25th Canadian Congress of Applied Mechanics (CANCAM 2015)

Progress in Canadian Mechanics Volume 1

London, Ontario, Canada
31 May – 4 June 2015

Editors:

J.M. Floryan
L. Jiang
O.R. Tutunea-Fatan

ISBN: 978-1-5108-3689-1

Printed from e-media with permission by:

Curran Associates, Inc.
57 Morehouse Lane
Red Hook, NY 12571

Some format issues inherent in the e-media version may also appear in this print version.

Printed by Curran Associates, Inc. (2017)

For permission requests, please contact Western University – Department of Mechanical and Materials Engineering at the address below.

Western University – Department of Mechanical and Materials Engineering
Spencer Engineering Building
London ON N6A 5B9
Canada

Phone: 519-661-2136
Fax: 519-661-3020

mme@uwo.ca

Additional copies of this publication are available from:

Curran Associates, Inc.
57 Morehouse Lane
Red Hook, NY 12571 USA
Phone: 845-758-0400
Fax: 845-758-2633
Email: curran@proceedings.com
Web: www.proceedings.com

CANADIAN CONGRESS OF APPLIED MECHANICS

CANCAM

CONGRÈS CANADIEN DE MÉCANIQUE APPLIQUÉE

Progress in Canadian Mechanics

Volume 1

Proceedings of the 25th Canadian Congress of
Applied Mechanics (CANCAM 2015)
London, Ontario, May 31 – June 4, 2015

Editors: **Floryan, J.M.**
Jiang, L.
Tutunea-Fatan, O.R.

PREFACE

These Proceedings contain all papers presented during the 25th Canadian Congress of Applied Mechanics held in London, Ontario, May 31 – June 4, 2015. We wish to acknowledge the work of the Organizing and Program Committees, as well as the guidance of the CANCAM Central Committee.

Canadian Congress of Applied Mechanics (CANCAM) has been held every two years since the inaugural gathering at the Université Laval in Québec City in 1967. This year Western University had had the privilege of hosting it for the second time, having previously held the event 30 years ago, in 1985, through the Department of Mechanical and Materials Engineering.

Founded in 1878 by Bishop Isaac Hellmuth of the Anglican Diocese of Huron, "The Western University of London Ontario" opened its doors to students for the first time in 1881 with four faculties – Arts, Divinity, Law and Medicine. In 1916, the current campus was purchased from the Kingsmill family. In 1923 the university was renamed as The University of Western Ontario and then in 2014 was, once again, rebranded as Western University. Since that first class graduated in 1883, the university has become a vibrant centre of learning. Through twelve faculties and three affiliated university colleges, Western delivers education to its 36,000-plus students through more than 400 specializations. It was at Western where Sir Frederick Banting rose from a restless sleep in 1920 and wrote out 25 words that led to his discovery of insulin. It was at Western where researchers led by Ivan Smith developed the world's first "cobalt bomb" to treat cancer in 1951.

The Faculty of Engineering has celebrated its 60th anniversary in 2014. The Faculty consists of four departments that offer nine accredited undergraduate programs. Presently, Western Engineering hosts 1,700 undergraduate students, 700 graduate students and 96 faculty members.

J.M. Floryan
General Chair, CANCAM 2015

CANCAM 2015 Committees

Organizing Committee

Honorary Chair: David Weaver (McMaster University)
General Chair: Jerzy M. Floryan (Western University)
Program Chair: Remus Tutunea-Fatan (Western University)
Paper Chair: Liying Jiang (Western University)

Program Committee

Martin Agelin-Chaab (University of Ontario Institute of Technology)
Samuel Asokanthan (Western University)
Evgueni Bordatchev (NRC Automotive & Surface Transportation London)
Ralph Buchal (Western University)
Ashraf El Damatty (Western University)
Louis Ferreira (Western University)
Clinton Groth (University of Toronto)
Horia Hangan (Western University)
Mona Hasanzadeh (Western University)
Mohammad Hossain (Western University)
Amir Khajepour (University of Waterloo)
Robert Klassen (Western University)
Bruce Minaker (University of Windsor)
Sushanta Mitra (York University)
Alireza Mohammadi (Western University)
Hadi Moradi (Western University)
Michael Naish (Western University)
Timothy Newson (Western University)
Aaron Price (Western University)
Eric Savory (Western University)
Kamran Siddiqui (Western University)
David Steinman (University of Toronto)
Pierre Sullivan (University of Toronto)
Mark Tachie (University of Manitoba)
Seyed Mohammad Taghavi (Université Laval)
Jeffrey Wood (Western University)
Xiaohua Wu (Royal Military College of Canada)
Jun Yang (Western University)
Serhiy Yarusevych (University of Waterloo)

CANCAM Central Committee

President: Clarence W. De Silva (University of British Columbia)
Past President: Sadik Dost (University of Victoria)
Vice-Predient: Jim Bugg (University of Saskatchewan)
Secretary: Arun Misra (McGill University)
Treasurer: Leszek Sudak (University of Calgary)
Member: Donald Bergstrom (University of Saskatchewan)
Member: J. Maciej Floryan (Western University)
Member: Remus Tutunea-Fatan (Western University)

Honorary Chair

David S. Weaver
Department of Mechanical Engineering
McMaster University
Hamilton, Ontario, Canada
weaverds@mcmaster.ca

Professor David S Weaver earned his BASc and MASc in Mechanical Engineering at the University of Toronto in 1964 and 1966 respectively, and his PhD in Civil Engineering (theoretical mechanics) at the University of Waterloo in 1969. Following 2 years as a Research Assistant Professor at Waterloo, he joined McMaster University as an Assistant Professor of Mechanical Engineering in 1971. He is currently Professor Emeritus at McMaster, where he continues to conduct research and supervise graduate students.

For nearly 50 years, Professor Weaver had conducted experimental and theoretical research in the field of flow-excited vibrations and acoustics. These problems typically involve high Reynolds number, unsteady flows over bluff bodies with moving boundaries and are, therefore, challenging to model both theoretically and experimentally. The research approach taken is to develop an understanding of the underlying physics of the fluid-structure interaction phenomena and then to create predictive tools for their prevention or avoidance in design. Specific applications range over a large variety of problems from the strumming of underwater sonar tow cables and the vibrations of large hydraulic control gates, to acoustic resonance in steam line relief valves and fluidelastic instability in nuclear steam generator tube arrays. Current research involves an experimental study of the loading on steam generator tubes during the rapid depressurization following a main steam line break in a nuclear plant. He has published several hundred technical papers and reports.

Professor Weaver has carried out contract research and consulting in the field of flow-induced vibrations for many companies around the world from Mitsubishi Heavy Industries in Japan to The Joint European Torus (JET) Fusion Energy Project in UK. He was invited to teach a short course in Flow-Induced Vibrations at Cranfield University (UK) for a number of years, and in Kyoto, Japan, by the Japanese Society for Mechanical Engineers. His work is widely used in the design of components against damaging flow-excited vibrations and has been recognized by many awards including the Engineering Applications of Mechanics Medal in 1990 and the CANCAM medal in 2013. He has been honored as a Fellow of ASME, CSME, EIC and The Canadian Academy of Engineering.

In terms of service, Professor Weaver served 3 terms as Chair of the Dept. of Mechanical Engineering at McMaster and was Assistant Dean of Engineering (External Affairs). He was a member of the NSERC Mechanical Engineering Grant Selection Committee 1991-1994, including 2 years as its Chair. He was a member of a number of NSERC award committees, the College of Reviewers for the Canada Research Chairs Program, and the Mechanical Engineering Panel of the Science Foundation Ireland Frontiers Research Program. He is an Associate Editor of the Journal of Fluids and Structures and has been on the organizing committee of innumerable international conferences in flow-induced vibrations and acoustics for 35 years. Amongst service on many technical societies and organizations, Professor Weaver is a member of the Canadian National Committee of The International Union of Theoretical and Applied Mechanics (IUTAM), including 6 years as a Canadian member of the IUTAM General Assembly. He is also a member of the Board of Directors of the CSME and served as its President in 1997-98.

A dedicated teacher, Professor Weaver has also been recognized for his outstanding contributions to engineering education, including Certificates of Excellence from the McMaster Students Union (MSU) in 1979 and 1998.

Plenary Speakers

Greg Kopp
Department of Civil and Environmental Engineering
The University of Western Ontario
London, Ontario N6A 5B9 Canada
gakopp@uwo.ca

Professor G.A. Kopp received a B.Sc.M.E. from the University of Manitoba in 1989, a M.Eng. from McMaster University in 1991 and a Ph.D. in Mechanical Engineering from the University of Toronto in 1995. Between 1995 and 1997 he held a NSERC Post-doctoral Fellowship in the Chemical Engineering Department at the Universitat Rovira i Virgili in Tarragona, Spain. He returned to Canada in the summer of 1997 to an appointment of Assistant Professor at the University of Western Ontario. He was promoted to Professor in the Department of Civil & Environmental Engineering in July 2007 (and has a cross-appointment in the Department of Mechanical & Materials Engineering). In January 2015 he was appointed Associate Dean (Graduate and Postdoctoral Studies) in the Faculty of Engineering. Dr. Kopp's expertise and research projects relate to mitigating damage to structures during extreme wind storms such as tornadoes and hurricanes. Details include model-scale wind tunnel and full-scale component test methods, field surveys of damage caused by tornadoes, building aerodynamics, wind effects on building component and cladding systems, tornado and thunderstorm winds, wind loads on solar arrays, the role of turbulence on wind loads, wind-borne debris, and turbulent shear flows.

Robert M. McMeeking
Tony Evans Professor of Structural Materials and Professor of Mechanical Engineering
University of California, Santa Barbara, U.S.A.
Sixth Century Professor of Engineering Materials, University of Aberdeen, Scotland
rmcm@engineering.ucsb.edu

Professor McMeeking completed his Ph.D. in solid mechanics at Brown University in 1976 and then spent 2 years at Stanford University as Acting Assistant Professor. He moved to the University of Illinois at Urbana-Champaign as Assistant Professor, and subsequently Associate Professor, in the Theoretical and Applied Mechanics Department. Professor McMeeking joined the University of California, Santa Barbara (UCSB) in 1985 as Professor of Materials and of Mechanical Engineering. He was Chair of the Department of Mechanical Engineering 1992-1995 and again during 1999-2003. He has published over 250 scientific papers on such subjects as plasticity, fracture mechanics, computational methods, glaciology, tough ceramics, composite materials, materials processing, powder consolidation and sintering, ferroelectrics, microstructural evolution, nanotribology, actuating structures, blast and fragment protection of structures, fluid structure interactions arising from underwater blast waves, the mechanics of the cell and its cytoskeleton, lithium-ion batteries and fuel-cells. Professor McMeeking is a member of the U.S. National Academy of Engineering, Fellow of both the U.K. Royal Academy of Engineering and the Royal Society of Edinburgh, and held a Humboldt Award for Senior Scientists in 2006 and again in 2013. He was given the Brown Engineering Alumni Medal in 2007, the 2014 William Prager Medal of the Society of Engineering Science, and the 2014 Timoshenko Medal of the American Society of Mechanical Engineers.

Stavros Tavoularis
Department of Mechanical Engineering
University of Ottawa
Ottawa, Ontario K1N 6N5 Canada
tav@genie.uottawa.ca

Professor Tavoularis received a Ph.D. from The Johns Hopkins University, Baltimore, USA. Since 1980, he has been a faculty member in the Department of Mechanical Engineering at the University of Ottawa, where he served terms as Department Chair and Director of the Ottawa-Carleton Institute for Mechanical and Aerospace Engineering. He is currently cross-appointed in Physics Department and has been an HPCVL - Sun Microsystems Research Chair in Computational Science and Engineering. As Director of the Fluid Mechanics Laboratory, he supervises a large team of graduate and undergraduate students, postdoctoral fellows and other researchers on a variety of experimental and computational projects concerning turbulence, turbulent mixing, vortex dynamics, aerodynamics, nuclear reactor thermalhydraulics, bio-fluid dynamics, and design of flow apparatus and instrumentation. His work has been funded by government organizations and private industries. Professor Tavoularis has been elected as a Fellow of the Canadian Academy of Engineering, a Fellow of the Engineering Institute of Canada, a Fellow of the Canadian Society for Mechanical Engineering and a Fellow of the American Physical Society and is a recipient of the George S. Glinski Award for Excellence in Research. He is the author of the graduate textbook *Measurement in Fluid Mechanics*, published by Cambridge University Press, and numerous research articles and reports.

Kon-Well Wang
Department of Mechanical Engineering
University of Michigan
Ann Arbor, MI, U.S.A.
kwwang@umich.edu

Kon-Well Wang is the Stephen P. Timoshenko Professor and Tim Manganello/BorgWarner Department Chair of Mechanical Engineering at the University of Michigan. He received his Ph.D. degree from the University of California at Berkeley in 1985, worked at the General Motors Research Labs as a Senior Research Engineer, and started his academic career at the Pennsylvania State University in 1988. During his Penn State years, Professor Wang has served as the William E. Diefenderfer Chaired Professor in Mechanical Engineering, Director of the Structural Dynamics and Controls Lab, Associate Director of the Vertical Lift Research Center of Excellence, and Group Leader for the Center for Acoustics and Vibration. He joined the University of Michigan in 2008.

Professor Wang's main technical interests are in adaptive structural systems and structural dynamics & controls. He has received various recognitions, such as the SPIE Smart Structures and Materials Lifetime Achievement Award, the ASME Adaptive Structures and Materials Systems Award, the ASME N.O. Myklestad Award, the ASME Rudolf Kalman Award, the ASME Adaptive Structures and Material Systems Best Paper Award, the NASA Tech Brief Award, and the SAE Ralph Teetor Award. He is a Fellow of the ASME, AAAS, and IOP. He is the Publication Executive of the ASME Design Engineering Division Executive Committee and the Chair of the ASME Mechanical Engineering Department Heads Executive Committee. Professor Wang has been the Chief Editor for the ASME Journal of Vibration and Acoustics. He is currently an Associate Editor for the Journal of Intelligent Material Systems and Structures and an Editorial Advisory Board Member for the Journal of Sound and Vibration.

Jun Yang
Department of Mechanical and Materials Engineering
The University of Western Ontario
London, ON N6A 5B9 Canada
jyang@eng.uwo.ca

Jun Yang is an associate professor in Department of Mechanical and Materials Engineering, and Biomedical Engineering Program, Western University (The University of Western Ontario). He received his Ph.D. degree in Mechanical Engineering from University of Alberta in 2004. He was a postdoctoral fellow in Mechanical/Biomedical Engineering from 2004 to 2005 at Georgia Institute of Technology. He joined Western University in September 2005. Dr. Jun Yang's research interests include micro/nanofabrication, surface engineering, additive manufacturing, printed electronics, flexible electronics, MEMS/NEMS/BioMEMS, lab-on-a-chip, sensors and actuators.

Dr. Yang was a recipient of 2006 PetroCanada Young Innovators Award, 2008 Canada Foundation for Innovation-Leaders Opportunity Fund (LOF) Award, 2009 Early Researcher Award, 2012 University's Faculty Scholar Award and 2015 Xerox UAC Award.

Table of Contents

Plenary Lecture

The Changing Face of Applied Mechanics 1967-2015 2
David S. Weaver

The Generation of Stress in the Storage Particles of Lithium-Ion Batteries 5
Robert M. McMeeking

Assessing Wind Speeds in Tornadoes from Observed Damage 6
Greg Kopp

Recent Advances in Adaptive Metastructures for Structural Dynamics Enhancement 7
Kon-Well Wang

Vortices 8
Stavros Tavoularis

Design and Micro/Nano-Engineering of Functional Materials, Surfaces and Devices: from Fundamentals to Applications 9
Jun Yang

Actuation and Smart Materials

Dynamic Analyses of a Viscoelastic Dielectric Elastomer Oscillator 11
Jianyou Zhou, Liying Jiang, Roger E. Khayat

Dynamic Analysis of a Vibration Shaker Based on a Dielectric Elastomer Actuator 15
Ujjaval Gupta, Hareesh Godaba, Jian Zhu

Durability of a New Generation of Basalt-FRP Reinforcing Bars 19
Mohamed Hassan, Brahim Benmokrane

Advanced Manufacturing

An Improvement of Fixtureless Inspection for Non-Rigid Parts Based on Filtering Sample Points 24
Sasan Sattarpanah Karganroudi, Jean-Christophe Cuillière, Vincent Francois, Souheil-Antoine Tahan

Nd:YAG Laser Surface Treatment of Stainless Steel 304 Sheets 28
Mitra Radmanesh, AmirKianoosh Kiani

Measurement Methods and Data Representations for Mechanical Part Refurbishment/Replacement 32
H.M.H. Bubshait, Allan D. Spence

Current State-of-the-Art in Modeling and Simulation of Laser Polishing Process 36
Shirzad Mohajerani, Evgueni Bordatchev, O. Remus Tutunea-Fatan

Development of Ultrasonic Surface Treatment Device 40
Yanjun Qian, H. J. Kwon

Human Head and Face Detection and Tracking Using Single Camera to Improve Human Robot Interaction............................44
Sumesh Kunwar, Meysar Zeinali.

Parameter-Driven Geometric Modeling of Retroreflective Features............................48
Sama Hussein, Benjamin Hamilton, Evgueni Bordatchev, O. Remus Tutunea-Fatan

Experimental Study of Touch Probing with a Five Axis Machine Tool............................52
Md Mizanur Rahman, J.R.R. Mayer

Ice-Microtome Design for Procurement and Crystal Analysis of Ice Thin Sections............................56
Steve Bruneau, Ashley Sullivan, Zetian Zhang, Bruce Colborne

Strategies in Single Point Inverted Cutting for Fabrication of Structured Surfaces............................60
Ben Hamilton, Sama Hussein, O. Remus Tutunea-Fatan, Evgueni Bordatchev

Aerodynamics

Analysis of Wind Turbine Stall Using Tuft Flow Visualization............................65
Tyler E. Gallant, David A. Johnson

CFD Modeling of the Wind Chamber Used for Radioactive Particle Resuspension Experiment............................69
Sharman Perera, Edward Waller, Paul Walsh, Andrew Doxtator, Ali Akhtar

Effects of Wind Tunnel Model End Conditions on Low Reynolds Number Airfoil Studies............................74
Mark Feero, Pierre Sullivan, Philippe Lavoie

Large-Eddy Simulation of Turbulent Round Jets in Crossflows............................78
Paul Ziadé, Pierre Sullivan

Measurement of Rotating Wind Turbine Blade Angle of Attack with a Five-Hole Pressure Probe............................82
Tyler E. Gallant, David A. Johnson

Wingtip Vortex Control Via Tip-Mounted Half-Delta Wings............................86
Ying Yu Su, Steve Choi, Tim Lee

Near Wake Flow Topology of Ahmed Bodies............................90
Graham Feltham, Alis Ekmekci

Vortex Shedding in the Wakes of Finned Cylinders............................96
Jeff McClure, Serhiy Yarusevych

Propagation of Fluctuations and Flow Separation on an Unsteadily Loaded Airfoil............................100
Andrew S. Tenney, Jacques Lewalle

Flow-Topology Around a Cylinder Submerged in Different Width Wake............................104
Md. Mahbub Alam, Longjun Wang, Yu Zhou

Biological Fluid Mechanics

The Effect of Sinus Geometry on the Turbulence Downstream of a Bileaflet Heart Valve............................109
Laura Haya, Stavros Tavoularis

PIV Measurement of Jets Through Model Mitral Valve with Different Suture Configurations.............113
Anna Chtchetinina, Lyes Kadem, Hoi Dick Ng

Ascertaining the Influence of Intraglottal Vortices upon the Dynamics of the Vocal Folds.............117
Brian Q. Kettlewell, Sean D. Peterson

A Computational Fluid Dynamics Model of Failing Fontan Circulation for Use in the Design of a Novel Treatment.............121
Matthew G. Doyle, Wanyu Zhang, Fabian Alvarado, Haruki Ide, Lars Grosse-Wortmann, Richard D. Weisel, Osami Honjo, Leonard W. Tse, Cristina H. Amon

Experimental Investigation of Hemodynamic Variations Inside Abdominal Aortic Aneurysm During Change of Posture.............125
Hani Abulkhair, Lyes Kadem

Non-Newtonian Blood-Mimicking Fluid for Particle Image Velocimetry.............129
Adam J Fortais, John R de Bruyn, Tamie L Poepping

Lumped Parameter Model of Flow Through the Carotid Bifurcation.............133
Onaizah Onaizah, Tamie Poepping, Mair Zamir

Computational Modeling of Microbubble Distribution in Microvascular Networks.............137
Daniel Goldman, Graham M. Fraser, Christopher G. Ellis

Evaluating Flow Disturbances in the Carotid Bifurcation.............140
Tamie L. Poepping, Amanda L. DiCarlo, Onaizah Onaizah, Mair Zamir

The Role of Heparin Sulphate in Endothelial Cell Leukocyte Attachment in Regions of Spatial Wall Shear Stress Gradients.............144
Scott Cooper, Richard L. Leask

Laser Doppler Velocimetry Measurements of Wall Shear Stress Under Steady and Pulsatile Flow Conditions in a Parallel Plate Flow Chamber.............147
Hamed Avari, Eric Savory

Biomaterials

On the Energetics of the Edge of an Open Lipid Bilayer.............152
Meisam Asgari, Aisa Biria

Study of Laser Adjustable Sutures for Trabeculectomy.............156
Kongying Xie, Husam Elkaseh, Eddie B. Moss, Robert J. Campbell, Yong Jun Lai

Flexible Armors Inspired from Natural Fish Scales and Osteoderms.............160
Roberto Martini, Ravi Chintapalli, Francois Barthelat

Doctor Blading Artificial Nacre and Bone.............161
Mohammad Mirkhalaf, Francois Barthelat

A Model for the Collagen Fibrils in Biological Soft Tissues.............162
Meisam Asgari, Hossein K. Heris, Luc Mongeau

Validation of a Novel Tonometry Device for the Developing World....166
Melanie Chin, Tim Newson, Gord Campbell

Biomechanics

Nonlinear Mechanical Response of Porcine Knee Joint....171
Marcel Rodriguez, LePing Li

Enhancing Biocompatibility of Grade 4-Titanium Using Laser Surface Texturing for Bone and Tissue Transplant Applications....174
Mitra Radmanesh, AmirKianoosh Kian

Contact Stress Analysis of Surface Guided Knee Implant Using Finite Element Modeling....177
Ida Khosravipour, Yunhua Luo, Shabnam Pejhan, Urs P. Wyss

Experimental Apparatus for Analysis of Bone Removal During Surgical Resurfacing Procedures....181
Jonathan Kusins, O. Remus Tutunea-Fatan, Louis Ferreira

Indicators of Motion Induced Interruption Occurrences in Heavy-Weather Sea Conditions....185
N.R. Bourgeois, R.G. Langlois

Boundary Layers

Fluctuating Pressures and Flow Characteristics Within Separation Bubbles on Surface-Mounted Prisms....190
Abul Fahad Akon, Gregory Alan Kopp

Numerical Simulation of Laminar Shear-Layer Reattachment in a Planar Sudden Expansion....194
Nima Moallemi, Joshua Brinkerhoff

Vortex Visualization and Characterization in the Laminar Separation Bubble over a NACA 0018 Airfoil....198
Andrew Lambert, Serhiy Yarusevych

Modeling of Dielectric Barrier Discharge Actuator for Airflow Control....202
Mohammadreza Ghazanchaei, Kazimierz Adamiak, G. S. Peter Castle

Numerical Investigation of Blockage Development in an Aeroengine Transonic Axial Compressor Stage....206
Shady Ali, Eric Savory, Chao Zhang

Combustion and Flames

Test Methodology for Comparing Wildland Fire Chemical Treatments....211
Shammawi A. Anderson, Andre G. McDonald

Modified Biodiesel Chemical Kinetics Based on Numerical Simulation in an Ignition Quality Tester....215
Mahmoud Elhalwagy, Chao Zhang

Buoyant Instabilities and Flame Spread in Wildland Fires: Implication of the Need for Scaling Instability Analysis....219
M.A. Finney, J.D. Cohen, J.A. Forthofer, S.S. McAllister, N.K. Akafuah, K. Saito, M.J. Gollner, D.J. Gorham

Structured Natural Convection 223
J.M. Floryan, M.Z. Hossain, A. Bassom

Computational Fluid Dynamics

Characteristics of New Mode of Instability in a Channel with Longitudinal Grooves........................228
A. Mohammadi, J. M. Floryan

Spectral Algorithm for Analysis of Flows in Vibrating Channels..232
S. Zandi, A. Mohammadi, J.M. Floryan

Three-Dimensional Spectrally-Accurate Algorithm for the Navier-Stokes Equations.............................237
Md. Sakib, J. M. Floryan

Hydrodynamic Study of Gas-Particle Interaction in an FCC Downer Reactor.......................................241
Tatiana López, Carlos A. Bustamante, César Nieto

Numerical Simulation of Displacement Flow of Viscoplastic Fluids..245
A. Eslami, K. Alba, I.A. Frigaard, S.M. Taghavi

Shear-Thinning Effects on Nanofiber Formation by Centrifugal Spinning..249
S. M. Taghavi, R.G. Larson

Two-Dimensional Hydrodynamics Study of FCC Riser Reactor by Eulerian-Eulerian Approach..........253
Natalia Gómez, Carlos A. Bustamante, César Nieto

Oil Flow Visualization of the Laminar Separation Bubble on a SD 7037 Airfoil.....................................257
Faegheh Ghorbanishohrat, David A. Johnson

Dispersed Phase Single-Particle Lagrangian and Correlation Dimensional Statistics in Kinematically Simulated Turbulence..261
S. A. Murray, M. F. Lightstone, S. Tullis

Numerical Investigation of Jet Pump with Twisted Tapes...265
Arpita Srivastava, Shaligram Tiwari, Mani Annamalai

CFD Simulation of Spray Cooling Systems...269
M. Jafari, R.M. Barron, R. Balachandar

Effect of Microstructured Roughness on Micro-Droplet Dynamics...273
S.M.Reza Attarzadeh, Moussa Tembely, Ali Dolatabadi

Simulation of Aysmmetric Turbulent Channel Flow with Roughness on One Wall................................277
Ming Teng, Donald J. Bergstrom

Spectral Algorithm for Linear Stability Analysis of Flow in Ribbed Annuli..281
H. V. Moradi, J. M. Floryan

Computational Solid Mechanics

Use of Rigid Bar Elements in Finite Element Analysis 286
Kevin Brooks, Nader Zamani

Simulating Debris Impacts into Tornado Saferooms with LS-DYNA 290
Yevgeniy Parfilko, Benjamin Varela

Transient Heat Flow Analysis Using FEA and Laplace Transform-Based Collocation Method 293
Amir Khalilollahi, Enayat Mahajerin, Gary Burgess

Structural Topology Optimization with Adaptive Mesh Refinement and De-Refinement 297
Alexandre Nana, Jean-Christophe Cuillière, Vincent François

On the Importance of Thermal Effects in Dislocation Dynamics Simulations 301
Oxana Skiba, Robert Gracie

A Phase-Field Modeling of the Microstructural Evolution in Ti-6Al-4V Alloy Under Mechanical Loading 305
Jianhui Zhai, Yu Su

A Novel Design of the Heat Echanger of an Aircraft's Environmental Controller 309
Negin Shahbazian, Mohammad Anisur Rahman, Kamran Behdinan

Redesign of a DC-Link Capacitor for an Aircraft Power and Thermal Management Controller 313
Mohammad Anisur Rahman, Negin Shahbazian, Kamran Behdinan

A Coupled Dynamic Simulation of Hydraulic Fracturing 317
Matin Parchei-Esfahani, Robert Gracie

Computational Modeling and Calibration of a 3D Metal Foam Heat Sink 321
Shahzada P. Aly, Khaled S. Al-Athel, Abul Fazal M. Arif, Javad Mostaghimi

Convection

Thermodynamics and Fluid Mechanics of Heavy Water and Supercritical Water 326
Robert L. Varty

Study of AL2O3/Water Nanofluid in a Multiport Slab Heat Exchanger 330
Mohammed Ismail, Shahram Fotowat, Amir Fartaj

Transient Heat Transfer Behavior in the Liquid Domain of a PCM During Phase Change 334
Mohammad Bashar, Kamran Siddiqui

Thermal Drift Due to Modulated Dual Heating 338
M. Z. Hossain, J. M. Floryan

Convective Heat Transfer Due to Multiple Circular Jets Impinging on a Large Cylinder 342
Chris Csernyei, Anthony G. Straatman

Drops, Bubbles

Computational Study of Multiphase Transport Through Fuel Cell Flow Channel....347
Abbas Ghasemi, Yanzhou Qin, Xianguo Li

Relation Between Paint Application Parameters and Droplet Diameter: A Way to Improve Efficiency and Reduce Air Emissions....351
Rajan Ray, Paul Henshaw

Experimental Investigation of Particles Turbulence in A High Reynolds Vertical Pipe Flow....355
Rouholluh Shokri, Sina Ghaemi, David S. Nobes, R. Sean Sanders

The Effect of Mesh-Type Bubble Breaker on Two Phase Vertical Co-Current Flow....359
Alan Kalbfleisch, Kamran Siddiqui

Characterization of the Two-Phase Flow Due to a Holed Bubble Breaker Inside an Effervescent Atomizer Using High-Speed Imaging....363
Mona Hassanzadeh Jobehdar, Kamran Siddiqui, Aly H. Gadallah, Wajid A. Chishty

Elasticity/Plasticity

Effect of Irradiation and Thermal Recovery on the Hardness of Inconel 800H and AISI 310 Alloys....368
H. Rajakumar, R.J. Klassen

Cyclic Plasticity Modelling of Wrought Magnesium Alloys....372
Ali Roostaei, Hamid Jahed

Plastic Deformation of Gold Micro Spheres....376
AZM Ariful Islam, V Bhakhri, Robert J. Klassen

Indentation-Induced Deformation Behaviour of Magnesium....380
V. Bhakhri, M. Haghshenas, R. Oviasuyi, R.J. Klassen

Plastic Buckling Paradox for Plates in Biaxial Loading....384
Suresh Shrivastava, Bob Munlemvo Mayemba

Experimental Methods

Electrical Resistivity of Sands....388
Zaid Al-Qaysi, Abouzar Sadrekarimi

Reynold Number Effect on Forward Facing Step in Pressure Gradient....392
Hassan Iftekhar, Weijie Shao, Martin Agelin-Chaab

Mixing Characteristics in Turbulent Orifice Jets....397
Yaw Y. Afriyie, Mark F. Tachie, Mikhail Koupriyanov, Tom Epp

Experimental and Numerical Study of Flow Characteristics Through a Gradual Contraction....401
Sarwesh Mali, Ali Nematollahi, Ebenezer E. Essel, Mark F. Tachie

Applications of Fiber Bragg Grating Sensors to Strain Measurements of AZ31B Extrusion in Rotating Bending Cyclic Tests....405
Bahareh Marzbanrad, Ehsan Toyserkani, Hamid Jahed

Flow Around 2D Bluff Bodies Mounted Close to a Floor....409
Hassan Iftekhar, Martin Agelin-Chaab

Experimental Investigation on Overpressures in a Scale Model of Sidedraw Silo During Emptying....413
Chenying Dong, Kun Pang, Fang Yuan, Liying Jiang

Design of Kolsky Bar to Characterize the Dynamic Response of Elastomer Under High Strain Rate....417
Muhammad Salman Chaudhry, Alex Czekanski, Roger Carrick

Temperature Measurements in Industrial Scale Spray Cooling Systems....423
M. Jafari, R. Balachandar, R.M. Barron

Numerical Investigation of the Bearing Capacity of Ring Foundations on Clay....427
Maha Al Massri, Timothy A. Newson

Fluid-Structure Interactions

Sloshing of Supported Shells Conveying Fluid....432
Michael P. Païdoussis, Kostas Karazis

Transient Loading on Nuclear Steam Generator Tubes During a Two-Phase Blowdown....436
Ouajih Hamouda, David S. Weaver, Jovica Riznic

Numerical Modeling of Tornadic Load on a Tall Building....440
Zoheb Nasir, Girma T. Bitsuamlak

Transition in Dynamical Behaviour of a Flexible Slender Cylinder Subjected to Axial Flow, with a Variable Support at the Upstream End....444
Mojtaba Kheiri

Fracture Mechanics

Shape Effect of Traction-Separation Law on Dynamic Fracture of X80 Pipeline Steel....449
Y.J. Wang, C.Q. Ru

Residual Stress State in Single-Edge Notched Tension Specimen Caused by the Local Compression Technique....451
Yifan Huang, Wenxing Zhou

Cohesive Zone Model for Temperature Dependent Fracture Analysis of Pipeline Steel....455
Sourayon Chanda, CQ Ru

Overcoming Brittleness Through Bio-Inspiration and Microarchitecture....459
Mohammad Mirkhalaf, Ahmad K. Dastjerdi, Francois Barthelat.

Simulation of Dent-Crack Defect in Pipeline Using FEA....460
Hossein Ghaednia, Sreekanta Das, Jamshid Zohrehheydariha

Understanding the Strain Distributions Caused by a Dent Defect....464
Hossein Ghaednia, Sreekanta Das, Jamshid Zohrehheydariha, Sara Kenno

Geophysical Fluid Dynamics

Examination of Helmholtz Vortex Laws Using Rankine Combined Vortices in Circular Cylinders....469
Josué Njock Libii

Spatio-Temporal Dynamics of a Street Canyon Flow from Combined Field and Wind Tunnel Experiments....474
Laurent Perret, Karin Blackman, Eric Savory

The Flow Field of an EF5 Tornado in a Supercell Windstorm....478
W.E. Lin, A. Ozyoruk, E. Savory, L.G. Orf

Evaluating the Performance of Inflow Boundary Condition Models for Rans Simulation of ABL Flow Over Complex Topography....482
Tibebu H. Birhane, Girma T. Bitsuamlak

Unsteady Rans Modelling of Density-Driven Experimental Thunderstorm Downbursts....487
Matthew Roberto, Chris Oreskovic, Eric Savory, Juliette Porto

Experimental Characterizaton of Flow Around an Escarpment....491
Ryan Kilpatrick, Kamran Siddiqui, Horia Hangan

Mechatronics

Optical Fiber Sensors for In-Situ Monitoring of Lithium Ion Batteries....496
Hamid Alemohammad, AbdulRahman Ghannoum, Liliana Zdravkova, Krishna Iyer, Patricia Nieva

A Novel Nonlinear Amplitude-Modulation Gyroscope Incorporating Internal Resonance....500
S. Amir Mousavi Lajimi, Navid Noori, Amr Marzouk, Behraad Bahreyni, Farid Golnaraghi

Scale Experimental Test Vehicle for Rollover Study....504
Paul White, Rhyse Maryniuk, Sobhan Sadri, Christine Wu

State of Charge Estimation of Li-ion Batteries Using the Dynamic 2nd-Order Smooth Variable Structure Filter....509
Hamed Hossein Afshari, Andrew Gadsden, Ryan Ahmed, Saeid Habibi

Development of Residential HVAC Control and Demand Management System Using Intelligent Strategy Planning Models....513
Nima Alibabaei, Alan S. Fung, Kaamran Raahemifar

Positioning Accuracy Recovery for RRR Compliant Mechanism System under PZT Actuator Failure....517
Xuejun Wang, Wenjun Zhang, Changli Liu

Micro- and Nano-fluids

Experimental Study on Enhancement of Heat Transfer Using Nanoparticles Based Fluids....522
Nabeel Al-Rawahi, Myo Tay Zar Myint, Joydeep Dutta

Combining Micromilling and Liquid-Based Solvent Bonding for Ultra-Rapid Fabrication of Plastic Microchips................526
Alwin M. D. Wan, Edmond W. K. Young

Design Considerations for On-Chip Biosensing in Microfluidic Cardiovascular Models..528
Jeremy Wong, Craig A. Simmons

A Novel Microfluidic Switch Design with Response Characteristics to High Acceleration..532
Weirong Nie, Huang Liu, Liying Jiang

Multibody/Vehicle Dynamics

Automated Design Synthesis of LQR Controllers for Road Vehicle Active Suspension Systems..537
Eungkil Lee, Yuping He

Symbolic Integration of Finite Element Models with Dynamic Multibody Systems...541
Yiteng Liang, Joydeep Banerjee, John McPhee, Asaad Sellmann, Orang Vahid Araghi

Acausal Modelling for Dynamic Multibody Simulation of Damped Three-Dimensional Flexible Beams.............................546
Joydeep Banerjee, John McPhee

Design and Analysis of an Interconnected Suspension for a Small Off-Road Vehicle...550
Bruce P Minaker, Zheng Yao

Viscoelastic Multibody Dynamics of Whiplash..554
Mohamed T.Z. Hassan, S.A. Meguid

Road Profile Recovery Using Onboard Acceleration Data...558
Fei Hu, Bruce P Minaker, Mehrzad Abdollahi

Multicomponent Materials

Numerical Prediction on Elastic Properties of Chopped Fiber Reinforced Composites...563
Hui Liu, Liying Jiang, Danielle Zeng, Yang Li

Design of Stiffness by Controlled Organization of Microstructure in an Elastomeric Nanocomposite...................................567
Peiying Jennifer Tsai, Suvojit Ghosh, Peidong Wu, Ishwar K. Puri

The Influence of Filler Volume Fraction on Sensitivity to Distributed Strain in Carbon Nanofiber/Polyurethane Nanocomposites...571
T. N. Tallman, K. W. Wang

Effect of Surface Treatment on the Strength and the Environmental Durability of Bond at the Metal-Polymer Composite Interface...575
Syeda Noor E Sumaiya, Shailesh Chekurthi, J. Raghavan, Sitarz Blaine, Brian D McMillan

Strain-Rate-Dependent Mixed Mode Failure Analysis of Unidirectional Carbon Fiber Reinforced Composite Laminates...579
Shailesh Chekurthi, J. Raghavan

Computational Homogenization of Metal Matrix Composites....583
M. U. Siddiqui, A. F. M. Arif

Nanostructures and MEMS

Multiscale Modelling of Thermo-Electro-Mechanical Behaviours of Gallium Nitride Nanowires: Size and Temperature Dependence of the Piezoelectric Potential....589
Jin Zhang, S. A. Meguid

Sound Absorption by Multilayer of Micro perforated Membranes....593
Xiaobing Cai, Jun Yang

Air Bubble Assisted Detection for Micro Mass Sensors....597
Jacky Chow, Yong Jun Lai

Laser Modification of Printed Graphene-Oxide (GO) and Graphene-Cellulose (G-CMC) Thin Films....601
Dogan Sinar, George K. Knopf, Suwas Nikumb

Remote Manipulation of 3D Printed Complex Magnetic Architecture after Surface Modification....605
Dongxing Zhang, Jun Yang

A Rapid Method for Preparation of Metal Coated 3D Printed Objects....608
Qiuquan Guo, Mingjun Hu, Xiaobing Cai, Jun Yang

A Multifunctional MEMS Sensor System for Brake System Applications....611
Rogelio Hernández Aguirre, Patricia Nieva, Amir Khajepour

Natural Frequencies of Double and Triple Walled Carbon Nanotubes Using Finite Element Formulation....615
Ishan Ali Khan, Seyed M. Hashemi

Friction Behaviour of Hexagonal-Shaped Graphene Islands Grown on Copper by Chemical Vapour Deposition (CVD)....619
Philip Egberts

Non-Newtonian Fluids

Displacement Flows in Slightly Converging Channels....621
R. Mollaabbaasi, K. Alba, S. M. Taghavi

A Microfluidic Method for Optical Studies of Early Stage Biofilm Response to Applied Shear Stresses....625
M. Parvinzadeh Gashti, J. Greener

Two-Layer Flow of Immiscible Non-Newtonian Fluids: Linear Stability Analysis....630
K. Alba, S.M. Taghavi

Effect of Nanoparticles on Rheological Properties of Herschel-Bulkley Fluids....634
R.Mollaabbasi, S.H. Noie

Magnetorheological Effects from Tracer Residence Time Distribution in Nanofluid Poiseuille Flows....638
F. Larachi, P. Hajiani

Porous Media

A Computational Model of Temperature Monitoring at a Leakage in a Leak Detection System of a Pipeline........................643
A. S. Bhuiyan, M. A. Hossain, J. M. Alam

Buoyancy-Driven Flows in a Confined Porous Medium........................648
Chunendra K. Sahu, M. R. Flynn

Carbon Sequestration Modelling Using Stabilized Finite Element Methods........................652
Chris Ladubec, Robert Gracie

Reaction-Driven Flow in a Porous Medium........................656
Jared Penney, Marek Stastna

Modelling Solids Separation in Rotating Belt Filters Using a Semi-Empirical Numerical Approach........................660
Christopher T. DeGroot

Renewable Energy Systems

A Numerical Study of Basement Heat Losses........................665
Lam Dang, Wey H. Leong

Seasonal Performance Investigation of a Dual Tank Solar-Assisted Heat Pump System........................669
Carsen J. Banister, Michael R. Collins

Optimized Load Allocation of Boilers Using Mixed Integer Non-Linear Programming for Reduced Natural Gas Consumption........................674
Altamash Ahmad Baig, Alan S. Fung

Improved Gasification Through Torrefaction of Biomass........................678
Prabir Basu, Manoj Kumar Raut, Bishnu Acharya

The Role of HRW Systems in Reducing Greenhouse Gas Emission and Energy Savings of Ontario-Based Plants................682
Afarin Amirirad, Alan S. Fung

A New Design for the Improvement of Uniform Cooling in Aircooled Li-ion Battery Stacks........................686
Soroush Arghavan, Mohammadhadi Akbari

An Air Source Heat Pump for Hydronic Home Heating, Application of an Air Based Photovoltaic/Thermal Collector.......691
Omid Nemati, Alan S. Fung

Comparison Analysis of Weather Normalization for Industrial Plants Using Prism and Excel........................695
Md Maniruzzaman Akan, Alan S Fung, Tamima Ahmed

Energy Benchmarking Analysis Based on Ventilation Consumption of Small and Medium Powder Coating Industries in the Greater Toronto Area (GTA)........................699
Tamima Ahmed, Alan S. Fung

Experimental Performance Study of a Vertical and a Horizontal Ground Loops Coupled to a Ground Source Heat Pump System........................703
Waleed S. Alzahrani, Alan S. Fung, Wey H. Leong

Turbulence

Boundary Layers Properties for Turbulent Taylor-Couettepoiseuille Flows in a Narrow-Gap Cavity....................................708
Poncet Sébastien, Viazzo Stéphane

Very-Lage-Scale Direct Numerical Simulation of the Canonical Flat-Plate Boundary Layer..712
Jean-Pierre Hickey, Xiaohua Wu

Turbulet Flow Behind a Partially Blocking Grid..716
Fama Fouladi, Paul Henshaw, David S-K. Ting

The Turbulent Structure of Inner Layer over Superhydrophobic Surface...720
Babak Vajdi Hokmabad, Sina Ghaemi

Effect of Aspect Ratio on the Near Field Dynamics of Free Surface Jet..724
Anuvrat Mishra, Mark F. Tachie, David C.S. Kuhn

Turbulence & Vortex Dynamics

Suppression of Vortex Shedding for Flow Past around a Circular Cylinder using Splitter Plate...729
Abdelkrem Eltaggaz

Quasi-Steady Hypothesis and Conductor Galloping..734
Karine Lavertu, André Laneville, Hachimi Fellouah

Large Eddy Simulation of Suspended Sediments in Turbulent Flow..738
M. Dallali, V.Armenio

Spatial Development of the Zone of Negative Turbulence Production in an Asymmetry Channel Flow................................742
Ali Esmaeilnejad, Sina Ghaemi

Vibration of Structures

Recreation of Random Vibration During Truck Transportation...747
Enayat Mahajerin, Gary Burgess

Fourth Order Approximate Solutions for the Undamped Free Vibration of Strongly Nonlinear Oscillators Subject to Initial Displacements..751
Vinicius B. F. Rabelo, M. L. Liu

Seismic Response Investigations of Nonlinear Single-Degree of-Freedom Systems Equipped with Gyromass Dampers...755
Reza Mirza Hessabi, Jiejing Zhou, Oya Mercan

High-Cycle Fatigue Testing on Aero-Engine Blade...759
Nima Bahrani, Kamran Behdinan

A Structured Approach for Solving the Inverse Eigenvalue Problem of a Beam with a Grounded Mass-Spring Attachement...763
Farhad Mir Hosseini, Patrick Dumond, Natalie Baddour

Stochastic Dynamics and Stability of a Class of Mems Gyroscopes....767
Soroush Arghavan, Mohamed Bognash, Samuel Asokanthan

Transient Response of a Sandwich Structure Damped with a Fibrous Core Material....771
Elsa Piollet, Guilhem Michon, Annie Ross

Vibration Control of Tall Buildings Using Aerodynamic Optimization....775
Ahmed Elshaer, Girma Bitsuamlak, Ashraf El-Damatty

Bouncing Dynamics Considerations for Micro Switch Design....779
Mohamed Bognash, Tianfu Wang, Samuel Asokanthan

Horizontal Models of Unbonded Fiber Reinforced Elastomeric Isolators....783
Niel Van Engelen, Michael Tait, Dimitrios Konstantinidis

Nonlinear Thermo-Electro-Mechanically Induced Vibration of FGPM Beams....787
M. Komijani, R.Gracie

The Inverse Eigenvalue Determinant Method for Jacobi and Tridiagonal Matrices....790
Patrick Dumond, Natalie Baddour

Vibration Analysis of Coupled Axial and Transverse Vibrations of Oil Well Drill Strings Without Effect of Drilling Mud....794
Mohammad Damshenas, Mohammad Taghi Ahmadian

Seismic Analysis for Composite Conical Tanks....800
Ahmed Elansary, Ashraf El Damatty

Low-Frequency Vibrations of Inhomogeneous Elastic Structures with Contrast Properties....804
Danila Prikazchikov, Olga Sergushova

Seismic Behaviour of Steel Liquid-Filled Conical Tanks Designed Under Hydrostatic Pressure....808
A. Musa, A.A. El Damatty

An Algorithm for Optimal Seat Selection for Heavy Mobile Machinery in Different Industrial Vibration Environments....812
Xiaoxu Ji, James P. Dickey, Tammy R. Eger

Asymmetric Influences on the Energy Dissipation Properties of a Metastable Module Excited Near Resonance....816
Narayanan Kidambi, R.L. Harne, K.W. Wang

Vibration Analysis of Drill String Under Inclined Drilling Including Stick-Slip Phenomenon with the Effect of Drilling Mud Injection....820
Mohammad Damshenas

Wind Engineering

Applicability of Quasi-Steady Theory on Building Point Pressures....827
Chieh-Hsun Wu, Gregory A. Kopp

Large Scale Particle Streak Velocimetry in the Windeee Dome....832
Dan-Stefan Parvu, Adrian Costache, Horia Hangan

Effect of Atmospheric Wind on a Ground Mounted Solar Panel Array....836
Chowdhury Jubayer, Horia Hangan

Aerodynamic and Acoustic Effects of a Serrated Trailing Edge on Wind Turbine Blades....840
Nicholas Tam, Kobra Gharali, David A. Johnson

Large Eddy Simulation of Wind-Induced Pressure on a Low Rise Building....844
B. Hajra, H. Aboshosha, G. T. Bitsuamlak

Author Index....848

Plenary Lectures

THE CHANGING FACE OF APPLIED MECHANICS 1967-2015

David S. Weaver
Department of Mechanical Engineering
McMaster University
Hamilton, Ontario, Canada
weaverds@mcmaster.ca

ABSTRACT

It is a great honour for me to have been chosen to be the Honourary Chair of this, the 25th CANCAM Conference. The first CANCAM was held at Universite Laval in 1967 and I must be one of the few remaining active individuals who attended that historic event. It was Canada's Centennial year and the conference attracted over 250 papers from around the world. The inaugural address was given by Maurice Roy, the President of the International Union of Theoretical and Applied Mechanics (IUTAM), and the list of general lecturers included Brian Spalding of Imperial College, a 'father' of CFD; Anatol Roshko, Theodore von Karman Professor of Aeronautics at Caltech; Hans Ziegler, Professor of Mechanics at the ETH in Zurich; and Clifford Truesdell, Professor of Rational Mechanics at Johns Hopkins University. It was an exciting and stimulating conference and its impact is still being felt now, nearly half a century later. That said, over these years, the discipline and its public visibility, or 'face', of applied mechanics has changed enormously and I wanted to take this opportunity to reflect on these changes.

In the 1950's and 60's, engineering education was expanding rapidly and being made much more mathematically rigorous. Graduate student numbers were much smaller in those days and, at many schools, courses in advanced mathematics were core and obligatory. Until the late 60's, digital computers were in their relative infancy and most calculations were done using simplified methods and slide rules. This may seem laughable now but these methods worked astonishingly well, producing such achievements as the SR-71 Blackbird Mach 3 spy plane which flew over the Soviet Union with impunity for nearly 30 years. Another notable achievement is our first generation of nuclear reactors, some of which are still operating now. This relatively simple approach had 2 distinct advantages: in order to develop a simplified but useful model, one had to understand the underlying physics sufficiently well to determine what could be ignored while still producing verifiable results; and one had to have some idea of what numbers to expect, given the nature of slide rule calculations. In other words, **one had to understand the limitations of the model and to be able to assess the correctness of the answers it produced.**

A significant aspect of the research in applied mechanics at the time of the first CANCAM in 1967 was the concentration on generic issues related to the physics and the mathematical or experimental tools, often with no specific practical application in mind. In the first issue (1956) of Clifford Truesdell's *Archive for Rational Mechanics and Analysis*, he wrote "Contributions for pure analysis are welcome; while they need not have physical application, when such application is possible, it should be developed". One could call the 50's and 60's the 'hay days' of applied mechanics. The motivation was discovery, there was enormous freedom to pursue the problems of greatest interest, and the work was considered to be 'fun' by its practitioners. In recognition of the importance of applied mechanics to modern engineering, Departments of Engineering Mechanics were established, either as independent departments or in conjunction with others.

As one might expect during the 'hay days' of mechanics, research funding was almost exclusively from the government and interaction between universities and industry was very rare. At that time, university research funding was the responsibility of the NRC, NSERC having not been created until 1978. A small group of us thought that university-industry collaboration could provide substantial mutual benefits and in 1972 organized a symposium on the engineering applications of mechanics, designed to stimulate university-industry interaction. On the occasion of the first symposium, a Dean of Engineering stated during a panel session that there were no industrial problems around which were worthy of an academic's attention. The rejoinder from an industry representative was that there were no academics capable of solving any real industry problems. This exchange was horrifying to those of us responsible for organizing the conference, but it did have the value of demonstrating the enormity of the gulf that existed between industry and the universities at the time, and the prejudices which would have to be overcome to develop effective collaborations. Of course, we were not the only ones who saw the potential benefits of encouraging more research in the industrial sector, and amongst other government policies, programs were developed to provide incentives for industrial collaboration with academia. Generally speaking, these programs were very successful. The conferences on Engineering Applications of Mechanics continued every 2nd year, alternating with CANCAM and emphasizing industrial participation, until 1994. By this time, university-industry collaboration was the norm and the 'applications oriented pendulum' had, in my opinion, actually swung too far. Non-government research funding in mechanical engineering was about 5 times the NSERC Discovery Grants funding, and it had become relatively difficult to attract research support for basic 'blue sky' research with no specific application in mind. I say that the pendulum had swung too far because I believe that there needs to be a reasonable balance between highly focussed applied research and curiosity driven basic research. If everyone is forced to do short term research for industry, who will develop the long term vision, make new discoveries, or study the problems affecting humanity with no immediate expectation of a financial profit for the sponsor?

Returning to the days of CANCAM 1967, digital computers were just starting to become effective tools for engineering calculations. While frustratingly slow and cumbersome by today's standards, they provided potential for the analysis of problems which could only have been dreamt about previously. Users had to write their own code, punch programs onto cards, and submit them for processing on a mainframe computer. Pick-up was usually the next day, if you were lucky. Initially, finite difference methods were generally used, being extensions of methodologies applied in the days of hand calculations. When finite element methods were being developed, there were often debates about which method was the 'best' and competitions were even held to establish the point. In one case, the competitors were some of the leading figures in computational mechanics. The posed problem involved the 3-D axial buckling of a thin triangular cylindrical shell with rounded corners and cut-out circles in the side walls. Interestingly, when the day of revelation came (a year later), it was a non-contest. The contestants got different answers for various reasons including making errors in the analysis so the methodologies could not be properly compared. Clearly, the profound benefits of computational mechanics did not come without its challenges in basic mechanics.

One of the great applied mechanicians of the day was George Carrier, T Jefferson Coolidge Professor of Applied Mechanics at Harvard University. His work exemplified the application of mechanics to the solution of practical problems. He had an intuitive ability to understand the basic physics of complex problems and to generate simple models which could be solved analytically, an ability which he exercised over his career with such skill that he was awarded the US National Medal of Science in 1990 by the then President George H W Bush. At the 6th US National Congress of Applied Mechanics at Harvard in 1970, George Carrier presented a paper entitled "*How much mathematics does it take to ruin a good engineer*". He was expressing his concern that there was a tendency to attack all engineering problems with the most powerful tools available, with too little thought being given to understanding the underlying physics, and to the precision of the answer required for a particular problem. My observation is that, over the intervening years, the situation has become worse, except that I would modify the question to read *How much computer power does it take to ruin a good engineer*? It seems that many engineers are inclined to replace thinking with the use of complex computer tools when a simpler approach would not only be more efficient but better engineering. The issue becomes potentially worse when sophisticated codes are actually required and the user has little understanding of the physics of the problem, little, if any, idea of the limitations of the code, and simply believes the results of the computations without question. This is the antithesis of the George Carrier approach and is encouraged in our engineering curricula when we compress courses in basic mechanics to make room for 'black box' approaches using commercial codes. It is not unusual for graduate students to use sophisticated commercial codes in their research without first taking advanced courses in the theory of elasticity, non-linear dynamics, continuum mechanics, fluid dynamics, stability theory, mathematical physics, numerical analysis, etc. The problem is exacerbated by code vendors who exaggerate the power of their products, or at least, fail to understand their limitations. Of course, this is a very complex problem. Students are expected to graduate with the ability to use modern engineering tools and it follows that room must be made in an already full Curriculum to accomplish this. If the length of an engineering program is not increased, something has to go. Unfortunately, it seems that the basics suffer in the process. At the same time, faculty typically warn students about the need to understand what is happening in the 'black box', but the practicality is that such full understanding is not always achieved. I am sure that many of us could tell stories of the sometimes disastrous effects produced.

As I prepared this presentation, I thought that a definition of *applied mechanics* might help me to make my points. My search was admittedly not exhaustive but nothing suitable was found. Thus, I have made up my own with all the potential criticism that could invite.

Applied Mechanics is a foundational discipline which studies physical phenomena, discovers the underlying physics, develops predictive models, and creates the mathematical and/or experimental tools required for their analysis.

It follows from this definition that applied mechanics is basic to most engineering disciplines and crosses many traditional discipline boundaries. The results of applied mechanics research touch nearly every physical aspect of our lives, making life easier, safer, more enjoyable, and in some cases, even possible. The benefits are all around us and are undeniable. The breadth of applied mechanics is amply demonstrated by the enormous variety of the research topics being reported at this 25th CANCAM.

So how has applied mechanics changed since CANCAM 1 at Laval? There have been many very positive developments. The knowledge generated in applied mechanics over the last 50 years has been truly astounding. The computational and experimental tools developed provide opportunities for studying, modelling, and analyzing physical phenomena which were beyond our imagination 50 years ago. Numerous engineering problems which could only be solved experimentally can now be solved numerically, thereby obsoleting a number of experimental methods and reducing design and development time substantially. Many routine engineering tasks can now be pretty much automated using computer codes, providing much improved efficiencies. There is no question that the broad field of applied mechanics is healthy and has impacted society in innumerable, very constructive ways. On the other hand, Departments of Mechanics have all but disappeared. All too often, some foundational courses in applied mechanics are no longer required or perhaps not even available, for graduate students, having given way to more highly focused and specialized ones concentrated on the use of computational tools. Conferences on applied mechanics have generally suffered from the tendency to organize conferences focused on sub-disciplines. An era of curiosity driven research with virtually no university-industry collaboration has given way to a time when it is difficult to fund research without having a fairly immediate industrial need. Thus, it appears as though applied mechanics has lost its 'public face', i.e., its broad recognition as an independent academic discipline, and its funding has largely become dependent on relatively short term return on investment.

What about the future? I do not see the issue of funding in applied mechanics being changed any time soon, certainly not in Canada as long as the present government is in power. The tendency towards increasing specialization and highly focused conferences seems inevitable and does have its definite advantages. However, the down-side is that the benefits of cross-fertilization may be lost, and the common phenomena and analysis tools are often rediscovered and re-invented. However, I see the greatest concern is the tendency to emphasize the use of sophisticated computer codes at the expense of foundational courses in applied mechanics. This can delude the practitioner into believing that the answers produced are correct, when there is inadequate understanding of the underlying physics and/or the limitations of the code. There are a couple of outstanding examples to illustrate this point. A number of small nuclear reactors which were designed and built for research in the days before digital computers and they have performed well for over half a century in some cases. An example is the AECL NRU which was producing the majority of the world's radio-pharmaceuticals until a few years ago. The intended replacement reactor was called the Maple Reactor and the project was cancelled after more than a decade of effort and more than a billion dollars of investment. The best computer codes available in the world failed to predict the positive power coefficient found during commissioning. Basically this means that an increase in core reactivity is associated with an increase in reactor power and continuous active control is required to maintain reactor stability. A similar story is found related to domestic nuclear power plants. The first generation steam generators were developed with relatively primitive computational tools compared to those currently available and performed mechanically very well for decades. Their replacements have been developed using much more powerful computer codes but some have tended to be more prone to early mechanical failures. The San Onofre Nuclear Generating Station (SONGS) was recently shut down permanently because the replacement steam generators failed, in one case, in less than 1 year of service. The structural dynamics and thermalhydraulics of modern nuclear steam generators are too complex to be modelled in detail by current codes and there is substantial dependence on empiricism and experience. A failure to understand completely the underlying physics and/or the limitations of the codes has had disastrous economic effects, happily without also creating a human safety issue. I hasten to add that decisions such as shutting down these reactors are complex and, in the end, involve many factors other than just the computer code deficiencies.

I will conclude by saying that the developments in applied mechanics over the last 50 years has been nothing short of spectacular, and that society has benefited enormously in the process. I am continually impressed with the seemingly intractable problems I saw as a student which can now be routinely solved. Automobiles can be designed and gotten it into production in less than 3 years. Modern aircraft can be built with a minimum of wind tunnel testing. Vehicles can be designed for crashworthiness, thereby improving human safety. The field of biomechanics has evolved with productive collaboration between medicine and engineering, to the great benefit in quality to human life. The list goes on and on. There is no doubt in my mind that applied mechanics will continue to prosper and to grow in importance as long as we do not lose sight of the foundations on which the discipline has been built.

THE GENERATION OF STRESS IN THE STORAGE PARTICLES OF LITHIUM-ION BATTERIES

Robert M. McMeeking
Tony Evans Professor of Structural Materials and Professor of Mechanical Engineering
University of California, Santa Barbara, U.S.A.
Sixth Century Professor of Engineering Materials, University of Aberdeen, Scotland
Email: rmcm@engineering.ucsb.edu

ABSTRACT

Models are developed for the transport of Li ions in the electrolyte of lithium ion batteries, their diffusion through storage electrode particles, and their kinetics through the surface of the particles between the electrolyte and the particles. As a consequence of the Li ion intercalating in the storage particles, their lattice swells, leading to elastic stress when the concentration of Li ions in the particles is not uniform. The models of transport are based on standard concepts for multi-component diffusion in liquids and solids, but are not restricted to dilute solutions, or to small changes in the concentration of the diffusing species. In addition, phase changes are permitted during mass transport as the concentration of lithium varies from the almost depleted state of the storage particle to one where the material is saturated with its ions. The elastic swelling and shrinkage may involve very large dilatations, which can be allowed for in the formulation of the model. Thus, the models can be suitable for storage particle, where the amount of Li can vary by large amounts depending on the state of charge, for staging as observed in the storage process in graphite, for the enormous swelling that takes place when silicon is used for storage, and for electrolytes in which the concentration of Li ions is high. The model is used to compute the processes of charging and discharging the battery to assess the parameters that influence the development of stress in the storage particles, and to deduce the likelihood of fracture of the storage particle material. The objective is to assess designs of porous electrode microstructures that permit rapid charging and discharging, but obviate the likelihood of fracture and other mechanical damage that limit the performance and reliability of the battery.

ASSESSING WIND SPEEDS IN TORNADOES FROM OBSERVED DAMAGE

Greg Kopp
Department of Civil and Environmental Engineering
The University of Western Ontario
London, Ontario N6A 5B9 Canada
gakopp@uwo.ca

ABSTRACT

Wind speeds in tornadoes are rarely measured directly. Rather, reported wind speeds are based on estimates from the observed damage. For example, if the roof structure blows-off of a house, the level of damage is categorized as EF-2 on the Enhanced-Fujita Scale, with wind speeds estimated to be in the range from 180-220 km/hr. This relationship between damage and wind speed has not been validated scientifically and is based primarily on expert opinion. It is of some importance to know the wind speeds associated with tornadoes, since critical infrastructure (such as nuclear power plants), is often designed for these events. In the talk, Prof. Kopp will discuss typical damage observations in Canada, full-scale and model-scale studies to determine structural capacities related to these observed failures, and wind tunnel studies to determine aerodynamic loads in tornadoes. Current knowledge gaps, and future research directions, will also be examined.

Proceedings of the 25th CANCAM
London, Ontario, Canada, May 31 – June 4, 2015

RECENT ADVANCES IN ADAPTIVE METASTRUCTURES FOR STRUCTURAL DYNAMICS ENHANCEMENT

Kon-Well Wang
Department of Mechanical Engineering
University of Michigan
Ann Arbor, MI, U.S.A.
kwwang@umich.edu

ABSTRACT

During the past couple of decades, due to the advances in materials, electronics, and system integration technologies, structural dynamics and controls researchers in various engineering disciplines have been investigating the feasibility of creating adaptive structures. The vision is to develop a multifunctional structural system that has various distributed and built-in autonomous abilities, such as vibration and stability controls, shape configuration and morphing, materials and mechanical property variations, energy harvesting and distribution, and health monitoring and self-healing. From a structural system point of view, one of the major challenges is on how to best synthesize the cross-field and local-global coupling characteristics of the various adaptive materials and elements to optimize the overall structure performance. In recent years, interesting approaches have been explored to achieve adaptive metastructures based on synergistic modular architectures, often observed in biological and atomistic systems or in artificial meta-materials. It is recognized that to achieve significant new advances in adaptive structural systems, researchers have to conduct even more cross talks with various disciplines. This presentation will review and discuss some of the recent interdisciplinary research efforts in synthesizing adaptive metastructures for structural dynamics enhancement.

Proceedings of the 25th CANCAM
London, Ontario, Canada, May 31 – June 4, 2015

VORTICES

Stavros Tavoularis
Department of Mechanical Engineering
University of Ottawa
Ottawa, Ontario K1N 6N5 Canada
tav@genie.uottawa.ca

ABSTRACT

Much of our insight into the structure and evolution of fluid flows has been acquired by consideration of vorticity, whose variation obeys universal physical laws and the corresponding mathematical equations. Essential elements of most fluid flows are isolated vortices, namely fluid structures that contain concentrated vorticity. It is well-known that vortex-induced motions and vortex interactions have profound effects on mass, heat and momentum transport as well as on forces on immersed solid walls. Following a brief review of some basic theoretical concepts concerning vorticity and vortices, this presentation will outline some findings of several experimental and analytical studies that have been conducted over the years in the Fluid Mechanics Laboratory of the University of Ottawa and demonstrate the significance vortices have in fundamental flow phenomena and in diverse environmental, industrial, aerospace and biological applications. Particular aspects of vortical flows that will be discussed include the following: the generation of horseshoe-shaped vortices as the dominant coherent structures in a turbulent shear flow, and their contribution to scalar transport; the triggering and evolution of "gap vortex streets" in flows through annular channels and the formation of "vortex networks" in nuclear reactor fuel channels ("rod bundles"); the transport of gas through sink-vortices in stratified gas-liquid flows; the properties and impact of wing-tip and LEX vortices; the succession of multi-type vortical structures in smokestack plumes subjected to cross-flows; and the characteristics of vortices in pulsatile blood flows in the left ventricle of the heart and past a mechanical aortic valve.

DESIGN AND MICRO/NANO-ENGINEERING OF FUNCTIONAL MATERIALS, SURFACES AND DEVICES: FROM FUNDAMENTALS TO APPLICATIONS

Jun Yang
Department of Mechanical and Materials Engineering
The University of Western Ontario
London, Ontario N6A 5B9 Canada
Email: jyang@eng.uwo.ca

ABSTRACT

Functions of materials, surfaces and devices are related to not only the chemical substances but also the structures. Study of structure-function-property relationships is always a central issue of mechanics. Recent advances in micro/nano-engineering technology enable researchers to design and fabricate materials, surfaces and devices with precisely controlled structures. These micro/nanostructured materials, surfaces and devices can achieve superior and even unprecedented properties and/or functions. For instance, artificially structured metamaterials present negative refractive index, mass density and dynamic modulus under certain circumstances; hierarchical micro/nanostructured surfaces appear as superoleophobicity which does not exist in nature. Nowadays these micro/nanostructured materials have been extensively used for energy, environmental, information technology (IT) and biomedical applications. In addition to applied research, micro/nano-engineering also acts as powerful approaches for fundamental studies. This presentation will review and discuss recent advances in micro/nano-engineering techniques and their applications in making functional materials, surfaces and devices.

Actuation and Smart Materials

DYNAMIC ANALYSES OF A VISCOELASTIC DIELECTRIC ELASTOMER OSCILLATOR

Jianyou Zhou, Liying Jiang*, and Roger E. Khayat
Department of Mechanical and Materials Engineering
The University of Western Ontario
London, Ontario N6A 5B9, Canada
jzhou252@uwo.ca, lyjiang@eng.uwo.ca*, and rkhayat@eng.uwo.ca

ABSTRACT

With tunable resonant frequency, DE-based oscillators have shown promise for a potential alternative to the traditional silicon-based devices in MEMs. The resonant and dynamic behavior of a DE oscillator is strongly influenced by the viscoelasticity of the elastomers. Based on the finite-deformation viscoelasticity model for dielectrics, this work presents dynamic analyses on a DE membrane oscillator with the consideration of the viscoelastic effect. Simulation results show a significant difference of the tuned frequency between a purely elastic DE oscillator and a viscoelastic DE one. This work can be claimed as a useful guideline for optimal design of viscoelastic DE-based oscillators.

KEYWORDS: Dielectric elastomers, oscillation, resonant frequency, viscoelasticity

INTRODUCTION

As a category of electroative polymers that possess large-deformation capability, dielectric elastomers (DEs) have been extensively explored and showed potential for a variety of applications such as soft robots, actuators, resonators, energy harvesters, MEMS devices and artificial muscles.[1]

Early studies on the DEs mainly focused on their electromechanical coupling behavior under the circumstance of quasi-static deformation. However, as DEs have been developed as resonators and oscillators in the recent years,[2] investigating the dynamic behavior of DEs has become very essential and useful to these particular applications. For a DE resonator or oscillator, oscillation can be generated when an alternating voltage is applied and its resonant frequency can be actively tuned by applying a static electric voltage to the DE. This property makes DE-based resonators and oscillators a potential alternative to the traditional silicon-based devices in MEMs since the resonant frequency of the silicon-based devices is basically fixed after fabrication.

Nevertheless, a recent study has proven that the dynamic performance of the DEs is strongly affected by their material viscoelasticity.[3] In fact, much effort has been devoted to achieving a better understanding on the viscoelastic behavior of the DEs, but until recently, a model that is capable of accounting for both the finite inelastic deformation and the electromechanical coupling of the DEs was proposed by Hong.[3] Based on the coupled field theory for elastic dielectrics by Suo et. al.[4] and the finite-deformation viscoelasticity theory by Reese and Govindjee,[5] Hong's model[3] claims to be able to adopt most finite-deformation constitutive relations and evolution laws of viscoelastic solids.

By adopting Hong's model[3] to investigate the viscoelastic effect of the DEs, this work aims to explain the complex interplay of the material viscoelasticity, the failure modes and the dynamic behavior of viscoelastic DE oscillators.

FORMULATION

The schematic of a DE oscillator with a commonly used configuration (i.e., the membrane configuration[6-7]) is shown in Fig. 1. Fig. 1(a) shows the undeformed state of a DE membrane coated with two compliant electrodes on its top and bottom surfaces. The dimensions of the undeformed DE membrane are L_1, L_2 and L_3, respectively. Then the undeformed membrane is pre-stretched in 1- and 2-direction,

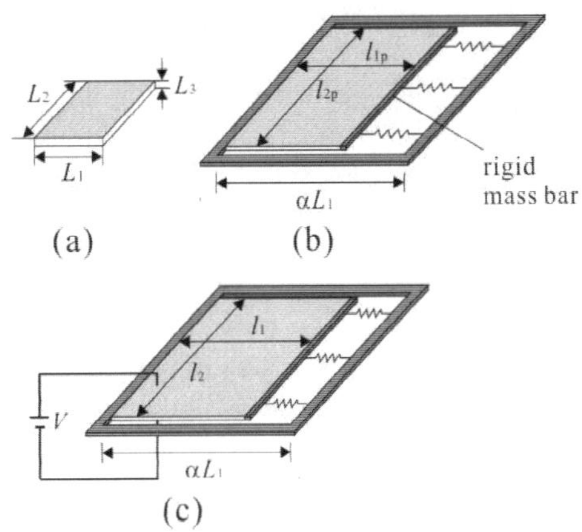

Figure 1: Schematics of a DE membrane oscillator: (a) undeformed state; (b) pre-stretch state; (c) current state

and bonded to a rigid frame and a rigid bar with its two edges along 2-direction. The rigid bar with a mass of m is then further connected to the other side of frame by the springs with an effective stiffness of k (Fig. 1(b)). The original length of each spring is βL_1 and the length of the rigid frame in 1-

direction is αL_1. In the pre-stretch state shown in Fig. 1(b), the dimensions of the membrane are l_{1p}, l_{2p} and l_{3p}, which gives the pre-stretch ratios as $\lambda_{1p}=l_{1p}/L_1$, $\lambda_{2p}=l_{2p}/L_2$ and $\lambda_{3p}=l_{3p}/L_3$, respectively. An electric voltage V is then applied to the electrodes of the membrane, which forces the membrane to deform to the current state with the dimensions of l_1, l_2 and l_3 (Fig. 1(c)). Bonded to the rigid frame and bar, the membrane is subjected to tensile forces P_1 and P_2 in 1- and 2-direction. Similarly, the rigid bar is under the tensile forces from the membrane ($-P_1$) and the springs (P_s). The deformation gradient of the elastomer in the current state is expressed as,

$$\mathbf{F}=\begin{pmatrix}\lambda_1 & 0 & 0\\ 0 & \lambda_2 & 0\\ 0 & 0 & \lambda_3\end{pmatrix}.$$

Following the work of Hong[3], the total deformation gradient can be further multiplicatively decomposed into an elastic part and an inelastic part, i.e., $\mathbf{F}=\mathbf{F}^e\mathbf{F}^i$, where

$$\mathbf{F}^e=\begin{pmatrix}\lambda_1^{\mathrm{e}} & 0 & 0\\ 0 & \lambda_2^{\mathrm{e}} & 0\\ 0 & 0 & \lambda_3^{\mathrm{e}}\end{pmatrix} \text{ and } \mathbf{F}^i=\begin{pmatrix}\lambda_1^{\mathrm{i}} & 0 & 0\\ 0 & \lambda_2^{\mathrm{i}} & 0\\ 0 & 0 & \lambda_3^{\mathrm{i}}\end{pmatrix},$$

and the superscripts "e" and "i" signify elastic and inelastic, respectively. Also, as commonly treated in the literature, the DE is assumed to be incompressible in this work, which gives that $\lambda_3=1/\lambda_1\lambda_2$, $\lambda_3^{\mathrm{e}}=1/\lambda_1^{\mathrm{e}}\lambda_2^{\mathrm{e}}$ and $\lambda_3^{\mathrm{i}}=1/\lambda_1^{\mathrm{i}}\lambda_2^{\mathrm{i}}$.

When a perturbation is applied to the membrane in the current state, the variation of the total Helmholtz free energy W of the DE induced by the small changes of the total stretch ratios (namely, $\delta\lambda_1$ and $\delta\lambda_2$), is equal to the work done by the tensile forces (P_1 and P_2) and the inertia force,[6-7] i.e.,

$$L_1L_2L_3\left(\frac{\partial W}{\partial\lambda_1}\delta\lambda_1+\frac{\partial W}{\partial\lambda_2}\delta\lambda_2\right)=P_1L_1\delta\lambda_1+P_2L_2\delta\lambda_2-L_2L_3\frac{L_1^3}{3}\rho\frac{d^2\lambda_1}{dt^2}\delta\lambda_1, \quad (1)$$

where ρ is the mass density of the DE membrane. Note that the work done by the inertia force in the 3-direction is neglected in Eq. (1) since $L_3 \ll L_1$.

Moreover, according to the work of Reese and Govindjee[5] and Hong[3], the total Helmholtz free energy can also be split into two parts: a non-equilibrium part W^{NEQ} and an equilibrium part W^{EQ}, which results in

$$W=W^{\mathrm{EQ}}(\lambda_1,\lambda_2,V)+W^{\mathrm{NEQ}}\left(\lambda_1^{\mathrm{e}},\lambda_2^{\mathrm{e}}\right). \quad (2)$$

Here, W^{NEQ} is a function of the elastic deformation, while W^{EQ} is related to both the total deformation and the applied voltage V. Following Huang and Suo,[8] W^{EQ} is in the form

$$W^{\mathrm{EQ}}=W_s(\lambda_1,\lambda_2)+\frac{\varepsilon\varepsilon_0}{2}\left(\frac{V}{L_3}\right)^2\lambda_1^2\lambda_2^2, \quad (3)$$

where $W_s(\lambda_1,\lambda_2)$ is the strain energy density function, ε_0 is the permittivity of the vacuum and ε is the relative dielectric constant. Since the non-equilibrium Helmholtz free energy is only related to the elastic deformation, we can assume that W^{NEQ} also takes the same form as the strain energy density function. For large deformation of the DE, the hyperelastic Gent model[9] is adopted here, which gives,

$$W_s=-\frac{G^{\mathrm{EQ}}J_{\mathrm{lim}}^{\mathrm{EQ}}}{2}\ln\left(1-\frac{\lambda_1^2+\lambda_2^2+\lambda_1^{-2}\lambda_2^{-2}-3}{J_{\mathrm{lim}}^{\mathrm{EQ}}}\right), \quad (4)$$

$$W^{\mathrm{NEQ}}=-\frac{G^{\mathrm{NEQ}}J_{\mathrm{lim}}^{\mathrm{NEQ}}}{2}\ln\left[1-\frac{\left(\lambda_1^{\mathrm{e}}\right)^2+\left(\lambda_2^{\mathrm{e}}\right)^2+\left(\lambda_1^{\mathrm{e}}\lambda_2^{\mathrm{e}}\right)^{-2}-3}{J_{\mathrm{lim}}^{\mathrm{NEQ}}}\right], \quad (5)$$

where G^{EQ} and G^{NEQ} are the equilibrium and non-equilibrium shear modulus, $J_{\mathrm{lim}}^{\mathrm{EQ}}$ is material constant determined by the total stretching limit and $J_{\mathrm{lim}}^{\mathrm{NEQ}}$ is determined by the stretching limit of the elastic part.

Combining Eqs. (1), (2), (3), (4) and (5), we can obtain

$$\frac{P_1}{GL_2L_3}=\frac{\chi J_{\mathrm{lim}}^{\mathrm{EQ}}\left(\lambda_1-\lambda_1^{-3}\lambda_2^{-2}\right)}{J_{\mathrm{lim}}^{\mathrm{EQ}}-\lambda_1^2-\lambda_2^2-\lambda_1^{-2}\lambda_2^{-2}+3}+\frac{L_1^2}{3G}\rho\frac{d^2\lambda_1}{dt^2}-\frac{\varepsilon\varepsilon_0}{G}\left(\frac{V}{L_3}\right)^2\lambda_1\lambda_2^2 +\frac{(1-\chi)J_{\mathrm{lim}}^{\mathrm{NEQ}}\left[\lambda_1\left(\lambda_1^{\mathrm{i}}\right)^{-2}-\lambda_1^{-3}\lambda_2^{-2}\left(\lambda_1^{\mathrm{i}}\right)^2\left(\lambda_2^{\mathrm{i}}\right)^2\right]}{J_{\mathrm{lim}}^{\mathrm{NEQ}}-\left(\frac{\lambda_1}{\lambda_1^{\mathrm{i}}}\right)^2-\left(\frac{\lambda_2}{\lambda_2^{\mathrm{i}}}\right)^2-\left(\frac{\lambda_1\lambda_2}{\lambda_1^{\mathrm{i}}\lambda_2^{\mathrm{i}}}\right)^{-2}+3}, \quad (6)$$

$$\frac{P_2}{GL_1L_3}=\frac{\chi J_{\mathrm{lim}}^{\mathrm{EQ}}\left(\lambda_2-\lambda_2^{-3}\lambda_1^{-2}\right)}{J_{\mathrm{lim}}^{\mathrm{EQ}}-\lambda_1^2-\lambda_2^2-\lambda_1^{-2}\lambda_2^{-2}+3}-\frac{\varepsilon\varepsilon_0}{G}\left(\frac{V}{L_3}\right)^2\lambda_2\lambda_1^2 +\frac{(1-\chi)J_{\mathrm{lim}}^{\mathrm{NEQ}}\left[\lambda_2\left(\lambda_2^{\mathrm{i}}\right)^{-2}-\lambda_2^{-3}\lambda_1^{-2}\left(\lambda_1^{\mathrm{i}}\right)^2\left(\lambda_2^{\mathrm{i}}\right)^2\right]}{J_{\mathrm{lim}}^{\mathrm{NEQ}}-\left(\frac{\lambda_1}{\lambda_1^{\mathrm{i}}}\right)^2-\left(\frac{\lambda_2}{\lambda_2^{\mathrm{i}}}\right)^2-\left(\frac{\lambda_1\lambda_2}{\lambda_1^{\mathrm{i}}\lambda_2^{\mathrm{i}}}\right)^{-2}+3}, \quad (7)$$

where $G=G^{\mathrm{EQ}}+G^{\mathrm{NEQ}}$, and $\chi=G^{\mathrm{EQ}}/G$ indicates the fraction of the time-independent polymer networks.

When the oscillator is modeled with the oscillation of the rigid bar, its equation of motion is expressed as

$$\frac{P_1}{GL_2L_3}-\frac{P_s}{GL_2L_3}+\frac{mL_1}{GL_2L_3}\frac{d^2\lambda_1}{dt^2}=0, \quad (8)$$

where $P_s=k(\alpha-\lambda_1-\beta)L_1$. Particularly, when the oscillator is static, the inertia terms in Eqs. (6) and (8) vanish.

It can be noticed from Eq. (6) that the inelastic stretch ratios are involved in Eq. (8). According to the work of Reese and Govindjee[5], these time-dependent inelastic stretch ratios satisfy the following evolution equation, i.e.,

$$-\frac{1}{2}\mathbf{F}\frac{d\left[(\mathbf{C}^i)^{-1}\right]}{dt}\mathbf{F}^{\mathrm{T}}\cdot\left(\mathbf{b}^{\mathrm{e}}\right)^{-1}=\gamma^{-1}:\boldsymbol{\sigma}^{\mathrm{NEQ}}, \quad (9)$$

where, $\mathbf{C}^{\mathrm{i}}=\left(\mathbf{F}^{\mathrm{i}}\right)^{\mathrm{T}}\mathbf{F}^{\mathrm{i}}$, $\mathbf{b}^{\mathrm{e}}=\mathbf{F}^{\mathrm{e}}\left(\mathbf{F}^{\mathrm{e}}\right)^{\mathrm{T}}$, $\boldsymbol{\sigma}^{\mathrm{NEQ}}=2\mathbf{F}^{\mathrm{e}}\frac{\partial W^{\mathrm{NEQ}}}{\partial\mathbf{C}^{\mathrm{e}}}\left(\mathbf{F}^{\mathrm{e}}\right)^{\mathrm{T}}$, $\mathbf{C}^{\mathrm{e}}=\left(\mathbf{F}^{\mathrm{e}}\right)^{\mathrm{T}}\mathbf{F}^{\mathrm{e}}$ and γ^{-1} is an isotropic rank-four mobility tensor that takes the form

$$\gamma^{-1} = \frac{1}{2\eta}\left(\mathbf{I}^4 - \frac{1}{3}\mathbf{I}\otimes\mathbf{I}\right), \quad (10)$$

where η is the shear viscosity, $\mathbf{I}^4$ is the fourth order symmetric identity tensor and $\mathbf{I}$ is the second order identity tensor. Therefore, $d\lambda_1^i / dt$ and $d\lambda_2^i / dt$ can be obtained by expanding Eq. (9).

It should also be mentioned that there are two typical failure modes to be avoided during the operation of the oscillator. First, to prevent the membrane from loss-of-tension, we need to ensure that $P_1 > 0$, $P_2 > 0$ and $P_s > 0$. Also, the applied voltage should be kept below the breakdown voltage of the DE, i.e., the applied voltage satisfies

$$\frac{V}{L_3}\sqrt{\frac{\varepsilon\varepsilon_0}{G}} < d\lambda_1^{-1}\lambda_2^{-1}, \quad (11)$$

where $d = E_{EB}\sqrt{\varepsilon\varepsilon_0 / G}$ and E_{EB} is the dielectric strength of the DE. For the simulation in the current work, it is ensured that those failure modes are avoided.

RESONANT BEHAVIOR OF THE DE OSCILLATOR

Noting that $\lambda_2 = \lambda_{2p}$, and substituting Eq. (6) into Eq. (8), we can obtain[7]

$$\frac{d^2\lambda_1}{dt^2} + g\left(\lambda_1, \lambda_1^i, \lambda_2^i, V^*\right) = 0 . \quad (12)$$

where $V^* = V\sqrt{\varepsilon\varepsilon_0 / G} / L_3$.

When the DE membrane is subjected to a small perturbation $\Delta(t)$ in 1-direction at time t, $\lambda_1(t) = \lambda_1' + \Delta(t)$, where λ_1' is the stretch ratio in 1-direction before the perturbation. By expanding function g into Taylor series up to the first order, Eq. (12) can be re-written as

$$\frac{d^2\Delta}{dt^2} + \frac{\partial g\left(\lambda_1', \lambda_1^i, \lambda_2^i, V^*\right)}{\partial \lambda_1}\Delta = 0 . \quad (13)$$

Therefore, the resonant frequency of the DE oscillator is determined as

$$\omega_n^2 = \frac{\partial g\left(\lambda_1', \lambda_1^i, \lambda_2^i, V^*\right)}{\partial \lambda_1} . \quad (14)$$

As mentioned above, the resonant frequency of a DE oscillator can be tuned by applying an electric voltage. Fig. 2 shows the resonant frequency tuning process (the change of the dimensionless resonant frequency $\Omega^* = \omega_n / \sqrt{\varphi}$ with time) of the DE oscillator. At time t =0s, the oscillator is fully relaxed in the pre-stretch state with λ_{1p}=1.5 and λ_{2p}=3. At time t =10s, an electric voltage is applied to the membrane at a loading rate of $r = dV^* / dt$ until the dimensionless voltage V^* reaches a pre-determined value of V_p= 0.15, which will be maintained afterwards. As shown in this figure, the resonant frequency drops when the voltage increases, and further decreases to a steady value after the voltage level is maintained. After the voltage is applied at 10s, the membrane deforms and starts to relax again until it reaches an equilibrium state where the inelastic stretch ratios equal to the total stretch ratios and the frequency becomes steady. Moreover, when the pre-determined voltage level V_p is fixed at 0.15, the resonant frequency decreases to the same steady value at the end of the tuning process regardless of the voltage loading rate r.

To demonstrate the significance of accounting for the material viscoelasticity of DE oscillators, a comparison of the frequency tuning between a purely elastic DE oscillator and a viscoelastic oscillator is presented in Fig. 3. It can be seen that the tuned resonant frequency of the purely elastic DE oscillator is much higher than that of the viscoelastic oscillator. Therefore, neglecting the viscoelastic effect of DE oscillators may lead to significant error in practical applications.

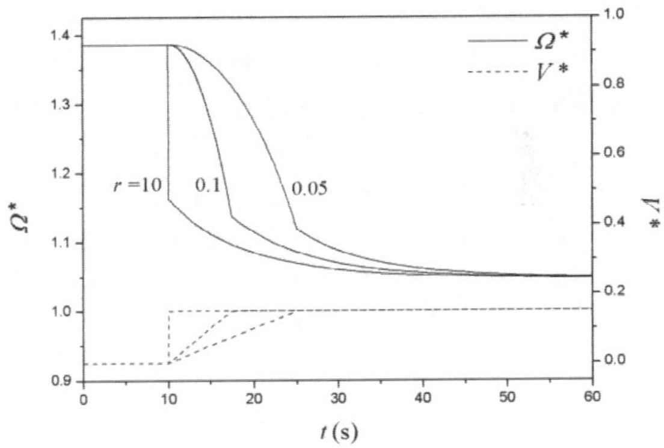

Figure 2: Frequency tuning process of the viscoelastic DE membrane oscillator

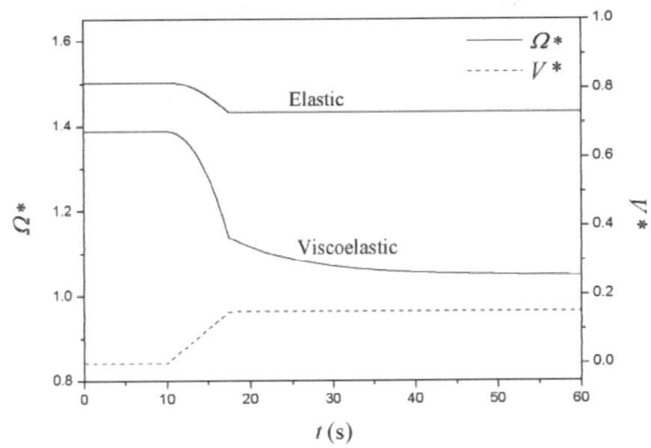

Figure 3: Comparison of the frequency tuning between a purely elastic DE oscillator and a viscoelastic DE oscillator

PARAMETRIC EXCITATION OF THE OSCILLATOR

To explore the dynamic behavior of the oscillator, the applied voltage V^* is now separated into two parts and takes the form

$$V^* = V_D^* + V_A^* \sin(\omega t), \tag{15}$$

where V_D^* is a static voltage, V_A^* is the amplitude of the periodic voltage and ω is the frequency of the excitation. According to Eqs (12) and (15), Fig. 4 depicts the variation of the total stretch ratio in 1-direction as a function of t when the excitation is applied with $\omega / \sqrt{\varphi} = 0.8$, $V_A^* = 0.1 V_D^*$ and $V_D^* = 0.15$. The irregular transient response to the excitation is mainly due to the material viscoelasticity. As the excitation continues, the membrane relaxes to an equilibrium state and a steady response is achieved after a short period.

In addition, to verify the resonant behavior of the DE oscillator in the Fig. 2, we plot the steady-state amplitude of the oscillation as a function of the dimensionless frequency of excitation in Fig. 5. It can be seen that the peak of the amplitude is at about $\omega / \sqrt{\varphi} = 1.05$, which is in agreement with the tuned resonant frequency in Fig. 2.

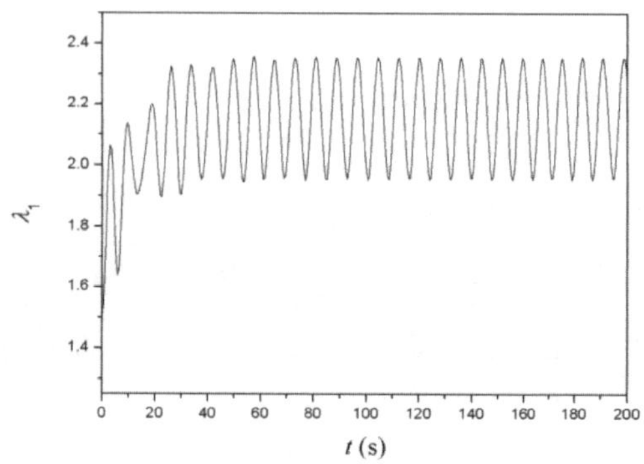

Figure 4: Excitation response of the viscoelastic DE membrane oscillator

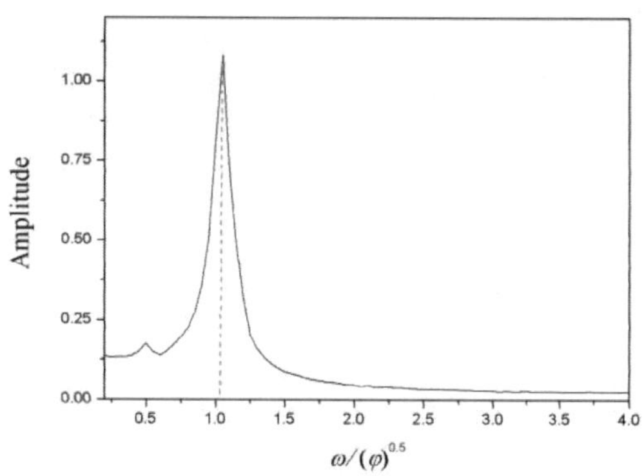

Figure 5: Amplitude of the oscillation as a function of the dimensionless frequency

CONCLUSIONS

By modeling the in-plane oscillation of a DE membrane oscillator, this work demonstrates the frequency tuning and the dynamic behavior of the oscillator with the consideration of the material viscoelasticity. It is found that the resonant frequency is insensitive to the loading rate of the applied voltage when the pre-determined voltage level is fixed. Also, by comparing the frequency tuning between an elastic and a viscoelastic DE oscillator, it is again proven that the material viscoelasticity exerts a strong effect on the dynamic performance of DE oscillators. The current work is expected to provide a better understanding on the dynamic characteristic of DE oscillators.

ACKNOWLEDGMENTS

This work is supported by Natural Sciences and Engineering Research Council of Canada (NSERC).

REFERENCES

1. Carpi, F., Rossi, D. D., Kornbluh, R., Pelrine, R., and Sommer-Larsen, P. 2008, *Dielectric Elastomers as Electromechanical Transducers: Fundamentals, Materials, Devices, Models and Applications of an Emerging Electroactive Polymer Technology*, Elsevier, Amsterdam.

2. O'Brien, B. M., Rosset, S., Shea, H. R., and Anderson, I. A. 2012, "Cutting the fat: artificial muscle oscillators for lighter, cheaper, and slimmer devices", *Proc. SPIE*, 8340, pp. 834008.

3. Hong, W. 2011, "Modeling viscoelastic dielectrics", *Journal of the Mechanics and Physics of Solids*, 59, pp. 637-650.

4. Suo, Z., Zhao, X., and Greene, W. H. 2008, "A nonlinear field theory of deformable dielectrics", *Journal of the Mechanics and Physics of Solids*, 56, pp. 467-486.

5. Reese, S., and Govindjee, S. 1998, "A theory of finite viscoelasticity and numerical aspects", *International Journal of Solids and Structures*, 35, pp. 3455-3482.

6. Li, T., Qu, S., and Yang, W. 2012, "Electromechanical and dynamic analyses of tunable dielectric elastomer resonator", *International Journal of Solids and Structures*, 49, pp. 3754-3761.

7. Zhou, J., Jiang, L., and Khayat, R. E. 2014, "Viscoelastic effects on frequency tuning of a dielectric elastomer membrane resonator", *Journal of Applied Physics*, 115, 124106.

8. Huang, R., and Suo, Z. 2012, "Electromenchanical phase transition in dielectric elastomers", *Proceedings of the Royal Society A*, 468, pp. 1014-1040.

9. Gent, A. N. 1996, "A New Constitutive relation for rubber", *Rubber Chemistry and Technology*, 69, pp. 59-61.

Dynamic analysis of a vibration shaker based on a dielectric elastomer actuator

Ujjaval Gupta, Hareesh Godaba, *Jian Zhu
Department of Mechanical Engineering,
National University of Singapore
9 Engineering Drive 1, Singapore
***mpezhuj@nus.edu.sg**

ABSTRACT

Dielectric elastomers are thin membranes made of soft active materials, capable of undergoing fast, reversible deformation when subject to voltage. In this paper we make a combined study of theory and experiments to study the dynamic response of a vibration shaker based on a dielectric elastomer actuator. The clamped configuration of dielectric elastomer made of natural rubber (Oppo band, Singapore) is prepared with prestretches. Subject to DC voltage, the dielectric elastomer deforms and the vibration shaker moves. Subject to AC voltage, the dielectric elastomer and the vibration shaker oscillate. The oscillation of the vibration shaker is measured experimentally. The theoretical model is developed, and the calculations are qualitatively consistent with the experimental data.

KEYWORDS: dielectric elastomer, dynamic analysis, haptic feedback

INTRODUCTION

A dielectric elastomer actuator is one class of soft active materials, and can deform in response to voltage. A thin layer of dielectric elastomer material, typically with a thickness ~ 100 μm, is sandwiched between compliant electrodes on both sides [1-2]. When the material is subject to voltage, positive charges accumulate on one electrode, while negative charges accumulate on the other electrode. Due to the mutual attraction of the opposite charges the elastomer experiences voltage-induced stress, known as the Maxwell stress, which squeezes the elastomer in thickness and expands it in area. Subject to DC voltage, the dielectric elastomer actuator can exhibit large voltage-induced deformation. When coupled with water, the actuator can achieve an area strain of 1165% [3]. When coupled with air, it can achieve an area strain of 1692% [4]. Dielectric elastomers are similar to natural muscles in the aspects of large deformation and high energy density, and are commonly referred to as "Artificial Muscles" [5].

When a dielectric elastomer is subject to AC voltage, it will oscillate with a frequency ranging from several Hz to several kHz. Due to large deformation and nonlinear electromechanical coupling, the vibration of dielectric elastomers is strongly nonlinear [6-7]. Dynamic applications of dielectric elastomers include soft pumps, soft speakers, soft motors, etc [8]. The first commercial application of dielectric elastomer actuators is a vibration shaker developed by Vivitouch [9], which can be attached to a gaming headset/console to provide haptic feedback to the user. In this application it is pertinent to tune the output displacement/force of the device which has a direct effect on the user's tactile sensation [9-10].

This paper presents dynamic analysis of a vibration shaker based on a dielectric elastomer actuator. The oscillation of the vibration shaker is measured experimentally. The theoretical model is developed to interpret the experiments.

EXPERIMENTAL SETUP

Experiments are conducted to measure the output displacement of the vibrating shaker. The setup is shown in Fig. 1. The voltage signal is programmed through Labview (BNC-2120 connector block, NI) and amplified using a high voltage amplifier (10/40A, Trek). Upon application of AC voltage, the dielectric elastomer membrane oscillates, producing displacement in the attached vibration shaker whose displacement is measured using a laser displacement sensor (ILD1700, MicroEpsilon).

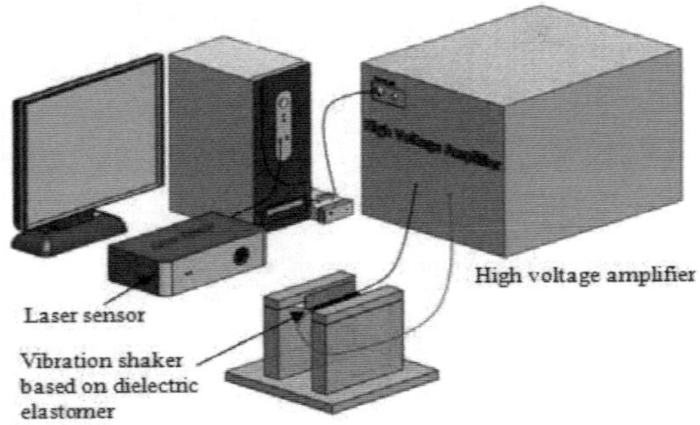

Fig. 1. Schematic of the experimental setup

Given the low viscoelastic effects combined with high fracture toughness and excellent fatigue strength of the natural rubber, Oppo Band, Singapore with a thickness (H = 0.48 mm) is used as a dielectric elastomer [11]. The vibration shaker based on a dielectric elastomer actuator is fabricated as follows. A piece of natural rubber (Oppo band, Singapore) is cut in dimensions $L_1 \times L_2 \times H$ as shown in the reference state (Fig. 2a). The membrane is in relaxed state without any loads applied. It is then prestretched mechanically in two in-plane directions (1 & 2) and the two ends are clamped as shown in Fig. 2b. Unless otherwise stated, the experiments are conducted on a membrane with fixed prestretches $\lambda_{1p} = 2$ and $\lambda_{2p} = 2$. An inertia mass is attached in the middle dividing the membrane into two parts *A* and *B*. The function of the mass is to create vibration upon application of voltage. The mass of the membrane is 0.375g. The vibration shaker is made of an acrylic plate of dimensions 91mm×4mm×21mm with a mass of 6.4g. The vibration shaker produces haptic feedback which can be felt by the user. In the undeformed state, parts *A* and *B* have equal length L_2, but widths L_{1A} and L_{1B}, respectively. The clamps are separated by a constant distance, $d = \lambda_{1p}(L_{1A} + L_{1B}) = 40mm$. When voltage is off (Fig. 2b), part *A* is smeared with carbon grease acting as electrodes while electrodes are absent in part *B*. Parts *A* and *B* have same length $\lambda_{2p}L_2$ = 100mm, and widths $\lambda_{1p}L_{1A}$ and $\lambda_{1p}L_{1B}$, respectively. When voltage is on (Fig. 2c), the top and bottom electrodes in part *A* are connected to a voltage source. Upon application of AC voltage, part *A* oscillates along direction 1, causing the mass to vibrate outputting force/displacement. In Fig. 2c, parts A and B have equal length $\lambda_{2p}L_2$, but have widths $\lambda_{1A}L_{1A}$ and $\lambda_{1B}L_{1B}$, respectively.

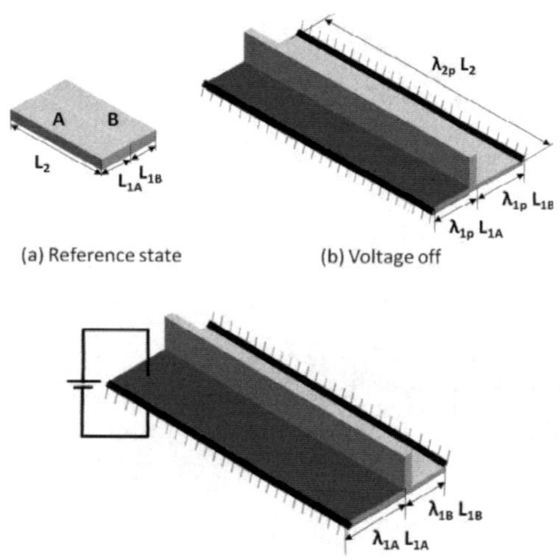

Fig. 2. Schematic of vibration shaker based on dielectric elastomer actuator

EXPERIMENTAL RESULTS

Figure 3a shows the displacement recorded by the laser displacement sensor. The membrane is subject to AC voltage with a constant amplitude ($V_{ac} = 7kV$) but a ramping frequency. Figure 3b plots the amplitude of oscillation as a function of the frequency of AC voltage. The amplitude of oscillation is defined as the difference between the maximum and minimum amplitude values for a given frequency. As shown in Fig. 3b, the curve reaches peak when the harmonic resonance is excited. When $V_{ac} = 7kV$, the largest amplitude of oscillation is about $2000\mu m$. The dielectric elastomer made of Oppo band, Singapore seems excellent in both actuation and energy harvesting due to its low viscoelasticity [11].

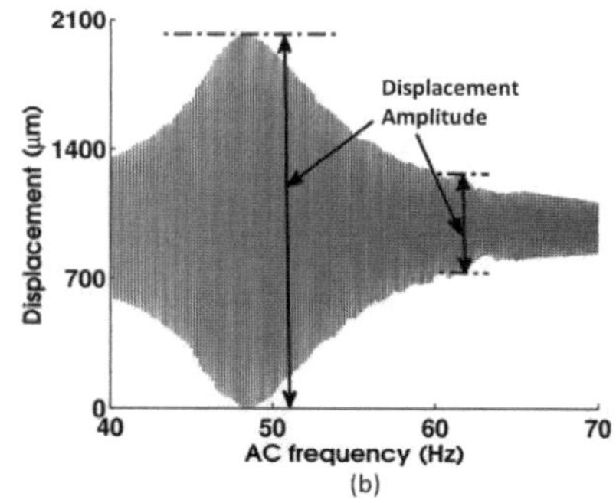

(a)

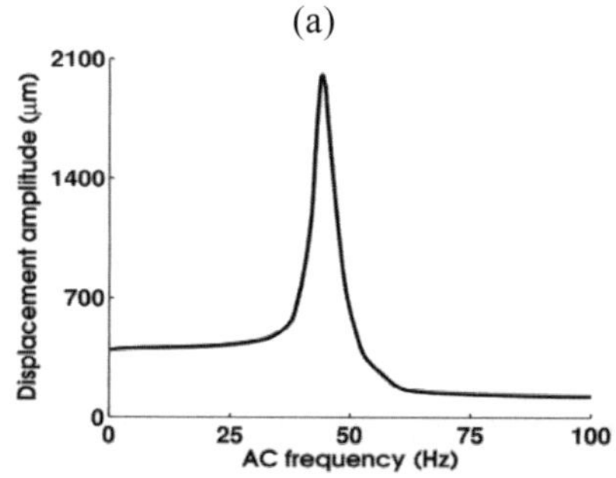

(b)

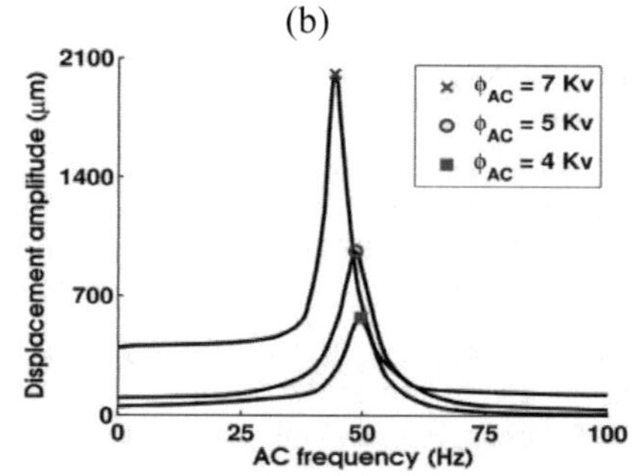

(c)

Fig. 3. (a) Displacement recorded by the laser sensor at $V_{ac} = 7kV$. (b) Amplitude of oscillation as a function of frequency of AC voltage at $V_{ac} = 7kV$. (c) Amplitudes of oscillation at different amplitudes of AC voltage.

Figure 3c shows the amplitudes of oscillation at several amplitudes of AC voltage. As we can see, as the amplitude of AC voltage increases, the resonant frequency of the system decreases. This can be understood as follows. When a dielectric elastomer membrane is subject to AC voltage $\Phi = V_{ac}\sin(\omega t)$, the voltage-induced Maxwell stress is given by $\varepsilon(\Phi/h)^2$, where h is the thickness of the membrane. The time-varying Maxwell stress consists of two terms, one is a constant term $\frac{1}{2}\varepsilon(V_{ac}/h)^2$ and the other is an alternating term $\frac{1}{2}\varepsilon(V_{ac}/h)^2\cos(2\omega t)$. The constant term has an effect similar to that of DC voltage [7, 12], which can reduce the tension in the membrane and thus reduce the natural freuency of the system. The alternating term is responsible for the excited vibration of the elastomeric membrane. When the excited frequency (2ω) is close to the natural frequency of the system, the harmonic resonance is excited.

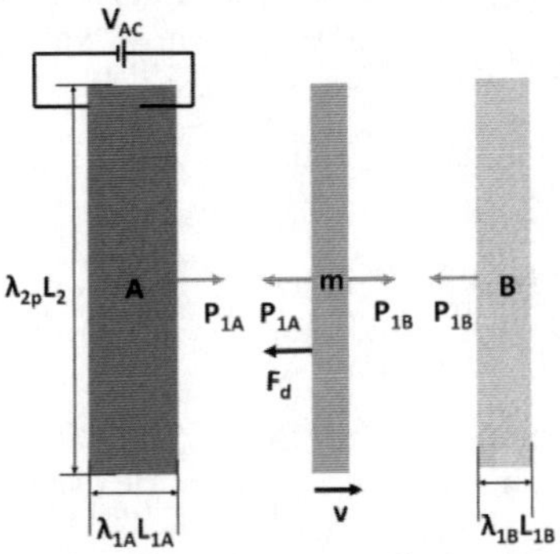

Fig. 4. Schematic of the forces applied to part *A*, part *B* and mass *m*

THEORETICAL MODELLING

We develop a theoretical model to study the dyanmic behaviour of the vibration shaker. Figure 4 illustrates the forces applied to part *A*, part *B* and the mass *m* along direction 1. In addition, a linear damping effect is taken into account [10], which assumes that the damping force applied to the vibration shaker is proportional to its velocity, given by $F_d = cL_{1A}\frac{d\lambda_{1A}}{dt}$, where *c* is the damping coefficient. Due to the low viscoelasticity of Oppo Band [11], viscoelastic effects have been ignored. For the vibration shaker *m*, the equation of motion is given by,

$$mL_{1A}\frac{d^2\lambda_{1A}}{dt^2} = P_{1B} - P_{1A} - cL_{1A}\frac{d\lambda_{1A}}{dt} \tag{1}$$

where λ_{1A} is the stretch of part *A* at the actuated state, P_{1A} and P_{1B} are the forces applied by parts *A* and *B*, respectively. Since the mass of the membrane (0.375g) is much smaller as compared to that of the vibration shaker (6.4g), inertia effects of the membrane are ignored. The equation of state for part *A* is given by

$$\lambda_{1A}\frac{P_{1A}}{L_2H} + \varepsilon\left(\frac{\Phi}{H}\right)^2\lambda_{1A}^2\lambda_{2p}^2 = \lambda_{1A}\frac{\partial W(\lambda_{1A},\lambda_{2p})}{\partial\lambda_{1A}} \tag{2}$$

where the first term represents the mechanical stress, the second term represents the Maxwell stress, and *W* represents the elastic free energy density of the elastomer [13]. Since voltage is absent in part *B*, the equation of state for part *B* is given by,

$$\lambda_{1B}\frac{P_{1B}}{L_2H} = \lambda_{1B}\frac{\partial W(\lambda_{1B},\lambda_{2p})}{\partial\lambda_{1B}} \tag{3}$$

In the experiments, the distance between the clamps is fixed, $d = \lambda_{1A}L_{1A} + \lambda_{1B}L_{1B} = \lambda_{1p}(L_{1A} + L_{1B})$. Since $L_{1A} = L_{1B}$, we obtain $\lambda_{1A} + \lambda_{1B} = 2\lambda_{1p}$. The elastomer is assumed to be incompressible, that is

$$\lambda_1\lambda_2\lambda_3 = 1 \tag{4}$$

where λ_1, λ_2 and λ_3 are the stretches along the width, length, and thickness directions, respectively. As the stretches observed experimentally are not very large, for simplicity, a neo-Hookean material model is employed with the elastic free energy density

$$W(\lambda_1,\lambda_2) = \frac{\mu}{2}\left(\lambda_1^2 + \lambda_2^2 + \lambda_1^{-2}\lambda_2^{-2} - 3\right) \tag{5}$$

where μ is the small-strain shear modulus of the material. Substituting (5) into (1), we obtain

$$\frac{d^2\lambda_{1A}}{dt^2} - \frac{L_2H\mu}{mL_{1A}}\left(2(\lambda_{1p} - \lambda_{1A}) + \lambda_{1A}^{-3}\lambda_{2p}^{-2} - (2\lambda_{1p} - \lambda_{1A})^{-3}\lambda_{2p}^{-2} + \frac{\varepsilon}{\mu}\left(\frac{\Phi}{H}\right)^2\lambda_{1A}\lambda_{2p}^2\right) + \frac{c}{m}L_{1A}\frac{d\lambda_{1A}}{dt} = 0 \tag{6}$$

The output displacement of the vibration shaker is

$$\Delta = L_{1A}(\lambda_{1A} - \lambda_{1p}) \tag{7}$$

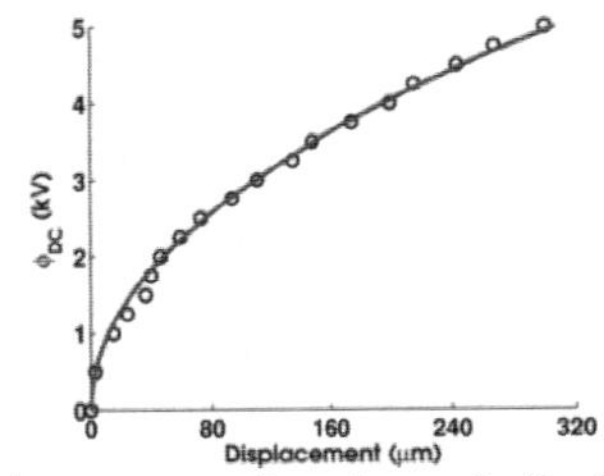

Fig. 5. DC voltage as a function of displacement of the vibration shaker

The static equilibrium of the vibration shaker is analyzed first, when part *A* is subject to DC voltage. Letting $d^2\lambda_{1A}/dt^2$=0 and $d\lambda_{1A}/dt$=0, we obtain the stretch λ_{1A} by inserting $\Phi = V_{dc}$ into Eq. (6). The output displacement is thus calculated by Eq. (7). Figure 5 plots the DC voltage as a function of the displacement of the vibration shaker. The blue curve represents the calculations, in which the permittivity $\varepsilon = 2.39\times10^{-11}F/m$ [11] and the shear modulus $\mu = 321kPa$ are used. The open circles represents the experimental data.

Using the above material parameters, we simulate the oscillation of the mass when it is subjected to AC voltage. Substituting $\Phi = V_{ac}\sin(\omega t)$ into Eq.(6), we obtain

$$\frac{L_2 H\mu}{mL_{1A}}\left(2(\lambda_{1p}-\lambda_{1A})+\lambda_{1A}^{-3}\lambda_{2p}^{-2}-(2\lambda_{1p}-\alpha\lambda_{1A})^{-3}\lambda_{2p}^{-2}+\frac{\varepsilon}{\mu}\left(\frac{V_{ac}}{H}\right)^2\frac{1-\cos(2\omega t)}{2}\lambda_{1A}\lambda_{2p}^2\right)$$
$$-\frac{d^2\lambda_{1A}}{dt^2}-\frac{c}{m}\frac{d\lambda_{1A}}{dt}=0 \qquad (8)$$

The time-dependent stretch $\lambda_{1A}(t)$ can be obtained by solving (8) for any initial conditions. Unless otherwise stated, we assume that the mass starts at rest with the initial conditions: $\lambda_{1A}\big|_{t=0}=\lambda_{1p}$, and $\left.\frac{d\lambda_{1A}}{dt}\right|_{t=0}=0$. Figure 6 shows the displacement amplitude as a function of the frequency of AC voltage when $V_{ac}=7kV$. The blue curve represents the theoretical calculations with $c/m=1.125\times10^{-3}\sec^{-1}$, and the black curve represents the experimental data (Fig. 3b). The calculations match the experiments in the frequency and the amplitude at the resonance.

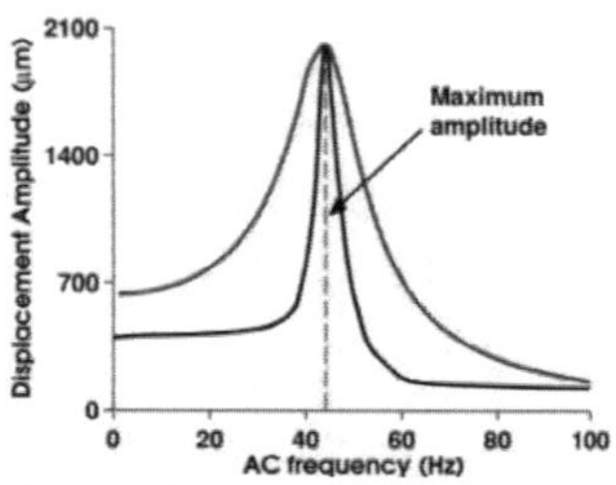

Fig. 6. Comparison of theoretical calculations (in blue) to experimental observations (in black).

CONCLUSION

In this paper we investigate dynamic behaviour of a vibration shaker based on a dielectric elastomer actuator. The vibrations are strongly nonlinear due to large deformation and nonlinear material properties. Oppo Band, Singapore is employed as the dielectric elastomer, and the membrane exhibits a large amplitude of oscillation. The oscillation of the vibration shaker is measured experimentally. The theoretical model is developed, which takes into account the damping effect. The theoretical results are qualitatively consistent with the experimental observations.

ACKNOWLEDGMENTS

U.G. and H.G. are supported by NUS Research Scholarships. J.Z. acknowledges support from MOE startup, Singapore (R-265-000-444-133) and MOE Tier 1, Singapore (R-265-000-497-112).

REFERENCES

1. Carpi, F., Bauer,S., and Rossi,D.De 2010,"Stretching dielectric elastomer performance", *Science*, 330(6012), 1759 -1761.
2. Pelrine, R., Kornbluh, R., Pei, Q.B. and Joseph, J. 2000, "High-speed electrically actuated elastomers with strain greater than 100%", *Science*, 287, 836-839.
3. Godaba, H., Foo, C.C., Zhang, Z.Q., Khoo, B.C., and Zhu, J. 2014, "Giant voltage-induced deformation of a dielectric elastomer under constant pressure", *Appl. Phys. Lett.*, 105, 112901.
4. Keplinger, C., Li, T.F., Baumgartner, R., Suo, Z.G., Bauer, S. 2012, "Harnessing snap-through inst. in soft dielectrics to achieve giant voltage-triggered deformation", *Soft Matter*, 8(2), 285-288 .
5. S, Ashley. "Artificial muscles." *Scientific American*, October 2003: pp. 52–59
6. Fox, J.W., and Goulbourne, N.C. 2008, "On the dynamic electromechanical loading of dielectric elastomer membrane", *J. Mech. Phys. Solids.*, 56(8), 2669-2686.
7. Zhu, J., Cai, S.Q., and Suo, Z.G. 2010, "Resonant behavior of a membrane of a dielectric elastomer", *Int. J. Solids Struct.*,47,3254-3262.
8. Carpi, F., Rossi, D. D., Kornbluh, R., Pelrine, R., Sommer-Larsen, P., 2008, *Dielectric elastomers as electromechanical transducers*, Amsterdam, The Netherlands: Elsevier.
9. Biggs,S.,and Hitchcock,R. 2010, "Artificial muscle actuators for haptic displays", *Proc. SPIE*, 7642, 76420I.
10. Gupta, U., Godaba, H., Zhao, Z.J., Chui, C. K., Zhu, J., 2015, "Tunable force/displacement of a vibration shaker driven by a dielectric elastomer actuator", *Extreme Mechanics Letters*, available online.
11. Kaltseis, R., Keplinger, C., Koh, S.J.A., Baumgartner, R., Goh, Y.F., Ng, W.H., Kogler, A., Trols, A., Foo, C.C., Suo, Z.G., Bauer. S. 2014, "Natural rubber for sustainable high-power electrical energy generation", *RSC Advances*, 4, 27905-27913.
12. Dubois, P., Rosset, S., Niklaus, M., Dadras, M., Shea, H. 2008, "Voltage control of the resonance frequency of DEAP membrane", *Microelectromech. S.*, 17, 1072.
13. Suo, Z.G. 2010, "Theory of DE", Acta Mech. Solida Sin., 23, 549.

DURABILITY OF A NEW GENERATION OF BASALT-FRP REINFORCING BARS

Dr. Mohamed Hassan
Department of Civil Engineering
Queen's University
Kingston, ON Canada
Mohamed.Hassan@USherbrooke.ca

Prof. Brahim Benmokrane
Department of Civil Engineering
University of Shebrooke
Sherbrooke, QC Canada
Brahim.Benmokrane@USherbrooke.ca

ABSTRACT

This study is a part of an ongoing durability research program on a new generation of basalt-fiber-reinforced polymer (BFRP) reinforcing bars for concrete structures. BFRP bars, namely, 12-mm-diameter BFRP deformed bars, was investigated. The BFRP specimens were exposed to an alkaline solution (pH=13) for 3 and 6 months at an elevated temperature (60°C) simulating a concrete environment in order to examine their durability performance. Thereafter, the physical and mechanical properties of the conditioned bars were assessed and compared to that of the unconditioned specimens. After 6 months of exposure, the BFRP bars exhibited average retentions of tensile strength, transverse shear strength, and interlaminar shear strength of 69%, 84%, and 88%, respectively. The strength reduction of the conditioned specimens may be attributed to the degradation of the fiber–resin interface as confirmed from the SEM and microstructural analyses. To promote this new material for concrete structures and inclusion in FRP standards and guides, more experimental tests targeting durability issues and full-scale structure testing are needed.

KEYWORDS: Basalt fibers; Durability; Physical and Mechanical Properties; Characterization; Strength retention.

INTRODUCTION

The use of basalt fiber–reinforced polymers (BFRPs) in construction applications is relatively new. This type of structural material is expected to provide performance comparable or superior to that of glass-fiber–reinforced polymers (GFRP), while being significantly cost-effective (Wang et al. 2012; Benmokrane et al 2015). Limited studies, however, have addressed the durability performance of BFRP bars under real and simulated harsh environmental conditions. An extensive research project is being conducted at the University of Sherbrooke to assess the short-term and long-term characteristics of BFRP bars as a preliminary step in introducing this new material into pilot projects and FRP design codes and material specifications. The project addresses a new generation of five different types of BFRP bars and one type of BFRP tendon.

RESEARCH SCOPE

This study aimed at assessing the characteristics of BFRP bars made with basalt fibers and their long-term durability through conditioning in an alkaline solution simulating a moist concrete environment at elevated temperature. The findings will contribute to integrating BFRP into FRP standards and guides, such as (ACI 2008) and (CSA 2010). It is worth mentioning that basalt-FRP bars have not yet been incorporated into design standards and specifications.

EXPERIMENTAL WORK

This study was carried out on a new generation of BFRP bars, namely, 12-mm-diameter BFRP deformed bars, as shown in Figure 1. The experimental program included three phases. Phase I focused on determining the physical properties of the BFRP bars. Phase II investigated short-term mechanical properties, including tensile strength, transverse shear strength and interlaminar shear strength. The test results served as a reference for residual strength after conditioning. Phase III included a durability study to assess the performance of BFRP bars immersed in an alkaline solution at an elevated temperature of 60°C for 3 and 6 months as per CSA (2010) and ACI (2008). The changes in the physical and mechanical characteristics were assessed by comparing the characteristics of the conditioned specimens to those of the reference bars from Phase I and II.

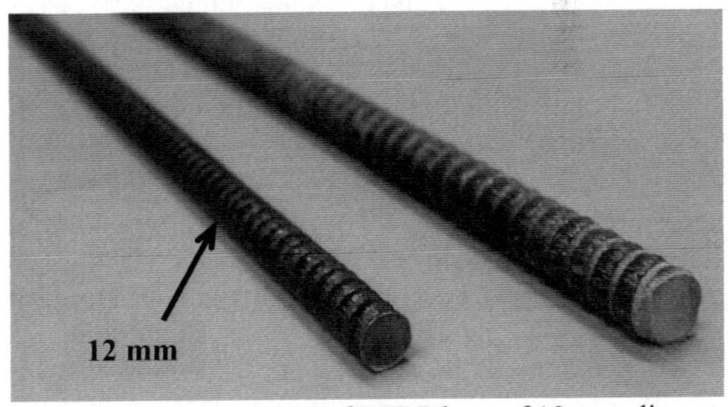

Figure 1: New generation of BFRP bars of 12 mm diameter

PHASE I: PHYSICAL CHARACTERIZATION

Physical and microstructural analyses were conducted on reference specimens (unconditioned) to investigate the material properties and microstructure according to CSA (2010) and ACI (2008). Table 1 presents results of the physical properties of unconditioned BFRP bars. The test results show that the tested BFRP bars had a fiber content of around 81% (by weight), a cure ratio around 98%, a glass-transition temperature (T_g) around 117°C, and moisture uptake at saturation of about 0.13%. The physical properties shown by the investigated BFRP bar of 12 mm diameter met the requirements of the requirements of CSA S807 (2010) and ACI 440.6-08 (2008).

Scanning-electron-microscope (SEM) micrographs at different magnifications taken of the reference specimens (see Figure 5) show some pores and voids in the coating layer on the external surface of the bars. No pores or defects such as debonding at the fiber–matrix interface were, however, observed in the core of the bars.

Table 1: Physical properties of the investigated BFRP bars

Property	12- mm control bar	12- mm conditioned bar (after 3 months)	Specified limit	
			ACI (2008)	CSA (2010)
Cross-sectional area (mm^2)	125	-	N/A	N/A
Fiber content by weight (%)	81.5	-	55% (by vol.)	70 (by wt)
Transverse CET ($\times 10^{-6}\,°C^{-1}$)	-	-	N/A	40
Density	2.1	-	N/A	N/A
Moisture uptake (%)	0.13	0.35	1	1.0 (D2); 0.75 (D1)
Cure ratio (%)	98.0	99.0	N/A	93 (D2); 95 (D1)
T_g (°C)	117.0	111.0 (run1) 119.0 (run 2)	100	80 (D2); 100(D1)

* D1 and D2 classifications can be found elsewhere in (CSA 2010).

PHASE II: MECHANICAL CHARACTERIZATION OF REFERENCE BARS

The mechanical characterization included testing of representative BFRP specimens to determine their tensile strength, transverse shear strength, and interlaminar shear strength.

TENSILE STRENGTH

Longitudinal tensile tests were carried out on 18 unconditioned specimens of three batches. The BFRP specimens' length as well as the length and diameter of steel anchors at both ends to be used for the tensile test were calculated according to ASTM D7205 (2011). Each specimen was instrumented with a linear variable differential transformer (LVDT) with a gauge length of 200 mm to record elongation during testing. The test was carried out with a 2000 kN Baldwin testing machine at a loading rate ranging from 250 to 300 MPa/min. The applied load and bar elongation were recorded with a data-acquisition system monitored by a computer.

All the specimens failed in a brittle manner with fiber rupture in the test length region (see Figure 2). The stress–strain curves were linear elastic up to failure. Table 2 gives the test results for the mechanical properties of the unconditioned BFRP bars. The test results indicate that the tensile strength of the 12 mm unconditioned BFRP bars was 1702 ± 41.0 MPa while the tensile modulus of elasticity was 62 GPa. The BFRP bars evidenced maximum strain at failure of 2.5%, which is higher than the minimum value of 1.2% (CSA 2010).

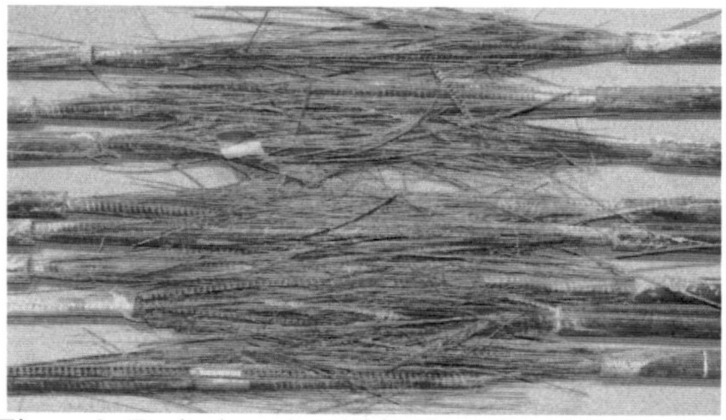

Figure 2: Typical mode of failure of the test specimens

TRANSVERSE SHEAR STRENGTH

The transverse shear tests were determined in accordance with ASTM D7617 (2011). A total of 24 unconditioned specimens (three batches) 200 mm in length were tested on an MTS 810 testing machine equipped with a 500 kN load cell. A controlled displacement rate of 1.5 mm/min was used during the test until specimen failure. The applied load was recorded with a data-acquisition system monitored by a computer. Table 2 shows the transverse shear strength of the unconditioned BFRP specimens. The average transverse shear strength of the BFRP specimens was 272 ± 18 MPa which indicates that the test results met the criteria as specified in (CSA 2010) (>160 MPa). Figure 3 shows a typical transverse shear mode of failure of the BFRP bars.

a) Control b) conditioned after 6 M

Figure 3: Typical mode of failure of transverse shear test

INTERLAMINAR SHEAR STRENGTH

The interlaminar shear tests were conducted in accordance with the test method in ASTM D4475 (2008). The interlaminar shear strength test was conducted on 24 unconditioned specimens (three batches). The tests were carried out using a 500 kN MTS 810 testing machine at a displacement-controlled rate of 1.3 mm/min. The test specimens were cut into lengths of 70 mm. The distance between supports in the test setup was set to four times the nominal bar diameter. All specimens failed typically in a horizontal shear plan. Figure 4 shows a typical interlaminar shear mode of failure of the specimens. The average interlaminar shear strength for the reference specimens was 68 ± 4.9 MPa.

a) Control

b) Conditioned after 6M

Figure 4: Typical mode of failure of interlaminar shear test

PHASE III: DURABILITY AND LONG-TERM PERFORMANCE ASSESSMENT STUDY

ENVIRONMENTAL CONDITIONING

Accelerated aging tests were conducted in accordance with ASTM D7705 (2012). The conditioning of the BFRP bars included the combined exposure to a harsh alkaline environment and elevated temperature. The alkaline solution was prepared to have a composition representative of the pore water inside portland-cement concrete, specifically, 118.5 g of $Ca(OH)_2$, 0.9 g of NaOH, and 4.2 g of KOH per liter of deionized water. The solution had a pH of 12.8 to 13.0, which is representative of a mature concrete pore solution. The BFRP bars were immersed in this solution at 60°C for conditioning periods of 3 and 6 months. The conditioning time started once the solution reached the prescribed temperature. The durability was assessed by comparing the physical and mechanical characteristics of the conditioned bars to those of the reference bars. In addition, the microstructure of the conditioned BFRP bars was investigated with SEM analysis to assess any changes or environmental degradation.

PHYSICAL CHARACTERIZATION OF CONDITIOINED BFRP BARS

Figure 5 shows the SEM micrographs of the outer surface and cross-section of the 12 mm diameter BFRP bars before and after conditioning at different magnifications. As shown in Figure 5, the BFRP bars after conditioning show no significant changes in the coating layer on the external surface of the bar compared with that of unconditioned specimens. However, the conditioned specimen seems to be affected by the accelerated aging, which show some debonding at the fiber/resin interface. This can explain the higher moisture uptake ratio measured at saturation. In fact, the moisture uptake ratio has increased from 0.13% before aging to 0.35% after aging in alkaline solution at an elevated temperature (60°C) (Table 2). This may also justify the interlaminar shear strength reduction after the accelerating aging, as indicates in the following section.

Table 2 presents the T_g values for the conditioned specimens after 3 months of exposure. Two scans were performed for the conditioned specimens. The T_g of conditioned specimen was slightly decreased by (6%) after the first scan with respect to the unconditioned specimen, this is an indication of plasticizing effect or chemical degradation. However, after the second scan, the T_g is in the same range as that of the reference specimen, this is indicative of a reversible plasticizing effect attributable to the moisture absorption.

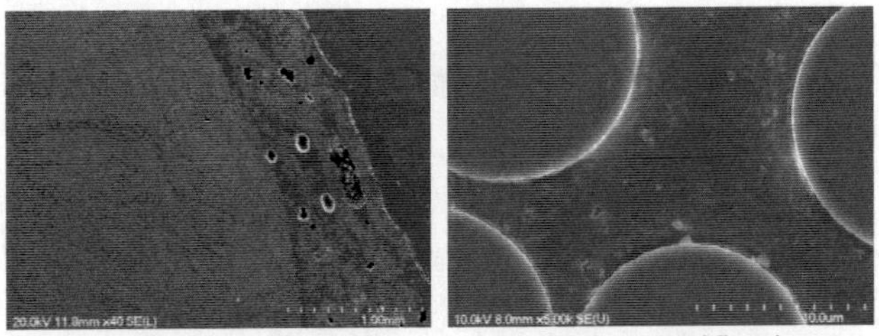

(a) Low magnification (b) High magnification

(i) Before conditioning

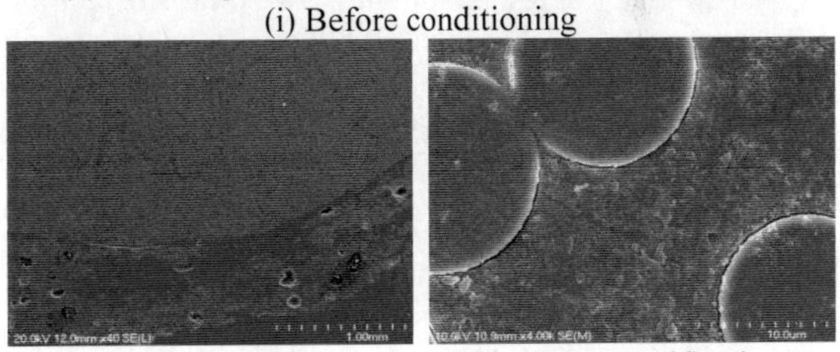

(a) Low magnification (b) High magnification

(i) After conditioning

Figure 5: Micrographs of the cross section of the reference and conditioned BFRP bars of the 12-mm diameter after 3 months at different magnifications

MECHANICAL PROPERTIES OF CONDITIOINED BFRP BARS

A total of 15, 24, and 24 conditioned specimens (three batches) were subjected to direct tensile, transverse shear, and interlaminar shear tests after conditioning for 3 and 6 months of exposure, respectively. Table 2 gives the mechanical properties of the conditioned BFRP bars. As shown in Table 2, the tensile strength of the conditioned specimens was affected by the accelerated aging: the tensile strength decreased as the immersion time. The unconditioned and conditioned specimens exhibited similar brittle manner mode of failure with fiber rupture in the test length region. The average tensile-strength test results obtained after conditioning for 3 and 6 months were 1260 ± 57.6 and 1171 ± 82.0 MPa while the corresponding values for the modulus of elasticity were 62.0 ± 1.1 and 61.9 ± 2.4 GPa, respectively. This indicates that the loss of tensile strength resistance after 3 and 6 months was around of 26% and 31%, respectively. There were no significant differences in the tensile modulus of elasticity of the aged BFRP specimens in comparison to the unconditioned bars. The strength reduction of the conditioned specimens may be attributed to the degradation of the fiber/resin interface, which reduced the stress transfer. Table 2 shows the retention of tensile strength, transverse shear strength, and interlaminar shear strength of the conditioned specimens.

Table 2: Retention and mechanical properties of the reference and conditioned BFRP bars

FRP Bar	f_{fu}	R	E_f	ε_u	τ_u	R	S_u	R
C-12	1706 ±40	-	62.1 ±5.6	2.52 ±0.2	272 ±18	-	68± 4.9	-
12-3M	1260 ±57	74	62.0 ±1.1	2.03 ±0.07	247 ±17	91	62± 4.8	91
12-6M	1171 ±82	69	61.9 ±2.4	1.79 ±0.17	240 ±16	84	58± 8.0	88

C-12 = 12 mm diameter (control bars); 12-3M= 12 mm diameter (bars conditioned for 3 months); f_{fu} = ultimate tensile strength (MPa); R = retention (%); E_f = ultimate tensile modulus of elasticity (GPa); ε_u = ultimate tensile strain (%); τ_u = ultimate transverse shear strength (MPa); and S_u = ultimate interlaminar shear strength (MPa).

The average test results for transverse shear strength for the 12 mm conditioned specimens after 3 and 6 months of exposure were 247 ± 17 and 240 ± 8 MPa, respectively, which indicates that the test results met the requirements of (CSA 2010) (>160 MPa). After 6 months of exposure, the retention of transverse shear strength was 84%. On the other hand, the average interlaminar shear strength of the conditioned specimens was 62 ± 5 and 58 ± 8 MPa after 3 and 6 months of exposure, respectively. Besides, the mode of failure of the transverse shear and interlaminar shear for all conditioned specimens was the same as compared to that of the unconditioned specimens (see Figure 3 and 4).

CONCLUSIONS

This study is a part of an ongoing durability research program at the University of Shebrooke on a new generation of BFRP reinforcing bars for concrete structure. An overview of the physical and microstructural analyses as well as a preliminary durability characterization of newly developed BFRP bars of 12 mm in diameter is presented. The tested specimens were exposed to an alkaline solution (pH=13) for 3 and 6 months at elevated temperature (60°C). Based on the test results, the following conclusions can be drawn:

1. The tensile strength of unconditioned and conditioned BFRP bars after 3 and 6 months of exposure was 1706, 1260, and 1176 MPa, while the modulus of elasticity and ultimate tensile strain was 62.1, 62.0, and 61.9 GPa and 2.52%, 2.01%, and 1.79%, respectively.
2. The tensile strength of the tested BFRP bars was significantly affected by the accelerated aging (reduced by 26% and 31%, respectively, after 3 and 6 months). There were no significant differences in the tensile modulus of elasticity of the aged specimens compared to that of the unconditioned bars.
3. After 6 months of exposure, the average retention of transverse shear strength and interlaminar shear strength of the 12 mm diameter BFRP bars was 84% and 88% in comparison to the unconditioned bars, respectively.
4. SEM micrographs of the conditioned BFRP bars seem to be affected by the accelerated aging, which show some debonding at the fiber/resin interface. However, there was no deterioration in the basalt fibers.
5. The newly developed 12 mm diameter BFRP bars satisfied the minimum requirements of (ACI 2008) and (CSA 2010) concerning their physical and mechanical properties for use as nonprestressed reinforcement for concrete structures.
6. To promote this new material for use in civil engineering for concrete structures and inclusion in FRP standards and guides, more experimental tests targeting durability issues and full-scale structure testing are needed.

ACKNOWLEDGMENTS

The authors wish to acknowledge the financial support of the Natural Sciences and Engineering Research Council of Canada (NSERC), and the Fonds québécois de la recherche sur la nature et les technologies (FQRNT) of Quebec.

REFERENCES

1. ACI Committee 440. 2008, "Specification for Carbon and Glass Fiber-Reinforced Polymer Bar Materials for Concrete Reinforcement (ACI 440.6M-08)", American Concrete Institute, Farmington, Hills, MI, 6 p.
2. ASTM D4475. 2008, "Standard Test Method for Apparent Horizontal Shear Strength of Pultruded Reinforced Plastic Rods by the Short Beam Method", American Society for Testing and Materials, Conshohocken, USA. 4p.
3. ASTM D7617. 2011, "Standard Test Method for Transverse Shear Strength of Fiber-reinforced Polymer Matrix Composite Bars", American Society for Testing and Materials, Conshohocken, USA, 12 p.
4. ASTM D7205. 2011, "Standard Test Method for Tensile Properties of Fiber Reinforced Polymer Matrix Composite Bars", American Society for Testing and Materials, Conshohocken, USA, 12 p.
5. Benmokrane, B., Elgabbas, F., Ahmed, E., and Cousin, P. 2015, "Characterization and Comparative Durability Study of Glass/Vinylester, Basalt/Vinylester, and Basalt/Epoxy FRP Bars", Journal of Composites for Construction, ASCE, ISSN 1090-0268/04015008(12).
6. Canadian Standards Association (CSA). 2010, "Specification for Fibre-Reinforced Polymers (CAN/CSA S807–10)", Rexdale, ON, Canada, 27 p.
7. Wang, X., Wu, G., and Wu, Z. (2012), "Tensile property of prestressing basalt FRP and hybrid FRP tendons under salt solution." Proc., 6th Int. Conf. on FRP Composites in Civil Engineering, International Institute for FRP in Construction, Kingston, ON, Canada.

Advanced Manufacturing

AN IMPROVEMENT OF FIXTURELESS INSPECTION FOR NON-RIGID PARTS BASED ON FILTERING SAMPLE POINTS

Sasan Sattarpanah Karganroudi sattarpa@uqtr.ca
Jean-Christophe Cuillière cuillier@uqtr.ca
Vincent Francois francois@uqtr.ca
Équipe de Recherche en Intégration Cao-CAlcul (ÉRICCA)
Université du Québec à Trois-Rivières
Trois-Rivières, Québec, Canada

Souheil-Antoine Tahan
Antoine.Tahan@etsmtl.ca
Laboratoire d'ingénierie des produits, procédés et systèmes (LIPPS)
École de Technologie Supérieure (ETS)
Montreal, Québec, Canada

ABSTRACT

Geometric inspection of non-rigid (flexible) parts in free-state can aid manufacturers to automatically and inexpensively improve quality by detecting defects on manufactured parts. Non-contact optic data acquisition devices (scanners) are used to measure point clouds on the boundary of manufactured part in free-state and compare it with its nominal computer aided design (CAD) model in a common coordinate system. In the case of non-rigid parts, deformation during free-state inspection, mostly due to weight and to residual stresses, causes specific problems since this deformation has to be distinguished from defects. In this study the *generalized numerical inspection fixture (GNIF)* method is applied to generate a prior set of corresponding sample points between CAD and scanned models. These points are then used to deform the CAD model to conform to scanned data, through finite element non-rigid registration. Corresponding sample points generated using GNIF are evenly distributed over both models. Therefore, some sample points may be located on defects, which makes that these defects cannot be measured accurately. In this paper we introduce a method, aimed at automatically removing these sample points, which is based on von Mises stress and curvature criteria. Once removed, these points are no longer used in non-rigid FEA registration, which makes that associated defects can be identified and measured accurately.

KEYWORDS: Geometric inspection, Non-rigid parts, von Mises stress, Discrete curvature, GNIF, FEA, Mesh modification.

INTRODUCTION

Competitive markets imply high quality standards for manufactured parts, which requires setting up fast and accurate geometric and dimensional measurement methods. Beholden to the improvements in computer graphics and optic scanners, manual and tactile methods of measurement and inspection have been replaced by Computer Aided Inspection (CAI) methods. Manufactured parts are automatically scanned as point clouds and this scanned data is compared with the nominal CAD model in a common coordinate system. This allows identifying and measuring defects on these parts. Non-rigid (flexible) manufactured parts, e.g. parts with a very thin thickness (if compared with the other dimensions) deform to an extent in free-state which is beyond the dimensional and/or geometrical tolerances on the designed drawing. Parts are only subjected to gravity during free-state inspection and, as shown in Fig. 1-b, a non-rigid part deforms due to its own weight and/or the release of residual stresses.

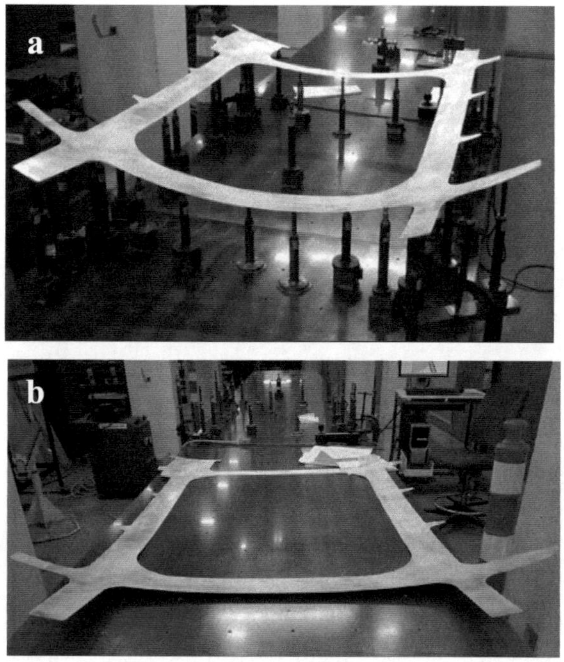

Figure 1: Compliance behavior of a non-rigid aerospace panel, a) constrained on its inspection fixture, b) in free-state inspection. [1].

Undesired dimensional and geometric deviation of non-rigid parts in free-state can be compensated by mounting these parts on special embedding fixtures. However, in most cases these fixtures are sophisticated and expensive increasing the lead time in the quality control process. Thus far, depending on the type of non-rigid parts, several virtual fixtureless inspection methods have been developed, among which methods introduced in [2-4]. In this study the generalized numerical inspection fixture (GNIF) method [5] is applied to find the intrinsic similarity between nominal CAD and scanned data in free-state. It is based on the assumption that deformation is isometric (deformation preserves geodesic distances). Based on this assumption and on fast marching algorithm for the calculation of geodesic distances on CAD and scan triangulations, GNIF generates corresponding sample points between the CAD and scanned models. In this purpose, generalized multidimensional scaling (GMDS) is a minimization algorithm that finds corresponding sample point on CAD model which has the minimum geodesic distortion respect to its sample point on scanned data. Corresponding sample points are used to deform the CAD model conforming to scanned data by means of FEA and introducing displacement vectors on the CAD sample points. Bringing the deformed CAD and scanned model in a common coordinate and making a comparison between them, defects will be detected and can be measured.

Since corresponding sample points are evenly distributed all over the CAD and scanned model, some samples can be located on defect areas. It causes inaccurate measurement of defects because sample points on a defect deform the CAD model to take the defect shape. In this paper, two filtration methods are explained to automatically remove sample points on defect areas. The first one is based on von Mises stress criterion and the other applies difference between discrete curvatures on CAD and scanned model. Since the removed sample points will no more be applied in FEA, the CAD model will deform to simulate only the elastic deformation of scanned part due to its flexibility in free-state but not the deviated defects. This improves the measurement of possible defects during inspection of manufactured parts.

FILTERING GNIF SAMPLE POINTS

The generated corresponding sample points by GNIF method are evenly distributed over the CAD and scanned model, and applying FEA with displacement vector on CAD sample points deforms the CAD model to conform to scanned data. Since displacing the CAD sample points on defect areas impels the CAD model to take the shape of defects on scanned data, it is required to estimate the defect areas beforehand and filter these CAD sample points from FEA boundary conditions for improving defect measurement.

Study of von Mises stress contour on the CAD model deformed by FEA applying all the generated sample points, clarifies that the existence of defects increase locally the stress value in the defect areas. Defining a threshold for the stress value provides an estimation for the position of defects, in other words, defects are located where the stress value is higher than the defined threshold value.

Meanwhile, comparing the curvature map between the CAD and scanned model shows that defects cause significant difference between the models curvatures. This can also be utilized as a criterion to detect defects and to filter sample points on the defect areas. For this purpose, the principal discrete curvatures on the CAD and scanned triangulation are calculated by Eq. (1) and Eq. (2) for maximum and minimum curvatures respectively.

$$K_1(p) = K_H(p) + \sqrt{K_H^2(p) - K_G(p)} \quad (1)$$

$$K_2(p) = K_H(p) - \sqrt{K_H^2(p) - K_G(p)} \quad (2)$$

Where K_H is the mean curvature and K_G is the Gaussian curvature. The mean and Gaussian curvatures are also discretely calculated by the Gauss-Bonnet scheme [6].

CASE STUDY

The case study herein is an aerospace aluminum panel of approximately 1.5 m length to 1 m width and 1 mm thickness. As shown in Fig. 1, the CAD model is schematically compared with the scanned mesh which is deformed (representing the free-state deformation of non-rigid part) and is included three defects which their height is measured as 1 mm for bumps No. 1 and 3, and also 1.5 mm for bump No. 2.

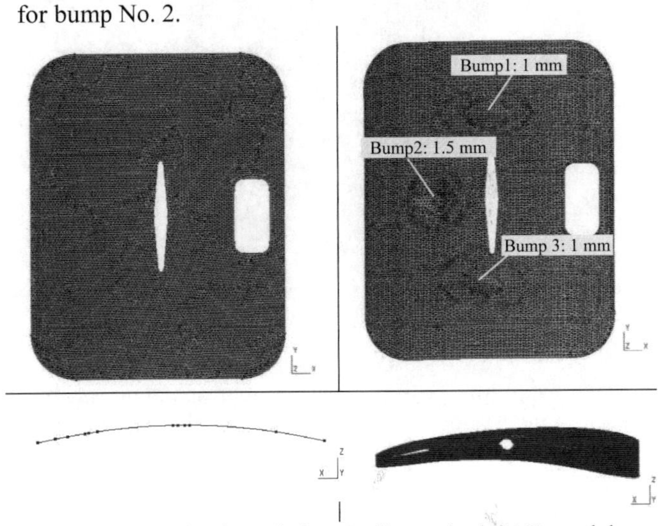

Figure 1: Mesh triangulation; Left) nominal CAD model, Right) scanned mesh (deformed and included three defects).

The corresponding sample points, which are generated by GNIF method as shown in Fig. 2, are inserted into the CAD mesh of the case study by incremental Delaunay triangulation method which is illustrated schematically in Fig. 3.

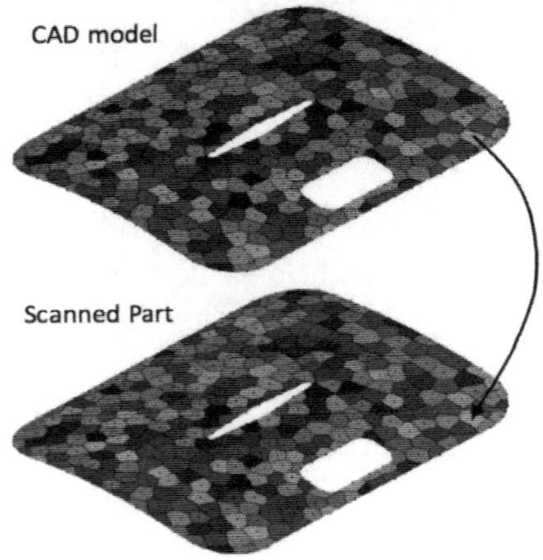

Figure 2: Corresponding sample points generated by GNIF method between the CAD and scanned models.

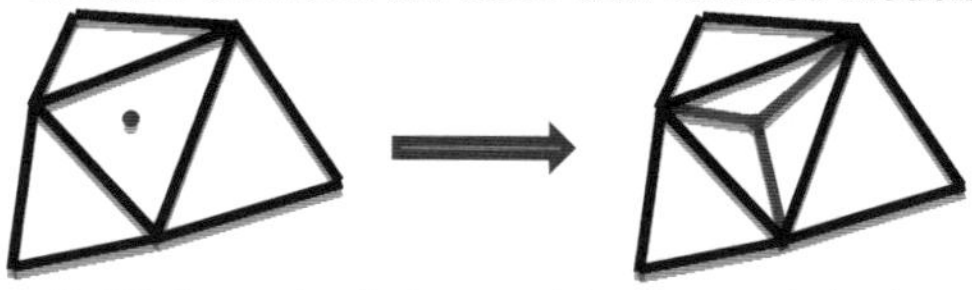

Figure 3: Violet point is inserted into a mesh by incremental Delaunay triangulation method.

The inserted sample points into the CAD mesh are evenly distributed all over the model which are depicted in Fig. 4 by green color for triangles sharing the sample points.

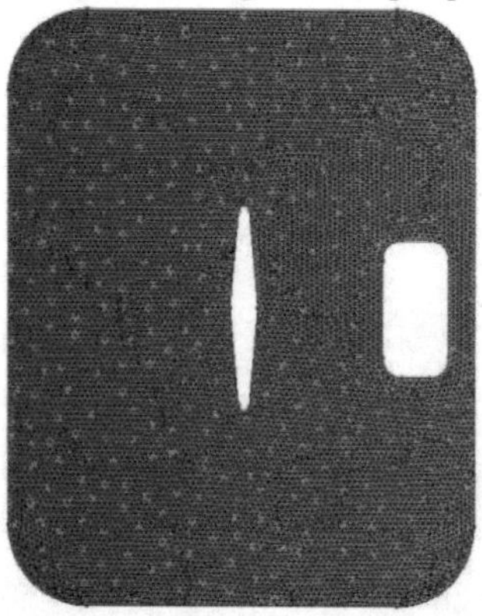

Figure 4: The inserted sample points are evenly distributed over the CAD mesh (green triangles).

Applying FEA to displace the sample points on the CAD model toward the position of their corresponding sample points on the scanned mesh, the CAD model is deformed to conform to scanned part. Bringing both deformed CAD and scanned model in a common coordinate and comparing the models, defects are measured as shown in Fig. 5. The defects are measured for bump No. 1 to 3 as 0.13 mm, 0.19 mm and 0.32 mm which has respectively 87%, 87% and 68% error comparing to the actual measured defects.

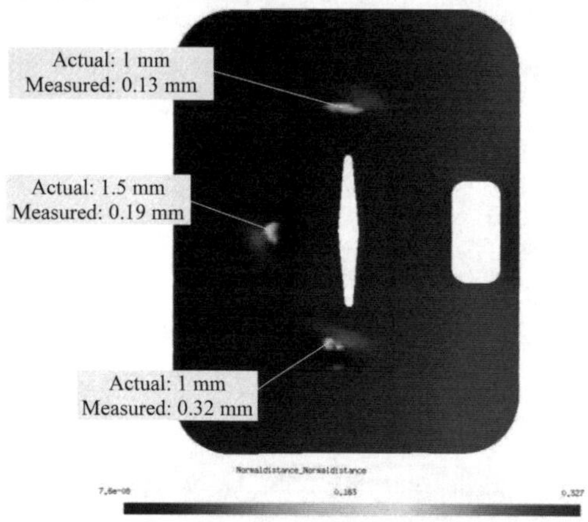

Figure 5: Defects measurement applying all sample points.

Comparing to the actual defect size, the measured defects include significant errors and it is required to filter the CAD sample points located in defect areas. The GNIF method has an inherent inaccuracy during generating corresponding sample points which causes high stress values distributed all over the deformed CAD model. However, as shown in Fig. 6, the von Mises stress value on defect areas are also locally higher than the other zones. The stress variation on defects are different from noise stresses caused by GNIF error. Determining the median value of von Mises stress range (3000 MPa) as threshold for this model, the defects are indicated on areas with von Mises stress higher than the threshold value.

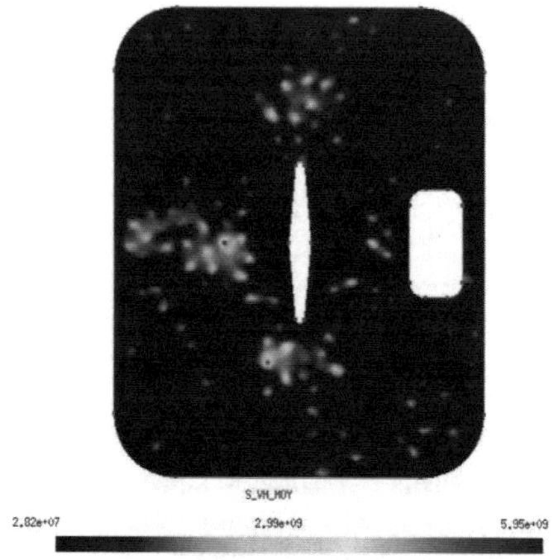

Figure 6: von Mises stress (range: 28 – 6000 MPa) for deformed CAD model applying FEA to conform to scan data.

Removing sample points where the stress value exceed the threshold value and applying displacement vectors by FEA on the remained sample points, improves the measurement of defects for bumps No. 1 to 3 as 0.22 mm, 0.45 mm and 0.67 mm which has respectively 78%, 70% and 33% error compared to actual defect sizes.

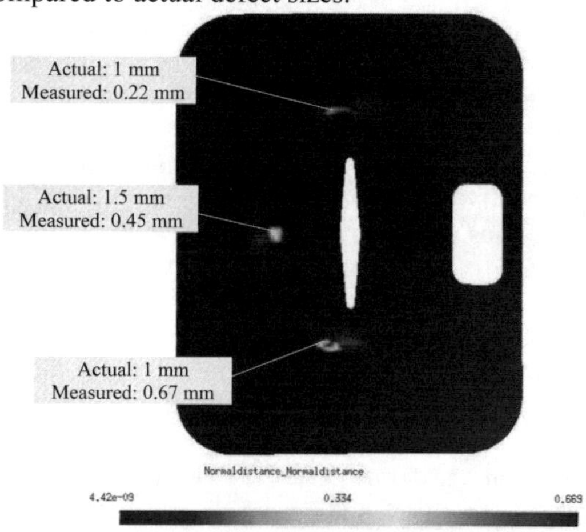

Figure 7: Defects measurement after filtering the sample points by von Mises stress criterion on defect areas.

Even though the filtration method by stress criterion improves the measurement of defects size, the defect measurement is more accurate as applying curvature criterion for sample point filtration. The difference of discrete principal curvatures between the CAD and scanned model is utilized to detect the defect areas. Fig. 8 illustrates, as an example, the discrete maximum curvature map calculated on the CAD and scanned mesh for the case study model. After calculating the difference of principal curvatures on CAD model compared to scanned data, the quarter of maximum value for curvature difference is applied as the threshold value.

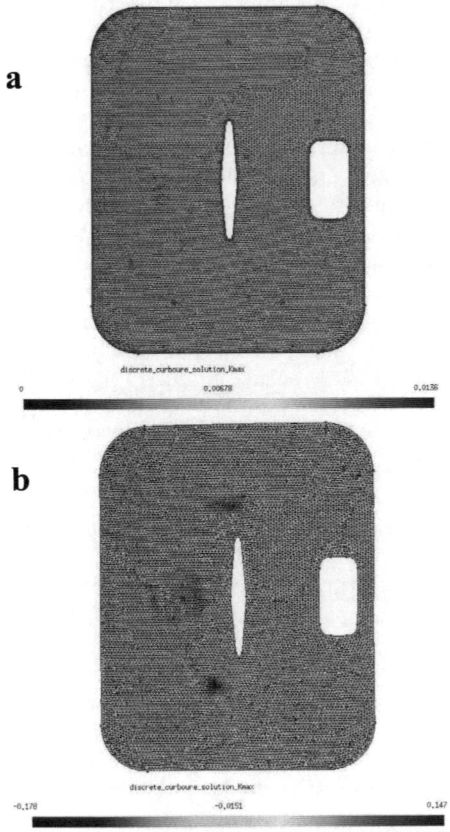

Figure 8: K_{max} curvature on a) the CAD mesh (range: 0. - 0.014 mm^{-1}), b) the scan data (range: -0.178 - 0.147 mm^{-1}).

Removing the corresponding sample points located on the CAD model which exceeds the threshold value provides a much more accurate defect measurement as shown in Fig. 9. The defects are measured for bumps No. 1 to 3 as 0.75 mm, 1.49 mm and 0.80 mm which has respectively 25%, 0.7% and 20% error comparing to actual defects size.

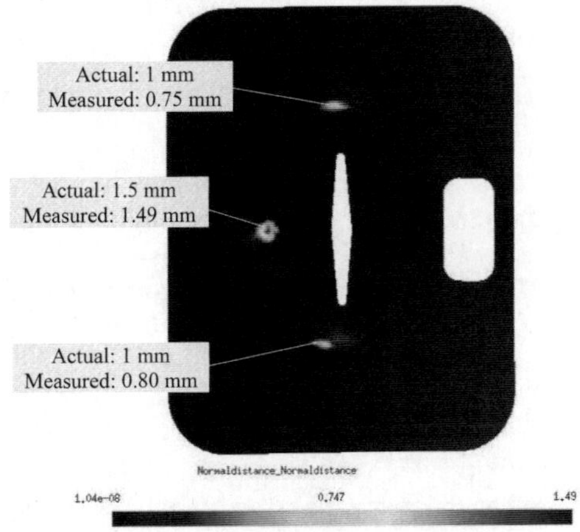

Figure 9: Defects measurement after filtering the sample points on defect areas by curvature criterion on defect areas.

CONCLUSIONS

In this study the fixtureless inspection of non-rigid parts applying the GNIF method is improved by augmenting the measurement accuracy of defects on the manufactured parts. Applying stress and curvature criterions, sample points associated to defect areas can automatically be removed which result in a deformation of CAD model to conform to scanned data except on defects. This development improves the GNIF method to automatically filter sample points and only apply those which are required to simulate the compliance deformation of part in free-state, and measure defects more accurately. The implementation of this method can be generalized by defining the threshold values automatically for stress and curvature criteria, and also the measurement accuracy can be improved by implementing successively the stress criterion after curvature criterion. Meanwhile, the method needs to be checked on cases which include global defects comparing to the defects used in the case study.

ACKNOWLEDGMENTS

The authors would like to thank the National Sciences and Engineering Research Council (NSERC), industrial partners, and Consortium for Aerospace Research and Innovation in Québec (CRIAQ) for their support and financial contribution.

REFERENCES

[1] G. N. Abenhaim, S. A. Tahan, A. Desrochers, and J.-F. Lalonde, "Aerospace Panels Fixtureless Inspection Methods with Restraining Force Requirements; A Technology Review," SAE Technical Paper 2013.

[2] A. Weckenmann and J. Weickmann, "Optical Inspection of Formed Sheet Metal Parts Applying Fringe Projection Systems and Virtual Fixation," Metrology and Measurement Systems, vol. 13, pp. 321-330, 2006.

Nd:YAG Laser Surface Treatment of Stainless Steel 304 Sheets

Mitra Radmanesh
Department of Mechanical Engineering
University of New Brunswick
Fredericton, New Brunswick, Canada
Mitra.radmanesh@unb.ca

AmirKianoosh Kiani
Department of Mechanical Engineering
University of New Brunswick
Fredericton, New Brunswick, Canada
a.kiani@unb.ca

ABSTRACT

Laser processing and laser surface texturing has become a growing topic of studies in various fields. It is essential to understand the laser irradiation mechanism in order for high quality results. Through this study, the effects of power and pulse duration, on surface structure of stainless steel type 304 were examined. The analyzed sample was prepared by using pulse durations of 20 ms, 10 ms, 5ms, and variety of pulsed energy. Looking at the variation of non-dimensional temperature along the surface and center of the sample, the effects of pulse duration and intensity (corresponding to voltage) were observed. Upon considering all the aspects of the irradiated spots, such as diameter of the heat affected area and surface patterns, advantages and disadvantages of short and long pulse durations are mapped out. Also, a method to obtain the ablation threshold of stainless steel is introduced, and a thorough analytical solution is obtained.

KEYWORDS: laser material processing, surface treatment, stainless steel, and ablation threshold

INTRODUCTION

In recent decades, the use of lasers in various fields of medical and engineering has increased, and modeling of laser material interactions has become an important topic of studies. Important contributions of lasers are surface annealing, surface treatment, and processing of all metals and alloys. Generally, laser processing of materials involves high concentrated heating of the targets using laser, in order to create several types of structural changes, such as amorphization, hardening, and surface patterning, in a fast and precise manner [1]. Lasers are able to deliver energy in a wide range, with very high precision in all aspects. In comparison to conventional methods, laser approach is very flexible and fast, making it particularly suitable for mass production, rapid prototyping, and custom-scale manufacturing of vast variety of applications such as micro welding, drilling, and cutting of metals and alloys [2]. In recent decades, there have been major improvements on laser systems to increase the interaction time (pulse duration) between the laser and the material; hence reducing the effects on the bulk material [3].

In order to fully understand the mechanism behind the laser-material interaction, understanding and controlling the effects of process parameters are very essential; hence choosing the most optimal pulse duration is influential in quality of the particular application [3]. This study has been conducted to identify the effects of key laser process parameters such as power and pulse duration on the structure and surface pattern of stainless steel. Also, a new method for calculation of ablation threshold has been introduced for stainless steel, which can be customized for other metals and alloys.

EXPERIMENTAL SETUP

In order to investigate the effects of power and pulse duration, and develop a new optimal map for processing of materials, an experimental design is composed. Stainless steel type 304 sample, with a thickness of 0.5 mm, and a pulsed Nd:YAG laser system with a peak power of 6 kW, and beam diameter of 20 um are chosen to conduct the experiment. For a wavelength of 1064 nm provided by this laser system, a coefficient of absorption of 0.3 is used for stainless steel in the calculations [4]. To alter the power of the system, the voltage of the laser system reducing in 25V intervals starting from 400V to 225V is used. This trend was applied to three different pulse durations of 20 ms, 10 ms, and 5 ms. This study provided a very confined and controlled experimental environment for the required observations. Upon completing the experiment, the effects of each of the parameters on the surface structure and heat-affected zones were fully examined.

METHODOLOGY

In order to proceed with the analytical examinations, the nondimensionalization method is introduced to track the changes of each parameter, by calculating the non-dimensional temperature at the surface and center [5]. To proceed with this method, a cylindrical symmetric flow without internal heat generation, and constant, isotropic material properties was assumed; from there a transient heat conduction equation and proper boundary condition, as shown in the following, were defined to develop the temperature history of the process to estimate the temperature at each point [6].

$$\frac{1}{r}\frac{\partial}{\partial r}\left(r\frac{\partial T}{\partial r}\right)+\frac{\partial^2 T}{\partial z^2}=\frac{1}{\alpha}\frac{\partial T}{\partial t} \quad (1)$$

where α is the thermal diffusivity, and is defined as $\alpha = k/\rho Cp$, with k being the thermal conductivity, and ρ the density, and C_p the specific heat.

$$\frac{\partial T}{\partial z}\Big|_{z=0} = -\frac{1}{k}\frac{\eta P(t)}{\pi r_0^2}\exp(-r^2/r_0^2) \quad (2)$$

where P is defined as the peak power delivered, η is the fraction of energy absorbed by the surface, and r_0 is the beam diameter (in this study the beam diameter is 20 um)

Taking into account the vast range of scale variation between the required parameters, it is more convenient to define a set of non-dimensional groups, and simplify the obtained equations. The customized non-dimensional variables are as follow [6]:

$$R = r/r_0 ; Z = z/r_0 ; \theta = T/T_{ref} ; \tau = 4\alpha t/r_0^2$$

where, R is the non-dimensional Temperature (It is assumed that reference temperature is the melting temperature of the material), Z is the non-dimensional depth, θ is the non-dimensional temperature and finally, τ is non-dimensional time for $t \le t_p$. These non-dimensional variables are then used to simplify Equations 1 & 2. Equation 3 displays the simplified boundary equation [7]:

$$\frac{\partial \theta}{\partial Z}\Big|_{Z=0} = -Q\exp(-R^2)\left[u(\tau) - u(\tau - \tau_p)\right] \quad (3)$$

where Q is the non-dimensional power and is defined as: $Q = \eta P/k\pi r_0 T_{ref}$

Through mathematical procedure, the differential equation of the redefined boundary condition was used to calculate the integral form of the temperature. Hence, the maximum surface temperatures of various pulse durations were obtained to show the non-dimensional temperature profiles under each condition. Equations 4, and 5 define temperature profiles at along radios and center respectively:

$$\theta(R, Z = 0, \tau_p) \approx \frac{Q}{\sqrt{\pi}}\arctan(\sqrt{\tau_p})\exp(-R^2) \quad (4)$$

$$\theta(0,0,\tau) = \frac{Q}{\sqrt{\pi}}\sqrt{\tau} \quad , \text{for } \tau < \tau_p \quad (5)$$

Comparing the temperature profiles at each case, consequently leads to the most optimal process parameters, useful in processing applications. An important theory to be considered when working with non-dimensional temperatures is the behavior of the temperature profile in cases of constant power and constant energy. When working with constant power, having a shorter pulse duration results in a lower non-dimensional temperature [5]. This is because, with a constant power being delivered by the laser, a shorter time results in less heat being accumulated, and hence lower non-dimensional temperature will be achieved. On the contrary, when using constant energy, the non-dimensional temperature is higher with shorter pulse durations. In other words, energy is delivered in a more concentrated manner with shorter pulse durations; hence higher non-dimensional temperatures are resulted [5].

RESULTS AND DISCUSSION

Multiple irradiated spots are created on the surface of the sample, upon completion of the experiment. Figure 1 displays the optical images taken of the fifth points of 20 ms, 10 ms, and 5 ms pulse durations. It can be seen that the size and surface texture of each spot changes based on the corresponding voltage and pulse duration.

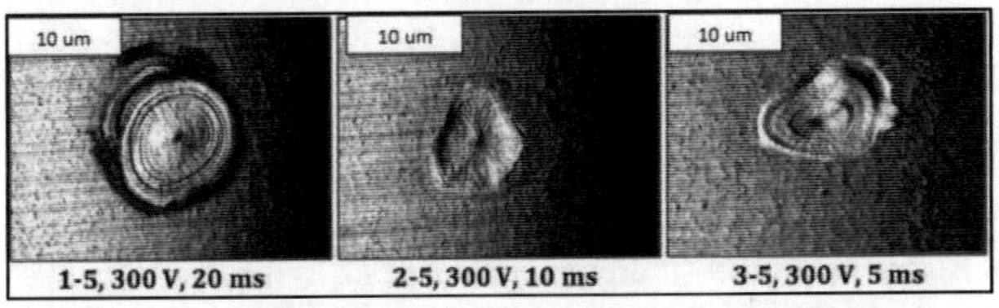

Figure 1: Optical Images of Irradiated Spots (5X Mag)

Creation of different zones on surface of each irradiated spot is due to the keyhole generation phenomenon that takes place during any laser surface texturing process [4]. Looking at Figure 2, the irradiated spot is almost consisted of three zones as displayed:

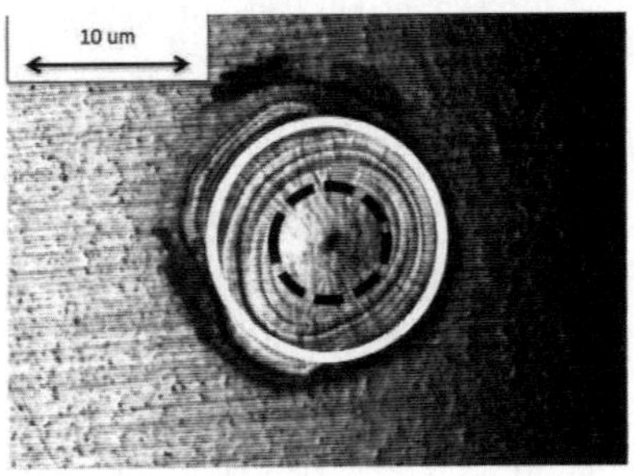

Figure 2: Surface Layers of Irradiated Spot

As it can be seen, the outer area, which is indicated with the solid circle, marks the secondary heat affected zone. The next layer shown with the dashed circle indicates the primary heat affected zone (HAZ), and is the center of the penetration zone. Lastly, the area appearing as waves on the surface of the irradiated spot signifies the molten area caused by the accumulation of heat on the surface of the work piece [4]. When processing materials using laser, controlling the size of each of these layers is an essential factor in the precision and quality of the final results. Figure 3 illustrates the variation of secondary HAZ based on intensity values, which are correspondence to the voltages used.

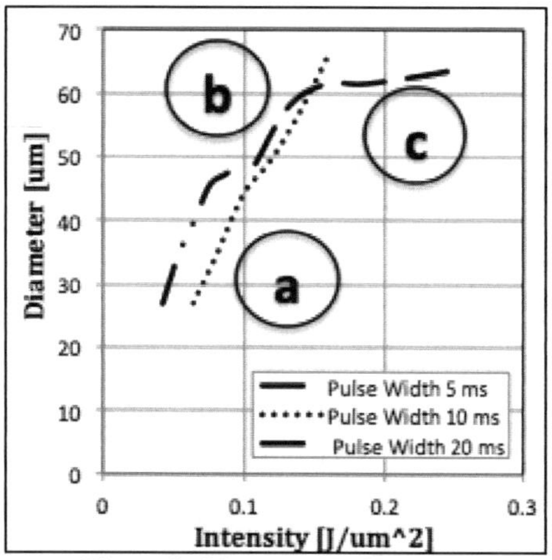

Figure 3: Diameter Vs. Intensity

This figure displays the trend of change in diameter according to the intensity. It is important to note that, in this study, intensity is obtained based on the energy delivered to the work piece. As expected the diameter of the HAZ increases with an increase in intensity. As shown, at a constant intensity (energy per area), the shorter pulse duration results in a larger outer diameter. In the case of a constant intensity, the amount of available energy is limited; hence, when using shorter pulse duration, the constant energy will be delivered in a more concentrated manner, since the process is not long enough for the energy to spread across the material. This then results in more heat accumulation, and therefore a larger HAZ diameter will result using shorter pulse duration [5].

After studying the secondary HAZ, in accordance to intensity and pulse duration, exploring it at various key points is necessary in understanding its behavior. Hence, the important zones at which significant changes occur in the behavior of outer diameter, are indicated on Figure 3. Starting with zone **"a"**, the trend of changes is as expected; the outer diameter increases with an increase in intensity, and shorter pulse duration results in a larger diameter value at a constant intensity. However, when reaching zone **"b"**, it can be seen the diameters starting to reach similar values, particularly for 10 ms and 20 ms pulse durations. At this point, effects of pulse duration on HAZ start to decrease as the intensity increases. This indicates the temperature of material is starting to reach the ablation temperature threshold, and hence material is reaching the saturation state. Laser ablation is generally defined as material removal in microscopic scale from the surface of the work piece due to significant changes in the state, where the ablated amount transforms into gas or plasma phase [8]. In case where laser radiation reaches a sufficiently high intensity, the surface of the material starts ionization, and transforms into a form of dense plasma. When the delivered energy by the laser is increased while the pulse duration is decreased, the accumulated heat will gradually exceed the breakdown threshold of the material. Thus, ablation is expected [8]. Hence, looking at zone **"c"**, it is clear the material has reached the saturation point, and the HAZ remains constant despite the changes in intensity. Reaching the saturation point suggests the material has exceeded its ablation threshold at this point, and as the intensity increases the material starts being evaporated, hence the diameter of the heat affected zone remains constant [9]. Determining the threshold for ablation temperature is very essential in laser processing of material. Depending on the application required, the intensity and pulse duration could be controlled to give the desired results in any of the presented zones. For lasers with Gaussian beam profile, if heat propagation effects are neglected, the future size of the ablated area can be calculated by [10,11]:

$$D^2 = 2\omega_O^2 \{\ln \phi_O - \ln \phi_{th}(N_s)\} \tag{6}$$

where, ω_O is the radius of laser spot, $\varphi_{th}(N_s)$ is the multi-pulse ablation threshold which depends both on the number of pulses and material and φ_O is the maximum laser flounce which is given by:

$$\phi_O = \frac{2E_{pulse}}{\pi\omega_O^2} \tag{7}$$

From Eq. 6 and 7, the one pulse ablation threshold of the stainless steel can be calculated by:

$$\phi_{th}(1) = 2E_{pulse} \Big/ \pi.\omega_O^2.e^{0.5(D/\omega_O)^2} \tag{8}$$

Figure 4 shows the computed results for $\varphi_{th}(1)$ at different pulse durations based on the analytical solution. As shown in Figure 4, the ablation threshold for one pulse decreases with decrease in the pulse duration. Also, the threshold nearly reaches saturation at diameter of 60 um which is in a good agreement with our experimental results shown in Figure 3.

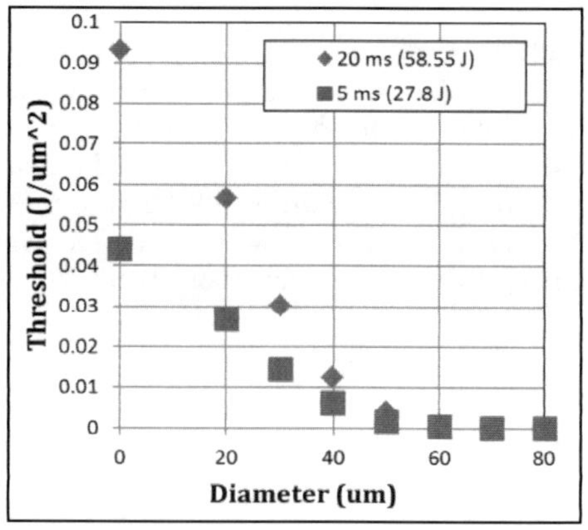

Figure 4: Computed Results for Ablation Threshold

Next, variation of non-dimensional temperature along the surface of the irradiated spot was examined. Figure 5, illustrates this variation. As shown, the non-dimensional surface temperature is at its maximum at the center of the irradiated spot, and gradually decreases along the radius of the spot, and eventually reaches a constant value at the room temperature. The three layers created on surface of the irradiated spot can easily be shown along these curves. As indicated in the figure by the lines on the curves, the middle zone corresponds to the molten area indicated earlier.

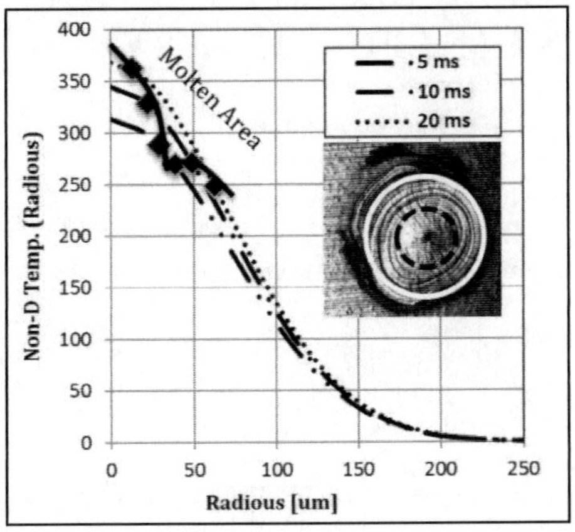

Figure 5: Temperature Trend Across the Surface

As shown in our experimental results, at shorter pulse durations, the secondary heat affected zone (molten area) is significantly smaller than the longer pulse durations; which can clearly be seen in Figure 5. The bottom dashed line, the indicator for pulse duration of 5 ms, has the smallest part in the middle section of the curves compared to the dashed and point lines. Since shorter pulse duration, deliver the energy in a more concentrated manner; they involve rapid energy deposition and creation of vapor and plasma phases, hence results in absence of molten area [12]. This reduces the effects of energy on the surrounding of the target area (heat accumulation). Hence, the non-homogenous aspect of energy deposition, which causes the wave-like patterns, is reduced. Hence, shorter pulse durations result in a better surface resolution, and smaller heat affected zone based on the application [12].

CONCLUSIONS

To conclude, a stainless steel type 304 sample was used to investigate the effects of pulse duration and power on the surface structure and patterns of materials, during laser processing. Also, a method for obtaining the ablation threshold of stainless steel for one pulse was introduced. After a thorough analysis, it was observed that pulse duration is an essential factor in controlling the quality of the process materials. In order to obtain a better surface resolution, with minimum damage to the surrounding areas of the irradiated spot, shorter pulse duration should be used. For applications involving surface, larger pulse durations are recommended due to their larger heat affected zone diameters. Further, the introduced method for the ablation calculation showed that the ablation threshold for one pulse decreases with decrease in the pulse duration, and the results were in a very close agreement to the obtained experimental results.

REFERENCES

[1] J. Majumdar and I. Manna, "Laser Material Processing," *International Materials Reviews,* vol. 56, pp. 341-388, 2011.

[2] M. Harooni, B. Carlson, B. R. Strohmeier and R. Kovacevic, "Pore formation mechanism and its mitigation in laser welding of AZ31B magnesium alloy in lap joint configuration," *Materials & Design, 58*, pp. 265-276, 2014.

[3] T. Ahmmed, E. J. Yang Ling, P. Servio and A. kietzig, "Introducing a New Optimization Tool for Femtosecond Laser-Induced Surface Texturing on Titanium, Stainless Steel, Aluminum and Copper," *Optics and Lasers in Engineering,* vol. 66, pp. 258-268, 2014.

[4] K. Kazemi and J. A. Goldak, "Numerical Simulation of Laser Full Penetration Welding," *Computational Materials Science,* vol. 44, pp. 841-849, 2009.

[5] A. Kiani, K. Venkatakrishnan and B. Tan, "Enhancement of the Optical Absorption of Thin-film of Amorphorized Silicon for Photovoltaic Energy Conversion," *Solar Energy,* vol. 85, pp. 1817-1823, 2011.

[6] P. Woodard and J. Druden, "Thermal Analysis of a Laser Pulse for Discrete Sopt Surface Transformation Hardening," *Applied Science,* vol. 85, pp. 2488-2496, 1998.

[7] A. Kiani, K. Venkatakrishnan and B. Tan, "Optimal Absorption Enhancement in 3D Silicon Oxide Nano-sandwich Type Solar Cell," *Optics Express,* vol. 22, 2013.

[8] S. Nolte, C. Momma, H. Jacobs and A. Tunnermann, "Ablation of Metals by Ultrashort Laser Pulses," *Optical Society of America,* vol. 14, pp. 2716-2722, 1997.

[9] L. A. Dobrzanski and A. Drygala, "Laser Processing of Multicrystalline Silicon for Texturization of Solar Cells," *Material Processing Technology,* vol. 191, pp. 228-231, 2007.

[10] K. Venkatakrishnan, P. Stanley, N.R. Sivakumar, B. Tan, L.E.N. Lim, *Appl. Phys. A* **77**, 655-658 (2003).

[11] A. Borowiec, H.K. Haugen, Appl. Phys. A 79, 521-529 (2004).

[12] M. Erdogan, B. Oktem, H. Kalaycioglu and S. Yavas, "texturing of Titanium (Ti6A14V) Medical Implant Surfaces With Mhz-Repetition-Rate Femtosecond and Picosecond Yb-Doped Fiber Lasers," *Optic Express,* vol. 19, pp. 10986-10996, 2011.

MEASUREMENT METHODS AND DATA REPRESENTATIONS FOR MECHANICAL PART REFURBISHMENT / REPLACEMENT

H.M.H. Bubshait
Mechanical Engineering
McMaster University
Hamilton, ON, CANADA
bubshahm@mcmaster.ca

Allan D. Spence
Mechanical Engineering
McMaster University
Hamilton, ON, CANADA
adspence@mcmaster.ca

ABSTRACT

Contact and non-contact measurement systems are used in a wide range of dimensional metrology applications. Since each system and associated Computer Aided Design (CAD) sample point representation has its own strengths and weaknesses, it is proposed to explore the possibility to combine these measurement systems to reap the best benefits of each system. The paper includes a brief introduction of the common contact and non-contact measurement methods, as well as a review of existing hybrid systems and CAD representations. An included example part measurement illustrates these ideas.

KEYWORDS:
CMMs, Hybrid Measurement, CAD Data Representation

INTRODUCTION

When an in-service component fails, significant down time losses may result. A broken combine means that a farmer cannot harvest crops. A broken aircraft component means a grounded plane. A navy ship may be vulnerable to enemy attack or storm if a critical part fails. A space station may no longer be able to sustain life if a critical component fails. In many cases the device or component is many years old, and CAD (Computer Aided Design) geometry files may not be available. The location may be geographically distant from a machine shop. Therefore, rapid methods for digitizing, conversion to CAD, and refurbishment or replacement using a minimum amount of material, and involving lay persons without narrow technological expertise, are desirable.

To geometrically represent a part, it must be digitized and the resulting sample points converted to a CAD representation that can as required be visualized, analyzed, and manufactured. Digitizing the part geometry can be accomplished via touch probe contact systems or non-contact devices such as laser scanners. Each of these measurement systems has its advantages and disadvantages. Thus, it is proposed to investigate complementary (best of both systems) solutions.

LITERATURE REVIEW

Capturing part geometry using laser scanners can be fast. Often, however, touch probe measurement is required to verify compliance with mating surface Geometric Dimensioning & Tolerancing (GD&T) specifications. For example, planes and holes require this more accurate measurement method. Laser scanner data is less accurate and cannot capture the hole locations. Moreover, the datum planes must be flat. To solve this issue, it is proposed to investigate a method that can combine the benefits of a portable CMM and laser scanner to achieve a hybrid or grafted part CAD model.

Carbone et al. [5] discussed combining a non-contact vision system with a contact CMM. The main idea is to capture the geometry of the part using the vision system to create a rough CAD model that can be used as a guideline for the CMM. After several iterations to reach the desired results, the model can be exported as IGES file to a CAD environment. Case studies are included for reconstruction of a door handle and turbine blade.

Huang and Qian [11] proposed a method to integrate laser scanners with CMM data acquisition. The laser scanner will be used to capture the desired area at first. Points with the maximum measurement uncertainties will then be determined. Next, the CMM is used to measure the points with uncertainties and the whole model will update dynamically as the CMM measurements are acquired.

A recent publication by Li et al. [12] proposed a new method based on competitive multi sensors approach. The method utilizes the competitive and complementary multi-sensors configurations to capture the geometry of the part. The process will start by capturing the whole geometry of the part using laser scanner. Intelligent feature recognition and segmentation algorithms will then be applied to the point cloud data. Finally, the touch probe will re-measure the features that need higher accuracy.

Liu and Wang [13] proposed a method to combine the data of laser scanner and CMM independently. That is, each measurement can be carried out separately. The laser scanner can be used to capture the overall geometry of the part resulting in a triangular mesh. The CMM measures the main features producing primitive shapes. Both data measurement sets will be registered into a common coordinate system.

The concentration of the current paper is to perform practical measurements using lay person level instruments, with the resulting CAD data merged or grafted so that subsequent optimization and 3D printing or other manufacturing is streamlined.

CONTACT AND NON-CONTACT DIGITIZING

TOUCH PROBE EQUIPPED COORDINATE MEASURING MACHINES

Conventional Coordinate Measuring Machine (CMM) systems use a touch probe to obtain Cartesian sample point coordinates on the subject part [3]. That is, under computer control, the touch trigger probe of the CMM will contact the surface of the part and register the x, y and z coordinates of the measured point. Typically the data is analyzed with respect to Geometric Dimensioning & Tolerancing (GD&T) specifications, and results are displayed to summarize measured sizes, distances, angles, form, etc.

Non-Cartesian CMMs include articulated arms. These systems have either 6 or 7 rotary degrees of freedom and are manually manipulated. The touch probe is rigid, and sample points are recorded when the operator holds the probe against the part surface and presses a trigger switch. Rotary encoder outputs are converted to Cartesian to determine the corresponding x, y, and z coordinates [8]. Compared to the conventional CMM, articulated arms are less accurate and highly dependent on the skills of the operator. On the other hand, the articulated arm is portable and versatile.

NON-CONTACT DIGITIZING

Although contact based systems are accurate to better than 10 μm, even with a continuous analog touch probe the method of data point collection is limited to a sequence of individual points at a rate no higher than a few hundred points per second. This is much too slow for large areas on complex surfaces. Hence non-contact digitizers such as laser scanners, which collect over 100,000 points per second, are attractive for these circumstances. Resolution and accuracy is in the 50 μm to 100 μm range, and measurement capability deteriorates on shiny and dark surfaces. A comparison of digitizing methods is reported in [1].

CAD REPRESENTATION

BOUNDARY REPRESENTATION

Representing a CAD model can be achieved by different methods. One such method is the boundary representation (Figure 1) [16]. The geometry of a model can be broken down to surfaces, curves and points. When these geometric elements are bounded, the model will have faces, edges and vertices. To complete the skin of the solid, the faces have to be bounded without any missing parts. This can be achieved by supplying the geometric model with topology information defining the relationship between the faces.

POINT CLOUD MESH

In non-contact digitizers, the geometric data is collected in the form of three dimensional points with varying densities. These points are unorganized and measurement motion sweeps overlap with each other. One of the approaches used to convert point cloud data to CAD is the polyhedral mesh [17] in which the shape of the object is approximated using a triangular mesh that is optimized by removing overlapping points. An example of a point cloud mesh can be seen in Figure 2.

In the ensuing CAD models, low density measurements from contact probes are normally represented by using Least Squares and Chebyshev feature fitting methods with substitute features such as planes, cylinders, etc. [14,15]. These surfaces are then combined into a standard B-rep model. Digitizer data for complex surfaces is represented using a mesh. Illustration of this procedure for a sample part is summarized below.

EXPERIMENTAL PART PROCESSING

To illustrate the workflow, a small (damaged) hose support bracket, for which no CAD model is available, was measured (Figure 3) using both a Creaform REVscan [7] laser digitizer with GeoMagic [10] software, and a FARO [9] articulating arm and BuildIT [4] software (Figure 4). Hole data was difficult to acquire due to shadows. To overcome this, the FARO arm was used. Created plane and cylinder (hole) data is shown in Figure 5.

To merge the data, the FARO measurements are used as orthogonal datum planes features (Figure 6). Using GeoMagic software, the point cloud mesh from REVScan is aligned with respect to these datum (Figure 7). At present the point cloud mesh is manually trimmed by the more accurate FARO hole features. A resulting 3D printed part is shown in Figure 8.

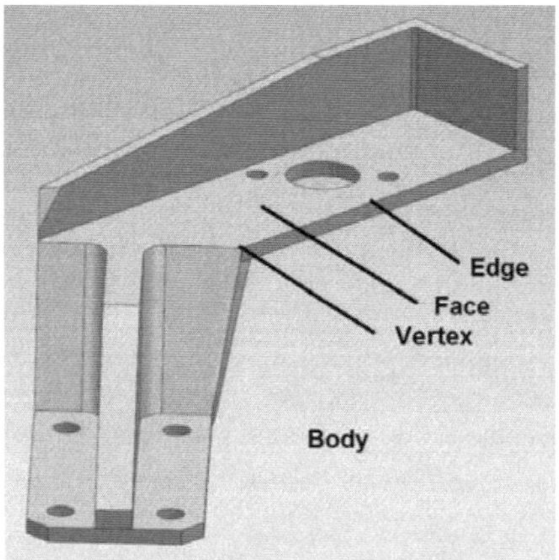

Figure 1: Boundary Representation

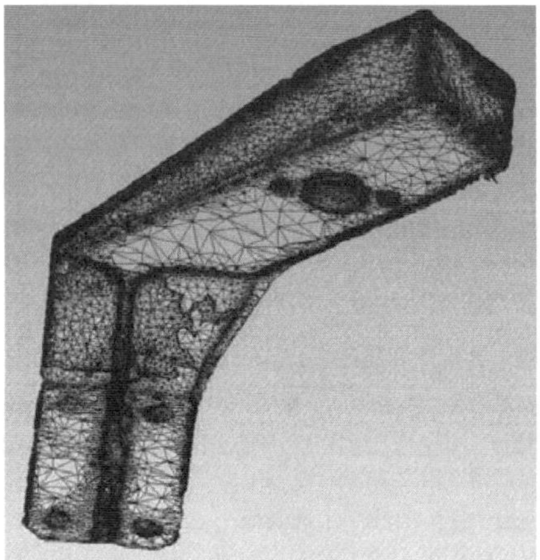

Figure 2: Triangular Mesh

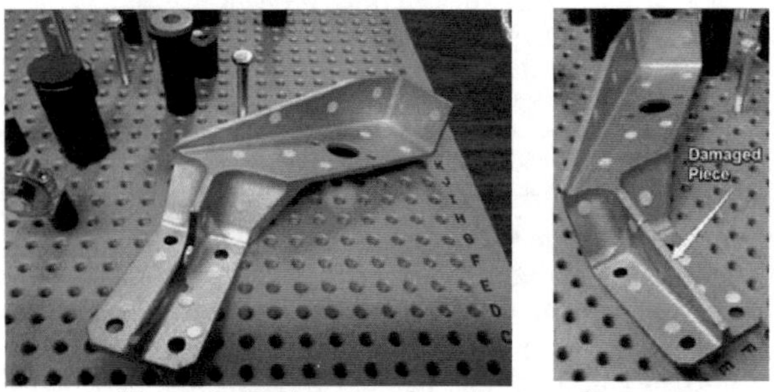

Figure 3: Broken hose bracket sample part

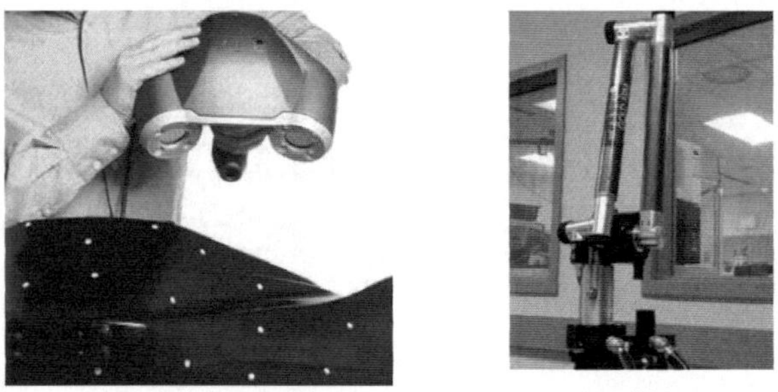

(a) (b)

Figure 4: (a) REVscan [7] (b) FARO Arm [9]

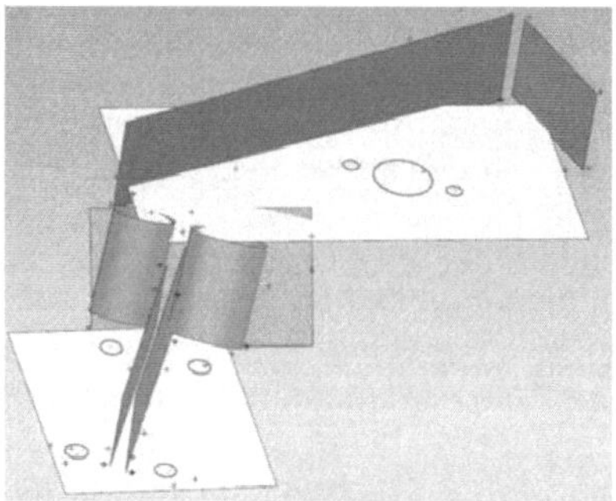

Figure 5: FARO Arm Measurements (features)

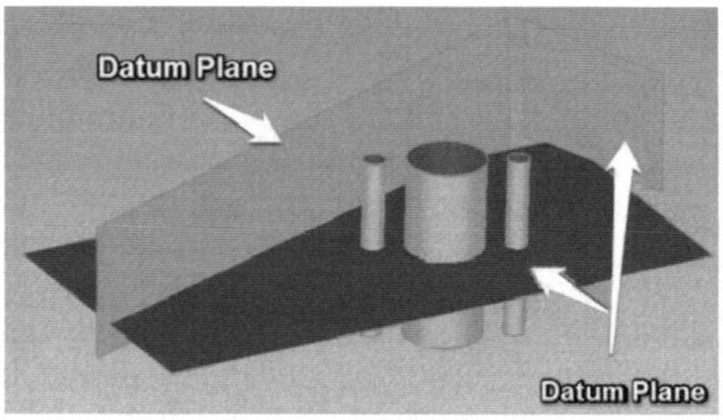

Figure 6: FARO datum planes

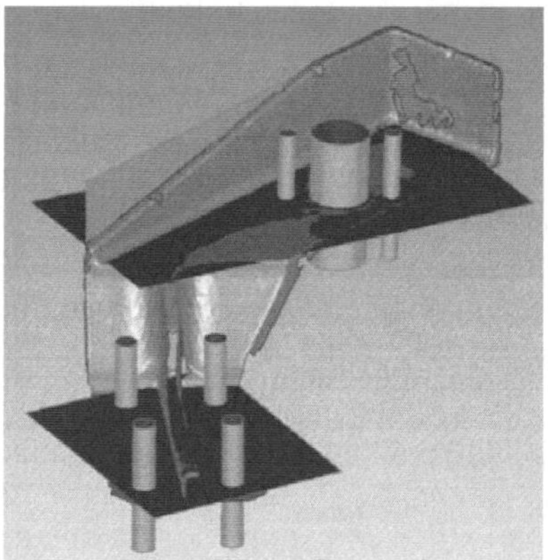

Figure 7: REVScan-FARO data alignment

TOPOLOGY OPTIMIZATION AND ADDITIVE MANUFACTURING

The purpose of topology optimization is to find the optimal distribution of structural material, and requires information about the loads, support conditions, feature locations and volume of the part [2]. It is most effective to integrate it with CAD to obtain a functional model for ultimate realization of the product by machining, additive manufacturing, etc. Additive manufacturing is defined by the American Society for Testing and Materials (ASTM) as "a process of joining materials to make objects from 3D model data, usually layer upon layer, as opposed to subtractive manufacturing methodologies" [6]. It is an ideal complement to topology optimization reverse engineering. Currently reported research performs these steps manually, and ongoing research will concentrate on automation of the workflow.

Figure 8: Resulting 3D printed part

CONCLUSIONS

There are several hybrid systems strategies that can be used to automate the process of data acquisition to capture the geometry of the part. Non-contact digitizers collect a large number of points quickly, but with limited ability to accurately measure holes and mating datum features. Contact systems complement these deficiencies. Grafting the data from both systems together provides advantages for downstream shape and topology optimization, and for subsequent 3D printing or other manufacturing methods. At present the grafting is a manual procedure, and additional work is needed to automate this for practical application.

ACKNOWLEDGMENTS

This work was financially supported by Saudi Arabian Government Scholarship Program administered by Saudi Arabian Cultural Bureau in Canada (Bubshait), and a Discovery Grant from the Natural Sciences and Engineering Research Council of Canada (Spence). BuildIT provided the FARO arm interface software.

REFERENCES

1. Barbero, B.R., and Ureta, E.S. 2011, "Comparative study of different digitization techniques and their accuracy" *Computer-Aided Design* 43(2), pp. 188-206.

2. Bendsøe, M.P., and Sigmund, O. 2004, *Topology Optimization*, 2nd Ed., Springer, Berlin.

3. Bosch, J.A., Ed. 1995, *Coordinate Measuring Machines and Systems*, CRC Press, Boca Raton, FL.

4. BuildIT Software and Solutions, Ltd., Montréal, Québec, www.builditsoftware.com.

5. Carbone, V. et al. 2001, "Combination of a Vision System and a Coordinate Measuring Machine for the Reverse Engineering of Freeform Surfaces" *Int. J. Adv. Manuf. Tech.*, 17(4), pp. 263-271.

6. Chua, C.K., and Leong, K.F. 2014, *3D Printing and Additive Manufacturing: Principles and Applications*, World Scientific Publishing Co., Singapore.

7. Creaform Inc., Lévis, Québec, www.creaform3d.com.

8. Desai, S., and Bidanda, B. 2006, "Reverse Engineering: A Review & Evaluation of Contact Based Systems", *Rapid Prototyping*, Springer, Berlin, pp. 107-131.

9. FARO Technologies Inc., Lake Mary, FL, www.faro.com.

10. GeoMagic 3D Systems, Cary, SC, www.geomagic.com.

11. Huang, Y., and Qian, X. 2007, "A Dynamic Sensing-and-Modeling Approach to Three-Dimensional Point- and Area-Sensor Integration." *ASME J. Manufacturing Sci. Eng.*, 129(3), pp. 623-635.

12. Li, F. et al. 2014, "Rapid and Accurate Reverse Engineering of Geometry Based on a Multi-Sensor System." *Int. J. Adv. Manuf. Tech.,* 74(1-4), pp. 369-382.

13. Liu, P. X., and Wang, Y. 2011, "Model Construction Based on CMM and Optical Scanning Data for Reverse Engineering". *Global Product Development*, Springer, Berlin. pp. 691-696.

14. Shakarji, C.M. 1998, "Least-Squares Fitting Algorithms of the NIST Algorithm Testing System", *J. NIST*, 103(6), pp. 633-641.

15. Shakarji, C.M., and Clement, A. "Reference Algorithms for Chebyshev and One-Sided Data Fitting for Coordinate Metrology" 2004, CIRP Annals 53(1), pp. 439-442.

16. Zeid, I. 2005, *Mastering CAD/CAM*, McGraw-Hill, Boston, MA.

17. Zhang, Y. F., Y. Wong, S., and Loh, H.T. 2008, "Relationship between Reverse Engineering and Rapid Prototyping", *Reverse Engineering,* Ch. 6, pp. 119-139.

CURRENT STATE-OF-THE-ART IN MODELING AND SIMULATION OF LASER POLISHING PROCESS

Shirzad Mohajerani
Mechanical and Materials Engineering
Western University
London, Ontario, Canada
smohaje@uwo.ca

Evgueni Bordatchev
National Research Council of Canada
London, Ontario, Canada
evgueni.bordatchev@nrc-cnrc.gc.ca

O. Remus Tutunea-Fatan
Mechanical and Materials Engineering
Western University
London, Ontario, Canada
rtutunea@eng.uwo.ca

ABSTRACT

Among the multitude of machining operations, polishing is a finishing process whose primary purpose is to significantly enhance the quality of the workpiece surface. Conventionally, most polishing methods such as grinding, lapping or chemical-mechanical planarization are based on the removal of a thin layer of material. Contrasting with them, laser polishing (LP) is a new and advanced non-contact finishing technique that could have an important effect on the surface roughness. Similar other machining operations, LP is associated with significant number of physical phenomena that coexist in the laser-material interaction zone. The primary objectives of this study were: i) to identify the nature of these overlapping phenomena and ii) to describe the progress made so far with respect to their theoretical and numerical analysis. To address these goals, a synthesis of the theories and governing equations used so far in modeling of the LP will be presented in this study. It is anticipated that this systemized approach will constitute a foundation for future studies attempting to optimize LP process parameters and to formulate predictions with respect of the quality of the post-LP surface while accounting for initial surface geometry and workpiece material properties.

KEYWORDS: Laser polishing, laser-material interactions, physical processes, modelling, simulation, review

INTRODUCTION

Laser polishing (LP) by the material re-melting is a relatively new finishing technology has been under extensive development under the past decade, mainly due to its high productivity as well as complete and facile integration with CNC technology. To support this, it could be mentioned that the CNC-driven LP rate could reach 1 min/cm^2 which strongly contrasts with 30 min/cm^2 that is typically associated with conventional manual polishing techniques applied on flat surfaces. Generally speaking, two types of LP processes are typically acknowledged in the literature, namely macro and micro LP, respectively. The distinction between them is typically related to laser beam diameter as well as re-melting depth. Laser macro-polishing is typically performed by means of high-power continuous-wave lasers that induce a melting depth between 20 μm and 200 μm on surfaces with poor initial surface roughness (S_a >1 μm) [1]. Following the initial melting, the relatively large layer of molten material solidifies in a configuration that is somewhat less dependent on its initial shape. By contrast, laser micro-polishing typically relies on pulsed laser radiation in the range of micro- or nanoseconds [2], or even picoseconds [3]. In this latter case, LP is based on the re-melting a relatively thin (< 5 μm) top layer of workpiece material by means of the laser radiation. In this case, the reduction of the surface asperities/roughness is caused by the redistribution of the molten material under the surface tension, a process followed by the rapid solidification of the redistributed material. A schematic of the LP process is depicted in Fig. 1.

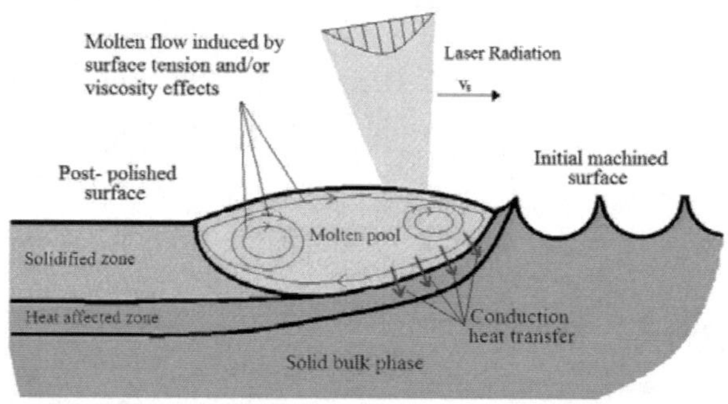

Figure 1: Laser polishing and its phenomena

The present study will focus further on the discussion of the progress made so far on the identification and modeling of the principal physical phenomena that were deemed so far responsible for rapid melting and solidification that occurs during LP. The mathematical models and governing equations of these phenomena will also be reviewed in order to ascertain future directions of investigation.

PHYSICAL PROCESSES DURING THE LP

It is well known that a broad palette of physical heat and mass transfer mechanisms underlie the LP process and their diversity significantly increases the complexity of any theoretical/numerical model to be developed for outcome prediction purposes. During the incipient phases of LP, the energy of the laser beam is transferred to the workpiece surface by means of the radiation and this in turn leads to the creation of a pool of molten material. The amount of energy delivered to the surface has to be precisely controlled in order to avert the onset of ablation that could induce vaporization of the workpiece material. This can be achieved through well-controlled adjustments of the laser power, working distance (*i.e.* the distance between the focal plane of the laser lens and workpiece surface), pulse duration and scanning speed of the laser beam. Following radiation, laser energy propagates through the workpiece material by means of conduction. The energy received from the beam induces increased oscillations of the atoms in workpiece material. The atoms eventually break off from their lattice and this will trigger the phase change associated with melting. It is presently assumed that conduction determines the travelling speed of the solid-liquid boundary as well as the geometry of the pool for molten material. The phenomenon of convection is typically accompanied by a transfer of the heat through the interface between the outer surface of the molten material pool and its surroundings. It is worthwhile to mention that some amount of the heat is lost through the convection to the surrounding area; however it is much lesser than the thermal propagation into the workpiece body during the rapid solidification process.

During LP (see Fig. 1), laser energy determines a continuous flow of the molten material of the surface peaks into the adjacent valleys and this in turn has an overall leveling effect on the polished surface. The primary driver of this modification is the surface tension caused by thermal gradients that are present in the molten material. If the gravity forces are neglected because of the small scale of the molten material pool [4], the flow of the molten material could be regarded as being controlled by the opposite effects of surface tension and viscosity. The time required for an asperity to be melted and then leveled out should be typically longer than the time required for heating and solidification of the asperity through thermal diffusion [5]. Another important requirement of LP demands that the depth of the molten material to be lower than the maximum valley of the surface. This is necessary to provide sufficient space and time to allow the flow of molten material to travel from peaks to valleys before becoming solid. As it shown in Fig. 1, a peripheral heat affected zone (HAZ) is inevitably produced around the molten material pool through conduction and propagation. As such, it would be reasonable to predict that the use of pulsed laser with short or even ultrashort pulse duration will yield narrow HAZs that are critically desirable in LP [6].

MODELING AND SIMULATION OF LASER POLISHING PROCESS

A review of the modeling and simulation efforts that were reported so far reveals that almost without any exception they involved discretizations in time domain performed by means of numerical methods such as finite elements and/or finite volume. To exemplify this, several models available in the surveyed literature will be presented further.

Dutto *et al.* [7] proposed a one dimensional unsteady model in order to find the melting threshold (energy per unit area applied before the start of the melting) and surface temperature during LP of silicon carbide. In this model, conduction was assumed as the prominent heat transfer mode. Other attempt was made by Shen *et al.* [8] who have suggested an analytical 1D model to calculate the temperature before melting as well as the liquid-solid boundary after melting. A different approach was proposed by Ramos *et al.* [9] who have suggested a thermo-physical model of LP by assuming an exponential decay of the surface height after solidification. In this case, surface tension gradient was regarded as the dominant factor influencing the flow in the liquid phase and the final surface height after solidification.

A modified fixed domain method has been used by Mai *et al.* [10] in order to trace the liquid-solid boundary in a molten pool dominated by 2D conduction and yielded during polishing of 304 stainless steel. The study relied on Stefan boundary equation to determine the exact solid-liquid boundary in the grid element. By contrast, Liu *et al.* [11] used Laplace transform to solve the heat transfer equation in conduction in order to allow them to forecast temperature gradient and solidification velocity during the irradiation of a magnesium alloy. Shao *et al.* [12] proposed an 1D unsteady model to correlate the melting depth and laser irradiation time. The study has made a number of simplifying assumptions with respect to the ypical geometry of the asperities, the dominant conduction mode of the heat, the uniform laser energy distribution in the spot area and concluded that nanosecond lasers are effective in polishing of metallic surfaces. Other researchers such as Yilbas *et al.* [13] utilized an enthalpy porosity model involving finite volume method to describe a mushy (*e.g.* partially solidified) zone and then predict the melt depth during the phase change.

More recently, Perry *et al.* [14] have investigated the effect of laser pulse duration and scanning speed during the pulsed laser micro-polishing (PLμP) of Ni by means of a two-step simulation. More specifically, during the first step, the melt depth and laser pulse duration have been estimated through a 1D transient finite element model. During the second step, the uneven initial surface was simulated as a combination of the Fourier components with different spatial frequencies. Perry *et al.* concluded that if melt duration is considerably longer than the damping time for a specific Fourier component, a significant reduction in the amplitude of the specific surface geometry component will occur.

Another effort in the simulation of the laser-material interaction was presented and validated by Sowdari *et al.* [15] whose enthalpy-based finite element model for temperature distribution and melt depth hypothesized that laser beam is stationary during irradiation. Along the same line of thoughts, Bohun *et al.* [15] developed a predictive model based on lubrication theory. The 1D unsteady model used to analyze the volumetric laser heating and LP of the glass revealed that for smoothening of the glass surface at specific height, the heat delivered by the laser should penetrate at least 50 times deeper.

In their comprehensive approach built on prior work [14], Vadali *et al.* [4] relied on a transient 2D axisymmetric heat transfer model to predict melt depth and duration. In this work, the surface of the molten material pool was modeled as an oscillating wave on which viscosity and surface tension act as damping forces and a critical frequency affecting the most significant amplitude reduction was determined analytically. Similar considerations were made by Ukar *et al.* [16] whose model was based on the low-pass filtering of the spatial frequency spectrum for initial topography followed by the reconstructions of the surface by means of the remaining frequencies.

A statistical approach in modeling and analysis of surface geometry during LμP was developed for the first time by Chow *et al.* [17]. In this study, the authors have considered LμP as a single-input (original surface)/single-output (polished surface) dynamic system and experimental investigations of the dynamic operator LμP were performed by means of transfer functions involving statistical metrics. Other attempts in the analysis of LμP for Ti alloys were made by Ma *et al.* [18] who relied on an axisymmetric numerical model to assess melt depth and surface profile after solidification, but omitted to consider the effect of initial surface geometry.

DISCUSSION

The studies referenced above seem to collectively suggest that while the available models are capable to efficiently simulate the temperature distribution during the heating of the solid phase, the profile and the flow of the molten material are insufficiently known and understood. Furthermore, the effect of the surface tension during solidification was not analyzed in details and it is absolutely clear that the complexity of the calculations involved in all proposed models requires significant assumptions that may dramatically restrict their validity. It is important to note that even though simplifying assumptions are the "necessary evil" of any numerical simulation, the decision with respect to the phenomena to be ignored and the ones to the accounted for is of paramount importance with respect to the reliability and feasibility of the developed model. Finally, the topography of the initial surface has to be considered in much greater details for more accurate predictions of the LP outcome.

All in all, LP process should be modelled as a complex thermo-physical system capable make predictions on the polished surface geometry by taking into account the actual process parameters, such as: laser wavelength, power, frequency, pulse duration, focal spot diameter, scanning speed, material, and initial surface geometry.

It is well known that material properties, such as density, conduction coefficient and absorption change during the solid-liquid material phase change. The absence of these properties models leads to the non-realistic and non-reliable simulation results. These properties are considered to remain constant during phase heating, the temperature distribution could be written as:

$$\frac{\partial T}{\partial t}+V_x\frac{\partial T}{\partial x}+V_y\frac{\partial T}{\partial y}+V_z\frac{\partial T}{\partial z}=\lambda\left(\frac{\partial^2 T}{\partial x^2}+\frac{\partial^2 T}{\partial y^2}+\frac{\partial^2 T}{\partial z^2}\right) \quad (1)$$

where T is the temperature, V_x, V_y and V_z are the velocity field components, and λ is the diffusivity:

$$\lambda = k/\rho c_p \quad (2)$$

where ρ, c_p and k are the density, heat capacity and thermal conductivity, respectively.

The pulsed laser radiation could be considered as one of the boundary condition for the Eq. (1). It could be modeled as a cylindrical shape and Gaussian distribution over time [3, 4, 18] as:

$$I(x,y,t)=I_0\exp\left[-2\left(x^2+y^2\right)/\left(D_B/2\right)^2-\kappa\left(t/\tau_p\right)^2\right] \quad (3)$$

where I_0 is the peak power reached at the center of the laser beam at $t=0$, D_B is the beam diameter, κ is a laser specific constant, and τ_p is the pulse width.

The initial surface roughness may affect the capability of the surface to absorb more laser radiation and to increase the depth of the molten material pool that should be considered in more details in our future studies [18].

For the solid-liquid boundary (between molten material pool and HAZ), Stefan condition can be expressed as:

$$\rho_s L\frac{\partial S(t)}{\partial t}=k_s\frac{\partial T_s(u,t)}{\partial u}-k_l\frac{\partial T_l(u,t)}{\partial u},\quad u=S(t) \quad (4)$$

where L is the melting latent heat of the material, u is the direction of melting, and $S(t)$ is the time-depended location of the solid-liquid boundary.

In order to find the velocity filed in Eq. (1), a momentum equation in the fluid phase – known as Navier-Stokes equation for non-compressible and Newtonian flows – should be solved by coupling with energy equation Eq. (1) in the molten material pool and considering surface tension as a boundary condition for the molten pool. As an example of in this regard, Vatsya and Nikumb [6] have proposed an axisymmetric model:

$$\frac{\partial V_z}{\partial r}+\frac{\partial V_r}{\partial z}-\frac{1}{\mu}P=\frac{1}{\mu_l}\frac{\partial \alpha}{\partial r}\quad,\quad z=0, \quad (5)$$

where V_z and V_r are the vertical and radial component of flow velocity, α is the surface tension, and the pressure value P is taken relative to the surrounding pressure. The relationship between the surface tension and the temperature can be described as:

$$\frac{\partial \alpha}{\partial r} = \frac{\partial \alpha}{\partial T}\frac{\partial T}{\partial r}, \qquad \frac{\partial \alpha}{\partial T} = const \tag{6}$$

CONCLUSIONS

Laser polishing – currently regarded as an innovative force-free polishing technique – can be effectively applied in different industries including tools and dies, automotive, aerospace as well as many others. The physical fundamentals of LP along with their modelling techniques have been reviewed in this study with a focus on identification of the principal physical phenomena underlying LP.

The diversity of thermo-physical phenomena involved in LP makes its simulation extremely difficult. The majority of models proposed so far seem rough and approximate primarily due to the simplifying assumptions made that most commonly ignore the effects of surface tension and/or initial surface topography. Based on this, it is believed that future simulation studies should focus on the development of a more profound understanding of these factors with a critical impact on the mechanics of the laser polishing.

ACKNOWLEDGMENTS

This paper is the result of collaboration between Western University (London, Ontario) and National Research Council (London, Ontario). Partial financial support was also provided by Natural Sciences and Engineering Research Council (NSERC) of Canada and AUTO21 Network of Centres of Excellence.

REFERENCES

1. Willenborg, E., 2011, "Polishing with laser radiation", In: *Tailored Light 2*, edited by Poprawe, R., Springer: Berlin/Heidelberg, pp. 196-202

2. Temmler, A., Willenborg, E., and Wissenbach, K., 2012, "Laser polishing", *Proc. of the SPIE*, 8243, #82430W

3. Bordatchev, E.V., Hafiz. A.M., Tutunea-Fatan, O.R., 2014, "Performance of laser polishing in finishing of metallic surfaces", *Int J Adv Manuf Tech*, 73, pp. 35-52

4. Vadali, M., Ma, C., Duffie, N.A., Li, X. and Pfefferkorn, F.E. 2012, "Pulsed laser micro polishing: Surface prediction model", *J. of Manuf. Proc.*. 14, pp. 307–15

5. Bohun, C.S., Cumberbatch, E., Fitt, A., Huang, H., Li, B., Soltannia, B., and Westbrook, R., 2011, "Modelling of CO2 Laser Polishing of Glass", *4-th Industrial Problem Solving Workshop*,, Montreal, Quebec, Canada

6. Vatsya, S.R. and Nikumb, S.K., 2003, "Modeling of fluid dynamical processes during pulsed-laser texturing of material surfaces", *Phisical Review B* 68, 035410

7. Dutto, C., Fogarassy, E., Mathiot, D. 2001," Numerical and experimental analysis of pulsed excimer laser processing of silicon carbide", *Applied Surface Science,* 184 (2001) 362–366

8. Shen, Z.H., Zhang, S.Y., Lu, J., and Ni, X.W. 2001, "Mathematical modeling of laser induced heating and melting in solids", *Optics & Laser Technology* , 33 (2001) 533–537.

9. Ramos, J.A., Bourell, D.L., and Beaman, J.J., 2003, "Surface Over-melt During Laser Polishing of Indirect-SLS Parts", *Mat Res Soc Symp Proc*, 758, pp. 53-61.

10. Mai, T.A., Lim, G.C. 2004, "Micromelting and its effects on surface topography and properties in laser polishing of stainless steel", *Journal of Laser Applications,* 15(4), pp. 221-8.

11. Liu, S.Y., Hu, J.D., Yang, Y. Guo, Z.X. and Wang, H.Y. 2005, "Microstructure analysis of magnesium alloy melted by laser irradiation", *Applied Surface Science* 252, pp. 1723–31.

12. Shao, T.M., Hua, M., Tam, H.Y. and Cheung, E.H.M., 2005, "An approach to modelling of laser polishing of metals", *Surface & Coatings Technology,* 197, pp.77-84.

13. Yilbas, B.S., Shuja, S.Z., Khan, S.M.A. and Aleem, A. 2009, "Laser melting of carbide tool surface: Model and experimental studies", *Applied Surface Science* 255, pp. 9396-403

14. Perry, T.L., Werschmoeller, D., Li, X., Pfefferkorn, F.E. and Duffie, N.A. 2009, "The Effect of Laser Pulse Duration and Feed Rate on Pulsed Laser Polishing of Microfabricated Nickel Samples", *J of Manufacturing Science and Engineering,* 131 / 031002-1

15. Sowdari, D. and Majumdar, P. 2010, "Finite element analysis of laser irradiated metal heating and melting processes", *Optics & Laser Technology*, 42, pp. 855-865.

16. Ukar, E., Lamikiz, A., Martínez, S., Tabernero, I. and López de Lacalle , L.N. 2012, "Roughness prediction on laser polished surfaces", *Journal of Materials Processing Technology,* 212, pp. 1305-13

17. Chow, M.T.C., Hafiz, A.M.K., Bordatchev, E.V., Tutunea-Fatan, O. R., and Knopf, G.K., 2010, "Experimental statistical analysis of laser micropolishing process", *Proc. of the Int'l Symp. on Optomechatronic Tech.*, Toronto, Canada, paper LPTI-1, 6 p.

18. Ma, Ch., Vadali, M., Duffie, N.A., Pfefferkorn , F.E and Li, X., 2013 , "Melt Pool Flow and Surface Evolution During Pulsed Laser Micro Polishing of Ti6Al4V ", *J of Manufacturing Science and Engineering,* 135 / 061023-1

Proceedings of the 25th CANCAM
London, Ontario, Canada, May 31-June 4, 2015

DEVELOPMENT OF ULTRASONIC SURFACE TREATMENT DEVICE

Yanjun Qian[1], H. J. Kwon[2],
Mechanical and Mechatronics Engineering
University of Waterloo
Waterloo, Ontario, Canada
moneyqyj813@gmail.com[1], hjkwon@uwaterloo.ca[2],

ABSTRACT

Surface mechanical properties, such as roughness and residual stress, are essential for tribological properties and fatigue life of mechanical structures. Recent research shows that using ultrasonic surface treatment (UST), it is possible to modify the surface topology into a designated pattern, and even make it a multifunctional surface. Inspired by the UST device that has been employed in surface enhancement practices, we are developing a piezoelectric UST device that may be much more efficient and versatile than other alternatives such as femtosecond laser or magnetostrictive method for patterning the material surface.

Keywords: Surface patterning, ultrasonic surface treatment

INTRODUCTION

The failures of the materials caused by repeated loading, corrosion, and wear have been big concerns to industries. Surface physical and chemical properties of mechanical components and parts play vital role in those failures.

In order to improve the surface property of the product against the failures, numerous methods of surface treatment have been proposed and put into practice. They can be varied by the nature of the treatment ways such as metallurgical, mechanical, and chemical or by adding a coating. One of the popular and effective treatment methods is a micro-scale mechanical process called ultrasonic surface treatment (UST).

Development of UST

In 1980s, some researchers found that mechanical properties and surface structure of the material could be altered by applying ultrasonic energy to the surface. They designed a device capable of performing ultrasonic shot peening on the object. After the treatment, the fatigue strength of the treated surface increased by 15-25%, which is caused by the residual compressive stress [1]. The distribution of dislocations and internal stresses in an ultrasonically strengthened layer is more uniform than in normal static deformation, leading to the improvement of fatigue performance.

In 2005, Pyoun et al [2] developed an experimental device to perform the ultrasonic cold forging technology (UCFT). Tests of the device on the automobile parts showed the improvement of surface hardness and surface roughness, which led to the performance increase in tribological behavior and service life [3]. Later on they continued to use the device to conduct experiments on various kinds of materials to determine the effectiveness of the ultrasonic nanocrystalline surface modification [4-7].

However, most of the studies described above are primarily focused on the change of mechanical properties such as strength and fatigue life. For the surface conditions, it is more focused on the depth of the nanocrystallized layer and the deformation state or cracks in the depth sections instead of much detailed surface topology studies. Not until the recent reports, they began to pay more attention to surface topology in detailed tribology studies [8].

New application of the UST

An application of femtosecond laser pulses has been recently reported that aims to create a multifunctional metal surface via surface modification [9]. The multifunctional surface exhibits combined effects of dramatically enhanced broadband absorption, superhydrophobicity, and self-cleaning. In this study they changed the topology of the material surface into a designated pattern that demonstrates the different properties. But due to the cost and time-consuming nature of the femtosecond laser pulses technique, it is hard to be introduced into massive production.

If such modification can be achieved by piezoelectric UST indenter, it should be a much more efficient and affordable method than the current femtosecond laser pulse device. So in this paper, we intend to design a piezoelectric UST device to properly modify the surface, and also provide a way to examine the surface pattern.

DEVICE AND METHODS

Structure of the ultrasonic indenter

The schematic of a typical horn-type piezoelectric ultrasonic indenter is shown in Fig. 1.

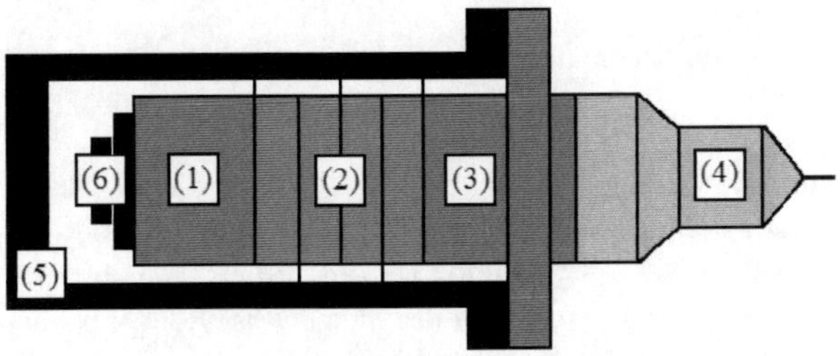

Figure 1: Structure of the ultrasonic indenter: (1) Back mass. (2) Piezoelectric ceramic stack. (3) Front mass. (4) Amplifier horn with a tip. (5) Back cover. (6) Preload bolt and nut.

The transducer comprises the piezoelectric ceramic stack, back mass, front mass and a horn [10]. The piezoelectric ceramic stack is clamped between back mass and front mass. A bolt is used to apply some tightening load to the structure in order to prevent the ceramics from failure.

Piezoelectric ceramic stack acts as the power source of the system. Responding to the high frequency input electrical signal, the ceramic stacks are designed to react in its thickness change, which can be observed as the ultrasonic oscillation induced by the piezoelectric effect. Since the amplitude of the oscillation generate by the ultrasonic transducer is inadequate for material processing, a horn with a certain profile is designed to magnify the amplitude at the processing end.

As for the surface treatment, the tip functions as the working part which makes direct contact with the processing surface. The material and the size for the tip should be carefully selected, because it will directly influence the outcome of the process.

In addition to the normal configuration, a back cover is introduced to serve as an interface to be coupled with other mechanical parts in our design, and it also provides some protections to the transducer.

Mechanism of the ultrasonic indenter

The shape of the tip at contact point can be simplified to semi-sphere (Fig. 2). The tip of the ultrasonic indenter will be reciprocating in Z direction at high frequency over 20KHz, so that it will strike more than several 10,000 times/sec on the moving surface. The striking force is composed of two parts, which are a preload P_s, which is applied to ensure the solid impacts of the actions and a dynamic load P_d sourcing from the high frequency oscillation.

The strikes will deliberately cause severe plastic deformation and thus change the topology on the surface. Desired pattern are carved by controlling the moving speed in X and Y directions.

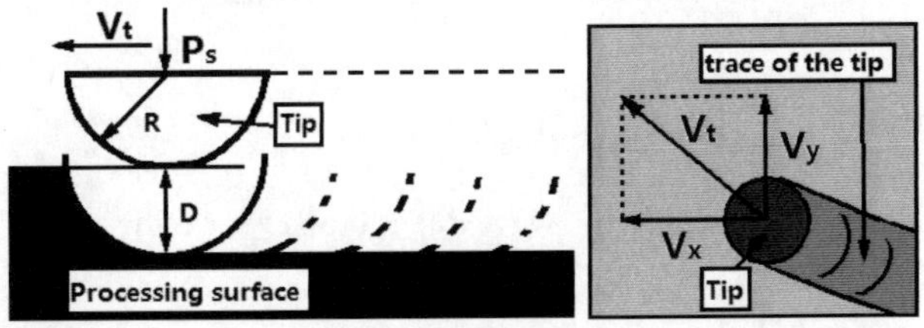

Figure 2: Mechanism of the ultrasonic indenter. Ps: Preload applied on the surface. R: Radius of the tip. D: Maximum displacement of the tip. Vt: Relative speed between the tip and the surface. Vx: Relative speed in x direction between the tip and the surface. Vy: Relative speed in y direction between the tip and the surface.

Analytical design of the ultrasonic indenter

Piezoelectric ceramic stack

The number of the ceramics involved in the stack is determined by the power of the system. It can be calculated with the Eq. (1).

$$n = \frac{W}{A \cdot V \cdot f} \tag{1}$$

where W is the needed power, A is the power capacity constant for the ceramics (usually 2~3 W/(cm^3·kHz) [11]), V is the total volume of the ceramics and f is the working frequency of the ceramic.

1. *Vibrator (i.e. back mass, ceramic stack and front mass)*

After the dimensions of the ceramic stack are set, the dimensions of other parts of the vibrator can also be determined through equivalent circuit network theory [11].

Usually, the total length of the transducer is multiples of the half wave length. This is basically a preliminary calculation for the transducer, so we will ignore the working load of the transducer and design the vibrator in the manner of two quarter wave length (Fig. 3).

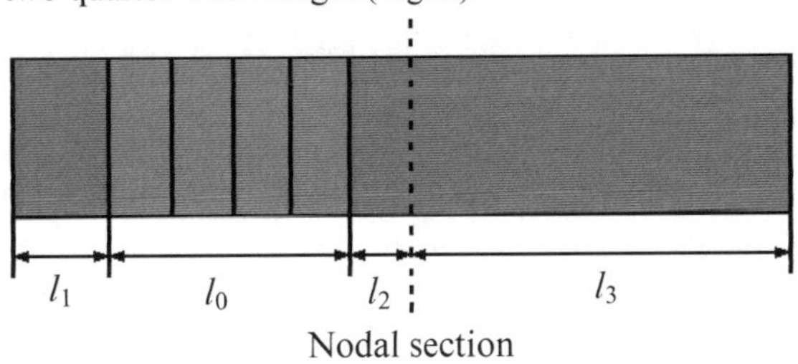

Figure 3: Simplified vibrator diagram: back mass (l_1), ceramic stack (l_0) and part of the front mass (l_2) on the left side of the nodal section forms one quarter wave length while the rest of the front mass (l_3) forms the other quarter wave length.

Derived from transmission line theory and the frequency equations, we can establish the relationship between l_1, l_2 and l_3 in Eq. (2) [11].

$$\frac{Z_1}{Z_0}\tan k_0 l_0 \tan k_1 l_1 + \frac{Z_1}{Z_2}\tan k_2 l_2 \tan k_1 l_1 + \frac{Z_0}{Z_2}\tan k_2 l_2 \tan k_0 l_0 = 1 \quad (2)$$

where k represents the wave length and Z is the wave impedance.

2. *Horn design*

The profile of the horn can be varied in accordance to the actual requirement. Stepped profile horn is widely used due to its relatively high amplification factor and machinability (Fig. 4) [12].

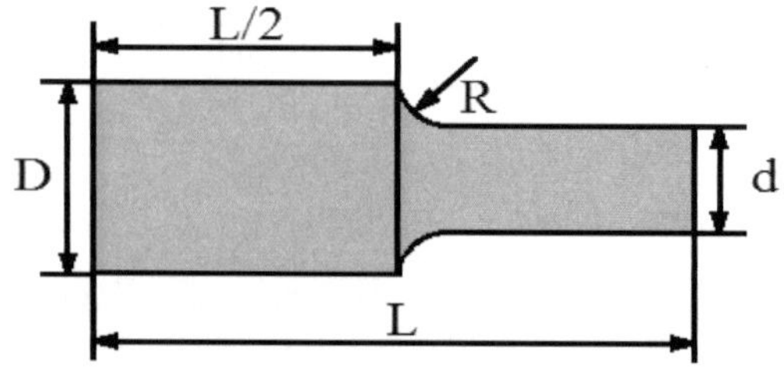

Figure 4: Typical stepped horn. R: Radius of the round transition. D and d: diameter of the input and output end. L: length of the horn.

As mentioned previously, the length of the whole transducer is usually multiples of the half wave length. So it is reasonable to design the horn into the length of half wave length of its material. To prevent the deformation stress piles up at the diameter drop section, it is recommended to design the both ends into the same length. Additionally, placing a round transition at the neck will be helpful to avoid excessive stress concentration. However, the radius of the round transition is certainly to affect the nature frequency of the structure and may be refined by various approaches such as analytical calculation or FEM analysis [13].

The input end diameter is determined by the previous structure, and the output end diameter depends on the desired displacement amplification ratio.

3. *Tip*

Tip comprises a body holder and a needle-like pin. The tip body part may be coupled to the horn using a screw joint, so that the tip can be considered as an attachment to the output end of the horn.

For simplicity, the pin part is neglected here, while the body uses a cylinder shape with diameter D_d and uses the same material as the horn. It has a total length l_b. Eq. (3) can be introduced under this circumstance to calculate the compensational length (deduction length of the output end of the horn) l_c [14].

$$l_c = l_b \frac{D_d^2}{d^2} \quad (3)$$

where d is still the diameter of the output end of the horn.

4. *Design summary:*

Based on the calculation, a simplified model of ultrasonic indenter is finished (Fig. 5).

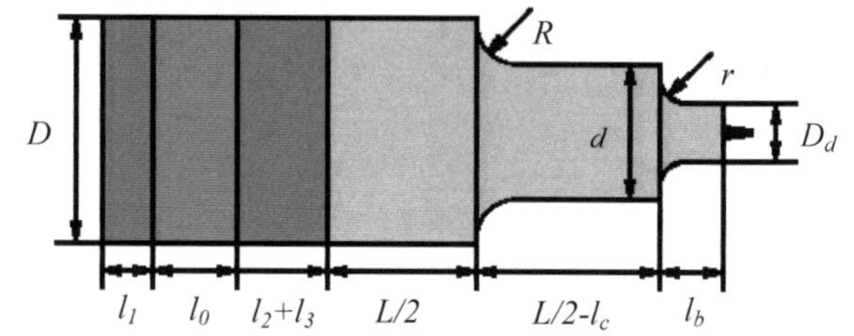

Figure 5: Simplified shape and dimensions of the ultrasonic indenter.

Requirements and Materials

The indenter should emit at least 500W power when working at the frequency at 20 kHz or a little bit higher.

PZT8 ceramic rings, whose density is 7.8 g/cm^3 and sound speed is 3100 m/s, are used as the material for the ceramic stack. The dimensional size is 38mm*15mm*5mm (outer diameter*inner hole diameter*thickness) for each ring.

In order to transmit as much energy to the front part as possible, the material for the back mass prefers higher acoustic impedance than the front part. On the other hand, the front mass and the horn should consider using materials with lower acoustic impedance and higher fatigue performance.

Take the synthesis of the conditions above into account, AISI 1045 steel is selected as the material for back mass while 7075-T6 Aluminum for front mass and horn. Their basic information is listed below in Table 1.

Table 1. Basic mechanical properties of the Materials

Material	Young's Modulus (GPa)	Density (g/cm^3)	Yield Strength (MPa)
AISI 1045	200	7.87	450
7075-T6	71.7	2.81	503

RESULTS AND CONCEPTS

Modelling of the ultrasonic indenter

After going through all the calculations, the simplified model is built in Solidworks (Fig. 6).

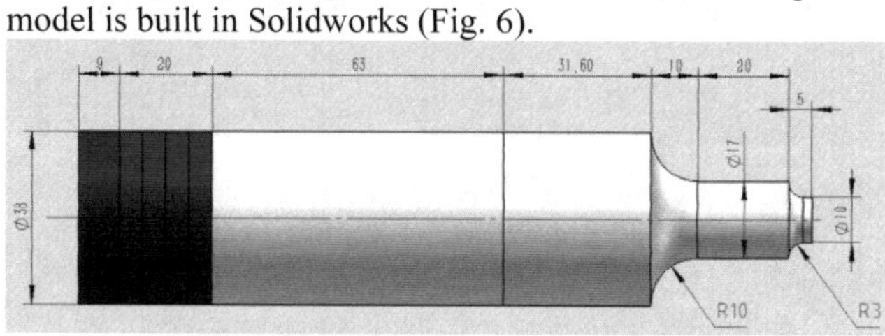

Figure 6: Initial ultrasonic indenter model.

The initial model for finite element analysis is ready. Based on the results of analysis and more detailed adjustments, the model will be modified accordingly. After trial and error, the final model will be determined for manufacture.

UST processing apparatus

Based on the concept of the working mechanism of the ultrasonic indenter, some peripheral devices are imported into the design of the apparatus (Fig. 7).

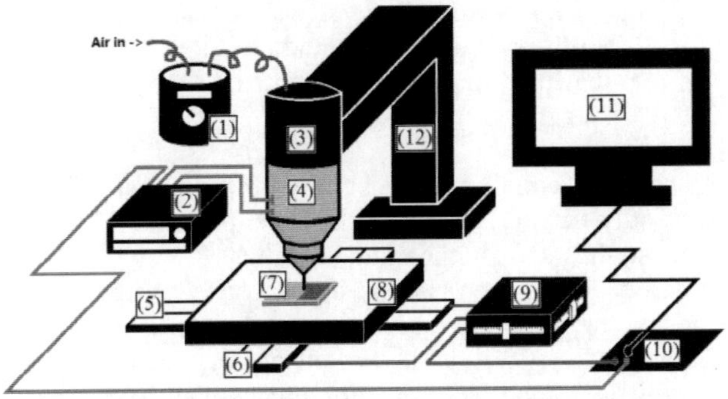

Figure 7: Schematic of the UST apparatus: (1) Compressed air regulator. (2) Ultrasound signal generator. (3) Air cylinder. (4) Ultrasonic indenter. (5) Linear guide for X direction. (6) Linear guide for Y direction. (7) Workpiece. (8) Stage. (9) XY stage controller. (10) Digital IO board. (11) Computer with LabVIEW programme. (12) Support.

The system is controlled by a digital input/output board via LabVIEW programme in PC. The ultrasound signal from the ultrasonic signal generator is transmitted to the ultrasonic indenter through wires. Compressed air is controlled by the regulator and injected into air cylinder to apply appropriate preload P_s.

Instead of moving the ultrasonic indenter to make the pattern, linear guides controlled by the XY stage controller are used to move the stage and the workpiece fixed on it.

Measurement solution

After being processed by the UST apparatus, the surface awaits the scanning electron microscope (SEM) to do a general scan and then atomic-force microscope (AFM) for micro-structures. The AFM system used in our laboratory is XE-100 from Park System Corp.

ACKNOWLEDGMENTS

The work was sponsored by Natural Sciences and Engineering Research Council of Canada (NSERC).

REFERENCES

1. Kulemin, A. V., V. V. Kononov, and I. A. Stebel'kov, 1981, "Increasing the fatigue strength of parts by ultrasonic surface treatment", *Strength of Materials,* 13.1: 76-81.
2. Pyoun, Y. S., et al., 2008, "Development of Evolutionary Cone Type LSD for SUV/RV Utilizing the Axiomatic Approach and the Ultrasonic Nanocrystal Surface Modification Technology", *International Journal of Automotive Technology* 9.1: 61-70.
3. Cherif, A., Y. Pyoun, and B. Scholtes., 2010, "Effects of Ultrasonic Nanocrystal Surface Modification (UNSM) on Residual Stress State and Fatigue Strength of AISI 304", *Journal of mMaterials Engineering and Performance* 19.2: 282-286.
4. Lee, Chang-Soon, et al., 2010, "Rolling Contact Fatigue Characteristics of SAE52100 by Ultrasonic Nanocrystal Surface Modification Technology", *International Journal of Modern Physics* B 24.15n16: 3065-3070.
5. Cho, In Shik, et al., 2011, "Wear Behavior of Cu–Zn Alloy by Ultrasonic Nanocrystalline Surface Modification", *Journal of Nanoscience and Nanotechnology* 11.7: 6443-6447.
6. Wu, Bo, et al., 2012, "Effect of Ultrasonic Nanocrystal Surface Modification on the Fatigue Behaviors of Plasma-nitrided S45C Steel", *Surface and Coatings Technology* 213: 271-277.
7. Yasuoka, Manabu, et al., 2013, "Improvement of the Fatigue Strength of SUS304 Austenite Stainless Steel Using Ultrasonic Nanocrystal Surface Modification" *Surface and Coatings Technology* 218: 93-98.
8. Amanov, Auezhan, Young-Sik Pyun, and Shinya Sasaki., 2014, "Effects of Ultrasonic Nanocrystalline Surface Modification (UNSM) Technique on the Tribological Behavior of Sintered Cu-based Alloy", *Tribology International* 72: 187-197.
9. A. Y. Vorobyev and C. Guo, 2015, "Multifunctional surfaces produced by femtosecond laser pulses", *Journal of Applied Physics*, 117, 033103
10. Iula, Antonio, et al., 2002, "Finite Element Three-dimensional Analysis of the Vibrational Behaviour of the Langevin-type Transducer", *Ultrasonics* 40.1: 513-517.
11. Lin Shuyu, 2004, *Theories and Designs of the Ultrasonic Transducer*, Science Press, ISBN 9787030134196 [in Chinese]
12. Lin Dzon-Mou., 1987, *Theories and Designs of the Ultrasonic Horn*, Science Press, ISBN 9787030000088 [in Chinese].
13. Nanu, A. S., MARINESCU, N. I., & Ghiculescu, D., 2011, Study on Ultrasonic Stepped Horn Geometry Design and FEM Simulation. *Nonconventional Technologies Reviews*, 4, 25-30.
14. Xueyan Cheng, 2005, *A New Type of Rotary Ultrasonic Composite Grinding Head and Serialization,* Tianjin University, Tianjin China.

HUMAN HEAD AND FACE DETECTION AND TRACKING USING SINGLE CAMERA TO IMPROVE HUMAN ROBOT INTERACTION

Sumesh Kunwar
School of Engineering
Laurentian University
Sudbury, Ontario, Canada
E-mail:skunwar@laurentian.ca

Meysar Zeinali
School of Engineering
Laurentian University
Sudbury, Ontario, Canada
E-mail:mzeinali@laurentian.ca

ABSTRACT

Human robot interaction has been a prime focus of researchers working on computer vision and robotics for the last few years. Human detection and human tracking is the preliminary and the most important step towards human robot interaction. In the current paper, a human detection and tracking system using image processing is presented by using a single camera. First an edge detector system is applied to acquire an edge image as pre-processing step. Then, all the segments which are linear or almost linear are removed. A curve fitting process is then applied to fit the ellipse representing a human head. The algorithm is implemented in MATLAB/SIMULINK environment and the experimental results obtained are promising.

KEYWORDS: Head detection, Ellipse fitting, Human Robot Interaction

INTRODUCTION

Human robot interaction (HRI) is frequent in the hybrid automated manufacturing (HAM) systems which are increasingly becoming accepted. The hybrid automated manufacturing systems can increase productivity of manufacturing systems by integrating a human's unique skills and a robot's capacities and accuracy to collaborate as intelligent entities [1]. As the use of robot in production and service sector continues to increase, the safety of people working along with a robot is of big concern.

Human robot interaction is a field of study dedicated to understanding, designing, and evaluating the robotic systems for use by or with humans [2]. The objective of using robots together with human is to alleviate the work process for human operators and to improve operations with the goal of increasing precision, shortening the product life cycle in manufacturing and reducing the risks in production [3].

Efficient HRI enables a mobile robot to learn its user's preference and behaviors over a period of time which is essential for effective and efficient use of service robots [4]. For the execution of natural HRI, a visual interference is crucial [5]. Considering the mobility of mobile robot, it is obvious that the robot is faced with various imaging conditions such as pose variations, illumination changes, and clustered backgrounds [5]. Detecting a human and tracking under complex conditions can be a challenging task. Human head or only face has been used to detect presence of human in many studies done in the past.

Face detection has been popular in the last few years mainly due to its application in security control, intelligent human machine interference, visual surveillance and content based retrieval [6]. Human face detection process can be classified as knowledge-based methods, invariant feature methods, template matching methods and appearance based methods [7]. Knowledge based face detection approach takes into consideration human knowledge of what a face looks like; a face consists of two ears, two eyes, a nose and a mouth [7].

In invariant feature based method, face detection problem is addressed by finding local geometric or appearance features which are invariant to pose, viewpoint or lighting conditions. [7]. Some of the early works on face detection were feature-based system in which distance and angles between fiducial points are compared [8]. In template based method, a template is formed based on the facial structure and the template is searched in the input image to locate any face present in the input image. The disadvantage of this method is that it isn't pose invariant.

Many popular modern face detection systems are based on appearance based method. In appearance based method, the method learns face model from a number of training samples to perform face detection [9]. A large number of training faces is required for the training phase and the detector works mostly for the frontal views. Thus, it can be seen that face detection, application can be limited because the frontal view not always available. Since human head can be detected from any orientation, thus detecting human head is more general and is not limited to any specific view, therefore it can be used to detect human.

Recently, some researchers have used the elliptical property of a human head to detect human. Zou et al. [10] have first isolated moving object and then they have used elliptical detection method to detect human face using ellipse fitting.

In this paper, a human detection algorithm based on detecting the features of human head is presented. To make detection system robust, the circular shape of human head is taken as the most important feature while detecting a human. The proposed system is computationally feasible and can be applied on application requiring real-time human detection system.

The remainder of this paper is organized as follows: the following section presents the problem statement and definition, after that human tracking and head detection methodology is presented, then the following section describes the experimental set up and results, and the last section concludes the article.

PROBLEM STATEMENT AND DEFINITION

While the detection of a human face for a human is trivial, the detection is more complicated for computers; as for the computer human face is a set of large array of numbers. The objective of a face detection system is to determine whether there are faces in an input image and to determine their respective location. This is an important preliminary step for a robot to acknowledge the presence of the human in the surrounding and workspace of a robot accordingly. Therefore, a human detection system needs to be very accurate and robust to the environmental conditions and other variables.

The trend in the robotic research community is to make a robot more intelligent by increasing its capacity to take decision based on the past knowledge and the information available at the present. Thus for a robot working side by side with human, adequate knowledge of the workplace where it is working should be given. It should be used in a way that it is safe for other instruments and human to be in the place where a robot is working. Machine vision helps the robot to be aware of its surrounding. Thus, a robot can acknowledge the presence of human with the help of machine vision. Human detection system uses machine vision to make a robot aware of the presence of human in its surrounding.

In daily life, real time information about its working environment is crucial for a robot. Thus, a face detection system should be able to work on real time. Various ways of face detection system are available. Some of them are computationally expensive and some of them are not applicable in real world scenario [9].

During the last few years, many researches were carried out on the detection of the human faces [10]. Mostly the researches were focused on the detection of the human face if images of frontal views were available. In many face detection system, the detection task is viewed as detecting an object. In those researches, the detection of human face could also be taken as a pattern recognition problem. To detect an object, a trainer is trained using a large number of face and non-face samples. The advantage of such a system is that they have high accuracy rate but they are however concentrated on detecting face having frontal views. In reality, people won't always be looking at the camera. Similarly, people with long beards, cap or big glasses could not be recognized by the system. For a robot working together with human, such type of error is not acceptable.

Some researchers have used triangular property of a face to localize the facial region. Lin and Fan [11] have introduced new criteria for localizing face. Their criteria for finding potential facial region is that the combination of two eyes and centre of the mouth forms an isosceles triangle whereas the combination of the centre of the ear, nose and eye forms a right angled triangle. The problem with this approach is that ear may be sometimes totally and partially occluded thus the system will fail to detect the person.

Some researches [9] have also used the skin color to detect people. However, the system is prone to make error under varying illumination conditions. Additionally, people have different color of faces; white, red, yellowish white, brown and black. While trying to detect these color, chances are high that different object having colors similar to the human face will be considered as the human face. Also, people detection becomes difficult when people's face is not properly visible to the camera.

Therefore, detecting people based on the shape of the human head can be a better solution to human detection problem. The shape of the human head is almost circular and this property can be used to detect the head. The advantage of such a system is that it is pose invariant, i.e. the system can detect human not only from frontal and profile views but from all directions. To make sure, that the tracking is done on human head, the aspect ratio of the human head and the average human head size are also taken into account.

HUMAN TRACKING AND HEAD DETECTION METHODOLOGY

In order to detect the shape of human head, first an edge detection algorithm is applied to the acquired images of the human in the field of view of a camera. Edge detection is one of the most frequently used operations in image processing. Most of the edge detection works on the simple idea of finding out pixel where the intensity values changes drastically. Edge in a grey image is found where intensity value changes drastically. For traditional edge detector, there is contradiction between ability to smooth noise and ability to locate the edge. Prewitt operator reaches extreme detection in the edge and makes noise smooth by grayscale difference of adjacent pixels from top to bottom, left and right [12].

After finding the edge as shown in Figure 3 in experimental results (i.e., next section), there are undesired edges. In order to eliminate the linear segments of the detected edges, a line detection method proposed and is being used. After eliminating those linear segments, image shown in Figure 4, was found, which can be used to find the elliptical object using a curve fitting method. Thus by applying suitable pre-processing steps, the number of coordinates in edge vector was considerably reduced. Working on fewer numbers of pixels will definitely require lesser time.

While trying to eliminate the lines, nonlinear segments were grouped into groups of pixel to construct the circular/elliptical curves similar to human head curves. In this paper, to eliminate the linear segments, first, pixels with similar pose coordinates were grouped into smaller groups, then the two point slope form was applied to find the slope of various segments, Figure 3. The angle between tangents of two consecutive segments is an important factor to identify whether two consecutive segments are linear or they are part of a curve. The slope of tangent line of each segment was calculated using two point method, equation (1).

$$m = \frac{y_2 - y_1}{x_2 - x_1} \qquad (1)$$

From the slope obtained from equation (1), the angle made by the two consecutive tangent lines can be obtained as:

$$\theta = \tan^{-1}(m) \qquad (2)$$

In Figure 1, it can be seen that the difference between angles made by tangents of segments and horizontal axis which are almost linear are small. In other word, the normal to tangent are almost parallel indicating that the segments can be considered linear segment. By choosing suitable angle as threshold ($\theta_{threshold}$), all the linear segments making angle less than the threshold can be omitted.

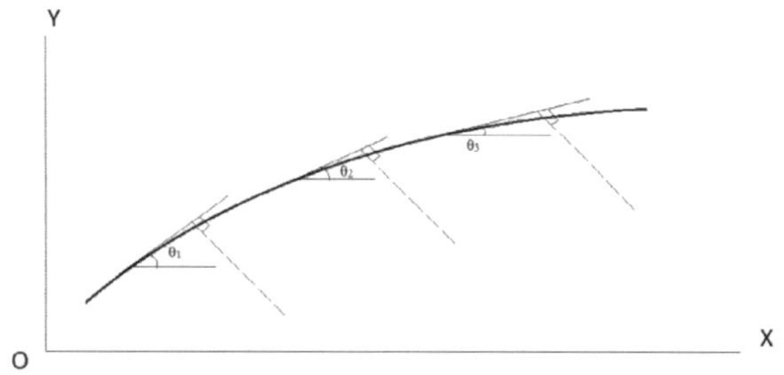

Figure 1: Line segments having angle less than threshold

Figure 2 illustrates the case that the detected edge is a curve and the angle difference made by the tangents lines and horizontal axis at different points of the curve is more than a threshold angle. The angle difference more than a defined threshold, indicates that the segment is curve and it can be part of human head curve, thus the segments are retained for further investigation.

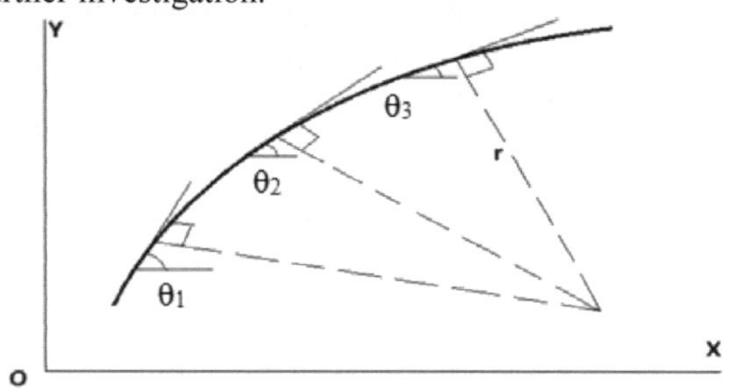

Figure 2: Line segments making more angle than threshold

In figure 2, the angle difference is greater than the threshold.

$$\theta_2 - \theta_1 = \theta_{threshold} \qquad (3)$$

and it can be seen that normals drawn from different tangent intersect at a point which gives the approximate center of the curve. The definition of the curvature of a curve can be used to find smooth curves fits to each curvilinear segment of the edge.

After finding the curves, ellipse fitting is done to locate the human head, since the average size of the human head is known. The general equation of the ellipse [9] is:

$$\left(\frac{x - x_o}{a}\right)^2 + \left(\frac{y - y_o}{b}\right)^2 = 1, \qquad (5)$$

Where x_o and y_o are the ellipse center and, a and b are the length of major and minor axis respectively.

The equation can also be represented in the polar form as:

$$x = x_o + \rho(\theta)\cos(\theta),$$

$$y = y_o + \rho(\theta)\sin(\theta) \qquad (6)$$

where, θ is the point angle of the edge point and

$$\tan(\theta) = \frac{(y - y_o)}{(x - x_o)} \quad \text{and}$$

$$\rho(\theta) = \frac{ab}{\sqrt{a^2 \sin^2 \theta + b^2 \cos^2 \theta}} \qquad (7)$$

Thus, by finding an ellipse whose aspect ratio is close to that of human head between 1.0 and 2.0 [10], human head can be localized. Then, the human motion can be tracked using object distance method recently proposed in Alizadeh, and Zeinali [13].

EXPERIMENTAL SETUP AND RESULTS

A number of simple pre-processing steps are essential for this human detection system. In this research, Logitech HD Webcam c615 with a resolution of 640 X 480 is used. For experimental purpose, human head is expected to be at the top most part of the image. The color space conversion is the initial step as edge detector works on the intensity value of the image. Then, the obtained gray image is processed to get an edge image as shown in Figure 3 by using built-in MATLAB edge detector function.

Figure 3: An edge image

After eliminating vertical and horizontal lines, figure similar to Figure 4 is obtained. For comparison purposes, Hough transform [10], originally used to detect straight lines, are used to detect circular arcs. However, the disadvantage of Hough algorithm is that it is computationally expensive. Thus, the main reason using proposed ellipse fitting method to track the human is that it is computationally inexpensive as compared to methods using Hough transform.

Figure 4: Image obtained after removing linear segments

After finding the elliptical arc representing human head, tracking process starts, tracking can be done as per the requirement of the system and based on the distance measurement method using single camera proposed by Alizadeh, and Zeinali [13]. In presence of other curves whose curvature is similar to human head, the program is making error. The problem will be solved by searching for skin color around the curve location.

Figure 5: Tracking in progress

CONCLUSIONS

In this paper, a method of human detection and tracking using single camera and the features of human head is proposed. This approach can detect human under various poses. By using the proposed method, along with the object distance measurement algorithm, it is possible to track human movements and to find its speed in real-time. Thus, the robot gets more knowledge about the human around its working environment. Therefore, a robot working together with human can be made safer and more reliable. The proposed method is applicable to areas including health care, hybrid automated manufacturing systems and other application areas where human robot interaction is needed.

REFERENCES

1. Lin C.T. and Wang. M.-J. 1999, "Human-robot interaction in an aircraft wing drilling system", *International Journal of Industrial Ergonomics*, 23(0) , pp. 83-94.
2. A. Goodrich, M. and C. Schultz. 2007, A., "Human-Robot Interaction: A Survey", *Foundations and Trends in Human-Computer Interaction*, 1(3), pp.203-275.
3. Šekoranja, B., Bašić, D., Švaco, M., Šuligoj, F. and Jerbić, B. 2014, "Human-Robot Interaction Based on Use of Capacitive Sensors", *Procedia Engineering*, 69(0), pp. 464-468.
4. Sekmen, A. and Challa, P. 2013, "Assessment of adaptive human–robot interactions", *Knowledge-Based Systems*, vol. 42, No. 0, pp. 49-59.
5. Lee, T., Park, S. and Park, M. 2006, "An effective method for detecting facial features and face in human–robot interaction", *Information Sciences*, 176(21), pp. 3166-3189.
6. Du, C.H., Zhu, H., Luo, L.M., Liu, J. and Huang, X.Y. 2013, "Face detection in video based on AdaBoost algorithm and skin model", *The Journal of China Universities of Posts and Telecommunications*, 20(1), pp. 6-9.
7. Yang, M.H., Kriegman, D.J. and Ahuja, N. 2002, "Detecting Faces in Images: A Survey", *IEEE Transactions on pattern analysis and machine intelligence*, 24(1) , pp. 34-58.
8. Zhang, G. and Wang, Y. 2011, "Robust 3D face recognition based on resolution invariant features", *Pattern Recognition Letters*, 32(7), pp. 1009-1019.
9. Zhang, C., and Zhang, Z. 2010, A survey of Recent Advances in Face Detection, *Technical report*, TR-2010-66, Microsoft Corporation, Redmond, Washington.
10. Zou, W., Li, Y., Yuan, K. and Xu, D. 2009, "Real-time elliptical head contour detection under arbitrary pose and wide distance range", *Journal of Visual Communication and Image Representation*, 20(3), pp. 217-228.
11. Lin, C. and Fan, K. 2001, "Triangle-based approach to the detection of human face", *Pattern Recognition*, 34(6), pp. 1271-1284.
12. Xishan, T. 2012, "A novel image edge detection algorithm based on Prewitt operator and wavelet transform", International journal of advancements in computing technology, 4(19), pp. 73-82.
13. Alizadeh, P. and Zeinali M. 2013, "A Real-Time Object Distance Measurement Using a Monocular Camera" Proceedings of *24th IASTED international Conference on Modelling and Simulation*, Alberta, Canada.

Proceedings of the 25th CANCAM
London, Ontario, Canada, May 31 – June 4, 2015

PARAMETER-DRIVEN GEOMETRIC MODELING OF RETROREFLECTIVE FEATURES

Sama Hussein
Mechanical and Materials Engineering
Western University
London, Ontario, Canada
sabdelw3@uwo.ca

Benjamin Hamilton
Mechanical and Materials Engineering
Western University
London, Ontario, Canada
bhamilt8@uwo.ca

Evgueni Bordatchev
National Research Council of Canada
London, Ontario, Canada
evgueni.bordatchev@nrc-cnrc.gc.ca

O. Remus Tutunea-Fatan
Mechanical and Materials Engineering
Western University
London, Ontario, Canada
rtutunea@eng.uwo.ca

ABSTRACT

Retroreflectors (RR) are passive optical components that reflect the incident light back to its original source. Because of this characteristic, RRs are used in a broad palette of applications ranging from metrology to traffic safety. Among the variety of possible geometries for RRs, there are three main types of inverted corner cube (ICC) shapes: rectangular, triangular and hexagonal. While all types of RRs are capable of fulfilling their primary retroreflection function, their optical performances tend to make them more or less suitable to a particular application. In this regard, the present study is primarily focused on the application of ICC RRs for automotive taillights. Non-sequential optical modeling of RRs has been used to outline their basic optical functionalities and to analyze the trajectory of the retroreflected light caused by the total internal reflection (TIR) phenomenon. Based on this, a parametric approach to define the principal geometric parameters of ICC RRs is proposed by considering three levels of design parametrization: internal, areal and directional. The present study opens up new optical methods to be used in optimization and control of the optical performance for ICC RRs in terms of the retroreflection efficiency and vertical/horizontal divergence.

KEYWORDS: Retroreflective microoptics, geometric modeling, optical analysis, parametric modeling

INTRODUCTION

Retroreflection is an optical phenomenon characterized by an intentional reflection of the incoming light back to its original source(s) with a controlled amount of scattering [1, 2]. This optical functionality is critically required in safety applications involved in automotive and surface transportation (*e.g.* roadside markers, lighting, safety jackets/signs, etc.). Generally speaking, three main types of RR geometries have been used so far in almost all automotive lighting applications: lens-and-mirror (also called cat's eye), corner cube (CC) and inverted corner cube (ICC). The basic functionality of these RRs is depicted in Fig. 1.

In the lens-and-mirror (cat's eye) RR (see Fig. 1a), the

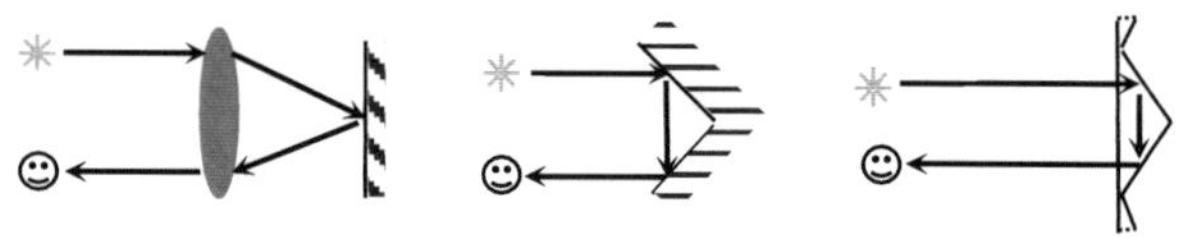

a) lens-and-mirror b) corner cube (CC) c) inverted CC

Figure 1: Principal retroreflector designs

incoming light ray is directed towards a reflective surface and reflected back through a lens such that it is returned in a direction parallel to that of the incident ray. The CC RR typically consists of two or more mutually orthogonal reflective surfaces that are also capable to return the incident ray in a parallel direction (Fig. 1b). However, no lens is required in this case since the return of the incident light is caused by the particularities of the RR geometry that speculates on the equivalence between incidence and reflection angles. CC RRs are known to have higher reflectivity than cat's eye designs although the latter ones can accept higher range of incident angles [3]. Finally, ICC RRs (Fig. 1c) are based on a different optical phenomenon called total internal reflection (TIR). Here, the incident ray is reflected at critical angle by an interface between two optical media with different indexes of refraction. This type of RRs is widely utilized in automotive lighting and it will constitute the main focus of the present study.

WORKING PRINCIPLES, FABRICATION AND TYPICAL DESIGNS

In general terms, the ICC RRs used in automotive rear and/or side lighting applications have to adequately and concurrently serve both safety and illumination purposes (Fig. 2). For this purpose, the typical design of an ICC RR is comprised of an array of elementary features like the one shown in Fig. 1c. This design ensures that the light reflected by the tailgate RR of the preceding car is returned to the succeeding vehicle, such that its driver is made aware of a preceding traffic hazard. On the other hand, the retroreflective element incorporates multiple illumination/signaling/brake warning functionalities that allow the driver of the preceding car to communicate his/her presence and/or actions/intentions. It is important to point out that in addition to the aforementioned traffic-related functions, rearlights are also required to meet aesthetic/styling requirements derived from their placement on the visible exterior of the automobile. However, this type of design constraints is not specifically considered within the limited scope of this work.

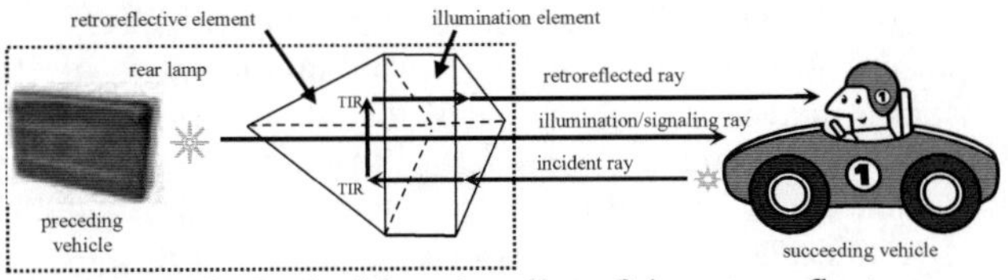

Figure 2: Optical functionality of the retroreflectors

To successfully meet its two principal functionalities most taillights incorporate both illumination and retroreflective elements. While the actual embodiment broadly varies among manufacturers, illumination is typically ensured by a cut-off piece/bezel forming the exterior surface of the lamp. By contrast, retroreflectivity is provided by means of the RR that is attached to the illumination system.

Commonly, each of the elementary RRs embedded in a rear lighting system has one ceiling/top incidence facet and at least two retroreflective facets. As a consequence, the incident light penetrates the optical body of the RR through the top facet and it is then reflected back at the interface formed between air and lamp material which constitutes the core of the TIR phenomenon. Because of this, the angle formed by the retroreflective surfaces represents one of the most critical elements of RR design and Fig. 3 clearly demonstrates that relatively small deviations from its optimal 90° value will cause dramatic decreases in the ratio between the reflected and incident power of the light beam.

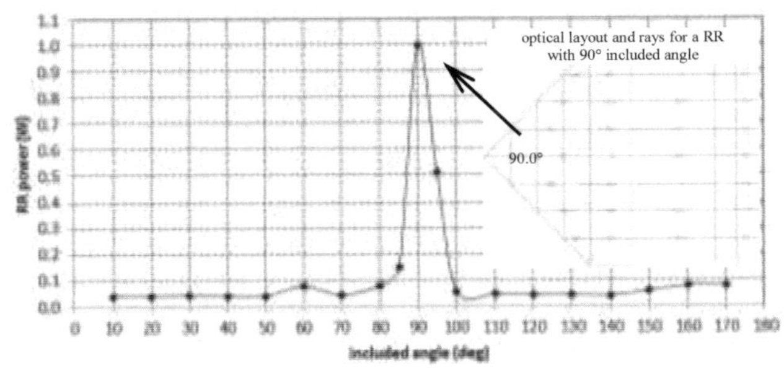

Figure 3: Correlation between the optical efficiency of the RR and its included angle

It is also necessary to reemphasize that the TIR associated with ICC RRs is caused by uncoated facets. The absence of metal-coated reflection facets that is characteristic to other types of RRs results in cost-effective components capable of repeatable and reliable functionality.

Conventionally, mass fabrication of RRs is a complex process comprised of several steps revolving around injection molding. As such, the optical performance of the RRs is tightly related to quality and accuracy of the master mold used. While newer approaches advocate for ultraprecise machining involving diamond flycutting/ruling or thin shims, state-of-the-art master fabrication continue to rely on traditional pin bundling techniques. According to them, molding of the RR geometry can be performed either directly on the forming ends on the pins or by fabricating an intermediary electroform on which RR will be subsequently molded. The main phases involved in the latter manufacturing path are illustrated in Fig. 3.

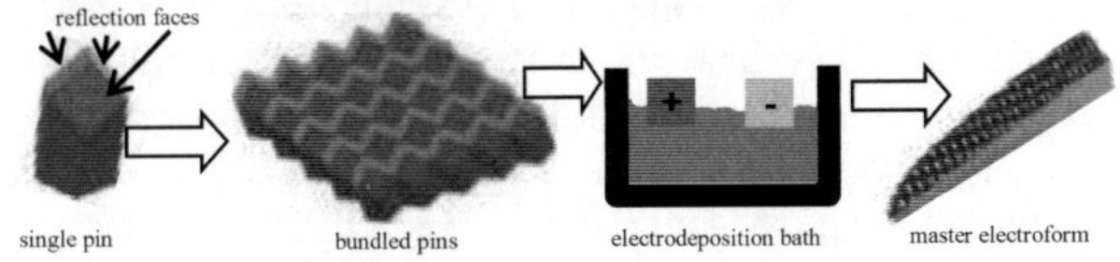

Figure 3: Phased fabrication of master electroforms

Typical pins are characterized by hexagonal cross sections and their forming tips have to replicate with high accuracy the "negative" of the intended RR geometry. Because of this, the pins involved in ICC RR fabrication are defined by three mutually orthogonal facets. In order to preserve the superior optical performance of the RR, the three retroreflective facets have to meet high optical quality requirements ($R_a < 10$ nm) while the angles between cannot deviate by more than 0.05° from the perpendicularity condition. At the next manufacturing step, hexagonal pins are assembled and secured together and their forming ends are then immersed into an electrodeposition bath that allows a continuous depositing of the nickel atoms. The master electroform that is generated at the end of the process becomes the key component of a core or cavity insert to be integrated in the mold to fabricate the final rear lighting product.

The vast majority of automotive taillights are fabricated from transparent polymeric materials such as polymethyl methacrylate (PMMA), acrylic or polycarbonate. All of these materials are capable to meet all optical requirements of the RRs, both from safety and illumination perspectives.

Depending on the practical application, automotive lighting generally relies on three typical RR designs: right triangular prism, and CC with triangular or hexagonal apertures. Their characteristic geometries are illustrated in Fig. 4.

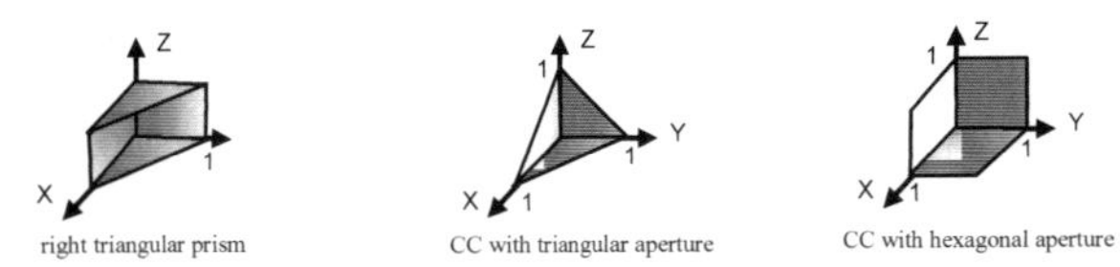

Figure 4: Typical RR geometries

The differences between these three designs reside in both the geometry of the elementary RR features as well as the way in which they are positioned and oriented in the overall array forming the structured surface. The optical performance of retroreflective component is tightly related to the reflective area of its facets and it can characterized by two parameters: i) retroreflection efficiency (RRE); and ii) divergence of the retroreflected light.

Retroreflection efficiency is defined as the ratio between the incident and reflected power of the light. It is obvious that while the target RRE is 100%, not the entire area of the RR contributes equally to the TIR phenomenon. Because of this, definitions of retroreflection area [4] and effective aperture [5] were introduced in the optical analysis/design of the RRs. Similar to the direct dependence between RRE and the optical performance, divergence represents – from a pure theoretical standpoint – a desirable safety trait of the RRs since it enhances their visibility by means of a wider observation angle. However, divergence can also be nothing else but a consequence of the manufacturing imperfections translated into angular deviations between the RR facets.

To illustrate this aspect, results from numerical simulations performed by means of Zemax OpticStudio Professional software (Fig. 5) clearly suggest that an error of ±0.5° between the orthogonal reflection facets does not significantly affect RRE values since 99.992% corresponds to 89.5°, same 99.992% is associated with 90.0° while 95.551% is determined by a 90.5° angle (see Fig. 5, top row). However, the retroreflective light pattern represented by the incoherent irradiance in W/cm^2 units (see Fig. 5, bottom row) will significantly affect the functionality of the RR, especially its divergence.

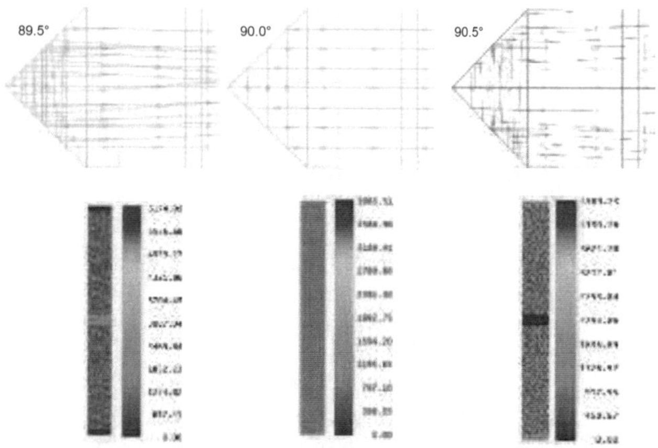

Figure 5: The effect of manufacturing defects on the optical performance of the RRs

Clearly, the attainment of the theoretical value of 90.0° for the included angle of a RR will ensure an ideal maximum RRE of 100% coupled with a zero divergence (Fig. 5, central column). However, as soon as the manufactured included angle departs from its prescribed nominal value, both RRE and divergence will be affected. On one hand, in case of an undersized included angle, some of the incident rays will deviate from their ideal parallel trajectories and will concentrate in the central region of the RR before reaching the observer (Fig. 5, left column). On the other hand, a significant majority of the incoming rays will diverge outward in case of an oversized included angle, the only exception being constituted by the central rays. While it might appear that the divergence represents an undesirable byproduct of the imperfections that are inherently associated with RR fabrication, it will be reiterated here that in fact a certain amount of divergence is highly desirable in a RR since it is associated with a wider observation angle.

PARAMETRIC MODELING FRAMEWORK

By taking into consideration the aforementioned typical designs, optical functionalities, and fabrication aspects, it seems that a parametric modeling approach will enable a facile modification of the RR design in an attempt to meet concurrent and sometimes contradictory design constraints. The proposed parametric approach will be able to establish a functional dependence between the input design parameters and the output metrics associated with the optical performance of the RRs, namely RRE and divergence. To illustrate this, the remainder of the study will exemplify the proposed approach in the context of triangular prismatic RR.

When it comes to the design of an automotive rear lighting component whose elementary RR consists of a right triangular prism, three different types of design parameters can be identified (Fig. 6).

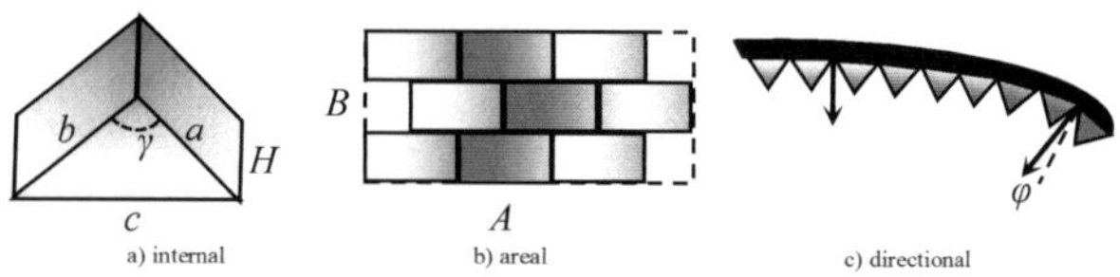

Figure 6: Types of parameters controlling the design of automotive rear lighting

As it can be seen in Fig 6a, if the two triangular bases of the right prism are right triangles ($\gamma = 90°$), then the geometry of the elementary RR is fully constrained by: the length of the base/hypotenuse (c), the length of the catheti (a, b) and width/height of the prism (H). As such, all other secondary design parameters can be calculated by means of well-known trigonometric formulas. Therefore, the proposed parametric modeling will assume a, b, c, and H as input parameters that exert a direct influence on the optical performance of the RR:

$$\max\{RRE, divergence\} = f(a,b,c,H) \quad (1)$$

In addition to the parameters that are internal to the elementary RR, areal parameters are used to control the pattern in which the base retroreflective elements are positioned and oriented on the structured surface. At this level, both elementary (a, b, c, H) and areal (A, B) parameters have to be included in the optimization of optical performance for a RR.

Finally, a higher (*e.g.* third) level of directional parametric modeling has to be considered when the elementary RRs are to be located on nonplanar/freeform surfaces, which constitutes in fact the most common instance in automotive lighting. In this case, an orientation angle φ is required to set the orientation of the base RR with respect to the surface normal passing through the point opposing the hypotenuse of the triangular base of the prism. Depending on the simplifying assumptions made, the optical analysis can be conducted either in planar or 3D scenarios. Moreover, depending on the particularities of the rear lighting application, it is usually necessary to direct the reflected light upward and toward the eyes of the observer (*e.g.* the driver of the succeeding car) as the taillights of the preceding vehicle are commonly located below the normal line of sight of the succeeding driver. Lastly, directional orientation plays a foreseeable important role on vertical and horizontal divergence, both with relevance in the traffic safety context.

It is well known that all automotive OEMs invest considerable efforts in making their end products more esthetically appealing while improving the overall aerodynamics of the car. Consequently, all externally mounted fixtures – including taillights – have to maintain their optimal functionality while matching an increasingly complex outer shape of the body. However, it is necessary to emphasize that conventional taillight fabrication techniques relying on pin bundling are faced with serious challenges when used on freeform surfaces alternating between quasi-planar and curved zones such is the case shown in Fig. 7.

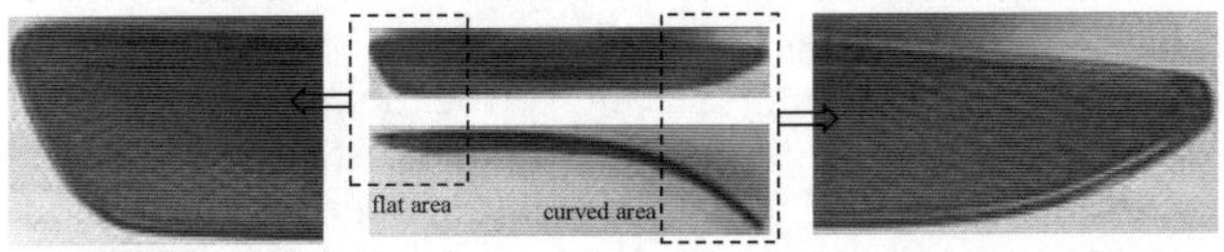

Figure 7: Representative sample of an automotive taillight

As it can be noticed from the extreme detail pictures, the geometry of the elementary RR features varies greatly between the quasi-flat and curved zones of the rear light cover. Because of this, significant differences between the forming ends of the pins used on the two primary zones have to exist in order to ensure uniform optical performances across the entire lighting component (Fig. 8). While alternate fabrication methods could involve tilted pins, their feasibility and reliability have not been tested so far.

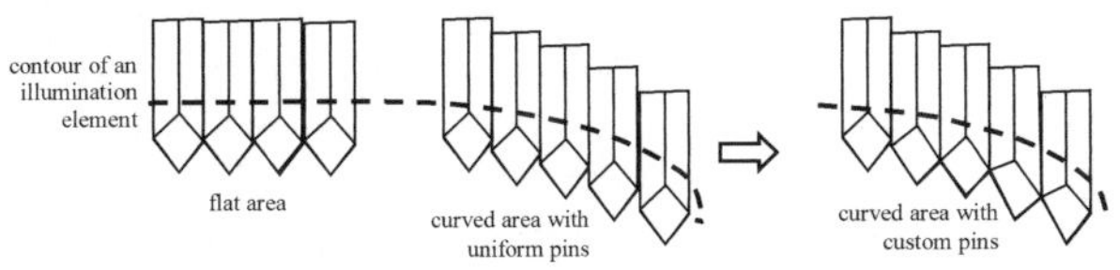

Figure 8: Correlation between forming end geometry and ICC RR feature positioning in an array

SUMMARY AND CONCLUSIONS

The present study has been centered on the analysis of a particular type of optical components involved in automotive lighting called retroreflectors. While there are three main types of RRs, ICC remains the design that is most commonly employed in automotive and safety applications, primarily due to its superior optical performance.

Inverted corner cube RRs are comprised of two main optical elements that serve simultaneously both illumination and retroflection purposes. On one hand, the illumination element directs the taillight to the surroundings for the purpose of signaling and/or warning. On the other hand, the retroreflective element redirects the light originating from the headlights of a succeeding vehicle back to its driver. This feature is associated with important traffic safety connotations.

Inverted corner cube RRs rely on the TIR phenomenon to redirect incident light in a direction that is opposite and parallel to the incident ray. To be able to efficiently exploit the TIR principles, ICC RR has to incorporate one "pass through" facet through which the light both enters and exits, and at least two mutually perpendicular reflection facets on which the light will be totally reflected. Despite the known superiority of ICC over other types of RRs, their overall prototyping and fabrication cost is high (~$2000/in^2) and is associated with long fabrication time (12-14 weeks per electroform).

To address some of the known drawbacks of the merely empirical approach constituting the core of the state-of-the-art technology, a parametric modeling framework capable to generically enclose all typical designs of ICC RRs is proposed. The proposed approach consists of three possible levels for design parametrization (internal, areal and directional) and opens up new investigational avenues in the parameter-driven optimization of the ICC RRs.

Future extensions of this work will be focused on the development of parametrizations for other RR designs, the ongoing goal being to establish a comprehensive platform capable to integrate geometry, optical performance as well as the fabrication of the retroreflective components both at macro and microscale levels.

ACKNOWLEDGMENTS

This study is a result of the collaboration between Western University and National Research Council (London, Ontario). Partial financial support was also provided by Natural Sciences and Engineering Research Council of Canada and AUTO21 Network of Centres of Excellence.

REFERENCES

1. Liepmann, T. W., 1994, "How retroreflectors bring the light back," *Laser Focus World*, 30, pp. 129-132.

2. Nilsen, R.B. and Lu, X. J., 2004, "Retroreflection technology", *Proceedings of SPIE*, 5616, pp. 47-60.

3. So, B.S., Jung, Y.H. & Lee, D.W., 2002, "Shape design of efficient retroreflective articles", *Journal of Materials Processing Tech*, 130, pp. 632-640.

4. Brinksmeier, E., Gläbe, R., and Flucke, C., 2008, "Manufacturing of molds for replication of micro cube corner retroreflectors", *Prod. Eng. Res. Devel.*, 2, 33-38.

5. Yuan, J., Chang, S., Li, S., and Zhang, Y., 2002, "Design and fabrication of micro-cube-corner array retro-reflectors", *Optics Communications*, 209, pp. 75-83.

Proceedings of the 25th CANCAM
London, Ontario, Canada, May 31 – June 4, 2015

EXPERIMENTAL STUDY OF TOUCH PROBING WITH A FIVE AXIS MACHINE TOOL

Md Mizanur Rahman
Department of Mechanical Engineering
Ecole Polytechnique de Montréal
Montréal, Québec, Canada
mdmizanur.rahman@polymtl.ca

J.R.R. Mayer
Department of Mechanical Engineering
Ecole Polytechnique de Montréal
Montréal, Québec, Canada
rene.mayer@polymtl.ca

ABSTRACT

The touch trigger probe plays an important role in modern metrology because of its robust and compact design with crash protection, long life and excellent accuracy. Aside from coordinate measuring machine (CMM), touch trigger probes are used for the accuracy assessment of multi-axis machine tools as an interchangeable device with the cutting tool. As a result, with the performance of the touch trigger probe is a matter of interest to the users. The touch trigger probe itself as well as the measuring surface, measuring environment etc. can contribute to the measurement inaccuracies. The paper presents the effect of surface irregularities, surface wetness due to cutting fluid, unidirectional repeatability of the probe and probing direction on probing accuracy on a machine tool.

KEYWORDS: Probing, machine tool, touch probe, repeatability, measurement inaccuracy

INTRODUCTION

Users of machine tools seek to integrate touch trigger probe within the machine tools. The purpose is use the machine tool not only for machining operation but also inspect the machined part throughout the machining operation. In general, a touch trigger probe provides a binary signal when contact occurs with the workpiece surface which is used to trigger readings of the X-, Y- and Z-axis positions. The touch trigger probe temporarily replaces the cutting tool at the machine spindle. On machine measurement of the part being machined allows to take corrective actions on the machine tool [1]. Thus the accuracy of the measuring device (probe) has significant influence on the measurement results.

Cauchick-Miguel et al. studied the factors that affects probe accuracy and assessed the influences of the approach direction, probe lobing, measuring speed, probe indexing angle and spring pressure which are responsible for the inaccuracies in the probing measurement by measuring a Rank Taylor Hobson standard hemisphere and a precision sphere [2]. Wozniak et al. considered probe pretravel variation as one of the major sources of probe inaccuracies. They experimentally demonstrated that, configuration of the probe, triggering force, stylus length and stiffness have influence on the probe pretravel variation [3]. Later on, they proposed a new technique for the verification of CMM touch trigger probe accuracy by measuring the distance between a reference point and the triggering point for different directions. The reference point is defined by the axis of rotation of the most stable axis of any rotary table [4]. By using the technique described in [4] they also showed that the measured object's shape, surface roughness and material stiffness contribute to the inaccuracies caused by a touch trigger probe [5].

Johnson et al. investigated the dynamic error characteristics of touch trigger probes on a CMM and found measurement speed, probe longitude, approach distance, probe latitude, stylus length/stylus tip diameter, probe orientation, operating mode (scanning and non-scanning), scan pitch, preload spring force (gauging force), probe type, and the surface approach angle had an important impact on probe accuracy. They also proposed some solution to avoid such error sources such as, optimizing the measurement speed along with a defined probe spring force and selection of probe type [6].

Fesperman et al. proposed an evaluation method for the standard of numerical control machines' measuring performance assessment furnished with touch trigger probe. Based on the designed tests, probing repeatability, two dimensional (2D) and three dimensional (3D) probing errors as well as the inaccuracies in distinguishing the WCS (workpiece coordinate system) in MCS (machine coordinate system) is also considered for this evaluation method [7]. This work aimed to create a baseline for on-machine measurement uncertainty budget data set [8].

Jankowski et al. built a portable setup by using a master artefact with an inner hemisphere for the indirect assessment of the accuracy of the touch-trigger probe. This setup is used to estimate the triggering radius variation in 2D and 3D measurements as well as unidirectional repeatability of a touch-trigger probe [9].

All the above works was conducted for CMM probing. On a machine tool additional factors must be considered. Machining process requires coolant liquid for the cutting process that makes the surfaces wet. Also the surface might not be smooth. Hence, this paper focuses on the external influences that can yield inaccuracies of the probing measurements on a machine tool. Surface irregularities, wetness effect, repeatability of the probe and equivalent probe lobing are experimentally evaluated.

MEASUREMENT METHODS AND RESULTS

For the tests, the table of a five axis WCBXFZYT horizontal machine tool is probed as shown is Figure 1 with a Renishaw machine touch trigger probe MP700 is used with a 100 mm stylus. Four different test procedures have been followed.
In procedure I, the unidirectional repeatability of the probe is evaluated by measuring a surface point 300 times in X-, Y-, Z-, 2D- (an inclined XY plane 45^0 with the horizontal axis) and 3D-directions (all the linear axes need to move simultaneously). The surface is cleaned and dried before the measurements commence.

Figure 1: Probing on a WCBXFZYT machine tool

Figure 2: Spindle with nozzle to spray cutting fluid holding the probe

In Procedure I, total time required for 300 probing with the respective direction is 1H45M approximately with the full retraction of the probe in Z-axis by 300mm after each probing. Figure 3 shows the presence of a drift for an unknow reason. Table 1 shows the results for the Procedure I (unidirectional repeatability of the probe).

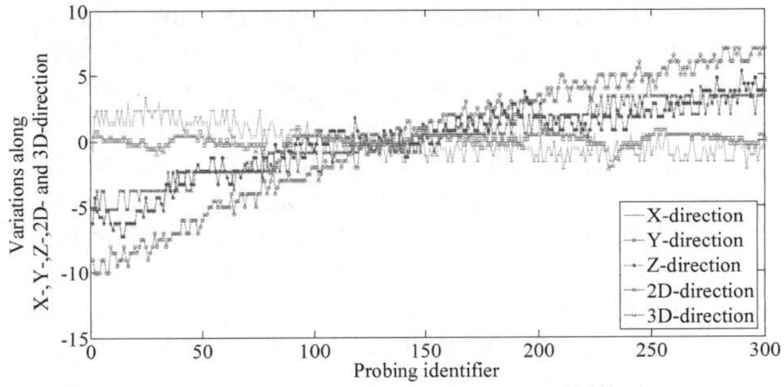

Figure 3: Procedure I (Unidirectional repeatability)

The drift could be responsible for the higher value range of Y-, Z- and 2D-directional measurements. The standard deviation of the measurement in 3D direction is unexpectedly lower and the reason is unknown.

Table 1: Range and standard deviation of the probing measurements for 300 repetitions

	X	Y	Z	2D	3D
Range (μm)	5.5	17.0	12.5	9.9	3.9
Standard deviation (μm)	1.2	4.9	2.7	2.5	0.6

To remove the effect of the drift on the statistical results, pooled standard deviation is calculated for these five directional measurement. For pooled standard deviation calculation, 12 small windows of 25 probings are defined. From these 12 populations, the pooled standard deviation is calculated by using the following formula:
If the populations are indexed as i=1, 2… m, then the pooled variance s_p^2 can be estimated by the mean of the variances s_i^2

$$s_{p,k}^2 = \frac{\sum_{i=1}^{m}(n_i-1)s_{i,k}^2}{\sum_{i=1}^{m}(n_i-1)}, \qquad (1)$$

where n_i is the sample size of population i and k refers to the X-, Y-, Z-, 2D- and 3D-directions. The pooled standard deviations are given in Table 2. Hence, a significant improvement in measurement repeatability is observed.

Table 2: Pooled standard deviation for a window of probing measurements

	X	Y	Z	2D	3D
Pooled standard deviations (μm)	0.63	0.59	0.63	0.64	0.46

In procedure II, surface wetness effect on probing is assessed by measuring a single surface point on Z- and 3D-directions. Procedure II includes four different approaches. For Approach A, a surface point on Z-axis is measured 600 times while for the first 100 probing the surface remain dry and clean, for the second 100 probing cutting fluid is sprayed manually on the surface to make it wet. In this manner the measurement continued for 600 probing. Approach B is similar to approach A but instead of manually wetting the surface the machines' own system (Figure 2) is used to apply the cutting fluid on the surface point. The cutting fluid passes through the machine spindle to the targeted surface area. In Approach C, the same surface point is measured 600 times without using cutting fluid and the surface is kept clean and dry. Finally, in Approach D a surface point is measured in a 3D pose of the machine tool for 600 times but with the same manner than Approach A. Figure 3 shows the measurement variation over 600 probing for Approach A, B, C and D. Total time required for the measurements of each Approach is 1H53M approximately.

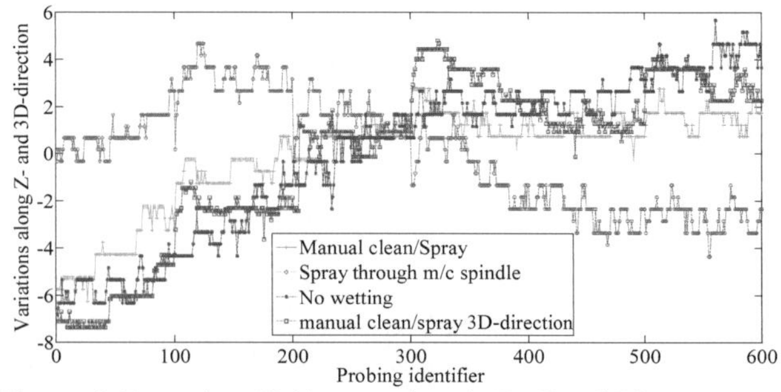

Figure 4: Procedure II (Approaches A, B, C and D)

Figure 4 shows that the drift contributes to the range and standard deviation for the measurements in Procedure II until 300 probing approximately and then the readings have a tendency to stabilize. Table 3 shows the range and standard deviation for Procedure II.

Table 3: Range and standard deviation for four different approaches in Procedure II

	Approach A	Approach B	Approach C	Approach D
Range (μm)	9.0	9.0	12.5	12.2
Standard deviation (μm)	2.1	2.3	3.4	3.5

Figure 4 shows that, the machine start stabilizing after 300 probing for approach A, C and D. As mentioned earlier, in Approaches A and D the surface was wetted by manually spraying the cutting fluid to the desired location and in Approach C the surface was dry and clean. But for Approach B, the cutting fluid sprayed through machine spindle as shown in Figure 2 and the machine seems still drifting which may be an implication of the machine instability due to cutting fluid passes through the spindle and causes thermal instability.

Procedure III is for surface irregularity assessment, a spatial grid of 2 by 2 mm with i ($i = 1$ to n where, n=25) regularly spaced points is measured along k–directions (k= X-, Y-, Z-, 2D- and 3D). The grid is measured $j = 1$ to m times with m=19 for the respective direction hence each point on the grid is revisited m times. Then the effect of surface irregularities is observed by calculating the mean of each point $\hat{P}_{i,k}$ over all m cycles for a particular direction k (k=X-, Y-, Z-, 2D- and 3D-direction). By doing this, the effect of repeatability and drift can be removed and the graphical representation of the surface irregularity effect is given in Figure 5. An indirect machine tool inspection technique like [10], measured surface can influence the overall measurement inaccuracies as measurements were conducted directly on the machine tool table hence the surface contribution is analyzed.

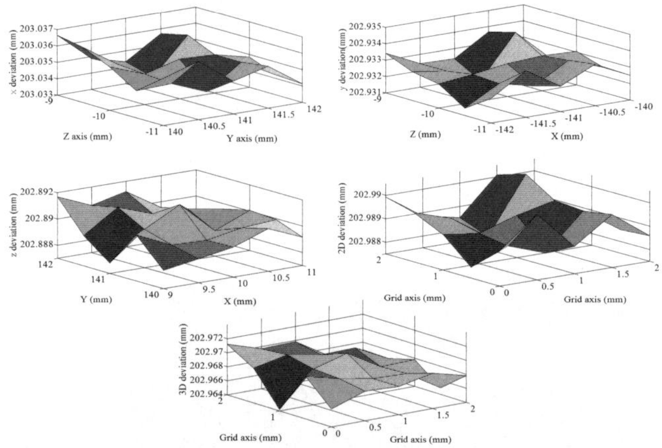

Figure 5: Effect of surface irregularities on probing of a 2 by 2 mm spatial grid along the X-, Y-, Z-, 2D- and 3D-directions

Grid measurement cycle start with measuring the spatial grid along the X-direction followed by the Y-, Z-, 2D- and 3D-directions. Table 4 shows the range of mean $\hat{P}_{i,k}$ and standard deviations for each grid point along the k directions. The mean of each grid point $\hat{P}_{i,k}$ over all m cycles for a particular direction k is,

$$\hat{P}_{i,k} = \frac{1}{m}\sum_{j=1}^{m} P_{i,j,k}; \qquad (2)$$

The standard deviation of the n=25 grid points means is then calculated for each respective direction.

$$S_k^2 = \sum_{i=1}^{n} \frac{(\hat{P}_{i,k} - \hat{T}_k)^2}{n-1}; \qquad (3)$$

where, $\hat{T}_k = \frac{1}{n}\sum_{i=1}^{n} \hat{P}_{i,k}$; (4)

Table 4: Range and standard deviation of the 5×5 spatial grid probing

	X	Y	Z	2D	3D
Range (μm)	2.9	2.7	3.9	2.5	7.0
Standard deviation (μm)	0.86	0.68	0.97	0.73	1.7

Table 4 gives the range and standard deviation of $\hat{P}_{i,k}$ shows the measurements variation along different directions due to surface irregularities. The probing on Z- and 3D-directions is largely affected by surface irregularities compared to X-, Y-, and 2D- directional measurements. As given in Table 4, Z and 3D probing has range and standard deviation of (3.9, 0.97) µm and (7.0, 1.7) µm respectively.

Procedure IV focused on the effect of the probe and machine. Systematic lobing of the probing results is estimated by probing a precision sphere with a sphericity of 1.6 µm (as measured on a CMM). The on machine tool probing is done every 30 degrees longitudinally and every 10 degree latitudinally. Then a sphere is fitted to the measurements. The maximum and minimum residuals, out of sphericity as well as the standard deviation of the residuals are 4.71, -3.37, 8.1 and 1.7 µm respectively.

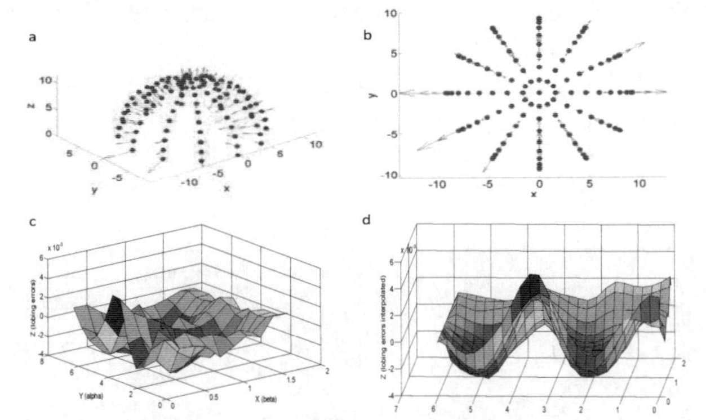

Figure 6: Residuals for a sphere probing (1000x) (a) and (b), probe lobing error (c) and interpolated lobing error (d)

Figure 6 (b) shows that, there is an elliptical trend over the residuals with two visible lobes also visible in Figure 6 (c and d). This error can be the combined effect of the probe, sphere and the machine itself.

CONCLUSIONS

Machine tool probing accuracy is affected by the probe, the machine as well as the measurement environment. Results show that, there is an unknown drift of 10 µm approximately for the first hour while the measurements are performed. Surface irregularities are the major contributor to the inaccuracy of the probing when the probe approaches in Z and 3D directions with a deviation range of 3.9 and 7 µm. But the drift might also contribute to the inaccuracies of the surface measured in different directions. Disregarding the drift, surface wetness does not have observable effect on probing measurements unless cutting fluid is supplied through the machine's system to wet the surface. Because when the cutting fluid passes through the machine spindle, it may change the thermal state of the spindle and cause possible alteration of the spindle position. Thus it can also affect the accuracy of the probing measurement. Repeatability of the probe causes significant effects on the measurements. The worst case is the repeatability for the Y-axis measurement with a standard deviation of 4.9 µm while the 3D measurement has unexpectedly lower standard deviation of 0.6 µm. Out of sphericity of the probing results was significant at 8.1 µm.

ACKNOWLEDGMENTS

This research work is funded by NSERC, CRIAQ, Pratt & Whitney Canada, Meloche Group and SONACA Montreal. The authors are very grateful for the experimental support of Guy Gironne and Vincent Mayer.

REFERENCES

[1] Guiassa, R., Mayer, J. R. R., Balazinski, M., Engin, S., and Delorme, F. E., 2014, "Closed door machining error compensation of complex surfaces using the cutting compliance coefficient and on-machine measurement for a milling process". *International Journal of Computer Integrated Manufacturing*, Volume 27, Issue 11, 2014

[2] Cauchick-Miguel, P. A., and Kings, T. G., 1998, "Factors which influence CMM touch trigger probe performance," *International Journal of Machine Tools and Manufacture*, 38(4), pp. 363-374.

[3] Wozniak, A., and Dobosz, M., 2005, "Factors influencing probing accuracy of a coordinate measuring machine," *IEEE Transactions on Instrumentation and Measurement*, 54(6), pp. 2540-2548.

[4] Dobosz, M., and Wozniak, A., 2005, "CMM touch trigger probes testing using a reference axis," *Precision Engineering*, 29(3), pp. 281-289.

[5] Wozniak, A., and Dobosz, M., 2005, "Influence of measured objects parameters on CMM touch trigger probe accuracy of probing," *Precision Engineering*, 29(3), pp. 290-297.

[6] Johnson, R. P., Qingping, Y., and Butler, C., "Dynamic error characteristics of touch trigger probes fitted to coordinate measuring machines," *Proc. IMTC/98 Conference. IEEE Instrumentation and Measurement Technology Conference.*18-21 May 1998, IEEE, pp. 1168-1172.

[7] Fesperman, R. R., Moylan, S. P., and Donmez, M. A., "Methods, practices, and standards for evaluating on-machine touch trigger probing of workpieces," *Proc. 25th Annual Meeting of the American Society for Precision Engineering, ASPE 2010*, October 31, 2010 - November 4, 2010, American Society for Precision Engineering, ASPE, pp. 309-312.

[8] ISO/DIS, 2011, "*Test code for machine tools-Part 10: Determination of measuring performance of probing systems of numerically controlled machine tools*," 230-10.

[9] Jankowski, M., Wozniak, A., and Byszewski, M., 2014, "*Machine tool probes testing using a moving inner hemispherical master artefact*," Precision Engineering, 38(2), pp. 421-427

[10] Rahman, M.M., Mayer, J.R.R., 2014, "*Five axis machine tool volumetric error prediction through an indirect estimation of intra and inter axis error parameters by probing facets on a scale enriched uncalibrated indigenous artefact*", Precision Engineering, doi:10.1016/j.precisioneng. 2014.10.010

Proceedings of the 25th CANCAM
London, Ontario, Canada, May 31 – June 4, 2015

ICE-MICROTOME DESIGN FOR PROCUREMENT AND CRYSTAL ANALYSIS OF ICE THIN SECTIONS

Steve Bruneau, Ashley Sullivan, Zetian Zhang, and Bruce Colborne
Faculty of Engineering and Applied Science
Memorial University of Newfoundland
St. John's, Canada

ABSTRACT

In this paper, a novel device called an ice-microtome is introduced to be used to produce thin sections of ice for the study of internal structure. The ice-microtome provides an easy and precise procurement of ice thin sections, thus making the further analysis more efficient and accurate. Important features of this device include a scissor-lift mechanism, oblique blade design and low cost. This device has been fabricated and tested, showing great functionality and portability.

INTRODUCTION

Analysis of ice thin sections is a frequently used method for examining the internal structure of ice. Within the STePS^2 research program at Memorial University the ice in question is used in laboratory investigations of high energy collisions between marine ice and steel structures. The study aims to develop validated practical tools that will permit the safe design and assessment of ships and offshore structures for arctic conditions [1]. Ice thin sections are acquired from the ice samples in damaged and undamaged zones to be examined in terms of their crystal structure to investigate properties such as crystal orientation, size and overall size distribution.

Thin Section Analysis

Once ice thin sections have been prepared, a set of two polarizers is used to acquire images of the sections. Two polarizing filters are arranged perpendicular to each other, with a light source installed underneath the lower polarizer. A polarizing filter only allows light waves of the same orientation to pass through, which means two perpendicularly positioned filters block visual light waves completely.

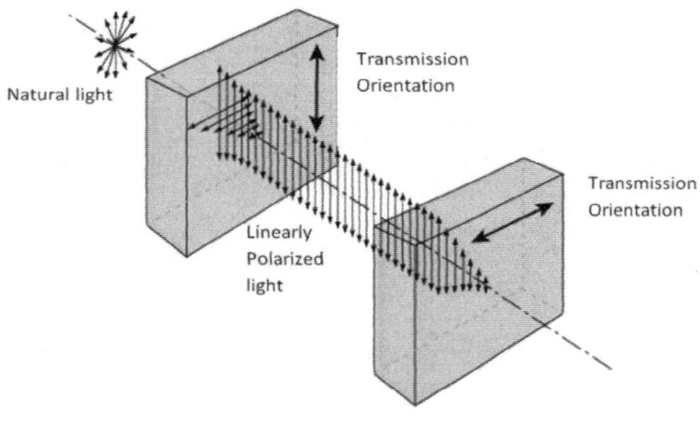

Figure 1: Perpendicular Polarizing Filters Blocks Visual Light [2]

As can be seen in Figure 1, the light source goes through the first polarizer and becomes linearly polarized light, which is totally blocked by the perpendicularly arranged second polarizer.

When an ice thin section is placed between the two polarizers, the polarized light is refracted to multiple directions due to the different orientations of individual ice crystals in the ice sheet. This technique reveals the otherwise transparent and indistinguishable crystal morphology.

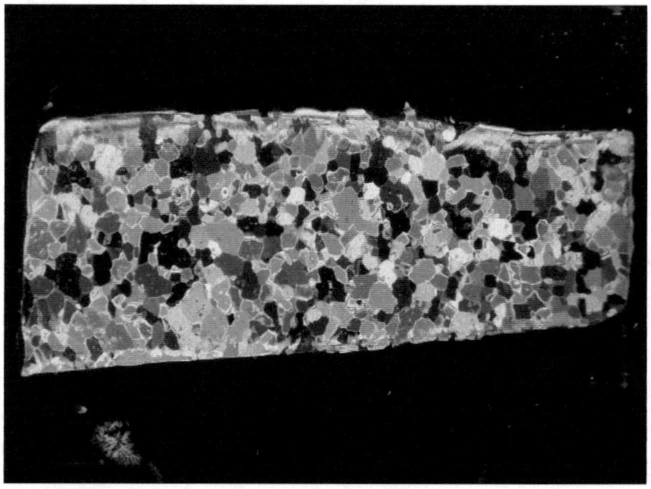

Figure 2: Polarized Ice Crystal Morphology [3]

Ice Crystal Counter

Once the pictures of polarized ice thin sections are obtained, a program may be used to analyze the images, one such, developed by the authors, is called Ice Crystal Counter. This program retrieves the number, density and average area of crystals in the section, it also levels the image horizontally and overlays a 1cm^2 grid on the image. While much is automated, many manual adjustment options are available to ensure accuracy of processing for each unique thin section image [4].

Thin Section Preparation and Rationale for Microtome

In the STePS^2 program crushed or uncrushed large scale (1m) ice samples are retrieved from an aluminum frame holder with the use of heating belts. The sample is then cut into large chunks with an electric chainsaw. These chunks are further cut into 6-10mm thick sections with the use of a typical woodworking band saw. These thick sections are finally trimmed to produce a relatively flat sample featuring the ice region under consideration. Ice samples are then mounted to a glass slide (approx. 100x150mm) by placing the flat ice surface onto a cold glass slide, applying pressure and then dropping small water droplets with a pipette on the perimeter of the section.

Previously, ice thin sections (0.5-1.5mm) were then procured by slowly melting down the thick ice specimen via manually rubbing the surface with a thick, warm aluminum plate [5]. However, this method needed improvement since melting creates a non-uniform surface of ice and a blurring of ice crystals. Thus, a new procedure of shaving ice thick sections with a microtome was introduced, which overcomes the flaws and tedium of the melting procedure.

IMPORTANT FEATURES

Scissor-Lift Microtome Design

The microtome design is based on a scissor-lift platform, which is moved vertically by a mechanism with the issue of linked, folding supports in an "X" pattern (scissor mechanism). The upward motion is achieved by applying pressure to the outside of the lowest set of supports, elongating the crossing pattern, and propelling the platform vertically. The contraction of the scissor action is mechanical via a lead-screw. This design achieves the desired action by a ½"-10, left-handed Acme platform nut coupled with a ½"-10, left-handed Acme threaded rod. This rod is controlled by a crank handle located on the back of the device and is secured by flange bearings on each end that allow the rod to rotate.

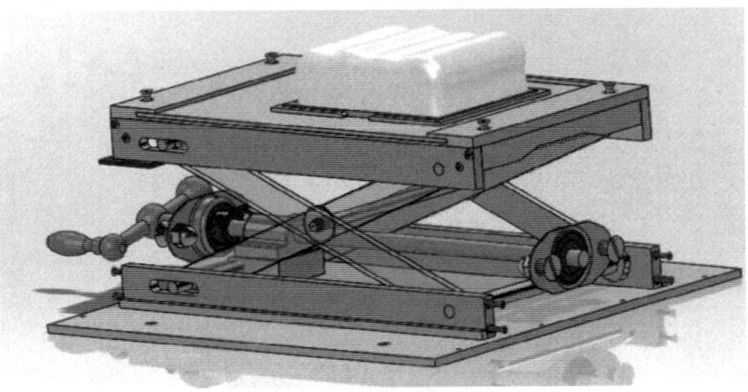

Figure 3: Isolated View of Scissor Lift Mechanism

Ease of Use

To minimize accidental impact of the shaving assembly with the main frame, rubber stops are located on the contacting points.

Figure 4: Rubber Stops

The shaving assembly is designed to ride along 3/4" diameter hardened shafts with four pillow block linear bearings that give a smooth and fast slide.

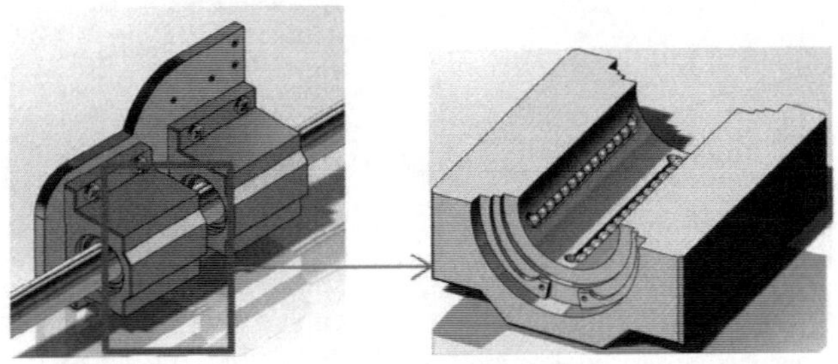

Figure 5: Pillow Block Bearing Mechanism

Footprint

Overall, the design was developed to have a small footprint allowing for ease of portability and use in confined locations. This type of device is normally used in a cold room environment where space is usually limited. The device was designed to be compact, having a footprint of 15"x14"x7.25".

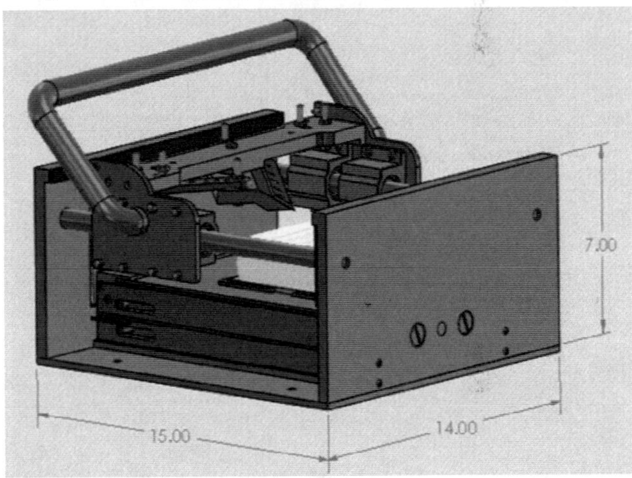

Figure 6: Footprint of the Microtome

PERFORMANCE EXPERIMENTS

Adjustable Blade Design for Experimental Purpose

In order to get the best performance of ice shaving, an experimental design with an adjustable blade was made to determine the ideal geometry of blade that will create the smoothest surface of ice. The adjustable design gives great control of yaw angle of the blade in order to study how the oblique angle affects the cutting performance. When talking about other parameters that affect cutting performance, in particular the rake angle, Gunay et al. [6] have discovered that the cutting resistance force decreases with increased rake angle.

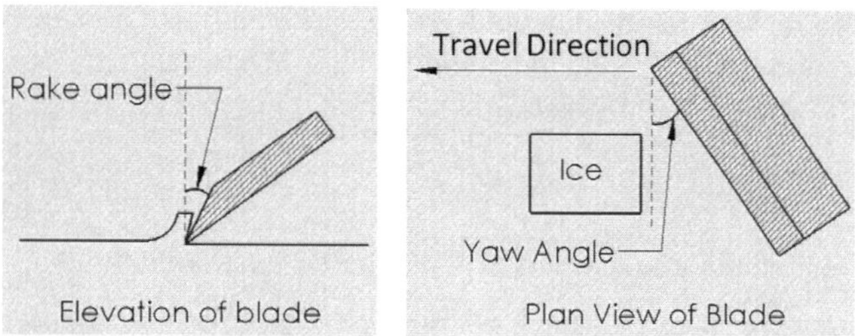

Figure 7: Rake Angle and Yaw Angle

With different yaw angle and rake angle of the blade, the different fluency on cutting and different smoothness of the shaved surface can be achieved. With the design shown in Figure 8, the blade can be adjusted from perpendicular to 45 degrees to the travel direction, with 6 choices of angle. The middle supporting bar provides the main driving and supporting force to the knife, while the two side arms fix the knife in the desired angle. The blade being used on this device has a tip angle of 30^0 and the aluminum plate that supports the blade needs a minimum 20^0 angle to provide the strength. As a result, the rake angle of the blade was fixed at 40^0 for this device.

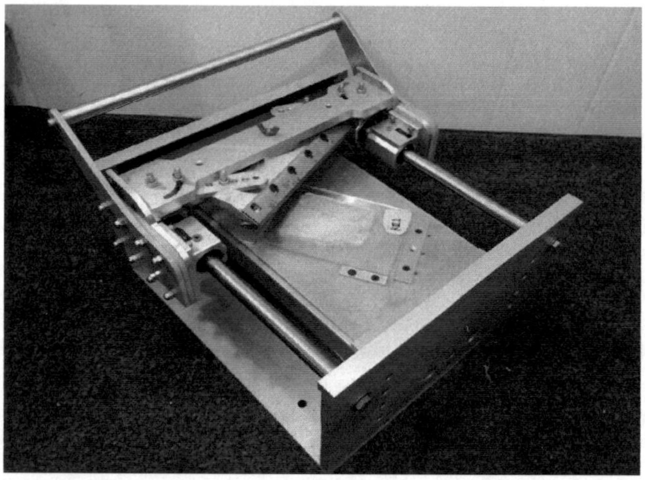

Figure 8: Adjustable Blade Design for Experimental Purpose

Expectations of the Experiment

A. Moufki et al. [7] have discovered that with an increase trend in the yaw angle, the lateral force will increase while the cutting force and thrust force will decrease, resulting in easier shaving. It was predicted that the bigger the yaw angle, the more fluent the shaving performance will be, and thus the smoother the ice surface will appear.

Experimental Settings

A set of experiments were conducted to test the performance of the microtome, based on the fluency of cutting and smoothness of shaved ice surface.

The blade yaw angle was set to be , 22.5, 3 , 37.5 and 45 to perform the shaving procedure in a cold room with -7C air temperature. A camera was used to record the shaving process and take pictures of the shaved surface. The fluency of cutting was determined based on the personnel's feeling and the video recorded. One other way to determine the fluency of shaving was by looking at the ripples formed parallel to the cutting edge on the ice surface due to intermittent shaving motion caused by the friction force and chatter.

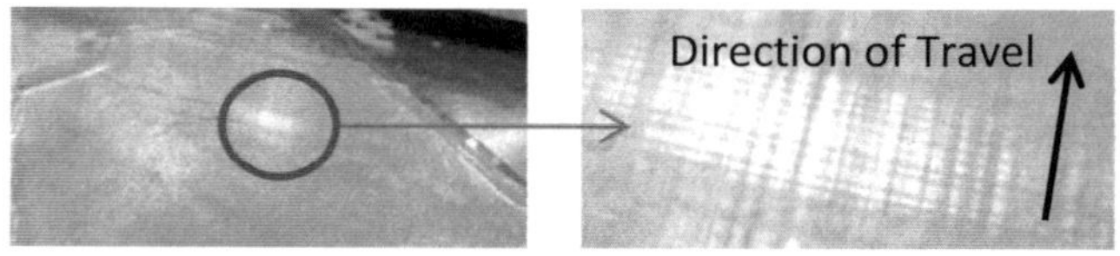

Figure 9: Ripples Developed Due to the Friction and Chatter

Experimental Results

Table 1: Experimental Results

	Fluency of shaving	Space between Ripples (parallel to cutting edge)
Orthogonal	Intermittent	0.0565 inch
22.5	Fluent	0.0705 inch
3	Fluent	0.0703 inch
37.5	Fluent	0.0698 inch
45	Fluent	0.0762 inch

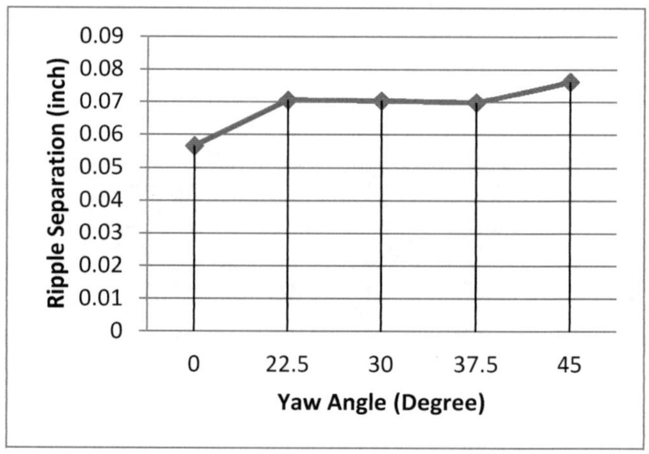

Figure 10: Ripple Separation Distance vs Yaw Angle

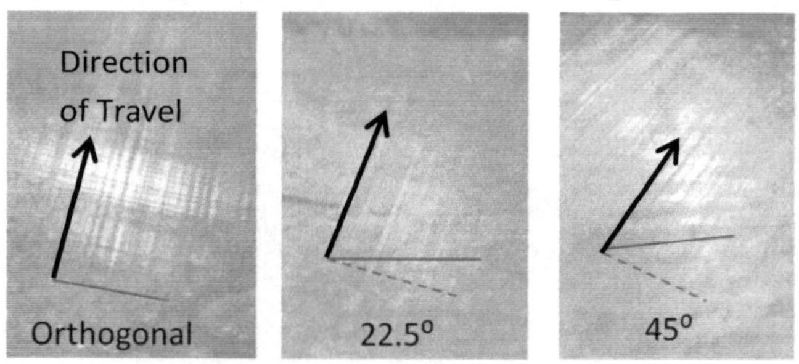

Figure 11: Comparison of Ripples Formed by Intermittent Shaving with Different Yaw Angles (Red Lines)

Comments

The experimental results agreed somewhat with the expected results. There was an obvious difference between orthogonal shaving (yaw = 0) and oblique shaving (yaw ≠), nevertheless the degree of yaw angle did not make a noticeable change to the performance. There were two kinds of ripples on the shaved surface of ice, one parallel to the traveling direction while another parallel to the cutting edge. The ripples parallel to traveling direction were caused by the bur on the blade edge, which can be minimized by sharpening the blade. The ripples parallel to the cutting edge were formed by the intermittent motion of shaving due to friction and chatter. When the blade is inclined at an angle to the traveling direction, the shaving process became much more fluent than orthogonal shaving, and the ripples parallel to cutting edge were reduced.

Final Design with Fixed Blade Angle

Based on the experimental results, a final design with fixed yaw angle was made. The yaw angle of blade was designed to be 22.5^0 and the rake angle is 40^0, which resulted in good travel range, fluent shaving performance and a smooth ice surface.

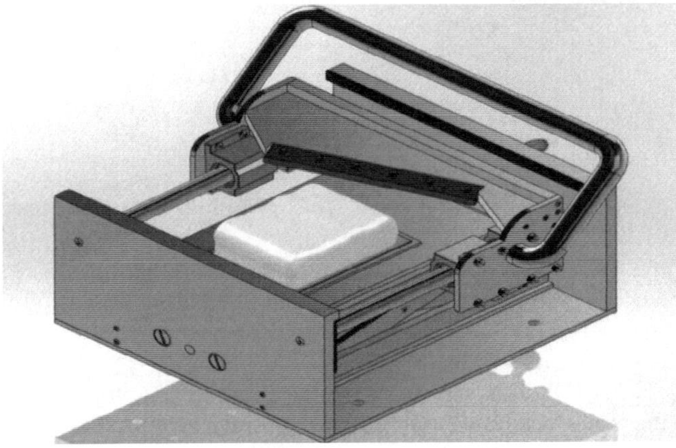

Figure 12: Final Design with Fixed Blade Yaw Angle

Cost

This device is mainly made from 6061 Aluminum alloy, which is widely used in the industry applications for its low density, corrosion resistance and low price. The very sharp and hard blade is a commercially available blade (DW7332), which is typically used in a wood planning device. The cost of the whole device is below $2000, which is very budget friendly, comparing to the commercially available microtomes that are around $10,000.

Safety Considerations

Working with microtomes, whether preparing or cutting a sample, is an acquired skill. Hands-on training is required from trained personnel before using a microtome. Because of the associated hazards, safety must be incorporated into every step of the process to keep fingers and hands protected. The blade is covered with a blade cover when not being used. Leather gloves are recommended and long handled brushes should be used when cleaning the device to avoid accidental contact with the blade.

CONCLUSIONS

In this paper, the entire procedure of ice thin section procurement and analysis has been introduced. The device that involved in the procurement process of ice called ice-microtome is designed and manufactured to shave ice samples into thin sections that is 0.5~1.5 mm thick. Several experiments have been carried on to test the performance of this microtome, based on the fluency of cutting and smoothness of shaved surface. The results of the experiments led to a final design of the microtome that has a 22.5-degree blade yaw angle and 40-degree rake angle. It was observed that the ice-microtome is a capable and inexpensive device to be used in ice thin section analysis procedure.

ACKNOWLEDGEMENTS

This research has been done under STePS2 project and was supported by: ABS, Atlantic Canada Opportunities Agency, BMT Fleet Technology, Husky Oil Operations Ltd, Research and Development Council, Newfoundland and Labrador and Samsung Heavy Industries.

REFERENCES

[1] Faculty of Engineering and Applied Science. (2012, April 10). *Preparing for harsh conditions* [online]. Available: http://www.engr.mun.ca/home/Steps2.php.

[2] R. Nave. (2015, Mar 10). *Crossed Polarizers* [online]. Available: http://hyperphysics.phy-astr.gsu.edu/hbase/phyopt/polcross.html

[3] Andyextance. (2014, May 10). *The urgent voice who refused to be silenced on climate danger* [Online]. Available: https://simpleclimate.wordpress.com/tag/antarctica/

[4] J. Sheppard, "Ice Crystal Counter User Manual," Memorial University of Newfoundland Sustainable Technology for Polar Ships and Structures, 2012, 1-7.

[5] A. Sullivan, "Influence of Ice Shapes on Indentation Loads," Memorial University of Newfoundland Sustainable Technology for Polar Ships and Structures, 2012, 1-28.

[6] A.K. Baldoukas, F.A. Soukatzidis, G.A. Demosthenous and A.E. Lontos, "Experimental investigation of the effect of cutting depth, tool rake angle and workpiece material type on the main cutting force during a turning process," Proceedings of the 3rd International Conference on Manufacturing Engineering (ICMEN), 1-3 October 2008

[7] A. Moufki, D. Dudzinski, A. Molinari, M. Rausch, "Thermoviscoplastic modelling of oblique cutting: forces and chip flow predictions," International Journal of Mechanical Sciences, Volume 42, Issue 6, June 2000, Pages 1205-1232, ISSN 0020-7403, Available: http://dx.doi.org/10.1016/S0020-7403(99)00036-3.

STRATEGIES IN SINGLE POINT INVERTED CUTTING FOR FABRICATION OF STRUCTURED SURFACES

Ben Hamilton
Mechanical and Materials Engineering
Western University
London, Ontario, Canada
bhamilt8@uwo.ca

Sama Hussein
Mechanical and Materials Engineering
Western University
London, Ontario, Canada
sabdelw3@uwo.ca

O. Remus Tutunea-Fatan
Mechanical and Materials Engineering
Western University
London, Ontario, Canada
rtutunea@eng.uwo.ca

Evgueni Bordatchev
National Research Council of Canada
London, Ontario, Canada
evgueni.bordatchev@nrc-cnrc.gc.ca

ABSTRACT

Corner cube (CC) constitutes the most commonly used geometry encountered in the design of retroreflectors (RRs). Due to their property to reflect back an incident beam, RRs are optical components that have found multiple industrial applications, including but not limited, to automotive lighting. Despite their ubiquity, CC retroreflective arrays cannot be generated by means of traditional multi-axis machine tools, primarily due to their kinematic limitations. However, while the alternate pin bundling techniques that were developed are fully capable to generate macroscale CC features, their applicability to microfeatures remains limited. To address this, the present study will examine the feasibility of ultraprecise single point inverted cutting techniques as a viable alternative for fabrication of microscale CC geometries.

KEYWORDS: retroreflector, cube-corner, micro-machining, monocrystalline, diamond

INTRODUCTION

Retroreflectors (RRs) are passive optical components that have the ability to reflect a light beam back to its source through a range of incident angles deviating from the normal axis. In general, retroreflectors can be divided into two primary groups: lens-and-mirror (also called "cat's-eye"), and triple mirrors with mutually perpendicular facets (also called "corner cube") [1, 2]. The latter type – that is geometrically equivalent to a cube-originating pyramid that is indented into a planar surface – is characterized by an optical efficiency that is superior to the former, primarily due to its considerably reduced amount of light scattering. While each type of RR has its set of applications, this paper will focus on the fabrication of corner cube (CC) RRs.

Depending on the relative positioning of the microfeatures within retroreflective pattern, there are two main types of CC RRs: with hexagonal and triangular apertures, respectively (Fig. 1). Among them, the hexagonal aperture is characterized by a higher efficiency at small incident angles [3]. Because of this, the remainder of this paper will focus on the manufacturing of hexagonal aperture RRs, although the core concepts to be presented can also be extrapolated to other types.

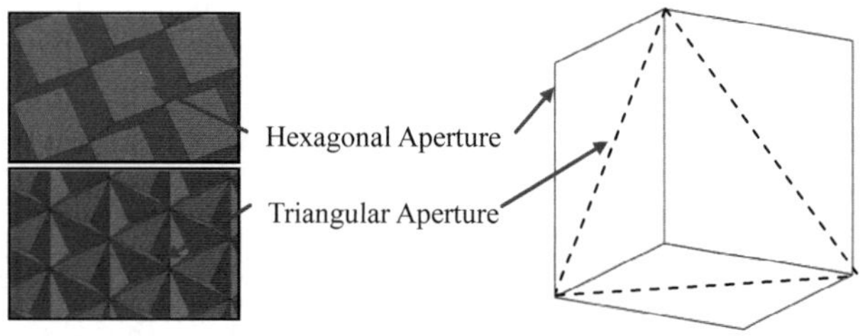

Figure 1: Detailed geometry of corner cube retroreflectors

STATE-OF-THE-ART IN THE FABRICATION OF RETROREFLECTIVE OPTICS

The present fabrication approaches pertaining retroreflective arrays can be divided into two primary categories: micro- and macroscale. While no formal threshold between them has been proposed so far, a characteristic dimension (C) of 500 μm could be considered as the dividing line between micro- and macrosize CC features [3, 4].

In general terms, macrofeatures are mass-produced by means of injection moulding. The tooling used is generated by means of a pin bundling technique involving pins with hexagonal cross sections and CC forming ends oriented in a consistent manner (Fig. 2) [5]. The pin bundling technique is one of the few possible options capable to overcome the kinematic limitations of the traditional multi-axis machine tools that make them incapable to generate the orthogonal geometry around the apex of the CC feature.

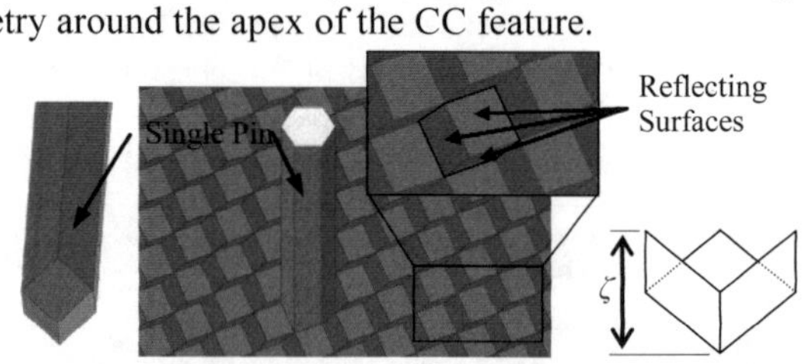

Figure 2: Fabrication of CC arrays through pin bundling

Despite its ubiquity, pin bundling remains a manual, error prone as well as time-consuming technique, especially when considering that as the characteristic geometry is being reduced in size, the number of individual facets increases in a quadratic fashion. For instance, a 10 cm^2 array with a 1000 μm characteristic dimension requires 578 pins, whereas the same array requires 2310 pins if the characteristic dimension (ζ in Fig. 2) is decreased to 500 μm. Given that many retroreflective arrays are required to cover more than 100 cm^2, combined with the fact that the present miniaturization needs of the industry often require thousands or even millions of pins, it becomes apparent that more versatile and efficient manufacturing approaches are needed. To address this, few alternative methods have been developed in an effort to replace the aforementioned pin bundling technique.

One such technique to be mentioned in this context is constituted by the microchiseling approach involving the use of an ultraprecise five-axis micromachining center along with a monocrystalline diamond insert (MCD) [4, 5]. Similar with other diamond cutting operations, while the quality of the resulting surfaces approaches the optical domain, the durability of the cutting tool remains a challenge, especially in case of metallic materials.

INVERTED CUTTING KINEMATICS AND TOOLING

Along the same lines, the primary goal of this study is to investigate additional micromachining pathways involving single point inverted cutting conceived as a combination between five-axis machining and diamond turning (Fig. 3).

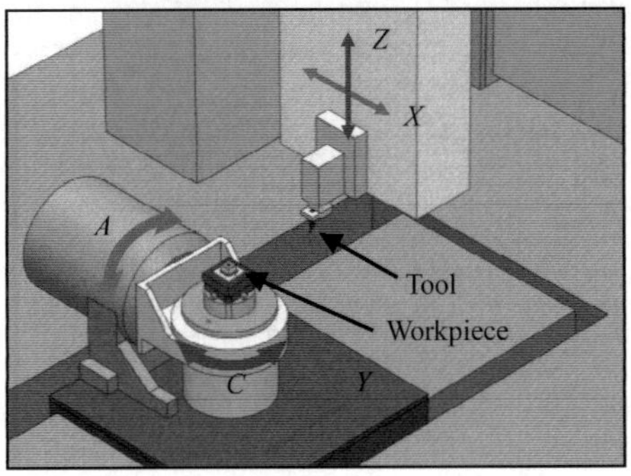

Figure 3: Single point inverted cutting kinematics

In the developed approach, the position of the diamond tool is determined by the two translational motions along *X* and *Z* whereas the workpiece can be rotated around *X* and *Z* axes (*Z*-axis being the primary rotation) and can be translated along *Y*-axis. The presence of the two rotational axes is required to enable an adequate alignment of diamond insert surfaces with the facet of the RR to be machined.

To generate the desired CC geometry, a V-shaped MCD insert with turning-resembling geometry was used (Fig. 4). As it can be noticed, the positioning of the rake and clearance faces is somewhat counterintuitive with respect to traditional turning operations, and that primarily to ensure an adequate generation of the CC facets.

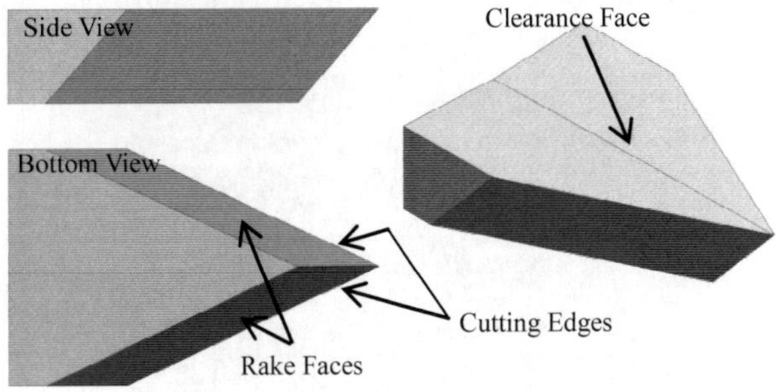

Figure 4: Schematic representation of the MCD insert

INVERTED CUTTING STRATEGIES

In order to ensure the desired levels of optical surface quality, the generation of the desired CC microfeatures can be conceived as a succession of roughing and finishing steps as it is unlikely that layers of materials thicker than 4-10 μm can be removed in the final passes.

Evidently, depending of size of the characteristic dimension, roughing operations might or might not be required. However, in a more conservative approach attempting to maximize the life of the tool as well as the quality of the machined surface, it can be ascertained that both type of operations will be required.

FINISHING APPROACHES

While most conventional depictions of the machining processes tend to describe them in the sequence in which they are being applied to the workpiece (*e.g.* roughing, finishing), in this study they will be presented in the reverse order, primarily for clarity reasons.

The principal assumption of the finishing approaches is that the geometry of the CC microfeatures has been already been brought to a "near net shape" that entails a minimal amount of finishing, a hypothesis that is required to ensure the desired optical surface quality of the machined CC facets.

Unidirectional Translational V-Motion

In this approach, the workpiece has to rotate -35.26° around the secondary *X*-axis in order to align the first unfinished facet with the vertical *ZX* plane of the machine tool. After that, the tip of the tool has to linearly trace the edge of the facet to be finished in a *–X/–Z* motion (motion 1 in Fig. 5). Once the tip reaches the apex of the cavity, the motion of the tool shifts to trace the adjacent facet edge (*-X/+Z* direction or motion 2 in Fig. 5). In this manner, a complete V-motion sequence generates one facet of the CC structure with hexagonal aperture.

Following the completion of one CC facet, the workpiece has to rotate 120° about the primary *Z*-axis in order to expose the next unfinished CC facet and then the process repeats once more for the third facet of the retroreflective feature. Finally, the complete array can be generated through successive shifts in the horizontal *XY*-plane of the described V-motion pattern.

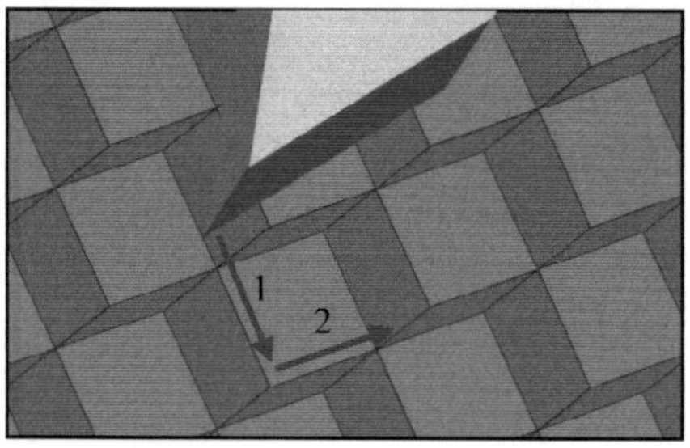

Figure 5: Unidirectional translational V-motion

Given the permanent contact between one particular cutting edge of the diamond insert and workpiece, this straightforward cutting approach has the potential to induce a premature and fast wear on the tool even though the underformed chip thickness is typically kept under 4 μm. Because of this constraint, increasingly deeper passes in *Z* direction are required in order to achieve the desired depth of the CC feature. The total length of the toolpath involved in the generation of one facet is around 3.5 times the characteristic dimension or 3.5ζ.

Alternating Translational V-Motion

To improve the aforementioned unidirectional approach, other cutting strategies can be imagined and one more or less evident choice would be to ensure a more even distribution of the wear on the tool by enforcing the contact on both cutting edges of the V-shaped diamond insert. In order to accomplish this, an alternating, rather than unidirectional V-motion would be preferable, as shown in Fig. 6.

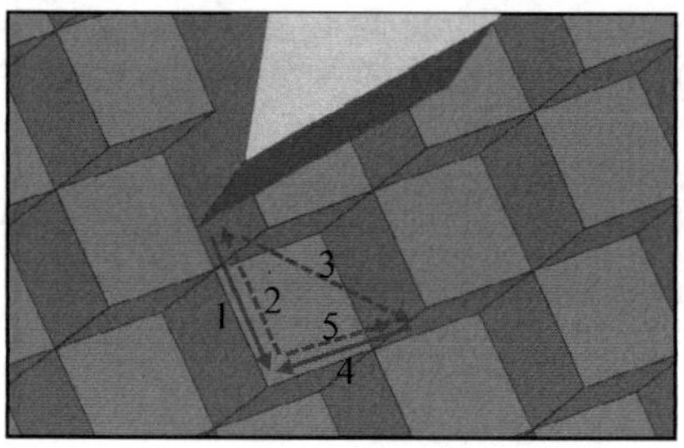

Figure 6: Alternating translational V-motion

While the overall wear progression rate on the tool will decrease, it can be noticed that the total tool path length increases to 8ζ, partly as a result of the additional ancillary positioning/non-cutting time required in this scenario. However, while a more in-depth investigation would be required in order to be able to present a documented conclusion on this topic, it could be speculated here that the overall machining time could still experience a reduction in this case, primarily due to the decreased number of tool changes/setup procedures which are time consuming [4].

Alternating Direct Plunge Motion

Another possible alternative makes use of a tool whose geometry matches that of the CC, such that instead of the previous 50° angle at the tip, the insert is characterized by a 90° opening angle. In this case, a simple alternating "in-and-out" motion along the –*Z* direction (and movements of *Z* and *Y* axes) would be sufficient to generate one of the CC facets (Fig. 7).

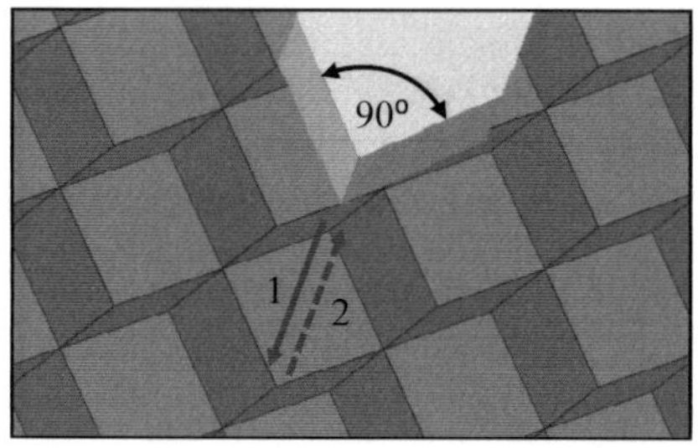

Figure 7: Alternating direct plunge motion.

After tool retraction (motion 2 in Fig. 7), the workpiece has to perform a 120° rotation about *C*-axis in order to expose the next facet to be cut, such that after three consecutive plunges, the entire CC geometry will be completed.

Since in this cutting scenario, the tool is concurrently engaged on both cutting edges at the same time, cutting forces are likely to increase. However, it remains to be determined if this increased tool load translates into premature tool wear and/or failure. Regardless of the actual tool durability outcome, this approach remains the most productive since it is characterized by the shortest tool path.

ROUGHING APPROACHES

These approaches are required in order to bring the CC geometry closer to its finished shape/size. While for small enough characteristic dimensions roughing could be deemed as unnecessary, this does not represent the most common scenario since the absence of roughing could translate in: i) premature tool failure associated with large material removal rates that are desirable for productivity/efficiency reasons; and ii) impossibility to guarantee optical surface quality at large material removal rates. Based on these considerations, two different roughing strategies will be detailed further.

Enlarging Pyramidal Recess

Depending on the geometry of the cutting tool, this approach would entail a gradual enlargement of the CC structure from center to periphery (Fig. 8). When each of the three facets delimiting the contour is completed, the tool had to perform an outward motion in order to advance to the next concentric contour. The amount of outward advancement is determined by the maximum allowable thickness of the layer that can be removed in one plunge motion.

Because of this, a 150 μm deep structure would translate into 18 layers of 4 μm thickness each [6], the last ones being the finishing passes. In the event that cutting inserts with both angles are available, a number of combinations between roughing and finishing passes can also be envisioned.

Figure 8: Enlarging pyramidal recess

Spherical Recess

An alternative to the pyramidal recess approach would involve the use of a ball-end mill for roughing to be followed by any of the aforementioned options for finishing (Fig. 9).

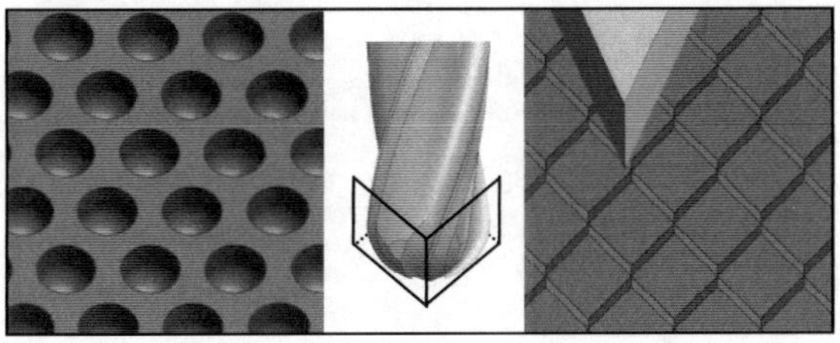

Figure 9: Spherical recess

While it is evident that in this case the amount of material to be removed in the initial finishing passes will be unevenly distributed compared to the prior roughing approach, this technique involving rotational tools is more productive. Furthermore, a certain tradeoff exists between the size of the end mill and the depth of the recess to be generated as larger tools are more efficient in the radial direction, but less effective with respect to the axial one. This translates into important finishing thickness non-uniformities in either of the two directions.

CONCLUSIONS

The need for smaller cube-corner retroreflective arrays has prompted for the investigation of more efficient processes capable to generate these arrays. Conventionally, these structured surfaces are fabricated by means of master inserts generated through pin-bundling techniques. However, as the size of the feature decreases into to the microscale domain, traditional fabrication techniques tend to become inefficient.

To circumvent the drawbacks of the traditional manufacturing techniques, novel techniques involving five-axis single point inverted cutting have been analyzed in the present study. Based on this, it becomes apparent that various combinations between different roughing and finishing strategies will have to be further investigated in order to determine the cutting parameters capable to generate optical surface quality in a repeatable and predictable manner.

Further research will be carried out towards the development of a cost-efficient mass-fabrication procedure capable to ensure high optical characteristics of the generated retroreflective microfeatures.

ACKNOWLEDGMENTS

This study is the result of the collaboration between Western University and National Research Council of Canada. Partial financial support was provided by Natural Sciences and Engineering Research Council (NSERC) and AUTO21 Network of Centres of Excellence.

REFERENCES

1. Davies, M.A., Evans, C.J., Patterson, S.R., Vohra, R., and Bergner, B.C. 2003, "Application of Precision Diamond Machining to the Manufacture of Micro-Photonics Components", *Proceedings of SPIE*, Vol. 5183, pp. 94-108.
2. Takeuchi, Y., Yoneyama, Y., Ishida, T., and Kawai, T. 2009, "6-Axis Control Ultraprecision Microgrooving on Sculptured Surfaces with Non-Rotational Cutting Tool", *CIRP Annals - Manufacturing Technology*, Vol. 58, pp. 53-56.
3. Jinghe, Y., Shengjiang, C., Sumei, L., and Yanxin, Z. 2002, "Design and Fabrication of Micro-cube-corner array Retro-reflectors", *Optics Communications*, Vol. 209, pp. 75-83.
4. Flucke, C., Glabe, R., and Brinksmeier, E. 2008, "Diamond Micro Chiseling of Molding Inserts for Optical Micro Structures", *Proceedings of the 23rd ASPE Annual Meeting and 12th ICPE*, Portland, USA, pp. 92-95
5. Brinksmeier, E., Reimer, O., Osmer, J., Twardy, S., and Rickens, K. 2009, "Machining Processes for Precision Molds", *CDROM Proceedings of the 24th Annual Meeting of the American Society for Precision Engineering*, Monterey, USA.
6. Brinksmeier, E., Glabe, R., and Schonemann, L. 2012, "Diamond Micro Chiseling of Large-scale Retroreflective Arrays", *Precision Engineering*, Vol 36, pp. 650-657.

Aerodynamics

Proceedings of the 25th CANCAM
London, Ontario, Canada, May 31 – June 4, 2015

ANALYSIS OF WIND TURBINE STALL USING TUFT FLOW VISUALIZATION

Tyler E. Gallant
Mechanical Engineering
University of Waterloo
Waterloo, On, CA
tgallant@uwaterloo.ca

Professor David A. Johnson
Mechanical Engineering
University of Waterloo
Waterloo, On, CA
david.johnson@uwaterloo.ca

ABSTRACT

The formation of stall over a wind turbine blade and the resulting flow field are not currently well understood despite their significant effects on turbine performance. Here, the study of the formation of stall on a blade as a function of the tip speed ratio, azimuthal position and yaw-offset position is presented. Torque generated by the turbine is also examined to observe the effect of stall on performance.

Results consistently showed that, as the upstream wind speed increased, the stall on the blade, as well as the torque generated, increased. With a 0° yaw-offset, the stall fraction (ζ) was found to be relatively consistent with azimuthal position. However, when the turbine was yawed ±15°, azimuthal variations in ζ were apparent. It was found that, as the blade rotated towards the oncoming wind, ζ decreased, and as the blade rotated away from the wind, ζ increased. It is posited that this variation is due to the changing velocity geometry and angle of attack throughout the turbine rotation.

KEYWORDS: Wind turbine, flow visualization, stall, aerodynamics, fluid dynamics

INTRODUCTION

One of the most abundant sources of energy from renewable resources is wind turbines. As demand for more efficient, reliable and quieter wind turbines increases, further research is required to understand the aerodynamics of wind turbine blades in all possible conditions. One aerodynamic condition that is not currently well understood occurs when a wind turbine blade stalls. Stall occurs when the air flow over the blade separates from the surface, a point which is related to the angle of attack of the blade, α, the magnitude of the relative wind and other parameters such as geometry.

One technique for studying stall over a wind turbine blade is tuft flow visualization. This technique consists of attaching short pieces of string or yarn to a surface immersed in a flow. Images of the tufts movement and orientation are recorded and provide an indication of the flow direction as well as flow separation. In the context of wind turbines, tufts are attached to the surface of the blade on the suction side of the airfoil. If a tuft at a particular location is aligned with the flow direction, it indicates that the flow at that point is attached to the surface. Similarly, if the tuft is not oriented with the flow, it indicates that the flow at that point has separated. An example image of tufts on a wind turbine blade immersed in air flow is provided in Figure 1.

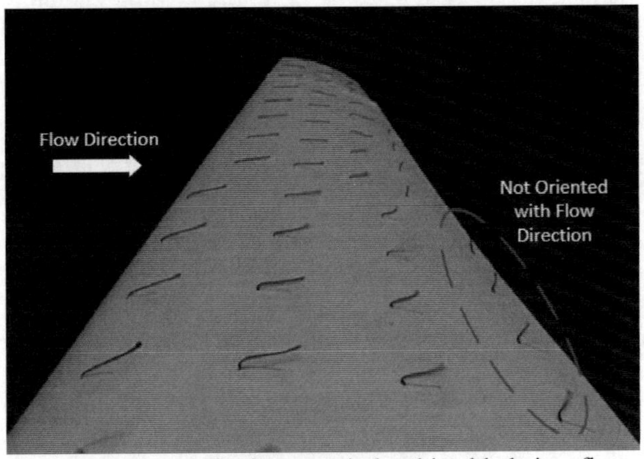

Figure 1: Image of tufts on a wind turbine blade in a flow

This study examines the formation of stall on a wind turbine blade as a function of the tip speed ratio λ, the azimuthal position, and the yaw-offset position. A secondary objective of the study is to examine the effect of stall formation on the shaft torque generated during operation.

PREVIOUS STUDIES

Several experimental studies examining the formation of stall on a wind turbine blade with tuft flow visualization have been published in the literature, including Eggleston and Starcher [1], Haans et. al. [2] and Maeda and Kawabuchi [3]. However, in these studies tuft images were analyzed manually by the researchers with unclear criteria for what tuft orientation was indicative of stall. This may have resulted in erroneous or inconsistent results, as well as significantly limited the amount of tuft video data which could be practically analyzed. These problems were recently addressed by Swytink-Binnema and Johnson [4] with the development of digital image processing software capable of analyzing large numbers of tuft images using strict guidelines for stall recognition.

They conducted tests using a 30 kW upwind wind turbine equipped with two 5 m long blades with a nominal 3°

pitch and no twist [4]. Stall was observed to increase as a function of the wind speed, and appeared to vary cyclically along the blade with the azimuthal position during testing [4]. Experiments were performed in an open environment with an unknown yaw-offset position, which the author suggested may be responsible for the azimuthal variations in stall [4]. They defined the stall fraction ζ as the percentage of tufts in the image indicating that the flow has separated.

EXPERIMENTAL SET-UP

All tests were conducted at the UW Wind Generation Research Facility. The facility is equipped with an open circuit wind tunnel with a fan area of 11.4 m by 7.6 m [7]. The fan bank is capable of generating winds up to 11 m/s [7], allowing for full-scale testing with the test turbine in a controlled, constant wind speed with known yaw conditions.

Tests were conducted using the University of Waterloo (UW) Wind Energy Group test turbine. The test turbine is equipped with a variable speed generator allowing for testing with set operational speeds up to 230 rpm [5]. Three 1.6 m long test blades designed and fabricated by Gertz [6] were installed on the test turbine. Each blade has a tapered chord and a twist of approximately 19°. The blades are designed to operate at 200 rpm in a 6.5 m/s wind, resulting in a design tip speed ratio of 5.15 [6]. An image of the test turbine with the Gertz blades installed is shown in Figure 2.

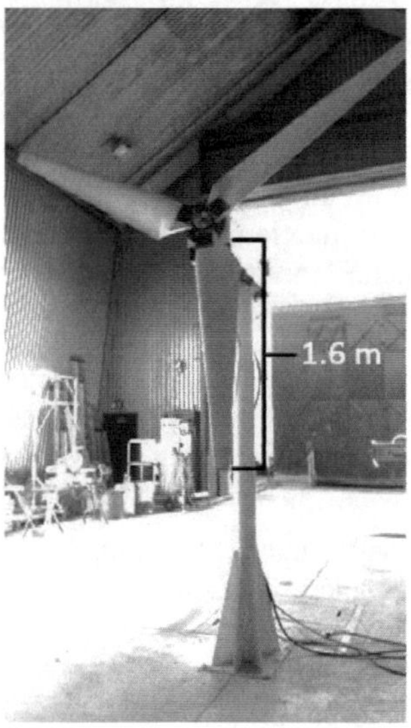

Figure 2: Image of Test Turbine with Gertz blades installed

After preliminary testing of tuft lengths and materials, it was decided to use 3 cm lengths of cotton yarn for the experiment. Approximately 50 tufts were distributed evenly over the blade, aligned parallel to the quarter chord line and spaced 4 to 8 cm apart. This was done to prevent tufts from becoming tangled with each other as well as to keep tufts separated in the camera images for post-processing.

To record the tufts movement during operation, a high definition (HD) video camera was attached to the blade root via an adhesive mount. Video was recorded at 1080p resolution at a rate of 60 frames per second. An image of the camera and tufts attached to the blade is provided in Figure 3.

Figure 3: Photograph of tufts and camera attached to blade

A Futek Rotary Torque Sensor (Model TRS605) [8] is installed on the main shaft of the turbine between the rotor and the gearbox. Torque measurements were made at a frequency of 720 Hz during testing.

The torque sensor is equipped with an optical encoder which is used to measure the azimuthal position of the test blade during rotation. All measurements of the azimuthal position using the optical encoder are recorded relative to the vertically downward orientation, which is set as 0°. Images from the video recording were synchronized with the optical encoder and torque measurements in post-processing.

Images recorded during the experiments were analyzed using a modified version of the image processing software reported in [4]. Modifications included the development of input parameters to account for the change in blades, an increase in the tolerance for tuft orientation indicating attached flow to ±20° to account for the curvature of the blade, and other minor changes.

Tests were conducted at three tip speed ratios, λ, defined as the ratio of tip speed to wind speed. Each test was repeated with a 0°, -15° and +15° yaw-offset.

RESULTS AND DISCUSSION

Ten minute videos (approximately 18,000 images each) of the tufts movements were processed using the modified image processing software. Example images of the software output at $\lambda = 6.3$ and $\lambda = 3.7$ are presented in Figure **4**. Tufts identified as being indicative of stall are highlighted in black.

Examining the frame captured while testing at the lower wind speed (Fig. 4(a)), $n = 40$ tufts have been recognized in the image with a stall fraction of $\zeta = 20\%$, meaning 20% of the tufts in the image indicate that the flow has separated. Except for one tuft at the leading edge, all of the tufts recognized as stalled are located on the trailing edge of the blade. As the wind speed is increased in Fig. 4(b), the output shows that the stall on the trailing edge has extended further along the chord, reaching approximately $\zeta = 40\%$ of the tufts. It is expected that the aerodynamic design of the blade (i.*e*. the tapered chord and twist) is responsible for maintaining the attached flow at the leading half of the blade.

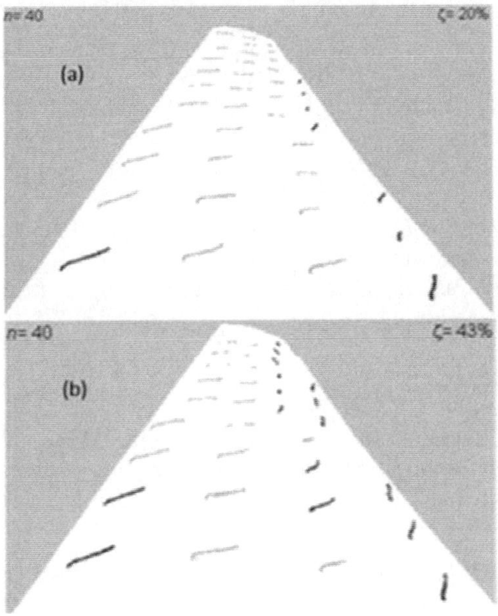

Figure 4: ζ for two cases (a) $\lambda = 6.3$ and (b) $\lambda = 3.7$. Tufts identified as being indicative of stall are highlighted in black

ζ and torque data were sorted into 72 azimuthal position bins 5° in size, averaged and plotted to see trends in the stall formation and performance. ζ results from tests conducted with a 0° yaw-offset are presented in Figure 5, and the corresponding torque results are presented in Figure 6.

As expected, ζ generally increased as the tip speed ratio decreased and wind speed increased. Small variations in ζ with azimuthal position are also observed at low wind speeds. However, at the highest wind speed, a significant decrease in ζ is observed between positions 180° and 320°.

An increase in the stall corresponds to an increase in the angle of attack and therefore, to a point, an increase in the lift. This is observed in the torque results shown in Figure 6. Between the lowest and highest wind speeds tested, the torque is increased by approximately 5 times. Therefore, even with the increase in ζ as observed in Figure 5, the torque performance of the turbine has only increased. Again, it is expected that the high performance of the blade at high wind speeds (lower λ) is due to the aerodynamic design.

It is noted that the torque measurements at 0° yaw-offset remain essentially constant at all azimuthal positions. This is to be expected given that the turbine is equipped with three blades, which distributes the loads evenly when exposed to a uniform flow field.

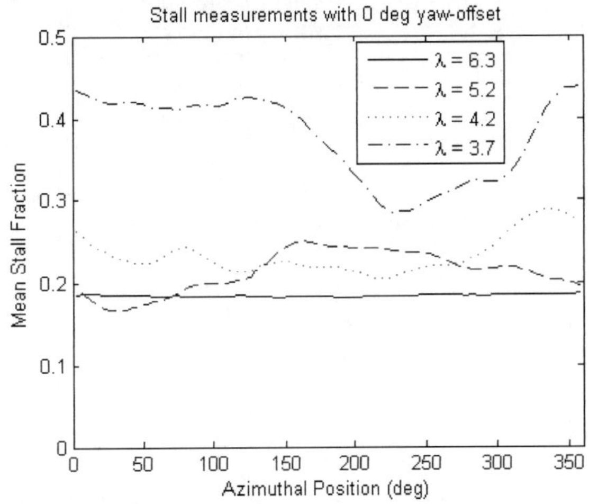

Figure 5: Stall fraction ζ with a 0° yaw-offset

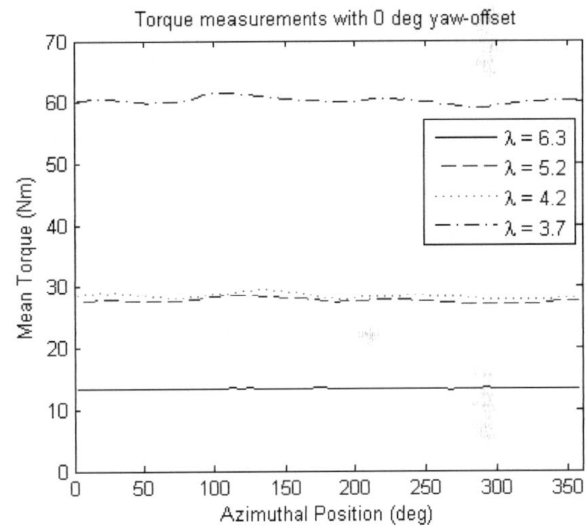

Figure 6: Torque measurements with a 0° yaw-offset

When the turbine is yawed ±15°, ζ changes significantly compared to the 0° yaw-offset case. Stall data was again sorted into azimuthal position bins and plotted for the +15° yaw case in Figure 7 and for the -15° yaw case in Figure 8. Note that the turbine rotates towards the wind at the upper part of the rotation when yawed +15°, and away from the wind at the upper part of the rotation when yawed -15°.

At +15° yaw, ζ measured increases at every wind speed tested relative to the zero-yaw case plotted in Figure 5. ζ is also observed to increase gradually with increasing wind speed. However, a consistent decrease in ζ is observed as the test blade travels through the upper part of the rotation in the $\lambda = 5.2$, 4.2 and 3.7 cases.

When the turbine is yawed -15° the measured ζ again increases with the increasing wind speed and decreasing tip speed ratio. However, the azimuthal variation in ζ increases significantly, with low ζ observed during the lower part of the blade rotation towards the wind and ζ as high as 80% at the upper positions rotating away from the wind.

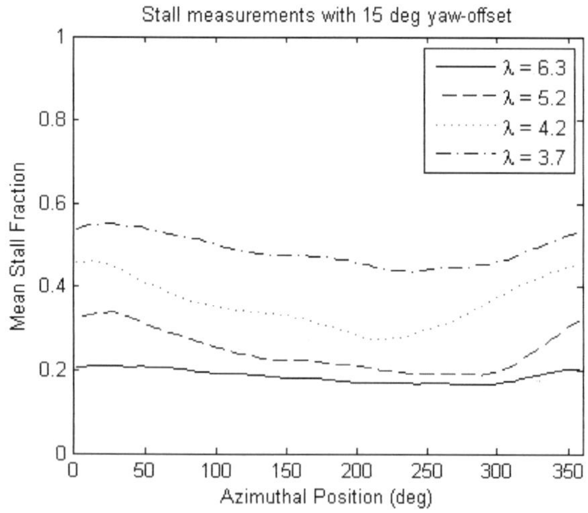

Figure 7: ζ versus azimuthal angle with a +15° yaw-offset

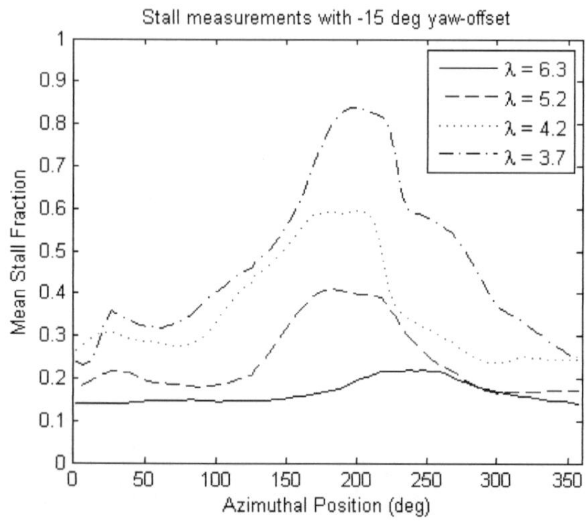

Figure 8: ζ versus azimuthal angle with a -15° yaw-offset

It is expected that the azimuthal variation in ζ observed in both yaw cases is due to the varying angle of attack throughout the rotation. As the blade travels towards the wind, the upstream wind vector is offset by the yaw angle such that the axial velocity on the blade is decreased and the tangential velocity is increased, resulting in a reduced angle of attack. When the blade travels away from the wind, the opposite happens and the angle of attack is increased. Because the stall on the blade is dependent on the angle of attack, the variations in the velocity geometry throughout the rotation is reflected in the measured ζ values.

CONCLUSIONS

Tuft flow visualization experiments were conducted using a full-scale test turbine with the objective of studying the formation of stall on a wind turbine blade as a function of the tip speed ratio, azimuthal position and yaw-offset position. Shaft torque measurements were also recorded to examine the effect of stall on turbine performance.

Results consistently showed that, as the wind velocity was increased and the tip speed ratio decreased, ζ and the shaft torque measured both increased. With a 0° yaw-offset, ζ and torque measurements remained essentially consistent with azimuthal position, except in the case of $\lambda = 3.7$.

However, when the wind turbine was yawed ±15°, it was found that as the blade rotated towards the wind, ζ on the blade decreased and as the blade rotated away from the wind, the stall on the blade increased. It is posited that the variation in ζ with azimuthal position is due to the varying velocity geometry and angle of attack through the rotation.

ACKNOWLEDGMENTS

The authors would like to thank Nigel Swytink-Binnema for his assistance with the image processing algorithm, and Wind Energy Group members and for support from the Natural Science and Engineering Research Council (NSERC).

REFERENCES

1. Eggleston, D. and Starcher, K. 1990, "A comparative study of the aerodynamics of several wind turbines using flow visualization," *J Sol Energy*, Vol. 112, no. 4, pp. 301-309.
2. Haans, W., Sant, T., van Kuik, G. and van Bussel, G. 2006, "Stall in yawed flow conditions: a correlation of blade element momentum predictions with experiments," in 44th AIAA Aero. Sci. Meeting, (Reno, Nevada).
3. Maeda, T. and Kawabuchi, H. 2005, "Surface pressure measurement on a rotating blade of field horizontal axis wind turbine in yawed condition," *JSME Int J B-Fluid T*, Vol. 48, no. 1, pp. 156-163.
4. Swytink-Binnema, N. and Johnson, D.A. 2014, *J Phys Conf. Series*, 524(1):012016. DOI:10.1088/1742-6596/524/1/012016.
5. Abdelrahman, A. 2014. "Development of a Wind Turbine Test Rig and Rotor for Trailing Edge Flap Investigation." M.A.Sc. thesis, UWaterloo, Canada.
6. Gertz, D. and Johnson, D.A. 2011. "An evaluation testbed for wind turbine blade tip designs–baseline case", *Int. J. Energy Research*, Vol. 35, 15, pp. 1360-1370.
7. "UW Live Fire Research Facility." UWaterloo Wind Energy Group. Available: windenergy.uwaterloo.ca. [Accessed Dec. 5, 2014].
8. Futek. "Futek Model TRS605: Non Contact Shaft to Shaft Rotary Torque Sensor W/ Encoder." Internet: http://www.futek.com/files/pdf/Product%20Drawings/trs605.pdf. 2014 [Oct. 3, 2014].

Proceedings of the 25th CANCAM
London, Ontario, Canada, May 31 – June 4, 2015

CFD MODELING OF THE WIND CHAMBER USED FOR RADIOACTIVE PARTICLE RESUSPENSION EXPERIMENT

Perera, Sharman
Faculty of Energy System and Nuclear Science
University of Ontario Institute of Technology
Oshawa, Ontario, Canada
Sharman.perera@uoit.ca

Waller, Edward
Faculty of Energy System and Nuclear Science
University of Ontario Institute of Technology
Oshawa, Ontario, Canada
Ed.waller@uoit.ca

Walsh, Paul
Department of Aerospace Engineering
Ryerson University
Toronto, Ontario, Canada
Paul.walsh@ryerson.ca

Doxtator, Andrew
Faculty of Energy System and Nuclear Science
University of Ontario Institute of Technology
Oshawa, Ontario, Canada
Andrew.doxtator@uoit.ca

Akhtar, Ali
Faculty of Energy System and Nuclear Science
University of Ontario Institute of Technology
Oshawa, Ontario, Canada
Ali.akhtar@uoit.ca

ABSTRACT

Commercially available CFD software, FLUENT, was successfully used to model the flow fields of the wind chamber used to conduct radioactive particle resuspension experiments at WIS facility in Germany using Reynolds-averaged Navier-Stokes equations (RANS). Time averaged continuity equation and momentum equations coupled with the Shear Stress Transport (SST) *k*-omega *(ω)* turbulence model equations were successfully solved simulating similar velocity profiles inside the wind chamber compared to experimental results.

KEYWORDS: Resuspension, CFD, Turbulent, Radioactive Rxperimental,

1. INTRODUCTION

Freshly settled radioactive particles that were released by a radiological dispersion device (**RDD**) can be resuspended back to the atmosphere by external forces acting on them due to air moving over these particles. This resuspension would spread contamination and impose health hazards to humans.

2. RESUSPENSION EXPERIMENTS

Resuspension experiments were conducted in a 10 m long, open-ended wind chamber that was built exclusively for the purpose of radioactive particle resuspension experiments. This wind chamber was setup inside Wehrwissenschaftliches Institut für Schutztechnologien (WIS) facility located in Munster, Germany. The WIS facility is equipped with an appropriate ventilation system to handle experiments with live agent radioactive materials.

Figure 1 shows the layout of this wind chamber experimental setup and the locations of measuring devices used during the resuspension experiment. The wind chamber was mapped prior to resuspension experiments using two sets of hotwire anemometers (Perera, Waller, & Akhtar, 2015).

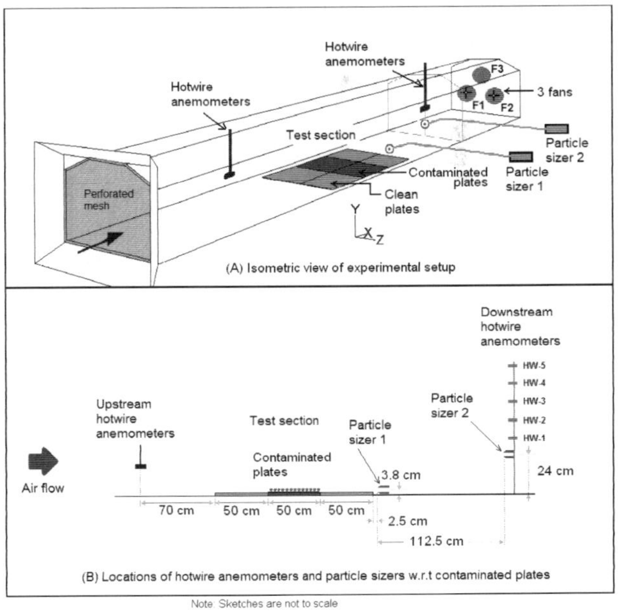

Figure 1: Layout of the experimental setup of WIS facility test chamber.

This wind chamber can generate winds up to 7 m/s at the center of the test section using 3 fans each moving 10,000 m^3 air per hour. Fans F1 and F2 were axial propeller fans and fan F3 was a ducted vent connected to a fan located 2m behind the wind chamber. Each fan has diameter of 40 cm and was mounted on the back of the 10 m long wind chamber to provide constant mass flux across the test section.

3. CFD Modeling of the WIS Facility Wind Chamber

The experimental wind chamber used to conduct particle resuspension studies at the WIS facility was very complex. A good understanding of CFD modeling and aerodynamics knowledge is required in order to simplify this complex geometry into a simple and solvable domain while also trying to capture all the important aerodynamic parameters of the resuspension process.

The CFD model of the WIS facility wind chamber was modeled in 2D due to symmetric nature of the flow patterns at the test section and the geometry.

3.1 MESH GENERATION OF 2D WIS FACILITY WIND CHAMBER

The two dimensional (2D) computational domain of the WIS facility wind chamber was generated using rectilinear mesh elements. A very fine mesh was generated close to the wall surfaces of the wind chamber to ensure that there were adequate number of nodes in the region of the viscosity dominated boundary layer to model turbulent properties accurately and also to gain the velocity profile close to the floor to model the resuspension process inside the laminar sub-layer region in future studies. The original mesh was refined until y+ of all wall surfaces were less than 2.5.

Figure 2 shows boundaries and the final 2D mesh of the selected areas of 2D WIS facility wind chamber computational domain.

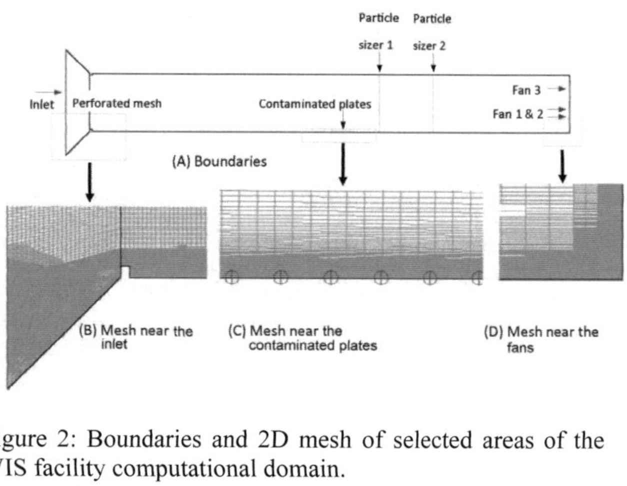

Figure 2: Boundaries and 2D mesh of selected areas of the WIS facility computational domain.

Figure 2-(A) shows the boundaries used in the computational domain. Figure 2-(B) shows the zoomed in 2D mesh close to the lower surface of the wind chamber intake nozzle, perforated mesh and the turbulators used to trip the boundary layer, Figure 2-(C) shows the zoomed in 2D mesh close to the contaminated plate. Figure 2-(D) shows the zoomed in 2D mesh close to the wall where the fans were mounted.

FLUENT solves the boundary layer either using wall functions when $y^+ > 30$ or resolving turbulence models all the way to wall surfaces when $1 < y^+ < 5$. It should be noted FLUENT recommends to use y^+ close to 1 (Fluent Inc., 2010). The original mesh of the 2D WIS facility wind chamber was developed in 1:1 scale using GAMBIT mesh generation software. The initial mesh had 132,905 nodes and the y^+ less than 1 for the ceiling and the floor, but the y^+ of the converging intake nozzle of the wind chamber were between 250 and 550 and the back walls of the wind chamber where the fans were mounted were between 5 and 500. Later this mesh was refined using a y^+ adaptation algorithm available in FLUENT software until the y^+ of all wall surfaces were less than 2.5. The corresponding mesh ended up with 428,213 nodes.

3.2 CFD MODELING OF THE 2D WIS FACILITY WIND CHAMBER FLOW FIELDS

The boundary layer of WIS facility wind chamber was tripped using turbulators located just after the honeycomb structure. Also the average Reynolds number based on hydraulic diameter ($\mathbf{Re}_L$) at the test section was calculated to be 5.6E5 at Standard Atmospheric Condition (SAC). The flow at the test section was turbulent since $\mathbf{Re}_L$ was bigger than 2000 (Biswas & Eswaran, 2002).

3.2.1 SELECTION OF APPROPRIATE CFD MODELS TO SOLVE THE 2D WIS FACILITY COMPUTATIONAL DOMAIN

Since the general flow in the test section is turbulent, the Reynolds-averaged Navier-Stokes equations (RANS) approach will be used to find the solutions of the computational domain. RANS approach uses time averaged fluctuating quantities to find the solutions of the 2D WIS facility computational domain.

Continuity equation and momentum equations coupled with the Shear Stress *Transport* (SST) *k*-omega *(ω)* turbulence model equations were solved to find the flow fields of the WIS facility computational domain. Here k represents the turbulence kinematic energy and ω represents the specific dissipation rate of turbulence.

The SST *k-omega* turbulence model was selected to model the turbulent transport because of its capabilities in modeling turbulence transport parameters down to the wall surfaces (Fluent Inc., 2010) where the initial resuspension process takes place. The Shear Stress Transport (SST) k-ω turbulence model uses both standard k-ω and k-ε models. The standard k-ω and k-ε are two famous turbulence models and information about these two models can be found in any turbulent flow fluid dynamics text books. In the SST k-ω turbulence model, the turbulence properties of the inner region of the boundary layer are found using standard k-ω model and rest of the flow turbulence properties are found using standard k-ε model. Historically, the SST k-ω turbulence model had proven to predict accurate solutions for wall bounded flows even in highly separated flow regions (Georgiadis & Yoder, 2006).

3.2.2 BOUNDARY CONDITIONS OF THE 2D WIS FACILITY COMPUTATIONAL DOMAIN

The boundaries of the computational domain of the 2D WIS facility were selected to represent the physical model of the experimental setup.

3.2.2.1 INLET

The inlet of the wind chamber was modeled in FLUENT as a **pressure-inlet** boundary condition.

The direction of the flow was selected normal to the inlet. Since the inlet of the wind chamber was made with 2.2 m x 2.2 m square cross section and it was opened to the WIS facility surrounding, the hydraulic diameter of 2.2 m and turbulent intensity of 1% was used.

3.2.2.2 PERFORATED MESH

The boundary condition of the perforated mesh was modeled in FLUENT as a **porous jump**. The pressure drop across a porous medium was modeled using Forchhimer equation as given below.

$$\frac{dp}{dx} = \frac{\mu}{K}u + \frac{1}{2}\rho C u^2 \qquad (1)$$

Here, K (m^2) is the permeability of the medium, u is superficial velocity, ρ is the density of air, C (1/m) is the pressure-jump coefficient (Fluent Inc., 2010). Superficial velocity through the porous jump boundary condition was calculated by dividing the volume flow rate through the wind chamber by the cross sectional area of the wind chamber where the perforated mesh was located.

The values of K and C used in Equation (1) were calculated modeling the perforated mesh inside a virtual wind chamber using FLUENT software for a range of velocities that WIS facility wind chamber will undergo during its normal operation (Sapounas, Campen, Wildschut, & Bot, 2010). Table 1 gives the values that were used in the 2D WIS facility wind chamber model as input parameters to model the perforated mesh boundary conditions found using FLUENT software.

Table 1: Porous jump boundary values used in FLUENT to model the perforated mesh used inside the WIS facility wind chamber.

Face Permeability (K) (m^2)	1.2E-8 (Sapounas, Campen, Wildschut, & Bot, 2010).
Porous Medium Thickness (m)	0.001 Thickness of the plate
Pressure-Jump Coefficient (C) (1/m)	3220 (Sapounas, Campen, Wildschut, & Bot, 2010).
Discrete Phase BC Type	Interior

3.2.2.3 FANS

Boundary conditions of the fans used in the WIS facility wind chamber were modeled in FLUENT as **exhaust-fan**. The mass flow rates of the fans in the CFD model were selected to match the WIS facility experimental model. Target mass flow rates of 5.22 kg/s and 2.61 kg/s were introduced to the fans F1&F2 and the fan F3 boundary conditions.

The pressure drop across the fans F1&F2 was used as zero since these fans were forcing the wind chamber air out to the WIS facility surroundings. The pressure drop across fan F3 was used as 37 Pa since this fan was forcing air through a 2m long duct and 90° elbow to WIS facility surrounding. The pressure drop across the duct and the elbow was calculated according to data found from ASHRAE hand book (ASHRAE handbook, fundamentals, 1997).

The ***intensity and length-scale*** option available in FLUENT was selected to introduce turbulence parameters of the backflow of fan 1&2 boundary condition and for fan 3, ***intensity and hydraulic diameter*** option was selected (Fluent Inc., 2010). Turbulent intensity of 20% and length-scale of 0.004 m (10% of blade span) were introduced to the fans F1&F2 and for the fan F3 turbulent intensity of 10% and hydraulic diameter of 40 cm were introduced as per the recommendations given by FLUENT software manual.

3.2.2.4 WALLS

No-slip boundary conditions were enforced at all walls of 2D CFD model of the WIS facility wind chamber. Since the majority of distances from the wall boundaries to centroids of the wall-adjacent cells in the computational domain were smaller than the peak to peak roughness height of the walls

(R_z=9.174 μm), it is meaningless to define the roughness parameters that will be used in FLUENT software to modify the active thickness of the laminar sub layer (Fluent Inc., 2010). In this situation FLUENT will solve the turbulent parameters all the way to the walls assuming all walls were smooth.

3.2.3 SOLVING GOVERNING EQUATIONS

The computational domain of 2D WIS facility wind chamber has 428,213 mesh nodes. The solution of this computational domain needs to be found using a solution algorithm that is capable of solving nonlinear governing equations utilizing available computer power at UOIT. Therefore, a pressure based, unsteady, implicit segregated solver was selected due to its lesser computer memory requirement compared to the pressure based coupled algorithm, but the rate of solution convergence for segregated solver is significantly lower (Fluent Inc., 2010).

The implicit time integration used here is unconditionally stable unlike explicit time integration where time step size is limited by the Courant-Friedrichs-Lewy (CFL) condition. For 2D computational domains, Courant number (**C**) is defined as seen in Equation (2).

$$\mathbf{C} = \frac{u\Delta t}{\Delta x} < 1, \qquad (2)$$

Here u is the velocity, Δt is time step size and Δx is length interval. The corresponding time step size for explicit time marching was calculated to be 2.9 ns (using u= 6.5 m/s and Δx =1.88E-8 m) compared to 10 μs time step used in implicit method to achieve solutions up to 1E-7 convergence criterion.

Since the computational domain of the WIS facility was made of structured rectilinear mesh, default Green-Gauss Cell Based method was used to calculate the convection terms, secondary diffusion terms and velocity derivatives of the governing equations (Fluent Inc., 2010).

3.3 CFD RESULTS OF 2D WIS FACILITY EXPERIMENT

The computational domain of the 2D WIS facility wind chamber flow fields were solved until 1E-7 residual convergence criterion for continuity, momentum and turbulence transport parameters were achieved.

Figure 3 shows the streamwise velocities at down steam hotwire locations found using experiments (circles) and superimposed CFD results along the vertical line connecting the tips of those hotwire anemometers (solid line).

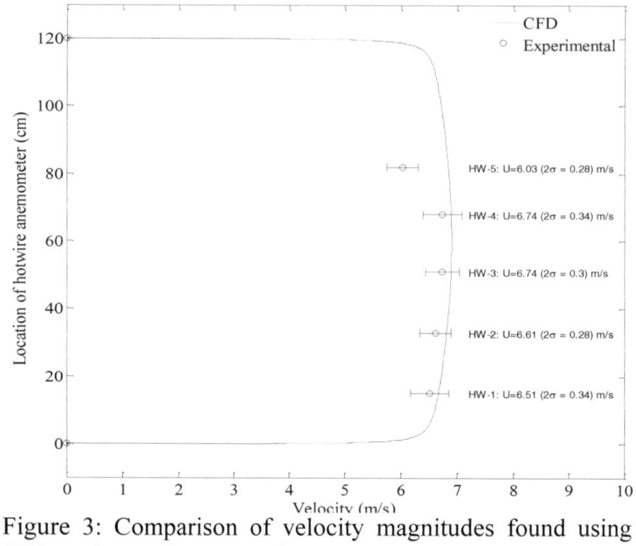

Figure 3: Comparison of velocity magnitudes found using CFD and experiments at downstream hotwire anemometers locations. CFD results were calculated across the horizontal line running through the tips of the Hotwire anemometers (**Re** = 5.6E5 at SAC).

Figure 3 shows that the 2D WIS facility CFD model generated using FLUENT successfully modeled the streamwise velocity components at the locations of the hotwire anemometers except at the HW-5 location. The HW-5 was located where the cross section of the wind chamber changes its shape from rectangular to trapezium. The transition of the shape was not introduced into the 2D WIS facility computational model to reduce the computational demand since the distance from the floor of the wind chamber to the locations was almost 8E5 times away compared to the height of the computational domain that will be used to model the resuspension process at the Laminar Sub Layer region.

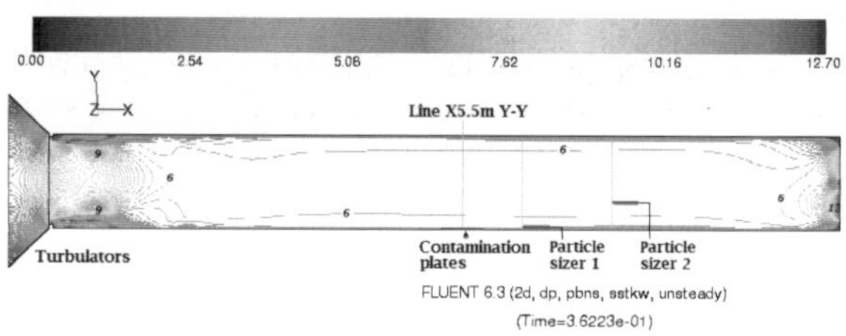

Figure 4: Velocity magnitude contours of 2D WIS facility computational domain. (**Re** = 5.6E5 at SAC).

According to CFD results, air from the WIS facility surrounding enters the wind chamber at a velocity of 5.9 m/s (at the center). Beyond this point velocity started increasing up to 7.4 m/s at the perforated mesh. As air turns the corner from converging section to settling chamber of the wind chamber, it goes over 10 cm tall turbulators further increasing its velocity up to maximum of 9 m/s due to the vortexes formed at the aft of the turbulators. After this point, flow inside the wind chamber started settling. The velocity vectors of selected profiles inside the wind chamber are shown in Figure 5.

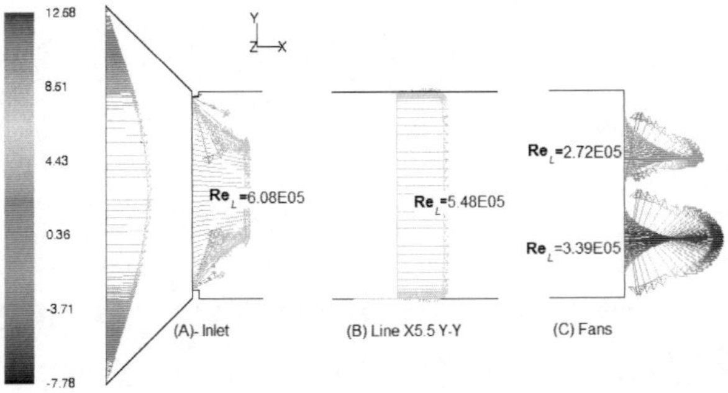

Figure 5: Velocity vectors, colored by x velocity, of selected profiles and corresponding maximum, hydraulic diameter based Reynolds number (**Re**$_L$) at SAC. Profile (A) shows the Inlet and Perforated plate, profile (B) shows the Line X5.5 Y-Y and profile (C) shows velocity vectors of fans.

As shown in Figure 5, the velocity vector of the inlet profile has a hyperbolic shape with a focus at 5.9 m/s. The velocity profile at the mid-section of the perforated plate has an almost linear velocity profile averaging at 7.4 m/s. The velocity vector profile at Line X5.5 Y-Y is almost linear across the test section except near the ceiling and the floor of the wind chamber. The velocity vectors patterns of Fans F1&F2 and Fan F3 show the fanning effect with maximum velocities of 12.6 m/s and 10.2 m/s respectively.

4. Conclusions

Commercially available CFD software, FLUENT, was successfully used to model the flow fields of the wind chamber used to conduct radioactive particle resuspension experiments at WIS facility in Germany.

Converged 2D CFD results provided the time averaged velocity profile close the wall surfaces where initial particle resuspension takes place.

Time averaged velocity with fluctuating velocity components will provide the instantaneous velocity profile required to model the coherent structure of the laminar sub-layer close to the center of the contaminated plate as the next stage of this project.

5. References

1. S. R. Perera, J. E. Waller and A. Akhtar, "Experimental Modeling of Wind-Driven Bin-by-Bin Resuspension Factors of Freshly Fallen Radionuclides After an Energetic Release From a adiological Dispersal Device," *ASME Journal of Nuclear Engineering and Radiation Science,* 2015.

2. Fluent Inc., "Fluent 6.3 documentation," April 2010. [Online]. Available: http://my.fit.edu/itresources/manuals/fluent6.3/help/. [Accessed 2010].

3. G. Biswas and V. Eswaran, Turbulent Flows, P. Kumar, R. Prasad and V. Gupta, Eds., New Delhi: Mehra,N.K., Narosa Publishing House, 2002.

4. N. J. Georgiadis and D. A. Yoder, "Evaluation of Modified Two-Equation Turbulence Models for Jet Flow Predictions," in *AIAA 2006-490*, Reno, Nevada, 2006.

5. A. A. Sapounas, J. B. Campen, J. Wildschut and G. P. A. Bot, "Energy Saviing During Bulb Storage Applying Modelling with Computational Fluid Dynamics," in *XVIIth World Congress of the International Commission of Agricultural Engineering (CIGR)*, Quebec, 2010.

6. ASHRAE handbook, fundamentals, 1997.

Proceedings of the 25th CANCAM
London, Ontario, Canada, May 31 - June 4, 2015

EFFECTS OF WIND TUNNEL MODEL END CONDITIONS ON LOW REYNOLDS NUMBER AIRFOIL STUDIES

Mark Feero
Institute for Aerospace Studies
University of Toronto
Toronto, Ontario, Canada
m.feero@mail.utoronto.ca

Pierre Sullivan
Mechanical and Industrial Engineering
University of Toronto
Toronto, Ontario, Canada
sullivan@mie.utoronto.ca

Philippe Lavoie
Institute for Aerospace Studies
University of Toronto
Toronto, Ontario, Canada
lavoie@utias.utoronto.ca

ABSTRACT

Wind tunnel experiments are used to investigate the influence of end effects on the flow over a conventional airfoil at low Reynolds number. When gaps are present between the model ends and the tunnel walls, significant spanwise variation is observed in the mean streamwise velocity in the wake. Minimizing the gaps improves the two-dimensionality of the flow, however the best results are achieved when end plates are used. The presence of gaps versus the use of end plates has a dramatic effect on the stall angle and the shape of the lift versus angle-of-attack curve.

Keywords: Low-speed aerodynamics, airfoil, end plates

INTRODUCTION

The study of airfoils operating at chord-based Reynolds numbers, Re_c, less than 5×10^5 has increased in recent years as a result of applications such as unmanned aerial vehicles (UAVs), micro air vehicles (MAVs) and wind turbine blades. Low Reynolds number flow on an airfoil is characterized by a boundary layer on the suction surface that is laminar upon separation and may reattach, forming a laminar separation bubble (LSB) [1]. Reattachment occurs due to a growth of instabilities in the separated shear layer that cause the boundary layer to transition to turbulence [2]. If the adverse pressure gradient is too severe, the flow will remain separated and a large unstable wake will develop.

Experimental investigations of airfoils at low Reynolds number provide unique challenges due to the sensitivity of the flow to the experimental setup and facility. Olson et al. [3] showed that the freestream turbulence intensity has a significant effect on the LSB characteristics of a SD7003 airfoil at $Re_c \approx 10^4$ due to the sensitivity of the laminar boundary layer to broadband excitation. Wang et al. [4] also demonstrated that freestream turbulence affected separation and reattachment characteristics and aerodynamic forces on a NACA 0012 airfoil in a similar manner to Reynolds number (i.e., increasing turbulence intensity or Reynolds number can delay stall). Boutilier and Yarusevych [5] investigated the effects of end plate spacing and test section blockage on a NACA 0018 airfoil at $Re_c = 10^5$. The authors showed that the use of end plates significantly improved the spanwise uniformity of both mean and fluctuating surface pressure when the spacing was at least 7 times the projected frontal thickness of the airfoil. Furthermore, they showed that while transition and reattachment locations were not significantly affected by solid blockages between 4% and 8%, a noticeable error was observed in the maximum lift.

The goal of the present study is to investigate the effect of end conditions on the spanwise uniformity of the flow and the lift coefficient determined from measurements at midspan. In particular, the spanwise uniformity of the flow over the airfoil with varying gaps at the model ends and with end plates will be compared.

EXPERIMENTAL SETUP

Experiments were performed in a low-turbulence recirculating wind tunnel with a rectangular test section that is 5 m long, 0.91 m wide and 1.22 m tall. The flow enters the test section after passing through seven screens and a 9:1 contraction. The freestream velocity in the test section is adjustable from approximately $U_\infty = 2.6$ m/s to 18 m/s and is monitored using a pitot-static tube located at the test section inlet. Over

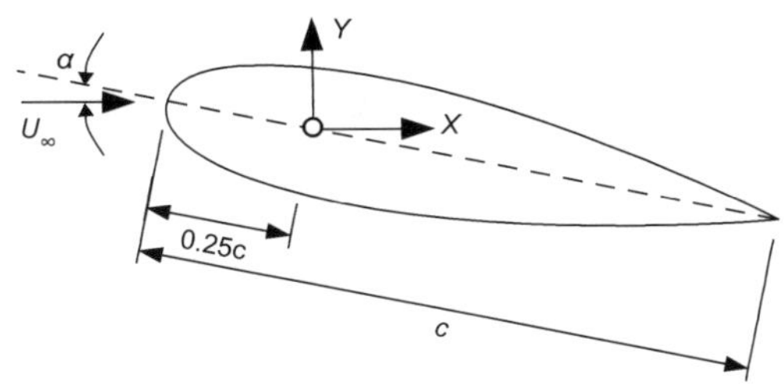

Figure 1: Description of the global coordinate system.

Three configurations with different end conditions were tested: (1) large gaps between the model ends and test section walls, (2) small gaps, and (3) end plates fixed to the model ends. In configuration (1), the model span was $w = 895$ mm such that there was a gap of approximately 7.5 mm ($0.025c$) between the model ends and the walls of the test section. The gaps were reduced in the second configuration to approximately 1.15 mm ($0.003c$) by attaching 6.35 mm thick end caps with the same airfoil profile, giving a model span $w = 908$ mm. The span was modified to $w = 885$ mm for configuration (3) to facilitate the installation of end plates on either end of the model. The inner edges of the end plates were approximately 15 mm from the tunnel walls, placing them outside the 12 mm wall boundary layer at the model leading edge. The aspect ratio of the model in these configurations varied only slightly from AR $= w/c = 2.95$ to 3.02 and it can be considered that AR ≈ 3. The end plates were designed according to Boutilier and Yarusevych [5] who performed a thorough review of the use of end plates on aerodynamic models, including cylinders and airfoils. The end plates were circular with a $2.25c$ diameter, extended $0.8c$ downstream of the trailing edge and had a 30° chamfer on the outer edge. The end plates added an additional 0.5% blockage.

A chordwise row of 64 pressure taps at midspan allowed measurement of the static pressure distribution on the airfoil surface. The surface pressure measurements were performed using a 64-port Scanivalve multiplexer connected to an MKS 226A Baratron pressure transducer with a dynamic range of ± 26.7 Pa and an accuracy of ± 0.08 Pa. Aerodynamic $\alpha = 0$ was obtained by adjusting the model angle until the pressure distribution on the upper surface matched the lower surface. Measurements of the mean streamwise velocity, U, in the airfoil wake were obtained by traversing a pitot-static tube connected to the same pressure transducer. Typical measurement uncertainties in U were less than 1%.

RESULTS

The severity of the end effects due to various model configurations can be assessed from the two-dimensionality of the flow field. Measurements of the mean streamwise velocity, U, in the YZ plane were performed for the three model configurations discussed in the preceding section. The measurement plane was located $2c$ downstream of the trailing edge. The angle-of-attack $\alpha = 10°$ was selected for configurations (1) and (2) since this corresponded to stalled flow. For configuration (3), both $\alpha = 0°$ and $\alpha = 12°$ were tested. In the former case, a laminar separation bubble forms on the airfoil, while the flow is stalled for the latter.

Figure 2a shows U/U_∞ measured in the YZ plane for configuration (1) with $0.025c$ gaps. It is immediately apparent from these results that not only is the flow 3D, it is asymmetric about the midspan. Towards the $Z/w = 0.5$ end of the model, the wake is narrower than at $Z/w = 0$ and the centre of the velocity deficit is shifted down. This shows that the flow has transitioned from separated to quasi-attached from $Z/w = 0$ to 0.33. This can be observed in more detail by comparing profiles of U/U_∞ at $Z/w = 0$ and $Z/w = 0.33$ (Figure 3). These measurements were acquired over a 10 s duration, and the upper portion of the velocity profile at $Z/w = 0.33$ has an apparent jaggedness despite the long sampling time. This suggests that the flow is likely alternating between an attached and separated state due to disturbances introduced into the boundary layer by the $0.025c$ gap. It is not clear why this only occurs on one end of the model, however it highlights the sensitivity of the flow at low Re_c.

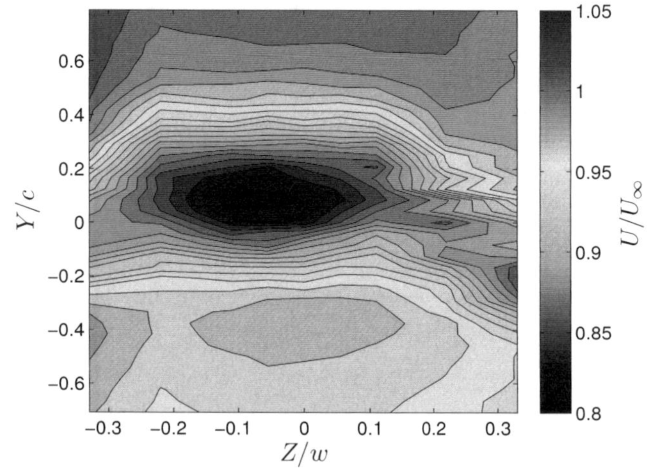

(a) $0.025c$ gaps

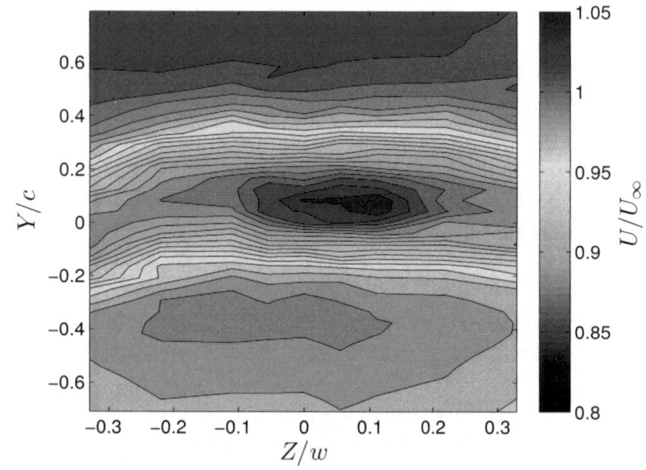

(b) $0.003c$ gaps

Figure 2: Streamwise velocity contours $2c$ downstream of the trailing edge for $\alpha = 10°$ and $Re_c = 10^5$.

The contour plot of U/U_∞ for configuration (2) with $0.003c$ gaps is shown in Figure 2b. With the size of the gaps reduced to less than 1% chord, the flow is separated over the entire measurement range and can be considered quasi-2D over at least 60% of the model span. The maximum spanwise deviation in U, defined as $|U(Y,Z) - U(Y,0)|/U_\infty$, is 6% for this configuration. However, it can still be observed that there is a slight asymmetry to the flow (i.e., the wake on the $Z/w = -0.5$ is shifted down slightly) and the maximum velocity deficit at a given Z location is larger near midspan.

The streamwise velocity contour plots for configuration (3) at $\alpha = 0°$ and $\alpha = 12°$ are shown in Figure 4. For these measurements, the spanwise range was increased to $-0.44 \leq Z/w \leq 0.44$ to capture more of the wake. The results show that for each flow regime (i.e., attached with a

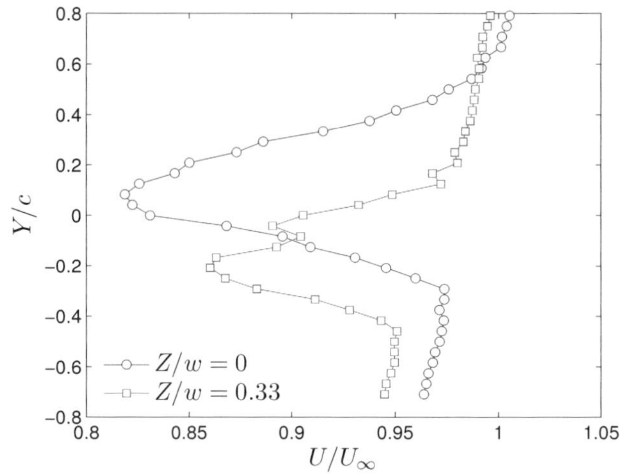

Figure 3: Comparison of mean U profiles at different spanwise locations for configuration (1).

LSB and stalled), the flow field in the wake is symmetric about $Z/w = 0$ and is two-dimensional over approximately 60% of the model span. Even in the remaining 28% span, the spanwise variation in U is no larger than 6%.

The dramatic consequence of end effects at low Reynolds number can be demonstrated by the variation in the lift coefficient C_L with α, where C_L is calculated from the surface pressure distribution at $Z/w = 0$. Figure 5 shows the lift curve for each of the three configurations, with the thin airfoil prediction $C_L = 2\pi\alpha$ included for reference. No wind tunnel corrections were applied since the drag force was not measured, and standard corrections are not considered valid in the post-stall regime. For configuration (1), the airfoil abruptly stalls at $\alpha = 6°$ and the lift is dramatically reduced to $C_L \approx 0$. Once the gap is minimized in configuration (2), the stall angle is increased to $\alpha = 9°$, yet C_L agrees well with configuration (1) for $\alpha \leq 5°$ and $\alpha \geq 9$. This shows that not only do the $0.025c$ gaps cause a spanwise variation in the flow, the disturbances propagate to midspan and cause the flow to prematurely stall. This result is not unexpected due to the highly sensitive nature of the laminar boundary layer present on the suction surface. With the end plates installed, the stall angle is increased slightly to $\alpha = 10°$ and a more typical lift curve is observed. That is, there is a plateau in C_L prior to stall. However, a linear increase in C_L at low angles-of-attack is not observed. This is likely a blockage effect, as further evidenced by the fact that C_L becomes larger than the thin airfoil value prior to stall for $5° \leq \alpha \leq 7°$. It is possible that wind tunnel corrections applied at pre-stall angles would amend this effect.

CONCLUSIONS

The influence of end effects on low Reynolds number airfoil wind tunnel experiments was investigated by considering

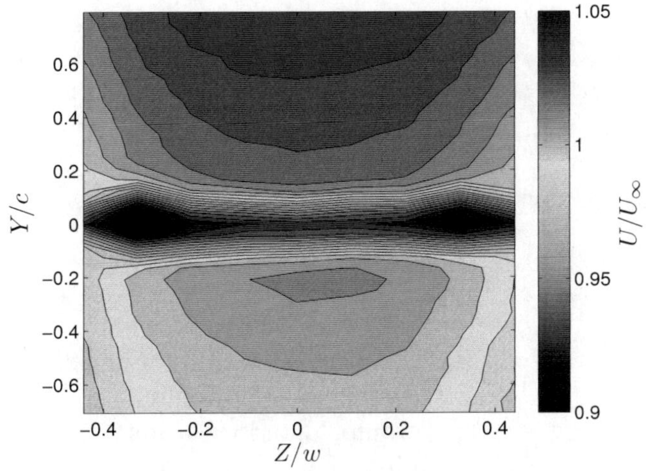

(a) $\alpha = 0^\circ$

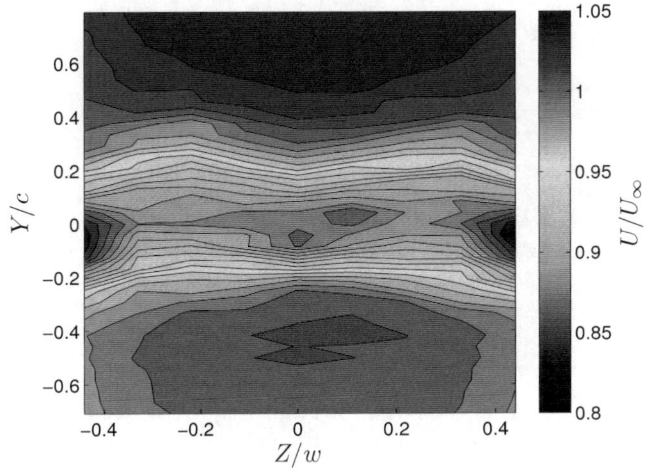

(b) $\alpha = 12^\circ$

Figure 4: Streamwise velocity contours $2c$ downstream of the trailing edge for $Re_c = 10^5$ with end plates installed.

three end configurations: significant gaps between the model and tunnel walls, minimal gaps, and end plates installed on the model ends. Planar measurements of the mean streamwise velocity in the wake of the airfoil were used to assess the spanwise uniformity of the flow. The results demonstrate that when $0.025c$ gaps are present, the flow varies considerably in the spanwise direction and even transitions from separated at midspan to quasi-attached near the model end. When the gaps were minimized to $0.003c$, the spanwise uniformity of the flow was increased and U varied within only 6% relative to midspan. Installing end plates led to further improvements, in particular the flow was sufficiently uniform to be considered two-dimensional over at least 60% of the model span. The variation in U in the outer portions near the end plates was no more than 6%.

The importance of end plates for low Reynolds number air-

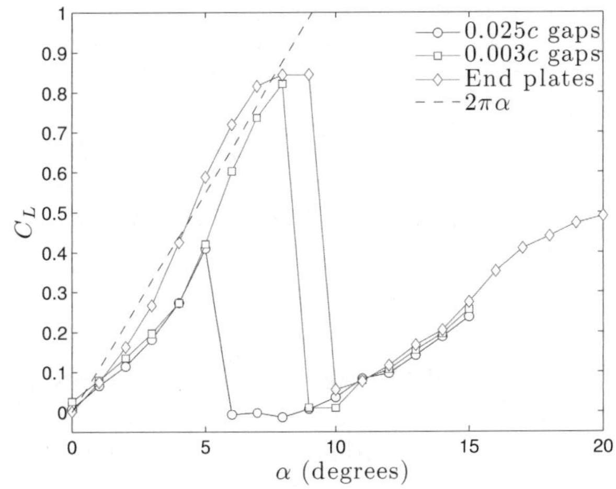

Figure 5: Lift coefficient curves at $Re_c = 10^5$ for each end configuration.

foil studies was clearly demonstrated in the comparison of the lift curve for each configuration. With end plates installed, a typical lift curve showing a plateau in C_L prior to stall at $\alpha = 10^\circ$ was observed (although blockage effects were likely present). However, the presence of $0.025c$ gaps caused the flow at midspan to stall abruptly and prematurely at $\alpha = 6^\circ$. With minimized gaps, this effect was reduced but premature stall still occurred at $\alpha = 9^\circ$. The results demonstrate that 2D airfoil studies at low Reynolds number should employ end plates to ensure the quality of the flow and to ensure consistency between experiments.

ACKNOWLEDGMENTS

The authors gratefully acknowledge support from the Natural Sciences and Engineering Research Council of Canada.

REFERENCES

[1] Lissaman, P. B. S. 1983, "Low-Reynolds-number airfoils", *Ann. Rev. Fluid Mech.*, Vol(15), pp. 223-239.

[2] Dovgal, A. V., et al. 1994, "Laminar boundary layer separation: instability and associated phenomena", *Prog. Aero. Sci.*, Vol(30), pp. 61-94.

[3] Olson, D. A., et al. 2013, "On the challenges in experimental characterization of flow separation over airfoils at low Reynolds number", *Exp. Fluids*, Vol(54), pp. 1-11.

[4] Wang, S., et al. 2014, "Turbulent intensity and Reynolds number effects on an airfoil at low Reynolds numbers," *Phys. Fluids*, Vol(26).

[5] Boutilier, M., Yarusevych, S. 2012, "Effects of End Plates and Blockage on Low-Reynolds-Number Flows Over Airfoils", *AIAA Journal*, Vol(50), pp. 1547-1559.

Proceedings of the 25th CANCAM
London, Ontario, Canada, May 31 - June 4, 2015

LARGE-EDDY SIMULATION OF TURBULENT ROUND JETS IN CROSSFLOWS

Paul Ziadé
Dept. of Mechanical & Industrial Engineering
University of Toronto
Toronto, Ontario, Canada
paul.ziade@mail.utoronto.ca

Pierre Sullivan
Dept. of Mechanical & Industrial Engineering
University of Toronto
Toronto, Ontario, Canada
sullivan@mie.utoronto.ca

ABSTRACT

Large-Eddy Simulation is performed for a series of jets in crossflows. Comparisons with experimental and DNS results show excellent agreement of velocity profiles and jet trajectories. The key vortical structures are reproduced by the computations exhibiting good quantitative and qualitative behaviour.
Keywords: Turbulence, Large-Eddy Simulation, Flow Control, Aerodynamics

INTRODUCTION

Separated flows occur in a wide range of engineering applications. Active flow control devices have received much attention in recent years and have been shown feasible for industrial applications[1]. One type of flow control device, a Synthetic Jet Actuator (SJA), has shown particular promise, especially for aerodynamic applications. SJAs produce zero-net-mass-flux oscillatory jets, or synthetic jets, which help to stabilize the boundary layer by adding or removing momentum to or from the boundary layer.

The overall goal of this project is to answer certain open questions relating to active flow control on airfoils using computational techniques and high performance computing. Firstly, there is no consensus on the optimal actuation frequency. In addition, the optimal location of actuation, whether upstream or downstream of the separation point, needs to be determined. Recent experimental work also suggests that the internal cavity shape can influence the performance of the synthetic jet[2]. In this project, detailed parametric studies will be performed using large-eddy simulation (LES) and detached-eddy simulation (DES) to shed some light on the above questions.

This paper presents preliminary computational results for a series of turbulent round jets in crossflows. This canonical flow is seen as a stepping stone towards the simulation of the more complicated synthetic jet actuator and the full three-dimensional airfoil with actuator. The physical characteristics of the jet in crossflow (JIC), previous experiments, computational studies, and a comparison and discussion will be presented.

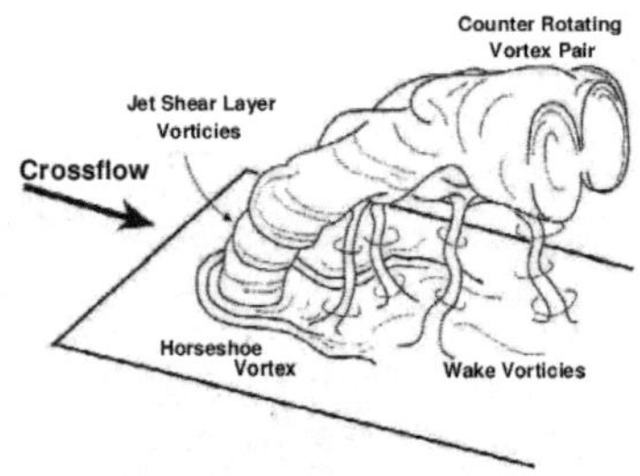

Figure 1: Schematic of typical JIC flow structure[4]

JIC STRUCTURE

A set of coherent structures appearing in the JIC has been cited by numerous researchers. The four well known vortex systems in these jets are shown in Figure 1. The most dominant vortex system is the counter-rotating vortex pair (CVP) initiated at the pipe wall and developing downstream of the jet exit. The mixing layer between the jet and crossflow forms the jet shear layer vortices. Due to a Kelvin-Helmholtz instability, the flow rolls up near the exit of the jet. The horseshoe vortex forms upstream of the jet and is largely due to the adverse pressure gradient as the crossflow approaches the jet. The wake vortices form downstream of the jet exit and often connect the wall to the CVP. It is widely accepted that the controlling parameter is the velocity ratio, $r = \overline{U_j}/U_e$, where $\overline{U_j}$ is the mean jet velocity and U_e is the freestream crossflow velocity[3].

STUDIES FOR COMPARISON & NUMERICAL METHODOLOGY

Two studies have been selected for this computational analysis. The first is a DNS simulation performed by Muppidi & Mahesh[4]. The main objective of their study was to compute the jet trajectories under different conditions. The second study is the experiments of Andreopoulos & Rodi[3]. Detailed three-sensor hot-wire probe velocity measurements

Description	Details	Re_D	Velocity Ratio	N-Cells	N-Procs/Cluster	Ref
Experimental	Steady Round JIC	20,500	0.5	5.6x10^6	64/BGQ	Andreopoulos
		41,000	1	5.6x10^6	64/BGQ	
		82,000	2	5.6x10^6	64/BGQ	
DNS	Steady Round JIC	1,500	1.52	3.2x10^6	64/GPC	Muppidi
		5,000	5.7	5.6x10^6	128/GPC	

Table 1: JIC studies considered for LES comparison

were taken for three cases of JICs. A summary of the different cases considered in this report is given in Table 1. The two DNS cases have a prescribed parabolic jet inlet profile. The crossflow boundary layer thickness is given at the upstream lip of the jet. For the lower velocity ratio simulations (r=1.52), the computational domain spans 48D in the streamwise direction, 24D in the wall-normal direction and 24D in the spanwise direction. The higher velocity ratio simulations (r=5.7) have a domain with dimensions 48Dx48Dx48D. For both simulations, the domain includes a pipe length of 10D.
A plug flow profile is produced at the inlet of the 12D-lendth pipe for the experimental study by Andreopoulos. The crossflow boundary layer is given 4D upstream of the jet exit. To initiate turbulence, the jet flow is tripped by a notch in the pipe wall. The upstream boundary is located 10D upstream of the jet exit, and the wall-normal dimension is 24D.
The dimensions of the LES computational set-up are exactly those of the corresponding studies. A structured hexahedral mesh is generated with mesh refinement near the pipe wall, bottom crossflow wall, and high-gradient areas near the jet-crossflow boundary and the downstream wake. The crossflow boundary layer at the domain inlet is determined using the Blasius analytical solution and mapped onto the computational mesh. Convective terms are discretized using second order upwinding and time integration is performed using a second order backward-implicit scheme. The discretized pressure-momentum equations are solved using the PISO algorithm on 64 to 128 processors using the Scinet supercomputer[5]. The filtered Navier-Stokes equations are given below:

$$\frac{\partial \overline{u_i}}{\partial t} + \frac{\partial}{\partial x_j}(\overline{u_i}\,\overline{u_j}) = -\frac{\partial \overline{p}}{\partial x_i} + \nu \frac{\partial}{\partial x_j}\left(\frac{\partial \overline{u_i}}{\partial x_j} + \frac{\partial \overline{u_j}}{\partial x_i}\right) - \frac{\partial \tau_{ij}}{\partial x_j}$$

The sub-grid scale (SGS) stress, τ_{ij}, is modelled using Yoshizawa's k-equation model[6].

$$\frac{\partial K}{\partial t} + \nabla \cdot (K\overline{\boldsymbol{u}}) = \nabla \cdot [(\nu + \nu_r)\nabla(K)] - \epsilon - \boldsymbol{\tau} : \overline{\boldsymbol{S}}$$

where,

$$\nu_r = C_k K^{1/2}\Delta,\ \epsilon = C_\epsilon K^{3/2}/\Delta, \text{and } \tau_{ij} = -2\nu_r \overline{S}_{ij}$$

Detached-eddy simulation (DES) is performed as well for comparison with the experiments according to the model proposed by Spalart et al.[7].

RESULTS - FLOW STRUCTURE

Coherent structures are identified using the Q-criterion[8]:

$$Q = \frac{1}{2}[|\boldsymbol{\Omega}|^2 - |\boldsymbol{S}|^2] > 0$$

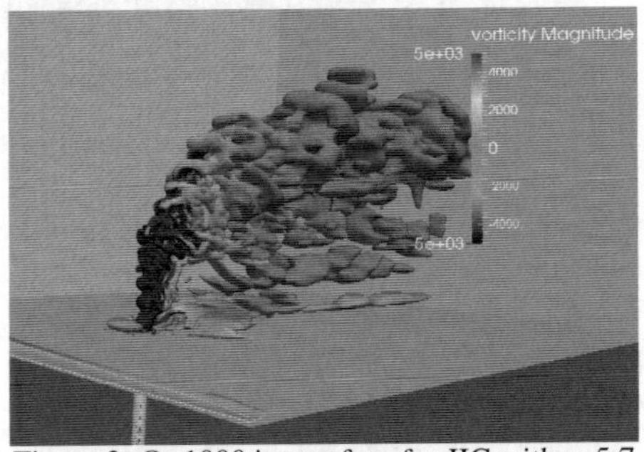

Figure 2: Q=1000 iso-surface for JIC with r=5.7

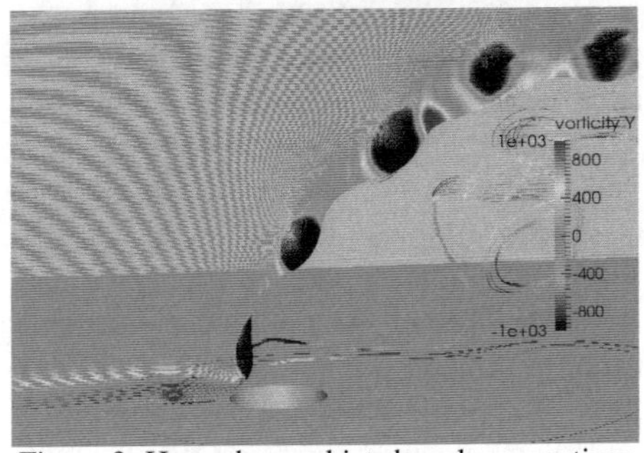

Figure 3: Horseshoe and jet shear layer vortices

where:
$\boldsymbol{\Omega} = \frac{1}{2}\left(\frac{\partial u_i}{\partial x_j} - \frac{\partial u_j}{\partial x_i}\right)$ is the vorticity tensor
$\boldsymbol{S} = \frac{1}{2}\left(\frac{\partial u_i}{\partial x_j} + \frac{\partial u_j}{\partial x_i}\right)$ is the rate of strain tensor.
According to the Q-criterion, a vortex is identified within a spatial region where the vorticity dominates the rate of strain. Figure 2 clearly shows a wide range of structures for Q=1000, including the jet shear layer vortices and the wake vortices. The horseshoe vortex can be seen upstream of the jet exit in Figure 3. In addition, the jet-crossflow interaction produces an unsteady shear layer at the interface which rolls up to form the jet shear layer vortices due to a Helmholtz instability. The CVP is shown in Figure 4, demonstrating its initiation very early in the flow domain.

RESULTS - JET TRAJECTORY

In the DNS study considered, the jet trajectory is defined as the mid-plane mean velocity streamline originating at the

Figure 4: Counter-rotating vortex pair - mean velocity streamlines

jet exit center. The statistics for the LES simulations were computed after 4 flow-through times to ensure that initial transients have left the domain. The jet trajectories are shown in Figure 5.
Excellent agreement between the DNS and LES is obtained. It can be seen that, as expected, the higher velocity ratio JIC penetrates deeper into the flow whereas the effect of the crossflow is felt very early on for the lower velocity ratio. Interestingly, the jet begins to bend inside the pipe at low velocity ratios, therefore emphasizing the need to model the pipe domain.

RESULTS - VELOCITY PROFILES

A comparison of the LES solution with the experimental velocity profiles of [3] is presented for r=0.5, 1, and 2 in figures 6-8. The wake behaviour is clearly visible, particularly for the higher velocity ratio. At lower velocity ratio, the crossflow fluid acts like a partial cover, quickly bending the jet and accelerating it to a higher velocity than U_e. It can be seen that at lower r, the flow quickly regains boundary- layer-like behaviour, whereas the wake persists for a longer distance at higher r. Very good agreement is reached between the experiment and computations, with DES showing more favourable results. Due to the higher operating Reynolds number, the proper computation of the pipe flow is critical to the success of the simulation, thus making DES a more effective choice with a less restrictive y^+ requirement than would be required for a wall-resolved LES.

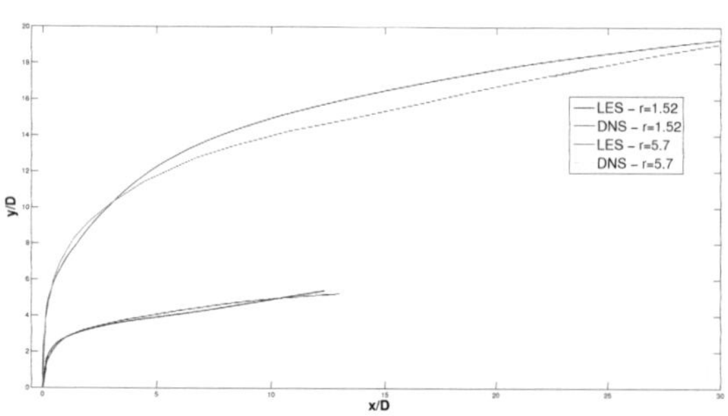

Figure 5: Mean velocity streamline trajectory

CONCLUSIONS

The jet in crossflow is discussed as a stepping stone towards the simulation of synthetic jet actuators in separated airfoils. LES and DES is performed for a series of jets in crossflows. Comparisons with experimental and DNS results show excellent agreement of velocity profiles and jet trajectories. The key vortical structures are reproduced by the computations. These results will be extended to oscillating conditions and presented at the upcoming congress.

ACKNOWLEDGMENTS

Computations were performed on the GPC and BGQ supercomputers at the SciNet HPC Consortium. SciNet is funded by: the Canada Foundation for Innovation under the auspices of Compute Canada; the Government of Ontario; Ontario Research Fund - Research Excellence; and the University of Toronto.

REFERENCES

[1] M. Amitay et al. Aerodynamic flow control over an unconventional airfoil using synthetic jet actuator. *AIAA Journal*, 39(3), 2001.

[2] M. Feero et al. Influence of cavity shape on synthetic jet performance. *Sensors and Actuators A:Physical*, 2014. Under second review.

[3] J. Andreopoulos & W. Rodi. Experimental investigation of jets in a crossflow. *J. Fluid Mech.*, 138:93–127, 1984.

[4] S. Muppidi & K. Mahesh. Study of trajectories of jets in crossflow using direct numerical simulation. *J. Fluid Mech*, 530:81–100, 2005.

[5] C. Loken et al. Scinet: Lessons learned from building a power-efficient top-20 system and data centre. *J. of Physics:Conference Series*, 256(1), 2010.

[6] A.Yoshizawa. A statistically-derived subgrid-scale kinetic energy model for the large-eddy simulation of turbulent flows. *J. Physical Society of Japan*, 54(8):2834–2839, 1985.

[7] P.R. Spalart et al. A new version of detached-eddy simulation, resistant to ambiguous grid densities. *Theor.Comput.Fluid Dyn.*, 20:181–195, 2006.

[8] J.C.R. Hunt et al. Eddies, streams, and convergence zones in turbulent flows. *Center for Turbulence Research Report CTR-S88*, 1988.

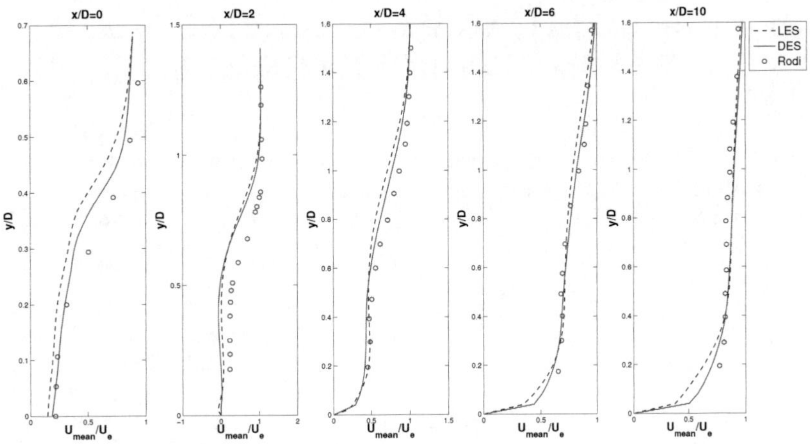

Figure 6: Mean streamwise velocity profiles for r=0.5

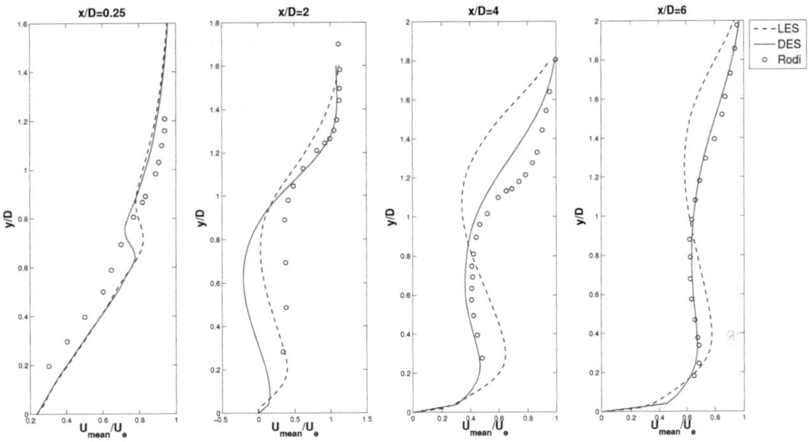

Figure 7: Mean streamwise velocity profiles for r=1

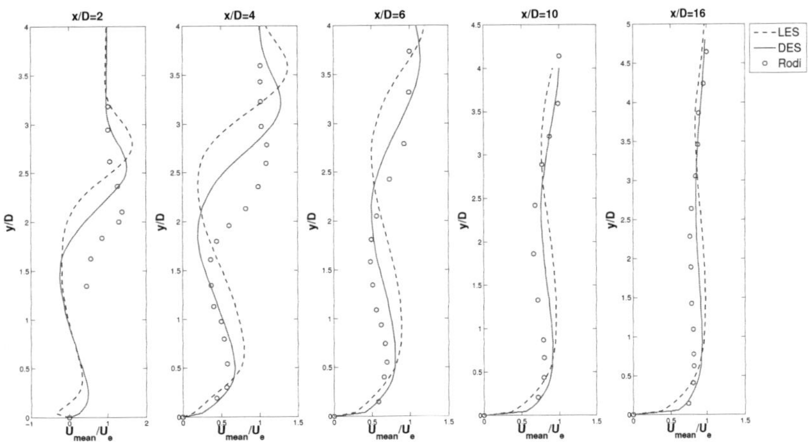

Figure 8: Mean streamwise velocity profiles for r=2

MEASUREMENT OF ROTATING WIND TURBINE BLADE ANGLE OF ATTACK WITH A FIVE-HOLE PRESSURE PROBE

Tyler E. Gallant
Mechanical and Mechatronics Engineering
University of Waterloo
Waterloo, ON, CA
tgallant@uwaterloo.ca

Professor David A. Johnson
Mechanical and Mechatronics Engineering
University of Waterloo
Waterloo, ON, CA
david.johnson@uwaterloo.ca

ABSTRACT

The torque generated by a wind turbine blade is dependent on the blade performance, which is itself a strong function of the angle of attack. Unfortunately, the wind approaching the blade slows relative to the upstream wind, meaning that the angle of attack at the turbine blade cannot be measured using instrumentation upstream. Here the angle of attack α and span-wise flow angle β are measured directly on the rotating blade using a five-hole pressure probe, and variations in the angles are observed as a function of azimuthal position and tip speed ratio.

Turbine performance was evaluated at the University of Waterloo (UW) Wind Generation Facility. Testing was conducted at three wind speeds, and all tests were conducted with a 0° yaw-offset. Results showed that the angle of attack at the probe location increased by over 20° between the low and high wind speeds. An increase in the angle of attack of approximately 5° was consistently observed at azimuthal positions between 100° and 130° as the blade approached the vertically upward position. Further testing is required to investigate possible causes of the azimuthal variation in angle of attack.

KEYWORDS: Wind turbine, angle of attack, five-hole probe, aerodynamics, fluid mechanics

INTRODUCTION

Experimental studies are often conducted in wind turbine design for potential design improvements in terms of efficiency, noise emissions, and safety. Conventionally during turbine testing, the wind speed approaching a wind turbine is measured upstream of the turbine via a separate instrumented tower or through instruments located on the turbine nacelle. However, such measurements do not account for the reduction of wind speed prior to the rotor caused by the interaction between the wind and the blades. Because of this, the wind speed at the blade is typically not known, making it difficult to estimate the true angle of attack at the blade. Because the performance of a wind turbine blade is highly dependent on the angle of attack, the error in this estimate may negatively affect the design of turbine blades specific to certain flow conditions.

Here a method is shown to measure how the local turbine blade flow angles, angle of attack α and span-wise angle β, vary with the azimuthal position during rotation. A secondary objective of the study was to observe how the angle of attack and span-wise angle vary with wind speed. Flow angles were measured directly at the blade using a five-hole pressure probe.

BACKGROUND

Several studies related to the performance of wind turbines have observed variations in the angle of attack with position. A flow visualization field study of stall formation over a wind turbine blade by Swytink-Binnema and Johnson [1] showed that the amount of stall on the blade increased as the blade was rotating towards the ground in comparison to when the blade was travelling away from the ground. The paper concluded that such variation may be either due to a yaw offset which was not measured or vertical wind shear causing a variation across the rotor plane [1]. Because the formation of stall over a blade is dependent on the angle of attack, it can also be concluded that the angle of attack varied with azimuthal position during testing [1].

A similar study by Maeda and Kawabuchi [2] examining stall over a wind turbine blade also resulted in observed variations in the angle of attack with azimuthal position. Their results showed that, with a zero degree yaw-offset, the angle of attack was at a maximum when the blade was oriented vertically upwards and at a minimum when the blade was oriented vertically downwards [2]. The authors state that the effect was likely due to the atmospheric boundary layer. With a 45° yaw offset, the angle of attack was observed to be greatest at azimuthal positions of 180° to 270° (the quadrant following vertically upwards) [2]. The angle of attack was therefore not only a function of wind speed, but also of azimuthal position and the yaw-offset.

Numerous wind turbine studies have used five-hole pressure probes to measure the angle of attack at the blade, including Maeda and Kawabuchi [2] and the NREL Unsteady Aerodynamics Experiments [3]. While many algorithms are present in the literature for calculating the flow angle given the distribution of pressures measured by the probe, here the calibration method and algorithm proposed by Paul et al. [4] was used.

Following this calibration method, probe orientations were divided into 5 zones, with each one corresponding to a maximum pressure measurement in one of the five probe holes. For each zone, pitch, span-wise, total and static pressure coefficients were calculated for each set of calibration pressure measurements. The calibration therefore resulted in a distribution of pitch, span-wise, static pressure and total pressure coefficients which correspond to specific pitch, α_p, and span-wise, β_p, angles relative to the probe. After accumulating test data, the same equations were used to calculate experimental pressure coefficients which were then related back to α_p and β_p relative to the probe through the calibration data.

Once the pitch and span angles relative to the probe were known, the measured angles could be related back to the angle of attack and span wise flow angle through equations reported by NREL [3]:

$$\propto = \operatorname{atan}\left(\frac{\cos \alpha_p * \cos(\beta_p + \varepsilon) * \sin\gamma + \sin\alpha_p * \cos\gamma}{\cos(\alpha_p) * \cos(\beta_p + \varepsilon) * \cos\gamma - \sin\alpha_p \sin\gamma}\right)$$

$$\beta = \operatorname{atan}\left(\frac{\cos\alpha_p * \sin(\beta_p + \varepsilon)}{\cos\alpha_p * \cos(\beta_p + \varepsilon) * \cos\gamma - \sin\alpha_p * \sin\gamma}\right)$$

Here, α is the angle of attack with respect to the blade, β is the span-wise flow angle with respect to the blade, γ is the probe angle offset in the pitch direction and ε is the probe angle offset in the span-wise direction. All angles are in units of degrees.

EXPERIMENTAL SET-UP

The experiment was conducted at the University of Waterloo (UW) Wind Generation Research Facility using the UW Wind Energy Group (WEG) test turbine. The facility is equipped with a fan bank 11.5 m by 7.5 m in size which is capable of generating wind speeds up to 11.5 m/s [5]. The fan bank was used to create constant velocity flows over the turbine. Tests were conducted at multiple tip speed ratios, defined in Eq. (1). Here, Ω is the rotational speed of the turbine, R is the radius of the turbine rotor and U_∞ is the upstream wind speed.

$$\lambda = \frac{\Omega R}{U_\infty} \qquad (1)$$

The turbine itself consisted of one modular 3D-printed test blade and two weighted rods which act as placeholders for the remaining blades. The blade is designed for a wind speed of 8.5 m/s and a rotational speed of 200 rpm, resulting in a design tip speed ratio of 4.2 [6]. The blade is 1.7 m long and has geometry based on the S833 airfoil with a constant 178 mm chord and 6° pitch [6]. The two weighted rods are used during testing to maintain the proper weight distribution for the turbine, as was done in reference [6]. An image of the turbine is provided in Figure 1.

Figure 1: UW Wind Energy Group Test Turbine [6]

As part of a separate research project, one blade section was 3D printed using the same aerodynamic design and material but with the addition of an opening for a five-hole pressure probe [7]. A schematic of the layout is shown in Figure 2. Given that original blade design was modular, the pressure probe blade section was installed in the blade without any modifications to the rotor.

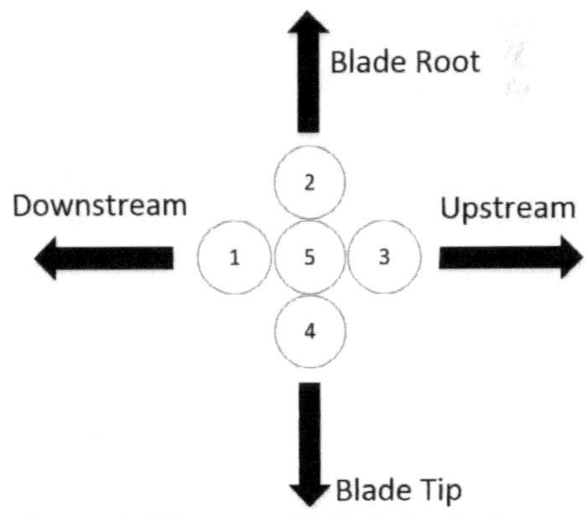

Figure 2: Diagram of 5-Hole Probe Layout

The probe was installed at 55% of the blade span from the root (i.e. $r/R = 0.55$), and is offset from the chord by approximately 15° to match the expected angle of attack at the radial location [6]. A labelled image of the five-hole probe at the leading edge of the blade is shown in Figure 3. Pressure measurements were sampled using an Arduino microcontroller installed in the blade section, which then transmitted data to acquisition software via Bluetooth. Further details are available in Moscardi [7].

Figure 3: Photograph of five-hole pressure probe located in the test blade

A Futek Rotary Torque Sensor (Model TRS605) [7] equipped with an optical encoder was installed between the rotor and gearbox of the wind turbine, allowing for measurements of azimuthal position during testing. All measurements were made relative to a 0° reference defined as the test blade being oriented vertically downward.

The test blade being used for the experiment is designed for a tip speed ratio, λ, of 4.2. This setting was taken as the base case, and the tunnel wind speed was decreased to provide higher tip speed ratio cases. Rotational speed and wind speed were measured to verify the tip speed ratios tested, and all results plotted are labelled with the measured tip speed ratios. Pressure data were organized into 36 azimuthal bins 10° in width. Each data set consisted of 5 minutes of measurements with a sampling rate of approximately 20 Hz, resulting in between 140 and 180 measurements sorted into each bin.

RESULTS AND DISCUSSION

Pressure coefficients were calculated according to the method outlined in Paul et al. [4] and were interpolated with calibration data to determine the pitch and span-wise angles relative to the probe (α_p and β_p, respectively). Using the equations reported by NREL [3] with the measured probe offset angles, the measured angle of attack and span-wise angles were calculated and are plotted in Figure 4 and Figure 5 respectively.

The angle of attack α was found to vary both with the tip speed ratio λ and the azimuthal position. As the wind speed is increased and the tip speed ratio decreases, the angle of attack increases from approximately 11 - 14° to as high as 22 - 26°. The increase in angle is due to changes in the upstream wind speed, resulting in a change in the direction of the relative velocity towards the upstream wind.

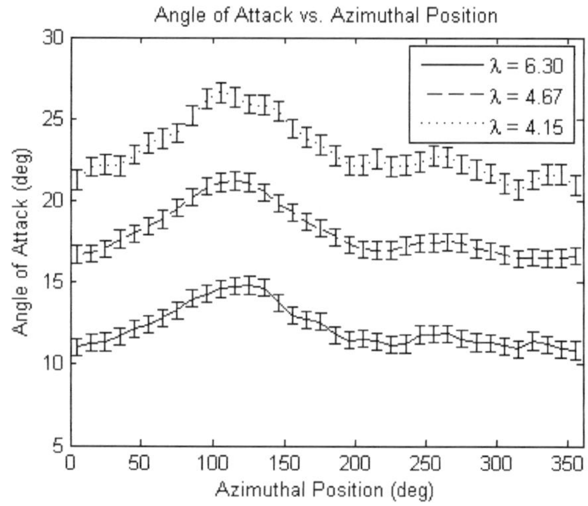

Figure 4: Average Angles of Attack vs. Azimuthal Position for three tip speed (λ) ratios

An increase in the angle of attack of about 5° is observed at azimuthal positions between approximately 100° and 130° in every test case, as the blade rotates towards vertically upward. Further examination of the raw data showed that the peak in the angle of attack is not caused by: a yaw condition caused by a crosswind from the fan bank; an inconsistent effect of the centrifugal force on the pressure transducers in the blade; or the calibration of the probe and coefficients given the raw pressure measurements. The peak occurs with a consistent magnitude and azimuthal location throughout all nine tests conducted.

It is posited that, given the similar azimuthal location of the increase in angle of attack to the increase in stall observed in yaw conditions by Swytink-Binnema and Johnson [1] and Maeda and Kawabuchi [2], the testing conditions are likely similar. Thus, it can be posited that the variation in the angle of attack with azimuthal position may be due to variations in the relative speed of the flow over the blade, potentially due to dynamic stall. Confirmation of such an anomaly requires further testing with the WEG test turbine.

Measurements of the span-wise flow angle were also sorted into 36 azimuthal position bins and averaged between three tests. Results are plotted against azimuthal position in Figure 5.

The span-wise angle of the flow relative to the blade was measured to be between 13° and 17° for all wind speeds tested. Some variation was observed with azimuthal position, with an increase in the span-wise angle occurring at azimuthal positions of approximately 180° to 240° as the blade passes the vertically upwards orientation. It is expected that the azimuthal variation in the span-wise flow angle is caused by the same aerodynamic effect causing the azimuthal variation in the angle of attack measurements.

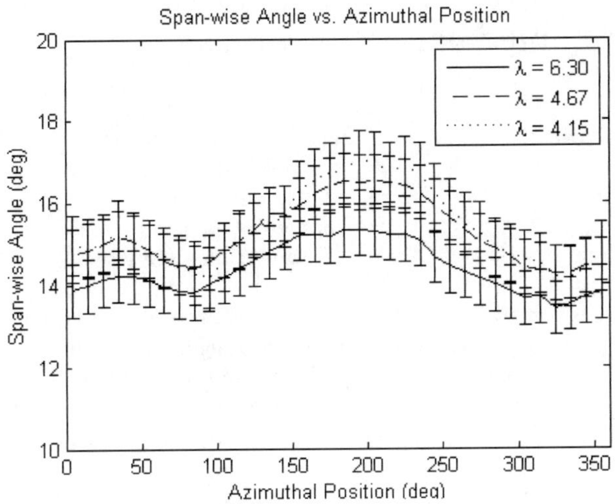

Figure 5: Average Span-wise Angles vs. Azimuthal Position for three tip speed (λ) ratios

CONCLUSIONS

Measurements have confirmed the expected increase in the angle of attack at the blade with wind speed. An average increase of over 10° was observed between the low and high wind speeds tested.

In all tests conducted, the angle of attack was observed to increase by as much as 5° between azimuthal positions of 100° to 130° as the blade rotated towards the upward vertical position. After reviewing the upstream wind measurements, the effect of the centrifugal force on the pressure measurements, and the raw probe pressure measurements, it was concluded that the peak in α is caused by aerodynamic effects such as dynamic stall. Comparison with yawed stall measurements obtained by Swytink-Binnema and Johnson [1] and Maeda and Kawabuchi [2] supports this possibility. Confirmation of such an anomaly requires further testing at the Wind Generation Facility, and an investigation of these results will be conducted as future work.

Future work will also include further testing with the five-hole pressure probe and test turbine at various tip speed ratios and yaw conditions.

ACKNOWLEDGMENTS

The authors would like to thank the Wind Energy Group at UW for their support with this project, specifically Andrea Moscardi for his assistance with the five-hole pressure probe and Nicholas Tam for his assistance with installing the turbine blades and for support from the Natural Science and Engineering Research Council (NSERC).

REFERENCES

1. Swytink-Binnema, N. and Johnson, D.A. 2014, "Novel Image Analysis Method for Blade Aerodynamic Performance on Operational Turbine" *J. Phys. Conf. Ser.* 524(1):012016.DOI:10.1088/1742-6596/524/1/012016 .
2. Maeda, T. and Kawabuchi, H. 2005, "Surface pressure measurement on a rotating blade of field horizontal axis wind turbine in yawed condition," *JSME International Journal Series B*, Vol. 48, no. 1, pp. 156-163.
3. Hand, M.M., Simms, D.A., Fingersh, L.J., Jager, D.W., Cotrell, J.R., Schreck, S., Larwood. S.M., 2001, "Unsteady Aerodynamics Experiment Phase VI: Wind Tunnel Test Configurations and Available Data Campaigns." NREL/TP-500-29955.
4. Paul, A.R., Upadhyay, R.R., Jain, A., 2011, "A novel calibration algorithm for five-hole pressure probe." *Int. J. Eng., Sci., Tech.*, Vol.3, no. 2, pp. 89-95.
5. "UW Live Fire Research Facility." University of Waterloo Wind Energy Group. Available: http://windenergy.uwaterloo.ca/facil_firelab.php. [Accessed Dec. 5, 2014].
6. Abdelrahman, A., 2014. "Development of a Wind Turbine Test Rig and Rotor for Trailing Edge Flap Investigation." MASc thesis, University of Waterloo, Canada.
7. Moscardi, A. 2014, "Five Hole Pressure Probe For Local Inflow Study on a Horizontal Axis Wind Turbine" MASc thesis University of Waterloo/Politecnio di Milano.
8. Futek. "Futek Model TRS605: Non-Contact Shaft to Shaft Rotary Torque Sensor W/ Encoder." Internet: http://www.futek.com/files/pdf/Product%20Drawings/trs605.pdf. 2014 [Oct. 3, 2014]

Proceedings of the 25th CANCAM
London, Ontario, Canada, May 31 – June 4, 2015

Wingtip vortex control via tip-mounted half-delta wings

Ying Yu Su
Department of Mechanical Engineering
McGill University
Montreal, Quebec, Canada
ying.su@mail.mcgill.ca

Steve Choi
Department of Mechanical Engineering
McGill University
Montreal, Quebec, Canada
sunghun.choi@mail.mcgill.ca

Professor Tim Lee
Department of Mechanical Engineering
McGill University
Montreal, Quebec, Canada
tim.lee@mcgill.ca

ABSTRACT

The control of the tip vortex, generated by a rectangular NACA 0012 wing, via tip-mounted half-delta wings (HDWs), of different slendernesses Λ, and root chords c_r, was investigated experimentally at Re = 2.45 × 10^5. The results show that the addition of HDWs consistently led to a diffused tip vortex. The degree of diffusion was, however, found to increase with decreasing Λ and c_r. HDWs with $c_r \leq$ 50% of the baseline-wing chord c caused a rapid diffusion of vorticity and rendered a weak circulation flow-like tip vortex, suggesting an enhanced wake-vortex decay and alleviation. The c_r = 0.5c HDW also produced an improved lift-to-drag ratio. A unique double-vortex pattern also exhibited downstream of the $c_r \leq$ 50%c HDW wings.

KEYWORDS: tip vortex modification, NACA0012 wing, slender, non-slender, delta wing, lift-induced drag

1. INTRODUCTION

The suppression of wingtip vortex and the reduction of lift-induced drag of aircraft wings have continuously posed a challenge to aerodynamicists and fluid dynamists. An excellent review on the airplane trailing vortices is given by [1]. Various passive wingtip vortex control techniques, for example, winglets, spoilers, stub/subwing, and porous tips and leading edges have thus been attempted by researchers elsewhere to modify the strength and structure of the tip vortices. More recently, an alternative wingtip vortex control was investigated by [2] through the addition of a slender full-chord half-delta wing mounted to the tip of a static rectangular NACA 0012 wing.

Reference [2] found that the tip vortex generated behind a rectangular NACA 0012 wing was greatly diffused and enlarged by a tip-mounted 65°-sweep half-delta wing (HDW), as a result of the breakdown of the HDW vortex developed on the upper surface of the half-delta wing. The HDW tip vortex control concept was inspired by the pioneering work of [3-5] in which the HDW was attempted to improve the wing lift generating capability. Lee and Pereira [2] further noticed that, in addition to the observed changes in the size and strength of the HDW-wing tip vortex, the lowered vorticity level also led to a much reduced lift-induced drag coefficient C_{Di} (= $D_i/\frac{1}{2}\rho_\infty u_\infty^2 S$). The lift-induced drag D_i was computed using the Maskell wake integral model [6-8].The reduced C_{Di} also translates into a smaller C_D, especially at high C_L range, compared to the baseline wing at the same lift condition. This HDW tip vortex control concept is, however, shadowed by the undesired HDW-induced increase in wing weight and bending moment.

The objective of this investigation was to examine the minimization of the above-mentioned undesired effects incurred by the full-chord HDW through the use of small-chord HDWs with $c_r \leq$ 0.5c at Re = 2.45 × 10^5. Small-chord HDWs of different slenderness Λ (= 50° and 65°) were also considered. Special emphasis was placed on the impact of the small-chord HDW, of different Λ, and c_r on the near-field tip vortex at x/c = 2.8 at different angles of attack α . The growth and development of the HDW vortex both along the tip and in the near field of the HDW wing, at α = 10°, was also acquired to better understand the impact of small-chord HDWs on the tip vortex. These tip-vortex flowfield measurements were also supplemented with wind-tunnel force balance measurements.

2. EXPERIMENTAL METHOD

The experiment was conducted in a subsonic wind tunnel with a test section of 90 cm × 120 cm × 270 cm at McGill University. A NC-machined semi-wing, of a NACA 0012 profile and a chord of c = 28 cm and a half-wing span of s = 50.8 cm, was used as the test model. Boundary-layer transition was fixed on the upper and the lower wing surfaces by using a trip strip (of a width of 5%c) located at 10%c downstream of the wing leading edge. The freestream velocity was fixed at u_∞ = 13.5 m/s, which rendered a chord Reynolds number of Re = 2.45 × 10^5. The square tip of the semi-wing model was also equipped with a half-delta wing (HDW), made of aluminum plate with a thickness of or t/c = 1.1%, of different sweeps Λ and root chords c_r.

The vortex flow structures both along and in the near field of the HDW wing configuration were obtained by using a miniature seven-hole pressure probe of an outside diameter of 2.8 mm. Eight Honeywell DC002NDR5 pressure transducers (seven for the probe and one for tunnel reference total pressure) were used. The pressure signals were sampled at 500 Hz with a sampling time of 5 seconds, and were recorded on a PC through a 16-bit A/D board. Probe traversing was achieved through a custom-built computer-controlled traversing system. For lift and drag measurements,

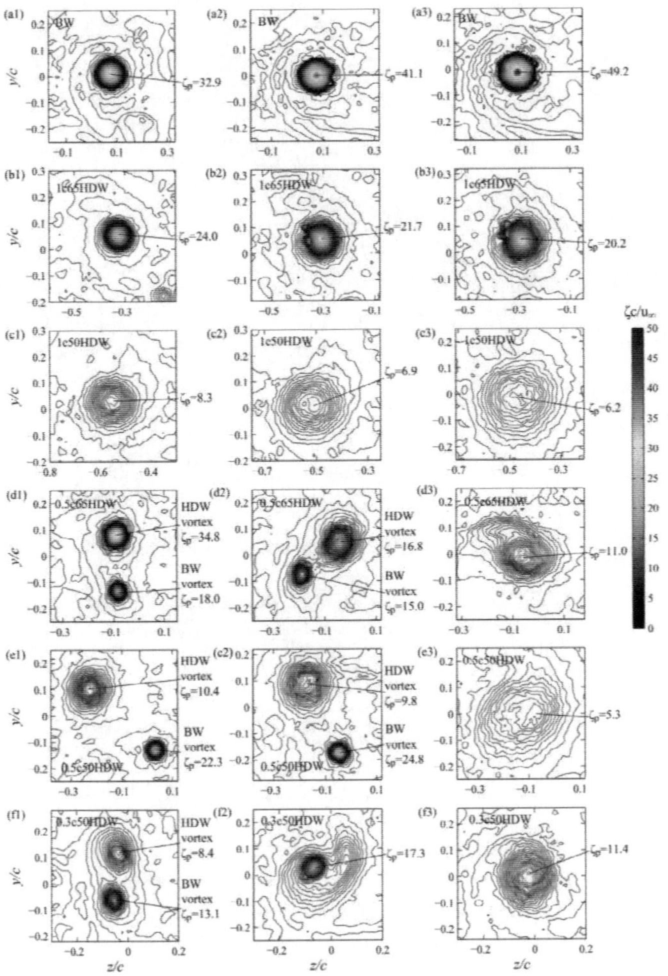

Figure 1 : Impact of Λ and c_r of the HDW on the normalized iso-vorticity contours of the tip vortex at x/c = 2.8 for α = 6°, 8° and 10°. (a1)-(a3): BW; (b1)-(b3): 1c65HDW; (c1)-(c3): 1c50HDW; (d1)-(d3): 0.5c65HDW; (e1)-(e3): 0.5c50HDW; and (f1)-(f3): 0.3c50HDW.

the wing model was mounted vertically on an external two-component force balance located outside the wind tunnel.

Each near-field data plane, covering up to 70% of the wing semi-span, had 2,400 measuring grid points with an increment of $\Delta y = \Delta z$ = 3.2 mm (or 1.15%c), except along the span of the wing where Δz = 6.4 mm. The lift-induced drag was computed, based on the vorticity distributions inferred from the measured *vw*-crossflow fields, via the Maskell model. The maximum uncertainty for the experimental data is as follows [10]: mean velocity 3.5%, vorticity component 8%, and vortex radius 4%. The maximum uncertainty for the lift and drag coefficient is ±0.025 and ±0.005, respectively. Note that for the HDW wing configuration, the total wing area S_{total} and total semi span were used in the normalization of the lift and drag forces and the lift-induced drag.

3. RESULTS AND DISCUSSION

3.1 Flow analysis for full-chord or c_r = c HDW

Figures 1b1-c3 show the impact of sweep or slenderness of the full-chord HDW on the normalized iso-vorticity ($\zeta c/u_\infty$) contours of the tip vortex at x/c = 2.8 for α = 6°, 8° and 10° (covering low-to-medium α regime). The HDW deflection was fixed at δ = 0°. The streamwise vorticity ζ (= $\partial w/\partial y - \partial v/\partial z$) was calculated from the crossflow (*vw*) measurements by using a central differencing scheme to evaluate the derivatives. The tip vortex generated behind the baseline wing (BW) is also included in Figs. 1a1-a3 for a direct comparison. Figures 1a1-a3 show that the BW tip vortex always remained concentrated and had the highest normalized peak vorticity ζ_p (= $\zeta_{peak}c/u_\infty$; indicative of vorticity level) compared to all the HDW wings tested in the present study. For the BW, ζ_p was also found to increase with α (up to the static-stall angle α_{ss}), as a result of the interaction and merging of the secondary vortices with the main tip vortex along the baseline-wing tip (for x/c < 1; Fig. 2a). The tip vortex continued to progress downstream of the BW trailing edge. For $x/c \geq 2.5$, the inner-flow region of the tip vortex was found to attain near axisymmetry (see [9,11] for details).

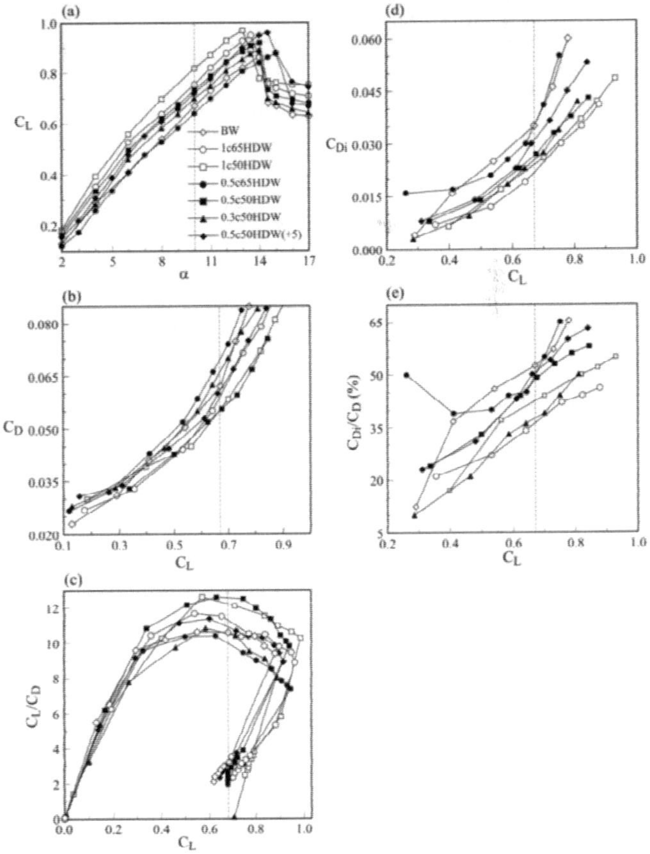

Figure 3 : Effect of slenderness and root chord of HDW on aerodynamic characteristics.

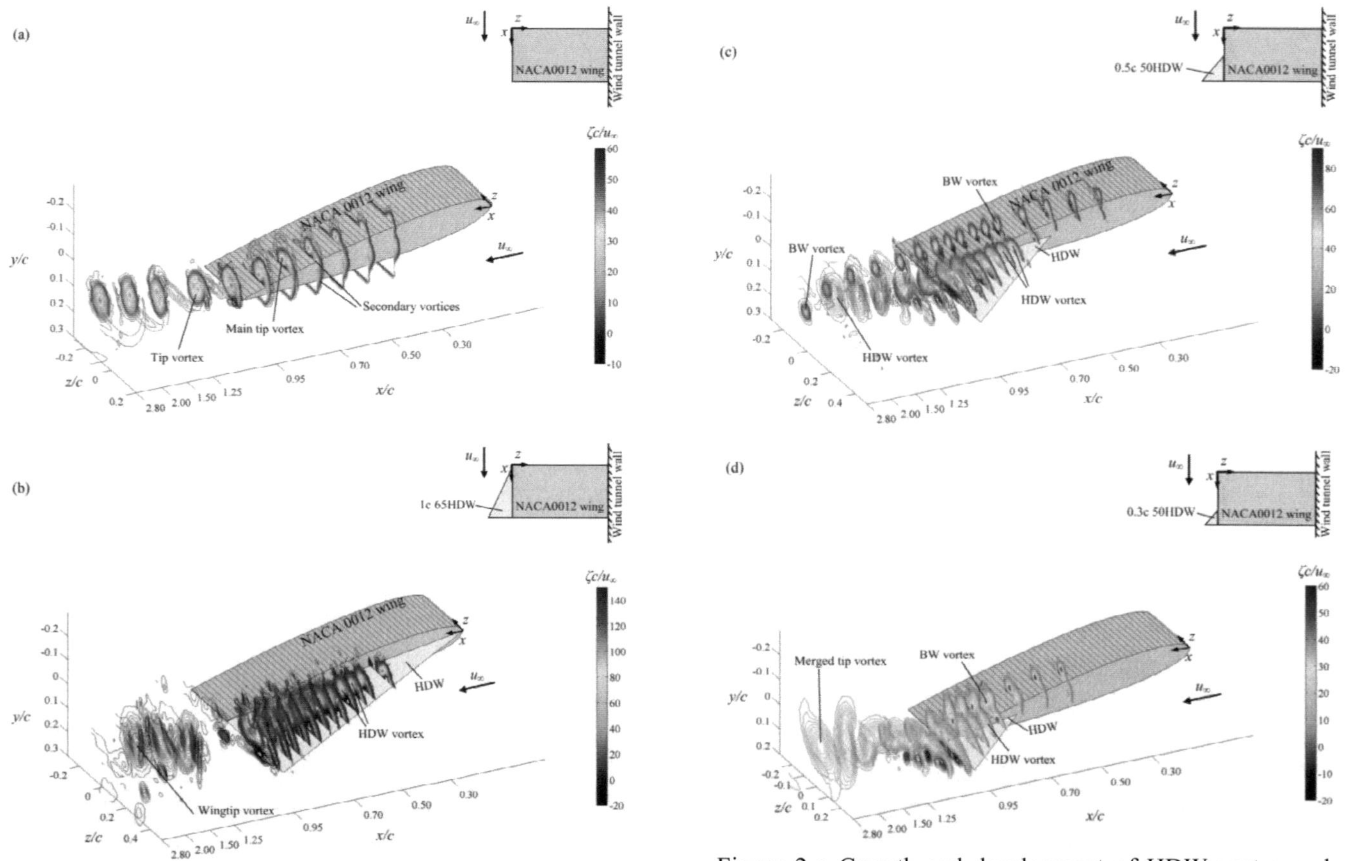

Figure 2 : Growth and development of HDW vortex and BW vortex on half-delta wing configuration. (a) BW at α = 10°, (b) 1c65HDW wing at α =10° (c) 0.5c50HDW at α = 10°, and (d) 0.3c50HDW at α = 10°.

Figures 1b1-c3 show that the addition of the full-chord HDWs, however, led to an enlarged and diffused tip vortex with a lower ζ_p compared to their BW counterpart. The degree of diffusion increased as Λ was decreased. The lowered vorticity level can be attributed to the HDW vortex breakdown on the slender half-delta wing surface. Figure 3b shows that at α = 10° the HDW vortex breakdown was observed between x/c = 0.85 and 0.9 (denoted by the large drop in the peak vorticity of the HDW vortex) on the 1c65HDW wing, resulting in the diffused tip vortex presented in Fig. 1b3.

For the nonslender HDW with Λ = 50° and c_r = c (designated as 1c50HDW), the tip vortex became circulation flow-like with a significantly enlarged size and lowered vorticity level or ζ_p compared to its slender counterpart (Figs. 1c1-c3). The presence of the circulation flow-like tip vortex can be attributed to the close proximity of the HDW vortex to the 1c50HDW surface, which, in turn, led to HDW vortex breakdown at $x/c \approx 0.5$ (not shown here). In short, despite the observed large impact of the 1c50HDW on the tip vortex, the addition was, however, accompanied by a large increase in S_{total} and AR, which can translate into an increased wing weight and bending moment. To relieve the HDW-induced large increase in S_{total} and AR, HDWs of smaller chord (i.e., $c_r \leq 0.5c$) were considered.

3.2 Flow analysis for small-chord or $c_r \leq 0.5c$ HDW

Figures 1d1-e3 show that the tip vortex generated by HDW wings with $c_r \leq 0.5c$ is characterized by a double-vortex pattern (i.e., the coexistence of a BW vortex, originating from the free tip portion of the rectangular wing, and a HDW vortex) at low α. Meanwhile, the HDW vortex moved toward and engulfed the BW vortex, causing a greatly enlarged and circulation flow-like tip vortex as α was increased. The nonslender HDW produced a more diffused and enlarged tip vortex compared to its slender counterpart. Typical formation and growth of the double vortex and their interaction along the 0.5c50HDW wing and in its near field are displayed in Fig. 2c.

For the 0.3c50HDW wing, the double vortex pattern also appeared and had a lower ζ_p and was closer (Figs. 2f1-f3) in comparison to its c_r = 0.5c counterpart for $\alpha \leq 6°$. The close proximity of the double vortex expedited the interaction and merging for $\alpha > 6°$. The growth, interaction and merging of the double vortex associated with the 0.3c50HDW wing for $0 < x/c \leq 2.8$ is illustrated in Fig. 2d.

3.3 C_L, C_D, C_L/C_D and C_{Di} analysis

Figure 3a shows that the presence of full-chord HDWs can produce an increased C_L compared to the BW. A 22.4% and 13.4% increase in C_L, for example, at $\alpha = 10^o$ (a representative medium-α regime) was obtained for the 1c50HDW and 1c65HDW wings, respectively, in comparison to the BW (with $C_L = 0.67$). The observed C_L increment can be attributed to the HDW-induced improved wing tip conditions. The force-balance measurements also show that full-chord HDWs also gave rise to a reduced C_D, especially for $C_L > 0.3$ regime (Fig. 3b), and subsequently an improved C_L/C_D ratio compared to the BW at the same lift condition (Fig. 3c). The 1c50HDW wing, however, produced a lower C_D compared to its slender counterpart.

For the $c_r = 0.5c$ nonslender HDW (i.e., 0.5c50HDW) wing, the C_L increment was consistently lower than its full-chord counterpart while the C_D reduction remained comparable. More importantly, the 0.5c50HDW wing was found to produce the best C_L/C_D performance among all HDWs tested (Fig. 5c). The cause for the observed C_D reduction can be examined via the lift-induced drag coefficient C_{Di}. Figure 3d shows that the C_{Di} of the 0.5c50HDW wing always had a lower-than-BW value, especially, for the high α or lift regime. Figure 3e further indicates that at higher α the total C_D is dominated by C_{Di}.

The effect of the HDW root chord on the aerodynamic characteristics was further investigated by the use of the 0.3c50HDW wing. A 10% increase in C_L and 23% decrease in C_{Di} at $\alpha = 10^o$ compared to the BW were obtained. The C_L/C_D, however, remained somewhat inferior to the BW value at the same lift condition. The 0.3c50HDW was also able to produce a reduced C_{Di} similar to that of the 0.5c50HDW wing (see Fig. 3d) and also a C_{Di}/C_D ratio comparable to the 1c65HDW wing (see Fig. 3e).

4. CONCLUSIONS

The tip vortex developed behind a rectangular NACA 0012 wing with tip-mounted HDWs, of different Λ and c_r was invetigated in a subsonic wind tunnel. The results show that, regardless of Λ and c_r the addition of HDWs always led to an enlarged and diffused tip vortex and increased C_L compared to the baseline wing. The near-field vortex flow behind the small-chord HDW, however, exhibited a double-vortex pattern, which interacted and merged with each other as it progressed downstream, forming a circulation-like flow. The interaction and merge of the double vortex was found to be expedited by upward HDW deflection. The smaller the c_r of the nonslender HDW the lower the lift-induced drag. $c_r = 0.5c$ nonslender HDW wing produced the largest lift-to-drag ratio among all the HDWs tested.

ACKNOWLEDGEMENTS

This work was supported by the Natural Science and Engineering Research Council (NSERC) of Canada.

NOMENCLATURE

AR	aspect ratio, $= b^2/S$
B,	wing span, main wing chord
C_D, C_{Di}	total and lift-induced drag coefficients
C_L	total lift coefficient
c_r	root chord of half-delta wing
D_i	lift-induced drag
Re	chord Reynolds number, $= cu_\infty/\nu$
S	wing surface area
s	half-wing span, $= b/2$
S_{BW}	baseline wing area
S_{HDW}	HDW surface area
S_{total}	$= S_{BW} + S_{HDW}$
$u,v,w,\ x,y,z$	axial, transverse and spanwise mean velocities and directions
u_∞	freestream velocity
α	angle of attack
Λ	sweep angle
ζ	mean streamwise vorticity
Γ	circulation or vortex strength

REFERENCES

1. Spalart, P.R., 1998, "Airplane trailing vortices," *Ann Rev Fluid Mech.*, 30, pp.107-138.
2. Lee, T., and Pereira, J., 2013, "Modification of static-wing tip vortex via a slender half-delta wing," *J. Fluids & Struc.*, 43, pp.1-14.
3. Staufenbiel, R., and Vitting, T., 1991, "On aircraft wake properties and some methods for stimulating decay and breakdown of tip vortices," *AGARD Conference Proceedings* 494, Neuilly Sur Seine, France.
4. Nikolic, V.R., 2005, "Movable tip strakes and wing aerodynamics," *J. Aircraft*, 42(6), pp.1418-1426.
5. Nikolic, V.R., 2011, "Optimal Movable Wing Tip Strake," *J. Aircraft*, 48(1), pp.335-341.
6. Maskell, E., 1973, "Progress towards a method for the measurement of the components of the drag of a wing of finite span," *RAE Technical Report*, 72232.
7. Brune, G.W., 1994, "Quantitative low-speed wake surveys," *J Aircraft*, 31(2), pp.249-255.
8. Kusonose, K., 1998, "Drag reduction based on a wake-integral method," AIAA-98-2723.
9. Birch, D., and Lee, T., 2004, "The structure and induced drag of a tip vortex," *J. Aircraft*, 41(5), pp.1138-1145.
10. Moffat, R.J., 1988, "Describing the uncertainties in experimental results," Exp Thermal & Fluid Sci., **1**, pp.3-17.
11. Lee, T., and Pereira, J., 2010, "On the nature of wake- and jet-like axial tip-vortex flow," *J Aircraft,* 47(6), pp.1946-1954.

Proceedings of the 25th CANCAM
London, Ontario, Canada, May 31 – June 4, 2014

NEAR WAKE FLOW TOPOLOGY OF AHMED BODIES

Graham Feltham
Aerospace Engineering
University of Toronto
Toronto, Ontario, Canada
graham.feltham@mail.utoronto.ca

Alis Ekmekci*
Aerospace Engineering
University of Toronto
Toronto, Ontario, Canada
ekmekci@ utias.utoronto.ca

ABSTRACT

The near wake flow topology of Ahmed bodies with back slant angles of 25, 30, and 35 degrees have been investigated using the techniques of surface-wire hydrogen bubble visualization and Particle Image Velocimetry (PIV). The results provide a new perspective into the flow behaviour in the near wake of these geometries and point out areas that require further study.

KEYWORDS: Ahmed body, wake flow, hydrogen bubble visualization, flow visualization, flow topology, PIV

INTRODUCTION

The flow topology at the rear of automobile shapes has been previously studied in great detail in order to determine the dominant structures in the near wake. The most common geometry used to study the rear wake flow region in automobiles is a simplified representation known as the Ahmed body. Ahmed et al. [1] found that by changing the rear slant angle, the lift and drag experienced by the body was altered as a result of a change in the flow topology at the rear of the body. Later investigations [2-6] refined the models of flow topology at the rear of the Ahmed body. Based on a detailed understanding of the flow topology, development of control methods to mitigate the pressure and vortex drag incurred by the flow structures have been a subject of intense research [7, 8].

The focus of the current investigation is to visualize and compare the flow structures forming in the near wake of Ahmed body geometries with different rear slant angles. A surface-based visualization technique is used along with PIV measurements. The results provide a new look into the near-wake flow topology of Ahmed bodies and identify areas that require further study.

EXPERIMENTAL SETUP AND TECHNIQUES

Experiments were conducted in a recirculating-type water channel, located at the University of Toronto Institute for Aerospace Studies. This facility provided a continuous and uniform flow along its 5-m-long test section, which had an area of 0.61 m x 0.68 m. The interior of the test section was visually accessible from all sides. Prior to entering the test-section, the flow passed through a chamber, composed of a set of honeycomb and screens, followed by a 6:1 contraction, which yielded a free-stream turbulence intensity of less than 0.3%.

Three Ahmed body geometries were investigated with back-slant angles of 25, 30, 35 degrees. The models were 3D printed using the SLA technique and were hand finished to produce smooth surfaces. The models were painted flat black and were mounted to an endplate with four stainless steel pins. The model position in the water channel and the model's ground clearance C are shown in Figure 1 with the ground clearance being identical to that used in Ahmed et al. [1]. The endplate was made from Plexiglas and was wrapped using flat black vinyl. The leading edge of the endplate had an elliptical profile (with a semi-major/semi-minor axis ratio of a/b = 4/3) to prevent flow separation upstream of the Ahmed models (this behavior was verified with a separate PIV investigation). The trailing edge of the endplate was angled at 10 degrees in order to correct for the circulation about the plate due to the blockage effects; this angle was carefully selected after a series of hydrogen bubble and PIV investigations. The blockage of the complete experimental setup was 5.4%.

Experiments were performed using surface-wire hydrogen bubble visualization and 2D-PIV techniques. Hydrogen bubble visualization experiments were performed at a flow speed of 0.668 m/s, resulting in a Reynolds number based on the length of the Ahmed body models of 196,000. PIV experiments were performed at a flow speed of 0.786 m/s, resulting in a Reynolds number of 230,000.

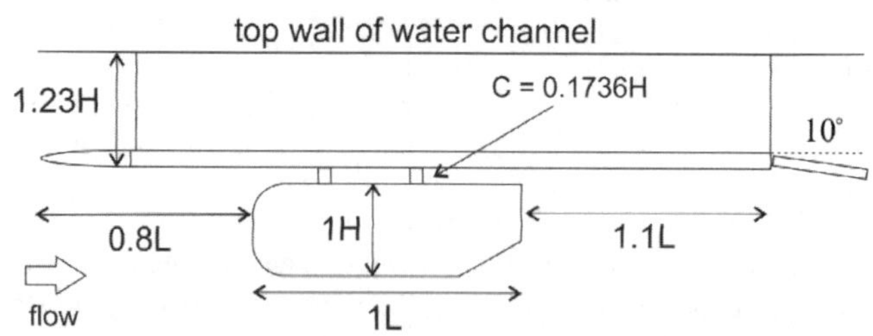

Figure 1: Experimental setup, L = 0.261 m, H = 0.072 m.

To perform the surface-wire hydrogen bubble visualization tests, platinum wires with a diameter of 100 μ were laid in position and tacked at several points to the surface of the models with cyanoacrylate glue. Sections of the wire where it was undesirable to generate hydrogen bubbles were covered by a layer of Kapton tape. To generate

* Corresponding author: Tel.: + 1 416 667 7855; fax: + 1 416 667 7799.
E-mail: ekmekci@utias.utoronto.ca (A. Ekmekci)

hydrogen bubbles at different locations around the model geometries, independent wire circuits were implemented (as shown in Figure 2). Single or multiple wire circuits could be energized using a DC power source at any time with a circuit selector board. The wire placement was selected such that the flow structures at the model centerline and at the model edge could be visualized. The electrical wires providing power to the visualization circuits were taped to the surface of the endplate to minimize their flow disturbance.

The presence of the platinum wires on the model surface is not expected to affect the behavior of the boundary layer or flow structures because most of the wires were located on the back faces of the models, which are regions of full or partial separation, and the boundary layers on the upper and side surfaces of the models were expected to be turbulent upstream of the wires as the flow separates and reattaches aft of the model nose (this behavior was confirmed during a separate PIV investigation).

Figure 2: Multiple wire circuits are implemented using 100-μ-diameter platinum wires at the back of the Ahmed body models in surface-wire hydrogen bubble visualization tests.

A dual-slit light box containing six 3-Watt-power LEDs was utilized in illumination of the hydrogen bubbles. The flow structures were photographed using a Samsung NX1000 camera with a Sigma 105 mm DG Macro lens and were videotaped using a Canon HFG10 video camera. The shutter speed of the camera was varied depending on the velocity of the flow in the region of visualization.

PIV experiments were performed using a NewWave Nd:YAG laser with a maximum energy output of 200 mJ/pulse and a wavelength of 523 nm. Hollow glass spheres with a mean diameter of 10 μ were used as seeding particles. The laser system was located below the channel and directed a ~3-mm-thick planar light sheet vertically toward the model through a TSI 90° mirror module. Data acquisition was performed at 14.5 Hz, which corresponded to the maximum operating frequency of the laser. For each model, 1000 image pairs were recorded, which corresponded to 69 s of continuous data acquisition. All data processing was carried out in INSIGHT 4G software supplied by TSI Inc. Further information on the PIV apparatus and technique are provided in Feltham [9].

25 DEGREES OF BACK SLANT ANGLE

In Figure 3, an instantaneous hydrogen bubble visualization image and a time-averaged PIV result, depicting the contour patterns of velocity magnitude and the streamline topology, in the plane coincident with the mid-plane of the Ahmed body with a back slant angle of 25 degrees are shown. The flow along the slant is attached and the existence of two large recirculation cores in the separated wake is evident. The rotational orientations of these cores are opposite and the upper core (A) is larger in size. From movies of the hydrogen bubble visualization, the upper core (A) is observed to deform and change in shape but is relatively stationary in location. The lower core (B) is periodically shed downstream. This topology is consistent with that of Ahmed et. al [1].

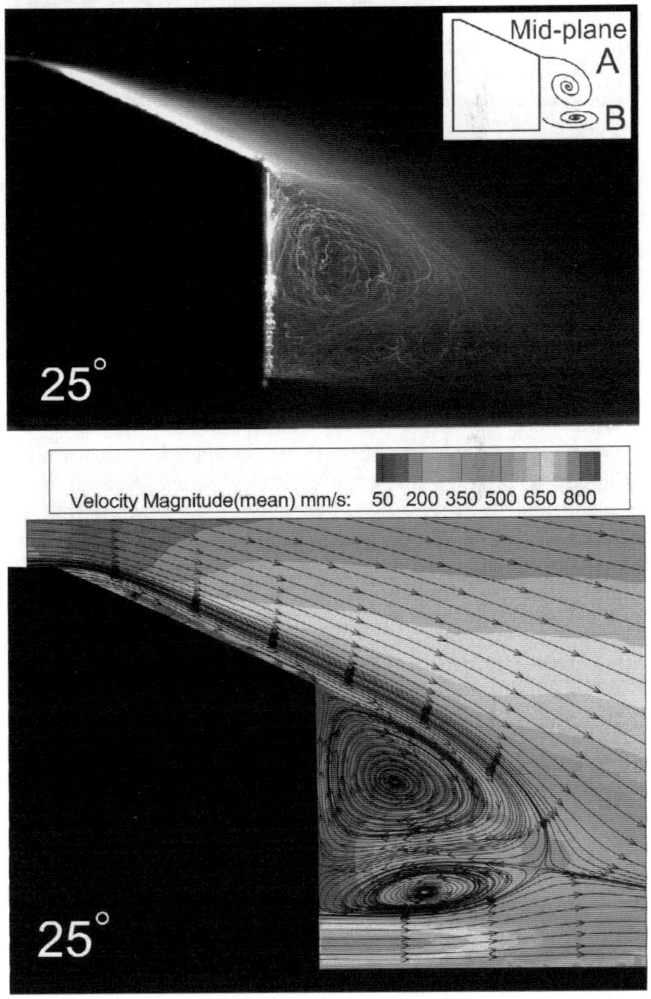

Figure 3: Instantaneous hydrogen bubble visualization (top) and time-averaged PIV (bottom) in the mid-plane of the near-wake for the 25 degrees of back slant angle.

An instantaneous image from the hydrogen bubble visualization in the plane coincident with the edge of the back slant with the 25 degrees is provided in Figure 4. The formation of a strong longitudinal vortex along the back slant is evident. This vortex persists into the wake behind the body. The angle of the vortex core is about 17 degrees. From movies of the hydrogen bubble visualization, the longitudinal vortex is observed to be stable and stationary in position.

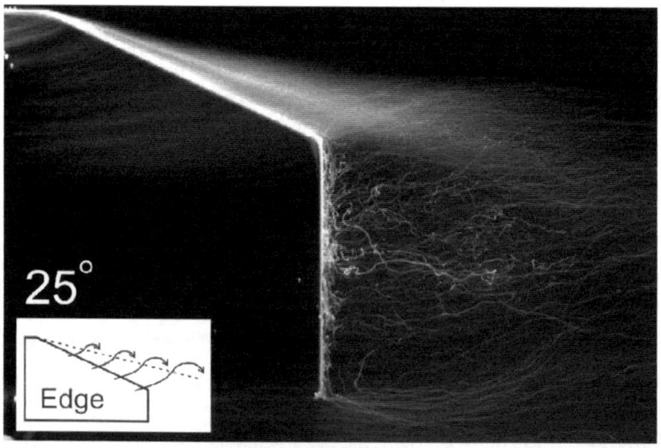

Figure 4: An instantaneous hydrogen bubble visualization image in the plane coincident with the edge of the back slant with an angle of 25 degrees.

30 DEGREES OF BACK SLANT ANGLE

Instantaneous hydrogen bubble visualization and time-averaged PIV patterns from the mid-plane of the Ahmed body with a back slant angle of 30 degrees are shown in Figures 5 and 6. This angle is commonly referred to as the "critical angle" [10] as the flow depicts one of two distinct flow modes, which leads to a sharp discontinuity in lift and drag. By comparing the results given in Figures 5 and 6, it is evident that two different flow modes are indeed observed in the present study at 30 degrees as explained below. During experiments, the starting condition of the channel (whether the speed was increased abruptly from rest or gradually in steps) was observed to determine the mode of the flow at this critical angle. Although the exact relationship between the starting conditions and the flow modes is left for a future study, the flow topology at each mode is investigated in the present study for the critical back slant angle.

During the first flow mode, movies of the hydrogen bubble visualization showed two flow states that were observed to occur interchangeably along the slant: at times the flow separated at the beginning of the slant and re-attached at the end of the slant (an instant of which is shown in Figure 5, first image from the top), while at other times the flow is observed to separate and remain separated along the slant (an instant of which is shown in Figure 5, second from the top). The re-circulation bubble (marked as C) which can be observed in the re-attached state of this mode appears to eventually shed into the wake in an unsteady manner. For this mode, during PIV experiments, however, the slant face of the Ahmed body was not within the field of view and hence it was not possible to study the separation/reattachment state along the slant from the present PIV results. The flow states,

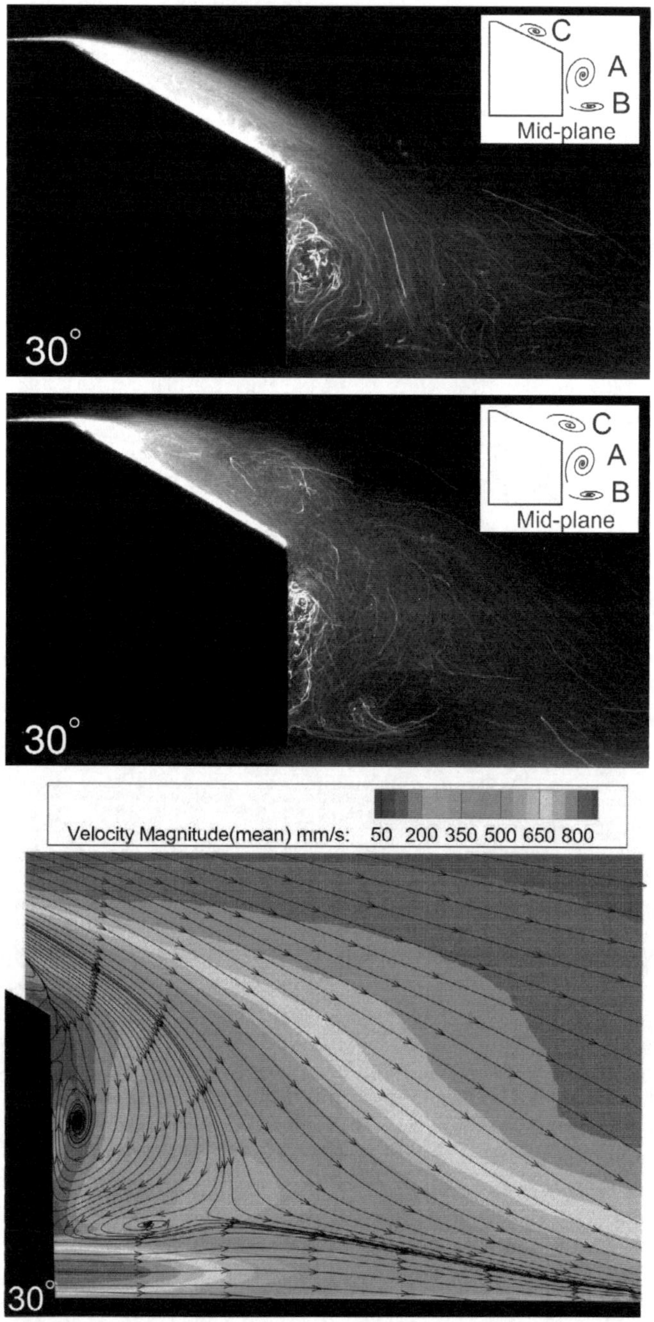

Figure 5: Instantaneous hydrogen bubble visualization (fist two images from the top) and time-averaged PIV result (bottom image) from the mid-plane of the near-wake for the 30 degrees of back slant angle during the first mode.

observed along the slant during the present hydrogen bubble visualization tests, for the first flow mode is interesting and therefore is identified as an area requiring further PIV analyses in future research. The near-wake topology forming right behind the rear of the Ahmed body during this first mode is found to show a distinctive flow pattern irrespective of the flow state observed along the slant in both the hydrogen bubble visualization and PIV results (see Figure 5). The size of the low-speed region in the near wake was the smallest observed throughout this study. The upper recirculation bubble (A) was directly adjacent to the rear face of the model and both recirculation bubbles (upper A and lower B) were smaller in size. These features altogether result in the highest lift and drag (in the time-averaged sense) in the Ahmed body [1] for this mode (this has been confirmed by the authors during a separate experiment using a two-component force balance and the current models).

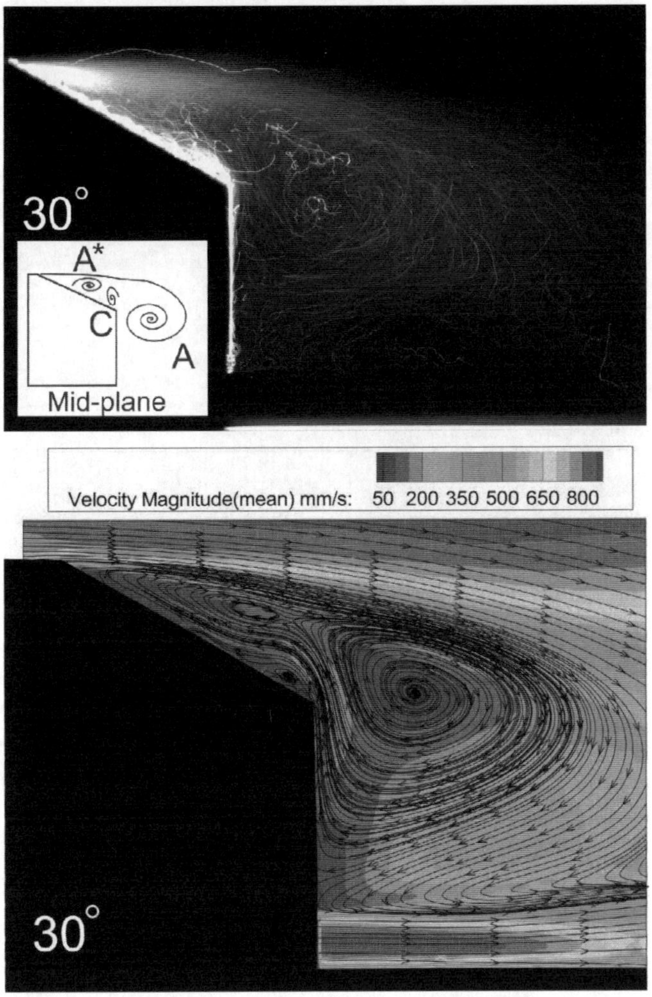

Figure 6: Instantaneous hydrogen bubble visualization (top) and time-averaged PIV (bottom) results in the mid-plane of the near-wake for the 30 degrees of back slant angle for the second flow mode (which depicts continual flow separation starting from the beginning of the slant).

During the second flow mode (shown in Figure 6), the size of the low-speed near wake region forming right behind the rear of the Ahmed body was significantly increased. The flow separated at all times at the beginning of the rear slant and a large recirculation bubble (A) dominated the near wake topology. The existence of a strong backflow along the slant and a small secondary recirculation bubble (C) is evident. The PIV result (bottom image in Figure 6) shows the existence of a smaller recirculation bubble (A*) with the same orientation as the large upper near wake bubble (A).

Instantaneous hydrogen bubble visualization images corresponding to each state in the plane coincident with the edge of the back slant are provided for the slant angle of 30 degrees in Figure 7. For the first flow mode, the formation of a longitudinal vortex along the back slant is evident (top image of Figure 7). This vortex is larger in size and displays a greater unsteadiness compared to that of the model with the 25 degrees slant angle (which was shown earlier in Figure 4). The angle of the vortex core is approximately 19.5 degrees. For the second flow mode, a longitudinal vortex is not observed (bottom of Figure 7).

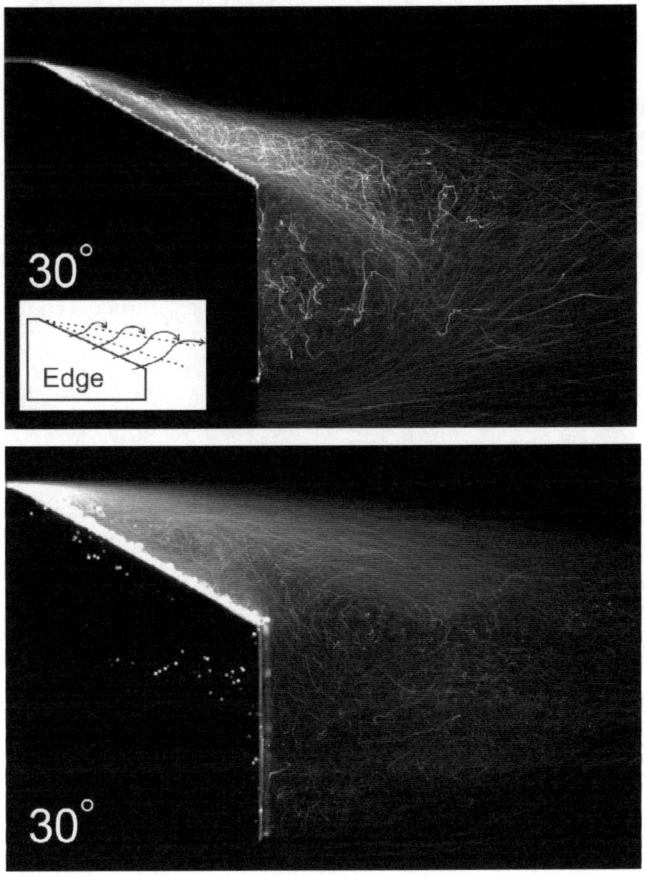

Figure 7: Instantaneous hydrogen bubble visualization images from the plane coincident with the edge of the back slant for the model with a back slant angle of 30 degrees. The top image corresponds to the first flow mode and the bottom image corresponds to the second flow mode.

35 DEGREES OF BACK SLANT ANGLE

An instantaneous hydrogen bubble visualization image and the time-averaged PIV result, depicting the contour patterns of velocity magnitude and streamline topology, in the plane coincident with the mid-plane of the Ahmed body with a back slant angle of 35 degrees are shown in Figure 8. The flow separates at the beginning of the slant and remains fully separated in the wake. As in the second mode of the 30 degrees slant angle where flow separation was observed starting from the beginning of the slant, the formation of a large recirculation bubble dominates the wake (A). The existence of a small counter-rotating recirculation bubble (B) was also observed (from the bottom image of Figure 8), whilst this was not observed for the slant angle of 30 degrees in the second mode. The two small recirculation structures observed along the slant for the 30 degree model in the second flow mode involving separated state (i.e., A* and C) are not found in the time-averaged PIV for 35 degrees, however the structure C was found to form sporadically in the movie of the hydrogen bubble visualization. Strong backflow along the slant is evident.

An instantaneous hydrogen bubble visualization image taken from the plane coincident with the edge of the back slant for the model with a back slant angle of 35 degrees is also shown in Figure 9. As in the separated state of the 30 degrees slant angle, there is no evident formation of a longitudinal vortex.

CONCLUSIONS

The near-wake flow topology of Ahmed bodies with back slant angles of 25, 30, and 35 degrees were investigated using the techniques of surface-wire hydrogen bubble visualization and Particle Image Velocimetry (PIV). The flow topology for the 25 degrees matched with that previously reported by Ahmed [1]. Two flow modes were observed for the "critical angle" of 30 degrees depending on the starting condition of the flow. The first mode depicted an unstable separation-reattachment phenomenon along the slant, which is identified as a subject for further systematic study. The second flow mode for the critical slant angle depicted a separated flow state starting from the beginning of the slant, which closely resembled the case observed for the 35 degrees slant angle. The dependency of the flow mode on the starting condition of the flow is a further area identified as a subject of future study.

ACKNOWLEDGMENTS

This research is supported by NSERC and Multimatic Inc. The authors would like to express gratitude towards Multimatic Inc. for their interest.

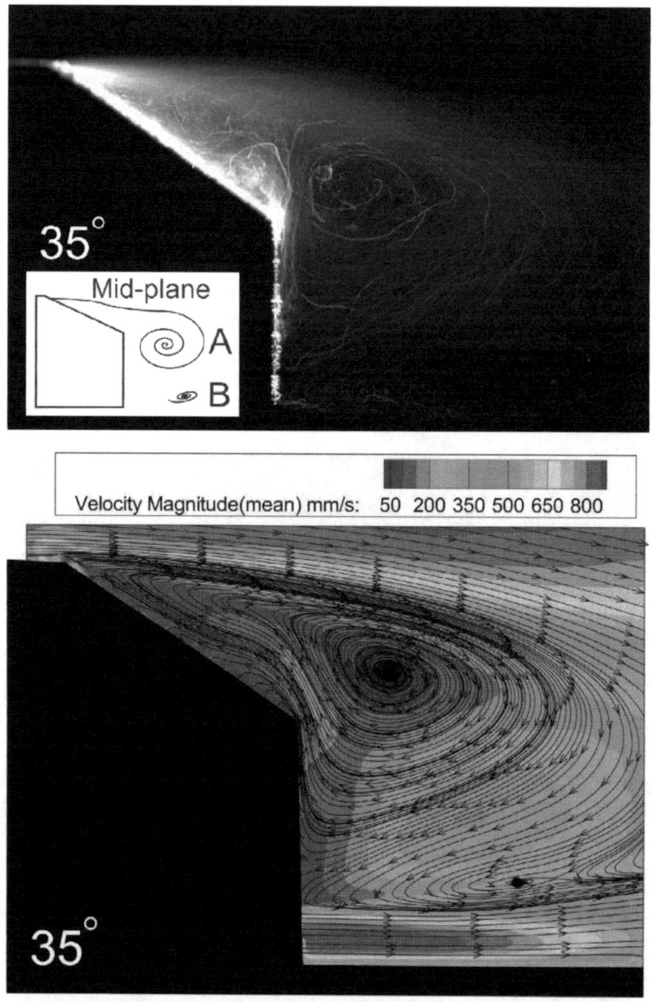

Figure 8: Instantaneous hydrogen bubble visualization (top) and time-averaged PIV (bottom) results in the mid-plane of the near-wake for the 35 degrees of back slant angle.

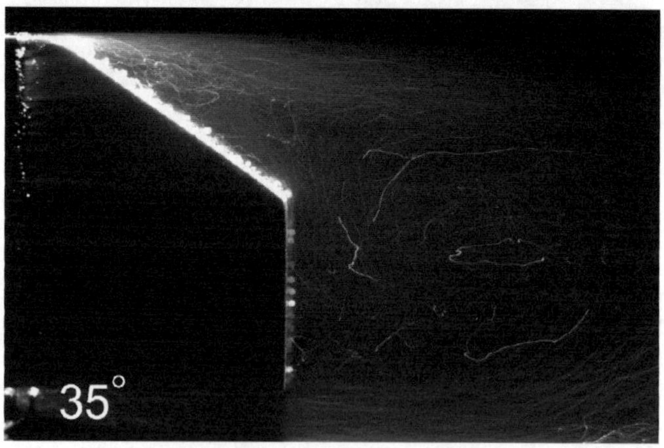

Figure 9: An instantaneous hydrogen bubble visualization image from the plane coincident with the edge of the back slant with a back slant angle of 35 degrees.

REFERENCES

1. Ahmed, S.R., Ramm, G., and Faltin, G. 1984, "Some Salient Features of the Time-Averaged Ground Vehicle Wake", *SAE Technical Paper Series*, 840300.

2. Vino, G., Watkins, S., Mousley, P., Watmuff, J., and Prasad, S. 2005, "Flow structures in the near-wake of the Ahmed model", *Journal of Fluids and Structures*, vol. 20, pp. 673-695.

3. Krajnovic, S., and Davidson, L. 2005, "Flow Around a Simplified Car, Part 1: Large Eddy Simulation", *Journal of Fluids Engineering*, vol. 127, pp. 907-918.

4. Krajnovic, S., and Davidson, L. 2005, "Flow Around a Simplified Car, Part 2: Understanding the Flow", *Journal of Fluids Engineering*, vol. 127, pp. 919-928.

5. Conan, B., Anthoine, J., and Planquart, P. 2011, "Experimental aerodynamic study of a car-type bluff body", Experiments in Fluids, vol. 50, pp. 1273-1284.

6. Wang, X., Zhou, Y., Pin, Y., and Chan, T. 2013, "Turbulent near wake of an Ahmed vehicle model", *Experiments in Fluids*, vol. 9, pp. 1490-1509.

7. Bruun, A., Wassen, E., Sperber, D., Nitsche, W., and Thiele, F. 2007, "Active Drag Control for a Generic Car Model", *Active Flow Control*, vol. 95, pp. 247-259.

8. Bruneau, C.H, Creuse, E., Depeyras, D., Gillieron, P., and Mortazavi, I. 2012, "Active and Passive Flow Control around Simplified Ground Vehicles", *Journal of Applied Fluid Mechanics*, vol. 5, pp. 89-93.

9. Feltham, G. 2013, "Flow topology around a simplified two-wheel landing gear with emphasis on the stagnation point," M.Sc. thesis, University of Toronto.

10. Hucho, W-H. H. 1998, *Aerodynamics of Road Vehicles 4ed.*, Society of Automotive Engineers Inc.,Warrendale.

VORTEX SHEDDING IN THE WAKES OF FINNED CYLINDERS

Jeff McClure
Department of Mechanical and Mechatronics Engineering
University of Waterloo
Waterloo, Ontario, Canada
jejmcclu@uwaterloo.ca

Serhiy Yarusevych
Department of Mechanical and Mechatronics Engineering
University of Waterloo
Waterloo, Ontario, Canada
syarus@uwaterloo.ca

ABSTRACT

Laser Doppler Velocimetry (LDV) and time-resolved, two-component Particle Image Velocimetry (PIV) measurements are performed in a water flume facility to investigate the flow development in the wake of uniform circular cylinders with circular fins. The effect of pitch ratio ($0.083 \leq c/D \leq 1.0$) is studied at a diameter ratio (D_f/D) of 2.0, and a fin thickness ratio (t/D) of 0.027. The tests are performed at a constant Reynolds number (Re_{Df}) of 3150, corresponding to the separated shear layer transition regime for uniform cylinders. The results indicate that von Karman vortex shedding occurs in the wake for all models. The boundary layer development on the fin's surface is shown to affect separation characteristics and, hence, vortex formation. A modified frequency scaling is proposed, which accounts for the added flow blockage created by the boundary layer growth on the fins.

KEYWORDS: Vortex shedding, bluff body aerodynamics, vortex dynamics, finned cylinders.

INTRODUCTION

Cylinders with fins are commonly encountered in heat exchanger devices due to their ability to enhance heat transfer characteristics[1]. The geometry of finned tube heat exchangers may differ significantly depending on the application. Critical to the design and performance of a heat exchanger are the mean and fluctuating loading on the tubes, as they relate to pressure losses and may lead to structural vibrations. The loading characteristics can be related to vortex shedding from finned tubes, which motivates the need to understand the unsteady vortex shedding phenomenon from such geometry.

The geometry of an isolated finned cylinder is shown in Figure 1. Cylinder with either solid or serrated fins affixed perpendicular or wound helically along the span have been shown to exhibit similar flow development over a wide range of fin thickness ratios (t/D), pitch ratios (c/D), diameter ratios (Df/D), and Reynolds numbers (Re_D)[2-6]. Previous studies report von Karman vortex shedding for the following practical range of parameters for finned tube heat exchangers: $0.0125 \leq t/D \leq 0.064$, $0.06 \leq c/D \leq 1.0$, $1.2 \leq Df/D \leq 2.0$, and $1.0 \times 10^3 \leq Re_D \leq 1.85 \times 10^5$. In some cases, more intense and coherent vortex shedding is observed from finned cylinders compared to that from a bare cylinder[3,4,6]. In contrast, for large fin thickness ($0.33 < t/D < 4$) and fin spacing within $2 < c/D < 9$, Nakamura and Igarashi[4] show that vortex shedding occurs in discrete cells between adjacent fin pairs.

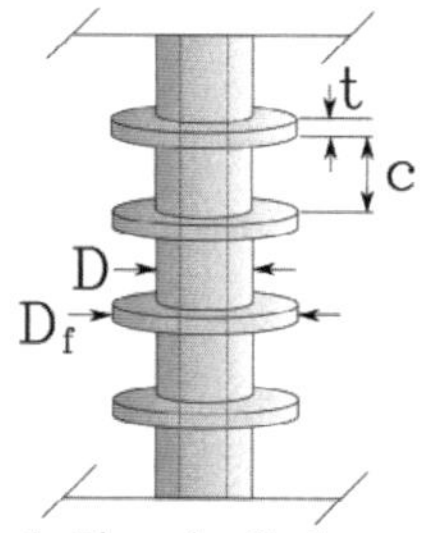

Figure 1: Finned cylinder geometry

Mair[1] introduced the definition of the effective diameter (Eq. 1) based on equating the volume of the finned cylinder to that of a uniform cylinder of diameter D_E. A number of previous studies on isolated cylinders of various fin geometry find that the Strouhal number based on the effective diameter remains relatively constant over a wide range of Re_D and geometrical parameters[2-4,6].

$$D_E = \frac{(c-t)D + tD_f}{c} \tag{1}$$

In contrast, Hamakawa et al.[5] suggest that, once a critical minimum c/D is reached for cylinders with serrated helical strakes, the Strouhal number decreases suddenly to a value expected for a uniform cylinder of diameter D_f rather than the effective diameter (D_E). In this case, flow separation occurs at the fin diameter, and it has been proposed to modify the definition of the effective diameter to $D_E = D_f$ for a specific pitch ratio and Reynolds number range ($c/D \leq 0.16$ and $76000 \leq Re_D \leq 190000$).

Previous studies have shown that the pitch ratio is the most critical parameter affecting the flow development over finned cylinders[2-4,6]. While vortex shedding was detected in the wake of a finned cylinder in most previous studies, the effect of fins on unsteady wake topology has not been investigated in detail, as most studies were confined to point measurements in the wake. The present work seeks to investigate spatio-temporal wake development of a finned cylinder.

EXPERIMENTAL METHODOLOGY

All experiments were performed in the water flume facility located at the University of Waterloo Fluid Mechanics Laboratory. A constant free-stream velocity of 86 *mm/s* was maintained in the test section. Three finned cylinder models of varying pitch ratio (c/D = 1.0, 0.33, 0.083) were investigated along with two uniform cylinder models whose diameters matched the base diameter and the fin diameter, corresponding to $Re_D = 1575$ and $Re_{Df} = 3150$, respectively.

Figure 2 shows the experimental configuration employed. Each model was affixed between 27 *cm* diameter endplates ($7D_f$). Time-resolved, planar Particle Image Velocimetry (PIV) measurements are obtained in horizontal and vertical planes using two high-speed 1024x1024 pixel Photron cameras and a high repetition rate Nd:YLF pulsed laser. The single frame images were acquired at 100 Hz, with 5457 images per set. The images were acquired and processed using Davis 8 software, using iterative multi-pass interrogation widows. For the final pass of image processing, 16x16 pixel interrogation windows were used with 75% overlap, yielding a vector pitch of $0.02D_f$. The flow was seeded with 10 μm diameter hollow glass spheres. Horizontal planar measurements (x-y planes) were obtained at the mid-span plane ($z = 0$) of each model, at the mid-pitch plane ($z = c/2D$) of each finned cylinder model. Vertical plane measurements (x-z planes) were obtained at a plane centred at half the effective diameter ($y = 0.5D_{eff}$). In both horizontal and vertical configurations, the two high speed cameras were offset from each other in the streamwise direction such that their image planes overlap by approximately 10%. Following image processing, the vector fields from each camera were joined based on the correlation of velocity vectors in the overlap region between the two cameras. Within the overlap region, a linear weighting function was employed.

Laser Doppler Velocimetry (LDV) measurements in the wake were acquired downstream of each model for spectral analysis. The measurement position was at x/D_{eff} = 0.5and y/D_{eff} = 5. The velocity data were resampled at 20 *Hz* using the sample-and-hold technique utilized by Adrian and Yao[6]. The minimum frequency resolution of the obtain velocity spectra was $0.002St_D$. Solid blockage of the models in the test section was less than 3%, and no blockage correction was applied to the measured shedding frequency.

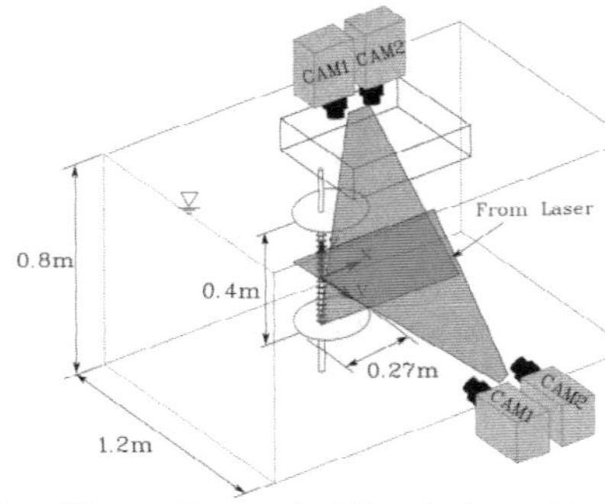

Figure 2: Water flume facility test section and PIV configuration

RESULTS

Analysis of the time-resolved PIV results indicated that von Karman vortex shedding occurs in the wake of each model investigated. However, notable changes in vortex shedding characteristics occur with changing fin pitch ratio. Figure 3 presents planar contours of mean streamwise velocity acquired at a mid-pitch plane, which illustrate that the addition of fins produces substantial changes in the extent of the vortex formation region. The vortex formation length is

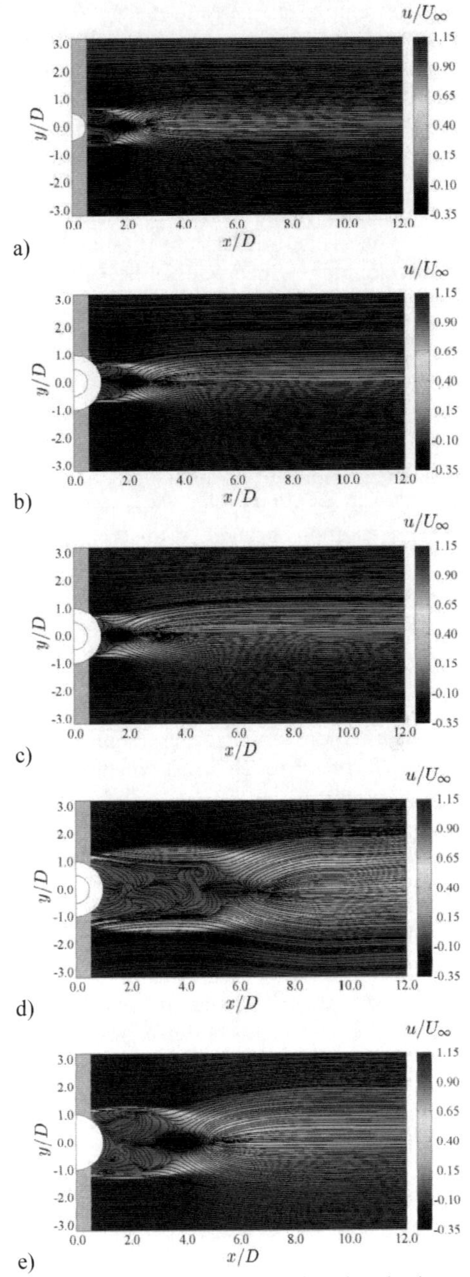

Figure 3: Mean flow fields and streamlines in a horizontal plane at the mid-span $z/D = 0$: a) uniform cylinder of D, b) $c/D = 1.0$, c) $c/D = 0.33$, d) $c/D = 0.083$, and e) uniform cylinder of D_f.

defined by the streamwise distance along the wake centerline to the peak in the streamwise velocity RMS, and the characteristic width of the vortex formation region is estimated using the transverse distance between two distinct peaks in streamwise velocity fluctuations which are pronounced at the end of the formation region. The trends in these two characteristic dimensions with pitch ratio are shown in Figure 4. For a uniform cylinder of base diameter (Fig. 3a), the roll-up of the mean streamlines and the downstream saddle point indicate the presence and extent of the recirculation region downstream of the cylinder. Adding fins with c/D = 1.0 (Fig. 3b) and c/D = 0.33 (Fig. 3c), produces a similar recirculation region to that of a uniform base cylinder, with a minor increase in both the formation length and wake width (Fig. 4). This indicates that flow surrounding the fins has a marginal effect on the flow development at the mid-pitch ($y = c/2D$) plane for sufficiently large pitch ratios. However, decreasing the pitch ratio to c/D = 0.083 (Fig. 3d) leads to drastic changes in the near wake development. Specifically, the vortex formation region enlarges significantly, extending up to x/D = 5. The formation length for this pitch ratio exceeds that for a uniform cylinder of fin diameter (Fig. 3e), while the transverse extent of the vortex formation region is approximately the same (Fig. 4). The observed significant changes in the scale of vortex formation region can be related to changes in the location of boundary layer separation. For pitch ratios of 1.0 and 0.33, the shear layer trajectories indicate that separation occurs from the base cylinder, leading to a similar sized formation region as that of a uniform cylinder of base diameter. In contrast, for a pitch ratio of 0.083, separation occurs at the fin diameter instead, leading to a significant increase in wake width (Fig. 4). However, while separation occurs at comparable positions for a pitch ratio of 0.083 and a uniform cylinder of fin diameter, the formation length differs significantly, which will be explored further through measurements in the vertical plane.

To study the effect fins on near wake development in more detail, velocity measurements were conducted in the vertical plane at y/D_{eff} = 0.5. The obtained mean streamwise velocity contours are presented in Fig. 5. For the larger fin pitch ratios, c/D = 1.0 (Fig. 5a) and c/D = 0.33 (Fig. 5b), low velocity streaks are introduced downstream of each fin due to the boundary layer growth on the fin surfaces. However, forc/D = 0.083 (Fig. 5c), a uniform region of reduced velocity is observed immediately downstream of the

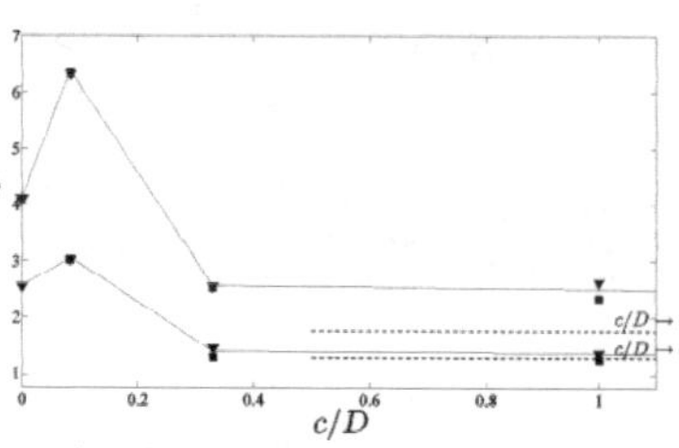

Figure 4: Formation length (L_f) and characteristic width of the formation region (w).

finned cylinder, similar to that for a uniform diameter cylinder (Fig. 5d). For such closely spaced fins, boundary layers growing between adjacent fins coalesce. This acts to displace the outer flow around the finned cylinder towards the outer fin diameter, leading to the observed changes in the flow separation. In this case, the finned cylinder essentially acts as a solid cylinder with bleed flow[8]. The bleed flow increases the base pressure[8], leading to a substantial enlargement of the vortex formation length compared to that expected for the solid model.

The Strouhal number data based on the effective diameter is plotted against the inverse pitch ratio (D/c) in Fig. 6a, along with the results of previous studies[2,3,6]. Overall, a poor collapse of the Strouhal number data with changing pitch ratio is observed. In particular, for the present study, there is a notable decrease of the Strouhal number with increasing the inverse pitch ration (i.e., with decreasing the fin pitch). This trend is attributed the boundary layer growth on the finned surfaces, which has been shown to significantly

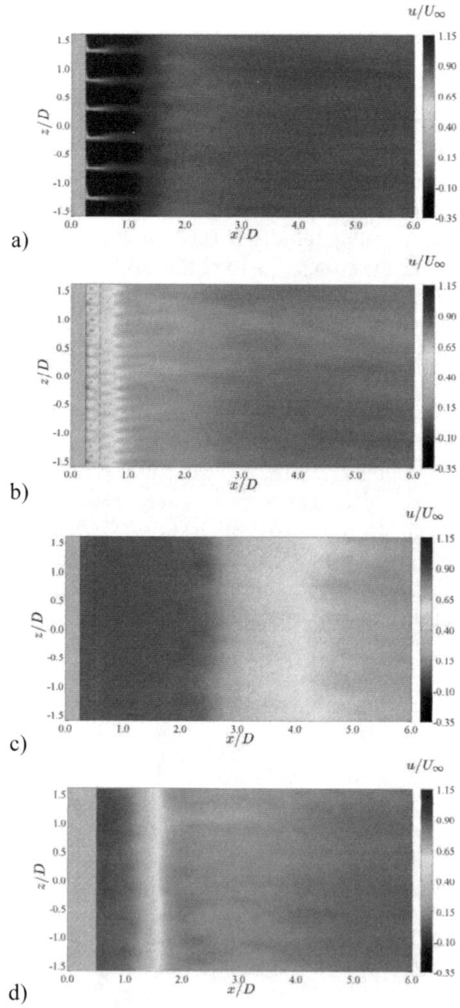

Figure 5: Mean streamwise velocity contours in a vertical plane at y/D_{eff} = 0.5 for a) c/D =1.0, b) c/D = 0.33, c) c/D = 0.083, and d) Uniform cylinder of D_f

affect the flow development for closely spaced fins (Figs. 3-4). While a similar decrease in Strouhal number is seen in other data, it is more pronounced for the present study due to the lower Reynolds number which produces thicker boundary layers on the fins. To account for the boundary layer growth on the fins, it is proposed to modify the effective diameter definition by introducing the effective fin thickness defined as $t^* = t + 2\delta^*$, where δ^* is the average boundary layer displacement thickness on the fin. The displacement thickness is estimated using Blasius' flat plate solution using fin parameters at $\frac{1}{2}D_f$. This modification increases the effective diameter, which accounts by the physical outer flow displacement due to the boundary layer growth. In the case where fin density is so high that the boundary layers coalesce ($\delta/c > 0.5$) the effective diameter is replaced with the fin diameter. This modified definition of the effective diameter (D_{eff}^*) now reflects the Reynolds number effect on flow development. The adjusted Strouhal data, is plotted against the boundary layer thickness normalized by fin pitch ratio in Fig. 6b, demonstrating a better collapse of the results and reduced scatter for all the data sets.

CONCLUSIONS

The effect of fin pitch ratio ($0.083 \leq c/D \leq 1.0$) on vortex development in the shear layer transition regime has been studied and compared to the flow development of uniform cylinders at $Re_D = 1575$ and $Re_{Df} = 3175$. For the conditions investigated, vortex shedding occurs in the wake of all models. The near wake development and the characteristics of the vortex formation region are not affected significantly by the presence of fins for high pitch ratios ($c/D \geq 0.33$). However, for the lowest pitch ratio 0.083, coalescence of the boundary layers on the fin surfaces drastically affect flow

a)

b)

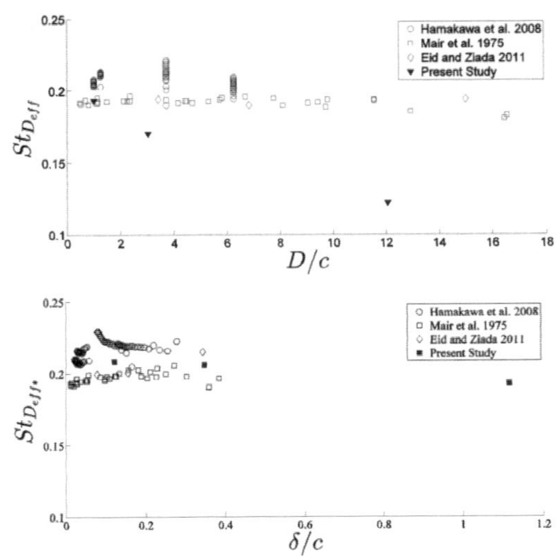

Figure 6: a) Strouhal number based on the effective diameter vs. inverse pitch ratio D/c, b) Strouhal number based on the modified effective diameter vs. δ/c.

development over the model and, hence, wake vortex formation region and shedding characteristics. In particular, the outer flow is directed around the outer surface of the fins away from the base cylinder leading to a significant enlargement of the vortex formation region. The wake development is found to be similar to that for a cylinder with a bleed flow, where flow between the fins increases the base pressure, leading to a significant enlargement in the vortex formation length. A modification of the effective diameter definition is proposed for the Strouhal number which accounts for the added flow blockage due to the fin boundary layer growth. The resultant Strouhal numbers exhibit significantly less scatter compared to those obtained based on the standard definition of the effective diameter.

ACKNOWLEDGMENTS

The financial support from the Natural Sciences and Engineering Research Council of Canada (NSERC) is gratefully acknowledged by the authors.

REFERENCES

1. Wang, C.C. and Chi, K.Y., 2000, "Heat transfer and friction characteristics of plain fin-and-tube heat exchangers, part I: new experimental data", *International Journal of Heat and Mass Transfer*, Vol(43), pp. 2681 – 2691.
2. Mair, W.A., Jones P.D.F, and Palmer R.K.W., 1975, "Vortex shedding from finned tubes", *Journal of Sound and Vibration*, Vol(39), pp. 293-296.
3. Eid, M. and Ziada, S., 2011, "Vortex shedding and acoustic resonance of single and tandem finned cylinders", *Journal of Fluids and Structures*, Vol(27), pp. 1035-1048.
4. Hamakawa, H. et al., 2008, "Vortex shedding from a circular cylinder with spiral fin", *Journal of Fluid Science and Technology*, Vol.(3), pp. 787-795.
5. Nakamura, H, and Igarashi, T., 2008, "Omnidirectional reductions in drag and fluctuating forces for a circular cylinder by attaching rings", *Journal of Wind Engineering and Industrial Aerodynamics*, Vol.(96), pp. 887-899.
6. Hamakawa, H., Fukano, T., Nishida, E., Aragaki, M., 2010, "Vortex shedding from a circular cylinder with fin", *AIAA Aeroacoustics Conference*, Maastricht, Paper AIAA-2001-2215.
7. Adrian R.J., Yao C.S., 1987, "Power spectra of fluid velocities measured by laser Doppler velocimetry", *Exp. Fluids,* Vol(5), pp.17–28.
8. Fransson, J.H.M, Konieczny, P., Alfredsson, P.H., 2004, "Flow around a porous cylinder subject to continuous suction or blowing", *Journal of Fluids and Structures*, Vol(19), pp. 1031-1048.

Proceedings of the 25th CANCAM
London, Ontario, Canada, May 31 - June 4, 2015

PROPAGATION OF FLUCTUATIONS AND FLOW SEPARATION ON AN UNSTEADILY LOADED AIRFOIL

Andrew S. Tenney
Mechanical and Aerospace Engineering
Syracuse University
Syracuse, NY, USA
astenney@syr.edu

Jacques Lewalle
Mechanical and Aerospace Engineering
Syracuse University
Syracuse, NY, USA
jlewalle@syr.edu

ABSTRACT

We analyze pressure data from 18 taps located along the surface of a DU-96-W180 airfoil in unsteady flow conditions. The conditions were set to mimic the flow conditions experienced by a wind turbine blade under unsteady loading to test and to quantify the effects of several flow control schemes. Here we are interested in the propagation of fluctuations along the pressure and suction sides, particularly in relation to the fluctuating separation point. An unsteady phase of the incoming fluctuations is defined using Morlet wavelets, and phase-conditioned cross-correlations are calculated. Using wavelet-based pattern recognition, individual events in the pressure data are identified with several different algorithms utilizing both the original time series pressure signals and their corresponding scalograms.

Keywords: wavelets, separation, turbulence, unsteady flow

INTRODUCTION

In the past decade, the global capacity for producing energy via wind power has increased tenfold, and is expected to double within the next five years. This growth has sparked a need for more efficient and robust turbine systems that are capable of producing energy in less than ideal conditions. In the case of large scale wind farms, only the first row of turbines are subjected to quasi-steady atmospheric flow unobstructed by upstream structures. The subsequent turbines are affected by the unsteadiness generated by this first row of turbines, and consequently suffer a performance drop.

In this study, we quantify the effects of unsteadiness on

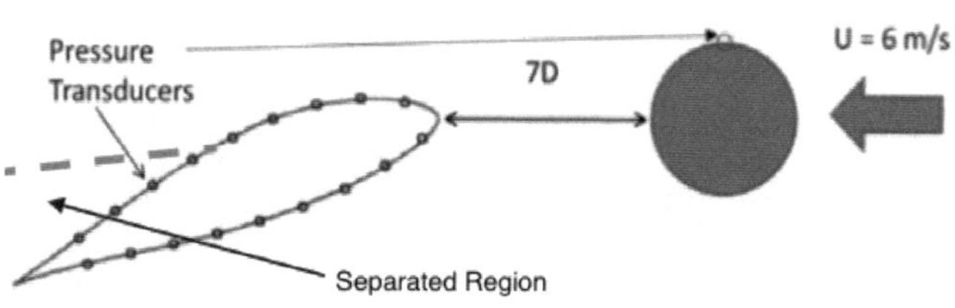

Figure 1: Experimental Setup

a common airfoil used in the design of turbine blades with the goal of testing and comparing several active flow control schemes. To attain the desired test conditions, a cylinder of diameter 203mm was placed 7 diameters upstream of a DU-96-W180 airfoil of chord length 242mm (Figure 1). The Reynolds number based on the cylinder diameter was 78,000, leading to a fully turbulent cylinder wake and a peak shedding Strouhal number of approximately 0.2 (corresponding to a dominant shedding frequency of 7Hz). [4]

METHODS: STATISTICAL

Time series pressure signals from 18 taps on the airfoil surface and 1 on the cylinder surface were analyzed. A continuous Morlet wavelet transform was utilized to represent the signals in time-frequency space. Mathematically, this transform takes the form

$$\tilde{u}_M = \int u(t')\Psi^*_M((t-t'),\omega)dt' \qquad (1)$$

where Ψ_M is the scaled Morlet wavelet [3]. Figure 2 details the two dimensional wavelet map or "scalogram" of a sample

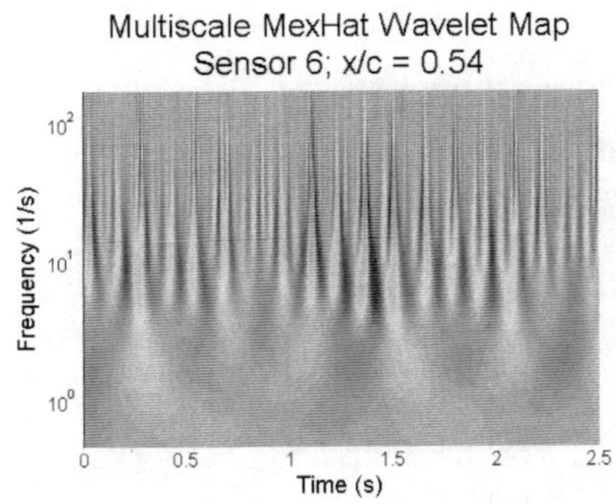

Figure 2: Wavelet map of time series pressure signal; Sensor 6; x/c = 0.54; Separated region

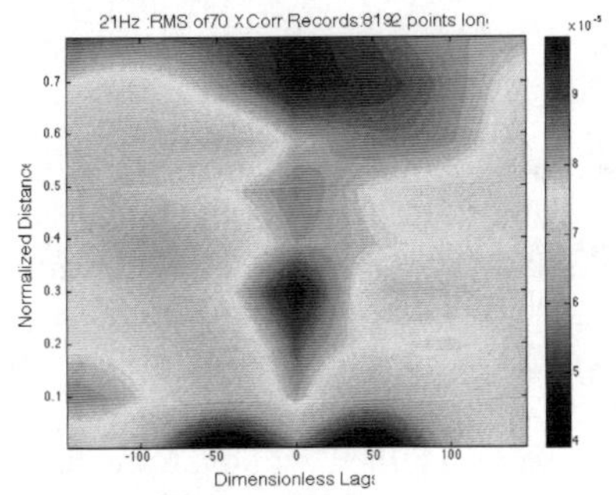

Figure 3: RMS of short-sample correlation maps showing variability in correlation maps as a function of time lag and distance along the airfoil surface (28Hz harmonic, 8192 point excerpts, 70 excerpts)

pressure signal.

We studied the effect of sample length on the convergence of the correlation maps to our accepted statistical mean values (based on using the entire 64000 point signals). The individually captured time series were divided into excerpts of 2^n (n = 12,13,14) points, and correlation coefficients of the band-pass filtered signal (fundamental frequency and 4 harmonics) from the leading edge and the downstream sensors were calculated. A three dimensional array of correlation values was then constructed using each correlation map as a new layer in the third dimension. The RMS of the resulting 1D vector at every lag and spatial location was computed and plotted (Figure 3). The result is a graphical representation of the variability in the correlation maps at all time lags and sensor locations. The power spectra of the band-pass filtered signals were also averaged over many signals to ensure statistical convergence.

The square norm of the two dimensional Morlet scalogram of a time series signal is a representation of the signal's energy distribution in the time-frequency domain in terms of the envelope of wave packets[5]. Taking a slice of this scalogram at any particular frequency represents the modulation of the signal at that frequency. We studied these modulations at the fundamental shedding frequency of 7 Hz, as well as 4 harmonics and two higer frequencies related to the turbulent scales in the separated region. These modulations were then treated as

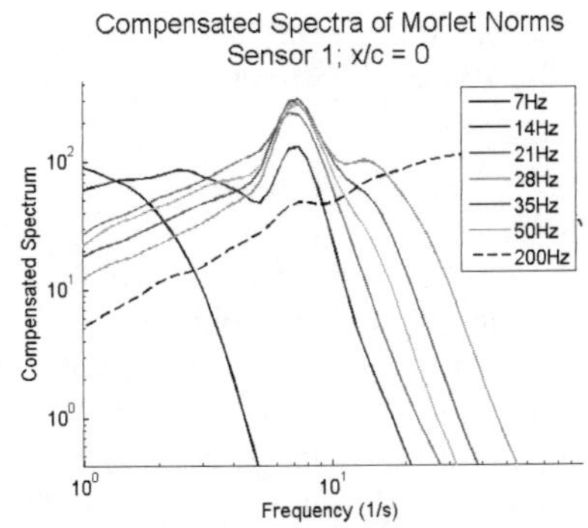

Figure 4: Compensated spectrum of the norm of 7Hz and its harmonics wavelet coefficients; sensor 1 along airfoil surface

individual time series signals and their power spectra were calculated and plotted to quantify similarities between the modulations (Figure 4). The peak in the power spectra shared by the last 3 harmonics confirms the qualitatively noted similarities in the modulations, and motivates an analysis through use of cross correlations.

Each harmonic in the signal captured by the sensor at the leading edge is correlated with its counterparts captured by the downstream transducers. Both positive and negative time lags are utilized to explore the possibility of upstream propagation of information along the airfoil surface (Figure 5).

METHODS: EVENT LEVEL

We identify individual structures in the flow by looking for their distinctive signatures propagating in the flow from one pressure sensor to the next. This is accomplished through the use of a pattern recognition algorithm originally developed to analyze axisymmetric jet flow. [6][7]

This algorithm most recently developed algorithm is the Time-Frequency-Lag or TFL algorithm. The two signals being compared are transofrmed using continuous wavelets, and are correlated at each frequency of interst over a series of plausible time lags relating to the timescales present in the flow. A Gaussian filter is then applied to these correlation vectors. The filtered vectors are then scanned for portions that meet a predetermined condition (one-fifth of the maximum vector magnitude in this case), which are then tagged a possible events. These lists are then sorted to eliminate events that have been counted more than once, or are clearly unphysical. [6][7]

This algorithm is used to identify events within signals captured simultaneously by consecutive pressure sensors. The resulting lists can then be binned according to lag and frequency to identify trends in the propagation of information at frequencies of interest. These trends give insight into the mechanismis behind the upstream and downstream propagation identified by the statistical methods.

Individual events, or wavepackets, can be clipped from the signals captured at each pressure tap, based on their time of occurance and the frequency of the events. These events can then be tracked as they propagate in the flow, while simultaneously moitoring changes in the amplitude and shape of the event as it travels from sensor to sensor.

RESULTS

The scalogram in Figure 2 shows a band of activity around the fundamental frequency of 7Hz. The pressure signal used to create this scalogram was collected by the transducer at x/c = 0.54 in the separated region. The activity in the 7Hz band highlights the dominance of this fundamental frequency even in the presence of turbulence. The intermittency at higher frequencies is characteristic of the pressure in turbulent flows.

Figure 5 is a graphical representation of the variability in the 2D correlation maps. The value at each point in the plane represents the RMS value of the 1D vector formed by the difference of 70 individual correlation maps from the statistical mean stacked in a 3D array. The smaller the RMS value at any point, the smaller the fluctuations in the correlation maps from the accepted mean at that lag and time. The large spike at a normalized distance of 0.3 indicates that the correlation maps tend to vary significantly from the mean value. This greater than average difference in signals being compared gives us a good estimate for the separation point, as the pressure difference across the shear layer is drastic and we would expect this greater level of variation.

Performing cross correlation calculations on simultaneously captured pressure signals allows us to see how fluctuations travel long the airfoil on a case by case basis. By averaging a large number of correlation maps, the dominant trends in the transmission of information along the airfoil surface can be observed and interpreted. The leftward leaning slope of the ridge denoting maximum correlation values in Figure 4 points to the possibility that information is traveling upstream in the region ahead of the separation point. The clear disjunction between the leftward leaning ridge and the rightward leaning ridge starting at a normalized distance of 0.3 is indicative of an abrupt change in flow conditions (ie. the point of flow separation). The identification of a chordwise location of the separation point allows us to define a specific set of sensors on which to focus our event level search while trying to identify the source of separation point fluctuations.

In analyzing individual cross correlations between sensors, it was determined that there were unexpected gaps within the 2 dimensional correlation maps at certain frequencies. These extended gaps indicate a phase difference between the two signals being compared of approximately $\frac{\pi}{2}$ rad or a zero phase angle for both signals for uncharacteristically long periods. These phase dislocations are in direct contrast with the observed regular oscillatory nature of the two signals being compared. One explanation for this in the case of sensors in the area where separation begins is the drastic change in the nature of the signals themselves. The turbulence within the separated region leads to less coherence and greater intermittency in the time series signals. In the region just upstream where we expect separation to occur, a short period of time where there is a phase discrepancy between two signals could signify the movement of the separation point to include the downstream

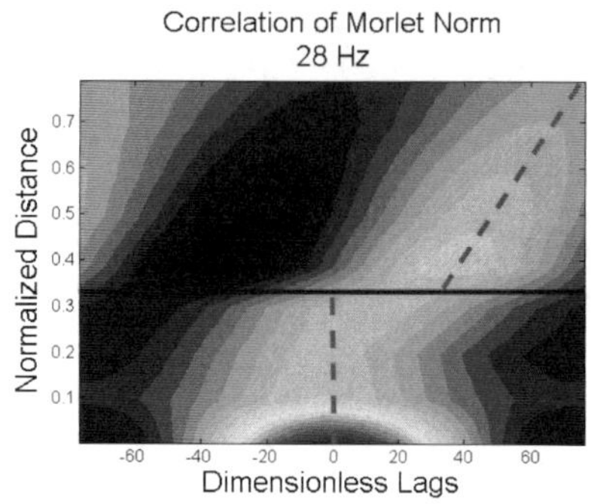

Figure 5: Averaged and normalized correlation map of Morlet norms. Red dashed lines represent ridges of maximum correlation, black line is approximate location of disjunction. (28Hz Harmonic)

sensor used in the cross correlation calculations. Another explanation of these phase dislocations could also be the irregularity of large scale structures travelling along the airfoil surfaces. For example, shedded vortices may not always travel along the same surface, resulting in phase gaps in the time series signals.

Indiviual events were identified within the pressure signals using the TFL algortihm (see METHODS: EVENT LEVEL). To verify that the majority of coherent events were occuring at the 7Hz fundamental and its harmonics, the pressure signals were band pass filtered between 6Hz and 40Hz and the TFL algorithm was applied. Upon this analysis, it was discovered that a strong majority of events do lie within the 7Hz and harmonic grequency bands, allowing us to transform the pressure signals using only those bands to speed up computation time.

Events were identified over all samples in the pressure data, and binned according to their lag and frequency. These bins were then normalized by their related frequency to give a more realistic picture of relative event counts. Figure 6 shows the distribution of events identified between sensor 2 and sensor 3 on the suction side of the airfoil where the flow is always attached. The events identified in the 7Hz band in this region are centered between 3 and 4 timsesteps, which is the approximate convective timescale for this flow. The event populaions in the higher frequency harmonics are associated with lags closer to zero, both positive and negative. The negative time lags are associated with upstream propagation of information, and the smaller time lags indicate a mechanism of propagation that is not explained by convection.

Individual events matched between several sensors can be clipped and compared to one another to monitor changes in wavepacket shape and amplitude as it travels from sensor to sensor. Figure 7 details an event captured in sensors 2, 3, and 4 as well as their propagation between these sensors. This event is one of many captured propagating upstream in the flow. As Figure 7 illustrates, the osciallations of the wavepacket may change in amplitude and shape somewhat as the event propagates, but the overall structure of the wavepacket as a whole is maintained. By continuing to study these wavepackets on

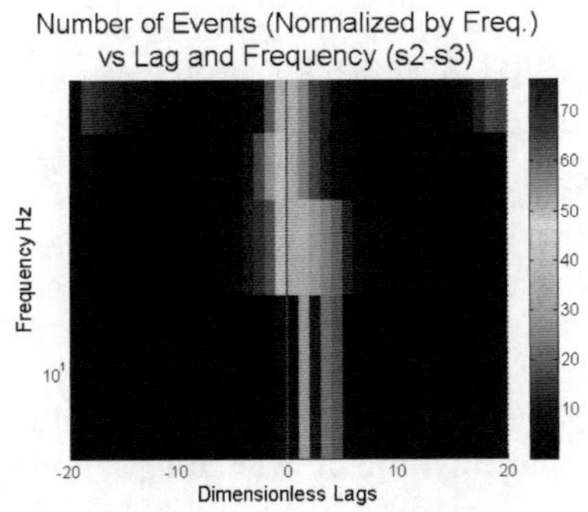

Figure 6: Populations of events identified vs. time and frequency (only 7hz fundamental and harmoncs up tp 35Hz; Normalized by frequency)

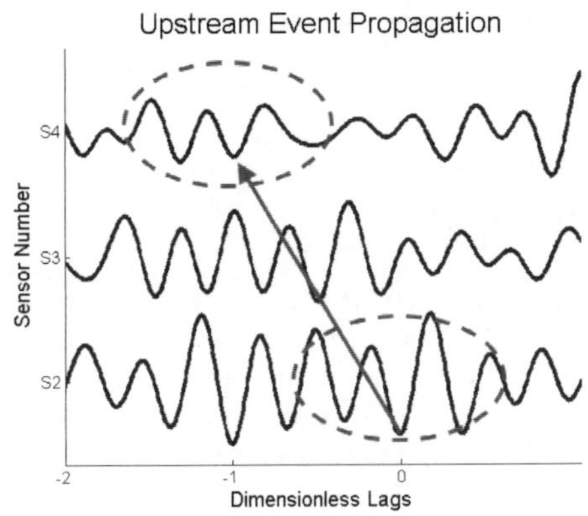

Figure 7: Propagation of individual event between sensors 2, 3, and 4.

an individual level, we may be able to better understand the nature of their formation, as well as the effects they have on the fluctuation of the separation point.

CONCLUSIONS

The increase in correlation coefficients at negative time lags suggests that information is being propagated upstream in the flow. This trend is observed at all harmonics of the dominant 7Hz frequency, and will be studied in more detail in future research. In addition, the high RMS values at the start of the separated region (see Fig. 3) correspond to a separation point which varies in time. The mechanisms of these fluctuations include upstream wave propagation in the shear layer, as well as convection of structures along the pressure surface of the airfoil and around the trailing edge. The fluctuation in separation point location will also introduce intermittency in the pressure signals captured by adjacent stations.

The event level analysis provides further insight into the nature of the flow itself, as well as some of the mechanisms for information transmission within the flow. At the fundamental frequency of 7Hz, convection along th airfoil surface is the main mode for the propagation of structures and information, while at the higher frequency harmonics, the timescales associated with the largest popuations of events indicate some other mode of transmission. In further analysis on the propagation of individual events, it may be possible to determine their origins, as well as the mechanisms that drive thier propagation upstream and at velocities not associated with convection. Going forward, PIV data captured in concurrence with the pressure data will be utilized to identify the cause of the fluctuating separation point and give more insight into how the 7Hz fundamental and its harmonics are affecting this process.

ACKNOWLEDGMENTS

The work of Dr. Guannan Wang and Zhe Bai are acknowledged as the basis and inspiration for this current work.

REFERENCES

[1] Bai, Zhe; Weekly Reports, 2013-2014

[2] Lewalle, J., Low, K. R., Glauser, M. N., Berger, Z. P., Correlations of Near- and Far-Field Acoustic Data in a High-Speed Jet, Seventh International Symposium On Turbulence and Shear Flow Phenomena (TSFP-7), 2011

[3] Lewalle, Jacques; Continuous and Orthogonal Wavelets, http://lcs3.syr.edu/faculty/lewalle/wavelets/wavelets.htm; accessed Oct. 2014

[4] Wang, Guannan; Experimental Investigation of the Active Flow Control on a Two-Dimensional Wind Turbine Airfoil, Syracuse University SURFACE, 2013

[5] Wang, G., Lewalle, J., Glauser, M., and Walczak, J., Investigation of the Benefits of Unsteady Blowing Actuation on a 2D Wind Turbine Blade, Journal of Turbulence, Volume: 14, Issue: 01, pages 165 - 189

[6] Kan, P., Lewalle, J., Gogineni, S.;Mach-number dependence of acoustic source properties in high speed jets - Part I: ensemble statistics of acoustically active regions, AIAA SciTech, National Harbor, MD, 2014

[7] Kan, P., Lewalle, J., Gogineni, S.;Mach-number dependence of acoustic source properties in high speed jets - Part II: event-based description, AIAA SciTech, National Harbor, MD, 2014

Proceedings of the 25th CANCAM
London, Ontario, Canada, May 31 – June 4, 2015

FLOW-TOPOLOGY AROUND A CYLINDER SUBMERGED IN DIFFERENT WIDTH WAKE

Md. Mahbub Alam[1,2]
[1]Institute for Turbulence-Noise-Vibration Interaction and Control, Shenzhen Graduate School, Harbin Institute of Technology, Shenzhen 518055, China
[2]Key Kab of Advanced Manufacturing and Technology, Shenzhen Graduate School, Harbin Institute of Technology, Shenzhen 518055, China
Email: alamm28@yahoo.com

Longjun Wang
Institute for Turbulence-Noise-Vibration Interaction and Control, Shenzhen Graduate School, Harbin Institute of Technology, Shenzhen 518055, China

Yu Zhou
Institute for Turbulence-Noise-Vibration Interaction and Control, Shenzhen Graduate School, Harbin Institute of Technology, Shenzhen 518055, China

ABSTRACT

Fluid dynamics around a circular cylinder (diameter D) placed in the wake of another of smaller diameter d is examined, including time-mean drag coefficient (C_D), fluctuating drag and lift coefficients (C_D' and C_L'), Strouhal number (St, based on D) and flow structures at the spacing ratio L/d = 1.0 ~ 8.0, where L is the distance from the wake generating (upstream) cylinder center to the forward stagnation point of the monitoring (downstream) cylinder. While d is varied as 8, 16, 24, 32 and 40 mm, D is fixed at 40 mm, corresponding to a diameter ratio d/D ranging from 0.2 to 1.0. The Reynolds number is kept constant at 4.27×10^4 based on D. A remarkable influence of the upstream cylinder diameter on C_D, C_D', C_L', St and flow structure around the downstream cylinder is observed. For each d/D, an increase in L/d leads to a transition from reattachment to coshedding flows. The critical spacing dividing the reattachment and coshedding flows appears at a larger L/d at smaller d/D. C_D, C_D' and C_L' at the critical spacing leap for d/D =1.0 - 0.4, but decline for d/D = 0.2. In the reattachment regime St(s) of the two cylinders are identical, regardless of d/D. On the other hand, in the coshedding regime the St(s) are identical for d/D = 1.0, 0.8 and 0.6 only, following the upstream cylinder St, in addition to a subharmonic lock-in for d/D = 0.6; the St(s) are however different for d/D = 0.4 and 0.2, the St of the upstream cylinder being much higher than that of the downstream cylinder. C_D in general increases with d/D. C_D' and C_L', on the other hand, generally wane and grow as d/D decreases from 1.0 to 0.6 and 0.4 to 0.2, respectively. While the former phenomenon is dominantly influenced by impaired vortices/shear-layer associated with smaller d/D, the latter by increased flow velocity in the gap.

KEYWORDS: Flow-induced forces, Strouhal number, tandem cylinders, bistable flow, wake

INTRODUCTION

Slender structures appear in groups in many engineering applications. Naturally, some structures in a group are submerged in the wake of the others. Two inline cylinders may be considered as the basic element of multiple structures, where the leeward cylinder is in the wake of the windward cylinder (Fig. 1). The flow around two tandem cylinders of identical diameters is in general classified into three major regimes [1]: (i) the overshoot regime ($0.5 < L/d < 1.0$; d is the upstream cylinder diameter, and L is the distance between the center of the upstream cylinder and the leading stagnation point of the downstream cylinder); (ii) the reattachment regime ($1.0 < L/d < 3.5$); (iii) the coshedding regime ($L/d >$ 3.5). There is a transition L/d range between the reattachment and coshedding regimes, where both reattachment and coshedding flows appear intermittently, switching from one to the other. The transition L/d is called critical or bistable flow spacing. Furthermore, the reattachment regime can further be divided into a number of flows depending on the reattachment position [2]. At the coshedding regime, it is well established that the two cylinders shed vortices separately at the same frequency [2-4]. For a given downstream cylinder, different cross sections of the upstream cylinder with the same characteristic width result in a different frequency of vortex shedding from the upstream cylinder, and the frequency of vortex shedding from the downstream cylinder modifies accordingly to lock-in, adjusting itself to that of the upstream one [5-7]. On the other hand, for a given upstream cylinder size, a change in the cross-section of the downstream cylinder does not influence the shedding frequency of the upstream cylinder [6]. It was concluded that the hydrodynamic stability of the flow around two tandem cylinders is predominantly controlled by the upstream cylinder, not by the downstream one.

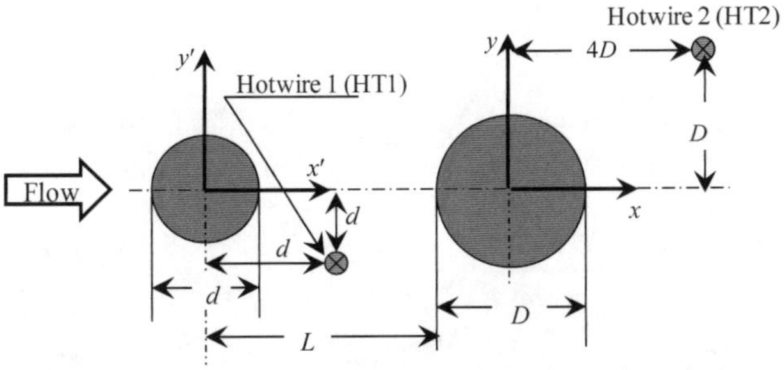

Fig. 1. Experimental setup and definition of symbols.

The reported studies were concerned with two cylinders of an identical diameter. There has not been extensive research on C_D, $C_D{}'$ and $C_L{}'$, St and flow around two cylinders of different diameters. The structures in a group or close proximity are not always of the same diameters. For example, a piggyback pipeline normally comprises of one (or more) small pipe (or pipes) and a large pipe. The small and the large pipes could either be in contact or separate, depending on design and installation requirements. The application demands engineers to know the answer of the following questions. What is the effect of the upstream cylinder diameter on C_D, $C_D{}'$, $C_L{}'$, St and wake of the downstream cylinder? How are these parameters dependent on L/d? What are the characteristics of fluid dynamics, as compared with those associated with two identical cylinders?

This work aims to examine the upstream cylinder diameter effect on C_D, $C_D{}'$, $C_L{}'$, St , critical spacing, and wake of the downstream cylinder at L/d = 1.0 – 8.0, covering all possible flow regimes. The upstream cylinder diameter (d) is varied, with the downstream cylinder diameter (D) unchanged, so that the ratio d/D varies from 0.2 to 1.0.

EXPERIMENTAL SETUP

Experiments were performed in a closed-circuit wind tunnel with a 5.5 m long test section of 0.8 m × 1.0 m. The downstream cylinder diameter D (= 40 mm) was kept fixed, and the upstream cylinder diameter d was changed as d = 8, 16, 24, 32 and 40 mm, resulting in a diameter ratio d/D = 0.2, 0.4, 0.6, 0.8, and 1.0, respectively. The spacing ratio L/d between the cylinders was systematically varied from 1.0 to 8.0. The freestream velocity U_∞ was 16 m/s, corresponding to $Re = 4.27\times10^4$ based on D. The turbulence intensity of the uniform flow was 0.45%. The maximum blockage is 4% and the aspect ratio of the cylinder is 20.

Fluid forces acting on the downstream cylinder are measured using a sectional load cell (65 mm long) installed at the midsection of the cylinder. Therefore, C_D, $C_D{}'$, and $C_L{}'$ are estimated from the load cell output. Fig.1 shows schematically the arrangement of the cylinders and definitions of coordinates (x', y') and (x, y) with the origins defined at the upstream and downstream cylinder centers, respectively. Two single tungsten hotwires, placed in the gap between the cylinders and behind the downstream cylinder, respectively, were used to measure the longitudinal velocity fluctuation u from which vortex shedding frequencies were extracted using FFT. Smoke-flow visualization technique was used to investigate the flow structure in the gap between two cylinders and behind the downstream cylinder. Pressure measurement on the periphery of the downstream cylinder also carried out, with 32 pressure taps connected to a pressure scanner.

WAKE AND STROUHAL NUMBER

The effect of d/D on flow around the around the downstream cylinder is shown in Figure 2 where the left column displays the photographs corresponding to reattachment flow and the right column to the coshedding flow. Clearly the upstream cylinder shear layers at d/D = 1.0 reattach on the downstream cylinder at L/d = 2.5 (Fig. 2a1) but roll up in the gap between the cylinders at L/d = 3.0 (Fig. 2a2). That is, the critical spacing lies between L/d = 2.5 and 3.0. Similarly, that at d/D = 0.8 and 0.6 nestles between L/d = 3.0 and 3.5 (Fig. 2b-c). d/D = 0.4 and 0.2, nevertheless, correspond to the critical spacing at 4.0 < L/d < 4.5 and 6.0 < L/d < 7.0, respectively (Fig. 2d-e). What is interesting here is that with a decrease in d/D from 1.0 to 0.2 the upstream cylinder wake width shrinks gradually; the shrinking is larger in the coshedding regime than the reattachment regime.

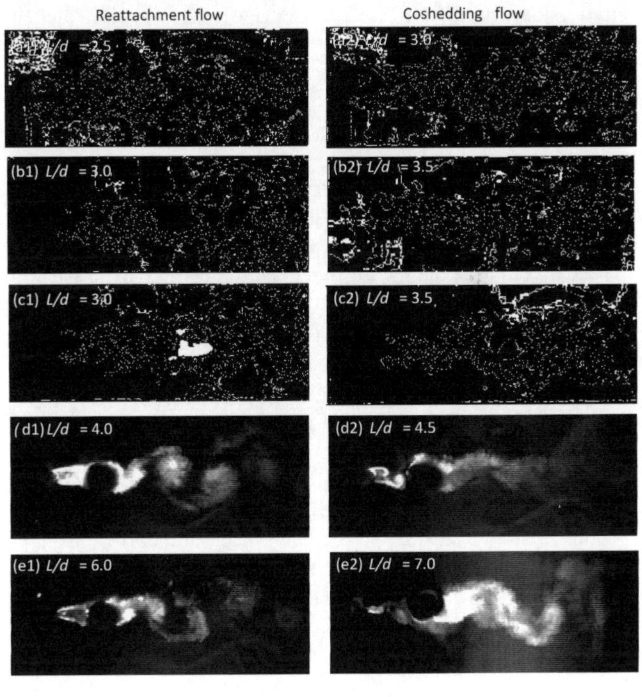

Fig. 2. Visualized flow structures at (a) d/D = 1.0, (b) d/D = 0.8, (c) d/D = 0.6, (d) d/D = 0.4, and (e) d/D = 0.2. Left- and right-column pictures represent reattachment and coshedding flows, respectively.

The transmutation of flow with d/D and L/d could be further discussed with the aid of power spectral density functions (Fig. 3) of fluctuating velocities obtained in the gap

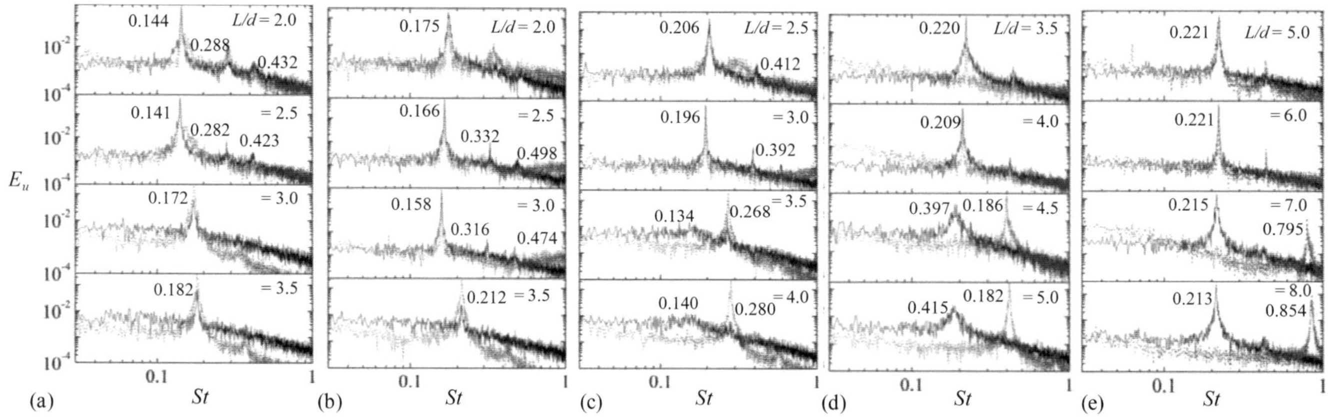

Fig. 3. Power spectra of hotwire (HT1 and HT2) signals: (a) $d/D = 1.0$, (b) $d/D = 0.8$, (c) $d/D = 0.6$, (d) $d/D = 0.4$, and (e) $d/D = 0.2$, at different L/d. Dotted line, HT1; solid line, HT2.

between the cylinders (HT1) and behind the downstream cylinder (HT2). St ($= f_v D/U_\infty$, f_v is the vortex shedding frequency) jumps from 0.141 to 0.172 between $L/d = 2.5$ and 3.0 for $d/D = 1.0$ and from 0.158 to 0.212 between $L/d = 3.0$ and 3.5 for $d/D = 0.8$. Apparently, the jump is due to a drastic modification of reattachment flow to coshedding flow when L/d is increased. While the higher St at the former d/D is smaller than the St (= 0.197) of an isolated cylinder, that at the latter is higher. In the coshedding flow (i.e., $L/d \geq 3.0$) for $d/D = 1.0$, albeit shedding vortices individually, the two cylinders have an identical St, slightly smaller than the isolated cylinder's. This is an established phenomenon for the same diameter tandem cylinders that the convective vortices from the upstream cylinder trigger the shedding from the downstream cylinder, making both St identical [6]. On the other hand, though having different diameters at $d/D = 0.8$, the two cylinders have identical St again, due to a lock-in phenomenon. One may arise a question, why is this St larger? Indeed, now the upstream cylinder diameter is small, $d = 0.8D$. The shedding frequency from the upstream cylinder is thus larger, given the same approaching flow velocity. St based on d can be calculated as $0.212 \times 0.8 = 0.170$ that is almost the same as that at $d/D = 1.0$. The downstream cylinder displays two St (say, 0.134 and 0.268 at $L/d = 3.5$) at $d/D = 0.6$ in the coshedding regime, where the upstream cylinder $St = 0.268$, suggesting both fundamental and subharmonic locks-in. At $d/D = 0.4$ and 0.2, though St(s) of two cylinders are identical in the reattachment regime, they are different in the coshedding regime. The fundamental St (upstream cylinder, downstream cylinder) = (0.268, 0.134) at $L/d = 3.5$ for $d/D = 0.6$, (0.397, 0.186) at $L/d = 4.5$ for $d/D = 0.4$, and (0.795, 0.215) at $L/d = 7.0$ for $d/D = 0.2$. Both St augment with d/D. While the augmentation of the upstream cylinder St with a decrease in d/D is due to a decrease in diameter, that of the downstream cylinder is caused by a larger flow velocity in the gap between the cylinders. When the upstream cylinder $St = 0.268$ at (d/D, L/d) = (0.6, 3.5), 0.397 at (d/D, L/d) = (0.4, 4.5), and 0.795 at (d/D, L/d) = (0.2, 7.0) are normalized by d instead of D, they become 0.161, 0.159, and 0.159, respectively, comparable to those at $d/D = 0.8$ and 1.0.

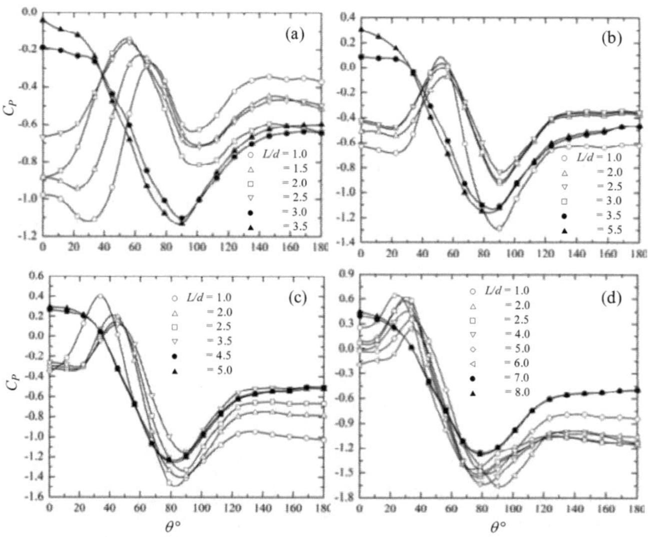

Fig. 4. Distribution of time-averaged pressure coefficient C_P along the surface of the cylinder: (a) $d/D = 1.0$, (b) $d/D = 0.6$, (c) $d/D = 0.4$, (d) $d/D = 0.2$.

CYLINDER SURFACE PRESSURE

Time-mean pressure coefficient C_p distributions around the surface of the cylinder at various L/d and d/D are shown in Fig. 4. C_p displays a peak at $\theta = 20°$-$70°$ depending on d/D for the reattachment flow regime and at $\theta = 0°$ for the coshedding flow regime. The peak at the reattachment flow regime represents the shear layer reattachment. C_p at the whole surface in the reattachment regime (open symbols) is negative for $d/D \geq 0.6$, even at the reattachment peak, while that for other d/D is positive at and near the reattachment. At a given L/d, C_p at the reattachment increases with a decrease in d/D. At the same time, the reattachment position moves toward the forward stagnation point. In the coshedding regime as well (solid symbols), a smaller d/D accompanies a larger C_p at the forward stagnation point, which further insinuates that the smaller the *d/D*, the greater the flow velocity in the gap between the cylinders. It is worth noting that when a reattachment flow modifies to a coshedding flow, C_P on the base of the cylinder decreases significantly for *d/D* = 1.0-0.6, declines a little for *d/D* = 0.4 and increases considerably for the other *d/D*.

FLOW-INDUCED FORCE

C_D, C_D' and C_L' variations with *L/d* shown in Fig. 5(a-c) reflect that they are strong functions of *L/d* and *d/D*, especially before the critical spacing. For *d/D* = 1.0 and 0.8, as *L/d* increases from 1.0 to 2.0, C_D, C_D' and C_L' all escalate. At the same *L/d* range, *d/D* = 0.6 and 0.2 have the opposite behaviors; C_D, C_D' and C_L' all drop. C_D, C_D' and C_L' decrease slightly for *d/D* = 1.0 and 0.8 between *L/d* = 2.0 and 2.5. It is remarkable that at the critical spacing C_D, C_D' and C_L' all jump for *d/D* = 1.0 - 0.4, but drop for *d/D* = 0.2. C_D in general enhances with a decrease in *d/D* except at *d/D* = 1.0, *L/d* = 2.0. C_D' and C_L', on the other hand, generally decrease with *d/D* decreasing from 1.0 to 0.6 and increase with *d/D* decreasing from 0.4 to 0.2. While the former occurrence is mostly dictated by vortices/shear-layer weakening with the decrease in *d/D*, the latter by a larger flow velocity in the gap. A flow map showing reattachment, bistable and coshedding flows is given in Fig. 5(d). The critical spacing associated with the possible bistable flow extends with a decrease in *d/D*.

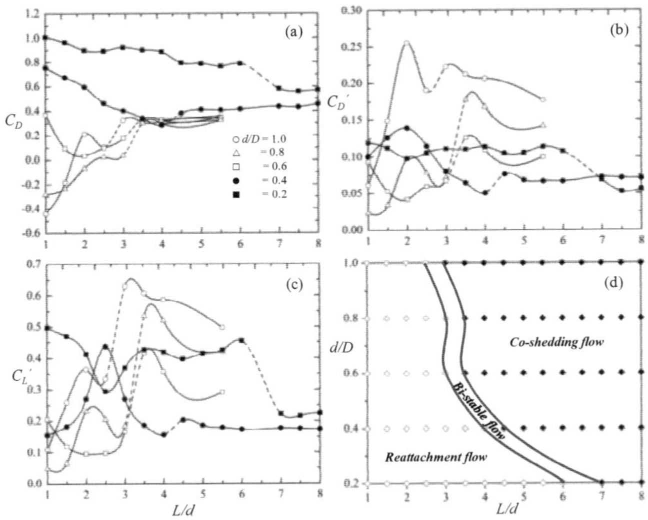

Fig. 5. Dependence on *L/d* of (a) C_D, (b) C_D', and (c) C_L'. (d) Flow map in *L/d* - *d/D* plane.

CONCLUSIONS

The main results of this study may be summarized as follows. *St* is identical for both cylinders in the reattachment regime regardless of *d/D*. In the coshedding regime, the downstream cylinder shedding frequency locks-in to the upstream cylinder for *d/D* = 1.0, 0.8 and 0.6. At *d/D* = 0.6, a lock-in of the downstream cylinder shedding with the subharmonic of the upstream cylinder shedding is also noticed. For the other two *d/D*, the upstream cylinder shedding frequency is much higher than the downstream cylinder; lock-in is not observed. The critical *L/d* dividing the reattachment and coshedding flows is larger at smaller *d/D*, nestling at 2.5 < *L/d* < 3.0, 3.0 < *L/d* < 3.5, 4.0 < *L/d* < 4.5, and 6.0 < *L/d* < 7.0 for *d/D* =1.0, 0.8-0.6, 0.4 and 0.2, respectively. At these critical spacings, while C_D, C_D' and C_L' leap for *d/D* =1.0 - 0.4, they fall for *d/D* = 0.2. C_D, C_D' and C_L' are very sensitive to *L/d* in the reattachment regime. C_D in general increases with *d/D* except at *d/D* = 1.0, *L/d* =2.0. C_D' and C_L' generally diminish and climb with *d/D* decreasing from 1.0 to 0.6 and 0.4 to 0.2, respectively. While the former behavior is mostly connected to vortices/shear-layer weakening with the decrease in *d/D*, the latter to a larger flow velocity in the gap.

ACKNOWLEDGMENTS

Alam wishes to acknowledge supports given to him from the Research Grant Council of Shenzhen Government through grants JCYJ20120613145300404 and JCYJ20130402100505796.

REFERENCES

[1] M.M., Zdravkovich, 1987, "The effects of interference between circular cylinders in cross flow", *Journal of Fluids and Structures*, Vol. 1, pp. 239–261.

[2] M.M., Alam, 2014, "The aerodynamics of a cylinder submerged in the wake of another", *Journal of Fluids and Structures*, Vol. 51, pp. 393-400.

[3] M.M., Alam, M., Moriya, K., Takai, H., Sakamoto, 2003, "Fluctuating fluid forces acting on two circular cylinders in a tandem arrangement at a subcritical Reynolds number", *Journal of Wind Engineering and Industrial Aerodynamics*, Vol. 91, pp. 139–154.

[4] S., Ishigai, E., Nishikawa, E., Nishimura, K., Cho, 1972, "Experimental study of structure of gas flow in tube banks axes normal to flow", *Bull. Japan Society of Mechanical Engineering*, Vol. 15, pp. 949–956.

[5] M.M., Alam, Y., Zhou, 2007, "Phase lag between vortex sheddings from two tandem bluff bodies", *Journal of Fluids and Structures*, Vol. 23, pp. 339–347.

[6] M.M., Alam, H., Sakamoto, Y., Zhou, 2005, "Determination of flow configurations and fluid forces acting on two staggered circular cylinders of equal diameter in cross-flow", *Journal of Fluids and Structures*, Vol. 21, pp. 363–394.

[7] M.M., Alam, Y., Zhou, 2008, "Strouhal numbers, forces and flow structures around two tandem cylinders of different diameters", *Journal of Fluids and Structures*, Vol. 24, pp. 505-526.

Biological Fluid Mechanics

THE EFFECT OF SINUS GEOMETRY ON THE TURBULENCE DOWNSTREAM OF A BILEAFLET HEART VALVE

Laura Haya
Mechanical Engineering
University of Ottawa
Ottawa, Ontario, Canada
Laura.Haya@uottawa.ca

Stavros Tavoularis
Mechanical Engineering
University of Ottawa
Ottawa, Ontario, Canada
Stavros.Tavoularis@uottawa.ca

ABSTRACT

The levels of velocity fluctuation and Reynolds stresses generated by a bileaflet mechanical heart valve inserted in a straight, axisymmetric channel and in an anatomical model of the aorta, and operating at physiological flow conditions, were measured using particle image velocimetry. In both channels, the strongest turbulence occurred at the interface between the lateral jets and the recirculating sinus flow, but the level of turbulence was significantly greater in the axisymmetric model. This was largely attributed to the straight walls, high height-to-width ratio, and sudden contraction of the axisymmetric sinus. It was concluded that the shape of the anatomical aortic sinus was effective in reducing turbulence.

KEYWORDS: Bileaflet mechanical heart valve, particle image velocimetry, Reynolds stress, turbulence, aorta.

INTRODUCTION

Bileaflet mechanical heart valves (BMHVs) are routinely implanted to replace diseased native aortic valves. Flow patterns behind BMHVs are non-physiological and are believed to be responsible for the observed high incidence of thrombosis and blood cell damage. Pathological thrombosis is initiated by platelets that have been damaged by high levels of viscous shear stress at the cellular level (Ge *et al.*, 2008), which in turn have been correlated strongly with high levels of turbulence (Antiga and Steinman, 2009).

The flow characteristics downstream of BMHVs have been studied extensively, both experimentally and numerically. In most of these studies, the valve was embedded in channels with drastically simplified geometries, rather than in models of physiological vessels, which are difficult to replicate experimentally and computationally expensive to simulate *in silico*. Several studies considered a BMHV inserted in a straight tube, with or without an axisymmetric expansion at its root to represent the Valsalva sinuses. Recent studies have shown that the characteristics of flows past BMHVs depend strongly on the geometrical features of the vessels in which they have been embedded (Grigioni *et al.* 2005; Borazjani *et al.* 2010). Therefore, it is of interest to investigate the flow fields past BMHVs in vessels that model as realistically as possible the human aorta and compare them to results obtained in simplified geometries.

In this work, the velocity fields past a BMHV mounted within an axisymmetric channel and an anatomical model of the aorta were mapped using particle image velocimetry. Measures of turbulence in the two fields, including velocity fluctuation levels and maximum Reynolds stresses, were compared.

APPARATUS AND EXPERIMENTAL PROCEDURES

The measurements were conducted in a mock circulation loop that closely reproduced physiological pressure and flow conditions in the aorta (Fig. 1). A 21 mm St. Jude Medical Regent™ BMHV (Fig. 2), having an outer diameter D = 22.6 mm, was mounted between an upstream test section and a downstream one, both cast from clear silicone. The upstream section consisted of a cylindrical channel, 25.4 mm in diameter and 140 mm long. The downstream section was either an axisymmetric channel or an anatomical model of the aorta (Fig. 3). The inlet of the axisymmetric channel consisted of a cylindrical sudden expansion, 31.8 mm in diameter, which formed a "pseudo sinus", 20.3 mm long. This geometry is similar to those used in previous studies (Dasi *et al.*, 2007; Yun *et al.*, 2014). The anatomical model (Elastrat, Geneva, Switzerland) comprised the geometry of the triple sinus at the aortic root, the coronary arteries, and the aortic arch with its branching arteries. The results reported in this article were obtained with all branching arteries clamped.

The valve was oriented with its line of symmetry transecting the non-coronary sinus and the left- and right-coronary sinuses situated on either side of the valve's lateral orifices. The coordinate origin was at the valve centre, at the downstream edge of the valve body, the z-axis was normal to the valve plane and positive in the downstream direction, and the x- and y- axes ran perpendicular and parallel to the valve's plane of symmetry, respectively. The aortic arch curved toward the positive x-direction, which corresponds to the anatomical left side. The working fluid was a mixture of water (47.4% by weight), glycerin (39.9%) and sodium iodide (15.7%). Its kinematic viscosity was 3.65 cSt, which approximates that of blood at physiological shear rates, whereas its refractive index matched that of silicone.

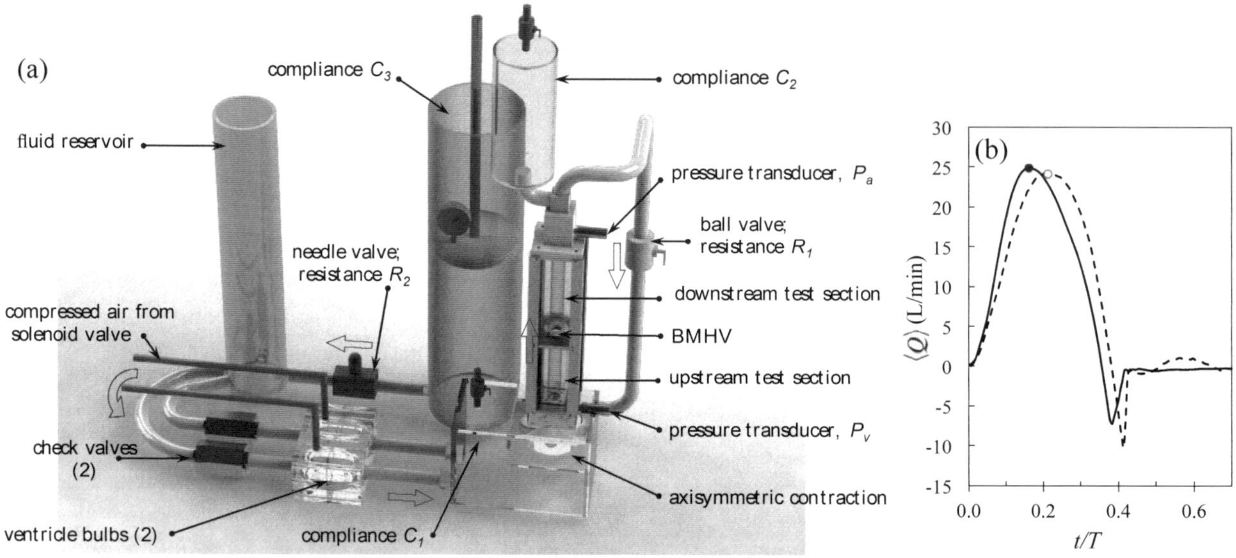

Figure 1: (a) Mock circulation loop; (b) bulk flow rate $\langle Q \rangle$ for measurements in the axisymmetric model (solid line) and the anatomical model (dashed line). Symbols mark the peak flow rates.

Figure 2: St. Jude Regent™ valve (BMHV).

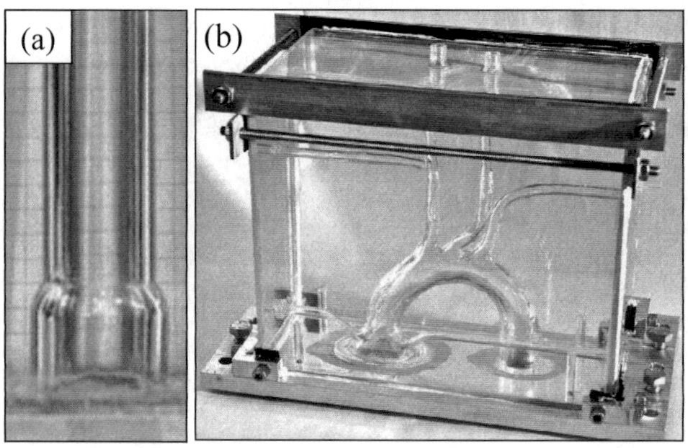

Figure 3: Axisymmetric (a) and anatomical (b) test sections.

The loop was operated at a frequency of 70.6 bpm (period T=850 ms), a bulk flow rate of 5.3 L/min and a peak flow rate of 24.9 L/min (Fig 1(b)). The pressure measured downstream of the BMHV was 75 mm Hg at the onset of forward flow (diastolic pressure) and 115 mm Hg at peak flow (systolic pressure).

Phase-locked planar PIV measurements were made downstream of the valve in the x-z and y-z centre-planes, illuminated by a double-pulsed laser sheet from a dual cavity Nd:YAG laser (120 Solo XT; ESI, NWR Division Sunnyvale, CA, USA). 1000 instantaneous maps were recorded at each of several key phases of the cycle, but only measurements at the peak flow phase are presented here.

Velocity vectors were calculated by applying cross-correlation algorithms on image pairs (Davis 8.1.4, LaVision). The vector resolution was 105 μm for the axi-symmetric section and 270 μm for the anatomical one.

The planar Reynolds stress tensor was calculated as

$$RS = \rho \begin{vmatrix} \langle u^2 \rangle & \langle uw \rangle \\ \langle uw \rangle & \langle w^2 \rangle \end{vmatrix}, \quad (1)$$

where ρ is the fluid density, u and w are the velocity fluctuations in the x- and z-directions, and angle brackets indicate phase-averages. The 2D coordinate-invariant maximum Reynolds stress was calculated as

$$RS_{max} = \frac{1}{2}(\sigma_{max} - \sigma_{min}), \quad (2)$$

where σ_{max} and σ_{min} are the maximum and minimum principal stresses of the Reynolds stress tensor (Ge *et al.*, 2007). Finally, a characteristic magnitude of velocity fluctuations on the x-z plane was calculated as

$$q' = \sqrt{\langle u^2 \rangle + \langle w^2 \rangle}\ . \quad (3)$$

RESULTS

Profiles of $q'$and the phase-averaged axial velocity $\langle W \rangle$ in the two test sections, in a direction perpendicular to the valve's symmetry plane at z/D=0.50 and 1.00 and during peak flow are shown in Fig. 4. The mean velocity profiles were not significantly different in the two channels, with the exception of an apparent shift of the edge of the jet towards the inner curve of the aortic arch.

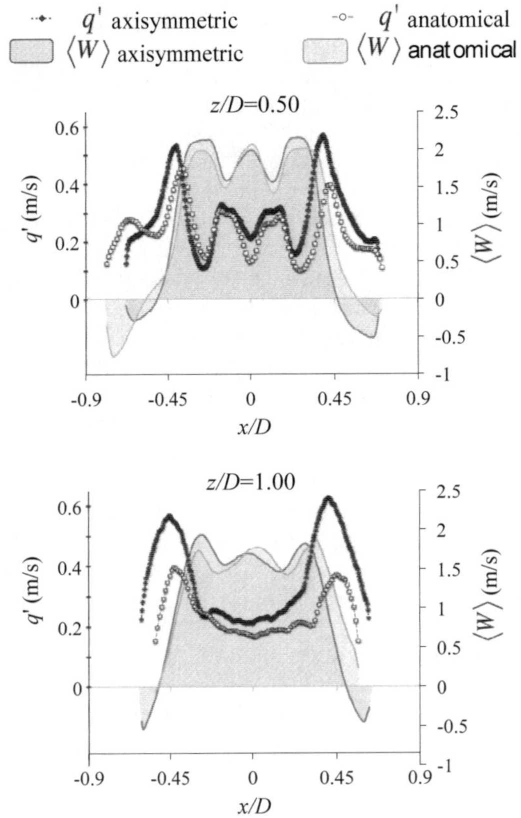

Figure 4: Velocity fluctuations q' at peak flow in the *x-z* centre-plane at $z/D = 0.5$ (top) and $z/D = 1.00$ (bottom).

In both cases, the velocity fluctuations were highest at locations where the mean velocity gradient was largest, most notably, at the interface between the lateral jets and the recirculation zones within the sinuses. In general, the velocity fluctuation levels were lower in the anatomical than the axisymmetric test section and this difference tended to be larger for the downstream profiles, for which the peak q' were roughly 50% higher in the axisymmetric channel.

Maps of the Reynolds stress metric RS_{max} for the two geometries are presented in Fig 5. The results on the *x-z* centre-plane are consistent with the velocity fluctuation profiles shown in Fig 4. In both geometries, the Reynolds stress was largest at the outside edges of the lateral jets and reached maxima near the channel contraction, at $z/D \approx 1.00$. In these jet edges, the shear stress level was much greater in the axisymmetric model and the difference grew with increasing distance from the valve, but, in the core of the channel, the levels of Reynolds stress were comparable for the two geometries at all downstream distances. In the *y-z* planes, the Reynolds stresses in the two channels were generally comparable, but also had the following minor differences. In the axisymmetric model, a region of moderately high stress occurred at the channel contraction, especially on the positive *y* side. The corresponding region

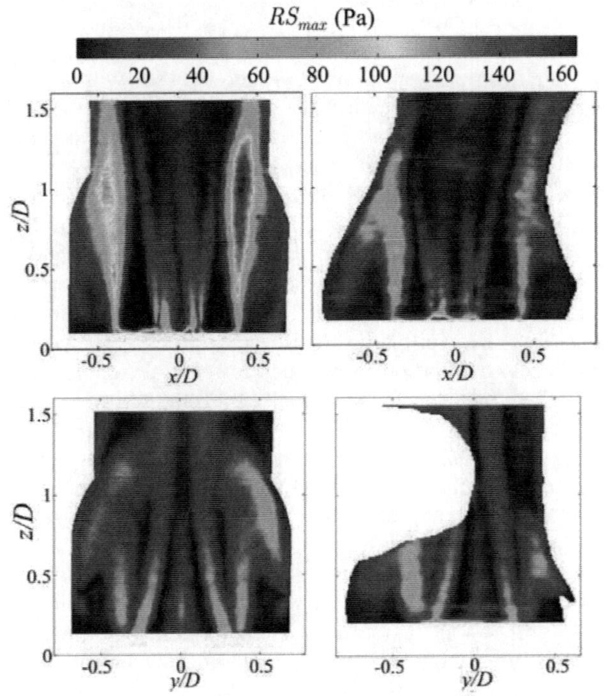

Figure 5: Maps of Reynolds shear stress RS_{max} in the *x-z* centre-planes (top) and the *y-z* centre-planes (bottom) in the axisymmetric (left) and anatomical (right) channels.

in the anatomical model was not as large, and exhibited lower levels of stress, likely because the sinus contracted more gradually in this case. In both geometries, regions of moderately high stress developed between the central jet and the recirculation zone in the sinus. In the anatomical model, the stress was higher than in the axisymmetric model on the negative *y* side of the jet, which was adjacent to the non-coronary sinus, but nearly imperceptible on the positive *y* side, where there was no sinus. The recirculating flow in the non-coronary sinus was much greater in the anatomical model compared with the axisymmetric one, which is the likely cause of the increased stress in this location.

DISCUSSION AND CONCLUSIONS

The previous results show that the turbulence produced by the valve in both the axisymmetric and the anatomical channels reached its highest levels at the interfaces of the lateral jets with the sinus recirculation zones, but these levels, as well as the corresponding global maximum, were significantly higher in the axisymmetric channel than in the anatomical one. As operating conditions were effectively the same for both cases, the differences in turbulence may be attributed largely to differences in the shapes of the two channels, particularly the fact that the axisymmetric sinus contracted rather abruptly, whereas the anatomical sinus contracted much more gradually. Sample instantaneous planar "streamlines" are shown in Fig. 6 for the two sinus geometries, in each case together with a "streamline" from phase-averaged velocities. In the axisymmetric sinus, the recirculating fluid was forced to rotate by 180° with a very short radius; bounded by the straight sinus wall, the downward flow was forced to align abruptly with the vertical direction, and to continue in this direction for the entire length of the sinus. On the contrary, the anatomical sinus permitted the fluid within it to rotate more gradually; at the top of the sinus, the recirculating fluid was forced to rotate by only about 135°, whereas, at the bottom of the sinus, where the fluid completed its rotation, the turning radius was relatively large. As a result of these geometrical differences, the vorticity was much lower in the anatomical sinus than the straight-walled sinus of the axisymmetric model.

The instantaneous planar streamlines presented in Fig. 6 reveal that, during peak flow in the axisymmetric sinus, the recirculation region contained several relatively small vortices, some of which approached the bottom of the sinus. Inspection of such streamlines at different phases of the cycle demonstrated that at the onset of flow acceleration, a single vortex was first formed at the bottom of the sinus, and was advected downstream, stretching towards the end of the sinus as the fluid continued to accelerate. Subsequently and under intense stretching in the narrow sinus, this vortex lost its coherence and broke into numerous smaller vortices, thus also generating strong local turbulence. In the anatomical model, however, sequences of instantaneous fields revealed that the sinus recirculation retained its coherence, even at peak flow; this is attributed partly to the gradual reorientation of the flow at the sinus top and partly to the relatively low height-to-width ratio of the anatomical sinus.

Compared to the straight, axisymmetric sinus, the anatomical sinus permitted a greater lateral spread of jets emanating from the valve. This moderated the rate of rotation of the fluid in the recirculation region that formed in the sinus, and also reduced the velocity gradient at the interface between the jet and the recirculation region; both

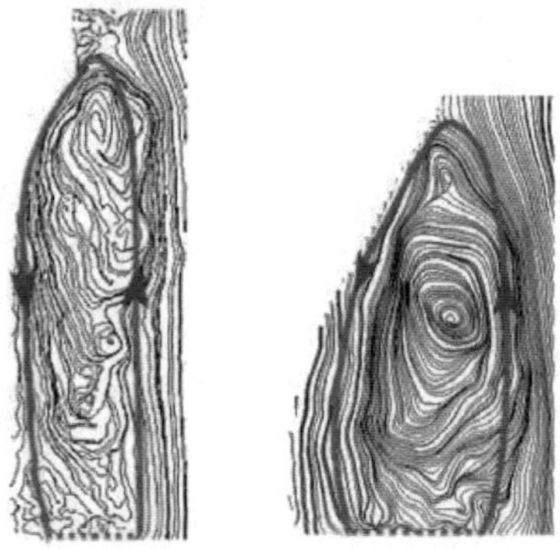

(a) (b)

Figure 6: Sample instantaneous planar "streamlines" in (a) the axisymmetric sinus and (b) the anatomical sinus. Red lines indicate "streamlines" of the phase-averaged velocity.

effects resulted in keeping turbulence at relatively low levels, when compared to those in the straight sinus. As suggested by Borazjani *et al.* (2010), secondary flow structures caused by the aorta curvature may have further contributed to the lateral spread of the jets by impeding their downstream progression.

The present results suggest that the testing of BMHVs in axisymmetric models of the aorta having a sudden expansion to represent the sinuses would produce turbulent stresses that would be significantly higher than those in anatomical models. Because the use of accurate values of local turbulent and viscous stresses is essential in the assessment of mechanisms of thrombosis and blood damage, this finding emphasizes the importance of accurate geometric modeling in bio-fluid dynamic studies.

An additional finding is that a sinus cavity with a relatively low height-to-width ratio and gradual geometrical changes engendered lower turbulence than a cavity with a high height-to-width ratio and abrupt changes in geometry. This finding can be applied to the design and implantation siting of prosthetic devices, as well as to the surgical reconstruction of the aortic root.

ACKNOWLEDGMENTS

The bileaflet valve was kindly donated by St. Jude Medical. Financial support was provided by the Natural Sciences and Engineering Research Council of Canada.

REFERENCES

1. Antiga, L. and Steinman, D. A., 2009, "Rethinking turbulence in blood", *Biorheology,* 46 (2), pp. 77–81.

2. Borazjani, I., Ge, L., and Sotiropoulos, F. 2010, "High-resolution fluid–structure interaction simulations of flow through a bi-leaflet mechanical heart valve in an anatomic aorta", *Ann Biomed Eng*, 38(2), pp. 326-344.

3. Dasi, L. P., Ge, L., Simon, H. A., Sotiropoulous, F., and Yoganathan, A. 2007, "Vorticity dynamics of a bileaflet mechanical heart valve in an axisymmetric aorta", *Phys Fluids,* 19(6), 067105.

4. Ge, L., Dasi, L. P., Sotiropoulous, F., Yoganathan, A., 2008, "Characterization of hemodynamic forces induced by mechanical heart valves: Reynolds *vs.* viscous stresses", *Ann Biomed Eng*, 36(2), pp. 276-97.

5. Grigioni, M., Daniele, C., Del Gaudio, C., Morbiducci, U., Balducci, A., D'Avenio, G., and Barbaro, V., 2005, "Three-dimensional numeric simulation of flow through an aortic bileaflet valve in a realistic model of aortic root", *ASAIO*, 51(3), pp. 176-183.

6. Yun, B. M., Dasi, L.P., Aidun, C.K., and Yoganathan, A.P., 2014, "Highly resolved pulsatile flows through prosthetic heart valves using the entropic lattice-Boltzmann method", *J Fluid Mech,* 754, pp. 122–160.

PIV measurement of jets through model mitral valve with different suture configurations

Anna Chtchetinina, Lyes Kadem and Hoi Dick Ng†
Department of Mechanical and Industrial Engineering
Concordia University
Montréal, Québec, Canada
†hoing@encs.concordia.ca

ABSTRACT

This paper presents an experimental study of jets emitted from flexible orifices of different configurations aimed at understanding the effects of a repair procedure for a mal-functioning mitral valve, known as edge-to-edge repair. This study takes the simple orifice geometries of no-clip, middle-clip and side-clip configurations. Time-resolved Particle Image Velocimetry (TR-PIV) is used for the flow measurement and the velocity data are analyzed to reveal different flow characteristic features once the jets interact.

KEYWORDS: Mitral valve, Particle Image Velocimetry (PIV), jets, orifices

INTRODUCTION

Mitral valve (MV) regurgitation, or simply mitral regurgitation (MR), is a MV deficiency whereby some of the blood from the left ventricle is being pushed back into the left atrium during one of the cardiac phases [1]. Such a condition is caused by imperfect valve closure. It can lead to many physical complications, varying from general chronic fatigue to lethal heart failure. For several common cases of MR, the most frequently used repair option is an Edge-to-Edge repair (ETER). The technique was developed by Alfieri et al. [2] to simplify the pre-existing methods. ETER is based on the approximation of the free edge of the leaflets at the site of regurgitation, and then placing a suture or a clip connecting two leaflets to reduce their height. The most common area of prolapse is associated with a central region of the MV, so the suture is often positioned in the center or slightly to the side of the MV opening, resulting in an artificial double-orifice MV (DOMV) [3, 4]. Therefore, such positions were chosen for the present investigation.

There already exist clinical trials using ETER to report its effect on the function of a MV and to provide an insight into the mechanical interaction of ETER with the leaflets, e.g., [5, 6]. However, further fundamental and detailed studies are needed to clarify its effect on the flow pattern and hemodynamics. One of the consequences of this repair technique is that a single jet flow through the valve is transformed into an offset double jet configuration. The resultant changes in flow pattern may pose serious danger to the entire cardiovascular system and hence, such knowledge is of significance for improvement of the existing repair techniques.

In this study, an experimental investigation of offset jets is carried out, aimed at understanding the effects of an ETER procedure on the flow pattern. This research takes on a simplified model by focusing on two jets of the same and then of different characteristics emitted from a flexible orifice closely resembling the physical properties of a real MV, and investigates their dynamics once interacting. The purpose of the study is to shed light on how two jets influence each other and to improve understanding of the physical processes governing the nature of the jet instability. The results of this study can also lead to possible implications on the design of MV repair techniques.

METHODOLOGY

Experiments are carried out to analyze the dynamics of liquid jet interactions generated from a pulsatile fluid flow to mimic the cardiovascular system. The experimental setup consists of a small diameter piston-cylinder device connected to a large transparent Plexigas water tank. The flow is driven by the piston-cylinder through the pump control system (SuperPump pump by ViVitro Systems Inc.). The piston strokes tested were L = 46 mm and 55 mm. A 25.5-mm flexible nozzle with various orifice configurations is attached at the end of the piston-cylinder before the flow emerges into the water tank. A schematic of the setup is given in Fig. 1. Three different orifice geometries are used to mimic different MV configurations: a) with no-clip (NC), D = 25.5 mm, b) middle-clip (MC), $D_L = D_R$ = 10.25 mm, and c) side-clip (SC) arrangement, D_L = 13 mm, D_R = 7.5 mm, respectively, see Fig. 2.

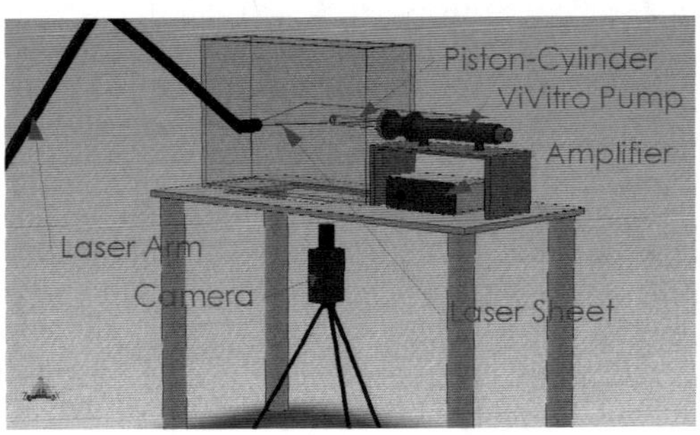

Figure 1: Experimental setup.

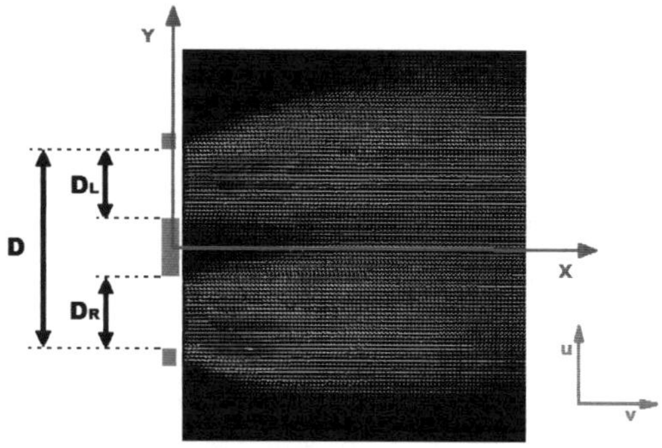

Figure 2: Coordinates for the orifice configuration.

The flow is measured using a Time-Resolved Particle Image Velocimetry (TR-PIV) system. The dual-frame CCD camera is located below the tank, directly underneath the cylinder exit, normal to the exit plane. A double-pulsed Nd-YAG laser sheet is aligned with the center of the MV flexible orifices, crossing orifices along the axis of vortex ring motion. Other details for the PIV conditions are listed in Table 1. Both image evaluation and initial vector field acquisition were processed digitally using commercially available software DaVis 7.2 by LaVision. All of the other vector and visual operations performed in this project were done using MATLAB R2008a.

Parameters used in the PIV measurement	
Number of images	350
Recording rate	350 Hz
Time step Δt	900 μs
Recording duration	1 s
Camera adjusted chip size	1632 x 1120 pixels
Picture size	127.103 x 87.2274 mm
Scale factor	0.0778816 mm/pixel
Seeding particles	10 μm silver-coated hollow borosilicate glass spheres with a density of 1.4 g/cm^3

Table 1. Parameters used for the PIV measurement.

RESULTS AND DISCUSSIONS

Figures 3 to 5 first show the average velocity field contours for different orifice configurations and piston stroke lengths L. For the NC configuration, Fig. 3 shows an area of a slow flow with two symmetrical recirculation regions on both edges of the orifice. The two vortex cores are approximately 1.6D apart radially from each other, and 0.24D axially away from the exit plain. Note that the downwards shift can possibly be explained by the fact that the nozzle is flexible, and does not always open in a perfectly straight manner.

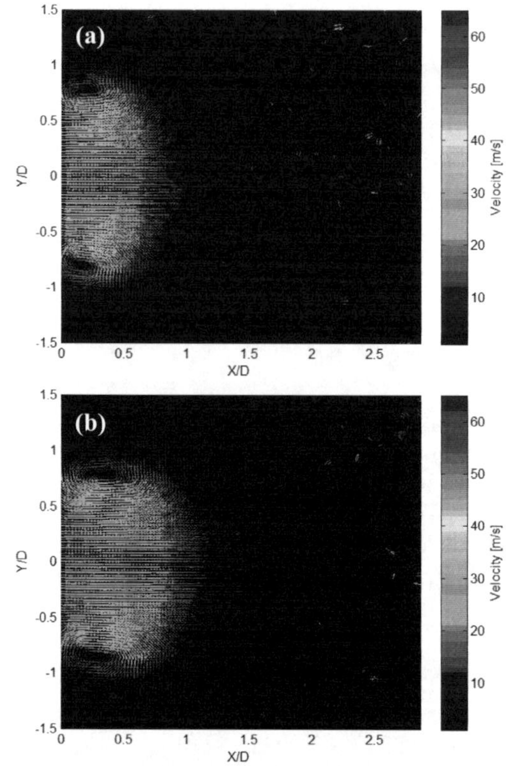

Figure 3: Average velocity field for the case with no clip and piston stroke a) L = 46 mm; and b) L =55 mm.

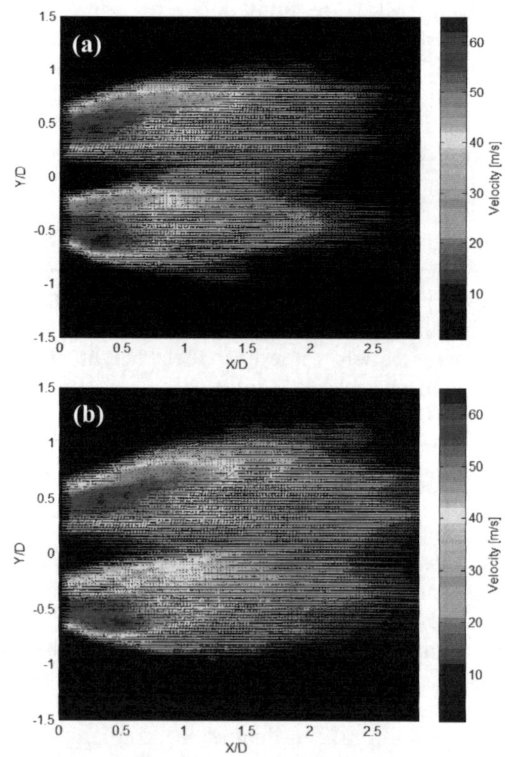

Figure 4: Average velocity field for the case with middle clip and piston stroke a) L = 46 mm; and b) L = 55 mm.

The mean flow for the MC configuration consists of two jets with an ill-defined core as shown in Fig. 4. Still, the overall shape of the mean flow resembles the expected characteristics of a parallel jet flow field first defined by Tanaka [7, 8]. One of the main features of the parallel jet flow is the entrainment of the surrounding fluid by turbulent jets in the region between the two jets, which causes a low pressure region near the nozzles, leading the jets to draw towards each other through mutual entrainment. In this experiment, the exit distance between the two jets is so small that the jets seem to reattach almost immediately after ejection. The two jets reach the combined point only at the very end of the tested field.

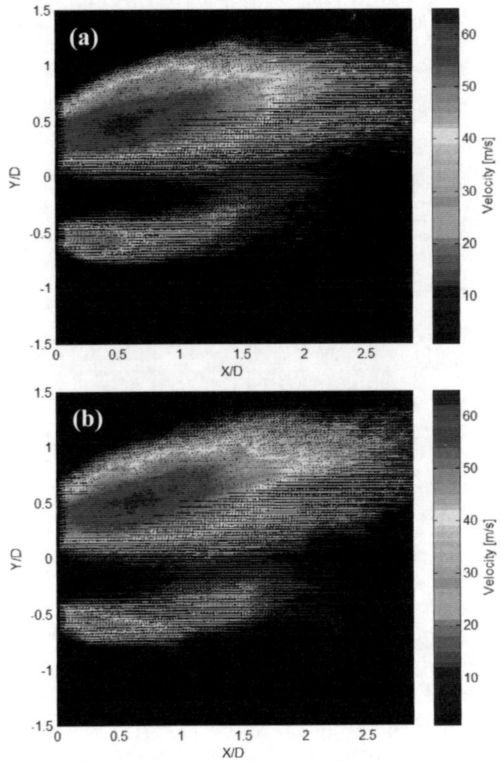

Figure 5: Average velocity field for the case with side clip and piston stroke a) $L = 46$ mm; and b) $L = 55$ mm.

From the present experiments, it is found that no combined region was fully attained even for the largest stroke tested for the SC case. Hence, the average velocity vector field in Fig. 5 represents mainly the merging region, while the combined region is not present for the examined flow duration. The interesting feature of uneven parallel jets is the attraction of the lower velocity jet towards the higher velocity one, creating an inclined jet profile favoring the stronger jet. This results from the entrainment rate of the faster velocity jet being higher than that of the slower jet. It should be noted that in this case the larger opening ejects a higher velocity jet. The flow is shifted upwards about the jet centerline for all cases. This is due to the fact that the two jets are ejected from a flexible double orifice nozzle, as opposed to two separate rigid nozzles.

Mean axial velocity profiles of the jets' interaction field at 0.25, 1.0, 1.75, and 2.5 X/D are also given in Fig. 6 for the piston stroke L = 55mm. The ordinate represents the radial dimension, or the position normalized using single orifice diameter D = 25.5 mm. The abscissa represents the velocity normalized with respect to the piston velocity V_p corresponding to the stroke of interest. At 0.25 X/D, the velocity profile of the MC configuration represents two peaks with almost the same local maximum in both jets. The velocity profile of the SC configuration also possesses two peaks: the large one for the primary jet and the smaller one for the secondary jet. It should be noted that the local maximum velocity of the primary jet is very similar to that of the MC one, with the latter being slightly faster. It is also found that for other stroke lengths tested the difference in Reynolds number produced by the piston does not affect the proportional difference between the two jets in the SC configuration near the exit. The distribution at 1.0 X/D widens radially for the MC case with the profile below the centerline being slightly slower than the upper one. Such discrepancy in the local maximum of the two peaks produced by two equivalent jets can be attributed to the flexible nozzle. The velocity segment between the two jets starts to flatten out as the two jets combine. In the SC configuration the proportional difference between the primary and the secondary jets increases. The primary local maximum for the SC configuration becomes slightly ahead of that for the MC configuration above the centerline. In addition, the local maximum of the secondary jet starts to diminish faster than the primary one. This can be attributed to the fact that the secondary jet is now 'feeding' the primary one. This makes the primary jet stronger, and hence, slower to diminish. The velocity profile widens radially only above the centerline or for the primary jet. The widening is also more significant than in the MC case. Further radial widening of the MC configuration takes place only in the top segment, and eventually the two peaks combine into one flat hump-like profile. For the SC configuration, the top section of the profile widens even further and the entire profile appears to shift upwards with the secondary peak being almost fully absorbed.

CONCLUSIONS

In this study, flow measurement of jets resulting from different orifice configurations to mimic various MV arrangements is carried out using time-resolved PIV. In general, the flow dynamics for NC case is dominated by the first two vortices caused by boundary layer roll-up around the orifice lips. In the MC case, the velocity profile just past the exit plain possesses two equivalent local peaks corresponding to the two equivalent jets. The overall profile begins to flatten and widen radially as the flow progresses, finally transforming into a combined hump-like profile at the end of the examined field for the larger piston strokes where the two jets merge at the very end of the tested field.

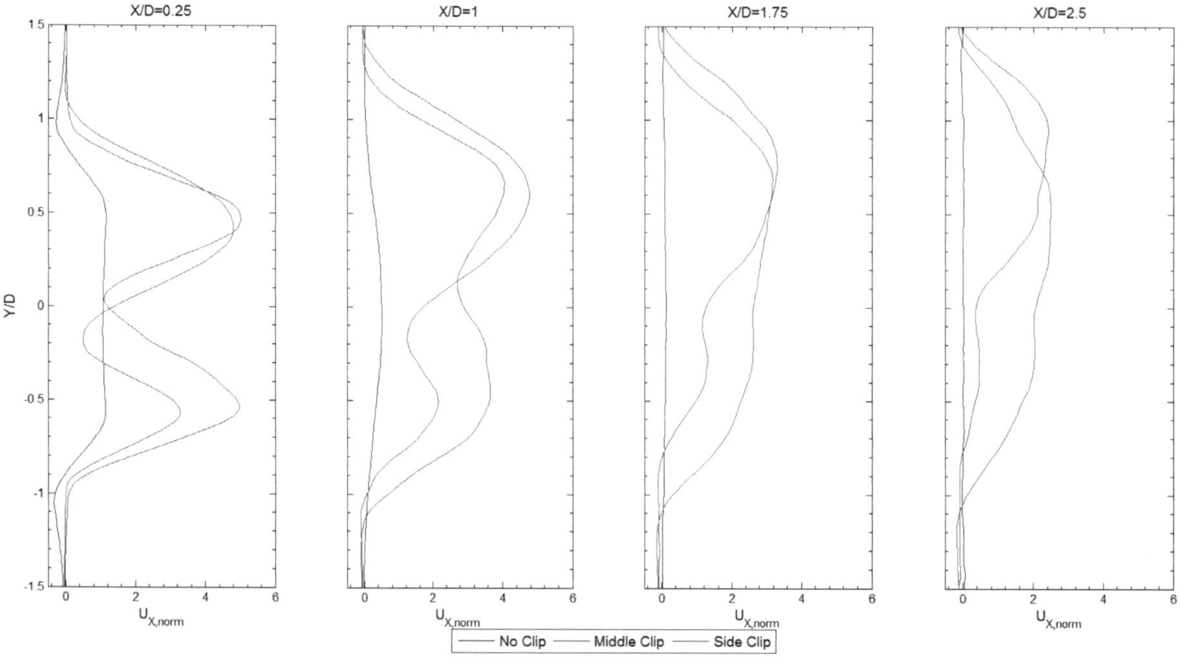

Figure 6: Distribution of mean axial velocity at 0.25, 1, 1.75, and 2.5 *X*/*D* for *L* = 55mm.

Overall, the visual properties of the SC case are very similar to the ones observed for MC. The prominent difference detected is the attraction of the smaller jet to the larger one. The initial flow of the SC case exhibits one larger peak of a primary jet and one smaller for the secondary jet, with the local maximum of the primary jet location equivalent to the initial maximum of the MC case. Additionally, the proportion between the two peaks of the SC case does not seem to be affected by the Reynolds number. However, as the flow progresses the proportional difference begins to increase as a result of the secondary jet being absorbed by the primary one. This leads to the primary jet's ability to keep its maximum almost equivalent to the MC configuration downstream. The additional mass entrained into the primary jet and the diminishing of the secondary one result into a shift of the overall profile upwards. By the end of the test section the secondary peak is nearly fully absorbed into the primary for the largest stroke tested. However, the fully combined region of the SC case was not reached in this experiment. With this time-resolved velocity data, the dynamic structures and energy distribution of these jets can be further analyzed using the Proper Orthogonal Decomposition (POD) method. Other turbulence quantities which can affect the hemodyanmics can as well be computed.

ACKNOWLEDGMENTS

This work is supported by the Natural Sciences and Engineering Research Council of Canada (NSERC) and the Canada Foundation for Innovation (CFI).

REFERENCES

1. Enriquez-Sarano, M., Akins, C.W., Vahanian, A. 2009, "Mitral regurgitation", Lancet, 373, pp. 1382-1394.
2. Alfieri, O., Maisano, F., De Bonis, M. et al. 2001, "The double-orifice technique in mitral valve repair: a simple solution for complex problems", J. Thorac. Cardiovasc. Surg., 122, pp. 674-681.
3. Maisano, F., Torracca, L., Oppizzi, M., et al. 1998, "The edge-to-edge technique: a simplified method to correct mitral insufficiency", J. Thorac. Cardiovasc. Surg., 13, pp. 240-246.
4. Maisano, F., Schreuder, J.J., Oppizzi, M., et al. 2000, "The double-orifice technique as a standardized approach to treat mitral regurgitation due to sever myxomatous disease: surgical technique", J. Thorac. Cardiovasc. Surg., 17, pp. 201-205.
5. Franzen, O., Baldus, S., Rudolph, V. et al. 2010, "Acute outcome of MitraClip therapy for mitral regurgitation in high-surgical-risk patients: emphasis on adverse valve morphology and severe left ventricular dysfunction", Eur. Heart J., 31, pp. 1373-1381.
6. Tamburino, C., Ussia, G.P., Maisano, F. et al. 2010, "Percutaneous mitral valve repair with the MitraClip system: acute results from a real world setting", Eur. Heart J., 31, pp. 1382-1389.
7. Tanaka, E. 1970, "The interference of two-dimensional parallel jets: 1st report, experiments on dual jet", Bull. JSME, 13(56), pp. 272-280.
8. Tanaka, E. 1974, "The interference of two-dimensional parallel jets: 2nd report, experiments on the combined flow of dual jet", Bull. JSME, 17(109), pp. 920-927.

Proceedings of the 25th CANCAM
London, Ontario, Canada, May 31 - June 4, 2015

ASCERTAINING THE INFLUENCE OF INTRAGLOTTAL VORTICES UPON THE DYNAMICS OF THE VOCAL FOLDS

Brian Q. Kettlewell
Department of Applied Mathematics
University of Waterloo
Waterloo, Ontario, Canada
brian.kettlewell@uwaterloo.ca

Sean D. Peterson
Department of Mechanical and Mechatronics Engineering
University of Waterloo
Waterloo, Ontario, Canada
peterson@mme.uwaterloo.ca

ABSTRACT

The rate at which the vocal folds close is generally understood to contribute to the intelligibility of speech, with rapid glottal closure increasing the high frequency content of the resulting acoustical speech signal, leading to a richer sounding voice. Recent studies have noted the formation and shedding of vortices within the glottis during the closing phase of the phonatory cycle, leading some to hypothesize that the low pressure at the centres of these intraglottal vortices may promote the rapid closing of the glottis. In the present investigation, we assess the relationship between vocal fold dynamics and intraglottal vortices via a potential flow model of the glottal flow coupled to a low-order body-cover model of the vocal fold tissues. The glottal flow is simulated with and without a pair of counter-rotating irrotational vortices, and the response of the tissue model is compared. The vortices are found to have minimal influence on the vocal fold dynamics due, in part, to limited interaction time with the tissue as they advect through the glottis.

Keywords: intraglottal vortices, vocal folds, human phonation, vortex dynamics

INTRODUCTION

Recent studies have demonstrated that coherent vortical structures form within the glottis when it is in a diverging configuration during the closing phase of the phonatory cycle, see for example [1]. However, the impact of these vortices on the vocal fold dynamics is debated. Khosla et al. [2] hypothesize that the roll-up of these vortices aids in rapidly drawing the glottis closed due to their low pressure cores, which alter the aerodynamic loading on the vocal fold surfaces. In a follow-up paper, Khosla et al. [3] correlate the presence of intraglottal vortices with increased magnitude of high frequency acoustic content, which is generally recognized as a major factor contributing to richness of the voice. These correlations do not, however, clearly articulate any causal mechanism relating the vortices to altered vocal fold dynamics.

To elucidate the link between the presence of a negative pressure induced at the vocal fold surfaces on the rate of glottal closure, Farahani and Zhang [4] performed simulations of a coupled fluid-structure interaction with a potential flow model of the glottal flow with a superposed *ad hoc* negative pressure correction at the medial surfaces of the vocal folds to account for the influence of intraglottal vortices. They performed simulations of the vocal fold dynamics with and without the pressure correction and found the influence of the pressure perturbation to be negligible in comparison with the recoil forces of the vocal fold tissues.

In this paper, we directly study the influence of intraglottal vortices on vocal fold dynamics using a potential flow model of the glottal airflow with advecting vortices coupled to a reduced order structural model of the vocal fold tissues. As opposed to an *ad hoc* pressure correction on the vocal folds to account for the presence of a vortex, herein we directly model the vortices and their kinematics in a fully coupled fluid-structure interaction simulation.

PROBLEM FORMULATION

The vocal folds have a high aspect ratio; consequently, a mid-membranous coronal section, as shown in Figure 1, may be taken as representative of the vocal folds. The figure contrasts a schematic of the layered structure of one vocal fold with a representative bar-plate body-cover mechanical model of the vocal fold tissues [5]. The boundary of the flow domain, which coincides with the medial vocal fold surface determined by the surface of the cover mass m, is approximated by a polygon. The cover mass is permitted to translate laterally and rotate about a nodal point with moment of inertia I_c. The two degrees-of-freedom of the cover capture the inferior-superior motion of the mucosal wave, while the body mass M accounts for the inertia of the thyroarytenoid muscle tissues.

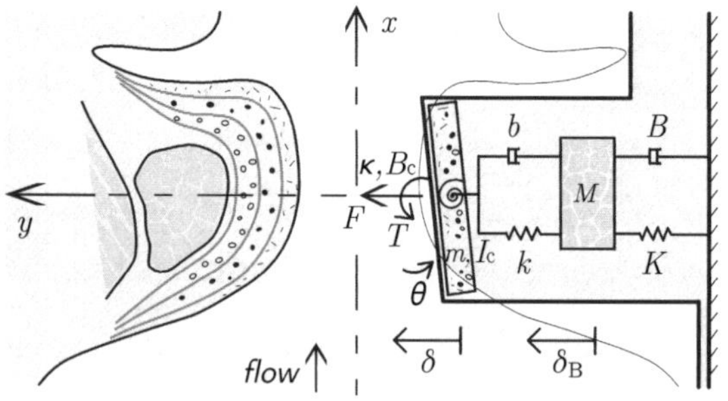

Figure 1: A mid-membranous coronal section of the larynx displaying the layered structure of vocal fold tissue with one vocal fold replaced by a representative mechanical model.

A Cartesian coordinate system is defined with the x-axis oriented along the glottal centreline and the origin located such that the y-axis intersects the nodal point of each vocal fold. The inferior margin of the vocal fold is located at $x = -l_{\text{node}}$, while the superior margin is located at $x = t_{\text{VF}} - l_{\text{node}}$, where t_{VF} is the total vocal fold thickness. Displacements δ and θ represent translational and angular displacement from equilibrium of the cover mass, and δ_{B} represents the displacement of the body mass. The geometry of the glottis is fully determined when subglottal and epilaryngeal channel widths h_{s} and h_{e} as well as the absolute positions of the equilibrium points of the cover are specified.

Conservation of linear momentum for the body and cover masses together with conservation of angular momentum for the cover yield the following coupled equations of motion

$$m\,\ddot{\delta} + b\left(\dot{\delta} - \dot{\delta}_{\text{B}}\right) + k\,(\delta - \delta_{\text{B}}) = F \tag{1}$$

$$M\,\ddot{\delta}_{\text{B}} + B\,\dot{\delta}_{\text{B}} + b\left(\dot{\delta}_{\text{B}} - \dot{\delta}\right) + k\,(\delta_{\text{B}} - \delta) + K\,\delta_{\text{B}} = 0 \tag{2}$$

$$I_{\text{c}}\,\ddot{\theta} + B_{\text{c}}\,\dot{\theta} + \kappa\,\theta = T, \tag{3}$$

in which the "dot" notation indicates a temporal derivative. The cover is viscoelastically coupled to the body mass via torsional elastic and damping elements κ and B_{c} and translational elasticity and damping k and b. K and B are the stiffness and damping coupling the body to the laryngeal cartilages. In the present study, we assume the vocal fold vibration to be symmetric, and thus we solve the dynamics for only one vocal fold.

The forcing functions, which appear on the right hand sides of (1) and (3), are due to the fluidic loading of the medial vocal fold surface due to the glottal airflow. F is the resultant force due to pressure resolved in the transverse direction, and T is a signed magnitude of the resultant pitching moment. With surface normal $\hat{n}$ directed into the glottis, we have

$$F = \left(\int_{\mathcal{C}(t)} -P\,(\vec{x}, t)\,\hat{n}\ L\,\mathrm{d}s \right) \cdot \hat{\jmath} \tag{4}$$

$$T = \pm \left\| \int_{\mathcal{C}(t)} \vec{r} \times \left(-P\,(\vec{x}, t)\,\hat{n} \right) L\,\mathrm{d}s \right\| \tag{5}$$

where $P\,(\vec{x}, t)$ is the fluid pressure at location $\vec{x} = (x, y)$ in the flow domain at time t, L is the vocal fold length into the page, $\mathcal{C}(t)$ is a curve which coincides with the medial vocal fold surface at time t, $\hat{\jmath}$ is a unit vector in the y direction, and $\vec{r}(t)$ is a position vector from the nodal point to some other location along $\mathcal{C}$. The pitching moment T is taken positive when it tends to increase θ.

The glottal airflow is modelled using potential flow, which assumes the fluid to be ideal; *i.e.*, incompressible, inviscid, and irrotational, except possibly at a finite number of isolated points. We note that the inviscid assumption is implied in (4) and (5). In potential flow, the fluid motion is completely governed by kinematics. Specifically, defining a complex variable $z = x + \mathrm{i}y$, where i is the imaginary number, we may define a complex potential $F(z) = \Phi + \mathrm{i}\Psi$, where Φ and Ψ are velocity potential and streamfunction, respectively. The fluid velocity, with x component u and y component v, is obtained via the conjugate of the derivative of the complex potential,

$$w(z) = \frac{\mathrm{d}F}{\mathrm{d}z} = u - \mathrm{i}v. \tag{6}$$

Both Φ and Ψ satisfy conservation of mass identically, and solving for either one completely determines the flow.

The flow domain is defined by a set of streamlines which trace the polygonal glottal geometry shown in Figure 1 and on the right side of Figure 2. To facilitate solution of the flow in this complicated geometry, we employ a Schwarz-Christoffel (SC) transformation to map the glottal geometry to an infinite horizontal strip, $S = \{\zeta \in \mathbb{C} \mid 0 < \mathfrak{Im}\,\zeta < 1\}$, where $\zeta = \xi + \mathrm{i}\eta$ is a complex variable defining position in the auxiliary plane, shown on the left in Figure 2. The SC map is given by

$$f\,(\zeta) = C_1 + C_2 \int^{\zeta} \exp\left(\frac{\pi}{2}\,(\alpha_- - \alpha_+)\,s \right) \cdot \prod_{k=1}^{n} \left(\sinh\left(\frac{\pi}{2}\,(s - \zeta_k) \right) \right)^{\beta_k} ds \tag{7}$$

in which the hyperbolic sine maps S to a slit half plane, the ends of the strip map to ∞, and $\exp\left(\frac{\pi}{2}\,(\alpha_- - \alpha_+)\,s\right)$ is inserted to account for the divergence angle $\alpha_\pm$ at each end. Parameter C_1 allows flexibility in selecting the lower limit of integration in (7), and C_2 accounts for the orientation of the polygonal image. For the polygonal region in the physical plane, with vertices z_k and corresponding exterior angles β_k, the SC map associates pre-vertices ζ_k according to $\zeta_k = f^{-1}\,(z_k)$.

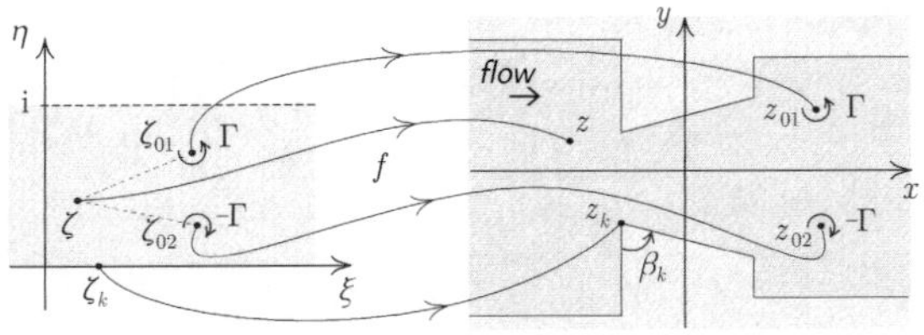

Figure 2: A schematic of the Schwarz-Christoffel mapping f between the horizontal strip of the auxiliary plane on the left and the flow domain within the physical plane on the right.

The solution of the flow within the auxiliary plane, where streamlines coincide with the boundaries of the flow domain, corresponds to a flow in a channel with static impermeable walls. Since the SC map preserves angles, wall-normal velocities vanish at the boundaries of the flow domain in the physical plane as well. Though the appropriate boundary condition in the physical plane requires that the wall-normal velocity of the fluid relative to the moving wall vanish, because the timescale of the vocal fold vibration is large relative to the convective timescale of the glottal flow, we make the assumption of a quasi-steady geometry. Consequently, solution to the glottal flow problem is obtained in the auxiliary plane, which is subsequently mapped to the instantaneous glottal geometry in the physical plane.

Superposition of vortices of strength $\Gamma_1 = \Gamma$ and $\Gamma_2 = -\Gamma$ located at ζ_{01} and ζ_{02} in the auxiliary plane (see Figure 2) together with a horizontal free stream results in the complex potential at a point $\zeta \in S$ given by

$$\widetilde{F}(\zeta) = V_s\, h_s\, \zeta + \sum_{k=1}^{2} \frac{\Gamma_k}{2\pi \mathrm{i}} \log \frac{\sinh\left(\frac{\pi}{2}\left(\zeta - \zeta_{0k}\right)\right)}{\sinh\left(\frac{\pi}{2}\left(\zeta - \overline{\zeta}_{0k}\right)\right)}, \qquad (8)$$

where the first term is the free stream contribution and the sum accumulates contributions from the two vortices present in the flow. The scaling of subglottal velocity V_s by the subglottal width h_s ensures that the subglottal velocity is correct in the transformed channel geometry in the physical plane.

Since an SC map entails the relation between the complex potential $\widetilde{F}(\zeta)$ in S and the desired complex potential in the physical plane, $F(z) = \widetilde{F}\left(f^{-1}(z)\right)$, the complex velocity in equation (6) may be recast in terms of $\zeta \in S$ as

$$w(z) = \frac{dF}{dz} = \frac{d\widetilde{F}}{d\zeta} \frac{1}{f'(\zeta)}, \qquad (9)$$

where $\zeta = f^{-1}(z)$ and f is the SC map capturing the geometry of the instantaneous glottal configuration. Consequently, via the complex potential (8) in S, where vortex locations $\zeta_{0k} = f^{-1}(z_{0k})$ are preimages of vortex locations in the physical plane, the desired velocity field in the physical plane may be obtained via (9) for the geometry captured by f.

The vortex positions are governed by

$$\dot{z}_{0k} = \overline{w^*(z_{0k})}, \qquad (10)$$

where the overbar indicates complex conjugation and the asterisk indicates that the velocity field is desingularized; *i.e.*, the velocity field with the contribution from the vortex at z_{0k} removed.

Finally, the unsteady Bernoulli equation for the pressure $P(z,t)$ may be written as

$$P(z,t) = P_s + \frac{1}{2}\rho\left(V_s^2 - w(z)\overline{w(z)}\right) - \rho\frac{\partial}{\partial t}\mathfrak{Re}F \qquad (11)$$

where ρ is the air density, P_s is subglottal pressure, and $\mathfrak{Re}F$ denotes the real part of the complex potential $F(z)$, namely $\Phi(z,t)$. Equation (11) determines a pressure field which ensures that conservation of momentum is satisfied for the velocity field $w(z)$ for the given geometry and vortex position.

SOLUTION PROCEDURE

The simulation employs a temporal marching scheme; at each timestep, the current state of the dynamical vocal fold model determines displacements from equilibrium and velocities of the masses. The displacements together with specified absolute equilibrium positions determine the current geometry. For the current glottal configuration, a polygonal region with vertices z_k and corresponding exterior angles β_k, the SC map f is obtained numerically. Together with the current vortical positions within the glottis, z_{0k}, the velocity field is computed from (9) via f and (8). Hence, the pressure distribution determined by (11) may be obtained with the unsteady term computed numerically using a second-order backward difference formulation given $F(z)$ and its recent time history as well as the dynamical state of the vocal folds. The pressure field is only required to be determined at a small number of control points along the medial vocal fold surface to allow numerical integration of (4) and (5). Once the forcing functions are determined, the next glottal configuration is obtained by simultaneously solving (1), (2), and (3) over one timestep with a fourth-order Runge-Kutta solver.

As for the pair of advecting vortices, disposed symmetrically about the glottal midline, they are initially placed at some upstream station and allowed to advect through the glottis at a rate to ensure that the pair arrives near the superior region of the glottis in the closing phase of each phonatory cycle. In each timestep, given the current vortex positions, the next vortex position is determined by evaluating (10) over one timestep using a first-order backward difference formulation. The speed of the vortices is determined by the vortex strength, the spacing and placement of the vortices, and the free stream velocity within the glottis. Once the vortices have advected sufficiently far downstream of the glottis, they are removed and replaced at their original upstream station and released so they arrive in the glottis in the closing phase of the next period of the phonatory cycle.

RESULTS AND DISCUSSION

Simulations are performed, one with vortices and one without. The grey band in the lower frame of Figure 3 indicates the range of pressures which prevail in the absence of vortices. Other pressure distributions show the perturbation of the pressure loading on the medial vocal fold surface as a function of axial station of the advecting vortex pair. The pressure on the medial surface of the vocal folds varies significantly near the inferior margin of the medial surfaces from one time to the next due to significant streamline curvature as the flow enters the glottis. Because these strong pressures are at the extreme of the vocal fold, they act on a longer moment arm relative to the nodal point leading to large pitching moments. Also shown in the upper frame is a representative glottal configuration with corresponding streamlines, which vary only slightly in the time required for vortices to advect from glottal entrance to exit.

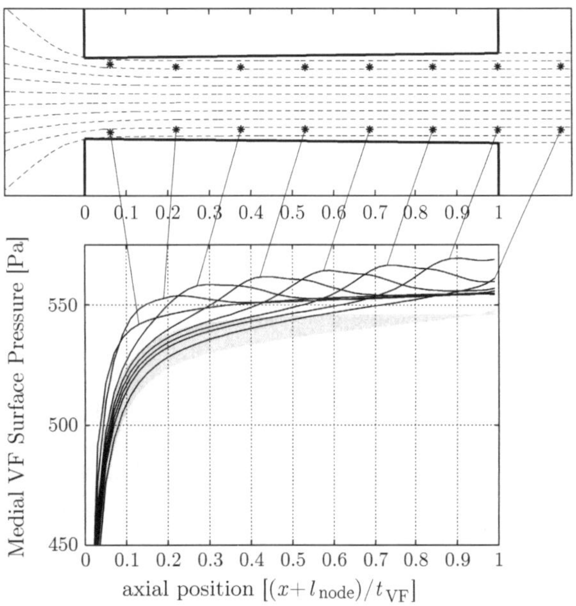

Figure 3: Medial vocal fold surface pressure distributions corresponding to axial positions of the intraglottal vortex pairs.

To quantify the effect of intraglottal vortices on the glottal waveform, Figure 4 overlays the derivatives of the glottal area waveforms of the baseline case without vortices and the case with vortex advection. The cases with and without vortices show qualitatively similar structures in the closing phases, which correspond to portions of Figure 4 in which the derivatives of the glottal area waveform are negative. Specifically, their magnitudes are similar during the closing phase, being nearly identical for many cycles, though they may differ by as much as 25% on occasion. Despite some cycle-to-cycle variation, Figure 4 indicates that advecting intraglottal vortices do not provide a mechanism which systematically causes the vocal folds to close more rapidly.

The simulation results show that the pressure perturbations due to the advecting vortices do not impact the vocal fold dynamics significantly. Although the pressure distribution on the medial surfaces of the vocal folds is perturbed by the advecting vortices, the effect is either short lived, weak, or both. Thus it may be concluded that intraglottal vortices, although they do modestly influence the dynamics of the vocal

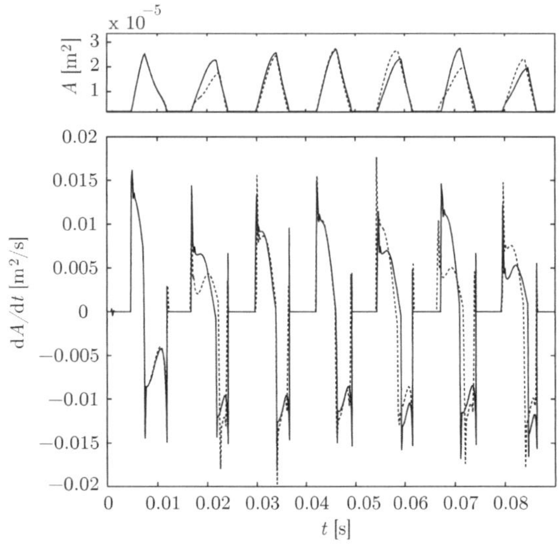

Figure 4: Glottal area waveforms (upper frame) and their temporal derivatives (lower frame). Solid traces correspond to the case with vortex advection, while the hatched traces correspond to the case without vortices.

folds, do not systematically modify the rate of closure of the vocal folds in the closing phase of the phonatory cycle. This study furnishes evidence against the hypothesis that intraglottal vortices augment the intelligibility of voiced speech due to increased high frequency content from a vocal tract excited by a more rapidly closing glottis.

REFERENCES

[1] Erath, B. D., Plesniak, M. W. 2010, "Viscous flow features in scaled-up physical models of normal and pathological phonation," *Int. J. Heat Fluid Flow*, 31(3), pp. 468-481.

[2] Khosla, S. M., Murugappan, S., Gutmark, E. 2008, "What can vortices tell us about vocal fold vibration and voice production," *Curr. Opin. Otolaryngol. Head Neck Surg.*, 16(3), pp. 183-187.

[3] Khosla, S., Murugappan, S., Paniello, R., Ying, J., Gutmark, E. 2009, "Role of vortices in voice production: Normal versus asymmetric tension," *Laryngoscope*, 119(1), pp. 216-221.

[4] Farahani, M. H., Zhang, Z. 2014, "A computational study of the effect of intraglottal vortex-induced negative pressure on vocal fold vibration," *J. Acoust. Soc. Am.*, 136(5), pp. EL369-EL375.

[5] Titze, I. R. 2002, "Regulating glottal airflow in phonation: Application of the maximum power transfer theorem to a low dimensional phonatio model," *J. Acoust. Soc. Am.*, 111(1), pp. 367-376.

A COMPUTATIONAL FLUID DYNAMICS MODEL OF FAILING FONTAN CIRCULATION FOR USE IN THE DESIGN OF A NOVEL TREATMENT

Matthew G. Doyle
Dept. of Mechanical and Industrial Engineering
University of Toronto
Toronto, ON, Canada
mg.doyle@utoronto.ca

Wanyu Zhang
Division of Engineering Science
University of Toronto
Toronto, ON, Canada
wy.zhang@mail.utoronto.ca

Fabian Alvarado
Dept. of Mechanical and Industrial Engineering
University of Toronto
Toronto, ON, Canada
fsalvara@ualberta.ca

Haruki Ide
Division of Cardiovascular Surgery
Hospital for Sick Children
Toronto, ON, Canada
haruki.ide@sickkids.ca

Lars Grosse-Wortmann
Division of Cardiology
Hospital for Sick Children
Toronto, ON, Canada
lars.grosse-wortmann@sickkids.ca

Richard D. Weisel
Division of Cardiac Surgery
Toronto General Hospital, PMCC, UHN
Toronto, ON, Canada
rweisel@uhnresearch.ca

Osami Honjo
Division of Cardiovascular Surgery
Hospital for Sick Children
Toronto, ON, Canada
osami.honjo@sickkids.ca

Leonard W. Tse
Division of Vascular Surgery
Toronto General Hospital, PMCC, UHN
Toronto, ON, Canada
leonard.tse@uhn.ca

Cristina H. Amon
Mechanical and Industrial Engineering & IBBME
University of Toronto
Toronto, ON, Canada
cristina.amon@utoronto.ca

ABSTRACT

Patients with failing Fontan circulation can often have elevated venous blood pressures and reduced cardiac output. There is currently no short-term non-invasive treatment to bridge these patients to transplant or further surgical intervention. It is the overall goal of this project to develop a cannula-pump system as this treatment option. As a first step, we have defined a representative geometry for the Fontan circulation along with appropriate physiological boundary conditions. We have also defined a lumped parameter model to couple to the three-dimensional geometry in a closed loop. Preliminary computational fluid dynamics simulation results are presented along with a description of the limitations of our study and our plans for future work.

KEYWORDS: biological fluid mechanics, design optimization, Fontan circulation, lumped parameter model, total cavopulmonary connection

INTRODUCTION

The Fontan procedure is the last in a series of surgeries that modify the vasculature in patients born with a single functioning heart ventricle. Anastomoses are created to connect the superior vena cava (SVC) and inferior vena cava (IVC) to the right pulmonary artery (RPA), the latter of which is done via a synthetic tube referred to as the extracardiac conduit (ECC). This results in blood flowing directly from the SVC and ECC into the left pulmonary artery (LPA) and RPA, bypassing the heart. This modified anatomy is referred to as the total cavopulmonary connection (TCPC).

The Fontan circulation can often result in elevated venous pressures and reduced cardiac output [1]. Patients diagnosed as having failing Fontan circulations and experiencing these dysfunctions may require additional surgical interventions or a heart transplant.

Currently, there is no short-term treatment to bridge failing Fontan patients to surgery or transplant, and it is the goal of this project to develop a device to provide such a treatment. Our device will consist of a cannula inserted into the TCPC that will take blood from the SVC and ECC to an external commercially available pump, and pump the blood back to the LPA and RPA.

Cardiac index (CI) is defined as the ratio of cardiac output to body surface area. Fontan pressure p_F is defined as the blood pressure measured within the ECC. In the context of this project, we will consider the Fontan circulation to be failing if $CI \leq 1.7\ (L/min)/m^2$ and $p_F \geq 18$ mm Hg. Our target values for the cannula design are $2.2 \leq CI \leq 2.4\ (L/min)/m^2$ and $10 \leq p_F \leq 12$ mm Hg. Note that CI and p_F are not independent parameters, and that a change in one will have an impact on the other.

The objective of this study is to define a representative computational model of blood flow in the TCPC along with appropriate boundary conditions to represent the rest of the circulatory system. Once these objectives are reached, a prototype cannula design will be added to the computational model for design optimization.

METHODS

We obtained angiograms of the TCPC in eight healthy (mean age 4 years) and two failing (mean age 5 years) pediatric Fontan patients treated at the Hospital for Sick Children. Under the assumption that the failing Fontan circulation is not caused by anatomical differences in the TCPC, we measured various dimensions from angiograms of both healthy and failing Fontan TCPCs to be used to calculate an average TCPC geometry. Figure 1 shows angiograms of two anterior-posterior views from a representative patient along with the relevant dimensions.

To define an average TCPC geometry, shown in Fig. 2, we assumed that all vessels lie in a single coronal plane, the tops of the LPA and RPA are aligned, and that all vessels except the ECC are cylindrical. The ECC was modelled as a curved pipe that varied in cross-section from a circle at its base to an ellipse at the anastomosis. The SVC and ECC diameters were calculated as the averages of measurements at three locations (B, C, and D; and I, J, K, respectively) along their lengths. The SVC and ECC lengths (A and H) were measured from their respective anastomoses to the base for the ECC and to the major bend for the SVC. The SVC and ECC anastomoses lengths (Q and S) were measured as the distances between the left and right endpoints of the SVC and ECC intersecting the RPA. The ECC curvature (R) was measured using an arc passing through the midpoints of three cross-sections along its length. The LPA and RPA diameters were calculated as the averages of measurements at the TCPC junction (O and F) and at the branch lobes (G and P). The lengths of the LPA and RPA (E and N) were defined as the centre to centre distance between the diameter measurement locations. The offset between the SVC and ECC (T) was measured as the distance between the centres of the two anastomoses.

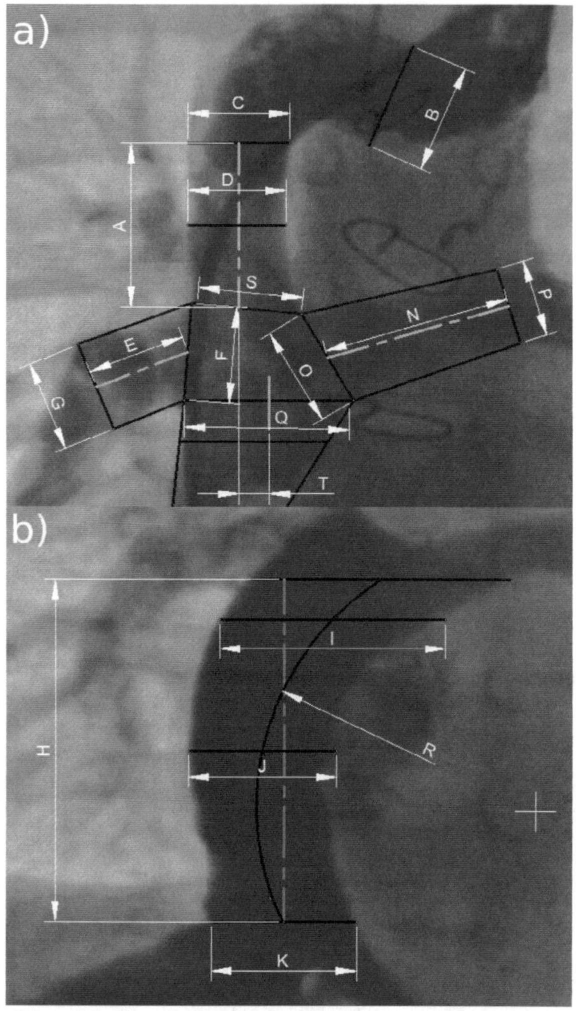

Figure 1: Representative angiograms used to measure dimensions of SVC, ECC, LPA, RPA, and anastomoses

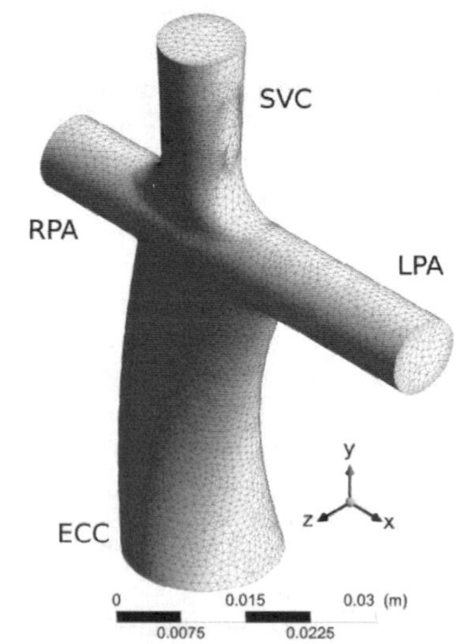

Figure 2: Average TCPC geometry and mesh

Computational fluid dynamics (CFD) simulations of flow in healthy and failing Fontan circulations were performed using the commercial software ANSYS Fluent v. 15 (ANSYS, Inc., Canonsburg, PA, USA). We assumed the flow to be laminar and steady, and blood to act as a Newtonian fluid with a density of $\rho = 1050$ kg/m^3 and a viscosity of $\mu = 0.00316$ kg/m·s [3].

Physiological volumetric flow rate boundary conditions were applied at the inlets. For the healthy Fontan circulation, $Q_{IVC} = 0.897$ L/min and $Q_{SVC} = 0.483$ L/min, and for the failing Fontan circulation, $Q_{IVC} = 0.585$ L/min and $Q_{SVC} = 0.315$ L/min. These values were calculated based on CI = 2.3 (L/min)/m^2 and 1.5 (L/min)/m^2, for healthy and failing cases respectively, a body surface area of 0.6 m^2, and a 65%/35% split of the flow to the IVC and SVC [4].

At the outlets, pressure boundary conditions of $p_{LPA} = p_{RPA} = 11$ mm Hg and 13 mm Hg were applied for healthy and mild failing Fontan circulations, respectively. The values were selected from Table 2 in Ref. [5] based on pulmonary vascular resistances of 3 mm Hg/(L/min) and 4 mm Hg/(L/min), respectively. The former was based on the value proposed by Ref. [6] and the latter was chosen to be greater than the former because pulmonary vascular resistance is known to increase in the failing Fontan circulation.

Because changes to the flow in the TCPC introduced by the cannula-pump system will result in changes throughout the rest of the circulation, it is necessary to couple the TCPC geometry with appropriate boundary conditions to represent the entire circulatory system. This is done by coupling the three-dimensional TCPC geometry to a zero-dimensional lumped-parameter model (LPM), defined using an analogy between electric circuits and fluid flow, for which vessel compliance C is analogous to capacitance, vessel resistance R is analogous to electrical resistance, flow rate Q is analogous to current, and pressure p is analogous to voltage.

Our LPM, shown in Fig. 3, is based on the model developed by Ref. [6]. We assumed the model to be steady-state using an approach define by Peskin [7]. Furthermore, we assumed that large systemic and pulmonary arteries and veins are compliance vessels with no resistance, and smaller vessels (i.e. capillary bed) are resistance vessels with no compliance [7]. Parameter values for our LPM are given in Table 1 for healthy and failing pediatric Fontan patients. The parameter K is a pump coefficient for the single ventricle and is defined as the product of the heart rate and the diastolic ventricular compliance. Note that we adjusted R_{LBB} and R_{UBB} to values that will lead to a physiological 65%/35% flow split.

The LPM was coupled to our CFD simulations using user-defined functions in Fluent. Iterations were performed between the CFD and LPM results until convergence was achieved between the two models.

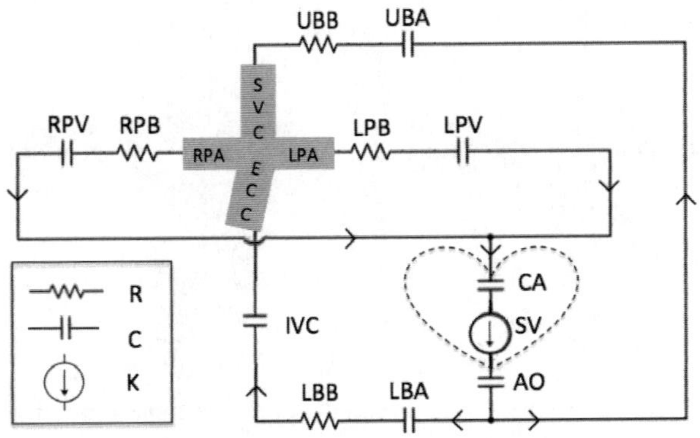

Figure 3: Lumped parameter model coupled to the TCPC

Table 1: Parameter value for lumped parameter model

Block	Parameter	Healthy	Failing
Upper body arteries (UBA)	C	1.0 ml/mmHg	1.0 ml/mm Hg
Upper body bed (UBB)	R	1.6 mmHg/(ml/s)	2.1 mm Hg/(ml/s)
Lower body arteries (LBA)	C	3.0 ml/mmHg	3.0 ml/mm Hg
Lower body bed (LBB)	R	2.9 mmHg/(ml/s)	3.9 mmHg/(ml/s)
Left pulmonary veins (LPV)	C	15 ml/mmHg	15 ml/mmHg
Left pulmonary bed (LPB)	R	0.132 mmHg/(ml/s)	0.480 mmHg/(ml/s)
Right pulmonary veins (RPV)	C	15 ml/mmHg	15 ml/mmHg
Right pulmonary bed (RPB)	R	0.132 mmHg/(ml/s)	0.132 mmHg/(ml/s)
Inferior vena cava (IVC)	C	15 ml/mmHg	15 ml/mmHg
Single ventricle (SV)	K	6.5 (ml/s)/mmHg	6.5 (ml/s)/mmHg
Common atrium (CA)	C	5.0 ml/mmHg	5.0 ml/mmHg
Aorta (AO)	C	1.0 ml/mmHg	1.0 ml/mmHg

RESULTS AND DISCUSSION

Figures 4 and 5 show pressure contours and velocity vectors, respectively, in a coronal plane at the centre of the TCPC geometry for both healthy and failing Fontan circulations without being coupled to the LPM. Consistent with our imposed boundary conditions and known physiology, pressures were higher and velocities were lower in the failing Fontan circulation compared to the healthy case.

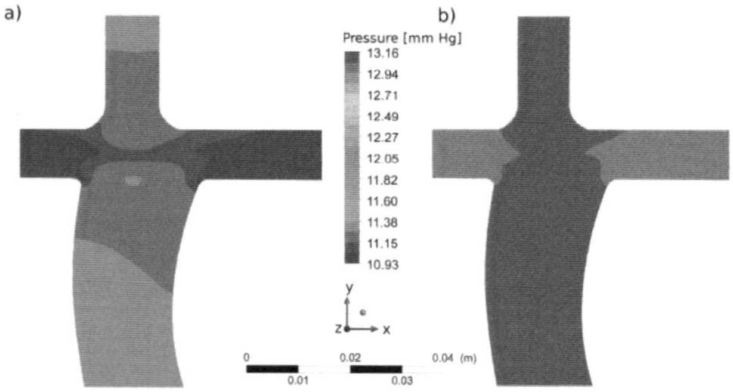

Figure 4: Pressure contours in the centre coronal plane for healthy (a) and failing b) Fontan circulations

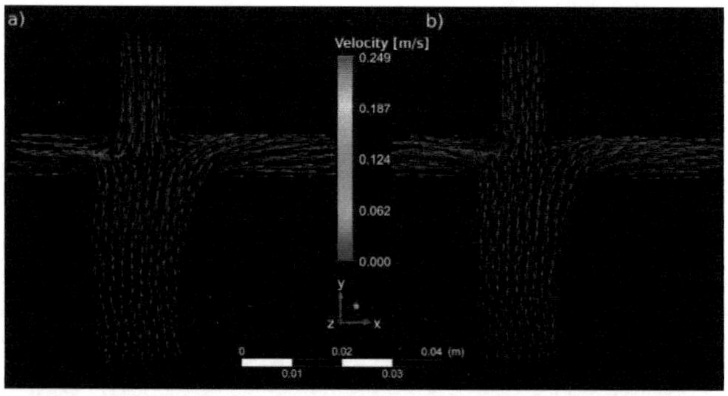

Figure 5: Velocity vectors in the centre coronal plane for healthy (a) and failing b) Fontan circulations

Our initial simulations were performed using several simplifying assumptions, the impact of which will be investigated in future versions of our model. We initially neglected out-of-plane geometric variations, which are expected to have an impact on the three-dimensionality of the flow in the TCPC. We will perform a sensitivity analysis including these variations, particularly for the ECC, which is known to deviate from a single coronal plane. We will try different inlet velocity profiles to allow for physiological mixing to occur in the TCPC.

Reynolds numbers in the SVC and ECC were 328 and 333, respectively, for the healthy case, and 214 and 217, for the mild failing case, which is consistent with our assumption of laminar flow.

While flow in the circulatory system is transient, the distance from the heart to the TCPC and the use of a continuous-flow pump connected to our cannula make the assumption of steady flow a reasonable first approximation. By making this assumption, we reduced the LPM from a system of differential equations to a system of algebraic equations, greatly simplifying the implementation of this model. However, transient effects may be important, especially during the adaptation of the circulatory system through the introduction of the cannula-pump system. These effects will be investigated in a future version of the model.

CONCLUSIONS

A representative TCPC geometry was defined for use in CFD simulations to study healthy and failing Fontan circulations and design a cannula-pump system for treatment of the failing Fontan circulation. Results from initial CFD simulations were presented without the presence of the cannula or the lumped parameter model boundary conditions, which will be added in future versions of our model.

Our next step is to complete the coupling of the CFD simulations with the LPM. This will be followed by the introduction of the cannula-pump system to the model, which will allow for cannula design optimization to be performed, with goal of developing a cannula-pump system as a short-term bridge to transplantation or surgical intervention in patients with failing Fontan circulations.

REFERENCES

1. Khambadkone S. 2008, "The Fontan Pathway: What's down the road?", *Annals of Pediatric Cardiology*, 1(2), pp. 83-92.
2. Laganà, K., Balossino, R., Migliavacca, F., Pennati, G., Bove, E.L., de Leval, M.R., and Dubini, G. 2005, "Multiscale modeling of the cardiovascular system: Application to the study of pulmonary and coronary perfusions in the univentricular circulation", *Journal of Biomechanics*, 38, pp. 1129-1141.
3. Cheng, Y., Oertel, H., and Schenkel, T. 2005, "Fluid-structure coupled CFD simulation of the left ventricular flow during filling phase", *Annals of Biomedical Engineering*, 33(5), pp. 567-576.
4. Cavalcanti, S., Gnudi, G., Masetti, P., Ussia, G.P., and Marcelletti, C.F. 2001, "Analysis by mathematical model of haemodynamic data in the failing Fontan circulation", *Physiological Measurement*, 22, pp. 209-222.
5. Migliavacca, F., Dubini, G., Bove, E.L., and de Leval, M.R. 2003, "Computational fluid dynamics simulations in realistic 3-D geometries of the total cavopulmonary anastomosis: The influence of the inferior caval anastomosis", *Journal of Biomechanical Engineering*, 125, pp. 805-813.
6. Pennati, G., Migliavacca, F., Dubini, G., Pietrabissa, R., and de Leval, M.R. 1997, "A mathematical model of circulation in the presence of bidirectional cavopulmonary anastomosis in children with a univentricular heart", *Medical Engineering and Physics*, 19(3), pp. 223-234.
7. Hoppensteadt, F.C., and Peskin, C.S. 2001, *Modeling and Simulation in Medicine and the Life Sciences, 2nd Ed.* Springer-Verlag, New York.

Proceedings of the 25th CANCAM
London, Ontario, Canada, May 31 – June 4, 2015

Experimental Investigation of Hemodynamic Variations inside Abdominal Aortic Aneurysm During Change of Posture

Hani Abulkhair
Mechanical Engineering department
King AbdulAziz University
Jeddah, Saudi Arabia
haboalkhair@kau.edu.sa

Lyes Kadem
Mechanical and Industrial Engineering
Concordia University
Montreal, Canada
kadem@encs.concordia.ca

ABSTRACT

Rupture of the abdominal aorta is the 10th largest killer among seniors in western countries. The rupture occurs after a dilation of the aorta in the abdominal region which forms what is called an Abdominal Aortic Aneurysm (AAA). Current treatment methods are surgical interventions which are decided based on the aneurysm size or its growth rate. Accelerating blood flow through the abdominal region has been hypothesized to limit thrombus formation in the AAA and as a potential consequence to reduce AAA growth rate. The change of the body posture has a substantial effect on the blood flow rate in the abdominal region which may be in favor of accelerating the blood flow rate. In this study, the effect of a regular change from sitting to standing on AAAs flow field will be experimentally investigated using time-resolved particle image velocimetry.

KEYWORDS: Abdominal aortic aneurysm, Hemodynamics, Thrombus formation, Particle Image Velocimetry.

INTRODUCTION

An abdominal aortic aneurysm (AAA) is a vascular disease that results from a dilation of the abdominal aorta. It is considered a health threat due its susceptibility to rupture. According to [2], rupture of the abdominal aorta is the 10th largest killer among seniors in western countries. Globally, 500,000 patients are affected by AAAs each year [1], besides the risks of rupture, the lumen inside an AAA promotes intraluminal thrombus (ILT) formation. Seventy five percent of the aneurysms are accompanied with ILT [3], [7]. This thrombus is found to have a strong relation with weakening the aortic wall where it works as a barrier between the aortic walls and the red blood cells that convey oxygen [7]. Besides that, they can breakdown and block the iliac arteries or other small arteries [10].

Usually aneurysms are detected by ultrasound imaging since the symptoms of this disease are not a clear indication of it. According to group health in USA [12], people are recommended for CT screening between 55 and 65 based on their gender, family and smoking history. Once the AAA is detected, the patient will be under annual screening until the aneurysm reached a critical diameter. Then, he/she will be advised to surgical intervention.

Interestingly, a recent study by [4] showed that frequent exercise activity might help in reducing the progression of AAAs. The study hypothesizes that increasing the level of activity during the day will influence hemodynamics in a way that reduces the progression of the aneurysm, and washes blood particles faster from the aneurysm in a way that the probability of thrombus formation is reduced. Due to the potential reduced ability of old people to perform exercise activities, the current study hypothesized that simple body movement can help these people to have the same beneficial effects on AAA flow field as performing regular exercise.

This study will investigate the effect of change of body posture effect on blood flow inside AAA. Due to lack of studies that discuss the temporal effect hemodynamics during the change of posture, the change from sitting to standing will be utilized to validate the current hypotheses. This change will be simulated and tested experimentally using Particle Image Velcimetry (PIV).

METHODS

A schematic experimental setup is shown in Figure 1a. This test section includes a Leeson centrifugal pump, a tank, a replica of the ascending and descending aortas, an AAA replica, a PIV (particle image velocimetry setup), valves, a flow meter and a pressure probe. A Leeson centrifugal pump sucks a blood analog (40% water and 60% glycerol in volume) from a tank and pumps it through the aortic arch. This solution has a density of 1100 kg/m^3 and viscosity of 3.7 cps. The pump is controlled by a custom-made LabVIEW program that generates a user-defined flow waveform. A mechanical heart valve, simulating a native aortic valve, is installed in the aortic position at the entrance of the ascending aorta model. The descending aorta includes renal arteries, the model of the AAA and two iliac arteries. The AAA model is installed in a transparent Plexiglas box which is filled with water and glycerol to minimize distortion.

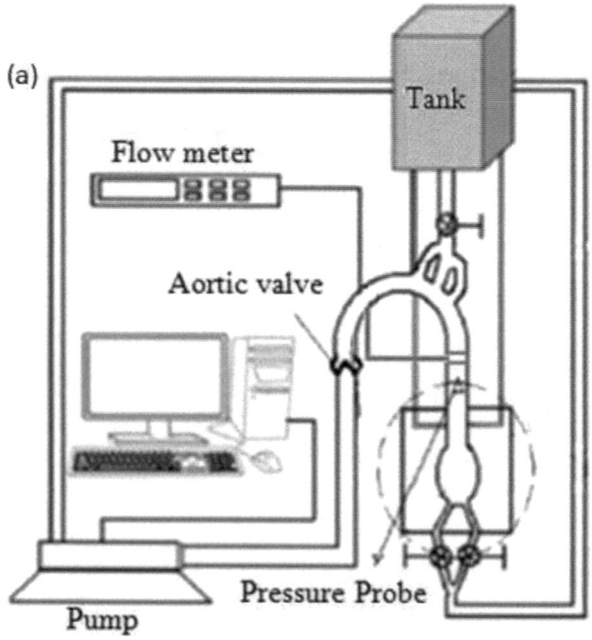

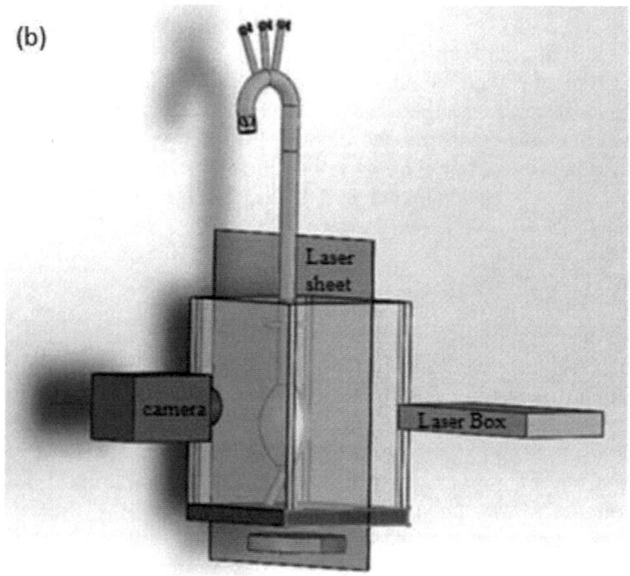

Figure 1: a) Schematic diagram of the test facility b) Schematic of PIV measurement

All the tests were performed at room temperature. Transonic Flow meter T201 was installed upstream of the renal arteries, and the flow was divided between the renal and iliac arteries evenly to reflect the actual resting condition, See, [8]. A Millar catheter pressure sensor was installed 10 cm above the aneurysm. Figure 2 shows the pressure and flow rates obtained during the actual measurements. The heart rate was fixed at 72 bpm which is equivalent to a period of 0.83 sec. The frequency parameter (Womersly number $= r\sqrt{\omega/v}$) was equal to 12.6, where v is the kinematic viscosity, $\omega = 2\pi/T$, T is the cycle period, and r is the radius of the abdominal aorta upstream of AAA. The maximum Reynolds number is about 1415 and mean Reynolds number about 568.

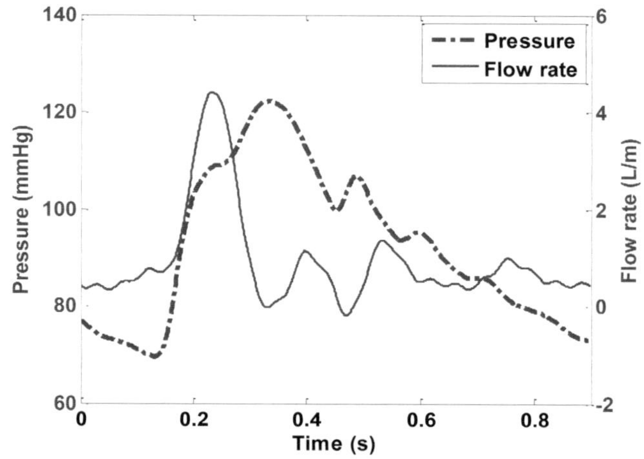

Figure 2: Pressure and flow rate waveforms during the cardiac cycle corresponding to the normal case

The PIV measurements were performed in the mid-plane of the anterior view as shown in Figure 1b. The seeding particles are PSP Polyamide particles, with a size of 50 μm and a density of 1030 kg/m^3. The difference in density does not have a strong impact in the velocity lag where the relaxation time was found to be equal to $4.4\mu s$. The time difference between the two laser pulses was optimized. Image processing was performed by fast Fourier transform based on the cross-correlation. The interrogation area was selected to be 16*16 pixels with 50% overlap which gives a spatial resolution of (0.732mm×0.732 mm). The overall uncertainty of PIV measurements due to magnification factor and image displacement is less than 5%. The peak locking effect was minimized by having image particles around 3 pixels.

The change of posture has been simulated experimentally to be compatible with the lumped parameter model introduced by [6]. Figures 3 & 4 show the variations in pressure and flow rate in aorta as a result of simulated change of posture from sitting to standing. Overall, 26 cycles are to simulate the transient effects due to change of posture.

In order to evaluate the potential advantage of change of posture on hemodynamics inside AAA, a custom-made Matlab code was used to track particles in the flow field. The number of particles leaving the AAA after N cycles is used to compare resting condition with change of posture condition.

RESULTS AND DISCUSSION

Figure 5 displays the velocity vectors at the peak systolic pressure for the normal case, while Figure 6 shows the velocity field corresponding to the peak systole of the cycle just following the transient effects due to change in posture (cycle number 24); see the green dot in Figure 3. The maximum velocity magnitude in both cases was found to be around 0.35 m/s at the entrance of AAA. By taking the whole domain of AAA, the general magnitude of the velocity field inside the domain of the aneurysm is higher in the change of posture case which means the overall particle residence time due to change in posture change should be reduced.

The magnified region of AAA in Figure 5, corresponding to the normal case, shows a tendency to form a circulation region which results in more stasis of the blood particles, and a non-optimal washout of AAA. For the same region AAA with the change of posture case, has been observed that the maximum velocity has increase to 0.15 m/s and the direction of the velocity vectors promote a better washout of AAA.

More specifically, after applying the particle tracking method, the advantage of the change becomes clearer. Figure 7 shows particles that left AAA region during the change of posture case while stayed in AAA region for the normal case. In vivo AAA, thrombosis are formed near the edges of the AAA are the main cause for hypoxia and arterial inflation.

CONCLUSION

AAA hemodynamics is a wide area of research. In this work, the effect of simple change of posture (from sitting to standing) has been evaluated experimentally inside AAA to evaluate the potential advantage of this method in reducing the progression of AAA. The results show some advantages when applying this method. If patient could perform this change of posture routinely, it might be a simple and helpful method for delaying the progression of early diagnosed AAAs. Finally, further studies have to specifically compare the benefits of repeated change in posture with moderate exercise.

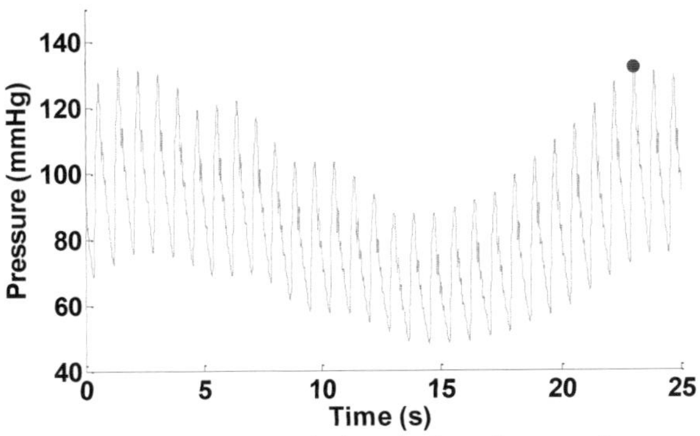

Figure 3: Pressure variation during change of posture

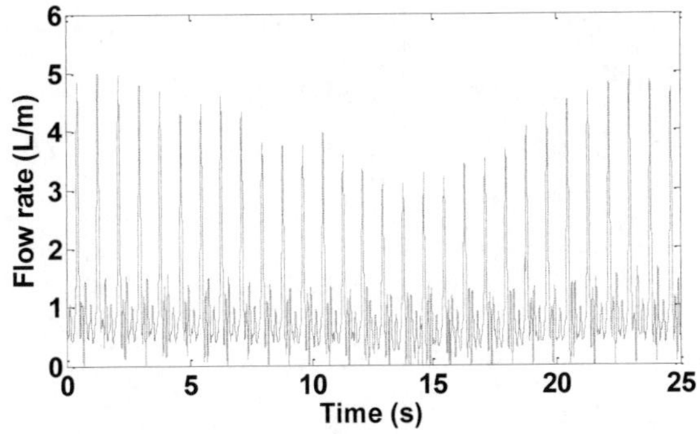

Figure 4: Flow rate variation during change of posture

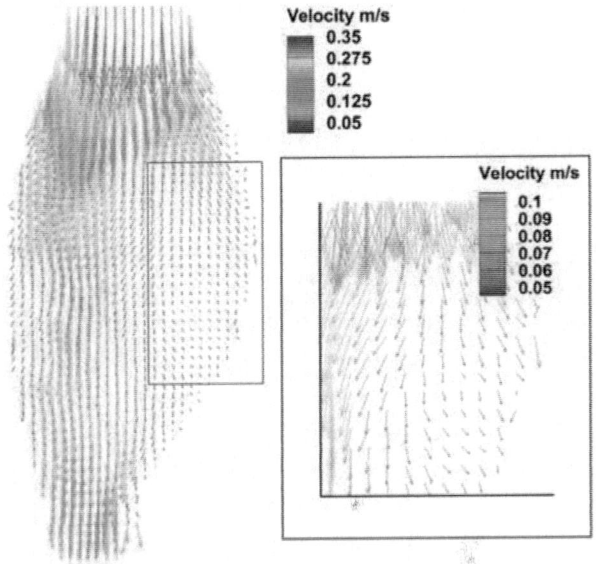

Figure 5: Velocity vectors at the peak of systolic pressure for the normal case

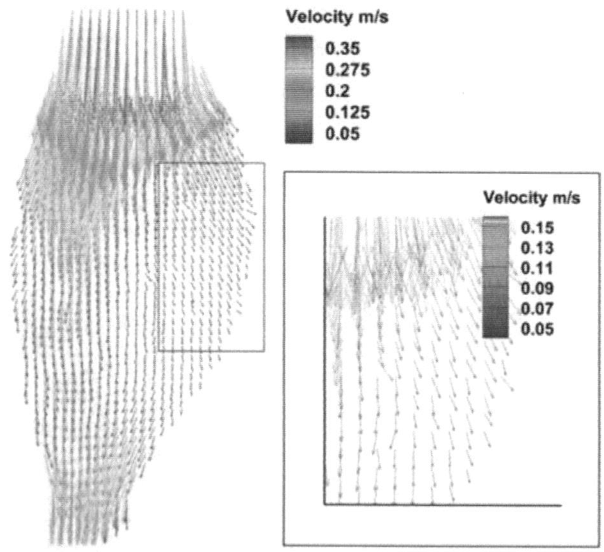

Figure 6: Velocity vectors at the peak of systolic pressure for the change of posture case

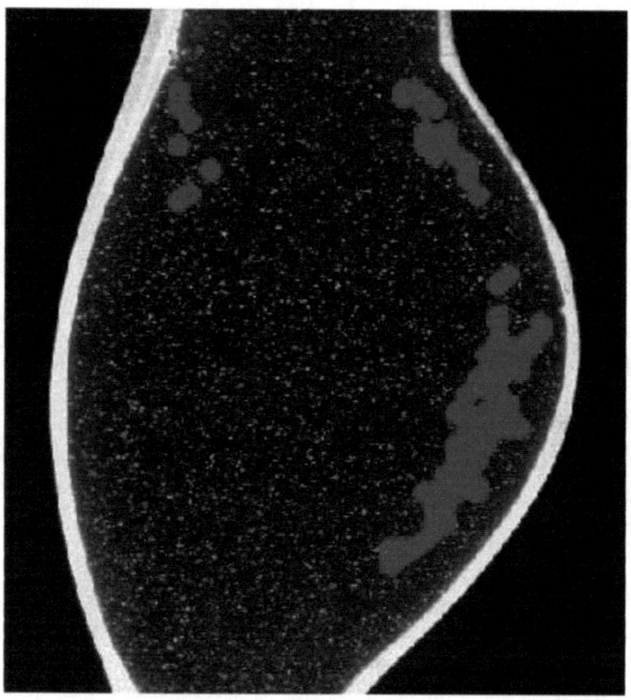

Figure 7: Location of particles that did not leave the aneurysm region during the normal case and left during the change of posture

REFERENCES

1. Bosch, J. L., Lester, J. S., McMahon, P. M., Beinfeld, M. T., Halpern, E. F., Kaufman, J. A., & Gazelle, G. S., 2001. "Hospital Costs for Elective Endovascular and Surgical Repairs of Infrarenal Abdominal Aortic Aneurysms 1". *Radiology*, 220(2), pp.492-497.
2. Zankl, a R., Schumacher, H., Krumsdorf, U., Katus, H. a, Jahn, L., & Tiefenbacher, C. P., 2007. "Pathology, natural history and treatment of abdominal aortic aneurysms". *Clinical research in cardiology : Official Journal of the German Cardiac Society*, 96(3), pp.140-151.
3. Sakalihasan, N., Limet, R., & Defawe, O. D., 2005. "Abdominal aortic aneurysm". *The Lancet*, 365(9470), pp.1577-1589.
4. Suh, G., Les, A. S., Tenforde, A. S., Shadden, S. C., Ryan, L., Yeung, J. J., & Cheng, C. P., 2012. "Quantification of Particle Residence Time in Abdominal Aortic Aneurysms Using Magnetic Resonance Imaging and Computational Fluid Dynamics". *Ann Biomedical Engineering*, 39(2), pp.864-883.
5. Lasheras, J. C., 2007. "The Biomechanics of Arterial Aneurysms". *Annual Review of Fluid Mechanics*, 39(1), pp.293-319.
6. Olufsen, M. S., Nadim, A., & Lipsitz, L. A., 2002. "Dynamics of cerebral blood flow regulation explained using a lumped parameter model". *American Journal of Physiology-Regulatory, Integrative and Comparative Physiology*, 282(2), pp.611-622.
7. Vorp, D. A., Raghavan, M. L., & Webster, M. W., 1998. "Mechanical wall stress in abdominal aortic aneurysm: influence of diameter and asymmetry". *Journal of Vascular Surgery*, 27(4), pp.632-639.
8. Moore, J. E., Ku, D. N., Zarins, C. K., & Glagov, S., 1992. "Pulsatile flow visualization in the abdominal aorta under differing physiologic conditions: implications for increased susceptibility to atherosclerosis. *Journal of biomechanical engineering*", 114(3), pp.391-397.
9. Chambers, D., Epstein, D., Walker, S., Fayter, D., Paton, F., Wright, K., & Michaels, J., 2009. 'Endovascular stents for abdominal aortic aneurysms: a systematic review and economic model". *Health Technology Assessment,* 13(48), pp.1-189.
10. www.patient.co.uk/health/abdominal-aortic-aneurysm (Accessed: 15 Jan 2015).
11. http://www.ghc.org. (Accessed: 15 Jan 2015).

Proceedings of the 25th CANCAM
London, Ontario, Canada, May 31 – June 4, 2015

Non-Newtonian Blood-Mimicking Fluid for Particle Image Velocimetry

Adam J Fortais, John R de Bruyn, Tamie L Poepping
Department of Physics and Astronomy
London, ON, Canada
poepping@uwo.ca

ABSTRACT

Laboratory studies in vascular flow models using optical techniques, such as particle image velocimetry (PIV), aim to accurately model blood flow scenarios. Most studies to date have used Newtonian blood-mimicking fluids, which fail to capture the non-Newtonian rheology of blood at shear rates less than approximately 100 s^{-1}. Using experimental data from the recent literature on blood rheology, an expression for the dynamic viscosity of blood as a function of shear rate was found. A non-Newtonian blood-mimicking fluid, consisting of 51.9 wt% water, 25.9 wt% NaI, 22.2 wt% glycerol, and 0.019 wt% XG, was formulated. The rheological behavior of this fluid was characterized using rotational rheometry, and its viscosity as a function of shear rate was found to be a good match to that of blood at shear rates above 1 s^{-1}, The refractive index of this fluid was adjusted to 1.4145±0.0005 to match that of polydimethylsiloxane (PDMS) vascular phantoms commonly used in PIV studies. The rheological and optical properties of the fluid make it an excellent candidate for flow studies in vascular models.

KEYWORDS: Rheology, non-Newtonian, blood mimicking fluid, BMF, blood analogue, hemodynamics, refractive index, viscosity, shear thinning, polydimethylsiloxane

INTRODUCTION

Laboratory vascular flow studies using PIV, or other optical techniques, require an optically transparent blood-mimicking fluid (BMF) that not only approximates the dynamic viscosity of blood but also matches the refractive index of the flow phantoms. Phantoms are commonly made from PDMS due to its optical clarity, availability, and ease of molding into the required complex geometries. A number of fluids have been used to simulate blood, but most of them have been Newtonian – meaning their viscosity, η, is independent of shear rate, $\dot{\gamma}$. For example, Yousif et al. {Yousif 2011} described a Newtonian fluid composed of glycerol, water and sodium iodide (NaI) designed to match the refractive index of PDMS vascular phantoms. Such fluids do not adequately mimic the shear-thinning behavior displayed by real blood at low shear rates, nor do they capture the behavior of blood in microcirculatory systems. As a result, experiments using Newtonian blood-mimicking fluids may not reproduce physiological flow behavior in complex flow conditions and geometries. Several groups have developed non-Newtonian BMF, however, none to date have simultaneously replicated physiological viscosity while maintaining an appropriate refractive index for performing PIV (Brooksheir et al. 1993, Gijsen et al. 1999, Walker et al. 2013).

Blood is a shear-thinning non-Newtonian fluid that approaches a constant viscosity of approximately 4 cP at high shear rates but has a significantly higher viscosity at low shear rates (Antonova et al. 2008, Baskurt et al. 2003, Bor-Kucukatay et al. 2008, Carrera et al. 2008, Galduróz et al. 2007, Merrill 1969). At low shear rates, RBC tend to aggregate into rouleaux – a second-order structure often visualized as a stack of coins. This structure can be broken up by sufficiently strong shear but will reform when the shear is reduced (Goldstone et al. 1970). This effect is facilitated by plasma proteins – predominantly fibrinogen – and is less prevalent in RBC-saline solutions (Goldstone et al. 1970, Rand et al. 1964). RBC aggregation is the dominant cause of the shear dependence of whole blood viscosity at low shear rates. It is commonly stated that blood has a yield stress, σ_y, (i.e. it does not flow unless the shear stress exceeds this value). Charm (Charm et al. 1964) reported a yield stress of 9±3 mPa in men and 25±5 mPa for women with non-Newtonian behavior observed only for $\dot{\gamma}$ < 40 s^{-1}. These reported values of the yield stress are orders of magnitude smaller than typical blood pressure. Furthermore, a fluid with a yield stress of 10 mPa would yield under its own weight if it was only 1 μm deep. Even if such a small yield stress exists in whole blood, it can have no physiological significance.

For the present study, the results of viscosity measurements on 683 blood samples compiled from six recent studies (Antonova et al. 2008, Amparo et al. 2008, Bor-Kucukatay et al. 2008, Carrera et al. 2008, Fehr et al. 2008, Galduroz et al. 2007) were analyzed to characterize the viscosity of blood as a function of shear rate. These data had been measured at shear rates between 0.01 and 250 s^{-1} and at a temperature of 37°C (normal body temperature) with either a cone-and-plate or a Couette (concentric cylinder) rheometer. Measurements were done using blood with a hematocrit level between 40 and 45%, either with or without anti-coagulant. These literature data for the viscosity of blood, with uncertainties provided in the original papers, are plotted in Figure 1.
The fluids in this paper were fitted to the Cross model,

$$\eta(\dot{\gamma}) = \eta_\infty + \frac{\eta_0 - \eta_\infty}{1+\tau\dot{\gamma}^{1-q}} , \qquad (1)$$

an empirical four-parameter function that assumes no yield stress. It describes a fluid with a Newtonian viscosity plateau at low shear rates and a second viscosity plateau at high shear rates. Here η_0 and η_∞ are the viscosities as $\dot{\gamma}$ approaches zero and infinity, respectively, while τ and q are empirical parameters related to the rate and degree of shear thinning.

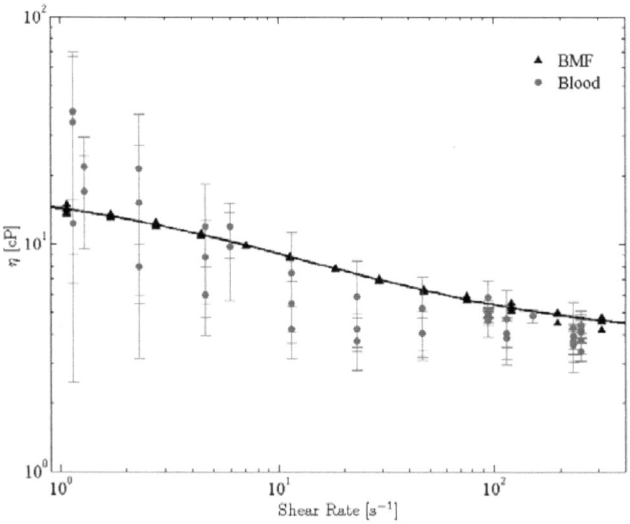

Figure 1 Viscosity as a function of shear rate for the literature whole-blood data and a BMF consisting of 51.9 wt% water, 25.9 wt% NaI, 22.2 wt% glycerol, and 0.019 wt% XG. The solid line is a fit of Equation (1) to the BMF data. Three BMF trials are displayed

In this paper, a non-Newtonian BMF that both simulates the macroscopic rheological properties of *in vivo* blood across a range of shear rates and is refractive-index-matched for use in a PDMS vascular phantom is described. Details are provided that should enable tailoring of this fluid for use in other systems, including flow models made from other materials.

METHOD

We investigated fluids composed of glycerol (G-3700, ACP Chemicals Inc, Montreal QC, CAN), distilled water, NaI (7920-1 Caldon Laboratories, Georgetown ON CAN), and xanthan gum (XG)(G1253 Sigma Aldrich, St. Louis MO, USA). XG is a water-soluble, long-chain polysaccharide with a molecular weight of several million. We found that very low concentrations of XG, between 0.002 wt% and 0.025 wt%, were sufficient to impart the desired degree of shear thinning to the fluids. Prospective BMF solutions were selected based on the degree to which they accurately matched the high-shear-rate viscosity of blood and their qualitative match to the low $\dot{\gamma}$ shear thinning of blood illustrated in Figure 1. A single fluid was then selected that best characterized the macroscopic rheological properties of blood within the region of testable shear rates while also matching the refractive index of PDMS.

A strain-controlled rotational rheometer with a concentric cylinder tool (MCR301, Anton Paar USA Inc, Ashland VA, USA) was used to measure viscosity as a function of shear rate. In practice, viscosities as low as 2 cP could be measured at shear rates down to 1 s^{-1}. Viscosity measurements could be performed for $\dot{\gamma}$ up to about 300 s^{-1}, at which point secondary flow effects became noticeable. The rheometer tool was temperature controlled, and all measurements were performed at 22°C. Mechanical degradation of the fluid was studied using a stress-controlled rotational rheometer with a 25-mm parallel-plate tool (AR 1500ex, TA Instruments USA, New Castle DE, USA). The gap between the plates was in the range 280-310 μm, and the shear rate was between 20 and 3000 s^{-1}.

Refractive indices were measured using an Abbé refractometer (33.45.58 3L, Bausch & Lomb, Rochester NY, USA) with an accuracy of 0.0005. The refractometer was calibrated with distilled water at room temperature. All measurements were performed at room temperature.

RESULTS

Figure 2 shows the viscosity as a function of shear rate for aqueous solutions containing a fixed glycerol concentration of 30 wt% and a range of XG concentrations. The degree of shear thinning increases as the concentration of XG is increased. Figure 2 also shows data for a fluid made by adding 20 g NaI to 100 g of 30 wt% glycerol and 0.025 wt% XG in water. The addition of NaI to the solution results in a noticeable decrease in the degree of shear thinning, making the viscosity curve for this fluid comparable to that of the 0.010% XG solution with no NaI. Other concentrations of NaI (10 wt % and 25 wt %, not shown) were also tested but had viscosity curves very similar to that of the 20% NaI solution plotted in Figure 2.

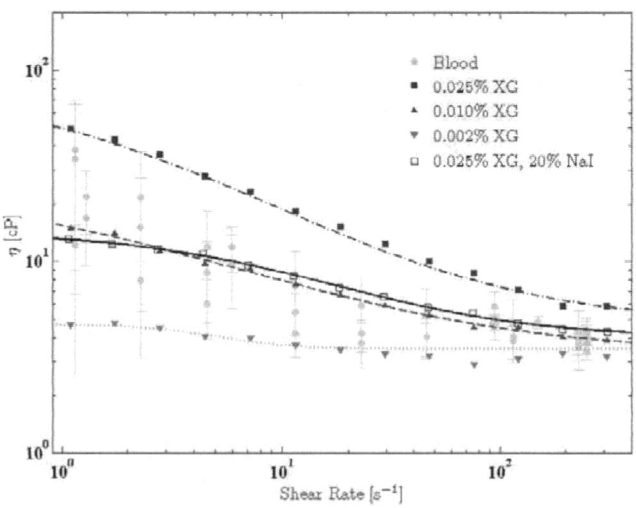

Figure 2 Viscosity as a function of shear rate for solutions of 30 wt% glycerol for different XG and NaI concentrations, as indicated in the legend, shown in comparison to literature whole-blood data. The curves are fits of the solutions data to Equation (1)

For each fluid tested, the Cross model, Eq. (1), was fit to the aggregate data from three independent sets of measurements of viscosity as a function of shear rate. The infinite shear rate viscosity η_∞ determined from the fits was compared to the glycerol concentration, *G,* for several XG concentrations, *X.* In all cases, η_∞ increased linearly with *G*. Least-squares fits of a straight line to the data for each value of *X* were made, and the corresponding fit parameters are given in Table 1. The uncertainties in the fit parameters are substantial because of the small number of data points however; these fits are simply intended to parameterize the results and to provide a means of estimating η_∞ for different fluid compositions.

Table 1 Parameters from straight-line fits to η_∞ as a function of glycerol concentration *G*, where m_1 is the slope and b_1 is the intercept.

XG Conc. *X* (wt%)	m_1 (cP/wt%)	b_1 (cP)
0	0.112	-1.00
0.002	0.145	-1.38
0.010	0.188	-2.36
0.025	0.226	-2.91

The dependence of η_∞ on *X* for several glycerol concentrations was measured, showing η_∞ increases with *X*, and, over the range of *X* studied here, can be described by the form

$$\eta_\infty = aX^b + c, \quad (2)$$

where *a*, *b*, and *c* are fit parameters as given in Table 2.

Table 2 Parameters from fit of η_∞ as a function of XG concentration, *X,* as given by Equation (2).

Glycerol Conc. *G* (wt %)	*a* (cP/wt%)	*b*	*c* (cP)
30	0.559	0.285	2.41
35	0.480	0.389	2.91
40	0.652	0.356	3.41
44	1.13	0.223	3.99

The refractive index *n* of the solution is important for PIV experiments using vascular phantoms. *n* as a function of glycerol concentration for fluids with no NaI and with 16.7% NaI added were measured, showing that *n* increases linearly with *G*. Straight lines were fit to the data, and the parameters are provided in Table 3.

Table 3 Parameters from straight line fits to the refractive index *n* as a function of glycerol concentration *G*, where m_2 is the slope and b_2 is the intercept.

NaI Conc. (wt %)	m_2 (wt%$^{-1}$)	b_2
0	0.00134 ± 0.00005	1.3300
16.7	0.00139 ± 0.00005	1.3550

The refractive index was then measured as a function of NaI concentration *N* for several concentrations of glycerol. In all cases, the slope of the data increases as *N* is increased. The data can be described by a quadratic equation,

$$n(N) = fN^2 + gN + h, \quad (3)$$

for each glycerol concentration *G*. The corresponding fit parameters are displayed in Table 4. The refractive index was independent of *X* for the very low XG concentrations used in this study.

Table 4 Parameters from fits of *n* as a function of glycerol concentration, *G*, to quadratic Equation (3).

Glycerol Conc. *G* (wt %)	*f* (10^{-5} wt%$^{-2}$)	*g* (wt%$^{-1}$)	*h*
0	2.16	0.00109	1.3340
30	1.25	0.00131	1.3720
32	1.05	0.00139	1.3739
35	1.12	0.00137	1.3780
42	0.382	0.00146	1.3872

Based on these results, a fluid formulation with composition 51.9 wt% water, 25.9 wt% NaI, 22.2 wt% glycerol, and 0.019 wt% XG was selected as most closely mimicking the shear-rate dependent viscosity of blood and, simultaneously, the refractive index of the PDMS phantoms commonly used in PIV studies. Three independent measurements of the viscosity of this fluid as a function of shear rate are shown in Figure 1. The data for our BMF agrees with the blood data within the error bars in the latter, although its viscosity may be systematically slightly higher than that of blood at intermediate shear rates. In particular, our BMF reproduces the shear thinning behavior of blood over the range of shear rates for which data are available, and it has the same high-shear-rate visocsity as blood. The BMF data were fit to Equation (1); this fit is also plotted in Figure 1. The corresponding fit parameters are displayed in Table 5 along with those for the blood literature data.

Table 5 Parameters from fits of η as a function of $\dot{\gamma}$ to Equation (1) for our BMF and the literature data for blood. Uncertainties are one standard deviation.

Fluid	η_∞ (cP)	η_0 (cP)	τ (s^{1-q})	*q*
BMF	4.0 ± 0.2	18.3 ± 0.7	0.37 ± 0.05	0.32 ± 0.03
Blood	4.4 ± 0.7	60 ± 3	1.4 ± 0.3	-0.3 ± 0.3

The refractive index of this fluid was measured to be n = 1.4145±0.0005, in agreement with the value of 1.4140±0.0008 previously specified by Yousif et al. (Yousif 2011) for the PDMS vascular phantom. Figure 3 shows an empty vascular phantom next to one filled with the non-Newtonian BMF. The latter is visibly transparent.

We verified that the viscosity of our BMF does not change as a result of sustained exposure to high shear, such as might occur in a pump. The viscosity of a sample of the fluid was measured as a function of $\dot{\gamma}$ before and after subjecting the fluid to a shear rate of approximately 3000 s^{-1} for one hour. The two viscosity curves agreed within experimental error, indicating that no mechanical degradation occurred.

Figure 3 PDMS vascular phantom with and without non-Newtonian BMF

The temperature dependence of the viscosity as a function of shear rate was measured at 20, 22 and 25°C (results not shown) and found to be consistent with the results of pure glycerol. Finally, fluids made with XG purchased from a bulk foods store (Bulk Barn) were also prepared and found to have properties that were not significantly different from those of fluids made using XG purchased from a chemical supply firm.

CONCLUSION

A non-Newtonian blood-mimicking fluid has been designed that matches the shear-rate dependence of the viscosity of human blood with a hematocrit between 40-45% at physiological temperature. The fluid is intended for PIV experiments on blood-mimicking flow, and its refractive index is matched to that of the PDMS commonly used to fabricate vascular phantoms. This fluid reproduces the rheological properties of blood more realistically than the Newtonian fluids used in most previous in vitro flow studies. By modifying its composition, the properties of the fluid can easily be modified to mimic those of blood at different hematocrit levels and for use in phantoms made from different materials.

ACKNOWLEDGMENTS

Financial support was provided by the Heart and Stroke Foundation of Ontario (Grant #T-6427) and the Natural Sciences and Engineering Research Council of Canada (Discovery Grant).

REFERENCES

Antonova, N., Zvetkova, E., Ivanov, I., and Savov, Y. (2008). Hemorheological changes and characteristic parameters derived from whole blood viscometry in chronic heroin addicts. Clin Hemorheol Microcirc *39*, 53–61.

Amparo,, V., Murado, J., Santaolaria, M., Simo, M., Mico, L., Calvo, J., Todoli, J., and Ricart, J.M. (2007). Haemorheological changes in patients with systematic lupus erythematosus do not seem to be related to thrombotic events. Clin Hemorheo Microcirc *38, 23-29.*

Baskurt, O.K., and Meiselman, H.J. (2003). Blood rheology and hemodynamics. N. York, ed. (c1974-: Stratton Intercontinental Medical Book Corporation), pp. 435–450.

Brookshier, K.A., and Tarbell, J.M. (1993). Evaluation of a Transparent Blood Analog Fluid: Aqueous Xanthan Gum/Glycerin. Biorheology *30*, 107–116.

Bor-Kucukatay, M., Keskin, A., Akdam, H., Kabukcu-Hacioglu, S., Erken, G., Atsak, P., and Kucukatay, V. (2008). Effect of thrombocytapheresis on blood rheology in healthy donors: Role of nitric oxide. Transfu Apher Sci *39*, 101–108.

Carrera, L.I., Etchepare, R., D'Arrigo, M., Vaira, S.M., Valverde, J., D'Ottavio, A.E., and Foresto, P. (2008). Hemorheologic changes in type 2 diabetic patients with microangiopathic skin lesions. Journal Diabetes Complications *22*, 132–136.

Charm, S. E., McComis, E., and Kurland, G. (1964). Rheology and structure of blood suspensions. J. Appl. Physiol. *19* 127-133.

Fehr, M., Galliard-Grigioni, K.S., and Reinhart, W.H. (2008). Influence of acute alcohol exposure on hemorheological parameters and platelet function in vivo and in vitro. Clin Hemorheo Microcirc *39*, 351–358.

Galduróz, J., Antunes, H.K., and Santos, R.F. (2007). Gender- and age-related variations in blood viscosity in normal volunteers: A study of the effects of extract of Allium sativum and Ginkgo biloba. Phytomedicine *14*, 447–451.

Gijsen, F., Van de Vosse, F., and Janssen, J. (1999). The influence of the non-Newtonian properties of blood on the flow in large arteries: steady flow in a carotid bifurcation model. J Biomech *32*, 601–608.

Goldstone, J., Schmid-Schönbein, H., and Wells, R. (1970). The rheology of red blood cell aggregates. Microvasc Res *2*, 273–286.

Merrill, E.W. (1969). Rheology of blood. Physiol Rev *49*, 863–888.

Rand, P.W., and Lacombe, E. (1964). Hemodilution, tonicity, and blood viscosity. J Clin Invest *43*. 2214-2226.

Walker, A.M., Johnston, C.R., and Rival, D.E. (2014). On the Characterization of a Non-Newtonian Blood Analog and Its Response to Pulsatile Flow Downstream of a Simplified Stenosis. Ann of Biomed Eng *42*, 97–109.

Yousif, M.Y., Holdsworth, D.W., and Poepping, T.L. (2011). A blood-mimicking fluid for particle image velocimetry with silicone vascular models. Exp Fluids *50*, 769–774.

LUMPED PARAMETER MODEL OF FLOW THROUGH THE CAROTID BIFURCATION

Onaizah Onaizah
Department of Medical Biophysics
University of Western Ontario
London, ON, Canada
oonaizah@uwo.ca

Tamie Poepping
Department of Physics and Astronomy
University of Western Ontario
London, ON, Canada
poepping@uwo.ca

Mair Zamir
Department of Applied Mathematics
University of Western Ontario
London, ON, Canada
zamir@uwo.ca

ABSTRACT

The carotid artery is one of the major supply routes of blood to the brain. Currently, it is unknown what impact local changes within the carotid artery have on global circulation and specifically on the pressure and flow relationship in the common carotid artery (CCA) to the internal carotid artery (ICA) and eventually into the cerebral circulation. We used a lumped parameter model matched to an in vitro flow loop to study the effects of local stenosis and stiffness changes in the carotid artery, together with a family of carotid artery phantoms incorporating a progressively increasing stenosis or varying compliance. We found that in general local stenosis had minimal impact on blood supply to the brain, up to a certain threshold. Statistically significant changes were seen in the mean ICA flow beyond a 70% stenosis ($p<0.05$) and also in the modeled resistance ($p<0.01$). Local stiffness of the carotid artery had a significant effect on the mean ICA flow ($p<0.05$) and was reflected in the compliance derived by the lumped parameter model ($p<0.001$).

KEYWORDS: carotid stenosis, carotid compliance, vascular impedance, lumped parameter model, hemodynamics

INTRODUCTION

The internal carotid artery (ICA) – a branch of the common carotid artery - is one of the major suppliers of blood to the brain. The carotid bifurcation is prone to plaque development that can alter local blood flow patterns, contributing to potential plaque rupture and blood clot formation that ultimately can result in ischemic stroke. These changes occur primarily due to age and disease and are influenced by many factors. While changes in local flow patterns have been widely studied using imaging techniques [1] and CFD [2], a comprehensive study of the effects of carotid artery disease (associated with the progression of stenosis and stiffness) on the global hemodynamics is not available.

A lumped parameter (LP) model has often been used to study the distal impedance of a vascular system. The LP model is essentially an electrical analogue to the fluid mechanical system where voltage and current are treated as pressure and flow rate to output an impedance of the associated circuit. Vascular impedance is an extension of vascular resistance as it provides a more complete relationship between pressure and flow. Vascular impedance takes into the account the more dynamic and pulsatile nature of the flow as well as the elastic nature of the arteries (vascular compliance) and the influences of inertia [3, 4, 5].

Studying how changes in the compliance impact the carotid artery can be of considerable significance. Waveform analysis and the use of the lumped parameter model is not a new concept in the study of blood flow. In fact, the original Windkessel LP model was first defined in 1899 by Otto Frank. It has since evolved and been used to study different arteries but a gap does exist in its application to cerebral flow [6] and carotid flow (leading to cerebral and extra-cranial flow).

The objective of this study was to implement an LP model matched to an experimental in vitro flow loop containing carotid phantoms. The LP model was validated using experimentally measured pressure and flow rate waveforms to determine the impedance of the system. The model was applied to study two cases, one of increasing stenosis and one of decreasing compliance, and each of their impacts on mean ICA flow. The aim of this project was to better understand the role of local vascular parameters on blood flow to the brain.

METHOD

A. Experimental

An experimental in vitro flow loop containing life-size carotid artery phantoms was used (Fig. 1a). The flow loop (Fig. 1b) consisted of a programmable positive-displacement pump [7] (CompuFlow 1000, Shelley Medical Imaging, London CAN) that circulated the blood mimicking fluid (BMF) [8] (dynamic viscosity of 4.31 ± 0.03 mPa.s) stored in the reservoir. The pump was programmed to deliver an idealized common carotid artery (CCA) flow rate waveform (cardiac cycle of 0.92 s) [9].

The flow loop incorporated a family of eccentrically stenosed (relative to distal ICA diameter as per the NASCET criteria – normal, 30%, 50% and 70%) carotid artery phantoms with an 8 mm CCA inner diameter (ID) that bifurcates into an internal carotid artery (ICA) and external carotid artery (ECA) with inner diameters of 5.5 mm and 4.5

mm, respectively. Phantoms of two different compliances were implemented. The phantoms were manufactured in house, and both phantoms were made from polydimethylsiloxane (PDMS) (Sylgard 184, Dow Corning). The block phantom consists of a hollow flow channel inside a block of PDMS, and the thin-walled phantom has a vessel of 1-mm thick PDMS wall mounted in a box filled with the BMF for the purpose of neutral buoyancy.

Three in-line electromagnetic flow meter probes (EP620 and EP625, Carolina Medical Electronics) are attached at the inlet of the CCA and outlets of the ICA and ECA. The custom-built flow meter sections include a hemostasis valve for inserting a pressure catheter (SPR 350S, size 5F, Millar Instruments, Inc., Houston, TX), which thus enables simultaneous measurements of the pressure and volumetric flow-rate waveforms.

B. Theoretical

A lumped parameter model was developed to match the bifurcating in vitro flow loop with an upper ICA and lower ECA branch, as depicted in Fig. 1c. The branches contain elements associated specifically with the resistance due to the stenosis (R_{ica}) and the compliance of the phantom (C_{ica} and C_{eca}); both of these parameters can play a significant role in carotid artery disease as they change the pressure and flow waveforms that ultimately reach the brain. The compliance (C_{ed}/C_{id}) and resistance (R_{ed}/R_{id}) of the peripheral tubing and that of the phantom are separate components in the model. Eqs. (1) and (2) show the impedance of the ICA and ECA branches of the LP model which are combined in Eq. (3) to calculate the total impedance of the system.

The pressure waveform measured at the CCA inlet of the in vitro flow loop was used as the input to the LP model. This waveform is broken into its Fourier components, and each component is divided by a frequency-dependent impedance (model) to generate flow rate frequency components that can be recombined to form a theoretical flow-rate waveform. The theoretical waveform is then matched to the experimentally measured flow-rate waveform by adjusting the impedance (R, L, C parameters). The parameter values that best match the theoretical flow-rate waveform to the experimental waveform are used to characterize the system impedance.

$$Z_{ica} = R_{ica} + i\omega L_{ica} + \frac{1}{i\omega C_{id} + i\omega C_{ica} + \frac{1}{R_{id}}} \quad (1)$$

$$Z_{eca} = R_{eca} + i\omega L_{eca} + \frac{1}{i\omega C_{ed} + i\omega C_{eca} + \frac{1}{R_{ed}}} \quad (2)$$

$$\frac{1}{Z_{total}} = \frac{1}{Z_{ica}} + \frac{1}{Z_{eca}} \quad (3)$$

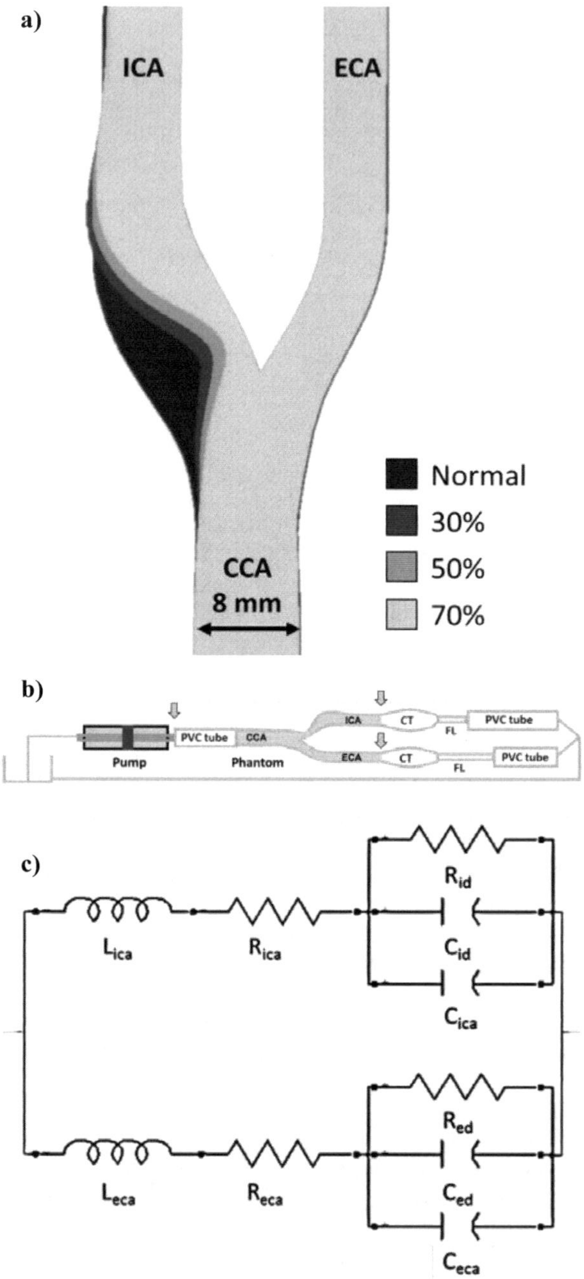

Figure 1: Family of phantom geometries with increasing plaque progression forming more severe stenosis overlaid on the normal (plaque-free) geometry. b) In vitro flow loop configuration used. Pressure and flow-rate waveforms were measured at the locations indicated by the arrows. c) 10-element LP model matched to in vitro flow loop shown above. R_{ed} and C_{ed} represent the resistance and compliance of elements located downstream from the ECA, C_{id} and R_{id} represent the resistance and compliance of elements located downstream from the ICA, and the remaining elements (R_{ica}, C_{ica}, L_{ica}) represent the resistance, compliance and inductance respectively for the ICA and ECA (R_{eca}, C_{eca}, L_{eca}) branches of the phantom.

RESULTS

The LP model was validated as shown in Fig. 2 where the measured and modelled flow-rate waveforms are plotted. Simultaneous measurement of pressure and flow-rate were made proximal to the bifurcation (at the inlet of the CCA). The downstream resistance (R_{ed} = 44.22 ± 0.5, R_{id} = 31.59 ± 0.35 mmHg/(mL/s)) and compliance (C_{ed} = C_{id} = 2.4 ± 0.05 x 10^{-3} mL/mmHg) were calculated beforehand by using a simpler flow loop configuration and held fixed while the new phantom components were allowed to vary. The results listed in Table 1 suggest that local changes happening within the carotid do not make a significant impact on the impedance as the R, C, and L parameter associated with the phantom are much smaller than the downstream parameters. The experimental and theoretical flow rate curves are well matched with a goodness of fit (R^2) of greater than 0.97.

Fig. 3 shows the mean flow through the ICA for progressively higher stenosis and decreasing compliance. In Fig. 3a, as the stenosis increases, the mean flow decreases with the biggest drop between the rigid 50% and 70% phantoms. The results are statistically significant for stenosis of 70% or greater ($p<0.05$). In Fig. 3b, as the compliance of the carotid artery phantom decreases, so does the mean flow through the ICA which is also statistically significant ($p<0.05$). In fact, it is important to note that the compliant 50% phantom actually has a higher mean flow through the ICA than the rigid normal phantom, which suggests that compliance plays quite a large role on blood supply to the brain.

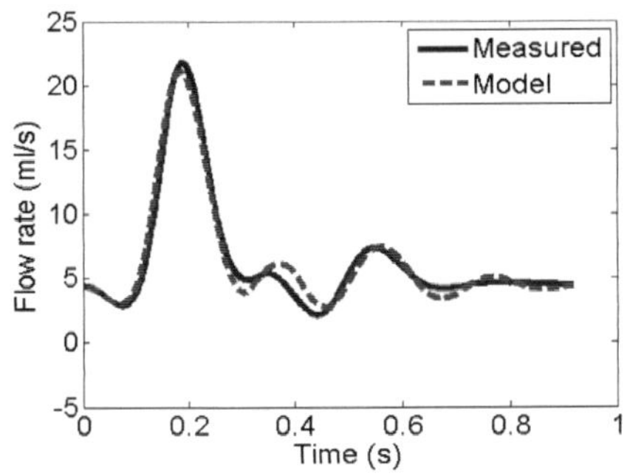

Figure 2: Measured and modelled waveforms for a rigid carotid phantom with 50% ICA stenosis. Waveforms were measured at the CCA inlet.

DISCUSSION

A waveform analysis using a lumped parameter model was performed to study blood supply to the brain via the carotid arteries. Carotid artery phantoms with increasing stenosis were used and the main finding was that local changes within the carotid artery were found to have minimal impact on the global hemodynamics up to a certain threshold (70% - severe stenosis) beyond which significant changes in blood supply were observed.

Table 1 Output parameter for all cases where the model flow-rate waveform is iteratively matched to the measured waveform and mean flow-rates are calculated from the waveforms.
*p<0.05, ** p<0.01, *** p<0.001; Units: [L: mmHg/(mL/s^2)] [R: mmHg/(mL/s)] [C: 10^{-3}*mL/mmHg] [Flow: mL/s]

	Rigid Normal	Rigid 30%	Rigid 50%	Rigid 70%	Compliant 50%
R_{ica}	0.44±0.06	0.92±0.06	1.30±0.14	2.56±0.14 **	0.44±0.06
C_{ica}	0.59± 0.01	0.41±0.01	0.58±0.11	0.35±0.04	1.86±0.04***
L_{ica}	0.29±0.01	0.30±0.01	0.30±0.01	0.32±0.01	0.30±0.01
R_{eca}	0.84±0.05	0.84±0.05	0.84±0.05	0.84±0.05	0.84±0.05
C_{eca}	0.59±0.01	0.41±0.01	0.58±0.11	0.35±0.04	1.86±0.04***
L_{eca}	0.29±0.01	0.30±0.01	0.30±0.01	0.32±0.01	0.30 ± 0.01
ICA Flow	3.74±0.10	3.72±0.07	3.70±0.09	3.56±0.08*	3.88±0.12*
CCA Flow	6.00±0.07	5.98±0.05	6.08±0.09	5.95±0.12	6.08±0.08

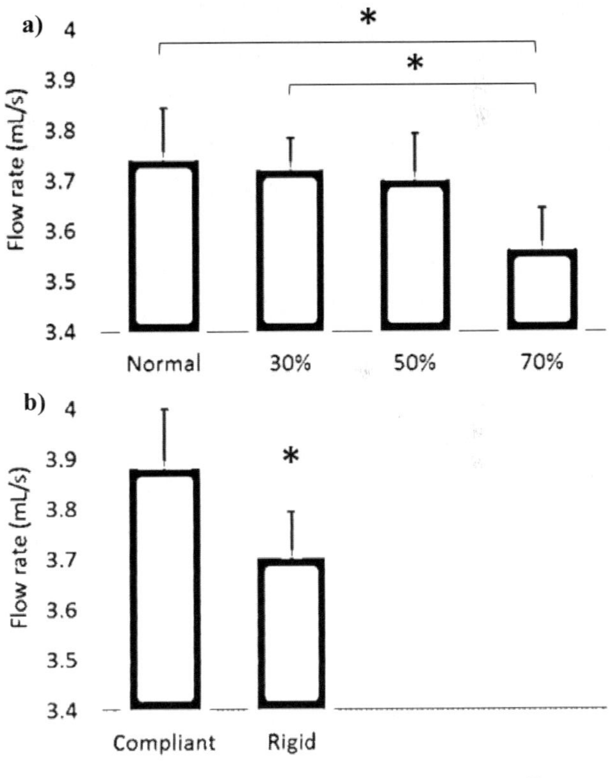

Figure 3: a) Effects of Obstructive Disease. Mean flow-rate (measured) through the ICA is shown for a family of 4 rigid phantoms corresponding to increasing ICA plaque formation from a normal (plaque-free) geometry up to a severe (70%) stenosis. b) Effects of Sclerotic Disease. Mean ICA flow-rate measured in a 50% stenosed compliant and rigid phantom.

Changes in local compliance were also studied and were found to have a significant impact on the global hemodynamics. This is of substantial importance since aging has a severe impact on compliance, it decreases dramatically (by approximately a factor of 4) between the ages of 25 and 75 [10]. Overall, the combination of an LP model and an experimental in vitro flow loop provide a powerful tool to study the effects of local changes within the carotid on blood supply to the brain.

Limitations of the study include the fact that the model does not account for wave reflections, and that the thin-walled compliant phantom, although more compliant than the rigid model, still has a relatively low compliance, and thus the effects of more physiologically realistic compliance could not be studied. Also, the in vitro flow loop incorporates downstream resistances and compliances that are based on the best available data which may provide a reasonable approximation.

CONCLUSIONS

A lumped parameter model was successfully designed and validated to match an experimental in vitro carotid artery flow loop and the model was then used to study how local changes in the carotid artery impact blood supply to the brain. The changes in the carotid artery due to stenosis (up to 70%) were found to have minimal impact on the downstream perfusion. However, the elasticity of the carotid artery had a significant impact on the vascular impedance and downstream perfusion.

ACKNOWLEDGMENTS

The authors would like to acknowledge Amanda Dicarlo for help with experiments, Hristo Nikolov for the phantom fabrication, and Brian Dalrymple and Frank Van Sas for machining and technical help. Financial support is acknowledged from the Ontario Ministry of Research and Innovation (Early Researcher Award), Heart and Stroke Foundation of Ontario (operating grant #T-6427), Natural Sciences and Engineering Research Council of Canada (Discovery Grant), O.O. was partially supported by a Canadian Institutes of Health Research Training Fellowship in Vascular Research.

REFERENCES

1. Kefayati, S., Milner, J.S., Holdsworth, D.W., Poepping, T.L. 2014, "In vitro shear stress measurements using particle image velocimetry in a family of carotid artery models: effect of stenosis severity, plaque eccentricity, and ulceration", PLoS One, Vol(9), pp. e98209.
2. Steinman, D.A., Poepping, T.L., Tambasco, M., Rankin, R.N., Holdsworth, D.W. 2000, "Flow patterns at the stenosed carotid bifurcation: effect of concentric versus eccentric stenosis", Ann Biomed Eng, Vol(28), pp. 415-23.
3. Tzeng, Y.C., Chan, G.S. 2011, "Unraveling the human cerebral circulation: insights from cerebral blood pressure and flow recordings", J Appl Physiol (1985), Vol(111), pp. 349-50.
4. Zhu, Y.S., Tseng, B.Y., Shibata, S., Levine, B.D., Zhang, R. 2011, "Increases in cerebrovascular impedance in older adults", J Appl Physiol (1985), Vol(111), pp. 376-81.
5. Zamir, M. 2012, *The Physics of Coronary Blood Flow*, Springer, New York.
6. Zuj, K. A. 2012, Analysis of Transcranial Doppler Ultrasound Waveform Morphology for the Assessment of Cerebrovascular Hemodynamics. Kinesiology. Electronic Theses and Dissertations (UW), University of Waterloo.
7. Holdsworth, D.W., Rickey, D.W., Drangova, M., Miller, D.J., Fenster, A. 1991, "Computer-controlled positive displacement pump for physiological flow simulation", Med Biol Eng Comput, Vol(29), pp. 565-70.
8. Yousif, M.Y., Holdsworth, D.W., Poepping, T.L. 2009, "Deriving a blood-mimicking fluid for particle image velocimetry in Sylgard-184 vascular models", Conf Proc IEEE Eng Med Biol Soc, 2009, Minneapolis, Minnesota, USA, pp. 1412-5.
9. Holdsworth, D.W., Norley, C.J., Frayne, R., Steinman, D.A., Rutt, B.K. 1999, "Characterization of common carotid artery blood-flow waveforms in normal human subjects", Physiol Meas, Vol(20), pp. 219-40.
10. Lanne, T., Hansen, F., Mangell, P., Sonesson, B. 1994, "Differences in mechanical properties of the common carotid artery and abdominal aorta in healthy males", J Vasc Surg, Vol(20), pp. 218-25.

Proceedings of the 25th CANCAM
London, Ontario, Canada, May 31 – June 4, 2015

COMPUTATIONAL MODELING OF MICROBUBBLE DISTRIBUTION IN MICROVASCULAR NETWORKS

Daniel Goldman
Department of Medical Biophysics
University of Western Ontario
London, ON, Canada
dgoldma2@uwo.ca

Graham M. Fraser
Department of Medical Biophysics
University of Western Ontario
London, ON, Canada
gmfraser@uwo.ca

Christopher G. Ellis
Department of Medical Biophysics
University of Western Ontario
London, ON, Canada
cgellis@uwo.ca

ABSTRACT

Contrast-enhanced ultrasound using microbubbles as the contrast agent is becoming an increasingly important medical imaging technique. Given the complexity of red blood cell distribution in the microvasculature, it is expected that microbubble distribution will also be complex. However, the details of microbubble distribution at the microvascular level have not been extensively studied. Therefore, we are developing a computational model of microbubble flow within individual microvessels, at diverging bifurcations, and within microvascular networks comprising both capillaries and several branching orders of arterioles. Intravital microscopy experiments are planned to determine bubble distribution properties for input into the model, and to verify the predictions of the model. In this paper we present the framework to be used for our computational model and present results for simplified cases of bubble distribution at diverging bifurcations.

KEYWORDS: CEUS, microbubble, microvascular, perfusion, mathematical model

INTRODUCTION

Contrast-enhanced ultrasound (CEUS) using encapsulated microbubbles has been applied to a number of clinical situations, including the response of microvascular blood flow distributions to increased plasma insulin [1]. This flow response is believed to be important for insulin utilization under normal conditions, and a key factor in the occurrence of insulin resistance in pre-diabetes and type 2 diabetes. In particular, CEUS has been used to suggest that capillary recruitment is induced by increased insulin under normal conditions, and that this recruitment is impaired in animals and humans that are insulin-resistant. CEUS measurements in skeletal muscle have also indicated that small increases in plasma insulin affect microvascular blood flow distribution, while larger increases are required to alter total blood flow. However, it has not been clear from these experiments exactly how insulin acts along the arteriolar tree or how it alters red blood cell (RBC) flow in capillaries.

We are currently exploring blood flow and microbubble distribution in microvascular networks and their alteration due to increased plasma insulin using two approaches. First, we have initiated intravital microscopy experiments to observe distribution of both RBCs and fluorescently labeled microbubbles of various sizes in skeletal muscle arteriolar trees and capillaries. Second, we are developing a computational model to calculate how specific changes in arteriolar diameters within an arteriolar tree will alter microbubble distribution at the capillary level, and how this will vary with bubble size, vessel diameter and hematocrit (RBC volume fraction).

In this paper we describe initial steps towards this computational model, including our existing two-phase (RBCs and plasma) steady-state blood flow model, a model of skeletal muscle microvascular network geometry, and simple models of bubble distribution at diverging arteriolar bifurcations.

MODELING METHODS

Blood flow. Our steady-state two-phase blood flow model, which has been previously described, is based on the model developed and validated in [2]. This model conserves total volume flow rate and RBC volume flow rate at each node connecting two or three vessel segments (assumed to be cylindrical), and includes an empirical function describing the nonlinear way in which RBCs distribute at diverging bifurcations. It also includes an empirical relation describing how blood viscosity varies as a function of vessel diameter and hematocrit.

Microvascular network geometry. Our microvascular network geometry is based on measurements of arteriolar trees in skeletal muscle, particular those recently published from in vivo experiments on the rat gluteus maximus [3].

Starting from a first-order arteriole with a diameter (D) of 100μm, our network bifurcates 8 times until reaching terminal arterioles with a mean diameter of 15μ□. Asymmetries in diameters and lengths of daughter vessels at each bifurcation are included to yield a realistic range of blood flows (Q) at each level in the network and, in particular, to match the power law log(Q) ~ 2.63*log(D) found in [3]. At the terminal end of each arteriolar flow path, it is assumed that blood flows into a group of capillaries whose number is proportional to the blood flow in the supplying terminal arteriole.

Radial bubble distribution. Encapsulated microbubbles are nearly rigid spheres and therefore at very low Reynolds number (Re<<1) should undergo minimal radial migration. However, in medium to large arterioles Re~1 and therefore some size- and Re-dependent variation in radial position is expected. We assume this variation depends on the ratio of vessel diameter to bubble diameter, D/d, and on the local Re=DV/ν_{blood}, where V is the mean blood velocity and ν_{blood} is the hematocrit-dependent blood viscosity. Based on theoretical and experimental studies [4,5], we are working to construct a radial bubble density function ϕ of the form

$$\phi(r) = \bar{\phi} \cdot f(r/R, r_*/R, \phi_{max}/\bar{\phi}, \phi_0/\bar{\phi}) \quad (1)$$

where $\bar{\phi}$ is the mean bubble density in the vessel being considered, R is the vessel radius, R-r_* is the bubble-free region near the vessel wall, ϕ_{max} is the maximum bubble density, and ϕ_0 is the centerline bubble density. The parameters r_*, ϕ_{max}, and ϕ_0 will depend on D/d and Re.

Bubble distribution at diverging bifurcations. Given the radial bubble density function above and assumed velocity profiles $u(r)$ in microvessels, the relative bubble flow rate into daughter branches at diverging bifurcations can be estimated. If we assume the flow from the parent vessel into the two daughter vessels is separated by a plane, then the location of this plane can be calculated knowing the daughter flow rates, which are given by our network flow model. The bubble flow rate to each daughter can then be estimated by integrating $\phi(r)\,u(r)$ over the respective portion of the parent vessel that supplies each daughter. In this way, we can estimate the bubble distribution at each bifurcation proceeding down the arteriolar tree, and obtain a bubble flow rate in each terminal arteriole. Since the radial distribution of bubbles is expected to differ from that of RBCs, which are highly deformable and generally larger than microbubbles used in CEUS, we expect the distribution of bubble flow rates in the terminal arterioles will differ from the distribution of RBC flow rates.

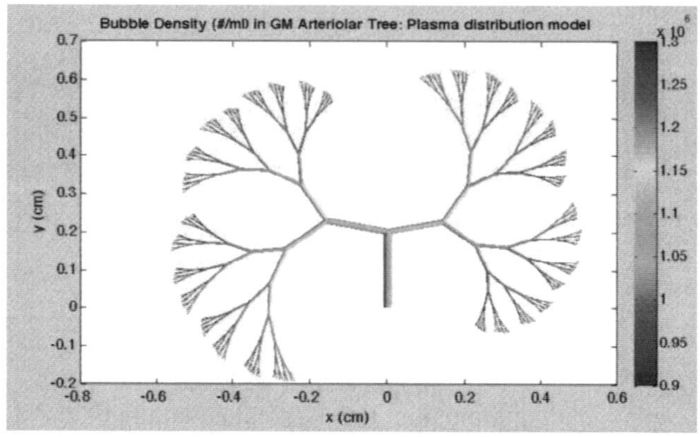

Figure 1: Bubble distribution for plasma model

RESULTS

We are still in the process of determining suitable radial bubble density functions $\phi(r)$, and therefore consider two simplified cases based on the calculations in our blood flow model. We first compute the RBC and plasma flow in the network considered, and fix the bubble supply rate into the first-order arteriole. This results in different initial bubble densities depending on the radial distribution model considered (since it affects the mean axial bubble velocity). We then implement our simplified bubble distribution models at diverging bifurcations to determine bubble flow and density throughout the network.

Figure 1 shows arteriolar bubble density (capillaries are not shown) for the case where microbubbles are assumed to distribute in the same way as plasma. Figure 2 shows the case where microbubbles are assumed to distribute in the same way as RBCs. The differences in the bubble densities in the two cases are a result of both the bubble distribution at diverging bifurcations and the velocities of the bubbles, which are faster in the RBC model than in the plasma model.

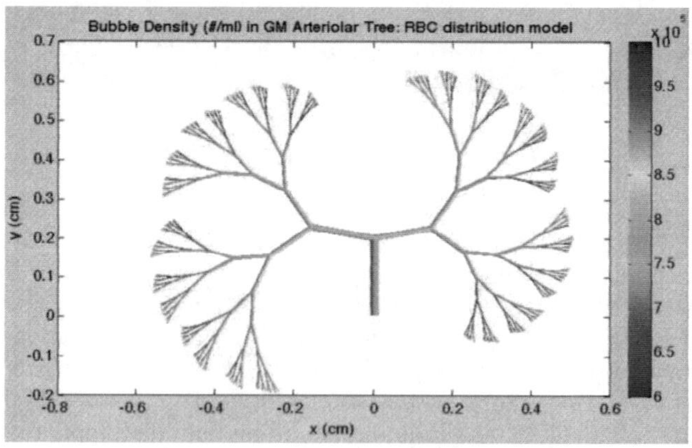

Figure 2: Bubble distribution for RBC model

CONCLUSIONS

We have presented a framework for computational modeling of microbubble distribution in microvascular networks. In the future, we plan to implement this approach using specific experiment-based functions in Eq. (1) to model radial bubble distribution and compare calculated RBC and bubble distributions in capillary beds for microbubble sizes in the range of 1-10μm.

We plan to validate our computational model by comparing our results to measurements made in rat skeletal muscle using intravital videomicroscopy with fluorescently labeled microbubbles.

Since the number of microbubbles used in CEUS imaging is relatively small (<1 per 1000 RBCs), our computational model will help interpret CEUS results on blood flow magnitude and, more importantly, distribution at the microvascular level. Distribution of not only blood flow as a whole but also of plasma and RBCs separately is crucial in understanding the transport to tissue of oxygen and other substances such as glucose, and our model will be used to improve the ability of CEUS to provide this type of information. The main application considered at present is in evaluating the blood flow response to insulin in healthy, pre-diabetic and type 2 diabetic patients, but applications to other diseases affecting the microvasculature are anticipated in the future.

ACKNOWLEDGMENTS

This work was supported by a Natural Sciences and Engineering Research Council (NSERC) grant (R4081A03) awarded to D.G.

REFERENCES

1. Barrett, E.J., Wang, H., Upchurch, C.T. & Liu, Z. (2011). Insulin regulates its own delivery to skeletal muscle by feed-forward actions on the vasculature. *Am. J. Physiol. Endocrinol. Metab.* 301, E252-E263.

2. Pries, A. R., Secomb, T. W., Gessner, T., Sperandio, M. B., Gross, J. F. & Gaehtgens, P. (1994). Resistance to blood flow in microvessels in vivo. *Circulation Research, 75*(5), 904-915.

3. Al-Khazraji, B. K., Novielli, N. M., Goldman, D., Medeiros, P. J. & Jackson, D. N. (2012). A simple streak length method for quantifying and characterizing red blood cell velocity profiles and blood flow in rat skeletal muscle arterioles. *Microcirculation, 19*(4), 327-335.

4. Choi, Y.-S. & Lee, S.-J. (2010). Holographic analysis of three-dimensional inertial migration of spherical particles in micro-scale pipe flow. *Microfluid Nanofluid.* 9, 819-829.

5. Bhattacharya, S., Gurung, D.K. & Navardi, S. (2013). Radial lift on a suspended finite-sized sphere due to fluid inertia for low-Reynolds-number flow through a cylinder. *J. Fluid Mech.* 722, 159-186.

EVALUATING FLOW DISTURBANCES IN THE CAROTID BIFURCATION

Tamie L. Poepping
Department of Physics & Astronomy
Western University
London, ON, Canada
poepping@uwo.ca

Amanda L. DiCarlo
Department of Physics & Astronomy
Western University
London, ON, Canada
adicarl3@uwo.ca

Onaizah Onaizah
Department of Medical Biophysics
Western University
London, ON, Canada
oonaizah@uwo.ca

Mair Zamir
Department of Applied Mathematics
Western University
London, ON, Canada
zamir@uwo.ca

ABSTRACT

Atherosclerosis in the carotid artery, which is a major supply route to the brain, is a primary risk factor for stroke or brain attack. Arterial compliance, a measure of the elasticity of blood vessels, is a common indicator of vascular disease and is known to decrease in association with other stroke risk factors such as age, diabetes, and high blood pressure. Decreased carotid compliance leads to changes in the flow and pressure waveforms and corresponding changes in the velocity patterns. Resulting hemodynamic parameters – such as increased shear stress and turbulence – are thought to play a key role in the processes of both plaque and blood clot formation. Complementary methods of numerical modelling and particle image velocimetry (PIV) in physical models serve to enable controlled studies of both global blood supply and local flow patterns as a function of various independent variables, including vessel compliance and stenosis severity. PIV measurements of velocity fields with high temporal and spatial resolution have been obtained in a family of carotid phantoms, including comparisons in a matched set with varying compliance. The reduced vessel compliance leads to higher peak pressures, peak velocities, and turbulence intensities. Regions of disturbed flow also see higher levels of turbulence intensity that also are sustained for a longer portion of the cardiac cycle. A lumped parameter model, matched to the existing in vitro flow loop, can be used to study the global effects due to local stenosis and stiffness changes in the carotid, as well as quantifying their relative contributions to the impedance of the system.

KEYWORDS: carotid stenosis, arterial compliance, vascular impedance, lumped parameter model, particle image velocimetry, turbulence, shear stress

INTRODUCTION

Atherosclerosis is a progressive arterial disease whereby lesions characteristically develop at localized 'atheroprone' regions of disturbed arterial flow, such as curved regions and bifurcations, resulting in a thickening and hardening of the artery wall. At these atherosusceptible regions, flow departs from pulsatile, unidirectional shear stress to create zones of flow separation, flow reversal, oscillating bi-directional shear stress, and turbulence. Hemodynamics also play a significant role in stroke risk, where thrombus formation may be accelerated in regions of slow or recirculating flow, high shear and increased turbulence, and where plaque rupture has been correlated with localized high shear stress regions. Flow disturbances have also been shown to contribute to the pathogenesis of many other clinical disorders, including aneurysms, aortic valve calcification, post-intervention intimal hyperplasia, and stent restenosis.

Arterial compliance decreases in conjunction with the presence of a number of known cardiovascular risk factors – including age, hypertension, and diabetes (Cheng et al. 2002) – and is associated with atherosclerosis at various sites in the arterial tree (van Popele et al. 2001). Furthermore, arterial compliance has been shown to be an independent predictor of ischemic stroke risk, transient ischemic attack, and even fatal stroke events (Sugioka et al. 2002; Laurent et al. 2003; Dijk et al. 2004; Tsivgoulis et al. 2006), separate from additional cardiovascular risk factors.

An in vitro facility incorporating a family of anthropomorphic phantoms was used to analyze the effect of compliance on carotid artery flow patterns, shear stress, and turbulence intensity experimentally within carotid artery phantoms using particle image velocimetry. A lumped parameter model enabled quantification of the impedance induced by increasing stenosis severity and decreasing compliance.

METHODS

A. Experimental Flow Model

An in vitro flow loop incorporating life-sized phantoms provides a facility for simulating and measuring physiological flow scenarios. At the carotid artery bifurcation, the common carotid artery (CCA) branches into the external carotid artery (ECA), supplying blood to the face and neck, and the internal carotid artery (ICA) supplying flow to the brain. A family of anthropomorphic carotid bifurcation models, based on in-vivo x-ray angiographic studies (Smith et al. 1996), includes geometries ranging from a disease-free ('normal') geometry to a severely stenosed (70% ICA diameter reduction), as depicted in Figure 1a. For each stenosis severity, there are two geometric configurations or eccentricities – based on concentric (axisymmetric) and eccentric (asymmetric) plaque deposition within the ICA sinus – and two different vessel compliances. Phantoms were manufactured from optically transparent polydimethylsiloxane (PDMS) (Sylgard 184, Dow Corning) using a lost-core method to generate a low-compliance block phantom, with a hollow flow channel inside a block of PDMS, and a thin-walled phantom, with a 1-mm thick PDMS vessel wall (Poepping et al. 2010). Physiologically realistic flows are generated using a programmable volumetric-displacement pump (Holdsworth et al. 1999)(CompuFlow 1000, Shelley Medical Imaging, London ON) to circulate a blood-mimicking fluid (BMF) with suitable dynamic viscosity of 4.31±0.03 mPa•s and a refractive index (n= 1.4140) closely matched to the PDMS models (Yousif et al. 2011). Downstream flow resistors, consisting of small diameter tubing, are used to mimic the downstream vascular bed and achieve an approximate flow division between the ICA and ECA of 2: 1 (Poepping et al. 2002); (Hoi et al. 2010). Custom CCA flow-rate waveforms can be generated, varying from one representing young healthy adults (Holdsworth et al. 1999) to one for older adults (Hoi et al. 2010). In-line pressure and flow meters enable co-localized measurement of experimental intraluminal pressure and volumetric flow-rate waveforms at each of the CCA inlet, and ICA and ECA outlets.

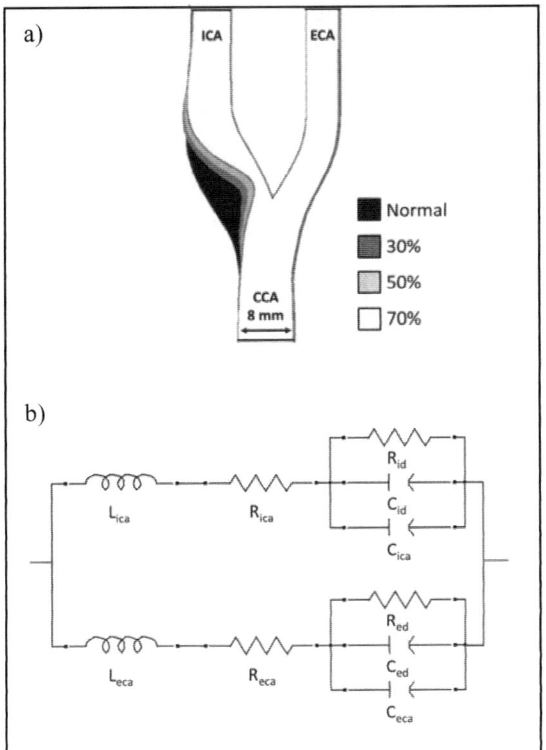

Figure 1: a) Schematic of the geometries for a family of carotid models with plaque formation progressing from a 'normal' or disease-free geometry to increasing stenosis severity in the ICA (measured clinically as a % constriction in inner diameter relative to the upstream ICA), and b) A 10-element circuit specifically designed to match the in-vitro flow system.

B. Lumped Parameter Model

A lumped parameter (LP) model was constructed to match the in-vitro flow facility, where voltage and current are treated as pressure and flow rate in an electrical analogue of the fluid mechanical circuit in order to estimate the impedance of the flow system. Vascular impedance considers both resistance and compliance (capacitance) of the pulsatile flow within the elastic arteries, as well as inertial (inductance) effects of the dynamic flow properties (Tzeng and Chan 2011; Zhu et al. 2011). The circuit, shown in Figure 1b, contains elements associated specifically with resistance due to the ICA stenosis (R_{ica}) and compliance of the carotid artery (C_{ica} and C_{eca}), which are key parameters in carotid artery disease. The compliance (C_{id}, C_{ed}) and resistance (R_{id}, R_{ed}) of components downstream from the ICA and ECA are considered separately and kept constant across the family of carotid models. Inertial effects are lumped into a single inertance for each of the ICA (L_{ica}) and ECA (L_{eca}). The impedance of the ICA (Eqn. 1) and ECA (Eqn. 2) branches in the model are combined to calculate the total impedance of the system (Eqn. 3) as follows:

$$Z_{ica} = R_{ica} + i\omega L_{ica} + \frac{1}{i\omega C_{id} + i\omega C_{ica} + \frac{1}{R_{id}}} \quad (1)$$

$$Z_{eca} = R_{eca} + i\omega L_{eca} + \frac{1}{i\omega C_{ed} + i\omega C_{eca} + \frac{1}{R_{ed}}} \quad (2)$$

$$\frac{1}{Z_{total}} = \frac{1}{Z_{ica}} + \frac{1}{Z_{eca}} \quad (3)$$

C. Particle Image Velocimetry Measurements

High-resolution particle images were obtained in the transparent PDMS carotid phantoms using a stereoscopic time-resolved PIV system previously described (Kefayati et al. 2014). The laser sheet was collimated to 1 mm thick in the region of interest. High-speed cameras were equipped with low-pass filters to eliminate reflections and isolate signal from fluorescent tracer particles (Fluostar). Volumetric data was acquired by adjusting the elevation of the phantom in 0.5 mm increments for a total of 15 horizontal planes. Thirty cardiac cycles were collected for each plane, where acquisition was synchronized via an external trigger signal generated by the pump. Double-frame exposure PIV images were collected at a frame rate of 100 Hz.

A root-mean-square mask was applied to raw particle images prior to velocity computation to eliminate non-flow regions. Unique masks were created for each horizontal image plane and each cardiac cycle time point and plane to account for possible wall motion in the compliant model. Velocity vectors were computed using a fast-Fourier transform cross-correlation multi-pass algorithm (DaVis 7.2, LaVision, Inc.) with decreasing window size from 64x64 pixels down to 16x16 pixels with 50% overlap on windows, resulting in approximately 0.3 mm in-plane (x and y) final vector spacing. A post-processing median filter and smoothing 3x3 kernel was applied once to resultant vector fields. All further post processing and computations were completed in Matlab.

Quantification of flow disturbances was based on the turbulence intensity metric. Velocity maps were first ensemble averaged at each phase point based on the following:

$$V_i = \frac{1}{N}\sum_{n=0}^{N-1} v_i(t+nT), \qquad i = (x,y,z)$$

where N represents the number of total cardiac cycles (N=30) and T is the cycle period (T=920ms). Next, the phase-averaged flow was separated from non-coherent fluctuations using a Reynolds decomposition method (Reynolds and Hussain 1972):

$$v_i'(t,n) = v_i(t,n) - V_i(t)$$

From these velocity fluctuations, the total turbulent intensity was estimated as follows:

$$TI_{TOT}(t) = \sqrt{(TI_x(t))^2 + (TI_y(t))^2 + (TI_z(t))^2}$$

$$TI_i(t) = \sqrt{\frac{\sum_{n=1}^{N} v_i'(t,n)v_i'(t,n)}{N-1}}, \qquad i = x,y,z$$

RESULTS

Waveform analysis using a lumped parameter model enabled estimation of the resistance associated with increasing ICA stenosis, which increased more than six fold from 0.44 to 2.6 mmHg/(ml/s) while progressing from normal (no plaque or stenosis) to severely stenosed. Increasing resistance due to carotid stenosis severity had minimal impact on the global hemodynamics (pressure and volumetric flow rate waveforms) that would supply blood to the brain. However, changes in local compliance had a significant impact on both the average flow rate through the ICA and the local flow waveforms.

Figure 2 shows phase-averaged velocity magnitude for both the thin-walled compliant vessel and rigid block model at peak systole. Maximum CCA inlet velocity occurs 10 ms later for the compliant vessel model compared to the rigid model as the compliance induces a phase delay. Higher velocities are observed overall in the block phantom, including higher maximum jet velocities compared to the compliant model. Both models exhibit a jet pattern in the ICA that begins at the stenosis throat, hugs the inner wall, crosses over and impinges on the outer wall. Two large recirculation zones exist on either side of the jet. A velocity vector map is shown for a cross-sectional slice located 12 mm distal to the apex in the ICA branch illustrating similar

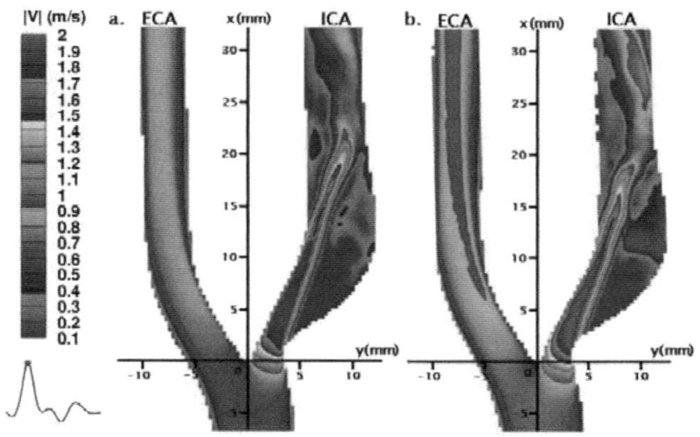

Figure 2: Flow visualization based on the central plane phase-averaged velocity magnitude at peak systole are shown for a compliant thin-vessel phantom (a) and rigid block phantom model (b)..

transverse flow patterns with a slight spiral forming in the jet of the rigid model.

Figure 3 shows turbulence intensity maps in the ICA. Maximum turbulence intensity occurred during the deceleration phase following peak systole when the flow destabilizes and the jet dissipated. The rigid model shows a higher level of disturbed flow that also spans a broader area. This region of elevated turbulence intensity occurs more proximal, or upstream, in the rigid model, and notably where blood potentially feeds back into the larger upstream recirculation zone. In vivo, platelets activated by high shear gradients are prone to preferentially adhere to the vessel wall in adjoining low shear regions. Thus increased turbulence in close proximity to the recirculation zone, as observed in the more rigid models, could lead to increased platelet aggregation and thrombus formation in these regions, as well as expose vulnerable plaques to highly disturbed flow.

While the PDMS carotid models used here incorporate idealized geometry with highly planar symmetry, and thus inducing flow symmetry, they provide a unique matched model family enabling well-controlled studies of the effects of various dependent variables, such as vessel compliance and stenosis severity, to be isolated. Also, due to manufacturing limitations, the carotid phantoms have a low vessel compliance compared to in vivo, but demonstrate the relative significance of compliance.

CONCLUSIONS

A matched model study was used to investigate the role of arterial compliance on the local hemodynamics in carotid artery phantoms with identical vessel geometry. Decreased compliance results in higher velocities, higher intraluminal pressures, and higher levels of flow disturbance – all of which suggest potentially increased stroke risk.

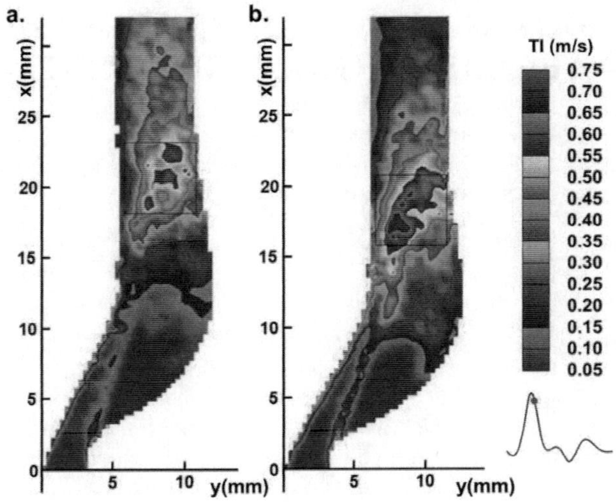

Figure 3: Turbulence intensity maps in the ICA of the compliant vessel (a) and rigid block phantom (b) for the deceleration phase immediately following peak systole.

ACKNOWLEDGMENTS

The authors acknowledge Hristo Nikolov for the phantom fabrication, Brian Dalrymple and Frank Van Sas for machining and technical help. Funding is gratefully acknowledged from the Canada Foundation for Innovation, Ontario Ministry of Research and Innovation, Heart and Stroke Foundation of Ontario (operating grant #T-6427) and Natural Sciences and Engineering Research Council of Canada (Discovery Grant). A.D. and O.O. were partially supported by a Canadian Institutes of Health Research Training Fellowship in Vascular Research.

REFERENCES

Cheng KS, Baker CR, Hamilton G, Hoeks AP, Seifalian AM. Arterial elastic properties and cardiovascular risk/event. European journal of vascular and endovascular surgery : the official journal of the European Society for Vascular Surgery 2002;24:383-97.

Dijk JM, van der Graaf Y, Grobbee DE, Bots ML, Group SS. Carotid stiffness indicates risk of ischemic stroke and TIA in patients with internal carotid artery stenosis: the SMART study. Stroke; a journal of cerebral circulation 2004;35:2258-62.

Hoi Y, Wasserman BA, Xie YJ, Najjar SS, Ferruci L, Lakatta EG, Gerstenblith G, Steinman DA. Characterization of volumetric flow rate waveforms at the carotid bifurcations of older adults. Physiol Meas 2010;31:291-302.

Holdsworth DW, Norley CJ, Frayne R, Steinman DA, Rutt BK. Characterization of common carotid artery blood-flow waveforms in normal human subjects. Physiol Meas 1999;20:219-40.

Kefayati S, Holdsworth DW, Poepping TL. Turbulence intensity measurements using particle image velocimetry in diseased carotid artery models: Effect of stenosis severity, plaque eccentricity, and ulceration. Journal of Biomechanics 2014;47:253-63.

Laurent S, Katsahian S, Fassot C, Tropeano AI, Gautier I, Laloux B, Boutouyrie P. Aortic stiffness is an independent predictor of fatal stroke in essential hypertension. Stroke; a journal of cerebral circulation 2003;34:1203-6.

Poepping TL, Nikolov HN, Rankin RN, Lee M, Holdsworth DW. An in vitro system for Doppler ultrasound flow studies in the stenosed carotid artery bifurcation. Ultrasound in Medicine and Biology 2002;28:495-506.

Poepping TL, Rankin RN, Holdsworth DW. Flow patterns in carotid bifurcation models using pulsed Doppler ultrasound: Effect of concentric vs. eccentric stenosis on turbulence and recirculation. Ultrasound in Medicine and Biology 2010;36:1125-34.

Reynolds WC, Hussain AKM. Mechanics of an organized wave in turbulent shear-flow: Theoretical Models and Comparison with Experiments. J Fluid Mech 54:263-8 DOI 10.1017/s0022112072000679 1972

Smith RF, Rutt BK, Fox AJ, Rankin RN, Holdsworth DW. Geometric characterization of stenosed human carotid arteries. Academic Radiology 1996;3:898-911.

Sugioka K, Hozumi T, Sciacca RR, Miyake Y, Titova I, Gaspard G, Sacco RL, Homma S, Di Tullio MR. Impact of aortic stiffness on ischemic stroke in elderly patients. Stroke; a journal of cerebral circulation 2002;33:2077-81.

Tsivgoulis G, Vemmos K, Papamichael C, Spengos K, Daffertshofer M, Cimboneriu A, Zis V, Lekakis J, Zakopoulos N, Mavrikakis M. Common carotid arterial stiffness and the risk of ischaemic stroke. European journal of neurology : the official journal of the European Federation of Neurological Societies 2006;13:475-81.

Tzeng YC, Chan GS. Unraveling the human cerebral circulation: insights from cerebral blood pressure and flow recordings. J Appl Physiol (1985) 2011;111:349-50.

van Popele NM, Grobbee DE, Bots ML, Asmar R, Topouchian J, Reneman RS, Hoeks AP, van der Kuip DA, Hofman A, Witteman JC. Association between arterial stiffness and atherosclerosis: the Rotterdam Study. Stroke; a journal of cerebral circulation 2001;32:454-60.

Yousif MY, Holdsworth DW, Poepping TL. A blood-mimicking fluid for particle image velocimetry with silicone vascular models. Experiments in Fluids 2011;50:769-74.

Zhu YS, Tseng BY, Shibata S, Levine BD, Zhang R. Increases in cerebrovascular impedance in older adults. J Appl Physiol (1985) 2011;111:376-81.

Proceedings of the 25th CANCAM
London, Ontario, Canada, May 31 – June 4, 2015

THE ROLE OF HEPARIN SULPHATE IN ENDOTHELIAL CELL LEUKOCYTE ATTACHMENT IN REGIONS OF SPATIAL WALL SHEAR STRESS GRADIENTS

Scott Cooper
Chemical Engineering
McGill University
Montreal, Quebec, Canada
scott.cooper@mail.mcgill.ca

Richard L. Leask
Chemical Engineering
McGill University
Montreal, Quebec, Canada
richard.leask@mcgill.ca

ABSTRACT

Atherosclerosis is an inflammatory disease of the endothelium which shows increased prevalence in areas of disturbed blood flow such as areas of wall shear stress (WSS) gradients. It is hypothesized that endothelial cells (ECs) sense and respond to WSS by mechanotransduction through the EC glycocalyx. Heparin sulphate (HS), a major component of the EC glycocalyx has been linked to mechanotransduction of uniform WSS. In this study, three dimensional models were used to impose WSS gradients on cultured ECs after degradation of a significant portion of HS from the glycocalyx. It was found that following degradation, ECs did not elicit the same pattern of morphological response as control cells. Furthermore, patterns of firm leukocyte attachment in the models were significantly affected by the degradation of HS. Finally, degradation of HS resulted in a significantly larger amount of firm leukocyte attachment in all conditions. These results support the hypothesis that HS in the EC glycocalyx plays a major role in the mechanotransduction of hemodynamic forces, helping us better understand the early development of atherosclerosis.

KEYWORDS: Mechanotransduction, endothelial cells, leukocyte attachment, wall shear stress gradients

INTRODUCTION

Atherosclerosis is focal in nature and is prevalent in areas of disturbed flow leading to the hypothesis that disturbed flow causes endothelial cell (EC) dysfunction initiating the disease (1). ECs, which line the vasculature, respond to the wall shear stress (WSS) imparted by the fluid flow through a process termed mechanotransduction which is still poorly understood (2, 3). Some key observations of EC response to disturbed flow are morphological rounding and increased occurrence of firm leukocyte attachment.

The glycocalyx is a layer of membrane-bound macromolecules lining the endothelium and has been found to extend on the scale of 0.5-4.5μm into the vessel, *in vivo* (4, 5). Being at the interface of the EC surface and fluid flow places the structure in a prime location to be involved in mechanotransduction. The most abundant component of this structure is heparin sulphate, accounting for 50-90% of glycosaminoglycans in the glycocalyx (6).

Using HS as a target to compromise the glycocalyx, in this study, we performed perfusion experiments in a three dimensional tissue culture model, containing a 50% asymmetric stenosis, Figure 1. Investigation into the morphological response and leukocyte attachment patterns in the regions of WSS gradients explored the role HS plays in the mechanotransduction of these forces.

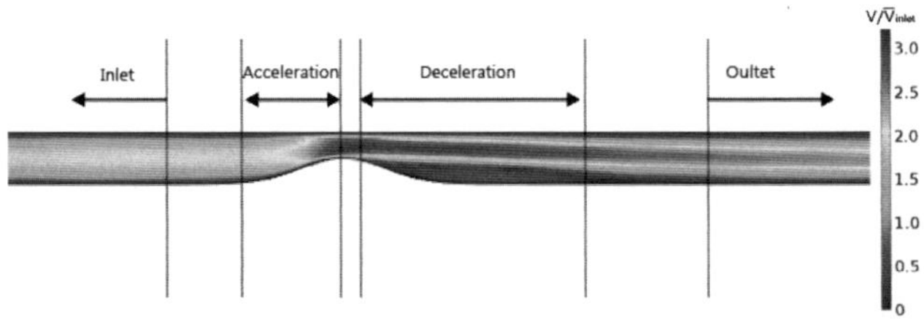

Figure 1: Comsol Simulation of the pre-shearing flow field in in the 50% asymmetric stenosis.

MORPHOLOGICAL RESPONSE

Prior to perfusion experiments, a degradation protocol was established of 2 hour incubation of human abdominal aortic ECs (HAAECs) with an 180 mU/mL solution of heparinase III resulting in a significant decrease in HS (33±4%, two-tailed t-test, $P<0.01$). Morphological response of the HAAECs to WSS gradients was investigated by performing 24 hour preshearing of cultured cells following degradation at an inlet WSS of 10 dyne/cm^2. Using the shape index (previously described by Nerem et al, (7)), the morphological response in the four regions of the model: inlet, acceleration (proximal side of stenosis), deceleration (distal side of stenosis) and outlet, were investigated, Figure 2.

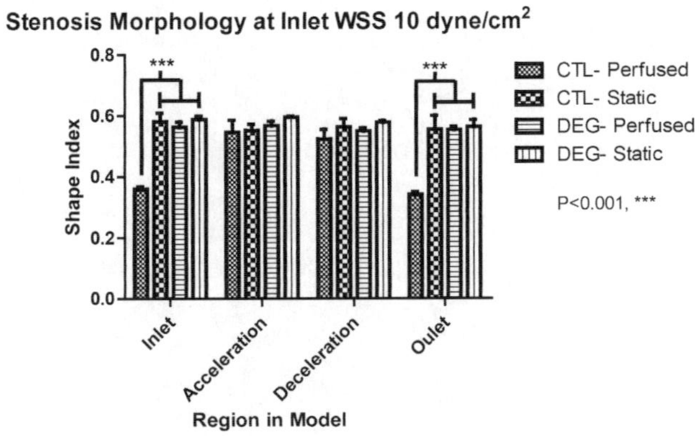

Figure 2: Average endothelial cell shape index in each region with and without HS degradation.

As shown in Figure 2, a decreased shape index in response to flow in the inlet and outlet regions of control cells was not observed following the degradation of HS.

LEUKOCYTE ATTACHMENT

Leukocyte adhesion to the endothelium was investigated by introducing neutrophil-like cells (ATRA stimulated NB4 cell line) and quantifying the number of firmly attached cells post-experiment. Cultured HAAECs were presheared for twenty four hours at an inlet WSS of 10 dyne/cm^2 following HS degradation. Immediately after preshearing, a solution of 1×10^6 NB4 cells/mL suspension was perfused through the models at an inlet WSS of 1 dyne/cm^2 and subsequently fixed with paraformaldehyde. The number of firmly attached cells was assessed, Figure 3.

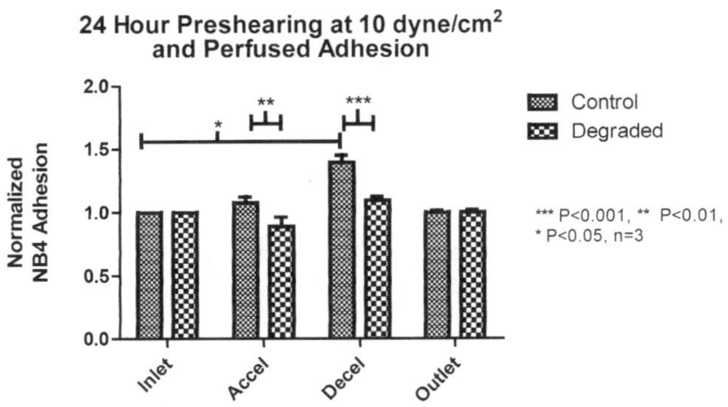

Figure 3: Leukocyte adhesion to each region with and without HS degradation.

As can be seen, the pattern of adhesion was statistically different between control and HS degraded models. Control cells showed an increase in firm attachment in the regions of WSS gradients (acceleration and deceleration regions) whereas there was no change in adhesion in any of the regions of the HS degraded models.

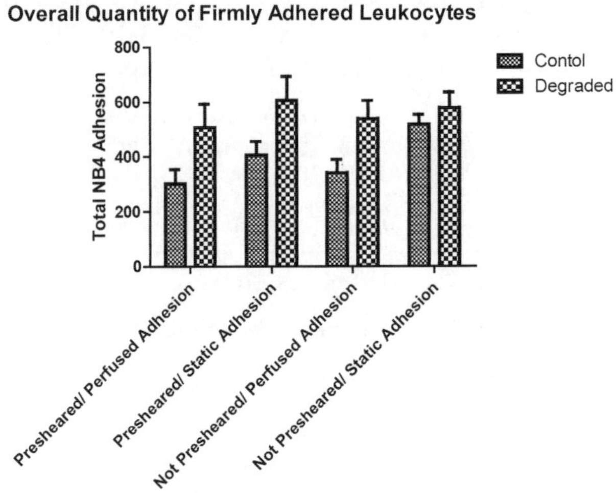

Figure 4: Overall leukocyte attachment for each flow condition with and without HS degradation.

The overall quantity of firmly attached leukocytes was also seen to be significantly higher following HS degradation as seen in Figure 4.

CONCLUSIONS

Heparin sulphate, a major component of the EC glycocalyx, has been suggested as a location for the mechanotransduction of WSS. Following degradation of a significant portion of this glycosaminoglycan, significantly different morphological responses and patterns of leukocyte attachment were found in HAAECs exposed to WSS gradients. Finally, it was also found that an increased in leukocyte adhesion occurred following HS degradation, suggesting the glycocalyx also plays an important role in firm attachment of these cells. These adverse findings help strengthen the hypothesis that the glycocalyx is in fact implicated in mechanotransduction of WSS in ECs.

ACKNOWLEDGMENTS

The authors would like to acknowledge the help by Karli McDonald and Daniel Burkat on this project. Funding was provided by CIHR MOP-119292.

REFERENCES

1. Lopez, A.D., et al., *Global and regional burden of disease and risk factors, 2001: systematic analysis of population health data.* Lancet, 2006. **367**(9524): p. 1747-57.

2. Mulivor, A.W. and H.H. Lipowsky, *Role of glycocalyx in leukocyte-endothelial cell adhesion.* Am J Physiol Heart Circ Physiol, 2002. **283**(4): p. H1282-91.

3. Giuffre, L., et al., *Monocyte adhesion to activated aortic endothelium: role of L-selectin and heparan sulfate proteoglycans.* J Cell Biol, 1997. **136**(4): p. 945-56.

4. Nieuwdorp, M., et al., *The endothelial glycocalyx: a potential barrier between health and vascular disease.* Curr Opin Lipidol, 2005. **16**(5): p. 507-11.

5. Coombe, D.R. and W.C. Kett, *Heparan sulfate-protein interactions: therapeutic potential through structure-function insights.* Cell Mol Life Sci, 2005. **62**(4): p. 410-24.

6. van Haaren, P.M., et al., *Localization of the permeability barrier to solutes in isolated arteries by confocal microscopy.* Am J Physiol Heart Circ Physiol, 2003. **285**(6): p. H2848-56.

7. Ihrcke, N.S., et al., *Role of heparan sulfate in immune system-blood vessel interactions.* Immunol Today, 1993. **14**(10): p. 500-5.

8. Nerem, R.M., M.J. Levesque, and J.F. Cornhill, *Vascular endothelial morphology as an indicator of the pattern of blood flow.* J Biomech Eng, 1981. **103**(3): p. 172-6.

LASER DOPPLER VELOCIMETRY MEASUREMENTS OF WALL SHEAR STRESS UNDER STEADY AND PULSATILE FLOW CONDITIONS IN A PARALLEL PLATE FLOW CHAMBER

Hamed Avari
Department of Mechanical & Material Engineering
University of Western Ontario
London, Ontario, Canada
havari@uwo.ca

Eric Savory
Department of Mechanical & Material Engineering
University of Western Ontario
London, Ontario, Canada
esavory@eng.uwo.ca

ABSTRACT

Most of the studies using Parallel plate flow chambers (PPFC), have relied on theoretical relations for estimating the wall shear stress (WSS) for steady flow only. In the current study, a laser Doppler velocimetry (LDV) method was used to quantify the flow in a PPFC (H = 1.8 mm × W = 17.5 mm, D_h = 3.26 mm, aspect ratio 9.72) under steady and laminar pulsatile flow conditions. No other literature was found on flow quantification using LDV in such a small-scale PPFC. The uniformity of the flow across the span of the channel, as well as the WSS assessment for all flow conditions, were measured with the uncertainties being within 16%. Comparison of the results with theoretical relations, as well as similar reported experiments, suggests good agreement with the differences <12%.

KEYWORDS: LDV, PPFC, steady, Pulsatile, wall shear stress, hemodynamics

I. INTRODUCTION

The pathophysiological effects of hemodynamics on Endothelial Cells (ECs) have been the subject of study for over half a century [1-5]. Due to the difficulties associated with conducting *in vivo* experiments in this field, *in vitro* experiments using parallel plate flow chambers (PPFC) have been widely reported [3, 4, 6]. Based on theoretical relations the wall shear stress (WSS, τ_w) for a steady flow in a rectangular channel for laminar flow is defined as

$$\tau_w = 6Q\mu/(WH^2) \quad (1)$$

where Q, μ, W and H are the volume flow rate, fluid dynamic viscosity, channel width and height, respectively [7]. Although this equation strictly applies to a channel flow with infinite aspect ratio (W/H=∞), when W/H>50, τ_w is within ±1% accuracy of that obtained from this equation [7]. Application of Eq. 1 has been reported in many research studies using PPFC [3, 6, 8, 4], although the effect of aspect ratio on the accuracy of WSS in those studies remains debatable. Eq. 1 is not applicable for predicting the WSS flow regimes for other than steady laminar (e.g., pulsatile). Additionally, the uniformity of the flow over ECs remains extremely under-reported in the literature even though it is a critical factor. Analytical solutions, as well as numerical simulations, have been applied to solve for pulsatile pressure-driven flows through vessels. No analytical solution is capable of solving a flow regime where flow become intermittently turbulent before re-laminarization, (a phenomenon that happens in stenosed arteries *in vivo*) followed by an unsteady recirculation region [9, 10]. Numerical methods depict substantial discrepancies from reported detailed experimental studies [11] due to the limitations of the applied turbulence models [10, 12] and, hence, experimental studies are essential for flow quantifications in such cases.

Among flow measurement techniques, laser Doppler velocimetry (LDV) has been reported in many studies for boundary layer measurements in a channel/tube flow due to its high frequency response (for pulsatile flow measurements) and spatial resolution to determine the near wall velocities. Very few studies were found in the literature where LDV was employed in a channel where H<60 mm. These studies included steady laminar [13, 14, 15], pulsatile laminar [16] or turbulent flows [17, 18]. The smallest and largest H among these studies was 7 mm [17] and 54 mm [16], respectively. LDV measurements become more challenging in a confined channel geometry, with the associate effects of the walls (e.g., light scatter) and limitations associated with the LDV measuring volume size and beam half-angle. The current paper reports on the use of LDV to experimentally quantify the WSS by measuring velocities in steady laminar and pulsatile flow regimes in a small-scale channel flow. The methods and challenges associated with measuring velocity in a small channel along with some results and discussion will be presented.

II. EXPERIMENTAL SETUP

In this section, the flow facility, instruments and measuring techniques and experimental conditions will be discussed.

A. INSTRUMENTS AND MEASUREMENT TECHNIQUES

A PPFC (manufactured from ULTEM 1000) was designed and developed for this study consisting of top and bottom cavities with the coverslip test section positioned on the bottom cavity and upstream and downstream flow development transition lengths of 251 mm and 94 mm, respectively. The channel test section has a L=22 mm with rectangular cross section (height, H=1.8 mm and width, W=17.5 mm). The working section is made from precision glass (Angstrom Precision Optics Inc., USA) for optical access. The flow was driven by a positive displacement pump (CompuFlow 1000 MR, Shelley Medical Imaging Technology, Canada). The flow loop also included a reservoir and a controlled heating unit to maintain the fluid temperature at 37 °C. A Dulbecco's Modified Eagle Medium (DMEM) based biological fluid was prepared and characterized for the viscosity and density to be 0.737±0.001 mPa.s and 997.96±0.12 kg/m^3, respectively.

An LDV system (with a customized setup) consisted of a 5 Watt water-cooled Argon-ion laser (COHERENT, USA), a 2-component (4 beams) mini-probe (TRx20, TSI, USA) with measuring control volume size of 68 μm × 64 μm × 1.1 mm (horizontal axis, vertical axis, length), a photo detector multiplier (PDM 1000, TSI, USA), a FSA 3500 signal processor (TSI, USA) and FlowSizer software (version 2.0.4) [38] was applied. A pair of shifted and unshifted blue beams with wavelengths λ= 488.0 nm were used to measure the u-velocity. The probe was mounted on a multi-axis traverse (THORLABS, USA) with positional accuracy of ±5 μm (in all three directions) to ensure accurate translation in all three directions. Typical data rates were 50-100 Hz near the wall and 500-2000 Hz (depending on the distance from the wall and flow velocity) further from the wall. All the acquired data were analyzed with MATLAB (Total Academic Headcount license, 2013a). A schematic of the LDV layout is depicted in Fig. 1. Titanium dioxide particles (Showa America, USA), with size distribution ranging from 0.24-0.47 μm (average of ≈0.36 μm) were used as the seeding particles. To determine the wall location in this study, the LDV probe was tilted by φ=4.5° (slightly greater than the beam half angle θ=3.9°) to allow for near wall measurements without the beams encountering the channel lower wall. By traversing the measuring volume from the lower channel wall upward, the voltage output of the PDM was monitored with an oscilloscope and the location of the maximum voltage monitored was assigned as being the location of the centre of the measurement volume (with a positional uncertainty of ±27 μm).

B. EXPERIMENTAL CONDITIONS

Flow quantification for a laminar and a laminar pulsatile (carotid waveform) were carried out. To assess whether uniform flow conditions existed in the test section, vertical and lateral profiles of the time-averaged streamwise velocity (U) component were measured at three different streamwise (A, B and C) and spanwise (D, E and F) locations in the test section. Fig. 1-B illustrates the schematic of the glass working section, along with the reference coordinate system and measurement locations. The Re for a channel flow is defined as $Re=U_b(2\delta)/\nu$ or $Re_o=U_o\delta/\nu$, where U_b, U_o, δ and ν are the channel flow bulk velocity (=Q/HW), centre-line maximum velocity, channel half height (δ=H/2) and kinematic viscosity of the fluid, respectively [19]. The flow is reported to be laminar for Re<1350. WSS was calculated as

$$\tau_w=\mu(\partial U/\partial y)_{@y=0} \quad (2)$$

The maximum uncertainties associated with y, U and τ_w (determined from the time-averaged mean velocity gradient) were found to be ±30 μm, ±7.0 % and ±15.4%, respectively.

III. RRESULTS AND DISCUSSION

A. STEADY LAMINAR FLOW

A laminar flow with Q=12.8 ml/s (Re=990) was quantified in the test section. To monitor the uniformity of the flow across the span of the channel, Fig. 2-A illustrates the time-averaged streamwise velocity U measured at locations A, B and C compared with the analytical solution from [7]

$$U(y,z) = \frac{\Delta PH^2}{8\mu L}\left(1-\frac{4y^2}{H^2}\right)-\frac{\Delta PH^2}{8\mu L}\sum_{n=0}^{\infty}\frac{32(-1)^n cosh\left(\frac{(2n+1)\pi z}{H}\right)\cos\left(\frac{(2n+1)\pi y}{H}\right)}{(2n+1)^3\pi^3\cosh\left(\frac{(2n+1)\pi W}{2H}\right)} \quad (3)$$

where U(y,z), ΔP, μ and L are the streamwise velocity at any specified y and z coordinate, the pressure loss along (x) the channel, dynamic viscosity and channel length, respectively. The results show that about 65-75% of the test section is exposed to a uniform flow.

To assess the WSS (τ_w), the time-averaged streamwise velocity U profiles at A, B and C were plotted vs. the analytical solution for the WSS derived from Eq. 2. A 2nd order polynomial curve fitting method was applied to estimate the gradient of the velocity profiles at the wall for both laminar and pulsatile flows. The average experimental (using Eq. 2) and analytical τ_w (derived from Eq. 3) were found to be 11.48±0.98 dyn/cm^2 and 10.45±0.37 dyn/cm^2, respectively, with the 9% difference being within experimental errors.

B. PULSATILE LAMINAR FLOW (CAROTID)

For a flow with imposed fluctuations, the instantaneous velocity u(t) can be Reynolds-decomposed [20] as

$$u = U + \tilde{u} + u'; \qquad <u>= U + \tilde{u} \quad (4)$$

where u, U, $\tilde{u}$, u' and <u> are the u-component instantaneous velocity, the global time-averaged mean velocity, the periodic bin-average velocity, the turbulent fluctuations of the velocity and phase-averaged velocity, respectively. Since the flow was laminar, the u' term was zero [21]. For each measurement location, 200 waveform cycles [22] of period (T) of 920 ms were averaged within 23 bins per cycle. A carotid pulse with f≈1.1 Hz was produced by the pump. The Local minimum and maximum Reynolds numbers were $R_{0\text{-}min}$≈100, $Re_{0\text{-}max}$≈560 and $Re_{0\text{-}mean}$≈282. The Womersley number (α), which is the ratio between the pulsation flow frequency with respect to viscous effects, is defined as $\alpha=\delta(\omega/\nu)^{0.5}$ [7] which returns α=2.76 for the current case. The normalized phase-averaged streamwise velocity $<u>^*$ at location E is depicted in Fig. 3-A for 2y/H=0.33, 0.66 and 1 where there are no significant phase shifts between these profiles. To examine the uniformity of the flow in the test section, the spanwise direction normalized phase-averaged velocity profiles $\langle u\rangle^*$ at $t_1^*=0.19$, $t_2^*=0.33$ and $t_3^*=0.85$ for locations A, B and C are illustrated in Fig. 3-B. The profiles that belong to the same phase at the three different locations compare well with each other. Examining all of the phases reveals that there is no significant difference in the width of the uniform region of the flow in the test section (about 60% of the channel width). The streamwise direction normalized phase-averaged velocity profiles $\langle u\rangle^*$ at $t_1^*=0.19$, $t_2^*=0.33$ and $t_3^*=0.85$ for locations D, E and F are presented in Fig. 3-C. Comparing the graphs corresponding to the same phase of the cycle but at different locations, shows that there are no significant spatial differences, implying no streamwise change at any given time in the test section. Comparison between temporal changes the streamwise velocity profiles shows that the profiles become more blunt at $0<t^*<0.33$, then at $0.33<t^*<0.63$, the profiles tend toward a parabolic profile at heights about $0.88<2y/H<1$ and again at $0.63<t^*<1$, they tend to be more blunt at $0.88<2y/H<1$. The distortion of velocity profiles is dependent upon the value of α [7]. As α increases (>1), inertial forces dominate and, hence, the profile becomes more blunt near the centre-line. A similar pattern is reported for a carotid waveform with $\alpha=4.7$ where a larger region near the centre ($\approx 30\%$ of the vessel diameter) has a blunt region [7].

The phase-averaged shear stresses $\langle \tau_w \rangle$ at locations D, E and F are depicted in Fig. 4. The WSS is positive, indicating a unidirectional WSS over the test section, and the minimum, maximum and mean WSS over the entire pulse was found to be 3.32, 6.02 and 4.82 dyne/cm^2 which are in the physiological range for a carotid pulse (4-15 dyne/cm^2 [7]).

IV. CONCLUSIONS

Laser Doppler Velocimetry (LDV) was applied to quantify the wall shear stress (WSS) in steady laminar and pulsatile laminar flows in a small scale channel flow (H=1.8 mm) which to the best of authors, knowledge is the only study at this scale using LDV. 2nd order polynomial curve fitting was applied to estimate the WSS, which was in good agreement ($\approx 9\%$ difference) with the analytical solution, for steady flow. Phase-averaging was employed to analyze the velocity profiles for WSS estimation for pulsatile flow condition and the velocity profiles showed Womersley number (α) dependent behavior, similar to the reported literature. The flow was found to be uniform over 60-70% of the channel span. The current experimental set-up design and development has been applied in other studies where the pulsatile flow becomes intermittently turbulent.

V. ACKNOWLEDGMENTS

The authors should like to thank Prof. Kem Rogers, the University Machine Services (UMS) at the University of Western Ontario and Mr. Marin Vratonjic for their input into the concept design and fabrication of the PPFC. This work was supported by the CFI Infrastructure Operating Fund and CIHR Strategic Training Program in Vascular Research at Robarts Research Institute, London, Ontario.

REFERENCES

[1] C. F. Dewey and S. R. Bussolari, "The dynamic response of vascular endothelial cells to fluid shear stress," *J. Biomech. Eng.*, vol. 103, pp. 177-185, 1981.

[2] S. R. Bussolari, F. Dewey and M. A. Gimbrone, "Apparatus for subjecting living cells to fluid shear stress," *Rev. Sci. Instrum*, vol. 53, no. 12, pp. 1851-1854, 1982.

[3] R. M. Nerem, M. J. Levesque and J. F. Cornhill, "Vascular endothelial morphology as an indicator of the pattern of blood flow," *J. Biomech. Eng.*, vol. 103, pp. 172-176, 1981.

[4] S. Chien, "mechanotransduction and endothelial cell homeostasis; the wisdom of the cell," *Am J Physiol Heart Circ Physiol*, vol. 292, pp. 1209-1224, 2007.

[5] J.-J. Chiu and S. Chien, "Effects of disturbed flow on vascular endothelium; Pathophysiological basis and clinical perspectives," *Physiol Rev*, vol. 91, pp. 327-387, 2011.

[6] B. R. Blackman, K. a Barbee and L. E. Thibault, "In vitro cell shearing device to investigate the dynamic response of cells in a controlled hydrodynamic environment," *Ann. Biomed. Eng*, vol. 28, no. 4, pp. 363-372, 2000.

[7] G. A. Truskey, F. Yuan and D. F. Katz, Transport phenomena in biological systems, NY: Pearson, 2009.

[8] R. G. Bacabaca, T. H. Smith, S. C. Cowinc, J. J. Van Lanoona, F. T. Nieuwstadtd, R. Heethaarb and J. Klein-Nulend, "Dynamic shear stress in parallel-plate flow chambers," *J. Biomech.*, vol. 38, pp. 159-167, 2005.

[9] M. A. Koo, J.-K. Kang, M. H. Lee, H. J. Seo, B.-J. Kwon, K. E. You, M. S. Kim, D. Kim and J.-C. Park, "Stimulated migration and penetration of vascular endothelial cells into poly (L-lactic acid) scaffolds," *Biomat. Res.*, vol. 18, no. 7, pp. 1-8, 2014.

[10] G. A. Truskey, A. Pinero and K. Barber, "Effects of recirculating flow on U-938 cell adhesion to human vein endothelial cells," *Am J Physiol Heart Circ Physiol*, vol. 275, pp. 591-599, 1998.

[11] F. Ghalichi , X. Deng , A. De Champlain , Y. Douville , M. King and R. Guidoin, "Low Reynolds number turbulence modeling of blood flow in arterial stenoses," *Biorheol.*, vol. 34, no. 4-5, pp. 281-94, 1998.

[12] N. Beratlis, E. Balaras, B. Parvinian and K. Kiger, "A Numerical and experimental investigation of transitional pulsatile flow in a stenosed channel," *J. Biomech. Eng.*, vol. 127, pp. 1147-1157, 2005.

[13] U. Piomelli and J. Liu, "Large-eddy simulation of rotating channels flows using a localized dynamic model," *Phys. Fluids*, vol. 7, no. 4, pp. 839-847, 1994.

[14] S. So, H. Morikita, S. Takagi and Y. Matsmotu, "Laser Doppler velocimetry of turbulent bubbly channel Flow," *Exp. Fluids*, vol. 33, pp. 135-142, 2002.

[15] J. M. Verhelst and F. Nieuwstadt, "Visco-elastic flow past circular cylinders mounted in a channel: experimental measurements of velocity and drag," *J. Nonnewton Fluid Mech.*, vol. 116, pp. 301-328, 2004.

[16] Y. Zhang and D. Che, "Turbulent statistics in a rectangular channel flow with one groove-roughened wall," in *Sixth International Conference on Fluid Mechanics*, Guangzhou, 2011.

[17] J. R. Castrejón-Pita, A. A. Castrejón-Pita, G. Huelsz and R. Tovar, "Experimental demonstration of the Rayleigh acoustic viscous boundary layer theory," *Phys. Rev.*, vol. 73, pp. 036601-1-5, 2006.

[18] C. Appel, J. Mantzaras, R. Schaeren, R. Bombach and A. Inauen, "Turbulent catalytically stabalized combustion of hydrogen/air mixtures in entry channel flows," *Combast. Flame*, vol. 140, pp. 70-92, 2005.

[19] D. Marx, Y. Auregan, H. Bailliet and J. Valiere, "PIV and LDV evidence of hydrodynamic instability over a liner in a duct with flow," *J. Sound Vib.*, vol. 329, pp. 3797-3812, 2010.

[20] S. B. Pope, Turbulent flows, Cambridge: Cambridge, 2000.

[21] W. C. Reynolds and A. K. F. M. Hussain, "The mechanics of an organized wave in turbulent shear flow; part 3, theoretical models and comparisons with experiments," *J. Fluid Mech.*, vol. 54, pp. 263-268, 1972.

[22] S.-C. Wu, "Phase-averaged method applied to periodic flow between shrouded co-rotating disks," *Int. J. Rot. Machin.*, vol. 8, no. 6, pp. 413-421, 2002.

[23] S. Kefayati, J. S. Milner, D. W. Holdsworth and T. L. Poepping, "In vitro shear stress measurements using particle image velocimetry in a family of carotid artery models: effect of stenosis severity, plaque eccentricity, and ulceration," *PLOS ONE*, vol. 9, no. 7, pp. 0098209-1-12, 2014.

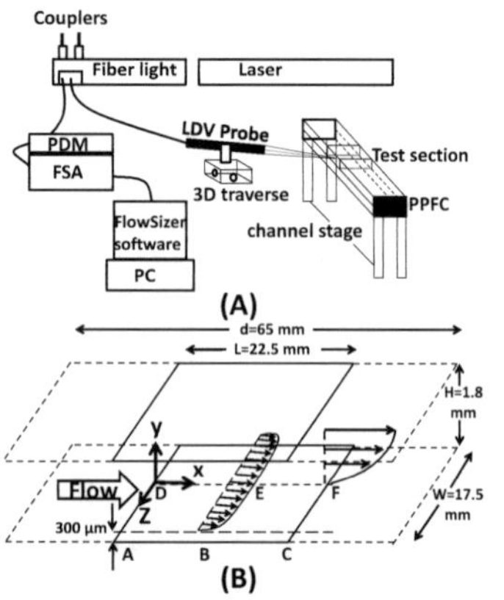

Figure 1 Schematic of the LDV set-up layout (A); Schematic of the test working section, coordinate system and measurement locations (B)

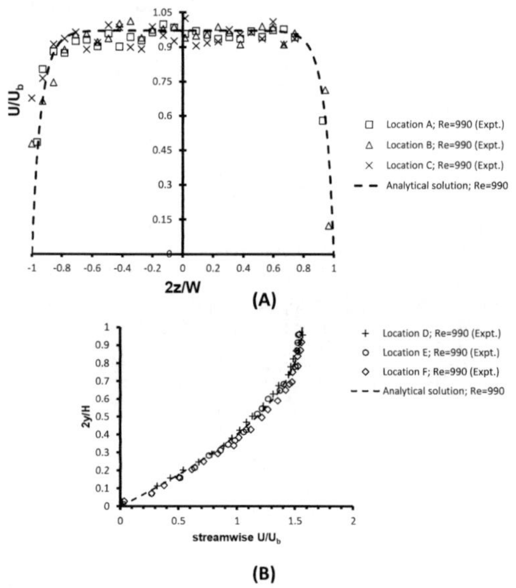

Figure 2 Comparison of streamwise direction time-averaged mean velocity U normalized by U_b in the span of the channel with the analytical solution from for -1<2z/W<1 at locations A, B and C for Re=990. The average deviation of the experimental data points from the analytical solution at locations A, B and C are ±2.4, ±2.9 and ±4.4% (A) Comparison of streamwise time-averaged mean velocity U normalized by U_b (bulk velocity) with the analytical solution from for 0<2y/H<1 at locations D, E and F for Re=990. The average deviation of the experimental data points from the analytical solution at locations D, E and F are ±1.8, ±3.2 and ±6.2% (B)

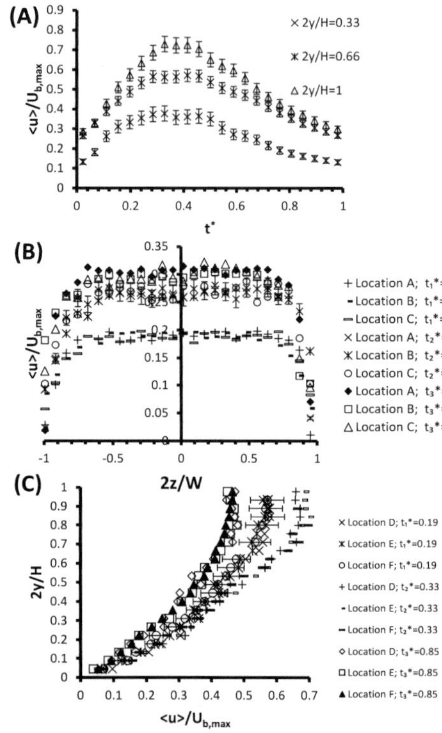

Figure 3 Phase-averaged streamwise velocity profiles <u> normalized by $U_{b,max}$ at 2y/H=0.33, 0.66 and 1 for location E (A); Streamwise direction phase-averaged velocity profiles <u> normalized by $U_{b,max}$ at locations A, B and C for the phases $t_1^*= 0.19$, $t_2^*= 0.33$ and $t_3^*= 0.85$ across the span of the channel -1<2z/W<1 for the carotid pulse (B); Streamwise direction phase-averaged velocity profiles <u> normalized by $U_{b,max}$ at locations D, E and F for the phases $t_1^*= 0.19$, $t_2^*= 0.33$ and $t_3^*= 0.8585$ across the channel half height 0<2y/H<1 for the carotid pulse (C) Note: error bars are indication of experimental uncertainty and are drawn on one data set only for the clarity of the graphs

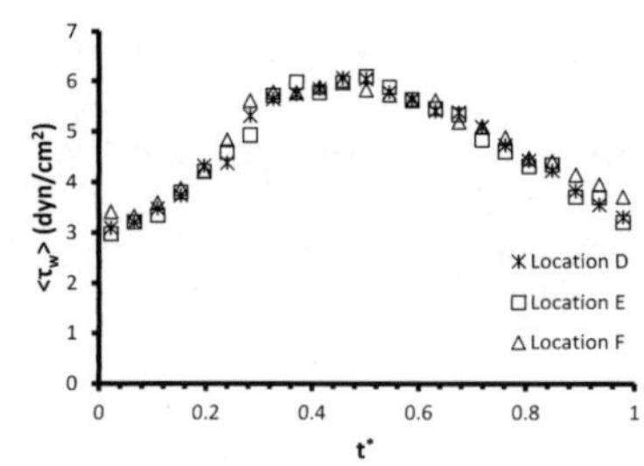

Figure 4 Phase-averaged WSS, $<\tau_w>$, vs. normalized time at locations D, E and F for the carotid pulse

Biomaterials

Proceedings of the 25th CANCAM
London, Ontario, Canada, May 31 - June 4, 2015

ON THE ENERGETICS OF THE EDGE OF AN OPEN LIPID BILAYER

Meisam Asgari
Department of Mechanical Engineering
McGill University
Montréal, QC, Canada
meisam.asgari@mail.mcgill.ca

Aisa Biria
Department of Mechanical Engineering
McGill University
Montréal, QC, Canada
aisa.biria@mail.mcgill.ca

ABSTRACT

The formation and growing of pores in a lipid bilayer have attracted considerable attention from an energetic point of view in recent years. Such pores permit targeted delivery of drugs and genes to the cell. Understanding the energy and elastic behaviour of a lipid-bilayer edge is significant for controlling the formation and growth of such pores. In the present study, the interactions in the molecular level are used to obtain the free energy of the edge of an open lipid bilayer. The resulted free-energy density includes terms associated with flexural and torsional energies of the edge, in addition to a line-tension contribution. The resulting free energy can be used to study the stability and equilibrium configurations of open lipid bilayers.

Keywords: Open lipid bilayer, Elastic free-energy, Molecular interactions, Differential geometry.

INTRODUCTION

A phospholipid molecule consists of a hydrophilic head and two hydrophobic tails. In an aqueous solution at sufficient concentrations, such molecules self-assemble into nanostructures such as lipid bilayers, to protect the tail from the solvent [1]. Lipid bilayers are the main constituents of cell membrane in most living organisms. They act as barriers for water-soluble molecules such as ions, and proteins, and provide the cell and its substructures with compartment and shape [2]. Lipid bilayers are composed of two leaflets of phospholipid molecules oriented transversely and set tail-to-tail.

The formation of an edge in lipid membranes results in exposer of the tail groups at the edge to water, which is energetically unfavourable. As a result, the molecules rapidly rearrange around the exposed edge, forming a semicylindrical rim. Such rearrangement is the source of a line energy at the edge. In order to eliminate this edge energy, lipid bilayers tend to form closed structures such as spheroids. [3]. Nevertheless, they may momentarily open due to various stimuli such as mechanical stresses and thermal instabilities. The formation of these transient pores is essential for regulation of PH, transmembrane electrochemical potential, and concentrations of different molecules in the cell. Additionally, transient open membranes are formed during electro-formation [4]. Recently, stabilizing pores and controling their size have been pursued by means of electric fields [5], sonication [6], and use of edge-active chemical agents [7]. The increasing progress in these techniques has attracted increasing attention to the study of the open lipid bilayers, including molecular dynamic simulations, as well as continuum mechanical treatment and numerical investigations of the equilibrium configurations.

Previous theoretical studies of the stability of pored membranes have mainly relied on constitutive assumptions for the edge, regardless of its flexural and torsional elasticity. For instance, Boal and Rao [8], and Capovilla *et al.* [9] considered the edge energy of an open lipid bilayer as a given constant. Tu and Ou-Yang [10] considered dependence of the edge energy on its geometry, namely geodesic and normal curvatures. Nevertheless, their assumptions on the form of the line energy have not been precisely justified.

May [11] obtained the line energy of a lipid bilayer edge through optimization of the lipid packing at the vicinity of the edge. He modeled the edge as a semicylindrical micelle, and took the free energy per molecule to depend upon the chain length of the molecules, their cross-sectional area, and the strength of the interactions of the molecules with each other and with the surrounding solution. Although successful in obtaining the line tension, that framework did not capture the bending and torsional energetics of the edge. The gap in the literature to successfully relate the macro-scale edge energy to its microstructure has motivated the current study.

A microphysical approach can be used to obtain the free-energy of a material. Using such an approach, Keller and Merchant [12] found the internal energy, surface tension, and bending energy of a liquid surface. As an application of the work of Keller and Merchant [12], the Canham–Helfrich free-energy density for a lipid bilayer was derived based on microphysical considerations [13]. Similarly, a model for the free energy of a wormlike micelle was derived [14].

In this study, we use the microphysical approach of Keller and Merchant [12] to investigate the elastic behavior of the edge of a lipid bilayer. Following May [11] and motivated by previous studies [15, 16, 17, 18], the edge is modeled as a semicylindrical surface. In addition, the phospholipid molecules comprising the edge are modeled as one-dimensional rigid rods of constant length, oriented perpendicular to the centerline of the edge. The applied framework enables us to extract the form of the free energy and the flexural and torsional moduli of the edge, based on the intermolecular energetic interaction between phospholipid molecules.

MATHEMATICAL PRELIMINARY

Consider a smooth, orientable, open surface $\mathcal{S}$ representing the open lipid bilayer, with boundary $\mathcal{C} = \partial\mathcal{S}$, as depicted in Figure 1. Let $\mathcal{C} = \{\boldsymbol{x} : \boldsymbol{x} = \boldsymbol{x}(s), 0 \leq s \leq L\}$, denote the arclength parametrization of the closed boundary curve $\mathcal{C}$. On denoting the differentiation with respect to the arclength s by a superposed dot, the unit tangent of $\mathcal{C}$ is introduced, in terms of the arclength parametrization $\boldsymbol{x}$, by $\mathbf{t} := \dot{\boldsymbol{x}}$. The orientation of $\dot{\mathbf{t}}$ is called the unit normal to the curve $\mathcal{C}$, and is denoted by $\mathbf{N}$. Further, $\kappa = |\dot{\boldsymbol{x}} \times \ddot{\boldsymbol{x}}|$ denotes the curvature of $\mathcal{C}$ at that point. For an arbitrary point on curve $\mathcal{C}$ at which $\kappa \neq 0$, the unit binormal vector is defined by $\mathbf{B} = \mathbf{t} \times \mathbf{N}$. The unit tangent $\mathbf{t}$, unit normal $\mathbf{N}$, and unit binormal $\mathbf{B}$ at each point of $\mathcal{C}$, form the Frenet frame $\{\mathbf{t}, \mathbf{N}, \mathbf{B}\}$ at that point. Further, the torsion τ of $\mathcal{C}$ is defined by $\dot{\mathbf{B}} = -\tau\mathbf{N}$.

On the boundary curve $\mathcal{C}$ of the surface $\mathcal{S}$, the unit normal to the surface is denoted by $\boldsymbol{n}$. Further, let $\boldsymbol{p}$ denote the unit vector in the tangent plane of $\mathcal{S}$ perpendicular to the unit tangent $\boldsymbol{t}$ while pointing outward. We call $\boldsymbol{p}$ the unit tangent-normal. The set of unit normal $\boldsymbol{n}$ to the surface $\mathcal{S}$ at $\mathcal{C}$, unit tangent-normal $\boldsymbol{p}$, and the unit tangent $\boldsymbol{t}$, which is $\boldsymbol{t} = \boldsymbol{n} \times \boldsymbol{p}$, form the oriented basis $\{\boldsymbol{t}, \boldsymbol{n}, \boldsymbol{p}\}$ on $\mathcal{C}$, known as the Darboux frame (Figure 1). Considering that the normal $\mathbf{N}$ and binormal $\mathbf{B}$ of the Frenet frame of $\mathcal{C}$ are also perpendicular to $\boldsymbol{t}$, they both lie in the plane spanned by the normal $\boldsymbol{n}$ and tangent-normal $\boldsymbol{p}$ of the Darboux frame. Therefore, they are related to $\boldsymbol{n}$ and $\boldsymbol{p}$ by $\mathbf{N} = (\cos\psi)\mathbf{n} - (\sin\psi)\mathbf{p}$, and $\mathbf{B} = (\sin\psi)\mathbf{n} + (\cos\psi)\mathbf{p}$, where ψ denotes the angle between the unit normal $\mathbf{N}$ of the curve $\mathcal{C}$ and the unit normal $\boldsymbol{n}$ to the surface $\mathcal{S}$. The derivatives of $\boldsymbol{t}$, $\boldsymbol{n}$, and $\boldsymbol{p}$ with respect to the arclength s along $\mathcal{C}$ are expressed as

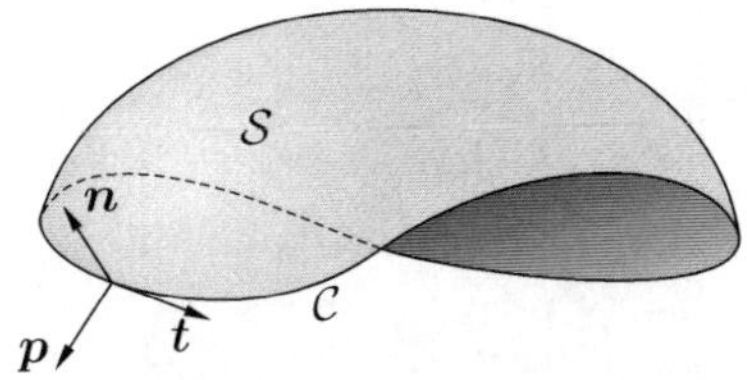

Figure 1: Mathematical identification of an open lipid bilayer as an open surface $\mathcal{S}$ with boundary $\mathcal{C} = \partial\mathcal{S}$ on which a Darboux frame has been shown.

$$\begin{Bmatrix} \dot{\boldsymbol{t}} \\ \dot{\boldsymbol{n}} \\ \dot{\boldsymbol{p}} \end{Bmatrix} = \begin{bmatrix} 0 & \kappa\cos\psi & -\kappa\sin\psi \\ -\kappa\cos\psi & 0 & \tau + \dot{\psi} \\ \kappa\sin\psi & -\tau - \dot{\psi} & 0 \end{bmatrix} \begin{Bmatrix} \boldsymbol{t} \\ \boldsymbol{n} \\ \boldsymbol{p} \end{Bmatrix}. \tag{1}$$

The quantity $\tau_g = \tau + \dot{\psi}$ is called the geodesic torsion of the curve $\mathcal{C}$ on the surface $\mathcal{S}$. The quantity $\kappa\cos\psi$ is the magnitude of the projection of the curvature κ of the curve $\mathcal{C}$ on the unit normal $\boldsymbol{n}$ to the surface $\mathcal{S}$, and is called the normal curvature κ_n of the curve $\mathcal{C}$ on $\mathcal{S}$. Further, the tangential (or geodesic) curvature κ_g of the curve $\mathcal{C}$ on the surface $\mathcal{S}$ is $\kappa\sin\psi$. Thus, the arclength derivatives of $\{\boldsymbol{t}, \boldsymbol{n}, \boldsymbol{p}\}$ in (1) are expressed as $\dot{\boldsymbol{t}} = \kappa_n\boldsymbol{n} - \kappa_g\boldsymbol{p}$, $\dot{\boldsymbol{n}} = -\kappa_n\boldsymbol{t} + \tau_g\boldsymbol{p}$, and $\dot{\boldsymbol{p}} = \kappa_g\boldsymbol{t} - \tau_g\boldsymbol{n}$.

UNDERLYING ASSUMPTIONS

The assumptions below are considered to model the edge of an open lipid bilayer:

- The phospholipid molecules comprising the edge are modeled as one-dimensional rigid rods with the same length a. The lipid molecules are assumed to be perpendicular to the centerline $\mathcal{C}$ [15, 17] (see Fig. 2b).
- The distribution of the molecules along $\mathcal{C}$ is denoted by the molecular density function ζ.
- The molecules at any cross-section of the edge have uniform angular distribution [15].
- The maximum curvature that the boundary $\mathcal{C}$ can support at each point is denoted by $1/\ell$.

Hence, the phospholipid molecules comprising the edge are allocated so that their hydrophilic parts lie on a thin semicylindrical surface as shown in Fig. 2 to form a core shielding the hydrophobic tails from the surrounding solution. The centerline of the edge is denoted by a boundary curve $\mathcal{C}$.

Consider a lipid molecule at the position corresponding to s on $\mathcal{C}$ with orientation θ measured from the corresponding tangent-normal $\boldsymbol{p}(s)$, as shown in Fig. 2. Let the director $\boldsymbol{d}(s,\theta)$ denote the orientation of this molecule. Such a director can be expressed as a linear combination $\boldsymbol{d}(s,\theta) = (\cos\theta)\,\boldsymbol{p}(s) + (\sin\theta)\,\boldsymbol{n}(s)$, where $\boldsymbol{p}(s)$ and $\boldsymbol{n}(s)$ denote the tangent-normal and the unit normal (to $\mathcal{S}$) at the position corresponding to s.

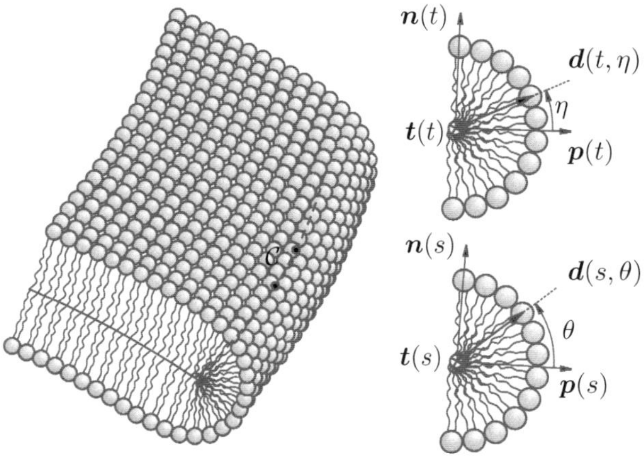

Figure 2: (a) A section of the edge of an open bilayer, (b) cross-sections of the edge at positions $\boldsymbol{x}(s)$ and $\boldsymbol{x}(t)$ with Darboux frame $\{\boldsymbol{t}, \boldsymbol{n}, \boldsymbol{p}\}$ at those positions.

FREE ENERGY DENSITY OF THE EDGE

Consider two molecules, with directors $\boldsymbol{d}$ and $\boldsymbol{e}$, located respectively at positions $\boldsymbol{x}$ and $\boldsymbol{y}$ interior to $\mathcal{C}$. Let the interaction energy (encompassing steric, electrostatic, and other relevant effects) between the molecules under consideration be denoted by $\Omega(\boldsymbol{x}, \boldsymbol{y}, \boldsymbol{d}, \boldsymbol{e})$. Similar to Keller and Merchant [12], we assume that the interaction energy between two molecules separated by more than a fixed cutoff distance δ vanishes. In the present setting, δ is required to be small relative to the minimum radius of curvature ℓ introduced in assumption 3, so that $\delta \ll \ell$, or equivalently, $\epsilon := \delta/\ell \ll 1$. Hereafter, we restrict attention to interaction energies $\Omega(\boldsymbol{x}, \boldsymbol{y}, \boldsymbol{d}, \boldsymbol{e})$ that are frame indifferent. It then follows that $\Omega(\boldsymbol{x}, \boldsymbol{y}, \boldsymbol{d}, \boldsymbol{e})$ may depend on the positions $\boldsymbol{x}$ and $\boldsymbol{y}$ and the directors $\boldsymbol{d}$ and $\boldsymbol{e}$ only through the length $|\boldsymbol{x}-\boldsymbol{y}|$ of the vector between $\boldsymbol{x}$ and $\boldsymbol{y}$, the dot products $(\boldsymbol{x}-\boldsymbol{y})\cdot\boldsymbol{d}$ and $(\boldsymbol{x}-\boldsymbol{y})\cdot\boldsymbol{e}$ formed by the directors and that vector, and the dot product $\boldsymbol{d}\cdot\boldsymbol{e}$ formed by the directors. Like Keller and Merchant [12], we assume that dependence of the interaction energy on the length of the relative position vector is scaled by the ratio ϵ and, that

$$\Omega(\boldsymbol{x}, \boldsymbol{y}, \boldsymbol{d}, \boldsymbol{e}) = 2\,\tilde{\Omega}(\epsilon^{-2}\boldsymbol{r}\cdot\boldsymbol{r}, \boldsymbol{r}\cdot\boldsymbol{d}, \boldsymbol{r}\cdot\boldsymbol{e}, \boldsymbol{d}\cdot\boldsymbol{e}), \quad (2)$$

with $\boldsymbol{r} = \boldsymbol{x}-\boldsymbol{y}$. The factor of two in the right-hand side of (2) is for simplifying subsequent calculations.

As a consequence of the foregoing discussion, the net free-energy Ψ_{net} of the edge can be expressed as

$$\Psi_{\text{net}} = \int_0^L \frac{1}{2}\Big(\int_{-\frac{\pi}{2}}^{\frac{\pi}{2}} \omega(s,\theta)\,\mathrm{d}\theta\Big)\zeta(s)\,\mathrm{d}s, \quad (3)$$

where

$$\omega(s,\theta) = \int_0^L\int_{\frac{-\pi}{2}}^{\frac{\pi}{2}} \Omega(\boldsymbol{x}(s), \boldsymbol{x}(t), \boldsymbol{d}(s,\theta), \boldsymbol{d}(t,\eta))\zeta(t)\,\mathrm{d}\eta\,\mathrm{d}t \quad (4)$$

is the free energy due to the interactions between the molecule with director $\boldsymbol{d}(s,\theta)$ at $\boldsymbol{x}(s)$ with all other molecules and where a factor of one-half compensates for the double counting of interactions arising from integrating over both s and t from 0 to L. From (3), the free-energy density Ψ at position $\boldsymbol{x}(s)$ on $\mathcal{C}$ is

$$\Psi = \frac{1}{2}\Big(\int_{\frac{-\pi}{2}}^{\frac{\pi}{2}} \omega(s,\theta)\,\mathrm{d}\theta\Big)\zeta(s). \quad (5)$$

The function ζ denotes the density of the lipid molecules at any point of the curve $\mathcal{C}$. Upon replacing $\omega(s,\theta)$ in (5), and substituting Ω by its equivalent from (2), and Taylor expanding the integrand of (5) with respect to ϵ up to the second derivative term, (5) becomes

$$\Psi = k_\circ + k_1\kappa_g^2 + k_2\kappa_n^2 + k_3\kappa_g + k_4\kappa_n + k_5\kappa_n\kappa_g + k_6\tau_g^2, \quad (6)$$

which includes a quadratic expression in terms of κ_g, κ_n, and τ_g. Also, $k_\circ$ and the coefficients k_1–k_6 in (6) are the material parameters entering the model. Such parameters are functions of the interaction potential $\bar{\Omega}$, and molecular distribution function ζ. The free-energy density in (6) may be simplified as

$$\Psi = k_\circ + k_1(\kappa_g - \kappa_{g\circ})^2 + k_2(\kappa_n - \kappa_{n\circ})^2 + k_5\kappa_n\kappa_g + k_6{\tau_g}^2, \quad (7)$$

so that $\kappa_{g\circ} = -k_3/2k_1$, and $\kappa_{n\circ} = -k_4/2k_2$ represent the spontaneous geodesic and normal curvatures of $\mathcal{C}$. Notice that $k_\circ$ is the standard line energy of the edge—the part which is independent of edge geometry—while the remaining terms show how Ψ depends on the geometry of the edge.

The first term $k_\circ$ on the right-hand side of (7) is insensitive to the shape of the boundary. Since the molecular distribution on the boundary has implicit dependence upon the ambient temperature and concentration, these effects may be encompassed in $k_\circ$ and in the moduli k_1–k_6.

CONCLUSIONS

A free-energy density function was derived for the edge of an open lipid bilayer, taking into account the interactions between its molecules. The final expression contains quadratic terms in geodesic curvature, normal curvature, and geodesic torsion of the boundary curve and a term including the multiplication of geodesic and normal curvatures. It also includes a term representing the line tension corresponding the boundary. The free-energy of the edge is the evidence of the excess energy due to the specific arrangement of the phospholipid molecules in the vicinity of the boundary of the open bilayer. Our study supplements the previous works in which the free energy due to the edge is assumed constant and, thus, independent of the geometry of the boundary.

The derivation of our model gives rise to integral representations for the elastic moduli present in the model. In particular, molecular-base origins of the spontaneous curvatures for the edge of a lipid bilayer are investigated. The calculations predict the existence of the spontaneous normal and geodesic curvatures for the centerline of the edge.

Inspired by investigations on rupturing soap films [19], Litster [20] developed a continuum model for the free-energy Ψ needed for opening-up of a pore in a lipid membrane in the form

$$2\pi r(\varrho - \frac{\gamma_s r}{2}), \qquad (8)$$

with ϱ the line energy of the edge and γ_s the interfacial surface tension. It is evident from (8) that transient pores with sizes less than a critical radius $r^* = \varrho/\gamma_s$ tend to reseal, while those having the size exceeding this critical radius, may grow indefinitely, leading to the rupture of the membrane. Nevertheless, it is well-accepted today (see for example the review by Jähnig [21]) that lipid membranes possess zero surface tension, by which the second term on the right hand of (8) vanishes. This transpires the agreement of the current result based on molecular interactions, with that obtained previously on continuum grounds. The increasing cost of generating a larger pore confirms the stability of a lipid membrane with respect to the fluctuations that might bring about transient pores.

ACKNOWLEDGMENTS

Financial support from National Institutes of Health (NIDCD) grant DC 005788 (PI: Luc Mongeau) is gratefully acknowledged.

REFERENCES

[1] Israelachvili, J. N. 2011, *Intermolecular and surface forces: revised third edition*, Academic Press.

[2] Luckey, M. 2014, *Membrane structural biology: with biochemical and biophysical foundations*, Cambridge University Press, New York.

[3] Seifert, U. 1997, "Configurations of fluid membranes and vesicles,"*Advances in physics*, Vol(46), pp. 13–137.

[4] Lasic, D. D. 1988, "The mechanism of vesicle formation,"*Biochemical Journal*, Vol(256), pp. 1.

[5] Tsong, T. Y. 1991, "Electroporation of cell membranes,"*Biochemical Journal*, Vol(60), pp. 297–306.

[6] Marmottant, P., Hilgenfeldt, S. 2003, "Controlled vesicle deformation analysis by single oscillating bubbles,"*Nature*, Vol(423), pp. 153–156.

[7] Fromherz, P., Röcker, C., Rüppel, D. 1986, "From discoid micelles to spherical vesicles. The concept of edge activity,"*Faraday Discussions of the Chemical Society*, Vol(81), pp. 39–48.

[8] Boal, David H., Rao, M. 1992, "Scaling behavior of fluid membranes in three dimensions,"*Physical Review A*, Vol(45), pp. R6947.

[9] Capovilla, R., Guven, J., Santiago, J. A. 2002, "Lipid membranes with an edge,"*Physical Review E*, Vol(66), pp. 021607.

[10] Tu, Z. C., Ou-Yang, Z. C. 2003, "Lipid membranes with free edges,"*Physical Review E*, Vol(68), pp. 061915.

[11] May, S. 2000, "A molecular model for the line tension of lipid membranes,"*The European Physical Journal E*, Vol(3), pp. 37–44.

[12] Keller, J. B., Merchant, G. J. 1991, "Flexural rigidity of a liquid surface,"*Journal of statistical physics*, Vol(63), pp. 1039–1051.

[13] Seguin, B., Fried, E. 2014, "Microphysical derivation of the canham–helfrich free-energy density,"*Journal of mathematical biology*, Vol(68), pp. 647–665.

[14] Asgari, M. 2015, "Elastic free-energy of wormlike micellar chains: theory and suggested experiments,"*arXiv preprint arXiv:1502.02338*.

[15] Jiang, F. Y., Bouret, Y., Kindt, J. T. 2004, "Molecular dynamics simulations of the lipid bilayer edge,"*Biophysical journal*, Vol(87), pp. 182–192.

[16] de Joannis, J., Jiang, F. Y., Kindt, J. T. 2006, "Coarse-grained model simulations of mixed-lipid systems: composition and line tension of a stabilized bilayer edge,"*Langmuir*, Vol(22), pp. 998–1005.

[17] Wohlert, J., Den Otter, W. K., Edholm, O., Briels, W. J. 2006, "Free energy of a trans-membrane pore calculated from atomistic molecular dynamics simulations,"*The Journal of chemical physics*, Vol(124), pp. 154905.

[18] Karatekin, E., Sandre, O., Guitouni, H., Borghi, N., Puech, P., Brochard-Wyart, F. 2003, "Cascades of transient pores in giant vesicles: line tension and transport,"*Biophysical journal*, Vol(84), pp. 1734–1749.

[19] Deryagin, B. V., Gutop, Y. V. 1962, "Theory of the breakdown (rupture) of free films,"*Kolloidn. Zh*, Vol(24), pp. 370–374.

[20] Litster, J. D. 1975, "Stability of lipid bilayers and red blood cell membranes,"*Physics Letters A*, Vol(53), pp. 193–194.

[21] Jähnig, F. 1996, "What is the surface tension of a lipid bilayer membrane?"*Physics Letters A*, Vol(71), pp. 1348.

Study of Laser Adjustable Sutures for Trabeculectomy

Kongying Xie[1], Husam Elkaseh[1], Eddie B. Moss[2], Robert J. Campbell[2] and Yong Jun Lai[1]

[1]Department of Mechanical and Materials Engineering
Queen's University,
Kingston, Ontario, Canada
lai@queensu.ca

[2]Department of ophthalmology
Hotel Dieu Hospital
Kingston General Hospital
Kingston, Ontario, Canada

ABSTRACT

This paper presents a novel design of suture which could enable length adjustment in post treatment of trabeculectomy surgery and avoid medical complications. The new suture is able to provide up to 4 mm elongation at extension load over 0.2N. Testing of these new sutures using cadaveric eyes has been conducted. It was found that lysing the central suture decrease IOP (intraocular pressure) by 36.8% and lysing one additional side of the device further lowered IOP by a median of 5.5%.

Keywords: micromilling, IOP, adjustable suture

1. INTRODUCTION

Glaucoma is the leading cause of irreversible blindness in the world. This disease is primarily mediated by high pressure within the eye. The most common surgical approach to treating glaucoma and preventing blindness involves the creation of a new fluid outflow pathway within the eye. This surgery is termed a 'trabeculectomy', and it involves the creation of a scleral flap (a guarding trap door of tissue) and a small opening (fistula) leading into the anterior part of the eye [1]. This scleral flap is has traditionally been secured via standard sutures, which are left slightly loose to allow the egress of the fluid from the eye. The ocular fluid (aqueous humour) is captured under the mucus membrane layer of the eye called the conjunctiva, and then reabsorbed by the body.

Unfortunately, this is a risky procedure [2]. The required suture tension to attain the desired pressure lowering within the eye is difficult to precisely gauge intraoperatively. Of particular concern, very serious consequences may result if the pressure in the eye is lowered too far.

Surgeons have developed a number of approaches to address this issue [3, 4]. In general, to avoid overly low pressures, sutures are tied somewhat tight during initial surgery. Surgeons then plan to release the sutures post operatively. This can be easily achieved in the clinic by applying laser to the suture, which then breaks. However, studies have shown that this is frequently associated with a precipitous drop in pressure, which may cause serious problems including severe hemorrhage within the eye and loss of vision.

In this paper, we present a new design which could allow surgeons to safely and incrementally increase outflow in an easy control manner by using existing facilities during the post-operative period [5].

2. DESIGN AND FABRICATION

2.1 Design

The adjustable sutures comprise a standard 10-0 nylon suture and an added portion in the shape of a ring (see Figure 1below).

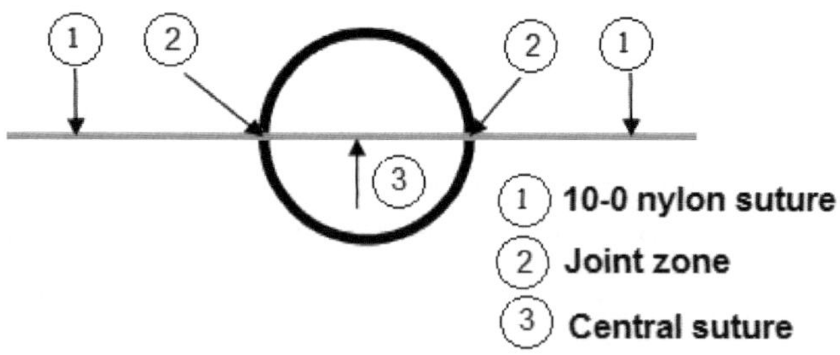

Figure 1. Schematic design of the new adjustable suture.

The suture spans across the diameter of the ring-shaped added portion forming a "ϕ" shaped device. The suture is connected with the added portion at the joint zone. When the suture is used in the trapdoor in trabeculectomy, the suture is generally tightened up and therefore it is straight under a tension force. When needed, a surgeon can fire a laser at the central suture and break it, and then the added portion will take the tension load. The added portion (the ring here) is able to provide some relaxation while carry the load (see Figure 2 below), and consequently it allows a faster flow from the trapdoor from trabeculectomy and lowers the IOP.

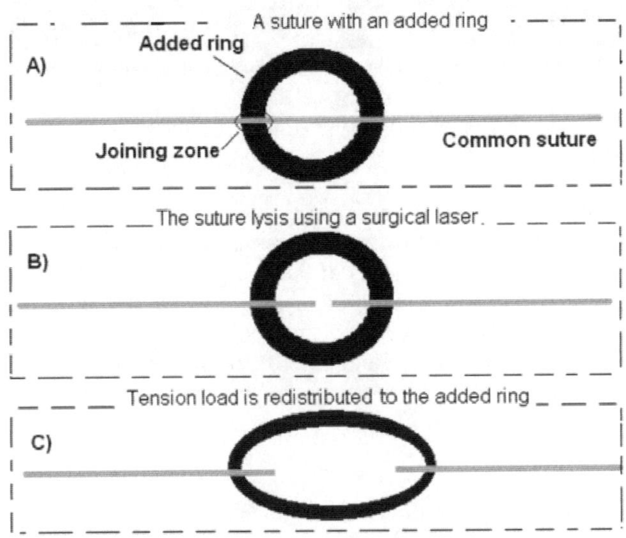

Figure 2. Schematic of working principle for the new suture.

As argon lasers are widely available in clinical settings and have a laser spot size of approximately 50 μm. In order to avoid misfire, the inner diameter of the added portion (ring) is at least 100 μm. The width of the ring can vary. In this case, the inner diameter is set as 1mm and the width of the ring is 250 μm.

In order to prototype this device, a Microlution™ high speed micromilling machine was used to develop micromolds for the ring and suture. A milling tool with 25 μm diameter is used with spindle speed at 80000rpm with a feed rate of 100mm/s. Once the mold is ready, a suture is place in the trench and a drop of liquid silicone elastomer (medical grade) from NuSil Silicone Technology (California, U.S.A.). mixed with its curing agent was deposited into the mold, which sits on a spin coater to remove the excess silicone. The mold was placed on a hot plate at 150°C for 10 minutes, then cooled to room temperature. The nylon suture was thus joined with the silicone ring tightly and released from the micromold.

3. TESTING

3.1 Strength testing

To examine the bonding strength of the joining zone, the diameter-spanning segment of the suture was cut and tensile testing was conducted for a set of 12 identical suture prototypes using a texture analyzer. Figure 3 shows the maximum allowable tension load that the new suture can carry varying from 0.264N to 0.152N (see A points in Figure 3), before one of lysed suture slips from the joint zone. Once the suture starts slipping, the load capable to hold drops down (see B points in Figure 3). As the load (reduced) continues, the lysed suture will slip out the joint zone and zero load can be carried. The linear elongations from the tension load can reach to over 4mm for the 12 tested sets. In clinical applications, the expected deformation (relaxation) that will be needed to induce the required reduction in resistance to aqueous outflow is expected to be below 1mm. The actual tightening load on a surgically tied suture is well below the critical threshold load where "pull out" occurs.

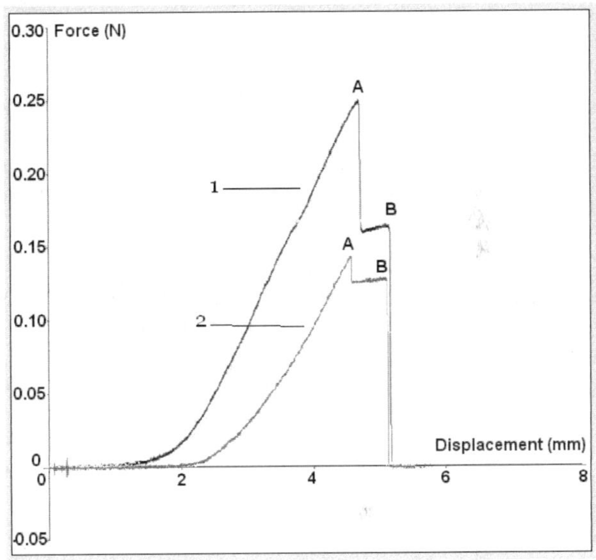

Figure 3: Allowable tension load vs. displacement (max. and min.) for the new suture. (only max. and min. tension load were selected, the rest results were between these two curves)

3.2 *Ex-vivo* testing

A triangle shaped a scleral flap (trap door) showed in Figure 4 below is tightened using the new sutures from this research. When central suture was lysed the tightness of the flap will relax and consequently let more flow out.

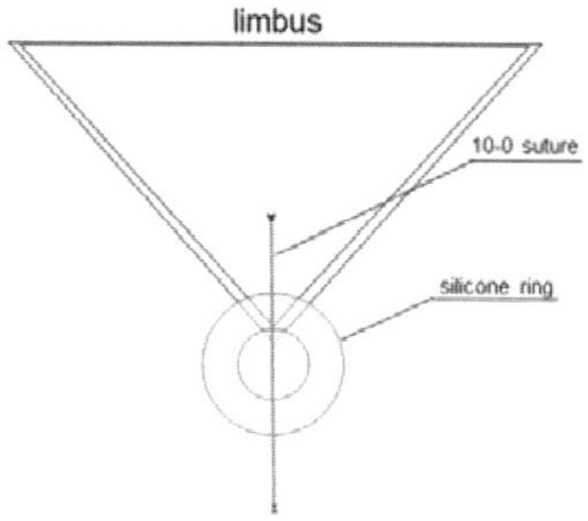

a)

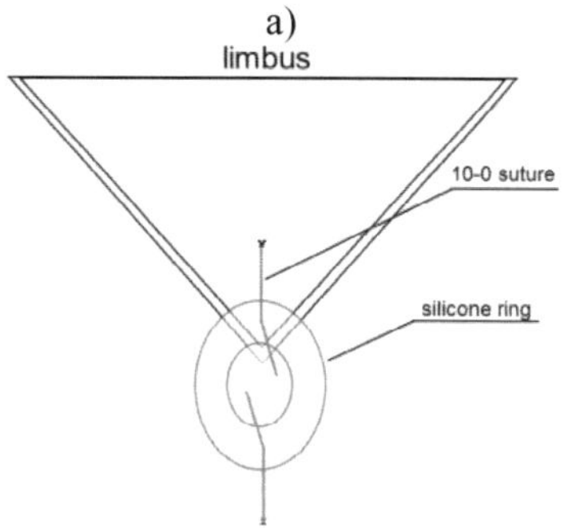

b)

Figure 4: Schematic of the new suture used in a scleral flap: a) tightened state, b) central suture is lysed.

Human cadaver eyes underwent trabeculectomy surgery with the prototype device was investigated. Eyes were perfused with an infusion pump and pressure within the eye was monitored (see Figure 5). To simulate in vivo performance, changes in pressure and outflow facility from baseline were measured while sequential clinically relevant device interventions were carried out. Specifically, flowrate was first adjusted to obtain steady-state conditions at a preset initial IOP, and three successive modifications to the surgical outflow pathway took place: 1) the central suture spanning the ring was lysed, 2) one side of the device was lysed to investigate the effect of increasing device deformability, and 3) the 10-0 suture was completely lysed.

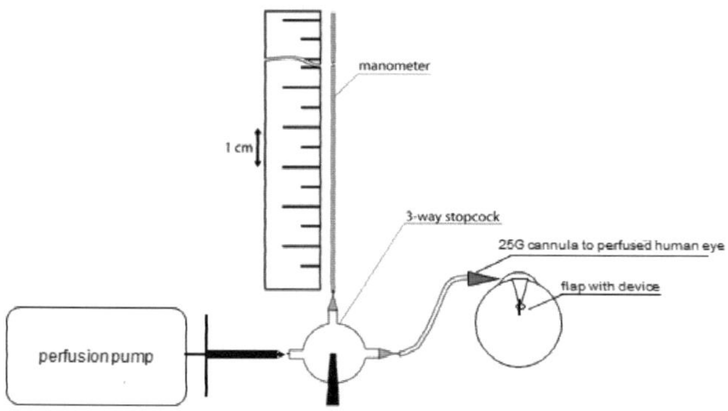

Figure 5. Schematic for test setup.

Figure 6 shows one of the new sutures used in the testing before and after its central suture was lysed by using clinical argon laser.

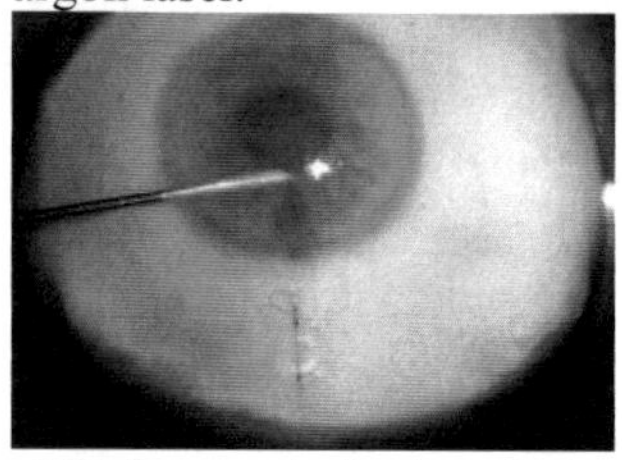

a) Before central suture lysis

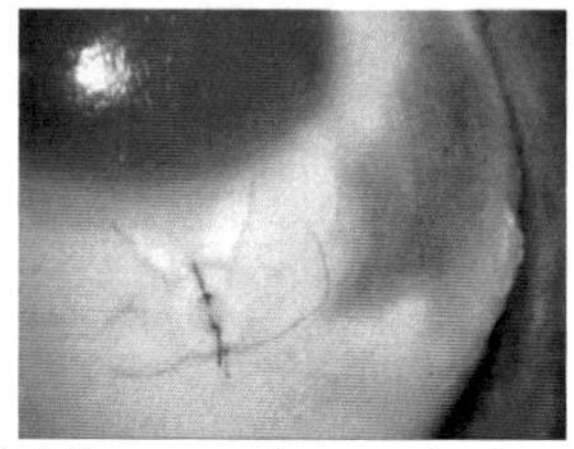

b) After central suture lysis

Figure 6. Picture of a new suture on a cadaveric eye during testing.

Results were obtained from four eyes. Perfusion rates of 0.6 to 4.8 ml/hr achieved baseline pressures that ranged from 23.3 to 28.7 mmHg. Lysis of the central suture produced a median IOP decrease of 36.8%. Lysing one side of the device further lowered IOP by a median of 5.5%. Complete suture lysis resulted in IOP ranging from 3.1 to 5.9 mmHg.

4. DISCUSSIONS

We found that the dimensions of the ring are difficult to choose due to practical constraints. A thicker ring can take higher load and more reliable, but is not desirable for clinical use. A wider ring can provide more stable surface contact with eye ball surface, but it becomes stiff due to large surface tension force. Optimizing the dimension of the ring is a part of the continuing research and we also explored some possibility of using adhesive to join the suture at the joint zone. Rather to add an elastomer ring to the suture, we attempt to loop the suture into the "Φ" shape and use adhesive to permanently keep the joint zone bonded.

A probe with 300μm in diameter with 45° slant angle at the tip was used to produce reliable size of adhesive droplets. The Figure 7 shows the droplet diameter can be reduced to around 100 μm which is a desirable size.

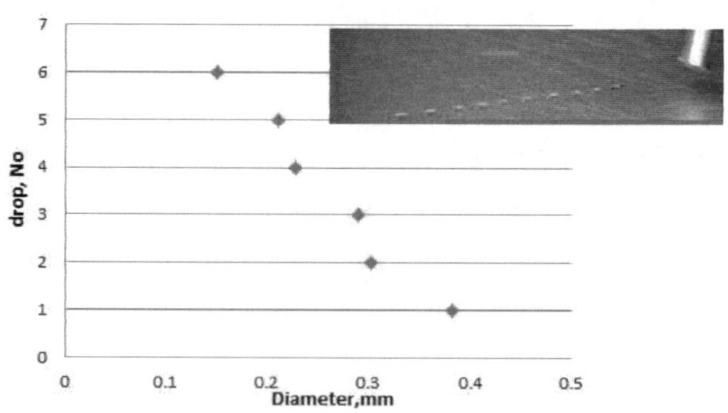

Figure 7. Adhesive droplet diameter using a probe with 300μm in diameter.

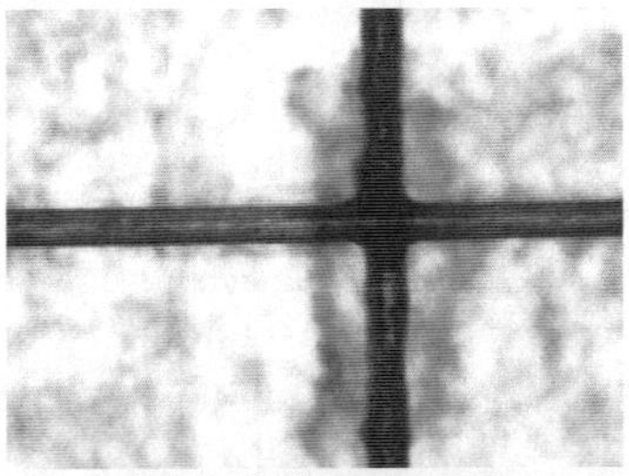

a) Bonded at crossing

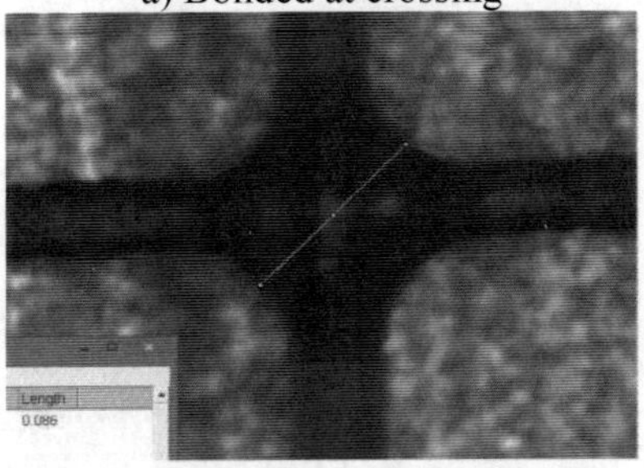

b) Closeup of joint zone (~90μm)

Figure 8. View of a joint zone from a looped suture using a UV adhesive droplet.

Figure 8 shows the crossing zone (joint zone) of a looped "Φ" shape suture, where a UV adhesive droplet with a diameter of 120 μm was landed and cured. The joint zone of the suture is found almost as strong as the suture itself. Due to its small joint zone size and easy processing, it holds a great potential to be produced in a larger scale.

5. CONCLUSIONS

This proof of principle experiment demonstrated that the suture-integrated ring device may incrementally lower IOP in perfused human eyes, by increasing facility of outflow in a regulated way. This concept may form the basis for the development of a new device capable of increasing the safety of trabeculectomy. Additional investigations will be needed to validate these preliminary results.

6. ACKNOWLEDGEMENT

The authors would like to thank Queen's SARC program, NSERC and CIHR for funding support. Also thanks go to Kingston General Hospital and Hotel Dieu Hospital for allowing access to their clinical facilities.

7. REFERENCES

[1]. Birchall W and Wells AP. The effect of sceral flap edge apposition on intraocular pressure control in experimental trabeculectomy. Clin Exp Ophthalmol 2008; 36:353-357.

[2]. Reibaldi A, Uva MG, and Longo A. Nine-year follow-up of trabeculectomy with or without low-dosage mitomycin-C in primary open angle glaucoma. Br J Ophthalmol 2008;92:1666-1670.

[3]. Aykan U et al. Laser suture lysis or releasable sutures after trabeculectomy. J Glaucoma 2007;16(2):240-245.

[4]. Macken P, Buys Y, and Trope GE. Glaucoma laser suture lysis. Br J Ophthalmol 1996;80:398-401.

[5]. Xie K, Campbell R, and Lai Y, Adjustable suture US patent 13/684,346 (pending)

FLEXIBLE ARMORS INSPIRED FROM NATURAL FISH SCALES AND OSTEODERMS

Roberto Martini
Mechanical Engineering Dept.
McGill University
Montreal, QC, Canada
roberto.martini@mcgill.ca

Ravi Chintapalli
Mechanical Engineering Dept.
McGill University
Montreal, QC, Canada
ravi.chintapalli@mail.mcgill.ca

Francois Barthelat
Mechanical Engineering Dept.
McGill University
Montreal, QC, Canada
francois.barthelat@mcgill.ca

ABSTRACT

Crocodiles, armadillo, turtles, fish and many other animal species have evolved flexible armored skins in the form of hard scales or osteoderms, which can be described as hard plates of finite size embedded in softer tissues. The individual hard segments provide protection from predators, while the relative motion of these segments provides the flexibility required for efficient locomotion. In this work, we duplicated these broad concepts in a bio-inspired segmented armor. Hexagonal segments of well-defined size and shape were carved within a thin glass plate using laser engraving. The engraved plate was then placed on a soft substrate which simulated soft tissues, and then punctured with a sharp needle mounted on a miniature loading stage. The resistance of our segmented armor was significantly higher when smaller hexagons were used, and our bio-inspired segmented glass displayed an increase in puncture resistance of up to 70% compared to a continuous plate of glass of the same thickness. Detailed structural analyses aided by finite elements revealed that this extraordinary improvement is due to the reduced span of individual segments, which decreases flexural stresses and delays fracture. This effect can only be however only be achieved if the plates are at least 1000 stiffer than the underlying substrate, which is the case for natural armor systems. Our bio-inspired system also displayed many of the attributes of natural armors: flexible, robust with "multi-hit" capabilities. This new segmented glass therefore suggests interesting bio-inspired strategies and mechanisms which could be systematically exploited in high-performance flexible armors. This study also provides new insights and a better understanding of the mechanics of natural armors such as scales and osteoderms.

KEYWORDS: Bio-inspired materials, flexible armor, laser engraving, thin films on soft substrates

Figure 1: Examples of segmented natural armor: (a) Fish (adapted from [4]); (b) Armadillo [9] (c) Crocodile

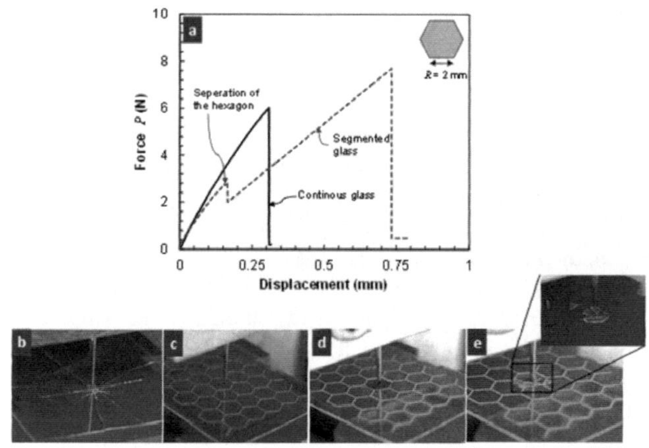

Figure 2: (a) Typical puncture force-displacement responses for a continuous glass plate and for segmented glass plate with R=1.5 mm (b) failure mode for continuous glass plate; (c)-(e) failure sequence for segmented glass plate with R=1.5 mm

DOCTOR BLADING ARTIFICIAL NACRE AND BONE

Mohammad Mirkhalaf
Mechanical Engineering Department
McGill University
Montreal, QC, Canada
mohammad.mirkhalaf@mail.mcgill.ca

Francois Barthelat
Mechanical Engineering Department
McGill University
Montreal, QC, Canada
francois.barthelat@mcgill.ca

ABSTRACT

The remarkable mechanical performance of biological materials such as bone, nacre, and spider silk stems from their staggered microstructure in which stiff and strong reinforcements are elongated in the direction of loading, separated by softer interfaces, and shifted relative to each other. This structure results in useful combinations of modulus, strength and toughness and therefore is increasingly mimicked in bio-inspired engineering composites. Here, we report the use of a simple and versatile technique based on doctor-blading to fabricate staggered composites of microscopic alumina tablets with high alignment in a chitosan matrix. Tensile tests on these nacre-like materials show that the modulus and strength of the composite films are enhanced by the incorporation of ceramic tablets, but only up to 15 % vol. after which all properties degrade. This phenomenon, also reported in the past for nacre-like materials obtained from different techniques, has been limiting our ability to produce large volumes of high-performance nacre-like materials. Examination of the structure of the films revealed that at lower tablet concentrations the tablets are well-aligned and well dispersed thorough the volume of the film. At 15% vol. and beyond, we observed tablet misalignment and clustering. In order to investigate the impact of these imperfections on material performance we developed large scale finite elements models representative of the structure of the composites films. These models show that mechanical performance significantly degrades with tablet misalignment, and especially at high tablet concentrations. The simulations along with the SEM images therefore quantitatively explain the experimental trends, e. g. the degradation of mechanical properties at high tablet contents, observed is most of experimental studies on similar materials. The doctor blading technique also paves the pathways towards the development of biomaterials or high-performance coatings.

KEYWORDS: Artificial nacre, bio-inspired materials, staggered composites, optimization, finite element modelling

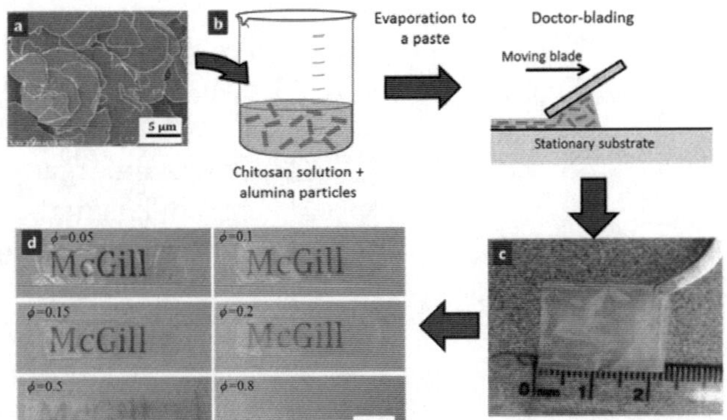

Figure 1: (a) SEM images of alumina tablets, showing their size and shape, (b) schematic of the fabrication process, (c) optical image of the film with 15 vol. % tablet concentration released from the substrate (d) optical images of the composite films with 5 vol. %, 10 vol. %, 15 vol. %, 20 vol. %, 50 vol. % and 80 vol. % ceramic concentrations (ϕ denotes the tablet volume concentration)

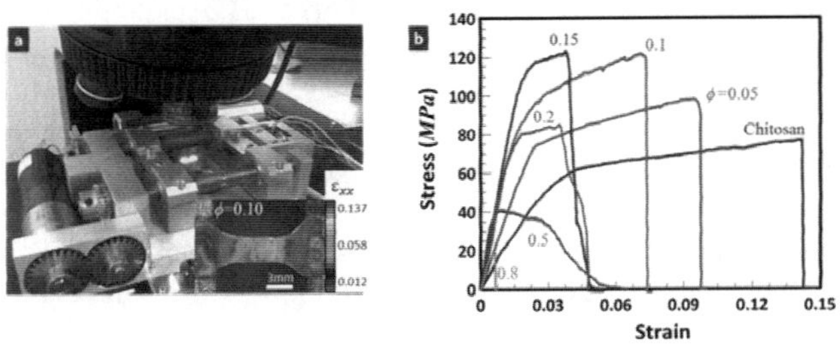

Figure 2: (a) Loading set up for the tensile tests. The distribution of the tensile strain for a dogbone sample with 10 vol. % tablet concentration just before failure is also shown, (b) tensile stress-strain curves of the composite films with different tablet concentrations.

Proceedings of the 25th CANCAM
London, Ontario, Canada, May 31 - June 4, 2015

A MODEL FOR THE COLLAGEN FIBRILS IN BIOLOGICAL SOFT TISSUES

Meisam Asgari, Hossein K. Heris, and Luc Mongeau
Department of Mechanical Engineering
McGill University, Montréal, QC, Canada
meisam.asgari@mail.mcgill.ca
hossein.khadiviheris@mail.mcgill.ca
luc.mongeau@mcgill.ca

ABSTRACT

Collagen fibers are the most abundant proteins in the extracellular matrix of biological soft tissue. They contribute to the tensile strength of the tissue, and its mechanical stability, cohesiveness, cellular migration regulation, and remodelling. These fibers have a characteristic D-banding periodicity of 67 nm. They are synthesized by the assembly of collagen molecules, which have a length of 300 nm and a diameter of 1.4 nm. In the present study, a continuum-mechanical model for the elasticity of wave-like structures similar in shape to collagen fibers observed in biological soft tissues was investigated. The structure of the collagen fiber at the micro-scale is idealized as intertwined helical fibrils. The model is based on the linear isotropic elasticity assumption. It accounts for the local shear, twist, stretch, and bending of the helical fibrils forming the collagen fiber. Key material parameters include the Young and shear moduli of the helical fibrils forming the collagen fiber. The model was used to quantify normal and shear stresses in the collagen fiber in response to bending deformations. Experimental procedures to obtain the elastic moduli of the fibrils are discussed.

Keywords: Collagen fibril, Elastic moduli, Differential geometry.

INTRODUCTION

Many biological soft tissues are comprised of extra-cellular matrix (ECM) embedded in a fluid substrate. The ECM includes different protein-made structures such as elastin and collagen fibers, which can be imaged using nonlinear laser scanning microscopy [1]. Collagen is one of the most abundant proteins in the ECM. It provides its structural strength. There are many applications for collagen such as in manufacturing cosmetics, glue and gelatin. It is used to produce biomaterials [2]. Various types of collagen have been identified in soft tissues. Among them, collagen type I is believed to be the most significant tension-bearing component of the ECM.

The structure of each collagen fiber incorporates a hierarchical arrangement. Laterally packed, triple-helical monomeric tropocollagen molecules with a thickness of 1.5 nm and a length of 300 nm [3], are organized in fibrillar units with a diameter of 4–5 nm, in turn forming larger micrometer sized fibrils. One feature of collagen structure is its D-banding periodicity of 67 nm, resulting from the quarter-staggered alignment of collagen chains during fibrillogenesis.

In the present study, it is hypothesized that the anisotropic behaviour of soft tissues might be caused by the anisotropy of the collagen fibers within the ECM [4]. The research objective was therefore to characterize macroscopic properties based on the characteristics at microscopic scales. To this end, tensile test with a microelectromechanical system [5], bending test on a suspended fiber by using atomic force microscopy (AFM) [2], and nano-indentation have been performed [6]. These efforts have yielded estimates of the elastic modulus of the collagen fibers in different tissues. However, previous studies have not taken into account the shear deformation of the collagen fibers as an independent material parameter.

Lanir [7], and Comninou and Yannas [8] assumed a sinusoidal crimp wave form, and derived a constitutive relation for the crimped collagen fibers. In that approach, the collagen fibers are wavy within the tissue at rest, and become straight when the tissue is stretched. This perspective reasonably agrees with the experimental observations. Another perspective for modeling the wavy structure of the collagen fibers is the zig-zag model based on kinematic linkages with rigid or flexible hinges and links forming the fiber [9]. However, such models do not account for the shear deformation in fibers [10].

The constitutive model of collagen fibers proposed by Freed and Doehring [10] assume each collagen fiber to form a cylindrical helix. This idea stems from the idea of Evans and Barbenel [11] and Yahia and Drouin [12] who observed planar and helical crimp patterns in mitral valve chordae tendineae taken from porcine. Prior to Freed and Doehring [10], Beskos and Jenkins [14] considered a constitutive model in which a cylindrical helix represents the wavy structure of the crimped collagen. In contrast to the model by Freed and Doehring [10], Beskos and Jenkins [14] assumed the collagen fibers to be inextensible, and thus, they found an infinite stiffness for the collagen fibers at full extension. The inextensibility of the collagen fibers in the tissue is not an accurate assumption [10].

A rope model for collagen has been investigated [3]. In such a model, the fibril-like structure of the collagen is assumed as a set of intertwined helical fibrils (see Figure 1). In previous studies of the rope model, the elasticity of the collagen fibers was characterized, regardless of the micro-scale modeling of the fibrils forming each fiber.

The objective of the present study is to derive a continuum-mechanical model to quantify the elastic moduli of the collagen fibrils forming each fiber within the framework of the rope model. Concepts of differential geometry of helical curves are employed. Micromechanical bending test is designed to be performed on a single fiber using atomic force microscopy (AFM). The three-point bending test for a clamped beam subject to a lateral point load in the middle is considered. For simplicity, we assume the fibers to be isotropic and linearly elastic. As a consequence of fitting the force-displacement curve to experimental data, the elastic constants including Young and shear moduli of the collagen fibrils forming the fiber can be determined.

GEOMETRY AND KINEMATICS

Let $\mathbf{x} = \mathbf{x}(s), 0 \le s \le L$ denote the arc-length parametrization of the curve $\mathcal{C}$. Using a prime to denote differentiation with respect to the arclength s, the unit tangent $\mathbf{t}$, unit normal $\mathbf{n}$, and unit binormal $\mathbf{b}$ of the Frenet frame $\{\mathbf{t}, \mathbf{n}, \mathbf{b}\}$ of $\mathcal{C}$ are given in terms of the arc-length parametrization $\mathbf{x}$ of $\mathcal{C}$ by $\mathbf{t} = \mathbf{x}'$, $\mathbf{n} = \mathbf{x}''/|\mathbf{x}''|$, and $\mathbf{b} = (\mathbf{x}' \times \mathbf{x}'')/|\mathbf{x}''|$. Moreover, the curvature κ and torsion τ of $\mathcal{C}$ are given by $\kappa = |\mathbf{t}'| = |\mathbf{x}''|$, and $\tau = \pm|\mathbf{b}'| = \mathbf{x}' \cdot (\mathbf{x}'' \times \mathbf{x}''')/|\mathbf{x}''|^2$. The curvature κ and torsion τ are two locally invariant quantities that characterize the space curve $\mathcal{C}$ up to a rigid translation in space. More specifically, the arc-length parametrization of a helix in $\mathbb{R}^3$ is expressed as

$$\mathbf{x} = r\cos(\chi s)\,\mathbf{e}_1 + r\sin(\chi s)\,\mathbf{e}_2 + h\chi s\,\mathbf{e}_3. \tag{1}$$

where $\chi = (r^2 + h^2)^{-1/2}$, with r and h denoting the radius and the pitch of the helix respectively. Further, $\kappa = r/(r^2 + \chi^2)$ and $\tau = h/(r^2 + \chi^2)$ denote the curvature and the torsion τ of the helix [13].

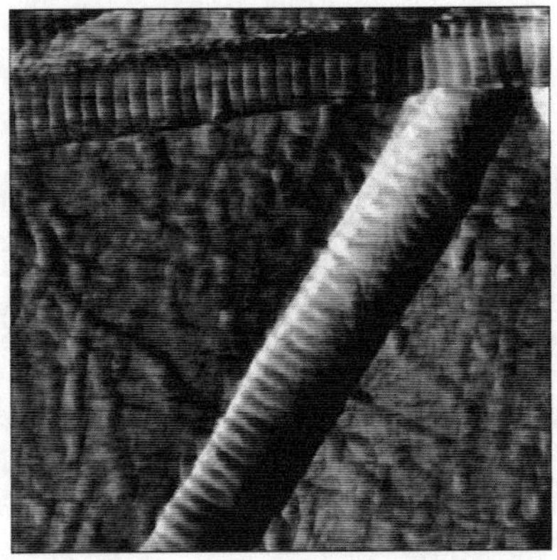

Figure 1: Pictures of two single collagen fibers, obtained using Atomic Force Microscopy (AFM). Geometric features such as the radius r and the pitch h of the helix are determined from these pictures. For a fiber comprised of a single fibril, $h \approx 67$ nm, and $r \approx 150$ nm.

STRAIN ENERGY OF A FIBRIL

Consider a single collagen fiber, freely-supported at the ends, and subjected to 3-point bending with a vertical point load F applied at the midpoint. Linear elasticity and isotropy are assumed. The shear force, $\mathbf{V}$, and the bending moment $\mathbf{M}$, at the cross-section of the beam located at x can be obtained from the general beam theory $\mathbf{V} = V\,\mathbf{e}_1 = \frac{F}{2}\,\mathbf{e}_1$ and $\mathbf{M} = (Fx/2 - FL/8)\,\mathbf{e}_2$, respectively. The variable L denotes the total length of the beam, and $-\mathbf{e}_1$ and $\mathbf{e}_3$ represent the directions of the lateral load, $\mathbf{F}$, and that of the beam in the undeformed state, respectively. At each point of the beam, the shear force $\mathbf{V}$, and the bending moment $\mathbf{M}$ are projected on the three local unit vectors forming the Frenet frame $\{\mathbf{t}, \mathbf{n}, \mathbf{b}\}$ of $\mathcal{C}$. The result of this projection for the local force, $\mathbf{V}_\ell$, is

$$\mathbf{V}_\ell = \bar{v}_1\,\mathbf{t} + \bar{v}_2\,\mathbf{n} + \bar{v}_3\,\mathbf{b}. \tag{2}$$

A similar projection for the bending moment, $\mathbf{m} = \bar{\mathbf{m}} + \tilde{\mathbf{m}}$, yields

$$\bar{\mathbf{m}} = \bar{m}_1\,\mathbf{t} + \bar{m}_2\,\mathbf{n} + \bar{m}_3\,\mathbf{b}, \tag{3}$$

with $\tilde{\mathbf{m}} = r\,\mathbf{n} \times \mathbf{V} = \tilde{m}_1\,\mathbf{t} + \tilde{m}_2\,\mathbf{n} + \tilde{m}_3\,\mathbf{b}$. Hence, the

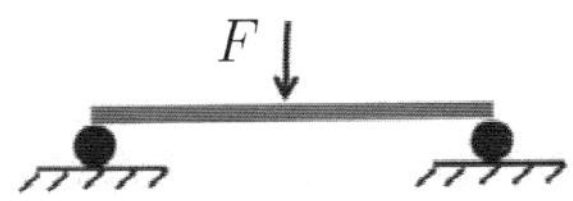

Figure 2: Three-point bending of a collagen fiber, with simply-supported edges. The probe of the AFM applied the force F at the midpoint.

Figure 3: Schematics of the microstructure of the beam. The beam represents a fiber comprised of two fibrils.

local shear force within the cross-section of the helix, with a cross-sectional radius r, is $V_\ell = \sqrt{\bar{v}_2^2 + \bar{v}_3^2}$. The local axial load is v_1. The local bending moment at the cross-section of the helix is

$$M_\ell = \sqrt{(\bar{m}_2 + \tilde{m}_2)^2 + (\bar{m}_3 + \tilde{m}_3)^2}, \tag{4}$$

and the torque is $\bar{m}_1 + \tilde{m}_1$. Having $\{\mathbf{e}_1, \mathbf{e}_2, \mathbf{e}_3\}$ as the right-handed coordinate system on the beam, the bending moment vector, $\mathbf{M}$, can be expressed as

$$\mathbf{M} = M\mathbf{e}_2 = \frac{F}{8}(4x - L)\,\mathbf{e}_2. \tag{5}$$

Consequently, (2), (3), and (5) yield

$$V_\ell = Vr\chi\sin(\chi s)\,\mathbf{t} + V\cos(\chi s)\,\mathbf{n} - Vh\chi\sin(\chi s)\,\mathbf{b}, \tag{6}$$

for the local force V_ℓ, and

$$\bar{\mathbf{m}} = \frac{Mr\cos(\chi s)}{\chi(h^2+r^2)}\,\mathbf{t} - M\sin(\chi s)\,\mathbf{n} - \frac{Mh\cos(\chi s)}{\chi(h^2+r^2)}\,\mathbf{b}, \tag{7}$$

and

$$\tilde{\mathbf{m}} = -\frac{Vrh\sin(\chi s)}{\chi(h^2+r^2)}\,\mathbf{t} - \frac{Vr^2\sin(\chi s)}{\chi(h^2+r^2)}\,\mathbf{b} = \tilde{m}_1\,\mathbf{t} + \tilde{m}_3\,\mathbf{b}, \tag{8}$$

for the local moments $\bar{\mathbf{m}}$ and $\tilde{\mathbf{m}}$ respectively. Consequently, the local axial force P, local shear force V_ℓ, local torque T, and local bending moment M_ℓ are obtained as

$$\left.\begin{aligned} P &= Vr\chi\sin(\chi s), \\ V_\ell &= \sqrt{(V\cos(\chi s))^2 + (Vh\chi\sin(\chi s))^2}, \\ T &= \frac{Mr\cos(\chi s)}{\chi(h^2+r^2)}, \\ &\text{and} \\ M_\ell &= \sqrt{M^2\sin^2(\chi s) + \left(\frac{Mh\cos(\chi s)}{\chi(h^2+r^2)} + \frac{Vr^2\sin(\chi s)}{\chi(h^2+r^2)}\right)^2}, \end{aligned}\right\} \tag{9}$$

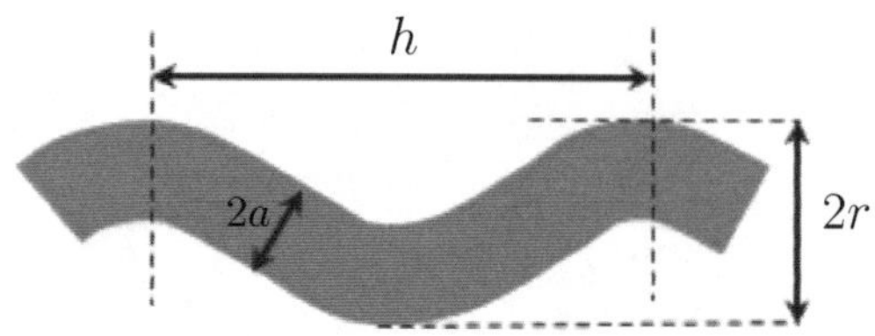

Figure 4: A helix representing a single collagen fibril. The pitch h, the radius r, and the thickness $2a$ of the helix are shown.

where $V = F/2$, $M = F(4x - L)/8$, and $x = r\cos(\chi s)$.

The strain energy $\mathrm{d}\Psi$ corresponding to an element $\mathrm{d}\ell$ of the helical fibril is the sum $\mathrm{d}\Psi = \mathrm{d}\Psi_{\text{axi}} + \mathrm{d}\Psi_{\text{bend}} + \mathrm{d}\Psi_{\text{twist}} + \mathrm{d}\Psi_{\text{shear}}$ where $\mathrm{d}\Psi_{\text{axi}} = P^2\,\mathrm{d}\mathcal{L}/2AE$, $\mathrm{d}\Psi_{\text{bend}} = M_\ell^2\,\mathrm{d}\mathcal{L}/2\,EI$, $\mathrm{d}\Psi_{\text{twist}} = T^2\,\mathrm{d}\mathcal{L}/2\,JG$, and $\mathrm{d}\Psi_{\text{shear}} = V_\ell^2\,\mathrm{d}\mathcal{L}/2AG$ are the strain energy portions due to the axial force P, bending moment M_ℓ, torsional moment (torque) T, and shear force V_ℓ, respectively. In these relations, E denotes the Young's modulus, and G represents the shear modulus of the fibril. The total strain energy, Ψ_{tot}, of a single fibril is obtained by integrating each term over the entire length $\mathcal{L}$ of the helical fibril.

PRILIMINARY NUMERICAL RESULTS

The strain energy of an elastic helical rod is used to find the deflection of the midpoint of the beam. In doing so, Castigliano's first theorem is employed. The strain energy Ψ of a single helix is obtained as

$$\Psi_{\text{tot}} = \int_{\bar{\mathcal{C}}} \mathrm{d}\Psi = \int_0^{\mathcal{L}} \left(\frac{P^2}{2AE} + \frac{V_\ell{}^2}{2AG} + \frac{M_\ell^2}{2EI} + \frac{T^2}{2JG}\right)\mathrm{d}\mathcal{L}, \tag{10}$$

where $\bar{\mathcal{C}}$ denotes the helical curve representing a fibril, $A = \pi a^2$ is the area of the cross-section of the helix, $I = \pi a^4/4$ and $J = \pi a^4/2$ denote the moment of inertia and the polar moment of inertia of the cross-section of the helix, with a being the radius of the cross-section of the helix.

Figure 5 shows the strain energy of the beam in terms of the applied force. As evident from Figure 5, the free energy of the helix increases as the ratio of the pitch h to the radius r is decreased. This is in agreement with previous models [10].

The deflection of the midpoint of the beam subjected to the point load F is thus $\delta_v = \partial\Psi_{\text{tot}}/\partial F$. The plot of the vertical deflection, δ_v, versus the applied force, F, for selected values of the geometric parameters of the helical strip, and chosen material parameters, as shown in Figure 6. Increasing the pitch, h, of the helix with respect to its radius, r, causes the deflection of the midpoint to decrease. This is intuitively correct, since in such case, the helical strip resembles a straight

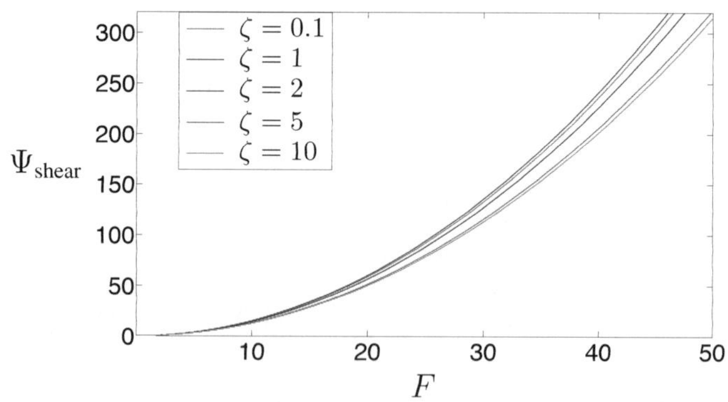

Figure 5: Plot of the shear strain energy in terms of the applied lateral force at the midpoint, for $\mathcal{L} = 10L$, and different values of $\zeta = h/r$, and $AE = AG = EI = JG = 1$.

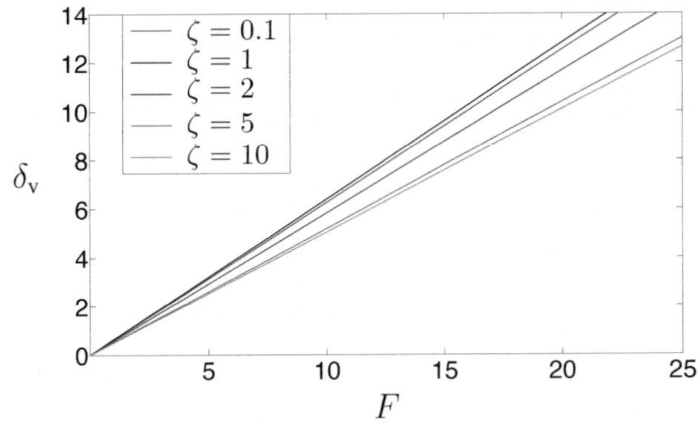

Figure 6: Plot of the deflection in terms of the applied lateral force at the midpoint, for $\mathcal{L} = 10L$, and different values of $\zeta = h/r$, and $AE = AG = EI = JG = 1$.

line, and thus it becomes stiffer in response to bending. This result agrees with previous observations [10].

CONCLUDING REMARKS AND FUTURE WORK

A continuum mechanical model of collagen fibers in biological soft tissues was developed based on linear isotropic elasticity. The model accounts for local bending, shear, twisting, and stretch of individual helical fibrils forming the collagen fiber. The idea stems from previously developed rope model for collagen fibers, in which the fiber is postulated to consist of a finite number of analogous intertwined helical fibrils. The model is novel because it characterizes the Young and shear moduli of the helices representing each collagen fibril forming the fiber. The derived model shows the sensitivity of the strain energy to the geometric features of the collagen fibrils forming the collagen fiber. For instance, fibrils with higher pitch or lower radius display more bending stiffness. Further, the strain energy resulted from the derived model is convex, which agrees with the previously developed mathematical models. Moreover, the strain energy stored in fibrils with larger pitch or smaller radius is less than that of the fibrils with smaller pitch or larger radius.

Single fibers spanning multiple channels in a substrate will be subjected to three-point bending tests using an AFM cantilever without a tip. The prevalent models are those for clamped or simply-supported beams loaded at the midpoint, in a way that the external point load remains vertical, while the beam behaves linearly elastic. The material parameters (i.e. Young and shear moduli) of collagen fibrils will be found by fitting the experimental force-displacement curve, obtained from 3-point bending of the collagen fibers, to the one resulted from the developed model.

ACKNOWLEDGMENTS

Financial support from CFI grant 229251, and National Institutes of Health (NIDCD) grant R01 DC005788 is gratefully acknowledged.

REFERENCES

[1] Miri, A. K., Heris, H. K., Tripathy, U., Wiseman, P. W., Mongeau, L. 2013, "Microstructural characterization of vocal folds toward a strain-energy model of collagen remodeling,"*Acta biomaterialia*, Vol(9), pp. 7957–7967.

[2] Yang, L., Fitie, C., van der Werf, K. O. Bennink, M., Dijkstra, P., Feijen, J. 2008, "Mechanical properties of single electrospun collagen type I fibers,"*Biomaterials*, Vol(29), pp. 955–962.

[3] Bozec, L., van der Heijden, G., Horton, M. 2007, "Collagen fibrils: nanoscale ropes,"*Biophysical Journal*, Vol(92), pp. 70–75.

[4] Wenger, M. P. E., Bozec, L., Horton, M. A., Mesquida, P. 2007, "Mechanical properties of collagen fibrils,"*Biophysical journal*, Vol(93), pp. 1255–1263.

[5] Eppell, S. J., Smith, B. N., Kahn, H., Ballarini, R. 2006, "Nano measurements with micro-devices: mechanical properties of hydrated collagen fibrils,"*Journal of the Royal Society Interface*, Vol(3), pp. 117–121.

[6] Radmacher, M. 2002, "Measuring the elastic properties of living cells by the atomic force microscope,"*Methods in cell biology*, Vol(68), pp. 67–90.

[7] Lanir, Y. 1983, "Constitutive equations for fibrous connective tissues,"*Journal of Biomechanics*, Vol(16), pp. 1–12.

[8] Comninou, M., Yannas, I. V. 1976, "Dependence of stress-strain nonlinearity of connective tissues on the geometry of collagen fibres,"*Journal of Biomechanics*, Vol(9), pp. 427–433.

[9] Kastelic, J., Palley, I., Baer, E. 1980, "A structural mechanical model for tendon crimping,"*Journal of Biomechanics*, Vol(13), pp. 887–893.

[10] Freed, A. D., Doehring, T. C. 2005, "Elastic model for crimped collagen fibrils,"*Journal of biomechanical engineering*, Vol(127), pp. 587–593.

[11] Evans, J. H., Barbenel, J. C. 1975, "Structural and mechanical properties of tendon related to function,"*Equine Veterinary Journal*, Vol(7), pp. 1–8.

[12] Yahia, L. H., Drouin, G. 1989, "Microscopical investigation of canine anterior cruciate ligament and patellar tendon: collagen fascicle morphology and architecture,"*Journal of Orthopaedic Research*, Vol(7), pp. 243–251.

[13] Millman, R. S., Parker, G. D. 1977, *Elements of differential geometry*, Prentice-Hall.

[14] Beskos, D. E., Jenkins, J. T. 1975, "A mechanical model for mammalian tendon,"*Journal of Applied Mechanics*, Vol(42), pp. 755–758.

VALIDATION OF A NOVEL TONOMETRY DEVICE FOR THE DEVELOPING WORLD

Melanie Chin
Medical Biophysics
Western University
London, Ontario, Canada
mchin43@uwo.ca

Dr. Tim Newson
Faculty of Engineering
Western University
London, Ontario, Canada
tnewson2@uwo.ca

Dr. Gord Campbell
The National Research
Council of Canada
London, Ontario, Canada
Gordon.Campbell@nrc-cnrc.gc.ca

ABSTRACT

Glaucoma is the second leading cause of irreversible blindness in the world, but if detected and treated early, prognosis is good. This makes it a suitable disease for widespread preventative screening programs. This investigation explores the feasibility of creating an inexpensive and portable glaucoma screening device that would be straight forward for any local health care professional to use and interpret. We propose the use of a relatively cheap direct-contact tonometer, which incorporates an internal snapdome mechanism that could provide rapid estimates of intraocular pressure. Initial testing indicates that the proposed device reliably buckles at the expected target force at an average of 197.265g ±5.3% (mean ± SD) for planar specimens. Our findings indicate the mechanism should also translate well to the more complex hemispherical ocular geometries.

KEYWORDS: glaucoma, intra-ocular pressure, tonometry, applanation, cornea, screening.

INTRODUCTION

Approximately 67 million people are blind worldwide and over two-thirds of them live in developing nations.[1] Glaucoma is the second leading cause of blindness in the world and is generally characterized by increased intraocular pressure (IOP). An elevated IOP often leads to irreversible damage of the optic nerve, which is normally responsible for transmitting images to the brain. If an elevated IOP persists, glaucoma can result in total, permanent loss of vision within a few years.[1]

In the developing world, glaucoma is often diagnosed late because it is typically painless and because central visual acuity is normally preserved until very late in the disease progression. However, if detected and treated early, vision loss can be minimized and patients have a much better prognosis. In most cases, glaucoma can be easily treatable with inexpensive topical eye drops or oral medication. For these reasons, glaucoma is an attractive disease to target for earlier diagnosis with screening programs.[1]

Conversely, in developed countries, glaucoma is not a leading cause of blindness since elevated intraocular pressures are detected by optometrists or ophthalmologists during routine testing.[2] The "tonometry" or "pressure testing" devices used in the developed world typically require a highly trained individual to operate and interpret the results. Additionally, the equipment can be quite expensive and is not readily mobile. As a result, tonometry methods commonly used in developed countries—like non-contact "air puff" tonometry or Goldmann slit-lamp tonometry—cannot easily be used to screen for glaucoma in the poorest and often widely dispersed regions of the developing world.[1,2]

However, portable tonometry devices like the Tono-pen™ (Reichert, Inc., Depew NY) have been developed as alternatives.[3] This device is a hand-held commercial applanation tonometer. It is portable, small, light, operates on batteries and uses a disposable latex tip cover to eliminate cross-contamination between eyes. The direct contact with the eye is fast and measurements can be done without anesthetic. Although the device is simple enough for most humanitarian workers to use, the steep cost (approximately $4000USD) makes it unrealistic for wide spread use.[3] Furthermore, the internal electronics may be very sensitive to environmental conditions like heat and high humidity, and can be susceptible to damage and have maintenance issues.

What is urgently needed is a very cheap, low-technology glaucoma screening tool for the general population that separates potentially healthy from unhealthy individuals not seeking treatment, which could produce rapid, approximate estimates of intraocular pressure. The aim of this project is to investigate the possibility of creating an inexpensive and portable glaucoma screening device that would be simple for any local health care professional to use and interpret.

THEORY

The gold standard test in tonometry is the Goldmann Applanation Tonometer (GAT), which determines IOP from the force required to "applanate" or flatten a specified area of the cornea.[4] The Imbert-Fick law states that the pressure in a spherical volume filled with liquid and surrounded by an infinitely thin membrane can be measured by the counter-pressure, which just flattens part of the membrane according to Eq. (1): [4]

$$IOP = F/A \qquad (1)$$

Where IOP is the estimate of intraocular pressure, F is the force necessary to applanate a constant area of the cornea, and A is the area of the cornea flattened by the indenter. The Maklakoff tonometer is an early example of an applanation tonometer that was widely used through the early part of the 20th century. This consisted of an elongated dumbbell shaped indenter placed on the eye and assesses the amount of corneal flattening produced by a known weight.

This project has investigated the feasibility of creating a Maklakoff style self-weight tonometer combined with a "snap dome" mechanism in contact with the cornea. This tonometry device would be based on a binary response that could be used to determine if an eye is beyond a risky threshold IOP. The indenter would applanate a portion of the cornea and then determine if 'snap-through buckling' is observed. Snap-through buckling is a phenomenon that occurs with shallow arches, thin domes and curved plates, when downward loading causes a complete reversal of curvature over all or part of the structure.[5] When the loads can no longer be supported, there is a bifurcation or 'snap-through' buckling that occurs as a sudden jump into another equilibrium state.[5]

Using Eq. (1), the force that would be required to cause a snap-through buckle at a particular intraocular pressure can be determined given a known area of the indentation of the corneal surface. A Snaptron Inc. snapdome would be selected for the tonometer such that low or average intraocular pressures (corresponding to a relatively soft eye) would not induce a buckle of the snapdome during indentation. Conversely, an elevated IOP—indicating an individual is hypertensive (with a correspondingly stiffer eye) and potentially at risk for developing glaucoma—would induce an observable buckle in the snap dome during indentation.

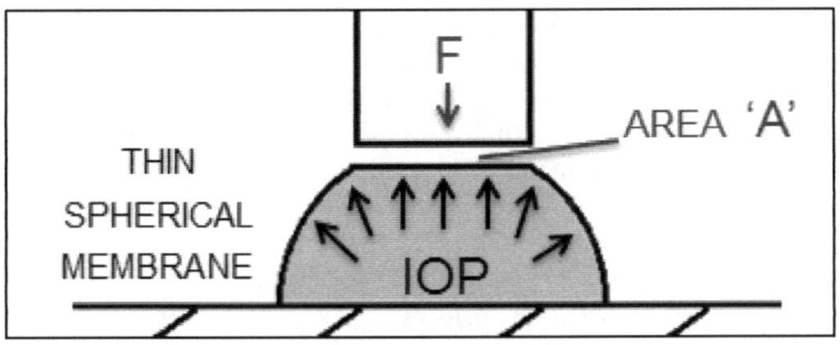

Figure 1 A force is applied using the cylindrical indenter head to a thin spherical membrane which applanates an area A. Using the Eq. (1), the intraocular pressure (IOP) can be calculated.

Thus, the aim this project is to investigate the feasibility of creating a tonometry device that incorporates a snapdome to provide a simple binary result for each test. If intraocular pressure is normal, there would be a "no snap condition" if intraocular pressure is high there would be a "snap condition" and the individual could be encouraged to seek further testing and professional care.

METHODS

A proof-of-principal version of the snap-dome indenter has been constructed and tested. Using the Versa Loader from ELE International, standard mechanical tests consisting of parallel plate uniaxial unconfined compression, rigid cylindrical indentation, and snap-dome indentation were conducted on samples with known stiffness made from planar rubber and on tissue mimicking phantoms made of polyvinyl alcohol cryogel (PVA-c). These model materials were selected to simulate the mechanical non-linearity of ocular tissue and to create simple and reproducible ocular models for an initial assessment of the capabilities of the snap dome indenter. Further tests on PVA-c and rubber spheres and hemisphere are being conducted, but are outside the scope of this paper.

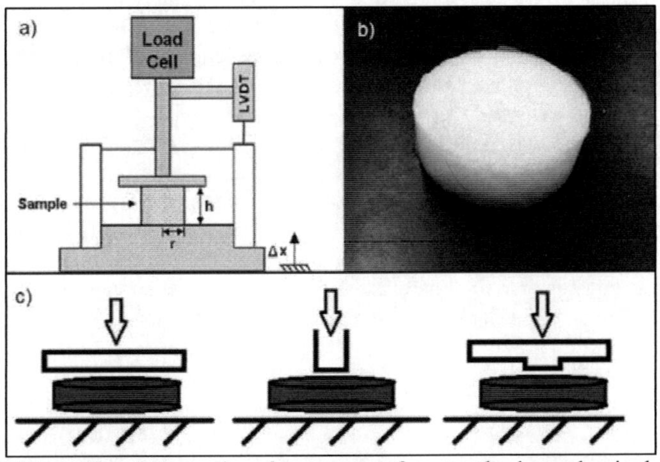

Figure 2a) Schematic of apparatus for standard mechanical testing. **b)** Photograph of PVA-c sample with diameter=35mm and height=12mm. **c)** Schematic of methods used to apply force to samples representing unconfined compression (UC), rigid indentation, and snapdome indentation shown respectively from left to right.

According to the Poisson's effect, when a material undergoes an axial load from one direction it has a tendency to expand in the plane direction perpendicular to the applied force, essentially exhibiting conservation of volume. Thus, it is important to correct for the increasing cross-sectional area of any sample undergoing an unconfined compression using the equations below to ensure the appropriate stress—and in turn Young's moduli—are being computed. The strain, ε can be computed as:[9]

$$Strain\ (\varepsilon) = \frac{\Delta L}{L_o} \tag{2}$$

Where ΔL is the change in length and L_o is the original length. Then, since cross-sectional area changes with deformation corrected area, A' can be calculated:

$$A' = \frac{A_o}{1 - \varepsilon} \tag{3}$$

Where A_o is the original cross-sectional area and ε is the strain determined in Eq. (2). Finally the stress, σ can be calculated from the applied force, F and the corrected area, A' found using Eq. (3):

$$Stress\ (\sigma) = \frac{F}{A'} \tag{4}$$

Young's modulus, E for a sample undergoing unconfined compression can then be determined by taking the slope of the line tangent to the stress-strain curve for a particular strain value. In contrast, the Young's modulus, E of a sample undergoing a plane ended cylindrical indentation can be computed according to the following relationship:[6]

$$E = \frac{(1 - v^2)}{2r} \times \frac{F}{\Delta L} \quad (5)$$

Where v is Poisson's ratio (assumed to be v=0.5 for these incompressible materials), F is the applied load, r is the radius of the cylindrical indenter, and ΔL represents the displacement of the body under the indenter. The use of this equation assumes that the sample approximates a semi-infinite medium under rigid cylindrical indentation.

One of the materials tested, polyvinyl alcohol cryogel (PVA-c), is a synthetic polymer with tissue-mimicking capabilities. It is created as a hydro-gel solution that undergoes freeze-thaw cycling (FTC) in a temperature chamber.[7] Increasing the PVA-to-water ratio or increasing the amount of FTCs will increase the stiffness of the material and allow for the simulation of different conditions such as a healthy eye, or a stiff eye experiencing a high IOP.[7]

RESULTS

Unconfined compression tests were initially conducted to create stress-strain curves to better understand the constitutive behavior of these materials and determine strain related stiffness changes, as exhibited in Fig. 3.

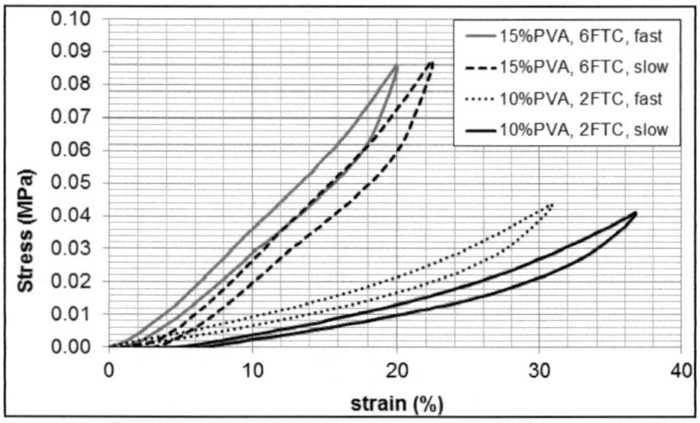

Figure 3 Stress-strain curves from the unconfined compression of PVA-c samples of varying stiffness at different loading rates.

Each curve in Fig. 3 shows the load-unload cycle for different PVA-c materials. The slope of the tangent at a specific strain represents the elastic Young's modulus, an intrinsic property of a material that relates to the material's stiffness. These materials generally increase in stiffness for a given strain level with increased loading rate (fast=2mm/min and slow=0.2mm/min) and for increased number of freeze-thaw cycles (2FTC vs. 6FTC). Similarly, the materials increase in stiffness for increased PVA concentrations (10% vs. 15%). As seen with other elastomers, the materials display stiffening with increased strains and unloading hysteresis due to entropy changes and stress relaxation.

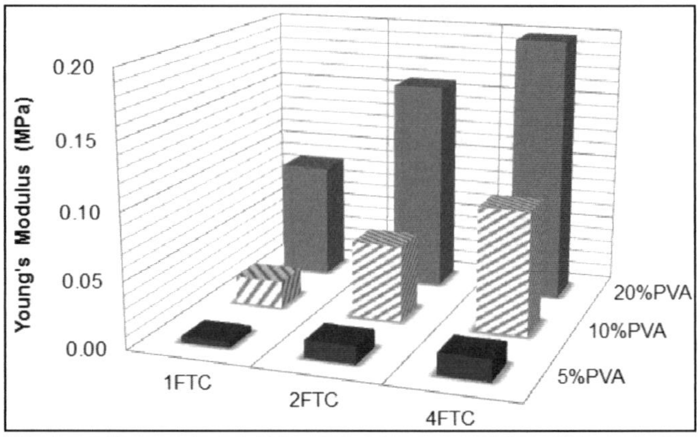

Figure 4 Comparison of Young's elastic moduli at 5% strain for various PVA-c samples.

Fig. 4 displays Young's moduli calculated for different materials based on Eq. 2. The figure shows a range of E values from 0.04 to 0.194 MPa at 5% strain giving stiffness values in the same range as experimental corneal testing.[10,11] Hence, Fig. 4 figure shows that the properties of the tissue mimicking polymer PVA-C can be altered to simulate various levels of tissue stiffness, covering healthy and diseased eye states.

In Fig. 5, the force-displacement curves for the unconfined compression, rigid indentation and snapdome indentation of a single PVA-c sample are shown. Initially, the slope of the curve corresponding with rigid indentation closely follows the slope of the curve corresponding with snapdome indentation, as the area of the indenter tip in contact with the material's surface is identical. Any variations are due to the compliance of the snapdome prior to buckling. A snap-through buckle occurs at ~1.7mm of displacement. Then, after >2mm of displacement, the slope of the snapdome indent curve more closely mimics that of unconfined compression but with an offset displacement. This is a result of the applied force spreading over a larger surface area surrounding the snap-dome tip which is similar to an unconfined compression (see Fig. 2c).

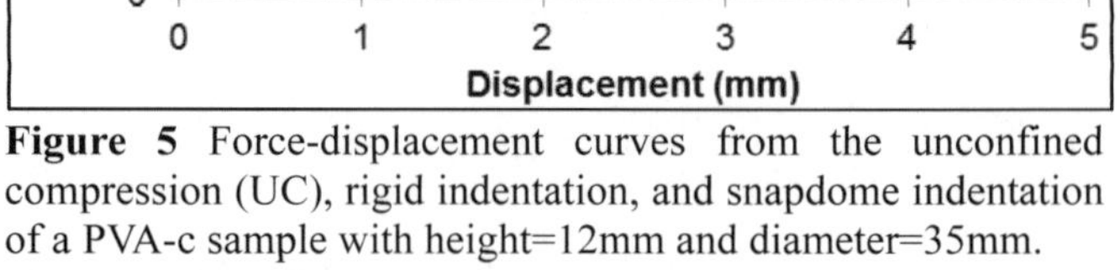

Figure 5 Force-displacement curves from the unconfined compression (UC), rigid indentation, and snapdome indentation of a PVA-c sample with height=12mm and diameter=35mm.

Fig. 6 shows the binary responses observed when using the snapdome indenter on various materials, with a positive snap condition observed when testing a stiff material (10%PVA, 4FTC), and no-snap observed when indenting less stiff materials (10% PVA, 2FTC and 10%PVA, 1FTC). In this figure, the positive snap condition is clearly visible as a discontinuity or sudden drop in the force displacement curve for the 10%PVA, 4FTC sample, in contrast to the smooth curves for the less stiff materials that do not induce a snap which suggests that the concept snapdome tonometer will identify elevated IOP as desired.

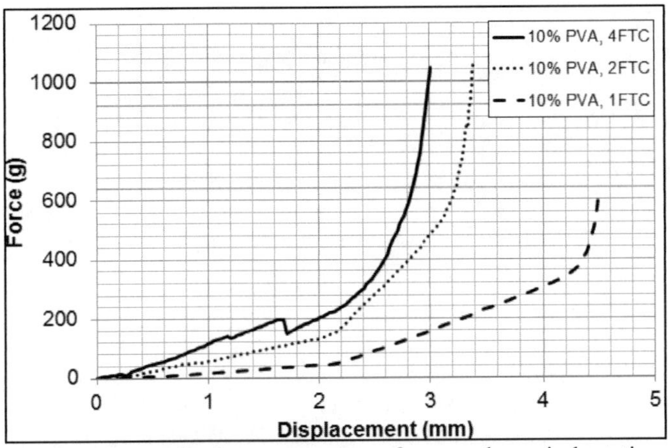

Figure 6 Force-displacement curves for snapdome indentation on PVA-c samples of decreasing stiffness.

Fig. 7 shows that when a material was stiff enough to induce a snap, the tip buckled consistently. The average trip force of 197.265g ±5.3% (mean ± SD) closely corresponds with the manufacturer's published trip force of 200g. This indicates the results are reliable and repeatable.

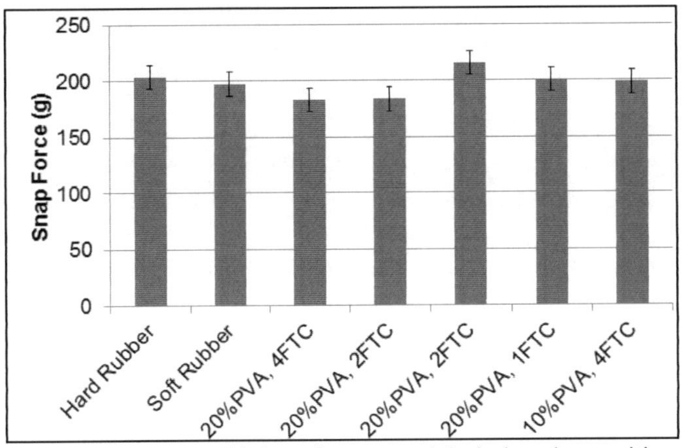

Figure 7 Force at which various materials induced a positive snap condition in the proposed snap dome indenter. Average trip force of 197.265g ±5.3% (mean ± SD). The hard rubber and soft rubber samples had shore durometer hardness (A) of 70 and 10, respectively.

DISCUSSION

This project has, generated promising results for the development and validation of an inexpensive and portable tonometry device that would be simple for local workers to operate and interpret. Initial testing of the prototype tonometer indicated that it functioned as predicted, repeatedly producing an observable snap against stiff materials, but no observable snap against less stiff materials. Further testing would be necessary to investigate how this mechanism would be transferrable to human eyes, effectively allowing for the differentiation between a stiff eye experiencing elevated IOP and a healthy eye. This would require exploration of the effects of material and geometric non-linearity, and would necessitate testing on spherical or hemispherical elastic models. Optimization of the shoulders next to the indenter tip of the snapdome device is require to ensure the snapdome is not engaged for higher displacements on curved surfaces.

Future work might include tests in a clinical setting, undertaking a comparative analysis of the snapdome tonometer to other available portable tonometry devices like the Tono-pen™. A comprehensive study should not only compare the accuracy and reliability of the proposed indenter against a device already available on the market, but also provide a cost analysis.

CONCLUSIONS

Initial testing indicates that the proposed snapdome tonometer repeatedly produces a positive result when used on stiff materials, and a negative result on less stiff materials. On planar materials, the snapdome reliably buckles at the expected target force at an average of 197.265g ±5.3% (mean ± SD). The results indicate that the response of the snapdome may translate well to the more complex ocular geometries. Further work is required to calibrate the device for specific IOP/stiffness combinations.

ACKNOWLEDGMENTS

Sincerest thanks Masters student Jordan Kiss for his support. We would like to recognize NSERC-USRA, The Schulich School of Medicine, Western Engineering and the NRC for providing funding, materials, and laboratory facilities necessary to make this investigation possible.

REFERENCES

1. Thomas, R., Sekhar, G. C., & Parikh, R. (2007). Primary angle closure glaucoma: a developing world perspective. *Clinical & experimental ophthalmology*, *35*(4), 374-378.
2. Leite, M. T., Sakata, L. M., & Medeiros, F. A. (2011). Managing glaucoma in developing countries. *Arquivos brasileiros de oftalmologia*, *74*(2), 83-84.
3. Wigers, A. Midelfart, A (1994). Clinical comparison of the ProTon and TonoPen tonometers with the Goldmann applanation tonometer. *British Journal of Ophthamology*. 78(12), 895-898.
4. Orssengo, G. J., & Pye, D. C. (1999). Determination of the true intraocular pressure and modulus of elasticity of the human cornea in vivo. *Bulletin of mathematical biology*, *61*(3), 551-572.
5. Karagiozova, D., Zhang, X. W., & Yu, T. X. (2012). Static and dynamic snap-through behaviour of an elastic spherical shell. *Acta Mechanica Sinica*, *28*(3), 695-710.
6. Hayes W, Keer L, Herrmann G, and Mockros L. (1972). A mathematical analysis for indentation tests of articular cartilage *J. Biomechanics*: 5,541–51
7. Hao Wang, W., Campbell, G (2008). Tissue Mimicking Artificial Human Intervertebral Disc Part II. *NRC. Controlled Technical Report* (2008).

Biomechanics

Proceedings of the 25th CANCAM
London, Ontario, Canada, May 31 – June 4, 2015

NONLINEAR MECHANICAL RESPONSE OF PORCINE KNEE JOINT

Rodriguez, Marcel
Department of Mechanical and Manufacturing Engineering
University of Calgary
Calgary, Alberta, Canada
mlrodrig@ucalgary.ca

Li, LePing
Department of Mechanical and Manufacturing Engineering
University of Calgary
Calgary, Alberta, Canada
Leping.Li@ucalgary.ca

ABSTRACT

The load response of articular cartilage to compression highly depends on the compression rate. This response has been extensively investigated in vitro (in tissue disc), but not sufficiently examined in situ (in intact joint). The influence of complex anatomy and interplay between tissues must be considered before in vitro studies can be used to understand joint mechanics. The objective of the present study is to quantify the compression-rate dependence in a knee joint. We are interested in the poromechanical response of the human knee, but used porcine joints as a first step. Results revealed a significant nonlinear compression-rate dependent load response of the knee joint. Reaction force varied two times between the smallest and fastest compression rates with a fluid load support of up to ~70%. Our study will provide more accurate understanding in load share between the fluid pressure and tissue matrices within the knee joint, and how the load share changes with compression rate that varies in daily activity.

KEYWORDS: Biomechanics, Joint Mechanics, Articular Cartilage, Meniscus, Compression-rate dependence.

INTRODUCTION

Cartilage response to load is extensively investigated both computationally and experimentally using relatively simple geometries, such cartilage discs. This relationship is characterized to be non-linear and viscoelastic by nature with a fluid load support of 90% when a uniform compression is applied rapidly [1, 2]. Several studies have revealed how the load is transmitted within the joint when certain flexion angles are used [3, 4]. It is also known that cartilage response is different when it is healthy or degenerated. In the former, the reaction force can be up to ten times greater between slow and fast compression [2, 5]. In the latter case these differences are not noticeable.

We wanted to evaluate the effect of compression rate on the reaction force for a given displacement in a healthy joint, as a first stage for further studies which will include degenerated cartilage and partial meniscectomy in the knee.

As a viscoelastic material, preconditioning influences mechanical response of tissues. It has been found that preconditioning to different strain than the test will produce a difference stress-strain response even in the case when same strain rate is used in both cases [6]. Also, when more than one test is performed with the same specimen, recovery time between tests must be sufficient. If this time is not long enough, tissues will remain partially deformed and stiffness for the next test will be affected.

The objective of the present study is to determine the nonlinear load response of the knee, especially the response on the compression-rate. This is a key factor to understand the mechanical response of a healthy and degenerated knee. Our long term goal is to obtain this response for the human knee joint, but porcine knee joints are used in this initial stage.

MATERIALS AND METHODS

Compression tests of porcine knees were performed on a Mini Bionix II (MTS, Minnesota). Fresh porcine stifle joints with intact joint capsule were purchased within 24h of slaughter. The ankle joint was mostly retained to keep the tibia and fibula together. The joints were kept in sealed bags with Phosphate Buffered Saline (PBS) solution and stored in a fridge at 4 ºC prior to testing next day. The femur was fixed in a hollow metal cylinder and the ankle joint was fixed in another cylinder using dental cement (Fastray™ LC, Bosworth®). Metal bolts were screwed in from the outer cylinder surface to the bone surface prior to cementing, in order to prevent possible sliding between cement and cylinder or bone. The femur was first cemented and the cylinder holding it was then attached to the load cell (10kN) fixing on the moving part of the MTS. In order to maintain the natural angle of the joint (~ 45º), the cylinder for the ankle joint was positioned appropriately and then fixed on the MTS base prior to cementing. Wet gauzes were used to cover the joint to keep it hydrated during the whole test (Fig. 1).

Firstly, a preconditioning was performed at 0.25 Hz and 400 μm amplitude for 30 cycles. The response tends to be repeatable after a few cycles (Fig. 2). Then an 800-μm ramp compression was applied at 6 rates: 10, 50, 100, 500, 1000 and 2000 μm/s, respectively, followed by a 20 min relaxation period. The joint was then unloaded and remained unloaded for 20 min to allow tissue recovery from fluid loss and deformation prior to next test. Knee compression and reaction force were monitored using the Station Manager® software installed on a computer which controls the MTS. Fourteen joints have been tested so far.

RESULTS

Compression-rate dependent load response was noteworthy: peak reaction force was almost doubled between the slowest and the fastest rate of compression (~55N @ 10 µm/s and ~95 N @ 2000 µm/s, Fig. 3). Relationship in reaction force in two consecutive compression rates is more noticeable when these rates are smaller (~55 @ 10 µm/s and ~70 N @ 50 µm/s) rather than higher (~92 @ 1000 µm/s and ~95 N @ 2000 µm/s). This can be explained as follows: when compression rate is increased, it approximates to an instantaneous compression when there is no time for fluid exudation and the fluid remains in cartilages. Mathematically, the reaction force of instantaneous compression is the maximum or asymptote of all compressions at different rates. When the compression rate is small, the force increases quickly with increased rate because it is far from the asymptote. When the compression rate is sufficiently great, the force slowly increases approaching the asymptote (Fig. 4).

After the initial ramp compression, reaction force decayed rapidly during the first 180s and continued to decrease monotonically (Fig. 5) as fluid exudates from the cartilages.

Equilibrium load for the test shown here is approximately 30% (~30/108) of the peak load. In our human knee joint computer modeling, the equilibrium load was ~33% of the peak load [7].

CONCLUSIONS

Reaction force highly depends on the rate of compression for most healthy pig knee joints. Conditions such as freezing and hydration were found to affect the results significantly.

The difference between the peak and equilibrium load is believed to contribute by the fluid pressure in cartilage and menisci. In our case, the fluid pressure can support 60-70% of the total loading in most pig knee joints, when the joint is subjected to a fast compression.

Waiting time between tests is also vital for repeatable results. It is necessary to ensure completely recovery of tissues prior to any rerun.

Our results indicate the influence of complex anatomy and interplay between tissues on the load support of the joint because up to 90% fluid pressure load support was found in a cartilage disk.

Our study will aid to understand the load share between the fluid pressure and tissue matrices within the healthy knee joint. We will use this approach to further study how the load share changes with compression rate for injured and degenerated knees.

ACKNOWLEDGMENTS

AIHS Interdisciplinary Team Grant in Osteoarthritis Trainee Studentship Award.

REFERENCES

1. Atheshian GA, Hung CT, 2005, "A theoretical solution for the frictionless rolling contact of cylindrical biphasic articular cartilage layers", *Clinical Orthopaedics & Related Research* 81, 81-90.
2. Li LP, Herzog W, 2005, "Electromechanical response of articular cartilage in indentation-considerations on the determination of cartilage properties during arthroscopy", *Computer Methods Biomechanics Biomedical Engineering* 8, 83-91.
3. Ahmed, AM, Burke DL, 1983. "In-vitro measurement of static pressure distribution in synovial joints—Part I: Tibial surface of the knee", *Journal of Biomechanical Engineering* 105, 216-225.
4. Thomas DB, Daniel TS, 1984. "In-Vitro Contact Stress Distribution on the Femoral Condyles", *Journal of Orthopaedic Research* 2,190-199.
5. Li LP, Buschmann MD, Shirazi-Adl A, 2003, "Investigation of mechanical behavior of articular cartilage by fibril reinforced poroelastic models", *Journal of Biomechanical Engineering* 125, 161-168.
6. Cheng S, Clarke E, Bilston L, 2009, "The effects of preconditioning strain on measured tissue properties", *Journal of Biomechanics* 42, 1360-1362.
7. Dabiri Y, Li LP, 2013, "Influences of the depth-dependent material inhomogeneity of articular cartilage on the fluid pressurization in the human knee", *Medical Engineering & Physics* 35, 1591-1598.

Figure 1. Stifle joint testing setup. Screws and dental cement (in blue) were used to attach the joint to metal cylinders that were attached to the MTS. Gauze was also used to keep the joint hydrated. The load cell (in black) is shown to attach to the actuator of the MTS.

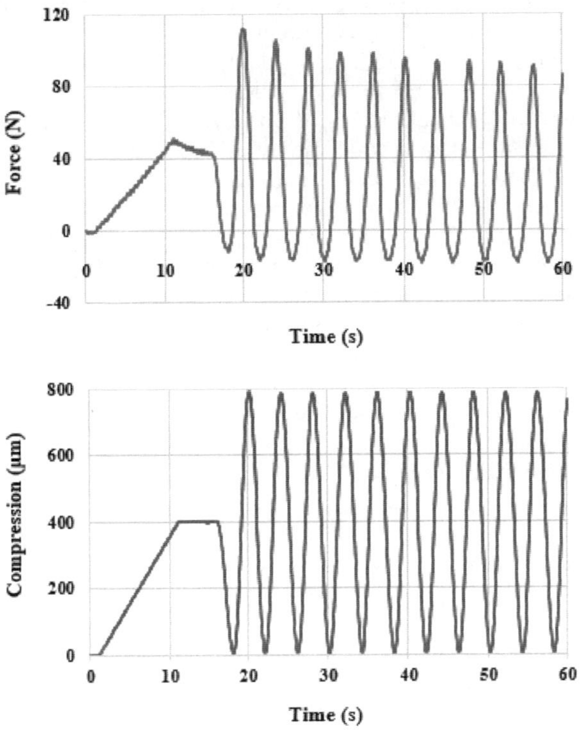

Figure 2. First ten cycles of knee joint preconditioning: reaction force tends to be stable as a function of time.

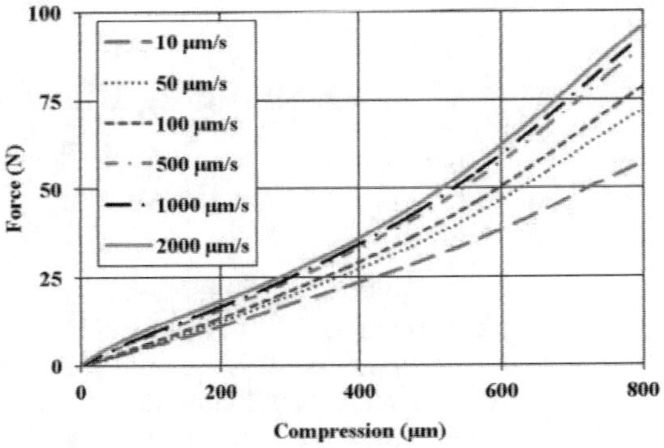

Figure 3. Load and compression relationship of the knee joint. Force nonlinearly increases with the increase of the compression.

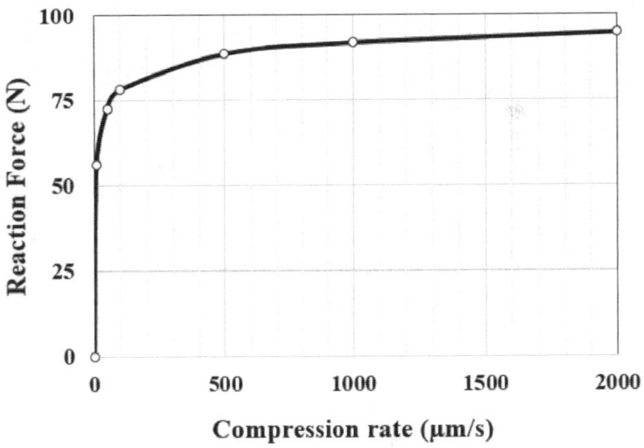

Figure 4. Reaction force as a function of compression rate, at 800 µm of compression

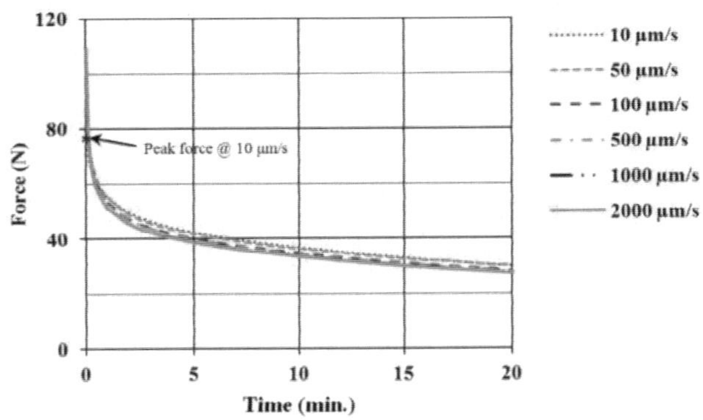

Figure 5. Relaxation of the joint at different compression rates. Reaction depends on compression rate but approaches to the same value (~30 N) at long time (a complete equilibrium was not reached).

Proceedings of the 25th CANCAM
London, Ontario, Canada, May 31 – June 4, 2015

Enhancing Biocompatibility of Grade 4-Titanium Using Laser Surface Texturing For Bone and Tissue Transplant Applications

Mitra Radmanesh
Department of Mechanical Engineering
University of New Brunswick
Fredericton, New Brunswick, Canada
Mitra.radmanesh@unb.ca

AmirKianoosh Kiani
Department of Mechanical Engineering
University of New Brunswick
Fredericton, New Brunswick, Canada
a.kiani@unb.ca

ABSTRACT

Enhancing the biocompatibility of wider range of materials is an essential step in improving the effectiveness of any bone and tissue implant applications. Through this study, the effects of laser surface texturing on the biocompatibility of grade 4 – titanium is investigated, by using simulated body fluid, energy dispersive x-ray, and x-ray diffraction. It was observed that, by producing various patterns on the surface of the titanium samples using a speed range of 50 mm/s, 100 mm/s, and 300 mm/s, the roughness of the surfaces of samples were varied. The variation of the surface roughness affected the cell adhesion rate to the surface of the titanium. Upon completion of the analysis, it is concluded that, using a lower processing speed, a lower frequency, and more exposure to SBF results in a better biocompatibility improvement as oppose to higher frequencies.

KEYWORDS: Osseointegration; Titania; Laser ablation; Surface modification; biomaterials

INTRODUCTION

In recent decades, biomedical engineering has become one of the most advanced fields in research and development societies. A significant contribution of this growing field is the bone and tissue transplant. This treatment is one of the most promising techniques in medical studies and orthopedic surgeries, with numerous advantages [1, 2]. In order for the organ replacement to take place, a scaffold system is required to gradually degrade and resorb as the tissue inside the body. The scaffold system generally provides the required support for cells to proliferate and attach to the structure, and eventually grows to the shape of the replaced bone. It is of utmost importance for the materials used in scaffold systems to have high biocompatibility to avoid problems such as infections, immune system rejections, and long periods of healing [2,3]. When implanted in the body, the only portion of the scaffold systems in touch with the cell is the surface of the structure. Hence, altering surface properties of the materials is one of the effective methods of increasing the biocompatibility of the materials used in structure of scaffolds [3, 4]. A new developing method for surface texturing in recent decades has been the laser surface texturing. This new technology enables the creation of patterns in scales as small as Micro/Nano on the surface of materials such as titanium and its alloys to increase the cell adhesion on the surface of these materials [5]. Further, despite the conventional methods, laser surface texturing offers better control and precision, wide range of selectivity, more feasibility, and is environmentally friendly [3, 6]. In comparison to conventional methods, this laser approach is more flexible and reduces processing time, making it particularly suitable for rapid prototyping and custom-scale manufacturing for a wide variety of applications such as in bio-NEMS and nano-biomaterials fabrication [7].

The purpose of this paper is to investigate the effects of laser surface texturing on the biocompatibility of grade 4 - titanium, by altering the surface roughness of the titanium samples. The biocompatibility of the processed samples is examined through the use of simulated body fluid (SBF), and measuring the composition of bone-like apatite layer by using energy dispersive x-ray (EDX) and x-ray diffraction (XRD).

EXPERIMENTAL SETUP

To study the biocompatibility of titanium, grade 4 thin sheets of titanium were used. First, the samples were irradiated with nanosecond pulses to create the predetermined patterns on the surface of titanium. The treated samples were then place into simulated body fluid for three days and seven days period. Experiments were completed using a Sol 10 W laser system, with a peak power of 50 kW, wavelength of 1064 nm, and a beam spot diameter of 8 mm. The pulse frequency of this laser system is in the range of 10 kHz to 100 kHz. In this experiment, the pulse repetition rate (frequency) was varied at 25 kHz, and 50 kHz; while the scanning speed was set to 50 mm/s, 100 mm/s, and 300 mm/s for each repetition rate, using the EZCAD© software. The laser power throughout the experiment was measured to be 8 W - 10 W, which was calculated using a power-meter prior to annealing of the samples. To proceed with the experiment, simulated body fluid was conceived to evaluate the biocompatibility of the textures created on the surface of the samples. Upon conceiving the simulated body fluid, processed samples were placed in the fluid for a set period of time to evaluate their biocompatibility. The obtained results are presented and discussed in the following sections.

RESULTS AND DISCUSSION

As explained previously, altering materials' surface topography properties can enhance their biocompatibility. In this experiment, increasing their surface roughness altered the surface topography of titanium samples. Horizontal patterns were generated on the surface of these samples, which increases the surface roughness by melting the surface in the location of the patterns. Generally, an increase in the roughness of the surface increases the cell adhesion rate by providing a larger surface area for the cells to attach themselves. The contact area increases by an increase in the deviation of the material surface; hence creating the patterns on the surface of the samples increases the average arithmetic deviation of the surface, which results in a higher average surface roughness [8, 9, 10]. This can be clearly seen in the samples resulted in this experiment.

First, the set of samples with the frequency of 25 KHz are examined.

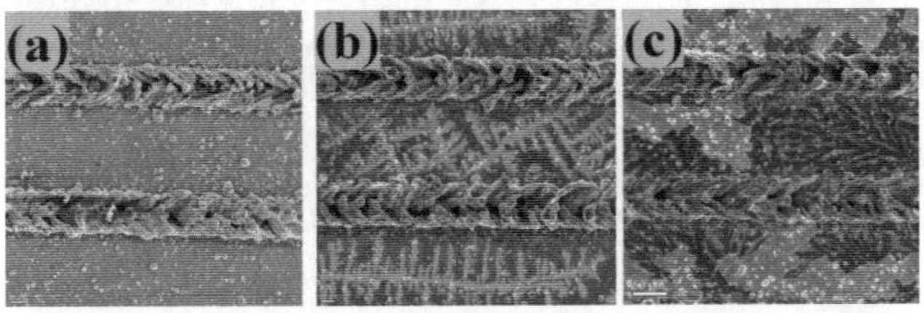

Figure 1: (a) No SBF; (b) 3 Days in SBF; (c) 7 Days in SBF

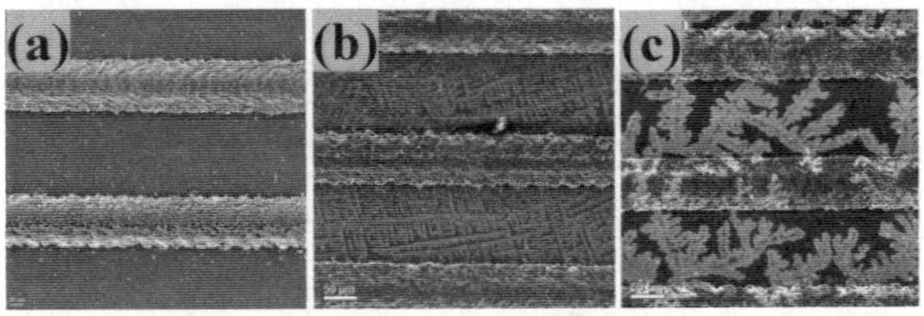

Figure 2: (a) No SBF; (b) 3 Days in SBF; (c) 7 Days in SBF

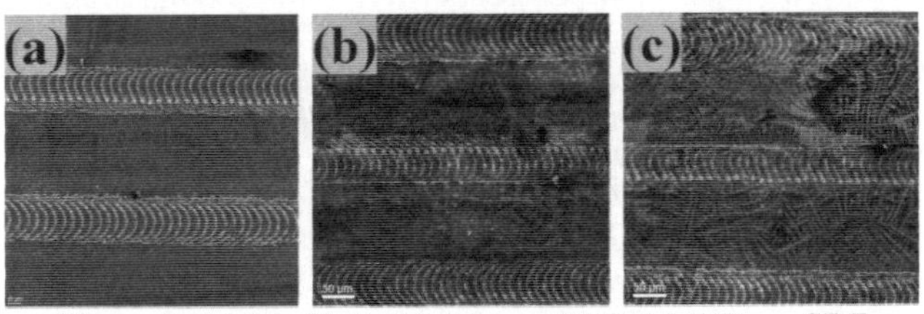

Figure 3: (a) No SBF; (b) 3 Days SBF; (c) 7 Days SBF

Fig. 1,2, and 3 display the samples prepared using processing speed of 50mm/s, 100 mm/s and 300 mm/s, respectively. Looking at pictures (b) and (c) in all three figures, it is clearly shown that bone-like apatite layers are deposited on the surface of these figures; the white layers shown on the surface of the figures in comparison to picture (a). The deposition of these bone-like apatite layers is an indicator of the biocompatibility of the processed samples. The more deposition of apatite layer on the surface of the samples, the more the biocompatibility of the sample has been improved. Hence, when comparing pictures (b) and (c) on each figure, it is clear that in picture (c), the samples with seven days of processing time in SBF, have more deposition than picture (b), since more time was given to the cells to attach to the surface of these samples.

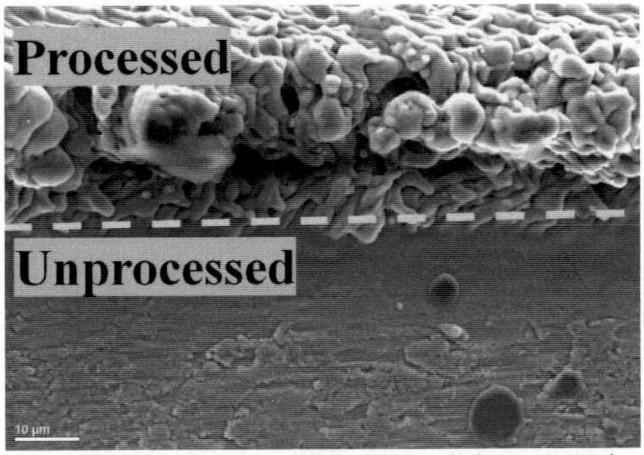

Figure 4: SEM micrograph of the Ti morphology comparing the different apatite-inducing abilities of the unprocessed area and the processed area after soaking in SBF.

This can be seen more clearly in Fig. 4; a closer illustration of the irradiated patterns of the 7 days processed samples in SBF. Figure 5 (a), (b), and (c) represent a closer illustration of samples with speeds of 50 mm/s, 100 mm/s, and 300 mm/s processing speeds respectively.

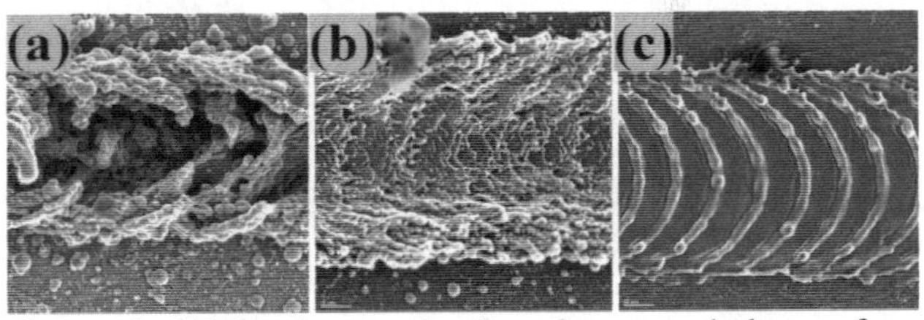

Figure 5: SEM micrograph of surface morphology of an processed Ti sample after 7 days soaking in SBF

As shown in Fig. 4, the bone-like apatite particles are distinctly more on top of the irradiated pattern in comparison to the regular surface of the titanium. It should be noted that, upon laser irradiation on the surface of the samples, the molten material pours around the indicated path and therefore, increases the roughness of the surrounding surfaces. Hence, the small depositions that can be seen surrounding the created patterns can be explained. This observation is in agreement with the fact that roughness increases the cell attachment rate to the surface of the materials. Also, looking closely at the samples in Fig. 5, it is distinctly shown Fig. 5 (a), the lowest processing speed (50 mm/s), resulted in more bone-like apatite deposition on the surface compared to Fig. 5 (b) and (c). This is due to the fact that, at lower processing speeds, more time is required to create the patterns, and hence more time for heat accumulation on the surface of the irradiated line, thus a higher amount of heat accumulates at that area. This in return results in a deeper penetration of the irradiated line, and producing higher roughness on the surface of the sample. In other words, lower processing speeds; increase the average roughness of the surface by creating larger irregularities upon melting the surface of the material [11,12].

EDX analysis of the composition of the apatite layer deposited after soaking in SBF for Ti indicates the presence of titanium, calcium, phosphorous, and oxygen, which corresponds to hydroxyapatite (HA) ($Ca_5H_2(PO4)_3{\cdot}OH$) as shown in Fig. 6. HA has a composition similar to the mineral phase of bone, and it has been demonstrated that bone-like HA possesses good osteoconductivity and has a high affinity for living bone cells.

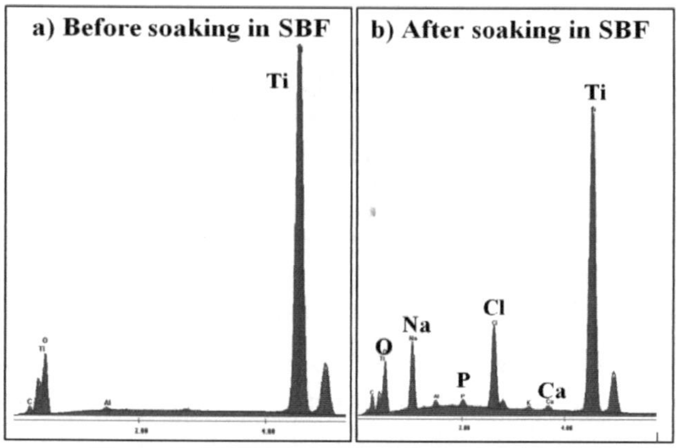

Figure 6: EDX Analysis - 100 mm/s Samples

The existence of bone-like apatite layer is further investigated by analyzing the samples using XRD; as indicated on Fig. 7, upon obtaining XRD analysis, traces of calcium was detected on the surface of the processed titanium samples after soaking in SBF which is in close agreements with the expectations.

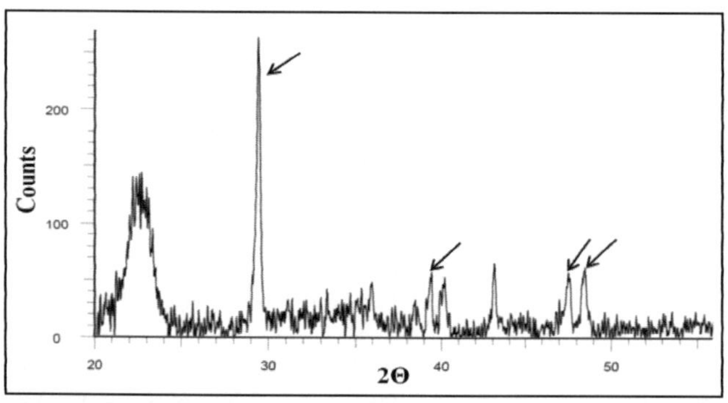

Figure 7: XRD diffraction patterns of processed Ti after soaking in SBF

Next section of the study examines the samples using frequency of 50 KHz.

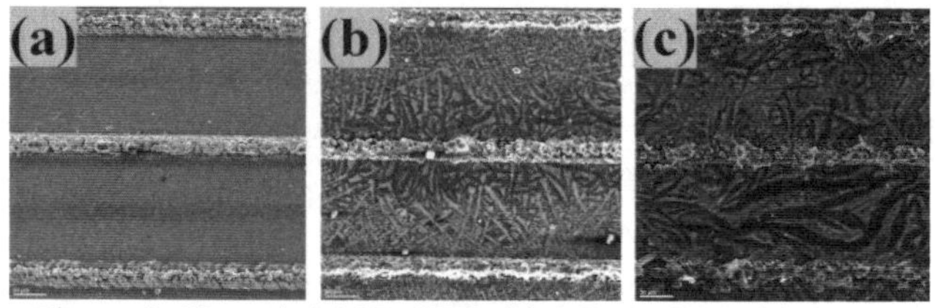

Figure 8: 50mm/s (a) No SBF; (b) 3 Days SBF; (c) 7 Days SBF

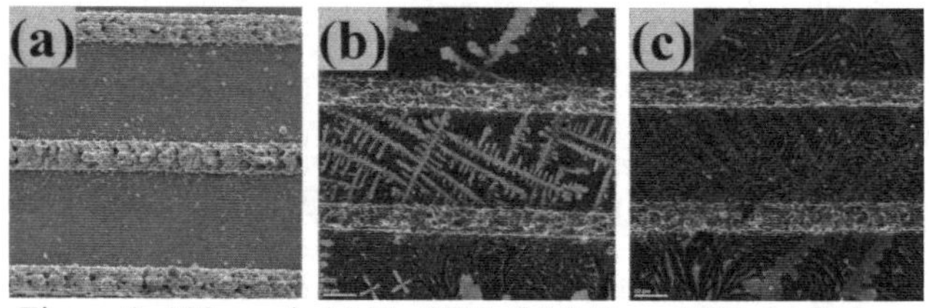

Figure 9: 100 mm/s (a) No SBF; (b) 3 Days SBF; (c) 7 Days SBF

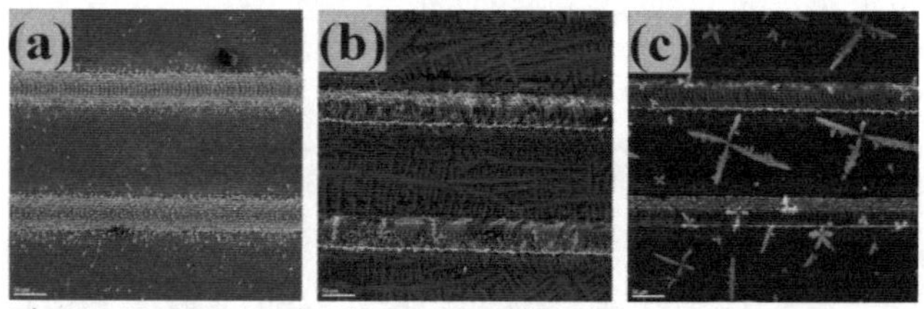

Figure 10: 300 mm/s (a) No SBF; (b) 3 Days SBF; (c) 7 Days SBF

Figures 8,9, and 10 represent the samples made using processing speed of 50 mm/s, 100 mm/s, and 300 mm/s respectively with frequency of 50 KHz.

Similar to the previous case, samples given seven days of processing on SBF show a higher increase in biocompatibility compared to pictures (b), the 3 days SBF processing samples, of all three figures. Pictures (a) in all three figures, show that an increase in the processing speed of irradiated patterns reduces the roughness of the surface, since they appear smoother as the speed increases. Hence, similar to the previous case, if samples with 300 mm/s were compared to samples with 50 mm/s and 100 mm/s on Figures 8, 9, and 10, the samples with lower speeds show more layers of deposition across their surfaces.

The closer magnification of samples, and the EDX and XRD analysis for this set of samples were not considered since due to the high frequency, the amount of deposition on the surface are relatively insignificant compared to the previous case.

An important fact that needs to be considered when comparing the three cases discussed, is the biocompatibility of samples with regards to the frequencies used in each case. Looking at Fig. 1, 2, 3, and 8, 9, 10 it is evidently shown that samples produced using a lower frequency, resulted in a higher amount of deposition on their surfaces.

Proceedings of the 25th CANCAM
London, Ontario, Canada, May 31 – June 4, 2015

CONTACT STRESS ANALYSIS OF SURFACE GUIDED KNEE IMPLANT USING FINITE ELEMENT MODELING

Ida Khosravipour
Mechanical Engineering
University of Manitoba
Winnipeg, Manitoba, Canada
khosravi@myumanitoba.ca

Yunhua Luo
Mechanical Engineering
University of Manitoba
Winnipeg, Manitoba, Canada
Yunhua.Luo@umanitoba.ca

Shabnam Pejhan
Mechanical Engineering
University of Manitoba
Winnipeg, Manitoba, Canada
pejhans@myumanitoba.ca

Urs P. Wyss
Mechanical Engineering
University of Manitoba
Winnipeg, Manitoba, Canada
Urs.Wyss@umanitoba.ca

ABSTRACT

After Total Knee Arthroplasty, contact stresses at the surface and stresses at the implant-cement-bone interface are directly related to the joint contact forces. These stresses are a major factor in wear and fatigue, and aseptic loosening. Implant contact stresses influence the wear and fatigue damage of the Ultra High Molecular Weight Polyethylene (UHMWPE) articulating surface, decreasing the longevity of the implant. In order to test the performance of the new surface guided knee implant design, finite element modeling is used to analyze the stresses of the design from extension to deep flexion.
KEYWORDS: knee arthroplasty, finite element modeling, knee implant, contact stress, surface guided

INTRODUCTION

In several previous studies finite element modeling has been used to estimate the stress distribution in implants during the gait cycle. Ishikawa et al. [1] developed a two-dimensional finite element model to evaluate the effect of contact stresses on the polyethylene tibial insert during a gait cycle. Balasubramanian et al. [2] used more realistic positions of total knee replacement (TKR) components in a gait. Fluoroscopy was performed on a patient who flexed the knee from 0 to 85 degrees. A two-dimensional FE model of the prosthesis in the sagittal plane was built and analysis was done based on the fluoroscopic data. Reeves et al. [3] used a two-dimensional sagittal plane model to examine the generation of plastic strains in the polyethylene tibial insert due to repetitive loading. Godest et al. [4] used an explicit finite element model to simulate the kinematics and the internal stresses in knee implants during a gait cycle. In the study, a three-dimensional finite element model was constructed to predict the motion, and the polyethylene stresses when a knee joint replacement is subjected to complex loading conditions. Miyoshi et al. [5] developed a three-dimensional finite element model of a total knee replacement's articulating surface to examine the component with and without a posterior cut-out slot. Beillas et al. [6] developed a generic finite element model of the lower body limbs by using data from two patients to validate a one legged forward hopping test. In this study, the proximal femur, distal tibia, foot and ankle were modeled by using rigid body elements while using shell elements for the surfaces. Harryson et al. [7] evaluated their patient specific designed knee implant by comparing the proposed smooth surface at the interface with a model of conventional five flat interface surfaces. Chandran and Amirouche [8] focused their 3-D analysis on a part of a specific knee design, a femoral cam-tibial post articulation in a posterior cruciate substituting TKR, and suggested various design parameters that would optimize femoral rollback. Tuncer et al. [9] conducted in vitro mechanical experiments on ten human cadaveric knees, and the results of bone strain measurements were used to validate four tibia and four femur models of the uni-compartmental knee replacement (UKR) implanted knee. Sensitivity assessments and convergence studies were conducted in the study to optimize modelling parameters. Kwon et al. [10] compared biomechanical specifications and stress results of fixed-bearing and mobile-bearing for uni-compartmental knee arthroplasty by using finite element method. They concluded that the in vitro wear of polyethylene inserts in knee arthroplasty devices depends on the magnitude, frequency, and orientation of kinematic inputs.

METHOD

Achieving an adequate and close to normal range of motion, and overall normal performance of the knee joint is the final goal of all knee replacement designs and surgeries. The design requirements for a preliminary conceptual design of a surface-guided knee by Amiri et al. [11] was investigated considering patient specific anatomy of the femur and tibia. The main goal of this research was to identify the design features that need to be optimized as they relate to the functionality of the implant by creating a CAD model of the implant. The 3D model of the knee geometry was created by segmentation of magnetic resonance imaging (MRI) data from the Osteoarthritis Initiative (OAI) data set. The MRI scans had a resolution of 384×384 pixels, with a pixel size of 0.7mm. In this study a finite element model would be used for the prediction of the contact stresses of tibiofemoral articulation of TKR design. The segmentation of the MRI data was accomplished using the thresholding function and LiveWire tool in Mimics (Materialise, Leuven, Belgium). The 3D models were exported into 3-matic v.8.0 (Materialise, Leuven, Belgium) to define the required functional parameters for the design step. The functional axes of femur and tibia, as well as intersection planes were defined through visible and easily defined landmarks on the femur and tibia, based on the data of previous research on knee alignment and functional axes [12]. The complexity of knee joint geometry which consists of continuous varying radii along the surface of each condyle has led to the development of customized knee implant designs. This study proposed a methodology to use guiding features that are incorporated into the shapes of the condyles and create the implant via extracting the geometrical and functional parameters from patient's MRI data of the bone and cartilage. Two knee implant models have been created based on two patients' MRI data [13].

This study includes the application of joint contact forces at three flexion angles (with the highest amount of load) followed by an analysis of the contact area, peak contact stress and von-Mises stresses in the tibial insert. It is suggested to model the femoral component as rigid body to reduce computational time since it is assumed to be made of the CoCr alloy which has modulus of elasticity far greater than that of the polyethylene insert. This assumption is in line with many of previous studies confirming that a rigid body model of CoCr alloy femoral part would be an acceptable assumption [14]. The material properties for the two parts of implant are provided in the Table 1. The polyethylene insert could be modeled as a deformable body.

Table 1: Material properties of Femoral and Tibial part

Component	Material	Young's modulus (MPa)	Poisson ratio	Density (kg/m^3)
Femoral part	CoCr alloy	193,000	0.29	8500
Tibial insert	UHMWPE	680	0.46	936

A surface to surface contact with finite sliding is defined between articular surfaces of the femoral and tibial component. The femoral articular surface is defined as the master surface, while the tibial insert articular surface is defined as the slave surface, indicating that the femoral articular surface is allowed to penetrate the tibial articular surface. This kind of contact formulation approximately resembles a deformable contact in the normal direction. The behavior of the contact in the normal direction is suggested to be defined with a pressure-over closure relation that is based on the elastic foundation principle. Only bone on bone forces can be used to determine the strength of an implant although it is a fact that even the bone-on-bone forces do not include co-contraction of muscles [15]. In this study axial and anterior posterior loads were applied at three different flexion angles (0 and 15 degrees of flexion in level walking and 60 degrees of flexion in stair ascending), without considering any other loads based on the data in the literature [15]. A body mass of 80 kg has been used, which is the main body weight reported by the study from which load data were used. In this study axial load equivalent to three times of the body weight is applied distributed on the upper surface of the femoral part, mimicking the bone-implant condition for a few degrees of flexion. The contact between the femoral and tibial parts of implant is modeled as pure penalty contact. The contact is defined as frictional with friction coefficient of 0.04 which is reported in literature to be in agreement with experimental tests [14, 16]. Hex-dominant meshing method is used for tibial component. The hexahedral elements were used due to their higher convergence rate. Articular surface of the femoral component is meshed with linear quadrilateral elements. Also, it's important to note that the surface to surface contact which is used for this project has no issue with quadrilateral element types. The element size of the tibia component is 0.74 mm. For the purpose of this study, contact stress analysis, fixed support has been assigned on the down surface of the tibia component. Thus, it is constrained in all directions at the distal end.

RESULTS

The outcome parameters have been derived for the evaluation of finite element analysis and comparing the model with previous studies. Mesh sensitivity was studied and the mesh refinement changed the predicted peak contact stress by less than 2.7%. Table 2 shows the peak von-Mises stresses for the two knee implant designs. Based on Ries et al. study [17], the upper and lower limits of polyethylene fatigue failure stress obtained from tensile testing are 32 MPa and 17 MPa, respectively. Since the yield properties of polyethylene have been shown to be similar in compression and tension, these limits are assumed as the criteria. All von-Mises stresses in these two designs are less than both upper and lower limits of polyethylene fatigue failure. Also, it should be noted that the peak von-Mises stress occurs 1-3 mm beneath the surface.

Table 2: Peak von-Mises stresses for knee implant 1 and knee implant 2

Flexion angle	0 degree of flexion	15 degrees of flexion	60 degrees of flexion
Knee Implant 1	13.5 MPa	16.8 MPa	14.9 MPa
Knee Implant 2	7.2 MPa	14.6 MPa	12.99 MPa

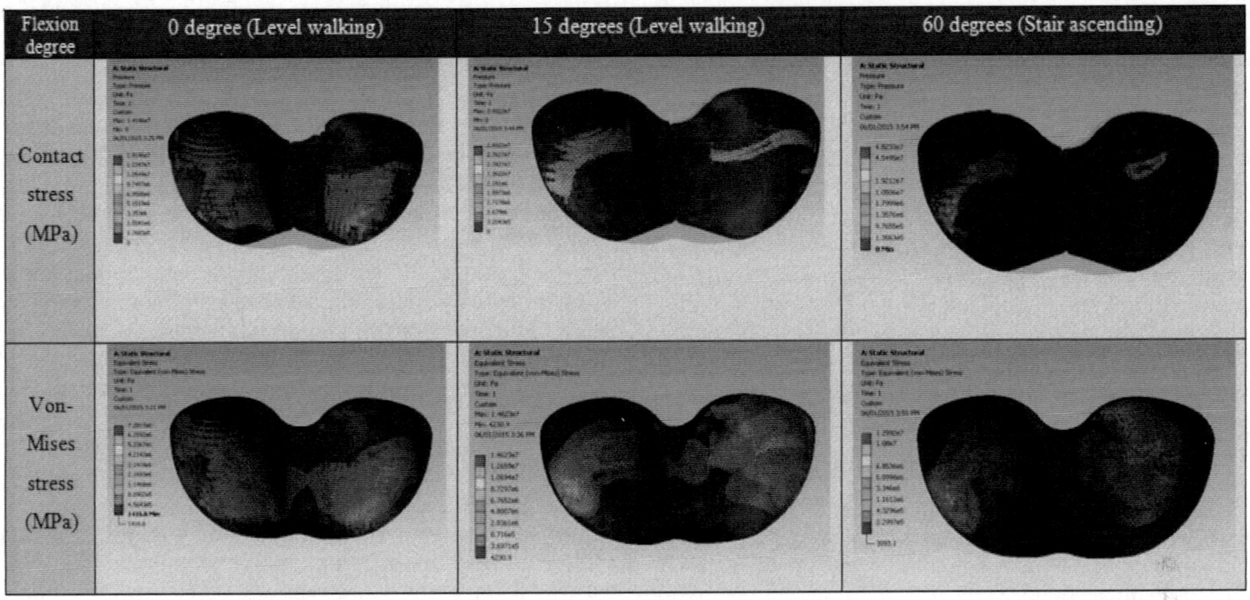

Figure 1: Contact pressure and von-Mises stress pattern for 0 degree and 15 degrees of flexion in level walking and 60 degrees of stair ascending in knee implant 2

The equivalent von-Mises stress, peak contact pressure and contact pressure distribution width on the insert part are determined and analyzed. Fig.1 shows the contact stress and von-Mises stress on the polyethylene tibia compartment at 0°, 15°, and 60 ° flexion of knee implant 2. As it can be seen from the results of the contact pressure distribution figures, the contact area for 15 and 60 degrees of flexion are smaller. This can lead to higher contact pressure on these models and the contact pressure is not distributed on all parts of the tibial surface. In these two models the peak contact pressure is on the lateral side of the tibia although the contact area is larger on the medial side.

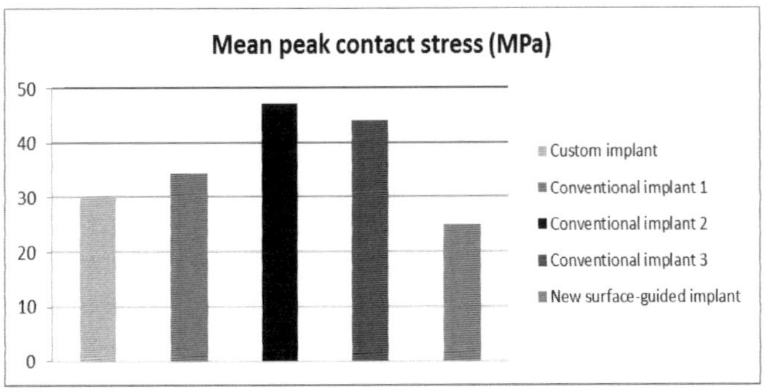

Figure 2: Mean peak contact stress comparison with the work of Heever et al.

For validation and also comparison with previous studies, work of Heever et al. [18] has been considered. In this work, 4 different models of implant have been studied and the contact stress of these models has been evaluated. In figure 2 contact stress of this new surface guided knee implant has been compared with these 4 models. The loading and boundary conditions for all these implants were the same.

CONCLUSIONS

In Total Knee Replacement the polyethylene stresses are dependent on the kinematics and on the design of the implant. The aim of this study was to examine whether the new surface guided knee implant can perform similar or better under loading conditions, and potentially wear. The most uniform stress distribution happens when the design is patient specific and also surface guided. This model has the potential of providing good contact stress distribution with less contact stress and less maximum von-Mises stress on tibia component which can reduce the possibility of surface fatigue and wear. The mean peak contact pressure of this surface guided knee implant is almost 10% lower than in previous studies. Another important aspect that has been evaluated in this study is the effect of different flexion degrees on the tibial component stress distribution. As it has been mentioned in previous studies [19], the maximum contact stress on tibio-femoral articulation happens during stair ascending with 60 degrees of flexion in knee implant. In addition, fatigue analysis could be done in order to evaluate the short and long term advantages of this new design. Cyclic stresses are required to be conducted to determine the wear, deformation and possible damages.

ACKNOWLEDGMENTS

The authors would like to gratefully thank DePuy for providing financial support.

REFERENCES

1.Ishikawa H., Yasuda K., Fujiki. H., "Contact analysis of ultrahigh molecular weight polyethylene articular plate in artificial knee joint during gait movement", Journal of biomechanical engineering, vol. 118(3), pp. 377-386, 1996.
2. Balasubramanian V, Shanbhag A, Li G, Rubash HE. "Effects of flexion on stresses in UHMWPE tibial liners" Transactions of the Annual Meeting- Orthopaedic Research Society, vol. 793, 1997.
3. Reeves E.A., Barton DC, FitzPatrick DP, Fisher J., "A two-dimensional model of cyclic strain accumulation in ultra high molecular weight polyethylene knee replacements", Proceedings of the institution of mechanical engineers, vol. 212, no. 3, pp. 189-198, 1998.
4. Godest A.C., Beaugonin M., Haug E., Taylor M., Gregson P.J., "Simulation of a knee joint replacement during a gait cycle using explicit finite element analysis,", Journal of biomechanics, vol. 35, pp. 267-275, 2002.
5. Miyoshia S., Takahashia T., Ohtania M., Yamamotoa H., Kameyamab K., "Analysis of the shape of the tibial tray in total knee arthroplasty using a three-dimension finite element model", Clinical Biomechanics, vol. 17, pp. 521-525, 2002.
6. Beillas P., Papaioannou G., Tashman S., Yang K.H., "A new method to investigate in vivo knee behavior using a finite element model of the lower limb", Journal of biomechanics, vol. 17, pp. 521-525, 2004.
7. Harrysson O.L., Hosni Y.A., Nayfeh J.F., "Custom designed orthopedic implants evaluated using finite element analysis of patient specific computed tomography data: femoral-component case study", BMC Musculoskeletal disorders, vol. 8, pp. 91-101, 2007.
8.Chandran N., Amirouche F., "Optimization of the posterior stabilizaed tibial post for greater femoral rollback after total knee arthroplasty- a finite element analysis", International orthopaedics (SICOT), vol. 33, pp. 687-693, 2009.
9. Tuncer M., Cobb J.P., Hansen U.N., Amis A.A., "Validation of multiple subject-specific finite element models of unicompartmental knee replacement", Journal of Medical Engineering & Physics, vol. 35, pp. 1457-1464, 2013.
10. Kwon O., Kang K., Son J., Kwon S., So S., Suh D., Choi Y., Kim H., Koh Y., "Biomechanical comparison of fixed- and mobile-bearing for unicompartmental knee arthroplasty using finite element analysis", Journal of orthopaedic research , vol. 10, pp. 338-346, 2013.
11. Amiri S., Cook D.V., Wyss U.P., " Conceptual design for condylar guiding features of a total knee replacement", Journal of Medical Devices, vol. 5: pp. 1-9, 2011.
12. Cook D.V., Scudamore A., Li J., Wyss U.P., T. Bryant, P. Costigan, "Axial Lower-limb Alignment: Comparison of Knee Geometry in Normal Volunteers and Osteoarthritis Patients", Osteoarthritis and Cartilage, vol. 5:pp. 39-47, 1997.
13. Pejhan S, Essex O, Bohm E., Brandt J., Wyss U., "Design features and kinematic behavior of a customized surface-guided knee implant", accepted in Orthopedic Research Society conference, 2015.
14. O'Brien, S., Luo, Y., Wu, C., Petrak, M., Bohm, E., & Brandt, J. M., "Prediction of backside micromotion in total knee replacements by finite element simulation"., Proceedings of the Institution of Mechanical Engineers, Part H: Journal of Engineering in Medicine, Vol. 226(3), pp. 235-245, 2012.
15. Wyss U.P., Costigan P.A., "Gait analysis Biomechanical Tool in Development of Artificial Joints", Morscher EW (ed) Endoprosthetics; Springer, pp. 103-115, 1995.
16. Halloran J.P., Easley S.K., Petrella A.J., Rullkoetter P.J., "Comparison of Deformable and Elastic Foundation Finite Element Simulations for Predicting Knee Replacement Mechanics", Journal of Biomechanical Engineering; 127: pp. 813-818, 2005.
17. Sathasivam S., Walker P.S., "Computer model to predict subsurface damage in tibial inserts of total knees", Journal of Orthopaedic Research, 16:pp. 564-571, 1998.
18. Van den Heever D.J., Scheffer C., Erasmus P., Dillon E., "Contact stresses in a patient-specific uni-compartmental knee replacement", Clinical Biomechanics, 26:pp. 159-166, 2011.
19. Van Den Heever D.J., Scheffer C., Erasmus P.J., Dillon E.M., "Development of a patient-specific femoralcomponent for unicompartmental knee replacement", 31st annual international conference of IEEE EMBS, 2009.

EXPERIMENTAL APPARATUS FOR ANALYSIS OF BONE REMOVAL DURING SURGICAL RESURFACING PROCEDURES

Jonathan Kusins
Mechanical and Materials Engineering
Western University
London, Ontario, Canada
jkusins@uwo.ca

O. Remus Tutunea-Fatan
Mechanical and Materials Engineering
Western University
London, Ontario, Canada
rtutunea@eng.uwo.ca

Louis Ferreira
Mechanical and Materials Engineering
Western University
London, Ontario, Canada
Louis.Ferreira@uwo.ca

ABSTRACT

The experimental quantification of the bone removal characteristics represents a desirable outcome from the perspective of design of control feedback loops involved in orthopaedic robotic systems. The current study is focused on the development of an experimental apparatus capable to assess the parameters associated with light bone removal operations, such as those involved in glenoid resurfacing performed in the context of shoulder arthroplasty. For this purpose, an orthopaedic burring tool has been instrumented with an accelerometer capable to monitor tool vibration as well as an infrared pyrometer capable to measure the superficial temperature attained by the bone as a result of passing of the cutting tool. In addition to this, a workpiece mount has been instrumented with a three-axis load cell in order to track the cutting forces associated with the bone removal process. The developed system allows a concurrent measurement of the three outputs associated with bone removal process (cutting force, vibration, and temperature) as a function of various machining-specific parameters such as cutting tool's size and type, rotary speed, feed rate, depth of cut as well as the orientation of the tool with respect to the workpiece. A representative sample of output variables is presented in this study as a demonstration of the capabilities of the developed device.

KEYWORDS: Joint replacement, bone resurfacing, bone removal parameters, experimental apparatus, superficial temperature, cutting forces, vibration amplitude

INTRODUCTION

In orthopaedics, resurfacing is a surgical procedure whose typical purpose to lightly alter the outer shape of the bone in order to facilitate the subsequent implantation operations involved in joint replacements. Similar to other bone machining instances, the need for improved accuracy and reliability has led to the introduction of robotic surgical systems. For instance, an autonomous burring robot Robodoc® (Curexo Technology Corporation, Fremont, California) was first introduced in a surgical setting in 1992 [1]. Currently, the Robotic Arm Interactive Orthopedic System (RIO®) (MAKO Surgical Corp., Fort Lauderdale, Florida) is a commercially available surgeon-guided robot used primarily for hip replacements and partial knee resurfacing [2].

It is important to note that the emergence of robotic systems provides an opportunity to implement closed-loop feedback control to improve the effectiveness of automated bone removal systems with respect to their manual counterparts. To address this aspect, several prior studies have attempted to characterize the bone removal process by means of the temperature generated at the surface of the osseous specimen. The interest in this particular bone removal process outcome stems in the fact that it is known that permanent thermal bone damage (*i.e.* necrosis) could be associated with temperatures in excess of 50°C, and this in turn can limit the post-surgery bone ingrowth [3]. Along the same lines, Denis *et al.* and Shin *et al.* relied on infrared pyrometers to measure the superficial temperature of freshly cut specimens [4, 5], while Sugita *et al.* developed a force-feedback control loop capable to monitor cutting forces and to select the optimal machining parameters from the perspective of a minimal thermal effect [6]. However, although temperature and cutting forces have been monitored in some of the prior studies, no attempts were made towards their integration with the dynamics of the bone removal process, which represents the primary goal of this work.

MATERIALS AND METHODS

Process Parameters and Measurands

An experimental apparatus was designed to quantify some of the characteristic parameters associated with bone removal process. For the purpose of this study, the following variables were regarded as the controlled inputs of the bone removal process:

- Rotational speed of the tool (rpm)
- Tool type
- Depth of cut (mm)
- Feed rate (mm/s)
- Cutting track overlap (%)
- Inclination angle (°)
- Tilt angle (°)

The measurands selected to evaluate the performance of the bone removal process were:

- Superficial temperature of the bone (°C)
- Three-axial cutting force components (N)
- Vibration amplitude (m/s^2)

Experimental Apparatus

The structure of the developed experimental apparatus consists of the following: cutting tool mount, workpiece clamp, and a servo-hydraulic actuator. More details on each of these components are provided further.

Tool Holding System

The Midas Rex® EHS Stylus® (Medtronic Inc, Minneapolis, Minnesota) was selected to provide the main rotational motion of the cutting tool. The rotational speed of the orthopaedic system ranges from 200 to 75,000 rpm and is controlled by an integrated power console supplied by Medtronic. The translational motion required between the cutting tool and the workpiece is supplied by means of an Instron® actuator to which the bone removal system has been securely attached. The orientation of the cutting tool with respect to the workpiece is determined by two different angles: inclination and tilt, respectively. While the tilt of the cutting tool with respect of the workpiece normal at the cutter contact point (*e.g.* the theoretical contact point between the tool and workpiece) is performed in a horizontal plane by means of the Instron kinematics, its inclination in the vertical plane enclosing workpiece normal passing through the cutter contact point is ensured by means of the circular pattern of holes depicted in Fig. 1. These holes allow an incremental change in the inclination of the tool between -45° and 45° (5° increments).

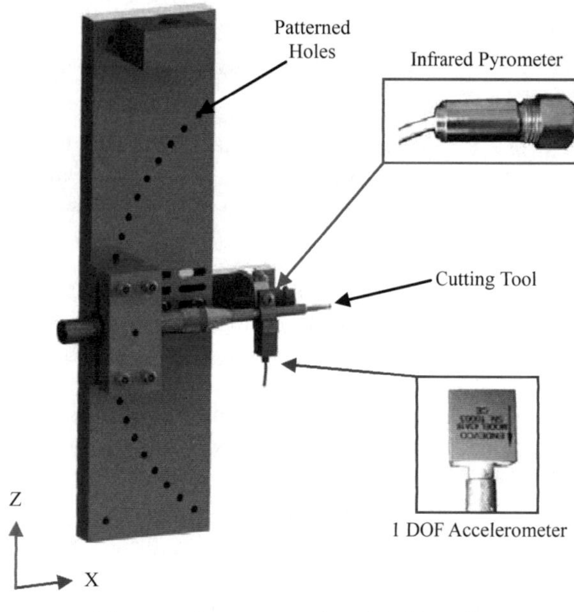

Figure 1: Instrumented tool holder with adjustable orientation

The pyrometer used to capture the post-cutting surface temperature was mounted in a position to ensure a direct line of sight to the machined track located immediately behind the cutting tool. Finally the single DOF accelerometer was mounted to capture the vertical (*e.g.* parallel to tool feed) vibrations of the entire tool holding system.

Workholding System

The fixture illustrated in Fig. 2 was designed to maintain the workpiece in a fixed spatial position with respect to the Instron frame. The primary component ensuring the location, support as well as clamping of the prismatic samples was a conventional tool shop vise. For the testing and validation purposes of the present study, cancellous-grade sawbone was chosen as the workpiece material since it ensures a uniform and consistent biomechanical structure [7]. In agreement with the overall context of the present study, the selected sawbone has a density of 20 lb/ft^3 which is comparable to that of a healthy joint bone [8]. The load cell placed between the workholding device and the two-axis stage was included specifically for the measurement of the three components of the cutting force.

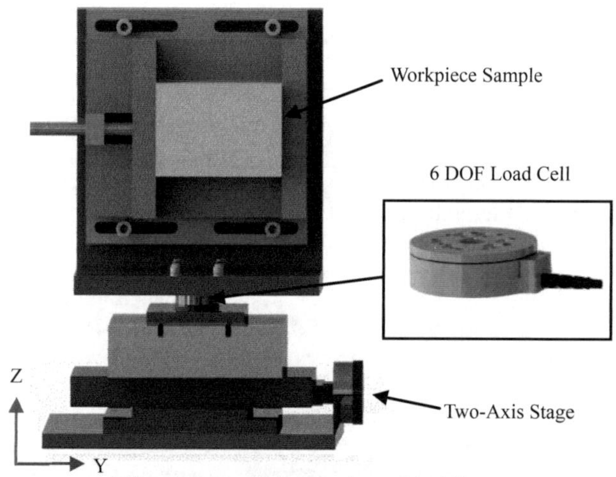

Figure 2: Instrumented workholding system with adjustable horizontal position

Data Acquisition System

Cutting Force Measurement

A three-axis load cell (Mini45, ATI Technologies, Markham, Ontario) with a resolution of 1/16 N was used to acquire the triaxial components of the cutting force. A double Butterworth low pass filter of 10 Hz was applied on the cutting force signal that was decimated at 1 kHz for post-processing purposes.

Vibration Measurement

A single-axis piezoelectric accelerometer (Endevco® model 42A16, Meggitt Sensing Systems, Fribourg, Switzerland) was used in dynamic data acquisition. The relevant technical specifications of the accelerometer are: ±50 g range, 100 mV/g sensitivity, and an amplitude response (±5%) of 1 to 10 kHz. The accelerometer was coupled with a power supply and signal conditioner (Model 4416B Endevco®). A double Butterworth low pass filter of 10 kHz was applied to the accelerometer. A Fast Fourier Transform (FFT) was applied to the vibration measurements to translate data from the time domain into the frequency domain.

Temperature Measurement

An infrared pyrometer (Micro-Epsilon, Model CT-SF02, Raleigh, North Carolina) was used for temperature measurement. The instrument comes with an accuracy of 1°C and resolution of 0.1°C. A CF02 lens was added to the pyrometer to reduce the spot size to 2.4 mm at a standoff distance of 30 mm. Although the ideal measurement of the temperature should happen at the cutter contact point, this optimal position is permanently occluded by the cutting tool itself. Therefore, the pyrometer was focused on an area located immediately behind the cutting tool (Fig. 3).

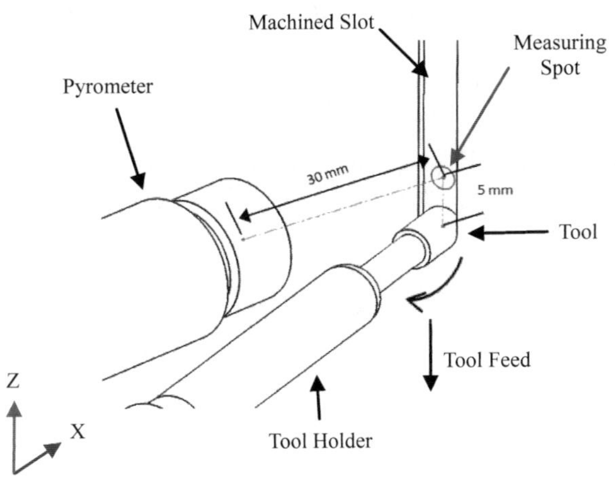

Figure 3: Temperature measurement setup

Given the quasi-stability of the temperature during cutting, data was decimated to 100 Hz for post-processing purposes.

Data Acquisition Unit

To ensure the correct time synchronization of all three measurands, all data collected by the sensors was supplied to an USB-6210 data acquisition unit (National Instruments Corporation, Austin, Texas). NI LabVIEW software was used for collection and post-processing of the data that was sampled at 25 kHz for all transducers.

RESULTS

The following parameters were used for a sample cut performed with the developed apparatus:

- Rotational speed of the tool: 45,000 rpm
- Tool type: spherical burring tool of 4 mm diameter
- Depth of cut: 0.5 mm
- Feed rate: 6 mm/s
- Cutting track overlap: 0%
- Inclination angle: 0°
- Tilt angle: 0°

Under these cutting conditions, a representative sample of pyrometer, load cell and accelerometer measurements is presented in Fig. 4. Depending on the status of the cutting tool as well as its relative position with respect to the workpiece, the plots indicate that the cutter could be either off, disengaged/idling or engaged in a cut. As such, all subsequent results will only be discussed in the context of the engaged state of the tool. Furthermore, during the bone removal phase associated with tool/workpiece engagement, it can be noticed that a stable cutting regime is attained when the tool has become fully engaged with the material and this should preclude the start and end portions of the cut when the tool is less than the diameter of the tool away from the material boundaries. Three independent cuts were performed to verify the repeatability of the developed system.

The analysis of the acquired data indicates that – under the tested cutting conditions – the superficial temperature of the workpiece raised to 31.9±0.18 °C, while the cutting forces generated in the process were less than 1 N: $F_x = 0.36 \pm 0.01$ N, $F_y = 0.18 \pm 0.02$ N, $F_z = -0.11 \pm 0.01$ N. The vertical accelerations generated during the process varied between +25 and -30 g (peak to peak values) with a root mean square value of 4.01±0.52 g. In frequency domain (FFT transform), peaks were recorded at 750 and 1500 Hz. In the disengaged state of the tool, vibration and cutting forces were identical before and after the cut, while the temperature dropped gradually after the cut before reaching the room temperature value of 22 °C.

DISCUSSION

Based on the preliminary investigations performed in the scope of this study, the developed experimental apparatus is capable to monitor the targeted outputs of the bone removal process. The data obtained was within the expected ranges, with vibration frequencies peaking at 750 and 1500 Hz, corresponding to tool's rotational speed (45,000 rpm) and the second harmonic of the base frequency, respectively. The measurement of the load cell was calibrated against an offline measurement performed by means of comparable static weights (data not shown), such that it is reasonable to expect that their values were also accurate, despite the relatively low net values of the cutting force that could be negatively affected by the noise. Further investigations will attempt to determine the accuracy of the temperature measurements that is expected to drop slightly due to the offset/trailing position of the measuring point with respect to the cutter contact point.

CONCLUSIONS

The present study outlines the developmental steps performed towards the development of an experimental apparatus capable to monitor concurrently several important output variables associated with bone removal. Future work will focus on the use of the device in investigation of the dependencies between input variables and output variables, as experienced during the light cuts associated with bone resurfacing procedures involved, for instance, in glenoid reaming performed in total shoulder arthroplasty.

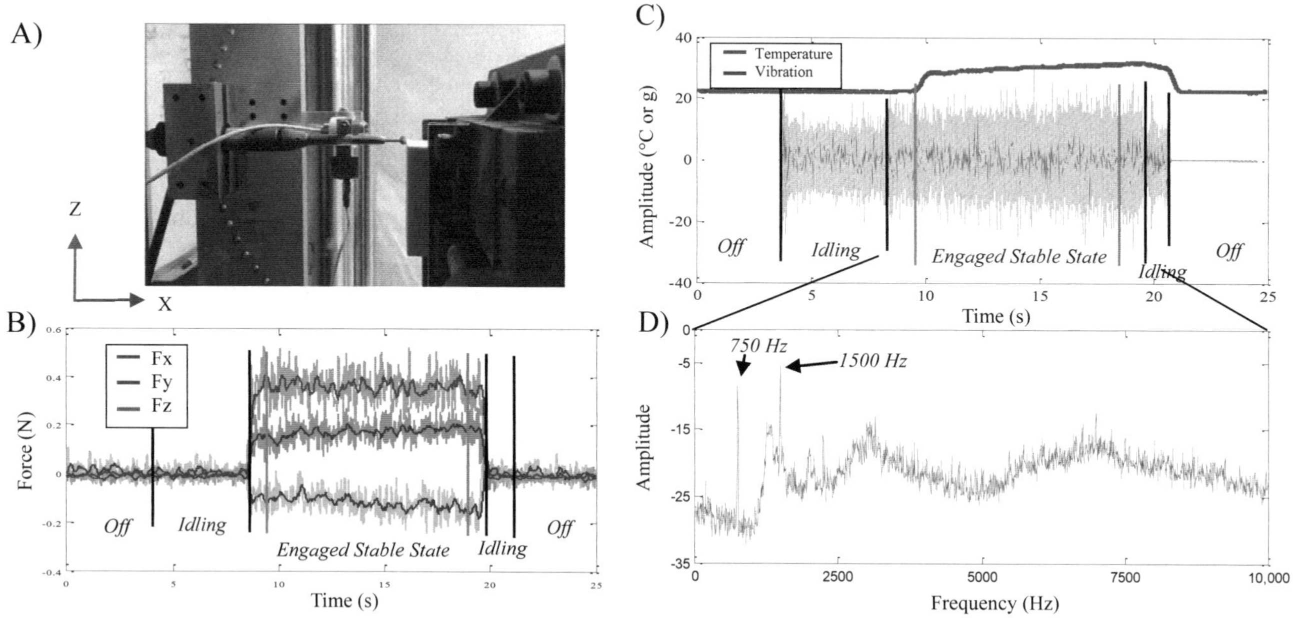

Figure 4: (A) Physical implementation of the experimental apparatus;
(B, C, D) Sample measurements for cutting force, temperature and vibration, respectively

In this regard, it is reasonable to expect that an in-depth understanding of the machining characteristics associated with joint replacement or resurfacing will facilitate optimal decisions with respect to an efficient use of the robotic and/or computer-driven bone removal solutions that are presently available.

ACKNOWLEDGMENTS

This work was supported in part by National Sciences and Engineering Research Council of Canada (NSERC), also through a Collaborative Research and Training Experience (CREATE) program on Computer-Assisted Medical Intervention (CAMI).

REFERENCES

1. A. P. Schulz, K. Seide, C. Queitsch, A. von Haugwitz, J. Meiners, B. Kienast, *et al.*, "Results of total hip replacement using the Robodoc surgical assistant system: clinical outcome and evaluation of complications for 97 procedures," *The International Journal of Medical Robotics and Computer Assisted Surgery,* vol. 3, pp. 301-306, 2007.
2. R. Tarwala and L. D. Dorr, "Robotic assisted total hip arthroplasty using the MAKO platform," *Current Reviews in Musculoskeletal Medicine,* vol. 4, pp. 151-156, 2011.
3. A. R. Eriksson and T. Albrektsson, "Temperature threshold levels for heat-induced bone tissue injury: A vital-microscopic study in the rabbit," *The Journal of prosthetic dentistry,* vol. 50, pp. 101-107, 1983.
4. H. C. Shin and Y. S. Yoon, "Bone temperature estimation during orthopaedic round bur burring operations," *Journal of Biomechanics,* vol. 39, pp. 33-39, 2006.
5. K. Denis, G. Van Ham, J. Vander Sloten, R. Van Audekercke, G. Van der Perre, J. De Schutter, *et al.*, "Influence of bone burring parameters on the temperature rise, burring forces and surface flatness in view of robot-assisted total knee arthroplasty," *International Congress Series,* vol. 1230, pp. 300-306, 2001.
6. N. Sugita, T. Nakano, Y. Nakajima, K. Fujiwara, N. Abe, T. Ozaki, *et al.*, "Dynamic controlled burring process for bone machining," *Journal of Materials Processing Tech,* vol. 209, pp. 5777-5784, 2009.
7. ASTM F1839-08, "Standard Specification for Rigid Polyurethane Foam for Use as a Standard Material for Testing Orthopaedic Devices and Instruments,", ASTM International, West Conshohocken, PA, 2012.
8. L. H. Frich, N. C. Jensen, A. Odgaard, C. M. Pedersen, J. O. Søjbjerg, and M. Dalstra, "Bone strength and material properties of the glenoid," *Journal of Shoulder and Elbow Surgery,* vol. 6, pp. 97-104, 1997.

INDICATORS OF MOTION INDUCED INTERRUPTION OCCURRENCES IN HEAVY–WEATHER SEA CONDITIONS

N.R. Bourgeois
Mechanical and Aerospace Engineering
Carleton University
Ottawa, ON, Canada
Nicholas.Bourgeois@carleton.ca

R.G. Langlois
Mechanical and Aerospace Engineering
Carleton University
Ottawa, ON, Canada
Robert.Langlois@carleton.ca

ABSTRACT

During a recent heavy-weather sea trial, human postural stability experiments were performed which included measurements of joint positions using a motion capture system, and foot pressures using instrumented footwear insoles. From these data, centre of mass, centre of force, and base of support parameters were determined. These data were then compared to observed occurrences of motion induced interruptions in order to determine if a simple correlation exists between them.

Keywords: Sea trial, human postural stability, centre of mass, base of support, motion induced interruption, MII

INTRODUCTION

Human postural stability experiments were carried out on the ship Canadian Forces Auxiliary Vessel (CFAV) Quest during the Q-348 heavy-weather sea trial from November 20th through November 28th, 2012 [1]. Participants took part in experimental sessions in which they were asked to maintain balance while performing a cognitive task, at various orientations with respect to the ship, and at various sea state conditions. The data measurement techniques used were unique in that they attempted to measure all of the sensory inputs that a person might use to maintain balance, including joint positions and orientations, and foot pressure/force data. This paper presents a short overview of how portions of that data have been post-processed, and an early look at various derived parameters that may be indicative of loss of balance events known as motion induced interruptions (MIIs) [2].

MOTION CAPTURE DATA

Motion tracking data was recorded using two Microsoft Kinect sensors and an iPi Studio software package. The software's proprietary algorithms map depth data recordings from the two sensors onto a model of a human skeleton. A sample of the skeleton model used is shown in Figure 1.

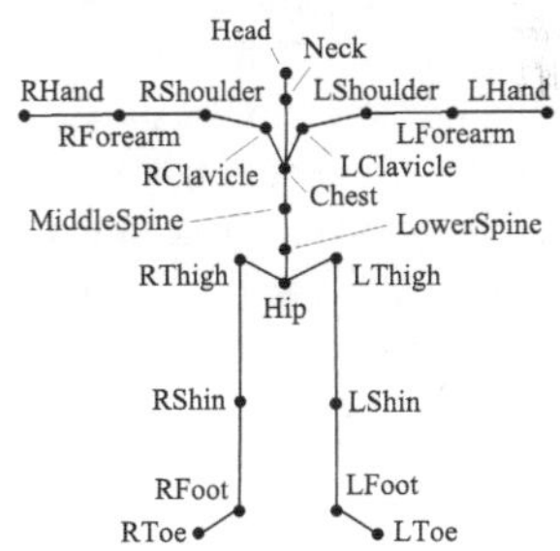

Figure 1: Skeletal structure output by iPi Studio software.

GLOBAL JOINT POSITIONS

The motion tracking data are stored in Biovision or BVH files, which is a file format designed to be used to record the motion history of a body with a rigid skeletal structure. The files begin with the initial positions of each joint and the degrees of freedom of motion available to each. Following that is a time history of joint angle 'channels', which are used to specify changes to the initial geometry over the course of the recording. This means that as long as the correct coordinate and rotation conventions are followed, it is possible to calculate the global position of each joint at any point in time.

The rotation convention used is XYZ Euler angles (Bryant angles), so the transformation between individual joint coordinate systems is defined as:

$$R = \begin{bmatrix} 1 & 0 & 0 \\ 0 & c_{\theta_x} & -s_{\theta_x} \\ 0 & s_{\theta_x} & c_{\theta_x} \end{bmatrix} \begin{bmatrix} c_{\theta_y} & 0 & s_{\theta_y} \\ 0 & 1 & 0 \\ -s_{\theta_y} & 0 & c_{\theta_y} \end{bmatrix} \begin{bmatrix} c_{\theta_z} & -s_{\theta_z} & 0 \\ s_{\theta_z} & c_{\theta_z} & 0 \\ 0 & 0 & 1 \end{bmatrix} \quad (1)$$

Rotations are defined as the relative angles between coordinate systems. A sample joint position calculation can be presented as:

$$r_D = R_{a/O}\{(r_{B/A})_a + R_{b/a}\{(r_{C/B})_b + R_{c/b}(r_{D/C})_c\}\} \quad (2)$$

where the uppercase subscripts indicate relative joint positions and the lowercase subscripts indicate which coordinate system they are defined in. While this is not a typical form for this type of calculation, it is computationally efficient if the global positions of intermediate joints are not required. Using this form, the global right ankle position can be calculated as:

$$\begin{aligned} r_{RFoot/inertial} = {} & r_{Hip} + R_{Hip}(r_{RThigh} \\ & + R_{RThigh}(r_{RShin} + R_{RShin}(r_{RFoot}))) \end{aligned} \quad (3)$$

where the subscripts indicate which initial position or recorded channel should be used.

CENTRE OF MASS POSITION

Of practical use is an estimate of the location of a person's centre of mass (CoM), which can be derived at each moment in time from the motion capture data. Approximate mass values and centre of mass locations of each individual body segment were calculated using proximal distances and mass fractions taken from literature [3]. Overall body CoM position was calculated by using a weighted average of all of the individual segment locations.

INSOLE DATA

Pressure-sensitive insoles were used to measure subject foot pressures during the sea trial experiments. A sample software view of one data frame is shown in Figure 2.

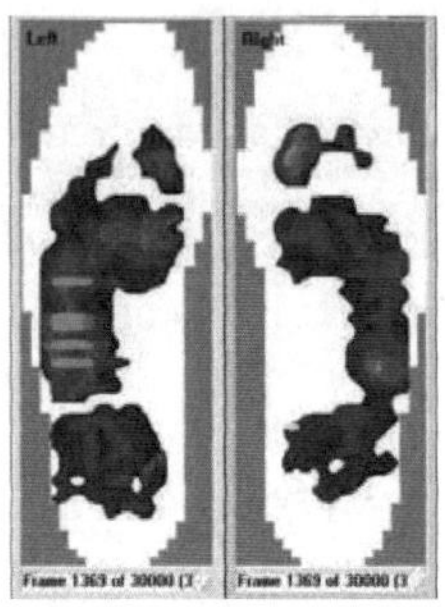

Figure 2: Screenshot of insole pressure recording software.

Recordings of this type of data were used in conjunction with the motion capture data to calculate individual base of support (BoS) boundaries for each participant at any point in time. The relationship between BoS and CoM is a common metric in postural stability studies [4]. The first step was to calculate a mask which would encompass the area beneath a subject's foot. The next step was to reduce that to an outline based on the outer limits of the mask. Sample results from each calculation step are shown in Figure 3.

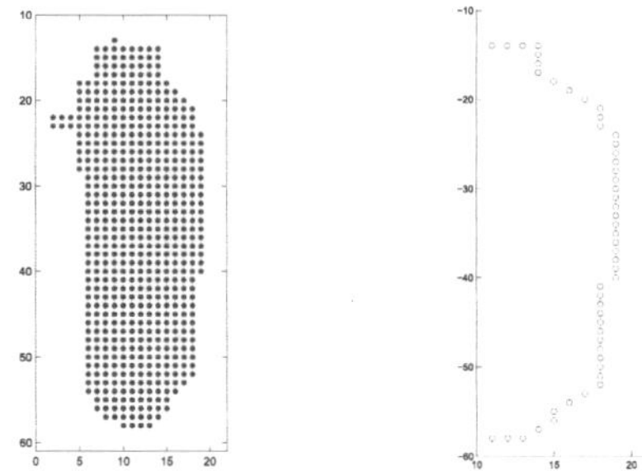

(a) Insole mask (b) Insole outline

Figure 3: Insole data processing steps.

Additional steps were carried out which included:

- Estimation of ankle locations based on ground-foot contact areas.
- Calibration of pressure measurements based on subject mass, subject and ship motions, and auxiliary force measurements made using a load cell located under one foot.
- Summation of foot pressures in order to determine the overall vertical force exerted by each foot.
- Matching of the insole's coordinate system to the motion capture's coordinate system and translation of insole data to ankle locations.
- Calculation of BoS boundaries that connect the left and right insole outlines.
- Calculation of total body centre of force (CoF) based on locations of individual foot forces and their relative magnitudes.

A sample result of all of these calculations is shown in Figure 4.

RESULTS

The objective of the analysis in this paper is to observe several metrics derived from the calculations performed in the previous sections in order to determine whether there are any trends that can indicate the onset of MIIs.

Shown in this section are two short data profiles of a single participant during a single experimental trial. The participant was instructed to stand facing at an angle of 90 degrees with respect to the ship centreline while performing a cognitive clipboard/tablet task, and while maintaining balance in heavy-weather sea conditions. The ship's motion was dominated by roll so this was the most difficult stance direction to maintain.

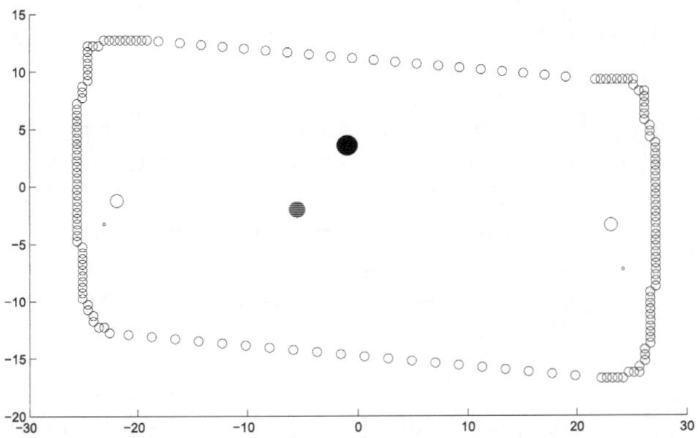

Figure 4: Base of support outline with ankle locations (small side circles), individual foot CoF locations (large side circles), body CoM (black and larger filled circle), and overall CoF (gray and smaller filled circle).

MEASUREMENT METRICS

The following metrics are presented in graphical form:

Right Foot Relative Position

During the experiments, participants were instructed to keep their left foot on top of a load cell plate. Because of this they naturally tended to correct stance imbalances by moving only their right foot. This means that relative motion of the right foot was a good indicator of the onset of and recovery from MIIs.

Centre of Mass to Base of Support

The shortest distance between the CoM and BoS was calculated and plotted.

Subject Reaction Moment

The subject reaction moment is the moment generated by the CoF about the CoM. It is calculated as the total vertical force measured by the insoles, multiplied by the distance between the CoM and CoF.

Base of Support Angle

Another useful metric for determining when an MII has occurred is to check if the CoM is no longer contained within the BoS. This can be calculated by adding together all of the angles created by two adjacent points of the BoS, and the point corresponding to the CoM. If the CoM is within the BoS, this sum should be equal to 360 degrees.

Ship Motions

There is undoubtedly a relationship between the magnitude of ship motions and occurrences of MIIs, although no research to date has been able to form a conclusive correlation. Plotted here are the two ship motions that would dominate a subject's response while standing at 90 degrees with respect to the ship centreline: roll angle and lateral acceleration.

SAMPLE DATA 1

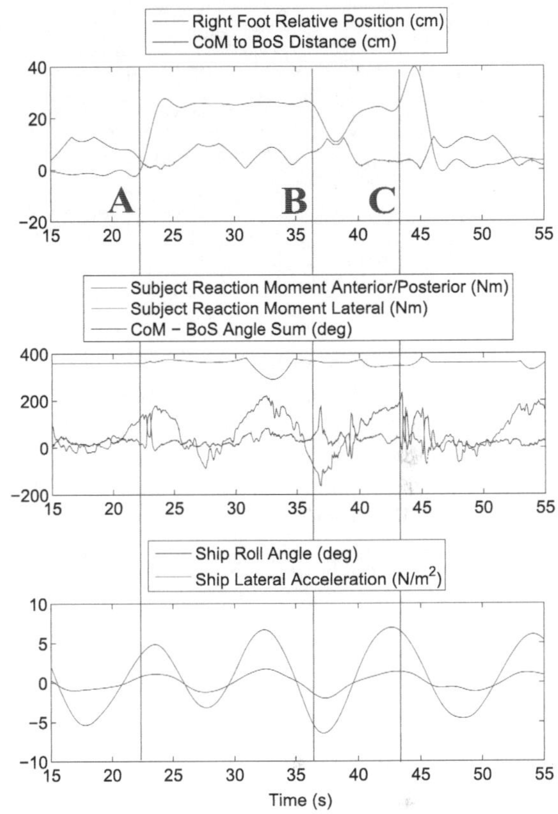

Figure 5: Measured and calculated postural stability data containing 3 MII events.

The letter labels on the graph indicate the occurrence of MIIs, which were identified manually from the motion capture recordings.

- A - Starting from neutral, subject steps back and decides to hold that position.
- B - Subject steps forward and then moves back to previous foot position.
- C - Subject steps even farther back, then returns to neutral.

SAMPLE DATA 2

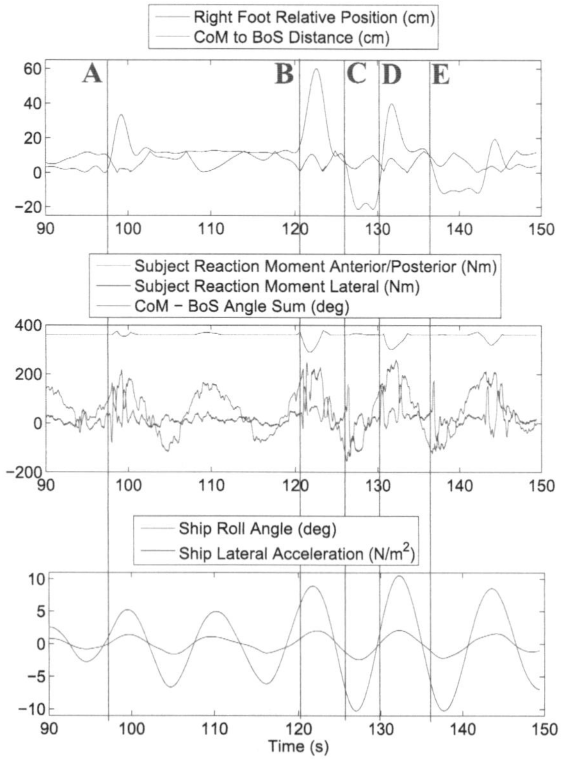

Figure 6: Measured and calculated postural stability data containing 5 MII events.

- A - Subject steps back and returns to neutral.
- B - Subject steps back and returns to neutral.
- C - Subject steps forward.
- D - Subject attempts to return to neutral but is forced to step back instead, then returns to neutral.
- E - Subject steps forward and holds before repositioning to neutral.

DISCUSSION

Several useful observations can be made from the two data sets presented, and from the process of generating them.

1. Foot position is a useful indicator of the onset of MIIs. An alternative approach that was tested was to only look at the total magnitude of the foot's movement. This was also useful for indicating the onset of MIIs, but had the disadvantage of not indicating the step direction. This is important information because frequently a participant would miss and step through the neutral position rather than returning to it due to unexpected ship motions.
2. The CoM - BoS angle sum is of questionable usefulness. While it does indicate the occurrence of MIIs, it is also subject to false positives.
3. While the reaction moment does not appear to provide information on the onset of MIIs, it does appear to display a trend on the recovery from MIIs. While it is fully expected for there to be a peak in lateral reaction moment due to one foot carrying all of the force while stepping, there is also a peak in the anterior/posterior direction.
4. There does not seem to be any definite trend between CoM - BoS distance and the occurrence of MIIs, although it has been observed in these data sets and others, that the distance is often close to its maximum (CoM centred) before the onset of an MII rather than at a minimum (CoM near the edge of BoS).
5. From these data sets there does appear to be some relationship between roll angle, lateral acceleration, and occurrence of MIIs, although as expected, it is not a simple relationship.

CONCLUSIONS

This paper examines the relationship between occurrence of motion induced interruptions (MIIs) and various measured and derived parameters that may have relevance to the study of human postural stability. Those parameters are mainly focused on the locations of body centre of mass (CoM) and centre of force (CoF), and their relative positions within the base of support (BoS). These measurements were experimentally derived from motion capture data and foot pressure insoles during a recent heavy-weather sea trial. Future work will improve on this result by comparing the results in more detail with specific published results and by including a dynamic model of the human participant in the analysis.

REFERENCES

[1] Bourgeois, N., Langlois, R. and Hunter, A. "Quest Q-348 sea trial: Human postural stability studies." *In "International Conference of Control, Dynamic Systems, and Robotics,"* Ottawa, Canada, 2014.

[2] Graham, R., Baitis, A. and Meyers, W. "On the development of seakeeping criteria." *Naval Engineers Journal*, Vol. 104, pp. 259–275, 1992.

[3] Winter, D. *Biomechanics and Motor Control of Human Movement, 4th Edition*. John Wiley and Sons Inc., Hoboken, New Jersey, 2009.

[4] Pai, Y.C. and Patton, J. "Center of mass velocity-position predictions for balance control." *Journal of Biomechanics*, Vol. 30, No. 4, pp. 347–354, 1997.

Boundary Layers

Proceedings of the 25th CANCAM
London, Ontario, Canada, May 31 – June 4, 2015

FLUCTUATING PRESSURES AND FLOW CHARACTERISTICS WITHIN SEPARATION BUBBLES ON SURFACE-MOUNTED PRISMS

Abul Fahad Akon
Dept. of Mechanical & Materials Engineering
University of Western Ontario
London, ON, Canada
aakon@uwo.ca

Gregory Alan Kopp
Boundary Layer Wind Tunnel Laboratory
University of Western Ontario
London, ON, Canada
gakopp@uwo.ca

ABSTRACT

Investigating fluctuating surface pressure fields underneath separation bubbles on sharp-edged, three-dimensional, surface-mounted, bluff bodies in turbulent boundary layer flows is important as these are related to the forces acting on buildings in the wind. The distributions of the fluctuating pressures are dependent on the turbulence properties in the upstream flow and are different from the pressures acting on bluff bodies in a uniform, smooth flow. The paper presents results from time-resolved Particle Image Velocimetry and synchronized pressure measurements taken above the roof of surface-mounted prisms.

KEYWORDS: Bluff-body aerodynamics, atmospheric turbulence, separated flows, surface-mounted prisms, wind loads

INTRODUCTION

In the area of bluff body aerodynamics, the study of the separating and reattaching flows around bluff bodies is important because of the large surface pressure fluctuations. These large fluctuating pressures cause large instantaneous uplifting loads on the roofs of buildings, which is the primary source of failures during windstorms. The pressure field and aerodynamic forces acting on a bluff body depend on the nature of the flow inside and around the separation bubble. Investigations show that the structure of the separation bubble, pressure fluctuations underneath the separation bubble, the length of the separation bubble and forces exerted on the bluff bodies vary with the turbulence properties in the free stream (Gartshore (1973), Hillier & Cherry (1981)).

Turbulence intensity and the turbulent scales (L_x) have large impacts on the separation bubble properties and the pressure fields on the surface under the separation bubble. Hillier & Cherry (1981) show that pressure fluctuations under the separation bubble for two-dimensional sharp-edged bluff bodies placed in uniform flow are strongly dependent on the free stream turbulence intensity. Similar results were found by Saathoff & Melbourne (1989, 1997). Fig. 1, which contains results from Saathoff & Melbourne (1997) and Cherry *et al.* (1984), shows that the fluctuating pressures under the separation bubble for two-dimensional sharp-edged bluff bodies are strongly dependent on the free stream turbulence intensities and length scales. Their results for two-dimensional bluff bodies in uniform upstream flows show an increase in the surface pressure fluctuations with increasing turbulence intensity and turbulent length scales in the free stream. In addition, the location of the maximum fluctuations is moved towards the separation point (leading edge) as turbulence in the free stream is increased. However, turbulence length scales do not alter the location of maximum pressure fluctuation. For smooth upstream cases, also investigated by Cherry *et al.* (1984) and Saathoff & Melbourne (1997), the maximum fluctuations are observed very close to the reattachment point, where the distance from the leading edge (separation point) to the mean reattachment length is defined as X_r.

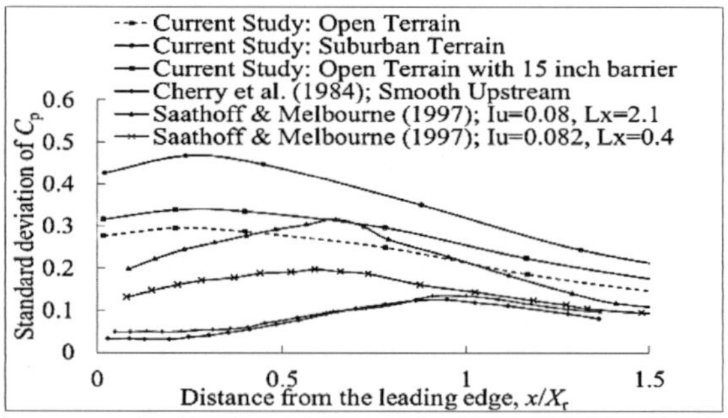

Figure 1: Distribution of fluctuating pressure coefficients under the separation bubble. (The results of Saathoff & Melbourne (1997) are regenerated)

Surface pressures for three-dimensional bluff bodies in boundary layer flows are also affected by the turbulence in the boundary layer (Castro & Robins (1977)). Simultaneous pressure and velocity field measurements on the surface of a gable-roofed building (Kopp *et al.*, 2012b) show that a vortex forms near the leading edge and convects downstream. These vortices are responsible for the highest instantaneous uplift.

In this paper, measurements of velocity and pressure fields obtained from Time-Resolved Particle Image Velocimetry measurements synchronized with surface pressure measurements for a surface-mounted rectangular prism are reported for three different upstream boundary layer conditions. Comparisons of the velocity and fluctuating pressure fields were made with velocity and fluctuating pressure fields of two-dimensional bluff body in uniform upstream flow.

EXPERIMENTAL APPROACH

The surface mounted prism used in the present experiment is a scale model of the TTU-WERFL test building. A row of 9 pressure taps from the top surface of the model were connected to the pressure scanners by a tubing system (Ho *et al.* (2005)). The model was placed on the turntable in the high-speed test section in the Boundary Layer Wind Tunnel II at the University of Western Ontario and was tested under three different upstream conditions, as indicated in Table 1. Two upstream conditions are "Open Terrain" and "Suburban Terrain", where the turbulence levels are altered with only a small change to the integral scales. The third upstream condition is the "Open Terrain" with a 15" barrier at the entrance of the test section, which increases the integral scale with only small changes in the overall turbulence level.

Table 1: Characteristics of the upstream conditions

Upstream Condition	Barrier	Streamwise Turbulence Intensity $[I_u]$	Length scale $[L_x/H]$
Open Terrain	15"	16.5	11
Open Terrain	0"	16.5	8
Suburban Terrain	0"	26	7

In order to capture the velocity field around the TTU building model simultaneously with pressure field measurements on the top surface of the model, the Time-Resolved Particle Image Velocimetry (TR-PIV) system was synchronized with pressure measurement system. Two fields of view, one capturing the flow upstream of the model and the other capturing the separated and reattaching flow at the top surface of the model, were chosen. The dimensions of the model, the co-ordinate systems used and the locations of the fields of view are shown in Fig. 2. The TR-PIV system is able to sample PIV velocity field at 500 Hz. Pressure data were sampled at 1108 Hz to avoid the aliasing the effects. Atomized olive oil particles were seeded in the flow and were illuminated by a double-head, diode-pumped Q-switched Nd:YLF laser operating at 1000 Hz. Two 1 Mb Photron FASTCAM-1024PCI CMOS cameras were used to take PIV images. A detailed description of the TR-PIV system can be found in Taylor *et al.* (2010). Pressure and velocity data were sampled for approximately 160 s. A time delay of 85μs was allowed between each of the image pairs to restrict the movement of the particles within 6 pixels. TSI Insight 4G, was used to find the velocity fields where FFT cross-correlation algorithm was used. An interrogation window size of 32x32 pixels with 50% overlap were used in the processing of velocity vectors. Global standard deviation filter, followed by local mean and median filters was used to identify the spurious vectors and these were replaced by interpolated vectors. Standard cross-correlation algorithms have a special uncertainty of less approximately 0.1 pixels (Huang *et al.* (1997)).

Pressure coefficients were recorded with reference to the pressure at the reference point near the centre of the wind tunnel and pressure coefficients referenced to the model height was calculated by Eqn. 1 expressed as,

$$C_{p,H} = C_{p,ref}(q_{ref}/q_H) = C_{p,ref}(U_{ref}/U_h)^2 \quad (1)$$

where, $C_{p,H}$ is pressure coefficients referenced to model height, $C_{p,ref}$ is the pressure coefficient referenced to the reference point, q_{ref} and q_H are the dynamic pressures at the reference height and at model height respectively, $U_{ref}$is the mean velocity at the reference point and U_h is the mean velocity at the model roof height.

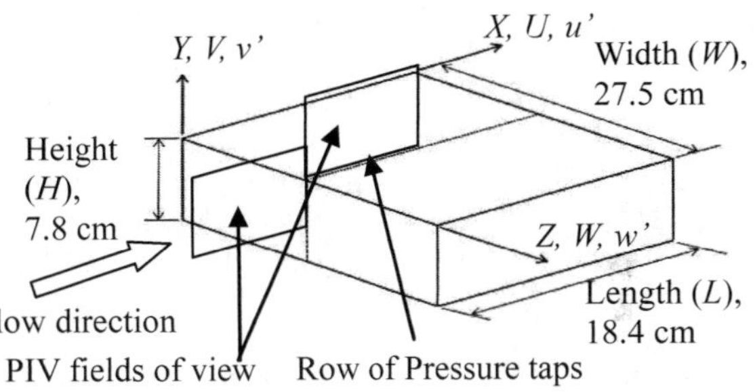

Figure 2: Schematic diagram of the model and the co-ordinate system.

RESULTS

The mean reattachment length (X_r) was found from the mean flow field on the top surface of the model. The mean reattachment point is the point where the gradient of the streamwise component of the velocity in vertical direction is zero. In Table 2 the mean reattachment lengths and the maximum height of the separation bubble corresponding to three upstream conditions are presented. Under the separation bubble the standard deviations of the fluctuating pressure coefficients for three different boundary layer upstream conditions are presented in Fig. 1. In the figure the results from Cherry *et al.* (1984) and Saathoff & Melbourne (1997) for two-dimensional sharp-edged bluff body placed in uniform upstream flow are also shown. For the present experiment, it is observed that increasing turbulence intensity and length scales increases the magnitude of the surface pressure fluctu-

Table 2: Mean reattachment lengths

Upstream Conditions	Barrier	Mean Reattachment Length $[X_r/H]$	Maximum Thickness of the Separation Bubble $[T_{b,max}/H]$
Open Terrain	15"	1.05	0.21
Open Terrain	0"	1.05	0.19
Suburban Terrain	0"	0.88	0.17

ations like the two-dimensional bluff bodies in uniform upstream flows. It is observed that for smooth uniform upstream case, the maximum fluctuations of surface pressures are very close to the reattachment point while for turbulent boundary layer cases, the maximum fluctuations are observed much closer to the leading edge.

In order to understand the effects of length scales in the free stream, the present experimental results were normalized by the model height turbulence intensities. The results are presented in Fig. 3. In Fig. 3 the vertical scale on the right shows the fluctuating pressures for smooth upstream case obtained by Saathoff & Melbourne (1997). From the figure, it is observed that, normalizing the fluctuating pressure by the corresponding turbulence intensities for the present experiment makes the fluctuating pressures of similar turbulence length scales in the upstream ("Open Terrain" and "Suburban Terrain") fall on the same curve. This indicates a linear relationship of turbulence intensity with the fluctuating pressure coefficients under the separation bubble, in agreement with the formulation proposed by Bearman & Morel (1983). However, when the upstream turbulence length scales are different, even though the model height turbulence intensity is the same ("Open Terrain" and "Open Terrain" with 15" barrier), differences in magnitude of the fluctuating pressure coefficients are observed. This strongly indicates that there are effects of upstream length scales on the fluctuating pressure coefficients. Same findings were observed for the turbulent uniform upstream data sets obtained by Saathoff & Melbourne (1997).

For the smooth upstream case of Saathoff & Melbourne (1997), the fluctuating pressure coefficients normalized by very low turbulence intensity show very large values compared to the present experimental results and the results for turbulent uniform upstream cases obtained by Saathoff & Melbourne (1997). The results of Cherry *et al.* (1984), when normalized by the turbulence intensities in the upstream (I_u=0.0007) also show very high values (1000 times higher than the present experimental results and are not shown on Fig. 3). This breakdown of results for smooth upstream case indicates that the pressure fluctuations for smooth upstream do not follow the Quasi-steady theory as all of the fluctuations are due to body-generated turbulence. On the other hand, for the turbulent upstream cases for Saathoff & Melbourne (1997) and also for the present experiment, in addition to the body generated turbulence, the flow around the bluff bodies are also affected by the turbulence in the upstream.

To understand the differences in the flow features in and around the separation bubble for the three dimensional surface mounted bluff bodies and two-dimensional bluff bodies in uniform upstream flows, comparisons of the fluctuating velocity fields were made for the present experimental results with the results obtained by Castro & Haque (1987). The results of Castro & Haque (1987) are for two-dimensional flat plate with a splitter plate placed normal to the uniform upstream flow of 0.25% turbulence intensity. These results are shown in the following figures (Figs. 4-5).

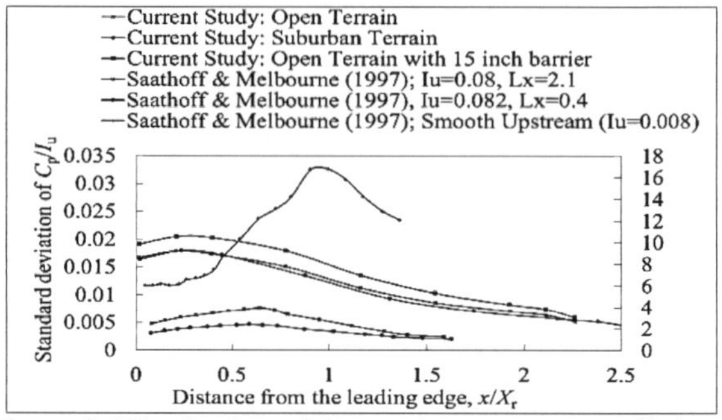

Figure 3: Distribution of fluctuating pressure coefficients under the separation bubble.

Fluctuating velocity field comparisons were made at three different axial locations of the separation bubble. The axial locations are $0.35X_r$, $0.68X_r$ and $1.02X_r$. For Castro & Haque (1987), the smooth upstream flow separates from the leading edge (0.5D above the splitter plate, here D is the thickness of the flat plate) of the flat plate and reattaches on the splitter plate. On the other hand, for the present experiment, the separation point and the reattachment point are vertically located at the same height. In order to compare the velocity fields in and around the separation bubble for two-dimensional bluff bodies in smooth upstream flow with three-dimensional bluff bodies in turbulent boundary layer flow, the vertical distance from the reattachment plane of Figs. 4-5 is presented normalized by the maximum thickness ($T_{b,max}$) of the separation bubble.

The vertical distribution of streamwise Reynolds normal stress (Fig. 4) for the present experiment shows that, by increasing the turbulence intensity in the free stream, the value of streamwise Reynolds normal stress increases and as the separated flow moves downstream while the region of fluctuations in streamwise velocity spreads vertically. In contrast, for the uniform upstream case (Castro & Haque (1987)), the streamwise Reynolds stress is low compared to the turbulent boundary layer case near the leading edge, it increases as separated flow moves downstream and at the reattachment point it is even slightly higher than the turbulent boundary layer upstream condition. The location of maximum streamwise velocity fluctuations always remain within the maximum height of the separation bubble for smooth flow over two-dimensional bluff bodies and for three-dimensional bluff body in boundary layer upstream, the location of maximum streamwise fluctuations moves beyond the maximum height of the separation bubble as the reattachment point is approached.

In contrast, the Reynolds normal stress in vertical direction (Fig. 5) for uniform upstream over two-dimensional bluff body is always higher than turbulent boundary flow cases. As the reattachment point approaches, the magnitude of vertical velocity fluctuations increases and further deviates from the turbulent boundary layer upstream case. Here also the maximum fluctuations of vertical velocity component are restricted within the maximum height of the separation bubble for two-dimensional bluff body in smooth flow. However, the magnitude of vertical velocity fluctuations is higher than streamwise velocity fluctuations for two-dimensional bluff body in smooth upstream. For the present experiment, even though the vertical velocity fluctuations spreads vertically in the similar pattern like the streamwise velocity fluctuations, the magnitude of vertical velocity fluctuations are much smaller than streamwise fluctuations.

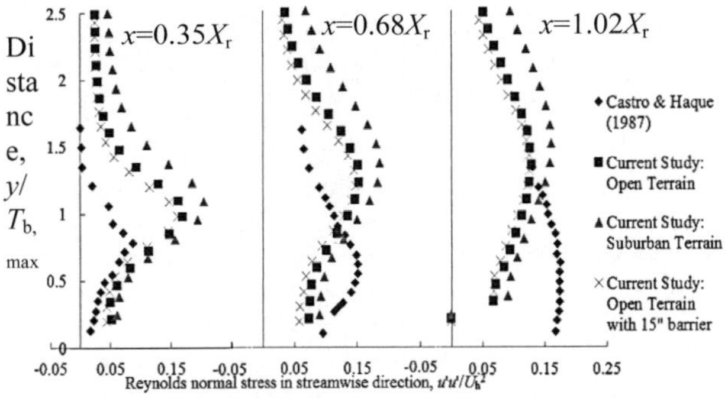

Figure 4: Distribution of streamwise Reynolds stress.

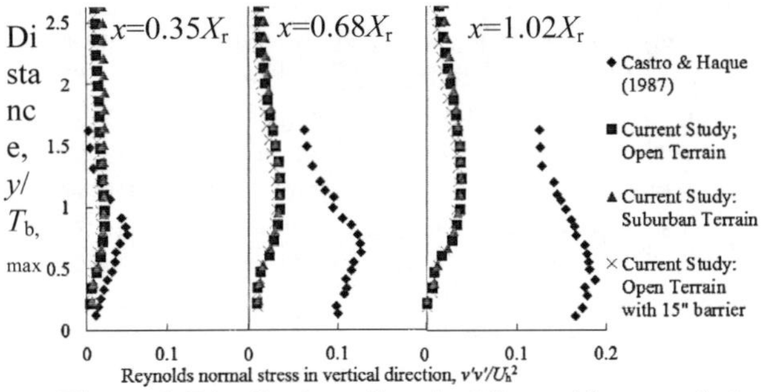

Figure 5: Distribution of vertical Reynolds normal stress.

CONCLUSIONS

From the observed results, the following conclusions can be drawn:

- Both the free stream turbulence intensity and length scale affects the fluctuating pressure coefficients on the surface under separation bubbles for three-dimensional, surface-mounted bluff bodies.
- The turbulence intensity in the upstream increases the streamwise Reynolds stress bounding the separation bubble, but does not have significant effects on the vertical Reynolds stress.
- The maximum velocity fluctuations for three dimensional bluff bodies in turbulent boundary layer upstream occur further away from the reattaching surface when compared to two-dimensional bluff bodies in uniform smooth upstream flows.
- Surface pressure fluctuations have important relationships with the Reynolds stresses closer to the surface. The full conference presentation will address these in detail.

ACKNOWLEDGMENTS

This work was funded by NSERC through the Discovery Grant program.

REFERENCES

1. Gartshore, I. S. 1973, "The effects of free stream turbulence on the grad of rectangular two-dimensional prisms", *Boundary Layer Wind Tunnel Laboratory Report*, University of Western Ontario, London, 4-73.
2. Hillier, R. and Cherry, N. J. 1981, "The effects of stream turbulence on separation bubbles", *Journal of Wind Engineering and Industrial Aerodynamics*, Vol(8), pp. 49-58.
3. Saathoff, P. J. and Melbourne, W. H. 1989, "The generation of peak pressures in separated/reattaching flow", *Journal of Wind Engineering and Industrial Aerodynamics*, Vol(32), pp. 121-134.
4. Saathoff, P. J. and Melbourne, W. H. 1997, "Effects of free stream turbulence on surface pressure fluctuations in a separation bubble", *Journal of Fluid Mechanics*, Vol(337), pp. 1-24.
5. Cherry. N. J., Hillier, R. and Latour, P. M. 1984, "Unsteady measurements in a separated and reattaching flow", *Journal of Fluid Mechanics*, Vol(144), pp. 13-46.
6. Castro, I. P. and Robins, A. G. 1977, "The flow around a surface-mounted cube in uniform and turbulent streams", *Journal of Fluid Mechanics*, Vol(79), pp. 301-335.
7. Kopp, G. A., Morrison, M. J. and Henderson, D. J. 2012, "Full-scale testing of low-rise, residential buildings with realistic wind loads", *Journal of Wind Engineering and Industrial Aerodynamics*, Vol(104-106), pp. 25-39.
8. Ho, T. C. E., Surry, D., Morrish, D. and Kopp, G. A. 2005, "The UWO contributions to the NIST aerodynamic database for wind loads on low buildings: Part1. Archiving format and basic aerodynamic data", *Journal of Wind Engineering and Industrial Aerodynamics*, Vol(93), pp. 1-30.
9. Taylor, Z. J., Gurka, R., Kopp, G. A. and Liberzon, A. 2010, "Long-Duration Time-Resolved PIV to Study Unsteady Aerodynamics", *IEEE Transactions on Instrumentation and Measurement*, Vol(59), pp. 3262-3269.
10. Huang, H., Dabiri, D. and Gharib, M. 1997, "On errors of digital Particle Image Velocimetry", *Measurement Science and Technology*, Vol(8), pp. 1427-1440.
11. Bearman, P. W. and Morel, T. 1983, "Effects of free stream turbulence on the flow around bluff bodies", *Progress in Aerospace Science*, Vol(20), pp. 97-123.
12. Castro, I. P. and Haque, A. 1987, "The structure of a turbulent shear layer bounding a separation region", *Journal of Fluid Mechanics*, Vol(179), pp. 439-468.

NUMERICAL SIMULATION OF LAMINAR SHEAR-LAYER REATTACHMENT IN A PLANAR SUDDEN EXPANSION

Nima Moallemi [1]
School of Engineering
University of British Columbia, Okanagan
Kelowna, BC, Canada

Joshua Brinkerhoff [2]
School of Engineering
University of British Columbia, Okanagan
Kelowna, BC, Canada
joshua.brinkerhoff@ubc.ca

ABSTRACT

This paper describes numerical simulations of laminar incompressible flow in a planar sudden expansion. Shear layer reattachment is studied at Reynolds numbers up to 3000 and expansion ratios in the range of 1.3 to 4.0. Correlations are developed for the non-dimensional reattachment length and the maximum velocity magnitude in the reverse-flow region in terms of Reynolds numbers and expansion ratios. The correlations and computed results achieve excellent agreement with published literature. The onset of asymmetry in the sudden expansion flow is also observed and values of the critical Reynolds numbers for the onset of asymmetry for various expansion ratios are identified.

KEYWORDS: computational fluid dynamics; sudden expansions; laminar flow; shear layer reattachment

INTRODUCTION

The flow of an incompressible fluid in a sudden planar expansion is representative of numerous flows in pipeline networks, combustion chambers, and fluidized beds. The heat-transfer coefficient and pressure drop are strongly affected by the length of a reversed-flow region immediately downstream of the expansion due to the high momentum exchange that occurs in that region. Devices requiring enhanced mixing of the flow may benefit from this momentum exchange and sudden expansions can be added to promote passive mixing. For instance, Ko and Sung [1] investigated mixing in a sudden expansion to enhance the mixing of air and fuel in a cylindrical combustion chamber. Design optimization of engineering devices in which sudden expansions are used to promote passive mixing requires empirical correlations that can accurately predict the reattachment length and the magnitude of flow reversal for a range of Reynolds numbers and geometrical configurations.

The primary parameters affecting sudden expansions are the flow Reynolds number based on the inlet channel height and maximum inlet velocity (Re) and the ratio of the channel height upstream of the sudden expansion to the height downstream, termed the expansion ratio (Er). Several studies [2-3] have found that the streamwise length of the reversed-flow region, termed the reattachment length, increases linearly with Reynolds number up to approximately $Re = 250$, which corresponds to a laminar flow regime, and then grows exponentially in the range $400 < Re < 1000$ as unsteady fluctuations develop within the shear layer. The maximum reattachment length occurs at about Re=1000, for which transition to turbulence is completed at the reattachment location [4]. An abrupt decrease in the reattachment length is observed for $1000<Re<8000$ as the transition location moves upstream towards the step. Finally, a slight increase in reattachment length is observed for $Re > 10^4$ due to transition to turbulent flow occurring upstream of the expansion.

Expansion ratio also has a strong impact on the flow development downstream of a sudden expansion [5-6]. These studies show that the reattachment length tends to increase with expansion for a given Reynolds number, with transition to turbulence occurring at successively lower Reynolds numbers as Er increases. As the Reynolds number increases at a given expansion ratio, numerous studies have observed an abrupt bifurcation in which reattachment occurs at different locations on the upper and lower channel walls. Early experimental studies showed that bifurcation occurs after a critical Reynolds number is reached [7-8], and that the critical Reynolds number decreases with increasing expansion ratio [9]. Battaglia and Papadoupoulos [10] showed that bifurcation develops as a two-dimensional (2D) instability mode even in three-dimensional geometries. Furthermore, because the 2D instability mechanism is very sensitive to the initial amplitude of disturbances present in the flow, Sanmiguel-Rojas et al. [11] concluded that identifying a critical Reynolds number for bifurcation onset is still challenging. As such, correlations predicting the critical Reynolds number for a range of expansion ratios are not available.

The present work focuses on solving the Navier-Stokes equations for transient flow in a planar sudden expansion for a Reynolds number range of Re=25-3000 and an expansion ratio range of Er=1.3-4.0. The shear layer reattachment and bifurcation phenomena are investigated using direct numerical simulation (DNS) to provide insight on the development of flow reattachment, maximum reversed-flow velocities, and the onset of bifurcation. This study aims to develop improved design correlations for these phenomena for a broad range of Reynolds numbers and expansion ratios. The study also serves to validate the present computational approach for follow-on simulations of 3D sudden expansions with transitional and turbulent conditions.

[1]Ph.D. student
[2]Assistant professor. Corresponding author.

NUMERICAL METHOD

A direct numerical simulation (DNS) was performed for laminar, incompressible flow in a 2D planar sudden expansion. The computational domain shown in Fig. 1 consists of two channels with heights of d and D mated together to produce a sudden expansion with a step-height of h. The lengths of the computational domain upstream and downstream of the sudden expansion are L_i=10h and L_o=60h, respectively. A steady, parabolic inflow boundary is imposed at the inlet boundary and a zero gauge static pressure and zero streamwise velocity gradient are imposed at the outlet boundary. All walls are represented by no-slip boundaries. The outlet boundary is placed far from the expansion to minimize outlet boundary effects on the region of interest.

Structured, non-uniform, orthogonal grids were mapped to the computational domain. The numbers of grid cells in the regions upstream and downstream of the expansion are given in Table 1. The cells are non-uniformly spaced in order to place a higher number of cells in the region near the no-slip walls and close to the sudden expansion. A grid-refinement study was performed by varying the total number of cells at 4 refinement levels between approximately 15,000 and 180,000. Using the reattachment length as a metric, it was confirmed that 100,000 cells is sufficient to guarantee grid independence for all Reynolds numbers and expansion ratios considered in this paper, and thus the grids described in Table 1 are conservatively refined.

OpenFOAM®, an open source computational fluid dynamics software package, was used to solve the time-varying incompressible Navier–Stokes equations via the finite volume method [12]. Second-order central differencing and implicit second-order Euler backward differencing are used to discretize the spatial and temporal derivatives, respectively. The time-step size in each simulation varies depending on the Reynolds number and grid size (Moallemi, 2014), with $\Delta t = 10^{-5}$ s being the smallest time-step size used. Air at 1 atm and 25°C is the working fluid. The simulations were initialized with a zero relative static pressure and an initial velocity equal to the mean velocity at the inflow boundary. The cases were run for approximately 130,000 time-steps to reach a statistically-steady state and then approximately 200,000 time-steps were collected for analysis. Each simulation was performed in a parallel-processing environment on 24 CPUs.

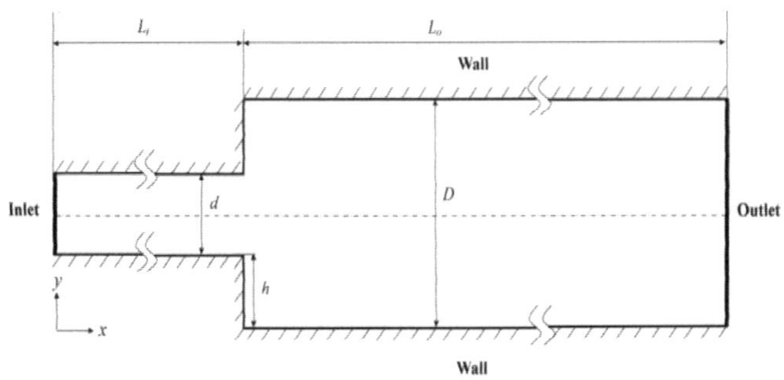

Figure 1: Schematic of the computational domain.

Table 1: Distribution of finite-volume grid nodes in the spatial grids of the present study.

Er	$(N_x \times N_y)_{Li}$	$(N_x \times N_y)_{Lo}$	*Number of cells*
1.3	100 × 120	600 × 200	132,000
1.7	100 × 90	600 × 210	135,000
2.0	100 × 80	600 × 240	152,000
2.6	100 × 62	600 × 262	163,400
3.2	100 × 54	600 × 274	169,800
4.0	100 × 40	600 × 280	172,000

RESULTS AND DISCUSSION

A total of 30 simulations were performed with Reynolds numbers between Re = 25-11,000 and with expansion ratios of Er = 1.3, 1.7, 2.0, 2.6, 3.2, and 4.0. Transition to turbulence was noted in some of these cases, but because the computational domain is two-dimensional and thus unable to accurately resolve the 3D fluctuations accompanying transition, the analysis in the present section is limited to the laminar regions of the simulation. The discussion of the results focuses first on the reattachment length, followed by the onset of asymmetry in the reattachment region via bifurcation, and finally the maximum velocity magnitude in the reversed-flow region.

Reattachment length

Figure 2 compares the computed reattachment lengths (normalized by the step-height, h) as a function of the Reynolds number for Er = 2.0 with published numerical and experimental results. The present results are in good agreement with published investigations for $Re \leq 1100$. Above this Reynolds number, transition to turbulence is initiated in the shear layer and the present 2D simulations begin to deviate from the published results. The excellent agreement for lower Reynolds numbers suggests that the 2D simulations are capable of capturing reattachment of the laminar shear layer. Although not shown, similar agreement is obtained between the present results and Back et al. [3] at Er = 2.6. Figure 3 plots the normalized reattachment length as a function of Reynolds number for expansion ratios of Er = 1.3-4.0. The maximum Reynolds number plotted in Fig. 3 for each expansion ratio value corresponds to the point where time-dependent flow is observed in the simulations and transition to turbulence is initiated. As will be shown later in this paper, this occurs at approximately three times the critical Reynolds number for bifurcation. Because bifurcation and the subsequent transition to turbulence occurs at successively lower Reynolds numbers as the expansion ratio increases, the Reynolds-number ranges are different in Fig. 3 for each value of the expansion ratio.

The following correlation for the normalized reattachment length is fitted to the computed results in Fig. 3:

$$\frac{L_r}{h} \qquad Re - \qquad Er \qquad Re\,Er \qquad Re\,Er \quad . \quad (1)$$

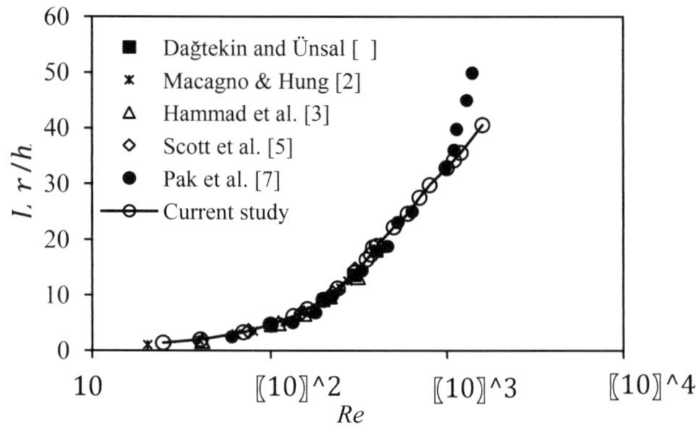

Figure 2: Reynolds number distribution of the reattachment length for Er = 2.0.

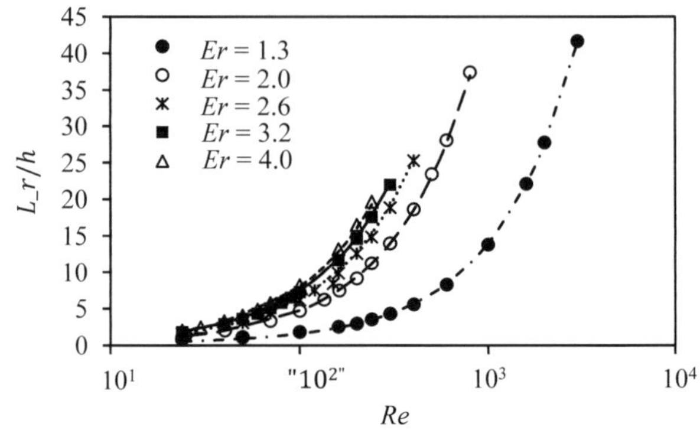

Figure 3: Reynolds number distribution of the reattachment length for various expansion ratios. Symbols denote numerical results and lines denote Eq. (1).

From Fig. 3, it is clear that Eq. (1) achieves excellent agreement with the computed results; the norm of the residual is $R^2 > 0.99$. The correlation is validated for expansion ratios in the range $1.3 < Er < 4.0$ and Reynolds numbers in the range $24 < Re < 3Re_{c,b}$, where $Re_{c,b}$ is the critical Reynolds number for the onset of bifurcation.

Onset of flow bifurcation

In the introduction, it was noted that sudden-expansion flows undergo a bifurcation in which the reattachment region becomes asymmetric and multiple recirculation zones form downstream of the expansion. Bifurcation is observed in the present results for all the expansion ratios under investigation. To provide a qualitative illustration of bifurcation, Fig. 4 plots the streamline patterns for Re = 200 and 500 at an expansion ratio of Er = 2.0. For the lower Reynolds number, the streamline pattern is symmetric and the reattachment length is equal on the upper and lower channel walls. For the higher Reynolds number, however, the symmetry-breaking bifurcation occurs and an asymmetric streamline pattern develops. In the bifurcation phenomena, the recirculation region shrinks on one wall and lengthens on the other, and a secondary recirculation region develops on the side with the smaller primary region. The critical Reynolds number for the onset of bifurcation, denoted $Re_{c,b}$, is tabulated in Table 2 for the expansion ratios studied. Figure 5 illustrates the variation in the reattachment lengths of the longer and shorter reattachment regions (L_{r1} and L_{r2}, respectively) in a case that bifurcates at a critical Reynolds number of $Re_{c,b}$ = 370. In most cases, the growth of L_{r1} with Reynolds number exactly follows the trend shown in Fig. 2 (and predicted in Eq. (1)) until about $3Re_{c,b}$, beyond which unsteadiness of the recirculation zone occurs and transition to turbulence rapidly ensues. An exception to this was in the Er = 1.3 cases; transition did not occur and L_{r1} followed the trend of Fig. 2 to the end of the computational domain.

Reversed-flow velocity magnitude

Mass and momentum conservation at the sudden expansion dictates that the sudden increase in the flow area is accompanied by a decrease in the average velocity and a corresponding increase in static pressure, yielding an adverse streamwise pressure gradient in the vicinity of the sudden expansion. The flow recirculation noted in Fig. 4 occurs in the low-momentum fluid near the wall in response to the local adverse pressure gradient. When sudden expansions are used as passive mixing devices, the maximum velocity magnitude in the reversed-flow region provides a convenient

Table 2: Critical Reynolds number for the onset of bifurcation in laminar flows in planar sudden expansions.

Er	$Re_{c,b}$
1.3	2860 ± 10
2.0	370 ± 10
2.6	160 ± 10
3.2	90 ± 10
4.0	70 ± 10

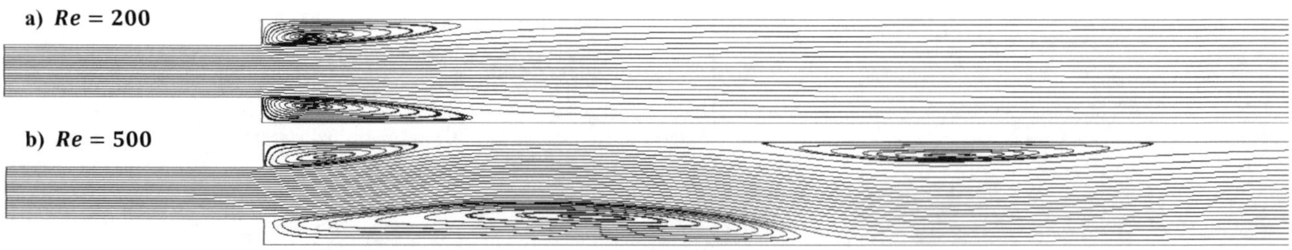

Figure 4: Streamline pattern near the expansion (zoomed view). (a) Symmetric reattachment for Re = 200 and Er = 2.0. (b) Asymmetric reattachment for Re = 500 and Er = 2.0.

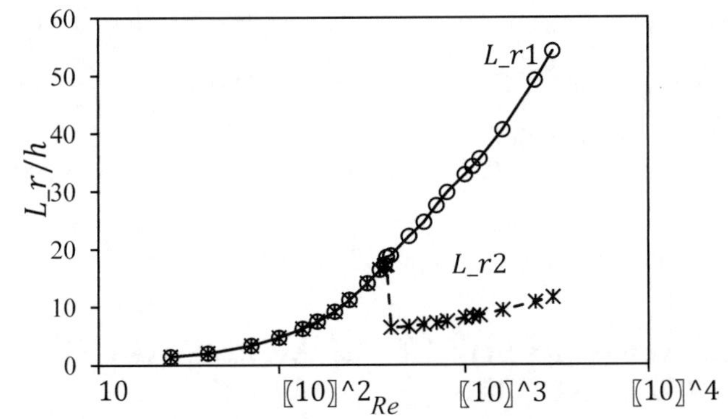

Figure 5: Reynolds number distribution of the reattachment location for a bifurcating case with $Er = 2.0$. The critical Reynolds number for bifurcation is $Re_{c,b} = 370$.

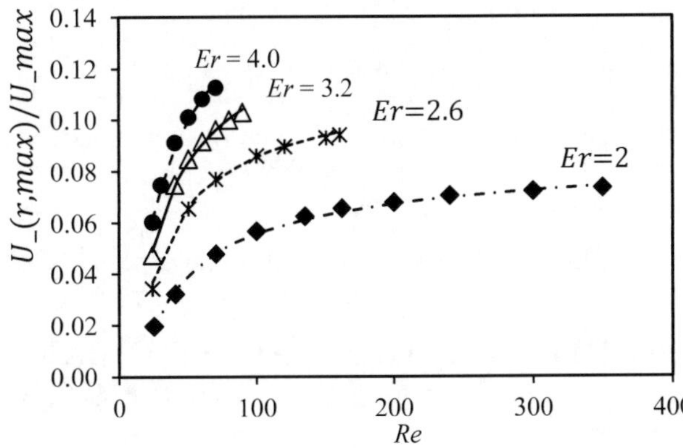

Figure 6: Reynolds number distribution of the reattachment length for various expansion ratios. Symbols denote numerical results and lines denote Eq. (2).

estimate of the level of shear stress and is thus a simple way of approximating the mixing effectiveness of the flow. Figure 6 plots the maximum velocity magnitude of the reversed flow ($U_{r,max}$) normalized by the maximum inlet velocity (U_{max}) for Er = 2.0-4.0. The results show that expansion ratio has a dominant effect on the maximum velocity in the reversed-flow region, and hence would likely exhibit a stronger influence on mixing effectiveness than Reynolds number within the laminar regime.

A correlation for predicting the maximum magnitude of the reversed velocity is given below:

$$\frac{U_{r,max}}{U_{max}} = \quad Er \quad \frac{Er}{Re^{\quad Er}} \quad \frac{Er}{Re^{\quad Er}}. \quad (2)$$

This correlation is valid for $\leq Re \leq Re_{c,b}$ and $2.0 \leq Er \leq 4.0$. Unlike Eq. (1), Eq. (2) cannot be applied for Reynolds numbers higher than $Re_{c,b}$ because once bifurcation occurs, very irregular velocities are observed in the various recirculation zones. Eq. (2) is accurate to within 5% of the computed results, as illustrated in Fig. 6.

CONCLUSIONS

A numerical analysis is performed of reattachment in a sudden expansion. The effect of Reynolds number and expansion ratio on the reattachment length, the onset of flow bifurcation, and the maximum velocity magnitude in the reversed-flow region are quantified. Correlations are proposed for the shear layer reattachment length and maximum reversed-flow velocity in terms of the Reynolds number and expansion ratio. The correlations achieve excellent agreement with the computed data and results from published literature. Critical Reynolds numbers for the onset of flow bifurcation are identified. The study facilitates optimization of sudden expansions and validates the numerical approach for follow-on 3D studies.

REFERENCES

1. Ko, S C and Sung, H J , "Large-eddy simulation of turbulent flow inside a sudden-expansion cylindrical chamber," *J. of Turbulence*, 3(N4) DOI: 10.1088/1468-5248/3/1/004.
2. Macagno, E.O., Hung, T-K 9 , "Computational and experimental study of a captive annular eddy", *J. of Fluid Mechanics*, 28(01):43-64
3. Hammad, K.J., Ötügen, M V , Arik, E B 999 , "A piv study of the laminar axisymmetric sudden expansion flow", *Experiments in Fluids*, 26(3), pp. 266–72.
4. Adams E, Johnston J 988, "Effects of the separating shear layer on the reattachment flow structure part 2: reattachment length and wall shear stress", *Experiments in Fluids*, 6, pp. 493–99.
5. Scott, P , Mirza, F , Vlachopoulos, J 98 , "A finite element analysis of laminar flows through planar and axisymmetric abrupt expansions", *Computers & Fluids*, 14(4), pp. 423-32.
6. Dağtekin, İ , Ünsal, M , "Numerical analysis of axisymmetric and planar sudden expansion flows for laminar regime", *International J. for Numerical Methods in Fluids*, 65(9), pp.1133-144.
7. Pak, B., Cho, Y.I., Choi, S.U.S 99 , "Separation and reattachment of non-Newtonian flows in a sudden expansion pipe fluid", *J. of Non-Newtonian Fluid Mechanics*, 37(2), pp. 175-99.
8. Cherdron, W , Durst, F , Whitelaw, J H 9 8, "Asymmetric flows and instabilities in symmetric ducts with sudden expansions", *J. of Fluid Mechanics*, 84(1), pp. 13-31.
9. Drikakis, D , 9 9 , "Bifurcation phenomena in incompressible sudden expansion flows", *Physics of Fluids*, 9(1), pp.76-87.
10. Battaglia, F , Papadopoulos, G , "Bifurcation characteristics of flows in rectangular sudden expansion channels" *J. of Fluids Engineering*, 128(4), pp. 671-679.
11. Sanmiguel- o as , E , Del Pino, C , Guti rrez-Montes, C. 2010, "Global mode analysis of a pipe flow through a : axisymmetric sudden expansion", *Physics of Fluids*, 22(7), pp. 071702.
12. Ferziger, J H , Perić, M , "*Computational methods for fluid dynamics*, Vol. 3, Springer, Berlin.

Proceedings of the 25th CANCAM
London, Ontario, Canada, May 31 - June 4, 2015

VORTEX VISUALIZATION AND CHARACTERIZATION IN THE LAMINAR SEPARATION BUBBLE OVER A NACA 0018 AIRFOIL

Andrew Lambert
Mechanical and Mechatronics Engineering
University of Waterloo
Waterloo, Ontario, Canada
arlamber@uwaterloo.ca

Serhiy Yarusevych
Mechanical and Mechatronics Engineering
University of Waterloo
Waterloo, Ontario, Canada
syarus@uwaterloo.ca

ABSTRACT

Coherent structures that form in a laminar separation bubble over a NACA 0018 airfoil are studied experimentally using time-resolved flow visualization and embedded microphone measurements. Flow visualization reveals the formation of periodic vortices in the aft portion of the separation bubble. Surface pressure measurements synchronized with the visualizations show that these structures produce distinct surface pressure fluctuations, whose character is similar to that expected from a continuous train of near-wall vortices in potential flow. These vortices may merge, which alters the periodic pressure signal as the vortices and hence pressure minima merge, making it possible to detect the passage of individual vortices as well as vortex merging based solely on the analysis of surface pressure signals. Analysis of the experimental results and a complementary potential flow model demonstrates that the merging process is initiated due to the induced orbiting motion of the neighboring vortices, which is promoted by the mean velocity gradient in the shear layer. The results provide insight into vortex dynamics within laminar separation bubbles and provide basis for future experiments based solely on surface pressure measurements.

Keywords: laminar separation bubble, boundary layer, transition, vortices, vortex merging, flow visualization

INTRODUCTION

Airfoil operation at chord Reynolds numbers below 500,000 is often accompanied by the formation of a Laminar Separation Bubble (LSB) over the airfoil's suction side. This LSB structure is formed as a consequence of laminar boundary layer separation, followed by flow transition and mean flow reattachment [1]. It can have a significant impact on airfoil performance [2], which has motivated active research efforts into its behaviour and characteristics.

Previous studies have found that transition of the separated shear layer is induced by the amplification of a band of disturbances [3]. The initial growth of perturbations is governed by a Kelvin-Helmholtz instability [4]. As the amplitude of disturbances increases, the separated shear layer can roll up into individual vortices. These vortices have been observed in PIV and DNS over airfoils [5], and may contribute to the increased momentum transfer which leads to reattachment.

Experimental studies on other geometries have linked the development of similar coherent structures to surface pressure fluctuations in near-wall flows. For example, strong minima in surface pressure signals were found to be related to the presence of vortices in flows over blunt-edged splitters [6] and backward-facing steps [7]. For airfoils in low Reynolds number flows, Boutilier and Yarusevych [8] have shown that such quantitative characteristics of the shear layer disturbances as the growth rate, convection speed, and frequency can be estimated using embedded microphones. In the present investigation, the merging of shear layer roll-up vortices is studied by performing novel time-resolved visualizations of the separation bubble on an airfoil, synchronized with time-resolved surface pressure measurements via embedded microphones.

EXPERIMENTAL SETUP

Experiments were performed in the University of Waterloo adaptive-wall wind tunnel. This open-return, suction type

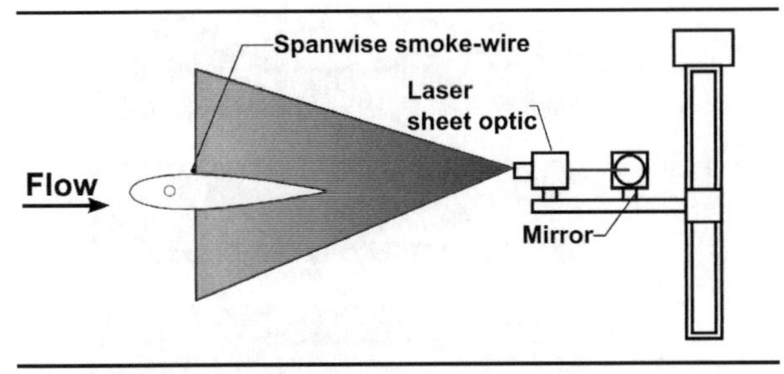

Figure 1: Schematic of airfoil and flow visualization equipment within the wind tunnel test section.

wind tunnel has four screens and a honeycomb flow conditioning element upstream of a 9:1 contraction, which results in a turbulence intensity of less than 0.2% for the conditions investigated. The test section is 6.5m long with a cross-section 0.61 by 0.89 metres. For this investigation, the walls were set to a geometrically straight configuration, with a constant cross-sectional area along the test section.

A schematic of the experimental setup is shown in Fig. 1. The aluminum NACA 018 airfoil model spans the test section, with a solid blockage ratio of under 5%. Fluctuating surface pressures are measured by 25 Panasonic WM-62C back-electret microphones embedded under the airfoils surface along its chord. The microphone signals are recorded through a National Instruments PCI-4472 data acquisition system that measures eight signals simultaneously. Further information on the airfoil model design and instrumentation can be found in Ref. 9.

For the time-resolved visualizations of the laminar separation bubble, the smoke-wire was placed within the separated shear layer across the span of the airfoil. Smoke was produced by applying a mixture of 70% glycerine 30% water to the wire, and evaporating it through electric heating. A light sheet from a Photonics Industries DM20-527 Nd:YLF laser illuminated the vertical plane aligned with the row of embedded microphones. High-speed image sequences were recorded with a Photron Fastcam SA4 camera at 5000 Hz at a resolution of 1024 by 512 pixels using a Nikkor 105 mm macro lens. The image acquisition was synchronized with measurements from five surface microphones sampled simultaneously at 40 kHz.

Results

All experimental results presented here were obtained at a chord-based Reynolds number of 100,000 and an angle of attack of 8°. At these conditions, a separation bubble forms on the suction side of the airfoil and produces distinct coherent structures. Flow development in the aft portion of the separation bubble is depicted in Figs. 2a-c. The presented sequence of images shows a streakline produced from the smoke wire which illustrates the amplification of periodic oscillations in the separated shear layer that leads to roll-up and shedding of vortices at x/c $\approx$ 0.26. To be able to gain insight into the development of these structures, it is of interest to establish a direct

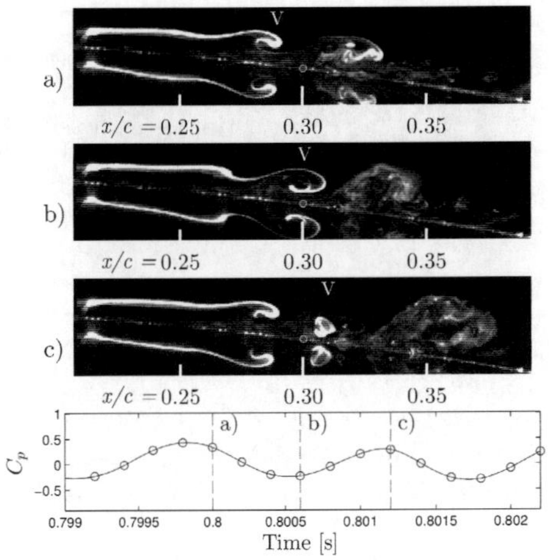

Figure 2: Pressure signals produced due to the passage of vortices over a microphone. The microphone location is marked by a red dot in the visualization images.

correlation between the passage of shear layer vortices and the pressure fluctuations they induce on the surface. Figure 2 presents a segment of surface pressure fluctuations measured at a microphone location when a vortex passes over a microphone port labeled by a red dot in the corresponding flow visualization images. As the vortex approaches the microphone port (Fig. 2a), the flow impinging on the surface leads to an increase in surface pressure a maximum in the pressure signal. When the vortex is above the microphone port (Fig. 2b), a pressure minimum is attained in the surface pressure signal. The pressure begins to recover as the vortex continues to move downstream (Fig. 2c). Thus, a train of shear layer vortices induces periodic surface pressure fluctuations at a given microphone location.

It is instructive to substantiate the observation made based on the comparison of flow visualizations and pressure signals with a simplified model. A train of vortices over a surface is modeled using mirror pairs of opposite-signed irrotational vortices, whose streamwise and wall normal spacing are modeled based on experimental flow visualization. The simulated pressure signature is obtained by calculating pressure variation along the dividing streamline, and the results are shown in Fig. 3. For this model, the strength of the vortices is selected such that the magnitude of the resulting surface pressure fluctuations approximately matches those measured experimentally. The results presented in Fig. 3a show that the region of low pressure surrounding the vortex core extends to the wall, leading to a pronounced surface pressure minimum (Fig. 3b). The train of shear layer vortices induces periodic surface pressure variations, with local pressure maxima attained between the neighboring vortices and pressure minima corresponding to the streamwise location of each vortex. The model results are in agreement with experimental observation (Fig. 2) and show that surface pressure fluctuations can be used to detect the streamwise position of vortices downstream of their formation, enabling the quantification of essential elements of

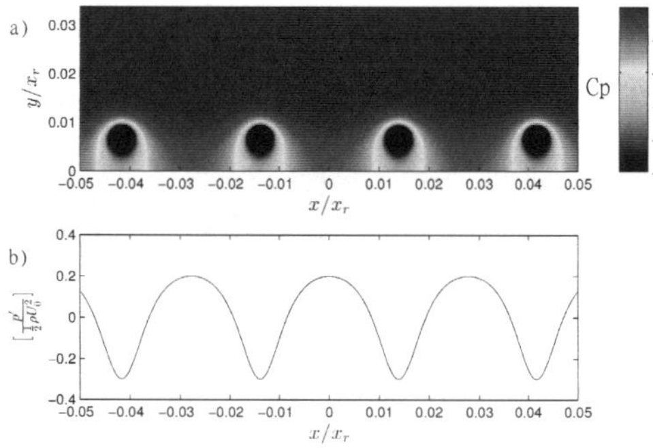

Figure 3: Inviscid model results for a vortex train passing over a surface: a) pressure contours and b) surface pressure variation.

vortex dynamics based solely on surface pressure measurements with the microphone array.

Analysis of flow visualizations showed that some shear layer vortices can merge in the separation bubble. Figure 4 illustrates the merging process in a sequence of high-speed visualization images. In this case, the merging occurs at approximately x/c=0.30. The vortex located just upstream of x/c=0.30 in Fig. 4a is captured merging with the following vortex, which is forming at approximately x/c=0.27 in Fig. 4a. It can be seen that the leading vortex is located closer to the wall and its convective velocity is lower than that of the upstream vortex (Figs. 4a-b). The leading vortex gradually slows down becoming nearly stationary, while the following vortex moves away from the surface and closes in on the leading vortex (Figs. 4b-c). The vortices begin to orbit as they approach each other (Fig. 4d-e) and the cores merge eventually (Fig. 4f). This evolution from orbiting to the merging

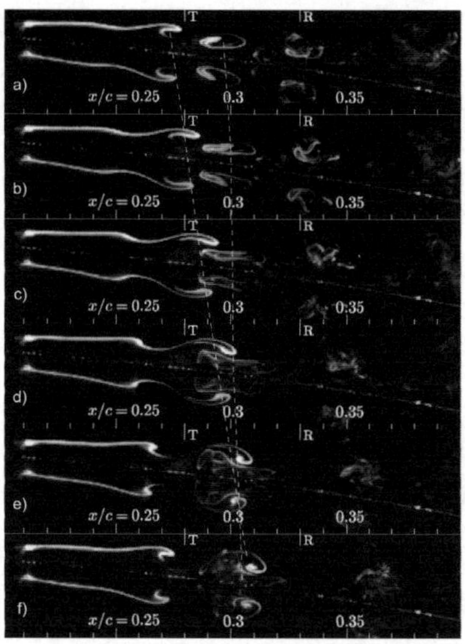

Figure 4: Visualization of vortex merging inside separation bubble. Images are at time intervals of 2 ms, with dotted lines indicating the progression of the cores of the vortices

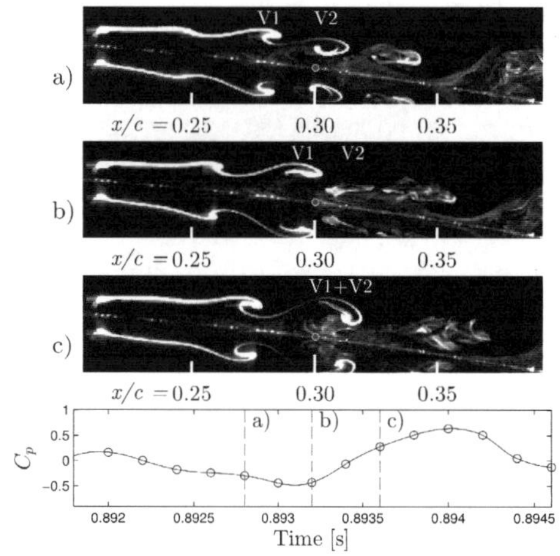

Figure 5: Pressure signals produced due to merging of two shear layer vortices. The microphone is shown by a red dot.

of vortex cores agrees with the stages of vortex merging described by Meunier et al.[10].

A vortex merging event with the corresponding surface pressure signals is shown in Fig. 3. The event causes a distinct change to the otherwise periodic pattern of surface pressure fluctuations. Specifically, when vortex V1 passes over the microphone port, no distinct surface pressure minimum is attained, which is attributed to the close proximity of the upstream vortex, V2 (Fig. 3a-b). As the vortices begin to orbit and merge, they instead produce a single minimum in the surface pressure signal at that location. Once the pair have passed the microphone port (Fig. 3c), the surface pressure begins to recover, resuming the trend expected for the continuous vortex street passing over the surface. The results suggests that, by detecting such events in surface pressure signals, the occurrence and location of vortex merging can be identified.

The initial vortex interaction and the associated pressure field response were simulated using an inviscid model. This simplified model relates the events within the flow to what is measurable at the surface, as well as gaining insight into the merging phenomenon. A pair of co-rotating vortices were modeled near the wall in the presence of uniform free stream with a circulation that gives the same rms of wall pressure fluctuations as measured experimentally. With an imposed mean velocity, the pair advects downstream against the direction of their self-induced velocity. A sequence of the pressure fields induced by the vortices are shown in Fig. 6. The simulation results show that, as expected, the vortices begin to orbit, with the leading vortex moving closer to the surface, while the following vortex moves away from the surface. As a result of the change of vortex positions near the wall, the convective velocity of the leading vortex decreases, while that of the following vortex increases. In a separation bubble, this effect is enhanced significantly due to the mean velocity increasing away from the wall. As a result, the the orbiting occurs faster in the experiment compared to the simulation (cf. Figs. 4 and 6). The simulated surface pressure signals essen-

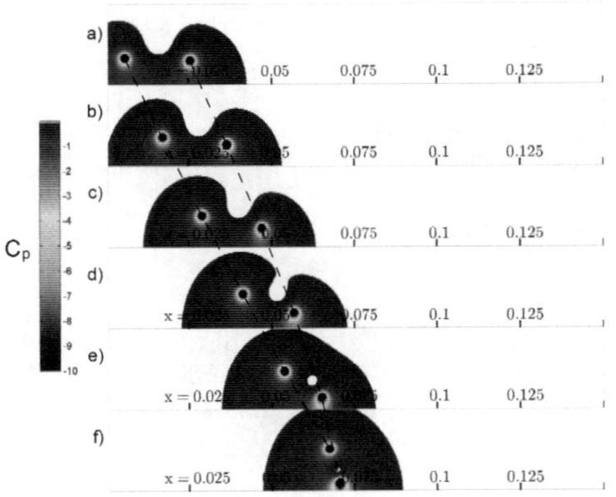

Figure 6: Inviscid simulation of vortex merging, with frames at intervals of 7.2 ms.

tially produce only one minimum in surface pressure when the streamwise spacing between the two vortices is reduced to approximately 70% of their initial streamwise spacing, (Fig 6 d) which agrees with the experimental results shown in Fig. 2. Comparing the experimental results and inviscid simulations, it can be concluded that the initial relative displacement between the adjacent vortices undergoing merging is similar to the induced velocity effect in inviscid simulations. However, this effect is further amplified in real flow due to the presence of the significant wall-normal velocity gradient, which accelerates the merging process. Indeed, the orbiting sequence preceeding merging in Fig. 4 occurs approximately three times faster than in inviscid simulations (Fig. 6). It should be noted that merging events were not observed periodically in this study, and they involved only a subset of shear layer vortices. This suggests that vortex merging in separation bubbles occurs due to perturbations in the vortex shedding process affecting the spacing between the consecutive vortices and/or their strength.

CONCLUSIONS

The development of coherent structures in a laminar separation bubble was investigated experimentally using high-speed flow visualizations synchronized with fluctuating surface pressure measurements. Wind tunnel tests were performed on a NACA 0018 airfoil at $Re_c = 100,000$ and $AOA = 8°$. The results show periodic shedding of shear layer vortices in the aft portion of the separation bubble. The passage of an individual vortex over an embedded microphone produces a distinct pressure minimum, and a sequence of these vortices produces a periodic signal, making it possible to detect and characterize these structures based on time-resolved surface pressure measurements. Some of the shear layer vortices merge downstream. When merging occurs, periodic surface pressure fluctuations are disrupted, as a single pressure minimum forms due to the passage of the merged vortex. A simplified potential flow model is considered to support the direct relation between the near-wall flow structures and the associated surface pressure fluctuations, as well as to gain insight into the merging process. It is shown that, while merging is initiated by the orbiting motion of the two neighboring vortices, it is accelerated significantly by the mean flow gradient normal to the wall. The results provide a foundation for characterizing the dynamics of coherent structures in laminar separation bubbles based on time-resolved surface pressure measurements.

ACKNOWLEDGMENTS

Thanks are extended to the Natural Sciences and Engineering Research Council of Canada (NSERC) and Bombardier Aerospace for funding this work. The authors also thank Mr. Thomas M. Kirk for his assistance with the experiments.

REFERENCES

[1] M. S. Selig, "Low Reynolds Number Airfoil Design Lecture Notes VKI Lecture Series Various Approaches to Airfoil Design," 2003.

[2] Lissaman, B. S. 1983, "Low-Reynolds-Number Airfoils," *Annu. Rev. Fluid Mech.*, Vol(15), pp. 223-239.

[3] Gaster, M. "The Structure and Behaviour of Laminar Separation Bubbles," Aeronautical Research Council, London

[4] Dovgal,A., Kozlov, V., and Michalke, A., 1994, "Laminar boundary layer separation: instability and associated phenomena," *Prog. Aerosp. Sci.*, Vol(30), pp. 61-94.

[5] O. Marxen, M. Lang, and U. Rist, "Vortex formation and vortex breakup in a laminar separation bubble," *J. Fluid Mech.*, pp. 5890, 2013.

[6] M. Kiya and K. Sasaki, "Structure of large-scale vortices and unsteady reverse flow in the reattaching zone of a turbulent separation bubble," *Journal of Fluid Mechanics*, 154, pp 463-491, 1985.

[7] L. M. Hudy, A. Naguib, and W. M. Humphreys, "Stochastic estimation of a separated-flow field using wall-pressure-array measurements," *Phys. Fluids*, vol. 19, no. 2, p. 024103, 2007.

[8] M. S. H. Boutilier and S. Yarusevych, "Separated shear layer transition over an airfoil at a low Reynolds number," *Phys. Fluids*, vol. 24, no. 8, pp. 08410508410523, 2012.

[9] R. Gerakopulos and S. Yarusevych, "Novel time-resolved pressure measurements on an airfoil at a low reynolds number," *AIAA J.*, vol. 50, no. 5, pp. 11891200, May 2012.

[10] P. Meunier, S. Le Dizes, and T. Leweke, "Physics of vortex merging," *Comptes Rendus Phys.*, vol. 6, pp. 431450, 2005.

[11] T. M. Kirk, "The Later Stages of Transition over a NACA0018 Airfoil at a Low Reynolds Number," University of Waterloo, 2014.

Proceedings of the 25th CANCAM
London, Ontario, Canada, May 31 – June 4, 2015

MODELING OF DIELECTRIC BARRIER DISCHARGE ACTUATOR FOR AIRFLOW CONTROL

Mohammadreza Ghazanchaei
Department of Electrical & Computer engineering
University of Western Ontario
London, ON, Canada
mghazanc@uwo.ca

Kazimierz Adamiak
Department of Electrical & Computer engineering
University of Western Ontario
London, ON, Canada
kadamiak@eng.uwo.ca

G. S. Peter Castle
Department of Electrical & Computer engineering
University of Western Ontario
London, ON, Canada
pcastle@eng.uwo.ca

ABSTRACT

In this study, a two-dimensional model of dielectric barrier discharge (DBD) is numerically simulated in order to study the ability of such system to actively modify low-velocity airflow along a flat plate. The investigated model involves mutual interaction between the electric field, space charge density and gas flow. A single species model including the effect of positive and negative ions has been adopted for this study. The DBD system consists of two electrodes, placed in a way to accelerate the tangential airflow very close to the wall. In order to improve our understanding of the physical mechanisms associated with the system, velocity profiles within the boundary layer of the flat plate are presented for a free air stream velocity up to a few meters per second. The effect of frequency and configuration on the system performance has been discussed. Effect of the DBD system on the mechanical drag is also studied. The results show the EHD actuator can cause drag reduction.

KEYWORDS: Electro-hydrodynamics, dielectric barrier discharge, airflow velocity, flow control, electric field

INTRODUCTION

When a high voltage is applied between two electrodes, ions are produced in air and they drift from the discharge electrode to the collecting one under the Coulomb forces. These ions exchange momentum with air molecules and induce a fluid motion, called electro-hydrodynamic flow (EHD). It has been proven that the glow discharge at atmospheric pressure using a dielectric barrier discharge can induce fluid flow and operate as an actuator for flow control. The plasma jet generated by a DBD actuator is able to alter the velocity inside the boundary layer and can modify the interaction between fluid and body surface without any mechanical moving parts. Such EHD phenomenon directly converts electric energy into kinetic energy with very small response time compared to the characteristic times of the fluid dynamics.

There are many articles with a wide review on AC surface DBD actuators applied to airflow control [1, 2]. The idea of using weakly ionized gas formed at the surface of a dielectric material for producing a wall jet close to the surface of an insulator was first demonstrated by Roth [3]. Later on both electrical and mechanical characteristics of surface DBD were presented and discussed by few research groups. However, because of its complicated physics, most of the investigations on DBD actuators are experimental. Focusing on the instantaneous evolution of the velocity at a given point, Forte [4] observed the non-stationary distribution of instantaneous velocity for different points. It has been shown that the actuator generates a pulsed induced flow at the same frequency as the frequency of high voltage. In their experiments the authors have shown that at points close to the exposed electrode negative corona discharge induces higher velocities as compared with the positive corona. It was already revealed that the structure of the plasma is not exactly the same during both half cycles [5]. However, just 10 mm downstream of the exposed electrode the experiment shows the same behavior during both positive and negative half cycles with very small fluctuation around a mean value. A reason for decreased difference between two cycles is that far enough from the exposed electrode, there is no effect of electrons and the net body force results from positive and negative ions with very small difference in their mobility.

Several authors tried to model the whole process taking into account the chemistry of discharges in various gases. Boeuf [6] showed that for a sinusoidal voltage, the discharge current consists of short pulses with large amplitude along with a low frequency current discharge of the corona type. However, the contribution of the low frequency current to the total force may be predominant in comparison with the pulse contribution because the force induced during the corona phases acts over a much larger volume and for a longer time. Likhanskii et al. [7] conducted a simulation taking into account the chemistry of the discharge including both positive and negative ions. Although they could model the whole process, they had to increase the level of frequency up to 100kHz in order to control the number of pulses happening during a half period of voltage. In most of existing experiments the working frequency for DBD actuators is between a few hundred hertz to a few kilo hertz. Therefore, such an approach can be useful for detailed study of some physical processes, but not as a design or optimization tool in engineering applications.

The authors interested in engineering applications of DBD often ignore the processes in the ionization layer, what is justified due to its small thickness, generally in the order of the corona electrode radius of curvature. Besides, it must be noted that the ionic wind in the active region is mainly caused by ions drift and the electrons have negligible effect due to their small mass compared to the ion mass. Thus, in this study, first we numerically simulate DBD assuming a 2D model in order to study the ability of the discharge to actively modify low-velocity airflow along a flat plate. The investigated model involves mutual interaction between the electric field, space charge density and gas flow. A single species model including positive and negative ions has been developed in this study for AC input voltage. Since the ionic wind in this case is mainly caused by ions, the effect of the electrons in the model has been ignored. The DBD system consists of two electrodes one placed on the insulating surface of the plate and exposed to high voltage and the other one buried inside the dielectric material and grounded. In this configuration, Fig.1, the ionic wind may be able to accelerate the tangential airflow very close to the wall. In order to improve our understanding of the physical mechanisms associated with the system, velocity profiles within the boundary layer of the flat plate are presented for a free air stream velocity up to a few meters per second. The effect of frequency level and geometry of the DBD on the system performance has been discussed. Effect of DBD on the mechanical drag is also studied. The results show that the EHD actuator causes drag reduction. This study could help researchers in optimization of the actuator.

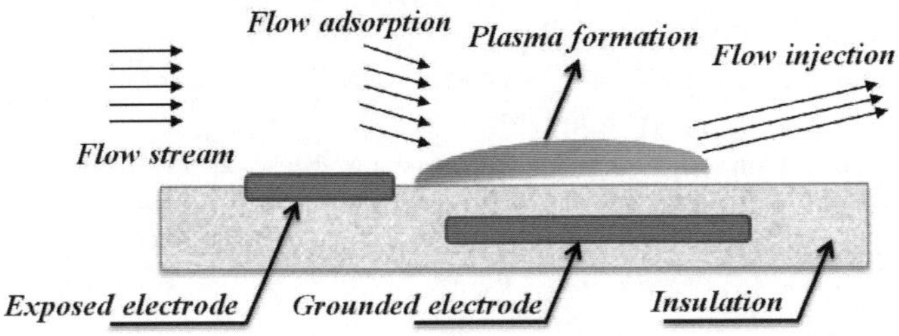

Figure 1: Idea of DBD actuator

GOVERNING EQUATIONS

FLUID FLOW MODEL

Detailed descriptions of the governing equations are given in [8]. The summary of the equations of motion for high enough Reynolds number and for the two-dimensional flow are:

$$\frac{\partial u}{\partial x}+\frac{\partial v}{\partial y}=0$$
$$\frac{\partial u}{\partial t}+u\frac{\partial u}{\partial x}+v\frac{\partial u}{\partial y}=-\frac{1}{\rho}\frac{\partial P}{\partial x}+\lambda\frac{\partial^2 u}{\partial y^2}+\frac{\rho_c}{\rho}E_x \tag{1}$$
$$-\frac{1}{\rho}\frac{\partial P}{\partial y}=0$$

where u and v are the velocity components in the x and y directions, respectively, ρ is the fluid density, μ is the dynamic viscosity, $\lambda(={}^{\mu}/_{\rho})$ is the fluid kinematic viscosity, E_x is the electric field strength in the x direction, and ρ_c is the net ion charge density. To solve the above system of equations it is necessary to specify initial and boundary conditions. At the plate surface there are no slip conditions.

$$u=0\ and\ v=0\ at\ y=0$$

At infinity, outside the boundary layer and away from the plate, the flow is undistorted

$$u\rightarrow U_\infty\ as\ y\rightarrow\infty$$

ELECTRICAL MODEL

In the model used in this paper, Poisson's equation governs the electric potential, and consequently the electric field intensity distribution for a given charge distribution [9]:

$$\nabla^2 V=-\frac{\rho_c}{\varepsilon} \tag{2}$$

The charge drift creates the electric current with a density defined as:

$$\vec{J}=\rho_c(\mu_e\vec{E}+\vec{u})-D\nabla\rho_c \tag{3}$$

where $\vec{J}$ is the electric current density, $\vec{E}$ is the electric field intensity, $\vec{u}$ is the gas velocity, ε is the gas permittivity, μ_e is the mobility of ions and D is the ions diffusion coefficient. The current density must satisfy the charge conservation equation:

$$\frac{\partial\rho_c}{\partial t}+\nabla\cdot\left(\rho_c(\mu_e\vec{E}+\vec{u})-D\nabla\rho_c\right)=0 \tag{4}$$

The problem of the corona discharge is governed by two partial differential equations with unknown electric potential and space charge density. The boundary conditions for the potential are straightforward: a given AC electric potential (V_{ap}) on the corona electrode and zero potential on the grounded plane [9]. However, formulation of proper boundary conditions for the space charge density is not so easy because the ionization layer has been neglected. One possible approach is to use Kaptzov's hypothesis which assumes that when the corona discharge occurs at some point on the electrode, the electric field at this point remains at the corona onset value. Peek's equation is used to find the onset level in air [10]:

$$E_0=3.1\times10^6\left(1+\frac{0.308}{\sqrt{R}}\right)\left[\frac{V}{m}\right] \tag{5}$$

where R is the electrode radius in cm.

The initial value for the space charge density has to be guessed or taken from the previous solution. After the problem is solved, the electric field on the electrode surface is compared with Peek's value and the electric charge updated. The following formula has been used to update the space charge density on the surface of corona electrode [10]:

$$\rho_c^{new}=\rho_c^{old}+\beta\times(E-E_0) \tag{6}$$

where β is an experimentally found constant and E is the actual value of the electric field.

An important point in simulating DBD is to include surface charge accumulation on the dielectric surface. This surface charge produces a field directed against the electric field produced by the voltage source to an extent that the ionization eventually stops. Actually, the surface charge accumulation plays a controlling role for the self-limiting feature of DBD preventing transition from a corona to a spark discharge. According to Maxwell equations, the tangential and normal components of the electrical intensity and electrical displacement at the interface between air and dielectric material must satisfy the following conditions [11]:

$$\vec{n}\times(E_d-E_a)=0 \tag{7}$$

$$\vec{n}\times(D_d-D_a)=\rho_s \tag{8}$$

In the above equations, ρ_s is the accumulated surface charge density, and the subscripts d and a, correspond to dielectric or air. The time derivative of surface charge density is related to the normal current density:

$$\frac{\partial\rho_s}{\partial t}=\vec{n}\cdot\vec{J} \tag{9}$$

The current density $\vec{J}$ is defined in Eq. (3), so, the surface charge accumulation can be calculated as:

$$\rho_s=\int J_n dt+C_0 \tag{10}$$

RESULTS AND DISCUSSION

ELECTRIC CURRENT

All simulations have been performed by utilizing a single surface DBD actuator immersed in air at atmospheric pressure. The electrodes are made of copper strips with a thickness of 0.04 mm. The upper electrode is 5 mm wide, and the lower one is buried and is 15 mm wide. A 5 mm thick insulator has been used as dielectric

material. A 4 mm horizontal gap between the two electrodes is assumed in basic configuration. Typical voltage and current waveforms are presented in Fig. 2 assuming a discharge for a frequency of 1 *kHz* and amplitude of 15 *kV*.
In this study, a formula for the discharge current for general electrode geometry is derived from the energy balance equation. With a few simplifications the current equation can be calculated as follow [12]:

$$I = \frac{1}{V_{ap}} \iiint \rho\mu \vec{E} \cdot \vec{E}_s dv \qquad (11)$$

The $\pi/2$ phase-shifted capacitive current has been neglected on this graph.

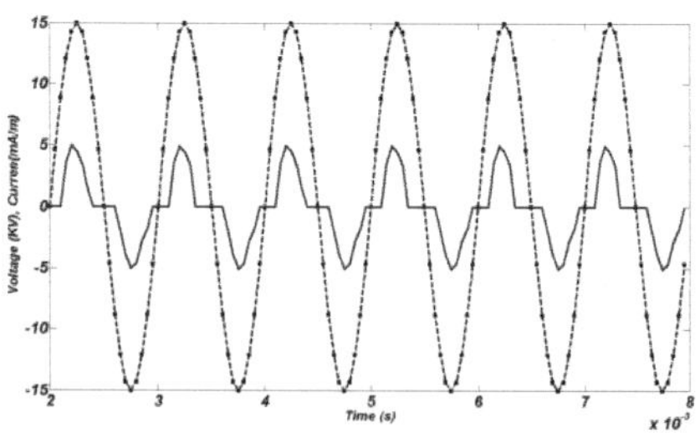

Figure 2: Voltage (kV) and current (mA/m) waveforms in DBD

As it is shown in Fig. 2 the chocking effect on current waveform happens and the current is not in phase with the applied voltage.

FLOW VELOCITY

In order to study the DBD effect, the velocity profile for inlet velocity equal to 1 m/s has been shown in Fig. 3 for systems with/without discharge.

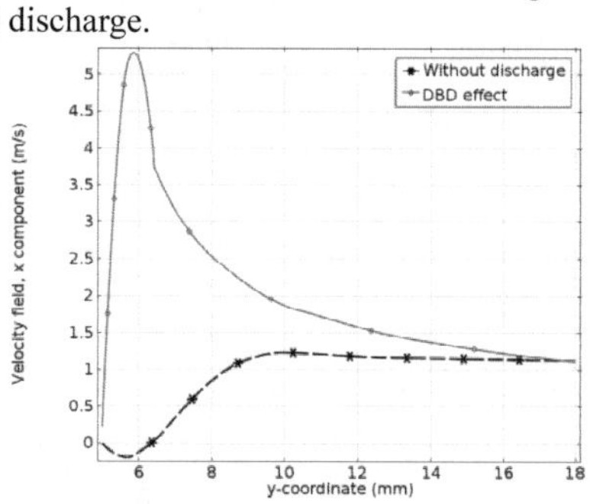

Figure 3: Velocity profile along a direction normal to the surface at X=5mm ahead of the exposed electrode

Fig. 4 shows the effect of the plasma generated EHD flow on the air flow in the boundary layer. The velocity streamlines are attracted towards the surface and form a wall jet. By using this feature, momentum would be injected in regions of boundary layer. Depends on the geometry and the configuration of the electrodes, the DBD has the contribution up to few meters per second on the velocity profile enhancement.

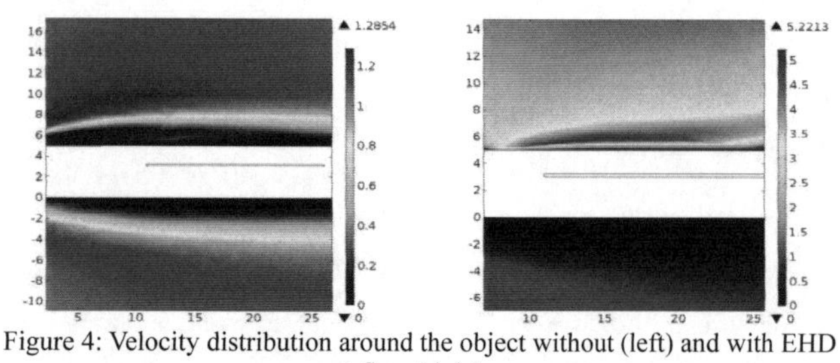

Figure 4: Velocity distribution around the object without (left) and with EHD flow (right)

Velocity profiles above a DBD actuator are shown in Fig. 5 for different X values. The higher velocity is reached at x closer to the exposed electrode. Away from the discharge electrode the value of the maximum velocity decreases but the point of maximum moves farther from the wall.

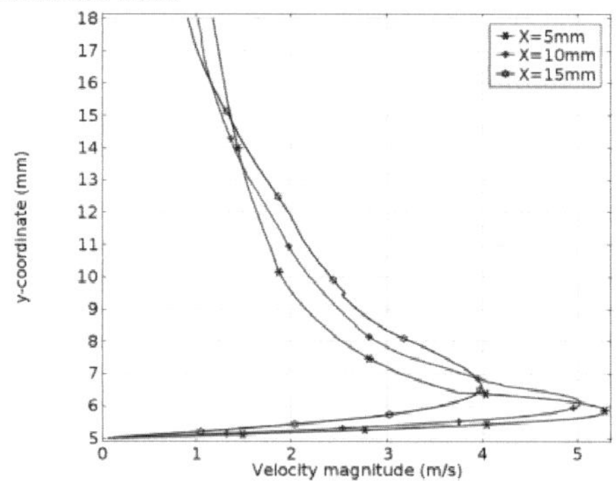

Figure 5: Velocity profiles above the DBD actuator for different distances from the discharge electrode

Time evolution of the horizontal component of the air velocity (u) is presented in Fig. 6 for two different points:

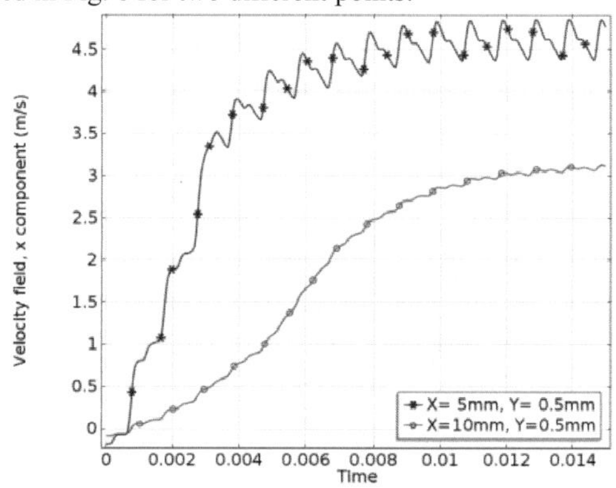

Figure 6: Time evolution of the x velocity component

The following can be observed from the calculated distributions:

- A pulsed induced flow is generated at the same frequency as the frequency of high voltage.
- Velocity fluctuations disappear at larger distances from the discharge electrode and the horizontal velocity stabilizes around a mean value.
- The velocity profile stabilizes after around ten cycles of the voltage period.

FORCE CALCULATION

The total force acting on air flow in the boundary layer resulted by corona discharge is an important parameters of the actuator. This force is the accumulation of the frictional drag force and the body force produced by the DBD system. For low-viscosity fluids total force per unit length is equal to [13]:

$$\sum F = \rho \int_0^\infty U_\infty (U_\infty - U) dy \; [N/m] \qquad (12)$$

Fig. 7 shows the total force over the flat plate for different input velocity levels. Simulation results confirm the reduction of the total force for the velocity levels up to a few meters per second. It seems that the drag reduction comes from stream wise momentum injected into the flow by DBD system.

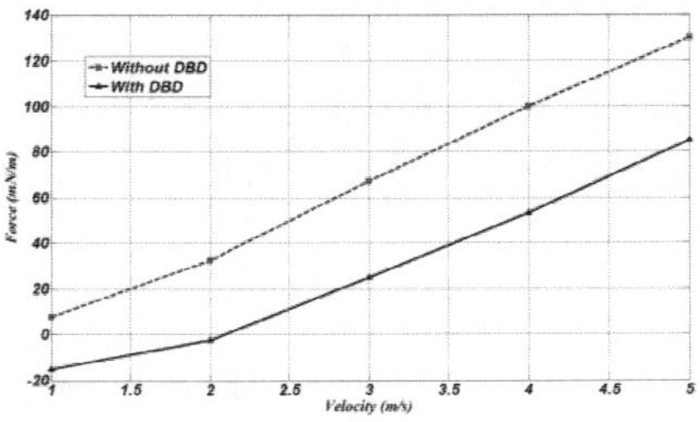

Figure 7: Total force calculation at X=5 mm ahead of exposed electrode versus free airflow velocity $\mathbf{U}_\infty$ in absence and in presence of the DBD

PARAMETRIC STUDY

There are few geometrical and physical features that have strong effects on actuator performance. Parametric study of those features would be useful for optimized design of the system. In this study the system was analyzed for different grounded electrode width, and for different frequency levels. Fig. 8 shows the plasma extension for two different electrode lengths. It has been found that the plasma extends up to the rightmost point of the grounded electrode on the surface. After that, the effect of DBD disappears.

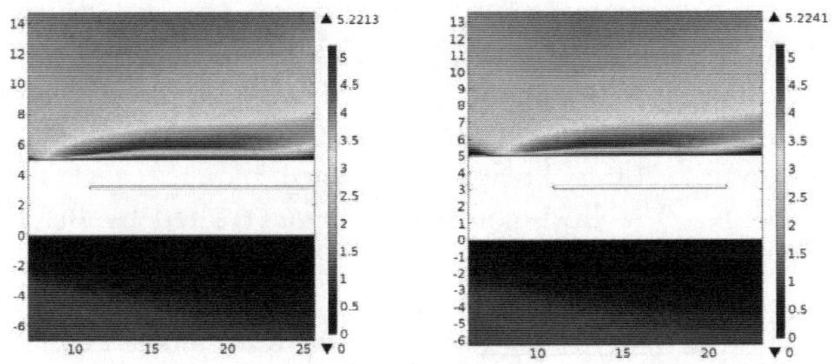

Figure 8: The effect of DBD on boundary layer for 15mm (left), and 10mm (right) electrode length

However, since the distance between the tip of the exposed electrode and the grounded electrode is constant, the onset voltage level, and consequently the maximum current is the same for both cases. Therefore, almost the same level of the electric energy is transferred to the air. This is confirmed by the results in Fig. 9, where, the evolution of the velocity for two different cases at the same point is shown: the narrower the grounded electrode, the greater the velocity magnitude.

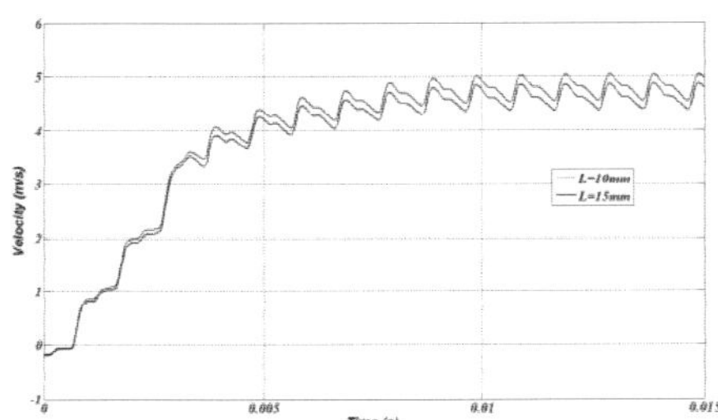

Figure 9: Velocity evolution for the same point and for two different widths of the ground electrode

Fig. 10 shows the evolution of the velocity at the same point for two different frequencies. It can be noted that at the lower frequencies the velocity reaches it quasi stationary form after smaller number of cycles, but the velocity fluctuations are much bigger compared to that at higher frequencies. Meantime, higher frequency results in a greater magnitude of the mean velocity.

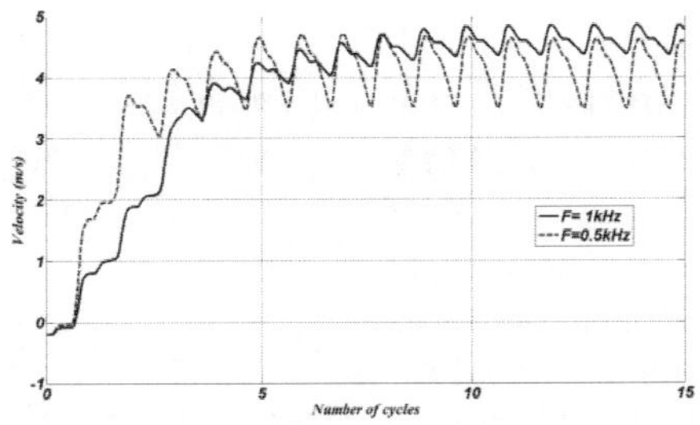

Figure 10: Velocity evolution at the same point for two different frequencies of the applied voltage

CONCLUSIONS

DBD seems to have a significant effect on the velocity profile in boundary layer. Even for very small electrical power consumption. Following conclusions could be noted from the performed simulations:

- Single species model provides the results with a good agreement with experiments published in literature.
- DBD system results in drag reduction up to air velocities which correspond to glow discharge regime.
- Extension and magnitude of the plasma depends on the configuration of the system.
- Higher voltage frequencies show better actuator performance with less velocity fluctuations.

REFERENCES

1. Benard, N and Moreau, E," Electrical and mechanical characteristics of surface AC dielectric barrier discharge plasma actuators applied to airflow control", Exp. Fluids 55:1846, November, 2014.
2. Corke, T., Enloe, C. and Wilkinson, S.," Dielectric Barrier Discharge Plasma Actuators for Flow Control", Annul. Rev. Fluid Mech., 42:505–29, 2010.
3. Roth J., Sherman M. and Wilkinson S., "Boundary layer flow control with a One Atmosphere Uniform Glow Discharge Surface Plasma". AIAA Paper 98-0328, Proc. of the 36th AIAA Aerospace Sciences Meeting & Exhibit Reno, NV, January 12-15, 1998
4. Forte, M, Jolibois, J., Pons, J., Moreau, E., Touchard, G., Cazalens M.," Optimization of a dielectric barrier discharge actuator by stationary and non-stationary measurements of the induced flow velocity: application to airflow control", Exp. Fluids, 43:917–928, 2007
5. Enloe, C., McLaughlin, T., VanDyken, R., Kachner K., Jumper, E. and Corke T.," Mechanisms and responses of a single dielectric barrier plasma actuator: plasma morphology", AIAA J, 42(3):589–594, 2004.
6. Boeuf, J., Lagmich, Y., Unfer, T., Callegari, T., Pitchford, L.," Electrohydrodynamic force in dielectric barrier discharge plasma actuators", J. Phys. D: Appl. Phys.40:652–662, 2007
7. Likhanskii, A., Shneider, M., Macheret, S. and Miles, R.," Modeling of dielectric barrier discharge plasma actuator in air", J. Phys. D: Appl. Phys., 103, 053305, 2008.
8. H. H. Schlichting,"Boundary layer theory", New York: McGraw-Hill, 2000.
9. Sattari, P. Castle, G. S. P. and Adamiak, K.," FEM FCT based dynamic simulation of corona discharge in point–plane configuration", IEEE Trans. Ind. Appl., Vol. 46, pp. 1699-1706, September 2010
10. Adamiak, K. and Atten, P.," Simulation of corona discharge in point-plane configuration", J. Electros., Vol. 61, pp. 85-98, 2004
11. Shang, J. S., Huang, P. G.," Modeling of ac dielectric barrier discharge", J. Appl. Phys., Vol. 107: 113302-7, June 2010.
12. Sato, N., Discharge current induced by the motion of charged particles", J. Phys. D: Appl. Phys., Vol. 13, pp. 3-6, 1980
13. Moreau, E., Léger, L. and Touchard, G.," Effect of a DC surface-corona discharge on a flat plate boundary layer for air flow velocity up to 25 m/s", J. Electros., Vol. 64, pp. 215–225, 2006

NUMERICAL INVESTIGATION OF BLOCKAGE DEVELOPMENT IN AN AEROENGINE TRANSONIC AXIAL COMPRESSOR STAGE

Shady Ali
Department of Mechanical and Materials Engineering
University of Western Ontario
London, Ontario, Canada
sali332@uwo.ca

Eric Savory
Department of Mechanical and Materials Engineering
University of Western Ontario
London, Ontario, Canada
esavory@uwo.ca

Chao Zhang
Department of Mechanical and Materials Engineering
University of Western Ontario
London, Ontario, Canada
czhang3@uwo.ca

ABSTRACT

Numerical simulations have been carried out to quantify the development of flow blockage along the rotor and stator of a transonic compressor stage using the boundary layer thickness deducted from the boundary layer theory. The radial profiles of blockage have been obtained at different axial planes starting from the rotor leading edge and ending with the stator trailing edge. Three operating points at design speed have been used in the analysis to identify the regions where blockage is sensitive to the reduction of the mass flow rate. These conditions correspond to the maximum efficiency, a high total pressure ratio point prior to the maximum pressure, presumably a pre-stall point, and the last numerically stable point which might correspond to a post stall point. The results show that the rotor casing region exhibits high radial blockage at all operating points due to the shock induced separation of the rotor casing boundary layer. At the stator while the flow travels along the passage, a considerable radial blockage growth is observed when the mass flow is reduced due to massive separations near the stator hub. The overall planar blockage along the compressor stage shows a major blockage growth at the stator trailing edge when approaching the reduced mass flow conditions. The results suggest that the contribution of stator stall in the compressor stalling process should not be disregarded.

KEYWORDS: Transonic compressor, Blockage, Stall, CFD

INTRODUCTION

Axial transonic compressors are core components in aero-engines nowadays due to their high pressure ratio per stage compared to axial subsonic compressors. The operating range of the whole engine is highly dependent upon the performance and stability of the compressor. When the engine runs at reduced mass flow rates or speeds, there is a risk that the compressor experiences stall and surge. Therefore, understanding of the stall physics is of great importance.

Blockage is a velocity deficit that reduces the effective flow area. Compressor stall initiation is linked to the growth of the low velocity regions. It is crucial to understand the sources of blockage losses in the compressor stage to optimize the design of axial compressors. The classical stall hypothesis, proposed by Emmons et al [1], linked the spike stall to the blockage growth by the separation of the suction side boundary layer and, hence, high flow incidence occurs. In transonic compressors, the blockage growth may be due to the casing boundary layer separation downstream of the shock [2]. On the other hand, Chen et al [3] excluded the existence of any boundary layer separation induced by the shock while operating near stall in a transonic rotor. They linked the spike stall to the vortex breakdown created by the shock/TLV interaction. The low momentum region created by the tip vortex breakdown has been observed in both subsonic and transonic compressors [4-5]. In transonic compressors operating at lower speeds, a secondary vortex along the blade suction surface may be another source of blockage [6]. All these blockage regions may exist in the same compressor; however, the link between stall and the growth of a specific blockage is not yet clear. The objective of this study is to identify and understand the mechanisms of blockage growth that may lead to stall in an axial transonic compressor stage.

QUANTIFICATION OF BLOCKAGE

The total blockage is the reduction in the effective flow area, which is given by [7].

$$B=1-\frac{Effective\ area}{Geometric\ area}=1-\frac{A-\int \delta^*(r)dr}{A}=\frac{\int \delta^*(r)dr}{A} \quad (1)$$

where A is the total area of the axial plane and $\delta^*(r)$ is assumed to be the integral axial mass flux deficit (ρu) [7]. Analogous to the displacement thickness of boundary layer theory, δ^* is obtained at each constant radial position using the following equation.

$$\delta^*(r)=\int_{\theta_{min}(r)}^{\theta_{max}(r)}\left(1-\frac{(\rho u)}{(\rho u)_{inviscid}}\right)rd\theta \quad (2)$$

For simplicity, the axial velocity component will be used in analysis at all sections, although this approach was applied only downstream the rotor [7]. This means that within the blade passage, the combined effect of blade boundary layers and any vortices or wakes will be included in the integration. To obtain the inviscid flow velocities, a defect region is calculated using radial and circumferential gradients of the axial mass flux.

$$Defect\ Region=\sqrt{\left(\frac{\partial(\rho u)}{\partial r}\right)^2+\left(\frac{\partial(\rho u)}{\partial(r\theta)}\right)^2}\geq Cutoff \quad (3)$$

$$Cutoff=\frac{C\times(\rho u)_{Inlet}}{Chord}$$

where *Cutoff* is proportional to the axial flux at the inlet of the rotor or stator divided by the rotor or stator chord at the tip, respectively. The predicted displacement thickness is not sensitive to the cutoff value since the gradients are much stronger in the defect region [7]. The value of C is taken as 2 [7]. The inviscid flux field $(\rho u)_{inviscid}$ has the same value as the local flux field (ρu) in the core region (i.e. the region outside the defect zone) which gives zero local blockage at these points. On the other hand, the inviscid flux field within the defect region is linearly extrapolated from the local velocity field at the core region.

GEOMETRY AND NUMERICAL METHOD

The compressor stage analyzed in this work is the first stage of an aeroengine multistage transonic compressor. The stage consists of inlet guide vanes (IGV), a rotor (R) and a stator (S). Steady state simulations have been performed using ANSYS CFX 14.5 [8]. The CFX solver is a finite volume, coupled, pressure based; multigrid solver [9]. The Rhie-Chow interpolation scheme [10] is used for pressure-velocity coupling to avoid decoupling of the co-located grid. For all transport equations, a high resolution advection scheme [11] that uses a non-linear blending function for each node is applied. Turbulence is simulated using the shear stress transport model (SST) [12] that switches between the k-ω model near the wall and the k-ε model away from the wall through a blending function.

At the preliminary investigation stage, it is more practical to simulate a single blade passage with periodic boundaries instead of the full wheel. The computational model is shown in Fig. 1. Two stator blade passages are simulated in order to reduce the pitch ratio at the rotor/stator interface to be used for future unsteady analyses. A large intake is mounted upstream of the inlet guide vanes and a variable area nozzle is mounted downstream the stator to avoid interactions between the blade rows and the effects of the boundary conditions at inlet and outlet [13]. It has been shown that including a variable area nozzle at the compressor exit improves the stability of the simulations near stall [13]. Total pressure and temperature are imposed at the inlet while a constant static pressure is imposed at the outlet. Different operating points are obtained by changing the exit area of the nozzle. All blades, hubs and casings are set as no-slip adiabatic walls. For the IGV/rotor and rotor/stator interfaces, the mixing plane approach is applied [14]. A circumferential averaging of the flow quantities is performed at the interface while meridional profiles are maintained between the stationary and rotary components.

The details of the grid resolution and topology are listed in Table 1 for coarse, medium and fine meshes. The O-J grid is used for the rotor and stator for a better mesh quality in the case of high stagger angles. Ansys Turbogrid 14.5 is used for the mesh generation. The average value of y^+ is maintained below 2 for the blades and 5 for the hubs and shrouds. Kroll et al [15] have shown that $y^+<5$ is sufficient for predictions of separation points and skin friction in transonic airfoils.

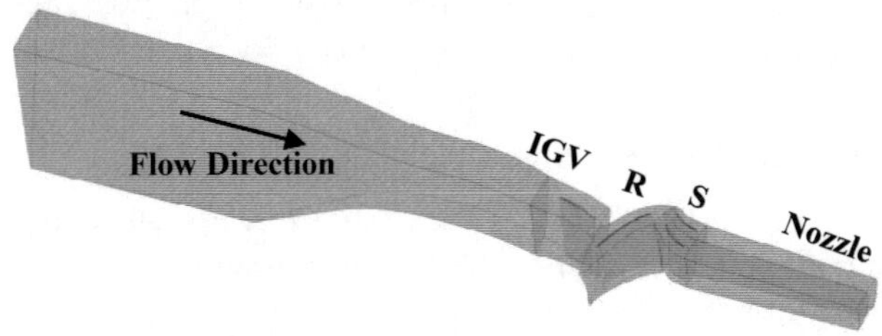

Figure 1: The Computational Domain

Table 1: Mesh Data

Component / Density	Intake	IGV	R	S	Nozzle
Coarse	143,450	182,707	364,569	306,236	70,602
Medium	1.08 M	1.30 M	2.50 M	1.11 M	258,400
Fine	3.56 M	11.3 M	18.7M	8.3 M	1.69 M
Topology	H	O-H	O-J	O-J	H

GRID INDEPENDENCE STUDY

Since all mesh densities show excellent agreement at point A (i.e. differences are below 1%), those results are not represented here. Significant differences between different mesh results are observed as the compressor operates near the reduced mass flow points. Point B (Fig. 3) is selected for comparison between meshes in terms of global performance (Table 2) and the local flow field (Fig. 2). Table 2 shows that the agreement between the medium and fine meshes is better than the agreement between the coarse and other mesh densities, particularly for the pressure ratio. The spanwise profiles of circumferential averaged pressure, temperature, axial velocity and flow angle are plotted downstream of both rotor and stator for all mesh densities in Fig. 2. As expected, the medium mesh profiles are closer to the fine results than the coarse mesh, particularly downstream of the stator. The medium mesh axial velocity and flow angle results show a good agreement with the fine mesh results. Therefore, the medium mesh is used for the analysis.

PERFORMANCE CHARACTERSITICS OF THE COMPRESSOR STAGE

The pressure and efficiency are plotted versus the flow coefficient in the characteristic curve for the three different meshes shown in Fig. 3. It is well known that the choking phenomenon takes place when the compressor mass flow rate is insensitive to the reduction of the static back pressure. This is obvious at the last two points where the compressor operates at its maximum flow rate. Researchers have often been concerned with reduced mass flow operating conditions on the left hand side of the characteristics (points B and C) since the compressor may experience the serious phenomena of stall and surge [3-7]. Massive separations and system vibrations resulting from stall may shorten the lifetime of compressors, therefore, it is crucial to understand the physics and mechanisms of stall in order to develop effective control systems.

BLOCKAGE DEVELOPMENT IN THE STAGE

The radial distribution of the flow blockage along the compressor stage is shown in Fig. 4. The distribution is plotted starting from the axial plane passing by the rotor tip leading edge. The profile at 130% of the rotor leading edge is located at midway between the rotor and the stator at the tip. The major contribution of blockage development is located at the rotor tip due to shock-induced casing boundary layer separation, which will be discussed later. For the rotor, the boundary layers on the suction and pressure sides make a small contribution to the blockage compared to the tip and hub regions. For the stator, the blockage grows significantly between 0% and 40% span because of the massive boundary layer separations at the suction side. The distribution of the total blockage across the stage shown in Fig. 5 suggests the importance of stator blockage as the compressor operates at off-design conditions. The drop in the blockage between 130% and 165% streamwise is due to mixing out of the rotor wake by the mixing plane approach.

SOURCES OF BLOCKAGE WITHIN ROTOR AND STATOR BLADE ROWS

The regions of strong velocity deficit can be clearly seen in Fig. 6 and Fig. 7. As the mass flow is reduced as in point C, the rotor shock interacts with the weak momentum flow at the tip resulting in strong separations at the leading edge plane which are depicted in Fig. 6. The low momentum fluid downstream the shock explains the blockage growth near the rotor tip region. For the stator, the suction side boundary layer separations play a major role in the flow blockage as shown in Fig. 7. The streamlines near the suction side show strong vertical flows which produce a region of negative flow downstream of the leading edge, as illustrated by the axial velocity contours.

Table 2: Grid Independence Tests

Variable	Coarse-Medium % Change	Coarse-Fine % Change	Medium-Fine % Change
PR	0.881	1.278	0.397
Efficiency	0.345	0.163	0.182

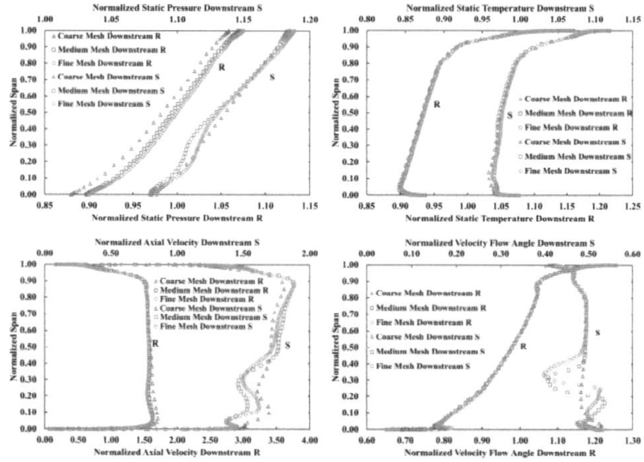

Figure 2: Grid Independence Test at Point B

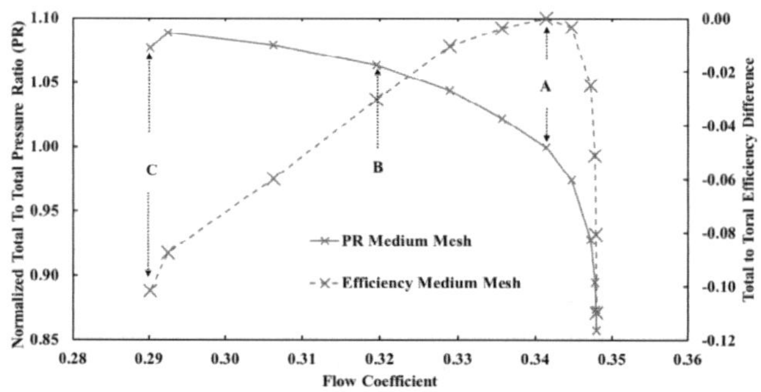

Figure 3: Performance Characteristics Curve

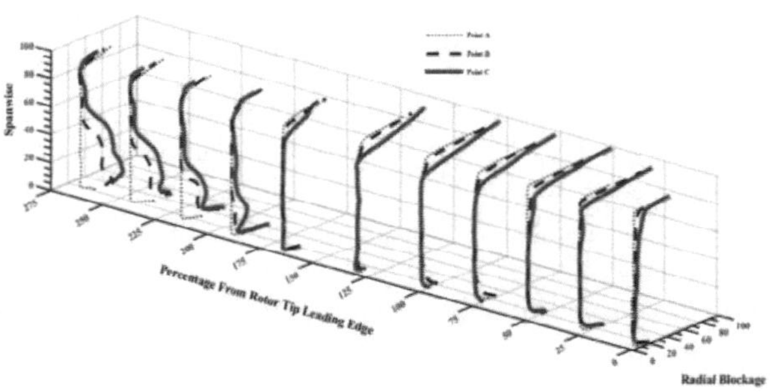

Figure 4: Radial Blockage Development along the Stage

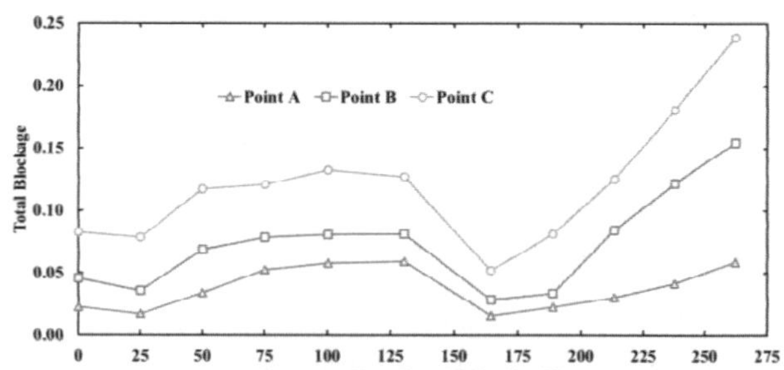

Figure 5: Total Blockage Development along the Stage

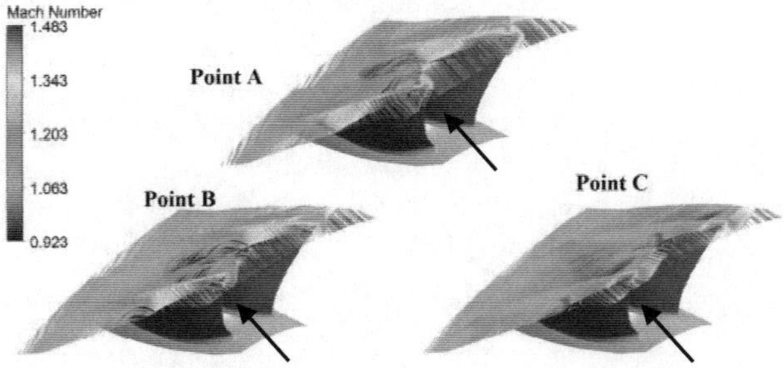

Figure 6: Rotor Casing Boundary Layer Streamlines

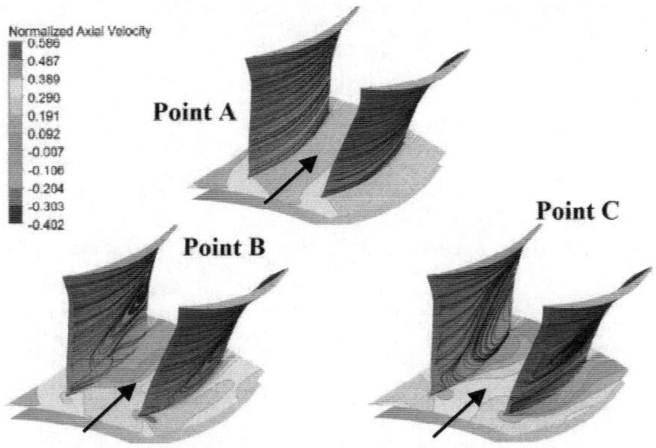

Figure 7: Stator SS Boundary Layer and Axial Velocity Contour at 10% Span

CONCLUSIONS

The blockage development has been numerically investigated in an aeroengine transonic axial compressor stage using the CFD commercial software ANSYS CFX. The boundary layer theory has been adopted to quantify the distribution of the blockage in the radial direction. The results indicate that the blockage is significant near the rotor tip and the stator hub due to the rotor shock-boundary layer interaction and the stator suction side boundary layer separation, respectively. The total blockage development suggests that stall might initiate at in the stator of the compressor stage examined here.

ACKNOWLEDGMENTS

This work was made possible by the facilities of the Shared Hierarchical Academic Research Computing Network (SHARCNET:www.sharcnet.ca) and Compute/Calcul Canada.

REFERENCES

[1] Emmons, H. W., Pearson, C. F., and Grant, H. P., "Compressor surge and stall propagation," *Trans. ASME*, vol. 77, pp. 455–469, 1955.

[2] Hah, C., Rabe, D., and Wadia, A., "Role of tip-leakage vortices and passage shock in stall inception in a swept transonic compressor rotor," in *Proceedings of ASME Turbo Expo, Vienna, Austria*, 2004.

[3] Chen, J.-P., Hathaway, M. D., and Herrick, G. P., "Prestall behavior of a transonic axial compressor stage via time-accurate numerical simulation," *Journal of Turbomachinery*, vol. 130, no. 4, pp. 1–12, 2008.

[4] Furukawa, M., Inoue, M., Saiki, K., and Yamada, K., "The role of tip leakage vortex breakdown in compressor rotor aerodynamics," *Journal of Turbomachinery*, vol. 121, no. 3, pp. 469–480, 1999.

[5] Yamada, K., Funazaki, K., and Sasaki, H., "Numerical investigation of relation between unsteady behavior of tip leakage vortex and rotating disturbance in a transonic axial compressor rotor," in *Proceedings of ASME Turbo Expo, Berlin, Germany*, 2008.

[6] Suder, K. L. and Celestina, M. L., "Experimental and computational investigation of the tip clearance flow in a transonic axial compressor rotor," *Journal of Turbomachinery*, vol. 118, no. 2, pp. 218–229, 1996.

[7] Suder, L., "Experimental investigation of the flow in a transonic , axial flow compressor with respect to the development of blockage and loss," PhD thesis, Department of Mechanical and Aerospace Engineering, Case Western Reserve University.

[8] "ANSYS CFX Version 14.5, Ansys Inc., Canonsburg, PA." 2012.

[9] Raw, M., "Robustness of coupled algebraic multigrid for the Navier-Stokes equations," in *AIAA 34th Aerospace Sciences Meeting and Exhibit, Reno, Nevada, USA*, 1996.

[10] Rhie, C. and Chow, W., "Numerical study of the turbulent flow past an airfoil with trailing edge separation," *AIAA journal*, vol. 21, no. 11, pp. 1525–1532, 1983.

[11] Barth, T. J. and Jespersen, D. C., "The design and application of upwind schemes on unstructured meshes," in *27th Aerospace Sciences Meeting, Reno, Nevada, USA*, 1989.

[12] Menter, F. R., "Two-equation eddy-viscosity turbulence models for engineering applications," *AIAA Journal*, vol. 32, no. 8, pp. 1598–1605, 1994.

[13] Vahdati, M., Sayma, a. I., Freeman, C., and Imregun, M., "On the use of atmospheric boundary conditions for axial-flow compressor stall simulations," *Journal of Turbomachinery*, vol. 127, no. 2, pp. 349–351, 2005.

[14] Denton, J. D., "The calculation of three dimensional viscous flow through multistage turbomachines," *Journal of Turbomachinery*, vol. 114, no. 1, pp. 18–26, 1992.

[15] Kroll, N., Schwamborn, D., Becker, K., Rieger, H., and Thiele, F., *MEGADESIGN and MegaOpt – German initiatives for aerodynamic simulation and optimization in aircraft design*, Vol. 107. Springer, 2007.

Combustion and Flames

Proceedings of the 25th CANCAM
London, Ontario, Canada, May 31 – June 4, 2015

TEST METHODOLOGY FOR COMPARING WILDLAND FIRE CHEMICAL TREATMENTS

Shammawi A. Anderson
Department of Mechanical Engineering
University of Alberta
Edmonton, AB, Canada
shammawi@ualberta.ca

Andre G. McDonald
Department of Mechanical Engineering
University of Alberta
Edmonton, AB, Canada
andre2@ualberta.ca

ABSTRACT

A simple and effective laboratory test methodology was developed for comparing the effectiveness of wildland fire chemicals in delaying the ignition time of vegetative fuels. The proposed test assembly consisted of an electric-powered radiant heater that was used to produce a uniform radiant thermal load, which allowed for a maximum incident heat flux of 90 kW/m^2 to ignite vegetative fuel samples. The samples (untreated, water treated, and foam treated) were mounted on to a load cell, rated at a maximum of 2 kg, to determine the transient mass loss during the combustion process. The time to glowing ignition was determined by visual observation, while the time to flaming ignition was determined by using transient mass loss data. A custom-made heat flux sensor was also used to validate the time to flaming ignition. Statistical *t*-test analysis was conducted on the time to flaming ignition to determine whether the results were statistically significant for the different chemical treatments. This indicated that the test methodology allowed for effective differentiation between the wildfire chemical treatments by comparing their mean ignition times. The narrow standard deviations of the average ignition times suggested that the test methodology was able to produce repeatable results. In addition to the ignition time, the methodology also provided useful quantitative data such as transient mass loss and heat release all in one test, which can be incorporated into various fire models. The test methodology may also be used to complement the US Forest Service Wildland Fire Chemicals Standard Test Procedures in which no vegetative fuel samples are used, to provide an indication of field performance of wildland fire chemicals.

KEYWORDS: Fire, Heat Flux, Ignition, Transient Mass Loss, Wild Fire Chemicals.

INTRODUCTION

Fire agencies worldwide apply millions of liters of fire suppression chemicals in an attempt to aid in the containment and extinguishing of wildfires. These chemicals have been proven to be effective firefighting tools for retarding the spread and lowering the intensity of wildfires [1]. The chemicals can mitigate fires directly (through treatment of the burning fuels) or indirectly (in advance of a fire front to create control lines or to reinforce constructed firelines in the unburned fuel).

Fire suppression chemicals include long-term fire retardants, which inhibit combustion even after vaporization of their water-based matrix, water enhancers, which lose their effectiveness after evaporation of their water content, firefighting foams, which form small bubbles when mixed with water, and wetting agents, which reduce the surface tension of water and increase its ability to spread and whet the fuel surfaces [1-2].

Various analytical techniques have been used to evaluate the performance of wildfire chemicals, such as gas chromatography, Fourier transform infrared spectroscopy, mass spectrometry, and thermal gravimetric analysis [2-7]. However, the small samples used and the rapid removal of pyrolysis or combustion products can lead to erroneous interpretations in terms of forest fuel flammability in real situations [2]. Therefore, the information provided by the analytical methods should be supported by other laboratory scale fire tests. There are a few laboratory-scale tests applicable for evaluating wildfire chemicals. The most widely used is the ASTM 1321 Standard Test Method for Determining Material Ignition and Flame Spread Properties [8]. The US Forest Service utilizes this standard to develop the Evaluation of Wildfire Chemicals Test Procedure [9]. These tests, however, can be costly and tedious. Also, vegetative samples are not used to provide an indication of field performance.

The mass loss cone calorimeter, a common bench-scale apparatus that uses oxygen consumption to obtain heat release rate, has also been used for fire retardancy measurements on wood products [10]. This device, apart from its high cost, has been used to measure energy release rates of ignited solid and thick materials such as building materials, plastics, wood boards, and wood products, rather than foliar (vegetative) samples.

To date, there is little public information on the relative effectiveness of the wildfire chemicals in relation to one another. A simple specifications sheet provided by the US Forest Service has to be supplied by the manufacturer to be eligible for listing on the Qualified Products List (QPL) [11-13]. Comparing the various types of suppression chemicals will aid in the strategic planning of combatting wildfires as well as reduce the immense costs necessary to contain them. The purpose of this study was to develop an appropriate laboratory scale test methodology for comparing the effectiveness of wildland forest fire chemicals that delay the ignition time of vegetative fuels.

EXPERIMENTAL METHODS

Apparatus Assembly

A test skid was fabricated to accommodate the test equipment and instrumentation as shown in Fig. 1.

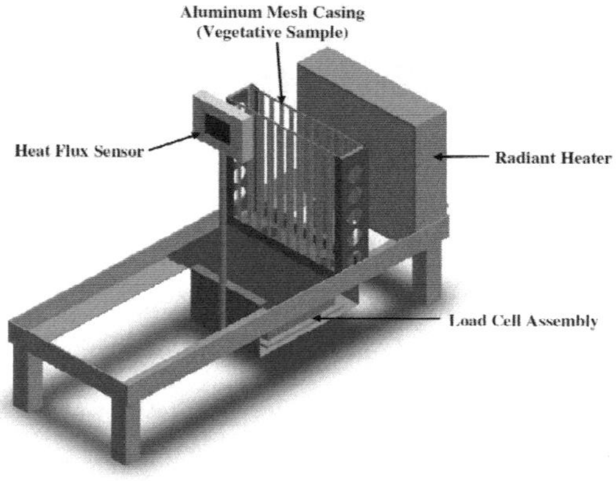

Figure 1: Radiant Heater Test Skid Assembly.

A single panel, 0.31 m x 0.31 m (12 in x 12 in) radiant heater (QH-121260, Omega Corp., Laval, QC, Canada) was installed. It was rated at 240 VAC, 1 phase, 1660 – 8640 watts, which allowed for incident heat fluxes of up to 90 kW/m^2, with a maximum su C. The heater was connected to a silicone-controlled rectifier (SCR) power controller (SCR39P- 24-040, Omega Corp., Laval, QC, Canada), a 1/32 DIN Temperature Controller (CN7533, Omega Corp., Laval, QC, Canada), and an AC Current Indicator (DP450-HACC, Omega Corp., Laval, QC, Canada), all of which were securely mounted inside a NEMA (National Electrical Manufacturers Association) 13 enclosure.

The radiant heater panel was securely mounted on to the skid in front of the vegetative fuel sample in an aluminum mesh cage (see Fig. 1). The vegetative fuel sample was mounted on to a load cell, rated for a maximum mass load of 2 kg (4.4 lbs.). The load cell assembly was capable of moving along the skid to vary the exposure energy as a function of the distance from the radiant heater (the stand-off distance). A custom-made robust heat flux sensor, was positioned directly behind the fuel sample. Details on the fabrication and testing of the heat flux sensor were provided by Sullivan and McDonald [14].

For each test, the results were collected using a stand-alone data acquisition and analysis system (DaqPROTM 5300, Fourier Systems Inc., Mokena, IL, USA). The data acquisition system used a 16-bit system with 8 channels for data collection. The temperature measurements were collected at a sampling rate of 1 Hz for the thermocouples from the heat flux sensor, as well as voltage measurements from the load cell.

Wildfire Chemicals

The treatments that were used in this study were water and foam. The foam chemical was Phos-Chek® WD 881 Class A Foam Concentrate (ICL Performance Products LP, St. Louis, Missouri, USA). The Phos-Chek WD 881 was mechanically blended with water at a concentration of 0.3% by volume to prepare a 0.5 liter (0.132 gallons) solution. The concentration of foam chosen was within the range of the mix ratio qualified by the US Forest Service Specifications (0.1–1%) [12].

Test Methodology

Lodge-Pole Pine was the vegetative fuel sample that was used to evaluate the performance of the specified chemical treatments. The pine samples were allowed to dry inside the laboratory over a 2-month period. The moisture content was measured, in accordance with the ASTM D4442 oven dry standard method, before the burn tests were conducted [15]. This was necessary to ensure that a "dry" sample was used (less than 5% moisture content) to reduce variability in the test results and provide an attainable and repeatable condition of the vegetative fuel material. The vegetative samples, weighing 75 g, were prepared from the pine branches and then sandwiched inside the 0.31 m x 0.31 m x 0.025 m (12 in x 12 in x 1 in) aluminum mesh casing. The treatments (water and foam) were applied to the vegetative fuel samples by dipping the entire aluminum mesh casing into a container containing the suppression material. The mesh casing containing the vegetative fuel samples was then mounted on the load cell, which was positioned at a stand-off distance of 0.102 m (4 in) from the radiant heater. Before each test, the samples were shielded from the heater by using an aluminum sheet C. Then, the shield was removed to expose the sample to the thermal load. This shielding process ensured that for each burn test that was performed, the vegetative fuel samples were consistently exposed to the same initial heat flux of 5 kW/m^2 at the 0.102 m (4 in) stand-off distance.

The time to glowing ignition was determined by visual observation and with a stopwatch. An exhaust fan in the test hood was powered on after the sample started to glow to reduce the accumulation of heavy smoke inside the space. The transient mass loss data from the load cell was used to determine the time to flaming ignition and the heat flux data was used for verification of that ignition time. A picture of the test during combustion is shown in Fig. 2.

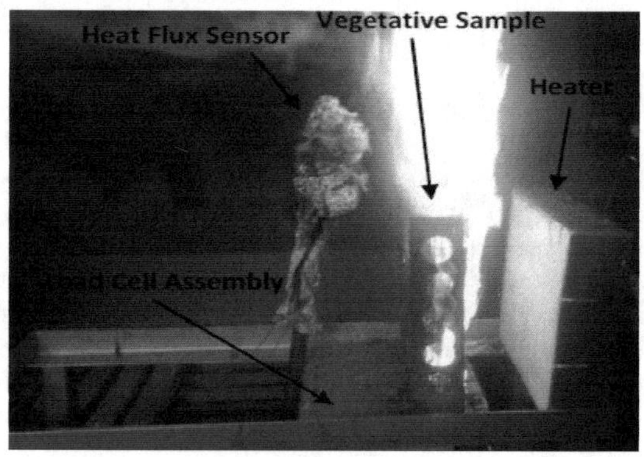

Figure 2: Burn Test during Combustion (Flaming of Vegetative Sample after Ignition).

RESULTS AND ANALYSIS

The mean ignition times, based on glowing ignition and flaming ignition, for each of the treatments that were explored are shown in Table 1, along with the standard deviation and number of tests (n) that were conducted. The glowing ignition time data that are presented in the table were gathered manually with the use of a stopwatch and visual observation. The data for the time to flaming ignition were determined from the mass loss and heat flux measurements.

Treatment	Ignition Time Glowing (sec.)	Ignition Time Flaming (sec.)
Untreated	162 ± 4 (n = 14)	184 ± 6 (n = 14)
Water Treated	182 ± 13 (n = 18)	209 ± 9 (n = 18)
Foam Treated	312 ± 21 (n = 11)	342 ± 44 (n = 11)

Table 1: Results from the Burn Tests

The results presented in Table 1 show the variations in the ignition time (both glowing and flaming ignition) for the different treatments, with the dry, untreated fuel serving as a control. There was a slight delay in the transition of the ignition time from glowing ignition to flaming ignition as shown in the table. As the vegetative fuel is heated by the radiant heater, off-gassing and charring is initiated. Glowing ignition is the solid phase combustion of the charred products, which is then immediately followed by flaming ignition (combustion of volatile gases). The transient mass loss and heat flux data were not able to identify the time to glowing ignition.

From the results, the most effective treatment that delayed ignition and extended the ignition time was the foam-based chemical. As expected, the untreated sample had the shortest ignition time. The foam-based chemical treatment produced the longest time to ignition because it lowers the surface tension of water, which aids in the wetting and saturation of the vegetative fuel samples with water.

Transient Mass Loss Data

Transient mass loss data was used to verify the time to ignition that initiated flaming of the fuel. A graph of the transient mass of an untreated spruce sample is shown in Fig. 3. The time at which flaming was initiated (184 seconds) is indicated by a sharp, nearly singular, decrease in the mass of the fuel as it is consumed during the burning process. Given the level of noise in the data, the glowing ignition time cannot be confidently identified using this method.

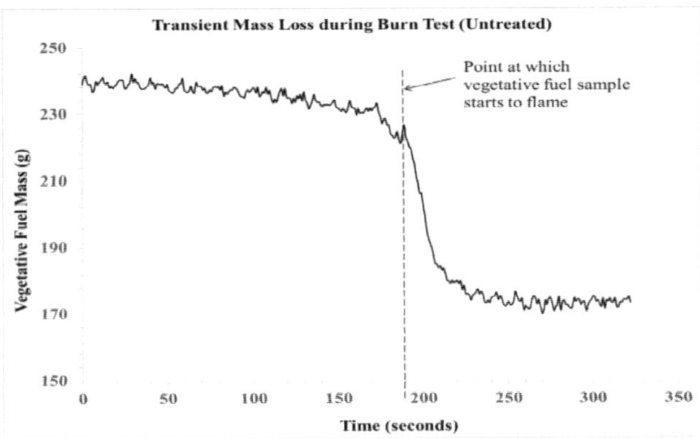

Figure 3: Mean Ignition Time of an Untreated Sample from Transient Mass Data

Heat Flux Data

The heat flux sensor used was developed by Sullivan and McDonald [14]. The incident heat flux on the surface of the sensor was determined by using Eqn. 1 [14]:

$$q''(t) = \frac{1}{\varepsilon}\left(\frac{k[T(d,t) - T_1(t)]}{(L-d)} + \sigma\varepsilon\left[T(d,t)^4 + T_1(t)^4 - 2T_\infty^4\right]\right), \quad (1)$$

where t is time, k is thermal conductivity, ε is emissivity, σ is the Stefan-Boltzmann constant, L is the thickness of the sensor, d is the depth into the sensor where a thermocouple is located, T_∞ is the surrounding ambient temperature, $T_1(t)$ is the backside temperature of the sensor, and $T(d,t)$ is the temperature at location d in the sensor.

Figure 4 shows a curve of the incident heat flux on the sensor that originated from the heated, untreated vegetative fuel. The curve shows that after approximately 182 seconds of heating, the heat flux originating from the fuel increases suddenly, which is indicative of ignition that caused flaming. As the sample burned, the heat flux increased to a maximum value, and decreased as the material was consumed by the fire. Similar interpretations of signals have been presented, where a relation between the patterns of voltage signals indicated the occurrence of a physical phenomenon or a change in a measured parameter [16].

Comparison of the ignition times obtained from the transient mass loss and heat flux data are presented in Table 2. The results show that the variation in the ignition times required for flaming of the fuel, as determined from data from the two methods, was less than 1%. The results from the two methods were also in accord with respect to the time at which the flame was extinguished and the burnt fuel began to cool. From Figs. 3 and 4 for the untreated sample, that time appears to be approximately 225 seconds after initial heating of the fuel. For the transient mass, the mass did not decrease further beyond 225 seconds and for the transient heat flux, the heat flux began to decrease from a maximum at approximately the same time. Based on all the samples that were studied, there was some variability in the maximum heat flux released due to non-uniformity and inhomogeneity of the composition of the vegetative fuel sample.

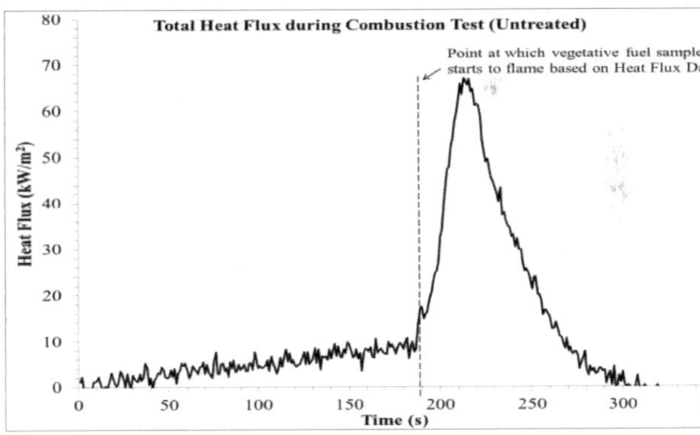

Figure 4: Mean Ignition Time of an Untreated Sample from Transient Heat Flux Data

Treatment	Flaming Ignition Time (sec) - Mass Loss	Flaming Ignition Time (sec) – Heat Flux
Untreated	184	182
Water	209	210
Foam	342	341

Table 2: Comparison of Flaming Ignition Time from the Mass Loss and Heat Flux Data

Statistical Analysis

A detailed statistical analysis, using a two sided t-test at a significance level of $\alpha = 0.05$ (conventional standard) [17], was conducted on the mean time to flaming ignition to determine if the results from the burn test were statistically significant for the different treatments. The results from the t-tests are shown in Tables 3 for the different treatments.

Treatment	Mean Ignition Time (s)	*t*-Value Calculated	*t*-Value Distribution Tables [17]
Untreated	184 ± 6 (n = 14)	9.1	2.8
Water	209 ± 9 (n = 18)		
Untreated	184 ± 6 (n = 14)	13.0	2.8
Foam	342 ± 44 (n = 11)		
Water	209 ± 9 (n = 18)	11.9	2.8
Foam	342 ± 44 (n = 11)		

Table 3: *t*-test Result of Flaming Ignition Times for Untreated & Water Treated, Untreated & Foam Treated, and Water & Foam Treated.

Based on the results from Tables 3, since the calculated *t*-values in each case is greater than the *t*-value from the distribution tables, it can be concluded that the mean ignition times are statistically different for the different treatments. This result suggests that the test methodology developed is effective in comparing different wildland forest fire chemicals.

CONCLUSIONS

The primary objective of this study was to develop a simple and effective laboratory-scale test methodology for comparing the effectiveness of various wildland forest fire chemicals. The main emphasis of the methodology was to determine the time to glowing and flaming ignition of the vegetative samples to compare the different treatments. It was possible to determine the time to flaming ignition from the transient mass loss and heat flux data however this was not the case for time to glowing ignition which was determined only visually. Comparing the results obtained from the transient mass loss and heat flux data, the time to flaming ignition is within < 1%, which validates the ignition time obtained. There was a slight delay in the ignition time transitioning from glowing ignition to flaming ignition due to the physical and chemical burning process of vegetative fuels. Statistical analysis using a two sided *t*-test, indicated that the results from the burn test were statistically significant for the different chemical treatments. This indicated that the test methodology allowed for effective differentiation between the wildfire chemical treatments by comparing their mean ignition times. The narrow standard deviations of the average ignition times suggested that the test methodology was able to produce repeatable results. In addition to the ignition time, the methodology also provided useful quantitative data on transient mass loss and heat release all in one test, which can be incorporated into various fire models. The test methodology may be used to complement the US Forest Service Wildland Fire Chemicals Standard Test Procedures in which no vegetative fuel samples are used to provide an indication of field performance of wildland fire chemicals.

REFERENCES

[1] Kalabokidis, K.D., Effects of Wildfire Suppression Chemicals on People and the Environment – A Review, Global Nest: The Int. J. Vol. 2, 2000, pp 129-137.

[2] Liodakis, S., Antonopoulos, I., T. Kakardakis, Evaluating the use of minerals as forest fire retardants, Fire Safety Journal 45 (2010) 98–105.

[3] Liodakis, S., Bakirtzis, D., Dimitrakopoulos, A.P., Auto ignition and thermogravimetric analysis of forest species treated with fire retardants, Thermochimica Acta 399 (2003) 31 – 42.

[4] Lyons, J.W., The Chemistry and Uses of Fire Retardants, RE Krieger Publishing Company, Malabar, FL, 1987.

[5] Pappa, A.A., Tzamtzis, N.E., Statheropoulos, M.K., Liodakis, S.E., Parissakis, G.K., A comparative study of the effects of fire retardants on the pyrolysis of cellulose and Pinus halepensis pine-needles, Journal of Analytical and Applied Pyrolysis 31 (1995) 85 – 100.

[6] Pouwels, A.D., Eijkel, G.B., Boon, J.J., Curie-point pyrolysis–capillary gas chromatography high resolution mass spectrometry of micro crystalline cellulose, Journal of Analytical and Applied Pyrolysis 14 (1989) 237 – 280.

[7] Wang, Q., Li, J., Winandy, E., Chemical mechanism of fire retardance of boric acid on wood, Wood Science and Technology 38 (2004) 375 – 389.

[8] ASTM E1321-13, Standard Test Method for Determining Material Ignition and Flame Spread Properties, ASTM International, West Conshohocken, PA, USA, 2013.

[9] United States Department of Agriculture (USDA) Forest Service, Wildland Fire Chemicals Standard Test Procedure, Section 2: Fire Tests, STP 2.1–2.2, available from: http://www.fs.fed.us/rm/fire/wfcs/tests/index.htm.

[10] Grexa, O., Lubke, H., Flammability parameters of wood tested on a cone calorimeter, Polymer Degradation and Stability 74 (2001) 427 – 432.

[11] Specification for Long Term Retardant, Wildland Firefighting, United States Department of Agriculture Forest Service: Specification 5100-340c, 2010.

[12] Specification for Fire Suppressant Foam (Class A. Foam) for Wildland Firefighting, United States Department of Agriculture Forest Service: Specification 5100-307a, 2010.

[13] Specification for Water Enhancers (Gels) for Wildland Firefighting, United States Department of Agriculture Forest Service: Specification 5100-306a, 2010.

[14] Sullivan, E., McDonald, A., "M Md d S Development for Measuring Energy Transfer from Wildland F ", I . J. W d d F , 3, 14, 5 – 1004.

[15] ASTM D4442-07, Standard Test Methods for Direct Moisture Content Measurement of Wood and Wood-Base Materials, ASTM International, West Conshohocken, PA, 2007.

[16] McDonald, A., Moreau, C., Chandra, S., Use of thermal emission signals to characterize the impact of fully and partially molten plasma-sprayed zirconia particles on glass surfaces, Surface and Coating Technology, 204, 2010, 2323 – 2330.

[17] D. C. Montgomery, Design and Analysis of Experiments, 8th Edition, John Wiley and Sons, Inc., Arizona, 2013, pp. 36-38, 686.

MODIFIED BIODIESEL CHEMICAL KINETICS BASED ON NUMERICAL SIMULATION IN AN IGNITION QUALITY TESTER

Mahmoud Elhalwagy
Mechanical & Materials Engineering
Western University
London, Ontario, Canada
melhalwa@uwo.ca

Chao Zhang
Mechanical & Materials Engineering
Western University
London, Ontario, Canada
czhang3@uwo.ca

ABSTRACT

Numerical methods are becoming a key factor in the development of combustion chemical kinetic mechanisms. However, limited applicability of CFD for detailed chemistry invoked reliance on volume independent simulations for mechanism validations. In the present study, a five-component biodiesel mechanism is constructed by adding five irreversible reaction steps to a surrogate mechanism. Tuning of the Arrhenius kinetics constants is achieved by CFD simulations of the combustion process in an ignition quality tester (IQT) to predict accurate values of Cetane numbers (CN). Physical properties relevant to the spray, atomization and vaporization processes are predicted for the five biodiesel components. The validation is conducted through the comparison to a recent engine experiment. CFD simulations are carried out using the CFD software, FORTE, with dynamic cell clustering for detailed chemistry simulations.

KEYWORDS: Biodiesel, Chemical Kinetics, Combustion, CFD, Ignition Quality Tester.

INTRODUCTION

Biodiesel fuels are methyl esters of fatty acids produced from vegetable oils and animal fats [1]. It is fairly accurate to model them as two saturated (single bonded) components namely, Methyl Stearate (MSTEA) $C_{19}H_{38}O_2$ and Methyl Palmitate (MPA) $C_{17}H_{34}O_2$, three unsaturated components known as Methyl Oleate (MEOLE) (1 double bond) $C_{19}H_{36}O_2$, Methyl Linoleate (MLINO) (2 double bonds) $C_{19}H_{34}O_2$ and Methyl Linolenate (3 double bonds) (MLIN1) $C_{19}H_{32}O_2$ [2]. Representation of chemical kinetics used in biodiesel combustion started with one step global reaction models in the same way implemented in KIVA codes [3]. Due to the reported inaccuracy of this technique, semi-detailed representation of biodiesel combustion using smaller molecules as surrogates like Methyl Butanoate (MB) $C_5H_{10}O_2$ [4] and Methyl Decanoate $C_{11}H_{22}O_2$ [5] was employed to increase the accuracy of the ignition and combustion simulations. Detailed mechanisms for the two surrogates have been combined and reduced to account for the biodiesel decomposition [6]–[8] giving better results however, proportions of the blend surrogates in the fuel have been determined empirically based only on experience and final validation. Additional drawbacks are the lack of proper presentation of biodiesel components' thermo-physical properties with higher focus on the chemistry and the absence of real biodiesel chemical constituents in the mechanisms with dependence only on the surrogates.

In the present work, five irreversible steps are added to the mechanism used in the work of Ismail et al.[7] , in which MB and MD mechanisms were combined. Basis of decomposition is the same as that of Ismail et al [7] and Yang et al [6] depending on the MB2D (i.e. unsaturated $C_5H_7O_2$ in MB mechanism) pathway to account for the unsaturated biodiesel components combustion. These steps decomposed the original five biodiesel constituents to the surrogates instead of specifying an empirically determined blend for the fuel. In the present study, the fuel's gravimetric decompositions are used directly in the simulations. The five proposed steps are:

MPA + O_2 => MD+MB+CH_2
MSTEA + O_2 => MPA+2CH_2
MEOLE + O_2 => MD+MB2D+3CH_2
MLINO + O_2 => MD+MB2D+C_3H_4
MLIN1 + O_2 => MD+MB2D+C_3H_2

Thermochemical data for the main five biodiesel constituents are taken from Saggese et al. [9]. The final reaction mechanism contains 115 species and 387 reactions.

Ignition quality tester ™ [10] is a device used to calculate the Cetane number of diesel-like fuels by evaluation of the autoignition delay in a constant volume spray chamber with standard dimensions. To obtain the Arrhenius coefficients for the irreversible steps, numerical simulations of the biodiesel combustion process is carried out in this standard chamber with real fuel like conditions as specified in the D6890- 13a ASTM report [10]. Relation between the auto-ignition delay (I.D.) defined by delay of the sudden pressure rise in the chamber and the derived CN is:

$$DCN = \frac{186.6}{I.D.(\text{m.sec.})} + 4.46, 3.3m.\text{sec} < I.D. < 6.4m.\text{sec}.$$

$$DCN = \frac{83.99}{(\text{I.D.} - 1.512)^{0.658}} + 3.547, \text{otherwise}$$

Cetane values for the five biodiesel components are taken from Westbrook et al [2]. Arrhenius constants are determined by matching the results of the predicted pressure rise point to the anticipated pressure rise point based on Cetane values.

THERMOPHYSICAL PROPERTIES

To perform CFD simulations, accurate values for the

variations of liquid density, liquid specific heat, vapor pressure, latent heat of vaporization, thermal conductivity, viscosity and surface tension with temperature are needed. Data compilations for biodiesel properties are rare. Only Methyl Oleate was reported [11]. In the present work, thermos-physical data are needed for all five biodiesel constituents for the engine applicable temperature range. Liquid density is based on recent experimental data [12] and the modified Rackette equation [13] outside of the experimental data range. Liquid specific heat is based on the modified Rowlinson method [14]. The rest of the physical properties required in addition to the critical temperatures for the five biodiesel components are obtained based on the methods recommended by An et al [15]. Implementation of these correlations to the numerical sub-models requires the results of physical properties predicted to be curve fitted to the function forms of the data compilation [11]. These functions are included in the fuel library of the FORTE CFD software [16]. Figure 1 shows the physical properties for the five Biodiesel components.

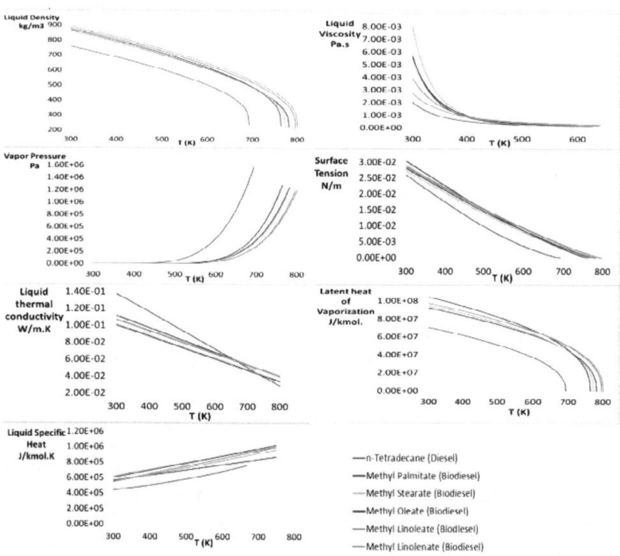

Figure 1 Biodiesel thermo-physical properties variation with temperature.

COMPUTATIONAL MODELS

The CFD simulation of the combustion process in the IQT is performed. The IQT is a 0.21 L chamber with a length of 17.65 cm and a diameter of 5.08 cm as reported by Bogin et al [17]. It is initially filled with air until fuel is sprayed into the chamber. Initial conditions are 861 K and 21 bar [10] and the nozzle diameter is taken as 700 microns [18]. Figure 2 depicts the IQT mesh created by the automatic mesh generation of FORTE software using the immersed boundary approach [19]. A global mesh size of 0.35, 0.2 and 0.15 cm is used. Mesh independence test is carried out showing that medium mesh size of 61728 cells can give the grid independent results. The mesh independence test and needle injection pattern are depicted in Figure 3.

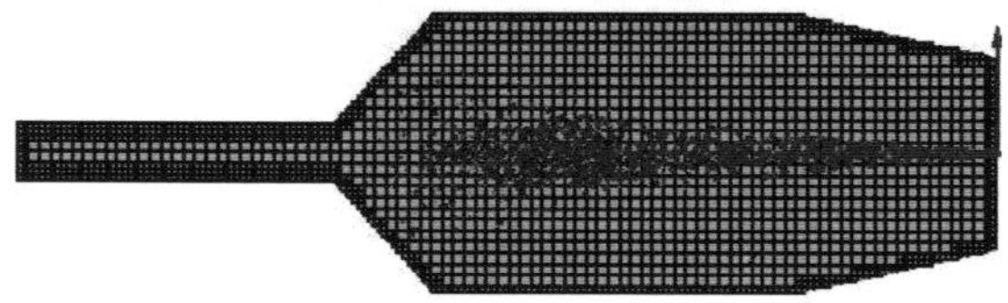

Figure 2 IQT mesh.

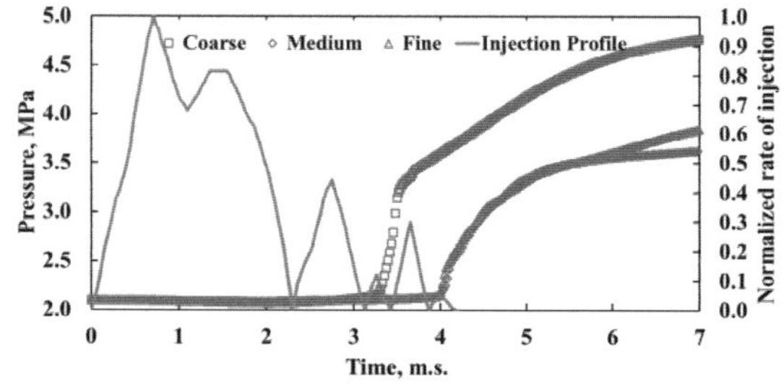

Figure 3 Mesh independence test and fuel injection rate.

Table 1 summary of CFD sub-models [16].

CFD Technique	Arbitrary Lagrangian-Eulerian ALE
Turbulence Model	RNG k-ε Model
Spray Tracking	Stochastic Lagrangian Tracking/ Monte Carlo Method
Spray Injection	Solid Cone Injection
Spray Break up	Hybrid KH-RT Model and Gas jet Model
Spray Collision	Radius of Influence
Combustion Model	Detailed chemistry with Dynamic cell clustering
Turbulence/Chemistry Interaction	Mixing Time scale Model
Heat Transfer Wall treatment	Improved law of the wall

In order to reduce the computational time, detailed chemistry with dynamic cell clustering (DCC) is utilized [20]. The threshold of the DCC is 10 K for temperature and 0.02 for equivalence ratio. Variable time step size is utilized with a maximum of 0.01 m.s. at very stable simulation time before spray and further into post combustion. Table 1 summarizes the different sub-models for the present simulations. Fuel mass injected is determined based on the same volume flow rate as in Bogin et al. [17] i.e. (mass injected of biodiesel/(78 mg/inj.) for diesel)) is proportional to the ratio of liquid densities at 861 K.

DETERMINATION OF ARRHENIUS CONSTANTS

The five biodiesel constituents are simulated in the IQT with tuning of the pre-exponential factor of their irreversible decomposition reactions. It is found that the rate of decomposition of the MPA and MSTEA saturated components depended on the high temperature combustion (specifically R1 and R2) instead of the MB2D pathway as concluded by Ismail et al [7]. In order to control the point of pressure rise to a realistically acceptable DCN, the constants tuning of R1 and R2 in table 2 is carried out. The rest of the unsaturated components namely MEOLE, MLINO and MLIN1 had their decomposition solely controlled by *A* constants of their irreversible decomposition steps since they highly depend on the pre-calibrated MB2D reaction pathway. Tables 2 and 3 represent the tuned Arrhenius constants and the corresponding DCN values based on CFD results, respectively. Ignition delays are estimated based on the time corresponding to the start of sharp pressure gradient. Figure 4 illustrates the CFD based ignition characteristic of different biodiesel constituents.

Table 2 Arrhenius Constants for modified reactions, Units are Mole,sec.,cm, Cal. and K

Reaction	*A*	b	E_a
MPA + O_2 => MD+MB+CH_2	5e+11	0	10500.0
MSTEA + O_2 => MPA+2CH_2	5e+11	0	0
MEOLE+O_2=> MD+MB2D+3CH_2	6.2e+9	0	10500.0
MLINO+O_2=>MD++MB2D+C_3H_4	1.7e+10	0	10500.0
MLIN1 + O_2 => MD++MB2D+C_3H_2	3.2e+9	0	10500.0
R1: C_7H_{16}+HO_2= $C_7H_{15\text{-}2}$+H_2O_2	8.5e+14	0	16950.0
R2: O+OH=O_2+H	5.5E+15	-0.5	0

Table 3 Cetane numbers and Ignition delays of Biodiesels.

Component	CN(Ref.2)	DCN (CFD)	I.D.,m.s.
MPA	86	**91.77**	2.44
MSTEA	101	**116.4**	2.15
MEOLE	59	**58.08**	3.48
MLINO	38	**37.3**	5.68
MLIN1	23	**22.95**	10.78

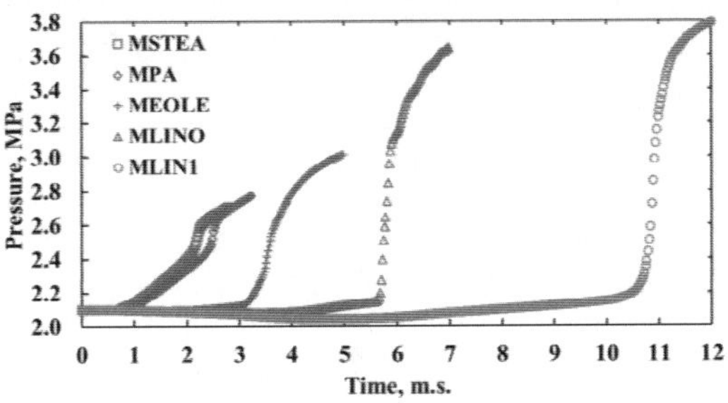

Figure 4 Pressure rise of different Biodiesel Constituents

The inverse correlation between the cetane number and the degree of unsaturation [21] in Biodiesel is well established from Figure 4 with higher ignition delay for highly unsaturated biodiesels. Figure 5 shows contours of the temperature variations for MPA (major saturated species), MEOLE (unsaturated) and MLIN1 (highly unsaturated) as representations of different behaviors of biodiesel constituents based on the number of double bonds in the parent fatty acid. The typical diesel cone behavior is noticed in MPA contours with combustion development depending on the spray characteristics and diffusion combustion propagation at the spray/air interface. The similarity is attributed to similar physical properties and the same chemical behavior dictated by single bonded straight chains as in diesel structure. In MEOLE and MLIN1, spray vaporization happen faster due to longer chemical delays. For MEOLE, the combustion starts at the end of the spray penetration length with similar to the premixed combustion behavior after that. MLIN1 is almost fully premixed combustion independent of spray. This behavior is dictated by the presence of double bonds in their fatty acids.

Experimental Validation

Experimental data of rape seed biodiesel RME combustion in a Volvo D12C heavy duty single cylinder direct injection diesel engine has been used [22] at 1000 RPM, 25% load, 25% EGR and two cases of injection starts - 3.3 and 2.3 CA ATDC to validate the modified mechanism. The injection duration and mass are 9.2° and 70 mg, respectively. Sizing and specifications of the engine can be found in [22]. RME is modelled as 0.599 MEOLE, 0.043 MPA, 0.014 MSTEA, 0.132 MLIN1 and 0.212 MLINO [2].

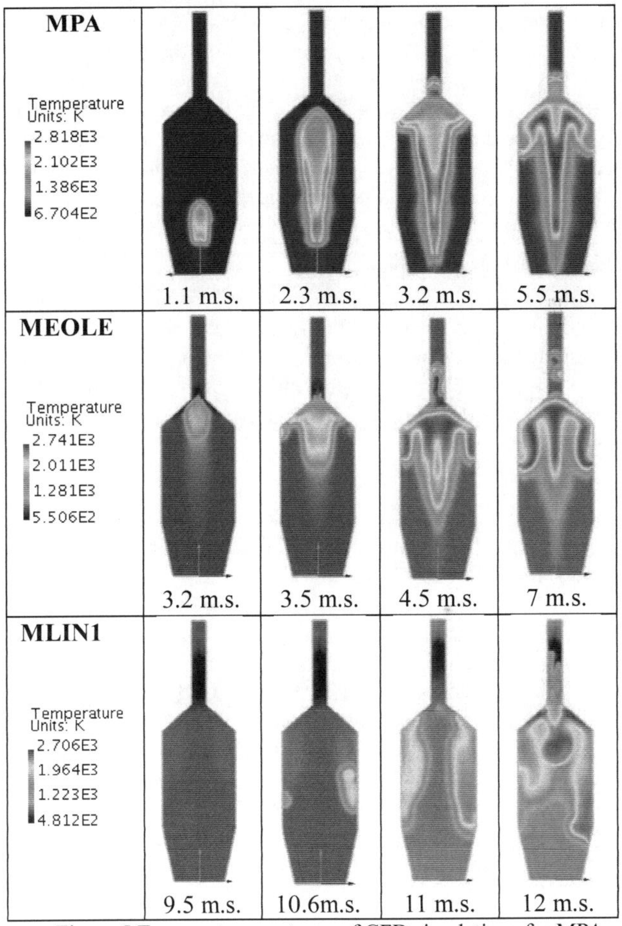

Figure 5 Temperature contours of CFD simulations for MPA, MEOLE and MLIN1 in the IQT.

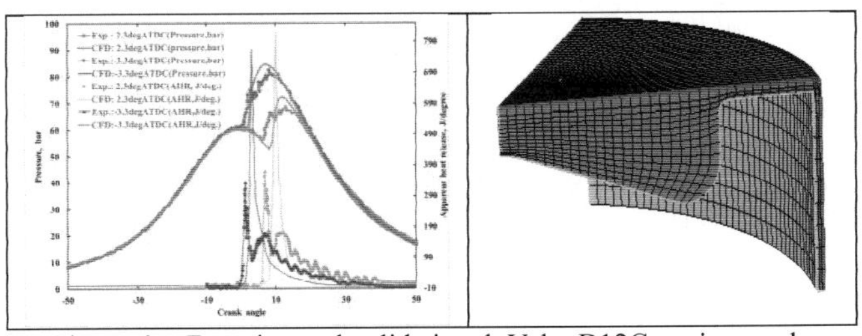

Figure 6 a. Experimental validation, b.VolvoD12C engine mesh.

FORTE sector mesh generator is utilized for a 72° sector (i.e. a five nozzles engine). Crevice volume is added to account for the engine blow by. A node count is of 16,200 at TDC. Figure 6 depicts the pressure-crank angle diagram for the current simulations. The present modification provided good agreement with the experimental data in terms of pressure and apparent heat release in the cylinder.

CONCLUSIONS

Biodiesel five-component mechanism has been proposed in the present paper. The mechanism is constituted by including 5 irreversible steps to existing reduced chemistry with CFD simulations in the IQT^{TM} as a guide for generating chemical kinetics constants of the proposed reactions. FORTE software has been used for CFD simulations to study the ignition and cetane number properties and match the behavior to experimental values. It was found that the longer chemical delay is correlated to the level of unsaturation of biodiesel constituents. The present mechanism also showed good accuracy in pressure and heat release predictions when compared to data from a heavy duty diesel engine.

ACKNOWLEDGMENTS

The authors would like to thank Prof. Valeri Golovitchev in the Chalmers institute of technology, Gotenberg, Sweden for providing engine data necessary for simulations. We also thank SHARCNET for providing high performance computing, NSERC for the financial support and Reaction Design for the software.

REFERENCES

[1] J. H. Van Gerpen, C. L. Peterson, and C. E. Goering, "Biodiesel : An Alternative Fuel for Compression Ignition Engines," no. 31, pp. 1–22, 2007.

[2] C. K. Westbrook, C. V. Naik, O. Herbinet, W. J. Pitz, M. Mehl, S. M. Sarathy, and H. J. Curran, "Detailed chemical kinetic reaction mechanisms for soy and rapeseed biodiesel fuels," *Combust. Flame*, vol. 158, no. 4, pp. 742–755, 2011.

[3] A. A. Amsden, P. J. O'Rourke and T. D. Butler, "KIVA-II: A Computer Program for Chemically Reactive Flows with Sprays". Report: LA-11560-MS. *Los Alamos National Laboratory*, Los Alamos, New Mexico, USA, 1989.

[4] S. Gaïl, S. M. Sarathy, M. J. Thomson, P. Diévart, and P. Dagaut, "Experimental and chemical kinetic modeling study of small methyl esters oxidation: Methyl (E)-2-butenoate and methyl butanoate," *Combust. Flame*, vol. 155, no. 4, pp. 635–650, Dec. 2008.

[5] K. Seshadri, T. Lu, O. Herbinet, S. Humer, U. Niemann, W. J. Pitz, R. Seiser, and C. K. Law, "Experimental and kinetic modeling study of extinction and ignition of methyl decanoate in laminar non-premixed flows," *Proc. Combust. Inst.*, vol. 32, no. 1, pp. 1067–1074, Jan. 2009.

[6] J. Yang, M. Johansson, and V. Golovitchev, "Engine Performance and Emissions Formation for RME and Conventional Diesel Oil: A Comparative Study," *ASME 2009 Intern. Combust. Engine Div. Spring Tech. Conf.*, pp. 125–134, 2009.

[7] H. Mohamed Ismail, H. K. Ng, S. Gan, T. Lucchini, and A. Onorati, "Development of a reduced biodiesel combustion kinetics mechanism for CFD modelling of a light-duty diesel engine," *Fuel*, vol. 106, pp. 388–400, Apr. 2013.

[8] H. An, W. M. Yang, a. Maghbouli, J. Li, and K. J. Chua, "A skeletal mechanism for biodiesel blend surrogates combustion," *Energy Convers. Manag.*, vol. 81, 2014.

[9] C. Saggese, A. Frassoldati, A. Cuoci, T. Faravelli, and E. Ranzi, "A lumped approach to the kinetic modeling of pyrolysis and combustion of biodiesel fuels," *Proc. Combust. Inst.*, vol. 34, no. 1, pp. 427–434, Jan. 2013.

[10] "Standard Test Method for Determination of Ignition Delay and Derived Cetane Number (DCN) of Diesel Fuel Oils by Combustion in a Constant volume chamber," *ASTM international* D6890-13b , 2013.

[11]T.E. Daubert, R.P. Danner, "Physical and thermodynamic properties of pure chemicals: data compilation". Washington, DC: *Taylor& Francis*; 1989.

[12] M. J. Pratas, S. Freitas, M. B. Oliveira, S. C. Monteiro, A. S. Lima, and J. a. P. Coutinho, "Densities and Viscosities of Fatty Acid Methyl and Ethyl Esters," *J. Chem. Eng. Data*, vol. 55, no. 9, pp. 3983–3990, Sep. 2010.

[13] E. Blangino, A. F. Riverós, and S. D. Romano, "Physics and Chemistry of Liquids : An Numerical expressions for viscosity , surface tension and density of biodiesel : analysis and experimental validation," no. November 2013, pp.37–41.

[14] A. Bondi,_"Estimation of the Heat Capacity of Liquids," *Industrial and Engineering Chemistry Fundamentals* **5**(4), 442-449, 1966.

[15] H. An, W. M. Yang, a. Maghbouli, S. K. Chou, and K. J. Chua, "Detailed physical properties prediction of pure methyl esters for biodiesel combustion modeling," *Appl. Energy*, vol. 102, pp. 647–656, Feb. 2013.

[16] Reaction Design Inc. , "FORTE, FOR-UG-40132-1307-1," ReactionDesign Inc., San Diego, CA, 2013..

[17] G. E. Bogin, A. DeFilippo, J. Y. Chen, G. Chin, J. Luecke, M. a. Ratcliff, B. T. Zigler, and A. M. Dean, "Numerical and experimental investigation of n-heptane autoignition in the ignition quality tester (IQT)," *Energy and Fuels*, vol. 25, pp. 5562–5572, 2011.

[18] C. V. Naik, K. Puduppakkam, E. Meeks, and L. Liang, "Ignition Quality Tester Guided Improvements to Reaction Mechanisms for n -Alkanes: n -Heptane to n -Hexadecane," *SAE Technical Paper* .Apr. 2012.

[19] L. Liang, A. Shelburn, C. Wang, D. Hodgson, and E. Meeks, "A New Automatic and Dynamic Mesh Generation Technique Based On Immersed Boundary Method," 2011.

[20] L. Liang, J. G. Stevens, and J. T. Farrell, "A Dynamic Multi-Zone Partitioning Scheme for Solving Detailed Chemical Kinetics in Reactive Flow Computations," *Combust. Sci. Technol.*, vol. 181, no. March 2015, pp. 1345–1371, 2009.

[21] G. Knothe, "Dependence of biodiesel fuel properties on the structure of fatty acid alkyl esters," *Fuel Process. Technol.*, vol. 86, no. 10, pp. 1059–1070, Jun. 2005.

[22] J. Yang, PhD thesis, Biodiesel Spray Combustion Modeling Based on a Detailed Chemistry Approach., Chalmers institute, Sweden, 2012.

BUOYANT INSTABILITIES AND FLAME SPREAD IN WILDLAND FIRES: IMPLICATION OF THE NEED FOR SCALING INSTABILITY ANALYSIS

M.A. Finney, J.D. Cohen, J.A. Forthofer, S.S. McAllister
US Forest Service, Missoula Fire Science Laboratory, Missoula MT 59808, USA
e-mail: mfinney@fs.fed.us

N.K. Akafuah, K. Saito
Dept. of Mechanical Engineering, University of Kentucky, Lexington, KY 40506, USA
e-mail; ksaito@uky.edu
M.J. Gollner, D.J. Gorham
Dept. Fire Protection Engineering, University of Maryland, MD 20742, USA
e-mail: mgollner@umd.edu

ABSTRACT

This paper reports scaling correlations on the wind tunnel laboratory scale flame spread experimental data using artificially made cardboard fuel beds, outdoor large scale wood crib experiments and observations made on the actual forest fires. Fine thermocouples were used to measure transient behavior of flame temperature during flame spread, a high speed camera captured dynamic behavior of spreading flames, and infrared thermography imaging technique to transient thermal images of the condensed fuel beds. As a result, we found all these data were correlated based on Strouhal-Froude numbers, where the average flame height, horizontal wind speed, and frequency of flame fluctuation were used.

Forest fire, flame spread, scaling, convective heat transfer

INTRODUCTION

Despite decades of research on wildland fires, a common understanding has not emerged as to how heat transfer processes are organized to produce fuel particle ignition and flame spread. This has left physical modelling of wildland fires without an experimental basis for the many assumptions, including radiation and convective heat transfer. Strong experimental evidence now suggest a secondary role for radiation in flame spread in fine fuel beds because 1) the optical attenuation through the fuel bed diminishes irradiance to particles, and 2) cooling by forced or natural convection efficiently prevents fine fuels from reaching ignition. Thus, we focus here on recent laboratory and field experiments designed to determine how convective heating occurs. We found clear evidence that non-steady heating by flame contact produces fuel particle ignition and considering scaling analysis to reveal the basic governing mechanisms on flame spread.

EXPERIMENTAL METHODS

Laboratory fire spread experiments were conducted in uniform fuel beds made of laser-cut cardboard. Cardboard and paper strips have been used previously (Emmons and Shen, 1971) and offer advantages of known homogenous properties such as density and customizable physical dimensions of discrete particles (length, surface area etc.). Fuel beds were burned in the wind tunnel at the Missoula Fire Sciences Laboratory. The wind profiles of the 3m cross-section have been described previously (Rothermel and Anderson 1966, Catchpole et al. 1998) and are laminar except along the bottom surface where an upstream trip-fence produces a turbulent boundary layer. Wind speeds were varied from 0.22 m s-1 to 2.3 m s-1 with relative humidity of approximately 25%.

Fuels for the laboratory burns were engineered from cardboard with a commercial CO_2 laser system to produce fuel elements that are connected at regular spacing along a common spine. The cards or "combs" could then be arranged in rows at various spacing to form a fuel bed with vertically standing particles (Figure 1). The cardboard used was brown "chip board" 1.27mm (0.05 inch) thick with approximately 60% recycled content. Fuel particles were created at different lengths and widths and arranged at different row spacing to achieve specific fuel bed properties. The laser cutter/engraver system was a Universal Laser Systems Inc. ILS12.150D model equipped with two 60W laser cartridges. The beams from both lasers were collimated for cutting. The table accommodates sheets of cardboard 0.61m X 1.22m (2 ft X 4 ft) so that multiple combs can be cut from the same sheet in one operation

Fuel beds constructed of these cardboard combs were 1.22m to 2.45m in width and 3.05m to 6.1m in length. The combs were supported and arranged on a foundation of cement-board strips (Hardy Board) 0.635cm x 5.08cm (1/4 in. x 2 in) each separated by a steel spacer 0.158cm x 2.54cm

(1/16 in x 1.0 in). The steel spacers rested on the floor to preserve a slot at the upper surface which pinched the spine of the fuel combs such that only the vertical tines were exposed (Figure 1). Tine lengths of 2.54cm (1in), 10.1cm (4in), 20.3cm (8in), and 35.6cm (14in) were used in the burns. The longitudinal spacing of the combs could be adjusted every 1.43cm (5/16in). To limit inflow to the combustion zone along the lateral edges during burning, the sides of the beds were lined with paper that was treated with the flame retardant Diammonium Phosphate ($(NH_4)_2HPO_4$). This technique was described by Byram et al. (1964), where fire retardant limits independent flaming combustion but allows the paper to burn in conjunction with the advancing fire front. The consumption of the paper sideliners at the trailing edge of the burning zone avoids channelling of air inflow to the rear of the fire which has been shown to affect fire spread on slopes (Smith 1992). Cutouts of the sideliner permitted filming of the ignition process within the fuel bed.

Field-scale fires were assessed to determine potential scaling relationships beyond those possible within the laboratory. At Fort Swift, Texas, grass 1 m deep was burned on separate plots with average winds of 1.7 m s-1 and 4.5 m s-1 and produced flames of approximately 3m tall. A rectangular wooden crib was also constructed of square ponderosa pine lath 0.025 m thick at 0.12x0.12 m square spacing with overall dimensions of (1.2m wide, 1m tall, and 17m long) and was burned to provide stationary flame source with approximately 6 m flames.

Figure 2. Picture of laser-cut cardboard laboratory fuel bed. Note that the scale is graduated in inches.

RESULTS AND DISCUSSION

The unprecedented uniformity of the fuel beds from engineered cardboard allowed us to observe flame structure and behaviour isolated from environmental heterogeneity. The first and most obvious characteristic of flame fronts in these flame sources is the peak and trough pattern of the flame edge (Figure 2). This geometry is similar to instabilities in boundary layer flow caused by Taylor-Görtler vortices, oriented in the flow direction, that arise when air inflow to the flame zone is lifted by buoyancy (Floryan 1991). From the literature (*e.g.* Sparrow and Husar 1969), studies of flow over heated plates and of boundary layer transition, pairing of counter-rotating Taylor-Görtler vortices form an alternating series of convergence zones in the transverse direction – with downward force creating the flame trough and up-ward convergence in flame peaks (Figure 2). The significance of these vortices is that they force flame down and forward into the fuel bed and advect secondary instabilities forward through the flame zone. Data from spreading wind tunnel fires, prescribed grassland burns, and stationary crib fires all reveal that the transverse wavelength of these vortices is about ½ the flame length (Figure 2D).

Figure 1. Flame structure in laboratory-scale cardboard fuel beds(2A) reveal the same sawtooth flame structure as grass fires (2B) and stationary crib fires (2C) because of Taylor- Görtler vortices. The transverse wavelength of these flame peaks app appears to be about ½ the flame

The second dynamic discovered in these experiments is a series of flame parcels (Figure 3a) which are first seen to originate near the back edge of the flame zone. These parcels remain coherent on video and move forward through the flame zone to burst forward and impinge fuel particles for some considerable distance ahead of the combustion interface. Thermocouple temperatures measured in the fuel beds (Figure 3b) fluctuate with predictable average frequencies that showed strong Strouhal-Froude scaling (Figure 3c) similar to the pulsing frequency from pool fires (Malalakasera 2006). This scaling, using the variables of wind speed and flame length, means the bursts frequency increases with wind speed and decreases with flame size. The origin of the bursts appears to be from instabilities arising at the upstream edge of the flame zone similar to Tollmien-Schlichting waves observed in boundary layer instabilities (Floryan 1991). These are visible first as transverse ridges and concave packets in the upper flame surface (Figure 3A). With downstream movement through the flame zone they are drawn into the circulations from Görtler vorticity and expelled forward past the leading flame edge into the fresh fuels ahead. This bursting frequency was recorded as temperature spikes on thermocouple arrays embedded in the fuel bed (Figure 3B) which demonstrates the coherency of the flame structures as they advect through the flame zone. Remarkably, data from all field sources suggests consistent St-Fr (Strouhal-Froude) number scaling as laboratory-scale fires (stationary or spreading) (Figure 3C). Instability associated with St number may require further clarification. Here we define St = *fL/U,* where *f* is the frequency associated with vortex shedding, *L* is the characteristic length (the average visible flame height) and *U* is the horizontal wind velocity. The frequency of vortex shading may have two different sources: wind share-driven and buoyancy force-driven, but we are not able to specifically identify which source and wait for future study.

flame behaviours in both studies suggests broad applicability across a wide range of wildland fire sizes and fuel (Adam et al., 2014). The strong dependency of intermittent convective heating to thermal behaviour and ignition of individual fuel elements to radiation and convection is revealed in two other studies (Cohen et al, 2014; English et al, 2014).

Quasi-periodic flame behaviour in spreading fires has been noticed previously for trench fires for pine needle beds, but the role of such non-steady flame impingement in igniting fuel particles and flame spread has not been considered (Atkinson et al. 1995, Woodburn and Drysdale 1998, and Dupuy et al. 2011). The Strouhal-Froude scaling suggested by our data is consistent with buoyant dynamics of stationary fire phenomena (Malalasekera et al. 1996) but new to spreading fires.

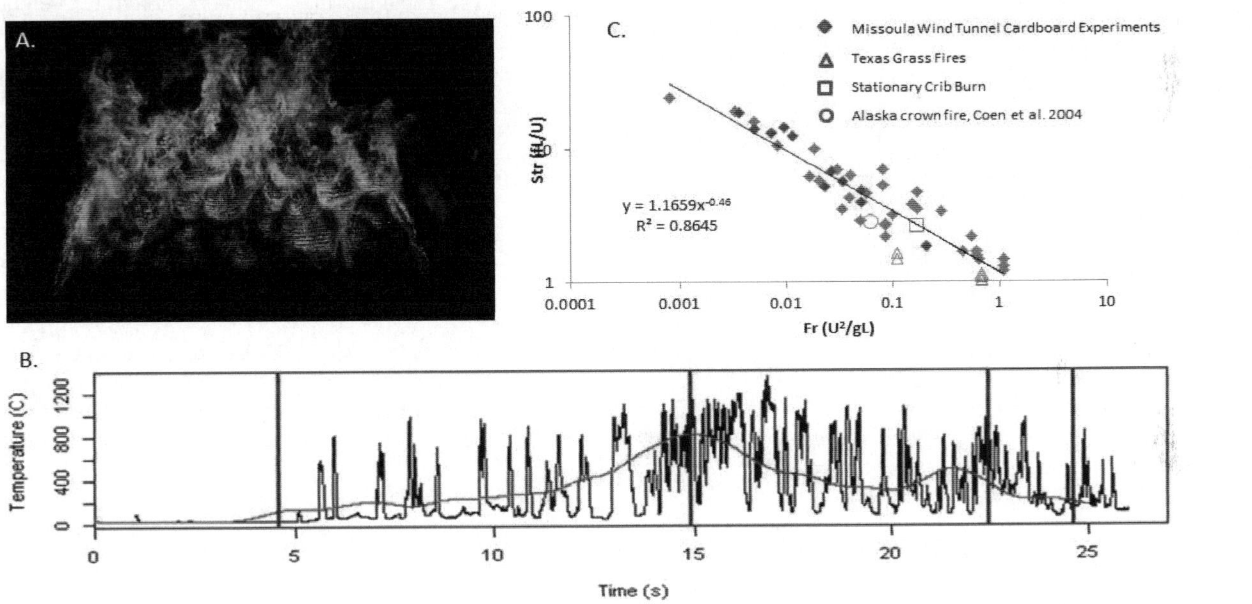

Figure 3. Instabilities in the flame zone are revealed in experimental fires burning in cardboard fuel beds as dish-shaped patterns originating near the rear edge of the flame zone (3A). These produce forward bursts of flame ahead of the combustion interface that result in a pulsing-type temperature signal on thermocouples (3B). Average burst frequencies (*f*) scale with a Strouhal-Froude correlation (St-Fr, 3C) using wind speed (*U*) and flame length (*L*) of laboratory and field-scale fires (see Figure 2).

This research reveals instabilities of the flame that are similar to buoyant dynamics of non-reacting flows (Floryan 1991). These instabilities appear responsible for creating flame intermittency at the leading edge of spreading fires and forcing flames downward into the fuel bed. Similar flame structure and intermittency was recently found using stationary gas flames, with and without forced flow (Gorham et al., 2014). The use of the St-Fr scaling of non-steady

These findings suggest that the difficulty of identifying an integral length scale for convection related to ignition (Anderson et al. 2010) comes from assuming away the time-dependency of particle heating. It also suggests by virtue of the inverse dependency of average pulse frequency on flame dimensions that slower buoyant dynamics of larger flame zones compensate for the increases in energy release and convective heating distance to avoid runaway spread. If buoyant instabilities are responsible for the flame behaviours and particle ignition, then it strongly suggests that laboratory-scale fire spread processes should extend readily to field proportions because flame temperature in diffusion flames (and thus buoyancy) remains approximately the same regardless of fire size.

At this time, our limited field-scale experiments and observations seem to be in agreement with this scaling. Much work is yet to be done to understand useful scaling relationships, but the ultimate goal is to someday incorporate these findings into practical tools for wildland fire managers.

CONCLUSIONS

We report here our current experimental data and new observations on the strong interaction of fire and wind in forest fires. Our findings suggest that the interaction of flow dynamics created between the fire generated buoyancy-driven force and the wind generated inertia-driven force wind flow has a profound influence in flame spread in forest fires, the mechanism has been ignored for decades under the preference of radiation as the dominant mechanism. The forest fuel particles are critically heated by the unsteady dynamic convective heat transfer process, but not by the normally assumed steady process. Even large particles in uniform fuel beds show a jump in heating temperature from flame contact as a necessary step for ignition, but not by a steady radiation heating by fire because optical attenuation drastically reduces the irradiance effect.

REFERENCES

Adam, B.A., JD. English, NK. Akafuah, K.Saito, M Finney, J Forthofer, and J Cohen, 2014. The Strouhal-Froude number scaling for wildland fire spread. Presented at the 7th International Conference on Forest Fire Research, Portugal.

Anderson, W., E. Catchpole and B. Butler, 2010. Convective heat transfer in fire spread through fine fuel beds. Intl. J. Wildl. Fire, 19:284-298.

Atkinson, G., D. Drysdale and Y. Wu, 1995. Fire driven flow in an inclined trench. Fire Safety Journal, 28:141-158.

Byram, G., H. Clements, E. Elliott and P. George, 1964. An experimental study of model fires. USDA Forest Service, Southeastern Forest Experiment, Technical Report No. 3., Macon, Georgia.

Catchpole, W., E. Catchpole, B. Butler, R. C. Rothermel, G. A. Morris and D. J. Latham. 1998. Rate of Spread of Free-Burning Fires in Woody Fuels in a Wind Tunnel. Comb. Sci. Tech., 145:1-37.

Coen, J. L., S. Mahalingam, and J. W. Daily, 2004. Infrared imagery of crown-fire dynamics during Frostfire. J. Appl. Meteor., 43:1241-1259.

Cohen, J.D. and M.A. Finney, 2014. Fine particle heating during experimental laboratory fires. Presented at the 7th International Conference on Forest Fire Research, Portugal.

Dupuy, JI, J. Marechal, D. Portier and J.-C. Valette. 2011. The effects of slope and fuel bed width on laboratory fire behavior," Intl. J. Wildl. Fire, 20: 272–288.

Emmons, H. and T. Shen. 1971. Fire spread in paper arrays. Proc. The Combust. Institute, 13: 917-926.

English, J.D., NK. Akafuah, Brittany A. Adam, M.Finney, J. Forthofer, J.Cohen, S. McAllister and K. Saito, 2014. Ignition behaviour of cardboard fuel particles. Presented at the 7th International Conference on Forest Fire Research, Portugal.

Floryan, J.M. 1991. On the Görtler instability of boundary layers. Progress in Aerospace Science, 28:235-271.

Gorham, DJ., R. Hakes, A.V. Singh, J. Forthofer, J. Cohen, S. McAllister, M.A. Finney and M.J. Gollner. 2014. Studying wildland fire spread using stationary fires. Presented at the 7th International Conference on Forest Fire Research, Portugal.

Malalasekera, W. M. G., H. K. Versteeg and K. Gilchrist. 1996 . A review of research and an experimental study on the pulsation of buoyant diffusion flames and pool fires. Fire and Materials, 20: 261–271.

Rothermel R.C. and H. Anderson. 1966. Fire spread characteristics determined in the laboratory. US Forest Service, Missoula MT.

Smith, D. 1992. Measurements of flame length and flame angle in an inclined trench," Fire Safety Journal, 18: 231-44.

Sparrow E. and R. Husar. 1969. Longitudinal vortices in natural convection flow on inclined plates. J. Fluid. Mech., 37: 251-255.

Woodburn, P., and D. Drysdale. 1998. Fires in inclined trenches: time-varying features of the attached plume. Fire Safety Journal, 31: 165-172.

STRUCTURED NATURAL CONVECTION

J.M. Floryan and M.Z. Hossain
Department of Mechanical and Materials Engineering
The University of Western Ontario
London, Ontario, Canada
mfloryan@eng.uwo.ca

A. Bassom
School of Mathematics and Statistics
The University of Western Australia
Crawley, Western Australia, Australia
andrew.bassom@uwa.edu.au

ABSTRACT

The term "structured convection" refers to convection driven by patterns of heating. Such patterns create horizontal pressure gradients that drive convection regardless of the magnitude of the heating. Choice of the heating pattern represents a control mechanism which may lead to a wide range of flow responses. Flow phenomena associated with the structured convection for the simplest case of heating described by one Fourier mode include an increase of the transverse heat flow and a reduction of the pressure losses in heated channels.

KEYWORDS: natural convection, patterns.

INTRODUCTION

The natural convection occurs in a horizontal layer through the buoyancy driven instability of the conductive state of equilibrium. Such convection appears only when the heating is applied from below, i.e. Rayleigh-Bénard (RB) convection [1-3]. Laminar RB convection has been the origin of much research on nonlinear dynamics and pattern formation [4]. Recent efforts have been focused on the explanation of the roles played by various boundary imperfections [5-6]. A large enough heating intensity leads to turbulent RB convection which has many practical applications in the area of heat transfer [7-8]. Current efforts are focused on the use of boundary imperfections with an appropriate spatial pattern in order to increase the relevant Nusselt number [9]. It is less well known that the instability may occur when the fluid is heated from above (anti-convection) as long as the system contains at least one interface [10-13]. Anti-convection may occur in fluids with arbitrary properties when combined with heat release/extraction at the interface [14-15].

Structured convection occurs in response to patterns of heating. The convection structure may be either directly related to the pattern of heating or could be dictated by the secondary states. The understanding of the structured convection is required for a reliable prediction of the local contaminant transport in urban environments where the form of the convection is dictated by the surface topography (building pattern) and thermally-relevant features of this topography like color variations (color patterns of roofs, streets and parks). Similar conditions can be encountered in rural environments where local circulation can be driven by different heating rates of patterns of forests and lakes, and can be significantly modified by the rural terrain topography. The associated problem of urban heat islands occurs due to human activity, e.g. patterns of dwelling provide patterns of heat sources that may drive local convective currents. The insulating effect of continents on the mantle convection within the Earth represents a spatially structured convection with heating from above [16]. Similar problems are found in geological applications where a system of fractures, leads and polynyas in sea ice leads to convection in both the ocean and atmosphere [17]. Other examples include patterns of local fires, patterns of computer chips, and so on. The important unifying features of all these problems is the existence of a certain spatially distributed pattern of external heating applied to the system of interest. It is the prediction of the response of flow systems exposed to such heating patterns that is the main interest in this study.

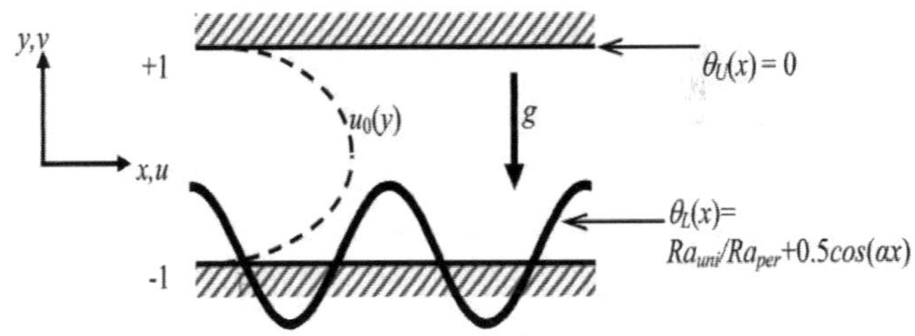

Figure 1. Sketch of the flow system.

PROBLEM FORMULATION

We shall use the model problem considered in the literature to illustrate effects of simple heating patterns [6, 18-20]. Consider the steady flow of a fluid confined in a channel bounded by two parallel walls extending to $\pm\infty$ in the x-direction. The bounding walls are placed at a distance $2h$ apart with the gravitational acceleration g acting in the negative y-direction, as shown in Fig.1. The flow is driven in the positive x-direction by a pressure gradient. The channel is filled with a Boussinesq fluid, i.e. the fluid is incompressible,

Newtonian with thermal conductivity k, specific heat c, thermal diffusivity $\kappa=k/\rho c$, kinematic viscosity ν, dynamic viscosity μ and thermal expansion coefficient Γ. The lower wall is subject to externally imposed heating which results in the walls' temperatures of the form

$$\theta_L(x)=\sum_{n=-1}^{n=+1}\theta_L^{(n)}e^{in\alpha x}, \qquad \theta_U(x)=0 \qquad (1a,b)$$

where θ denotes the difference between the actual temperature and the temperature of the upper wall $\theta = T - T_U$, T denotes the absolute temperature, $\theta_L^{(n)}=\theta_L^{(-n)*}$ are the reality conditions, stars denote complex conjugates, $\lambda=2\pi/\alpha$ is the wavelength of the heating, and subscripts L and U refer to the lower and upper walls, respectively. Mode zero in Eq.(1a), i.e., $\theta_L^{(0)}$, describes the uniform heating while the remaining modes describe the spatially varying heating. The conductive temperature field θ_0 is described by

$$\frac{\partial^2\theta_0}{\partial x^2}+\frac{\partial^2\theta_0}{\partial y^2}=0 \qquad (2)$$

subject to boundary conditions (1) and its solution has the form

$$\theta_0(x,y)=\frac{Ra_{uni}}{Ra_{per}}\hat{\theta}_0(y)+\hat{\hat{\theta}}_0(x,y) \qquad (3a)$$

where $\hat{\theta}_0(y)=\frac{1}{2}(1-y)$, (3b)

$$\hat{\hat{\theta}}_0(x,y)=\hat{\hat{\theta}}_0^{(1)}(y)e^{i\alpha x}+CC= \frac{1}{8}\left[\frac{cosh(\alpha y)}{cosh(\alpha)}-\frac{sinh(\alpha y)}{sinh(\alpha)}\right]e^{in\alpha x}+CC \qquad (3c)$$

and CC stands for the complex conjugate. Two temperature scales have been used. The uniform temperature component $\hat{\theta}_0$ has been scaled with the difference T_{uni} between the mean temperature of the lower wall and the temperature of the upper wall. The complete conductive temperature θ_0, as well as its periodic part, have been scaled with the amplitude of the temperature variations along the lower wall T_{per} resulting in the temperature distribution at this wall in the form $\theta_L(x)=Ra_{uni}/Ra_{per}+1/2cos(\alpha x)$. The channel half height h has been used as the length scale. The Rayleigh numbers based on the uniform and spatial modulation temperature scales have been defined as $Ra_{uni}=g\Gamma h^3T_{uni}/\nu\kappa$ and $Ra_{per}=g\Gamma h^3T_{per}/\nu\kappa$, respectively.

The velocity and pressure fields in the absence of heating have the form

$$\mathbf{v}_0(x,y)=[u_0(y),\ 0]=[1-y^2,\ 0],$$
$$p_0(x,y)=-2x/Re \qquad (4)$$

where $\mathbf{v}_0=(u_0,v_0)$ denotes the velocity vector scaled with the maximum of the x-velocity U_{max}, p_0 stands for the pressure scaled with ρU_{max}^2 and the Reynolds number is defined as $Re=U_{max}h/\nu$.

The heating produces convection and the resulting changes in the flow and temperature fields can be represented as

$$u_T(x,y)=Re\,u_0(y)+u_1(x,y),\quad v_T(x,y)=v_1(x,y),$$
$$\theta_T(x,y)=Pr^{-1}\theta_0(x,y)+\theta_1(x,y), \qquad (5)$$
$$p_T(x,y)=Re^2p_0(x)+p_1(x,y).$$

In the above, (u_T,v_T), p_T and θ_T denote the complete velocity, pressure and temperature fields, respectively, and (u_1,v_1), p_1 and θ_1 denote the velocity, pressure and temperature field modifications caused by the heating, respectively. The complete velocity vector and the velocity modifications have been scaled using the convective velocity scale $U_v=\nu/h$ where $U_{max}/U_v=Re$, the pressure modifications have been scaled using ρU_v^2 and $T_v=T_{per}\nu/\kappa$ has been used as the convective temperature scale, where $T_v/T_{per}=Pr$ defines the Prandtl number. The field equations for the flow and temperature modifications have the form

$$(Re\,u_0+u_1)\frac{\partial u_1}{\partial x}+Re\,v_1\frac{du_0}{dy}+v_1\frac{\partial u_1}{\partial y}= -\frac{\partial p_1}{\partial x}+\frac{\partial^2 u_1}{\partial x^2}+\frac{\partial^2 u_1}{\partial y^2}, \qquad (6a)$$

$$(Re u_0+u_1)\frac{\partial v_1}{\partial x}+v_1\frac{\partial v_1}{\partial y}=-\frac{\partial p_1}{\partial y}+\frac{\partial^2 v_1}{\partial x^2}+\frac{\partial^2 v_1}{\partial y^2} +Ra_{per}\theta_1+Ra_{uni}Pr^{-1}\hat{\theta}_0+Ra_{per}Pr^{-1}\hat{\hat{\theta}}_0, \qquad (6b)$$

$$(Re u_0+u_1)\left(\frac{\partial\hat{\hat{\theta}}_0}{\partial x}+Pr\frac{\partial\theta_1}{\partial x}\right)+ v_1\left(-\frac{Ra_{uni}}{2Ra_{per}}+\frac{\partial\hat{\hat{\theta}}_0}{\partial y}+Pr\frac{\partial\theta_1}{\partial y}\right)=\frac{\partial^2\theta_1}{\partial x^2}+\frac{\partial^2\theta_1}{\partial y^2}, \qquad (6c)$$

$$\frac{\partial u_1}{\partial x}+\frac{\partial v_1}{\partial y}=0. \qquad (6d)$$

The spatially modulated heating generates convection regardless of its intensity. It is assumed that the uniform heating is weak enough that it is unable to create convection on its own; it contributes to the overall motion through Eq.(6c). The term associated with the uniform heating in Eq.(6b) contributes only to the formation of the vertical pressure gradient.

One is interested in the determination of the heat transfer characteristics across the channel as well as the mean pressure modification p_1 associated with the imposed heating. It is assumed that the flow rates through the heated and isothermal channels are the same, i.e., the solution must satisfy the flow rate constraint of the form

$$Q = \int_{-1}^{1} u_T dy = \int_{-1}^{1} (Re u_0 + u_1) dy = 4 Re/3. \quad (7)$$

The problem formulation is closed by specifying the no-slip, the no-penetration and the thermal boundary conditions at the bounding walls which have the form

$$u_1(\pm 1) = 0, \quad v_1(\pm 1) = 0, \quad \theta_1(\pm 1) = 0. \quad (8)$$

The solution of (6)-(8) results in the determination of the velocity, temperature and pressure fields.

DISCUSSION

Figure 2 illustrates variations of the average Nusselt number Nu_{av} as a function of the heating wave number α and the heating intensity Ra_{per} when the fluid layer is exposed to a combination of the spatially periodic and the uniform heating in such a way so that the ratio of the relevant Rayleigh numbers remains constant, e.g. $Ra_{uni}/Ra_{per} = 0.5$. The forced convection is characterized by a very small Reynolds number $Re = 1$. It can be seen that the convection generates a net heat flow between the walls for all Rayleigh numbers. The magnitude of heat flow increases proportionally Ra_{per} with the most effective heating wave number corresponding to $\alpha \approx 1$.

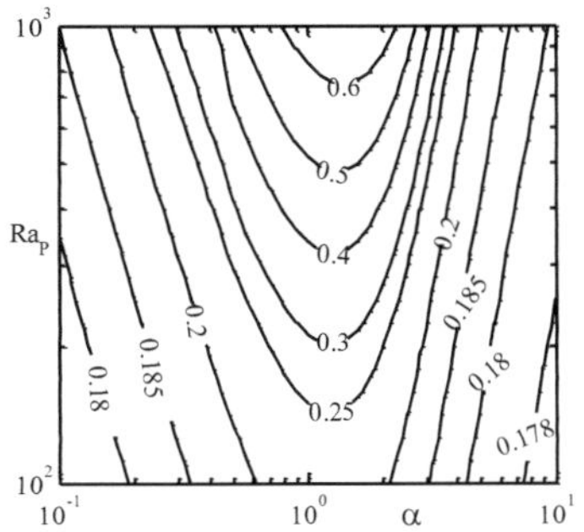

Figure 2. Variations of the average Nusselt number Nu_{av} as a function of α and Ra_{per} for $Re = 1$, $Ra_{uni}/Ra_{per} = 0.5$.

Figure 3 demonstrates that a significant reduction of the pressure losses associated with the maintenance of the forced convection can be achieved by the same type of heating as used in Fig.2. This reduction is associated with the horizontal variations of the buoyancy effects [18,19]. Figure 4 illustrates emergence of the secondary states when the uniform heating component becomes too large. These states are driven either by the RB effect associated with the uniform heating or by the spatial parametric resonance effects associated with the spatially modulated heating [20]. A proper combination of both types of heating can be used for an earlier initiation of the buoyancy driven instabilities and, thus, can be used for an intensification of the heat transfer associated with the natural convection.

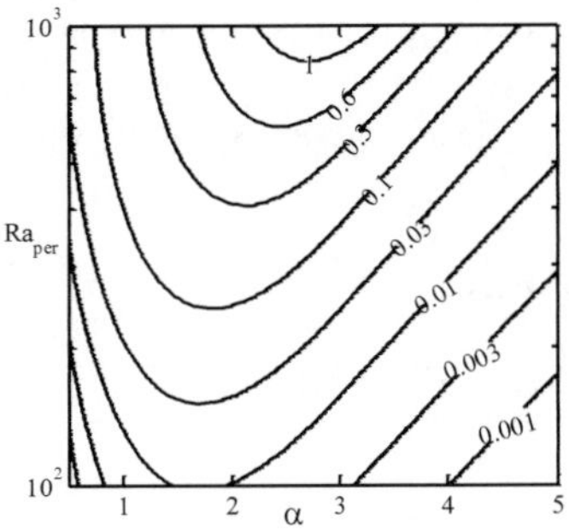

Figure 3. Variations of the pressure gradient correction A/Re as a function of α and Ra_{per} for the uniform heating corresponding to $Ra_{uni}/Ra_{per} = 0.5$ for $Re = 1$.

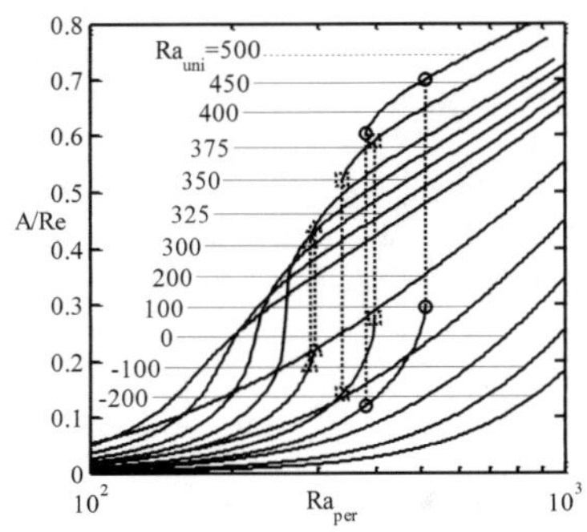

Figure 4. Variations of the pressure gradient correction A/Re as a function of Ra_{per} for $\alpha = 2$ and $Re = 1$ for different levels of the uniform heating measured either in terms of Ra_{uni}.

Convection associated with the long wavelength heating is intriguing as it locally breaks the naturally occurring long convection rolls into secondary forms at locations corresponding to the temperature peaks, as illustrated in Fig.5. The analytical structure of such convection and the relevant mechanisms remain to be studied.

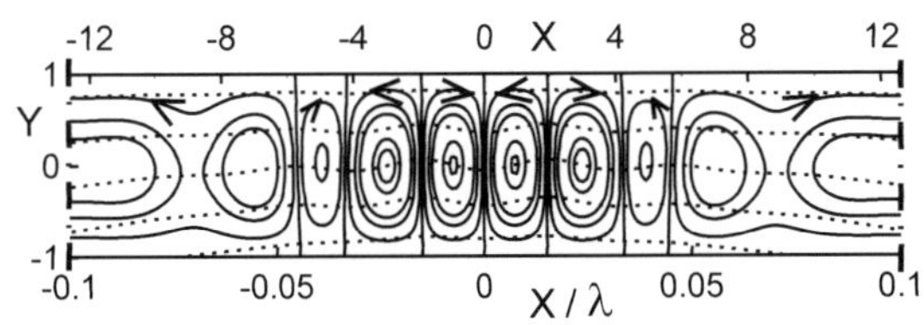

Figure 5. Flow structures for heating with $Ra_{per} = 440$, $Ra_{uni} = 0$, and $\alpha = 0.05$.

CONCLUSIONS

An analysis of convection driven by a spatially distributed heating applied at the lower wall of a horizontal slot has been presented. This form of the heating represents one of the simplest heating patterns. It has been shown that convection driven by this heating leads to the net heat transfer across the layer even for very small heating rates and reduces the pressure losses for flows driven through the slot by a pressure gradient.

ACKNOWLEDGMENTS

This work has been carried out with the support of the NSERC of Canada

REFERENCES

1. Bénard, H., 1900, Les Tourbillons Cellulaires dans une Nappe Liquide, *Revue Générale Science Pure et Applique Revue Gen. Sci. Pure Appl.*, vol.11, pp.1261-1271.

2. Chandrasekhar, S., 1961, *Hydrodynamic and Hydromagnetic Stability* (Oxford University Press).

3. Rayleigh, J.W.S., 1916, On convection currents in a horizontal layer of fluid when the higher temperature is on the under side, *Philosophical Magazine,* vol.32, pp.529-546.

4. Bodenschatz, E., Pesch, W., and Ahlers, G., 2000, Recent Developments in Rayleigh-Bénard Convection, *Ann. Rev. Fluid Mech.*, vol.32, pp. 709-778.

5. Freund, G., Pesch, W. and Zimmermann, W., 2011, Rayleigh-Bénard convection in the presence of spatial temperature modulations, *J. Fluid Mech.*, vol.673, pp.318-348.

6. Hossain, M.Z. and Floryan, J.M., 2013, Instabilities of natural convection in a periodically heated layer, *J. Fluid Mech.*, vol.733, pp.33-67.

7. Ahlers, G., Grossmann S. and D. Lohse, 2009, Heat ransfer and large scale dynamics in turbulent Rayleigh-Bénard Convection, *Rev. Mod. Phys.*, vol.81, pp.503-538 .

8. Lohse, D. and Xia, K.Q., 2010, Small-Scale Properties of Turbulent Rayleigh-Bénard Convection, *Annu. Rev. Fluid Mech.* vol. 42, pp.335-364.

9. Shiskina, O. and Wagner, C., 2011, Modelling the influence of wall roughness on heat transfer in thermal convection, *J. Fluid Mech.*, vol. 686, pp.568-582.

10. Welander, P., 1964, Convective instability in a two-lwyer fluid heated uniformly from above, Tellus, vol. 16, pp.349-358.

11. Gershuni, G.Z. and Zhukhovitskii, E.M., 1980, Instability of a system of horizontal layers of immiscible fluids heated from above, *Fluid Dyn.*, vol.15, pp.816-822.

12. Perestenko, O.V. and Ingel, L.K.H., 1990, The occurrence of moist "anticonvection" in a water-air system, *J. Fluid Mech.,* vol. 287, pp. 1-20.

13. Nepomnyashchy, A.A. and Simanovskii, I.B., 1990, Onset of convection due to heating from above and heat release at an interface, *Fluid Dyn.,* vol. 25, pp. 340-343.

14. Nepomnyashchy, A.A., Simanovskii, I.B. and Braveman, L.M., 2000, Anticonvection in system with heat release at the interface, *Phys. Fluids*, vol. 12, 1129-1132.

15. Merkt, D. and Bestehorn, M., 2012, Pattern formation in anticonvective systems, *Fluid Dyn. Res*., vol.44, 031413.

16. Lenardic, A., Moresi, L., Jellinek, A.M. and Manga, M., 2005, Continental insulation, mantle cooling, and the surface area of oceans and continents, *Earth Planet. Sci. Lett.*, vol. 234, pp.317-333.

17. Marcq, S., Weiss, J., 2012 Influence of sea ice lead-width distribution on turbulent heat transfer between the ocean and the atmosphere, *Cryosphere,* vol.6, pp.143–156.

18. Hossain, M.Z., Floryan, D. and Floryan, J.M., 2012, Drag reduction due to spatial thermal modulations, *J. Fluid Mech.*, vol.713, pp. 398-419.

19. Floryan, Daniel, and Floryan, J.M., 2015, "Drag Reduction in Heated Channels", *J. Fluid Mech.,* vol. 765, pp. 353-395.

20. Hossain, M.Z., and Floryan, J.M., 2015, Mixed Convection in a Periodically Heated Channel, *J. Fluid Mech.*, vol. 768, pp.51-90.

Computational Fluid Dynamics

CHARACTERISTICS OF NEW MODE OF INSTABILITY IN A CHANNEL WITH LONGITUDINAL GROOVES

Alireza Mohammadi
Department of Mechanical & Materials Engineering
University of Western Ontario
London, Ontario, Canada
amoham69@alumni.uwo.ca

J.M. Floryan
Department of Mechanical & Materials Engineering
University of Western Ontario
London, Ontario, Canada
mfloryan@eng.uwo.ca

ABSTRACT

It is known that grooves which are parallel to the flow direction, i.e. longitudinal grooves, may either stabilize or destabilize the travelling wave instability in a pressure-gradient-driven flow in a channel depending on the groove wave number. These waves reduce to the classical Tollmien-Schlichting (TS) waves in the smooth channel limit. It is shown that another class of travelling wave instability exists if grooves with sufficiently high amplitude and proper wavelengths are used. It is demonstrated that the new instability is driven by inviscid mechanisms, with the disturbance motion having the form of a wave propagating in the streamwise direction with the phase speed about four times larger than the TS waves speed and with its streamwise wavelength being about twice the spanwise groove wavelength. The instability motion is concentrated mostly in the middle of the channel and has a primarily planar character, i.e. the dominant velocity components are parallel to the walls. A significant reduction of the corresponding critical Reynolds number can be achieved by increasing the groove amplitude. This mode reduces to the highly attenuated Squire mode in the smooth channel limit.

KEYWORDS: instability, transition to turbulence, channel flow, distributed surface roughness

INTRODUCTION

It has been known since the original work of Reynolds [1] that surface roughness plays an important role in the laminar–turbulent transition. This problem has most frequently been studied in the context of the identification of conditions when the presence of roughness can be ignored, i.e. when the wall can be viewed as hydraulically smooth. The question of roughness effects in turbulent flows has been studied in the context of drag determination starting with Hagen [2] and Darcy [3]; see [4, 5] for reviews.

The most promising method for the mathematical description of the hydrodynamic properties of a surface with an arbitrary topography relies on the reduced-order geometry model [6]. The geometric properties are categorized by projecting the surface geometry onto a convenient functional space, e.g. Fourier space, with the expectation that only a few leading Fourier modes representing the topography matter as far as hydrodynamics are concerned. This technique permits the identification of features of the topography that have a decisive influence on the flow response, with irrelevant details removed from consideration. Indeed, it has been demonstrated that, in many instances, it is sufficient to use only the leading Fourier mode to capture the main physical processes with accuracy sufficient for most applications [7].

The primary objective of this work is to carry out a systematic analysis of the effects of the small-amplitude, longitudinal, two-dimensional grooves of various shapes on the stability of pressure-gradient-driven flows in a channel. The particular focus of this work is the search for instability modes other than the TS waves. The new results will expand the understanding of the effects of distributed surface roughness and how such roughness is able to influence the laminar–turbulent transition and, thus, complement the results available for the same grooves placed transversely with respect to the flow [7]. Because of the drag reducing capabilities of the long wavelength grooves, special attention is paid to the flow stability in the presence of such grooves.

FLOW IN A CHANNEL WITH LONGITUDINAL GROOVES

Consider steady and fully-developed flow through a channel with longitudinal grooves of arbitrary geometry and driven by a constant pressure gradient along the x-direction (see Fig. 1). The grooves geometry can be expressed in terms of Fourier expansions of the form

$$y_L(z) = -1 + \sum_{m=-N_A}^{m=N_A} H_L^{(m)} e^{im\beta z}, \quad y_U(z) = 1 + \sum_{m=-N_A}^{m=N_A} H_U^{(m)} e^{im\beta z} \tag{1a,b}$$

,

where the subscripts L and U refer to the lower and upper walls, respectively, β is the groove wave number, $\lambda_z=2\pi/\beta$ denotes the groove wavelength, N_A represents the number of Fourier modes required to describe the geometry, $H_L^{(m)} = H_L^{(-m)*}$ and $H_U^{(m)} = H_U^{(-m)*}$ express the reality conditions, and stars indicate the complex conjugates. Our interest is in the analysis of effects of flow modulations and, thus, we assume that the mean openings of the grooved and the reference smooth channels are the same, i.e. $H_L^{(0)} = H_U^{(0)} = 0$.

The driving mechanisms and the groove geometry do not depend on the x-coordinate and, thus, the fluid movement is governed by the simplified x-momentum equation and boundary conditions of the form

$$\frac{\partial^2 u_B}{\partial y^2} + \frac{\partial^2 u_B}{\partial z^2} - Re\frac{dp_B}{dx} = 0\,,\, u_B(y_L,z) = 0\,,\, u_B(y_U,z) = 0 \qquad \text{(2a–c)}$$

.

The fixed volume flow rate constraint of the form of

$$Q = \lambda_z^{-1} \int_{z=0}^{z=\lambda_z} \int_{y=y_L(z)}^{y=y_U(z)} u_B(y,z)\,dy\,dz = \frac{4}{3} \qquad (3)$$

has been chosen as the closing condition. This constraint states that the flow rates in the grooved and smooth channels are the same. The pressure gradient that is required to maintain the same flow rate in the grooved channel represents a useful measure for assessing the ability of the grooves to reduce drag.

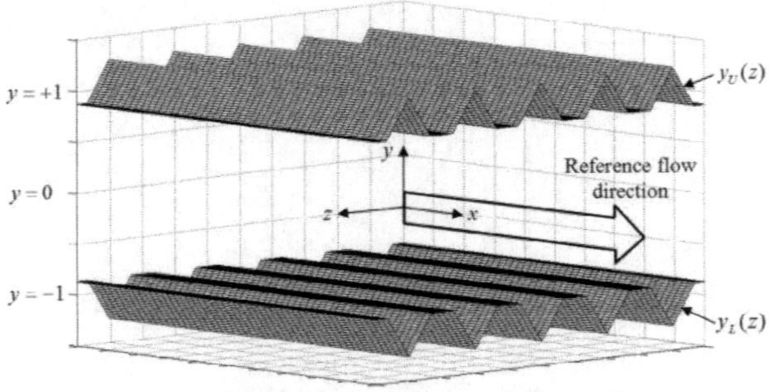

Figure 1: Sketch of the flow configuration.

The properties of the mean flow are illustrated for a simple groove geometry of the form

$$y_L(z) = -1 + S\cos(\beta z)\,, \qquad y_U = 1\,, \qquad \text{(4a,b)}$$

i.e. the lower wall is fitted with sinusoidal grooves while the upper wall is smooth. Figure 2 illustrates variations of the pressure gradient corrections as a function of β and S, and demonstrates that grooves of sufficiently long wavelength reduce the pressure gradient required to drive the flow.

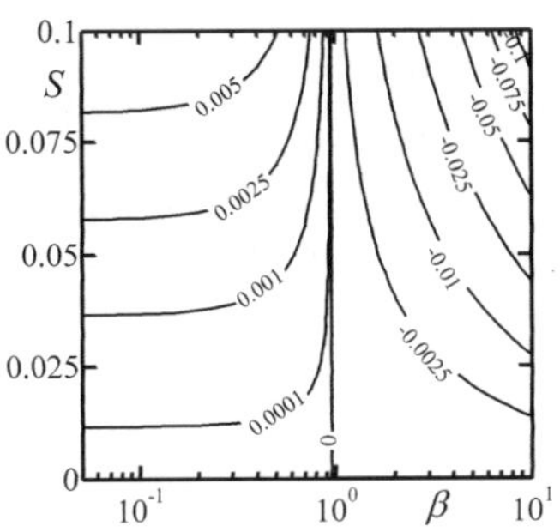

Figure 2: Variations of the pressure gradient correction Re dp_1/dx in a channel with geometry defined by Eq. (4) as a function of the groove wave number β and the groove amplitude S.

LINEAR STABILITY ANALYSIS

The linear stability analysis permits identification of conditions that signal the beginning of transition from the laminar to the turbulent state. In particular, the study of stability properties of the flow modified by the long wavelength grooves is important due to the drag reducing abilities of such grooves; it is of interest to determine the maximum value of the Reynolds number which guarantees that the flow remains laminar. We shall assume that the disturbance level is low and, thus, the asymptotic stability can predict the onset conditions for secondary states. A high level of disturbances requires analysis of the transient growth [8] as well as information about the initial structure of the disturbance field.

The governing equations for the linear stability analysis can be expressed in terms of the continuity and the vorticity transport equations of the form

$$\nabla \cdot \boldsymbol{V} = 0\,, \qquad \text{(5a)}$$

$$\frac{\partial \boldsymbol{\omega}}{\partial t} - (\boldsymbol{\omega} \cdot \nabla)\boldsymbol{V} + (\boldsymbol{V} \cdot \nabla)\boldsymbol{\omega} - \frac{1}{Re}\nabla^2 \boldsymbol{\omega} = 0\,, \qquad \text{(5b)}$$

where $\boldsymbol{V}$ and $\boldsymbol{\omega} = \nabla \times \boldsymbol{V}$ are the velocity and vorticity vectors, respectively [6]. The flow quantities are expressed as a superposition of the basic flow and three-dimensional disturbances, i.e.

$$\boldsymbol{V}(x,y,z,t) = \boldsymbol{V}_B(y,z) + \boldsymbol{V}_D(x,y,z,t)\,,$$

$$\boldsymbol{\omega}(x,y,z,t) = \boldsymbol{\omega}_B(y,z) + \boldsymbol{\omega}_D(x,y,z,t)\,. \qquad \text{(6a,b)}$$

In the above, subscripts B and D denote the basic flow and the disturbance field, respectively, and $\boldsymbol{V}_D = (u_D, v_D, w_D)$ and $\boldsymbol{\omega}_D = (\xi_D, \eta_D, \phi_D)$ stand for the disturbance velocity and vorticity vectors, respectively. Substitution of (6) into (5), subtraction of the basic flow and linearization lead to the disturbance equations of the form shown in [9] and subject to the homogeneous boundary conditions of the form

$$\boldsymbol{V}_D(x,y_L,z,t) = 0, \qquad \boldsymbol{V}_D(x,y_U,z,t) = 0\,. \qquad \text{(7a,b)}$$

Coefficients appearing in (7) are functions of the (y, z)-coordinates and, therefore, the solution can be written as

$$\boldsymbol{V}_D(x,y,z,t) = \boldsymbol{G}_D(y,z)\, e^{i(\delta x + \mu z - \sigma t)} + c.c.\,,$$

$$\boldsymbol{\omega}_D(x,y,z,t) = \boldsymbol{\Omega}_D(y,z)\, e^{i(\delta x + \mu z - \sigma t)} + c.c.\,, \qquad \text{(8a,b)}$$

where δ and μ denote the real wave numbers in the x- and z-directions, respectively, $\sigma = \sigma_r + i\sigma_i$ is the complex amplification rate, σ_i is the rate of growth of disturbances, σ_r is the frequency of disturbances and $c.c.$ refers to complex conjugates. Amplitude functions $\boldsymbol{G}_D(y,z)$ and $\boldsymbol{\Omega}_D(y,z)$ are periodic functions of z and, thus, can be expressed in terms of Fourier expansions of the form

$$\boldsymbol{G}_D(y,z) = \sum_{n=-\infty}^{n=+\infty} [g_u^{(n)}(y), g_v^{(n)}(y), g_w^{(n)}(y)] e^{in\beta z} + c.c.\,, \qquad \text{(9a)}$$

$$\boldsymbol{\Omega}_D(y,z) = \sum_{n=-\infty}^{n=+\infty} [g_\xi^{(n)}(y), g_\eta^{(n)}(y), g_\phi^{(n)}(y)] e^{in\beta z} + c.c.\,. \qquad \text{(9b)}$$

Substitution of (9) into (8) leads to the disturbance velocity and vorticity components of the form

$$\boldsymbol{V}_D(x,y,z,t) = \sum_{n=-\infty}^{n=+\infty} [g_u^{(n)}(y), g_v^{(n)}(y), g_w^{(n)}(y)] e^{i[\delta x + (\mu + n\beta)z - \sigma t]} + c.c.\,, \qquad \text{(10a)}$$

$$\boldsymbol{\omega}_D(x,y,z,t) = \sum_{n=-\infty}^{n=+\infty} [g_\xi^{(n)}(y), g_\eta^{(n)}(y), g_\phi^{(n)}(y)] e^{i[\delta x + (\mu + n\beta)z - \sigma t]} + c.c.\,. \qquad \text{(10b)}$$

A system of linear homogeneous ordinary differential equations for $g_v^{(n)}(y)$ and $g_\eta^{(n)}(y)$ is obtained by substituting (10) into (5) and separating Fourier modes. The complete problem represents an eigenvalue problem which is solved using the method explained in [9].

RESULTS

We rely on the reduced-order geometry model and thus consider sinusoidal grooves placed at the lower wall with the channel geometry described by Eq. (4). The full spectrum for this channel for Re = 6000, S = 0.035, β = 0.7 for "two dimensional" disturbances with δ = 0.3, μ = 0 is displayed in Fig. 3. The unstable eigenvalue is located at the tip of the P branch (centre modes) and its location is strongly affected by the groove amplitude. This mode connects to the Squire mode in the limit of $S\rightarrow 0$ as demonstrated in Fig. 4. The disturbance velocity field displayed in Fig. 5 shows that the instability motion is concentrated in the widest channel opening. While the grooves provide a rather small change in the geometry of the lower wall, the zone of the most rapid instability motion is located far away from the wall and leads to a rapid increase of the intensity of the spanwise motion (see Fig. 5).

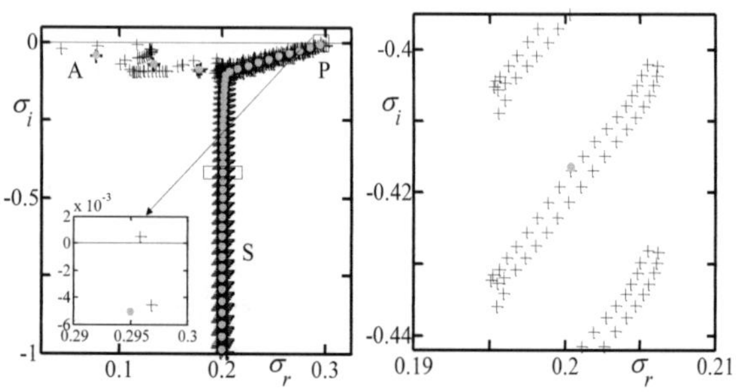

Figure 3. Disturbance spectrum for flow in a channel with geometry defined by Eq. (4) with β = 0.7 and S = 0.035 for Re = 6000 and disturbances with δ = 0.3, μ = 0. Black + identify elements of the spectrum. The Squire spectrum for the smooth channel is given for comparison purposes and is identified using grey dots. Figure 3b displays enlargement of the box shown on the branch S of Fig. 3a. Width of branch S is dictated by the number of Fourier modes used in the solution.

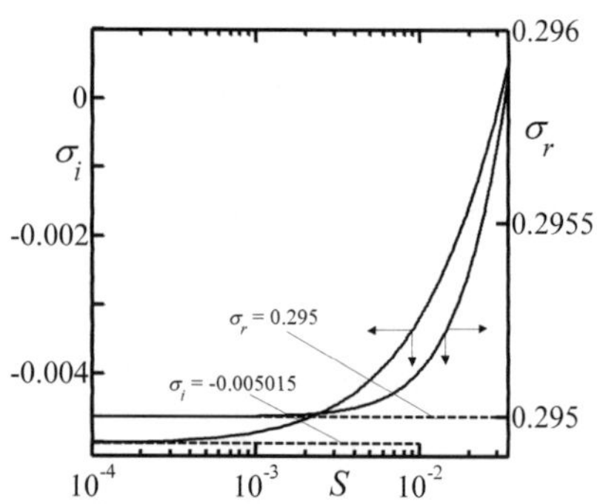

Figure 4. Variations of the growth rate σ_i and the frequency σ_r of the "two-dimensional" disturbances (μ = 0) as a function of the groove amplitude S for the groove geometry described by Eq. (4) with β = 0.7 for the flow Reynolds number Re = 6000 and disturbances with the streamwise wave number δ = 0.3. Dashed lines identify the least stable Squire mode with δ = 0.3 in the smooth channel.

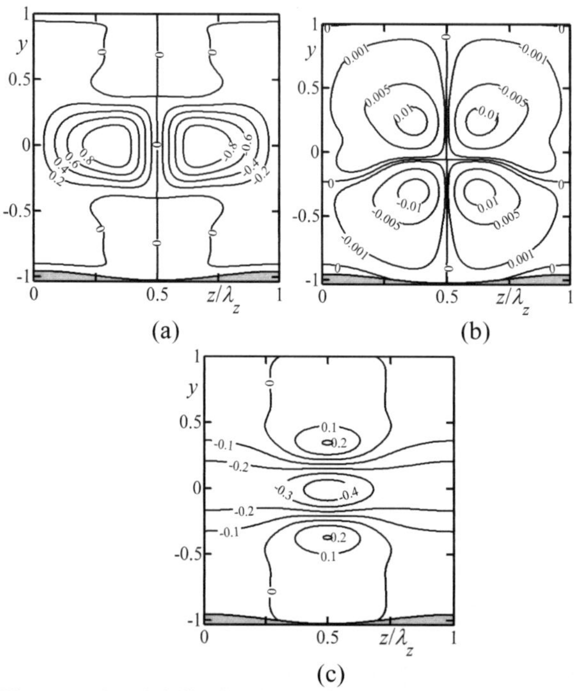

Figure 5. Spatial distributions of the streamwise u_D (Fig. 5a), normal v_D (Fig. 5b) and spanwise w_D (Fig. 5c) disturbance velocity components at (x,t) = (0,0) normalized with condition $\max_{y_L\le y\le y_U,\,0\le z\le\lambda_z}|u_D(0,y,z,0)|=1$ and determined for the same conditions as in Fig. 3 for the new mode.

We shall now address the question of the origin of the new instability. It is known that the TS instability is driven by friction and a sufficient increase of Re will lead to its eventual stabilization. This process is illustrated in Fig. 6a displaying variations of the amplification rate σ_i as a function of δ for increasing Re. Initial destabilization is followed by stabilization when Re attains sufficiently large magnitude. Figure 6b displays results of a similar study for the new mode and demonstrates that the amplification rate becomes Re-independent once Re becomes large enough. Figure 7 displays the distribution of the mean flow y-vorticity component, which can be viewed as a measure of the "driving force", as well as the distribution of the dissipation function $\Phi_B = (\frac{\partial u_B}{\partial y})^2 + (\frac{\partial u_B}{\partial z})^2$, which can be viewed as a measure of the "opposing force". Quotations are used here to underline the qualitative character of these terms. The "driving force" increases monotonically in the downward direction while the dissipation function is smallest around the channel mid-line with local minima in the widest and narrowest channel openings and increases with distance away from the channel center. The optimal conditions for the initiation of the instability are on both sides of the widest channel opening slightly below the channel mid-line as this is where the ratio of the "driving" and "opposing forces" is the highest.

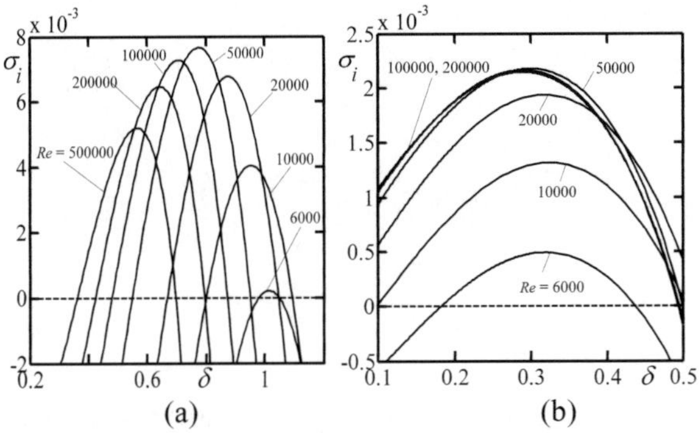

Figure 6. Variations of the growth rate σ_i of the "two-dimensional" disturbances (μ = 0) as a function of the streamwise wave number δ for flow in a channel with geometry described by Eq. (4) with $S = 0.035$, $\beta = 0.7$ for the TS waves (Fig. 6a) and the new mode (Fig. 6b).

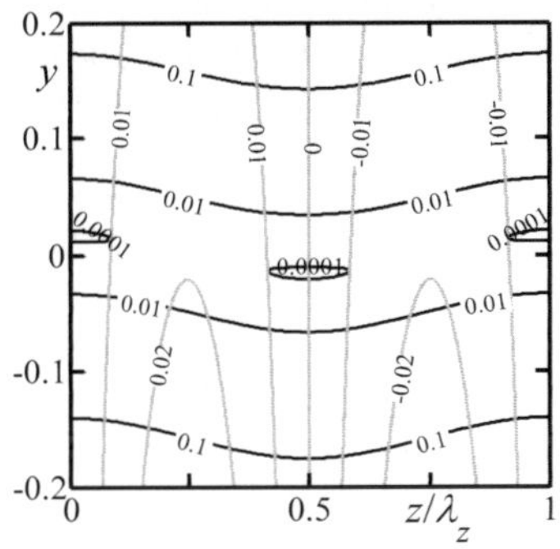

Figure 7. Distributions of the mean flow dissipation function Φ_B (black lines) and the mean flow normal vorticity component $\eta_B = \partial u_B/\partial z$ (grey lines) in the middle of the channel with geometry described by Eq. (4) with $S = 0.035$ and $\beta = 0.7$.

CONCLUSIONS

A new type of disturbances unstable in the range of Reynolds numbers of interest has been identified. These disturbances have the form of travelling waves but with characteristics different from the TS waves. The new mode grows only in the presence of grooves with the proper wave number and the proper amplitude; it becomes highly attenuated in the absence of grooves.

It has been shown that the new mode is driven by the inviscid instability mechanism associated with both the vorticity dynamics and the spatial parametric resonance. The "driving force" is associated with the spanwise gradients of the streamwise velocity component which are essential for both mechanisms. Reduction of the groove wave number reduces spanwise velocity gradients due to smaller geometry gradients and, thus, decreases the "driving force" and leads to the stabilization of this mode. An increase of the groove wave number leads to the stream lift up above the grooves and reduction of the flow modulations and, thus, a reduction of the "driving force" and flow stabilization. As a result, the new mode is able to grow only in a well defined range of the groove wave numbers.

ACKNOWLEDGMENTS

This work has been carried out with support from the Natural Sciences and Engineering Research Council (NSERC) of Canada.

REFERENCES

1. Reynolds, O. 1883, "An experimental investigation of the circumstances which determine whether the motion of water shall be direct or sinuous, and of the law of resistance in parallel channels", *Philos. Trans. R. Soc. London*, **174**, 935–982.

2. Hagen, G. 1854, "Über den Einfluss der Temperatur auf die Bewegung des Wassers in Röhren", *Math. Abh. Akad. Wiss. Berlin* **17**, 17–98.

3. Darcy, H. 1857 *Recherches expérimentales relatives au mouvement de l'eau dans les tuyaux*. Mallet-Bachelier.

4. Jiménez, J. 2004 Turbulent flows over rough walls. *Ann. Rev. Fluid Mech.* **36**,173–196.

5. Herwig, H., Gloss, D. & Wenterodt, T. 2008 A new approach to understanding and modelling the influence of wall roughness on friction factors for pipe and channel flows. *J. Fluid Mech.* **613**, 35–53.

6. Floryan, J. M. 1997 Stability of wall-bounded shear layers in the presence of simulated distributed surface roughness. *J. Fluid Mech.* **335**, 29–55.

7. Floryan, J. M. 2007 Three-dimensional instabilities of laminar flow in a rough channel and the concept of hydraulically smooth wall. *Eur. J. Mech. B/Fluids* **26**, 305–329.

8. Szumbarski, J. & Floryan, J. M. 2006 Transient disturbance growth in a corrugated channel. *J. Fluid Mech.* **568**, 243–272.

9. Moradi, H. V. & Floryan, J. M. 2014 Stability of flow in a channel with longitudinal grooves. *J. Fluid Mech.* **757**, 613–648.

Spectral Algorithm for Analysis of Flows in Vibrating Channels

S. Zandi szandi@uwo.ca
A. Mohammadi amoham69@alumni.uwo.ca
J.M. Floryan mfloryan@eng.uwo.ca
Department of Mechanical and Materials Engineering
The University of Western Ontario
London, Ontario, Canada

ABSTRACT

A spectral method for analysis of flows in channels with vibrating walls has been developed. Difficulties associated with irregularities of the boundaries have been overcome by employing the immersed boundary conditions method (IBC). The algorithm relies on the Fourier expansions in the flow direction and the Chebyshev expansions in the transverse direction. Various tests confirm the spectral accuracy of the algorithm.

KEYWORDS: spectral discretization, vibrating walls, rough boundaries, immersed boundary conditions method.

INTRODUCTION

The "immersed boundary conditions" (IBC) method, which uses spectral discretization, has been developed for two-dimensional flow problems in domains bounded by vibrating boundaries. The shapes of physical boundaries are described using Fourier expansions in the flow direction. Chebyshev expansions have been used for discretization in transverse direction. Galileo transformation has been employed in order to convert this unsteady problem into a steady one.

PROBLEM FORMULATION

Consider steady, two-dimensional flow of a fluid confined in a channel bounded by two parallel walls extending to $\pm\infty$ in the X-direction and placed at a distance $2h$ apart as shown in Fig.1. The flow is driven in the positive X-direction by a pressure gradient resulting in the velocity and pressure fields, and the flow rate of the form

$$\boldsymbol{v}_0(X,Y) = (1-Y^2, 0),\ p_0(X,Y) = -2X/Re,\quad \Psi_0 = Y - \frac{Y^3}{3} + \frac{2}{3},\ Q_0 = \frac{4}{3}$$

(1)

where $\boldsymbol{v}_0 = (u_0, v_0)$ denotes the velocity vector scaled with the maximum of the X-velocity u_{max}, p_0 stands for the pressure scaled with ρu_{max}^2 where ρ stands for the density, Ψ_0 stands for the stream function, Q_0 denotes the flow rate, the Reynolds number is defined as $Re = u_{max}h/\nu$ where ν stands for the kinematic viscosity, and h has been used as the length scale. The flow is modified by imposing wall vibrations in the form of travelling waves with known amplitudes. The resulting time-dependent channel geometry is described as

$$Y_U(t,X) = 1 + h_U(X-ct) = 1 + \sum_{n=-N_A, n\neq 0}^{n=N_A} H_U^{\langle n\rangle} e^{in\alpha(X-ct)},$$

(2a)

$$Y_L(t,X) = -1 + h_L(X-ct) = -1 + \sum_{n=-N_A, n\neq 0}^{n=N_A} H_L^{\langle n\rangle} e^{in\alpha(X-ct)}$$

(2b)

where h_U and h_L are known, the subscripts L and U refer to the lower and upper walls, respectively, N_A denotes the number of Fourier modes required to describe the shape of the wave, c denotes the wave phase speed, and $H_U^{\langle -n\rangle} = H_U^{\langle n\rangle *}$ are the reality condition with * denoting the complex conjugate. Our interests are in the waves that do not affect the mean channel opening and, thus, coefficients of modes zero in the Fourier expansions (2) have been set to zero.

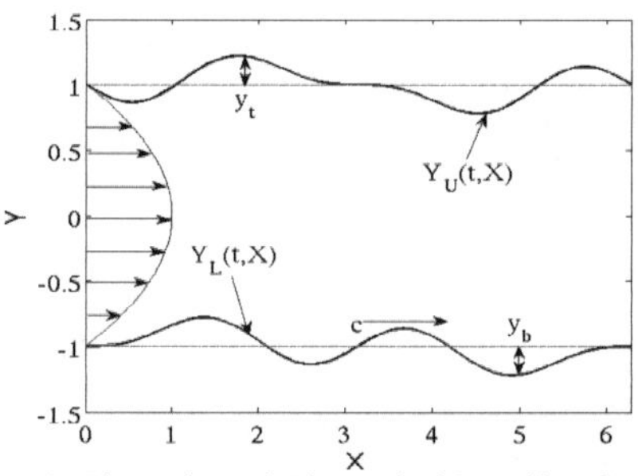

Figure 1. Flow domain bounded by vibrating walls.

We shall represent all flow quantities as sums of the reference flow and the vibration-induced modifications, i.e.

$$u_T(t,X,Y) = u_0(Y) + u_1(t,X,Y),\ v_T(t,X,Y) = v_1(t,X,Y)$$
$$p_T(t,X,Y) = p_0(X) + p_1(t,X,Y). \tag{3}$$

In the above, (u_T, v_T) and p_T denote the complete velocity and pressure fields, respectively, and (u_1, v_1) and p_1 denote the velocity and pressure field modifications caused by the vibrations, respectively. Substitution of (3) into the Navier-Stokes and continuity equations and use of (1) lead to the field equations for the flow modifications of the form

$$\frac{\partial u_1}{\partial t} + (u_0 + u_1)\frac{\partial u_1}{\partial X} + v_1\frac{du_0}{dY} + v_1\frac{\partial u_1}{\partial Y} = -\frac{\partial p_1}{\partial X} + Re^{-1}\left(\frac{\partial^2 u_1}{\partial X^2} + \frac{\partial^2 u_1}{\partial Y^2}\right),$$

(4a)

$$\frac{\partial v_1}{\partial t} + (u_0 + u_1)\frac{\partial v_1}{\partial X} + v_1\frac{\partial v_1}{\partial Y} =$$

$$-\frac{\partial p_1}{\partial Y} + Re^{-1}\left(\frac{\partial^2 v_1}{\partial X^2} + \frac{\partial^2 v_1}{\partial Y^2}\right), \quad (4b)$$

$$\frac{\partial u_1}{\partial X} + \frac{\partial v_1}{\partial Y} = 0 \quad (4c)$$

subject to boundary conditions of the form

$$u_T\left(t, X, Y_U(t,X)\right) = 0,$$
$$v_T\left(t, X, Y_U(t,X)\right) = \frac{\partial Y_U(t,X)}{\partial t} = -ch'_U(X-ct), \quad (4d)$$
$$u_T\left(t, X, Y_L(t,X)\right) = 0,$$
$$v_T\left(t, X, Y_L(t,X)\right) = \frac{\partial Y_L(t,X)}{\partial t} = -ch'_L(X-ct) \quad (4e)$$

where h/u_{max} has been used as the time scale, ∂ denotes partial differentiation, subscripts t, X and Y denote the arguments of partial differentiations and prime denotes differentiation with respect to the argument. We are interested in the determination if the surface vibrations can lead to the reduction of the pressure gradient required to maintain the specified flow rate. Accordingly, we impose the mass flow rate constraint of the form

$$Q(t,X)|_{mean} = \left(\int_{Y_L(t,x)}^{Y_U(t,x)} u(t,X,Y)\,dY\right)\Big|_{mean} = \frac{4}{3}, \quad (4f)$$

i.e. the net flow rate through the vibrating channel at any X-location must be the same as through the stationary reference channel.

SOLUTION METHOD

Introduction of the Galileo transformation of the form

$$y = Y, \; x = X - ct \quad (5)$$

converts the unsteady problem (4) into the steady problem of the form

$$(u_0 + u_1 - c)\frac{\partial u_1}{\partial x} + v_1\frac{du_0}{dy} + v_1\frac{\partial u_1}{\partial y} = -\frac{\partial p_1}{\partial x} + Re^{-1}\left(\frac{\partial^2 u_1}{\partial x^2} + \frac{\partial^2 u_1}{\partial y^2}\right), \quad (6a)$$

$$(u_0 + u_1 - c)\frac{\partial v_1}{\partial x} + v_1\frac{\partial v_1}{\partial y} = -\frac{\partial p_1}{\partial y} + Re^{-1}\left(\frac{\partial^2 v_1}{\partial x^2} + \frac{\partial^2 v_1}{\partial y^2}\right), \quad (6b)$$

$$\frac{\partial u_1}{\partial x} + \frac{\partial v_1}{\partial y} = 0, \quad (6c)$$

$$y = y_U(x): \quad u_1 = -u_0, \quad v_1 = -ch'_U(x), \quad (6d)$$
$$y = y_L(x): \quad u_1 = -u_0, \quad v_1 = -ch'_L(x), \quad (6e)$$

$$Q(x)|_{mean} = \left(\int_{y_L(x)}^{y_U(x)} u_T(x,y)\,dy\right)\Big|_{mean} = \frac{4}{3} \quad (6f)$$

where the locations of the boundaries are given as

$$y_U(x) = 1 + h_U(x) = 1 + \sum_{n=-N_A, n\neq 0}^{n=N_A} H_U^{\langle n\rangle} e^{inax}, \quad (7a)$$
$$y_L(x) = -1 + h_L(x) = -1 + \sum_{n=-N_A n\neq 0}^{n=N_A} H_L^{\langle n\rangle} e^{inax}. \quad (7b)$$

The reader may note that $Q(x)$ is a periodic function of x and its mean value corresponds to the 0th mode of its Fourier expansion. Continuity equation can be satisfied identically by introducing functions Ψ_T, Ψ_1 defined as

$$u_T = u_0 + u_1 = \frac{\partial \Psi_T}{\partial y} = \frac{d\Psi_0}{dy} + \frac{\partial \Psi_1}{\partial y},$$
$$v_T = v_1 = -\frac{\partial \Psi_1}{\partial x}, \qquad \Psi_T = \Psi_0 + \Psi_1. \quad (8)$$

Use of (8) and elimination of pressure lead to the flow problem of the form

$$-Re^{-1}\nabla^2(\nabla^2\Psi_1) + (u_0 - c)\frac{\partial}{\partial x}\nabla^2\Psi_1 - \frac{d^2u_0}{dy^2}\frac{\partial \Psi_1}{\partial x} = -\left(\frac{\partial \Psi_1}{\partial y}\frac{\partial}{\partial x} - \frac{\partial \Psi_1}{\partial x}\frac{\partial}{\partial y}\right)\nabla^2\Psi_1, \quad (9)$$

$$y = y_U(x): \quad \frac{\partial \Psi_1}{\partial y} = -u_0, \quad \frac{\partial \Psi_1}{\partial x} = ch'_U(x), \quad (10a\text{-}b)$$

$$y = y_L(x): \quad \frac{\partial \Psi_1}{\partial y} = -u_0, \quad \frac{\partial \Psi_1}{\partial x} = ch'_L(x), \quad (11a\text{-}b)$$

$$Q(x)|_{mean} = \left[(\Psi_0 + \Psi_1)|_{y_U(x)} - (\Psi_0 + \Psi_1)|_{y_L(x)}\right]\Big|_{mean} = \frac{4}{3}. \quad (12)$$

Condition (11b) can be written in a different form by noting that variations of Ψ_T along the lower wall can be expressed as

$$d\Psi_{T,L} = \left(\frac{\partial \Psi_T}{\partial x}dx + \frac{\partial \Psi_T}{\partial y}dy\right)\Big|_{y_L(x)} = c\,\frac{dy_L}{dx}dx. \quad (13)$$

Integration along the wall results in

$$\Psi_{T,L}(x) = c[y_L(x) - y_L(x_0)] \quad (14)$$

where the constant of integration has been selected by assuming that $\Psi_{T,L}(x_0) = 0$ with x_0 representing an arbitrary point on the wall. Similar analysis carried out for the upper wall leads to an alternative form of (10b), i.e.

$$\Psi_{T,U}(x) = c[y_U(x) - y_L(x_0)] - 2c + \frac{4}{3} \quad (15)$$

where the constraint (12) has been used in order to determine integration constant. Equations (9), (10a), (11a), (14) and (15) represent an alternative problem formulation. This particular formulation is better suited for the analysis while (9)-(12) is better suited for the IBC method implementation.

DISCRETIZATION OF THE FIELD EQUATIONS

We seek a spectrally accurate numerical solution of the above problem with the channel geometry expressed by (7). The immersed boundary conditions (IBC) method is used to deal with the complexities associated with the irregular geometry. In this method, a fixed computational domain with the physical domain immersed in its interior is used and the flow boundary conditions are replaced with constraints. The size of the computational domain in the y-direction cannot be smaller than $-1-y_b < y < 1+y_t$ where y_t, y_b denote locations of extremities of the upper and lower walls, respectively (see Fig.1). The locations of these extremities determine the minimum y-expanse of the computational domain which needs to be used and transformation of the form

$$\hat{y} = 2\left[\frac{y-(1+y_t)}{y_t+y_b+2}\right]+1 \tag{16}$$

maps this domain into $[-1,1]$ to facilitate the use of the standard form of Chebyshev polynomials. The flow problem expressed using $\hat{y}$ rather than y coordinate has the form

$$-Re^{-1}\left(\Gamma^4\frac{\partial^4\Psi_1}{\partial\hat{y}^4}+2\Gamma^2\frac{\partial^2}{\partial x^2}\left(\frac{\partial^2\Psi_1}{\partial\hat{y}^2}\right)+\frac{\partial^4\Psi_1}{\partial x^4}\right)+ (u_0-c)\frac{\partial}{\partial x}\left(\frac{\partial^2\Psi_1}{\partial x^2}+\Gamma^2\frac{\partial^2\Psi_1}{\partial\hat{y}^2}\right)-\Gamma^2\frac{d^2u_0}{d\hat{y}^2}\frac{\partial\Psi_1}{\partial x}= -\Gamma\frac{\partial}{\partial\hat{y}}\left(\frac{\partial\widehat{u_1u_1}}{\partial x}+\Gamma\frac{\partial\widehat{u_1v_1}}{\partial\hat{y}}\right)+\frac{\partial}{\partial x}\left(\frac{\partial\widehat{u_1v_1}}{\partial x}+\Gamma\frac{\partial\widehat{v_1v_1}}{\partial\hat{y}}\right), \tag{17a}$$

$$\hat{y}=\hat{y}_U(x):\ \frac{\partial\Psi_1}{\partial\hat{y}} = -\Gamma^{-1}u_0 \tag{17b}$$

$$\Psi_{T,U}(x) = c\Gamma^{-1}[\hat{y}_U(x)-\hat{y}_L(x_0)]-2c+4/3 \qquad \frac{\partial\Psi_1}{\partial x} = c\Gamma^{-1}\frac{d\hat{y}_U(x)}{dx}, \tag{17c}$$

$$\hat{y}=\hat{y}_L(x):\ \frac{\partial\Psi_1}{\partial\hat{y}} = -\Gamma^{-1}u_0, \tag{17d}$$

$$\Psi_{T,L}(x) = c\Gamma^{-1}[\hat{y}_L(x)-\hat{y}_L(x_0)], \qquad \frac{\partial\Psi_1}{\partial x} = c\Gamma^{-1}\frac{d\hat{y}_L(x)}{dx}, \tag{17e}$$

$$Q(x)|_{mean} = \left[(\Psi_0+\Psi_1)|_{\hat{y}_U(x)}-(\Psi_0+\Psi_1)|_{\hat{y}_L(x)}\right]\Big|_{mean} = 4/3, \tag{17f}$$

$$\hat{y}_U = \sum_{n=-N_A}^{N_A} A_U^{\langle n\rangle}e^{in\alpha x} \quad \text{with} \quad A_U^{\langle 0\rangle} = 1-\Gamma y_t,\ \ A_U^{\langle n\rangle} = \Gamma H_U^{\langle n\rangle} \ \text{ for } n\neq 0 \tag{17g}$$

$$\hat{y}_L = \sum_{n=-N_A}^{N_A} A_L^{\langle n\rangle}e^{in\alpha x} \quad \text{with} \quad A_L^{\langle 0\rangle} = 1+\Gamma(-2-y_t),\ \ A_L^{\langle n\rangle} = \Gamma H_L^{\langle n\rangle} \ \text{ for } n\neq 0 \tag{17h}$$

where $\Gamma = \frac{d\hat{y}}{dy} = \frac{2}{y_t+y_b+2}$ and $A_U^{\langle n\rangle} = A_U^{\langle -n\rangle *}$, $A_L^{\langle n\rangle} = A_L^{\langle -n\rangle *}$ represent the reality conditions. We shall keep two forms of (17c) and (17e) as use of boundary conditions in the form of derivatives is more efficient computationally while specification of values of Ψ_T along the walls is more convenient for analytic solutions and enforcement of the flow rate constraint.

The x-periodicity of the solution domain suggests expressing the unknown Ψ_1 as a Fourier expansion of the form

$$\Psi_1(x,\hat{y}) = \sum_{n=-\infty}^{\infty}\Phi^{\langle n\rangle}(\hat{y})e^{in\alpha x} \approx \sum_{n=-N_M}^{N_M}\Phi^{\langle n\rangle}(\hat{y})e^{in\alpha x}, \tag{18}$$

where $\Phi^{\langle n\rangle}$ are the modal functions satisfying the reality conditions of the form $\Phi^{\langle n\rangle} = \Phi^{\langle -n\rangle *}$ and N_M is the number of Fourier modes used in the solution.

The discretization of the field equation begins with substitution of (18) into (17a). The resulting equation can be expressed in terms of a system of modal equations if the nonlinear terms on the right hand side of (17a) can be expressed as Fourier expansions. Such expansions can be written in the following forms

$$[\widehat{u_1u_1},\widehat{u_1v_1},\widehat{v_1v_1}](x,\hat{y}) \approx \sum_{n=-N_M}^{N_M}\left[\widehat{uu}^{\langle n\rangle},\widehat{uv}^{\langle n\rangle},\widehat{vv}^{\langle n\rangle}\right](\hat{y})e^{in\alpha x}. \tag{19}$$

Substitution of (19) into (17a) and separation of Fourier modes lead to the modal equations of the form

$$-Re^{-1}[\Gamma^4D^4-2\Gamma^2(n\alpha)^2D^2+(n\alpha)^4]\Phi^{\langle n\rangle}+ in\alpha(u_0-c)[\Gamma^2D^2-(n\alpha)^2]\Phi^{\langle n\rangle}-in\alpha\Gamma^2D^2u_0\Phi^{\langle n\rangle} = -\Gamma D\left[in\alpha\,\widehat{uv}^{\langle n\rangle}+\Gamma D\widehat{uv}^{\langle n\rangle}\right]+ in\alpha\left[in\alpha\,\widehat{uv}^{\langle n\rangle}+\Gamma D\widehat{vv}^{\langle n\rangle}\right] \tag{20}$$

where $-N_M < n < N_M$, $D = d/d\hat{y}$. These equations are coupled explicitly through the nonlinear terms and coupled implicitly through the boundary conditions which shall be discussed later.

Differential system (20) needs to be converted into an algebraic system before the actual computations can begin. Discretization begins with representing $\Phi^{\langle n\rangle}$ in terms of Chebyshev expansions of the form

$$\Phi^{\langle n\rangle}(\hat{y}) = \sum_{k=0}^{\infty}G_k^{\langle n\rangle}T_k(\hat{y}) \approx \sum_{k=0}^{N_T-1}G_k^{\langle n\rangle}T_k(\hat{y}) \tag{21}$$

where T_k denotes the Chebyshev polynomial of kth order, $G_k^{\langle n\rangle}$ denotes the unknown coefficients of the expansion and N_T denotes the number of Chebyshev polynomials retained in the solution. The nonlinear terms on the right hand side of (20) have to be expressed in terms of similar expansions, i.e.

$$[\widehat{uu}^{\langle n\rangle}, \widehat{uv}^{\langle n\rangle}, \widehat{vv}^{\langle n\rangle}](\hat{y}) = \sum_{k=0}^{\infty}[G\widehat{uu}_k^{\langle n\rangle}, G\widehat{uv}_k^{\langle n\rangle}, G\widehat{vv}_k^{\langle n\rangle}]T_k(\hat{y}) \approx \sum_{k=0}^{N_T-1}[G\widehat{uu}_k^{\langle n\rangle}, G\widehat{uv}_k^{\langle n\rangle}, G\widehat{vv}_k^{\langle n\rangle}]T_k(\hat{y})$$
(22)

where $G\widehat{uu}_k^{\langle n\rangle}, G\widehat{uv}_k^{\langle n\rangle}, G\widehat{vv}_k^{\langle n\rangle}$ are considered to be known. The reader may note that these coefficients need to be recomputed at the beginning of the each iteration. We shall address this issue later in the presentation. The derivatives of $\Phi^{\langle n\rangle}(\hat{y})$ are evaluated using formula of the form

$$D^n\Phi^{\langle n\rangle} = \sum_{k=0}^{N_T-1} G_k^{\langle n\rangle} D^n T_k(\hat{y}).$$
(23)

Similar formulae are used for the $\hat{y}$ derivatives of $\widehat{uu}^{\langle n\rangle}, \widehat{uv}^{\langle n\rangle}, \widehat{vv}^{\langle n\rangle}$. Substitution of (21)-(22) into (20) and use of (23) results in the modal equations of the form

$$\sum_{k=0}^{N_T-1}\{-Re^{-1}[\Gamma^4 D^4 T_k - 2\Gamma^2(n\alpha)^2 D^2 T_k + (n\alpha)^4 T_k] + in\alpha[\Gamma^2(u_0 - c)D^2T_k - (n\alpha)^2(u_0-c)T_k] - in\alpha\Gamma^2(D^2u_0)T_k\}G_k^{\langle n\rangle} = \sum_{k=0}^{N_T-1}[-in\alpha\Gamma G\widehat{uu}_k^{\langle n\rangle} DT_k - \Gamma^2 G\widehat{uv}_k^{\langle n\rangle} D^2 T_k -$$

na2GuvknTk+inaΓGvvknDTk. (24)

The Galerkin procedure is used to extract algebraic equations for $G_k^{\langle n\rangle}$. This procedure involves projection of the residuum on the basis functions and setting it to zero and, in practical terms, involves multiplication of (24) by $T_j(\hat{y})$, $j = [0, N_{T-1}]$ and integration with the weight factor of $\hat{\omega} = (1-\hat{y}^2)^{-1/2}$. This process results in

$$\sum_{k=0}^{N_T-1}\{-Re^{-1}[\Gamma^4\langle T_j, D^4T_k\rangle - 2\Gamma^2(n\alpha)^2\langle T_j, D^2T_k\rangle + (n\alpha)^4\langle T_j, T_k\rangle] + in\alpha[\Gamma^2\langle T_j, (u_0-c)D^2T_k\rangle - (n\alpha)^2\langle T_j,(u_0-c)T_k\rangle] - in\alpha\Gamma^2\langle T_j,(D^2u_0)T_k\rangle\}G_k^{\langle n\rangle} = -in\alpha\Gamma\sum_{k=0}^{N_T-1}G\widehat{uu}_k^{\langle n\rangle}\langle T_j, DT_k\rangle - \Gamma^2\sum_{k=0}^{N_T-1}G\widehat{uv}_k^{\langle n\rangle}\langle T_j, D^2T_k\rangle - (n\alpha)^2\sum_{k=0}^{N_T-1}G\widehat{uv}_k^{\langle n\rangle}\langle T_j, T_k\rangle + in\alpha\Gamma\sum_{k=0}^{N_T-1}G\widehat{vv}_k^{\langle n\rangle}\langle T_j, DT_k\rangle,$$
(25)

where $\langle\ \rangle$ denotes the inner product.

DISCUSSION OF RESULTS

We shall now investigate accuracy of the method for a sinusoidal wave propagating along the lower wall resulting in the channel geometry of the form

$$y_U = 1,\ y_L = -1 + A\cos(\alpha x).$$
(26)

The overall accuracy depends on the accuracy of the enforcement of boundary conditions [5]. We shall use two types of quantities to illustrate this error. The first one involves evaluation of u_T and $v_T + ch'_L(x)$ along the lower wall. Since these quantities are supposed to be zero, their values, denoted as u_L and $v_L + ch'_L$, provide a direct measure of the error. Figure 2 illustrates distributions of u_L and $v_L + ch'_L$ for the Reynolds number $Re = 5$, the phase speed $c = 1.3$, the wave amplitude $A = 0.05$ and the wave number $\alpha = 5$ determined using $N_M = 20$ Fourier modes and $N_T = 80$ Chebyshev polynomials. It can be seen that the largest error occurs at the widest channel opening. The second quantity involves L_∞ norms of u_L and $v_L + ch'_L$ defined as

$$\|u_L(x)\|_\infty = \sup|u_L(x, y_L(x))|, \quad (27a)$$
$$\|v_L(x) + ch'_L(x)\|_\infty = \sup|v_L(x, y_L(x)) + ch'_L(x)|, \quad (27b)$$

where $0 \le x \le 2\pi/\alpha$. Figure 3 displays variations of these norms as functions of the total number of Fourier modes N_M used in the calculations for the same conditions as in Fig.2 except for the wave number $\alpha = 1$ and with different wave amplitudes A determined using $N_T = 80$ Chebyshev polynomials. It can be seen both norms decrease exponentially with an increase of the number of Fourier used, which demonstrates the spectral convergence of the algorithm.

CONCLUSIONS

Description and validation of a spectral algorithm suitable for analysis of flows through vibrating channels have been presented. The flow boundary conditions have been enforced on the vibrating walls using the immersed boundary conditions (IBC) method. The spatial discretization is based on the Fourier expansions in the flow direction and Chebyshev expansions in the transverse direction. Various tests confirm that the algorithm provides spectral accuracy.

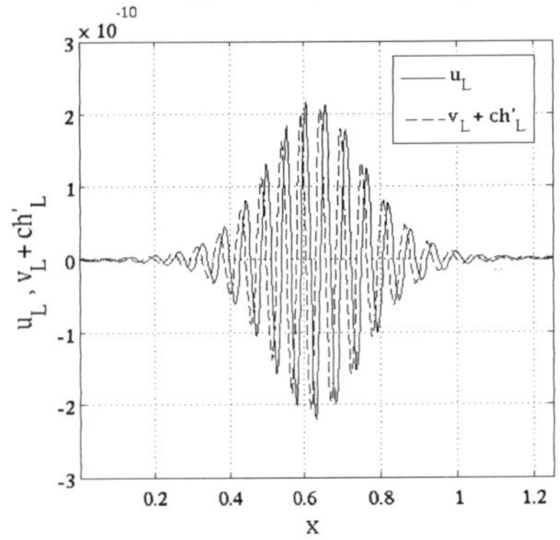

Figure 2. Distributions of u_L and $v_L + ch'_L$

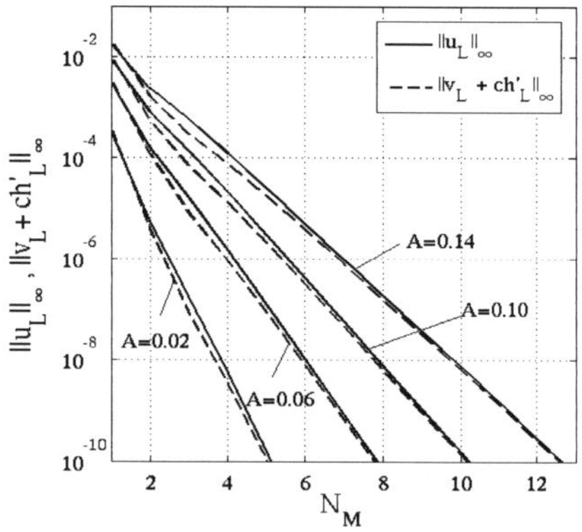

Figure 3. Variations of $\|u_L\|_\infty$ and $\|v_L + ch'_L\|_\infty$

ACKNOWLEDGMENTS

This work has been carried out with support from the Natural Sciences and Engineering Research Council (NSERC) of Canada.

REFERENCES

1. Cabal, A., Szumbarski, J., Floryan, J.M. 2001, "Numerical Simulation of Flows over Corrugated Walls", *Comp. & Fluids,* (30), 753-776.
2. Floryan, J.M. 1986, "Conformal-mapping-based coordinate generation for flows in periodic configurations", *J. Comp. Phys.,* (62), 221-247.
3. Floryan, J.M., Zemach, C. 1987, "Schwarz-Christoffel transformations - a general approach", *J. Comp. Phys.,* (72), 347-371.
4. Floryan, J.M., Zemach, C. 1993, "Schwarz-Christoffel methods for conformal mappings of regions with a periodic boundary", *J. Comp. App. Math.*, (46), 77-102.
5. Husain, S.Z., Floryan, J.M. 2008, "Gridless Spectral Algorithm for Stokes Flow with Moving Boundaries", *Computer Methods in Applied Mechanics and Engineering,* (198), 245-259.

THREE-DIMENSIONAL SPECTRALLY-ACCURATE ALGORITHM FOR THE NAVIER-STOKES EQUATIONS

Md. Sakib **J. M. Floryan**
msakib@uwo.ca **mfloryan@eng.uwo.ca**
Department of Mechanical and Materials Engineering
The University of western Ontario
London, Ontario, Canada

ABSTRACT

In this work a spectrally-accurate algorithm for three-dimensional flows in domains bounded by rough walls is presented. The algorithm is based on the velocity-vorticity formulation and uses the Immersed Boundary Conditions concept. Discretization is based on the Fourier expansions in the two periodic directions and the Chebyshev expansion in the transverse direction. Special solver that takes the advantage of the structure of the relevant coefficient matrix has been developed. Numerical tests confirm the spectral accuracy.

KEYWORDS: Spectral method, immersed boundary conditions method, rough boundaries, three-dimensional channel flow.

INTRODUCTION

The main difficulty associated with the numerical simulations of flows in domains bounded by rough walls is the modeling of the effect of surface roughness. Methods based on the mapping of the irregular boundaries to a rectangular domain are highly accurate but the coefficient matrices associated with these methods are non-sparse and the computational cost is high [1]. In addition, such mappings are available only for a limited class of geometries.

A recent development in the modeling of rough walls is the immersed boundary (IB) methods proposed by Peskin in the context of cardiac dynamics [2]. The common limitations of the IB methods are the low spatial accuracy as they are based on either the low-order finite-difference, or the finite-volume or the finite-element techniques [3]. Szumbarski and Floryan proposed an alternative version for the IB method, referred to as the Immersed Boundary Conditions (IBC) method [4], which takes the advantage of the spectral discretization to provide the highest possible spatial accuracy. This algorithm has been extended to time dependent problems and problems with moving boundaries, and its accuracy has been improved through the use of the over-determined formulation. Fernandez et al [5] applied the IBC method to three-dimensional problems governed by Laplace equation. The present work describes the extension of the IBC method to three-dimensional problems governed by the steady Navier-Stokes equation.

PROBLEM FORMULATION

The flow problem of interest consists of a three-dimensional viscous steady flow in a channel bounded by two rough walls. The geometry of the walls is described by the following relations

$$y_U(x,z) = 1 + \sum_{n=-\infty}^{n=+\infty}\sum_{m=-\infty}^{m=+\infty} H_U^{(n.m)} e^{i(n\alpha x + m\beta z)}, \quad (1)$$
$$y_L(x,z) = -1 + \sum_{n=-\infty}^{n=+\infty}\sum_{m=-\infty}^{m=+\infty} H_L^{(n.m)} e^{i(n\alpha x + m\beta z)} \quad (2)$$

where $H_U^{(n,m)} = H_U^{(-n,-m)*}$ and $H_L^{(n,m)} = H_L^{(-n,-m)*}$ and the stars denote the complex conjugates. The walls are periodic and extend to $\pm\infty$ in the x- and z-directions, with $\lambda_x = 2\pi/\alpha$ and $\lambda_z = 2\pi/\beta$ denoting the wave lengths (Fig. 1).

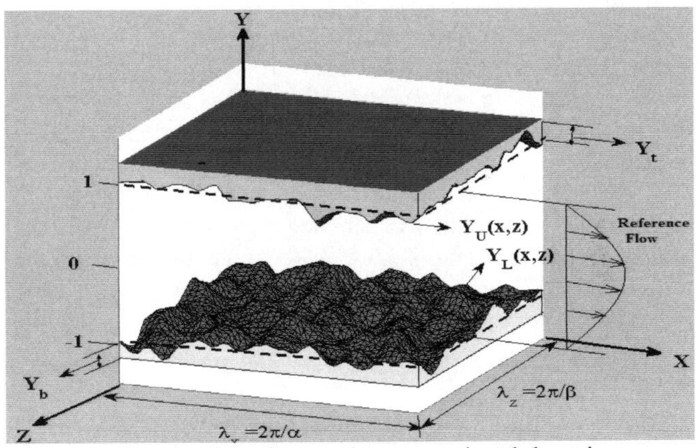

Figure 1: Sketch of the flow and computational domains.

Taking the flow between two parallel plates located at $y = \pm 1$ as the reference flow and eliminating the pressure from the Navier-stokes equations, two coupled equations for the normal-to-the-wall components of the velocity (v_1) and the vorticity (η) can be derived. This form of the governing equations, referred to as the velocity-vorticity formulation, can be written as

$$-\frac{1}{Re}\nabla^2(\nabla^2 v_1) + \left(u_0\frac{\partial}{\partial x} + w_0\frac{\partial}{\partial z}\right)(\nabla^2 v_1) - \frac{\partial^2 u_0}{\partial y^2}\frac{\partial v_1}{\partial x} - \frac{\partial^2 w_0}{\partial y^2}\frac{\partial v_1}{\partial z} = N_v, \quad (3)$$

$$-\frac{1}{Re}\nabla^2\eta + u_0\frac{\partial\eta}{\partial x} + w_0\frac{\partial\eta}{\partial z} + \frac{\partial v_1}{\partial z}\frac{\partial u_0}{\partial y} - \frac{\partial v_1}{\partial x}\frac{\partial w_0}{\partial y} = N_\eta \tag{4}$$

where

$$N_\eta = \frac{\partial H_w}{\partial x} - \frac{\partial H_u}{\partial z}, \tag{5}$$

$$N_v = \frac{\partial}{\partial y}\left(\frac{\partial H_u}{\partial x} + \frac{\partial H_w}{\partial z}\right) - \left(\frac{\partial^2}{\partial x^2} + \frac{\partial^2}{\partial z^2}\right)H_v, \tag{6}$$

$$H_u = \frac{\partial}{\partial x}(u_1u_1) + \frac{\partial}{\partial y}(u_1v_1) + \frac{\partial}{\partial z}(u_1w_1), \tag{7}$$

$$H_v = \frac{\partial}{\partial x}(u_1v_1) + \frac{\partial}{\partial y}(v_1v_1) + \frac{\partial}{\partial z}(v_1w_1), \tag{8}$$

$$H_w = \frac{\partial}{\partial x}(u_1w_1) + \frac{\partial}{\partial y}(v_1w_1) + \frac{\partial}{\partial z}(w_1w_1). \tag{9}$$

In the above, subscripts 0 and 1 refer to the reference flow and flow modifications due to the presence of the roughness, respectively.

The boundary conditions come from the no-slip and no-penetration conditions at walls and can be written in the case of the upper wall in the following form

$$u_1(y_U(x,z)) = -u_0(y_U(x,z)), \tag{10}$$

$$v_1(y_U(x,z)) = 0, \tag{11}$$

$$w_1(y_U(x,z)) = -w_0(y_U(x,z)). \tag{12}$$

Conditions for the lower wall are similar. The form of these conditions expressed in terms of the wall-normal velocity and vorticity is presented in the next section.

DISCRETIZATION METHOD

To implement the spectral discretization we need a regular computational domain extending over one period in the x- and z-directions and extending in the y-direction far enough so that the flow domain is completely submerged inside the computational domain. Figure 1 illustrates the form of the computational domain. Since the standard definition of the Chebyshev polynomials will be used, the y-extent of the computational domain needs to be mapped into (-1, 1). The required mapping is of the form

$$\hat{y} = [y - (1 + y_t)]\Gamma + 1 \tag{13}$$

where $\hat{y} \in \langle -1, 1\rangle$, $\Gamma = 2/(2 + y_t + y_b)$ is a constant and y_t and y_b stands for the maximum deviations of rough walls from 1 and -1, respectively. Fourier expansions are used for discretization in the x- and z-directions. The solution takes the following form

$$f(x,\hat{y},z) \approx \sum_{n=-N_M}^{n=+N_M}\sum_{m=-M_M}^{m=+M_M} f^{(n,m)}(\hat{y})e^{i(n\alpha x + m\beta z)} \tag{14}$$

where N_M and M_M denote the number of Fourier modes in the x- and z-directions, $f^{(n,m)} = f^{(-n,-m)^*}$, stars denote the complex conjugates, $f(x,\hat{y},z)$ represents the velocity components as well as the normal-to-wall vorticity component, and $f^{(n,m)}$ represent the modal functions for these components. Substitution of Eqs (23)-(26) into (8)-(9) results in the following modal form of the governing equations

$$\left[\frac{D_{nm}^2}{Re} + in\alpha(-u_0D_{nm} + D^2u_0) + im\beta(-w_0D_{nm} + D^2w_0)\right]v^{(n,m)} = -\left(\hat{N}_v\right)^{(n,m)}, \tag{15}$$

$$\left(\frac{D_{nm}}{Re} - in\alpha u_0 - im\beta w_0\right)\eta^{(n,m)} + (in\alpha Dw_0 - im\beta Du_0)v^{(n,m)} = -\left(\hat{N}_\eta\right)^{(n,m)} \tag{16}$$

where

$$D^q = \Gamma^q\frac{d^q}{d\hat{y}^q}, \tag{17}$$

$$D_{nm} = D^2 - k_{nm}^2, \tag{18}$$

$$k_{nm}^2 = n^2\alpha^2 + m^2\beta^2 \,. \tag{19}$$

Using the modal form of the continuity equation and applying the definition of the wall-normal vorticity, we obtain for $(n,m) \neq (0,0)$

$$u^{(n,m)} = \frac{1}{k_{nm}^2}\left[in\alpha D(v^{(n,m)}) - im\beta\eta^{(n,m)}\right], \tag{20}$$

$$w^{(n,m)} = \frac{1}{k_{nm}^2}\left[im\beta D(v^{(n,m)}) + in\alpha\eta^{(n,m)}\right]. \tag{21}$$

Use of (20)-(21) combined with (14) leads to the following form of the boundary conditions for the upper wall

$$\sum_{\substack{n=-N_M \\ (n,m)\neq(0,0)}}^{n=N_M}\sum_{m=-M_M}^{m=M_M}\frac{1}{k_{nm}^2}\left[in\alpha D(v^{(n,m)})(\hat{y}_U(x,z)) - im\beta\eta^{(n,m)}(\hat{y}_U(x,z))\right]e^{i(n\alpha x+m\beta z)} + u^{(0,0)}(\hat{y}_U(x,z)) = -u_0(\hat{y}_U), \tag{22}$$

$$\sum_{n=-N_M}^{n=N_M}\sum_{m=-M_M}^{m=M_M} v^{(n,m)}(\hat{y}_U(x,z))e^{i(n\alpha x+m\beta z)} = 0, \tag{23}$$

$$\sum_{\substack{n=-N_M \\ (n,m)\neq(0,0)}}^{n=N_M}\sum_{m=-M_M}^{m=M_M}\frac{1}{k_{nm}^2}\left[im\beta D(v^{(n,m)})(\hat{y}_U(x,z)) + in\alpha\eta^{(n,m)}(\hat{y}_U(x,z))\right]e^{i(n\alpha x+m\beta z)} + w^{(0,0)}(\hat{y}_U(x,z)) = -w_0(\hat{y}_U). \tag{24}$$

The removal of the mode (0, 0) of the u- and w-velocity components requires development of special procedures.

In the case of three-dimensional flows, we need two constraints to close the system of equations, namely, either the constant pressure gradient or the constant volume flow rate in both the x- and z- directions. In the context of the velocity-vorticity formulation, the constraints enter the system of governing equations for $u^{(0,0)}$ *and* $w^{(0,0)}$. These additional governing equations replace the equations for mode (0, 0) of the wall-normal velocity and vorticity as $\eta^{(0,0)}$ and $v^{(0,0)}$ can be shown to be zero.

Substitution of the Fourier expansions into the x- and z-component of the Navier-stokes equations and setting

$(n,m) = (0,0)$ provides us with the governing equations for $u^{(0,0)}$ and $w^{(0,0)}$ of the following form

$$\frac{1}{Re}D^2u^{(0,0)} = e'_x + D\langle\widehat{u_1v_1}\rangle^{(0,0)}, \qquad (25)$$

$$\frac{1}{Re}D^2w^{(0,0)} = e'_z + D\langle\widehat{v_1w_1}\rangle^{(0,0)}, \qquad (26)$$

where e'_x and e'_z are the modifications of the pressure gradients in the x- and z-direction due to the presence of the roughness Taking the Chebyshev expansions of modal functions and applying the Galerkin projection method, the system of equations can be written as follows

$$\sum_{k=0}^{NT}\Big[\Big(\frac{1}{Re}\langle T_j, D^4T_k\rangle - \frac{2k_{nm}^2}{Re}\langle T_j, D^2T_k\rangle + \frac{k_{nm}^4}{Re}\langle T_j, T_k\rangle - in\alpha\langle T_j, u_0D^2T_k\rangle + in\alpha k_{nm}^2\langle T_j, u_0T_k\rangle + in\alpha\langle T_j, D^2u_0T_k\rangle - im\beta\langle T_j, w_0D^2T_k\rangle + im\beta k_{nm}^2\langle T_j, w_0T_k\rangle + im\beta\langle T_j, D^2w_0T_k\rangle\Big)\hat{v}_k^{(n,m)}\Big] = \sum_{k=0}^{k=NT}\Big\{\big[-in\alpha(G_{uv})_k^{(n,m)} - im\beta(G_{vw})_k^{(n,m)}\big]\langle T_j, D^2T_k\rangle + \big[n^2\alpha^2(G_{uu})_k^{(n,m)} + 2nm\alpha\beta(G_{uw})_k^{(n,m)} + m^2\beta^2(G_{ww})_k^{(n,m)} - k_{nm}^2(G_{vv})_k^{(n,m)}\big]\langle T_j, DT_k\rangle + \big[-ink_{nm}^2\alpha(G_{uv})_k^{(n,m)} - -imk_{nm}^2\beta(G_{vw})_k^{(n,m)}\big]\langle T_j, T_k\rangle\Big\}, \quad 0 \le j \le NT-4, \qquad (27)$$

$$\sum_{k=0}^{k=NT}\Big(\frac{1}{Re}\langle T_j, D^2T_k\rangle - \frac{k_{nm}^2}{Re}\langle T_j, T_k\rangle - in\alpha\langle T_j, u_0T_k\rangle - im\beta\langle T_j, w_0T_k\rangle\Big)\hat{\eta}_k^{(n,m)} + \big(in\alpha\langle T_j, Dw_0T_k\rangle - im\beta\langle T_j, Du_0T_k\rangle\big)\hat{v}_k^{(n,m)} = \sum_{k=0}^{k=NT}\Big\{\big[-in\alpha(G_{vw})_k^{(n,m)} + im\beta(G_{uv})_k^{(n,m)}\big]\langle T_j, DT_k\rangle + \big[nm\alpha\beta\beta^2(G_{ww})_k^{(n,m)} - nm\alpha\beta\beta^2(G_{uu})_k^{(n,m)} + (n^2\alpha^2 - m^2\beta^2)\beta^2(G_{uw})_k^{(n,m)}\big]\langle T_j, T_k\rangle\Big\}, \quad 0 \le j \le NT-2 \qquad (28)$$

where Eqs (58)-(59) provide the algebraic equations for $N_M \le n \le N_M$, $M_M \le m \le M_M$, and $(n,m) \ne (0,0)$.

$$\frac{1}{Re}\sum_{k=0}^{k=NT} u_k^{(0,0)}\langle T_j, D^2T_k\rangle = e'_x\langle T_j, T_0\rangle + \sum_{k=0}^{k=NT}(G_{uv})_k^{(0,0)}, \quad 0 \le j \le NT-2, \qquad (29)$$

$$\frac{1}{Re}\sum_{k=0}^{k=NT} w_k^{(0,0)}\langle T_j, D^2T_k\rangle = e'_z\langle T_j, T_0\rangle + \sum_{k=0}^{k=NT}(G_{vw})_k^{(0,0)}, \quad 0 \le j \le NT-2. \qquad (30)$$

Substitution of the Chebyshev expansions and use of the Fourier expansions for values of the Chebyshev polynomials and their derivatives evaluated along the wall, and for values of u_0 *and* w_0 evaluated along the wall, leads to the final form of the boundary conditions:

$$\sum_{\substack{p=-N_M\\(p,q)\ne(0,0)}}^{p=N_M}\sum_{q=-M_M}^{q=M_M}\sum_{k=0}^{k=NT}\frac{1}{k_{pq}^2}\Big[ip\alpha\hat{v}_k^{(p,q)}(EU)_k^{(n-p,m-q)} - iq\beta\hat{\eta}_k^{(p,q)}(CU)_k^{(n-p,m-q)}\Big] + \sum_{k=0}^{k=NT}\hat{u}_k^{(0,0)}(CU)_k^{(n,m)} = -\big(\widehat{U}_0^U\big)^{(n,m)}, \qquad (31)$$

$$\sum_{\substack{p=-N_M\\(p,q)\ne(0,0)}}^{p=N_M}\sum_{q=-M_M}^{q=M_M}\sum_{k=0}^{k=NT}\hat{v}_k^{(p,q)}(CU)_k^{(n-p,m-q)} = 0, \qquad (32)$$

$$\sum_{\substack{p=-N_M\\(p,q)\ne(0,0)}}^{p=N_M}\sum_{q=-M_M}^{q=M_M}\sum_{k=0}^{k=NT}\frac{1}{k_{pq}^2}\Big[iq\beta\hat{v}_k^{(p,q)}(EU)_k^{(n-p,m-q)} + ip\alpha\hat{\eta}_k^{(p,q)}(CU)_k^{(n-p,m-q)}\Big] + \sum_{k=0}^{k=NT}\widehat{w}_k^{(0,0)}(CU)_k^{(n,m)} = -\big(\widehat{W}_0^U\big)^{(n,m)}, \qquad (33)$$

where $-N_M \le n \le N_M$ and $-M_M \le m \le M_M$. In the above, the coefficients $CU, EU, \widehat{U}_0^U, \widehat{W}_0^U$ are the Fourier expansion coefficients of $T_k(\hat{y}_U)$, $DT_k(\hat{y}_U)$, $u_0(\hat{y}_U)$, *and* $w_0(\hat{y}_U)$ respectively. Use of identical expressions, with L replacing U, provides the boundary conditions at the lower wall.

SOLUTION PROCESS

The system (27)-(30) along with the boundary conditions (31)-(33) represents a coupled nonlinear system of algebraic equations which is solved iteratively. All nonlinear terms are placed in the right hand side, and are computed using the information available from the previous iteration. A specialized direct solver has been used in order to take advantage of the particular structure of the left hand side matrix associated with the IBC method. Husain et al [6] have shown that this solver has substantial benefits regarding both the memory and the speed of execution in case of two dimensional problems. We have developed a modified version of the solver adapted to the coefficient matrix of the three-dimensional problem.

PERFORMANCE OF THE ALGORITHM

To examine the performance of the algorithm a model problem is introduced. In this problem, the upper wall remains flat and the shape of the lower wall is described by one Fourier mode, i.e.

$$y_L = -1 + y_b sin(\alpha x)cos(\beta z), \; y_U = 1. \qquad (34)$$

All tests have been carried out using the following parameters:

$$y_b = 0.1, \; \alpha = \beta = 1, \; Re = 5, \; e'_x = 0, \; e'_z = 0 \; . \qquad (35)$$

Since it is known that the maximum error occurs at the boundary, the following error measure has been used for testing

$$\big\|u_{er,BC}\big\|_\infty = \sup\,\big|u_{er,BC}\big|, \qquad (36)$$

where $0 \le x \le \frac{2\pi}{\alpha}$ *and* $0 \le z \le \frac{2\pi}{\beta}$.

Test discussed below have been carried out with NT = 40 Chebyshev polynomials which is sufficient to reduce the error of the Chebyshev expansions below the machine error. The remaining error is associated with the Fourier expansions. Results displayed in Fig. 2 demonstrate that the error decreases exponentially with an increase of the number of Fourier modes used and, thus, the algorithm provides the spectral accuracy.

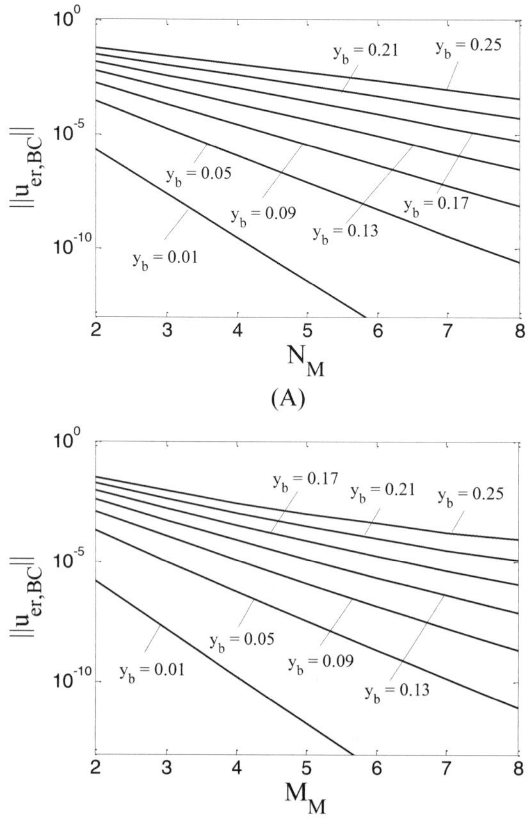

Figure 2: Variation of $\left\|u_{er,BC}\right\|_\infty$ as a function of the number of Fourier modes using in the computations for $\alpha = \beta = 1$, $Re = 5$. The presented results have been determined with NT = 40 and $N_M = 10$ (Fig.2A) and $M_M = 10$ (Fig.2B).

Figure 3 displays spectrum of the error at the lower wall. It can be seen that the first (N_M, M_M) modes have been eliminated which is consistent with the construction of the boundary conditions.

CONCLUSIONS

A gridless, spectrally accurate algorithm has been developed for simulations of three-dimensional viscous flows in domains bounded by rough walls. The algorithm is based on the concept of Immersed Boundary Conditions where the flow domain is immersed inside a fixed computational domain. The algorithm uses Fourier expansions in the parallel-to-the-wall directions and the Chebyshev expansions in the wall-normal direction. Various tests demonstrate that the algorithm delivers spectral accuracy.

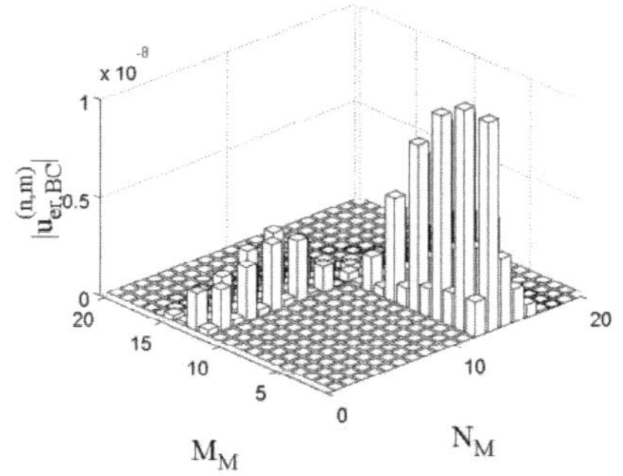

Figure 3: Fourier spectra for the u-velocity component evaluated at the lower wall. The presented results are for NM = MM = 10, $\alpha = \beta = 1$, $Re = 5$.

ACKNOWLEDGMENTS

This work has been carried out with support from the Natural Sciences and Research Council (NSERC) of Canada.

REFERENCES

1. Husain, S. Z., Floryan, J. M. 2008, "Immersed Boundary Conditions Method for Unsteady Flow Problems described by the Laplace Operator", *Int. J. Num. Meth. Fluid Mech.*, Vol(56), pp. 1765-1786.
2. Peskin, C. S. 1982, "The Fluid Dynamics of Heart Valves: Experimental, Theoretical and Computational Methods", *Ann. Rev. Fluid Mech.*, Vol(14), pp. 235-259.
3. Mittal, R., Iaccarino, G. 2005, "Immersed Boundary Methods", *Ann. Rev. Fluid Mech.*, Vol(37), pp. 239-261.
4. Szumbarski, J., Floryan, J. M. 1999, "A Direct Spectral Method for Flows over Rough Boundaries", *J. Comp. Phys.*, Vol(153), pp. 378-402.
5. Fernandez, D. C. D. R., Husain, S. Z., Floryan, J. M. 2011, "Immersed Boundary Conditions Method for Heat Conduction Problems in Slots with Time Dependent Geomery", *Int. J. Numer. Meth. Fluids.*, Vol(67), pp. 478-500.
6. Husain, S. Z., Floryan, J. M. 2013, "Effective Solvers for the Immersed Boundary Conditions Method". *Computers and Fluids,* Vol(84), pp.127-140.

HYDRODYNAMIC STUDY OF GAS-PARTICLE INTERACTION IN AN FCC DOWNER REACTOR

Tatiana López
Grupo de Energía y Termodinámica, Facultad de Ingeniería Mecánica, Escuela de Ingenierías
Universidad Pontificia Bolivariana
Medellín, Colombia
tatiana.lopezmo@upb.edu.co

Carlos A. Bustamante
Grupo de Energía y Termodinámica, Facultad de Ingeniería mecánica, Escuela de Ingenierías
Universidad Pontificia Bolivariana
Medellín, Colombia
carlos.bustamante@upb.edu.co

César Nieto
Grupo de Investigación en Ingeniería Aeroespacial, Facultad de Ingeniería Aeronáutica, Escuela de Ingenierías
Universidad Pontificia Bolivariana
Medellín, Colombia
cesar.nieto@upb.edu.co

ABSTRACT

Hydrodynamics of gas-solids in a downer reactor is studied by solving mass and momentum conservation equations. A 2D Eulerian-Eulerian computational fluid dynamics (CFD) model based on the kinetic theory of granular flow with a κ-ε realizable turbulence model is used. Particle-wall interaction is estimated by the Jackson and Johnson boundary condition model while particle-gas momentum interaction is obtained by using the Gidaspow approach. Computational results for radial distribution of solid volume fraction and solid velocity are compared with experimental measurements available in literature showing good agreement.

KEYWORDS: FCC, multiphase flow, downer reactor, hydrodynamics.

INTRODUCTION

In a downer reactor, gas and solids flow co-currently in the direction of gravity. This characteristic implies several advantages over other types of circulating fluidized beds, such as the riser reactor which is widely used in fluid catalytic cracking processes. Some of these advantages are uniform gas and solids flow structures (radial and axial), reduced residence time and improved contact between phases. Downer reactors have been studied for a long time but not general agreement about its design has yet been reached (1). As for the hydrodynamic behavior inside the downer reactor, several authors have worked in cold models. Wang et al (2) named three different zones along the downer length corresponding to the gas-solids flow behavior: the first and second acceleration zones and the constant velocity zone. Zhang et al (3)(4) worked on a 0.1 m downer laboratory setup where they ran several experiments varying the gas velocity and the solids flux. Their results clearly showed the three development zones inside the dower as well as the near plug behavior that makes a uniform radial profile of solids holdup. In the experimental study done by Zhao and Takei (5), they had an especial distributor where air entered axially and radially into the downer. That distributor proved to be better than those used in previous studies for the case in Which the solid distribution near to the entrance of the downer is uniform. Several others studies similar to those mentioned can be found in the literature showing experimental behavior clearly depends on inlet radial solid distribution, gas velocity and solid mass flux. Cheng et al. (6), worked on a 3D CFD code based on the kinetic theory coupled with turbulence models of both gas and particle phase. The results showed good agreement with the experimental data of a pilot-scale downer setup, it is also noted that a dense-ring structured was present and sustain in the fully developed region. Although other CFD models are reported in the literature showing good agreement with experimental data (7)(8), they do not present the complete set of parameters and equations.

In the present work, the hydrodynamics study was made by using CFD commercial software. The main aim of the work was to validate the computational model so it could be used later for improving downer design before a cold model is constructed. Only the result of the model validation is presented here since the other part of the work is currently being developed.

MATHEMATICAL MODEL

A two phase Eulerian-Eulerian model based on the kinetic theory of granular flow (KTGF) was used for simulating two-dimensional and isothermal gas-solid flow inside a downer reactor. The conservation equations solved are shown below:

Continuity (q^{th} phase)

$$\frac{\partial}{\partial t}\left(\alpha_q \rho_q\right) + \nabla \cdot \left(\alpha_q \rho_q \vec{v}_q\right) = 0, \tag{1}$$

where $\vec{v}_q$ is the velocity, α_q is the volume fraction and ρ_q is the density of the phase q, namely f for gas phase and s for solid phase.

Momentum (gas phase)

$$\frac{\partial}{\partial t}(\alpha_f \rho_f \vec{v}_f) + \nabla \cdot (\alpha_f \rho_f \vec{v}_f \vec{v}_f) = -\alpha_f \nabla p + \nabla \cdot \bar{\bar{\tau}}_f + \alpha_f \rho_f \vec{g} + K_{sf}(\vec{v}_s - \vec{v}_f), \quad (2)$$

with $\vec{g}$ as the gravity vector and $\bar{\bar{\tau}}_f$ is the gas phase stress-strain tensor.

$$\bar{\bar{\tau}}_f = \alpha_f \mu_f (\nabla \vec{v}_f + \nabla v_f^T) + \alpha_f \left(\lambda_f - \frac{2}{3}\mu_f\right) \nabla \cdot \vec{v}_f \bar{\bar{I}} \quad (3)$$

Here μ_f and λ_f are the shear and bulk viscosity of the fluid phase, K_{sf} is an interaction force between phases, and p is the pressure shared by all phases.

Momentum (solid phase)

$$\frac{\partial}{\partial t}(\alpha_s \rho_s \vec{v}_s) + \nabla \cdot (\alpha_s \rho_s \vec{v}_s \vec{v}_s) = -\alpha_s \nabla p - \nabla p_s + \nabla \cdot \bar{\bar{\tau}}_s + \alpha_s \rho_s \vec{g} + K_{sf}(\vec{v}_f - \vec{v}_s), \quad (4)$$

where p_s is the solids pressure, which is obtained from the KTGF in terms of the coefficient of restitution for particle collisions, the radial distribution function and the granular temperature.

$$\bar{\bar{\tau}}_s = \alpha_s \mu_s (\nabla \vec{v}_s + \nabla v_s^T) + \alpha_s \left(\lambda_s - \frac{2}{3}\mu_s\right) \nabla \cdot \vec{v}_s \bar{\bar{I}}$$

Here μ_s and λ_s are the granular and bulk viscosity of the solid phase. Additionally, for a binary system it is clear that $K_{sf} = K_{fs}$.

The other equations, parameters and expressions needed for the mathematical model completion are mentioned in **Table *1*** which also contains their corresponding value or reference where it was reported.

Table 1 Model parameters.

KTGF parameters	Specularity coefficient	0.0003
	Restitution coefficient between particle and wall	0.9
	Restitution coefficient between particles	0.95
Multiphase model	Drag closure, granular viscosity and granular conductivity	(9)
	Bulk viscosity, solids pressure, radial distribution	(10)
	Frictional viscosity	(11)
	Solids volume fraction	0.0125

COMPUTATIONAL MODEL

The geometry and working parameters of the downer reactor were taken from literature in order to have reference values for comparison. The computational simulation has been performed using the commercial software Ansys Fluent 15. ***Table 2*** shows the parameters used in the simulation. After a mesh independency study in a two-dimensional domain, the mesh selected was a structured grid with 62720 elements.

Table 2 Geometrical and computational parameters.

Downer system specifications (3)	Reactor diameter [m]	0,1
	Reactor length [m]	9.3
	Gas density [kg/m^3]	1,2
	solids density [kg/m^3]	1500
	Particles diameter [μm]	67
Computational model	Turbulence model	K-ε realizable
	Solver	2D planar, unsteady
	Pressure velocity coupling	SIMPLE
	Spatial Discretization scheme	Second order UpWind
	Time step [s]	0.001
	Simulation time [s]	40
	Inlet granular temperature	0.1
	Wall boundary condition	(12)

NUMERICAL RESULTS

In order to assure similarity between the experimental procedure and computational model, several simulations were performed by varying solid velocity and keeping constant the gas velocity until the reported average solid flux through a surface normal to flow inside reactor was reached. The following computational results and its corresponding experimental values, taken from (3,4), are for a gas velocity of 10.2 m/s and a solid velocity of 10.92 m/s which corresponds to a mass flux of 205 kg/m^{2}s.

Initially, a base case simulation was ran for the downer reactor according to the parameters presented in **Table *1***. Then, a sensibility analysis was made with parameters such as the granular temperature, restitution coefficient between particle and wall and between particles, and the specularity coefficient, in order to obtain the final combination of parameters presented in **Table 2** whose corresponding results show the best matching to the reference experimental measurements. In **Figure *1*,** the radial distribution of local particle velocity is presented for two different downer heights. Both the experimental and the computational results share the same tendency with the highest value towards the center of reactor and high gradients near to the wall. It was found that the variation of the KTGF parameters have little to none influence on the velocity radial profile and tendency.

As mention in the literature (3)(1), a positive axial pressure gradient is presented in most of the downer height (**Figure *2***) given by the almost constant velocity of solids and gas during their displacement through the downer. Gas velocity is constant because the solids volume fraction is low. In the solid case, the central region particles quickly reach the terminal velocity. Therefore, the potential energy of the mixture is transformed into kinetic energy in the first section of the downer until the constant velocities are reached and then it is converted into flow work allowing a pressure increment. Some of the initial potential and kinetic energy is lost due to friction

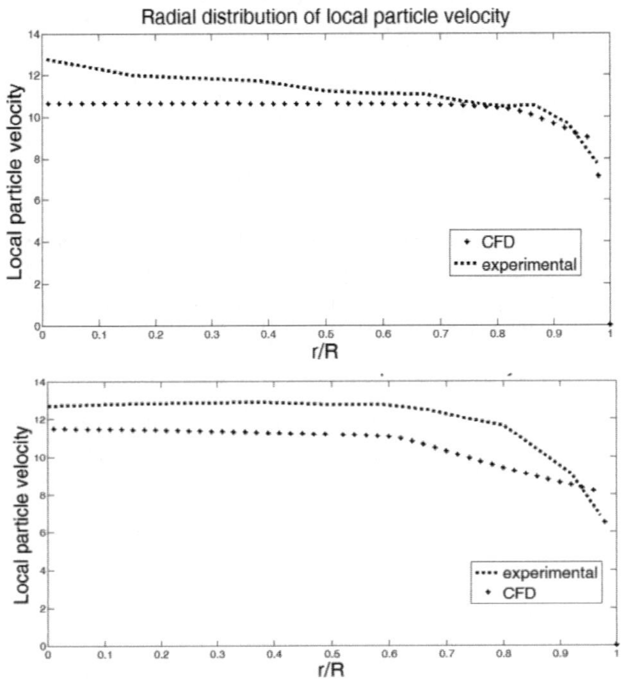

Figure 1: Radial distribution of local particle velocity for a downer height of 2.112 (up) and 8.056 m (down).

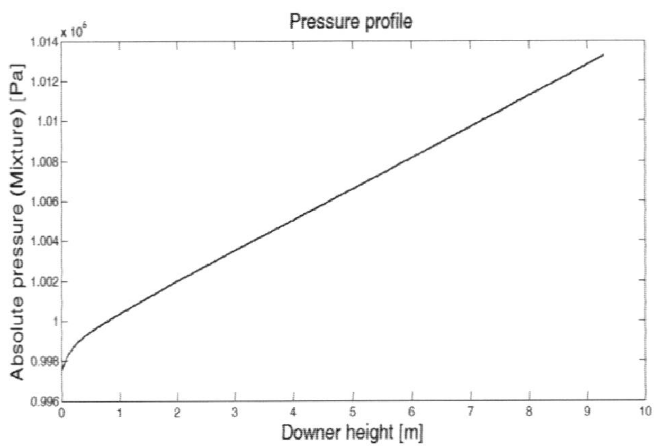

Figure 2: Pressure inside the reactor.

The radial distribution for the local solids volume fraction is presented for the same downer heights in ***Figure 3***. Close to the downer entrance the experimental results shows a volume fraction peak near to the wall that is also present in the computational results but the difference increases towards the center of the downer. Whereas near to the bottom of the downer, the experimental results shows a flatter profile while the CFD simulation keeps the peak. This discrepancy may be mostly due to the inlet radial distribution of solids in the experimental model. Zhang et al (3) use a solids- gas distributor with a special configuration which causes a high solid volume fraction near to the reactor wall while almost zero values towards the center. As it is well known (5)(3)(13), solid distribution inside downer reactor strongly depends on inlet configuration. Although computational results were obtained by assuming uniform inlet solid distribution, the volume fraction range agrees to the experimental data. It was also found that the solids holdup peak in the radial profile was influenced by the KTGF parameters. Depending on the combination of the aforementioned parameters the results will be a higher or lower peak, also it can lead to the displacement of the peak away from the wall and, in cases where the specularity coefficient is low enough, it can result in a double peak occurrence. For all the combinations tried in this study the solids holdup peak was always present.

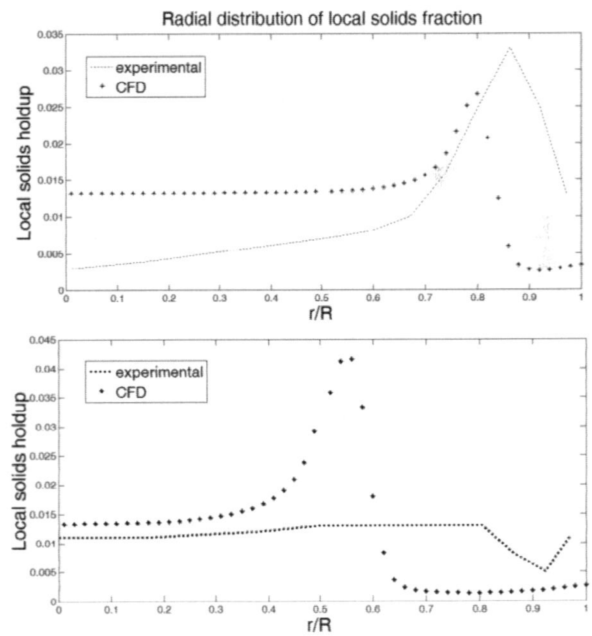

Figure 3: Radial distribution of local solid fraction for a downer height of 2.112 and 8.056 m, respectively.

CONCLUSIONS

An Eulerian-Eulerian approach based on the KTGF computational model for the gas-solids flow into a downer reactor has been presented. The working parameters have been adjusted and compared to an experimental work reported in the literature. The results showed good agreement in terms of range and tendency. It was found that the interaction of the solids in the zone near the wall is very critical for the flow development, and strongly depends on the arrangement made at the entrance of the downer. For a better match between the computational and experimental work, the geometry and configuration of the solids and gas distributor should be copy as close as possible. Since this is an ongoing work, this topic would be addressed in a later stage. Besides, the computational model is acceptable and it is going to be adapted to a 3D geometry where a better match between results is expected.

ACKNOWLEDGMENTS

The present work is being developed with the financial support provided by the Colombian Department of Sciences, Technology, and Innovation (COLCIENCIAS), the Empresa Colombiana de Petróleos (ECOPETROL) and The Universidad Pontificia Bolivariana.

REFERENCES

1. Zhu J, Yu Z, Jin Y. Cocurrent downflow circulating fluidized bed (downer) reactors—a state of the art review. Can J … [Internet]. 1995 [cited 2014 Feb 27];73.

2. Wang Z, Bai D, Jin Y. Hydrodynamics of cocurrent downflow circulating fluidized bed (CDCFB). Powder Technol [Internet]. 1992 Jun;70(3):271–5.

3. Zhang H, Zhu J, Bergougnou M. Hydrodynamics in downflow fluidized beds (1): solids concentration profiles and pressure gradient distributions. Chem Eng Sci [Internet]. 1999 [cited 2014 Aug 4];54:5461–70.

4. Zhang H, Zhu J. Hydrodynamics in downflow fluidized beds (2): particle velocity and solids flux profiles. Chem Eng Sci [Internet]. 2000 [cited 2014 Aug 4];55:4367–77.

5. Zhao T, Takei M. Discussion of the solids distribution behavior in a downer with new designed distributor based on concentration images obtained by electrical capacitance tomography. Powder Technol [Internet]. Elsevier B.V.; 2010 Feb [cited 2014 Feb 27];198(1):120–30.

6. Cheng Y, Guo Y, Wei F, Jin Y, Lin W. Modeling the hydrodynamics of downer reactors based on kinetic theory. Chem Eng Sci [Internet]. 1999 Jul;54(13-14):2019–27.

7. Samruamphianskun T, Piumsomboon P, Chalermsinsuwan B. Computation of system turbulences and dispersion coefficients in circulating fluidized bed downer using CFD simulation. Chem Eng Res Des [Internet]. Institution of Chemical Engineers; 2012 Dec [cited 2014 Jan 28];90(12):2164–78.

8. Peng G, Dong P, Li Z, Wang J, Lin W. Eulerian simulation of gas–solid flow in a countercurrent downer. Chem Eng J [Internet]. Elsevier B.V.; 2013 Aug [cited 2014 Aug 4];230:406–14.

9. Gidaspow D, Bezburuah R, Ding J. Hydrodynamics of circulating fluidized beds, kinetic theory approach. Fluidization VII, 7th engineering foundation conference on fluidization. 1992. p. 75–82.

10. Lun CKK, Savage SB, Jeffrey DJ, Chepurniy N. Kinetic theories for granular flow: Inelastic particles in Couette flow and slightky inelastic particles in a general flow field. J Fluid Mech. 1984;140:223–56.

11. Schaeffer DG. Instability in the evolution equations describing incompressible granular flow. J Differ equations. 1987;66:19–50.

12. Johnson PC, Jackson R. Frictional-Collisional constitutive relations for granular materials, with application to plane shearing. J Fluid Mech. 1987;176:67–93.

13. Cheng Y, Wei F, Guo Y, Jin Y. CFD simulation of hydrodynamics in the entrance region of a downer. Chem Eng Sci [Internet]. 2001 Feb;56(4):1687–96.

Proceedings of the 25th CANCAM
London, Ontario, Canada, May 31 - June 4, 2015

NUMERICAL SIMULATION OF DISPLACEMENT FLOW OF VISCOPLASTIC FLUIDS

A. Eslami
Department of Chemical Engineering
Laval University
Quebec, QC, Canada
Ali.Eslami.1@ulaval.ca

K. Alba
Department of Engineering Technology
University of Huston
Huston, TX, USA
kalba@uh.edu

I.A. Frigaard
Department of Mechanical Engineering
University of British Columbia
Vancouver, BC, Canada
frigaard@math.ubc.ca

S.M. Taghavi
Department of Chemical Engineering
Laval University
Quebec, QC, Canada
Seyed-Mohammad.Taghavi@gch.ulaval.ca

ABSTRACT

The displacement flow of a viscoplastic fluid by a Newtonian fluid is investigated numerically in an inclined 2D channel. The fluids are considered to be miscible. We focus on a situation where the displacing Newtonian fluid is heavier than the displaced viscoplastic one; therefore, the configuration considered is density-unstable. We demonstrate that static residual layers of the displaced fluid are found at near-horizontal inclinations. On the other hand, unstable flows and irregular interfacial patterns are observed at higher inclinations. In addition, we show how the velocity of the displacing penetrating font varies with the imposed flow velocity.

Keywords: Displacement flow, viscoplastic fluid, static layer

INTRODUCTION

Many industrial processes require the removal of a gelled material from a duct. For example, we can mention oil & gas well cementing, waxy crude oil pipeline restarts, biomedical applications (mucus, biofilms), cleaning of equipment, and food processing. An import issue in these processes is the presence of the residual deposits. Although turbulent flows can be found in some situations, process limitations can often enforce the flows to be laminar. In the present computational study, we focus on this laminar-imposed-flow case, which is of relevance to many industrial applications.

Previously, we have studied the effects of circular pipe inclination on the displacement of a yield-stress fluid by a Newtonian fluid of higher density experimentally [1,2], which covered various inclination angles from near-horizontal to fully vertical. For brevity, we refer to these papers for a more detailed introduction and review of previous work. The main observation made in our previous works was that a strong yield-stress fluid in displacement flows can lead to the presence of two main regimes that we called center-type and slump-type displacements [1,2]. In the current work, we would like to study the effects of the presence of a yield stress for displacement flows that take place in a 2D channel geometry computationally. The novelty of the current study firstly is that through a computational approach, the displacement flow problem may now be studied over a broader range of parameters compared to those used in experiments. Secondly, more detailed information on the flow concentration and velocity fields have been obtained that can lead scientists better understand the underlying mechanism of the displacement flow problem. Thirdly, the current numerical approach sets the basis to model more complicated flows than 2D such as pipes or other 3D geometries.

GOVERNING EQUATIONS

We study the displacement of a Bingham fluid (fluid 2) by a Newtonian one (fluid 1) along a channel of width $\hat{D}$ .The fluids are miscible and have different densities. We choose Cartesian coordinates $(\hat{x}, \hat{y})$ with $\hat{x}$ representing the streamwise direction.[1] The dimensionless equations of motion are:

$$[1+\phi At]Re\left[\frac{\partial}{\partial t}\mathbf{u}+(\mathbf{u}.\nabla)\,\mathbf{u}\right] = -\nabla p+\nabla.\tau+\frac{\phi Re}{Fr^2}\mathbf{e}_g, \quad (1)$$

$$\nabla.\mathbf{u} = 0, \quad (2)$$

$$C_t+\mathbf{u}\cdot\nabla C = \frac{1}{Pe}\nabla^2 C. \quad (3)$$

where $\mathbf{u}$ denotes the velocity, p is the pressure, τ shows the deviatoric stress. Here $\mathbf{e}_g = (\cos\beta, -\sin\beta)$ and the function $\phi(C) = 2C - 1$ interpolates linearly between -1 and 1 for $C \in [0, 1]$. No slip conditions are satisfied at the walls, the heavy fluid enters fully developed (plane Poiseuille profile) at $x = -L/4$ and outflow conditions are applied at $x = 3L/4$. In order to make these equations dimensionless, we have used the following scaling:

$$\mathbf{x}=\frac{\hat{\mathbf{x}}}{\hat{D}},\quad t=\frac{\hat{V}_0}{\hat{D}}\hat{t},\quad \mathbf{u}=\frac{\hat{\mathbf{u}}}{\hat{V}_0},$$
$$p=\frac{\hat{D}}{\hat{\mu}_1\hat{V}_0}\hat{p},\quad \tau=\frac{\hat{D}}{\hat{\mu}_1\hat{V}_0}\hat{\tau} \quad (4)$$

where $\hat{\mu}_1$ denotes the viscosity of the displacing fluid, which is Newtonian. The dimensionless groups that appear in (1-3) are the Reynolds number Re, densimetric Froude number Fr, channel inclination β, Atwood number At and Péclet number Pe. We denote the density of the displacing fluid by $\hat{\rho}_H$ and that of the displaced fluid by $\hat{\rho}_L$. The Atwood number is defined as $At = (\hat{\rho}_H - \hat{\rho}_L)/(\hat{\rho}_H + \hat{\rho}_L)$, representing a dimensionless density difference. For this study $At > 0$ since $\hat{\rho}_H > \hat{\rho}_L$ (heavy fluid displacing the light fluid in the downwards direction). Our simulations are all performed for small At, the significance of which is that a Boussinesq approximation is valid. Briefly, this means that density differences can significantly affect buoyancy forces, captured by the densimetric Froude number, $Fr = \hat{V}_0/\sqrt{At\hat{g}\hat{D}}$, but do not significantly affect the acceleration of the individual fluids. Here $\hat{g}$ is the gravitational acceleration, $\hat{V}_0$ is the mean imposed velocity. The Reynolds number is defined as $Re = \hat{V}_0\hat{D}/\hat{\nu}$, where $\hat{\nu}$ is defined using the mean density $\hat{\rho} = (\hat{\rho}_L + \hat{\rho}_H)/2$ and the viscosity of fluid 1, $\hat{\mu}_1$. Finally, the Péclet number $Pe = \hat{V}_0\hat{D}/\hat{D}_m$ (here $\hat{D}_m$ is the molecular diffusivity), and typically we have $Pe \gg 1$. This suggests that on the timescale of interest, molecular diffusivity does not play a major role in the flows studied.

The Newtonian displacing fluid has the constitutive law of $\tau(\mathbf{u}) = \dot{\gamma}(\mathbf{u})$ with $\dot{\gamma} = \nabla\mathbf{u} + (\nabla\mathbf{u})^T$. The displaced fluid is assumed to be a Bingham fluid, which presents a yield stress effect. For this model, we have:

$$\tau_2(\mathbf{u}) = m\left[1+\frac{B}{\dot{\gamma}(\mathbf{u})}\right]\dot{\gamma}(\mathbf{u}) \Leftrightarrow \tau_2(\mathbf{u}) > mB, \quad (5)$$
$$\dot{\gamma}(\mathbf{u}) = 0 \Leftrightarrow \tau_2(\mathbf{u}) \le mB \quad (6)$$

where the second invariants, $\dot{\gamma}(\mathbf{u})$ and $\tau_2(\mathbf{u})$, are defined by:

$$\dot{\gamma}(\mathbf{u}) = \left[\frac{1}{2}\sum_{i,j=1}^{2}[\dot{\gamma}_{ij}(\mathbf{u})]^2\right]^{1/2},$$
$$\tau_2(\mathbf{u}) = \left[\frac{1}{2}\sum_{i,j=1}^{2}[\tau_{2,ij}(\mathbf{u})]^2\right]^{1/2}. \quad (7)$$

The viscosity ratio m is:

$$m \equiv \frac{\hat{\mu}_2}{\hat{\mu}_1} \quad (8)$$

where $\hat{\mu}_2$ is a viscosity scale for fluid 2 (displaced) and Bingham number B can be introduced as:

$$B \equiv \frac{\hat{\tau}_Y}{\hat{\mu}_1[\hat{V}_0/\hat{D}_0]}. \quad (9)$$

Fig. 1 shows the flow geometry and notation used in the current study. Equations of motion are discretised using a mixed finite element, finite volume method. The Navier-Stokes equations are solved using Galerkin finite element method. The divergence-free condition is enforced by an augmented Lagrangian technique [3]. A fixed time step is used for the Navier-Stokes equations, advancing from time step N to $N+1$. The convective velocity is approximated at time step N

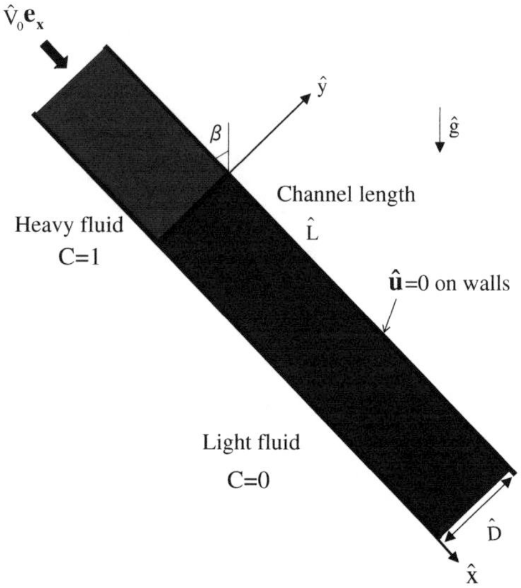

Figure 1: Schematic view of the numerical domain used.

[1] In this paper, we adopt the convention of denoting dimensional quantities with the ^ symbol and dimensionless quantities without.

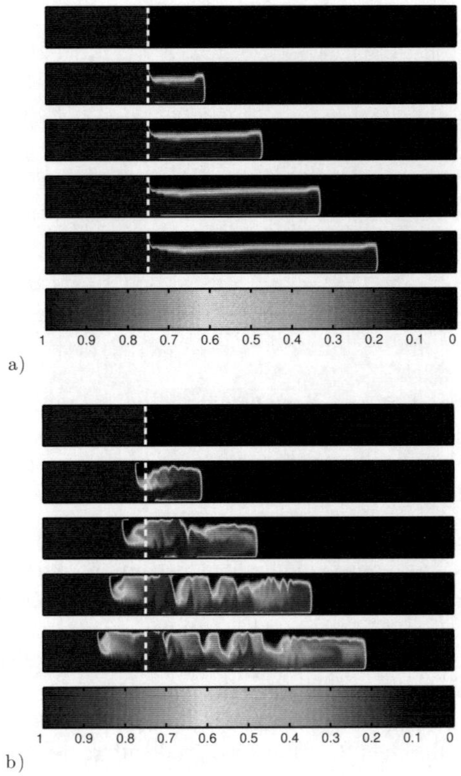

Figure 2: Concentration maps of the numerical simulation as times $t = [0,\ 11.03,\ 22.06,\ 33.09,\ 44.13]$, run for $At = 0.01$, $m = 1$, $B = 40$, $Re = 400$ and $Fr = 0.485$ for a) $\beta = 80°$ and b) $\beta = 60°$. In this figure $L = 105$. The last image at the bottom of the figure is the colourbar of the concentration values.

while the linear spatial derivatives of the velocity are approximated implicitly at time step $N + 1$. The pressure is approximated at time step $N + 1$ (semi-implicit method regarding the nonlinear terms). More details of the computational methodology is given in [4]. To check the accuracy of the code, various simple test problems have been implemented.

We are mostly interested in predicting global features of the displacement flow. For the meshes in most of the computations we have used 28 cells across the channel of height 19.05 (mm), refined slightly towards the walls, and 400 cells along the $\hat{L} = 2$ (m) length of the channel. The initial interface between the two fluids in most cases was placed at a distance $L/4$ from the left side of the computational domain where $L = \hat{L}/\hat{D} \approx 10^2$ is the dimensionless channel length (see also Fig. 1).

At $t = 0$, the two fluids are separated by an imaginary gate valve located at $x = 0$. As time progresses, due to the imposed mean velocity, the heavy layer penetrates through the light layer and displaces it. The streamwise distance of the tip of the advancing displacing front from the gate valve increases with time.

RESULTS

In this section we present our computational results. Fig. 2 shows the concentration maps of the numerical simulation as time evolves, run for $At = 0.01$, $m = 1$, $B = 40$, $Re = 400$ and $Fr = 0.485$. The only difference between the Fig. 2(a) and Fig. 2(b) is that the channel is less inclined with respect to vertical in the latter. Fig. 2(a) presents a qualitatively stable flow where the yield stress of the displaced fluid has been able to damp interfacial instabilities. If this simulation is run with the same parameter set except with the displaced fluid being Newtonian ($B = 0$), the interface between the two fluid will become unstable (see for example [5]). Fig. 2(a) also shows that the displaced fluid layer close to top wall becomes static after the displacing front passes underneath. Fig. 2(b) illustrates a completely different behavior. The interface between the two fluids in this figure is unstable leading to appearance of irregular interfacial patterns. Also note that there is stronger *back flow* in this case compared to $\beta = 80°$ case, meaning that the displaced fluid has largely moved upstream of the gate valve location. The reason is that due to the fact that the channel is closer to the vertical in $\beta = 60°$ case, the driving buoyant forces in the streamwise direction are larger.

Fig. 3 shows the spatiotemporal diagram of the depth-averaged concentration values, $\bar{C}_y$ for the same simulation as Fig. 2. As it can be observed in Fig. 3(a), the flow is stable over a long time. Displacing front seems to stably move with a constant speed. The displaced flow trailing front only slightly moves backward. Fig. 3(b) shows that the flow is unstable. The displacing fluid leading front has a nearly constant speed. However, the trailing front initially moves backward but stops for the duration of the simulation.

In Fig. 4 we show the velocity vector fields for the same simulations as Figs. 2,3. Fig. 4(a) confirms that the flow is nearly stable and that the displaced fluid layer close to top wall is entirely static. We observe the fluid within the displaced layer has a zero velocity. Fig. 4(b) image confirms that the flow is unstable; however, the flow velocity becomes close to zero locally and static segments of the displaced layer are found at least for short periods of time.

Fig. 5 depicts the speed of the leading front $\hat{V}_f$ versus $\hat{V}_0$ for $At = 0.01$, $m = 1$ and $B = 40$ for $\beta = 80°$ (black) and $\beta = 60°$ (red). By increasing the mean flow speed $\hat{V}_0$, the front velocity $\hat{V}_f$ should also naturally increase as suggested by the figure. The data in Fig. 5 also imply that the leading front velocity increases nearly linearly with mean imposed velocity. Last observation in Fig. 5 is that $\hat{V}_f$ is slightly larger when the channel is more inclined with respect to vertical. It seems as if the flow instability in the case of $\beta = 60°$ leads to the displacing front deceleration.

CONCLUSIONS

In this paper, we have considered an interesting displacement flow where a heavier Newtonian fluid displaces a lighter viscoplastic fluid. To do so, we have performed computational simulations of miscible displacement flows in an in-

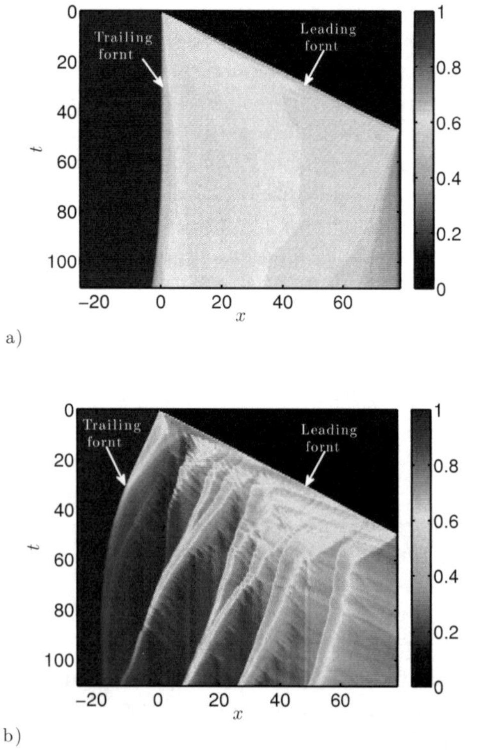

Figure 3: Spatiotemporal diagram of the depth-averaged concentration values, $\bar{C}_y$ for the same simulation as Fig. 2.

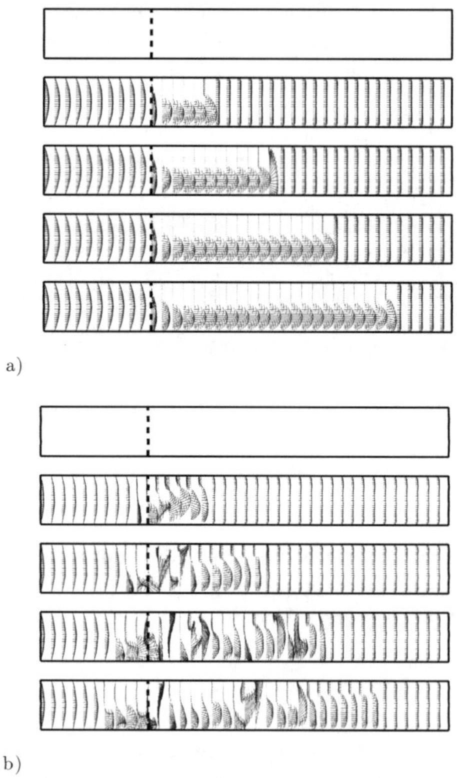

Figure 4: Velocity vector fields for the same simulations as Fig. 2.

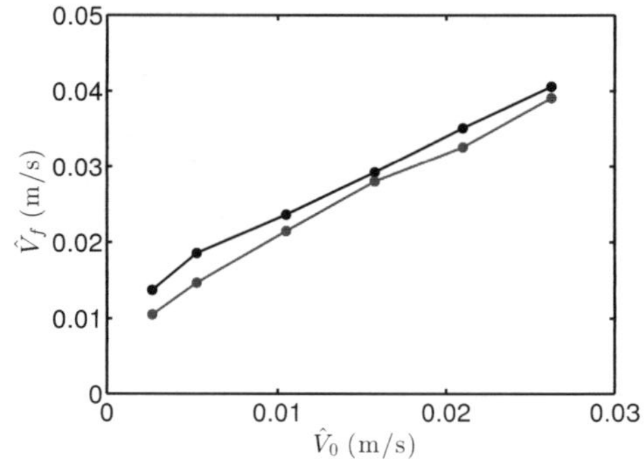

Figure 5: Speed of the leading front $\hat{V}_f$ versus $\hat{V}_0$ for $At = 0.01$, $m = 1$ and $B = 40$ for $\beta = 80°$ (black) and $\beta = 60°$ (red).

clined channel. Our results show that the yield stress, in some cases, can damp the interfacial instabilities and that truly static layers of the displaced fluid can be found for certain flow parameters. Displacement flows at lower angles with the respect to vertical can be more unstable. In this case, irregular interfacial patterns between the two fluids form. In addition, it has been found that the leading front velocity is slightly larger when the channel is more inclined with respect to vertical.

ACKNOWLEDGEMENT

The research has been enabled in part by support provided by Calcul Québec, WestGrid and Compute Canada/Calcul Canada (www.computecanada.ca).

REFERENCES

[1] Taghavi, S.M., Alba, K., Moyers-Gonzalez, M., and Frigaard, I.A. 2012 "Incomplete fluid-fluid displacement of yield stress fluids in near-horizontal pipes: experiments and theory," *J. non-Newt. Fluid Mech.*, Vol(167-168), pp. 59-74.

[2] Alba, K., Taghavi, S.M., de Bruyn, J.R., and Frigaard, I.A. 2013 "Incomplete fluid– fluid displacement of yield-stress fluids. part 2: Highly inclined pipes," *J. non-Newt. Fluid Mech.*, Vol(201), pp. 80-93.

[3] Glowinski, R., and Le Tallec, P. 1989 "Augmented Lagrangian and operator-splitting methods in nonlinear," *SIAM*.

[4] Alba, K., Taghavi, S.M., and Frigaard, I.A. 2014 "Miscible density-unstable displacement flows in inclined channels," *Phys. Fluids*, Vol(26), pp. 122104.

[5] Taghavi, S.M. and Alba, K., and Frigaard, I.A. 2012 "Buoyant miscible displacement flows at moderate viscosity ratios and low Atwood numbers in near-horizontal ducts," *J. Fluid Mech.*, Vol(69), pp. 404-418.

SHEAR-THINNING EFFECTS ON NANOFIBER FORMATION BY CENTRIFUGAL SPINNING

S. M. Taghavi
Department of Chemical Engineering
Laval University
Quebec, QC, Canada
Seyed-Mohammad.Taghavi@gch.ulaval.ca

R.G. Larson
Department of Chemical Engineering
University of Michigan
Ann Arbor, MI, USA
Rlarson@umich.edu

ABSTRACT

We extend a regularized thin fiber model to determine the steady fiber velocity and diameter of a shear-thinning viscous fiber emerging from a nozzle rotating about an axis in the presence of centrifugal, inertial, surface-tension and viscous forces of arbitrary magnitudes. We focus on the case where the rotation is fast and fiber is viscous, as it is of interest in practical situations. We find that the shear-thinning effects can significantly modify the trajectory, the speed, and the radius of the fiber as it moves away from the center for the rotation.

Keywords: Nanofiber, shear-thinning, regularized model

INTRODUCTION

Nanofibers are important materials with enormous applications thanks to their high surface-area-to-volume ratio. They can also provide improved thermal and mechanical properties. Nanofibers are produced through many methods, a recent of which is through applying a centrifugal force (centrifugal spinning) [1]. In this process a nozzle or orifice rotates at high speed around its axis of symmetry and produces high volumes of polymeric nanofibers. To apply this simple process, a polymer solution or melt should be loaded into a specially designed spinneret that is capable of rotating the fluid at a high speed. Figure 1 illustrates a schematic view of the centrifugal spinning process. Due to the centrifugal force applied by rotation, the polymeric fluid is forced through the orifices of the spinneret producing continuous polymer jets. As it can be seen in the figure, the trajectory of each polymer jet is curved due to the rotational forces. The fibers are stretched during the process and they are then collected as a fine web on the collector positioned away from the center of rotation.

Marheineke *et al.* [2] derived an asymptotic "string" model for the dynamics of curved viscous inertial Newtonian fibers subject to a surface tension. The string model is a zeroth-order slender body model, whose slenderness parameter is the ratio of fiber radius to radial position of the orifice. This formulation can provide relatively simple model to describe the fiber behavior. However these types of models suffer from limitations when viscous forces dominate the flow at short distances from the orifice. Arne *et al.* [3] showed that the string model has no physically relevant stationary solution when $ReRb^2 < 1$, with Reynolds number Re (expressing the ratios between inertia and viscosity) and Rossby number Rb) (expressing the ratios between inertia and rotation). The regime of small Rb and Re is, however, the regime of high spinning speeds and viscous fibers; therefore, it is the most important regime of greatest interest in production of nanofibers in centrifugal spinning. This limitation of the string model also effectively exists when the model includes shear-thinning effects. We have recently proposed a regularized thin fiber (string) model that overcomes past numerical limitations and for the first time allows determination of the steady fiber velocity and diameter of a (semi-infinit) Newtonian viscous fiber emerging from a nozzle rotating about an axis in the presence of centrifugal, inertial, and viscous forces of arbitrary magnitudes [4]. Here, we extend this model to include realsitic shear-thinning effects. These effects have been only studied for the regime of large Rb and Re [5].

GOVERNING EQUATIONS

We consider a spinneret of radius $\hat{r}_0$ rotating about its axis with constant angular velocity $\hat{\Omega}$ (in a counter-clockwise di-

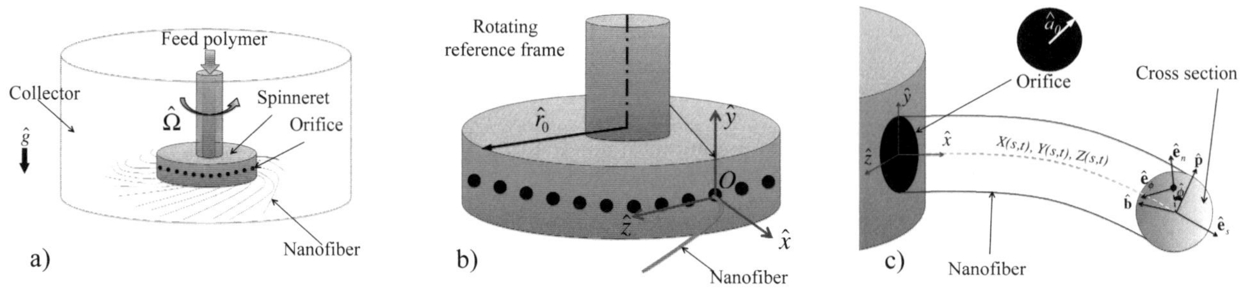

Figure 1: a) Schematic view of nanofiber formation through centrifugal spinning; b) Spinneret and rotating Cartesian reference frame; c) Closer view of the orifice and depiction of the jet with curvilinear coordinate system.

rection) along its axis of symmetry. In practice, hundreds of liquid jets typically emerge from the nozzles of the spinneret. However, here for the purpose of simplicity, we consider only one nozzle and neglect the effects of the jets on one another. We also assume that the orifice is circular. Typically the radius of the orifice of the nozzle, $\hat{a}_0$ and the radius of the liquid jet, $\hat{a}$ are much smaller than the radius of the spinneret, $\hat{r}_0$. We also introduce dimensionless fiber radius as $R = \hat{a}/\hat{a}_0$. [1] For simplicity, we here neglect the effects of gravity, heat transfer, mass transfer, and air drag.

We consider a frame of reference with coordinate system $(\hat{x}, \hat{y}, \hat{z})$, which rotates with the angular speed $\hat{\Omega}$; therefore the orifice from which the jet emerges is fixed (see figure 1b). For a rapidly rotating spinneret, we assume that inertial forces due to the rotation dominate gravitational body forces acting in the direction of the negative $\hat{y}$-axis so that the fiber is moving primarily in the plane perpendicular to the axis of rotation ($\hat{x}, \hat{z}$-plane). Extending the work of Wallwork *et al.* [6] and Parau *et al.* [7], we deploy a curvilinear coordinate system $(\hat{s}, \hat{n}, \hat{\phi})$ where $\hat{s}$ is the arclength along the fiber centerline of the jet from the orifice, and $(\hat{n}, \hat{\phi})$ are cross-sectional plane polar coordinates in the radial and azimuthal directions. The associated unit vectors can be shown by $\hat{\mathbf{e}}_s$, $\hat{\mathbf{e}}_n$, and $\hat{\mathbf{e}}_\phi$, respectively (see figure 1c). Extensive discussion and proper derivation for a similar case are explained further in Decent *et al.* [8].

Having its origin in the centre of the circular orifice (denoted by $\hat{O}$), the position of the centerline of the jet can be described by Cartesian coordinates $(X(s,t), 0, Z(s,t))$, where s is the arclength along the centerline of the jet made dimensionless with $\hat{r}_0$, and X and Z are $\hat{x}$ and $\hat{z}$ positions made dimensionless in the same way, and t is time made dimensionless with $\hat{r}_0/\hat{u}_0$. In this coordinate system the $\hat{x}$-axis is directed normal to the surface of the spinneret in the initial direction of the jet and the $\hat{z}$-axis is orthogonal to the $\hat{x}$-axis in the plane of the centerline of the jet. Positive values of $\hat{z}$ are in the direction opposite to the motion of the spinneret. Following Wallwork *et al.* [6], the unit vectors in this coordinate system can be calculated by a principal normal vector $\hat{\mathbf{p}}$ and a binormal vector $\hat{\mathbf{b}}$ to the centreline (see Decent *et al.* [8] for details). The choice of the coordinate system is motivated by Wallwork *et al.* [6], Parau *et al.* [7], and Uddin *et al.* [5].

Using the velocity vector $\hat{\mathbf{u}} = \hat{u}\hat{\mathbf{e}}_s + \hat{v}\hat{\mathbf{e}}_n + \hat{w}\hat{\mathbf{e}}_\phi$, where $\hat{u}$, $\hat{v}$ and $\hat{w}$ are the tangential, radial and azimuthal velocity components relative to the centerline of the jet respectively, the equations of motion can be written as:

$$\hat{\rho}\left(\frac{\partial \hat{\mathbf{u}}}{\partial \hat{t}} + \hat{\mathbf{u}}.\nabla\hat{\mathbf{u}}\right) = -\nabla\hat{p} + \nabla.\hat{\mathbf{T}} - 2\hat{\boldsymbol{\omega}} \times \hat{\mathbf{u}} - \hat{\boldsymbol{\omega}} \times (\hat{\boldsymbol{\omega}} \times \hat{\mathbf{r}}), \tag{1}$$

$$\nabla.\hat{\mathbf{u}} = 0, \tag{2}$$

where $\hat{t}$ is time, $\hat{\rho}$ is the constant density of the fluid, $\hat{p}$ is the pressure within the jet and $\hat{\boldsymbol{\omega}} = \left(0, \hat{\Omega}, 0\right)$ is the angular velocity vector of the spinneret. To include the shear-thinning effects, we rely on a simple power-law fluid model for the constitutive equation ($\hat{\mathbf{T}} = \hat{\eta}\hat{\dot{\gamma}}$, where $\hat{\eta}$ is the apparent viscosity defined as $\hat{\eta} = \hat{\mu}_0\hat{\dot{\gamma}}^{n-1}$, where $\hat{\dot{\gamma}}$ is the shear rate, $\hat{\mu}_0$ is the fluid constancy, and n is the power-law index). Here we focus on the steady-state solutions of the governing equations, in which all the variables are only functions of the arclength, s. Assuming that jet is a long, slender, thin object with the aspect ratio $\epsilon = \hat{a}_0/\hat{r}_0$, one can expand velocities and pressure in Taylor series in ϵn and R, X, Z in asymptotic series in ϵ, and at zeroth order in ϵ arrive at the so-called "string" equations, where the filament diameter is so thin compared to other length scales that the fiber radius is vanishing compared to other length scales [5]:

$$-3\left(\frac{\left|\sqrt{3}u_s\right|^{n-1}u_s}{u}\right) + \frac{\mathrm{Re}}{2We\sqrt{u}}\left(\frac{u_s}{u}\right) + \mathrm{Re}\,u_s - \frac{\mathrm{Re}\left[(X+1)X_s + ZZ_s\right]}{Rb^2} = 0, \tag{3}$$

$$\begin{aligned}\delta\kappa_{sss} &= \left(\frac{(7-n)}{2}\left|\sqrt{3}u_s\right|^{n-1} - \mathrm{Re}\,u + \frac{\mathrm{Re}}{We\sqrt{u}}\right)\kappa \\ &+ \left(\frac{-2\,\mathrm{Re}}{Rb} + \frac{\mathrm{Re}\left[ZX_s - (X+1)Z_s\right]}{Rb^2 u}\right) \\ &- (n-1)\left|\sqrt{3}u_s\right|^{n-1}\kappa_s,\end{aligned} \tag{4}$$

$$Z_{ss} = -X_s\kappa, \tag{5}$$

$$X_s^2 + Z_s^2 = 1, \tag{6}$$

[1] In this paper, we adopt the convention of denoting dimensional quantities with the ^ symbol (e.g. the orifice radius $\hat{a}_0$) and dimensionless quantities without.

where all the parameters are of leading order in ϵ, the subscript s represents a derivative with respect to s, and κ is the generalized curvature of the liquid jet. Equation (3) represents the linear momentum balance in the axial direction, equation (4) is the equation of motion in the radial direction forming a solvability condition [5,7], equation (5) is the relation for the generalized curvature, and equation (6) is the arclength condition. To obtain these dimensionless equations, the velocity u has been non-dimensionalised with respect to the jet exit velocity u_0. The dimensionless parameters in these equations are the Rossby number $\mathrm{Rb} = \hat{u}_0/\hat{r}_0\hat{\Omega}$, the Reynolds number $\mathrm{Re} = \hat{\rho}\hat{u}_0^{2-n}\hat{r}_0^n/\hat{\mu}_0$ and the Weber number $We = \hat{\rho}\hat{u}_0^2\hat{a}/\hat{r}_0$. See [5] for more details on the derivation. Note that in Eq. (4), δ is introduced as a regularization coefficient, which is zero in lowest order in ϵ in the derivation of the equations of the usual string model (e.g. see [5]).

Physically the regularization coefficient represents the order of the thickness of the boundary layer close to the orifice, which is expected to have little effect much further away from the orifice. For a very thin jet, the layer is accordingly very small; therefore the solution away from the orifice becomes independent of the choice of δ. In our computations, we have in fact observed that fiber trajectory, speed and radius are independent of the choice of the regularization coefficient as long as δ is kept small (typically $\delta \leq 0.001$).

The upstream boundary conditions for equations (3-6) are $X(0) = Z(0) = Z_s(0) = \kappa(0) = 0$, and $u(0) = X_s(0) = 1$. For small Rb, in the domain between $s = 0$ and $s = L$ we treat the problem as a boundary value problem with the end boundary conditions of $\kappa_s(L) = \kappa_{ss}(L) = 0$ and $u_s(L)/u(L) = 0$. The former boundary condition keeps the curvature bounded within the domain; the latter is equivalent to a downstream condition that the viscous stress there be zero (i.e., stress-free dynamic boundary condition at a certain length). In this domain, we have set up systems of nonlinear equations using a Runge-Kutta collocation method combined with a Newton method. In order to validate our model, we compared various results including the fiber radius and the trajectory of the centerline of the steady jet (Z, X) with those of Parau *et al.* [7] for $\mathrm{Re}\,\mathrm{Rb}^2 > 1$ as well as with Cosserot rod results of Arne *et al.* [3] for $\mathrm{Re}\,\mathrm{Rb}^2 < 1$, and both showed excellent agreement.

RESULTS

For simplicity we have included our main results in one figure, i.e., Fig. 2, where we show the results of model simulation for a shear-thinning fluid, all plotted for the same parameters. Fig. 2(a) shows that steady centerline (Z, X) versus s, where it is evident that the shear-thinning properties can modify the trajectory of the fiber even over a short length. When the fluid is more shear-thinning, the fiber turns inwards further as the spinneret rotates. Fig. 2(b) plots the steady fiber speed and shows that the fiber speed increases with increasing the shear-thinning properties (i.e., decreasing the power-law index n). Fig. 2(c) shows the steady fiber radius and demonstrates that the fiber with the lowest n thins more rapidly. Variation of u_s/u versus s is plotted in Fig. 2(d) showing that the shear-thinning effects can modify u_s/u. Fiber angles α for $n = 1$, $n = 0.75$ and $n = 0.5$ are plotted in Fig. 2(e). This figure shows that the fiber angles, which are more or less the same for small s, becomes different at longer distances, with the more shear-thinning fluid having the largest (negative) angle. The steady fiber curvature κ versus s plotted in Fig. 2(f) shows the shear-thinning effects only slightly affect the fiber curvature.

CONCLUSIONS

We have developed a regularized string model to predict the fiber trajectory and diameter as a function of distance from the spinneret for centrifugal spinning with centrifugal, inertial, surface-tension and shear-thinning viscous forces of arbitrary magnitudes. The fiber trajectory, radius, speed, angle, and curvature can be found through our model. We have found that when the shear-thinning effects are higher (i.e., the power-law index n is lower), the fiber thins faster.

REFERENCES

[1] Padron, S., Fuentes, A., Caruntu, D. and Lozano, K. 2013 "Experimental study of nanofiber production through forcespinning," *J. Appl. Phys.*, Vol(113), pp. 024318.

[2] Marheineke, N., and Wegener, R., 2009 "Asymptotic model for the dynamics of curved viscous fibres with surface tension," *J. Fluid Mech.*, Vol(622), pp. 345-369

[3] Arne, W., Marheineke, N., Meister, A. and Wegener, R. 2010 "Numerical analysis of Cosserat rod and string models for viscous jets in rotational spinning processes," *Math. Models Methods Appl. Sci.*, Vol(20), pp. 1941-1965.

[4] Taghavi, S.M., and Larson, R. G. 2014 "Regularized thin-fiber model for nanofiber formation by centrifugal spinning," *Phys. Rev. E*, Vol(89), pp. 023011 (Erratum *Phys. Rev. E* 89, 059903 (2014)).

[5] Uddin, J., Decent, S.P., and Simmons, M.J.H. 2008, "Non-linear waves along a rotating non-Newtonian liquid jet," *Int. J. Eng. Sci.*, Vol(46), pp.1253-1265.

[6] Wallwork, I.M., Decent, S.P., King., A.C., and Schulkes, R.M.S.M. 2002, "The trajectory and stability of a spiralling liquid jet. Part 1. Inviscid theory," *J. Fluid Mech.*, Vol(459), pp. 43- 65

[7] Parau, E.I., King., Decent, S.P., Simmons, M.J.H., Wong, D.C.Y., and King. A.C. 2007 "Nonlinear viscous liquid jets from a rotating orifice," *J. Eng. Math.*, Vol(57), pp. 159-179.

[8] Decent, S.P., King., A.C., and Wallwork, I.M. 2002, "Free jets spun from a prilling tower," *J. Eng. Math.*, Vol(42), pp. 265-282.

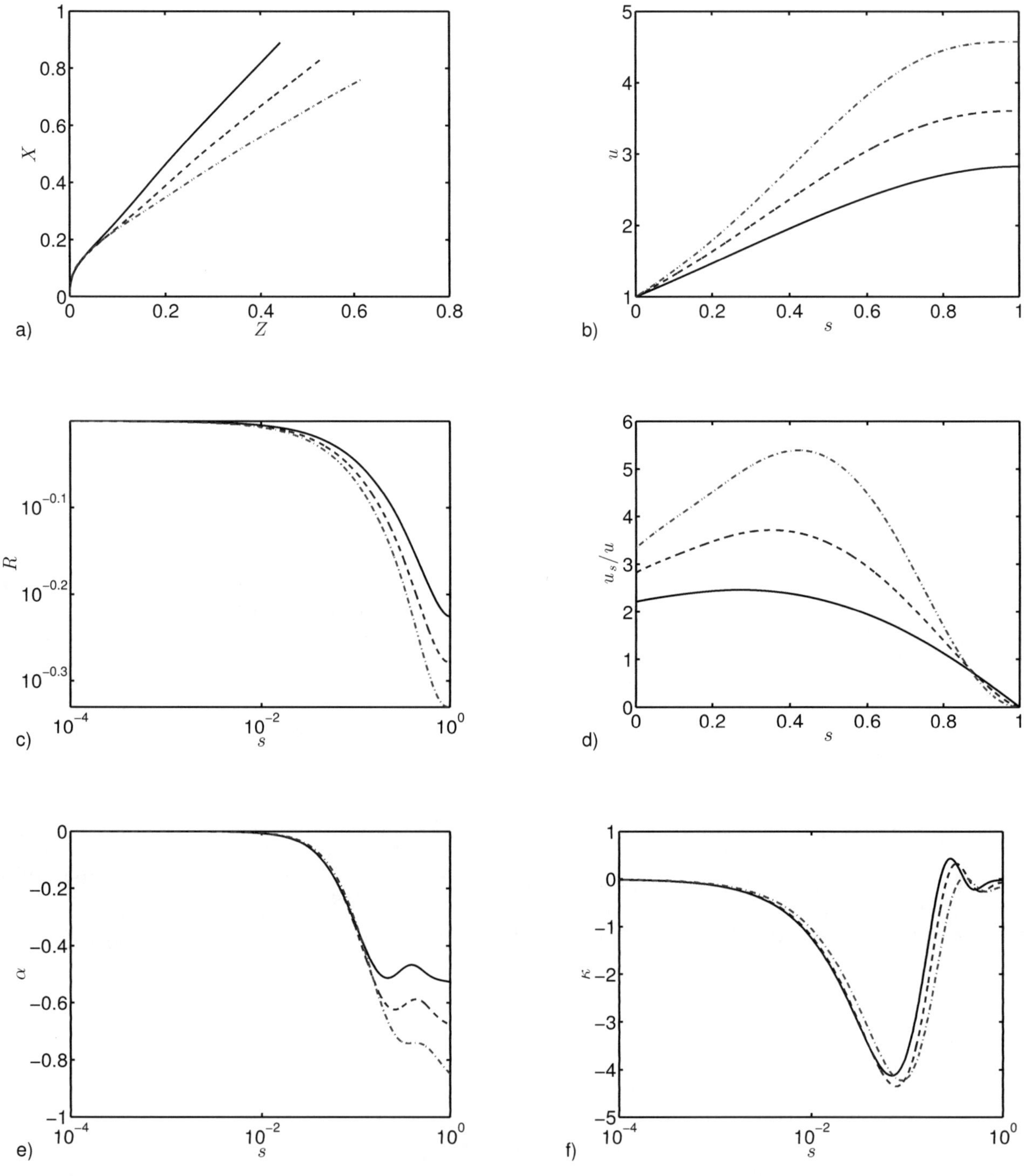

Figure 2: The results of model simulation for a shear-thinning fluid: (a) The steady centerline (Z, X); (b) Steady fiber speed u versus s, (c) Steady fiber radius R versus s; (d) Variation of u_s/u versus s; (e) Steady fiber angle α versus s; (e) Steady fiber curvature κ versus s. Parameters are $Rb = 0.1$, $Re = 0.1$, $We = 10$, $\delta = 0.001$, $L = 1$ for $n = 1$ (solid line), $n = 0.75$ (dashed line), and $n = 0.5$ (dashed-dotted line).

TWO-DIMENSIONAL HYDRODYNAMICS STUDY OF FCC RISER REACTOR BY EULERIAN- EULERIAN APPROACH

Natalia Gómez
Grupo de Energía y Termodinámica, Facultad de Ingeniería Mecánica, Escuela de Ingenierías
Universidad Pontificia Bolivariana
Medellín, Colombia
natalia.gomezv@upb.edu.co

Carlos A. Bustamante
Grupo de Energía y Termodinámica, Facultad de Ingeniería Mecánica, Escuela de Ingenierías
Universidad Pontificia Bolivariana
Medellín, Colombia
carlos.bustamante@upb.edu.co

César Nieto
Grupo de Investigación en Ingeniería Aeroespacial, Facultad de Ingeniería Aeronáutica, Escuela de Ingenierías
Universidad Pontificia Bolivariana
Medellín, Colombia
cesar.nieto@upb.edu.co

ABSTRACT

A study of two-dimensional two-phase flow in a fluid catalytic cracking (FCC) riser reactor by means of computational fluid dynamics (CFD) simulations is presented. Mass and momentum conservation equations are solved in addition to turbulence (k-e model) and the kinetic theory of granular flow (KTGF) equations, for the gas phase and the catalyst particles. Both phases are modeled by an Eulerian approach, for which several models for the momentum interaction between phases are considered (the Gidaspow approach and Syamlal-Obrien drag models) in order to analyze its effect over particle distribution. The results obtained for the radial distribution of solid velocity and solid volume fraction are compared with experimental results reported by others authors showing a similar behavior. As expected for the solid volume fraction values a higher concentration towards the axis of the reactor is reported. The values obtained in the simulation are closer to the experimental data as the steady state is attained.

KEYWORDS: FCC, Riser, hydrodynamics, two-phase flow, volume fraction.

INTRODUCTION

Riser reactors for fluid catalytic cracking (FCC) process has been widely used in last fifty years for processing crude oil in products, especially gasoline. Therefore, the knowledge of hydrodynamic behavior inside the reactor has an important role on the improvement of performance, product quality and, in consequence, operational cost.

This work presents a isothermal hydrodynamics study on a laboratory scale FCC riser reactor by employing a CFD commercial software. A two dimensional Eulerian-Eulerian model based on kinetic theory of granular flow (KTGF) is used, and results obtained are compared with experimental measurements which are reported and used by other authors (3,4,5,7)). Also, different drag models (8,9) are employed and their results compared in order to choose the one whose results are closer to the experimental behavior.

The velocity results obtained with all of the models have very similar trend, with low velocity in the reactor walls and maximum value in the central area of the reactor. As expected, the volumetric fraction radial distribution shows higher concentration of particles near the walls where solid velocity is smaller.

This work seeks to validate the ideal computational model for this type of reactor, and to select the drag model suitable for the particular case.

MATHEMATICAL MODEL

A two phase Eulerian-Eulerian mathematical model that describes the behavior of gas flow and solid particles (catalyst) based on mass and momentum conservation equations, turbulent (mixture) and KTGF in two-dimensional isothermal riser reactor is solved.

Continuity

$$\frac{\partial}{\partial t}(\alpha_q\rho_q) + \nabla\cdot(\alpha_q\rho_q\vec{v}_q) = 0 \qquad (1)$$

Where q is the corresponding phase, namely g for gas phase and s for solid phase.

Gas phase momentum

$$\frac{\partial}{\partial t}\left(\alpha_g\rho_g\vec{v}_g\right) + \nabla\cdot\left(\alpha_g\rho_g\vec{v}_g\vec{v}_g\right) = -\alpha_g\nabla p + \nabla\cdot\bar{\bar{\tau}}_g + \alpha_g\rho_g\vec{g} + K_{sg}\left(\vec{v}_s - \vec{v}_g\right) \qquad (2)$$

Solid phase momentum

$$\frac{\partial}{\partial t}\left(\alpha_s\rho_s\vec{v}_s\right) + \nabla\cdot\left(\alpha_s\rho_s\vec{v}_s\vec{v}_s\right) = -\alpha_s\nabla p - \nabla p_s + \nabla\cdot\bar{\bar{\tau}}_s + \alpha_s\rho_s\vec{g} + K_{sg}\left(\vec{v}_g - \vec{v}_s\right) \qquad (3)$$

Solid pressure and solid stress tensor are computed by the KTGF model in terms of radial distribution (1), frictional pressure (2) and granular temperature. In the next section, drag models for calculating the interaction between phase (Ksg) are presented.

COMPUTATIONAL MODEL

A two dimensional model, with two inlets and outlets placed at both sides of reactor in order to emulate the axisymmetric behavior observed in the three dimensional experiments, is built to simulate the FCC reactor hydrodynamics under the operational parameters reported in (3). Geometrical and physical parameters used during the simulation are shown in Tabla 1.

The computational simulation is performed by using the commercial software Ansys Fluent 15. After mesh independency study, a mapped mesh of 26050 square elements, with the highest elements density at the particle inlet and the reactor outlet zones, was used.

Tabla 1. Geometrical parameters

Diameter (m)	0.2
Height (m)	14.2
Solid density (Kg/m^3)	1700
Gas density(Kg/m^3)	1.2
Particle diameter (μm)	55
Volume fraction (particle)	0.4
Velocity inlet gas (m/s)	5.2

InTable 2, it is presented the computational parameters used during the simulation regarding coupling scheme, spatial discretization aspects and values used for the temporal discretization implicit scheme.

Table 2. Computational parameters

Pressure-Velocity coupling Scheme	Phase coupled SIMPLE
Spatial discretization:	
Momentum, Turbulent kinetic energy, turbulent dissipation rate	*Second order UpWind*
Valume fraction	*Quick*
Transcient formulation	*Second order UpWind*
Time step	0.0005
Total simulation time	30

The Interaction between phases (K_{sg}) in the different drag models are expressed in terms of volume fraction, particle based Reynolds, particle diameter, terminal velocity correlation for the solid phase and velocity magnitude.

The drag model and their equations used for the research are presented below:

Gidaspow

$$C_D = 150\frac{\varphi_s^2 \mu_{Gg}}{(1-\varphi_{sg})d_p^2} + \frac{7}{4}\frac{\varphi_s \rho_g |U_r|}{d_p}$$

Syamlal Obrien

$$C_D = \left(\frac{0.63}{v_{r,s}} + \frac{4.8}{\sqrt{v_{r,s} Re_s}}\right)^2$$

Schiller Naumann

$$C_D = \frac{24}{Re_p}(1 + Re_p^{0.687})$$

RESULTS

Figure 1 presents a comparison between the experimental results reported for volume fraction (3)-(4) and the computational results for the different drag models. These results are compared at the same height of 3.9 m from the gas inlet (3), and correspond to the average value over the period between 20 and 30 seconds of physical time in computational simulation.

The volume fraction behavior shows, for all of the models, a similar behavior at the central area of the reactor, however the distribution of particles on the walls is considerably different for the Syamlal Obrien models, since concentration is zero at wall and increases to a maximum concentration of 0.23 at a 0.05 m radio. Conversely, Gidapow model results are closer, both in trend and numerical values, to the experimental measurements, where the maximum concentration is around 0.08 m radius of the reactor central zone; this behavior shows a high concentration of particle in the wall area and lower in the central zone of reactor.

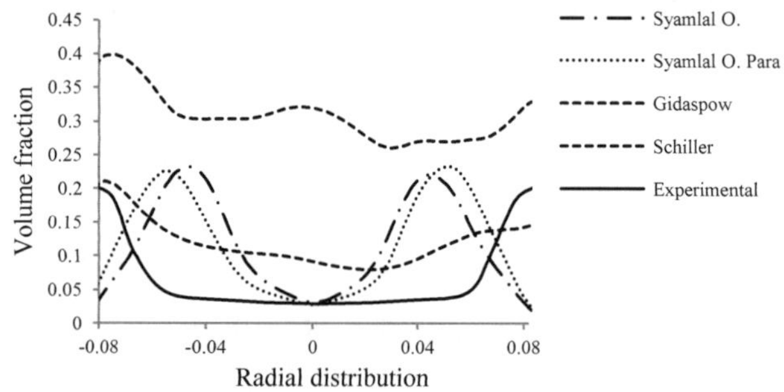

Figure 1. Computational and experimental volume fraction.

Error! Reference source not found. presents RMS error in volume fraction with respect to the experimental measurements, being the Gidaspow model result the one with the lower value.

Tabla 3. RMS Error Drag model whit experimental volume fraction

RMS Error	Syamlal Obrien	Syamlal Obien Para	Gidaspow	Schiller
Volume fraction	0.129	0.123	0.061	0.225

The computational results of solid mass flux are compared with the experimental measurements, at the same height 3.9 m and in the same period of physical time. Figure 2 shows de solid mass flux behavior in the radial position of reactor. Clearly, the Gidaspow model is closer to the experimental measurements. The maximum value in the experimental measurement is around 800 kg/m^2s, while the Gidaspow model are below the 500 kg/m^2s in most of the interval considered which is close to the value reported in (4). Schiller Naumann model has a similar trend, but the values in some points almost triple the value of the experimental measurement.

The Syamlal Obrien models have a different behavior and solid mass flux values: as it can be visualized in Figure 2, a maximum value at a location 0.05 m and minimum values on the walls and the central area.

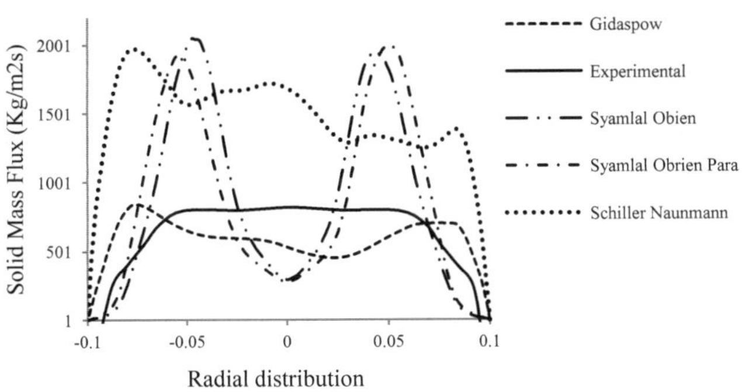

Figure 2. Computational and experimental Solid Mass Flux.

Figure 3 presents the radial distribution of particle velocity obtained by computational simulation with different drag models. The results show that in all models the behavior is similar, with velocities equal to zero at the wall and an increment towards reactor center where the maximum value is reached, which is similar to the reported experimental trend.

However, the velocity values obtained by using the Syamlal models are considerably larger than Gidaspow and Schiller Naumann models

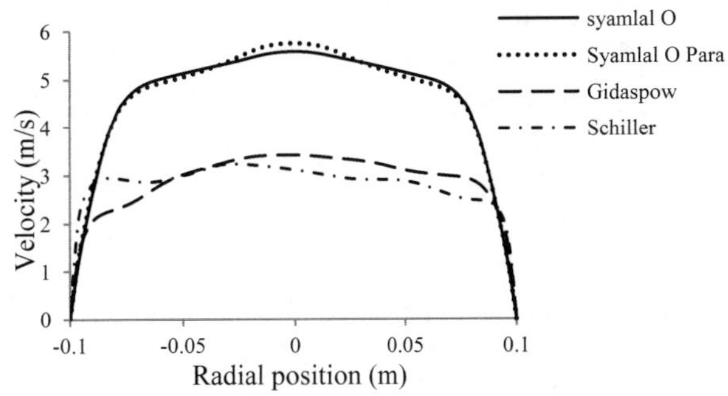

Figure 3. Particle velocity in radial position.

According to the computational simulation, the Gidaspow model presents a similar behavior in comparison to the experimental measurements and the other authors (4,5) and, in consequence, this is the model that best suits and predicts the hydrodynamic behavior inside the reactor, and it can be used for future work related to 3D simulations..

Figure 4 presents a solid volume fraction contour in the bottom zone, obtained by using the Gidaspow model. It can be observed that the higher concentration of particles occurs in a specific range of radial distance, being frequent the formation of solid high concentration zones towards the wall, as it is presented in **Error! Reference source not found.**

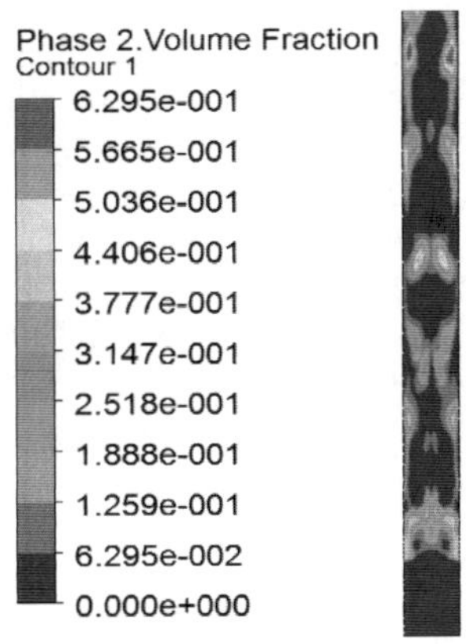

Figure 4. Volume fraction particle phase contour

It is important to note that the geometry built is an adaptation of the experimental model (3D) to two dimensions affecting the simulated hydrodynamic behavior by generating a kind of bubbling in the central zone of the reactor, which could be caused by the additional particle inlet and pressure outlet.

Notation

g	Gas phase subscript
s	Solid phase subscript
α_q	Volume fraction of phase q

ρ_q	Density of phase q
$\vec{v}_q$	Velocity of phase q
$\bar{\bar{\tau}}_q$	q- phase stress tensor
$\vec{g}$	gravity
p	Mixture pressure
P_s	Solid pressure
K_{sg}	Interaction between phases
C_D	Drag coefficient
$v_{r,s}$	Terminal velocity correlation for the solid phase
Re_p	Particle basedReynolds

CONCLUSION

A two-dimensional computational model of a laboratory scale FCC riser reactor is simulated. Two phase eulerian-eulerian approach was used to model catalyst particle and gas hydrodynamics inside reactor as well as several drag models (momentum interaction between phases). The best match to the experimental measurements is obtained by using the Gidaspow drag model. Although the velocity values attained by the other models are not the expected, the tendency of Gidaspow, Syamlal Obrien and Schiller Naumann models are similar.

In the case of volume fraction, values are overestimated in a wide range of radial distance, except when employing Gidaspow model. By using Gidaspow scheme, the behavior is similar to experimental with higher solid concentration towards the walls and smaller values close to the reactor center. Given the complexity of the phenomena involved, this is a first approach to achieve a suitable simulation of momentum interaction inside riser reactor. Future work will be focused on simulating 3D reactor geometries in order to analyze industrial situation. .

ACKNOWLEDGMENTS

This work was made possible by the support of the research through innovation projects and improvements program in Colombia that promotes education and development of the country Colciencias, the Universidad Pontificia Bolivariana and the Ecopetrol

REFERENCES

1. Lun CKK, Savage SB, Jeffrey DJ, Chepurniy N. Kinetic theories for granular flow: Inelastic particles in Couette flow and slightky inelastic particles in a general flow field. J Fluid Mech. 1984;140:223–56.

2. Schaeffer DG. Instability in the evolution equations describing incompressible granular flow. J Differ equations. 1987;66:19–50.

3. T. Knowlton, D. Geldart, J. Master, D. King. Comparison of CFB hydrodynamic model PSRI challenge problem: VIII fluidization conference, Tour-France, 1995

4. Arastoopour, H. (2001). Numerical simulation and experimental analysis of gas/solid flow systems: 1999 Fluor-Daniel Plenary lecture. *Powder Technology*, *119*(2-3), 59–67. doi:10.1016/S0032-5910(00)00417-4

5. Shah, M. T., Utikar, R. P., Pareek, V. K., Tade, M. O., & Evans, G. M. (2014). Effect of Closure Models on Eulerian-Eulerian Gas-Solid Flow Predictions in Riser. *Powder Technology*. doi:10.1016/j.powtec.2014.08.064

6. Gidaspow D, Bezburuah R, Ding J. Hydrodynamics of circulating fluidized beds, kinetic theory approach. Fluidization VII, 7th engineering foundation conference on fluidization. 1992. p. 75–82.

7. Shah, M. T., Utikar, R. P., Pareek, V. K., Tade, M. O., & Evans, G. M. (2014). Effect of Closure Models on Eulerian-Eulerian Gas-Solid Flow Predictions in Riser. Powder Technology. doi:10.1016/j.powtec.2014.08.064

8. Gómez, L. C., & Milioli, F. E. (2004). A NUMERICAL SIMULATION ANALYSIS OF THE EFFECT OF THE INTERFACE DRAG FUNCTION ON CLUSTER EVOLUTION IN A CFB RISER GAS-SOLID FLOW, 21(04), 569–583.

9. Benzarti, S., Mhiri, H., & Bournot, H. (2012). Drag models for Simulation Gas-Solid Flow in the Bubbling Fluidized Bed of FCC Particles. Waset.org, 61, 1138–1143. Retrieved from http://waset.org/journals/waset/v61/v61-210.pdf

Proceedings of the 25th CANCAM
London, Ontario, Canada, May 31 – June 4, 2015

OIL FLOW VISUALIZATION OF THE LAMINAR SEPARATION BUBBLE ON A SD 7037 AIRFOIL

Faegheh Ghorbanishohrat
Mechanical Engineering Department
University of Waterloo
Waterloo, ON, Canada
fghorban@uwaterloo.ca

David A. Johnson
Mechanical Engineering Department
University of Waterloo
Waterloo, ON, Canada
da3johns@uwaterloo.ca

ABSTRACT

In recent years, expanding utilization of small wind turbines, unmanned air vehicles (UAV), and micro air vehicles (MAV) has attracted a lot of attention in order to increase their performance. Considering the important role that flow separation and reattachment plays on the performance of airfoils especially at low Reynolds numbers (*Re*), this study is focused on identifying the laminar flow separation and reattachment over the SD7037 airfoil using the surface oil flow visualization technique. The experiments were conducted at three different *Re* equal to 32000, 38000 and 43800 with angle of attack (α) varying between 1 and 11 degrees. The results show that as *Re* is increased separation is delayed and the reattachment point moves toward the leading edge. Also there is a reduction in the length of the bubble. On the other hand by increasing the angle of attack, the laminar separation bubble moves toward the leading edge of the airfoil.

KEYWORDS: Oil flow visualization, Laminar Separation Bubble, Low Reynolds number flows, Small wind turbine

INTRODUCTION

For devices that work in low Reynolds number (*Re*) flows, usually a laminar boundary layer remains attached to the surface until sufficient adverse pressure gradient occurs and the flow detaches from the surface. Due to an unstable separated boundary layer small vortices are formed and transition occurs. Sometimes these created disturbances are strong enough to overcome the adverse pressure gradient and reattach the flow to the surface as a turbulent boundary layer [1, 2]. Figure 1 shows the profile of a laminar separation bubble (LSB) by Horton [3].

The SD7037 airfoil is a low *Re* airfoil. Surface oil flow visualization for this airfoil is reported in McGranahan and Selig [4] for three *Re* $=2\text{x}10^5$, $3\text{x}10^5$, and $5\text{x}10^5$. They mentioned that the time to do similar tests for *Re* of $1\text{x}10^6$ was very long and all details would be lost. For this particular airfoil even at the highest *Re* (Re=$5\text{x}10^5$) transition occurs in the presence of a LSB and this bubble remains for all angles of attack until the airfoil stalls. Similar work has been done by Genc et al. [5,6] that estimated the features of the LSB for a NACA2415 airfoil by surface oil flow visualization and they measured the pressure distribution over the airfoil at *Re* = $2\text{x}10^5$, $3\text{x}10^5$ and $1\text{x}10^6$.

Review of previous works [1, 2, 7, and 8] also reveals that the features of the bubble, i.e. length of the bubble, may vary with changes in *Re*, airfoil angle of attack (α) or unsteady flow field around the airfoil. Therefore determining the LSB parameters for a specific airfoil over an applicable range of *Re* and α, in both static and dynamic cases is critical before using that airfoil. By performing a complete series of experimental tests to cover steady and unsteady flows the importance of this phenomenon on the performance of wind turbine can be found.

The purpose of this research is determination of the characteristics of the LSB, location and length, for the SD7037 airfoil at different *Re* and α in static cases. This airfoil was chosen because it is popular for sailplanes and small wind turbines [9]. To achieve this goal surface oil flow visualization has been used.

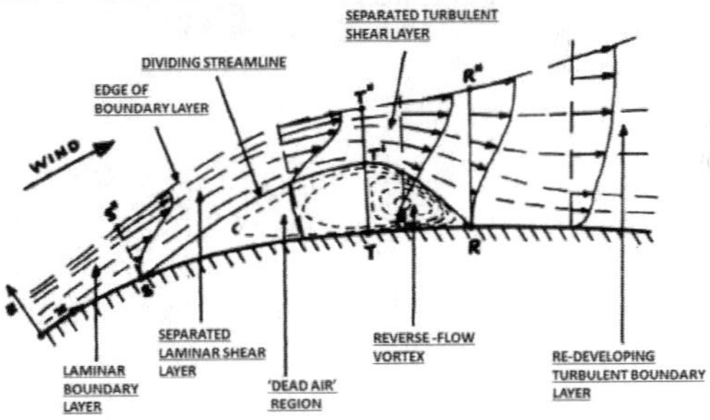

Figure 1 Profile of Laminar Separation Bubble [3]

EXPERIMENTAL APPARATUS AND METHOD

All tests were conducted in a closed return subsonic wind tunnel located at the University of Waterloo. The test section of the tunnel has dimensions of 152.4 mm×152.4 mm× 450 mm with the maximum turbulence intensity of the tunnel less than 1%.

The airfoil models of the SD7037 airfoil have a chord length (*c*) of 25.7 mm to avoid flow blockage at high angles of attack. To check the repeatability of experiments two models of same airfoil were fabricated. Both airfoil models were designed from the same software directly to a 5 axis CNC machine with a tolerance of 0.025 mm and were manufactured from a solid aluminum blank although the process of preparation made slight geometric differences that are difficult to quantify. These uncertainties have an effect on the roughness of the surface of the airfoil. This roughness has two effects; first, it aids in turbulence development and secondly changes the skin friction. Therefore using two models will give us an estimation of this uncertainty. Both airfoils were polished and anodized black so a white pigment was used for oil flow visualization. The blockage ratio for SD7037 airfoil at zero angle of attack is 2.9%. To change the angle of the attack of the airfoil a servo motor with 22.2 counts per degree of shaft rotation was used to position the airfoil [10]. The setup is shown in Fig. 2.

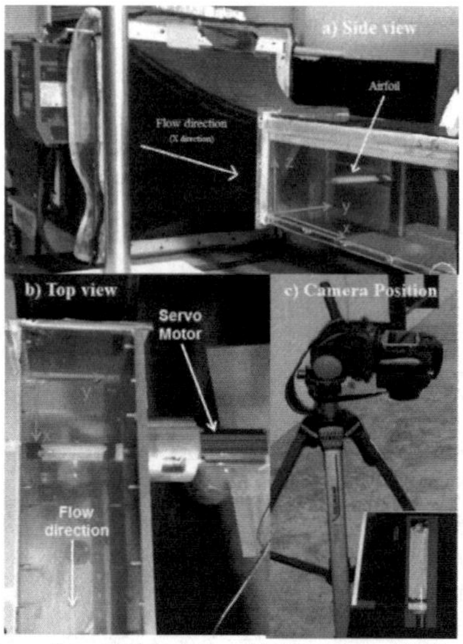

Figure 2 Experimental setup, a) Side view b) Top view, and c) Camera positions for taking images.

In surface oil flow visualization, the upper surface of the airfoil was painted with a pigmented oil mixture. The process requires placing the airfoil in the running wind tunnel and waiting for a few minutes so the flow pattern and shape of the streaklines on the surface are stabilized and observable. After some time running the wind tunnel, depending on the mixture of oil, free stream velocity, and size of the model, the oil mixture has dried and the model can be taken out for photographing. Many different mixtures were evaluated during the development of this study. The final oil mixture used in this study contained mineral oil, (Monarch Oil #70), titanium dioxide and a very small amount of linseed oil (the ratio of mineral oil to titanium dioxide to linseed oil was 15:15:4) and was painted on the surface of the airfoil with a thin and smooth paintbrush, Fig. 3. The mineral oil had a density of 0.842 to 0.8639 kg/L at 15°C. It should be noted that to reduce the oil viscosity, for each 8 m/s velocity decrement, 1 droplet of linseed oil was added to the mixture. In this study after 5 to 10 minutes of running the wind tunnel, the oil mixture had dried and the model was taken out for photographs. A Nikon D70 camera with a Nikkor 50 mm F/1.8 lens was used with the resolution of 3008×2000 pixels. The field of view for all the images was fixed at 422mm×277mm.

In this research, for one of the airfoil models the experiments were conducted for three chord *Re*, $3.2\text{x}10^4$, $3.8\text{x}10^4$, and $4.38\text{x}10^4$ that are based on the airfoil chord as stated by Eq. (1)

$$Re = \frac{\rho u c}{\mu} \quad (1)$$

Where ρ and μ are air density and viscosity, u is free stream velocity, and c is the airfoil chord. For Re=$4.38\text{x}10^4$ the experiment was repeated using the second airfoil model. The angle of attack was increased from 1° to 11° in 1° steps.

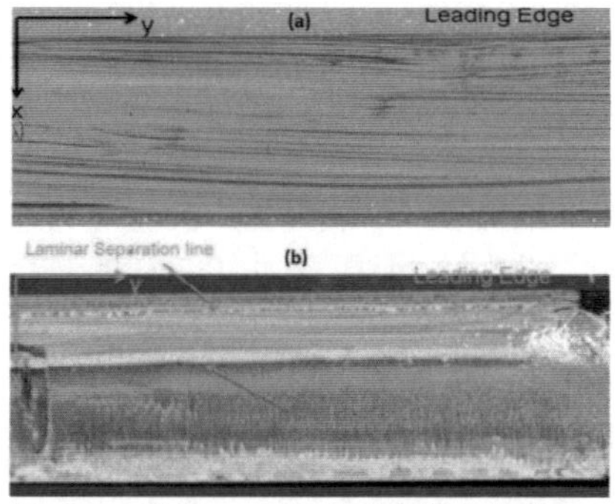

Figure 3 a) Surface of the painted airfoil a) before turning on the wind tunnel b) after turning on the wind tunnel

The uncertainty of the experiments depends on the uncertainty in the measuring devices, repeatability, linearization, changes in density and viscosity due to temperature, position of the airfoil (adjustments with the servo motor), etc. Pressure changes in the wind tunnel have a negligible effect on *Re*. The uncertainty for the *Re* in this experiment is about 5.4%.

RESULTS

In this work we take advantage of the surface oil flow visualization technique to determine the features of the LSB. For every freestream velocity and angle of attack the oil mixture was uniformly distributed on the surface of the airfoil as a thin layer. During tests the wind tunnel was running for about 10 minutes. After turning off the wind tunnel, images were captured. Using this method, laminar, transition and turbulent regions on the surface are observable.

For the first model of the SD7037 airfoil at Re=$4.38\text{x}10^4$ at all designated angles of attack, the experiment was repeated 3 or 4 times on different days to make sure that the results are repeatable. Repeating these tests allowed calculation of the uncertainty of the results due to the unsteadiness. Figure 4 shows these results. The distance of separation (x_s) and reattachment (x_r) points from the leading edge are divided by the chord. As can be seen in this graph the separation points move from 0.6c to 0.04c. Along with the separation point, the reattachment point changes from 0.9c to 0.3c. Comparing these numbers simply indicates that the reattachment point moves faster toward the leading edge than the separation point therefore the length of the bubble is reduced by increasing the angle of attack. Also the results of the second airfoil model are compared with these results. As can be seen almost all the results of separation and reattachment points for the second model are found to be located very close to the same points of the first model. Except for two points in the graph of the second model other points do not exceed the uncertainty range. In the second model the separation point moves about 50% along the chord from 1° to 11° and the reattachment point moves about 60% toward the leading edge that causes a reduction in the length of the bubble by increasing the angle of attack similar to the first airfoil model.

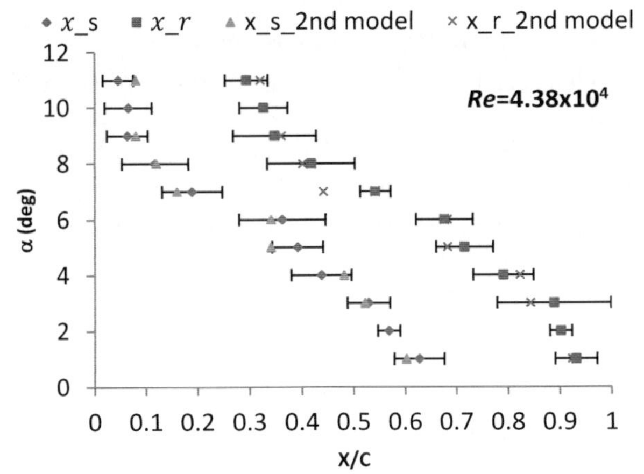

Figure 4 Position of separation and reattachment points with changing angle of attack at Re =4.38x10^4.

These results reveal that the error due to the airfoil manufacturing, sanding and painting process is in the range of the uncertainty of the experiment and is acceptable.

Figure 5 shows some images of surface oil flow visualization that were performed in this study. By looking at these results movement of the separation and reattachment with increasing α is clear but also there are more interesting features in these images. For α > 5° the difference between the accumulation line and the reattachment line is very clear. For 7° < α < 10°, the reattached flow at higher Re is turbulent and attached to the surface so that the streaklines were fully visible. For lower Re the flow barely attaches to the surface. At α < 3° laminar separation occurred at Re=3.2x10^4 but the disturbances were not strong enough to eliminate reverse flow and it reattaches to the surface. On the other hand at α =11° for the lower Re, it seems that an unstable LSB was burst or in other words "bursting the bubble" defined in previous work as a sudden increase in the length of the bubble in the streamwise direction causing an intense change in pressure distribution over the surface of the airfoil. The location and size of the LSB has effects on the bubble bursting phenomenon [11] while in higher Re flow it was still attached. In Fig. 6 the characteristics of the LSB, and separation and reattachment points, for different Re are presented briefly. The major difference can be seen between Re=4.38x10^4 and 3.2x10^4 while there is not a significant difference for Re=4.38x10^4 and 3.8x10^4. The difference in the reattachment point for these Re are more visible than the separation point. This could be explained by the fact that the flows are laminar until the separation point, but in the reattachment point momentum transfer occurs due to turbulent mixing and higher Re will aid momentum transfer.

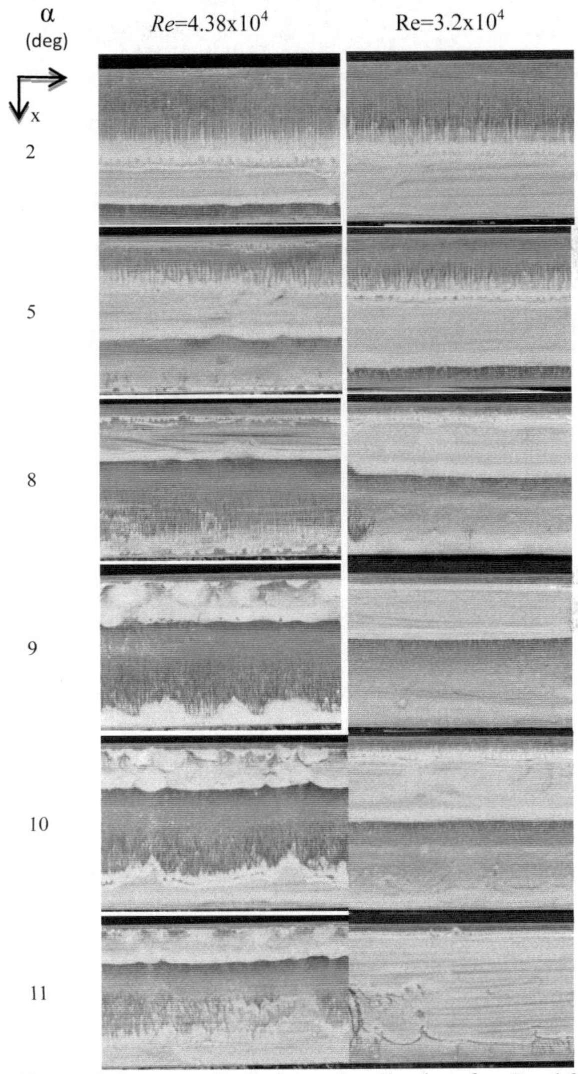

Figure 5 Images of oil flow visualization for Re=4.38x10^4 and 3.2x10^4 at different α

Figure 7 summarizes the results of this study by presenting the changes in the length of the LSB at different α and Re. As can be seen for the lowest Re the difference between minimum lengths of LSB occurs at 9° and the maximum length of bubble is at 4° at 0.14*c*. But for the highest *Re* in the majority of α, the length is about 0.30*c*. In other words, for higher *Re* the rate of change of the length of the LSB is slower than the lower *Re*.

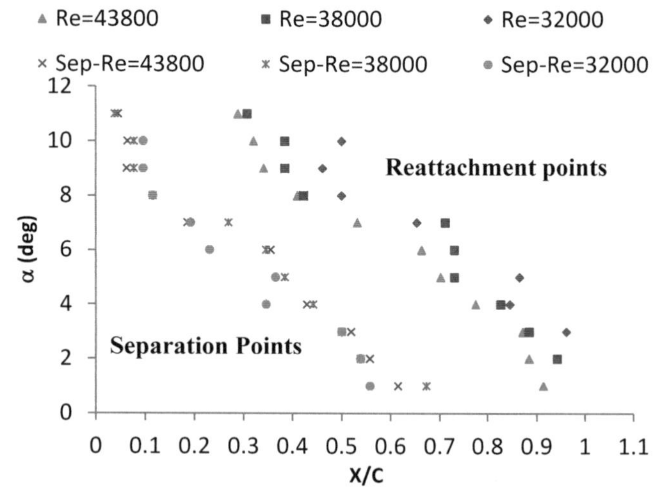

Figure 6 Position of separation and reattachment points on the surface of the airfoil by changing *Re*

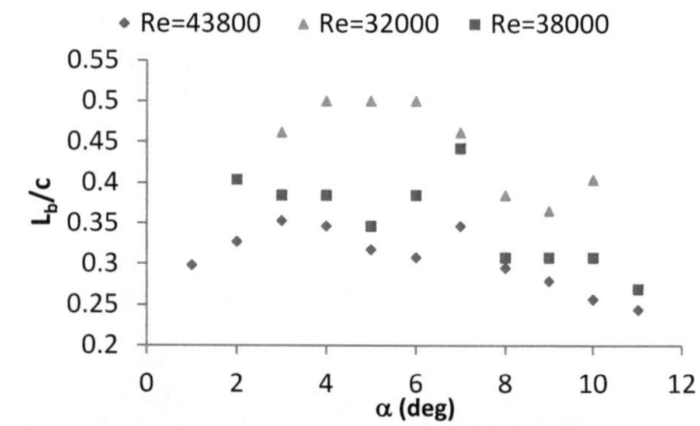

Figure 7 Changes of the length of the LSB by changing α and *Re*

CONCLUSIONS

In this study surface oil flow visualization for a SD7037 airfoil has been completed using a mixture of mineral oil, TiO_2, and linseed oil in the ratio of 15:15:4 respectively at three *Re* (4.38×10^4, 3.8×10^4, and 3.2×10^4) for range of α between 1° and 11° to characterize the location and length of the LSB. The results showed that with increasing *Re* the separation point goes toward the trailing edge and the reattachment point goes toward the leading edge and therefore the length of the bubble is decreased. By increasing α the length of the bubble decreases and moves in the direction of the leading edge. Because of the small chord length of the airfoil, the run time was about 10 minutes per data point therefore it provided a significant advantage in the ability to utilize the surface oil visualization technique for *Re* even lower than 3×10^4 which can be useful in small wind turbines, MAVs and UAVs.

ACKNOWLEDGMENTS

The authors would like to thank NSERC, and the Wind Energy Group at UWaterloo for their support with this project.

REFERENCES

1. Lissaman P. B. S., 1983, "Low-Reynolds-Number Airfoils", *Ann Rev Fluid Mech.15*-223-39.

2. Jahanmiri M., 2011, "Laminar separation bubble: Its Structure, Dynamics and Control", *Research Report*, ISSN 1652-8546, Chalmers University of Technology.

3. Horton H., 1968, "Laminar Separation Bubbles in Two and Three Dimensional Incompressible Flow", Department of Aeronautical Engineering, Queen Mary College, University of London, London.

4. McGranahan B. D., and Selig, M. S., 2003, "Surface Oil Flow Measurements on Several Airfoils at Low Reynolds Numbers", *21st AIAA Applied Aerodynamics Conference*, 2003-4067.

5. Genc, M., Serdar, Karasu, I., and Acikel, H. Hakan, 2012, "An Experimental Study on Aerodynamics of NACA2415 aerofoil at Low Reynolds Numbers", *Experimental thermal and fluid science*, 39, 252-264.

6. Genc, M., Serdar, Karasu, I., Acikel, H. Hakan, and Akpolat, M., Tugrul, 2012, "Low Reynolds Number Flows and Transition", *InTech*.

7. Gharali K., Johnson D., 2013, "Reduced frequency effects on Laminar Separation Bubbles", *10th International Symposium on PIV*, Delft, The Netherlands.

8. O'Meara M. M., Mueller, T. J., 1987, "Laminar Separation Bubble Characteristics on an Airfoil at Low Reynolds Numbers", *AIAA Journal*, Vol. 25, No 8, Aug. issue, pp 1033-1041.

9. Lyon Ch. A., Broeren A. P., Giguere P. , Gopalarathnam A., and Selig, M. S. , 1997 ,"Summary of Low-Speed Airfoil Data" , Dept Aero Eng, U Illinois Urbana-Champaign, SoarTech Publications, Virginia Beach, Virginia, Volume 3.

10. Gharali K., 2013, "Pitching airfoil study and freestream effects for wind turbine applications", *Ph.D. thesis*, University of Waterloo.

11. Haggmark C., 2000, "Investigation of disturbances developing in a laminar separation bubble flow", Technical reports from Royal Institute of Technology Department of Mechanics.

Proceedings of the 25th CANCAM
London, Ontario, Canada, May 31 - June 4, 2015

DISPERSED PHASE SINGLE-PARTICLE LAGRANGIAN AND CORRELATION DIMENSIONAL STATISTICS IN KINEMATICALLY SIMULATED TURBULENCE

S. A. Murray, M. F. Lightstone, S. Tullis

McMaster University
Hamilton, Ontario, Canada

ABSTRACT

Kinematic simulation (KS) is a means of reducing the solution of the Navier-Stokes equations to a manageable number of degrees of freedom, and has attracted attention due to the promise of bringing turbulence simulation within the reach of computational tractability. Such models have also been applied in particle-laden flows, due to their ability to enforce spatial organization of the fluid velocity field. A critical evaluation of KS is presented; in particular, we examine its ability to reproduce single-particle Lagrangian statistics and the desired spatial organization of an ensemble of inertial particles. Some computational results are presented, in which Lagrangian particles are transported alternatively by (1) turbulence generated by direct numerical simulation (DNS) of the incompressible Navier-Stokes equations, and (2) KS. We find that KS reproduces clustering of inertial particles and the intermittent simulated turbulent velocity signal, but not to the same extent as is seen in the DNS.

Keywords: Particle-laden flows, turbulence, RANS modelling, kinematic simulation, direct numerical simulation, computational fluid dynamics

INTRODUCTION

Particle-laden flows are a class of two-phase flows in which one of the phases is a continuously connected fluid, and the other phase is comprised of small, immiscible, typically dilute particles dispersed throughout the fluid. The modelling of turbulent particle-laden flows can be applied to a tremendous range of phenomena of environmental and industrial interest, such as the dispersion of pollutants in the atmosphere, fluidization and turbulent mixing in combustion processes, deposition of medicinal aerosols in the extrathoracic airway, processing and transport of liquid/solid nuclear waste and water droplet formation within clouds. Many physical/chemical processes depend upon the local concentration of the dispersed species, hence the development of accurate computational models of these processes depends on the ability to reproduce particle clustering and dispersion. Even single phase turbulent flows pose enormous challenges for researchers; often many simplifying assumptions are required to make meaningful predictions. This difficulty is exacerbated with the addition of a particle phase, since a balance must be sought in simulating the turbulence in a manner which is computationally tractable, but does not sacrifice the detailed correlated information "seen" by the dispersed particles.

If the particles are treated in the Lagrangian frame of reference, and if the only fluid information available is the time-averaged statistics (as in the Reynolds-averaged Navier-Stokes (RANS) context), one can stochastically generate a history of correlated fluid velocities seen by individual particles, in a manner which reflects these statistics - this information is then used to integrate the (Lagrangian) momentum equation for an individual particle. Such methods are known as stochastic separated flow (SSF) models. However a significant pitfall of this approach is the inability to reproduce structure; this shortcoming is manifested chiefly in two ways; from the perspective of the single-particle Lagrangian statistics (the velocity history seen by individual particles is necessarily Markovian, whereas in real turbulence it is not); and also the method could fail to reproduce the preferential concentration of an ensemble of inertial particles. This is because particles could "see" very different fluid velocities, even if they are very close together - which of course is highly unphysical. This motivates the need for a method which uses more information regarding how real turbulence is known to behave, namely, its spatial/temporal organization.

One method of enforcing spatial structure of the turbulence is to stochastically generate an entire flow field, also in a manner which reflects the expected second-order Eulerian statistics - this is the idea behind kinematic simulation (KS), a method which has its foundation in the work of [1], and has attracted much attention from both pure and applied fluid dynamicists.

KS is naturally alluring as a modelling tool due to its ability to reproduce sensible information on the turbulence structure, with a vast reduction in degrees of freedom relative to DNS. Indeed in various studies it has been applied to particle-laden flows for cases in which the Reynolds number would be out of reach for DNS [2, 3, 4]. Apart from KS as a modelling tool, significantly more effort has been devoted to critically analyzing the ability of KS to reproduce statistics relating to structure in a variety of scenarios - in particular the properties of *pairs* of fluid tracers or inertial particles, due to the close link between these statistics and the concentration variance of passive scalars [5].

In this investigation, the *single-particle* Lagrangian statistics are studied (in particular, dispersion, defined as a particle's departure from its initial position, averaged over the particle ensemble). Such statistics have traditionally been overlooked, since they are thought to be in the domain of simpler SSF approaches. However; SSF approaches are far from perfect; they are fundamentally limited by their sensitivity to the estimate of Lagrangian integral timescale (defined in the next section), and are typically informed by a great deal of empiricisms. Moreover, many pairwise particle statistics hinge upon accurate prediction of the Lagrangian integral timescale. Also clustering of inertial particles cannot be reproduced by SSF approaches. If particle-laden flow models can more sensibly synthesize turbulence *from the perspective of dispersed particles*, the door is opened to simulation techniques with broader scopes of applicability.

METHODOLOGY

An individual particle is modelled in the Lagrangian frame of reference according to the appropriately simplified [6] momentum equation:

$$\frac{d}{dt}\mathbf{u}_p(t) = \frac{1}{\tau_p}\left(\mathbf{u}_f(\mathbf{x}_p(t), t) - \mathbf{u}_p(t)\right) \tag{1}$$

where $\mathbf{u}_p(t)$ is the velocity of an individual particle, $\mathbf{u}_f(\mathbf{x}_p(t), t)$ is the fluid velocity at the particle location, $\mathbf{x}_p(t)$, and τ_p is the particle relaxation time. This quantity represents, loosely speaking, how quickly a particle reacts to changes in fluid velocity; heavy particles are associated with high τ_p values, and light particles with low ones. A parameter of interest is the *Stokes number*, defined as:

$$St = \frac{\tau_p}{\tau_\eta} \tag{2}$$

where τ_η is the Kolmogorov microscale, that is, the smallest length scale of the turbulent cascade. Clustering is known to occur maximally for $St \sim 1$ particles [7]. The interface between the (modelled or directly simulated) flow field and the particle motion is given by the fluid velocity at the particle location $\mathbf{u}_f(\mathbf{x}_p(t), t)$. Here one-way coupling is assumed, that is, the particle dynamics are influenced by the fluid dynamics, and not vice versa.

The turbulent fluid velocity, $\mathbf{u}_f$, can be given by DNS, or it can be *kinematically simulated* at each particle location at a given time t, by summing randomly oriented sines and cosines of varying spatial scales, and whose amplitudes are determined by an input Eulerian energy spectrum (given either empirically, or simulated in a manner consistent with Kolmogorov scale relations). A commonly applied formulation [8] is:

$$\mathbf{u}_f(\mathbf{x}, t) = \sum_{n=1}^{N} \mathbf{a}_n \cos(\boldsymbol{\kappa}_n \cdot \mathbf{x} + \omega_n t) + \mathbf{b}_n \sin(\boldsymbol{\kappa}_n \cdot \mathbf{x} + \omega_n t) \tag{3}$$

where the $\boldsymbol{\kappa}_n$ are randomly sampled vectors, whose orientations are uniformly distributed over a sphere. Also $\mathbf{a}_n$ and $\mathbf{b}_n$ are chosen with random orientations, but under the constraint that their orientations be perpendicular to $\boldsymbol{\kappa}_n$:

$$\mathbf{a}_n \cdot \boldsymbol{\kappa}_n \text{ and } \mathbf{b}_n \cdot \boldsymbol{\kappa}_n \text{ for all } n \tag{4}$$

Also $|\boldsymbol{\kappa}_n| = \kappa_n$, where κ_n is the sequence of N wavenumbers for which the spectrum is given. Consequently $E(\kappa_n)$ is the sequence of corresponding energy density values. The magnitudes of the vectors $\mathbf{a}_n$ and $\mathbf{b}_n$ should be chosen such that the correct amount of energy for each particular mode is injected:

$$|\mathbf{a}_n|^2 = |\mathbf{b}_n|^2 = 2E(\kappa_n)\Delta\kappa_n \tag{5}$$

where $\Delta\kappa_n$ is the interval width associated with the wavenumber κ_n, and is given as:

$$\Delta\kappa_n = \begin{cases} \frac{\kappa_2 - \kappa_1}{2} & \text{if } n = 1 \\ \frac{\kappa_{n+1} - \kappa_{n-1}}{2} & \text{if } 2 \leq n \leq N-1 \\ \frac{\kappa_N - \kappa_{N-1}}{2} & \text{if } n = N \end{cases} \tag{6}$$

the unsteadiness of the KS field is enforced via the scale-dependent ω_n term in equation (3), whose interpretation will be discussed in further detail shortly.

In much particle-laden flows literature, the aim is to reproduce dispersion, defined as $\langle (x_p(t+\tau) - x_p(t))^2 \rangle$ (the brackets $\langle\cdot\rangle$ indicate an average over the particle ensemble). Here, the non-boldface x_p represents a single component of the vector $\mathbf{x}_p$. Since the scope of this investigation is isotropic turbulence, the statistics of interest are the roughly the same over all components. Related is the particle velocity autocorrelation:

$$R_L(\tau) = \frac{\langle u_p(\mathbf{x}_p(t), t) u_p(\mathbf{x}_p(t+\tau), t+\tau) \rangle}{\sqrt{\langle (u_p(\mathbf{x}_p(t))^2 \rangle} \sqrt{\langle u_p(\mathbf{x}_p(t+\tau), t+\tau)^2 \rangle}} \tag{7}$$

The relationship between dispersion and autocorrelation is underpinned by the Lagrangian integral timescale, defined as:

$$T_L = \int_0^\infty R_L(\tau)\, d\tau \tag{8}$$

Also of great interest is the ability of KS to reproduce the clustering of the particle phase - of chief importance is the correlation dimension, which, loosely speaking, describes the extent to which the particle phase is space-filling. More rigorously, it is given by:

$$\mathcal{D}_2 = \lim_{r \to 0} \frac{\log(P_2(r_i))}{\log(r_i)} \tag{9}$$

where $P_2(r_i)$ is the probability of finding two particles within r_i of one another. For randomly distributed particles, $\mathcal{D}_2 = 3$, and for particles on a plane, $\mathcal{D}_2 = 2$. For inertial particles in real turbulence, $\mathcal{D}_2$ is expected to settle between 2 and 3, with the smallest values occurring for $St \sim 1$ particles.

The unsteadiness term in equation (3), ω_n, is somewhat open to interpretation. Initially it has been informed by heuristic arguments about motions at a given scale; in particular, it should represent some sort of *turnover time* of eddies corresponding to that scale:

$$\omega_n = \lambda \sqrt{\kappa_n^3 E(\kappa_n)} \tag{10}$$

where $\lambda \sim 1$. It can also be more rigorously defined according to how quickly the field should decorrelate at a fixed position (as in [8]), yielding:

$$\omega_n = \lambda u_{rms} \kappa_n \tag{11}$$

A pervasive theme in KS literature, however, is that even with the Eulerian properties of a field enforced, sensible Lagrangian behaviour of the simulated flow is far from guaranteed. The simple explanation is that a *kinematic* field is not *dynamic*, and the statistical hallmarks of transport of small eddies by large ones will necessarily be lost. The scope of this brief communication is to address the following question: in view of single-particle Lagrangian and correlation dimensional statistics of dispersed particles being of fundamental importance, to what extent are they sacrificed in the name of preserving certain Eulerian features of the KS flow? While this question is addressed to a certain extent in the literature [9, 10], it is not enough to simply capture inertial range scale relations; in practice, the utility of KS *as a model* (for single or pairwise particle dynamics) hinges upon whether it can reproduce T_L, and/or whether it can reproduce particle structure.

To provide a benchmark for comparison of the KS results, particle-laden DNS were carried out. The DNS employed a pseudospectral, parallelized solver for the forced, incompressible Navier-Stokes equations on a periodic domain with 256^3 collocation points. Although one is restricted to low Reynolds numbers with DNS, there exists the advantage that the energy spectrum, $E(\kappa)$, can be evaluated from the DNS, and can in turn be used as input for KS, offering a robust evaluation of the ability of KS to capture (or not capture) the physics of real turbulence.

For this study, a particle-laden DNS was carried out for $St \sim 0.75$ particles, with Taylor-scale Reynolds number ~ 90. The output spectrum was in turn used as input for a variety of KS, also for $St \sim 0.75$. The effect of unsteadiness

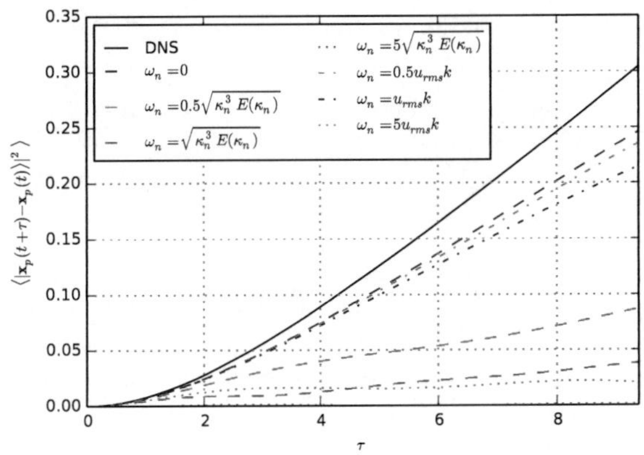

Figure 1: Dispersion of $St \sim 0.75$ inertial particles in DNS and KS

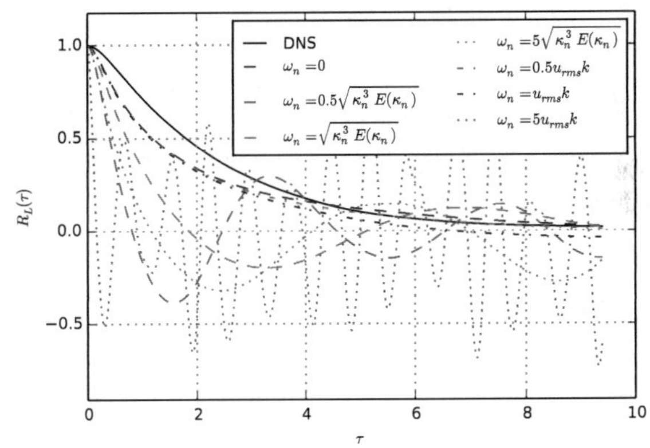

Figure 2: Particle velocity autocorrelation

was examined, by testing a variety of unsteadiness parameters: $\lambda = 0, 0.5, 1$ and 5, under either unsteadiness formulation (equations (10) and (11)). For these tests, gravity was neglected.

RESULTS

Broadly speaking, the imposed unsteadiness undermines sensible Lagrangian behaviour (for either fluid tracers or inertial particles), and indeed the greater the unsteadiness (controlled by greater λ, under either the formulation), the greater the extent of underprediction of dispersion (see Fig. 1). This shortcoming is manifested also in the particle velocity autocorrelation (Fig. 2), which, by equation (8), corresponds to underpredicted T_L. For the highest λ cases, the particles "see" almost exclusively the contribution from a single mode, suggesting the particles simply become stuck in orbits (and hence exhibit minimal dispersion). Indeed the greater the variation of a field, the more quickly a particle is thrust into a new "eddy", and the more quickly the velocity signal decorrelates.

A similar shortcoming is seen in the prediction of particle

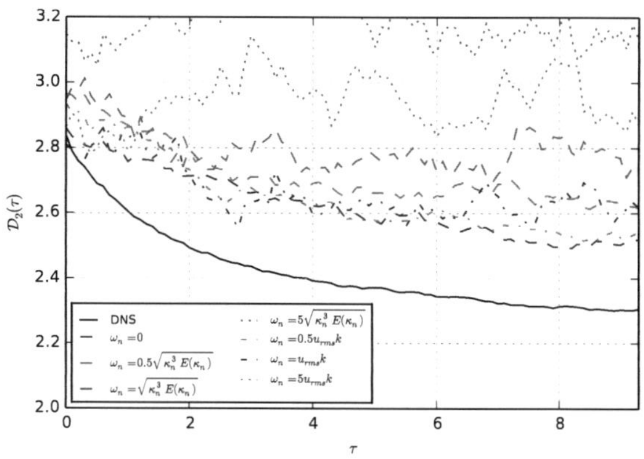

Figure 3: Correlation dimension

clustering: the higher λ (under either unsteadiness formulation), the higher D_2 (corresponding to diminished clustering) - see Fig. 3. Indeed clustering is contingent upon the persistence of coherent structures in the turbulence with which particles can interact and resonate; the more rapidly the field exhibits temporal fluctuations, the more this persistence is compromised.

CONCLUSIONS

KS is moderately capable of reproducing a variety of statistics of interest in Reynolds number, particle-laden, homogeneous, isotropic turbulence. While unsteadiness may restore the scale relations of the Eulerian one-point, two-time structure function, this occurs to the detriment of sensible single-particle Lagrangian statistics, which, in many cases, are of fundamental importance to modellers. The preferred strategy for kinematically simulating these statistics is to employ a steady field. In the literature, steady KS has already been suggested as a subgrid scale complement [9, 4], though for particle-laden flows, the justification of KS as a subgrid model, vs. *no subgrid model at all*, is unclear [11]. In the future, the applicability of a simulated spectrum (informed only by the RANS parameters of turbulent kinetic energy and dissipation) will be explored.

ACKNOWLEDGMENTS

We acknowledge the financial assistance of NSERC for this project, Compute Canada and Shared Hierarchical Academic Research Network (SharcNET) for the provision of high-performance computing facilities and support, Dr. Nicholas Kevlahan (Department of Mathematics & Statistics at McMaster University) for the contribution of his FORTRAN DNS solver distribution, and former research group members Matthew Cernick and Cathy Strutt for their work on the particle tracking utilities.

REFERENCES

[1] Robert H Kraichnan. Diffusion by a random velocity field. *Physics of Fluids*, 13(1):22–31, 1970.

[2] P Flohr and JC Vassilicos. A scalar subgrid model with flow structure for large-eddy simulations of scalar variances. *Journal of Fluid Mechanics*, 407:315–349, 2000.

[3] A ElMaihy and F Nicolleau. Investigation of the dispersion of heavy-particle pairs and Richardson's law using kinematic simulation. *Physical Review E*, 71(4):046307, 2005.

[4] M. A. I. Khan, Luo. X.Y., F.C.G.A. Nicolleau, P.G. Tucker, and G.L. Iacono. Effects of LES sub-grid flow structure on particle deposition in a plane channel with a ribbed wall. *International Journal for Numerical Methods in Biomedical Engineering*, 26:999–1015, 2010.

[5] PA Durbin. A stochastic model of two-particle dispersion and concentration fluctuations in homogeneous turbulence. *Journal of Fluid Mechanics*, 100(02):279–302, 1980.

[6] V. Armenio and V. Fiorotto. The importance of the forces acting on particles in turbulent flows . *Phys. Fluids*, 13(8):2437–2440, 2001.

[7] Jérémie Bec, Massimo Cencini, and Rafaela Hillerbrand. Heavy particles in incompressible flows: the large stokes number asymptotics. *Physica D: Nonlinear Phenomena*, 226(1):11–22, 2007.

[8] DR Osborne, JC Vassilicos, and JD Haigh. One-particle two-time diffusion in three-dimensional homogeneous isotropic turbulence. *Physics of Fluids (1994-present)*, 17(3):035104, 2005.

[9] JCH Fung, JCR Hunt, NA Malik, and RJ Perkins. Kinematic simulation of homogeneous turbulence by unsteady random fourier modes. *Journal of Fluid Mechanics*, 236(1):281–318, 1992.

[10] N. A. Malik and J. C. Vassilicos. A Lagrangian model for turbulent dispersion with turbulent-like flow structure: Comparison with direct numerical simulation for two-particle statistics. *Physics of Fluids*, 11:1572–1580, June 1999. doi: 10.1063/1.870019.

[11] Matthew J Cernick, SW Tullis, and MF Lightstone. Particle subgrid scale modelling in large-eddy simulations of particle-laden turbulence. *Journal of Turbulence*, 16 (2):101–135, 2015.

NUMERICAL INVESTIGATION OF JET PUMP WITH TWISTED TAPES

Arpita Srivastava
Department of Mechanical Engineering
Indian Institute of Technology
Chennai, Tamil Nadu, India
arpita.sri.lko@gmail.com

Shaligram Tiwari
Department of Mechanical Engineering
Indian Institute of Technology
Chennai, Tamil Nadu, India
shaligram@gmail.com

Mani Annamalai
Department of Mechanical Engineering
Indian Institute of Technology
Chennai, Tamil Nadu, India
mania@iitm.ac.in

ABSTRACT

Jet pump is used to transfer momentum from a high velocity primary stream to a secondary stream that gets entrained. It is packaged with the advantage of geometrical simplicity without any moving parts. It comprises mainly of a nozzle, mixing chamber and diffuser. It is used in numerous applications, one such being desalination systems, where it helps in creating vacuum inside the flash chamber without employing conventional oil vacuum pump. Aim of the present study is to carry out three-dimensional numerical investigations of two phase flow jet pump in presence of twisted tapes under optimized conditions of the parameters such as upstream and downstream pressures, primary jet mass flow rate, entrained secondary stream mass flow rate, geometry of the ejector, etc. Investigations of swirling turbulent jet flows are carried out using suitable turbulent model in ANSYS FLUENT. This investigation helps in predicting the flow behavior at the exit of nozzle with respect to different nozzle profiles causing dispersion in jet and also due to introduction of the swirl. At first, the three-dimensional numerical computations have been carried out to validate the existing literature without introduction of swirl in primary stream. Further investigations are carried out to identify the effect of tape with respect to the case without tape. Water and air have been used as fluids in the primary and secondary streams respectively. Nozzle profiles selected is conical. The mean diameter of nozzles are kept to be 4mm and 6mm. Results in present study suggest that the momentum exchange achieved by the swirl causes significant enhancement efficiency of the jet pump.

KEYWORDS: Jet pump, nozzle profile, two phase flow, twisted tape.

INTRODUCTION

The ejectors employed in jet pumps causes a complex flow physics and hence it is preferably modeled in 3D incorporating CFD which gives us the advantage over neglecting 1D assumptions and predicted very near to actual situation. Several investigations have been reported for enhancing its efficiency. Many researchers carried out valuable experiments to determine the conditions at which improvements can be achieved. Pfliedderer [1] proposed that the pressure ratio and air flow rate are independent of each other. Martinelli et al. [2] reported that increase in air flow rate at secondary inlet with an increase in flow rate of water at the primary side is due to increased pressure at primary motive fluid. Many researchers have performed studies on multi hole nozzle. Decay of primary jet flow as well as length of potential core depends on how the phenomenon of momentum exchange occurs between two fluids stream in mixing zone near to primary nozzle exit. Sharma [9] in his work experimentally verified that, for lower area ratio, nozzles having elliptical profile are capable of producing higher efficiency then nozzles with circular or conical profile. He experimented with different area ratio of nozzle. The geometrical design for the present study is taken from Sharma's experiment. Hansen and Kinnavy [5] performed experiments on ejectors keeping different area ratio and then studied different parametric effect. It was concluded that area ratio has major role to play in ejector efficiency than any other parameter.
Abdus Samad [10] investigated the influence of the introduction of swirl low performance of an ejector and concluded that it enhance the jet breakup resulting in higher suction rates at different optimized swirl angles.

GOVERNING EQUATION AND NUMERICAL TREATMENT

Governing equations solved are the mass and momentum conservation equations. For steady state, incompressible flow with constant viscosity, these equations can be expressed as

$$\frac{\partial u_i}{\partial x_i} = 0 \qquad (1)$$

$$u_i \frac{\partial u_j}{\partial x_i} - \frac{\partial}{\partial x_j}\left[(\nu + \nu_t)\frac{\partial u_j}{\partial x_j}\right] + \frac{1}{\rho}\frac{\partial P}{\partial x_j} = 0 \qquad (2)$$

In these equations, u_i refers to the mean velocity along x axis, P the static pressure and ν the kinematic viscosity of fluid.
CFD helps to formulate the required governing equations involved in fluid flow into numbers such as to get a final desired description of the complete flow field under observation numerically. The observation and results obtained out of CFD are near to actual flow and it has been proved in large number of applications. Riffat et *al.* [6] have employed CFD tools and application to determine the optimum design of the ejector. Not only this, he examined the various performance characteristic obtained during

simulation to compare using primary nozzles of various geometries with different types of working fluid. Later, the performance of ejector in a compressible flow was studied by varying the nozzle position with respect to mixing chamber by Riffat and Omer [7] for a refrigeration system using CFD model. Gas powered Jet pump performance also depends on its diffuser design which was analyzed by Neve [11] numerically using CFD tools.

Ouzzane and Aidoun [12], investigated on ejector and put forward one-dimensional flow model involved in their studies. Bartosiewicz [8] performed numerical analysis on the ejector with working fluid as air, using CFD software FLUENT. They validated the existing CFD results obtained with the experimental investigations made. Further these studies were extended to optimize the turbulent model to capture more physical realistic situation occurring in flow field. Many studies has proposed that introduction of swirl in the primary jet may increase the overall efficiency of jet pump by improving the entrainment of secondary fluid.

STATEMENT OF THE PROBLEM

The present numerical validation studies have conducted with the results of Sharma's experimental work [9]. Problem is being defined numerically considering it to be a steady flow, incompressible situation. The pressure-based steady flow Navier-Stokes solution's algorithm has been followed to reach out the solution.

The efficiency of the jet pump was calculated by considering upstream pressure, pressure at diffuser, secondary and flow ratio.

$$\eta = \frac{M(P_{down}-P_{\sec})}{(P_{up}-P_{down})} \quad (3)$$

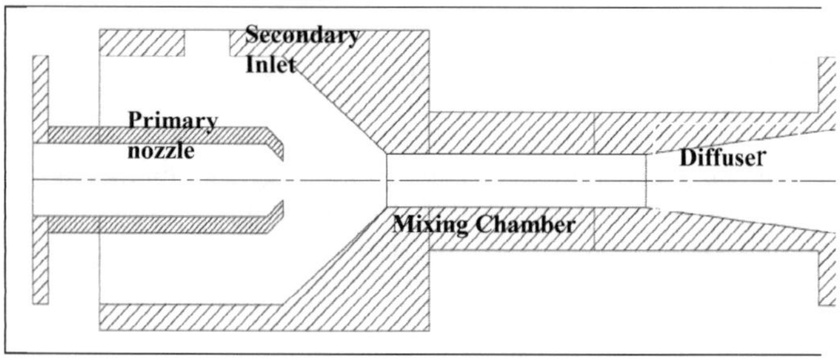

Figure1: Schematic cross section of jet pump

It excludes the effect of pipes and connecting elements used in the experimental setup. The steady state of the jet pump is modeled maximum evacuation of air in the vacuum chamber is achieved. The 3-D geometry is simulated to ensure that maximum physical turbulent flow field can be captured and studied. The major details of the dimensions of the ejector employed are given in Table 1. The profile of the nozzle is changed keeping other geometrical parameters as the same. For validation, nozzle of circular, elliptical and conical profiles, with 4mm nozzle diameter, are compared and found that elliptical nozzle has better efficiency than others.

Table 1.Design parameters of nozzle

Notation	Name	Dimension
D_o	Nozzle diameter	4 mm and 6mm
D_{suc}	Suction chamber diameter	21mm
D_{mt}	Mixing tube diameter	10mm
D_{st}	Inlet diameter of supply tube	21mm
L_{mt}	Length of the mixing tube	265mm
S	Distance between nozzle and mixing tube	31mm
L_d	Length of the diffuser	135mm
D_d	Diameter of the diffuser at outlet	21mm
θ	Diffuser semi cone angle	2°30'

This geometry is termed as the 'base geometry' in this study as shown in figure 1. For the studies with twisted tape in conical nozzle, the diameter of the mixing tube (D_{mt}) and the diffuser element downstream of the mixing tube were kept constant.

For the present study CFD software ANSYS Fluent is used which incorporates a finite volume code for the preset simulations. This code solves the discretized equations in a segregated manner, with SIMPLE algorithm. The first-order upwind scheme is taken for momentum, volume fraction, turbulent Kinetic energy and turbulent dissipation rate discretisation. The solutions were assumed to have converged for the residual level of 10^{-4} for continuity, x-velocity, and y velocity and 10^{-6} for k -epsilon.

Creation of the geometries was carried out in Solid works and then imported to ANSYS Workbench for meshing of the computational domains. To ensure better insight view of fluid flow and mixing grid independent study was carried out for ejector with conical nozzle incorporating twisted tape. For performing the grid study, geometries with mesh size of 167231cells, 257111 cells, 430017 cells 707808 cells, 864906 cells, 1072648 cells and 1755598 cells were numerically analyzed. Minimum orthogonal quality was kept at 0.5.The grid size was optimized by head difference. It was observed that for the geometries with higher mesh size, variation in head difference was almost negligible. Since after 7.8lakhs grid size there was no such significant difference in pressure contours this was fixed as grid size for all the further simulations which include swirl generators also. Boundary conditions used for each simulation are known static absolute pressure. Turbulent intensity was selected 5 % and respective hydraulic diameters were used at each of the flow boundaries.

DESCRIPTION

To get the swirling flows single and double twisted tapes, as shown in figures 2 to 3, has been used in figure 4. The twisted-tapes are characterized by the twist ratio defined by the ratio between the tape turn length of 180° along its axis and the tube diameter. The length of twisted tape is 40mm and width is 18mm.Thickness of the tape used are 1mm and

1.5mm Single Twisted tapes, used in current study, are 90 degree tape turn with twist ratio 1.11.
Similarly double twisted tapes are also used in primary flow such as two tapes right angled to each other are twisted at 90 degree.

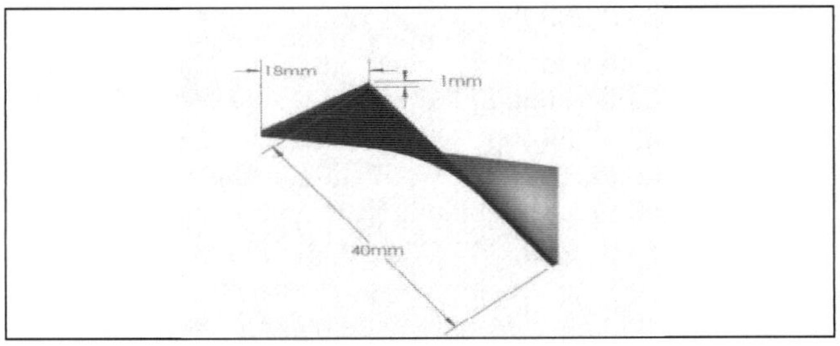

Figure 2: Single twisted tape

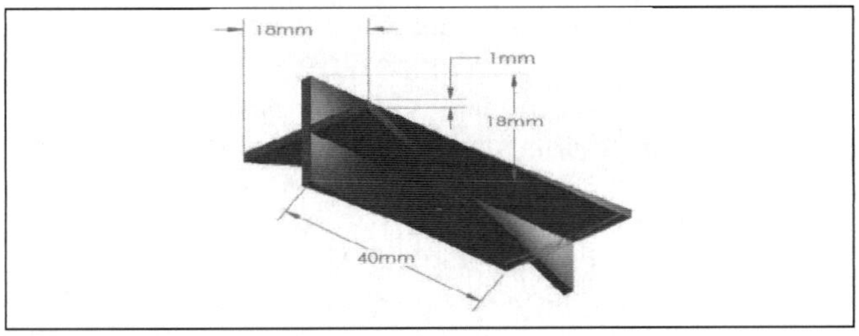

Figure 3: Double Twisted Tape

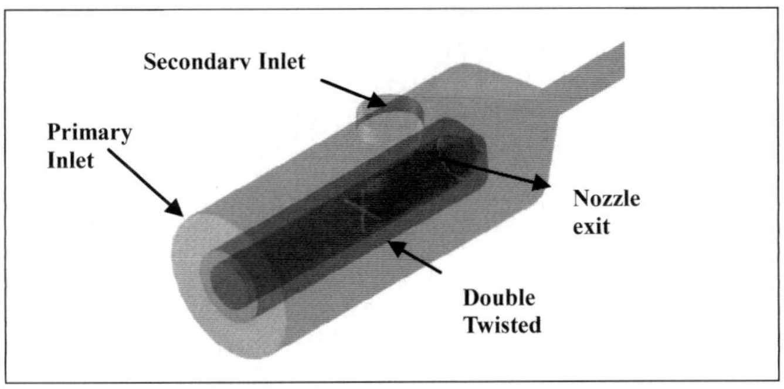

Figure 4: Pictorial representation of double twisted tape incorporated in primary nozzle

RESULTS AND DISCUSSION

On performing numerical studies, the ejector incorporating double twisted tape is found to be highly efficient comparatively to other ejector combination. It is also found that single twisted tape has the lesser efficiency than double twisted tape but better than without tape. The Table 2.shows the comparative parametric studies carried out along with performance obtained. The twisted tape studies are performed for higher area ratio in conical nozzle as for higher diameter of nozzles are found more efficient than smoother profiled nozzle by Sharma [9]. The velocity, turbulent kinetic energy and vorticity magnitude graphs have been plotted against the length of the ejector, as shown in figures 5 to 7.The velocity contours in figure 8 shows that the maximum velocity was attained in the conical nozzle with double twisted tape having thickness 1.5mm. Efficiency of this arrangement is increased by 10% when compared with nozzle, without any tape.

Table 2: Comparison of various operating parameters and performance

Parameters	Without tape	Single twisted tape		Double twisted Tape	
		1mm	1.5mm	1.5mm	1mm
Water flowrate Q_w(lps)	1	1	1	1	1
Air Flowrate Q_a(lps)	1.6	1.63	1.63	1.8	1.72
Flow Ratio	1.7	1.63	1.63	1.8	1.72
Upstream Pressure(Bar absolute) (P_{up})	8.9	8.9	8.9	8.9	8.9
Downstream Pressure(Bar Absolute) (P_{down})	1.81	1.81	1.81	1.81	1.81
Suction Pressure(Bar Absolute) (P_{sec})	1.01306	1.01311	1.01311	1.01306	1.01306
Efficiency η (in %)	18.4	18.4	18.4	20.4	19.5

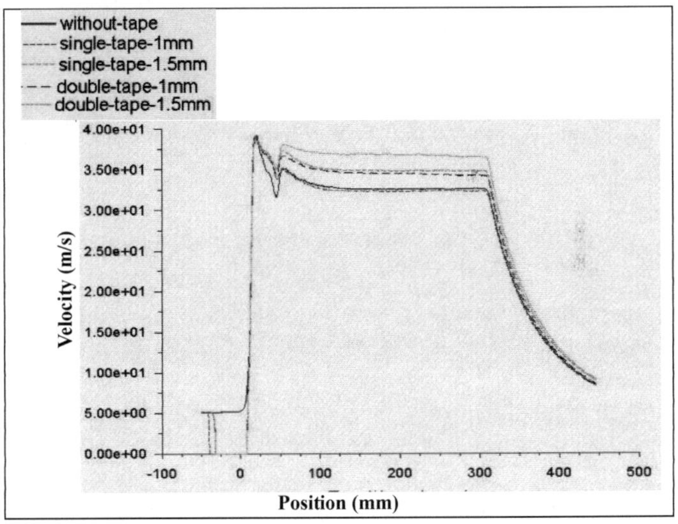

Figure 5: Velocity profiles for jet pump with different twisted tape

CONCLUSIONS

It has been observed that there is a change in flow behavior after the nozzle when twisted tapes are inserted in the upstream primary fluid and more pressure drops after the nozzle is observed due to high vorticity magnitude. This enhances the volume flow rate of the secondary air which ultimately results in better entrainment.

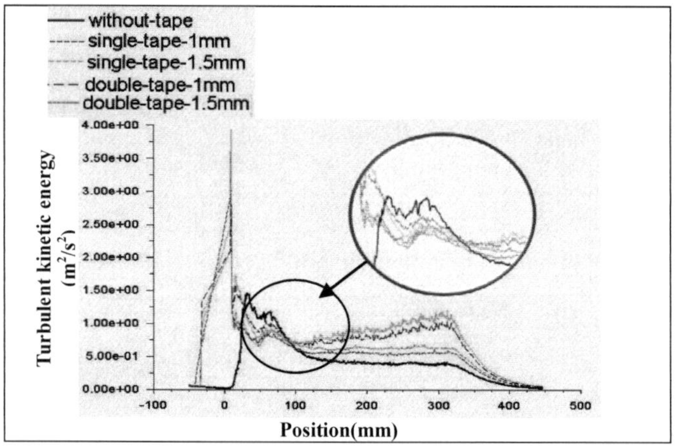

Figure 6: Turbulent kinetic energy profiles for jet pump with different twisted tape

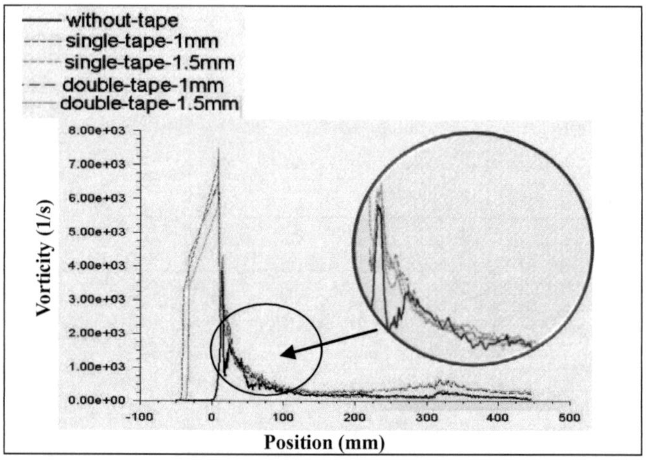

Figure 7: Vorticity magnitude profiles for jet pump with different twisted tape

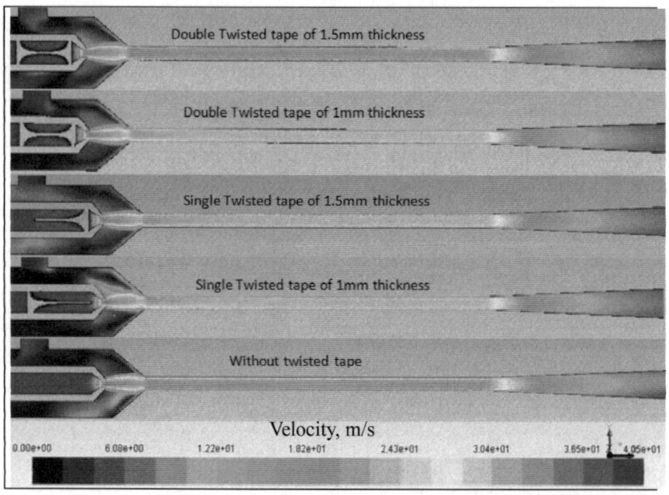

Figure 8: Velocity contours at the mid plane for different Jet pumps

REFERENCES

1. Pfleiderer, C., "Experiments on jet pump for its performance", C. Zeit, *VDI*, 58, 965 &1011.
2. Martinelli, R.C., Boelter, L. M. K., Morrin, E. H., "Theoretical and experimental analysis of ejectors Trans. *ASME*, 66, pp. 139-151.
3. Senthil Kumar, R., Mani, A., Kumaraswamy, S. 2004, "Selection of Pumps for Vacuum Desalination System Utilizing Ocean Thermal Energy, *31st National Conference on Fluid Mechanics and Fluid Power*, Vol. 1, pp. 409–416.
4. Senthil Kumar, R., Mani,A., Kumaraswamy, S., 2007, "Experimental Investigation on Two-Phase Jet Pump used in Desalination System " , *Desalination,* 204, pp. 437-447.
5. Hansen, A. G., and Kinnavy, R. 1965, "The design of water jet pumps part I – experimental determination of optimum design parameters". *ASME Paper* 65-WA/FE-31.
6. Riffat, S. B., Gan, G., and Smith, S., 1996, Computational fluid dynamics applied to ejector heat pumps, *Journal of Applied Thermal Engineering,* 16, 291-297.
7. Riffat, S. B., and Omer, S. A. 2001, CFD modeling and experimental investigations of an ejector refrigeration system, *Int. Journal on Energy Research,* 25, 115-128.
8. Bartosiewicz, Y., Aidoun, Z., Mercadier, Y., 2006, "Numerical assessment of ejector operation for refrigeration applications based on CFD*", Applied Thermal Engineering*, n 26, pp.604–612.
9. Sharma, V. Kumar., Kumaraswamy, S., Mani, A. 2012, "Effect of Various Nozzle Profiles on Performance of a Two Phase Flow Jet Pump". *Int. Journal of Mechanical and Aerospace Eng,6,* p.136-142.
10. Samad, A., Omar, R., Hewakandamby, B., Lowndes I., and Short, G. 2012, "Swirl Induced Flow Through a Venturi-Ejector," *ASME 2012 Fluids Engineering Division Summer Meeting (FEDSM2012),* Puerto Rico, USA.
11. Neve, R. S. 1993, "Computational fluid dynamics analysis of diffuser performance in gas-powered jet pumps", *Int. Journal on Heat and Fluid flow*, 14, 401-407.
12. Ouzzane, M., and Aidoun, Z. 2003, Model development and numerical procedure for detailed ejector analysis and design, *Journal of Applied Engineering*, 23, 2337-2351.

Proceedings of the 25th CANCAM
London, Ontario, Canada, May 31 – June 4, 2015

CFD SIMULATION OF SPRAY COOLING SYSTEMS

M. Jafari
Mech., Auto. & Materials Engg.
University of Windsor
Windsor, ON, Canada
jafari@uwindsor.ca

R.M. Barron
Mech., Auto. & Materials Engg.
University of Windsor
Windsor, ON, Canada
az3@uwindsor.ca

R. Balachandar
Mech., Auto. & Materials Engg.
University of Windsor
Windsor, ON, Canada
rambala@uwindsor.ca

ABSTRACT

In this study, an efficient approach is devised to analyze several important aspects of the spray cooling process. A computational fluid dynamics (CFD) code is used to solve continuity, momentum and energy equations coupled with a Lagrangian-Eulerian solver capable of simulating droplets as well as thin fluid film. The model is validated against experimental data for spray impacting on a circular flat surface and the predicted heat transfer coefficient is in good agreement with the experiments. A liquid film is generated as a result of the spray impact and heat transfer occurs between the liquid and solid. The effects of mass flux and spray Reynolds number on the spray behaviour and thermal characteristics have been studied.

KEYWORDS: Spray cooling, heat transfer, fluid film, computational fluid dynamics

INTRODUCTION

The impingement of sprays onto dry and wet walls and the associated heat transfer occurs in many engineering applications. These applications include internal combustion engines, gas turbines, spray drying, spray coating and spray cooling. The fluid dynamics and heat transfer characteristics of liquid films created by spray impingement are very complex and determining the underlying physics requires fundamental studies. The complex physics involves simultaneous interaction of spray droplets, a thin film surface and spray droplet evaporation, nucleation, convection and condensation. An additional complication in understanding spray cooling heat transfer mechanisms is the small scales at which it occurs. This is the main reason that there are few reliable models which are able to describe the outcome of spray impact and the corresponding heat transfer with good accuracy for different regimes.

Computational modeling can be an effective analysis and design tool for spray equipment manufacturers and process engineers. Depending on the application, modeling can also be very complex, requiring accurate physical models to represent the effects of droplet momentum, film thickness, gravity, surface tension, surface characteristics and phase change.

The overall objective of this work is to develop a better understanding of the hydrodynamics and heat transfer associated with the spray impact onto a heated surface. In order to simulate the spray impinging on heated flat and curved surfaces, a model is developed herein and simulated using commercial CFD software. Based on previous research it is known that, as the droplets impinge onto a heated surface, a thin fluid layer is formed. Research has suggested that the best way to investigate spray cooling is to specify the existence of this thin layer and study the interaction between the layer and the impinging spray droplets.

In this study, an efficient and practical approach is devised for considering critical aspects of the spray cooling process. The computational fluid dynamics (CFD) methodology includes numerous droplets and is designed to predict the spray-wall impact outcome based on reliable correlations. Even though, as a numerical model, it is not an exact representation of the interaction between the spray and the liquid layer, it provides an acceptable picture of the transport phenomena. The STAR-CCM+ CFD code (CD-adapco, 2012) has been used to solve continuity, momentum and energy equations coupled with a Lagrangian-Eulerian solver capable of simulating the droplets motion as well as the thin fluid film that forms on the surface.

Since the main concern in any cooling method is the heat flux and surface temperature, the heat transfer coefficient (HTC) and temperature distribution on the hot surface are the main parameters to be studied. The experiments of Karwa et al. (2007) were selected as the benchmark for validating this CFD methodology. Their experiments were performed at steady state for a water spray with temperature of 25°C impinging on a flat circular plate. The hot plate surface temperature was maintained below 95°C.

NUMERICAL MODEL

The computational domain and boundary conditions are shown in Fig. 1. A uniform mass flow rate is maintained throughout the impingement process, a solid cone spray model is assumed for the injector and a Rosin-Rammler model is used for the spray droplet size distribution. A uniform heat flux is imposed on the top surface of the impingement plate and the skirt surface of the plate is assumed to be an adiabatic wall. The turbulent flow is modeled using RANS equations with k-ω SST, but the fluid film that forms on the surface is assumed to be laminar. The water has constant physical properties throughout the process. The modified model proposed by Bai et al. (2002) was selected as the spray impact model since it is the most widely used model for this type of application. Since the Weber number in our case is low, the TAB breakup model is a reasonable choice. The TAB model was developed by O'Rourke and Amsden (1987) using the Taylor analogy to determine when and how the droplets break up under the action of aerodynamic forces.

Three cases with the same type of nozzle (cone spray) were chosen from the experiments of Karwa et al. (2007) for validation. The CFD results were obtained under various operating conditions of the spray system (different cases), with a spray nozzle distance to the plate H ranging between 18 and 29 mm, and Reynolds number varying between 162 and 230. In the cases considered, the nozzle supply pressure changes between 1.41 bars (Case 1) and 2.94 bars (Case 3).

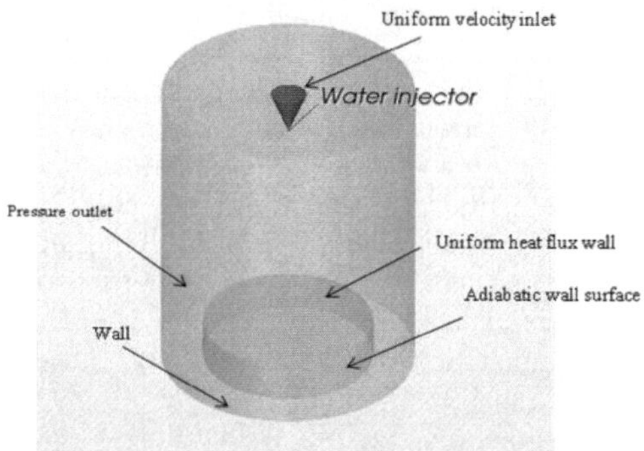

Figure 1: Computational domain and boundary conditions

Table 1 Operating parameters from Karwa et al. (2007) (Unijet #1 nozzle, orifice diameter $d_0 = 510$ μm)

Case #	Mass flux, G (kg/m² s)	SMD d_{32} (μm)	Droplet velocity (m/s)	Re	We
Case 1	6.21	149.2	16.8	162	590
Case 2	7.32	137.1	19.9	191	760
Case 3	8.78	123.4	24.3	230	1020

RESULTS

This section provides results for the cases outlined in Table 1, which are based on the experiments by Karwa et al. (2007). The effect of mass flux on the droplet flow field, film thickness, surface temperature and HTC and the Nusselt number (Nu) variation with spray Reynolds number are investigated. The average mass flux is used to define the spray Reynolds number Re.

The variation of HTC with the flow rate is shown in Fig. 2, indicating good agreement with the results of Karwa et al. (2007). These results are also in agreement with Oliphant et al. (1998) who used air-assist spray impingement cooling to show that the heat transfer coefficient is strongly dependent on mass flux in the non-boiling regime, which was also reported by Wang et al. (2011) and Lin et al. (2004).

Figure 3 presents the CFD results obtained at the operating flow conditions of the spray system (Table 1). In addition to the experimental measurements, Karwa et al. (2007) have developed a correlation for Nusselt number based on the spray Reynolds number, given by

$$Nu = 20.344\,\mathrm{Re}^{0.6589} \quad (1)$$

Wang et al. (2011) have also introduced a relevant correlation, expressed as

$$Nu = 0.1275\,\mathrm{Re}^{0.9322}\,\xi^{2.2485} \quad (2)$$

where $\xi = \frac{T_{wall}}{T_{boiling} - T_e}$ and T_{wall} is the wall temperature, $T_{boiling}$ is the boiling temperature of the working fluid and T_e is the environment temperature. As illustrated in Fig. 3, at the larger Reynolds numbers the present CFD model slightly underpredict the Nusselt number in comparison to the measurements of Karwa et al. (2007), while their correlation eqn. (1) slightly overestimates Nu. The correlation of Wang et al. (2011), eqn. (2), significantly overestimates Nu at larger Re. The CFD results for the Nusselt number are within 10% of the experimental data of Karwa et al. (2007).

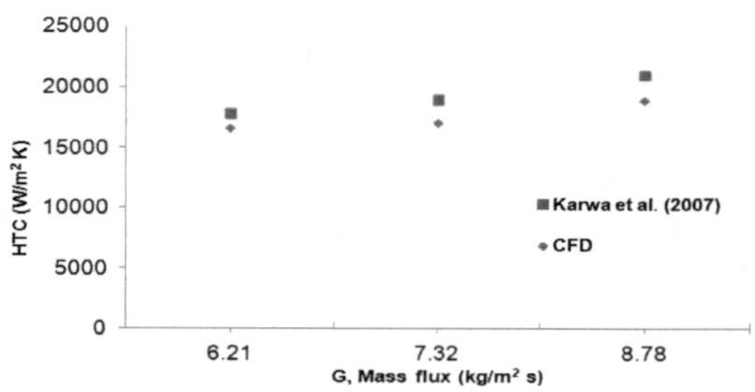

Figure 2: Heat transfer coefficient versus flow rate

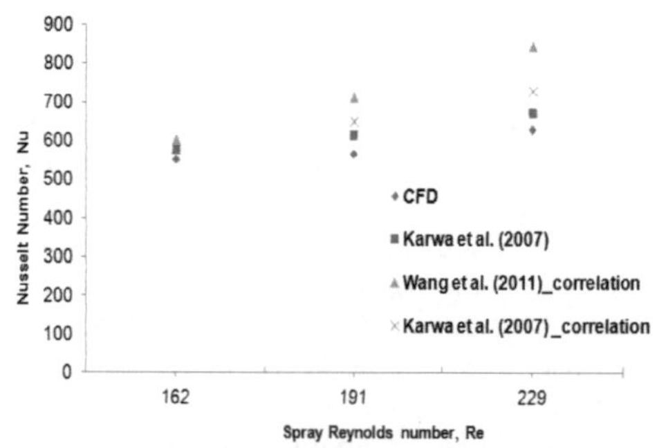

Figure 3: Comparison of predicted Nusselt number with Karwa et al. (2007) and Wang et al. (2011)

Based on the results of the simulation, good agreement is observed for the HTC values for these cases involving spray impact and fluid film formation over a flat solid surface. The estimated value of HTC lies between 16591.84 to 19170.04 W/m^2 K. It is also observed that as the mass flux increases, the HTC increases, consistent with the findings of Freund et al. (2007). To obtain an acceptable value for the HTC, a "quasi-steady" condition must be reached. Even though the spray parameters such as the droplet diameter and velocities vary, droplet mass flux has been found to be the dominant factor that influences heat transfer.

Figure 4 shows the film thickness for the three cases. In order to maintain the full coverage of the target plate by spray droplets, Case 3 must have a larger spray angle than the other cases. Case 3 involves higher droplet velocities and mass flux, while the droplet sizes are the smallest for this case. This leads to higher lateral velocity and ultimately a thinner fluid film. A thinner film leads to a smaller boundary layer thickness and higher heat removal rate.

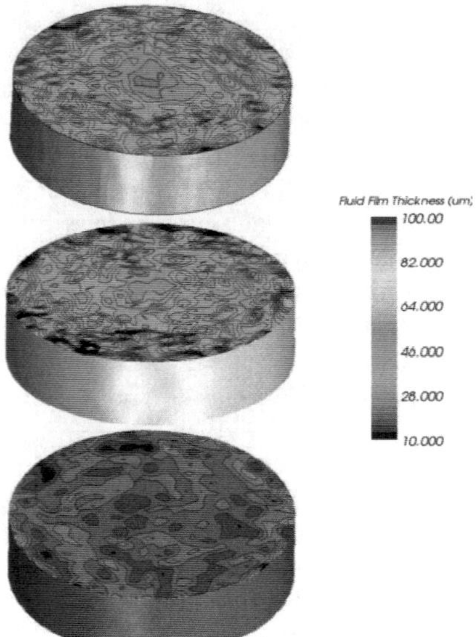

Figure 4: Film thickness on the solid surface at 15 *msec*; Case 1 (top), Case 2 (middle) and Case 3 (bottom)

Figure 5 shows the velocity vectors of the spray droplets for all three cases. The spray cone angle and the distance from the spray nozzle to the target surface vary between these cases. This distance is the smallest for Case 3, leading to the formation of large vortical structures near the plate which promotes a higher HTC.

Figure 6 shows that the HTC decreases in the radial direction from the centre to the edge of the metal target plate. At higher radii (further from the centre of the circular plate) the film decelerates and both the film and thermal boundary layer become thicker, resulting in a drop in HTC. Another factor contributing to the lower HTC could be that the liquid coolant's temperature has already increased as it reaches the edges of the plate, as seen in Fig. 7. Furthermore, as the droplet density increases due to deceleration along the plate, the droplet–film heat transfer increases, thereby reducing the surface temperature.

Freund et al. (2007) reported that the peak HTC values around the centre of the plate increase as the flow rate increases. The heat transfer rate in the central region is high, but as the fluid moves out radially along the surface, the fluid loses its momentum and the heat transfer coefficient decreases, as is observed in the current study. This pattern also matches the results obtained by Pautsch (2004), who found the highest performance of heat removal at the centre with the lowest performance of the heat removal occurring at the edges. As the spray flow rate increases, droplet momentum introduces more agitation to the liquid, generating more mixing in the flow. This enhances the heat transfer locally, and increases the HTC.

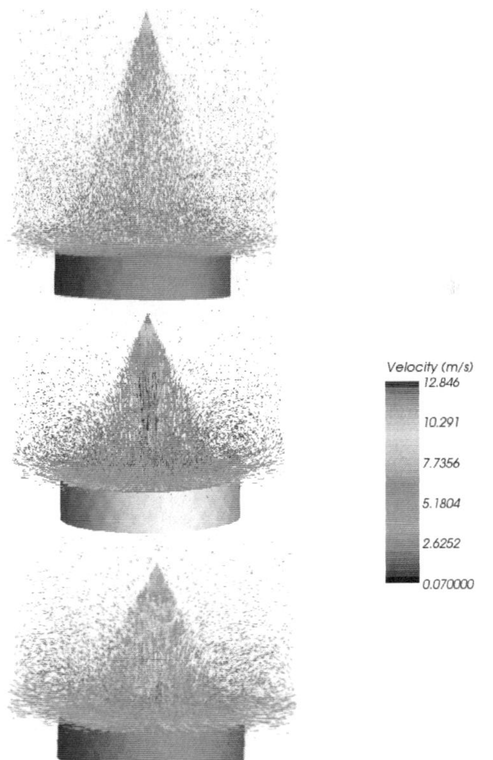

Figure 5: Spray droplet velocity vectors at 15 *msec*; Case 1 (top), Case 2 (middle) and Case 3 (bottom)

The behaviour of the circumferentially averaged temperature of the hot plate as a function of radius is shown in Fig. 7. The temperature distribution of the heated surface is non-uniform. By increasing the mass flow rate, higher temperature reduction rate can be achieved at the centre of the plate. A reduction of approximately 4°C in the mean surface temperature is observed when the flow rate increases from 6.21 to 8.78 kg/m^2s. It is worth noting that the surface temperature in the present study maintained a level below 95°C in agreement with the experimental evidence. Therefore, it can be concluded that evaporation does not have a significant impact on the heat transfer and the majority of heat is transferred by single phase convection to the water droplets and film formed on the hot surface. i.e., the main sources of heat transfer during the cooling process can be expressed as

$$\dot{Q}_{in} = \dot{Q}_{drop} + \dot{Q}_{film} \quad (3)$$

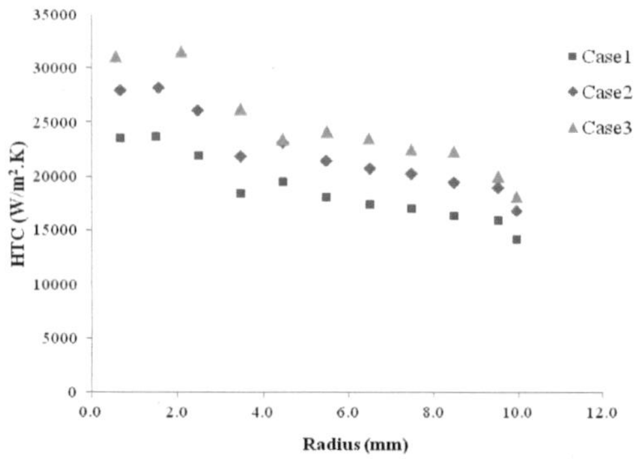

Figure 6: Heat transfer coefficient distribution versus surface radius

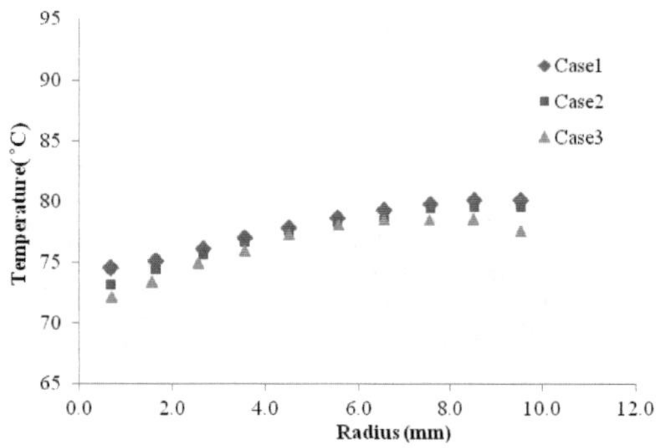

Figure 7: Temperature versus surface radius

Due to the interaction of the phenomena mentioned above and the inability to only control droplet velocity or mass flux, it is difficult to study the influence of a single parameter on the heat transfer. Mudawar and Estes (1996) argued that the mass flux, not the droplet velocity, was the dominant factor for heat transfer performance. As a result of increased mass flux, the combined effect of mixing and unsteady boundary layer evolution leads to enhanced heat transfer rates. The level of enhancement depends on the boost in mass flux, characterized by the number of droplets exiting the nozzle per second, their velocity and diameter.

CONCLUSIONS

A finite volume CFD method which includes Lagrangian spray modeling, fluid film development and conjugate heat transfer has been implemented to simulate spray cooling conditions. The model considers the interaction between the spray and the thin film that forms on the target surface, and the flow of heat from one to the other. It also considers heat transfer in the fluid as well as in the solid. The CFD results are in good agreement with the available test data and correlations, demonstrating that the method employed in this study could be successfully utilized to analyze realistic spray cooling applications.

REFERENCES

1. Bai, C., Rusche, H., Gosman, A., "Modeling of gasoline spray impingement", *Atomization and Sprays*, 12:1-27, 2002.
2. CD-adapco, STAR-CCM+ User's Manual, Version 7.06, 2012.
3. Fruend, S., Pautsch, A.G., Shedd, T.A., Kabelac, S., "Local heat transfer coefficients in spray cooling systems measured with temperature oscillation IR thermography", *Int. J. Heat Mass Transfer*, 50:1953-1962, 2007.
4. Karwa, N., Kale, S.R., Subbarao, P.M.V., "Experimental study of non-boiling heat transfer from a horizontal surface by water sprays", *Expt. Thermal and Fluid Science*, 32:571–579, 2007.
5. Lin, L., Ponnappan, R., Yerkes, K., Hager, B., "Large area spray cooling", *Proc. 42nd AIAA Aerospace Sciences Meeting and Exhibit*, Reno, NV, 2004.
6. Mudawar, I., Estes, K.A., "Optimizing and predicting CHF in spray cooling of a square surface", *J. Heat Transfer*, 118(3):672-679, 1996.
7. Oliphant, K., Webb, B.W., McQuay, M.Q., "An experimental comparison of liquid jet array and spray impingement cooling in the non-boiling regime", *Expt. Thermal and Fluid Science*, 18:1-10, 1998.
8. O'Rourke, P.J., Amsden, A.A., "The TAB method for numerical calculation of spray droplet breakup", *SAE Technical Paper* 872089, 1987.
9. Pautsch, A.G., Heat Transfer and Film Thickness Characteristics of Spray Cooling with Phase Change, *Master's Thesis*, Univ. of Wisconsin - Madison, 2004.
10. Wang, Y., Liu, M., Liu, D., Xu, K., "Heat flux correlation for spray cooling in the non-boiling regime", *Heat Transfer Engineering*, 32(11):1075–1081, 2011.

Proceedings of the 25th CANCAM
London, Ontario, Canada, May 31 - June 4, 2015

EFFECT OF MICROSTRUCTURED ROUGHNESS ON MICRO-DROPLET DYNAMICS

S.M.Reza Attarzadeh*, Moussa Tembely, and Ali Dolatabadi
Department of Mechanical Engineering
University of Concordia
Montreal, Quebec, Canada
E-mails: s_attarz@live.concordia.ca
moussa.tembely@gmail.com, ali.dolatabadi@concordia.ca

ABSTRACT

Understanding the various stages of droplet dynamic on superhydrophobic textured surface is of interest for many industrial applications such as aerospace industries. In this paper, we investigate numerically the dynamic of micro-droplet impact on a flat and textured substrates at different surface roughness comparable to the size of the droplet, which is similar to real scenario of impacting cloud particles (20 μm) over the structure of the aircraft. The volume of fluid (VOF) method coupled with a dynamic contact angle model is used to simulate the impact of a 20 μm droplet with the velocity of 1.6 m.s^{-1} on a flat surface, and two micro-textured surfaces with all having intrinsic contact angle of 111°. The simulation successfully captured the different stages of the droplet dynamic ranging from the impact, spreading phase, to the detachment which is mainly due to the effect of entrapped air beneath the droplet. It is also observed that increase in the surface roughness result in an increase of droplet mobility and thus to a higher apparent contact angles. In addition, the shape of the textured surfaces on the contact time is also highlighted.

Keywords: Superhydrophobic, Volume of Fluid, Textured surface, Micro-droplet, Roughness

INTRODUCTION

Surfaces with designed microscopic roughness possess remarkable non-wetting properties, and extensive attention is drawn recently by the micro and nano-technology research communities [1, 2, 3] into development of the materials, aiming at finding new ways of controlling solid surface-fluid interface properties. Although according to the Young's equation, a flat surface with a contact angle approaching 180° could theoretically be possible, no physical evidence have been reported that demonstrates this situation. Theoretical models describing the influence of surface roughening with respect to the wetting properties of the surface have been introduced earlier either following the (i)Wenzel [4] , and (ii) Cassie-Baxter [5] models. In Wenzel it is assumed that the liquid impales the roughness elements, and follows the surface topography, resulting in higher surface wettability due to the increase in contact area [4] :

$$\cos\theta_W = R\cos\theta_e \quad (1)$$

where R is the ratio of the actual area of the solid/liquid interface to the normally projected area, while θ_e is the equilibrium or static contact angle. In the Cassie-Baxtar model [5], air entrapped beneath the droplet inside the groove features form a composite hydrophobic surface, holding the droplet on the tip of the pillars and resulting in a larger contact angle,

$$\cos\theta_C = -1 + f(\cos\theta_e + 1) \quad (2)$$

where f is the solid-liquid fraction ratio. A large number of experiments and theoretical investigations on droplets-surface interaction has been reported, however they are based on smooth substrate with intrinsic contact angle [7, 8]. However, the outcome of droplet impact on textured surface is different from that on smooth surface, due to the impact force, the liquid penetrates into the micro-cavities, and exhibits different wetting scenario on textured surface. Deng et al. [13] discussed different types of pressures and wetting states experienced by the impacting droplets on textured surface. However, as his experiment was conducted for droplet size of 1mm, there is no evidence indicating that the dynamic of droplet would be the same for micron size drops when the size of the droplet is almost comparable with the size of the pillars. Unfortunately due to some uncertain factors, such as, penetration depth, geometrical configuration of recoiling droplet and the complicated viscous dissipation rate, a direct theoretical and experimental investigation of the bounding behavior of the droplet on textured surface is difficult. Therefore, the numerical simulation can overcome these difficulties and provide more detailed information under varied initial and operating conditions.

NUMERICAL METHODOLOGY

Volume of Fluid (VOF) [6] interface capturing method conjugated with OpenFoam 2.1.1, is used to investigate the effect of micro-structured roughness on spreading and depinning of micro-droplet. The liquid-air interface is tracked by the VOF formulation. In VOF model [6], tracking of the interface is modeled by solving continuity equation for one of the two phases in each computational cell at every time steps. The governing equations for continuity and momentum are,

$$\vec{\nabla}.\vec{V} = 0 \tag{3}$$

$$\frac{\partial \rho \vec{V}}{\partial t} + \vec{\nabla}.(\rho \vec{V}\vec{V}) = -\vec{\nabla} p^* + \vec{\nabla}.(\mu.\nabla \vec{V}) + \vec{F_b} \tag{4}$$

Where t is time, $\vec{V}$ is the velocity vector, $p^* = p - \rho g$ is the modified pressure, $\vec{F_b}$ is the body force, and α is the volume fraction which is used to calculate the viscosity(μ), and density(ρ), as weighted averages based on the distribution of the liquid fraction,

$$\begin{aligned} \rho &= \alpha \rho_l + (1-\alpha)\rho_g \\ \mu &= \alpha \mu_l + (1-\alpha)\mu_g \end{aligned} \tag{5}$$

where the subscripts l and g denote the liquid and gas phases, respectively. In VOF method the phase fraction scalers function α, admit value between 0 and 1, and $\alpha = 0$ corresponds liquid phase, $\alpha = 1$ for gases, and $0 < \alpha < 1$ where the interface is located. As the interface evolves with the motion of fluid, it is essential that α satisfies the conservation equation while advecting with the flow,

$$\frac{\partial \alpha}{\partial t} + (\vec{V}\vec{\nabla})\alpha + \vec{\nabla}.(\vec{V_r}\alpha(1-\alpha)) = 0 \tag{6}$$

where $\vec{V_r}$ is the vector of relative velocity, designated as the compression velocity introduced by Rusche [12] to facilitate a sharp interface between phases,

$$\vec{V_r} = k_c \vec{n} \max \frac{|\vec{n}.\vec{V}|}{|S|^2} \tag{7}$$

where k_c is an adjustable coefficient used to adjust the amount of compression, and S, the cell surface area vector. The interface unit normal $\vec{n}$ is computed by taking the gradient of smoothed volume fraction α at the cell faces,

$$\vec{n} = \frac{\vec{\nabla}\alpha}{|\vec{\nabla}\alpha| + \delta} \tag{8}$$

where δ is a small number in order to stabilize the calculation in regions outside the transition region where $|\vec{\nabla}\alpha| \to 0$. In this study, $\delta = 10^{-8}$ is used for the entire simulations. The main benefit of such formulation is in the possibility of capturing the sharp interface region. The continuum surface force method [14] is used to model surface tension as a body force $\vec{F_b}$ that acts only on interfacial cells,

$$\vec{F_b} = \sigma k \vec{\nabla}\alpha \tag{9}$$

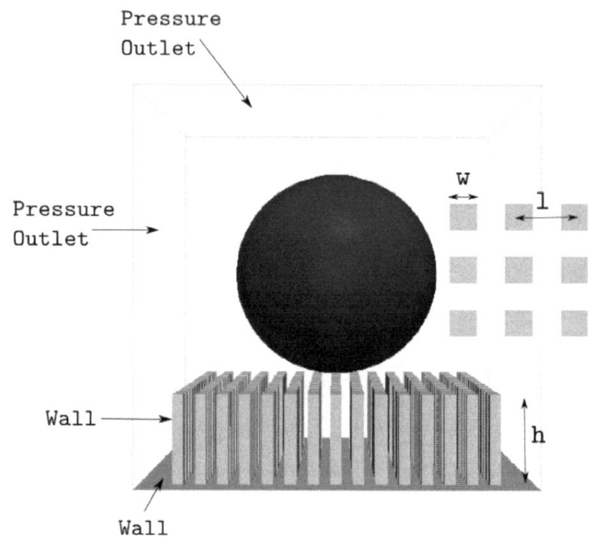

Figure 1: 3D computational domain, droplet and textured surface

where σ is the surface tension, and k is the mean curvature of the free surface, calculated by

$$k = -\vec{\nabla}.\left(\frac{\vec{\nabla}\alpha}{|\vec{\nabla}\alpha|}\right) \tag{10}$$

The pressure-Implicit with splitting operators (PISO) scheme in transient flow is used for the pressure-velocity coupling. The wettability effects at the substrate are taken into account by using the dynamic contact angle which is assumed to be a function of the contact line velocity. The Kistler's correlation [15] is used to calculated the dynamic contact angle in each time step,

$$\theta_d = f_H[Ca + f_H^{-1}(\theta_e)], \tag{11}$$

where f_H is the Hoffman's function and is defined as,

$$f_H = \arccos\{1 - 2\tanh[5.16[\frac{x}{1 + 1.31x^{0.99}}]^{0.706}]\} \tag{12}$$

in the above equation the equilibrium contact angle θ_e is replaced by either the advancing contact angle, θ_A, or the receding contact angle, θ_R, depending on the direction of the velocity vector at the contact line or the static contact angle, θ_s, if the contact line velocity is zero. It is worth mentioning that the no-slip models such as Kistler's correlation [15] have been extensively used to investigate droplet dynamics [9, 10, 11]. In addition, it is shown that this model is capable to capture the relevant physics of droplet-substrate interaction and provides good agreement with experiment.

The computational domain size of 0.034×0.034×0.040 mm is used for the 3D simulation of the textured surface (Figure 1) with w, l, and h correspond to pillars width, the pitch between two adjacent pillars, and the height of the pillars, respectively.

A mesh independence test is performed for all textured models. The domain is discretized by about 3.3 million, 7.8 million, and 15.6 million Hexahedral mesh with local refinement beneath the droplet and the interfacial cells. As the change of interface evolution in 7.8 million and 15.6 million mesh was minimal, the domain composed of 7.8 million mesh is applied for this study. The cells are refined so to have 120 cells per diameter of the droplet before impingement. In this study, $2\mu m$ pitch (l) is taken consistent through all simulations. In order to compare the results with the data obtaining from experiments, it is assumed the average surface roughness of $1.9\mu m$ as a reference, hence width (w) and height (h), $1\mu m$ and $8\mu m$ respectively is obtained according to the surface roughness.

RESULTS AND DISCUSSION

The effect of textured roughening was examined by considering the impact of 20μm diameter water droplet onto a series of micro-textures (pillars) of square pattern in the atmospheric condition. Figure 2 illustrating the time-laps sequence of droplet impingement on a (a) smooth substrate, (b) textured surface with pillars width of 0.5μm, and (c) textured surface with pillars width of 1μm respectively. As shown in the Figure 4-a the droplet is not repelled from the substrate due to the higher contact area with the droplet. However detachment is encountered in textured surfaces at different contact times, depending on the topographical characteristics of the surfaces. It worth noting that the repellency of the droplet by adding the pillars on surface is well captured by the model, in agreement with experimental result of manufacturing superhydrophobic surface by texturing. It was observed the solid fraction beneath droplet has an important influence on penetration depth of the micro-droplet. As the pitch (l) is kept constant in all simulations, hence, the effect of pitch on dynamic of droplet, as an important parameter is not individually evaluated.

Figure 3 provides a quantitative comparison between the microdroplets spreading factor, as a function of time. It is shown that droplet impinging on the surface with smaller pillar width (w) significantly affect droplet dynamics in terms of both the spreading diameter and the contact time (Figure 3). Increasing projected solid fraction by two times results in reducing contact time about 20%, and the maximum spreading diameter approximately increases by 15%. Due to the smaller projected solid fraction in Figure 2-b, the droplet penetrates before it bounces off the surface. The penetrated volume, geometrically preserves its spherical shape, due to the effect of surface tension. It is observed that smaller solid fraction induces higher penetration depth of the droplet, as well as higher wetting area of the surface.

Experimentally it is difficult to determine the exact location of droplet impingement, specially for droplets within 20-100 μm size range, therefore a quantitative study was carried out to examine contact time for different possible impact points on the textured surfaces (Figure 4). Black dots in the figure 4, representing the exact location of the impingement

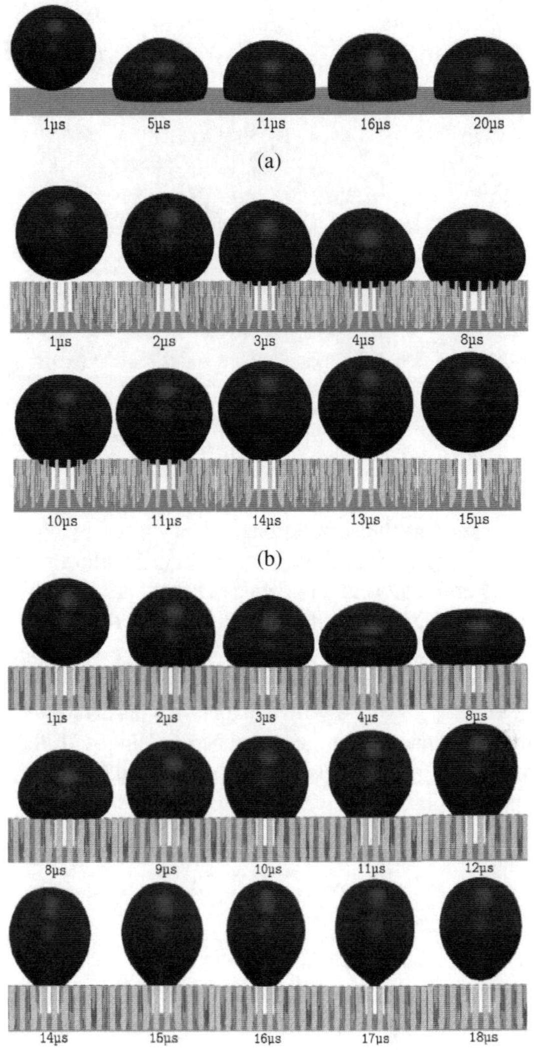

Figure 2: Micro-droplet impact on flat and textured surface with intrinsic contact angle 111°. (a) Flat substrate, (b) Pillars $w = 1\mu$m, and (c) Pillars $w = 0.5\mu$m

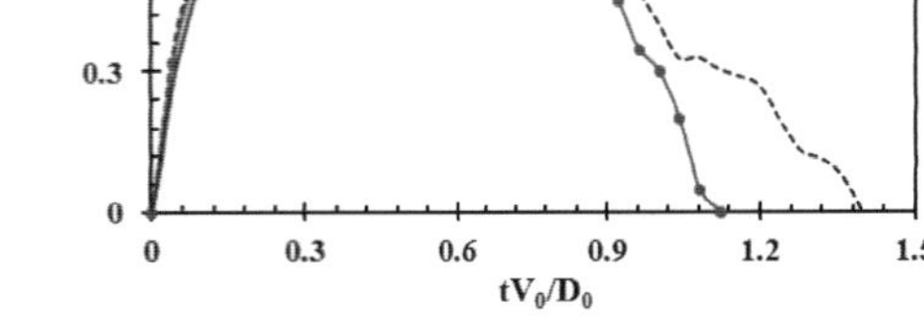

Figure 3: Numerical results of the 20 μm microdroplet spreading factor evolution at 1.6 m.s^{-1} on both smooth and textured surfaces ($\theta_e = 111°$).

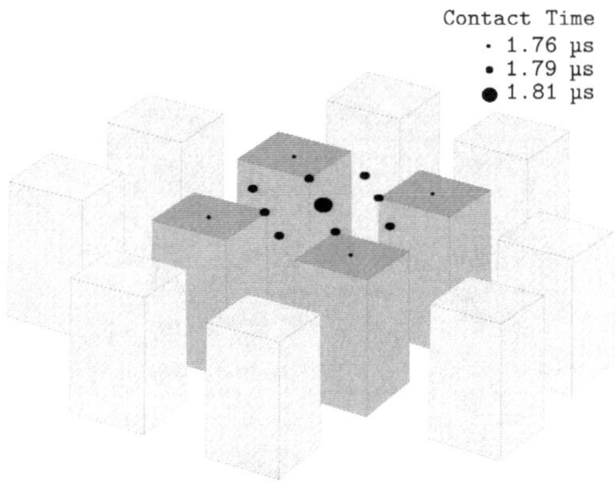

Figure 4: Contact time at different impact locations

that is set in the simulation. Based on the results, the maximum contact time ΔT_{max} is observed when the droplet impacts at the gap centers between the pillars, and the least contact time, ΔT_{min}, when the droplet hits the center of a pillar upon impact. Nevertheless, contact time was changed only by 3%. Although the effect of the impact point on the the contact time may not be significant for a single droplet, such effect may become very important when dealing with the impact of a spray on textured surfaces, as required by many industrial applications.

CONCLUSIONS

The numerical simulation of the impact of micro-droplets on various textured surfaces is performed using a volume of fluid method along with a dynamics contact angle model to describe the droplet-substrate interaction. The simulation successfully captured the dynamic of impinging micro-droplet on textured surface with designed surface roughness, comparable to the size of the droplet. The model captured the effect of entrapped air, which increased the apparent contact angle, and lower surface energy as a result of mechanical roughening. In addition, the simulation indicates that superhydropbobicity can be affected by both the pillars design, and the droplet impact position. The pillars width is found to control both the contact time and the maximum spreading diameter of the droplet, making it one the interesting design parameter for textured surfaces.

ACKNOWLEDGMENTS

The authors gratefully acknowledge the financial support from Natural Sciences and Engineering Research Council of Canada (NSERC), Consortium de Recherche et dInnovation en Aerospatiale au Quebec (CRIAQ), and Bombardier Aerospace and Pratt & Whitney Canada.

REFERENCES

[1] Shang, HM., Wang, Y., A.2005, Thin Solid Films, 472(12):3743.

[2] Sun, T., Feng, L., A.2005, Acc Chem Res, 38:644.

[3] Feng, L., Song Y.,A.2003, Angewandte Chem Int Ed, 42(7):8002

[4] Wenzel RN.,A.1936, Ind Eng Chemistry, 28:98894.

[5] Cassie, A., Baxter S.A.1944, Trans Faraday Soc. 1944;40:54651.

[6] Hirt, C. W., Nichols, B. D., A.1981, " Volume of Fluid (VOF) Method for the Dynamics of Free Boundaries", J. Comput. Phys. 39, 201225.

[7] Gunjal, P. R., Ranade, V. V., Chaudhari, R. V.,A.2005, "Dynamics of Drop Impact on Solid Surface". AIChE J., 51, 5978.

[8] Mitra, S., Sathe, M., Doroodchi, E., Utikar, R., Shah, M., Pareek, V., Joshi, J. B., Evans, G., A.2013, "Droplet Impact Dynamics on A Spherical Particle", Chem Eng Sci, 100, 105119.

[9] Sikalo, S., Wilhelm, H., Roisman, I.V., Jakirli, S., Tropea, C., A.2005, "Dynamic contact angle of spreading droplets", Phys. Fluids, 062103. doi:10.1063/1.1928828

[10] Roisman, I. V., Opfer, L.,Tropea, C.,Raessi, M., Mostaghimi, J. and Chandra, S., A.2008, "Drop Impact onto a Dry Surface: Role of the Dynamic Contact Angle", Colloids and Surfaces A: Physicochemical and Engineering Aspects. 322, no. 13

[11] Malgarinos, I., Nikolopoulos, N., Marengo, M., Antonini, C., Gavaises, M., A.2014. "VOF simulations of the contact angle dynamics during the drop spreading". Adv. Colloid Interface Sci. 212, 120.

[12] Rusche, H., A.2002 "Computational fluid dynamics of dispersed two-phase flows at high phase fractions", Ph.D. dissertation, Imperial College, London, England.

[13] Deng, T., Varanasi, K. K., Hsu, M., Bhate, N., Keimel, C., Stein,J., Blohm, M., A.2009, "Nonwetting of Impinging Droplets on Textured Surfaces". Appl. Phys. Lett, 94, 133109.

[14] Bussmann, M., Mostaghimi, J., and Chandra, S., A.1999,On a three-dimensional volume tracking model of droplet impact, Phys.Fluids, 11, 14061417.

[15] S. F. Kistler, A.1993 "Hydrodynamics of wetting", in Wettability, edited by J. C. Berg, Marcel Dekker, New York, 1993, 311429.

[16] Kwon, D., Lee, S. J.,A.2012, "Impact and Wetting Behaviors of Impinging Microdroplets on Superhydrophobic Textured Surfaces" Appl. Phys. Lett. 100, 171601.

Proceedings of the 25th CANCAM
London, Ontario, Canada, May 31 – June 4, 2015

SIMULATION OF AYSMMETRIC TURBULENT CHANNEL FLOW WITH ROUGHNESS ON ONE WALL

Ming TENG
Mechanical Engineering Department
University of Saskatchewan
Saskatoon, Saskatchewan, Canada
mit738@mail.usask.ca

Donald J. Bergstrom
Mechanical Engineering Department
University of Saskatchewan
Saskatoon, Saskatchewan, Canada
don.bergstrom@usask.ca

ABSTRACT

The finite volume method was used to solve the Reynolds Averaged Navier-Stokes equations for fully-developed asymmetric turbulent channel flow with one smooth and one rough wall. The flow over the rough wall was modeled using a mixing length model which was modified to include the effects of surface roughness. The results were compared to two direct numerical simulation studies, one of which considered sand-grain type roughness while the other considered two-dimensional rib roughness. The eddy viscosity model was more successful in predicting the mean velocity profiles for the sand-grain roughness than for the rib roughness. This is due to the challenge of using a single roughness parameter such as equivalent sand-grain roughness to describe the effects of the two-dimensional ribs.

KEYWORDS: Asymmetric turbulent channel flow, rough wall turbulence, eddy viscosity model

INTRODUCTION

As compared to laminar flow, the most essential and practical characteristic of turbulence lies in the enhanced mixing and transport, which it facilitates. For a rough surface, the turbulent transport is further enhanced over that for a smooth wall. Motivated by this characteristic, fully turbulent flows over rough walls have been investigated for some time, mostly through experiments. Among many different methods employed to create surface roughness on a wall, two-dimensional ribs aligned transverse to the flow and sand-grain roughness are among the most popular options. However, there are fundamental differences between these two geometries. For example, based on the spacing of the roughness elements, two-dimensional ribs can produce either d-type or k-type-roughness. Unlike two-dimensional ribs, sand-grain roughness, usually simulated by distributions of minute spherical elements, represents a much more homogeneous distribution of roughness on the wall. For an asymmetric turbulent channel flow, the geometry and type of roughness are known to significantly influence the magnitude and location of maximum stream-wise velocity and other characteristics of the velocity profile.

There are different methods for simulating turbulent flows. For industrial applications, the most popular approach is to solve the so-called Reynolds Averaged Navier-Stokes (RANS) equations. The RANS equations efficiently describe the spatial variations of the mean or time-average quantities, and include the dominant effects of the turbulent Reynolds stresses.

Due to the inherent nonlinearity of the RANS equations relating to the convective terms and also the Reynolds stress relations, their solution typically requires use of numerical methods, more specifically computational fluid dynamics (CFD). There are several methods widely employed in CFD, including the Finite Volume Method (FVM), Finite Difference Method (FDM), and Finite Element Method (FEM). In this study, the FVM is employed, since it has been found to be a robust and efficient approach for the discretization of the relevant transport equations.

For the prediction of turbulent flows using a RANS approach, a so-called turbulence model is required to close the mean transport equations by relating the turbulent fluxes to the mean velocity field. A hierarchy of models has been developed of increasing complexity. For the two-dimensional channel flows considered in the present study, an eddy viscosity model is adopted. More specifically, an eddy viscosity model formulation is tested which has the capability to include the effects of surface roughness. Krogstad (1991) developed an eddy viscosity model closure for prediction of boundary layer flows over walls with surface roughness.

In the present paper, fully developed turbulent channel flow with roughness on one wall is predicted using the eddy viscosity model of Krogstad (1991). The presence of one smooth and one rough wall creates an asymmetric mean velocity profile, with the peak velocity shifted toward the smooth wall. Two different roughness geometries are considered: sand grain and rib roughness. In both cases, the roughness is incorporated using the concept of equivalent sand grain roughness, which is correlated with the roughness shift observed in the mean velocity profile. The study demonstrates the capability of this specific turbulence model to capture the effects of roughness on the mean velocity field and wall shear stress. Not surprisingly, it turns out that the rib roughness is not as well modeled as the sand grain roughness.

MATHEMATICAL MODEL AND NUMERICAL METHOD

The present study uses the finite volume method to solve the RANS equation for fully developed turbulent flow in a two-dimensional channel with one smooth and one rough wall. The wall normal velocity is identically zero and the flow is driven in the stream-wise (x) direction by a constant pressure gradient. The mean momentum equation is closed using an eddy viscosity model based on the mixing length formulations described below.

Mixing length model

In order to incorporate the effect of turbulence, the discrete transport coefficients for diffusion include both the laminar and turbulent contributions. The latter is typically dominant and given by a mixing length model, i.e.

$$\nu_t = {l_m}^2 \left|\frac{\partial U}{\partial y}\right| \tag{1.1}$$

For a smooth-wall flow, the mixing length l_m is calculated using a van Driest damping function given by:

$$l_m = \kappa y \left[1 - exp\left(-\frac{y^+}{A^+}\right)\right] \tag{1.2}$$

where $\kappa = 0.41$ is the von Kármán constant.

Modification of damping function for roughness

For a rough-wall flow, the mixing length takes the following form:

$$l_m = \kappa y F \tag{1.3}$$

where F is a damping function modified for surface roughness.

The following damping function was first proposed by van Driest (1956):

$$F = 1 - \exp\left(-\frac{y^+}{A^+}\right) + \exp\left(-\frac{y^+}{A^+}\frac{R^+}{k^+}\right) \tag{1.4}$$

where A^+ and R^+ are model constants. In the damping functions, k^+ is the dimensionless roughness height and y^+ is the dimensionless wall-normal distance defined, respectively, as:

$$k^+ = k * u_\tau / \nu,\ y^+ = y * u_\tau / \nu \tag{1.5}$$

where u_τ is the local friction velocity.

As pointed out by Sivykh (1984), significant under-prediction of the velocity shift, ΔU^+, was generated by equation (1.4) for large roughness elements when F is limited to be less than or approximately equal to one. To correct this deficiency, Krogstad (1991) proposed the following formulation:

$$F = 1 - \exp\left(-\frac{y^+}{A^+}\right) + \exp\left(-\frac{y^+}{A^+}\left(\frac{R^+}{k^+}\right)^{\frac{3}{2}}\right)\sqrt{1 + \exp\left(-\frac{R^+}{k^+}\right)} \tag{1.6}$$

Krogstad (1991) showed that the added square root term effectively increased the slope by the necessary amount for large values of k^+. The values of the constants involved in the damping function (1.6) were set to $A^+ = 26$ and $R^+ = 70$ and as such calibrated for the onset of completely rough flow (Rotta, 1962).

The discrete momentum equation using the mixing length model described above was discretised using the finite volume method. The discrete equations were solved on a non-uniform grid consisting of 100 control volumes. Note that the wall normal distance y^+ was measured from the wall which characterized the local near-wall region.

RESULTS

The first test case was the direct numerical simulation (DNS) of Miyake *et al.* (2000), which modeled sand-grain roughness. Two types of roughness were considered in their study: a sparse distribution (case a) and dense distribution (case b). In both cases, the mean roughness height was modelled to have a Gaussian distribution. To simulate these two cases, the current study employed the equivalent sand grain roughness values of k_s^+= 27 and 86.2 for the sparse case and dense case, respectively. These values were determined from the velocity shift of each velocity profile calculated by Miyake et al. (2000). The equivalent sand grain roughness was calculated from the measured velocity shift, ΔU^+, using the Prandtl-Schlichting relation, i.e.

$$\Delta U^+ = \frac{1}{\kappa} ln k_s^+ - 3.37 \tag{1.7}$$

Table 1. Test case 1: flow parameters.

	k_s^+	Y_{max}/H	$Re_{\tau r}$	ΔU^+
C 1	dense	1.24H	576.9	7.5
S 1	86.2	1.29H	558	7.65
C 2	sparse	1.06H	465	3.5
S 2	27	1.08H	476	3.5

Based on the comparisons presented in Table 1, the results of the eddy viscosity model were generally in good agreement with the simulations of Miyake *et al.* (2000). The eddy viscosity model was able to include the effects of roughness for both the dense and sparse roughness distributions. The eddy viscosity model tended to over-predict the location of the peak velocity relative to the rough wall.

The mean velocity profiles normalized using inner coordinates are presented in Figure 1. The sparse and dense roughness distributions correspond to transitionally and fully rough flows, respectively. Note that the smooth wall profile is not affected by the roughness on the opposite wall, even though due to roughness the peak velocity is displaced toward the smooth wall.

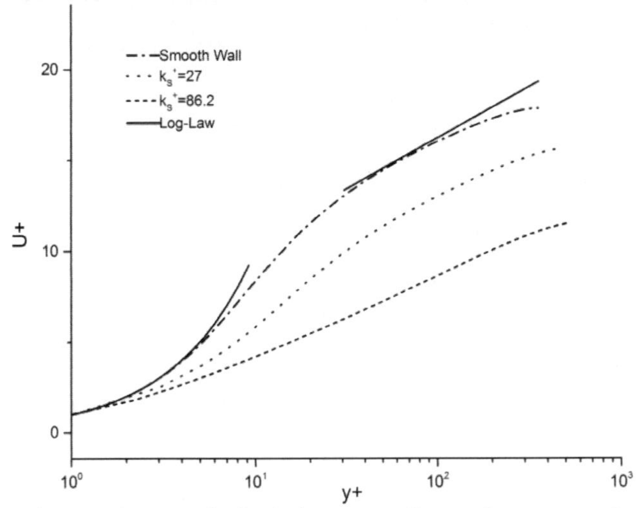

Figure 1. Mean velocity in inner coordinates for test case 1

Figure 2 presents the mean stream-wise velocity profiles predicted for the two rough flows plotted in outer coordinates, i.e. *y/2H*. For comparison with the data of Miyake *et al.* (2000), the mean velocity was normalized by the mean friction velocity. Overall, the agreement is good, with a maximum deviation of less than 7 %.

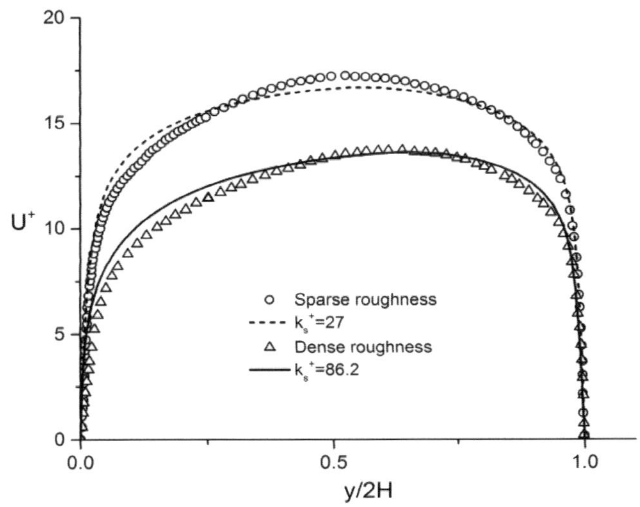

Figure 2. Mean velocity in outer coordinates for test case 1

The relatively good agreement demonstrates that the eddy viscosity model is well suited for predicting the effect of surface roughness due to sand grain type roughness.

The second test case considered prediction of the rough-wall flows studied both experimentally and numerically by Burattini *et al.* (2008). In their study, the roughness was created by two-dimensional ribs, and three different flow cases were considered. Note that whereas the DNS of Burattini *et al.* (2008) characterized the flow in terms of the dimensionless rib height k^+, the eddy viscosity model used an equivalent sand grain roughness k_s^+ based on the roughness shift. Using the measured values of the roughness shift, ΔU^+, the corresponding equivalent sand grain roughness values were determined to be k_s^+ = 51.4, 137, and 224.1, respectively, as shown in Table 2.

Table 2. Test case 2: flow parameters.

	Re_b	$Re_{\tau r}$	$k^+ or\ k_s^+$	Y_{max}/H
DNS 1	2800	260	26	1.22
DNS 2	6900	634	63	1.27
DNS 3	12000	1094	109	1.27
S 1	2760	265	51.4	1.22
S 2	6800	676	137	1.34
S 3	11500	1120	224.1	1.38

Figure 3 presents the predictions for the mean stream-wise velocity profiles using inner coordinates. The last two flow cases clearly represent fully rough flow. Again, the smooth wall case is in close agreement with the canonical log-law for a smooth wall.

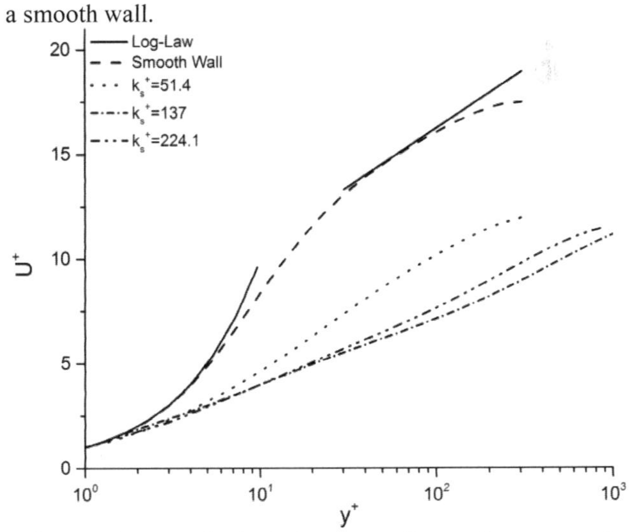

Figure 3. Mean velocity in inner coordinates for test case 2

Figure 4 presents the predictions for the mean velocity profiles using outer coordinates. Note that the velocity profiles for DNS 2 and DNS 3 are shifted upward by 0.5 and

1 unit, respectively. For case 2, the eddy viscosity model is less able to capture the features of the mean velocity profile on the rough wall. The eddy viscosity model results in an over-prediction of the mean velocity profile on the rough wall. On the other hand, the predictions generally agree with the results of Burattini *et al.* (2008) in terms of the magnitude of the Reynolds numbers based on the bulk velocity and the friction velocity on the rough wall.

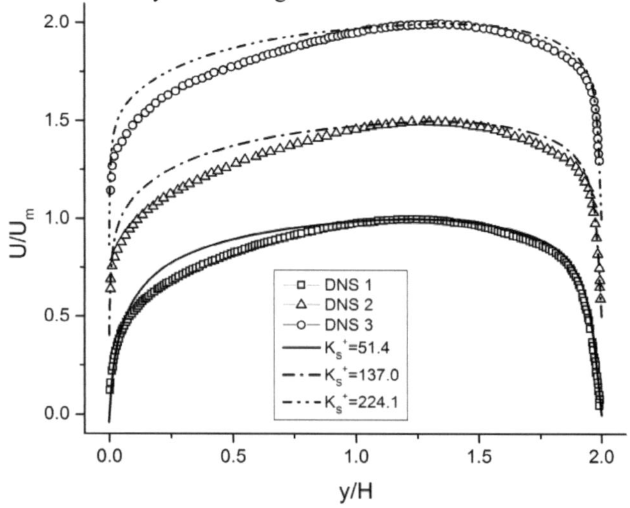

Figure 4. Mean velocity in outer coordinates for test case 2

One possible reason for the significant disagreement on the rough wall observed in Figure 4 would be that the eddy viscosity model does a poor job of recreating the form drag due to the two-dimensional ribs. Equivalent sand-grain roughness apparently is not an effective parameter for characterizing this type of roughness geometry.

CONCLUSIONS

The eddy viscosity model developed by Krogstad (1991) is used to study asymmetric turbulent channel flow, where one wall is smooth and the other is rough. The finite volume method is used to discretise the RANS equations, which are solved for two different test cases: the sand grain roughness by Miyake *et al.* (2000) and the two-dimensional rib roughness by Burattini *et al.* (2008). Satisfactory agreement was obtained for the transitionally and fully rough flows considered by Miyake *et al.* (2000) using a sand grain roughness. On the other hand, the eddy viscosity model was less successful at predicting the rib-roughness flow of Burattini *et al.* (2008). This suggests that although the equivalent sand grain roughness has proven to be a useful approach for characterizing wall roughness in engineering calculations, as a single parameter it may not be effective in modeling roughness with very different geometric characteristics, such as rib-roughness.

ACKNOWLEDGEMENTS

This paper is a result of an introductory research project performed by the first author (MT) while employed as an undergraduate summer research student investigating numerical models for predicting rough-wall turbulent flow. Funding for this project was provided by NSERC via the research grant of the second author (DJB).

REFERENCES

Burattini, P., Leonardi, S., Orlandi, P. and Antonia. R. 2008, "Comparison between experiments and direct numerical simulations in a channel flow with roughness on one wall", *Journal of Fluid Mechanics*, Vol. 600, pp. 403-426.

Krogstad, P. -Å. 1991, "Modification of the van Driest damping function to include the effects of surface roughness", *AIAA Journal*, Vol. 29, pp. 888-894.

Miyake, Y., Tsujimoto, K. and Yasuaki A. 2000, "A DNS of a turbulent flow in a rough-wall channel using roughness elements mode", *JSME international Journal*, Vol. B 43 (8), pp. 233-242.

Rotta, J. C. 1962, "Turbulent boundary layers in incompressible flow", *Progress in Aeronautical Sciences,* Pergamon, Oxford, Vol. 2, pp. 1-219.

Sivykh, G. F. 1984, "Turbulent viscosity for the analysis of an incompressible boundary layer on a rough surface", *Journal of Applied Mechanical and Technical Physics*, Vol. 25, pp. 226-232.

Van Driest, E. R. 1956, "On turbulent flow near a wall", *Journal of Aeronautical Science*, Vol. 23, pp. 1007-1011.

Proceedings of the 25th CANCAM
London, Ontario, Canada, May 31 – June 4, 2015

Spectral Algorithm for Linear Stability Analysis of Flow in Ribbed Annuli

H. V. Moradi **J. M. Floryan**
hvafadar@uwo.ca **mfloryan@eng.uwo.ca**
Department of Mechanical and Materials Engineering
The University of Western Ontario
London, Ontario, Canada

ABSTRACT

A spectrally accurate algorithm for linear stability analysis of flows in annuli bounded by walls with axisymmetric ribs subjected to arbitrary three-dimensional disturbances has been developed. It is assumed that the flow environment is quiescent and, thus, the system evolution is well described by the asymptotic stability. Immersed boundary conditions method (IBC) has been adopted to overcome difficulties associated with irregularities of the geometry. Various tests confirm the spectral accuracy of the algorithm.

KEYWORDS: spectral discretization, immersed boundary conditions method, linear stability analysis.

INTRODUCTION

Spectral methods provide the lowest error for spatial discretization of the field equations but are generally limited to solution domains with regular geometries. The first spectrally accurate implementation of the immersed boundary concept was developed by Szumbarski and Floryan [7] and is referred to as the Immersed Boundary Conditions (IBC) method in the rest of this discussion. This method relies on a purely formal construction of boundary constraints in order to generate the required closing relations. The construction relies on the representation of the physical boundaries in the spectral space and nullifying the relevant Fourier modes. The attractiveness of this concept is associated with the precise mathematical formalism and high accuracy. The method has been implemented to study unsteady problems, moving boundary problems involving Laplace, biharmonic and Navier-Stokes operators [2-4].

PROBLEM FORMULATION

The geometry of the annulus and structure of the flow field are described using the cylindrical system of coordinates with z, r and θ describing the axial, radial and circumferential directions. Steady, pressure-gradient-driven axial flow of viscous incompressible fluid in an annulus extending to $\pm\infty$ in the z-direction is considered. The annulus is modified by introducing axisymmetric surface ribs of arbitrary cross-section (see Figure 1). The geometry of the cylinders is described using Fourier expansions of the form

$$r_{in}(z) = R_1 + \sum_{l=-N_A}^{N_A} H_{in}^{(l)} e^{il\alpha z}, \quad r_{out}(z) = 1 + R_1 + \sum_{l=-N_A}^{N_A} H_{out}^{(l)} e^{il\alpha z} \tag{1}$$

where $H_{in}^{(l)} = H_{in}^{(-l)*}$ and $H_{out}^{(l)} = H_{out}^{(-l)*}$ are the reality conditions, stars denote the complex conjugate, R_1 stands for the average radius of the inner cylinder, subscripts "*out*" and "*in*" denote the outer and inner cylinders, respectively, N_A is the number of Fourier modes required for description of the geometry and the average annulus opening L has been used as the length scale. The ribs are periodic in the z-direction with wavelength $\lambda = 2\pi/\alpha$.
Flow in a smooth annulus with walls located at $r = R_1$ and $r = 1 + R_1$ serves as the reference flow (annular Poiseuille flow; APF) with the solution given in [6]. The maximum of the axial velocity component U_{max} is used as the velocity scale, ρU_{max}^2 is used as the pressure scale, ρ stands for the density and the Reynolds number is defined as $Re = U_{max}L/\nu$ where ν denotes the kinematic viscosity.

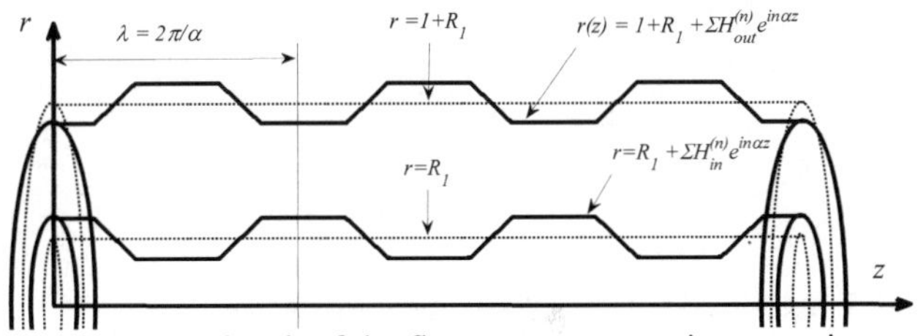

Figure 1. Sketch of the flow geometry - axisymmetric annulus with transverse ribs of arbitrary cross-section. Dash lines represent the smooth reference annulus.

Flow in the ribbed annuli is represented as a superposition of the APF and modifications, i.e.

$$\vec{V}_2(z,r) = [u_2(z,r), v_2(z,r)] = \vec{V}_0(r) + \vec{V}_1(z,r) = [u_0(r), 0] + [u_1(z,r), v_1(z,r)], p_2(z,r) = p_0(z) + p_1(z,r) \tag{2}$$

where $\vec{V}$ denotes the velocity vector, and subscripts 0, 1, 2 denote the reference flow, the rib-induced modifications and the complete flow quantities, respectively. Substitution of (2) into the axisymmetric form of the Navier-Stokes and continuity equations, introduction of the Stokes stream function defined as

$$u_1 = \frac{-1}{r}\frac{\partial \psi_1}{\partial r}, \quad v_1 = \frac{1}{r}\frac{\partial \psi_1}{\partial z} \tag{3}$$

and elimination of pressure lead to a single forth-order partial differential equation for ψ_1 of the form

$$\left[\frac{E^4}{Re} - u_0\frac{\partial E^2}{\partial z} + \left(D^2u_0 - \frac{Du_0}{r}\right)\frac{\partial}{\partial z}\right]\psi_1$$
$$= r\left[\frac{\partial}{\partial z}\left(\frac{\partial\widehat{v_1v_1}}{\partial r} + \frac{\partial\widehat{u_1v_1}}{\partial z}\right) - \frac{\partial}{\partial r}\left(\frac{\partial\widehat{u_1v_1}}{\partial r} + \frac{\partial\widehat{u_1u_1}}{\partial z}\right)\right] + \frac{\partial\widehat{v_1v_1}}{\partial z} + \frac{\widehat{u_1v_1}}{r} - \frac{\partial\widehat{u_1v_1}}{\partial z} \quad (4)$$

where hat denotes velocity products, $D \equiv d/dr$ and operator E^2 is of the form

$$E^2 = \frac{\partial^2}{\partial r^2} + \frac{\partial^2}{\partial z^2} - \frac{1}{r}\frac{\partial}{\partial r}. \quad (5)$$

The boundary conditions are the no-slip and no-penetration conditions at the inner and outer cylinders which can be expressed in terms of the stream function of the form

$$\frac{1}{r}\frac{\partial\psi_1}{\partial r} = u_0(z,r), \quad \frac{1}{r}\frac{\partial\psi_1}{\partial z} = 0 \quad \text{at} \quad r = r_{in}(z),$$
$$\frac{1}{r}\frac{\partial\psi_1}{\partial r} = u_0(z,r), \quad \frac{1}{r}\frac{\partial\psi_1}{\partial z} = 0 \quad \text{at} \quad r = r_{out}(z). \quad (6)$$

Spectrally accurate solution of (4) subjected to the boundary relations (6) was discussed in details in [5]. The flow field is expressed as

$$\vec{V_2}(z,\bar{r}) = \vec{V_0}(\bar{r}) + \vec{V_1}(z,\bar{r}) = [u_0(\bar{r}), 0] + \sum_{n=-N_M}^{N_M}\left[f_u^{(n)}(\bar{r}), f_v^{(n)}(\bar{r}), 0\right]e^{ina z} \quad (7)$$

LINEAR STABILITY ANALYSIS

The analysis begins with the momentum and continuity equations written in terms of primitive variables of the form

$$\frac{\partial\vec{V}}{\partial t} + (\vec{V}.\vec{\nabla})\vec{V} = -\vec{\nabla}p + \frac{1}{Re}\nabla^2\vec{V}, \quad (8)$$
$$\vec{\nabla}.\vec{V} = 0, \quad (9)$$

where $\vec{V} = (u, v, w)$ denotes the velocity vector with components in the (z, r, θ) directions, respectively, and p stands for the pressure. All quantities are expressed as the superposition of the unsteady, three-dimensional disturbances and the mean part. Subtraction of the mean parts and linearization lead to the disturbance equations of the form

$$\vec{\nabla}.\vec{V_3} = 0, \quad (10)$$
$$\frac{\partial\vec{V_3}}{\partial t} + (\vec{V_2}.\vec{\nabla})\vec{V_3} + (\vec{V_3}.\vec{\nabla})\vec{V_2} = -\vec{\nabla}p_3 + \frac{1}{Re}\nabla^2\vec{V_3} \quad (11)$$

where subscripts 2 and 3 refer to the mean flow and the disturbance fields, respectively and subject to the no-slip and no-penetration boundary conditions at the inner and outer cylinders of the form

$$\vec{V_3} = 0 \text{ at } r = r_{in}(z) \text{ and } r = r_{out}(z). \quad (12)$$

Disturbance flow quantities can be expressed in the form

$$\left[\vec{V_3}(z,r,\theta,t), p(z,r,\theta,t)\right] \approx \sum_{n=-N_N}^{N_N}\left[g_u^{(n)}, g_v^{(n)}, g_w^{(n)}, g_p^{(n)}\right]e^{i[(\delta+n\alpha)z+M\theta-\sigma t]} + c.c \quad (13)$$

Substitution of (13) into (10) and (11), separation of Fourier modes, and after rather lengthy algebra, result in a set of two equations for each Fourier mode n of the form

$$\mathcal{L}_{OS}^{(n)}G_v^{(n)} + \mathcal{C}_{OS}^{(n)}\Omega^{(n)} = \sum_{m=-N_J}^{N_J}\left(E_\Omega^{(n,m)}\Omega^{(n-m)} + E_v^{(n,m)}G_v^{(n-m)}\right) \quad (14)$$
$$\mathcal{L}_{SQ}^{(n)}\Omega^{(n)} + \mathcal{C}_{SQ}^{(n)}G_v^{(n)} = \sum_{m=-N_J}^{N_J}\left(F_\Omega^{(n,m)}\Omega^{(n-m)} + F_v^{(n,m)}G_v^{(n-m)}\right), \quad (15)$$

where $\mathcal{L}_{OS}^{(n)}$, $\mathcal{L}_{SQ}^{(n)}$, $\mathcal{C}_{OS}^{(n)}$, and $\mathcal{C}_{SQ}^{(n)}$ are the Orr-Sommerfeld, Squire and coupling operators defined as

$$\mathcal{L}_{OS}^{(n)} = (-i\sigma + it_nu_0)\mathcal{T}^{(n)} - irt_nk_n^2D\left(\frac{Du_0}{rk_n^2}\right) - Re^{-1}\mathcal{T}^{(n)}\mathcal{T}^{(n)}, \quad (16)$$
$$\mathcal{C}_{OS}^{(n)} = 2Mt_nRe^{-1}\mathcal{T}^{(n)},$$
$$\mathcal{L}_{SQ}^{(n)} = (-i\sigma + it_nu_0) - Re^{-1}\mathcal{S}^{(n)},$$
$$\mathcal{C}_{SQ}^{(n)} = -\frac{iMDu_0}{r^3k_n^2} - \frac{2Mt_n}{r^4k_n^4Re}\mathcal{T}^{(n)}, \quad (17)$$
$$\mathcal{T}^{(n)}(r) = rk_n^2D\left(\frac{D}{rk_n^2}\right) - k_n^2, \quad (18)$$
$$\mathcal{S}^{(n)}(r) = \frac{1}{r^3k_n^2}D(r^3k_n^2D) - k_n^2. \quad (19)$$

In the above,

$$G_v^{(n)}(r) = -irg_v^{(n)},$$
$$\Omega^{(n)}(r) = \frac{ig_\eta^{(n)}}{rk_n^2} = \frac{rt_ng_w^{(n)} - Mg_u^{(n)}}{r^2k_n^2} \quad (20)$$

and $g_\eta^{(n)}(r)$ stands for modal function of radial vorticity. Terms grouped on the left hand side of (14) and (15) represent a 6th order system for mode n while terms on the right hand side provide the inter-modal coupling associated with the ribs. The complete problem consists of $2(2N_N + 1)$ systems.

NUMERICAL DISCRETIZATION OF THE FIELD EQUATIONS

We wish to determine spectrally accurate solution of the problem described by Eqs (14) and (15). Because of the irregularities of the boundaries, the immersed boundary conditions (IBC) method is used to enforce the boundary conditions. Another alternative is offered by the boundary conditions transfer procedure but this option has not been pursued due to its accuracy limitations [1]. Figure 1 illustrates the computational domain which is bounded by $-R_{in}+R_1$ from below and by $1+R_1+R_{out}$ from above in the radial direction, where R_{in} and R_{out} denote locations of the rib extremities. The radial domain is mapped onto $[-1,1]$ in order to use the standard definition of Chebyshev polynomials. Transformation used in this study has the form

$$\begin{aligned}\bar{r}=\Gamma(r+C),\Gamma=2(1+R_{in}+R_{out})^{-1},\\ C=0.5(R_{in}-R_{out}-2R_1-1),\end{aligned} \tag{21}$$

Fourier series representing the vertical velocity and vertical vorticity are truncated at N_N, i.e.

$$\begin{aligned}&[v_D(z,r,\theta,t),\eta_D(z,r,\theta,t)]\approx\\ &\sum_{m=-N_N}^{m=+N_N}\left[i\frac{G_v^{(m)}(r)}{r},irk_m^2\Omega^{(m)}(r)\right]e^{i[(\delta+m\alpha)z+M\theta-\sigma t]}\\ &+c.c.\end{aligned} \tag{22}$$

resulting in $2(2N_N+1)$ systems of type (14) and (15). The modal functions $G_v^{(m)}(\bar{r})$ and $\Omega^{(m)}(\bar{r})$ are expressed in terms of Chebyshev expansions of the form

$$\left[G_v^{(m)}(\bar{r}),\Omega^{(m)}(\bar{r})\right]\approx\sum_{k=0}^{N_E}\left[G_{k,v}^{(m)},G_{k,\Omega}^{(m)}\right]T_k(\bar{r}) \tag{23}$$

where T_k denotes the Chebyshev polynomial of the kth order, and $G_{k,v}^{(m)}$ and $G_{k,\Omega}^{(m)}$ stand for the unknown coefficients of the expansions. Chebyshev collocation technique based on the Gauss-Chebyshev-Lobatto points is used to form a system of algebraic equations. Points at the edges of the computational domain and immediately adjacent to the edges are omitted to provide space for the boundary conditions. The boundary conditions (12) are implemented using the immersed boundary conditions concept. The resulting algebraic system is posed as a general eigenvalue problem for σ.

DISCRETIZATION OF THE BOUNDARY CONDITIONS

Methods used for the discretization of the boundary conditions at the inner and outer cylinders are identical and, thus, the following presentation is focused on the axial velocity component of the disturbances at the inner cylinder only. Axial velocity component for the inner cylinder expressed in terms of the radial disturbance velocity and the radial disturbance vorticity have the following form

$$\begin{aligned}u_3[z,\bar{r}_{in}(z),\theta,t]=\sum_{n=-N_N}^{N_N}\left(-\frac{\Gamma t_n\bar{D}G_v^{(n)}}{fk_n^2}-M\Omega^{(n)}\right)\\ \times e^{i[(\delta+n\alpha)z+M\theta-\sigma t]}=0,\end{aligned} \tag{24}$$

Use of (22) and (23) brings (24) to the following form

$$\begin{aligned}\sum_{n=-N_N}^{N_N}\sum_{k=0}^{N_T}\left(-\frac{\Gamma t_n\bar{D}T_kG_{k,v}^{(n)}}{fk_n^2}-MT_kG_{k,\Omega}^{(n)}\right)\\ \times e^{i[(\delta+n\alpha)z+M\theta-\sigma t]}=0,\end{aligned} \tag{25}$$

The above relations require evaluation of the Chebyshev polynomials and their derivatives along the cylinder surface. The resulting values represent periodic functions of z and, thus, can be expressed using Fourier expansions of the form

$$\begin{aligned}T_k[\bar{r}_{in}(z)]&=\sum_{m=-N_S}^{N_S}(w_{in})_k^{(m)}e^{im\alpha z},\\ \bar{D}T_k[\bar{r}_{in}(z)]&=\sum_{m=-N_S}^{N_S}(d_{in})_k^{(m)}e^{im\alpha z}\end{aligned} \tag{26}$$

where $N_S=N_TN_A$. Substitution of (26) into (25) leads to boundary condition of the form

$$\begin{aligned}\sum_{n=-N_S-N_N}^{N_S+N_N}\sum_{m=-N_N}^{N_N}\sum_{k=0}^{N_T}\Bigg[-\frac{\Gamma t_mG_{k,v}^{(m)}(d_{in})_k^{(n-m)}}{fk_m^2}\\ -MG_{k,\Omega}^{(m)}(w_{in})_k^{(n-m)}\Bigg]e^{i[(\delta+n\alpha)z+M\theta-\sigma t]}=0,\end{aligned} \tag{27}$$

Separation of Fourier modes result in the boundary relations whose enforcement is equivalent to the enforcement of the flow boundary conditions and whose form is suitable for numerical implementation. These boundary relations are of the form

$$\begin{aligned}\sum_{m=-N_N}^{N_N}\sum_{k=0}^{N_T}\Gamma t_mG_{k,v}^{(m)}\sum_{p=-N_N-N_S}^{N_N+N_S}(d_{in})_k^{(p-m)}F_{u,m}^{(n-p)}\\ +\sum_{m=-N_N}^{N_N}\sum_{k=0}^{N_T}MG_{k,\Omega}^{(m)}(w_{in})_k^{(n-m)}=0,\end{aligned} \tag{28}$$

where

$$\frac{1}{fk_m^2}=\frac{1}{ft_m^2+M^2/f}\approx\sum_{p=-N_R}^{N_R}F_{u,m}^{(p)}e^{ip\alpha z} \tag{29}$$

DISCUSSION OF RESULTS

We shall limit ourselves to study a smooth annulus whose inner cylinder is modified by introducing axisymmetric ribs with sinusoidal shape. The geometry of the annulus can be expressed in the form

$$r_{in}(z) = R_1 + S_{in}cos(\alpha z), r_{out}(z) = 1 + R_1 \quad (30)$$

where α and S_{in} denote wavenumber and amplitude of the ribs, respectively. The solution accuracy depends on the convergence rates of the Fourier and Chebyshev expansions used in (22) and (23) as well as on the accuracy of the discretization of the boundary conditions. Exponential convergence of the solution for the disturbance flow field can be demonstrated by introducing Chebyshev norm for the axial component of the disturbance velocity in the form

$$\left\|\Phi_u^{(n)}(\bar{r})\right\|_\omega = \left\{\int_{-1}^{1} g_u^{(m)}(\bar{r}) g_u^{(m)*}(\bar{r})\omega(\bar{r})d\bar{r}\right\}^{1/2} \quad (31)$$

where $\omega(\bar{r}) = 1/\sqrt{1-\bar{r}^2}$ and star denotes the complex conjugate. Figure 2 demonstrates that the convergence of the expansion for the disturbance velocity field is exponential but its rate slows down but for larger α.

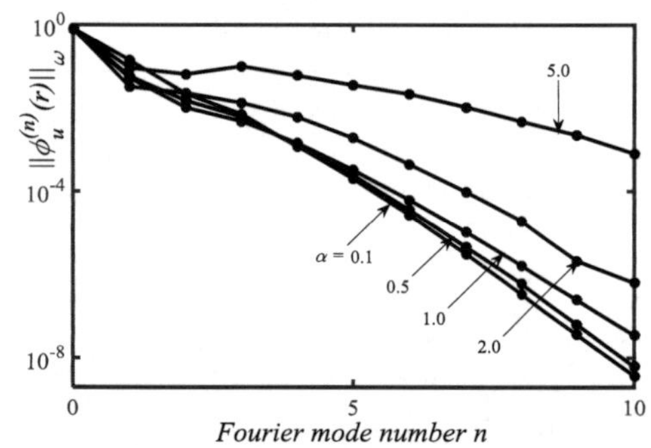

Figure 2. Variations of the disturbance Chebyshev norm $\left\|\Phi_u^{(n)}(r)\right\|_\omega$ as a function of the number of Fourier modes N_N used in the solution for the rib geometry described by Eq. (30) with $R_1 = 1$ and $S_{in} = 0.015$, for $Re = 15000$ and disturbances with $\delta = 1.6$ and $M = 2$.

Results presented in Figure 3 can be used to select numerical parameters which guarantee the required absolute accuracy for eigenvalues. They can be determined with a four digit accuracy using $N_E = 50$ Chebyshev polynomials and $N_N = 5$ Fourier modes. Table 1 presents detailed results of numerical convergence studies which can also be used for selection of numerical parameters which guarantee the desired absolute accuracy.

CONCLUSIONS

Linear stability analysis of APF modified by the axisymmetric ribs has been considered. Spectrally accurate algorithm based on Fourier and Chebyshev expansions have been developed. The flow boundary conditions have been enforced on the vibrating walls using the immersed boundary conditions (IBC) method. Gauss-Lobatto collocation points have been used to form a system of equations. The resulting algebraic system is posed as a general eigenvalue problem. Various tests confirm that the algorithm provides spectral accuracy.

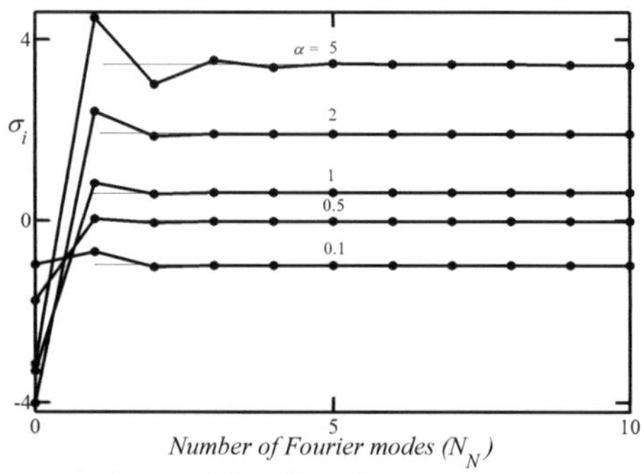

Figure 3. Variations of the disturbance growth rate $\sigma_i \times 10^3$ as a function of the number of Fourier modes N_N used in the solution for the rib geometry described by Eq. (30) with $R_1 = 1$ and $S_{in} = 0.015$, for $Re = 15000$ and disturbances with $\delta = 1.6$ and $M = 2$.

ACKNOWLEDGMENTS

This work has been carried out with support from the Natural Sciences and Engineering Research Council (NSERC) of Canada.

REFERENCES

1. Cabal, A., Szumbarski, J., Floryan, J.M. 2001 Numerical Simulation of Flows over Corrugated Walls. *Comp. & Fluids* **30**, 753-776.
2. Husain, S. Z. Floryan, J. M. 2008 Gridless Spectral Algorithm for Stokes Flow with Moving Boundaries. *Computer Methods in Applied Mechanics and Engineering* **198**, 245-259.
3. Husain, S. Z., Floryan, J. M. 2010 Spectrally-accurate algorithm for moving boundary problems for the Navier-Stokes equations. *J. Comp. Phys.* **229**, 2287–2313.
4. Husain, S. Z., Floryan, J. M. 2007 Immersed boundary conditions method for unsteady flow problems described by the Laplace operator, *Int. J. Numer. Meth. Fluids* **56**, 1765–1786.
5. Moradi, H. V., Floryan, J. M. 2012 Algorithm for analysis of flows in ribbed annuli. *Int. J. Numer. Meth. Fluids* **68**, 805–838.
6. Moradi, H. V., Floryan, J. M. 2013a Flows in annuli with longitudinal grooves. *J. Fluid Mech.* **716**, 280–315.
7. Szumbarski, J., Floryan, J. M. 1999 A direct spectral method for determination of flows over corrugated boundaries, *J. Comput. Phys.* **153**, 378-402.

Computational Solid Mechanics

Proceedings of the 25th CANCAM
London, Ontario, Canada, May 31 – June 4, 2015

USE OF RIGID BAR ELEMENTS IN FINITE ELEMENT ANALYSIS

Mr. Kevin Brooks
Department of Mechanical, Automotive, and Materials Engineering
University of Windsor
Windsor, Ontario, Canada
brook111@uwindsor.ca

Dr. Nader Zamani
Department of Mechanical, Automotive, and Materials Engineering
University of Windsor
Windsor, Ontario, Canada
zamani@uwindsor.ca

ABSTRACT

This paper describes the behaviour of the smooth and rigid connection properties in finite element calculations. To make discussion easier, the effects of these connection types are analysed using a simple case study involving a cantilever beam. There would seem to be some confusion regarding the application of smooth and rigid connection properties which leads to improper use. Many finite element analysis programs have smooth and rigid connection elements for calculations, however they are sometimes labelled under different names and are not well documented [1][2][4]. This paper is attempting to provide useful information for end users.

KEYWORDS: FEA, Catia, Connections, Rigid Bar Element

INTRODUCTION

Connecting multiple parts in Finite Element Analysis (FEA) is necessary for analysing many complicated parts and assemblies. For multiple parts to interact, there must be a defined connection between them. Depending on the nature of the connection, this definition may include the introduction of an element. This is the case for the rigid and smooth connection properties in Catia, which are the focus of this paper. The rigid and smooth connection properties create a null-length rigid bar element at the centroid of the connected regions which is used in computing the responses.

RIGID CONNECTION

The rigid connection connects the nodes of connected regions to a null-length rigid bar which acts to control connected regions by preventing motion relative to the bar [2]. A bi product of the rigid connection's behaviour is that the connected faces cannot deform relative to the rigid bar. Thus, the faces must maintain the original shape. This property can change the behaviour of the interacting regions resulting in properties one might expect from an infinitely rigid interface, such as induced stresses to counteract the effect of the Poisson ratio. This is analogous to using a rigid bar to connect the connected regions. This behaviour will be explained later.

SMOOTH CONNECTION

The smooth connection is essentially the opposite of the rigid connection. While it looks the same, connecting nodes to a null-length rigid bar, with the smooth connection each of the nodes acts as a controlling element for the rigid bar [2]. This way, the smooth connection allows the connected faces to deform freely under the applied loading. The smooth connection's rigid bar movement is governed simply by the motions of the connected nodes, taking the average movements of the nodes as being the movement for the rigid bar element. The smooth connection behaviour will be explained later.

Figure 1 illustrates how two regions can be connected using a single reference element. For the smooth connection, the motion of the "rigid bar" is mostly governed by the motion of the "connected nodes", whereas for the rigid connection, the motion of the "connected nodes" is governed by the "rigid bar" [2].

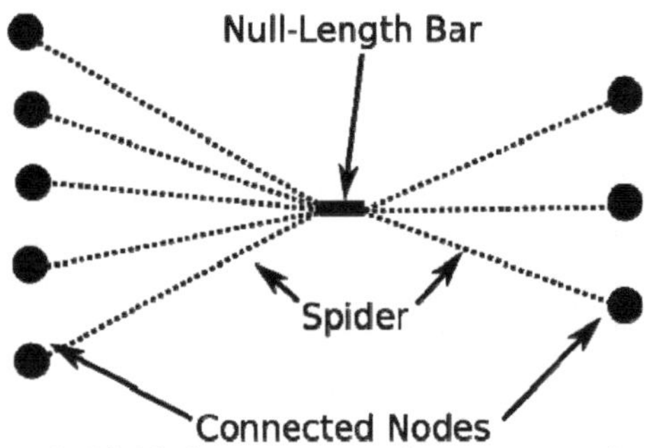

Figure 1: Rigid element connecting two sets of nodes

The graphic in figure 1 is a very good representation of connections using rigid elements in general. A benchmark analysis will now be used to demonstrate the key differences in the different connections. In addition to testing smooth and rigid connections, a fastened connection will be used to compare results.

BENCHMARK ANALYSIS

For this benchmark analysis, a 30-cm long square cantilever beam with side lengths 5 cm will be tested under conditions of axial, bending, and torsional loads using Catia V5's Finite Element Analysis workbench. Figure 2 shows the primary points of interest. To reduce calculation time, the region between the dashed outlines will be replaced with 1-dimensional elements for testing the smooth and rigid connections. For this analysis, all solid elements are parabolic tetrahedral elements of side length 4 mm. The 1-dimensional elements are linear with a length of 1 mm. The fastened connection will use a model with all solid elements.

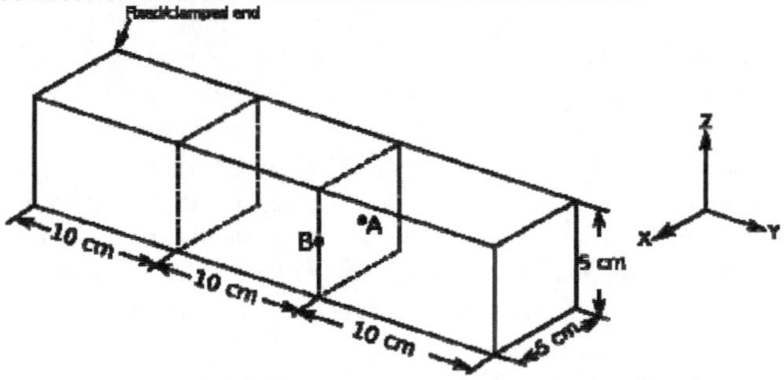

Figure 2: Rigid element connecting two sets of nodes

For this benchmark analysis, the stresses will be considered. The deflections are found to be very similar to each other and to the theory, so they will not be included. There will be 3 cases considered: axial compression, bending, and torsion. The axial case will consider a uniform load of 250 N in compression applied uniformly over the end of the beam. The bending case will have the 250 N load applied on the end of the beam in the negative Z-direction. Finally, there will be a torsional load of 200 N.m applied along the axis of the beam. The summary of the stresses are listed in Tables 1 through 3. The theoretical stress for the axial case is found using Eq. (1)

$$\sigma = P / A \tag{1}$$

where σ is the axial stress, P is the axial load, and A is the cross-sectional area [3].

The theoretical maximum shear stress is found using Eq. (2)

$$\tau = \frac{3F}{2A} \tag{2}$$

where τ is the shear stress, F is the bending load, and A is the cross-sectional area [3].

The theoretical torsional stress is found using the mathematical theory of elasticity for torsion, Eq. (3)

$$\tau_{max} = \frac{M}{\mu b^3} \tag{3}$$

where τ_{max} is the maximum shear stress (in the middle of one of the side faces), M is the applied moment, μ is a constant depending on geometry, and b the beam cross-section side length [3]. These equations will be used to calculate the von Mises (VM) stress for comparison purposes.

Connection type	VM Stress (Catia, kPa)	VM Stress (Theory, kPa)
Fastened	104.5	≈ 100 kPa
Rigid	40.362	≈ 100 kPa
Smooth	219.511	≈ 100 kPa

Table 1: Axial stress results

Connection type	VM Stress (Catia, kPa)	VM Stress (Theory, kPa)
Fastened	264.838	259.808
Rigid	56.8203	259.808
Smooth	235.235	259.808

Table 2: Bending stress results

Connection type	VM Stress (Catia, kPa)	VM Stress (Theory, kPa)
Fastened	13054.7	13323.468
Rigid	10987.9	13323.468
Smooth	16066.2	13323.468

Table 3: Torsion stress results

In tables 1 and 2, we note that the results for the rigid and smooth connections are very different from each other. Catia also returns the stress tensors, which will help to analyse the nature of the stresses attained.

The fastened connection is exactly as expected, showing an axial stress of 103.6 kPa in compression for the axial compression case and about 144.824 kPa in YZ shear for the bending case. These values closely relate to the theoretical stress values that were used to calculate the VM stress. The similarity in values for the VM stress reflects this.

According to the stress tensor for axial compression, the rigid connection has an axial stress of about 85.7 kPa in compression, and there is an additional longitudinal stress of 45.3 kPa in compression. The longitudinal stresses are a result of the Poisson ratio of the material. Since the rigid connection prevents any deformation of the interface, the beam cannot deform longitudinally, so there has to be an induced stress to counteract the response of the beam to an axial load. With the Poisson ratio of 0.346, the longitudinal stress can be reasonably expected. The rigid connection seems to be most useful in cases where connected bodies are to interact as though the interface is entirely rigid.

The smooth connection has an axial stress tensor component of 191.5 kPa in compression for the axial case. This value is not realistic for anything. Upon inspection, the connected faces with the smooth connection appear to have a lot of noise. If we refine the mesh on the connected face, the average stress of the smooth connection approaches a similar value to the fastened connection, and the stress distribution becomes more uniform. It would appear as though the smooth connection is subject to some level of noise at the interface where it is applied, so some mesh refinement may be necessary.

The bending stresses appeared to have some variability in the stress values returned. Most importantly, the rigid connection returned a stress of only 56.8203 kPa. Elsewhere along the beam, we find that the influence of the rigid connection can result in a VM stress as high as 530 kPa. If we run the analysis again setting the Poisson ratio to zero, we get a VM stress for the rigid connection of 213.843 kPa, which is significantly closer to the theoretical stress. This suggests that the Poisson ratio is once again influencing the results attained by the use of the rigid connection.

In all cases, the stress values for the torsional load were reasonably close to the theoretical value. The torsional load seems to be transmitted in a similar way for all three connection types. Since there is not a significant difference associated with the torsional load, there will be no further inspection of its stress values.

APPLICATIONS

Some applications of the smooth and rigid connections will now be discussed. The analysis of the stresses under different loads is beneficial for getting an idea of the interactions caused by the smooth and rigid connections, however they do not suggest any particular application. Here, the nature of the connections will be considered and linked with certain industrial requirements for structural analysis.

APPLICATIONS – RIGID CONNECTION

The most obvious case for the rigid connection is in connecting two parts where the materials have significantly different stiffness. For example, when an aluminium part connects to a steel part of comparable size, the steel would seem to be rigid with respect to the aluminium. So, if a part containing a large, nearly rigid section is being analysed, the nearly rigid section may likely be replaced with a rigid connection without much loss of accuracy.

Another application for the rigid connection may be when a part is being attached with a cable in tension. If the cable is to be replaced with 1-dimensional elements and is being attached to a part somehow, the cable may be attached to the part using a rigid connection. Care should be taken, however, to apply the rigid connection appropriately. If the rigid connection is applied only to where the cable is connecting to the part, then the results should be as expected. If, however, the connection is applied further outside the region that the cable is expected to interact on, the results will be skewed. An example of this would be applying the connection over an entire face while it is only expected to act over a small part of the face. It is important to realize that using this connection type for a cable is only really accurate for cables under tension.

Additionally, for a cable, the degrees of freedom transmitted by the connection can be modified such that the cable behaves the way it is intended to. If the cable is welded to the part, then it will transfer moments and rotations. However, if the cable is tied to the part, it may be more appropriate to disable the transfer of moments as the cable is free to rotate relative to the attached surface. The interface in this case would be expected to have a constant geometry, otherwise the rigid connection may give erroneous results. A similar use for this connection would be to approximate a fastened connection between two incompatible mesh types, such as fastening 1-dimensional elements to 2-dimensional elements.

Finally, another application for the rigid connection may be in connecting two parts with a low elastic modulus with a bonding agent that has a high elastic modulus. In such a situation, it may be reasonable to approximate the bonding agent using a rigid connection. This would simplify the problem while providing reasonably accurate results. Since no part is truly rigid, it should be noted that the rigid connection is only an approximation and not exact. It should also be noted that the method by which something is being connected changes the applicability of the rigid connection. For example, if a steel part is glued to an aluminium part where the glue has a low stiffness, then the interface would no longer seem rigid and thus approximating the interface with a rigid connection would be erroneous.

APPLICATIONS – SMOOTH CONNECTION

The smooth connection is useful under conditions where the connected interface is expected to deform under the applied loading. This is useful when two parts are connected in a way that would typically merit a rigid connection, however the part being replaced has a relatively low stiffness.

One example of this would be if two parts are connected with a bonding agent that has a relatively low elastic modulus. An example of such would be two copper pipes being welded with tin. To approximate the response of this connection, a smooth connection can be used to simplify the problem. This would be especially useful if the connected parts have different mesh types, for example, beam elements connecting to shell elements. It should be noted, however, that the smooth connection doesn't have any material. Instead, the smooth connection treats the interface as having no resistance to the change of the connected regions. The only major effect of the smooth connection is to keep the connected regions separated by the specified distance.

In terms of applicability, the smooth connection should not be used in cases where deflection is important. This is because the connection will not account for the elasticity of the material being replaced. The smooth connection is useful for approximating the stresses in the part, however the deflections will likely be different.

CONCLUSIONS

Essentially, the rigid and smooth connection properties can be simplified to a few simple cases. For the rigid connection, applications include any time that a structure with a high stiffness relative to the connected elements is to be neglected. In that case, the assumption made by using the rigid connection is that the parts are somehow fastened such that the connected regions do not deform relative to the connection.

The smooth connection is useful under the opposite conditions. Where a structure with a high stiffness is being considered and interacts with a structure with low stiffness, the structure with low stiffness can be replaced with a smooth connection. Again, the assumption made is that the parts are fastened. In the case of the smooth connection, the regions are allowed to deform, which assumes that the structures with a smooth connection applied are of sufficiently high stiffness that the region with the smooth connection does not have a significant influence on the reaction of the beam to the applied loading. Essentially, the smooth connection makes the assumption that the region being replaced will transmit loads, however the stiffness of the connected regions are so high that they act as though there is nothing resisting their movement.

In both cases, it is important to note that the rigid and smooth connections do not account for deflections into the connection. Thus, the region which is being replaced will maintain a constant distance between the regions being connected.

REFERENCES

1. MSC Software. 2004, *MSC.Nastran 2004 Reference Manual*, web, mscsoftware.com.
2. Dassault Systems. 2014, *Catia Users Manual.*
3. Beer, Johnson, DeWolf, Mazurek. 2012, *Mechanics of Materials, Sixth Edition*, McGraw-Hill, New York.
4. N.G. Zamani. 2012, *Catia V5 FEA Tutorials*, SDC Publications.

SIMULATING DEBRIS IMPACTS INTO TORNADO SAFEROOMS WITH LS-DYNA

Yevgeniy Parfilko
Department of Mechanical Engineering
Rochester Institute of Technology
Rochester, New York, USA
yvp4331@rit.edu

Benjamin Varela, Ph.D.
Department of Mechanical Engineering
Rochester Institute of Technology
Rochester, New York, USA
bxveme@rit.edu

ABSTRACT

Structures are typically designed to withstand two types of loading: distributed loads such as wind loads or high-energy hard impacts such as steel projectiles. Saferooms that protect against severe storms such as tornadoes are affected by a third type of loading - soft impacts caused by windborne debris. In this paper, impacts between a wood missile and a reinforced-concrete wall are simulated in LS-DYNA. Parameters that were used in calibrating the simulation to match test results are described. The model is then extended to a tornado saferoom design and evaluated against multiple soft impacts. Damage is measured and characterized by an internal material model output variable, as well as by measuring the erosion and deformation of the structure. Additionally, the effect of repeated impacts into the same point is simulated to determine whether a saferoom that has taken damage from a storm will maintain structural integrity. The simulations show that soft impacts typically cause the most damage to the exterior surface of the concrete at the point of impact, with comparatively little spallation effects. This differs from the plug-formation seen in hard impacts. Repeated impacts result in accumulated damage that eventually results in perforation. These simulations are used to make predictions about the number of consecutive storms a saferoom can withstand before being replaced.

KEYWORDS: Tornadoes, Impact Analysis, Finite Element, LS-DYNA, Safe-Rooms, Concrete

INTRODUCTION

Despite technological advances, severe storms claim hundreds of lives every year. The current standard for the design of storm shelters is published by the National Storm Shelter Association and outlines the minimum impacts that storm shelters must withstand [1]. A reliable damage prediction model will aid in improving the design of tornado saferooms and similar protective structures.

During a storm, these structures may be subject to three types of loading: wind loads, hard impacts and soft impacts. Most saferooms can withstand the wind loads and are not subject to hard impacts such as steel ballistic projectiles. Soft impacts, however, are common because they are caused by low-density and low-speed projectiles that deform significantly on impact with a target. They do not cause catastrophic failure, but may still affect the structural integrity of the concrete, thus understanding the scope of the damage is vital for saferooms that must withstand multiple storms.

This paper first compares empirical formulas for hard and soft impacts with the material model used in simulation. The second section examines simulations to measure damage from impacts into protective concrete structures. The third section examines how damage is accumulated during repetitive impact. The results of these simulations are then generalized in a model to predict the damage done to a structure based on the wind speed.

Simulations in this paper were carried out in LS-DYNA, an explicit dynamics FEA software that is the industry standard for high-deformation impacts. The concrete model used was the RHT concrete model, developed at the Ernst-Mach institute in 1991.

METHODS

Many protective structures rely on thick reinforced concrete to prevent perforation from missiles [6] [4]. Several empirical formulas have been developed with the purpose of predicting the missile penetration, most notably the NDRC formula for ballistic missile penetration [7]. In 1978, Kar [5] modified the NDRC formula to account for softer projectiles. The formulas are shown in Figure 1. The NSSA Standard gives speeds for tornado-generated debris ranging from 50 to 100 mph, or 22 to 45 m/s. Table 1 gives debris speeds based on the Fujita-scale rating of the tornado.

The test detailed in the NSSA Standard was recreated with a finite element model. LS-DYNA material 272 (RHT

Concrete) was used for the slab and material 143 was used to model anisotropic wood. The RHT concrete model is a continuous surface cap model that incorporates a crush pressure to limit the amount of compressive strain the material can withstand [2]. This allows accurate simulation of damaged and deformation for impact loads and combined loads.

Impact simulations were run over a range of missile velocities. Penetration data was collected by tracking nodal deformations in the impact region. The maximum nodal deformation in the direction orthogonal to the concrete surface was selected as the representative data point for the trial.

For comparison, predicted penetration curves were numerically derived from the NDRC and Kar formulas.

NDRC FORMULA

$$G\left(\frac{x}{d}\right) = \frac{180}{(f_c')^{0.5}}\, N\, D\, d^{0.2} \left(\frac{V_0}{1000}\right)^{1.8}$$

KAR FORMULA

$$G\left(\frac{x}{d}\right) = \frac{180}{(f_c')^{0.5}} \left(\frac{E}{E_m}\right)^{1.25} \frac{W\, N_2}{D\, d^{1.8}} \left(\frac{V_0}{1000}\right)^{1.8}$$

CONSTRAINTS

$$G\left(\frac{x}{d}\right) = \begin{cases} \left(\frac{x}{2d}\right)^2 \; for \; \frac{x}{d} < 2 \\ \frac{x}{d} - 1 \; for \; \frac{x}{d} \geq 2 \end{cases}$$

Figure 1: NDRC and Kar formulas with the nondimensional penetration constraints. The Kar formula uses scaling factor based on the missile's modulus of elasticity relative to steel.

Design Wind Speed	Missile Speed and Shelter Impact Surface
130 mph	80 mph Vertical Surfaces 53 mph Horizontal Surfaces
160 mph	84 mph Vertical Surfaces 56 mph Horizontal Surfaces
200 mph	90 mph Vertical Surfaces 60 mph Horizontal Surfaces
250 mph	100 mph Vertical Surfaces 67 mph Horizontal Surfaces

Table 1: ICC/NSSA Standard speeds for a 15-lb sawn lumber 2x4 missiles based on wind speed. Categories correspond to EF2, EF3, EF4 and EF5 tornadoes.

RESULTS

Several simulations were run with varying missile properties and concrete stiffness to reflect variation in environmental conditions. The penetration in the presence of environmental conditions showed a variance of ±0.5 mm. In comparison, penetration from impact at 45 m/s was around 5 mm.

In the Figure 3 below, the NDRC and Kar formulas are plotted with dotted lines and effectively represent upper and lower bounds for penetration depth. During an F2 or F3 tornado, the depth is close to Kar formula predictions. In an F4 or F5 tornado, the penetration increases nonlinearly and instead approaches the NDRC formula.

Figure 2: Visualization of the FEA test. The slab dimensions are 1200mm x 1200mm x 150mm. The lumber is US standard 15-lb 2x4 in.

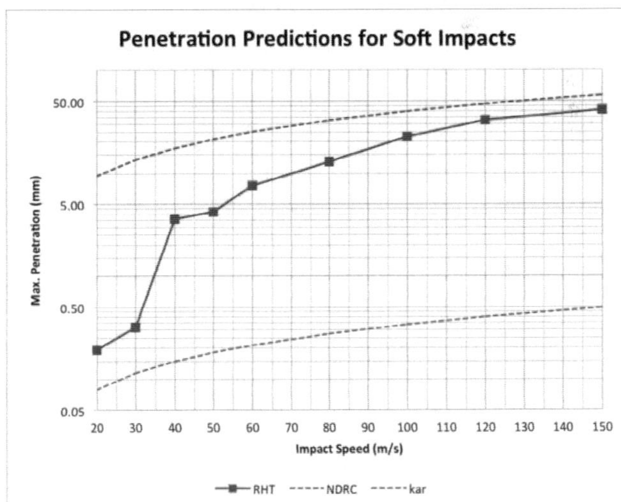

Figure 3: Frontal impact penetration test plotted in reference to empirical equations.

Because of the variance due to environmental condition variation, the penetrations given by the Kar formula can be considered insignificant. The impacts near the NDRC curve, however, do cause measureable damage.

For a frontal impact, nodal deformation roughly followed a bell-curve shape around the impact zone. Plastic deformation occurred in a zone that extended roughly to twice the diameter of the missile, for missiles whose diameter was less than the concrete thickness.

Additional tests were carried out to examine the damage along a free surface. Deformations that were more than a diameter-length away from the edge caused similar deformations to the frontal impact. Significant damage, seen in Figure 4, occurred near the corners when the footprint of the missile was not fully supported by the edge.

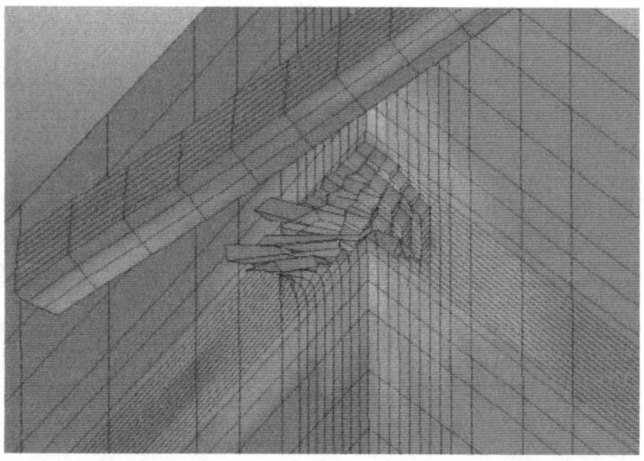

Figure 4: Visulization of a corner impact test (top) and a cross section of a vehicle impact test (bottom) into a concrete wall. Pressure distribution is shown in the color scale.

The frontal and corner impact tests revealed that presence and location of rebar had no effect on the shape or size of the damaged area for soft impacts. To observe how rebar affected the failure of concrete in the storm shelter, a heavy deformable mass, acting as a vehicle, was impacted into a model of an actual tornado saferoom. The rebar prevented the full penetration of the wall, and the concrete in the mid-plane of the wall experienced failure first. Spallation occurred only if the mass made contact with the rebar elements.

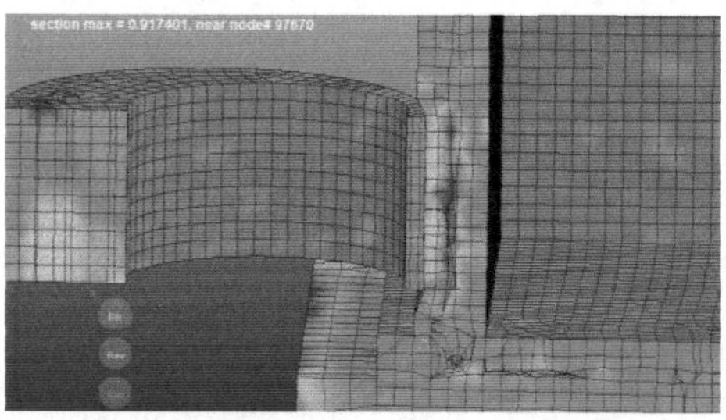

Figure 5: Cross section of a vehicle impact test on a concrete wall. Pressure distribution is shown in the color scale, with red indicating pressure above 34.5 MPa (compaction failure).

CONCLUSIONS

Simulations have shown that concrete exhibits a nonlinear response to soft impacts, with a sharp increase in penetration depth around 40 m/s impact speeds. From these results, we can make important generalizations that will allow us to solve the problem of characterizing soft impacts. Foremost, damage from these impacts is local, with no affect on the reinforcement and interior walls. This validates that inhabitants are safe from shrapnel due to spallation.

The damage zone can be bounded by twice the frontal area of the impact missile, and twice as deep as the penetration depth. This means that in order to weaken the concrete, the cumulative theoretical penetrations must add up to half the distance between the exterior surface and the reinforcement. This approach is suggested as a means of providing a lower bound on the accumulated damage before failure, and may be improved upon by physical test data. However, it is valid only in cases where the theoretical plug diameter is larger than the distance between reinforcing bars [8]. It is hoped that these observations will enable a better understanding of accumulated damage from wind impacts, and allow reliable damage modeling and assessment of protective reinforced concrete structures.

ACKNOWLEDGMENTS

The authors would like to thank OZ Saferooms for their support of this project.

REFERENCES

[1] American National Standrad. 2008. ICC/NSSA standard for the design and construction of storm shelters. Washington, DC: ICC.

[2] Borvall, T. et. al. The RHT concrete model in LS-DYNA. in: Proceedings of the 8th European LS-DYNA Users Conference, Strasbourg, 2011.

[3] Brannon, R.M. Survey of Four Damage Models for Concrete. Sandia Report, 2009. SAND2009-5544.

[4] James, R.J. Impact of High Velocity Objects Into Concrete Structures - Methodology and Application. Proceedings of IMECE'03, 2003.

[5] Kar, A.K.. Local Effects of Tornado-generated Missiles, J. of Struct. Div. ASCE, Vol. 104, No. ST5, pp. 809-816, May 1978.

[6] Murthy et.al. Impact analysis of Concrete Structural Components. Defense Science Journal 2010, 60, 3, p.307-319.

[7] Teland, J.A. A Review of Empirical Equations for Missile Impact Effects on Concrete. Norwegian Defense Research Establishment, 1998. FFI/RAPPORT-97/05856.

[8] Yankelevsky, D.Z. Local Response of Concrete Slabs to Low Velocity Missile Impact. Int. J. of Impact Engineering, 1997, 19, 4, p. 331-343.

TRANSIENT HEAT FLOW ANALYSIS USING FEA AND LAPLACE TRANSFORM-BASED COLLOCATION METHOD

Amir Khalilollahi
Dept. of Mechanical Engineering
Pennsylvania State University- Erie,
Erie, PA, USA
amir-k@psu.edu

Enayat Mahajerin
Dept. of Mechanical Engineering,
Saginaw Valley State University,
University Center, MI, USA
mahajeri@svsu.edu

Gary Burgess
School of Packaging,
Michigan State University,
E. Lansing, Michigan, USA
burgess@msu.edu

ABSTRACT

Finite Element Analysis (FEA) and the Laplace Transform-Based Fundamental Collocation Method (FCM) are applied to regions of arbitrary shapes subjected to arbitrary initial and mixed type boundary conditions. In the FEA process the time derivative is replaced with finite difference method (FCM) and resulting the time dependent global equations are solved incrementally using the initial conditions. The FCM approach is applied in the Laplace transform domain temperatures are obtained in the Laplace domain. Subsequent inversion technique is used to retrieve the time domain solution. To assess applicability and accuracy of these methods, they are applied to typical transient heat flow cases for which exact solutions can be obtained by separation of variables.

KEYWORDS: Finite Element Analysis (FEA), Fundamental Collocation Method (FCM); Finite Difference Method (FDM), Transient Heat Flow, Laplace Transform.

INTRODUCTION

The problem of transient heat diffusion in two dimensions [1] can be handled using Finite Element Analysis (FEA) code [2] and the Fundamental Collocation Method (FCM) [3]. The FEA solution is achieved by developing models in ANSYS®. The solution starts from the governing differential equation of the transient heat conduction and proceeds to building thermal element matrices. The element matrices are assembled and the corresponding time-dependent global matrix is constructed which is solved incrementally using the specified initial and boundary conditions. The time-dependent FEA approach assumes that conditions at time t have been computed and then they are used to determine temperatures at t+Δt. In a case with a spatially variable initial and boundary condition, an executable script was developed in ANSYS® APDL.

The FCM approach is applied in the Laplace transform domain and subsequent inversion technique is used to retrieve the time domain solution. This process may also be limited in accuracy. The FCM algorithm incorporated here is applied directly in the time domain where a Green's function solution to the governing partial differential equation is readily available.

The purpose of this paper is to demonstrate the applicability of both FEA and Collocation FCM methods to two transient heat conduction scenarios, and to investigate the closeness of their solutions.

MATHEMATICAL MODELING

The time-dependent temperature field, T(x,y,t), in a general two dimensional homogeneous, isotropic region R with its boundary ∂R is governed by [1]

$$k\nabla^2 T = \rho C_p \frac{\partial T}{\partial t} \quad for\ x, y\ \ in\ \ R \tag{1}$$

$$T(x, y, 0) = 0 \quad in\ \ R \tag{2}$$

$$aT_o + b\frac{\partial T}{\partial n} = c \quad on\ \ \partial R \tag{3}$$

where k=thermal conductivity, ρ=mass density, C_p=specific heat, $\alpha=k/\rho C_p$=thermal diffusivity, $\nabla^2 = \partial^2/\partial x^2+\partial^2/\partial y^2$ is the Laplace operator, T_0=specified initial temperature distribution, n= outward normal to the boundary, and a, b, c are specified functions of position

The FEA Model:

The FEA model for Equation (1) is

$$C\dot{T} + KT = F \tag{4}$$

where K, T, and F are the global conductance matrix, nodal temperature column vector and the nodal force vector and are the same as in the steady state FEA formulation. The C matrix, is known as the global *capacitance* matrix. Using finite difference approximation of the time derivative, Equation (1) becomes

$$C\left[\frac{T^{t+\Delta t}-T^{t}}{\Delta t}\right]+K\left[\theta T^{t+\Delta t}+(1-\theta)T^{t}\right]=F \tag{5}$$

where $0\leq\Theta\leq 1$.

Using the initial conditions, $T(t=0)=T_0$, temperatures at $t+\Delta t$ can be computed and the results for $T(t+\Delta t)$ are used to obtain $T(t+2\Delta t)$, and so on...as time progresses. The ANSYS models utilized 8-node-Quad elements and were finely meshed to render results with minimal error (thermal energy error <0.5% in ANSYS®).

The FCM Model:
Laplace Transform of Equation (1) is taken,

$$k\nabla^2 T(x,y,s)-\rho C_p s\overline{T}(x,y,s)\ =0 \tag{6}$$

where,

$$\overline{T}(x,y,s)\ =\int_0^{\infty} e^{-st}T(x,y,t)dt \tag{7}$$

The Laplace Transform of the boundary condition in Equation (3) is

$$a\overline{T}+b\frac{\partial\overline{T}}{\partial n}\ =\frac{c-aT_o}{s} \quad on\ \partial R \tag{8}$$

In the FCM approach [5, 6], there are NB chosen boundary points at which boundary conditions will be satisfied by collocation. This is done by placing NB point heat sources outside the region, usually on a congruent boundary, located at a specified distance DS away from the boundary of the region. Strengths, W, of the heat sources are adjusted such that Equation (6) is satisfied. For each value of Laplace argument s, there will be NB linear equations in the NB unknown source strengths, W's. In matrix form obtain

$$D\,W=f \tag{9}$$

where D is the NB x NB "*influence matrix*", *W* is a column vector containing the strengths of the heat sources, and *f* is a column vector containing the prescribed boundary conditions. Once W's are found, temperatures in the s-domain are computed using collocation. Numerical inversion of the s-domain results are used to obtain the t-time domain. In this case, comparing with the FEA matrices, fewer equations are used and as a result, dimension of matrix D in Equation (9) is much smaller than the FEA matrices.

EXAMPLES
Example 1: A rectangular plate of uniform thickness (h= 1 cm, A=20 cm, B=10 cm) with a thermal diffusivity of α=14.1 mm^2/sec is subjected to the boundary and initial conditions specified in Figure 1. Results are summarized in Figures 2-4. Table 1 shows comparison of FEA with FCM. Overall the results are in good agreement, although FCM results deviate for locations away from the origin.

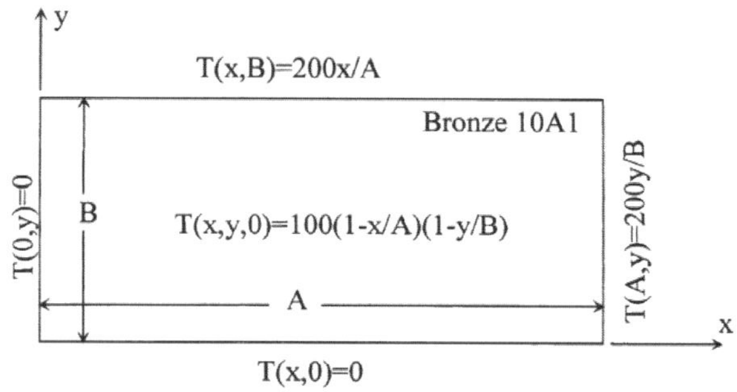

Figure1: Geometry and conditions, Example 1.

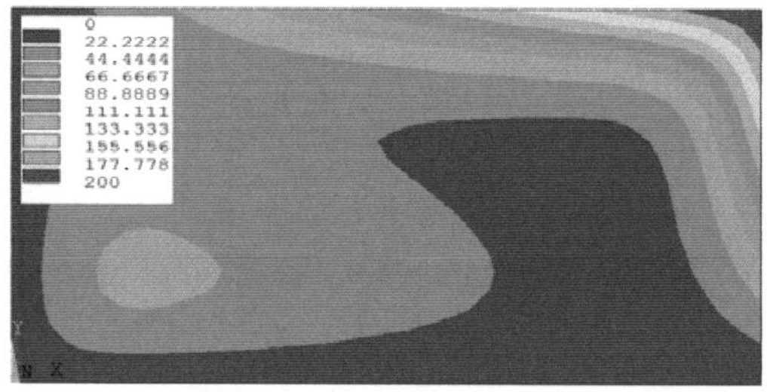

Figure 2: Temp contour plot (t=10 s)

Figure 3: Temp contour plot (t=30 s)

Figure 4: Temp contour plot (t=100 s)

Example 2.
A quarter circular plate of radius R_o = 50 mm and uniform thickness, h=1 cm is made of a material with thermal diffusivity α=10 mm^2/s, subjected to the initial and the

boundary conditions shown in Figure 5. Available steady state theoretical solutions are computed for this problem and presented in Table 2.

X,Y mm	t =10 s	t = 30 s	t = 100 s
25,50	46.42/45.83/ 45.50	24.01/24.40/ 25.55	11.44/11.63/ 11.68
50,100	25.08/25.14/ 25.29	29.70/29.30/ 29.52	44.51/43.03/ 43.16
75,150	26.81/27.09/ 27.75	75.51/70.67/ 70.57	110.67/106.08/ 106.14

Table 1. Comparison of Temperature: FCM/Exact/FEA

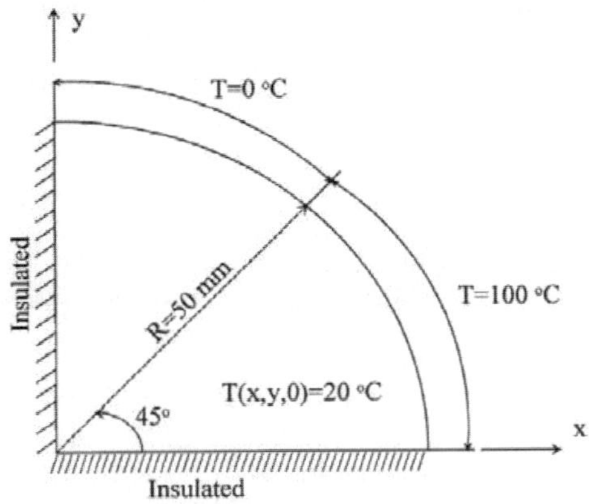

Figure 5: Geometry and conditions, Example 2.

Figure 6: Temp contour plot, steady state)

r/R, θ	Theory	FCM, t=1000s	FEA
.5,0	65.59	65.60	65.59
.5,22.5	61.49	61.48	61.49
.5,45	50	49.89	50
.9,67.5	9.28	8.52	9.28
.9,22.5	90.72	91.43	91.71

Table 2. Comparison of Temperature (Steady State)

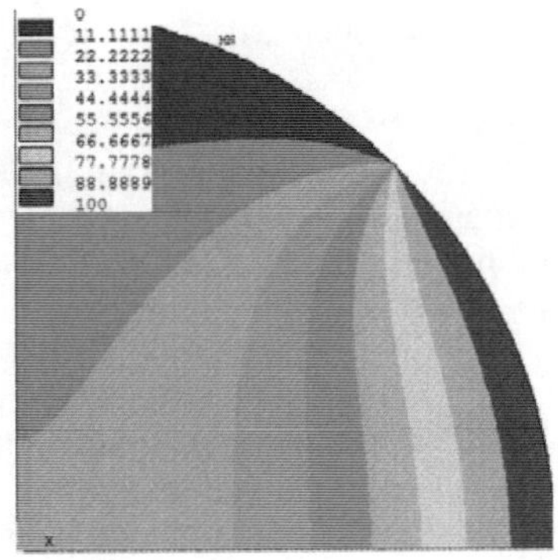

Figure 7: Example 2 (t=20 sec.)

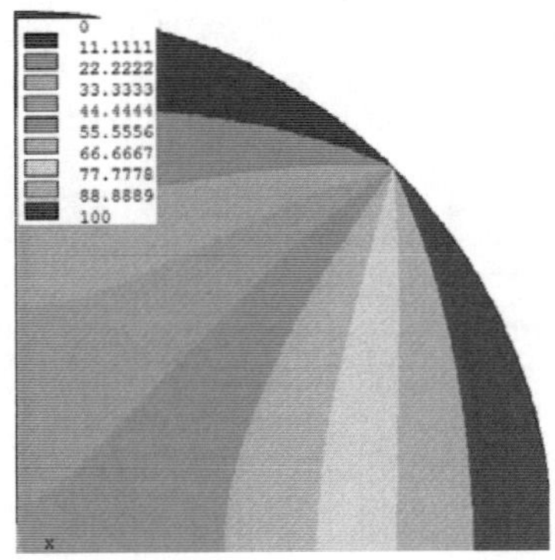

Figure 8: Example 2 (t=100 sec.)

r/R, θ	FCM	FEA
.5,0	34.80	39.53
.5,22.5	32.31	36.85
.5,45	21.35	28.88
.9,67.5	4.65	6.06
.9,22.5	87.28	84.33

Table 3. Comparison of Temperature (t=20 s)

r/R, θ	FCM	FEA
.5,0	61.59	62.20
.5,22.5	57.50	58.09
.5,45	45.90	46.61
.9,67.5	7.24	8.62
.9,22.5	90.64	90.05

Table 4. Comparison of Temperature (t=100 s)

FCM model of this plate included 16 boundary points with the external heat sources located outside the boundary at a distance DS=10 mm away from the actual boundary. Figures 6-8 and Table 2-4 show the corresponding FEA and FCM results. It is discerned that the FCM results agree more with FEA's at longer times as it is evident in those tables.

Heat Flow

The total heat flow rate across all boundaries of the rectangle plate (example 1) is shown in Figure 9, while the heat rate across cooling boundaries (the isothermal bottom and left edges) is shown in Figures 10. It is interesting to note a minimum value for the total heat rate around 67 s.

Similarly the heat rate across isothermal boundaries in Example 2 is presented in Figure 11. The rates are based on the unit value for conductivity of materials and the thickness of 1 cm in both examples. It is noted that the heat flow from the cold boundary (T=0 °C, θ=45-90°) is negative as heat exits from the domain, but the absolute value is used to demonstrate that the heat crossing each boundary will eventually approach to the same value (≈1.78 W, t= 150 s).

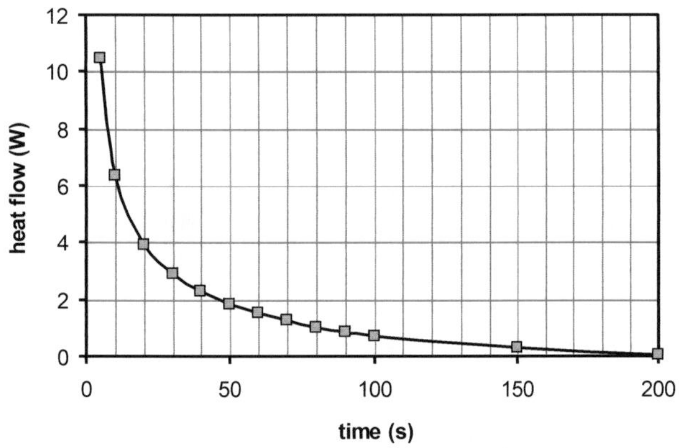

Figure 9: Heat Flow (Example 1).

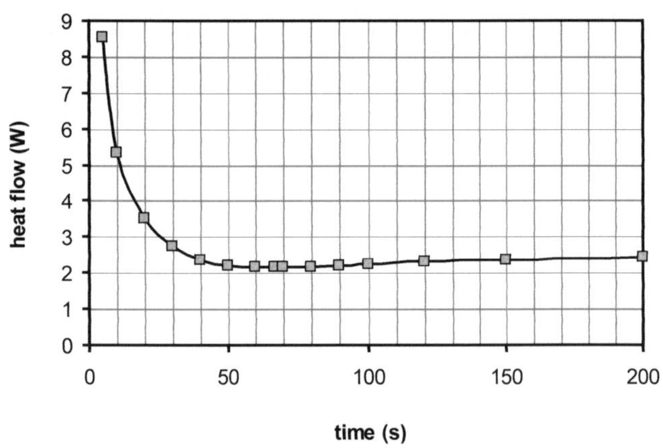

Figure 10: Heat Flow (Example 1).

CONCLUSIONS

This paper has presented comparison and applicability of FEA and the Laplace transform-based Fundamental Collocation Method (FCM) for solving transient heat conduction problems. General two-dimensional regions of arbitrary shapes are used to inspect applicability and accuracy of these methods. The results for the example problems demonstrate that both methods are suitable for regions of arbitrary shapes and mixed type boundary conditions. However, the FEA requires using commercial tools such as ANSYS®. The FCM can be applied to the heat flow problems independently because it requires a relatively small matrix size and therefore, fewer computations.

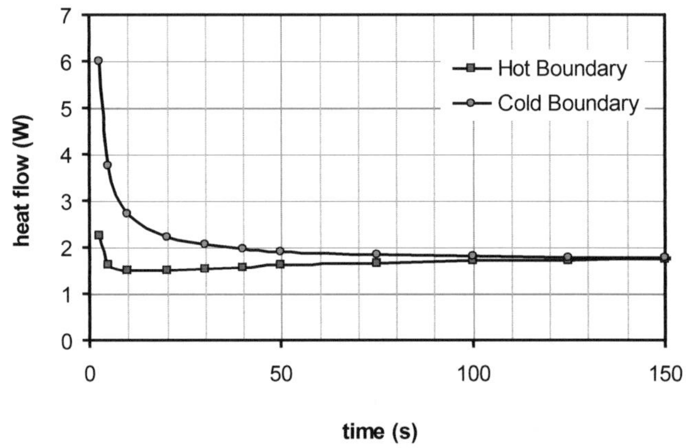

Figure 11: Cross boundary heat flow (Example 2)

REFERENCES

1. Incropera, F. P. and Dewitt, D. P. *"Introduction to Heat Transfer,"* J. Wiley, 2011.
2. Logan, D. L., 2012 *"A first Course in the Finite Element Method,"*, Cengage Learning,.
3. Burgess, G. and Mahajerin, E., 1985,"On the Numerical Solution of Laplace's Equation Using Personal Computers," *International Journal of Mechanical Engineering Education*, 2(1) pp 45- 54,.
4. ANSYS 2015, ANSYS, Inc. 2600 ANSYS Drive, Canonsburg, PA 15317,USA
5. Burgess, G., and Mahajerin, E., "Transient Heat Flow Analysis Using the Fundamental Collocation Method," *International Journal of Applied Thermal Engineering*, Vol. 23, pp 893-904 (2003).
6. Mahajerin, E and Burgess, B., "A Laplace Transform Based Fundamental Collocation Method for Two-Dimensional Heat Flow,*"International Journal of Applied Thermal Engineering,* Vol. 23, pp 101-111 (2003).

Proceedings of the 25th CANCAM
London, Ontario, Canada, May 31 – June 4, 2015

STRUCTURAL TOPOLOGY OPTIMIZATION WITH ADAPTIVE MESH REFINEMENT AND DE-REFINEMENT

Alexandre Nana nana@uqtr.ca
Jean-Christophe Cuillière cuillier@uqtr.ca
Vincent François francois@uqtr.ca
Équipe de Recherche en Intégration CAO-Calcul (ÉRICCA)
Université du Québec à Trois-Rivières
Trois-Rivières, Québec, Canada

ABSTRACT

This paper presents an adaptive mesh refinement and de-refinement strategy for topology optimization of continuum structures. Using an *h*-adaptivity scheme, the proposed mesh sizing function generates coarser elements inside solid and void material and smaller elements at the solid-void interface. This proposed framework also introduces an adaptation of the popular heuristic filter technique to take into account mesh adaptations. Using the SIMP optimization scheme, we obtain the same mesh-independent and checkerboard-free design as the optimal design on uniformly dense meshes, but with reduced numbers of elements and a more accurate description of structural boundaries.

KEYWORDS: topology optimization, SIMP, refinement, de-refinement, filtering, boundary.

INTRODUCTION

Topology optimization of continuum structures [1] is a powerful tool that becomes an essential step in product development and design. It consists in automatically calculating the optimal distribution of a given volume of material in a design domain under constraints. It is important in practical engineering since it achieves significantly superior designs than shape optimization or conventional trial and error approaches. One of the most widely used topology optimization techniques is the Solid Isotropic Material with Penalization (SIMP) [1].

Globally, the SIMP method consists in determining whether or not a point of a design domain should be solid material or void. Volume gain and, by the way, the final volume of void is prescribed as an input, and it is kept constant throughout optimization iterations. In its classical implementation, the raw SIMP outcome consists on a relative density distribution $\boldsymbol{\rho}(\boldsymbol{x}, \boldsymbol{y}, \boldsymbol{z})$.

By using 3D unstructured meshes, our team recently illustrated the fact that, on the same optimization problem, with the same input data and with different mesh sizing, the SIMP result can be very different [2]. This suggests that SIMP results might be improved by mesh adaptations. Also, although most topology optimization methods are performed from uniform and dense meshes, yet, high-accuracy is only necessary within localized regions.

Improvements of topology optimization results are necessary when one seeks for high quality solutions and high accuracy in the description of boundaries, especially when other transformations based on the optimal design are to follow. Few approaches have been proposed to incorporate mesh adaptations in the optimization process.

Maute and Ramm [3] proposed to extract an iso-density line of the result, refine the mesh, and consider this result as a new design problem with the same boundary conditions and loads. Several optimization cycles are implemented, until the desired quality in the result is reached. Costa and Alves [4] refined a given triangular coarse mesh after a fixed number of optimization steps. After each refinement step, a constrained Laplacian smoothing is performed to improve the quality of finite elements. Roman Stainko [5] was among the pioneers to consider only remeshing the solid-void interface and to apply the approach to tetrahedral meshes. He introduced a filter radius to control the number of elements around the solid-void interface. Wang et al [6] introduced the AMR (Adaptive Mesh Refinement) concept to achieve the design that would be obtained from an uniformly fine mesh. The AMR is carried out continuously throughout the optimization process and de-refinement is performed only in void regions.

In most of the aforementioned methods, either only 2D structures are treated, or both solid and boundary elements are refined. Moreover, refinement and de-refinement are not implemented simultaneously. More recently, Wang et al. [7, 8] proposed a refinement technique for topology optimization based on independent error control for displacement and density fields. However, they used non-conventional finite elements (so-called *level-one mesh incompatibility* [6]) which can be computationally more expensive and less general.

ADAPTIVE TOPOLOGY OPTIMIZATION PROCESS

The objective of the work presented here is simultaneously refining and de-refining specific regions of finite element meshes used along topology optimization iterations to improve resolution in the definition of structural boundaries, at a lower computational cost. This section introduces the main idea and the basic principles that are inherent to the proposed refinement and de-refinement strategy.

1) The topology optimization method

The SIMP optimization scheme is based on an artificial solid isotropic material with power-law penalization of intermediate densities to steer the optimal solution to a discrete 0-1 design. The detailed steps of the implementation of the SIMP method can be found in [9, 10]. In its classical implementation, the relative density $\rho(x,y,z)$ is related to the distribution of a virtual elastic modulus according to the penalization law:

$$\widetilde{\boldsymbol{E}}(\boldsymbol{x},\boldsymbol{y},\boldsymbol{z}) = \boldsymbol{E}\cdot(\boldsymbol{\rho}(\boldsymbol{x},\boldsymbol{y},\boldsymbol{z}))^{\boldsymbol{p}} \quad \textbf{(1)}$$

Where p is a penalization coefficient ($\boldsymbol{p}=\boldsymbol{3}$ in this work). In the following equations, as for the elastic modulus $\tilde{E}(x,y,z)$, all objects that are affected by the relative density field $\rho(x,y,z)$ are noted using a $\sim$. The optimization problem in terms of minimizing the compliance $\tilde{C}$ while constraining with a volume fraction f is formulated as:

$$\begin{cases} \boldsymbol{min}\ \widetilde{\boldsymbol{C}} = \{\widetilde{\boldsymbol{U}}\}^{t}\cdot[\widetilde{\boldsymbol{K}}]\cdot\{\widetilde{\boldsymbol{U}}\} \\ \boldsymbol{s.t.} \begin{cases} [\widetilde{\boldsymbol{K}}]\cdot\{\widetilde{\boldsymbol{U}}\} = \{\boldsymbol{F}\} \\ \dfrac{\widetilde{\boldsymbol{V}}}{\boldsymbol{V_d}} = \boldsymbol{f} \\ \widetilde{\boldsymbol{E}}(\boldsymbol{x},\boldsymbol{y},\boldsymbol{z}) = \boldsymbol{E}\cdot(\boldsymbol{\rho}(\boldsymbol{x},\boldsymbol{y},\boldsymbol{z}))^{\boldsymbol{p}} \\ \boldsymbol{0} < \boldsymbol{\rho_{void}} \leq \boldsymbol{\rho} \leq \boldsymbol{1} \end{cases} \end{cases} \quad (1)$$

Where $\tilde{V}$ is the material volume once $\rho(x,y,z)$ is applied, V_d the design domain volume, $[\tilde{K}]$ the global stiffness matrix, $\{\tilde{U}\}$ and $\{F\}$ the global vectors of respectively the displacement and the external load. The threshold $\rho_{void}(= 0{,}001$ in practice) is the lower bound of the relative density used to avoid numerical singularities.

The SIMP process converges (an optimal design is assumed to be achieved) if the relative difference of the compliance values between two successive iterations is less than a threshold Δ_i. Moreover, as explained with details in [2], deriving an optimized shape from raw SIMP results (a relative density distribution $\rho(x,y,z)$) is based on a density threshold ρ_{min}. The optimized shape is made with design material for which $\rho(x,y,z) > \rho_{min}$.

2) A revisited filtering scheme

To obtain checkerboard free designs and reduce mesh dependence of SIMP solutions, it is of common usage to apply filters on the compliance sensitivity and on the relative density. A commonly used filter has been proposed by Sigmund and Petersson [11]. It modifies the sensitivity of a specific element e by the weighted average of the element sensitivities in its neighborhood. This neighborhood, is based on a fixed length r_{min} which is the radius of the sphere centered at the centroid of e. Thus, is doesn't change with mesh adaptations. All the elements which center is inside the sphere are identified as neighboring elements that influence the sensitivity analysis and the relative density of e.

In this work, we consider that two elements are adjacent if they share a face, an edge or a node, and we use the first adjacent elements to a given element for filtering both sensitivity and relative density (see Figure 1).

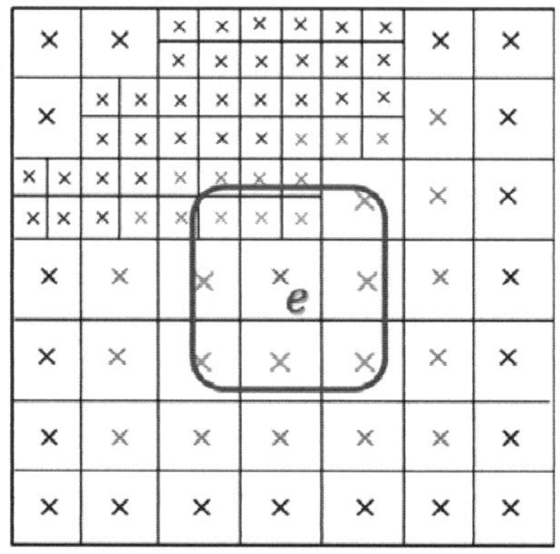

Figure 1: First (green) and second (pink) layers of adjacent elements to element e considered in the compliance's sensitivity and the relative density filtering.

3) Detection of structural boundaries

Our idea towards automatically localizing the boundaries of the optimal shape is based on the fact that boundaries of the optimal shape can easily be identified from the relative density gradient $\overrightarrow{\boldsymbol{grad}}(\boldsymbol{\rho})$. This relative density gradient is used here as a tool to rigorously detect the regions where either refinement or de-refinement should be performed. Indeed, as we will see in the example below, the relative density gradient value is closed to 0 (almost no variation of the relative density) inside void and fully solid regions, and reaches a maximum value at the solid-void interface (structural boundaries of the optimal design).

4) Adaptive mesh refinement and de-refinement strategy

The proposed mesh refinement and de-refinement strategy is based on a mesh sizing function that follows an exponential law to insure that the prescribed element size never reaches null values. The targeted element size distribution $\boldsymbol{d_j(x,y,z)}$ at refinement and de-refinement step j is obtained from the distribution $\overrightarrow{grad}\left(\rho_{j-1}(x,y,z)\right)$ using:

$$\begin{cases} \boldsymbol{d_j(x,y,z)} = \boldsymbol{d_g} & \boldsymbol{for}\ \boldsymbol{j}=\boldsymbol{0} \\ \boldsymbol{d_j(x,y,z)} = \boldsymbol{E_{nm}} e^{-\mu\left\|\overrightarrow{grad}\left(\rho_{j-1}(x,y,z)\right)\right\|} & \boldsymbol{for}\ \boldsymbol{j} > 0 \end{cases} \quad \textbf{(3)}$$

E_{nm} stands for the maximum element size allowed and it is usually higher than the initial uniform mesh size d_g. μ is a coefficient that controls refinement and de-refinement severity. The objectives underlying this adapted mesh sizing function is generating coarser elements inside solid and void material and finer elements at the solid-void interface.

Our refinement and de-refinement scheme is based on the following guidelines:

- The mesh sizing function used for the first SIMP optimization (step 0) is constant. It is noted $d_0(x,y,z) = d_g$.
- $d_j(x,y,z)$ should be maximum and higher that the initial mesh size d_g where the relative density gradient, at step $(j-1)$, noted $\overrightarrow{grad}\left(\rho_{j-1}(x,y,z)\right)$, is close to 0.
- $d_j(x,y,z)$ should be minimum where $\overrightarrow{grad}\left(\rho_{j-1}(x,y,z)\right)$ is maximum.
- $d_j(x,y,z)$ should vary smoothly enough so that a reasonable element quality is guaranteed.

NUMERICAL EXAMPLE: A CONNECTING ROD

The method presented here is implemented in our research platform. This platform integrates geometric modelling, mesh generation, FEA and optimization capabilities. In this example, Young's modulus is 69 GPa and Poisson's ratio is 0,33. The adaptive topology optimization process starts with an initial CAD model on which material data, boundary conditions and loads are applied (Figure 2a). The SIMP result presented in Figure 2b is obtained using an initial uniform mesh with $\boldsymbol{d_g = 4,5\ mm}$ without adaptive refinement. The volume fraction imposed is $\boldsymbol{f = 0,3}$ and the derived relative density gradient is shown in Figure 2c.

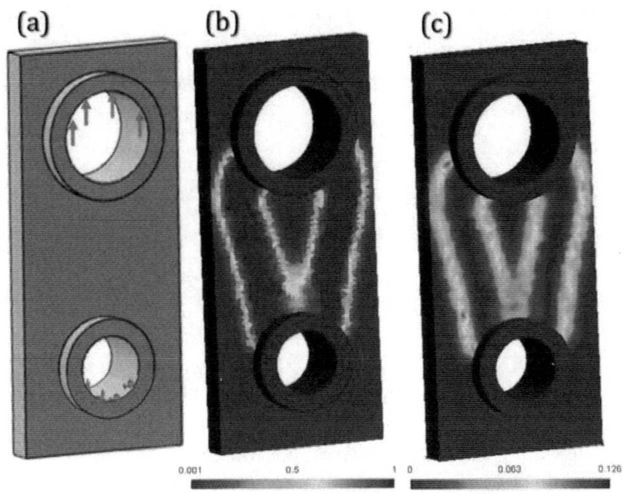

Figure 2: (a) connecting rod model, loads and boundary conditions, (b) SIMP result using an uniform mesh ($\boldsymbol{d_g = 4,5\ mm}$) and (c) relative density gradient derived.

Figure 3 introduces mesh adaptations applied on the case introduced in Figure 2. Figure 3a displays a 2D view of Figure 2b. Figure 3b, 3c and 3d show SIMP results after the first, second and third mesh adaptations. In this case, $\boldsymbol{E_{nm} = 9\ mm}$ and $\boldsymbol{\mu = 11}$.

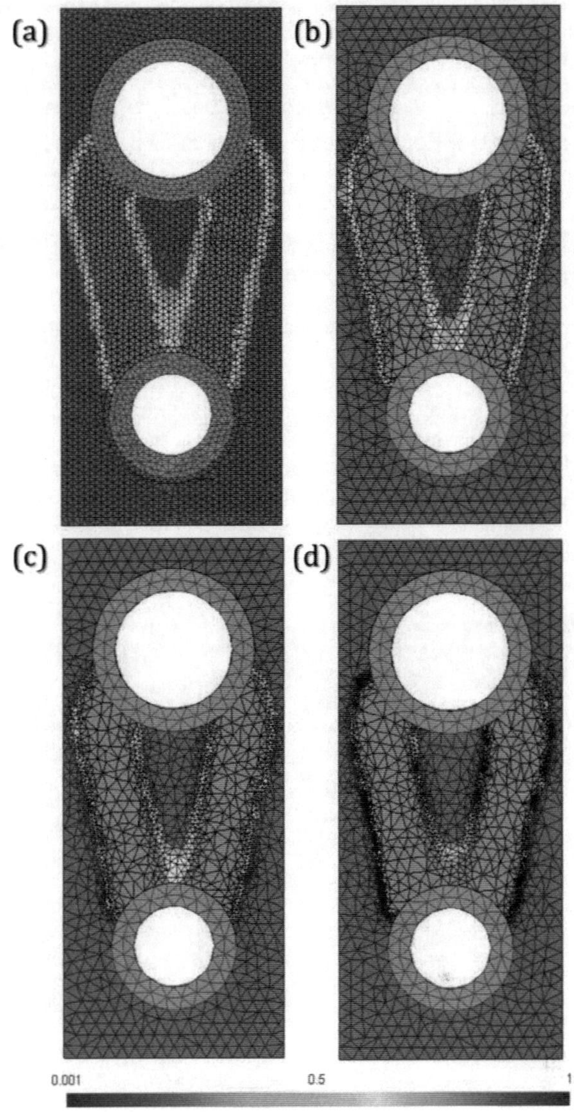

Figure 3: SIMP results using (a) an initial uniform mesh with $\boldsymbol{d_g = 4,5\ mm}$, (b), (c) and (d) after the first, second and third mesh adaptation with $\boldsymbol{E_{nm} = 9\ mm}$ and $\boldsymbol{\mu = 11}$.

Minimum element sizes obtained are $\boldsymbol{2,40\ mm}$, $\boldsymbol{0,65\ mm}$ and $\boldsymbol{0,30\ mm}$ after the first, second and third mesh adaptations. From Figure 3 and the statistics given in Table 1 it is clear that a higher resolution of the structural boundaries and a lower final compliance are achieved along mesh adaptations. Compliance finally converges after 65 iterations to $0{,}042\,J$ with $\boldsymbol{\Delta_i = 0,001\%}$. As suggested by [8], since the actual interest is on the final solution, the convergence criterion is gradually decreased from a relatively loose one to a stricter one along mesh adaptations. Furthermore, Figure 3 also shows that after the first adaptation, new elements are added only at the structural boundaries (at the solid-void interface) while the rest of the mesh remains constant.

	Nb_{iter}	Nb_{tetra}	SIMP Conv. criter. (%)	Compliance (Joules)
Initial mesh	11	114 544	1,000	0,051
1st adapt.	14	33 014	0,100	0,043
2nd adapt.	18	45 524	0,010	0,042
3rd adapt.	22	100 720	0,001	0,042
Total	65			

Table 1: Number of elements and SIMP method statistics.

This example demonstrates the ability of the proposed approach to generate symmetrical designs with a high-quality boundary description using asymmetric and unstructured tetrahedral meshes. The final density distribution after three mesh adaptations, the rough optimal shape using $\rho_{min} = 0{,}5$ and the smooth shape derived based on the algorithm presented in [12] are shown in Figure 4.

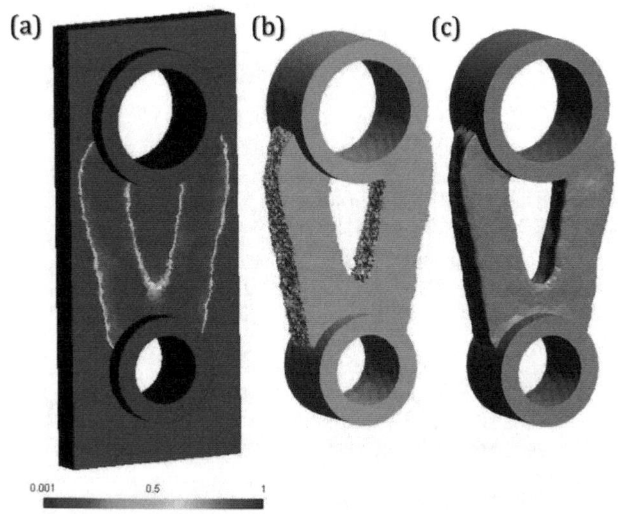

Figure 4: Final (a) density distribution and optimal shapes (b) before and (c) after the smoothing.

CONCLUSION

In this paper we propose an automated mesh refinement and de-refinement strategy for topology optimization methods to improve quality of the optimal solution and accuracy in the description of structural boundaries. This is achieved by simultaneous mesh refinements and de-refinements that are performed from a mesh sizing function. This function is based on the relative density gradient. Due to high-variations in mesh sizing, classical schemes for filtering the sensitivity and the relative density need to be revisited. The approach presented here is quite promising since it can be applied successfully to many optimization problems and extended to other topology optimization schemes.

ACKNOWLEDGMENTS

This study was carried out as part of a project supported by research funding from the Québec Nature and Technology Research Fund and by the Natural Sciences and Engineering Research Council of Canada (NSERC).

REFERENCES

1. Bendsoe, M.P. and O. Sigmund, *Topology optimization - Theory,Methods and Applications*. 2nd ed. 2003, Berlin: Springer. 370.
2. Cuillière, J.-C., V. François, and J.-M. Drouet, *Automatic mesh generation and transformation for topology optimization methods.* Computer-Aided Design, 2013. **45**(12): p. 1489-1506.
3. Maute, K. and E. Ramm, *Adaptive topology optimization.* Structural Optimization, 1995. **10**(2): p. 100-112.
4. Arantes Costa Jr, J.C. and M.K. Alves, *Layout optimization with h-adaptivity of structures.* International Journal for Numerical Methods in Engineering, 2003. **58**(1): p. 83-102.
5. Stainko, R., *An adaptive multilevel approach to the minimal compliance problem in topology optimization.* Communications in Numerical Methods in Engineering, 2006. **22**(2): p. 109-118.
6. Wang, S., E. de Sturler, and G.H. Paulino, *Dynamic adaptive mesh refinement for topology optimization.* arXiv preprint arXiv:1009.4975, 2010.
7. Wang, Y., Z. Kang, and Q. He, *An adaptive refinement approach for topology optimization based on separated density field description.* Computers and Structures, 2013. **117**: p. 10-22.
8. Wang, Y., Z. Kang, and Q. He, *Adaptive topology optimization with independent error control for separated displacement and density fields.* Computers and Structures, 2014. **135**: p. 50-61.
9. Sigmund, O., *A 99 line topology optimization code written in Matlab.* Struct. Multidisc. Optim., 2001. **21**: p. 120-127.
10. Cuillière, J.C., V. Francois, and J.M. Drouet, *Towards the Integration of Topology Optimization into the CAD Process.* Computer-Aided Design and Applications, 2014. **11**(2): p. 120-140.
11. Sigmund, O. and J. Petersson, *Numerical instabilities in topology optimization : A survey on procedures dealing with checkerboards, mesh-dependencies and local minima.* Structural Optimization, 1998. **16**: p. 68-75.
12. Chen, C.-Y. and K.-Y. Cheng, *A direction-oriented sharpness dependent filter for 3D polygon meshes.* Computers & Graphics, 2008. **32**(2): p. 129-140.

Proceedings of the 25th CANCAM
London, Ontario, Canada, May 31 - June 4, 2015

ON THE IMPORTANCE OF THERMAL EFFECTS IN DISLOCATION DYNAMICS SIMULATIONS

Oxana Skiba
Department of Civil and Environmental Engineering
University of Waterloo
Waterloo, Ontario, Canada
oskiba@uwaterloo.ca

Robert Gracie
Department of Civil and Environmental Engineering
University of Waterloo
Waterloo, Ontario, Canada
rgracie@uwaterloo.ca

ABSTRACT

Dislocations motion in crystalline solids under high strain rate plastic deformation causes significant heat generation. Such thermo-mechanical phenomenon should be taken into account when modeling dislocation dynamics. In addition, the increase in temperature affects dislocation drag and the hardening behavior of the material. A transient coupled two-dimensional thermo-mechanical dislocation dynamics model is developed for high strain rate crystalline solids plastic deformation. The model is based on the eXtended Finite Element Method (XFEM), which captures the dislocation slip through the Heaviside step enrichment function. The coupled governing and constitutive equations are presented. The XFEM approximations for the displacement and temperature fields are given. The methodology to incorporate the change in dislocations drag with temperature is shown. An example which illustrates the thermal effects in a coupled system is provided. Results illustrate that thermal effects are important to dislocation dynamics simulations.

Keywords: Dislocation Dynamics, XFEM, plasticity, thermo-mechanical, transient

INTRODUCTION

When dislocations move under high strain rate deformation they generate heat. Since the heat generation rate is much higher than the heat transfer rate it causes strong local heating and softening of the material. Therefore, in modelling this phenomenon an additional equation of heat conduction alone with Fourier law must be added to the system of the equations [1]. Further, the experiments show that dislocation drag coefficient linearly increases with the temperature [2]. In case of the temperature rises that are typical for the shear bands, such dislocation drag increase can be significant. Therefore, underestimated dislocation drag can lead to the inaccuracies in predicted material behaviour. It is especially important when the strain rate is high, since the effect of dislocation drag increases with increase of strain rate [3]. Dislocation dynamics models are widely used and provide a means to study plastic deformation at the length scale between molecular dynamics models and crystal plasticity models [4, 5].

The current work proposes a coupled thermo–mechanical dislocation dynamics (DD) model that captures the heat generation effect and incorporates heat conduction. This model is based on the eXtended Finite Element Method (XFEM). The methodology was successfully used to model DD [6, 7]. Thus, the XFEM–DD does not require the discontinuity created by dislocations to conform to the mesh and so allows to model DD without remeshing.

GOVERNING AND CONSTITUTIVE EQUATIONS

Consider the domain Ω bounded by Γ, as shown in Fig.1. The domain contains n_d dislocations (surfaces of discontinuity). The internal surface of discontinuity α is represented by Γ_d^α. This surface is the active part of the slip plane. The core region Ω_d^α is the region, where the strain energy is unbounded. We denote $\Gamma_d = \cup_\alpha \Gamma_d^\alpha$ and $\Omega_d = \cup_\alpha \Omega_d^\alpha$, where α is from 1 to n_d. Therefore, the strain energy in Ω / Ω_d is bounded. The set of the equations that govern the deformation and incorporate

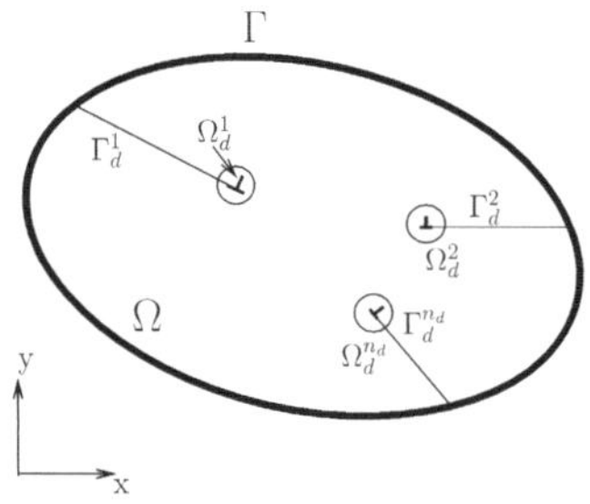

Figure 1: Domain definition and notations. Modified from [7].

heat conduction is the following:

$$\nabla \cdot \boldsymbol{\sigma} + \mathbf{g} = 0, \tag{1}$$

$$C_p \rho_T \frac{\partial \boldsymbol{\Theta}(\mathbf{x}, t)}{\partial t} = \mathbf{k} \Delta \boldsymbol{\Theta}(\mathbf{x}, t) + S(\mathbf{x}, t), \tag{2}$$

where $\boldsymbol{\Theta} = T(\mathbf{x}, t) - \Theta_0$ is the absolute temperature change from a stress free reference temperature Θ_0 at the point $\mathbf{x} \in \Omega$ at the moment t. Here, $\rho_T > 0$ is the mass density, $\mathbf{k}$ is the matrix of coefficients of thermal conductivity, and $C_p > 0$ is the specific heat capacity. Body heat source function $S(\mathbf{x}, t)$ is a linear combination of dirac-delta functions that is defined in Ω and changes with time t:

$$S(\mathbf{x}, t) = \sum_{\alpha=1}^{nd} s_\alpha \delta(\mathbf{x} - \mathbf{x}_{s_\alpha}) \tag{3}$$

The strength of the heat source s_α is the heat generated when a dislocation α has moved and is calculated as a rate of work done by the driving force on a dislocation, which is known as Peach–Koehler force. The position of the heat source corresponding the the dislocation α is $\mathbf{x}_{s_\alpha}$. The coupled problem is complete when appropriate boundary conditions are imposed as well. Here, we emphasize, that the system is also subjected to internal boundary conditions:

$$[|\mathbf{u}|] = \mathbf{b}^\alpha \text{ on } \Gamma_d^\alpha \tag{4}$$

where $\mathbf{u}$ is the displacement, and $\mathbf{b}^\alpha$ is the Burgers vector of dislocation α.
The constitutive equation for the stress is:

$$\boldsymbol{\sigma} = \mathbf{C} : \boldsymbol{\varepsilon} - \boldsymbol{\alpha}\Theta, \tag{5}$$

where $\boldsymbol{\alpha}$ are stress–temperature(thermo–mechanical) coefficients that are related to coefficients of thermal expansion through the tensor of the elastic moduli, $\mathbf{C}$.

XFEM APPROXIMATIONS

In XFEM approach to DD, a standard local displacement approximation is extended by the discontinuities that dislocations create. In our model we consider edge dislocations. Their geometry is defined by two affine functions, $f^\alpha(\mathbf{x}) = \alpha_0 + \alpha_i \mathbf{x_i}$ and $g^\alpha(\mathbf{x}) = \beta_0 + \beta_i \mathbf{x_i}$, in a way that the intersections of these functions define the location of the dislocation core. The active part of the glide plane is defined by $f^\alpha(\mathbf{x}) = 0,\ g^\alpha(\mathbf{x}) \leq 0$, see Fig.2. The displacement approximation for a domain with n_d edge dislocations and Burgers vectors $\mathbf{b}^\alpha$ can be written as:

$$\mathbf{u}^h(\mathbf{x}) = \sum_{I \in S} N_I(\mathbf{x}) \mathbf{d}_I + \sum_{\alpha=1}^{n_d} \mathbf{b}^\alpha \sum_{J \in S^\alpha} N_J(\mathbf{x})[H(f^\alpha(\mathbf{x})) - H(f^\alpha(\mathbf{x}_J))], \quad \forall \mathbf{x} \in \Omega \tag{6}$$

where N_I are the standard finite element shape functions, $\mathbf{d}_I$ are the nodal degrees of freedom, and $\mathbf{x}_J$ are the coordinates of node J. The set of all nodes is denoted by S. The set of enriched nodes is denoted by S^α. The enriched nodes are the nodes of elements that are completely cut by the active part of the glide plane glide plane as shown in Fig.2. The Heaviside

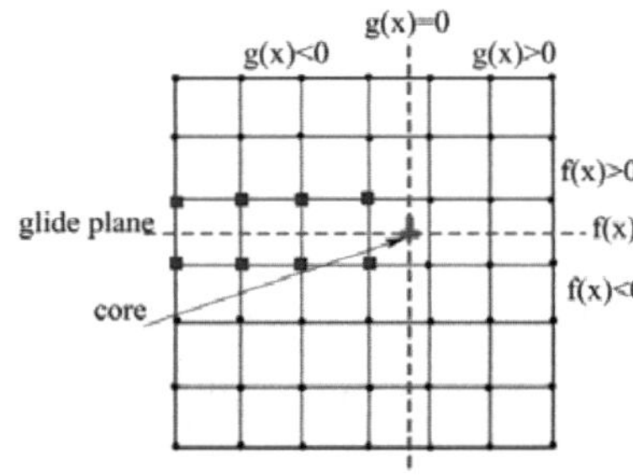

Figure 2: Illustration of the dislocation geometry and enriched nodes. Dark squares represent nodes in S^α. [7].

step function is given by:

$$H(z) = \begin{cases} 1 & \text{if } z > 0 \\ 0 & \text{otherwise} \end{cases} . \tag{7}$$

The temperature field is approximated by:

$$\boldsymbol{\Theta}^h(\mathbf{x}, t) = \sum_{I \in S} \bar{N}_I(\mathbf{x}) \Theta_I(t), \quad \forall \mathbf{x} \in \Omega \tag{8}$$

where, $\bar{N}_I$ are the standard finite element shape functions, and $\Theta_I(t)$ are time-dependent nodal temperature changes.
Since Burgers vector, $\mathbf{b}^\alpha$, is prescribed for each dislocation, the discontinuous enrichment in Eq. (6) does not lead to additional degrees of freedom. We use the Crank-Nicolson scheme to integrate the equation Eq. (2) in time.

THERMAL EFFECT ON DISLOCATION DRAG

In this work the velocity $\mathbf{v}^\alpha$ of dislocation α is calculated from a phenomenological equation of motion:

$$\mathbf{B} \cdot \mathbf{v}^\alpha = \mathbf{F}^\alpha \tag{9}$$

where $\mathbf{F}^\alpha$ is the Peach–Koehler force acting on dislocation α, and $\mathbf{B}$ is the tensor of the dislocation drag. The motion of the dislocations is restricted to the glide plane, so that $\mathbf{B} = B\mathbf{b} \otimes \mathbf{n}$. Here, $\mathbf{n}$ is the normal to the glide plane and B is the drag coefficient.

In order to incorporate the effect of mobility change with temperature, we represent the drag coefficient of each dislocation α, which is the inverse of the dislocation mobility, as:

$$B_\alpha = B_1\Theta_\alpha(t) + B_0;\ B_1 > 0 \qquad (10)$$

where $\Theta_\alpha(t)$ is the temperature rise from the reference temperature, Θ_0, at the location of dislocation α at moment t, and B_0 is the drag of the dislocation corresponding to the reference temperature. The parameter B_1 determines the rate of growth of each dislocation drag with increase of the temperature. Variation of parameter B_1 allows the effect of the temperature through the drag coefficient on material response to be studied. Notice, that each dislocation in the model is subject to a different drag depending on the temperature at the dislocation core.

The suggestion of Eq.(10) is based on the phonon drag mechanism. Such mechanism linearly relates the drag coefficient to the temperature [2]. Moreover, several studies have been done validating linear approximation of the dislocation drag with temperature, for example [8].

NUMERICAL EXAMPLE

First, the temperature rise distribution over the domain is presented in this example. Second, the effect of dislocation drag increase with temperature on material response is illustrated. A body with dimensions $2L \times W$, where $L = 0.75\mu m$, with the tensile axis aligned with the x_1 direction is considered, see Fig.3. Tension is imposed by prescribing loads at $x_1 = 2L$ and putting rollers along $x_1 = 0$. Copper is the material used, which has the following properties: thermal conductivity k = 385 W/(m K); density $\rho_T = 8960$ kg/m^3; mass specific heat capacity: $C_p = 385$ J/(kg K); elastic modulus E =117 $\cdot$ 10^9 Pa (N/m^2); Poison's ratio ν= 0.36; stress–temperature coefficients $\alpha = 17 \cdot 10^6$ 1/K.

A Frank-Read source mechanism is used to nucleate new dislocations. The sources are randomly distributed on the slip planes with a density $\rho_{src} = 125\mu m^{-2}$. The nucleation strength, τ_{nuc}, is randomly assigned to each source using a Gaussian distribution with an average strength of $50MPa$ and standard deviation of $1MPa$. The distance between two dislocations is taken to be L_{nuc} such that the attractive force is balanced by the internal stress in the absence of an electric field. The nucleation time for the sources, t_{nuc}, is taken to be $2.5ns$. Annihilation distance is $L_{an} = 8.5nm$. One slip system at an angle of $\alpha = 60°$ to the positive x_1 axis is considered. The slip planes are $100b$ apart from each other, where $b = 0.25nm$ is the magnitude of the Burger's vector of the edge dislocations. The material is dislocation and stress free at the beginning of simulation. The surface is flux free. The initial temperature is set to $T_0 = 290$ K. The temperature change on the boundary was fixed to $\bar{\Theta} = 0$ K for all the simulations, which represents the Dirichlet boundary conditions.

First, in order to demonstrate the heat generation phenomenon, the temperature distribution over the domain at a strain of 0.002 and loading rate of $\dot{\sigma} = 0.24$ MPa$\cdot ns^{-1}$ is shown in Fig.4. The most heated area, up to 600 K, is localized at the center of the domain, which corresponds to the area of the highest dislocations activity.

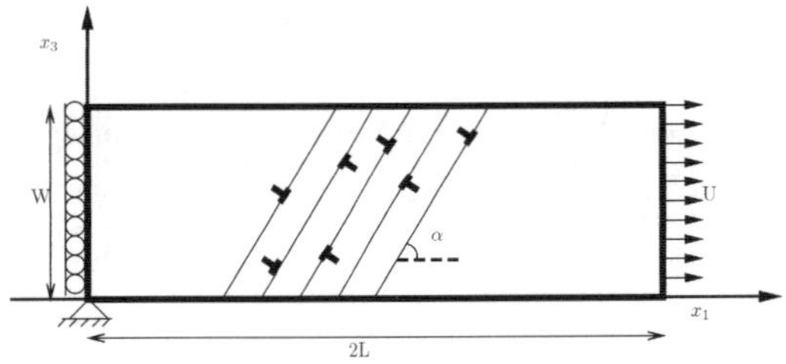

Figure 3: Single crystal specimen. [7].

Second, to study the role of the temperature dependence of the dislocation drag coefficient, B_1 in Eq.(10) is taken to be 0, $0.5 \cdot 10^{-7}$, 10^{-7} $(Pa \cdot s)/K$. The choice of the range of values of B_1 is based on the experimental work in [10], where dislocation drag is measured as a function of temperature. Two sets of simulations were performed at different loading rates, one at a loading rate of $\dot{\sigma} = 0.12$ MPa$\cdot ns^{-1}$ (yielding a strain rate of $\dot{\varepsilon} = 5.6 \cdot 10^3 s^{-1}$ at $\varepsilon = 0.002$) and one at $\dot{\sigma} = 0.24$ MPa$\cdot ns^{-1}$ ($\dot{\varepsilon} = 1.1 \cdot 10^4 s^{-1}$ at $\varepsilon = 0.002$). The results are shown in Fig.5 and Fig.6. In order to demonstrate the underestimated softening behavior, the response of the material without thermocoupling is also presented in both figures.

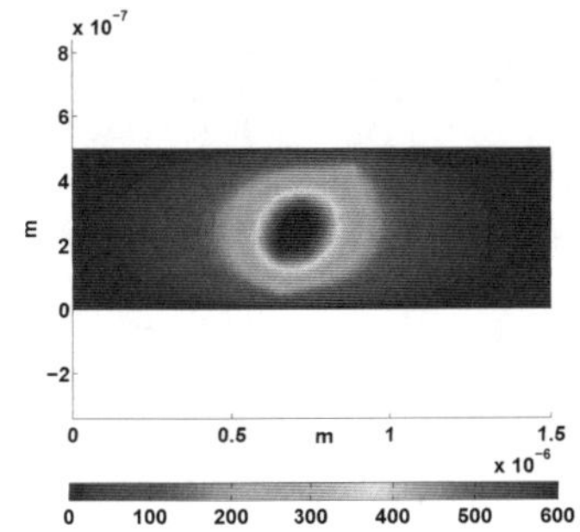

Figure 4: Temperature rise distribution over the domain. Temperature change, Θ, is fixed on the boundary to 0 K. [11]

The results show that the temperature dependence of dislocation drag does not significantly effect the stress-strain response of the material at a lower loading rate. However, it is not the case for the higher loading rates. The thermal hardening is observed with the increase of dislocation drag at the higher loading rates. The larger B_1, the more drag a dislocation will experience at a given temperature, leading to lower

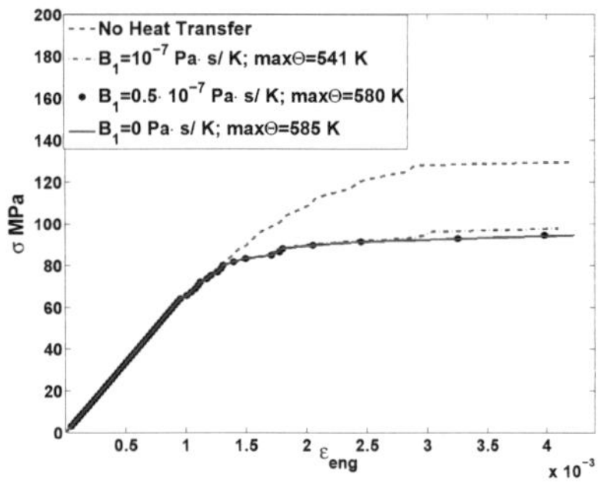

Figure 5: The stress-strain curve for the initial drag $B_0 = 10^{-4} Pa \cdot s$ at the loading rate $\dot{\sigma}$=0.12 MPa $\cdot ns^{-1}$. [11]

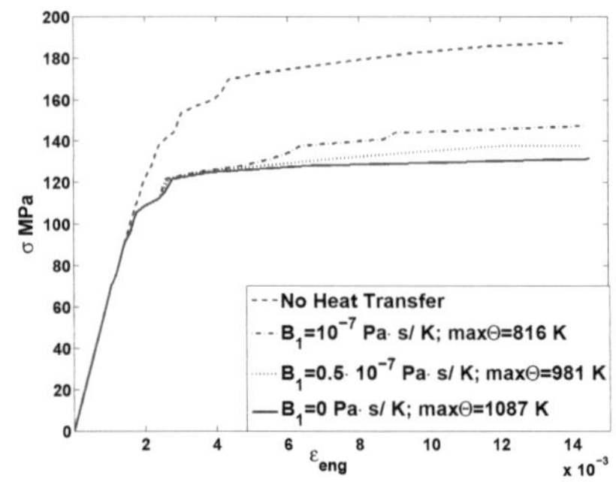

Figure 6: The stress-strain curve for the $B_0 = 10^{-4} Pa \cdot s$ at the loading rate $\dot{\sigma}$=0.24 MPa $\cdot ns^{-1}$. [11]

dislocation velocities. The lower dislocation velocities result in less heat generation for the same Peach–Koehler force. The maximum temperature rise for each simulation is also reported in Fig.5 and Fig.6, where it can be observed that increase in B_1 leads to more hardening and a lower maximum temperatures.

The highest degree of softening occurs at $B_1 = 0$, when the dislocation drag does not change the velocity of the dislocation. Our model is in qualitative agreement with the results obtained by Zirelli et al. in [3], where they showed that the effect of constant dislocation drag is negligible at strain rate of $10^3 s^{-1}$ and is significant at a strain rate of $2 \cdot 10^3 s^{-1}$.

CONCLUSIONS

A transient thermo-mechanical dislocation dynamics model based on the eXtended Finite Element method was developed to simulate the response of the crystalline solids under high strain rate plastic deformation. The study case demonstrated that dislocation movement under high strain rate deformation can cause significant local heating, which leads to the softening of the material. The hardening of the material was observed due to temperature dependent dislocation drag. The simulations showed that the effect of temperature dependent drag increases with deformation rate. The developed model provides a framework for further investigation of thermo–mechanical phenomenon of high strain rate plastic deformation.

ACKNOWLEDGMENTS

We gratefully acknowledge the support from a Discovery Grant from the Natural Science and Engineering Research Council of Canada and a start-up grant from the University of Waterloo.

REFERENCES

[1] Ghoniem, N., Tong, S., and Sun, L. 2000, "Parametric dislocation dynamics: a thermodynamics-based approach to investigations of mesoscopic plastic deformation," *Physical Review B*, Vol(61), pp. 913.

[2] Kocks, U., Argon, A., and Ashby, M. 1975, "Progress in materials science," *Pergamon Press: Oxford*, Vol(19), pp.1–129.

[3] Zerilli, F., Armstrong, R. 1992, "The effect of dislocation drag on the stress-strain behavior of FCC metals," *Acta Metallurgica et Materialia*, Vol(40), pp.1803–1808.

[4] Benzerga, A., Shaver, N. 2006, "Scale dependence of mechanical properties of single crystals under uniform deformation," *Scripta materialia*, Vol(54), pp. 1937-1941.

[5] Monnet, G., Vincent, L., and Devincre, B. 2013, "Dislocation-dynamics based crystal plasticity law for the low-and high-temperature deformation regimes of bcc crystal," *Acta Materialia*, Vol(61), pp. 6178–6190.

[6] Gracie, R., Ventura, G., and Belytschko, T. 2007, "A new fast finite element method for dislocations based on interior discontinuities," *International Journal for Numerical Methods in Engineering*, Vol(69), pp. 423–441.

[7] Skiba, O., Gracie, R., and Potapenko, S. 2013, "Electromechanical simulations of dislocations," *Modelling and Simulation in Materials Science and Engineering*, Vol(21), 16 pp.

[8] Kubin, L. 2013, *Dislocations, mesoscale simulations and plastic flow*, Oxford University Press, Oxford.

[9] Giessen, E., Needleman, A. 1995, "Discrete dislocation plasticity: a simple planar model," *Modelling and Simulation in Materials Science and Engineering*, Vol(3), pp.689-735.

[10] Jassby, K., Vreeland, T. 1970, "An experimental study of the mobility of edge dislocations in pure copper single crystals," *Philosophical Magazine*, Vol(21), pp.1147–1168.

[11] Skiba, O., Gracie, R., "Thermomechanical dislocation dynamics model for high strain rate plastic deformation," *Submitted for publication Sept 2014*.

A PHASE-FIELD MODELING OF THE MICROSTRUCTURAL EVOLUTION IN Ti-6Al-4V ALLOY UNDER MECHANICAL LOADING

Jianhui Zhai
Department of Mechanics
School of Aerospace Engineering
Beijing Institute of Technology
Beijing, 100081 China

Yu Su
Department of Mechanics
School of Aerospace Engineering
Beijing Institute of Technology
Beijing, 100081 China
Email: adamyusu@bit.edu.cn

ABSTRACT

In this work, a phase-field model is proposed to study the phase transformation in Ti-6Al-4V alloys under mechanical loading. The titanium alloy undergoes stress- and temperature-induced phase transformation, during which the crystalline structure evolves from the body centered cubic (BCC) β-phase to the orthorhombic α''- phase. In the proposed model, a spatially and temporally continuous scalar order parameter is employed to characterize the state of the crystalline structure. The effect of thermo-mechanical coupling is taken into account in the model. The presented approach is implemented in a finite element procedure to solve selected sample problems for the phase transformation in Ti-6Al-4V alloys.

KEYWORDS: Phase field, Ti-6Al-4V alloy, microstructural evolution, adiabatic heating, dynamic loading

INTRODUCTION

Ti-6Al-4V titanium alloy is widely employed in aircraft industries due to its remarkable mechanical and chemical properties. Upon deformation, the stress- and temperature-induced phase transformation may occur within the Ti-6Al-4V alloy. The crystalline structure can evolve from the body centered cubic (BCC) β-phase to the orthorhombic α''-phase. As the microstructure characteristics of the material significantly affect the overall mechanical behavior, it is worth of the effort to study the phenomenon of the phase transformation in the Ti-6Al-4V titanium alloy and summarize the effect of such process on the mechanical behavior upon mechanical loading.

The crystalline microstructural transformation in Ti-6Al-4V titanium alloys is a complex process with thermo-mechanical coupling. The phase transformation itself is an irreversible process with crystallite geometric change, leading to the phase transformation strain. Additionally, during the dynamic loading process, such as high strain-rate impact and high speed machining, the adiabatic heating has to be taken into account. Thus the process of heat conduction within the material is deemed as another key feature for the problem. A phase-field model is developed in this paper in order to study the microstructural evolution process. The phase-field method is widely employed to simulate the solid-state microstructural evolution, such as the process of martensitic phase transformation in shape memory alloys[1] and polarization switching in ferroelectric crystals[2,3]. In this work, we refer to the approach by Grandi[1] and propose a phase-field model, which is consistent to the second law of thermodynamics. The proposed model will be utilized to simulate the stress- and temperature-induced phase transformation in Ti-6Al-4V alloy. Three sample problems are given to demonstrate the applicability of this approach.

THE PHASE-FIELD MODEL

We define a spatially and temporally continuous phase field or order parameter, φ, to describe the state of the crystalline structure in the material. In this model, φ =0 indicates a complete occupancy of β-phase in an infinitely small region, whereas, φ =1 indicates a complete occupancy of the α''- phase. Therefore the value of φ at any moment and position indicate the density of the α''- phase in that limiting volume. We define the Landau–Ginzburg–Devonshire free energy density function, $\hat{\psi}$, in Gibbs free energy form, whose value relies on the stress tensor ($\boldsymbol{\sigma}$), absolute temperature (θ), the α''- phase density (φ) and its gradient ($\nabla\varphi$), as

$$\hat{\psi} = -\frac{\rho c}{2\theta_{ref}}\theta^2 - \frac{1}{2}\boldsymbol{\sigma}:\boldsymbol{\lambda}:\boldsymbol{\sigma} + \frac{1}{2}\kappa\,|\nabla\varphi|^2 + L[\theta_0\,\mathrm{F}(\varphi) + [\mathrm{b}(\theta_A - \theta) - \frac{\alpha}{L^2}\boldsymbol{\sigma}^2]\mathrm{G}(\varphi)] \quad (1)$$

The first term in Eq. (1) represents the thermal energy, the second term is the elastic energy, the third term is the gradient energy, and the last term is the Landau-type energy. The parameter c is specific heat, θ_{ref} denotes the reference temperature, $\boldsymbol{\lambda}$ represents the compliance modulus tensor, κ is the interface energy parameter, and L, α, and θ_0 are material constants. In addition, the two functions **F** and **G** are defined by:

$$F(\varphi) = \frac{1}{2}\varphi^4 - \frac{2}{3}\varphi^3 \tag{2}$$

and

$$G(\varphi) = \begin{cases} 0, & (\varphi < 0) \\ \frac{1}{2}\varphi^4 - \frac{4}{3}\varphi^3 + \varphi^2, & (0 \le \varphi \le 1) \\ \frac{1}{6}, & (\varphi > 1) \end{cases} \tag{3}$$

These two functions of φ are plotted in Fig. 1(a) and (b). The Landau-type energy is plotted in Fig. 2(a) and (b) for selected temperatures and selected equivalent stress, respectively.

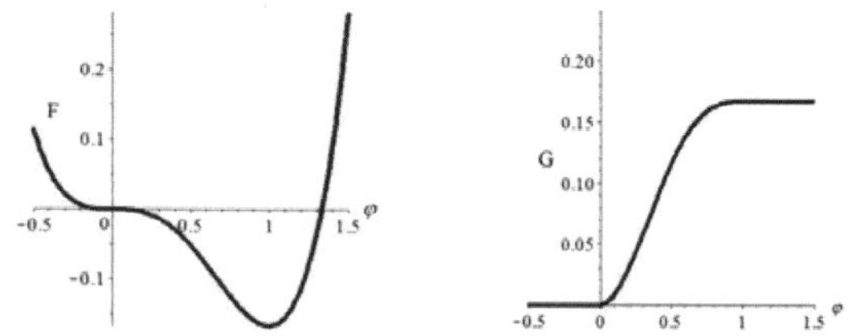

Figure 1: The function: (a) $\mathbf{F}(\boldsymbol{\varphi})$ and (b) $\mathbf{G}(\boldsymbol{\varphi})$

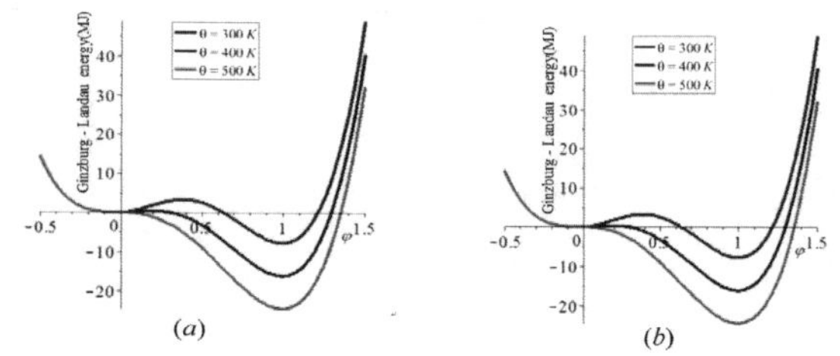

Figure 2: The Landau-type energy density at (a) selected temperatures ($\boldsymbol{\sigma}$=200 MPa), and (b) selected equivalent stresses ($\boldsymbol{\theta}$=300 K)

It should be noticed that a small perturbation needs to be added to function **F** in order to numerically initiate the phase transformation. As it can be seen in Figs. 2, the Landau-type energy density has local minima only at $\varphi = 0$ or at φ=1 for any value of the stress and temperature.

With the Ginzburg-Landau free energy density, we now obtain the constitutive relation through the partial differential of the free energy with respect to the total strain, as

$$\boldsymbol{\varepsilon} = -\frac{\partial \psi}{\partial \boldsymbol{\sigma}} = \boldsymbol{\lambda} : \boldsymbol{\sigma} + 2\alpha \frac{\boldsymbol{\sigma}}{L} G(\varphi) \cdot \tag{4}$$

Similarly, the entropy can be obtained by taking the partial differential of the free energy density with respect to the temperature, as

$$\eta = -\frac{\partial \hat{\psi}}{\partial \theta} = \frac{\rho c}{\theta_{ref}} \theta + bLG(\varphi) \cdot \tag{5}$$

In phase-field method, the time-dependent Ginzburg-Landau (TDGL) kinetic equation governs the evolution of the microstructure by

$$\tau\dot{\varphi} = -\frac{\partial \hat{\psi}}{\partial \varphi} + \nabla \cdot \left(\frac{\partial \hat{\psi}}{\partial \nabla \varphi}\right), \tag{6}$$

where τ is the mobile coefficient. By substituting Eq. (1) into Eq. (6), one is able to obtain the explicit governing equation of the evolution.

For heat conduction, the first law of thermodynamics requires that the following equation must be satisfied:

$$\dot{e} = P_m^i + P_\varphi^i - \nabla \cdot \boldsymbol{q} + r \tag{7}$$

where e is the internal energy density, $\boldsymbol{q}$ is the heat flux, r is the external heat source, and P_m^i and P_φ^i are the internal mechanical power and internal configurational force power, respectively.

In Eq. (7), the heat flux complies with the heat conduction law, as

$$\boldsymbol{q} = -k\nabla\theta \quad (k > 0), \tag{8}$$

where k is positive thermal conductivity. The specific form of the internal energy density is given by

$$e = \hat{\psi} + \boldsymbol{\sigma} : \boldsymbol{\varepsilon} + \theta\eta \cdot \tag{9}$$

The internal mechanical power and internal configurational force power are given by

$$P_m^i = \boldsymbol{\sigma} : \dot{\boldsymbol{\varepsilon}} \tag{10}$$

and

$$P_\varphi^i = \tau\dot{\varphi}^2 + \hat{\psi}_{,\varphi}\dot{\varphi} + \kappa\nabla\varphi \cdot \nabla\dot{\varphi}, \tag{11}$$

respectively.

By referring to Eqs. (1), (4) and (5), one is able to obtain the explicit formulas for Eqs. (9) and (10). Thus the first law of thermodynamics can be explicitly obtained by substituting Eqs. (9)-(11) into Eq. (7).

So far, the three fundamental equations for the studied problem, namely the linear momentum balance equation, the TDGL kinetic equation, and the equation for the first law of thermodynamics are obtained and presented in Eq. (12), as

$$\begin{cases} \rho\ddot{\boldsymbol{u}} = \nabla \cdot \boldsymbol{\sigma} + \rho \boldsymbol{f} \\ \tau\dot{\varphi} = -L\{\theta_0 F(\varphi) + [\mathrm{b}(\theta_A - \theta) \\ \qquad - \alpha\boldsymbol{\sigma} : \boldsymbol{\sigma} / L^2]\mathrm{G}(\varphi)\} + \kappa\Delta\varphi \\ \rho c / \theta_{ref}\theta\dot{\theta} + bL\theta G'(\varphi)\dot{\varphi} - \tau\dot{\varphi}^2 = k\Delta\theta + r \ . \end{cases} \tag{12}$$

It should be noticed that Eq. (12) must be satisfied for the studied problem. Finite element formulas can be developed based on Eq. (12) and the corresponding boundary problem is ready to be numerically solved by a finite element procedure or a commercial finite element solver such as ABAQUS with user defined subroutines (UEL).

SAMPLE RESULTS AND DISCUSSIONS

In this section, a developed two-dimensional (2D) finite element procedure based on Eq. (12) is employed to investigate several sample problems. Demonstration of the capability of the proposed phase-field approach is provided in capturing the stress- and temperature-induced phase transformation in Ti-6Al-4V titanium alloys. The values of parameters are listed in Table 1. E and v are elasticity modulus and poisson ratio, respectively.

Parameter	Value	Unit
τ	2.8×10^{6}	$Pa \cdot s$
L	10^{6}	$Pa \cdot K^{-1}$
κ	0.05	N
α	2×10^{-3}	-
b	1.0	-
θ_A	523.5	K
θ_0	253.15	K
θ_{ref}	298	K
ρ	4420	$Kg \cdot m^{-3}$
c	560	$J \cdot Kg^{-1} \cdot K^{-1}$
k	7.2	$W \cdot m^{-1} \cdot K^{-1}$
E	114×10^{9}	Pa
v	0.342	-

Table 1: Values of the model parameters

As schematically illustrated in Fig. 3, a 2D titanium alloy occupies a rectangle region, Ω, with dimension of 25×5 mm^2. This piece of alloy is under static uniaxial mechanical loading along the length direction.

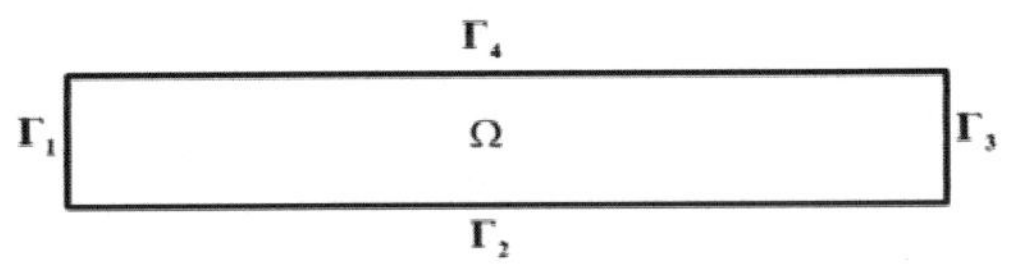

Figure 3: A schematic of 2D Ti-6Al-4V alloy

The boundary conditions for the temperature, the order parameter, and the mechanical part in the problem are generally given by

$$\nabla\theta(x,y,t)\cdot \boldsymbol{n}\,|_{\partial\Omega} = g(x,y), \tag{13}$$

$$\nabla\varphi(x,y,t)\cdot \boldsymbol{n}\,|_{\partial\Omega} = 0, \tag{14}$$

$$u_y(x,y,t)\,|_{\Gamma_2} = 0,\ u_x(x,y,t)\,|_{\Gamma_3} = a(t), \tag{15}$$

and

$$u_x(x = 10mm, y, t) = 0, \tag{16}$$

where the vector $\boldsymbol{n}$ is the outward normal to the boundary and the profile of the prescribed displacement function $a(t)$ is illustrated in Fig. 4. u_x and u_y are the displacements in x (horizontal) and y (vertical) directions, respectively. For this study we ignore the heat convection, i.e., $g(x, y) = 0$.

The initial conditions for temperature and the order parameter are determined to be

$$\theta(x,y,t=0) = \theta_0(x,y) \tag{17}$$

and

$$\varphi(x,y,t=0) = \varphi_0(x,y), \tag{18}$$

respectively. The values of $\theta_0(x,y)$ and $\varphi_0(x,y)$ are 298K and 0, respectively.

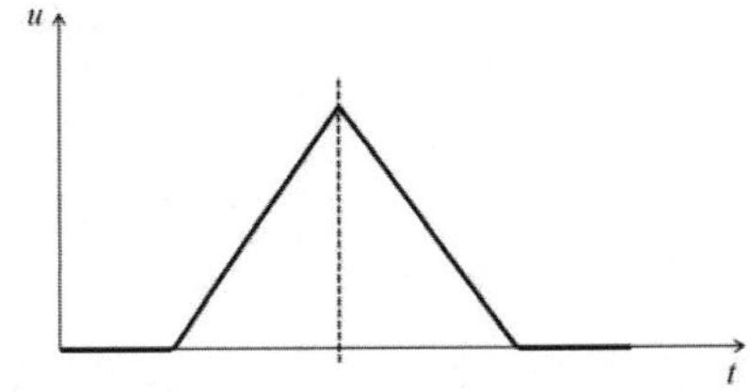

Figure 4: A schematic of the loading history

The computed distribution of φ is plotted for the initial state, the intermediate loading state, and the final loading state in Figs.5(a)-(c), respectively. As it is observed in the plots, the entire material is in β-phase during the stress-free state. As the loading starts to increase slowly, a part of the alloy starts to undergo microstructural transformation from the β-phase to the α"- phase. A finite-thickness interface exists between the two phases. The existence of the interface is attributed to the gradient energy term in the Ginzburg-Landau free energy density in Eq. (1).

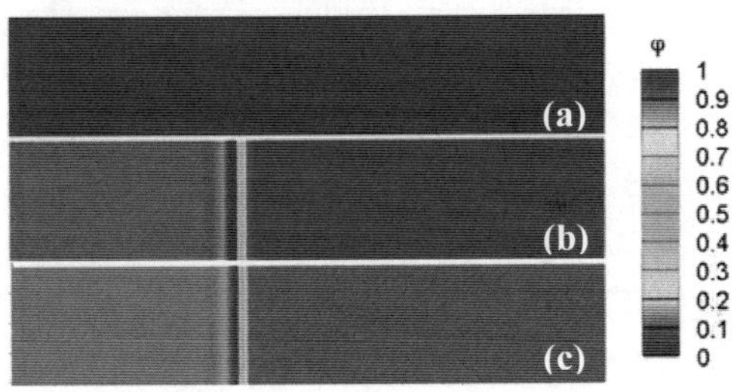

Figure 5: Tension-induced phase transformation: (a) the initial state; (b) the intermediate state, and (c) the final state

In the next, we are examining the stress-induced phase transformation under applied shear stress. In order to do that, we adopted a squared titanium alloy (5mm × 5mm) as schematically shown in Fig. 6(a). Mechanical load is applied to the boundary of the material to generate simple shear deformation. The boundary conditions for temperature and the order parameter are identical to the case of the first sample problem. The applied mechanical boundary conditions and constraints are given by

$$u_x(x,y,t)\,|_{\Gamma_2} = 0\ \&\ u_x(x,y,t)\,|_{\Gamma_4} = b(t), \tag{19}$$

and

$$u_y(x,y,t) = 0, \tag{20}$$

where the profile of the prescribed displacement function $b(t)$ is illustrated in Fig. 6(b).

The numerically obtained distribution of the order parameter is shown in Figs. 7(a) and (b) for the initial state and the state upon simple shear loading, respectively. It is evident that the proposed model is able to mimic the phase transformation induced by shear stress.

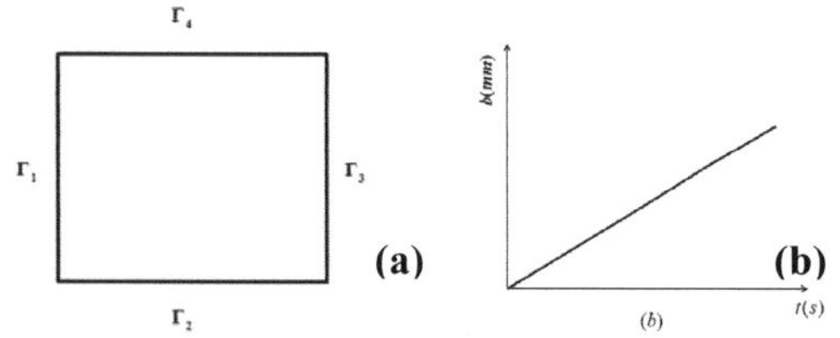

Figure 6: Simple shear problem: (a) A schematic of 2D Ti-6Al-4V alloy; (b) a schematic of the loading history

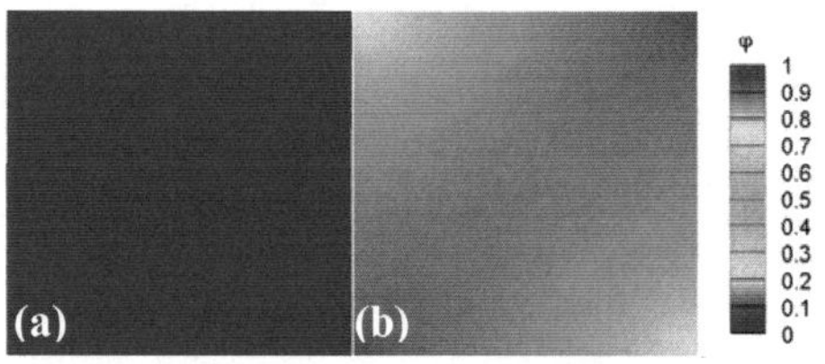

Figure 7: Simple shear-induced phase transformation: (a) the initial state; (b) after the shear loading

At last, we demonstrate the temperature-induced in β-phase transformation in the material. The geometry of the problem is identical to that of the first sample problem except that a higher temperature region occupies the center of the area as illustrated in Fig. 8.

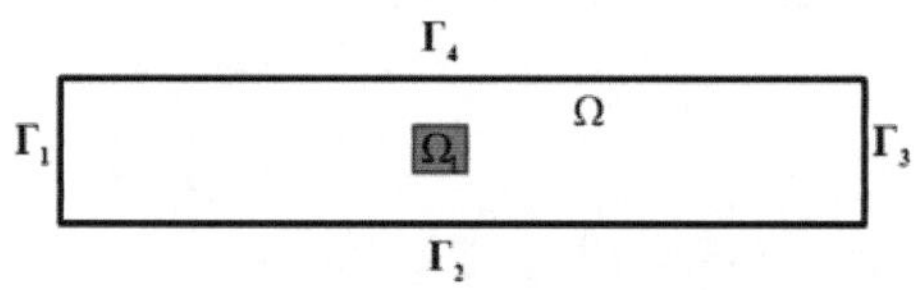

Figure 8: A schematic of 2D Ti-6Al-4V alloy with a higher temperature region located at the center of the area

The boundary conditions for temperature and the order parameter are identical to the case of the first sample problem. The mechanical boundary conditions are given by

$$u_x(x,y,t)|_{\Gamma_1} = 0 \ \& \ u_y(x,y,t)|_{\Gamma_2} = 0 \cdot \quad (21)$$

(22)

The initial condition for the temperature is assigned as

$$\theta(x,y,t=0)|_{\Omega} = \theta_0(x,y) \quad (23)$$

and

$$\theta(x,y,t=0)|_{\Omega_1} = \theta_1(x,y), \quad (24)$$

whereof $\theta_0(x,y)$=298 K and $\theta_1(x,y)$=900 K

The distribution of the order parameter is computed and plotted in Fig. 9(a)-(c) corresponding to the initial state, the intermediate state and the last state, respectively. It is clear that, due to the heat conduction, the initial lower temperature region got heated up and the rise of temperature causes microstructural evolution. The temperature difference leads to the interface of the microstructure as illustrated in Fig. 9(c).

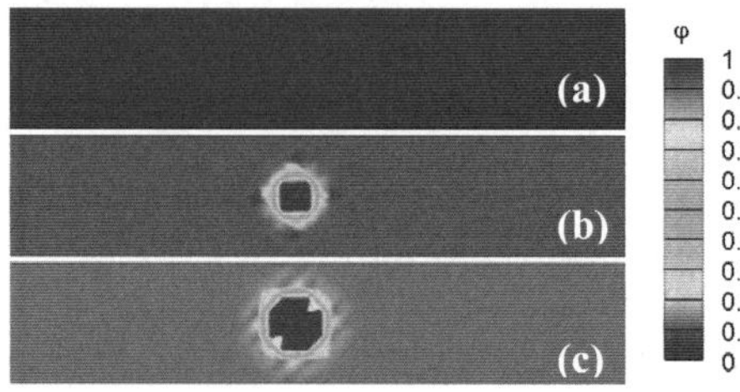

Figure 9: Temperature-induced phase transformation: (a) the initial state; (b) the intermediate state, and (c) the final state

CONCLUSIONS

In this work a continuum-based phase-field model is developed to study the microstructural evolution in Ti-6Al-4V titanium alloy under mechanical loading. The phase transformation from β crystalline structure to α'' structure is simulated by a finite element procedure based on the proposed model. Three sample problems are respectively solved and it is demonstrated that the proposed approach can be used to study the phase transformation processes adequately.

ACKNOWLEDGMENTS

The authors thank the support from National Natural Science Foundation of China (Grant No. 11132011), and BIT International Research Collaboration Project (Grant No. GZ2014015101).

REFERENCES

1. Grandi, D., Maraldi, M., and Molari, L. 2012, "A macroscale phase-field model for shape memory alloys with non-isothermal effects: influence of strain rate and environmental conditions on the mechanical response", *Acta Materialia*, Vol (60), pp. 179-191.
2. Su, Y., Liu, N. and Weng, G.J. 2015, "A phase field study of frequency dependence and grain-size effects in nanocrystalline ferroelectric polycrystals", *Acta Materialila*, Vol (87), pp. 293-308.
3. Su, Y. and Landis, C.M. 2007, "Continuum thermodynamics of ferroelectric domain evolution: theory, finite element implementation, and application to domain wall pinning", *Journal of the Mechanics and Physics of Solids*, Vol (55), pp. 280-305.

Proceedings of the 25th CANCAM
London, Ontario, Canada, May 31 – June 4, 2015

A NOVEL DESIGN OF THE HEAT ECHANGER OF AN AIRCRAFT'S ENVIRONMENTAL CONTROLLER

Negin Shahbazian and Mohammad Anisur Rahman
Mechanical and Industrial Engineering
University of Toronto
Toronto, Ontario, Canada
negin.shahbazian@mail.utoronto.ca
anis.rahman@mail.utoronto.ca

Professor Kamran Behdinan
Mechanical and Industrial Engineering
University of Toronto
Toronto, Ontario, Canada
behdinan@mie.utoronto.ca

ABSTRACT

Redesign of a heat exchanger chassis of an aircraft's environmental controller was carried out using finite element analysis (FEA) to reduce overall cost and weight of the end unit. Seven design alternatives were proposed to replace the current design and were evaluated using a decision matrix. Selected design involved introduction of bolted flanges into the chassis. Von-Mises stresses in the flanges and shear and tensile forces in the bolts were obtained by FEA using ANSYS. Flanges were further modified using safety margins and weight as optimization factors. The optimized flange contains a fluid passage with 0.08in wall thickness reduction and 0.14in passage length reduction. Furthermore, the overall weight was reduced by 0.026kg. After optimization, the margin of safety was improved from 0.041 to 0.580 for maximum operating pressure, and from -0.297 to 0.067 for burst pressure. The novel design accomplished weight and cost reduction along with improving the margins of safety.

KEYWORDS: Heat exchanger, bolted flange, FEA, optimization, margin of safety

INTRODUCTION

Design and manufacturing industries are rapidly moving towards employing novel manufacturing techniques that are more efficient and cost effective, and improve the product quality. In order to accomplish that, products must be designed to achieve satisfactory functionality, safety, and quality. At the same time minimize the complexity, time consumption, and risk in the manufacturing and assembly processes. The leading companies must invest sufficient resources in research and development in order to maintain their leading status within the relevant industry by offering the optimal balance between price and product quality. Furthermore, it is important to improve the overall value of an existing product through a novel product design approach.

To accomplish lower product cost for the environmental controller in hand in advance of a production ramp-up planned in approximately two years time, a feasibility study was initiated to understand the technical viability of a number of redesign options that were under consideration.

In an endeavor to support the cost reduction goal, authors have considered reengineering the housing chassis of the controller unit in order to reduce the manufacturing cost as well as the cost of raw materials. The housing chassis in hand is a complex assembly that contains a heat exchanging system in its walls. The walls of the chassis contain fluid passages integrated in them for cooling purposes. The fluid cavities are generated by vacuum brazing lids onto the base panel of each chassis wall. In order to reduce the overall cost of the chassis, eight design alternatives were proposed and investigated in this research in order to replace the current in production design. The main objective of this research was to redesign the housing chassis to employ less expensive manufacturing processes as well as less complicated and cost efficient assembly techniques. Additionally, the novel design considered minimizing the cost of raw materials. A critical factor considered in the redesign of this chassis was weight. Since weight plays a crucial role in aerospace applications, the new design aimed to be lighter or the same weight as the current design in production.

INVESTIGATED DESIGN ALTERNATIVES

Eight design options including the current in production design were proposed, investigated, and compared against one another. The first design option was the current vacuum brazed design. The second design alternative was the modification of the existing chassis with only some of the panels redesigned. The third proposed option was using gaskets and bolted joints to connect the lids to the walls to eliminate the vacuum brazing process. This design was prototyped for one out of the six chassis panels with the most complicated fluid passage shape. After running tests (including cycling and burst pressure tests) on the prototype, the panel marginally passed the burst pressure test. The test results were not satisfactory due to the customer's demand for higher margins. Furthermore, a hybrid design option, which is a combination of the gasket and bolted walls and the current brazed walls, was investigated. The hybrid method considered using gaskets only in two of the chassis walls with the shortest fluid passage length and least complicated passage shapes. The next option studied was embedding tubes in the walls instead of forming the fluid passages by connecting the walls and lids. Furthermore, electron beam (EB) welding was studied as a replacement for the current vacuum brazing technique. Additionally, bonding using no gaskets was investigated.

The eight design alternatives were ranked based on six different criteria, cost, weight, design effort, design risk, manufacturing risk and quality, and reliability risk. The mentioned criteria were ranked based on their importance to customer. A weighted decision matrix was generated for all the design options. The sum of the points for each design was calculated to determine the design with the highest points.

The most favorable design alternative was design with some panels modified. This design corresponds to modifying three of the panels in the chassis in a way that the machining time and effort required is significantly lowered compared to that of the current in production design.

DESIGN WITH SOME PANELS MODIFIED

The chassis contains six panels. Three of the adjacent panels as shown in Fig. 1 contain inlet outlet flanges that are machined on the panel. These flanges drove the need for thick sheets of raw material stock and significant machining time. This design alternative focuses on machining those flanges separately and joining them to the relative panels using bolts. Therefore, thinner sheets of raw material can be used, which allows for reducing the stock cost, the machining time, and labor cost.

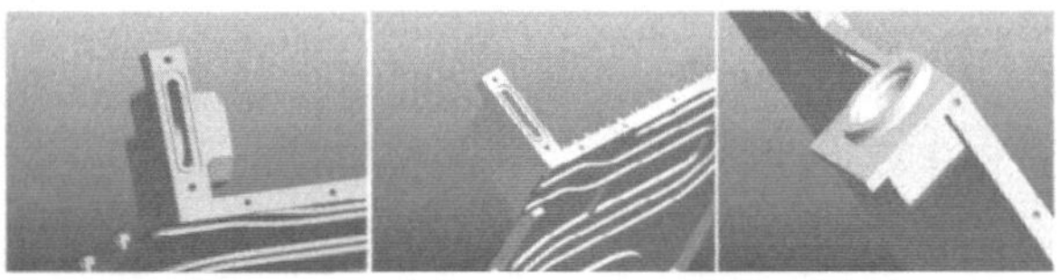

Figure 1: Three flanges machined on panels.

FINITE ELEMENT ANALYSIS AND MODEL

A 3D model of the design was constructed using Creo Elements/Pro software and an FEA was performed using ANSYS Workbench 15.0 platform as a static structural analysis model. Mesh sensitivity tests were run on the fluid channels and flanges. A hex dominant mesh was chosen for flanges and a uniform mesh transitioning was generated from one section to the next. To set up the model, constant fluid pressure equal to the maximum operating pressure of the heat exchanger was applied normal to the internal faces of fluid cavities. FEA results showed that the highest deformation and von-Mises stresses in the chassis were found in one of the flanges, which became the main area of focus. In application of maximum operating pressure, the flange marginally passed the allowable yield stress. The ratio of burst pressure to maximum operating pressure values is known to be 1.666. By incorporating that ratio the resulting von-Mises stresses in the application of burst pressure were calculated. Stresses in this case exceeded the allowable ultimate tensile stress leading in further design modifications.

REPLACING THE SHARP EDGE WITH A CHAMFER

The highest stress value in flange was found on a sharp edge inside the fluid cavity. In order to reduce the equivalent stresses on the sharp edge, the edge was replaced by a 30° chamfer. The von-Mises stresses on the chamfered edge were decreased by 27% compared to that of the sharp edge. Fig 2 shows the contours of maximum equivalent stresses on the flange after applying the chamfer.

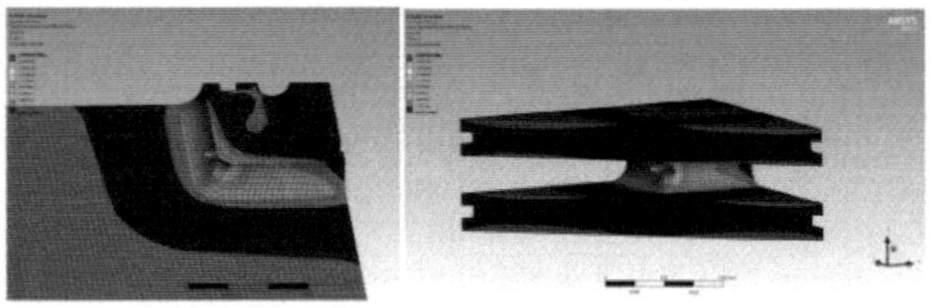

Figure 2: Contours of von-Mises stresses on chamfered flange.

FURTHER DESIGN MODIFICATIONS

The primary drivers of stress were identified to be the wall thickness (t) and the free length of the passage (L). Therefore, these two parameters were chosen as variables in the design optimization. Wall thickness was decreased by 0.050 and 0.075 inches and the free length of fluid passage was decreased by 80% and 60% by adding material inside the fluid passage. Four different CAD models for the mentioned dimension changes were generated and the FEA results were recorded. The von-Mises stress values were then plotted in two separate graphs for L and t. Using analysis and curve fitting the data, functions for von-Mises stresses in each curve in the form of a polynomial were obtained. Two equations for the normalized maximum equivalent stresses with respect to Δt and ΔL are as follows:

$$\sigma = 267093\Delta t^2 + 39755\Delta t + C_0 \quad (1)$$

$$\sigma = 13736\Delta L^2 + 22583\Delta L + C_0 \quad (2)$$

where σ is the normalized von-Mises stress and C_0 is a constant corresponding to the normalized von-Mises stress in the current flange dimensions. Combining Eq. (1) and Eq. (2) results in an equation, which corresponds to the changes in wall thickness and passage length simultaneously.

$$\sigma = 267093\Delta t^2 + 39755\Delta t + 13736\Delta L^2 + 22583\Delta L + C_0 \quad (3)$$

After performing several iterations on Δt and ΔL values and using Eq. (3), different value points for the combination of (Δt, ΔL, σ) were obtained. Furthermore, the margins of safety for all the calculated combinations were calculated using the following formula [1]:

$$MS = \left|\frac{\sigma_{all}}{\sigma_{act}}\right| - 1 \quad (4)$$

where MS is the margin of safety and σ_{all} and σ_{act} correspond to the allowable stress and the actual stress respectively. The MS values are presented in Fig. 3.

		Δt										
		0.000	-0.0075	-0.015	-0.0225	-0.03	-0.0375	-0.045	-0.0525	-0.06	-0.0675	-0.075
ΔL	0.000	0.041	0.064	0.086	0.107	0.125	0.140	0.153	0.164	0.171	0.176	0.177
	-0.070	0.180	0.211	0.239	0.265	0.289	0.310	0.327	0.341	0.351	0.357	0.358
	-0.140	0.345	0.384	0.421	0.456	0.487	0.515	0.538	0.557	0.570	0.578	0.580
	-0.210	0.537	0.589	0.639	0.685	0.727	0.764	0.796	0.821	0.839	0.850	0.854
	-0.280	0.763	0.832	0.898	0.960	1.017	1.068	1.112	1.146	1.172	1.187	1.192
	-0.350	1.024	1.115	1.203	1.288	1.366	1.436	1.497	1.546	1.582	1.604	1.610
	-0.420	1.320	1.440	1.559	1.673	1.781	1.879	1.964	2.033	2.084	2.115	2.124
	-0.490	1.645	1.802	1.960	2.114	2.261	2.396	2.515	2.613	2.686	2.730	2.743
	-0.560	1.983	2.184	2.389	2.593	2.790	2.974	3.138	3.274	3.377	3.439	3.458
	-0.630	2.305	2.555	2.812	3.072	3.327	3.569	3.787	3.970	4.109	4.194	4.220
	-0.700	2.573	2.866	3.172	3.486	3.797	4.096	4.368	4.600	4.777	4.886	4.919

Figure 3: Safety margin values for maximum operating pressure.

The margin of safety for the ultimate tensile stress in application of burst pressure was calculated by the same method used for yield margin of safety. Calculations showed dimension combinations with negative MS values in some areas. Hence, the corresponding dimensions were unacceptable for the design.

WEIGHT OPTIMIZATION

Moving from the initial dimensions toward the new proposed dimensions, the wall thickness was decreased while the free length of passage was decreased as well. The former caused reduction in weight while the latter introduced new material inside the fluid cavity walls resulting in weight increase. Hence, a weight optimization was performed to find the lightest part with the best yield and UTS margins.

Similar to the previous section, the weight versus different thickness values and the weight versus different passage lengths were plotted. Furthermore, by curve fitting the data, a polynomial equation was obtained for each case based on the value t and L being the variables. Since three points were used in the curve fitting, the equations for weight would give us a rough estimate of the actual weight values in various dimension adjustments. Additionally, these equations were converted to equations using Δt and ΔL as variables. Eq. (5) and (6) correspond to the normalized weight values, Wt as a function of Δt and ΔL respectively.

$$Wt = -4.7359\Delta t^2 + 2.2606\Delta t + 1 \quad (5)$$

$$Wt = -0.1002\,\Delta L^2 - 0.2784\Delta L + 1 \quad (6)$$

Combining the two polynomials formed a general equation for weight using Δt and ΔL as variables as can be seen in Eq. (5).

$$Wt = -4.7359\Delta t^2 + 2.2606\Delta t - 0.1002\,\Delta L^2 - 0.2784\Delta L + 1 \quad (5)$$

Different iterations of Δt and ΔL were put in this polynomial to calculate weight values. Using the analysis from previous section, the weight values that correspond to the negative margin of safety for UTS were eliminated. Furthermore, Fig. 4 was generated demonstrating the ΔWt or change in the weight values (change with respect to the initial design with Δt and ΔL equal to zero).

		Δt										
		0.000	-0.01	-0.02	-0.02	-0.03	-0.04	-0.05	-0.05	-0.06	-0.07	-0.08
ΔL	0.000	No	No	No	No	No	No	No	No	No	No	No
	-0.070	No	No	No	No	No	No	No	No	No	No	No
	-0.140	No	No	No	No	-0.006	-0.009	-0.012	-0.016	-0.019	-0.023	0.026
	-0.210	0.009	0.006	0.003	[illegible]	-0.003	-0.006	[illegible]	[illegible]	[illegible]	-0.020	-0.024
	-0.280	0.012	0.009	0.006	0.003	[illegible]	-0.004	[illegible]	[illegible]	[illegible]	-0.017	-0.021
	-0.350	0.014	0.011	0.008	0.005	0.002	[illegible]	-0.004	-0.008	-0.011	-0.015	-0.018
	-0.420	0.016	0.014	0.011	0.008	0.005	0.001	-0.002	-0.005	-0.009	-0.012	-0.016
	-0.490	0.019	0.016	[illegible]	[illegible]	[illegible]	0.003	[illegible]	-0.003	-0.007	-0.010	-0.014
	-0.560	0.021	0.018	[illegible]	[illegible]	[illegible]	0.005	0.002	[illegible]	-0.005	-0.008	-0.012
	-0.630	0.023	0.020	0.017	0.014	0.011	0.007	0.004	[illegible]	-0.003	-0.006	-0.010
	-0.700	0.024	0.021	0.018	0.015	0.012	0.009	0.006	0.002	-0.001	-0.005	-0.008

Weigth Reduction
Weigth Increase

Figure 4: Values of actual weight change with respect to different Δt and ΔL values.

In Fig. 4, the ΔWt values that indicate negative UTS margins of safety were marked as No. The rest are values with positive margins of safety. A black curve drawn on the table shows the approximate location of ΔWt values equal to zero. The right side of the curve corresponds to the negative ΔWt values, which means there is reduction in part weight compared to the initial design. The left side of the curve shows the areas with positive change of weight meaning that the weight has increased. Therefore, the optimal design is chosen from the right side of the curve. The cell with the largest absolute value on the right shows a part with highest weight reduction, which has a positive margin, both for yield and ultimate tensile stress. The largest value for |ΔWt| on the right side is equal to 0.026kg which corresponds to a design with ΔL = −0.140 and Δt = −0.08 inches.

In the aerospace applications weight is an extremely crucial factor. The proposed dimensions along with providing satisfactory margins of safety decrease the weight by 26 grams. This weight reduction is considered significant in airframe parts with the size of the flange in hand and it is highly desirable for the aircraft manufacturers.

EVALUATION OF BOLTED JOINTS

The bolts were modeled as circular beams in the FEA. Beam probes were mounted on every single beam in order to obtain results of the pressure loading. These results include values of axial force on the bolt, bolt torque, shear force at I and J the two ends of the bolt, and moment at the I and J ends. The flange is bolted to the chassis panels using seven bolts. There are two different types of bolts used in this part. Fig. 5 shows the position of different bolts in this part. Bolt number 1, 2, and 3 are flat countersunk head screws. While bolt number 4, 5, 6, and 7 are hex head screws.

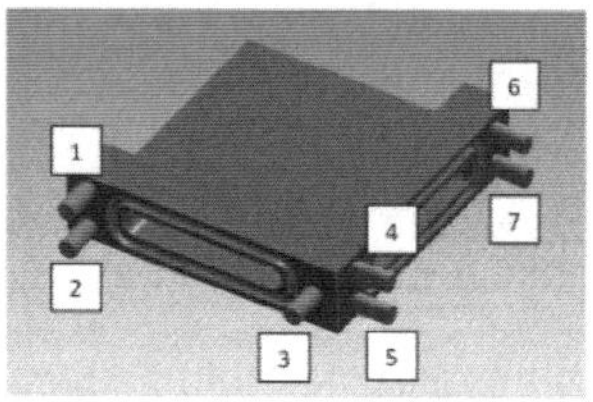

Figure 5: Position of different bolt numbers in the flange.

NASA Technical Memorandum document [2] presents different bolt configurations and the method to calculate bolt and joint stiffness of each configuration. Since there are no nuts used in this design, some of the mentioned configurations are well matched with the bolt types used in this part. Using the methods presented in this document, the stiffness of bolt and joint, K_B and K_J are calculated for both bolt types. Furthermore, the ratio of joint to bolt stiffness was calculated for both cases. These ratios are 3.93 and 4.68 for the hex head and countersunk flat head bolts respectively. Using the load obtained from FEA on each bolt, the margin of safety for each bolt was calculated [3]. The values are presented in Fig. 6. Evaluation of the bolt loads in the joints indicated sufficient margin.

Bolt Type	Bolt Number	MS
Countersunk Head	1	6.018389155
	2	4.925776095
	3	4.962593022
Hex Head	4	6.083427462
	5	7.211567816
	6	4.595368972
	7	5.899342086

Figure 5: MS values for all the bolt types and bolt numbers.

CONCLUSIONS

In summary, due to the growth of design and manufacturing industries, it is sometimes favorable for the companies to replace the current in production designs with more effective novel design alternatives. This research proposed a novel design for a heat exchanger of an aircraft's environmental controller to replace the current design and maintain competitive in the aerospace industry. The objective of this design proposal was to reduce the end unit cost and overall part weight, and improve the safety margins.

Seven alternatives to the current design were proposed and ranked based on six criteria using a decision matrix. One of the alternatives was selected and studied. By performing FEA on the CAD models of the proposed design and verifying the FEA results with the hand calculations, areas of high stress were identified and modified. The dimensions were adjusted to obtain optimal weight and safety margins.

The proposed design with its original dimensions was analyzed for the given loads. Consequently, areas of negative margin of safety were identified and the design was further modified. The final optimized design improved the margin of safety for the maximum operating pressure from 0.041 to 0.580 and for the burst pressure from -0.297 to 0.067. Additionally, the optimized design weighs less by 0.026 kg compared to the current design. This is considered to be a significant weight reduction value in aerospace applications for parts with similar weight and size.

The novel design significantly reduced the required machining time resulting in significant cost take-out in manufacturing. Furthermore, it reduced the overall amount of raw materials resulting in significant cost saving.

ACKNOWLEDGMENTS

The authors would like to acknowledge the funding and support provided by the University of Toronto Institute for Multidisciplinary Design and Innovation (UT-IMDI), NSERC, OCE, and Honeywell Aerospace Canada for the completion of this work. They further extend their gratitude to the industry advisors Mark Phillips and Vahe Gharakhanian at Honeywell Aerospace Canada for their continued support and cooperation through various stages of this project.

REFERENCES

1. N. T. Standard, NASA-STD-5020, *Requirements For Threaded Fastening Systems In Space-flight Hardware*. Washington, DC: National Aeronautics and Space Administration, 2012.

2. J. A. Chambers, "Preloaded joint analysis methodology for space flight systems," *NASA Technical Memorandum*, no. 106943, pp. 13–15, 1995.

3. R. C. Rice, D. of Transportation, F. Administration, and O. of Aviation Research, *Metallic Materials Properties Development and Standardization (MMPDS)*. Washington, DC: U.S. Department of Transportation, Federal Aviation Administration, 2003.

REDESIGN OF A DC-LINK CAPACITOR FOR AN AIRCRAFT POWER AND THERMAL MANAGEMENT CONTROLLER

Mohammad Anisur Rahman And Negin Shahbazian
Mechanical & Industrial Engineering
University of Toronto
Toronto, Ontario, Canada
anis.rahman@mail.utoronto.ca ,
negin.shahbazian@mail.utoronto.ca

Professor Kamran Behdinan
Mechanical & Industrial Engineering
University of Toronto
Toronto, Ontario, Canada
behdinan@mie.utoronto.ca

ABSTRACT

DC –Link Capacitor used in power converters as an energy buffer is one of the key components of aircraft Power and Thermal Management Controller (PTMC). Several factors in order to reduce the cost of the PTMC have been identified, including the integration of the DC-link capacitors, DC bus bars and Snubber caps. The key factor of this redesigned DC-link capacitor is to justify its structural functionality subjected to random vibration loading.

In this research, Pro/Engineer, commercial Computer Aided Design (CAD) software, is utilized to develop the CAD model of redesigning DC-link capacitor integrated with DC bus bars and Snubber caps. Finite Element Method (FEM) using ANSYS is used to predict the life cycle of the structure (redesign DC link capacitor integrated with DC bus bar and Snubber caps) subjected to random loading.

KEYWORDS: DC-Link Capacitor, PTMC, Fatigue Life, FEM, ANSYS, Pro/Engineer

INTRODUCTION

Aircraft Power and Thermal Management Controller (PTMC) is an integrated system, which in combination with the aircraft vehicle management system (VMS), performs aircraft functions traditionally reserved for the auxiliary power unit (APU), the environmental control system (ECS) and the backup power system.

In order for the Canadian aerospace industry to maintain its contribution to the country's economy, Canadian aerospace companies must remain competitive in the global aerospace market and constantly maintain their high quality, reliable and first to market products.

The PTMC consists of three major components, the chassis or housing, the High Power Switch Module (HPSM) and the control electronics which consist of multiple circuit card assemblies (CCA's).

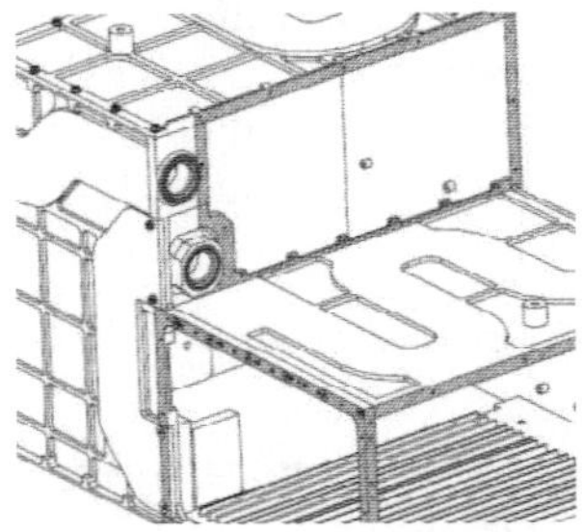

Figure 1: Housing of PTMC

Among the three major components of PTMC, HPSM is the most costly and challenging component due to its volume, weight and assembly into the housing of PTMC. The authors reached to reduce the cost of PTMC by reducing cost of HPSM.

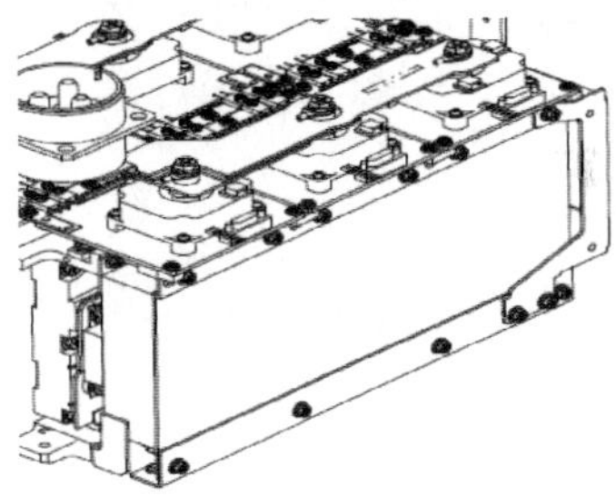

Figure 2: High Power Switch Module (HPSM)

This research primarily focused on reducing the cost of the DC-link capacitors, bus bars and Snubber caps which are the parts of HPSM. Several opportunities for cost takeout had been identified, including the integration of some of the aforementioned subassemblies. The secondary objective of the research was to reduce the volume of the HPSM. This makes it easier to assemble in the chassis and reduce end unit labor cost.

In this research, numerical method was used to predict the life cycle of the structure (redesign DC link capacitor integrated with DC bus bar and Snubber caps) due to random loading.

Finite Element based tools for fatigue life prediction are now widely available. The basic aim of such tools is to enable fatigue life calculations to be done at the design stage of a development process. Nearly all structures or components have traditionally been designed using time based structural and fatigue analysis methods [1]. However, by developing a frequency based fatigue analysis approach [2], the true composition of the random stress or strain responses can be retained within a much optimized fatigue design process.

In this research, frequency based fatigue technique was used because this can yield many advantages, the most important being, (i) an improved understanding of system behavior, (ii) the capability to fully include the true structural behavior rather than a potentially inadequate simplified version and (iii) a more computationally efficient fatigue analysis procedure.

METHODOLOGY

The first step of this research was to develop part models of the various components of proposed DC link capacitor, DC bus bar and Snubber caps. After developing all part models, an assembly model of High Power Switch Module (HPSM), which consists of two modified DC link capacitors, four modified DC bus bars, six modified Snubber caps, one existing Heat Exchanger and six existing Integrated Bipolar Gate Drivers, was developed by using Pro/Engineer.

Figure 3: Existing models of the DC-Link Capacitor, Snubber, mounting bracket and DC Bus bar

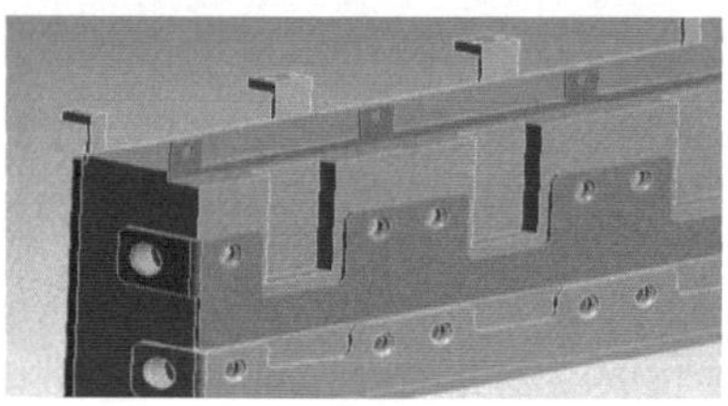

Figure 4: Proposed CAD model of the DC-link capacitor integrated with DC bus bars and Snubber caps

The second step of this research was to generate the mesh (a discretized geometry which is used for calculation) of the assembly model of HPSM which was used for the numerical simulation. The Finite Element Method (FEM), which is a computational simulation technique, particularly good at solving complex structural problems, was used. The idea was to discretize the assembly model (HPSM) into regions, each of which has many smaller elements and nodes (intersections of mesh elements). The material properties for each component into the assembly model were defined in terms of elasticity and density and then the theoretical vibrational behavior of the assembly model was solved through matrix manipulations. This allowed the primary Eigen frequencies of a structure to be determined, as well as the visualization of the mode shapes [3]. After modal analysis, the input Power Spectral Density (PSD) and Damping ratio were specified to calculate the von-Mises stresses at critical region of every component of redesigning DC link capacitor integrated with DC bus bar and Snubber caps.

The commercial software package ANSYS was used for mesh generation, obtaining natural frequencies and stresses (von-Misses). In this research, weakest component was identified on the basis of computing von-Mises stresses and Cumulative Damage Index (CDI) of that component was also calculated. On the basis of calculating CDI, life cycle of redesign DC link capacitor integrated with the DC bus bar and Snubber caps was predicted.

FINITE ELEMENT MODEL (FEM) OF HPSM

The analysis was performed using the Finite Element (FE) Method. The FE program, ANSYS, Rev. 15.0, was used to solve and post-process the analysis. The FE model was generated using the ANSYS Workbench module by importing solid models from Pro/Engineer, a CAD/CAM, feature based solid modeling program. The CAD model was simplified for the purpose of analysis. Following were the assumptions of the HPSM assembly model:

The DC-Link Capacitor integrated with DC bus bars and Snubber caps, peripheral heat sinks and other components, where applicable, were modeled with solid elements. The CCAs were modeled with shell elements representing the printed wiring board (PWB) with their nominal thickness.

For dynamic analysis, modal damping ratio of 5% was used for all components of the model. The damping ratio was specified as a material parameter in the ANSYS software.

A web (or spoke) made of beam elements was connected to the fastener holes. The center of the web of the corresponding fastener holes was connected by a beam element representing the fastener. In the FE model, the fastener is considered for the purpose of load transfer between components and therefore is characterized to be rigid and of negligible mass.

In general, where necessary, the densities of structural components were adjusted to reflect the measured or estimated design weight. The total mass of DC-Link

Capacitor integrated with DC bus bars and Snubber caps was calculated by using the ANSYS FE model.

MESH OF THE MODEL

The mesh quality and size were chosen to be uniform throughout the model with Tet10 (Tetrahedral), Hex20 (Hexahedral) and Wed15 elements. Figure 5 shows the mesh model of HPSM.

Figure 5: Mesh Model of the HPSM Assembly

The figure 6 shows that the natural frequency of the first mode is not too much sensitive with the number of elements of the model. In this research, a fairly large FE model having 138315 elements and 282080 nodes was used which necessitated using a powerful computing machine to obtain the desired solution in a reasonable amount of time.

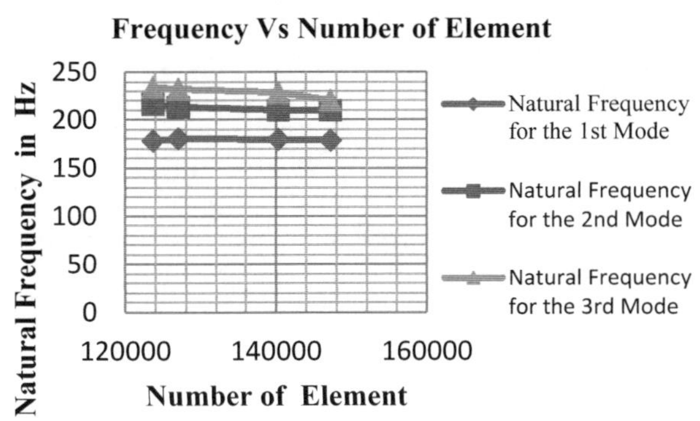

Figure 6: Natural Frequency Vs Number of Element Curve

FATIGUE ANALYSIS

Analysis for random load is a two-step process. In the first step, a modal (Eigenvalue) analysis was performed to obtain natural frequencies and modes of the structure. Table 1 presents the first six natural frequencies of the structure.

In the second step, the input PSD (power spectrum density) spectra (given by the customer) and damping ratio were specified and a spectrum analysis was performed.

Number of Mode	Frequency (Hz)
1	178.79
2	214.77
3	232.76
4	257.18
5	275.18
6	277.16

Table 1: First six natural frequencies of the HPSM assembly model

In this analysis, the duration for the random vibration test representing one life of the controller was 8 hours (given by the customer) along each axis. Analysis was performed in each orthogonal direction. Numbers of service cycles at various levels were calculated by using the following equations:

$$M = f \times t \times C \quad (1)$$

Where

f = frequency (Hz , cycles per second)
t = test duration (hours)
C = conversion factor (3600 seconds/hour)

To obtain the number of cycles at the 1σ, 2σ and 3σ stress levels, the probability associated with each level was multiplied by the total number of cycles (M).

$$\text{Number of service cycle at } 1\sigma \text{ level}, n_1 = M \times 0.683 \quad (2)$$
$$\text{Number of service cycle at } 2\sigma \text{ level}, n_2 = M \times 0.271 \quad (3)$$
$$\text{Number of service cycle at } 3\sigma \text{ level}, n_3 = M \times 0.0433 \quad (4)$$

Table 2 represents the number of the service cycle (n) at various stress levels in the natural frequency of the first mode.

	Stress level: σ_1	Stress level : σ_2	Stress level : σ_3
Probability	0.683	0.271	0.0433
Duration(hrs), t	8	8	8
Factor (Sec/hr), C	3600	3600	3600
Natural Frequency (Hz), f	178.79	178.79	178.79
No. of Service Cycle (n)	3516871	1395420	222958

Table 2: No. of Service Cycle (n) at Various Levels

Computed von-Mises stresses of the weakest component of the model due to the force along x-direction shows in figure 7.

Figure 7: Max. Von-Mises stress (3992.2 psi) on weakest component of the model

Equivalent Von-Mises stresses for fatigue were calculated after getting maximum Von-Mises stress (3992.2 psi) for the weakest component. This was done by simply multiplying the maximum Von-Mises stresses with corresponding casting factor, fitting factor and stress level factor. By using S-N data for Al 5052, numbers of cycles, which could cause failure, were obtained corresponding to equivalent Von-Mises stresses for fatigue at various stress levels.

In this research, Palmgren-Miner rule [4] was used to calculate the damage fraction at any stress level σ_i which is linearly proportional to the ratio of the number of cycles of operation to the total number of cycles that produces failure at that stress level, that is

$$D_i = \frac{n_i}{N_i} \quad (5)$$

Then, a total damage can be defined as the sum of all the fractional damages over a total of k blocks,

$$D = \sum_{i=1}^{k} \frac{n_i}{N_i} \quad (6)$$

Cumulative Damage Index in X direction is shown in Table 3.

	Stress level: σ_1	Stress level: σ_2	Stress level: σ_3
Equivalent Von Mises Stress for fatigue (SEQVf) (psi)	4591.03	9182.06	13773.09
Number of cycles which could cause failure, N (from S-N curve based on SEQVf)	5.00E+08	5.00E+08	5.00E+08
Number of Service cycles, n	3516871	1395420	222958
Damage Index (DI) = n/N	0.0070	0.0028	0.0004
Cumulative Damage Index (CDI), $D = \sum_{i=1}^{3} \frac{n_i}{N_i}$	0.01		

Table 3: Cumulative Damage Index (0.01) at X direction ≤ 1

Similarly, Cumulative Damage Indices of the weakest component along other orthogonal directions were also calculated.

CONCLUSION

In this analysis, the weakest component of the modified DC-link Capacitors was identified and Cumulative Damage Index (CDI) at all orthogonal directions for that component was calculated. Calculated CDI along the X, Y and Z directions were respectively 0.01, 0.01 and 0.01. Based on CDI, it can be concluded that modified DC-link Capacitor will survive at least 2 life cycles per customer requirement. Therefore, the structural functionality of the redesigned DC-link capacitor integrated with DC bus bar and Snubber caps due to random loading is satisfactory.

ACKNOWLEDGMENTS

The authors would like to acknowledge the funding and support provided by the University of Toronto Institute for Multidisciplinary Design and Innovation (UT-IMDI), NSERC, OCE, and Honeywell Aerospace Canada for the completion of this work. They further extend their gratitude to the industry advisors Mark Phillips and Vahe Gharakhanian at Honeywell Aerospace Canada for their continued support and cooperation through various stages of this project.

REFERENCES

1. Da Yu , Abdullah Al-Yafawi, Tung T. Nguyen, Seungbae Park and Soonwan Chung, A. 2011, "High-cycle fatigue life prediction for Pb-free BGA under random vibration loading", Microelectron Reliab, Vol 51, pp. 649–656.

2. N.W.M. Bishop and Hu Zhihua, A. 1991, The Fatigue Analysis of wind Turbine Blades Using Frequency Domain Techniques, European Wind Energy Conference, EWEC 91, Amsterdam, pp. 246-250.

3. Bo Yuan, Limei Xu, Jianfeng Ren, Dagui Huang, A. 2008 "Vibration Suppression Optimization of Electronic Apparatus Rack using Design of Experiment Method" Asia International Symposium on Mechatronics, pp. 27-31.

4. M. A. Miner, A.1945, "Cumulative Damage in Fatigue", Transactions of the ASME, Journal of Applied Mechanics, vol. 67, pp. 159.

A COUPLED DYNAMIC SIMULATION OF HYDRAULIC FRACTURING

Matin Parchei-Esfahani
Civil and Environmental Engineering
University of Waterloo
Waterloo, ON, Canada
mparchei@uwaterloo.ca

Robert Gracie
Civil and Environmental Engineering
University of Waterloo
Waterloo, ON, Canada
rgracie@uwaterloo.ca

ABSTRACT

A two-dimensional dynamic model is presented to simulate the Hydraulic Fracturing (HF) process using the eXtended Finite Element Method (XFEM). The fracture is assumed to be induced hydraulically in a homogenous and isotropic elastic reservoir by an incompressible Newtonian fluid. Darcy's law is assumed to be valid for the flow within the fracture network; hence, the Navier-Stokes equations are further reduced to the Reynolds lubrication equations. Such assumptions for the flow are valid for a range of shale gas and oil reservoirs. An explicit dynamic XFEM model is used to simulate the fractured rock formation, and the fluid domain is discretized using a one-dimensional Eulerian finite element mesh. A sequential coupling strategy is employed for the hydro-mechanical interaction between the fractured reservoir and the injected fluid.

KEYWORDS: hydraulic fracturing, Reynolds lubrication equation, Extended Finite Element Method, fluid-solid interaction

INTRODUCTION

Hydraulic fracturing is the most common technique for reservoir stimulation in oil and gas industry. This technique has also become prevalent for fracturing impermeable rock masses to dispose wastes or sequestrate carbon dioxide, geothermal heat production, and measuring in situ stresses in geomechanical studies.

In its most simplified form, hydraulic fracturing involves coupling of at least three different physical processes: 1) the injection and flow of a fluid into/through a solid rock mass; 2) the deformation of the solid matrix; and 3) and the propagation of fractures [1]. Such an idealized model can be utilized for real time design optimization as the HF process goes on and as updated information comes from monitoring and measurements. The current model and implementation builds upon two dimensional mathematical models are developed by Barani et al. [2] and Khoei et al. [3] in which quasi-static and dynamic propagation of a hydraulically induced fracture is presented in a partially saturated porous medium. These early models, however, have mainly considered crack propagation in porous media rather than hydraulically induced fractures.

Fully coupled models have also been of interest. Studies by de Borst et al [4], and Mohammadnejad and Khoei [5, 6], considered fully coupled simulations for hydraulic fracture propagation in porous media. The former developed a quasi-static model while the latter considered dynamic crack propagation. Comprehensive studies on different coupling strategies have been published by Gordeliy and Peirce [7, 8, 9] and Kim et al. [10, 11]. Stability and convergence rate of different coupling schemes are discussed in details in these studies. Three-dimensional models for non-planar hydraulic fractures are also available in the literature [12].

In this article, we present a dynamic model to take into account the coupling of the three aforementioned processes. Dynamic fracture propagation is modeled using the XFEM, which takes advantage of local enrichment functions to represent fracture discontinuities. We also utilized a sequential coupling strategy in which the coupled system of nonlinear equations is decoupled at each time step and solved independently for a single coupling parameter based on the estimated values for the other parameters. This strategy substantially reduces computational costs compared to a fully coupled solution as there is no need for inverting huge asymmetric matrices.

In the following sections, we first introduce the mathematical representation of the physical problem. The development of the discrete equations is presented in details in the second section. The third section provides a short discussion on the coupling strategy, and the paper concludes by presenting a numerical example to illustrate validity of the model.

PHYSICAL MODEL

Consider a homogenous and isotropic elastic rock formation. Assume also that the solid matrix is of negligible permeability and that the fluid-formation interaction is exclusively along the fracture boundaries. Thus, no leak-off flow is considered from the fracture network into the rock matrix, and interaction forces are only applied on crack faces. Moreover, the process is assumed to take place under isothermal conditions so that the variation of temperature-dependent properties such as bulk density and fluid viscosity of the formation and the injected fluid can also be considered negligible.

With the above assumptions, the linear momentum balance for the solid phase of the model is given as

$$\nabla \cdot \boldsymbol{\sigma}_s + \rho_s \mathbf{b} = \rho_s \ddot{\mathbf{u}} \tag{1}$$

where $\boldsymbol{\sigma}$ is the Cauchy stress tensor, $\mathbf{b}$ is the body force per unit mass, ρ_s is the mass density, and $\ddot{\mathbf{u}}$ is the solid matrix acceleration vector. Subscript s indicates values corresponding to the solid matrix.

Assuming small deformations and brittle fracture conditions, the rock formation can be considered to behave linearly elastic according to Hooke's law

$$\boldsymbol{\sigma}_s = \mathbf{C} : \boldsymbol{\varepsilon}_s(\mathbf{u}) - \boldsymbol{\sigma}_0 \tag{2}$$

where $\mathbf{u}$ is the displacement vector, $\boldsymbol{\sigma}_0$ is the state of confining stresses, $\mathbf{C}$ is the fourth order elasticity tensor, and $\boldsymbol{\varepsilon}_s(\mathbf{u})$ is the symmetric part of the deformation gradient tensor, i.e.,

$$\boldsymbol{\varepsilon}_s(\mathbf{u}) = \frac{1}{2}\left(\nabla \mathbf{u} + \nabla \mathbf{u}^{\mathrm{T}}\right) \tag{3}$$

The governing equation for fluid flow within the fracture network is the Reynolds equation, which is valid provided that the fluid flow is linear, laminar, and planar, and the fracture aperture is small compared to the fracture length and the fracture walls are smooth (low relative roughness). The Reynolds equation for one-dimensional flow along a straight fracture is

$$\dot{w} = \frac{1}{12\mu}\frac{\partial}{\partial s}\left(w^3 \frac{\partial p}{\partial s}\right) + Q(t)\delta(s) - \frac{2C_L}{\sqrt{t - t_0(s)}} - 2S_0\delta(t - t_0(s)) \tag{4}$$

where s is the coordinate defined along the fracture/flow path, p is the fluid pressure, w and $\dot{w}$ indicate the fracture aperture, and its derivative with respect to time (i.e., the fracture opening velocity), μ is the Newtonian fluid viscosity, $Q(t)$ is the source injection rate, C_L is leak-off coefficient with dimension $[LT^{-1/2}]$, $\delta(.)$ is the Dirac delta function, t_0 represents the leak-off initiation delay factor, and S_0 is the spurt which represents a sink surface or the initial leak-off from the fracture faces before complete formation of the filter cake at fracture walls [1].

The third and the forth terms on the right hand side of (4) represent the leak-off flow from the fracture to the solid matrix. Considering the assumption of negligible permeability for the rock formation in our model, there would be no leak-off flow to the solid matrix, and therefore, the last two terms of (4) vanish. Equation (4) then reduces to

$$\dot{w} = \frac{1}{12\mu}\frac{\partial}{\partial s}\left(w^3 \frac{\partial p}{\partial s}\right) + Q(t)\delta(s) \tag{5}$$

which is the governing equation for the one-dimensional fluid flow in a hydraulically induced fracture with no leak-off from the fracture surfaces.

Now that the governing equations for the solid and fluid phase of the model have been developed, we need to discuss the solution strategy. Equations (1) and (5) must be solved simultaneously for the fluid pressure, p, and fracture aperture, w. The effect of fluid pressure on the solid matrix is taken into account as a natural boundary condition on the fracture surfaces.

DISCRETE MODEL

A schematic of the solid matrix domain Ω is depicted in Fig. 1. Ω is bounded by boundary $\Gamma = \Gamma_t \cup \Gamma_c \cup \Gamma_u$ such that $(\Gamma_t \cup \Gamma_c) \cap \Gamma_u = \phi$, where Γ_t is the portion of the external boundary where traction are applied, Γ_c is the internal boundary (crack surfaces) where fluid pressure is applied, and Γ_u is the portion of the boundary where displacements are prescribed. The variational form of (1) over domain Ω is

$$\int_\Omega \boldsymbol{\varepsilon}_s(\mathbf{u}) : \mathbf{C} : \delta\boldsymbol{\varepsilon}_s(\mathbf{u}) d\Omega - \int_\Omega \boldsymbol{\sigma}_0 : \delta\boldsymbol{\varepsilon}_s(\mathbf{u}) d\Omega + \int_\Omega \rho_s \ddot{\mathbf{u}} \cdot \delta\mathbf{u} d\Omega = \int_\Omega \rho_s \mathbf{b} \cdot \delta\mathbf{u} d\Omega + \int_{\Gamma_t} \mathbf{t}_t \cdot \delta\mathbf{u} d\Gamma + \int_{\Gamma_c} \mathbf{t}_c \cdot \delta\mathbf{u} d\Gamma \tag{6}$$

where $\mathbf{t}_t$ and $\mathbf{t}_c$ are external tractions applied on Γ_t and Γ_c respectively, and δ representing a small admissible variation of the field of interest.

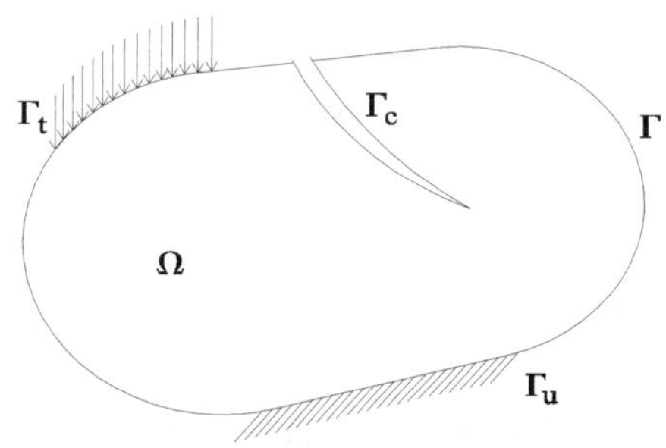

Figure 1: schematic of the domain of the problem

We use an enriched XFEM approximation for each component of the displacement field

$$u = \sum_{j=1}^{n} N_j \left(u_j + \sum_{k=1}^{m} \Psi_k a_{jk}\right) \tag{7}$$

where u is a component of the displacement field $\mathbf{u}$, u_j are the nodal displacement components, a_{jk} are the enrichment degrees of freedom, N_j represent the standard finite element shape functions and Ψ_k are the XFEM enrichment functions. Substituting (7) into the variational formulation presented in (6), the discrete form of the governing equation can be derived

$$\mathbf{M}_s \ddot{\mathbf{u}} + \mathbf{K}_s \mathbf{u} - \mathbf{f}_0^{\mathrm{int}} = \mathbf{f}_t^{\mathrm{ext}} + \mathbf{f}_c^{\mathrm{pres}} \tag{8}$$

where $\mathbf{M}_s$ and $\mathbf{K}_s$ are the mass and stiffness matrices of the solid matrix, respectively. $\mathbf{f}_0^{\mathrm{int}}$ is the internal force vector due to the initial stress state in the solid matrix, $\mathbf{f}_t^{\mathrm{ext}}$ is the external force vector due to external traction applied on the traction boundary Γ_t and body forces, and $\mathbf{f}_c^{\mathrm{pres}}$ is due to the fluid pressure acting on the fracture surfaces. For a detailed review of the enrichment functions and discrete equations, the reader is referred to [13].

The variational form (5) over the flow domain (i.e. the crack boundary Γ_c) is

$$\int_{\Gamma_c} \frac{\partial p}{\partial s}\cdot\frac{w^3}{12\mu}\delta\left(\frac{\partial p}{\partial s}\right)d\Gamma = \int_{\Gamma_c}\frac{\partial w}{\partial t}\delta p d\Gamma + \int_{\Gamma_c} Q(t)\delta(s)\delta p d\Gamma + \bar{q}_w\delta p \qquad (9)$$

where $\bar{q}_w$ is the injection flux from the wellbore, and δ is an admissible variation of the field of interest. The finite element approximation for the pressure field along the fracture is taken as

$$p = \sum_{i=1}^{n} N_i p_i \qquad (10)$$

in which p_i are the nodal values of fluid pressure, and N_i are linear finite element shape functions. Substituting (10) into (9) yields the discrete equation for the fluid

$$\mathbf{K}_f\mathbf{p} - \mathbf{f}_w = \mathbf{f}_Q + \mathbf{f}_q \qquad (11)$$

where $\mathbf{p}$ is the fluid pressure, $\mathbf{K}_f$ is the conductivity matrix, $\mathbf{f}_w$ is the change in flow rate corresponding to the variation of fracture aperture, $\mathbf{f}_Q$ is the inflow rate due to point source Q, and $\mathbf{f}_q$ is the inflow rate due to injection from the wellbore. Eqs. (8) and (11) are solved based on the coupling strategy discussed in the following section.

COUPLING STRATEGY

The choice of the coupling strategy is a crucial step in solving the set of discrete equations with coupled unknown parameters. While a fully-coupled strategy may lead to a more accurate solution, the higher computational costs of such a method can make it inapplicable especially to problems dealing with large domains or numerous coupling parameters. A detailed discussion on different coupling strategies, the accuracy and the efficiency of each method is presented in [7-11].

In this study, we employed a sequential coupling strategy to solve the set of discrete equations (8) and (11). Assuming an initial pressure estimation, the crack aperture is calculated from (8) using an explicit central difference time integration scheme. The resulting displacements are then applied as an initial estimation in (11) to solve for the fluid pressure in the next time step. It is shown in [9] that this coupling strategy converges faster and is more accurate compared to the method starting with an initial fracture.

NUMERICAL EXAMPLE

The eXtended Leak-Off Test (XLOT) is a reliable technique for measuring in-situ stresses with the assistance of hydraulic fracturing. Fluid is injected to an isolated section of the wellbore with a constant rate to open a natural fracture until it becomes unstable. This pressure is known as break-down pressure which propagates the fracture to some distance away from the wellbore and the pressure remains almost constant at this stage. For more details about the XLOT technique the reader is referred to [14].

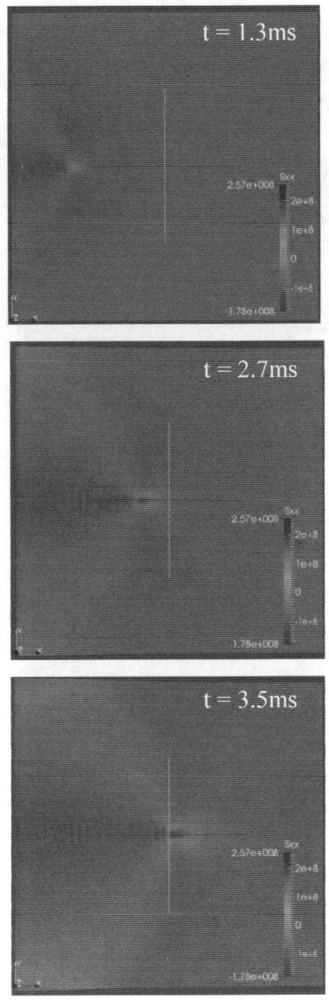

Figure 2: σ_{xx} at different time steps

In order to simulate the unstable crack propagation during the break-down stage of an XLOT, we assume that a vertical natural fracture with length $L_0 = 0.7m$ exists at the edge of a homogeneous and isotropic $10m$ by $10m$ domain which represents the formation with physical properties

$\rho_s = 2800 kg/m^3$, $E = 20GPa$, and $\nu = 0.35$. We also assume that at the instant of instability, bottom hole pressure is $p_0 = 80MPa$ and the crack propagates with a speed close to the limit fracture speed which is the Reighley wave speed, c_R. Let crack propagation speed be $v_f = 0.95c_R \approx 0.87c_s$. where c_s is the shear wave speed. Given the values of E and ν for the formation, the shear wave speed is calculated as $c_s = 1.62km/s$, and therefore, fracture propagation speed becomes $v_{cr} = 1.42km/s$. Fig. (2) and Fig. (3) show the changes in the stress field around the crack on a plane normal to the fracture plane at different time steps after the crack becomes instable.

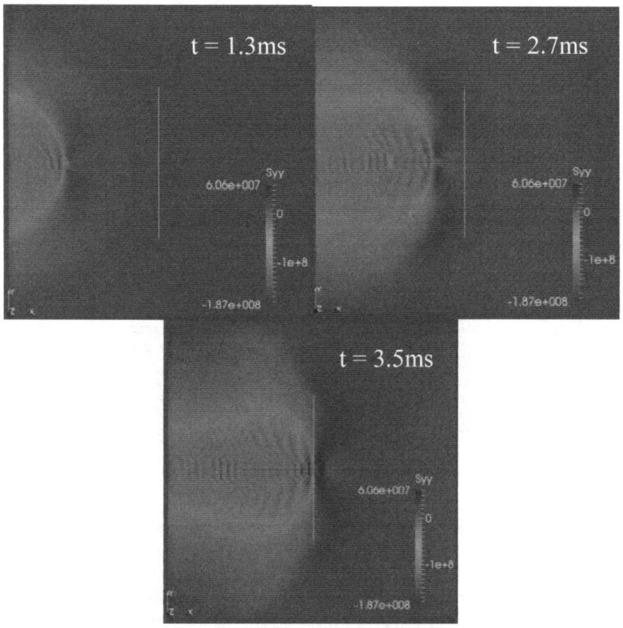

Figure 3: σ_{yy} at different time steps

CONCLUSIONS

An explicit XFEM model is successfully implemented to model dynamic propagation of hydraulic fractures. The sequential coupling strategy provides an acceptable rate of convergence, and sufficient accuracy for the results. The model is capable of being applied to more complex hydraulic fracturing problems which is the subject of an ongoing research.

REFERENCES

1. Adachi, J., Siebrits, E., Peirce, A. & Desroches, J. 2007, "Computer simulation of hydraulic fractures", International Journal of Rock Mechanics and Mining Sciences, vol. 44, no. 5, pp. 739-757.
2. Barani, O.R., Khoei, A.R. & Mofid, M. 2011, "Modeling of cohesive crack growth in partially saturated porous media; A study on the permeability of cohesive fracture", International Journal of Fracture, vol. 167, no. 1, pp. 15-31.
3. Khoei, A.R., Barani, O.R. & Mofid, M. 2011, "Modeling of dynamic cohesive fracture propagation in porous saturated media", International Journal for Numerical and Analytical Methods in Geomechanics, vol. 35, no. 10, pp. 1160-1184.
4. De Borst, R., Réthoré, J. & Abellan, M.-. 2006, "A numerical approach for arbitrary cracks in a fluid-saturated medium", Archive of Applied Mechanics, vol. 75, no. 10-12, pp. 595-606.
5. Mohammadnejad, T. & Khoei, A.R. 2013, "Hydro-mechanical modeling of cohesive crack propagation in multiphase porous media using the extended finite element method", International Journal for Numerical and Analytical Methods in Geomechanics, vol. 37, no. 10, pp. 1247-1279.
6. Mohammadnejad, T. & Khoei, A.R. 2013, "An extended finite element method for hydraulic fracture propagation in deformable porous media with the cohesive crack model", Finite Elements in Analysis and Design, vol. 73, pp. 77-95.
7. Gordeliy, E. & Peirce, A. 2013, "Coupling schemes for modeling hydraulic fracture propagation using the XFEM", Computer Methods in Applied Mechanics and Engineering, vol. 253, pp. 305-322.
8. Gordeliy, E. & Peirce, A. 2013, "Implicit level set schemes for modeling hydraulic fractures using the XFEM", Computer Methods in Applied Mechanics and Engineering, vol. 266, pp. 125-143.
9. Gordeliy, E. & Peirce, A. 2015, "Enrichment strategies and convergence properties of the XFEM for hydraulic fracture problems", Computer Methods in Applied Mechanics and Engineering, vol. 283, pp. 474-502.
10. Kim, J., Tchelepi, H.A. & Juanes, R. 2011, "Stability and convergence of sequential methods for coupled flow and geomechanics: Fixed-stress and fixed-strain splits", Computer Methods in Applied Mechanics and Engineering, vol. 200, no. 13-16, pp. 1591-1606.
11. Kim, J., Tchelepi, H.A. & Juanes, R. 2011, "Stability and convergence of sequential methods for coupled flow and geomechanics: Drained and undrained splits", Computer Methods in Applied Mechanics and Engineering, vol. 200, no. 23-24, pp. 2094-2116.
12. Gupta, P. & Duarte, C.A. 2014, "Simulation of non-planar three-dimensional hydraulic fracture propagation", International Journal for Numerical and Analytical Methods in Geomechanics, vol. 38, no. 13, pp. 1397-1430.
13. Moes, N., Dolbow, J., & Belytschko, T. 1999, " A finite element method for crack growth without remeshing", International Journal for Numerical Methods in Engineering, vol. 46, pp. 131-150.
14. Kunze, K.R. & Steiger, R.P. 1992, "Accurate in-situ stress measurements during drilling operations", *Proceedings - SPE Annual Technical Conference and Exhibition*, pp. 491.

COMPUTATIONAL MODELING AND CALIBRATION OF A 3D METAL FOAM HEAT SINK

Shahzada P. Aly
Mechanical Department
KFUPM
Dhahran, Saudi Arabia
pamir254@homail.com

Khaled S. Al-Athel
Mechanical Department
KFUPM
Dhahran, Saudi Arabia
kathel@kfupm.edu.sa

Abul Fazal M. Arif
Mechanical Department
KFUPM
Dhahran, Saudi Arabia
afmarif@kfupm.edu.sa

Javad Mostaghimi
Mechanical and Industrial Engineering
University of Toronto
Toronto, Ontario, Canada
mostag@mie.utoronto.ca

ABSTRACT

Metal foams are those porous materials that have a very high percentage of void spaces at macroscopic level. Depending on the manufacturing technique they are either produced as open or closed cells. Metallic foams have taken over significant amount of applications due to their improved thermal and mechanical properties. Due to their complex geometry, as an easy alternative, researchers usually prefer computational models based on common techniques. An innovative approach has been undertaken to capture the three-dimensional (3D) optimized finite element model of open-cell metal foams. Computed tomography (CT) has been utilized to extract extreme precise 3D surface model and Computer Aided Design (CAD) software have been employed to "clean" the surface model for finite element (FE) analysis. The viability of this model has been demonstrated here as their use in heat sinks and is calibrated against experimental results. The validated and calibrated model is then used for assessing different scenarios via simulations.

INTRODUCTION

Metal foams promise attractive applications as their thermal and mechanical properties can be controlled by altering three main factors during their production; base metal, foaming process and heat treatment conditions [1]. They mostly retain the physical properties of their parent metal. The two parameters, porosity (percentage) and pore density (PPC-pores per cm or PPI-pores per inch) are generally used to describe metal foams. Thus a lot of research has been carried out to find their thermo-mechanical behavior solely relying on these two parameters [2]. Thermal conductivity of metal foams is also affected by the porosity and material type. Improved manufacturing techniques in the recent decade have enhanced the versatile nature of open-cells, such as their high thermal conductivity, lower densities, higher surface area to volume fraction and the astonishing ability to readily mix with the flowing fluid makes them a splendid candidate for thermal applications like heat exchangers, fuel cells and efficient heat dissipaters in electronic cooling [3], [4]. Metal foams thus increase the heat transfer rate but at an expense on additional pressure drop in the flow stream [5]. But the trade-off of this drawback is reduced size of heat exchangers and heat sinks.

Due to foams' highly random structure, scientists have only been able to approximate their physical behavior [6], [7]. Thus exact real life models are needed to predict the true thermo-mechanical behavior as simplified models cannot do justice with the accuracy of the results. Initially to model the 3D structure of metal foams, researchers worked on the concept unit cell approach, e.g. Kelvin and Weaire–Phelan (WP) structures [8]. Weaire–Phelan model is superior to Kelvin model, but due to simplicity and reasonable accuracy of Kelvin's, researchers still prefer it for modeling. Others like Sihn *et al*, represented their unit cell by subtracting four spheres from all corners of single tetrahedral structure [9]. Playing with the size of these spheres they varied shape, size and dispersion of pores.

Straatman *et al,* extended the previous mentioned approach by subtraction a sphere from the middle of cuboid, that was a representation of single pore of metal foam [10]. Druma *et al*, found the use of body centered cubic (BCC) structures as a good means of idealizing open cell honeycomb metal foams [11]. This was done by extracting nine same size balls from a cube.

Whereas for closed-cell three dimensional modeling, Voronoi tessellation method and node-bond model is generally used [12]. More or less, variations of these cited unit cells are still in practice for 3D modeling of foams.

Ramirez *et al,* presented and tested a 3D model that was generated using a micro-CT scan [13]. The model showed more accuracy than previous models presented in the literature as it was based on an actual metal foam sample that was scanned using a micro-CT. It is important to note that their sample was smaller than the unit dimension used to define the PPI which might cause some errors if the model is used in large scale problems or applications.

METAL FOAM 3D MODELLING

Highly complex and irregular structure of metal foams, poses a challenge if an accurate 3D model is to be acquired. It's very difficult if not impossible to model it using CAD or FE software. So for such type of structures, reverse engineering techniques like computed tomography (CT) scan can be used to generate 3D models.

Aluminum (6101) alloy with 93% porosity and pore density of 4 pores per centimeter (~10 PPI), obtained from "Goodfellow" has been used for scanning [14]. CT scan was performed on Skyscan 1172 and the sample size was 25x25x6.35 mm. This size was chosen so that minimum of twice the PPC unit length (1 cm) of sample is obtained, both horizontally and vertically. X-ray source was set at 40 kV and 250 μA. By keeping the step size of 0.7° (deg), total 512 slices were captured. Scanning trajectory was complete 360° for thorough capture of sample. As our sample was low density so no filter was used. Total scan time was approximately 15 minutes.

For Skyscan 1172 model, sample size cannot exceed 18 mm in height, whereas our sample is about 25 mm in height and width. Thus a total of 2 scans were performed i.e. one for lower and one for upper half of height. Later these two scans were conjoined together by the software in reconstruction phase.

After the scan these image slices are used to generate a "raw" 3D reconstruction. The word raw here suggests that this 3D model is only good for visual aid, but not for FE analysis. For FE analysis this model needs to be free from any errors in geometry. So initial 3D rendering is done by "Skyscan NRecon" software that comes with this CT scanner model. The setting of parameters at this phase greatly effects the final quality of final 3D model, so great emphasis is placed on an optimized set of variables. Through trial and error, different values of parameters were set for filters to get least noise and geometric errors in cleaned images. Also no density threshold value was given as the interest is in capturing the sample as a whole.

These cleaned images were than used for constructing the 3D model in "Skyscan CTan". This 3D model was saved in .stl format as it provides length units (i.e. mm, cm etc.) of model, thus depicting the true size of sample.

This raw 3D model is generally quite rough and an ample amount of image processing is required before it can be imported to FE tools. So for "cleaning" and "healing", "Geomagic Studio" was used. Geomagic highlights all the susceptible issues with the geometry, which were taken care of by its specialized editing tools. Another very important service rendered by Geomagic is to significantly reduce the number of surfaces, thus decreasing the overall size of model without compromising its accuracy and shape details. This in return significantly reduces the computational resources for FE analysis. The final Geomagic cleaned file is ready for a FE analysis but it is only a surface model. "Solidworks" is used to convert this surface model to volume model.

Figure 1 shows the steps for 3D modeling of metal foam. Numbering in the figure are: **(1)** Aluminum alloy metal foam sample, **(2)** Skyscan NRecon, **(3)** CTvol, **(4)** Geomagic unclean raw geometry, **(5)** Geomagic cleaned raw geometry, **(6)** Geomagic clean 3D surface model and **(7)** SolidWorks 3D volume model.

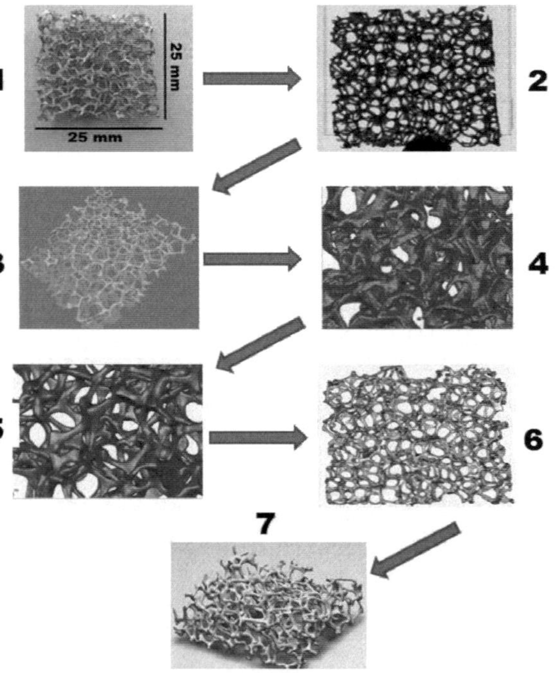

Figure 1: 3D modeling steps for metal foam

METAL FOAM HEAT SINK EXPERIMENT

Parameters already available for simulation are shown in table 1. Metal foams retain these parameters from their parent metal and are the same specified values by the manufacturer. The heat transfer coefficient (*h*), have to be obtained from experiments. In this section, we perform experiments to calibrate the *h* value for the simulations.

Table 1: Heat sink material properties

Material	Aluminum 6101
Density (Kg/m³)	2700
Thermal conductivity (W/m °C)	218
Coef. of thermal expansion (1/°C)	23.58×10^{-6}
Specific heat (J/g °C)	0.895

Figure 2 shows the experimental setup. The single fin heat sink sample was placed inside a wind tunnel with room temperature of 24°C and power input of 1.3W (approx.). For first 60 min temperature was recorded for fan-off conditions (natural convection) and after that for further 30 min, temperature readings were taken for fan-on (forced convection), with wind speed of 1.6 m/s inside tunnel.

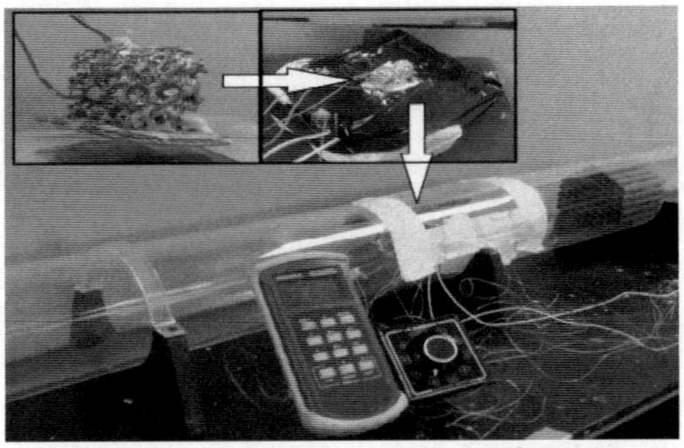

Figure 2: Heat sink experimental setup

Temperature was monitored by T-type thermocouples and "Omega HH506RA" was used for temperature data logging. Temperature variations for heat sink over a span of 90 min are shown in figure 3. As shown, temperature readings were taken of heat source, fin top and fin base.

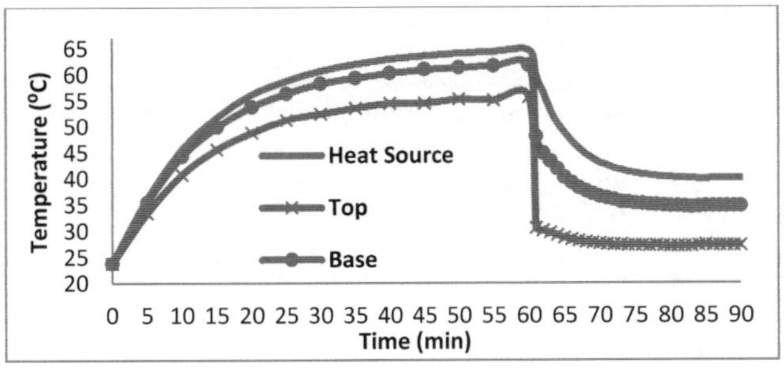

Figure 3: Temperature variation of metal foam heat sink

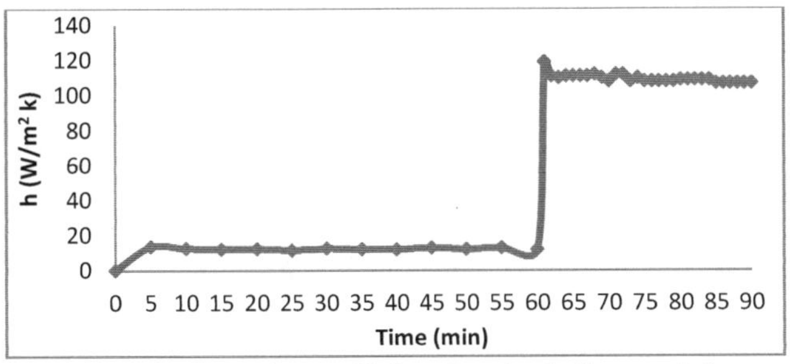

Figure 4: Heat transfer coefficient for single fin metal foam – fan-off and fan-on conditions

From experimental data (Figure 3) we can obtain temperature distribution of heat sink at any time interval. The model is exact replica of the real sample, so base temperature from any instance in Figure 3 was imposed as an input to the base of heat sink model. Thermal analysis was run and 'h' was calibrated such that temperature distribution of simulation matches to that of experiment for that particular instant. This way we acquired the *h* value for that specific moment only. The procedure was repeated for different time intervals of entire time span of fan-off and fan-on phase. Figure 4 shows the calibrated value of *h* as a function of time during fan-on and fan-off. The figure also enlightens the fact that *h* value for fan off is approximately 12.5 W/m^2K and that of fan-on is approximately 109.5 W/m^2K.

FINITE ELEMENT METAL FOAM HEAT SINK SIMULATIONS

After calibrating the 3D model and getting the values of *h*, this model can be used to simulate a variety of cases. To demonstrate this, transient thermal analysis of one and three fins 4 PPC metal foam heat sinks are compared under same conditions. Figures 5 and 6, show temperature distributions of one and three fin 4 PPC heat sink models, after a full cycle of fan-off and fan-on.

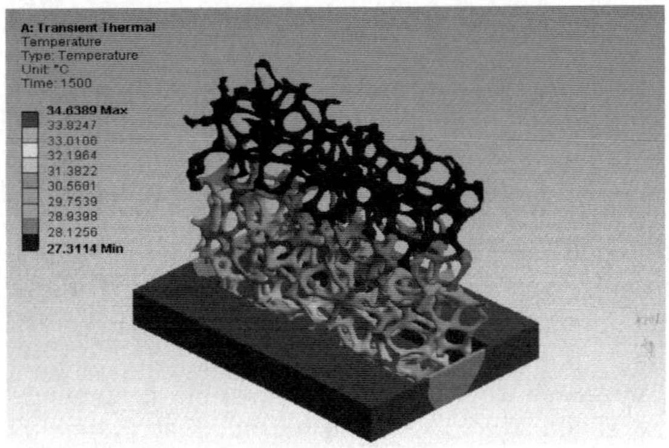

Figure 5: Temperature profile of 1 fin metal foam heat sink

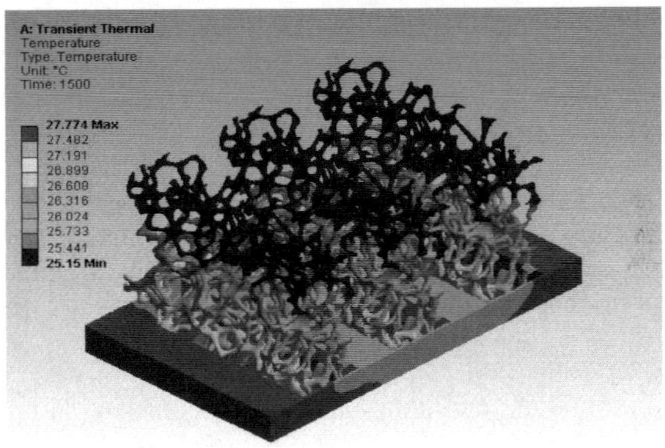

Figure 6: Temperature profile of 3 fins metal foam heat sink

Experimental data was only used to determine the value of 'h for our model. Mancin *et al.* [15] showed that *h* does not depend on imposed heat flux, therefore for sake of comparison only, for both one and three fins 4 PPC heat sink models, heat of 1W has been provided to the base. Other boundary conditions for both cases are also same.

Figure 7 shows the transient analysis simulations run for both one and three fin cases over a span of 1500 s (25 min). As shown, maximum temperatures reached by 1 fin and 3 fins during fan-off (natural convection) are 82.88 °C and 44.49 °C respectively. Also, *h* value for first 900 s (15 min) is that of fan-off (h=12.5) and for remaining 600 s (10 min) is of fan-on (h=109.5) respectively (calibrated in Figure 4).

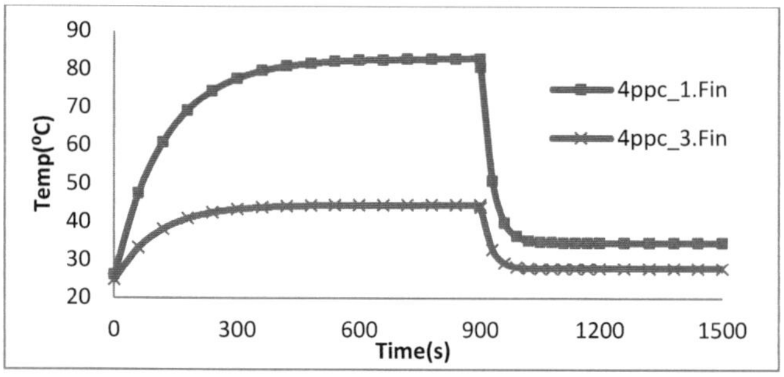

Figure 7: Transient analysis of 1 and 3 fins heat sinks

Simulations were run using ANSYS software and the solid model was meshed using tetrahedral elements.

CONCLUSION

This work presents a methodology for developing a realistic and accurate 3D computational model of open-cell metal foams. CT scan was used to capture the precise geometry of metal foam. This paper discussed the optimized process of obtaining a clean workable FE volume (solid) model.

Experiment was run on single fin 4 PPC metal foam heat sink to obtain *h* values. It can be seen that value of *h* is almost consistent at fan-off and fan-on conditions. This *h* value can be used for FE simulations of different scenarios.

The cleaned model was used in Ansys to develop one and three fins 4 PPC metal foam heat sinks. These two cases were run in simulations. The effect of changing the number of fins for same heat input and same environmental conditions is clearly observed from temperature distribution of both simulations. Figure 7 shows that even for fan-off (natural convection) conditions, base temperature is lower in 3 fins. Thus indicating that higher the number of fins, lower is the temperature increase and hence less thermal stresses. Also it takes less time for 3 fins to reach steady-state, in both fan-off and fan-on conditions. Size scaling of this very same model can be used for varying pore density (PPC) for different scenarios, without the need of taking scan for different samples.

ACKNOWLEDGEMENT

The authors would like to acknowledge the support of King Fahd University of Petroleum and Minerals through DSR project # SB131002.

REFERENCES

[1] M. Khairul Alam and M. S. Anghelescu, "Analysis of Deformation and Residual Stresses in Composites Processed on a Carbon Foam Tooling," *J. Compos. Mater.*, vol. 43, no. 19, pp. 2057–2070, 2009.

[2] S. Mancin, C. Zilio, A. Cavallini, and L. Rossetto, "Heat transfer during air flow in aluminum foams," *Int. J. Heat Mass Transf.*, vol. 53, no. 21–22, pp. 4976–4984, 2010.

[3] X.-H. Han, Q. Wang, Y.-G. Park, C. T'Joen, A. Sommers, and A. Jacobi, "A Review of Metal Foam and Metal Matrix Composites for Heat Exchangers and Heat Sinks," *Heat Transf. Eng.*, vol. 33, no. February, pp. 991–1009, 2012.

[4] I. Ghosh, "How Good Is Open-Cell Metal Foam as Heat Transfer Surface?," *J. Heat Transfer*, vol. 131, no. 10, p. 101004, Oct. 2009.

[5] F. Topin, J. P. Bonnet, B. Madani, and L. Tadrist, "Experimental analysis of multiphase flow in metallic foam: Flow laws, heat transfer and convective boiling," *Adv. Eng. Mater.*, vol. 8, no. 9, pp. 890–899, 2006.

[6] T. J. Lu, H. a. Stone, and M. F. Ashby, "Heat transfer in open-cell metal foams," *Acta Mater.*, vol. 46, no. 10, pp. 3619–3635, 1998.

[7] J. Banhart, "Manufacture, characterisation and application of cellular metals and metal foams," *Prog. Mater. Sci.*, vol. 46, pp. 559–632, 2001.

[8] a. Kopanidis, a. Theodorakakos, E. Gavaises, and D. Bouris, "3D numerical simulation of flow and conjugate heat transfer through a pore scale model of high porosity open cell metal foam," *Int. J. Heat Mass Transf.*, vol. 53, no. 11–12, pp. 2539–2550, 2010.

[9] S. Sihn and A. K. Roy, "Modeling and prediction of bulk properties of open-cell carbon foam," *J. Mech. Phys. Solids*, vol. 52, pp. 167–191, 2004.

[10] A. G. Straatman, N. C. Gallego, Q. Yu, L. Betchen, and B. E. Thompson, "Forced Convection Heat Transfer and Hydraulic Losses in Graphitic Foam," *J. Heat Transfer*, vol. 129, no. 9, p. 1237, Sep. 2007.

[11] A. Druma, M. K. Alam, M. Anghelescu, C. Druma, and B. Maruyama, "Three Dimensional Modeling of Carbon Foams," in *Heat Transfer, Part A*, 2005, vol. 2005, pp. 209–211.

[12] a. Liebscher, C. Proppe, C. Redenbach, and D. Schwarzer, "Stochastic multiscale modeling of metal foams," *Procedia IUTAM*, vol. 6, pp. 87–96, 2013.

[13] L. E. Murr, S. M. Gaytan, F. Medina, E. Martinez, J. L. Martinez, D. H. Hernandez, B. I. Machado, D. a. Ramirez, and R. B. Wicker, "Characterization of Ti-6Al-4V open cellular foams fabricated by additive manufacturing using electron beam melting," *Mater. Sci. Eng. A*, vol. 527, no. 7–8, pp. 1861–1868, 2010.

[14] "http://www.goodfellow.com/." .

[15] S. Mancin, C. Zilio, A. Diani, and L. Rossetto, "Experimental air heat transfer and pressure drop through copper foams," *Exp. Therm. Fluid Sci.*, vol. 36, pp. 224–232, 2012.

Convection

Proceedings of the 25th CANCAM
London, Ontario, Canada, May 31 – June 4, 2015

THERMODYNAMICS AND FLUID MECHANICS OF HEAVY WATER AND SUPERCRITICAL WATER

Robert L. Varty
Consultant in Engineering Science
Edmonton, Alberta, Canada
robvarty@telus.net

ABSTRACT

Information about the thermodynamic properties and the fluid-dynamic properties of water is used in the analysis and the design of nuclear reactors for electrical power generation. In this study, the properties of ordinary water and heavy water in subcritical, critical and supercritical thermodynamic states were considered. Significant differences exist between the properties of ordinary water and heavy water, so calculations for heavy-water flows should be based on information for heavy water.

KEYWORDS: Steam properties, viscosity, heavy water, supercritical water, convection.

INTRODUCTION

Water is used as a working fluid in most of the thermal power plants that generate electricity on a large scale. Water is abundant, non-toxic and non-flammable; it flows easily and it transfers heat well.

Pure water consists of molecules of hydrogen oxide (H_2O). The chemical element hydrogen has three isotopes: protium (^1H), deuterium (^2H or D) and tritium (^3H or T). Protium and deuterium occur naturally in significant quantities. Tritium is radioactive and it is produced terrestrially in nuclear reactors. The chemical element oxygen has three stable (non-radioactive) isotopes: ^{16}O, ^{17}O and ^{18}O. These isotopes occur naturally in significant quantities. The terrestrial supply of ordinary water (hydrogen oxide) contains a small proportion of heavy water (deuterium oxide, D_2O). This proportion is approximately one part heavy water to 7000 parts of ordinary water. Industrial supplies of heavy water have been extracted from ordinary water by an exchange process. This process is based upon the behaviour of deuterium in a mixture of liquid hydrogen oxide and gaseous hydrogen sulphide (H_2S) at high temperatures, and at low temperatures [1].

The nuclei of the hydrogen atoms in a water molecule can slow down ("moderate") efficiently the fast neutrons from the fission of uranium nuclei. The slow neutrons are required to stimulate the fission of additional uranium nuclei, so that a nuclear reactor continues to supply heat in a power plant. A deuterium nucleus can absorb a neutron less readily than a protium nucleus. Hence, heavy water is a better moderator than ordinary water ("light water") for nuclear reactors that use uranium-dioxide fuel with natural (0.7% ^{235}U) uranium or low-enriched (1% to 2% ^{235}U) uranium.

The Canada Deuterium Uranium (CANDU) reactors are nuclear-fission reactors that have heavy-water moderator and uranium-based fuel. The primary coolant is separated from the moderator by the pressure tubes that convey the coolant past the nuclear fuel. A CANDU pressurized heavy-water reactor (CANDU PHWR) has heavy-water coolant. A recent version of this reactor type operates with an outlet temperature of 310 °C and an outlet pressure of 10.0 MPa (absolute) [2]. An Advanced CANDU Reactor (ACR) with light-water coolant has been designed; it would operate with an outlet temperature of 319 °C and an outlet pressure of 11.2 MPa (absolute) [3]. A CANDU supercritical-water-cooled reactor (CANDU SCWR) has been proposed; it would operate with an outlet temperature of 625 °C and an outlet pressure of 25.0 MPa (absolute) [4].

Ordinary water and heavy water differ significantly in their molar masses: hydrogen oxide has $M = 18.015\,\mathrm{g/mol}$ and deuterium oxide has $M = 20.028\,\mathrm{g/mol}$. This mass difference leads directly to a significant difference in the density ρ. The molecular-mass difference contributes to significant differences in thermodynamic properties, and a significant difference in the viscosity μ [5].

The future development of nuclear reactors for electrical power generation will benefit from new information about the thermodynamic properties and the fluid-dynamic properties of water. Information about heavy water and supercritical water will be relevant to the Canadian nuclear-power industry.

THERMODYNAMICS AND FLUID MECHANICS

The theory of thermodynamics can be based on a set of postulates about a thermodynamic system and its fundamental relation [6]. The present study is based on this formulation of thermodynamics.

The primary fundamental relation of a simple thermodynamic system with one chemical component is the function $S(U, V, N)$, where S is the entropy, U is the internal energy, V is the volume, and N is the number of moles of substance. The equilibrium state of a system is determined by the entropy-maximum postulate. Lower-case letters are used to denote specific (per unit mass) quantities, so $s = S/m$,

$u = U/m$, $v = V/m$, and m represents the mass of a system.

The thermodynamic potentials are derived from the function $S(U, V, N)$ by inversion and Legendre transformations. The thermodynamic potentials contain individually the complete thermodynamic information about a system. The relevant thermodynamic potentials are the specific Helmholtz potential f, the specific enthalpy potential h, and the specific Gibbs potential g.

The relevant intensive parameters are the absolute temperature T, the absolute pressure p, and the chemical potential μ'. For a system with one chemical component, the specific Gibbs potential is proportional to the chemical potential.

The thermodynamic system is a sample of water in the present study. The reader may be familiar with steam tables that contain numerical information about the thermodynamic properties of water. Such tables can be calculated from an equation for f. The coefficients for such an equation can be determined from experimental data for measurable quantities, such as pressure and enthalpy change.

The theory of fluid mechanics has a provision for the resistance of a fluid to motion. This provision can involve Newtonian viscous behaviour, with a coefficient of viscosity μ. For a one-directional flow, the shear stress is $\tau = \mu \cdot du/dy$, where u is the velocity and y is the transverse position. The viscosity μ is a function of the absolute temperature T and the absolute pressure p. For a typical liquid, μ decreases rapidly with increasing T. For a low-pressure gas, μ increases with increasing T. For a typical fluid, μ increases with increasing p [7]. A sample of water exhibits Newtonian viscous behaviour.

PROPERTIES OF WATER AND STEAM

The International Association for the Properties of Water and Steam (IAPWS) has published thermodynamic information about ordinary water [8] and heavy water [9]. The IAPWS has published thermodynamic data for ordinary water and heavy water at their thermodynamic critical points [10]. The IAPWS has published fluid-dynamic information about ordinary water [11] and heavy water [12]. These publications, and the current location of the IAPWS secretariat, can be obtained on the Internet at http://www.iapws.org/.

The general IAPWS formulations for ordinary water [8] and heavy water [9] are based on the specific Helmholtz potential f. The formulation for ordinary water represents stable fluid states for temperatures from −22 °C to 1000 °C (251 K to 1273 K), and for absolute pressures from 0.0002 MPa to 1000 MPa. The formulation for heavy water represents stable fluid states for temperatures from 3.80 °C to 527 °C (276.95 K to 800 K), and for absolute pressures from 0 MPa to 100 MPa.

The liquid phase of water and the vapour phase of water can coexist. Thermodynamic equilibrium requires equality of the chemical potential of the two phases, so there is equality of specific Gibbs potential of the two phases. This equality determines a liquid-vapour coexistence curve on a graph of p versus T. One terminus of this curve is the triple point, where the solid, liquid and vapour phases coexist. The other terminus of this curve is the critical point, where the liquid and vapour phases are indistinguishable. This curve corresponds to the "saturated" states, where the liquid and vapour phases are in equilibrium. The triple points of ordinary water and heavy water are described in Table 1. The critical points of ordinary water and heavy water are described in Table 2.

Table 1: Triple points of ordinary water and heavy water [8, 9].

Property	Ordinary water	Heavy water
Temperature (°C)	0.01	3.80
Absolute temperature, T (K)	273.16	276.95
Absolute pressure, p (MPa)	0.0006117	0.0006601

Table 2: Critical points of ordinary water and light water [8, 9].

Property	Ordinary water	Heavy water
Temperature (°C)	373.946	370.697
Absolute temperature, T (K)	647.096	643.847
Absolute pressure, p (MPa)	22.064	21.671

CRITICAL PHENOMENA AND FLUID PROPERTIES

The transition between the liquid phase of water and the vapour phase of water is a first-order phase transition. This transition is related closely to the thermodynamic stability of the system (a sample of water). The van der Waals model of a fluid [6] can be derived with statistical mechanics. The function $g(T, P)$ can be determined for the van der Waals model, and this function can be used to study thermodynamic stability [6]. The consideration of stability accounts for the difference in the values of a thermodynamic property (other than g) between the liquid and vapour phases, at a given temperature and pressure.

The phenomenon of critical opalescence was observed experimentally in the 1860s. Critical opalescence is an optical effect that results from large-scale density fluctuations in a fluid near its critical point [6]. These density fluctuations are very effective at scattering light, so a transparent fluid becomes highly reflective at temperatures and pressures near the critical-point values.

On a graph of p versus v, the isotherm with the critical temperature T_c has a point of inflection at the critical pressure p_c. This inflection is predicted by the van der Waals model, and it is indicated by experimental data. This inflection suggests that the isothermal compressibility approaches infinity near the critical point. Thus, small statistical fluctuations in pressure can produce large fluctuations in specific volume v and density ρ ($\rho = 1/v$).

A supercritical thermodynamic state has a temperature greater than the critical temperature and a pressure greater than the critical pressure. A subcritical thermodynamic state has a temperature less than the critical temperature or a pressure less than the critical pressure; an inclusive "or" applies. The critical thermodynamic state has a temperature equal to the critical temperature and a pressure equal to the critical pressure; it is represented in thermodynamic-state space by the critical point.

Basic thermodynamic properties of ordinary water and heavy water at the critical points are shown in Table 3. The liquid-vapour coexistence curve for ordinary water is close to the liquid-vapour coexistence curve for heavy water, on a graph of p versus T. There are significant differences in the termini of the coexistence curves, as described in Tables 1, 2 and 3.

Table 3: Thermodynamic properties of ordinary water and heavy water at the critical points [10].

Property	Ordinary water	Heavy water
Temperature ($^\circ$C)	373.946	370.697
Absolute temperature, T (K)	647.096	643.847
Absolute pressure, p (MPa)	22.064	21.671
Specific volume, υ (m^3/kg)	0.003106	0.002809
Density, ρ (kg/m^3)	322	356

VISCOSITY AND DENSITY

The fluid-dynamic properties of water include the viscosity μ and the density ρ. The IAPWS has published formulations for the viscosity of ordinary water [11] and heavy water [12]. The formulation for ordinary water includes stable fluid states for temperatures from the melting temperature to 900 °C (1173.15 K), and for absolute pressures from the triple-point pressure to 300 MPa. The formulation for heavy water includes stable fluid states for temperatures from the melting temperature to 502 °C (775 K), and for absolute pressures from 0 MPa to 100 MPa. The melting temperature depends on the absolute pressure for each formulation. The viscosity is expressed as a function of absolute temperature and density, $\mu(T, \rho)$, for each formulation.

An example of the variation of viscosity with temperature is shown in Table 4. The substance is heavy water, in its liquid phase. The absolute pressure is $p = 10.0$ MPa. The density is $\rho \approx 1100$ kg/m^3. These conditions are relevant to a CANDU PHWR.

Table 4: Example of the viscosity of heavy water [12].

Temp. ($^\circ$C)	Abs. temp., T (K)	Viscosity, μ (μPa · s)
25.0	298.15	1091.0
100.0	373.15	332.3
160.0	433.15	201.5
300.0	573.15	94.3

CRITICAL-POINT BEHAVIOUR

The theory of the first-order phase transition is not valid at, or near, a critical point. A valid theory of the second-order phase transition in a small neighbourhood of a critical point was achieved in the 1970s. The valid theory of critical phenomena incorporates a correct treatment of the correspondence between thermodynamics and statistical mechanics [6].

Experimental data for specific heat capacity at constant volume, c_υ, versus temperature have strongly non-linear behaviour for temperatures near the critical temperature, for both ordinary water and heavy water. The IAPWS formulations for ordinary water and heavy water represent thermodynamic properties and viscosity well for both subcritical states and supercritical states, except for thermodynamic states in a small neighbourhood of a critical point.

Thermodynamic properties can have very large values at non-critical points in thermodynamic-state space. An example of this type of phenomenon occurs with the specific heat capacity at constant pressure, c_p, of ordinary water. At a given pressure greater than the critical pressure, a graph of c_p versus T has a sharp peak at a pseudocritical temperature T_{pc} greater than the critical temperature [13]. A case of this example occurs at $p = 24.5$ MPa: $T_{pc} \approx 658$ K and $c_p \approx 87$ kJ/(kg · K). In comparison, the critical point occurs at $p = 22.064$ MPa and $T_c = 647.096$ K. A constant-pressure system exhibits a capacity to receive a large amount of heat as its temperature is increased near its pseudocritical point. This example is relevant to a proposed CANDU SCWR [14].

CONCLUSION

Ordinary water and heavy water are used as coolants for nuclear-fission reactors. Liquid-vapour phase transitions and critical-point behaviour are explained by theories based on thermodynamics and statistical mechanics. Thermodynamic properties and viscosity are known well for subcritical states and supercritical states, except in a small neighbourhood of a critical point.

Significant differences exist between the properties of ordinary water and heavy water. Therefore, calculations for heavy-water flows should be based on information about heavy water, rather than easy-to-find information about ordinary water.

Present water-cooled nuclear-fission reactors do not operate near the critical points of their coolants during normal operation. Future reactors with supercritical-water coolant would operate near the critical points of their coolants during transient processes that would likely occur during normal operation.

ACKNOWLEDGEMENTS

The author thanks members of the author's family for their tolerance of the research activity, and for their patience during the publication of the research.

REFERENCES

1. "Nuclear power technology" 2005, in Considine, G.D., editor, *Van Nostrand's Encyclopedia of Chemistry*, 5th ed., John Wiley & Sons, Inc., Hoboken, New Jersey, pp. 1101–1124.

2. Candu Energy Inc. 2012, *Enhanced CANDU 6 Technical Summary*, Mississauga, Ontario.

3. Atomic Energy of Canada Limited 2010, *ACR-1000 Advanced CANDU Reactor Technical Summary*, Mississauga,

Ontario.

4. Gupta, S., and Pioro, I. 2014, "An overview of supercritical water-cooled reactor (SCWR) concepts", *Bull. Can. Nucl. Soc.*, 35(3), pp. 25–30.

5. Poling, B.E., Prausnitz, J.M., and O'Connell, J.P. 2001, *The Properties of Gases and Liquids*, 5th ed., The McGraw-Hill Companies, Inc., New York.

6. Callen, H.B. 1985, *Thermodynamics and an Introduction to Thermostatistics*, 2nd ed., John Wiley & Sons, Inc., New York.

7. White, F.M. 1991, *Viscous Fluid Flow*, 2nd ed., McGraw-Hill, Inc., New York.

8. The International Association for the Properties of Water and Steam 2009, *Revised Release on the IAPWS Formulation 1995 for the Thermodynamic Properties of Ordinary Water Substance for General and Scientific Use*.

9. The International Association for the Properties of Water and Steam 2005, *Revised Release on the IAPS Formulation 1984 for the Thermodynamic Properties of Heavy Water Substance*.

10. The International Association for the Properties of Water and Steam 1992, *IAPWS Release on the Values of Temperature, Pressure and Density of Ordinary and Heavy Water Substances at their Respective Critical Points*.

11. The International Association for the Properties of Water and Steam 2008, *Release on the IAPWS Formulation 2008 for the Viscosity of Ordinary Water Substance*.

12. The International Association for the Properties of Water and Steam 2007, *Revised Release on Viscosity and Thermal Conductivity of Heavy Water Substance*.

13. Yamagata, K., et al. 1972, "Forced convective heat transfer to supercritical water flowing in tubes", *Int. J. Heat Mass Transfer*, 15, pp. 2575–2593.

14. Pioro, I.L., and Duffey, R.B. 2007, *Heat Transfer and Hydraulic Resistance at Supercritical Pressures in Power-Engineering Applications*, ASME Press, New York.

STUDY OF AL2O3/WATER NANOFLUID IN A MULTIPORT SLAB HEAT EXCHANGER

Mohammed Ismail
Mechanical Automotive and Materials Engineering
University of Windsor
Windsor, Ontario, Canada
ismailf@uwindsor.ca

Shahram Fotowat
Mechanical Automotive and Materials Engineering
University of Windsor
Windsor, Ontario, Canada
fotowat@uwindsor.ca

Amir Fartaj
Mechanical Automotive and Materials Engineering
University of Windsor
Windsor, Ontario, Canada
fartaj@uwindsor.ca

ABSTRACT

Miniature heat exchanger comprising narrow size flow channels is becoming popular in automotive and many other industrial applications. The main focus of the researchers is to implement minichannels due to enhanced heat flux, larger heat transfer area density, and energy efficiency compared with conventional one. The use of nanofluids has also recently become the core interest of many researchers due to its elevated heat transfer performance. In the present study, heat transfer and flow characteristics of water and Al2O3/water nanofluid are numerically investigated as a homogeneous single phase fluid in a multiport slab minichannel heat exchanger (MICHX). In order to cool the liquid, cold air at a constant mass flow rate of 0.507kg/s and temperature of 14°C is allowed to pass through the finned passages of the heat exchanger core in a cross-flow orientation. The liquid at different Reynolds number ranging from 300 to 1850 is forced to flow through the channels of the MICHX. The volumetric concentration of the nanofluid is varied from 1.0% to 4.0%, while the inlet temperature of liquid is kept constant at 76°C. In this study, liquid-side pressure drop, heat transfer co-efficient, and Nusselt number are computed.

KEYWORDS: Multiport slab, nanofluids, numerical simulation, minichannel heat exchanger, multiport.

INTRODUCTION

The heat removal or temperature control is considered as a vital issue in many critical applications involving high heat fluxes. Nanofluids are potential heat transfer fluids which can be applied in many devices for better performances. It can be used for a wide range of industries, such as electronics, transportations, HVAC as well as process heating and cooling to energy conversion [1] Nanofluids are manufactured by dispersing nanometer-sized (usually <100nm) materials including nanoparticles, nanotubes, nanofibers, and nanorods in the base fluids [2]. Nanofluid possesses enhanced thermophysical properties including thermal conductivity, viscosity, and thermal diffusivity compared with those of base fluids such as water, ethylene glycol, and engine oil.

The stability of nanofluid is one of the most key issues in heat transfer performance, and it is a big challenge to achieve the anticipated stability. In recent years, nanofluids have attracted more and more attention [3]. The other important parameters which effect the heat transfer and flow behaviors of nanofluids include the nanoparticles material, volume concentration, size, shape, temperature, and additives as well as the base fluid [4] In the past several years, numerous investigations on convective heat transfer of nanofluids were performed experimentally and numerically [5-8]. The enhanced convective heat transfer in nanofluids has been reported by some studies. For the heat transfer in nanofluids, it is also found that there are inconsistencies in the reported results published by some researches. The heat transfer in nanofluids is found inconsistent in their proposed results. More study is needed to find out the relationships between the enhancement and the nanoparticle concentration, temperature, and non-dimensional numbers such as Reynolds number and Nusselt number [1]. The purpose of this study is to investigate the cooling performance of Al2O3/water in a multiport slab MICHX.

NUMERICAL METHOD

The numerical simulation is conducted in a 100mm×2mm×304mm MICHX with 68 extruded circular channels. The Model as well as specifications of MICHX considered for the current study is presented in Fig. 1 and Table 1. Three-dimensional double precision pressure-based simulations are performed using ANSYS FLUENT, a finite volume method (FVM) based commercial CFD code. The Navier-Stokes and energy equations are used to solve the convective heat transfer process for steady state incompressible fluid flow. Temperature dependent thermophysical properties are used for all liquids.

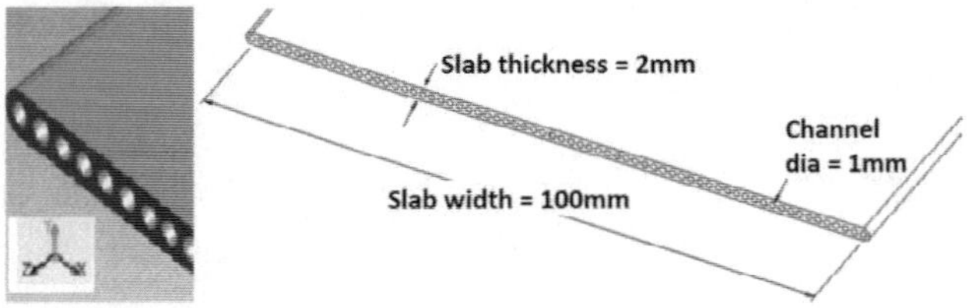

Figure 1: Dimensions of circular channels and the MICHX slab used in numerical simulation

Table 1. Details of MICHX

Parameters	Specifications
MICHX Material	Al alloy 3003
Diameter of circular channel	1mm
Number of channels in a slab	68
Distance between channel centers	1.463mm
Slab length in x direction,	100mm
Slab thickness in y direction	2mm
Slab width in z direction	304mm
Inlet and outlet diameter of header	4.76mm

The mass flow rate and temperature are specified at both liquid and air inlets, outflow conditions are specified for the fluid outlet, and type 1 thermal boundary condition is specified at the walls of test chamber. Standard k-ε model is used in order to account the turbulence effects. For the flow and energy variables, the sum of the respective normalized absolute residuals in each control volume is reduced to 10^{-4} and 10^{-9} respectively, when the consecutive iteration is stopped.

MODEL VALIDATION

In order to ensure the accuracy and consistency of the model, numerical results are verified by checking the mass balance (MB) and heat balance (HB) as shown in Fig. 2. The error in MB and HB are found to be ±0.08% and ±1.01% respectively, which are very reliable.

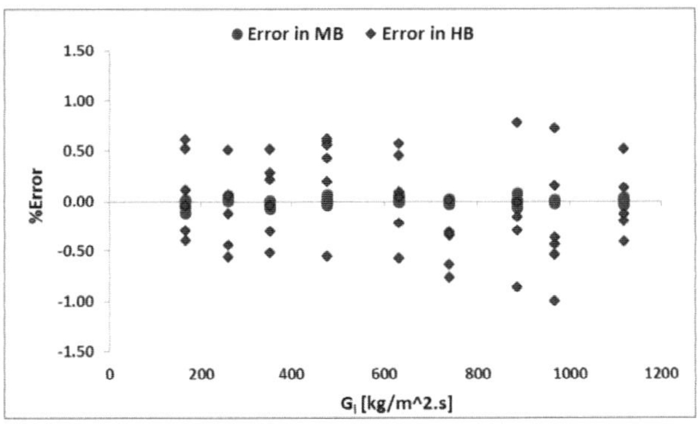

Figure 2: Error in mass balance and heat balance for different mass velocity

A set of simulation results is also compared with experimental data [10] for the air-side outlet temperature and the surface temperature as well as the water-side outlet temperature and pressure drop. Simulation results show a very good agreement with experimental data. The details are previously published [9] and is not repeated here. Therefore, the results obtained from the simulation are reliable and can be used to investigate fluid flow and heat transfer characteristics of similar MICHX.

RESULTS AND DISCUSSIONS

Liquid velocity, temperature, and Nusselt number

The variations of stream-wise local centerline velocity, mean surface temperature, fluid mean temperature, and Nusselt number for different volume concentrations are shown in Figs. 3, 4, 5, and 6 respectively. The mass velocity of 1117kg/m^{2}s is kept constant for all fluids. In the top multi-port slab, the nanofluids travels 304mm along the channels inside the MICHX slab.

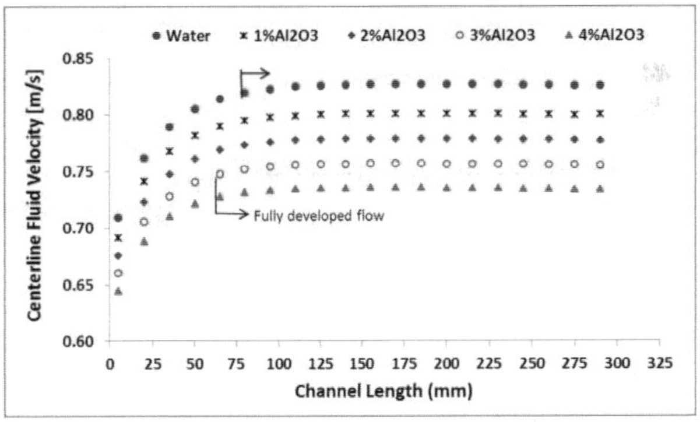

Figure 3: Variations of local centerline fluid velocity along the channel length for diff. vol. concentration of nanofluid for a specified mass velocity of 1117 kg/m^{2}s.

Fig. 3 shows that the nanofluids reach the fully developed region sooner than water. It is also observed that the higher concentration fluid becomes fully developed earlier than lower concentration fluid.

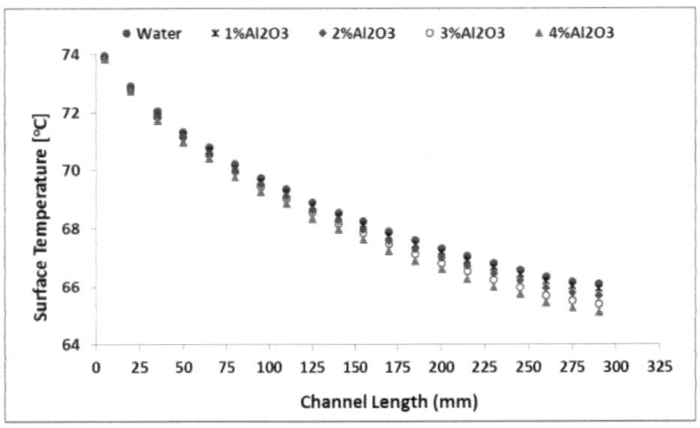

Figure 4: Variations of local surface temperature along the channel length for diff. vol. concentration of nanofluids for a specified mass velocity of 1117 kg/m^{2}s.

From Fig. 4, it is observed that the surface temperature decreases along the length of the channel. It is also shown that the nanofluid is enhancing the mean surface temperature drop along the channel compare to water and this temperature drop increases with increase of the nanofluids concentration.

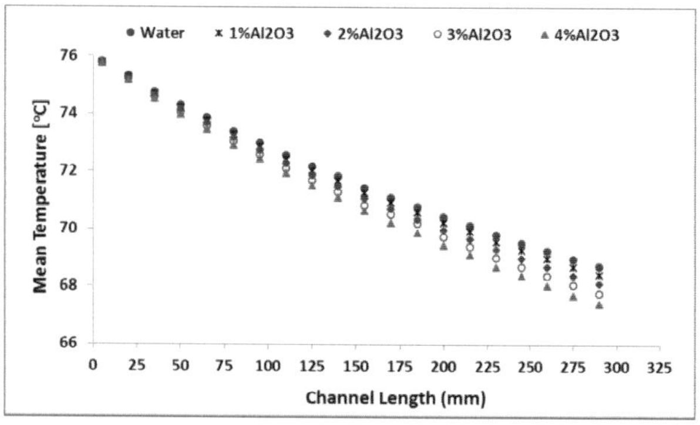

Figure 5: Variations of local fluid mean temperature along the channel length for diff. vol. concentration of nanofluids for a specified mass velocity of 1117 kg/m²s.

Fig. 5 shows that the local mean temperature of liquids decreases almost linearly through the channel length. From the Figure, it is also obvious that the nanofluid is enhancing the mean surface temperature drop along the channel compare to water and this temperature drop increases with increase of the concentration.

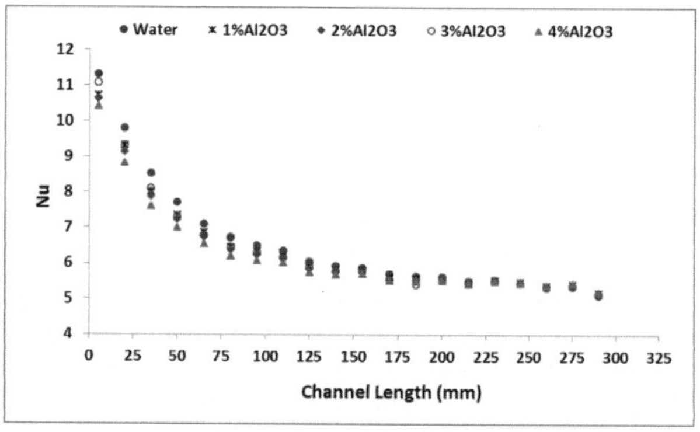

Figure 6: Variations of local Nusselt number along the channel length for diff. vol. concentration of nanofluids.

Fig. 6 is presenting the local Nusselt number of different liquids through the channel length. The Nu is observed lower for the nanofluids compare to the water, and the difference is higher for higher nanoparticles concentration.

Heat transfer rate and hot fluid side pressure drop

This study is conducted at constant temperatures and mass flow rates of cold air flow, 14°C and 0.507 kg/s accordingly as the boundary conditions for numerical simulation. For the liquid-side condition, inlet temperature is kept constant at 76°C, and the fluids are computed according to the variations of Re from 300 to 1850 through the minichannels. The dimensionless temperature is defined as $\theta = \Delta T_l / \Delta T_{max}$ to investigate the liquid-side temperature drop over maximum fluids temperature differences.

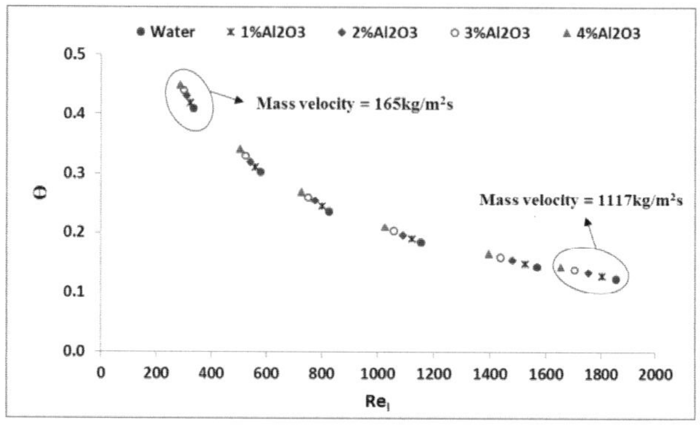

Figure 7: Variations of dimensionless temperature for different Reynolds number

Fig. 7 illustrates that θ decreases, i.e. temperature drop decreases with the increase of liquid-side Re. Furthermore, it is found that for a constant mass flow rate, θ is higher for nanofluids compare to the water even for higher Re. Also for higher concentration, θ is higher which means higher liquid-side temperature drop. Moreover, the rate of decreasing θ is lower for higher Re.

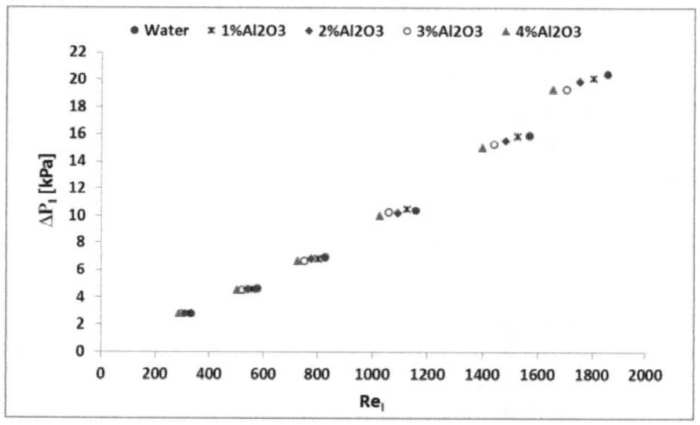

Figure 8: Effect of volume concentrations of Al2O3 nanoparticles in liquid-side pressure drop for diff. Re.

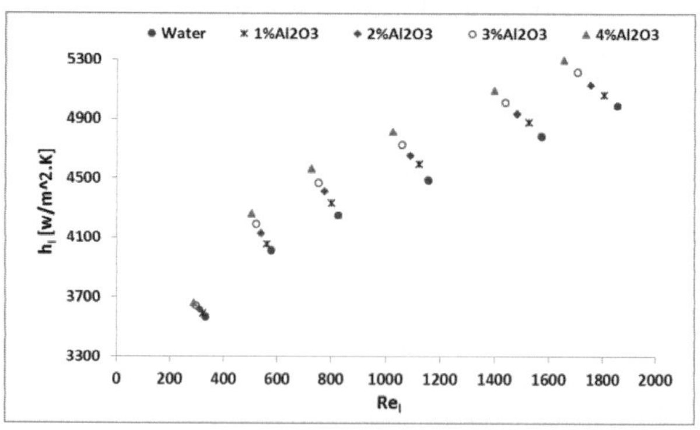

Figure 9: Effect of volume concentrations of Al_2O_3 nanoparticles in heat transfer coefficient for diff. Re

Fig. 8 shows that for a specified mass flow rate, the pressure drop (ΔP) in the MICHX core is insignificant for all fluids at lower Re; however, a noticeable pressure drop is observed at higher Re. It is also found that the ΔP increases with the increase of Re. For a specified Re, the pressure drop is increased by 10-12% with the increase of nanofluid volume concentration up to 4%.

Fig. 9 indicates that the heat transfer coefficient of fluids, h_f increases by the increase of Re_f. The result also shows the nanofluids has higher heat transfer coefficient compare to water and nanofluids with higher concentration has higher heat transfer coefficient for the same mass flow rate even with the lower Re_f. The enhancement of the heat transfer coefficient is found 7-9.5% by using nanofluids.

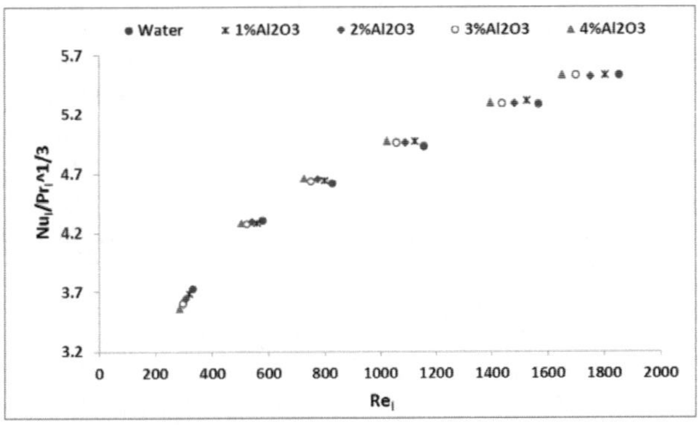

Figure 10: Effect of volume concentrations of Al_2O_3 nanoparticles on liquid-side Nusselt number for diff. Re.

Fig. 10 demonstrates the relation between Nu and Re. Each group of data represents same mass flow rate for all fluids. It is clear from the graph that the Nu is higher with higher Re. It is also illustrating that the Nu does not change significantly with the change of concentration for same mass flow rate.

CONCLUSIONS

In the current study, numerical simulation is used to evaluate the performance of nanofluids with water using a slab MICHX, and the following results are concluded.

1-Even though the pressure drop, ΔP in the MICHX core is negligible for all fluids at lower Re, a significant pressure drop is observed at higher Re. It is also found that the ΔP increases with the increase of Re as well as with the increase of nanofluid volume concentration.

2-The dimensionless temperature, θ decreases with the increase of liquid-side Re which implies higher temperature drop as well as higher heat transfer rate through the channel. For a constant mass flow rate, θ is higher for higher concentration of nanofluids compare to the water. The rate of decreasing θ is lower for higher Re.

3-The nanofluids has higher heat transfer coefficient compare to water and higher concentration possesses higher heat transfer coefficient for the same mass flow rate. Nanofluids, improved the heat transfer coefficient by 7.0-9.5% with the cost of 10-12% pressure drop.

4-Nu is higher for water compare to nanofluids for the same mass flow rate in the MICHX. It is also found that for same mass flow rate, the Nu does not change considerably with the change of concentration.

ACKNOWLEDGMENTS

The authors would like to thank the Natural Sciences and Engineering Research Council of Canada and the University of Windsor for the financial support of the research.

REFERENCES

1. Pang, C., Lee, J.W., and Kang, Y.T. 2015, "Review on combined heat and mass transfer characteristics in nanofluids", *Int. J. Thermal Sci.*, Vol(87), pp. 49-67.
2. Choi, S.U.S. and Eastman, J.A. 1995, "Enhancing thermal conductivity of fluids with nanoparticles", *Argonne National Laboratory*, Argonne, IL, USA.
3. Yu, W. and Xie, H. 2012, "A Review on Nanofluids: Preparation, StabilityMechanisms, and Applications", *Journal of Nanomaterials*, Vol(2012), pp. 1-12.
4. Lomascolo, M., Colangelo, G., and Milanese, M., and Risi, A. 2015, "Review of heat transfer in nanofluids: Conductive, convective and radiative experimental results", *Renewable and Sustainable Energy Reviews,* Vol(43), pp.1182–1198.
5. Ray, D., Das, D., and Vajjha, R. 2014, "Experimental and numerical investigations of nanofluids performance in a compact minichannel plate heat exchanger", *Int. J. of Heat and Mass Transfer*, Vol(71), pp. 732–746.
6. Akbaridoust, F., Rakhsha, M., Abbassi, A. Saffar-Avval, M. 2013, "Experimental investigation on the convective heat transfer of nanofluid flow", *International Journal of Heat and Mass Transfer,* Vol(58), pp. 480–491.
7. Pandey, S.D. and Nema, V.K. 2012, "Experimental analysis of heat transfer and friction factor of nanofluid corrugated plate heat exchanger", *Experimental Thermal and Fluid Science*, Vol(38), pp. 248–256.
8. Wen, D. and Ding, Y. 2004, "Experimental investigation into convective heat transfer of nanofluids at the entrance region under laminar flow conditions", *Int. J. of Heat and Mass Transfer*, Vol(47), pp. 5181–5188.
9. Ismail, M., Fartaj, A, and Karimi, M. 2013, "Numerical Investigation on Heat Transfer and Fluid Flow Behaviors of Viscous Fluids in a Minichannel Heat Exchanger", *Numerical Heat Transfer, Part A*, Vol(64), 1-29.
10. Khan, M.G. and Fartaj, A. 2011, "A Review on Microchannel Heat Exchangers and Potential Applications", *Int. J. Energy Res.*, Vol(35), pp. 553–582.

TRANSIENT HEAT TRANSFER BEHAVIOR IN THE LIQUID DOMAIN OF A PCM DURING PHASE CHANGE

Mohammad Bashar
Mechanical and Materials Engineering
University of Western Ontario
London, Ontario, Canada
mbashar5@uwo.ca

Kamran Siddiqui
Mechanical and Materials Engineering
University of Western Ontario
London, Ontario, Canada
ksiddiqui@eng.uwo.ca

ABSTRACT

An experimental study was conducted to investigate the convection heat transfer process in a phase change material (PCM) during its phase transition. Paraffin wax was used as the PCM. A horizontal heated tube near the bottom of a narrow rectangular shaped storage enclosure acted as the heat source. The behavior of the heat transfer process in the liquid domain and the movement of the interface lying between the solid and the liquid phase were quantified by means of a temperature recording system and an imaging technique. Three different input heat loads with average heat flux of 5630 W/m^2, 8670 W/m^2 and 10050 W/m^2 respectively were used to see the effects of changing heat flux. The results show that the behavior and motion of the convective cell immensely influences the heat transfer rate as well as the formation and nature of the interface structure. Also the changing volume of the melted domain during the melting process significantly effects the convective heat transfer coefficient.

KEYWORDS: PCM, thermal energy storage, natural convection, convective heat transfer coefficient, interface.

INTRODUCTION

At present, one of the top priorities in the energy sector is to reduce consumption of depleting fossil fuels, which also has detrimental effects on the environment. A great emphasis is given on the utilization of renewable energy resources as well as the energy recovery from wasted heat generated in industrial, commercial and residential processes. Several protocols have been signed in the international arena to limit greenhouse gas emissions and reduce global warming. To implement the conditions proposed in these forums and to fulfill the requirements, it is imperative to make substantial changes in the energy production patterns and their utilization. Maximize the use of renewable energy by mixing them with the conventional energy resources and minimizing the heat waste by using different heat recovery mechanisms are by far the best options. Since most of the renewable energies and wasted heat from various sectors are intermittent in intensity and nature, they do not always match with the demand. Hence, storing the thermal energy when it is surplus and utilizing it at a later period when there is a demand, is an optimal solution.

Thermal energy storage systems using phase change material (PCM) as the storage medium take the advantage of the latent heat of the material to increase the energy storage density. These systems have many engineering applications including building climate control, electronic equipment cooling and heating and food process industry. Among the most distinctive characteristics of PCMs, their high energy density and isothermal nature during any thermal energy change process are most important. Despite many advantages, most of the potential PCMs have one major disadvantage, their 'low thermal conductivity'. Thus, designing a PCM thermal storage remains a major challenge to the engineers to ensure an effective heat transfer to and from the PCM materials, which greatly influences the efficiency and performance of the storage [1]. The heat transfer during the solid-liquid phase change is a crucial process that involves many variables and has complexity which is non-linear. The heat transfer during solid-liquid phase change is an active area of research.

There are large numbers of PCMs with a wide range of melting/solidification temperatures, making them suitable for different applications [2]. A comprehensive review on thermal energy storage (TES) using solid-liquid phase change materials reflects diverse aspects of these materials, different heat transfer parameters and applications [3]. An inherent feature in a phase change process is its solid-liquid interface, which not only physically separates the two phases but also isolates these two zones that have significantly different thermo-physical properties. The thermal energy absorbed or released across this interface, which is considered as the dynamic zone, determines the rate and manner in which the interface would move, which in turn greatly affects the heat exchange between the two phases. Since the physical movement of the interface is the result of the whole process, such problem is also called moving boundary problem [4].

Such problems are also called in the literature as the conjugate heat transfer problem [5]. For accurate analysis of this transient heat transfer process, the conduction in the solid part and the convective heat transfer in the adjacent liquid part is required to be analyzed separately and later coupled or conjugated [6]. Such type of complex coupling, non-linearity in the governing equations, and the complicacy in geometries involved with the moving boundary have imposed major challenges for analytical and computational investigations of this problems, forcing to seek experimental approach [7]. Accurate prediction of the convection heat transfer is an essential part of the thermal design process in all industrial applications [8]. Betrand et al. [9] compared different physical models and numerical procedures for relatively simpler problems and discussed where the phase change in the form of melting is driven by laminar thermal convection in the melt.

The present study is focused on a comprehensive experimental investigation of the transient heat transfer process during the solid-liquid phase change in a compact PCM thermal energy storage system. An imaging technique was used to capture the phase change process and track the dynamics of the solid-liquid interface. A set of thermocouples were also embedded to quantify the heat transfer in the liquid domain.

EXPERIMENTAL SETUP

For the purpose of this experimental study, a small rectangular shaped enclosure was built as the thermal energy storage as shown in Fig. 1. The inner dimensions of the enclosure were 50 mm (length) × 65 mm (height) × 12 mm (width). A U-shaped block made from an 8 mm thick aluminum (6061T6) plate was used as the main structural part of the enclosure. Two plain glass sheets (5mm thick) formed the front and back walls and two pieces of 2 mm thick rubber gaskets placed between the glass and the aluminum block prevented leak and absorbed thermal expansions. The glass panes facilitated the optical access for visualization of the melting process and uninterrupted recording of the images during the melting process. The enclosure top was left open to room atmosphere to accommodate thermal expansion of PCM. A 4.76 mm OD (3/8") brass tube was passed through the PCM enclosure near the bottom acted as the heat source. Hot and metered air was used as the heat transfer fluid (HTF) passing through the tube. Brass tube and other areas of the enclosure, except the glass panes, were insulated to prevent heat losses.

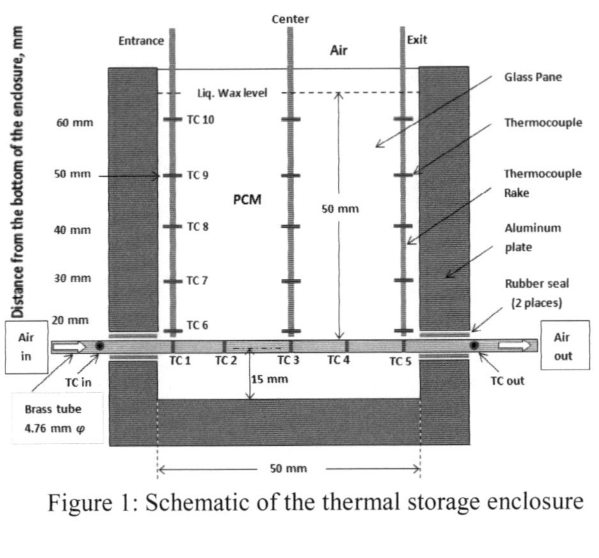

Figure 1: Schematic of the thermal storage enclosure

The schematic of the complete experimental setup is shown in Fig. 2. The hot air serving as the heat transfer fluid was supplied from a compressed air reservoir via a pressure regulator to an air heater (Farnam, CC-A15), capable of controlling the temperature precisely. The air flow rate was measured using an inline flow meter upstream of the heater. Commercially available Paraffin wax, named as 'Polyfin' (melting temperature 55°C) was used as the PCM. Small amount of tracer particle was added to the wax to help quantifying the convective flow behavior and interface movement of the molten liquid.

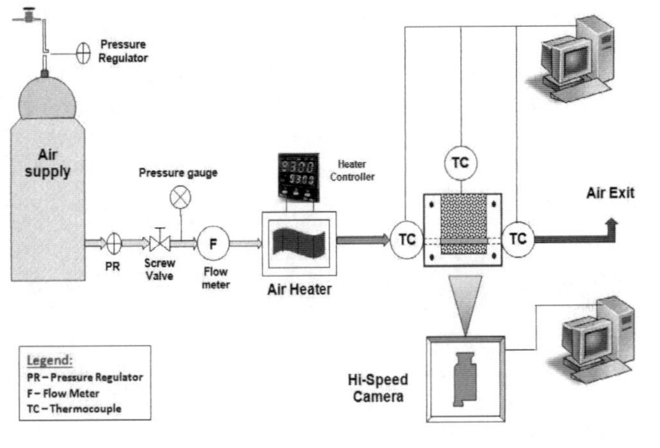

Figure 2: Schematic of the experimental set up

Three different input heat fluxes were considered by varying the temperature of the inlet air to the storage enclosure. The temperature of the HTF passing through the chamber was continuously measured at the inlet and outlet of the tube. The surface temperature of the tube inside the chamber was measured at five different locations. The temperature of the PCM was also measured at five vertical locations within the PCM domain using a rake with 'T' type thermocouples (denoted as TC) fixed on it. By changing the position of the rake in three axial locations, temperatures were recorded near the entrance, center and exit of the storage. The measurements at each of these axial locations were conducted in separate set of experiments under identical conditions. For each experiment, TCs were positioned with the TC holder in one of these three positions to record the temperature, which was termed as local parameters. Local data from all three stations were averaged to get the bulk or global parameters.

The temperature data from all these 12 thermocouples were recorded via a data acquisition card (NI-9211) controlled by the LabVIEW software. During a given experiment, the PCM was continuously heated by the HTF passing through the bottom horizontal tube, to the desired temperature. A high resolution 4 Mega pixel CCD camera (Flare 4M180-CL) was used to capture images of the phase change process and track the solid-liquid interface and the flow structure of the melted wax. The camera was connected to an image acquisition system (IOI DVR core) to record images. The camera was controlled via the DVR Core View software installed on a PC. For each run, images were captured until the wax was completely melted and temperatures were stabilized. The camera frame rate was set between 40 and 45 frames per second for different experimental runs. Different in-house Matlab codes were used for the image processing of the acquired data. The codes were used to detect and track the interface and calculate the melted fraction of the wax and interface velocity. The temperature data were used to compute heat transfer rate into the PCM and other related parameters including the heat transfer coefficient.

RESULTS AND DISCUSSIONS

Three different input heat loads with an average heat flux of 10050 W/m^2, 8670 W/m^2 and 5630 W/m^2 respectively were used in the experiment. The narrow width of the enclosure was chosen in order to simulate a 2D liquid flow pattern inside the chamber once the PCM was melted.

At the beginning, the thermal energy transferred to the PCM via the imbedded horizontal tube was through conduction since the PCM was initially at the room temperature. As the PCM in the vicinity of the heat source reached melting temperature, the phase change started. In the melted liquid domain, natural convection was induced forming local convective cells. This caused non-uniformity in the displacement of the melting front.

The time series of the PCM temperature at four vertical locations, when the TC rake was in the center position in the storage domain, are plotted in Fig. 3. The temperature of the PCM in the vicinity of the horizontal heat source showed quicker rise and stabilized at the melting temperature as expected. Other locations (denoted by the increasing TC numbers) followed similar pattern.

The interface between the solid and liquid phases was tracked by processing of the recorded images, as mentioned earlier. Fig. 4 shows an example image along with the detected interface with the red color. The solid-liquid interface rose within the domain with time due to the progression of the melting process. Time history of the interface positions is plotted in Fig. 5. The results indicate the nonlinear nature of the interface throughout the melting process. The time history of the recorded coordinate of the interface was used to compute local instantaneous interface velocity along the front at a given input heat flux. The spatially-averaged interface velocity at a given time was calculated by averaging the local interface velocity at different axial locations. The results indicate some fluctuations in the instantaneous interfacial velocity but overall, the interface velocity remains relatively uniform throughout the melting process at a given input heat flux. The time-averaged interface velocity over the melting phase is plotted versus the input flux in Fig. 6. As expected, the interface velocity monotonically increased with an increase in the heat flux. The results also indicate that the relationship between the average interface velocity and the input heat flux is relatively linear.

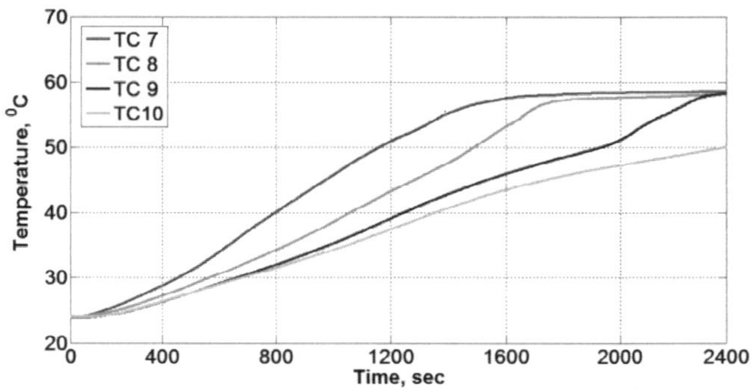

Figure 3: Time-Temperature profile during PCM melting (For vertical position of each thermocouple see Fig. 1)

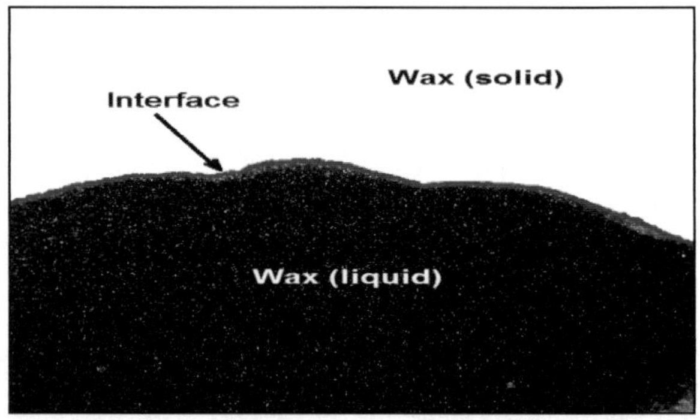

Figure 4: Instantaneous position of the solid-liquid interface

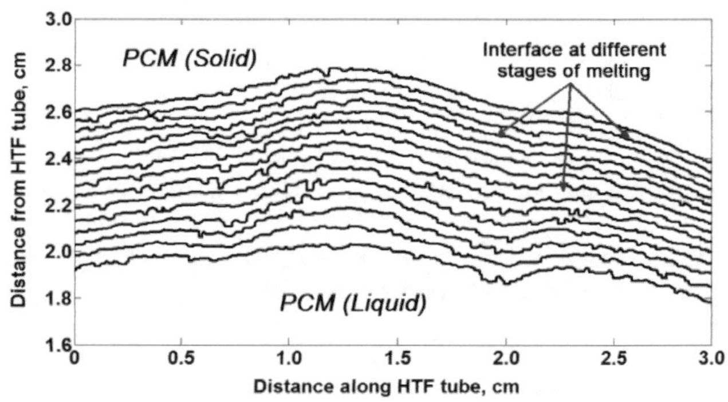

Figure 5: Time history of the solid-liquid interface positions

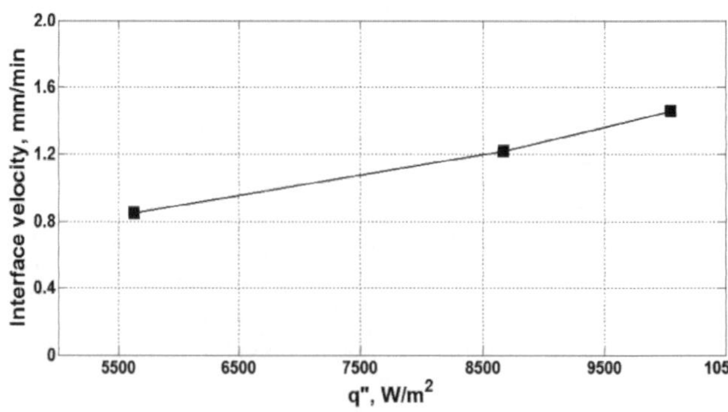

Figure 6: Interface velocity (V_{IF}) with different heat flux

The temperature data inside the PCM and from the tube surface, and the input heat flux allowed computing the local heat transfer coefficient within the melted domain in a transient manner. That is, the heat transfer coefficient was computed at different fractions of the melted region based on the locations of embedded thermocouples within the PCM domain. The local heat transfer coefficient was computed using the Newton's law of cooling,

$$q = hA_T(T_S - T_B) \quad (1)$$

Where, h is the heat transfer coefficient, A_T is the tube outer surface area, T_S is the tube surface temperature and T_B is the bulk liquid temperature in Eq. (1). The heat transfer coefficients were computed locally in the entrance, middle and exit regions of the storage at different melted fractions. The spatially-averaged heat transfer coefficient values at different melted fractions are plotted in Fig. 7 for the three input heat fluxes. The results show that the heat transfer coefficient decreased with an increase in the liquid fraction. A plausible explanation for this trend is that the convective cells formed in the liquid phase are stronger when the liquid fraction is low, which was also observed in the recorded image sequences. This strong convective current lead to higher heat transfer coefficients when the melted region was smaller. The results also show that at a given liquid fraction, the heat transfer coefficient increased with the heat flux, as expected.

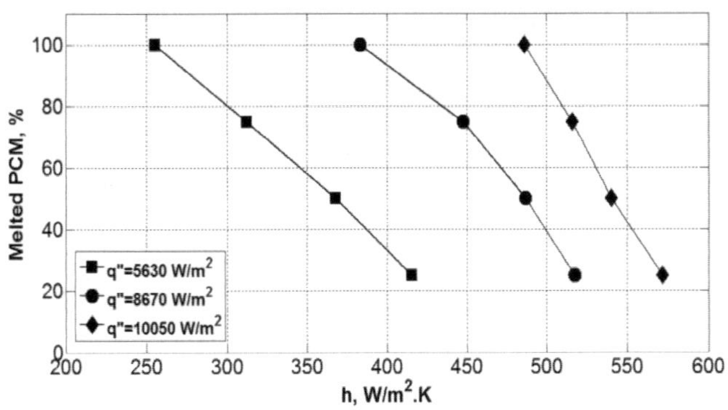

Figure 7: Convective heat transfer coefficient with melting

CONCLUSIONS

The phase change process in a PCM storage system have been studied experimentally using a temperature recording system and an imaging technique, where a horizontal heated tube placed near the bottom of the enclosure acted as the heat source. Results show that the interface moved in a nonlinear fashion although the moving interface showed nearly a uniform velocity along the length of the storage enclosure. The magnitude of the interface velocity increased almost monotonically with an increase in the input heat flux. The results also show that the heat transfer rate was significantly affected by the convective motions in the melted phase of the PCM wax. Results also show that the local heat transfer coefficient decreased with an increase in the liquid fraction, which is likely due to the weakening of the convective cells.

REFERENCES

1. Basak, T., Roy, S., and Balakrishnan, A. 2006, "Effects of thermal boundary conditions on natural convection flow within a square cavity", *International Journal of Heat and Mass Transfer,* Vol (49), pp. 4525–4535.
2. Sharma, A., Tyagi, V., Chen, C., and Buddhi, D. 2009, "Review on thermal energy storage with phase change materials and Applications", *Renewable and Sustainable Energy Reviews*, Vol (13), pp. 318–345.
3. Zalba, B., Marin, J., Cabeza, L. and Mehling, H. 2003, "Review on thermal energy storage with phase change: materials, heat transfer analysis and applications", *Applied Thermal Engineering*, Vol (23), pp. 251-283.
4. Hu, H., and Argyropoulos, S. 1996, "Mathematical modelling of solidification and melting: a review", *Modelling Simul. Mater. Sci. Eng.* Vol (4), pp. 371–396.
5. Fortunato, B., Camporeale, S., Torresi, M., Albano, M. 2012, "Simple Mathematical Model of Thermal Storage with PCM", *AASRI Procedia*, Vol (2), pp. 241- 248.
6. Viskanta, R., 1983, *Phase Change Heat Transfer, in Solar Heat Storage: Latent Heat Materials*, CRC Press, Boca Raton, Fl, USA.
7. He, B., Martin, V., and Setterwall, F. 2004, "Phase transition temperature ranges and storage density of paraffin wax phase change materials", *Energy,* Vol (29) 1785–1804.
8. Xu, R. and Naterer, G. 2004, "Controlled interface acceleration in unidirectional solidification", *Int. J. of Heat and Mass Transfer*, Vol (47), pp. 4821–4832.
9. Betrand, O. et al. 1999, "Melting driven by natural convection", International Journal of Therm. Science, Vol (38), pp. 5-26.

THERMAL DRIFT DUE TO MODULATED DUAL HEATING

M. Z. Hossain
Dept. of Mechanical and Materials Engineering
University of Western Ontario
London, ON, Canada
mhossa7@uwo.ca

J. M. Floryan
Dept. of Mechanical and Materials Engineering
University of Western Ontario
London, ON, Canada
mfloryan@eng.uwo.ca

ABSTRACT

Convection in a horizontal slot subject to spatially modulated heating applied at the upper and lower walls is considered. It is demonstrated that a wide variety of convection patterns can be generated by changing the relative position of both heating patterns. A significant intensification of convection, compared to convection resulting from heating applied at one wall only, results if there is no phase shift between the two patterns, while a significant reduction of convection results from the phase shift corresponding to the half-heating-wavelength. The system generates a non-zero mean shear stress at each wall for all phase shifts except shifts corresponding to half of and one full heating wavelength. This effect, which is generated within one convection cell, gives rise to a global effect which may lead to a thermally-induced drift of the walls if such a drift was allowed.

KEYWORDS: Natural convection, periodic heating, wall shear stress.

INTRODUCTION

Natural convection due to spatially inhomogeneous heating creates horizontal density variations which lead to motions frequently referred to as horizontal convection [1-2]. These gradients create different vertical pressure distributions at different horizontal locations which, in turn, create horizontal pressure gradients. The resulting configurations are statically unstable resulting in the fluid movement regardless of the intensity of the heating. This motion can be generated by heating applied from below as well as from above [3]. There are an infinite number of possible heating patterns but a systematic study and categorization of their effects is not yet available. A comprehensive analysis of the simplest pattern consisting of sinusoidal heating with an arbitrary wavelength has been given only recently [4] and has led to the concept of structured convection. A proper combination of structured convection and the axial pressure gradient leads to drag reduction in heated channels through an effect similar to the super-hydrophobic effect, and is referred to as the Super-Thermo-Hydrophobic effect [5-6].

The convection in question affects contaminant transport in urban environments as local heating rates are determined by the dissimilar thermal properties of roofs, streets and parks. It affects rural environments where local circulation is driven by variations in the heating rates of forests, fields and lakes. The insulating effect of continents on the mantle convection within the Earth represents a spatially structured convection with heating from above [7]. Spatial variability can also result from the use of mixed insulating and conducting boundary conditions [8]. One can further look at systems of localized fires and prediction of their propagation [9], systems of computer chips, thermal patterning in micro-fluidic devices and so on.

The main objective of the present analysis is to study the properties of natural convection driven by two heating patterns, one applied at the upper wall and one at the lower wall. We focus our attention on the simplest patterns characterized by a single wave number, two amplitudes and a phase shift between them.

PROBLEM FORMULATION

Consider fluid confined between two parallel walls extending to $\pm\infty$ in the x-direction as shown in Fig.1. The gravitational acceleration g acting in the negative y-direction. The fluid is incompressible and Newtonian with thermal conductivity k, specific heat c, thermal diffusivity $\kappa = k/\rho c$, kinematic viscosity ν, dynamic viscosity μ, thermal expansion coefficient Γ and variations of the density ρ that follow the Boussinesq approximation. All material properties are evaluated at the reference temperature T_R. The lower and upper walls are subject to periodic heating with phase difference Ω between them, resulting in the walls' temperatures of the form

$$\theta_L(x) = \cos(\alpha x)/2\,, \qquad \theta_U(x) = \cos(\alpha x + \Omega)/2 \qquad (1)$$

where subscripts L and U refer to the lower and upper walls, respectively, θ_L denotes the relative temperature of the lower wall scaled with the amplitude of its peak-to-peak variations $T_{p,U}$, i.e., $\theta_U = (T - T_R)/T_{p,U}$, θ_U denotes the similarly defined relative temperature of the upper wall, T denotes the absolute temperature, $\lambda = 2\pi/\alpha$ is the wavelength of the heating and the half-channel-height h has been used as the length scale.

The horizontal temperature gradients lead to natural convection and the resulting temperature field can be represented as

$$\theta(x,y) = Ra_{p,L}\,\theta_{0,L}(x,y) + Ra_{p,U}\,\theta_{0,U}(x,y) + \theta_1(x,y) \qquad (2)$$

where $\theta_{0,L}$ stands for the conductive temperature field associated with the lower wall heating, $\theta_{0,U}$ stands for the conductive temperature field associated with the upper wall heating, θ_1 denotes the temperature modifications associated with convection and θ stands for the complete temperature field scaled with $\nu\kappa/(g\Gamma h^3)$ as the temperature scale. $Ra_{p,L} = g\Gamma h^3 T_{p,L}/(\nu\kappa)$ and $Ra_{p,U} = g\Gamma h^3 T_{p,U}/(\nu\kappa)$ are the lower and upper periodic Rayleigh numbers expressing the intensity of the heating applied at these walls.

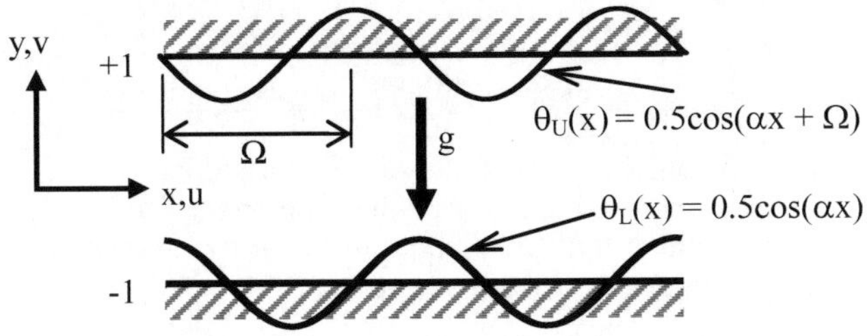

Figure 1: Sketch of the system configuration.

The field equations for the flow field and the temperature modifications have the form

$$u\frac{\partial u}{\partial x} + v\frac{\partial u}{\partial y} = -\frac{\partial p}{\partial x} + \nabla^2 u, \tag{3a}$$

$$u\frac{\partial v}{\partial x} + v\frac{\partial v}{\partial y} = -\frac{\partial p}{\partial y} + \nabla^2 v + Pr^{-1}\theta_1 + Ra_{p,L}Pr^{-1}\theta_{0,L} + Ra_{p,U}Pr^{-1}\theta_{0,U}, \tag{3b}$$

$$u\left(Ra_{p,L}\frac{\partial \theta_{0,L}}{\partial x} + Ra_{p,U}\frac{\partial \theta_{0,U}}{\partial x} + \frac{\partial \theta_1}{\partial x}\right) + v\left(Ra_{p,L}\frac{\partial \theta_{0,L}}{\partial y} + Ra_{p,U}\frac{\partial \theta_{0,U}}{\partial y} + \frac{\partial \theta_1}{\partial y}\right) = Pr^{-1}\nabla^2\theta_1, \tag{3c}$$

$$\frac{\partial u}{\partial x} + \frac{\partial v}{\partial y} = 0 \tag{3d}$$

where ∇^2 denotes the Laplace operator. The boundary conditions consist of the no-slip, the no-penetration and the thermal boundary conditions of the form

$$u(\pm 1) = 0, \quad v(\pm 1) = 0, \quad \theta_1(\pm 1) = 0. \tag{4}$$

METHOD OF SOLUTION

To solve the system (3), we define the stream function $\psi(x,y)$ in the usual manner as, $u = \partial\psi/\partial y$, $v = -\partial\psi/\partial x$, and eliminate the pressure terms. The resulting system of equations are transformed into Fourier space (in x-direction) using the following expansions,

$$\psi(x,y) = \sum_{n=-\infty}^{n=+\infty} \varphi^{(n)}(y)\, e^{in\alpha x}, \quad \theta_1(x,y) = \sum_{n=-\infty}^{n=+\infty} \phi^{(n)}(y)\, e^{in\alpha x} \tag{5}$$

To obtain the numerical solution, the expansions (5) have been truncated after N_M terms. The y-discretization method uses the Chebyshev Collocation technique based on N_T collocation points [10]. The resulting nonlinear system of algebraic equations is solved using an iterative technique combined with under-relaxation. The number of collocation points and the number of Fourier modes used in the solution were selected through systematic numerical tests so that the flow quantities of interest were determined with at least six digits accuracy.

After computing of the velocity and temperature fields, the wall shear stresses acting on the fluid at the upper (τ_U) and lower (τ_L) walls are calculated as

$$\tau_U = \left.\frac{\partial u}{\partial y}\right|_{y=1}, \quad \tau_L = -\left.\frac{\partial u}{\partial y}\right|_{y=-1}, \tag{6}$$

and their average values can be expressed as

$$\tau_{av,L} = \frac{1}{\lambda}\int_0^{\lambda}\left(-\left.\frac{\partial u}{\partial y}\right|_{y=-1}\right)dx = -\left.\frac{du^{(0)}}{dy}\right|_{y=-1},$$

$$\tau_{av,U} = \frac{1}{\lambda}\int_0^{\lambda}\left(\left.\frac{\partial u}{\partial y}\right|_{y=1}\right)dx = -\left.\frac{du^{(0)}}{dy}\right|_{y=1}. \tag{7}$$

DISCUSSION OF RESULTS

The flow patterns created by the different phase shifts between the upper and lower heating patterns are displayed in Fig.2. Convection consists of counter-rotating rolls of equal strength when the hot spots are above each other when $\Omega = 0$ with columns of heated fluid centered around the hot spots and moving upwards and columns of cooled fluid centered around the cold spots and moving downwards. Fluid elements rotate inside cells and periodically come closer to the upper or lower walls, facilitating the net heat transfer between the walls. Use of the phase difference of $\Omega = 3\pi/4$ results in rolls of unequal strength tilted in the negative x-direction. The upward moving heated fluid concentrates in the tilted columns connecting the upper and lower hot spots, and the cooled fluid moving downwards concentrates in tilted columns connecting the cold spots. Further increase of the phase shift leads to an increase of the tilting and, eventually, to a division of rolls into two layers at $\Omega = \pi$ with half of the fluid permanently trapped in the lower section of the slot and the remaining fluid trapped in the upper section. The upper and lower rolls rotate in the opposite directions resulting in the formation of two layers of heated and cooled fluid, with the zones of heated fluid in the lower layers shifted by half of a wavelength in the horizontal directions with respect to the heated zones in the upper layer. The location of the warmer fluid corresponds to the upward fluid motion in both layers. Further increase of Ω results in reversed changes to what has been described above; the form of these changes can be deduced from the horizontal periodicity conditions. The case of $\Omega = 5\pi/4$ is shown for illustration purposes.

The distributions of the wall shear stress acting on the fluid for conditions taken from Fig.2 are illustrated in Fig. 3. These distributions have interesting features depending on how the averages are taken. When $\Omega = 0$, equal and opposite stresses act at the upper and lower walls resulting in zero net stress at each vertical slot section as well as zero mean stress at each wall when averaged over the complete heating wavelength or zero mean stress for both walls when averaged over half of the heating wavelength. This symmetry is lost when Ω increases, resulting in the appearance of a non-zero net stress at each vertical section as well as in a non-zero mean stress at each wall when averaged over one heating period. The latter stress acts in the negative direction at the lower wall and in the positive direction at the upper wall. Distributions of both stresses become identical for $\Omega = \pi$ resulting in a non-zero net stress acting at each vertical slot section and a non-zero mean stress at each wall when averaged over half of the heating wavelength. Since each half gives an opposite direction of the mean shear, their sum gives zero average shear over the full heating wavelength. Further increase of the phase shift till $\Omega = 2\pi$ results in similar variations of shear but with the directions of shear reversed when compared with their directions for $0 < \Omega < \pi$; details of these changes can be deduced from Fig.3 with the help of the periodicity condition.

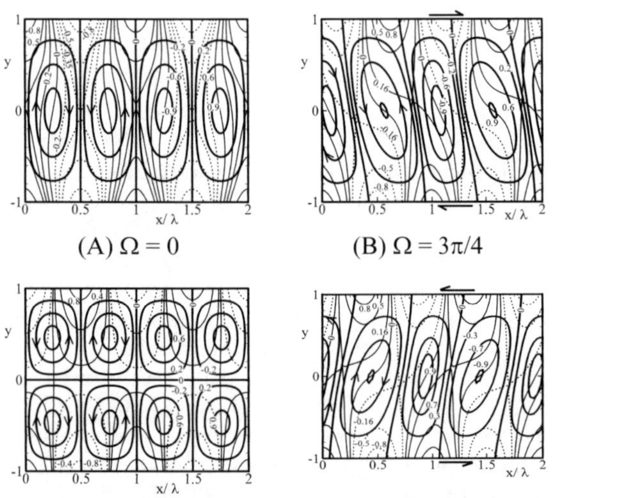

(A) $\Omega = 0$ (B) $\Omega = 3\pi/4$

(C) $\Omega = \pi$ (D) $\Omega = 5\pi/4$

Figure 2: Flow and temperature patterns for $Ra_{p,L} = Ra_{p,U} = 200$, $Pr = 0.71$, $\alpha = 1.6$ and $\Omega = 0, 3\pi/4, \pi, 5\pi/4$, respectively. Thick solid lines correspond to streamlines, and thin solid and dashed lines correspond to the positive and negative isotherms, respectively. Arrows show direction of the mean stress acting on the fluid.

Variations of the average shear (over one heating period) acting at each wall as a function of Ω are depicted in Fig. 4 for $Ra_{p,L} = Ra_{p,U} = 200, 400, 1000$. At the lower wall the mean shear increases slowly from $\tau_{av,L} = 0$ at $\Omega = 0$, reaches a maximum for $\pi/2 < \Omega < \pi$, drops down to zero at $\Omega = \pi$, changes sign and follows a similar pattern for $\pi < \Omega < 2\pi$ dropping again down to zero at $\Omega = 2\pi$. The distribution of $\tau_{av,L}$ is anti-symmetric with respect to $\Omega = \pi$. The mean shear at the upper wall has exactly the same magnitude and distribution but the opposite direction. Increase of $Ra_{p,L} = Ra_{p,U}$ results in much larger magnitudes of $\tau_{av,L}$ and $\tau_{av,U}$ with the maxima moving closer to $\Omega = \pi$ and the appearance of very large gradients of $d(\tau_{av,L})/d\Omega$ and $d(\tau_{av,U})/d\Omega$ in the vicinity of $\Omega = \pi$.

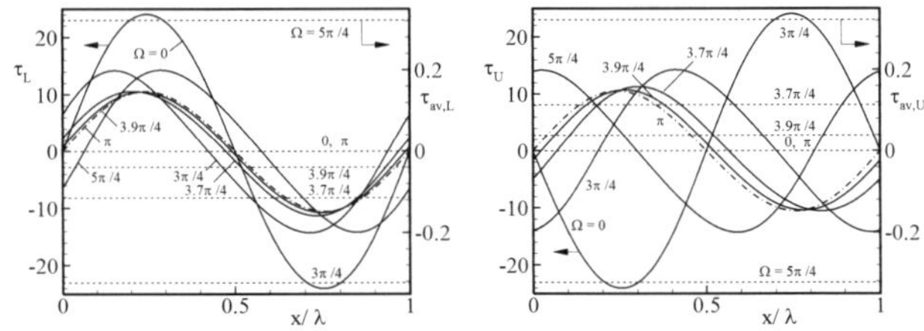

Figure 3: Distributions of the shear stress and its average taken over one heating wavelength acting on the fluid at the lower (left) and upper (right) walls for conditions corresponding to Fig.2. Dashed line corresponds to $\Omega = \pi$.

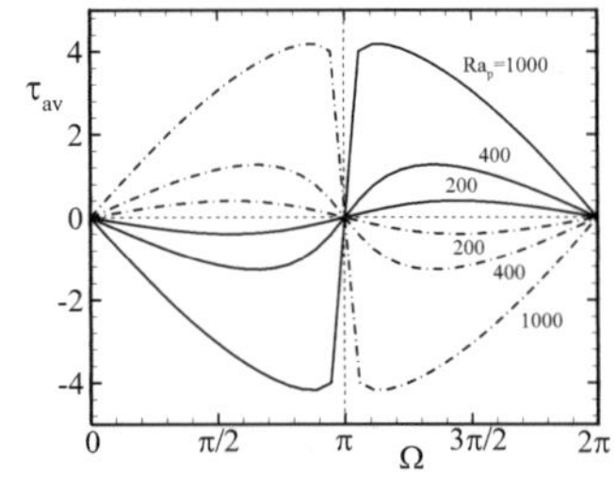

Figure 4: Variations of the average shear stress acting on the fluid at the lower (solid lines) and the upper (dashed lines) walls, as functions of the phase shift Ω for $Ra_{p,L} = Ra_{p,U} = 200, 400, 1000$, and $Pr = 0.71$, $\alpha = 1.6$.

Existence of a non-zero mean shear leads to the generation of a net force acting on each wall. This is an interesting effect as convection within one cell is able to generate an overall effect reaching beyond this cell. This force, summed over several convection cells, could add up to a sizeable magnitude and could lead to a change in the relative positions of both walls assuming that the walls are allowed to move and assuming that the heating patterns are fixed with respect to each wall during such movement. We shall refer to this effect as thermally-induced drift. The directions of the forces acting on the walls are such that the phase shift $\Omega = \pi$ corresponds to an unstable equilibrium and the phase shifts $\Omega = 0, 2\pi$ correspond to stable equilibria. Assuming that drift is allowed, the system will eventually end up in a position where the hot spots are located above each other, i.e. $\Omega = 0, 2\pi$, with the final flow topology illustrated in Fig.2A. Further interesting system properties can be identified by allowing stretching of the boundaries, e.g. boundaries formed by stretchable membranes. For example, when $\Omega = \pi$, membranes on both walls are pulled in the same direction within each half heating wavelength while for $\Omega = 0$ they are pulled in the opposite directions.

The effect of changing the heating wavelength on the average shear is presented in Fig. 5 for $Ra_{p,L} = Ra_{p,U} = 200$. The average shear acting at the lower wall has two extrema at

$\alpha \approx 1.6$, one for $\Omega \approx \pi/2$ and another for $\Omega \approx 3\pi/2$. These extrema have the same magnitude but opposite signs which account for the change in the direction of rotation of the convective roll. The magnitude of $\tau_{av,L}$ decreases to zero when $\alpha \to 0$ and $\alpha \to \infty$; its variations are anti-symmetric with respect to $\Omega = \pi$ for all α's and values are negative when $0 < \Omega < \pi$ and positive $\pi < \Omega < 2\pi$. The average shear acting at the upper wall has the same distribution but the opposite direction.

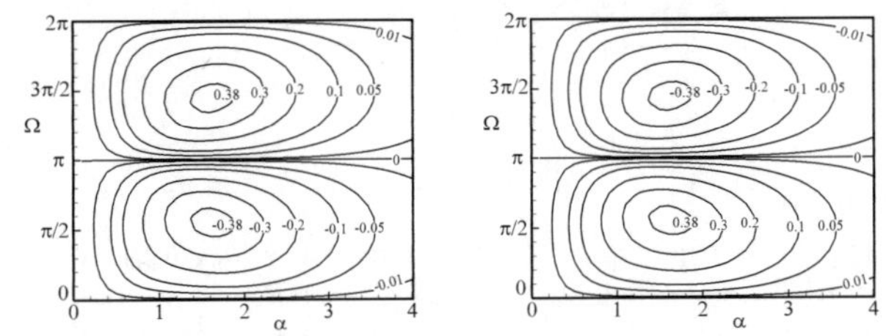

Figure 5: Variations of the average wall shear stress acting on the fluid at the lower (left) and upper (right) walls as functions of α and Ω for $Ra_{p,L} = Ra_{p,U} = 200$, $Pr = 0.71$.

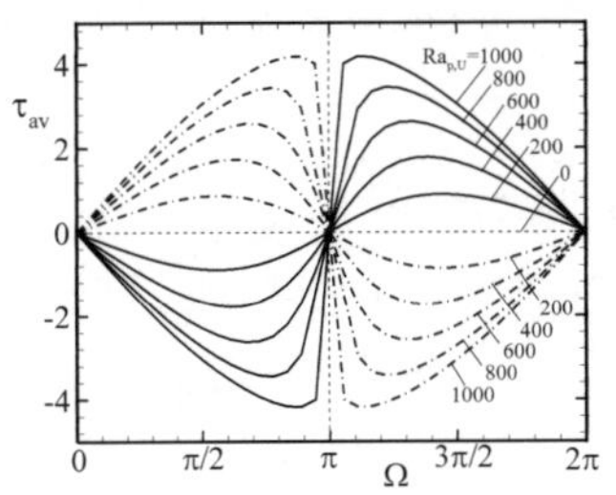

Figure 6: Variations of the average shear stress acting on the fluid at the lower (solid lines) and the upper (dashed lines) walls, as a function of the phase shift Ω for $Ra_{p,L} = 1000$, $Pr = 0.71$, $\alpha = 1.6$ and $Ra_{p,U} = 0, 200, 400, 600, 800, 1000$.

The property of the system for different heating intensities applied at each wall is illustrated in Fig. 6. In this case, the lower wall has constant heating corresponding to $Ra_{p,L} = 1000$ while heating at the upper wall gradually increases. The average wall shear stress is zero at both walls when only one wall is heated. Its magnitude increases as $Ra_{p,U}$ increases, the increase is the same at both walls and values of τ_{av} reach the maximum when the heating intensities at both walls are the same. The phase shift, which produces the highest τ_{av}, changes from $\Omega \approx \pi/2$ at the lowest $Ra_{p,U}$ considered to $\Omega \approx 7\pi/8$ when $Ra_{p,L} = Ra_{p,U} = 1000$. Since the distribution of τ_{av} as a function of Ω is anti-symmetric, a similar shift occurs when the phase shift changes from $\Omega \approx 3\pi/2$ to $\Omega \approx 9\pi/8$.

CONCLUSIONS

Convection in a horizontal slot subject to periodic heating applied at the upper and lower walls has been analyzed. The upper heating has been shifted with respect to the lower heating by a phase shift Ω. A significant change in the convection pattern has been observed as Ω varies from 0 to 2π. This pattern generally consists of one layer of pairs of counter-rotating rolls but morphs into two layers for $\Omega \approx \pi$. Changes of Ω result in significant variations in the intensity of convection, with the weakest convection corresponding to $\Omega = \pi$ and the strongest convection taking place for Ω closed to 0 and 2π. This convection creates a wall shear stress with a non-zero mean for all phase shifts except $\Omega = 0, \pi, 2\pi$. The mean shear, when summed over several heating wavelengths, may lead to the generation of a sizeable force which could lead to a drift in the relative position of both walls. This effect, which is referred to as thermally-induced drift, illustrates the ability of single convection cells to create a global effect.

ACKNOWLEDGMENTS

This work has been carried out with the support of NSERC of Canada.

REFERENCES

1. Hughes, G.O., and Griffiths, R.W., 2008, "Horizontal convection", *Annu. Rev. Fluid Mech.*, vol 40, pp. 185-208.
2. Winters, K.B., and Young,W.R., 2009, "Available potential energy and buoyancy variance in horizontal convection", *J. Fluid Mech.*, vol 629, pp. 221-230.
3. Hossain, M.Z., and Floryan, J.M., 2014, "Natural convection in a fluid layer periodically heated from above" *Phys. Rev. E*, Vol 90, 023015.
4. Hossain, M. Z., and Floryan, J. M., 2013, "Heat transfer due to natural convection in a periodically heated slot", *ASME J. Heat Transfer*, Vol 135, 022503.
5. Hossain, M.Z., Floryan, D., and Floryan, J.M., 2012, "Drag reduction due to spatial thermal modulations", *J. Fluid Mech.*, Vol 713, pp. 398-419.
6. Floryan, D., and Floryan, J.M., 2015, "Drag reduction in heated channels", *J. Fluid Mech.*, Vol 765, pp. 353-395.
7. Lenardic, A., Moresi, L., Jellinek, A.M., and Manga, M., 2005, "Continental insulation, mantle cooling, and the surface area of oceans and continents", *Earth Planet. Sci. Lett.*, Vol 234, pp. 317-333.
8. Ripesi ,P., Biferale, L., Sbragaglia , M., and Wirth, A., 2014, "Natural convection with mixed insulating and conducting boundary conditions: low- and high-Rayleigh-number regimes", *J. Fluid Mech.*, Vol 742, pp. 636-663.
9. Finney, M.A., Cohen, J.D., McAllister, S.S., and Jolly, W.M., 2012, "On the need for a theory of wildland fire spread", *Int. J. of Wildland Fire*, 22(1) 25-36.
10. C. Canuto, M.Y. Hussaini, A. Quarteroni, T.A. Zang, 2006, *Spectral Methods: Fundamentals in Single Domains*, Springer-Verlag, Berlin Heidelberg.

CONVECTIVE HEAT TRANSFER DUE TO MULTIPLE CIRCULAR JETS IMPINGING ON A LARGE CYLINDER

Chris Csernyei
Mechanical & Materials Engineering
Western University
London, Ontario, Canada
ccsernye@uwo.ca

Anthony G. Straatman
Mechanical & Materials Engineering
Western University
London, Ontario, Canada
astraatman@eng.uwo.ca

ABSTRACT

Through the utilization of ANSYS Fluent 13.5, simulations to investigate the effects of convective heat transfer due to multiple circular jets on a large cylinder have been undertaken. The importance of this study stems from the real-world application of convective cooling of rotary cement kilns using large axial fans. The operating conditions were validated through small scale simulations, which were then applied to the full scale model. Effects on heat transfer from the cylinder due to a variation in the number of jets, the jet position, and the jet velocity for a fixed cylinder-to-jet nozzle separation distance (y/d) were studied. In the simulations performed, the use of symmetry to model a greater region was implemented. In regards to general locations of the jets, one was located vertically beneath the kiln with the other angled 45° from the vertical plane, and forming a 90° angle with the cylinder surface. Jet velocity (V_J) was modeled at 5, 10, and 15 m/s. The resulting influences on velocity profile, heat transfer distribution and global Nusselt number are displayed. It was observed that the global Nusselt number increased nearly linearly with increasing jet velocity, and that the case where the vertical and angled fans were positioned at the same axial position produced the highest overall heat transfer

KEYWORDS: Impinging jet, convection, CFD

INTRODUCTION

Rotary cement kilns are utilized in the cement industry for the production of cement clinker. A simple schematic of the layout of a rotary cement kiln is shown in Figure 1. Tilted at an angle of 2 - 5° with respect to horizontal, and operating at a slow rotating speed of approximately 5 rpm, the raw feed (typically limestone, silica, aluminum and iron oxide, Mujumdar [2006]) travels through the cylindrical kiln due to gravity/tumbling. The kiln is comprised of a steel shell for strength and rigidity, lined on the inner surface with refractory brick to enhance the thermal resistance, and to isolate the steel shell from the high temperature process taking place inside. Raw feed enters the kiln at one end, and the fuel (petroleum coke), which is combusted in a burner located approximately 1m into the kiln, enters at the opposite end.

The limestone is first calcined in the decomposition region of the kiln, near the raw feed entrance. Solid-solid reactions begin to occur after the decomposition of limestone into calcium oxide and carbon dioxide completes.

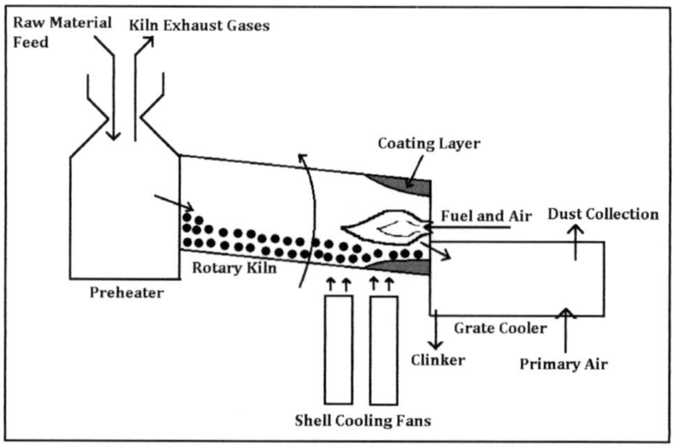

Figure 1: Schematic of a Rotary Cement Kiln

Near the burner at the opposite end of the kiln, the temperature of the exhaust gases can reach upwards of 2200 K, and the reacting raw material can reach as high as 1900 K. The solids begin to melt in this region allowing for solid-liquid reactions to occur. As the liquid solidifies, it forms tiny balls referred to as clinker, as well as forming a relatively thin coating layer over the inner refractory of the kiln. The formation of this coating layer is essential for the longevity of the rotary kiln as it adds to the thermal resistance of the kiln wall.

To reduce the shell temperature of the kiln and promote the formation of this coating layer, kiln shell cooling fans are often used. The highest heat transfer occurs in the burner region of the kiln where the tumbling particle bed, which occupies about 10-15% of the cross-sectional area, makes contact with the refractory. Due to the rotation of the kiln, the particle bed is skewed to one side (see Fig. 2), and in many installations, two different sized fans are used to promote heat transfer. One set of fans is typically positioned vertically beneath the kiln; however, it is offset from the kiln centerline to accommodate the skewed particle bed. A second set is angled 45° from vertical, while still impinging with the kiln outer surface at an angle of 70° - 90°. The orientations of these fans can also be seen in Fig. 2.

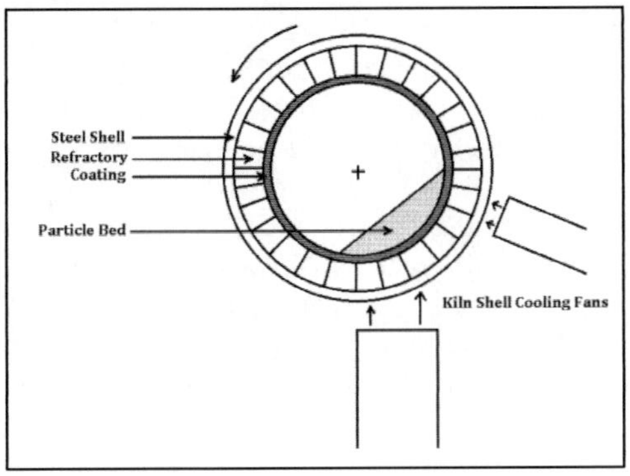

Figure 2: Cross-section of Rotary Cement Kiln

Simulations have been carried out to study the convective cooling of the rotary cement kiln. In these simulations, a constant surface temperature was assumed, enabling a considerable reduction in the size of the computational domain. The focus of the present work is on how convective heat transfer from the surface of the shell is impacted by fan position and flow speed.

MODELLING OF CYLINDER FOR MULTIPLE IMPINGING CIRCULAR JETS

A segment of the studied rotary cement kiln was chosen for these simulations. A representation of the kiln region being studied can be seen in Fig. 3. The rotary kiln is 4.8m in diameter and 75m in length. To reduce simulation run time, the complete kiln was reduced to a segment of length 0.89m, as shown in Figs. 3&4. This length was appropriate to represent an approximately repeating pattern along the kiln axis in terms of fan spacing. With this segment length, six different geometric conditions could be modelled with only the positions and number of jets along the segment being adjusted. The geometry shown in Fig. 3 and Fig. 4 correspond to a 2:1 ratio of vertical to angled fans, and a fan diameter ratio of 29:15. In each of the six cases studied, the vertical jet was bisected by one symmetry plane. The y/d spacing of the vertical jet remained constant for each case. This jet was located directly beneath the kiln and also offset by a distance of the jet's radius. Orientation of the second jet varied from being centered in the domain, to being offset away from the vertical fan. In all cases, a repeating axial pattern could be preserved without changing the width of the segment. These angled fans were positioned to be closer to the kiln shell wall, in accordance with a typical cement plant installation.

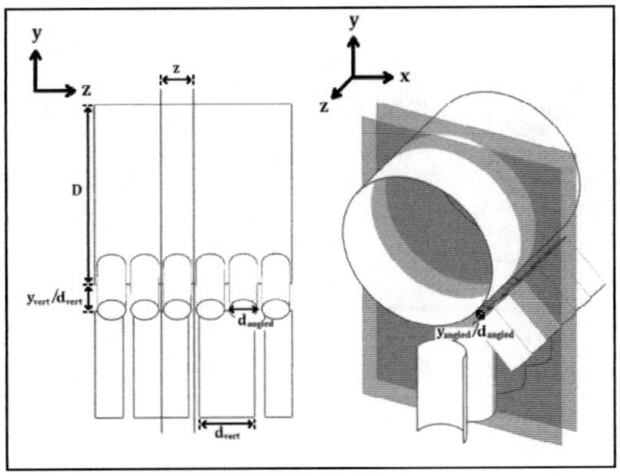

Figure 3: Segmentation of Rotary Cement Kiln

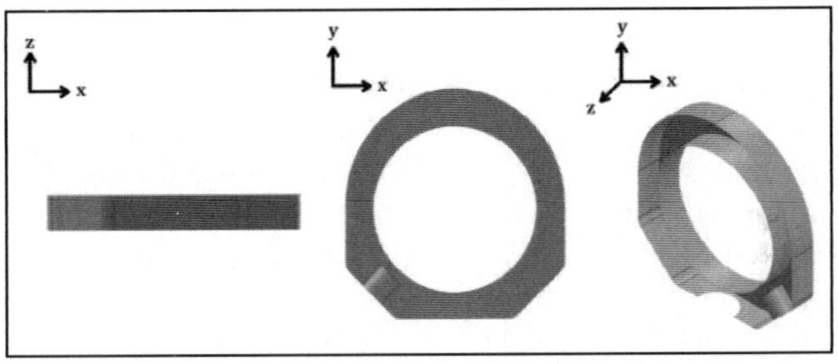

Figure 4: Reduced geometry as extracted from Fig. 3.

The model geometry was designed and modified using SolidWorks 2013 and imported into ANSYS ICEM for mesh generation. Naming of each geometry was determined based on the positioning of the jets.

MESH GENERATION AND CFD MODEL SETUP

Mesh refinement and grid independence was performed prior to selecting an appropriate mesh size for the cases being studied. A tetrahedral mesh was generated for each case.

A finer mesh size was used for the cylinder surface to accommodate for rapid variations in heat transfer in regions of jet impingement. Additionally, a finer mesh was used at the duct inlet, with refinement around the duct wall to account for the entrainment in this region. Element count for the cylinder surface was approximately 60,000 elements, and duct outlet element counts ranged from 2000 – 4000 elements depending on the size of the duct. The majority of the elements were located in the internal body of the domain in order to capture eddies formed due to the collision of air streams. An overall grid size of approximately 1,000,000 elements was used to obtain solutions that were grid independent to 5% in terms of overall heat transfer. Surface meshes for the geometry shown in Fig. 4 can be seen in Fig. 5.

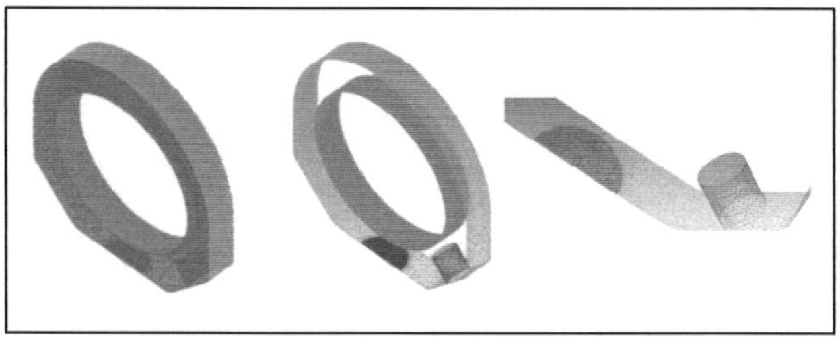

Figure 5: Isometric view (left), isometric without symmetry planes (middle), and close up of vertical and angled jet (right).

The cylinder was modeled as an isothermal stationary wall with a constant temperature of 473K for each case. This was based on a sample of data taken for average shell temperature of a typical kiln. Flow at the jet inlets was set using a velocity inlet condition, with a radially uniform velocity for both the vertical and angled jet set as the same value. Turbulent intensity for each jet was set to 10% to accommodate the large diameter of the jets and distance from the fan to the jet outlet. A study was performed to show the variation in turbulent intensity from 5% to 10% resulted in a percent difference of <5% for the overall heat transfer on the kiln wall. The outer surface of the domain was modeled as a constant-static-pressure boundary to account for outflow and backflow. The pressure for this condition was set as ambient static pressure [Zuckerman 2007], with temperature set to an ambient temperature of 298K. Symmetry conditions were imposed on the two axial planes. CFD simulations were performed using ANSYS FLUENT 13.5. The energy model and k-ε turbulence model with standard wall-functions were utilized. Second-order upwind differencing was used to model advection in all transport equations. Simulation time averaged between 4 – 7 hours per case using a 4 core Intel Core i7-3820 3.60 GHz CPU with 16 GB of RAM running Windows 7 Professional.

RESULTS AND DISCUSSION

Post-processing the data using ANSYS FLUENT resulted in an analysis of the total surface heat transfer (W/m^2). An overall heat transfer coefficient was produced according to Eq. (1),

$$h_{global} = \frac{q''}{[T_{cyl} - T_J]} \quad (1)$$

where q'' is the heat flux, T_{cyl} is the static temperature of the cylinder at 473K, and T_J is the temperature of the jets which were set as an ambient 298K. A global Nusselt number was also determined from the heat transfer coefficient according to Eq. (2),

$$Nu_{global} = h_{global}\left(\frac{D}{k}\right) \quad (2)$$

where, D is the diameter of the cylinder in meters, and k is the thermal conductivity of the air.

The predicted trend in global heat flux (Fig. 6) shows that heat transfer increases in cases where the effect of the angled jet was greater, as might be expected. To show the impact of the fan flow rate, results of global Nusselt number (Nu_{global}) were plotted against the respective jet velocity in Fig. 6. Due to the difference in characteristic diameter for the two jets, a comparison against Reynolds number is not given. Respective Reynolds numbers for each jet can be seen in Table 1. The jet Reynolds number was determined according to Eq. 3,

$$Re = \frac{V_J d}{\nu}, \quad (3)$$

where V_J is the jet velocity, d is the characteristic diameter of the jet, and ν is kinematic viscosity of air.

Table 1: Jet Reynolds numbers at various air stream velocities

Air Stream Velocity (m/s)	Vertical Jet	Angled Jet
5	412,000	213,000
10	823,000	426,000
15	1,236,000	640,000

Flow from each jet was determined to be highly turbulent. Results of the global Nusselt number of the kiln are shown to increase approximately linearly increasing jet velocity.

Figure 6 also shows that Nu_{global} increases approximately 6% with the addition of a third jet and approximately 12 % with the addition of a fourth jet. In addition, changing the position of the vertical jet from directly underneath the cylinder to an offset position results in a difference of <1%.In all the cases where the vertical jet was centered, Nu_{global} was slightly higher than when the jet was offset; however, this result is more pronounced in the offset angled jet cases. A difference of approximately 4% was seen between the centered vertical jet case and the offset vertical jet case. A comparison between the flow fields at the symmetry plane intersecting the angled jet can be seen in Figure 7. In the case where the vertical jet is offset, the flow field from the angled jet is not as pronounced around the cylinder; indicating that the offset jet is actually deflecting away the air stream which is reducing the heat transfer effect of the angled jet.

Increasing the jet velocity of both jets produced a corresponding increase in Nu_{global}. Analysis beyond 15 m/s was not performed in this study as jet airflow speeds of above 15 m/s are typically not utilized in operation.

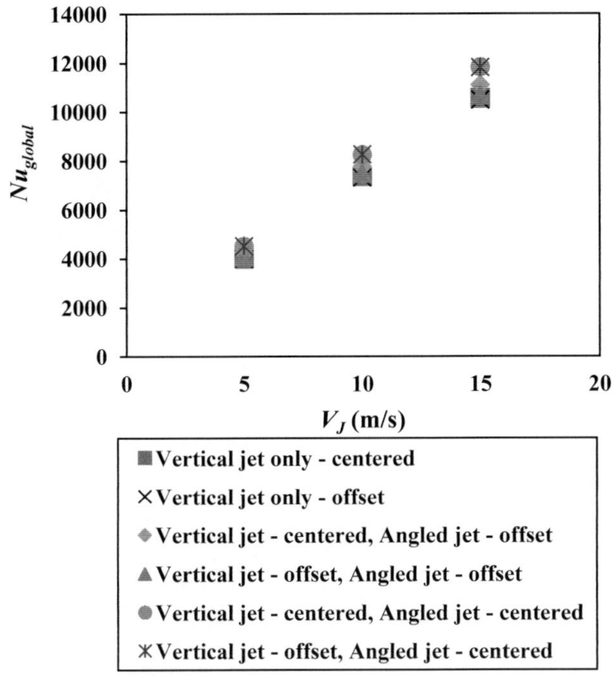

Figure 6: Global Nusselt number for cylinder segment with changing jet velocity

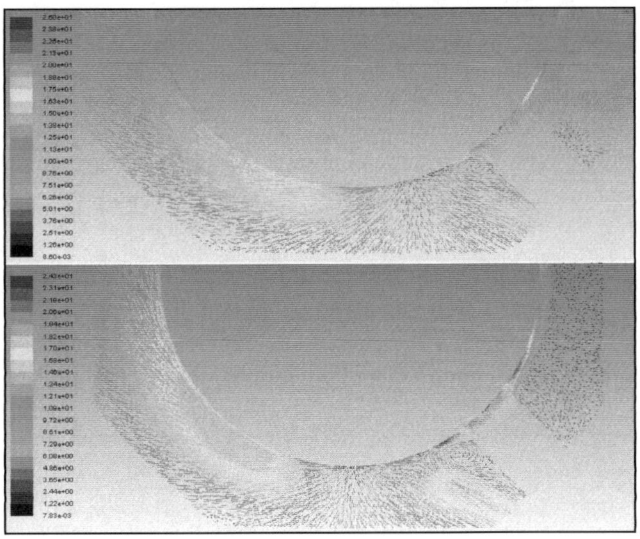

Figure 7: Velocity flow field for 3 jets with offset vertical jet (top) and 3 jets with centered vertical jet (bottom)

Figure 8 shows an isometric view of the total surface heat transfer contours of the cylinder for geometry shown in Figures 4 and 5. Typical heat transfer contours indicate that the area of the largest heat transfer is in the stagnation region on the cylinder where the jet is impinging.

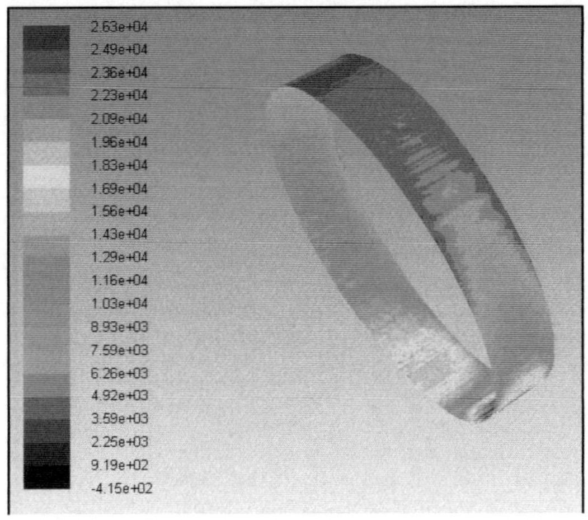

Figure 8: Isometric view of total surface heat transfer (W/m^2) for the centered vertical jet with a centered angled jet case

CONCLUSIONS

A CFD study has been done to study the convective heat transfer due to multiple large jets impinging on an isothermal cylinder. Increasing the airstream velocity of the jets resulted in a corresponding (and expected) increase in overall heat transfer. An almost negligible variation in heat transfer was observed when changing the position of the large vertical jet from centered beneath the kiln to a position offset from center. The case where the vertical and angled fans were centered produced the highest global heat transfer.

ACKNOWLEDGMENTS

The authors gratefully acknowledge the financial support from MITACS Canada.

REFERENCES

1. Li, S.-Q., Ma, L.-B., Wan, W., and Yao, Q., 2005, "A mathematical model of heat transfer in a rotary kiln thermo-reactor", *Chemical Engineering and Technology*, Vol 28, pp. 1480 - 1489

2. Mujumdar, K., and Ranade, V. 2006, "Simulation of rotary cement kiln using a one-dimensional model", *Chemical Engineering Research and Design*, Vol 84, pp. 165 – 177

3. Tawfek, A. 1999, "Heat transfer due to a round jet impinging normal to a circular cylinder", *Heat and Mass Transfer*, Vol 35, pp. 327 – 333

4. Zuckerman, N., and Lior, N. 2007, "Radial Slot Jet Impingement Flow and Heat Transfer on a Cylindrical Target", *Journal of Thermophysics and Heat Transfer*, Vol 21, pp. 548 – 561

Drops, Bubbles

Proceedings of the 25th CANCAM
London, Ontario, Canada, May 31 – June 4, 2015

COMPUTATIONAL STUDY OF MULTIPHASE TRANSPORT THROUGH FUEL CELL FLOW CHANNEL

Abbas Ghasemi
Department of Mechanical and Mechatronics Engineering, University of Waterloo, 200 University Avenue West, Waterloo, Ontario, Canada N2L 3G1
Email: a6ghasem@uwaterloo.ca

Yanzhou Qin
Department of Mechanical and Mechatronics Engineering, University of Waterloo, 200 University Avenue West, Waterloo, Ontario, Canada N2L 3G1
Email: y27qin@ uwaterloo.ca

Xianguo Li
Department of Mechanical and Mechatronics Engineering, University of Waterloo, 200 University Avenue West, Waterloo, Ontario, Canada N2L 3G1
Email:xianguo.li@uwaterloo.ca

ABSTRACT

Liquid flow into the gas flow channel similar to that of fuel cell condition is simulated by using the volume of fluid (VOF) method. The temporal evolution of the liquid into the air stream in the channel is analyzed. The transient development of the issued liquid flow before its impingement on the channel wall is similar to a laminar jet slightly deflected by the air cross flow. The formation and evolution of a liquid wall-jet is observed upon the liquid impinging on the channel wall. Finally in the post-impingement period, the growth of the instabilities on the liquid surface breaks the liquid core down to smaller chunks of liquid, and a slug region is formed in the leading tip of the liquid which grows while traveling to the downstream of the channel.

KEYWORDS: Volume of fluid, Channel flow, liquid jet, break-up.

INTRODUCTION

Efficient performance of the proton exchange membrane (PEM) fuel cells is important for their application as clean power generation devices. Optimum performance of a PEM fuel cell depends on various parameters. One of the important parameters is the water flow condition in gas flow channels [1]. Therefore, various experimental [2-5] and numerical techniques have been [6-8] utilized to study the different aspects of the two-phase channel flow.

In the case of lower liquid flow rates, droplet formation and its dynamics play an important role in the channel flow field and pressure distribution [9]. Droplet deformation and transport, is influenced by the gas flow regime and the channel wettability as well as droplet interaction with the channel surface. Obviously, higher inertia provided by larger gas velocities contributes to the water removal [9]. In addition, at higher contact angles between the droplet and channel surface droplet has less interfacial interaction with wall and can be removed more easily, but droplet tends to have a higher blockage of the channel for the gas flow [9]. Furthermore, droplet detachment and deformation can also depend on channel flow and temperature conditions [10]. Another important factor in the liquid transport is the channel configuration being straight, parallel or serpentine shape [11, 12]. In the case of faster droplet growth or lower gas velocities slug flow formation is also observed in the channels [13, 14].

In most of the previous studies, liquid was introduced in the form of a droplet in the middle of channel or seating on a channel wall. In addition, there is a need for the study of liquid emergence form an orifice while subjected to the gas flow. This can provide information on the liquid jet formation, transition and consequent formation of droplets due to break-up. Therefore, in the present study, a straight section of the channel is considered. A relatively high rate of liquid is assumed to emerge continuously from a single pore on the top side of the gas flow channel. The objective is to study the evolution of the liquid jet into the channel under the influence of the gas phase flow inside the channel. Computational fluid dynamics provides detailed temporally and spatially resolved information on the multiphase flow evolution inside the channel. A volume of fluid (VOF) multiphase approach enables us to track the liquid-gas interface during the solution of the Navier-Stokes equation. Laminar regime governs the gas and liquid flow inside the chamber.

FORMULATION / COMPUTATIONAL APPROACH

Computational domain used for the present study is shown in Fig. 1. Liquid water enters normal to the top wall of the square channel from a square shape opening. Entering liquid is subjected to cross air flow from the channel entrance. Dimensions of the computational domain shown in Fig. 1 are D = 0.1 mm, W = 0.4 mm and L = 4 mm.

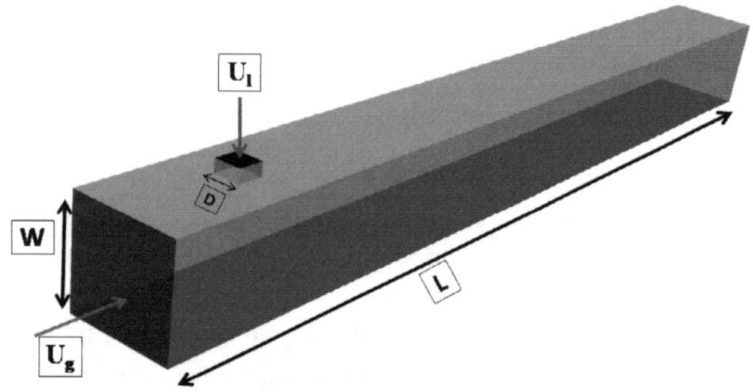

Figure 1: Computational domain

Liquid and air velocities are taken as U_l = 12 m/s and U_g = 10 m/s, respectively, corresponding to the inlet liquid Reynolds number ($Re = U_lD/\upsilon_l$) in the order of 1400, inlet liquid Weber number ($We = \rho_lU_l^2D/\sigma$) of 100, and the inlet gas Reynolds number ($Re = U_gW/\upsilon_g$) of 430. The surface tension, liquid kinematic viscosity and air kinematic viscosity are shown as σ, υ_l and υ_g respectively.

In order to solve for the multiphase flow dynamics a volume of fluid (VOF) approach is implemented using CONVERGE CFD tool. In the VOF method the void fraction α allows for the definition of liquid or gas regions, where $\alpha = 0$ characterizes a cell filled only with liquid and $\alpha = 1$ only gas. Cells with $0 < \alpha < 1$ contain both liquid and gas. In order to capture the liquid/gas interface, a high resolution interface capturing (HRIC) method [15] is used. The solution method based on the void fraction is governed by the following equations for the conservation of mass and momentum:

$$\nabla . u = 0 \quad (1)$$

$$\rho\left[\frac{\partial u}{\partial t} + u.\nabla u\right] = -\nabla p + \mu\nabla.[\nabla u + (\nabla u)^T] + \sigma\kappa\delta_s n \quad (2)$$

where u, p and κ are the velocity vector, pressure, and interface curvature, respectively. The values for the density (ρ) and dynamic viscosity (μ) are defined as:

$$\rho = \alpha\rho_g + (1-\alpha)\rho_l \quad (3)$$

$$\mu = \alpha\mu_g + (1-\alpha)\mu_l \quad (4)$$

In order to account for the advection of the liquid/gas interface, the conservation equation needs to be solved for the void fraction:

$$\frac{\partial \alpha}{\partial t} + u.\nabla\alpha = 0 \quad (5)$$

The computational domain is discretized using a finite volume approach generating Cartesian grid. Velocity inlet is implemented for inlet liquid and gas boundary conditions. All the walls were treated with no-slip condition and outflow conditions were applied to the flow exit region. A second order accurate Crank-Nicholson time marching scheme is used with the time step $\Delta t = 1 \times 10^{-8}\ s$. This choice of time step is to maintain solution stability according to Courant–Friedrichs–Lewy (CFL) criterion. Convection CFL number $CFL_u = \frac{U_{max}\Delta t}{\Delta}$ based on maximum velocity in the domain (U_{max}) and minimum local grid size (Δ) is evaluated as 0.07.

RESULTS AND DISCUSSION

In this section the temporal evolution of the liquid is investigated before, during and after the liquid jet impingement on the channel wall.

In Fig. 2, liquid jet evolution in the central plane of the channel is shown. Liquid is issued from the square shape opening at time t = 0 s. During the development of liquid into the ambient air it is subjected to a cross air flow entering the channel. In this pre-impingement phase the liquid is behaving similar to a laminar jet in cross flow. However, the deflection of the jet axis subjected to the cross flow is not very significant in this phase due to the high inertia of the liquid jet. In addition, the relative gas/liquid velocity creates Kelvin-Helmholtz instabilities causing the interface deformations at the shear layer. The tip of the liquid jet develops a rather round shape in this phase due to the surface tension effect. During t = 40-60 μs, the impingement process starts as the tip of the liquid jet starts to touch the channel wall.

Figure 2: Liquid development into the channel prior to impingement.

During the impingement process shown in Fig. 3, the evolution of the liquid jet is influenced by the no-slip condition at the wall as well as the wall adhesion forces. It should be noted that, wall surface is assumed to have hydrophobic characteristics with the contact angles of 140 degrees. During t = 70-100 μs, a wall-jet region is formed near the wall. This wall-jet region develops until t = 110 μs when its tip starts to detach from the wall. Another observation is the shedding of the liquid phase in the downstream of the liquid jet column. This is due to the separation of cross flow gas phase in the downstream of the liquid column.

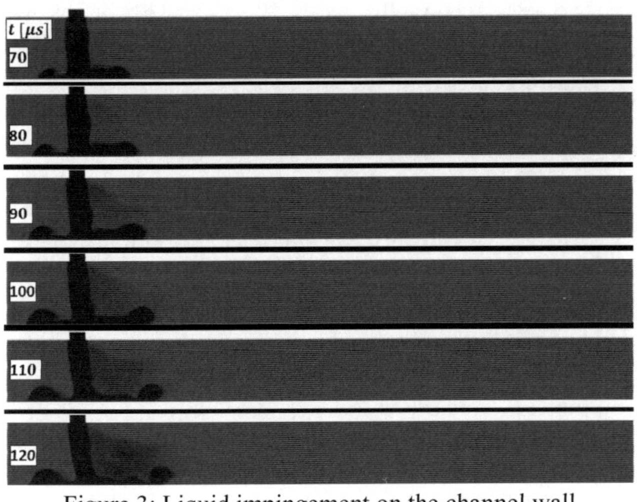

Figure 3: Liquid impingement on the channel wall

In the post-impingement period shown in Fig. 4, the detached liquid in the downstream of the impingement zone evolves into the channel. Due to the interaction of the liquid phase with the wall as well as the air flow in the channel instabilities grow in the downstream of the impingement zone. This is accompanied with meandering and deformation in the gas liquid interface. This causes some patches of liquid to detach from the main body of the liquid. It should be noted that liquid break-up can be a function of many different parameters. If liquid issues from the pores at higher velocities, higher Weber numbers are achieved. It means that inertial forces can be more dominant compared to surface tension and viscous forces. In addition, higher air flows can intensify the cross flow effect in destabilizing the liquid core. Consequently, the liquid phase can break up faster and finer droplets are generated.

It can be seen at the leading tip of the liquid that a slug region is formed. This slug region grows and travels downstream of the channel with increasing time. In addition, the slug flow region grows while traveling downstream. This can create blockage in the channel and result into a significant pressure drop, which is equivalent to a considerable increase in the requirement for the pumping power for the air flow in the channel.

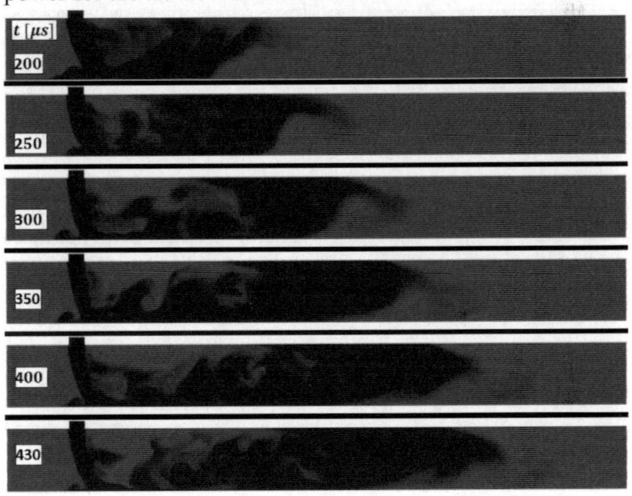

Figure 4: Liquid development into the channel after the wall impingement.

CONCLUSIONS

Computational fluid dynamics (CFD) is used to study the development of liquid water issued into the channel flow similar to that occurs in a PEM fuel cell. Entering liquid is subjected to the air cross flow entering the channel. Both liquid water and air enter the channel with low Reynolds numbers corresponding to the laminar flow regime. Interaction of the liquid with air is calculated using a volume of fluid (VOF) multiphase approach. Evolution of liquid into the air channel is temporally studied in three periods: the development of the liquid issuing from the top wall into a transient laminar jet affected by the cross flow air; the impingement of the liquid jet on the channel wall; and the formation and evolution of a liquid wall jet on the channel wall surface. Finally, in the post-impingement period, downstream evolution of the liquid into the air channel is tracked temporally. It is shown that the growth of the instabilities eventually breaks the liquid core down to smaller chunks of liquid. In addition, a slug region is formed at the leading tip of the liquid causing blockage in the channel.

ACKNOWLEDGMENTS

This work is financially supported by the Natural Sciences and Engineering Research Council of Canada through a Discovery Grant, the Government of Ontario through the 3rd Round of Ontario-China Research and Innovation Fund, and MITACS through an Accelerate Cluster funding.

REFERENCES

1. Anderson, R., Zhang, L., Ding, Y., Blanco, M., Bi, X. and Wilkinson, D.P. 2010, "A critical review of two-phase flow in gas flow channels of proton exchange membrane fuel cells"*Journal of Power Sources,* Vol(195), pp. 4531–4553.

2. Zawodzinski, J.T.A., Derouin, C., Radzinski, S., Sherman, R.J., Smith, V.T., Springer, T.E. and Gottesfeld, S. 1993,"Water uptake by and transport through Nafion 117 membranes" *Journal of Electrochemical Society,* Vol(140), pp. 1041–1047.

3. Zawodzinski, T.A., Davey, J., Valerio, J. and Gottesfeld, S. 1995, "The water content dependence of electro-osmotic drag in proton-conducting polymer electrolytes" *Eletrochimica Acta,*Vol(40), pp. 297–302.

4. Motupally, S., Becker, A.J., and Weidner, J.W. 2000,"Diffusion of water in Nafion 115 membranes",*Journal of Electrochemical Society,* Vol(147), pp. 3171–3177.

5. Choi, P. and Datta, R. 2003 "Sorption in proton-exchange membranes, an explanation of Schroeder's paradox", *Journal of Electrochemical Society,* Vol(1500), pp. 601–607.

6. Pasaogullari, U. and Wang, C.Y., 2004,"Two-phase transport and the role of micro-porous layer in polymer electrolyte fuel cells", *Electrochimica Acta,* Vol(49), pp. 4359–4369.

7. Satija, R., Jacobson, D.L., Arif, M. and Werner, S.A. 2004, "In situ neutron imaging technique for evaluation of water management systems in operating PEM fuel cells", *Journal of Power Sources*, Vol(129), pp. 238–245.

8. Chen, K.S., Hickner, M.A. and Noble D.R. 2005,"Simplified models for predicting the onset of liquid water droplet instability at the gas diffusion layer/gas flow channel interface", *International Journal of Energy Research,*Vol(29), pp. 1113–1132.

9. Mondal, B., Jiao, K. and Li, X. 2011,"Three-dimensional simulation of water droplet movement in PEM fuel cell flow channels with hydrophilic surfaces", *International Journal of Energy Research*, Vol(35), pp. 1200–1212.

10. Cho, S.C., Wang, Y. and Chen, K. S. 2012, "Droplet dynamics in a polymer electrolyte fuel cell gas flow channel: Forces, Deformation and detachment. II: Comparisons of analytical solution with numerical and experimental results", *Journal of Power Sources*, Vol(210), pp. 191–197.

11. Ryan Y.D., Zhang, A.L., Bi X. and Wilkinson, D.P. 2013, "Simulations of two-phase flow distribution in communicating parallel channels for a PEM fuel cell", *International Journal of Multiphase Flow,* Vol(52), pp. 35–45.

12. Jiao, K., Zhou, B., Quan, P., 2006,"Liquid water transport in parallel serpentine channels with manifolds on cathode side of a PEM fuel cell stack", *Journal of Power Sources,* Vol(154), pp. 124–137.

13. Colosqui, C.E., Cheah, M.J., Kevrekidis, I.G. and Benziger, J.B. 2011, "Droplet and slug formation in polymer electrolyte membrane fuel cell flow channels: The role of interfacial forces", *Journal of Power Sources,* Vol(196),pp. 57– 68.

14. Zhang, F., Yang, X. and Wang, C. 2006,"Liquid Water Removal from a Polymer Electrolyte Fuel Cell",*Journal of Electrochemical Society,*Vol(153), pp. 225-232.

15. Waclawczyk, T. and Koronowicz, T., 2006, "Modeling of the wave breaking with CICSAM and HRIC high-resolution scheme", *Proceedings of the European Conference on Computational Fluid Dynamics*, Egmondaan Zee, The Netherlands, pp.1-19.

RELATION BETWEEN PAINT APPLICATION PARAMETERS AND DROPLET DIAMETER: A WAY TO IMPROVE EFFICIENCY AND REDUCE AIR EMISSIONS

Rajan Ray
Department of Civil & Environmental Engineering
University of Windsor
Windsor, Ontario, Canada
rayr@uwindsor.ca

Dr. Paul Henshaw
Department of Civil & Environmental Engineering
University of Windsor
Windsor, Ontario, Canada
henshaw@uwindsor.ca

ABSTRACT

The automotive coating sector is always a key concern for both automotive and environmental engineering. Appearance is the most important factor in customer's initial satisfaction, while a good coating is a requirement for better lifespan of the vehicle. In respect of environmental engineering, coating is responsible for more than 90% of volatile organic compounds (VOC) in the automotive industry. A proper investigation of different types of coatings and the effect of the variation of application parameters can help the industry. It will also help to better understand the relationship between application and finish, which will eventually optimize the coating application and decrease the VOC emission.

KEYWORDS: Coatings, Fluid flow, Spray, Atomization.

INTRODUCTION

Different types and layers of coatings are applied in the automotive coating industry. Although a variety of coating application processes is available, the basis of all coatings remains the same concept: a film-forming material is applied on the surface of a body. In automotive paint shops primer, basecoat and clear coat are typically used as coating layers. The current trend is that basecoats contain metallic flakes to enhance sparkle. The clear coat is a protective layer that enhances gloss.

In the automotive paint industry, appearance and cost efficiency are the two key aspects which drive improvements in technology. Appearance is directly related to automotive sales and customer satisfaction. In spite of recent advances, the industry is still fighting with the third aspect: controlling the VOC emissions, which has been a concern for decades.

Surface coating is still the most significant source of VOC emission from the automotive industry. The solvents used in the coating can easily evaporate in the air during coating application and curing. US EPA has a standard that maximum 1.40 kg of VOC/ L of applied coating solids (ACS) and 1.47 kg of VOC/ L of ACS for primer and top coat is allowable for automotive paint industry (Subpart MM, 2005). The UNECE 1999 Protocol (Gothenberg Protocol) states that the VOC emission limit values for automotive surface coating of new car plants should be 45 g/m^2 (Ozone Protocol, 1999). According to Canadian National Pollutant Release Inventory (NPRI) 2003 report, the Canadian paint and solvent sector accounts for around 18%% (323000 tonnes) of total VOC emitted in 2012 (Environmental Canada, 2014). Almost all of this VOC comes from automotive surface coating.

A Phase Doppler Anemometer (PDA) system was used for particle diameter measurement in the spray from a Rotary Bell Atomizer (RBA) in the current study. A RBA is a special type of paint applicator where paint is pumped into the interior of a cup (the "bell") rotating at thousands of revolutions per minute (krpm). The paint forms ligaments at the edge of the cup, and these subsequently form droplets. RBAs are normally used in high volume production environments like the automotive industry where its superior transfer efficiency and spray pattern consistency result in a smooth finish. A typical bell applicator also has a cone of shaping air, blown in the same direction as the paint, to contain the spray. RBAs produce fine droplets, and so the spray must be charged electrostatically, in order to reach the electrically-grounded target. In a PDA system two out-of-phase laser beams create an interference pattern where they cross. Particles which pass through the measurement volume reflect the interference pattern, which is shifted, depending on the velocity of the particle. When the droplet is small and spherical, the light refracts from it, and the extent of refraction is a function of particle size. Thus the PDA is used to calculate the size, velocity and volume flux of spherical particles, droplets or bubbles suspended in gaseous or liquid flows. A PDA and RBA are shown in Figure 1.

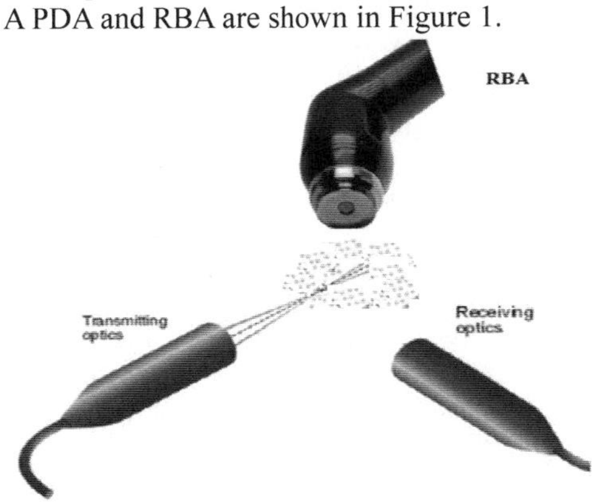

Figure 1: Schematic experimental diagram of PDA with the RBAA proper investigation which correlates various

paint application parameters to paint flux can increase the efficiency of the spray while decreasing the VOC emissions in the automotive paint shop. The Bell Speed (BS) and Flow Rate (FR) of the paint to the RBA are two major operational controlling factors. A detailed investigation of these factors on different types of coatings can provide a better picture of coating evaporation which is important to optimize the use of coatings applied in the industry.

This research investigated the effects of BS and FR on particle Sauter mean diameter (D_{32}) of water and clear coat sprayed from a rotary bell atomizer, measured using a Phase Doppler Anemometer (PDA). It also strives to calculate the effects of these parameters on metallic paint. A Rotary bell atomizer (RBA) guided with a robotic arm was used to atomize the fluid. Water was used as the fluid initially to establish a method to investigate these effects, a method which was used here for clear coat and metallic paint.

LITERATURE REVIEW

Frazer R.P. *et al.* conducted a range of investigations in this area in 1962 to better realize the formation of liquid sheets during atomization. A spinning cup similar to the current rotary bell was used in an air stream released vertically with the fluid spray. They figured out that the initial drop size from the breaking of the liquid sheet was a function of sheet thicknesses. They did not find any significant impact of air containment on atomization. However, they rationalized that a good finish is highly dependent on the sheet uniformity and critically dependent on application parameters like the spinning rate of the cup, flow rate and viscosity of the fluid.

A direct relation between bell speed, flow rate and mean drop size was established by Corbells *et al.* (1991). A higher bell speed, resulting in higher centrifugal and Coriolis forces accounted for a lower mean drop size, while a higher flow rate accounted for higher mean drop size when the bell speed was lower than 20 krpm. If the flow rate increases with a constant bell speed, the average drop size increases. But a higher bell speed with a constant flow rate decreases the average drop size and increases the droplet number.

Di Domenico and Henshaw (2012) established a full factorial two-level DOE matrix to determine the relationship between the paint finish appearance and four variable RBA parameters: flow rate, bell speed, shaping air and electrostatic potential. The appearance was categorized by evaluating the distinctness of image, surface waviness, and multi-angle color. It was hypothesized that a change of particle diameter due to the change in these important factors played an important role in the evaporation of the paint droplets in-flight, and hence affected the appearance of the cured paint.

There is a direct relation between the droplet size, surface area and evaporation. The higher the drop size, the lower the total surface area (for a given volume of paint) and the lower the total evaporation. Zhentao W. (2009) used a PDA and four types of circular nozzle to atomize a water jet sprayed upwards at a constant pressure and temperature. They found that with increasing axial distance, the droplet diameter increased and the particle velocity decreased.

Metallic solvent borne and water borne paint sprayed from a RBA was used to measure the particle size characteristics by using a shadowgraphy technique (Basu *et al.,* 2010). A CCD camera was used to calculate the particle size. For bell speeds within a 30-50 krpm range, the mean particle size (D_{32}) decreased with an increasing bell speed. The shaping air was found insignificant for 30-50 krpm range bell speed.

In a previous investigation using water spray (Ray *et al.*, 2015), the PDA was used to show the effect on evaporation of water by varying BS, FR, and SA. Electrostatic charging was not used. All other variables like atmospheric pressure and temperature were kept constant. The effect of shaping air was found to be statistically insignificant. The total evaporation (cm^3/s) between 22.5 and 37.5 cm from the atomizer was calculated from point measurements of the droplet flux. It was found that the droplet diameter decreased and hence the evaporation increased with higher bell speed or lower flow rate.

EXPERIMENTAL SETUP

Fluid (water, clear coat or base coat) was fed into a high speed bell atomizer (ABB RB1000) which was attached to a computer guided robotic arm system (ABB IRB-5500X). A PDA system (Dantec Dynamic A/S, Denmark) was used to measure the particle diameter, velocity, volume flux, and particle size distribution of the spray. The PDA system had receiving and transmitting optics inside the spray booth. The control and processing unit outside the booth consisted of a computer and the BSA flow software package. A nebulizer (Airlife-002002) was used to align the PDA and robotic arm initially. The spray booth was maintained with a constant temperature of 24° C (75±1°F) and 65±2% humidity. The booth had a constant and uniform, mostly vertical, downward air velocity of 2.1 m/s. But the air had a tendency to flow towards the outer wall of the booth. In order to minimize this effect, the PDA was placed at the center of the booth. Ray *et al.* (2015) described the experimental setup used for this current investigation.

Measurements were taken along a traverse through the spray in an x-y plane at 22.5 cm along the axis of the bell cup rotation (z-axis). The points in the spray were sampled in 2.5 cm radial intervals by moving the robot arm holding the RBA relative to the fixed PDA probe and receiver. The shaping air was set at a constant 200 standard liters per minute (L/min), bell speed was set to 30, 40 and 50 thousand rotations per minute (krpm) and the flow rate was varied between 100, 200 or 300 cubic centimeters per minute (cm^3/min). The electrostatic charge was applied and varied for the base coat and clear coat. No electrostatic charge was used when using only water. The electrostatic change was varied between 60, 70 or 80 kV for clear coat. Only 70 kV electrostatic charge was used for base coat.

The PDA system was set to acquire data for a maximum of 10,000 particle counts at each position or to acquire for 5 seconds, whichever came first. This ensured a reasonable maximum number of counts at the most distal point in the spray where particle counts were lower, while not overloading the system at the center of the spray where a higher particle density existed. However, the system can only determine a statistically correct diameter or velocity with a minimum 1500 particle counts. These limitations confined the spray's realistic range up to 7.5 cm for water and 12.5 cm for clear coat. It is important to note that particle counts beyond this range were low and represents less than 4% and 7% of the total flux for water and clear coat, respectively (Ray et al., 2015). Similarity in the x and y directions was assumed. Figure 1 shows the experimental setup in the spray booth for the current testing method.

Each test was repeated 20 times for water and 3 times for clearcoat and metallic paint to estimate the variance and the variance was multiplied by 1.96 to calculate 95% confidence limit of the (Dantec Dynamics, 2006; Albrecht H.-E. *et al.*, 2003). This was used to calculate the uncertainty of the PDA measurement. The static repeatability of the positioning robot is 0.15 mm (ABB, 2009). It was included to calculate the total uncertainty of the test. Ray et al.(2015) describes a detailed calculation of uncertainity.

RESULTS AND DISCUSSIONS

Particle mean diameter decreased with increasing bell speed and decreasing axial distance when water was used as a spray fluid at 300 cm^3/s flow rate (Figure 2). Figure 3 shows the same trend while using clear coat. It is noted that the initial decrease in mean diameter is steep while using water where a modest decrease was found in the case of clear coat.

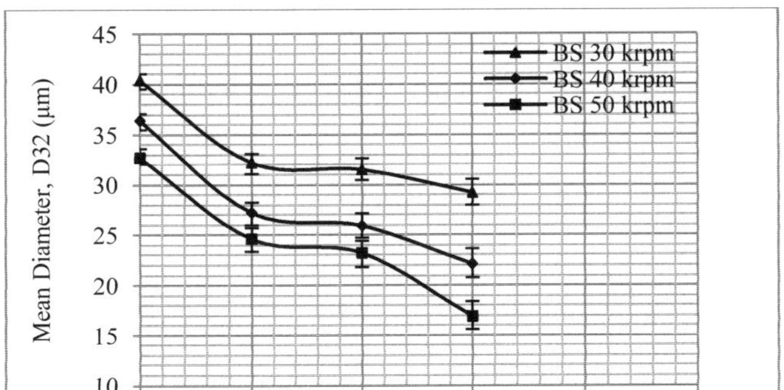

Figure 2: Mean diameter, D_{32} (µm) with increasing Bell speed for 300 cm^3/s flow rate for water

At higher bell speed, the initial sheet thickness is less, which causes a smaller initial drop diameter (Frazer *et al.*, 1962). At the same time, the fluid particles have a higher velocity. This causes two effects: the droplets travel faster in space which decreases the exposure to unsaturated air and hence evaporation, but the increased velocity means that there is less flight time for evaporation. Ultimately, the combined effect increases the evaporation and hence decreases the drop diameter. Our findings support those from previous researchers.

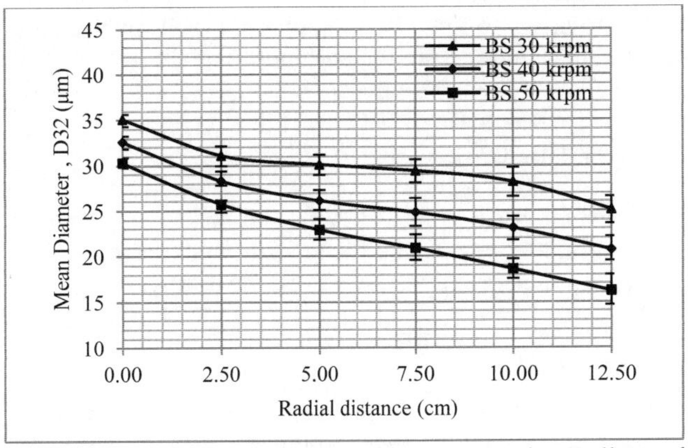

Figure 3: Mean diameter D_{32} (µm) with increasing Bell speed for 300 cm^3/s flow rate of clear coat

This research attempted to figure out if the same trend exists for basecoat. Figure 4 show that the results for metallic basecoat followed the same trend as clear coat and water. However, these results should be used with caution. The basecoat droplets have a high opacity and do not refract the light so well. With water and clear coat, the validation was more than 90% for all cases; However, it was less than 65% for base coat. According to Dantec Dynamic, the provider of PDA system, experimental values with less than 80% validation are statistically unreliable (Dantec Dynamics, 2006). Because of this uncertainty, the flux measurement for the basecoat was considered invalid, and the evaporation rate between x-y planes was not calculated.

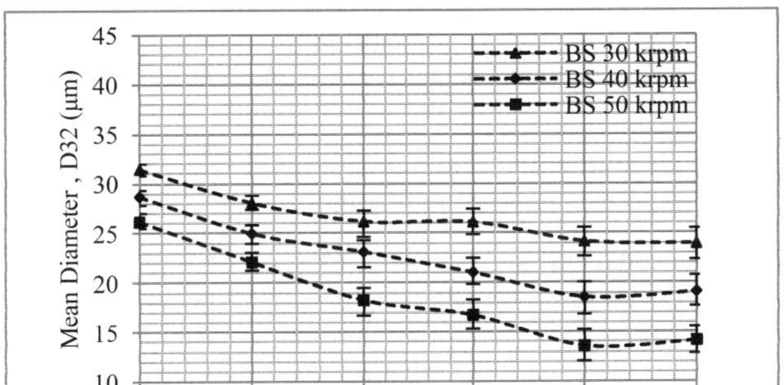

Figure 4: Mean diameter D_{32}(µm) with increasing Bell speed for 300 cm^3/s flow rate of base coat (metallic, brilliant black)

CONCLUSIONS

The trend of mean diameter decreasing with increasing bell speed that was found in water and clear coat was also found for basecoat. However, the measurements with metallic-pigmented basecoat had a low statistical viability, and were deemed unsuitable for calculating the flux through parallel planes, which is required for calculating the evaporation. Future plans include repeating the test with a non-metallic ("straight shade") basecoat.

ACKNOWLEDGMENTS

University of Windsor/Chrysler Canada Automotive Research and Development Center provided the necessary training, equipment and facilities to do the testing.

REFERENCES

1. Albrecht H.-E., Heinz-Eberhard (2003). Laser doppler and phase doppler measurement techniques. Springer, Berlin; New York

2. ABB K. K., 2009, Installation and Operation Manual RBB 1000 Series, RB-1000WSC Rev-01. *shimada shi*, Japan.

3. Air Pollutant Emission Summaries and Historical Emission Trends, 2014 Preliminary Version Release, *Environment Canada*, 2014.

   http://www.ec.gc.ca/indicateurs-indicators/default.asp?lang=en&n=64B9E95D-1

4. Basu K. S., Zhou J., Harding A., Moncier J., Williams C., Baker L., MacCreight K. (2010) "Effect of atomization and rheology control additives on particle size and appearance of automotive coatings". *ISCST* 2010.

5. Corbeels P. L., D. W. Senser, A. H. Lefebvre: Atomization characteristics of a high-speed rotary-bell paint applicator. *Atomization and Sprays*, 1992, 2, 87-89.

6. Dantec Dynamics, 2006, BSA Flow software V.4.10, *Installation and User guide*, Denmark.

7. Di Domenico J. and Henshaw P. (2012) The Effects of Base Coat Bell Application Parameters on Elements of Appearance for an Automotive Coatings Process, *JCTR.*

8. Fraser, R.P., Eisenklam, P., Dombrowski, N., and Hasson, D. 1962. Drop formation from rapidly moving liquid sheets. *Am. Inst. Chem. Eng. J.,* 8(5), 672–680.

9. Ozone, UNECE, 1999 Gothenburg Protocol to Abate Acidification, Eutrophication and Ground-level Ozone, *Ozone Protocol, 1999.*

   http://www.unece.org/env/lrtap/multi_h1.htm

10. Ray, R., Henshaw, P., Biswas, N. & Sak C., 2015 Evaporation of Water Spray from a Rotary Bell Atomizer using a Phase Doppler Anemometer, *Atomization and Sprays* (Accepted).

11. Subpart MM, US EPA, Title 40, Part 60, Standards of Performance for Automobile and Light Duty Truck Surface Coating Operations, *Subpart MM, 2005.*

12. Zhentao W., 2009. Experimental Investigation on Spray Characteristics of Aerodynamic Cooling Nozzle for Metal Cutting. In Proceedings of the 2009 International Conference on Energy and Environment Technology - Volume 03 (ICEET '09), Vol. 3. *IEEE Computer Society, Washington, DC, USA*, 323-326.

Proceedings of the 25th CANCAM
London, Ontario, Canada, May 31 – June 4, 2015

EXPERIMENTAL INVESTIGATION OF PARTICLES TURBULENCE IN A HIGH REYNOLDS VERTICAL PIPE FLOW

Rouholluh Shokri
Department of Chemical and Materials Eng.
University of Alberta
Edmonton, AB, Canada, T6G 2V4
shokri@ualberta.ca

Sina Ghaemi
Department of Mechanical Eng.
University of Alberta
Edmonton, AB, Canada, T6G 2G
ghaemi@ualberta.ca

David S. Nobes
Department of Mechanical Eng.
University of Alberta
Edmonton, AB, Canada, T6G 2G
dnobes@ualberta.ca

R. Sean Sanders
Department of Chemical and Materials Eng.
University of Alberta
Edmonton, AB, Canada, T6G 2V4
ssanders@ualberta.ca

ABSTRACT

Particle-laden turbulent flows are highly complicated due to the poorly understood interactions between the fluid turbulence and the dispersed particles. A high-Reynolds-number experimental investigation of the turbulent motion of solid particles in liquid flow at an unprecedented Reynolds number of Re = 320,000 is conducted in a vertically upward section of a closed slurry pipe loop. The solid particles are 0.5, 1, and 2 mm glass beads with volumetric concentration of 0.1, 0.4 and 0.8 percent, respectively. Measurements of the turbulent motion of the solid particles are conducted using particle tracking velocimetry (PTV) technique while the surrounding liquid phase is measured by PIV technique. Results show that the mean velocity of the glass beads is slower than the liquid phase at the central region of the pipe while it is faster in the near wall region. In the core of the flow larger particles have higher streamwise turbulent fluctuations $\langle u^2 \rangle$ while the smaller ones have a higher $\langle u^2 \rangle$ near the wall. The particles have much higher radial fluctuation than the fluid which is intensified by increase of the size of particles. Measurement of particle concentration also shows that larger 2 mm particles tend to accumulate in core region and follow a linear distribution.

KEYWORDS: turbulence, fluctuation, vertical pipe, particle-laden flow, PIV/PTV

INTRODUCTION

The presence of small solid particles affects the transport of mass, momentum, and heat in high Reynolds number turbulent pipe flows operating in a variety of engineering systems. The motion of solid particles and their interaction with the fluid turbulence introduces a complicated system which is a function of Re and Stokes number of the particles, particle/fluid density ratio, flow direction, and solid phase concentration. This complexity has limited analytical models and numerical simulations of particle-laden flows to simplified conditions and relatively low Re numbers. The higher Re numbers are accessible by experimental investigation although according to Balachandar & Eaton[1] measurement in turbulent particle-laden flows is still limited to Re number of 30,000 far lower than most industrial applications.

In order to develop a good understanding of such a complex particulate systems, the relative turbulent motions of the carrier and solid phases have been investigated in the literature. Kulic et al.[2] and Varaksin et al.[3] illustrated that for small particles in the range of 50 to 70 µm in a downward gas flow at Reynolds number less than 15,000, streamwise turbulence intensity of particles is higher than the single phase. However the radial turbulence intensity of particles is lower that the single phase flow. Kussin and Sommerfeld[4] measured particle turbulence intensities in particulate gas flow in a horizontal pipe with different wall roughness up to Reynolds numbers of 57,000 and showed that wall roughness significantly impacts the turbulence intensity of the particles. Caraman et al.[5] reported the turbulent statistics for 60 µm glass beads in a downward gas flow at Reynolds number of 5,300. They found out that the particles had higher streamwise fluctuating velocities than the gas and the fluctuations in normal direction to the flow direction were almost the same for both phases. Fluctuating velocities of particles in normal direction to the flow were much higher than the fluid fluctuations. Kameyama et al.[6] conducted some experiments for 0.625 mm glass beads in both upward and downward water flows at Re=19,500 and showed that both radial and axial turbulence intensity fluctuations of particles are equal to or higher than the water in both up/downward flow directions. Aside from the limited low Reynolds numbers (lower than 57,000) experimental investigations on the particle turbulent motions, the experimental data on turbulent statistics of the particles in liquid flows are extremely scarce as most of the investigations are particle-laden gas flows. Further experimental investigation is needed to obtain high Re data and scrutinize the solid phase turbulence in liquid phase (low ρ_P/ρ_f).

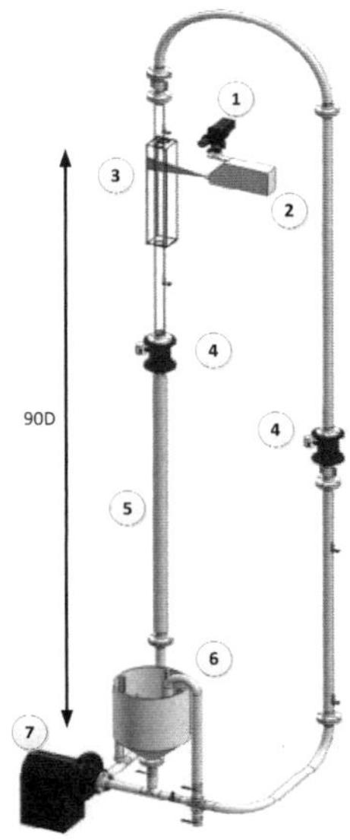

Figure 1: Schematic of the experimental setup consisting of (1) camera (2) laser, (3) acrylic pipe and viewing box, (4) magnetic flow meters, (5) double pipe heat exchanger, (6)-Feeding tank, (7) and centrifugal pump.

The current experimental investigation provides detailed characterization of turbulent motion of solid particles dispersed in water in a 2 inch upward flow at *Re* of 320,000. Apart from the industrial application of vertical flows such as well drilling and hydraulic hoisting, the interactions between fluid turbulence and particles are not additionally complicated by having gravity creating highly skewed concentration over the cross section. Glass beads were used as the particulate phase with diameters of 0.5, 1 and 2 mm and volumetric concentrations of 0.1, 0.2, and 0.8 percent respectively. A combined PIV/PTV (Particle Image/Tracking velocimetry) technique is applied for simultaneous measurement of turbulent statistics of both phases as detailed in the subsequent sections.

EXPERIMENTS

A schematic of the closed slurry loop is shown in fig.1. The loop operates using a centrifugal pump controlled by a variable frequency drive while the fluid temperature is held constant at 25°C using a heat exchanger. Measurements are conducted in the upward flow pipe section with pipe diameter of D=50.6 mm. A transparent test section made from acrylic is located 90D after the lower bend providing sufficient length to reach a fully developed flow condition. In order to minimize image distortion due to the curvature of the pipe wall, a rectangular acrylic box filled with water is placed around the test section. The particulate phase consists of glass beads with nominal average diameter of 0.5, 1 and 2 mm with volumetric concentrations of 0.1, 0.4 and 0.8 %, respectively. Glass beads have the density of 2500kg/m^3 resulting in ρ_p/ρ_f =2.5 where ρ_p and ρ_f are particle and fluid density, respectively. Average bulk velocity is 5.74 m/s which correspond to Reynolds number (*Re*) of 320,000.

A planar Particle Image/Tracking Velocimetry (PIV/PTV) technique is employed to capture the motion of both liquid and particulate phases. Flow is seeded by 18 μm hollow glass tracers with density of 600 kg/m^3 (60P18 Potters Industries). PIV Images are captured by a CCD camera (Imager Intense, Lavision) with 1376×1040 pixel resolution which has a physical pixel size of 6.45×6.45 μm. The required PIV illumination is provided by an Nd:YAG laser (Solo III-15, New Wave Research) which creates a sheet with a thickness of about 1 mm. For each set of experiments, 10,000 pairs of double-frame images are acquired using Davis 8.2 (LaVision). Magnification and spatial resolution of the imaging system is set at 0.27 and 42.6 pixel/mm, respectively.

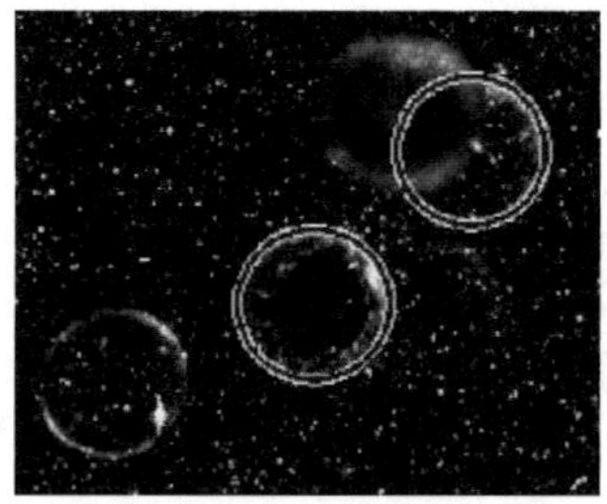

Figure 2: Blown up section of an image containing seeding tracers and 2mm glass beads. In-focus glass beads identified with blue circles were detected for PTV analysis.

The images capture both the large glass beads and the PIV tracers. The large glass beads are detected using "imfindcircle" function in MATLAB[7] which is based on Hough transform for detection of circular objects. The script requires the range of acceptable particle radius (set to ±%20 of average particle radius) and also a gradient-based threshold for edge detection as input parameters. The latter is based on the high intensity gradient at the sharp boundary of in-focus particles while the out-of-focus particles have a smooth gradient. Two different low and high gradient-based thresholds are considered for edge-detection. The low threshold is applied to detect and mask out all the in-focus and out-of-focus particles from both frames for PIV analysis of the liquid phase. However a higher threshold is applied to only detect the in-focus particles for the PTV process (fig.2). By having the centroid location and radius of the in focus particles, displacement (i.e. velocity) of the detected glass beads are measured in the PTV algorithm developed in MATLAB. In the PIV process, cross-correlation method with 32×32 pix^2 window size is applied to the double frame images to obtain the instantaneous velocity field of liquid phase.

RESULTS AND DISCUSSION

The radial direction is indicated by r starting from the centre of the pipe r=0 (r/R=0) towards the pipe wall located at r=25.3 mm (r/R=1). The symbols (U, V) and (u, v) are average velocity and fluctuating velocities in the streamwise and radial direction, respectively. As shown in fig.3, the mean velocity of the glass beads is slightly slower than the carrier phase in the core region of the flow (r/R<0.85). This velocity lag is higher for the larger particles due to their higher weight (gravitational force) at the pipe centreline. As shown the difference between the average velocity of the particles and the liquid phase velocity becomes smaller by approaching the wall. At around r/R=0.85 called "crossing point", the velocities of liquid and solid phases become almost identical. Beyond r/R>0.85, the particle velocity becomes larger than the liquid velocity. Therefore, the relative velocity of the particles and the fluid in the near wall region is in the opposite direction of that in the core flow. It is also observed that the larger particles have the largest velocity in the near wall region (r/R>0.85). This phenomena was reported by others[6,8,9] and is attributed to the "slip boundary condition" of the particles at the wall[8]. Particles after colliding with the wall bounce off to the core of the flow while preserving most of their streamwise momentum. Moreover high velocity particles are transported from core of the flow to the wall region by their lateral fluctuations. Since these particles have high response time (τ_p), they don't quickly respond to the surrounding liquid gradient field and do not effectively decelerate in this region. Particles with larger size have longer response time, therefore, larger particles acquire higher velocity in the near wall region.

Fig.4 illustrates that particles have a larger $<u^2>$ than the carrier phase almost everywhere across the pipe radius which is in agreement with other works[2-6]. The streamwise turbulent intensity is larger for the larger particles at the pipe centreline while the smaller ones have a higher intensity near the wall. Caraman et al.[10] provided a graph of $<u^2>$ for 60 and 90 μm particles showing that smaller particles have a larger $<u^2>$ in the near wall region. Since particles have higher inertia than the fluid element they can move further in the flow field while keeping their initial inertia. This leads to particles with larger axial fluctuations than the liquid phase. On the same ground, larger turbulence production of 2mm particles is expected due to their higher inertia. However, production of axial turbulence for 0.5 mm particles is larger near the wall which can be attributed to the larger mean velocity gradient of 0.5 mm particles in this region. The radial fluctuations of particles are observed to be higher than the fluid in fig.5. The 2mm glass beads also demonstrated larger radial fluctuation than the rest. The results are in agreement with other works such as Kameyama et al.[6] and Boree et al.[10].

In order to investigate the sources of particle fluctuations in the radial direction, the cross section of the pipe is divided in to two sections: core region and near wall region. In the core region, in addition to the fluid turbulence, particles are subject to the lift force which acts towards the pipe centre and particle-particle interactions which is a dispersive force and scatters the particles towards the pipe wall. The interaction of the 2 mm particles with the fluid turbulence is lower than the other smaller particles because of its higher Stokes number. However, the other two sources work stronger in favour of 2mm particles thus they have much higher radial turbulence in the core of the flow. In the near wall region, fluid turbulence and particle-particle interaction are negligible. In this region, the lift force is reversed due to larger velocity relative to the liquid phase[9] and pushes the particles towards the wall. Therefore, the reverse lift force and the particle-wall collision are regarded as the main sources for the particles radial fluctuations which again are stronger for 2 mm particles.

Concentration profiles of the glass beads (fig.6) are obtained from PTV data using the number of particles (N_p) at any radial position, normalized by the total number of detected particles (N_{total}) across the radius. Concentration profiles for 0.5 and 1 mm particles are almost constant in core of the flow (r/R<0.7) and then it rapidly declines in the near wall region. This trend was observed by Kameyama et al.[6] for 625 μm glass beads in the upward liquid flow. However the concentration profile for 2 mm particles is linearly declining from pipe centreline towards the pipe wall. The same trend has been seen by Hosokawa and Tomiyama[11] for 2.5 and 4mm ceramic particles in the upward water flow.

These trends show that there is a lift force that pushes the particles away from the wall and it is more effective in case of 2mm particles. This lift force accumulates the particles in the core of the flow. It is clear that particles are subject to the liquid velocity gradient from pipe centre (r/R=0) to crossing point (r/R=0.85) which causes particles to spin and creates a lift force towards the pipe centreline. At the core region of the flow (below r/R=0.7) the velocity gradient becomes smaller and the magnitude of the lift force declines. This explains the flat concentration profiles for 0.5 and 1 mm in the core of the flow. Because of the large size of 2mm particles, this force is still effective even below r/R=0.7 and keeps pushing the particles towards the centreline which creates a centre-peaked profile for 2mm particles.

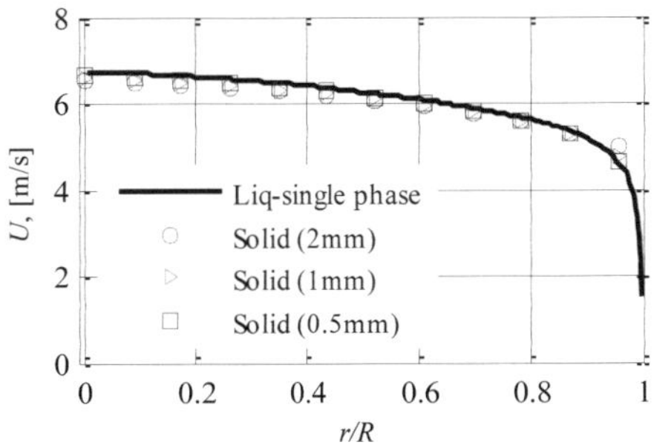

Figure 3: Mean velocity profiles of the particles and the liquid phase (unladen-fluid).

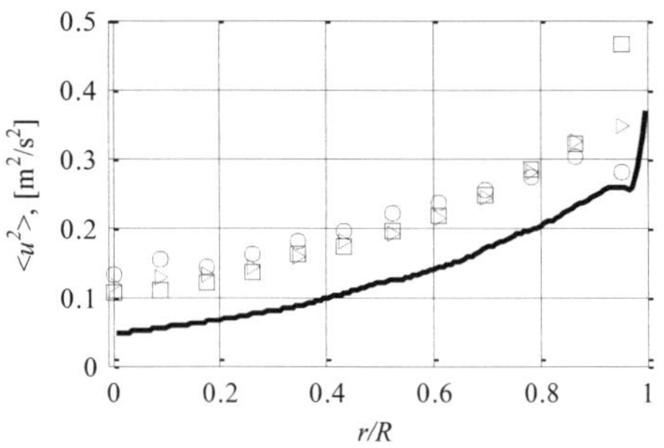

Figure 4: Streamwise fluctuating velocity profiles (symbols similar to fig.3)

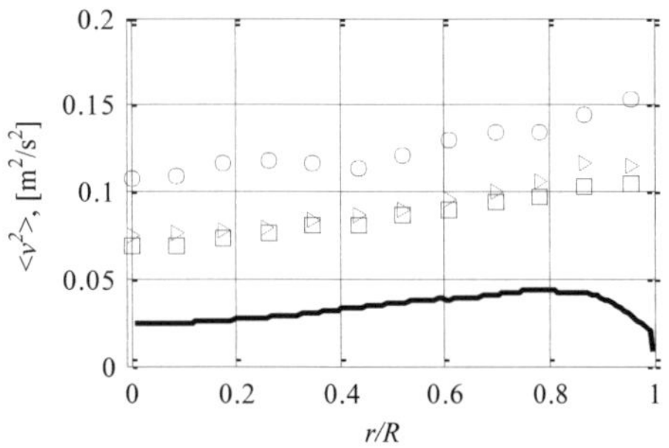

Figure 5: Radial fluctuating velocity profiles (symbols similar to fig.3)

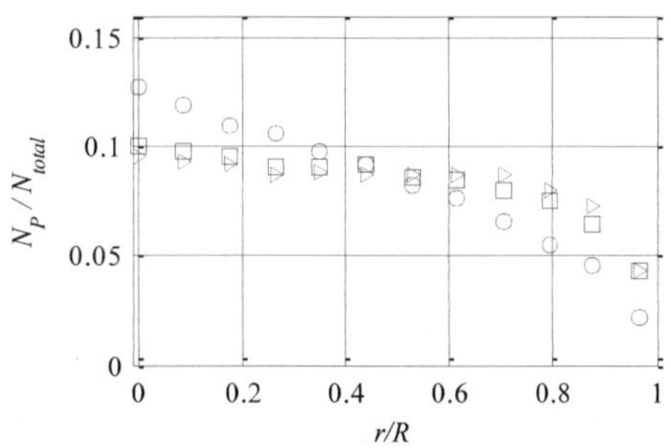

Figure 6: Number density distribution of particles (symbols similar to fig.3)

CONCLUSIONS

The turbulent motions of solid particles have been measured in upward solid/liquid flows at a high unprecedented Reynolds number of 320,000. Results show that the 0.5, 1 and 2 mm glass beads lag behind the fluid in the core of the flow (*r/R*<0.85) because of the gravity, however they move faster than the liquid phase in the near wall region (*r/R*>0.85). The turbulent fluctuations of the particles in both streamwise and transverse directions are larger than the carrier phase. Axial fluctuations are larger for 2mm particles at the pipe centreline while the 0.5 mm particles acquire the largest streamwise fluctuations near the wall. Particles tend to accumulate in the core region because of lift force which is more vigorous in case of 2 mm particles, creating a linear concentration profile with a centreline peak.

ACKNOLEDGMENTS

Authors would like to thank NSERC Industrial Research Chair in Pipeline Transport Processes for the financial support of this work.

REFERENCES

1. Balachandar, S. & Eaton, J. K. 2010, "Turbulent dispersed multiphase flow", *Annu. Rev. Fluid Mech.,*Vol(42), pp. 111–133.
2. Kulick, J. D., Fessler, J. R. & Eaton, J. K. 1994, "Particle response and turbulence modification in fully developed channel flow", *J. Fluid Mech.*,Vol(227), pp. 109–134.
3. Varaksin, A. Y., Polezhaev, Y. V & Polyakov, A. F. 2000, "Effect of particle concentration on flucctuating velocity of the disperse phase for turbulent pipe flow", *Int. J. Heat Fluid Flow*, Vol(21). pp. 562–567.
4. Kussin, J. & Sommerfeld, M. 2000, "Experimental studies on particle behaviour and turbulence modification in horizontal channel flow with different wall roughness", *Exp. Fluids,* Vol(33), pp. 143–159 .
5. a raman, , or e , J. & Simonin, O. 2003, "Effect of collisions on the dispersed phase fluctuation in a dilute tube flow: Experimental and theoretical analysis", *Phys. Fluids,*Vol(15), pp. 3602–3612.
6. Kameyama, K., Kanai, H., Kawashima, H. & Ishima, T. 2014, "Evaluation of particle motion in solid-liquid two-phase pipe flow with downward/upward flow directions", *17th Internatonal Symposium on Applications of Laser Techniques to Fluid Mechanics,* Lisbon, Portugal, pp. 1–15.
7. MATLAB R2013a., The MathWorks Inc., Natick, Massachusetts, United States.
8. Tsuji, Y., Morikawa, Y. & Shiomi, H. 1984, "LDV measurements of an air-solid two-phase flow in a vertical pipe" *J. Fluid Mech.*, Vol(139), pp. 417-434 (1984).
9. Lee, S. L. & Durst, F. 1982, "On the motion of particles in turbulent duct flows", *Int. J. Multiph. Flow,* Vol(8), pp.125–146.
10. or e , J. & Caraman, N. 2005, "Dilute bidispersed tube flow: Role of interclass collisions at increased loadings", *Phys. Fluids,* Vol(17), pp. 055108–1–9.
11. Hosokawa, S. & Tomiyama, A. 2004, "Turbulence modification in gas–liquid and solid–liquid dispersed two-phase pipe flows", *Int. J. Heat Fluid Flow,*Vol(25), pp. 489–498.

THE EFFECT OF MESH-TYPE BUBBLE BREAKER ON TWO PHASE VERTICAL CO-CURRENT FLOW

Alan Kalbfleisch
Mechanical and Materials Engineering
Western University
London, Ontario, Canada
akalbfl@uwo.ca

Kamran Siddiqui
Mechanical and Materials Engineering
Western University
London, Ontario, Canada
ksiddiqui@eng.uwo.ca

ABSTRACT

An experimental investigation of the influence of a mesh-type bubble breaker on the two-phase flow in a vertical co-flowing column was experimentally investigated using high speed imaging. A range of gas-to-liquid flow rates ratio (GLR) was considered that cover the three two-phase flow regimes from bubbly to annular flows. The results show that the mesh bubble breaker effectively reduces the initial bubble size by 50-70% in all three flow regimes. The results also show that the bubble breaker delays the transition of flow regime from bubbly to slug to higher GLRs, in particular at low liquid flow rates.

KEYWORDS: Two Phase Flow, Bubble Break Up, Mesh Bubble Breaker, Vertical Tube Absorber

INTRODUCTION

Bubble column reactors and vertical tube absorbers are used in a variety of mechanical and chemical industrial processes due to their high rates of heat and mass transfer [1-2]. Due to these qualities, the use of vertical tube absorbers has been proposed for small scale heat absorption refrigeration systems [3]. Although these systems can be designed for residential applications, the large size of the absorbers makes them undesirable for commercial use. The large gas-liquid flow rates ratio (GLR) used by these systems can result in slug or annular flow, decreasing the interfacial area and hence, limiting the mass transfer between the phases [4-6]. It has been proposed that mesh bubble breakers can be used to reduce the bubble size or prevent the transition from bubbly flow to slug or annular flow. Several studies have shown that the size of bubbles generated by a nozzle in vertical co-flow can be predicted and controlled based on the flow rates and the geometry of the system [7-9]. Using a mesh bubble breaker has also shown predictable results for the bubble size at low GLRs in a vertical tube absorber [10]. The current study expands on the study conducted by Gadallah and Siddiqui [10] by considering the effect of a mesh type bubble breaker over a larger range of GLRs. It is predicted that by using a mesh type bubble breaker in a two phase vertical co-flow, the initial interfacial area can be reduced allowing for the design of more compact vertical tube absorbers.

EXPERIMENTAL SETUP AND PROCEDURE

An experimental apparatus was assembled to study the two-phase co-flow in a vertical tube using high speed imaging. The images were then used to estimate the size of the bubbles before and after the mesh bubble breaker to analyze the effect of the mesh bubble breaker on a two-phase vertical co-flow. A schematic of the experimental apparatus is shown in Figure 1.

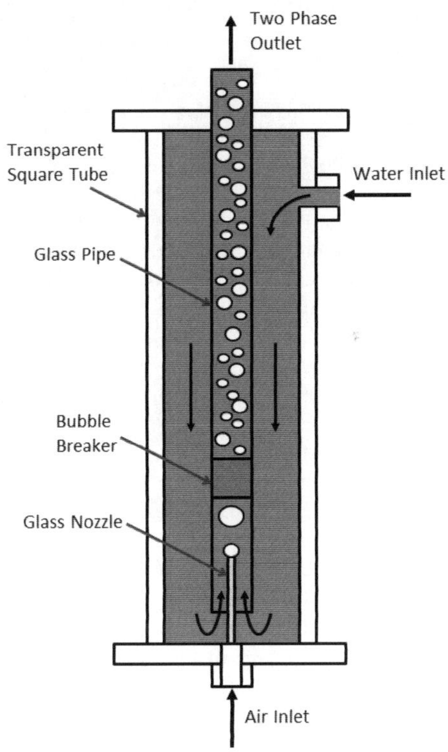

Figure 1: Schematic of the experimental setup.

The apparatus used in this study was modelled based on the apparatus used by Gadallah and Siddiqui [10]. The vertical absorber was made of a glass tube (D_i=15mm, D_o=19mm and L=350 mm). Air and water were used as gas and liquid phases, respectively. To avoid image distorted due to the surface curvature of the glass tube, the absorber was placed inside a square acrylic tube with water in between the space. Water at room temperature flowed through the inlet on the side of square tube. The water was channeled to the entrance of the vertical absorber (see figure 1). A vertical nozzle made of a glass tube (D_i=2.5mm, D_o=3.2mm) was placed at the bottom of the apparatus. Air at room temperature was supplied to the nozzle to generate bubbles into the liquid co-flow. The liquid and gas flow rates were separately controlled using manual valve flow meters. The experiments were conducted over a range of liquid and gas flow, $_l$=0.63-3.8 L/min and $_g$=0.2-2.5 L/min, respectively. The corresponding GLR range was 0.05-4.0. A mesh bubble breaker, shown in Figure 2, was placed in the absorber, one inch downstream from the tip of the glass nozzle.

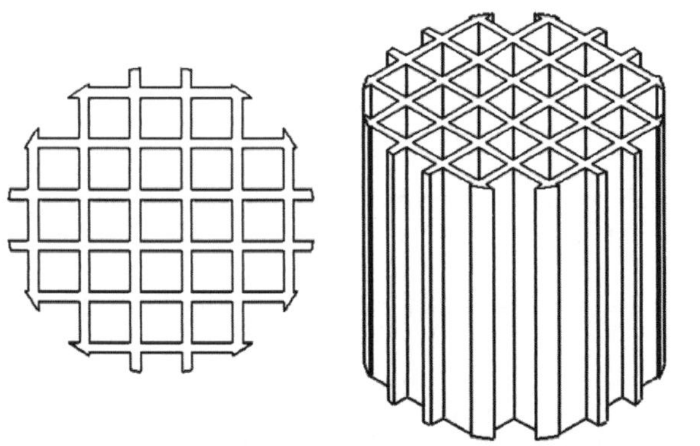

Figure 2: Bubble breaker model top view and isometric view.

The bubble breaker used in this study was designed using 3D CAD modeling software and manufactured using a MakerBot 3D printer. The bubble breaker used a square pore opening with sides 2mm × 2 mm. The wall thickness of the mesh structure was 0.5mm. The height of the bubble breaker was 15mm. The outer edges of the breaker were profiled to provide a circular shape to insert the breaker in the circular absorber as a press fit.

To capture the flow dynamics and the bubble size, a Photron Fastcam high speed camera was used. The apparatus was backlit using a halogen lamp and a diffuser screen. This ensured a clear contrast between the gas and liquid phases to facilities image processing. Greyscale images were captured at a framerate of 1000 fps for 5 seconds for each flow configuration. A Matlab algorithm was used to automatically detect and measure the size of the bubbles generated by the nozzle (upstream of the bubble breaker). The bubbles were observed to have an elliptical shape and the diameter of each bubble was measured as the average of the maximum height and maximum width. For the detection of bubbles generated by the bubble breaker, another interactive Matlab algorithm was used which allowed the user to manually detect each bubble. This interactive approach was used due to the large overlap of bubbles in the images. The algorithm then computed the diameter of each detected bubble. To determine the mean bubble size of the nozzle-generated bubbles, 50 bubbles at each flow condition were measured, while for the breaker-generated bubbles 500 bubbles were measure.

RESULTS AND DISCUSSION

The images captured by the high speed camera were used to investigate the effect of the bubble breaker in bubbly, slug and annular flow regimes. Figure 3 qualitatively shows the three flow regimes in the absence and presence of the bubble breaker.

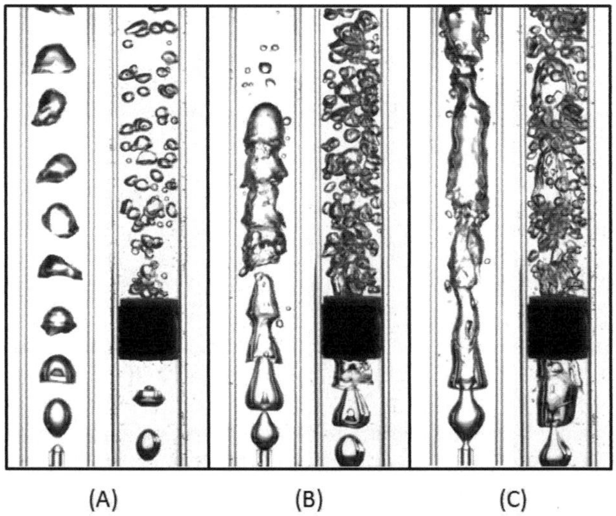

Figure 3: Comparison of two phase vertical co-current flow with and without the use of a bubble breaker for (a) bubbly flow (GLR=0.2), (b), slug flow (GLR=0.8), and (c) annular flow (GLR=1.0). The bubble breaker is shown as the black square in the images.

Figure 3(a) compares the two phase flow with and without a bubble breaker in the bubbly flow regime. As seen in the Figure, the bubble breaker is effective in reducing the bubble size of a single bubble in the bubbly flow regime. After passing through the bubble breaker, the single large bubble is broken up into numerous small bubbles of similar sizes and dispersed over the cross section of the vertical tube. In the slug flow regime, the images in Figure 3(b) show the formation of a chain of slug bubbles, which in the absence of the bubble breaker, coalesce as they advect downstream. However, the bubble breaker prevents the large bubbles from coalescence and transforms the flow regime into the bubbly mode by breaking the large slug bubbles in numerous small bubbles. Since the slug flow regime exists at GLRs greater than that of the bubbly flow regime, the number of broken bubbles is larger than that for the bubbly flow regime. This also increases the chance of bubble coalescence, which increases the size of bubbles as they advect downstream of the bubble breaker. The bubble breaker is also effective at breaking up annular bubble as shown in Figure 3(c). As figure shows, the annular bubble is broken into a combination of bubble slugs in the center of the tube and small bubbles in the peripheral region and in the wake of the slugs. The detailed analysis of the image data indicate that (not shown here) with a further increase in the gas flow rate, the bubble slugs eventually transform into a full annular flow in the center of the tube after the breaker and smaller bubbles in relatively large quantities still exist in the peripheral region. The results in Figure 3 clearly demonstrate that in all three flow regimes, the bubble breaker was effective in breaking the bubbles and increasing the interfacial surface area between the liquid and gas phases compared to the corresponding cases without a bubble breaker.

The results in Figure 3 also indicate that the bubble breaker delays the transition of flow regime from bubbly to slug to higher GLRs. To quantify this effect, the GLR range for transition from bubbly to slug flow was identified at a given liquid flow rate both in the presence of absence of the bubble breaker. The results are presented in Figure 4. The lower bound of the interval represents the highest GLR at which fully bubbly flow regime exists while the upper bound of the interval represents the GLR at which the flow regime is completely transitioned to the slug flow.

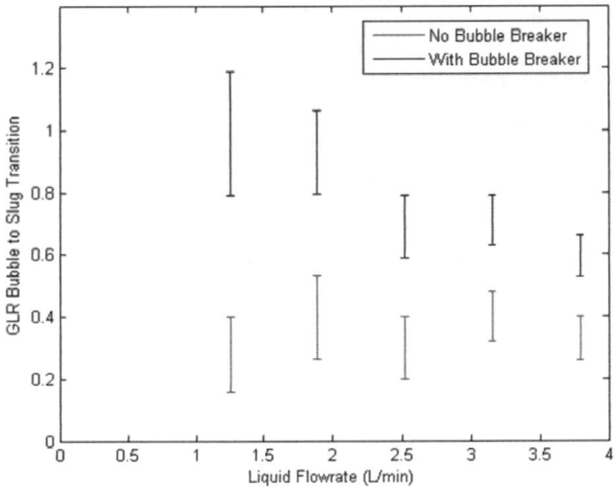

Figure 4: GLR range at which the flow transitions from bubble to slug flow with and without a bubble breaker present.

The figure shows that at the lowest liquid flow rate ($_L$=1.3LPM) without the bubble breaker, the transition starts at the GLR of about 1.8, and fully slug flow regime establishes at a GLR of 0.4. When the bubble breaker was present, the flow was still observed to be in the bubbly regime at a GLR of 0.8. This result bodes well for use in vertical tube absorbers that require high GLRs as the bubbly flow can aid in the enhanced gas-liquid mass transfer. As the liquid flowrate increases, the transition range in the presence of the bubble breaker decreases and get closer to that for the case without a bubble breaker. These results show that the bubble breaker is more effective at lower liquid flow rates.

In the study performed by Gadallah and Siddiqui [10], it was found that the bubbles generated by the bubble breaker remains almost the same over the studied GLR range. It should be noted that their study was focused only on the two-phase flow in the bubbly flow regime. The images data recorded in the present study were used to determine if this trend remains true as the two-phase flow passes through the bubbly flow regime. The mean bubble diameter of the entire range of GLRs at which a bubbly flow regime was present is shown in Figure 5.

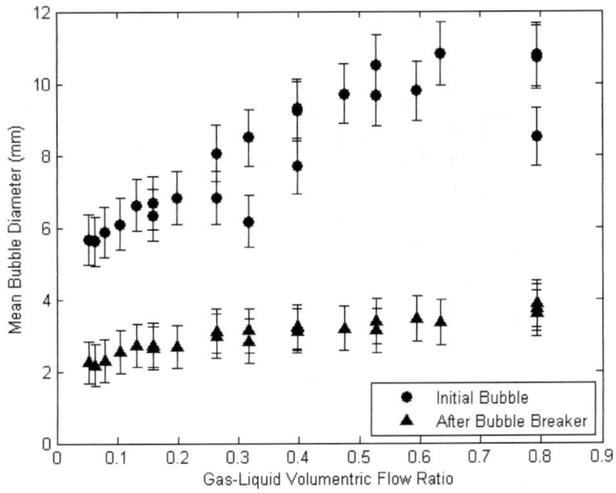

Figure 5: Diameter of initial nozzle generated bubbles and bubbles generated by the bubble breaker.

Figure 5 presents the mean bubble diameter before and after the bubble breaker over the given range of GLRs. The error bars represent the measurement uncertainty due to pixel size calibration and the bubble size estimation. The results in Figure 5 clearly show that the bubble breaker is effective at reducing the size of the bubbles present in the flow. The bubble size reduction for the entire range of GLR was found to be within about 50-70%. The bubble size reduction with reference to the initial bubble size was found to increase generally with an increase in the GLR.

The results in Figure 5 also show that in the absence of the bubble breaker, the bubble size in general, increased with an increase in the GLR, as expected. However, in the presence of the bubble breaker, a slight increase in the mean bubble diameter with GLR was observed. These results are consistent with those of Gadallah and Siddiqui [10]. The increase in the average bubble size after the breaker with an increase in GLR was found to be due to the increase in the bubble coalescence. Immediately after leaving the bubble breaker, it was observed that the bubbles were similar in size and independent of GLR. However, at higher GLRs, the bubbles quickly coalesced causing the presence of a number of similar-sized large bubbles. As the GLR further increased, the number of initial small bubbles decreased and the number of large coalesced bubbles increased. The large bubbles varied in size; the largest of the coalesced bubbles eventually formed slugs at the transition GLR. This result can be seen from the plot of the bubble histograms for three GLRs shown in Figure 6.

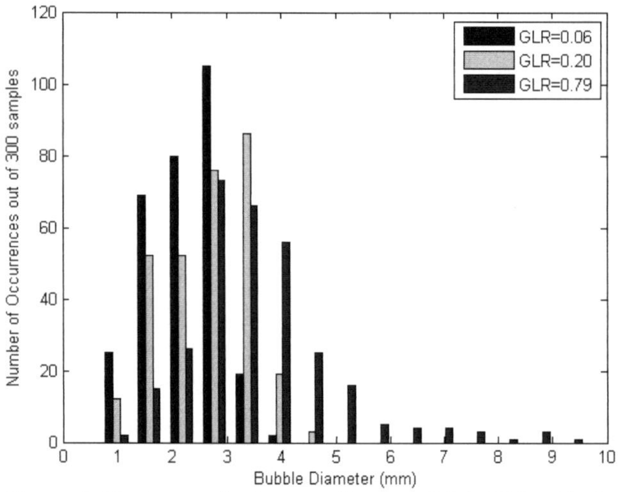

Figure 6: Histogram for low, medium and high GLRs. Each case shown had a bubble sample size of 300.

The histograms in Figure 6 show how coalescence affects the average bubble size. Although the peak bubble size in all three cases remains constant, larger bubbles are seen in the high GLR case. The presence of small bubbles generated by the bubble breaker decreased as the GLR increased as more bubbles coalesced to form larger bubbles. In terms of mass transfer in gas-liquid absorbers, these results indicate that the overall size of the absorber cannot be predicted by the average size of the bubbles after the bubble breaker. The large bubbles present at GLRs close to the transition point will be the limiting factor in absorber length as they will take the longest time to be fully absorbed into the liquid.

CONCLUSIONS

The role of a mesh bubble breaker was investigated in a two phase vertical co-current flow column. The GLR range considered in the study cover all three two-phase flow regimes from bubbly to annular flow. The results demonstrated that the mesh bubble breaker effectively breaks large bubbles into smaller ones. A decrease in the bubble diameter by 50-70% was observed over the given GLR range. The increase in the average bubble size with GLR was found to be relatively small in the presence of the bubble breaker compared to that without a breaker. Although the average bubble size in the tube after the bubble breaker increased, the peak bubble size remained almost the same; the increase in the average bubble size was primarily due to coalescence occurring downstream of the bubble breaker. It was also found that the bubble breaker delays the transition of flow regime from bubbly to slug to higher GLRs in particular at low liquid flow rates.

REFERENCES

1. Kantarci, N., Borak, F., and Ulgen, K.O. 2005, "Bubble column reactors", *Process Biochemistry*, 40, pp. 2263-2283.
2. Duduković M.P., Larachi F., Mills P.L. 1999 "Multiphase reactors – revisited" *Chemical Engineering Science*, 54, pp. 1975–1995.
3. Fernandez-Seara, J., Sieres, J., Rodriguez, C., and Vazquez, M. 2005, "Ammonia-water absorption in vertical tubular absorbers", *International Journal of Thermal Sciences*, 44, pp. 277-288.
4. Deckwer, W.D., Louisi, Y., Zaidi, A., and Ralek, M. 1980, "Hydrodynamic Properties of the Fischer-Tropsch Slurry Process", *Industrial & Engineering Chemistry Process Design and Development*, 19(4), pp 699-708.
5. Salman, W., Gavriilidis, A., Angeli, P., 2006, "On the formation of Taylor bubbles in small tubes", *Chemical Engineering Science,* 61, pp 6653-6666
6. Chilton, T.H, and Colburn A.P., 1934, "Mass Transfer (Absorption) Coefficients", *Industrial and Engineering Chemistry,* 26(11), pp 1183-1187.
7. Sada, E., A. Yasunishi, S. Katoh, and M. Nishioka, 1978 "Bubble Formation in Flowing Liquid" *Canadian Journal of Chemical Engineering,* 56, pp. 669-672.
8. Oğuz, H.N., Prosperetti, A. 1993, "Dynamics of bubble-growth and detachment from a needle" *Journal of Fluid Mechanics,* 257, pp. 111–145.
9. Terasaka, K., Tsuge, H. Matsue, H. 1999 "Bubble formation in cocurrently upward flowing liquid" *Canadian Journal of Chemical Engineering,* 77(3), pp. 458–464.
10. Gadallah, A.H., and Siddiqui, K. 2014, "Bubble size reduction using honeycomb monolith breaker in co-current upward flowing liquid", CSME International Congress 2014, Toronto, Canada.

Proceedings of the 25th CANCAM
London, Ontario, Canada, May 31 – June 4, 2015

Characterization of the two-phase flow due to a holed bubble breaker inside an effervescent atomizer using high-speed imaging

Mona Hassanzadeh Jobehdar
Department of Mechanical and Materials Engineering, University of Western Ontario, London, ON, Canada, N6A 5B9

Kamran Siddiqui
Department of Mechanical and Materials Engineering, University of Western Ontario, London, ON, Canada, N6A 5B9

Aly H. Gadallah
Department of Mechanical and power Engineering, Tanta University, Tanta, Egypt

Wajid A. Chishty
Institute for Aerospace Research National Research Council Canada Ottawa, ON, Canada, K1A 0R6

ABSTRACT

An experimental study was conducted to investigate two-phase flow characteristics inside and outside an effervescent atomizer. The influence of a new bubble breaker with different configurations (size and number of holes) over a range of gas to liquid flow rates ratios (GLRs) from 0.001 to 0.022 were considered. To visualize and quantify bubble and spray characteristics, high-speed imaging was used. The results showed that the two-phase flow characteristics inside and outside the effervescent atomizer are strongly influenced by the diameter and numbers of holes of the bubble breaker.

KEYWORDS: Bubble breaker, effervescent atomizer, high-speed imaging, bubble size, droplet size

INTRODUCTION

Effervescent atomizer is a type of twin-fluid atomizers in which a gas is bubbled into a liquid stream, which produces fine spray by shattering gas bubbles at the atomizer exit [1]. Some of applications where effervescent atomization technique was used include gas turbines [2, 3], internal combustion engines [4], furnaces and burners [5], and pharmaceutical sprays [6].

In an effervescent atomizer bubbles are formed by the injection of gas into the liquid stream through the aerator tube holes [7, 8]. The bubble-liquid mixture then passes through a mixing chamber before exiting from the orifice. A high-pressure drop at the exit orifice causes the gas bubbles to expand suddenly and shatter. This transforms the liquid into fine ligaments and then small droplets [8]. The spray quality which determines the performance of an atomizer is influenced by the internal two-phase flow [9]. The internal gas-liquid flow normally exists in one of the three flow regimes; bubbly flow, slug flow and annular flow [7, 10, 11]. The relative flow rates of gas and liquid, typically expressed as gas to liquid flow rates ratio (GLR) has a major impact on the internal two-phase flow regimes and the external spray [7, 8, 12-14].

The spray droplet size is a key parameter that determines the spray quality. It has been reported that the atomizer internal geometry [13, 14] and operating condition [12-14] influence the droplet size. An increase in GLR has also shown to decrease the droplet size [8, 12-14]. The spray steadiness is an important consideration is practical applications. An earlier work has reported that an internal bubbly flow at the exit orifice produces steadier spray than an intermittent internal flow of gas-liquid [15]. Hence, it is desirable to maintain bubbly flow regime inside the atomizer with uniform small-size bubbles. This can be achieved by using a bubble breaker that fragments large bubbles and slugs into smaller bubbles. Different types of bubble breakers were implemented inside effervescent atomizers [11]. Porous media and perforated inserts were mainly used as bubble breakers in the effervescent atomizer are. In the present study, a new type of bubble breaker in an effervescent atomizer is investigated.

EXPERIMENTAL SETUP AND TECHNIQUES

The schematic of the experimental setup is shown in figure 1. Water and air were used as liquid and gas phases, respectively and their flow rates were measured using rotameters installed upstream of the inlet valves. A range of liquid and gas flow rates were considered in this study, which covered the GLR (by mass) range from 0.001 to 0.022. To enable direct visualization of the internal two-phase flow, an effervescent atomizer with full optical access was designed and fabricated. To investigate the impact of size and number of holes of the bubble breaker on the bubbles fragmentation and consequently the spray quality, five bubble breaker inserts with different hole-configurations were considered. All breakers were cylindrical in shape and were made from acrylic. To study the effect of breaker hole size, three breakers were built with the hole-diameter of 2 mm, 3.18 mm and 5 mm, hereinafter referred to as small, medium and large, respectively. In addition, three breakers with one, three and five holes with the hole-diameter of 3.18 mm, 1.83 mm and 1.42 mm, were considered to study the influence of the number of breaker holes. Hereinafter, they are referred to as one, three and five, respectively.

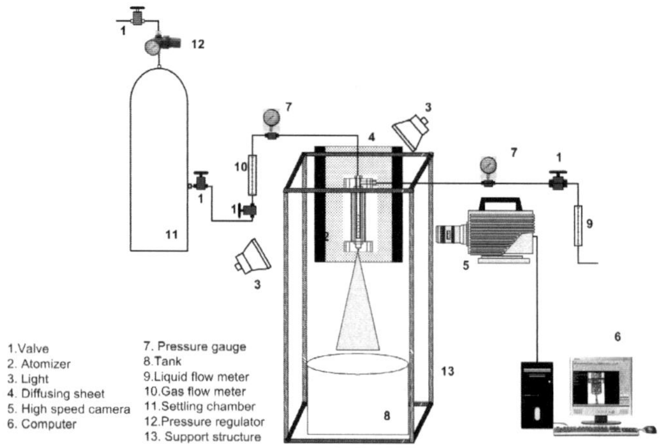

Figure 1: Schematic of the experimental setup

An imaging system comprised of a high-speed camera (Photron SA5) with a 60 mm lens was used to capture the images of the internal two-phase flow. Back-lit shadowgraphy technique using a 500W halogen lamp and a diffusion screen was implemented to improve the image quality. At each GLR, 14,000 images at a rate of 20,000 frames per second were captured. The same high-speed camera but with a 12X zoom lens was used to capture the images of the spray droplets. Due to a smaller field of view and higher frame rate, a powerful light source (continuous Diode-Pumped Solid-State laser) was used for back-lit shadowgraphy. The spray measurements were made at the location of z =150 mm downstream from the exit orifice where the spray was fully developed. To break any interference effects in the spray and randomize the spray pattern, 10 sets of measurements were made at different times for each GLR. In each set, 5000 spray images at a rate of 150,000 frames per second were captured. An in-house algorithm in the Matlab environment was used to measure the internal bubble size and spray droplet characteristics.

RESULTS AND DISCISSION

The results are presented in sub-sections describing separately the impact of single-hole diameter and multi-hole of the bubble breaker on the two-phase flow behavior inside and outside an effervescent atomizer.

SINGLE-HOLE BUBBLE BREAKERS

Figure 2 qualitatively shows the effect of bubble breaker's single-hole diameter at three different GLRs. It is evident from the figure that the size of the breaker hole has an impact on the bubble breakup. At the low GLR of 0.001 (Figure 2a), results show that for all three sizes of the breaker hole, the flow regime upstream of the breaker is bubbly flow and the upstream large bubbles are fragmented into smaller-size bubbles as they pass through the breakers. An overall assessment is that a decrease in the size of the breaker hole leads to the generation of smaller bubbles in numerous quantities. A reduction in the cross-section area of the flow and increased in the local velocity and the velocity gradients. This consequently increased the shear force exerted by the liquid on the approaching bubbles which caused the bubbles to stretch or elongate and eventually break.

The size of bubbles upstream of the bubble breaker increased with an increase in the GLR increased, which increased the size of broken bubbles inside the bubble breaker (see Figure 2b). The effectiveness of the smaller bubble breaker hole is still evident at this GLR. As the GLR further increased, the bubbles upstream of breaker are still present but this effect diminished with an increase in the breaker-hole diameter (see Figure 2c). Inside the breaker, the bubble breakup is still observed in the entrance region of the bubble breaker. However, as the size of breaker hole increased to medium size, the unsteady slug-annular flow is observed inside the breaker hole. An unsteady annular flow inside the bubble breaker is observed as the size of the breaker-hole further increased (large diameter case).

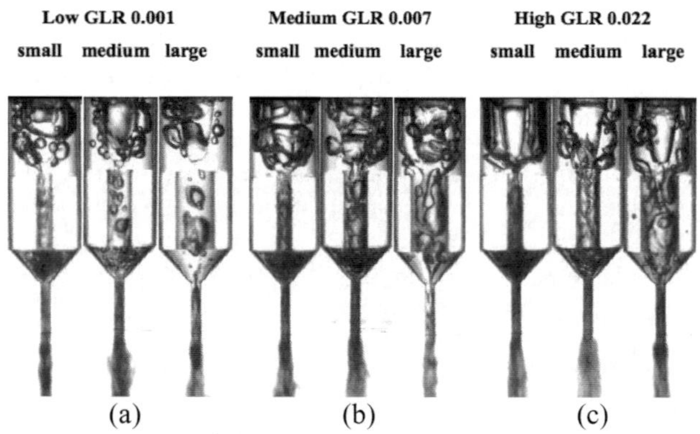

Figure 2: Effect of bubble breaker hole diameter at different GLRs (a) GLR=0.001, (b) GLR=0.007, (c) GLR=0.022.

Figure 3(a) shows the mean bubble size for the three breaker-hole diameters at different GLRs. The mean bubble size without a breaker is also presented for reference and comparison. The results clearly demonstrate that at all GLRs, the new bubble breakers generate smaller bubbles. Although the mean bubble size generated by the breakers increased with an increase in GLR, as expected, the rate of increment of the bubble size is significantly less than that for the case without a bubble breaker. The average size of bubbles inside the bubble breakers is 24%, 34% and 61% smaller than the mean bubble size without a breaker at low, medium and high GLRs, respectively. Figure 3(b) show that for all cases, the droplet size decreased with an increase in GLR, which is consistent with previous studies [12-14]. The results also show that the atomizer with the new bubble breaker produced better quality of the spray (smaller droplet size) compared to the case without a bubble breaker, over the given GLR range.

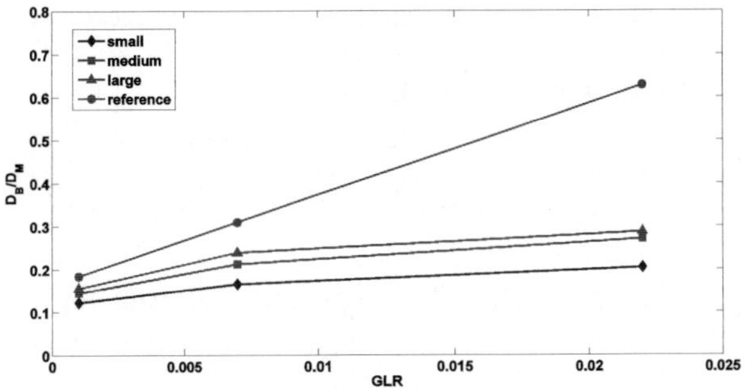

(a)

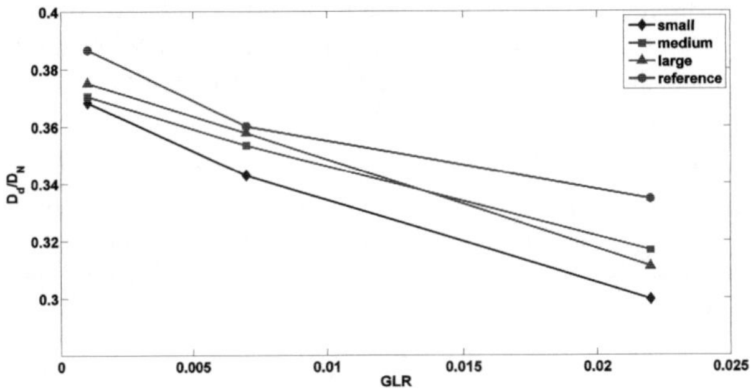

(b)

Figure 3: (a) Normalized mean bubble diameter (D_B/D_M) inside the bubble breakers and (b) Normalized mean droplet diameter (D_d/D_N) versus GLR. D_M is the inner diameter of the mixing chamber (D_M=9.6mm), and D_N is the inner diameter of the exit orifice (D_N=1.27mm). Error bars (based on the standard error of the mean) are smaller than the size of the symbols.

MULTI-HOLE BUBBLE BREAKERS

The effect of number of breaker holes and the holes configuration on the bubble fragmentation is qualitatively shown in Figure 4 at different GLRs. The flow upstream of the breaker was in the bubbly for all three breakers at the low GLR of 0.001 (see Figure 4a) but the size of the upstream bubbles was influenced by the configuration of the bubble breaker. Smallest upstream bubbles were observed for the single-hole breaker, whose size increased for the five-hole breaker and the largest bubbles were formed upstream of the three-hole breaker. This trend could be due to the reason that for the single-hole and five-hole breakers, a hole was in the middle and aligned with the conical base of the aerator tube. This caused strong velocity gradients due to flow streamlining and hence, effectively broke upstream bubbles, while, no hole in the three-hole breaker was in the middle.

The results also show that the multi-hole breaker is more effective in breaking the bubbles and generating smaller bubbles inside the breaker holes. The size of bubbles upstream of the breaker increased with an increase in GLR, which increased the size of the bubbles inside the bubble breaker as well (see Figure 4b). Multi-hole breakers are still found to be more effective. As the GLR further increased, the influence of multi-hole configurations on the upstream bubble breaker tends to diminish and slug-annular flow is observed inside all bubble breakers (see figure 4c).

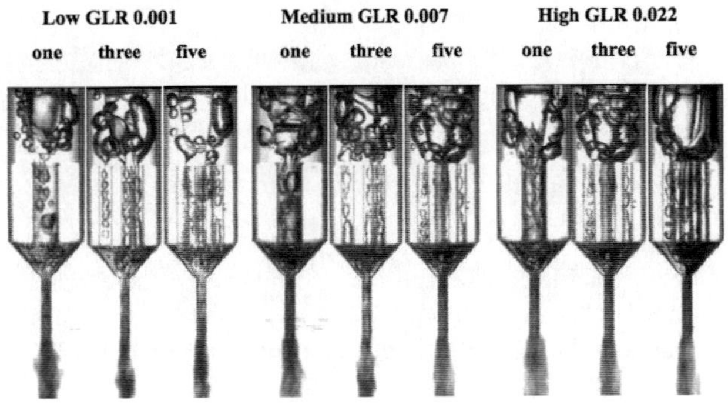

(a) (b) (c)

Figure 4: Figure Effect of number of holes in the bubble breakers at different GLRs, (a) GLR=0.001, (b) GLR=0.007, (c) GLR=0.022.

Figure 5(a) shows the mean bubble size over the given range of GLR for different bubble breaker configurations. The results show that the bubble size inside the breaker increased with an increase in GLR, as expected. The multi-hole breakers produce smaller bubbles compared to the single-hole breaker with an average size about 30%, 15% and 20% smaller than that in the single-hole breaker at GLRs of 0.001, 0.007 and 0.022, respectively. The mean droplet size for different breaker configurations over the given range of GLR is shown in figure 5(b). A distinct effect of single- and multi-hole bubble breakers on the mean droplet size is observed. The results show a monotonic decrease in the droplet size with an increase in the number of breaker holes over the entire range of GLRs. In comparison with the no breaker case, the results show that an atomizer with multi-hole bubble breakers produced on average, 8% smaller droplets than the atomizer without a breaker over the given GLR range. The results also show that the effect of the number of bubble breaker hole on the spray quality becomes more significant with an increase in GLR.

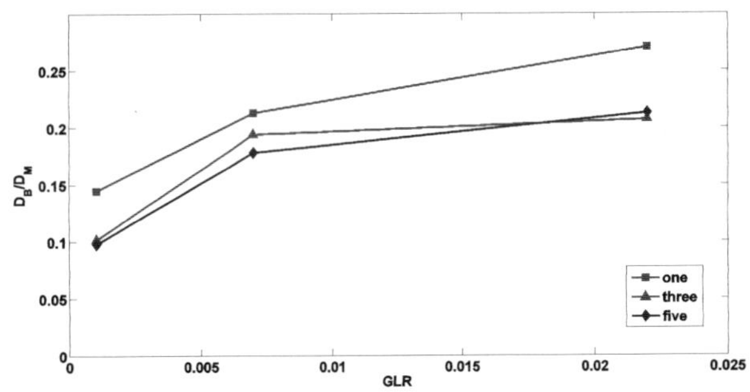

(a)

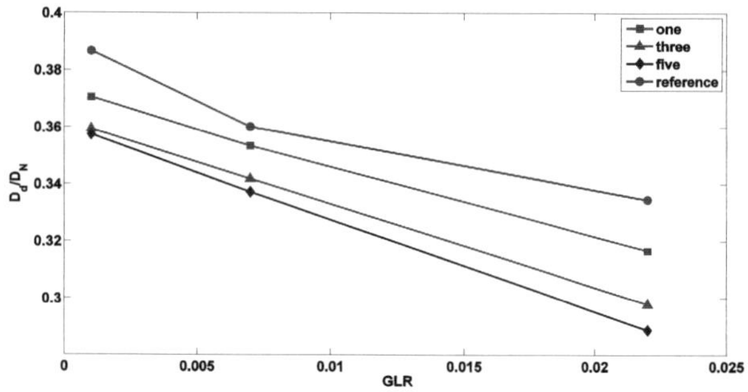

(b)

Figure 5: (a) Normalized mean bubble diameter (D_B/D_M) inside the bubble breakers and (b) Normalized mean droplet diameter (D_d/D_N) versus GLR. D_M=9.6mm and D_N=1.27mm. Error bars (based on the standard error of the mean) are smaller than the size of the symbols.

CONCLUSIONS

An experimental study was conducted to study the influence of a new type of bubble breaker over a range of GLRs using high-speed imaging technique. The results show that the new type of bubble breakers reduces the bubble size by approximately 15% to 70% over the entire range of GLRs. The results also illustrate that the size of single-hole breaker has strong impact on bubble fragmentation and spray droplet size inside and outside the atomizer, respectively. It was also observed that the number of breaker holes (same total flow area) has an impact on the internal and external two-phase flow in the atomizer. The multi-hole breakers compared to the single-hole breaker with the same total flow area were found to generate smaller fragmented bubbles at different GLRs. It was also illustrated that with an increase in the number of breaker holes the droplet size decreases monotonically over the given range of GLR. It is concluded that compared to the atomizer without the bubble breaker, the atomizer with the bubble breaker exhibits more effective atomization. However, the bubble breaker inside the atomizer increase flow losses and hence higher pressure drop.

ACKNOWLEDGMENTS

The authors would like to thank Western university and National Research Council Canada (NRC) for their support.

REFERENCES

1. Sovani, S. D., Sojka, P. E., Lefebvre, A. H., "Effervescent atomization" Progress in Energy and Combustion Science, 2001: 27, 483-521.
2. Wang, X. F., Chin, J.S., Lefebvre, A. H., "Influence of gas injector geometry on atomization performance of aerated-liquid nozzles." International Journal of Turbo & Jet-Engines, 1989: 6, 271-280.
3. Chin, J. S., Lefebvre, A. H., "A design procedure for effervescent atomizers." Journal of Engineering for Gas Turbines Power, 1995: 117, 266-271.
4. Rehman, S., Zaidi, K., "Design and fabrication of the high pressure effervescent spray combustion system ", International Journal of Advance Research and Innovation, 2014: 1, 72-77.
5. Sovani, S. D., Chou, E., Sojka P. E., Gore, J. P., Eckerle, W. A., Crofts, J. D., "High pressure effervescent atomization: effect of ambient pressure on spray cone angles." Fuel, 2000: 80, 427-35.
6. Petersen, F. J., Worts, O., Schaefer, T., Sojka, P. E., "Effervescent atomization of aqueous polymer solutions and dispersions." Pharmaceutical Development and Technology, 2001: 6, 201-210.
7. Roesler, T. C., Lefebvre, A. H., "Photographic studies on aerated-liquid atomization, combustion fundamentals and applications", Proceedings of the Meeting of the Central States Section of the Combustion Inst., April 1988, Indianapolis, U.S.
8. Lefebvre, A. H., "Atomization and sprays" Hemisphere Publishing Corporation, New York, USA, 1988.
9. Lin, K. C., Kennedy, P. J., Jackson, T. A., "Structures of internal flow and the corresponding spray for aerated-liquid injectors", 37th AIAA/ASME/SAE/ASEE Joint Propulsion Conference, (American Institute of Aeronautics and Astronautics) AIAA 2001-3569, 8–11th July, Salt Lake City, Utah, USA, 2001.
10. Kleinstreuer, C., "Two-phase flow theory and application." New York: Taylor & Francis 2003.
11. Sutherland, J. J., Sojka, P. E., Plesniak M. W., "Entrainment by ligament-controlled effervescent atomizer-produce spray", International Journal of Multiphase Flow, 1997: 23(5), 865-884.
12. Huang, X., Wang, X., and Liao, G., "Visualization of two phase flow inside an effervescent atomizer." Journal of Visualization, 2008: 11, 299-308.
13. Jedelsky, J., Jicha, M., Slama, J., Otahal, J., "Development of an effervescent atomizer for industrial burners" Energy Fuels, 2009: 23, 6112-6130.
14. Jobehdar, M. H., "Experimental Study of Two-Phase Flow in a Liquid Cross-Flow and an Effervescent Atomizer" PhD Thesis, Department of Mechanical and Materials Engineering, University of Western Ontario, 2014.
15. Maldonado, S., Fleck, B., Heidrick, T., Amirfazli, A., Chan, E.W., Knapper, B., "Development of an experimental method to evaluate the stability of gas–liquid sprays", Atomization and Sprays 2008: 18, 699–722.

Elasticity/Plasticity

Proceedings of the 25th CANCAM
London, Ontario, Canada, May 31 – June 4, 2015

EFFECT OF IRRADIATION AND THERMAL RECOVERY ON THE HARDNESS OF INCONEL 800H AND AISI 310 ALLOYS

H. Rajakumar
Dept. of Mechanical and Materials Engineering
University of Western Ontario
London, Ontario, Canada
hrajakum@uwo.ca

R.J. Klassen
Dept. of Mechanical and Materials Engineering
University of Western Ontario
London, Ontario, Canada
rjklasse@uwo.ca

ABSTRACT

The proposed Canadian Gen-IV supercritical water cooled nuclear reactor (SCWR) has in-core components that will be exposed to very high levels of neutron flux and temperature. It is therefore important to understand the effect of concurrent irradiation hardening and thermal recovery on the mechanical properties of these components. We present here the results of indentation hardness tests performed on two alloys, Inconel 800H and AISI 310, which are being considered as candidate in-core materials for the Canadian Gen-IV SCWR. The alloys were irradiated with 8 MeV Fe^{4+} ions to various levels, up to 15 displacements per atom (dpa), to simulate neutron irradiation damage and then subjected to isothermal anneals at 400°C and 500°C to induce thermal recovery of the irradiation damage.

The indentation hardness of both alloys increased by 20 to 40% with relatively small levels of ion irradiation corresponding to about 1 dpa and followed a power-law dependence upon dpa. The irradiation hardening in both alloys recovered rapidly with annealing and the time t_c for complete recovery also displayed a power-law dependence upon dpa.

KEYWORDS: irradiation hardening, thermal recovery, indentation hardness, ion irradiation, nuclear materials.

INTRODUCTION

A primary challenge associated with the Canadian proposed Gen. IV SCWR is the selection of suitable alloys for the nuclear fuel cladding and pressure tube liner components [1-2]. The alloys must be capable of enduring high levels of neutron irradiation in supercritical water at temperature up to about 750°C. Two alloys that are currently being considered for this application are AISI 310 and Inconel 800H (Table 1). To date little is known of the effect of prolonged high-temperature neutron irradiation on the mechanical properties of these alloys.

It is well known that neutron irradiation has a significant hardening effect on crystalline material and this is the result of the generation of neutron-induced crystal defects in the material [3]. The effect of neutron irradiation on the hardening of common nuclear materials, such as iron and zirconium alloys, is now quite well characterized over the 250 to 350°C temperature range typical of current water-cooled nuclear reactors. The rate of neutron irradiation hardening above this temperature will, presumably, be affected by faster rates of thermal recovery of the irradiation damage however this has yet to be quantified.

Recently there is increased use of high energy (E > 1MeV) self-similar ion irradiation to invoke irradiation hardening in test samples at a rate that is orders of magnitude faster than in-reactor neutron irradiation [4-8]. This allows much more efficient studies to be undertaken of the effect of irradiation hardening on the mechanical properties of in-reactor materials. The depth of penetration of high energy heavy ions into a crystalline solid is only several micrometers; therefore, the use of ion irradiation to simulate neutron irradiation hardening must be accompanied with a nano-mechanical testing technique capable of sampling the hardness of this small region.

This paper reports some results from our study of the effect of high energy Fe^{4+} irradiation on the hardness of the AISI 310 and Inconel 800H alloys under temperature conditions of 400 and 500°C. We perform room temperature indentation hardness tests on both alloys exposed to various levels of Fe^{4+} irradiation and then expose these samples to isothermal heating, at 400 and 500°C, for various lengths of time to measure the rate of recovery of the indentation hardness.

PROCEDURE

The chemical composition of the AISI 310 and Inconel 800H alloys is given in Table 1. Polished surfaces, 12.7mm x 12.7mm, were prepared from each material by conventional mechanical grinding and polishing techniques. The final polishing step was performed with an aqueous suspension of 0.05 μm Al_2O_3 abrasive particles. The as-polished samples were annealed for 20 minutes in a protective atmosphere at 1000°C to remove residual cold-work from the grinding/polishing steps. The samples were then lightly polished with the 0.05 μm Al_2O_3 suspension to remove any surface oxidation resulting from annealing. The microstructure of the alloys consists of large equiaxed grains between 50 and 100 μm diameter.

The polished samples were irradiated with 8.0 MeV Fe^{4+} ions at the Tandetron Ion Accelerator facility at the Western University (http://www.isw.physics.uwo.ca/). The samples were masked such that different regions of the polished surface received different levels of ion irradiation. The SRIM software (http://www.srim.org/) was used to calculate the ion penetration depth and the ion dosage necessary to invoke specific levels of irradiation damage, in units of displacements per atom (dpa), was calculated using the Kinchin-Pease model. Regions of the AISI 310 samples were exposed to ion dosage levels corresponding to damage levels from 0.01 to 10dpa while regions of the Inconel 800H samples were irradiated to levels from 0.10 to 15dpa damage.

Multiple indentation hardness tests were performed on each dpa region of the ion-irradiated samples. The tests were performed with a diamond Berkovich indenter at a constant ratio of indentation loading rate to load of $\dot{F}/F = 0.05\,\text{sec}^{-1}$. Multiple partial unloadings were performed during each indentation test to allow determination of the indentation hardness at various indentation depths during each test. The indentation plastic depth h was calculated at each partial unloading using the Doerner-Nix equation [9].

The indentation hardness H was calculated as

$$H = \frac{F}{A(h)} \tag{1}$$

where $A(h)$ is a high-order polynomial function describing the dependence of the projected indentation contact area upon the indentation depth h. The function $A(h)$ was determined from calibration tests performed by indentation of fused silica.

The indented ion-irradiated samples were then annealed at 400 and 500°C for various lengths of time from 1 to 100 minutes. The samples were removed from the furnace after each annealing time and multiple indentation hardness measurements were performed using the method described above. Approximately ten indentations, each containing nine partial unloading, were performed at each test condition.

Table 1. Chemical composition, in weight percent, of the two alloys in this study.

	C	Al	Si	P	S	Ti	Cr	Mn	Fe	Ni	Cu	Mo
AISI 310	0.08	-	0.75	0.05	0.03	-	25.00	2.00	50.35	20.50	0.50	0.75
Inconel 800H	0.08	1.40	1.00	0.05	0.02	1.40	21.00	1.50	39.50	33.00	0.75	-

RESULTS AND DISCUSSION

Figure 1 shows indentation force versus depth curves for the AISI 310 alloy in the non-irradiated and the ion-irradiated (8 dpa) conditions. The material clearly becomes hardened as a result of ion irradiation. The hardness of both the alloys displayed an indentation depth effect with higher hardness when the indentation depth was small (Figure 2). This is consistent with the well known indentation depth dependence of hardness that is generally observed for ductile materials [10].

Although the ion irradiated material displayed higher hardness than the non-irradiated material at all indentation depths, the increase in hardness was greatest for indentations of depth less than about 500 nm (Figure 2). This is likely due to the geometry of the test system as shown in Figure 3. Since the hardness reflects the average stress necessary to invoke plastic strain in a finite volume beneath the indenter, only indentations of depth less than that in which the plastic zone does not extend beyond the ion penetration depth will reflect the true effect of irradiation damage on the hardness.

Johnson proposed the following equation for the radius c of a hemi-spherical plastic zone beneath a conical indenter of flank angle β and width a (Figure 3)

$$\frac{c}{a} = \left[\frac{E^*\tan\beta}{6\sigma_y(1-\upsilon)} + \frac{2}{3}\left(\frac{1-2\upsilon}{1-\upsilon}\right)\right]^{\frac{1}{3}} \tag{2}$$

where σ_y is the flow stress and ν is the Poisson's ratio of the indented material while E* is the reduced elastic modulus of the indenter/substrate system [11]. Substituting into this equation values representative of our test material and indenter: $\beta = 20^{o}$, $\nu = 0.36$, $E^* = 88$GPa, and $\sigma_{y(\text{irradiated})} = 600$MPa we find that radius c of the plastic zone is about 1.7, 3.3, and 8.3 μm for indentation depths h of 200nm, 400nm, and 1.0μm respectively. This relation, $c \approx 8h$, is close to the $c = 12h$ calculated, using elastic-plastic finite element analyses, for pyramidal indentations of the same geometry as that used in this study [12, 13]. Since the calculated depth of the ion-irradiation hardening in our study is about 2 μm, and based upon the analysis above, we will assess the effect of ion-irradiation on the hardness by considering the hardness data from indentation depths of 200 and 400nm.

The average non-irradiated indentation hardness, at depths of 200 and 400nm, was 3.6 and 3.1GPa for the AISI 310 and 4.1 and 3.7GPa for the Inconel 800H alloys. Figure 4 shows the effect of increasing Fe^{4+} ion irradiation damage, in units of dpa, on the average hardness of both alloys. The average indentation hardness was determined from multiple indentations at each indentation depth, 200nm and 400nm, and each dpa level. The scatter in the data in Figure 4 indicates the variability in the hardness arising primarily as a result of surface roughness.

The following power-law functions for hardness H, in units of GPa, were obtained by fitting to the data in Figure 4(a) for the 200nm indentations.

$$H_{200nm\,\text{AISI}\,310} = 3.6 + 1.6(dpa)^{0.09} \tag{3}$$

$$H_{200nm\,\text{Inconel800H}} = 4.1 + 2.5(dpa)^{0.02} \tag{4}$$

One can see that the two alloys show very similar irradiation hardening characteristics with rapidly increasing hardness after small amounts of irradiation damage such that H increases by 20 to 40% after about 1.0 dpa of irradiation damage. These hardness values, and the extent of irradiation hardening at low dpa levels, are very consistent with previously reported findings for the ion irradiated AISI 316 alloy [14].

Figure 5 indicates how the average room-temperature indentation hardness H, for 200nm deep indentations, decreases as a function of annealing time, t, at 400°C. The hardness displays a logarithmic dependence upon annealing time. Curves of $H = H_{0dpa} + At^n$ were fit to the data for each dpa level and these equations were used to determine the time t_c required for complete recovery of the irradiation damage, i.e. the annealing time needed for $H = H_{0dpa}$.

Figure 6 depicts a logarithmic plot of t_c versus dpa for data obtained from tests performed, primarily at annealing temperature of 400°C. Linear regression analysis of the data indicate the following functional relationship between t_c, in units of minutes, and irradiation damage (dpa)

$$t_c = 574(dpa)^{1.26} \quad (4)$$

The data suggest that the rate of thermal recovery of ion-induced crystallographic damage follows a single trend for both alloys and both indentation depths.

Included in Figure 6 are data from 400nm depth hardness tests performed on the AISI 310 alloy at 500°C. These data lie on the same trend as the 400°C data indicating that the rate of irradiation damage recovery is not significantly greater, within the precision limits of our measurement techniques, at 500°C than at 400°C.

The time for complete recovery of irradiation damage at 400°C and 500°C is quite short for these alloy. For example, Equation (4) predicts that 1dpa of irradiation damage, which is in the order of magnitude of the annual accumulated irradiation damage of in-core components such as fuel cladding, is recovered in about 10 hours at 400°C or 500°C.

CONCLUSIONS

In this study we performed 8.0 Mev Fe^{4+} ion irradiation, to levels up to 15 displacements per atom (dpa), followed by isothermal annealing at 400°C and 500°C to assess for the AISI 310 and Inconel 800H alloys: i) the dependence of indentation hardness upon ion irradiation induced crystal damage and ii) the thermal recovery rate of the irradiation damage.

The average indentation hardness of both alloys increased very rapidly, by about 20 to 40%, with relatively small levels of ion irradiation corresponding to about 1 dpa. Our measured average hardness of the non-irradiated and the ion-irradiated material is similar, both in magnitude and dependence upon irradiation damage, to what was previously reported for similar alloys [14]. Similar power-law functions of H(dpa) were determined for both alloys indicating that their hardness displayed similar dependence upon irradiation damage.

We observed that the ion-irradiation hardening in both alloys recovered rapidly with annealing at 400°C and 500°C. Differences in the rate of recovery with annealing temperature from 400°C to 500°C could not be detected for the AISI 310 alloy. The time t_c for complete recovery of the irradiation damage was found to increase with the level of dpa and a power-law function of t_c(dpa) was obtained that predicted t_c for both alloys at 400°C and for AISI 310 alloy at 500°C.

Our findings suggest that in the case of high-temperature in-core applications, typical of those being proposed for fuel cladding and pressure tube liners in Canada's Gen-IV SCWR, both AISI 310 and Inconel 800H alloys will undergo at least 20% increase in hardness as a result of low levels of neutron irradiation damage typical of what could be expected within one year in core. This increase will however will negated by very rapid concurrent thermal recovery. Our data suggest that the recovery rate is sufficiently rapid that no net irradiation hardening will occur. We are currently continuing our studies to more accurately quantify the rates of irradiation hardening and thermal recovery in these alloys.

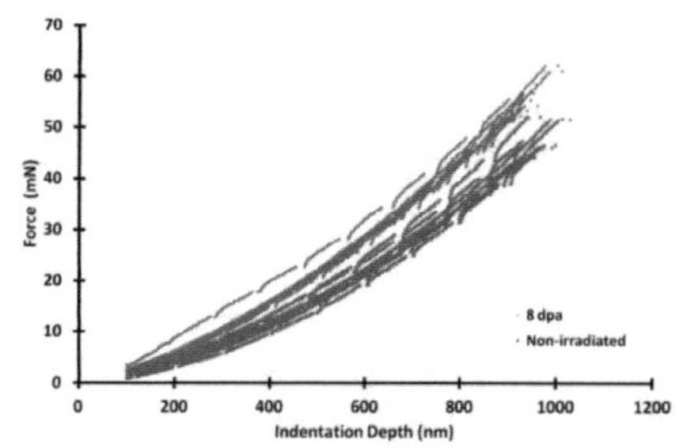

Figure 1: Indentation force versus depth from multiple tests performed on the AISI 310 alloy in the non-irradiated and the ion-irradiated (8 dpa) conditions.

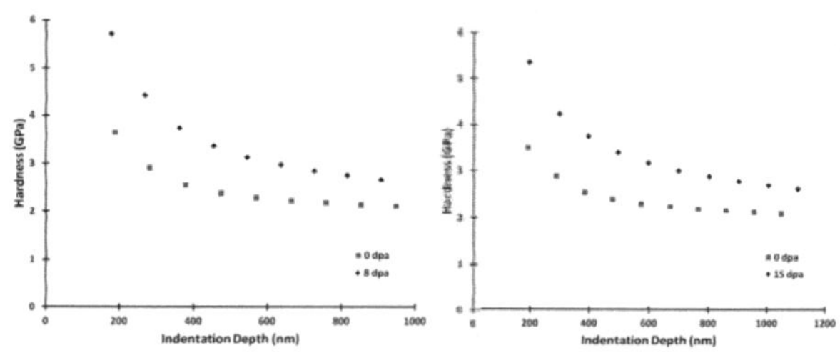

Figure 2: Indentation hardness versus depth for: (left) the AISI 310 alloy in the non-irradiated and the 8 dpa ion-irradiated conditions and (right) the Inconel 800H alloy in the non-irradiated and the 15 dpa ion-irradiated conditions.

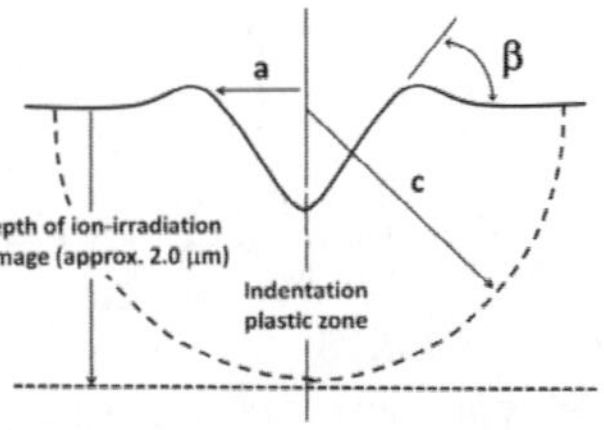

Figure 3. Schematic illustration of an indentation and its associated hemispherical plastic zone of radius c. In our study, the indentations must be made at a small enough depth that c is less than the ion irradiation depth.

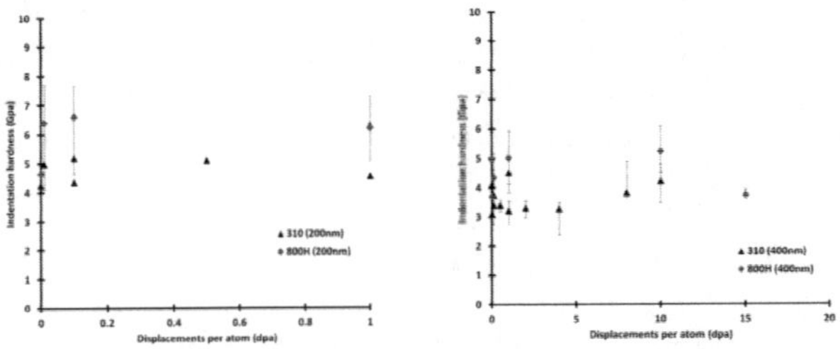

Figure 4: Average indentation hardness versus Fe^{4+} ion irradiation damage in the AISI 310 and Inconel 800H alloys. The 200nm deep indentations (left) show increased hardness compared to the 400nm deep indentations (right).

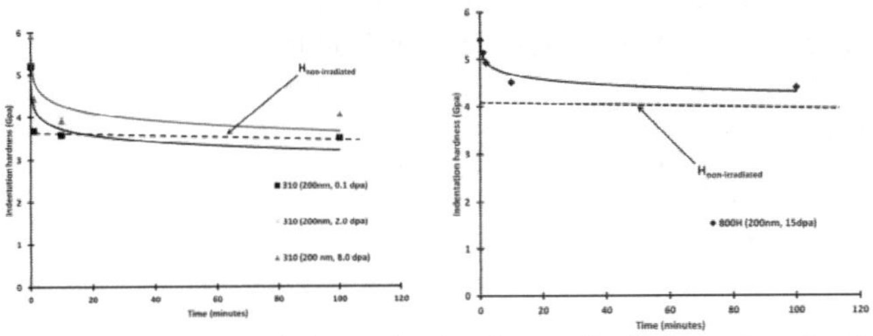

Figure 5: Average indentation hardness, for indentation depth of 200 nm, versus 400°C annealing time for (left) AISI 310 and (right) Inconel 800H samples. The solid curves indicated power-law functions fitted to the data.

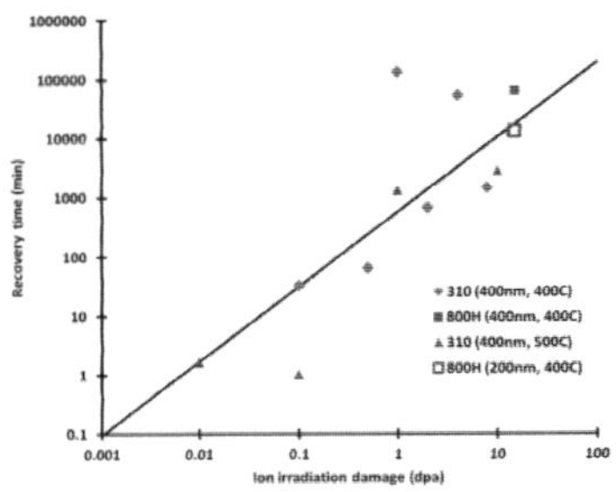

Figure 6. Time t_c required to completely recover the irradiation hardness, by annealing at 400°C or 500°C, of the AISI 310 and Inconel 800H alloys versus dpa.

ACKNOWLEDGMENTS

We wish to acknowledge the financial support for this research from the Natural Science and Engineering Research Council of Canada. The technical assistance of Mr. Jack Hendriks at the Tandetron Ion Accelerator facilities at the University of Western Ontario is also gratefully acknowledged. The many helpful discussions concerning the findings of this study with members of the Canadian Gen. IV Materials Research group, coordinated by Dr. David Guzonas, are also gratefully acknowledged.

REFERENCES

1. Khartabil, H.F., Duffey, R.B., Spinks N., Diamond, W. 2005, Proceedings of ICAPP '05, Seoul, Korea, May 15-19, 2005, Paper 5564.
2. U.S. DOE Nuclear Energy Research Advisory Committee and the Generation IV International Forum; GIF-002-00, 2002.
3. Bruemmer, S.M., Simonen, E.P., Scott, P.M., Andersen, P.L., Was, G.S., Nelson, J.L. 1999, Journal of Nuclear Materials, Vol (274), pp. 299.
4. Dutton, R., Lim, C.S. 1986, Canadian Metallurgical Quarterly, Vol (25), pp. 169.
5. Radjabov, D. 1991,Vacuum, Vol (42) pp. 163.
6. Romanov, I.G., Tsariova, I.N. 2003 Vacuum, Vol (68) pp. 213.
7. Li, N., Martin, M.S., Anderoglu, O., Misra, A., Shao, L., Wang, H. , Zhang, X. 2009, Journal of Applied Physics, Vol (105), pp.123522.
8. Hosemann, P., Dai, Y., Stergar, E., Leitner, H., Olivas, E., Nelson, A.T., Maloy, S.A. 2011, Experimental Mechanics, Vol (55), pp. 1095.
9. Doerner, M.F., Nix, W.D. 1986 Journal of Materials Research, Vol (1), pp.601.
10. Nix, W.D., Gao, H. 1998, Journal of Mechanics and Physics of Solids. Vol (46) pp.411.
11. Johnson, K.L. 1985 "Contact Mechanics", Cambridge Press, Cambridge, United Kingdom.
12. Murakami, Y., Itokazu, M. 1997, Int. J. Solids Structures Vol (34) pp. 4005.
13. Larsson, P.-L., Giannakopoulos, A. E., Soderlund, E., Rowcliffe, D. J., Vestergaard, R. 1996, Int. J. Solids Structures Vol (33) pp.221.
14. Hunn, J.D., Lee E.H., Byun, T.S., Mansur, L.K. 2000, J. of Nuclear Materials, Vol (282) pp.131.

CYCLIC PLASTICITY MODELLING OF WROUGHT MAGNESIUM ALLOYS

Ali Roostaei
Mechanical & Mechatronics Department
University of Waterloo
Waterloo, Ontario, Canada
aaroostaei@uwaterloo.ca

Hamid Jahed
Mechanical & Mechatronics Department
University of Waterloo
Waterloo, Ontario, Canada
hjahed@uwaterloo.ca

ABSTRACT

Wrought magnesium alloys show unusual characteristics during axial cyclic plastic straining, including asymmetric yielding and direction-dependent behaviour while their shear behaviour is fairly symmetric. In this paper, a cyclic plasticity model is proposed which is able to capture these unique features. The proposed model is based on an anisotropic modification of Ziegler's hardening rule, which is done by the introduction of plastic modulus matrix. The components of this matrix explicitly correlate with the material properties in different directions. Uniaxial tension-compression and pure shear cyclic tests are employed to calibrate the model parameters. The model is, then, verified using proportional and non-proportional multiaxial cyclic tests on two types of wrought magnesium alloys. It is concluded that the model predictions are in good accordance with the experimental results.

KEYWORDS: Wrought magnesium alloys, Asymmetric yielding, Directional anisotropy, Cyclic plasticity, Multiaxial loading

INTRODUCTION

In recent years, the automotive industries' commitment to reduce vehicle's weight, and thereby fuel consumption, has brought attention to light metals. Magnesium (and its alloys), being the lightest commercially available structural metal (with the density of 1.75 gr/cm^3), has recently been the focal point for this increased interest.

Currently, magnesium alloys are widely being applied to automotive components used in vehicle interior, housings and trim parts. Extending current application of magnesium alloys to load-bearing components such as suspension arms is feasible through comprehensive understanding of their plastic behaviour. Fatigue modelling is also necessary to estimate the component's life under cyclic loading conditions, which, in turn, requires an accurate constitutive cyclic plasticity model.

Owing to the strong propensity of wrought magnesium alloys to mechanical twinning at room temperature, they exhibit exceptional macroscopic plastic characteristics such as yield asymmetry, plastic flow asymmetry and directional anisotropy. Therefore, an accurate material model should be able to take these unusual plastic features of wrought magnesium alloys into account. This paper deals with the development of such material model (in continuation of an earlier work by Noban et al. [1]), its calibration using experimental results and finally preliminary verification of the proposed model.

PROPOSED FORMULATION

As one of the essential elements of continuum-based plasticity models, a yield function should be defined. For the current investigation, the well-known isotropic von Mises yield criterion has been incorporated:

$$\phi = (\boldsymbol{S} - \boldsymbol{A}):(\boldsymbol{S} - \boldsymbol{A}) - \frac{2}{3}Y^2 = 0 \qquad (1)$$

where $\boldsymbol{S}$ and $\boldsymbol{A}$ is the stress deviator and back-stress deviator tensors, respectively, and Y is the uniaxial yield stress. The yield surface in Eq. (1), initially centered at the origin, evolves only by kinematic hardening (represented by back stress deviator tensor, $\boldsymbol{A}$), i.e., it translates in the stress space without changing its shape and size.

With the purpose of describing the unusual cyclic plastic behaviour of magnesium alloys, an anisotropic generalization of Ziegler's hardening rule is proposed for the evolution of back stress tensor:

$$d\boldsymbol{\alpha} = \sqrt{\frac{3}{2}}[H]\,\boldsymbol{m}\,dp \qquad (2)$$

where $\boldsymbol{m}$ gives the evolution direction of yield surface according to Ziegler's rule and dp is the equivalent plastic strain increment. The matrix $[H]$ in Eq. (2) contains the hardening or plastic moduli. Its matrix form can be exploited to make explicit reference to the properties of the material along different directions.

Consistency condition in conjunction with associated flow rule and von Mises yield function, can be employed to give the plastic modulus function:

$$h_p = \sqrt{\frac{3}{2}}\boldsymbol{n} : [H]\boldsymbol{m} \qquad (3)$$

where $\boldsymbol{n}$ is the unit outward normal tensor to the yield surface defined by:

$$\boldsymbol{n} = \frac{S-A}{|S-A|} = \sqrt{\frac{3}{2}}\frac{S-A}{Y} \qquad (4)$$

Eq. (3) allows for calibrating the components of the $[H]$ matrix independently, corresponding to material properties in different directions.

MODEL CALIBRATION

In order to calibrate the model parameters for different types of magnesium alloys, we need to find proper mathematical functions to fit experimental plastic moduli curves exhibited by material during simple uniaxial tension-compression as well as pure shear cyclic tests. Under uniaxial cyclic tension-compression along 1-direction $\tilde{n}$ and $\tilde{m}$ can be written in Voigt notation:

$$\tilde{n}^T = \left[\sqrt{2/3} \quad -\sqrt{1/6} \quad -\sqrt{1/6} \quad 0 \quad 0 \quad 0\right] \quad (5)$$

$$\tilde{m}^T = [1 \quad 0 \quad 0 \quad 0 \quad 0 \quad 0] \quad (6)$$

where superscript "T" refers to the transpose of a vector quantity. Substituting Eqs. (5) and (6) into Eq. (3) gives the uniaxial plastic modulus:

$$h_p^{Uniaxial} = \sqrt{\frac{3}{2}} n_{11} H_{11} m_{11} = H_{11} \quad (7)$$

Eq. (7) suggests that an empirical fit function can be used to represent the evolution of experimental plastic modulus with progression of uniaxial cyclic plastic straining along 1-direction.

Furthermore, to account for yield asymmetry, a general function for the plastic modulus in 1-direction, Eq. (7), is assumed to take the following form:

$$H_{11} = [H_0 + \Delta H_L \eta(n_{11}) + \Delta H_U \eta(-n_{11})]|m_{11}| \quad (8)$$

where subscripts "L" and "U" denote functions for loading and unloading reversals, respectively, H_0 is a material parameter assumed equal to the magnitude of plastic modulus upon onset of yielding and η is a Heaviside step function defined as follows:

$$\eta(x) = \begin{cases} 1 & x \geq 0 \\ 0 & x < 0 \end{cases} \quad (9)$$

A typical uniaxial tension-compression stabilized hysteresis loop for AZ31B Mg alloy extrusion is shown in Fig. 1. All cyclic tests on AZ31B extrusion being presented hereafter were carried out by Albinmousa et al. [2, 3].

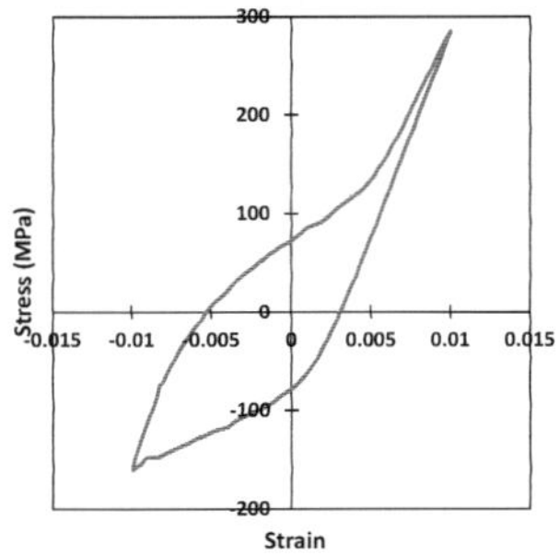

Figure 1: Typical uniaxial stabilized hysteresis for AZ31B Mg alloy extrusion at strain amplitude of 1% [2].

To be able to clearly see the evolution of plastic modulus with further plastic straining, stress-plastic strain curves for tensile and compressive reversals are separated and illustrated in Fig. 2.

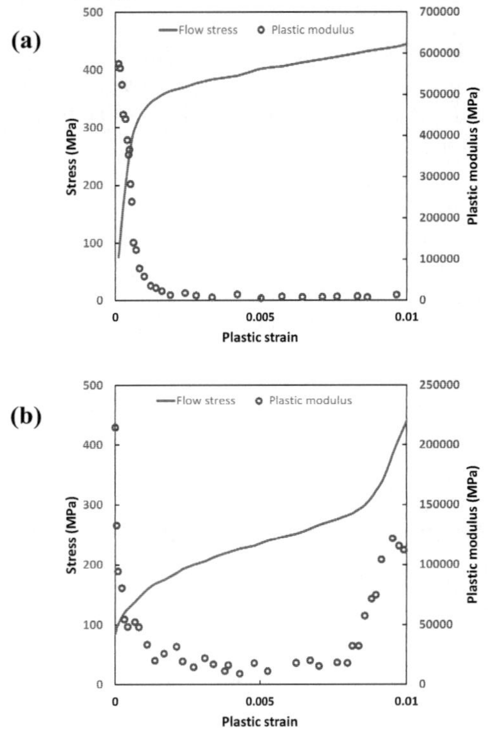

Figure 2: Stresses and plastic moduli of AZ31B Mg alloy extrusion as a function of absolute plastic strain for (a) compressive and (b) tensile reversals.

With regard to the compressive reversal, Fig. 2(a), a single-sigmoid function is sufficient for fitting experimental plastic modulus curve, whereas, a double-sigmoid function is needed to capture two peaks observed in Fig. 2(b). In this study, hyperbolic tangent functions of the following form have been utilized to capture different features of plastic modulus curves:

$$\Delta H_{L,U} = \frac{h_{L,U}}{2}\left(1 + tanh\left(\frac{\sigma_{11} - z_{L,U}}{d_{L,U}}\right)\right) \quad (10)$$

where $h_{L,U}$, $z_{L,U}$ and $d_{L,U}$ are material parameters to be calibrated based on the experimental results and σ_{11} is the component of stress in 1-direction.

Under pure shear condition, simplifying Eq. (3) leads to:

$$h_p^{Shear} = \sqrt{\frac{3}{2}} H_{12} \quad (11)$$

where h_p^{Shear} is the experimental plastic modulus during cyclic pure shear loading in 12-direction. Unlike the axial behaviour, the shear behaviour of magnesium alloys can be approximated to be symmetric. In addition, it is safe to assume Masing behaviour for magnesium alloys under pure shear loading. Taking these assumptions into consideration, a

Ramberg-Osgood power relation is proposed as fitting function for experimental shear plastic modulus:

$$H_{12} = 2^{(1-n_s)} n_s K_s \left(\frac{2^{(n_s-1)}\Sigma_{12}}{K_s}\right)^{\frac{n_s-1}{n_s}} \left(\sqrt{2}|m_{12}|\right) \quad (12)$$

where n_s and K_s are cyclic strain hardening exponent and cyclic strength coefficient in shear, respectively, and Σ_{12} is the translated shear stress component in 12-direction given by:

$$\Sigma_{12} = \left||Max(\sigma_{12})| + Sign(n_{12})\sigma_{12}\right| \quad (13)$$

in which $Sign(n_{12})$ gives the sign of the component of unit outward normal vector in 12-direction. The Ramberg-Osgood coefficients can be extracted from the plot of cyclic shear stress-plastic strain in a logarithmic scale.

Using the uniaxial cyclic tension-compression tests as well as cyclic shear tests at various strain amplitudes, the model parameters in Eqs. (10) and (12) are calibrated for AZ31B Mg alloy extrusion and tabulated in Table 1.

Table 1: Parameters for AZ31B Mg alloy extrusion.

Param.	First term	Second term
h_L	-247000	$1.05\times10^9\varepsilon_{11,max}{}^2 - 2.58\times10^7\varepsilon_{11,max} + 229500$
z_L	$5\times10^6\varepsilon_{11,max}{}^2 - 95000\varepsilon_{11,max} + 360$	$-2.9\times10^6\varepsilon_{11,max}{}^2 + 57916\varepsilon_{11,max} - 157.5$
d_L	80	20
h_U	596000	0
z_U	$-2.9\times10^6\varepsilon_{11,max}{}^2 + 57916\varepsilon_{11,max} - 262.5$	0
d_U	-50	0
Shear	$K_s = 384.27\ MPa,\ n_s = 0.3077$	

Employing constants from Table 1, the cyclic uniaxial and torsional tests used to fit the model parameters are reproduced and superposed to the experimental results for AZ31B extrusion in Fig. 3. Three different strain amplitudes are selected for each loading condition.

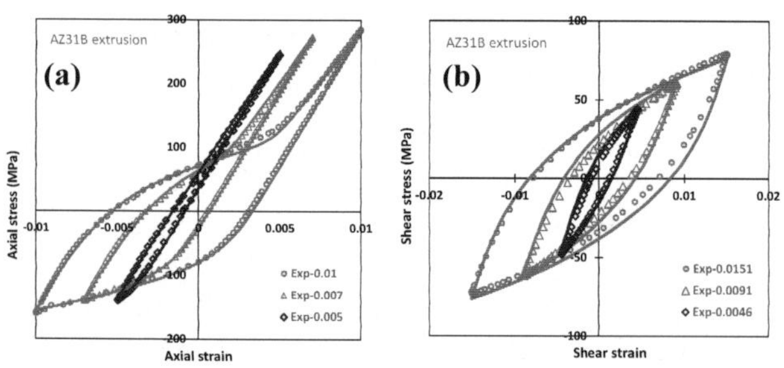

Figure 3: Comparisons of measured [2] and predicted stresses for AZ31B during cyclic (a) uniaxial pull-push and (b) pure torsional loading.

According to Fig. 3, the measured and predicted hysteresis loops are in good agreement, showing proper fitting. Significantly, a smooth transient behaviour from elastic to plastic region and tensile reversal with upward concavity are correctly captured.

PRELIMINARY VERIFICATION

It is hypothesized that the share of each direction properties in overall multi-directional loading response is incorporated into the proposed hardening rule, and thus, the cyclic plastic behaviour of the material during combined complex loading paths can be predicted by the model with calibrated parameters.

For the purpose of initial verification, multi-axial axial-torsional cyclic test results on AZ31B magnesium alloy extrusion under different loading conditions are reproduced form literature. Axial and shear response predictions under proportional (in-phase) and non-proportional (90° out-of-phase) multi-axial loading conditions are compared with the measured responses in Figs. 4 and 5.

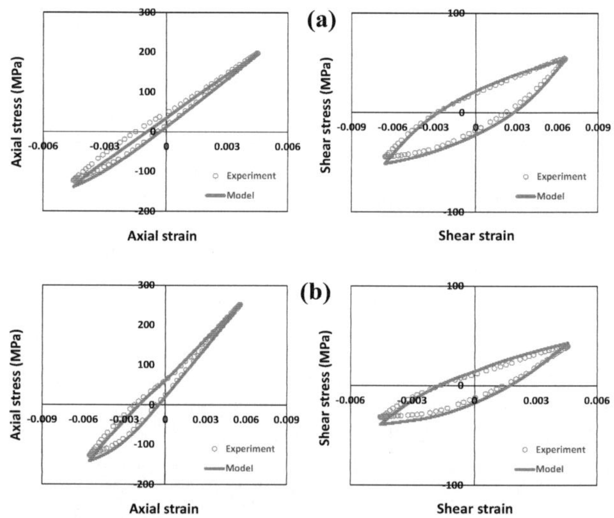

Figure 4: Experimental results [3] vs. predictions during proportional in-phase axial-torsional cyclic tests on AZ31B Mg alloy: (a) $\varepsilon_a = 0.45\%, \gamma_a = 0.65\%$ and (b) $\varepsilon_a = 0.55\%, \gamma_a = 0.45\%$.

Following the same calibration procedure using the experimental cyclic test results of Zhang et al. [4] on AZ61A magnesium alloy extrusion, the model parameters were extracted. Then, predictions from model were compared with the experimental results under various loading conditions and strain amplitudes. These comparisons are depicted in Figs. 6 and 7.

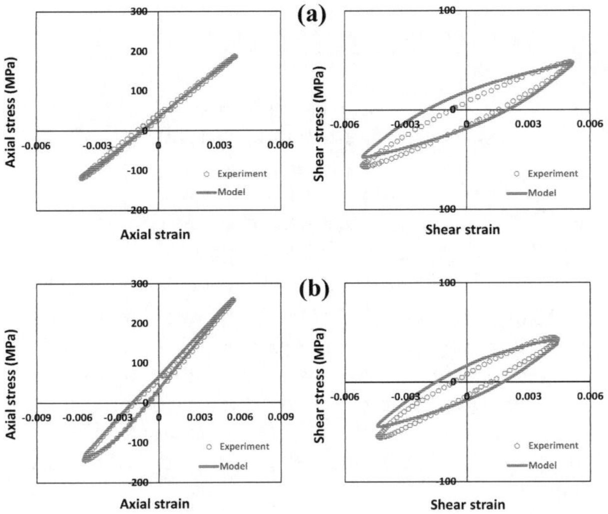

Figure 5: Experimental results [3] vs. predictions during non-proportional 90° out-of-phase axial-torsional cyclic tests on AZ31B Mg alloy: (a) $\varepsilon_a = 0.38\%, \gamma_a = 0.51\%$ and (b) $\varepsilon_a = 0.55\%, \gamma_a = 0.44\%$.

As can be seen in Figs. 4, 5, 6 and 7, though the model parameters were obtained from uniaxial test results, the predictions under multi-axial loading are in good agreement with the measurements. This implies the ability of the developed model in accommodating combined multi-axial loading paths as well as capturing unusual plastic behaviour of Mg alloys.

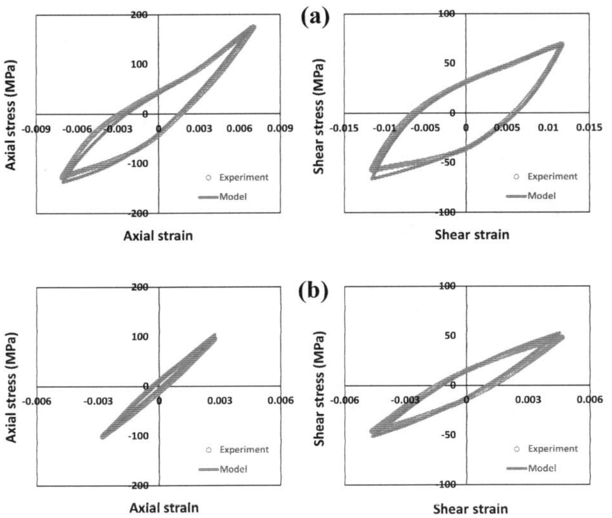

Figure 6: Experimental results [4] vs. predictions during proportional in-phase axial-torsional cyclic tests on AZ61A Mg alloy: (a) $\varepsilon_a = 0.70\%, \gamma_a = 1.16\%$ and (b) $\varepsilon_a = 0.27\%, \gamma_a = 0.46\%$.

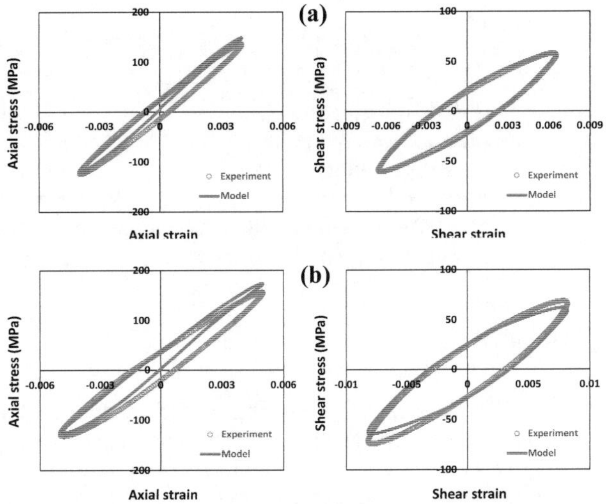

Figure 7: Experimental results [4] vs. predictions during non-proportional 90° out-of-phase axial-torsional cyclic tests on AZ61A Mg alloy: (a) $\varepsilon_a = 0.40\%, \gamma_a = 0.65\%$ and (b) $\varepsilon_a = 0.50\%, \gamma_a = 0.82\%$.

CONCLUSIONS

Ziegler's hardening rule is further extended to accommodate unusual hardening behaviour of wrought magnesium alloys.
Matrix form of the developed hardening modulus function allows for explicit reference to material behaviour along different directions.
A hyperbolic tangent sigmoid function and a Ramberg-Osgood power law function are employed to represent plastic modulus evolution during axial and shear loading, respectively.
Model prediction under complex multiaxial loading agrees well with the experimental results.

REFERENCES

1. Noban, M., Albinmousa, J., Jahed, H., and Lambert, S., 2011, "A Continuum-Based Cyclic Plasticity Model for AZ31B Magnesium Alloy under Proportional Loading", *Procedia Engineering, Vol.* 10, pp. 1366–1371.
2. Albinmousa, J., Jahed, H., and Lambert, S., 2011, "Cyclic Axial and Cyclic Torsional Behaviour of Extruded AZ31B Magnesium Alloy", *International Journal of Fatigue,* Vol. 33(11), pp. 1403–1416.
3. Albinmousa, J., Jahed, H., and Lambert, S., 2011, "Cyclic Behaviour of Wrought Magnesium Alloy under Multiaxial Load", *International Journal of Fatigue,* Vol. 33(8), pp. 1127–1139.
4. Zhang, J., Yu, Q., Jiang, Y., and Li, Q., 2011, "An Experimental Study of Cyclic Deformation of Extruded AZ61A Magnesium Alloy", *International Journal of Plasticity,* Vol. 27(5), pp. 768–787.

PLASTIC DEFORMATION OF GOLD MICRO SPHERES

AZM Ariful Islam
Mechanical and Materials Engineering
Western University
London, Ontario, Canada
aislam32@uwo.ca

V Bhakhri
Mechanical and Materials Engineering
Western University
London, Ontario, Canada
vbhakhr@uwo.ca

Robert J. Klassen
Mechanical and Materials Engineering
Western University
London, Ontario, Canada
rjklasse@uwo.ca

ABSTRACT

Size and strain-rate dependence of plastic deformation of Au microspheres was investigated at room-temperature using micro-compression testing. The contact yield stress was observed to decrease from 865 MPa for 0.75 μm to 210 MPa for 6 μm diameter spheres. The apparent activation volume, V^*, associated with the rate dependent plastic deformation remained constant at 4-5b^3 for 0.75 and 1 μm spheres over strains up to 20% whereas it displayed an increasing trend from 10 to 38b^3 for 3 and 6 μm diameter specimens. These findings suggest that the operative deformation mechanism was specimen size dependent. The larger spheres deformed by a mechanism of dislocation-obstacle limited glide whereas, the smaller spheres deformed by a dislocation-starvation mechanism.

KEYWORDS: Au microsphere, plastic deformation, size effect, finite element simulation, Activation volume.

INTRODUCTION

The topic of mechanical behavior of materials in small volumes has been of tremendous interest to the scientific community for over past decades [1]. In initial studies, the measured hardness of common materials displayed an inverse scaling with the indentation depth and this was explained in terms of hardening associated with the formation of geometrical necessary dislocations (GND) [2]. Recently, micro-compression tests on focused ion beam (FIB) machined nano- and micro-pillars have shown that single crystalline metals containing dislocations also size dependent yield stresses even in the absence of strain gradients [3][4]. These high strengths were associated with the dislocation starved conditions that prevail in small volumes, causing the operative deformation mechanism to be limited solely by the dislocation nucleation process rather than dislocation-obstacle interaction. While all of these studies are consistent with the tenet that smaller is stronger, a unified plasticity mechanism encompassing all deformation length scales has yet to be established [5].

In this work, we have investigated the size and strain-rate dependence of the mechanical behavior of metallic nano-/micro-spheres. For this purpose, Gold (Au) spheres of four sizes (0.75 μm to 6 μm diameters) were fabricated and subsequently, micro-compression tests were performed at different loading rates. Finite Element (FE) analysis was performed to estimate the stress-strain distribution in the compressed microspheres. Finally, strain rate dependence of the estimated stress on the different size microsphere was investigated to determine the operative deformation mechanism.

MATERIALS AND METHODS

Sample Preparation

Gold microspheres were fabricated on a (0001)-oriented sapphire (α-Al_2O_3) surface. An A8 PMMA positive photoresist layer was spin-coated on the O_2 plasma cleaned sapphire surface and e-beam lithography was used to project a grid of 1.5 to 12 μm diameter circles upon the photoresist. After the exposed photoresist was removed, a polycrystalline Au film of 100 to 600 nm thickness was deposited by e-beam evaporation. The remaining photoresist was then removed and the resulting Au cylinders, of 1.5 to 12 μm diameter and 0.1 to 0.6 μm height, were annealed in-vacuum at 1000°C. During annealing the Au de-wetted from the sapphire substrate and took the form of facetted single-crystal spheres of 0.75 μm to 6 μm diameter (Fig. 1). The facets were {111} close packed planes [6]. About 100 Au microspheres were fabricated for each diameter.

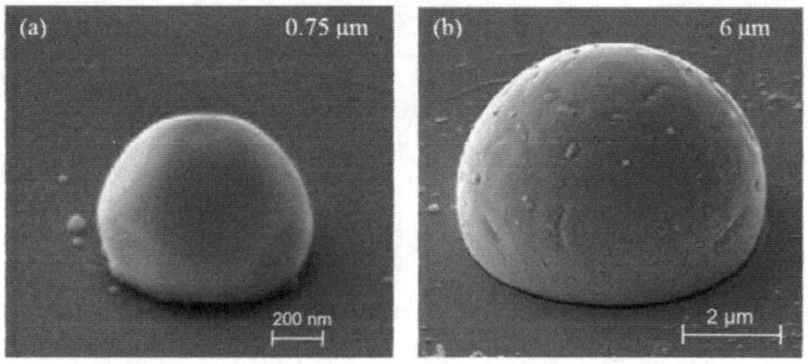

Figure 1: SEM images of as fabricated: (a) 0.75 μm and (b) 6 μm diameter Au microspheres

Microsphere Compression

The Au microspheres were compressed with an instrumented NanoTest platform by Micro Materials Ltd (Wrexham, UK) fitted with a 9 μm diameter diamond flat punch indenter. About 60 compression tests were performed for four different sized spheres. The compression tests were carried out by loading the indenter at a constant-rate to a maximum depth of about 20% of the initial sphere diameter. Different loading rates, ranging from 0.01 to 1 mN/sec, were used to study the strain-rate dependence of plastic deformation of the spheres.

Finite Element Simulations

FE simulations of the sphere compression were performed using the mesh shown in Fig. 2. The shape and dimension of the microsphere were obtained from measurements made from SEM images. In the simulation, the flat-punch indenter and substrate were assumed to be analytical rigid and frictionless surface-to-surface contact condition was defined at the flat-punch/sphere and the substrate/sphere interfaces. The microsphere was meshed with 4-node linear axisymmetric quadrilateral stress elements.

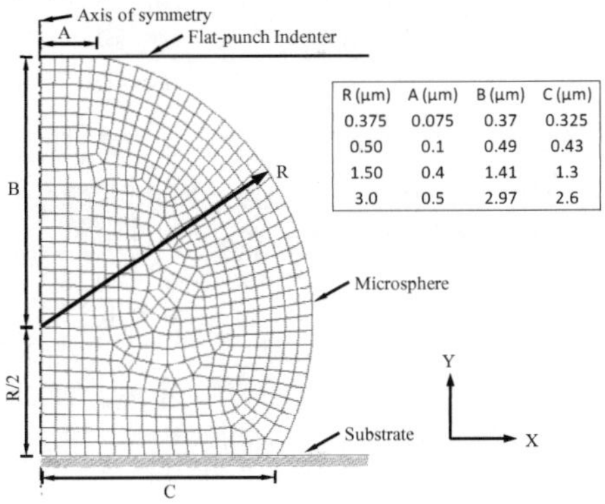

Figure 2: Schematic representation of the FE mesh used in this study. The dimension of the different Au microspheres analyzed is shown in the insets.

The compression process was simulated by imposing downward displacement of the nodes at the flat punch/sphere interface (Region A in Fig. 2). All simulations were done by defining multiple analysis steps. Isotropic elastic properties of the gold, E = 79 GPa and ν = 0.44, were used. The force-depth response obtained from the simulation was matched with the experimental force-depth response curve for the tested microspheres, by adjusting the multi-linear isotropic plastic flow properties of Au in the FE model [7]. In total, 12 models were created for four different diameter microspheres (0.75 to 6 μm) compressed at three loading-rates.

RESULTS

Experimental Microcompression Tests

Figure 3 shows the experimental force-displacement (*F-h*) curves for the smallest, 0.75 μm, and largest, 6 μm, diameter microspheres compressed at loading rates of 0.01 to 1.0 mN/sec. The deformation force F for the small 0.75 μm diameter spheres shows a strong dependence upon the loading rate.

Besides loading rate dependence, discrete load jumps in the *F-h* curves were observed for the tested spheres and the 0.75 μm diameter spheres displayed a higher frequency of these jumps.

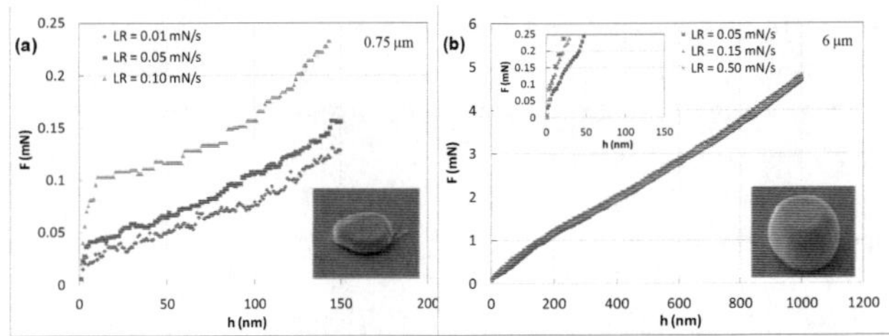

Figure 3: Experimentally obtained force F versus displacement h response for (a) 0.75 μm and (b) 6 μm diameter Au microspheres loaded at three different rates.

Similar load-jump features were previously reported in nanoindentation and nano-/micro-pillar compression studies and were associated with the initiation of dislocation nucleation/avalanche events in nano-sized crystalline samples [4][5][8]. Included in the insets in Fig. 3 are SEM images of the deformed 0.75 μm and 6.0 μm diameter spheres. Multiple slips lines corresponding to {111}<110> slip systems can be seen.

The apparent contact stress corresponding to the start of plastic yielding of the Au microspheres was expressed as the initial yielding force divided by the area of the {111} facet at the top of the sphere. In this way the contact stress of the smallest, 0.75 μm, diameter sphere was estimated to be between 670 MPa and 2 GPa while the largest, 6 μm, diameter sphere was between 210 MPa and 350 MPa over the three loading rates tested.

Stress-strain Curve from FE Simulation

The stress-strain distribution in the compressed spheres was analyzed using FE simulations described earlier. The average Von-Mises stress-strain data, collected from the nodes along the axis of symmetry of the model, were recorded for each load-depth increment. Figure 4 shows the average Von-Mises stress versus Von-Mises strain curves for the 0.75 μm and 6μm diameter microspheres compressed at the three loading rates.

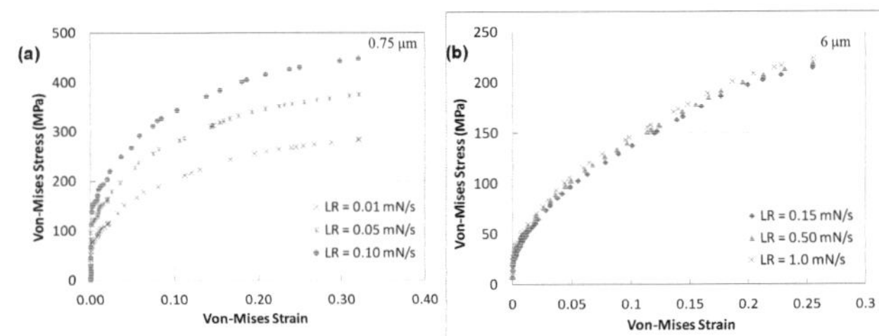

Figure 4: Von-Mises stress verses strain responses for the (a) 0.75 μm and (d) 6 μm diameter Au microspheres obtained from FE analyses performed at three loading rates.

The yield stress increased with increasing loading rates for all size microspheres; however, the yield stress of the smallest spheres displayed significantly greater loading rate dependence. The Von-Mises stress values corresponding to 20% Von-Mises strain for 0.75 μm spheres, increased from 275 MPa to 425 MPa with increasing loading rate from 0.01 to 0.10 mN/s whereas, for largest 6 μm diameter spheres, the equivalent stress increased from 215 MPa to 225 MPa. Since the strain rate increases with loading rate, this indicates that the strain-rate dependence of yield stress is increased with decreasing sphere diameter.

DISCUSSION

Besides enhanced strain-rate sensitivity with decreasing sphere size it is clear from Fig. 4 that, for a particular amount of strain, smaller spheres display a higher flow stress than large diameter spheres and this is consistent with previously reported length scale-dependence of plasticity [1][5][9].The strain, strain rate, and temperature dependence of the flow stress σ of plastically deforming metals is often described in terms of the following Arrhenius relationship [10];

$$\sigma = A\varepsilon^{n}\dot{\varepsilon}^{m}\exp\left[-\frac{Q}{RT}\right] \tag{1}$$

Where ε is strain, $\dot{\varepsilon}$ strain rate, A is a material constant, n is the strain hardening exponent, m is the strain rate sensitivity, Q is the activation energy, T is the absolute temperature and R is the universal gas constant. From this expression, at a constant $\dot{\varepsilon}$ and T, the strain dependence of stress provides the information on the strain hardening coefficient, n. Similarly, for a particular ε and T, the strain-rate dependence of stress yields the strain rate sensitivity, m. Figure 5a shows the strain dependence of the flow stress at strain rates of 3.5×10^{-2} sec^{-1}, 1.4×10^{-1} sec^{-1} and 2.6×10^{-1} sec^{-1} at 295k for 0.75 μm diameter spheres. The calculated values of n were 0.30, 0.27, and 0.26 respectively. Similarly, for 0.75 μm diameter microspheres, the calculated values of m, corresponding to four constant ε values of 0.05, 0.10, 0.15, 0.20, were 0.30, 0.28, 0.26, and 0.24 (Fig. 5b).

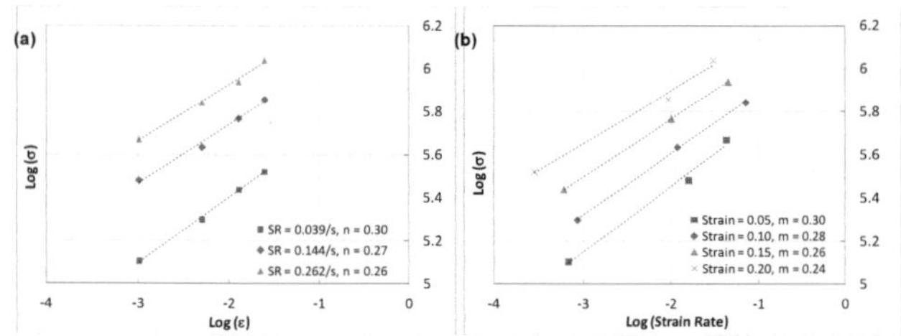

Figure 5: (a) Log(σ) vs. Log (ε) plots showing the effect of strain rate on n and (b) Log (σ) vs. Log (strain rate) plots showing the effect of strain on m for 0.75 μm Au spheres.

Figures 6 shows n (plotted against $\dot{\varepsilon}$) and m (plotted against ε) for all the microsphere sizes tested in this study. The strain-hardening exponent n for the larger, 6 and 3 μm diameter spheres ranged between 0.40-0.50 and showed clear strain-rate dependence. These values are consistent with those reported in the literature for bulk soft FCC metals [10]. The calculated n values for the smaller diameter Au microspheres (0.75 to 1.0 μm diameter) were, however, lower, approximately 0.3, and were significantly less dependent upon the strain-rate. These noticeable differences between the smaller and larger diameter spheres indicate that the smaller spheres deform differently than the larger spheres, which undergo plastic deformation similar to bulk metals.

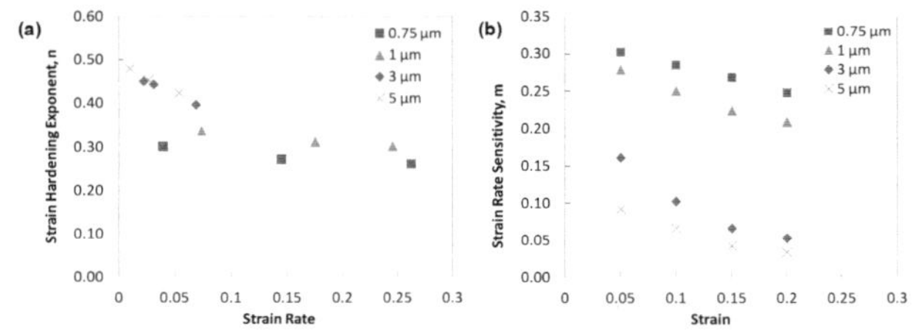

Figure 6: (a) Strain hardening exponent, n and (b) Strain rate sensitivity, m for all diameters Au microspheres tested in this study

The values of m, Fig. 6 (b), increased with decreasing sphere diameters indicating that specimens became more rate-sensitive with decreasing size. For the largest diameter 6 μm sphere m decreased from 0.08 to 0.03 as the ε increased from 0.05 to 0.20. The m value of 0.03 at is typical of Bulk FCC metals. This is another indication that the large diameter Au microspheres deformed in similar manner to bulk Au [11]. The smaller diameter spheres (0.75 to 1.0 μm) displayed significantly higher m value ranging between 0.2-0.3.

Metals deform plastically due to the motion of line defects, i.e. dislocations. Mobile dislocations in pure metals are frequently obstructed by various obstacles, such as point defects and other dislocations and these barriers are overcome by thermal activation. The volume swept by the hindered dislocations during activation is known as activation volume, V. The relationship between the flow stress and the strain rate or temperature has been used previously to quantify apparent activation energy and V^* [12]. These fundamental deformation rate-controlling parameters are used to predict the underlying operative dislocation-obstacle interaction mechanism controlling the plastic deformation process.

Using the strain rate sensitivity values m, V^* can be calculated using the following equation [13];

$$V^* = \sqrt{3}kT/m\sigma \tag{2}$$

Where k is the Boltzmann constant.

Figure 7 shows the calculated V^* versus strain for the Au microspheres in this study. The plotted V^* are normalized with respect to b^3, where b= 0.298 nm is the Burgers vector of Au. The V^* remained independent of strain during compression, with approximately a constant value of 4-5b^3, for smaller diameter, 0.75 μm and 1 μm, spheres whereas, for larger diameter, 3 μm and 6 μm spheres, the estimated V^* increased from 10b^3-38b^3, with increasing strain. The low V^* values, 4-5b^3, for smaller diameter spheres in this work are similar to previously reported estimations on FCC metals by Mook and Gerberich [14], and are characteristic of dislocation nucleation events during plastic deformation. These findings are also consistent with frequent discrete load jumps observed in F-h curves for 0.75 μm and 1 μm spheres (Fig. 3a) and suggest that plastic deformation for smaller diameter spheres is controlled by dislocation nucleation mechanism. This notion is further supported by high contact stress estimations (~865MPa) for the smaller spheres. These contact stress values close in magnitude to the ideal shear strength of Au (~800MPa) calculated by Ogata et al. [15]. On the other hand, the larger diameter, 3 µm and 6 µm, spheres displayed smoother *F-h* curves (Fig. 3b) and deformed at much lower contact stresses (340-210 MPa). Also, the V* magnitude ($10b^3$-$38b^3$) and its increasing trend with strain are typical of the operation of a dislocation-obstacle interaction type of deformation rate controlling mechanism typical of that common in bulk ductile metals [12].

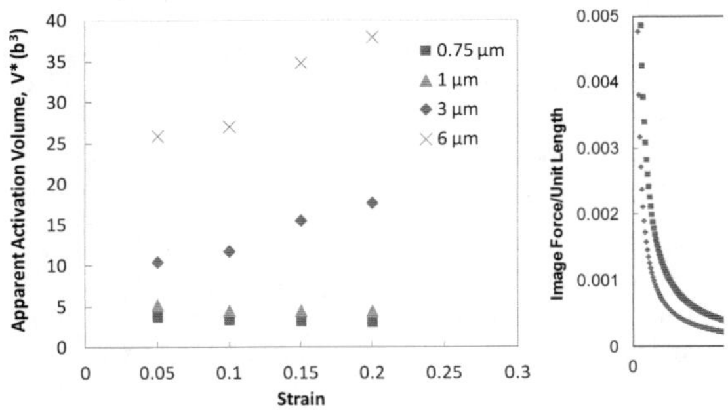

Figure 7: The change in apparent activation volume V^* with strain for different diameter Au microspheres.

CONCLUSIONS

In this study the dependence of the plastic deformation of Au microspheres upon sphere diameter (0.75-6 µm) and strain-rate at ambient temperature is investigated using micro-compression testing. The operative deformation rate-controlling mechanism was found to be size-dependent. The larger, 3 µm and 6 µm diameter, spheres deformed by a dislocation-obstacle interaction mechanism typical of bulk FCC metal specimens whereas plasticity in the smaller 0.75 and 1 µm diameter, spheres was controlled by a dislocation-starvation mechanism. These findings are particularly important in case of emerging nano/micro-technologies, making use of materials at ever decreasing scales.

ACKNOWLEDGMENTS

Authors wish to acknowledge NSERC for providing financial support for this work.

REFERENCES

1. Mook W. M., Niederberger C., Bechelany M., Philippe L., and Michler J., Feb. 2010, "Compression of freestanding gold nanostructures: from stochastic yield to predictable flow", *Nanotechnology*, Vol. 21, pp. 055701–9.
2. Nix W. and Gao H., 1998, "Indentation size effects in crystalline materials: a law for strain gradient plasticity", *J. Mech. Phys. Solids*, Vol. 46, pp. 411–425.
3. Greer J. R., Oliver W. C., and Nix W. D., Apr. 2005, "Size dependence of mechanical properties of gold at the micron scale in the absence of strain gradients", *Acta Mater.*, Vol. 53, pp. 1821–1830.
4. Greer J., 2006, "Bridging the gap between computational and experimental length scales: a review on nano-scale plasticity", *Rev. Adv. Mater. Sci*, Vol. 13, pp. 59–70.
5. Nix W. D., Greer J. R., Feng G., and Lilleodden E. T., Feb. 2007, "Deformation at the nanometer and micrometer length scales: Effects of strain gradients and dislocation starvation", *Thin Solid Films*, Vol. 515, pp. 3152–3157.
6. Mordehai D., Lee S.-W., Backes B., Srolovitz D. J., Nix W. D., and Rabkin E., Aug. 2011, "Size effect in compression of single-crystal gold microparticles", *Acta Mater.*, Vol. 59, pp. 5202–5215.
7. Gibson R. F., Jang H.-K., Simkhada S., Yu Q., Kim H.-I., and Suhr J., Sep. 2014, "Use of Nanoindentation, Finite Element Simulations, and a Combined Experimental/Numerical Approach to Characterize Elastic Moduli of Individual Porous Silica Particles", *Part. Sci. Technol.*
8. Zaiser M., Schwerdtfeger J., Schneider A. S., Frick C. P., Clark B. G., Gruber P. A., and Arzt E., Oct. 2008, "Strain bursts in plastically deforming molybdenum micro- and nanopillars", *Philos. Mag.*, Vol. 88, pp. 3861–3874.
9. Greer J. and Nix W., Jun. 2006, "Nanoscale gold pillars strengthened through dislocation starvation", *Phys. Rev. B*, Vol. 73, pp. 245410–6.
10. Kelly A. and Nicholson R., 1972, *Strengthening methods in crystals*, Wiley, Halstead Press Division.
11. Dao M., Lu L., Shen Y. F., and Suresh S., Dec. 2006, "Strength, strain-rate sensitivity and ductility of copper with nanoscale twins", *Acta Mater.*, Vol. 54, pp. 5421–5432.
12. Kocks U. F., Argon A. S., and Ashby M. F., 1975, *Thermodynamics and kinetics of slip*, Pergamon Press.
13. Wang Y., Hamza A., and Ma E., Jun. 2006, "Temperature-dependent strain rate sensitivity and activation volume of nanocrystalline Ni", *Acta Mater.*, Vol. 54, pp. 2715–2726.
14. Mook W. M., Lund M. S., Leighton C., and Gerberich W. W., Oct. 2008, "Flow stresses and activation volumes for highly deformed nanoposts", *Mater. Sci. Eng. A*, Vol. 493, pp. 12–20.
15. Ogata S., Li J., Hirosaki N., Shibutani Y., and Yip S., Sep. 2004, "Ideal shear strain of metals and ceramics", *Phys. Rev. B*, Vol. 70, pp. 104104–7.

INDENTATION-INDUCED DEFORMATION BEHAVIOUR OF MAGNESIUM

V. Bhakhri
Mechanical & Materials Engineering
The University of Western Ontario
London, ON, Canada
vbhakhr@uwo.ca

M. Haghshenas
Mechanical & Mechatronics Engineering
University of Waterloo
Waterloo, ON, Canada
mhaghshe@alumni.uwo.ca

R. Oviasuyi
NOVA Chemicals Corp.
Calgary, AB, Canada
roviasuyi@novachem.com

R.J. Klassen
Mechanical & Materials Engineering
The University of Western Ontario
London, ON, Canada
rklasse@uwo.ca

ABSTRACT

Dual-stage, constant loading-rate followed by constant-load stage, pyramidal indentation experiments were performed to investigate the strain-rate (10^{-5}-10^{-1} sec^{-1}) and temperature (295K-573K) dependence of commercially pure magnesium (Mg). The estimated total activation energy, Q (0.69-1.1eV), and apparent activation volume, V*(17-28b^3), over the tested temperature range indicated that plastic deformation is controlled by Cross-slip mechanism. The combined results from the present study and previously reported high strain-rate findings[1] confirm that operative deformation mechanism in Mg over a wide strain-rate, 10^{-5}-150 sec^{-1}, and temperature, 298K to 773K, range remains the same, i.e. Cross-slip, under the multi-axial stress state invoked by a pyramidal indenter.

KEYWORDS: Micro-indentation; Magnesium; Thermal Activation; Cross-slip

INTRODUCTION

Pure magnesium and magnesium alloys have found applications in areas as diverse as; electronics, automotive and more recently in the armour and ballistic applications[2][3]. For most of these applications requiring high specific strength, various magnesium alloy systems are being developed. This in turn has fuelled efforts to mechanically characterize the hexagonally closed pack (HCP) material system under various loading, strain-rate and temperature conditions. Much of the existing knowledge on mechanical behaviour of Mg has been retrieved from the uni-axial, tensile/compressive, testing of single and poly-crystalline specimens[4]. These investigations revealed that there are multiple deformation modes operative depending upon the loading direction, grain size and temperature. The slip (primary and secondary) and twinning occur at room and elevated temperatures. Primary slip takes place on the (0001) basal plane, along <1120> most closely packed direction at room temperature, while secondary slip occurs in the <1120> direction on the {1010} prism face planes. On the other hand, primary twinning takes place across the {1012} planes, whereas secondary twinning occurs on the {3034}. Slip can also be activated along <1120> direction on {1011} pyramidal planes and the twinning occurs across the {1013} planes.

Besides this information from the uni-axial stress state experiments, various studies employed instrumented indentation testing to investigate the mechanical behavior of Mg[1][2][5][6][7]. In particular, Somekawa and Schuh[5][6][1] used dynamic indentation experiments to explore the mechanical response of coarse and fine grained pure Mg and its alloys. Authors studied their high strain-rates (0.02-150 sec^{-1}) behavior at ambient and elevated temperatures (up to 773K) during multi-axial stress state invoked by pyramidal indenters and concluded Cross-slip to be dominant deformation-rate controlling mechanism. Despite several investigations on mechanisms of plastic deformation in Mg, its deformation behavior is still under discussion. Considering the versatility of its application, it is sensible to mechanically characterize the behavior of pure Mg across a broad range of strain-rate and temperature, and deduce information on underlying deformation mechanisms. In this work, we report the results from two stage, constant loading-rate and constant-load dwell, indentation experiments performed on a commercially pure Mg at 295K to 573K. The indentation stress variation with strain-rate (10^{-5}-10^{-2} sec^{-1}) and temperature is interpreted in terms of a thermally activated dislocation flow model to obtain fundamental deformation parameters; apparent activation volume, V*, and total activation energy, Q, of the deformation-rate controlling obstacles. The combined results from the present study and works by Somekawa and Schuh[1] provides the information on operative deformation mechanisms in Mg over a wide strain-rate, 10^{-5}-150 sec^{-1}, and temperature, 295K to 773K range.

EXPERIMENTAL PROCEDURE

Polycrystalline commercially pure casting magnesium (99.69 *wt%* Mg and 0.31 *wt%* Zn) was used in the as-received condition. The specimen was first mechanically polished to 0.05μm level surface roughness finish, followed by chemical polishing to obtain deformation-free surface.

Dual stage, constant loading-rate stage followed by constant-load stage, indentation tests were performed at 295K to 573K using NanoTest platform (Micro Materials Ltd., Wrexham, UK). During a typical indentation test, a Berkovich indenter was loaded at a rate of 25 mN/s until a pre-defined maximum indentation depth of ~10 μm was reached. The indentation load was then held constant for 30, 450, 900, and 1800 seconds during which the indentation depth, *h*, was recorded at half-second intervals. Subsequently, the indenter was unloaded at original rate to 10% of the maximum load and held for another 120 sec for thermal drift correction. For high temperature experiments, both specimen and the indenter tip were heated to test temperatures value using separate heaters to minimize the temperature gradient across specimen surface-indenter tip contact. This arrangement minimized the thermal drift contribution to the measured depth signal during and indentation experiment. The measured thermal drift increased from 0.01nm/sec at 295K to 0.04nm/sec at 573K.

RESULTS & DISCUSSION

Figures1 shows an optical micrograph of the indentation made at 573K. The area around the indent revealed a coarse-grained polycrystalline structure with average grain size of ~80 µm. The grain-boundary features around the indent at 573K depict either de-cohesion or thermal etching at elevated temperature.

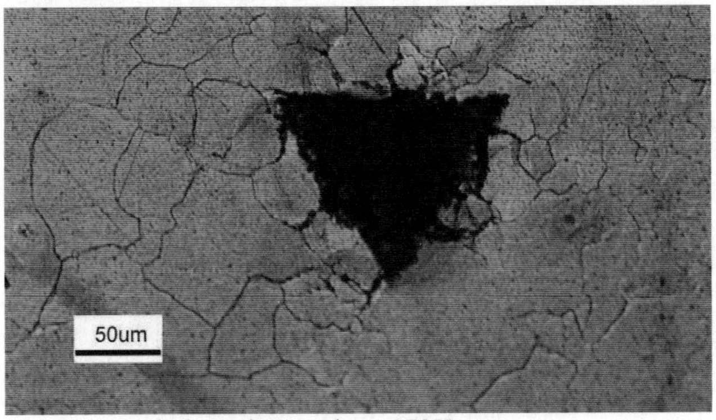

Figure 1: An indentation made at 573K.

The representative Force (P)-Displacement (h) curves obtained from the constant loading-rate and constant-load stages of indentation tests performed at 295K, 423K and 573K are shown in Fig. 2. During constant-loading rate stage, the indenter penetrated into the material to ~10µm depth, followed by a Constant-load stage, where the indenter load P was made to dwell for 900 seconds. The indentation depth increased with increasing time during dwell (Fig2b), due to the material creep under constant-load condition. The load required to reach 10µm depth decreased with increasing temperature, whereas the amount of creep increased as the temperature increased from 295K to 573K.

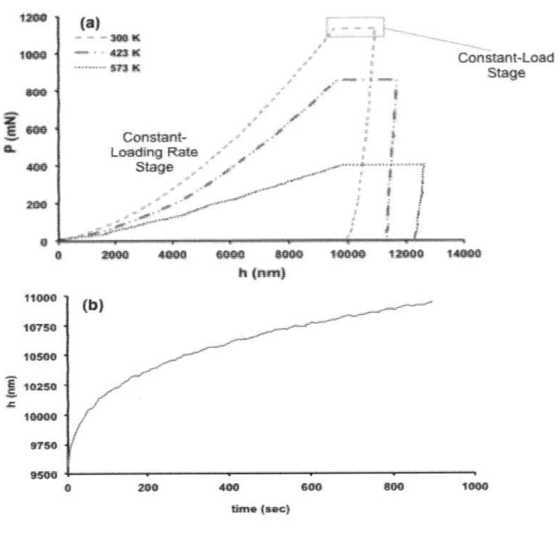

Figure 1: **(a)** Indentation load *P* versus indentation depth *h* for representative indentations made at 295-573K; **(b)** Indentation depth against the holding time of 900 sec during Constant-load stage at 295K.

The indentation stress, σ, and corresponding Shear stress, τ, can be calculated using indentation test data (*P*, *h*, *t*) by following expressions;

$$\sigma = \frac{P}{24.5h^2}, \quad \tau = \frac{\sigma}{3\sqrt{3}}, \qquad (1)$$

Here $24.5h^2$ is the indentation contact area function for a Berkovich indenter, and the normal to multi-axial stress conversion factor $3\sqrt{3}$ is based on the Von-Mises yield criterion.

Similarly, Indentation Strain-rate, $\frac{d\varepsilon}{dt}$, and Shear Strain-rate, $\frac{d\gamma}{dt}$, can be calculated using indentation depth, h, and time, t, data as;

$$\frac{d\varepsilon}{dt} = K\left(\frac{dh}{dt}\frac{1}{h}\right) \quad \frac{d\gamma}{dt} = \sqrt{3}\left(\frac{d\varepsilon}{dt}\right) \qquad (2)$$

where, K is a indenter geometry constant , and $\sqrt{3}$ is again a normal to multi-axial strain-rate conversion factor based on Von-Mises yield criterion.

The deformation strain-rate, $\frac{d\gamma}{dt}$, resulting from the mobile dislocation density, ρ_m, moving at a velocity, *v*, on a slip plane under the influence of shear flow stress, τ , is given by the following classical Orowan's equation[8];

$$\frac{d\gamma}{dt} = \rho_m b v \qquad (3)$$

where the velocity of moving dislocations, v , can be estimated using Boltzmann probability function;

$$v = \vartheta b\left[exp\left(-\frac{(\tau_p-\tau)V^*}{kT}\right)\right] \qquad (4)$$

where ϑ is the attempt frequency and τ_p is the lattice resistance or Peierls' stress for pure metals, i.e. the stress needed to make the dislocations move in the absence of any thermal energy. V^* is the activation volume, *T* is the absolute temperature and *k* is the Boltzmann's constant. Combining these equations and solving for the applied shear stress the expression can be simplified to;

$$\tau = \tau_p + \frac{kT}{V^*} ln\left[\frac{d\gamma}{dt}\right] - \frac{kT}{V^*} ln[\rho_m b^2 \vartheta] \qquad (5)$$

From this expression, a plot of τ against the natural logarithm of $\frac{d\gamma}{dt}$, at a particular temperature, should result in a linear relationship with slope kT/V^*. Therefore, V^* can be determined through the obtained slope. Similarly, intercept of the plot of shear flow stress, τ, versus temperature for a particular $\frac{d\gamma}{dt}$ provides information on the Peierls' stress, τ_p, and the slope of this curve can be used to determine the mobile dislocation density using the activation volume estimated earlier. Using the information on V^* and τ_p from these two plots, one can determine the activation energy Q of the deformation rate controlling obstacles as $Q = \tau_p V^*$. Both Q and V* are signatures of the operative deformation mechanism during plastic deformation process.

Figure3 shows the strain rate dependence of τ for commercially pure magnesium tested in this study. τ increased from 68 ($\pm$6) MPa to 102 ($\pm$11) MPa as the strain-rate increased from 10^{-5} sec^{-1} to $2x10^{-2}$ sec^{-1} at 295K. As can be noticed from the Fig. 3, the flow stress values, at a particular strain-rate, dropped with increasing temperature. However, it displayed similar strain-rate dependence regardless of test temperatures. The dropping τ with increasing temperature indicates the ease of dislocation motion on slip planes resulting from additional thermal energy at higher temperatures. As described earlier in the text, the slope of Fig. 3 provides the information on an activation parameter, V^*. The calculated values for V^* at tested temperatures, 295K to 573K, ranged between $17b^3$ to $28b^3$. Also included in Fig. 3(a) are the high strain-rate (10^{-2} to 150 sec^{-1}) data at 295K by Somekawa and Schuh[1] [7]. Our results from the low strain-rate (10^{-5} to 10^{-2} sec^{-1}) experiments in the present study align very well with the high strain-rate data from dynamic indentation testing by the authors. Also, a similar range of $15\text{-}45b^3$ for the estimated V* values from these high strain-rate experiments has been reported, indicating that similar deformation mechanism in pure Mg is operative over strain-rates ranging from 10^{-5}-150 sec^{-1}.

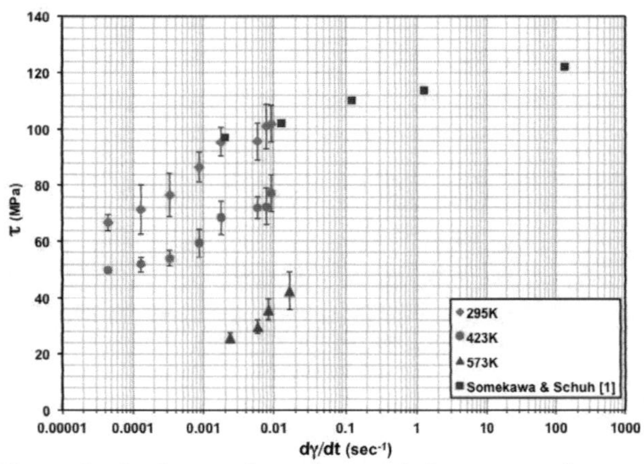

Figure 3: Strain-rate dependence of shear stress for tested coarse-grained Mg at 295K to 573K. Also included in the plot, are the room temperature high strain-rate data (10^{-2} to 150 sec^{-1}) by Somekawa & Schuh [7].

The temperature dependence of τ in Fig. 4 shows that τ decreases from 100 ($\pm$10) MPa at 295K to 30 ($\pm$10) MPa at 573K. The model expression in Eq. 5 is fitted (solid lines) for different strain-rates at variable temperatures to the data obtained in this work. Here the value of Burger's vector, b=0.321nm, for Mg was used and mobile dislocation density, ρ_m, and the attempt frequency, ϑ were used as adjustable parameters. The extrapolation of these trends to 0K provides the Perierls' stress *i.e.* the shear stress to move at dislocation in the absence of thermal energy. The Perierls' stress value obtained from Fib. 4 is of the order of 176 MPa. This value is also fairly close to 146 MPa provided in Deformation Mechanism Maps[9]. In the tested polycrystalline pure Mg, the activation energy, Q, of the deformation rate controlling obstacles ranged between 0.69 eV to 1.1 eV for tested temperature range of 295K to 573K.

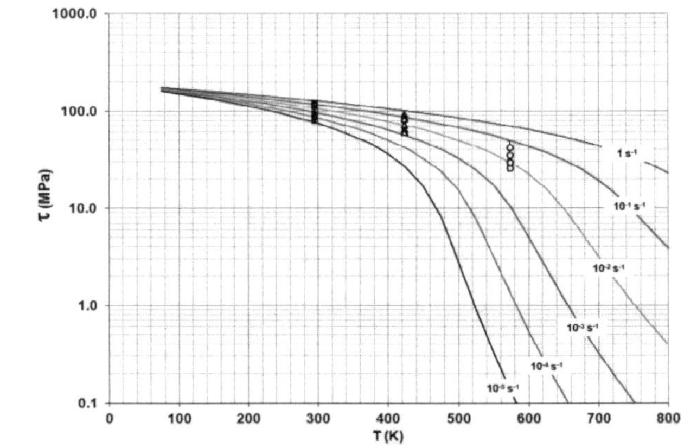

Figure4: Temperature dependence of coarse-grained Magnesium along with the predictions (solid lines) by Eq.5 for strain-rates ranging from 10^{-5} to 1 sec^{-1}.

The estimated apparent activation volume in present study ($17\text{-}28b^3$) and in the works by Schuh and Shomekawa[1] ($15\text{-}45b^3$) are similar and orders of magnitude lower than previously reported V* values from macro-scale uniaxial testing of coarse-grained materials (*i.e.* $1000b^3$ [7]). This difference emphasizes that the deformation in material plastic zone around a pyramidal indenter is entirely different from the one induced under uni-axial stress condition. It has been demonstrated that the plastic deformation of magnesium is significantly influenced by grain boundaries. The basal plane acts as a dominant dislocation slip system in single-crystalline Mg at room temperature. Whereas, the non-basal dislocations can be activated in the vicinity of grain boundaries, due to the necessity of stress/strain compatibility, in polycrystalline Mg[10]. The other mode of low-temperature and high-strain-rate plasticity, twinning, is also affected by grain boundaries in Mg; deformation twins preferentially nucleate at grain boundaries and grow to accommodate plastic deformation due to the lack of available activated slip systems[11]. The Cross-sectional observations of the deformed microstructures beneath dynamic indentations made in pure coarse grain Mg at 295K confirmed the presence of {10-12} type of twins[1]. These twins were proposed to form at early stages of indentation to accommodate the induced strains in the absence of favorable slip systems. With the progression of deformation, the twin-boundaries promote cross-slip in a manner similar to grain-boundaries due to strain compatibility constraints. The Q (0.69-1.1eV) and V* ($17\text{-}28b^3$) values obtained in the present work are consistent with Cross-slip as a dominant deformation rate controlling mechanism across 10^{-5}-10^{-2} sec^{-1}

strain-rate range at 295-573K.

The grain-boundary activity around the indent made at 573K in Fig.1 indicates that there can be another competing mechanism operative besides Cross-slip at this temperature. The estimated Q=1.1eV at 573K is also consistent with the slip assisted Grain-boundary sliding mechanism (Q=~1.0eV)[12]. This mechanism is considered to operate by plastic compatibility conditions in the presence of plastic strain anisotropy of Mg. Nevertheless, from the experimental data in Fig.3, similar strain-rate dependence of τ over tested temperature range (295K-573K), indicates Cross-slip to be a dominant rate controlling mechanism.

CONCLUSIONS

Dual-stage, constant loading-rate followed by constant load, pyramidal indentation experiments were performed to investigate the strain-rate (10^{-5}-10^{-1} sec^{-1}) and temperature (300K-573K) dependence of commercially pure magnesium (Mg). The estimated total activation energy, Q increased from 0.69eV at 300K to 1.1eV at 573K. Similarly, the apparent activation volume, V*, also displayed an increasing trend from $15b^3$ at 300K to $40b^3$ at 573K. Both Q and V* parameters indicated that plastic deformation is controlled by Cross-slip mechanism over the tested strain-rate and temperature range. The combined results from the present study and previously reported high strain-rate findings[1] confirm that operative deformation mechanism in Mg over a wide strain-rate, 10^{-5}-150 sec^{-1}, and temperature, 298K to 773K range remains the same, i.e. Cross-slip, under the multi-axial stress state invoked by a pyramidal indenter.

ACKNOWLEDGMENTS

Authors wish to thank NSERC for providing financial support for this work.

REFERENCES

[1] Somekawa H, Schuh CA. Nanoindentation behavior and deformed microstructures in coarse-grained magnesium alloys. Scr Mater 2013;68:416–9.

[2] Wang CL, Mukai T, Nieh TG. Room temperature creep of fine-grained pure Mg: A direct comparison between nanoindentation and uniaxial tension. J Mater Res 2009;24:1615–8.

[3] Sklenička V, Pahutová M, Kuchařová K, Svoboda M, Langdon TG. Creep processes in magnesium alloys and their composites. Metall Mater Trans A 2002;33:883–9.

[4] Agnew SR, Yoo MH, Tomé CN. Application of texture simulation to understanding mechanical behavior of Mg and solid solution alloys containing Li or Y. Acta Mater 2001;49:4277–89.

[5] Somekawa H, Schuh CA. High-strain-rate nanoindentation behavior of fine-grained magnesium alloys. J Mater Res 2012;27:1295–302.

[6] Somekawa H, Schuh CA. Effect of solid solution elements on nanoindentation hardness, rate dependence, and incipient plasticity in fine grained magnesium alloys. Acta Mater 2011;59:7554–63.

[7] Catoor D, Gao YF, Geng J, Prasad MJNV, Herbert EG, Kumar KS, et al. Incipient plasticity and deformation mechanisms in single-crystal Mg during spherical nanoindentation. Acta Mater 2013;61:2953–65.

[8] Kocks UF, Argon AS, Ashby F. Thermodynamics and Kinetics of Slip. Pergamon Press; 1975.

[9] Frost HJ, Ashby F. Deformation-mechanism maps: the plasticity and creep of metals and ceramics. Pergamon Press; 1982.

[10] Koike J, Kobayashi T, Mukai T, Watanabe H, Suzuki M, Maruyama K, et al. The activity of non-basal slip systems and dynamic recovery at room temperature in fine-grained AZ31B magnesium alloys. Acta Mater 2003;51:2055–65.

[11] Lubarda VA, Meyers MA, Vo O. The onset of twinning in metals: a constitutive description 2001;49:4025–39.

[12] Koike J. Enhanced deformation mechanisms by anisotropic plasticity in polycrystalline Mg alloys at room temperature. Metall Mater Trans A 2005;36:1689–96.

PLASTIC BUCKLING PARADOX FOR PLATES IN BIAXIAL LOADING

Suresh Shrivastava
Dept. Civil Engineering and Applied Mechanics
McGill University
Montreal, Quebec, Canada H3A 2K6
email: suresh.shrivastava@mcgill.ca

Bob Munlemvo Mayemba
Dept. Civil Engineering and Applied Mechanics
McGill University
Montreal, Quebec, Canada H3A 2K6

ABSTRACT

Exact plastic bifurcation buckling analysis of simply supported rectangular plates, bonded to elastic foundation, and subjected to uniform biaxial stress is presented. In particular, plastic buckling paradox is examined for three cases of loading: (1) uniaxial compression, (2) equibiaxial compression, and (3) equibiaxial compression-tension.

KEYWORDS: Plastic Buckling Paradox, Bifurcation, Biaxial Loading, Incremental and Deformation Theories of Plasticity, Isotropic Strain Hardening.

1. INTRODUCTION

Plastic buckling of rectangular plates in uniform uniaxial compression has been investigated by many researchers. Here, the bifurcation buckling under uniform biaxial loading is presented. Additionally, the plate is bonded to an elastic foundation. Exact analyses are performed for the two rival theories of strain-hardening plasticity, J_2 deformation and J_2 incremental, both obeying the J_2 (Mises) yield criterion.

Experiments favour the incremental theory as the correct phenomenological theory of plasticity [1]. However, for bifurcation of plates in uniaxial compression, it has been well known, since Handleman and Prager [2], that the incremental theory gives bifurcation loads which are much higher than the experimental buckling loads. In contrast, the predictions from the deformation theory are in safe-side agreement with experiments. This situation, wherein, seemingly, a correct theory yields the wrong result, and an incorrect one gives the right result, has been dubbed as the plastic buckling paradox.

2. CONSTITUTIVE RELATIONS

The strains are small; the plastic part is volume preserving. The pre buckling strains are uniform ($\varepsilon_{xx}^0, \varepsilon_{yy}^0$) where ε_{xx}^0 is always compressive, but ε_{yy}^0 can be tensile. Loading is such that $\sigma_{yy}^0 = \alpha\sigma_{xx}^0$ where α is a number $-1 \le \alpha \le 1$. Numerically, $P_2 = \sigma_{yy}^0 t = \alpha\sigma_{xx}^0 t = \alpha P_1$.

In bifurcation, the stresses change from σ_{ij}^0 to $\sigma_{ij}^0 + \sigma_{ij}$ where $\sigma_{ij}(\sigma_{xx}, \sigma_{yy}, \sigma_{xy})$ are stress increments, non-proportional to σ_{ij}^0. They are related to the strain increments ε_{ij} arising from a state of yield. For the J_2 deformation theory of strain-hardening plasticity, these relations are [3]:

$$\sigma_{xx} = B'\varepsilon_{xx} + C'\varepsilon_{yy}, \sigma_{yy} = C'\varepsilon_{xx} + D'\varepsilon_{yy}, \sigma_{xy} = 2F'\varepsilon_{xy} \text{ where}$$

$$F' = \frac{E}{(2+2\nu+3e)}, C' = \frac{E\{2(1+\alpha^2)(\lambda-1+2\nu)+\alpha(5+3e-5\lambda-4\nu)\}}{\Delta},$$
$$B' = \frac{E\{(\lambda+3+3e+4\lambda\alpha(\alpha-1)\}}{\Delta}, D' = \frac{E\{(\alpha^2(\lambda+3+3e)-4\lambda(\alpha-1)\}}{\Delta},$$
$$\Delta = (1+\alpha^2-\alpha)\{\lambda(5+3e-4\nu)-(1-2\nu)^2\}+3\alpha(1-\lambda+e)(1-2\nu).$$

The parameters $\lambda = E/E_t, e = E/E_s - 1$ where $E_t = d\sigma/d\varepsilon$ and $E_s = \sigma/\varepsilon$ are the tangent and secant moduli, obtained from a compression or tension test of the material, at the effective stress level σ_{eff}^0, which for the biaxial state of stress $\sigma_{yy}^0 = \alpha\sigma_{xx}^0$, is

$$\sigma_{eff}^0 = \sqrt{3J_2} = \sqrt{(\sigma_{xx}^0)^2 - \sigma_{xx}^0\sigma_{yy}^0 + (\sigma_{yy}^0)^2} = \sqrt{1+\alpha+\alpha^2} \times \sigma_{xx}^0$$

Moduli of the J_2 incremental theory are obtained by substituting $e = 0$ in the above, whereas those for the standard elastic behaviour are obtained for $e = 0$ and $\lambda = 1$. Thus, the constitutive relations, as written, incorporate the two plasticity theories, and also the elastic behaviour. Winkler foundation, of modulus k, offers reaction proportional to the plate deflection.

3. GOVERNING EQUATION AND ITS SOLUTION

Figure 1 shows the plate configuration. The plate length, width, and thickness are $a, b,$ and t respectively. The four edges are considered simply supported. As noted before, the pre-bifurcation loading is $P_1 = \sigma_{xx}^0 t$ in the longitudinal direction, and biaxial σ_{cr} and $P_2 = \sigma_{yy}^0 t = \alpha\sigma_{xx}^0 t$ in the transverse direction. Foundation is shown by springs.

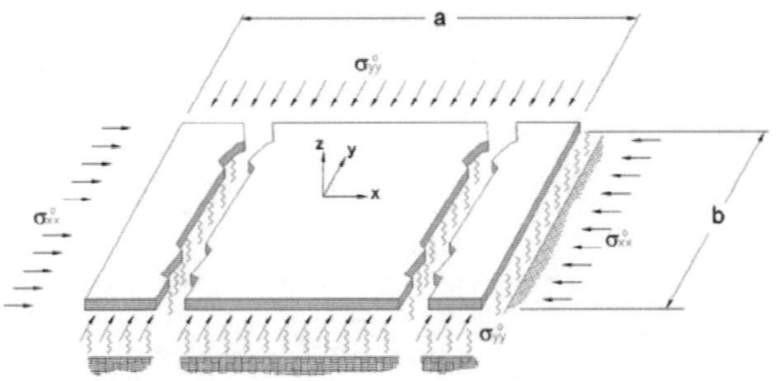

Fig. 1: Plate Configuration

At bifurcation, the z displacement w and the strains ε_{ij} emerge. Following Kirchhoff, they are

$$\varepsilon_{xx} = -zw_{xx}, \varepsilon_{yy} = -zw_{yy}, \varepsilon_{xy} = -zw_{xy}.$$

The principle of virtual work

$$\int_v (\sigma_{ij}\delta\varepsilon_{ij} + kw\delta w - P_1 w_x \delta w_x - P_2 w_y \delta w_y)\delta V = 0,$$

yields the following equation of equilibrium

$$\frac{\partial^2 M_{xx}}{\partial x^2} + \frac{\partial^2 M_{yy}}{\partial y^2} + \frac{\partial^2 M_{xy}}{\partial x \partial y} + P_1(w_{xx} + \alpha w_{yy}) + kw = 0$$

of the buckled plate, where $M_{xx} = \int \sigma_{xx} z dz$ etc. are the moment resultants per unit length.

Shanley's concept of plastic bifurcation under increasing load is held valid, and hence no unloading needs to be considered. The above loading relations are applicable throughout the plate thickness, and the moment resultants can be written as

$$M_{xx} = Bw_{xx} + Cw_{yy}, M_{yy} = Cw_{xx} + Dw_{yy}, M_{xy} = 2Fw_{xy},$$

$$B = B't^3/12, C = C't^3/12, D = D't^3/12, F = F't^3/12.$$

The governing equation is then:

$$B\frac{\partial^4 w}{\partial x^4} + 2(2F + C)\frac{\partial^4 w}{\partial x^2 \partial y^2} + D\frac{\partial^4 w}{\partial y^4} + P_1\left(\frac{\partial^2 w}{\partial x^2} + \alpha\frac{\partial^2 w}{\partial y^2}\right) + kw = 0\cdot$$

For simple supports $w = M_{xx} = 0$ at $x = 0, a,$ and $w = M_{yy} = 0$ at $y = 0, b,$ the solution is

$$w(x, y) = A_{mn}\sin\frac{m\pi x}{a}\sin\frac{n\pi y}{b},$$ and correspondingly,

$$P_1 = \pi^2 \frac{B\dfrac{m^4}{a^4} + 2(2F + C)\dfrac{m^2 n^2}{a^2 b^2} + D\dfrac{n^4}{b^4} + k}{\dfrac{m^2}{a^2} + \alpha\dfrac{n^2}{b^2}}$$

where m, n are the number of half waves of the bifurcated plate. A_{mn} is an arbitrary deflection at the centre of the plate.

4. RESULTS FOR THREE PARTICULAR CASES

Results are given for only the no foundation cases, $k = 0$. For full results, the reader may consult [3].

(1) Uniaxial Compression, $\alpha = 0$. The moduli of the constitutive relations reduce to

$$F' = \frac{E}{(2+2\nu+3e)}, C' = \frac{2E(\lambda - 1 + 2\nu)}{\Delta}, B' = \frac{E(\lambda + 3 + 3e)}{\Delta}, D' = \frac{4E\lambda}{\Delta},$$

$$\Delta = \lambda(5 + 3e - 4\nu) - (1 - 2\nu)^2.$$

For the lowest critical load, $n = 1$ and for long (say $a/b \geq 5$) plates, $m^2 = \frac{a^2}{b^2}\sqrt{\frac{D}{B}}$. This then gives

$$P_{1cr} = \pi^2 \times \frac{2\sqrt{BD} + 2C + 4F}{b^2}, \text{ or } \sigma_{xcr} = \frac{\pi^2 t^2}{12b^2}(2\sqrt{B'D'} + 2C' + 4F')\cdot$$

We can now calculate the bifurcation stresses σ^d_{xcr} for the deformation theory, and σ^i_{xcr} for the incremental theory $(e = 0)$ for plates of the same material and same dimensions. Although not indicated, in all cases, λ and e are different for the two stress levels of the two theories.

(2) Equibiaxial Compression $\alpha = 1$. In this case

$$F' = \frac{E}{(2+2\nu+3e)}, C' = \frac{E(1 - \lambda + 4\nu + 3e)}{\Delta}, B' = D' = \frac{E(\lambda + 3 + 3e)}{\Delta},$$

$\Delta = (\lambda + 1 - 2\nu)(2 + 2\nu + 3e)$, and additionally $2F' + C' = D'$.

For this case $P_{1cr} = \pi^2 B\left(\frac{m^2}{a^2} + \frac{n^2}{b^2}\right)$, or $P_{1cr} = \pi^2 B\left(\frac{a^2 + b^2}{a^2 b^2}\right)$ since one must have $m = n = 1$ for minimum P_{1cr}.

The bifurcation stress for the deformation theory is

$$\sigma^d_{xcr} = \frac{(\lambda + 3 + 3e)}{(\lambda + 1 - 2\nu)(2 + 2\nu + 3e)}\frac{\pi^2 E t^2}{12}\frac{(a^2 + b^2)}{a^2 b^2},$$

and for the incremental theory, it is

$$\sigma^{i}_{xcr} = \frac{(\lambda+3)}{(\lambda+1-2\nu)(2+2\nu)}\frac{\pi^2 E t^2}{12}\frac{(a^2+b^2)}{a^2b^2}.$$

(3) Equibiaxial Compression-Tension $\alpha=-1$. We find that

$$F'=\frac{E}{(2+2\nu+3e)}, C'=\frac{E(9\lambda-9+12\nu-3e)}{\Delta}, B'=D'=\frac{E(9\lambda+3+3e)}{\Delta},$$

$\Delta = 3(2-2\nu+e)(3\lambda-1+2\nu)$, and

$$P_{1cr}=\frac{B\frac{m^4}{a^4}+2(2F+C)\frac{m^2}{a^2b^2}+\frac{Bn^4}{b^4}}{\frac{m^2}{a^2}-\frac{n^2}{b^2}}.$$ For the minimum P_{1cr},

$n=1$, and for long plates $\frac{\partial^2 P_{1cr}}{\partial m^2}=0$,

$$m^2=\frac{a^2}{b^2}+\sqrt{\frac{2a^4}{b^4}\frac{(B+2F+C)}{B}}$$, and accordingly,

$$\sigma_{1cr}=\frac{2\pi^2 E t^2}{12b^2}\{B'+2B'F'+C'+\sqrt{2B'(B'+2F'+C')}\}.$$

5. COMPARISON OF THE NUMERICAL RESULTS

Numerical results are obtained for plates with $a/b=5$ and different b/t ratios. The material is Al alloy 24S-T3, with a tension test $\sigma-\varepsilon$ curve modeled by $\varepsilon=\frac{\sigma}{E}+0.002(\frac{\sigma}{300})^7$ with $E=76{,}530$ MPa and $\nu=0.33$. The experimental validity is up to 310 MPa, with plasticity setting around 160 MPa. Graphs of σ^{d}_{xcr} and σ^{i}_{xcr} versus t/b have been drawn up to $\sigma^{d}_{xcr}=\sigma_{eq}/\sqrt{1+\alpha^2-\alpha}$, σ_{eq} reaching the experimental limit ≈ 310MPa.

The ratio $\sigma^{i}_{xcr}/\sigma^{d}_{xcr}$ is $371/300=1.24$ for the $\alpha=0$ case, $337/300=1.12$ for the $\alpha=1$ case, and $267/179=1.49$ for the $\alpha=-1$ case.

Figure 2 gives the graphs. The paradox is (i) evident in the uniaxial case, (ii) quite muted for the equibixaxial compression, and (iii) reappearing, rather strongly, for the equibiaxial compression-tension case.

6. CONCLUSIONS

It is clear that the isotropic hardening is not a good model for bifurcation, except perhaps for the symmetric case of equibiaxial compression. A model, which takes into account the stress induced anisotropy in the evolution of the yield surface, might resolve the paradox.

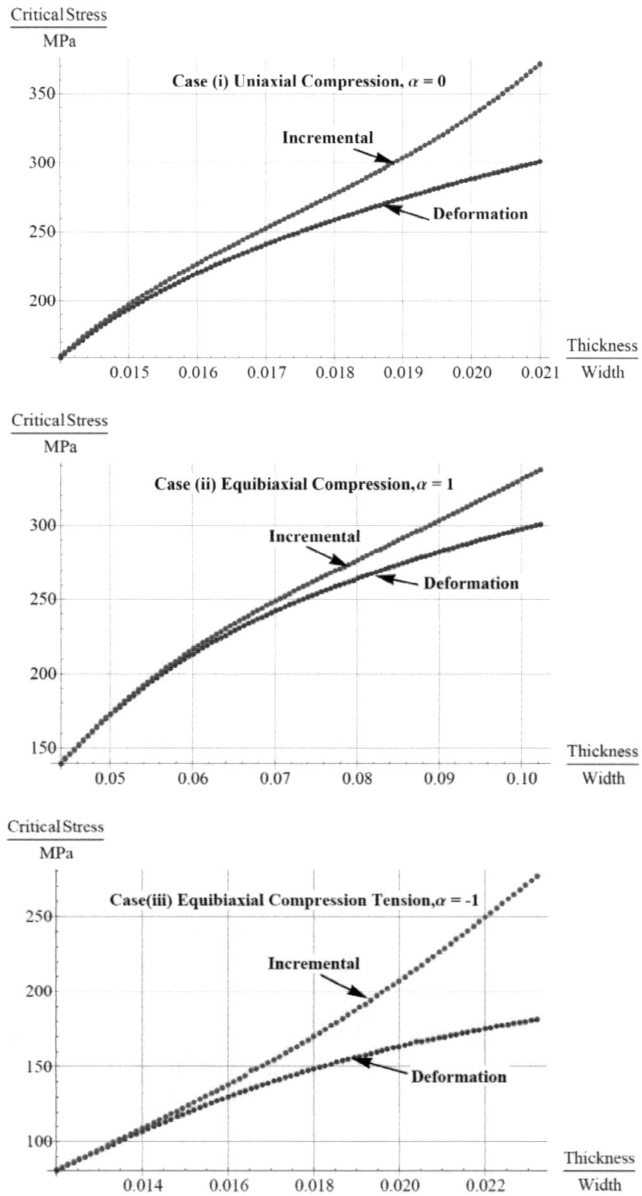

Fig. 2 Critical Stresses Predicted by the Two Theories

REFERENCES

1. G. H. Handelman and W. Prager, 1949, Plastic Buckling of a Rectangular Plate under Edge Thrusts, NACA Report 946.

2. R. Hill, 1950, The Mathematical Theory of Plasticity, Oxford University Press, London.

3. Bob Munlemvo Mayemba, 2012, Plastic Buckling of Rectangular Plates Bonded to Elastic Foundation in Biaxial Loading, M. Eng. Project, McGill University.

Experimental Methods

ELECTRICAL RESISTIVITY OF SANDS

Zaid Al-Qaysi
Dept. of Civil and Environmental Eng.
Western University
London, Ontario, Canada
zalqaysi@uwo.ca

Abouzar Sadrekarimi
Dept. of Civil and Environmental Eng.
Western University
London, Ontario, Canada
asadrek@uwo.ca

ABSTRACT

The difficulty of sampling unconsolidated saturated cohesionless soils for assessing their in-situ porosity (void ratio) due to sample disturbance and expensive methods of obtaining undisturbed field samples has increased the need for finding inexpensive methods for estimating the in-situ porosity. The approach of using geophysical techniques by measuring the electrical resistance of cohesionless soil can be used to define many geotechnical parameters and physical properties of sand without the need to obtain field samples, thereby minimizing some of the effects of soil disturbance. This paper presents an experimental testing program for investigating the effects of pore water salinity, soil fabric, void ratio, and fines content on electrical resistivity of saturated sands. The results show that the electrical resistivity of saturated sands decreases with increasing void ratio, increasing electrolyte concentration, and decreasing fines content at a certain void ratio; sand fabric and gradation have relatively minor influences. Empirical correlations are developed for estimating sand porosity.

KEYWORDS: porosity, sand, silt content, electrical resistivity, geophysical methods.

INTRODUCTION

The behavior of cohesionless soils is largely controlled by their density and porosity. Therefore, determination of in-situ porosity and thus density of sands is essential for replicating the in-situ shearing behavior and predicting field strength and liquefaction susceptibility behavior. These parameters in the top 10 m of a seabed are particularly important for the design of most offshore structures (e.g. oil platforms, wind turbine foundations, oil pipelines). However, direct measurement of the in-situ porosity and density of cohesionless soils is challenging due to the difficulties in obtaining undisturbed samples for laboratory testing and the susceptibility of cohesionless soil samples to disturbance caused by borehole excavation, sampling, during transportation, and sample extrusion and handling. In particular, saturated sands are notoriously difficult to sample without disturbance. These challenges have increased the need of finding more reliable and cost-effective in-situ testing methods. Geophysical techniques, including soil electrical resistivity measurement, can be used to define many geotechnical parameters and physical properties of sand without obtaining field samples, and thus minimize soil disturbance effects.

The objective of this study is to establish a relationship between electrical resistivity and porosity of cohesionless soils.

THEORETICAL BASIS

Electrical resistivity, ρ (Ohm·m) of a material is a measure of how well the material retards the flow of an electrical current. As particles (composed of for example quartz, feldspar, or carbonates) of a cohesionless soil are often non-conductive and act as insulators, the flow of an electrical current through a saturated granular soil occurs primarily through the pore water. Therefore, conductivity, volume, and the distribution of the pore water control the bulk electrical resistivity of sands (ρ_b). Accordingly, ρ_b can be directly related to the pore water resistivity (ρ_f) and pore volume (characterized by soil porosity n) using the following general empirical relationship (Winsauer et al., 1952):

$$\frac{\rho_b}{\rho_f} = an^{-m} \tag{1}$$

Where exponent m and coefficient a are fitting parameters which depend on pore volume tortuosity and soil particle cementation. These parameters can be determined from laboratory calibration experiments. The ratio ρ_b/ρ_f is known as the "formation factor" (*FF*) which is employed to remove the effects of changes in pore water resistivity (as sand particles are insulators) and geometrical characteristics of the measuring probe from the electrical resistivity of a saturated sand. Similar to ρ_b, *FF* is an intrinsic property of a soil which depends on the volume and the geometry of soil pore spaces.

EXPERIMENTAL PROGRAM

The experiments of this study were designed to determine the effects of fines content, sand fabric, salt concentration and void ratio on the electrical resistivity of sands as discussed in the following paragraphs.

Sample preparation
Sand samples were prepared by moist tamping in a cylindrical acrylic chamber with an internal diameter and

height of 114.4 mm and 250 mm, respectively. The samples were then saturated by filling them with water. This method led to more uniform samples and repeatable test results.

Materials

Two gradations of Ottawa sand (called "Barco 32" and "Barco 71") as well as a local silty sand (with a fines content of 11%) from the Boler Mountain (London, Ontario) were used in the experiments of this investigation. The particles of Barco 32 and Barco 71 sands are purely quartz with rounded to subrounded shapes, while the Boler Mountain sand is composed of subrounded to subangular particle shapes with mixed mineralogy of quartz, feldspar and carbonates. Barco 71 sand was mixed with different amounts of quartz silt particles to produce different silty sands and investigate the effect of fines content (FC). Figure 1 shows the gradations of these materials.

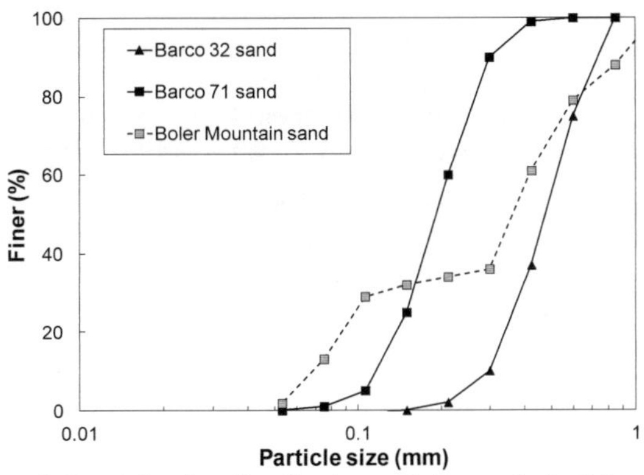

Figure 1: Particle size distributions of the sands in this study

Electrical resistivity measurement

A portable electrical resistivity probe was used to measure the electrical resistance (R) as well as the temperature of the soil water system. This probe introduces an electrical current (I) into the soil sample at a certain depth through two electrodes, and measures the potential voltage difference (V) between them. The average electrical resistance of the soil sample around the electrodes is then calculated from Ohm's law ($R = V/I$). These data are transferred to and displayed on a handheld Personal Digital Assistant (PDA). Soil electrical resistivity (ρ_b), a fundamental soil property, is obtained by multiplying the measured resistance (R) by the geometrical factor, K of the probe ($\rho_b = K \times R$) which depends on the size and the separation of the electrodes. An average geometrical factor of 0.872 was determined for the probe by calibration in an ionic buffer solution of 3g/L salt (NaCl) with a known resistivity of 0.052 Ohm.m. All experiments were performed at a controlled laboratory temperature of 20°C and in a non-metallic (acrylic) cylinder as metal could interfere with the measurements. The acrylic container had an indside diameter of 114.4 mm and a height of 250 mm. The probe is made of four 55-mm long stainless electrodes which are arranged in an annular pattern of 35 mm in diameter.

RESULTS AND DISCUSSION

The results of the experiments are presented below. For each experiment, sample void ratio and porosity was determined based on the weight of the soil used for sample preparation and sample volume, while ρ_b was measured using the electrical resistivity probe.

Effect of electrolyte concentration

Electrical resistivity of the pore water (ρ_f) depends on the electrolyte concentration as a result of pore water salinity, and thus at a certain void ratio the electrical resistance of a saturated cohesionless soil would become a function of the amount of salt in the pore fluid. Figure 2 presents the electrical resistance of saturated Barco 71 sand samples with different amounts of pore water salinity expressed in grams of dissolved salt per 1 liter of distilled water. Based on this figure, salt content of the pore fluid electrolyte has a profound impact on the electrical resistance of soil as R decreases with increasing salt concentration of the pore water at any certain void ratio.

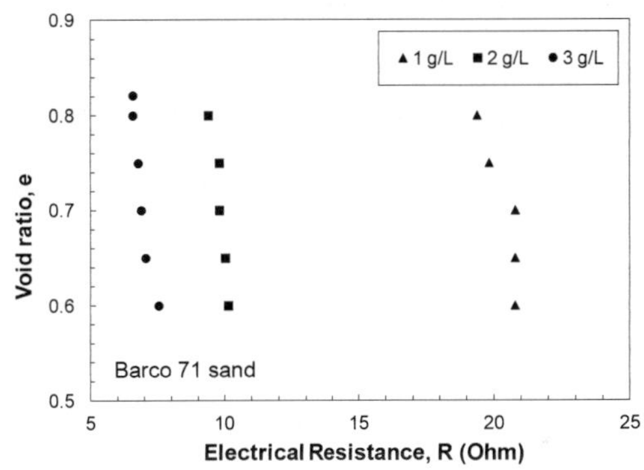

Figure 2: Effect of pore water salinity on electrical resistance of saturated Barco 71 sand samples

Figure 2 further indicates that soil electrical resistance is inversely related to its void ratio. As void ratio increases, the electrical resistance decreases and the electrical conductivity increases. Increasing of the electrical conductivity is due to the increase of the conducting pore water volume among the sand particles and the ability of the electrolyte water to conduct electricity, reducing the electrical resistivity of the saturated sand.

Effect of particle size distribution

Figure 3 compares the electrical resistances of the different sands tested in this study, which clearly indicates increasing soil electrical resistance with increasing of the amount of fines. Increasing fines content (FC), even at the same void ratio, reduces the number and the volume of connections among the pores of a sand and thus the electrical connectivity and conductivity of the pore fluid. Despite differences in particle size distributions (see Fig. 1), the Barco 32 and Barco 71 sands exhibit nearly similar electrical resistance at

FC = 0%, or the Boler Mountain and Barco 71 sands with about 10% fines contents show similar electrical resistances. Hence, the effect of variations in particle size distribution on the electrical resistance of sands seems to be secondary to the effect of FC. A similar observation was made by Keller and Frischkecht (1966).

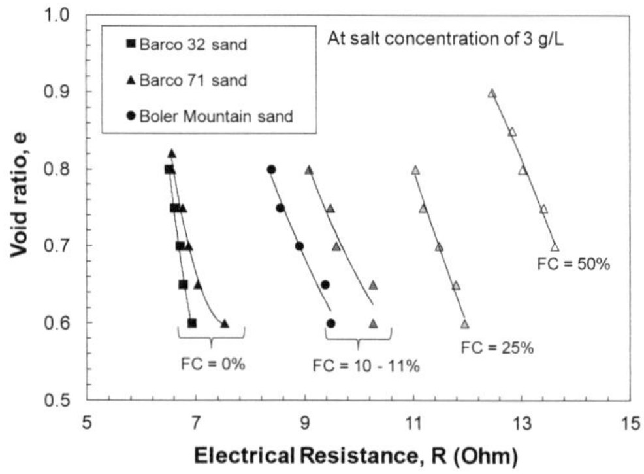

Figure 3: Effect of sand particle size distribution and fines content (FC) on electrical resistance with a pore water salt concentration of 3 g/L

Effect of specimen preparation method

In order to investigate the effect of sand fabric on the electrical resistivity of saturated sands, a number of experiments were conducted on samples prepared by air and water pluviation methods. These methods resemble sand fabrics formed by natural aeolian and fluvial process (e.g. in rivers and streams), respectively (Vaid and Eliadorani, 1998). Air pluviated specimens were prepared by raining sand particles into the cylindrical mold through air. The void ratio and density of the samples prepared by air pluviation was varied by changing the free-fall height and thus the depositional velocity of the particles. For preparing very loose samples, sand particles were rained with nearly zero drop height. A similar procedure was used to prepare water pluviated specimens by raining sand particles through water. However, as the particles reach a terminal velocity after a certain drop height in water, different void ratios were produced by tamping the sides of the specimen mold and thus densifying the sand sample. Figure 4 shows the results of these experiments. Electrical resistances of the samples prepared by the pluviation methods fall nearly on the same trendline, despite the larger void ratios of the water pluviated samples. However, the electrical resistances of the moist tamped samples are slightly greater which is likely because of the comparatively isolated pores of the honeycomb moist tamped sand fabric.

Estimation of porosity

To develop a practical relationship, the effects of electrolyte salinity and probe geometry are removed by presenting the results in terms of formation factor (*FF*) and porosity (*n*) in Figure 5. Electrical resistivity data for several other sands (platy sand, sand with 1% shell content, rounded sand, and sand and gravel) with different particle size distributions and particle shapes (Arulmoli et al., 1985; Jackson et al., 1978) are also included in this figure. According to Figure 5, *FF* generally increases with decreasing sand porosity (or increasing density) due to the smaller water saturated pore volume of dense sands. According to these data, variations in gradation, mean particle size, and particle shape have little influence on the trend of data. The most significant changes in *FF* are produced by changes in FC, while the scatter at certain FC could be because of variations in sand gradation and method of sample preparation.

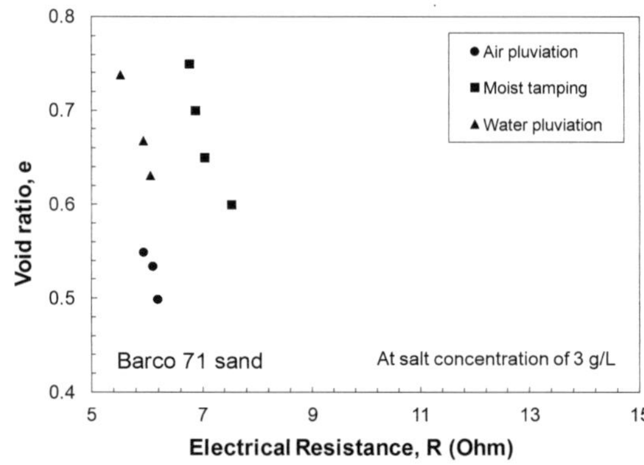

Figure 4: Effect of specimen preparation method on electrical resistance of sand

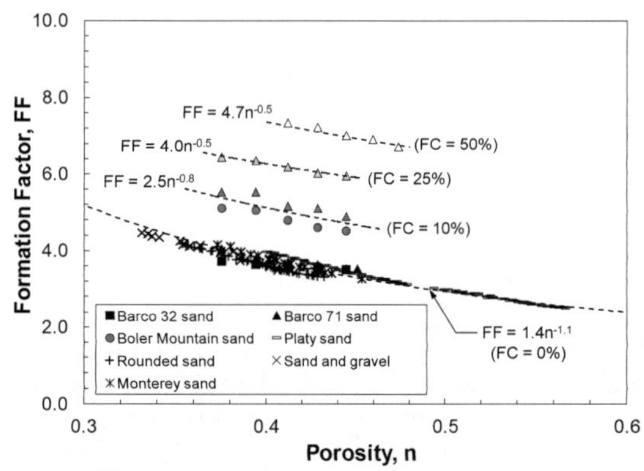

Figure 5: Effect of porosity on *FF* for the cohesionless soils tested in this study as well as for several other sands (Arulmoli et al., 1985; Jackson et al., 1978)

Data at certain FC are also curve fitted with Equation (1) and the fitting parameters (*a, m*) are shown in Figure 5. In contrast to other studies, rather than particle shape (Jackson et al., 1978) or cementation (Archie, 1942), Figure 5 indicates that the parameters of Equation (1) largely depend on FC. Coefficient *a* and the exponent *m* respectively increase and decrease with increasing FC. Fine particles increase the tortuosity and thus the length of the path that an electrical current must follow as well as constricting the openness of these flow channels. Therefore FF increases with increasing FC. As the experiments of this study were performed at small stress levels (< 5 kPa) the effect of stress level cannot be inferred from these results. However, based on past experimental observations (Hulbert et al., 1982; Lee et al., 2008), *FF* varies with stress to the extent that soil porosity is altered by stress. In other words, at a certain porosity stress does not appear to affect *FF*.

FIELD APPLICATIONS

Site specific calibration is always recommended for using electrical resistivity measurements for estimating sand porosity. Yet in the absence of such data, the correlations presented in Figure 5 are suggested to estimate the in-situ density of sands. This method can be applied to the electrical resistivity measurements of a resistivity cone penetrometer (RCPTu) to identify soil stratigraphy as well as soil porosity. This will provide an additional level of redundancy to the existing empirical methods of estimating the void ratio (i.e. porosity) of cohesionless soils from CPTu. In addition to RCPTu, electrical resistivity tomography could be used for obtaining three dimensional resistivity images of the subsurface and thus a more extensive characterization of soil porosity. In particular, an initial surface electrical resistivity surveying prior to drilling and sampling would help to locate loose and potentially liquefiable and problematic cohesionless soils and assist in selecting strategic borehole locations and sampling prior to drilling. Therefore, an integrated geophysical, geologic, and drilling study would provide a more rapid, cost effective, and accurate understanding of the subsurface.

Since changes in electrical resistivity result from changes in pore water volume, the application of this method would require saturated soil conditions. Offshore and seabed sediments would thus be the ideal conditions for the application of this method as they are naturally saturated with ionized sea water. In particular, the in-situ resistivity of pore water (ρ_f) – which is needed to calculate *FF* - can be readily measured just before the probe penetrates into the sediment without taking water samples. Although some other geophysical techniques like the nuclear magnetic resonance method, gamma-ray absorption techniques or X-ray tomography allow high resolution porosity determination, these methods involve highly complex and expensive instrumentation. Whereas electrical resistivity measurement and interpretation is fast, reliable, and low-cost.

CONCLUSIONS

The results of this study show that the electrical resistivity of cohesionless soils decreases with increasing pore water salinity, and porosity while soil fabric, particle size distribution and shape of the particles have negligible effects on electrical resistivity for clean sands. However, the amount of fine particles (< 0.075 mm) has a profound impact on the resistivity of a saturated cohesionless soil to an electrical current. Accordingly, a number of correlations are developed between electrical resistivity and porosity of cohesionless soils for certain fines contents. Measurement of void ratio and in-situ density of cohesionless soils could be difficult due to sample disturbance and the inherent variability of in-situ soils.

REFERENCES

1. Archie, G.E., 1942, "The electrical resistivity log as an aid in determining some reservoir characteristics" *Transactions of the American Institute of Mining and Metallurgical Engineers Petroleum Division*, Vol(146), pp. 54 - 62.
2. Arulmoli, K., Arulanandan, K., Seed, H.B., 1985, "New method for evaluating liquefaction potential", *Journal of Geotechnical Engineering*, Vol(111), pp. 95-114.
3. Erickson, S.N., Jarrard, R.D., 1998, "Porosity/formation-factor relationships for siliciclastic sediments from Amazon Fan", *Geophysical Research Letters*, Vol(25), pp. 2309 - 2312.
4. Hulbert, M.H., Bennett, R.H., Lambert, D.N., 1982, "Seabed geotechnical parameters from electrical conductivity measurements", *Geo-Marine Letters*, Vol(2), pp. 219 - 222.
5. Jackson, P.D., Taylor-Smith, D., Stanford, P.N., 1978, "Resistivity-porosity-particle shape relationships for marine sands", *Geophysics*, Vol(43), pp. 1250 - 1268.
6. Jones, P.H., Buford, T.B., 1951, "Electric logging applied to ground-water exploration", *Geophysics*, Vol(16), pp. 115 - 139.
7. Keller, G.V., and Frischknecht, F. C. , 1966, *Electrical methods in geophysical prospecting*, Pergamon Press, New York.
8. Lee, C., Lee, J.-S., Lee, W., Cho, T.-H., 2008, "Experiment setup for shear wave and electrical resistance measurements in an oedometer", *Geotechnical Testing Journal(ASTM)*, Vol(31), pp. 149 - 156.
9. Vaid, Y.P., Eliadorani, A., 1998, "Instability and liquefaction of granular soils under undrained and partially drained states", *Canadian Geotechnical Journal*, Vol(35), pp. 1053-1062.
10. Winsauer, W.O., Shearin, H.M., Masson, P.H., Williams, M., 1952, "Resistivity of brinesaturated sands in relation to pore geometry", *Bulletin of the American American Association of Petroleum Geologists*, Vol(36), pp. 253 - 257.

REYNOLD NUMBER EFFECT ON FORWARD FACING STEP IN PRESSURE GRADIENT

Hassan Iftekhar
Automotive, Mechanical and Manufacturing Engineering
University of Ontario Institute of Technology
Oshawa, ON, Canada
hassan.iftekhar@uoit.ca

Weijie Shao
Automotive, Mechanical and Manufacturing Engineering
University of Ontario Institute of Technology
Oshawa, ON, Canada
weijie.shao@uoit.ca

Martin Agelin-Chaab
Automotive, Mechanical and Manufacturing Engineering
University of Ontario Institute of Technology
Oshawa, ON, Canada
martin.agelin-chaab@uoit.ca

ABSTRACT

This paper reports an investigation of the effects of Reynolds numbers on turbulent FFS flows in adverse pressure gradient (APG). Three Reynolds numbers of $Re_h = U_m h/v$ of 1600, 3200, and 4800 were created for this study using diverging channels, where U_m is the maximum approach velocity, h is the step height, and ν is the kinematic viscosity of water. A PIV technique was used to conduct the measurements at several locations. The results indicate that Re_h has no significant effect on the mean flow. The mean flow characteristics such as the mean reattachment length (X_r) obtained from APG-4 showed no significant Re_h dependence. However, the APG-0 case showed an increase in X_r as the Re_h increased to 4800.

KEYWORDS: ***Forward facing-step, adverse pressure gradient, turbulent flows, particle image velocimetry***

INTRODUCTION

Flow separation and recirculation caused by blunt obstacles occur in diverse industrial applications. For example, flows over obstacles are encountered in diffusers, heat exchangers and combustors. After separation, the flow usually reattaches and in the reattachment region, the shear layer may be overwhelmingly distorted so that the usual thin layer approximations are not valid. In this regard, separated and reattached turbulent flows can be used as prototypical flows studying complex turbulent shear flows.

A wide range of simple laboratory geometries such as the backward-facing step (BFS) [1], forward-facing step (FFS) [2-3], transverse ribs [4] and blunt plate [5] have been used to generate separation and reattachment. A detailed summary of previous studies on separated reattached flows are summarized by Shah and Tachie [6] and Weijie and Agelin-chaab [7] and is not repeated here.

Figure 1 shows typical features of FFS flows including the two recirculating regions and the reattachment length (X_r), which is the distance from the leading edge to the reattachment point as indicated in the figure. After reattachment the flow redevelops downstream.

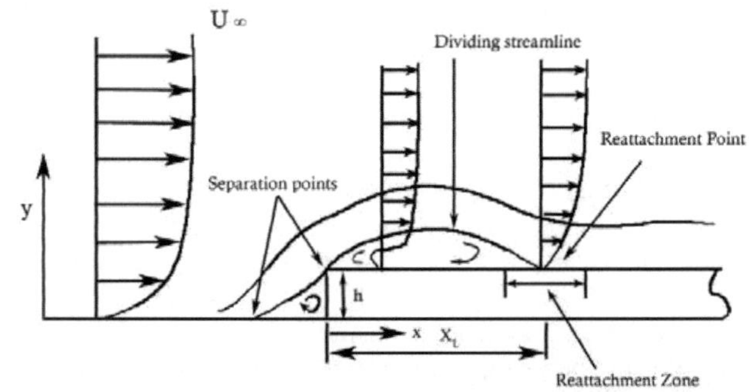

Figure 1: Mean flow features of a forward facing step [8].

Tachie et al. [3] studied mean velocity and turbulence intensity on a FFS in an open channel flow at the following Reynolds numbers: $960 \leq Re_h \leq 1890$. Their results indicated that the mean velocity profiles at 50 step heights downstream collapsed reasonably with those obtained upstream of the FFS. However, the profiles of the turbulence intensities did not collapse satisfactorily even at 100 step heights downstream.

Turbulent flows with pressure gradient also play an important role in many engineering applications, such as in diffusers and nozzles or diverging/converging channels. Prior research [9], demonstrated that adverse pressure gradient (APG) considerably slows the mean flow close to the wall and thereby thickens the boundary layer. APG also increases the turbulence intensities and Reynolds shear stress compared with values reported in zero pressure gradient (ZPG) turbulent boundary layers. Aubertine and Eaton [10] studied mild APG turbulent boundary layers and found that the displacement thickness, momentum thickness, and the Reynolds number based on momentum thickness increased in the diverging section. However, Skåre and Krogstad [11] examined strong APG and concluded that although the Reynolds stress distribution across the boundary layer is different from ZPG turbulent boundary layers, the stress ratios were nearly independent of pressure gradient. A detailed review of previous APG studies can be found in [12-13]. On the other hand, Shah and Tachie [6] studied flow over a transverse square rib in adverse pressure gradient using particle image velocimetry (PIV). It was observed that the flow dynamics in the upper boundary layer of the separated region were insensitive to pressure gradient but significant differences existed in the lower boundary layer. For example, APG modified the mean velocity and turbulence quantities across a larger portion of the channel. Furthermore, APG produced higher mass and momentum deficits as well as Reynolds stresses than in ZPG. More recently Weijie and Agelin-chaab [7, 14] reported the combined effects of APG and FFS on the redeveloping turbulence structures after reattachment. In the present study, the effect of Reynolds number on the mean flow and Reynolds stresses is examined.

EXPERIMENTAL SETUP AND MEASUREMENT PROCEDURE

Experimental Set-up

The experiments were performed in a re-circulating closed water tunnel with a test section that was 2500 mm long, 200 mm wide and 200 mm deep. The test section of the channel was constructed with clear acrylic to facilitate optical access. Figure 2 shows a schematic of the test section with the associated dimensions and coordinate system used. Note that $x = 0$ corresponds to the leading edge of the step and $y = 0$ is at the top surface of the step. A FFS of height $h = 6$ mm and length 1300 mm was used to induce flow separation. To minimize three-dimensionality of the flow over the step, the aspect ratio (AR = W/h) = 31 was used which is significantly larger than the minimum value of AR = 10 required to establish a two-dimensional mean flow at the mid-span of the test section [15]. The adverse pressure gradients were created

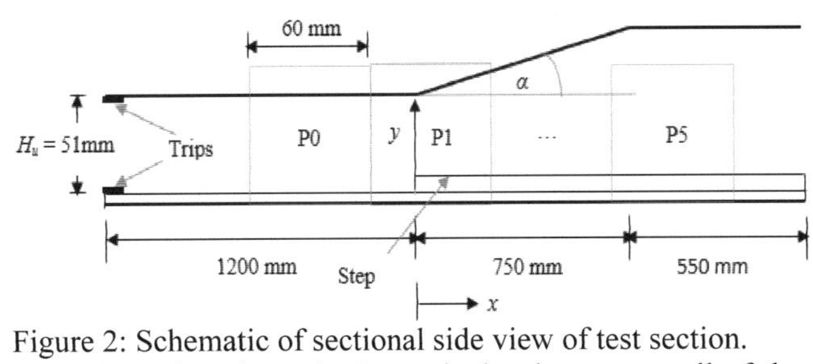

Figure 2: Schematic of sectional side view of test section.

using diverging channels. In particular the upper wall of the channel was diverged at an angle (α) to the horizontal as shown in Fig. 2. Three different angles were studied: α = 0°, 2° and 4° (hereafter denoted by APG-0, APG-2 and APG-4, respectively).

Measurement Procedure

A PIV technique was used to conduct the field velocity measurements. The flow was seeded with poly methyl methacrylate particles coated with a fluorescent dye, Rhodamine B. The particles have mean diameter and density of 10 μm and 1.19 g/cm^3, respectively. The fluorescent dye absorbs light at a wavelength of 550 nm and emits orange light at a wavelength of 590 nm. An Nd-YAG, 120 mJ/pulse laser of 532 nm wavelength was used to illuminate the flow field. The laser sheet was located at the mid-span of the test section for all measurements. A 12-bit high-resolution digital camera that uses a CCD with 2048 pixels × 2048 pixels and a 7.4 μm pixel pitch was used to image the flow field. The measurements were made at a field of view of 60 mm × 60 mm. A 32 pixels × 32 pixels interrogation window with 50% overlap was used to process the 4000 instantaneous image pairs resulting in a vector spacing of 0.47 mm. The uncertainty in the mean velocities and Reynolds stresses at 95% confidence level was estimated to be ±2% and ± 10% of the stream wise velocity, respectively [16-17].

Prior to conducting measurements, data were obtained at -20h (upstream of the step) to characterize the nature of the approach flow. The Reynolds numbers based on the step height (h) were Re_h (= $U_m h/\nu$) = 1600, 3200, and 4800; where U_m is the maximum approach velocity and ν is the kinematic viscosity of water.

RESULTS AND DISCUSSION

Mean Flow

The reattachment lengths (X_r) in the recirculation region was determined as the average of where the dividing streamline $\psi = 0$ and the 50% forward fraction line reattaches to the top surface of the step after separation. The X_r obtained from experiments for each test condition are listed in Table 1. The error in estimating X_r is 0.1h. The data in the table indicate that as the Re_h increased to 4800 for the APG-0, then there is an increase in X_r. This is expected since the high Re_h flow is more energetic to travel longer distances before reattaching. However, the dramatic increase by about 30% for Re_h = 4800 is not understood yet. When the APG is increased as in the case of APG-4, there is no increase in the X_r values. This implies the significant increase in pressure gradient basically nullified the trend observed in APG-0.

The profiles of stream wise mean velocities at three different locations over the step in the near field are reported in Fig.3. Note that all quantities in this and subsequent figures are made dimensionless using free stream velocity U_m of the approach flow which was measured at x/h = -20 upstream of the leading edge of the step.

Table 1: Reattachment lengths with Reynolds number.

Test	Reynolds number	Reattachment length (X_r^*)
APG-0	1600	1.7
APG-0	3200	1.8
APG-0	4800	2.3
APG-4	1600	2.1
APG-4	3200	2.2
APG-4	4800	2.2

Also, (a) to (c) represent APG-0 and (d) to (f) represent APG-4. Figure 3 shows that the overall mean velocities for both APG-0 and APG-4 collapsed beyond x/h = 2 indicating no significant Reynolds number effects on the mean flow beyond the recirculation bubble. This is interesting since the reattachment length of APG-4 is similar to $2h$. However, at x/h = 2 for APG-0, the velocity profiles are larger for higher Reynolds numbers, which is also consistent with the reattachment lengths reported in Table 1.

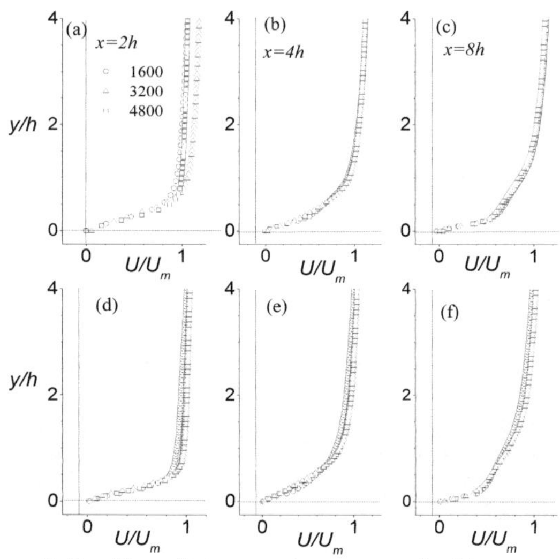

Figure 3: Profiles of stream-wise mean velocities.
(a)- (c) is APG-0 and (d) – (f) is APG-4.

The profiles of the Reynolds stresses are used to examine the evolution of the turbulent quantities with the Reynolds numbers. Profiles of Reynolds stresses are reported in Figs. 4-6. In general the turbulent quantities are limited to the wall (y/h < 1.0) and increase with Re_h. The profiles show highest peak values at x = $2h$ but decreased downstream as expected. Although, as the flow evolved downstream, the region with high turbulence spreads outward to y/h > 1.0. For APG-0, the streamwise Reynolds normal stresses (<*uu*>) for Re_h = 1600 and 3200 appear to collapse at x/h = 4 but not Re_h = 4800. For APG-4 on the other hand, none of the profiles collapsed. This inconsistency in the <*uu*> values in the region reported is not understood. In addition, both pressure gradient cases showed double peaks at x/h = 8 consistent with the shear layer in that region but single peaks at x/h = 4 and earlier. Also, the peak values in both cases are similar. The trends observed for <*vv*> are similar to those described for <*uu*> and will not be repeated, except that the values are smaller; about half of those for <*uu*>. This is expected since the flow is driven more in the stream-wise direction than the wall-normal. The profiles of the Reynolds shear stresses (<*uv*>) are much smaller but also show less significant variation between them at x/h= 8 for the normal stresses discussed earlier. This may suggest that the influence of the Re number is limited to the region x/h < 8.

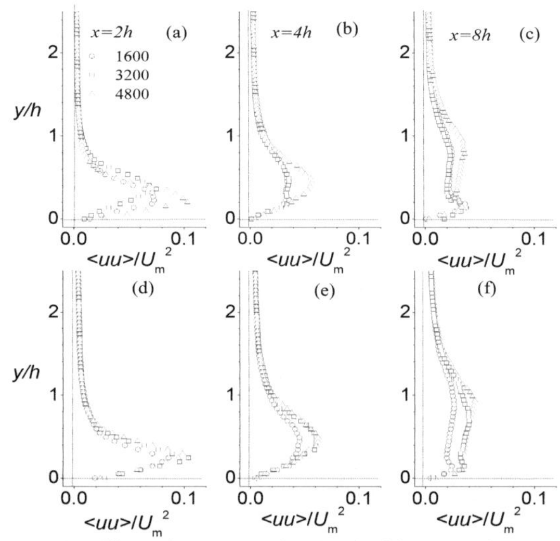

Figure 4: Profiles of stream-wise Reynolds normal stresses.
(a)- (c) is APG-0 and (d) – (f) is APG-4.

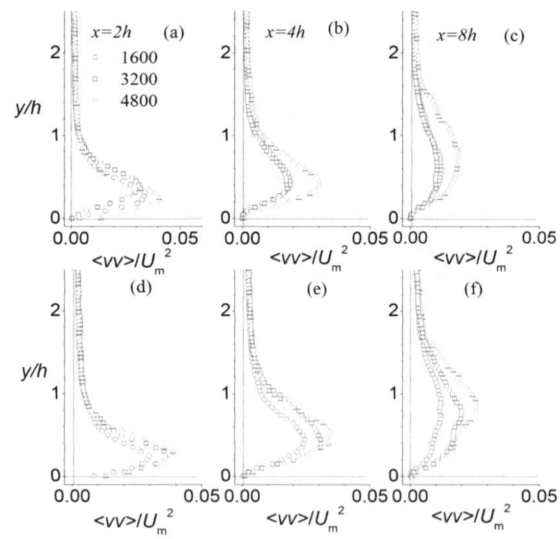

Figure 5: Profiles of wall-normal Reynolds normal stresses.
(a)- (c) is APG-0 and (d) – (f) is APG-4.

CONCLUSIONS

This paper reports an investigation of the effects of Reynolds numbers on turbulent FFS flows in adverse pressure gradient (APG). Three Reynolds numbers of Re_h = 1600, 3200, and 4800 were created for this study. Diverging channels were used to simulate the APG conditions and a PIV technique was used to conduct the measurements at several locations.

The results indicate that Re_h has no significant effect on the mean flow. The mean flow characteristics such as the mean reattachment length (X_r) obtained from APG-4 showed no significant Re_h dependence. However, the APG-0 case showed increase in X_r as the Re_h increased to 4800. Although the Reynolds stresses were limited to the wall region, $y/h < 1.0$, they appeared to be strongly dependent on Re_h and in fact, do increase slightly with Re_h irrespective of pressure gradient. The variation in the Reynolds shear stresses due to Re_h is the least.

This physical insight and data reported are useful for assessing the ability of turbulence models to reproduce the behaviour of complex flows.

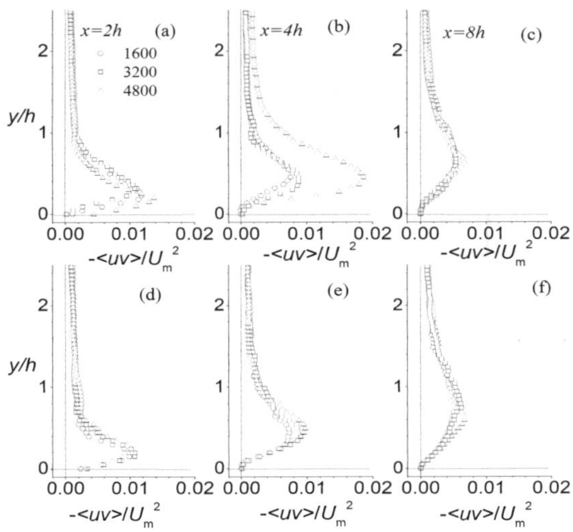

Figure 6: Profiles of Reynolds shear stresses.
(a)- (c) is APG-0 and (d) – (f) is APG-4.

ACKNOWLEDGMENTS

Financial support by the Natural Sciences and Engineering Research Council of Canada for this work is gratefully acknowledged by the second author. The authors would like to thank Dr. M. F. Tachie for the permission to use his lab and Mr. E. K. Essel for his help with the data acquisition.

REFERENCES

[1]. Bradshaw, P. and Wong, F. Y. F., 1972, "The Reattachment and Relaxation of a Turbulent Shear Layer," *J. Fluid Mech.*, 52, pp. 113-135.

[2]. Abu-Mulaweh, H. I., 2005, "Turbulent Mixed Convection Flow over a Forward-Facing Step - The Effect of Step Height," *Int'l J. Thermal Sci.*, 44, pp. 155-162.

[3]. Tachie, M. F., Balachandar, R. and Bergstrom, D.J., 2001, "Open Channel Boundary Layer Relaxation Behind a Forward Facing Step at Low Reynolds Numbers," *J. Fluids Eng.*, 123, pp. 539-544.

[4]. Agelinchaab, M. & Tachie, M. F. 2008 PIV Study of Separated and Reattached Open Channel Flow over Surface Mounted Blocks. *ASME Journal of Fluids Engineering*, Vol. 130, No. 6, pp. 061206.

[5]. Kiya, M. and Sasaki, K., 1983, "Structure of a Turbulent Separation Bubble*," J. Fluid Mech.*, 137, pp. 83-113.

[6]. Shah, M. K. and Tachie, M. F., "Flow Relaxation Past a Transverse Square Rib in Pressure Gradients", *AIAA Journal*, Vol.46, No. 7 (2008), pp. 1849-1863.

[7]. Shao, W. J., and Agelin-chaab, M., "The structure of forward facing step flows in adverse pressure gradient", *ASME, FEDSM2014-21357*, 2014, Chicago, Illinois, USA.

[8]. Sherry, M., Lo Jaconoa, D., and Sheridana, J., 2010, "An experimental investigation of the recirculation zone formed downstream of a forward facing step," *Journal of Wind Engineering and Industrial Aerodynamics*, 98 (120), pp. 888-894.

[9]. Kim, D. S., White, B. R., Ayala, A. and Bagheri, N., 1999, "Higher-Order Turbulence Products of Velocity and Temperature for Adverse Pressure Gradient Boundary Layer Flows," *Proceedings of Heat Transfer and Fluid Mechanics Institute*, pp. 125-132.

[10]. Aubertine, C. D. and Eaton, J. K., 2005, "Turbulence Development in a Non-Equilibrium Turbulent Boundary Layer with Mild Adverse Pressure Gradient," *J. Fluid Mech.*, 532, pp. 345-364.

[11]. Skåre, P. E., and Krogstad, P.-Å., 1994, "A Turbulent Equilibrium Boundary Layer Near Separation," *J. Fluid Mech.*, 272, pp. 319-348.

[12]. Dengel, P., and Fernholz, H. H., 1990, "An Experimental Investigation of an Incompressible Turbulent Boundary Layer in the Vicinity of Separation," *J. Fluid Mech.*, 212, pp. 615–636.

[13]. Angele, K. P. and Muhammad-Klingmann, B., 2006, "PIV Measurements in a Weakly Separating and Reattaching Turbulent Boundary Layer," *Eur. J. Mech. B/Fluids*, 25, pp. 204-222.

[14] Shao, W. J., and Agelin-chaab, M., "Forward facing step flows in adverse pressure gradient", *CSME*, 2014, Toronto, Ontario, Canada.

[15]. De Brederode, V., and Bradshaw, P., 1972, "Three-dimensional flow in nominally two dimensional separation bubbles flow behind a rearward-facing step," Imp. *Coll. Aeronaut.* Rep. 72–19.

[16]. Coleman, H., and Steele, W., 1995, "Engineering application of experimental uncertainty analysis," *AIAA J*, 33, pp. 1888–1895.
[17]. Forliti, D. J., Strykowski, P. J., and Debatin, K., 2000, "Bias and precision errors of digital particle image velocimetry," *Exp. Fluids*, 28, pp. 436–447.

MIXING CHARACTERISTICS IN TURBULENT ORIFICE JETS

Yaw Y. Afriyie & Mark F. Tachie
Department of Mechanical Engineering
University of Manitoba
Winnipeg, MB, Canada, R3T 5V6

Mikhail Koupriyanov & Tom Epp
Price Industries Limited
Winnipeg, MB, Canada, R2K 3Z9

ABSTRACT

An experimental investigation was conducted to study the effect of slot geometry on turbulent free orifice jets. Particle image velocimetry (PIV) technique was used for the velocity measurements in the near and developing regions of the jets ($0 \leq x/d \leq 21$), where d is the equivalent diameter of the nozzle. These measurements were performed for different orifice nozzles which include round, square, rectangle and ellipse. The Reynolds number based on equivalent diameter was kept constant by maintaining the exit velocity of all the jets at approximately 0.6 m/s ($Re_d = 5400$). The non-circular nozzles show faster decay of centerline velocity, greater spread rate and higher turbulent intensities. This was especially the case in the rectangular and elliptical nozzles with an aspect ratio of 2.

KEYWORDS: Free jet, Orifice nozzle, Axis switching, Mixing.

INTRODUCTION

Turbulent free jets have diverse engineering and industrial applications, for example in heating, ventilation and air conditioning (HVAC) systems and heat transfer augmentation. Also because of their relative simplicity compared with wall-bounded turbulent shear flows, they are used as prototypical flows for exploring turbulent flow physics. As a result, turbulent free jets have received a lot of research attention. Majority of the fundamental research on free jet flows has focused on plane and circular nozzles. A summary of such studies can be found in literature (e.g. Abdel-Rahman, 2010). Non-circular nozzles (e.g. rectangular, elliptical and triangular) have received some research attention recently because they provide enhanced mixing between the jet flow and its surroundings compared with circular jets. A summary of studies on non-circular nozzles is also available in literature (Gutmark & Grinstein, 1999).

Although, there has been substantial research on jet flows, majority of these studies have not examined the effect of nozzle geometry for a wide range of shapes at identical initial conditions; apart from the study of Mi & Nathan, (2010), who used hot-wire anemometry (HWA) to study the effect of nozzle geometry on turbulent statistical properties on nine different orifice plate nozzles. This makes a direct comparison of nozzle geometry effects problematic, due to the influence of initial conditions on the development of the jet flow.

The objective of the study is to investigate the effect of nozzle geometry on the mixing characteristics of different orifice plate nozzles using particle image velocimetry (PIV). The exit geometries to be studied include a round (RD), square (SQ), rectangular (Rect) and elliptical (Elp) nozzles.

EXPERIMENTAL SETUP

The experiments were conducted in an open recirculating water channel with dimensions 2500 (length) × 200 (width) × 200 (height) mm. The side walls of the channel were made from smooth acrylic to facilitate optical access. A series of steel plates and honey combs are fitted within the settling chamber to condition the flow before entering the contraction section. The presence of the contraction causes the flow to accelerate thereby reducing the turbulence levels of the flow.

The nozzle assembly consists of a dam assembly and nozzle plates. The nozzle plates were made of stainless steel of thickness 1 mm. The nozzle assembly was mounted 150 mm from the inlet of the water channel. The dam assembly was tight-fitted onto the sides of the open channel. The circular nozzle is of diameter 9 mm and all other nozzle shapes had the same opening area as the circular nozzle. In the case of the rectangular and elliptical nozzles, measurements were made in both the major and minor axis planes (denoted as maj and min respectively). The Reynolds number based on equivalent diameter (d), which is defined as the diameter of a circle with the same opening area as the non-circular nozzle, and jet exit velocity (U_j) was kept constant at $Re_d = 5400$.

PIV SYSTEM AND MEASUREMENT PROCEDURE

A high resolution PIV system was used to obtain velocity measurements in the streamwise-wall-normal plane (x-y). The flow was seeded using 10 μm silver coated hollow glass spheres having a specific gravity of 1.4. The flow field was illuminated with a New Wave Solo Nd:YAG double-pulsed laser that emits green light up to a maximum of 120 mJ/pulse. The laser sheet was aligned with the mid-span of the channel, and a 12-bit CCD camera with 2048 pixel × 2048 pixel array and 7.4 μm pixel pitch was used to image the flow field. The field of view was set to 65 mm × 65 mm. Detailed velocity measurements were made in 3 planes, extending from the nozzle exit to x/d = 21. Based on a convergence test, a sample size of 6000 instantaneous image pairs was acquired in each plane to calculate the ensemble statistics reported subsequently. The interrogation window sizes were set to 32 pixels × 32 pixels with 50 % overlap. The spacing between adjacent vectors was 0.057d.

Following the methodology of Coleman & Steele, 1995 and Forliti et. al., 2000, the measurement uncertainties in the mean velocity, turbulence intensities and Reynolds shear stress at 95 % confidence level was estimated to be ±2 %, ±7 % and ± 10 % respectively.

RESULTS AND DISCUSSION

MEAN VELOCITY AND TURBULENT INTENSITY

Figure 1 shows the mean velocity contours normalized by the jet exit velocity and the equivalent diameter in the near field (i.e. $U^* = U/U_j$, $x^* = x/d$ and $y^* = y/d$). For all the nozzles, the jet velocity increases from the exit till it reaches its maximum some distance from the exit due to the vena contracta effect, before decaying with streamwise distance. For the elliptical and rectangular nozzle (not shown), as the jet evolves, there is initially more spreading in the major axis compared with the minor one and this is seen by the thicker initial jet thickness; but at some point, the axes switch over, and there is more spreading in the minor axis plane. This phenomenon is known as axis switching, and the switch over location is estimated as $x^* = 1.25$ and $x^* = 2.3$ for the elliptical and rectangular nozzles respectively. This phenomenon is the reason why the rectangular and elliptical jets exhibit better mixing qualities as compared to their square and round counterparts respectively.

The normalized streamwise turbulent intensity (u_{rms}^*), also normalized by U_j and d is shown in Figure 2. The local maxima values are observed to occur on either side of the jet centerline. Since intense mixing between the core jet and the ambient fluid takes place in the initial shear layer, it is expected that the peak values will occur in that region. The turbulent intensities then decay with streamwise distance. The maximum value for the elliptical nozzle occurs in the major plane, and is 50 % larger than that for the round nozzle. This implies greater mixing in the elliptical nozzle compared with the round nozzle. The peak u_{rms}^* value for the rectangle nozzle (not shown) also occur in the major plane, and is 34 % larger than that for the square nozzle. The maximum values in the minor axis planes of both the elliptical and rectangular nozzles (not shown) are comparable with the reference square and round nozzles respectively.

The contours of the Reynolds shear stress (not shown) show a positive value above and negative one below the jet centerline immediately after the nozzle exit for the elliptical and rectangular nozzles. There is then a change in sign, i.e. further downstream, the positive values are found below and the negative ones found above the jet centerline. There is no such change in sign observed for the round and square nozzles. The change in sign observed occurs after ($x^* = 2$) in both the elliptical and rectangular nozzles. The highest value of the Reynolds shear stress is found in the rectangular case and is almost twice as large as the peak in the case of the square jet.

MEAN VELOCITY DECAY AND SPREAD RATE

Figure 3 shows plots for both the decay and spread of the jets considered in the study. The elliptical and rectangular nozzles show the most rapid decay of the centerline velocity with streamwise distance. The jet spread is usually quantified using the half-velocity width. This is defined as the distance from the jet centerline to the point where the local streamwise mean velocity has decayed to 50 % of the corresponding mean velocity. In this study, other velocity widths were examined in addition to the half-velocity width. These jet velocity widths are defined as the distance from the jet centerline to the point where the local streamwise velocity has decayed to 90 %, 75 %, 25 % and 10 %. These jet velocity widths for each nozzle are plotted against streamwise distance and fitted to the linear equation:

$$\frac{y_b}{d} = K_s\left(\frac{x}{d} + C_s\right) \qquad (1)$$

where y_b refers to the various velocity widths (i.e. $y_{0.9}$, $y_{0.75}$, $y_{0.5}$, $y_{0.25}$ and $y_{0.1}$), K_s is the spread rate and C_s is the geometric virtual origin (as shown in Figure 3b). The spread rate for each jet width is then plotted against its corresponding jet width as shown in Figure 3c. In the case of the rectangular and elliptical nozzles, the geometric mean (avg) was used to facilitate comparison with the other nozzles. The plot shows that for each jet width considered, the spread rate for the rectangular nozzle is the highest. Also, the least spread occurs in the round and square nozzles. Finally the difference in spread rates becomes more significant as one moves further away from the jet centerline. For example, the spread rate at jet width $y_{0.1}$ for the rectangular nozzle is 63 % greater than that of the square nozzle, while the spread rate of the elliptical nozzle at the same jet width location is 38 % greater than the round nozzle.

CONCLUSIONS

PIV technique was used to investigate the effect of nozzle geometry on different orifice plate nozzles at identical initial conditions. The decay, spread rates, and shear layer values of the streamwise turbulent intensity and Reynolds shear stress suggest that free jets issuing from rectangular and elliptical nozzles show superior mixing capacity compared with jets from the other nozzles studied.

ACKNOWLEDGMENTS

The financial support of Price Industries Limited and the Natural Sciences and Engineering Research Council of Canada (NSERC) is appreciated.

REFERENCES

1. Abdel-Rahman, A. (2010). A review of effects of initial and boundary conditions on turbulent jets. *WSEAS Transactions on Fluid Mechanics*, *5*(4), 257–275. Retrieved from http://www.wseas.us/e-library/transactions/fluid/2010/52-334.pdf

2. Coleman, H., & Steele, W. (1995). Engineering application of experimental uncertainty analysis. *AIAA Journal*, *33*(10), 1888 – 1896. Retrieved from http://arc.aiaa.org/doi/abs/10.2514/3.12742

3. Forliti, D. J., Strykowski, P. J., & Debatin, K. (2000). Bias and precision errors of digital particle image velocimetry. *Experiments in Fluids*, *28*(5), 436–447. doi:10.1007/s003480050403

4. Gutmark, E., & Grinstein, F. (1999). Flow control with noncircular jets 1. *Annual Review of Fluid Mechanics*, *31*, 239–272. Retrieved from http://www.annualreviews.org/doi/abs/10.1146/annurev.fluid.31.1.239

5. Mi, J., & Nathan, G. J. (2010). Statistical Properties of Turbulent Free Jets Issuing from Nine Differently-Shaped Nozzles. *Flow, Turbulence and Combustion*, *84*, 583–606. doi:10.1007/s10494-009-9240-0

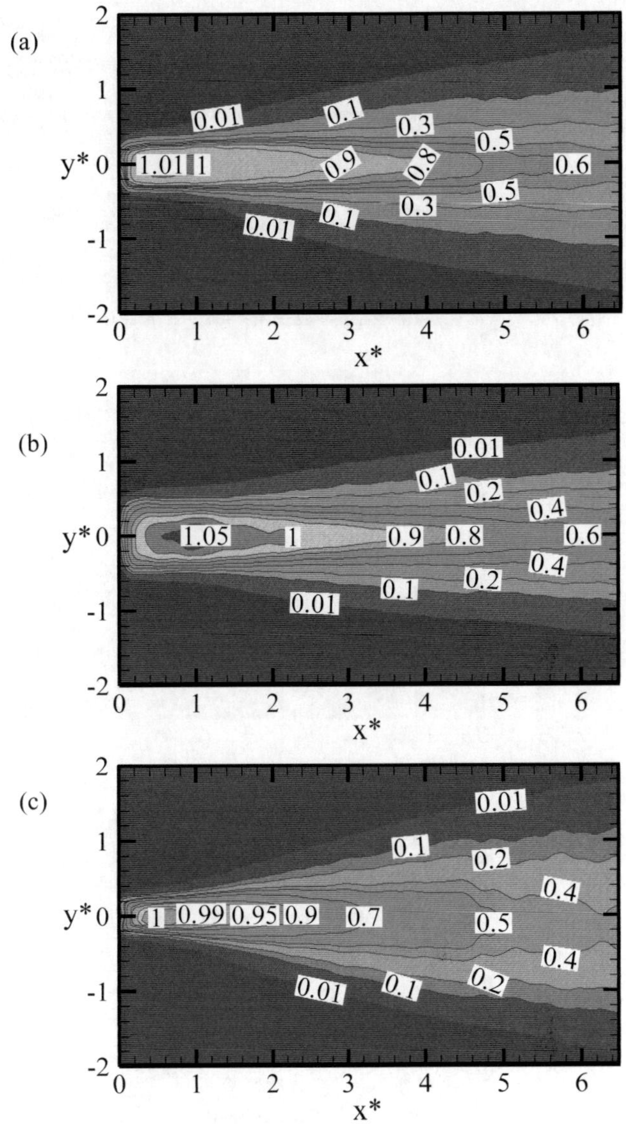

Figure 1: Normalized streamwise velocity (U^*) contour plots of (a) round nozzle, and the ellipse in the (b) major axis plane and (c) minor axis plane.

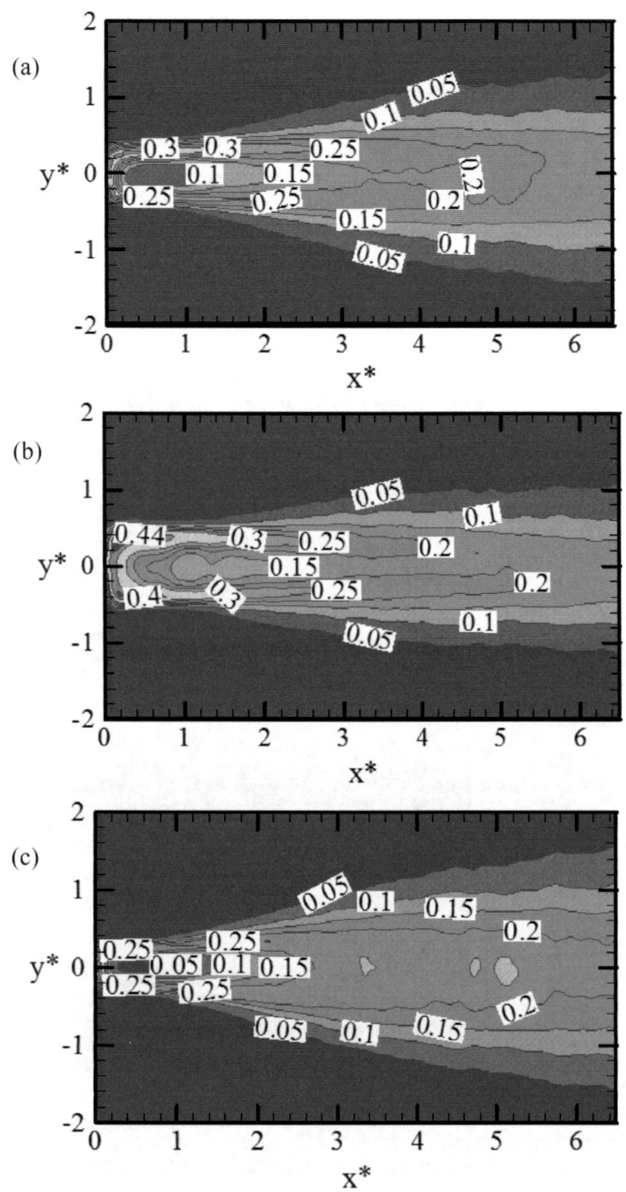

Figure 2: Normalized streamwise turbulent intensity (u_{rms}^*) contour plots of (a) round nozzle, and the ellipse in the (b) major axis plane and (c) minor axis plane.

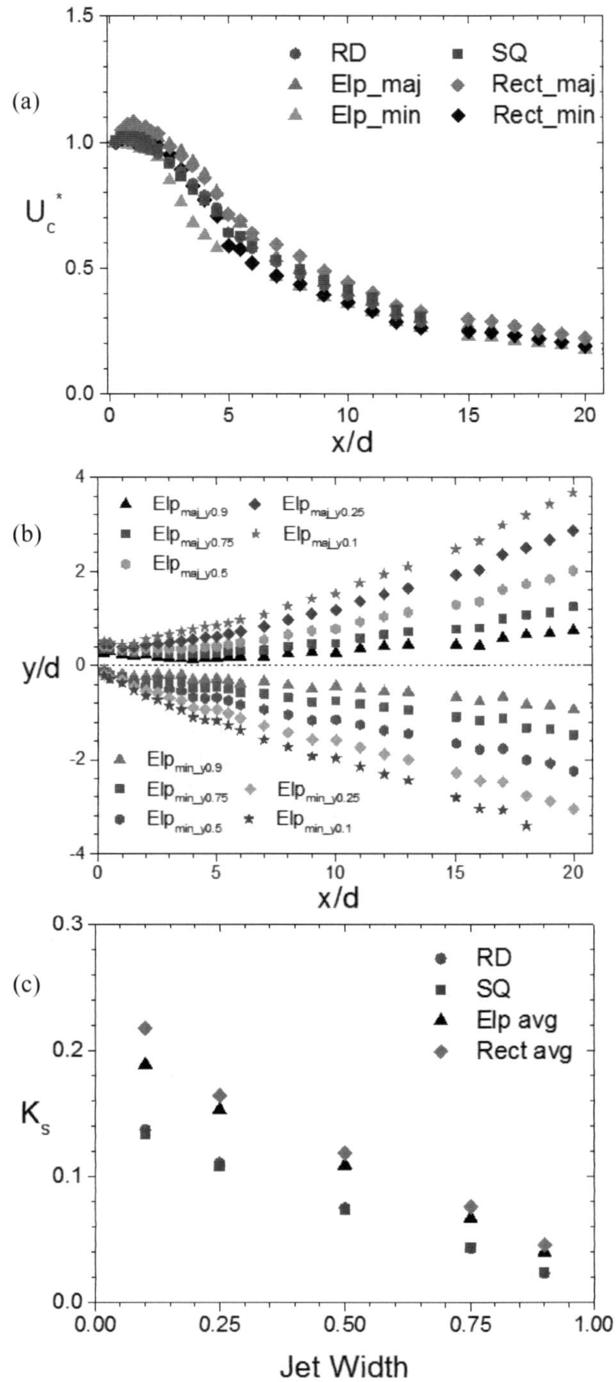

Figure 3: Plot of (a) mean velocity decay with streamwise distance, (b) spread of jet with streamwise distance and (c) variation of the spread rate (K_s) with jet velocity width.

EXPERIMENTAL AND NUMERICAL STUDY OF FLOW CHARACTERISTICS THROUGH A GRADUAL CONTRACTION

Sarwesh Mali, Ali Nematollahi, Ebenezer E. Essel and Mark F. Tachie
Department of Mechanical Engineering
University of Manitoba
Winnipeg, MB, R3T 5V6, Canada
Mark.Tachie@umanitoba.ca

ABSTRACT

Turbulent flow in a hopper with a gradual contraction is studied experimentally and numerically. The velocity measurements were performed using a high resolution particle image velocimetry (PIV) system. The measurements were performed for two Reynolds numbers, Re_D = 49,000 and 23,000. The steady and isothermal Reynolds-Averaged Navier-Stokes (RANS) equations and two turbulence models, standard $k-$epsilon and $k-$omega based Shear Stress Transport (SST) models, were used to predict the flow characteristics. The mean velocities predicted by the two turbulence models are in satisfactory agreement with those measured by PIV, however only $k-$omega based shear stress transport model is able to reasonably predict mean velocities near the wall.

KEYWORDS: PIV, RANS, CFD simulation, turbulence models, contraction, hopper

INTRODUCTION

The flow field within gradual and sudden contractions is of great importance in engineering and industrial applications. Detailed investigations of mean velocity profiles and pressure losses through a sudden contraction were performed by Sanchez et al. [1] using experimental and numerical methods. Also, detailed investigations of Reynolds number effects on flow through gradual contraction were performed by Prinos and Goulas [2]. Flows through hoppers are another configuration of flow through a gradual contraction which is of industrial significance. A hopper is a mechanical device with a converging section attached to a pipe. They are used to collect and facilitate the flow of substances entering through their larger opening and exiting through the attached pipe with a smaller opening. Depending on the specific applications, hoppers may have round, pyramidal, conical or square taper surface connecting the fluid entry and exit sections. In spite of their diverse applications, the flow characteristics through hoppers have not been studied in detail. Towards this end, the University of Manitoba and Rimer Alco North America (RANA), a leading manufacturer of mine refuge stations in North America, initiated a collaborative research program. The objective of this research is to conduct velocity measurements to improve our understanding of flow characteristics through a conical hopper. It is acknowledged that a better understanding of the flow characteristics through these kind of geometries is critically important to the development of accurate numerical predictive models for both velocity and pressure fields. In this paper, experimental results obtained in turbulent flow through a hopper using a particle image velocimetry (PIV) system is used to validate the numerical results obtained using commercial CFD software ANSYS® CFX 15.0.

EXPERIMENTAL PROCEDURE

The experimental test facility is shown in Figure 1. A part of the hopper (hereafter referred to as the test section) is made up of acrylic material to facilitate optical access as shown in Figure 2(a). The facility consists of a flow conditioning unit, a conical hopper supplied by RANA, circulating pump and piping, supporting framework, laser and CCD camera. The flow conditioning unit is 667 *mm* in length, 309 *mm* in width and 300 *mm* in height. The unit contains a series of sieves to reduce turbulence and straighten the flow prior to entering the test section. The test section has dimensions of 495 *mm* in length (L) and 195 *mm* in width (W) at the inlet section, and 215 *mm* in height (H). The inlet section of the test section transitions into a circular cross-section at the exit plane with a diameter, D = 82 *mm*. The resulting contraction ratio is 18. However, during the fabrication of test section the polyurethane adhesive used to seal the exit connecting to a pipe prevented optical access at the exit. The Cartesian coordinate system adopted has the origin of the streamwise distance ($x = 0$) at the inlet of the test section and the origin of the wall normal distance ($y = 0$) at the mid-section of the test section. The experiments were conducted at two flow rates 0.0032 m^3/s and 0.0015 m^3/s, which results in Reynolds numbers based on the maximum exit velocities (U_{max} = 0.59 *m/s* and 0.28 *m/s*) and exit diameter of the hopper (D = 0.082 m), Re_D = 49,000 and Re_D =23,000, respectively. The two flow rates were chosen to investigate possible effects of Reynolds number on the flow characteristics. For brevity, the experiments conducted at Re_D = 49,000 and 23,000 are hereafter referred to as Q1 and Q2, respectively.

A PIV system was used to conduct velocity measurements in the $x-y$ plane located at the mid-span of the test section. PIV is a non-intrusive velocity measurement technique that provides whole-field velocity information. In the present study, the working fluid (water) was seeded with 10 μm silver coated hollow glass spheres having specific gravity of approximately 1.1 g/cm^3. The size of the particles was chosen so as to ensure that they follow the flow faithfully. An Nd-YAG double pulsed laser (120 *mJ/pulse*) was used to illuminate the flow field and the scattered light from the tracer particles was captured using a high resolution 2048 pixels × 2048 pixels, 12 bit CCD camera. The images captured were stored on a hard disk. The field of view of the camera was set to 108 *mm* × 108 *mm* in the *x-y* plane. Four different overlapping measurement planes (P1 - P4) were used to map out the transparent quarter of the flow domain as shown in Figure 2(b). Based on a convergence test, a sample size of 6000 instantaneous image pairs was acquired in each plane and was post processed using the adaptive correlation option of commercial software Dynamic Studio v3.41 developed by Dantec Dynamics. The interrogation area size for the adaptive correlation was set to 64 pixels × 64 pixels with 50% overlap in both the *x* and *y* directions. A comprehensive in-house MATLAB - C++ code was developed and used to post-process the velocity data to obtain mean velocities.

Figure 1: Experimental setup showing the test facility and the components of the PIV setup

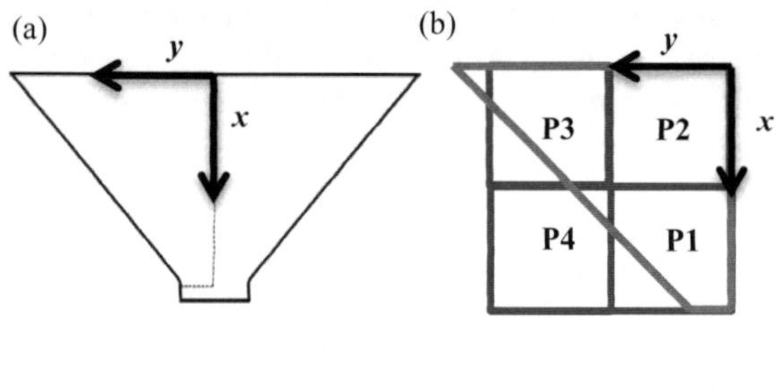

Figure 2: Schematic drawings of (a) hopper test section (b) measurement planes (P1 - P4)

NUMERICAL PROCEDURE

In this section, the governing equations for the three-dimensional, steady, incompressible, turbulent flow are presented with the standard $k - \varepsilon$ [3] and $k - \omega$ based SST [4] models. The grid generation is described briefly together with appropriate boundary conditions for both turbulence models. The Reynolds-Averaged Navier-Strokes (RANS) equations for continuity and momentum conservation are given by equations (1) and (2), respectively. The two equation models assumed an eddy viscosity relationship for the Reynolds stresses as in equation (3) below

$$\frac{\partial U_j}{\partial x_j} = 0 \tag{1}$$

$$\frac{\partial}{\partial x_j}(\rho U_j U_i) = -\frac{\partial P}{\partial x_i} + \frac{\partial}{\partial x_j}\left(\mu \frac{\partial U_i}{\partial x_j} - \overline{\rho u_i u_j}\right) \tag{2}$$

$$-\overline{\rho u_i u_j} = \mu_t \left(\frac{\partial U_i}{\partial x_j} + \frac{\partial U_j}{\partial x_i}\right) - \frac{2}{3}\rho \delta_{ij} k \tag{3}$$

where μ_t is the eddy viscosity and δ_{ij} is the Kronecker delta. In the solution of the model equations, the pressure is the thermodynamic pressure plus $\frac{2}{3}\rho\delta_{ij}k$.

The $k - \varepsilon$ model solved the following two equations for turbulent kinetic energy, k and its dissipation rate ε:

$$\frac{\partial}{\partial x_j}(\rho U_j k) = \frac{\partial}{\partial x_j}\left(\left(\mu + \frac{\mu_t}{\sigma_k}\right)\frac{\partial k}{\partial x_j}\right) + P_k - \rho\varepsilon \tag{4}$$

$$\frac{\partial}{\partial x_j}(\rho U_j \varepsilon) = \frac{\partial}{\partial x_j}\left(\left(\mu + \frac{\mu_t}{\sigma_\varepsilon}\right)\frac{\partial \varepsilon}{\partial x_j}\right) + \frac{\varepsilon}{k}(C_{\varepsilon 1} - \rho C_{\varepsilon 2}\varepsilon) \tag{5}$$

where the production term P_k is:

$$P_k = \mu_t \left(\frac{\partial U_i}{\partial x_j} + \frac{\partial U_j}{\partial x_i}\right)\frac{\partial U_i}{\partial x_j} \tag{6}$$

and the eddy viscosity is computed using $\mu_t = \rho C_\mu k^2/\varepsilon$. The $k - \varepsilon$ model constants are $C_{\varepsilon 1} = 1.44, C_{\varepsilon 2} = 1.92, \sigma_k = 1.0, \sigma_\epsilon = 1.3$ and $C_\mu = 0.09$.

The equations solved for the $k - \omega$ based SST model are:

$$\frac{\partial}{\partial x_j}(\rho U_j k) = \frac{\partial}{\partial x_j}\left(\left(\mu + \frac{\mu_t}{\sigma_{k2}}\right)\frac{\partial k}{\partial x_j}\right) + P_k - \rho\beta^* k\omega \tag{7}$$

$$\frac{\partial}{\partial x_j}(\rho U_j \omega) = \frac{\partial}{\partial x_j}\left(\left(\mu + \frac{\mu_t}{\sigma_{\omega 2}}\right)\frac{\partial \omega}{\partial x_j}\right) + (1 - F_1)2\rho\frac{1}{\sigma_{\omega 2}\omega}\frac{\partial k}{\partial x_j}\frac{\partial \omega}{\partial x_j} + \alpha_2\frac{\omega}{k}P_k - \rho\beta_2\omega^2 \tag{8}$$

where $\alpha_2, \sigma_{\omega 2}, \sigma_{k2}, \beta_2, \beta^* and\ \alpha_1$ are constants with values 0.44, 10.856, 1.0, 0.0828, 0.09 and 0.31, respectively. All other symbols have their usual meaning and blending functions, F_1:

$$F_1 = \tanh(arg_1^4) \quad (9)$$

with,

$$arg_1 = \min\left(\max\left(\frac{\sqrt{k}}{\beta^*\omega y}, \frac{500v}{y^2\omega}\right), \frac{4\rho k}{CD_{k\omega}\sigma_{\omega 2}y^2}\right), \text{ and}$$

$$CD_{k\omega} = \max\left(2\rho\frac{1}{\sigma_{\omega 2}\omega}\frac{\partial k}{\partial x_j}\frac{\partial \omega}{\partial x_j}, 1.0 \times 10^{-10}\right) \quad (10)$$

The distance of the first nodal point away from the wall is y and v is the kinematic viscosity. Also, the formulation of the eddy viscosity is given by

$$\mu_t = \frac{\rho\alpha_1 k}{\max(\alpha_1\omega, SF_2)} \quad (11)$$

with $F_2 = \tanh(arg_2^2)$ and $arg_2 = \max\left(\frac{2\sqrt{k}}{\beta^*\omega y}, \frac{500v}{y^2\omega}\right)$, where F_2 is a blending function similar to F_1 and S is an invariant measure of the strain rate, given by

$$S = \sqrt{\left(\frac{\partial U_j}{\partial x_i} + \frac{\partial U_i}{\partial x_j}\right)\frac{\partial U_i}{\partial x_j}} \quad (12)$$

The boundary conditions for the solution domain are as follows: at the inlet a constant mass flow rate, $\dot{m} = 3.15\ kg/s$ is specified corresponding to Q1. In addition, at the inlet the relative turbulence intensity is set to 5% based on the experimental results. At the outlet, the average static pressure is set to zero. Also, the outlet was specified at a distance of 4 m (48 D) downstream from the exit. This distance was chosen to ensure that the fully developed turbulent flow condition is achieved at the outlet. A no-slip boundary condition was prescribed at the test section boundaries. In the $k - \varepsilon$ model a scalable wall-function was used. A near wall treatment with automatic switch from wall-function to a low Reynolds number formulation based on grid refinement was employed for the SST model. For both turbulence models, $y^+ < 1$ was maintained. Double precision was used in the computations and the solution was considered converged when the normalized RMS residual of each of the discretized equations was less than 1.0×10^{-6}. The numerical results were considered to be grid independent by using 4 computational grids (1,866,000 elements, 3,837,910 elements, 5,299,553 elements and 6,011,243 elements). No significant difference was observed between the finest two grids and hence the second last grid (5,299,553 elements) was used for all the subsequent calculations. For the sake of brevity no grid independent results are presented here.

EXPERIMENTAL AND NUMERICAL RESULTS

Figure 3 shows the contour plots of mean velocities obtained from the PIV measurements. The contour plots obtained in the four planes of measurement were combined to obtain a whole field visualization of the flow field in the quarter part of the test section. The plots of streamwise mean velocity for the two flow rates are presented in Figure 3. In this and subsequent plots, the velocity data are normalized by the corresponding maximum streamwise velocity, U_{max} and the streamwise and wall-normal distances are normalized by the hopper exit diameter, D. For each test case, the streamwise mean velocity accelerates with increasing streamwise distance until the maximum velocity occurs near the exit of the test section. The acceleration of the flow with increasing streamwise distance is caused by the gradual contraction of the test section. The flow pattern near the exit of the test section is qualitatively similar for the two flow conditions investigated. The contour plots of the wall-normal mean velocity for Q1 and Q2 are shown in Figure 4. As can be seen, the maximum wall-normal velocity occurs close the wall and near the exit of the test section. At the centerline of the test section, the wall-normal velocity is identically zero. The maximum wall-normal velocity for Q1 is 10% higher than that of Q2. The gradual transition from a wider cross-sectional area at the inlet of the hopper to a smaller circular cross-section at the exit plane caused a gradual increase in the mean velocities with increasing streamwise distance.

Figure 5 shows the contour plots of the mean velocities for Q1 obtained using $k - \omega$ based SST model. The contour plots obtained using $k - \varepsilon$ model are not shown because they were qualitatively similar. The contours of mean velocities are comparable with the experimental data. Experimental and numerical results are compared at $x/D = 2.1$ location near the contraction exit as shown in Figure 6. In particular, the streamwise and wall-normal mean velocities measured at this location are compared with those predicted by the turbulence models. This location was chosen instead of the contraction exit because the curvature of the contraction prevented PIV from obtaining accurate velocity measurements near the wall. The streamwise and wall normal mean velocity profiles at this x/D location for experimental and numerical models closely follow each other. Both the turbulence models are able to predict the shape of the velocity profiles along the center of contraction, however close to the wall only $k - \omega$ based SST model is able to capture the mean velocities going to zero. This is because $k - \omega$ based SST model uses enhanced wall function and therefore it is able to capture the mean velocities going to zero near the wall. The prediction of the correct values of pressure with a turbulence model near the wall is based on accurate calculation of the mean velocities near the wall. Hence, the modelling of the near wall velocities is crucial for the calculation of pressure.

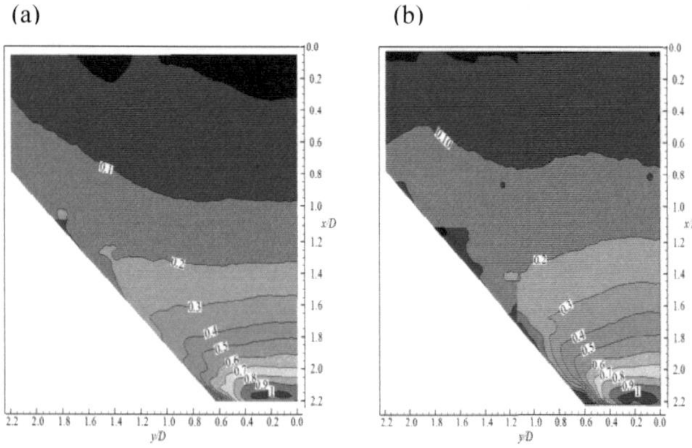

Figure 3: (a) Q1 (b) Q2 Streamwise mean velocity

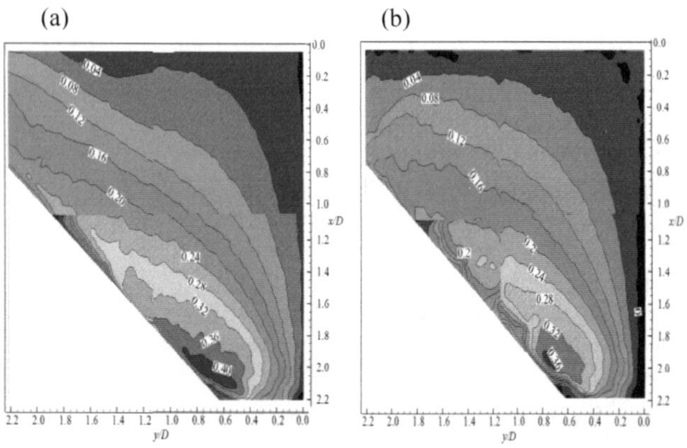

Figure 4: (a) Q1 (b) Q2 Wall normal mean velocity along y

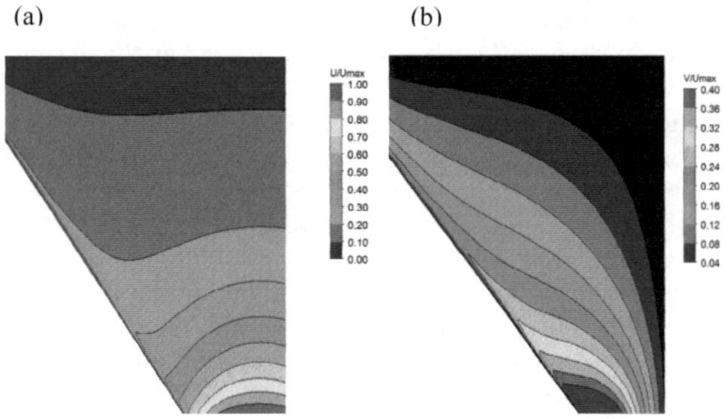

Figure 5: (a) Streamwise (b) Wall normal mean velocity for Q1 obtained using $k-\omega$ based SST model

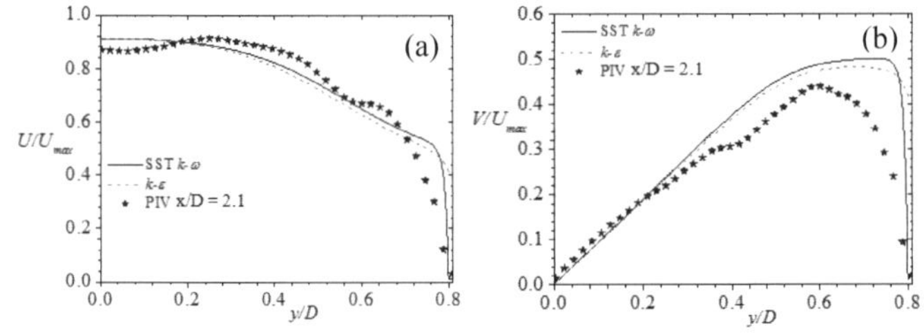

Figure 6: (a) Streamwise mean velocity (b) Wall normal mean velocity along y at $x/D = 2.1$

CONCLUSIONS

A high resolution PIV system was used to perform detailed velocity measurements within a conical hopper to better understand the flow characteristics. The experiments were conducted at Reynolds number based on the maximum exit velocity and exit diameter of the hopper, $Re_D =$ 49,000 and 23,000 to investigate the effects of Reynolds number on the flow characteristics. The results showed a weak dependence of mean flow characteristics for the two Reynolds numbers. Also, numerical computations were performed using $k-\varepsilon$ and $k-\omega$ based SST models in ANSYS® CFX 15.0. The results showed that both models were able to reasonably predict the mean velocities, however only $k-\omega$ based SST model captured velocities near the wall.

ACKNOWLEDGMENTS

The financial support provided by RANA and NSERC Engage grant is gratefully acknowledged.

REFERENCES

1. Sanchez, F.P., et.al., 2010, "Experimental and numerical study of turbulent Newtonian flow through an axisymmetric sudden contraction", *13th Brazilian congress of thermal Sciences and Engg.*, Dec 5-10, Uberlandia, Brazil
2. Prinos, P., Goulas, A., 1993, "Flow characteristics in a pipe with a gradual contraction", *J. of hydraulic research.* Vol(31), pp. 587-600.
3. Launder, B.E., Spalding, D.B., 1994, "The numerical computation of turbulent flows", *Comput. Methods Appl. Mech. Eng.*, Vol(3), pp. 269-289.
4. Menter, F.R., 1994, "Two-equation eddy-viscosity turbulence models for engineering applications", *AIAA J.*, Vol(32), pp. 37-40.

Proceedings of the 25th CANCAM
London, Ontario, Canada, May 31 – June 4, 2015

APPLICATIONS OF FIBER BRAGG GRATING SENSORS TO STRAIN MEASUREMENTS OF AZ31B EXTRUSION IN ROTATING BENDING CYCLIC TESTS

Bahareh Marzbanrad
Mechanical and Mechatronic Engineering
University of Waterloo
Waterloo, ON, Canada
bmarzban@uwaterloo.ca

Ehsan Toyserkani
Mechanical and Mechatronic Engineering
University of Waterloo
Waterloo, ON, Canada
ehsan.toyserkani@uwaterloo.ca

Hamid Jahed
Mechanical and Mechatronic Engineering
University of Waterloo
Waterloo, ON, Canada
hjahed@uwaterloo.ca

ABSTRACT

Fiber Bragg Grating (FBG) sensor was employed for measuring the strain induced during load-control rotating bending cyclic tests on AZ31B extrusion magnesium alloy specimens. The mid portion of a FBG sensor with a 5 mm grating was adhered to the AZ31B testing samples and cured under UV light. The instrumented AZ31B samples were then used in rotating bending tests at different load amplitudes from 2.5 to 6.5 Nm. Any changes in the grating periods due to the fiber deformation shift the wavelength of the reflected light passing through the fiber. The shift in wavelength was used to calculate the strain. The strain measurement of the FBG was verified by comparing its results with strain measurement using an extensometer in a uniaxial push-pull test. This technique enabled the measurement of strain amplitude evolution with respect to the number of cycles in the rotating bending cyclic test.

KEYWORDS: AZ31B, rotating bending test, strain measurement, Fiber Bragg Grating sensor.

INTRODUCTION

Fiber optic sensors were designed and fabricated in the 1960's. These sensors have been utilized in a wide range of applications such as aerospace, marine, electric power generation, oil and gas industry, nuclear industry, and biomedical apparatus [1]. The fiber optic is a hair-thin filament of fused silica consists of two parts, the core and cladding. Usually light travels inside the core, because the refractive index of the core is higher than the cladding. This difference generates due to doping the core by germanium. One type of fiber optic sensor is Fiber Bragg Grating, which is employed to measure strain and temperature. The Bragg grating are written into the germanium doped core of 4 to 9 μm resulting in the high intensity and narrow wavelength-band peaks. Any stretching of the gratings leads to a shift in the wavelength of these peaks. The deformation of the fiber can happen with mechanical strain or temperature changes [2]. Therefore, the wavelength shift represents strain and/or temperature change. This sensor has been employed for strain/deformation measurement in different applications. For example, the FBG sensors are resistant to corrosion and harsh environment and hence, they are utilized to measure the deformation of materials in these media [3]. Fatigue crack growth rate was monitored successfully by mounting a FBG sensor in front of the crack tip [4]. In other research, Flat-clad FBG sensor has been used to characterize low cycle fatigue deformation behavior of extruded AZ31B magnesium alloy [5].
Optimum engineering design to achieve the highest possible efficiency has encouraged the automotive and aerospace manufacturers to adopt light materials instead of conventional structural materials, such as steel and aluminum [6]. Magnesium is the lightest available commercial metal with HCP crystal structure that has high strength to weight ratio, and density of 1.738 gr/cm^3. Its density is 75% less than steel and about 35% less than aluminum. Magnesium and its alloys also have good castability, weldabilty, damping capacity, electrical and thermal conductivity [7]. Wrought magnesium alloys, such as extrusion and rolled AZ31B, have superior mechanical properties than cast magnesium alloys. As a result, these alloys are preferred for applications in load bearing components, especially in automotive and aerospace industries. Therefore, research on the mechanical behaviour of these materials is essential, especially because the behaviour of wrought magnesium alloys is not symmetric under tension and comparison [8]. The asymmetric behavior exists because of different deformation mechanism of magnesium under tensile and compressive load. It is reported that when load applying in extrusion/rolling direction, the dominant deformation mechanism of the AZ31B magnesium alloy is twinning under compression, while detwinning and slip are dominant deformation mechanism under tension [9]. Hence, characterization and modeling of the fatigue behavior are crucial for design and development of the mechanical components made of magnesium alloys. For this purpose, stress-control or strain-control cyclic tests are usually performed to evaluate the fatigue behaviour of materials [8-

12]. However, many engineering components are subjected to cyclic bending moment or load-control cyclic condition. The Rotating Bending Machine (RBM) is a simple machine that is usually employed to extract stress versus number of cycles until failure of the sample (S-N curve). In these experiments, applied bending moment in the elastic region is translated to normal stress through elastic flexural formula. Recently, the elastic-plastic behaviour of the AZ31B extrusion was modeled to investigate the cyclic stress-strain hysteresis in rotating bending test [13]. Obtaining the cyclic deformation hysteresis is required to estimate the fatigue behaviour of materials. However, the evolution of strain of a magnesium sample with respect to the number of cycles during the rotating bending experiment has not been reported yet.

In the present study, first, the accuracy of the strain measurement by the FBG sensor in elastic and plastic regimes was investigated and the results compared with a standard extensometer during uniaxial stress-control test. Second, the evolution of strain of AZ31B samples during rotating bending tests was measured using the FBG sensors.

MATERIALS AND EXPERIMENTS

The AZ31B magnesium alloy samples were cut from an air-quenched extrusion section, which was extruded with an exit speed of 50.8 mm/s and temperature between 360 and 382°C [8]. Shape and dimensions of the samples are shown in Figure 1.

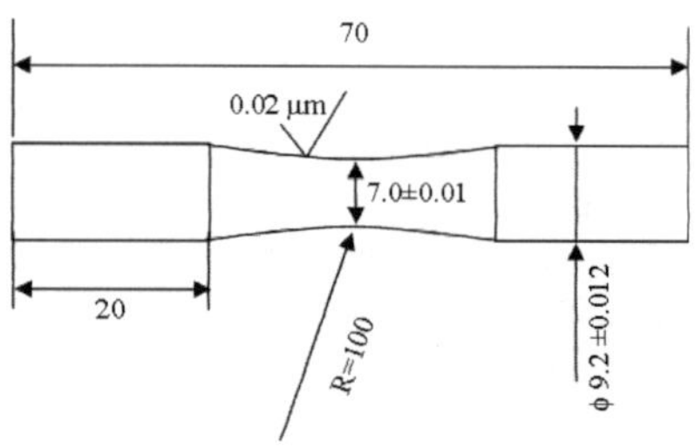

Figure 1: Rotating bending sample geometry, (dimensions: mm)

The FBG sensor was adhered to the sample using a commercial adhesive, Norland Optical Adhesive 61. For this purpose, surface of the sample was gently cleaned with isopropanol alcohol. The polyamide coat of the FBG sensor was removed by furniture stripper. The fiber was laid on the sample along the longitudinal direction of the sample such that the middle of the 5 mm long FBG sensor was fixed on the middle of the sample, where it had the minimum cross section. The glue was applied on the grated portion of the fiber using a needle and cured by the UV light irradiation for 20 minutes.

The data interrogator system of the FBG sensor used in this study had a sampling rate of 2 Hz, while the rotation speed of the RBM was in the range of 60 to 100 Hz. Therefore, the RBM machine was modified by retrofitting a stepper motor in parallel to the main drive of the RBM. This motor was utilized to rotate the sample with a speed as low as 0.001 Hz to enable the FBG data recording matched with the sampling rate of the FBG interrogator. The strain was measured as a function of cycles during rotating bending experiments with the load of 2.5, 3.5, 4.5, 5.5 and 6.5 Nm. The tests were carried out at a frequency of 20 Hz, except for the cycles in which strain was measured.

The mechanical strain can be calculated by the following equation:

$$\varepsilon = \frac{\Delta\lambda}{k \times \lambda_0} \qquad (1)$$

where $\Delta\lambda$ is the wavelength shift, λ_0 is the initial wavelength and k is a gage factor, which is equal to 0.78 for the fiber used [14].

RESULTS AND DISCUSSION

The effect of curing process on the sensor and the substrate was studied first. Heat Generated during the curing process of the adhesive can warm up the sample. Since the temperature may affect the microstructure of samples, temperature change of the sample during the curing process was investigated. For this purpose, a thermocouple was attached to the sample and the temperature change was monitored during the curing process. The thermal mismatch of magnesium and epoxy due to different coefficients of thermal expansion caused residual stress and consequently a mean strain in the vicinity of the attached sensor. This residual stress and mean strain must be considered for future strain measurements. To demonstrate these two phenomena, the temperature and wavelength change of the sample was recorded. Figure 2 shows the temperature history during the curing process, with the maximum temperature of 87°C (labeled by a). This figure also depicts the wavelength change of the FBG sensor during curing. The FBG wavelength after curing and cooling to the room temperature is less than the initial wavelength of the sensor (labeled by b). It means that a residual stress was produced in the sample after curing. In this case, the compressive strain of the sample was measured to be equal to 0.06 %, which translates into 25 MPa compressive residual stress in the sample.

The FBG strain measurements due to the application of cyclic load were first verified in a uniaxial push-pull test performed on an Instron servo-hydraulic testing machine. An FBG sensor was adhered to the sample and this sample was tested by the Instron machine while the strain was measured by the FBG sensor and an extensometer, simultaneously. Figure 3 depicts the results of this stress control cyclic experiment with stress amplitude of ±110 MPa. Based on this experiment, the FBG sensor and the extensometer follow the same curve. Therefore, FBG provided accurate measurements both in the linear elastic and nonlinear plastic regime up to a measured strain of 1.3%.

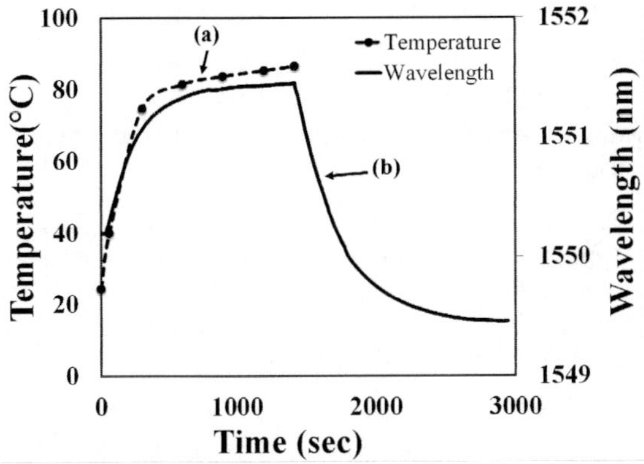

Figure 2: The temperature change during curing (a), and The wavelength changes within the curing process and cooling down to the room temperature (b).

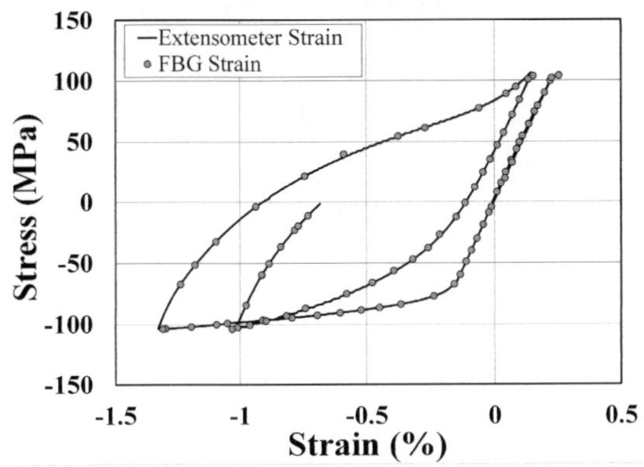

Figure 3: Stress-strain response for AZ31B extrusion; the strain was measured by FBG sensor and extensometer at the stress amplitude of ±110 MPa.

Figure 4 shows the recorded strain (obtained from an FBG sensor) versus the position of the sensor for four different bending moments: 2.5, 3.5, 4.5 and 5.5 Nm. In one full revolution of the sample, the FBG experiences maximum tension (when it's at the bottom) and maximum compression (when it's at the top). Depending on the initial location of FBG sensor before starting the test, the FBG may go through compression or tension first. In the case of 2.5 and 3.5 Nm tests, the FBG sensor was under the compression initially. The two other experiments started when the FBG sensor was at the bottom of the sample and experienced tensile stress at the beginning of the test.

The 2.5 and 3.5 Nm bending moments reveal the elastic deformation due to the symmetrical strain fluctuation. Higher bending moments caused plastic deformation that is evidenced by asymmetric strain curves. This behaviour is observed because the yield strength of AZ31B under compression is less than the one under tension.

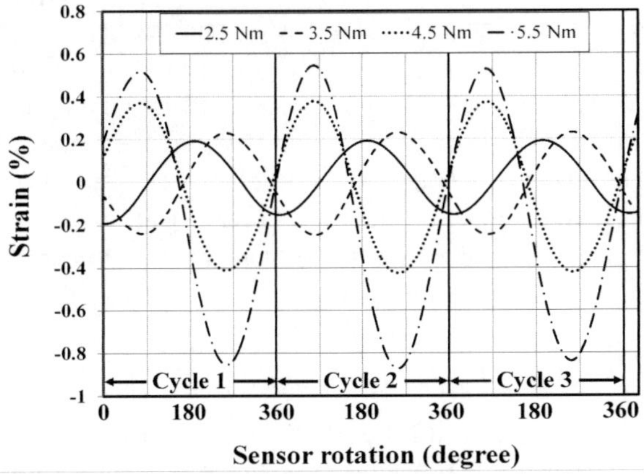

Figure 4: Strain history measured by FBG in the first few cycles under 2.5, 3.5, 4.5, and 5.5 Nm bending moments.

Figure 5 shows the peak and valley strain of AZ31B magnesium alloy sample which tested under 6.5 Nm bending moment up to 5000 cycles. The presented data demonstrate the change of strain amplitude of the sample under cyclic test. Stain amplitude of the sample in the second cycle is a little bit larger than the first cycle. From the cycle number of 3 to 2000, the strain amplitude as well as peak and the valley strain reduced. Albinmousa [8] reported that AZ31B extrusion has a considerable plastic strain recovery. Such behaviour is also observed in the present study as depicted in figure 5. From cycle number of 2000 to 5000, the strain amplitude is almost constant which can be concluded that the cyclic behaviour of the sample is stable. It was reported that during a stress control uniaxial cyclic test of AZ31B magnesium alloy with mean stress equal to zero, the peak and valley strain showed the same trend [10-12,15].

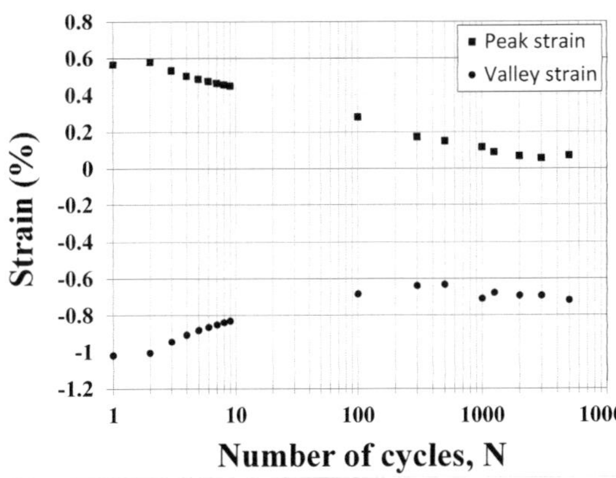

Figure 5: Curves of peak and valley Strains versus number of cycles, M=6.5 Nm

Based on that study, the sample became stable after a few cycles. However, in the load control rotating bending experiments, more cycle was required to reach the stable condition. Moreover, the comparison between the peak and the valley strains confirms the asymmetric behavior of the AZ31B under 6.5 Nm, which is compatible with the presented data in Figure 4. Also, there is a noticeable negative mean normal strain of -0.4% in the stabilized cycles. This is another indication of AZ31B asymmetric behaviour which is due to lower yield values in compression, and hence larger compressive strains, as compared to yield value in tension.

CONCLUSION

This research presents the strain evolution of the samples during the rotating bending experiment that is measured by the Fiber Bragg Grating sensor. The experiments were done on AZ31B extrusion magnesium alloy. The main outcomes of this research are summarized as follows:

1) An adhesion curing process was developed for successful attachment of FBG to Mg substrate.
2) Accuracy of FBG sensor strain measurements was verified through a push-pull test and the use of a calibrated axial extensometer.
3) The FBG sensor precisely measured strains up to 1.3%. The strain measurements were over elastic and plastic regimes of AZ31B. Strains larger than 1.3% were not examined in this study.
4) A residual stress was observed in the sample after curing of the epoxy. This was induced because of the difference between coefficient of thermal expansion of the sample and cured adhesive. This residual stress and its related strain were considered in the further strain measurements.
5) Change of the peak and valley strain with respect to cycles of the sample during rotating bending was obtained. This experiment revealed that under the load control rotating bending, the strain of the sample becomes stable after 2000 cycles.
6) A large negative mean strain is developed in the load control cyclic test of magnesium AZ31B. This can be attributed to the asymmetric behaviour of wrought magnesium alloys.

REFERENCES

1. Udd, E. and Spillman, W.B., 2011," Fiber Optic Sensors: An Introduction to Engineer and Scientists", 2nd edition, John Wiley & Sons.
2. Dockney, J. M., and Tatam, R., 1996, "Simultaneous independent temperature and strain measurement using in-fiber Bragg grating sensors", Electron. Lett., 32 (12), pp. 1133–1134.
3. Wnuk, V. P, Mendez, Ferguson, S. and Graver, T., 2005," Process for mounting and packaging of Fiber Bragg Grating strain sensors for use in harsh environment application, Smart structures conference, SPIE paper 5758-6
4. Brotzu, A., Felli, F., Paolozzi, A., Caputo, L., Passeggio, F. and Gaete, S., 2008," Use of Fiber Optic sensors for monitoring crack growth in fatigue and corrosion-fatigue tests", Adv. Mater. Res., 38, pp. 155-160
5. Gu, X, Li, C., Feng, A., and Chen, D, 2011, " Application of flat-clad optical Fiber Bragg Grating sensor in characterization of asymmetric fatigue deformation of extruded magnesium alloy", IEEE sensor J., 11 (11), pp. 3042-3046
6. Eliezer, D., Aghion, E. and Froes, F. H., 1998, "Magnesium science, technology and application" Adv. Perform. Mater. , 5, pp. 201-212
7. Mirza, F.A., Chen, D. L., 2013, "Fatigue of Magnesium alloy in: Aerospace materials handbook", CRC Press RES, Taylor & Francis, New York
8. Albinmousa, J., Jahed, H. and Lambert, S., 2011, "Cyclic behaviour of wrought magnesium alloy under multiaxial load" International Journal of fatigue, 33, pp. 1127-1139.
9. Kleiner, S. and Uggowitzer, P. J., 2004, "Mechanical anisotropy of extruded Mg–6% Al–1% Zn alloy", Mater. Sci. and Eng. A., 379, pp. 258-263
10. Hasegawa, S., Tsuchida, Y., Yano, H., and Matsui, M., 2007, "Evaluation of low cycle fatigue life in AZ31 magnesium alloy", Int. J. Fatigue, 29, pp. 1839-1845
11. Huppmann, M., Lentz, M., Brommelhoff, K., and Reimers, W., 2010, " Fatigue properties of the hot extruded magnesium alloy AZ31", Mat. Sci, and Eng. A., 527, pp. 5514-5521
12. Kang, G., Yu, C., Liu, Y. and Quan, G., 2014, " Uniaxial ratcheting of extruded AZ31 magnesium alloy: effect of mean stress", Mat. Sci, and Eng. A., 607, pp. 318-327
13. Kalatehmollaei, E. Mahmoudi-Asl H. and Jahed, H., 2014, An asymmetric elastic-plastic analysis of load-controlled rotating bending test and its application in the fatigue life estimation of wrought magnesium AZ31B", Int. J. Fatigue, 64, pp33-41
14. Julich, F., Aulbach, L., Wilfert, A., Kratzer, P., Kuttler, R. and Roths, J., 2013, "Gauge factors of Fiber Bragg Grating strain sensors in different types of optical fibers, Meas. Sci. Technol., 24, pp. 094007 (1-7)
15. Shiozawa, K., Kitajima, J., Kaminashi, T., Murai,T. and Takahashi,T, 2011, "Low-cycle fatigue deformation behaviour and evaluation fatigue life on extruded magnesium alloys" Procedia Eng. , 10, pp 1244-249.

Proceedings of the 25th CANCAM
London, Ontario, Canada, May 31 – June 4, 2015

FLOW AROUND 2D BLUFF BODIES MOUNTED CLOSE TO A FLOOR

Hassan Iftekhar
Automotive, Mechanical and Manufacturing Engineering
University of Ontario Institute of Technology
Oshawa, ON, Canada
hassan.iftekhar@uoit.ca

Martin Agelin-Chaab
Automotive, Mechanical and Manufacturing Engineering
University of Ontario Institute of Technology
Oshawa, ON, Canada
martin.agelin-chaab@uoit.ca

ABSTRACT

This study reports turbulent flows around 2D bluff bodies using particle image velocimetry (PIV). The bluff bodies used have profiles that resemble those of pickup trucks. A Reynolds number of Re_h ($=U_m h/\nu$) = 9000 was used for this study, where h is the maximum model height, U_m is the maximum approach velocity, and ν is the kinematic viscosity of water. The results indicate that the structure of the vortices behind the bluff bodies strongly depend on the cavity dimensions. Also the turbulence kinetic energy is reduced when the cavity length is increased but increased when the cavity depth or base height is increased. The Reynolds shear stress profiles show similar trends described, although the profiles are approximately anti-symmetric about the y-axis in conformity with the shear layer.

KEYWORDS: ***Bluff bodies, turbulence, particle image velocimetry, water tunnel***

INTRODUCTION

Turbulent flows around bluff bodies have diverse engineering application and as a result have been studied quite extensively. A wide variety of bluff bodies exist including complex structures such as space launchers and road vehicles. Recently, Wolf [1] conducted an experimental study on the aerodynamics of space launchers at subsonic conditions using a simplified model. The study revealed that the fluctuating field was comprised of multiple phenomena and not just vortex shedding alone. Flows around road vehicles are complicated by the ground effect and this makes them uniquely different [2]. In attempt to understand the physics of aerodynamics of complex bluff bodies such as road vehicles, simplified geometries have also been used in the past. The most comprehensive review of previous studies can be found in Hucho [3]. Subsequent studies reported by Balkanyi [4], Khalighi et al. [5] and Ha et al. [6] were more concerned with how drag reduction devices affect the aerodynamics of the near wake region.

Turbulent flows in these bluff bodies contain recirculating flow regions, unsteady vortex shedding and other complex interactions that pose a challenge to some experimental and numerical techniques [7]. Therefore, there is a need to provide comprehensive experimental data on the turbulence structures to improve our physical understanding of the flow and for the development and validation of turbulence models.

As reported earlier in [8], the bluff shape used in the current study is a complex two-dimensional (2D) bluff body that resembles the profile of a pickup truck. As it has been reported, the flow dynamics around pickup trucks are extremely complex mainly due to the intricate interaction of the separated flow and the bed, bed walls and tailgate with the separated shear layers formed behind the cabin [9]. It should be emphasized that the current study is not an investigation of the turbulent flow around a pickup truck. Rather, this is a fundamental study of turbulent flow around a complex 2D asymmetric bluff body with a side view that resembles the side profile of a pickup truck. The goal of the study is to provide fundamental insight to further the understanding of the flow structure in these 2D complex bluff body geometries. The high quality data will also be useful for validating turbulence models.

EXPERIMENTAL SETUP AND MEASUREMENT PROCEDURE

Experimental Set-Up: The experiments were performed in a re-circulating closed channel water tunnel with a test section that was 2500 mm long, 200 mm wide and 200 mm deep. A provisional test section made from 6 mm thick acrylic plate was inserted into the main channel. The insert was 2500 mm long, 188 mm wide and 188 mm deep, and its base was tightly screwed onto the floor of the main channel. Figure 1 is a schematic of the test model and the test section. The test model was also made of acrylic with a width of 187 mm that spans the entire width of the test section, and has a maximum height of 12 mm. Each test model was held firmly in position by screwing it to the side walls of the test section insert at 1250 mm downstream of the inlet as shown in Fig. 1. Note that the three test models with different dimensions (i.e., lengths, l, and cavity base height, t, as indicated in Fig. 1a) were used for this study. The models are denoted as L_1 with l = 23 mm, t = 8 mm; L_{1B} with l = 23 mm, t = 10 mm; and L_2 with l = 30 mm, t = 8 mm. The gap of $0.5h$ was maintained for all the tests. As shown in Fig. 1, $x = 0$ is at the trailing end of the test model, $y = 0$ is at the underside of the test model and $y' = 0$ is at the floor of the test section.

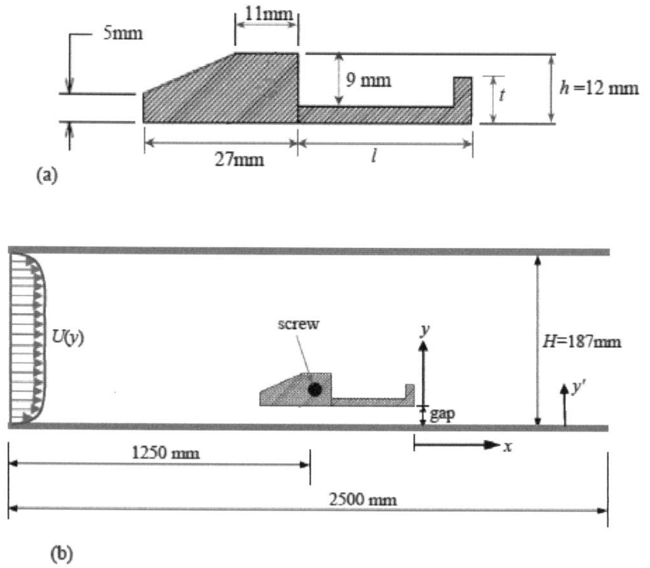

Figure 1: Schematic of the test model and its arrangement inside the test section. (a) Test model, and (b) test model inside test section.

Measurement procedure: A PIV technique was used to conduct the velocity measurements. The flow was seeded with poly methyl methacrylate particles coated with a fluorescent dye, Rhodamine B. The particles have mean diameter and density of 10 μm and 1.19 g/cm^3, respectively. The fluorescent dye absorbs light at a wavelength of 550 nm and emits orange light at a wavelength of 590 nm. The particle time constant is 7μs giving a Stokes number of approximately 0.001 based on the Kolmogorov time-scale. This small number confirms the particles ability to faithfully follow the flow. An Nd-YAG, 120 mJ/pulse laser of 532 nm wavelength was used to illuminate the flow field. The laser sheet was first located at the mid-span of the test section for the measurements. A 12-bit high-resolution digital camera (Dantec Dynamic HiSense 4M camera) that uses a CCD with 2048 pixels × 2048 pixels and has a 7.4 μm pixel pitch was used to image the flow field. The measurements were made at a field of view of 74 mm × 74 mm. The instantaneous images were processed using the adaptive correlation option of Dynamic Studio developed by Dantec Dynamics. A 32 pixels × 32 pixels interrogation window with 50% overlap was used, resulting in a vector spacing of 0.58 mm. Four thousand (4000) instantaneous image pairs were used to compute the mean velocity and turbulent statistics reported subsequently. Measurement uncertainty analysis was undertaken following the AIAA standard described by Coleman and Steele [10]. The uncertainty in the mean velocities at 95% confidence level was estimated to be ±2% of the streamwise velocity. The uncertainties in the turbulence kinetic energy and Reynolds shear stress are estimated ± 10 % of the peak values.

Test Conditions and Flow Qualification: Prior to conducting measurements around the bluff bodies, data were obtained at -25h (upstream of the leading edge of the test model) to characterize the nature of the approach flow. The maximum mean velocity was U_m = 0.75 *m/s* while the Reynolds number based on the test section height (H) and maximum model height (h) were Re_H = 141,000 and Re_h = 9,000, respectively. Note that $Re_H = U_mH/\nu$ and $Re_h = U_mh/\nu$, where ν is the kinematic viscosity of water. This was the highest Re_H attainable for this setup without vibration in the water tunnel. The blockage ratio defined as the ratio of the maximum cross-sectional area of the test model to the test section's cross-sectional area was about 6%. Data obtained at mid-plane of the test section and 20% (of the test section width) offset the mid-plane confirmed that the flow is not two-dimensional. The length and velocity scales used in the study are the maximum bluff body height (h = 12 mm) and maximum approach velocity (U_m = 0.75 m/s).

RESULTS AND DISCUSSION

Velocity magnitude, vectors and streamlines: These quantities were obtained with the vectors superimposed to reveal some of the qualitative features of the mean flow pattern around the test model. The approximate location of the test model is shown in each figure. The streamlines (Fig. 2) revealed that two distinct clockwise rotating vortices were formed inside the cavity (vortex to the left) and behind the model (vortex to the right). The latter was not reported in a previous three-dimensional (3D) study [9]. This suggests that either the spatial resolution in that study was unable to resolve the flow structures or the wake structure in the 3D model does not have the second vortex due to the side walls. The streamlines also indicate that the increasing the cavity length as in L_2 does not have any significant impact on the vortices. However, increasing the cavity deep or back height as in L_{1B} significantly changes the vortices in both inside and behind the cavity as shown. The velocity vectors and magnitudes show no significant obvious differences in the flow patterns especially away from the bluff bodies. The velocity profiles will be used to explain in more details.

Mean velocity profiles: To complement the qualitative information shown in Fig. 2, profiles of the streamwise mean velocities at selected locations are shown to reveal the effect of the different bluff body geometries on the near wake region where the flow is most complex. The profiles in Fig. 3 show that the ambient fluid is entrained as the flow evolved in the wake region and the maximum reverse velocity increased with the cavity depth. For example, the magnitude of the maximum reverse velocity at $x = 0.5h$ is $(U/U_e)_{max} \approx$ 12% for Test L_{1B} and 9% for Tests L_1 and L_2. The corresponding values at $x = 1.0h$ and $2.0h$ are $(U/U_e)_{max} \approx$ 15% and 0% for Test L_{1B}, respectively. For Tests L_1 and L_2 there is no reverse flow beyond $x = 0.5h$ because of their weak recirculation regions. Note that the profiles do not collapse even at $x = 2.0h$.

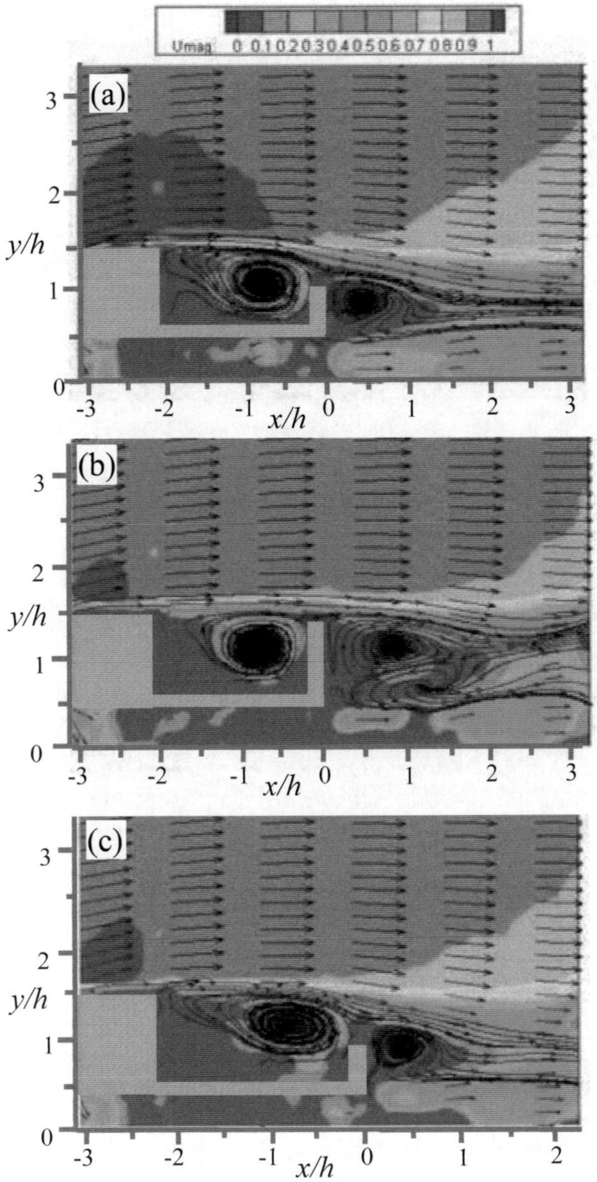

Figure 2: Contours of velocity magnitudes, vectors, and streamlines. (a) L_1, (b) L_{1B} and (c) L_2. U_{mag}= *Velocity magnitude.*

Turbulence kinetic energy and Reynolds shear stresses: Figures 4 and 5 contain profiles of the normalized turbulence kinetic energy $\{tke^* = (u^2+v^2)/U_m^2\}$, and Reynolds stress ($uv^*$ $= uv/U_m^2$) across the wakes at the selected locations indicated. The tke^* profiles (Fig. 4) have distinct double peaks as expected because the presence of vortices and the nature of the shear layer at those locations. The double peaks became broader with downstream distance as mixing and entrainment increases. The peaks for L_1 and L_2 considerably decayed at $x = 2.0h$. It can be observed that the turbulence kinetic energy is reduced when the cavity length is increased as in L_2 but increased when the cavity depth or base height is increased as in L_{1B}. The Reynolds shear stress profiles (Fig. 5) show similar trends described, although the profiles are approximately anti-symmetric about the y-axis in conformity with the shear layer

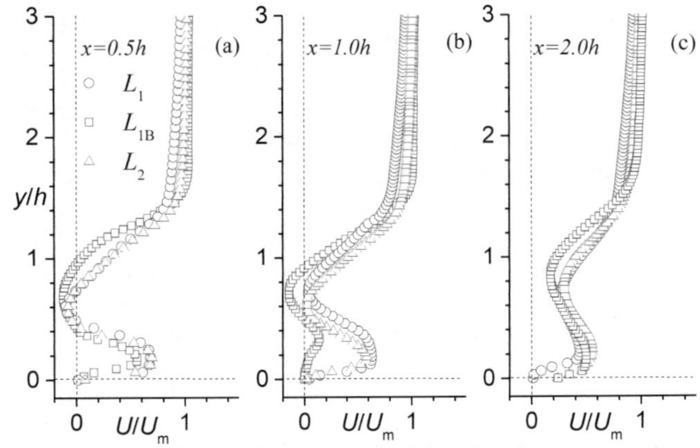

Figure 3: Profiles of mean velocities in the streamwise direction.

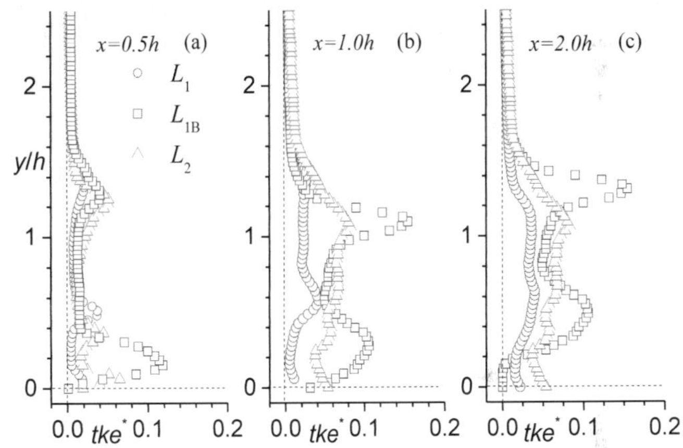

Figure 4: Profiles of turbulence kinetic energy.

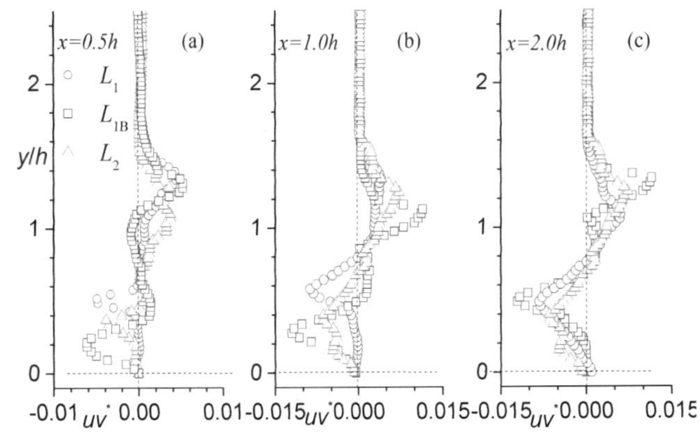

Figure 5: Profiles of Reynolds shear stresses.

Production of turbulence kinetic energy and Reynolds shear stresses

The general trends observed for the profiles of tke^* and uv^* above can be explained by the turbulence production. Since a planar PIV was employed in the present study, the production terms in the transport equations for the turbulent kinetic energy, P_k, and Reynolds shear stress, P_{uv}, were estimated as follows: $P_k = [-uv\partial U/\partial y + -uv\partial V/\partial x + u^2\partial U/\partial x + v^2\partial V/\partial y]$ and $P_{uv} = [-uv\partial V/\partial y + -uv\partial U/\partial x + u^2\partial V/\partial x + v^2\partial U/\partial y]$. Typical normalized profiles of P_k and P_{uv} are shown in Figs. 6 and 7, respectively, where ($P_k^* = P_k \times h/U_m^3$ and $P_{uv}^* = P_{uv} \times h/U_m^3$). Because $-uv\partial U/\partial y$ is the most dominant term there are negative and positive peaks in P_k^* which is consistent with the Reynolds stresses due to the lower and upper shear layers. In addition, the profiles for the test models L_{1B} are much higher while the others are comparable. Furthermore, the peak values increase downstream which is consistent with the trends observed in the Reynolds stresses discussed earlier. The P_{uv}^* profiles in Fig. 7 are similar to the trends described for P_k^* in Fig. 6. However, the trends in the profiles in Fig. 7 are anti-symmetry about the y-axis in a fashion similar to the Reynolds shear stresses but with opposite signs due the significant contribution from the term $-uv\partial V/\partial y$.

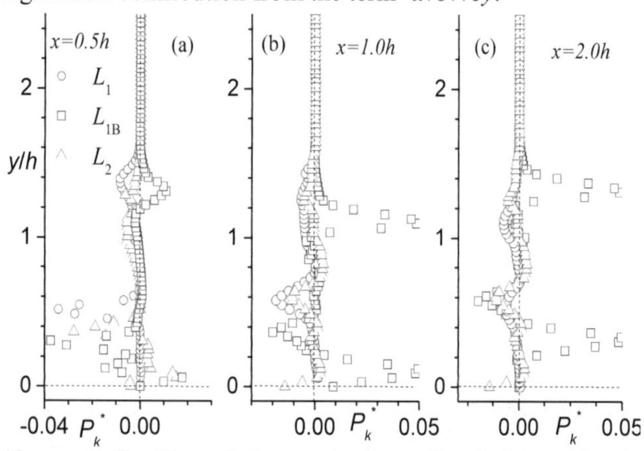

Figure 6: Profiles of the production of turbulence kinetic energy, P_k.

CONCLUSIONS

The effect of geometry on a turbulent flow around complex 2D bluff was studied using a high resolution PIV. It was observed that the sizes of the vortices inside the cavities are dependent on the cavity dimensions. In addition, the structure of the vortices behind the bluff bodies strongly depend on the cavity dimensions. It can be observed that the turbulence kinetic energy is reduced when the cavity length is increased as in L_2 but increased when the cavity depth or base height is increased as in L_{1B}. The Reynolds shear stress profiles show similar trends described, although the profiles are approximately anti-symmetric about the y-axis in conformity with the shear layer.

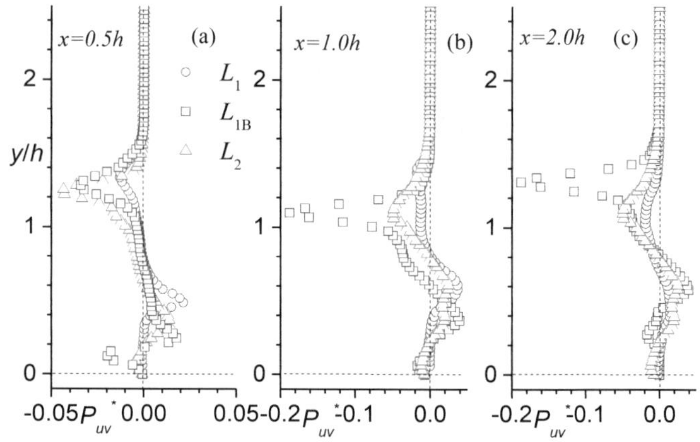

Figure 7: Profiles of the production of Reynolds shear stresses, P_{uv}.

ACKNOWLEDGMENTS

The author would like to thank Dr. M.F. Tachie for the use of his lab facilities and Messrs. J. Tsikata and E. Essel for their help with data acquisition and processing.

REFERENCES

[1] Wolf, C.C., 2013, "The Subsonic Near-Wake of Bluff Bodies," *PhD Dissertation RWTH Aachen University, Germany.*

[2] Bearman, P. W., 1980, "REVIEW—Bluff body flows applicable to vehicle aerodynamics," *Journal of Fluids Engineering*, 102, pp. 265- 274.

[3] Hucho, W. H., 1998, "Aerodynamics of road vehicles," *SAE International, Warrendale, Pa.*

[4] Balkanyi, S. R., 2001, "An experimental investigation of drag reduction devices for blunt based road vehicles," *MS thesis (2001), Delft University of Technology.*

[5] Khalighi, B. Zang, S. Koromilas, C. Balkanyi, S. Bernal, L.P. Iaccarino, G. and Moin, P., 2001, "Experimental and computational study of unsteady wake flow behind a body with a drag reduction device," *SAE Paper* 2001-01-1042.

[6] Ha, J., Jeong, S. and Obayashi, S., 2010, "Flow characteristics of a pickup truck with regard to the bed geometry variation," *Proceedings of the Institution of Mechanical Engineers, Part D, Journal of Automobile Engineering,* 224, pp. 881–891.

[7] Holloway, S. Leylek, J.H., York, W.D. and Khalighi, B., 2009, "Aerodynamics of a pickup truck: Combined CFD and experimental study," *SAE Int. J. Commercial Vehicles*, 2, pp. 88-100.

[8] Agelin-Chaab, M., 2014, "Structure of turbulent flows over two-dimensional bluff bodies inspired by a pickup truck geometry." Int. J. Heat and Fluid Flow, 50, pp. 417-430.

[9] Al-Gami, A. Bernal, L. Khallghi, B., 2003, "Experimental investigation of the near wake of a pickup truck," *SAE World Congress,* 2003-01-0651, Detroit, USA.

[10] Coleman, H., Steel, W., "Engineering application of experimental uncertainty analysis." *AIAA J.* 33, pp. 1888-1895.

EXPERIMENTAL INVESTIGATION ON OVERPRESSURES IN A SCALE MODEL OF SIDEDRAW SILO DURING EMPTYING

Chenying Dong, Kun Pang, Fang Yuan*
School of Civil Engineering and Architecture
Henan University of Technology
Zhengzhou, Henan, China
***fangyuan64@163.com**

Liying Jiang
Department of Mechanical and Materials Engineering
The University of Western Ontario
London, Ontario, Canada
lyjiang@eng.uwo.ca

ABSTRACT

The dynamic pressure on silo walls during emptying is related to the way of emptying and flow pattern of the storage materials. Therefore, it has always been an important and fundamental issue to guarantee the safety of storage facilities through improving the way of emptying and controlling the flow pattern. In this paper, three silo models with different emptying processes, including sidedraw emptying with and without chutes and bottom center discharging, are tested. Both static pressure and the dynamic pressure are measured by the sensors distributed along the silo circumference at different depth. The maximum overpressure coefficient for different models is assessed, which reaches the peak value at a particular depth for the point farthest away from the discharging port. It is concluded that the sidedraw silo with chutes is a better choice if the overpressure phenomenon is the concern.

KEYWORDS: Scale model of silo, Sidedraw chutes, Silo emptying, Discharge tube, Overpressure

INTRODUCTION

Silos with special sidedraw chutes designed by the GSI company have been widely adopted in North America, which have threefold advantages. Firstly, the design can make use of the self-weight of stored material to flow automatically and orderly in separated layers from top to bottom. Secondly, the way of discharging can effectively avoid serious engineering accidents possibly caused by mass-flow. Thirdly, it can load and unload directly even without using other conveyors or equipments. Regardless of these benefits, the load is eccentric in the emptying (or discharging) process of such silo structures. Therefore, how to ensure the safety in the emptying process of the silos has become the most concerned issue in recent years, which has attracted considerable attention from researchers for understanding the emptying process.

It has been confirmed that the dynamic pressure generated in the discharging process is bigger than the static one [1-2]. For example, Reimbert [1] carried out a few tests on large silos and the results showed that the coefficients of overpressure are very variable. The maximum coefficients varied from 1.76 to 2.39. The eccentric emptying has also been investigated in deferent kind of silos [3-4].The current work will focus on the eccentric emptying of sidedraw silo with chutes. The lateral pressure and flow pattern of such a silo will be investigated by model tests.

EXPERIMENTS

Based on the real silo, a scale of 1/ 20 model was made with the height of 1100mm, the inner diameter of 500mm, and the wall thickness of 5mm. The sidedraw chutes location and the position of the discharge tube with a 45° inclined angle are shown in Fig.1(a). In order to observe the flow of the storage materials clearly, the silo wall is made of Plexiglas. It should be mentioned that the silo has a height to diameter ratio of 2.2, which belongs to a deep silo according to Chinese Silo Code [5].

DYB-3B resistance strain earth gauges are used to measure the pressure values at different positions on the sidewall. The data collection instrument is S3886 resistance

Zhengzhou Academician workstation. *(ZHENG TECHNOLOGY PLAN (2013)No.4:131PYSGZ205).*

strain gauge, which can collect the data automatically for further data analysis by using appropriate software. The Fujian Pingtan standard sands are used as storage materials.

The objectives of the experiment includes: (1) test the dynamic pressure on the silo walls during emptying; (2) observe the flow pattern of the stored materials; (3) explore the lateral pressure distribution and the location of the maximum pressure value; (4) investigate the influence of eccentric discharging problems.

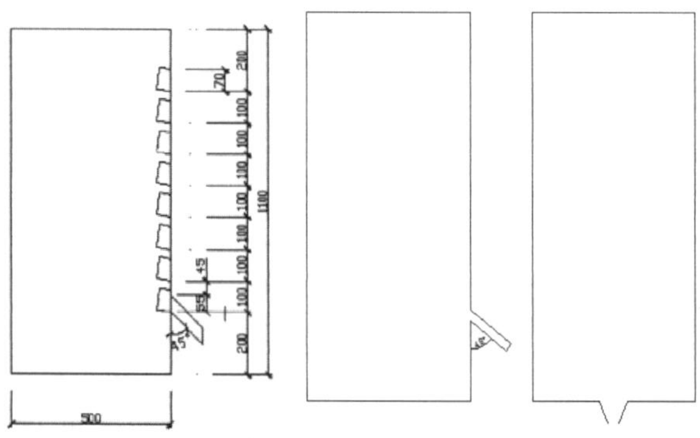

(a) (b) (c)

Figure 1: The scale models of silo with different discharge ways

The distribution of the measuring points is shown in Fig. 2, which are classified into three columns, named as A, B and C. Along each column there are seven measuring points at different depth of the silo. When the sensors are placed at these pre-determined positions, then the silo is filled with sand and the strain gauge is connected to the computer for measuring.

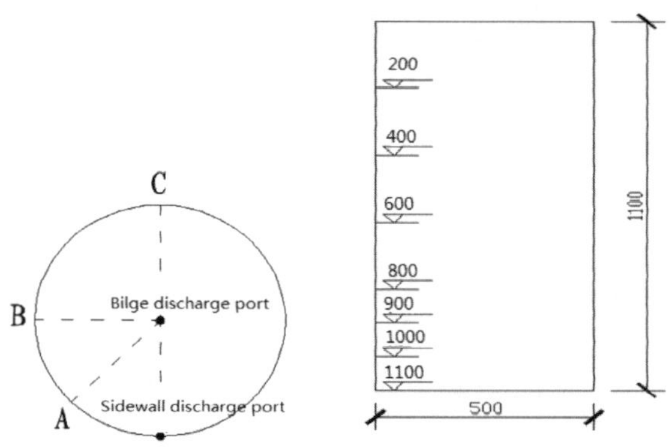

Figure 2: Sensors arrangement

The static pressures are taken first. Then the dynamic pressures are measured in three emptying processes: (1) the sidedraw emptying with chutes as shown in Fig.1(a); (2) the sidedraw emptying (without chutes) as shown in Fig.1 (b); (3) the bottom center emptying as shown in Fig.1(c).

(1) Static lateral pressure measurement: Pour the sand into the silo as uniformly as possible and measure the data after the sand get dense automatically by its own gravity. Several tests are taken to get the average values for the static pressures.

(2) Dynamic pressure measurement: Measure the lateral pressure at different positions on the wall during the emptying process. Set the strain gauge to record the strain every second at each measuring point. The pressures at the measuring points will be determined from the calibrated pressure-strain equation. The flow pattern of stored materials in the emptying process will also be recorded by cameras. Three emptying models as shown in Fig. 1 will be tested.

RESULTS OF EXPERIMENTS ON THE SCALE SILO

(1) The test of static pressures

After several tests the static pressures at different measuring points are calculated by average. It is found that the values of the static pressures are in good agreement with the code [5], with deviation in the normal range of experimental error.

(2) The test of dynamic pressures

(a) The sidedraw discharge (without chutes)

In the process of sidedraw emptying, the pressure variation with time for the measuring points at the depth of 0.6m, 0.9m and 1.1m of the columns A and C are shown in the Fig.3 and Fig.4, respectively. As shown in Fig. 3, the pressure at the measuring points of column A near the sidewall discharge port keeps almost constant until a sudden jump up at the both the depth of 0.6 m and 0.9 m, while a jump down at the 1.1 m depth. During the emptying process, the pressure has a fluctuating variation. Until near the end of emptying, the pressure at 1.1m depth has a second increase then drops down, while the pressure at the other depths decays to zero. This is due to the dead space formed by the sidedraw. For the measuring points of column C located farthest away from the sidewall discharge port, the pressure variation trend is different from that of the measuring points of column A as is shown in Fig.4. For example, the pressure at depth of 0.6m drops gradually, while the pressure at depth 0.9m and 1.1 m suddenly drop down at the end of emptying. It is obvious that the asymmetric load or eccentric load does exist.

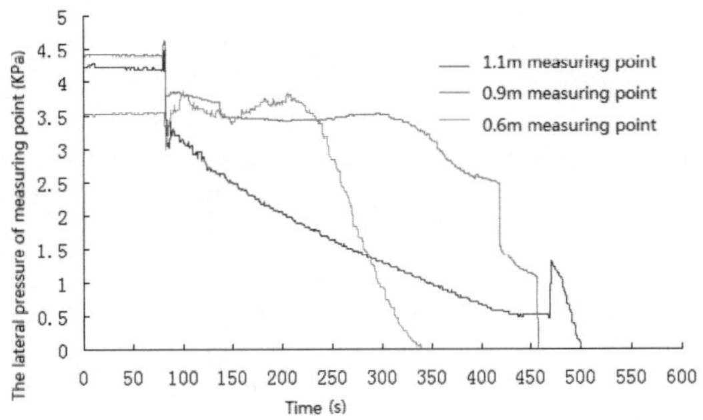

Figure3:The lateral pressures of column A in the sidedraw emptying

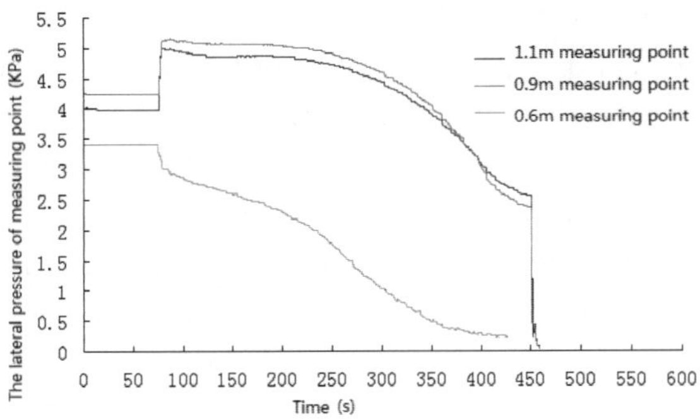

Figure4: The lateral pressures of column C in the sidedraw emptying

(b) The sidedraw emptying with chutes

The pressure curves for sidedraw silo with chutes measured by column A and C sensors are presented by Fig.5 and Fig.6, respectively.

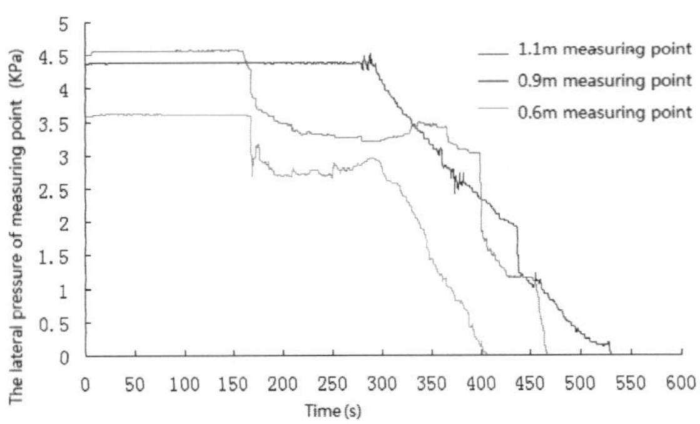

Figure5: The lateral pressures of column A in the sidedraw emptying with chutes

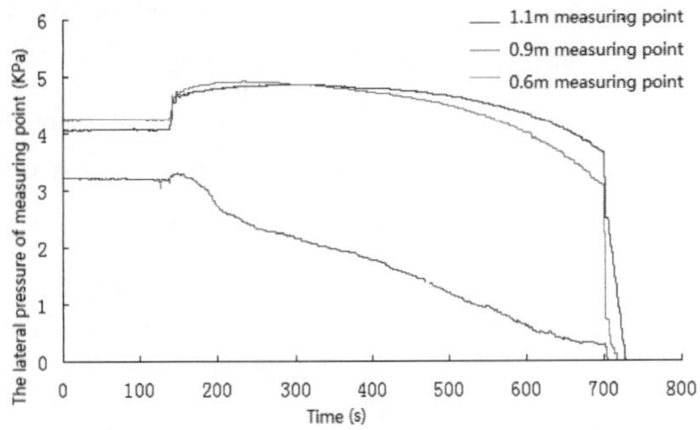

Figure 6: The lateral pressures of column C in the sidedraw emptying with chutes

It is found that the pressure variation at the measuring points of column C is similar for both silos with and without chutes. However, pressure variation trend for the points of column A which are located nearest to the discharge port is quite different for these two cases. It means that the chutes have more significant influence for the dynamic pressure at the position near the discharge port.

(c) The bottom center emptying

In the bottom centre emptying, the pressures variation with time at the depth of 0.6m, 0.9m and 1.1m of the column A is shown in the Fig.7, which is different from the other two emptying processes. However it is similar to the other two cases that the maxim pressure in the depth of 0.9m is bigger than those of 0.6m and 1.1m.

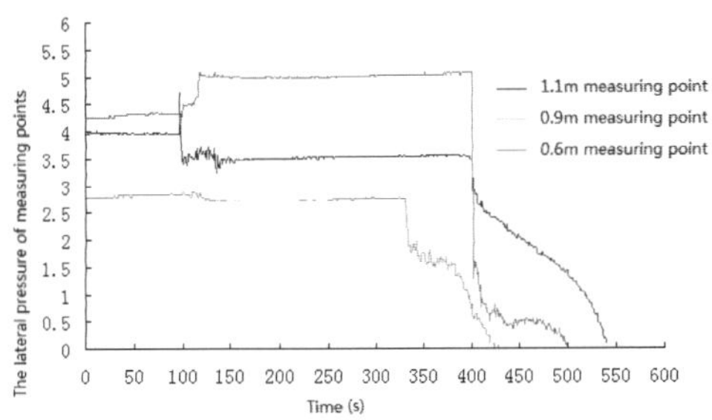

Figure 7: The lateral pressures of group A in the bottom center emptying

OVERPRESSURE AND FLOW PATTERN

In order to compare the dynamic lateral pressure and the static lateral pressure of the different silo models, the overpressure coefficient is introduced, which is defined as the maximum dynamic lateral pressure divided by the static pressure. The overpressure coefficients along the depth of the silo measured by columns A and C sensors for the silos with and without chutes are compared in Fig.8, which shows a similar trend. It is observed that the overpressure coefficient of column C is larger than that of column A in general, except some close points. It is also found that the peak value of the overpressure coefficient of the sidedraw silo without chutes is relatively higher. For example, the maximum overpressure coefficient of sidedraw one is 1.29, which occurs at the approximate depth of 0.8m. The maximum overpressure coefficient reaches the maximum value at both the depth of 0.2m and 0.8m for the measuring points of column C with a value of 1.25 for the silo with chutes.

For the center emptying, shown in Fig.8, the overpressure phenomenon also occurs in the process of the discharging with a maximum value of 1.32 occurring at the depth of 0.8m for both columns A and C.

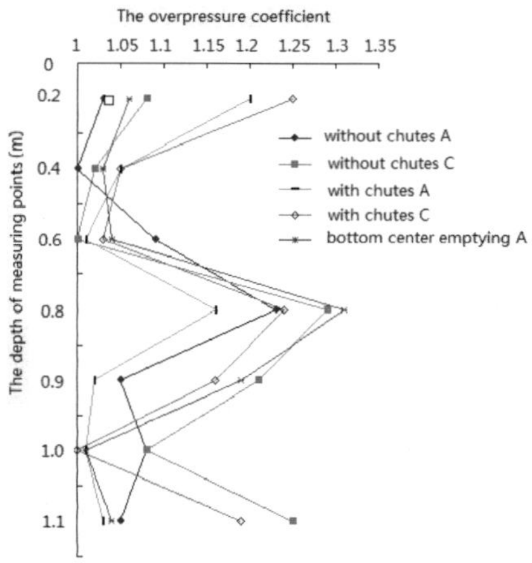

Figure 8: The overpressure coefficient of the three emptying process

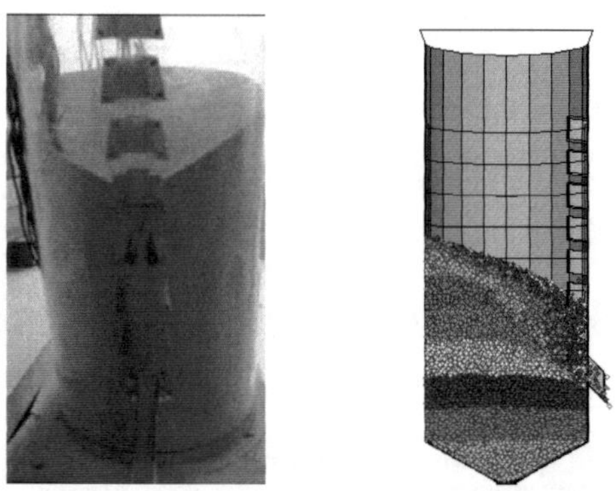

Figure 9: The flow pattern

Two flow patterns are show in Fig.9. The left One is taken by camera and the other one is by numerical simulations using PFC3D（Particle Flow Code in 3 Dimensions. It can be seen that the numerical simulation results are similar to the camera image.

CONCLUSION

Based on the experimental data and analysis performed on the silo models with different emptying processes, it is concluded that:

(1) The dynamic pressure in the sidedraw silo with chutes depends on the position along the side wall, with smaller value at the points near the discharging port.

(2) The increase of the dynamic pressure in the sidedraw silo with chutes is relatively small, in other words, the overpressure phenomenon is alleviated for such kind of silos. The maximum overpressure coefficient occurs at the depth of 0.2m and 0.8m at the points farthest away from the discharging port along the silo circumference.

(3) Considering the overpressure phenomenon, the sidedraw silo with chutes is a better choice than both the sidedraw silo without chutes and the bottom center emptying silo. In practice, the speed of discharging is slow and the overpressure and the eccentric problem are under control.

REFERENCES

1. M.& A. Reimbert. 1976, *Silos Theory and Practice*. Trans Tech Publications, Clausthal.

2. Su L X. 1998, “Mechanics analysis of over pressure to the silo wall during discharge”, *Journal of Zheng zhou Grain college*, 19(4), pp15-20.

3. Zhang S S and Yuan F. 2009, PFC3D simulation on dynamic lateral pressure of large-diameter squat silo centrifugal model. *Journal of Henan University of technology*, 30(5), pp92-96.

4. Song Y H. 2011, *Study on dynamic pressure and eccentric discharge in squat silo* [D]. Henan University of Technology, Zhengzhou.

5. Cui Y R, etc. 2003, *Reinforced Concrete Design Code（GB50077-2003）* [S]. China Planning Press, Beijing.

Proceedings of the 25th CANCAM
London, Ontario, Canada, May 31 – June 4, 2015

DESIGN OF KOLSKY BAR TO CHARACTERIZE THE DYNAMIC RESPONSE OF ELASTOMER UNDER HIGH STRAIN RATE

Muhammad Salman Chaudhry
Lassonde School Of Engineering
York University
Toronto, Ontario, Canada
salman01@yorku.ca

Dr. Alex Czekanski
Associate Professor
York University
Toronto, Ontario, Canada
alex.czekanski@lassonde.yorku.ca

Roger Carrick
(Designer in Residence)
York University
Toronto, Ontario, Canada
carrick@yorku.ca

ABSTRACT

The need for dynamic characterization of elastomers under high strain rate is increasing with the increase of wide variety of dynamic application of elastomers. The most common apparatus used to study the dynamic behavior of materials in the strain rate regimes of 100 – 1000/s is the Kolsky bar apparatus. Kolsky bar needs to be modified when it is used to characterize low strength materials such as elastomers. However, as there are no standard guidelines for the design and construction of Kolsky bar, this paper presents an overall design requirements like bar and specimen, considerations and cautions that needs to be taken into account when designing and building such an apparatus for elastomer. For e.g. one of the requirement of a valid Kolsky bar is that the specimen attains uniform stress distribution. In the next half of this article a design for such a Kolsky bar apparatus is given, which is currently under construction

KEYWORDS: Kolsky bar, dynamic characterization, high strain rate testing, elastomers

INTRODUCTION

Low impedance materials such as elastomers are finding new applications in automotive, biomedical, aerospace etc. An example of a dynamic application of elastomers is shock absorbers. For such dynamic applications researchers cannot rely on properties determined under static conditions. Accurate models and techniques are needed to characterize elastomers in the dynamic range. Hyperleastic materials like elastomers also show dependency on the strain rate at which the loading is applied. In order to generate strain rate dependent constitutive models for these materials, testing at different strain rate is required. This allow researchers and designers to accurately predict the materials response.

The most common method to characterize materials at a strain rate of > 100/s is by using a Kolsky bar apparatus. The Kolsky bar is commonly used to dynamically characterize materials such as, metals [1], ceramics [2], foam [3], composites [4] and "smart materials" [5]. The Kolsky bar is an ideal choice to characterize the stress strain response of material as a function of strain rate. A family of curves can then be used to generate dynamic models to accurately predict performance in impact applications. A conventional Kolsky bar apparatus needs to be modified to test soft materials. The characteristic of soft materials are low stiffness, and low elastic wave velocities. Considerations such as impedance mismatch, uniform response of the specimens have to be taken into account, which could result in the distorted and weak signal. To account for these effects modifications such as using sensitive force transducers, low impedance bar materials, hollow rods [6], and pulse shaping techniques [7] have been investigated by researchers. This paper will examine these techniques as well as other important design parameters and requirements. Finally a design for dynamic characterization of elastomers is presented based the modification needed for its use with soft materials.

CONVENTIONAL KOLSKY COMPRESSION BAR

A Kolsky bar setup is the most commonly used high strain rate testing equipment to determine the dynamic response of various engineering materials. Although there is no standard design, most researchers use apparatus which includes the following components [1].

- Two long cylindrical elastic bars (incident, transmission).
- A striker launching mechanism (usually achieved by a gas gun)
- Force/Strain measuring sensors
- Bearing and base setup, for the axial alignment of bars.
- Data acquisition system

This general Kolsky setup is modified in different ways to allow for the testing of a wide variety of materials (Fig.1). In a Kolsky bar, a sample is sandwiched between the incident and transmission bar. An elastic compressive stress wave is generated, when the striker bars impacts the incident bar. This stress wave travel through the incident bar, upon reaching the incident bar and specimen interface, part of the wave is reflected and some part of it is transmitted through the specimen into the transmission bar (Fig.2). The pulse profile for the incident, transmitted and reflected wave is recorded as a function of time in both of the bars by strain gauges. This pulse profile is further analyzed to determine the dynamic stress strain response of the specimen [8].

The traditional choice of the bar materials is usually a high-strength structural material such as Steel (200 GPa), Aluminum (70-90 GPa) or Nickel Alloy. These high strength materials are used because the yield strength of the bar

determines the maximum attainable stress. Some researchers have also used a choice of low strength materials for bars such as magnesium (45 GPa) or polymeric bars (<20 GPa) [9], [10], [11]. Low strength bars provides a high signal-to-noise ratio and facilitates high-resolution dynamic testing.
Ideally the striker bar is of the same diameter and material as of the incident and transmission bar [1]. The impacting velocity and the length of striker dictates the loading time of the pulse. This loading time is critical when testing low impedance materials. It has to be greater than the time taken for the specimen to reach stress equilibrium [12]. Under equilibrium, the amplitude of the reflected wave is directly proportional to maximum achievable strain rate and similarly the amplitude of the transmitted pulse is directly proportional to the stress developed within the specimen [1].

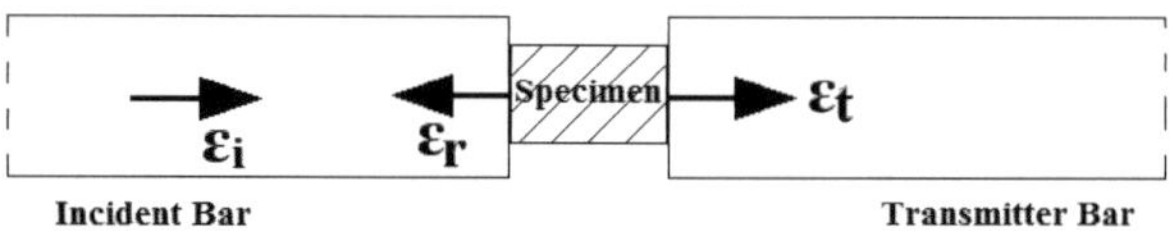

Figure 2: Schematic showing the incident, reflected and transmitted pulse

THEORY OF THE KOLSKY BAR

Working principle of the Kolsky bar is based on the one dimensional propagation of the compressive wave through the bars. This one dimensional compressive stress wave is related to the velocities that the particles experience at the either end of the specimen by the following equation.

$$v_1 = C_b(\varepsilon_i - \varepsilon_r) \quad (1)$$

$$v_2 = C_b(\varepsilon_t) \quad (2)$$

Where ε_i, ε_r, ε_t, are the measured strain pulses recorded from the strain gages. C_b is the elastic wave speed of the bar material. Using these equations the relation for averaged engineering strain integrated over the pulse duration t in specimen having the length L_s is

$$\varepsilon = \int_0^t \frac{(v_1 - v_2)}{L_s} dt = \frac{C_b}{L_s}\int_0^t (\varepsilon_i - \varepsilon_r - \varepsilon_t)dt \quad (3)$$

(3)

And the stresses at both of the interfaces of the specimen can be determined from

$$\sigma_1 = \frac{A_b}{A_s} E_b(\varepsilon_i + \varepsilon_r) \quad (4)$$

$$\sigma_2 = \frac{A_b}{A_s} E_b \varepsilon_t \quad (5)$$

A_b and A_s are the bar and specimen cross section area and E_b being the Young's Modulus. One of the assumption that is made while determining the stress-strain response, is that the specimen is in a state of equilibrated stress ($\sigma_1 = \sigma_2$). Making this assumption we can arrive to following conclusion.

$$\varepsilon_r + \varepsilon_i = \varepsilon_t \quad (6)$$

Thus equation (3) and (5) can be solved to yield the strain rate, strain and stress profile in the specimen.

$$\sigma = \frac{A_b}{A_s} . E_b \varepsilon_t \quad (7)$$

$$\dot{\varepsilon} = -2\frac{C_b}{L_s}\varepsilon_r \quad (8)$$

$$\varepsilon(t) = -2\frac{C_b}{L_s}\int_0^t \varepsilon_r \, dt \quad (9)$$

GENERAL DESIGN OF KOLSKY BAR

The design of the Kolsky bar apparatus is centered on the choice of specimen size and the max desired strain rate. The strain in the incident pulse can be determined by the impact velocity of the striker by the following relation.

$$\varepsilon_i = \frac{V}{2C_B} \quad (10)$$

Here V denotes the striker impact velocity. Using the stress equilibrium assumption on both the surface of specimen and by substituting (10) and (8) in (7) the overall specimen characteristic response equation is found (11).

$$\sigma_s = E_b\left(\frac{A_b}{A_s}\right)\left(\frac{V}{2C_b}\right) - E_b\left(\frac{A_b}{A_s}\right)\left(\frac{L_s}{2C_b}\right)\dot{\varepsilon}_s \quad (11)$$

This expression relates the stress that can be achieved in the specimen as a function of strain and striking velocity. A particular striking velocity is chosen to achieve the desired strain rate (12) and stress (13)

$$\dot{\varepsilon}_{s(max)} = \frac{V}{L_s} \quad (12)$$

$$\sigma_s = E_b\left(\frac{A_b}{A_s}\right)\left(\frac{V}{2C_b}\right) \quad (13)$$

Higher levels of stress can be achieved by increasing the impact velocity or by increasing the area mismatch Ab/As. And similarly reducing specimen length and increasing the impact velocity would increase the maximum attainable strain rate.

When choosing the bar and specimen dimensions certain design requirement have to be met for a valid Kolsky bar. L_b/D_b ratio of at least 20 is required to allow one dimensional wave propagation. Typically this ratio is in an order of 100 [13]. The diameter of the bar should be around 2 to 4 times the diameter of the specimen. Choosing a large ratio for D_b/D_s is desirable for achieving high stress but can result in punch problems. This ratio should be kept such that the bar surface should not act like a plane for the specimen. L_s/ D_s is usually is selected in between 0.6 to 1 [13], this L_s /D_s ratio also depends on the type of material being tested. Soft materials have shown dependency on the length of the specimen which could result in axial and radial inertia. The striker bar is made up of the same material and diameter as of the incident and transmitter bars to avoid wave dispersion.

KOLKSY BAR TESTING FOR LOW IMPEDANCE MATERIALS

Uniform deformation and Inertial effects

When designing a Kolsky bar for soft materials, some consideration have to be taken into account. As discussed in the previous section, the measurement are based on the assumption of uniform deformation in the specimen. Due to the low strength, stiffness and acoustic impedance of soft materials, their mechanical response depends upon the compressive stress wave profile. Due to the low wave speed in specimen it takes non-zero time to achieve stress equilibration. The specimen can be considered to be in a state of stress equilibrium by allowing several reverberations of the stress wave inside the specimen. It requires almost 3 times the time taken for the wave to pass the specimen to achieve stress in equilibrium [14]. The time for a one way trip of stress wave through he specimen can be calculated by (14)

$$T = \frac{L_s}{C_s} \quad (14)$$

Where L_s is the length of the specimen and C_s is the characteristic elastic wave velocity. As it can be seen from (14) the length of the specimen is the most important variable in achieving a uniform state of stress in the specimen. Reducing the length of the specimen would help achieve this state quickly but on the contrary it has been observed that a small value of L_s could result in the increased frictional effect between the specimen bar interfaces, creating distortion in the recorded signal. The non-uniform deformation also results in the axial and radial inertia which is evident when testing very soft materials like elastomeric foams and gels. The inertial and radial effect due to impact loading is studied in detail by Song *et al* 2007. They investigated that this radial inertia is not only dependent on the loading conditions, but also on the specimen geometry and density. These effects cannot be ignored when optimizing the specimen dimensions.

In addition high strain rates can also produce inertia induced stress, especially when the specimen is going through large deformation. This inertia induced stress can greatly affect the strain profiles recorded by the strain gauges [15].

Low Transmitted Signals

Another problem associated with testing softer materials using the Kolsky bar experiment is the very low amplitude of the transmitted pulse [16]. Because of the impedance mismatch between the specimen and bar material, almost all of the stress wave is reflected back and only very small portion of it is transmitted. This weak signal becomes difficult to measure by regular strain sensor and more responsive and sensitive strain gages are required. One of the way to address this problem is to us low impedance bars, for e.g. polymer bars. Using polymer bars allows for greater wave dispersion and spectral analysis techniques are required to filter out the dispersion effect. Another technique of using a hollow transmitter bar also helped increase the amplitude of the transmitted signal up to several times [17]

Loading Profile

Elastomers also show dependency on the applied strain rate, generating different stress - strain responses for different loading rates. Therefore when testing elastomers, care must be taken to ensure that the stress pulse profile is achieved at a constant strain rate with low rate of initial loading. A low initial rate of loading is required because the acceleration induced inertia is inevitable in the initial period of loading, when reaching a particular strain rate. A common way of avoiding this acceleration induced inertia is a constant loading pulse with extended rise time.

Specimen Size effect

Specimen size needs to be properly optimized to mitigate the effects of associated radial and axial inertial effects, achieving uniform deformation and to attain the desired strain rate. The specimen diameter has to be smaller than the bar diameters. Upon the assumption that the maximum diameter during deformation does not exceed the bar diameter and with the choice of desirable strain, the required diameter for the specimen can be calculated by (15).

$$D_s = D_b\sqrt{1-\varepsilon} \qquad (15)$$

For example with the bar diameter of 20 mm and specimen diameter of 8 mm the maximum engineering strain that can be achieved is 0.7. After the selection of the specimen diameter, appropriate length of the specimen can be selected, keeping in mind the dependency of inertial effect and the stress equilibration on the length of the specimen. Thinner specimen helps in achieving a state of equilibrated stress quicker [18], but also produces visible frictional effects.

A specimen can be considered to be in equilibrated stress by comparing the stress profile on both of the faces of the specimen. Previous researchers have determined that equilibrium is reached if the ratio of the difference between the stress from either end '$\Delta\sigma$' the mean stress in the specimen 'σ_m' is less than 0.05 [19]

$$R(t) = \left|\frac{\Delta\sigma}{\sigma_m}\right| \leq 0.05 \qquad (16)$$

Song *et al* 2004 investigated that a thin specimen maybe one of the requirements in Kolsky bar experiment but it is not sufficient to achieve equilibrium. Specimen design can also help avoid inertia in radial direction. It was also investigated that this inertia is maximum at the center and diminishes at the edges. So using a hollow specimen could also help address the problem [3].

KOLSKY BAR DESIGN AND SETUP

This section describes the guidelines for designing a modified Kolsky bar apparatus to characterize behavior of elastomer when subjected to high strain rates between 10^2/s and 10^4/s. For the experimental design of the Kolsky bar apparatus the specimen geometry was optimized to be $D_s = 8$ mm and $L_s = 1.6$ mm. The length and the diameter of the bars are $L_b = 2000$mm and $D_b = 20$ mm. The selected values satisfy the design parameter ratios that were discussed in the earlier section. We get $L_b/D_b = 100$, $D_b/D_s = 2.5$ and $L_s/D_s = 0.2$. It can be noted that the L_s/D_s ratio is different then what was discussed in the previous section (between 0.6 - 1). The specimen geometry was optimized carefully by taking into account the effect of L_s on the uniform deformation, and equilibration of stress as discussed in the previous section.

Using (15) the choice of D_b and D_s could achieve strain level up to 0.8. The striker diameter D_{st} is kept equal to the incident and the transmitter rods to avoid wave dispersions. The length of striker L_{st} = 500 mm. Relatively longer L_{st} is to account for stress equilibration, as the loading pulse duration needs to be 3 times more than the time it takes the wave to travel back and forth inside the specimen. By using the relation$T = \frac{2L_{st}}{C_b}$, it can be seen that the choice of L_{st} would generate a loading pulse of duration 200 μs. Which is more than enough for the specimen to attain uniform stress distribution and respond according to it.

From the available choices of bar material. Aluminum 6061 Anodized (Igus®) was selected. It is apparent form (7) that to increase the transmitted strain either Young's modulus E_b or A_b/A_s must be reduced. This is the main reason for choosing Aluminum as bar material as compared to other higher strength material bar that are reportedly used before in Kolsky setups. Other option for bar material includes polymeric bars. The use of these low strength bars provides a sufficiently high signal-to-noise ratio but this low impedance mismatch between the specimen and bar will delay the dynamic equilibrium process. Moreover, the use of polymeric bars also increase wave dispersion and attenuation. It is recommended that striker, transmitter and incident bar are made up of the same material. Because of impedance mismatch the transmitted wave is very weak. To address this problem a hollow transmitter bar was chosen. As demonstrated by *Chen* (1998), hollow transmitter bars can be used to increase the amplitude of the transmitted wave. When using hollow bars modification needs to be made in (9). Accounting for the change in the cross-sectional area of the bars with the use of a hollow transmission bar, *Chen* formulated this relation to be

$$\varepsilon = \frac{C_b}{L}\left(1 - \frac{A_h}{A_b}\right)\int_0^t \varepsilon_i dt - \frac{C_b}{L}\left(1 + \frac{A_h}{A_b}\right)\int_0^t \varepsilon_r dt \qquad (17)$$

To ensure uniform deformation, a pulse shaping technique is adopted. A pulse shaper is attached in between the incident and striker bar impacting surface. Control and smoothness of the loading pulse can be achieved by using a pulse shaper as it absorbs the high frequency components of the pulse. Different choices of pulse shapers are used by researchers, such as polymer disks (Chen, 1998), Copper disks (Mutter, 2010) and tissue paper. Different geometries of pulse shapers can be tested to generate the desired pulse profile.

Another very important modification made to ensure specimen state of equilibrium is to use Piezoelectric Quartz crystals as force transducers on either side of specimen bar interface. Quartz crystal are chosen as force transducers because of the similar impedance of quartz and aluminum. Quartz crystals have been used in the Kolsky bar setup by Chen *et al* (1999) to measure the relatively weak transmitted signal, by sandwiching the crystal in the middle of the transmitter bar, to measure the weak transmitted signal. These quartz crystal as force transducers can also be attached to either side of the specimen to record the force profile. This force profile can be analyzed to ensure the stress equilibrium condition in the specimen (16), which is one of the requirements for a valid Kolsky bar experiment [8]. Casem *et al* (2005) thoroughly studied the inertial effects of quartz transducers in the Kolsky bar. *Casem* reported that using a single quartz crystal could induce inertial effects and instead three quartz crystal should be used to reduce this effect by a factor of 50.

To record the strain profile from the bars, semiconductor strain gages are recommend over foil gauge. Semiconductor strain gages have a gage factor of 140 which is approximately 70 times larger than the foil gages. Highly sensitive semi-conductor gages are needed to measure the very weak transmitted signal.

Alignment of the bars

A very important constraint in successfully conducting the Kolsky bar experiment is the alignment of bars. It is known that misalignment could cause distortion in the recorded signal which in some cases cannot be differentiated from the actual strain signals. Because the goal is to record strain of micro levels, caution must be taken when setting up the apparatus. Kareem *et al* (2012) has investigated in detail and quantified the effect of misalignment on the recorded stress strain response of the specimen. It was experimentally verified that in order to minimize this distortion the straightness of bars has to be at least 0.08mm per 205 m, centerless grinded to a target diameter with tolerance of ± 0.025mm and finally perfectly perpendicular lubricated end face with tolerance of ± 0.03 degrees.

Gas gun Setup

As discussed before, the maximum achievable strain rate depends upon the impact velocity of the striker. Most common way of accelerating the striker is by using a compressed gas gun. The proposed design of the gas gun is a two staged gas gun. The main supply of high pressure gas or a compressor is used to fill the reservoir. For the reservoir, 1 L sampling cylinder of a pressure rating of up to 3000 psi was selected. Pressure is released using a solenoid valve, with a pressure rating of 750 psi. As the velocity of striker depends upon the initial gas pressure, choosing a high pressure reservoir would generate a wide range velocities in the striker. Having control over a wide range of velocities is also needed because the striking velocity relates to the strain rate achieved in the specimen. Using this system, it was calculated that striker could be accelerated to values up to 80 m/s upon exiting the barrel.

CONCLUSION AND FUTURE WORK

This paper presents a description of modifications to a standard Kolsky bar apparatus required to characterize elastomers under high strain rate, including a sample apparatus design. When testing low strength materials such as elastomers using Kolsky bar, it is very important to make sure that the stress is equilibrated and the specimen goes through uniform deformation. Some consideration have to be made regarding the axial and radial inertia induced due to acceleration. To account for these effects the specimen geometry need to be properly optimized. When designing a Kolsky bar experiment, trade-offs have are made between the max desirable strain rate and stress induced in the specimen. The striker dictates the lading time of the pulse. A longer loading pulse is required to make sure the specimen has reached the state of equilibrated stress. This papers is the first in a series of more to come. The objective of study is the static and dynamic characterization and control of elastomers in a multi-body system. The next step in this study is to validate the results of Kolsky bar apparatus using commercially available Finite-Element-Modelling software.

I. References

[1] G. T. G. III, "Classic Split-Hopkinson Pressure Bar Testing," in *ASM Handbook: Mechanical Testing and Evaluation*, 2000, pp. 1027-1036.

[2] R. W. Chen, "Dynamic Compressive Behaviour of a Glass ceramic Under Lateral Confinement," *J. Mech. Physical Solids,* vol. 45, pp. 1303-1328, 1997.

[3] B. Song, M. Forrestal and W. Chen, "Dynamic and quasi-static propagationof compaction waves in a low-density epoxy foam," *Experimental Mechanics,* vol. 46, pp. 127-136, 2006.

[4] J. T. L. Ninan and C. Sun, "Use of Split Hopkinson Pressure bar for testing off-axis composites," *Int Journal of Impact Engineering,* vol. 25, pp. 291-313, 2001.

[5] W. Chen, Q. Wu, J. Kang and N. Winfree, "Compressive superelastic behavior of a NiTi shape memory alloy at a strain rates of 0.001/s to 750/s," *Int. J. Solids Struct.,* vol. 38, pp. 8989-8998, 2001.

[6] W. Chen, F. Lu and B. Zhou, "A Quartz-cryatl-embedded Split Hopkinson Pressure Bar for Soft Materials," *Experimental Mechanics,* vol. 40, no. 1, pp. 1-6, 2000.

[7] R. Naghdabadi and J. A. M.J. Ashrafi, "Experimental and numerical investigation of pulse-shaped split," *Materials Science and Engineering,* vol. 539, no. A, pp. 285-293, 2012.

[8] K. Ramesh, "High Strain Rate and Impact Experiments," in *Springer Handbook of Experimental Solid Mechanics*, J. W. N. Sharpe, Ed., Springer, 2008, p. Part D 33.

[9] G. Gray and Z. Zhao, "On the Use of SHPB Techniques to Determine the DYnamioc Behavior of Materials in the Range of Small Strains," *Int. J. Solids Struct,* pp. 3363-3375, 1996.

[10] G. Gray, L. Rota and H. Zhao, "Testing Viscous Soft Materials at Medium and High STrain Rates," *Constitutive Relation in High/Very High Strain Rates,* pp. 25-32, 1996.

[11] G.T. Gray III, W.R. Blumenthal, C.P. Trujillo and R.W. Carpenter II, "Infuence of Temerature and Strain Rate on the Mechacnical BEhavior of Adiprene-L100," *J. Phys,* pp. 523-528, 1997.

[12] X. Wu and D. Gorham, "Stress Equilibrium in the Split Hopkinson bar test," *J. Phys,* vol. IV, pp. 91-96, 1997.

[13] P. Follansbee, "The Hopkinson Bar, Mechanical Testing," *ASM Handbook,* pp. 198-203, 1985.

[14] E. D. H. Davies and S. C. Hunter, "The Dynamic Compression Testing Of Solids By the Method Of The Split Hopkinson Pressure Bar," *Jouranl of Mechanics and Physical Solids,* vol. 11, pp. 115-179, 1962.

[15] B. Song, Y. Ge, W. W. Chen and T. Weerasooriya, "Radial Inertia Effects in Kolsky Bar Testing of Extra-soft Speicmens," *Experimental Mechanics,* vol. 47, pp. 659-670, 2007.

[16] W. Chen and B. Song, Split Hopkinson (Kolsky) Bar: Deign, Testing and Applications, 233 Spring Street, New York: Springer Science+Business Media, LLC, 2011.

[17] W. Chen, B. Zhang and M. J. Forrestal, "A Split Hopkinson Bar Technique for Low-impedance Materials," *Experimental Mechanics,* p. 5, 1998.

[18] B. Song and W. Chen, "Dynamic Stress Equilibration in Split Hopkinson Pressure Bar Tests on Soft Materials," *Society for Experimental Mechanics,* vol. 44, no. 3, pp. 300-312, 2004.

[19] Ravichandran and Subhash, "Critical appraisal of limiting strain rates for compression testing of ceramics in a split Hopkinson pressure bar," *J. Amer. Ceram. Soc,* vol. 77, pp. 263-267), 1994.

[20] W. C. Weinong and S. Bo, Dynamic Characterization of Soft Materials, A. Shukla, D. R. Yapa and R. Guruswami , Eds., (Springer Science+Business Media, LLC, 233 Spring Street, New York,: Springer New York Dordrecht Heidelberg London, 2010.

[21] S. Huang, "Dynamic Testing of Soft and Ultra-Soft Materials," 2009.

[22] J. A. Bryson, "Impact Response of Polyurethane," 2009.

[23] G. Gary, J. Klepaczko and H. Zhao, "Generalization of Split Hopkinson Bar Technique to use Viscoelatic Materials," *Int. J. Impact Eng,* pp. 529-530, 1995.

[24] R. Thevamaran and C. Daraio, "An Experimental Technique for the Dynamic Characterization of Soft Complex Materials," *Society for Experimental Mechanics,* 2014.

[25] Mutter and N. J., "Characterization of Dynamic and Static Mechanical Behavior of Polyetherimide," 2012.

[26] M. L. Shawn, "Design of a Split Hopkinson Bar Apparatus for use with Fiber Reinforced Composite Materials," 2012.

[27] B. Song, K. Connelly, J. Korellis, W.-Y. Lu and B. R. Antoun, "Improved Kolsky-bar design for Mechanical

Characterization of Materials at High Strain Rates," *Measurement Science and Technology,* 2009.

[28] D. Casem, T. Weerasooriya and P.Moy, "Inertial Effects of Quartz Force Transducers Embedded in a Split Hopkinson Pressure Bar," *Society for Experimental Mechanics,* vol. 45, no. 4, 2005.

[29] H. Liao, W. Chen and V. Chiarito, "Mechanical Response of a Soft Rubber Compound Compressed at Various Strain Rates," *Mechanics of Time-Dependent Materials,* 2013.

[30] M. A. Kariem, J. H. Beynon and D. Ruan, "Misalignment effect in the split Hopkinson pressure bar technique," *Internatioal Journal of Impact Engineering,* vol. 47, pp. 60-70, 2012.

[31] B. Song and W. W. Chen, "One-Dimensional Dynamic Compressive Behavior of EPDM Rubber," *Journal of Engineering Materials and Technology,* vol. 125, 2003.

[32] C. W.W. and S. Bo, Split Hopkinson (Kolsky) Bar: Design, Testing and Aplications, 233 Spring Street, New York: © Springer Science+Business Media,, 2011.

[33] S.Sharma, V. Chavan, R. Agrawal, R. K. R.j. Patel and J. Chakravartty, "Split-Hopkinson Pressure Bar: An Experimental Technique for High Strain Rate Tests," 2011.

[34] S. B., W. Chen, Y. Ge and W. T, "Dynamic and Quasi-static compressive response of porcine muscle," *J. Biomech,* vol. 40, pp. 2999-3005, 2007.

TEMPERATURE MEASUREMENTS IN INDUSTRIAL SCALE SPRAY COOLING SYSTEMS

M. Jafari
Mech., Auto. & Materials Engg.
University of Windsor
Windsor, ON, Canada
jafari@uwindsor.ca

R. Balachandar
Mech., Auto. & Materials Engg.
University of Windsor
Windsor, ON, Canada
az3@uwindsor.ca

R.M. Barron
Mech., Auto. & Materials Engg.
University of Windsor
Windsor, ON, Canada
rambala@uwindsor.ca

ABSTRACT

This study provides details of the measurements performed on the spray cooler system of a tomato juice processing plant. The cans containing hot fluid pass through a series of spray nozzles. Cold water from the nozzles impinges on the cans, cooling them in the process. The collected test data includes fluid temperature at different stages of the cooling period using a portable computer chip called an iButton®. The trend of the recorded temperature drop is similar to the theoretical predictions from Newton's Cooling Law (NCL). This suggests that NCL could be used to obtain a reasonable estimate of the juice temperature in the intermediate stages of the cooling process. Considering the simplicity of NCL, it performs reasonably well, and can be used when other more refined tools (such as experiments and/or simulation) are not available and a quick estimate of the juice temperature is required.

KEYWORDS: Spray cooling, spin cooler system, temperature measurements, iButton®

INTRODUCTION

The spin cooler system is mainly used as part of the juice manufacturing process to cool down the juice. Once the cans are filled with hot juice, they are sealed and moved through a series of water spray nozzles for cooling. The spin cooler consists of this series of stationary nozzles and a flat conveyor system. These nozzles spray water droplets onto the hot cans as they pass along the conveyor. A significant amount of water and energy is usually consumed in the cooling process. A typical nozzle used in one of the lines in the cooling system is depicted in Fig. 1 and its associated parameters are listed in Table 1. A single nozzle spray system was set up in the lab, as shown in Fig. 2, to become familiar with the spray characteristics before designing the experiments to be carried out in the industrial facility.

After the filling and seaming process, the hot tomato juice cans (with juice temperature ≈ 92 °C) are carried in an upright position along an outdoor conveyor for 4 - 5 minutes, exposed to the outside environment. During this period, due to the air convection, the can temperature drops by approximately 3 to 4 °C. At the end of this upright conveyor, the cans fall on their side into a water bath for a short distance, emerging onto the conveyor that moves under the spray nozzles.

Figure 1: Nozzle used in spray cooler (Spraying Systems Co®)

Table 1: HH-SQ 3/4" NPT nozzle parameters

Nozzle orifice diameter (mm)	Spray angle (°)	Approximate pressure (psi)	Empirical value for droplet diameter (μm)
6.756	68	30	0.452

Figure 2: Image of the spray produced at the lab

There are between 57 to 76 nozzles, approximately uniformly spaced, mounted along the horizontal water supply pipe to spray water droplets onto the hot surface of the cans. The cylindrical tomato juice cans used in the experiments varied in size, with 0.54 L, 1.36 L and 2.84 L cans referred to as small, medium and large, respectively. The height from nozzle to impact surface also varied, depending on the production line and can size. The conveyor belt creates an axial rotation of the can in the (nearly) horizontal plane of the conveyor, inducing mixing inside the can by agitation, and thus increases the rate of heat transfer from the centre (i.e., slowest cooling point) of the can. The volume of the slowest cooling zone (SCZ) is thus reduced with the help of rotation. In general, it is claimed in most cases that the quality of agitated products is by far superior to those processed by the static method (Coles et al., 2003).

INITIAL DATA COLLECTION

Several different measurements were made at the tomato juice processing plant to collect important data, including tomato juice temperature at different stages and conveyor speeds. In the first set of measurements, the cans were removed from the line and opened. Using a thermometer, juice temperatures were recorded at three locations in each can, near the bottom, middle and top portions. Measurements were taken at four locations along the conveyor, designated as A, B, C and D. The average time for a can to travel the whole length of the conveyor was obtained from three runs at 276 s., 193 s. and 218 s. The time variation between the runs is because the cans, although they continuously rotate on the conveyor, may slip on the belt and not move towards the exit region. This occurs if there is a break in the stream of cans exiting the bath; a can is push towards the exit by cans behind it. The first set of data is summarized in Tables 2 – 6.

As can be seen in Table 2, there are two columns indicated by T_{bb} and T_{ab}, representing the juice temperature before the can enters the bath and after exiting the bath, respectively. The bath water temperature was in the range of 28.9-29.5 °C. The temperature of the juice was measured at the three positions in the can as described above and averaged as shown in Table 2. It is interesting to note that the temperature is significantly higher in the top portion of the can when it enters the bath, but is much more uniform at the end of the bath. This occurs because the can is in an upright position until it enters the bath and any mixing in the juice is the result of weak natural convection currents. As the can enters the bath it falls to its side and rolls towards the bath exit, experiencing strong forced mixing.

Table 2: Juice temperatures for the bath section

	T_{bb}			T_{ab}		
Position in can	Trial 1 (°C)	Trial 2 (°C)	Ave. Value	Trial 1 (°C)	Trial 2 (°C)	Ave. Value
Bottom	77-84	75	80	81-82	88-89	85
Middle	91	91	91	89-91.5	92	90
Top	96	96	96	93	90	91

In Tables 3 to 6, T_A, T_B, T_C and T_D represent the juice temperature as the cans passed sections A, B, C and D, respectively. Four lines of tomato juice cans moved under the cold water spray along the conveyor. Since the nozzles were expected to create an axisymmetric spray pattern, the two inside lines were considered together (inner), as were the two outside lines (outer). Tables 3, 4 and 5 show that the cans are not all cooled at the same rate. In particular, the inner rows tend to cool more slowly than the outer rows up to location A, followed by a reversed trend to the end of the conveyer. It was visually observed that the outer cans were impacted by a larger amount of water and it is postulated that a thin liquid film forms on the surface of these cans, causing a reduction in the heat transfer coefficient as they are subjected to more water spray.

The cans reach the end of the conveyer and move out of the spray zone into a common area at location D, so inner and outer row cans are indistinguishable. The data in Table 6 illustrates that the juice temperature has dropped to about 46 °C, and the temperature is uniform throughout the can. This temperature meets the industry packaging criteria.

Table 3: Juice temperatures at location A

	T_A	
Location in can	Outer row (°C)	Inner row (°C)
Bottom	64.95	70.8
Middle	68.4	75.5
Top	66	75.5
Ave. Value	66.45	73.93

Table 4: Juice temperatures at location B

	T_B	
Location in can	Outer row (°C)	Inner row (°C)
Bottom	60.5	56.1
Middle	58.2	59.1
Top	58.2	55.7
Ave. Value	58.96	57

Table 5: Juice temperatures at location C

	T_C	
Location in can	Outer row (°C)	Inner row (°C)
Bottom	48.5	51.1
Middle	50.6	46.5
Top	47.1	42.7
Ave. Value	48.7	46.8

Table 6: Juice temperatures at location D

	T_D		
	Outside the spray zone (°C)		
Location in can	Trial 1	Trial 2	Ave. Value
Bottom	48.7	44.9	46.8
Middle	45.8	45	45.4
Top	46	46.1	46.05
Ave. Value	46.8	45.3	46.1

ANALYTICAL METHOD

A simple model for the cooling of objects is Newton's Cooling Law (NCL), which states that the rate of change of temperature in a cooling body is proportional to the temperature difference between the body and its surroundings. Such behaviour has been observed in many laboratory experiments, leading to a wide acceptance of this approach. The solution of the resulting differential equation provides the temperature

$$T(t) = T_s + (T_0 - T_s)e^{kt} \quad (1)$$

where k is the decay rate, T_s is the temperature of the surroundings and T_0 is the initial juice temperature before entering the spray zone. Assuming k is constant for the whole process, it can be calculated from

$$e^k = \left(\frac{T_1 - T_s}{T_0 - T_s}\right)^{\frac{1}{t_1}} \quad (2)$$

where T_1 is the juice temperature at the exit section (D) and t_1 is total time taken for a can to travel along the full length of the conveyor.

The NCL predictions and a sample comparison to field data are shown in Fig. 3. It is worthwhile noting that the reduction in the measured and predicted juice temperature follows the same trend as the can rolls along the conveyor.

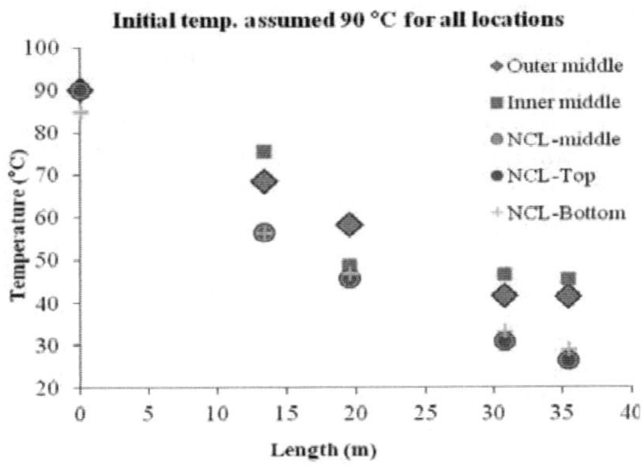

Figure 3: Comparison of measured juice temperature at different locations with Newton's cooling law

Errors in the temperature measurements may occur due to the intrusive presence of the thermometer and possible mixing before an accurate local measurement is recorded. The placement of a thermometer to record the temperature at various positions in the can during cooling also disturbs the flow patterns. Furthermore, it is difficult to measure the temperature at the SCZ because this is a region of the fluid which continually rolls and moves during cooling.

ENHANCED DATA COLLECTION

To eliminate some of the issues mentioned above, the temperature measurement approach was changed to the use of iButtons® (Maxim integrated$_{TM}$). An iButton® is a portable computer chip enclosed in a 16 mm thick stainless steel casing with 512 bytes of general memory plus 64 bytes of calibration memory. iButtons® have a sampling rate from 1 second up to 273 hours with an accuracy around ±0.5°C and are highly resistant to environmental hazards such as dirt, moisture and shock. Measured temperatures of tomato juice with the iButtons® inside the small, medium and large cans versus conveyor length and travel time are shown in Figs. 4-9. Gathering of field measurement data was done on two different days, day 1 and day 2. All the data for small, medium and large cans was compared with NCL using Eqns. (1) and (2). As mentioned earlier, there are four lines of cans moving on the conveyor. In Figs. 4-9, O.L, O.R, I.L and I.R stand for outer left and right, inner left and right, respectively.

Figure 4 shows the temperature variation versus length of the conveyor and Fig. 5 illustrates the temperature variation versus total travel time, for small cans. Close to the end of the line, it is clear that the tomato juice temperature becomes nearly constant. This is mainly due to the lower temperature difference between the impingement water temperature and tomato juice temperature. This suggests that fewer nozzles or less mass flux is needed close to the exit.

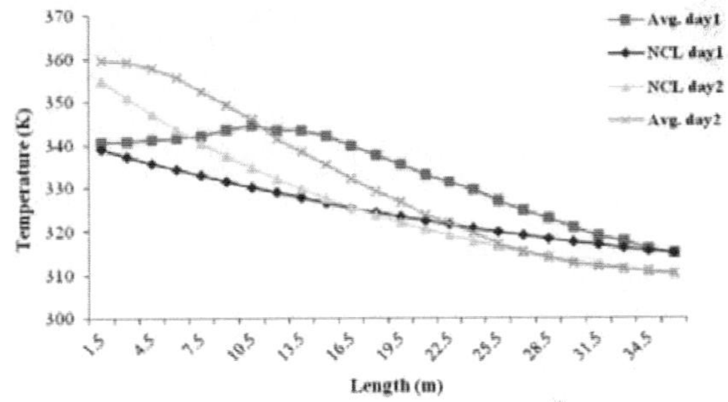

Figure 4: Temperature variation of tomato juice along the conveyor length (can size = 0.54 L)

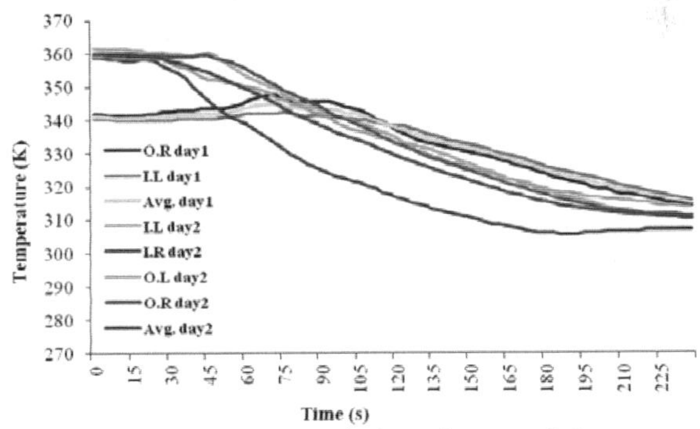

Figure 5: Temperature variation of tomato juice versus traveling time (can size = 0.54 L)

As shown in these figures, the graph approximately follows an exponential curve. It is clear from Figs. 4, 6 and 8 that there is reasonable agreement between the NCL and measured data.

Figure 7 demonstrates that the juice temperature on the second day reached the exit temperature of the can on the first day in half the travel time. Hence, time can be saved and efficiency of the conveyor can be increased, provided the cans are continually pushed towards the exit. Large can data is displayed in Figs. 8 and 9.

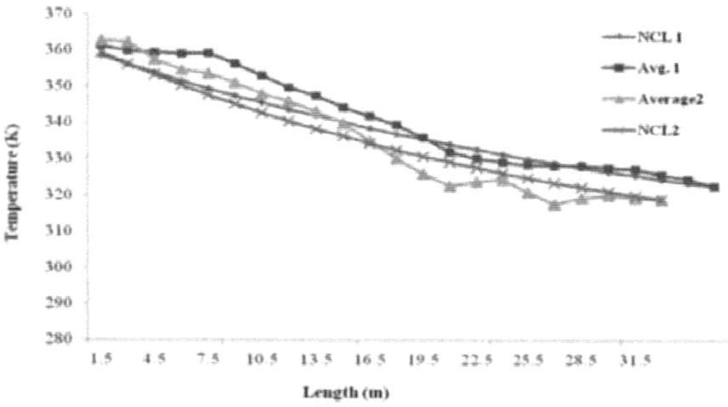

Figure 6: Temperature variation of tomato juice along the conveyor length (can size = 1.36 L)

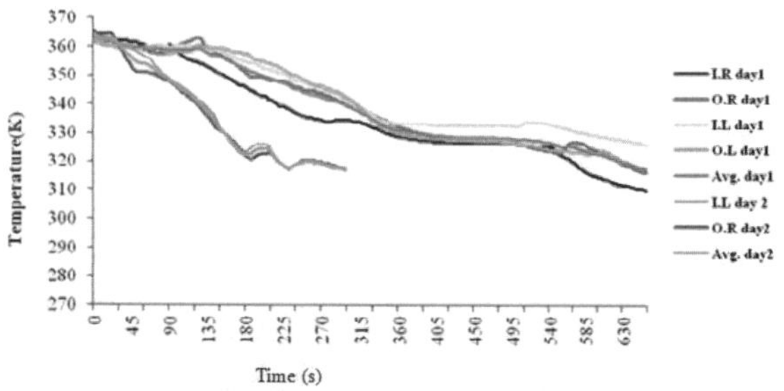

Figure 7: Temperature variation of tomato juice versus traveling time (can size = 1.36 L)

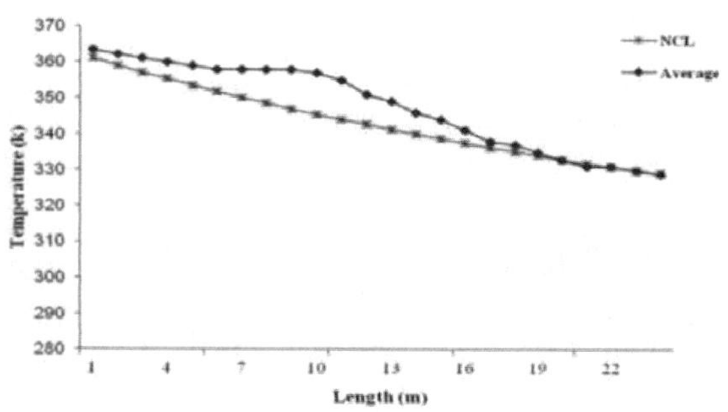

Figure 8: Temperature variation of tomato juice along the conveyor length (can size = 2.84 L)

CONCLUSIONS

There is a good agreement between juice temperatures recorded at a tomato juice processing plant using the iButton® and the theoretical predictions by Newton's Cooling Law. This suggests that NCL could be used as a preliminary tool to predict the temperature in the intermediate stages of the cooling process to obtain a rough estimate of the cool-down behaviour.

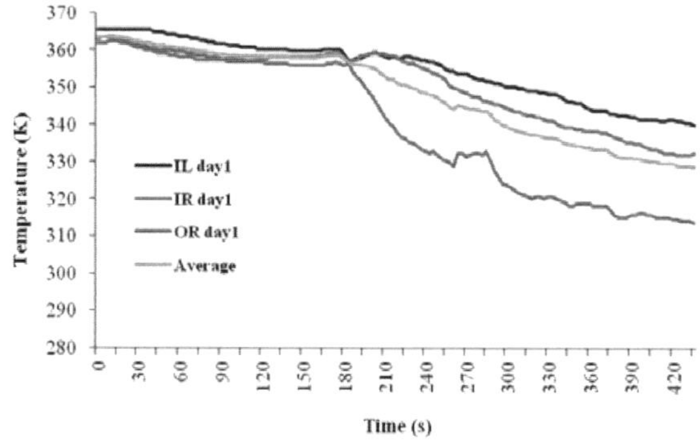

Figure 9: Temperature variation of tomato juice versus traveling time (can size = 2.84 L)

Considering the simplicity of NCL, it performs reasonably well. Based on the graphs provided, it can be concluded that fewer nozzles or less mass flux is needed close to the exit because of the smaller temperature difference between the impingement water temperature and tomato juice temperature. In order to reduce the volume of the SCZ (Slow Cooling Zone), tomato juice cans spin while moving along the conveyor. In addition, by spinning, more uniform tomato juice temperature can be achieved. Due to difficulty in measuring the temperature at the SCZ, prediction of temperature distribution during the cooling process using CFD modeling could be a good step towards understanding the heat transfer between the juice and the spray. The simulation would provide more insight into the temperature gradient and SCZ size inside the can and their dependence on the rotational speed of the cans.

ACKNOWLEDGMENTS

The authors gratefully acknowledge Heinz Canada for granting access to their facility in Leamington, ON and for technical support. This research was supported by NSERC through an ENGAGE Grant.

REFERENCES

1. Spraying Systems Full Jet Product Catalog, www.spray.com
2. Coles, R., McDowell, D., Kirwan, M.J. (eds), “Food Packaging Technology”, CRC Press, 2003.

Proceedings of the 25th CANCAM
London, Ontario, Canada, May 31 – June 4, 2015

Numerical Investigation of the Bearing Capacity of Ring Foundations on Clay

Maha Al Massri
Civil and Environmental Engineering
University of Western Ontario
London, Ontario, Canada
malmass@uwo.ca

Prof. Timothy A. Newson
Civil and Environmental Engineering
University of Western Ontario
London, Ontario, Canada
tnewson2@uwo.ca

ABSTRACT

The bearing capacity of ring foundations on drained soils has been addressed occasionally in the literature. However, little research has been conducted on the bearing capacity of ring foundations for undrained soils. This paper presents the results of a numerical investigation of the performance of ring footings on soft clay soils. The study has concentrated on the effect of the footing diameter ratios on the undrained bearing capacity factor (N_c). The results are compared to some of the available research in the literature. The study has been conducted using finite element analysis with the software PLAXIS. The variation of (N_c) is shown to vary with geometric properties. The research has both practical application for design and academic interest related to the interaction of closely spaced foundations of different types.

KEYWORDS: Ring Footing, Undrained, Numerical, Finite Elements, Foundations, Bearing Capacity.

INTRODUCTION

Bearing capacity theory for shallow foundations was developed in 1943 by Karl Terzaghi "the father of soil mechanics". His famous equation for determining bearing capacity for the general shear failure case is shown in equation *(1)* below.

$$q_{ult} = c\,N_c + q\,N_q + 0.50\,\gamma\,B\,N_\gamma \qquad (1)$$

Where q_{ult} is the ultimate bearing capacity, (γ) is the soil unit weight, (c) is soil cohesion, (B) is the footing width, and $(N_c,\ N_q \text{ and } N_\gamma)$ are bearing capacity factors, which are functions of the material properties and geometry.

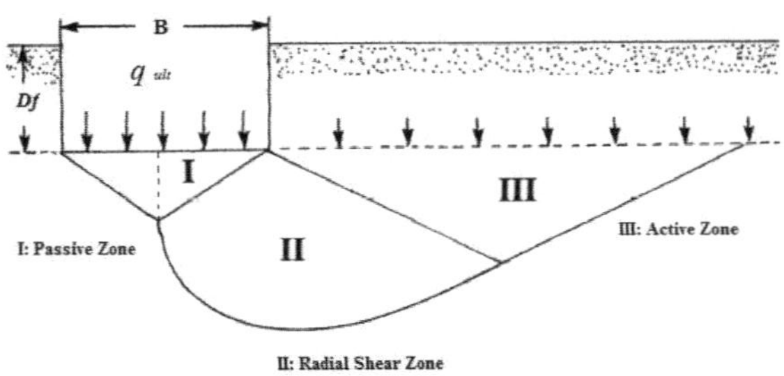

Fig. 1: General shear failure for a continous strip footing on a homogeneous isotropic soil (Terzaghi, 1943)

This bearing capacity equation was created for a rigid continuous (two dimensional, plane strain) rough based footing on a homogeneous isotropic soil as shown in *(Fig.1)*. Much research has been conducted to calculate bearing capacity of different types of foundations, such as circular and rectangular footings for various soil conditions. Only limited work has been done for the bearing capacity of drained ring foundations (eg. Zhu and Fanyu, 1998; Laman and Yildiz, 2003; Boushehrian and Hataf, 2009) and no analysis to date for undrained ring foundations.

For a foundation on the surface of an undrained soft clay $(\varphi'=0°)$ with no embedment $(D_f=0)$ bearing capacity is calculated using equation *(2)*.

$$q_{ult} = c\,N_c \qquad (2)$$

Where (c) is the undrained shear strength and the value of (N_c) for a surface strip footing was found to be $(2+\pi)$ by Prandtl (1920).

The equivalent value for (Nc) for a circular footing (the axisymmetrical case) has been found to range from 5.55 to 5.95 for smooth and rough bases respectively (Gourvenec and Mana, 2011). The interaction of closely spaced strip footings on clay has been investigated by a number of researchers: a) theoretically by Stuart (1962) who studied the interference effect of two strip footings on their ultimate bearing capacity; b) experimentally using small-scale model tests (eg. Das and Larbi-Cherif, 1983; Kumar and Saran 2003); c) by using upper bound limit analysis in combination with finite elements (eg. Kumar and Kouzar 2008). Further research was done by Martin and Hazell in (2005) who investigated the bearing capacity of parallel strip footings on clay by using the method of characteristics. Since this is an analog of the ring footing problem, this has been used to validate some of the solutions obtained during our study.

METHODS

The investigations have been based on two main geometric forms: *i)* modeling two parallel continuous strip footings with the same width (B), and a specified space between the two footings *(S)*; edge-to-edge spacing as shown in (*Fig.2*). *ii)* a ring footing having the same width as the strip footing

(B) and with the same spacing as the strip footing *(S)* as shown in (*Fig. 3*). Both models were investigated for the same soil parameters, considering isotropic homogeneous soil, with no embedment and for the undrained soil conditions *($\varphi'=0°$)* and undrained shear strength *(C_u)*. The two strip footings are assumed to be rigidly connected, such that they move down together without any horizontal displacement or rotation. In both cases, the strip and ring footing have a rough surface between the soil and the edge of the footing. For the strip footing a parametric study was conducted to investigate the effect of spacing ratio *(S/B)* on the bearing capacity of the soft clay soil. The same models were transformed to the axisymmetric case, which is a ring footing and the numerical simulation was repeated keeping all dimensions and soil parameters the same.

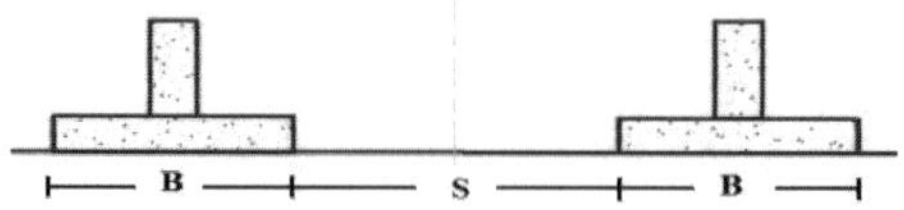

Fig. 2: Two Parallel continuous strip footings on the surface of undrained soft clay.

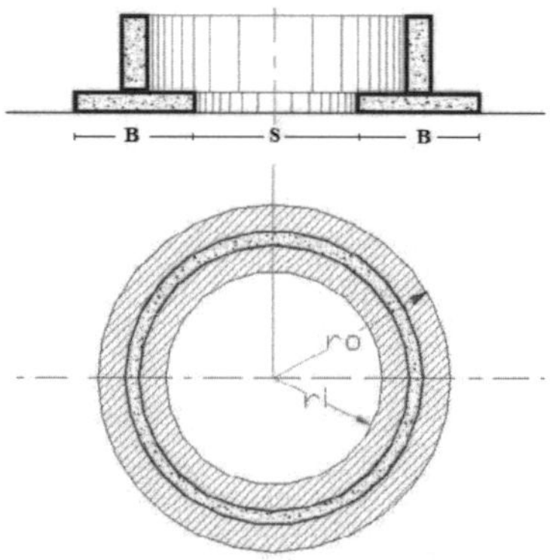

Fig. 3: Ring Footing on the surface of undrained soft clay.

A typical mesh is shown in *(Fig. 4)* with the boundary conditions and fixities. Analyses were repeated with increasing mesh refinement until converged results were found. A typical mesh had 5612, 4th order 15 node triangular elements. The soil was modeled with isotropic elasticity and a perfectly plastic Tresca failure criteria with undrained shear strength of $C_u=20$ *kPa*, soil unit weight $\gamma=\gamma_{sat}=20$ *kN/m*^{3}, modulus of elasticity $E=3000$ *kN/m*^{2} and Poisson's ratio $\upsilon=0.485$. The loading effect of the foundation was modelled with uniform vertical nodal displacements, simulating a rigid foundation of semi-width *(B/2)*.

NUMERICAL RESULTS ANALYSIS

Initially the strip footing results were plotted against the Martin & Hazell (2005) numerical results to compare the findings for the bearing capacity of parallel strip footings on a homogeneous clay for a rough footing as shown in (Fig.5).

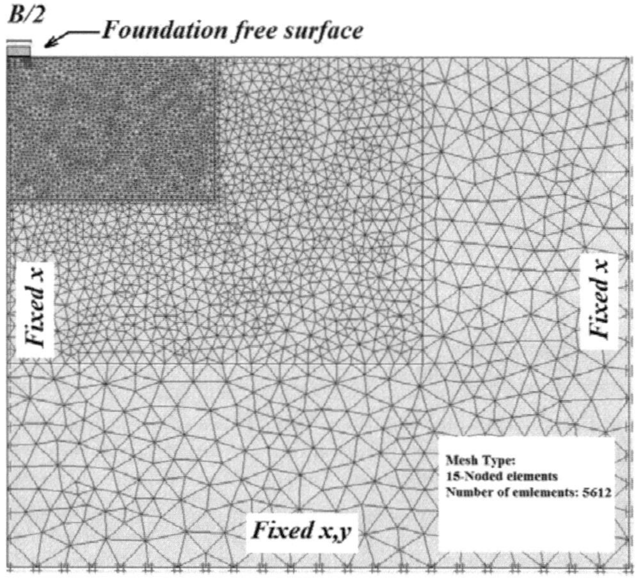

Fig. 4: Typical mesh showing elements, nodes and fixities.

The efficiency term as presented in Martin & Hazell, (2005) refers to the ratio of the overall (group of footings) bearing capacity to the sum of the individual (isolated footings) bearing capacities. The numerical results for the two parallel strip footings show a good agreement between the findings of Martin & Hazell (2005) and PLAXIS results as shown in *(Fig.5)*. The numerical analysis confirms that peak efficiency occurs at (*S/B =0.15)* and does not exceed 1.07 for two parallel strip footings, which agrees with Mandel (1963).

Another investigation was done to explore the effect of the variation of *(S/B)* on the bearing capacity factor *(Nc)*. The results are presented in *(Fig.6)* for strip a footing and *(Fig .7)* for a ring footing. In each case the results have been compared with the benchmark *(Nc)* value of a single footing. The ring footing shows less improvement over the *(Nc)* value for a circular footings. Since there is less consensus regarding the value of *(Nc)* for a circle, (this can be as high as 6.2, Meyerhof, 1963), then this may represent an efficiency of <1.0 for the ring footing. Also there does not seem to be such a clear peak at S/B≈0.15 for the ring footing case.

The failure mechanisms for the two parallel strip footings are shown in *(Fig.8)*. These illustrate the effect of varying *(S/B<0, 0.15 and 0.4)* on the failure surfaces. Increasing spacing of the footing *(S/B)* shows a transition from a Prandtl mechanism with a single wedge beneath the foundation *(Fig.8(a))* to an extended Prandtl mechanism, where the wedge extends beyond the edge of the footing *(Fig.8(b))*. Finally a pseudo-Hill type mechanism is seen, where the symmetry and secondary wedge and radial fans extends beyond the edge of the footing towards the line of symmetry. These mechanisms are a reflection of the reducing interaction between the two foundations and eventually when *(S/B>0.8)* the footings behave independently.

The failure mechanism of a ring footing is presented in *(Fig.9)* for *(S/B=0.4)*. Comparing the mechanism for *(S/B>0.4)* with the strip footing shows a similar shape, but with a relatively larger radial fan and a smaller passive wedge. A size comparison with the foundation geometry *(B)* shows a smaller mechanism (also seen for S/B=0). Since the lateral extent of the mechanism is smaller for the axisymmetrical case, the reduced interaction at lower *(S/B)* values is expected. The bearing capacity must also take into account two effects: greater capacity of a circular footing for a given *(φ)* and reduced *(φ)* of the soil in axisymmetry compared to the plane strain case *(eg. Meyerhof, 1963)*.

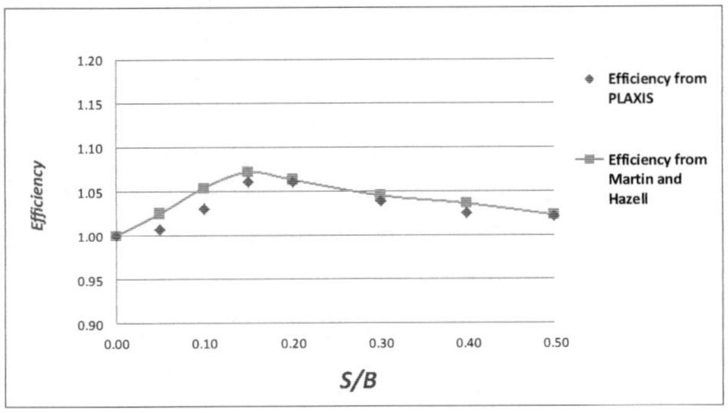

Fig. 5: Results for two parallel strip footing from PLAXIS and the Martin & Hazell (2005) study.

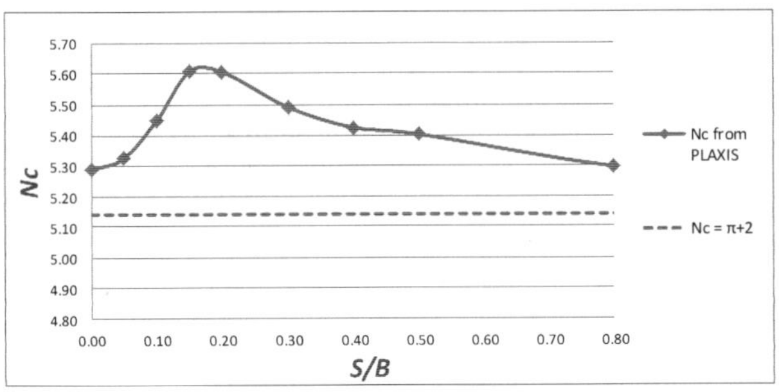

Fig. 6: Results for two parallel strip footings showing the effect of variation *(S/B)* on the bearing capacity factor *(Nc)*.

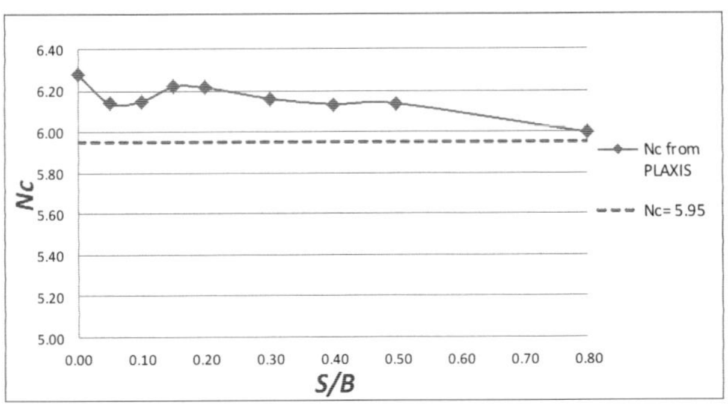

Fig. 7: Results for a ring footing showing the effect of variation *(S/B)* on the bearing capacity factor *(Nc)*.

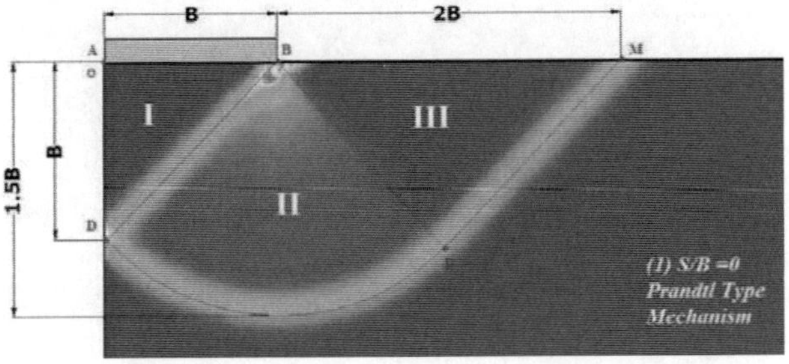

(a): Case (1) *(S/B=0)*: Prandtl Type Mechanism, similar to Terzaghi's mechanism with *(φ'=0°)*, single wedge beneath the footing *(I)*, with radial shear fan *(II)* and another passive wedge adjacent to footing *(III)*.

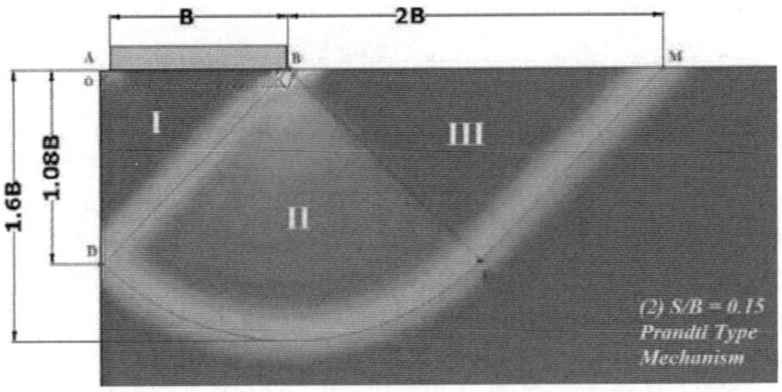

(b): Case (2) *(S/B=0.15)*: Prandtl Type Mechanism, similar to case (1) , but with wedge zone *(I)* extending further into gap between footing, extending range of mechanism and increasing bearing capacity factor.

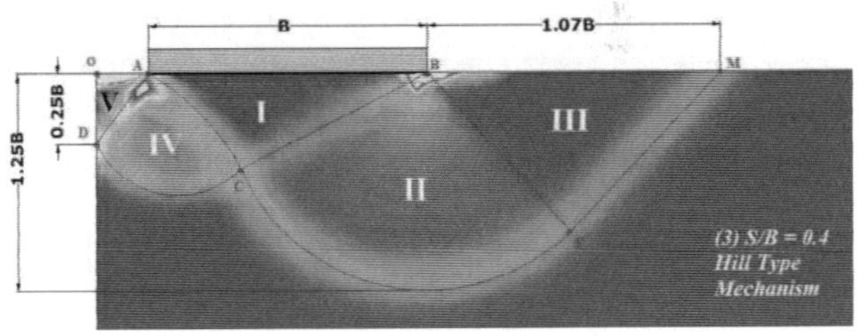

(c): Case (3) *(S/B=0.4)*: The mechanism is similar to a Hill type mechanism with two wedges *(I)* beneath the footings (due to symmetry) with mirror wedges about the line of symmetry. A third wedge is extended upwards between the two footings *(V)* and an additional radial shear fan *(IV)* occurs in addition to *(II)*. Again the larger volume of soil involved increases the bearing capacity.

Fig. 8: Incremental strain contours and assumed failure mechanisms for two parallel strip footings showing the effect of variation *(S/B=0, 0.15 and 0.4)* on the responses.

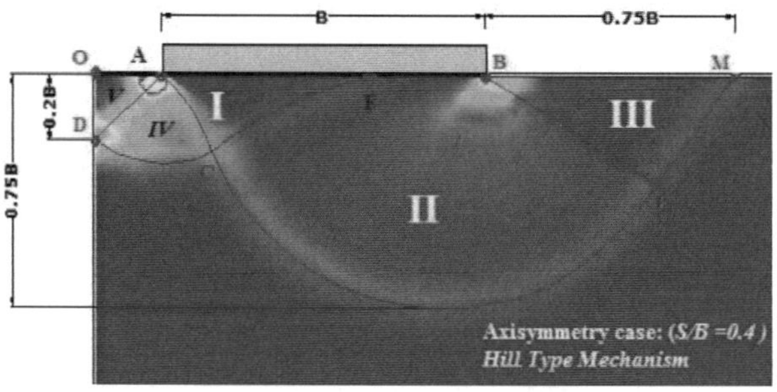

Fig. 9: Incremental strain contours and assumed failure mechanism for the axisymmetric case. Hill type mechanism with smaller wedges *(I, III and V)* beneath and to the sides of the foundation, and relatively larger radial shear fan for zone *(II)*.

CONCLUSIONS

This paper has presented a numerical investigation for the vertical bearing capacity of parallel strip footings and a ring footing on soft clay. Both footings were studied under undrained soil conditions. The results have been obtained using the finite element method and have been confirmed by comparing the results with the work of Martin and Hazell (2005). The generally smaller extent of the axisymmetric soil failure mechanism shows less interaction and lower efficiencies than occur for comparable *(S/B)* ratios for the plane strain footing case. Further work needs to be conducted for soils with varying soil strength with depth to generalize the results.

ACKNOWLEDGMENTS

Recognition goes to Western Engineering for providing funding, materials, and the laboratory facilities that made this investigation possible. The funding of NSERC for this work is also acknowledged.

REFERENCES

1. Hazell, E., Martin, C., 2005, "Bearing capacity of parallel strip footings on non-homogeneous clay", *Proceedings of the International Symposium. on Frontiers in Offshore Geotechnics*, Perth, WA, Australia, pp. 19-21.
2. Mabrouki, A., Benmeddour, D., Frank, R., and Mellas, M. June 2010, "Numerical study of the bearing capacity for two interfering strip footings on sands", *Computers and Geotechnics*, Vol (37)4, pp.431-439
3. Gourvenec, S., Salliyil Kodakkattu Mana, D. 2011, "Undrained vertical bearing capacity factors for shallow foundations", *Geotechnique Letters*, Vol (1)4, pp. 101-108.
4. Meyerhof, G.,. 1963, "Some Recent Research on the Bearing Capacity of Foundations", *8. Canadian Geotechnical Journal*, Vol (1)1 , pp. 16-26.
5. Karaulov, A., 2006, "Experimental and theoretical research on the bearing capacity of ring-foundation beds", *Soil Mechanics and Foundation Engineering*, Vol(43), pp. 37-40.
6. Das B. M., Larbi-Cherif S., 1983, "Bearing capacity of two closely-spaced shallow foundations on sand", *Soils and Foundations*, Vol(23)1, pp. 1-7.
7. Kumar J. and Kouzer, 2008, "Bearing capacity of two interfering footings", *International Journal for Numerical and Analytical Methods in Geomechanics*, Vol(32), pp. 251-264.
8. Stuart J.G., 1962, "Interference between foundations, with special reference to surface footings in sand", *Geotechnique*, Vol(12)1, pp. 15–22.
9. Kumar A. and Saran S., 2003, "Closely spaced footings on geogrid-reinforced sand", *Journal of Geotechnical and Geoenvironmental Engineering*, Vol(129)7, pp. 660–664.
10. Boushehrian, J., Hataf, N. 2003, "Experimental and numerical investigation of the bearing capacity of model circular and ring footings on reinforced sand", *Geotextiles and Geomembranes*, Vol(21)4, pp.241 - 256.
11. Laman, M., Yildiz, A. 2003, "Model studies of ring foundations on geogrid-reinforced sand", *Geosynthetics International*, Vol(10)5, pp.142 - 152.
12. Zhu, Fanyu . 1998, *Centrifuge modelling and numerical analysis of bearing capacity of ring foundations on sand*, Memorial University of Newfoundland (Canada).

Fluid-Structure Interactions

Proceedings of the 25th CANCAM
London, Ontario, Canada, May 31 – June 4, 2015

SLOSHING OF SUPPORTED SHELLS CONVEYING FLUID

Michael P. Païdoussis
Department of Mechanical Engineering
McGill University
Montreal, Quebec, Canada
michael.paidoussis@mcgill.ca

Kostas Karazis
Fuel Engineering Mechanics
AREVA Inc.
Lynchburg, Virginia, USA
kostas.karazis@areva.com

ABSTRACT

This study investigates the nonlinear dynamics of thin tubes with clamped ends subjected to axial internal fluid flow and examines the dynamical behaviour after the onset of instability. Experimental evidence shows that the tube loses stability by a dynamic divergence phenomenon, which is well modelled by the type of sloshing mechanism described in the recent work by Heil & Boyle [3] and Heil & Hazel [2]. A geometrically nonlinear model is developed to investigate the nature of the response of a tube in axial internal flow. The nonlinear shell model is coupled with a linear fluid-flow model and their interaction is analyzed using bifurcation theory. The numerical results show that the tube loses stability by static divergence, corresponding to dynamic divergence in the experiments, exhibiting large hysteresis for the full range of flow velocities used in this study.

KEYWORDS: flow-induced oscillation, dynamic divergence, sloshing, theory, experiments, shells, axial flow

INTRODUCTION

The dynamics of a cylindrical shell conveying incompressible fluid was initially investigated in the study by Païdoussis & Denise [9]. Their results showed that for shells with supported ends, theory predicted loss of stability by static divergence, while in the experiments flutter, rather than divergence, was observed. This paradox was finally resolved in the first decade of 2000 by the work of [7, 8], showing that the observed 'flutter' was in fact a manifestation of *dynamic divergence*. Recently, Jensen & Heil [6] presented for the first time the so-called *sloshing mechanism* for self-excited oscillations, which was later studied further by [3, 5, 12-14]. An excellent review is provided in [2]. A simple description of the sloshing mechanism is provided by [2, 5]. The oscillatory wall motion displaces fluid periodically from the collapsible part of the system into the rigid upstream and downstream tubes, as shown in Figure 1(a). If the amplitude of this "sloshing" is larger upstream than downstream, as shown in the second panel of the figure, there will be a net influx of kinetic energy; if this exceeds the energy lost by dissipation, the oscillation will grow. It is shown in [6] that the requisite upstream/downstream flow asymmetry can be achieved by making the downstream rigid section longer than the upstream one. Heil & Waters [4] have investigated whether and how the 2-D mechanism can be adjusted for 3-D oscillations. Some important differences are found but not so important that the mechanism in [6] would need to be adjusted. It was found that it can operate efficiently if (a) the undeformed cross-section of the tube is noncircular, e.g. elliptical, or (b) the initially axisymmetric cylindrical tube has buckled nonaxisymmetrically; these two situations are referred to as *scenario a* and *scenario b* by [2].

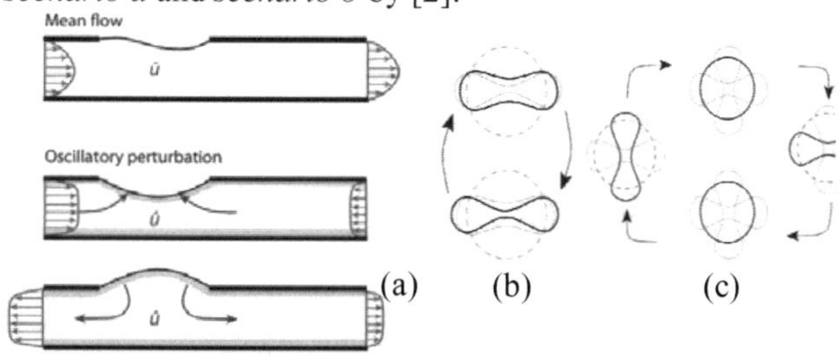

Figure 1: Illustration of the sloshing mechanism. (a) At sufficiently high frequency, the sloshing flows have a blunt inviscid core, with thin Stokes layers near the walls (Heil & Hazel 2011); (b) "Type A" and (c) "Type B" oscillations; [3]; cf. Weaver & Païdoussis [11]

Scenario (a) in [2] is studied by [12]. A tube of elliptical cross-section is considered in a Starling resistor set-up. The work on wall-driven oscillations developed by [14] is extended to deal with self-excited motions, which necessitates that the oscillation profile and frequency be evaluated as part of the solution, rather than being imposed, and use is made of the rational tube law of [13]. In [2], scenario (b) involves an initial nonaxisymmetrically buckled tube and it has been studied by [3]. In contrast to the work related to scenario (a), in this case solutions have been obtained numerically. At onset, the tube is subjected to self-excited small-amplitude "Type A" oscillations, but at higher flows the oscillation grows until the system traverses the statically unstable axisymmetric state, giving rise to "Type B" oscillations. As shown in Figure 1, Type A and Type B oscillations are the same as the 'flapping' and 'flutter' in [11]. A link can also be made to the combined experimental and theoretical studies by [7, 8, 10] on cylindrical shells with supported ends containing flow. Theory predicts loss of stability by a strongly subcritical static divergence. This is confirmed by experiments with relatively stiff tubes. More flexible elastomer tubes, on the other hand, display a loss of stability by large-amplitude *oscillations*, determined to be a manifestation of a dynamic divergence [10] which look just like the "Type B" oscillations depicted in Figure 1(c).

It is conjectured that the onset of large-amplitude Type B oscillations may be predicted by computing the threshold of static divergence in a cylindrical tube by the analytical theory of [7, 8]. In fact, this is the rationale for this paper: to explore the possibility, and degree of success, of predicting self-excited oscillations of cylindrical shells conveying fluid by predicting the onset of static divergence which would lead to dynamic divergence; moreover, to do this utilizing the available analytical model for static divergence of [8]. In the process, the available experimental data from the same study, as well as new ones, are reanalyzed to suit the purposes of the present study.

Specifically, this paper presents experimental and theoretical results on the stability of circular cylindrical shells clamped at both ends subjected to internal fluid flow. The experimental results resemble the Type B oscillations discussed in [3], suggesting loss of stability by dynamic divergence with large deformation amplitudes. The theoretical fluid-structure interaction model employs asymmetric and axisymmetric modes in the modal solution expansion and geometrical structural nonlinearities, predicting loss of stability by divergence. The extracted critical experimental points for L/R=2 shells are compared with the numerical stability solution branches, showing a good agreement between theory and experiments.

NUMERICAL MODEL

The final fluid-structure interaction numerical model utilizes a strong coupling between the structural and fluid models. The structural model consists of a thin circular cylindrical shell of length L, mean radius R, and thickness h such that $h/R \ll 1$. The upstream (x=0) and downstream (x=L) ends of the shell are assumed to be clamped. The origin of the cylindrical coordinate system, (O;x,y,z), is positioned at the centre of the upstream end of the shell, as shown in Figure 2. The shell is assumed to be of homogeneous, isotropic elastic material of Young's modulus E and Poisson's ratio v. Along the middle surface of the shell the displacement components are denoted by u, v and w, in the axial, circumferential and radial direction, respectively. The structural model is based on the Donnell nonlinear equations for shallow shells. The equation of equilibrium in the finite deflection theory is given by

$$D\nabla^4 w + ch\dot{w} + \rho h\ddot{w} = -p + \frac{\partial^2 F}{R\partial x^2} + \left(\frac{\partial^2 F}{\partial y^2}\frac{\partial^2 w}{\partial x^2} - 2\frac{\partial^2 F}{\partial x\partial y}\frac{\partial^2 w}{\partial x\partial y} + \frac{\partial^2 F}{\partial x^2}\frac{\partial^2 w}{\partial y^2}\right),$$

where F is the in-plane Airy stress function satisfying the following compatibility equation:

$$\frac{\nabla^4 F}{Eh} = -\frac{\partial^2 w}{R\partial x^2} - \frac{\partial^2 w}{\partial x^2}\frac{\partial^2 w}{\partial y^2} + \left(\frac{\partial^2 w}{\partial x\partial y}\right)^2, \qquad (1)$$

where $D = Eh^3/12(1-v^2)$ is the flexural stiffness of the shell, c the structural damping coefficient, and p the transmural pressure acting on the surface of the shell. The overdot represents a time derivative. The biharmonic operator is defined as $\nabla^4 = [\partial^2(\ )/\partial x^2 + \partial^2(\ )/\partial y^2]^2$.

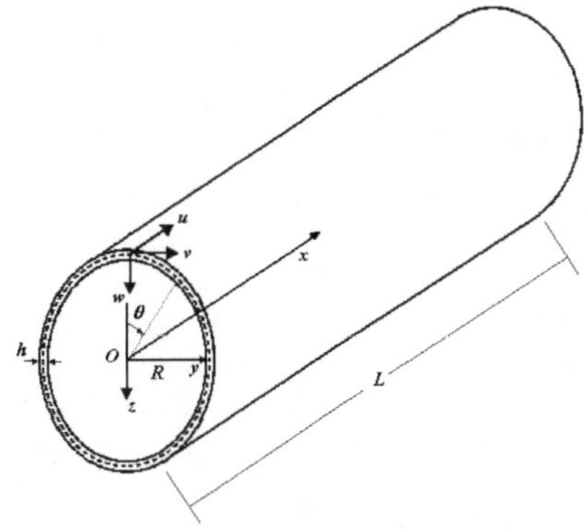

Figure 2: Shell geometry under consideration

The boundary conditions, at the clamped ends are

$$u = v = w = 0 \text{ and } \partial w/\partial x = 0 \text{ at } x=0 \text{ and } x=L. \qquad (3)$$

The assumed radial deformation mode for the nonlinear analysis of the clamped-clamped thin cylindrical shell is

$$w(x,y,t) = \sum_{m_i,i=1}^{M}\sum_{n=1}^{N}\left(A_{m,n}(t)\cos\left(\frac{ny}{R}\right) + B_{m,n}(t)\sin\left(\frac{ny}{R}\right)\right)\varphi_{m_i} + \sum_{m_{2i-1},i=1}^{M} A_{m_{2i-1},0}(t)\,\varphi_{m_{2i-1}} \qquad (4)$$

where φ_m are the eigenfunctions of a clamped-clamped beam, and M and N the upper limits of the summations. The amplitude functions $A_{m,n}(t), B_{m,n}(t)$ (asymmetric modes) and $A_{m_{2i-1},0}(t)$ (axisymmetric modes), are unknown time functions.

In the case of fluid-structure interaction the perturbation pressure p in equation (1) must be defined. The shell is assumed to be completely filled with internally flowing fluid. It is assumed that the fluid is inviscid, isentropic and the flow irrotational. In addition, a linear fluid-dynamic model is assumed to be sufficient to describe the flow structure. Using the irrotationality assumption, $\nabla \times V = 0$, the potential function, for the unsteady flow (unsteady because of shell motions), can be expressed as a combination of the undisturbed axial flow velocity U and the unsteady perturbation potential Φ associated with the motions of the shell. The perturbation velocity potential must satisfy the Laplace equation in cylindrical coordinates:

$$p = -\rho_F(\partial\Phi/\partial t + U\partial\Phi/\partial x), \qquad (5)$$

where ρ_F is the density of the fluid.

The assumed solution for the velocity potential is found using the separation of variables method for all spatial dimensions and time component. The fluid boundary conditions are:

at r=0 the flow is finite,

$$\text{at } r=R \Rightarrow \frac{\partial\Phi}{\partial r} = \left(\frac{\partial w}{\partial t} + U\frac{\partial w}{\partial x}\right), \qquad (6)$$

where the second is the impermeability condition, applied at r=R. For the specific case described here it should be mentioned that the flow velocity perturbation solution is calculated for a fluid in a cylindrical domain that has infinite length within an infinite length shell that is supported periodically. This assumption necessitates that the shell deformation w is defined for any $x \in (-\infty,\infty)$ [8]. Assuming a Fourier series of an orthogonal set of functions for the radial displacement w and applying the fluid boundary conditions given in equation (6) along with the method of separation of variables the final expression for the velocity potential Φ is given by

$$\Phi = \sum_m \sum_n \frac{\gamma_m(r)}{\gamma_m'(r)|_{r=R}} \left[\frac{\partial w_{m,n}}{\partial t} + U \frac{\partial w_{m,n}}{\partial x}\right] \quad (7)$$

where $\gamma_m(r)$ is the fluid flow solution in the radial direction that involves the modified Bessel functions of the first and second kind [8]. The Galerkin method is used to discretize the final form of equation (1) into a set of ordinary differential equations. These equations are integrated numerically using an arc-length continuation scheme embedded in the AUTO software [1]. In the present analysis the modal solution, $w(x,y,t)$, is expanded into an eight-degree-of-freedom (8-dof) model with four asymmetric and four axisymmetric modes.

EXPERIMENT

The studies by [7, 8] investigated the dynamical behaviour of thin cylindrical shells clamped at both ends and subjected to axial flow. In particular, for elastomer shells subjected to internal axial flow the results showed that the structure lost stability by dynamic divergence, while for the internal water-flow experiments using aluminum shells the structure lost stability by static divergence. Here, only a brief summary of the experimental set up used with the silicone based shells in internal air flow, described in [7], is provided. These experiments were performed using silicone-rubber circular cylindrical shells with clamped ends. The material properties and geometry for these shells are: Young's modulus: $2.82\text{x}10^5$ Pa; shell density: $1.16\text{x}10^3$ kg/m^3; Poisson's ratio 0.47; L/R=2; and internal shell radius= 0.023545 m. A detailed description of the manufacturing process for making shells and their material properties may be found in [10].
The experimental apparatus is shown in Figure 3. A number of experiments were conducted with internal air-flow and different shell lengths. Here we present results for L/R=2.

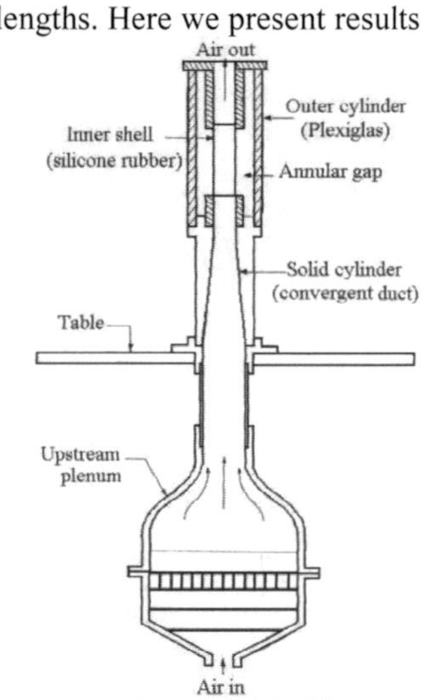

Figure 3: Apparatus for internal air-flow experiments [7]

The compressed air (entering at the bottom of the apparatus) flows through a honeycomb and screens in the lower part of the apparatus and follows the convergent duct section to reach the shell. In all the experiments a static pressure hole (pressure relief valve) ensures that the transmural pressure between the internal channel (volume of air inside the shell) and the annular space provided by the plexiglas cylinder, such that $P_{\text{ann}} = P_{\text{in}}(0)$. To record the response of the shell with increasing flow velocity an analog non-contacting laser sensor system was used. Results for L/R=2, shown in Figure 4, indicate that the shell lost stability by dynamic divergence, with a circumferential wavenumber n=4, once the flow reached the critical value for the onset of instability. As the shell lost stability by divergence with a very large amplitude, the pressure build-up upstream of the shell forced the shell to buckle in an antisymmetric fashion, and this phenomenon was repeated dynamically, so that the observed alternation of buckled shapes produces a "breathing" type oscillation, essentially indistinguishable from flutter similar to the Type B oscillations discussed in [10].

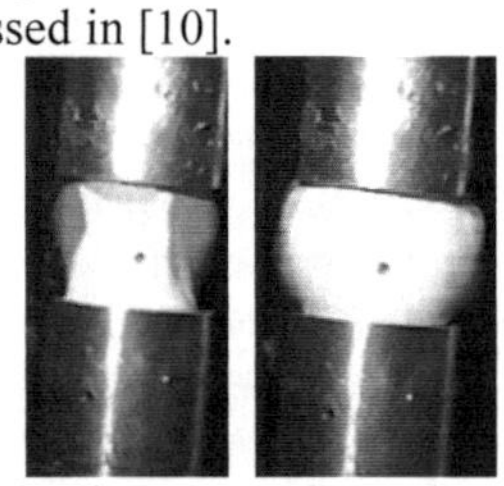

Figure 4: Dynamic divergence for a clamped silicone shell with internal air-flow, L/R=2 and n=4.

Because of the fast nature of the dynamic divergence response the amplitude of the shell during this phenomenon was not measured. However, using the left image from Figure 4 along with image digitization software an estimation of the shell amplitude at that instant is possible. Obviously, the shell amplitude values therein are only estimates and serve to provide a perspective of the large reduction in the uniform/circular cross-sectional flow area of the shell at instability. The substantial reduction in the uniform/circular flow area during the onset of instability is an essential feature of the sloshing mechanism. The digitization software revealed the maximum amlpitudes from both sides of the image: i) maximum right-hand-side shell amplitude: 0.01096 (m); ii) maximum left-hand-side shell amplitude: 0.007305 (m).

COMPARISON OF RESULTS

The original 8-dof theoretical model is reduced to two asymmetric and four axisymmetric modes and has been compared with experimental results for the elastomer shell clamped at both ends, L/R=2, and subjected to internal air-flow. The dimensionless flow velocity is $V = U/[\pi^2 \sqrt{D/\rho h}/L]$. Theoretical results for the internal air-flow/elastomer shell system with a small transmural pressure ΔP_{tm}=40 Pa, without companion-mode participation, indicate shell loss of stability by divergence (no oscillations). Using the shell displacement derived from the digitized image of Figure 4 (left image) and the recorded air flow velocity for the onset of dynamic divergence, a comparison between these experimental results and the theoretical model is attempted in Figure 5. Because the air-flow velocity reported in this figure corresponds to the critical flow at the onset of the instability and the reported tube deformations are for the maximum amplitude, it is expected that the correct air-flow velocity corresponding to this shell amplitude would be slightly higher. Therefore, the extracted experimental point and the numerical results may be considered to be in good agreement.

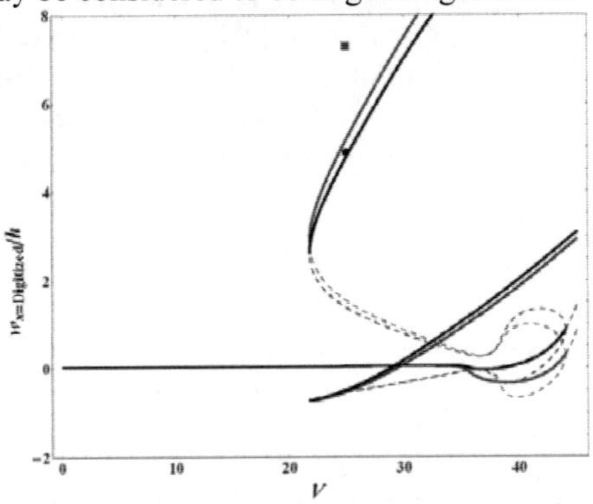

Figure 5: Comparison of extracted experimental and theoretical results for a clamped elastomer shell in axial internal air-flow, for n=4 and ΔP_{tm}=40 Pa. ■ is the experimental point extracted from the right-hand side in Figure 4 (left image) at x=0.02637 m; ● is the experimental point extracted from the left-hand side in Figure 4 (left image) at x=0.02852 m; solid and dotted lines represent the stable and unstable numerical solutions for x=0.02852 m (black lines) and x=0.02637 m. (blue lines).

The numerical model predicts only a static instability indicating that the dynamic divergence observed in the experiments is associated with Type B oscillations related to the upstream pressure build-up and relief due to the large deformation of the shell occurring at instability.

CONCLUSIONS

In this study, a set of experimental results on the stability of elastomer tubes with internal axial flow have been presented. The tube lost stability by dynamic divergence resembling the Type B oscillatory scenario discussed in [10], related to the sloshing phenomenon of [3]. Further evidence of the nonexistence of flutter type of instability for supported tubes in axial flow was supplied by numerical results obtained from a nonlinear fluid-structure interaction model. A comparison between the theoretical results and results extracted from digitized images taken from experiments (during instability) are in very good agreement for the amplitude and flow values related to the tube instability. The numerical model results revalidate the stability analysis for the onset of dynamic divergence based on the experimental observations for elastomer shells in internal flow presented in [7, 8]; they can be used to further support the loss of tube stability by Type B oscillations via the sloshing mechanism described in [3].

ACKNOWLEDGMENTS

The support of the Natural Sciences and Engineering Research Council (NSERC) of Canada is gratefully acknowledged.

REFERENCES

1. Doedel, E.J., Champneys, A.R., Fairgrieve, T.F., Kuznetsov, Y.A., Sandstede, B. and Wand, X. (1998) *AUTO 97: Continuation and Bifurcation Software for Ordinary Differential Equations (with HomCont).* Concordia University, Montreal, Canada.
2. Heil, M. & Hazel, A.L. 2011. Fluid-structure interaction in internal physiological flows. *Annual Review of Fluid Mechanics* **43**, 141-162.
3. Heil, M., Boyle, J., 2010. Self-excited oscillations in three-dimensional collapsible tubes: simulating their onset and large-amplitude oscillations. *Journal of Fluid Mechanics* **652**, 405-426.
4. Heil, M., Waters, S.L., 2006. Transverse flows in rapidly oscillating elastic cylindrical shells. *Journal of Fluid Mechanics* **547**, 185-214.
5. Heil, M., Waters, S.L., 2008. How rapidly oscillating collapsible tubes extract energy from a viscous mean flow. *Journal of Fluid Mechanics* **601**, 199-227.
6. Jensen, O.E., Heil, M., 2003. High-frequency self-excited oscillations in a collapsible-channel flow. *Journal of Fluids Mechanics* **481**, 235-268.
7. Karagiozis, K.N., Païdoussis, M.P., Misra, A.K., Grinevich, E., 2005. An experimental study of the nonlinear dynamics of cylindrical shells with clamped ends subjected to axial flow. *Journal of Fluids and Structures* **20**, 801-816.
8. Karagiozis, K.N., Païdoussis, M.P., Amabili, M., Misra, A.K., 2008. Nonlinear stability of cylindrical shells subjected to axial flow: Theory and experiments. *Journal of Sound and Vibration* **309**, 637-676.
9. Païdoussis, M.P., Denise, J.P., 1972, Flutter of thin cylindrical shells conveying fluid. *Journal of Sound and Vibration* **20**, 9-26.
10. Païdoussis, M.P., 2014. Fluid– Structure Interactions: Slender Structures and Axial Flow, 2nd edition. London, Academic Press
11. Weaver, D.S., Païdoussis, M.P., 1977. On collapse and flutter phenomena in thin tubes conveying fluid. *Journal of Sound and Vibration* **50**, 117-132
12. Whittaker, R.J., Heil. M., Jensen, O.E., Waters, S.L., 2010a. Predicting the onset of high-frequency self-excite oscillations in elastic-walled tubes. *Proceedings of the Royal Society* A **466**, 3635-3657
13. Whittaker, R.J., Heil, M., Jensen, O.E., Waters, S.L., 2010b. A rational derivation of a tube law from shell theory. *Quarterly Journal of Mechanics and Applied Mathematics* **63**, 465-496
14. Whittaker, R.J., Waters, S.L., Jensen, O.E., Boyle, J., Heil, M., 2010c. The energetic of flow through a rapidly oscillating tube. Part 1. General theory. *Journal of Fluid Mechanics* **648**, 83-121

TRANSIENT LOADING ON NUCLEAR STEAM GENERATOR TUBES DURING A TWO-PHASE BLOWDOWN

Ouajih Hamouda
Mechanical Engineering
McMaster University
Hamilton, ON, Canada
hamoudo@mcmaster.ca

David S. Weaver
Mechanical Engineering
McMaster University
Hamilton, ON, Canada
weaverds@mcmaster.ca

Jovica Riznic
Operational Engineering Assessment
Canadian Nuclear Safety Commission
Ottawa, ON, Canada
jovica.riznic@cnsc-ccsn.gc.ca

ABSTRACT

A main concern of nuclear power plant safety is to ensure the containment of radioactive materials produced by nuclear fission. This research investigates a hypothetical scenario involving the sudden rupture of the main steam pipe connecting the steam generators and steam turbines. A rapid 'blowdown' is produced in which the pressurised water in the steam generator is boiled off in a few seconds. Should the resulting transient loading on the steam generator tubes lead to their rupture, radioactive materials may be released beyond the reactor containment. The goal of this research is to improve our understanding of the effects of the postulated accident on the transient loading of nuclear steam generator tubes. The analysis deals with complex rapid transient two-phase flow dynamics and fluid-structural loading processes. A better understanding will enable the development of improved design tools to ensure steam generator safety. This paper presents a description of the experimental facility and experimental results that will ultimately provide guidance for the development of predictive modelling tools.

KEYWORDS: Steam generator tubes, rapid blowdown, transient two-phase flow, fluid-structure interaction

INTRODUCTION

A sudden rupture of the main steam line between a nuclear steam generator and the turbo-generators producing electricity results in a 'blowdown'. As a result, the pressurised water in the steam generator is rapidly discharged as a flashing steam-water mixture. This postulated nuclear power plant accident, referred to as a 'Main Steam Line Break (MSLB)', is the subject of this experimental research. In particular, it is desired to investigate the transient loading of the steam generator tubes during this simulated event.

The steam generator tubes represent the boundary between the irradiated primary side coolant and the secondary side coolant. Since the primary reactor coolant pressure is higher than the secondary coolant, any leakage from the tubes might possibly lead to the escape of radioactive materials in the primary side coolant from reactor containment. The loss of the structural integrity of this boundary is therefore unacceptable. Degraded tubes are especially vulnerable, for which sufficient design margins against failure must be ensured.

In the event of a MSLB accident, the rapid blowdown would produce a high-velocity steam-water flow across the tubes in the U-bend region. The loading on the tubes is potentially dangerous due to the substantial rise in the mass flow rate. The US NRC (Nuclear Regulatory Commission) reports that existing nuclear codes may not be suited to the simulation of transient two-phase three-dimensional cross-flow induced forces on tube bundles [1]. To the authors' knowledge, this phenomenon has never been thoroughly investigated experimentally, and the transient loading therefore remains difficult to predict with precision. The objective of this research is to address the potential for steam generator tube failure due to the transient loading.

EXPERIMENTAL FACILITY

An experimental model facility was designed and built to identify the important thermal-hydraulic phenomena during the simulated blowdown loading of steam generator tubes. Dynamic pressure and temperature measurements were provided, as well as two-phase high-speed video recordings. Direct dynamic load cell measurements of the transient drag loads were also acquired. An illustration of the purpose-built experimental facility used for the steam generator blowdown simulations is shown in Fig. 1.

The design basically consists of 3 main components: a static pressurised fluid reservoir with a flow area of about 186cm^2, a sectional model of a steam generator tube bundle, and a large downstream vacuum reservoir with a 60:1 expansion ratio, ensuring that the pressure drop across the tube bundle is not controlled by the rising downstream pressure. A rupture disc, which opens completely in a few milliseconds, separates the pressurised and vacuum reservoirs. Visual recording of the transient flow using high-speed imaging is possible through sight glasses located above and below the tube bundle.

The working fluid is refrigerant R-134a, which boils at approximately standard temperature and pressure and provides for reasonable scaling to full-scale conditions. R-134a has a relatively low vapour pressure, significantly reducing the cost of the experimental facility and enabling consistent fluid temperatures to be established without using any elaborate apparatus. This also allows thin rupture discs to

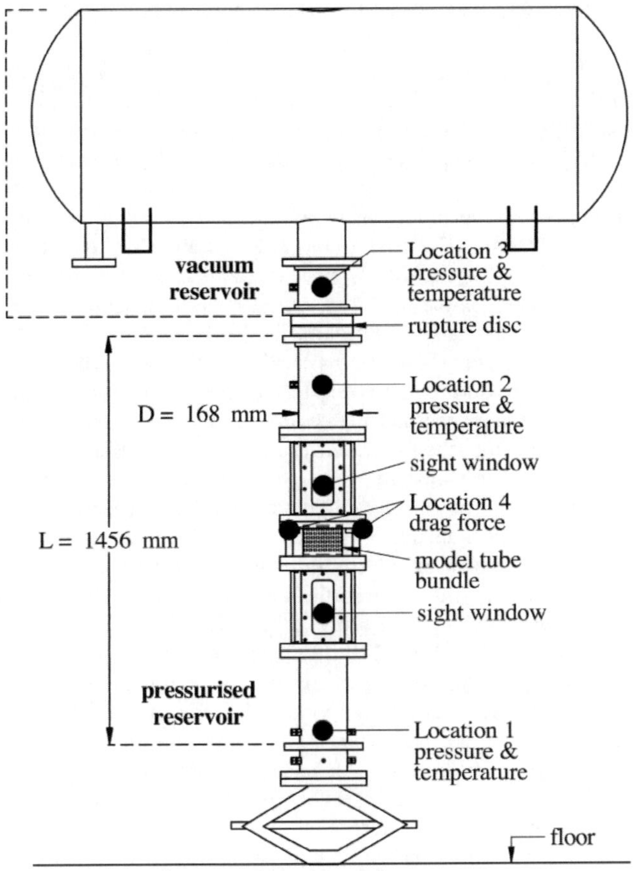

Figure 1: Experimental apparatus

be used, which in this study have always provided a full exit flow area from the pressurised reservoir.

The experiments are initiated by inserting the rupture disc and purging the system to a vacuum. The fluid reservoir is then charged with the desired volume of liquid R-134a. The final pressure required to burst the rupture disc is produced in a controlled manner using a N_2 gas cylinder and accumulator system that avoids R-134a contamination. The transient temperatures and pressures are measured along the vertical pipe axis at 3 locations, and the dynamic load is measured at the tube bundle.

The requirement for precise measurements in a violent explosive environment in which the main phenomena of interest occur in fractions of a second presented many technical challenges. These were overcome following a rigorous instrumentation and data collection system development program. Measurement reliability was validated for all frequency ranges of interest in this study, with associated uncertainties of ±3kPa for pressure and ±0.03kN for load measurements.

GENERAL DESCRIPTION OF THE PHENOMENA

An investigation of the thermal-hydraulic phenomena is essential for the proper identification of the attendant effects on the transient two-phase loading of a tube bundle in cross-flow. A set of transient pressure measurements at the 3 points along the pipe is provided in Fig. 2. The results were obtained from an experiment that was performed with 51% of the pressurised reservoir initially filled with liquid R-134a. The measurement points below the rupture disc are initially submerged in subcooled liquid (location 1) and saturated vapour (location 2). The liquid is initially at 584kPa and 18.4°C, and the vapour is at 571kPa and 18.8°C. The point above the rupture disc (location 3) is initially in a vacuum.

The phenomena shown in Fig. 2 were typically observed to occur in all of the two-phase blowdown tests performed. The pressure amplitudes and event durations varied from test to test due to changes in the initial conditions and the reservoir and test section geometries. Generally, the transient pressure traces show distinct features that can be split into 3 main segments. These take place in Fig. 2 between 0 – 10ms, 10 – 600ms, and from 600ms onwards, where $t = 0$ represents the instant of rupture disc opening.

When the rupture disc opens, the fluid is suddenly exposed to the downstream reservoir and accelerates upwards rapidly. In the first 10ms, pressure waves propagate at the speed of sound, originating at the point of the rupture disc. The pressure signals indicate a rapid change in pressure upon the arrival of the pressure waves. The pressurised region experiences a reduction in pressure due to the passage of a rarefaction wave, and the vacuum region increases in pressure. The wave propagation timings in the signals are consistent with expected theoretical propagation velocities based on the initial thermodynamic conditions.

Following the initial 10ms, once the pressure at the bottom of the reservoir is low enough, a period of vigorous vapour generation begins. The two-phase fluid mixture is simultaneously discharged from the pressurised reservoir. Vapour formation below the tube bundle (location 1) produces an increase in pressure and fluid discharge through the tube bundle causes a reduction in pressure. Eventually, a net 'quasi-steady' pressure is established, which remains more or less constant at about 400kPa until most of the liquid inventory is depleted. At this point in time, the rate of vapour generation cannot be sustained and the pressure drops towards equilibrium. Friction along the pipe above the tubes (locations 2 & 3) produces a small pressure gradient that diminishes as the fluid velocity decreases. At about 600ms,

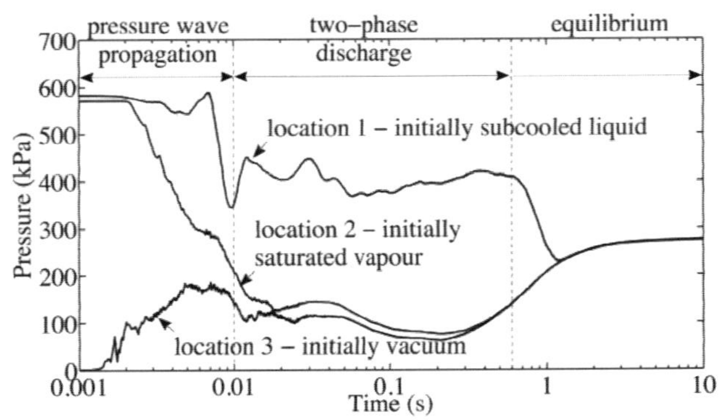

Figure 2: Transient blowdown pressure

the pressures everywhere in the pipe begin to equalise, signalling the end of the transient.

In an operating steam generator, the secondary side pressure would drop rapidly following a Main Steam Line Break from the operating pressure of 4.69MPa towards ambient atmospheric pressure conditions of 0.1MPa. The flashing steam-water mixture is simultaneously discharged through the break area. The pressurised water, initially at 260°C before it transitions to vapour, remains at a higher temperature than the saturation temperature corresponding to the local pressure conditions. Therefore, the liquid would temporarily be in a superheated 'meta-stable' state.

Similar non-equilibrium thermodynamic phenomena were observed in the present experiments. Analysis of the fluid thermodynamics was made possible by measurements from rapid response thermocouples. Generally, the temperatures of the liquid and vapour phases during the transient were observed to be different. In Fig. 3, the saturation pressure corresponding to the measured temperature is compared to the actual pressure measured at the locations above and below the tube bundle.

The liquid below the tube bundle retains its initial temperature during the initial phase of the transient. After a time period of about 110ms, the temperature drops towards the saturation temperature. The physical explanation for this behaviour is that the thermocouple junction is initially in contact with superheated liquid, which is in a non-equilibrium 'meta-stable' thermodynamic state. Once the liquid in the thermocouple vicinity begins to boil off, the temperature drops and approaches the temperature of the vapour being generated by the flashing liquid.

As soon as the two-phase mixture is discharged through the tube bundle, the temperature above the tubes rapidly proceeds towards the saturation temperature. The agreement between the measured and calculated pressures at this measurement location for practically the entire duration of the transient further confirms that the rapid depressurisation process of the liquid below the tubes occurs under a significant departure from thermal phase equilibrium. The physical modelling of such transient two-phase phenomena is very difficult and beyond current numerical modelling capabilities.

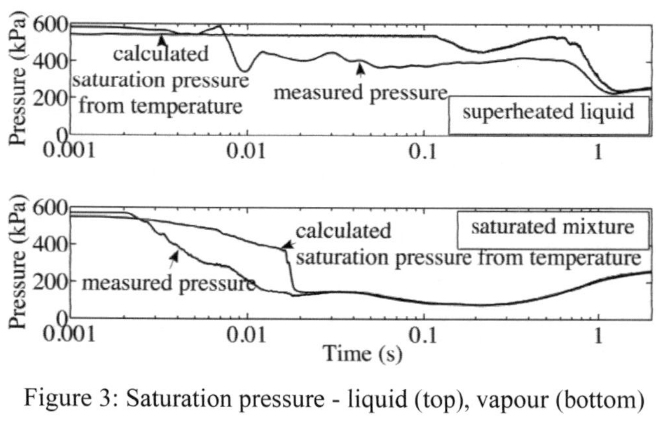

Figure 3: Saturation pressure - liquid (top), vapour (bottom)

VISUAL BLOWDOWN FLUID DYNAMICS

Synchronised high-speed imaging provides a basis for comparing the observed two-phase flashing flow mechanisms with the physical sensor measurements. A high-speed image sequence captured during the experiment presented above is shown in Fig. 4. The images were taken through the lower window below the tube bundle at a frame rate of 3000fps. The first frame at t = 0 shows a quiescent fluid before it is rapidly depressurised. The features present in the images shown in Fig. 4 were typically observed in all of the tests, and can be summarised as follows.

Vapour formation begins around the time when the measured pressure in the liquid has decreased to its minimum value. The dark regions in the frame at t = 0.1s represent vapour. The interface between the liquid and vapour phases reduces light transmission through the windows resulting in the vapour being discerned as darker. The main vapour generation mechanisms observed were the activation and growth of bubbles either at solid surface boundaries or pre-existing in the liquid. Bubble growth proceeds from the outside walls towards the centre of the pipe.

The velocity of the flashing vapour phase initially seems to have a vertical component towards the rupture disc at the top, as well as a radial component towards the undisturbed superheated liquid in the centre. At t = 0.2s, vapour bubbles and two-phase plumes can be seen originating from both the side steel walls of the reservoir and the steel base of the reservoir below the window.

The growth of the vapour phase proceeds very rapidly, with bubble coalescence producing continuous two-phase plumes that accelerate towards the vacuum reservoir. At 0.3s after disc rupture, the images show a dense two-phase concentration suggestive of a high void fraction, which develops early in the transient. The subsequent images continue to display vigorous boiling and a high void fraction until most of the liquid inventory in the reservoir is depleted.

TRANSIENT TUBE LOADING ANALYSIS

There has been much work in investigating the drag of a tube in a bundle subjected to cross-flow [2, 3]. The hydraulic drag is generally formulated as a function of pressure drop and empirical drag coefficients. These depend on the spacing

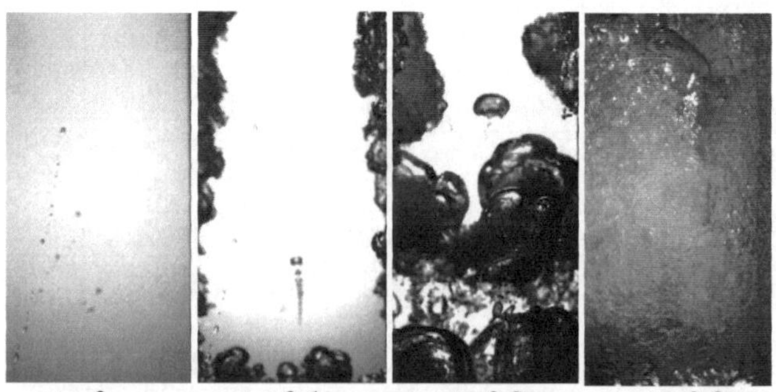

t = 0s **t = 0.1s** **t = 0.2s** **t = 0.3s**

Figure 4: High-speed blowdown flow visualisation

between the tubes, the number of transverse rows, and the fluid properties. By employing an experimentally determined empirical drag coefficient in this study, the fluid drag loading on the tube bundle, which depends on the number of rows, the flow discharge, and the tube bundle geometry, can be related to the transient pressure drop across the tube bundle.

There are several predictive models in the literature dealing with steady-state pressure drop through a tube bundle for a single-phase incompressible fluid. The pressure drop of a two-phase fluid flowing through a tube bundle in cross-flow is caused primarily by the form drag pressure losses. The pressure drop of an external flow past a bluff body, such as a bank of tubes, is related to the flow dynamic head according to Eq. (1)

$$\Delta p = \frac{1}{2}\rho u^2 \xi z \tag{1}$$

where Δp is the pressure drop, ρ is the density, u is the mean velocity in the tube bundle, ξ is a steady-state pressure drop coefficient dependent on the tube bundle geometry and the Reynolds number, and z is the number of rows of tubes.

The accuracy of two-phase tube bundle pressure drop correlations in the literature is limited by the range of mass flow rate, fluid properties, and tube bundle geometries used in the experiments for which the models were developed. The predictive capabilities of presently available models are not fully resolved. The prediction of the two-phase pressure drop across tube bundles therefore continues to be the focus of much experimental and theoretical research [4].

In Fig. 5, the transient drag load measured on the tubes is shown along with the pressure loading measured across the tube bundle, computed by means of an empirical drag coefficient. It is demonstrated that the form drag losses dominate the transient pressure drop beginning at about 30ms until the blowdown transient is complete. Both the pressure and the drag load are observed to attain a maximum of about 8kN between 150 – 450ms. Thus, a relationship between the pressure drop and the tube loading can be developed through an empirical drag coefficient. This is done by assuming flow-averaged densities and velocities, resulting in a constant two-phase mass flow rate through the tubes during this 'quasi-steady' period of the transient discharge.

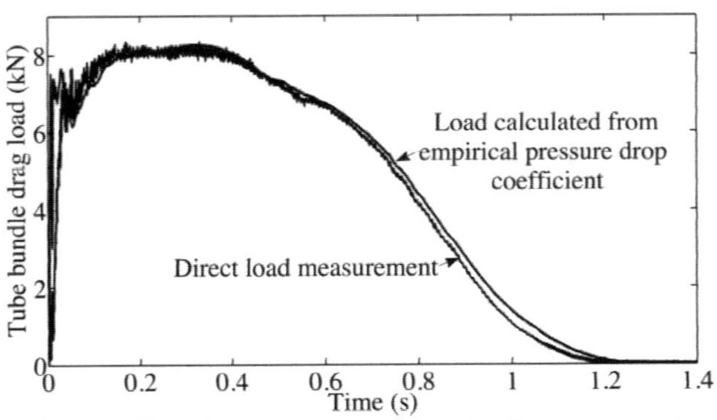

Figure 5: Transient pressure and drag load measurements

The transient distribution of the load shown in Fig. 5 is perfectly consistent with that of the pressure drop measured across the tube bundle. The transient drag force, F_D, is related to the pressure drop, Δp, according to Eq. (2)

$$F_D = C_D A_H \Delta p \tag{2}$$

where C_D is the drag coefficient and A_H is the area of the flow channel. By correlating the drag load and the pressure drop for 6 rows of tubes, the drag coefficient was empirically determined to be 1.41. Hence, the dynamic load exerted on the tube bundle by a rapidly depressurising two-phase flow can be predicted once the pressure drop is known.

CONCLUSIONS

An experimental investigation was performed in which the transient loading on steam generator tube banks was measured during a simulated steam generator rapid depressurisation. Pressure and temperature measurements and high-speed visualisations provided a valuable blowdown database for understanding the nature of the physical phenomena. The results provided insights into the development of a transient flashing two-phase flow through a sectional steam generator tube bundle. The pressure drop measured across the tube bundle was found to correlate very well with the tube bundle drag loading measurements. The tube loads can be estimated once the pressure drop is known.

ACKNOWLEDGEMENTS

The authors gratefully acknowledge the financial support of the Canadian Nuclear Safety Commission (CNSC) and its permission to publish this work.

REFERENCES

1. Reichelt, E., 2009, "Resolution of Generic Safety Issue 188: steam generator tube leaks or ruptures concurrent with containment bypass from Main Steam Line or Feedwater Line breaches", Office of Nuclear Regulatory Research – US NRC, NUREG-1919.
2. Shah, R. K. and Sekulic, D. P. 2003, *Fundamentals of Heat Exchanger Design*, John Wiley and Sons, Hoboken, New Jersey.
3. Zukauskas, A. 1987, "Heat transfer from tubes in cross-flow", *Advances in Heat Transfer*, Vol(18), pp. 87-159.
4. Ribatski, G. and Thome, J. R. 2007, "Two-phase flow and heat transfer across tube bundles – a review", *Heat Transfer Engineering*, Vol(28), Iss(6), pp. 508-524.

NUMERICAL MODELING OF TORNADIC LOAD ON A TALL BUILDING

Zoheb Nasir
Civil and Environmental Engineering/
WindEEE Research Institute
Western University
London, Ontario, Canada
znasir@uwo.ca

Dr. Girma T. Bitsuamlak
Civil and Environmental Engineering/
WindEEE Research Institute
Western University
London, Ontario, Canada
gbitsuam@uwo.ca

INTRODUCTION

Tornado is one of the most devastating natural peril. Each year 800-1000 tornadoes forms on average in North America alone, causing close to 80 fatalities and 1500 injuries, and property damage worth of a billons (on average). In 2014 alone, for instance, 584 were confirmed cases of Tornado in North America causing 51 fatalities. Various research is reported in literature that attempts to researchers and to analyse the origin and the flow structures of tornadoes of different intensities. Due to the higher swirling velocity near the ground and limitation in proper engineering instruments, core analysis of tornadic flow structure has been mostly limited to scaled-down laboratory based and numerical modeling. Some of the previous work focusing on assessing the impact of tornadic flow over engineering structures include: Jischke and Light (1983), Bienkiewicz and Dudhia (1993) and Mishra et al. (2003) measured the surface pressure distribution and flow structure experimentally over scaled down building models in the presence of tornado like vortex and straight line winds and reported that the pressure distribution is quite different among the two wind systems. Wang et al. (2001) and Fouts et al. (2003) reported the rapid change in pressure distribution with respect to the core position. Sarkar et al. (2006) and Sengupta et al. (2006) analysed the force coefficients on a cubic model. Hangan and Kim (2008) simulated tornadoes of different intensity in a laboratorical model. Haan et al. (2010) analyzed the impact of tornadic load and atmospheric boundary layer in a low rise building with the gable roof. Yang et al. (2011) analysed the impact of tornadic load over a tall building for a laboratory based model. Natarajan and Hangan (2012) analysed the impact of translation and surface roughness over a tornadic structure. Some of the numerical modeling includes the following: Selvam and Millet (2003) analysed the impact of tornadic load over a cubic building using large eddy simulation. They obtained that, translating tornadoes have higher force coefficients on the side walls and roof of the building than a stationary wind. More recently Refan et al. (2014) developed a consistent scaling method through the mini WindEEE Dome experiments and field measurements. The same technique was adopted recently by Nasir et al. (2014), to analyse the difference in pressure distribution and flow structure over a short rectangular bluff-body for different locations with respect to the core center. They obtained from their study that, suction pressure is higher for core center location of the bluff-body and it keep decreasing as it moves farther away from the core center. The present study extends the same approach to assess the tornado induced tall building aerodynamics.

NUMERICAL SETUP

A 3D computational modeling was adopted for numerical simulation using Reynolds Averaged Navier-Stokes (RANS) equations and for turbulence model Reynolds Stress Model (RSM) was used. The computational domain (CD) used here was adopted from Purdue tornado simulator which was modified from the Ward's tornado simulator (Church et al., 1979). However, for full scale analysis the numerical model was scaled up about 4000 times. This scaling factor was adopted by matching the location of maximum tangential velocity point of a numerically simulated tornado and an actual tornado (Refan et al., 2014). The CD is cylindrical in shape. The bottom wall part is velocity inlet and the upper wall part is shear free wall (see fig. 1). The top surface is a pressure outlet.

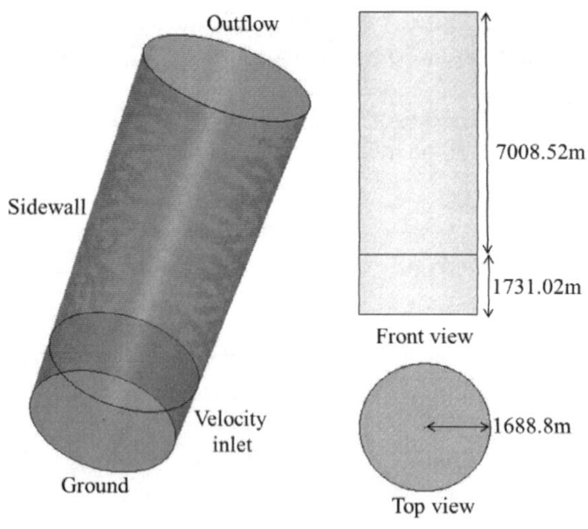

Figure 1: Details of the computational domain.

At the inlet, the flow velocity has two components (radial and tangential) in order to produce the swirling effect. The equation for the radial and tangential velocity components are as follows:

$$V_r = V_1 \times \left(\frac{z}{z_1}\right)^{1/7} \quad (1)$$

$$V_t = \frac{2H_o}{R_o} \times S \times V_r \quad (2)$$

where, V_r and V_t are the radial and tangential components of the velocity at z height, respectively. V_1 and z_1 are the reference velocity and height respectively. Based on the experimental research of baker (1981) on the laboratory scaled model, reference velocity and height are chosen 33.3m/sec and 106m respectively which corresponds to F2 tornado. H_o and R_o are the inlet height (1731.02m) and radius (1688.8m) of the domain, respectively (see fig. 1). S is the swirl ratio. In this analysis, a swirl ratio value of 0.4 was used which is the representation of F2 tornado in actual case (Refan et al., 2014) A standard tall building based on the Commonwealth Advisory Aeronautical Research Council (CAARC) is used (Dagnew and Bitsuamlak, 2014) widely used to validate new wind tunnels and CFD models for synoptic wind loads is used (Dagnew and Bitsuamlak, 2014). The CAARC building is rectangular in shape with dimensions 30.48m (width) by 40.72m (length) by 188.82m (height).

RESULTS AND DISCUSSION

Pressure coefficient distribution

Tornado induced pressure coefficients of a tall building, placed at three different locations with respect to the core center has been studied. These locations are at the core center (location of maximum ground suction), core radii (location of maximum tangential velocity) and outside core (location where suction is close to zero and tangential velocity is quite constant to some lower value) (fig. 2a). Since the building has a rectangular cross-section, the analysis was performed for two different orientation. The first one where the longer dimension (length) is perpendicular to the north direction and in the second case where the longer dimension (length) is parallel to the north direction. Pressure coefficients have been evaluated along the faces of the building. In this context mean pressure coefficient Cp was defined as,

$$C_P = \frac{P - P_0}{0.5\rho U_0^2} \quad (3)$$

where, P is the absolute pressure of the surface, P_0 is the atmospheric pressure, ρ is the density of air and U_0 is the reference velocity, which is maximum tangential velocity at the building height but in absence of the bluff-body.

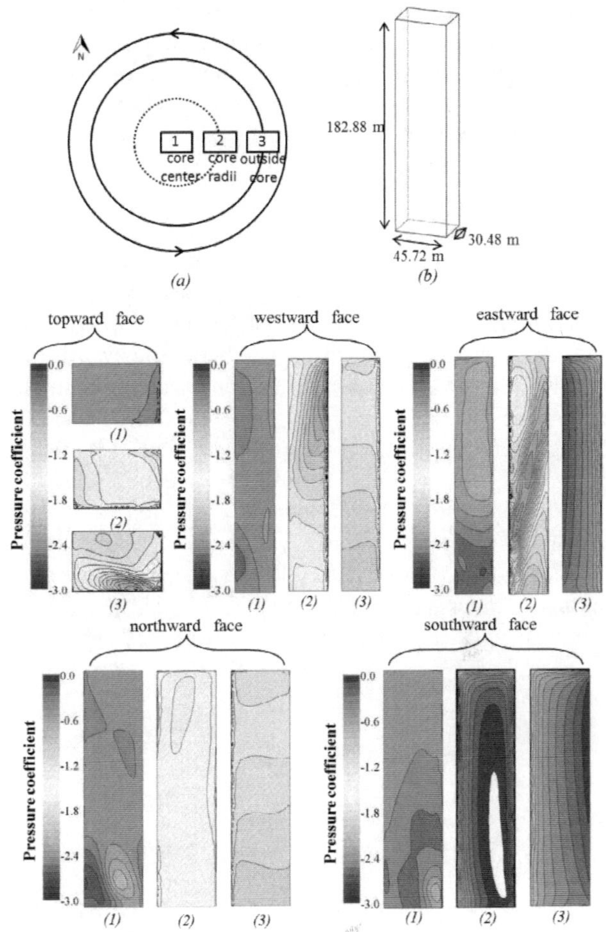

Figure 2: Mean pressure coefficient (Cp) distribution where wider surface faces N-S direction.

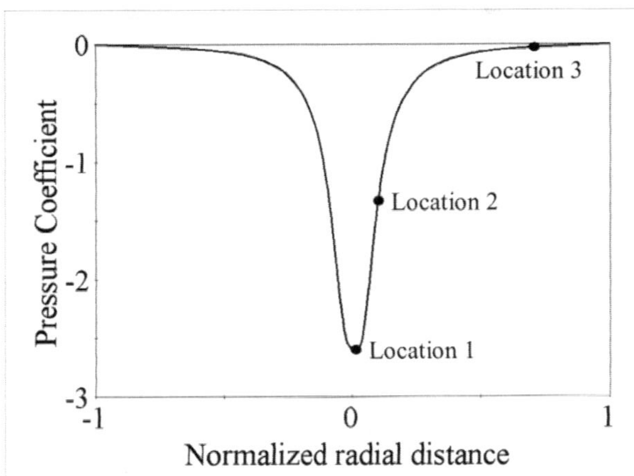

Figure 3: Mean Cp distribution on the floor (ground).

As shown in fig. 2, all the faces of the building are under suction for all three locations of tornado with respect to the building considered in the present study. However, suction is higher for location 1 which is at the core center (see fig. 2). This can be explained by ground pressure distribution of tornadic flow in the absence of building. Due to the high angular momentum of the swirling of the flow near the ground, it creates a high vertical pressure gradient which in turn creates high suction. Suction is higher at the core center and keep decreasing as it moves away from the core center (see fig. 3).

For building location 2, which is at the core radii, suction is minimum on the southward face which is due to the anticlockwise swirling of the present tornadic flow (see fig. 2b) which creates inward flow towards the southward face of the building (see fig. 4). In line with the pattern of ground suction, the suction pressure at the ground becomes less dominating for location 2 compared to location 1 (see figs 2 and 3). Thus the magnitudes of the *Cp* are still primarily dominated by the global suction that developed to high angular momentum of tornado but the local redistribution (contours) of the *Cp* is affected by the interaction with the building. This is more evident on the *Cp* distribution on so south facing wall as well as the east and west walls (fig. 2).

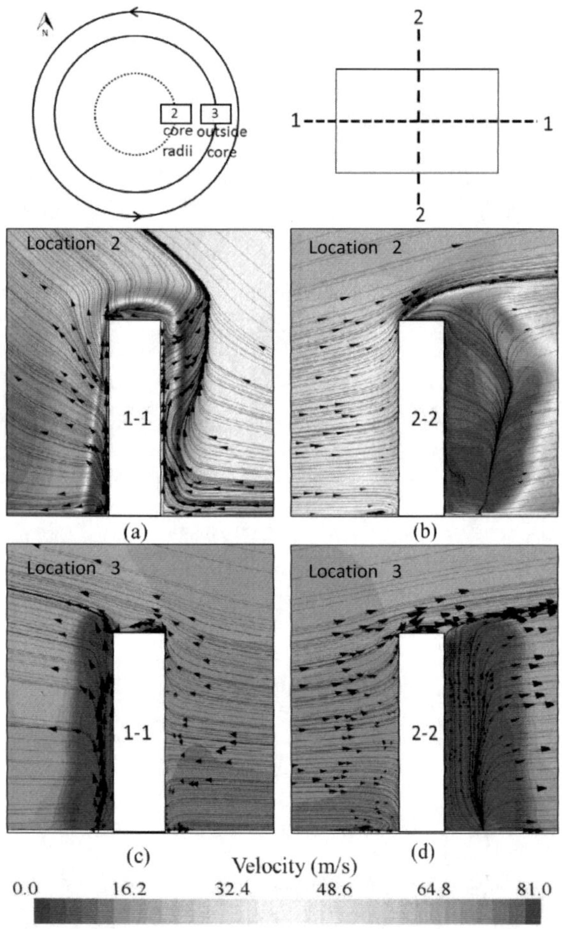

Figure 4: Mean velocity contour and velocity streamlines: (a) and (b) at core radii and (c) and (d) outside core.

A location 3, which is outside core and quite far from the core center. Suction is much less for southward and eastward face in comparison to northward and westward face. This is due to the tangential and radial component of the flow whose direction is toward the southward and eastward face of the building respectively (see fig. 4c and 4d). Also the ground suction is much less for this location compared to the other locations (fig. 3). So the bluff body aerodynamics (interaction between the wind and the building) becomes more dominating compared to the other locations (see fig. 3).

Similarly, for the second orientation where the longer dimension (length) is parallel with the north direction, pressure coefficients were measured for all three locations described earlier (also shown in fig. 5a).

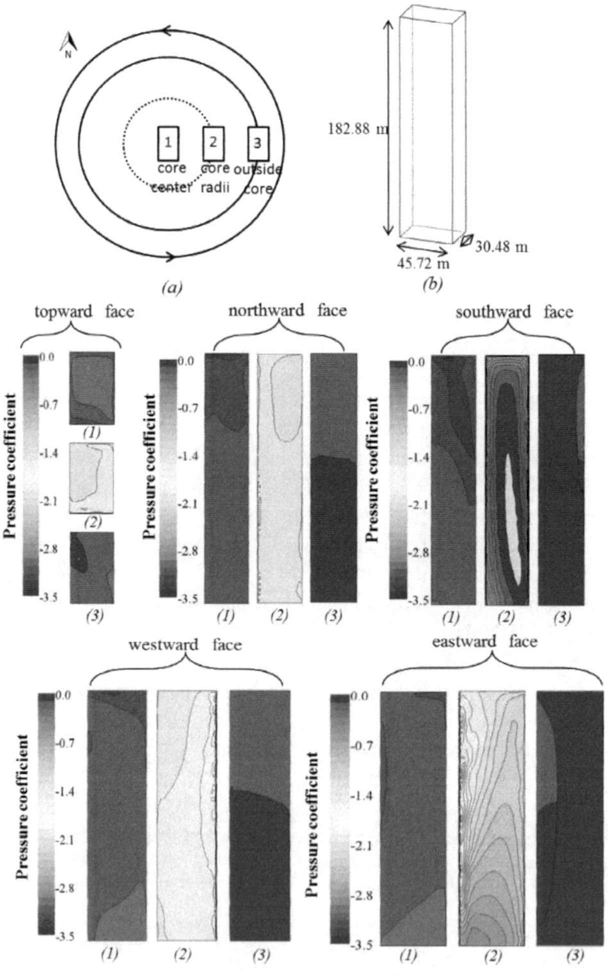

Figure 5: Pressure coefficient (*Cp*) distribution where wider surface faces E-W direction.

Suction is higher at the core center compare to the other locations (see fig. 5) similar to the previous orientation. For location 2, similar to the previous orientation suction is minimum at southward face compare to the other faces. Near the ground suction is less compare to the top section of the building due to the inward radial flow.

For location 3, suction is minimum at most of the portion of the southward and eastward faces. For northward and westward faces suction is higher at the top part of the than the bottom part (see fig. 5).

CONCLUSION

Here are some conclusions that are drawn from this study,

- Irrespective of the orientation of the building, suction is higher for core center location compare to other locations.
- For core radii location, suction was lesser in southward face compare to other faces of the building because of the anticlockwise rotation of the tornadic flow which provides a flow towards the southward face of the building and counteract the suction pressure due to the angular momentum of the flow.
- For outside core location, suction is lesser in southward and eastward face due to the tangential and radial component of the flow towards the faces respectively.
- At the core the angular momentum of the tornado dominates the aerodynamics, at locations far from the core center the wind and buildings affects the pressure distribution although the angular momentum still dictates the magnitude of the pressure coefficients.

REFERENCES

1. Bienkiewicz, B., Dudhia, P., 1993: Physical modeling of tornado like flow and tornado effects on building loading. *Proc. 7th U.S. National Conf. on Wind Engineering,* UCLA, Los Angeles, 95-106.
2. Church, C., Burgess, D., Doswell, C., Davies-Jones, R., 1993: *Tornado: Its structure, dynamics, predictions and hazards*. Washington, D.C., USA, American Geophysical union.
3. Dagnew, A., Bitsuamlak, G.T. 2014: Computational evaluation of wind loads on standard tall building using a large eddy simulation. *Wind & Structures*, 18(5), 567-598.
4. Fouts, L., James, D.L., Letchford, C.W., 2003: Pressure distribution on a cubical modeling tornado like flow. *Proc. 10th Intl Wind Engineering Conf.,* Texas, Tech. Univ., Lubbock, Tex.
5. Haan, F.L., Jr., Sarkar,P.P., Gallus, W.A., 2008: Design, Construction and Performance of a Large Tornado Simulator for Wind Engineering Applications. *Engineering Structures*, v. 30 1146-1159
6. Hangan, H., Kim, J-D., 2008: Swirl ratio effects on tornado vortices in relation to the Fujita scale. *Wind and Structures,* **11-4,** 291-302.
7. Jischke, M.C., Light, B.D., 1983. Laboratory simulation of tornadic wind loads on a rectangular model structure. In: Proceedings of Sixth International Conference on Wind Engineering, vol. 1, Australia & New Zealand.
8. Jischke, M.C., Light, B.D., 1983: Laboratory simulation of tornadic wind loads on a rectangular building. *J. Wind. Eng. Ind. Aerody.,* **13 (1-3),** 274-282.
9. Mishra, A.R., James, D.L., and Lecthcford, C.W., 2003: Comparison of pressure distribution on a cubical model in boundary layer and tornado like flow fields. *Proc. Americas Conference on Wind Engineering* , Baton Rouge, LA, USA.
10. Nasir, Z., Bitsuamlak, G., Hangan, H., 2014: Computational modeling of tornadic load on a bluff-body, 6th International Symposium on Computational Wind Engineering, Hamburg, Germany, June 8-12, 2014
11. Natarajan, D., Hangan, H., 2012: Large eddy simulation of translation and surface roughness effects on tornado like vortices. *J. Wind Eng. Ind. Aerodyn.* **104-106,** 577-584.
12. Refan, M., Hangan, H., Wurman, J., 2013: Reproducing tornadoes in laboratory using proper scaling. *J. Wind Eng. And Industrial Aerodynamics,* **135,** 136-148..
13. Sarkar, P.P., Haan, F.L., Balaramudu, V., Sengupta, A., 2006: Laboratory simulation of tornado and microburst to assess wind loads on buildings. *Proc. ASCE Structures Congress,* ASCE, Reston, Va.
14. Selvam, R.P., Millet, P.C., 2003: Computer modeling of tornado forces on a cubic building using large eddy simulation.*J. Arkansas Academy of Sci.* **57,** 140-146.
15. Sengupta, A., Haan, F.L., Sarkar, P.P., Balaramudu, S.V., 2006: Transient loads on buildings in microburst and tornado winds. *Proc. The fourth International Symposium on Comp. Wind Engr. (CWE2006),* Yokohama, Japan.
16. Wang, H, Letchford, J, D, Snow, R.J., 2001: Development of a prototype tornado simulator for the assessment of fluid-structure interaction. *Proc. 1st Americas Conference in Wind Engineering,* Clemson, SC.
17. Yang, Z., Sarkar, P., Hu, H., 2011: An experimental study of high-rise building in tornado-like winds. *J. Fluids and Structures,* v. 27 471-486.

TRANSITION IN DYNAMICAL BEHAVIOUR OF A FLEXIBLE SLENDER CYLINDER SUBJECTED TO AXIAL FLOW, WITH A VARIABLE SUPPORT AT THE UPSTREAM END

Mojtaba Kheiri
Department of Mechanical Engineering
McGill University
Montreal, Quebec, Canada
mojtaba.kheiri@mail.mcgill.ca

ABSTRACT

In this paper, the equation of motion and boundary conditions for a flexible cylinder supported only by a translational and a rotational spring at the upstream end and subjected to axial flow is presented. The equation of motion is solved numerically via Galerkin's method for a system in which the translational spring is infinitely stiff, thus acting as a pin. The dynamics of such a system with the rotational spring of an average stiffness is described briefly. Moreover, for two pinned-free cylinders of different lengths, additionally supported at the pinned end by a rotational spring, the variation of the lowest two critical flow velocities and the modal shapes of the unstable modes are shown as a function of the rotational stiffness.

Keywords: Pinned-free cylinder, divergence, flutter, yawing

INTRODUCTION

Since the study by Hawthorne [1] in 1961 on the dynamics of the Dracone, a flexible sausage-like container towed behind a small vessel, the dynamics and stability of flexible slender structures subjected to axial flow have been studied extensively. Païdoussis [2] has undertaken a comprehensive review of these studies.

It is now well-known that a flexible cylinder (cantilevered or with both ends supported) in axial flow, at sufficiently high flow velocities, may be subject to divergence (buckling) and to oscillatory instabilities (flutter). A towed flexible cylinder, on the other hand, is generally prone to both rigid-body and flexural instabilities, the former occurring at relatively low towing speeds, while the latter develop at higher towing speeds.

Some new theoretical/experimental developments have taken place in the recent decade in connection with the study of the dynamics of flexible slender structures in axial flow. For example, Modarres-Sadeghi et al. [3] developed a nonlinear model for an extensible flexible cylinder subjected to axial flow and analyzed the dynamics of such a system. Also, in a paper by de Langre et al. [4], it was shown that, contrary to previous predictions made via simplified models (refer to [2]), divergence and flutter may arise for thin, very long cylinders, provided that the free downstream end of the cylinder is well-streamlined.

Recently, a nonlinear model was developed for the dynamics of towed flexible cylinders [5]; also, in a two-part paper [6, 7], Kheiri et al. investigated the dynamics of long pipelines towed underwater with a configuration used in deep ocean storage of liquid CO_2.

In this paper, a theoretical model is presented for the dynamics of a flexible, slender cylinder flexibly supported (i.e. spring-supported) at its upstream end and free downstream. It is, in fact, of fundamental importance to study the transition in dynamical behaviour between a well-supported cylinder (e.g. clamped-free) and one supported in a flimsy way, which represents an intermediate but interesting state, approaching a free-free system. Here, the primary focus is on the study of the dynamics of a pinned-free flexible cylinder, additionally supported at the pinned end by a variable rotational spring.

THEORY

See Figure 1 which shows a flexible cylinder of length L and cross-sectional area A (i.e., $A = \pi D^2/4$, where D is the cylinder diameter). The cylinder is supported at one end by a translational and a rotational spring, the stiffnesses of which are, respectively, k_0 and c_0, and is free at the other. At the free end, the cylinder is fitted with a rigid tapering end-piece of length ℓ ($\ell << L$). The body is immersed in an incompressible fluid of density ρ flowing with uniform velocity U parallel to the x-axis, which coincides with the position of rest of the body. Except for the tapering end-piece, the body is of constant mass per unit length m and flexural rigidity EI. It is fairly reasonable to assume that the axial strain in the centreline of the cylinder is negligible, i.e. the *inextensibility* condition. It is further assumed that the body is neutrally buoyant, thus the forces due to gravity and buoyancy do not come into play, and the motions are considered to take place within the (x, y)-plane, which for the sake of simplicity is assumed to be horizontal.

The equation of motion for the cylinder shown in Figure 1

is obtained using the extended Hamilton's principle. The kinetic energy of the system is comprised of the kinetic energy of the cylinder and that of the end-piece. The potential energy of the system, on the other hand, in absence of gravity forces, includes the strain energy due to cylinder bending and potential energy of the springs. Also, the virtual work is obtained by considering fluid-dynamic forces acting on the cylinder and the tapering end-piece. These forces include F_L and F_N which are the viscous forces per unit length in the longitudinal and normal direction, respectively, and F_A which is the inviscid force per unit length in the transverse direction. For the sake of brevity, expressions for F_L, F_N and F_A are not given in this paper; instead, the reader is referred to [2, Chapter 8].

Over the end-piece, the inviscid hydrodynamic force is fF_A per unit length, where f accounts for departures from slender-body theory over the end-piece (see ref. [2]); the parameter f is normally between 0 and 1: for a perfectly streamlined end-piece, $f \to 1$, whereas for a perfectly blunt end-piece, $f \to 0$. In addition to F_A, F_L and F_N, possible boundary layer separation due to the shape of the end-piece may generate a form drag or pressure drag in the longitudinal direction; this force is represented by F_{TD}.

Thus, the expressions of kinetic and potential energies of the system and virtual work associated with the fluid-dynamic forces are inserted in the extended Hamilton's principle formulation. After considerable manipulation and recourse to integration by parts several times the following equation is obtained:

$$
\begin{aligned}
&\Big\{ \big[m + M + (m + fM)s_e\bar{\delta}(s-L)\big]\ddot{y} \\
&+ \big[2MU + fMUs_e\bar{\delta}(s-L)\big]\dot{y}' \\
&+ \big[\tfrac{1}{2}\rho DUC_N(1+\bar{s}_e\bar{\delta}(s-L)) + \tfrac{1}{2}\rho DC_D(1+\bar{s}_e\bar{\delta}(s-L)) \\
&- fMU\bar{\delta}(s-L)\big]\dot{y} + \big[\tfrac{1}{2}\rho DU^2C_N(1+\bar{s}_e\bar{\delta}(s-L)) \\
&- \tfrac{1}{2}\rho DU^2C_T(L+\bar{s}_e)\bar{\delta}(s) - fMU^2\bar{\delta}(s-L) \\
&- \tfrac{1}{2}\rho D^2U^2C_{TD}\bar{\delta}(s)\big]y' + k_0\bar{\delta}(s)y + EIy'''' \\
&+ \big[MU^2 - \tfrac{1}{2}\rho DU^2C_T(L+\bar{s}_e - s) - \tfrac{1}{2}\rho D^2U^2C_{TD}\big]y''\Big\}\delta y \\
&+ c_0\bar{\delta}(s)y'\delta y' = 0, \qquad (1)
\end{aligned}
$$

where $M = \rho A$; s_e and $\bar{s}_e$ are quantities which depend on the geometry of the end-piece; C_T and C_N are friction coefficients in the tangential (longitudinal) and normal direction, respectively; C_D is the (linearized) normal form-drag coefficient, and C_{TD} is the form-drag coefficient for the tapering end-piece; for more details regarding these parameters, see refs. [2, 5]. Also, $\bar{\delta}(s)$ is the Dirac delta function; s is the curvilinear coordinate along the centreline of the cylinder; δy is the virtual displacement in the y-direction; the *overdot* denotes a time derivative, and $(\,)' = \partial/\partial s$.

Since the effects of the end-springs and the end-piece have been taken into account in the equation of motion (i.e. the terms having $\bar{\delta}(s)$ or $\bar{\delta}(s-L)$ in equation (1)), the boundary conditions

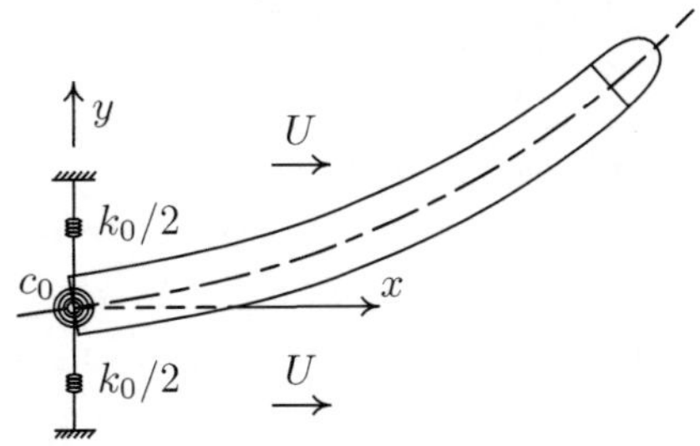

Figure 1: A flexible cylinder subjected to axial flow and supported only at the upstream end by a translational and a rotational spring the stiffnesses of which are represented, respectively, by k_0 and c_0.

take the following simple form:

$$EIy''' = EIy'' = 0, \qquad \text{at } s = 0 \text{ and } s = L, \qquad (2)$$

which are, in fact, the boundary conditions of a free-free Euler-Bernoulli beam.

ANALYSIS

The equation of motion, after being rendered dimensionless, is transformed into a set of second-order ODEs via the Galerkin method. In Galerkin's expansion, the dimensionless transverse displacement η (i.e. $\eta = y/L$) is written as $\eta(\xi,\tau) = \sum_{j=1}^{N}\phi_j(\xi)q_j(\tau)$, where $\phi_j(\xi)$ represent the free-free Euler-Bernoulli beam eigenfunctions, and $q_j(\tau)$ are the corresponding generalized coordinates; also, N denotes the number of eigenfunctions (modes) used in the expansion, and ξ and τ are the dimensionless distance and time, respectively.

Thus, the equation of motion, after applying the Galerkin method, takes the following form:

$$\mathbf{M}\ddot{\mathbf{q}} + \mathbf{C}\dot{\mathbf{q}} + \mathbf{K}\mathbf{q} = \mathbf{0}, \qquad (3)$$

where $\mathbf{q}$ is the vector of the generalized coordinates; $\mathbf{M}$, $\mathbf{C}$ and $\mathbf{K}$ correspond to the mass, damping and stiffness matrices, respectively.

By letting $\dot{\mathbf{q}} = \mathbf{u}$, equation (3) is reduced to first-order or state-space form which is then transformed from the time domain to the frequency domain by assuming a harmonic solution (i.e., $\{\mathbf{u}\ \mathbf{q}\}^{\mathrm{T}} = \{\bar{\mathbf{u}}\ \bar{\mathbf{q}}\}^{\mathrm{T}}\mathrm{e}^{\mathrm{i}\omega\tau}$). Generally complex values of ω ($\omega \equiv \mathcal{Re}(\omega) + \mathrm{i}\,\mathcal{Im}(\omega)$) are then obtained via an eigenvalue-problem solver, and their evolution with a system parameter, e.g. the dimensionless flow velocity u, is plotted in an Argand diagram. In the Argand diagram, the abscissa and the ordinate correspond to the real and imaginary parts of ω, ($\mathcal{Re}(\omega)$ and $\mathcal{Im}(\omega)$), respectively. It is noted that $\mathcal{Re}(\omega)$ corresponds to the dimensionless oscillation frequency, while $\mathcal{Im}(\omega)$ is related to damping, i.e. the damping ratio is $\zeta = \mathcal{Im}(\omega)/\mathcal{Re}(\omega)$. Thus, $\mathcal{Im}(\omega) > 0$ indicates stability, while $\mathcal{Im}(\omega) < 0$ means instability; more specifically, instability materializes by the crossing of ω locus from the positive to the negative half-plane in the Argand diagram.

It is fairly obvious that the translational spring acts as a pin if $\kappa_0 \to \infty$, where $\kappa_0 = k_0L^3/EI$ is the dimensionless translational stiffness; in fact, letting, for example, $\kappa_0 = 10^{10}$ in

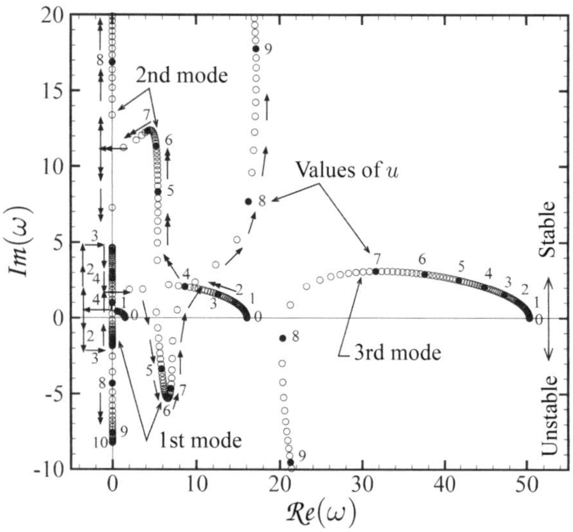

Figure 2: Argand diagram of the complex frequencies, ω, of the lowest three modes of a typical pinned-free cylinder with a rotational spring of stiffness $\kappa_0^* = 1$ at the pinned end and with a tapered free end ($f = 0.7$) in axial flow, as a function of u. Arrows show the evolution of the first mode and two headed arrows the evolution of the second mode.

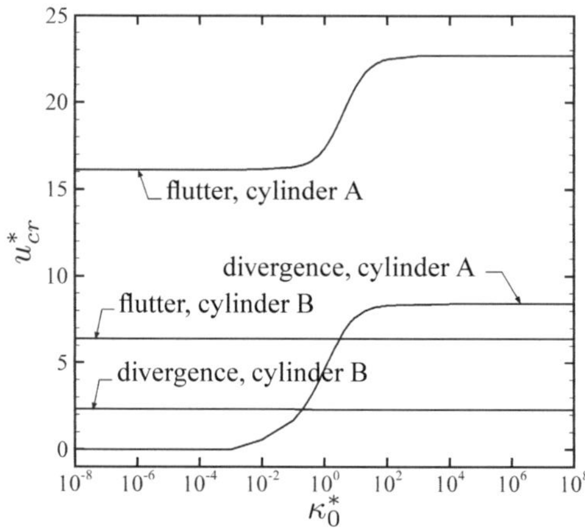

Figure 3: Variation of the lowest two critical flow velocities for static and dynamic instabilities of cylinders A and B as a function of κ_0^*.

the numerical solution may reasonably simulate a pinned end. As mentioned previously, the focus in this paper is on the study of the dynamics of a pinned-free flexible cylinder, additionally supported at the pinned (upstream) end by a rotational spring of stiffness κ_0^* ($\kappa_0^* = c_0 L/EI$ is the dimensionless rotational stiffness). It is interesting to see how the dynamical behaviour of the system changes as κ_0^* is varied.

The dynamics of a typical pinned-free cylinder, additionally supported at the pinned end by a rotational spring of stiffness $\kappa_0^* = 1$ and fitted at the free end with a fairly well streamlined end-piece ($f = 0.7$) is illustrated in Figure 2. More specifically, this figure shows the evolution of the lowest three complex frequencies of the system as the dimensionless flow velocity u (i.e. $u = (M/EI)^{1/2}UL$) is varied. As seen from the figure, all modes are stable at low flow velocities. For sufficiently high flow velocities, the frequency of the first and second modes successively become purely imaginary, bifurcating on the $Im(\omega)$-axis. The arrows just off the $Im(\omega)$-axis identify the points belonging to the loci of these modes evolving on the $Im(\omega)$-axis. The frequency associated with one branch of each of these two modes eventually vanishes altogether, indicating the onset of static divergence (buckling).

The first divergence occurs at $u \simeq 1.2$, in the first mode, followed at $u \simeq 7.7$ by divergence in the second mode. But before the second-mode divergence occurs, the negative branch of the first mode becomes positive at $3.9 < u < 4.0$, and at a slightly higher u, this branch and the other positive branch of the first-mode locus coalesce and leave the $Im(\omega)$-axis; shortly after, at $u \simeq 4.4$, the first-mode locus crosses from the positive to the negative half-plane, where $\mathcal{R}e(\omega) \neq 0$, indicating the onset of first-mode flutter. Before this flutter occurs, the system experiences a short interval of stability (i.e. $3.9 < u < 4.4$). Finally, third-mode flutter occurs at $u \simeq 8.0$.

Figure 3 shows the variation of the lowest two critical flow velocities for static and dynamic instabilities of two pinned-free cylinders (i.e. cylinders A and B), additionally supported at the pinned end by a rotational spring of stiffness κ_0^*, as a function of κ_0^*. The two cylinders have the same typical (dimensionless) parameters, except that cylinder B is more slender than cylinder A, i.e. $\varepsilon_B = 10\varepsilon_A$, where $\varepsilon = L/D$ is the slenderness ratio. The cylinders are fitted with tapering ends with $f = 0.7$. It is noted that the values of critical flow velocities in Figure 3 have been normalized such that they do not depend on L and D.

As seen from the figure, cylinder A becomes more stable as κ_0^* takes on large values, whereas it is less stable as κ_0^* takes on very small values. This is indeed very much expected since increasing κ_0^* means a stronger upstream support, which delays instabilities, and vice versa. In contrast, u_{cr}^* for cylinder B is independent of κ_0^*. In other words, increasing the stiffness of the rotational spring does not improve stability of cylinder B. It is also seen that for very low values of κ_0^*, cylinder A loses stability earlier than cylinder B, whereas for large values of κ_0^*, cylinder B becomes unstable first.

As also shown in Figure 3, for cylinder A, as κ_0^* takes on very small values (i.e. the system approaches a pinned-free cylinder), u_{cr}^* for the first instability approaches zero. This, in fact, corresponds to a rigid-body instability, called *stationary yawing*, which according to linear theory, manifests itself as a divergent non-oscillatory deviation from the position of rest, with or without flexing (for details refer to [2, Chapter 8]). This instability has been referred to as 'divergence' in Figure 3 for the sake of convenience.

Figure 4 shows the modal shapes of the unstable modes of cylinders A and B for different values of κ_0^* (i.e. $\kappa_0^* = 10^{-8}$, 1, 10 and 10^8). Subfigures (a)-(d) show the stationary yawing/divergence mode shape of cylinder A, while sub-

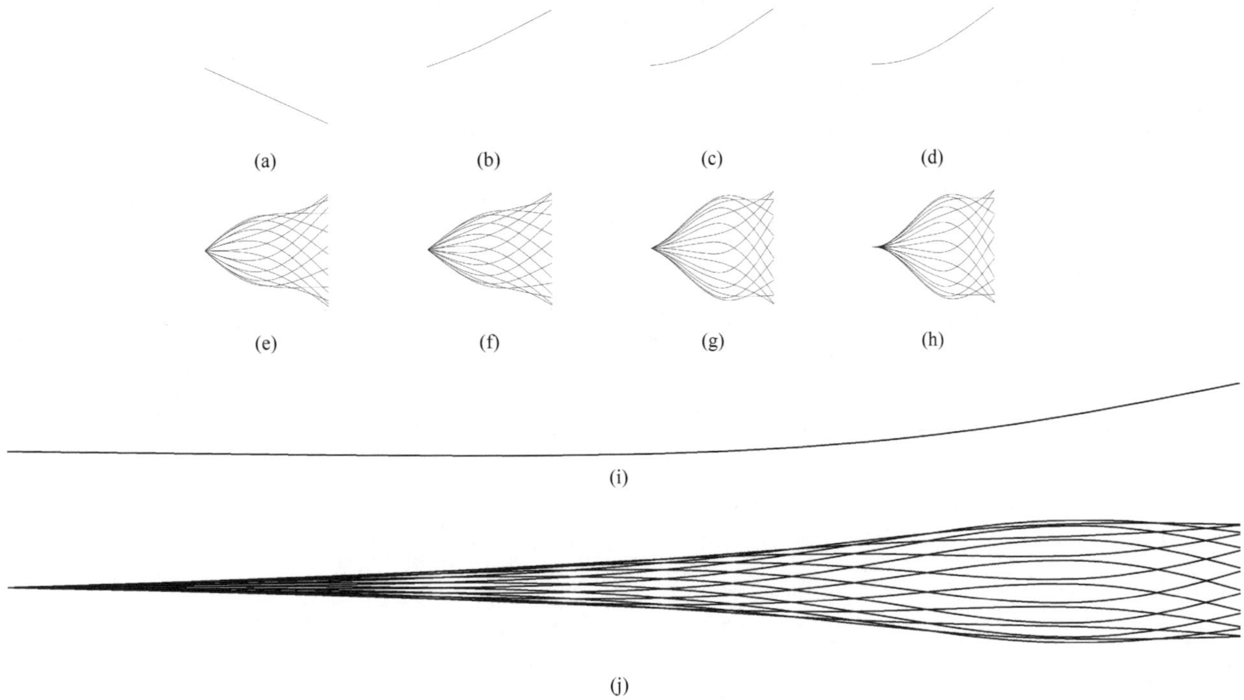

Figure 4: Modal shapes of the unstable modes of cylinders A and B for different values of κ_0^*. (a-d) Stationary yawing/divergence mode shape of cylinder A for $\kappa_0^* = 10^{-8},\ 1,\ 10$ and 10^8, respectively. (e-h) Flutter mode shape of cylinder A for several instants in a cycle of oscillation obtained at $u = 1.01u_{cf}$, for $\kappa_0^* = 10^{-8},\ 1,\ 10$ and 10^8, respectively. (i) Divergence mode shape of cylinder B. (j) Flutter mode shape of cylinder B for several instants in a cycle of oscillation obtained at $u = 1.01u_{cf}$.

figures (e)-(h) show a cycle of flutter motion of cylinder A, reconstructed at $u = 1.01u_{cf}$, where u_{cf} represents the threshold of flutter. It is seen from these figures that, as κ_0^* is increased from 10^{-8} to 10^8, the modal shape of the first unstable mode changes gradually from a skewed rigid body to a rigid body involving flexing and then to a buckled shape. Transition in the flutter mode shape as κ_0^* is varied is also evident from the figures, especially over the first half of the cylinder length. On the other hand, subfigures (i) and (j) show, respectively, the divergence mode shape and a cycle of flutter motion of cylinder B for $\kappa_0^* = 1$. In fact, it was found that the modal shapes of the unstable modes of cylinder B are independent of κ_0^* and therefore, they are given in Figure 4 for a single value of κ_0^*. Unlike the modal shapes of cylinder A, those of cylinder B show that the deformation is limited to a small region close to the downstream end of the cylinder; this dynamical feature of very slender cylinders was discovered previously (see, e.g. [4, 6, 7]).

CONCLUSIONS

In this paper, the equation of motion for a flexible, slender cylinder supported only by a translational and a rotational spring at the upstream end and subjected to axial flow was presented. For a typical cylinder (i.e. not a very slender cylinder) with a very stiff translational spring, it was found that at low flow velocities, stationary yawing or divergence may occur, depending on the rotational stiffness, while at higher flows, flutter may arise; moreover, the critical flow velocities for these instabilities and the modal shapes were found to be dependent on the rotational stiffness for a typical cylinder, while for a very slender cylinder, they were independent of the rotational stiffness.

REFERENCES

[1] Hawthorne, W. R. 1961, "The early development of the Dracone flexible barge," *Proceedings of the Institution of Mechanical Engineers*, Vol(175), pp. 52-83 .

[2] Païdoussis, M.P. 2004, *Fluid-Structure Interactions: Slender Structures and Axial Flow. Volume 2*, Elsevier Academic Press, Oxford.

[3] Modarres-Sadeghi, Y., Païdoussis, M. P. and Semler, C. 2005, "A nonlinear model for an extensible slender flexible cylinder subjected to axial flow," *Journal of Fluids and Structures*, Vol(21), pp. 609-627.

[4] de Langre, E., Païdoussis, M. P., Doaré, O. and Modarres-Sadeghi, Y. 2007, "Flutter of long flexible cylinders in axial flow," *Journal of Fluid Mechanics*, Vol(571), pp. 371-389.

[5] Kheiri, M., Païdoussis, M. P. and Amabili, M. 2013a, "A nonlinear model for a towed flexible cylinder," *Journal of Sound and Vibration*, Vol(332), pp. 1789-1806.

[6] Kheiri, M., Païdoussis, M. P., Amabili, M. and Epureanu, B. I. 2013b, "Three-dimensional dynamics of long pipes towed underwater. Part 1: The equations of motion," *Ocean Engineering*, Vol(64), pp. 153-160.

[7] Kheiri, M., Païdoussis, M. P., Amabili, M. and Epureanu, B. I. 2013c, "Three-dimensional dynamics of long pipes towed underwater. Part 2: Linear dynamics," *Ocean Engineering*, Vol(64), pp. 161-173.

Fracture Mechanics

Proceedings of the 25th CANCAM
London, Ontario, Canada, May 31 – June 4, 2015

SHAPE EFFECT OF TRACTION-SEPARATION LAW ON DYNAMIC FRACTURE OF X80 PIPELINE STEEL

Y.J. Wang
Department of Mechanical Engineering
University of Alberta
Edmonton, Alberta, Canada
yo7@ualberta.ca

C.Q. Ru
Department of Mechanical Engineering
University of Alberta
Edmonton, Alberta, Canada
cru@ualberta.ca

ABSTRACT

In this paper, the shape effect of traction-separation law (TSL) of a previously developed four-parameter cohesive zone model (CZM) is studied. The two key parameters of TSL: the maximum traction and the cohesive energy (defined by the area under the curve of TSL) are determined from the uniaxial true stress-strain curve (SS curve). Different shapes of TSL are defined by the other two parameters, separations S_0 and S_m, which correspond to the beginning and ending of fracture and determine the stiffness and softening shape of TSL, respectively. A uniaxial tensile test and a drop-weight tear test are simulated by Abaqus to examine the influence of TSL's shape on the nominal SS curve and the load-displacement curve. Results showed that the shape effect is important for high speed dynamic fracture although it can be ignored for quasi-static fracture of pipeline steels.

KEYWORDS: Dynamic fracture, cohesive zone model, traction-separation law, stress-strain curve, shape effect.

INTRODUCTION

Cohesive zone model (CZM) have been applied to dynamic fracture of ductile materials [1, 2]. A key problem of CZM is to determine the parameters of the traction-separation law (TSL). It is widely assumed that the most important two parameters are the maximum traction and the area under the TSL (called "cohesive energy"). However, the two key parameters alone cannot determine a complete TSL curve, at least one more parameter (for example, separation S_0 corresponding to the beginning of fracture) is needed to form a bi-linear TSL. For a linear-exponent TSL such as a previous model developed in Ref. [1, 2], two more parameters (S_0 and S_m, which correspond to the beginning and ending of fracture and determine the stiffness and softening shape of TSL, respectively) are needed. Therefore, it is of great interest to study TSL's shape effect on fracture of ductile materials such as X80 pipeline steel.

The two key TSL parameters have been studied for decades. It is now widely believed that the two key parameters largely determine fracture behavior of a ductile material and the shape effect defined by other extra parameters of TSL is usually ignorable. The present work aims to investigate this important issue for high speed dynamic fracture of pipeline steels.

TRACTION-SEPARATION LAW

The TSL used by Ren & Ru [1] with four-parameter (K, S_0, S_m, and α) can be simplified as

$$T = (1-D)KS \tag{1}$$

where, K is the scalar elastic stiffness, T is the traction, S is the separation, and D is a damage scalar which can be expressed as

$$D = \begin{cases} 0 & S < S_0 \\ 1-\dfrac{S_0}{S}\left[1-\left(\dfrac{S-S_0}{S_m\text{-}S_0}\right)^{\alpha}\right] & S \geq S_0 \end{cases} \tag{2}$$

where, S_0 and S_m are the separations at which the cohesive zone starts being damaged and fully fractured, respectively, and α is an exponent which controls the damage evolution.

The area under the TSL curve is the cohesive energy, which can be expressed as:

$$J_c = \left(1-\frac{1}{\alpha+1}\right)KS_0S_m + \left(\frac{1}{\alpha+1}-\frac{1}{2}\right)KS_0^{\ 2} \tag{3}$$

The determination of the maximum traction $T_{\max}$ and the cohesive energy J_c will be discussed in the next section. Obviously, with given $T_{\max}$ and J_c, the parameter S_0 controls the stiffness $K = T_{\max} / S_0$, and the parameter S_m is related to the damage evolution parameter α, and

$$S_0 \leq \frac{2J_c}{T_{\max}}, S_m > \frac{2J_c}{T_{\max}} \tag{4}$$

DETERMINATION OF THE TWO KEY COHESIVE ZONE MODEL PARAMETERS ($T_{\max}$ AND J_c)

1) Determination of the maximum traction

The maximum traction of TSL can be considered equal to the critical true stress at the fraction point. The true SS curve after necking can be derived by [3]

$$\sigma = \frac{\sigma_u}{2}\left[(1+\varepsilon-\varepsilon_u)+\frac{\varepsilon^{\varepsilon_u}}{\varepsilon_u^{\varepsilon_u}}\right] \quad (5)$$

where, σ_u and ε_u are the true stress and true strain at the necking point. The fracture point can be set as $\varepsilon_f = 0.25$ for this X80 pipeline steel, and the true stress at the fracture point is $\sigma_t^f = 944.05\text{MPa}$.

The critical stress can be calculated from [4]

$$\sigma_c = \sigma_{eq} + \sigma_m \quad (6)$$

where, σ_{eq} is the Von Mises equivalent stress, σ_m is the hydrostatic stress. For the one dimensional problem, $\sigma_{eq} = \sigma_t^f$, $\sigma_m = \sigma_t^f/3$, and thus σ_c=1258.73MPa, which gives the maximum traction T_{max} of TSL.

2) Determination of the cohesive energy

The cohesive energy is the area under the TSL, which is equal to the critical J-integral. A flat tensile specimen is modeled by Abaqus. The J-integral can be outputted by Abaqus directly, and the value corresponding to the beginning of fracture is selected, which is $J_c = 68.87\,\text{KJ/m}^2$ for X80 pipeline steel.

SHAPE EFFECT OF TRACTION-SEPARATION LAW

1) Simulation of an uniaxial tensile test

A simple 3D model with a cohesive layer laid on the tip of a small crack is simulated. Linear, exponential and softening controlled by Eq. (2) are used in the simulation. The nominal SS curves obtained from the simulations with and without CZM are plotted in Fig. 1, with a comparison to test result [1].

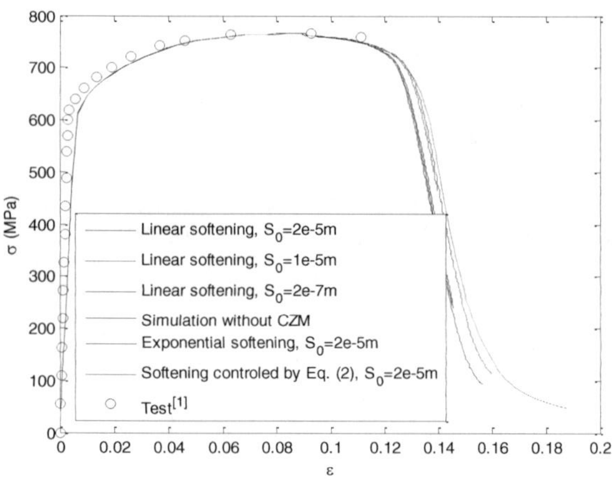

Figure 1: Nominal SS curve for different shapes of TSL

It can be seen that, although the results of exponential softening and the softening controlled by Eq. (2) are much closer to the simulation result without CZM, the shape effect of TSL can be ignored for quasi-static fracture.

2) Simulation of a drop-weight tear test

A standard DWTT simulation is used to examine the shape effect of TSL for high speed dynamic fracture. The rate-dependent constitutive model developed in Ref. [2] is used. In the steady-state, the two key CZM parameters are almost twice of the parameters obtained for quasi-static fracture. The load-displacement curves for different shapes of TSL are shown in Fig. 2. It can be seen that the shape of TSL has a significant effect on the simulation result for high speed dynamic fracture, especially at the beginning of fracture.

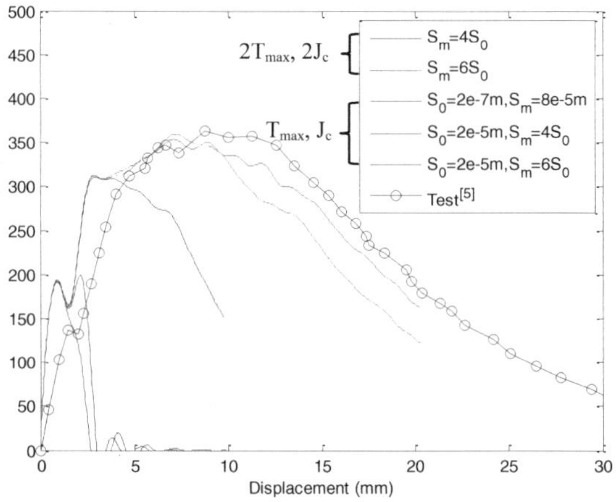

Figure 2: Load-displacement curves for different shapes of TSL

CONCLUSIONS

The present work studied the shape effect of traction-separation law (TSL) on dynamic fracture of pipeline steels. Our simulations indicated that the shape effect of TSL is important for dynamic fracture, although it can be ignored for quasi-static fracture.

ACKNOWLEDGMENTS

This work was supported by TransCanada Pipelines Ltd. and the NSERC (Natural Science & Engineering Research Council) of Canada (CRD program CRDPJ 399497).

REFERENCES

1. Ren, Z. J, & Ru, C.Q. 2013, "Numerical investigation of speed dependent dynamic fracture toughness of line pipe steels", *Engineering fracture mechanics*, Vol(99), pp. 214-222.

2. Yu, P. S. & Ru, C. Q. 2015, "Strain rate effects on dynamic fracture of pipeline steels: Finite element simulation", *International Journal of Pressure Vessels and Piping*, Vol(126-127), pp. 1-7.

3. Ling, Y. 1996, "Uniaxial true stress-strain after necking", *AMP Journal of Technology*, Vol(5), pp. 37-48.

4. Argon, A., Im, J. and Safoglu, R. 1975, "Cavity formation from inclusions in ductile fracture", *Metallurgical Transactions A*, Vol(6), pp. 825-837.

5. Shim, D.J., Wilkowski, G., Duan, D.M., & Zhou, J. 2010, "Effect of fracture speed on ductile fracture resistance: part 1—experimental", The 8th International Pipeline Conference, Calgary, Canada.

Proceedings of the 25th CANCAM
London, Ontario, Canada, May 31 – June 4, 2015

RESIDUAL STRESS STATE IN SINGLE-EDGE NOTCHED TENSION SPECIMEN CAUSED BY THE LOCAL COMPRESSION TECHNIQUE

Yifan Huang
Department of Civil & Environmental Engineering
Western University
London, Ontario, Canada
yhuan95@uwo.ca

Wenxing Zhou
Department of Civil & Environmental Engineering
Western University
London, Ontario, Canada
wzhou@eng.uwo.ca

ABSTRACT

Three-dimensional (3D) finite element analyses (FEA) are performed to simulate the local compression (LC) technique on the clamped single-edge notched tension (SE(T)) specimens. The analysis includes three types of indenters which are single pair of cylinder indenters (SPCI), double pairs of cylinder indenters (DPCI) and single pair of ring indenters (SPRI). The opening residual stress distribution in the uncracked ligament of the specimen is evaluated. The outcome of this study can facilitate the use of LC technique on SE(T) specimens.

KEYWORDS: Residual Stress, Single-edge Notched Specimen, Local Compression, Finite Element Analysis

INTRODUCTION

The fracture toughness resistance curve, i.e. the *J*-integral resistance (*J-R*) or crack tip opening displacement resistance (*CTOD-R*) curve is an important input in the engineering critical assessment (ECA) of steel structures such as pressure vessels and energy pipelines. The resistance curve is typically obtained from deeply-cracked single-edge bend (SE(B)) or compact tension (C(T)) specimens as standardized in ASTM E1820-13 [1] and BS7448-4 [2]. Recently, the use of the non-standard single-edge (notched) tension (SE(T) or SENT) specimen to determine the resistance curve has gained significant research interests [e.g. 3, 4] in the energy pipeline industry largely as a result of the development of the strain-based design and assessment methodologies. Figure 1 schematically shows the configuration of a typical SE(T) specimen including the width (W), thickness (B), crack length (a) and the daylight distance (distance between the clamped surfaces, H) of the specimen.

Most engineering manufacturing and structural construction process inevitably introduce the residual stress that can be retained in the engineering components. One example is the residual stress caused by the thermal cycles during the welding procedures remains in both weldments and the fusion region [5]. It is well-known that the residual stress has a non-negligible impact on the fracture toughness of the material/structural component [5]. The fracture toughness corresponding to the opening mode (mode I) is largely impacted by the opening residual stress, $\sigma_{yy(R)}$. In general, the tensile opening residual stress promotes the crack growth whereas the compressive opening residual stress retards the crack growth [5]. As a result, a non-uniform distribution of residual stress along the crack front yields non-uniform crack growth, i.e., a curved crack front. To ensure the accuracy of the experimentally determined *J*(*CTOD*)-*R* curve, testing standards (e.g. [1, 2]) usually specify acceptable levels of the crack front curvature for both the fatigue pre-crack and final crack fronts for the SE(B) and C(T) specimens.

For specimens containing notches or cracks located at weldments or the heat-affected zone, the residual stress in the uncracked ligament can be high and non-uniformly distributed along the crack front. In such cases, treatments are required to mitigate the residual stress prior to the fatigue pre-cracking. The local compression (LC) technique [6-8] is one of the mechanical treatment techniques that produce additional residual stresses to mitigate the existing residual stresses. Many experimental and numerical studies [6, 7, 9, 10] have been carried out to investigate the applicability of the LC technique for SE(B) and C(T) specimens. The focus of these studies is SE(B) and C(T) specimens with $a/W \geq 0.3$ and $0.25 \leq B/W \leq 1$. To our best knowledge, the applicability of the LC technique for SE(T) specimens has not been numerically investigated. Motivated by this observation, we carry out the three-dimensional (3D) finite element analyses (FEA) to investigate the opening residual stress state in the SE(T) specimen ($a/W = 0.25$, $B/W = 1$) caused by different LC techniques in the present study.

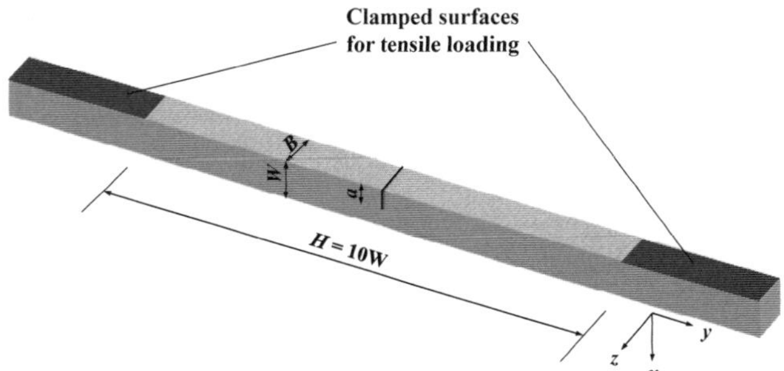

Figure 1: Schematic of SE(T) specimen

THE LOCAL COMPRESSION TECHNIQUE

Single Pair of Cylinder Indenters (SPCI)

Figure 2(a) illustrates the schematics of the typical LC method. As shown in the figure, a single pair of cylinder indenters (SPCI) is used to apply the out-of-plane (i.e., z direction) compression on the notched specimen. The indenter has a radius of R. δ_x and δ_y, denote the distances between the center of the indenter and notch tip in x and y direction, respectively. For SPCI, $\delta_y = 0$. As specified in Annex C of the standard of fracture toughness testing of welds ISO15653 [8], the 88% to 92% of the ligament in front of the machined notch shall be compressed (i.e., $\delta_x = (88\% \sim 92\%)(W - a) - R$). The indentation is controlled by the plastic strain, which shall not exceed 0.5%B on each side.

Double Pairs of Cylinder Indenters (DPCI)

Mahmoudi et al. [10] developed an LC technique by using double pairs of cylinder indenters (DPCI) (see Fig. 2(b)). It is reported in [10] that DPCI can produce either tensile or compressive opening residual stress depending on δ_x and δ_y. It was found that initial plastic strain adjacent to the crack tip leads to lower failure ductility [10]. When causing similar level of the opening residual stress, DPCI leads to smaller plastic strains near the crack tip than SPCI, which can reduce the plastic strain impacts on the toughness.

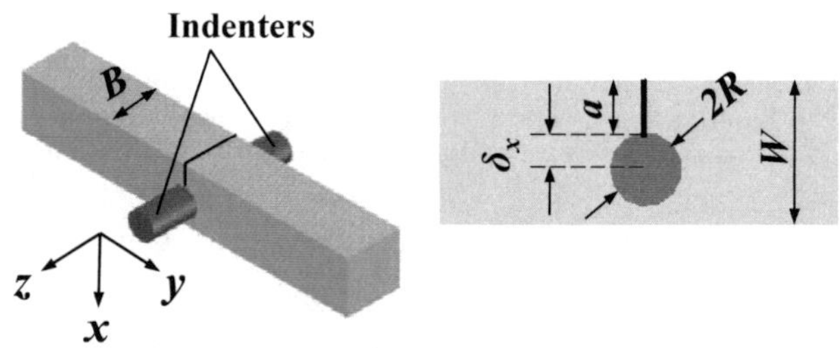

(a) Single pair of cylinder indenters (SPCI)

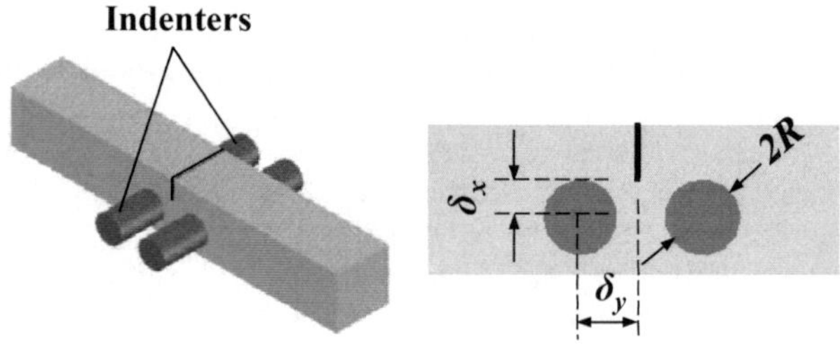

(b) Double pairs of cylinder indenters (DPCI)

Figure 2: The local compression technique

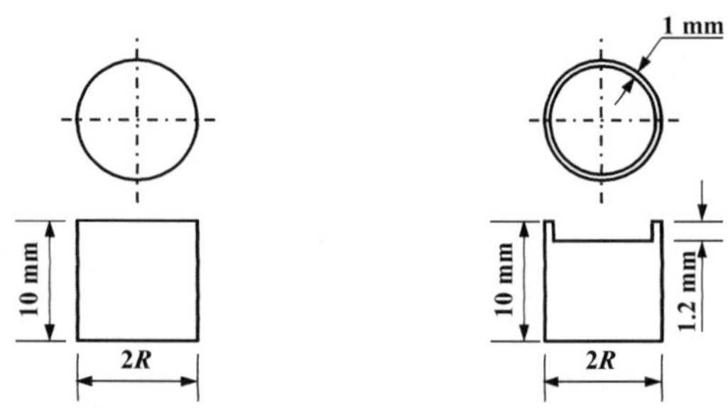

(a) Cylinder indenter (b) Ring indenter

Figure 3: Configurations of the indenters

Single Pair of Ring (SPR) Indenters

Lim et al. [11] suggested using a single pair of ring indenters (SPRI) to produce compressive residual stress near the crack tip. Based on the experiments results for the thin plate ($B/W = 0.04$) SE(B) specimens reported in [11], SPRI can effectively retard the fatigue crack growth.

NUMERICAL ANALYSES

The FEA code ABAQUS® [12], was employed to simulate the LC technique. Three-dimensional models of SE(T) specimens with $W = 20$ mm, $H/W = 10$, $a/W = 0.25$ and $B/W = 1$ were prepared for FEA. Three types of indenters, i.e., SPCI, DPCI and SPRI, were considered in the analysis. The configurations of the cylinder and ring indenters are shown in Figs. 3(a) and 3(b), respectively, with the radius of the indenters $R = 5$ mm. For SPCI, three indenter positions (i.e., $\delta_x = 0$, 0.5R and 1.0R) are considered whereas only one indenter position is considered for DPCI (i.e., $\delta_x = 0$ and $\delta_y = 1.2R$) and SPRI (i.e., $\delta_x = 0$).

Because of symmetry, only one quarter of the specimen with appropriate constraints imposed on the remaining ligament was modelled. A typical quarter-symmetric 3D model has 15 layers equally-distributed over the half thickness (B/2). The total number of elements is approximately 30,400 in a typical specimen. The 8-node 3D brick elements with 2×2×2 integration (i.e. C3D8R) were used. A typical FE model is shown in Fig. 4. A machine notch tip with a radius ($r = 0.1$ mm) as suggested in E1820 [1] was incorporated in the FE model (see Fig. 4). All of the FE models have 50 focused annular rings around the notch tip with 20 elements in each ring. The minimum in-plane dimension of the 1st ring is about 0.2r. A rigid and frictionless contact between specimen and indenter was assumed. The static load was applied based on the displacement control condition. Displacements were applied on a reference point associated with the rigid indenter. The applied displacement firstly increases to the target indentation values and then decreases to zero (i.e., unloading). In this study, two loading cases (i.e., 0.5% and 1%B total indentation on each side) were considered.

The finite-strain formulation was employed in FEA as it can more accurately simulate the deformed configuration of the specimen. The analyses utilized an elastic-plastic constitutive model with the J_2 flow theory. The true stress (σ) and true strain (ε) relationship of the material is characterized as follows:

$$\frac{\varepsilon}{\varepsilon_0} = \begin{cases} \dfrac{\sigma}{\sigma_{YS}} & ,\varepsilon \le \varepsilon_0 \\ \left(\dfrac{\sigma}{\sigma_{YS}}\right)^n & ,\varepsilon > \varepsilon_0 \end{cases} \tag{1}$$

where σ_{YS} is the yield stress; ε_0 ($\varepsilon_0 = \sigma_{YS}/E$) is the yield strain, and n denotes the strain hardening exponent. The yield stress, elastic modulus and Poisson's ratio and strain hardening exponent were assumed to be 510 MPa, 207 GPa, 0.3 and 13, respectively.

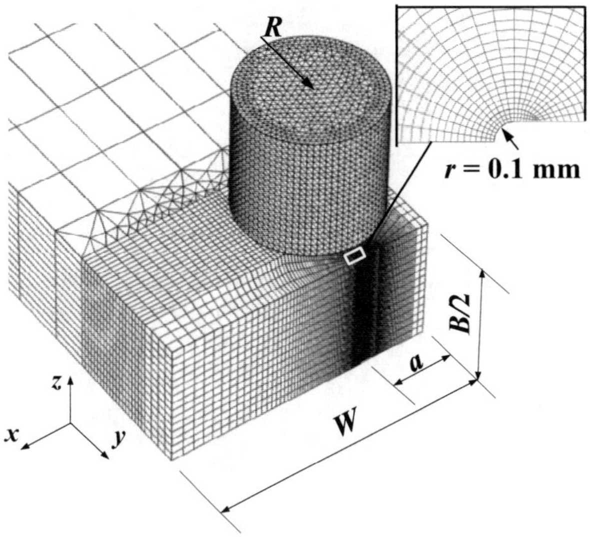

Figure 4: Typical FE model with DPCI

RESULTS AND DISCUSSIONS

The opening residual stress, $\sigma_{yy(R)}$, in the uncracked ligament was output from FEA. Figure 5 shows the distribution of $\sigma_{yy(R)}$ (normalized by σ_{YS}) at the crack tip (i.e., $x = 0$) along the crack front, where $z = 0$ and $B/2$ denote the mid-plane and free surfaces of the specimen, respectively. Positive and negative values of $\sigma_{yy(R)}/\sigma_{YS}$ represent tensile and compressive stresses, respectively. It is observed from Fig. 5(a) that the indentation caused by SPCI produces tensile stress near the specimen free surfaces (i.e., approximately $|2z/B| \geq 0.8$) and relatively uniform compressive stress at $|2z/B| < 0.8$. In contrast, indentation caused by SPRI yields compressive stress near the free surfaces and tensile stress at $|2z/B| \leq 0.5$. On the other hand, DPCI yields consistent compressive $\sigma_{yy(R)}$ along the crack front. Figures 5(a) and 5(b) suggeste that when increasing the indentation from 0.5%B to 1%B, the curves in Fig. 5(a) basically moves down by about $0.5\sigma_{yy(R)}/\sigma_{YS}$. The residual stress associated with SPRI and DPCI in the region $|2z/B| \leq 0.5$ become more uniform when the indentation is increased to 1%B.

Figure 6 shows the distribution of $\sigma_{yy(R)}$ at the mid-plane (i.e., $z = 0$) ahead of crack tip. It is observed that $\sigma_{yy(R)}$ is compressive right at the crack tip regardless of the choice of indenters and indentation magnitude. As x increases from 0 to $0.02(W - a)$, the values of $\sigma_{yy(R)}/\sigma_{YS}$ associated with SPRI and DPCI rapidly increases from -1 to 1 – 1.8 and decreases to around zero as x further increases. It is found that at the region $x/(W - a) \geq 0.2$, $\sigma_{yy(R)}$ associated with SPRI is

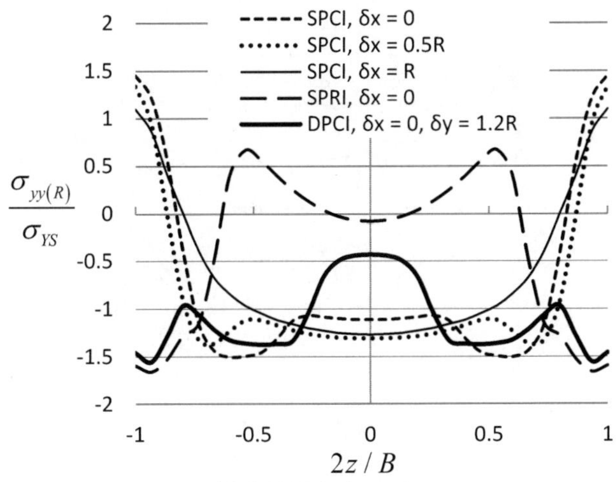

(a) 0.5%B indentation

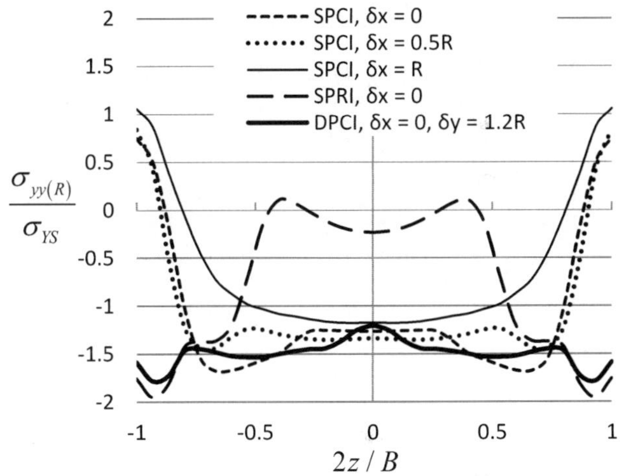

(b) 1%B indentation

Figure 5: Distribution of $\sigma_{yy(R)}$ along the crack front ($x = 0$)

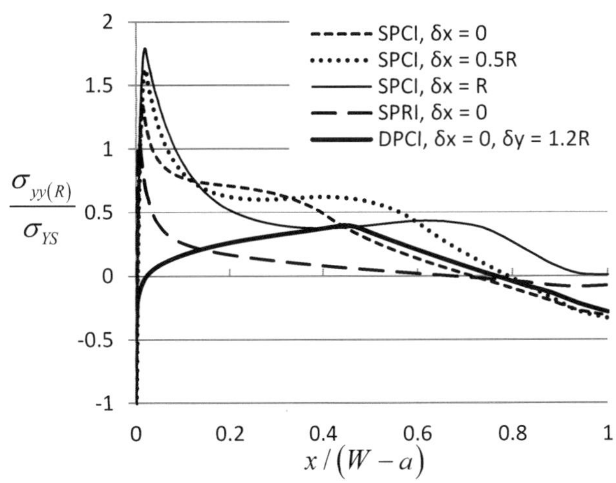

(a) 0.5%B total indentation

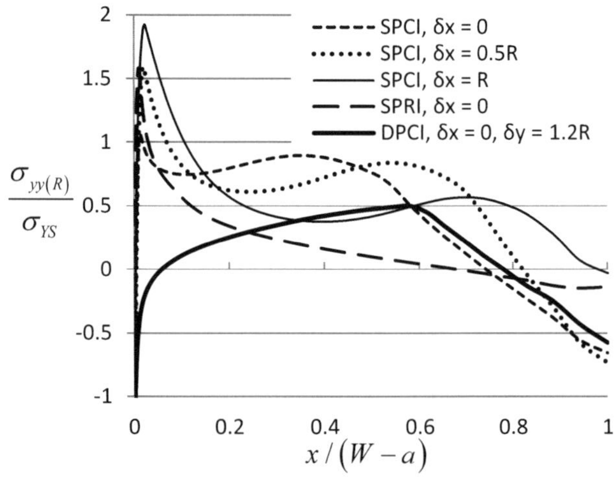

(b) 1%B total indentation

Figure 6: Distribution of $\sigma_{yy(R)}$ ahead of crack tip ($z = 0$)

relatively small ($\leq 0.2\sigma_{YS}$). On the other hand, the values of $\sigma_{yy(R)}/\sigma_{YS}$ associated with SPRI slowly increases from -0.5 to 0.5 as x increases from 0 to $0.5(W-a)$, and decreases back to -0.5 as x further increases.

Both Figs. 5 and 6 suggest that the distribution of $\sigma_{yy(R)}$ in the uncracked ligament is dependent on the configuration and position of the indenters, which agrees with the findings in the literature [10] with respect to LC technique on C(T) specimens. Once the existing residual stress state in the specimen is given, using the LC method with choices of appropriate indenters and indentation can effectively mitigate the existing residual stress in SE(T) specimen.

CONCLUSIONS

Three-dimensional finite element analyses have been performed on clamped SE(T) specimens to evaluate the residual stress state introduced by the local compression method. The analysis results suggest that the distribution of the opening residual stress in the uncracked ligament is dependent on the configuration and position of the indenters. Based on the given existing residual stress state, appropriate LC technique can be used to mitigate the existing residual stress in SE(T) specimen.

ACKNOWLEDGMENTS

This study is supported by TransCanada Corporation and the Natural Sciences and Engineering Research Council of Canada (NSERC) through the Collaborative Research and Development (CRD) program and by the NSERC-RIEM program. The authors gratefully acknowledge Dr. Henryk Pisarski in TWI Ltd. and Mr. James Gianetto in the Materials Technology Laboratory of Natural Resources Canada, for their discussions related to this study.

REFERENCES

1. ASTM. 2013, *ASTM E1820-13: Standard Test Method for Measurement of Fracture Toughness*, ASTM International, West Conshohocken, PA

2. BSI. 1997, *BS 7448: Part 4. Method for Determination of Fracture Resistance Curves and Initiation Values for Stable Crack Extension in Metallic Materials*, British Standards Institution, 389 Chiswick High Road, London.

3. Huang, Y., and Zhou, W. 2014, "Investigation of plastic eta factors for clamped SE(T) specimens based on three-dimensional finite element analyses", *Engineering Fracture Mechanics*, Vol(132), pp. 120-135.

4. Huang, Y., and Zhou, W. 2014, "*J-CTOD* relationship for clamped SE(T) specimens based on three-dimensional finite element analyses", *Engineering Fracture Mechanics*, Vol(131), pp. 643-655.

5. Zerbst U., Ainsworth R. A., Beier H. T., et al. 2014, "Review on fracture and crack propagation in weldments–A fracture mechanics perspective", *Engineering Fracture Mechanics*, Vol(132), pp. 200-276.

6. Dawes, M. 1971, "Fatigue pre-cracking weldment fracture-mechanics specimens", *Metal Construction and British Welding Journal*, Vol(3), pp. 61-65.

7. Towers, O. L., and Dawes, M. G. 1985, "Welding Institute research on the fatigue precracking of fracture toughness specimens", *Elastic-Plastic Fracture Test Methods: The User's Experience, ASTM STP 856*, pp. 23-46.

8. ISO. 2010, *ISO15653: Metallic materials-Method of test for the determination of quasistatic fracture toughness of welds*, International Organization for Standardization.

9. Hill, M. R., and Panontin, T. L. 1999, "Effect of residual stress on brittle fracture testing", *Fatigue and Fracture Mechanics, ASTM STP 1332*, Vol (29), pp. 154-175.

10. Mahmoudi, A., Truman, C., and Smith, D. 2008, "Using local out-of-plane compression (LOPC) to study the effects of residual stress on apparent fracture toughness", *Engineering Fracture Mechanics*, Vol(75), pp. 1516-1534.

11. Lim, W.-K., Song, J.-H., and Sankar, B. V. 2003, "Effect of ring indentation on fatigue crack growth in an aluminum plate", *International journal of fatigue*, Vol(25), pp. 1271-1277.

12. ABAQUS/Standard Version 6.12-2. 2013, *Analysis User's manual*, Dassault Systémes Simulia Corporation., Providence, RI.

Proceedings of the 25th CANCAM
London, Ontario, Canada, May 31 – June 4, 2015

COHESIVE ZONE MODEL FOR TEMPERATURE DEPENDENT FRACTURE ANALYSIS OF PIPELINE STEEL

Sourayon Chanda
Department of Mechanical Engineering
University of Alberta
Edmonton AB, Canada
Email: sourayon@ualberta.ca

CQ Ru
Department of Mechanical Engineering
University of Alberta
Edmonton AB, Canada
Email: cru@ualberta.ca

ABSTRACT

The present paper gives a cohesive zone model (CZM)-based finite element analysis to study the effects of temperature variation on dynamic fracture characteristics of pipeline steel (PS). Numerical simulation of drop weight tear test (DWTT) has been performed to analyze the temperature dependent dynamic fracture of PS. The key for such a model is to develop an appropriate temperature dependent traction-separation law (TSL) for the cohesive surfaces. The most significant CZM parameters which define TSL are the maximum traction and the area under the curve (called "fracture energy"). The maximum traction at any temperature in the TSL can be correlated to the maximum fracture stress at that specific temperature. In this regard, a modified Johnson-Cook constitutive model has been implemented to obtain true stress strain curves for PS at any temperature. Further, the area under the TSL equals the fracture energy of the material at that temperature. Based on these correlations, a temperature dependent CZM is presented to simulate dynamic fracture behavior of PS at varying temperature. The finite element modeling for the present problem has been done using Abaqus/CAE 6.13.

KEYWORDS: dynamic fracture, cohesive zone model, pipeline steel, traction separation law, temperature effect.

INTRODUCTION

Pipelines used for transportation of oil/gas by long distances often face different environmental conditions throughout the year in different parts of the world. Especially in northern Canada, surface pipelines run over a wide range of temperature varying from about -50°C in the winters to +35°C or higher in the summers. This could initiate a number of long-lasting impacts on the fracture behavior of pipelines and reduce service life of pipelines. In this regard, a complete analysis of temperature dependent dynamic fracture of pipelines is crucial. The present research aims to meet this need by developing a temperature-dependent CZM to study temperature dependency of dynamic fracture characteristics of PS.

A temperature dependent CZM has been developed to numerically simulate drop-weight tear test (DWTT) at different temperatures. To this end, the most challenging task is to develop a reasonable temperature-dependent TSL of CZM. The temperature dependency of CZM parameters is expected to depend on many factors such as maximum stress, fracture energy, separation, rate of separation, strain rate, damage evolution. However many researchers have suggested that the maximum traction and area under the TSL defined by fracture energy are the two most important parameters to define TSL. Based on this assumption, in the present analysis the primary focus is on temperature dependent variation of the maximum traction and the fracture energy. The present work is a continuation of previous finite element models by Ren & Ru [1] and Yu & Ru [2].

CONSTITUTIVE MODEL FOR TEMPERATURE DEPENDENCY OF PIPELINE STEEL

The present analysis requires an analytical model that could reproduce temperature dependent stress-strain curves (SSC) for PS for a wide range of temperature. Johnson and Cook [3] reported such a strength model for metals to correlate true stress with strain. The above model has been extended to low temperature in the present problem by a simple temperature-dependent modification term. Without considering strain rate effect, the modified Johnson and Cook model which can be used for both high and low temperatures can be expressed as:

$$\sigma = \left(A + B\varepsilon^n\right)\left(1 - \lambda\left|T^*\right|^m\right) \qquad (1)$$

where σ is the plastic stress, ε is the equivalent plastic strain $T^* = T_i - T_{room} / T_{melt} - T_{room}$ is the non-dimensionalized temperature, T_i is the actual temperature under consideration, T_{room} is the room temperature, T_{melt} is the melting point temperature of steel, which is taken as 1500°C for PS and λ is defined as:

$$\lambda = \begin{cases} T_i/\left|T_i\right| & \text{if } T_i \neq T_{ref} \\ 1 & \text{if } T_i = T_{ref} \end{cases} \qquad (2)$$

There are four material constants in Eq. (1), where A is equal to the yield stress for the material at the T_{room}. The other three constants B, n and m can be obtained by matching with true SSCs for PS at different temperatures[1,4]. The modified J-C model reproduces fairly accurate match with the true SSCs obtained from tests as seen in Figure 1. The PS considered for this research is API X80.

Table 1. Value of material constants for modified Johnson-Cook strength model for PS (obtained by matching with test results [1,4])

A	B	n	m
550 N/mm^2	702.45 N/mm^2	0.3646	0.81

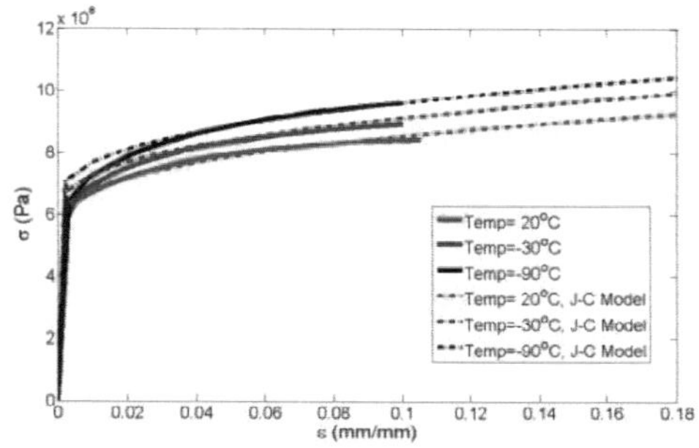

Figure 1. Modified J-C model which can reproduce true stress-strain data for different temperatures. [1,4]

In order to determine the maximum fracture stress and its dependency on temperature, first the temperature dependency of maximum fracture strain needs to be obtained. A number of engineering SSCs for PS [5–7] were analyzed which allow us to get a direct relationship between ultimate strength of PS σ_U with fracture strain ε_f, as shown in Figure 2. Finally, based on the known values of σ_U for PS at different temperatures, a relationship between the maximum fracture strain and non-dimensional temperature has been obtained as shown:

$$\varepsilon_f = -2\left(T^*\right)^2 + 0.41T^* + 0.25 \tag{3}$$

From Eq. (1) and Eq. (3) the maximum true stress at any temperature can be obtained.

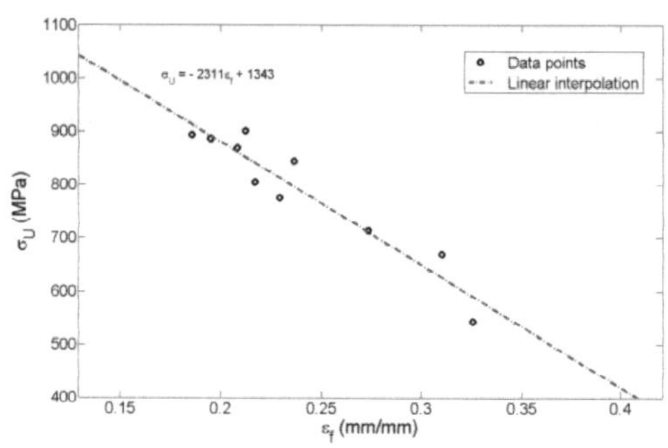

Figure 2. Relationship between ultimate strength of PS and fracture strain

To account for the voids generated in PS after damage initiation, a damage factor D^* [8] needs to be implemented to obtain the maximum fracture stress at failure. The maximum stress can, hence, be expressed as:

$$\sigma_D = \frac{\sigma}{1-D^*} \tag{4}$$

where σ_D is the true stress using damage factor, σ is the true stress without considering damage and D^* is the damage factor which varies from zero to $D^*_{\max}(\approx 0.3)$ for PS.

TEMPERATURE DEPENDENT COHESIVE ZONE MODEL

One of the primary shortcomings of LEFM for ductile materials such as PS is the stress singularity at the crack tip. CZM approach provides a more realistic scenario by considering cohesive surfaces, such that the stress remains finite at all locations including the crack tip. In this model, failure is characterized by complete separation of the two cohesive surfaces. The TSL provides a relation between traction and separation at the cracking surface of the cohesive zone. For mode I fracture as in the present case, only the normal components of traction and separation are considered for the two potential crack faces. To this end, the TSL can mathematically be expressed in terms of a scalar function as follows [1,2]:

$$T = (1-D)K \cdot S \tag{5}$$

where T is the traction stress in the normal direction, S is the corresponding separation, K is the scalar elastic stiffness and D is damage scalar which obeys the following evolution law [1]:

$$D = \begin{cases} 0, & \text{if } S < S_0 \\ 1 - \dfrac{S_0}{S}\left[1 - \left(\dfrac{S - S_0}{S_m - S_0}\right)^{\alpha}\right], & \text{if } S \geq S_0 \end{cases} \tag{6}$$

where S_0 is the separation at which damage initiates at the cohesive zone, $S_m = 4S_0$ [1,2] is the separation at which material is completely fractured, and α is a constant which controls the damage evolution.

In order to find the TSL for a material at any temperature, the key parameters are:

1. Maximum traction

The maximum traction of TSL for PS at a temperature can be taken equal to the maximum true stress of PS at that particular temperature. Thus, using Eq. (1) and Eq. (4), the maximum traction $T_{\max}$ can be expressed as:

$$T_{\max} = \frac{\left(A + B\varepsilon_f^{\,n}\right)\left(1 - \lambda\left|T^*\right|^m\right)}{1 - D^*_{\max}} \tag{7}$$

In terms of the known maximum traction, the scalar elastic stiffness K can be calculated as:

$$K = T_{\max}/S_0 \tag{8}$$

2. Fracture energy

The area under the TSL is equal to fracture energy at any particular temperature. Based on temperature dependency of fracture energy of a material, they can be divided into three regions, namely, brittle region, brittle-to-ductile transition region and ductile region. With extensive literature survey, the lower shelf critical fracture toughness and the upper shelf fracture energy was taken as 20 kJ/m^2 and 70 kJ/m^2 respectively, and the brittle-to-ductile transition was assumed to occur between -80°C and -130°C. Fig. 8 shows the variation of critical fracture energy with temperature considered. In this regard, an attempt has been made to describe the variation of fracture energy with temperature using a simple equation:

$$J_{Ic} = J_{Ic,ref} + C_1 \tanh\left[C_2\left(T_i - T_{ref}\right)\right] \quad (9)$$

where J_{Ic} is the critical fracture energy at any temperature T in kJ/m^2, T_{ref} is the reference temperature, taken equal to -105°C, $J_{Ic,ref}$=45 kJ/m^2 is the critical fracture energy at T_{ref}, C_1=25 kJ/m^2 is dependent on the upper shelf and lower shelf J_{Ic} values, and C_2=1/20. An illustration of Eq. (9) can be obtained from Figure 3. From the above equation, we can obtain the assumed value of fracture energy at any temperature.

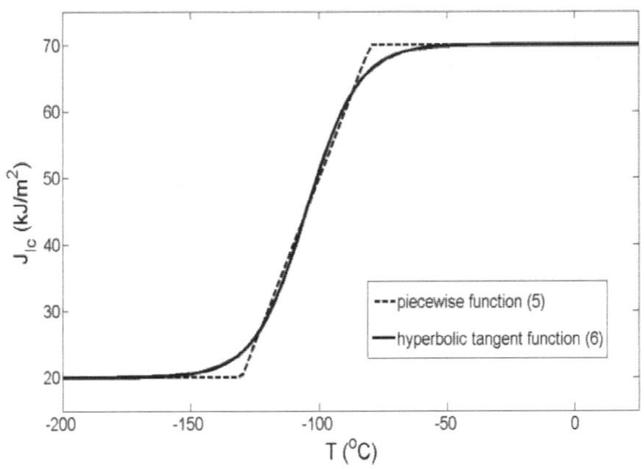

Figure 3. Critical fracture energy with temperature

3. Damage evolution

The damage evolution of the TSL is given by the constant α as explained earlier. With known T_{max} for a particular temperature, the area under the curve of TSL can be matched up to the critical fracture energy J_{Ic} with appropriate adjustment of α iteratively.

In the present model, to account for the temperature effect in the TSL, we shall assume that the separation at which damage initiates remains constant with change in temperature, equal to that of Ren & Ru's experimentally verified model [1]. Based on the above mentioned procedure, the TSL can be obtained for PS at any temperature. Figure 4 graphically shows the TSL for three different temperatures. The CZM-based finite element analysis has been conducted for these three temperatures.

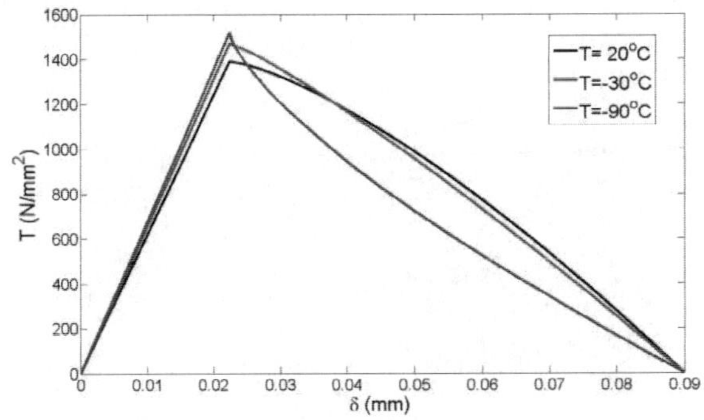

Figure 4. Traction separation law for three different temperatures

NUMERICAL SIMULATIONS

Based on the above mathematical model, the finite element model of DWTT has been done in Abaqus/CAE 6.13. The dimensions of the specimen, hammer, and support have been taken same as that in actual DWTT reported by Shim et. al. [9] The hammer velocity has been kept constant in all three cases and is equal to 7.53 m/s. The hammer weight considered is 780.19 kg. Simulations have been done for three temperatures, namely 25°C, -30°C and -90°C.

Figure 5 shows the load displacement curves for all three cases. The load-displacement curve changes as temperature reduces, suggesting that that fracture will occur earlier as the temperature decreases. Further, in the brittle-to-ductile transition zone, the maximum contact force in much less than that seen for the other two higher temperatures where PS is still in the ductile region. This indicates that significant differences may be noticed in the behavior of PS due to temperature changes. It can also be noticed that the load displacement curves does not vary significantly upto a displacement of 2mm, for all the cases.

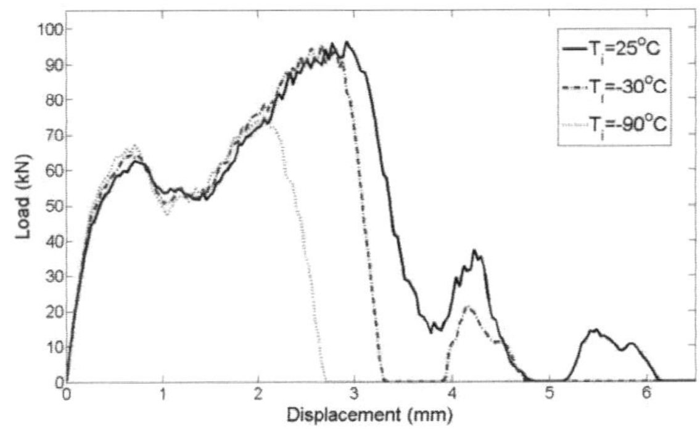

Figure 5. Temperature dependent load-displacement curves from finite element simulations

Figure 6 shows the the initiation of Mode I fracture for all the three cases considered. This figure also displays the mesh of the specimen near the crack tip.

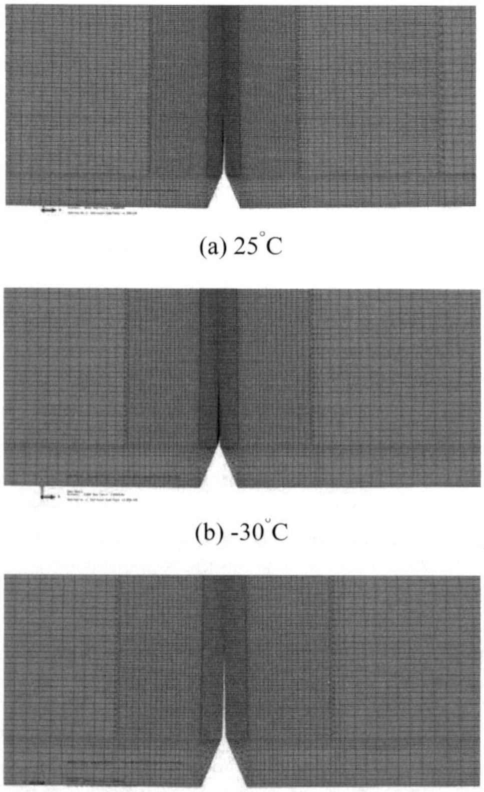

(a) 25°C

(b) -30°C

(c) -90°C

Figure 6. Initial of crack from in the DWTT specimen

CONCLUSIONS AND FUTURE WORK

The present work presents a cohesive zone model (CZM)-based finite element model to study the temperature effects of pipeline steel (PS). On the mathematical front, the model is based on a modified form of Johnson and Cook model to obtain the true SSCs at any temperature. Through proper identification of CZM parameters, a well-defined model has been proposed to generate temperature dependent traction-separation law (TSL) for the cohesive surfaces. After obtaining TSL at different temperatures, temperature dependent performance of PS can be obtained in greater detail. Presently, the simulations have been done for low temperatures only.

Based on the present model, a detailed finite element analysis will be performed to obtain the temperature dependency of dynamic fracture of PS. Numerical simulations will also be performed for high temperatures upto +50°C. With a systematic study of temperature variation of crack tip opening angle, crack tip opening displacement, speed of crack propagation, energy release rate, crack length etc. during fracture, we aim to present a complete analysis of the temperature dependency of dynamic fracture of PS. Further, a temperature-dependent strain rate-dependent finite element model will be developed for a more holistic analysis.

ACKNOWLEDGMENTS

The present research is supported by NSERC (Natural Science and Engineering Research Council) of Canada (CRD program CRDPJ 399497) and TransCanada Pipelines Ltd. The valuable suggestions of Dr. YJ Wang, University of Alberta are much appreciated.

REFERENCES

1. Ren, Z. J. & Ru, C. Q. Numerical investigation of speed dependent dynamic fracture toughness of line pipe steels. *Eng. Fract. Mech.* **99,** 214–222 (2013).

2. Yu, P. S. & Ru, C. Q. Strain rate effects on dynamic fracture of pipeline steels: Finite element simulation. *Int. J. Press. Vessel. Pip.* **126-127,** 1–7 (2015).

3. Johnson, G. & Cook, W. Fracture characteristics of three metals subjected to various strains, strain rates, temperatures and pressures. *Eng. Fract. Mech.* **21,** (1985).

4. Xu, J., Zhang, Z. L., Østby, E., Nyhus, B. & Sun, D. B. Effects of Temperature and Crack Tip Constraint on Cleavage Fracture Toughness in the Weld Thermal Simulated X80 Pipeline Steels. in *Proceediings 20th Int. Offshore Polar Eng. Conf.* **7,** 163–169 (2010).

5. Qiu, B. W. Research on Ductile Fracture of X80 Pipeline Steel and Finite Element Simulation [in Chinese]. (2010).

6. Wilkowski, G., Rudland, D., Shim, D., Brust, F. & Babu, S. *Advanced Integration of Multi-scale Mechanics and Welding Process Simulation in Weld Integrity Assessment.* (2008).

7. Rudland, D. . *et al.* Experimental investigation of CTOA in linepipe steels. *Eng. Fract. Mech.* **70,** 567–577 (2003).

8. Lemaitre, J. A Continuous Damage Mechanics Model for Ductile Fracture. *J. Eng. Mater. Technol.* **107,** 83–89 (2014).

9. Shim, D.-J., Wilkowski, G., Duan, D.-M. & Zhou, J. Effect of Fracture Speed on Ductile Fracture Resistance - Part 1: Experimental. in *Proc. 8th Int. Pipeline Conf.* (2010).

Proceedings of the 25th CANCAM
London, Ontario, Canada, May 31 – June 4, 2015

OVERCOMING BRITTLENESS THROUGH BIO-INSPIRATION AND MICROARCHITECTURE

Mohammad Mirkhalaf
Mechanical Engineering Dept.
McGill University
Montreal, QC, Canada
mohammad.mirkhalaf@mail.mcgill.ca

Ahmad K. Dastjerdi
Mechanical Engineering Dept.
McGill University
Montreal, QC, Canada
ahmad.khayer@mail.mcgill.ca

Francois Barthelat
Mechanical Engineering Dept.
McGill University
Montreal, QC, Canada
francois.barthelat@mcgill.ca

ABSTRACT

The fracture of highly mineralized natural materials such as bone, teeth or seashells is largely controlled by the interfaces they contain. These interfaces, relatively weak, deflect and guide cracks into configurations which eventually impede their propagation. As a result, weaker interfaces turn brittle minerals into tough materials which can deform and absorb energy from impacts. To explore these concepts in synthetic materials we used a three-dimensional laser-engraver to carve arrays of microcracks with well-defined geometries and toughness within the bulk of borosilicate glass slides. The microcracks, positioned along specific surfaces within the material, coalesce upon application of an external load and guide large cracks following the concept of "stamp holes". Using this approach we engraved curved interfaces with re-entrant features ahead of an incident crack in glass, defining jigsaw-like building blocks. After initial crack propagation along the weak interface, the jigsaw features produced a tremendous amount of toughness by pull out and interlocking, dissipating energy by friction. This powerful nonlinear failure mechanism, inspired by natural nacre, amplified the toughness of glass 100 times. Infiltrating the interfaces with polyurethane generated additional toughening mechanisms by ligament bridging, similar to the mechanics of proteins in hard biological materials. This additional step led to a material made of about 95% of glass but 200 times tougher, as stiff as plain glass but displaying non-linear deformations and failing at strains in excess of 3%. This bio-inspired glass not only shows how the powerful toughening mechanisms observed in nature can be harnessed in synthetic materials, but it also showcase a new approach to making bio-inspired composite materials by "fabricating" weak interfaces within stiff materials rather than assembling stiff building blocks within soft matrices.

KEYWORDS: Glass, bio-inspired materials, toughening mechanisms.

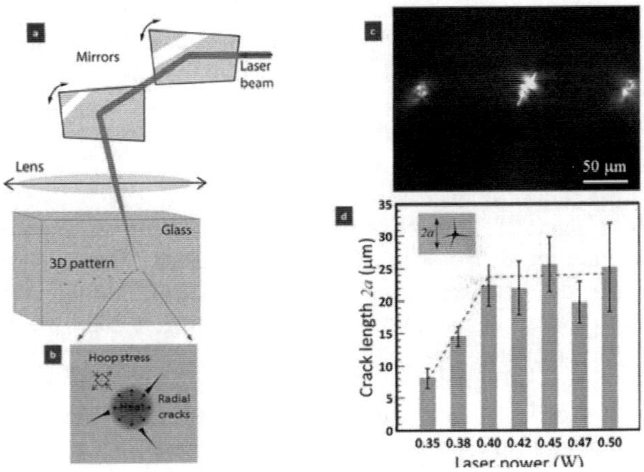

Figure 1: (a) Three-dimensional laser engraving setup. A laser beam is aimed and focused with high accuracy in the bulk of a block of glass. (b) The absorption of the energy provided by the laser raises the temperature locally, which generates hoop stresses and microcracking around the focal point. (c) Optical image of a line of micro-defects engraved within glass. (d) The size of the micro-defects can be controlled by adjusted the power of the laser.

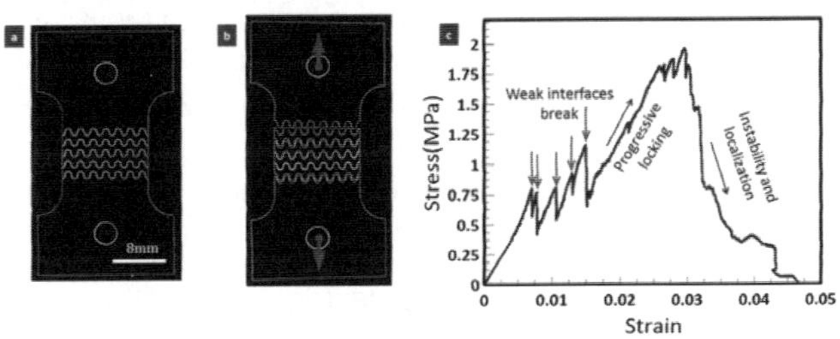

Figure 2: (a) initial and (b) most-mortem sample. (c) Typical stress-strain curve for engraved glass with five parallel jigsaw-like sutures. The arrows indicate the drops in stress corresponding to the sequential fracture of the five engraved interface.

SIMULATION OF DENT-CRACK DEFECT IN PIPELINE USING FEA

Hossein Ghaednia
Department of Civil and Environmental Engineering
University of Windsor
Windsor, Ontario, Canada
ghaedni@uwindsor.ca

Sreekanta Das
Department of Civil and Environmental Engineering
University of Windsor
Windsor, Ontario, Canada
sdas@uwindsor.ca

Jamshid Zohrehheydariha
Department of Civil and Environmental Engineering
University of Windsor
Windsor, Ontario, Canada
zohreh@uwindsor.ca

ABSTRACT

Experimental testing is the most conventional and reliable way to study the behavior of pipe while subjected to concentrate load and other loads. However, it is impossible to obtain all the information required for a thorough understanding from the experimental data. Experimental testing is also expensive and time consuming. Hence, it is not viable to consider full-scale tests for a wide range of test parameters. An alternative method to study and predict the behavior of such specimens is to use numerical tools such as finite element analysis (FEA) method. In this study, numerical modeling technique considering both material and geometric nonlinearity was employed to simulate the behavior of the dented pipes with a crack. In this study, developing of three-dimensional finite element pipe specimens were discussed and validated for determining structural behavior and burst strength of a pipe with the dent-crack defect. The load-deformation behavior, distributions of important strains and burst strength of defected pipes subject to this type of lateral denting load obtained from finite element analysis is discussed in this paper.

KEYWORDS: Dent-crack defect, FEA, oil and gas pipeline,

INTRODUCTION

Excavation operation often cause surface defects such as dent, gauge, crack, and/or their combinations in the field oil and gas pipeline. These defects, often called mechanical damages, and it is widely accepted that mechanical damage is the most common cause for pipeline failure in services [1]. It has been reported that the failure of oil and gas transmission pipelines resulting from mechanical damages ranges from 55% in the USA to around 70% in Europe [2-4].

The excavation in and around a field buried pipeline can also, cause a combination of dent, crack, and metal loose (gauges) in a small area on a pipe surface. Many investigations have been carried out to understand the effect of metal lose on pipeline such as gauges or cracks to predict the burst strength of pipeline analytically and experimentally [5]. Detailed guidelines on predicting the burst strength of such pipes are available in various pipeline design and maintenance codes and standards [6-8].

Several other studies completed to determine the behavior of pipe with single defect such as a dent or a crack or a gouge and the burst strength when such a single defect forms [9-12]. Also pipes with dent-gouge defect was subjects of many previous studies [13]. However, review of literatures available in the public domain reveals that no studies were undertaken to determine the burst strength of dented pipe that has developed crack in the dent. Such a combined defect is called dent-crack defect in this study. Therefore, this study was completed at the Centre for Engineering Research in Pipelines (CERP), University of Windsor to numerically predict the burst strength of NPS30 X70 pipe when a dent-crack defect develop on their surface. The numerical model was validated using full-scale test data [14, 15]. This paper discusses the results obtained from this research.

FINITE ELEMENT MODELING AND ANALYSIS

In this study, numerical modeling technique considering both material and geometric nonlinearities was used to simulate the behavior of the pipe with a dent-crack defect. Commercially available general purpose finite element analysis code, ABAQUS/Standard version 6.13 distributed by SIMULIA [16] was used to model the pipe behavior. The objective of developing the finite element model was to determine the burst strength of NPS30 X70 grade pipe used in this study when a crack-dent defect has developed on the pipe surface.

A full-length FE model, with the axial length of about $3D_0$, was adopted. Appropriate boundary conditions were applied along all edges of the full-length model. The model was created to closely reflect the pipe specimen and loading steps which was tested in previous study completed by CERP [14, 15]. The pipe and its defect specification are presented in Table 1.

Table 1: Pipe specification

Steel Grade	Outer Diameter	Thickness	Crack length	Crack depth	Dent length	Internal pressure
X70	762 mm	8.5 mm	100 mm	2 mm	100 mm	550 psi

Figure 1 shows a typical mesh used in numerical modelling. Vicinity of dented area was modeled with solid element, C3D8R (8-node linear brick), with reduced integration through the thickness. Minimum of four elements were used through the wall thickness. However, the 10-node quadratic (second order) tetrahedron solid element, C3D10, was used in and around the dent where geometry was complex and stress concentration was expected to be high (Figure 1a). Notch tip was modeled using Solid element, C3D20R (Figure 1b). The indenter and the T-shaped supports were modeled as discrete rigid body using 4-node rigid quadrilateral element, R3D4, because these supports and the indenter were relatively much more rigid than the pipe specimen and the deformations in these supports and the indenter were negligible. Mesh convergence studies were performed to obtain optimum sizes of various elements used in the FEA model.

a) Meshed pipe

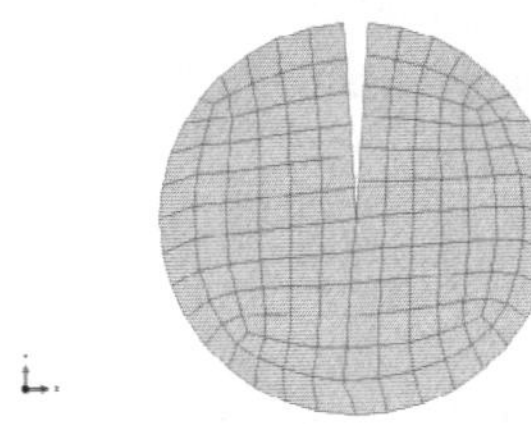

b) Mesh distribution around crack tip

Figure 1: Mesh distribution

The internal pressure of 550 psi was applied to the inner surface of the pipe wall. In the tests, very thick (12 mm) hemispherical steel end caps were used at both ends of the pipe specimen to hold the water and to apply internal pressure. Hence, in the FEA model, these end caps were modeled as an elastic-plastic steel with modulus of elasticity of 200 GPa.

The critical J integral (J_{1c}) can be used to predict the crack initiation in a pipe with dent-crack defect. Also, it can be assumed that a failure occurs when the maximum equivalent plastic (MEP) strain at any integration point reaches the maximum equivalent plastic strain at rupture obtained from material tensile tests. As result, the rupture strain can be conservatively considered as unit [17, 18]. In the FE model, it was assumed that a failure occurs when the J-integral (J) at any integration point around crack tip reaches the maximum value of 1.15 J_{1c} which is equivalent to plastic strain of unit. The J_{1c} for the pipe material obtained from fracture toughness tests in accordance with the ASTM E1820 [19]. The critical J integral (J_{1c}) was found to be 136 N/mm.

VALIDATION OF FE MODELS

The results obtained from the numerical analyses were compared with the test results of the test specimen from the previous study that has been done by CERP. Figure 2 shows the comparison of the load-deformation behavior obtained from the numerical models and test specimen in the laboratory. From Figure 2 it can be found that there is a good agreement between the experimental result and the numerical predictions.

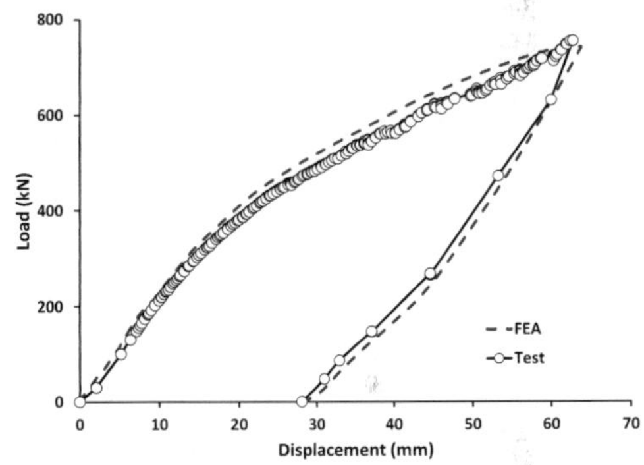

Figure 2: Comparison between experimental and numerical load-deformation

The comparison between the strain distribution obtained from the experimental result and finite element model in the circumferential and longitudinal directions are shown in Figure 3a and 3b. One of the objective of developing the finite element model was to determine the strain distributions under the indenter which is impossible to obtain otherwise from the experimental data. From Figure 3, it can be observed that there is a reasonably good agreement between the strain data obtained from strain gauges in the test specimen and those obtained from the numerical model. From Figure 3, it can also be observed that the maximum strain in the circumferential direction is tensile strains and it is 0.68%. In the longitudinal direction, there is no compressive strain as well and maximum tensile strains is 0.81%.

From Figures 3 it can be observed that maximum strain is not located inside the dented area. For example, maximum strain in circumferential direction occurs at 200 mm away from dent center, and for longitudinal direction, maximum strain is found 325 mm away from the dent center.

The burst strength of the pipe specimens in FE model was predicted using J_{1c} which was obtained from fracture toughness tests [19]. Crack initiation and burst strength of the specimen are shown in Figure 4. The J_{1c} for crack initiation is assumed to be 136 N/mm and the $1.15J_{1c}$ for the burst pressure prediction is assumed to be 156.4 N/mm. Figure 4 also shows the burst strength of specimen obtained from the test and FEA.

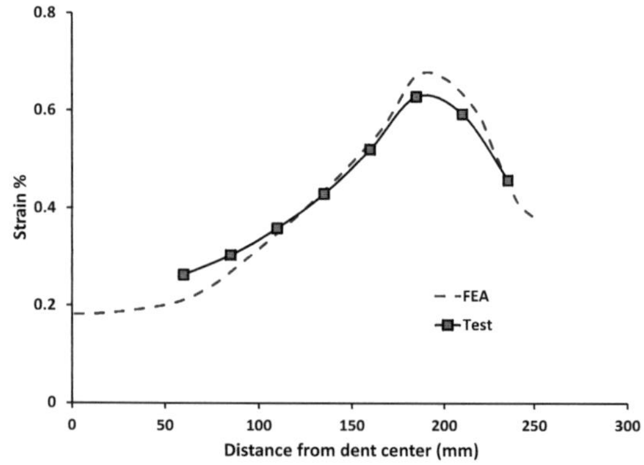

a) Circumferential strains

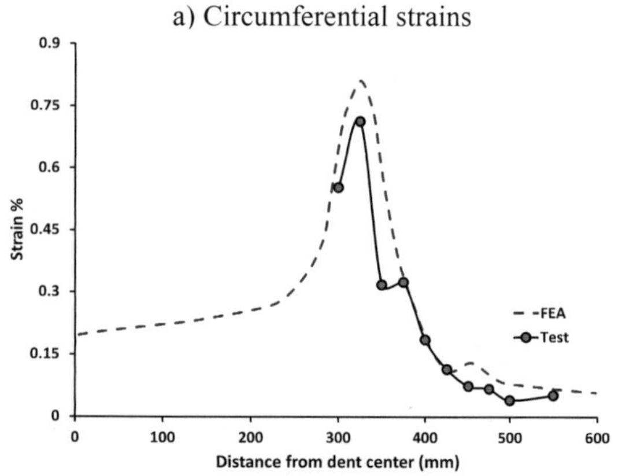

b) Longitudinal strains

Figure 3: Comparison between experimental and numerical strain distributions

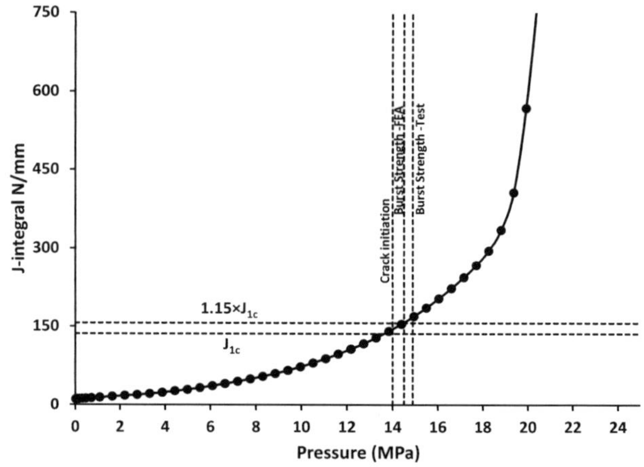

Figure 4: J-integral-pressure curve

The value of yield pressure (p_y) for NPS30 X70 grade pipe specimen used in this study is 1837 psi (12.67 MPa). For this specimens which had crack length of 100 mm, crack depth of 2 mm, and 4% of dent depth, the burst strength was 2160 psi (14.89 MPa). In other words, it can be said that the burst strength of this pipe specimen was found to be $1.17p_y$. Also from Figure 4, the predicted burst strength was found to be 2110 psi (14.55 MPa). This figure indicates that the burst pressure obtained from FEA is slightly conservative.

CONCLUSIONS

The following conclusions are made in this paper based on the finite element analyses completed in this study. Hence, these conclusions are limited to the pipe specimen, test method, defect type, and numerical study of this research work.

The maximum value of strains in both longitudinal and circumferential direction occurred outside the dented region. Also it was found for this specific indenter size and shape, there was no compressive strain in and around dented area.

Finite element model showed good agreement with test results in load-deformation behavior, and strain distribution.

FEA model predicts the burst strength well though the prediction is slightly conservative.

ACKNOWLEDGMENTS

The authors acknowledge the financial support received from the Natural Sciences and Engineering Research Council of Canada (NSERC) located in Ottawa, ON, Canada and TransCanada Pipelines located in Calgary, AB, Canada. The authors also acknowledge the support received from Center for Engineering Research in Pipelines (CERP) located in the University of Windsor.

REFERENCES

1. Cosham, A. and P. Hopkins, *The effect of dents in pipelines-guidance in the pipeline defect assessment manual.* International Journal of Pressure Vessels and Piping, 2004. **81**(2): p. 127-39.
2. Smith, R.B. and D.N. Gideon, *STATISTICAL ANALYSIS OF DOT-OPSO DATA.* Proceedings, Annual Symposium - Society of Flight Test Engineers, 1979: p. D. 1-D. 9.
3. Wang, K.C., et al., *The Effect of Mechanical Damage on Fracture Initiation in Line Pipe: Dents*. 1982: Energy, Mines and Resources Canada, CANMET, Energy Research Program.
4. Zarea, M.F., et al. *Numerical models for static denting and dynamic puncture of gas transmission linepipe and their validation*. 1996.
5. Staat, M. and V. Duc Khoi, *Limit analysis of flaws in pressurized pipes and cylindrical vessels. Part II: Circumferential defects.* Engineering Fracture Mechanics, 2013. **97**: p. 314-33.
6. CSA, *Z662-07: Oil and gas pipeline systems*. 2007, Canadian Standard Association: Mississauga, ON, Canada.
7. ASME, *B31G-1991 (R2004): Manual for determining remaining strength of corroded pipelines: supplement to B31 code-pressure piping*. 2004, ASME International: New York, NY, USA.
8. DNV, *OS-F101: Submarine Pipeline Systems*. 2012, Det Norske Veritas.
9. Allouti, M., et al., *Study of the influence of dent depth on the critical pressure of pipeline.* Engineering Failure Analysis, 2012. **21**: p. 40-51.
10. Blachut, J. and I.B. Iflefel, *Collapse of pipes with plain or gouged dents by bending moment.* International Journal of Pressure Vessels and Piping, 2007. **84**(9): p. 560-71.
11. Ghaednia, H., et al. *Out-of-roundness in NPS30 X70 pipes subjected to concentrated lateral load.* in *Proceedings of the Biennial International Pipeline Conference, IPC.* 2014.
12. Silva, J., H. Ghaednia, and S. Das. *Fatigue life assessment for nps30 steel pipe*. in *2012 9th International Pipeline Conference, IPC 2012, September 24, 2012 - September 28, 2012*. 2012. Calgary, AB, Canada: American Society of Mechanical Engineers.
13. Allouti, M., C. Schmitt, and G. Pluvinage, *Assessment of a gouge and dent defect in a pipeline by a combined criterion.* Engineering Failure Analysis, 2014. **36**: p. 1-13.
14. Ghaednia, H., et al., *Pressure tests on 30-in diameter X65 grade pipes with dent-crack defects.* The Journal of Pipeline Engineering, 2013. **12**(1): p. 7.
15. Ghaednia, H., et al. *Effect of dent depth on the burst pressure of NPS30 X70 pipes with dent-crack defect.* in *Proceedings of the Biennial International Pipeline Conference, IPC*. 2014.
16. SIMULIA, *Analysis User's Manuals*. Vol. 1-4. 2013, Rising Sun Mills, Providence, RI, USA: Dassault Systèmes Simulia Corp.
17. Das, S., J.J.R. Cheng, and D.W. Murray, *Prediction of fracture in wrinkled energy pipelines subjected to cyclic deformations.* International Journal of Offshore and Polar Engineering, 2007. **17**(3): p. 205-212.
18. Das, S., J.R. Cheng, and D.W. Murray, *Behavior of wrinkled steel pipelines subjected to cyclic axial loadings.* Canadian Journal of Civil Engineering, 2007. **34**(5): p. 598-607.
19. ASTM, *E1820-13: Standard Test Method for Measurement of Fracture Toughness*. 2013, ASTM International: West Conshohocken, Pennsylvania, USA.

UNDERSTANDING THE STRAIN DISTRIBUTIONS CAUSED BY A DENT DEFECT

Hossein Ghaednia
Department of Civil and Environmental Engineering
University of Windsor
Windsor, Ontario, Canada
ghaedni@uwindsor.ca

Sreekanta Das
Department of Civil and Environmental Engineering
University of Windsor
Windsor, Ontario, Canada
sdas@uwindsor.ca

Jamshid Zohrehheydariha
Department of Civil and Environmental Engineering
University of Windsor
Windsor, Ontario, Canada
zohreh@uwindsor.ca

Sara Kenno
Department of Civil and Environmental Engineering
University of Windsor
Windsor, Ontario, Canada
kenno2@uwindsor.ca

ABSTRACT

Structural integrity of oil and gas transmission pipelines is often threatened by external interferences such as concentrated lateral loads and as a result, a failure of the pipeline may occur due to "mechanical damages". Sometime, this load may not cause immediate rupture of pipes; rather form a dent which can reduce the pressure capacity of the pipeline. This kind of defect is a matter of serious concern for the pipeline operator since a rupture or a leak may occur. Accordingly, an extensive experimental study was completed at the Centre for Engineering Research in Pipelines (CERP), University of Windsor on 30 inch (762 mm) diameter and X55 grade pipes with D/t of 90. The aim of this research is to examine the influence of dent depth on strain distributions of dented pipe. The distributions of important strains of pipes subject to this type of lateral denting load obtained from experimental study is discussed in this paper.

KEYWORDS: Dent defect, dent depth, energy pipe

INTRODUCTION

External interference causes various defects in pipelines and this significantly affects the transportation of oil and gas. Corrosion, crack, puncture, dent, gouge, and combination of such damages from a variety of external interferences are some common examples of surface damage in pipelines. Gouges, dents, cracks, and punctures that form in the pipe wall as a result of contact and/or impact from foreign objects are often referred to as mechanical damage. Mechanical damage of oil and gas pipelines is referred to as the major cause of failure of pipelines in service and this damage may result in loss of product, explosions, fire, human and/or animal casualties, and pollution. It has been reported that the failure of oil and gas transmission pipelines resulting from mechanical damages ranges from 55% in the USA to around 70% in Europe [1-4].

A dent is an inward permanent plastic deformation of the pipe wall which causes a gross distortion of the pipe cross section[5]. A dent also causes stress and strain concentrations and a local reduction in the pipe diameter. Dent defect in energy pipelines has been a major concern for pipeline operators. Therefore, a significant number of investigations were completed to understand the effect of dents of various shapes and sizes on the behavior of energy pipes under monotonic and cyclic pressures. Many researches on other pipeline defects were completed as well [6-9]. It has been found and generally accepted that dent defect is not a threat to the integrity of pipe structure under monotonic pressure load when the dent depth is up to 6% of the diameter of the pipe [10-12]. However, depth is not always the most useful parameter for identifying whether a dent is a threat to pipeline integrity or not. The use of depth alone can result in unnecessary excavations of many dents which may not necessarily be harmful to the structural and operational integrity of the pipeline. On the other hand, it is also possible that a dent with a small dent depth could be a threat to the structural integrity of the pipeline if the change in curvature in the dent is high. Hence, strain values and their distributions are the appropriate and more relevant criteria for judging the severity and acceptability of a dent defect [13, 14]. Therefore, the distributions of strain in the pipe wall due to the formation of dent are of particular interest for this study. This study was being undertaken at the Centre of Engineering Research in Pipeline (CERP), University of Windsor. This paper discusses the results obtained from this study.

EXPERIMENTAL WORK

Test specimen

The purpose of conducting full-scale tests was to determine experimentally the critical strain values around dented area. Therefore, the size and material properties of pipe specimens were selected in such a way that represents the properties of pipes typically used in the oil and gas pipeline industry. Two specimens were fabricated from pipe with outer diameter of 762 mm, thickness of 8.5 mm, and conforming to grade, API 5L X55 and X70 steel pipes [15]. Most of the pipelines used in the current practice have a diameter to thickness (D/t) ratio ranging from 30 to 100. The D/t of the specimens used in this work was about 90. The length of each pipe specimen was 2500 mm. Two ends of the specimen were welded to semi-spherical steel caps of 12 mm thick for holding the water pressure. Three tensile specimens were cut from the virgin pipe section and the standard tensile tests were carried out to determine the mechanical properties of the pipe material in accordance with ASTM E8 [16]. The properties of the material which were obtained from the tensile test are presented in Table 1.

Table 1: Material properties

Material type	Yield strength	Tensile strength	Modulus of elasticity
X55	345 MPa	508 MPa	205 GPa
X70	540 MPa	620 MPa	204 GPa

The main parameter of the study was: the dent depth. The dent depth was varied from 2% to 4%. The internal (operating) pressure was hold on 550 psi (3.8 MPa) during indentation process. Details of the test specimens are given in Table 2 and Figure 1 shows the test specimen and test setup.

Table 2: Pipe specification

Specimen	Dent length	Dent depth	Internal pressure
X55-1	100 mm	28 mm (4%)	550 psi
X55-2	100 mm	15 mm (2%)	550 psi
X70-3	100 mm	28 mm (4%)	550 psi

Figure 1: Test setup

Test setup

Figure 1 shows the test setup. The pipe specimens were placed on three thick steel supports. Denting load was applied on the top surface of the pipe wall using a universal loading actuator. The total and plastic deformations of the pipe due to application of denting load was measured using linear variable differential transformers (LVDTs). Two LVDTs were used. The LVDTs were located at the top of actuator to acquire downward displacement of the indenter. Hemispherical steel end caps were welded at the end of each specimen to pressurize the pipes using a hydrostatic pump. The dent depth varied for different specimens. Specimens were instrumented with strain gauges to monitor and record the variations of strain around the dent. Two different lines of strain gages consisting of eight to ten strain gauges on each line were installed. The layout of the strain gauges is shown in Figure 2.

Test Procedure

For all three specimens same loading procedure was used. First, the pipe specimen was filled with water and pressurized using a hydrostatic pump. An internal pressure of 550 psi (0.3py) was applied in all specimens. The pressure, p_y, is the water pressure required for yielding of pipe material in the circumferential direction. Next, a monotonically increasing denting load was applied using the displacement control method while keeping the level of internal pressure unchanged. The internal pressure simulates operating pressure of the pipeline. The denting load was applied in several loading and unloading steps. After completion of each load step, that is, after complete removal of denting load, the internal pressure in the pipe specimen was reduced to zero. The objective was to obtain strain data when the pipe is completely unloaded. For all the specimens, the test was carried out and controlled to produce a dent with permanent depth of 2% and 4% of the outer diameter of the pipe.

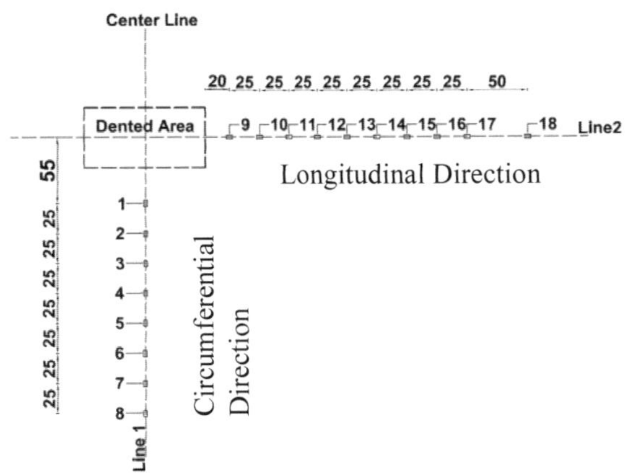

Figure 2: Strain gauge layout

TEST RESULTS

Strain data obtained from the strain gauges installed in the circumferential direction (Line 1 in Figure 2) is presented in Figure 3a. From the strain gauge readings, it is observed that as the dent depth increases the values of maximum circumferential strains increase as well. The highest circumferential strain for specimens X55-1 and X55-2 after unloading was 1.65% and 1.30%, respectively. The maximum values of tensile circumferential strain were obtained from the strain gauges 6 and 4, which were located at 180 and 130 mm away from the center of the indenter along the perimeter of the pipe (Figure 2), respectively. Hence, internal pressure not only affect the amount of maximum strain, but also change the location of that.

Figure 3a also compares strain distributions in the circumferential direction for specimens, X55-1 and X70-3, with the same operating pressure but different steel grade. In specimen X70-3, maximum tensile circumferential strain was 1.37% obtained from strain gauge 4, which was located 130 mm away from the center of the indenter along the perimeter of the pipe. However, for specimen X55-1, maximum tensile strain was 1.65% at strain gauge 6, which was located 180 mm away from the indenter center along the perimeter of the pipe (Figure 2). Hence, the specimen with the weaker material (X55-1) was subjected to a higher strain. This strain occurred at a further distance from the center of indenter than the maximum strain in specimen with the stronger material (X70-3).

Figure 3b shows the distributions of strains in the longitudinal direction and it shows similar trends as in Figure 3a. For specimens X55-1 and X55-2 maximum strain was recorded at strain gauge 11 (Figure 2), which was located at 120 mm away from the center of indenter. The maximum strain recorded was 2.88% for X55-1 and 1.82% for X55-2. It should be noted that all the strain values recorded from the test specimens are tensile.

For specimens X55-1 and X70-3 maximum strain was recorded at strain gauges 11 and 10, respectively (Figure 2), which were located at 120 and 95 mm away from the center of indenter. The maximum strain recorded was 2.88% for X55-1 and 1.92% for X70-3. It should be noted that all the strain values recorded from the test specimens are tensile. It is worth noting, maximum circumferential and longitudinal tensile strain in both specimens X55-2 and X70-3 are almost the same, however, the dent depth in X55-2 (2%) is half of the dent depth in X70-3 (4%).

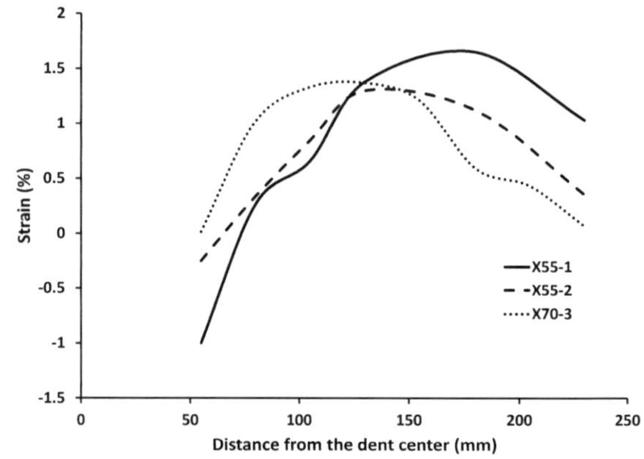

a) Circumferential strains

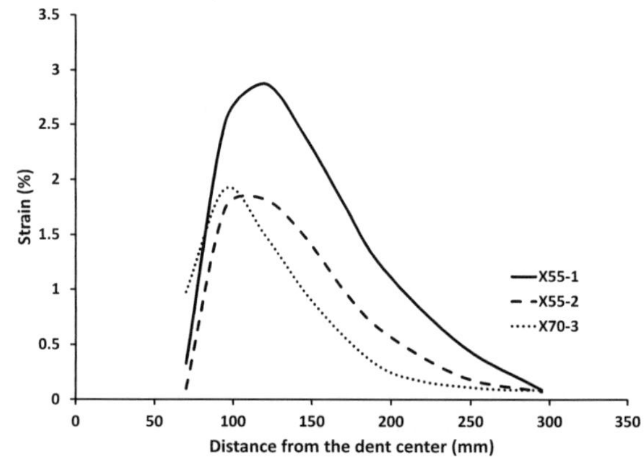

b) Longitudinal strains

Figure 3: Strain distributions

CONCLUSIONS

The following conclusions are made based on the tests presented in this paper. Hence, the conclusions are limited to the findings associated with specific test specimens presented in this paper.

Strain distributions in the dented region are highly dependent on the dent depth and steel grade. Magnitude of the maximum strains in the dented region is also influenced by the both the material properties and the dent depth.

Evaluation of severity of a dented pipeline solely based on dent depth is not appropriate. Rather, critical strain values must be included in the dent assessment criteria.

ACKNOWLEDGMENTS

The authors acknowledge the financial support received from the Natural Sciences and Engineering Research Council of Canada (NSERC) located in Ottawa, ON, Canada and TransCanada Pipelines located in Calgary, AB, Canada. The authors also acknowledge the support received from Center for Engineering Research in Pipelines (CERP) located in the University of Windsor.

REFERENCES

1. Smith, R.B. and D.N. Gideon, *STATISTICAL ANALYSIS OF DOT-OPSO DATA.* Proceedings, Annual Symposium - Society of Flight Test Engineers, 1979: p. D. 1-D. 9.
2. Wang, K.C., et al., *The Effect of Mechanical Damage on Fracture Initiation in Line Pipe: Dents*. 1982: Energy, Mines and Resources Canada, CANMET, Energy Research Program.
3. Zarea, M.F., et al. *Numerical models for static denting and dynamic puncture of gas transmission linepipe and their validation*. 1996.
4. Lancaster, E.R., *Burst pressures of pipes containing dents and gouges.* Proceedings of the Institution of Mechanical Engineers, Part E: Journal of Process Mechanical Engineering, 1996. **210**(1): p. 19-27.
5. Cosham, A. and P. Hopkins, *The effect of dents in pipelines-guidance in the pipeline defect assessment manual.* International Journal of Pressure Vessels and Piping, 2004. **81**(2): p. 127-39.
6. Das, S., J.R. Cheng, and D.W. Murray, *Behavior of wrinkled steel pipelines subjected to cyclic axial loadings.* Canadian Journal of Civil Engineering, 2007. **34**(5): p. 598-607.
7. Das, S., J.R. Cheng, and D.W. Murray. *Fracture in wrinkled linepipe under monotonic loading*. in *2002 4th International Pipeline Conference*. 2002. American Society of Mechanical Engineers.
8. Silva, J., H. Ghaednia, and S. Das. *Fatigue life assessment for nps30 steel pipe*. in *2012 9th International Pipeline Conference, IPC 2012, September 24, 2012 - September 28, 2012*. 2012. Calgary, AB, Canada: American Society of Mechanical Engineers.
9. Ghaednia, H., et al. *Out-of-roundness in NPS30 X70 pipes subjected to concentrated lateral load.* in *Proceedings of the Biennial International Pipeline Conference, IPC*. 2014.
10. ASME, *B31.4: Pipeline Transportation Systems for Liquids and Slurries*. 2012, ASME International: New York, NY, USA.
11. CSA, *Z662-07: Oil and gas pipeline systems*. 2007, Canadian Standard Association: Mississauga, ON, Canada.
12. Md Rafi, A.N., et al., *Revisiting ASME strain-based dent evaluation criterion.* Journal of Pressure Vessel Technology, Transactions of the ASME, 2012. **134**(4).
13. Gao, M., et al. *Strain-Based Models for Dent Assessment: A Review*. in *2008 7th International Pipeline Conference*. 2008. American Society of Mechanical Engineers.
14. Baker, M., *Department of Transportation, Office of Pipeline Safety, TTO Number 10, Integrity management program–dent study.* Delivery Order DTRS56-02-D-70036, final report, 2004.
15. API, *Spec 5L: Specification for Line Pipe*. 2012, American Petroleum Institute: Washington, DC, USA.
16. ASTM, *E8/E8M-11: Standard Test Methods for Tension Testing of Metallic Materials*. 2011, ASTM International West Conshohocken, Pennsylvania, USA.

Geophysical Fluid Dynamics

Proceedings of the 25th CANCAM
London, Ontario, Canada, May 31 – June 4, 2015

EXAMINATION OF HELMHOLTZ VORTEX LAWS USING RANKINE COMBINED VORTICES IN CIRCULAR CYLINDERS

Josué Njock Libii
Engineering Department
Indiana Univ.-Purdue Univ. Fort Wayne
Fort Wayne, IN, USA
Libii@ipfw.edu

ABSTRACT

Rankine combined vortices are created in water contained in circular cylinders that are vertical. The objective is to use such vortices for an experimental examination of Helmholtz vortex laws. Ten identical cylinders are used. Each cylinder has an orifice of different size at its bottom through which water can exit under two different circumstances: under the sole action of gravity or under the combined action of gravity and a vertical vortex created by stirring the water inside the cylinder about the vertical axis of the cylinder. Changing the size of the exit orifice changes the flow rate out of the cylinders; it also causes the streamlines followed by the exiting fluid to experience different amounts of constriction near the exit orifice. Vorticity meters are designed and used to measure vorticity at the free surface of the water in the tank and at points below that surface. It is found that vorticity increases in flow regions where streamlines are either constricted, or stretched, or constricted and stretched, as predicted by Helmholtz vortex laws.

KEYWORDS: Vorticity, Helmholtz vortex laws, Rankine combined vortex, conservation of angular momentum in a vortex, use of vorticity meters

INTRODUCTION

There are many kinds of vorticies and the literature on this subject is both mature and deeply rooted in the mathematical modelling of inviscid flows[Lamb, Batchelor, Wang, Saffman, Shtern, Lautrup]. We focus on a very common and well known case: that of the formation of a vortex near a small orifice that drains water from, say, a tank or bathtub, under the influence of gravity. This vortex is ordinarily explained using the principle of the conservation of angular momentum of incompressible fluid particles undergoing rotational motions. According to this principle, as fluid particles move closer to the axis of rotation, their radial positions decrease while their angular velocities increase in such a way that the product of the radial position and the speed of the particle remains constant throughout the motion.

As a first approximation, this phenomenon is modelled as a Rankine-combined vortex, which is the superposition of two plane elementary flows: an inviscid sink and an irrotational vortex that extends to infinity [Lamb]. Physically, this combined vortex consists of an inner cylindrical core of finite radius $r = R$, that is in rigid-body rotation and is surrounded by a potential flow with a circumferential velocity that is proportional to $1/r$, where r is the radial distance measured from the axis of rotation.

In the actual sink vortex, however, the radial flow is more complicated than is indicated by an inviscid sink. Complication arises for two reasons: Whereas the Rankine model is two dimensional and flow stays in one plane, in practice, flow is three dimensional and fluid is forced into the exit orifice from three different places: directly by the pull of gravity, by suction of water by the sink effect but there is also flow to the outlet from inside the viscous boundary layer that is formed at the bottom of a tank. Khoo et al.(1997) observed the detailed structure of the boundary layer beneath a Rankine-like vortex.

Noguchi, Yukimoto, Kimura and Niino (2003) studied the structure and instability of a sink vortex. They used a cylindrical tank made of plexiglass with a diameter of 48 cm, that could rotate with various speeds about the central axis. An exit orifice of diameter of 2.5 cm was centrally located at the bottom of the tank. Two parameters were used: the flow rate out of the tank and the rate of rotation of the rank about its vertical axis. The authors reported data collected at a height of 10 cm above the exit orifice using Particle Image Velocimetry (PIV) and Laser Induced Fluorescence (LIF). They tested flows generated at two rotation rates of 0.4 rad/s and 0.2 rad/s and two flow rates of approximately 34 cm^3/s and 160 cm^3/s . This process yielded four sets of data, one for each combined pair of parameters: RS (0.4 rad/s, 160 cm^3/s), Rs (0.4 rad/s, 34 cm^3/s), rS(0.2 rad/s, 150 cm^3/s), and rs(0.2 rad/s, 34 cm^3/s).

These authors reported the following: whether observed flow was similar to the Rankine combined vortex depended upon the magnitude of the exit flow rates. At a low flow rate (around 34 cm^3/s), experimental results showed that the conservation of angular momentum in the radial direction did not hold and, as a consequence, experimental results departed considerably from what was expected from the Rankine combined vortex. However, at a flow rate around160 cm^3/s, results were similar to what is predicted by the Rankine combined vortex. The streamlines obtained by sketching the instantaneous flow field revealed Rankine-like vortex that was almost steady. However, detailed observations demonstrated that the position of the axis of the vortex fluctuated about the axis of rotation of the tank. Ten different positions of the center of the vortex were determined during a complete revolution of the tank. The time elapsed between locations was constant and equal to 1/10 of the duration of the rotation of the tank. A sketch of the trajectory of the center of the vortex showed that, as it moved about, it traced a broken line of very irregular shape. Analysis of the coordinates of successive positions of the center of the vortex leads one to believe that the net displacement of the center of the vortex over a complete cycle is very small, causing one to assume that that center almost returns to its original position after one complete revolution of the tank.

In this paper, we fix the amount of vorticity that is given to the flow but vary the flow rates through the vortex and investigate the effect of the magnitude of the flow rate on the behavior of vorticity in a Rankine combined vortex.

We use Rankine combined vortices created in water contained in circular cylinders that are vertical. Each cylinder has an orifice at its bottom through which water can exit under either the sole action of gravity or under the combined actions of gravity and a vertical vortex created by stirring the water inside the cylinder about the vertical axis of the cylinder. Ten identical cylinders are tested; each cylinder is fitted with an exit orifice of a different diameter. Changing the size of the exit orifice changes the flow rate and causes the streamlines followed by the exiting fluid to experience different amounts of constriction near the exit orifice. Vorticity meters are designed and used to measure vorticity at the free surface of the water in the tank and at points below that surface, just above the exit orifice.
The tests measured

1. The effect of vorticity on the rate of efflux from the tank and on the time it took for a fixed amount of water to exit the tank; and
2. The effect of the size of the exit orifice on the magnitude of vorticity at the surface of the tank and below the surface of the tank.

Flow patterns associated with different exit orifices are studied. Collected data are used to infer the extent to which vortex lines are stretched based upon the measured increases in vorticity.

HELMHOLTZ'S VORTEX LAWS

Kelvin's circulation theorem states that the time rate of change of the circulation, Γ_C , associated with a closed curve C that is always made up of the same fluid particles is determined by the sum of the torques produced by all the forces that act on the fluid [NCFMF]. Thus, when pressure, body, and viscous forces create torques in the fluid, then, the time rate of change of the circulation around curve C is given by Eq. (1)

$$\frac{D\Gamma_C}{Dt} = -\oint \frac{dp}{\rho} + \oint \vec{G}\cdot\vec{dr} + \oint \frac{\mu}{\rho}\nabla^2\vec{V}\cdot\vec{dr} \qquad (1)$$

$$\oint \frac{dp}{\rho} = torques\ due\ to\ pressure\ forces$$

$$\oint \vec{G}\cdot\vec{dr} = torques\ due\ to\ body\ forces$$

$$\oint \frac{\mu}{\rho}\nabla^2\vec{V}\cdot\vec{dr} = torques\ due\ to\ viscous\ forces$$

When all factors that produce torques are either absent or negligible, then, the three Helmholtz's vortex laws hold, namely: 1) Vortex line cannot end in the fluid. They either form closed loops or end at a fluid boundary, and the circulation is the same for every contour that encloses a specific vortex line; 2) A fluid line that, at any instant of time, coincides with a vortex line will coincide with a vortex line forever. Therefore, vortex lines are said to be frozen to the fluid; and 3) On a vortex line of fixed identity, the ratio of the vorticity (ω) to the product of the fluid density (ρ) with the length (l) of the line remains constant as time proceeds ($\frac{\omega}{\rho l} = constant$). Thus, if the vortex line is stretched, the vorticity increases [NCFMF].

This paper presents the results of an experimental examination of Helmholtz vortex laws using Rankine combined vortices created in water contained in circular cylinders that are vertical. Each cylinder has an orifice at its bottom through which water can exit under the sole action of gravity or under the combined actions of gravity and a vertical vortex created by stirring the water inside the cylinder about the vertical axis of the cylinder. Ten identical cylinders are tested with each cylinder having been fitted with an exit orifice of a different diameter. Changing the size of the exit orifice changes the flow rate out of the cylinder; it also causes the streamlines followed by the exiting fluid to experience different amounts of constriction near the exit orifice. Vorticity meters are designed and used to measure vorticity at the free surface of the water in the tank and at points below that surface, just above the exit orifice.

TESTING APPARATUS

Figure 1 below shows a Solid Edge drawing of the model used to construct the testing apparatus and setup. It consists of a tank in the shape of an inverted bottle, the upper portion of which is a circular cylinder that is open to the atmosphere and the lower part of the tank is a funnel. A bottle with a capacity of 2 liters is securely fastened to a stable stand by means of a hanger. Below the stand is a catch tank that is used to collect and recycle water during experiments. Nine different caps had been 3-D printed so they could be threaded, one at a time, onto the mouth of the bottle in order to vary the exit flow rate from the tank. The outside diameter of each cap was the same as that of the mouth of the bottle (0.87 in) but each cap has a central orifice with a different diameter. The diameters of the central orifices varied with the cap and they increased from 10% to 100% of the diameter of the mouth of the bottle in steps of 10%, Figure 2. This made it possible to have ten different openings to test for.

EXPERIMENTAL PROCEDURE

Before the start of experiments, a selected cap is affixed to the mouth of the bottle with the exit orifice being closed. The bottle is filled with water and the water is allowed time to come to rest from the effects of the filling process. A stirrer is used to stir the water inside the bottle to add a specific amount of vorticity to it. Stirring causes the free surface of water in the tank to change its shape from a horizontal plane to that of a paraboloid of revolution with an upward facing concavity
[Njock Libii]. Once a fixed amount of vorticity has been added, the stirrer is removed and the free surface is allowed to become horizontal again. As soon as this happens, a vorticity meter is inserted into the rotating water and placed just below the free surface of water, Figure 3, and the central orifice is unplugged, allowing the rotating water to exit the tank; during this whole time the rotation of the vorticity meter is videotaped and saved for analysis at a later time. Once completed, a similar experiment is done with one single modification: the depth of insertion of the vorticity meter is changed. Once those two trials have been completed, a new cap is used and the process is repeated. This continues until all ten caps have been tested. In the data reported in this article, vorticity is measured at the free surface and also on a surface that is 6 inches below the free surface.

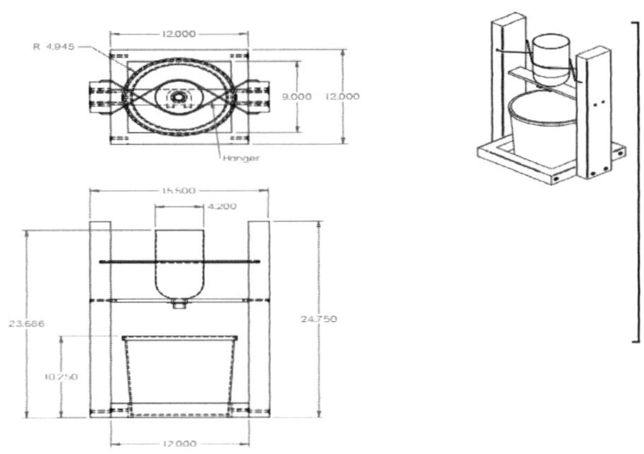

Figure 1: The Experimental Station

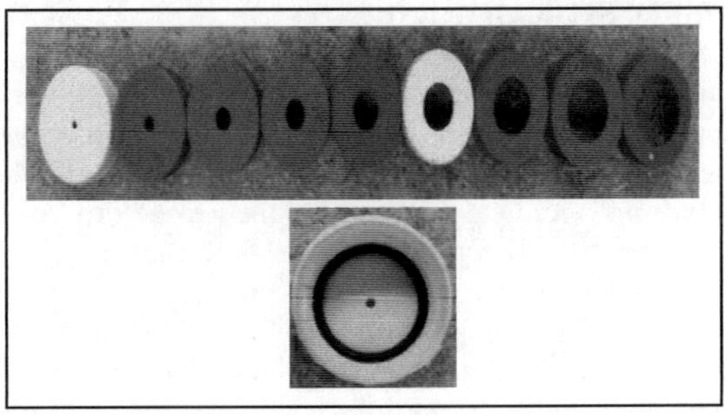

Figure 2: 3-D printed caps with central orifices

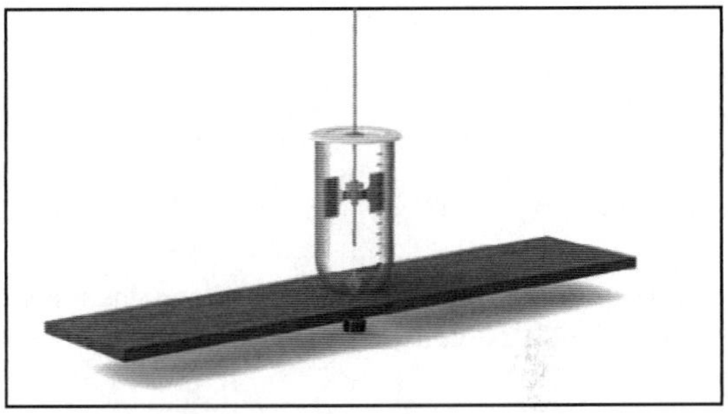

Figure 3: Vorticity meter inside the tank

RESULTS

For each of the ten caps, values of vorticity were measured continuously at the free surface and at a surface located 6 inches below the free surface. Measurements were taken with flow exiting the tank until the water level fell below the cylindrical portion of the tank. Values obtained at the free surface were averaged by dividing the total number of rotations of the vorticity meter by the total time it took the fluid to be evacuated from the cylindrical portion of the tank. Averages were calculated using the data obtained below the free surface similarly. Both averages were plotted in Figure 4. It can be seen from that Figure that submerged vorticities were larger than those measured at the free surface.

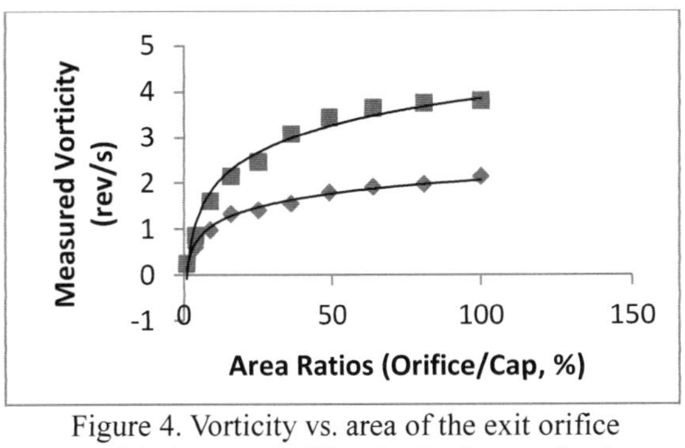

Figure 4. Vorticity vs. area of the exit orifice
Red (submerged); Blue (at the free surface)

DISCUSSION OF RESULTS

Once the exit orifice is opened and fluid starts to exit the tank, fluid particles must necessarily find their way towards the exit orifice. This causes the streamlines along which they move to converge towards the exit orifice. Such convergence brings fluid particles closer to the axis of rotation than before, thereby, reducing the radius r.

In the ideal Rankine combined vortex, the angular momentum of fluid partices is conserved. Thus, for a fluid particle that is rotating with a angular speed ω while it is located at a distance r from the axis of rotation, the conservation of its angular momentum requires that

$$\omega r^2 = constant.$$

It follows that a reduction in the radius r must be accompanied by an increase in the rate of rotation ω, which is what is observed in Figure 4. Thus, Figure 4 is in qualitative agreement with the conservation of angular momentum.

However, quantitative agreement is another matter. The rotating particle is on the periphery of a circle of radius r and area

$$A = \pi r^2$$

Hence, the conservation of angular momentum can be written using the area A as

$$\frac{\omega A}{\pi} = constant$$

Our data indicate that, in the cases tested, $\frac{\omega A}{\pi}$ is not a constant but is affected by the change is the area A. Thus, we write

$$A d\omega = C dA \quad (2)$$

After integration, this leads to

$$\omega = C_1 \ln(A) + C_2 \quad (3)$$

Note that, in our measurements, ω is an averaged vorticity over the time it took to empty the tank.

Here C_1 and C_2 are constants that depend upon the intial vorticity given to the fluid before flow started, the cross-sectional area of the central orifice, and that of the cap. Note that these quantities do not depend upon time because the vorticity calculated is not an instantaneous quantity. The data of Figure 4 follow a logarithmic trend, as shown by the fitted curve on that Figure.

The mechanism at work has been explained in the following manner in the literature [NCFMF): A vertical column of fluid that was initially on the axis of the tank is a vortex line. When flow is initiated in the vertical direction by opening the exit orifice, that column is stretched downwards towards the exit orifice, which increases vorticity below the free surface. This follows from the third vorticity law of Helmholtz, according to which the ratio of the vorticity (ω) to the product of the fluid density (ρ) with the length (l) of the line remains constant on a vortex line of fixed identity as time proceeds ($\frac{\omega}{\rho l} = constant$) [NCFMF].

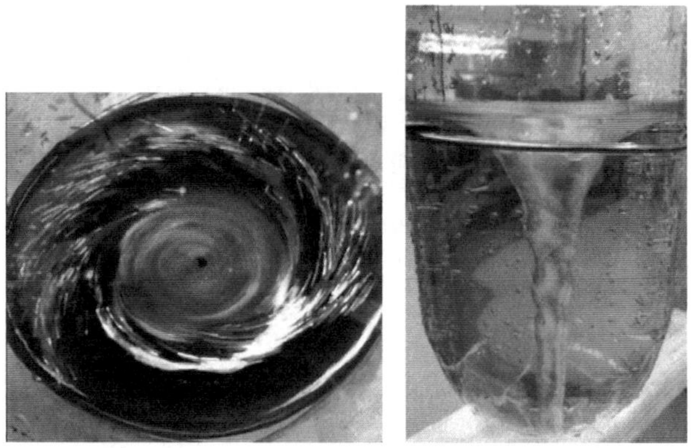

Figure 5: Top view (left) and side view of flow patterns

CONCLUSIONS

Vorticity was added to water in a cylindrical container and the rotating fluid so created was suddenly allowed to start draining through an exit orifice located at the bottom of the tank. When vorticity at the free surface was compared to submerged vorticity during the whole draining process, it was found that the latter was always greater than the former for each of the ten orifice sizes that were tested. The proposed explanation is that this is due to the fact that initially circular vortex lines are stretched as they move towards the exit orifice, causing their vorticity to increase.

Observation of the structure of the vortex core showed that changing the size of the exit orifice changed not only the volume flow rate out of the tank, which was expected, but also the flow patterns inside the tank; these changes affected the behavior of vorticity, Figure 5. Three distinct groups of flow patterns were observed, depending on the relative size of the exit orifice:

Group1: low flow rates: $0 < \frac{A_{exit}}{A_{cap}} \leq 0.2$;

Group 2: medium flow rates: $0.2 < \frac{A_{exit}}{A_{cap}} \leq 0.7$; and

Group 3: large flow rates: $0.7 < \frac{A_{exit}}{A_{cap}} \leq 1$.

At low flow rates, the difference between free surface vorticity and submerged vorticity was the smallest of the three. A large flow rates, the difference between free surface vorticity and submerged vorticity was the largest of the three. A medium flow rates, that difference was between the largest and the smallest.

ACKNOWLEDGMENTS

Pictures and data used in this paper were submitted by a variety of students who tested Rankine combined vortices while working on projects designed by the author for the Fluid Mechanics Laboratory at his home institution. They include Dustin Bruntz, Curtis Coverstone, Zachary Anderson, Josh Weaver, Julian Velazquez, and Austin Swihart.

REFERENCES

1. Lamb, H. 1993, *Hydrodynamics*, Cambridge University Press, Cambridge P. 23.
2. Lugt, H.J. 1997, *Vortex Flow in Nature and Technology*, John Wiley & Sons, New York.
3. Khoo,B.C., Yeo, K.S., and Lim, D.F. 1997, "The axisymmetric boundary layer beneath a Rankine- like vortex", *Experiments in Fluids*, 22, 300-311.
4. *Noguchi, T., Yukimoto, S., Kimura, R., and Niino, H. 2003,* "Structure and instability of a sink vortex", *4th Pacific Symposium on Flow Visualization and Image Processing (PSFVIP-4),* Chamonix, France, pp. 1- 6.
4. Shtern, V., Borissov, A.,and Hussain, F. 1997,"Vortex sinks with axial flow: Solution and applications", Physics of Fluids, 9 (10), , pp. 2941-2959.
5. Batchelor, G.K.1967, *An introduction to Fluid Dynamics*, Cambridge University Press, Cambridge.
6. Saffman, P.G. 1992, *Vortex Dynamics*, Cambridge University Press, Cambridge.
7. Wang, C.Y.1991, "Exact solutions of the steady-state Navier-Stokes equations", *Ann. Reviews Fluid Mechanics*, 23, 159.
8. National Committee for Fluid Mechanics Films (NCFMF), 1972, *Illustrated Experiments in Fluid Mechanics*, The MIT Press, Cambridge, pp. 63-74.
9. Lautrup, B. 2011, *Physics of Continuous Matter: Exotic and Everyday Phenomena in the Macroscopic World*, second edition, CRC Press, Boca Raton, pp. 462-478.
10. Njock Libii, J. 2009, "Using a rotating column of liquid on a recycled record player to measure the acceleration of gravity", *World Transactions on Engineering and Technology Education*, Vol (7), No.2, pp. 151-156.

Proceedings of the 25th CANCAM
London, Ontario, Canada, May 31 - June 4, 2015

SPATIO-TEMPORAL DYNAMICS OF A STREET CANYON FLOW FROM COMBINED FIELD AND WIND TUNNEL EXPERIMENTS.

Laurent Perret and Karin Blackman
LHEEA UMR CNRS 6598
LUNAM Université, Ecole Centrale de Nantes
Nantes, France
laurent.perret@ec-nantes.fr, karin.blackman@ec-nantes.fr

Eric Savory
Dept of Mech and Mat Eng
University of Western Ontario
London, Ontario, Canada
esavory@eng.uwo.ca

ABSTRACT

The present work investigates the dynamics of the flow in a simple street canyon using combined data from both full-scale and 1:200 scale wind tunnel experiments. It is shown how the application of a Stochastic Estimation (SE) method, using spatially well-resolved wind tunnel Particle Image Velocimetry (PIV) measurements may be used to predict full-scale flow dynamics from a limited number of sensors in an entire canyon cross-section (six sonic anemometers located within the canyon in the present case). The main features of the flow, such as the presence of the shear layer that develops over the canyon and the intermittent ejection and penetration events across the canyon opening, were well predicted by the SE. This combination of SE and wind tunnel PIV measurements may be used to optimize the number and location of wind velocity measurement locations in future street canyon field measurement campaigns, as well as for spatially extrapolating instantaneous flow features measured by the limited number of sonic anemometers.

Keywords: Street canyon, wind tunnel, PIV, field experiment, Stochastic Estimation

INTRODUCTION

Being able to model and predict the flow dynamics in the urban canopy is of crucial importance for understanding street-scale ventilation and scalar dispersion processes that have direct impact on pedestrian comfort and air quality. The street canyon configuration is usually used as a simpler model to investigate the characteristics of the flow existing in one or several neighboring streets within an urban area. The configuration for which the upstream wind is perpendicular to the street axis is of particular interest as it is the case in which the ventilation of the street is the most limited. Studies from the literature focusing on this configuration are based either on full-scale field experiments or reduced-scale modeling in wind-tunnels [2]. The former usually suffer from a poor spatial description of the flow due to the difficulty of implementing a large number of velocity sensors, whereas the latter offer the advantage of a well-controlled environment and the possibility of performing spatially-dense measurements to obtain a highly detailed stastistical description of the flow. However, there have been very few studies attempting to compare both approaches and assess the validity of wind-tunnel results (see [2] for a detailed review) and, to the authors' knowledge, none combine both experimental strategies to model and investigate the dynamics of the street canyon flow with the ambient wind perpendicular to the street axis.

The present study builds upon the previous work of [2] to demonstrate the potential of combining spatialy dense wind-tunnel measurements with sparse *in-situ* data via the use of Stochastic Estimation (SE) to obtain a dynamical model of the full-scale flow in a complete cross-section of the canyon.

STOCHASTIC ESTIMATION

Stochastic Estimation was first introduced by [1] as a mathematical approximation of conditional averages in order to identify coherent structures in turbulent flows. This method was further extended to estimate the time-evolution of the velocity field by using as a condition the time history of an instantaneous signal such as the velocity [3] or the pressure

[5] measured simultaneously at a few points. Mainly used in its linear version, it has been shown that its quadratic form is more appropriate when pressure signals are used as time-varying conditions [5]. [6] also recently showed the benefit of using quadratic Stochastic Estimation over its linear version for the prediction of the flow field inside an open-cavity when employing velocity reference signals. In this section, only the basic principles of the Stochastic Estimation are outlined. For an extensive description, the reader is referred to the work of [1], [3] and [4]. The Stochastic Estimation $\tilde{g}(x,t)$ of the conditional average $\langle g(x,t)|E(t)\rangle$ of a given quantity $g(x,t)$ under the condition $E(t)$ known at N_{ref} points of reference (where the $\langle\rangle$ denotes the ensemble-average operation) consists in approximating the conditional average by its Taylor's expansion truncated to a given order:

$$\langle g(x,t)|E(t)\rangle = \sum_{k=1}^{N_{ref}} A_k(x)E_k(t) + \sum_{k,l=1}^{N_{ref}} B_{kl}(x)E_k(t)E_l(t) + ...$$

In order to find the unknown coefficients $A_k(x)$, $B_{kl}(x)$, ..., the mean square error $e = \langle(\tilde{g}(x,t) - g(x,t))^2\rangle$ between the estimated and the true signals is minimized with respect to $A_k(x)$, $B_{kl}(x)$ by setting the derivatives of e with respect to the unknown coefficients to zero. When only the linear terms are retained in the Taylor's expansion, the approximation is referred as to the Linear Stochastic Estimation (LSE). The quadratic Stochastic Estimation (QSE) takes into account both the linear and quadratic terms from the Taylor's expansion of $\tilde{g}(x,t)$. It must be noted here that the coefficients$A_k(x)$, $B_{kl}(x)$ only depend on non-conditional statistics which need to be computed only once.

In the present study, the QSE is employed to estimate the ith instantaneous velocity component $u_i(x,y,z,t)$ at different locations (x,y,z) from the measurement of the two-component velocity field at N_{ref} locations:

$$\tilde{u}(x,y,z,t) = \sum_{j=1}^{N_c}\sum_{l=1}^{N_{ref}} A^i_{jl}(x)u_j(x_l,y_l,z_l,t) + \sum_{j,l=1}^{N_c}\sum_{m,n=1}^{N_{ref}} B^i_{jlmn}(x)u_j(x_m,y_m,z_m,t)u_j(x_n,y_n,z_n,t)$$

where N_c is the number of velocity components used as reference signals and (x_l,y_l,z_l) are the coordinates of the N_{ref} points where the reference signals are measured. This approach allows for the construction of a model for the time-evolution of the instantaneous flow in the plane containing the 6 sonic anemometers, which is based on the wind tunnel data and driven by the instantaneous reference velocity signals from the field experiment.

EXPERIMENTAL SETUP

In this section, only a brief description of the experimental setup is given, the reader being referred to the work of [2] for more details. During the field campaign, the 24m-long canyon was made from shipping crates, giving an aspect ratio of width to height, $W/h = 0.70$ and length to height, $L/h = 4.62$, and was located in a parking lot surrounded by woods and fields with some buildings nearby, outside of the city of Nantes. Within the canyon, six sonic anemometers were employed to measure the 3-component velocity in a cross-section of the canyon. They were aligned with the axis of the canyon, located 12.00 m from the canyon ends and spaced 0.80 m apart in the streamwise direction with three at a height of $z/h = 0.38$ (S14, S15, S16) and another three at $z/h = 0.77$ (S11, S12, S13) from the ground (figure 1a). A seventh sonic anemometer was located on a mast at 10.00 m height (S10), $x/h = 1.66$ and $y/h = 4.94$ away from the centre of the canyon, to determine the characteristics of the oncoming flow. All of the sonic anemometers have an acquisition frequency of 20 Hz and measure U, V, W velocity components and temperature..

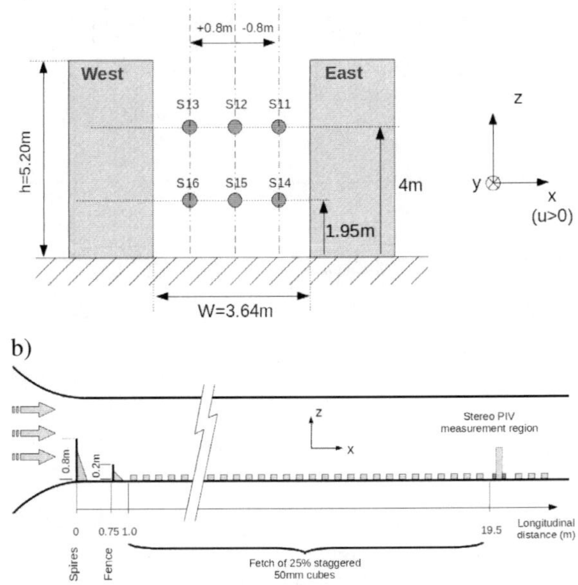

Figure 1: a) elevation of the canyon of the field experiment showing the sonic anemometer locations (○) b) wind tunnel setup.

The wind tunnel experiment was conducted in the low-speed, suck-down boundary layer wind tunnel in the Laboratoire de recherche en Hydrodynamique, Energetique et Environnement Atmospherique (LHEEA) at Ecole Centrale de Nantes (figure 1b), which has working section dimensions of 2 m (width) x 2 m (height) x 24 m (length). Flow measurements were conducted using stereoscopic PIV to record 5,000 three-component velocity fields. The upstream roughness used in the current case consisted of 50 mm cubes arranged in a staggered array with $\lambda_p = 25\%$ and the canyon had the same aspect ratio as the field experiment. The wind tunnel boundary layer was scaled using ESDU and was found to best match a

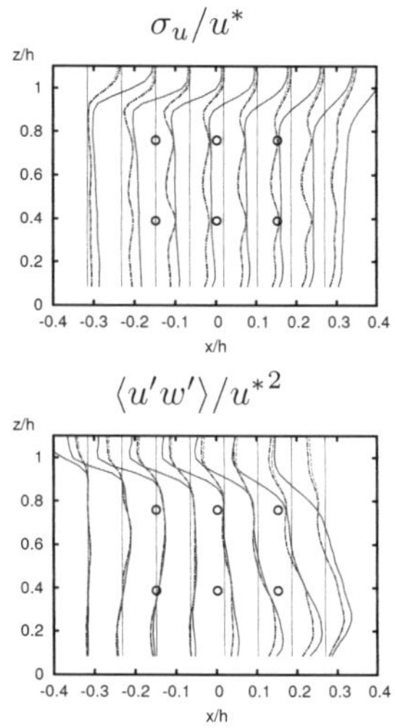

Figure 2: Statistics of the original (red) velocity field from the wind tunnel experiment and of its quadratic (blue) and linear (green) estimates using PIV reference data at the corresponding sonic anemometer locations (○).

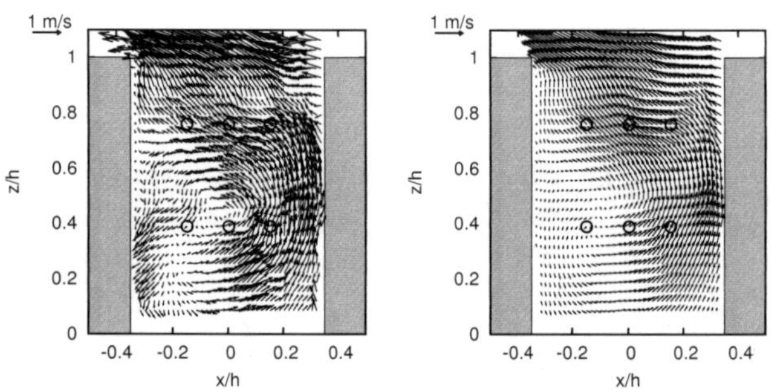

Figure 3: Comparison of (left) the instantaneous velocity field measured by PIV in the wind tunnel model and (right) the QSE velocity field using 6 reference points (○) located at the corresponding sonic anemometers in the field experiment (only the fluctuations are shown). The reference velocity signals are from the PIV measurements.

suburban terrain at a scale of 1:200. A detailed comparison of the statistics (including, mean velocity, Reynolds stress tensor, spectra and two-point correlations) of the flows in near-neutral stability conditions when the wind is perpendicular, modeled both in the wind-tunnel and *in-situ* can be found in [2]. These authors demonstrated that the wind tunnel model reproduces well the flow statistics and the organization of the main turbulent structures. Based on that, the spatially well-resolved wind tunnel data combined with the Stochastic Estimation (SE) method are used here to spatially extrapolate the sparse data from the field experiment.

WIND TUNNEL RESULTS

In the present study, both LSE and QSE have been used to derive a model for the estimation of both the longitudinal

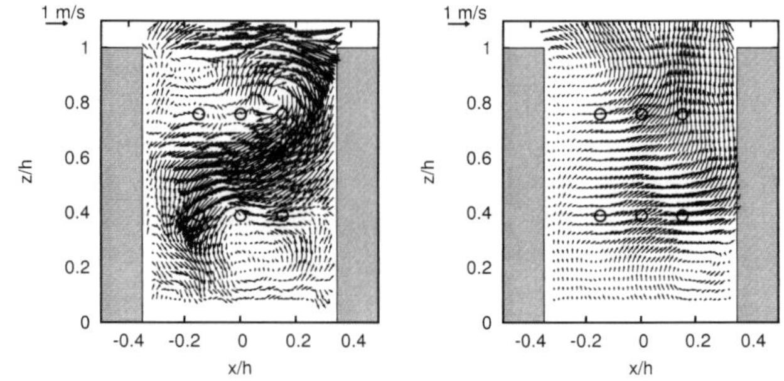

Figure 4: Examples of two different instantaneous velocity fields obtained by QSE using the 6 simultaneous measurements by sonic anemometers from the field experiment (○) and the QSE coefficient determined from the wind tunnel experiment (only the fluctuations are shown).

and vertical components at locations corresponding to the PIV vector fields, using both the longitudinal and vertical components known at the reference location of the sonic anemometers in the field experiment. The two approaches have first been tested using the PIV data as reference signals to obtain both a linear and a quadratic estimation of the original 5,000 PIV velocity fields. The statistics of the estimated PIV data are presented in figure 2. As already shown in previous work from the literature, the SE performs very well at the locations of the reference points and tends to under-estimate turbulent fluctuations when locations far from the reference points are considered, as a direct result of the drop of the correlation level when the distance to the reference points increases. Nevertheless, both the LSE and QSE are able to reproduce the main statistical features of the flow, especially the presence of the shear layer that develops at the top the upstream building. The use of the QSE leads to an improvement of the prediction of the level of both the turbulent fluctuations and the Reynolds shear stress within and at the top of the canyon. The same improvement has been found by [6] in their study of the flow dynamics inside an open cavity. The comparison between original and estimated instantaneous vector fields (figure 3) shows that the QSE is able to reproduce the large-scale motions in the canyon, in particular those responsible for strong ejections, penetrations and recirculation of fluid in the canyon, as well as some smaller-scale structures.

SPATIO-TEMPORAL ESTIMATION OF FIELD DATA

Based on the above results, the QSE model obtained from the PIV database has been used with velocity time-series from the sonic anemometers as reference signals to estimate the corresponding instantaneous velocity field in the entire canyon cross-section. The results presented here are based on a single 45 min period. The database obtained consists of 54,000 bi-dimensional, two-component vector fields with the same temporal frequency as the original field measurements (20 Hz). Examples of the instantaneous vector fields of the velocity fluctuations are presented in figure 4. In this case,

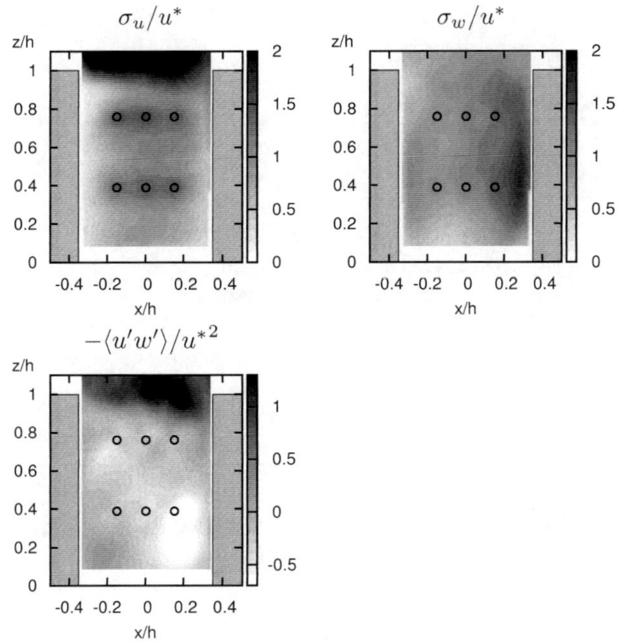

Figure 5: Statistics of the velocity field obtained by QSE using the 6 simultaneous measurements by sonic anemometers from the field experiment (o) and the QSE coefficient determined from the wind tunnel experiment.

the mean velocity flow field obtained from the wind tunnel database, properly scaled by the ambient wind velocity [2], is used to reconstruct the complete velocity field. The QSE flow field shows the presence of large-scale motions consisting of intermittent ejections and penetrations of fluid across the canyon opening and recirculation of fluid within the canyon accompanied by smaller-scale vortices. The intermittent presence of the flapping shear layer developing from the upstream canyon obstacle is also well-reproduced (not shown here). The corresponding statistics presented in figure 5 show good agreement with those obtained from the wind tunnel experiment. In particular, the high level of σ_u inside the shear-layer, the high level of σ_w along the vertical wall of the downstream building and the magnitude of the Reynolds shear stress at the canyon opening are well-reproduced.

CONCLUSIONS

To the authors' knowledge, the results presented here constitute the first attempt to combine spatially well-resolved PIV data obtained from a wind tunnel experiment with time-resolved but sparse velocity measurements from a field experiment representing the same street canyon configuration. At this stage, because no measurements were available except for those from the 6 reference sonic anemometers used to drive the QSE model, no extensive validation of the obtained dynamics is possible. However, the fact that the wind tunnel flow is properly scaled (geometrically and dynamically) with the field experiment flow and that one- and two-point statistics of both flows are in reasonable agreement, gives confidence in the present results for the prediction of the flow inside the canyon. Even if further work is needed to assess the quality of the model and to take into account the spanwise flow in the field canyon found present in most of the analyzed periods, the use of Stochastic Estimation with combined data from wind tunnel and field experiments offers great potential for the prediction of the ventilation inside the street canyon and for the design of field experiments for which the optimum location of sensors can be studied beforehand, depending on the objectives of the particular study.

ACKNOWLEDGMENTS

The authors should like to thank Mr Thibaut Piquet for his technical support during the experimental program and both the Region de Pays de La Loire and the Ontario Graduate Scholarship Program for providing funding.

REFERENCES

[1] Adrian, R.J., 1977 "On the role of conditional averages in turbulence theory". In *Turbulence in Liquids, Proc of the Fourth Biennial Symp*, Rolla, Mo., 22-24 Sep 1975, (Princeton, N.J: Science Press), pp. 323-332.

[2] Blackman, K., Perret, L., Savory, E. and Piquet, T., 2015 "Field and wind tunnel modeling of an idealized street canyon flow", *Atmospheric Environment*, Vol(106), pp. 139-153.

[3] Bonnet, J.P., Cole, D.R., Delville, J., Glauser, M.N. and Ukeiley, L.S., 1994, "Stochastic estimation and proper orthogonal decomposition: complementary techniques for identifying structure". *Exp Fluids*, Vol(17), pp. 307-314.

[4] Guezennec Y.G., 1989, "Stochastic estimation of coherent structures in turbulent boundary layers", *Phys Fluids A*, Vol(1), pp. 1054-1060.

[5] Murray, N.E. and Ukeiley, L.S., 2007, "Modified quadratic stochastic estimation of resonating subsonic cavity flow", *J Turbulence*, DOI: 10.1080/14685240701656121.

[6] Pey, Y.Y., Chua, L.P. and Siauw, W.L., 2012, "Stochastic estimation of cavity flow field", *World Acad Sci Eng Tech*, Vol(6), pp.11-26.

THE FLOW FIELD OF AN EF5 TORNADO IN A SUPERCELL WINDSTORM

W.E. Lin, A. Ozyoruk and E. Savory
Dept. of Mechanical & Materials Engineering
Western University
London, Ontario, Canada
wlin26@uwo.ca, aozyoruk@uwo.ca, esavory@uwo.ca

L.G. Orf
Dept. of Earth & Atmospheric Sciences
Central Michigan University
Mount Pleasant, Michigan, U.S.A
leigh.orf@cmich.edu

ABSTRACT

The flow field of an EF5 tornado arising within a numerical simulation of a supercell storm is discussed with respect to idealized tornado models in the literature. The size of the simulation domain was 120 km by 120 km in plan and 20 km in height but the most significant tornadic flow was located within a 1 km radius of the tornado centre and at the closest computational grid point from the surface (15 m above ground level). Significant deviations from axisymmetry arise from the variable alignment of the rotational and translational components of wind speed, as well as from flow structure complexities inferred from the vertical velocity field.

KEYWORDS: EF5 tornado, wind storm, supercell, numerical data, velocity fields

INTRODUCTION

Although it may be possible for a tornado-like flow to abruptly form in seemingly calm conditions at surface height (cf. discussion of landspouts [1]), the majority of damaging tornadoes are expected to form within larger convective weather systems such as a supercell windstorm. For instance, the 127 tornadoes of the 3-4 April 1974 outbreak occurred during a 24-hour duration and covered 571 km^2 within thirteen American states and one Canadian province; five tornadoes (4 %) warranted the most severe rating on the Fujita scale (F5) and had a cumulative path length of 32 km and a damage area of 0.2 km^2, twenty-three tornadoes were rated as F4 (18 %; 212 km; 5.3 km^2) and the remainder were weaker than F4 (2,996 km; 566 km^2) [2]. In a revisiting of the 1974 data, the total number of observed tornadoes was increased to 148 with thirty F4 and F5 tornadoes [3]. This extreme weather event was certainly extraordinary, yet the 27 April 2011 outbreak of 199 tornadoes (including four F/EF5 and eleven F/EF4 tornadoes, where the Enhanced Fujita (EF) scale has been adopted as an improved proxy for wind speed from damage assessment) [4] provided evidence that the speculated once-per-century return period for the 1974 outbreak [2] could be overly conservative. Moreover, the 2011 outbreak was generated by parent storms ranging from quasi-linear convective systems to long-lived discrete supercell storms.

Permanent residential structures of current typical construction can provide adequate shelter during tornadoes of up to EF3 strength. Re-inforced rooms or underground shelters can save lives in stronger tornadoes but widespread implementation by permanent dedicated shelters is presently unlikely due to financial cost (estimated at $44,000,000 per life saved) [5]. In want of great advances in construction materials science and structural design, evacuation before the arrival of a severe tornado and the siting of vulnerable infrastructure at a safe distance from persistent tornado tracks, if such paths do actually exist, are viewed as the only foreseeable possible measures for reducing EF4 and EF5 tornado losses. Both of these research directions require advancing the understanding of the wind field, in these less frequent but more devastating events, beyond idealistic axisymmetric models of a tornado-like vortex.

An isolated tornado has been modelled as a Rankine combined vortex retarded by friction at the surface [6, 7]. Wind speed profiles of Ekman-layer type within a boundary layer adjacent to the surface were smoothly matched to Rankine type profiles at greater height. This theoretical solution described an axisymmetric and steady flow where the tangential, radial and vertical wind speed components attained their respective maximum values at a prescribed radius from the tornado centre.

Laboratory experiments have been conducted within cylindrical vortex chambers [8 after 9, 10]. The volume flow rate through an outlet at the cylinder top was generated by a fan to model the tornado updraft. The subcloud layer and circulation around this updraft were represented by inflow and by angular momentum imparted by a rotating screen around the circumference of the Ward chamber. Numerical simulations have been performed with the enclosed and rotating domain of the Fiedler chamber to model the tornado updraft. The Fiedler chamber height represented the tropopause height, which is typically 10 km or more above ground level (AGL). Buoyancy per unit mass was imposed within the Fiedler chamber, instead of flow conditions being imposed at open boundaries as done with the Ward chamber.

These previous approaches represent a solid foundation upon which to understand an idealized tornado. To design structures to better withstand intense tornadoes, it is necessary to examine a tornado generated within its native environment of a supercell windstorm. The objective of the present work is to assess how the flow of an EF5 tornado arising within a numerical simulation of a supercell storm compares to idealized models.

BLUE WATERS/CM1 NUMERICAL SIMULATION

A tornado, rated EF5 on the Enhanced Fujita scale based upon mobile Doppler radar measurements, touched down on the northern outskirts of Oklahoma City on 24 May 2011 and left a damage path of 105 km through the communities of Calumet, El Reno, Piedmont and Guthrie in sequence [11]. As part of an ongoing research program on supercell tornadogenesis and tornado maintenance [12], the CM1 numerical model (version 16) [13] with microphysics and 20,000 cores of the Blue Waters supercomputer were utilized to simulate the storm. The size of the simulation domain was 120 km by 120 km in plan and 20 km in height, as shown in Fig. 1, which corresponded to a computational grid of 2,200 points by 2,200 points in plan and 380 points in height. The horizontal grid spacing was 30 m and isotropic within an inner domain of 60 km by 60 km in plan by 10 km in height. Outside of this inner domain, the grid spacing was increased. The vertical grid spacing was 30 m and isotropic with the closest grid point to the surface at 15 m AGL.

A 5th-order advection scheme was utilized with Smagorinsky turbulence closure and Morrison two-moment microphysics. A free-slip surface boundary condition was applied. The supercell was initiated using an updraft nudging (w-forcing) technique. The environmental base state was horizontally homogeneous and initialized from a 1-hour forecast sounding (Rapid Update Cycle from the National Oceanic and Atmospheric Administration/National Centers for Environmental Prediction) located off the right flank of the Calumet-El Reno-Piedmont-Guthrie supercell. The level of free convection (LFC) and the LFC equilibrium level were 895 m AGL and 13,198 m AGL, respectively. Following parcel theory, between these two heights, a parcel of air gained kinetic energy as it rose as is characteristic of an unstable atmosphere. The convective available potential energy (CAPE), available to a rising parcel, was 4.3 kJ/kg.

The resulting supercell lasted several hours with an EF5 tornado track of 105 km length by 3 hours of model time. The simulated tornado was on the ground for approximately the same duration as the observed tornado. The simulated storm was similar to the observed storm in terms of track length, direction, intensity, and tornado morphology, based upon renderings of the simulated moisture content and storm chaser footage. To the best of our knowledge, this is the first simulation of a supercell producing a long-track EF5 tornado.

The full-cloud simulation described above was run without including the effects of surface friction. Rather than analyze the tornado in the free-slip simulation for this work, we experimented with including surface friction. A simulation where surface roughness is applied for the entire simulation resulted in a supercell that did not create a tornado. This is not surprising, as simulations of deep moist convection are highly sensitive to initial conditions. A simulation was conducted where CM1 was restarted from (free-slip) checkpoint files at a time before tornado formation occurred. CM1 was modified such that the mature supercell passed over a region of gradually increasing surface roughness. This simulation did result in a strong tornado occurring over a region of nonzero surface roughness, but one that was weaker and shorter-lived than the free-slip simulation. The modified simulation, which includes surface friction only once the storm has matured, is analyzed below.

Figure 1: Northward elevation view of the precipitation isosurfaces in the simulated domain

SIMULATED PRESSURE PERTURBATION AND WIND VELOCITY

Contours of the pressure perturbation from the ambient baseline state gave an indication of the location of the tornado core. The largest pressure deficit during the simulation was 9.85 kPa less than ambient pressure and this occurred at the closest grid height above the surface (15 m AGL) at 6,133 seconds into the simulation; the point of largest pressure deficit will be referred to as the tornado centre and its spatial position changed during the 300-second interval when the tornado was at or near peak intensity. As shown in Fig. 2a, the significant pressure perturbation was confined within an approximate radius of 1 km around the tornado centre. Fig. 2b is an enlargement of this area of intense cyclonic wind velocities. Within this region, a significant non-uniformity was evident with the largest horizontal wind speeds appearing in the quadrant southeast of the tornado centre and the smallest horizontal wind speeds appearing in the quadrant northwest of the tornado centre. The former were approximately four times larger in magnitude than the latter.

The largest westward wind speed was 61 m/s, which occurred at a point 270 m north and 30 m west of the tornado centre, at 285 m AGL and at 6,135 s into the simulation. The largest eastward wind speed at t = 6,135 s was 83 m/s, which occurred at a point 150 m directly south of the tornado centre and at 15 m AGL. The largest pressure deficit at this instant of simulation time was 9.77 kPa at 15 m AGL.

The largest eastward wind speed was 94 m/s, which occurred at a point 150 m directly south of the tornado centre, at 15 m AGL and at 6,157 s into the simulation. The largest westward wind speed at t = 6,157 s was 57 m/s, which occurred at a point 210 m directly north of the tornado centre and at 285 m AGL. The largest pressure deficit at this instant of simulation time was 9.15 kPa at 15 m AGL; the tornado centre was displaced 90 m westward and 40 m southward from its position at 22 s earlier in the simulation.

The largest southward wind speed was 60 m/s, which

occurred at a point 90 m north and 180 m west of the tornado centre, at 405 m AGL and at 6,138 s into the simulation. The largest northward wind speed at t = 6,138 s was 73 m/s, which occurred at a point 30 m north and 150 m east of the tornado centre and at 135 m AGL. The largest pressure deficit at this instant of simulation time was 9.58 kPa at 15 m AGL; the tornado centre was displaced only one grid spacing (30 m) westward from its position at 3 s prior.

The largest northward wind speed was 88 m/s, which occurred at a point 150 m south and 180 m east of the tornado centre, at 15 m AGL and at 6,170 s into the simulation. The largest southward wind speed at t = 6,170 s was 58 m/s, which occurred at a point 120 m south and 210 m west of the tornado centre and at 375 m AGL. The largest pressure deficit at this instant of simulation time was 8.61 kPa at 885 m AGL; the tornado centre was also displaced 90 m westward from its position at 35 s earlier.

The horizontal translational speed of the tornado relative to the ground was 18 m/s towards the northeast. Separated into components in the cardinal directions, this wind speed was equivalent to 14 m/s northward and 12 m/s eastward. Thus, the translational component constituted 16 % of the largest northward wind speed and 13 % of the largest eastward wind speed during the 300 s interval near peak wind intensity.

In Fig. 3, the near-surface horizontal wind speed vectors in the 2-km-by-2-km area of the tornado core are presented, with the corresponding vertical wind speeds indicated by coloured contours. The vertical wind speeds were within a range of -5 m/s < w < 5 m/s and, thus, their magnitudes were clearly at least an order less than those of the largest horizontal wind speeds in the tornado core. The presence of regions of negative vertical velocity bely a simple view of a tornado as a helical flow pattern where near-surface winds converge at a tornado centre and corkscrew upwards.

It has been proposed that the pressure deficit at the tornado centre can induce a sinking flow there, with rising flow at greater radial distance from this central axis, i.e. a circulation bubble in the 'vortex breakdown' regime [14]. A conceptual drawing of circulation in a vertical cross-sectional plane through the tornado axis showed a double-cell flow pattern, where there is symmetry about the tornado axis with a cell comprised of sinking and rising air. This symmetry was clearly lacking from Fig. 3, which shows a lobe of rising flow immediately southwest of the tornado centre (near x = 8.4 km and y = -12.9 km) and a lobe of sinking flow immediately northeast of the tornado centre (near x = 8.7 km and y = 12.7 km). The wind velocities from the numerical simulation give an indication of air currents twisting downwards, as well as air currents twisting upwards about the tornado centre. Regions of sinking and rising air in close proximity to each other appear to curl into the tornado centre from greater radial distances and various cardinal directions (e.g. the band directly north of the tornado centre at y = 12.2 km). These flow features are consistent with the conceptual picture of a tornado in a supercell (cf. Figure 4.3 of [1] and Figure 5 of [8]) where baroclinically-generated tubes of horizontal vorticity are believed to stream into the tornado core, tilt upwards and become vertical vorticity tubes that play a role in tornado formation and maintenance.

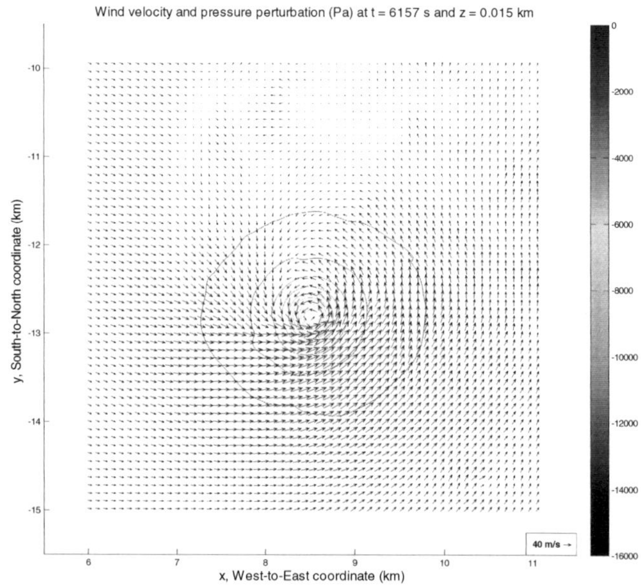

Figure 2a: Horizontal plane of pressure perturbation contours and velocity vectors at 15 m AGL (every third vector shown)

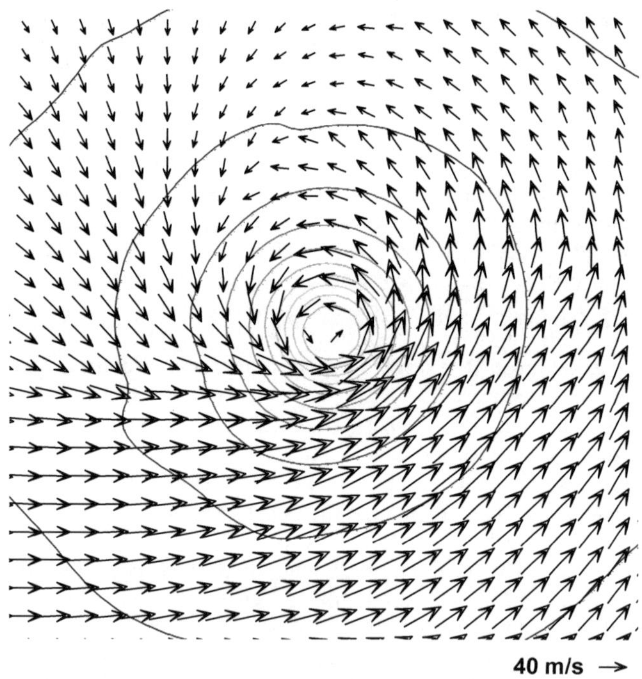

Figure 2b: Enlarged portion of Figure 2a showing a one-kilometre radius around the pressure minimum

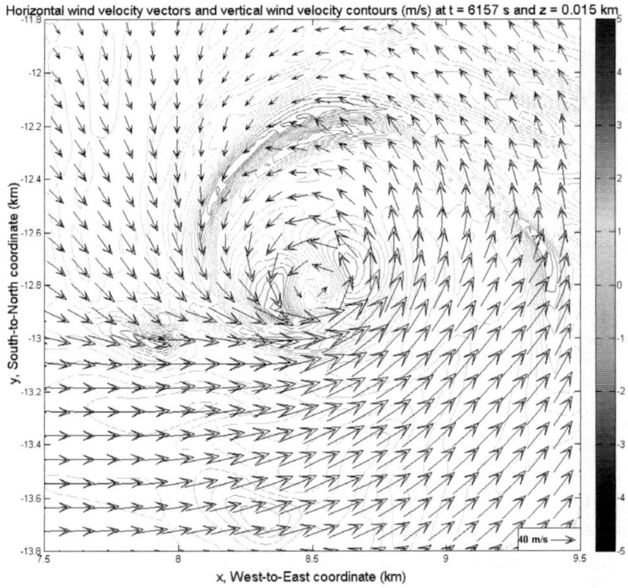

Figure 3: Horizontal plane showing horizontal wind velocity component vectors and vertical wind velocity contours

CONCLUSIONS

The numerical simulation of a supercell wind storm yielded an EF5 tornado of ~1 km radius at peak intensity. Looking in the direction of tornado translation, the largest horizontal wind speeds were to the right of the axis of rotation of the tornado winds, where the rotational and translational wind velocity components were cumulative. The smallest horizontal wind speeds occurred to the left of the axis of rotation where these components were in opposition. The largest horizontal wind speeds were about four times larger in magnitude than the simultaneously-occurring smallest horizontal wind speeds across the tornado core. Idealized tornado models ignore this lack of axisymmetry of the near-surface horizontal wind velocity. The vertical wind speeds at the tornado core were at least an order of magnitude less than the corresponding horizontal components. Considering all aspects of this analysis, the time-dependent and three-dimensional turbulent wind field of the tornado within the numerically-simulated supercell is notably more complex than the idealized models proposed in the literature.

ACKNOWLEDGMENTS

Financial support was provided by Hydro One. The numerical data was produced by the Blue Waters sustained-petascale computing project at the National Center for Supercomputing Applications at the University of Illinois at Urbana-Champaign, with support by the National Science Foundation and the state of Illinois.

REFERENCES

1. Grazulis, T.P. 2001, *The Tornado: Nature's Ultimate Windstorm*, University of Oklahoma Press, Norman.
2. Fujita, T.T. 1976, "Recent concepts of tornado winds", *Second U.S.A.-Japan Research Seminar on Wind Effects on Structures*, Kyoto, Japan, September 1974, pp. 3-14.
3. Corfidi, S.F., Weiss, S.J., Kain, J.S., Corfidi, S.J., Rabin, R.M., and Levit, J.J. 2010, "Revisiting the 3-4 April 1974 super outbreak of tornadoes", *Weather and Forecasting*, 25(2), pp. 465-510.
4. Knupp, K.R., Murphy, T.A., Coleman, T.A., Wade, R.A., Mullins, S.A., Schultz, C.J., Schultz, E.V., Carey, L., Sherrer, A., McCaul Jr., E.W., Carcione, B., Latimer, S., Kula, A., Laws, K., Marsh, P.T., and Klockow, K. 2014, "Meteorological overview of the devastating 27 April 2011 tornado outbreak", *Bulletin of the American Meteorological Society*, 95(7), pp. 1041-1062.
5. Simmons, K.M., and Sutter, D. 2012, "The 2011 tornadoes and the future of tornado research", *Bulletin of the American Meteorological Society*, 93(7), pp. 959-961.
6. Kuo, H.L. 1971, "Axisymmetric flows in the boundary layer of a maintained vortex", *Journal of Atmospheric Sciences*, 28(1), pp. 20-41.
7. Wen, Y.-K., and Ang, A.H.-S. 1975, "Tornado risk and wind effects on structures", *Fourth International Conference on Wind Effects on Buildings*, Heathrow, U.K., pp. 63-74.
8. Rotunno, R. 2013, "The fluid dynamics of tornadoes", *Annual Review of Fluid Mechanics*, 45, pp. 59-84.
9. Ward, N.B. 1972, "The exploration of certain features of tornado dynamics using a laboratory model", *Journal of Atmospheric Sciences*, 29(6), pp. 1194-1204.
10. Fiedler, B.H. 1994, "The thermodynamic speed limit and its violation in axisymmetric numerical simulations of tornado-like vortices", *Atmosphere-Ocean*, 32(2), pp. 335-359.
11. National Climatic Data Center, *Storm events database*, http://www.ncdc.noaa.gov/stormevents, 2011.
12. Orf, L., Wilhelmson, R., and Wicker, L.J. 2014, "A numerical simulation of a long-track EF5 tornado embedded within a supercell", *AMS Special Symposium on Severe Local Storms*, Atlanta, U.S.A.
13. Bryan, G.H., and Fritsch, J.M. 2002, "A benchmark simulation for moist nonhydrostatic numerical models", *Monthly Weather Review*, 130(12), pp. 2917-2928.
14. Davies-Jones, R.P. 1985, "Tornado dynamics" in *Thunderstorms: Morphology and Dynamics*, edited by E. Kessler, University of Oklahoma Press, Norman.

EVALUATING THE PERFORMANCE OF INFLOW BOUNDARY CONDITION MODELS FOR RANS SIMULATION OF ABL FLOW OVER COMPLEX TOPOGRAPHY

Tibebu H. Birhane
Department of Civil and Environmental Engineering
University of Western Ontario
London, ON, Canada
tbirhane@uwo.ca

Girma T. Bitsuamlak
Department of Civil and Environmental Engineering
University of Western Ontario
London, ON, Canada
gbitsuam@uwo.ca

ABSTRACT

Wind profiles based on constant shear stress assumption are commonly used as an inflow for computational simulation of atmospheric boundary layer (ABL) flow to estimate topographic induced wind speed-up. However, this assumption which leads to the Von Karman log-law velocity profile is valid for the lower region (around the bottom 30m thick layer) of a neutral ABL associated with large-scale strong wind storms. However, the upper part of the ABL flow is characterized by a variable shearing stress which is influenced by surface friction and the rotation of the Earth. RANS simulations based on $k-\varepsilon$ turbulence model were conducted to assess the impact of this assumption in calculating orographic induced wind speed-up. The computational simulations were carried out for Askervein Hill, whose field measurement data is available in literature. Wind speed-up results obtained from numerical simulations are compared with that of field measurements and boundary layer wind tunnel simulations from literature.

KEYWORDS: Topography, CFD, RANS, wind speed-up, Askervein hill

INTRODUCTION

Abrupt changes in the local topography causes wind speed-up which results in an increment of wind loading on structures constructed over such sites. The significance of wind speed-up due to abrupt changes in the local topography in evaluating wind loads have been investigated by several researchers since 1970's. A detailed summary review is given by Bitsuamlak et al. (2004). Following these studies, several national wind loading standards and codes of practice such as the National Building Code of Canada (NBCC 2005); the ASCE standard (ASCE/SEI 7-10); the Australian/New Zealand Standard (AS/NZS 1170.2:2002) and the European standard (EN 1991-1-4:2005) provide simplified formulations in the form of speed-up factors for generic topographic shapes including isolated hills, ridges or escarpments. However, these standards do not include provisions for wind speed-ups and the associated turbulence characteristics in rapidly varying terrain. For more complex situations, the practitioner is referred to either physical simulations in a boundary-layer wind tunnels (BLWT) or numerical simulations using Computational fluid dynamics (CFD).

For reliable CFD simulation, estimated "fully turbulent profiles" are required to be specified as the inlet boundary located at standard distance upstream of the region of interest (COST Action 732, 2007). Various empirical, semi-empirical, and theoretical formulas have been derived to represent the variation of wind velocity with height. The most familiar forms are the logarithmic and the exponential profiles. In CFD simulations, the logarithmic profile has been used most widely because of its consistency with the standard wall law for solid wall boundaries implemented in CFD codes. However, these models rarely produce equilibrium ABL flow parameters incident to plane of interest within the computational domain especially for a small upstream fetch distance (Hargreaves and Wright 2007). ESDU (1982) specifies that the log-law profile holds true near the ground surface up to height of approximately 30m above the ground, and wind has to traverse for at least 30km fetch of a particular terrain to develop an equilibrium boundary layer with the underlying ground surface roughness.

Near the ground surface, the shearing of ABL flow is predominantly caused by form drag due to surface roughness (Davenport 1960). This creates eddies within the ABL which intern induces shear stress on the mean flow. In eddy viscosity turbulent transport models such as $k-\varepsilon$, this turbulent shear stress within the computational domain is evaluated in terms of the mean velocity gradient via a turbulent viscosity $\left(-\rho\overline{u'w'} = \mu_T\left(\partial u/\partial z\right)\right)$ (Jones and Launder 1972). In which, the local value of turbulent viscosity is determined from the solution of transport equations for the turbulent kinetic energy (k) and turbulent dissipation rate (ε) as $\mu_T = c_\mu \rho k^2/\varepsilon; c_\mu = 0.09$. And the local mean strain rate is evaluated from solutions of mean momentum and continuity equations. As it is easy for practicing computational wind engineer to choose inlet profiles consistent with standard computational models instead of modifying the closure coefficients implemented within the model, testing of commonly used inlet profiles with standard closure coefficients is found relevant. Hence the main objective of this computational study is to compare predictions based on most often used inflow models (Richards and Hoxey 1993, Yang et al. 2009) with that of ESDU provisions. For this purpose, the Askervein '83 full scale test run TU30-B data and the standard $k-\varepsilon$ turbulence model without any modification on the closure coefficients are used in this computational study.

SPATIAL SIZING AND DISCRETIZATION OF THE COMPUTATIONAL DOMAIN

The Askervein hill is found near Scotland with its summit located at around 57.189^0 Northing and -7.375^0 Easting (Taylor and Teunissen 1987). Based on this global position, the digital elevation data defining the surface profile of the terrain was extracted from ASTER GDEM worldwide Elevation Data using Global Mapper and exported as STL file format. The hill has roughly an elliptical shape with major and minor axis of around 2km and 1km respectively. The size of computational domain was set as 6000mx6000mx800m. The height of the computational domain was determined based on the gradient height with an estimate of 775m using ESDU provision. Preliminarily grid size estimates were made based on the sensitivity studies carried out by Abdi and Bitsuamlak (2014) and Castro et al. (2003). However, new discretization scheme comprising of prismatic layers of cells fitting the ground surface profile and polyhedral cells in the core region was implemented as shown in Fig.1. In the vertical direction, the lower 40m layer of the computational domain was discretized using 10 layers of prismatic cells (with lateral size of 10m) and thickness of varying geometrically from 0.6m to size of 10m with stretching factor of 1.25. The core of the computational domain was discretized using polyhedral cells growing in every direction towards the top boundary.

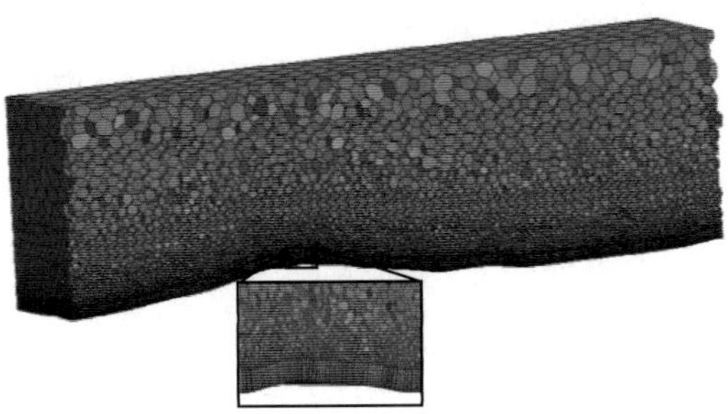

Figure 1: Discretized computational domain

SPECIFICATION OF TURBULENT BOUNDARY LAYER INFLOW

In the computational wind engineering community, the most widely used inflow profile is the model proposed by Richards and Hoxey (1993) which was derived from $k-\varepsilon$ transport equations based on the constant shear stress $\left(-\rho\overline{u'w'} = \mu_T\left(\partial u/\partial z\right) = \rho u_*^2\right)$ assumption over the height of the boundary layer. Numerical studies showed that this model doesn't produce an equilibrium boundary layer throughout the computational domain for the standard $k-\varepsilon$ model (Blocken et al. 2007, Hargreaves and Wright 2007, Yang et al. 2009). Recently, Richards and Norris (2011) presented the modified version as:

$$u = \frac{u_*}{k}\ln\left(\frac{z}{z_0}\right); k = \frac{u_*^2}{\sqrt{C_\mu}}; \varepsilon = \frac{u_*^3}{kz} \tag{1}$$

where κ is the von Karman constant to be taken as 0.433 and z_0 is the roughness length. As shown in the Fig. 2, the above formulation specifies a constant value for turbulent kinetic energy over the entire depth of the ABL. As an amendment for the decay of turbulent kinetic energy with height, Yang et al. (2009) proposed a modification for Richards and Hoxey (1993) inlet profiles given by:

$$\begin{aligned} u &= \frac{u_*}{k}\ln\left(\frac{z+z_0}{z_0}\right); \\ k &= \frac{u_*^2}{\sqrt{C_\mu}}\sqrt{C_1\ln\left(\frac{z+z_0}{z_0}\right)+C_2}; \\ \varepsilon &= \frac{u_*^3}{k(z+z_0)}\sqrt{C_1\ln\left(\frac{z+z_0}{z_0}\right)+C_2} \end{aligned} \tag{2}$$

where C_1 and C_2 are constants determined via experimental data fitting. Based on the standard $k-\varepsilon$ closure coefficients $\left(C_\mu = 0.09\right)$, they specified the curve fit constants as $C_1 = -0.55$ and $C_2 = 5.21$. However, for better predictions of turbulent kinetic energy throughout the computational domain, Yang et al. (2009) recommend to use constants derived based on modified $k-\varepsilon$ closure coefficients.

Mean wind speed, turbulent kinetic energy and dissipation rate profiles based on the above two models are plotted with those evaluated based on (ESDU 1982, ESDU 1985) provisions as shown in Fig. 2. The plots were produced using the friction velocity $u_* = 0.577$ and roughness length $z_0 = 0.02$ based on the Askervein '83 full scale (FS) test run TU30-B measurements at the reference site (Taylor and Teunissen 1987, Mickle, Cook et al. 1988) . The ESDU mean velocity profile is consistent with the full scale measurement for the lower regions of the ABL, whereas the deviation in the upper part may be associated the non-equilibrium condition due to roughness change in the upstream of the reference site(Mickle et al. 1988). The above two models misses the local peak kinetic energy near the ground surface which is the characteristics of the inner turbulent boundary layer(Wilcox 1998).

surface roughness parameter in wind simulation (Davenport 1960) defines surface roughness as "... is neither the shielding due to individual obstacles nor the orographic effects influencing the airflow in mountain regions, but the cumulative statistical drag effect of many obstructions on the wind" . Hence in this computational study, the undulations associated with landform of the bare ground are explicitly simulated as rough wall boundaries, whereas the surface mounted obstacles are indirectly accounted by specifying the roughness height in the standard wall function formulation (Blocken et al. 2007). Pursuant to the formulation of CFD

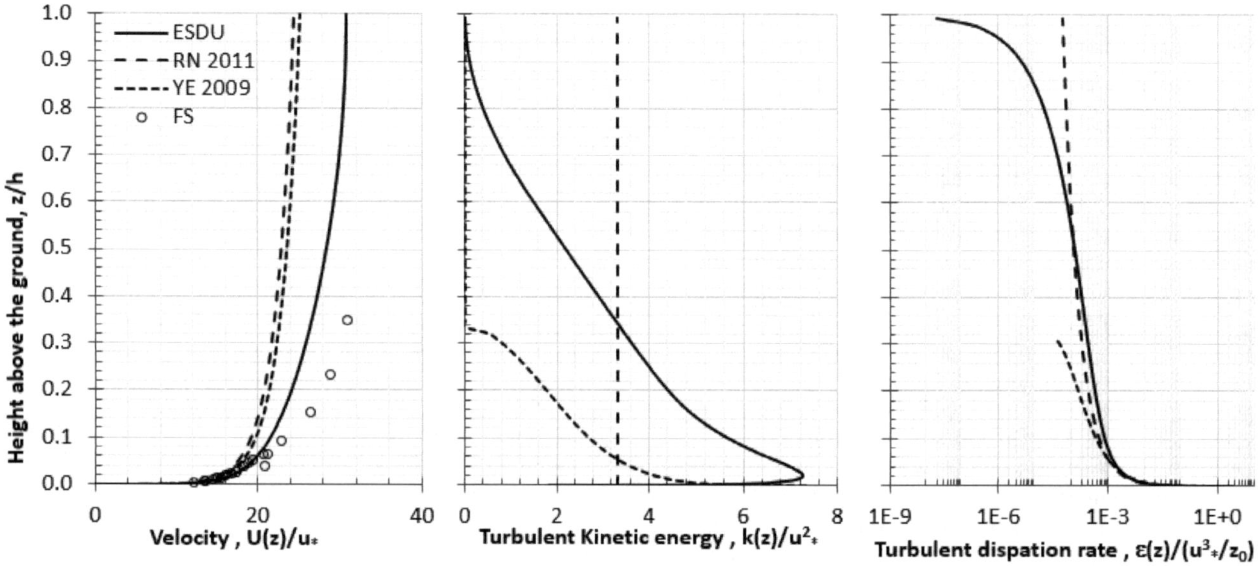

Figure 2: Inflow boundary conditions: FS = full scale measurements, ESDU = Engineering Science Data Unite, RN 2011 = Richards and Norris 2011, and YE 2009 = Yang et al. 2009

SPECIFICATION OF GROUND SURFACE BOUNDARY CONDITION

For flows bounded by smooth walls, viscous shear stress is dominant very close to the boundary (Van Driest 2012) in which the effective height of roughness elements is less than the viscous sublayer $\left(z^+ = zv/\mathrm{u}_* \leq 5\right)$. For high Reynolds Number ABL flow over a rough surface, the form drag from ground roughness is predominant (Van Driest 1956, Davenport 1960). Accordingly, the zero mean velocity boundary condition near the ground surface is generally considered to be at a distance equivalent to the roughness length $\left(z_0\right)$ above the ground (ESDU 1982, Richards and Norris 2011) to model ABL for wind engineering practice. For values of $z_0 = 0.02$ and $u_* = 0.57$ considered in this case study, this location $\left(z_0^+ = z_0 v/\mathrm{u}_* = 760\right)$ lies considerably above the viscous sublayer and the buffer sublayer $\left(z^+ = zv/\mathrm{u}_* \leq 30\right)$. Regarding the concept of

code by STAR-CCM+, no-slip boundary condition due to a rough wall boundary can be modeled using a standard log-low profile of the form $\left[u(z) = (u_*/\kappa)\ln(Ez/Cr)\right]$ where E and C are wall constants and equal to 9.0 and 0.253 for rough walls, and r is the effective roughness height to be specified depending on the roughness of the ground within the computational domain. As a layer of the first grid cells near the ground was placed at $z = 0.6m$ higher than the roughness length $\left(z_0 = 0.02\right)$, and the wind profile follows the log low close to the surface $\left[u(z) = (u_*/\kappa)\ln(z/z_0)\right]$, the effective roughness height was specified as $r = z_0\left(E/C\right)$ with default values of E and C defined above.

DISCUSSION OF SIMULATION RESULTS

The modifications in wind flow are usually reported as a factor called wind speed-up which represents the relative increase/decrease in mean wind speed with respect to the oncoming wind speed (mean) that is not affected by the terrain, measured at a similar height from the local ground. This non-dimensional parameter is evaluated using

$$\Delta s(z) = \frac{U(z) - U_i(z)}{U_i(z)}$$

where $U(z)$ is the computed mean horizontal velocity within the computational domain, and $U_i(z)$ is the inlet velocity, and z is the height measured from the local ground. Figure 3 shows the vertical profiles of wind speed-ups above the summit of the hill, HT (see Fig. 4) predicted using models using a realizable $k-\varepsilon$ turbulence model and the above three inlet velocity profile models proposed by Richards and Norris 2011 (represented as RN 2011), Yang et al. 2009 (represented as YE 2009) and ESDU . The computation was carried out without modifying the standard closure coefficients. The Figure shows the relative deviations of the computed results from that of full scale measurements (Mickle et al. 1988) and wind tunnel predictions (Teunissen et al. 1987). The speed-up predictions based on the three inflow models are in good agreement with that of the experimental observations for the upper section of the boundary layer. Whereas for the lower region, better wind speed-up estimates close to the experimental predictions were obtained using $k-\varepsilon$ turbulence model without any modification made on the standard closure coefficients.

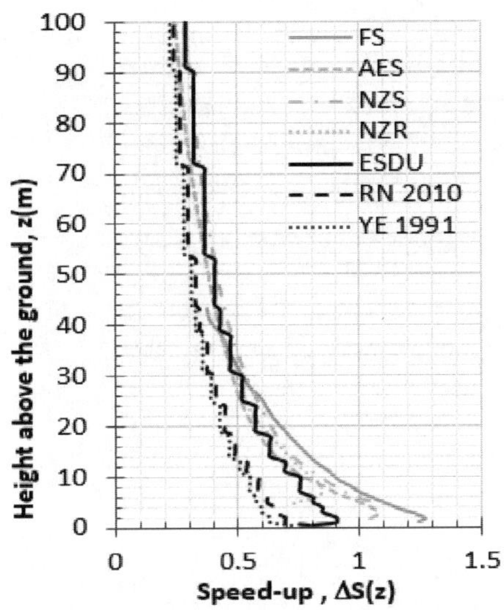

Figure 3: Vertical profile of wind speed-up over the Askervein hill summit: the broken grey lines represent experimental predictions (Teunissen et al. 1987) and solid grey line represents full-scale observation (Mickle et al. 1988)

Fig. 4 shows contour plots of wind speed-up over the surface defined by an offset of 10m from the ground surface, the values were calculated based on ESDU Velocity profile. The Figure shows the contour of wind speed-ups induced by the complex terrain where the hot red color shows the maximum wind speed-up and the dark blue shows the opposite. It can be inferred from the figure that building structures over the ridges could have a favorable effect in case of harnessing wind power or may have unfavorable effect associated with the overloading of structures such as towers of electric transmission lines.

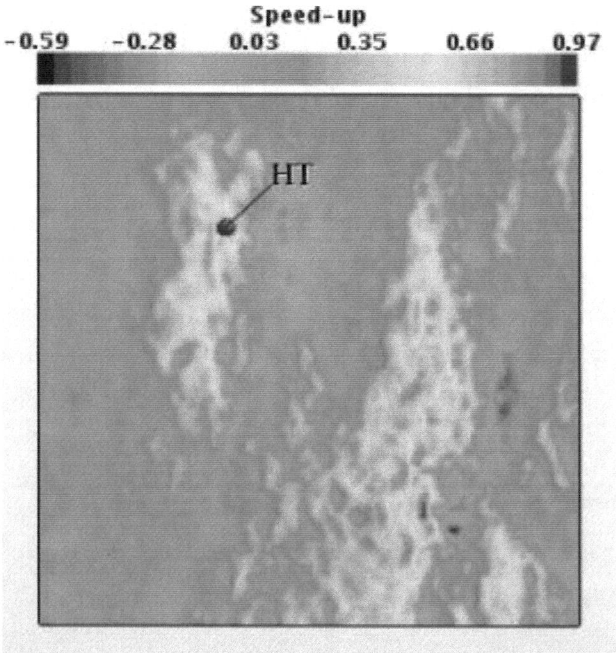

Figure 4: Wind speed-up contour over the Askervein terrain at distance of 10m from the nearest ground surface

ACKNOWLEDGMENTS

All of the computations have been performed using SHARCNET high performance computing facilities.

REFERENCES

1. Abdi, D. S. and G. T. Bitsuamlak (2014). "Wind flow simulations on idealized and real complex terrain using various turbulence models." Advances in Engineering Software 75: 30-41.
2. Bitsuamlak, G., T. Stathopoulos and C. Bédard (2004). "Numerical Evaluation of Wind Flow over Complex Terrain: Review." Journal of Aerospace Engineering 17(4): 135-145.
3. Blocken, B., T. Stathopoulos and J. Carmeliet (2007). "CFD simulation of the atmospheric boundary layer: wall function problems." Atmospheric environment 41(2): 238-252.
4. Castro, F. A., J. Palma and A. S. Lopes (2003). "Simulation of the Askervein Flow. Part 1: Reynolds Averaged Navier–Stokes Equations (k∈ Turbulence Model)." Boundary-Layer Meteorology 107(3): 501-530.
5. COST Action 732 (2007). Best practice guideline for CFD simulation of flows in the urban enviroment. Brussels, Belgium, COST office.
6. Davenport, A. G. (1960). "Rationale for determining design wind velocities." ASCE Journal of the Structural Division 86(5): 39-68.
7. ESDU (1982). Strong winds in the atmospheric boundary layer. Part 1: mean-hourly wind speeds. London, ESDU International Ltd. Item No. 82026.
8. ESDU (1985). Characterstics of atmospheric turbulence near the ground. Part II: single point data for strong winds (neutral atmosphere). London, ESDU International Ltd. Item No. 85020.
9. Hargreaves, D. and N. G. Wright (2007). "On the use of the k–ε model in commercial CFD software to model the neutral atmospheric boundary layer." Journal of Wind Engineering and Industrial Aerodynamics 95(5): 355-369.
10. Jones, W. and B. Launder (1972). "The prediction of laminarization with a two-equation model of turbulence." International journal of heat and mass transfer 15(2): 301-314.
11. Mickle, R., N. Cook, A. Hoff, N. Jensen, J. Salmon, P. Taylor, G. Tetzlaff and H. Teunissen (1988). "The Askervein Hill Project: Vertical profiles of wind and turbulence." Boundary-Layer Meteorology 43(1-2): 143-169.
12. Richards, P. and R. Hoxey (1993). "Appropriate boundary conditions for computational wind engineering models using the k-ϵ turbulence model." Journal of wind engineering and industrial aerodynamics 46: 145-153.
13. Richards, P. and S. Norris (2011). "Appropriate boundary conditions for computational wind engineering models revisited." Journal of Wind Engineering and Industrial Aerodynamics 99(4): 257-266.
14. Taylor, P. A. and H. W. Teunissen (1987). "The Askervein Hill project: Overview and background data." Boundary-Layer Meteorology 39(1-2): 15-39.
15. Teunissen, H., M. Shokr, A. Bowen, C. Wood and D. Green (1987). "The Askervein Hill project: wind-tunnel simulations at three length scales." Boundary-layer meteorology 40(1-2): 1-29.
16. Van Driest, E. (1956). "On turbulent flow near a wall." J. Aerospace Sci. 23: 1007-1011.
17. Wilcox, D. C. (1998). Turbulence modeling for CFD, DCW industries La Canada, CA.
18. Yang, Y., M. Gu, S. Q. Chen and X. Y. Jin (2009). "New inflow boundary conditions for modelling the neutral equilibrium atmospheric boundary layer in computational wind engineering." Journal of Wind Engineering and Industrial Aerodynamics 97(2): 88-95.

UNSTEADY RANS MODELLING OF DENSITY-DRIVEN EXPERIMENTAL THUNDERSTORM DOWNBURSTS

Matthew Roberto[1], Chris Oreskovic[1], Eric Savory[1], Juliette Porto[2]
[1] Dept. of Mech. and Mat. Engineering, University of Western Ontario, London, Ontario, Canada
[2] Institut National Polytechnique de Grenoble, Grenoble, France
mrober85@uwo.ca, coresko2@uwo.ca, esavory@uwo.ca, juliette.porto@ense3.grenoble-inp.fr

ABSTRACT

An Unsteady Reynolds-Averaged Navier-Stokes (URANS) simulation of a physical thunderstorm downburst experiment has been conducted utilizing a two-dimensional axi-symmetric domain. Two experimental set-ups have been modelled in ANSYS ICEM and Fluent, in order to investigate the implications of variations in the physical study domain. It is suggested from these preliminary results that the inclusion of drop cylinder geometry has some impact on the magnitude of the peak outflow wind speeds. Simulations, matching the configuration of an experimental, are performed and analyzed in terms of the areas of interest to the design of wind resistant structures. Four cases are run, covering two different density differences as well as cases where a drop cylinder wall is included and cases where no wall is included. The effect of the initial density difference on the radial outflow velocity is analyzed, as is the effect of the inclusion of a drop wall. The results are finally compared to the experimental data where it is found that the agreement with the numerical results is good, particularly when the drop wall is included.

KEYWORDS: Thunderstorm downbursts, density-driven model, computational fluid dynamics, wind engineering

INTRODUCTION

Downbursts are large masses of air that descend from the lower atmosphere and impinge on the ground, resulting in large radial outflows of strong winds. These intense winds have enormous potential to damage ground structures, such as power transmission lines [1]. Downbursts are density driven events, which result from thermodynamic and microphysical processes present in the formation of thunderstorms, such as the formation of rain, snow, hail etc. [2]. However, due to the downburst's strong temporal dependence, spatial variability and unpredictability, capturing field data of these events is particularly challenging. Additionally, downburst wind profiles are very different from those of everyday synoptic winds [3], and a greater understanding of these events would greatly benefit the design of ground structures. In the area of wind engineering, physical studies and numerical modelling of downburst phenomena has primarily focused on the impulsively-driven impinging circular jet model (IJ) [4]. However, this model is an idealized 'pipe in the sky' simplification of the real world phenomenon, which ignores all the physics present in actual downbursts, and introduces an unrealistic forcing of an impulsively-started jet not observed in nature. More recently, an experiment at the University Western Ontario has attempted to model downbursts by using a density-driven model [5]. This experimental set up aims to capture some of the realistic buoyancy driven phenomenon present in natural downbursts. However, due to the difficulty in experimental repeatability, numerical simulation of the experimental set up is critical to determine the ideal, or most realistic, experimental conditions. In order to achieve this, a number of unsteady RANS numerical simulations were conducted in which various experimental conditions were modified to investigate the effect that this would have on the reliability of the results.

DETAILS OF THE NUMERICAL MODEL

The two-dimensional axi-symmetric numerical model used for this work utilized an Unsteady Reynolds Averaged Navier Stokes Equations (URANS) model, with a shear stress transport k-ω turbulence model, and the Eulerian multiphase equations to capture two-phase flow, utilizing a second-order upwind scheme. The governing equations for the simulation are described fully in [5]. The commercial solver package ANSYS FLUENT was used in conjunction with ANSYS ICEM for all meshing. Two unique geometries were created, the first replicated the experimental set up of [6], including drop cylinder geometry. The second included no drop cylinder. Three meshes were then created, a fine, medium and coarse, in order to establish a grid independent solution, which can be observed in Figure 1 by plotting the velocity magnitude along a vertical line at r=0.01875m through the whole domain at t=0.775s. The percent change in velocity at vertical position of 0.144m shows 7.915% between the coarse and medium and 0.3727% between the medium and fine indicating a negligible difference for the latter. For this

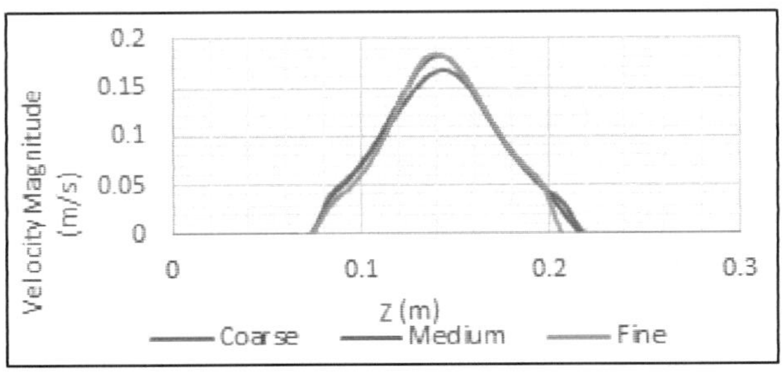

Figure 1 - Mesh independence velocity profile at r=0.01875m

reason, the medium resolved grid was used for all analysis. The model domain encompasses 0.316mx0.3m, used to replicate the experimental set up. A constant horizontal grid spacing of Δx=1.264mm, and a stretched vertical grid spacing was used to accurately resolve the flow closest to the ground, resulting in a total number of grid points of 62500 without a drop cylinder, and 61136 with the drop cylinder. A pressure outlet boundary condition was employed for the right and top boundary, while a wall boundary was used for the bottom and drop cylinder. This experiment is primarily axi-symmetric, and, therefore, an axis-symmetrical boundary condition was utilized for the left boundary, in order to reduce computational time. The computational domain meshing is identical for each case (wall and no wall). Figure 2. shows a more detailed schematic of the computational domain and the coordinate system presented in this paper.

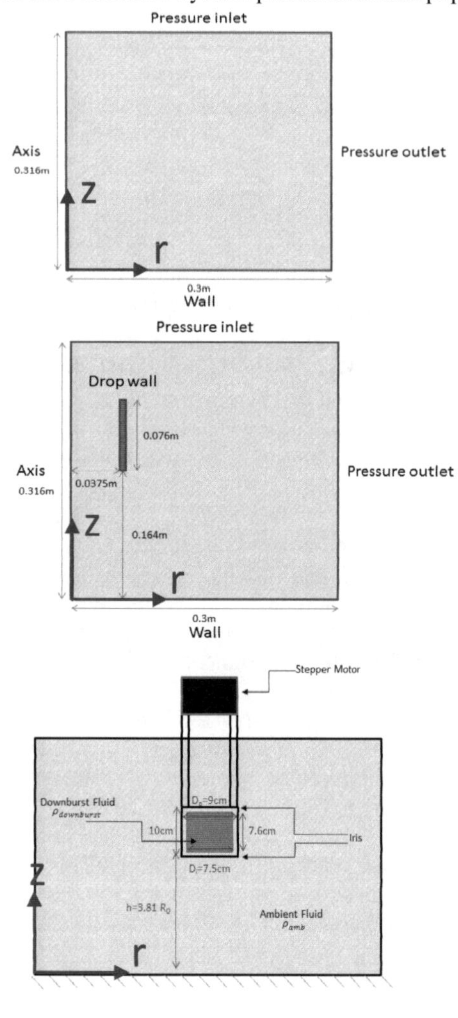

Figure 2 - Numerical and Experimental Domain

ANALYSIS OF NUMERICAL RESULTS

Simulations were run at two different values of Δρ/ρ; density #1 corresponds to a Δρ of 3% and density #2 corresponds to a Δρ/ρ of 4%, being representative of values for real events. In addition to the differing density cases, simulations were also run with and without a drop wall, which is a vertical wall separating the dense fluid from the ambient fluid at its initial location (and represents the physical cylinder which contains the fluid in a physical experiment).

Figure 3 shows an example of the evolution of the 4% Δρ/ρ simulated downburst over its whole life cycle from t=0 s through to t=3 s, with 0.5 second increments. Both the case with and without the drop wall are shown.

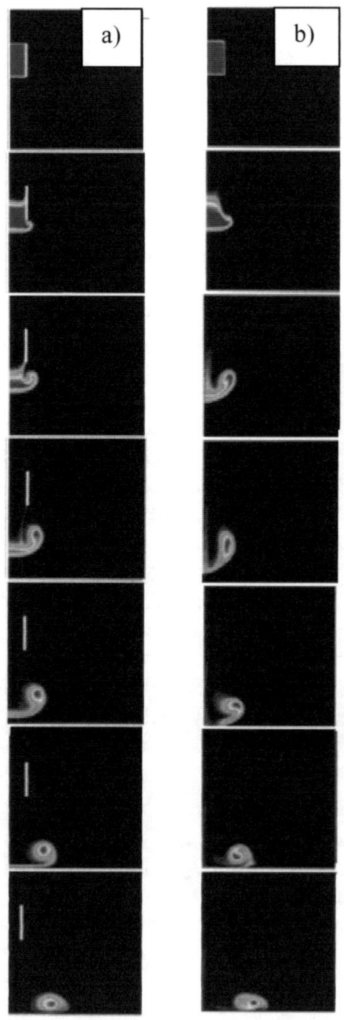

Figure 3 - 4% density downburst; a) with drop wall b) without drop wall

One particular area of interest for the simulation of thunderstorm downbursts is the maximum radial velocity. This peak corresponds to the maximum horizontal near-ground velocity that would be experienced by a structure. The maximum radial velocity is plotted in Figure 4 and it can be seen that for both the wall and non-wall cases, the case with the higher initial density difference experiences an earlier absolute maximum radial peak, with the 3% case lagging behind. Further analysis into the physical location of the absolute maximum radial velocity reveals that even though the time at which these peaks occur is different for different density trials, the radial position at which this happens is the same. For the cases without the drop wall this occurs at about 4.17D_i away from the axis. For the cases with the drop wall this occurs at about 4.15D_i away from the axis. Comparing the cases with the drop wall to those without the drop wall shows that the cases with the drop wall tend to experience higher radial velocities after impingement with the ground plane as well as a peak which occurs earlier in time than their counterparts with the drop wall. The curves for the wall cases also tend to be smoother, and lack the secondary peak that occurs around 0.9 s in the no-wall cases.

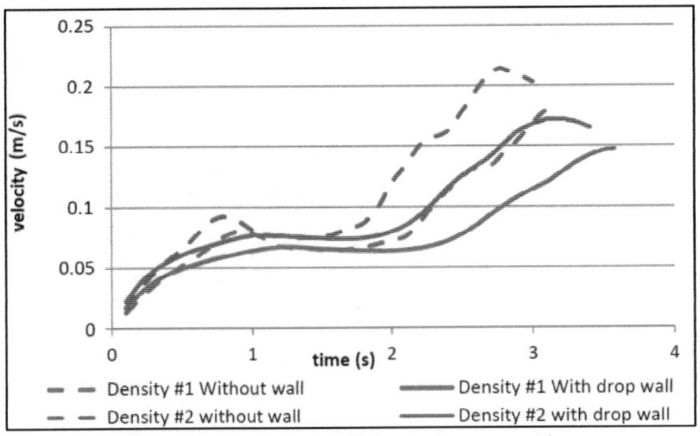

Figure 4 - Maximum radial velocity versus time

COMPARISON TO LITERATURE AND EXPERIMENTS

The four numerical trials are plotted against the matching experimental work, as well as against several experimental and numerical works from the literature. In these plots the height, radial position, and time are all scaled by non-dimensional parameters.

These parameters, developed by [7], allow scaling between model sized events and full scale downbursts. The length scale, R_0, which is the equivalent spherical diameter of the volume of dense fluid, is defined as shown in Eq. (1), where Q is the volume of dense fluid contained within the release cylinder. [8] also defines a time scale (Eq. (2)), where ρ is the ambient fluid density and $\Delta\rho$ is the density difference between the downburst and the ambient fluid. A velocity scale can then be defined by dividing the length scale by the time scale (Eq. (3)).

$$R_0 = \left(\frac{3Q}{4\pi}\right)^{1/3} \quad (1)$$

$$t_0 = \left(\frac{R_0\rho}{g\Delta\rho}\right)^{1/2} \quad (2)$$

$$V_0 = R_0/t_0 \quad (3)$$

Figure 5 shows a plot of the plot of the current height of the downburst front versus time from the time of release until the point where the downburst impinges on the ground. Downbursts 1 and 2 refer to the experimental trials [6]. They show good agreement with the values reported in the literature by Lundgren [7], which was a similar density-driven experimental trial, and will be the basis for comparing and validating the numerical results presented in this paper. The density difference in the experiments, as well as [7], was 3.3%. The 3% and 4% wall cases can be seen to agree quite well with the experimental results. The numerical curve is slightly more linear than in the experiments (the experimental curve starts flatter and ends steeper), but the overall trend matches extremely well. As expected, the higher density case has a steeper slope and impinges on the ground plane earlier than the lower density case. The cases without the wall both exhibit a steeper slope and quicker fall off than their walled counterparts. This may be due to the lack of initial friction at the interface between the dense fluid and the drop wall, leading to a slower acceleration of the fluid as it leaves the drop cylinder.

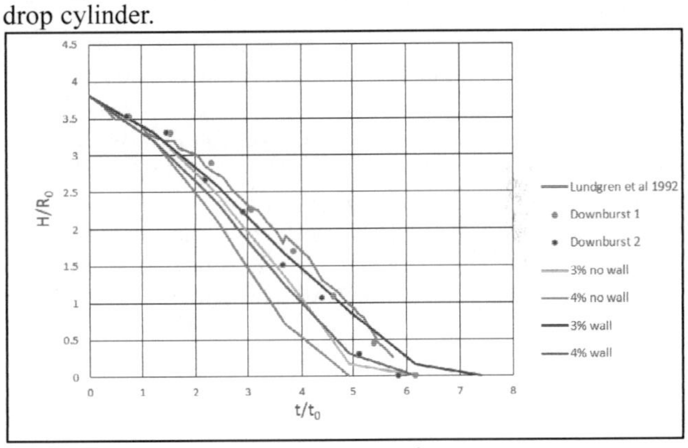

Figure 5 - Height of downburst front versus time

In Figure 6, the maximum radial position is plotted versus time, again non-dimensionalized. Similarly to the previous figure the experimental results (downburst 1 and 2) match quite well with experimental results reported in the literature [7, 8]. In addition to experimental work, this figure also makes comparisons to two numerical cooling source models [9, 10]. Those two numerical studies only report data in the post-impingement region. In both, it can be seen that there is a slightly higher slope than that exhibited by the experimental results. In the case of [10], the higher roughness value of the ground plane leads to a sudden reduction in radial velocity once the downburst vortex has travelled far enough from the point of impingement. Comparing the four trials conducted in this paper to the experimental results, it is obvious that there is a notable difference in the width of the downburst generated shortly after release of the dense fluid. This could be indicative of earlier vortex formation in the numerical model, or failure to include some aspect of the release physics such as the gate in the CFD model. After this initial period, which lasts until about $t/t_0 = 2.5$, the slope of the numerical results matches closely with the experiments. Again, the no-wall cases experience a steeper slope and spread farther from the axis than those cases with the drop wall.

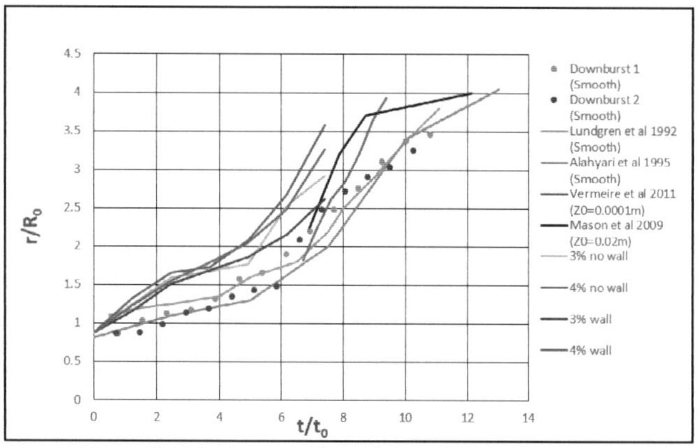

Figure 6 - Maximum radius of downburst versus time

CONCLUSIONS

A numerical simulation of a thunderstorm downburst was performed. This simulation was a two-dimensional URANS model which used a density-driven approach to represent the physics of a real downburst. The simulated domain represented a cylindrical volume of dense fluid being released into a larger volume of less dense ambient fluid. The domain of the event, including the fluid densities, domain size, and boundary conditions was designed to match an experimental study [6].

The simulation was analyzed for parameters of interest such as the maximum radial velocity for various cases with two different density differences (3% and 4%) and two different experimental conditions (with drop wall and without drop wall). The addition of the drop wall is found to result in a lower peak velocity which occurs later in time. Similarly, the trials with a smaller density difference also result in lower peak velocities and a later peak velocity time. Comparison of these numerical results to the experimental work showed very good agreement in terms of the evolution of large scale mean flow structures, particularly when comparing the plot of the position of the height of the downburst front. There was some initial disagreement in the evolution of the maximum downburst radius, but after $t/t_0 = 2.5$ the slope was well matched with the experimental data.

ACKNOWLEDGMENTS

The present work is supported by the Natural Sciences and Engineering Research Council of Canada (NSERC) and the Ontario Graduate Scholarship (OGS) of the Ontario Ministry of Training Colleges and Universities.

REFERENCES

1. Abd-Elaal, E.S., Mills, J. E. & Ma, X. 2013, "A coupled parametric-CFD study for determining ages of downbursts through investigation of different field parameters", *J Wind Eng Ind Aero*, 123, pp. 30–42.
2. Kim, J. and Hangan, H. 2007, "Downburst versus boundary layer induced wind loads for tall buildings", *Wind Struct.*, 10, pp. 481-494.
3. Orf, L., Oreskovic, C., Savory, E. & Kantor, E. 2014, "Circumferential analysis of a simulated three-dimensional downburst-producing thunderstorm outflow", *J Wind Eng Ind Aero*, 135, pp. 182-190.
4. Kim, J. & Hangan, H. 2007, "Numerical simulations of impinging jets with application to thunderstorm downbursts", *J Wind Eng Ind Aero*, 95, pp. 279–298.
5. ANSYS® Academic Research, Release 16.0, Help System, Coupled Field Theory Guide, ANSYS, Inc.
6. Roberto, M., Savory, E. 2014, "Experimental Simulation of Thunderstorm Downbursts", The Canadian Society for Mechanical Engineering International Congress 2014, Toronto, Canada.
7. Lundgren, T.S., Yao, J. and Mansour, N.N. 1992, "Microburst modelling and scaling", *J. Fluid Mech.*, 239, pp. 461-488.
8. Alahyari, A.A, and Longmire, E.K. 1995, "Dynamics of laboratory simulated microbursts", Ph.D. Thesis, University of Minnesota.
9. Vermeire, B.C., Orf, L.G., and Savory, E. 2011, "Improved modelling of downburst outflows for wind engineering applications using a cooling source approach", *J. Wind Eng. Ind. Aero.*, 99, pp. 801-814.
10. Mason, M.S., Wood, G.S., and Fletcher, D.F. 2009, "Numerical simulation of downburst winds", *J. Wind Eng. Ind. Aero.*, 97, pp. 523-539.

Proceedings of the 25th CANCAM
London, Ontario, Canada, May 31 – June 4, 2015

EXPERIMENTAL CHARACTERIZATON OF FLOW AROUND AN ESCARPMENT

Ryan Kilpatrick
WindEEE Research Institute/Department of Mechanical and Materials Engineering, University of Western Ontario, London, ON, Canada, N6A 5B9
rkilpatr@uwo.ca

Kamran Siddiqui
WindEEE Research Institute/Department of Mechanical and Materials Engineering, University of Western Ontario, London, ON, Canada, N6A 5B9
ksiddiq@uwo.ca

Horia Hangan
WindEEE Research Institute/Civil and Environmental Engineering
Western University
London, Ontario, Canada
hmhangan@uwo.ca

ABSTRACT

An experimental study on characterizing the mean and turbulent flow structure in the vicinity of a complex topographic feature is reported. Particle image velocimetry (PIV) was used to study the flow characteristics in the vicinity of the escarpment using a 1:100 scale model in a wind tunnel. Measurements were conducted along the Transect B 270° wind direction at a Reynolds number of 55,000. Mean streamwise velocity results showed good agreement with the measurements obtained from Cobra probe measurements as well as with a previous PIV study of the same escarpment.

Mean velocity was found to be accelerated on the top of the escarpment. A sharp increase in turbulence intensity in both the streamwise and vertical directions, as well as TKE was observed in the region just above the edge of the escarpment, as well as in the mid to lower section upstream of the escarpment.

KEYWORDS: Complex topography, wind energy, wind tunnel simulation, Bolund.

INTRODUCTION

Over the past few decades, wind turbines have emerged as a clean source of electrical power generation using wind energy. Multiple turbines are typically installed within close proximity and described as a wind farm. The most common type of wind turbine configuration is the horizontal-axis wind turbine (HAWT) in which the turbine blades rotate about a horizontal axis. As the wind industry matures, the focus turns to optimizing the efficiency of new wind farms, on the basis of kWh of electricity produced per dollar invested. This becomes particularly important as the very best sites for wind farms are used up and the use of more marginal terrain becomes necessary for continued growth of the industry.

One aspect of wind farms that is not fully understood is the interaction of the turbines with their local topography and with each other, the so-called wake interference. This is evidenced by the consistent underperformance of wind farms compared to their design capacity [1]. While considerable work has been done on wake characterization and interference, there is a scarcity of studies on how wind conditions upstream of a wind turbine are influenced by local topography and what effect this has on available wind energy, turbulence, wind loading and resulting long term performance of the turbine.

The present study is focused on an experimental investigation of the effects of topography on the near-surface flow characteristics using a scaled model in a wind tunnel setup.

THE MODEL

The topographic terrain considered in this study is the Bolund peninsula, a small hill located near the Risø Campus of Denmark Technical University (DTU). The Bolund hill has been the focus of much attention over the past several years, due to its topographical similarity to a typical wind turbine site in complex terrain. One of the motivations for studying this feature is that a better understanding of the flow around complex topography such as the steep escarpment present at Bolund should allow for improvements to predictive models for wind flow over various types of terrain. For example, current commercial linearized models have been shown to overestimate the speed up over the tops of steep ridges, leading to overly optimistic predictions of wind power generation [2]. A 1:100 scale model was made out of foam for the wind tunnel testing.

EXPERIMENTAL SETUP

The experiment was conducted in Tunnel 1 at the Boundary Layer Wind Tunnel Laboratory at Western University in London Ontario. The tunnel is an open circuit type with a length of 33 m and has the cross-section of 2.4 m (width) × 2.15 m (height) at the test section. To simulate a particular upstream boundary layer profile, the wind tunnel is equipped with pneumatic controlled roughness elements as well as spires and trips. In the present setup, a smooth boundary layer profile was produced hence, no roughness elements were used. The Reynolds number for the tests, using the maximum height of the island (11.73 m in full-scale) as the characteristic length, was 55,000.

Particle Image Velocimetry (PIV) technique was used to measure two-dimensional velocity fields in the vicinity of the escarpment. 3-component Series 100 Cobra probes (CP) were also used to measure the upstream flow velocity and turbulence at several locations close to and above the escarpment for comparison purposes. A schematic of the PIV setup is shown in Fig. 1, and a photograph of the setup is shown in Fig. 2. The model was aligned in the tunnel such that the streamwise direction matched the full-scale wind direction of 270° as described by Berg *et al.* [2].

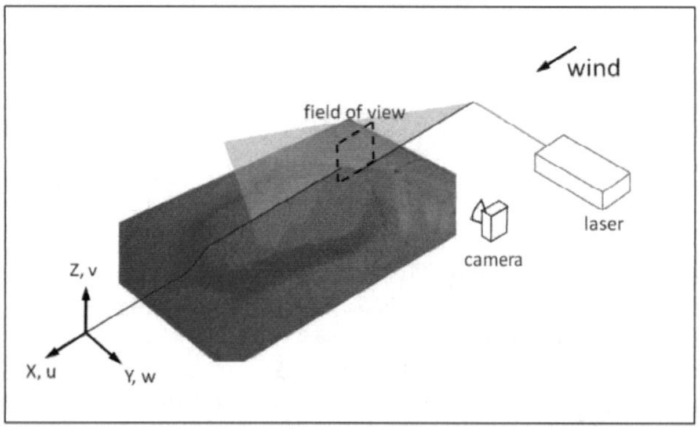

Figure 1: Schematic of PIV setup

Figure 2: Experimental setup

For this study, the frame of view was limited to the region of the escarpment, a single window measuring 21.6 cm x 21.6 cm. The PIV system comprised a pulsed 200 mJ dual-cavity Nd:YAG laser system (New Wave Research PIV Gemini) to illuminate the measurement region. The time difference between laser pulses was set at 100 μs. The tunnel was seeded with fog particles generated by Ultratec's Power Fog Industrial 9D theatrical fog system. A 4 megapixel CCD camera (Flare 4M180-CL) with 2048 × 2048 pixel resolution, equipped with a 75 mm lens was used to capture images. The camera was connected to an image acquisition system (IO Industries DVR Express Core). Images were acquired at a rate of 30 frames per second for 5 minutes i.e. 9000 PIV images that produced 4500 velocity fields.

The velocity field was obtained by cross-correlating the interrogation region in the first image with the search window in the second image, of a given PIV image pair. The size of the interrogation and search regions were set to 32×32 pixels and 64×64 pixels respectively. A 50% window overlap was used to obtain the nominal velocity resolution of 16.74 mm × 16.74 mm. An in-house code was utilized in a commercial image processing software heurisko® (AEON Verlag & Studio GmbH & Co. KG) to compute velocity fields. Just upstream of the escarpment, a reflection or other issue with the laser light sheet formation caused a concentrated area of bad vectors. A mask was defined to exclude these bad vectors from subsequent analyses. Spurious vectors were identified and then corrected using a local median test developed by Siddiqui *et al.* [3]. The local coordinate system for the scaled-model testing was transformed to the corresponding full-scale using the longitudinal and vertical co-ordinates of the actual Bolund Island, as given in Berg *et al.* [2].

The incoming wind profile was measured using Cobra probes at a location roughly 1.1 m upstream of the escarpment model. Fig. 3 shows the streamwise velocity and turbulence intensity profiles.

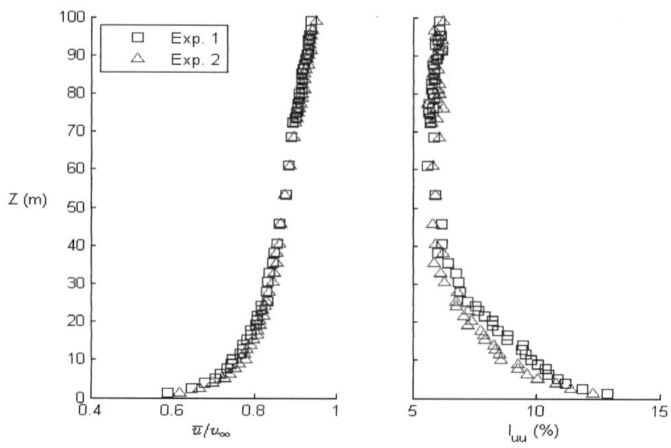

Figure 3: Upstream velocity and turbulence profiles (full-scale height is shown).

From the data points within the logarithmic overlap layer, the friction velocity and roughness length were estimated to be $u_* = 0.21$ and $z_0 = 3.9 \times 10^{-4}$ m, respectively. The roughness length compares reasonably well with the full-scale value of 6.0×10^{-4} described by Bechmann et al. [5], however friction velocity is considerably lower than the full-scale value of 0.47. This issue will be examined in further experiments.

RESULTS

Fig. 4 shows a sample instantaneous velocity field, the mean velocity field and a turbulent velocity field. The results in Fig. 4 highlight some key characteristics of the flow near the escarpment. The instantaneous and mean velocity fields show that the flow slows considerably as it approaches the base of the cliff and speeds up as it approaches and passes over the top. No recirculation zone is visible in the mean plot, however negative flow is intermittently observed in the instantaneous velocity fields, which is consistent with the observations by Yeow *et al.* [4] who used a 1:115 model of the Bolund Hill and tested at Reynolds numbers of 41,500 and 80,000 using PIV. Several vortical structures are visible in the turbulent velocity field.

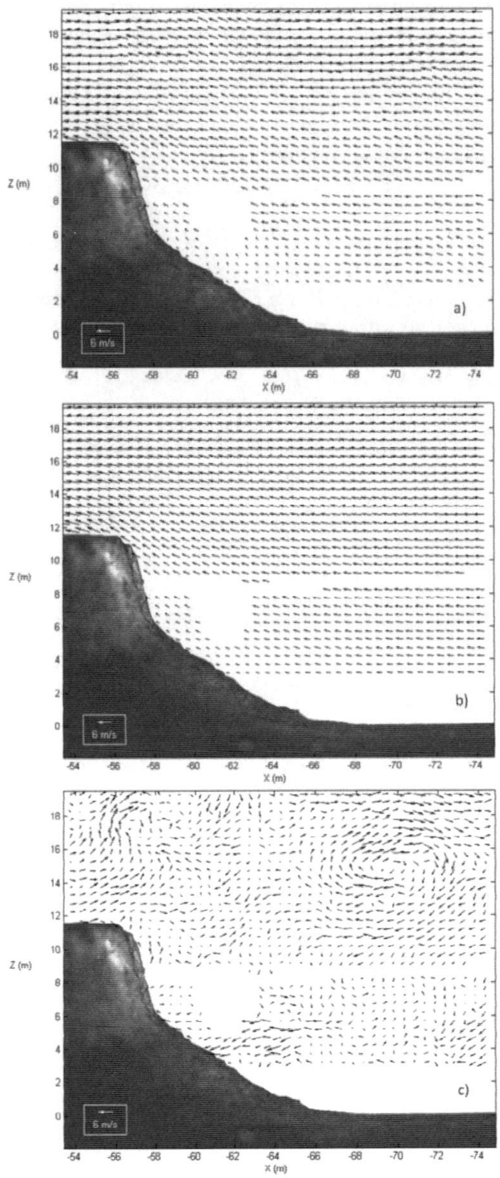

Figure 4: Velocity vector plots for a) Instantaneous b), mean and c) turbulent fields in full-scale co-ordinates

Mean streamwise velocity profiles at different streamwise locations are plotted in Fig. 5 to show the influence of local topographic features on the magnitude and vertical variation of the mean flow as the flow moves over the escarpment. The mean velocity data from the Cobra probe are also plotted for comparison. The results show that the escarpment locally influenced the mean velocity. Just above the escarpment, in the furthest downstream profile, emergence of a "nose" in the profile is observed, where the mean velocity speeds up to a value very close to the freestream velocity. Strong agreement is observed between PIV and Cobra probe data.

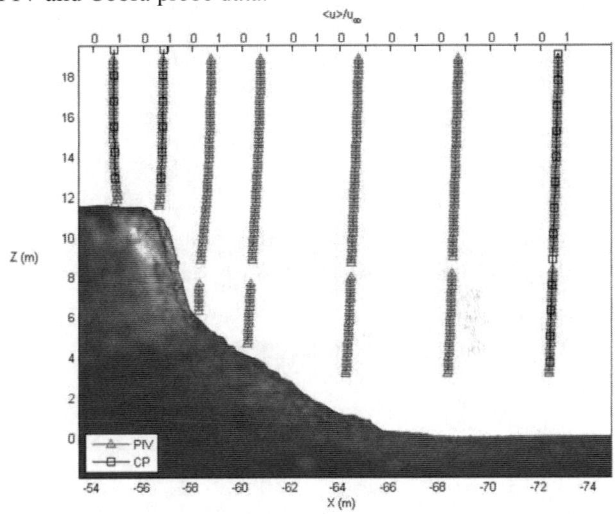

Figure 5: Streamwise velocity profiles above and upstream of the escarpment. Mean value of three horizontally adjacent columns used to generate PIV profiles.

To illustrate the spatial two-dimensional structure of the mean and turbulent flow, contour maps of the normalized mean streamwise velocity, streamwise and vertical turbulence intensities, and turbulent kinetic energy (TKE) are presented in Fig. 6. The mean velocity maps clearly demonstrate the flow acceleration over the escarpment top. The mean velocity contour map is in good agreement with that of Yeow *et al.* [5] for the same region. Results also indicate very high values of turbulence intensities and TKE at the same location just above the escarpment as well as in the mid to lower section immediately upstream of the escarpment.

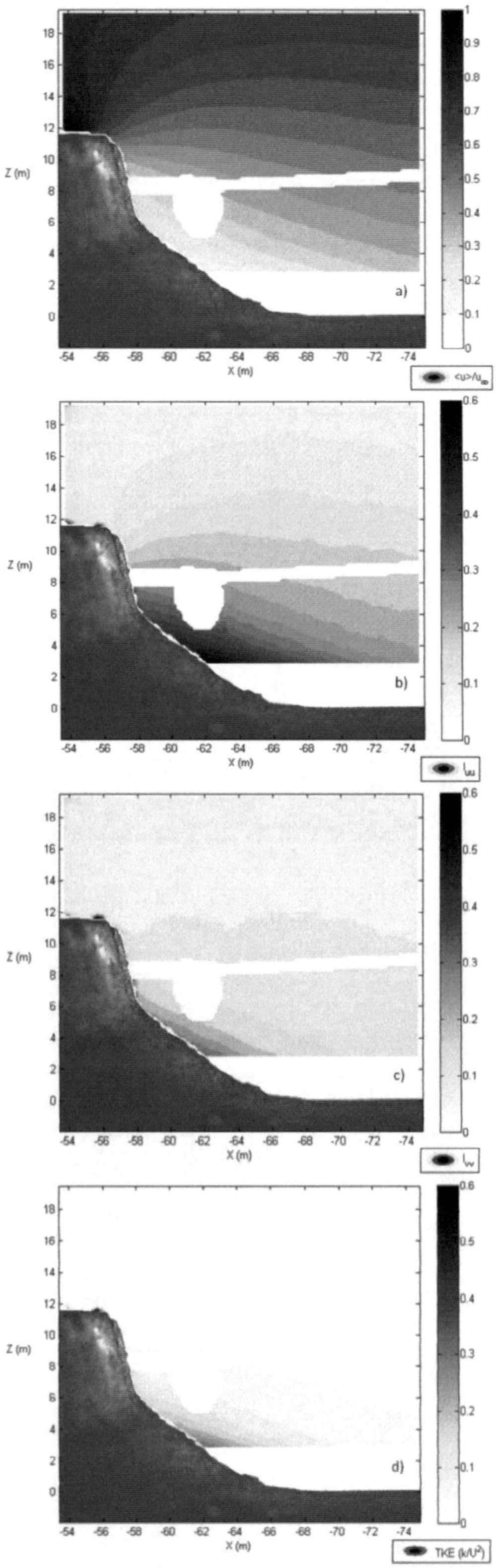

Figure 6: Contour plots of a) streamwise velocity, b) streamwise and c) vertical turbulence intensity, d) normalized TKE

CONCLUSIONS

The flow in the vicinity of the escarpment at Bolund hill model was studied by using particle image velocimetry in a wind tunnel. Tests were conducted along the 270° wind direction at a Reynolds number of 55,000. Mean streamwise velocity obtain from PIV data showed good agreement with that obtained from the Cobra probe as well as with the previous wind tunnel PIV study of Bolund Island conducted by Yeow *et al.* [4].

A sharp increase in turbulence intensity in both the streamwise and vertical directions, as well as TKE, was observed in the region just above the edge of the escarpment as well as in the mid to lower section upstream of the escarpment.

Future work will entail comparing these results with full scale field data, and to conduct further PIV experiments at different wind directions, with a larger scale model, a larger field of view and higher resolution cameras. The new WindEEE wind research facility will allow for a larger model of the island; tests are being planned for the 1:25 scale.

ACKNOWLEDGMENTS

The authors would like to acknowledge Natural Sciences and Engineering Research Council (NSERC) and University of Western Ontario for the support.

REFERENCES

1. Parvu, D. 2014, "Experimental Modeling of Wind Farm Topography and Canopy", Western University.
2. Berg, J. et al. The Bolund Experiment, Part I: Flow Over a Steep, Three-Dimensional Hill. Boundary-Layer Meteorology 2011, 141:219–243.
3. Siddiqui, M. H. K., M. R. Loewen, C. Richardson, W. E. Asher and A. T. Jessup, 2001: Simultaneous particle image velocimetry and infrared imagery of microscale breaking waves, Phys. Fluids, 13, 1891–903.
4. Yeow, T.S. et al. Reproducing the Bolund experiment in wind tunnel. Wind Energy 2013.
5. Bechmann A. et al. The Bolund Experiment, Part II: Blind Comparison of Microscale Flow Models. Boundary-Layer Meteorology 2011, 141:245–271.

Mechatronics

Proceedings of the 25th CANCAM
London, Ontario, Canada, May 31 – June 4, 2015

OPTICAL FIBER SENSORS FOR IN-SITU MONITORING OF LITHIUM ION BATTERIES

Hamid Alemohammad[1,*], AbdulRahman Ghannoum[1,2], Liliana Zdravkova[1], Krishna Iyer[1], Patricia Nieva[1], Aiping Yu[2], and Amir Khajepour[1]
[1]Department of Mechanical and Mechatronics Engineering
[2]Department of Chemical Engineering
University of Waterloo
Waterloo, Ontario, Canada
[*]shalemoh@uwaterloo.ca

ABSTRACT

This paper presents an embedded fiber optic sensor technology for in-situ monitoring of the performance of Lithium Ion Battery (LIB) cells used in electric vehicles. This type of monitoring system, which relies on the direct sensing of the opto-chemical properties of battery cells, can address the exiting challenges related to the estimation of State of Charge (SOC) and State of Health (SOH) in LIB cells. The sensor design specifications are derived from comprehensive optical characterization of commercial and in-house built LIB cells. The paper elaborates on the electrochemistry and opto-chemical properties of the LIB cells, sensor design and the development of a proprietary optical signal interrogation system. The experimental results showing the performance of the sensing system and the numerical techniques used for the estimation of SOC and SOH will also be presented.

KEYWORDS: Electric Vehicles, Lithium-Ion Battery Cell, Monitoring, Embedded Optical Fiber Sensor, State of Charge

INTRODUCTION

Given the ongoing global initiatives for promoting electric mobility, Lithium Ion Batteries (LIBs) have been strong candidates for use in electric vehicles due to their high energy density [1]. According to the International Energy Agency (IEA), the global number of electric vehicles is projected to reach 20M by the year 2020 [2]. The BCC Research market study speculates the LIB global market to reach $13.4B in 2018 from $6.5B in 2008 [3]. Electric vehicles are powered by a large number of LIB cells connected in series. The cells managed by a control system that manipulates the charging and discharging processes based on the estimation of the State of Charge (SOC), State of Health (SOH), failure status, and life expectancy of individual cells. The SOC estimations are currently achieved by measuring the external voltage and electric current of individual cells and running system identification and state estimation models [4]. However, complexity arises in measuring the voltage and current of individual cells connected in series with varying voltage reference levels. The SOC estimation is also based on integrating electric current data, which is highly dependent on the measurement accuracy and is susceptible to error accumulation. The real-time measurements are also affected by uncertain drive cycles [5][6]. These uncertainties have resulted in applying large safety factors in the design LIB cells. This makes the cells larger and heavier and increases negative environmental impacts when battery cells are disposed and recycled at the end of their effective life cycle. To increase the accuracy of the SOC estimation, it is imperative to implement in-situ measurement techniques using embedded sensors operating on the basis of direct interaction with the battery electrochemistry. Fiber optic sensors are viable candidates for this application due to their unique advantages which include immunity to electromagnetic interference and Ohmic loss, decent resistance to chemicals and corrosive environments, high signal-to-noise ratio (SNR), and long-term durability.

This paper presents, for the first time, the development of an electro-optical monitoring system with embedded optical fibers in LIB cells for in-situ monitoring of the cells during charge and discharge cycles. The optical sensor, with direct interaction with the electrochemistry of battery cells, can detect the changes in opto-chemical properties of the battery cell, which is not achievable with exiting current and voltage measurement techniques. The optical sensor data can be manipulated in battery management system for more accurate estimation of SOC and SOH in LIB cells.

ELECTROCHEMISTRY OF LIB CELLS

Rechargeable LIB cells consist of anode and cathode electrodes isolated by a non-conductive separator soaked in electrolyte. The most common intercalation anode material is graphite. Graphite has the advantages of relatively high capacity (372 mAh/g), high reversible lithium ion intercalation, good electrical conductivity, simple production process, and low cost [7]. The cathode material may be composed of different lithium compounds such as lithium iron phosphate ($LiFePO_4$), lithium manganese oxide ($LiMn_2O_4$) or lithium cobalt oxide ($LiCoO_2$). Orthorhombic $LiFePO_4$, commonly used in rechargeable LIB cells, is an inexpensive, environmentally benign, and abundant cathode material with a theoretical capacity of 170 mAh/g [8][9][10]. The operation of the LIB cell is based on the reversible intercalation of lithium ions in the layers of graphite; i.e., the graphite cathode is lithiated/de-lithiated during the battery charging/discharging cycle. Graphite anodes in LIB cells have ordered structures by the periodic arrangement of intercalated lithium ion layers within graphite sheets. The level of intercalation is classified by the number of graphite layers between adjacent layers of Li-ions: Stage I (LiC_6), stage II (LiC_{12}), stage III (LiC_{18}), and stage IV (LiC_{24}). The schematic of graphite intercalation in LIB cells is shown in Figure 1.

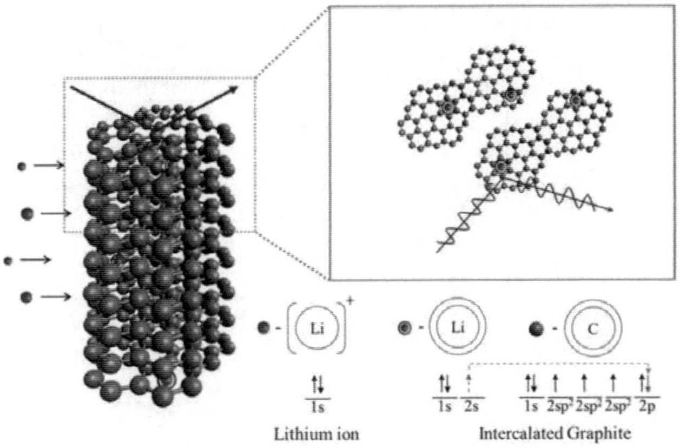

Figure 1: Intercalation of graphite anode in LIB cells and photon interaction with lithiated graphite

OPTICAL CHARACTERISTICS OF GRAPHITE ANODE

Pure graphite has a relatively low free carrier concentration at room temperature. The intercalation of lithium ions into graphite causes it to ionize due to the transfer of electrons from its valance band. This increases the free carrier concentration and affects the electrical, magnetic and optical properties of graphite [11]. The optical properties of lithiated graphite can be explained by the Drude model. The optical reflectivity of lithiated graphite exhibits characteristic Drude edge at the plasma frequency ω_p. As the frequency of the photons decrease below ω_p, the reflectivity increases, which is typical of metallic free carrier absorption [11]. The process of lithium intercalation increases the carrier density, which shifts ω_p to higher frequencies or smaller wavelengths. The photon interaction with lithiated graphite is also shown in Figure 1.

Stages I, II, and III of lithiated graphite can be identified by their visible colors, which have been identified as bright-yellow, red, and blue, respectively. Stage IV and higher are identified with a color that approaches the black color of pristine graphite [12]. The changes in the graphite anode color at different stages of lithiation and SOC are shown in Figure 2. In order to get a deeper insight into the optical characteristics of graphite anode, we conducted reflectance spectroscopy of graphite anode at different charge levels (i.e., different SOC levels). The results, as shown in Figure 3, indicate that the state of lithiation or SOC can be correlated to the optical reflectance of graphite. These results make the graphite anode ideal to be used as the basis for the design of an optical sensor to detect the different stages of lithiation in LIB cells which leads to a direct sensing method for the estimation of SOC.

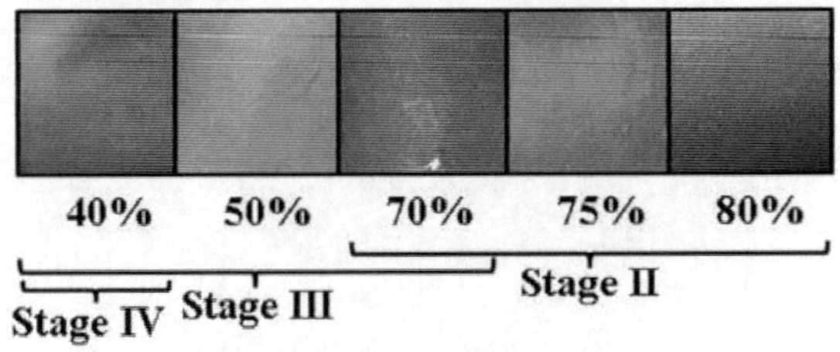

Figure 2: Colorimetry analysis of graphite anode at different charge levels or lithiation stages

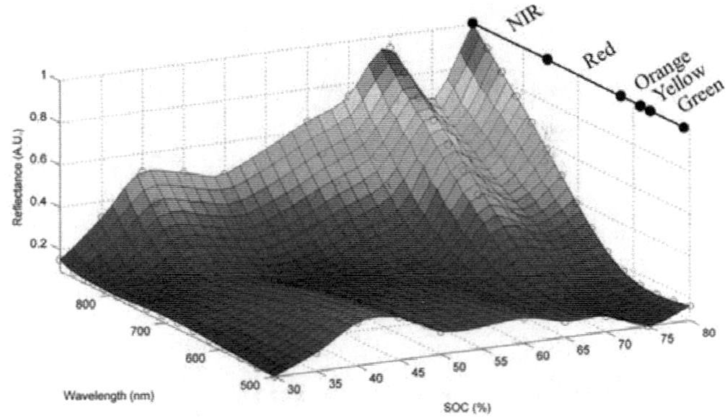

Figure 3: Reflectance spectroscopy measurements of graphite anode in LIB cells at different charge levels

EMBEDDED OPTICAL FIBER SENSOR

Given the optical characteristics of the graphite anode, we developed an in-situ measurement technique using embedded optical fiber sensors for real-time monitoring and a more accurate estimation of SOC and SOH in LIB cells. The sensing system is composed of a proprietary designed fiber optic sensor embedded inside an LIB cell and a miniaturized optical signal interrogation (OSI) system. Optical fiber has the

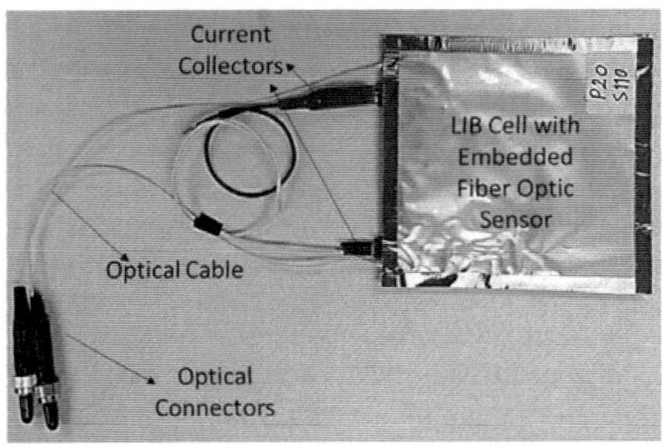

Figure 4: LIB cell with embedded optical fiber sensor

advantage of immunity to electromagnetic interference, robustness to corrosive chemicals, and near-zero Ohmic loss which makes it a viable solution for sensing inside battery cells. The sensor is interrogated in the transmission mode with optical intensity that is modulated by the state of lithiation in graphite. We have also developed an in-house integration procedure to embed optical fibers during the cell fabrication. Figure 4 shows an LIB cell with the embedded fiber optic sensor.

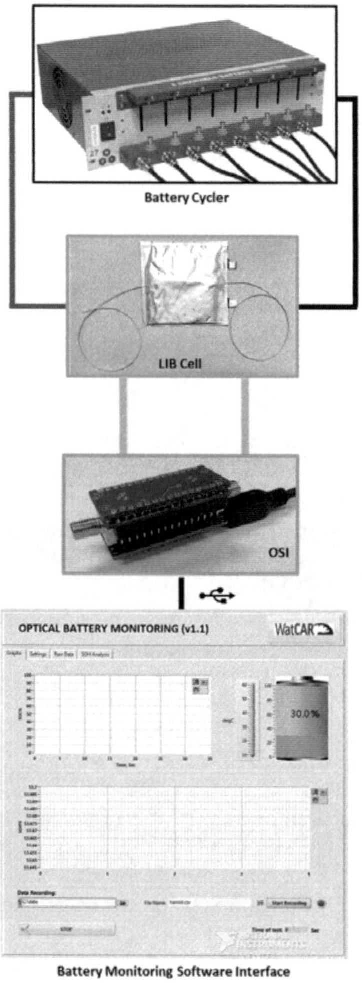

Figure 5: Experimental setup for battery testing

EXPERIMENTAL SETUP

The experimental setup for testing and validating battery cells and embedded optical fiber sensor, shown in Figure 5, consists of a battery cycler which is programmed to charge and discharge the battery cell at specific current rates. The cycler has the capability to measure the cell output voltage and current during cycling. We have developed a proprietary and miniaturized optical signal interrogation (OSI) system. The OSI system is composed of a broadband light source, light source driver, photo detector, signal amplifier, A/D convertor, and micro controller. The OSI system has a dual-purpose USB connection to power up the opto-electronic circuit and trasmit the measurement data. A customized data acquisition and analytical software is also developed to acquire the measurement data at a rate of up to 50 Hz.

RESULTS AND DISCUSSION

Figure 6 shows the experimental results obtained from cycling an LIB cell with the embedded optical fiber sensor. The graphs show the variations of the cell output voltage and the optical sensor signal. The cell voltage range corresponds to 0% to 100% SOC; cut-off voltages of 3.7 V and 2.5 V are used for 100% and 0% SOC, respectively. As indicated in the figure, two trends are observed in the optical measurement data: (1) the fast cycling trend indicating the battery charge/discharge cycles (2) the slow declining trend indicating the state of health related to battery aging.

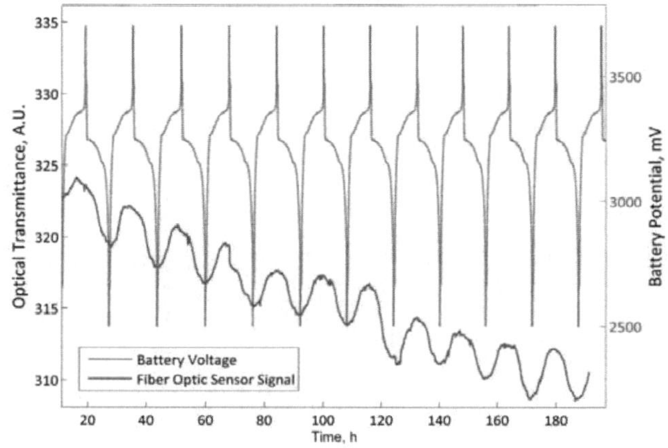

Figure 6: Output voltage and optical sensor data during LIB cell cycling

The fast and slow optical measurement data are decoupled and plotted in Figure 7. The fast trend in the optical signal follows the battery charge and discharge cycles. The shape of this graph is dependent on the internal electrochemistry of the battery and the direct interaction of the optical waves with the photo-chemistry of the cell. The changes in the photo-chemical properties of the battery cell at different stages of lithiation (i.e., different charge levels) modulate the guided waves in the fiber optic sensor. The state of charge in the LIB cell can be further estimated by demodulating the optical sensor data. Compared to the existing SOC estimation methods, the optical fiber sensing is more accurate as it directly measures the changes in the electrochemistry of the cell which is not completely detectable with external voltage and current measurements. Additionally, this technique is not prone to the measurement error accumulation.

We have developed an identification algorithm to estimate the level of charge by analyzing only the optical fiber sensor signal, as shown in Figure 8. The measurement data were obtained from a battery cell that went through several charge/discharge cycles while the optical sensor data were captured at 1 Hz. The data was post-processed and analyzed to optimize a linear auto-regressive (ARX) model for charge estimation (Figure 8). The algorithm has the capability to estimate the battery charge without using current or voltage measurement data. The goal is to analyze the optical sensor data as a direct measurement technique for more accurate estimation of SOC and SOH in LIB cells. Further analytical studies are in progress to improve the performance of the estimation algorithm.

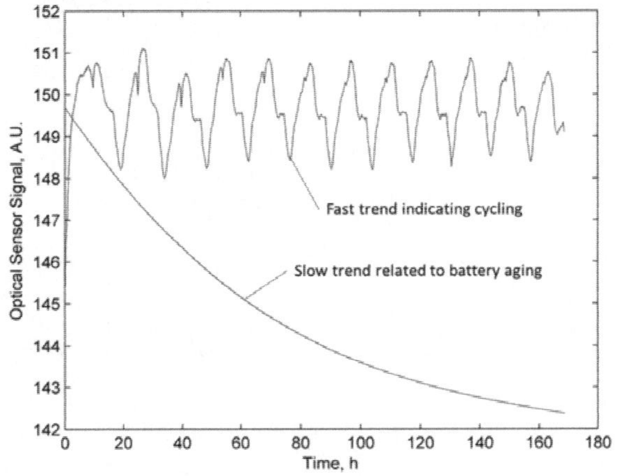

Figure 7: Decoupling of the slow and fast trends in the optical sensor measurement data

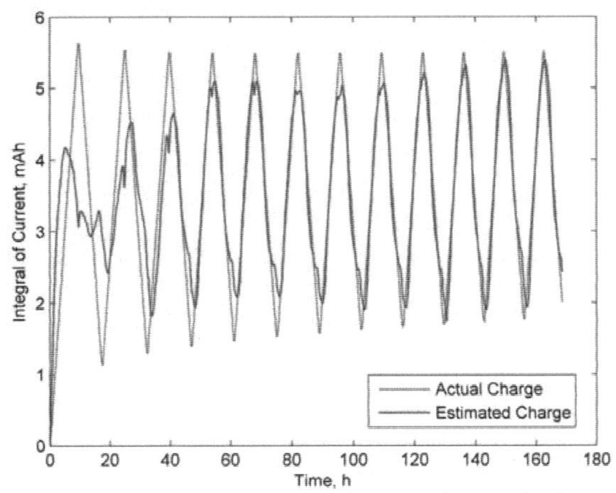

Figure 8: Estimation of battery charge using optical sensor data and an ARX model

CONCLUSIONS

This paper elaborated on a novel sensing system for monitoring of the performance of LIB cells. The monitoring system consisting of a proprietary designed optical fiber sensor and an in-house built optical sensor interrogation system, was tested in LIB cells made of graphite anode and $LiPO_4$ cathode. The system presented in this paper is the first in-situ monitoring system for rechargeable batteries. The experimental results indicate changes that can be correlated to the SOC and SOH of the battery cells. An estimation algorithm is under development to correlate the optical signal to the battery state of charge.

ACKNOWLEDGEMENT

The authors would like to acknowledge the funding for this research work provided by the "Green Intelligent Transportation Systems (GITS)" – Ontario Research Fund, Canada, the "Next Generation Electric Vehicles: Development of Key Technologies and Full Vehicle Testing" – Automotive Partnership Canada, and General Motors.

REFERENCES

[1] Gu, Y. J., Zeng, C. S., Wu, H. K., Cui, H. Z., Huang, X. W., Liu, X. B., Wang, C. L., Yang, Z.-N., and Liu, H. 2007, "Enhanced cycling performance and high energy density of $LiFePO_4$ based lithium ion batteries", *Materials Letters*, Vol(61), pp. 4700–4702.

[2] Global EV Outlook, *International Energy Agency*, April 2013.

[3] Lithium Batteries: Markets and Materials, *BCC Research*, October 2013.

[4] Wang, L. Y., Polis, M. P., Yin, G. G., Chen, W. , Fu, Y., and Mi, C. C. 2012, "Battery Cell Identification and SOC Estimation Using String Terminal Voltage Measurements", *IEEE Trans. Veh. Technol.*, Vol(61), pp. 2925–2935.

[5] Stuart, T. A., and Zhu, W. 2011, "Modularized battery management for large lithium ion cells", *J. Power Sources*, Vol(196), pp. 458–464.

[6] Xiong, R., He, H., Sun, F., Liu, X., and Liu, Z. 2013, "Model-based state of charge and peak power capability joint estimation of lithium-ion battery in plug-in hybrid elect vehicles", *J. Power Sources*, Vol(229), pp.159–169.

[7] Armstrong, M. J., O'Dwyer, C., Macklin, W. J., and Holmes, J. D. 2014, "Evaluating the performance of nanostructured materials as lithium-ion battery electrodes", *Nano Res.*, Vol(7), pp. 1–62.

[8] Takahashi, M., Tobishima, S., Takei, K., and Sakurai, Y. 2002, "Reaction behavior of $LiFePO_4$ as a cathode material for rechargeable lithium batteries", *Solid State Ion.*, Vol(148), pp. 283–289.

[9] "Nanophosphate® Basics: An Overview of the Structure, Properties and Benefits of A123 Systems' Proprietary Lithium Ion Battery Technology", *A123 Systems, Inc.*

[10] Chung, S. Y., Bloking, J. T., and Chiang, Y. M. 2002, "Electronically conductive phospho-olivines as lithium storage electrodes", *Nat. Mater.*, Vol(1), pp. 123–128.

[11] Dresselhaus, M. S., and Dresselhaus, G. 2002, "Intercalation compounds of graphite", *Adv. Phys.*, Vol(51), pp. 1–186.

[12] Pfluger, P., Geiser, V., Stolz, S., and Güntherodt, H. J. 1981, "Aspects of Alkali Metal Interclation and Deintercalation in Highly Oriented Pyrolytic Graphites", *Synth. Met.*, Vol(3), pp. 27–39.

Proceedings of the 25th CANCAM
London, Ontario, Canada, May 31 - June 4, 2015

A NOVEL NONLINEAR AMPLITUDE-MODULATION GYROSCOPE INCORPORATING INTERNAL RESONANCE

S. Amir Mousavi Lajimi, Navid Noori, Amr Marzouk, Behraad Bahreyni, Farid Golnaraghi
Mechatronic Systems Engineering, Simon Fraser University, B.C., Canada V3T 0A3
samirm@sfu.ca, samousav@uwaterloo.ca

ABSTRACT

We are presenting the design and the preliminary numerical and experimental analyses of two mismatched Coriolis vibratory gyroscopes incorporating nonlinear modal interaction. A novel double-H design includes two clamped-clamped beams and a suspended mass in the middle connected to the base beams via four short cantilevers. Another design is a T-shaped gyro including a primary doubly-clamped beam and a secondary sense beam. A combination of analytical, finite element, and experimental analyses are employed to study the characteristics of the nonlinear gyro. The drive mode matches the structure's second mode, while the sense mode matches the fundamental mode of the structure. Our preliminary study indicates that the bandwidth and the sensitivity of the rotation rate sensor are improved by employing the nonlinear modal interaction.

Keywords: Gyroscope, rate sensor, internal resonance, sensitivity, bandwidth, noise, saturation

INTRODUCTION

A conventional Coriolis vibratory gyro (CVG) refers to a micro- or macro-mechanical structure where the Coriolis effect transfers some energy from a primary oscillation mode to a secondary oscillation mode. The system used as a sensor, includes a primary (drive) mode excited by an external source and a secondary (sense) mode activated by the rotation rate-dependent terms. To operate a CVG in open-loop mode and amplitude-modulation scheme, the amplitude of the sense mode (pickoff) is used to identify the rotation rate [1]. Another operation mode for vibratory gyroscopes is the force-to-rebalance, or closed-loop, mode where the sense amplitude is estimated by an external signal proportional to the rotation rate [1].

The designs of rotation rate sensors (gyroscopes) are evolving to offer higher resolution, larger sensitivity, better performance and a lower fabrication cost. Beam-based gyroscopes offer a relatively simple structure and have been in-use for a long time [2]. Gyroscopes suffer from the low sensitivity and low mechanical scale factor. To overcome these difficulties, researchers have investigated beam-mass systems both experimentally and analytically [3, 4]. Furthermore, to operate the beam-mass gyroscope using frequency-modulation scheme the closed-form equation relating the angular rotation rate and the modal frequencies has been obtained [5, 6] and to exploit nonlinearities for higher performance, the nonlinear response has been characterized [7].

To increase the sensitivity of an amplitude-modulation gyro, the frequencies of the sense and drive modes are matched [8]. However, the bandwidth and other performance parameters are affected adversely and maybe improved by other arrangements of the modes [9]. The other possibility is to use different mode shapes of a structure and operate based on the features of nonlinearly coupled systems to increase the bandwidth and the sensitivity as it has been shown for resonators [10]. Furthermore, a parametrically resonating gyroscope offers a higher performance [11]. Recently, Golnaraghi et al. [12] revealed an application of internal resonance in designing gyroscopes with larger bandwidths and higher sensitivities [12].

In this work, we propose methods for the design and analysis of an amplitude-modulation gyro which is excited at the frequency of the drive mode being twice the frequency of the sense mode. The first and second frequencies are internally coupled because of the structure's design and offer a higher sensitivity and a larger bandwidth in the sensed response. The finite element analysis is performed in Ansys Mechanical APDL 14.5 package using static, modal, harmonic, and transient analyses. The initial designs are implemented in CoventorWare package and fabricated using a SOIMUMPs process [13]. Two designs are used to study two-to-one internally coupled gyros where the first one, a double-H gyro, is a novel design and the second one, a T-shaped structure, was previously studied as a resonator. A macro-scale model of the T-shaped structure is built and tested on a rate table and the frequency-response curves are obtained. Using a lumped-mass model, a mathematical model of the system is developed and solved numerically to obtain the saturation curves.

DESIGNS and METHODS

The schematic of the two micro-gyro designs with various configurations of the drive and sense electrodes are provided in Figs. 1 and 2. In Fig. 1, exciting the base beams along Y-axis, the Coriolis effect induces a response in the X-direction for a single-axis gyro rotating about the Z-axis (out-of-plane). It is worth mentioning that the same structure may be used for in-plane rate measurements placing the drive electrodes along the Z-axis under the drive beam on the Si substrate. The double-H design includes a mass to further amplify the sense signal. The second design is a T-shaped structure previously used as a resonator in [10], see Fig. 2.

Both designs are initially analyzed using MEMS CAD software CoventorWare and ANSYS 14.5 Workbench and fabricated using a standard SOIMUMPs process. Scanning electron microscope (SEM) images of the fabricated devices are presented in Figs. 1(a) and 2(a) obtained using a FEI Nova NanoSEM 430 SEM System. The structures are designed such that the modal frequency of the second mode is twice the modal frequency of the first mode interacting with each other through nonlinear modal interaction (internal resonance). The pair of sense electrodes provide the required means for differential sensing of the sense displacement. For the double-H gyro, Fig. 1, the drive electrodes operate in 180 degrees phase difference to increase the strength of actuation. Proper instrumentation of the experimentations are shown in Figs. 1(b) and 2(b).

To properly identify the nonlinear dynamics of the system, a macro-scale model of the T-shaped sensor is built and tested on a rate table, see Fig. 3. The parameters of the gyroscope include a base doubly-clamped beam of length 160.20mm, thickness 0.47mm, and width 15mm, and a secondary beam of length 80mm, thickness 0.203mm, and width 15mm. The actuation and sensing mechanisms include two pairs of PSI-5H4E piezoelectric ceramic transducers and a pair of LTS 15/03 and 15/12 laser displacement sensors. The device is mounted on an Ideal Aerosmith 1621-200A-TL rate table and operated and controlled through a customized MATLAB Simulink code and a dSpace data acquisition system.

A simplified mathematical model based on lumped-mass model, Fig. 4, of the T-shaped design, Fig. 2, is used to derive the equation of motion in the form

$$[M]\begin{Bmatrix}\ddot{r}_1\\ \ddot{\theta}_2\end{Bmatrix}+[C]\begin{Bmatrix}\dot{r}_1\\ \dot{\theta}_2\end{Bmatrix}+[K]\begin{Bmatrix}K_1\\ K_2\end{Bmatrix}=\begin{Bmatrix}F_r+F_e\\ F_\theta\end{Bmatrix} \quad (1)$$

where

$$[M]=\begin{Bmatrix}M_1+M_2 & -M_2r_2\sin(\theta_2)\\ -M_2r_2\sin(\theta_2) & M_2r_2^2\end{Bmatrix}$$

$$[C]=\begin{Bmatrix}\frac{c_1}{L_1^2-r_1^2} & 0\\ 0 & c_2\end{Bmatrix},\quad [K]=\begin{Bmatrix}\frac{\arcsin(\frac{r_1}{L_1})}{L_1\sqrt{1-\frac{r_1^2}{L_1^2}}} & 0\\ 0 & \theta_2\end{Bmatrix}$$

$$\begin{aligned}F_r =&M_2r_2\left(\dot{\theta}_2^2+\Omega^2\right)\cos(\theta_2)+2M_2r_2\Omega\cos(\theta_2)\dot{\theta}_2\\ &+(M_1+M_2)r_1\Omega^2\\ F_e =&a_e\sin(f_e\,t)\\ F_\theta =&-M_2r_2r_1\Omega^2\sin(\theta_2)-2M_2r_2\Omega\cos(\theta_2)\dot{r}_1\end{aligned}$$

where parameters and variables are introduced in Fig. 4. The mathematical model, Eq. (1), is used to numerically study the behaviour of the system and to derive a reduced-order model based on a two-variable perturbation method [14].

RESULTS

An Ansys Mechanical APDL code is developed to study the static, modal, harmonic, and transient responses of the structures. To perform the finite element analysis of the micro T-shaped gyro in a reasonable amount of time, the quality factor is set to 100 and a two-dimensional model of the structure, Fig. 2(a), is studied in Ansys Mechanical using Plane 182 2D structural solid elements. For a T-shaped gyro with a primary 388μm long clamped-clamped beam with width 20μm and a secondary 145.5μm long beam with width 8μm, the time-histories of the drive and sense displacements are computed using transient analysis and shown in Fig. 5. The energy is transferred from the drive beam (mode) to the sense beam

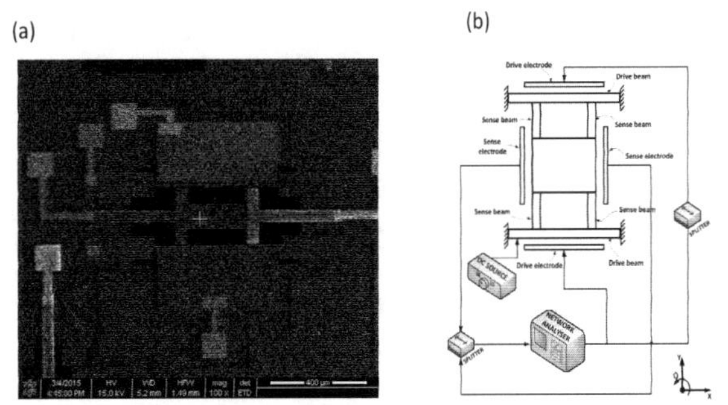

Figure 1: (a) A SEM image of the fabricated double-H shaped gyro and (b) the schematic of the gyroscope and the test setup.

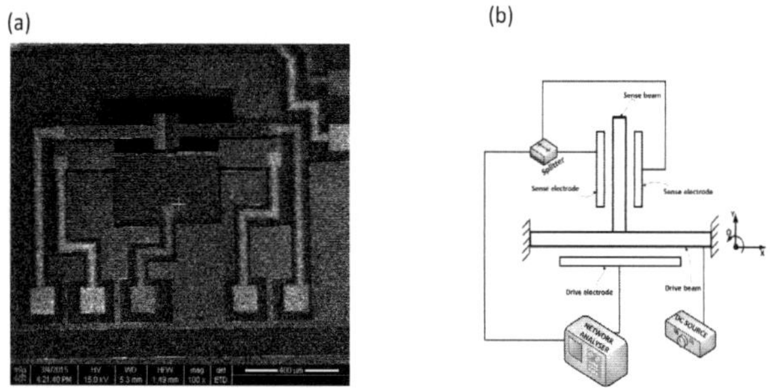

Figure 2: (a) A SEM image of the fabricated T-shaped gyro and (b) the schematic of gyroscope and the test setup.

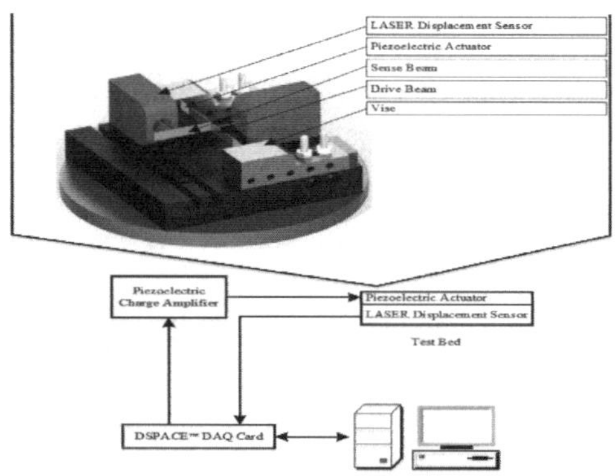

Figure 3: Experimental test setup for the T-shaped gyro. The structure is placed on the table inside the rate table chamber.

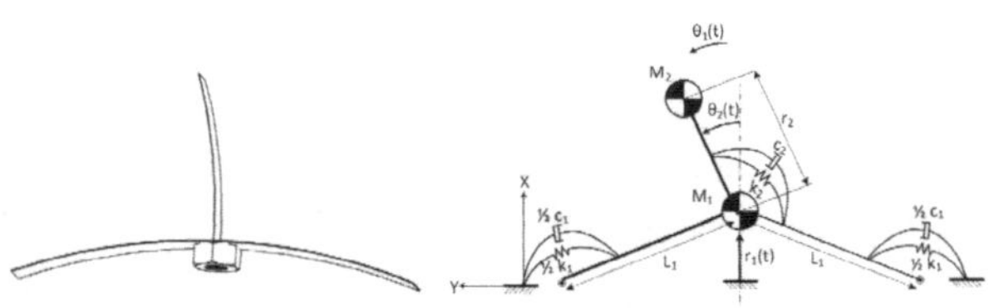

Figure 4: A lumped-mass model of the T-shaped gyroscope.

(mode) through internal resonance and the Coriolis effect giving rise to the large amplitude of the sense beam. A fast Fourier transform (FFT) of the sense and drive time-histories are given in Fig. 6 indicating the frequency components of the time-responses. The excitation frequency (the second peak) in the drive direction transforms into the primary frequency (the first peak) in the sense direction through the internal resonance and the Coriolis effect.

In Fig. 7, the experimental frequency-response curves for macro-structure of Fig. 3 are presented. Repeating the experiment for a set of excitation amplitudes, a family of frequency-response curves are obtained showing an increase in the bandwidth and the sense amplitude. For small excitation amplitudes, the response is similar to linear systems, however for large excitation amplitudes a flat region appears in the frequency-

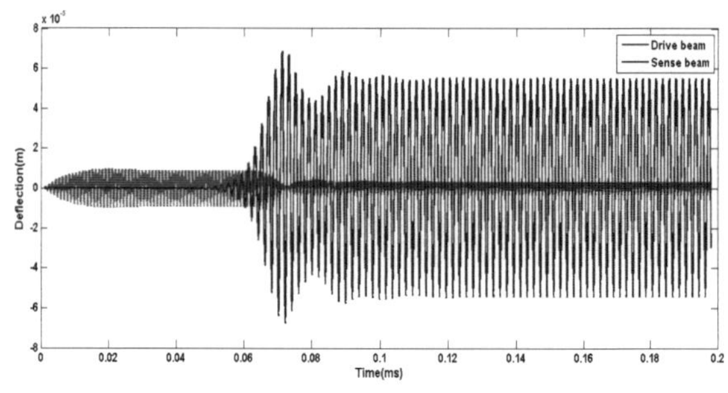

Figure 5: Displacement time histories of the drive and sense beams for micro T-gyro ($\Omega = 10°/s$).

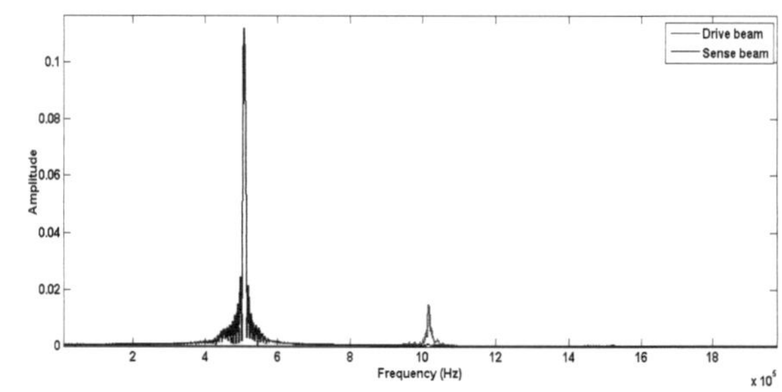

Figure 6: Fast Fourier transforms (FFT) of the displacement time histories for micro T-gyro ($\Omega = 10°/s$).

response curves offering a higher stability and a lesser sensitivity to where the measurements are taken along the frequency-response curve.

A numerical solution of the equation of motion, Eq. 1, is computed for an increasing amplitude of excitation by direct integration and plotted in Fig. 8. The sense mode starts to grow after a certain threshold of the excitation amplitude is passed where the drive amplitude reaches to a limit and remains unchanged. The saturation phenomena appears in systems having internally coupled modes of vibration. Similar results are obtained using a two-variable perturbation method and experimental analysis.

CONCLUSIONS

The coupled-system results in a multi-frequency response in the sense direction of the proposed novel Coriolis vibratory gyros. Initial results show improvements in the sensitivity and the bandwidth of the open-loop gyroscope. Using internal resonance offers a method to overcome the complexities of a closed-loop gyro. Employing the two-to-one internal resonance phenomenon, a wide bandwidth high amplitude response in the sense direction is achieved improving the stability and the sensitivity of the device. Furthermore, separating the drive and sense mode frequencies (for example the drive frequency is twice bigger than the sense frequency-*i.e* a 2:1 internal resonance), the sensed signal could be filtered for fre-

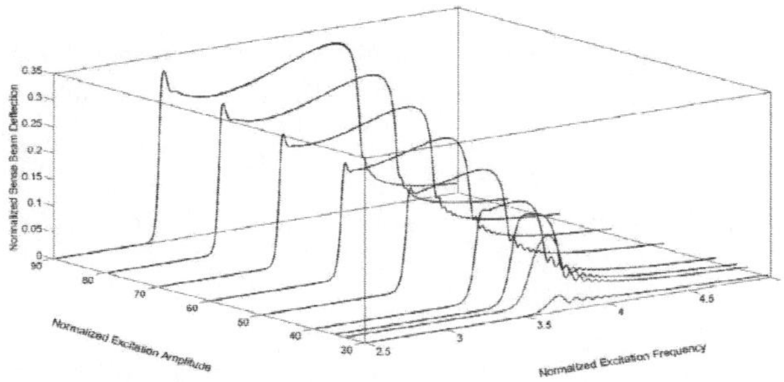

Figure 7: Experimental frequency-response curves of the sense beam for macro T-gyro showing an increase in the bandwidth of the response.

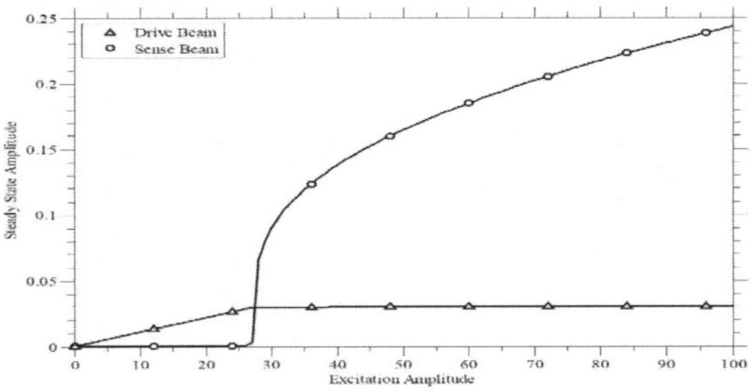

Figure 8: Numerical force-response curves of the sense and drive modes for macro T-gyro showing the nonlinear saturation of the drive amplitude.

quencies up to near the frequency of the drive mode reducing the noise effects around the natural frequency of the sense mode. It is expected to see a considerable improvement in the long term stability of the sensor by reducing the effects of the electronic noise on the readout circuit. Further investigations are underway to properly characterize the system response, examine the possibility of using the method to improve the performance of the amplitude-modulation gyroscopes, and to examine the possibility of using a closed-loop to control and manage the effects of nonlinearities.

REFERENCES

[1] Curey, R., Ash, M., Thielman, L., and Barker, C., 2004. "Proposed ieee inertial systems terminology standard and other inertial sensor standards". In Position Location and Navigation Symposium, pp. 83–90.

[2] O'Connor, J. M., and Shupe, D. M., 1983. Vibrating beam rotating sensor. U.S. Patent 4 381 672.

[3] Yang, H., Bao, M., Yin, H., and Shen, S., 2002. "A novel bulk micromachined gyroscope based on a rectangular beam-mass structure". *Sensors and Actuators A: Physical,* ***96***(2), pp. 145–151.

[4] Lajimi, S. A. M., Abdel-Rahman, E., and Heppler, G., 2014. "Modeling and sensitivity analysis of a mems vibratory rotation rate sensor". In Proceedings of SPIE 2014, Sensors and Smart Structures Technologies for Civil, Mechanical, and Aerospace Systems 2014, Vol. 9061, pp. 90612I–90612I–11.

[5] Lajimi, S. A. M., Heppler, G. R., and Abdel-Rahman, E., 2014. "The application of a new beam-rigid body mems gyroscope in the frequency-modulation mode". In Nano/Micro Engineered and Molecular Systems (NEMS), 2014 9th IEEE International Conference on, pp. 586–591.

[6] Lajimi, S., Heppler, G., and Abdel-Rahman, E., 2014. "A parametric study of the response of a beam-rigid-body microgyroscope". In ASME 2014 International Mechanical Engineering Congress and Exposition, American Society of Mechanical Engineers, pp. V010T13A052–V010T13A052.

[7] Lajimi, S., Heppler, G., and Abdel-Rahman, E., 2014. "Nonlinear dynamics of a beam-rigid body microgyroscope". In ASME 2014 International Design Engineering Technical Conferences and Computers and Information in Engineering Conference, American Society of Mechanical Engineers, pp. V008T11A036–V008T11A036.

[8] Zaman, M. F., Sharma, A., Hao, Z., and Ayazi, F., 2008. "A mode-matched silicon-yaw tuning-fork gyroscope with subdegree-per-hour allan deviation bias instability". *Microelectromechanical Systems, Journal of,* ***17***(6), pp. 1526–1536.

[9] Acar, C., and Shkel, A., 2008. *MEMS vibratory gyroscopes: structural approaches to improve robustness.* Springer.

[10] Vyas, A., Peroulis, D., and Bajaj, A. K., 2009. "A microresonator design based on nonlinear 1: 2 internal resonance in flexural structural modes". *Microelectromechanical Systems, Journal of,* ***18***(3), pp. 744–762.

[11] Oropeza-Ramos, L. A., Burgner, C. B., and Turner, K. L., 2009. "Robust micro-rate sensor actuated by parametric resonance". *Sensors and Actuators A: Physical,* ***152***(1), pp. 80–87.

[12] Golnaraghi, F., Bahreyni, B., Marzouk, A., Sarrafan, A., Lajimi, S. A. M., Pouyanfar, O., and Noori, N. Vibratory gyroscope utilizing nonlinear modal interaction. Filed on May 8, 2015, Application Number: 62159054.

[13] Cowen, A., Hames, G., Monk, D., Wilcenski, S., and Hardy, B. "Soimumps design handbook".

[14] Golnaraghi, M., Moon, F., and Rand, R., 1989. "Resonance in a high-speed flexible-arm robot". *Dynamics and Stability of Systems,* ***4***(3-4), pp. 169–188.

SCALE EXPERIMENTAL TEST VEHICLE FOR ROLLOVER STUDY

Paul White
Electrical/Computer Eng.
University of Manitoba
Winnipeg, MB, Canada
umwhi225@myumanitoba.ca

Rhyse Maryniuk
Mechanical Eng.
University of Manitoba
Winnipeg, MB, Canada
ummarynr@myumanitoba.ca

Sobhan Sadri
Mechanical Eng.
University of Manitoba
Winnipeg, MB, Canada
umsadri@myumanitoba.ca

Christine Wu
Mechanical Eng.
University of Manitoba
Winnipeg, MB, Canada
Christine.Wu@umanitoba.ca

ABSTRACT

The Scale Experimental Test Vehicle (SETV) is an autonomous scale vehicle developed to study the rollover dynamics of vehicles. This paper presents an overview of the hardware, controllers and software which make the SETV a mechatronic system. The SETV consists of an off-the-shelf scale vehicle, which has been modified with a roll safety structure. Inertial sensors measure linear accelerations and angular velocities of the sprung mass, while the rotational velocities of the wheels are measured by encoders. The SETV can be driven autonomously or by a user. In autonomous mode, a steering profile is used to drive the SETV at a predefined longitudinal velocity. A Graphical User Interface (GUI) allows a user to remotely monitor the sensor data, and load steering profiles for autonomous operation. The performance of the SETV in autonomous mode is demonstrated in an experiment.

Keywords: Scale vehicle, Rollover, Mechatronic system, PID controller, Computer software

I. INTRODUCTION

Vehicle rollover can cause serious injury and loss of life. In 2010 alone, more than 7,600 people lost their lives in rollover crashes in the USA [1]. To decrease rollover accidents, research on the vehicle roll dynamics and driver assistant systems is required, but examination of the research findings on a full-size vehicle is expensive and dangerous. Moreover, the number of facilities that perform rollover tests on full-size vehicles is limited. Fortunately, literature demonstrates that scale-model testing is a suitable alternative for full-size testing [2]. Furthermore, there are some advantageous for testing scale vehicles rather than full-size ones. For example, scale size vehicles are more durable and their maintenance is cheaper. Testing on a scale vehicle is more repeatable and less dangerous.

There is other research interested in using scale vehicles as experimental test beds. Sampei, et al. [3] used an articulated scale vehicle to examine a proposed path tracking control. Naoki, et al. [4] developed a four-wheel steering and four-wheel drive laboratory scale model vehicle as an experimental setup for testing their controller. Brennan and Alleyne presented the Illinois Roadway Simulator (IRS) in [5] as a scaled test bed to study vehicle dynamics and control. O'Brien, et al. [6] developed the Scale-Model Testing Apparatus, the simplified IRS, as a platform for their vehicle control projects. Travis, et al. [7] employed a scale vehicle in their study on rollover propensity. Verma, et al. [8] developed a scale vehicle to capture the longitudinal and powertrain dynamics of a full-size high-mobility multipurpose wheeled vehicle. Phanomchoeng and Rajamani used a $1/8^{th}$ scale vehicle in [9] to evaluate the newly proposed rollover index.

In this paper, the SETV developed at the University of Manitoba is presented. The objective for developing this SETV was to equip an off-the-shelf scale vehicle with sensors to use as a safe and economical test bed for rollover study. The SETV is equipped with an automatic driving mode to repeat a test any number of times. To increase the durability, the SETV is retrofitted with a roll prevention structure. The remainder of this paper is organized as follows. In Section II, an overview of the SETV systems is given. Section III provides an overview of the controllers used in SETV. The software architecture of SETV is described in Section IV. In Section V, an experiment test is performed to evaluate the performance of the SETV in the autonomous mode. Section VI includes the conclusion and future work.

Figure 1: SETV prototype

II. SYSTEM DESCRIPTION

This section serves as an overall description of the SETV, illustrating the basic hardware modules and outlining how the system is to be used. Figure 1 shows the completed SETV prototype. The SETV platform is based off the HPI racing $1/5^{th}$ scale Baja 5B SS which has been converted to its electric equivalent, the Baja 5B Flux. The test vehicle is designed for off road use and is capable of speeds up to 100 km/h. Therefore, it is well suited for rollover studies. The vehicle is driven by a Castle Creations $2028 - 780$ KV Brush Less DC (BLDC) motor and Mamba XL2 speed controller, which deliver up to 10 hp when powered by a pair of 4-cell Lithium Polymer (LiPo) batteries in series configuration. This powertrain configuration is ideal because the test vehicle can be tested extensively indoors.

The roll safety structure is made of VexRobotic steel hardware. This structure protects the electronics and it prevents the SETV from completely rolling over by means of outrigger wheels installed on the sides.

Figure 2 shows the electronic components of the SETV. An embedded computer (Phidgets 1073) runs the main software that communicates with the various sensor boards (Phidgets 1044, 1047, Bourns EMS 22 Q). Additionally, a servo motor controller (Phidgets 106) allows the computer to control the SETV's speed and steering angle. A USB flash drive can be attached to the embedded computer to log telemetry data, which can be viewed or analyzed on a computer at a later time. A wireless router allows a user to connect a device such as a laptop or tablet, which runs the user interface software to configure system settings. The user can also use this interface to upload steering profiles in csv format. All electronics are powered by a 4 cell LiPo battery through a voltage regulator.

To address safety of the hardware and surroundings, a hardware based manual override switch, shown in Figure 3, has been implemented. Using a switch built into the stock R/C remote, the user can switch between manual control and automatic control. This can allow the user to take control if the vehicle is approaching a dangerous situation, such as an inattentive bystander.

III. CONTROL SYSTEMS

The SETV's control systems include an open-loop steering controller and a closed loop longitudinal speed controller. The objective of the former is to implement a predefined steer profile, and the objective of the latter is to reach a preset longitudinal speed and to maintain it during the whole maneuver.

A. Open-Loop Steering Control

For the sake of simplicity, a simple open-loop steering controller was developed to adjust the front wheel angle. Assuming a simple bicycle model, the steering angle would be

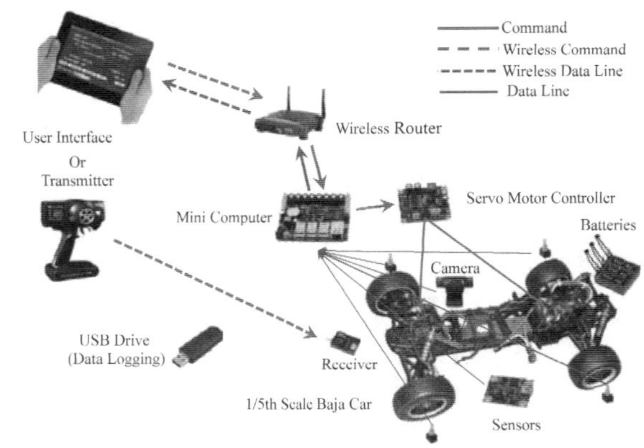

Figure 2: SETV electronics description

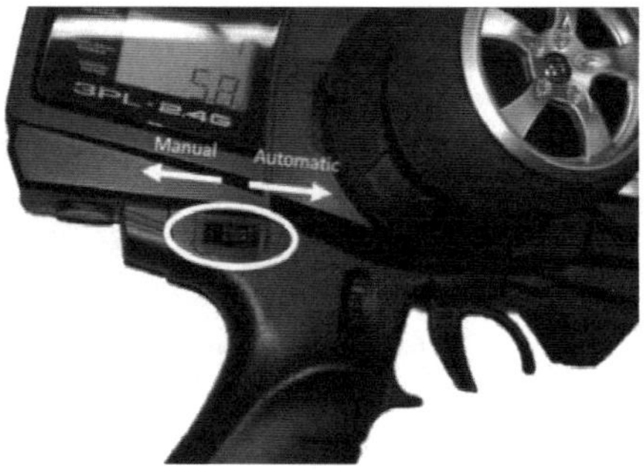

Figure 3: Manual override switch

the average of the left and right wheel, as illustrated in Figure 4.

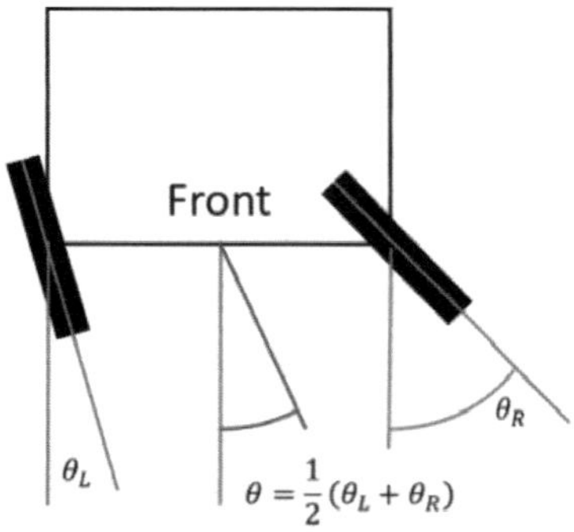

Figure 4: Open-loop steering model

To set the steering angle, θ, we had to specify the range of pulse-widths that the motor supported, and measure the range of steering angles that the maximal and minimal pulse widths caused. Assuming a linear correspondence between steering angle and servo angle, the steering angles can be interpolated. This was verified by setting the steering angle in software and measuring the resulting wheel angle while the vehicle was mounted on a test rig.

The open-loop controller was chosen because of its simplicity. The steering servo includes a controller that ensures that the motor position is accurate and steady.

B. Closed Loop Speed Control

In order to ensure accurate speed control of the vehicle, a closed loop speed controller was implemented. The vehicle system was modeled considering a resistive force linearly proportional to the vehicle velocity shown as the following differential equation:

$$m\dot{v}(t) = u(t) - bv(t) \tag{1}$$

where m is the vehicle mass, $\dot{v}$ is the vehicle acceleration, u is the input drive force at the wheels, and b is resistive force coefficient. The value for b was calculated by summing the resistive forces produced by rolling resistance (approximated as a linear function of velocity) and aero drag and approximating it as a linear function over the operating velocity. The results from the approximation are shown in Figure 5.

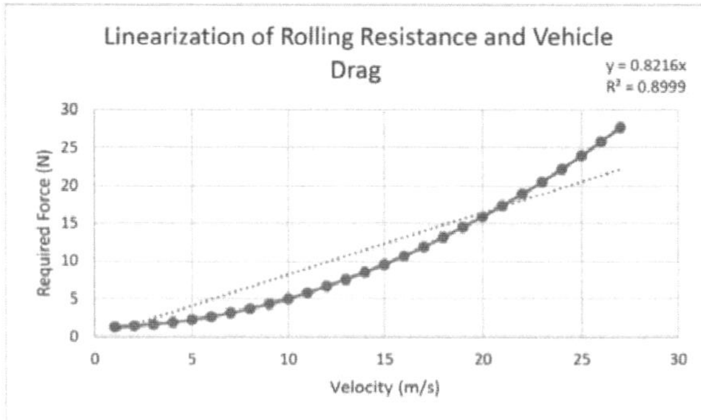

Figure 5: Determination of resistive coefficient

Closing the loop and applying PID control the transfer function for the system results in the block diagram shown in Figure 6.

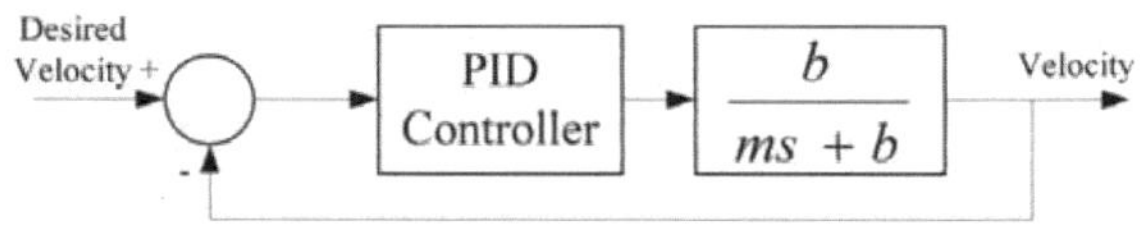

Figure 6: Block diagram of the closed loop speed controller

The controller needed to satisfy the following requirements:

- The system must be underdamped, this response resulted in some overshoot, but produced the fastest acting system.
- Less than 10% overshoot.
- The speed signal sent to the speed controller cannot be greater than the vehicle maximum speed.
- The system must be as simple as possible.

The controller was designed with the following constraints, which resulted in the system performance shown in Figure 7 to a unit step input. This response has a rise time of 2.68 seconds and 10% overshoot.

IV. SYSTEM SOFTWARE

The system software is composed of three modules: backend, user interface, and a web server.

A. Backend

The backend component runs on the embedded computer and communicates with the sensing/actuating hardware. It also performs data logging. This module is built using the

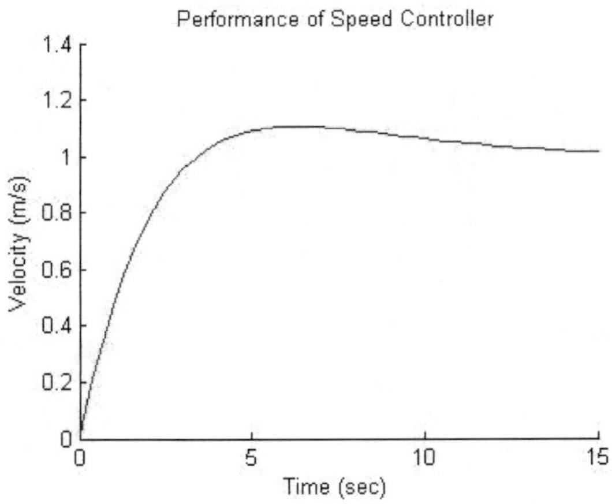

Figure 7: Response of the closed loop system to the unit step input

Java programming language. The Java language was selected due to its cross-platform compatibility. This means that if the vehicle requirements were to change, (e.g. support a much higher sampling rate, or perform computationally complex processing) the existing system software would simply need to be extended and could easily be loaded onto a more powerful computer. Since Java can run well on Linux as well as Windows, the software is not tightly coupled to a particular platform. It is also a popular learning language, so future students working on this project will likely be able to extend it. The backend is constructed in a very modular style, so custom controllers can be programmed and used without needing to fundamentally change the software.

B. User Interface

A user interface runs on a client device, such as a users tablet or laptop PC. The user interface is presented in the form of a web page, as illustrated in Figure 8. This means that it is entirely platform-independent, and should run in any device equipped with a modern web browser. We tested using the Google Chrome web browser (version 36.0.1985) since this browser is available on most major platforms. The user interface provides real-time graphical feedback from each of the sensors on the vehicle, as well as a live video feed if a webcam is present. The user can also install steering profiles (as described in Section II-B) to the vehicle and start/terminate profile execution.

C. Web Server

A web server runs on the embedded computer and interfaces between the user interface running on the client and the backend component running on the embedded computer. The bulk of this functionality was provided by the Jetty 9.2.0 server library. The web server is also implemented using the Java programming language.

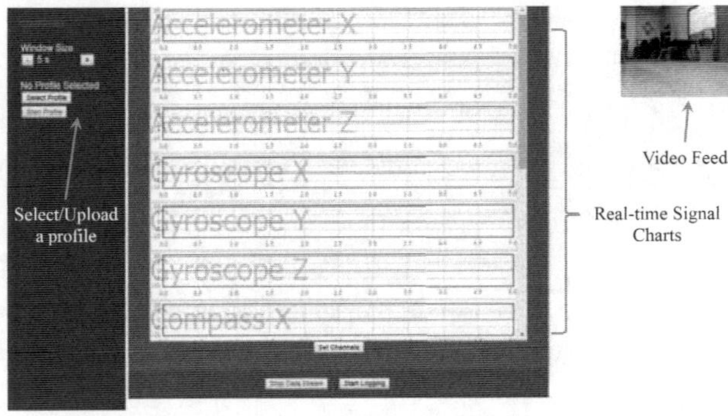

Figure 8: Screenshot of user interface

V. Performance Evaluation

The SETV is capable of monitoring the parameters described in Table 1.

Table 1: SETV parameters

Parameter	Units
Linear acceleration (3 axes)	$G's$
Angular velocity (3 axes)	$Degrees/s$
Wheel angular velocity (each wheel)	$Degrees/s$
Longitudinal velocity	m/s

These parameters are sampled at a frequency of approximately 14 Hz (70 ms sample interval). Each of these parameters can also be displayed in real-time in the UI. As mentioned before, the SETV is also capable of automatic navigation. The software parses the steering profile and sends commands to the steering and speed servos at a frequency of 5 Hz (or at 200 ms intervals).

To demonstrate the performance of the SETV, a maneuver is examined in autonomous mode. In this maneuver, the SETV accelerates to the target longitudinal velocity of $3\ m/s$, and then it steers 5 degrees to the right follows by 11 degrees to the left while maintaining the target velocity. In Figure 9, the longitudinal velocity of the SETV, the steer profile, and the yaw angle of the vehicle are plotted. From this figure, it can be seen that when the speed reaches to the $3\ m/s$, the steering controller steers the SETV to match the predefined steer profile while the speed controller keeps the speed constant. The yaw angle graph extracted from measured data by the IMU shows that in this manoeuvre the SETV (after reaching the target velocity) turned slightly to the right (negative yaw angle) then 1.75 circles on its left. This plot shows that the SETV executed the preset maneuvers. However, the open-loop steering controller does not appear to be correctly calibrated.

VI. CONCLUSIONS

In this paper, the development of the SETV at the University of Manitoba is presented. This mechatronic system is

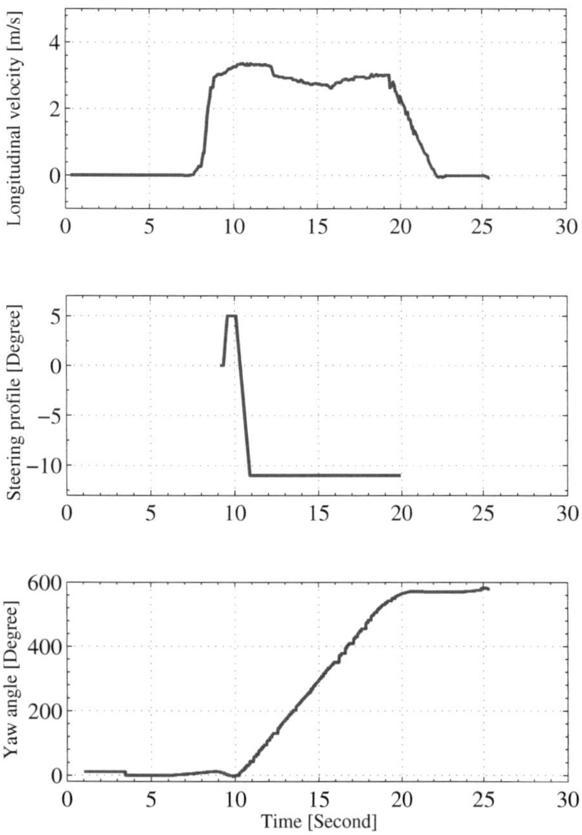

Figure 9: Longitudinal velocity, steering profile, and the yaw angle graphs of the SETV running in the autonomous mode

developed for studying the rollover dynamics of the vehicles. The components of the SETV are introduced and the software architecture is described. The autonomous performance of the SETV is demonstrated during a predefined maneuver. The experimental results verify that the whole system works properly. Improving the steering system with a closed loop controller to guarantee the precise steering in rollover maneuvers is considered as future work.

ACKNOWLEDGMENTS

The authors would like to acknowledge Jimmy Chiu and Ambarish Goswami from Honda Research Institute USA, California for their report on building a scale vehicle test platform. The authors would also like to thank Ahmad Byagowi from the University of Manitoba for his support with Ardu IMU, which was used to generate the 3rd graph in Figure 9.

References

[1] SafeCar national highway traffic safety administration (nhtsa). http://www.safercar.gov/Vehicle+Shoppers/Rollover/Fatalities. Accessed: 2015-02-24.

[2] E. Buckingham. On physically similar systems; illustrations of the use of dimensional equations. *Phys. Rev.*, 4:345–376, Oct 1914.

[3] M. Sampei, T. Tamura, T. Kobayashi, and N. Shibui. Arbitrary path tracking control of articulated vehicles using nonlinear control theory. *Control Systems Technology, IEEE Transactions on*, 3(1):125–131, Mar 1995.

[4] Naoki Matsumoto and M. Tomizuka. Vehicle lateral velocity and yaw rate control with two independent control inputs. In *American Control Conference, 1990*, pages 1868–1875, May 1990.

[5] S. Brennan and A. Alleyne. The illinois roadway simulator: a mechatronic testbed for vehicle dynamics and control. *Mechatronics, IEEE/ASME Transactions on*, 5(4):349–359, Dec 2000.

[6] Jr. O'Brien, R.T., J.A. Piepmeier, P.C. Hoblet, S.R. Burns, and C.E. George. Scale-model vehicle analysis using an off-the-shelf scale-model testing apparatus. In *American Control Conference, 2004. Proceedings of the 2004*, volume 4, pages 3387–3392 vol.4, June 2004.

[7] W.E. Travis, R.J. Whitehead, D.M. Bevly, and G.T. Flowers. Using scaled vehicles to investigate the influence of various properties on rollover propensity. In *American Control Conference, 2004. Proceedings of the 2004*, volume 4, pages 3381–3386 vol.4, June 2004.

[8] R. Verma, D. Del Vecchio, and H.K. Fathy. Development of a scaled vehicle with longitudinal dynamics of an hmmwv for an its testbed. *Mechatronics, IEEE/ASME Transactions on*, 13(1):46–57, Feb 2008.

[9] G. Phanomchoeng and R. Rajamani. New rollover index for detection of tripped and un-tripped rollovers. In *Decision and Control and European Control Conference (CDC-ECC), 2011 50th IEEE Conference on*, pages 7440–7445, Dec 2011.

STATE OF CHARGE ESTIMATION OF LI-ION BATTERIES USING THE DYNAMIC 2ND-ORDER SMOOTH VARIABLE STRUCTURE FILTER

Hamed Hossein Afshari
Mechanical Engineering Department
McMaster University, Hamilton, ON, Canada
afsharh@mcmaster.ca

Andrew Gadsden
Mechanical Engineering Department
University of Maryland, Baltimore County, Maryland, USA, gadsden@umbc.edu

Ryan Ahmed
Mechanical Engineering Department
McMaster University, Hamilton, ON, Canada
mohamerm@mcmaster.ca

Saeid Habibi
Mechanical Engineering Department
McMaster University, Hamilton, ON, Canada
habibi@mcmaster.ca

ABSTRACT

This paper presents a new approach for the robust state of charge (SOC) estimation of Lithium-ion (Li-ion) batteries. In this approach, the novel dynamic 2nd-order smooth variable structure filter (dynamic 2nd-order SVSF) applies to a Li-ion battery and generates robust SOC estimation under uncertain conditions. The dynamic 2nd-order SVSF is a model-based robust state estimation method that benefits from the robustness and chattering suppression of second-order sliding mode systems. To study the performance of Li-ion batteries, it is necessary to accurately estimate their SOC as a function of the operating time. The SOC estimation may be negatively affected by some factors including modeling imperfections, parametric uncertainties, and measurement noise relevant to the battery setup. To overcome or at least reduce effects of such factors on the SOC estimation, the robust state estimation is recommended that is insensitive to a wider range of noise and uncertainties. In this study, the robust characteristic of the dynamic 2nd-order SVSF method helps to accurately estimate the SOC of a Li-ion battery cell under uncertain and noisy conditions. The Li-ion cell is modeled using a simple first-order R-RC equivalent circuit model. The SOC estimation results are then compared to ones obtained by the extended Kalman filter (EKF) as a sub-optimal state estimator.

KEYWORDS: State of charge estimation, Li-ion battery, the 2nd-order smooth variable structure filter, robustness

INTRODUCTION

The Li-ion batteries have been extensively used in electric vehicles because of their high energy density, durability, safety, lack of hysteresis, and slow loss of charge when not in use. In order to improve their performance and increase their safety and efficiency, accurate management, monitoring, and control are required. For the battery management system, it is necessary to accurately estimate the SOC of the battery as a function of the operating time. The SOC presents the current battery capacity as a percentage of the maximum capacity. It has the same role as a fuel gage used in vehicles with an internal combustion engine. However note that in spite of the fuel gage that directly measures the amount of fuel in a fuel tank, the SOC cannot be measured. The SOC needs to be estimated using state and parameter estimation methods that extract the real-time value of the SOC based on the indirect, inaccurate and uncertain sensor measurements [1].

The SOC estimation may be affected by some factors including modeling imperfections, parametric uncertainties, and measurement noise. Noise and perturbations inherently exist in the measurement process, and caused by instruments and environmental factors. System uncertainties are usually due to inaccuracy in modeling the process, discretization error, and small variations of physical parameters from their nominal values. The Kalman filter is the most well-known method for state estimation that provides optimal state estimates by minimizing the state error covariance [2]. The main concern with the Kalman filter is that it is mainly designed based on an exact knowledge of the system's model with known parameters. In real applications, however, there may be considerable uncertainties about the model structure, physical parameters, level of noise, and initial conditions. These uncertainties may significantly degrade the Kalman filter's performance. To overcome or at least alleviate the effects of such factor on the SOC estimation, robust state estimation is proposed in which the main objective is to design a fix filter that is insensitive to a wider range of noise and modeling uncertainties. The main robust state estimation methods found in the literature are the robust Kalman (or H_2) filter [3], the H_∞ filter [4], and the smooth variable structure filter (SVSF) [5].

This paper presents a new approach for the robust SOC estimation of Li-ion batteries using the novel dynamic 2nd-order SVSF method [6]. The dynamic 2nd-order SVSF is a new model-based robust state estimation method that benefits from the robustness and chattering suppression characteristics of the second-order sliding mode systems. This method is used in this study in order to estimate the SOC of an experimental Li-ion battery cell under uncertain and noisy conditions. The Li-ion battery cell is modeled using a first-order R-RC equivalent circuit model. The SOC estimation results are then compared to ones obtained by the EKF method.

THE DYNAMIC 2ND-ORDER SVSF METHOD [6]

The dynamic 2nd-order SVSF is formulated in a predictor-corrector form. It has two main steps including the prediction and update. In the prediction step, the *a priori* state estimate is predicted using knowledge of the system prior to step *k*. In the update step, the calculated *a priori* estimate is refined into the *a posteriori* state estimate. The corrective gain of the dynamic 2nd-order SVSF pushes the innovation sequence and its first difference to zero in a finite time. This method applies to systems with a linear state and a linear measurement model. In order to describe this method, assume a stochastic system with a linear state model as:

$$x_{k+1} = \hat{F} x_k + \hat{G} u_k + w_k, \quad (1)$$

where $x_k \in \mathbb{R}^{n\times 1}$ is the state vector, $u_k \in \mathbb{R}^{p\times 1}$ is the control vector, and $z_k \in \mathbb{R}^{m\times 1}$ is the measurement vector. The measurement model is also linear and given by:

$$z_{k+1} = \hat{H} x_{k+1} + v_{k+1}, \quad (2)$$

where $z \in \mathbb{R}^{m\times 1}$ is the measurement vector, $v \in \mathbb{R}^{m\times 1}$ is the measurement noise, and $\hat{H} \in \mathbb{R}^{m\times n}$ is a measurement matrix.

The dynamic 2nd-order SVSF is presented as follows [6]:

i. Prediction of the *a priori* state estimate based on the system's state model as:

$$\hat{x}_{k+1|k} = \hat{F}\hat{x}_{k|k} + \hat{G}u_k. \quad (3)$$

where $\hat{F}$ is an estimate of the state model F. The *a priori* state estimate is calculated using the previous *a posteriori* state estimate $\hat{x}_{k|k}$. The *a priori* estimate of the measurement vector $\hat{z}_{k+1|k}$ is calculated using the estimated state vector and the measurement model as:

$$\hat{z}_{k+1|k} = \hat{H}\hat{x}_{k+1|k}, \quad (4)$$

where $\hat{H}$ is an estimate of the measurement model H.

ii. Calculation of the *a posteriori* and *a priori* measurement error, $e_{z_{k|k}} \in \mathbb{R}^{m\times 1}$ and $e_{z_{k+1|k}} \in \mathbb{R}^{m\times 1}$ respectively as:

$$e_{z_{k|k}} = z_k - \hat{H}\hat{x}_{k|k}, \quad (5)$$

$$e_{z_{k+1|k}} = z_{k+1} - \hat{H}\hat{x}_{k+1|k}. \quad (6)$$

iii. Calculation of the corrective gain for the dynamic 2nd-order SVSF $K_{k+1} \in \mathbb{R}^{n\times 1}$ as a function of the *a priori* and the *a posteriori* measurement errors as follows:

$$K_{k+1} = \hat{H}^{-1}\left[e_{z_{k+1|k}} - (\gamma + \Lambda_{k+1})e_{z_{k|k}} + \gamma\Lambda_{k+1}e_{z_{k-1|k-1}}\right]. \quad (7)$$

where $\gamma = Diag(\gamma_{ii}) \in \mathbb{R}^{m\times m}$ is a diagonal matrix with positive entries such that $0 < \gamma_{ii} < 1$. It represents the convergence rate corresponding to each entry.

iv. Update the *a priori* state estimate into the *a posteriori* state estimate $\hat{x}_{k+1|k+1}$ such that:

$$\hat{x}_{k+1|k+1} = \hat{x}_{k+1|k} + K_{k+1}. \quad (8)$$

In the dynamic 2nd-order SVSF method, the vector of sliding variables $S \in \mathbb{R}^{m\times 1}$ is defined as [6]:

$$S_k = e_{z_{k|k}}, \quad (9)$$

Following (7), the corrective gain represents a second-order Markov process and updated using the measurement error values at different time steps. The main advantage of the dynamic 2nd-order SVSF over other approaches is the use of a cut-off frequency coefficient within the corrective gain formulation. The cut-off frequency coefficient is assigned to each measurement that filters out unwanted chattering effects. This coefficient is formulated into the filter by defining a new dynamic sliding mode manifold that is given by:

$$\sigma_k = \Delta S_k + C S_k \quad (10)$$

where $C \in \mathbb{R}^{m\times m}$ denotes the manifold's cut-off frequency. Since the sliding variable is equal to $S_k = e_{z_{k|k}}$, the difference of the sliding variable is obtained by: $\Delta S_k = e_{z_{k|k}} - e_{z_{k-1|k-1}}$. Hence, by defining the sliding manifold as $\sigma_k = \Delta S_k + CS_k$ and proving the filter stability about it, it is ensured that the measurement error and its difference are decreasing in time.

THE LI-ION BATTERY MODELING

The experimental setup of the Li-ion battery is designed and built in the Centre for Mechatronics and Hybrid Technologies at McMaster University. The dynamics of the battery cell is simply modeled using a first-order R-RC equivalent circuit model. The first-order R-RC model is the simplest equivalent circuit model in which the battery dynamics are modeled using three elements including: 1) the open circuit voltage (OCV), 2) Internal resistance, and 3) capacitors [7]. Figure 1 shows a first-order R-RC model that is used in this study for modeling the Li-ion battery cell.

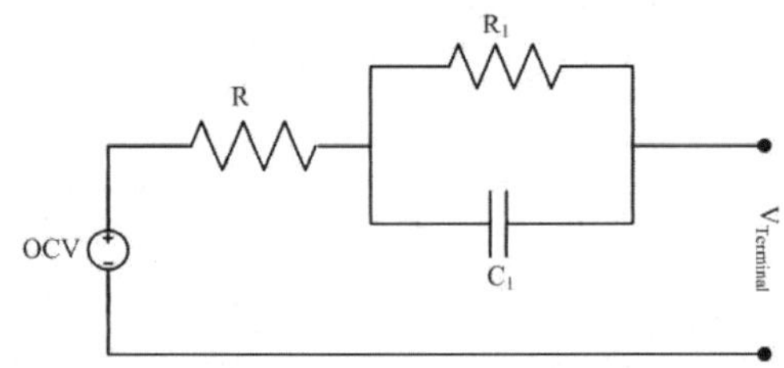

Figure 1. The first-order R-RC model of a battery cell

The first-order R-RC model in the state space from is [7]:

$$\begin{bmatrix} V_{k+1} \\ z_{k+1} \end{bmatrix} = \begin{bmatrix} 1 - \dfrac{\Delta t}{R_1 C_1} & 0 \\ 0 & 1 \end{bmatrix} \begin{bmatrix} V_k \\ z_k \end{bmatrix} + \begin{bmatrix} \dfrac{\Delta t}{C_1} \\ -\dfrac{\eta_i \Delta t}{C} \end{bmatrix} i_k, \quad (11)$$

$$y_k = OCV(z_k) - Ri_k - V_k, \quad (12)$$

where V_k is the voltage across the capacitor C_1, z_k is the state of charge, C is the battery nominal capacity, R is the battery ohmic resistance, and η_i is the cell Columbic efficiency. In addition, R_1C_1 represents the polarization time constant, and $OCV(z_k)$ represents the open circuit voltage as a function of the state of charge. The first-order R-RC model has two state variables that are the state of charge z_k and the voltage V_k. The input to the system is the current i_k, and the output y is the terminal voltage V_T. In this study, values of the physical parameters including R, R_1, C_1 are obtained using the Genetic algorithm optimization technique. The *OCV* is also formulated as a 9th-order polynomial function of the *SOC*. Note that its ten coefficients are calculated by firstly averaging the $OCV(z_k)$ curve for charging and discharging and then approximating it using a 9th-order polynomial function. Figure 2 presents a typical $OCV(z_k)$ of the under studied Li-ion polymer cell during a charging cycle.

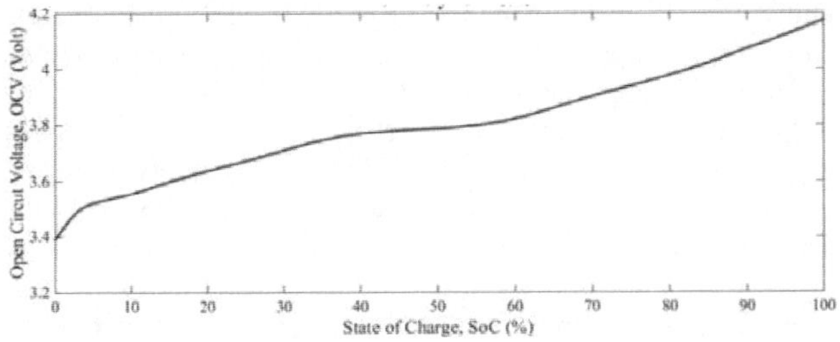

Figure 2. Experimental profile of OCV in terms of SoC

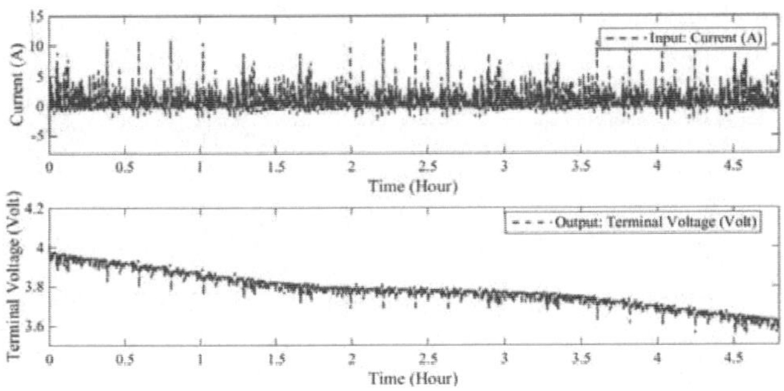

Figure 3. Profiles of the input current calculated from a UDDS cycle and the measured terminal voltage as the cell output

Figure 3 presents profiles of the input current i_k and the output terminal voltage V_T . The input profile is directly converted from a velocity profile of a mix of three benchmark driving schedules, namely: an Urban Dynamometer Driving Schedule (UDDS), a light duty drive cycle for high speed and high load (US06), and a Highway Fuel Economy Test (HWFET) [8]. The UDDS driving cycle is used to describe city driving conditions, and used to replicate the average speed, idle time, and number of stops for an average North American driver. The US06 cycle is a high acceleration, aggressive driving cycle, and the HWFET describes highway driving conditions with speeds below 60 miles/hour [8]. The three above mentioned driving cycles are shown in Figure 4. The pack current profiles from these driving cycles are scaled down to the cell-level while ignoring cell-to-cell balancing.

In order to generate the current profile from the velocity profile, a mid-size battery electric vehicle (BEV) model as shown in Figure 5 was modified from an existing hybrid vehicle model [9]. The BEV model consists of a lithium-ion battery pack, a vehicle speed controller, vehicle dynamic model, DC electric motor, and DC-DC converter. The model was developed in Simulink using Simscape library components. The driving range of the BEV is approximately 200 km per full charge.

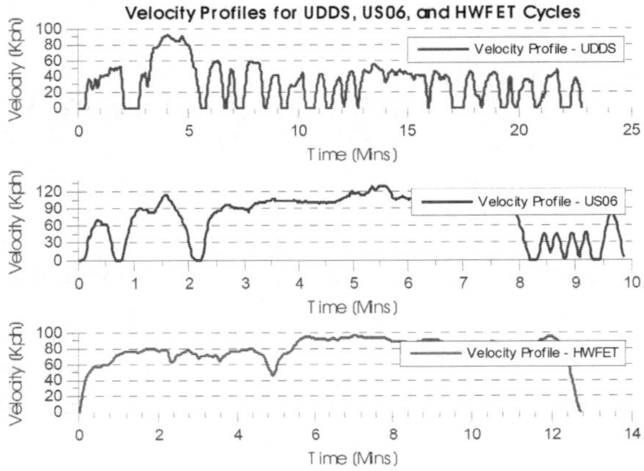

Figure 4. Velocity profiles for the UDDS (upper), US06 (middle), and HWFET (lower) Cycles [8]

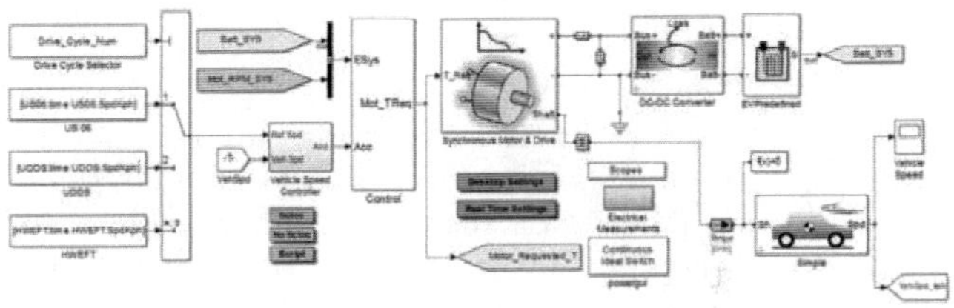

Figure 5. All-electric mid-size sedan simulation model in Simscape (Adopted from [9])

The parametric vector that needs to be optimized using the Genetic algorithm is: $\theta = [R, R_1, C_1]$. Modeling and estimation tasks are performed using the MATLAB-SIMULINK real-time environment. Figure 6 shows a block-diagram of the Li-ion battery model in SIMULINK. Table 1 also presents numeric values of the physical and optimized parameters.

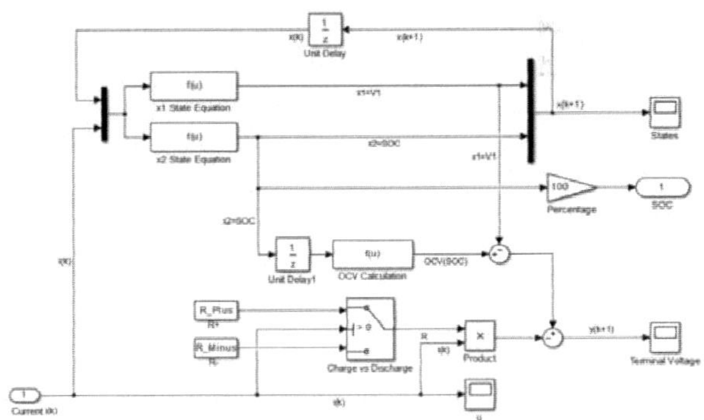

Figure 6. The Li-ion battery model in SIMULINK

Table 1: Numeric values of physical and optimized parameters

Parameter	Numeric Value
Nominal capacity, *C*	19440 (A.s)
Cell Columbic efficiency, η	1
Modeling capacity, C_1	3860.14 (A.s)
Modeling resistance, R_1	0.0049 (Ω)
Internal Resistance, *R*	0.0096 (Ω)
Initial state of charge	80 %
Sampling time, Δt	0.1 (s)

SOC ESTIMATION USING THE DYNAMIC 2ND- SVSF

In this section, the dynamic 2nd-order SVSF and the extended Kalman filter (EKF) methods are applied for the SOC estimation based on the provided Li-ion model and the captured input-output UDDS data where there exist noise and modeling uncertainties. The SOC estimation results are then compared with the actual SOC data generated from the experimental setup model. Comparison are made in terms of the root mean square error (RMSE). Figure 7 presents a block-diagram of the dynamic 2nd-order SVSF method built in SIMULINK. For the dynamic 2nd-order SVSF method, the convergence rate γ is set to 0.5, and the cut-off frequency matrix Λ is designed to be a diagonal matrix with elements equal to 0.4. For the EKF, the process noise covariance Q and the measurement noise covariance R are respectively set to:

$$Q = \begin{bmatrix} 10^3 & 0 & 0 \\ 0 & 10^4 & 0 \\ 0 & 0 & 5\times10^3 \end{bmatrix}, \quad \mathrm{R} = \left[3\times10^{-1}\right]. \tag{13}$$

Figure 8 compares the SOC estimation profile using the dynamic 2nd-order SVSF method with the actual profile. Table 2 also compares the RMSE values obtained by the EKF and the dynamic 2nd-order SVSF method. It demonstrates the higher accuracy of the dynamic 2nd-order SVSF over the EKF method under an uncertain experimental condition.

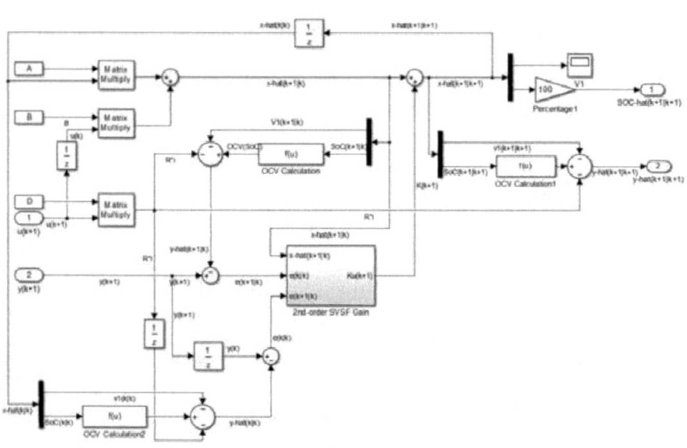

Figure 7. The dynamic 2nd-order SVSF in SIMULINK

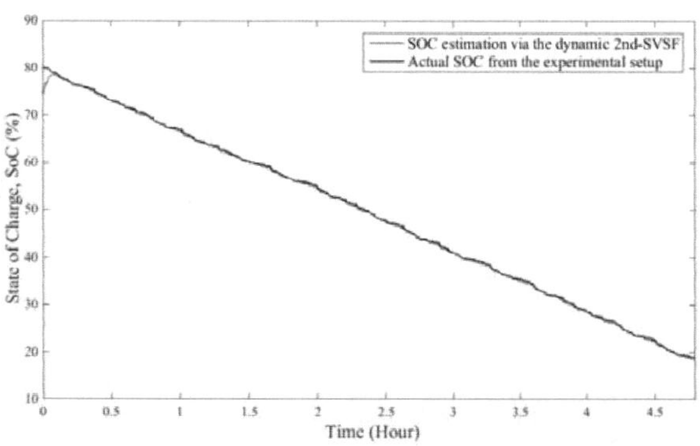

Figure 8. Actual and estimated SOC estimation profiles

Table 2: RMSE values obtained by different estimators

Estimator	RMSE for SOC Estimation
Extended Kalman filter	0.24
Dynamic 2nd-order SVSF	0.20

CONCLUSIONS

This paper introduces a new SOC estimation method that benefits from the dynamic 2nd-order SVSF as a robust state estimation technique. This method applies to an experimental Li-ion battery setup and the SOC estimation results are then compared with ones obtained by the EKF method. The input-output data set is provided from a mix of real-world, benchmark driving cycle. Estimation results show that the SOC estimation obtained by the dynamic 2nd-order SVSF precisely follows the actual SOC profile. The dynamic 2nd-order SVSF also produces a more accurate SOC estimate with a smaller RMSE in comparison to the EKF SOC estimate.

REFERENCES

1. Farag, M. S., Ahmed, R., Gadsden, S. A., Habibi, S., and Tjong, J., 2012, "A comparative study of Li-ion battery models and nonlinear dual estimation strategies", *IEEE 2012 Transportation Electrification Conference and Expo (ITEC)*, Dearborn, MI, USA, pp. 1-8.
2. Grewal, M. S., and Andrews, A. P., 2001, *Kalman Filtering: Theory and Practice Using MATLAB*, 2nd ed., John Wiley& Sons, Inc.
3. Xie, L., Soh, C., and Souza, C. E., 1994, "Robust Kalman filtering for uncertain discrete-time systems", *IEEE Transaction on Automatic Control,* 39(6), pp.1310-1314.
4. Zames, G., 1981, "Feedback and optimal sensitivity: model reference transformations, multiplicative seminorms and approximate inverse", *IEEE Transactions on Automatic Control,* 26(2), pp. 301-320.
5. Habibi, S., 2007, "Smooth variable structure filter", *Proceedings of the IEEE,* 95(5), pp. 1026-1059.
6. Afshari, H. H., and Habibi, S., 2014, "Dynamic 2nd-order smooth variable structure filter based on a dynamic sliding mode manifold", *IEEE Transactions on Control Systems Technology,* (Submitted).
7. Plett, G. L., 2004, "Extended Kalman filtering for battery management systems of LiPB-based HEV battery packs: part 2: modeling and identification", *Journal of Power Sources,* 134(2), pp. 262-276.
8. "http://www.epa.gov/nvfel/testing/dynamometer.htm," U.S. EPA. [Online].
9. S. Miller, "Hybrid-Electric Vehicle Model in Simulink" http://www.mathworks.com/matlabcentral/fileexchange/28441-hybrid-electric-vehicle-model-in-simulink 10 Aug 2010 (Updated 17 Apr 2014) [Online].

DEVELOPMENT OF RESIDENTIAL HVAC CONTROL AND DEMAND MANAGEMENT SYSTEM USING INTELLIGENT STRATEGY PLANNING MODELS

Nima Alibabaei
Mechanical Department
Ryerson University
Toronto, Ontario ,Canada
nima.alibabaei@ryerson.ca

Alan S. Fung
Mechanical Department
Ryerson University
Toronto, Ontario ,Canada
alanfung@ryerson.ca

Kaamran Raahemifar
Electrical Department
Ryerson University
Toronto, Ontario ,Canada
kraahemi@ryerson.ca

ABSTRACT

Residential houses can play significant role in managing the network energy system. Since houses have the potential of storing thermal energy, they present great opportunity for managing/controlling electric demand during ahead peak hours. Demand managing could be achieved by taking advantage of advanced predictive strategy planning models implemented on the Heating and Ventilating Air Conditioning (HVAC) system which is the largest energy consumer in residential houses. The main purpose of this study is to design a grid-friendly smart residential house. To this end, three different strategy planning models (SPMs) including: Load Shifting (LSH), Smart Dual Fuel Switching System (SDFSS), and LSHSDFSS, as the integration of fuel switching and load shifting strategy planning models are developed. A TRNSYS-Matlab co-simulator is developed in order to facilitate the implementation of different SPMs to the HVAC system of the case study house. The behavior of the HVAC system is analyzed in-depth numerically using each of the strategy planning models.

KEYWORDS: Load Shifting, Fuel Switching, Predictive Strategy Planning Models, Predictive HVAC Behavior, HVAC Demand Control, Residential Houses, Building Simulation

I. INTRODUCTION

Buildings must be seen as significant elements of the larger, dynamic energy system. This means when working on improving buildings' performance, the impact of buildings behavior on the network energy system should be considered; this includes criteria such as distribution networks of electricity, district heating, and natural gas consumption [1]. In this context, timing is essential, particularly in the case of electricity, which cannot be readily stored. In other words, time of consumption, as well as consumption itself, matter when it comes to improving buildings' performance [1].

In this study, attention is paid to the electric network and its peak loads. Peak power is used only a small fraction of the time, but the utility has to deliver it, even if it is more polluting and expensive to do. Hence, to be a grid-friendly consumer, we should avoid pulling energy from the utility grid during peak times [2]. This could be achieved by having smart buildings with control of Strategy Planning Models (SPMs). In addition to supporting the local grid, lower overall energy cost could be achieved for the homeowners with such smart building controller [2]. Among different strategy planning models, the ones concentrate on HVAC load are the most useful ones since HVAC system consumes the most part (40%) of electrical demand of a typical house [3].

Different methods are used to control the HVAC system. In developed/advanced controller, a model of the system (building and its HVAC) and the forecasted future weather condition are used to determine the best set of control operations [3]. In these control models, the prediction of the future state of the system is based on system model and generation of control vector signals as the output of them, drives the system towards desired state [4]. In this study, an accurate predictive control model is designed by using two powerful simulators. The overall conception of the building and its HVAC system is modeled using TRNSYS simulation environment while optimization of control strategies is performed by using Matlab. The main purpose of this article is to design a grid-friendly smart residential house controller. To this end, three different strategy planning models are developed in order to manage the hourly load demand of HVAC system for the next 24 hours. The operation of the HVAC system is analyzed in-depth using each strategy planning models.

This article is consisted of six sections. In Section II, different formats of forecast weather dataset is described. In Section III, the behavior of HVAC system is predicted using TRNSYS with forecast weather data. In Section IV, Smart Dual Fuel Switching System (SDFSS) is described as a smart control strategy for minimizing the HVAC system energy cost. In Section V, HVAC Load Shifting Strategy Planning Model (LSH-SPM) is presented. In Section VI, an advanced control model (LSHSDFSS) is developed by integrating both SDFSS and LSH-SPM models. Section VII describes the conclusion of the study.

II. WEATHER OPERATIONAL AND FORECAST INFORMATION

Deterministic and Probabilistic are two different methods used for generating predicted data. Each method has its own advantages [5]. Deterministic forecast is based on one scenario as the result of one or two predictive models. The data accuracy in this method is very high and this predicted data is usually used for short-term control process. Probabilistic forecast is the result of a group (sometime 21) of models produce a range of forecast data. This method is suitable for planning beyond tomorrow [5-6]. GRIB2 is the most common format used for presenting the forecast dataset. However, it should be retrieved into a standard numerical format to be useable in TRNSYS and Matlab.

Based on the nature of the project (accurate short-term forecast weather), High Resolution Deterministic Prediction System (HRDPS) source (bandage 2.5km) (http://dd.weather.gc.ca/model_hrdps/east/grib2/12) from the Canadian Meteorological Centre (CMC) web site is selected as the forecast dataset. In order to collect this information off of the database, wget was installed on the computer and a system command was called from within Matlab to collect the necessary data using wget [5]. Temperature (as the most important parameter which has impressive impact on the house energy system) for the next 24 hours is obtained by filtering this data. Figure 1 illustrates the grid format of the GRIB2 data. In this study, the TRNSYS program models the house [10] that is located at Kortright Centre for Conservation in Vaughan, Ontario, for this reason the grid element of (210, 490), corresponding to the location of the house, is selected to collect data from. In order to create an interface between Matlab and the GRIB2 data that is collected, the NCToolbox is installed and the ncgeodataset function is used.

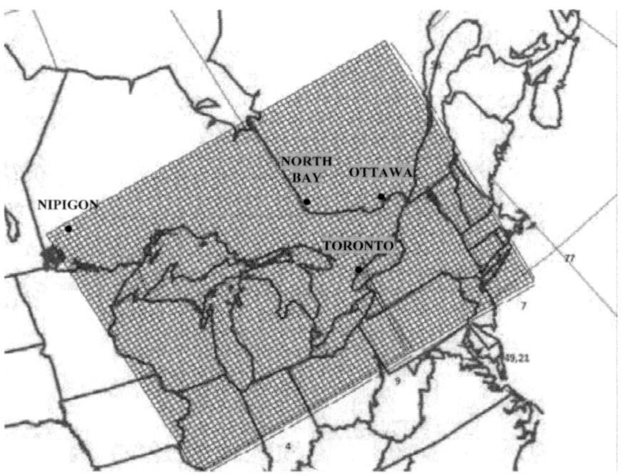

Figure 1: GRIB2 grid for eastern portion of Ontario

III. FORECASTING 24 HOURS-AHEAD HVAC SYSTEM ELECTRICAL DEMAND BASED ON WEATHER FORECAST DATASET (WFD)

A. House Description

The twin archetype sustainable houses (ASH), were constructed by the Toronto and Region Conservation Authority (TRCA) [8-10]. These twin-houses demonstrate sustainable housing technologies through experimentation and research and are one of the first Canadian projects to achieve a LEED for Homes Platinum Certification [8]. House A, which is made to have an air-tight building envelope according to the standards of ASHRAE 90.1, is selected in this project as the case study for testing different SPMs.

B. TRNSYS Model

TRNSYS is a transient system energy modeling software designed to solve complex energy system problems [9]. The House A TRNSYS model, developed in [9] research work, is used in this project for testing different strategy planning models. According to a study on various energy modeling programs [11], TRNSYS is reasonably powerful in terms of HVAC system modeling.

C. Methodology

The sequence of the study starts with downloading and processing of forecast weather data, running House A TRNSYS model with forecast data, implementing strategy planning models on TRNSYS system by generating operational command matrix, and lastly post processing the generated data. Figure 2 demonstrates the framework of the house energy simulation system is made by making powerful linkage between different but complementary software.

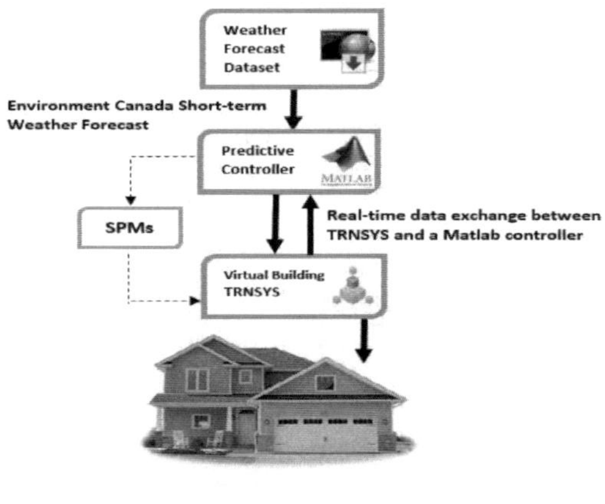

Figure 2: Framework of the house energy simulation system

Matlab is used in this study to control and manage each sequence, drive each process on time, make linkage between different programs, and store or call the required data.

D. Process Step Time

Initially, all the simulation was done based on one hour time step. But the obtained results were not very accurate since the events were taking place during given hours could not be monitored/processed by the control algorithm. In order to solve the problem, 5-minute time step was determined to be suitable after a few trials.

E. Electricity Prices

The electricity prices used in this project are 11.70 ¢/kWh, 15.40 ¢/kWh and 18.00 ¢/kWh respectively, for off-peak, mid-peak and peak hours, estimated by considering the Ontario Energy Board (OEB) Time-of-Use (TOU) prices. The model is run with January 4th, 2015 weather forecast data. The Daily Electricity Cost (DEC) of the HVAC system for this sample day is \$3.61. It should be taken into consideration that no strategy planning model is run yet.

IV. THE REDUCTION OF HVAC SYSTEM FUEL COST BY DESIGNING SMART DUAL FUEL SWITCHING STRATEGY PLANNING MODEL (SDFSS-SPM)

House A, has a natural gas mini boiler and an electric two-stage variable capacity ASHP to generate warm air through the Air Handling Unit (AHU) to meet the space heating demand. Based on an optimum scenario, at each time step, cheaper fuel (electricity or natural gas) would be selected and accordingly related system (ASHP or min boiler) would be set up to meet the space heating demand. In this study, a Labview program is developed to determine the hourly switching points as well as to drive each system towards the minimal cost while maintaining comfort level.

A. Calculating the fuel price consumed by ASHP

Outdoor temperature and electricity price are two essential parameters affect the ASHP energy cost. Outdoor temperature has direct impact on ASHP Coefficient of Performance (COP). Figure 3 shows the experimentally validated House A COP curve [10]. After determining COP and electricity price for the given hour, the cost of unit energy produced by ASHP is calculated by using Equation (1):

$$\text{ASHP electricity cost for preparing one unit of heat energy (\$/kwh)} = \frac{\text{Electricity price (¢/kWh)}}{COP*100} \qquad (1)$$

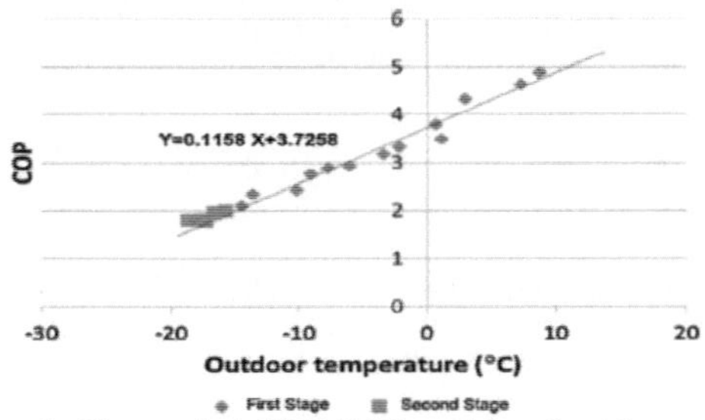

Figure 3: House A ASHP COP validated with outdoor tempreture [10]

B. Calculating Mini Boiler consumed fuel price

Natural gas price (41.60 ¢/m^3 estimated based on OEB prices) is constant for all hours. Thus, mini boiler efficiency is the only variable used to estimate the cost of each unit of thermal energy produced by the boiler. The efficiency of the boiler is determined based on the flow rate of water (load percentage) circulating through the boiler. Figure 4, illustrates the mini boiler efficiency curve provided by the manufacturer [12]. The fuel cost of Mini Boiler (for preparing one unit of heat energy) is calculated by using Equation (2):

$$\text{Mini boiler natural gas cost for preparing one unit of heat energy (\$/m}^3) = \frac{\text{Natural Gas price (\$/m}^3)}{Efficincy * 10.3} \qquad (2)$$

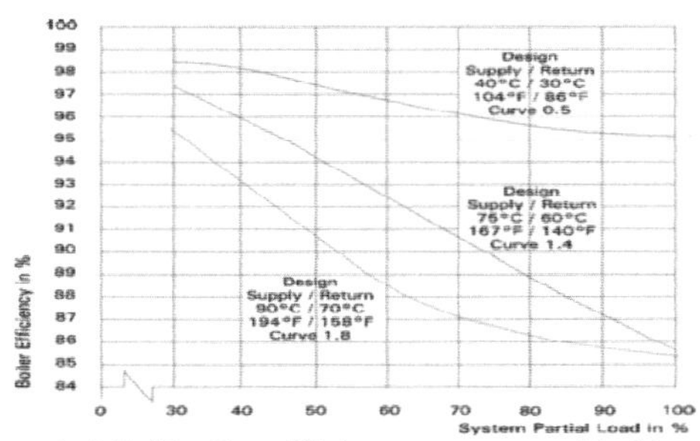

Figure 4: Mini boiler efficincy curve provided by the manufacturer [12]

C. Using SDFS system as a strategy planning model

Figure 5 demonstrates the simple flowchart view of SDFSS strategy planning model used for determining the switching points on different hours.

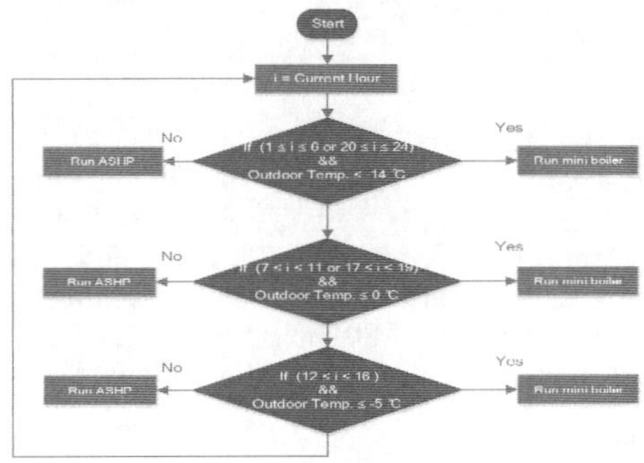

Figure 5: SDFSS strategy planning model

The Daily Fuel Cost (DFC) of the HVAC system after using (SDFSS-SPM) is found to be \$3.39.

D. Calculating the impact of SDFSS-SPM on DFC at different outdoor tempretures

Matlab model is run for Outdoor Temperatures between 0°C and -20°C. The predicted cost reduction occurs after using SDFSS-SPM is illustrated in Figure 6.

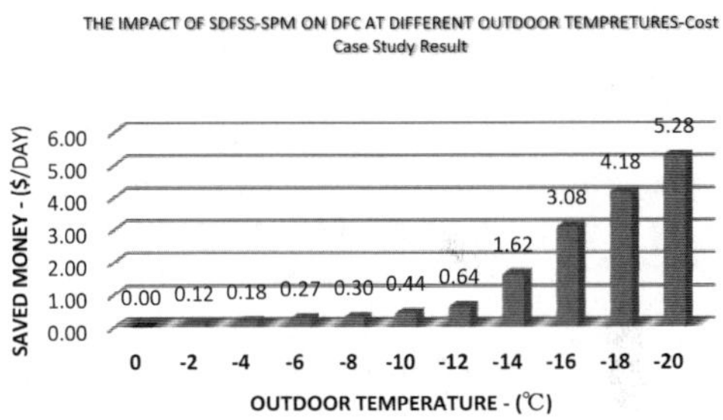

Figure 6 : Saved money occurred by using SDFSS-SPM at House A during different outdoor tempreture

V. MANAGING HVAC DEMAND BY USING LOAD SHIFTING STRATEGY PLANNING MODEL (LSH-SPM)

Occupant thermal comfort is the most important issue that should be taken into consideration in designing LSH-SPM model. Based on the ASHRAE Standard [13], the indoor temperature in the winter season should be kept between 20°C and 24°C for thermal comfort.

A. Methodology

Figure 7 demonstrates the framework used in this study for implementing LSH-SPM.

B. Pre-heating starting time

There will be no opportunity for shifting the HVAC load to off/mid hours without storing thermal energy with pre-heating process. If pre-heating process was started sooner than its optimum time, more electricity would be consumed

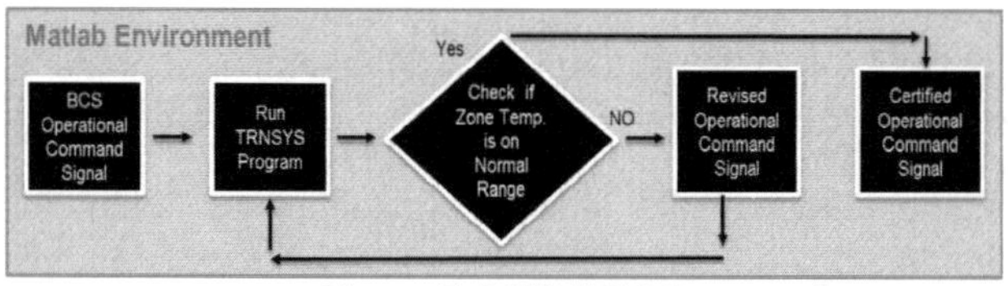

Figure 7: LSH-SPM framework

and room temperature would exceed the ASHRAE Standard range. If pre-heating was started later than its optimum time, the stored heat energy would not be sufficient to fully support LSH-SPM during peak load hours. In this study, House A TRNSYS model is run at each outdoor temperature to seek the optimum starting time. $4.01 is estimated for the DEC of the HVAC system in the sample day after running LSH-SPM.

C. Calculating the impact of LSH-SPM on DEC at different outdoor tempretures

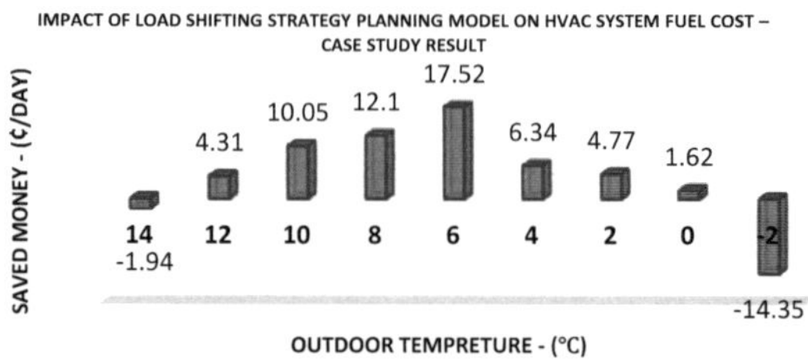

Figure 8 illustrates the saved money bars as the result of LSH-SPM on House A HVAC system.

Figure 8: Saved money occurred by using LSH-SPM at House A during different outdoor tempreture

VI. DEVELOPING LSH-SPM BY INTEGRATING SMART DUAL FUEL SWITCHING STRATEGY PLANNING MODEL (LSHSDFSS-SPM)

Under this scenario, SDFSS-SPM and LSH-SPM models are combined together to construct a developed intelligent strategy planning model, named LSHSDFSS-SPM. This model takes the advantages of load shifting and fuel switching system together. The DFC of the HVAC system after using this model is founded to $3.73. The result of different SPMs for House A in the sample day is presented and compared in Figure 9.

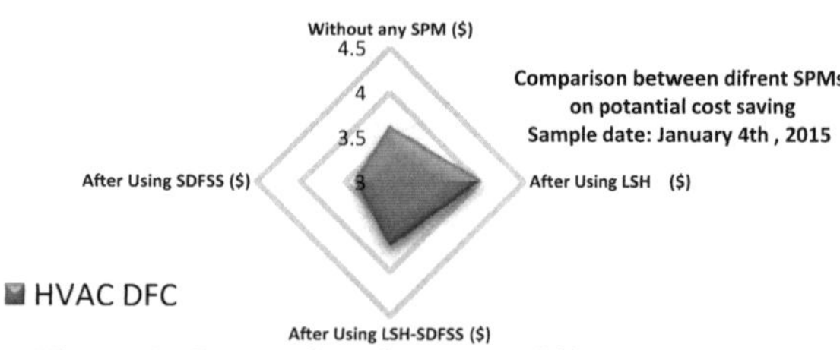

Figure 9: Comparison between different SPMs on potantial cost saving

VII. CONCLUTION

Since Residential houses have the potential of storing thermal energy, they present great opportunity for managing/controlling their demand by using controller based on predictive smart strategy planning models. SDFSS-SPM could notably reduce the HVAC energy cost. As the outdoor temperature gets colder, higher energy cost saving is achieved. LSH-SPM was the second strategy planning model implemented in this study. A smart method was used to determine the best HVAC starting time for pre-heating the house. LSHSDFSS-SPM was the third strategy planning model examined. This model took the advantage of the combined load shifting and fuel switching strategies together into a single system.

ACKNOWLEDGEMENT

The authors would like to acknowledge the financial support from the Smart Net-Zero Energy Buildings Research Network (SNEBRN), Toronto Hydro, MITACS Accelerate Ontario, CUE and TRCA.

REFERENCES

[1] Gruber, J.K., Jahromizadeh, b., Prodanovic, M., A. 2014, "Application-oriented modelling of domestic energy demand", *Electrical Power and Energy Systems,* Vol (61), pp. 656–664.

[2] Di Giorgio, A., Liberati, F., J. 2014, "Near real time load shifting control for residential electricity prosumers under designed and market indexed pricing models", *Applied Energy,* Vol (128), pp. 119–132.

[3] Afram, A., Janabi-Sharifi, F., F. 2014, "Theory and application of HVAC control systems – A review of model predictive control (MPC)", *Building and Environment,* Vol (72), pp. 343-355.

[4] Oldewurtela, F., Parisio, A., et al., A. 2012, "Use of model predictive control and weather forecasts for energy efficient building climate control", *Energy and Buildings,* Vol (45), pp. 15–27.

[5] Poulin, L., A. 2013, "Weather Forecast Data An Important Input into Building Management Systems", *National Prediction Operations Division, ICEBO 2013,* Montreal, Canada.

[6] Kim, S., Augenbroe, G., M. 2012, "Using the National Digital Forecast Database for model-based building controls", *Automation in Construction,* Vol (27), pp. 170–182.

[7] Poulin, L., A. 2013, "An NWP forecast information matrix in support of renewable energy applications", *CMOS 2013,* Saskatoon, Canada.

[8] Zhang, D., Rupayan, B., Fung, A., J. 2010, "Development of Monitoring System for the Sustainable Archetype House at Kortright Centre", *1st International High Performance Buildings Conference,* Indiana, USA.

[9] Safa, A., Fung, A., Kumar, R., F. 2015, "Performance of two-stage variable capacity air source heat pump: Field performance results and TRNSYS simulation", *Energy and Buildings,* Vol (94), pp. 80-90.

[10] Dembo, A., Fung, A.S., Ng, K.L.R., Pyrka, A., A. 2010, "The archetype sustainable house: inves- tigating its potentials to achieving the net-zero energy status based on the results of a detailed energy audit", *Proceedings of the 1st International High Performance Buildings Conference*", pp. 1–8.

[11] Zhang, D., Barua, R., Fung, A.S., M. 2011, " TRCA-BILD archetype sustainable house-overview of monitoring system and preliminary results for mechanical systems", *ASHRAE Trans,* vol(117-2), pp. 597–612.

[12] VIESMANN, Technical Data Manual, Gas-fired wall-mounted condensing boiler 12 to 285 MBH (3.5 to 83.5 kW), Vitodens 200-W Series B2HA.

[13] ASHRAE 2009, ASHRAE Handbook-Fundamentals, *American Society of Heating, Refrigeration and Air-Conditioning Engineers,* Inc., Atlanta, GA.

Proceedings of the 25th CANCAM
London, Ontario, Canada, May 31 – June 4, 2015

Positioning Accuracy Recovery for RRR Compliant Mechanism System under PZT actuator Failure

Xuejun Wang
Department of Mechanical Engineering
East China University of Science and Technology
Shanghai, China
wxjecust@gmail.com

Wenjun Zhang
Department of Mechanical Engineering
University of Saskatchewan
Saskatoon, Canada
wjz485@mail.usask.ca

Changli Liu
Department of Mechanical Engineering
East China University of Science and Technology
Shanghai, China
Clliu@ecust.edu.cn

ABSTRACT

This paper aims to address how the RRR mechanism recovers from the certain PZT failure to maintain its original positioning accuracy. The post-failure workspace analysis is analyzed, and different recovery strategies are proposed for different configurations of PZT failure, in particular, based on if the target position is included in post-failure workspace or not. The idea of recovery strategies are demonstrated via the experiment. This study provides the basic idea to the design stage of the entire compliant mechanism system, in which redundancy should be designed so that the system has the capability to maintain the original desired performance, and facilitates the compliant mechanism systems to keep functioning from unanticipated damages, to the sense of resilience.

KEYWORDS: Positioning Recovery, RRR mechanism, PZT failure, Resilience

INTRODUCTION

RRR mechanism is a parallel compliant mechanism driven by three PZT actuators; it has been widely in many applications such as cell manipulation, chip assembly and surgery (see Figure 1). Great efforts have been made in its kinematics [1] [2], dynamics [3] [4] and control [5] [6] and etc. to achieve the expected performance. Practically, the PZT actuators are assembled in the slots by the prestress that shown in Fig.1, hence, PZT may drop with respect to the decrease of the prestress or PZT actuator itself fails to work, which can however eventually leads the whole system to dysfunction, in this way, how to make the systems continue to work will greatly improve the resilience [7-9] of the RRR mechanism system, particularly this is significant important in the situations where the high accuracy and reliability are tremendously demanded. In this paper, how to make the failed system continue to work as expected is detailed discussed. One benefit of this study is such that the original desired performances are largely maintained the other benefit is that the "life expectancy" of the compliant mechanism systems will be extended in a post-failure configuration, all these will not only benefit the function but also the cost.

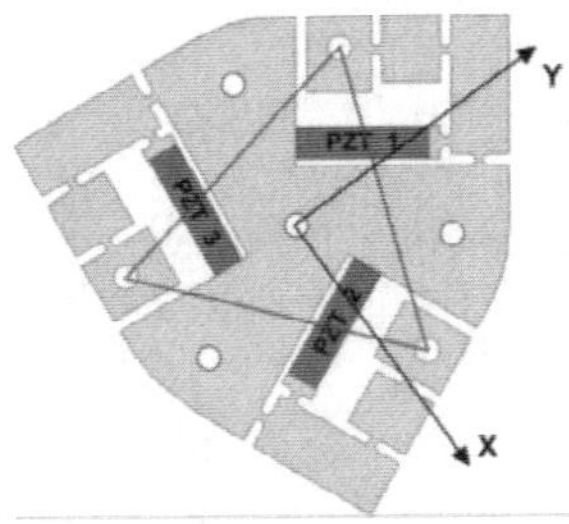

Figure 1: RRR mechanism

Resilient robot is currently a hot research field as the resilient robot can remain its original functions when encountering the unanticipated damages [10], basically, the uniqueness of such a robot is developed with the self-reconfiguration in the software and hardware bases and these two simultaneously, the resilient robot open the eyes for more powerful and reliable mechanical system, thus, the resilient robot concept provides us with fundamental idea to get a more reliable RRR mechanism. In the case of RRR mechanism, presence of PZT failure will compromise the positioning accuracy of RRR mechanism and further lead to unexpected performance, which is addressed in this paper to how to make a resilient RRR mechanism system in terms of positioning accuracy.

Based on the idea of resilient robot, the general methodology of the positioning recovery of RRR mechanism in the case of PZT failure basically has two levels: the first level is to get the desired point covered in the post-failure workspace, which is studied in next section. There are mainly two configurations of post-failure workspace because of different PZT actuator failures, the two configurations comes from the situation that if the target position is covered by the post-failure workspace or not. As such, the manipulation similar to self-configuration of resilient robot aforementioned need to be determined based on two configurations, in particular, the self-configuration in our case means rotating the whole RRR mechanism system, and this is demonstrated in this paper. The second level to coordinate the rest working PZT actuators for precise positioning, this is exactly based on the accurate kinematics.

Precise kinematic analysis underlies the accurate positioning for RRR mechanism, and is consists of forward and inverse kinematics, the inverse kinematics is easy to be done while the forward one is quite complex [11], which is not conducive to the controlling of micro-motion of the end-effector. Then an efficient way Constant-Jacobian Matrix method [12] was proposed to bridge PZT actuators' elongations and motion of end-effector, it greatly facilitate the kinematics analysis and the positioning control as well. Typically, the positioning was achieved by the PID closed loop control, and the Constant-Jacobian Matrix method was turned out to be accurate when compared to the experimental results [13].

POST-FAILURE WORKSPACE ANALYSIS

The original workspace projection in x-y plane [14] is presented as Figure 2 presented in blue. The workspace decreases when certain PZT fails to work, Figure 3 shows the post-failure workspaces of RRR mechanism under dysfunction of single PZT actuator PZT1, PZT2 and PZT3 respectively, the green area represents the post-failure workspace.

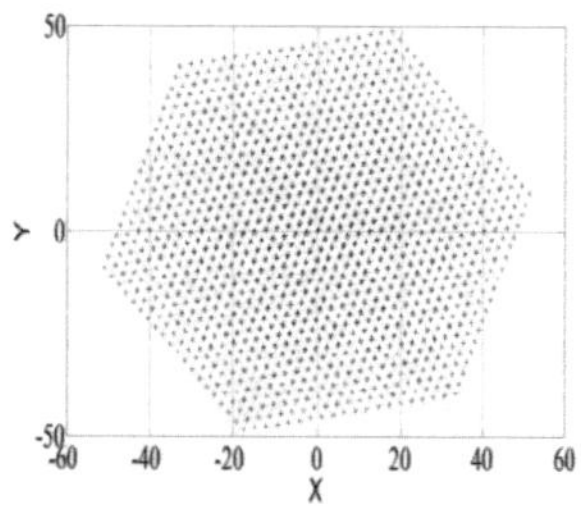

Figure 2: workspace of RRR mechanism

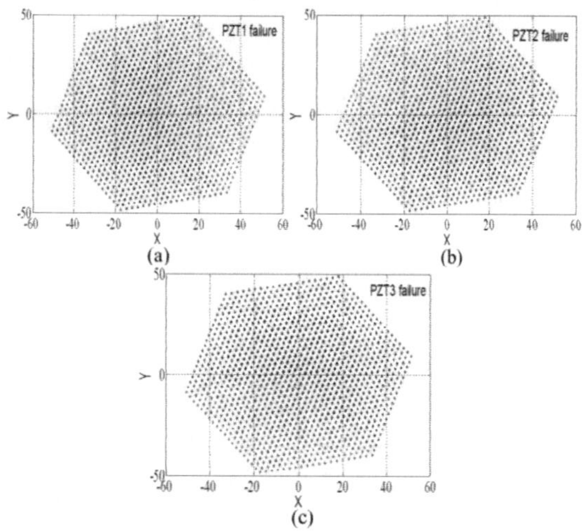

Figure 3: Post-failure Workspaces of RRR mechanism (In green) with different individual PZT failure. (a) is post-failure when PZT1 fails; (b) is post-failure when PZT2 fails; (c) is post-failure workspace when PZT3 fails

It is noted that the workspace decreases dramatically with PZT actuator failure, the post-failure workspaces under these three configurations are circumferential axisymmetric distributed, and all these three sub-workspaces constitute the whole original workspace. The post-failure workspace rotates by 120 degrees in the counterclockwise direction with respect to failures of different PZT actuators i.e., with the configurations of PZT1, PZT2 and PZT3 failures showed in Fig. 3, the corresponding post-failure workspaces are 120 degree difference. These interesting findings lay the foundation of positioning recovery in this study.

WORKSPACE RECOVERY

The post-failure workspace analysis above shows that the original workspace with PZT failure tremendously decreases, extension of the work range of PZTs may be regarded as one strategy to deal with that, but it turns out to be unfeasible, Figure 4 presents the post-failure workspace shown in green with work range increase of functioning PZTs, the post-failure workspace is presented in contrast with the original workspace. It turns outs that the original workspace cannot be recovered as the overlapping area remains the same.

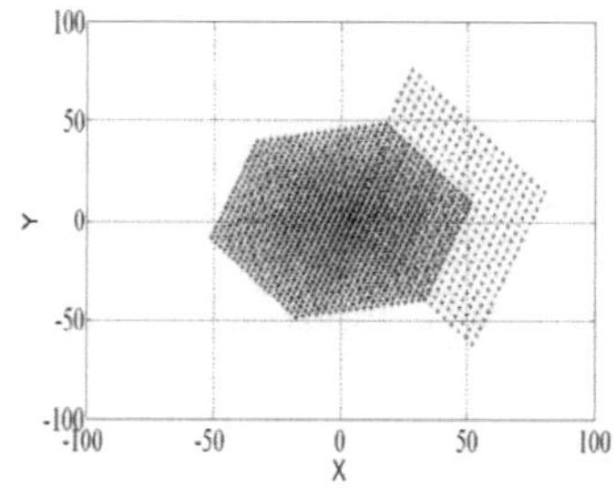

Figure 4: post-failure workspace with the work range increase of rest PZTs

However, according to the post-failure workspace analysis before, to achieve the other regions of original workspace while excluded by the post-failure workspace, we may rotate the post-failure workspace along the Z axis to the place where the workspace is demanded, that means the whole RRR mechanism should be rotated, the illustration is given in Figure 5.

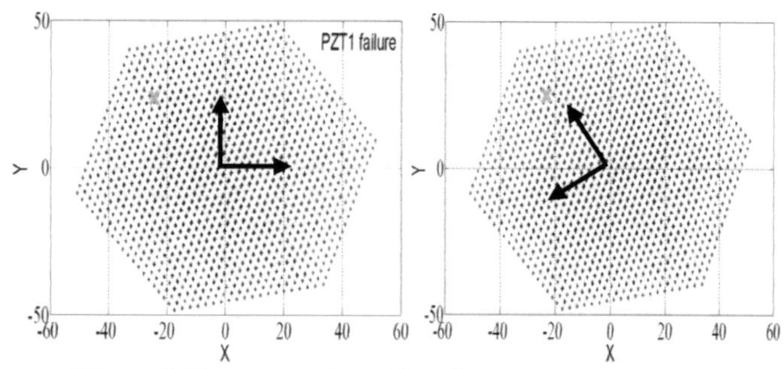

Figure 5: Demonstration of workspace recovery

In Figure 5, the cross point highlighted in (a) is assumed to be the target position, whereas the point is out of post-failure workspace with PZT1 failed, here both the target point and workspace are discussed in context of absolute coordinate. In order to let the post-failure workspace cover the target point, the whole RRR mechanism system then is rotated along the Z axis by 120 degrees, this leads to the post-failure workspace equivalent of that when PZT2 fails so that the target point can be achieved as displayed in (b), the principle here is available for the other diverse configurations of PZT failure.

POSITIONING ACCURACY RECOVERY

There are basically two configurations for positioning recovery, one is that the target position is in the post-failure workspace, the other is the target position is excluded by the post-failure workspace. In the first case, the target point is included in post-failure workspace. For RRR mechanism, the relationship between the elongations of PZT actuators and the motion of end-effector for RRR mechanism can be described as:

$$\mathbf{D} = \mathbf{JL} \tag{1}$$

Where $\mathbf{D}$ is the vector of end-effector motion, $\mathbf{L}$ represents the PZT actuator elongation, and $\mathbf{J}$ is 3x3 Constant-Jacobian Matrix, which in our case is:

$$\mathbf{J} = \begin{pmatrix} 0.451068716 & -0.083929543 & 0.125735391 \\ 3.888749854 & -0.097869666 & -4.155380077 \\ 0.001527845 & 0.002950417 & 0.018476359 \end{pmatrix} \tag{2}$$

According to [16], the coupling of three PZTs can be neglected, in this way, the Jacobian matrix is regarded to remain the same even certain of PZT doesn't work. Suppose one of the three PZT actuators i in RRR mechanism fails to work, the relationship then can be rewritten as:

$$\mathbf{D}^{i} = \mathbf{JL}^{i} \tag{3}$$

It is known from Eq. (1) that if one PZT fails to work, the rest PZT actuators can't fulfill the desired output of the end-effector in all three translations simultaneously. Practically, people mainly care about the plane motion of the device, hence, only plane motion is considered, then the updated input of the remaining PZT actuators are determined based on Eq. (3).

In the case of that the target point is excluded by the post-failure workspace. The first step is to rotate the RRR mechanism so that the desired point can be covered as described before; the next step is to update the input of the working PZT actuators. Usually, the desired point is considered in the absolute coordinate system, to simplify the kinematic analysis here, the absolute coordinate is decomposed to the local coordinate as shown in Figure 5, where the Constant-Jacobian Matrix remains constant for RRR mechanism. Here the situation in Figure 4 is cited as an example for demonstration, target point in (a) is (x, y) is not in the post-failure workspace, and then the decomposition of the point in the local coordinate (x', y') then is:

$$\begin{pmatrix} x' \\ y' \end{pmatrix} = \begin{pmatrix} \cos\theta & -\sin\theta \\ \sin\theta & \cos\theta \end{pmatrix} \begin{pmatrix} x \\ y \end{pmatrix} \tag{4}$$

Then the updated PZT actuators' elongations are determined by the Constant-Jacobian Matrix in the local coordinate system.

EXPERIMENTAL DEMONSTRATION

Figure 6 is the experimental set-up for the demonstration, the RRR mechanism are manufactured by wire-electrode cutting, and is mount on the platform that can be rotated, the RRR mechanism is driven by three PZT actuators (AE0505D16F, Thorlab) attached with strain gauges for PZT actuator controlling, and motion of the end-effector are measured by three eddy-current transducers (3330 XL, Bently Nevada), we display the realization of maintaining positioning accuracy under PZT actuator failure in the context of the aforementioned two configurations respectively, PZT actuator 1 is assumed to be failed in this experiment.

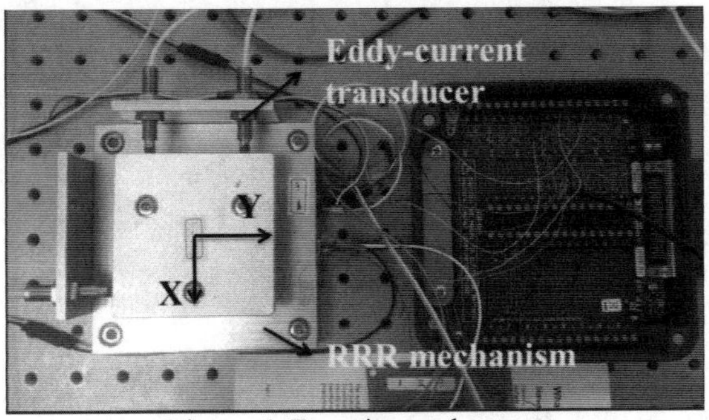

Figure 6: Experimental set-up

In this experiment, we set two target points for the two configurations, Table 1 and 2 exhibits the target position and experiment results as well as elongations of PZT actuators. With recalculation of PZT actuators' elongation and rotation of RRR mechanism, the original desired positioning can be done; this greatly highlights the feasibility of the proposed recovery strategy. The error involved in this experiment mainly attributed to the PZT controlling and the error induced by RRR mechanism rotation.

Table1
Positioning recovery under configuration 1

	Target point	Experimental results
Elongation(μm)	$P_0(3.466,-1.487)$	$P_1(3.158,-1.731)$
PZT1	3.7357186	0
PZT2	6.4048779	5.956413155
PZT3	6.4903018	5.913703199

Table2

Positioning recovery under configuration 1

	Target point	Experimental results
Elongation(μm)	$Q_0(-0.2018,3.831)$	$Q_1(-0.1717,2.973)$
PZT1	6.4048779	0
PZT2	6.4903018	1.216635477
PZT3	3.7357186	8.903806725

CONCLUSIONS

This paper investigates the positioning accuracy recovery for RRR mechanism system from PZT failure. Different recovery strategies are proposed based on the post-failure workspace analysis, and experimental demonstration results shows that the idea of positioning recovery presented in this paper are quite feasible. A better understanding of the resilience is promoted based on the resilience study in terms of RRR mechanism, while it doesn't loss generosity for the resilience realization on other compliant mechanism system. More importantly, this study also highlights the significance of system design for the resilience.

ACKNOWLEDGMENTS

The authors are grateful to the National Natural Science Foundation of China (NSFC, Grant No. 51175179), and the Fundamental Research Funds for the Central Universities for providing financial support for this work. The author (WJ Zhang) would also like to acknowledge a partial support from NSERC (Grant No. 247091) through a Discovery Grant for his involvement in this research.

REFERENCES

1. Yong, Y., Lu T. and Handley D., 2003. "Loop Closure Theory Deriving Linear and Simple Kinematic Modal for a 3-DOF parallel Micromanipulator", *In: Proceedings of SPIE*, Vol(5276), pp. 57-66.

2. Yong, Y., and Lu T., 2009. "Kinetostatic modeling of 3-RRR compliant micro-motion stages with flexure hinges", *Mechanism and machine theory*, Vol(44), pp.1156-1175.

3. Tian, Y., Shirinzadeh B., Zhang D., 2010. "Design and dynamics of a 3-DOF flexure-based parallel mechanism for micro/nano manipulation", *Microelectronic Engineering*, Vol(87), pp.230-241.

4. Yu, Y., Du, Z., Yang, J., and Li, Y. 2011, "An Experimental Study on the Dynamics of a 3-RRRFlexible Parallel Robot", *In: Proceedings of IEEE Transactions on Robotics*, Vol(27), pp. 992-997.

5. Lu, T., Handley, D., and Yong, Y. 2004, "Position Control of a 3 DOF Compliant Micro-Motion Stage", *In: Proceedings of IEEE Conference on Control, Automation, Robotics and Vision*, pp.1273-1278.

6. Acer, M., and Sabanovic, A. 2012, "Sliding-mode control of a flexure based mechanism using piezoelectric actuators", *Advanced Motion Control (AMC), 2012 12th IEEE International Workshop on*, pp.1-6.

7. Zhang, W., and Van, L. 2011, "Toward a resilient manufacture system", *CIRP Annals*, Vol(60), pp.469-472.

8. Zhang, W., and Lin, Y. 2010, "On the principle of design of resilient systems-application to enterprise information systems", *Enterprise Information Systems*, Vol(4), pp.99-110.

9. Sun, Z., Yang, G., Zhang, B. and Zhang W. 2011, "On the concept of resilient machine", *In: Proceeding of IEEE Conference on Industrial Electronics and Application, pp.357-360.*

10. Zhang, W., and Lin, Y. 2010, "Resilient information system: concept and principle", *Enterprise Information Systems*, Vol(4), pp. 26-40.

11. Huang, Z. and Fang, Y. 1996, "Kinematic characteristics analysis of 3DOF in-parallel actuated pyramid mechanism", *Mechanism and Machine Theory*, Vol(31), pp.1009-1018.

12. Zhang, W., Zou, J., Waston, L., and Zhao, W. 2002, "The Constant-Jacobian Method for Kinematics of a Three- DOF Planar Micro-Motion Stage", *Journal of Robotics systems,* Vol(19), pp.63-72.

13. Handley, D., Lu, T., and Yong, Y. 2004, "Workspace investigation of a three DOF compliant micro-motion stage", *In: Proceedings of the IEEE International Conference on control, automation, robotics and vision,* pp.1279-1284.

14. Daniel, C., Handley D., Zhang, W., Lu, T., and Zhao, W. 2002, "Multiple degree of freedom compliant mechanism possessing nearly uncoupled dynamics: experimental findings", *In: Proceedings of SPIE*, Vol(4936), pp. 168–176.

15. Zhang, W., and Lin, Y. 2010, "Resilient information system: concept and principle", *Enterprise Information Systems*, Vol(4), pp. 26-40.

Micro- and Nano-fluides

EXPERIMENTAL STUDY ON ENHANCEMENT OF HEAT TRANSFER USING NANOPARTICLES BASED FLUIDS

Nabeel Al-Rawahi
Mechanical and Industrial Engineering Department
College of Engineering, Sultan Qaboos University
Alkhoud, Muscat, Sultanate of Oman
alrawahi@squ.edu.om

Myo Tay Zar Myint
Sultan Qaboos University
Alkhoud, Muscat, Sultanate of Oman
myomyint@squ.edu.om

Joydeep Dutta
Sultan Qaboos University
Alkhoud, Muscat, Sultanate of Oman
dutta@squ.edu.om

ABSTRACT

Experimental study on enhancement of heat transfer using zinc oxide (ZnO) and aluminum oxide (Al_2O_3) nanoparticles based fluids is presented in this paper. A heat exchanger system was designed and fabricated to carry out the study. Nanofluids were prepared with different concentrations of nanoparticles in aqueous media. The nanoparticles and the fluid mixture were characterized using standard materials characterization techniques like X-ray diffraction, Scanning Electron Microscopy (SEM), Transmission Electron Microscopy (TEM), Zeta Potential, and Particle Size Analysis. The focus of this study is the stability of suspension of the nanoparticles in the base fluid and the heat transfer rate.

Compared to deionized water, nanofluids with particles concentrations of 0.02 wt %, results in about 30% enhancement in heat transfer efficiency. It was found that for 0.02 wt % ZnO or Al_2O_3 nanoparticles yielded the maximum heat transfer efficiency of nanofluids. The stabilization of the nanofluids could be further improved by stabilizing the nanoparticles with a surfactant (0.5 % wt of Sodium Dodecyl Sulfate (SDS)).

KEYWORDS: Heat Transfer, nanoparticles, nanofluid, stabilizing agents.

INTRODUCTION

The extensive requirements in heat transfer applications of fluids are encouraging to improve this important process continuously. There is a rise in the demand for transferring heat at higher rates from smaller areas or across a lower temperature gradient. As a result, over the last two decades different techniques for the enhancement of heat transfer have been reported [1].

Nanofluids have special properties that make them potentially useful in various applications of heat transfer. These applications include fuel cells, nuclear reactor coolant, vehicle thermal management, domestic refrigerator, chiller, heat exchanger, engine cooling, space technology, and in boiler flue gas temperature reduction [1]. Nanofluids provide enhanced thermal conductivity and increased convective heat transfer coefficient compared to the base fluid. It is critical to understand the performance of nanofluids in order to decide about their suitability for applications that utilize convective heat transfer.

It is known that thermal capability of traditional fluids can be improved by the inclusion of metallic or metal oxide particle additives. This has been practiced over more than a century by using bigger sized particles. However, because of their big size, any such suspensions lead to rapid settling of the particles. In addition, these particles can cause erosion in pipe walls and erode the pumps and other equipments. One benefit of the nanoparticles is that they increase the thermal conductivity coefficient and also convection heat transfer coefficient with much less pressure loss compared to traditional mixtures. It is therefore expected that the use of nanofluids in industries will solve problems related to equipments and thermal fluids[2].

Several studies have been conducted to investigate the effect on the thermal conductivity resulting from using nanoparticles in fluids (nanofluids). These include numerical [3], theoretical [4 -6], and experimental [7] studies. However, the exact cause of the high heat transfer rates in nanofluids are not yet identified. Keblinski et al. [4] studied the enhancement in heat transfer efficiency arising from Brownian motion of nano-scaled particles, formation of liquid layer at liquid/particle interface ballistic phonon effects and effects of nano particle clustering. They concluded that "*key factors in understanding thermal properties of nano-fluids are the ballistic phonon induced effects, rather than diffusive heat transfer, combined with direct or fluid mediated clustering effects that provide paths for rapid heat transport*". On the other hand, Vadasz et al. [5] investigated other mechanisms and concluded that "*hyperbolic heat conduction could have be the source of the excessively high effective thermal conductivity*" of the nanofluids. It should be noted that using nanofluids result in higher pressure drop which could be a concern in some systems. The advantage of using nanofluids with respect to heat transfer should take into account the increase in the required pumping energy. [8]

In this study, the relationship between the heat transfer rate and the suspension properties of nanoparticles in a base fluid is studied experimentally using commercially available nanoparticles of ZnO and Al_2O_3 with different types of stabilizing agents dispersed in an aqueous media using an in-house designed and fabricated heat exchanger system.

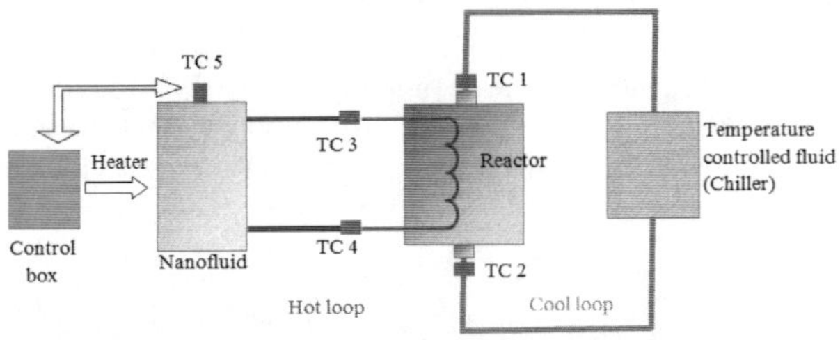

Figure 1: Schematic representation of the designed heat exchanger

Figure 2: Disassembled parts of the reactor

EXPERIMENTAL SETUP

In this study, a heat exchanger system was first designed and built. Then, different types of nanofluids were prepared. The nanofluids are classified as different % weight of nanofluids zinc oxide (ZnO) and alumina (Al_2O_3) nano particles capped by different % weight of stabilizing agents.

The schematic diagram of the designed heat exchange system is shown in Fig. 1. A simple reactor was designed and built consisting of a copper coil and a housing made up of polyvinyl chloride (PVC). The reactor is where the heat exchange process takes place. 8.6 m long copper tube used of 0.63 mm outer diameter and 0.5 mm inner diameter was used.

A glass container was used as a compartment for storage of the Nanofluid. It contains the immersion heating element (ca. 800 W) that is controlled by a programmable temperature controller (LT 400) via a solid state relay. The glass container was designed to facilitate easy filling and draining of fluids.

The response time of the thermocouple was determined by using two different temperatures of the water media. The temperature of the first media was kept at 24 ^{o}C while the second media was put in a controlled temperature of 40 ^{o}C. The volume of the media was measured to be 800 ml. The response time of the thermocouple for this volume was found to be ~ 3 seconds. Fig. 3 shows the schematic diagram of temperature control system. The temperature of nanofluid was controlled by PID controller (LT 400) connected to a thermocouple (K-type). Solid state relay with a maximum current rating of 40 A was used to supply alternating current (240 V) to the immersion heater (800 W) located in the nanofluid chamber. From this control circuit, the ramping rate of the temperature of the nanofluid can be controlled from room temperature to boiling point of the fluid. LabView program was utilized to establish a communication between NI-USB 6009 (National instrument) and thermocouple.

Different concentration of nanoparticles were used using stabilizing agents (also called capping agents) like polyvinyl alcohol (PVA), 3-Trimethoxysilyl propylamine (TMSP), and sodium dodecyl sulfate (SDS). Using different concentrations of the capping agents enables us to observe the role of both stabilization of the colloidal suspension and improvement in heat transfer efficiency in the nanoparticles .

Commercial ZnO nanoparticles used in the experiments are with diameters in the order of 10-30 nm while, Al_2O_3 nanoparticles were in the order of 50nm. Scanning Electron Microscopy (SEM) of both nanoparticles (ZnO and Al_2O_3) were observed. However, it was difficult to estimate the particle sizes from the micrographs due to the limitation of the resolution of the equipment. Therefore, high resolution TEM images were obtained as shown in Fig. 4 to estimate the average particle sizes. It was found that most of the ZnO particles were between 10 -5. nm in size, while most of Al_2O_3 particles were found to be between 10 – 30 nm in size.

The nanofluids were prepared at room temperature (~ 25 ^{o}C) following the sequence given below.

(1) 1, 0.5, 0.1, 0.02, 0.01 % wt of nanoparticles were prepared by dispersing 10, 5, 1, 0.2 and 0.1 g of nanoparticulate materials in 1000 ml of deionized water, respectively.
(2) The suspension was then stirred for 10 minutes in a magnetic stirrer and sonicated for 10 minutes at room temperature.
(3) All prepared fluids were then stirred for 48 hrs at room temperature.

These steps are shown in Fig.5.

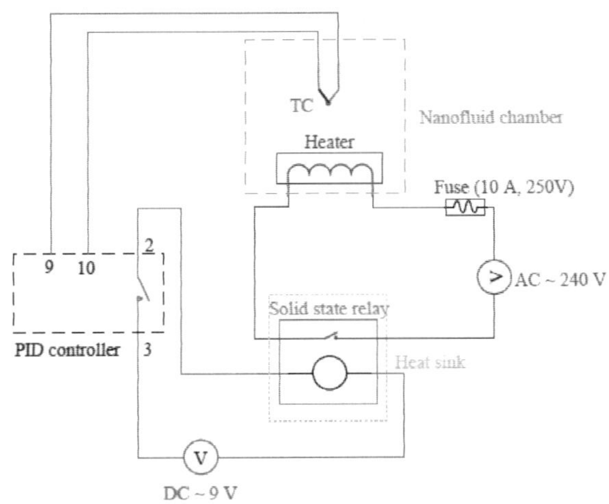

Figure 3: Schematic diagram of temperature control loop for the nanofluid chamber

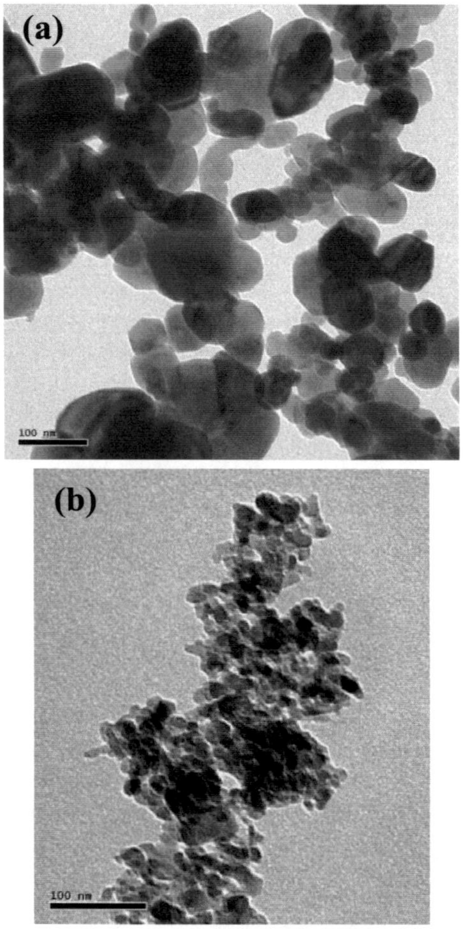

Figure 4: TEM image of (a) 0.02 % wt ZnO (b) 0.02 % wt Al_2O_3 nanofluids.

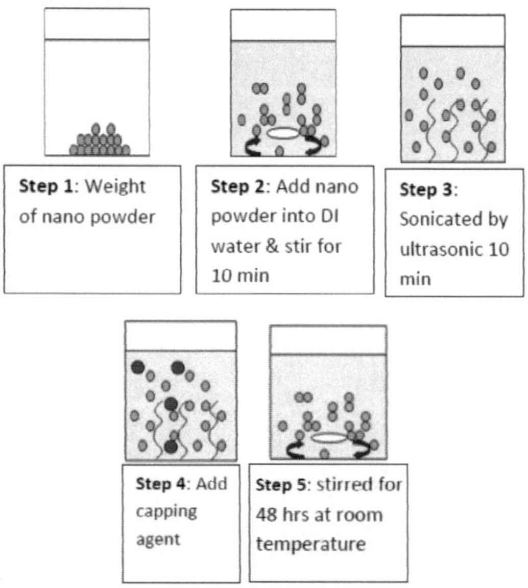

Figure 5: Steps of nanofluids preparation.

RESULTS AND DISCUSSION OF RESULTS

The fabricated heat exchanger system shown above was used with fixed flow rates of 10 ml/min. Heat transfer amount was calculated using the convection heat transfer equation

$$Q = H.A.(T_{out}-T_{in}) \qquad (1)$$

where: Q is the heat transfer amount in W, H is convection heat transfer coefficient for the medium which was taken to be = 50 W/m^{2}C. The typical value for water ranges from 20 - 100 W/m^{2}C [9]. A is the cross sectional area (internal) where the heat transfer occurs ~ 0.017 m^2 (calculated from the system design).

$$\text{Efficiency} = (Q_{nanofluid} - Q_{water})/\ Q_{nanofluid} \qquad (2)$$

The effect of using nanofluids that were prepared with different wt % of nanoparticles is shown in Fig. 6 and Fig. 7, which show the heat transfer achieved in each case. The amount of heat transfer reached a maximum value with 0.02 wt %. The increase in the heat transfer reached 29%. Other experimental results available in the literature show also an increase in the heat transfer efficiency with nanofluids. Xinyu Wu et al. [10] reported an increase in heat transfer coefficient of nanofluid up to 15.8% using Al_2O_3 nanoparticles with water.

The heat transfer decreases as the concentration of the of ZnO or Al_2O_3 nanoparticles in the base fluid increases (to 1 wt%) or decreases (to 0.01 wt%) from this optimum value. One of the possible reasons is the agglomeration of nanoparticles in the higher concentration of nanofluids. This results in a reduction of the surface area of heat conduction pathway [11].

It was also found that as the concentration of the capping agent increased, the heat transfer amount decreased. This is because the capping agent absorbs some of the heat and act as resistance to transfer the heat. In addition the capping agent is attached to the nano particle and covers part of its surface area which results in smaller particle surface area.

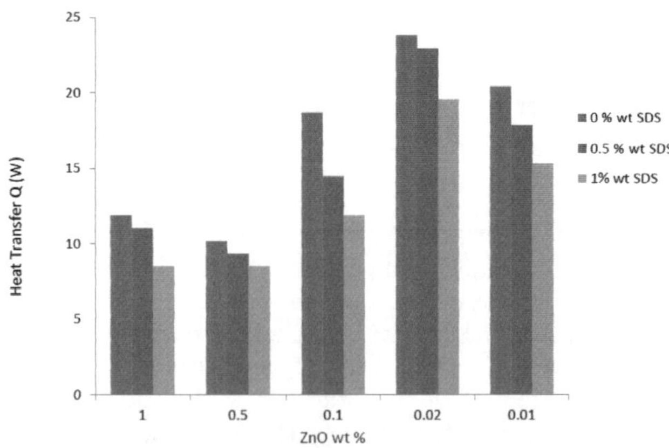

Figure 6: The effect of different ZnO nanofluid concentration and SDS percentage on the heat transfer.

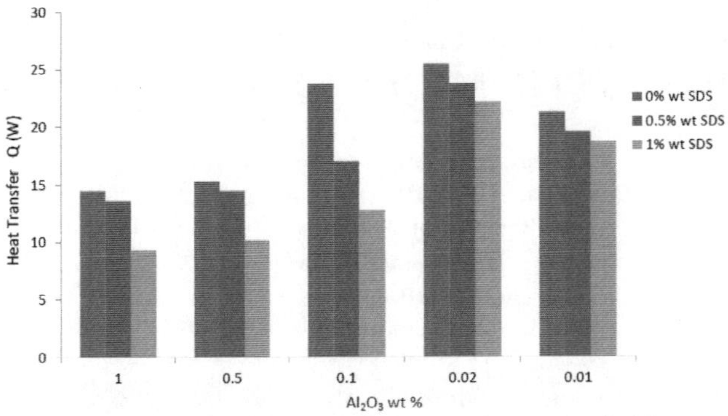

Figure 7: The effect of different Al_2O_3 nanofluid concentration and SDS percentage on the heat transfer.

CONCLUSIONS

In this study, a fit for purpose heat exchanger system was fabricated and used to measure and investigate the maximum heat transfer amount that can be achieved from different nanofluids. The set up consists of a copper coil and the housing made up of polyvinyl chloride (PVC).

Over thirty samples of nanofluids (different concentrations: 1, 0.5, 0.1, 0.02, 0.01 wt%) were prepared and studied in detail to correlate the suspension properties to the heat transfer rate. Commercially available nanoparticles of zinc oxide (ZnO) and aluminum oxide (Al_2O_3) were used to prepare water based nanofluids with sodium dodecyl sulfate (SDS), polyvinyl alcohol (PVA) and 3-Trimethoxysilyl propylamine (TMSP) as capping agents.

In conclusion, the results show an enhancement of upto 29 % in net heat transfer when using nanofluids with particle concentration of 0.02 wt% compared to the pure water (the base fluid). While the heat transfer using pure water was 17W, the maximum heat transfer amount provided by the nanofluid was 25.5 W. It was found that 0.02 % wt Al_2O_3 + 0.5 % wt SDS nanofluid and 0.02 % wt ZnO + 0.5 % wt SDS nanofluid gave the maximum heat transfer amount with better stability.

ACKNOWLEDGMENTS

The authors would like to acknowledge the funding provided by the Research Council of Oman through the Nanotechnology Chair in Sultan Qaboos University. The authors would like also to acknowledge the work done by: Nasser Al-Kindi, Mariam Al-Bahri. Salim Al-Riyami, and Juma Al-Maskry in setting and running some of these experiments.

REFERENCES

1. Paul, G., Chopkar, M., Manna, I., and Das, P.K. 2010, "Techniques for measuring the thermal conductivity of nanofluids: A review", *Renewable and Sustainable Energy Reviews*, Vol(14(7)), pp. 1913-1924.
2. Yu, W., et al. 2008, "Review and comparison of nanofluid thermal conductivity and heat transfer enhancements", *Heat Transfer Engineering*, Vol(29(5)), pp. 432-460.
3. Nimmagadda, R., Venkatasubbaiah, K. 2015, "Conjugate heat transfer analysis of micro-channel using novel hybrid nanofluids (Al_2O_3 + Ag/Water)", *European Journal of Mechanics B/Fluids*, Vol(52), pp. 19–27.
4. Keblinski, P., Phillpot, S.R., Choi, S.U.S., and Eastman, J.A. 2001, "Mechanisms of heat flow in suspensions of nano-sized particles (nanofluids)", *International Journal of Heat and Mass Transfer*, Vol(45(4)), pp. 855-863.
5. Vadasz, J. J., Govender, S., and Vadasz, P. 2005, "Heat transfer enhancement in nano-fluids suspensions: Possible mechanisms and explanation", *International Journal of Heat and Mass Transfer*, Vol((48,(13)), pp. 2673-2683.
6. E. Yamada, T. Ota, Effective thermal conductivity of dispersed materials, Heat Mass Transf. 13 (1980) 27-37.
7. Heris, S. Z., Nassan, T. H., Noie, S.H., Sardarabadi, H., and Sardara, M. 2013, "Laminar convective heat transfer of Al2O3/water nanofluid through square cross-sectional duct", *International Journal of Heat and Fluid Flow*, Vol(44), pp. 375-382.
8. Ferrouillat, S., Bontemps, A., Ribeiro, J. P., Gruss, J. A., and Soriano O. 2011, "Hydraulic and heat transfer study of SiO2/water nanofluids in horizontal tubes with imposed wall temperature boundary conditions", *International Journal of Heat and Fluid Flow*, Vol(32), pp. 424–439.
9. Yang, Y., et al. 2005, "Heat transfer properties of nanoparticle-in-fluid dispersions (nanofluids) in laminar flow", *International Journal of Heat and Mass Transfer*, Vol(48(6)), pp. 1107-1116
10. Wu, X., Wu, H., Cheng, P.,2009, "Pressure drop and heat transfer of Al_2O_3-H2O nanofluids through silicon microchannels", *J. Micromech. Microeng.* Vol(19(10)), pp. 1–11.
11. Witharana, S., Palabiyik, I., Musina, Z., and Ding, Y. 2013, "Stability of glycol nanofluids — The theory and experiment," *Powder Technology*, Vol(239), pp. 72–77.

Proceedings of the 25th CANCAM
London, Ontario, Canada, May 31 – June 4, 2015

COMBINING MICROMILLING AND LIQUID-BASED SOLVENT BONDING FOR ULTRA-RAPID FABRICATION OF PLASTIC MICROCHIPS

Alwin M. D. Wan
Dept. Mechanical & Industrial Engineering
University of Toronto
Toronto, ON, Canada
awan@mie.utoronto.ca

Edmond W. K. Young
Dept. Mechanical & Industrial Engineering
University of Toronto
Toronto, ON, Canada
eyoung@mie.utoronto.ca

ABSTRACT

We discuss experimental results from an alternative, unconventional procedure for fabricating microfluidic chips and devices in thermoplastics that reduces design-to-prototype turnaround times from ~1 week to less than an hour. The workflow consists of (1) computer-aided design, (2) CNC micromilling, and (3) liquid-based solvent bonding. Micromilling creates microchannels with higher surface roughness compared to hot embossing, but solvent bonding generates significantly higher bond strength compared to thermal diffusion bonding. Solvent concentration can be tuned to optimize bond strength and microchannel quality.

KEYWORDS: microfluidics, microfabrication, thermoplastics, milling, bonding

INTRODUCTION

Thermoplastics are excellent materials for fabrication of microfluidic chips because they are optically transparent, durable, and are compatible with both rapid prototyping and high volume fabrication [1]. Despite their potential for widespread usage in microscale technologies, the adoption of thermoplastics for microfluidic application in both academia and industry has been somewhat hindered due to remaining challenges in microfabrication. In particular, plastic microfabrication continues to require significant startup and development costs, including the need for expensive equipment, laborious mold-making processes, and long development times during the iterative design phase [2]. Recently, methods to achieve affordable rapid prototyping of plastic microfluidic devices were developed by combining low-cost epoxy mold-making, hot embossing, and thermal diffusion bonding processes, all in a streamlined workflow [3]. This procedure overall takes less than one week from computer-aided design (CAD) to fabricated microchip, including ~4-5 days for epoxy mold fabrication, and only ~20-30 min for plastic device fabrication, thereby making it comparable to rapid prototyping by soft lithography methods, and significantly faster than conventional hot embossing processes. In this work, we offer an alternative to hot embossing and thermal bonding procedures that shortens the design-to-prototype turnaround time even further by using micromilling and liquid-based solvent bonding. The result is an overall workflow that can be reduced to less than one hour (Figure 1).

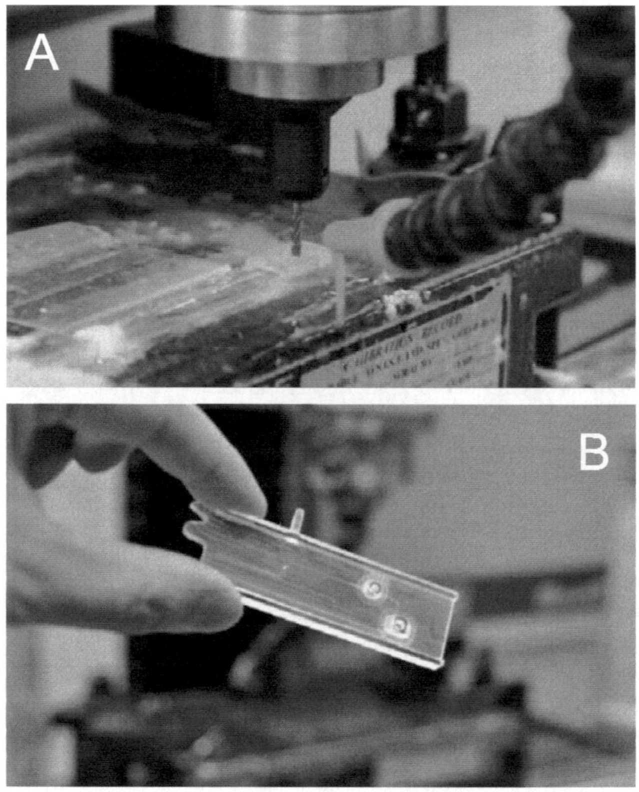

Figure 1: Micromilling (A) can be used to fabricate plastic microfluidic devices with complex 3D geometries (B), and do so with design-to-prototype turnaround times < 1 hour.

MATERIALS AND METHODS

Micromilling. Thermoplastic microfluidic devices were fabricated by micromilling with a Tormach PCNC 770 vertical milling machine (Tormach, Waunakee, WI, USA). Microfeatures on devices were modeled with Solidworks (Dassault Systemes, France), and the computer numerical control (CNC) program was generated in SprutCAM software (SprutCAM, Russia). Devices were milled into either poly(methylmethacrylate) (PMMA) or polystyrene

(PS) using different endmills ranging in diameter from 0.005" (127 μm) to 0.020" (508 μm).

Solvent Bonding. Varying concentrations of acetone mixed in deionized (DI) water were applied between the two bonding surfaces of plastic sheets or "layers" containing milled microfeatures. Layers were aligned, acetone solution was applied to surfaces by dispensing via eye-dropper or micropipette, and pressure (~5 psi) and heat (40°C) were applied for 20 min to achieve bonding at the interface.

Bond Strength Characterization. Bond strength can be quantified by forcing a wedge between bonded layers, separating the layers at the interface, and measuring the crack propagation length L created. From this method, a surface energy representing bond strength can be calculated, as previously shown by Maszara et al. [3]:

$$\gamma = \frac{3}{8}\frac{Et^3y^2}{L^4} \quad (1)$$

where γ represents the specific surface energy in [J/m^2]; E is the elastic modulus in [N/m^2]; t is the sheet thickness; and y is the half-thickness of the wedge used to separate the layers (Figure 2).

Microfeature quality characterization. After micromilling and solvent bonding, devices were sectioned across microchannel features, and imaged by brightfield microscopy (EVOS FL-Auto, EVOS, Life Technologies, Burlington, ON, Canada) to visualize mirochannel quality in terms of cross-sectional geometry achieved compared to cross-sectional geometry desired.

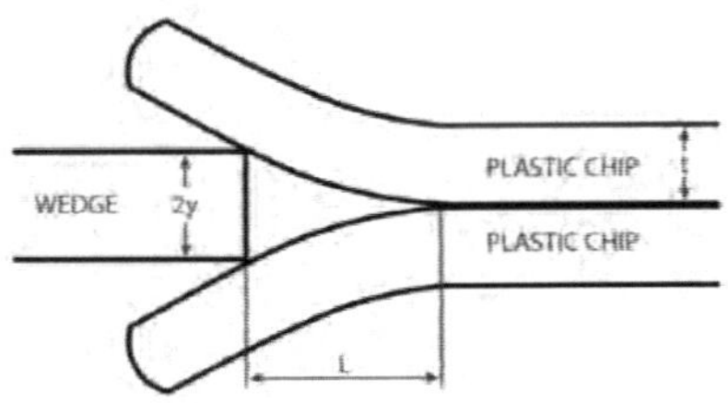

Figure 2: Cross-sectional side view of bond strength test.

RESULTS AND DISCUSSION

Bond Strength. Plastic layers were aligned and bonded by different concentrations of acetone diluted in DI water. Layers were then separated by the crack propagation length method (Figure 2), and the specific energy calculated to represent bond strength. For both PMMA and PS, increasing the acetone concentration from 50% to 100% improved bond strength by two orders of magnitude (Figure 3, A-B). However, this strength increase came at the cost of channel deformation due to polymer swelling near the microchannel surfaces (Figure 3, C-E, PMMA). Using PMMA as an example, 60-75% acetone concentration demonstrates good bond strength and high quality surfaces (without swelling), and should be used instead of >80%, when microfeature quality begins to deteriorate.

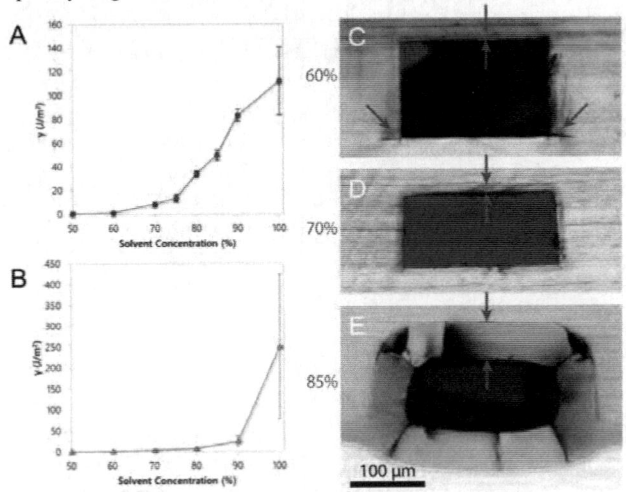

Figure 3: (A-B) Bond strength (specific energy, γ) versus acetone concentration for (A) PMMA and (B) PS. Error bars represent one standard deviation for n = 3. (C-E) Cross-sections of PMMA microchannels bonded with varying concentrations of acetone. Polymer swelling at high acetone concentrations leads to significant deformation.

CONCLUSIONS

Combining micromilling and liquid-based solvent bonding achieves high-quality microfluidic devices with high bond strength and minimal channel deformation under optimal solvent concentrations. Micromilling helps reduce design-to-prototype turnaround times from ~1 week to < 1 h, and can thus be very useful during the design iteration phase.

ACKNOWLEDGMENTS

This work was partially funded by Grand Challenges Canada (#S5-0391), an NSERC Discovery Grant (#436117-2003), and an NSERC Engage Grant (#455285-13).

REFERENCES

[1] Becker, H., and Gartner, C. 2000, "Polymer microfabrication methods for microfluidic analytical applications", *Electrophoresis*, 21(1), pp. 12–26.\

[2] Young, E.W.K., Berthier E., Guckenberger, D.J., Sackmann, E., Lamers, C., Meyvantsson, I., Huttenlocher, A, and Beebe, D.J. 2011, "Rapid prototyping of arrayed microfluidic systems in polystyrene for cell-based assays", *Analytical Chemistry*, 83(4), pp. 1408–17.

[3] Maszara, W., Goetz, G., Caviglia, A., and Mckitterick., J. 1988, "Bonding of silicon-wafers for silicon-on-insulator", *J. Applied Physics*, 64(10), pp. 4943–4950.

DESIGN CONSIDERATIONS FOR ON-CHIP BIOSENSING IN MICROFLUIDIC CARDIOVASCULAR MODELS

Jeremy Wong
Institute of Biomaterials & Biomedical Engineering
University of Toronto
Toronto, ON, Canada
jer.wong@mail.utoronto.ca

Craig A. Simmons
Dept. of Mechanical & Industrial Engineering
Institute of Biomaterials & Biomedical Engineering
University of Toronto
Toronto, ON, Canada
c.simmons@utoronto.ca

ABSTRACT

Microfluidic technologies are well-suited for modeling the vascular system and vascularized tissues, and therefore have excellent potential to improve the physiological relevance of in vitro systems for studying vascular functions. Emerging developments in microfluidic vascular models include the integration of on-chip sensing systems. In this talk, we describe our on-going efforts to this end, with particular focus on design considerations to implement on-chip biosensors to measure protein secretion by cells in response to microenvironmental cues.

KEYWORDS: microfluidics, hemodynamics, endothelial cells, biosensor, microenvironment

INTRODUCTION

The cells within blood vessels, including endothelial cells, mural cells, and circulating blood cells, function in a complex microenvironment that includes three-dimensional, soft extracellular matrices, heterotypic cell-cell interactions, and dynamic mechanical loading due to hemodynamic forces. However, traditional in vitro systems used to study vascular physiology and biology rarely recapitulate these features: in most cases, cells are grown under static conditions on the rigid, two-dimensional surfaces of microwell plates or at best, in macroscale flow chambers that incorporate flow-induced shear stresses. Beyond their limited physiological fidelity, these systems often require large amounts of expensive reagents and large numbers of precious cells, are low-throughput, and can be cumbersome. These deficiencies limit the novelty and translatability of the biological information that can be generated with traditional platforms, with important implications both to fundamental biological investigations and to applications such as drug screening.

To address the limitations of current in vitro systems, several groups have turned to microfluidic technologies, citing numerous putative advantages including:

- low Reynolds number flows with predictable fluid physics and transport, enabling precisely controlled flow environments;
- reduced reagent consumption and cell requirements, enabling improved throughput and reduced costs;
- substantial flexibility in design and fabrication, enabling physiological geometric features that are difficult to implement in traditional systems;
- the ability to readily incorporate biological hydrogels and other physiologically-relevant biomaterials into the platform; and
- the potential to integrate assays and sensors directly "on-chip".

With these advantages in mind, a wide variety of microfluidic systems have been developed to study various vascular functions and phenomena, often yielding improved models and novel insights (reviewed by Wong et al. (2012)).

Much current research and development in this field aims to extend existing platforms and broaden applications. Here we discuss our efforts in these directions, including development of a microfluidic platform to model vascularized tissues and design issues associated with our on-going efforts to incorporate on-chip biosensors to monitor cell responses to microenvironmental stimuli.

MEMBRANE MICROFLUIDIC PLATFORM

We developed a bilayer membrane microfluidic device to model various vascular functions (Fig. 1). The device consists of upper and lower microchannels separated by a polyethylene terephthalate, track-etched porous membrane with 1 to 8 μm diameter pores (depending on the intended application). The upper channel is fabricated in polydimethylsiloxane (PDMS) using standard lithography. The bottom PDMS channel is formed via squeeze fabrication with aluminum moulds (Moraes et al., 2009). The membrane is bonded between the two PDMS channels with a thin mortar layer of uncured PDMS via stamping (Young et al., 2010). The bottom channel is then bonded to a plasma-treated glass No. 1.5 coverslip. Tubing is inserted into the inlet and outlet of the top and bottom microchannels, enabling controlled fluid flow in both layers. In a typical application, the membrane is coated with an extracellular matrix protein and seeded with a monolayer of endothelial cells.

We have applied this device to measure endothelial permeability (Young et al., 2010), to model monocyte transmigration across the endothelium (Srigunapalan et al., 2011), and to model endothelial paracrine regulation of mural cells (Cheng et al., 2013). In the latter application, a photopolymerizable gelatin methacrylate (gel-MA) hydrogel containing mural cells was incorporated into the bottom channel as a physiologically-relevant subendothelial 3D matrix (Fig. 2). This arrangement – an endothelial layer overlying a 3D extracellular matrix with embedded tissue-specific cells – can be used to model a wide range of vascularized tissues, including liver, blood-brain barrier, and tumors, among others.

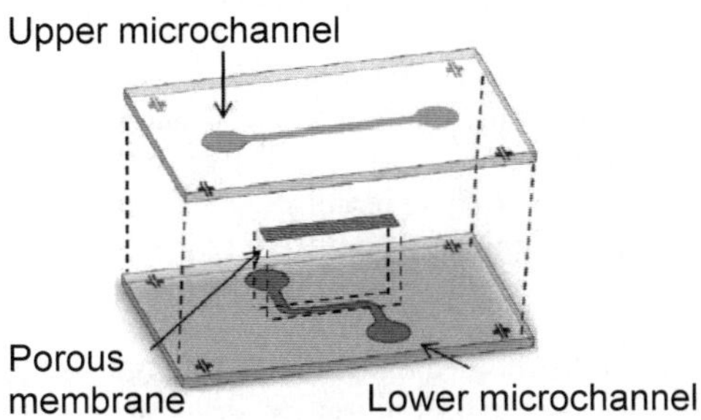

Figure 1: Exploded view schematic of the membrane microfluidic device showing the upper and lower layers sandwiching a porous membrane. After Srigunapalan et al. (2011).

ON-CHIP BIOSENSING

Although on-chip assays and sensing are touted as significant advantages of microfluidics, there are very few examples of microfluidic vascular models that integrate sensors on-chip. Such a development would enable, for example, continuous or real-time measurement of proteins secreted by the endothelium in response to shear stress, drug candidates, or other microenvironmental stimuli. A challenge with on-chip biosensing in a dynamic fluidic environment, particularly for proteins that may be expressed at low levels, is how to design the platform to maximize sensing sensitivity and performance.

To determine design guidelines for the microdevice and integrated biosensor, we are modeling transport of endothelial secreted proteins in the membrane microfluidic system. This builds on our previously published analytical model of transport of fluorescently-tagged proteins across an endothelialized membrane (Young et al., 2010). We consider the case shown in Fig. 2a of free flow in the upper channel (simulating the flow of blood in the lumen), and transport of an analyte as a dilute species through the membrane, modelled as a porous medium, to the bottom surface of the lower microchannel where the fluid is considered static and an integrated biosensor is most likely to be located due to fabrication and transport considerations. Our goal is to design a membrane microfluidic device that enables cells in the upper channel to be subjected to a broad range of physiological and pathophysiological dynamic shear stresses, while enabling sensitive protein quantification via on-chip biosensing in the lower channel. The current on-going computational modeling studies are being used to guide the design of the microchannels and biosensor, including channel dimensions, sensor positioning, and general measurement strategies in addition to assessing performance of a given design in meeting the desired functionality.

NUMERICAL MODELLING TO GUIDE MICRODEVICE DESIGN FOR ON-CHIP BIOSENSING

The microdevice is modelled in COMSOL Multiphysics, with typical membrane dimensions, porosity, and permeability as determined previously (Young et al., 2010). In initial simulations, a cell layer on the upper surface of the membrane (i.e., endothelial cells and/or adherent monocytes) is modelled as a constant source of 1 nM secreted tumour necrosis factor (TNF)-α, a cytokine of relevance to vascular function. The diffusivity of TNF-α was based on that used by Huang et al. (2012).

Endothelial cells in the upper channel are to be subjected to fluid flow-induced shear stress approximated as:

$$\tau = \frac{6\mu Q_t}{w_t h_t^2} \quad \text{Eq.1}$$

where μ is the viscosity of the cell culture medium; w_t and h_t

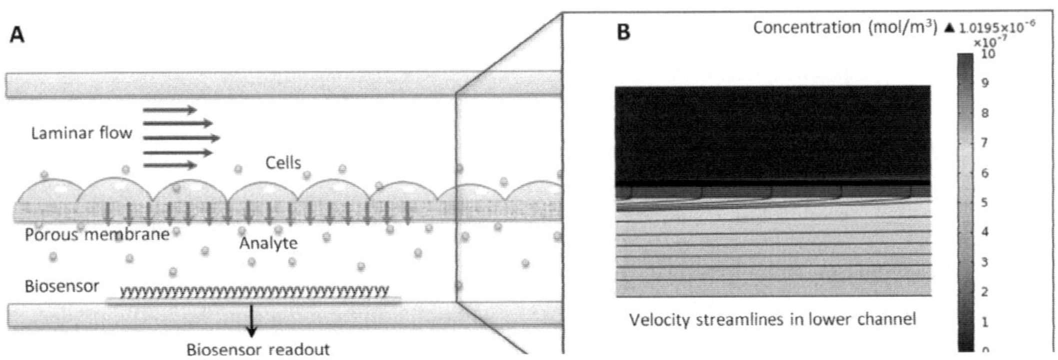

Figure 2: Membrane microfluidic device with on-chip biosensing. A) Schematic of upper and lower channels in the device: cells in the upper channel are subjected to flow-induced shear stress while the lower channel contains the biosensor. B) COMSOL-simulated concentration profile showing transport of the analyte to the lower channel along velocity streamlines.

are the top (upper) channel width and height, respectively; and Q_t is the upper channel flow rate. Flow in the upper channel is generated by applying a pressure gradient across the channel length. This also leads to a pressure difference between the upper and lower channels, across the porous membrane, which we call the transmembrane pressure, ΔP_T. Mass transport to the surface of the biosensor located in the lower channel is critically dependent on ΔP_T. For simplicity, we assume that the membrane is centered lengthwise at the middle of the upper channel. The transmembrane pressure at the midpoint of the membrane for a rectangular channel is then given by:

$$\Delta P_T = P_{t,m} - P_{b,m} = \frac{6\mu L_t Q_t}{w_t h_t^3} - P_{\text{hydrostatic}} \qquad \text{Eq.2}$$

where L_t is the upper channel length, and the pressure in the bottom (lower) chamber $P_{b,m} = P_{hydrostatic}$, determined by the hydraulic head in the tubes in the inlet and outlet of the lower channel under static conditions. Fig. 2b shows a concentration species, here TNF-α, being transported from the surface of the membrane down into the lower channel where it accumulates, driven by the transmembrane pressure. The velocity streamlines indicate that, given a high enough flow rate and corresponding pressure in the upper channel, the hydrostatic pressure in the lower channel inlet/outlet is overcome and fluid exits the lower channel. Such convection-dominated mass transport is significantly more rapid than relying purely on diffusive transport.

In order to obtain convection-dominated mass transport across the membrane, the fluid pressure in the upper channel must be greater than the fluid pressure in the bottom channel due to hydrostatic pressure. Setting Eq. 2 to zero and re-arranging the terms, we obtain an expression for the upper channel flow rate required for $\Delta P_T = 0$ at the midpoint of the membrane for a given set of channel dimensions:

$$Q_{\Delta P_T=0} = P_{\text{hydrostatic}} \frac{w_t h_t^3}{6\mu L_t} \qquad \text{Eq.3}$$

Flow rate $Q_t > Q_{\Delta P_T=0}$ thus leads to $\Delta P_T > 0$. Substituting Eq. 3 into the expression for shear stress (Eq. 1), we obtain an expression for the shear stress at the upper channel flow rate required for $\Delta P_T = 0$:

$$\tau_0 = P_{\text{hydrostatic}} \frac{h_t}{L_t} \qquad \text{Eq.4}$$

It is critical to design for $Q_{\Delta P_T=0}$ and τ_0 since upper microchannel parameters play an important in controlling both mass transport to the lower channel via the transmembrane pressure and the range of shear stress applied to the cells located in the upper channel. Ideally, transmembrane pressure should be maximized to drive mass transport to the lower channel to enable secreted protein biosensing, but at low enough $Q_{\Delta P_T=0}$ to achieve physiological shear stress, τ_0. Fig. 3 shows a plot of τ_0 (Eq. 3)

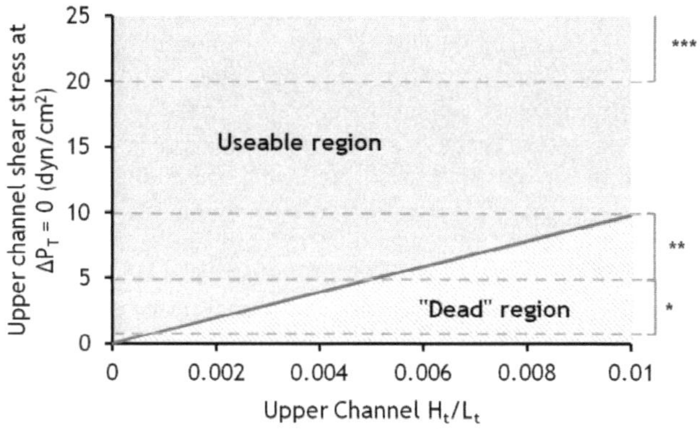

Figure 3: Shear stress in upper channel for different upper channel height-to-length ratios such that $\Delta P_T = 0$. The condition for mass transport to the lower channel $\Delta P_T > 0$ is met is only in the useable region, enabling both biosensing and shear stress application device functionality.

for different upper channel height-to-length ratios, and is divided into two regions: a useable shear stress region where conditions for mass transport to the biosensor in the lower channel are met, and a dead shear stress region where conditions for mass transport to the biosensor are not met and is thus a region where device functionality is limited. The plot also demonstrates that if an appropriate upper channel height-to-length ratio h_t/L_t is selected, physiological levels of vascular shear stress may be generated from low 1-5 dyn/cm^2 post-capillary venule shear stress (Boisseau et al., 2005) to higher 40-60 dyn/cm^2 arterial/capillary shear stress (Wong et al., 2012; Boisseau et al., 2005).

Mass transport efficiency was also considered via simulation of TNF-α transport through a subconfluent aortic endothelial monolayer, as determined by Young et al. (2010). Fig. 4a shows the simulated amount of time, i.e., the time constant, required for the concentration of analyte to reach 63% of the final saturating concentration C_{sat} (here 1 nM) in the region of interest for increasing transmembrane pressure values as well as for different lower channel heights h_b. The region of interest is defined as the surface of the lower channel located under the half of the porous membrane experiencing the largest transmembrane pressures, which corresponds to the half nearest to the upper channel inlet. Indeed, transport of the analyte to the bottom surface of the bottom channel is non-uniform along the length of the membrane, with the distribution and rate of accumulation demonstrating different sensitivities to channel dimensions and transmembrane pressure. For example, the time constant for analyte accumulation in the bottom channel is non-linearly dependent on the transmembrane pressure (Fig. 4a). Lowering microchannel height also significantly increases the rate of analyte accumulation in the bottom channel at lower transmembrane pressures and thus decreases the time constant. At a transmembrane pressure of zero, despite

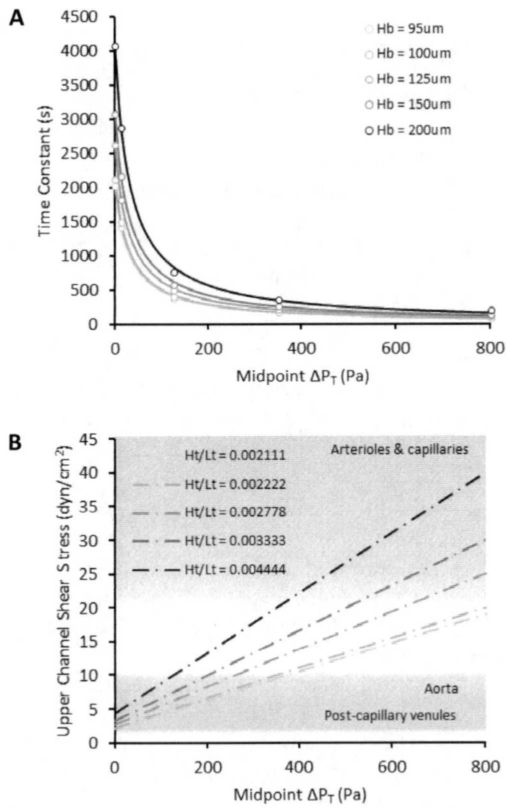

Figure 4: A) Device mass transport effectiveness at different ΔP_T for different bottom channel heights. The time constant is defined as the time necessary to attain 63% of the saturating concentration, here 1 nM. The data points on the plot were determined using the COMSOL model while the lines represent curve fits between data points. B) Shear stresses in upper channel for ΔP_T in panel A remain in physiologically-relevant range.

observing analyte transport into the lower channel, transport across the membrane is purely diffusive, which leads to impractical time constant values that are not conducive to real-time or continuous sampling biosensing. However, as transmembrane pressure increases, time constant values decrease to values that allow for continuous sampling to be performed at time intervals on the order of minutes. The time constant values also trend towards convergence for different lower channel heights at high transmembrane pressures. Fig. 4b also shows that predicted upper channel shear stresses remain within the wide range of interest in vascular biology, demonstrating that both application of physiological shear stress to cells and effective mass transport to the lower channel are possible within the device. By adjusting the upper channel height-to-length ratio, it is possible to modify the range of shear stresses achievable for a given range of upper channel flow rates. The transmembrane pressures and shear stresses shown in Fig. 4 do not represent upper limits for either quantity. Rather, the upper limits of transmembrane pressure and shear stress depend only on physical pumping capabilities.

CONCLUSIONS

Microfluidic technologies offer great potential as in vitro models of complex physiological systems, like the vasculature. The utility of these systems for a wide variety of applications can be improved by the incorporation of on-chip biosensing. For detection of cell-secreted proteins in dynamic flow environments, like the vasculature, careful consideration must be given to platform and sensor design to maximize sensing sensitivity and robustness. The preceding analysis establishes preliminary guidelines for the selection of design parameter values that are critical to shear stress generation and mass transport, including upper and lower channel dimensions and upper channel flow rates, such that on-chip measurements of flow-dependent protein production may be performed. In on-going work, we are testing the effects of a broader range of design parameters on sensor performance and incorporating additional model features, including improved models of the cellular-secreted analyte source and of the biosensor, considering sensor dimensions and binding kinetics. Design principles suggested by our modelling efforts will be implemented and tested experimentally to define system designs that meet the multiple criteria and constraints imposed by our application.

ACKNOWLEDGMENTS

Microfluidics work in the Simmons Lab is supported by CIHR, NSERC, HSF, and the CRC in Mechanobiology.

REFERENCES

1. Wong, K.H.K., Chan, J.M., Kamm, R.D., and Tien, J. 2012, "Microfluidic models of vascular functions", *Annual Review of Biomedical Engineering*, 14, pp. 205-230.

2. Moraes, C., Sun, Y., and Simmons, C.A. 2009, "Solving the shrinkage-induced PDMS alignment registration issue in multilayer soft lithography", *Journal of Micromechanics and Microengineering*, 19, 065015.

3. Young, E.W., Watson, M.W., Srigunapalan, S., Wheeler, A.R., and Simmons, C.A. 2010, "Technique for real-time measurements of endothelial permeability in a microfluidic membrane chip using laser-induced fluorescence detection", *Analytical Chemistry*, 82, pp. 808–816.

4. Srigunapalan, S., Lam, C., Wheeler, A.R., and Simmons, C.A. 2011, "A microfluidic membrane device to mimic critical components of the vascular microenvironment", *Biomicrofluidics*, 5, 13409.

5. Chen, M.B., Srigunapalan, S., Simmons, C.A. 2013, "A 3D microfluidic platform incorporating methacrylated gelatin hydrogels to study physiological cardiovascular cell–cell interactions", *Lab on a Chip*, 13, pp. 2591-2598.

6. Huang, N-T., Chen, W., Oh, B-R., Cornell, T.T., Shanley, T.P., Fu, J., and Kurabayashi, K. 2012, "An integrated microfluidic platform for in situ cellular cytokine secretion immunophenotyping", *Lab on a Chip*, 12, pp. 4093-4101.

7. Boisseau, M.R. 2005, "Roles of mechanical blood forces in vascular diseases. A clinical overview", *Clinical Hemorheology and Microcirculation*, 33, pp.201-207.

A NOVEL MICROFLUIDIC SWITCH DESIGN WITH RESPONSE CHARACTERISTICS TO HIGH ACCELERATION

Weirong Nie*, Huang Liu
Mechanical Engineering School
Nanjing University of Science and Technology
Nanjing, Jiangsu, China
njniewr@hotmail.com*, 448470991@qq.com

Liying Jiang
Department of Mechanical and Materials Engineering
The University of Western Ontario
London, Ontario, Canada
lyjiang@eng.uwo.ca

ABSTRACT

This paper proposes a novel microfluidic acceleration switch model with excellent response characteristics to high acceleration over 500g. The microfluidic switch works on the principles of micro fluidic dynamics and communicating vessels. Brine is selected as the function-fluid due to its conductivity, anti-dispersion capacity, and non-toxicity. Fluidic volume motion equation in the micro channel has been derived by modifying the Bernoulli equation with the incorporation of the brine viscous force. The structure conception and the theoretical model has been validated by a three-dimensional simulation model through FLUENT6.3 with VOF-CSF module and post-processed by TECPLOT360. The dynamic responses of the switch are studied under impulse acceleration with amplitude from 300g to 1000g. The threshold acceleration of the simulation model is between 470g to 480g, which is very close to the theoretically predicted threshold 476.9g. It is shown that the bigger acceleration amplitude leads to the shorter response time. Fluid dispersion problem is avoided in all the working processes.

KEYWORDS: Microfluidic, acceleration switch, communicating vessels, anti-dispersed

INTRODUCTION

Micro acceleration switch, as one kind of micro-volume inertial devices, is designed to sense the acceleration, which will touch the electrodes when the acceleration exceeds the threshold of the device.[1]Such kind of switches have been widely used in aeronautics and astronautics, automobile industry, medical apparatus and instruments, weapon systems etc.[2] Microfluidic acceleration switch is a relatively new type of acceleration switch, which replaces the traditional elastic solid electrode with the moving droplet in the micro-channel to sense the acceleration and make the signal electrodes connected. In conventional microfluidic acceleration switch, fluid dispersion problem often emerges when it suffers high acceleration loading, [3] which is believed to attribute to the fact that the whole droplet in the channel moves in the same direction with the loaded acceleration signal. The moving droplet in the micro-channel can be treated as a single slug liquid column. In order to maintain the integrity of the droplet, there must be strong enough pressure acting on the air interface. However, when the whole droplet moves in the same direction with an acceleration load, the droplet will disperse and mix with the air once the air pressure and surface tension aren't strong enough, leading to the fluid dispersion problem. [5] This situation makes the microfluidic acceleration switch lack of the performance with higher threshold.

To solve this dispersion problem, we designed a microfluidic acceleration switch by combining the communicating vessels, capillary valve and contact electrodes, which can automatically equalize the pressure of the liquid column. No matter which direction the liquid column is moving, there always exists back pressure acting on the liquid column. This back pressure is caused by the input acceleration, which is a negative feedback in the control mechanism. The higher of input acceleration, the stronger back pressure is exerted. The strong negative feedback can greatly improve the anti-dispersion capability of the function fluid. Thus the designed acceleration switch is characterized with high anti-dispersion capability, easy threshold adjustment, reliable electrical connection and small structure size.

STRUCTUREDESIGN

Fig.1 shows the diagram of the microfluidic acceleration switch, which consists of two main components, base plate and cover plate. The base plate is made of glass, and the gold electrodes are evaporated and developed on its bonding side. The cover plate is made of PDMS and embedded with rectangle cross-section micro groove and micro valve structure. A closed micro-channel is formed by the bonding the base plate and the cover plate along their bonding sides. Surface modification will be conducted on the base plate and the cover plate to form hydrophobic surfaces to the brine. A volume of brine is injected into the micro-channel to form a liquid column.

The microfluidic acceleration switch working principle is shown in Fig 2. The closed micro-channel consists of valve 1, channel 1, valve 2 and channel 2, which form U-shape communicating vessels. Valve 1 and valve 2 are also called capillary valve since they are designed with very tiny cross-section and specific transition shape so that the capillary force is sharply magnified when fluid passes through the valve. This enlarged capillary force has resistant effect to the brine column when it flows through the capillary valves. To go through the capillary valve, the brine column must overcome the resistance from capillary force.

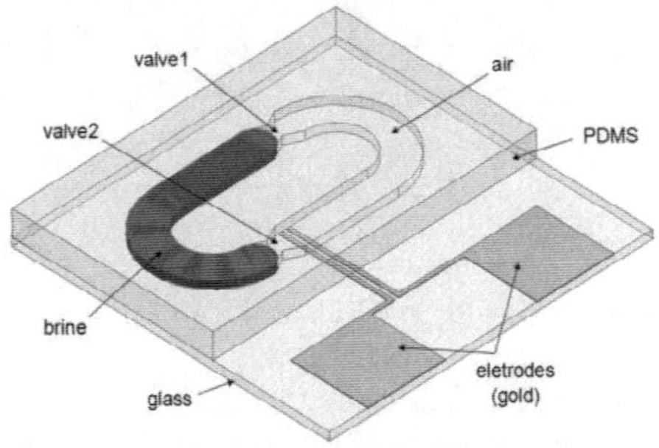

Fig 1. Three-dimensional diagram of the microfluidic acceleration switch

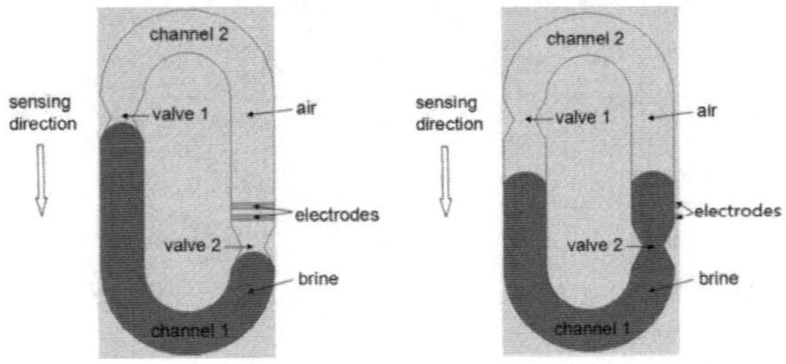

(a) off-state (b) on-state

Fig 2. The microfluidic switch working schematics

At the initial stage, the brine stays only in channel 1 and the switch is in off-state, as shown in Fig.2(a). The brine column will have the tendency to flow through valve 2 when the acceleration is loaded in the sensing direction of the switch. If the acceleration is strong enough, the brine column will pass through valve 2 and flow into channel 2. Once the brine column flows into channel 2, the brine column height in channel 1 and channel 2 will be equal according to the communicating vessels principle. This situation leads to the connection of the electrodes since the brine submerges the electrodes. The switch will thus be hold in on-state as shown in Fig.2(b).

THEORETICAL MODELING

For a single slug fluid flow in the micro-channel, when the fluid column length is much larger than the channel cross-sectional dimension, the microfluidic state can be described as non-compressible layered viscous flow[4] [8]. It is known that the Bernoulli equation can be applied to express the motion of non-viscous fluid in micro-channels. Considering the scale effect of the microfluidic devices, viscous force cannot be ignored, so it is essential to revise the Bernoulli equation considering the viscous force. With the two interfaces between the slug fluid and air labeled as A and B, the transient revised (or modified) Bernoulli equation with the consideration of the viscous force for the flow from A to B is expressed as

$$\frac{1}{2}\rho q_A^2 + p_A + \rho a h_A = \frac{1}{2}\rho q_B^2 + p_B + \rho a h_B + \rho\int_A^B \frac{dq}{dt} ds + p_f \quad (1)$$

Where ρ is the density of the fluid; q is the average velocity of the fluid cross-section; q is the flow rate ; p is the pressure on the fluid slug side; h is the height of interface with respect to the reference plane; a is the acceleration load; s is axial length of channel and p_f is revised viscous force. The subscripts "A" and "B" represent the sides A and B, respectively.

The value of viscous force p_f is influenced by the fluid viscosity μ , flow rate q, size and shape of the channel cross-section represented by the size coefficient k and the cross-section shape coefficient f. The revised viscous pressure can be expressed as

$$p_f = \mu\int_A^B kfq ds \quad (2)$$

The size coefficient k represents the effect of the characteristic scale of the micro-channel, which is a constant and can be calculated by the fitting method. The cross-section shape coefficient f is a factor representing the effect of cross-section shape with a dimension of L^{-2} . According to the dimensionless principle, both transient and steady flows have the same cross-section shape coefficient.

For a rectangular cross-section channel defined by width w and depth b , when the layered effect of the viscous fluid is considered, the average flow rate of the steady flow in the channel is expressed as empirical formula[9]

$$q = \frac{w^2}{12\mu}\frac{dp}{ds}\left[1 - \frac{192w}{\pi^5 b}\tanh\left(\frac{\pi b}{2w}\right)\right] \quad (3)$$

Where dp/ds is flow channel pressure drop on the steady flow. Correspondingly, the viscous pressure between interface A and B can be also determined as

$$p = \int_A^B \frac{dp}{ds} ds = \mu\int_A^B \frac{12\pi^5 bq}{w^2\left[\pi^5 b - 192w\tanh\left(\frac{\pi b}{2w}\right)\right]} ds \quad (4)$$

Moreover, by comparison of Eq.(2) with Eq.(4), the integral factor in Eq.(2) of size coefficient k , flow rate q and cross-section shape coefficient f can be expressed as

$$kfq = \frac{12\pi^5 bq}{w^2\left[\pi^5 b - 192w\tanh\left(\frac{\pi b}{2w}\right)\right]} \quad (5)$$

A simplified physical model of the microfluidic acceleration switch is demonstrated in Fig 3, with the red part representing the brine and the blue part representing the air. The depth of the micro channel is T, the width of the interface A is w_1 , and the width of interface B is w_2 . Ignoring the effect of the interface shape due to its micro area, the shape of interfaces A and B is simplified as a plane.

Due to the micro channel hydrophobicity, the capillary force is directed to the brine.

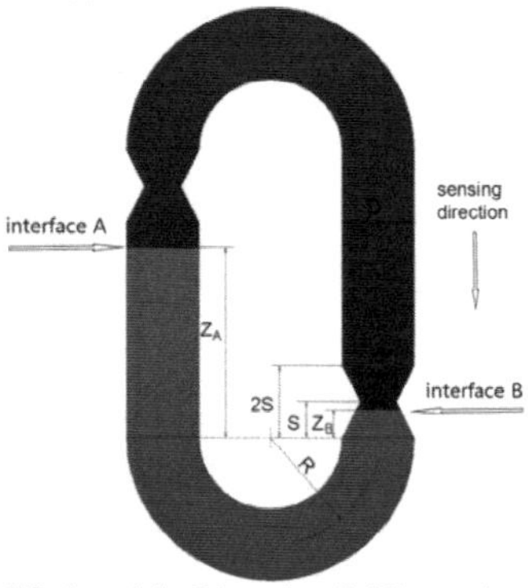

Fig 3. Simplified model of the microfluidic acceleration switch

According to Eq. (1), the brine flow from interface A to interface B can be described by

$$\frac{1}{2}\rho_1 q_A^2 + p_A + \Delta p_A + \rho_1 a z_A =$$

$$\frac{1}{2}\rho_1 q_B^2 + p_B + \Delta p_B + \rho_1 a z_B + \rho_1 \int_A^B \frac{dq}{dt} ds + \mu_1 \int_A^B k q_1 f ds \tag{6}$$

Where ρ_1 is the density of the brine, μ_1 is viscosity of the brine, and z_A, z_B represent the height of the brine on the A and B sides. p_A, p_B is pressure on the air side, $\Delta p_A, \Delta p_B$ is capillary force.

Similarly, the air flow from interface B to interface A can be expressed as

$$\frac{1}{2}\rho_2 q_B^2 + p_B + \rho_2 a z_B = \frac{1}{2}\rho_2 q_A^2 + p_A + \rho_2 a z_A +$$

$$\rho_2 \int_B^A \frac{dq}{dt} ds + \mu_2 \int_B^A k q_2 f ds \tag{7}$$

Where ρ_2 is density of the air, and μ_2 is viscosity of the air. Considering the density and viscosity of the air is much smaller than the brine, the effect of air can be ignored with assuming $\rho_2 = 0$ and $\mu_2 = 0$, leading to $p_B = p_A$ under this flow condition.

According to Eq.(6), the motion equation of the microfluidic acceleration switch can be expressed as

$$\rho_1 \int_A^B \frac{dq}{dt} ds = \rho_1 a (z_A - z_B) + (\Delta p_A - \Delta p_B) + \frac{1}{2}\rho_1 (q_A^2 - q_B^2) - \mu_1 \int_A^B k q_1 f ds \tag{8}$$

Without considering the effects of contact angle hysteresis, the capillary force of interface A and B can be expressed as[5]

$$\Delta p_A = \frac{2\sigma (w_1 + T)\cos(\pi - \theta)}{T w_1} \tag{9}$$

$$\Delta p_B = \frac{2\sigma D \cos(\pi - \theta) + 2\sigma T \cos(\pi - \theta + \alpha)}{T w_2} \tag{10}$$

Where θ is the contact angle, and α is angle of channel shave. The angle of channel shave α has the different value, for example, $\alpha = 0$ for straight channel, $\alpha < 0$ for shrinkage channel and $\alpha > 0$ for expansion channel. Here α is given the designed value of 26.6° .

The maximum capillary pressure appears when interface B reaches the neck of valve 2, while brine suffers maximum resistance. Switch static threshold of input acceleration load can be expressed as

$$g_{threshold} = \frac{\Delta p_B\big|_{w_2 = d} - \Delta p_A}{\rho_1 (z_A - S)} \tag{11}$$

STRUCTURE AND MODEL VALIDATION

To verify the structure concept and the theoretical model, a three-dimensional simulation model of the microfluidic acceleration switch has been built. The structure parameters of the simulation model are shown in Table 1.

Table 1. Structure parameters of the simulation model

Name	Value
R	150um
D	100um
d	50um
T	100um
S	50um
Brine's Density	1000kg/m^3
Brine's Viscosity	0.001
Air's Density	1.225kg/m^3
Air's Viscosity	0.000018
Surface Tension	0.075
Contact Angle	120°

FLUENT6.3 was adopted to simulate the dynamic response of the model. Its VOF-CSF module was used to describe the motion of the two brine-air interfaces A and B. The effect of contact angle hysteresis is not considered. The impulse acceleration is loaded on the sensing direction of the switch, and the amplitude of the acceleration ranges from 300g to 1000g. TECPLOT360 was used to post-process the simulation data. The working animations and dynamic response curves were generated. By analyzing all of the working processes, no fluid dispersion phenomenon was found. It means that the introduction of the communicating vessels structure can greatly improve the anti-dispersion ability of the microfluidic acceleration switch. The switch threshold of impulse acceleration is between 470g to 480g, which is very close tothe theoretical calculation threshold of 476.9g. When the microfluidic acceleration switch is subjected to the acceleration load with amplitude over than the threshold, the brine fluid will pass through valve 2 and the electrodes will be connected. The bigger acceleration loading, the shorter response time in which fluid flows into channel 2 and submerges the electrodes.

Fig 5(a) shows the displacement-time curves of the interface B at acceleration 500g theoretically and in simulation respectively, the curves are matched very well.

Fig.5(b) shows the average displacement-time curves of the interface B, which are corresponding to acceleration amplitude of 300g, 400g, 500g, 700g, 900g and 1000g, respectively. The displacement-time response curves with acceleration amplitude of 300g and 400g are under displacement 100 μm, which is the set height of valve 2. It means that the brine cannot pass through valve 2 to connect the electrodes under those loads. However, the curves corresponding to the acceleration amplitude of 500g, 700g, 900g and 1000g all reach over than 100 μm after a time period, indicating that the brine flows through the capillary value 2 resulting in the connection of the electrodes. Thus, the switch will transfer from an off-state to an on-state when it suffers an acceleration with amplitude larger than a threshold, while the switch is still kept in the original off-state when it experiences an acceleration with amplitude lower than the threshold.

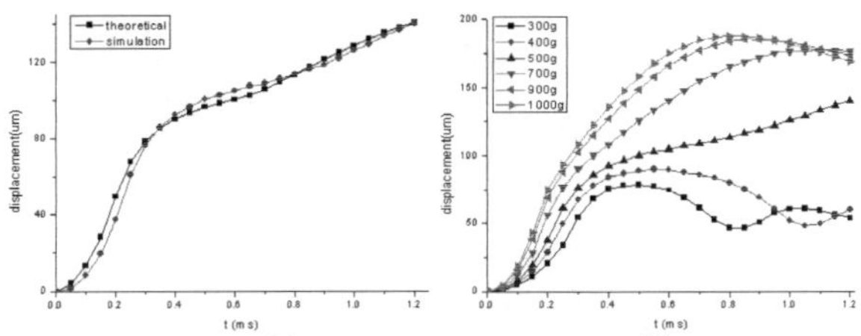

(a) (b)

Fig. 5 Displacement-time response curves

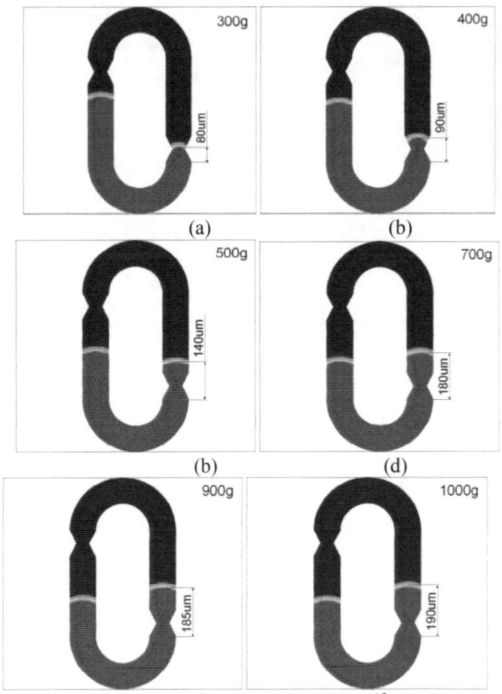

(a) (b) (b) (d) (e) (f)

Fig 6. The contours of the microfluidic switch

With different acceleration loads, interface B will reach its different maximum displacements, which are shown in Fig 6. It can be observed that the contours of the microfluidic column have different position corresponding to different accelerations. It indicates that the switch has the ability of distinguishing the acceleration amplitude if the electrodes arrays are set at special position.

CONCLUSIONS

This paper presents a newtype microfluidic inertial switch with working capacity over 500g high acceleration, which could be manufactured with low cost by MEMS technology due to its simple structure. The design is based on the microfluidic mechanics and communicating vessels principle. In the design process, Bernoulli equation is modified by introducing viscous force to solve the fluid dispersion problem of non-compressible layered viscous flow. The design model is validated with the modeling and simulation of the microfluidic acceleration switch. Simulation results also show that the theoretically designed model is applicable when the acceleration amplitude is bigger than the calculated threshold. The communicating vessels structure design can greatly help to improve the anti-dispersion ability of the function fluid. This novel of microfluidic acceleration switch can be used as an one-shoot trigger switch with characteristics of stable connection and hold-on.

ACKNOWLEDGMENTS

This research work is supported by China National Nature Science 51475245.

REFERENCES

1. Liu Tingting, Su Wei, Yang Tao, and Xu Yuan, "Vibration interference analysis and verification of micro-fluidic inertial switch", *AIP Advances* 4, 031313.2014.
2. Kwanghyun Yoo, Usung Park, Joonwon Kim, "Development and characterization of a novel configurable MEMS inertial switch using a microscale liquid-metal droplet in a microstructured channel", *Sensors and Actuators A,* 166 (2011), 234-240.
3. Hansang Cho,Ho-Young Kim, Ji Yoon Kang,Tae Song Kim, "How the capillary burst microvalve works", *Journal of Colloid and Interface Science*, 306 (2007) 379-385.
4. Ho-Young Kim, Heon Ju Lee and Byung Ha Kang, "Sliding of Liquid Drops Down an Inclined Solid Surface", *Journal of Colloid and Interface Science*, 247(2002), 372-380.
5. Yanying Feng, Zhaoying Zhou, Xiongying Ye, Jijun Xiong,"Passive valves based on hydrophobic microfluidics", *Sensors and Actuators A*, 108 (2003) 138-143.
6. Tzong-Shyng Leu, Pei-Yu Chang, "Pressure barrier of capillary stop valves in micro sample separators", *Sensors and Actuators A*, 115 (2004) 508-515.
7. Alain Gliere, Cyril Delattre, "Modeling and fabrication of capillary stop valvesfor planar microfluidic systems", *Sensors and Actuators A*, 130-131 (2006) 601-608.
8. Abraham E. Dukler, Matin G. Hubbard, "A Model for Gas-liguid slug flow in horizontal and near horirontal tubes". DOI:10.1021/i160056a011*Ind. Eng. Chem., Fundmn*, Vol(14), No.4,1975 337-347
9. Yongxiang Lu, "Hydraumatic and Pneumatic Technical Mannual",China Machine Press, 2002. 20-21

Multibody/Vehicle Dynamics

Proceedings of the 25th CANCAM
London, Ontario, Canada, May 31 – June 4, 2014

Automated Design Synthesis of LQR Controllers for Road Vehicle Active Suspension Systems

Eungkil Lee* and Yuping He

Department of Automotive, Mechanical and Manufacturing Engineering
University of Ontario Institute of Technology, Oshawa, Ontario, Canada
**Corresponding Author: eungkil.lee@uoit.ca*

ABSTRACT

This paper presents an automated design method for road vehicle active suspension controllers. In order to improve ride quality of road vehicles, controllers based on the Linear Quadratic Regulator (LQR) technique have been explored for active suspension systems of road vehicles. In conventional designs of LQR controllers for road vehicle active suspension systems, the weighting factors of the corresponding performance index are determined using trial and error method; the governing equations of motion for vehicle models are derived by hand. This is a tedious and time consuming process. To address this problem, an automated design method is proposed: the vehicle system model is generated using a multibody system software package, a genetic algorithm is utilized to search optimal weighting factors for the LQR controller, and the design optimization is implemented via parallel computing. To demonstrate the effectiveness of the proposed design method, it is applied to the LQR controller design for active suspensions of a road vehicle represented by a quarter-vehicle model. The simulation results show that the resultant LQR controller derived by the proposed method outperforms its counterpart based on conventional trial and error method in terms of displacement and vertical acceleration of sprung mass and suspension deflection. Moreover, the parallel computing technique significantly reduces numerical optimization time.

KEYWORDS: automated design, multibody system modelling, genetic algorithm, LQR controller, parallel computing, active suspensions, quarter-vehicle model.

1. INTRODUCTION

In active suspension controller design, a road vehicle model is usually generated by hand, and the control variables are tuned manually [1]. In order to relieve manual efforts of designers from the tedious and time consuming process, an automated design approach has been reported, and its effectiveness was demonstrated via numerical simulations [2]. However, in the LQR controller design reported in [2], the weighting factors of the corresponding performance index were normalized based on a reference design. In other words, the weighting factors were predetermined prior to the design optimization. Thus, whether the performance of the resultant LQR controller is optimal is questionable. Moreover, to ensure the robustness of the reported design method, a genetic algorithm (GA) was introduced, but the GA significantly degraded the computation efficiency of the model-based design method [2].

In order to address the problems arisen from the design method reported in [2], the present research proposes a new automated design method for road vehicle active suspension controllers: the vehicle system model is generated using a multibody system software package, EoM [3], a GA is utilized to search optimal weighting factors for the LQR controller, and the design optimization is implemented via parallel computing. To demonstrate the effectiveness of the proposed design method, it is applied to the LQR controller design for active suspensions of a road vehicle represented by a quarter-vehicle model. This paper only concentrates on optimization of control weighting factors. Optimization of suspension parameters is already considered in [2].

The rest of the paper is organized as follows. In section 2, the quarter-vehicle model is automatically generated using EoM and the model is validated via the hand-derived counterpart. In section 3, the LQR controller is introduced. The automated design of the LQR controller is described, and the parallel optimization implementation is outlined in section 4. In section 5, selected numerical simulation results are presented and discussed. Finally, conclusions are drawn in section 6.

2. VEHICLE SYSTEM MODELLING AND VALIDATION

The quarter-vehicle model is shown in Figure 1. And the two degrees of freedom (DOF) model can be described in terms of governing equations of motion as follow:

$$m_b\ddot{z}_2 = -k_s(z_2 - z_1) - c_s(\dot{z}_2 - \dot{z}_1) - U \quad (1)$$

$$m_w\ddot{z}_1 = -k_t(z_1 - q) + k_s(z_2 - z_1) + c_s(\dot{z}_2 - \dot{z}_1) + U \quad (2)$$

where m_b is the sprung mass, m_w is the unsprung mass, z_2 is the displacement of the sprung mass, z_1 is the displacement of the unsprung mass, q is the disturbance of the road, k_s and c_s are the coefficients of the spring stiffness and damping, respectively, k_t is the stiffness of the tire, and U is the control force.

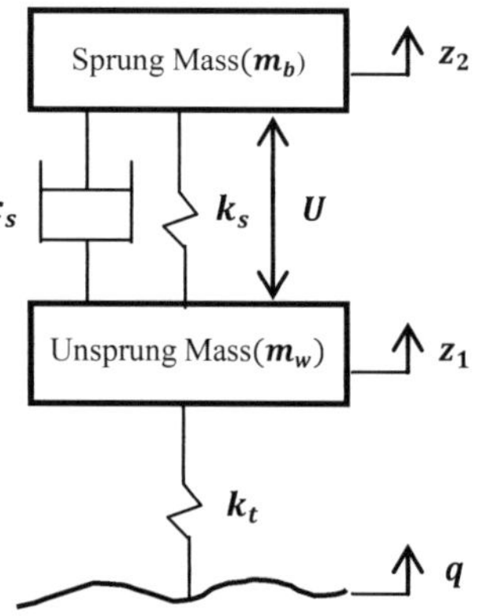

Figure 1 Quarter-car suspension model

Assume that the state variables x are defined as $[\dot{z}_2, \dot{z}_1, z_2, z_1]$. Thus, the state-space equation can be rewritten as,

$$\begin{aligned} \dot{x} &= Ax + B_c U + Bq \\ y &= Cx + DU \end{aligned} \tag{3}$$

where,

$$A = \begin{bmatrix} -c_s/m_b & c_s/m_b & -k_s/m_b & k_s/m_b \\ c_s/m_w & -c_s/m_w & k_s/m_w & -(k_t + k_s)/m_w \\ 1 & 0 & 0 & 0 \\ 0 & 1 & 0 & 0 \end{bmatrix}$$

$$B_c = \begin{bmatrix} -1/m_b \\ 1/m_w \\ 0 \\ 0 \end{bmatrix}, \quad B = \begin{bmatrix} 0 \\ k_t/m_w \\ 0 \\ 0 \end{bmatrix}$$

$$C = \begin{bmatrix} -c_s/m_b & c_s/m_b & -k_s/m_b & k_s/m_b \\ 0 & 0 & 1 & 0 \\ 0 & 0 & 1 & -1 \\ 0 & 0 & 0 & 1 \end{bmatrix}$$

$$D = \begin{bmatrix} -1/m_b \\ 0 \\ 0 \\ 0 \end{bmatrix} \tag{4}$$

Parameters of quarter-vehicle model are selected as: $m_b = 504.5kg$, $m_w = 62kg$, $k_s = 13100\ N/m$, $c_s = 400\ Ns/m$, $k_t = 252000\ N/m$ [4].

EoM software is designed to run in MATLAB. The software package is used to generate the quarter-vehicle model for automatic design process. To validate the vehicle model derived from EoM, this model is compared with its counterpart generated by hand. The selected simulation results based on the two models are shown in Figure 2, which indicate that the dynamic responses of the two models under the excitation of road irregularities are identical.

The road profile $q(t)$ representing a single road bump, which acts as disturbance, is given by the following equation:

$$q(t) = \begin{cases} 0.25(1 - \cos(8\pi t)) & 0 \le t \le 0.25 \\ 0 & otherwise \end{cases} \tag{5}$$

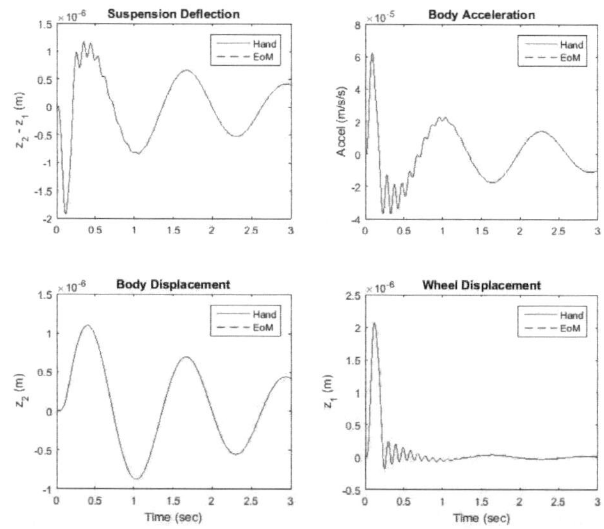

Figure 2: Dynamic responses of the quarter-vehicle models generated by EoM and derived by hand

3. LQR Controller Design

The LQR controller is designed for the active suspension system of the vehicle represented by the quarter-vehicle model shown in Figure 1. Consider a state variable feedback regulator for the system expressed as

$$u(t) = -Kx(t) \tag{6}$$

where K is the feedback control gain matrix. It has great influence on the control effects. The optimization procedure is to obtain optimal K value which minimizes the performance index **J** [5]. The **J** represents by

$$\mathbf{J} = \int_0^\infty (x^T(t)Qx(t) + u^T(t)R\,u(t))\,dt \tag{7}$$

where $u(t)$ is control input, and Q and R are weighting matrices.

The control gain matrix K is represented by

$$K = R^{-1}B^TP \tag{8}$$

where the matrix P is found from the Algebric Riccatiu Equation (ARE) given as

$$A^TP + PA - PBR^{-1}B^TP + Q = 0 \tag{9}$$

In this paper, ***lqr*** function in MATLAB is used to determine K matrix. For the LQR controller, the weighting matrices Q and R have significant influence on the performance of the LQR controller. In general, these are determined using trial and error method. However, using trial and error method to find optimal weighting matrices is time consuming, and the system may not be tuned to its optimum. The proposed design optimization method will be used to determine the ideal weighting matrices in order to ensure that the resultant LQR controller achieve optimal performance.

4. PROPOSED AUTOMATED DESIGN OPTIMIZATION

In the proposed automated design method, a GA is used as the search algorithm. In order to improve vehicle's ride quality and enhance handling performance, the active suspension controller is designed to achieve the minimum vertical acceleration and displacement for the sprung mass, displacement for the unsprung mass and suspension deflection under a given road irregularity excitation. To this end, the design variables are defined as q_1, q_2, q_3, q_4 *and* r, which constitute the weighting matrices as

$$Q = \begin{bmatrix} q_1 & 0 & 0 & 0 \\ 0 & q_2 & 0 & 0 \\ 0 & 0 & q_3 & 0 \\ 0 & 0 & 0 & q_4 \end{bmatrix}, \quad R = [r] \tag{10}$$

The design optimization is to find optimal values of the design variables in order to minimize the following objective function

$$f_{obj} = \frac{\varphi_1}{\varphi_1^p} + \frac{\varphi_2}{\varphi_2^p} + \frac{\varphi_3}{\varphi_3^p} + \frac{\varphi_4}{\varphi_4^p} \tag{11}$$

where $\varphi_1, \varphi_2, \varphi_3$ and φ_4 are the Root Mean Square (RMS) of the vertical acceleration of sprung mass, displacement of sprung mass, suspension deflection and displacement of unsprung mass with LQR controller respectively. $\varphi_1^p, \varphi_2^p, \varphi_3^p$ and φ_4^p are the RMS of the passive suspension system relative to $\varphi_1, \varphi_2, \varphi_3$ and φ_4 respectively. As shown in equation (11), the objective function is normalized by the RMS of the corresponding passive suspension system [6].

It is well-known that design optimization based on GA involve a significant number of cost function evaluation. In the design case concerned, the optimization includes a huge number of dynamic simulations of the vehicle model and performance measure evaluations. The computation efficiency of the automated design process is of importance. To ensure high computation efficiency of the automated design, the optimization is implemented on the Master/Slave Parallel Computer System in the Multidisciplinary Vehicle Systems Design (MVSD) laboratory at the University of Ontario Institute of Technology (UOIT) [7].

5. SIMULATION RESULTS

To demonstrate the effectiveness of the proposed design method, the LQR controller is designed by determining the weighing factors based on two approaches: 1) trial and error method, call LQR, and 2) the proposed automated design method, namely GA-LQR.

Table 1 offers the resulting weighting factors derived by the two methods, i.e. trial and error method and the proposed method.

Table 1: The weighting factors determined by the two methods

Design Parameters	GA-LQR	LQR	Bounds
q_1	9.55×10^6	6×10^5	$[0,\ 10^7]$
q_2	0.024	30	$[0,\ 10^7]$
q_3	9.62×10^6	5×10^4	$[0,\ 10^7]$
q_4	4.04×10^6	2×10^6	$[0,\ 10^7]$
r	0.0103	0.01	$[0,\ 10]$

The road disturbance is simulated by a single road bump as equation (5). As shown in Figures 3, 4 and 5, the resulting controller design based on GA-LQR outperforms that derived from LQR in terms of the vertical acceleration and displacement of the sprung mass and suspension deflection.

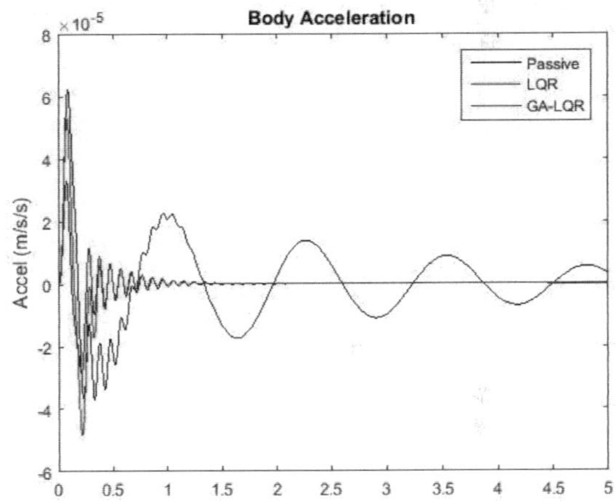

Figure 3: Body Acceleration

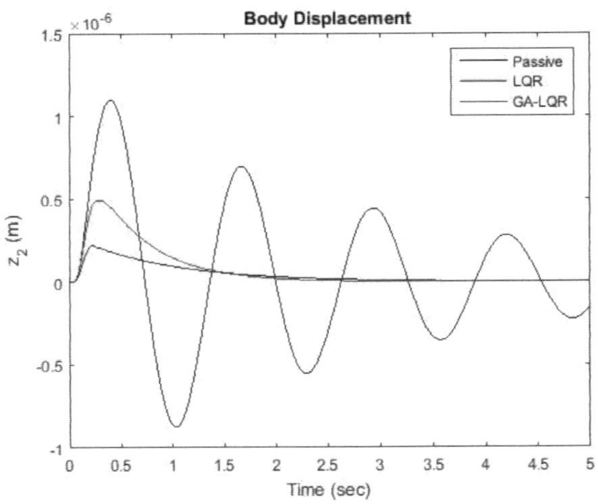

Figure 4: Body Displacement

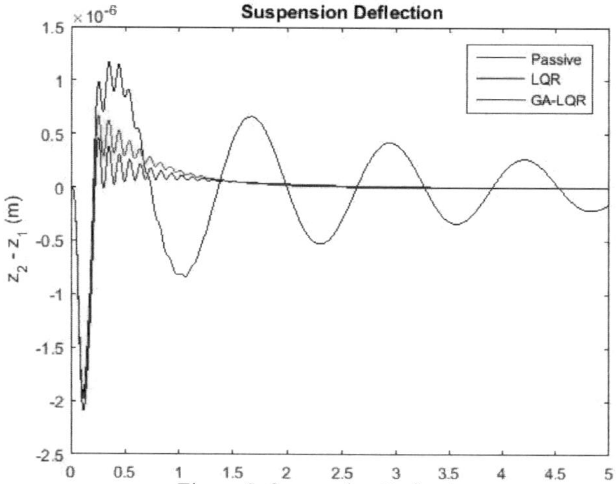

Figure 5: Suspension Deflection

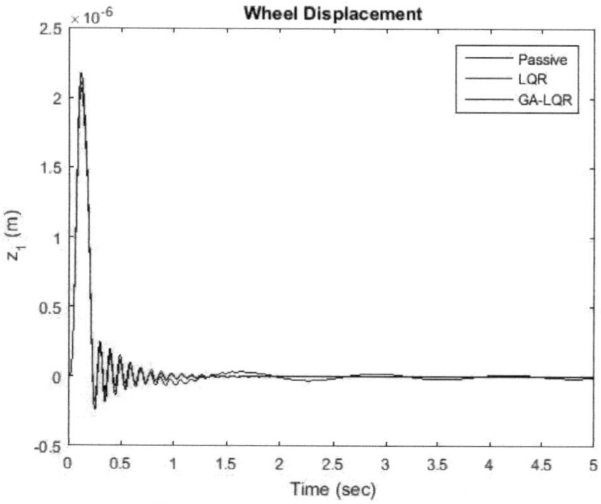

Figure 6: Wheel Displacement

As shown in Figure 6, displacement of the unsprung mass with GA-LQR gives similar amplitude with LQR and passive but settling time is faster than passive.

Table 2 shows the simulation times for the optimization using the GA based on single desktop computer and on the parallel computing system, respectively. It is clear that the parallel computing method is much faster than the regular method.

Table 2: Simulation Time Comparison

Type	Simulation Time
GA	648.731 seconds
GA-Parallel (4 workers)	186.183 seconds

6. Conclusions

This paper proposes an automated design method for controllers of road vehicle active suspensions. With the proposed method, the vehicle modeling and controller design variable selection are automatically implemented by a multibody system software package and a parallel genetic algorithm technique, respectively. Simulation results demonstrate that the proposed method is feasible, effective, computational efficient. This method can be applied to develop real-world vehicle active safety systems, e.g. electronic stability control, torque vectoring, active steering, etc.

Reference

1. Y. He and J. McPhee, "A Review of Automated Design Synthesis Approaches for Virtual Development of Ground Vehicle Suspensions", SAE Paper, 2007-01-0856.
2. Y. He and J. McPhee, "Multidisciplinary Design Optimization of Mechatronic Vehicles with Active Suspensions", Journal of Sound and Vibration, Vol. 283, 2005, pp. 217-241.
3. B.P. Minaker and R.J. Rieveley, "EoM User Guide", University of Windsor Vehicle Dynamics and Control Research Group.
4. K.D, Rao and S. Kumar, "Modelling And Simulation Of Quarter Car Semi Active Suspension System Using LQR Controller", Proc. of the 3rd Int. Conf. on Front. of Intell. Comput. (FICTA) 2014– Vol. 1, Advances in Intelligent Systems and Computing 327.
5. A. Agharkakli, U.S. Chavan and S. Phvithran, "Simulation And Analysis Of Passive And Active Suspension System Using Quarter Car Model For Non-uniform Road Profile", International Journal of Engineering Research and Applications (IJERA), Vol. 2, Issue 5, September- October 2012, pp.900-906.
6. G. Zhang, Z. Fang and L. Shu, "GALQR Optimal Control Method and Applying In The Active Suspension System, School of Mechatronic Engineering, Northwestern Polytechnical University.
7. M. Islam, Y. He, and T. Hu, "A Design Optimization Method for Multi-Trailer Articulated Heavy Vehicles with Active Safety Systems", Proceedings of 23rd International Symposium on Dynamics of Vehicles on Roads and Tracks, 2013, 10 pages.

SYMBOLIC INTEGRATION OF FINITE ELEMENT MODELS WITH DYNAMIC MULTIBODY SYSTEMS

Yiteng Liang, Joydeep Banerjee, John McPhee
Systems Design Engineering
University of Waterloo
Waterloo, Ontario, Canada
Y24liang@uwaterloo.ca

Asaad Sellmann
Mechanical Engineering
RWTH Aachen University
Aachen, Germany
asaad.sellmann@rwth-aachen.de

Orang Vahid Araghi
MapleSoft
Waterloo, Canada
ovahid@maplesoft.com

ABSTRACT

With advances in newer lightweight materials, engineering simulation increasingly requires integration of flexible bodies with multibody systems. Although considerable success has been made by using numerical methods, it still cannot meet the demand for high efficiency and accuracy in the presence of nonlinear dynamics. Symbolic computation shows a strong capability of generating highly-optimized simulation codes for mechatronic systems with rigid bodies. This paper will focus on the fundamentals of symbolic integration of finite element models with multibody system dynamics.

KEYWORDS: Multibody systems, dynamics, flexible bodies, symbolic computation

1. INTRODUCTION

Since the Finite Element Method (FEM) became widely applied to software-based simulation, engineering design has become more accurate and efficient. Although certain software packages such as Nastran and Abaqus show the versatility of FEM, its potential is restrained by numerical computation. The advent of symbolic computation enables faster simulation code, for which the development of symbolic FE modeling is at the initial stages (Shi [10] and Richard [11]). The integration of these flexible-body FE-models is expected to become a powerful and efficient simulation tool in the future. This paper will introduce the fundamentals of symbolic FEM by analyzing a 3D Bernoulli-Euler cantilever beam with beam elements. The results will be compared between different approaches using symbolic computation.

2. FLEXIBLE BEAM

2.1 3-D Beam Elements

Three assumptions are applied in the 3D beam element according to Astley [1]. First, the classical Bernoulli-Euler beam model assumes that the internal energy of the beam is entirely due to bending strains and stresses, so it neglects transverse shear deformations. In this case, the cross-sections remain plane and will be always perpendicular to the longitudinal axis during deformation. Second, for small deformations, axial deformation, bending, and twist can be decoupled and looked at separately. The last one is that the twist is regarded free, and the Saint-Venant method (Cook [2]) is used to establish the torsional stiffness.

Generally, a spatial beam element has 6 degree-of-freedoms (DoFs) (3 displacements and 3 rotations in Cartesian coordinate system) at the end of each node. At local frame of reference, as shown in Fig. 1.1, the origin is located at the center of mass with x-axis along the neutral axis. At node 1, there are three displacement vectors denoted by u_1, u_2, u_3 and three rotation vectors denoted by u_4, u_5, u_6 respectively (Cook [2] and Panzer [6]). It is identical for node 2.

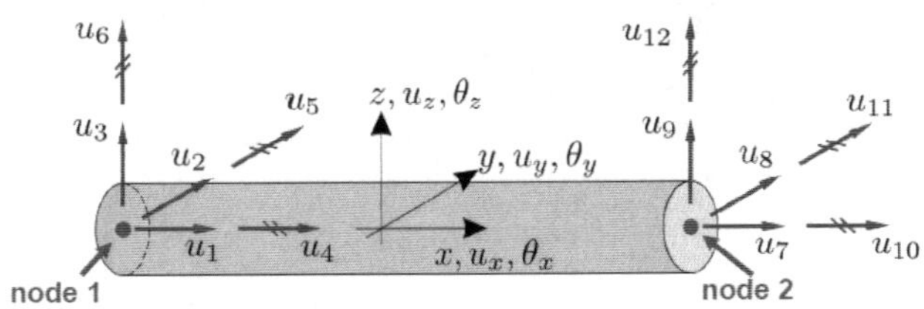

Figure 1.1: DoFs of a 3D beam element in local coordinates (Stærdahl [3])

Classical deformations of node 1 are shown in Fig. 1.2. The DoFs of one node can be clearly observed from the six deformation figures.

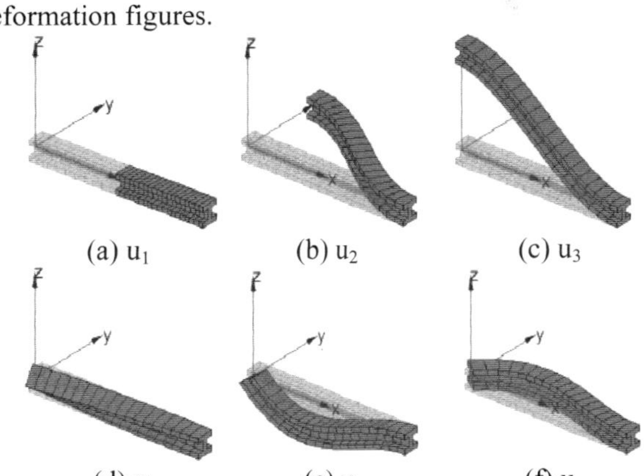

(a) u_1 (b) u_2 (c) u_3

(d) u_4 (e) u_5 (f) u_6

Figure 1.2: Deformation figures of node 1. (a), (b), and (c) indicate displacements of node 1 along x-, y-, and z-axis directions. (d), (e), and (f) indicate rotations of node 1 along x-, y-, and z-axis directions. (Stærdahl [3])

2.2 Element Matrices

An entire beam can be discretized into finite and continuous beam elements. The length of each element depends on the customized total number of elements assuming homogeneous discretization. Since nodal displacements are determined by initial and boundary conditions, it is critical to set up appropriate shape functions to trace any arbitrary point along the beam during deformation. Functions used for interpolation of the dependent variable and its derivative can be referred to (Cook [2]). As stated in the assumptions, the 12 DoFs can be divided into three groups in terms of axial deformation, bending, and twist. Furthermore, we assume that deformations are small in comparison to the geometry of the body at the beginning of the research. Therefore, the axial and bending deformation can be decoupled. The element frame of reference (xyz) is fixed to the rigid body rotation of the entire beam, so there is no rotation of the frame of reference in the local scale. The shape functions are also correspondingly grouped into three sets. The relationship between nodal and internal displacements is indicated by the general equation

$$\boldsymbol{u}(x)_{1X6} = \boldsymbol{N}_{12x6}\boldsymbol{u}_{i\,1x12} \tag{2.1}$$

where $\boldsymbol{N}$ is the shape function matrix connecting the nodal displacement matrix to the general displacement matrix and the numerical subscripts indicate dimensions. (Note that the bold $\boldsymbol{N}$ means matrix while the N_is below are equations) Each row of $\boldsymbol{N}$ corresponds to a specific deformation. The axial deformation (u_1, u_7) is assumed to be linear, and the corresponding shape function is shown below in matrix form

$$\boldsymbol{u}_x(x) = [N_1\ N_7]\begin{bmatrix}u_1\\u_7\end{bmatrix} = \begin{bmatrix}-\frac{1}{L}(x-x_7) & \frac{1}{L}(x-x_1)\end{bmatrix}\begin{bmatrix}u_1\\u_7\end{bmatrix} \tag{2.2}$$

where L is the length of an element. Note that the zeros in $[N_1\ N_7]$ and the remaining elements of $\begin{bmatrix}u_1\\u_7\end{bmatrix}$ are omitted.

Shape functions related to bending are polynomial functions. For bending in the x-y plane, the equation is

$$\boldsymbol{u}_y(x) = [N_2 N_6 N_8 N_{12}][u_2 u_6 u_8 u_{12}]^T \tag{2.3}$$

where N_2, N_6, N_8, and N_{12} are $1-3\frac{x^2}{L^2}+2\frac{x^3}{L^3}$, $x\left(-1+2\frac{x}{L}-\frac{x^2}{L^2}\right)$, $\frac{x^2}{L^2}\left(3-2\frac{x}{L}\right)$, and $\frac{x^2}{L^2}\left(3-2\frac{x}{L}\right)$ respectively. It is the same shape function for bending in the x-y plane, relating u_3, u_5, u_9, u_{11} to $\boldsymbol{u}_z(x)$

$$\boldsymbol{u}_z(x) = [N_3 N_5 N_9 N_{11}][u_3 u_5 u_9 u_{11}]^T \tag{2.4}$$

where $[N_3 N_5 N_9 N_{11}]$ is equal to $[N_2 N_6 N_8 N_{12}]$.

A linear approximation is used for the torsion between the nodes, so the shape function is identical to that of the axial deformation

$$\boldsymbol{\theta}_x(x) = [N_4\ N_{10}]\begin{bmatrix}u_4\\u_{10}\end{bmatrix} = \begin{bmatrix}-\frac{1}{L}(x-x_{10}) & \frac{1}{L}(x-x_4)\end{bmatrix}\begin{bmatrix}u_4\\u_{10}\end{bmatrix} \tag{2.5}$$

The shape functions for rotational coordinates about y- and z-axis are just derivatives of u_z and u_y

$$\begin{bmatrix}\boldsymbol{\theta}_y(x)\\\boldsymbol{\theta}_z(x)\end{bmatrix} = \begin{bmatrix}\frac{du_z}{dx}\\\frac{du_y}{dx}\end{bmatrix} \tag{2.6}$$

Before constructing the stiffness matrix, the strain-displacement matrix, $\boldsymbol{B}$, has to be derived separately for each of the three cases

$$\boldsymbol{B}_1 = \frac{dN_i}{dx}, \quad \boldsymbol{B}_2 = \frac{d^2N_j}{dx^2} \tag{2.7}$$

where N_i, and N_j are specific rows (or row) of the shape function matrix corresponding to the category of deformation when composing the partial stiffness matrices.

The entire stiffness matrix is composed of three parts, each corresponding to a different shape function matrix extracted from $\boldsymbol{N}$. First of all, the axial deformation stiffness matrix is obtained by

$$\boldsymbol{K}_a = \int_0^L \boldsymbol{B}_1{}^T AE\boldsymbol{B}_1 dx \tag{2.8}$$

The bending stiffness matrix in x-z and x-y planes are obtained by

$$\boldsymbol{K}_{by} = \int_0^L \boldsymbol{B}_2{}^T EI_y\boldsymbol{B}_2 dx \tag{2.9}$$

and

$$\boldsymbol{K}_{bz} = \int_0^L \boldsymbol{B}_2{}^T EI_z\boldsymbol{B}_2 dx \tag{2.10}$$

The torsional stiffness matrix is similar to that of the axial deformation because of linear variation

$$\boldsymbol{K}_t = \int_0^L \boldsymbol{B}_1{}^T GI_x\boldsymbol{B}_1 dx \tag{2.11}$$

The local stiffness matrix, $\boldsymbol{K}_e$, is the sum of (2.8)-(2.11)

$$\boldsymbol{K}_e = \boldsymbol{K}_a + \boldsymbol{K}_{by} + \boldsymbol{K}_{bz} + \boldsymbol{K}_t \tag{2.12}$$

The local mass matrix, $\boldsymbol{M}_e$, is obtained by

$$\boldsymbol{M}_e = \int_0^L \boldsymbol{N}^T A\rho\boldsymbol{N} dx \tag{2.13}$$

where A and ρ are the cross-section area and element density, respectively. According to the assumption, they are constant during deformation (Cook [2], Ottosen [9], and Bathe [3]).

2.3 Dynamic Equations

The dynamic equations can be written in the matrix form:

$$\boldsymbol{M}\ddot{\boldsymbol{u}}_i + \boldsymbol{K}\boldsymbol{u}_i = \boldsymbol{f}_i \tag{2.14}$$

where $\ddot{\boldsymbol{u}}_i$ is the nodal point accelerations (i.e., is the second time derivative of $\boldsymbol{u}_i$), and $\boldsymbol{f}_i$ is the nodal force vector. Both of them are time-dependent (Bathe [3]).

In measured dynamic responses of structures, it is observed that energy is dissipated during vibration, which in vibration analysis is usually taken account of by introducing velocity-dependent damping forces. $\dot{\boldsymbol{u}}_i$ is a vector of the nodal point velocities (i.e., the first time derivative of $\boldsymbol{u}_i$), and the equilibrium equations are modified to

$$\boldsymbol{M}\ddot{\boldsymbol{u}}_i + \boldsymbol{C}\dot{\boldsymbol{u}}_i + \boldsymbol{K}\boldsymbol{u}_i = \boldsymbol{f}_i \tag{2.15}$$

where $\boldsymbol{C}$ is the damping matrix; i.e., formally,

$$\boldsymbol{C} = \int_0^L \boldsymbol{N}^T A\kappa\boldsymbol{N} dx \tag{2.16}$$

where κ is the damping parameter of the element.

3. INTEGRATION WITH MULTIBODY SYSTEMS

Analyses of multibody systems (MBS) and finite element models are both well-established practices in computational engineering. Simulation of MBS is mainly used for analyses of mechanisms consisting of rigid components connected with joints. The mathematical description often results in a system of differential-algebraic equations (DAEs). With advancing complexity of developed systems, the goals of computational engineering are increasingly shifted to coupled problems.

There have been some established algorithms for integration of FEM with MBS. The classical and simplest method is to partition the coordinates into reference and modal subparts. One key assumption at the initial stage of development is that the inertia does not change with body deformations, i.e. the generalized $\boldsymbol{M}$ and $\boldsymbol{K}$ matrices are not a function of the model coordinates. The simplest partition for Eq. (1.14) is

$$\begin{bmatrix} \boldsymbol{M}_{rr} & \boldsymbol{M}_{rm} \\ \boldsymbol{M}_{mr} & \boldsymbol{M}_{mm} \end{bmatrix} \begin{bmatrix} \ddot{\boldsymbol{u}}_r \\ \ddot{\boldsymbol{u}}_m \end{bmatrix} + \begin{bmatrix} \boldsymbol{K}_{rr} & \boldsymbol{K}_{rm} \\ \boldsymbol{K}_{mr} & \boldsymbol{K}_{mm} \end{bmatrix} \begin{bmatrix} \boldsymbol{u}_r \\ \boldsymbol{u}_m \end{bmatrix} = \begin{bmatrix} \boldsymbol{f}_r \\ \boldsymbol{f}_m \end{bmatrix} \quad (3.1)$$

In this equation, subscripts r and m refer, respectively, to reference and modal coordinates. The dynamic coupling between the dynamic motion and the elastic deformation is represented by the matrices with subscripts rm and mr. (Shabana, [4])

4. SYMBOLIC PROGRAMMING

Essentially based on symbolic computation, symbolic programming subtly avoids common shortcoming that normally belongs to scientific computing relying on numerical computation. Symbolic computation uses variables, mathematical formulas, symbols such as π and ∞, and mathematical functions to realize simplification of expressions. It is not necessary to assign values to variables throughout the calculation unless and until a numeric approximation is needed. For instance, $\frac{\pi}{4}$ can remain as $\frac{\pi}{4}$ throughout the computation, resulting in an exact result by eliminating accumulated round-off errors. In addition, the symbolic simplification of expression enables higher efficiency and makes it easy to sweep parameters, optimize values, and study the behavior of the system. More benefit of symbolic computation can be found in Boyle [8].

Building FE-models involves encoding a number of large and complex matrices, but symbolic programming allows developers to just create matrices only containing variables and symbols. Matrices such as $\boldsymbol{K_e}$ and $\boldsymbol{M_e}$ in Eq. (2.12) and (2.13) can be directly input and ready for calculation. Not until the end when obtaining certain numeric results by solving the DAEs, do the matrices have to be evaluated. Symbolic programming is also able to make the FE-model easy to be manipulated. A detailed application of symbolic programming on FE-models can be referred to in Portela [5].

5. FLEXIBLE CANTILEVER BEAM

5.1 Parameters

In this section, an example of cantilever beam model with customizable beam component based on symbolic programming will be shown. This model enables the user to modify beams with flexible properties similar to the general FE approach. Variables such as length, cross-sectional area, density, Young's modulus, shear modulus, and second moment of cross-sectional area can be customized. However, the original beam component does not incorporate a damping factor, which leads to infinite oscillations upon excitation. Therefore, an additional damping matrix was developed to take damping into account。

The parameters of the beam component in this scenario are recorded in Table 5.1.

Table 5.1: The parameters of the beam component

Tip force (F)	-5t, (t ≤ 4s) -20 N (4 < t ≤10s)
Length (L)	0.5 m
Cross-Section (A)	0.00005 m^2
Density (ρ)	7850 kg/m^3
Young's Modulus (E)	2.1 E11 N/m^2
Poisson Ratio (ν)	0.3
Shear Modulus (G)	80769230769.231 N/m^2
Second Moment of Area about y (I_y)	4.17 E-10 m^4
Second Moment of Area about z (I_z)	1.04 E-10 m^4

The model will be created using Maple (a general-purpose symbolic computation environment) and MapleSim (a multi-domain physical model environment built on Maple's technology). Maple and MapleSim software are sold by Maplesoft, which also developed a FEA mathematical model to create custom components based on matrices from a finite element Maple script for a user-defined beam. The flexible beam component in MapleSim generates symbolic equations, and it serves the purpose of creating a beam component that can be integrated with multibody simulations. It is used to verify the results from our model. To verify these approaches, especially with regards to the newly-added damping effects, we define a scenario in which each approach (damped, undamped, and FEA worksheet) is tested and the results are compared against each other.

5.2 Matrix Set up

The matrices of interest in this case are the ($\boldsymbol{K}$) stiffness, ($\boldsymbol{M}$) mass, and ($\boldsymbol{C}$) damping matrices. Following Eq. (2.1)-(2.13), the final element stiffness and mass matrices can be obtained. Panzer [6] has given the detail of the matrices including a shearing factor, which has not been included here.

Depend on the number of elements, the size of the matrices varies. As the number go es up, the matrix size rises. It follows the superposition law (Panzer [6]) when adding up the matrices. After all of the matrices and the nodes are returned and meshed respectively, a group of dynamic DAEs will be generated when being combined with Eq. (2.14) (or Eq. (2.15) if damping is considered). The DAEs are, in tensor form,

$$M_{ij}\ddot{u}_j + C_{ij}\dot{u}_j + K_{ij}u_j = f_j\,,\ i,j = 1 \dots n \qquad (5.1)$$

where n is the total number of DoFs. As a cantilever beam, the first six DoFs will be permanently set to zeros. For example, the displacement vector becomes

$$u_j = [0,0,0,0,0,0,u_7,u_8,\dots,u_n]^T \qquad (5.2)$$

The force is applied at the tip of the beam to create the greatest excitation of the structure and to cause stress in every part of the beam. In this case, the force vector is

$$f_j = [0,\dots,0,F,0,0,0]^T \qquad (5.3)$$

All of the deflections are initially zeros. By solving Eq. (5.1) with the known boundary and initial conditions, the node deflection will be obtained.

5.3 Tip Deflection

The tip deflection is tracked, and Fig. 5.1 shows a plot of the tip deflection over the course of 10 seconds. The results of all three approaches are plotted to emphasize the similarities of their characteristic shape. The blue (thin) line represents the undamped flexible beam, which shows the strongest oscillation during the whole analysis. This is best visible from 4 seconds onwards to the end where the blue line encases both the red (thick) line and the green dots. The red line shows the flexible beam component with additional damping. As shown in Fig. 5.2 the oscillation has significantly decreased. Very similar to the results marked by the red line are the results for the flexible beam created with the Maplesoft FEA worksheet represented by the green dots. In all three cases the average tip deflection after 10 seconds is around 0.038 m.

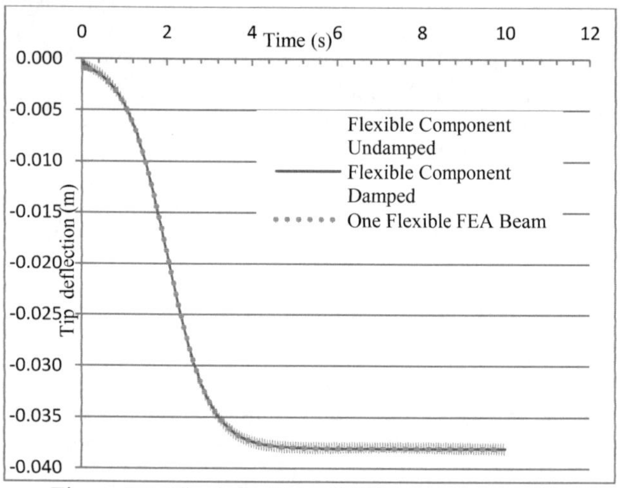

Figure 5.1: Tip deflection of different beam models

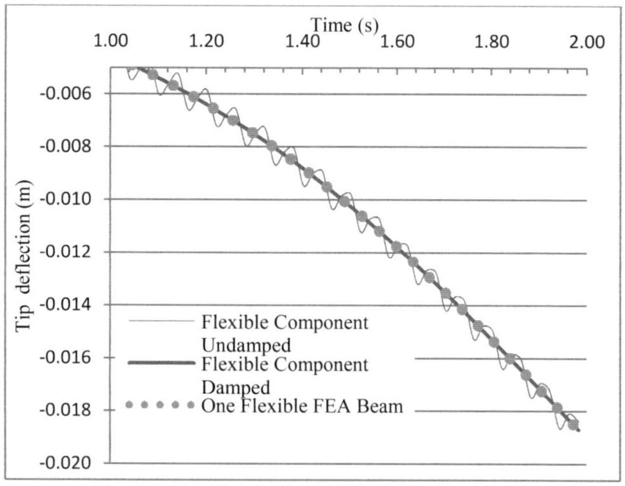

Figure 5.2: Zoom-in of the deflection curve

5.4 Convergence Study

Based on the convergence study and its plot given as Fig. 5.3, we can conclude that using more than 200 elements will not lead to a gain in accuracy for our model.

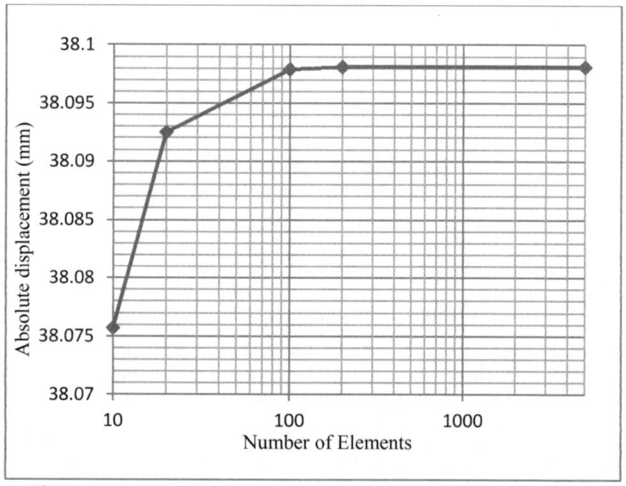

Figure 5.3: Tip displacement versus number of elements

6. CONCLUSIONS

The demands on simulations of dynamic systems has increased throughout the past years. In the past researchers had to ignore certain dynamics due to flexible bodies; today they can incorporate them into a dynamic multibody system. Although symbolic integration is still at the initial stage, it shows a strong capability of efficiently building large matrices and DAEs, and generating highly-optimized simulation code. Flexible beam models, due to their simplicity, have been developed first. In future research, rotary inertia relative to bending will be included in the mass matrix. Axial and bending deformation will be coupled. Torsional conditions will be tested as the next object, and the beam model will be used for the flexible half shafts in a longitudinal vehicle dynamics model. Creation of a symbolic plate model is another goal of our future work. The overall objective of this research is to develop simulation framework to include flexible bodies of arbitrary 3D geometries into multibody simulation. Time, cost, and adaptability of strictly numerical approaches and symbolic approaches will be compared after the model has been well developed and verified.

ACKNOWLEDGMENTS

This research was supported by NSERC, MapleSoft, and Toyota. This financial support is gratefully acknowledged.

REFERENCES

[1] Astley, R.J., 1992, *Finite Elements in Solids and Structures – An Introduction*, Chapman & Hall, London

[2] Cook, R.D., Malkus, D.S., Plesha, M.E., Witt, R.J., 2002, *Concepts and Applications of Finite Element Analysis*, John Wiley & Sons, New York

[3] Bathe K.J., 2003, *Finite Element Procedures*, Prentice Hall, New Jersey

[4] Shabana, A. A., Bauchau, O. A., & Hulbert, G. M., 2007, "Integration of large deformation finite element and multibody system algorithms", *Journal of Computational and Nonlinear Dynamics*, 2(4), pp. 351-359

[5] Portela, A. & Charafi, A., 2002, *Finite Elements Using Maple*, Springer, Germany

[6] Panzer, H., Hubele, J., Eid, R., & Lohmann, B., 2009, "Generating a parametric finite element model of a 3D cantilever Timoshenko beam using MATLAB", *Lehrstuhl für Regelungstechnik, Technische Universität München, Technical Reports on Automatic Control.*

[7] Stærdahl, J.W., Clausen, J., Andersen, L., n.d., *Finite element method II* [PDF document]. Retrieved from http://www.wind.civil.aau.dk/

[8] Boyle, A., Caviness, B.F., 1988, *Future Directions for Research in Symbolic Computation*, Society for Industrial and Applied Mathematics, Philadephia

[9] Ottosen, N. S., Petersson, H., & Saabye, N., 1992, *Introduction to the Finite Element Method.* Prentice Hall International.

[10] Shi, P., McPhee, J., & Heppler, 2001, "A deformation field for Euler-Bernoulli beams with applications to flexible multibody dynamics", *Multibody System Dynamics*, Springer Netherlands, 5(1), pp. 79-104

[11] Richard, M., Bouazara, M., & Therien, J., 2011, "Analysis of multibody systems with flexible plates using vibrational graph-theoretic methods", *Multibody System Dynamics*, Springer Netherlands, 25, pp. 43-63

Proceedings of the 25th CANCAM
London, Ontario, Canada, May 31 - June 4, 2015

ACAUSAL MODELLING FOR DYNAMIC MULTIBODY SIMULATION OF DAMPED THREE-DIMENSIONAL FLEXIBLE BEAMS

Joydeep Banerjee
Systems Design Engineering
University of Waterloo
Waterloo, Ontario, Canada
jbanerjee83@gmail.com

John McPhee
Systems Design Engineering
University of Waterloo
Waterloo, Ontario, Canada
mcphee@uwaterloo.ca

ABSTRACT

Flexible beams and plates are routinely encountered in practical multibody systems. Formulations based on the virtual work principle can be used to model flexible bodies and have been implemented in commercial software like MapleSim. In this paper, the derivation of the damped vibration of a flexible beam will be presented that can be used to simulate beams in the context of multibody systems under the influence of internal dissipative forces. The formulation will be used to derive the symbolic governing equations for the system. The integration of the model into a multibody system will also be demonstrated to illustrate the practicality of the implementation.

Keywords: Flexible beams, Multibody system dynamics, Virtual work formulation, Symbolic programming.

INTRODUCTION

Software tools for multibody simulations use different approaches to model flexible bodies. For example, MSC-ADAMS uses what is known as the Model Neutral Format to import modal analysis data for flexible bodies [1]. This method is not suitable for generation of symbolic governing equations. In contrast, MapleSim uses a virtual work based graph-theoretic approach to generate the governing equations for systems involving flexible multibody elements [9].

Graph-theoretic methods use linear graphs to keep track of the topology of the system and the connections between the through variables and the across variables associated with the systems [2]. For rigid multibody system simulations, the through variables are the forces and torques in the system and the across variables are the translational (and rotational) displacements, velocities, and accelerations.

For flexible multibody simulations, Shi et al. [6], have developed a modified graph-theoretic framework, where the virtual work done by the forces and torques of the system are also considered as through variables. Using this method, Shi et al. have developed models for different flexible multibody systems [5]. Richard and Bouazara [3] have used similar methods to model vibrating beams and shell elements.

Currently, the flexible beam element in MapleSim, which is based on the formulation by Shi and McPhee [6], does not include damping (external and internal) and as a result is not suitable for simulations where the damping forces affect the dynamics significantly. For example, since a vibrating beam in MapleSim keeps on vibrating for ever, it is not possible to use it for structural simulations which damps out vibrations internally. Similarly, it is also not possible to simulate the realistic swing of a flexible golf shaft since the flexible beam element does not include aerodynamic drag terms.

This paper will present a modification to the derivation by Shi et al. to include a strain rate dependent damping to the flexible beam element. This will allow users to create acausal models of multibody systems with flexible beam elements and generate symbolic governing equations, which can be used to develop optimized simulation codes for real time applications and control systems development.

THEORETICAL BACKGROUND

Shi et al. defined virtual work terminal equations to describe the behaviour of different components of multibody systems (e.g., rigid/flexible bodies, joints, forces etc.) [5].

Equation (1) shows the virtual work (VW) terminal equation for flexible bodies, where W is the virtual work done by the forces associated with the body, $\mathbf{f}_b$ is the applied body force, γ is the mass per unit volume, the quantities $\boldsymbol{\epsilon}$ and $\boldsymbol{\sigma}$ are components of the strain and stress tensors, and $\mathbf{r}$ is the displacement of the differential volume dV.

$$\delta W = -\int_V \delta\varepsilon^T \boldsymbol{\sigma} dV + \int_V \delta\mathbf{r}^T \left(\mathbf{f}_b - \gamma\ddot{\mathbf{r}}\right) dV \quad (1)$$

In the formulation by Shi et al., the governing equations for the flexible multibody system are generated by combining the VW terminal equations with the graph-theoretic topological equations. The objective of this presented work is to modify the formulation so that the implemented algorithm for equation generation results in the governing equations for a damped Rayleigh beam. In the next section equation (1) will be modified to include the effect of internal damping.

VIRTUAL WORK FORMULATION

Shi et al. [7] have proposed a second-order deformation field for Rayleigh beam elements which can generate elastic rotation matrices that satisfy the principles of orthogonality.

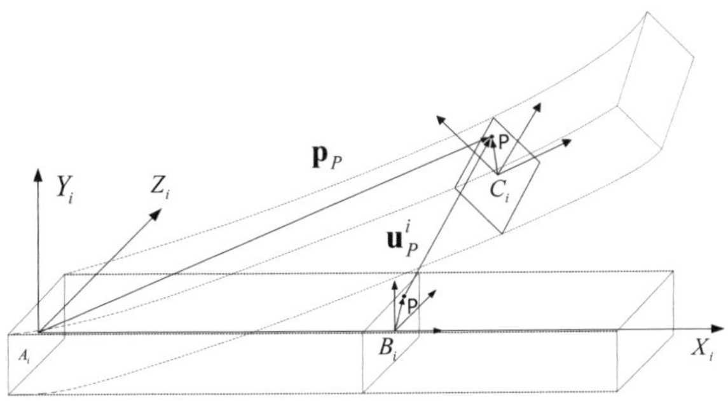

Figure 1: Displacement field for the flexible beam

In this approach the beam's elastic configuration is defined in terms of the axial deformation u, the two lateral bending v and w, and the torsional deformation ϕ, all of which are functions of both position x and time t. The effect of rotational inertia is included, but the shear deformation due to bending and warping due to torsion are neglected. Also it is assumed that the beam experiences small strain, but possibly large deflections. The deflection of the point P (i.e. $\mathbf{u}_P^i$ (figure 1) in the frame $\{X_i, Y_i, Z_i\}$) is given by equation (2).

$$\mathbf{u}_P^i = \begin{bmatrix} \psi_1 \\ v + \left(-\frac{v'^2}{2} - \frac{\phi^2}{2}\right) y + \left(-\phi - \frac{v'w'}{2}\right) z \\ w + \left(\phi - \frac{v'w'}{2}\right) y + \left(-\frac{w'^2}{2} - \frac{\phi^2}{2}\right) z \end{bmatrix} \quad (2)$$

$$\psi_1 = \begin{pmatrix} u - \int_0^x \left(v'^2/2 + w'^2/2\right) d\xi \\ -\left(-v' + u'v' - w'\phi\right) y \\ +\left(-w' + u'w' + v'\phi\right) z \end{pmatrix} \quad (3)$$

The deformation field from equation (2) can be used to determine the elements of strain tensors for the beam using nonlinear strain displacement relationships [4]. For a Rayleigh beam, the relevant quantities are ϵ_{xx} (the normal strain in the X_i direction), ϵ_{xy} (parallel to $[X_i, A_i, Y_i]$ plane), and ϵ_{xz} (parallel to $[X_i, A_i, Z_i]$ plane). The other strain components are considered negligible.

Using the strain-displacement relationships in (2), one can obtain the expressions for the strain components in terms of the elastic variables as shown in equation (4).

$$\begin{aligned} \varepsilon_{xx} &= H_{11} + yH_{12} + zH_{13} + H_{14} \\ \varepsilon_{xy} &= zH_{23} \qquad \varepsilon_{xz} = yH_{32} \end{aligned} \quad (4)$$

Where the following quantities shown in equation (5) are defined in terms of the elastic variables. In equation (4), the terms higher than second-order strains are ignored to reduce complexity without the loss of accuracy.

$$\begin{aligned} H_{11} &= u' + u'^2 \big/ 2 \\ H_{12} &= -v'' + u''v' - w''\phi \\ H_{13} &= -w'' + u''w' + v''\phi \\ H_{14} &= \left(a_1 y^2 + a_2 z^2 + a_3 yz\right)/2 \\ H_{32} &= \left(\phi'/2 + (v''w' - v'w'')/4\right) \quad H_{23} = -H_{32} \\ a_1 &= \left(v''^2 + \phi'^2\right) \quad a_2 = \left(w''^2 + \phi'^2\right) \quad a_3 = 2v''w'' \end{aligned} \quad (5)$$

To include the effect of internal damping, a Kelvin–Voigt type material behaviour is assumed. The stress-strain equation is shown in equation (6), where ζ, ξ are the damping factors, E is the Young's modulus, and G is the shear modulus.

$$\boldsymbol{\sigma} = \begin{bmatrix} \sigma_{xx} \\ \sigma_{xy} \\ \sigma_{xz} \end{bmatrix} = \begin{bmatrix} E\,\varepsilon_{xx} + \xi\,\dot{\varepsilon}_{xx} \\ 2G\,\varepsilon_{xy} + 2\zeta\,\dot{\varepsilon}_{xy} \\ 2G\,\varepsilon_{xz} + 2\zeta\,\dot{\varepsilon}_{xz} \end{bmatrix} \quad (6)$$

It is also worthwhile to mention that the first variation of the strains can be written in a form shown in equation (7).

$$\begin{aligned} \delta\varepsilon_{xx} &= \delta H_{11} + y\delta H_{12} + z\delta H_{13} + \delta H_{14} \\ \delta\varepsilon_{xy} &= z\delta H_{23} \quad \delta\varepsilon_{xz} = y\delta H_{32} \end{aligned} \quad (7)$$

Since material damping is a purely internal phenomenon, only the VW by the internal forces need to be considered. Using equation (6) in equation (1), equation (8) is obtained.

$$\begin{aligned} \delta W &= \int_V \delta\boldsymbol{\varepsilon}^T \boldsymbol{\sigma} dV = E \int_V \delta\varepsilon_{xx}\varepsilon_{xx} dV \\ &+ 4G \int_V \left(\delta\varepsilon_{xy}\varepsilon_{xy} + \delta\varepsilon_{xz}\varepsilon_{xz}\right) dV \\ &+ \xi \int_V \delta\varepsilon_{xx}\dot{\varepsilon}_{xx} dV + 4\zeta \int_V \left(\delta\varepsilon_{xy}\dot{\varepsilon}_{xy} + \delta\varepsilon_{xz}\dot{\varepsilon}_{xz}\right) dV \end{aligned} \quad (8)$$

If the cross-section of the beam is assumed to have two planes of symmetry, equation (8) can be simplified into equation (9).

$$\begin{aligned} \delta W &= E\int_0^L \left(\chi_1 + \frac{\chi_2}{2} + \frac{\chi_3}{4}\right) dx + 4G\int_0^L \chi_4 dx \\ &+\xi\int_0^L \left(\eta_1 + \frac{\eta_2}{2} + \frac{\eta_3}{4}\right) dx + 4\zeta\int_0^L \eta_4 dx \end{aligned} \tag{9}$$

Where

$$\begin{aligned} \chi_1 &= \begin{pmatrix} A\delta H_{11}H_{11} + J_{zz}\delta H_{12}H_{12} \\ +J_{yy}\delta H_{13}H_{13} \end{pmatrix} \\ \chi_2 &= \begin{pmatrix} \delta H_{11}\left(a_1 J_{zz} + a_2 J_{yy}\right) \\ +H_{11}\left(\delta a_1 J_{zz} + \delta a_2 J_{yy}\right) \end{pmatrix} \\ \chi_3 &= \begin{pmatrix} b_1 J_{zzzz} + b_2 J_{zzzy} + b_3 J_{zzyy} \\ +b_4 J_{zyyy} + b_5 J_{yyyy} \end{pmatrix} \\ \chi_4 &= \delta H_{23} H_{23} J_{yy} + \delta H_{32} H_{32} J_{zz} \end{aligned} \tag{10}$$

$$\begin{aligned} \eta_1 &= \begin{pmatrix} A\delta H_{11}\dot{H}_{11} + J_{zz}\delta H_{12}\dot{H}_{12} \\ +J_{yy}\delta H_{13}\dot{H}_{13} \end{pmatrix} \\ \eta_2 &= \begin{pmatrix} \delta H_{11}\left(\dot{a}_1 J_{zz} + \dot{a}_2 J_{yy}\right) \\ +\dot{H}_{11}\left(\delta a_1 J_{zz} + \delta a_2 J_{yy}\right) \end{pmatrix} \\ \eta_3 &= \begin{pmatrix} c_1 J_{zzzz} + c_2 J_{zzzy} + c_3 J_{zzyy} \\ +c_4 J_{zyyy} + c_5 J_{yyyy} \end{pmatrix} \\ \eta_4 &= \delta H_{23} \dot{H}_{23} J_{yy} + \delta H_{32} \dot{H}_{32} J_{zz} \end{aligned} \tag{11}$$

$$\begin{aligned} b_1 &= \delta a_1 a_1, \quad b_2 = \left(\delta a_1 a_3 + \delta a_3 a_1\right) \\ b_3 &= \left(\delta a_1 a_2 + \delta a_2 a_1 + \delta a_3 a_3\right) \\ b_4 &= \left(\delta a_3 a_2 + \delta a_2 a_3\right), \quad b_5 = \delta a_2 a_2 \end{aligned} \tag{12}$$

$$\begin{aligned} c_1 &= \delta a_1 \dot{a}_1, \quad c_2 = \left(\delta a_1 \dot{a}_3 + \delta a_3 \dot{a}_1\right) \\ c_3 &= \left(\delta a_1 \dot{a}_2 + \delta a_2 \dot{a}_1 + \delta a_3 \dot{a}_3\right) \\ c_4 &= \left(\delta a_3 \dot{a}_2 + \delta a_2 \dot{a}_3\right), \quad c_5 = \delta a_2 \dot{a}_2 \end{aligned} \tag{13}$$

The terms like J_{zzzz}, and J_{zz} are the fourth and second moments of area respectively. The moments of areas with odd powers are zero due to the assumed symmetry of cross-section.

SHAPE FUNCTIONS AND ELASTIC COORDINATES

MapleSim uses a Rayleigh—Ritz type approach for flexible multibody systems. In this method, the deformation variables u, v, w, and ϕ are approximated over the entire length of the beam using a variable separable form shown in equation (14).

$$\begin{aligned} u &= \sum_{n=0}^{n_u} u_n(t)\, T_n(x) \quad & v &= \sum_{n=0}^{n_u} v_n(t)\, T_n(x) \\ w &= \sum_{n=0}^{n_u} w_n(t)\, T_n(x) \quad & \phi &= \sum_{n=0}^{n_u} \phi_n(t)\, T_n(x) \end{aligned} \tag{14}$$

The terms $T_n(x)$ are the shape functions that satisfy the boundary conditions and are usually a series of polynomials (Taylor, Legendre, or Chebyshev) [8]. Using an user-specified number of shape functions, equation (14) can be substituted into (9), resulting in a set of ordinary differential equations in terms of the variables u_n, v_n, w_n and ϕ_n which are known as elastic coordinates. The total number of variables and hence the number of generated equations depend on the number of elastic coordinates selected for the process.

RESULTS

The presented formulation was implemented in MapleSim, which uses symbolic graph-theoretic formulation for flexible multibody simulations. To validate the implementation, a simple cantilever beam was simulated to vibrate under its own weight. The beam properties are given in equation (15).

$$\begin{aligned} L &= 1\,m & A &= 7.8537 \cdot 10^{-5}\,m^2 \\ \rho &= 2700\,Kg/m^3 & J &= 4.909 \cdot 10^{-10}\,m^4 \\ E &= 72 \cdot 10^9\,Pa & G &= 27 \cdot 10^9\,Pa \\ \zeta &= 3 \cdot 10^6\,Pa \cdot s & \xi &= 5 \cdot 10^6\,Pa \cdot s \end{aligned} \tag{15}$$

The simulation results from MapleSim were compared to the numerical results obtained by solving the partial differential equation governing the dynamics of a cantilever beam with identical properties. Figure 2 shows the congruity of the two methods for an undamped beam and a damped beam, which demonstrates the validity of the implementation.

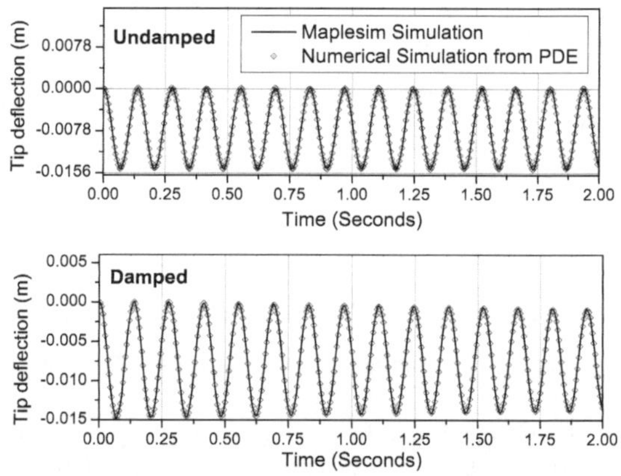

Figure 2: Validation of MapleSim Implementation

The MapleSim implementation was simulated assuming a planar beam and three elastic coordinates for the shape functions. The Cash-Karp 45 solver was used with error tolerance of $\epsilon = 0.00001$. For the PDE, a centred implicit integration scheme was used with spatial discritization of $0.01\,m$ and temporal discretization of $0.0001\,s$.

The same beam element was used in a three-dimensional spin-up beam simulation, where it was fixed to a revolute joint located at the origin with the vertical Y direction as its revolute axis. The beam's angular speed was ramped up from $0\,rad/s$ to $30\,rad/s$ in a $3\,s$ window, starting at $t = 1\,s$. The total simulation was carried on for $t = 5\,s$. The deflection of the beam tip in the vertical direction (i.e. $-Y$) was measured along with the torque required to drive the revolute joint). A fixed step RK4 solver was used for the simulation with timestep $h = 0.0001s$.

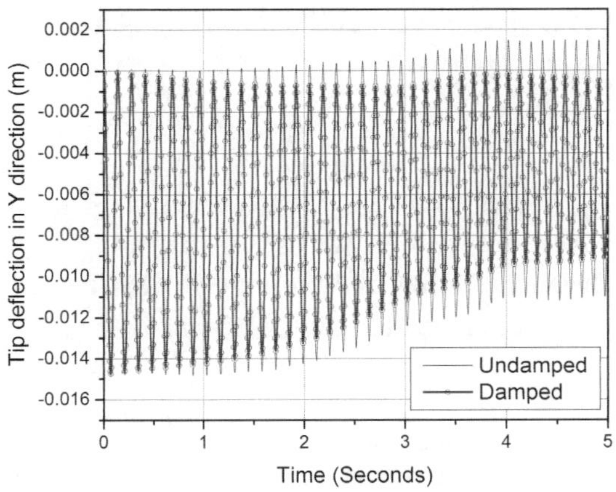

Figure 3: Tip delfection for damped and undamped beam

Figures 3 and 4 show the differences between the response from a damped and an undamped system, and demonstrate the applicability of the presented derivation for multibody simulation.

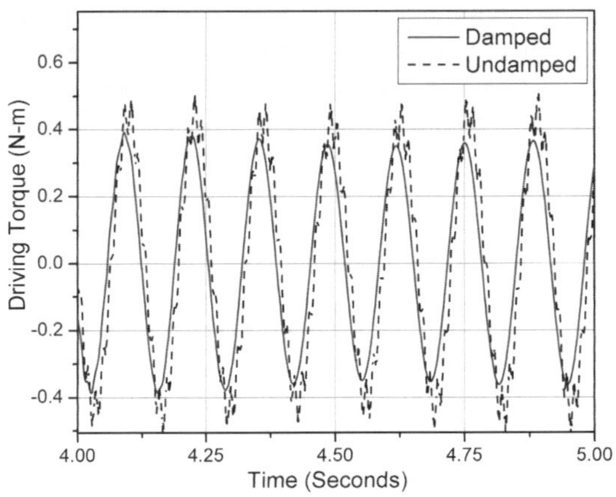

Figure 4: Driving torque for damped and undamped beam

CONCLUSIONS

This paper presents a virtual work based model for flexible beam elements with internal damping. The formulation has been implemented in MapleSim to enable users to include the damped beam in multibody systems in an acausal fashion. The method is an extension of the work by Shi and McPhee [6] and has been tailored to include strain rate dependent dissipation terms.

The presented model implementation uses a physics based approach (as opposed to proportional damping) to quantify the internal dissipation in beams. Although for regular components of automotive systems, the structural damping in itself is not very significant, it is still worthwhile to include the damping term due to the unrealistic nature of the undamped simulations. Also, this presented model can be tuned for engineered materials, where the equivalent model parameters can be estimated using parameter identification techniques.

Software simulation has demonstrated the effectiveness of the model and the requirement of the damping term for realistic simulation. Future research efforts will be directed toward the identification of damping parameters for accurate structural simulation.

ACKNOWLEDGMENTS

The authors would like to acknowledge the support from Dr. Chad Schmitke of Maplesoft and NSERC Canada.

REFERENCES

[1] Chaesil Kim, Yoon Gyo Jung, Soo Yong Cho, and Hoon Hyung Jung. Vibration control simulation for a multi-body high speed flexible rotor model using a phase adjusting method. In *ICCAS-SICE, 2009*, pages 2286–2290. IEEE, 2009.

[2] John J. McPhee. On the use of linear graph theory in multibody system dynamics. *Nonlinear Dynamics*, 9:73–90, 1996. 10.1007/BF01833294.

[3] Marc J. Richard and Mohamed Bouazara. Graph-theoretic approach for the dynamic simulation of flexible multibody systems. *Advances in Mechanical Engineering*, 4(530132):12, 2012.

[4] Irving H Shames and Clive L Dym. *Energy and finite element methods in structural mechanics*. Hemisphere Publishers, 1985.

[5] Pengfei Shi. *Flexible multibody dynamics, a new approach using virtual work and graph theory*. PhD thesis, University of Waterloo, 1998.

[6] Pengfei Shi and John McPhee. Dynamics of flexible multibody systems using virtual work and linear graph theory. *Multibody System Dynamics*, 4(4):355–381, 2000.

[7] Pengfei Shi, John McPhee, and Glenn Heppler. A deformation field for euler–bernoulli beams with applications to flexible multibody dynamics. *Multibody System Dynamics*, 5(1):79–104, 2001.

[8] Pengfei Shi, John McPhee, and Glenn Heppler. Polynomial shape functions and numerical methods for flexible multibody dynamics. *Mechanics of structures and machines*, 29(1):43–64, 2001.

[9] Wen J Sun, Hui Yan, Yong Ming Gao, and Hang Yin. Dynamics simulation of six-dof flexible robot arm based on maplesim. *Computer Technology and Development*, 10:054, 2011.

DESIGN AND ANALYSIS OF AN INTERCONNECTED SUSPENSION FOR A SMALL OFF-ROAD VEHICLE

Bruce P Minaker PhD PEng†
Department of Mechanical,
Automotive, & Materials Engineering
University of Windsor, Windsor ON N9B 3P4
Email:bminaker@uwindsor.ca

Zheng Yao
Department of Mechanical,
Automotive, & Materials Engineering
University of Windsor, Windsor ON N9B 3P4
Email:yao113@uwindsor.ca

ABSTRACT

The paper describes the design and multibody dynamic analysis of a mechanically interconnected suspension, as applied to a small off-road vehicle. Interconnected suspensions use some sort of connection between the axles of a vehicle in order improve ride quality or vehicle handling. In principle, the connection may be hydraulic, pneumatic, or mechanical, but for installation in a typical passenger car, a mechanical connection would likely be impractical due to weight and complexity. In this paper, the vehicle in question is the University of Windsor SAE Baja off-road competition vehicle, and novel mechanical design is proposed. A multibody dynamic analysis is performed on the proposed design using the EoM open source multibody software developed by the University of Windsor Vehicle Dynamics and Control research group in order to assess any potential performance improvements.

KEYWORDS: Vehicle dynamics, interconnected suspension, multibody dynamics

INTRODUCTION

The conventional vehicle suspension system in use today has one flexible connection between each wheel and the vehicle chassis, typically a mechanical coil spring with a hydraulic damper in parallel. In addition, many vehicles will add an additional flexible coupling, called a sway bar, or more precisely, an anti-roll bar, between the left and right hand sides of the front axle, or the rear axle, or sometimes both axles. This is typically done using a torsional spring, and serves two purposes. First, if the main springs of the vehicle are chosen sufficiently soft enough to allow good ride quality, they will often not be firm enough to control body roll during cornering. Excessive body roll can have adverse affects on tire grip, and can contribute to increased load transfer, or, in an extreme scenario, rollover. It is best avoided if possible. Secondly, the ratio of the front to rear roll stiffness can be modified using the anti-roll bar. It is important to note that the anti-roll bar stiffness has a small effect on lateral weight transfer, but a significant effect on the roll stiffness ratio of the front and rear axles. The roll stiffness ratio will influence the handling properties of the vehicle, with a high front roll stiffness contributing to an understeering and stable response; conversely, a high rear roll stiffness will contribute to an oversteering and very responsive, but potentially unstable, vehicle behaviour. For this reason, most passenger vehicles will use a front axle mounted anti-roll bar.

†corresponding author

When selecting the spring stiffnesses during a vehicle design, it is quite common to use a 'bounce-pitch-roll' model, where the vehicle is treated as a single rigid body with three degrees of freedom, and each wheel is permitted an additional vertical motion, giving a seven degrees-of-freedom model. When a modal analysis on this type of model is performed, a there are a number of interesting results. Typically, the modes will be divided into two groups based on their frequency. Three low frequency motions, typically between 0.5 and 2 Hz, are expected, where the response is largely chassis motion in the three bounce, pitch, and roll directions. Additionally, there are four high frequency motions, between roughly 8 and 12 Hz, that consist of mostly of wheel motion, referred to as 'wheel-hop'. The high frequency modes do not align to each wheel, but rather form a set of four coupled motions. These are typically 1) all wheels moving in phase (bounce), 2) front and rear out-of-phase (pitch), 3) left and right out-of-phase (roll), and 4) diagonally opposite pairs out of phase (twist or warp).

The desired characteristics of the three primary chassis motions will generally dictate the bounce, pitch, and roll stiffnesses, e.g., high roll stiffness is desired for the resistance to lateral loads during cornering, low bounce stiffness provides good ride quality, etc. The warp stiffness is often ignored, as conventional suspension design does not allow for individual selection of the stiffness in each of these wheel hop modes. However, there is no need for a high warp stiffness, as it is only encountered when the road or track is uneven. In fact, by reducing the warp stiffness, it is possible to reduce the stiffness of a single wheel deflection, and the overall response of the suspension to rough terrain can be improved. However, in order to decouple these motions, an interconnected suspension is required.

A number of researchers have discussed the concept of decoupling the bounce, roll, pitch, and twist stiffnesses by using interconnected suspensions. Papers by Zapletal[1] in 2000, and

Buj[2] in 2002 lay out many of the basic ideas, and describe the advantages of interconnected suspensions, particularly for racing applications. Zapletal was awarded a patent on a 'Balanced Suspension System' in 2004.

Nevertheless, there are very few instances of production passenger vehicles using interconnected suspensions, other than the typical anti-roll bar. They include the Citroën 2CV, which employed a mechanical interconnection of the main suspension coil springs on each side of the vehicle. The springs were arranged such that they lay in the floor of the vehicle and were oriented longitudinally. British Leyland marketed vehicles using the Moulton Hydragas interconnected suspension, which functioned similarly to the Citroën system, and was primarily intended for improvement of ride quality by coupling front and rear suspensions, particularly for vehicles with short wheelbases. Typically, passengers find pitch motions to be more uncomfortable than bounce motions, which partially explains why a longer wheelbase vehicle tends to have a better ride quality. Linking the front and rear suspensions so that they tend to move in opposing directions generates a similar effect. In 2003, Rideout and Anderson published a model of the Moulton Hydragas interconnected suspension[3].

Kinetic Pty Ltd and Tenneco Automotive implemented a hydraulic interconnected suspension that was utilized by both Citroen in the World Rally Car series, and by Mitsubishi in the Dakar Rally event. This implementation provided such an advantage that in 2006, the organizers of both events banned interconnected suspensions. The Lexus GX470 and Toyota Land Cruiser featured similar systems, under the trademark Kinetic Dynamic Suspension System. In 2005 and 2006, Wilde et. al. published studies on the performance of the Kinetic™system[4, 5], particularly as a means for avoiding rollover in sport utility vehicles.

A number of other researchers have continued to explore the benefits of coupled suspensions, with the primary focus on the hydraulic style of implementation, e.g., Smith and Walker[6], Mavroudakis and Eberhard[7], and Cao et. al.[8].

SAE BAJA

The SAE Baja is an annual design challenge event hosted by SAE International, as a part of their Collegiate Design Series, where teams of undergraduate and graduate engineering students design, analyze, build, test and compete in small off-road style race cars. It is very popular amongst Canadian and U.S. based engineering schools, with around 100 teams competing. The University of Windsor has been fielding an entry continuously since 2000. One of the challenges in the SAE Baja event is the design of a suspension system suitable for performance and driver comfort over very rough terrain. The University of Windsor SAE Baja car uses a double A-arm style front suspension, and independent trailing arm rear suspension. This paper discusses a proposal for a novel Baja vehicle design incorporating an interconnected suspension. The vehicle design is developed using the 3D CAD software CATIA, and analyzed using

Figure 1: CATIA model of Baja car showing interconnection

the Equations of Motion (EoM) multibody dynamics software, to evaluate the effect of the interconnection.

The design is based on a coupling of the anti-roll bar between the front and rear suspensions. Typically, the anti-roll bar will be mounted to the chassis using two bushings, and the two ends of the bar will be attached to the left and right side suspension arms. Deflection of the suspensions in opposing directions results in a twist of the anti-roll bar, and a corresponding resistance force. Deflection of the suspensions in the same direction simply rotates the bar freely in its mounting bushings. In this variation of the design, the anti-roll bar is not mounted directly to the chassis, but rather to a pair bell-cranks that are in turn mounted to the chassis. The bar sits in a location very near to that of a conventional front mounted anti-roll bar. The bell-cranks that carry the anti-roll bar are driven by deflection of the rear suspension. As a result, if the suspension motion is such that that axles are rolling in the same direction, as they would during a cornering maneuver, then the mounts move so that twist of the anti-roll bar is increased. If the suspension motion is such that the axles are rolling in opposing directions, as they would when a single wheel strikes a bump, the mounts move so that the twist of the anti-roll bar is reduced. An image of the CATIA model is shown in Figure 1.

DYNAMIC MODEL

The dynamic model of the vehicle is an extension of the one described in [9]. It has thirteen degrees of freedom, and is composed of twenty rigid bodies. The degrees of freedom are all six motions of the chassis, plus four suspension motions, two front wheel rotations, and the rear axle rotation. The bodies modelled are the chassis, each of the four wheels, an upper and lower A-arm and upright for each front suspension, a trailing arm for each rear suspension, and a drive axle coupling the rear wheels together. There are four bell-cranks that are added to drive the motion of the anti-roll bar, and two bodies that are used to represent the mass of the anti-roll bar.

The total number of constraints in this model is 107, applied through 22 connectors. Each of the upper and lower A-arms is connected to the vehicle body using a revolute joint; they can rotate around the longitudinal axis. Two ball joints connect the upper and lower A-arms to the upright. The front wheel is attached to the upright through a revolute joint around the lateral axis. Finally, the tie rod (modelled as a massless rigid link) connects the upright to the vehicle body. For the rear suspension, one revolute joint connects the swing arm to the chassis, allowing only rotation around the lateral axis. Similar to the front wheel, the rear wheel is attached to the swing arm and can also only rotate around the lateral axis. Finally, the axle is attached to the rear wheels using a constant velocity (CV) joint at each end, with one end also restricting lateral motion of the axle.

The mechanical interconnection starts by mounting each of four bell-cranks to the chassis with a revolute joint. Each of the rear bell cranks is connected to a rear trailing arm by a push-rod modelled as a massless rigid link, and to the front bell-crank on the same side, by another massless rigid link. Each of the front bell-cranks carries a small body representing the mass of the anti-roll bar; these small bodies are constrained to their respective bell-cranks by a revolute joint. Each of the anti-roll bar bodies is then driven by a short massless rigid link connected to a front lower control arm, representing the physical drop-link.

The two bodies that represent the anti-roll bar are connected with a torsional spring running laterally across the chassis with a stiffness of 2000 Nm/rad. The front and rear shock absorbers are all modelled as springs with stiffness, damping, and non-zero free length. The spring stiffness is set 20000 N/m, with a damping value of 100 Ns/m, at both the front and the rear. Each tire is modelled as a bushing, which works like a spring with no free length. The tire has stiffness in the vertical direction, and damping in both horizontal directions, to represent a linear force vs. slip tire model. The vertical stiffness is 75000 N/m and the horizontal damping is 5000 Ns/m at all four tires.

The weight of all the bodies is used to compute the preload in all connectors, and the associated tangent stiffness matrix. The tire-road contact is defined as the periphery of a disk, in the plane of the wheel. The effect of the contact geometry, although simple, is important, as it also factors into the tangent stiffness matrix. As a result, the natural tendency of the tire to 'fall over' side-to-side, and not front-to-back, is captured. The equations are generated around a constant forward speed of 5 m/s, with the associated angular velocity of the wheels and axle, to capture the gyroscopic effects.

Vehicle Configurations

In order to assess the contribution of the coupled suspension system, the vehicle was analyzed in three configurations. First, the analysis was done with no interconnection, and simply the four main suspension springs in place. In the second configuration, the interconnection system was added, with the intent of significantly increasing the roll stiffness while simultaneously reducing twist stiffness, and having a minimal effect on roll stiffness distribution. Finally, a third configuration with a traditional front mounted anti-roll bar was added. In the third configuration, the torsional stiffness of the anti-roll bar was twice that in the second configuration, in order to maintain a comparable increase in roll stiffness.

FORMULATION OF THE EQUATIONS OF MOTION

The equations of motion of the vehicle are formed using the EoM software. The EoM code, described in [9], has been implemented in MATLAB®/Octave and has successfully solved a number of benchmark problems. It is capable of analyzing three dimensional multibody linear systems with many degrees of freedom. It generates the equations of motion in linear descriptor state space form. Once in this form, a number of mathematical analyses are possible with relative ease.

SYSTEM RESPONSE

A total of five sensors were fitted to each configuration of the virtual vehicle. Three sensors were fitted at the right front wheel, to measure the absolute vertical motion of the chassis, the relative vertical motion between the chassis and the wheel, and the relative vertical motion between the wheel and the ground, i.e., the tire compression. The remaining two sensors were the right rear tire compression and the roll angle of the chassis. The motion was driven by two actuators. The first actuator was placed between the front tire and the road, to simulate road disturbances, and the second was a pure rolling moment acting on the chassis.

Modal Analysis

The first analysis was to determine the natural frequencies, damping ratios, time constants and wavelengths. There are three oscillatory modes, 15 damped modes, and eight rigid body modes. The results show that the three expected chassis oscillatory modes are present, but that the wheel hop modes are overdamped. For the baseline, the frequencies are 1.80 Hz for bounce, 2.69 Hz for pitch, and 2.82 Hz for roll. The results for the configurations with modified roll stiffness are almost identical. The bounce and pitch frequencies are unchanged, but the roll frequencies are 3.65 Hz and 3.51 Hz respectively. The increase in roll frequency shows both methods effectively increase roll stiffness. In all cases, the frequencies are relatively high, due to the light weight of the Baja vehicle in comparison to a normal passenger car.

Frequency Response

Next, the models were analyzed using a frequency response approach. Results for the right front corner of the baseline model are shown in Figure 2. Results for the other configurations are quite similar in appearance. The plots show that at low frequencies, the disturbances under the tire are mostly absorbed by motion of the chassis. As the frequency increases, the chassis motion increases as well, to values larger than the size of the disturbance. In the midrange of frequencies, the motion of the chassis starts decreasing, and the disturbance is absorbed by relative motion of the suspension. Finally, at high frequencies, the chassis and suspension motion decrease, and the disturbance is absorbed by tire deflection.

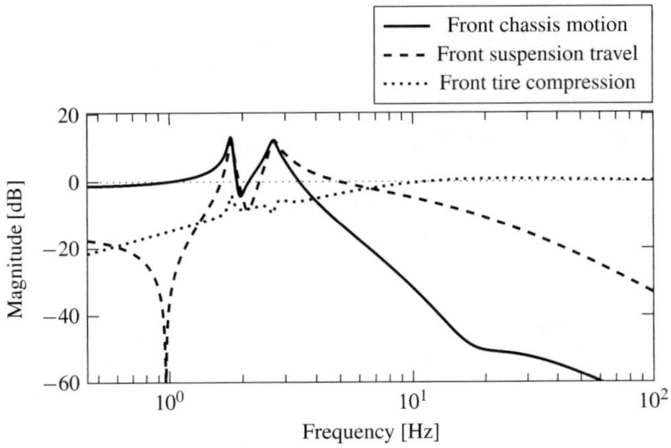

Figure 2: Frequency response: Response of vehicle at right front corner to wheel bump

Steady State Transfer Functions

Finally, a steady state transfer function analysis was also completed. The results for the baseline vehicle configuration are shown in Table 1, the interconnected configuration is shown in Table 2, and the traditional anti-roll bar configuration is shown in Table 3. The tables give the input actuators in columns, and the sensor output in rows. The bottom right entry in each table shows the ratio of chassis roll angle to applied roll moment, i.e, the inverse of the roll stiffness. The results agree with the eigenvalue analysis showing increased roll stiffness in the second and third configurations. The ratio of front tire compression to rear tire compression in response to roll moment is an indicator of the roll stiffness distribution. The tables show that the baseline and the interconnected configurations maintain close to the same ratio, while the third, as it has the anti-roll bar on the front axle, has a significant shift in roll stiffness distribution.

Table 1: Baseline steady state responses

G(s)	Front wheel bump [m]	Roll moment [kNm]
Front chassis motion [m]	8.2630×10^{-1}	8.0107×10^{-2}
Front suspension travel [m]	-1.6163×10^{-1}	6.6402×10^{-2}
Front tire compression [m]	-1.2069×10^{-2}	1.3705×10^{-2}
Rear tire compression [m]	1.3437×10^{-2}	1.2755×10^{-2}
Roll sensor [rad]	5.4523×10^{-1}	1.2174×10^{-1}

Table 2: Interconnected configuration

G(s)	Front wheel bump [m]	Roll moment [kNm]
Front chassis motion [m]	8.2437×10^{-1}	4.9675×10^{-2}
Front suspension travel [m]	-1.6338×10^{-1}	3.8838×10^{-2}
Front tire compression [m]	-1.2250×10^{-2}	1.0837×10^{-2}
Rear tire compression [m]	1.3224×10^{-2}	9.3950×10^{-3}
Roll sensor [rad]	5.4230×10^{-1}	7.5494×10^{-2}

Finally, the front chassis motion and front suspension travel in response to front wheel bump give an indication of the individual corner stiffness. A comparison of all three configurations shows that the interconnected system actually increases suspension travel and reduces chassis motion in response to wheel bump, when compared to the baseline, while the conventional anti-roll bar system system increases chassis motion and decreases suspension travel.

Table 3: Conventional front anti-roll bar configuration

G(s)	Front wheel bump [m]	Roll moment [kNm]
Front chassis motion [m]	8.9328×10^{-1}	5.4176×10^{-2}
Front suspension travel [m]	-9.7262×10^{-2}	4.1536×10^{-2}
Front tire compression [m]	-9.4567×10^{-3}	1.2640×10^{-2}
Rear tire compression [m]	2.4008×10^{-2}	8.6140×10^{-3}
Roll sensor [rad]	6.4719×10^{-1}	8.2356×10^{-2}

CONCLUSIONS

Based on the results seen in the analysis, it appears that the interconnected suspension should offer a performance advantage compared to a conventional anti-roll bar configuration. The results confirm that such a system is capable of simultaneously increasing roll stiffness while reducing twist stiffness, and consequently decreasing individual wheel suspension stiffness. Additional non-linear time history simulations of the vehicle motion in various configurations conducted using a commercial software are planned for further investigation.

REFERENCES

[1] Zapletal, E., 'Balanced Suspension', *SAE Technical Paper 2000-01-3572*, 2000.

[2] Buj, J.F., 'Integral Suspension System for Motor Vehicles Based on Passive Components', *SAE Technical Paper 2002-01-3105*, 2002.

[3] Rideout, G., Anderson, R.J., 'Experimental Testing and Mathematical Modeling of the Interconnected Hydragas Suspension System', *SAE Technical Paper 2003-01-0312*, 2003.

[4] Wilde, J.R., Heydinger, G.J., Guenther, D.A., , Mallin, T.P., Devenish, A.M., 'Experimental Evaluation of Fishhook Maneuver Performance of a Kinetic Suspension System', *SAE Technical Paper 2005-01-0392*, 2005.

[5] Wilde, J.R., Heydinger, G.J., Guenther, D.A., 'ADAMS Simulation of Ride and Handling Performance of the Kinetic™Suspension System', *SAE Technical Paper 2006-01-1972*, 2006.

[6] Smith, M.C., Walker, G.W., 'Interconnected Vehicle Suspension', *Proc. IMechE.*, vol. 219, Part D: J. Automobile Engineering, pp. 295–307, 2005.

[7] Mavroudakis, B., Peter Eberhard, P., 'Mode Decoupling in Vehicle Suspensions Applied to Race Cars', *Proceedings of the III European Conference on Computational Mechanics Solids, Structures and Coupled Problems in Engineering*, 2006.

[8] Cao, D., Rakheja, S., Su, C.Y.,'Roll and Pitch-Plane Coupled Hydro-Pneumatic Suspension', *Vehicle System Dynamics*, vol. 48, no. 3, pp. 361–386, 2010.

[9] Minaker, B.P., Yang, X., Li, S, 'Design Optimization of an SAE Baja Vehicle Using The EoM Open Source Multibody Dynamics Code', *Proceedings of The Canadian Society for Mechanical Engineering International Congress*, 2014.

VISCOELASTIC MULTIBODY DYNAMICS OF WHIPLASH

Mohamed T.Z. Hassan
Mechanical & Industrial Engineering
University of Toronto
Toronto, Ontario, Canada
mhassan@mie.utoronto.ca

S.A. Meguid
Mechanical & Industrial Engineering
University of Toronto
Toronto, Ontario, Canada
meguid@mie.utoronto.ca

ABSTRACT

Whiplash is one of the most common injuries in car accidents. The kinematics of the head and neck during rear impacts are important for assessment of possible trauma. In the current model the head and the neck are modeled as rigid bodies connected through 2-DOF joints. The model is validated using experimental results. Results show that the neck undergoes extension followed by the rebound phase. The head reaches its maximum displacement and rotation at around 150 milliseconds after the impact.

KEYWORDS: Biomechanics, whiplash, rigid body, Multibody dynamics, Lagrange principle

INTRODUCTION

According to the National Highway Traffic Safety Administration (NHTSA) in the US about 806,000 whiplash injuries occur yearly costing over $9 billion [1]. In Japan, about 50% of low impact speeds car accidents results in neck injuries. According to the statistics [2] about 95% of rear-end collisions results in minor injuries of which approximately 80% were concentrated on the neck. For drivers suffering from Whiplash-Associated Disorders (WAD), rear impacts cause some 60% of the injuries, while frontal and side impacts cause 27% and 16%, respectively [2]. Whiplash patients suffer from headache, neck pain, limited neck motion, visual impairment, weakness and dizziness [3-4].

Neck injury criteria are developed to assess whiplash injuries. Two of the most common criteria are the Neck Injury Criterion (NIC) [5] and Intervertebral Neck Injury Criterion IV-NIC [6]. These criteria evaluate the possibility of whiplash injury using the head velocity and acceleration, and the intervertebral rotation. Therefore, it is of great importance to capture the kinematics of the head and neck during rear-impacts

One of the first models was developed in 1968 by Martinez et al. [7] to simulate the response during whiplash using rigid body dynamics. Some models segmented the human body and neck into rigid segments which are connected with revolute joints having rotational spring and damper [8-9]. A mathematical model of the head and neck was developed by De Jager [10] in which the rigid head and vertebrae are connected by linear viscoelastic discs, non-linear viscoelastic elements, frictionless facets and Hill-type muscles. Esat et al. [11] modeled the vertebrae as rigid bodies which are connected by linear viscoelastic intervertebral discs and nonlinear viscoelastic ligaments. The muscle elements behave actively and passively.

Models in which the human neck is segmented into rigid bodies and are connected through revolute joints are based on the theory that the motion of two vertebrae relative to each other can be modeled as rotation around an instantaneous axis of rotation (IAR). However, during crashes the IAR changes its position moving upwards compared to the location of IAR during normal motion [12]. Therefore, limiting the motion between two adjacent vertebrae to pure rotation may not be accurate and will affect the resulting kinematics.

The objective of this study is to develop a computationally rigorous and efficient rigid body model to determine the kinematics of the passenger's head and neck during rear impacts. In this model, both rotational and extensional degrees of freedom are considered.

MULTIBODY DYNAMICS MODEL

In the current study, the head, the cervical spine (C_1-C_7) and the first thoracic vertebra T_1 are modeled as rigid bodies. The mass, moment of inertia and location of the center of mass (COM) for each segment is shown in Table 1 [13]. The initial position of COM of each segment, the global axes, and the local axes attached to each body are shown in Fig. 1.

Each two adjacent bodies are connected using a 2 DOF joint allowing in-plane rotation around the Z-axis and translation in the direction of the local y-axis of the inferior segment. The planar joints have damping and stiffness coefficients for each DOF. The linear and rotational stiffness of each intervertebral level was obtained from the study by Shea et al. [14]. Since no data were found for the damping coefficient for an entire intervertebral segment, the damping coefficients of intervertebral discs will be used instead. The linear and rotational damping coefficients of the intervertebral disc were obtained from study by van Lopik et al. [13]. The stiffness and damping coefficients are shown in Table 2.

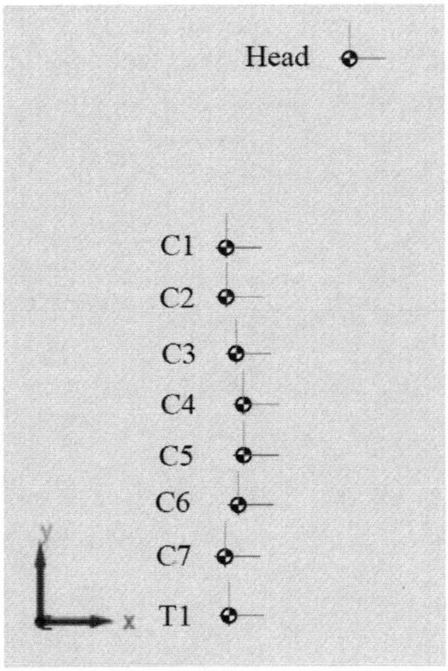

Figure 1: Initial position of the rigid bodies

Table 1: Mass, moment of inertia and location of COM of the vertebrae [13]

	Mass (kg)	I_{xx} (x10^{-4} kg.m^2)	I_{yy} (x10^{-4} kg.m^2)	I_{zz} (x10^{-4} kg.m^2)	I_{xy} (x10^{-4} kg.m^2)	X_g (mm)	Y_g (mm)
T_1	-	-	-	-	-	0	0
C_7	0.22	2.2	2.2	4.3	-	-1.3	19.7
C_6	0.24	2.4	2.4	4.7	-	3.1	36.9
C_5	0.23	2.3	2.3	4.5	-	4.9	53.6
C_4	0.23	2.3	2.3	4.4	-	4.8	70.5
C_3	0.24	2.4	2.4	4.6	-	2.5	87.5
C_2	0.25	2.5	2.5	4.8	-	-0.7	106.5
C_1	0.22	2.2	2.2	4.2	-	-0.7	123
Head	4.69	181	236	173	71	40	186

Table 2 Stiffness and damping coefficients [13-14]

Rotational stiffness (N.m.deg^{-1})	Rotational damping (N.m.s/deg)	Linear stiffness (x10^5 N.m^{-1})	Linear damping (N.s.m^{-1})
2.3	0.0262	11.14	1000

The system under analysis is a multibody dynamic system which can be analyzed using Lagrange's principle:

$$\frac{d}{dt}\left(\frac{\partial L}{\partial \dot{q}_k}\right) - \frac{\partial L}{\partial q_k} + \frac{\partial f}{\partial \dot{q}_k} = Q_k \quad (1)$$

where L is the Lagrangian given by Eq. (2), $\dot{q}_k$ is the generalized velocities, q_k is the generalized coordinates, f is the dissipation function given by Eq. (3), and Q_k is the generalized forces.

$$L = T - V \quad (2)$$

where T is the kinetic energy of the system given by Eq. (4) and V is the potential energy of the system given by Eq. (5).

$$f = \frac{1}{2}\sum_{i=1}^{8} c_i \dot{\theta}_{i,r}^2 \quad (3)$$

$$T = \sum_{i=1}^{8} I_i \dot{\theta}_i^2 + \sum_{i=1}^{8} m_i v_i^2 \quad (4)$$

$$V = \frac{1}{2}\sum_{i=1}^{8} k_i \theta_{i,r}^2 \quad (5)$$

where I_i is the mass moment of inertia of the body, m_i is the mass of the body, k_i is the joint stiffness, c_i is the joint damping coefficient, $\dot{\theta}_i$ is the absolute angular velocity, v_i is the absolute linear velocity, $\theta_{i,r}$ and $\dot{\theta}_{i,r}$ are the angle and angular velocity relative to the body below, respectively. The change in potential energy due to change of body height is neglected and only potential energy due to joint stiffness is considered.

The position of a body w.r.t to the local axes of the body below can be given by Eqs. (6) and (7).

$$x_i = l_i \sin(\theta_{i,r}) \quad (6)$$

$$y_i = l_i \cos(\theta_{i,r}) + y_{i,r} \quad (7)$$

where l_i is the initial distance between two adjacent bodies, $y_{i,r}$ is the relative displacement of a body in the y-direction w.r.t the body below. The position of a body w.r.t the global axis is given by Eq. (8).

$$\begin{bmatrix} X_k \\ Y_k \end{bmatrix} = \left(\sum_{i=1}^{k} [M_i]\right) \begin{bmatrix} x_k \\ y_k \end{bmatrix} \quad (8)$$

where M_i is the transformation matrix given by:

$$[M_i] = \begin{bmatrix} \cos(\theta_{i,r}) & \sin(\theta_{i,r}) \\ -\sin(\theta_{i,r}) & \cos(\theta_{i,r}) \end{bmatrix} \quad (9)$$

The linear velocity v_i is a function in the time derivatives of X_k and Y_k.

Substituting in Eq. (1) will provide 16 differential equations which can be solved to obtain θ and y for each body.

The model is developed using MATLAB Simulink. Fig.2 shows C_1 and C_2 vertebrae connected using planar joints. The joint is provided with stiffness and damping coefficients. Each body and joint is connected to a sensor to capture the position, velocity and acceleration. The impact is applied at the T_1 vertebra which is connected to the ground through a prismatic joint allowing movement in the x-direction only (not shown). The model is validated by comparing the results of the model with results obtained experimentally on cadavers [15].

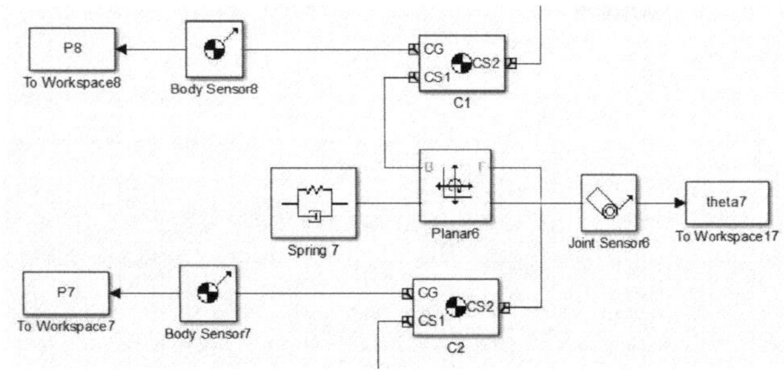

Figure 2: Simulink model

RESULTS & DISCUSSIONS

The impact applied to the T_1 vertebra is in the form of velocity profile in the positive x-axis direction as shown in Fig. 3. This velocity profile is the same one used in the experimental work [15].

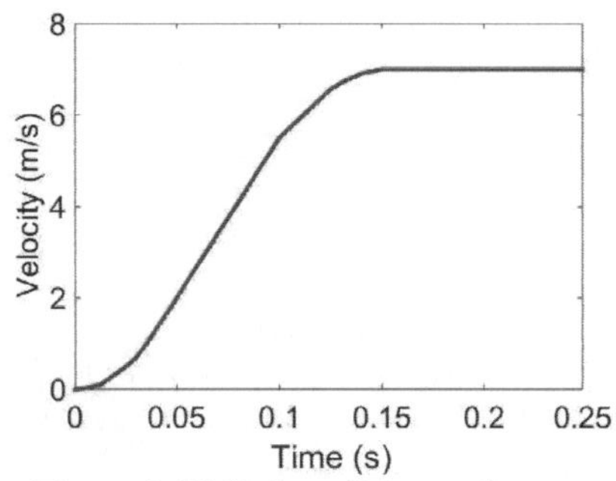

Figure 3: Velocity of T_1 vertebra

The displacement of the head with respect to T_1 vertebra is shown in Fig. 4. The model response is similar in trend to the experimental results for the horizontal displacement. The maximum horizontal displacement about 200 mm and is reached after about 150 ms from impact followed by the rebound phase. The vertical head displacement of the model is comparable to the experimental results. However, the magnitude and time of maximum displacement of the model is greater than that from expeimental results [15].

The total head rotation during impact is shown in Fig.5. Again, the model provides a trend similar to the experimental result. Again, the value and time of maximum head rotation of the model is greater than that of the experimental results.

The difference between the model's response and the experimental results can be related to the damping coefficients used in the study. The damping coefficients used for both linear and rotational DOF are those of the intervertebral discs only without taking into account the effect of the other tissues in the neck like ligaments and muscles. Incorporating the effect of these tissues will increase the damping of the model which will damp the response of the head.

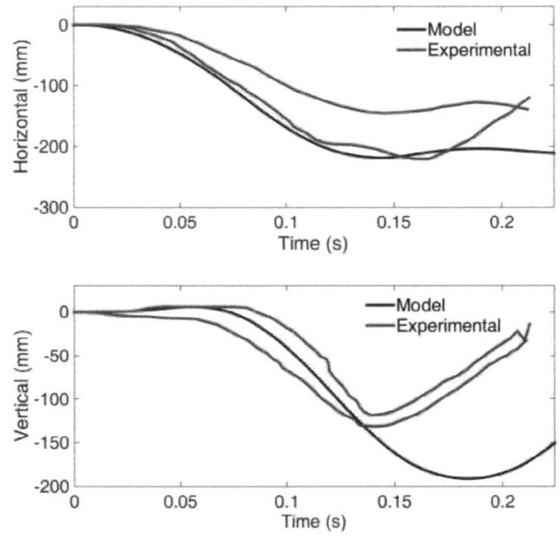

Figure 4: Head displacement with respect to T_1 vertebra

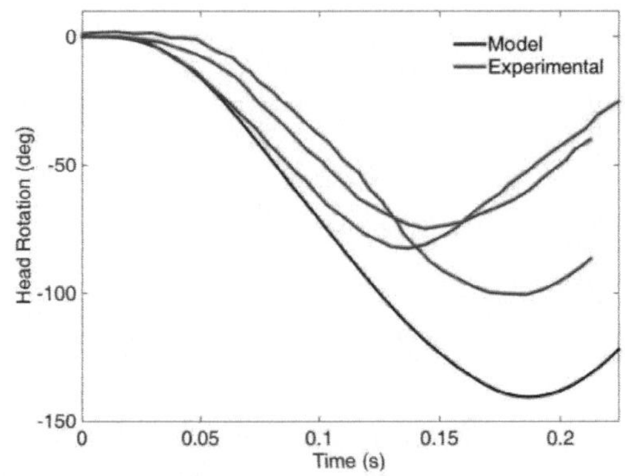

Figure 5: Head rotation

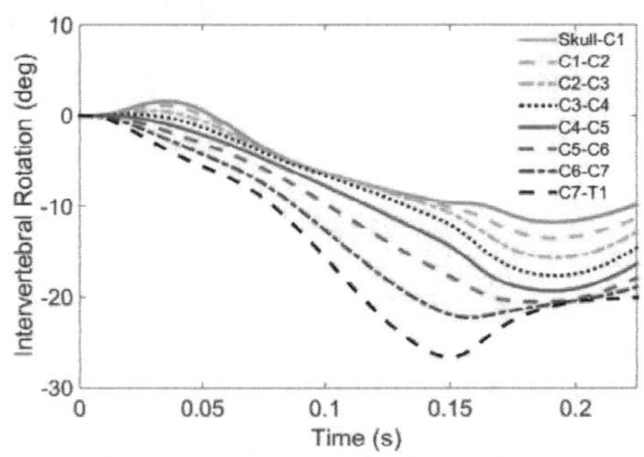

Figure 6: Intervertebral rotation

The intervertebral rotation (rotation between two adjacent vertebrae) of the model is shown in Fig.6. Positive rotation indicates flexion (forward rotation), while negative rotation indicates extension (backward rotation). During the first milliseconds the upper cervical spine (Skull-C_2) undergoes flexion for the first 50 milliseconds followed by extension, while the remaining part of the spine (C_2-T_1) undergoes extension only. This behavior produces the well-known S-shape of the spine during the first milliseconds of whiplash as reported by Grauer et al. [16]. The maximum intervertebral rotation (extension) is reached during the first 150-200 milliseconds followed by the rebound phase. The largest intervertebral rotation was recorded for the lower part of the cervical spine.

CONCLUSIONS

The model developed in this study captures the kinematics of the head and neck during rear impacts. After impact the neck undergoes extension until the head reaches its maximum position followed by the rebound phase of the head. The results of the current model show that whiplash injury is predicted to occur within the first 150-200 milliseconds after impact. During this time the maximum head displacement and rotation are recorded.

Accurate damping coefficients for the cervical spine segments incorporating the effect of all tissues instead of intervertebral disc's damping coefficients will enhance the response and accuracy of the model.

ACKNOWLEDGMENTS

The Authors gratefully acknowledge the kind support of Qatar National Research Fund (QNRF) under the National priority research program.

REFERENCES

1. Kuppa S., Saunders J., Stammen J., MalloryA., "Kinematically Based Whiplash Injury Criterion". s.l. : National Highway Traffic Safety Administration .
2. Ono K., Kanno M. 1996, "Influences of the Physical Parameters on the Risk to Neck Injuries in Low Impact Speed Rear-End Collisions", *Accident Analysis and Prevention*, Vol(28), pp. 493-499.
3. Hai-bin C., Yang K., Zheng-guo W. 2009, "Biomechanics of Whiplash Injury", *Chinese Journal of Traumatology*, Vol(12), pp. 305-314.
4. Siegmund G. P., Winkelstein B. A., Ivancic P. C., Svensson M. Y., Vasavada A. 2009, "The Anatomy and Biomechanics of Acute and Chronic Whiplash Injury". *Traffic Injury Prevention*, Vol(10), pp. 101-112.
5. Boström O., Svensson M. Y., Aldman B., Hansson H. A., Håland Y., Lövsund P., Seeman T., Suneson A., Säljö A., Örtengren T. 1996, "A New Neck Injury Criterion Candidate-Based on Injury Findings in the Cervical Spinal Ganglia After Experimental Neck Extension Trauma", *International IRCOBI Conference on the Biomechanics of Impact*, Dublin, Ireland.
6. Panjabi M. M., Ito S., Ivancic P. C., Rubin W. 2005, "Evaluation of the Intervertebral Neck Injury Criterion Using Simulated Rear Impacts". *Journal of Biomechanics*, Vol(38), pp.1694-1701.
7. Martinez J. L, Garcia D. J. 1968, "A Model for Whiplash". *Journal of Biomechanics*, Vol(1), pp. 23-32.
8. Garcia T., Ravani B. 2003, "A Biomechanical Evaluation of Whiplash Using a Multi-Body Dynamic Model", *Journal of Biomechanical Engineering*, Vol(125), pp. 254-265.
9. Himmetoglu S., Acar M., Bouazza-Marouf K., Taylor A. 2009, "A Multibody Human Model ofr Rear-Impact Simulation". *Proceedings of the Institution of Mechanical Engineers Part D: Journal of Automobile Engineering*, Vol(223), pp. 623-638.
10. De Jager M. K. J. 1996, "Mathematical Head-Neck Models for Acceleration Impacts". *(PhD Thesis)*.
11. Esat V., Acar M. 2007, "A Mulit-body Model of the Whole Human Spine For Whiplash Investigations", *20th Enhanced Safety of Vehicles Conference: Innovations for Safety: Opportunities and Challenges*, Lyon, France.
12. Kaneoka K., Ono K., Inami S., Hayashi K. 1999, "Motion Analysis of Cervical Vertebrae Druing Whiplash Loading", *Spine*, Vol(24), pp. 763-770.
13. Van Lopik D. W., Acar M. 2007 "Development of a Multi-body Computational Model of Human Head and Neck", *Proceedings of the Institution of Mechanical Engineers, Part K: Journal of Multi-body Dynamics*, Vol(221), pp. 175-197.
14. Shea M., Edwards W. T., White A. A., Hayes W. C. 1991, "Variations of Stiffness and Strength Along the Human Cervical Spine", *Journal of Biomechanics*, Vol(24), pp. 95-107.
15. Meyer F., Bourdet N., Deck C., Willinger R. 2004, "Human Neck Finite Element Model Development and Validation against Original Experimental Data". *Stapp Car Crash Journal*, Vol(48), pp.177-206.
16. Grauer J. N., Panjabi M. M., Cholewicki J., Nibu K., Dvorak J. 1997, "Whiplash Produces an S-Shaped Curvature of the Neck With Hyperextension at Lower Levels", *Spine*, Vol(22), pp. 2489-2494.

Proceedings of the 25th CANCAM
London, Ontario, Canada, May 31 - June 4, 2015

ROAD PROFILE RECOVERY USING ONBOARD ACCELERATION DATA

Fei Hu[†] **Bruce P Minaker PhD PEng**
Department of Mechanical,
Automotive, & Materials Engineering
University of Windsor, Windsor, ON N9B 3P4
Email:hu112@uwindsor.ca

Mehrzad Abdollahi PhD
Load Analysis Group,
Chrysler Technical Center
FCA US LLC, Auburn Hills, MI 48321-8004
Email:mehrzad.abdollahi@fcagroup.com

ABSTRACT

For the purpose of studying vehicle stability and suspension performance, road profiles are an important element that should be considered. Various methods have been developed to measure road surfaces. This paper develops a fast, computationally efficient, and economical method to obtain road profile data only using accelerometers. Two sensors are mounted on a vehicle to measure both wheel and body acceleration in the vertical direction. This approach is based on a dynamic model with an output tracking feedback control. Simulation results show that this method recovers road input with high accuracy.

KEYWORDS: Road profile, Dynamic model, Tracking controller, Accelerometer, Ride comfort

INTRODUCTION

Poor road profiles cause vehicle vibration and is one of the most important factors that should be considered in order to study vehicle stability and suspension performance, such as ride comfort or structural durability. Therefore, methods to accurately measure road profiles are very desireable.

There are a number of solutions that have been proposed to obtain road profiles, including physical or laser based measurement, and camera based techniques, typically for the purpose of road serviceability, survey, or maintenance. Many studies have been conducted with these types profilometers. One of the the first is the inertial profiler proposed by the General Motors Research Laboratories in 1964 [1, 2]. This instrument used a vehicle mounted accelerometer to collect data. However, it was found that inertial profilers can only measure and record road profile at certain speeds, e.g., between 16 and 112 km/h. In addition, this method depends heavily on the sensor location, because it is difficult to ensure a visual connection between the instrument and the ground. In [3], Pong has also found some drawbacks and limitations of its capabilities. Another example is a longitudinal profile analyser with a mounted laser sensor, developed by the Road and Bridges Central Laboratory in France [1, 4]. This system is a trailer towed mechanism that directly contacts the road surface while moving at constant speed. The limitation is that this device cannot be integrated in ordinary passenger cars. This instrument sometimes produces biased or corrupted measures as well.

The use of wheel force transducers to calculate the effective road profile is very popular in industry, but these are expensive, heavy enough to influence the suspension performance, and the procedure takes a long time. A more analytical approach is to use a dynamic model and an 'observer', a numerical technique that attempts to force the model states to converge to the real states. The sliding mode concept is quite popular in this area, e.g., [1, 5, 6]. Gonzalez et al. proposed a road roughness estimate method based on vehicle accelerometers in [7]. This method shows a relationship between the power spectral densities of the road surface and the vehicle accelerations by a transfer fuction. However, road profiles are only accurately estimated at a constant vehicle velocity, and a known profile is required. A similar road estimator based on the Fourier transform of the road is developed by Hong et. al. in [8]. In [9], the road estimator can be used at variable velocity but prior information on the road is required. In [10], a vehicle mounted system combining laser displacement sensors, accelerometers, and a gyroscope is developed. An algorithm based on the double integration of the acceleration with respect to time is designed to recover road profiles. However, this instrument is complicated and expensive.

In recognition of these disadvantages, a method using only accelerometer measurements is proposed as an alternative. Accelerometers are small, cost-efficient sensors commonly used to measure vibrations, with operating principles that are typically based on the properties of piezo-electric materials. They are increasingly commonly found in a wide range of equipment, even smartphones. In this approach, two (or more) sensors are mounted on a vehicle to measure both body and wheel accelerations in the vertical direction. A data recovery algorithm is developed using a dynamic model (quarter car) that is then coded and tested in MATLAB®/Octave. In order to develop the method, a complex multibody nonlinear quarter car model is simulated using Altair MotionView®, with the linear data recovery algorithm running in parallel.

[†] corresponding author

THEORETICAL BACKGROUND

A. *Profile*

A profile is a two-dimensional slice of a road surface, taken along an imaginary line. As shown in Figure 1, a profile consists of longitudinal and lateral profiles. This research is focused on the longitudinal profile as it shows the design grade, roughness, and texture [2, 11].

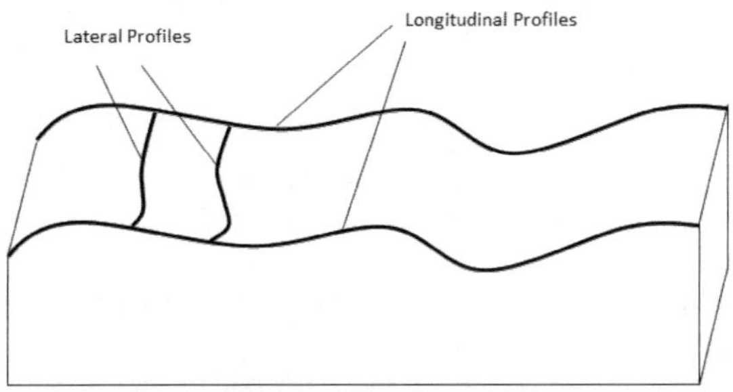

Figure 1: Road profile

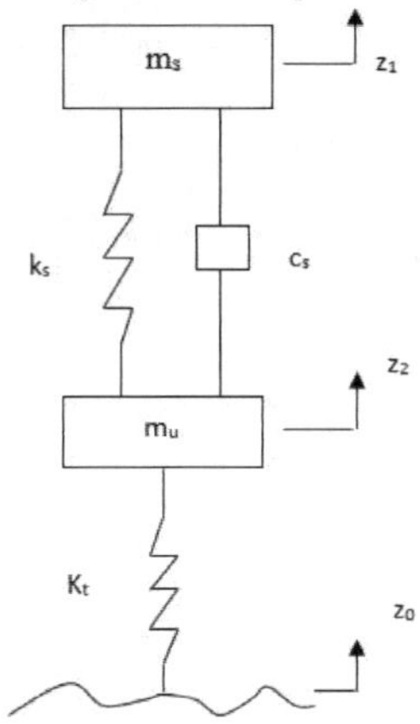

Figure 2: A quarter car model

B. *Vehicle modeling*

To describe the vertical dynamics of a vehicle, the well known quarter car model, shown in Figure 2 is considered. The model consists of a single sprung mass (the vehicle body, powertrain, passengers, cargo, etc.) connected to an unsprung mass (wheel and axle assembly). For simplicity, pitch and roll motions are not considered. The suspension between the sprung mass and unsprung mass is modeled as a passive viscous damper and linear coil spring. The tire is modeled as a simple linear spring without damping. The omission of tire damping is quite common in the quarter car model, as tires are generally surprisingly lightly damped; the tire is more accurately considered as an air spring than a rubber bushing.

Using Newton's second law, the equation of motion for the sprung mass and the unsprung mass are respectively:

$$m_s\ddot{z}_1 + c_s(\dot{z}_1 - \dot{z}_2) + k_s(z_1 - z_2) = 0 \quad (1)$$

$$m_u\ddot{z}_2 + c_s(\dot{z}_2 - \dot{z}_1) + k_s(z_2 - z_1) + k_t z_2 = k_t z_0 \quad (2)$$

where m_s and m_u are respectively the mass of the quarter car body and the wheel, k_s is the stiffness of the spring incorporated in the suspension system, k_t is the tire stiffness, c_s is the damping coefficient of the shock absorber, z_1 and z_2 are the position of the sprung mass and the unsprung mass respectively, z_0 is the displacement of the road surface. By denoting

$$\mathbf{z} = \begin{Bmatrix} z_1 \\ z_2 \end{Bmatrix}, \ \mathbf{u} = z_0 \quad (3)$$

the two equations can be combined into a vector equation.

$$\mathbf{M}\ddot{\mathbf{z}} + \mathbf{L}\dot{\mathbf{z}} + \mathbf{K}\mathbf{z} = \mathbf{F}\mathbf{u} \quad (4)$$

where

$$\mathbf{M} = \begin{bmatrix} m_s & 0 \\ 0 & m_u \end{bmatrix} \ \mathbf{L} = \begin{bmatrix} c_s & -c_s \\ -c_s & c_s \end{bmatrix}$$
$$\mathbf{K} = \begin{bmatrix} k_s & -k_s \\ -k_s & k_s + k_t \end{bmatrix} \quad \mathbf{F} = \begin{bmatrix} 0 \\ k_t \end{bmatrix} \quad (5)$$

The vector equations of motion can be recast in a first order form.

$$\begin{Bmatrix} \dot{\mathbf{x}} \\ \mathbf{y} \end{Bmatrix} = \begin{bmatrix} \mathbf{A} & \mathbf{B} \\ \mathbf{C} & \mathbf{D} \end{bmatrix} \begin{Bmatrix} \mathbf{x} \\ \mathbf{u} \end{Bmatrix} \quad (6)$$

The following definitions are required.

$$\mathbf{x} = \begin{Bmatrix} \mathbf{z} \\ \dot{\mathbf{z}} \end{Bmatrix}, \ \mathbf{y} = \ddot{\mathbf{z}} \quad (7)$$

$$\begin{bmatrix} \mathbf{A} & \mathbf{B} \\ \mathbf{C} & \mathbf{D} \end{bmatrix} = \left[\begin{array}{cc|c} \mathbf{0} & \mathbf{I} & \mathbf{0} \\ -\mathbf{M}^{-1}\mathbf{K} & -\mathbf{M}^{-1}\mathbf{L} & \mathbf{M}^{-1}\mathbf{F} \\ \hline -\mathbf{M}^{-1}\mathbf{K} & -\mathbf{M}^{-1}\mathbf{L} & \mathbf{M}^{-1}\mathbf{F} \end{array}\right] \quad (8)$$

In this model, the input $\mathbf{u}$ is the road displacement and the output $\mathbf{y}$ is sprung and unsprung mass accelerations. These equations of motion can be produced by hand, or for similar but more mechanically complex models, e.g., when including a suspension system, by using the equation of motion generator software 'EoM' that is developed by the University of Windsor Vehicle Dynamics and Control research group.

The error $\mathbf{e}$ is defined as the difference between the measured accelerations $\mathbf{r}$, and those predicted by the model.

$$\mathbf{e} = \mathbf{r} - \mathbf{y} \quad (9)$$

Next, a 'controller' that can drive the error between this acceleration and the measured acceleration to zero is introduced.

$$\dot{\mathbf{u}} = -\mathbf{G}\mathbf{u} + \mathbf{H}\mathbf{e} \quad (10)$$

The vehicle model and the controller can be joined into a combined model. Figure 3 shows the block diagram of the system. The selection of $\mathbf{G}$ and $\mathbf{H}$ are based on allowed convergence time and quality of fit respectively. If the $\mathbf{G}$ and $\mathbf{H}$ are chosen such that the system is stable and has the appropriate time response, once the measured acceleration is given as the

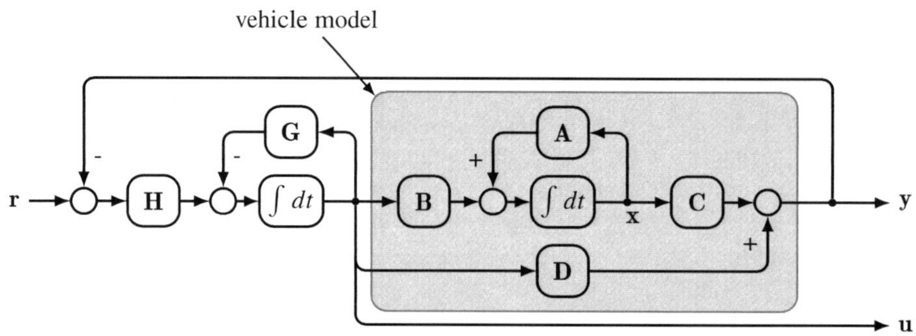

Figure 3: System block diagram

input of the system, the road displacement can be obtained from the output.

$$\begin{bmatrix} \mathbf{A}' & \mathbf{B}' \\ \mathbf{C}' & \mathbf{D}' \end{bmatrix} = \left[\begin{array}{cc|c} \mathbf{A} & \mathbf{B} & \mathbf{0} \\ -\mathbf{HC} & -\mathbf{G}-\mathbf{HD} & \mathbf{H} \\ \hline \mathbf{0} & \mathbf{I} & \mathbf{0} \end{array}\right] \tag{11}$$

$$\mathbf{x}' = \begin{Bmatrix} \mathbf{x} \\ \mathbf{u} \end{Bmatrix},\ \mathbf{u}' = \begin{Bmatrix} \mathbf{r} \end{Bmatrix},\ \mathbf{y}' = \begin{Bmatrix} \mathbf{u} \end{Bmatrix} \tag{12}$$

Of course, the real suspension will not be modeled perfectly using a linear quarter car model, and as a result there will be some discrepancy between the real road input and the recovered road input. However, the model used must be computationally simple and efficient enough that it can run in real time on low cost hardware. There is a tradeoff increased computation effort from model complexity, and the fidelity of the end result.

SIMULATION RESULTS AND DISCUSSION

To show the efficiency of the proposed method, some simulations were carried out using MATLAB® and Altair MotionView®. In order to test the algorithm, the road profile shown in Figure 4 is used as a baseline to compare the recovered road profile. The profile contains content that is spread over a span of wavelengths, with the amplitude decreasing with the wavelength. Table 1 shows the parameter values that are used in the quarter car model.

Using typical values for the vehicle parameters, and a value of $\mathbf{G} = 2$, giving the recovery algorithm a time constant of 0.5 seconds, several optimization trials using both genetic algorithm (GA) and *fminsearch* (Nelder-Mead method) function in MATLAB® have been done to select the **H**. in order to optimize the quality of the fit. Figure 5 shows the body and wheel accelerations in the vertical direction, which are generated from a linear quarter car model. These data are then used as the input of the recovery system. When using the accelerations as input, the recovered road profile compares almost exactly with the original road profile. The difference between the recovered and measured road profile is shown in Figure 6. The maximum value of the difference is below 10^{-4} m. When compared to the road profile, this result shows very high accuracy.

Of course, when using the same linear model to both generate the accelerations, and recover the displacement input, one would expect very good agreement. In order to verify the result in a more realistic test, a corresponding multibody model of the vehicle suspension is built in Altair MotionView®. As shown in Figure 7, two quarter car models are connected by a differential equation (controller), which represent a corner model of the test vehicle and a recovery quarter car model respectively. The model incorporates nonlinear kinematics due to the geometry of the suspension mechanism, and a non-linear damper. Nonlinearities in the tire are ignored in the results presented, but are source of future study. The **H** values in the recovery algorithm were also modified when adapting to the nonlinear model. With the linear data recovery algorithm running in parallel in the MotionView environment, the road profile can be also recovered. The accuracy is not as good as the result from the linear model, but it is still a promising achievement, which suggests that the algorithm will provide acceptable results when implemented in a real environment.

Table 1: Quarter car model parameter values

Parameter	Value	Unit
m_s	450	kg
m_u	61.8	kg
k_s	14130	N/m
c_s	3000	N s/m
k_t	150000	N/m
H	[43742 61937]	–

CONCLUSIONS AND FUTURE WORK

Road profile data plays a key role in assessing vehicle suspension performance. It becomes increasingly important to develop methods collect road profiles precisely. This paper presents a simple and economical method of recovering road profile effectively from a data recovery algorithm by using acceleration data. Results show that this approach can recover road profile with both linear and nonlinear vehicle parameters.

Numerous uncertainties still exist and are avenues of future study. These include methods for determination of the **G** and

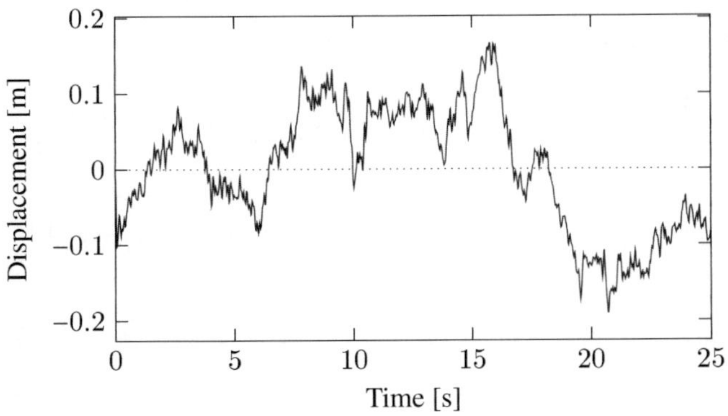

Figure 4: Road profile input

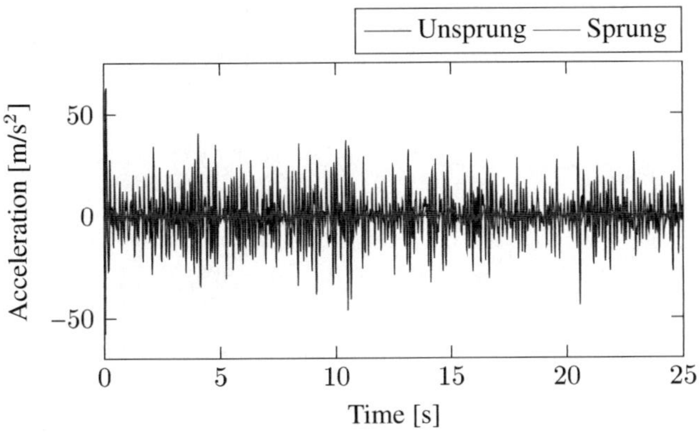

Figure 5: Vertical sprung and unsprung accelerations

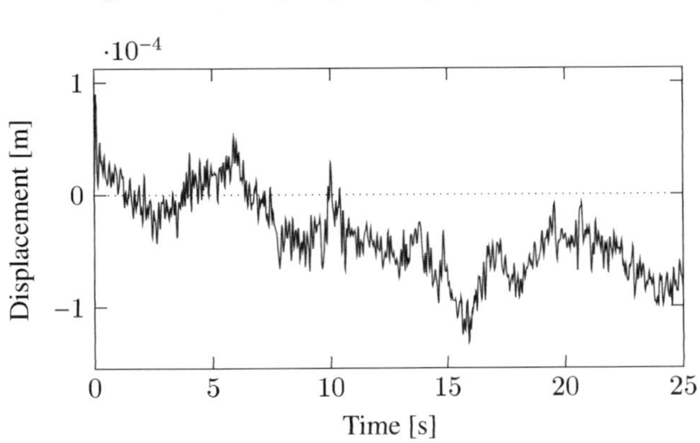

Figure 6: Difference between input and recovered road profile

H values independently of the road profile, the extension of the recovery model to half car or full car, the selection of the model parameters to best represent nonlinear vehicle dynamics, exploring the possibility of including nonlinearities in the vehicle model, and finally the effects of measurement noise on the results should all be considered.

REFERENCES

[1] Imine, H., Delanne, Y., M'Sirdi, N.K., 'Road profile input estimation in vehicle dynamics simulation', *Vehicle System Dynamics* Vol. 44, No. 4, 2006, pp. 285-303.

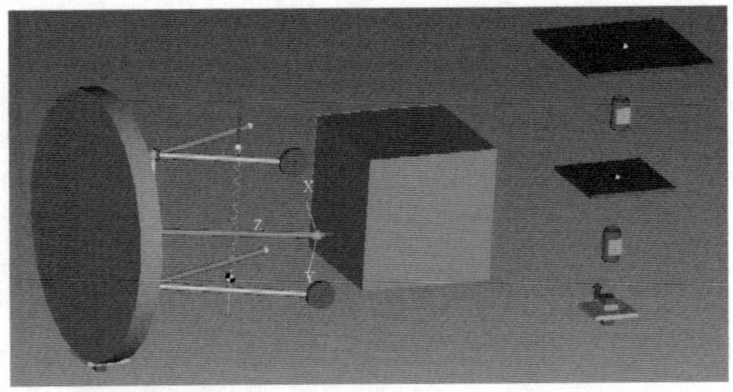

Figure 7: Recovery model in Altair MotionView®

[2] Karamihas, S. M., Sayers, M. W., 'The little book of profiling', *Ann Arbor: Transportation Research Institute, University of Michigan*,1998.

[3] Pong, M. F., 'The development of an extensive-range dynamic road profile and roughness measuring system', *Doctoral Dissertation, Pennsylvania State University*, 1992.

[4] Doumiati, M., et al., 'Estimation of road profile for vehicle dynamics motion: experimental validation', *American Control Conference (ACC)*, IEEE, 2011.

[5] McCann, R., Nguyen, S., 'System identification for a model-based observer of a road roughness profiler', *Proc of Region 5 Technical Conference*, IEEE, 2007.

[6] Rabhi, A., et al., 'Second order sliding mode observer for estimation of road profile', *Proc of Workshop on Variable Structure Systems VSS*, IEEE, 2006.

[7] Gonzalez, A., et al., 'The use of vehicle acceleration measurements to estimate road roughness', *Vehicle System Dynamics* Vol. 46, No. 6, 2008, pp. 483-499.

[8] Hong, K. S., Sohn, H. C., Hedrick, J. K., 'Modified skyhook control of semi-active suspensions: A new model, gain scheduling, and hardware-in-the-loop tuning', *Journal of Dynamic Systems, Measurement ,and Control*, V01. 124, No. 1, 2002, pp. 158-167.

[9] Heyns, T., Heyns, P. S., Villiers, J. P., 'A method for real-time condition monitoring of haul roads based on Bayesian parameter estimation', *Journal of Terramechanics* Vol. 49, 2012, pp. 103-113.

[10] Yang, D., Yi, H., Lian, X., 'Research on rapid measurement of medium short wave longitudinal road profiles', *International Conference on Electrical and Control Engineering (ICECE)*, IEEE, 2010.

[11] Reza-Kashyzadeh, K., et al., 'Investigating the effect of road roughness on automotive component', *Engineering Failure Analysis* Vol. 41, 2014, pp. 96-107.

Multicomponent Materials

NUMERICAL PREDICTION ON ELASTIC PROPERTIES OF CHOPPED FIBER REINFORCED COMPOSITES

Hui Liu, Liying Jiang*
Department of Mechanical and Materials Engineering
University of Western Ontario
London, Ontario, Canada
hliu389@uwo.ca, *lyjiang@eng.uwo.ca

Danielle Zeng, Yang Li
Research and Innovation Center
Ford Motor Company
Dearborn, Michigan, United States
dzeng@ford.com, yli197@ford.com

ABSTRACT

A RVE model with embedded solid elements is proposed to predict the equivalent elastic properties of chopped fiber reinforced composites. The major merit of the embedded element technique lies in the fact that the meshes of matrix and fibers could be created independently, which will significant improve the overall computation efficiency. The approach to determine the equivalent material properties of the heterogeneous RVE is illustrated. The proposed model based on the embedded element technique is verified by comparing the simulation results with the FE direct mesh analysis. In addition, mesh sensitivity for the computation model has been analyzed for the accuracy purpose.

KEYWORDS: Chopped fiber reinforced composite; Elastic properties; Embedded element technique; Strain energy equivalence; Representative volume element (RVE).

INTRODUCTION

As well known, chopped fiber reinforced composites have extensive applications particularly in automobile industry due to the low costs and ease of manufacture. Such materials usually lack information of detailed material properties and performance characterization, which are essential for their full potential applications. During the past decades, researchers have attempted to develop both theoretical and numerical approaches to predict their effective mechanical properties with lower cost compared to experimental tests. Generally, those approaches can be classified into two main categories, i.e., micromechanics modeling and representative volume element (RVE) / finite element analysis [1, 2]. The micromechanics models include the Eshelby's equivalent inclusion model [3], the self-consistent model [4], the Mori-Tanaka model [5], the bounding model [6], the Harpin-Tsai equation [7], and the shear lag model [8] among others. However, those models are limited since they cannot capture the micro-mechanical stress-strain state of materials realistically. Alternatively, the RVE method has been pursued by researchers to simulate the mechanical behaviors of the composites. In the RVE analysis, periodic boundary conditions are usually employed to eliminate the false boundary effects and improve the computational accuracy [2].

To impose the periodic boundary conditions on the RVE with FE direct meshes, the elements and nodes on opposite faces of the RVE are required to be exactly same. In addition, the grids on the interfaces between the matrix and fibers should be also coincident. Therefore it is very complex to produce these direct meshes, which consequently causes extreme computation expenses in the preprocessing of the FE analysis. In the current work, an embedded element technique is employed to achieve this purpose easily. With this technique the matrix and fibers in the RVEs can be meshed independently. Thus a structured mesh can be used for the matrix phase and free meshes can be employed for the fibers. Then the whole RVE analysis will be completed automatically by using the Python script in ABAQUS/CAE. In the literature, the embedded beam elements have been used to simulate the fibers in the composites [9]. It is found that the embedded beam elements work well for the composites with random distributed fibers. However, they seem not valid for the composites with aligned fibers since the model fails to capture the fiber in-plane deformation and the longitudinal shear. Therefore a RVE model with embedded solid elements is proposed here to predict the elastic properties of the composites.

BASIC COMPUTATIONAL PRINCIPLES

Homogenization technique based on the strain energy equivalence is employed in the RVE analysis to determine the effective elastic properties of the chopped fiber reinforced composites. The strain energy equivalence requires

$$\overline{\sigma_{ij}\varepsilon_{ij}} = \overline{\sigma}_{ij}\overline{\varepsilon}_{ij} \quad (1)$$

where σ_{ij} and ε_{ij} are stress and strain tensors, respectively, with the over bar representing the volume average.

For general anisotropic material, the constitutive equations are expressed as,

$$\overline{\sigma}_{ij} = C_{ijkl}\overline{\varepsilon}_{kl} \quad (2)$$

C_{ijkl} is the effective elastic tensor of the composite to be determined, which has 21 independent constants as shown below (in Voigt notation),

$$\boldsymbol{C} = \begin{bmatrix} C_{11} & C_{12} & C_{13} & C_{14} & C_{15} & C_{16} \\ C_{21} & C_{22} & C_{23} & C_{24} & C_{25} & C_{26} \\ C_{31} & C_{32} & C_{33} & C_{34} & C_{35} & C_{36} \\ C_{41} & C_{42} & C_{43} & C_{44} & C_{45} & C_{46} \\ C_{51} & C_{52} & C_{53} & C_{54} & C_{55} & C_{56} \\ C_{61} & C_{62} & C_{63} & C_{64} & C_{65} & C_{66} \end{bmatrix}, C_{ij} = C_{ji} \quad (3)$$

For the cases when the magnitudes of the non-diagonal components in rows and columns 4-6 of the stiffness matrix $\boldsymbol{C}$ (Eq. 3) are much smaller than the values of the other components, the composite can be regarded as orthotropic materials with the number of the independent material constants reduced to 9, i.e.,

$$\boldsymbol{C} = \begin{bmatrix} C_{11} & C_{12} & C_{13} & 0 & 0 & 0 \\ C_{21} & C_{22} & C_{23} & 0 & 0 & 0 \\ C_{31} & C_{32} & C_{33} & 0 & 0 & 0 \\ 0 & 0 & 0 & C_{44} & 0 & 0 \\ 0 & 0 & 0 & 0 & C_{55} & 0 \\ 0 & 0 & 0 & 0 & 0 & C_{66} \end{bmatrix}, C_{ij} = C_{ji} \quad (4)$$

The periodic boundary conditions for the displacement is described by

$$\boldsymbol{u}(\boldsymbol{x}+\boldsymbol{L}) = \boldsymbol{u}(\boldsymbol{x}) + \boldsymbol{\varepsilon}^0 \cdot \boldsymbol{x} \quad \forall \boldsymbol{x} \in \partial V \quad (5)$$

where $\boldsymbol{u}$ is the nodal displacement vector; $\boldsymbol{L}$ represents the dimensions of RVE; $\boldsymbol{\varepsilon}^0$ denotes the specified strain. More details can be found in the reference [2].

In the embedded element technique, the meshes of the fibers are embedded in the meshes of the matrix, while the material defined for the matrix element is not replaced by the material defined for the fiber element. To remove the additional stiffness, the Young's modulus E_f of the fiber embedded in the matrix is replaced by

$$E'_f = E_f - E_m \quad (6)$$

where E_m is the Young's modulus of the matrix [9].

SIMULATION RESULTS AND VALIDATIONS

In the simulation, the material constants and geometric parameters for the RVE analysis listed in Table 1 are taken for case studies.

Table 1. Computational parameters

Young's modulus of fiber, E_f	75 GPa
Poisson's ratio of fiber, μ_f	0.25
Young's modulus of matrix, E_m	1.6 GPa
Poisson's ratio of matrix, μ_m	0.35
Diameter of fiber, d_f	25 μm

A RVE with random oriented chopped fibers in the x-y plane as shown in Fig. 1 is created to investigate the mesh sensitivity of the embedded element technique, for which the fiber volume fraction and aspect ratio are set as 10% and 10, respectively.

Figure 1. RVE including chopped fibers which random distributed in *x-y* plane

Firstly, the equivalent stiffness matrix of this RVE is determined by using the direct FE mesh analysis, which is given below (GPa)

$$C = \begin{bmatrix} 3.869 & 1.806 & 1.546 & 0.0187 & -0.00036 & -0.00379 \\ 1.806 & 3.891 & 1.546 & -0.0252 & -0.00353 & -0.00444 \\ 1.546 & 1.546 & 3.003 & 0.00176 & 0.00276 & 0.00005 \\ 0.0187 & -0.02520 & 0.00176 & 1.06020 & -0.00330 & 0.00185 \\ -0.00036 & -0.00353 & 0.00276 & -0.00330 & 0.73276 & 0.00058 \\ -0.00379 & -0.00444 & 0.00005 & 0.00185 & 0.00058 & 0.73189 \end{bmatrix} \quad (6)$$

It is found that the magnitudes of the non-diagonal elements in the rows and columns 4-6 are much smaller than those of the other components. Thus the RVE can be approximated as orthotropic material and the equivalent

Young's moduli and Poisson's ratios can be calculated naturally.

The equivalent moduli of the RVE based on the embedded solid elements are calculated and plotted in Fig. 2 to check the matrix element size sensitivity when the element size is fixed as $C_f/10$ with C_f being the perimeter of the fiber cross-section. The corresponding moduli calculated from Eq. (6) based on the FE direct mesh are also provided for computation. With the matrix element size varying from $2d_f$ to $0.2d_f$, it is observed that the results from the embedded element technique converge to the ones of the direct mesh method when the matrix element size is small enough (smaller than $d_f/4.5$ for example).

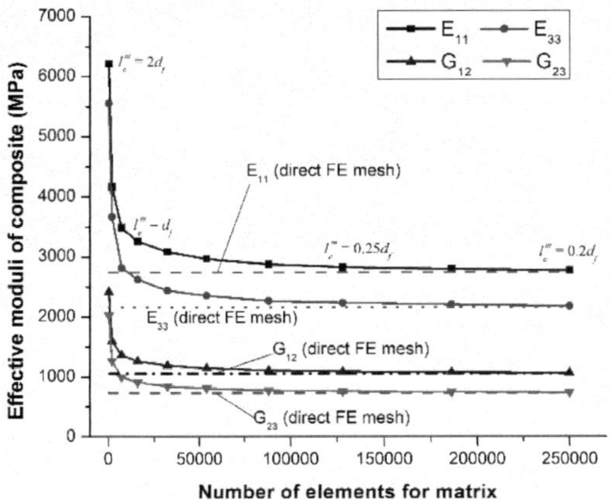

Figure 2. Variation of RVE effective moduli with matrix element size

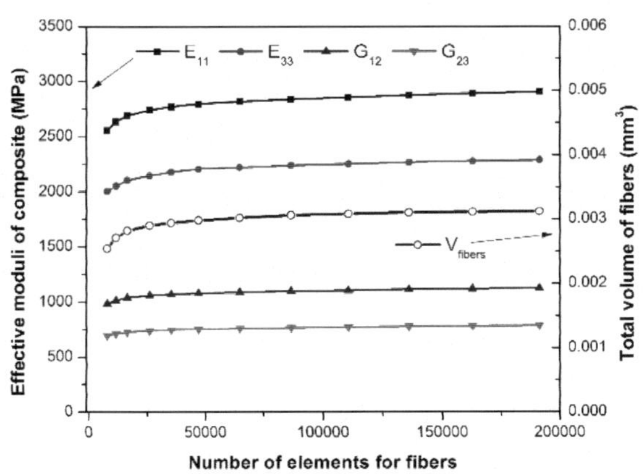

Figure 3. Variation of the effective moduli with the decreasing of the fiber element size

The mesh sensitivity of the fibers is also examined as plotted in Fig. 3 when the matrix element size is fixed as $d_f/4.5$. Similarly, the results converge to the constant values as predicted from the direct mesh analysis by decreasing the fiber element size. The variation of the total volume of all fibers with the decreasing of the fiber element size in this figure clearly demonstrates that the volume approaches its real value with the refining of the fiber meshing. It is concluded from Figs. 2 and 3 that the proposed solid element technique is validated by the FE direct mesh analysis when the meshing is sufficiently refined.

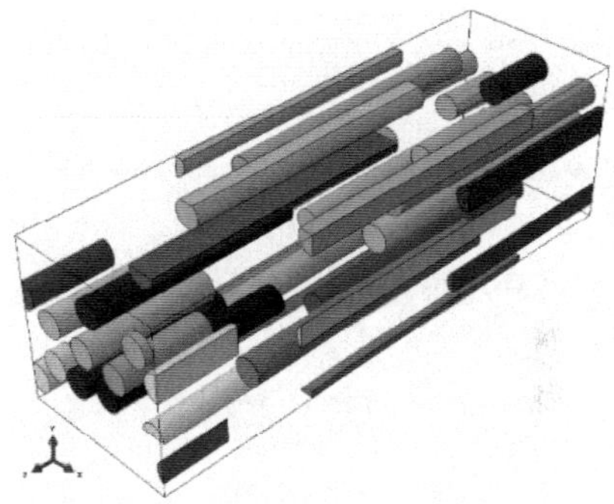

Figure 4. RVE with aligned chopped fibers

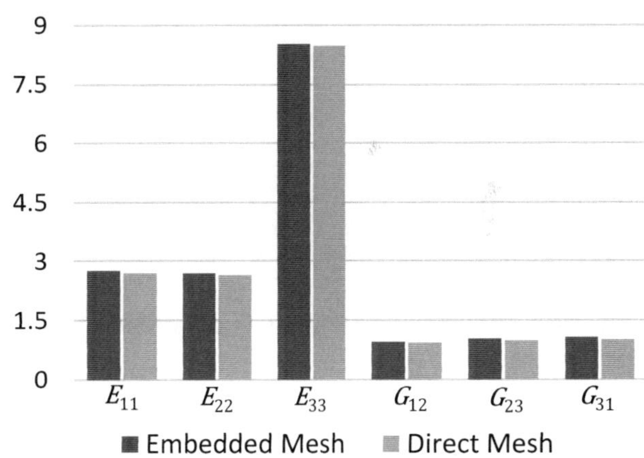

Figure 5. Comparison of the equivalent moduli calculated based on the embedded and direct meshes (Unit: GPa)

Another case study is performed for the RVE embedded with aligned chopped fibers as shown in Fig. 4, where the fiber volume fraction and aspect ratio are fixed as 20% and 10, respectively. Such a RVE can be also approximated as an orthotropic material due to the random distribution of the fibers within the transverse plane. Both the embedded mesh and direct mesh are generated for this RVE. In the embedded mesh technique the matrix and fiber element sizes are selected as $0.2d_f$ and $0.1C_f$. Computational results for the material constants are compared in Fig. 5, which shows excellent agreement between these two methodologies.

CONCLUSIONS

In this work, a RVE model with embedded solid elements is presented to predict the equivalent elastic properties of the chopped fiber reinforced composites numerically. In the embedded element technique, the matrix and fibers can be meshed independently. Thus it is more convenient to produce the meshes and impose the periodic boundary conditions on the RVE. Comparing with the results from the direct FE analysis, it is concluded that the developed RVE model with embedded solid elements is valid as long as the meshing is sufficiently refined. Obviously, the computational efficiency is significantly improved by avoiding the extreme computation expenses suffered by the direct FE mesh analysis.

ACKNOWLEDGMENTS

This work was supported by Ford University Research Program.

REFERENCES

1. Tucker III C.L., and Liang E., 1999, "Stiffness predictions for unidirectional short-fiber composites: Review and evaluation", *Compos. Sci. Technol.*, 59, 655-671.
2. Ostoja-Starzewski M., 2006, "Material spatial randomness: From statistical to representative volume element", *Probabilist. Eng. Mech.*, 21, 112-132.
3. Eshelby J.D., 1957, "The determination of the elastic field of an ellipsoidal inclusion, and related problems", *P. Roy. Soc. Lond. A Mat.*, 241, 376-396.
4. Chou T.W., Nomura S., and Taya M., 1980, "A self-consistent approach to the elastic stiffness of short-fiber composites", *J. Compos. Mater.*, 14, 178-188.
5. Mori T. and Tanaka K., 1973, "Average stress in matrix and average elastic energy of materials with misfitting inclusion", *Acta Metall. Mater.*, 21, 571-574.
6. Hashin Z. and Shtrikman S., 1963, "A variational approach to the theory of the elastic behavior of multiphase materials", *J. Mech. Phys. Solids*, 11, 127-140.
7. Halpin J.C. and Kardos J.L., 1976, "The Halpin-Tsai equations: A review", *Polym. Eng. Sci.*, 16, 344-352.
8. Cox H.L., 1952, "The elasticity and strength of paper and other fibrous materials", *Brit. J. Appl. Phys.*, 3, 72-79.
9. Lu Z., Yuan Z., and Liu Q., 2014, "3D numerical simulation for the elastic properties of random fiber composites with a wide range of fiber aspect ratios", *Comp. Mater. Sci.*, 90, 123-129.

DESIGN OF STIFFNESS BY CONTROLLED ORGANIZATION OF MICROSTRUCTURE IN AN ELASTOMERIC NANOCOMPOSITE

Peiying Jennifer Tsai
Mechanical Engineering
McMaster University
Hamilton, Ontario, Canada
tsaip@mcmaster.ca

Suvojit Ghosh
Engineering Physics
McMaster University
Hamilton, Ontario, Canada
sghosh@mcmaster.ca

Peidong Wu
Mechanical Engineering
McMaster University
Hamilton, Ontario, Canada
peidong@mcmaster.ca

Ishwar K. Puri
Mechanical Engineering
McMaster University
Hamilton, Ontario, Canada
ikpuri@mcmaster.ca

ABSTRACT

Recent developments enable *in situ* fabrication of microstructure in a liquid prepolymer by field-directed assembly of dispersed magnetic nanoparticles (MNPs). This allows enforcing a desired pattern in the microstructure of an elastomer-MNP nanocomposite. The geometry of the microstructure dictates the bulk properties of such composites. This presents an opportunity for on-demand control of material properties. Herein, we probe the use of such capabilities to control mechanical stiffness. Specifically, we examine the influence of organizing the dispersed particles into linear chains by simulating tensile tests using finite element methods. We find that such organization can lead to ~100 times enhancement in stiffness compared to a comparable material with randomly distributed particles. The result shows promising prospects for the next generation of additive manufacturing systems.

KEYWORDS: nanocomposite, magnetic nanoparticle, material design, programmable matter

INTRODUCTION

Composite materials comprising magnetic nanoparticles (MNPs) dispersed in an elastomeric matrix have gained considerable attention in the last decade. Such materials are synthesized by, e.g., solution-casting the polymers after dissolving them in a ferrofluid [1,2]. In such a process, an external magnetic field can be used to organize the MNPs into a desired microstructure in the solution [3]. Subsequently, the solvent is evaporated and the polymer is cured under exposure to the field. Thus, the microstructure is arrested in the surrounding matrix and preserved even once the field is removed.

Bulk properties of a composite material is strongly dependent on the geometry of its microstructure. Thus, the ability to enforce a desired pattern on the microstructure *in situ* enables a 'material printer'; beginning with the same polymer-MNP solution, it is possible to vary the properties of the final material by determining a suitable magnetic field *a priori* [4–7]. Various material properties can be controlled using this method, e.g., mechanical [6–10], magnetic [4,8], thermal [11] and electrical [12].

The simplest microstructure that can be assembled by magnetic fields are linear chains. Geometry and spacing of such chains depend on the externally applied magnetic field [3]. Herein, we study the influence of forming such chains of particles on the stiffness of the composite material. This is accomplished by simulating uniaxial tensile tests on various particle sizes and chain orientations using a commercial finite element package, Abaqus™. We select polydimethylsiloxane (PDMS) as our polymer matrix and iron as the core of the dispersed MNPs. The choice is motivated by the popularity of such materials: PDMS finds wide-ranging applications in, e.g., biomedical devices [13], gaskets and cookware. Iron is a popular material for MNPs.

We study various parameters describing the geometry of the linear chains, e.g., volume fractions (ψ) of the filler particles and orientation of the chains relative to the loading direction. The results are compared with pure PDMS and with PDMS composites comprising randomly distributed particles. We show that by organizing filler particles into linear chains, the stiffness of the composite can be enhanced by up to 100 times in situations when the loading occurs parallel to the chain direction. The stiffness-enhancing effect of such alignment increases with ψ. This creates the basis for designing the stiffness of materials by selection of suitable magnetic field while the composite material is being synthesized.

MATERIALS AND METHODS

PDMS, a rubber-like material, was modeled using a hyperelastic model [14]. The filler particles modeled as linear elastic. For simplicity, we assumed perfect bonding at the interface. The latter ensured continuity of stress and strain between matrix and MNPs.

Cross-linking in PDMS occurs through the hydrosilylation reaction wherein linear chains of PDMS react with a cross-linker yielding a 3D elastomer matrix [15,16]. The ratio between the two determines the mechanical properties of the finished material. We assumed a 10:1 (w/w) ratio. The resulting material is rubbery and can be described using a hyperelastic constitutive model assuming that the time scale of loading is less than relaxation time of the material. The most used model for rubbery materials is proposed by Arruda and Boyce [14]. It describes material loading as a strain energy potential. The strain is derived from the virtual chain microstructure in rubbers and rubber-like materials. In order to adopt this model for PDMS, a general compressibility term can be introduced [17] into the strain energy potential as

$$U=\mu\left\{\frac{1}{2}\left(\bar{I}_1-3\right)+\frac{1}{20\lambda_m^2}\left(\bar{I}_1^2-9\right)+\frac{11}{1050\lambda_m^4}\left(\bar{I}_1^3-27\right)+\frac{19}{7000\lambda_m^6}\left(\bar{I}_1^4-81\right)+\frac{519}{673750\lambda_m^8}\left(\bar{I}_1^5-243\right)\right\}+\frac{1}{D}\left(\frac{J_{el}^2-1}{2}-\ln J_{el}\right) \quad (1)$$

where U, μ, $\bar{I}_1$, λ_m, J_{el}, and D are the strain energy potential, shear modulus, 1st deviatoric strain invariant, chain locking stretch, elastic volume ratio, and compressibility parameter, respectively. For incompressible materials, $D=0$, and when the material coefficients are expressed as $\mu=nk\Theta$ and $\lambda_m=\sqrt{N}$, Eq. (1) becomes the equation (21) in Ref. [14]. This model was adopted for simulating the loading on PDMS. The various parameters, obtained by evaluating mechanical test data [18] using Abaqus™, are $\mu=0.169$ MPa, $\lambda_m=1.23$, and $D=0$.

We considered the filler particles to be iron, with Young's modulus E of 211GPa, density of 7.874 g/cm^3, and Poisson's ratio of 0.29.

In simulating the composite materials, the concept of a unit cell was employed [19]. A unit cell is a small volume of a material whose behavior represents the behavior of a larger material volume. Such a method can be employed when the material is expected to display some periodicity. It enables low computational cost. The method requires the use of periodic boundary conditions (PBCs). Herein, we used PBCs with translation and reflection symmetry under uniaxial loading, and further reduced the computational volume to 1/8 of a unit cell. The traction parts of boundary conditions are satisfied automatically when the variational principles based finite element methods are used [20]. Thus, only displacement boundary condition constraints were applied on the symmetry faces in our simulations. A large number of validation steps were used to ensure that the data from such a simulation matched that of a larger material. This was accomplished by comparison of data from the aforementioned simulation method with that from a much larger material block.

The stress-strain curves describe the mechanical behavior of a material. We defined the macroscopic nominal stress (Σ_n) and strain (E_n) as the following:

$$\Sigma_n=\frac{\text{total axial reaction force on top surface}}{\text{initial top surface area}}$$
$$\mathrm{E}_n=\frac{\text{displacement of top surface along } x_2\text{-axis}}{\text{initial length along } x_2\text{-axis}} \quad (2)$$

where x_2 is the direction of uniaxial loading. The stress-strain curves shown here are the macroscopic nominal stress-strain curves defined in Eq. (2).

RESULTS AND DISCUSSIONS

Herein, we probe the dependence of mechanical stiffness on microstructure geometry by observing the variation in the bulk Young's modulus with parameters describing chain geometry, notably (1) volume fraction of filler particles in the composite material (ψ) and (2) orientation of the chain with the loading direction. The observations are obtained from simulated uniaxial tensile loading. The following labels are used in describing the data: 'axial', 'lateral', 'random', and 'net'. In 'axial', the chains are parallel to the loading direction; in 'lateral', the chains are normal to the loading direction; in 'net', the chains are aligned at ±45° to the loading direction; in 'random', the particles are randomly distributed in the matrix. The ψ is varied between 0.77% and 34.22% by varying the unit cell dimensions. For the 'random' materials, the ψ is varied by varying the sphere radius between 0.0569 μm and 0.499 μm while keeping the simulated material volume at $1\ \mu\text{m}\times 1\ \mu\text{m}\times 1\ \mu\text{m}$, and the number of spheres within a 1/8 unit cell is ten. Five realizations of each volume fraction were obtained by 5 sets of coordinates for the 10 particles. Each set was obtained from a random number series subject to constrains that ensured that particles remained within the bounds of the 1/8 unit cell.

The stress-strain curves of the 'axial', 'lateral' and 'random' configurations are shown in **Error! Reference source not found.**. The 'net' configuration is omitted in this figure for brevity; it does not show significant difference from the 'lateral' configuration. All configurations show a stiffening with ψ, as is expected from any composite material with a stiffer dispersed phase. However, the effect is significantly more dramatic in the 'axial' configuration. Also, all configurations show strain hardening, but the effect is most pronounced in 'axial' samples. Similar trends of strengthening, but with smaller magnitudes, were observed for lateral models.

The observations above are explained well by considering the geometry of the heterogeneities in mechanical properties within the composite. Because the MNPs can be considered rigid relative to the PDMS matrix, the deformation due to loading is largely distributed in the elastomer matrix. Most of this load is borne by deformation in the elastomer between two consecutive particles along the loading direction. The elastomer available to bear such load is more constrained by the adjacent rigid particles in the 'axial' configuration than in any other configurations. Thus, for the same load, the elastomer in such a position in an 'axial' sample experiences a significantly higher stress, and thus presents a much higher stiffness. This is a direct consequence of the strain-hardening in rubbers. In contrast, in the 'lateral' configuration, the elastomer bearing this load has less constraint, yielding much lower stress, and in effect, much lower stiffness.

The effective Young's moduli was next calculated from the slope of stress-strain curve at 0.06 and 0.5 strain. This is presented as a function of ψ in **Error! Reference source not found.** for the various configurations we investigate. Fig. 2(a) shows that for all values of ψ, the 'axial' configuration is stiffer than any of the others. Even for extremely low filler concentration of $\psi = 0.77\%$, the Young's modulus of the 'axial' sample is 15.2 MPa while in lateral and random models, they are 0.9 MPa. The ratio of the two moduli increases with ψ until asymptoting at a value greater than 100 for $\psi > 12.3\%$. Thus, considering a ψ of, e.g., 0.2, we can control the stiffness by more than two orders of magnitude by simply having a magnetic field present during the curing of PDMS.

The difference between the 'lateral' and 'random' is much less significant, but becomes pronounced at higher ψ

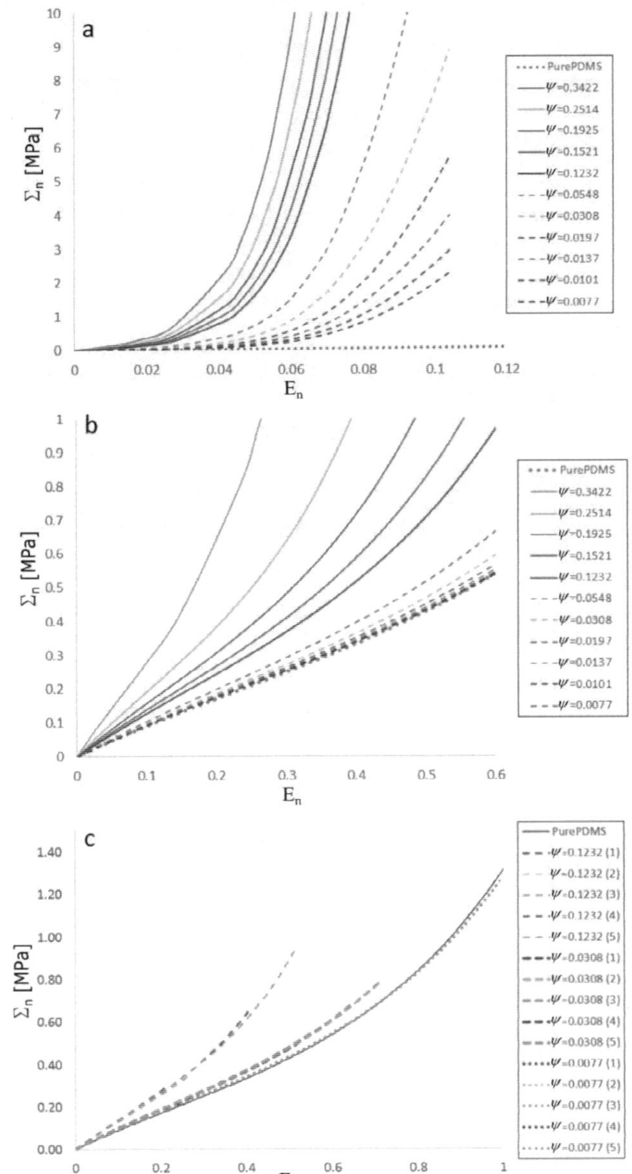

Figure 2: Simulated tensile test of composite samples with particles aligned in linear chains in parallel (a) or perpendicular (b) to the loading direction or randomly distributed (c). The chains parallel to the loading direction show a significantly higher stiffness compared to other arrangements of identical volume fraction.

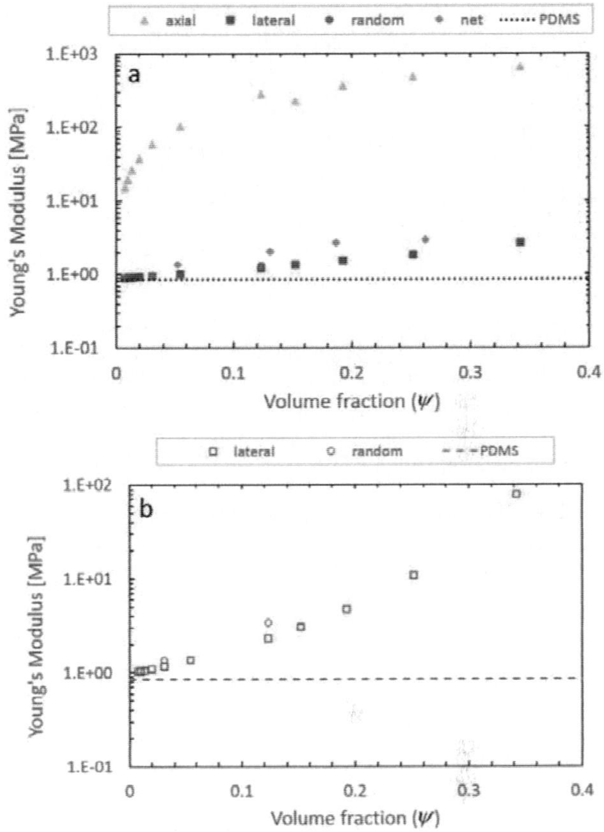

Figure 1: Young's modulus at 0.06 (a) and 0.5 (b) strain vs. ψ. At 0.06 strain, the Young's modulus of filler particles parallel to the loading direction (axial) is significantly higher than other arrangements. It increases sharply in the region $\psi < 10\%$. At 0.5 strain, the difference of modulus between random and lateral patterns increased with ψ.

as seen in Fig. 2(b). This demonstrates that the control on stiffness using this idealized technique is only possible when considering loading in a direction parallel to the chains of the filler particles.

CONCLUSIONS

We investigate that capability of enforcing a desired stiffness in a PDMS-MNP nanocomposite material by means of exposing the liquid prepolymer to a magnetic field during curing. We show that the stiffness of such a composite can be enhanced by almost 100 times by using a strong magnetic field which assembles the filler MNPs into linear chains. The stiffening effect, however, is only applicable when the loading occurs in the direction of the chains. Such findings have promising prospects in the context of additive manufacturing.

ACKNOWLEDGMENTS

The work was supported by a Discovery Grant from the Natural Sciences and Engineering Research Council of Canada (NSERC).

REFERENCES

1. Abramchuk, S., Kramarenko, E., Grishin, D., Stepanov, G., Nikitin, L. V., Filipcsei, G., Khokhlov, A. R., and Zrínyi, M., 2007, "Novel highly elastic magnetic materials for dampers and seals: Part II. Material behavior in a magnetic field," *Polym. Adv. Technol.*, Vol(18), pp. 513–518.

2. Abramchuk, S., Kramarenko, E., Stepanov, G., Nikitin, L. V., Filipcsei, G., Khokhlov, A. R., and Zrínyi, M., 2007, "Novel highly elastic magnetic materials for dampers and seals: Part I. Preparation and characterization of the elastic materials," *Polym. Adv. Technol.*, Vol(18), pp. 883–890.

3. Ghosh, S., and Puri, I. K., 2013, "Soft polymer magnetic nanocomposites: microstructure patterning by magnetophoretic transport and self-assembly," *Soft Matter*, Vol(9), p. 2024.

4. Fragouli, D., Buonsanti, R., Bertoni, G., Sangregorio, C., Innocenti, C., Falqui, A., Gatteschi, D., Cozzoli, P. D., Athanassiou, A., and Cingolani, R., 2010, "Dynamical formation of spatially localized arrays of aligned nanowires in plastic films with magnetic anisotropy," *ACS Nano*, Vol(4), pp. 1873–1878.

5. Jestin, J., Cousin, F., Dubais, I., Ménager, C., Schweins, R., Oberdisse, J., and Boué, F., 2008, "Anisotropic reinforcement of nanocomposites tuned by magnetic orientation of the filler network," *Adv. Mater.*, Vol(20), pp. 2533–2540.

6. Robbes, A. S., Cousin, F., Meneau, F., Dalmas, F., Boué, F., and Jestin, J., 2011, "Nanocomposite materials with controlled anisotropic reinforcement triggered by magnetic self-assembly," *Macromolecules*, Vol(44), pp. 8858–8865.

7. Varga, Z., Filipcsei, G., and Zrínyi, M., 2006, "Magnetic field sensitive functional elastomers with tuneable elastic modulus," *Polymer*, Vol(47), pp. 227–233.

8. Ruiz, M. M., Mietta, J. L., Soledad Antonel, P., Pérez, O. E., Martín Negri, R., and Jorge, G., 2013, "Structural and magnetic properties of Fe2−xCoSmxO4—nanoparticles and Fe2−xCoSmxO4—PDMS magnetoelastomers as a function of Sm content," *J. Magn. Magn. Mater.*, Vol(327), pp. 11–19.

9. Han, Y., Hong, W., and Faidley, L. E., 2013, "Field-stiffening effect of magneto-rheological elastomers," *Int. J. Solids Struct.*, Vol(50), pp. 2281–2288.

10. Evans, B. a, Fiser, B. L., Prins, W. J., Rapp, D. J., Shields, A. R., Glass, D. R., and Superfine, R., 2012, "A Highly Tunable Silicone-Based Magnetic Elastomer with Nanoscale Homogeneity.," *J. Magn. Magn. Mater.*, Vol(324), pp. 501–507.

11. Dallas, P., Georgakilas, V., Niarchos, D., Komninou, P., Kehagias, T., and Petridis, D., 2006, "Synthesis, characterization and thermal properties of polymer/magnetite nanocomposites," *Nanotechnology*, Vol(17), pp. 2046–2053.

12. Fragouli, D., Das, A., Innocenti, C., Guttikonda, Y., Rahman, S., and Liu, L., 2014, "Polymeric Films with Electric and Magnetic Anisotropy Due to Magnetically Assembled Functional Nano fibers," *ACS Appl. Mater. Interfaces*, Vol(6), pp. 4535–4541.

13. Mata, A., Fleischman, A. J., and Roy, S., 2005, "Characterization of polydimethylsiloxane (PDMS) properties for biomedical micro/nanosystems.," *Biomed. Microdevices*, Vol(7), pp. 281–93.

14. Arruda, E., and Boyce, M., 1993, "A three-dimensional constitutive model for the large stretch behavior of rubber elastic materials," *J. Mech. Phys. Solids*, Vol(41), pp. 389–412.

15. Owen, M., Clarson, S., and Semlyen, J., 1993, *Siloxane polymers*, PTR Prentice Hall, Englewood Cliffs, NJ.

16. Esteves, A. C. C., Brokken-Zijp, J., Laven, J., Huinink, H. P., Reuvers, N. J. W., Van, M. P., and de With, G., 2009, "Influence of cross-linker concentration on the cross-linking of PDMS and the network structures formed," *Polymer*, Vol(50), pp. 3955–3966.

17. Simulink, *Abaqus 6.12 user manual.*

18. Zang, J., Zhao, X., Cao, Y., and Hutchinson, J. W., 2012, "Localized ridge wrinkling of stiff films on compliant substrates," *J. Mech. Phys. Solids*, Vol(60), pp. 1265–1279.

19. Koplik, J., and Needleman, A., 1988, "Void growth and coalescence in porous plastic solids," *Int. J. Solids Struct.*, Vol(24), pp. 835–853.

20. Li, S., 2008, "Boundary conditions for unit cells from periodic microstructures and their implications," *Compos. Sci. Technol.*, Vol(68), pp. 1962–1974.

Proceedings of the 25th CANCAM
London, Ontario, Canada, May 31 – June 4, 2015

THE INFLUENCE OF FILLER VOLUME FRACTION ON SENSITIVITY TO DISTRIBUTED STRAIN IN CARBON NANOFIBER/POLYURETHANE NANOCOMPOSITES

T. N. Tallman
Department of Mechanical Engineering
University of Michigan
Ann Arbor, MI, USA
ttallman@umich.edu

K. W. Wang
Department of Mechanical Engineering
University of Michigan
Ann Arbor, MI, USA
kwwang@umich.edu

ABSTRACT

Nanocomposites have unprecedented potential as smart, self-sensing materials for strain detection and tactile sensing. Filler volume fraction is an important consideration in the development of strain-sensitive nanocomposites because of its influence on sensitivity and dispersion uniformity. The influence of filler volume fraction is herein studied for its effect on the sensitivity of distributed strain detection via electrical impedance tomography in carbon nanofiber/ polyurethane nanocomposites. It is found that lower volume fractions display the greatest sensitivity to strain; however, they also show the greatest variation in conductivity change indicating less uniform nanofiller dispersion.

KEYWORDS: nanocomposite, piezoresistivity, electrical impedance tomography, structural health monitoring, self-sensing

INTRODUCTION

Because their electrical conductivity depends on their mechanical state, nanocomposites have tremendous potential as high-sensitivity strain gauges [1-7] and for integrated damage detection [8-20]. Key to this potential is piezoresistivity. Piezoresistivity is rooted in the dependence of nanocomposite conductivity on well-connected nanofiller networks. That is, electrons traverse through nanofiller networks by tunneling between sufficiently proximate nanofillers. Strain that alters the connectivity of the network will also change the conductivity distribution of the nanocomposite, and damage that severs the network will manifest as a conductivity loss in the region of the damage.

Leveraging this for integrated damage detection has been studied by resistance change methods [11-14] and tomographic methods [15-20]. Resistance change methods measure the resistance between electrodes before and after damage and locate damage by interpolating resistance changes. Resistance change methods are computationally inexpensive, but require large electrode arrays. Tomographic methods such as electrical impedance tomography (EIT), conversely, require fewer electrodes located only along the periphery of the structure. EIT is more adept at spatially resolving damage or strain, but it is also more computationally expensive. Compared to damage detection, however, less work has been done to investigate either resistance change or tomographic methods for strictly strain detection.

In addition to strain sensing for structural health monitoring (SHM) applications, highly compliant nanocomposites have potential for tactile sensing wherein it is necessary to spatially resolve points of contact. This capability is of keen interest to a growing number of robotic and biomedical applications. Locating contact within flexible planar skins has been approached by incorporating a sensing medium into a compliant matrix such that a grid is formed either by the sensing medium [21] or by line electrodes sandwiching the sensing medium [22]. Pressure-induced capacitance changes are then measured at the grid points so that pressure fields can be imaged by interpolating capacitance changes between measurement points. Despite the success of this approach, an important limitation is the dependence on a grid of sensors which is costly and complex to manufacture.

In light of the preceding discussion, we identify a novel approach for advancing the state of the art – employing EIT for strain sensing and tactile imaging. We herein explore this by examining the ability of EIT to accurately locate strain-induced conductivity changes and the influence of filler volume fraction on strain sensitivity in highly flexible carbon nanofiber (CNF)/polyurethane (PU) nanocomposites.

ELECTRICAL IMPEDANCE TOMOGRAPHY

EIT is a noninvasive method of imaging an internal conductivity distribution. Procedurally, the domain to be imaged is line with electrodes as shown in Figure 1. Current is injected between the first electrode pair and the resulting voltage measured between electrode pairs not actively involved in the current injection. The current injection is then moved around the domain such that every electrode pair receives an injection while the boundary voltages are continually collected.

EIT functions by minimizing the difference between a vector of the previously described boundary voltages and an analytical operator, known as the forward operator, which predicts the same boundary voltages as shown in Eq. (1).

$$\hat{\sigma} = \text{argmin}_{\sigma} \|\boldsymbol{V}_m - \boldsymbol{F}(\sigma)\|^2 \quad (1)$$

Here, $\boldsymbol{V}_m$ is the vector of experimentally collected voltages, $\boldsymbol{F}(\sigma)$ is the analytical operator that predicts the boundary voltages, and $\hat{\sigma}$ is a conductivity distribution satisfying the minimization. Next, perform a Taylor series expansion on $\boldsymbol{F}(\sigma)$ about some initial conductivity estimate, σ_0, and retain only the linear terms as shown in Eq. (2).

$$\boldsymbol{F}(\sigma) \approx \boldsymbol{F}(\sigma_0) + \frac{\partial \boldsymbol{F}(\sigma_0)}{\partial \sigma}(\sigma - \sigma_0) \quad (2)$$

In order to proceed, we must next consider the forward operator. For steady-state diffusion, the relationship between currents and voltages is governed by Laplace's equation as shown in Eq. (3).

$$\boldsymbol{\nabla} \cdot \sigma \boldsymbol{\nabla} \phi = 0 \quad (3)$$

This is subjected to the complete electrode model boundary condition as shown in Eq. (4). This boundary condition assumes the electrodes are perfect conductors and that there is a voltage drop between the domain and the electrodes due to imperfect contact. Conservation of charge is also enforced by Eq. (5).

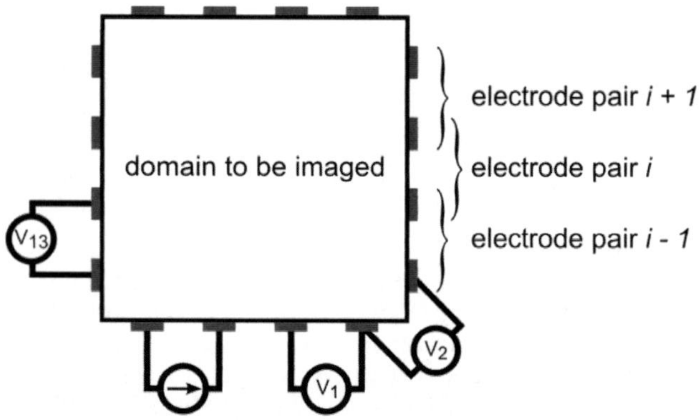

Figure 1. Representative EIT injection schematic shown with 16 electrodes.

$$\sigma \boldsymbol{\nabla} \phi \cdot \boldsymbol{n} = \frac{1}{z_l}(V_l - \phi) \quad (4)$$

$$\sum_{l=1}^{L} \int_{E_l} \sigma \boldsymbol{\nabla} \phi \cdot \boldsymbol{n} \, \mathrm{d}S_l = 0 \quad (5)$$

In the preceding, ϕ is the domain solution, $\boldsymbol{n}$ is the outward pointing normal vector, V_l is the voltage on the lth electrode, z_l is the contact impedance between the lth electrode and the domain, E_l is the length of the lth electrode, and L is the total number of electrodes.

The equations governing the forward operator are most expeditiously solved by finite element discretization as shown in Eq. (6).

$$\begin{bmatrix} \boldsymbol{A}_M + \boldsymbol{A}_Z & \boldsymbol{A}_W \\ \boldsymbol{A}_W^T & \boldsymbol{A}_D \end{bmatrix} \begin{bmatrix} \boldsymbol{\Phi} \\ \boldsymbol{V} \end{bmatrix} = \begin{bmatrix} \boldsymbol{0} \\ \boldsymbol{I} \end{bmatrix} \quad (6)$$

$$A_{Z\,ij} = \sum_{l=1}^{L} \int_{E_l} \frac{1}{z_l} w_i w_j \, \mathrm{d}S_l \quad (7)$$

$$A_{W\,li} = -\int_{E_l} \frac{1}{z_l} w_i \, \mathrm{d}S_l \quad (8)$$

$$A_D = \text{diag}\left(\frac{E_l}{z_l}\right) \quad (9)$$

In Eq. (6), $\boldsymbol{A}_M$ is the standard diffusion stiffness matrix, and the remaining terms account for the additional degrees of freedom belonging to the electrodes and the contribution of contact impedances. w_i is the ith finite element interpolation function used on the electrodes.

Having discretized the forward operator, we substitute Eq. (2) into Eq. (1) as shown in Eq. (10).

$$\hat{\boldsymbol{\sigma}} = \text{argmin}_{\sigma} \|\boldsymbol{V}_e - \boldsymbol{J}\Delta\boldsymbol{\sigma}\|^2 \quad (10)$$

Here, we have employed the substitutions $\boldsymbol{V}_e = \boldsymbol{V}_m - \boldsymbol{F}(\boldsymbol{\sigma}_0)$, $\boldsymbol{J} = \partial \boldsymbol{F}(\boldsymbol{\sigma}_0)/\partial \boldsymbol{\sigma}$, and $\Delta\boldsymbol{\sigma} = \boldsymbol{\sigma} - \boldsymbol{\sigma}_0$ and boldfaced quantities that have become vectors due to discretization. Recovering $\hat{\boldsymbol{\sigma}}$ from Eq. (10), however, still is not straightforward because of the severe rank deficiency of $\boldsymbol{J}$. Tikhonov regularization is therefore employed to update $\hat{\boldsymbol{\sigma}}$ iteratively as $\hat{\boldsymbol{\sigma}}_{n+1} = \hat{\boldsymbol{\sigma}}_n + \Delta\boldsymbol{\sigma}$ where $\Delta\boldsymbol{\sigma}$ is found as shown in Eq. (11).

$$\Delta\boldsymbol{\sigma} = (\boldsymbol{J}^T\boldsymbol{J} + \alpha^2 \boldsymbol{L}^T\boldsymbol{L})^{-1}\boldsymbol{J}^T\boldsymbol{V}_e \quad (11)$$

This iterative update continues until the error is acceptably minimized. In this research the discrete Laplace operator is used for the regularization term, $\boldsymbol{L}$.

EXPERIMENTAL PROCEDURES

CNF/PU specimens are produced by colleagues at Penn State University (see acknowledgements) using ReoFlex 20 PU and Pyrograf III-PR-24-XT-HHT CNFs. Measured amounts of PU and CNFs are combined to produce composites at 7.5, 10.5, 12.5, and 15% filler volume fraction.

Figure 2. Experimental EIT setup including glass marbles to generate a strain field to be imaged by EIT.

Distributed strain is induced in 2.54 cm × 2.54 cm CNF/PU samples by resting a 1.2 kg mass atop three bonded glass marbles as shown in Figure 2. Spherical glass marbles are used because they are non-conductive and their curvature does not cut into the soft CNF/PU. Three marbles are used to demonstrate the ability of EIT to clearly differentiate between multiple points of contact. A 28-electrode system is used to image strain-induced conductivity changes via an in-house EIT routine.

Electrodes are attached to the CNF/PU composites by first applying seven evenly spaced 1.59 mm patches of colloidal silver paste (TedPella 16032) per side. The paste is allowed to dry for at least one hour before jumper wires are pushed lightly into the patches and then glued to an acrylic (PMMA) base. An additional drop of silver paste is then applied where the jumper wire touches the originally applied patch to ensure good electrical contact. A Keithley 6221 current source is used to supply 2.5 mA DC injections between electrodes. Voltages are measured using two 16-channel National Instruments 6368-PXIe data acquisition cards for 10 s at 128 Hz. Data is collected via an in-house LabView code and smoothed using a moving average of half-width 256.

DISTRIBUTED STRAIN DETECTION VIA EIT

As shown in Figure 3, EIT accurately captures three distinct points of contact due to the spherical indenters for each volume fraction. Larger changes in conductivity are observed for lower volume fractions implying that they are more sensitive to the imposed strain fields. However, the EIT image produced for 7.5% filler volume fraction has a region in which the conductivity change is markedly larger. This is speculated to be due to non-uniform nanofiller dispersion and a region of lower nanofiller density in the vicinity of the region in question. Conversely, as the filler volume fraction increases, such deviations are less pronounced. Conductivity increases in the region of the distributed load indicating that the nanofillers are becoming closer together thereby decreasing the tunneling resistance felt by electrons. Additionally, the compression increases the density of the nanofiller network thereby increasing the number of viable tunneling junctions. These factors combine resulting in a net increase in conductivity.

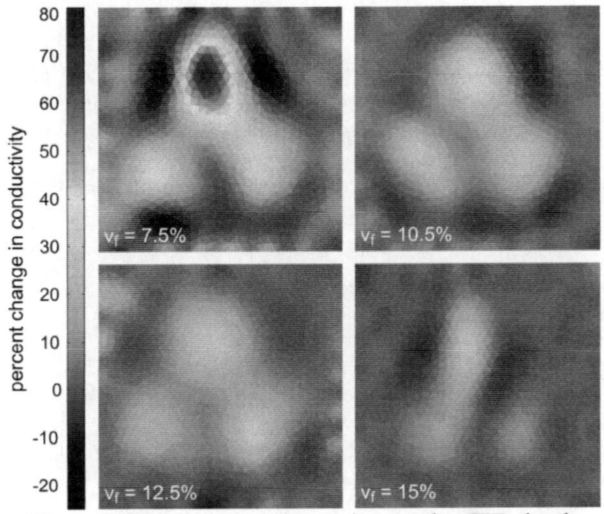

Figure 3. Distributed strain sensing results. EIT clearly captures three points of contact for each volume fraction.

SUMMARY AND CONCLUSIONS

Herein we have investigated employing CNF/PU nanocomposites and EIT for distributed strain sensing. Conductivity changes induced by bonded glass marbles are successfully imaged by EIT, and the point of contact of each marble is clearly discerned. Furthermore, it is found that sensitivity to distributed strain increases with decreasing filler volume fraction. However, the lowest volume fraction also shows the greatest variation in conductivity change potentially indicating less uniform dispersion. Nonetheless, all volume fractions accurately locate strain-induced conductivity changes thereby demonstrating the potential of distributed strain sensing via nanocomposites and EIT.

ACKNOWLEDGMENTS

This research is partially supported by the US Army Research Office (W911NF-10-1-00267) and the National Science Foundation (CMMI-DS-1232436).

The authors gratefully thank Dr. Sila Gungor and Professor Charles E. Bakis of the Pennsylvania State University for producing the CNF/PU.

REFERENCES

1. Hu N, Masuda Z, Yamamoto G, Fukunaga H, Hashida T, Qiu J. 2008, "Effect of fabrication process on electrical properties of polymer/multi-wall carbon nanotube

nanocomposites", *Composites Part A: Applied Science and Manufacturing*, Vol(39), pp. 893-903.

2. Hu N, Karube Y, Yan C, Masuda Z, Fukunaga H. 2008, "Tunneling effect in a polymer/carbon nanotube nanocomposite strain sensor", *Acta Materialia*, Vol(56), pp. 2929-2936.

3. Hu N, Karube Y, Arai M, Watanabe T, Yan C, Li Y, et al. 2010, "Investigation on sensitivity of a polymer/carbon nanotube composite strain sensor", *Carbon*, Vol(48), pp. 680-687.

4. Kang I, Schulz MJ, Kim JH, Shanov V, Shi D. 2006, "A carbon nanotube strain sensor for structural health monitoring", *Smart Materials and Structures*, Vol(15), pp. 737.

5. Loh KJ, Kim J, Lynch JP, Kam NWS, Kotov NA. 2007, "Multifunctional layer-by-layer carbon nanotube-polyelectrolyte thin films for strain and corrosion sensing", *Smart Materials and Structures*, Vol(16), pp. 429-438.

6. Loh KJ, Lynch JP, Shim BS, Kotov NA. 2008, "Tailoring piezoresistive sensitivity of multilayer carbon nanotube composite strain sensors", *Journal of Intelligent Material Systems and Structures*, Vol(19), pp. 747-764.

7. Tallman T, Wang KW. 2013, "An arbitrary strains carbon nanotube composite piezoresistivity model for finite element integration", *Applied Physics Letters*, Vol(102), pp. 011909.

8. Gao L, Chou TW, Thostenson ET, Zhang Z. 2010, "A comparative study of damage sensing in fiber composites using uniformly and non-uniformly dispersed carbon nanotubes", *Carbon*, Vol(48), pp. 3788-3794.

9. Gao L, Chou TW, Thostenson ET, Zhang Z, Coulaud M. 2011, "In situ sensing of impact damage in epoxy/glass fiber composites using percolating carbon nanotube networks", *Carbon*, Vol(49), pp. 3382-3385.

10. Thostenson ET, Chou TW. 2006, "Carbon Nanotube Networks: Sensing of Distributed Strain and Damage for Life Prediction and Self-Healing", *Advanced Materials*, Vol(18), pp. 2837-2841.

11. Naghashpour A, Hoa SV. 2013, "A technique for real-time detection, location and quantification of damage in large polymer composite structures made of electrically non-conductive fibers and carbon nanotube networks", *Nanotechnology*, Vol(24), pp. 455502.

12. Naghashpour A, Hoa SV. 2015, "A technique for real-time detecting, locating, and quantifying damage in large polymer composite structures made of carbon fibers and carbon nanotube networks", *Structural Health Monitoring*, Vol(14), pp. 35-45.

13. Zhang D, Ye L, Wang D, Tang Y, Mustapha S, Chen Y. 2012, "Assessment of transverse impact damage in GF/EP laminates of conductive nanoparticles using electrical resistive tomography", *Composites: Part A*, Vol(43), pp. 1587-1598.

14. Viets C, Kaysser S, Schulte K. 2014, "Damage mapping of GFRP via electrical resistance measurements using nanocomposite epoxy matrix systems" *Composites: Part B,* Vol(65), pp. 80-88.

15. Hou TC, Loh KJ, Lynch JP. 2007, "Spatial conductivity mapping of carbon nanotube composite thin films by electrical impedance tomography for sensing applications", *Nanotechnology*, Vol(18), pp. 315501.

16. Loh KJ, Hou TC, Lynch JP, Kotov NA. 2009, "Carbon nanotube sensing skins for spatial strain and impact damage identification", *Journal of Nondestructive Evaluation*, Vol(28), pp. 9-25.

17. Loyola BR, Briggs TM, Arronche L, Loh KJ, La Saponara V, O'Bryan G, et al. 2013, "Detection of spatially distributed damage in fiber-reinforced polymer composites", *Structural Health Monitoring*, Vol(12), pp. 225-239.

18. Tallman TN, Gungor S, Wang KW, Bakis CE. 2014, "Damage detection and conductivity evolution in carbon nanofiber epoxy via electrical impedance tomography", *Smart Materials and Structures*, Vol(23), pp. 045034.

19. Tallman TN, Gungor S, Wang KW,BCE. 2015, "Damage Detection via Electrical Impedance Tomography in Glass Fiber/Epoxy Laminates with Carbon Black Filler", *Structural Health Monitoring*, Vol(14), pp. 100-109.

20. Kim B, Lu Y, Kim T, Han JW, Meyyappan M, Li J. 2014, "Carbon Nanotube Coated Paper Sensors for Damage Diagnosis", *ACS Nano*, Vol(8), pp. 12092-12097.

21. Lipomi DJ, Vosgueritchian M, Tee BCK, Hellstrom SL, Lee JA, Fox CH, et al. 2011, "Skin-like pressure and strain sensors based on transparent elastic films of carbon nanotubes", *Nature Nanotechnology*, Vol(6), pp. 788-792.

22. Mannsfeld SCB, Tee BCK, Stoltenberg RM, Chen CVHH, Barman S, Muir BVO, et al. 2010, "Highly sensitive flexible pressure sensors with microstructured rubber dielectric layers", *Nature Materials*, Vol(9), pp. 859-864.

Proceedings of the 25th CANCAM
London, Ontario, Canada, May 31 – June 4, 2015

EFFECT OF SURFACE TREATMENT ON THE STRENGTH AND THE ENVIRONMENTAL DURABILITY OF BOND AT THE METAL – POLYMER COMPOSITE INTERFACE

Syeda Noor E Sumaiya, Shailesh Chekurthi, J. Raghavan
Composite Materials and Structures Research Group, Department of Mechanical Engineering
University of Manitoba
Winnipeg, Manitoba, Canada
nooresus@myumanitoba.ca
chekurts@myumanitoba.ca
Raghavan.Jayaraman@umanitoba.ca

Sitarz Blaine, Brian D McMillan
Standard Aero
Winnipeg, Manitoba, Canada
Blaine.Sitarz@StandardAero.com
Brian.McMillan@StandardAero.com

ABSTRACT

Effect of surface treatment on the strength and the environmental durability of bond between a glass fiber filled epoxy rub strip (Henkel Hysol EA 9891RP) and the aluminum (6061 T6) casing of an aircraft engine was studied using single lap shear test (ASTM D1002), wedge crack test (ASTM D3762), and FEM analysis. The optimal bonding conditions found from this study – room temperature curing at 103 kPa and bond thickness in the range of 0.15 – 0.25 mm - were used. The effect of four surface treatments, Untreated, Phosphoric Acid Etching (PAA), Pasa-Jell 105 (PJ), and Alodine 1132 was studied. In addition, the influence of Cytec's BR 127 primer, applied after the above surface treatments, was studied. The lap shear strength varied (9.7 to 13.2 MPa) with surface treatment in the following order - PJ < PAA <Alodine< PAA +BR127 <Untreated. The crack growth rate at 25°C and 95% RH decreased with various surface treatments in the following order - Untreated > PAA >Alodine> PJ > PAA +BR127. The crack growth rate during thermal cycling between +40°C and -50°C was lower than the crack growth at 40°C and 95% RH highlighting the deleterious effect of moisture. Thermal stress at the interface between the engine casing and the composite rub strip during thermal cycling was determined using ABAQUS 6.12 to be 10.5 MPa. This suggests the lap shear strength of the interface corresponding to the optimal surface treatment (PAA+BR127) should not degrade, due to moisture and thermal cycling, to a value less than 10.5 MPa. This is currently under investigation.

KEYWORDS: Metal-Composite bonding, Lap-shear, Surface treatment, Bond Durability, Finite element analysis

INTRODUCTION

Adhesive bonding of metals in general and aluminum in particular has been extensively studied in the past. The effect of various surface treatments on the strength and durability of the adhesive bond in assembled aluminum structures is relatively well understood [1]. Bonding between metals and polymer composites have also been studied in the past. These studies, related to aerospace applications, have used adhesives that cure at high temperature [2, 3]. Applicability of this knowledge to bonding of Aluminum to polymer composite at room temperature is unknown and hence, is the focus of this study. A polymer composite rub-strip is bonded to the inside surface of the aluminum casing of an engine to prevent the blades from coming in contact with the casing. While this rub-strip is bonded at high temperature and pressure by OEMs, room temperature bonding is used during repair and overhaul. Hence, the effect of various surface treatments on the interfacial shear strength and the environmental durability of bond between the rub strip and the aluminum, formed at room temperature, were studied experimentally and by numerical analysis.

EXPERIMENTAL DETAILS

Al (6061- T6) panels (152mm x 100 mm x1.6mm), purchased from a local supplier, were cleaned with water jet and subjected to following three surface treatments.

1. Untreated: Water-jet cleaning was followed by mechanical abrasion using 120 grit Alumina cloth. The panels were wiped clean with acetone.
2. Pasa-Jell 105(PJ) Treatment: Pasa-Jell 105 was applied to one end of the untreated panels, identified for bonding immediately after degreasing, as per manufacturer's guidelines. After 15 minutes, the treated surface was rinsed with distilled water and dried at room temperature.
3. Phosphoric Acid Anodizing (PAA): Phosphoric Acid anodizing of untreated panel was done in 85% phosphoric acid solution at 10 dc volts as per ASTM D 3933 for 20 – 25 minutes. It was a single-rack process where pH of the solution was maintained at 4.2. Panels were tested for water break after anodizing and dried at 60°C in an oven. Some panels were brush coated with Cytec's BR 127 primer, dried in air for 30 minutes, and cured at 120°C.
4. Alodine Treatment: TOUCH-N-PREP ALODINE 1132pen was used to coat the untreated Aluminum panel surface with Alodine, followed by air drying at room temperature.

Surface treated panels were bonded in single-lap shear configuration, within 30 minutes of treating the surface using Henkel Hysol's EA 9891RP. A mold was machined and used to bond single lap shear (SLS) panels of size 152 mm x 100 mm. The bonding was done at room temperature using a pressure of 103kPa. The length of the bond was 25 mm and the thickness of the bond varied in the range of 0.15 – 0.25 mm. A batch of SLS test coupons using untreated 6061–T6 aluminum (water cleaned, abraded and acetone wiped) was also prepared, tested and compared with the results from surface treated SLS coupons.

Wedge crack (WC) test coupons were manufactured using 6061-T6 aluminum panels (152 mm x 152 mm x3.2mm). Cleaning and surface treatment procedures, identical to those used in SLS panels, were used to prepare the WC panels. A mold was machined and used to bond these panels with a bond thickness of 0.3 mm. The WC panels were bonded using Henkel Hysol's EA 9891RP at room temperature using a pressure of 103 kPa. A batch of WC test coupons using untreated 6061–T6 aluminum (water cleaned, abraded and acetone wiped) was also prepared, tested and compared with the results from surface treated WC coupons.

All bonded test panels were cut using a band saw to yield test coupons, 25 mm in width, and tested using Instron's screw driven 5500R test frame. Special grips were machined to ensure alignment of the load axis with the center-line of the bond to avoid bending. SLS test coupons were tested as per ASTM D1002 using a cross-head rate of 0.05 inch/min. The edges of the WC coupons were polished to enable measurement of crack length. Wedges, prepared as per ASTM D3762, were used to generate the starter crack within the WC test coupons. Subsequently, these test coupons were aged in a humidity chamber at 40°C and 95% RH and the crack length was measured, using optical microscope, as a function of time. Additional WC test coupons were subjected to thermal cycling between +40°C and -50°C and the crack growth was monitored as a function of time/cycle. Four SLS specimens and five WC specimens were prepared and tested from each treatment.

FINITE ELEMENT ANALYSIS

The thermal stresses introduced, at the rub-strip – casing interface, during thermal cycling were analyzed using ABAQUS 6.12. A cross section of the aircraft engine casing was modeled using ABAQUS CAE. The aluminum and the polymer composite were modeled using 2-dimensional planar quadrilateral elements (CPE4) and the interface was modeled using contact elements. A fine mesh was used at the interface location shown in Fig.1, where maximum interfacial shear stresses were observed during preliminary analysis.

Cure shrinkage (-0.45%) and coefficient of thermal expansion of the composite were determined experimentally using Thermomechanical Analyzer (TMA). Young's modulus of the composite was measured by uniaxial testing of bulk specimen. Mechanical properties used in the analysis are listed in Table 1.

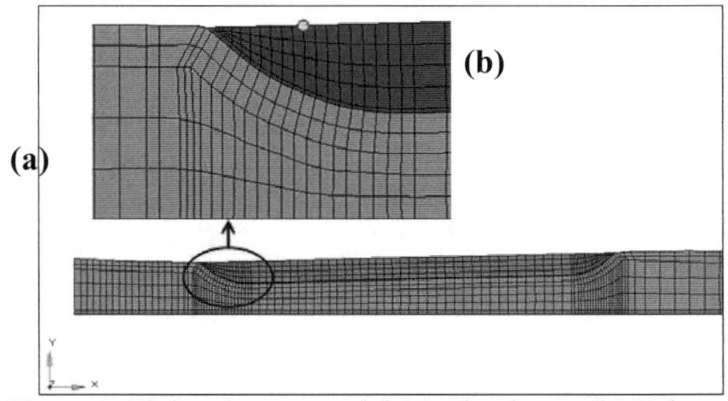

Figure 1: Finite element model (a) Aluminum, (b) Rub-strip

Component	Young's Modulus (E) GPa	Coefficient of thermal expansion (α) $\mu m.m^{-1}.K^{-1}$
Aluminum	69	23
Rub strip	0.63	83.26 (213 to 303 K) 129.4 (303 to 333 K)

Table 1: Mechanical properties of materials

To simulate the thermal cycling, the thermo-mechanical analysis was done inthree steps. Cure shrinkage was induced in the rub-strip as volume shrinkage boundary condition in the initial step. The FEM model with these cure induced pre-stresses was used as an initial condition for analysis during subsequent Heat-up (24°C to 50°C) and Cool-down (24°C to -50°C) steps.

RESULTS AND DISCUSSION

A comparison of interfacial strength measured using SLS test coupons is provided in Fig. 2.

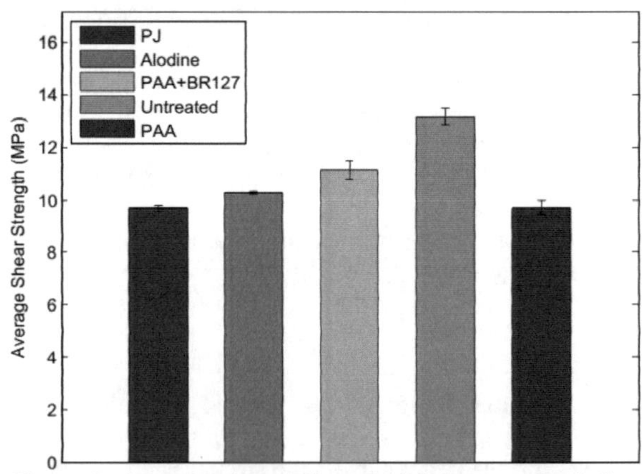

Figure 2: A comparison of interfacial shear strength for different surface treatments

The maximum interfacial strength (13.2 MPa) was measured with mechanically abraded and untreated SLS test coupons, while the surface treated coupons registered lower values. The interfacial shear strength varied (9.7 to 13.2 MPa) with surface treatment in the following order - PJ < PAA <Alodine< PAA +BR127 <Untreated. The interfacial strength for PJ, Alodine, PAA+BR127, Untreated and PAA are 9.67±0.12 MPa, 10.29±0.04 MPa, 11.13±0.35 MPa, 13.17±0.32 MPa and 9.70±0.29 MPa, respectively. The fracture surface revealed a mixed mode (cohesive in composite and adhesive at the interface) failure as shown in Fig. 3. Hence, the measured shear strength is an apparent value rather than a value corresponding to complete interface failure. Quantification of the two modes of failure may aid in understanding the variation in the interfacial strength with surface treatments. Use of primer marginally increased the strength of PAA coupons.

Figure 3: Fracture surface of a SLS coupon

The crack length in WC test coupons increased with aging time and reached a plateau value. The maximum crack length (corresponding to the plateau value beyond ~500 hours) during aging is compared in Fig.4 for various surface treatments. The crack growth was least in PAA+BR127 test coupons and maximum in untreated coupons. Alodine and PA resulted in similar crack growth – higher than PAA and slightly lower than PAA+BR127. In all these specimens (except PAA+BR127), the starter crack tip was close to the interface. For PAA+BR127, the starter crack tip was within the composite, which migrated to the interface during aging. Hence, one can conclude that these results represent the environmental durability (i.e. resistance to moisture), which increased in the following order- Untreated < PAA <Alodine / PJ < PAA +BR127.The average crack length for Untreated, PAA, Alodine, PJ and PAA + BR127 are 17.28±4.29 mm, 11.56±4.57 mm, 9.09±1.74 mm, 9.72±4.72mm and 7.47±2.99 mm, respectively. The primer increased the durability marginally. Results from thermal cycling are compared in Fig.5.

Humidity was not controlled during cycling between+40°C and -50°C. These results correspond to 72 cycles (288 hours) and the test is still continuing. Crack growth during thermal cycling was lower than the crack growth during aging in moisture, highlighting the deleterious effect of moisture on joint durability. The average crack length for PAA + BR127 and Untreated samples are 2.26±1.67 mm and 3.62±1.75 mm, respectively. The beneficial effect of primer is evident again; PAA+BR127 coupons registered marginally lower crack growth than untreated. The crack growth was interfacial.

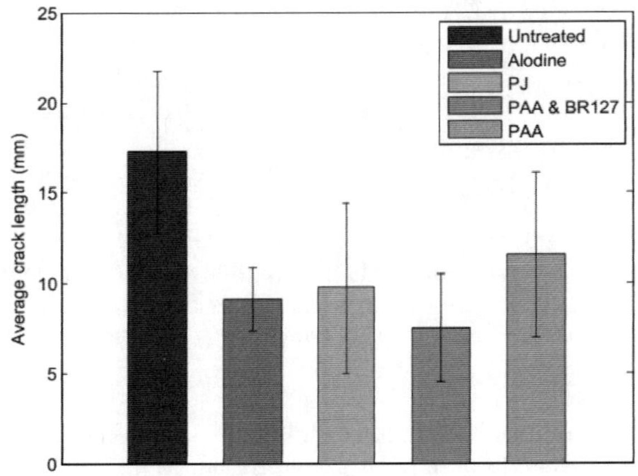

Figure 4: Average wedge crack growth for WC test coupons aged at 40°C and 95% RH

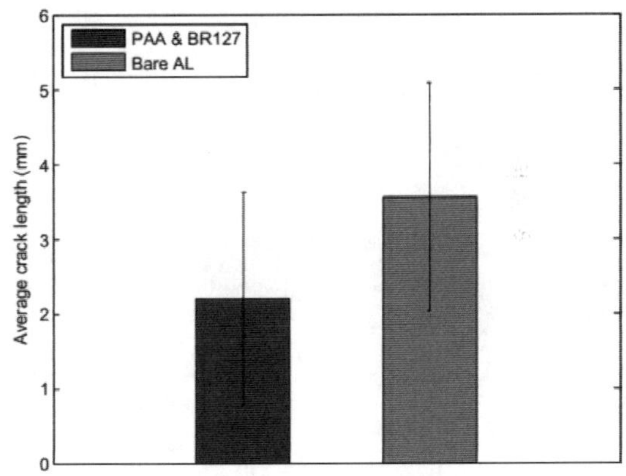

Figure 5: Average wedge crack growth for thermally cycled WC test coupons

The difference in interfacial strength could not be conclusively related to the differences in surface treatments since the fracture mode was not interfacial but mixed. Hence, the reason for the observed difference is not known at this time. Published interfacial strength for aluminum (PAA+B127 treated) bonded with 120°C curing adhesive is ~ 42 MPa [4]. Since the failure is not interfacial, such high values correspond to shear strength of the adhesive. Hence, substantially lower apparent interfacial shear strength values, measured in this study, correspond to the brittleness of the composite.

In contrast, the durability test results are conclusive. Surface treatments improve the durability; however, the difference in durability (aging in moisture) among the three treatments is marginal.

Some published studies concluded that PAA+FPL (Forest Products Laboratory primer) resulted in better durability than Pasa-Jell+BR127 at 50°C and95% RH for Al-2024 [5]. Others observed Pasa-Jell to perform better than PAA when both are coated with BR127 and bonded with 120°C curing film adhesive [6]. Fredell [7] evaluated the durability of Alodine 1200 and Pasa-Jell 105 with and without primer using a structural adhesive, at 35°C and 99% RH and found Alodine 1200 to result in a better durability than Pasa-Jell 105 and that primer improved the durability.

In this study, the effect of primer on durability was marginal. While previous studies used a high temperature curing adhesive to bond primed Aluminum, this study used a room temperature curing composite and this is believed to be the reason for the marginal beneficial effect of the primer. Further study is underway.

The predicted shear stress distribution at the end of the heating cycle from RT to +50°C is plotted in Fig. 6. A maximum shear stress of 10.5 MPa, at the casing – rub-strip interface (near the curved section in the figure), was observed. Similar results were obtained during cool down to -50°C.

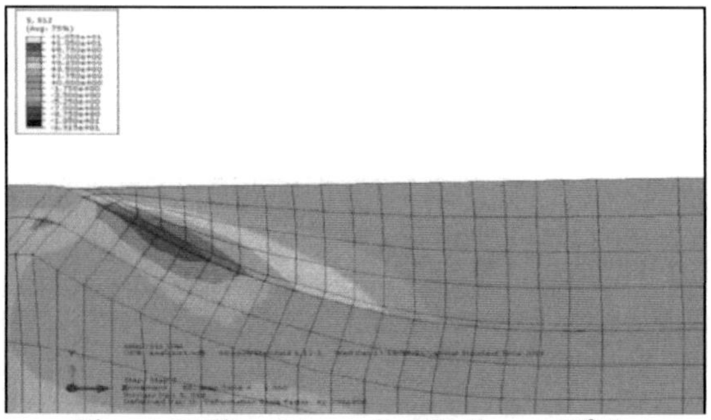

Figure 6: Shear stress distribution at 50°C.

This results suggest that only untreated and PAA+BR127 treatments would likely to result in intact bond during the first thermal cycle. Crack growth during subsequent thermal cycling could be predicted using the current FEM model provided the degradation in interfacial strength with thermal cycles is known. This study is currently underway. It can be concluded that shear strength of the interface corresponding to the optimal surface treatment should not degrade, under environmental conditions, to a value below the limiting value of 10.5 MPa. Else, interfacial crack will grow by debonding of the entire interface and result in lifting of the rub-strip as shown in Fig. 7, which would cause interference with the turbine blades.

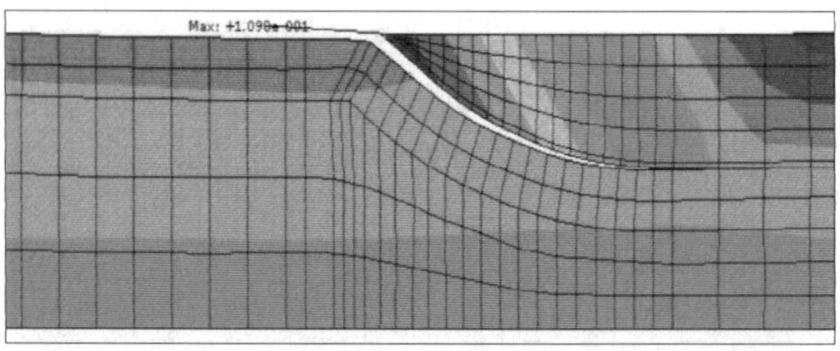

Figure 7: Predicted displacement plot at the end of Cool-down load step after crack propagation

CONCLUSIONS

The lap shear strength varied (9.7 to 13.2 MPa) with surface treatment in the following order - PJ < PAA <Alodine< PAA +BR127 <Untreated. However, the crack growth rate at 40°C and 95% RH decreased with various surface treatments in the following order - Untreated > PAA >Alodine / PJ > PAA +BR127. The crack growth rate during thermal cycling between +40°C and -50°C was lower than the crack growth at 40oC and 95% RH highlighting the deleterious effect of moisture. The beneficial effect of BR127 primer was marginal, which is believed to be due to room temperature curing of the composite.

ACKNOWLEDGMENT

The authors would like to acknowledge the financial support from NSERC Engage Grant and in-kind support from Standard Aero.

REFERENCES

[1] A. Higgins, "Adhesive bonding of aircraft structures," *International Journal of Adhesion and Adhesives,* pp. 367-376, 2000.

[2] E. M. Petrrie, "Adhesive bonding of aluminum alloys," *Metal Finishing,* vol. 105.2, pp. 49-56, 2007.

[3] E. M. Petrie, "Adhesives for the AssembLy of Aircraft Structures and Components: decades of performance improvement, with the new applications of the horizon,"*Metal Finishing* vol. 106.2, pp. 26-31, 2008

[4] J. J. Stayrook, D. L. Bellevou, R. T. Manson, A. A. Martinelli, S. Ridpath, J. R. Arnold and G. B. Gaskin, "Performance of Solvent and Water Based Primers in Epoxy Adhesive Systems,"*International SAMPE Symposium and Exhibition (Proceedings)* vol. 44, pp. 548-562, 1999

[5] A. Rider and . D. Arnott, "Boiling water and silane pre-treatment of aluminum alloys for durable adhesive bonding," *International Journal of Adhesion and Adhesives,* vol. 20, pp. 209–222, 2000.

[6] J. Zhang , X. Zhao, . Y. Zuo, J. Xinog and Zhange, "Effect of Surface Pretreatment on Adhesive Properties of Aluminum alloys," *Journal of Material Science and Technology,* vol. 24, pp. 236-240, 2008.

[7] S. Fredel, "Damage tolerant repair techniques for pressureised aircraft fuselages," Wright-Patterson Lab, Ohio, 1994.

Proceedings of the 25th CANCAM
London, Ontario, Canada, May 31 – June 4, 2015

STRAIN-RATE-DEPENDENT MIXED MODE FAILURE ANALYSIS OF UNIDIRECTIONAL CARBON FIBER REINFORCED COMPOSITE LAMINATES

Shailesh Chekurthi
Composite Materials and Structures Research Group, Department of Mechanical Engineering
University of Manitoba
Winnipeg, Manitoba, Canada
chekurts@myumanitoba.ca

J. Raghavan
Composite Materials and Structures Research Group, Department of Mechanical Engineering
University of Manitoba
Winnipeg, Manitoba, Canada
Raghavan.Jayaraman@umanitoba.ca

ABSTRACT

Rate dependent mixed-mode failure of unidirectional carbon fiber reinforced epoxy (Hexcel's G30-500/F263-7) laminates is studied using off axis laminate (0°, 10°, 15°, 30°, 45°, 60°, 75° and 90° degrees) and constant strain rate tests , at three strain rates (10^{-3}, 10^{-4} and 10^{-5} s^{-1}) in the temperature range 24°C to 270°C. Total fracture energy based failure criterion [1] is modified using critical stored elastic energy [2, 3], to account for the effect of viscoelasticity on fracture. Critical energy to failure in longitudinal normal, transverse normal and shear modes are determined using the procedure developed in Ref. 2 and experimental results from constant strain rate tensile (ASTM D3039) - and V-notch rail shear (ASTM D7078) testing of [0], [90] and $[0/90]_s$ laminates. The predicted failure envelope for off-axis laminates, at various strain rates and temperatures are in good agreement with experimental strength values.

KEYWORDS: Polymer matrix composites, Mixed-mode failure, Rate-dependent failure, Failure criterion

INTRODUCTION

Properties (modulus and strength) of a unidirectional polymer composite are dependent on strain rate due to visco-elasticity of the thermoset polymer matrix [2, 3]. In addition, its strength under uniaxial loading is different from multi-axial loading due to multiple failure modes contributing to the final failure. Multi-axial testing of unidirectional composites is normally accomplished using off-axis testing coupons [1-5] or filament wound tubes [6] or bi-axial test frames, and this results in data limited to one of the two stress quadrants. Sandhu [1] developed an energy-based failure criterion to predict tensile failure of off-axis test coupons. This study was focused on failure due to combined effect of transverse tensile (σ_{22}) and in-plane shear (τ_{12}) stresses. Kawai and Saito [4] and Daniel et al. [5] used uniaxial compression testing of off-axis test coupons to study the combined effect of transverse compressive and in-plane stresses on failure strength. In this second stress quadrant, significant interaction between these two stresses was observed i.e. shear stress at failure, in presence of transverse compressive was more than the in-plane shear strength obtained from uni-axial loading. While Kawai and Saito [4] used a modified Tsai-Wu model to predict the failure, Daniel et al. [5] developed a criterion based on dominant failure mode in this stress quadrant. All of these above studies are limited to small strain rate and temperature window, where the visco-elastic effect is not significant. This study expands the strain rate range by testing at a wide range of temperatures. Moreover, the Sandhu's model [1] is modified to account for the effect of visco-elasticity and develop a failure criterion for strain rate dependent mixed mode failure.

FAILURE CRITERION

Sandhu's failure criterion for failure of unidirectional laminate under on-axis and off-axis loading is expressed as:

$$\frac{\int_0^{\varepsilon_1}\sigma_{11}d\varepsilon}{\int_0^{\varepsilon_u}\sigma_{11}d\varepsilon}+\frac{\int_0^{\varepsilon_2}\sigma_{22}d\varepsilon}{\int_0^{\varepsilon_u}\sigma_{22}d\varepsilon}+\frac{\int_0^{\gamma_{12}}\tau_{12}d\gamma}{\int_0^{\gamma_u}\tau_{12}d\gamma}=1 \qquad -(1)$$

The non-linearity in the transverse normal and in-plane shear responses of the unidirectional composite was taken into account by assuming piecewise linear behavior and using small stress increments in simulation. The total energy for failure used in this criterion includes the energy dissipated during loading owing to the visco-elasticity of the thermoset matrix. This dissipated energy is subtracted from the total energy to determine the stored elastic energy at failure (i.e. critical stored energy (W_c)), using the model and procedure developed by Raghavan and Meshii [2, 3]. Using this, the Sandhu's criterion is modified to develop a critical stored energy criterion for mixed mode failure as:

$$\frac{W_{11}}{(W_{11})_C}+\frac{W_{22}}{(W_{22})_C}+\frac{W_{12}}{(W_{12})_C}=1 \qquad -(2)$$

where, W_{jj} (j = 1, 2) is the stored elastic energy in an off-axis laminate at any applied stress given by,

$$W_{jj}=\frac{[\sigma^*]^2}{2E_V}+\frac{[\sigma_i]^2}{2E_R} \qquad -(3)$$

where, σ^*, σ_i, E_V and E_R are model parameters determined experimentally as per procedure provided in Ref. 2 and 3.

$$[W_{jj}]_C=\frac{[\sigma_u^*]^2}{2E_V}+\frac{[\sigma_i^u]^2}{2E_R} \qquad -(4)$$

where, σ_u^* and σ_i^u are experimentally determined model parameters that correspond to off-axis failure.

Predictions using eqn. (2) are compared with experimental results and predictions using Sandhu's criterion, to evaluate the suitability of this new criterion to predict failure under multi-axial loading (stress quadrant I) over a wide range of temperatures and strain rates.

EXPERIMENTAL DETAILS

Hexcel Corporation's G30-500/F263-7 unidirectional tape prepeg was used to manufacture off-axis and $[0/90]_s$ laminates. The prepeg was manually stacked on a tool plate (bottom half of a mold), as per desired off-axis orientation, stacking sequence and thickness of the laminate. The stack of prepeg was vacuum bagged, covered by the top half of the mold and cured by holding and heating the mold between the plattens of a 50 ton Carver hydraulic hot press. The temperature of the platens was increased at 3°C / min to the cure temperature of 180°C and held at the temperature for 2 hours. Vacuum was applied to the laminate inside the vacuum bag until 80°C, when vacuum was cut and pressure was applied. Nitrogen gas was channeled into the space above the vacuum bag within the top half of the mold to simulate the autoclave pressure of 85±5Psi on the staked laminate. Laminate was post-cured in an oven at 230°C, to complete the cross linking reaction. Woven carbon fiber epoxy laminates, used as end tabs, were manufactured using a cure procedure similar to that for the unidirectional laminates. End tabs were bonded to the laminates using structural adhesive AF-31 supplied by 3M, at 2 tons pressure and 180°C, using the hydraulic press. A diamond saw cutter was used to cut the laminate into test coupons with desired dimensions. The edges of all coupons were ground using silicon carbide papers and polished using 6 μm alumina powder slurry to eliminate any edge damage. Strain gauges of type WK-06-125TM-350 and WK-06-250BG-350, bonded using Mbond 610 as per procedure suggested by Vishay Micro measurements, were used in measuring the strain.

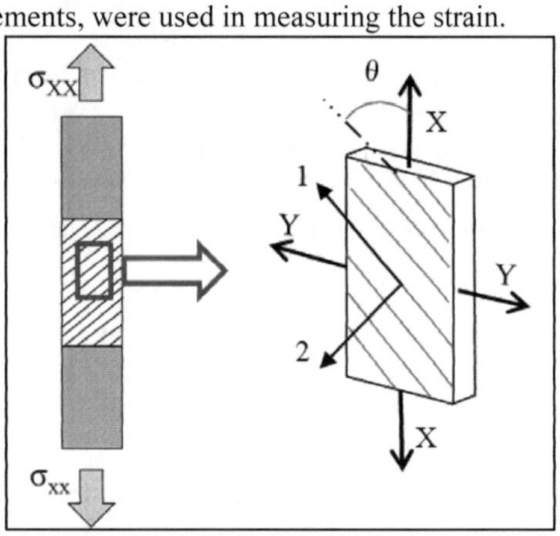

Figure 1: Off axis laminate coordinate reference system

Tensile and rail shear tests were carried out using Instron's servo electric 8562 and screw driven 5500R universal testing machines. These test frames were equipped with high temperature ovens, capable of maintaining required temperature within ±1°C. Samples were equilibrated at a required temperature for an hour, before testing. All measurements during testing were recorded using National Instrument's SCXI 1000 DAQ coupled with LabVIEW program.

Constant strain rate tensile tests were performed on off-axis laminates (0°, 10°, 15°, 30°, 45°, 60°, 75° and 90°) as per ASTM D3039. Samples were tested using strain rates of 10^{-3} s^{-1}, 10^{-4} s^{-1} and 10^{-5} s^{-1} and constant temperatures of 24°C, 80°C, 120°C, 160°C, 200°C, 245°C and 275°C at each strain rate. Sample geometry with global and principal coordinate reference systems is shown in Fig. 1. The $[0]_6$ and $[90]_{10}$ coupons yielded results for longitudinal normal and transverse properties in the principal material coordinates.

V-notched rail shear testing of $[0/90]_{4s}$ laminates was done as per ASTM D7078 to determine in-plane shear properties. Shear experiments were performed at strain rates and temperatures used for tensile testing. Grips manufactured by Wyoming Test Fixtures Inc. were used for this experiment. Test coupon geometry, gripped coupon and failure mode after testing are shown in Fig. 2.

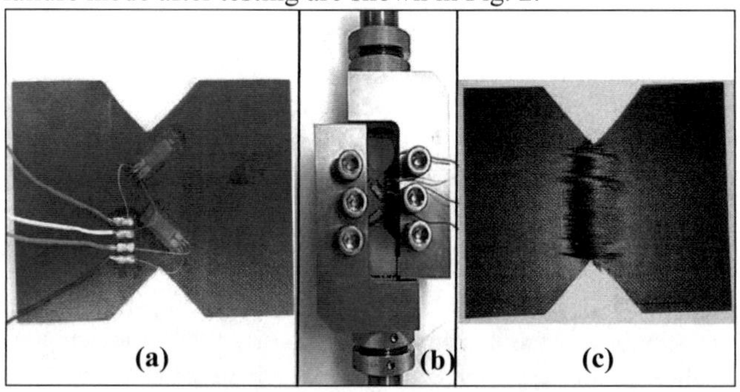

Figure 2: V-notch Rail Shear experiment details. (a) Sample geometry (b) Gripped sample (c) Failure mode

SIMULATION DETAILS

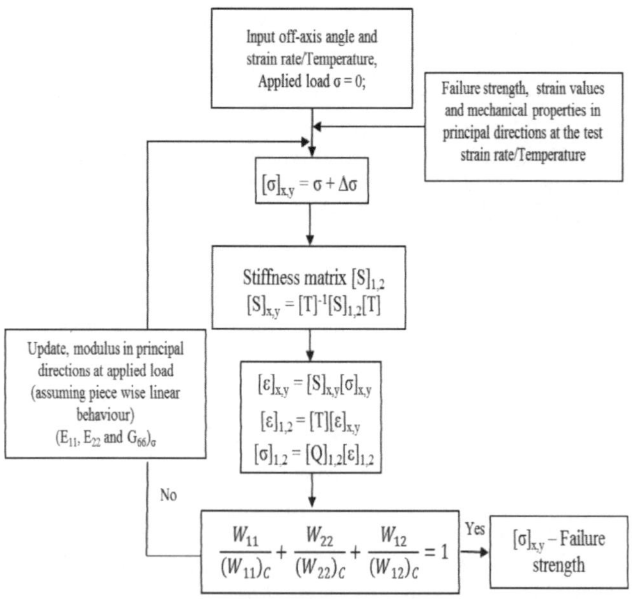

Figure 3: Flow chart for failure envelope prediction using critical stored elastic energy based fracture criterion.

The simulation procedure is detailed in Fig. 3.

RESULTS AND DISCUSSSION

Stress-strain plots for longitudinal normal, transverse normal and shear modes of loading at 24°C and at all strain rates are provided in Fig. 4. As expected, $[0]_6$ laminate did not exhibit significant effect of strain rate and temperature on modulus and strength. However, significant non-linearity was observed in the transverse and shear response, with increase in temperature or decrease in strain rate, since these responses are controlled by matrix properties. Average values for the measured modulus and strength along three principal directions are listed in Table1, for three strain rates and test temperatures of 24°C and 200°C. Decrease in modulus with decrease in strain rate and increase in temperature can be observed.

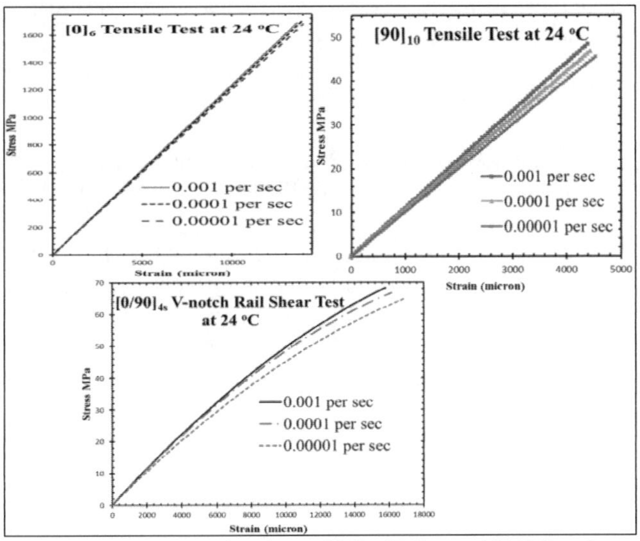

Figure 4: Stress-strain plots of $[0]_6$, $[90]_{10}$ and $[0/90]_{4s}$, for three strain rates at 24°C

	Strain Rates					
	10^{-3} s^{-1}		10^{-4} s^{-1}		10^{-5} s^{-1}	
	24°C	200°C	24°C	200°C	24°C	200°C
E_{11} (GPa)	118.5	115.8	118.5	114.8	118.2	114.4
E_{22} (GPa)	10.55	9.50	10.1	9.12	10.01	9.35
G_{66} (GPa)	5.7	4.1	5.6	4.1	5.3	3.8
σ^u_{11} (MPa)	1701	1681.7	1692.7	1666.2	1687.97	1661.7
σ^u_{22} (MPa)	48.4	41.65	44.72	41.1	44.12	36.5
τ^u_{66} (MPa)	68.24	51.8	66.63	50.8	64.98	46.40

Table 1: Experimentally measured in-plane properties for two temperatures

Fig. 5 compares the failure modes in off-axis laminates and similar failure was noticed at all test temperatures and strain rates. Fiber breakage is clearly the mode of failure in $[0]_6$ laminate. In $[90]_{10}$ laminates, matrix failure occurred in a plane parallel to the fiber axis, and perpendicular to load, indicating pure transverse mode of failure. In rail shear tested coupons; the failure was in the matrix parallel to the fibers as observed in Fig. 2(c), indicating shear mode of failure. In off-axis laminates, matrix failure in a plane parallel to the fiber axis of the sample was observed. Shear and transverse modes of failure are dominant in these fiber orientations, leading to mixed mode failure.

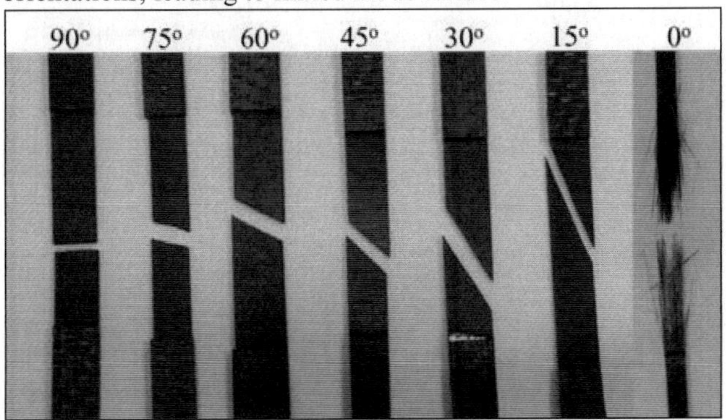

Figure 5: Failure modes in off-axis laminates tested at 10^{-3} s^{-1} and 24°C

Tensile strength, predicted using equations (1) and (2), are compared with experimental values for various off-axis laminates in Fig. 6. Predictions are in good agreement with the off-axis strength in transverse stress dominated region ($60<\theta<90$) within 1% error. However, the critical stored elastic energy criterion (eqn. (2)), overestimated the strength values by 12%, 9% and 16% for off axis angles 15°, 30° and 45° respectively. Eqn. (1) overestimated the values by 24%, 20% and 28%; and Tsai-Hill theory overestimated the strength values by 35%, 32% and 41% for 15°, 30° and 45° off-axis laminates respectively.

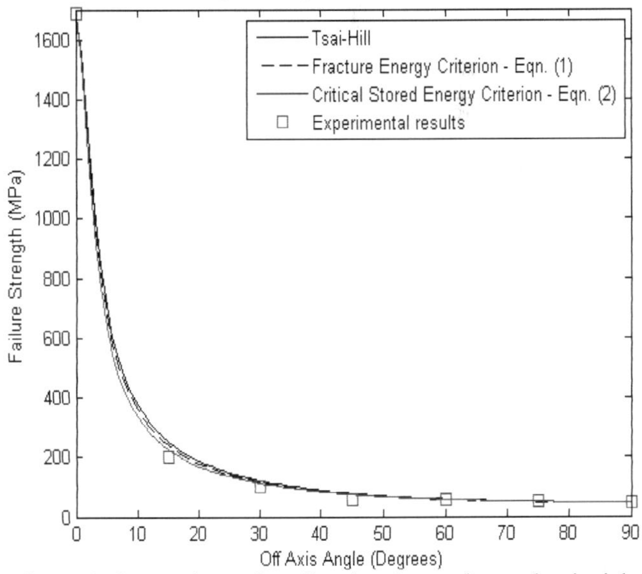

Figure 6: Comparison of predicted and experimental uniaxial strength for 10^{-3} s^{-1} and 24°C

Overall, critical stored elastic energy based criterion yielded better failure strength predictions when compared to fracture energy based criterion and Tsai-Hill theory. Failure strength predictions for two temperatures are compared with experimental values in Fig. 7 for a strain rate of 10^{-4} s^{-1}.

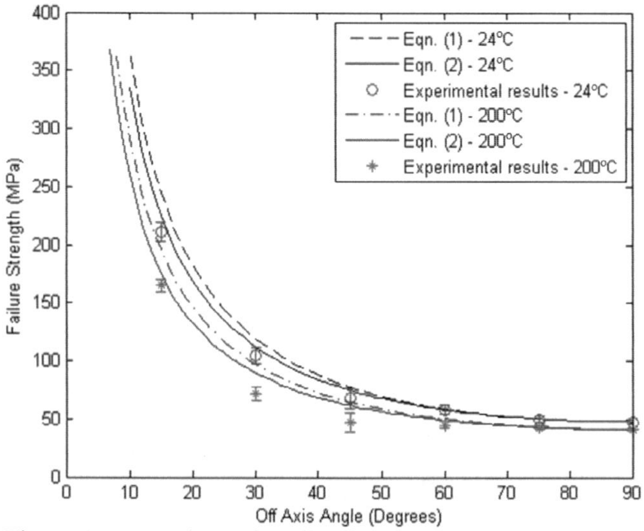

Figure 7: Comparison of predicted and experimental uniaxial strength for two temperatures at 10^{-4} s^{-1}

Eqn. (2) overestimated the strength values by 8%, 7% and 12% for off-axis angles of 15°, 30° and 45° respectively at 200°C. The predictions for 200°C are better than that for 24°C. Predictions using Sandhu's criterion (eqn. (1)) are significantly improved by deducting the dissipated energy from the total fracture energy.

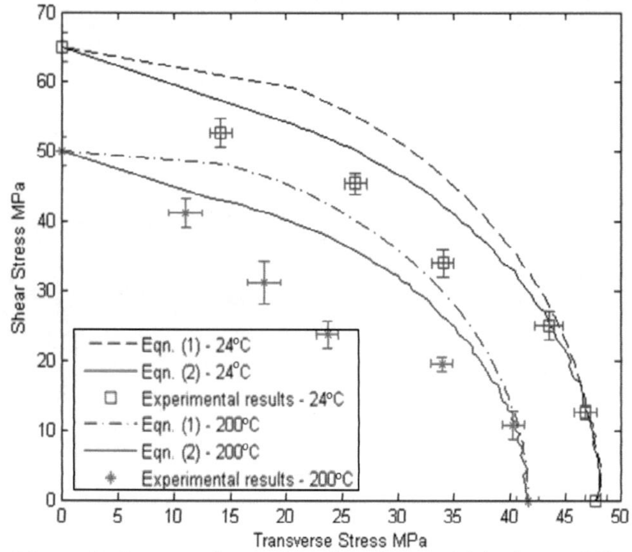

Figure 8: A comparison of predicted (eqn. (1) and eqn. (2)) failure envelope to the experimental transverse normal and shear stresses at 10^{-5} s^{-1}.

The results in Fig. 7 are re-plotted in Fig. 8 using shear and transverse normal stresses at failure to highlight the mixed mode of failure in off-axis test coupons. Eqn. (2) yields a better prediction and the observed difference between the predicted and experimental trends in Fig. 8 is under investigation.

CONCLUSIONS

Failure envelope for off axis laminates is predicted using critical energies for fracture in longitudinal normal, transverse normal, and shear modes, determined using on-axis tensile tests and V-notch rail shear tests at three strain rates and different temperatures. Predictions based on critical stored elastic energy criterion are better than that using total fracture energy based Sandhu's criterion and Tsai-Hill criterion.

ACKNOWLEDGMENTS

This work was supported by Discovery Grant from NSERC (Natural Sciences and Engineering Research Council of Canada). The authors gratefully acknowledge the donation of material by Hexcel Corporation.

REFERENCES

[1] R. Sandhu, "Nonlinear behaviour of unidirectional and angle ply laminates," *Journal of Aircarft,* pp. 104-111, 1975.

[2] J. Raghavan and M. Meshii, "Creep rupture of polymer composites," *Composites Science and Technology,* no. 57, pp. 375-388, 1997.

[3] J. Raghavan and M. Meshii, "Prediction of creep rupture of unidirectional carbon fiber reinforced polymer composite," *Material Science and Engineering,* no. 197, pp. 237-49, 1995.

[4] M. Kawai and S. Saito, "Off-axis strength differential effects in unidirectional carbon/epoxy laminates at different strain rates and predictions of associated failure envelopes," *Composites: Part A,* no. 40, pp. 1632-49, 2009.

[5] I. Daniel, B. Werner and J. Fenner, "Strain-rate-dependent failure criteria for composites," *Composites Science and Technology,* no. 71, pp. 357-364, 2011.

[6] E. Shin and K. Pae, "Effects of Hydrostatic Pressure onIn-Plane Shear Properties of Graphite/Epoxy Composites," *Journal of Composite Materials,* vol. 26, no. 6, pp. 828-868, 1992.

COMPUTATIONAL HOMOGENIZATION OF METAL MARIX COMPOSITES

M. U. Siddiqui
Mechanical Engineering department
King Fahd University of Petroleum and Minerals
Dhahran, Saudi Arabia
musiddiqui@kfupm.edu.sa

A. F. M. Arif
Mechanical Engineering department
King Fahd University of Petroleum and Minerals
Dhahran, Saudi Arabia
afmarif@kfupm.edu.sa

ABSTRACT

Computational homogenization provides an excellent tool for the design of composite materials. In the current work, a computational homogenization methodology is presented that is capable of estimating the elastic and rate-independent plastic constitutive behavior of metal matrix composites using finite element models of representative volume elements (RVEs) of the composite material. For this purpose, methodologies for the generation of RVEs and its size determination and the homogenization techniques are presented. Using the homogenized material response, effective constitutive models of the composite materials have been determined.

KEYWORDS: Metal matrix composite, homogenization, computational homogenization, representative volume element

INTRODUCTION

Various types of homogenization methods have been extensively employed in to determine the mechanical behaviors of heterogeneous materials by taking into account the length scale difference between the macrostructure and the individual microstructure components. These homogenization schemes fall into two major categories: mean-field homogenization and computational homogenization. Although mean-field homogenization schemes are much more computationally efficient than the computational homogenization schemes, homogenization in nonlinear regimes using mean-field approaches is clearly more cumbersome. In recent years, significant emphasis has been laid on computational homogenization techniques which for finding the material response are essentially solutions of finite element boundary value problem.

Several authors have presented homogenization methodologies for the estimation of the properties of the composite. Cho et al. [1] used computational homogenization to determine the thermoelastic properties for regular and random microstructures. They generated computation two phase RVEs using the random morphology description function (RMDF) and used the developed computational RVEs to determine the homogenized properties, study the effect of variations in particle shapes and orientation on the orthotropic properties and to study the response of a beam composed of a dual phase material. Galli et al. [2], [3] used computational homogenization for studying the macroscale behavior of particle-reinforced metal matrix composites (PRMMCs). Their work comprises a methodology for the generation of computational RVEs and the homogenization of the thermoelastic and plastic properties of the composites.

Selection of representative volume elements for computational homogenization has a huge impact on the accuracy of results. Several authors have focused solely on understand the relationship between the prediction accuracy and the determination of correct RVE size based on application [4]–[6]. Once the RVE has been developed, it can then be used to study the mechanical behavior of the material at the microscale[7]–[9] or can be used in a multiscale approach to include the microstructural effects at the macroscale[2], [10], [11].

The objective of the current work is the formulation of a methodology for carrying out computational homogenization of metal matrix composites. In this regard, methodologies for the generation of representative volume element, its size determination and then using the generated RVEs for the determination of elastic and elastoplastic properties of metal matrix composites are presented in the following sections.

HOMOGENIZATION METHODOLOGY

The homogenization problem consists of a representative volume element of the heterogeneous material. The RVE contains all constituent phases of the material and the size of the RVE is statistically determined to accurately represent the overall macroscopic response of the material. The constitutive behaviors of all phases present in the RVE are explicitly defined. Periodic boundary conditions are applied to the microscale model using constraint equations. The steps involved in the methodology are RVE generation, FE model generation, RVE size determination and homogenization steps.
The homogenized elasticity matrix of the heterogeneous material, given by equation (1), is calculated by applying six different loads (three in case of two-dimensional model) to the microscale model. Each load vector, shown in equation(2), provides one column of the tangent modulus by the homogenization of the stress field using equation(3). The

elements of C are then calculated by dividing σ_{ij}components with ε_o.

$$\bar{C} = \frac{d\sigma}{d\varepsilon} = \begin{bmatrix} C_{11} & C_{12} & C_{13} & C_{14} & C_{15} & C_{16} \\ C_{21} & C_{22} & C_{23} & C_{24} & C_{25} & C_{26} \\ C_{31} & C_{32} & C_{33} & C_{34} & C_{35} & C_{36} \\ C_{41} & C_{42} & C_{43} & C_{44} & C_{45} & C_{46} \\ C_{51} & C_{52} & C_{53} & C_{54} & C_{55} & C_{56} \\ C_{61} & C_{62} & C_{63} & C_{64} & C_{65} & C_{66} \end{bmatrix} \quad (1)$$

$$\bar{\varepsilon} = \begin{Bmatrix} \varepsilon_o \\ 0 \\ 0 \\ 0 \\ 0 \\ 0 \end{Bmatrix}, \begin{Bmatrix} 0 \\ \varepsilon_o \\ 0 \\ 0 \\ 0 \\ 0 \end{Bmatrix}, \begin{Bmatrix} 0 \\ 0 \\ \varepsilon_o \\ 0 \\ 0 \\ 0 \end{Bmatrix}, \begin{Bmatrix} 0 \\ 0 \\ 0 \\ \varepsilon_o \\ 0 \\ 0 \end{Bmatrix}, \begin{Bmatrix} 0 \\ 0 \\ 0 \\ 0 \\ \varepsilon_o \\ 0 \end{Bmatrix}, \begin{Bmatrix} 0 \\ 0 \\ 0 \\ 0 \\ 0 \\ \varepsilon_o \end{Bmatrix} \quad (2)$$

whereϵ_o is a small strain increment.

$$\bar{\sigma}_{ij} = \frac{1}{V}\int_V \sigma_{ij} dV \quad (3)$$

The calculation of the plastic properties requires the determination of the stress $\sigma_{y.loc}$ at which localized yielding starts and the stress strain curve. The yield stress $\sigma_{y,loc}$ needs to be iteratively determined by applying the correct strain load on the RVE that results in the beginning of yield. The resulting total strain and elastic strain components are homogenized using equations (4) and(5) respectively. The plastic strain components is calculated using equation (6). If the effective plastic strain is lower than a threshold value, the homogenized effective stress in the RVEis the yield stress of the composite. It is important to note that the plastic strain components cannot be homogenized directly.

$$\bar{\varepsilon}_{ij} = \frac{1}{V}\int_V \varepsilon_{ij} \quad (4)$$

$$\bar{\varepsilon}_{e,ij} = \frac{1}{V}\int_V \varepsilon_{e,ij} \quad (5)$$

$$\varepsilon_{p,ij} = \varepsilon_{ij} - \varepsilon_{e,ij} \quad (6)$$

Once the stress and strain at the localized yielding point are known, the remaining stress-strain curve can be determined by applying incremental load to the RVE. The points on the curve can also be fitted to a standard hardening function and its parameters determined.

RVE GENERATION AND SIZE ESTIMATION

The most common algorithm used for the generation of computational composite microstructures is the Random Sequential Adsorption (RSA) algorithm [12].
In the present work, a methodology similar to the Random Sequential Adsorption (RSA) algorithm is presented. It is capable of placing two different phases, an inclusion and porosity, of different average sizes in a matrix. The inputs to the algorithm include the target weight fraction of inclusions, target volume fraction of porosity, average inclusion size, target number of inclusions, target number of pores and matrix and inclusion densities. Using the inputs, the target inclusion volume fraction, RVE volume, RVE edge length and the average pore size are calculated. The algorithm starts by randomly placing the first inclusion particle in the matrix with a radius equal to the average inclusion radius. The remaining particles are sequentially placed. To determine the radius of a newly placed particle, its distance from all previously placed particles is calculated to determine its maximum allowable radius and a random radius is generated. The inclusion particles are placed until the volume fraction of the inclusions is equal to the desired value and the number of particles do not exceed the target number of inclusion particles by more than five percent. Three example RVEs generated using the algorithm implemented in MATLAB are shown in Figure 1.
A methodology to statistically determine the size of the RVE was presented by Gitman et al. [6] and has been applied here to determine the required RVE sizes for the determination of the elastic and inelastic properties of a composite. The methodology is based on the determination of a variation coefficient, known as the *chi-square criterion* for each RVE realization using equation (7). Once the *chi-square criterion* is below a threshold value the RVE size is converged. A threshold of 0.01 was selected in the present study.

$$\chi^2 = \frac{\sum_{i=1}^{n}\left(a_i - \langle a \rangle\right)^2}{\langle a \rangle} \quad (7)$$

Where a_i is the property under consideration for realization i, n is the total number of realizations of the same size and $\langle a \rangle$ is the mean value of the property under consideration.

MODEL VALIDATION

For validation of the methodology for elastic properties, results of Gudlur et al. [13]were selected. In their work, Gudlur et al. used object oriented finite element method code (OOF2) to determine the homogenized properties using actualmicrostructures of the composite. They selected four 50 µm x 50 µm cross sections of the SEM images of the composite and solved for the properties using OOF2. The results of Gudlur et al.'s work is given in Table 1 along with the results obtained using the current methodology. As can be seen from Table 1, the model shows good agreement with the results of Gudlur et al. In fact, the results of Gudler et al. show a greater scatter between different RVEs than the current work since they did not carry out a study to determine the required size of the RVE.

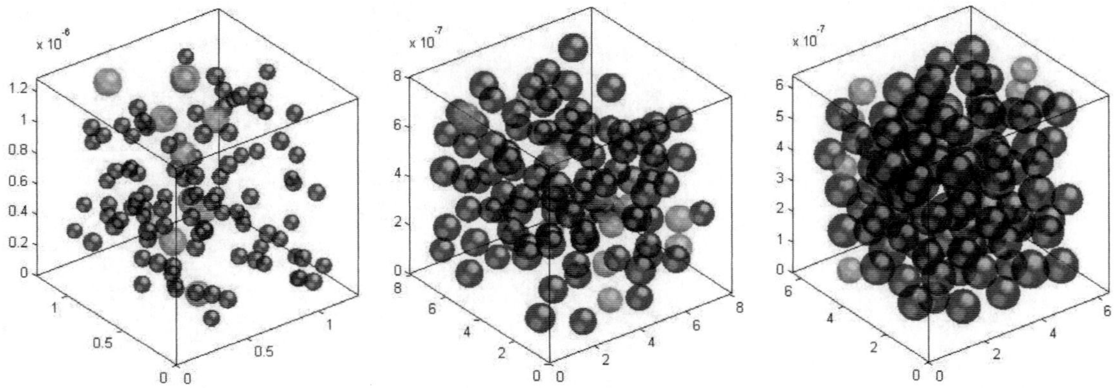

(a) (b) (c)

Figure 1. Example RVEs with 1% porosity and (a) 2.5% reinforcement (b) 10% reinforcement and (c) 20% reinforcement

Table 1. Validation of Elastic Property Homogenization.

RVE No.	Inclusion VF%	Porosity VF%	E [GPa] Gudlur et al.[13]	E [GPa] Current Work	Error
1	20.2	0.4	83.28	85.82	3.05%
2	21.7	0.1	86.67	87.16	0.56%
3	20.9	0.7	83.70	86.74	3.63%
4	21.2	0.0	88.05	92.26	4.78%

For the validation of plastic property homogenization, the homogenized stress-strain response of Al-Cu-Mg(2080)-10% SiC_p-T8 was predicted using the proposed methodology and compared with the experimental stress-strain response of the composite determined by Chawla and Shen [14]. The matrix was model as elastoplastic and the inclusions were modeled as elastic. The elastic modulus and the yield stress of the matrix were determined from the experimental stress-strain curve of the matrix provided by Chawla and Shen [14] and were found to be 79.3 GPa and 485.8 MPa respectively. The experimental hardening behavior of the matrix was also used in the simulation. The elastic modulus of the inclusion and its Poisson's ratio were 415 GPa and 0.225 respectively. The validation results are shown in Figure 2 and show good agreement between the homogenization and experimental results.

CASE STUDY A: ELASTIC PROPERTIES OF ALUMINUM-AL_2O_3 COMPOSITE

In order to further validate the model and to study the effect inclusion volume fraction on the macroscale elastic properties, Aluminum-Al_2O_3 composites with Al_2O_3inclusion volume fractions of 1%, 2.5%, 5%, 10%, 15%, 20% and 25% were modeled. A total of 300 inclusions were introduced in the matrix. The porosity fraction was set at 0. The homogenized elastic modulus was compared to those predicted by the Mori Tanaka model and the self-consistent model. The rule of mixtures upper and lower bounds were also calculated. The results are summarized in Table 2 and the comparison is presented in Figure 3.

The results show that the computational homogenization approach presented in the current work and self-consistent mean-field homogenization show very close agreement. In comparison, the Mori Tanaka approach under predicted the elastic modulus of the composite at higher inclusion volume fractions.

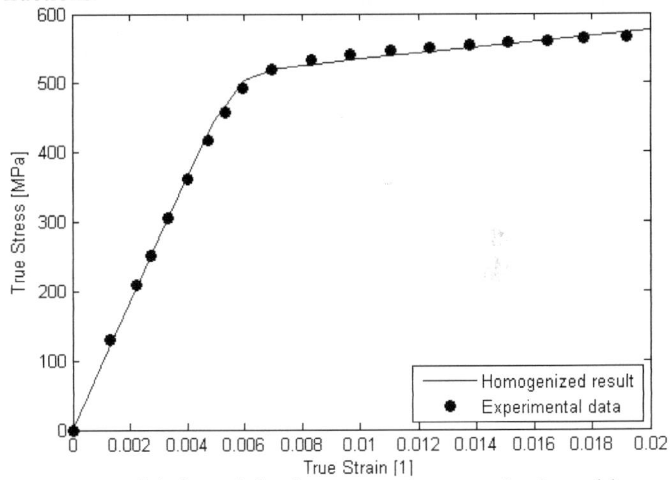

Figure 2. Validation of plastic property homogenization with experimental data [14].

Table 2. Elastic properties for Alumina-Aluminum composite

Inclusion volume fraction	E [GPa]	ν
1%	65.42	0.348
5%	70.49	0.342
10%	76.26	0.337
15%	82.93	0.330
20%	90.23	0.324
25%	98.22	0.316

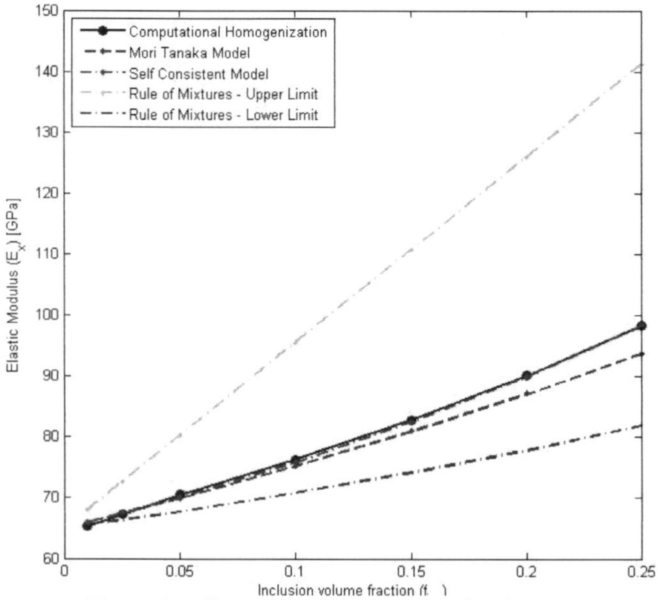

Figure 3. Effect of inclusion volume fraction.

CASE STUDY B: ELASTOPLASTIC PROPERTIES OF Al-Cu-Mg(2080) – SiC_p-T8 COMPOSITE

For the case of Al-Cu-Mg(2080)- SiC_p-T8 composite, the effect of inclusion volume fraction on the stress-strain response of the composite was studied. The stress-strain curves are shown in Figure 4. Using the results of computational homogenization, bulk yield stress and parameters of a hardening rule can also be determined. Here, the parameter of a Voce-type hardening law, given by equation (8) are determined using a Trust-Region optimization algorithm in *MATLAB*. The parameters of the hardening law are shown in Table 3.

$$\sigma_y = \sigma_o + \left(\sigma_A - \sigma_o\right)\left(1 - \frac{e^{-\varepsilon_p}}{\varepsilon_o}\right) + m\varepsilon_p \tag{8}$$

CONCLUDING REMARKS

In the current work, a homogenization methodology was presented for the estimation of the elastoplastic material properties of metals matrix composites.

- As part of the methodology, a methodology for the generation of RVEs for metal matrix composites was formulated that is capable of generating RVEs with spherical inclusions and pores.
- A methodology to determine the minimum RVE size required for correct property estimation was also presented.
- Computational homogenization for the estimation of elastic and rate-independent plastic properties of metal matrix composites was presented and validated.
- The effect of inclusion volume fraction on the bulk material properties of the composite was investigated and, effective material models were determined that can represent the bulk constitutive response of the composites.

Table 3. Voce-type hardening law parameters for Al-Cu-Mg(2080) - SiC_p-T8 composite.

Inclusion volume fraction	σ_0 [MPa]	σ_A [MPa]	m[GPa]	ε_0
1%	485	489.5	1.295	0.001625
5%	488	497.3	1.976	0.002778
10%	497.6	513.6	2.823	0.004156
15%	507.8	532.2	4.180	0.004711
20%	518.9	553.5	5.673	0.005551
25%	531.7	579.5	8.485	0.006016

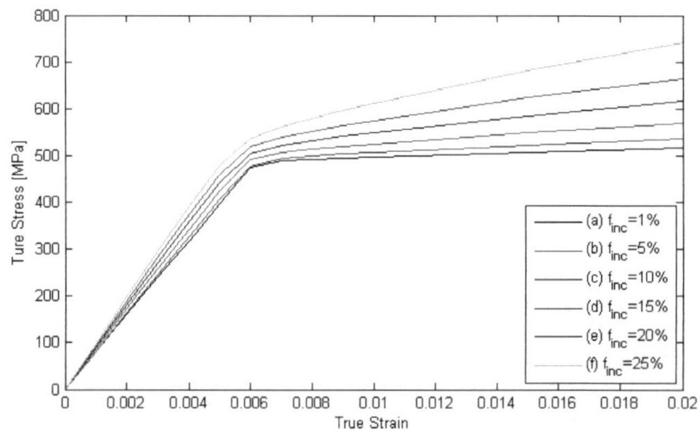

Figure 4. Effect of Inclusion volume fraction on stress-strain response of Al-Cu-Mg(2080) - SiC_p-T8 composite.

ACKNOWLEDGMENTS

The authors would like to acknowledge the support provided by King Abdulaziz City for Science and Technology (KACST) through the Science and Technology Unit at King Fahd University of Petroleum & Minerals (KFUPM) for funding this work through project 12-NAN2374-04 as part of the National Science, Technology and Innovation Plan.

REFERENCES

[1] J. R. Cho, Y. J. Kang, K. Y. Jeong, Y. J. Noh, and O. K. Lim, "Homogenization and thermoelastic analysis of heterogenous materials with regular and random microstructures," *Compos. Part B Eng.*, vol. 43, no. 5, pp. 2313–2323, Jul. 2012.

[2] M. Galli, J. Botsis, and J. Janczak-Rusch, "An elastoplastic three-dimensional homogenization model for particle reinforced composites," *Comput. Mater. Sci.*, vol. 41, no. 3, pp. 312–321, Jan. 2008.

[3] M. Galli, J. Cugnoni, and J. Botsis, "Numerical and statistical estimates of the representative volume element of elastoplastic random composites," *Eur. J. Mech. - A/Solids*, vol. 33, pp. 31–38, May 2012.

[4] Z. Shan and A. Gokhale, "Representative volume element for non-uniform micro-structure," *Comput. Mater. Sci.*, vol. 24, pp. 361–379, 2002.

[5] M. Stroeven, H. Askes, and L. J. Sluys, "Numerical determination of representative volumes for granular materials," *Comput. Methods Appl. Mech. Eng.*, vol. 193, no. 30–32, pp. 3221–3238, Jul. 2004.

[6] I. M. Gitman, H. Askes, and L. J. Sluys, "Representative volume: Existence and size determination," *Eng. Fract. Mech.*, vol. 74, no. 16, pp. 2518–2534, Nov. 2007.

[7] Y. Liu and X. Chen, "Evaluations of the effective material properties of carbon nanotube-based composites using a nanoscale representative volume element," *Mech. Mater.*, vol. 35, pp. 69–81, 2003.

[8] Z. Li and P. Steinmann, "RVE-based studies on the coupled effects of void size and void shape on yield behavior and void growth at micron scales," *Int. J. Plast.*, vol. 22, no. 7, pp. 1195–1216, Jul. 2006.

[9] B. R. Kim and H. K. Lee, "An RVE-based micromechanical analysis of fiber-reinforced composites considering fiber size dependency," *Compos. Struct.*, vol. 90, no. 4, pp. 418–427, Oct. 2009.

[10] P. Raghavan and S. Ghosh, "Concurrent multi-scale analysis of elastic composites by a multi-level computational model," *Comput. Methods Appl. Mech. Eng.*, vol. 193, no. 6–8, pp. 497–538, Feb. 2004.

[11] C. Toulemonde, R. Masson, and J. El Gharib, "Modeling the effective elastic behavior of composites: a mixed Finite Element and homogenisation approach," *Comptes Rendus Mécanique*, vol. 336, no. 3, pp. 275–282, Mar. 2008.

[12] M. Rintoul and S. Torquato, "Reconstruction of the Structure of Dispersions," *J. Colloid Interface Sci.*, vol. 186, pp. 467–76, 1997.

[13] P. Gudlur, A. Muliana, and M. Radovic, "The effect of microstructural morphology on the elastic , inelastic , and degradation behaviors of aluminum – alumina composites," *Mech. Res. Commun.*, vol. 57, pp. 49–56, 2014.

[14] N. Chawla and Y.-L. Shen, "Mechanical Behavior of Particle Reinforced Metal Matrix Composites," *Adv. Eng. Mater.*, vol. 3, no. 6, pp. 357–370, Jun. 2001.

Nanostructures and MEMS

MULTISCALE MODELLING OF THERMO-ELECTRO-MECHANICAL BEHAVIOURS OF GALLIUM NITRIDE NANOWIRES: SIZE AND TEMPERATURE DEPENDENCE OF THE PIEZOELECTRIC POTENTIAL

Jin Zhang
Mechanical and Industrial Engineering
University of Toronto
Toronto, Ontario, Canada
jinzhang@mie.utoronto.ca

S. A. Meguid
Mechanical and Industrial Engineering
University of Toronto
Toronto, Ontario, Canada
meguid@mie.utoronto.ca

ABSTRACT

A multiscale modelling approach has been proposed to study the piezoelectric potential generated in the deformed gallium nitride (GaN) nanowires (NWs). In this study, the material properties of GaN NWs are determined using molecular dynamics simulations and are found to depend on the NW's diameter and the working temperature. The obtained material properties are then used in the finite element analysis to determine the piezoelectric potential of GaN NWs. The obtained results show that the material properties are temperature- and size-dependent. Specifically, a core-surface model has been developed for NWs to reveal the physical origin of the size-dependence of the material properties. The results further show that the potential of the NW can increase by up to ten times that of its bulk counterpart due to the small-scale effect. In addition, the potential of a NW is also found to be sensitive to its working temperature.

KEYWORDS: piezoelectric potential, gallium nitride, nanowire, multiscale modelling, small-scale effect

INTRODUCTION

Recently, the wurtzite family materials especially the zinc oxide (ZnO) and gallium nitride (GaN) have received more attention due to their good coupling of piezoelectric and semiconducting properties. Specifically, the one dimensional (1D) ZnO and GaN nanostructures [e.g. nanowires (NWs)] have been used as basic building blocks for fabricating innovative devices, such as nanogenerators, piezoelectric field-effect transistors, piezoelectric diodes, piezoelectric chemical sensors, and piezo-phototronic devices [1]. One principle of the application of ZnO and GaN nanostructures is the transformation between the mechanical force and the piezoelectric potential. Thus, it is of great interest to accurately predict the piezoelectric potential in the NWs.

Extensive review of the literature indicates that the classical continuum mechanics together with bulk material properties was directly used as a cost-effective method [2, 3] for the analysis of nanoscopic systems. The obtained results play an important role in interpreting the experimental data and capturing the major parameters that control the overall structural response of piezoelectric NWs. Nevertheless, in contrast to their bulk counterparts, nanostructures usually exhibit size-dependent material properties, which is a unique physical phenomenon for nanoscopic structures and thus limits the usefulness of classical continuum theories.

In this paper, we have determined the material properties, including the elastic constant, piezoelectric coefficient and dielectric constant of GaN NWs using molecular dynamics simulations (MDSs). The obtained results show that these material properties are all temperature- and size-dependent. Specifically, a core-surface (CS) model has been developed for NWs to reveal the physical origin of the size-dependence of the material properties. Subsequently, the obtained material properties were employed to predict the piezoelectric potential of a GaN NW through the finite element method (FEM). The size and thermal effects are found to have significant influence on the piezoelectric potential of GaN NW.

DETERMINATION OF NANOSCOPIC MATERIAL CONSTANTS USING MDS

In this section, MDSs were employed to calculate the equivalent elastic, piezoelectric and dielectric properties of GaN NWs. In this paper, we consider the most common GaN NW whose growth direction is along the [100] crystalline direction. The NWs have hexagonal cross-sections with a six-fold symmetry and lateral surfaces {100} as shown in Fig. 1a. The length L of the NWs is fixed at 30 nm, whereas the radius a of the hexagonal cross-section is allowed to change to study size-dependent material properties of NWs.

A classical MDS method was employed in this work and the NVT ensemble (constant volume and temperature) was employed. The interactions between Ga-Ga, N-N, and Ga-N are described by the modified Stillinger-Weber (SW) potential [4], which is of the following form

$$U = \sum_i \sum_{j>i} \varphi_2(r_{ij}) + \sum_i \sum_{j\neq i} \sum_{k>j} \varphi_3(r_{ij}, r_{ik}, \theta_{ijk}) \cdot \quad (1)$$

Here U is the atomic energy, which is composed of a two-body term φ_2 and a three-body term φ_3. r_{ij} (or r_{ik}) is the distance between two atoms in a cell. θ_{ijk} is the angle between r_i and r_k subtended at atom j as vertex. The SW potentials have been used to reproduce bulk structures and mechanical properties; they have been successfully employed to evaluate the elastic properties, melting behavior and buckling of single-crystal GaN NWs [4, 5]. These calculations have demonstrated that the empirical SW potentials for GaN can be employed to study the mechanical properties of single-crystal GaN structures. In addition, the potentials can handle dangling bonds, wrong bonds, and excess bonds in bulk GaN very well. Therefore, these potentials should be reliable to study the mechanical characteristics of GaN NWs.

As a quasi 1D nanostructure, the elastic, piezoelectric and dielectric properties of GaN NWs in the axial direction (c-axis) are of interest and can be respectively characterized by the axial elastic constant c_{33}, piezoelectric coefficient e_{33}, and dielectric constant k_{33}. At the beginning of all simulations of these material properties, the equilibrium of the initialized structure was achieved corresponding to the lowest energy of the NWs (100 ps was used in this work). After the full relaxation, different treatments were applied to calculate the material properties (see Figs. 1b and 1c) and are discussed briefly below. Here all MDSs were conducted using LAMMPS [6] without periodic boundary conditions.

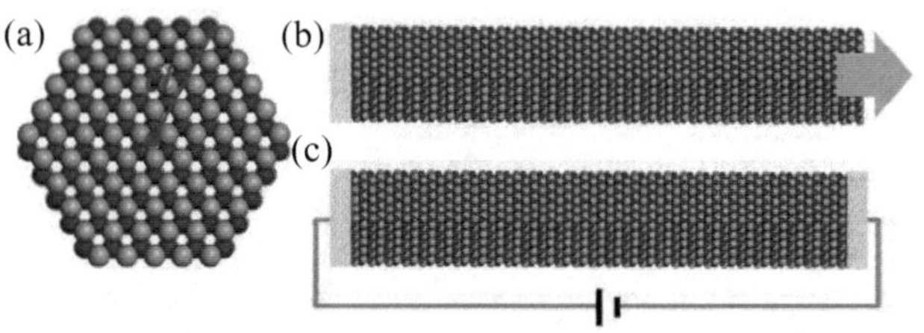

Figure 1: Molecular representation of the cross-section of a GaN NW. Simulation setups for the measurement of (b) the elastic property, and (c) the piezoelectric and dielectric properties of GaN NWs.

The elastic constant c_{33}: To determine the elastic constant c_{33}, one end of the NW was pulled along axial direction while the other end was fixed (Fig. 1b). The deformation of this NW was then measured by the axial strain λ_3 and also generated a tensile stress σ_3 in NWs. For small strain case, i.e., $\lambda_3 \leq 0.01$ considered in the present study, c_{33} can be obtained from the slope of the linear σ_3 - λ_3 curve, since $c_{33} = \partial\sigma_3/\partial\lambda_3$ for structures under small deformation.

The piezoelectric coefficient e_{33}: To determine the piezoelectric coefficient e_{33}, we fixed the two ends of the NWs and an electric field E_3 was applied to the NWs along the axial direction (Fig. 1c) and generated an external force F_3 on ion i. Finally, the NW was relaxed again to reach a new equilibrium state. The piezoelectric coefficient e_{33} can then be calculated as the negative slope of the resulting σ_3 - E_3 curve, since it is defined as $e_{33} = -\partial\sigma_3/\partial E_3$ [7].

The dielectric constant k_{33}: The dielectric constant k_{33} can be defined as $k_{33} = k_0\ (1+\chi_{33})$ [8] with $\chi_{33} = (\partial P_3/\partial E_3)/k_0$ being the electric susceptibility of the material. Here k_0 is the vacuum permittivity and P_3 is the axial polarization density. It is noted that P_3 is mainly determined by the polarization due to the relative displacement between Ga and N atoms, since the polarization between the nucleus and electron cloud is negligible [8]. To obtain χ_{33} and k_{33}, we fixed the two ends of the NWs after the initial relaxation, and applied an electric field E_3 to the NWs in the axial direction (Fig. 1c). Following this, the NW was relaxed again to reach a new equilibrium state. P_3 of this new equilibrium state was obtained according to the above definition. Finally, k_{33} can thus be identified from the slope of the obtained linear P_3 - E_3 curve.

NANOSCOPIC MATERIAL CONSTANTS

The elastic constant c_{33}, piezoelectric coefficient e_{33}, and relative dielectric constant ε_{33} ($\varepsilon_{33} = k_{33}/k_0$) obtained from the above approaches under the room temperature (300 K) are respectively plotted in Figs. 2a-2c (solid circles) as a function of the inverse of the radius $1/a$. It is seen from Figs. 2a-2c that the material properties of GaN NWs depend on the nanoscopic dimensions of the NWs. For example, in Figs. 2a-2c, c_{33} and ε_{33} of the NWs respectively increase by 71% and 56% as their a increases from 1 nm to 3.5 nm. Similarly, e_{33} is found to decrease by 46%. Similar tendencies of these materials properties are also found in the experimental studies of GaN NWs; see, e.g., Refs. [9-11].

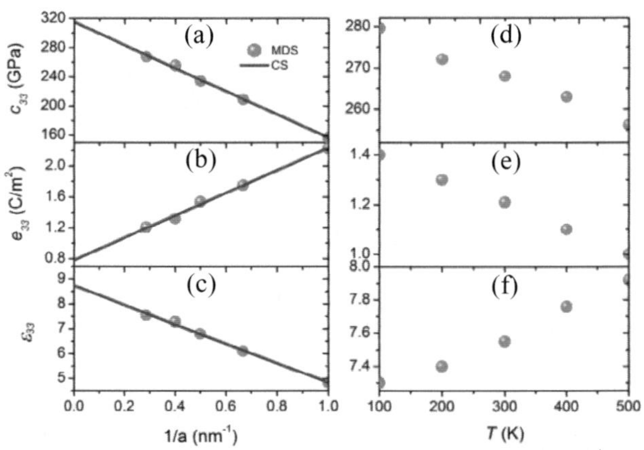

Figure 2: (a) The elastic constant c_{33}, (b) piezoelectric coefficient e_{33}, and (c) relative dielectric constant ε_{33} as a function of the inverse of the radius $1/a$, of GaN NWs under, room temperature. (d) c_{33}, (e) e_{33}, and (f) ε_{33} of GaN NWs with $a = 3.5$ nm as a function of the temperature T,

To understand the size-dependent material properties of NWs, we will introduce a core-surface (CS) model. It is known that the reduction in the size of materials to the nanoscale will enhance the influence of thin surface layers, where atoms experience structure reconstruction. In general, the surface reconstruction will lead to a distinct surface layer that is different to its bulk. For example, after full relaxation the potential energy of the NW's surface is about 23% greater than that of its bulk counterpart (see Fig. 3a). The difference of the physics between the surface layer and the bulk part will result in distinct material properties in layer and bulk, and is responsible for the size-dependent material properties observed at the nanoscale [12]. Inspired by this idea, a CS model was developed to characterize the size-dependent material properties observed in NWs [12], where a NW is modelled as a composite beam consisting of the core section of the bulk material and the surface layer with two distinct properties (see Fig. 3b). Following Ref. [13], we can obtain the equivalent material properties of the NW from the CS model as follows,

$$\rho_{33} = \rho_{33}^{b} + \frac{4\sqrt{3}}{3}\frac{\rho_{33}^{s}}{a}, \tag{2}$$

where ρ_{33}^{b} and ρ_{33}^{s} are the bulk and surface elastic constant, piezoelectric coefficient, or relative dielectric constant with $\rho \in (c, e, \varepsilon)$. According to Eq. 2, a linear curve fitting to the data was used in the MDSs depicted in Figs. 2a-2c. It is noted that the bulk values c_{33}^{b}, e_{33}^{b} and ε_{33}^{b} obtained from the present MDS-based CS model are 315 GPa, 0.78 C/m^2 and 8.7 and are in good agreement with the experimental and *ab initio* efforts (311 GPa, 0.73 C/m^2 and 9.7) described in Ref. [8].

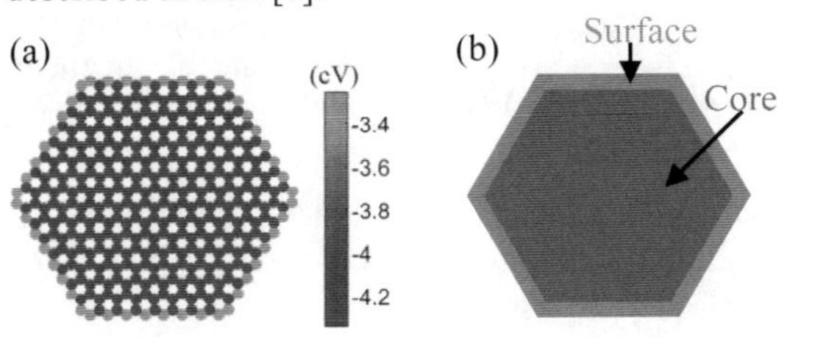

Figure 3: (a) The potential energy distribution of the cross-section of the GaN NW after the full relaxation. (b) An equivalent core-surface model of the GaN NW.

In the above studies, we have limited our attention to room temperature. Let us now extend the study to account for the effect of temperature. This will enable us to examine thermal effects on the material properties of GaN NWs. Conceptually, the temperature reflects the kinetic energy of atoms. At non-zero temperatures, the atoms of GaN crystals vibrate due to thermal oscillations around their equilibrium positions. The MDS is a valuable tool for investigating dynamical processes at non-zero temperatures.

c_{33}, e_{33} and ε_{33} of GaN NWs with the temperature T rising from 100 to 500 K have been measured and plotted in Figs. 2d-2f. Here, the radius of the NWs is 3.5 nm. It is known that higher temperature will induce higher kinetic energy of each atom of GaN, which will further result in easier slipping and elongation of GaN NWs. As a result, c_{33} and e_{33} of NWs in Figs. 2d and 2e are found to decrease with increasing T, while ε_{33} in Fig. 2f increases with increasing T. Similar tendencies of these materials constants are also found in the bulk materials of GaN and ZnO [7, 14].

PIEZOELECTRIC POTENTIAL OF GAN NWS

In this section, the material properties measured above are employed to study the piezoelectric potential generated in GaN NWs under compression. Here, as shown in Fig. 4a, a GaN NW is fixed and grounded at the substrate. Then, an axial pressure P is applied to the free surface of the NW to produce the piezoelectric polarization in the GaN NW. Theoretically, to obtain the piezoelectric potential, one should solve the following constitutive relations [2]:

$$\square_{\square} = c_{\square\square}\square_{\square} \square e_{\square k}E_k,$$
$$D_i = e_{i\square}\square_{\square} + k_{ik}E_k, \tag{3}$$

where $c_{\alpha\beta}$, $e_{\alpha k}$ and k_{ik} are the linear elastic constant, the piezoelectric coefficient and the dielectric constant. D_i and σ_α are the electric displacement and stress vectors. Incorporating constitutive equation (3) with the mechanical equation, geometrical compatibility equation, and Gauss equation in Ref. [2], we have calculated the piezoelectric potential in the deformed NW from the FEM, which was conducted using the commercial code ANSYS. In this process, SOLID5 element was selected to describe the NWs and 12000 elements were chosen after conducting element convergence tests.

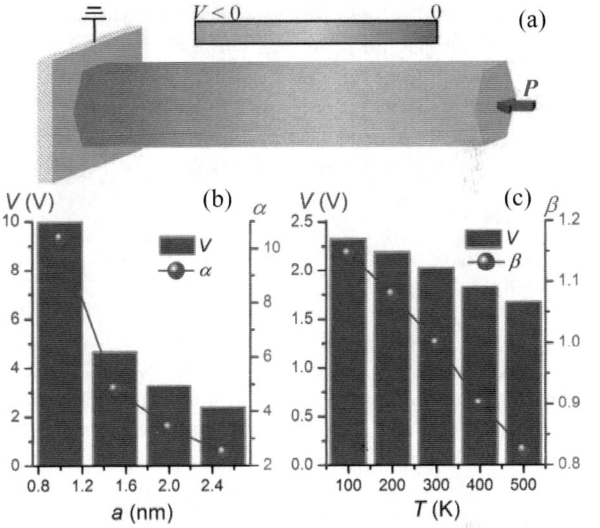

Figure 4: (a) Piezoelectric potential distribution of an NW subjected to the uniaxial compression. (b) The piezoelectric potential V and potential ratio α as a function of diameter a. (c) V and potential ratio β as a function of the temperature T.

The electric potential distribution in a deformed GaN NW obtained from the FEM calculation is plotted in Fig. 4a. It is found that when a compression force is uniformly applied on the free surface of the NW, it creates a negative potential between the NWs and the substrate. Fig. 4b shows the piezoelectric potential V of NWs as a function of the radius a, which decreases from 2.5 to 1 nm. Here P is taken as 1 GPa and the temperature is taken as $T = 300$ K. It is seen from the figure that the small-scale effect becomes more significant on V of the NW with smaller a. To qualitatively measure the small-scale effect on V, we have defined the potential ratio $\alpha = V/V_0$. Here, V_0 is the piezoelectric potential calculated based on bulk material properties and thus excludes the small-scale effect. α is then calculated in Fig. 4b with a decreasing from 2.5 to 1 nm. In this process, α is found to be raised from 2.5 to about 10.4, which denotes that the small-scale effect may increase V by up to about ten times. Based on these results, we come to the conclusions that the small-scale effect is substantial on the piezoelectric potential and thus has to be taken into consideration in the design of nanodevices. In other words, the bulk material properties will greatly underestimate the piezoelectric potential of GaN NWs.

In the above study, we have examined the piezoelectric potential of GaN NW under room temperature. Now, we will turn to the thermal effect on the piezoelectric potential. To this end we have calculated V as a function of T in Fig. 4c. Here, P is also taken as 1 GPa and a is fixed at 3.5 nm. Since the material properties of GaN NWs are temperature-dependent, the piezoelectric potential of them are accordingly found to strongly depend on the temperature and decrease with increasing temperature in Fig. 4c. To qualitatively measure the thermal effect on V, we have introduced another potential ratio $\beta = V/V_1$, where V_1 is the piezoelectric potential calculated based on the material properties of GaN at the room temperature. In Fig. 4c we can see that β decreases from 0.82 to 1.15 as T increases from 100 to 500 K. This observation suggests that the performance of GaN NW-based piezotronic nanodevices can be improved by decrease the working temperature of them.

CONCLUSIONS

In this paper, a multiscale modelling approach was performed to study the piezoelectric potential of GaN NWs. The material properties were measured based on MDSs and found to depend on the temperature and NW's diameter. The size-dependence of material properties was well explained by a CS model. The obtained material properties are then employed to the FEM analysis of the piezoelectric potential of a GaN NW. The results show that the potential in the NW can be up to ten times higher than its bulk counterpart due to the small-scale effect. In addition, a low working temperature can increase the potential of a GaN NW. The results may provide important guidance in rational design, fabrication, and use of GaN NW-based optoelectronic and electronic devices.

ACKNOWLEDGMENTS

This work was supported by the Natural Sciences and Engineering Research Council of Canada (NSERC) and the Discovery Accelerator Supplement.

REFERENCES

1. Wang, Z. L. 2012, "Progress in piezotronics and piezo-phototronics", *Advanced Materials*, Vol(24), pp. 4632-4646.
2. Gao Y. F., and Wang Z. L. 2007, "Electrostatic potential in a bent piezoelectric nanowire. The fundamental theory of nanogenerator and nanopiezotronics", *Nano Letters*, Vol(7), pp. 2499-2505.
3. Romano, G., Mantini, G., Di Carlo, A., D'Amico, A., Falconi, C., and Wang, Z. L. 2011, "Piezoelectric potential in vertically aligned nanowires for high output nanogenerators", *Nanotechnology*, Vol(22), pp. 465401.
4. Wang, Z. G., Zu, X. T., Yang, L., Gao, F., and Weber W. J. 2007, "Atomistic simulations of the size, orientation, and temperature dependence of tensile behavior in GaN nanowires", *Physical Review B*, Vol(76), pp. 045310.
5. Bernal, R. A., Agrawal, R., Peng, B., Bertness, K. A., Sanford, N. A., Davydov, A. V., and Espinosa, H. D. 2011, "Effect of growth orientation and diameter on the elasticity of GaN nanowires. A combined in situ TEM and atomistic modeling investigation", *Nano Letters*, Vol(11), pp. 548-555.
6. Plimpton, S. 1995, "Fast parallel algorithms for short-range molecular dynamics", Journal of Computational Physics, Vol(117), pp. 1-19.
7. Zhang, J., Wang, C. Y., Chowdhury, R., and Adhikari, S. 2013, "Size- and temperature-dependent piezoelectric properties of gallium nitride nanowires", *Scripta Materialia*, Vol(68), pp. 627-630.
8. Zhang, J. 2014, "Small-scale effect on the piezoelectric potential of gallium nitride nanowires", *Applied Physics Letters*, Vol(104), pp. 253110.
9. Nam, C. Y., Jaroenapibal, P., Tham, D., Luzzi, D. E., Evoy, S., and Fischer J. E. 2006, "Diameter-dependent electromechanical properties of GaN nanowires", *Nano Letters*, Vol(6), pp. 153-158.
10. Minary-Jolandan, M., Bernal, R. A., Kuljanishvili, I., Parpoil, V., and Espinosa, H. D. 2012, "Individual GaN nanowires exhibit strong piezoelectricity in 3D", *Nano Letters*, Vol(12), pp. 970-976.
11. Yang, Y., Guo, W., Wang, X., Wang, Z., Qie, J., and Zhang, Y. 2012, "Size dependence of dielectric constant in a single pencil-like ZnO nanowire", *Nano Letters*, Vol(12), pp. 1919-1922.
12. Zhang, J., Wang, C. Y., and Bowen, C. 2014, "Piezoelectric effects and electromechanical theories at the nanoscale", *Nanoscale*, Vol(6), pp. 13314-13327.
13. Dai, S. X., Gharbi, M., Sharma, P., and Park, H. S. 2011 "Surface piezoelectricity: Size effects in nanostructures and the emergence of piezoelectricity in non-piezoelectric materials", *Journal of Applied Physics*, Vol(110), pp. 104305.
14. Hill, N. A., and Waghmare, U. 2000 "First-principles study of strain-electronic interplay in ZnO: Stress and temperature dependence of the piezoelectric constants", *Physical Review B*, Vol(62), pp. 8802-8810.

Proceedings of the 25th CANCAM
London, Ontario, Canada, May 31 – June 4, 2015

SOUND ABSORPTION BY MULTILAYER OF MICROPERFORATED MEMBRANES

Xiaobing Cai
Department of Mechanical and Materials Engineering
Western University
London, Ontario, Canada
caixiaobing11@gmail.com

Jun Yang
Department of Mechanical and Materials Engineering
Western University
London, Ontario, Canada
jyang@eng.uwo.ca

ABSTRACT

Micro-perforated membranes have applications in various fields, such as serving as filters in microfluidics, as masks in micro-surface patterning, and as functional surfaces for hydrophobic or hydrophilic purposes. In this paper, we report that micro-perforated membranes exhibit super sound absorbing capabilities by providing abundant occurrence of airflow frictions within a specified sectional area during the penetration of sound waves. Theoretical analysis has been conducted to predict the sound absorption by multilayer of micro-perforated membranes comprising different pore sizes. Increased sound absorption performance of the micro-perforated membranes has been achieved at low frequencies. Based on microfabricaiton technology such as perforated membranes have been fabricated, with their sound absorption properties being measured.

KEYWORDS: sound absorption, micro-perforated membrane, transfer matrix method, photolighography

INTRODUCTION

Sound absorbing materials are widely served for noise mitigating as the latter could have detrimental effects on human's health and behaviour, especially for those under long term exposure, which commonly take places in downtown areas, airport vicinities and constructions sites and so on. So far, there are abundant of sound absorbing materials come into market every year and cover the usage of various situations in daily life. However, in a lot of circumstances, sound absorbing materials are still reluctant to meet people's demand. For example, aircraft passengers and crew still suffer a lot during every flight, as light and efficient sound absorbing materials are still beyond reach for airplane manufactures [1-2]. Automobile industries are still working hard to produce a thin and high effective noise absorber for sound management in car interior [3]. Although significant progress has been achieved in the past decades, currently existed sound absorbing materials are still not competent enough to eliminate the low frequency noise [4]. For particularly the vehicle manufacturing departments, efficient canceling the noise inside autos or fuselages would be a significant factor concerning the success of their business in the coming decades.

Sound absorbing materials depend on their microstructure to dissipate the noise energy. It is well accepted that friction between the micro pores and airflows driven by sound pressure is the dominant cause of sound energy damping [5], thus by strictly control the movement of airflow it is possible to achieve adaptable sound absorbing [4]. Thus the most observable sound absorbing materials are in form of porous foams, plates and wools, which are generally 'engineered structural materials' that their performance fully governed by microstructures. Scientists and engineers always make their best efforts to propose and produce porous materials with delicate microstructures to expect an exceptional sound absorbing performance [6].

However, due to the limitation on manufacturing method for highly architectural microstructure, most of sound absorbing materials are produced in an under-controlled way, which greatly impair their functions to reach a maximum level. It is easy to observe that a sound absorbing foam having lots of irregular pores inside even it is prepared by using latest technology. The failure in precise control in foam manufacturing makes it rarely performs better than glass fibre, which is generally considered as a good sound absorbing candidate. While benefits a lot from its long fibres' vibrating that greatly magnifies its sound dissipation, glass fibre also has an extremely disorganized structure that greatly limits its efficiency [7].

Due to the extremely complicated microstructures, the sound absorbing materials are inherently three dimensional objects that less likely to be realized by using even the ever existing state-of-the-art micro fabricating technologies. Fortunately, with the profound microfabrication technologies, such as photolithograpy and e-beam, which are typically proficient in producing precise membranes, multilayer micro-perforated membranes can be readily produced, as shown in figure 1. Each layer of the micro-perforated membrane has full controlled pore size and porosity, as well as the pattern of the pores. This multilayer structure can either stacked layer by layer or each held by a supporting structure, which is not shown here.

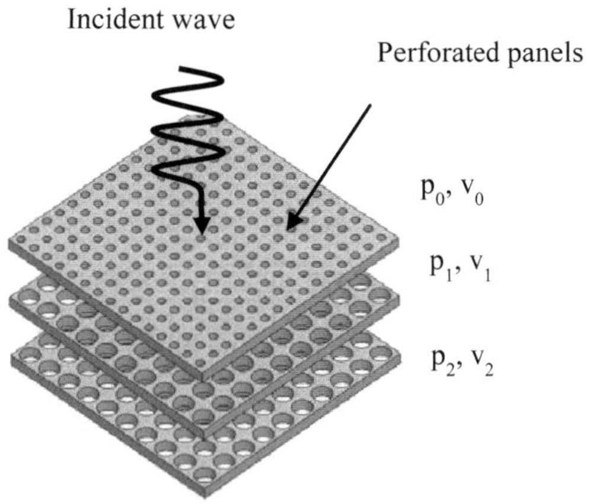

Figure 1: multi-layer of perforated panels with different pore size and porosity.

THEORETICAL ANALYSIS

Theoretical study is employed for qualitative analysis. Figure 2 illustrates a common noise control sketch that incident acoustic waves impinge on a sound absorbing material backed a hard wall. Consider the material (saturated by air) to be uniform, its acoustic properties can be generally characterized by bulk wave impedance Z_a and propagation constant k_a, thus the wave equations inside this medium can be expressed by [4]:

$$p(z,t) = \hat{p}e^{-ik_a x}e^{i\omega t}, \tag{1}$$

$$\langle p(z,t)\rangle_x = Z_a \langle \upsilon_z(z,t)\rangle_x, \tag{2}$$

where $\hat{p}$ is the amplitude of the pressure at $z=0$, υ is the air molecular velocity and $\langle\cdots\rangle$ represents averaging over an sectional area perpendicular to the propagation direction. Employing some boundary conditions, the surface resistance of bulk material before a hard wall can be readily found:

$$Z_1 = Z_a \frac{\sinh(bd)\cosh(bd) - i\sin(ad)\cos(ad)}{\cosh^2(bd) - \cos^2(ad)}, \tag{3}$$

in which a and b are the dimensionless real part and imaginary part of k_a. The surface resistance is related to the sound absorbing coefficient by:

$$\alpha = \frac{4Z_1' Z_0}{(Z_1' + Z_0)^2 + Z_1''^2}. \tag{4}$$

These equations could give qualitative results on how well a material performs, for instance, based on these equations, the dependence of sound absorbing coefficient of a material on its airflow resistance and thickness can be easily found. The maximum absorption usually happens when material thickness is about one quarter of the incident wavelength that air particle velocity becomes maximum and frictions are strongest, this is also the reason it is hard to design a sound absorbing material even thinner.

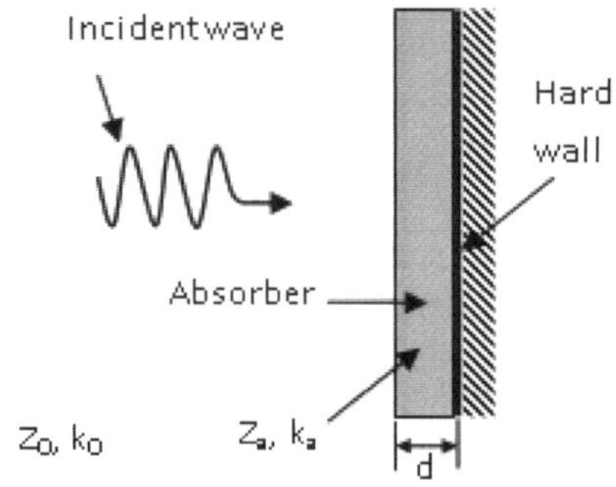

Figure 2: configuration of a sound absorbing material placed in front of a hard wall

For a real sound absorbing material, its impedance and propagation constant are strictly determined by its microstructure, e.g. the pore size, tortuosity and airflow resistance. Generally, impedance Z_a and propagation constant k_a are related to the medium's bulk density ρ_a and compressibility C_a by [1]:

$$Z_a = [\rho_a / C_a]^{1/2} \text{ and } k_a = i\omega(\rho_a C_a)^{1/2}. \tag{5}$$

In theoretically study, the bulk density and compressibility are usually converted from the averaged density and compressibility of a single pore

$$\rho = \rho_x \frac{\sigma}{\sigma_x}, \quad C = \Omega C_x \tag{6}$$

where σ and Ω is the porosity and airflow resistance of the bulk medium, ρ_x, C_x and σ_x is the average density, compressibility and airflow resistance of a single pore. For some materials with regular structures, such as straight holes, the last three parameters can be theoretically found. More complicated porous material that usually contains hundreds of thousands of pores and are of difference shapes and sizes, it is impracticable to perform a theoretically analysis.

With the advancement of computer technology, simulation on sound absorption by complicated porous materials becomes available. By taking the viscothermal properties of the air into account, and combining with Newton Fourier ideal gas model and heat conduction law, linearized time harmonic Navier Stokes equations can be established to replace classical acoustic Helmholtz equations that govern acoustic waves. Thus energy loss due to frictions between air flow and skeleton of the porous material can be simulated, as well as the losses induced by pressure variations. For normal incidence, sound absorption coefficient can be conveniently calculated by using the complex pressure data retrieved from two different points before the facing of a porous material, exactly the same as that in impedance tube measurement.

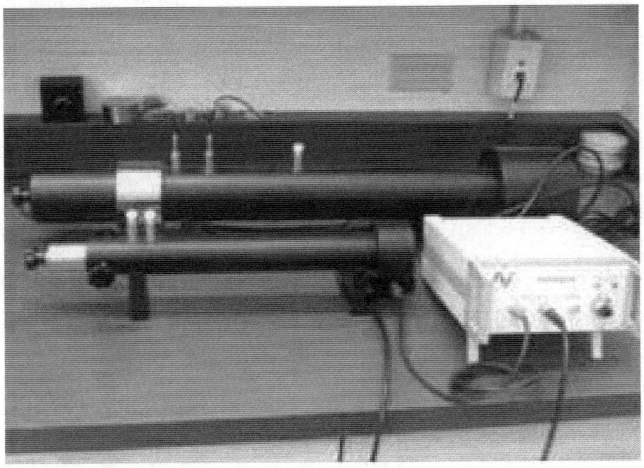

Figure 3: Sound impedance tube for standard material sound absorption coefficient measurement.

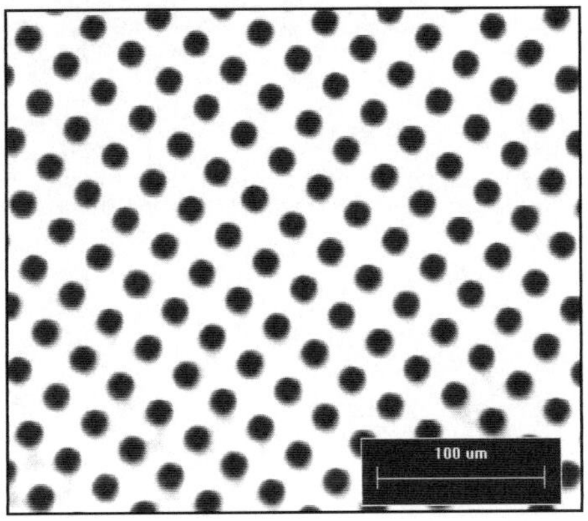

Figure 4: the picture of micro-perforated membrane

RESULTS AND DISCUSSION

The experiments concerned in this project are generally charactering the sound absorbing efficiency of samples with micro-perforated layers, fabricated by photolithography method. An impedance tube device has been setup for fulfilling this task based on standard method for measuring sound absorption coefficient of normal incidence, as shown in figure 3. Samples been tested are required to be well mounted into the tube at the far end and maintain flat surfaces. Modulated sounds are generated by a source loud-speaker and transit along inner wall of the tube to impinge onto the sample and the back hard wall if not been fully absorbed. The reflected waves thus superpose with the incident wave in front of the sample. By in situ taking the acoustic pressure at two separated locations along the tube from the source side and plus post data processing, this device gives sound absorbing coefficient of a material from frequency of 100Hz to 6300Hz, well covers the low frequencies.

Samples with delicate structures will be designed at first according to the principals addressed above, like soft frame vibration, tortuosity magnification and so on. Figure 4 shows a picture of one layer of the micro-perforated membrane fabricated by photolithography method. As shown by figure 4, The pores in this membrane have uniform sizes and distributed homogeneously, with pores size of 15 microns and porosity around 25%.

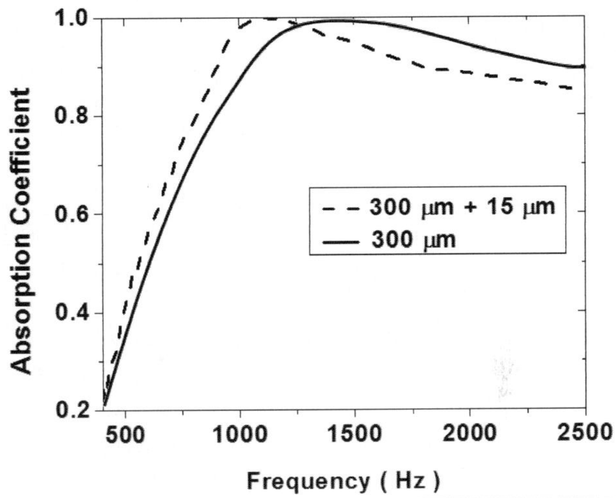

Figure 5: Measured results of sound absorption coefficients of multilayer micro-perforated membranes

Figure 5 shows the sound absorbing coefficient of a multilayer micro-perforated membrane system with two layers separated by a supporting layer containing pore size of 300 microns. A significant blue shift of peak absorption frequency can be found as after the micro-perforated membrane of 15 microns pores have been placed on the front. of same thickness, on the price of scarifying high frequency performance. This enhancement is induced by the high resistance of the facing, as the micro-holes make friction of airflow fiercer.

CONCLUSIONS

The application of micro-perforated membrane fabricated by photolithography technology to sound absorption has been proposed. The measured sound absorption coefficient show that the micro-perforated membranes have a capability of providing high sound impedance, based on what the sound absorption peak frequency can be reduced to lower frequency band, thus help to improve the performance of the original structure. The contribution of sound impedance to the sound absorption coefficient has been studied, and provides analytical support of the design of micro-perforated membrane.

ACKNOWLEDGMENTS

This work was supported by Natural Science and Engineering Research Council of Canada (NSERC) and Research Accelerator Grant of The University of Western Ontario.

REFERENCES

1. Wang, X. and Lu, T. J. 1999, "Optimized acoustic properties of cellular solids", J. Acoust. Soc. Am. Vol(106), pp. 756-765.

2. Han, F., Seiffert, G., Zhao, Y., and Gibbs, B. 2003, "Acoustic absorption behaviour of an open-celled aluminium foam", J. Phys. D. Appl. Phys. Vol(36), pp. 294-302.

3. Hakamada, M., Kuromura, T., Chen, Y., Kusuda, H. and Mabuchi, M. 2006, "High sound absorption of porous aluminum fabricated by spacer method", Appl. Phys. Lett. Vol(88), pp. 254106-1-3.

4. Mohrova, J. and Kalinova, K. 2012, "Different Structures of PVA Nanofibrous Membrane for Sound Absorption Application", J. Nanomater. Vol(2012), 1-4.

5. Mei, J., Ma, G., Yang, M., Yang, Z., Wen, W., and Sheng, P. 2012, "Dark acoustic metamaterials as super absorbers for low-frequency sound", Nat. Commun. Vol(3), pp. 756-1-7.

6. Schaedler, T., Jacobsen, A., Torrents, A., Sorensen, A., Lian, J., Greer, J., Valdevit, L., and Carter, W. 2011, "Ultralight metallic microlattices", Science Vol(334), pp. 962-965

7. Lee, D., and Kwon, Y. 2004, "Estimation of the absorption performance of multiple layer perforated panel systems by transfer matrix method", J. Sound Vib. Vol(278), pp. 847-860.

Proceedings of the 25th CANCAM
London, Ontario, Canada, May 31-June 4, 2015

AIR BUBBLE ASSISTED DETECTION FOR MICRO MASS SENSORS

Jacky Chow
Department of Mechanical and Materials Engineering
Queen's University,
Kingston, Ontario, Canada
chow@me.queensu.ca

Yong Jun Lai
Department of Mechanical and Materials Engineering
Queen's University,
Kingston, Ontario, Canada
lai@queensu.ca

ABSTRACT

This paper investigates the usage of air bubbles to create an air ambient environment for micro inertial sensors which are used for in-situ bioanalytes detection. Air bubbles are created by using direct injection and two phase fluid mixer. The feasibility of this concept has been proved by the results show the detection results are significantly improved with air bubble assistance.

Keywords: microsensor, mass sensor, air bubble

1. INTRODUCTION

Micromechanical resonators that are fully immersed in liquids during application face unique challenges when used for in-situ analyte detection. Virtual mass effect [1], viscous damping [2] and fluidic cell resonances [3] are among known factors that can potentially degrade the mass responsivity (Hz/g), quality factor and signal-to-noise (S/N) ratio of the overall biosensing instrument. As a result, the ability to detect analyte adsorption by measuring changes in resonant frequencies often diminish in liquids compared to actuation in more ideal environments such as in gases or in vacuum.

Biological analytes in their native form in nature, or in a state following standard biomedical sample preparation protocols, are often found suspended in liquids. Consequently, sensor-liquid contact is often unavoidable. The degradation of micromechanical detector performance is often unavoidable and expected when in-situ and real-time biodetection capabilities are required in application. For applications where continuous real-time biosensing is not a requirement, it has been common to mechanically characterize resonators under the most ideal conditions available, such as in atmospheric air or in vacuum, in order to obtain the most ideal spectral response of the resonator for mass detection.

However, real-time and in-situ biodetection in liquids remain an indispensable capability of biosensor with important implications in many scientific and engineering applications, such as providing insights for in-vitro studies, and continuous analyte monitoring in environmental applications. To work around limitations of liquid-mode operation while retaining continuous monitoring capabilities in real-time, some researchers have proposed novel solutions, such as the 'hollow' microchannel resonators [4] that are mechanically excited in a vacuum while liquid analyte media is continuously flowed internally inside the resonator, or the liquid-air resonator [5] with one interface exposed to air while the other interface is exposed to analyte media.

Another potential engineering solution to achieving high sensitivity continuous detection capabilities of resonators being proposed herein is to temporarily change the environment around the resonator to circumvent the need to perform mass detection procedures in liquids. For example, a gas or a fluid can be periodically applied to the sensor in controlled time-steps to control when analyte exposure occurs and when mechanical detection is performed.

2. METHODS

We will explore the strategy of using air bubbles to temporarily engulf otherwise liquid immersed resonators for the duration of the mechanical characterization procedure to enhance the overall detection sensitivity. Following mechanical characterization procedures, the air bubble can be promptly eliminated from the environment to re-establish liquid contact and analyte exposure to the sensor.

There are many techniques for generating gas bubbles in fluids in the viscinity of a MEMS structure. This can include thermal evaporation based techniques of the surrounding fluid [6], or the direct introduction of the gas phase. This paper will focus the study on methods based on introducing gases into the detection environment. Within this field of study, there are two main approaches: (i) direct injection of gas bubbles, or (ii) transport of gas bubbles to the environment. While any gas may equally be applicable, in this study we will focus on the use of air.

2.1 Direct Air Bubble Injection

Direct injection of an air bubble in the vicinity of the resonator is one method of establishing a gas phase to engulf the resonator. To do so may involve having the package of the sensor be designed with purpose-built air handling channels near the resonator as well as liquid serving channels for bioanalyte sample delivery. One advantage of direct injection is that the air bubbles can be introduced on-demand.

In this study a packaged device comprising a micromechanical clamped-clamped resonator with air and fluid handling channels was fabricated from the MetalMUMPs process[7] and used to demonstrate the presented principle. The resonator is a silicon nitride microstructure with overall dimensions of 100μm in length and 15μm in width with an internally embedded polysilicon electrode. The polysilicon electrode is 5μm in width and runs along the entire length of the resonator. The overall cross sectional thickness of the silicon nitride portion of the resonator is 0.7μm thick. The resonator is released from the substrate using a wet etch. A 20μm thick nickel layer is electroplated on top of the substrate plane to serve as external electrodes and for patterning the fluidic channels. It is noted that an underpass for the air channel allows air to inject from directly under the beam. Figure 1 shows the overall fabricated device with the air and liquid handing microchannel, the microresonator (in the center of the image) and with external electrodes (E).

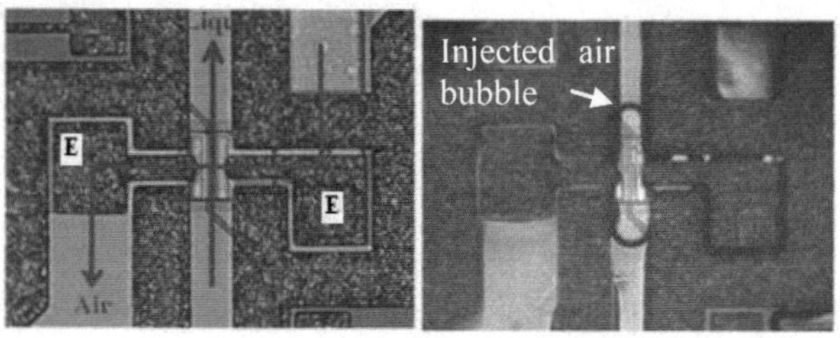

a) b)

Figure.1 a) Schematic MetalMUMPs device with direct air injection; b) Air bubble injected device

A top cover to be bonded on top of the microchannel is micromachined out of a 1.9mm thick cast piece of PMMA with high optical quality surfaces. The top cover has fluidic tube connectors directly machined into body of the part. Figure 2a shows a fluidic tube connector that is directly machined into the PMMA sheet by first drilling a 600μm diameter through hole and then milling away surrounding material to create a pillared tube with 50μm tube wall thickness and 1mm in length. This connector allows for liquid and air tight connections to be made with standard 0.64mm inner diameter 22 gauge PDMS tubing (Dow Corning Silastic Tubing). The top cover is then bonding to the MetalMUMPs chip by first applying a thin layer of UV curable glue onto a glass slide, and then imprinting a thin layer of glue over the top surface of the nickel layer on the MetalMUMPs chip. The connector is then manually positioned over the chip where it is placed on a hot plate at 473K for 10 seconds. The device is then placed under UV irradiation for 10 minutes. Figure 2b shows the MetalMUMPS chip with the bonded PMMA top cover.

(a) **(b)**

Figure 2. (a) Machined connector in PMMA and (b) image of packaged device.

The packaged chip is then secured and the PDMS tubes are then connected for air and liquid handling. DI water and air can be manually injected into the chamber using dedicated micro syringes attached to the PDMS tubes.

2.2 Transporting Air Bubbles

The previous section focused on directly injecting air bubbles in the vicinity of the resonator to change the environment from a liquid to a gas. An alternative technique for introducing gas to the sensor for higher sensitivity detection is to generate gas bubbles in an upstream process and then transport the gas bubbles downstream to the sensor within the confines of a microchannel. A microchannel two-phase mixer (see Figure 3) is situated upstream of the device capable of generating various sizes of gas bubbles. The two phase mixer can be used to generate periodic gas periodically that creates an alternating state between liquid and gas in a microchannel, which allows for a systematic detection scenario of introducing analytes for the sensor, and a high sensitivity detection phase in gas.

For the two-phase micromixer, a 300μm width and 300μm deep microchannel is fabricated by first machining patterns into PMMA. The mixer junction is a T-shaped junction and long channel lengths are added to allow for developed flows to be formed. The mixer comprises of two inlet ports and one outlet port. The fluidic tube connector for connecting 22 gauge PDMS tubes (inner diameter of 640 μm) described in the previous section, was also machined into another piece of PMMA. Then a mixture of 30% acetone and 70% ethanol was mixed. Bonding of the two PMMA components was accomplished by applying this mixture onto the surface of the PMMA part which will interface with the other PMMA part. Then the second PMMA part to be bonded was positioned in place and held overnight to dry. Figure 3a shows the two machined pieces of the microchannel, and Figure 3b shows the bonded part. Similarly, a PMMA package kit (see Figure 4a) was fabricated using micromilling machine and held a testing chip with a sensor (Figure 4b).

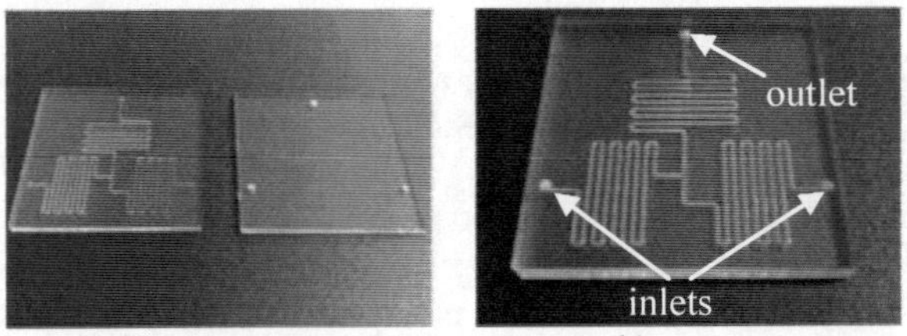

a) b)

Figure 3. a) Two machined components of the PMMA two-phase mixer, b) The bonded mixer.

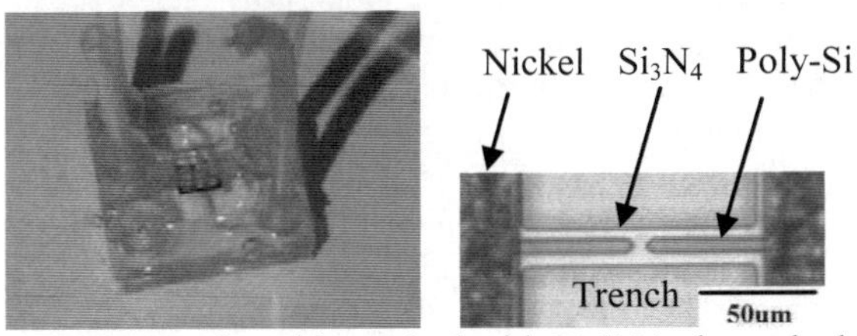

Figure 4. a) A testing chip packed with PMMA channels, b) a silicon nitride clamped-clamped beam.

A syringe pump held two syringes containing air and liquid (E.coli sample) respectively, connected to the microchannel mixer inlets. The outlet of the mixer was then connected to the inlet of the packaged fluidic-MEMS device. Figure 5a shows the image captured of the mixing at the T-shaped junction, and 5b is simulation at the T-junction of the mixer (using Comsol™ Two-Phase Flow module). Figure 5c shows an image of a PDMS tube section demonstrating the liquid plugs generated by the mixer.

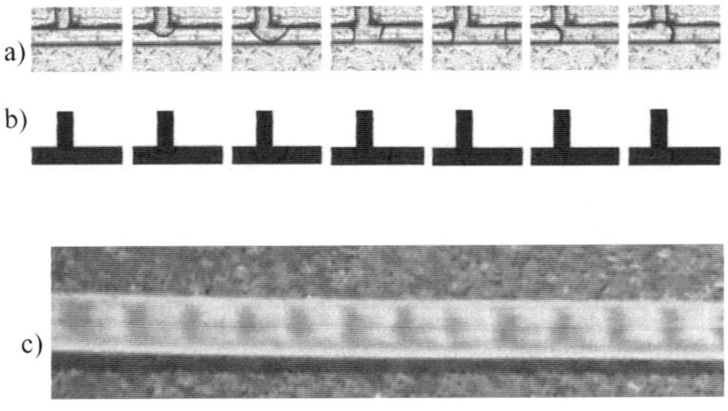

Figure 5. a) Optical micrographs captured of mixing action at the junction, b) COMSOL simulation of liquid-air mixing (blue for DI water, red for air), c) Liquid plugs generated by the two-phase mixer.

3. TESTS

3.1 Direct injected air bubble

The clamp-clamp beam on the packaged chip described in section 2 was excited to investigate its response. The device was first characterized in air before the packaging assembly in air. After packaging the device with the top cover assembly, DI water was injected into the chamber and then the device was mechanically characterized. The device is then characterized again after an air bubble is injected into the device and allowed to settle to 1 minute. Figure 6 show the results of the experiment by comparing the first mode response of the beam under the various scenarios. It can be observed that fully immersing the beam in the liquid resulted in the expected degradation of signal quality (amplitude, frequency and quality factor). Actually its quality factor decreases in the order of 100 in liquids. On the other hand, excitation of the beam in the air bubble was able to fully restore the response of the beam to the state it was in ambient air environments. The air bubble can be quickly redrawn and collapsed back through the air channel to allow liquid-structure contact to resume.

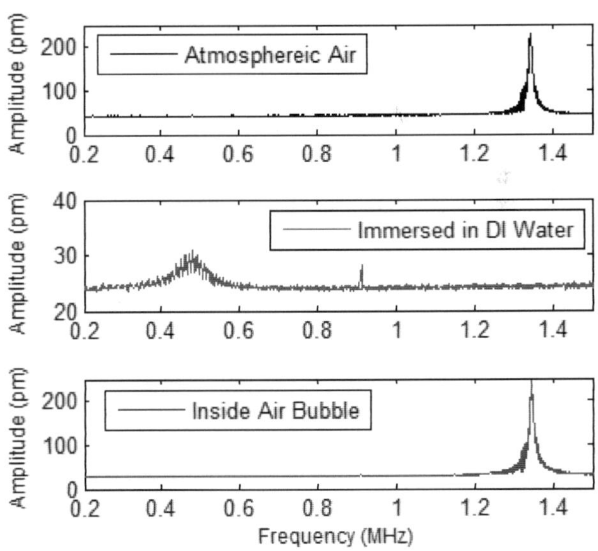

Figure 6. Comparing frequency/amplitude response of beam in dry atmospheric air, inside DI water and inside air bubble.

3.2 Transport air bubbles

The beam was then functionalized by passing poly-L-lysine solution into the chamber and allowed to settle for 30 minutes. The chamber and device was then washed by continuous flow of DI water for 10 minutes at a flow rate of 5ml/hr.

E. coli sample solution at a concentration of 10^5cell/ml was loaded onto one syringe of upstream of the two-phase mixer. Both syringe pumps were turned on simultaneously to allow the liquid-air mixer to generate liquid plugs upstream of the MEMS resonator, and then finally delivered to the MEMS resonator. The flow rate of the pump was set at 0.5ml/hr. E.coli bacteria were trapped to the sensing beam by using positive DEP generated by the excitation signal from the embedded electrodes. Since the fluid-gas mixer occur in discrete phases and are periodic, and it takes approximately 30 sec and 25 sec respectively for a fluid segment and air bubble to pass the chamber where the sensor locates. 6 total measurements were conducted. Each measurement was conducted one at the end of each fluid exposure phase before the transition to capture the end state response of the beam in the two different fluids of air and liquid. As such, 3 measurements of the 6 total will be made in liquid and 3 measurements will be made in air alternatively.

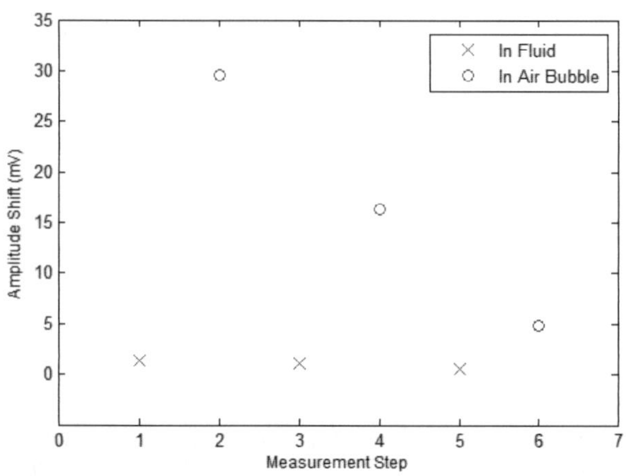

Figure 7. Graph of amplitude shift measured in different steps corresponding to periodic change of resonator from being in air bubble to in liquid.

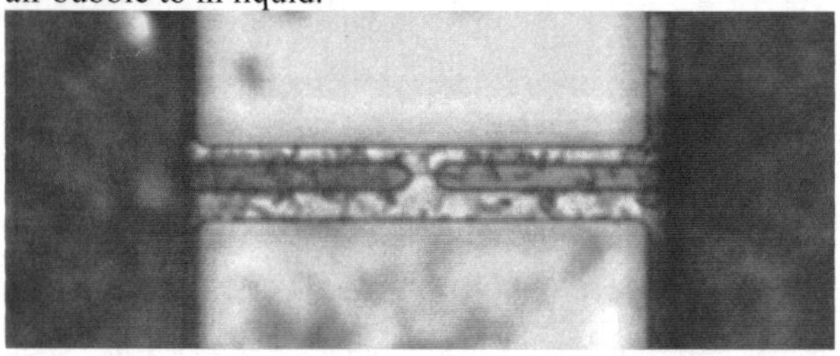
Figure 8. An optical micrograph of the device taken in dry air after the experiment showing collection of E. coli.

4. SUMMARY

This paper reported techniques to deliver air bubbles to the micro resonators as a method to temporarily change the excitation environment for the resonator and allow for higher mass sensitivity measurements to be conducted. Specifically, this paper focused on techniques that allow specific (pure) gases to be introduced. The main advantage of the presented direct injection method of introducing an air bubble is that it can be performed on demand. This allows the collection and in-situ detection of the experiment to continue, and when a higher sensitivity detection mode is needed, the liquid flow can be interrupted temporarily to allow the gas bubble to be injected into the device and a quick measurement to be taken. One drawback of this technique is that dedicated air handling channels are needed to be considered and fabricated onto the device before the strategy can be implemented, which may take more space and increase the complexity of the overall sensor.

The other technique was using an upstream gas-liquid mixer to generate liquid plugs that enable periodic delivery of air bubbles and liquid plugs to the sensor within a microchannel. It has been demonstrated that liquid plugs generated by a two-phase liquid-gas mixer can be used to generate periodic gas bubbles that can then be delivered to a MEMS device downstream. The periodic nature of the gas bubble delivery allows a reliable and consistent measurement to be conducted. In-situ detection can occur inside the liquid phase, while the gas phase can allow a temporarily higher sensitivity detection stage. The disadvantage of this method compared with the direct injection method is the lack of immediate on-demand control of when an air bubble assisted detection is to take place, since mixing is always done upstream and delivered later to the sensor in the present demonstration.

5. REFERENCES

[1] Braun, T, Barwich, V, Ghatkesar, MK, Bredekamp, AH, Gerber, C, Hegner M, Lang, HP. (2005). Micromechanical mass sensors for biomolecular detection in a physiological environment. Physical Review E 72, 031907.

[2] Sader, JE (1998). Frequency response of cantilever beams immersed in viscous fluids with application to the atomic force microscope. Journal of Applied Physics 84(1), 64-76.

[3] Kokavecz, J. and Mechler A. (2007). Investigation of fluid cell resonances in intermittent contact mode atomic force microscopy. Applied Physics Letters 91(2), 023113

[4] Burg, TP, and Manalis, SR. (2003). Suspended microchannel resonators for biomolecular detection. Applied Physics Letters 83(13), 2698-2700.

[5] Park, J, Nishida, S, Lambert, P, Kawakatsu, H., and Fujita, H. (2011). High-resolution cantilever biosensor resonating at air-liquid in a microchannel. Lab on a Chip 11, 4187.

[6] Liao, KM, Chen, RS, and Chou B. (2006). A novel thermal-bubble-based micromachined accelerometer. Sensors and Actuators A: Physical 130-131, 282-289.

[7] Cowen, A, et al, MetalMUMPs Design Handbook revision 1.0 MEMSCAP.

LASER MODIFICATION OF PRINTED GRAPHENE-OXIDE (GO) AND GRAPHENE-CELLULOSE (G-CMC) THIN FILMS

Dogan Sinar, George K. Knopf
Dept. of Mechanical and Materials Engineering
The University of Western Ontario
London, Ontario, Canada
dsinar@uwo.ca, gknopf@eng.uwo.ca

Suwas Nikumb
Automotive and Surface Transportation
National Research Council of Canada
London, Ontario, Canada
Suwas.Nikumb@nrc-cnrc.gc.ca

ABSTRACT

Although conductive polymer (PEDOT:PSS) and metal-based (Au, Ag) inks have very high conductivity, these materials are often very expensive and require unique ink formulations for different resistance values. As an alternative, two types of nontoxic water soluble graphene (G) derived inks are introduced in this paper. The first type of ink involves mixing hydrophilic graphene-oxide (GO) particles with polyethylene glycol and dionized (DI) water. The second type of ink is created by using a hydrophilic cellulose derivative, *Carboxymethyl Cellulose* (CMC), to help suspend the naturally hydrophobic G sheets in DI water. Once deposited on the substrate the electrical conductivity of both types of inkjet-printed films can be "tuned" using laser assisted thermal reduction. The process increases the conductivity of the printed GO film by removing oxygen molecules and the G-CMC film by decomposing the cellulose stabilizer. An experimental investigation shows that the resistivity of the low conductive GO film can be varied between $\sim\infty$ kΩ/sq. to 2 kΩ/sq., while the resistivity of the G-CMC films can be reduced to ~100 Ω/sq. Further tuning of laser's peak power intensity density by adjusting the beam spot sizeand power output allowed for partial ablationmachining of parts ofthese thin films without damaging the underlying substrate. Limited only by the wavelength of the laser (775 nm) and resolution of the positioning system (1-2 μm), it was possible to machine very fine features (< 10 μm) on the GO traces.

KEYWORDS: Graphene, carboxymethyl cellulose, graphene oxide, laser assisted thermal reduction, laser micromachining

INTRODUCTION

Flexible electronic circuitry is an emerging technology that will significantly impact the future of healthcare and medicine, food safety inspection, environmental monitoring, and public security. Recent advances in printing technology (screen, gravure, inkjet) and electrically conductive inks have enabled simple electronic circuitry to be fabricated on mechanically flexible polymers, paper, and bioresorbable silk. Although useful, screen and gravure printing lacks the flexibility of drop-on demand and inkjet printing. Inkjet printing uses very little material and can accurately deposit thin homogeneous films with high precision on a variety of rigid and flexible substrates. These material deposition systems permit the creation of very fine features in the range of 25μm. Furthermore, inkjet printing allows for a truly digital manufacturing process from CAD circuit designs to final printed product.

Metal based inks provide the highest conductivities for thin film electrodes and traces. On the other hand, particulate metal inks commonly require high sintering temperatures that can damage most flexible substrates [1]. In addition, use of dangerous chemicals, risk of oxidation of metal particles, and cost of precious metals are of concern in the case of industrial scale manufacturing [2]. Conductive polymers have been investigated as alternative material. But, low conductivity (usually < 300 S•cm^{-1}) [3] and carrier mobility [4] combined with low thermal and chemical stability have limited the commercial applications of conductive polymers.

Prior research has shown that low-cost graphene, and its derivative formulations, can also be used to create suitable conductive inks. Graphene (G) is a one atom thick layer composed of carbon atoms arranged in a hexagonal lattice forming a material which shows unique electrical properties and excellent mechanical, thermal, and chemical stability [5]. In contrast to G, graphene-oxide (GO) is a non-conductive, oxidized derivative but it can be heat treated to restore its conductivity. GO is shown to be highly hydrophilic and can form dispersions easily. Although oxygen-reduced graphene-oxide (rGO) is not as conductive as pristine G, its conductivity and band gap energy [6] can be tuned to a desired value by altering the level of oxidation. In other words, printed GO can be reduced to create 2D resistors for thin film devices while pristine G films can be used as conductive traces.

The following section describes the chemical synthesis of both graphene-oxide (GO) and graphene-cellulose (G-CMC) water based inks, and discusses the inkjet system used to print variable conductive microtraces on flexible substrates. An experimental study of tuning the electrical conductivity of rGO and G-rCMC films is then summarized. Finally, concluding comments and a brief discussion on future work is provided.

CHEMICAL SYNTHESIS AND INKJET PRINTING OF TUNABLE GO AND G-CMC INKS

A. Graphene-oxide (GO) Ink

GO ink was prepared by mixing dry GO powder with DI water at a concentration of 5 mg/mL. After 0.5 hours of sonication to properly disperse GO in water, 0.1 mg/mL PEG 400 was added and the mixture was sonicated for 2 more hours. PEG adsorbs on GO sheets and reduces the formation of coffee ring effects.

B. Graphene-cellulose (G-CMC) Ink

Aqueous graphene based ink was prepared by utilizing a cellulose derivative called *Carboxymethyl Cellulose* (CMC). In its natural state G possesses a very weak electrostatic charge (< 24 mV). As a result, aqueous dispersions of pristine G quickly agglomerate. Sodium salt of CMC is highly soluble in water. CMC chain has hydrophobic and hydrophilic functional sites which allow adsorption on G sheets. This adsorption allows G sheets to be stabilized in water by electrostatic and steric forces. To achieve formation of a proper dispersion, first 1.5 mg/mL CMC was dissolved in DI water under magnetic stirring. Once all CMC dissolved, 0.5 mg/mL G powder was added. The mixture was bath sonicated for 8 hours to reduce lateral dimensions of particles.

Initially, G sheets have lateral dimensions of up to 10 μm which could compromise inkjet printing by causing clogging of nozzle exit. Sonicated mixture was centrifuged for 30 minutes at 3500 G to remove large particles. Presence of excess CMC in the dispersion negatively affects film characteristics. To remove excess CMC, dispersed particles were salted out with 40% NaCl and centrifuged at 12000 G. After salt loaded water decanted, coagulated G particles were reconstituted in DI water and vacuum filtered twice to wash off left over salt. Finally, dry film on filter was reconstituted and sonicated for an hour before printing.

C. Inkjet Printing Water-based Graphene Derived Inks

Simple traces of GO and G-CMC were inkjet printed on flexible substrates using an Epson R2000 photo printer. A thick enough and homogenous film was achieved between 10 to 20 prints (Fig. 1). Using a profilometer, at 20 prints, film thickness was measured to be ~1μm. Conductivity of G-CMC films increase substantially after annealing process. CMC efficiently decomposes around 320°C. A 40 min. long annealing process at 320°C yields traces with resistivity of $\sim 6.6 \times 10^{-3}$ Ω-cm. This temperature can be achieved in a convection oven or by localized heating of a laser beam to decompose adsorbed cellulose. Efficient thermal removal of oxygen containing moieties from GO surface also required temperatures of 300°C or above [7]. As a result, both films can be printed the similar substrates and thermally treated with the same laser setup.

However, unlike G-CMC films, printed GO traces were dried around 210°C for 30 minutes in a convection oven before laser reduction. This allows GO traces to harden and have with a uniform 10 MΩ resistance (20 x 2 mm traces) before starting beginning laser reduction. It was observed that reduction of GO traces takes more exposure in comparison to G-CMC and provide a very large spectrum to adjust resistivity. To investigate this phenomenon, a femtosecond (120 fs) Ti: Sapphire laser source at 775nm wavelength was used. Printed samples were placed on an X-Y positioning table with +/-1 μm resolution. Traces were scanned with the laser beam at different output power (LP), feed rate (FR), and number of passes (NP) (Fig. 2).

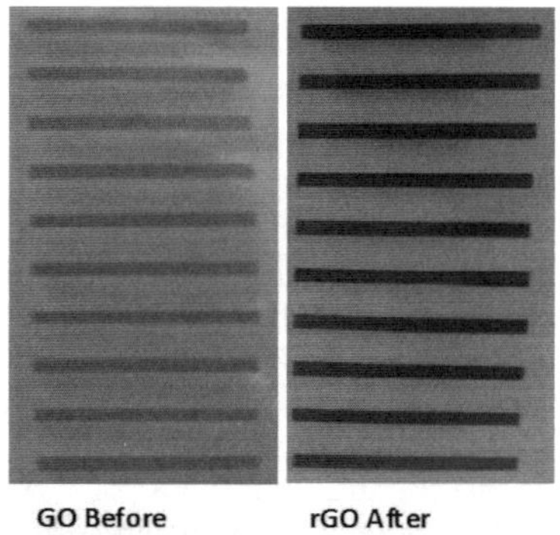

Figure 1: Deposited GO film strips before annealing (left) and the same rGO film strips after annealing (right).

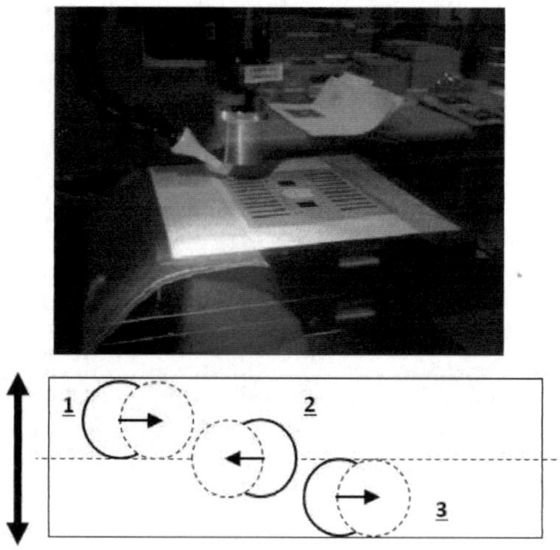

Figure 2: Experimental laser setup (top). Overlapping laser beam paths as beam travels across film (bottom).

EXPERIMENTAL RESULTS

Several tests were performed with different power levels and material feed rates to investigate the impact of laser beam exposure on the film sheet resistance (Rs). The experimental power levels (LP) were set to: 18-19, 20-21, 23-24, 24-26, and 27-29 mW; and the material feed rates (FR) ranged from 1000, 1200, 1800 and 2000 mm/min. The average laser power was 25mW at a spot size of ~ 220μm. The exposure time was controlled by adjusting the number of passes over the target surface (NP). The NP values were set to 6, 8, 10,

12, 14, 16, 18 and 20. More than one pass was necessary to observe a reduction in lower LP and/or high FR combinations. Predictable results are possible because laser systems provide precision heating. For example, reducing LP or increasing FR decreased heating and lowered the reduction in the film resistance. Due to the Gaussian shape of the laser beam, a threshold LP must also be reached prior to initiating the reduction process. Beyond this LP threshold, increasing LP impacts the instantaneous heating while changes to the FR affects the exposure time. The FR is determined by the speed and accuracy of the micro positioning system. To cover the desired surface on the film, the pulsating laser spots arewere overlapped (Fig. 2)

The laser reduction process resulted on further darkening of the annealed film along the beam pathway. The film thickness was measured using KLA Tencor P-10 surface profiler. For these surface profilometry readings, the GO films were printed on high quality microscope slides. The profilometry results show a weak connection between film thickness and double exposed areas (Fig. 3). This suggests that the material loss is minimal during the process. The sheet resistance (Rs) of the rGO film was determined by measuring bulk resistance and calculating resistivity using known dimensions. Fig. 4 shows the sheet resistance (R_S) of traces with increasing NP. As expected, rGO traces on polyimide substrates produce high Rs values for low NP.

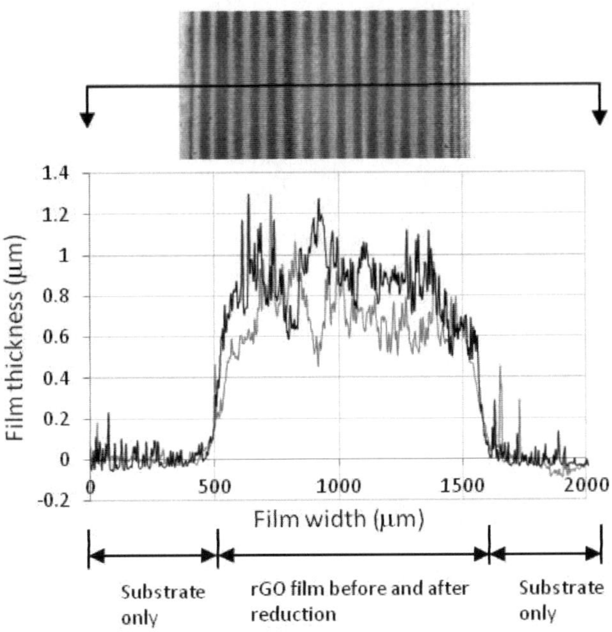

Figure 3: Film thickness profiles before (dark line) and after reduction (gray line) over the same cross-section. Laser process parameters: LP= ~19 mW, FR=1000 and NP=20. Variations in film thickness are due to laser spot overlap, instrumentation noise, and limitations of inkjet printer.

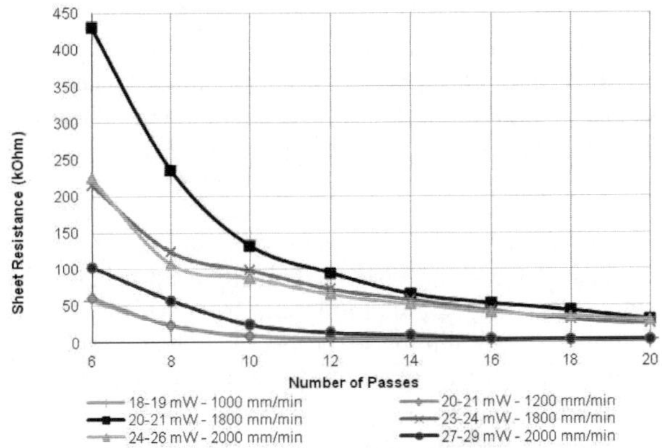

Figure 4: Sheet resistance values (Rs) for rGO traces after multiple laser passes (NP). The curves correspond to different PO and FR parameters.

One of the shortcomings of the inkjet printing method is the relatively large feature size. Small features (~25μm) can be achieved using specialized material deposition systems. Unfortunately, these systems tend to be expensive and place rigid restrictions on ink properties such as surface tension, viscosity, and particle size. Achieving homogenous and continuous traces with very small width can be problematic as blurring or "bleeding" of separate traces into each other. In contrast, post-processing using laser material removal allows local sections of the rGO and G-rCMC films to be accurately removed or ablated. Fig. 5 is an example of simple electrodes micromachined onto rGO film using 0.3mW LP and 100 mm/min FR.

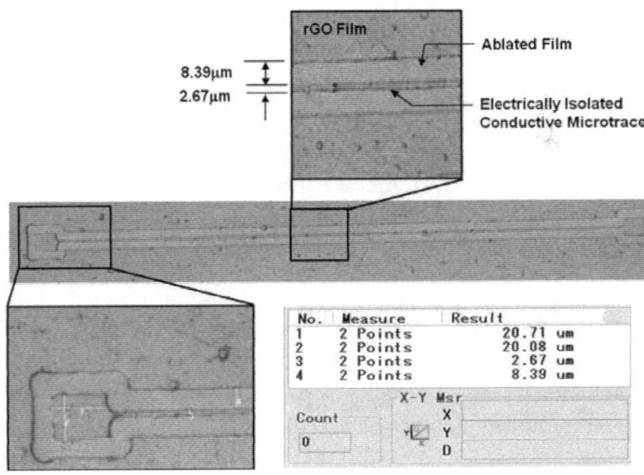

Figure 5: Microelectrode trace created by ablating the rGO film around the conductive line. The ablated regions have zero conductivity (R→∞). The width of the central conductive microtrace is 2.6μm.

Due to precision control over the power density of theproduced by a femtosecond pulsed laser, the underlying substrates does not experience any thermal damage during film ablation. Furthermore, the heated affected zone (HAZ) around the machined edges produced by the femtosecond pulsed laser is very smallsmall. In contrast, the HAZ produced during ablation by a continuous laser is substantially larger. For the femtosecond laser and above machining parameters, the smallest feature created on the rGO film was a 2.67µm wide conductive trace.

A series of tests were also performed to investigate the size of the HAZ produced during ablation. Fig. 6 shows simple ablated lines on the rGO film produced by different diameter beams. Deviations in the uniformity of the ablated lines were less than 2% for single pass machining for films. Furthermore, the HAZ was below 1% of ablated width for all traces.

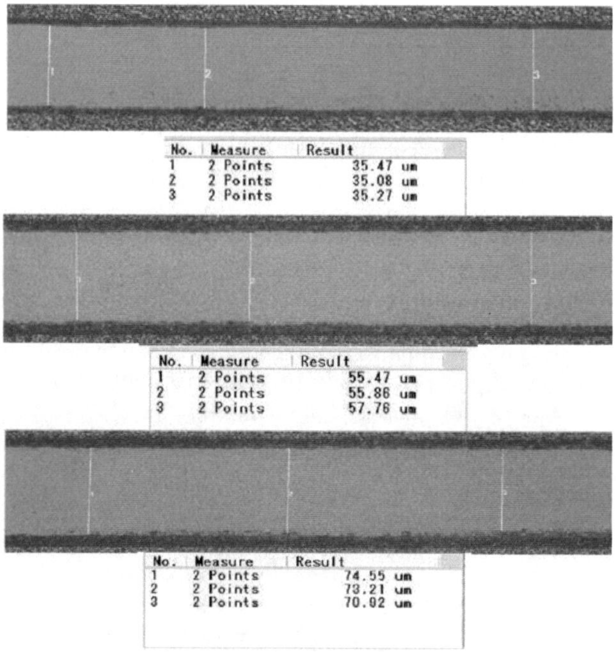

Figure 6: Ablated lines on the rGO film where the darker edges define the HAZ. The beam diameter has been changed for the three samples shown. In addition, LP and FR are adjusted to compensate for lower power intensity.

Finally, the mechanical properties of inkjet printed and laser modified GO and G-CMC thin films were assessed. Although metal based thin films provide excellent electrical performance, these materials are prone to cracks if severely bent. Preliminary experiments demonstrated that both thermally modified rGO and G-rCMC films have excellent mechanical stability. Mechanical flexing of both types of films showed little to no change in either the mechanical or electrical properties after the samples were folded in half 100 times. For example, the measured resistance values for rGO changed by only 0.22% when film was bent. No visible flaking of the thin film or separation from the substrate was observed. However, the films were susceptible to scratching by metal tools. Similar results were observed for G-rCMC films.

CONCLUSIONS

If inkjet printing is to become a viable method of electronic circuit fabrication then it is essential that the technology support various types of modifiable conductive inks. Modern electronic devices contain a large variety of discrete components including highly conductive micro-traces, resistors, inductors, and semiconductors with adjustable band gaps. The ability to print and then locally "tune" the conductivity, or resistivity, of rGO and G-rCMC features and microtraces provides a novel approach to creating more complex electronic circuitry on non-silicon substrates. These laser modified G-rCMC printed traces are currently being used to create Interdigitated Capacitors (IDC) for chemical sensing applications [8].

ACKNOWLEDGMENTS

This work has been supported by the Natural Sciences and Engineering Research Council of Canada (NSERC) and the National Research Council of Canada (NRCC).

REFERENCES

1. Perelaer, J., Hendriks, C.E., Laat, A., and Schubert, U. 2009, "One-step inkjet printing of conductive silver tracks on polymer substrate", *Nanotechnology*, 20, 165303 – 165308.
2. Suganuma, K. 2014, *Introduction to Printed Electronics*, Springer Science, New York.
3. Jonsson, S. K. M., Birgerson, J., Crispin, X., Greczynski, G., Osikowicz, W., Denier van der Gon, A. W., Salaneck, W. R., and Fahlman, M. 2003, "The effects of solvents in the morphology and sheet resistance in poly(3,4-ethylenedioxythiophene)-polystyrenesulfonic acid (PEDOT-PSS) films", *Synthetic Metals*, 139, 1-10.
4. Sirringhaus, H., Kawase, T., Friend, R.H., Shimoda, T., Inbasekaran, M., Wu W., and Woo, E.P. 2000, "High-resolution inkjet printing of all-polymer transistor circuits", *Science*, 290, 2123-2126.
5. Geim, A.K., 2009, "Graphene: status and prospects", *Science*, 324, 1530 – 1534.
6. Acik, M. and Chabal, Y. 2013, "A review on thermal exfoliation of graphene oxide", *J. of Mat. Sci. Res.*, 2(1), 101 – 112.
7. Huh, H.S. 2011, *Thermal Reduction of Graphene Oxide*, InTech.
8. Sinar, D. and Knopf, G. 2014, "Printed graphene interdigitated capacitive sensors on flexible polyimide substrates", *14th IEEE Int. Conf. on Nanotechnology*, Toronto, Canada, 538 – 542.

REMOTE MANIPULATION OF 3D PRINTED COMPLEX MAGNETIC ARCHITECTURE AFTER SURFACE MODIFICATION

Dongxing Zhang
Department of Mechanical and Materials Engineering, The University of Western Ontario, London, Ontario N6A 5B9, Canada
Email: dzhan43@uwo.ca

Jun Yang
Department of Mechanical and Materials Engineering, The University of Western Ontario, London, Ontario N6A 5B9, Canada
Email: jyang@eng.uwo.ca

ABSTRACT

Three-dimensional magnetic robot was fabricate by mixing directly a Br vinyl-terminated initiator in the commercial UV cured resin followed by 3D printing and subsequent surface modification via atomic transfer radical polymerization and electroless plating. Through the magnetic manipulation experiment, magnetic devices by initiator-integrated 3D printing were proven to be effective and may have a wide range of potential applications including targeted therapy, material removal, remote sensing, and as controllable structures, and they can be used in a number of systems in the human body .

KEYWORDS: 3D printing, magnetic robot, surface modification, magnetic manipulation

INTRODUCTION

Structures with magnetic property can be employed in a variety of applications[1,2]. Especially under the control of the magnetic field, this kind of robots will enable minimally invasive access to many hard-to-reach locations which shows a potential application in medical therapy[3]. Based on this, the magnetic robots can be used to perform tasks which are currently difficult or impossible, and even some tasks not yet conceived[4].

Recently, three dimensional (3D) printing technology has become a prevalent method to fabricate customized and complex structures which is often difficult or even impossible to be built by traditional fabrication technologies. Therefore, it provides an effective method to fabricate magnetic robots with complex geometry. Herein, we demonstrate the use of a new strategy to make magnetic structures/materials in which complex geometries are created by 3D printing and magnetic function is achieved by growing polyelectrolyte brushes via atomic-transfer radical polymerization (ATRP)[5-7]and subsequent ELP of magnetic elements[8-10].

EXPERIMENTAL SECTION

Architectures Created by 3D Printing. 5% 2-(2-Bromo-isobutyryloxy)ethyl methacrylate (BrMA) was mixed into the commercial 3D printing resin and stirred adequately. Then, 3D models are designed trough SOLIDWORKS and digitally sliced into a series of 2D layers with a thickness of 50 μ m, then the liquid resin was scanned by the ultraviolet laser light line by line from bottom to top until the whole 3D object is formed. After this, the printed structures needed rinsed ultrasonically in ethanol for 5 min to remove the absorbed resin and other small molecular monomers, then dried by nitrogen flow.

Next, the 3D printed initiator-integrated samples were immersed into the resulting solution of following mixture for 4 h at 60°C: 2-(methacryloyloxy)ethyl-trimethylammonium chloride (METAC) / methanol = 9:1 in weight was dissolved and degassed for 20 min by passing dry N_2 stream through the solution as it is stirred. Then 2, 2-Dipyridyl (0.48 g) and CuBr (0.12 g) were added into the resulting solution under the protection of N_2. Then the resulted sample was immersed into a 5 mM $(NH_4)_2PdCl_4$ aqueous solution for 15 min. Then, to grow the magnetic coating on the surface of the resulted sample, Ni-coated structures were prepared in this case. The electroless plating was performed in a plating bath consisting of 40 g/L of $Ni_2SO_4 \cdot 5H_2O$, 20 g/L of sodium citrate, 10 g/L of lactic acid, and a dimethylamine borane (DMAB) aqueous solution was prepared separately by 1 g/L of DMAB in water. Then a final Ni-plating solution for eletroless plating (ELP) was prepared for a 4:1 volumetric proportion of nickel-to-reductant stocks.

Characterization. The morphology of the structures was observed by SEM in a Zeiss Ultra 55. Substrates before and after magnetic coating were detected comparatively.

Magnetic manipulation. After electroless plating a layer of Ni coating on the surface of the 3D printing sample, a helix structure with magnetic coating was done for Ni acting is as a magnetic material. By applying a rotational magnetic field, the helix part can rotate in a plastic tube with scale bar. Thus, the non-uniform magnetic field emanating from rotating electromagnet coils can be utilized to provide wireless propulsion. Using this technique, a fundamental axial control strategy can be used to control the robot located along the rotation axis of the electromagnet coils.

RESULT AND DISCUSSION

As shown in the Fig. 1, metallic layer was successfully deposited on the surface of the 3D printed structures. Because of the polyelectrolyte brushes are capable of being nano-thick platforms to induce metal electroless plating metal layers are deposited with excellent adhesion to the substrate.

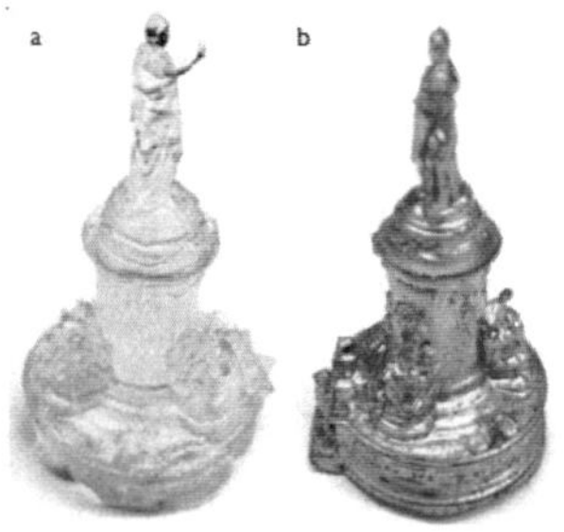

Figure 1: 3D printed sculpture integrated with initiator before (a) and after (b) magnetic coating process.

More evidences can be found in the SEM images below, it is clear that there is an obvious change of the surface before (Fig. 2a) and after (Fig. 2b) the Ni ELP process. Through the ion exchange, $PdCl_4^{-2}$ moieties were immobilized onto the surface of the sample. Then, the $PdCl_4^{-2}$ moieties-loaded samples can be coated uniformly with metallic layer.

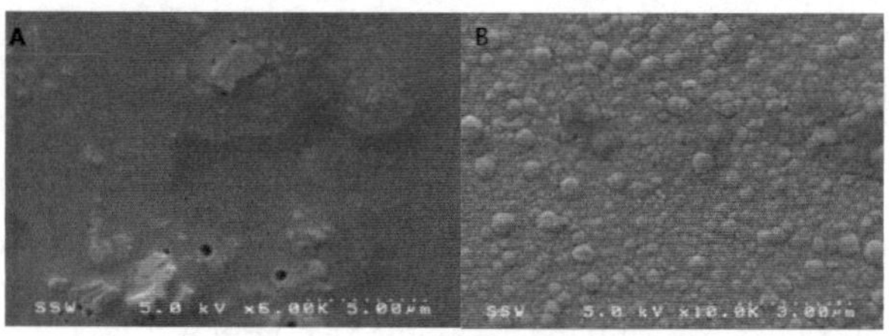

Figure 2: SEM images show the surface change before (a) and after (b) magnetic coating process

It's clear that surface-initiated ATRP based on the initiator of Br was processed to grow PMETAC brushes resulting in a dense coating on the surface, and all these evidences state that the surface-initiated ATRP works successfully to graft a uniform and dense magnetic coating on the sample.

To test the magnetic property under a real magnetic environment, a helix structure with magnetic coating was manipulated by applying a rotational magnetic field. Because of the sum of applied nonfluidic forces F and torques τ that are linearly related to its linear-axial velocity v and rotational velocity ω, rotation and propulsion of magnetic ellipsoid along their axes is well achieved. As shown in Fig. 3, under a rotate magnetic field, structures with magnetic coating was successfully rotated along with the field. Because of the combined action of the nonfluidic force and torque, rotation was effectively converted into linear motion (Fig. 3).

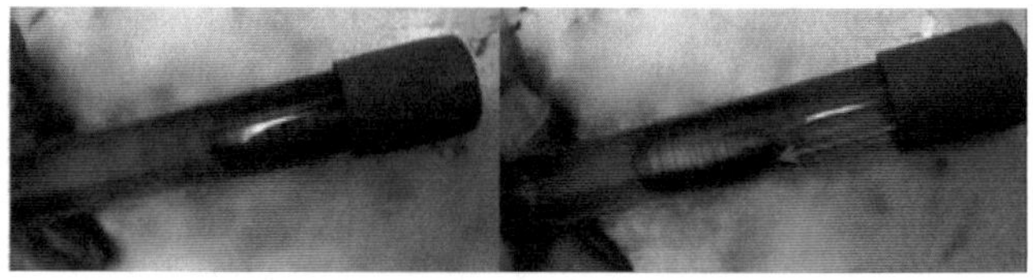

Figure 3: Magnetic manipulation of 3D printed ellipsoid (10mm length, 5mm diameter), rotated axially along the electromagnet (around 80mT in field strength), and a linear motion was realized

CONCLUSIONS

A new approach was demonstrated to fabricate magnetic architecture with complex geometries. By directly mixing a Br-containing vinyl-terminated initiator into photo-curable resin, 3D printed metallic structures were achieved via the surface-initiated ATRP and the subsequent polyelectrolyte-brush-assisted ELP of metals. Through the magnetic manipulation experiment, magnetic devices by initiator-integrated 3D printing were proven to have the potential to be used for a wide range of applications including targeted therapy [11] (e.g. drugs, hyperthermia), material removal [12] (e.g. ablation), remote sensing [13], and as controllable structures (e.g. stents)[14], and they can be used in a number of systems in the human body (e.g. circulatory system, urinary system, central nervous system)[15]. Moreover, by introducing etching technique, it was also capable of fabricating ultra-lightweight cellular metallic structures. Therefore, the capability of 3D printing is extended greatly by combining the merits of structure design, surface modification and polymer-assisted ELP. And it can also be adopted in the multi-scale printing technology, which can become a universal, robust, and cost-effective approach to make the 3D printing more practical in areas of electronics, integrated circuits, catalyst supports, and others.

REFERENCES

1. Nelson, B. J., Kaliakatsos, I. K., & Abbott, J. J. 2010, "Microrobots for minimally invasive medicine", *Annual review of biomedical engineering*, 12, pp. 55-85.
2. Zhang, L., Peyer, K. E., & Nelson, B. J. 2010, "Artificial bacterial flagella for micromanipulation", *Lab on a Chip*, *10*(17), pp. 2203-2215.
3. Fountain, T. W., Kailat, P. V., & Abbott, J. J. 2010, "Wireless control of magnetic helical microrobots using a rotating-permanent-magnet manipulator", *Robotics and Automation (ICRA), 2010 IEEE International Conference on* . IEEE, Alaska, USA, pp. 576-581.
4. Honda, T., Arai, K. I., & Ishiyama, K. 1996, "Micro swimming mechanisms propelled by external magnetic fields", *Magnetics, IEEE Transactions on*, *32*(5), pp. 5085-5087.
5. Liu, X., Wu, Y., Gao, Y., Wang, J., Li, Z., Han, J., ... & Ma, H. 2012, "Tuning the surface chemistry of iPDMS for

improved protein microarray performance", *Journal of Materials Chemistry*, *22*(13), pp. 6327-6334.
6. Li, Z., Zhang, S., Zhang, P., Yang, D., Jin, G., & Ma, H. 2012, "Surface initiated polymerization from integrated poly (dimethylsiloxane) enables crack - free large area wrinkle formation", *Polymers for Advanced Technologies*, *23*(9), pp. 1240-1245.
7. Wu, Y., Huang, Y., & Ma, H. 2007, "A facile method for permanent and functional surface modification of poly (dimethylsiloxane)", *Journal of the American Chemical Society*, *129*(23), pp. 7226-7227.
8. Wang, X., Hu, H., Shen, Y., Zhou, X., & Zheng, Z. 2011, "Stretchable conductors with ultrahigh tensile strain and stable metallic conductance enabled by prestrained polyelectrolyte nanoplatforms", *Advanced Materials*,*23*(27), pp. 3090-3094.
9. Guo, R., Yu, Y., Xie, Z., Liu, X., Zhou, X., Gao, Y., ... & Zheng, Z. 2013, "Matrix - Assisted Catalytic Printing for the Fabrication of Multiscale, Flexible, Foldable, and Stretchable Metal Conductors", *Advanced Materials*, *25*(24), pp. 3343-3350.
10. Azzaroni, O., Fonticelli, M. H., Benítez, G., Schilardi, P. L., Gago, R., Caretti, I., ... & Salvarezza, R. C. 2004, "Direct Nanopatterning of Metal Surfaces Using Self - Assembled Molecular Films", *Advanced Materials*, *16*(5), pp. 405-409.
11 . Edwards, J. C., Szczepański, L., Szechiński, J., Filipowicz-Sosnowska, A., Emery, P., Close, D. R., ... & Shaw, T. 2004, "Efficacy of B-cell–targeted therapy with rituximab in patients with rheumatoid arthritis", *New England Journal of Medicine*, *350*(25), pp. 2572-2581.
12. Charles, F. 2007, "System and techniques for magnetic manipulation of internal organs during minimally invasive surgery", *U.S. Patent Application 11/750,670*.
13. McNaughton, B. H., Anker, J. N., & Kopelman, R. 2005, "Magnetic microdrill as a modulated fluorescent pH sensor", *Journal of Magnetism and Magnetic Materials*, *293*(1), pp. 696-701.
14. Liu, G. L., Lu, Y., Kim, J., Doll, J. C., & Lee, L. P. 2005, "Magnetic Nanocrescents as Controllable Surface - Enhanced Raman Scattering Nanoprobes for Biomolecular Imaging", *Advanced Materials*, *17*(22), pp. 2683-2688.
15. Gillies, G. T., Ritter, R. C., Broaddus, W. C., Grady, M. S., Howard III, M. A., & McNeil, R. G. 1994, "Magnetic manipulation instrumentation for medical physics research", *Review of Scientific Instruments*, *65*(3), pp. 533-562.

A RAPID METHOD FOR PREPARATION OF METAL COATED 3D PRINTED OBJECTS

Qiuquan Guo, Mingjun Hu, Xiaobing Cai, Jun Yang*
Mechanical & Materials Engineering
Western University
London, Ontario, Canada
jyang@eng.uwo.ca

ABSTRACT

The prevail applications of three-dimensional printing (3DP) requires the selection of materials to be more flexible. However, in 3DP industry, plastic is still the dominate material for building up structure or devices. In this study, we present a facile method to fabricate various metallic structures by applying a polymer coating which can effectively capture catalyst for the following electroless plating of metal. Moreover, the method is also capable of fabricating metal coated 3D printed microchannel structure. By combining the merits of 3D printing in structure design with polymer-assisted electroless plating of metals, the developed technique is a facile, cost effective and fast approach to obtain metal coated 3D printed objects, which could greatly extend the capability of 3D printing technology and increase the application area as well, like in areas of electronics, moulding and others.

KEYWORDS: 3d printing, polymer coating, electroless plating, metal coating, silane coupling agent

INTRODUCTION

3D printing (3DP), also known as additive manufacturing or rapid prototyping, has been revolutionizing multiple industrial sectors all over the world, such as healthcare and biomedical, electronics, aerospace, military and defence, automotive and consumer products (e.g., jewelry, toys, fashion, etc.). 3DP is becoming a powerful manufacturing technology for creating sophisticated, bespoke and low-cost objects that, if possible to manufacture at all, were traditionally made by complicated procedures and facilities.

3D printing technology represents a group of different additive manufacturing methods [1]. Seven different categories of printing process are adopted by ASTM International Committee F42, including polymerization, material jetting, binder jetting, material extrusion, powder bed fusion, sheet lamination and direct energy deposition. Over last decade, various 3D printing applications have been developed, such as tissue and scaffold engineering [2-4], microfluidics [5], chemical synthesis [6], composites [7, 8], and electronics [9, 10]. Although the 3D printing technology has revolutionized the traditional manufacturing method, 3D printing technology also needs to overcome a number of technical barriers, among which developing functional printing materials is highly demanded.

As 3DP is becoming more prevalent, more diverse objects or products are being produced. Especially, those loading-bearing components, which require the parts to have necessary mechanical strength to function properly and safely, exert great challenge on 3DP. Among the wide spectrum of applications, 3D metallic objects have great advantages compared to other plastic objects, thanks to their strength, electronic and thermal properties, and anti-wearing characteristics. However, the plastic materials is still the dominated printing material in the market, which the 3D metal printing, like selective laser sintering (SLS), is very expensive.

In this study, we demonstrate a fast method of rendering the 3D printed object into metallic coating with the polymer-assisted electroless plating process. In most scenarios, only the surface needs to be coated with metal to realize desired functions. Instead of obtaining pure metal object, the polymer core and metal coated object is developed using this method. The central strategy is to apply a polymer coating on the 3D printed object, which can ready capture the catalyst for the subsequent electroless plating process. In addition, the polymer coating can be applied selectively on the 3D printed object, like microfluidic channel, to achieve metallic coating at desired locations.

EXPERIMENTALS

Materials

The Asiga 3D printer (based on Digital Laser Printing) was used and 3D printing resin base was purchased from Asiga as well. All other chemicals were obtained from Sigma Aldrich and solvents were purchased from Caledonlabs. All chemicals were of analytical grade.

Preparation of coating

The silane coupling agents were used as polymer coating, because the silane molecule has bi-functional groups that will react with two different phases, the printed object and the catalyst, thus function as a bridge between them. Two silanes were used, one is (3-Aminopropyl)trimethoxysilane (APS) and the other is (3-Mercaptopropyl)trimethoxysilane (MPS).

Both of them have alkoxy (-OCH3) group which will bond to the surface of 3D printed object. The NH2-terminus from APS and SH-terminus from MPS will capture the catalyst molecule. Isopropyl alcohol (IPA) was used as solvent. Firstly, 5 mL of APS was added into 40 mL of IPA, then 4 mL of MPS was added into the mixed solution. After mixing the solution thoroughly, 1 mL of H2O was added. The mixed solution was placed under ambient condition for at least 2 hours before the coating was applied.

For the case of coating the whole surface of 3D printed object, the sample was dipped into the solution and shaken to allow all the surface getting contact with the solution. After 30 sec, the sample was taken out and blow to dry. The sample was left at open air condition for another 10 min, then placed into the oven at 120°C for 20 min. For selective coating, only the part needs to be coated was fluxed with prepared silane solution.

Catalyst loading

Palladium(II) acetate was used as the catalyst for the following electroless plating process. 2.5 mg/mL of palladium acetate acetone solution was prepared. After the sample was taken out from the oven and cool down to ambient temperature, the sample was immersed into the palladium acetate solution for ~10 min to immobilize the catalyst on the surface. Then the sample was rinsed with water thoroughly and dried.

Electroless plating

In order to obtain the metallic coating, the sample was immersed into the house-made plating bath to obtain. In this study, copper (Cu) was used as an example. If desired, other metal, like nickel (Ni), can also be applied. The Cu plating bath containing a 1:1 mixture of freshly prepared solutions A and B. Solution A consists of 12 g/L NaOH, 13 g/L CuSO4·5H2O and 29 g/L potassium sodium tartrate. Solution B is 9.5 mL/L HCHO in water.

Characterization

The morphology of the resultant metallic architectures was investigated using a Hitachi S-4500 field emission scanning electro-microscope using a 5 kV accelerating voltage.

RESULTS AND DISCUSSIONS

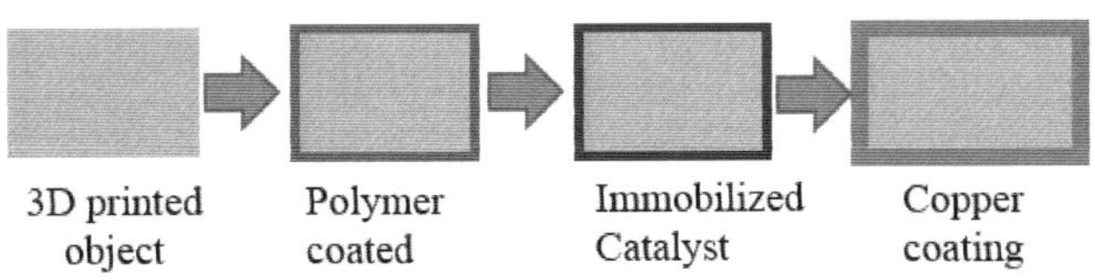

Figure 1. The schematic of the process for applying copper coating on 3D printed object

Figure 1 illustrates a typical process for applying copper coating on a 3D printed object. First, a desired structure was printed with any 3D printer, then silane coupling agent was coated on the surface. Thanks to the bi-functional property of silane molecule, the alkoxy group is readily bond to the 3D printed surfaces, while the other functional group is used to immobilize catalyst for the following electroless plating of copper. Finally, a metal coated 3D printed part is achieved by immersing the sample into electroplating solution. The thickness of the copper coating can be controlled by electroless plating time.

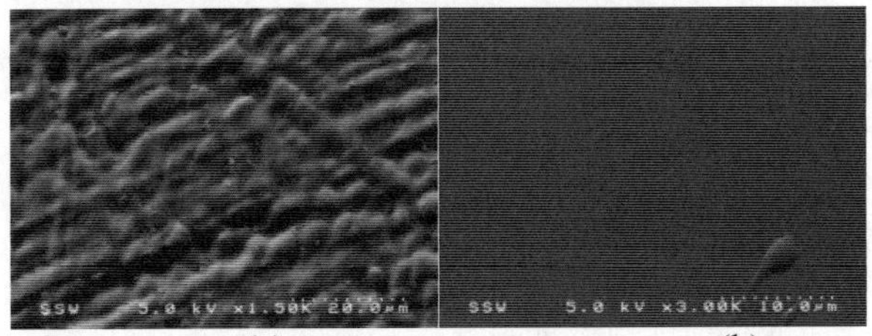

(a) (b)

Figure 2. (a) 3d printed surface; (b) after applying silane coating

It is worth to note that during 3D printing process, the object is printed under layer-by-layer manner, therefore, the surface is often rough, even with the best resolution printer. The plausible thing is that the silane coupling agent can range from several ten nanometers to several microns depending on the concentration being used. With easy control, the surface roughness can be greatly decreased. Figure 2 shows a typical location which was smoothed by applying the coating.

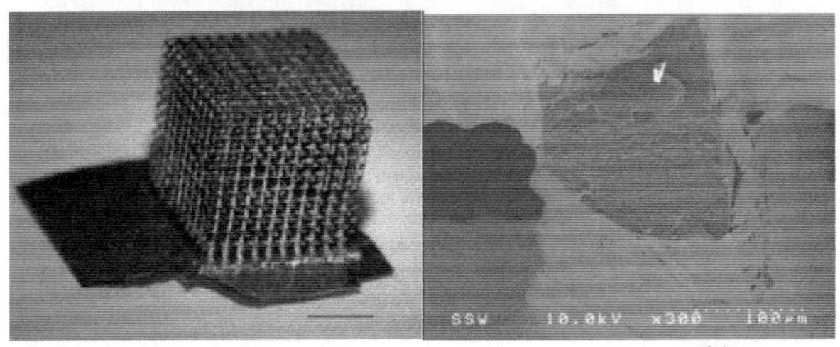

(a) (b)

Figure 3. (a) Copper coated 3D printed object, scale bar: 5 mm; (b) SEM image showing the cross section of the coating

Figure 3 shows a copper coated microlattice structure after ~1 hr electroplating time. The structure was cut and SEM was used to observe the cross section of coating. It is observed that around 200 nm thick copper layer was uniformly coated on the 3D printed object.

Recently, ultralight metallic cellular structure was developed, which opens the door for a number of potential application, especially in aerospace industry. As observed from Figure 3b, we can see that the copper has formed a continuous coating on the polymer template. The metallic cellular structure can be readily achieve by dissolving away the polymer structure if desired. The removing of polymer structure has been discussed in our previous work [11]. Briefly, 3 M NaOH aqueous solution was used and the sample was immersed in the solution for 24 hrs at 60°C.

CONCLUSIONS

In this study, we have demonstrated a rapid method for making metallic coating on the 3D printed object. Taking advantage of the bi-functional property of silane coupling agent, both the silane molecule and catalyst were strongly immobilized on the surface, allowing the subsequent growing of copper through electroless plating process. In addition, this method is able to selectively coat a partial of the surface if needed. For the microfluidic channel, it only requires to flux the silane solution into the channel and coat the channel surface with metal.

The developed method is universal for not only 3D printed object. We have also tried on polydimethylsiloxane (PDMS) surface and we could also get copper coated PDMS. Therefore, the developed technique can be leveraged to a number of other surfaces which requires metal coating.

ACKNOWLEDGMENTS

This work was supported by Natural Science and Engineering Research Council of Canada (NSERC), the Canada Foundation for Innovation (CFI), and Research Accelerator Grant of the University of Western Ontario.

REFERENCES

1. Pham, D.T. and R.S. Gault, *A comparison of rapid prototyping technologies.* International Journal of Machine Tools & Manufacture, 1998. **38**: p. 1257-87.
2. Villar, G., A.D. Graham, and H. Bayley, *A Tissue-Like Printed Material.* Science, 2013. **340**(6128): p. 48-52.
3. Duarte Campos, D.F., et al., *Three-dimensional printing of stem cell-laden hydrogels submerged in a hydrophobic high-density fluid.* Biofabrication, 2013. **5**(1): p. 015003.
4. Fedorovich, N.E., et al., *Evaluation of photocrosslinked lutrol hydrogel for tissue printing applications.* Biomacromolecules, 2009. **10**: p. 1689-1696.
5. Kitson, P.J., et al., *Configurable 3D-Printed millifluidic and microfluidic 'lab on a chip' reactionware devices.* Lab Chip, 2012. **12**(18): p. 3267-71.
6. Symes, M.D., et al., *Integrated 3D-printed reactionware for chemical synthesis and analysis.* Nature Chemistry, 2012. **4**(5): p. 349-354.
7. Melcher, R., et al., *Fabrication of Al2O3-based composites by indirect 3D-printing.* Materials Letters, 2006. **60**(4): p. 572-575.
8. Wu, C., et al., *3D-printing of highly uniform CaSiO3 ceramic scaffolds: preparation, characterization and in vivo osteogenesis.* Journal of Materials Chemistry, 2012. **22**(24): p. 12288.
9. Lopes, A.J., E. MacDonald, and R.B. Wicker, *Integrating stereolithography and direct print technologies for 3D structural electronics fabrication.* Rapid Prototyping Journal, 2012. **18**(2): p. 129-143.
10. Sun, K., et al., *3D Printing of Interdigitated Li-Ion Microbattery Architectures.* Advanced Materials, 2013: p. n/a-n/a.
11. Wang, X., et al., *Initiator integrated 3D printing enabling complex metallic architectures.* ACS Applied Materials & Interfaces, 2013.

Proceedings of the 25th CANCAM
London, Ontario, Canada, May 31 – June 4, 2015

A MULTIFUNCTIONAL MEMS SENSOR SYSTEM FOR BRAKE SYSTEM APPLICATIONS

Rogelio Hernández Aguirre[1], Patricia Nieva[2*], Amir Khajepour[3]
Department of Mechanical and Mechatronics
University of Waterloo
Waterloo, Ontario, Canada
r3hernan@uwaterloo.ca[1], pnieva@uwaterloo.ca[2*], a.khajepour@uwaterloo.ca[3]

*Corresponding author

ABSTRACT

In this study, a new multifunctional MEMS (Micro Electro Mechanical System) sensor system for the simultaneous measurement of pressure and temperature is being developed. The multifunctional sensor will be used in a new brake-by-wire system to provide brake force to each wheel individually. The new sensor system will monitor the pressure the temperature of the brake fluid to aid in the automatic control of the brake system. While numerous approaches have been taken to address the control and monitoring of the pressure in a brake system, no MEMS sensor system has yet been reported that can carry out real-time measurements of brake system of both the pressure and temperature simultaneously with minimal intrusion. The multifunctional MEMS pressure and temperature sensor is mounted in a single package, thus reducing, cost and size, making it ideal for its installation on the wheel cylinder with less penetration and thus reduced disturbance to the brake fluid environment.

KEYWORDS: MEMS, pressure sensor, temperature sensor, sensor integration, multifunctional sensor system

INTRODUCTION

The brake system is one of the most important subsystems in a vehicle. Over the years, it has drastically changed due to the progress in sensor technology, control systems, hydraulic and electrical systems, and the increased importance of safety conditions. The increase in the use of electrical components in braking systems has led to the development of what are called "brake-by-wire systems". One of the most innovative characteristics of these systems is the capacity to control the braking pressure of each wheel separately from the other wheels. The process of providing the brake force for each wheel independently through the vehicle control system as opposed to the driver input is called differential braking [1]. To monitor and control the different braking pressures on the car, such system requires pressure and temperature readings for each wheel brake. Current sensing systems installed in the brake system use a single pressure sensor for the whole brake system, and are designed in such a way that only one sensor can be installed in its main cylinder [2]. Moreover, these current systems monitor only the pressure but not temperature. While numerous MEMS approaches have been taken to address the control and monitoring of the pressure in a brake system [2], no sensor system has yet been reported that can carry out real-time measurements of brake system's pressure and temperature. The current paper investigates a solution that employ MEMS technology for the integration of pressure and temperature sensors as a component of an Electro Hydraulic Braking [EHB] system.

WORKING CONDITIONS AND SELECTION OF TECHNOLOGY

In a representative brake system, there are three phases during the braking process; apply, hold, and release. The brake pressure goes up when the driver increases the force exerted to the pedal, stays constant while the pedal is hold in the same position and finally drops when brake pedal is released. During the hold phase, the pressure can reach up to about 6 MPa. Table 1 shows the pressure specifications of the Electro Hydraulic Brake (EHB) system being developed in our lab. The working conditions inside the wheel cylinder take place at relatively high temperatures in an alkaline chemical environment. Temperatures up to 120°C [3], and pH ~ 11 are due primarily to the glycol-ether based composition of the brake fluid [4]. This high temperature can cause vaporization of the brake fluid, which is a problem because vapor is highly compressible relative to liquid, and therefore negates the hydraulic transfer of braking force.

Table 1: Working conditions of the sensor

Parameter	Value
Pressure Range	0.1 to 4 MPa
Temperature Range	-20 to 120 °C
Chemical environment	pH ~11

The braking performance of a vehicle can be significantly affected by the temperature rise in the brake components. High temperature during braking may cause brake fade, premature wear, bearing failure, thermal cracks, and thermally-excited vibration [5]. Moreover, braking distance is directly affected by the brake system temperature, while higher the temperature of the brake system, the longer the braking distance. Operational data from the brake system is typically taken at the main cylinder leading to one single pressure signal for the whole brake system [2]. To successfully deliver a constant and reliable signal with reliable information from all the cylinders of the brake system, the sensor system must be (1) compatible with the current sensor fittings used in hydraulic systems; (2) resistant to harsh environments; (3) capable to measure both signals using only one sensor penetration; (4) compatible with automotive electrical conditions; and (5) low cost. Having both transducers on the same packaging reduces the need of two different holes for sensor penetration. Using MEMS technology reduces the size of the package, allows for electrical compatibility with automotive electrical conditions and decreases the cost of the sensor system.

Considering that the pressure inside the wheel cylinder is the same due to the incompressibility of the brake fluid, and that the temperature does not have an spatial gradient due to the high conductivity of the caliper material and the very large contact area between the fluid and the caliper [3], it is possible to use a discrete approach using two different and independent MEMS transducers close to each other and in the same packaging. MEMS silicon-based pressure sensors are commercially available for many applications such as automotive, biomedical, and aerospace [6]. Piezoresistive MEMS pressure sensors are very common and have been chosen for this application over MEMS capacitive sensors due to their robustness, linearity, simple measurement circuitry and immunity to electromagnetic noise. However, difficulties in temperature compensation and fabrication control affect sensor offset and performance. Thus, a MEMS temperature sensor collocated next to the MEMS pressure sensor should provide temperature readings as well as the means for temperature compensation.

For the first prototype, a commercial MEMS piezoresistive pressure sensor manufactured by BCM sensors was selected. The temperature measurement is performed using a commercial resistance temperature detector (RTD) manufactured by OMEGA. This type of sensor measures temperature by correlating the resistance of the RTD element with temperature and it is known to be one of the most accurate and used MEMS temperature sensors [7]. Not only does it provide good accuracy, but it also provides excellent stability and repeatability.

PACKAGING AND CHEMICAL RESISTANCE TO THE HARSH ENVIRONMENT

To prevent sensor malfunction or degradation due to chemical attack, MEMS devices exposed to chemical environmental conditions must be protected against the media with a protective coating applied to the exposed area. Brake fluid is hygroscopic - that means it attracts and soaks up water .While this directly increases the boiling point of the brake fluid and reduces the viscosity of the brake fluid, the humidity levels in the brake fluid can damage the sensor systems. Zongyang Zhang et al [8] packaged an engine oil MEMS pressure sensor inside a metal chamber filled with silicon oil and covered by a corrugated diaphragm. This protected their sensor while keeping almost the same pressure stimulus due to the flexibility of the corrugated diaphragm. Although this option accomplishes the objective of protection, the main drawbacks are that it needs expensive packaging equipment and it is very difficult to keep the silicon oil clean before putting it inside the metal chamber.

Parylene coating has been extensively used in electronic components [9]. Several types of Parylene materials with slightly different chemical structures are available for different applications. The benefits of Parylene coating include room temperature conformal deposition, pinhole-free, excellent adhesion to most substrates, high moisture impermeability, chemical resistance and it has a maximum operating temperature of 450°C. This research explores the possibility of using Parylene as a coating material for high temperature operation in alkaline environments. The Parylene layer is designed to offer (1) dielectric protection; (2) protection of the wire bonding (electrical connections); (3) protection from the corrosive medium (brake fluid); and (4) protection from the moisture in the brake fluid.

PROTOTYPE FABRICATION

The prototype will be fabricated gluing both sensors on the same substrate or header. A commercial Transistor Outline (TO) Header was selected and purchased from Spectrum Semiconductors. This TO Header includes the lead pins needed to make the electrical connections to the sensors. Figure 1 shows the MEMS pressure sensor and the RTD sensor glued to the surface of the TO Header using ABLEBOND® 84-1LMI, a die attach adhesive designed for microelectronic chip bonding. This adhesive is resistant enough to support the specified high pressure and high temperature.

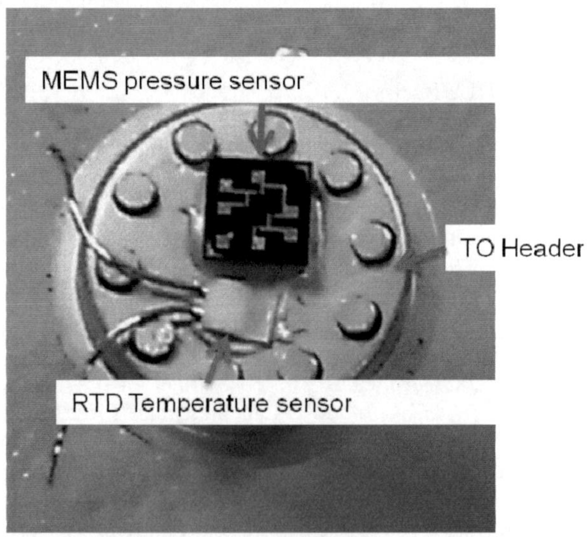

Figure 1: TO substrate with the MEMS and the RTD sensor glued on it.

Once the adhesive is cured, the leads of the RTD are solder to the TO Header lead pins using paste SynTECH-LF Lead Free. The MEMS pressure sensor is then wire bonded to its respective TO lead wires. Figure 2 shows the MEMS piezoresistive pressure sensor electrically connected to the TO header through wire bonding.

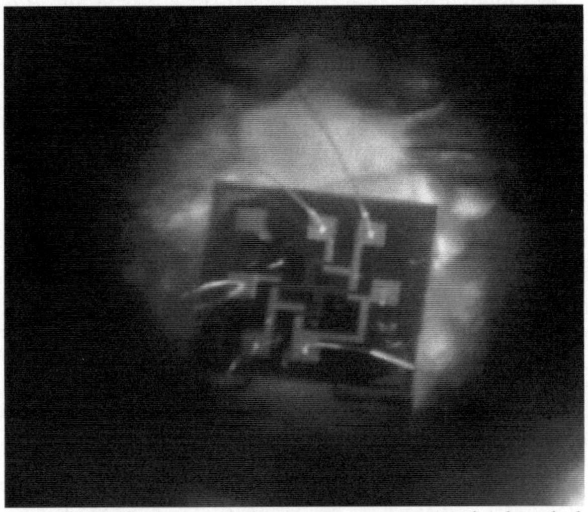

Figure 2: MEMS piezoresistive pressure sensor wire bonded to the TO header.

After the connections are finished, the parylene protection layer will be deposited over the sensors using a vapor deposition process at room temperature. The three-step deposition process includes: (1) Vaporization of solid dimmers to gaseous ones; (2) Pyrolysis of gaseous dimmers to monomers; and (3) Parylene deposition (see Figure 3) [10]. Parylene is a gaseous monomer which builds from the surface outward. This produces a pinhole free (pinhole free @ .6 micrometer or higher) conformal layer of uniform thickness. Surface tension causing the pulling away of sharp edges, bridging or thin out when conventional coatings are used in the curing process does not occur with the parylene process. Due to the Parylene coating is a room temperature process and there is no need a high cure temperature cycle, thus there is no thermal or mechanical stress.

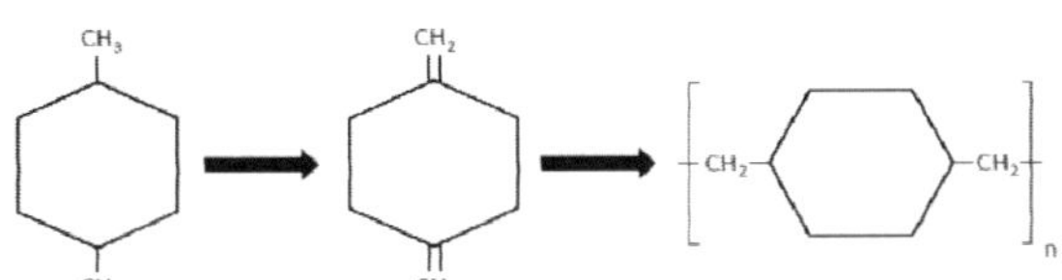

Figure 3: Three-Step Vapor Deposition Process of Parylene with chemical chains.

The last packaging step consists of building an interface of the sensor system to the brake system. Such interface should be compatible with the current fittings of the brake system. For this purpose, a commercial 5/16 to ¼ 18 NPT fitting will be used as a metal shell. The TO base and metal shell will be joined forming a sealed chamber using resistance welding, which will seal the chamber by joining the steel of the metal shell to the KOVAR alloy of the TO substrate [11]. Figure 4 shows the proposed final step on the packaging of the sensor system.

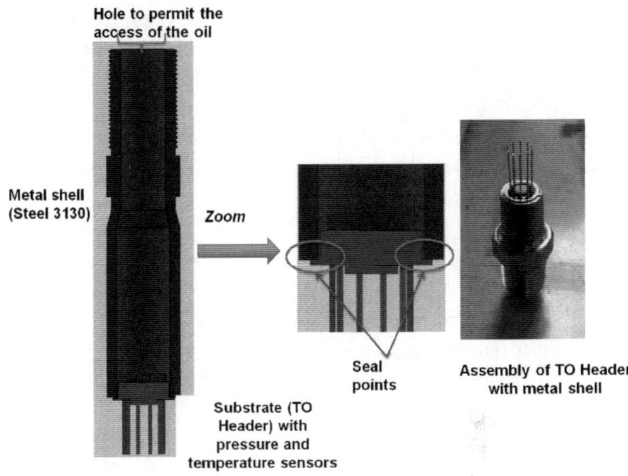

Figure 4: Schematic of the final packaging (left); excerpt showing details of the sealing points (center); and physical sample (right).

DISCUSSION OF FUTURE WORK

While the adhesion of parylene will protect the sensor system from the conditions of the brake fluid, also will affect the performance of the temperature and pressure sensor. Adding an extra layer of polymer on the surface of the RTD sensor will create an extra thermal mass which will be located between the sensing media and the RTD. This potentially can affect the response time and sensitivity of the sensor and should be considered for the final sensor system design. With regard to the pressure sensor, the adhesion of the parylene over the deflective diaphragm, will make it more rigid, and will cause a dampening effect, this directly will affect the stress response of the diaphragm and hence the sensor sensitivity. The sensor system is being modeling using COMSOL Multiphysics to account for the effect of the packaging on the performance of the sensor.

CONCLUSIONS

A new multifunctional MEMS sensor system for brake-by-wire systems has been proposed. The process to select the technology for the multifunctional sensor system has been reviewed and packaging steps for the prototyping of the sensor have been described. One of the main advantages of the proposed sensor system will reside in having both sensors in the same packaging and bonded to the same substrate. Using MEMS technology ensures reduction of size and cost. Reduction of size will reduce disturbance to the wheel cylinder and the brake system in general. Reducing sensor penetration will also decreases installation cost and improve life cycle.

ACKNOWLEDGMENTS

The authors would like to thank Natural Sciences and Engineering Research Council (NSERC) of Canada, and the CONACYT (National Council of Science and Technology of Mexico) for their financial support. They would also like to acknowledge CMC Microsystems for the provision of products and services that facilitated this research, including CAD tools and design methodology, fabrication services and engineering services. Este investigación se ha realizado con el apoyo de beca de la Secretaría de Educación Pública y del Gobierno Mexicano.

REFERENCES

1. V. Milanes, C. Gonzalez, J.E. Naranjo, 2010, "Electro-Hydraulic Braking System for Autonomous Vehicles", *International Journal of Automotive Technology*, Vol (11),pp.89-95
2. Kant, Bernhard. "Sensotronic Brake Control (SBC)." In Automotive Mechatronics, edited by Konrad Reif, 412–15. Bosch Professional Automotive Information. Springer Fachmedien Wiesbaden, 2015. http://link.springer.com/chapter/10.1007/978-3-658-03975-2_19.
3. Adrian Neys, "In Vehicle Brake System Temperature Model", Master's Thesis, Chalmers University of Technology, 2012. Retrieved from http://publications.lib.chalmers.se/records/fulltext/162888.pdf
4. International Society of Automotive Engineers, 2012, "Motor vehicle Brake Fluid", *SAE Specifications SAE J1703*.
5. Lee K, "Numerical prediction of brake fluid temperature rise during braking and heat soaking.", 1999, SAE, International Congress and Exposition Detroit, Michigan, 1-4 March. URL: http://delphi.com/pdf/techpapers/1999-01-0483.PDF
6. Eaton, W. P., and J. H. Smith. "Micromachined Pressure Sensors: Review and Recent Developments." Smart Materials and Structures 6, no. 5 (October 1, 1997): 530.doi:10.1088/0964-1726/6/5/004..
7. Santos, E. J P, and IB. Vasconcelos. "RTD-Based Smart Temperature Sensor: Process Development and Circuit Design." In 26th International Conference on Microelectronics, 2008. MIEL 2008, 333–36, 2008. doi:10.1109/ICMEL.2008.4559289.
8. Zhang, Zongyang, Chaojun Liu, Zhimin Wan, Gang Cao, Yun Lu, Bin Song, and Sheng liu."Optimization of Packaging Process of Piezoresistive Engine Oil Pressure Sensor." In 2010 11th International Conference on Electronic Packaging Technology High Density Packaging (ICEPT-HDP), 1362–65, 2010. doi:10.1109/ICEPT.2010.5582815.
9. Mohammadi, Abdolreza R., Chad P. J. Bennington, and Mu Chiao. "Development of a Combined Piezoresistive Pressure and Temperature Sensor Using a Chemical Protective Coating for Kraft Pulp Digester Process Monitoring." Journal of Micromechanics and Microengineering 21, no. 1 (January 1, 2011): 015009. doi:10.1088/0960-1317/21/1/015009.
10. Licari, James J. "5 - Reliability Assurance and Testing." In Coating Materials for Electronic Applications, edited by James J. Licari, 387–436. Materials and Processes for Electronic Applications. Norwich, NY: William Andrew Publishing,2003. http://www.sciencedirect.com/science/article/pii/B9780815514923500076.
11. Li, Xin, Qin Liu, Shixin Pang, Kaixian Xu, Hui Tang, and Chensong Sun. "High-Temperature Piezoresistive Pressure Sensor Based on Implantation of Oxygen into Silicon Wafer." Sensors and Actuators A: Physical 179 (June 2012): 277–82. doi:10.1016/j.sna.2012.03.027.

NATURAL FREQUENCIES OF DOUBLE AND TRIPLE WALLED CARBON NANOTUBES USING FINITE ELEMENT FORMULATION

Ishan Ali Khan
Department of Aerospace Engineering
Ryerson University
Toronto, ON, Canada
ishanali.khan@ryerson.ca

Seyed M. Hashemi
Department of Aerospace Engineering
Ryerson University
Toronto, ON, Canada
smhashem@ryerson.ca

ABSTRACT

Finite Element formulation is developed for double walled (DWCNTs) and triple walled carbon nanotubes (TWCNTs) to find their natural frequencies. Carbon nanotubes are modelled as Euler Bernoulli beams with the coupling factor as van der Waal's interlayer coefficient *c*. The residuals and weighting functions are introduced as per Galerkin weighted residual method for finding the element matrices, which are then assembled to find system's global stiffness and mass matrices. The linear eigenvalue problem, obtained after the application of boundary conditions, is then solved using a MATLAB code to derive the system natural frequencies, validated against the data available in literature.

KEYWORDS: carbon nanotubes, Euler Bernoulli beam, finite element modelling, linear eigenvalue problem, natural frequencies

INTRODUCTION

Carbon nanotubes have been under extensive research, since their discovery in 1991, due to the unique properties they possess. In recent years, the focus has been to understand their behavior, along with a lot of nano-sized structures including but not limited to nanoshells, nanocomposites etc, under a broader umbrella of nanomechanics. There has been extensive study on finding the vibrations of carbon nanotubes by modelling them as continuum models and by using atomistic approach. In continuum models carbon nanotubes have been studied as shells – owing to the tubular geometry - and as beams. Studying carbon nanotubes as continuum models take all the factors that affect the behavior of the carbon nanotubes into consideration by simple formulas. Demetris *et al* [1] considered carbon nanotubes as Euler Bernoulli beams, coupled by the van der Waals' forces, and found the natural frequencies for various boundary conditions using Bubnov – Galerkin and Petrov - Galerkin approximate approaches. The present study undertakes the Finite Element Method (FEM) approach, to preserve the generality of the method, using Galerkin's weighted residuals, to model carbon nanotubes as Euler Bernoulli beams for double walled carbon nanotubes (DWCNTs) and triple walled carbon nanotubes (TWCNTs). Taking the governing equations provided by literature and modelling Galerkin's weighted residual method by using hermite cubic interpolation functions, we obtain element stiffness and mass matrices. Eigenvalue value is obtained by different boundary conditions to the matrices – solution of which gives the natural frequencies.

THEORY

The governing equations for DWCNTs are given by [2]:

$$E_1 I_1 \frac{\partial^4 w_1}{\partial x^4} + \rho_1 A_1 \frac{\partial^4 w_1}{\partial t^2} = c(w_2 - w_1)$$

$$E_2 I_2 \frac{\partial^4 w_2}{\partial x^4} + \rho_2 A_2 \frac{\partial^4 w_2}{\partial t^2} = -c(w_2 - w_1) \qquad (1)$$

and for TWCNTs [2]:

$$E_1 I_1 \frac{\partial^4 w_1}{\partial x^4} + \rho_1 A_1 \frac{\partial^4 w_1}{\partial t^2} = c_1(w_2 - w_1)$$

$$E_2 I_2 \frac{\partial^4 w_2}{\partial x^4} + \rho_2 A_2 \frac{\partial^4 w_2}{\partial t^2} = c_2(w_3 - w_2) - c_1(w_2 - w_1)$$

$$E_3 I_3 \frac{\partial^4 w_3}{\partial x^4} + \rho_3 A_3 \frac{\partial^4 w_3}{\partial t^2} = -c_2(w_3 - w_2) \qquad (2)$$

where w_j is the transverse displacement, E_j is the Young's Modulus, I_j is the moment of inertia, A_j is the cross-sectional area of the tube, ρ_j is mass density and c is the van der Waal's interaction coefficient between the carbon nanotubes. $j=1,2$ for the DWCNTs and $j=1,2,3$ for TWCNTs with $j=1$ denoting the inner most tube, $j=2$ being the outer tube for DWCNT and middle tube for TWCNT and $j=3$ representing the outer most tube.

Assuming simple harmonic motion, with amplitudes of displacements represented by Wj, the Galerkin's weighted residual formulation is used on both the sets of equations with δW_j as the weighting functions with $j=1,2$ for DWCNTs:

$$\int_0^L (E_1 I_1 W'''' \delta W_1 + (c - \rho_1 A_1 \omega^2)\, W_1 \delta W_1)\, dx - \int_0^L c W_2 \delta W_1 = 0$$

$$\int_0^L (E_2 I_2 W'''' \delta W_2 + (c - \rho_2 A_2 \omega^2)\, W_2 \delta W_2) dx - \int_0^L c W_1 \delta W_2 = 0 \qquad (3)$$

and $j=1,2,3$ for TWCNT:

$$\int_0^L (E_1 I_1\, W''''\delta W_1 + (c_1 - \rho_1 A_1 \omega^2)\, W_1 \delta W_1)\, dx$$

$$- \int_0^L c_1\, W_2 \delta W_1 = 0$$

$$\int_0^L (E_2 I_2\, W''''\delta W_2 + (c_1 + c_2 - \rho_2 A_2 \omega^2) W_2\, \delta W_2) dx - \int_0^L c_2\, W_3 \delta W_2$$

$$- \int_0^L c_1\, W_2 \delta W_2 = 0$$

$$\int_0^L (E_3 I_3\, W''''\delta W_3 + (c_2 - \rho_3 A_3 \omega^2)\, \delta W_3) dx - \int_0^L c_2\, W_2 \delta W_3$$

$$= 0 \qquad (4)$$

Integrating the above sets of equations with parts and letting the boundary terms vanish with the application of system boundary conditions (e.g., $W_j = \delta W j = W j' = \delta W j' = 0$, for fixed B.C., and $W_j = \delta W j = 0$, for pinned B.C., …), the discretization of the system integral equations then leads to, the following sets of element integral equations for DWCNTs:

$$W_{w_1}^k = \int_0^{l^k} \delta W_1''(E_1 I_1 W_1'')dx + \int_0^{l^k} \delta W_1''(c - \rho_1 A_1 \omega^2)\frac{W_1^3}{6}$$

$$- \int_0^{l^k} \delta W_1 \frac{c W_2^2}{2} = 0$$

$$W_{w_2}^k = \int_0^{l^k} \delta W_2''(E_2 I_2 W_2'')dx + \int_0^{l^k} \delta W_2''(c - \rho_2 A_2 \omega^2)\frac{W_2^3}{6}$$

$$- \int_0^{l^k} \delta W_2 \frac{c W_1^2}{2} = 0 \qquad (5)$$

and the following expressions for TWCNTs:

$$W_{w_1}^k = \int_0^{l^k} \delta W_1''(E_1 I_1 W_1'')dx + \int_0^{l^k} \delta W_1''(c - \rho_1 A_1 \omega^2)\frac{W_1^3}{6}$$

$$- \int_0^{l^k} \delta W_1 \frac{c W_2^2}{2} = 0$$

$$W_{w_2}^k = \int_0^{l^k} \delta W_2''(E_2 I_2 W_2'')dx$$

$$+ \int_0^{l^k} \delta W_2''(c_1 + c_2 - \rho_2 A_2 \omega^2)\frac{W_2^3}{6}$$

$$- \int_0^{l^k} \delta W_2 \frac{c_2 W_3^2}{2} - \int_0^{l^k} \delta W_2 \frac{c_1 W_2^2}{2} = 0$$

$$W_{w_3}^k = \int_0^{l^k} \delta W_3''(E_3 I_3 W_3'')dx + \int_0^{l^k} \delta W_3''(c_2 - \rho_3 A_3 \omega^2)\frac{W_3^3}{6}$$

$$- \int_0^{l^k} \delta W_3 \frac{c_2 W_2^2}{2} = 0 \qquad (6)$$

In the above set of equations, we have:

$$W_i = \langle N_i(x_i)\rangle \{W_n\} \qquad (7)$$

where $\langle N_i(x_i)\rangle$ are the cubic shape functions of the beam elements given by:

$$N_1 = \frac{1}{L^3}(2x^3 - 3x^2 L + L^3);\ N_2 = \frac{1}{L^3}(x^3 L - 2x^2 L + xL);$$

$$N_3 = \frac{1}{L^3}(-2x^3 + 3x^2 L);\ N_4 = \frac{1}{L^3}(x^3 L - x^2 L^2); \qquad (8)$$

These shape functions are also used to approximate the virtual displacements:

$$\delta W(x) = \langle N_i(x_i)\rangle \{\delta W_n\} \qquad (9)$$

Using the equations defined in (7), (8) and (9), equations (5) can be represented in form of uncoupled matrices as:

$$W_{w_1}^k = \begin{pmatrix} \frac{12E_1I_1}{L^3} & \frac{6LE_1I_1}{L^3} & \frac{-12E_1I_1}{L^3} & \frac{6LE_1I_1}{L^3} \\ \frac{6LE_1I_1}{L^3} & \frac{4L^2E_1I_1}{L^3} & \frac{-6LE_1I_1}{L^3} & \frac{2L^2E_1I_1}{L^3} \\ \frac{-12E_1I_1}{L^3} & \frac{-6LE_1I_1}{L^3} & \frac{12E_1I_1}{L^3} & \frac{-6LE_1I_1}{L^3} \\ \frac{6LE_1I_1}{L^3} & \frac{2L^2E_1I_1}{L^3} & \frac{-6LE_1I_1}{L^3} & \frac{4L^2E_1I_1}{L^3} \end{pmatrix}^{*} \begin{Bmatrix} W_{11} \\ \theta_{11} \\ W_{12} \\ \theta_{12} \end{Bmatrix} +$$

$$\begin{pmatrix} \frac{156A_1\rho_1L}{420} & \frac{22L^2A_1\rho_1}{420} & \frac{54A_1\rho_1L}{420} & \frac{-13L^2A_1\rho_1}{420} \\ \frac{22L^2A_1\rho_1}{420} & \frac{4L^3A_1\rho_1}{420} & \frac{13L^2A_1\rho_1}{420} & \frac{-3L^3A_1\rho_1}{420} \\ \frac{54A_1\rho_1L}{420} & \frac{13L^2A_1\rho_1}{420} & \frac{156A_1\rho_1L}{420} & \frac{-22L^2A_1\rho_1}{420} \\ \frac{-13L^2A_1\rho_1}{420} & \frac{-3L^3A_1\rho_1}{420} & \frac{-22L^2A_1\rho_1}{420} & \frac{4L^3A_1\rho_1}{420} \end{pmatrix}^{**} \begin{Bmatrix} W_{11} \\ \theta_{11} \\ W_{12} \\ \theta_{12} \end{Bmatrix} +$$

$$\begin{pmatrix} 156c & 22Lc & 54c & -13Lc \\ 22Lc & 4L^2c & 13Lc & -3L^2 \\ 54c & 13Lc & 156c & -22Lc \\ -13Lc & -3L^2 & -22Lc & 4L^2c \end{pmatrix}^{***} \begin{Bmatrix} W_{11} \\ \theta_{11} \\ W_{12} \\ \theta_{12} \end{Bmatrix} -$$

$$\begin{pmatrix} 156c & 22Lc & 54c & -13Lc \\ 22Lc & 4L^2c & 13Lc & -3L^2 \\ 54c & 13Lc & 156c & -22Lc \\ -13Lc & -3L^2 & -22Lc & 4L^2c \end{pmatrix}^{****} \begin{Bmatrix} W_{21} \\ \theta_{21} \\ W_{22} \\ \theta_{22} \end{Bmatrix}$$

$$W_{w_2}^k = \begin{pmatrix} \frac{12E_2I_2}{L^3} & \frac{6LE_2I_2}{L^3} & \frac{-12E_2I_2}{L^3} & \frac{6LE_2I_2}{L^3} \\ \frac{6LE_2I_2}{L^3} & \frac{4L^2E_2I_2}{L^3} & \frac{-6LE_2I_2}{L^3} & \frac{2L^2E_2I_2}{L^3} \\ \frac{-12E_2I_2}{L^3} & \frac{-6LE_2I_2}{L^3} & \frac{12E_2I_2}{L^3} & \frac{-6LE_2I_2}{L^3} \\ \frac{6LE_2I_2}{L^3} & \frac{2L^2E_2I_2}{L^3} & \frac{-6LE_2I_2}{L^3} & \frac{4L^2E_2I_2}{L^3} \end{pmatrix}^{*} \begin{Bmatrix} W_{21} \\ \theta_{21} \\ W_{22} \\ \theta_{22} \end{Bmatrix} +$$

$$\begin{pmatrix} \frac{156A_2\rho_2L}{420} & \frac{22L^2A_2\rho_2}{420} & \frac{54A_2\rho_2L}{420} & \frac{-13L^2A_2\rho_2}{420} \\ \frac{22L^2A_2\rho_2}{420} & \frac{4L^3A_2\rho_2}{420} & \frac{13L^2A_2\rho_2}{420} & \frac{-3L^3A_2\rho_2}{420} \\ \frac{54A_2\rho_2L}{420} & \frac{13L^2A_2\rho_2}{420} & \frac{156A_2\rho_2L}{420} & \frac{-22L^2A_2\rho_2}{420} \\ \frac{-13L^2A_2\rho_2}{420} & \frac{-3L^3A_2\rho_2}{420} & \frac{-22L^2A_2\rho_2}{420} & \frac{4L^3A_2\rho_2}{420} \end{pmatrix}^{**} \begin{Bmatrix} W_{21} \\ \theta_{21} \\ W_{22} \\ \theta_{22} \end{Bmatrix} +$$

$$\begin{pmatrix} 156c & 22Lc & 54c & -13Lc \\ 22Lc & 4L^2c & 13Lc & -3L^2 \\ 54c & 13Lc & 156c & -22Lc \\ -13Lc & -3L^2 & -22Lc & 4L^2c \end{pmatrix}^{***} \begin{Bmatrix} w_{21} \\ \theta_{21} \\ w_{22} \\ \theta_{22} \end{Bmatrix} - \begin{pmatrix} 156c & 22Lc & 54c & -13Lc \\ 22Lc & 4L^2c & 13Lc & -3L^2 \\ 54c & 13Lc & 156c & -22Lc \\ -13Lc & -3L^2 & -22Lc & 4L^2c \end{pmatrix}^{****} \begin{Bmatrix} w_{11} \\ \theta_{11} \\ w_{12} \\ \theta_{12} \end{Bmatrix} \quad (10)$$

The stiffness, mass, semi-mass and coupling terms in Equation (10) are represented by the superscript *, **, *** and **** respectively. This form of equation (10) remains the same for TWCNTs, not represented here for the brevity, except for the middle tube having coupling terms from both inner and outer tubes. It should also be noted that a DWCNT has 8 DOF - 2 transverse displacements and 2 slopes for both ends of each carbon nanotube, and a TWCNT has 12 DOF in the same manner, as shown in Figure 1.

The element stiffness and mass matrices in the coupled form are represented as K_{elem} and M_{elem}, respectively. A FEM code in MATLAB would give the linear eigenvalue problem as:

$$M_{elem} = \begin{bmatrix} \frac{156A_1\rho_1L}{420} & \frac{22L^2A_1\rho_1}{420} & 0 & 0 & \frac{54A_1\rho_1L}{420} & \frac{-13L^2A_1\rho_1}{420} & 0 & 0 \\ \frac{22L^2A_1\rho_1}{420} & \frac{4L^3A_1\rho_1}{420} & 0 & 0 & \frac{13L^2A_1\rho_1}{420} & \frac{-3L^3A_1\rho_1}{420} & 0 & 0 \\ 0 & 0 & \frac{156A_2\rho_2L}{420} & \frac{22L^2A_2\rho_2}{420} & 0 & 0 & \frac{54A_2\rho_2L}{420} & \frac{-13L^2A_2\rho_2}{420} \\ 0 & 0 & \frac{22L^2A_2\rho_2}{420} & \frac{4L^3A_2\rho_2}{420} & 0 & 0 & \frac{13L^2A_2\rho_2}{420} & \frac{-3L^3A_2\rho_2}{420} \\ \frac{54A_1\rho_1L}{420} & \frac{13L^2A_1\rho_1}{420} & 0 & 0 & \frac{156A_1\rho_1L}{420} & \frac{-22L^2A_1\rho_1}{420} & 0 & 0 \\ \frac{-13L^2A_1\rho_1}{420} & \frac{-3L^3A_1\rho_1}{420} & 0 & 0 & \frac{-22L^2A_1\rho_1}{420} & \frac{4L^3A_1\rho_1}{420} & 0 & 0 \\ 0 & 0 & \frac{54A_2\rho_2L}{420} & \frac{13L^2A_2\rho_2}{420} & 0 & 0 & \frac{156A_2\rho_2L}{420} & \frac{-22L^2A_2\rho_2}{420} \\ 0 & 0 & \frac{-13L^2A_2\rho_2}{420} & \frac{-3L^3A_2\rho_2}{420} & 0 & 0 & \frac{-22L^2A_2\rho_2}{420} & \frac{4L^3A_2\rho_2}{420} \end{bmatrix}$$

$$K_{elem} = \begin{bmatrix} \frac{12E_1I_1}{L^3}+\frac{156Lc}{420} & \frac{6LE_1I_1}{L^3}+\frac{22L^2c}{420} & -\frac{156Lc}{420} & -\frac{22L^2c}{420} & -\frac{12E_1I_1}{L^3}+\frac{54Lc}{420} & \frac{6LE_1I_1}{L^3}-\frac{13L^2c}{420} & -\frac{54Lc}{420} & \frac{13L^2c}{420} \\ \frac{6LE_1I_1}{L^3}+\frac{22L^2c}{420} & \frac{4L^2E_1I_1}{L^3}+\frac{4L^3c}{420} & -\frac{22L^2c}{420} & -\frac{4L^3c}{420} & -\frac{6LE_1I_1}{L^3}+\frac{13L^2c}{420} & \frac{2L^2E_1I_1}{L^3}-\frac{3L^3c}{420} & -\frac{13L^2c}{420} & \frac{3L^3c}{420} \\ -\frac{156Lc}{420} & -\frac{22L^2c}{420} & \frac{12E_2I_2}{L^3}+\frac{156Lc}{420} & \frac{6LE_2I_2}{L^3}+\frac{22L^2c}{420} & -\frac{54Lc}{420} & \frac{13L^2c}{420} & -\frac{12E_2I_2}{L^3}+\frac{54Lc}{420} & \frac{6LE_2I_2}{L^3}-\frac{13L^2c}{420} \\ -\frac{22L^2c}{420} & -\frac{4L^3c}{420} & \frac{6LE_2I_2}{L^3}+\frac{22L^2c}{420} & \frac{4L^2E_1I_1}{L^3}+\frac{4L^3c}{420} & -\frac{13L^2c}{420} & \frac{3L^3c}{420} & -\frac{6LE_2I_2}{L^3}+\frac{13L^2c}{420} & \frac{2L^2E_2I_2}{L^3}-\frac{3L^3c}{420} \\ -\frac{12E_1I_1}{L^3}+\frac{54Lc}{420} & -\frac{6LE_1I_1}{L^3}+\frac{13L^2c}{420} & -\frac{54Lc}{420} & -\frac{13L^2c}{420} & \frac{12E_1I_1}{L^3}+\frac{156Lc}{420} & -\frac{6LE_2I_2}{L^3}-\frac{22L^2c}{420} & -\frac{156Lc}{420} & \frac{22L^2c}{420} \\ \frac{6LE_1I_1}{L^3}-\frac{13L^2c}{420} & \frac{2L^2E_1I_1}{L^3}-\frac{3L^3c}{420} & \frac{13L^2c}{420} & \frac{3L^3c}{420} & -\frac{6LE_1I_1}{L^3}-\frac{22L^2c}{420} & \frac{4L^2E_1I_1}{L^3}+\frac{4L^3c}{420} & \frac{22L^2c}{420} & -\frac{4L^3c}{420} \\ -\frac{54Lc}{420} & -\frac{13L^2c}{420} & -\frac{12E_2I_2}{L^3}+\frac{54Lc}{420} & -\frac{6LE_2I_2}{L^3}+\frac{13L^2c}{420} & -\frac{156Lc}{420} & \frac{22L^2c}{420} & \frac{12E_2I_2}{L^3}+\frac{156Lc}{420} & -\frac{6LE_2I_2}{L^3}-\frac{22L^2c}{420} \\ \frac{13L^2c}{420} & \frac{3L^3c}{420} & \frac{6LE_2I_2}{L^3}-\frac{13L^2c}{420} & \frac{2L^2E_2I_2}{L^3}-\frac{3L^3c}{420} & \frac{22L^2c}{420} & -\frac{4L^3c}{420} & -\frac{6LE_2I_2}{L^3}-\frac{22L^2c}{420} & \frac{4L^2E_1I_1}{L^3}+\frac{4L^3c}{420} \end{bmatrix}$$

$$\det\left(K_{global} - \omega^2 M_{global}\right) = 0 \quad (11)$$

with K_{global} and M_{global} being the assembled matrices. Element mass and stiffness matrices for TWCNTs are not presented for the sake of brevity.

The linear eigenvalue problem is solved, for natural frequencies taking:

$L = 14e - 9,$

$\rho_1 = \rho_2 = 2.3\frac{g}{cm^3},$

$E_1 = E_2 = 1\,TPa,$

$r_{1,i}(inner\ radius, inner\ tube) = 0.18nm,$

$r_{1,o}(outer\ radius, inner\ tube) = 0.52nm,$

$r_{2,i}(inner\ radius, outer\ tube) = 0.53nm,$

$r_{2,o}(outer\ radius, outer\ tube) = 0.87nm,$

$c = 71.11GPa,$ following Xu *et al.* [3].

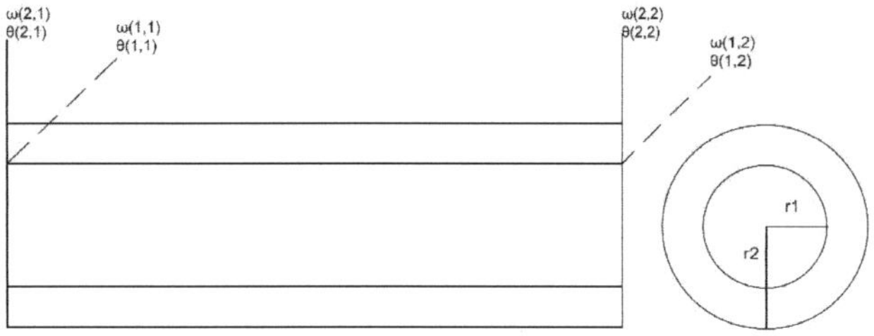

Figure 1 – Schematic Diagram of a DWCNT *(i,j)=(CNT,end)*

RESULTS

The natural frequencies for DWCNTs and TWCNTs obtained for different boundary conditions, are tabulated below. Finite element analysis on 10 elements, present below, is in agreement with the date available in literature.

Boundary Conditions		Xu *et al* [3] *(10^{12} rads/s)*	Elishakoff *et al* [1] *(10^{12} rads/s)*	Present Study *(10^{12} rads/s)*
Simply Supported Condition	ω_1	0.46	0.4673	0.467
	ω_2	7.71	7.8852	7.885
Clamped Clamped Condition	ω_1	1.06	1.049	1.057
	ω_2	7.75	7.924	7.925
Clamped - Free Condition	ω_1	0.17	0.167	0.166
	ω_2	7.7	7.87	7.876

Table 1: Natural Frequencies of DWCNTs

Boundary Conditions		Elishakoff *et al* [1] *(10^{12} rads/s)*	Present Study *(10^{12} rads/s)*
Simply Supported Condition	ω_1	0.648	0.6479
	ω_2	5.953	5.953
	ω_3	9.642	9.642
Clamped Clamped Condition	ω_1	1.448	1.459
	ω_2	6.046	6.048
	ω_3	9.706	9.707
Clamped - Free Condition	ω_1	0.223	0.2311
	ω_2	5.933	5.934
	ω_3	9.628	9.628

Table 2: Natural Frequencies of TWCNTs

CONCLUSION

Finite element modelling of double walled carbon nanotubes and triple walled carbon nanotubes is presented by taking Galerkin's method of weighted residual. The corresponding linear eigenvalue problem is solved to get the natural frequencies for clamped-clamped, clamped-free and simply supported boundary conditions, all found to be in agreement with the values present in the literature.

REFERENCES

[1] I. Elishakoff et al, Carbon Nanotubes and Nanosensors - Vibrations, Buckling, and Ballistic Impact, London: ISTE Ltd, 2012.

[2] C. Q. Ru, "Elastic Models for Carbon Nanotubes," in *Encyclopedia of Nanoscience and Nanotechnology*, H. S. Nalwa, Ed., American Scientific Publishers, 2004, pp. 731-744.

[3] K. Y. Xu et al, "Vibration of double- walled carbon nanotube aroused by nonlinear intertube van der Waals forces," *Journal of Aplied Physics,* vol. 99, pp. 064303-1 to 064303-7, 2006.

Proceedings of the 25th CANCAM
London, Ontario, Canada, May 31 - June 4, 2015

FRICTION BEHAVIOUR OF HEXAGONAL-SHAPED GRAPHENE ISLANDS GROWN ON COPPER BY CHEMICAL VAPOUR DEPOSITION (CVD)

Philip Egberts
Department of Mechanical and Manufacturing Engineering
Univeristy of Calgary
Calgary, Alberta, Canada
philip.egberts@ucalgary.ca

ABSTRACT

Single asperity friction experiments on graphene, as well as other atomically thin films, have shown that single layers of atoms can dramatically reduce friction when compared to the substrate material. In particular, graphene has been shown to be a remarkable (dry) lubricant on a number of surfaces including copper, nickel, silicon, and mica. However, because few layer graphene (FLG) films have atomic dimensions in one direction their mechanical properties are not isotropic. In this study, we examine high-quality chemical vapour deposited (CVD) graphene on polycrystalline surfaces using silicon atomic force microscopy (AFM) tips. Precise control over the deposition conditions allowed for growth of isolated, single-domain graphene islands that contain a very low defect density and are approximately 5-40 μm across. Compared to the surrounding copper surface, friction is reduced by a factor ranging from 1.5 to 7 over loads from the adhesive minimum up to 80 nN. Friction measurements also show a layer-dependence on friction, resulting from the formation of pucker around the sliding tip. Furthermore, substantial hysteresis in the load dependence of friction is observed when scanning on an individual graphene island while contact between the AFM tip and graphene film is maintained throughout the experiment. The observed hysteresis depends on sliding time and maximum applied normal force, indicating that there is a relatively weak adhesive interaction between graphene and the copper foil and that friction depends on the sliding history of the graphene-tip contact.

Keywords: atomic force microscope (AFM), graphene, copper, nanotribology, chemical vapor deposition (CVD)

ACKNOWLEDGMENTS

This work was supported by the National Science Foundation (NSF) under Awards NSF/MRSEC (No. DMR11-20901) & NSF/ENG (CMMI-1068741). Use of University of Pennsylvania Nano/Bio Interface Center instrumentation is acknowledged, which is funded through NSF Grants No. NSEC DMR08-32802, DBI-0721913, and DMR-0425780. The optical micrograph and Raman spectroscopy data in the Supporting Information were collected by Sung Ju Hong, Prof. Yung Woo Park research group at Seoul National University. P. Egberts would like to acknowledge financial support from the Natural Sciences and Engineering Council (NSERC) of Canada. Also, the support in creating the analysis scripts from Joel Lefever and James Hilbert as well as insightful discussions with Dr. Graham Wabiszewski are acknowledged. G. H. Han and A. T. C. Johnson acknowledge support from the Nano/Bio Interface Center, NSF Grant No. NSEC DMR08-32802.

REFERENCES

[1] Egberts, P., Han, G.H., Liu, X.Z., Johnson, A.T.C, and Carpick, R.W. 2014, "Frictional behavior of atomically thin fheets: Hexagonal-shaped graphene islands grown on copper by chemical vapor deposition", *ACS Nano*, 8(5), pp. 5010-5021.

Non-Newtonian Fluids

Proceedings of the 25th CANCAM
London, Ontario, Canada, May 31 - June 4, 2015

DISPLACEMENT FLOWS IN SLIGHTLY CONVERGING CHANNELS

R. Mollaabbaasi
Department of Chemical Engineering
Laval University
Quebec, QC, Canada
roozbeh.mollaabbasi.1@ulaval.ca

K. Alba
Department of Engineering Technology
University of Houston
Houston, TX, USA
kalba@uh.edu

S. M. Taghavi
Department of Chemical Engineering
Laval University
Quebec, QC, Canada
Seyed-Mohammad.Taghavi@gch.ulaval.ca

ABSTRACT

We consider displacement flows in non-uniform plane 2D channel geometry (i.e., a slightly converging channel in this paper) while the two fluids are miscible and buoyancy is significant. We assume that the channel is oriented close to horizontal. We employ a classical lubrication/thin-film approximation to simplify the governing equations to finally furnish a semi-analytical solution for the flux functions. We then numerically demonstrate how the interface between the heavy and the light fluids propagates in time. It is shown that the non-uniformity of the channel geometry can significantly change the shape of the interface and its front speed.

Keywords: Lubrication model, non-uniform geometry, buoyancy, viscous flows, interface

INTRODUCTION

In this paper we consider miscible heavy-light displacement flows along a non-uniform plane 2D channel of width $\hat{D}_0 + \tan\alpha\hat{x}$ (considering negative gradient $\alpha < 0$ & and also assuming $|\alpha| \ll 1$) in the large Péclet number limit, $Pe \gg 1$, (where $Pe = \hat{D}_0\hat{U}_0/\hat{D}_m$, with $\hat{U}_0$ being the mean displacement velocity and $\hat{D}_m$ the molecular diffusivity). This is a limit where, without a flow instability, the two fluids do not have enough time to mix on timescales of experimental interest. In this case, a reasonable approximation is to model the two fluids via a kinematic equation instead of a concentration-diffusion-equation (CDE) approach.

It has been demonstrated that the high Pe regime inevitably approaches the zero surface tension immiscible limit, ($Pe \to \infty$), if the flow remains stable [1]. In the limit of large Péclet number displacement flows, we consider the flows with the following characteristics: (1) buoyancy is a significant force in driving the fluid motion; (2) the flow occurs in a channel that is approximately horizontal; and (3) perhaps more interestingly, the channel is slightly converging (i.e., the channel upper wall is slightly angled with respect to the lower wall). We additionally assume that for sufficiently small density differences there exists a viscous regime in which inertial effects are relatively small so that the viscous regime persists for at least small Reynolds numbers, Re $(= \hat{\rho}\hat{U}_0\hat{D}_0/\hat{\mu})$, where $\hat{\rho}$ is the fluid density and $\hat{\mu}$ is the fluid viscosity.

Our work is one of the first efforts to study the effect of non-uniformity on the displacement flow (at least for miscible fluids). Our motivation comes from recent interesting studies [2,3] which have considered immiscible Newtonian displacements in a *tapered* rectilinear Hele-Shaw cell (HSC) by introducing a small depth gradient. By crafting a slightly converging/diverging HSC, these studies have shown that the flow instabilities (e.g., *viscous fingering*) can be stabilized through a variable transverse curvature mechanism. Note that studying the flow in a HSC provides a comfortable framework to analyze porous media flow [4]. In this work, for a simple case of a 2D channel flow, we would like to analytically understand how the displacement flow will be modified if the channel upper wall has a small (converging) angle with respect to the lower wall.

TWO-FLUID DISPLACEMENT FLOWS IN A NEAR-HORIZONTAL SLOT

We consider a two-dimensional region between two parallel plates, separated by a distance $\hat{D}$, that are oriented at an angle $\beta \approx \pi/2$ with respect to the vertical. The two fluids have the same viscosity but different densities. The slot is initially filled with fluid 2; fluid 1 is injected at $\hat{x} \to -\infty$ with a mean velocity $\hat{U}_0$. Cartesian coordinates $(\hat{x}, \hat{y})$ are as shown in Fig. 1. To simplify the derivation of our analytical model, we concentrate on the case of Newtonian fluids but we will also present the results for both Newtonian and shear-thinning fluids. The reader may refer to [5] regarding the formulation of the non-Newtonian problem in uniform channel geometry. Although the fluids are miscible, we consider the large Péclet number limit in which no significant mixing occurs over the timescales of interest. The dimensionless equations of motion, valid within each fluid region $\Omega_k,\ k = 1, 2$, are:

$$\phi_k \mathrm{Re}\left[\frac{\partial u}{\partial t} + u\frac{\partial u}{\partial x} + v\frac{\partial u}{\partial y}\right] = -\frac{\partial p}{\partial x} + \frac{\partial}{\partial x}\tau_{k,xx} + \frac{\partial}{\partial y}\tau_{k,xy} + \phi_k\frac{\cos\beta}{St}, \quad (1)$$

$$\phi_k \mathrm{Re}\left[\frac{\partial v}{\partial t} + u\frac{\partial v}{\partial x} + v\frac{\partial v}{\partial y}\right] = -\frac{\partial p}{\partial y} + \frac{\partial}{\partial x}\tau_{k,yx} + \frac{\partial}{\partial y}\tau_{k,yy} - \phi_k\frac{\sin\beta}{St}, \quad (2)$$

$$\frac{\partial u}{\partial x} + \frac{\partial v}{\partial y} = 0. \quad (3)$$

Here u, v denote the streamwise and depthwise velocity components respectively, p is the pressure, and $\tau_{k,ij}$ represents the ij-th component of the deviatoric stress in fluid k. The parameter $\phi_1 = 1$, and the three dimensionless parameters appearing above are the density ratio ϕ_2, the Reynolds number, Re, and the Stokes number, St, defined as:

$$\phi_2 = \phi = \frac{\hat{\rho}_2}{\hat{\rho}_1}, \quad \mathrm{Re} = \frac{\hat{\rho}_1\hat{U}_0\hat{D}}{\hat{\mu}_1}, \quad St = \frac{\hat{\mu}_1\hat{U}_0}{\hat{\rho}_1\hat{g}\hat{D}^2}. \quad (4)$$

Here $\hat{\mu}_1$ is the viscosity of fluid 1 and $\hat{g}$ is the gravitational acceleration. In order to derive (1)-(3) we have scaled distances using $\hat{D}_0$, velocities with $\hat{U}_0$, time with $\hat{D}_0/\hat{U}_0$, and pressure and stresses with $\hat{\mu}_1\hat{U}_0/\hat{D}_0$.

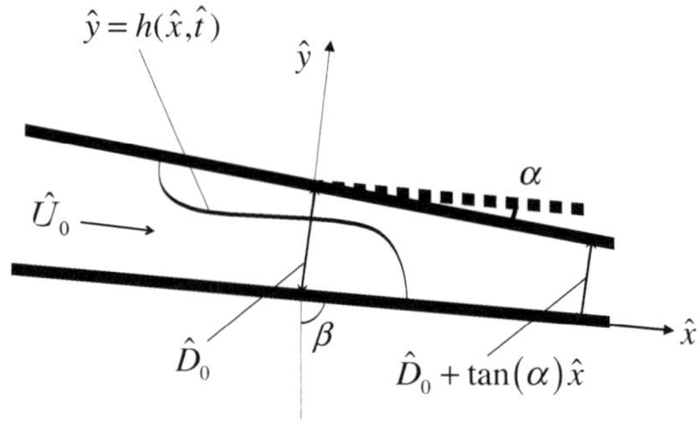

Figure 1: Schematic of displacement geometry. In this figure α is negative.

On the walls of the slot the no-slip condition is satisfied. Due to the scaling adopted, we have $\int_0^1 u\ dy = 1$ in each cross-section. The slot is assumed to be infinite in x, with the interface between fluids initially localised close to $x = 0$. We shall consider flows with buoyancy, in which the heavier fluid lies at the bottom of the slot. The fluid layers are separated by an interface ($y = h(x,t)$) which is assumed to be single-valued at each x. Across the interface, velocity and stress are continuous. The interface is simply advected with the flow, satisfying the kinematic condition.

Based on the small parameter of the problem, δ, which is defined as $\delta = \frac{\hat{D}_0}{\hat{L}}$ (with $\hat{L}$ being a characteristic length scale), we re-scale as follows

$$\delta x = \xi, \qquad \delta t = T, \qquad \delta p = P, \qquad v = \delta V,$$

and arrive at the following reduced system of equations, in each fluid region $\Omega_k,\ k = 1, 2$:

$$\delta\phi_k Re\left[\frac{\partial u}{\partial T} + u\frac{\partial u}{\partial \xi} + V\frac{\partial u}{\partial y}\right] = -\frac{\partial P}{\partial \xi} + \frac{\partial}{\partial y}\tau_{k,\xi y} + \phi_k\frac{\cos\beta}{St} + O(\delta^2),$$

$$\delta^3\phi_k Re\left[\frac{\partial V}{\partial T} + u\frac{\partial V}{\partial \xi} + V\frac{\partial V}{\partial y}\right] = -\frac{\partial P}{\partial y} - \delta\phi_k\frac{\sin\beta}{St} + O(\delta^2),$$

$$\frac{\partial u}{\partial \xi} + \frac{\partial V}{\partial y} = 0.$$

To aid interpretation of our model results, note that the time and length variables, (T, ξ), are related to the dimensional time and length by:

$$\frac{|\hat{\rho}_1 - \hat{\rho}_2|\hat{g}\sin\beta\hat{D}_0^3}{\hat{\mu}_1\hat{U}_0}\xi = \hat{x}, \qquad \frac{|\hat{\rho}_1 - \hat{\rho}_2|\hat{g}\sin\beta\hat{D}_0^3}{\hat{\mu}_1\hat{U}_0^2}T = \hat{t} \quad (5)$$

As mentioned before, we have used $\hat{D}_0/\hat{U}_0$ to scale $\hat{t}$, which is the usual convective timescale based on the mean velocity and $\hat{D}_0$. Therefore, the scale related to the *slow time* variable,

T, corresponds to the time taken to travel the characteristic spreading length $\hat{L}$ at mean velocity $\hat{U}_0$.

We now consider the limit $\delta \to 0$ with Re fixed:

$$0 = -\frac{\partial P}{\partial \xi} + \frac{\partial}{\partial y}\tau_{k,\xi y} + \chi \frac{\phi_k}{|1-\phi|}, \quad (6)$$

$$0 = -\frac{\partial P}{\partial y} - \frac{\phi_k}{|1-\phi|}, \quad (7)$$

where $\chi = \cot \beta / \delta$. The parameter χ measures the relative importance of the slope of the channel to the slope of the interface, in driving buoyancy related motions. We wish to consider channels that are close to horizontal, where the slopes of both the channel and the interface may be of comparable importance. Thus, we assume χ to be $O(1)$, i.e., we consider inclinations $\beta = \pi/2 + O(\delta)$. For $\chi > 0$ the slope of the channel is "downhill", in the direction of the flow, and for $\chi < 0$ the flow is "uphill". Note that for larger χ the model does not necessarily break down, but effectively we have chosen the wrong scaling as the effect of the channel slope is dominant.

We integrate (7) across both fluid layers to obtain the pressure:

$$P(\xi, y, T) = P_0(\xi, T) + \chi \frac{\phi_H}{|1-\phi|}\xi - \frac{\phi_H}{|1-\phi|} y \qquad y \in [0, h], \quad (8)$$

$$P(\xi, y, T) = P_0(\xi, T) + \chi \frac{\phi_H}{|1-\phi|}\xi - \frac{\phi_H - \phi_L}{|1-\phi|} h - \frac{\phi_L}{|1-\phi|} y \qquad y \in [h, 1+\alpha\xi], \quad (9)$$

where $P_0(\xi, T)$ is defined by:

$$P_0(\xi, T) = P(\xi, 0, T) - \chi \frac{\phi_H}{|1-\phi|}\xi,$$

with $\phi_H = \hat{\rho}_H / \hat{\rho}_1$ for the heavier fluid, $\phi_L = \hat{\rho}_L / \hat{\rho}_1$ for the lighter fluid. On substituting into (6), we arrive at:

$$0 = -\frac{\partial P_0}{\partial \xi} + \frac{\partial}{\partial y}\tau_{H,\xi y}, \quad y \in (0, h), \quad (10)$$

$$0 = -\frac{\partial P_0}{\partial \xi} + \frac{\partial}{\partial y}\tau_{L,\xi y} - \chi + \frac{\partial h}{\partial \xi}, \quad y \in (h, 1+\alpha\xi). \quad (11)$$

For now we assume that the solution of this problem may be computed and we note that the dependence of u on (ξ, T) enters only via $h(\xi, T)$, which satisfies

$$\frac{\partial h}{\partial T} + u\frac{\partial h}{\partial \xi} = V. \quad (12)$$

Combining the kinematic equation 12 with the divergence free constraint in the usual manner leads to the equation:

$$\frac{\partial h}{\partial T} + \frac{\partial}{\partial \xi} q(h, h_\xi) = 0, \quad (13)$$

where $q(h, h_\xi)$ is defined as:

$$q(h, h_\xi) = \int_0^h u(y, h, h_\xi)\, dy. \quad (14)$$

The remainder of our study concerns the behaviour of the solutions to the system (13) & (14).

As boundary conditions, we have that

$$h(\xi, T) \to 1+\alpha\xi, \text{ as } \xi \to -\infty; \quad h(\xi, T) \to 0, \text{ as } \xi \to \infty, \quad (15)$$

as the channel is assumed full of pure fluid 1 and fluid 2 at the two ends of the channel. As initial conditions we note that an initial profile in the unscaled variables $h(x, t=0) = h_0(x)$ is transformed to $h(\xi, T=0) = h_0(\xi/\delta)$. Since h_0 should be compatible with the far-field conditions we have that as $\delta \to 0$,

$$h(\xi, 0) \to 1 + \alpha\xi - H(\xi), \quad (16)$$

where $H(\xi)$ is the usual Heaviside function. In other words, in terms of ξ, the initial change in h is localised to $\xi = 0$.

The kinematic condition (13) can be discretized in the conservative form (second order in space and first order explicit in time) and afterwards integrated numerically by using a Van Leer flux limiter scheme. We will present the results of the numerical solution in the following section.

RESULTS

The results of the computed interface heights obtained from the numerical solution of equation (13) are given in this section. It was found that the long time behaviour of the interface is not sensitive to the initial condition of the problem, which we have taken as a linear function of ξ: typically $h(\xi, T=0) = -\xi + 0.5$.

Fig. 2 shows an example of the interface evolution with time for the case of $\alpha = 0$, i.e., a straight channel. This result confirms our pervious findings presented in [5]. We observe that the interface develops quickly into a slumping profile but after sufficiently long-time the shape of the interface remains self-similar. Over longer times, the solution generally has two segments: an advancing front moving at constant speed and a region at the top which is *stretched* with the top of the interface simply not moving.

Fig. 3 illustrates the interface evolution with time for the case of $\alpha = -0.01$, i.e., a slightly converging channel. Comparing this figure with Fig. 2, one observes that shape of the interface has significantly changed. The interface height, which seems to be constant at longer times in Fig. 2, decreases as the interface evolves. It is very interesting to note that the front seems to be progressively faster as it advances in a narrower channel made by the upper and lower walls. The frontal region also seems to accelerate and the front velocity is higher in this figure than that in Fig. 2. Note that we always limit the length of the channel so that the upper wall never contacts the lower wall. This is required to conform to the exit condition. For example for $\alpha = -0.01$, we limit the length so that $\xi < 99$.

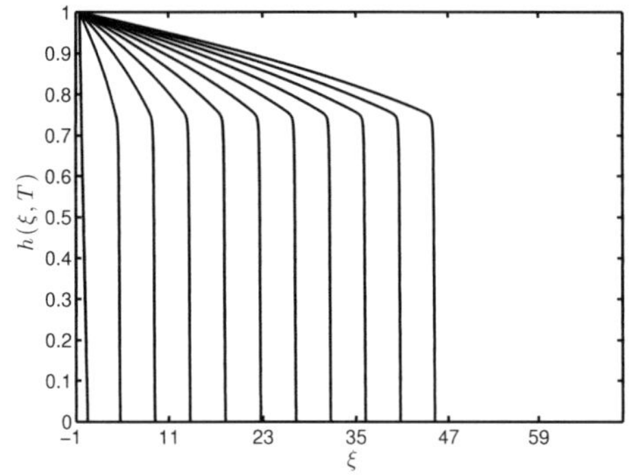

Figure 2: Example of the interface evolution in Newtonian displacements flow with $\alpha = 0$, $\chi = 0$ for $T = 0, 4, 8, ..., 32, 36, 40$.

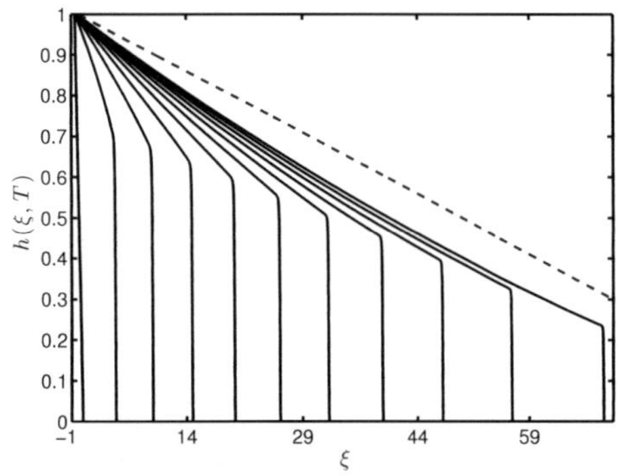

Figure 3: Example of the interface evolution for the same parameters as in Fig. 2 except $\alpha = -0.01$. The dashed line shows the position of upper wall.

Fig. 4 shows the interface evolution with time for the case of $\alpha = -0.01$, i.e., a slightly converging channel, for a non-Newtonian shear-thinning fluid. The derivation of the semi-analytical model for the non-Newtonian fluids has not been shown in this paper. However, the formulation for the similar problem of uniform channel geometry is available in [5], where the shear-thinning effects can be understood through a power-law index n. Based on Fig. 4, the non-uniformity of the geometry seems to have the same qualitative effect on the shear-thinning displacement as observed for the Newtonian displacement flows. The only difference is that in the case of a shear-thinning displacing fluid, the displacement front does not reach the upper wall of the channel as opposed to the Newtonian case.

CONCLUSIONS

A displacement flow of heavy fluid pushing a light fluid in a slightly converging plane 2D channel geometry was studied.

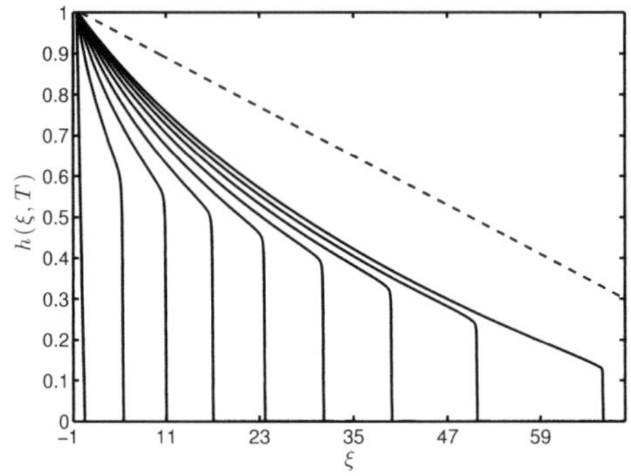

Figure 4: Example of the interface evolution for the same parameters as in Fig. 3 except that the displacing fluid is replaced by a shear-thinning one with power-law index $n = 0.25$. The dashed line shows the position of upper wall. Note that in this figure $T = 0, 4, 8, ..., 24, 28, 32$.

We assumed that the channel was oriented close to horizontal. The two fluids considered were miscible but could not mix over the time-scale of our interest due to the high Péclet number assumption. We have relied on a classical lubrication/thin-film approximation to simplify the governing equations and find a semi-analytical solution to the problem. The numerical simulation results show that the flow in the converging channel geometry can be significantly different than that in the straight channel. In particular, the shape of the interface and the frontal region speed are modified. This can have a great influence on studying displacement flows since most of the previous works in this area have considered uniform geometries. However, the geometries where these flows frequently occur usually have at least a degree of non-uniformity. Our results show that even a very small degree of non-uniformity can modify the displacement flow considerably.

REFERENCES

[1] Petitjeans, P., and Maxworthy, T. 1996 "Miscible displacements in capillary tubes. Part 1. Experiments," *J. Fluid Mech.*, Vol(326), pp. 37-56.

[2] Al-Housseiny, T.T., Tsai, P.A., and Stone, H.A. 2012, "Control of interfacial instabilities using flow geometry," *Nature Phys.*, Vol(8), pp. 747-750.

[3] Dias, E.O., and Miranda, J.A. 2013, "Taper-induced control of viscous fingering in variable-gap Hele-Shaw flows," *Phys. Rev. E*, Vol(87), pp. 053015.

[4] Homsy, G.M. 1987 "Viscous fingering in porous media," *Annu. Rev. Fluid Mech.*, Vol(19), pp. 271-311.

[5] Taghavi, S.M., Seon, T., Martinez, D.M., and Frigaard, I.A. 2009 "Buoyant-dominated slumping displacement flows in horizontal channels: the viscous limit," *J. Fluid Mech.*, Vol(639), pp. 1-35.

Proceedings of the 25th CANCAM
London, Ontario, Canada, May 31 – June 4, 2015

A MICROFLUIDIC METHOD FOR OPTICAL STUDIES OF EARLY STAGE BIOFILM RESPONSE TO APPLIED SHEAR STRESSES

M. Parvinzadeh Gashti
Département de Chimie
Université Laval
Québec, Québec, Canada
mazeyar.parvinzadeh-gashti.1@ulaval.ca

J. Greener
Département de Chimie
Université Laval
Québec, Québec, Canada
jesse.greener.@chm.ulaval.ca

ABSTRACT

Shear stresses are applied against growing biofilms in microchannels by a flowing nutrient solution. At certain times, the biofilms respond by moving in the direction of the flow. The displacement velocity of visibly identifiable biofilm segments and their changing optical density was monitored with optical microscopy and quantitatively analyzed by a semi-automated particle tracking routine. The method is used to reveal subtle differences in the response of cultivated biofilms based on their growth times and location in the channel. Following the demonstration of the approach, we discuss the potential of a system that employs highly-paralyzed microchannels and automated data acquisition for high-throughput measurements of biofilm displacement in different hydrodynamic and chemical environments.

KEYWORDS Biofilms, viscoelastic, non-Newtonian, shear stress, optical microscopy, video analysis, particle tracking.

INTRODUCTION

Biofilms are natural viscoelastic substances consisting of bacteria and a surrounding extra-cellular polymeric substance (EPS). It is a broad class of biomaterials that is well-represented in nature, ranging from slimy coatings on rocks in rivers to oral-dwelling variants. They are believed to represent the first multi-cellular life-form on Earth, and continue to dominate most ecosystems. Biofilms are also interesting from the perspective of new bio-technical applications, because they are in essence natural, self-repairing catalytic materials, which operate at ambient conditions. Some technological applications include biodegradation of toxic materials and biocatalysis of a range of chemical reactions.[1] One of our interests is in their ability to catalyse redox reactions, thereby opening the way for completely natural biofuel cells.[2]

In order to integrate biofilms into system-level bio-tech devices and have them faithfully accomplish a desired function, more must be done to understand and control their properties. Great advances have been made in understanding the molecular biology of biofilms in the last 5-10 years, yet there remains a strong need for quantitative physical characterisation of biofilms. However, biofilms are highly complex, heterogeneous materials, with properties that change in space and time. Therefore, they present a unique challenge to analytical scientists. In particular, biofilms are known to be extremely responsive to their hydrodynamic environment. Most biofilms require at least some tangential flow against their surface to facilitate transport of reagents and by-products through the biofilm-liquid interface. One of the most important roles of a biofilm is its ability to resist detachment from a surface due to resulting shear forces, which is driven by the mechanical properties of the EPS. Thus, biofilms in a flow-through bio-reactor should be sufficiently strong to resist wash-out under moderately high flow conditions. In addition, the mechanical properties of the biofilm serve as indicators of other biochemical and biophysical properties such as cross-linking density and porosity, respectively.

A previous study applied known shear stresses against several different mature biofilms to measure their compliance (measured strain/applied shear stress). It was observed that biofilm properties transitioned from elastic to viscous after a certain time of exposure to shear stresses above a critical threshold.[3]

In this proof-of principle study, we demonstrate a microfluidic platform to study the response of very young biofilms to well-defined shear stresses. This is accomplished by flowing the nutrient solution through a microfluidic growth channel during the first 70-150 hours of cultivation of a biofilm. The displacement of identifiable biofilm segments was monitored using an optical microscope to create high-resolution videos followed by a semi-automated particle tracking analysis to quantify the biofilm displacement in time and space. Furthermore, by integrating an automatic microscope stage, we were able to rapidly collect multiple time-series micrographs in parallel to simultaneously probe biofilm mechanical response to a range of hydrodynamic and chemical cultivation conditions.

MATERIALS

Photoresists (SU8 3000, Microchem, USA), sealing glass (Fisher Scientific, USA), tubing (Hamilton, Canada), 30 and 60 mL polypropylene syringes (BD, USA), tryptone (Sigma-Aldrich, USA), yeast extract (Sigma-Aldrich, USA), sodium chloride (Caledon Laboratories Ltd, Canada). Sterile distilled water for making media, *Pseudomonas sp.* as the bacteria used (gifted by G. Wolfhardt, Ryerson University, Canada).

APPARATUS

Acquisition of optical micrographs was accomplished with a microscope (IX73, Olympus, Canada) with 2x objective in transmission mode with an automatic stage (MS2, Applied Scientific Instruments MS2, USA). Both image acquisition and stage positioning were controlled by a customized macro within the imaging software environment (Image ProPlus 7.0, Media Cybernetics, USA). Image analysis was conducted using Fiji bundle for ImageJ and the particle tracking algorithm MTrackJ.[4] Optical density (OD) measurements were acquired from pixel intensities after a calibration process using target with known OD values. Numerical simulations were conducted using COMSOL multiphysics (v4.2) using physics for "Laminar Flow" and "Transport of Dilute Species" and free tetrahedral meshing with "fine" resolution.

A multi-channel microfluidic flowcell was fabricated by casting polydimethyl siloxaine (PDMS) against a template mould, which consisted of micro pattered photoresist features (SU8-3000) on a silicon wafer support substrate (FlowJEM, Canada). The resulting patterned PDMS was then sealed by a microscope slide after plasma activation (PCD-001 Harrick Plasma, Ithaca, NY, USA). The height, width and length of the microchannel were 400μm, 3000μm and 25mm, respectively. Liquid was injected a six-syringe pump (NE-1600, New Era, USA). Figure 1 shows an image of the 6 channel microfluidic device and peripheral setup.

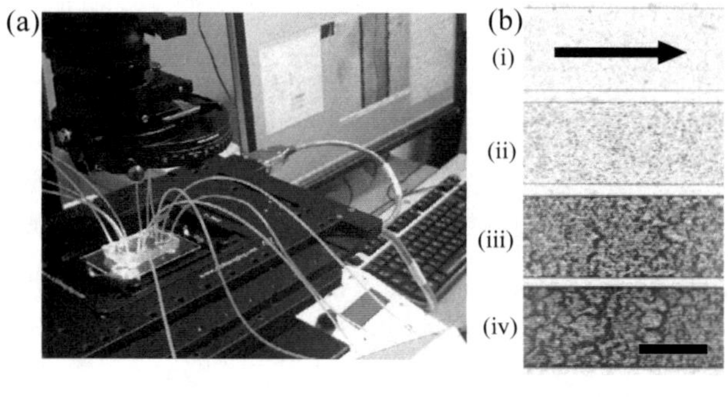

Figure 1(a) A six-channel microfluidic flow cell connected to syringe pump, is mounted on an automatic microscope stage. (b) Images of growing biofilm acquired at t=4h, 22h,34h, 52h (i-iv, respectively) after inoculation. Arrow represents the direction of flow of nutrient solution. Scale bar is 1500μm.

INOCULATION PROCESS

Colonies of *Pseudomonas sp.* were cultured at 4°C for 3 days in an LB nutrient Petri dish. Pre-cultured bacteria were used as inoculum for bacteria seeding. The bacteria colonies were scraped from the surface of LB plate and transferred to a tube containing 3mL of a modified LB medium, which consisted of 0.1% tryptone, 0.05% yeast extract and 0.1% sodium chloride. The tube was incubated for 18h on an orbital shaker at 30°C, at 300rpm. The tube connectors and 6 parallel microfluidic channels were disinfected by flowing 70% ethanol for 2h and then sterile distilled water (autoclaved) for 1h with flow rate of $2mL \cdot h^{-1}$. Inoculation was achieved by flowing bacteria pre-culture into the channels for 30min at $0.2mL \cdot h^{-1}$. After inoculation, the channels were exposed to a fresh LB nutrient with a known flow rate. It was this flowing solution that was responsible for applying a shear stress against the growing biofilm.

RESULTS AND DISCUSSION

Time-lapse videos: After inoculation the flowing nutrient solution produced a shear stress against all walls of the microchannel. The effect on growing biofilms was (a) the displacement of biofilm segments in time and (b) an increase in their optical density (OD), both of which were captured by time-lapse videos. Depending on the experimental setup, images were acquired every 30 mins or 60 mins for 65 hours to create the time lapse videos. Each 8-bit, 5 megapixel image was acquired using same parameters (exposure time, illumination intensity, gain). Figure 2 shows an example of four sequential frames from a data video between times 29h and 32h after inoculation. Due to the low-magnification used in these studies, the biofilm could only be observed after the aggregation of sufficient quantities of biomaterial resulted in segments with relatively high OD. Two such segments are highlighted in Fig. 2 using colour-coded arrows. Software tracking of these and other biofilm segments enabled us to monitor the evolution of their position and OD in time.

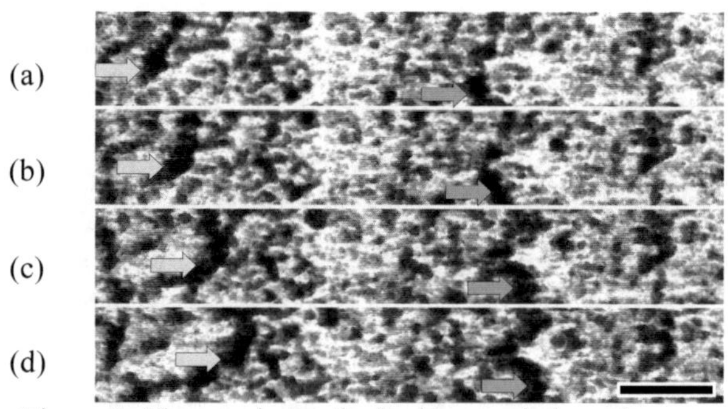

Figure 2: Time-series optical micrograph from the microchannel center at (a) 29h, (b) 30h, (c) 31h and (d) 32h after inoculation. Yellow and green arrows highlight the displacement of two optically dense biofilm segments in time. Scale bar length is 700μm.

Estimating height of tracked biofilm segments: As will be discussed shortly, it is important to know the height of the biofilm segments in order to understand the local hydrodynamic properties near them. The use of a confocal imaging instrumentation is one way to directly measure the three-dimensional structure of a biofilm. However, as this is expensive, specialized equipment we present instead a method that is compatible with regular two-dimensional images generated from optical microscopes. In this approach, we take advantage of the optical density (OD) of a biofilm, which is known to be directly proportional to the total biomass though which illuminating light must pass.[5] Assuming a constant biofilm density, OD values can be used as accurate estimates for the local biofilm height (h) given the right calibration procedure. We calibrated h from OD measurements using a two-point calibration method. First, OD was measured at the beginning of the experiment before any biofilm had accumulated (h_{min}=0µm). Secondly, OD values were measured in the channel corners (beside the vertical side-wall) at the end of the experiment, when we were sure that biofilm has formed a continuous sheet from the bottom of the microchannel to the top with a maximum height of h_{max}=400µm. In the present example, a linear calibration curve was generated from these two measurements, represented by

$$h = 160.84 \cdot OD - 42.30 \text{ (µm)} \quad (1)$$

Figure 3 shows the average OD for all tracked biofilm segments to generate Fig. 3 ($\widetilde{OD}_{track}$) and the average changes to the OD of the entire channel ($\widetilde{OD}_{chan}$). Comparison of the two curves reveals that at the end of the time-series measurements the average biofilm h=200µm, whereas the tracked biofilm segments were approximately h=280µm. This is because we tracked biofilm segments with larger OD, due to practical advantages of following them throughout the time series micrographs. Measurements of $\widetilde{OD}_{chan}$ had a relatively high standard deviation (shown by error bars in Fig. 3) compared to $\widetilde{OD}_{track}$ (not shown), this is because $\widetilde{OD}_{chan}$ encompasses a wider range of biofilm environments. The $\widetilde{OD}_{track}$ was approximately one standard deviation greater than the channel-wide measurements.

Effect of biofilm height on local shear stresses: It is important to account for the effect that the biofilm itself has on the local hydrodynamic environment due to the effect of constriction of channel free space on flow velocity. This effect has been shown to be very strong in microfluidic channels because growing biofilms can easily restrict the already limited channel space, which under a constant volumetric flow rate, can in turn cause rapid increases to shear stress applied to the biofilm.[6] We used knowledge of h, from the previous section, to estimate the local constriction in the free space for the liquid media to flow above individual tracked biofilm segments. The effect of this constriction in the headspace above tracked biofilm segments was accomplished using COMSOL, where the volumetric flow rate of a portion of the nutrient solution flowing above a particular track is constricted into a headspace that is equal to: *headspace*(µm)= 400-h. The result of this simple approach predicts strong, near exponential increases in shear stress when h is large. The differences in shear stresses due to this effect was 2.3mPa at t=0h and 6.4mPa at t=65h, approximately a three-fold increase.

Biofilm displacement and velocity: The displacement of tracked biofilm segments and their changing OD was measured by particle tracking. Note, we refer to an individual tracked biofilm segment as a "track" from this point forward. In the early times after inoculation, growth had a tendency to expand the clusters in all directions. We measured the expansion in both the downstream and upstream directions by tracking the displacement of the clusters' leading edges, d_{down} and d_{up}, respectively, during this period. The net displacement downstream, d_{net}, was calculated

$$d_{net} = d_{down} - d_{up} \quad (7\text{h} \le t \le 26\text{h}) \quad (2a)$$

$$d_{net} = d_{down} \quad (t > 26\text{h}) \quad (2b)$$

This was repeated for each of the tracks and the average value of d_{net} was calculated. The average velocity of all measured tracks in each 1h measurement period was therefore calculated

$$\tilde{v}_{net} = \left(\sum_i \frac{d_{net,i}}{n} \right) / t \quad (3)$$

Where the summation on the right determines the average net displacement for n tracks. At its peak (t=20h) d_{up} (2.5µm) was approximately 50% of d_{down} (5µm). After t=26h, d_{up} was much smaller than d_{down} and did not significantly contribute to d_{net}. Therefore, we used Eq. (2b) to calculate d_{net} for t>26h.

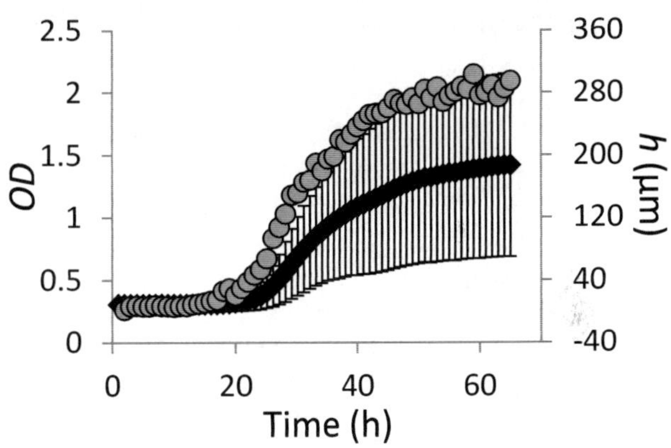

Figure 3: Results from measurements of average OD and corresponding average biofilm h as a function of time for entire channel area (black diamonds) and for the segments of the biofilm that were tracked (red circles). Error bars for the $\widetilde{OD}_{chan}$ were generated from +/- the standard deviation.

Wall effects: Flow profiles of the aqueous nutrient solution in microchannels differed depending on the position within the channel. Near to the vertical side-walls, flow velocity, velocity gradients and the resulting local shear stresses, were lower due to increased liquid/surface interactions. According to COMSOL flow simulations, the average shear stress resulting from the flowing nutrient solution (viscosity of 1 mPa·s) within a distance of 150µm from the microchannel wall was approximately a factor 2 less than those in the centre of the channel (Fig. 4). Therefore, we expected an observable difference in the track velocities near the channel vertical walls compared to its center portions. Figure 4 compares the $\tilde{v}_{net}$ values for tracks within 150μm of microchannel vertical walls to those in the center of the channel at different times following inoculation. A comparison of the results demonstrates three points. First, until t=31h, $\tilde{v}_{net}$ is approximately the same for all tracks. Secondly, after t=31h, $\tilde{v}_{net}$ of tracks in the center of the channel continues to increase until reaching a maximum at t=34h, while $\tilde{v}_{net}$ rapidly decreases for tracks near the vertical side-walls. Therefore, maximum $\tilde{v}_{net}$ is reached at earlier times near the side-walls (i.e., in the microchannel corners). Lastly, the maximum $\tilde{v}_{net}$ is lower near microchannel corners ($\tilde{v}_{net}$=44.3μm·h^{-1}) compared to the center portion ($\tilde{v}_{net}$=64.3 μm·h^{-1}).

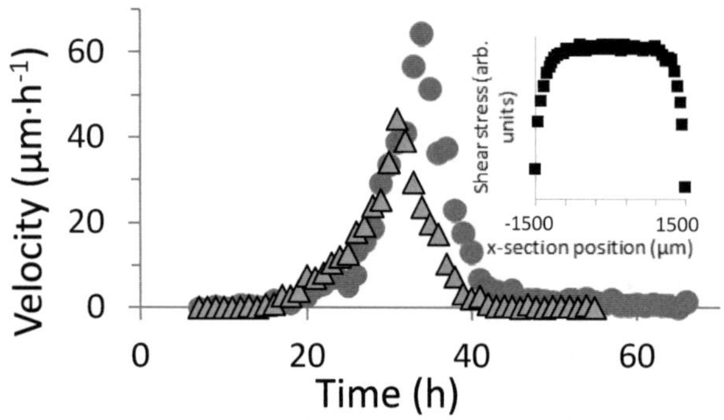

Figure 4: Time-dependent $\tilde{v}_{net}$ of tracks near the microchannel vertical side-walls (green triangles) and in the center (blue circles). LB medium had c_{tryp}=0.1% and Q_{nutr}=0.2mL·hr^{-1}. Each point represents the average from 15 different tracks. Inset shows profile for the change in applied shear stress for different positions along the cross-section.

Streamers: Other interesting details that were observed included the development of so-called "streamers", which are biofilm segments that have an upstream point of attachment, but are otherwise not connected to a surface. Since they are suspended in the channel interior, they are known to have a disproportionately high impact on the overall resistivity to flow. This type of formation has only ever been observed in microchannels which contain sharp bends in their flow path.[7]

PARALLEL STUDIES AND FUTURE STEPS

It is known that nutrient solutions with high concentrations of carbon source and low flow rates cause low density ("fluffy") biofilms. Therefore, preliminary results using automated parallel measurements included three nutrient solutions with different tryptone concentrations (c_{tryp}=0.02%, c_{tryp}=0.1% and c_{tryp}=0.18%), each flowing at Q_{nutr}=0.3 mL·h^{-1} and Q_{nutr}=0.6 mL·h^{-1}. In all cases, initial displacements started from small nucleation points on the horizontal channel wall, leading to small streaks down the length of the channel. In addition, all experiments at Q_{nutr}=0.3 mL·h^{-1} showed a rapid constriction of the channel starting from the vertical side walls, however, the time that this started was earlier and the rate of growth was faster as c_{tryp} was increased. Moreover, this effect was strongly reduced for experiments run at Q_{nutr}=0.6 mL·h^{-1}. In all cases, biofilms in the center portion of the channel did not stop moving as we observed in Fig.4. Therefore, it appears that above Q_{nutr}=0.2 mL·h^{-1}, the behaviour of the biofilm is fundamentally different. A full analysis of these results along with complementary experiments is underway for future publications. Other factors which are known to affect biofilm stability will also be studied. These include: (i) nutrient solution salinity (known to create more adherent biofilms), (ii) the presence of EPS destabilizers (eg, divalent cation chelators, cationic peptides and mucolytic agents, such as acetylcysteine) and (iii) constant vs. time-varying shear stress.

Future development plans include implementation of flowtemplated growth approach, pioneered by our group, which limits growth to one (horizontal) wall only.[8] This will ensure that all observed bio-density originates from one biofilm only. It also gives the ability to grow biofilms in center of the channel, thereby avoiding complex and variable hydrodynamic conditions in the channel corners.

CONCLUSION

This work opens the way for rigorous studies of young biofilms using video imaging and particle tracking to observe time-varying velocity and optical density of biofilm segments. Control of hydrodynamic and chemical conditions in microchannel complemented the platform's analytical capacity. Optical density measurements were used to estimate the height of the biofilms, thus enabling accurate estimations of the local shear stresses that were applied to the biofilms. We note that applied shear stress in this study were between 3 to 6 orders of magnitude lower than those required to cause mature biofilms to undergo viscous flow.[3] Preliminarily, we attribute this to age-related differences in mechanical properties.

REFERENCES

1. Halan, B., Buehler, K., Schmid, A., 2012, "Biofilms as living catalysts in continuous chemical syntheses", *Trends Biotechnol.*, 30(9), pp. 453–465.
2. Logan, B.E., Hamelers, et al., 2006, "Microbial Fuel Cells: Methodology and Technology.'' *Environ. Sci. Tech.*, 40(17), pp. 5181-5192.
3. Shaw, T., Winston, M., Rupp, C. J., Klapper, I., Stoodley, P., 2004, "Commonality of Elastic Relaxation Times in Biofilms", *Phys. Rev. Lett.*, 93(9), pp. 098102.
4. Schindelin, J., Arganda-Carreras, I., Frise, E., 2012, "Fiji: an open-source platform for biological-image analysis", *Nature Methods*, 9(7), pp. 676-682.
5. Bakke, R., Kommedal, R., Kalvenes, S., 2001, "Quantification of biofilm accumulation by an optical approach", *J. Microbiol. Methods*, 44(1), pp. 13-26.
6. Kim, J., Kim, H. S., Han, S., Lee, J. Y., Oh, J. E., Chung, S., Park, H. D., 2013, "Hydrodynamic effects on

bacterial biofilm development in a microfluidic environment", *Lab Chip*, 13(10), pp. 1846–1849.
7. Rusconi, R., Lecuyer, S., Stone, H. A., 2010, "Laminar flow around corners triggers the formation of biofilm streamers", *J R Soc Interface.*, 7(50), pp. 1293–1299.
8. Babaei, N., Safdar, M., Wolfaardt, G., Greener, J., 2014, "Micropatterned biofilm formations by laminar flow-templating", *Lab Chip*, 2014, 14(15), pp. 2666–2672

Proceedings of the 25th CANCAM
London, Ontario, Canada, May 31 - June 4, 2015

TWO-LAYER FLOW OF IMMISCIBLE NON-NEWTONIAN FLUIDS: LINEAR STABILITY ANALYSIS

K. Alba
Department of Engineering Technology
University of Houston
Houston, TX, USA
kalba@uh.edu

S.M. Taghavi
Department of Chemical Engineering
Laval University
Quebec, QC, Canada
Seyed-Mohammad.Taghavi@gch.ulaval.ca

ABSTRACT

Thin-film flow of two layers of non-Newtonian fluids are modeled in this paper. The model accounts for the weak inertial effects. The fluids are assumed to be *immiscible* i.e. an interfacial tension has been included in the momentum equations. Following a *weighted residual* approach, the momentum equations are reduced to a single equation relating the buoyancy parameter and interface height. The model is then closed by introducing the kinematic condition at the interface. A linear stability analysis is carried out for a two-layer parallel flow finding perfect agreement with the Orr-Sommerfeld results. The effect of fluids immiscibility (quantified through a Weber number) on the stability neutral curves is studied for a range of flow parameters. For the parameter range studied, it is generally found that the interfacial tension has a stabilizing effect on the flow. The model developed can be used for analyzing both displacement and co-extrusion flows.

Keywords: Thin-film flow, immiscible fluids, Herschel-Bulkley model

INTRODUCTION

Our motivation for this work comes from our recent and ongoing studies of density unstable buoyant miscible displacement flows along inclined pipes and channels [1, 2]. Whereas displacement flows in vertical pipes and channels may exhibit some degree of symmetry, once any significant inclination from vertical (β) is introduced, there is a strong tendency for the fluid layers to stratify, with the heavier fluid layer advancing along the lower side of the pipe or channel, i.e. as a gravity current superimposed on the mean flow. In [2], considering flows at small Atwood number ($At = (\hat{\rho}_H - \hat{\rho}_L)/(\hat{\rho}_H + \hat{\rho}_L)$) in duct inclinations that are close to horizontal, we have observed many flows for which viscous effects dominate in balancing buoyancy and for which the fluid streams remain laminar and separated over relatively long times. In the same studies, we have also observed flows in which inertia is significant leading to local instability and mixing at the interface. Taghavi *et. al.* [2] presents a comprehensive description of the parameter space of viscous and inertial displacement flows, for iso-viscous Newtonian fluids at high inclinations. The same overall description appears, at least qualitatively valid, if we consider fluids with modest viscosity ratios and shear-thinning effects, see [3].

The aim of our study is to develop tools appropriate for the identification and analysis of instability in the flows mentioned above and over a range of different fluids. In particular, we are interested in non-Newtonian fluids described by the Herschel-Bulkley fluid model. These fluids have a yield stress, below which the rate of strain is zero, and also exhibit shear-thinning behaviour. The common situation is the one in which the interface elongates during the displacement, due to both the mean flow and buoyancy. We are interested in exploiting the large aspect ratio of the flow to reduce the dimensionality of the problem. Two-layer thin film/lubrication models have been developed by [1, 2] and used to predict the front velocity in displacement flows for both pipes and channels. However, inertia is absent at leading order in such models and must be included to analyze the instability. More recently, we have studied two-layer thin-film flow of non-Newtonian fluids including weak-inertial terms assuming miscible fluids [4].

The methodology used is based on a *weighted residual* approach introduced in [5] for a single fluid. This methodology has been extended to two layer thin film flows in channels and pipes by [6, 7] and used to study channel co-extrusion flows by [8]. In [2], we have used the derivation of [6] in working with iso-viscous Newtonian fluid displacements. Here we first develop a general framework for two layer flows of non-Newtonian fluids within a plane channel assuming that the fluids are *immiscible*. The formulation used is very similar in essence to that in [4] except immiscible effects are included. In the second part, we consider the stability of these flows, first comparing against the long wavelength Orr-Sommerfeld results for parallel flows. Note that although our interest is in studying displacement flows, the formulation developed in this paper can be easily used for similar two-layer flows e.g. co-extrusion flows.

TWO-LAYER FLOW IN A 2D CHANNEL

The starting point for the model is a fully 2D plane channel immiscible displacement flow. In general, it will be assumed that the displacing fluid (fluid H) is heavier than the displaced (fluid L), so that the configuration is density unstable (see figure 1). The Navier-Stokes equations are made dimensionless using the channel height $\hat{D}$ as length-scale and the mean displacement velocity $\hat{V}_0$ as velocity scale. The model equations are:

$$[1 \pm At]\,[\mathbf{u}_t + \mathbf{u}\cdot\nabla\mathbf{u}] = -\nabla p + \frac{1}{Re}\nabla\cdot\tau \pm \frac{1}{Fr^2}\mathbf{e}_g, \quad (1)$$

$$\nabla\cdot\mathbf{u} = 0, \quad (2)$$

$$h_t + \mathbf{u}\cdot\nabla(h-y) = 0. \quad (3)$$

Here $\mathbf{e}_g = (\cos\beta, -\sin\beta)$ in directions (x,y) and the $\pm$ refers to the heavy/light fluid layers, respectively. The mean static pressure gradient has been subtracted from the pressure before scaling. The interface height is denoted by $y = h(x,t)$. For $t > 0$, no slip boundary conditions are satisfied at the solid walls and outflow conditions imposed at the channel exit. At the inflow, pure heavy fluid is pumped at mean velocity $\hat{V}_0$. Apart from the angle of inclination β and Atwood number, $At = (\hat{\rho}_H - \hat{\rho}_L)/(\hat{\rho}_H + \hat{\rho}_L)$, two other dimensionless parameters appearing in (1) are the Reynolds number, Re, and the (densimetric) Froude number, Fr:

$$Re \equiv \frac{\hat{V}_0\hat{D}}{\hat{\nu}}, \quad Fr \equiv \frac{\hat{V}_0}{\sqrt{At\hat{g}\hat{D}}}. \quad (4)$$

Here $\hat{\nu}$ is defined using the mean density $\hat{\rho} = (\hat{\rho}_H + \hat{\rho}_L)/2$ and a viscosity scale that will be derived from the rheological properties of the pure displacing fluid (fluid H). We see that for small At the flow is essentially governed by the 3 parameters β, Re & Fr. A dimensionless combination of these that arises naturally in thin layer flows is

$$\chi = \frac{2Re\cos\beta}{Fr^2}, \quad (5)$$

which represents the balance of axial buoyancy stresses and viscous stresses due to the imposed flow.

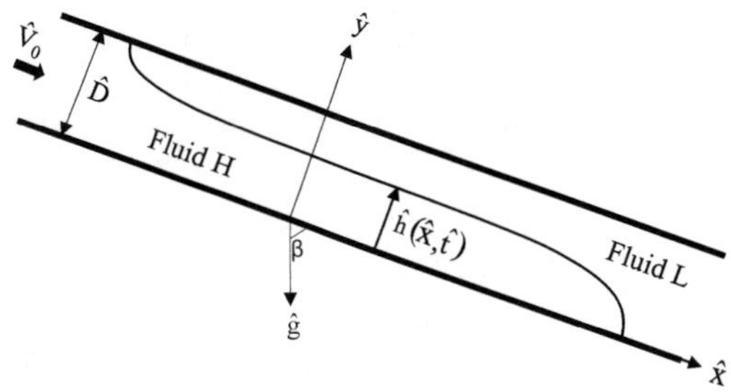

Figure 1: Schematic of the two-layer flow studied. Dimensional notations are used in the figure. The interface shape is illustrative only.

CONSTITUTIVE EQUATION

The fluids are assumed to be Herschel-Bulkley fluids, which incorporates also the simpler Bingham, power law and Newtonian models. Herschel-Bulkley fluids are described by three (dimensional) parameters namely a fluid consistency $\hat{\kappa}$, a yield stress $\hat{\tau}_Y$ and a power law index, n. Dimensionless constitutive laws for pure fluid k are:

$$\dot{\gamma}(\mathbf{u}) = 0 \Longleftrightarrow \tau_k(\mathbf{u}) \le B_k, \quad (6)$$

and

$$\tau_{k,ij}(\mathbf{u}) = \left[\kappa_k\dot{\gamma}^{n_k-1}(\mathbf{u}) + \frac{B_k}{\dot{\gamma}(\mathbf{u})}\right]\dot{\gamma}_{ij}(\mathbf{u}) \quad (7)$$

when $\tau_k(\mathbf{u}) > B_k$. Here, the strain rate tensor has components:

$$\dot{\gamma}_{ij}(\mathbf{u}) = \frac{\partial u_i}{\partial x_j} + \frac{\partial u_j}{\partial x_i}, \quad (8)$$

and the norms of these tensors, $\dot{\gamma}(\mathbf{u})$ and $\tau_k(\mathbf{u})$, are defined by:

$$\dot{\gamma}(\mathbf{u}) = \left[\frac{1}{2}\sum_{i,j=1}^{2}[\dot{\gamma}_{ij}(\mathbf{u})]^2\right]^{1/2},$$

$$\tau_k(\mathbf{u}) = \left[\frac{1}{2}\sum_{i,j=1}^{2}[\tau_{k,ij}(\mathbf{u})]^2\right]^{1/2}. \quad (9)$$

Here, subscripts $k = H, L$ are used to distinguish the fluids. The parameter $\kappa_H = 1$ and κ_L is the viscosity ratio m, equal to:

$$m \equiv \frac{\hat{\mu}_L}{\hat{\mu}_H} = \frac{\hat{\kappa}_L[\hat{V}_0/\hat{D}]^{n_L-1}}{\hat{\kappa}_H[\hat{V}_0/\hat{D}]^{n_H-1}}, \quad (10)$$

where $\hat{\mu}_L$ is a viscosity scale for fluid 2 (the less dense fluid). Note that in the case of 2 Newtonian fluids, $\hat{\mu}_k = \hat{\kappa}_k$. The Bingham numbers B_k are defined as:

$$B_k \equiv \frac{\hat{\tau}_{k,Y}}{\hat{\kappa}_H[\hat{V}_0/\hat{D}]^{n_H}}. \quad (11)$$

WEIGHTED RESIDUAL MODEL

The kinematic condition at the interface reads

$$\frac{\partial h}{\partial t}+U_0\frac{\partial h}{\partial x}+\frac{\partial h}{\partial x}\int_0^h\frac{\partial U_0}{\partial h}\,\mathrm{d}y+\frac{\partial b}{\partial x}\int_0^h\frac{\partial U_0}{\partial b}\,\mathrm{d}y=0, \quad (12)$$

where U_0 is leading order velocity defined by a system analogous to that of the parallel flow solution [4].

Following the same procedure as given in [4] for thin-film flows, one can arrive at the depth-averaged momentum equation given below

$$0=A^h(h,b)\frac{\partial h}{\partial t}+A^b(h,b)\frac{\partial b}{\partial t}+C^h(h,b)\frac{\partial h}{\partial x}+C^b(h,b)\frac{\partial b}{\partial x}$$
$$+\delta S(h,b)\left(\chi(1-\tan\beta\frac{\partial h}{\partial x})+W\frac{\partial^3 h}{\partial x^3}-b\right). \quad (13)$$

where $\delta \ll 1$ is the channel aspect ration. Note that the coefficients A^k & C^k are the same as those given in [4]. Here $W=\sigma/\hat{\rho}\hat{\nu}\hat{V}_0$ is the Weber number. In (13), b is a buoyancy parameter and h is the interface height; see Fig. 1. Basically the only difference with [4] is that in driving (13), a jump in hydrostatic pressure due to the interfascial tension has been included which reflects as $W\frac{\partial^3 h}{\partial x^3}$ term in (13).

STABILITY OF TWO-LAYER FLOWS

One of the main motivations for using the weighted residual approach is that it has proven effective in predicting long wavelength instabilities for weakly inertial thin film flows [5] and multi-layer flows [6, 4]. Here we explore whether this effectiveness carries over to multi-layer non-Newtonian fluid flows in the presence of an interfacial tension. The governing equations of the weighted residual method are (12) and (13), which model evolution of (h,b). To compare with a more classical Orr-Sommerfeld approach [4] we carry out a linear stability analysis of a uniform steady solution: $(h_0,b_0)=$ constant. A constant uniform solution of (12) and (13) is also a two-layer parallel solution of the full Navier-Stokes system, as has been described in [4]. Any $h_0\in[0,1]$ may be considered, but steady constant b_0 must satisfy:

$$b_0=\chi. \quad (14)$$

Recall that χ gives the ratio of viscous forces to buoyancy forces (along the channel).

As is usual, we suppose a linear perturbation:

$$h=h_0+h', \quad b=b_0+b', \quad (15)$$

and substitute (15) into (12) and (13), retaining only linear terms:

$$\frac{\partial h'}{\partial t}+\left(U_0+\int_0^{h_0}\frac{\partial U_0}{\partial h}\,\mathrm{d}y\right)\frac{\partial h'}{\partial x}+$$
$$\left(\int_0^{h_0}\frac{\partial U_0}{\partial b}\,\mathrm{d}y\right)\frac{\partial b'}{\partial x}=0, \quad (16)$$

$$A^h\frac{\partial h'}{\partial t}+A^b\frac{\partial b'}{\partial t}+C^h\frac{\partial h'}{\partial x}+C^b\frac{\partial b'}{\partial x}-$$
$$\delta S\chi\tan\beta\frac{\partial h'}{\partial x}+\delta SW\frac{\partial^3 h'}{\partial x^3}-\delta S b'=0, \quad (17)$$

Note that all algebraic functions above are evaluated at the steady $(h,b)=(h_0,b_0)$. We now assume a modal form for the linear perturbations, periodic in x, so that:

$$h'=\overline{h}e^{\mathrm{i}\alpha(x-ct)}, \quad b'=\overline{b}e^{\mathrm{i}\alpha(x-ct)}, \quad (18)$$

where $\overline{h}$ and $\overline{b}$ are constants. Substituting (18) into (16) and (17) leads to a dispersion relation that is quadratic in c:

$$\begin{vmatrix} \left(U_0+\int_0^{h_0}\frac{\partial U_0}{\partial h}\,\mathrm{d}y\right)-c & \left(\int_0^{h_0}\frac{\partial U_0}{\partial b}\,\mathrm{d}y\right) \\ \mathrm{i}\alpha(C^h-cA^h-\delta S\chi\tan\beta-\delta S\alpha^2 W) & \mathrm{i}\alpha(C^b-cA^b)-\delta S \end{vmatrix}=0 \quad (19)$$

If the imaginary part of c is positive, the flow is linearly unstable. Although we can find c for any wave number α, for the values we have tested the sign of the imaginary part of c is determined by the long wavelength limit, and in any case we wish to benchmark the solution against the long-wavelength Orr-Sommerfeld solution. The long wavelength analysis ($\alpha\to 0$) is made by expanding c, $\overline{h}$ and $\overline{b}$ with respect to $\alpha \ll 1$:

$$c=c_0+\alpha c_1+..., \quad \overline{h}=\overline{h}_0+\alpha\overline{h}_1+..., \quad \overline{b}=\overline{b}_0+\alpha\overline{b}_1+...$$

This gives at leading order $\overline{b}_0=0$ and:

$$c_0=U_0+\int_0^{h_0}\frac{\partial U_0}{\partial h}\,\mathrm{d}y. \quad (20)$$

We note that c_0 is real. At first order we find that $c_1=c_{1,R}+\mathrm{i}c_{1,I}$ is purely imaginary, determined by:

$$c_{1,I}=\frac{(C^h-c_0A^h-\delta S\chi\tan\beta-\delta S\bar{W})}{\delta S}\int_0^{h_0}\frac{\partial U_0}{\partial b}\,\mathrm{d}y. \quad (21)$$

Note again that this quantity is independent of δ since the coefficients A^k & C^k are linear in δ. Also note that it is assumed $\bar{W}=\alpha^2 W\sim O(1)$ to ensure the presence of capillary effects in $c_{1,I}$.

The validity of the weighted residual model in predicting the onset of stability was successfully tested upon Orr-Sommerfeld analysis in our previous study for the case of miscible fluids [4]. Figure 2 shows the variation of the first order growth rate, c_1, with n_H for $\bar{W}=1$, $Re=1$, $n_L=1$, $B_L=0$, $B_H=0.05$, $m=1$, $h_0=0.5$ and $\chi=0$. As can be seen from the figure, there is complete agreement found between the Orr-Sommerfeld [4] and weighted residual model equation (21) results.

We now move to explore the effect of interfacial tension on flow stability pictures for other governing parameters. Figure

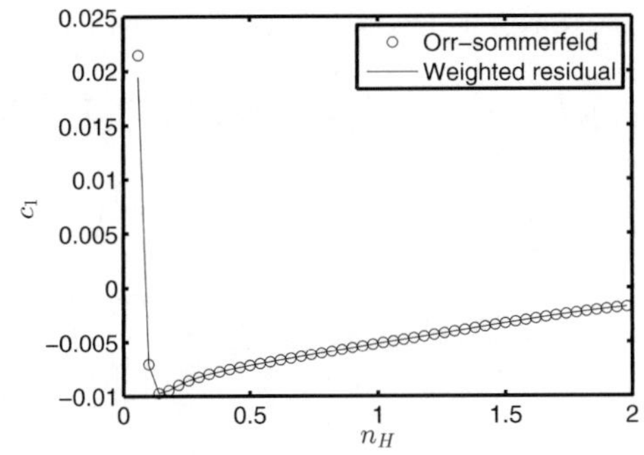

Figure 2: Comparison of the first order growth rate, c_1, in the long-wave limit: weighted residual and Orr-Sommerfeld analysis $\bar{W} = 1$, $Re = 1$, $n_L = 1$, $B_L = 0$, $B_H = 0.05$, $m = 1$, $h_0 = 0.5$ and $\chi = 0$ with n_H varying.

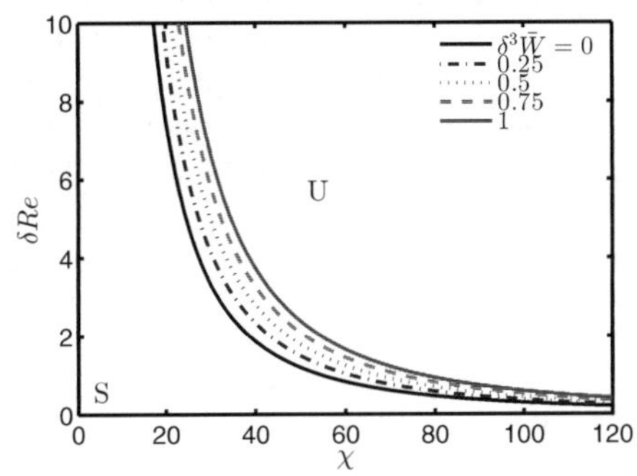

Figure 3: Marginal stability curves for the long-wave limit for $h_s = 0.5$ and $m = 1$ and different values of $\bar{W}$. Labels U and S represent unstable and stable flows respectively.

3 shows the marginal stability curves in the plane of $(\chi, \delta Re)$ assuming two iso-viscous Newtonian fluids. It can be observed from Fig. 3 that the interfacial tension has a stabilizing effect on the flow for the given range of parameters. The neutral stability curves are shifted upward with by increasing the Weber number, making the unstable regions smaller. A detailed look at the effects of interfascial tension on the flow stability for both Newtonian and non-Newtonian fluids requires further study which is not in the scope of the current manuscript.

CONCLUSIONS

In this paper an analytic framework based on a weighted residual method has been developed for immiscible fluids in a channel. The fluids assumed are Herschel-Bulkley. The model gives a first order accurate approximation to the interface thickness and buoyancy parameter. A conventional stability analysis is then carried out for our weighted residual evolution equations and compared with the corresponding Orr-Sommerfeld results, in the long wavelength limit. The growth rate values are predicted accurately by the weighted residual method. For the case of two iso-viscous Newtonian fluids, the interfacial tension was found to have a stabilizing effect on the flow. Given the model developed and tested, the next step can be to investigate the coupled effects of non-Newtonian fluids, buoyancy and interfacial tension on the stability of the two-layer thin-film flows.

REFERENCES

[1] Taghavi, S.M., Seon, T., Martinez, D.M., and Frigaard, I.A. 2009 "Buoyancy-dominated displacement flows in near-horizontal channels: the viscous limit," *J. Fluid Mech.*, Vol (639), pp. 1-35.

[2] Taghavi, S.M., Alba, K., Seon, T., Wielage-Burchard, K., Martinez, D.M., and Frigaard, I.A. 2012 "Miscible displacement flows in near-horizontal ducts at low Atwood number," *J. Fluid Mech.*, Vol (696), pp. 175-214.

[3] Taghavi, S.M., Alba, K., and Frigaard, I.A. 2012 "Buoyant miscible displacement flows at moderate viscosity ratios and low Atwood numbers in near-horizontal ducts," *Chem. Eng. Sc.*, Vol (69), pp. 404-418.

[4] Alba, K., Taghavi, S.M., and Frigaard, I.A. 2013 "A weighted residual method for two-layer non-Newtonian channel flows: steady-state results and their stability," *J. Fluid Mech.*, Vol (731), pp. 509-544.

[5] Ruyer-Quil, C., and Manneville, P. 2000 "Improved modeling of flows down inclined planes," *Eur. Phys. J. B*, Vol (15), pp. 357-369.

[6] Amaouche, M., Mehidi, N., and Amatousse, N. 2007 "Linear stability of a two-layer film flow down an inclined channel: A second-order weighted residual approach," *Phys. Fluids*, Vol (19), pp. 084106-14.

[7] Mehidi, N., and Amatousse, N. 2009 "Modélisation d'un écoulement coaxial en conduite circulaire de deux fluides visqueux," *C.R. Mecanique*, Vol (337), pp. 112-118.

[8] Alba, K., Laure, P., and Khayat, R.E. 2011 "Transient two-layer thin-film flow inside a channel," *Phys. Rev. E*, Vol (84), pp. 026320-14.

EFFECT OF NANOPARTICLES ON RHEOLOGICAL PROPERTIES OF HERSCHEL-BULKLEY FLUIDS

R.Mollaabbasi
Department of Chemical Engineering
University of Laval
Quebec, Quebec, Canada
Email: roozbeh.mollaabbasi.1@ulaval.ca

S.H. Noie
Department of Chemical Engineering
Ferdowsi University of Mashhad
Mashhad, K. Razavi, Iran
Email: noie@um.ac.ir

ABSTRACT

In recent decade, there are lots of efforts on energy saving and increasing efficiency of equipment in all industries. One of these issues is in drilling industries. Drilling mud is used to carry the rock excavated by the drill bit up to the surface and also to cool the drill. Changing the rheological properties of the Newtonian base fluid to Herschel-Bulkley fluid increases the efficiency of drilling mud for carrying the rock excavated. Also, adding the metallic nanoparticles can reinforce cooling ability of fluids. The effects of adding nanoparticles on rheological properties of Herschel-Bulkley fluids are investigated experimentally. Experiments show adding nano-particles increases the thermal conductivity of base fluids and decreases the power index of Herschel-Bulkley fluids. Pressure drop of base fluid increases less than 10% by increasing the nano-particle concentration.

KEYWORDS: Rheological Property, Herschel-Bulkley, Non-Newtonian Fluid, Drilling mud, Pressure drop

INTRODUCTION

The demand of energy has increased its cost in the world due to the growth of population. It has forced researchers to focus their attention on the ways of energy saving. The heat transfer property of fluids is one aspect of this issue. Researchers try to enhance heat transfer and decrease heat resistance. Non-Newtonian drilling is an applicable rule.
Drilling mud has several roles during the operation such as: removing, releasing, and suspending cutting from a well, controlling the formation of pressure, sealing, minimizing damages, lubricating beside of cooling. The water-based drilling mud is one of the most commonly used drilling fluids in industry and changing its rheological properties to Herschel–Bulkley fluid aids the transfer of cuttings from the bottom of the well to the surface [1]. The Eq. (1) explains relation of shear stress and shear rate of Herschel–Bulkley fluids[2]:

$$\tau = \tau_y + k\left(\frac{\partial u}{\partial y}\right)^n \quad (1)$$

τ is the shear stress (Pa), τ_y is the yield stress(Pa), k is the consistency index (Pa s^{-n}) and $\left(\frac{\partial u}{\partial y}\right)$ is the shear rate(1/s).
Suspending the metal nano-particles in a base fluid is one of the best methods for increasing the heat transfer. Increasing the thermal conductivity of base fluid is the main concept of this technique [3]. Nanoparticles have advantages over micro sized ones, such as: improving heat transfer and stability, reducing the pumping power and pressure drop [4].
In this research, the effects of adding nano-particles on rheological properties of Herschel-Bulkley fluids are investigated experimentally.

EXPERIMENTAL

We designed and made a flow loop (Fig.1). The gear pump provides the flow with the maximum capacity of 6 L/min. The copper pipe (140-cm long and 1 cm in diameter) is the most important part of the setup. This pipe is surrounded by three electrical heaters to supply the heat with a maximum power of 200W. Ceramic coat insulate the pipe. The condenser cools the hot fluid. Eight PT100-type thermocouples located on the pipe wall measure the temperature. A manometer and a flow meter measure the fluid pressure drop and volumetric flow rate respectively.

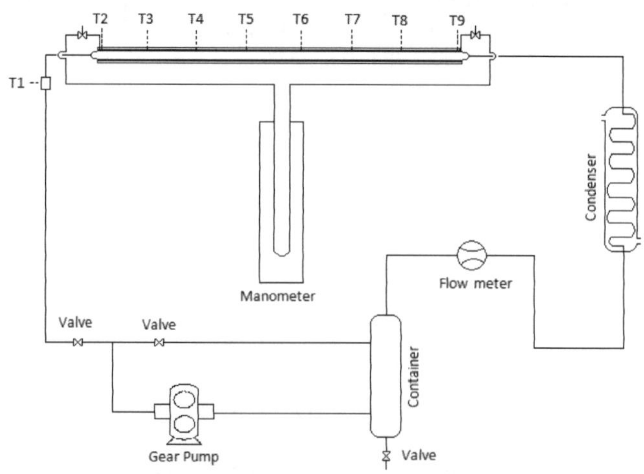

Figure 1: Schematic of flow loop

An aqueous polymer solution of 0.2% w/w polyacrylic acid (Aldrich) with molecular weight 3.2E+6 g/mol is used as a

Herschel–Bulkley fluid[5]. TiO_2 [6], γAl_2O_3 [7] and CuO [8] were used as nanoparticle additives (Particle size<20nm). The thermal conductivity of γAl_2O_3, TiO_2 and CuO are 46, 8.7 and 69 W/m K, subsequently. The nano-particle suspensions are prepared in 0.05, 0.1, 0.3 and 0.6 %v/v according to additive limitation in drilling mud [9]. The rheological properties of fluids are measured by Brookfield Rheometer (DV-III™ Ultra Rheometer). We repeated the experiments three times. The error was less than 8%.

RESULTS AND DISCUSIONS

PAA (polyacrylic acid) solution with different concentrations of Al_2O_3, TiO_2, and CuO nanoparticles is a Herschel–Bulkley fluid (Fig. 2). In Herschel–Bulkley fluid the strain experienced by the fluid is related to the stress in a complicated, non-linear way (Eq. (1)). Suspended particles in the fluid migrate from the high shear rate region to the low one [8]. So, the nanoparticle concentration and the apparent viscosity of the nano-fluid decrease near the wall and provides a smaller boundary-layer thickness.

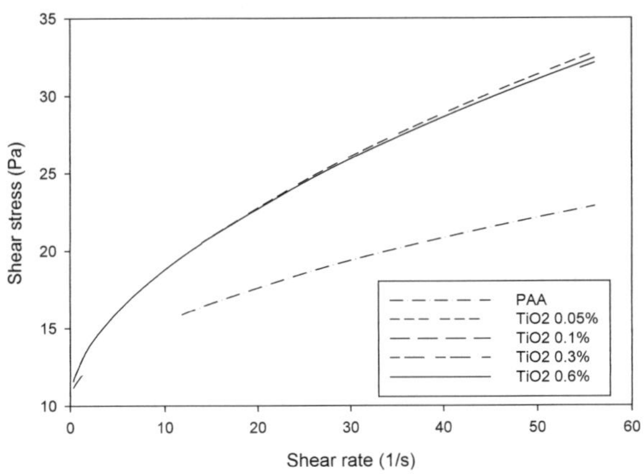

Figure 2(a)

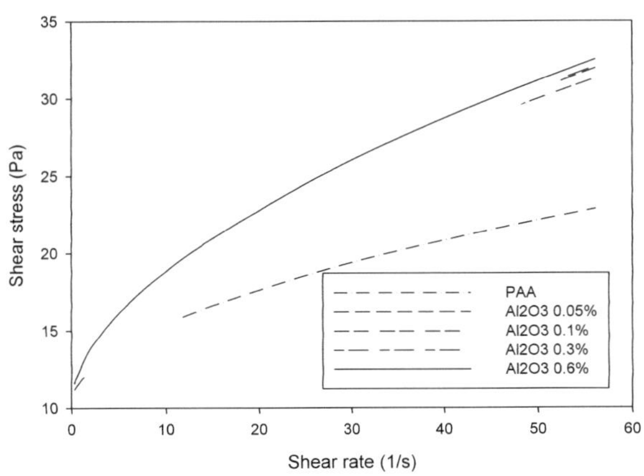

Figure 2(b)

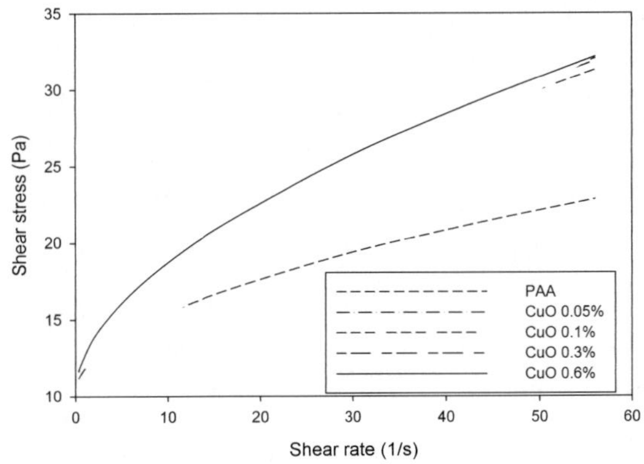

Figure 2(c)

Figure 2: Effect of nano-particles, TiO_2(a), Al_2O_3(b) and CuO(c) on rheological property of PAA at 35°C

In order to characterize the rheological properties of the base fluid and Herschel–Bulkley nano-fluids, we investigate the effect of adding nanoparticles on the consistency index, yield stress and power index of the PAA solution.

Fig.3 represents the consistency index (k) of the nano-fluids as a function of the particle concentration. Increase of nano-particle concentration increases consistency index of Al_2O_3, TiO_2, and CuO nano-fluids 40% in average (Fig. 3).

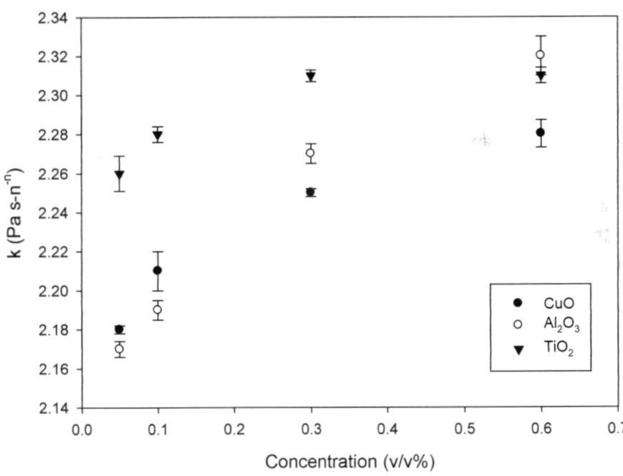

Figure 3: Effect of nano-particles on consistency index in Hersche-Bulkley model (Pa s^{-n}) of nano-fluids at 35°C

Fig.4. shows the effect of adding nanoparticles on the yield stress of the Herschel–Bulkley fluid. In this figure, the yield stress of a pure PAA corresponds on the zero concentration of nanoparticles. The adding of nanoparticles increases the yield stress of the Herschel–Bulkley fluid primary, but when the concentration increases more than 0.1%, the increasing concentration of nanoparticles does not exert any effect on the yield stress, and the graphs form a plateau. It shows that adding nanoparticles does not eliminate the yield stress of the base fluid but it reinforces it around 12% on average.

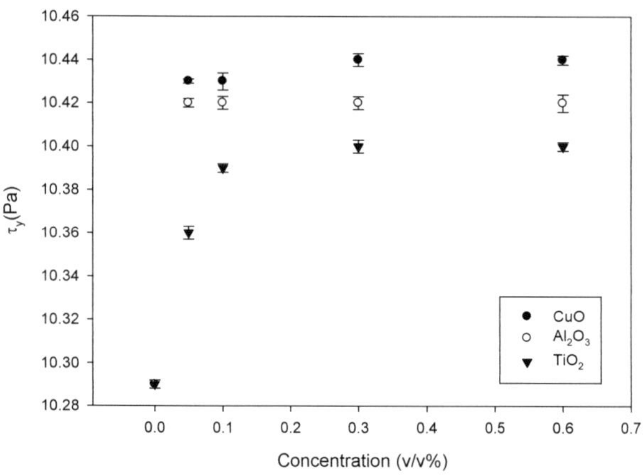

Figure 4: Effect of nano-particles on yield stress in Hersche-Bulkley model (Pa) of nano-fluids at 35°C

Fig.5. shows the power-law index of the nano-fluids as a function of the particle concentration. Adding nano-particles decreases the power-law index of the pseudo plastic fluids[10]. This effect is 6% in the Herschel-Bulkley fluids in average.

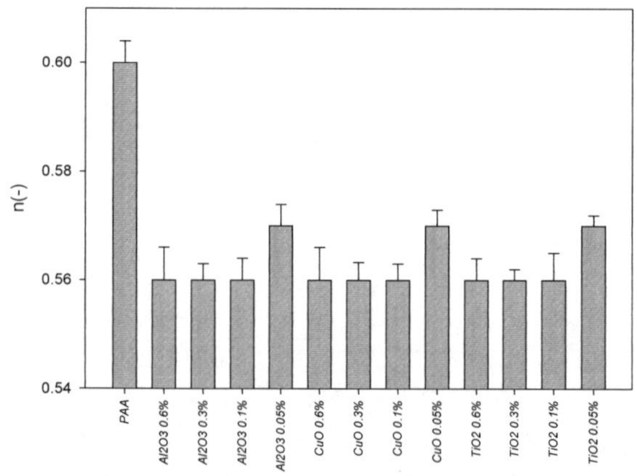

Figure 5: Effect of nanoparticles on power index in Hersche-Bulkley model (-) of nano-fluids at 35°C

The power-law index of CuO, Al_2O_3 and TiO_2 nano-fluids generally decreases with increasing the particle concentration. This reinforces the convective heat transfer of PAA solution 14-23%[11].

The thermal conductivity of a nano-fluid is calculated by[12]:

$$\lambda_{nf} = [\frac{\lambda_s + 2\lambda_{bf} + 2(\lambda_s - \lambda_{bf})(1+\varepsilon)^3 \varphi}{\lambda_s + \lambda_{bf} + (\lambda_s - \lambda_{bf})(1+\varepsilon)^3 \varphi}]\lambda_{bf} \qquad (2)$$

Where, λ is thermal conductivity (W/m.K), φ is the particle volume concentration (Dimensionless) and ε is the ratio of the nano layer thickness to the original particle radius. Normally $\varepsilon = 0.1$ is used to calculate the thermal conductivity of the nano-fluid. The subscribes s, bf and nf shows solid, base fluid and nano-fluid respectively.

Fig.6. shows the effect of nanoparticles on thermal conductivity of Herschel-Bulkley fluid. This shows that adding these metal nano-particles increase thermal conductivity of base fluid but it is not so much. According to Eq. (2), the particle volume concentration has an important role in the thermal conductivity of a nano-fluid. In these experiments the particle volume concentration is less than 1%.

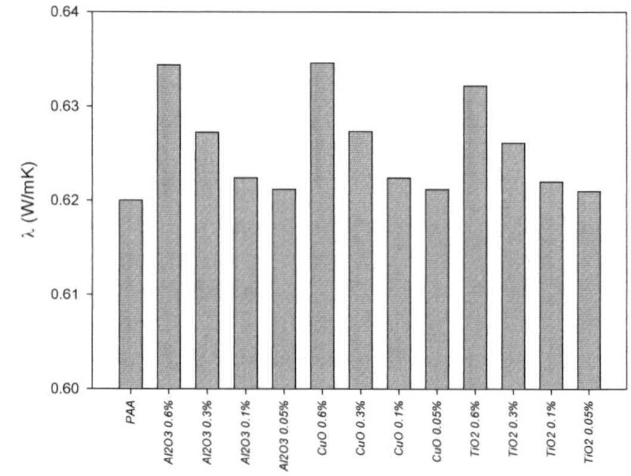

Figure 6: Effect of nano-particle's concentration on thermal conductivity of PAA at 30°C

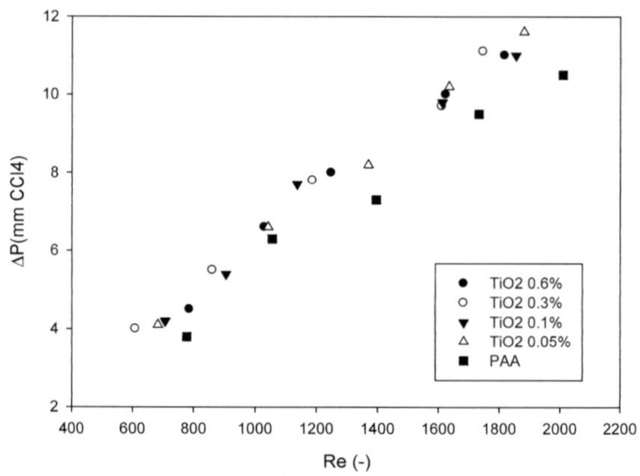

Figure 7(a)

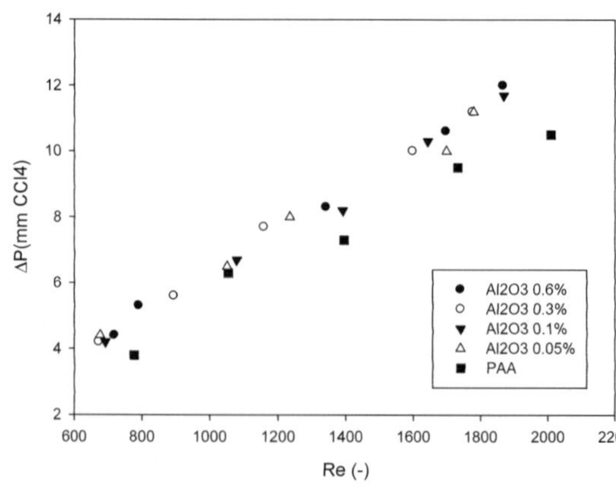

Figure 7(b)

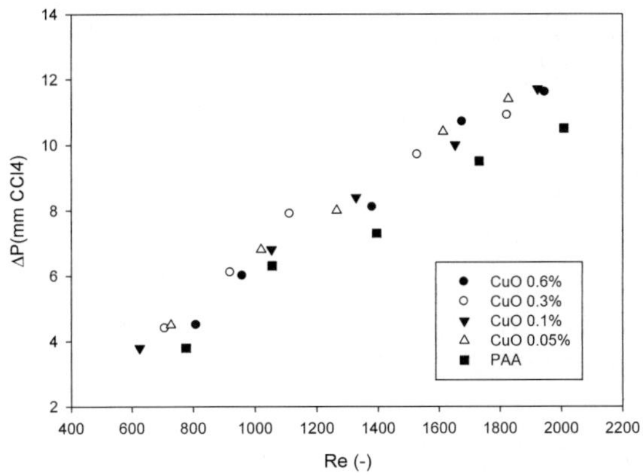

Figure 7(c)

Figure 7: Effect of nanoparticles TiO_2(a), Al_2O_3(b) and CuO(c) on pressure drop of nano-fluids at different Re number for laminar regime

The pressure drop in circular pipes is calculated by using Darcy-Weisbach equation[2]:

$$\Delta P = f\frac{L}{D}.\frac{\rho u^2}{2g} \quad (3)$$

In this equation, L and D are the length and diameter of the pipe (m) respectively, ρ is density (Kg/m^3), u is velocity (m/s). In a laminar flow (Re<2300) the friction factor (f) calculates by f=64/Re. According to this equation with increasing the velocity of fluid and increasing the Re number in the pipe, the pressure drop will be increased. This phenomenon is obvious in all of fluids at Figure 7.

There is a slight increasing in the pressure drop of the nano-fluids as compared with the pure PAA, at the same Re number. Adding nano-particles increases the average density of base fluid and increases the pressure drop of fluid in the pipe according the concentration of nanoparticles less than 10% in our experiments. Experiments show that Eq. (3) can predict pressure drop of these nano-fluids with error less than 4% in laminar regime.

CONCLUSION

Thermal conductivity of nano-fluids is a function of empirical shape factor, ratio of the nano layer thickness to the original particle radius and type of the particles. Adding nanoparticles increases the thermal conductivity of Herschel-Bulkley fluids and it depends on thermal conductivity of nanoparticles and will be increased by increasing solid concentration. Increasing Al_2O_3, TiO_2 and CuO nano particle concentration decreases the power index of Herschel-Bulkley fluid 6%, also increases consistency index and yield stress of base fluid 40% and 12% respectively. Adding CuO and Al_2O_3 nanoparticles more than 0.1% v/v does not have any special effect on the yield stress in the Herschel-Bulkley model but adding TiO_2 nanoparticles increases the yield stress in the Herschel-Bulkley model. Adding nanoparticles increases the average density of base fluid and increases the pressure drop of fluid in the pipe less than 10%. Darcy-Weisbach equation can predict pressure drop of Herschel-Bulkley fluids in laminar regimes.

REFERENCES

1. Committee, A.S.S., *Drilling Fluids Processing Handbook*2006: Elsevier.
2. Tropia, C., et al., *Handbook of Experimental Fluid Mechanics*2007: Springer.
3. Maxwell, J.C., *Electricity and magnetism*1873, Oxford,UK: Clarendon Press.
4. Choi, S.U.S., *Nanofluid technology: current status and future research*, in *Second Korean-American science and engineering association research trend study project review and the Korea U.S technical conference on strategic technologies.*1998: Vienna. p. 22-24.
5. Peixinho, J., C. Desaubry , and M. Lebouche *Heat transfer of a non-Newtonian fluid (Carbopol aqueous solution) in transitional pipe flow.* International journal of heat and mass transfer, 2008. **51**: p. 198–209.
6. Murshed , S.M.S., K.C. Leong , and C. Yang, *Enhanced thermal conductivity of TiO2-water based nanofluids.* Int. J. Thermal. Science., 2005. **44**: p. 367-373.
7. Fotukian , S.M. and M. Nasr Esfahany *Experimental investigation of turbulent convective heat transfer of dilute γ-Al2O3/water nanofluid inside a circular tube.* Int. J. Heat Fluid Flow, 2010. **31**: p. 606-612.
8. Heris , S.Z., S.G. Etemad, and M. Nasr Esfahany *Experimental investigation of oxide nanofluids laminar flow convective heat transfer.* Int. Commun. Heat Mass Transfer, 2006. **33**: p. 529-535.
9. Garg , P., et al., *An experimental study on the effect of ultrasonication on viscosity and heat transfer performance of multi-wall carbon nanotube-based aqueous nanofluids.* Int. J. Heat Mass Transfer 2009. **52**: p. 5090-5101.
10. Etemad, S.G., *The effect of sharp corners of channels on heat transfer of Non-Newtonian fluids.*, in *Proceeding of the 1th heat transfer conference*1998: Kyongju,Korea. p. 103.
11. Mollabbasi, R., S.H. Noie, and S.Z. Heris, *Experimental investigation of the Effect Exerted by Nanoparticles on the Heat transfer Coefficient of Herschel–Bulkley Fluids.* Heat Transfer Research, 2014. **45**(6).
12. YU , W. and S.U.S. Choi, *The role of interfacial layers in the enhanced thermal conductivity of nanofluids: a renovated Maxwell model.* J. Nanoparticle. Research., 2003. **5**: p. 167–171.

MAGNETORHEOLOGICAL EFFECTS FROM TRACER RESIDENCE TIME DISTRIBUTION IN NANOFLUID POISEUILLE FLOWS

F. Larachi, P. Hajiani
Department of Chemical Engineering
Laval University
Québec, Québec, Canada

ABSTRACT The influence of magnetic-field dependent viscosity (rotational viscosity) on molecular transport of species in dilute nanofluids has been studied using a Taylor dispersion capillary tube where suspended magnetic nanoparticles (MNPs) have been subjected to both magnetic field and low *Re* shear flow field. Axial dispersion has been quantified from residence time distributions (RTD) and tracer injection tests for rotating magnetic field (RMF), oscillating magnetic field (OMF) and static magnetic field (SMF). For RMF, an increase in lateral mixing is observed whereas no significant effect is detected for OMF. In SMF, the lateral mixing mechanism is retarded by magnetically locked MNPs. These findings highlight the effect of rotational viscosity on diffusion of other species hosted in dilute nanofluids and point to attractive applications to engineering fields where transport phenomena are central. Analysis of RTD breakthrough times enabled magnetically-induced non-Newtonian laminar velocity profiles to be reconstructed. It suggests that (magnetic field-free) parabolic velocity profiles evolve towards flattened and protruded shapes, respectively, in RMF and SMF. These results confirm that magnetically-excited MNPs may be considered as a potentially appealing tool to mediate molecular transport phenomena at the nanoscale such as in nano/microfluidic systems.

KEYWORDS: Magnetic nanoparticle, magnetic field, mixing, Taylor dispersion, residence time distribution

INTRODUCTION

Stable colloidal suspensions of single-domain magnetic nanoparticles (MNP) in nonmagnetic liquids exhibit fascinating rheological properties under magnetic fields (1). A magnetic field exerts a magnetic torque on the magnetic moment of MNPs to orient them toward its own direction (2). This magnetic body torque is often opposed by Brownian collisions and flow-field frictional torque. In sheared nanofluid flow, an external magnetic field forces MNPs to spin asynchronously relative to the local vorticity of the fluid, resulting in a directional modification of viscosity. This behavior, known as rotational viscosity, was first observed by McTague (3).

Rotational viscosity has been studied under three types of magnetic fields, static (SMF), oscillating (OMF) and rotating (RMF) (1,2,4). SMF rotational viscosity arises when the external magnetic field is not parallel to local fluid vorticity. While the fluid viscid nature forces MNPs to rotate in shear flows, the normal-to-vorticity component of magnetic field vector hinders the dipoles free rotation, thus leading to a greater rate of hydrodynamic energy dissipation around MNPs (2). This yields an inflated viscosity as observed by McTague (3) and explained theoretically by Shliomis (5).

Rotational viscosity under OMF was investigated experimentally by Bacri *et al.* (6) and theoretically by Shliomis and Morozov (7). Under OMF, MNPs try to chase the magnetic field via synchronized rotation due to periodic alternation of magnetic field. Subjecting a nanofluid laminar Poiseuille flow to low-frequency coaxial OMF (oscillating at frequency Ω_{OMF}) such that $\Omega_{OMF}\tau_B < 1$ ($\tau_B=3\eta_0 V_h/k_B T$ is Brownian relaxation time) partially impedes the azimuthal spin of MNPs and results in a positive rotational viscosity (8). Under high-frequency OMF such that $\Omega_{OMF}\tau_B > 1$ the magnetic field amplifies azimuthal MNP spin. Accordingly, electromagnetic energy is transferred to the fluid as kinetic energy, thereby enhancing flow nearby MNPs. This effect may be viewed as if the apparent nanofluid viscosity has decreased tantamount to negative rotational viscosity (6,8,9).

Rotational viscosity under RMF was first observed experimentally by Moskowitz and Rosensweig (10) in the form of nanofluid entrainment by the field originating from the average rotation of MNPs relative to their embedding fluid. The velocity profile of this macroscopic motion in RMF was studied experimentally and theoretically by Chaves *et al.* (11). The ferrofluid rheological properties under RMF have been obtained from measurement of torques of a submerged spindle. Based on this method, Rosental *et al.* (12) and Rinaldi *et al.* (13) detected negative and positive rotational viscosity with co-rotating and counter-rotating magnetic fields with respect to spindle resistive torques. Interestingly, under RMF, no minimum field frequency is required to achieve negative rotational viscosity in contrast with Bacri *et al.* (6) experiments reported under OMF.

In this contribution, we perform Taylor dispersion tests in a capillary tube and determine the axial dispersion coefficient (D_{ax}) as a means to investigate liquid-phase mass transfer in the presence of magnetically excited MNPs interacting with shear flow. Taylor dispersion is a phenomenon originating from combination between axial convective displacement of a tracer slab under laminar Poiseuille flow and its diffusive migration in radial direction (14). Hence in a given experimental condition, any phenomenon that alters molecular diffusion could be detected indirectly through its effect on D_{ax}. Residence time distribution measurements are thus performed using a Taylor dispersion capillary cell from which axial dispersion coefficients are obtained. MNPs in our Poiseuille experiments were excited successively by RMF, OMF and SMF to probe their influence on the lateral molecular transport mechanism in the capillary flows.

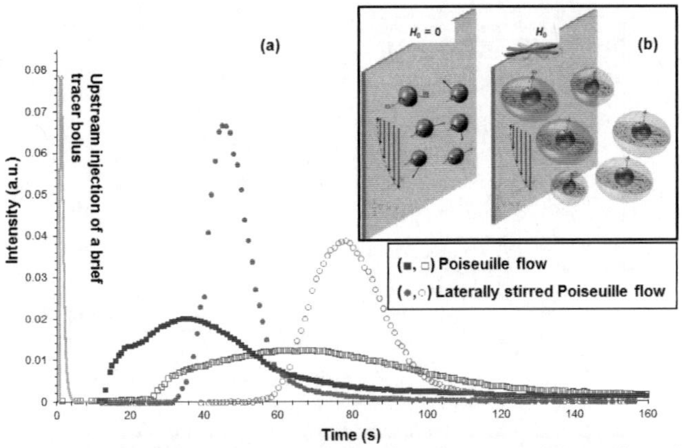

Figure 1 Nanofluid Taylor dispersion for pure ($H_0 = 0$) and magnetically-stimulated (H_0 = 36.5 kA/m) Poiseuille flow: Time evolution of tracer conductivity by upstream (full symbols) and downstream (empty symbols) electrodes (a); Orientation of nanoparticles parallel to fluid vorticity for H_0 = 0 and forced reorientation of MNP magnetic moments under rotating magnetic fields (b)

EXPERIMENTAL

MNP suspension Dilute aqueous suspensions of ferrite (Fe_3O_4) MNPs are prepared from commercial ferrofluids (Ferrotec EMG 705, 4.2%v/v). The magnetic core median diameter, d_c, is estimated to be 16.0 nm using an alternating gradient magnetometer. Suspensions with finer nanoparticles are also prepared through centrifugation of the as-received ferrofluids at 20,000 rpm for 4 h. After the magnetic properties of the centrifuged product are characterized, the desired particle volume fraction is achieved *via* dilution.

Magnet A tubular two-pole three-phase magnet was designed with bore dimensions as 55 mm height and 45 mm inner diameter. The magnet assembly consists of three identical coil pairs, spatially shifted from each other by 120° in azimuthal direction. Each coil is energized separately or coupled to other coils to generate rotating (RMF), oscillating (OMF) and static (SMF) magnetic fields with up to 50 mT intensity at the center axis. The magnet is cooled with a water-cooling jacket.

RTD test A 1 mm I.D. glass capillary tube, equipped with a T-shaped injection site and two conductivity electrodes, 3-cm apart, is introduced either vertically (for RMF and OMF tests), or horizontally (for SMF) at the bore center of a vertically aligned magnet. The flow of the dilute nanofluid is maintained using a syringe pump for volumetric flow rates corresponding to Reynolds numbers ranging from Re = 0.3 to 1.3. At t = 0, a small volume of salt tracer (ca. 0.5 μl) was instantly injected into the flow upstream of the two electrodes to allow determination of the nanofluid residence time distribution (RTD) by means of the so-called Aris' double detection method. The tracer subsequent spread under the joint actions of advection and diffusion is assessed in terms of an axial dispersion coefficient estimated by fitting to the measured exit (probe #2) signal, a computed exit signal obtained by convolution of measured inlet (probe #1) signal and an open-open boundary RTD axial dispersion model.

RESULTS & DISCUSSION

Taylor dispersion in RMF Fig. 1 illustrates the nanofluid tracer responses registered by the two downstream conductivity probes without (H_0 = 0) and with rotating magnetic field (H_0 = 36.5 kA/m, f = 50 Hz). An identical volumetric flow rate corresponding to pure laminar flow (Re ~ 1) was imposed for both instances. Three features emerge upon application of a rotating magnetic field in comparison to the case H_0 = 0: i) the signals' full-width at half maximum (FWHM) noticeably shrunk; ii) both probes' signal tails were significantly shortened; iii) both probes' signal breakthrough times were delayed. The first two features are interpreted to result from an enhanced transverse mixing of the tracer across the Poiseuille flow streamlines whereas the third one is ascribed to a flattening (or nose retraction) of velocity profile reminiscent of plug flow. Fig. 1b outlines a plausible mechanism the rotating magnetic field might have triggered whereby the observed RTD features are possibly altered by induced rotations of the nanoparticles in the plane transverse to the advective axial flow. The nanoparticles, as the magnetic field is off, are only allowed to spin collinearly to fluid vorticity; while, by being continuously randomly disoriented by Brownian agitation, their magnetic moments linger aimless (Fig. 1b). The FWHM, tail and breakthrough time of the tracer spread obeys classical Poiseuille flow in which, provided a fully developed flow is met, can be anticipated from the sole knowledge of the parabolic velocity profile and transverse molecular diffusion (Fig. 1a, H_0 = 0). The nanofluid state, on enabling a rotating magnetic field, shall be markedly altered by the tendency of the nanoparticle magnetic moments to seek, driven by incipient magnetic torques, spontaneous alignments with the magnetic vector field. In accordance with this latter's direction (Fig. 1b), the nanoparticles become confronted to two orthogonal rotational degrees of freedom. However, provided a magnetic field of sufficient strength energizes the nanofluid, the magnetic torques would build up to overcome both the shear-induced hydrodynamic torque and the Brownian collisions that erratically knock the nanoparticles. The ability to individually harness the nanoparticles to rotate in a direction nearly normal to vorticity brings about nanoconvective currents in the capillary transverse plane that otherwise would have been the private fiefdom of molecular diffusion should the magnetic field have been turned off. The view that the magnetic torque in the presence of a transverse rotating magnetic field sets into motion the nanoparticles is to our opinion the mechanism that subtends the tracer stimulated transverse mixing as revealed by the morphological changes of the nanofluid RTD under RMF (Fig. 1a).

Fig. 2 illustrates how the axial dispersion coefficient is dampened by increasing the magnetic field strength or frequency (Fig. 2a) or by increasing the volume fraction of magnetic cores in the nanofluid or the nanoparticle size (Fig. 2b). The mixed zones, represented in the form of mixing spheroids in Fig. 1b, are expected to swell and/or grow under the increase of one of the four operating variables until a plateau is reached which corresponds to an overlap of the mixed zones altogether.

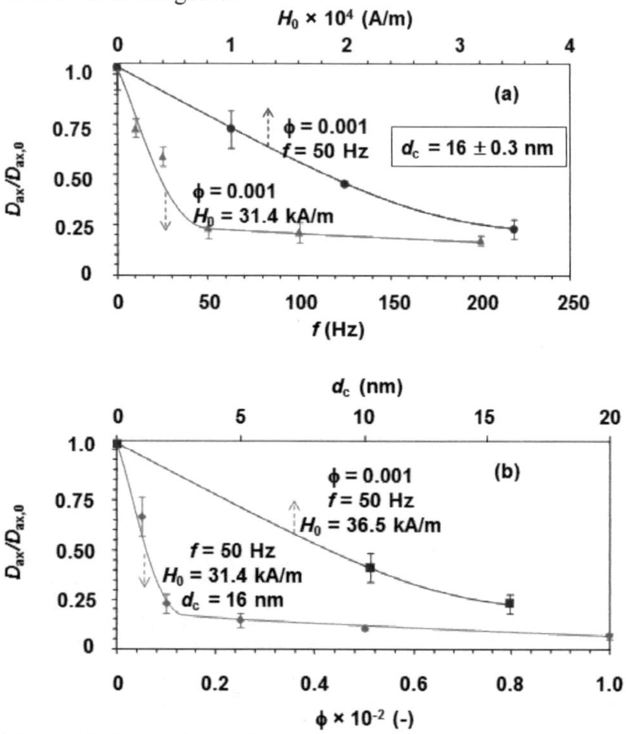

Figure 2 Axial dispersion ratio as a function of (a) RMF strength and frequency, and (b) MNP concentration/size

Taylor dispersion in OMF RTD tests were also performed to investigate OMF effect on excited MNP behavior at low-*Re* laminar flow. For the sake of comparability, OMF had rms field intensity equal to RMF strength and same frequency. The axial dispersion coefficients for a range of liquid flow rates were estimated from moment analysis of convoluted data acquired from conductivity electrodes' responses. Fig. 3 presents the Péclet number (UL/D_{ax}) versus *Re* number under equal strength of OMF and RMF (H_0 = 31.4 kA/m) and different frequencies. Subject to various OMF frequencies, UL/D_{ax} does not change significantly relative to magnetic field-free tests. This is unlike under RMF where axial dispersion decreases substantially. Hence, $D_{ax}/D_{ax,0}$ is almost constant under OMF indicating no pronounced lateral mixing effect in the capillary. In contrast to RMF, the very same MNPs will not give rise to lateral mixing under equally strong OMF. This observation raises the question of OMF ability to bring MNPs in gyration similar to what occurs under RMF.

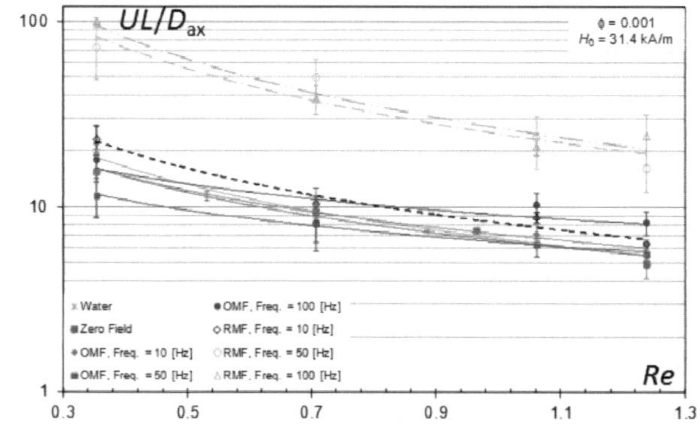

Figure 3 Péclet number *vs*. *Re* number in OMF and RMF

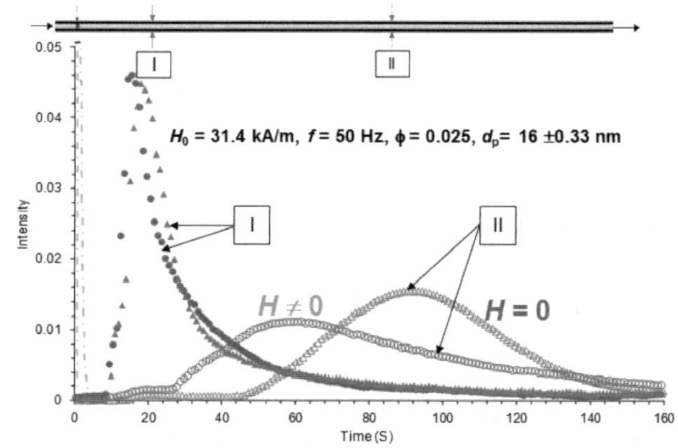

Figure 4 Nanofluid Taylor dispersion for pure (H_0 = 0) and SMF-stimulated Poiseuille flow

Taylor dispersion in SMF Fig. 4 shows RTD impulse responses with and without axial SMF. Signals from upstream electrodes are very similar whilst, downstream signals show marked deviations. The resemblance of inlet signals in presence and absence of SMF can be attributed to the short distance (1 cm) between tracer injection and first electrode. The 1st-electrode skewed tracer responses are ascribed to imperfect tracer injections that are well accounted for in the convolution integral for the estimation of axial dispersion coefficients. The outlet signal under SMF is wider and its breakthrough time occurs earlier than without magnetic field. Unlike in RMF, this observation demonstrates that axial dispersion inflated even further by SMF-excited MNPs interacting with shear flow.

MNPs alter laminar velocity profile Fig. 1a provides a crucial clue as the separation between the breakthrough times at the two electrodes amplifies with increased nanomixing whilst the mean residence time is kept identical. The breakthrough time features the fastest centerline tracer with shortest residence time (τ_{min}). The τ/τ_{min} ratio declines from 2 to 1 for a velocity profile evolving from parabolic to a completely flattened velocity-homogenized one. This ratio can be used to assess how far velocity profiles flatten due to nanomixing. By drawing an analogy with non-Newtonian power-law radial velocity profiles, $u(r)$, one has:

$$u(r)=\frac{3n+1}{1+n}U\left(1-\left(\frac{r}{a}\right)^{1+1/n}\right) \quad (1)$$

where the power-law index, n, varies between 1 (parabolic profile) and 0 (flat profile) and is expressed as a function of τ_{min}/τ according to:

$$n=\frac{1-\tau_{min}/\tau}{3\tau_{min}/\tau-1} \quad (2)$$

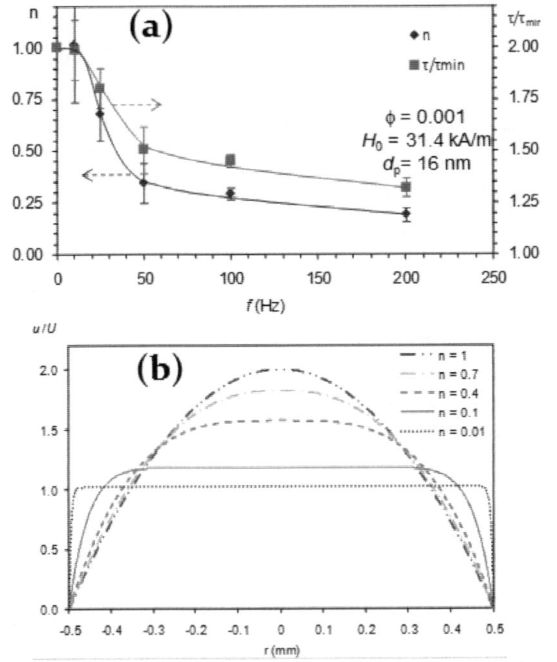

Figure 5 Power-law index n and τ/τ_{min} ratio as a function of (a) magnetic field frequency, (b) Simulated laminar flow velocity profile as a function of power-law index

Fig. 5a plots τ/τ_{min} and n versus RMF frequency. Sensitivity of τ/τ_{min} and n to frequency dwindles after 50 Hz. Unlike concentration homogenization, velocity homogenization (i.e., reduction of n) via nanomixing is more demanding and requires larger minimal H_0 and ϕ. This is illustrated in Fig. 5b which shows that the computed velocity profiles flatten as n decreases (from 1 to 10^{-2}) to echo increased nanomixing intensities. A similar analysis can be conducted in the case of the viscothickening effects objectified in the case of SMF.

CONCLUSION

Experimental works to elucidate the effect of magnetically-excited MNPs on molecular transport mechanism in nanofluids is lacking. By studying axial dispersion in dilute nanofluids flowing through a Taylor capillary cell under moderate strength and low frequency RMF, OMF and SMF, we investigated the mass transfer phenomena induced by the interactions of excited MNPs with Poiseuille flow fields. Our study demonstrates that MNPs in RMF and SMF manipulate tracer lateral diffusion rate in capillary whilst particle excitation by OMF has no significant impact. Furthermore, we correlated the shape of nanofluid laminar velocity profile in capillary to the breakthrough (minimum) residence time from RTD tests. Using a single parameter model, we evidenced that nanofluid laminar velocity profiles may change in SMF or RMF.

ACKNOWLEDGEMENTS The Canada Research Chair "Green processes for cleaner and sustainable energy" is gratefully acknowledged.

REFERENCES

(1) S. Odenbach, Magnetoviscous effects in ferrofluids, Springer, Berlin, 2002.

(2) R. E. Rosensweig, Ferrohydrodynamics, Dover Publications, Mineola, N.Y., 1997.

(3) J. P. McTague, Magnetoviscosity of magnetic colloids, J. Chem. Phys., 51 (1969) 133.

(4) S. Rhodes, X. W. He, S. Elborai, S. H. Lee, M. Zahn, Magnetic fluid behavior in uniform DC, AC, and rotating magnetic fields, J. Electrostat, 64 (2006) 513-519.

(5) M. I. Shliomis, Effective viscosity of magnetic suspensions, Sov. Phys. Jetp-Ussr, 34 (1972) 1291.

(6) J. C. Bacri, R. Perzynski, M. I. Shliomis, G. I. Burde, Negative-viscosity effect in a magnetic fluid, Phys. Rev. Lett., 75 (1995) 2128-2131.

(7) M. I. Shliomis, K. I. Morozov, Negative viscosity of ferrofluid under alternating magnetic-field, Phys. Fluids, 6 (1994) 2855-2861.

(8) A. Zeuner, R. Richter, I. Rehberg, Experiments on negative and positive magnetoviscosity in an alternating magnetic field, Phys. Rev. E., 58 (1998) 6287-6293.

(9) K. R. Schumacher, I. Sellien, G. S. Knoke, T. Cader, and B. A. Finlayson, Experiment and simulation of laminar and turbulent ferrofluid pipe flow in an oscillating magnetic field, Phys. Rev. E., 67 (2003) 026308.

(10) R. Moskowitz, R. E. Rosensweig, Nonmechanical Torque-Driven Flow of a Ferromagnetic Fluid by an Electromagnetic Field, Appl. Phys. Lett., 11 (1967) 301-303.

(11) A. Chaves, C. Rinaldi, S. Elborai, X. He, M. Zahn, Bulk flow in ferrofluids in a uniform rotating magnetic field, Phys. Rev. Lett., 96 (2006) 194501.

(12) A. D. Rosenthal, C. Rinaldi, T. Franklin, and M. Zahn, Torque measurements in spin-up flow of ferrofluids, J. Fluid Eng-T Asme, 126 (2004) 198-205.

(13) C. Rinaldi, F. Gutman, X. W. He, A. D. Rosenthal, M. Zahn, Torque measurements on ferrofluid cylinders in rotating magnetic fields, J. Magn. Magn. Mater., 289 (2005) 307-310.

(14) H. Brenner, D. A. Edwards, Macrotransport Processes, Butterworth-Heinemann, Boston, 1993.

Porous Media

Proceedings of the 25th CANCAM
London, Ontario, Canada, May 31 - June 4, 2015

A COMPUTATIONAL MODEL OF TEMPERATURE MONITORING AT A LEAKAGE IN A LEAK DETECTION SYSTEM OF A PIPELINE

A. S. Bhuiyan
Department of Mathematics and statistics
Memorial University of Newfoundland
St.John's, NL, Canada
masb80@mun.ca

M. A. Hossain, J. M. Alam
Department of Mathematics and statistics
Memorial University of Newfoundland
St.John's, NL, Canada
mah782@mun.ca, alamj@mun.ca

ABSTRACT

Monitoring the surrounding pipeline temperature through fibre optic sensors is an efficient technology for detecting leakage of oil and gas from pipelines. This article presents a computational model of temperature monitoring in such a leak detection system. A longitudinal cross section of an idealized pipe-in-pipe flow-line has been simulated, where a localized thermal source is placed on the lower boundary of the computational domain. The heat transfer rates and temperature profiles of the localized heat source have been investigated because a detailed understanding of the excess temperature surrounding a pipeline is essential to design efficient leak detection systems based on the active thermometry method. The excess temperature profile from a localized heat source is studied for various physical conditions. More effective results of the rate of heat transfer have been discussed, which are suggestive for controlling leakages in the pipelines. The present investigation would be useful for leak detection system.

Keywords: stable thermal gradient, localized heat source, wavelet method, leak detection system, pipe line monitoring

INTRODUCTION

Leakages from pipelines usually produce local thermal or hydraulic anomalies at the vicinity to the pipeline which could be measured using modern fibre optic sensor technologies from a central station that is located several kilometres away from the site of the leakage. Commerical operators require to install a leak detection system (LDS) in pipelines that transport oil/gas or other hazardous fluids [6, 20]. Active thermometry of localized heat source is one approach that is used in Leak detection system (LDS). An optical sensor cable is installed parallel to the pipeline on its entire length, which measures the temperature profile in the vicinity of the pipeline. In a state-of-the-art pipe-in-pipe (PIP) technology, the flow line is insulated – as depicted schematically in figure, 1 (insulation). One may idealize such a complicated flow line by an axial cross section (see figure 1 (insulation)) so that the complete leakage phenomena may be modelled based on pressure and temperature anomalies at the vicinity of the pipeline. An effective computational model of natural convection heat transfer and fluid flow from a localized source in a porous medium can be used to investigate how various physical conditions of an insulator may affect surrounding thermal and pressure anomalies in case a leak occurs. Such a model would also provide useful feedback to optimize the cost of insulation in a PIP system. For instance, in the Canadian Arctic offshore oil companies transport oil/gas at a temperature that is relatively higher than that of the surrounding ocean in the vicinity of the pipeline. Clearly, the natural convection heat transfer is the dominant phenomena that requires further investigation in the context of leakage monitoring in pipelines.

Considering the insulation as a porous medium, a number of numerical [15, 17, 18] as well as experimental [19] studies are available; most of such studies adopted a uniformly heated bottom wall. Note that such a study employing discrete or a localized heat source on the boundary is limited [4, 8, 17]. Recently, Oztop *et al.* [9] reported that both the local inclination angle of the pipeline and the sinusoidal heating at the bottom wall have significant impact on heat transfer. Khatib *et al.* [8] mentioned that an increase of the stable wall thermal gradient may keep the upper surface thermal energy independent of the

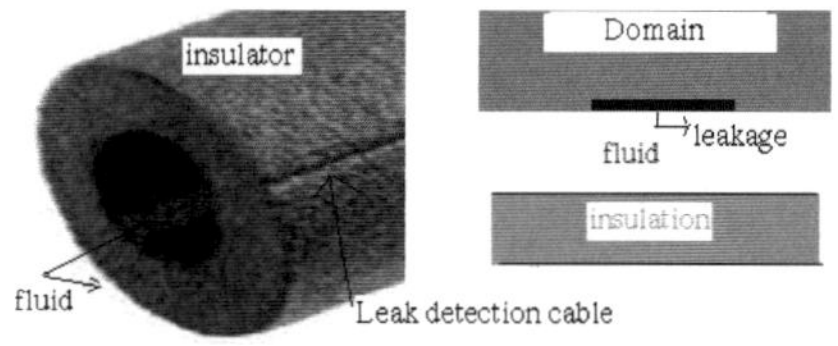

Figure 1: Cross section of a pipe.

Rayleigh number. Saied *et al.* [12] observed an increase of the heat transfer rate with an increase of the length of the heat source. Cotter and Michael [11] investigated the geometrical constraint on the natural convection heat transfer in recently produced crude oil. However, the leakage size in a pipeline is typically unknown and non-uniform, and a detailed characteristics of the aspect ratio effect on LDS is not fully understood (*e.g.* see [5]). In this brief report, we present primary results of a novel computational model that can efficiently analyse various physical conditions, and estimate necessary parameters in order to optimize a leakage detection technique.

FORMULATION OF THE PROBLEM

A fluid saturated porous medium with a localized heat source (see figure, 1 (domain)) has been considered for the present numerical simulations. The flow is presumed to be laminar, incompressible, and in a local thermodynamic equilibrium. The two vertical and upper walls are thermally insulated. The excess temperature vanishes on the lower boundary except a length of L_h, where a localized heat source is applied. The present mathematical model is similar to that of Sivasankaran and Younghae [15] except the momentum and energy equations have been extended;[*i.e.*]

$$\left.\begin{aligned}\frac{\partial u}{\partial t}+u\frac{\partial u}{\partial x}+v\frac{\partial u}{\partial y}&=-\frac{1}{\rho}\frac{\partial p}{\partial x}+\nu\left(\frac{\partial^2 u}{\partial x^2}+\frac{\partial^2 u}{\partial y^2}\right)\\&+[\beta\rho g\left(\theta-\theta_0\right)\sin\alpha-\frac{\nu}{K}\epsilon u],\end{aligned}\right\} \quad (1)$$

$$\left.\begin{aligned}\frac{\partial v}{\partial t}+u\frac{\partial v}{\partial x}+v\frac{\partial v}{\partial y}&=-\frac{1}{\rho}\frac{\partial p}{\partial y}+\nu\left(\frac{\partial^2 v}{\partial x^2}+\frac{\partial^2 v}{\partial y^2}\right)\\&+[\beta\rho g\left(\theta-\theta_0\right)\cos\alpha-\frac{\nu}{K}\epsilon v],\end{aligned}\right\} \quad (2)$$

$$\frac{\partial \theta}{\partial t}+u\frac{\partial \theta}{\partial x}+v\frac{\partial \theta}{\partial y}+[\frac{\partial \bar{\theta}}{\partial y}v]=\kappa\left(\frac{\partial^2 \theta}{\partial x^2}+\frac{\partial^2 \theta}{\partial y^2}\right). \quad (3)$$

Here, u and v are the velocity (m/s) components in horizontal and vertical direction respectively, (x, y) are reference coordinates, p is the pressure (pa), ν is kinematic viscosity (m^2/s), θ is the excess temperature (K), θ_0 is the temperature (K) at a reference state, α is an angle between the horizontal domain and the horizontal axis, K is permeability (m^2) of the medium, g is the acceleration (m^2/s) due to gravity, κ is thermal diffusivity (m^2/s), β is thermal expansion coefficient

Ra	10^3		10^4		10^5	
	Present	D. & T.	Present	D. & T.	Present	D. & T.
vorticity	1.957423	2.06900	3.659917	3.951325	5.345340	5.921375
Nu	0.1533277	0.148605	0.2373662	0.295132	0.3684891	0.643594

Table 1: Comparison of present results with the benchmark results of Dubois and Touzani(D. & T.) [16] in the air in presence of stable thermal gradient

Ra	de V. Davis(1983)	Ho and Lin(1997)	Shu and Wee(2002)	S.Sivasankaran(2011)	Present
10^3	1.117	1.118	1.118	1.117	1.118
10^4	2.238	2.248	2.244	2.247	2.243
10^5	4.509	4.528	4.518	4.521	4.523
10^6	8.817	8.824	8.822	8.806	8.823

Table 2: Comparison of present values of rate of heat transfer with the benchmark results of S. Sivasankeran *et al.* [15] and others.

(1/K), $\frac{\partial \bar{\theta}}{\partial y}$ is a prescribed vertical rate of variation of θ, t is time (s), L is horizontal length (m) of the domain, H is vertical length (m) of the domain, L/H is aspect ratio, ρ is fluid density. The boundary conditions are, $u = 0$, $\frac{\partial v}{\partial x} = 0$, $\frac{\partial \theta}{\partial x} = 0$ at $x = \pm L/2$, and $v = 0$, $\frac{\partial u}{\partial y} = 0$, $\frac{\partial \theta}{\partial y} = 0$ at $y = H$, and $u = 0, v = 0, \theta = 1,$ if $x \in [-L_h/2, L_h/2]$; $\theta = 0$, elswhere. The dimensionless parameters are the Prandtl number, $Pr = \frac{\nu}{\kappa}$, the Rayleigh number, $Ra = \frac{\beta L^3 g \Delta\theta}{\nu\kappa}$, the Darcy number, $Da = \frac{K}{L^2}$, and the Nusselt number, $Nu = -\frac{1}{L}\int_{-L/2}^{L/2}\frac{\partial \theta}{\partial y}\left(x, y = 0\right)dx$.

Numerical Method and Validation

The numerical method is based on the multiscale wavelet decomposition, and is documented in refs [1–3]. The present work adopts a Deslauriers-Dubuc wavelet system of order 6 that exactly reproduces polynomial of order 5. This leads to a a discretization error $\mathcal{O}(\Delta x^6)$ (see [3]). The code was previously validated with a number of heat transfer and fluid flow simulations (see [1–3]). To asses the numerical model development in the present study, we have carried out comparisons with previous benchmark simulations that are dynamically similar with the present work on a different application. Comparison results with 4 other benchmark heat transfer simulations have been summarized in table1 and 2. An excellent agreement has been observed.

RESULTS AND DISCUSSION

Unless it is stated otherwise, a mesh of uniform resolution (257×129) has been used. The resolutions have been altered occasionally to enhance the accuracy of the simulations for certain geometrical and physical situations. Fluid saturated insulator and porosity $\epsilon = 0.39$ in favor of fibrous and granular insulation at the insulator, *e.g.* Bagchi and Kulacki [4], have been used in all of our simulations. For computational and modelling convenience, we use the parameter Ra that represents the ratio of buoyancy force to thermal diffusivity. The present idealized paper is more consistent with Zhao *et al.*

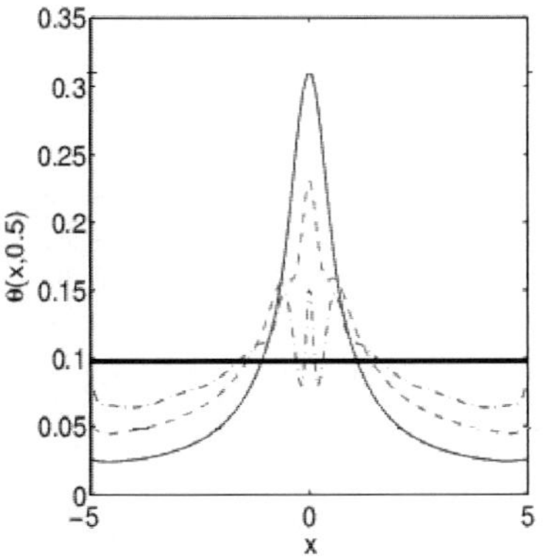

Figure 2: Excess temperature profiles for a fixed horizontal line at $y = 0.5\text{m}$ at different distance in the horizontal insulated domain for $Ra = 10^3$ (solid line), 10^4 (dot-dashed line), 10^5 (dashed line) while $Da = 10^{-6}$. Sensor cable showing as a thick line at $y = 0.5\text{m}$.

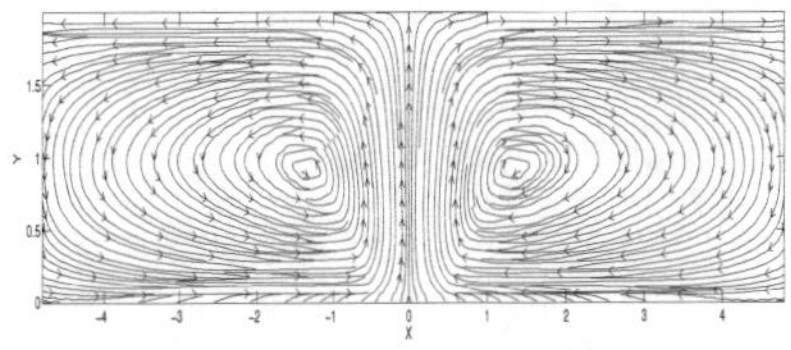

Figure 3: Stable stream lines for localized heat source of size 1 m and $Ra = 10^3$.

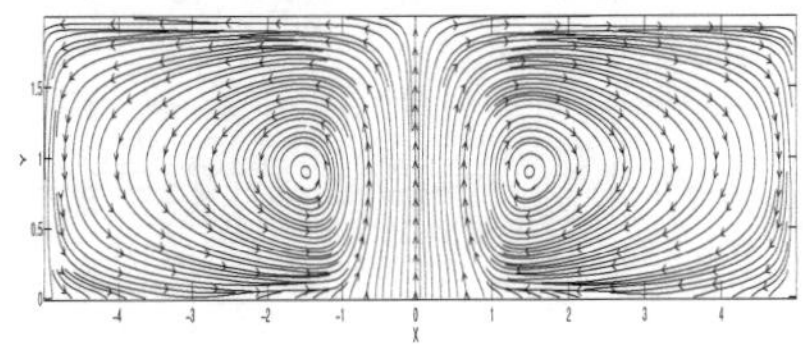

Figure 4: Stable stream lines for localized heat source of size 2 m and $Ra = 10$.

[19][*e.g.* a domain (7 m × 1 m)]. The initial temperature is 293.15 (K), and remains a constant.

Excess temperature variation

Since the heat source is introduced near $x = 0$ on the bottom boundary, the buoyancy force is imbalanced with the thermal diffusivity, and the localized heat profile would signal a leak through the fibre optic cable. Figure 2 presents a horizontal profile of temperature along a horizontal line at $y = 0.5$ m, where we placed the sensor cable, for three values of Ra. Clearly, the temperature maximum at $x = 0$ decreases with an increase of Ra. Eq (3) shows that this is an effect of the presence of the thermal gradient, which we have also confirmed by turning off the thermal gradient term in eq (3) (plots are not shown to save space). Clearly, the horizontal convection introduced by the thermal gradient has a significant impact on the strength of the signal received by the sensor. Spatial heterogeneity of the excess temperature profile is localized above the heating source because the buoyancy force is imbalanced by the thermal diffusivity. However, at the other part of the domain, buoyancy force is balanced by the inertial as well as the thermal diffusivity. The excess temperature is stationary at the 4.5 m difference from the left boundary of the domain at 12000 s, which is depicted in the figure 2. There is an excess temperature profiles are shown between the 4.5 m to 5.5 m distance from the left boundary. However, this temperature is not always excess enough to give a signal to the sensor due to the balance between buoyancy force and thermal diffusivity. More thermal diffusion comparing to buoyancy force may cause the excess temperature at the leakage and as a result of sending signals in the active thermal monitoring system. Hence, it is clear that the balance between buoyancy force and thermal diffusivity is another important factor in LDS. But, this two physical parameters depend on the physical as well as mechanical parameters: size and position of localized heat source, sizes of the insulator domain, permeability of the insulator domain etc.. Hence, now discussion about the mentioned parameters.

Controling proposals of excess temperature

We have placed our localized heat source at the center of the bottom wall. Our main motivation in this section is to make a balance between buoyancy force and thermal diffusivity in the insulator so that we could minimize excess temperature. Beyond this balance condition, the excess temperature will produce and give signal to the sensor to repair at that place. Flow responses for different sizes, $L_h = 1$ m, $L_h = 2$ m is exemplified in the figure 3 and 4 for a fixed value of permeability of the insulator, $Da = 1.4 \times 10^{-6}$. Flows have been simulated at single and double heat sources for different values of Ra. From the comparison figure 3 and 4, it is clear that buoyancy force is balanced by the heat source: there is a stationary circular flow in the middle of the domain from the heat source of size, $L_h = 1$ m with high values of $Ra = 10^3$. However, same patterns have shown for heat source of size, $L_h = 2$ m with $Ra = 10$. Not only Ra, but also time step is a factor in this case. We have used time step $\Delta t = 30$s for $Ra = 10^3$ and $\Delta t = 18$s for $Ra = 10$ in order to the numerical stability and faster convergence. It is evident that the length of the localized heat source and Ra has an inverse relation. Moreover, the effects of the localized heat source size on the convective heat transfer is shown in the figure 5. The influences have been revealed for different values of Ra and localized heat sources with $Da = 1.4 \times 10^{-6}$. It shows that the heat transfer rates are very high for lower values of Ra with the other same parameters. Moreover, the higher size of the heat source creates a high rate of the heat transfer because there is a much dissipation near to the longer heat source. Finally, the heat transfer rates and temperature profiles disclose that the heat transfer rates increases with the increases of the size of the insulator domain.

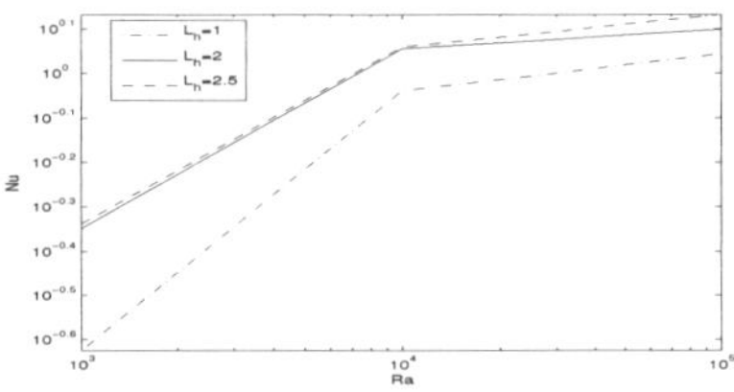

Figure 5: Comparison between the rate of heat transfer, Nu and Ra for different localized heat source sizes $L_h = 1$ m, $L_h = 2$ m and $L_h = 2.5$ m.

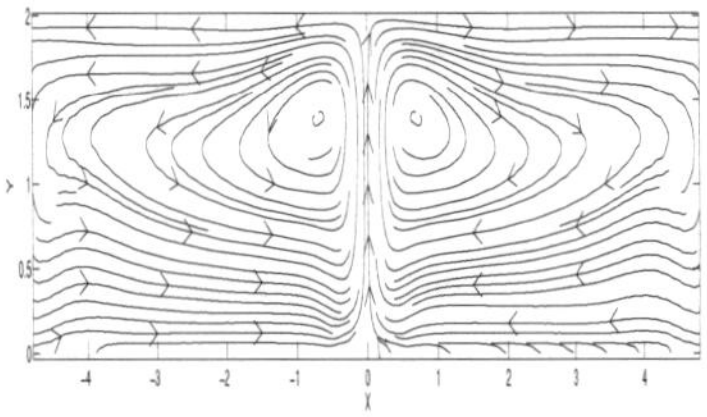

Figure 6: Stable thermal stratification at the core of the insulated domain for permeability 10^{-2} with $Ra = 10^4$.

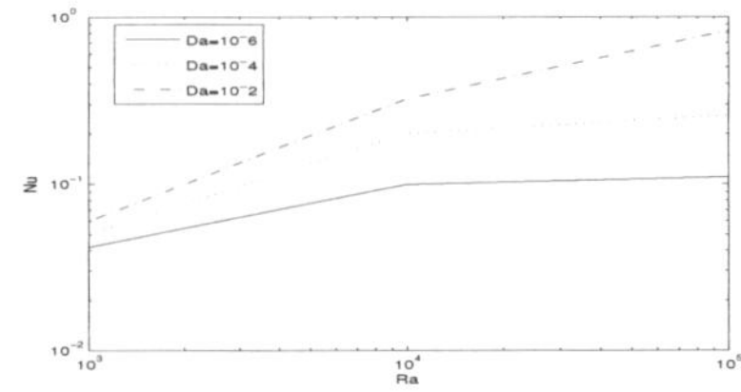

Figure 7: Nusselt number-versus-Ra relation for different permeabilities.

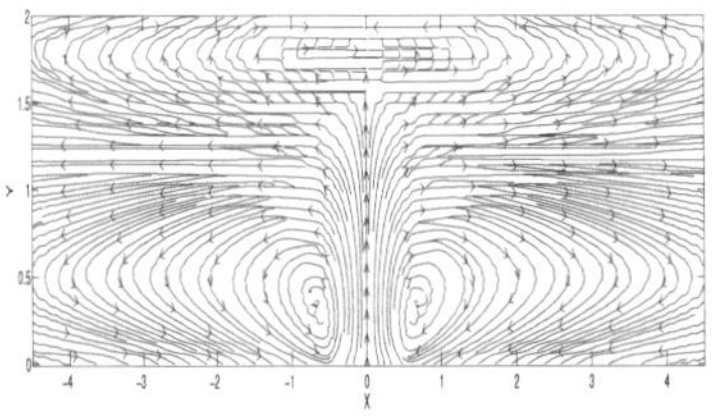

Figure 8: Stream lines at the insulation domain for heat source size 1 m, $Da = 1.4 \times 10^{-6}$, $Ra = 10^5$ in presence of the stable thermal gradient.

The permeability of the medium is important in the temperature monitoring, calculation because heat transfer is different for various permeability of the medium. In our simulation, we have used parameter, Da, to see the effects of permeability that comparing to the size of the insulator. The streamlines are shown in high to low permeability of the medium. There is an insulation drag in the heat transfer from the heat source and the heat transfer is given by the conduction dominated phenomena. The flow structure is circulated in near to the heat source for low Da, $Da = 10^{-2}$ and get chunk at the adiabatic wall and then travels in the direction to the cold wall. The fluid moves along the colder places before fill the entire domain at the core of the enclosure. There is a stable thermal stratification at the centre of the cavity. The thermal flow field is controlled by the convective heat transfer. Nusselt number is an increasing function of Da. The increase of Nusselt number is much for higher values of, $Da = 10^{-2}$. While there is a monotonic behaviours for the Da values of 10^{-4} to 10^{-6} in presence of $10^3 \leq Ra \leq 10^5$, aspect ratio 5 and localized heat source of size 1 m which is shown in the figure 7. In contemplation of physical views that more permeability of a medium has less temperature and heat transfer rate for considering the constant size of the particles in the insulation domain. At the bottom wall, heat transfer phenomenon is almost same for the both the increasing values of Da and Ra with the same heat source size and the horizontal position of the insulation domain. A stronger convective flow triumph in the cavity and the centre of the rotation moves to the opposite adiabatic wall, while strong velocity boundary layer produces at the near to the heat source which indicates that strong heat transfer is continuing in the domain. As a result, excess temperature profiles is not enough strong to give signal at the sensor for repair.

The stable thermal gradient is also important to make a balance between buoyancy force and thermal diffusivity at the domain. To see the influence of linear thermal gradient, equation (3), we have used horizontal rectangular both the air and porous compound with localized heat source on the bottom wall. At the porous domain, although the maximum and minimum value of the velocity, temperature and vorticity are decreased, the bottom wall heat transfer rates are increased. At higher values of Ra with the geometrical variable aspect ratio 5, horizontal heat source of size 1 m, Da=1.4×10^{-6}, the larger domain is influenced by the convective flow so that the streamlines move toward the upper wall in the presence of the stable thermal gradient. However, the streamlines indicates that the temperature field at the centre of the domain is stable for $Ra = 10^4$ in the presence of thermal gradients, while there is a circulatory flow performed at the center of the domain. Moreover, for the high values of Ra, $Ra = 10^5$, the flow becomes stronger and there is an interesting phenomena shows that the bicircular buoyant convective flow have been raised to a tricircular flow structure which is above far from the heat source. Hence, the effects of the size of the domain, for excess temperature at the leakage in LDS, are balanced by the fixed value of Ra in presence of the mentioned other corresponding parameters.

CONCLUSIONS

Our aim has been to find the excess temperature profiles for localized heat source at the leakage in an insulated monitoring pipe. From these excess temperature profiles, a sensor cable collects the signal and send to the main monitor. As a result, it is very easy to find the leak and fix it. Moreover, we also summarise some suggestions to control this excess temperature by dint of the condition of buoyancy force and thermal diffusivity at the insulator. Not only the size of the localized heat source, but also the size of the pipe as well as the permeability of the insulator at a leakage in the pipeline, is the cause of excess temperature. In a granular and fibrous insulator, spatial heterogeneity of excess temperature profiles is localized above the heating source. This excess profiles may be controlled by the surrounding balance of buoyancy force, inertia force as well as thermal diffusivity. The horizontal convection introduced by the thermal gradient has a significant impact on strength of the signal received by the sensor. Buoyancy force is balanced by the different sizes of the heat sources. The above conditions on buoyancy force, inertial force and thermal diffusivity in an insulator at the monitoring pipe may give us more flexibility and minimize cost of the leak detection system. We need to see the larger intensity of localized heat sources which may create irregular flow behaviours or turbulent flow at the localized heat source and, leak detection system in the monitoring pipe which are our future work.

REFERENCES

[1] Alam J. M., "A multiscale modeling study for the convective mass transfer in a subsurface aquifer", *Heat and Mass Transfer, DOI 10.1007/s00231-014-1490-2*, (2015).

[2] Alam J. M., Nicholas K.-R. Kevlahan, Oleg V. Vasilyev & Zahangir Hossain, "A Multiresolution Model for the Simulation of Transient Heat and Mass Transfer, Numerical Heat Transfer", *Part B: Fundamentals: An International Journal of Computation and Methodology* 61(3) (2012) 147–170.

[3] Alam J., Walsh R., Hossain M.A. & Rose A., "A computational methodology for two-dimensional fluid flows", *Int. J. Numer. Meth. Fluids*, 75(12) (2014) 835–859.

[4] A. Bagchi, F. A. Kulacki, "Natural convection in fluid-superposed porous layers heated locally from below", *Int. J. Heat Mass Transfer* 54 (2011) 3672–3682.

[5] A. Pregelj, M. Drab, "Leak detection methods and defining the sizes of leaks", *The 4th International Conference of Slovenian Society for Nondestructive Testing, Application of Contemporary Nondestructive Testing in Engineering*. 4(2) (1999).

[6] API 1130: "Computational Pipeline Monitoring for Liquid Pipelines", *2nd Edition, American Petroleum Institute (November, 2002)*.

[7] F. O. dos Santos, "Model development and numerical simulation for detectng pre-existing leaks in liquid pipeline", *22nd International Congress of Mechanical Engineering (COBEM 2013)*, (November 3-7, 2013).

[8] G. El-Khatib, V. Prasad, "Effects of Stratification on Thermal Convection in Horizontal Porous Layers With Localized Heating From Below", *J. Heat Transfer* 109 (1987) 683–687.

[9] H. F. Oztop, Khaled. Al-Salem, Y. Varol, I. Pop, M. Firat, "Effects of inclination angle on natural convection in an inclined open porous cavity with non-isothermally heated wall", *Int. J. Num. Meth. Heat and fluid flow* 22(8) (2012) 1053–1072.

[10] J. V. N. de Sousal, C. H. Sodre, A. G. B. de Limal, S. R F. Neto, "Numerical Analysis of Heavy Oil-Water Flow and Leak Detection in Vertical Pipeline," *Scientific research*, 3 (2013) 9–15.

[11] M. A. Cotter, E. C. Michael, "Transient cooling of petroleum by natural convection in cylindrical storage tanks-II. Effect of heat transfer coefficient, aspect ratio and temperature-dependent vis- cosity", *Int. J. Heat Mass Transfer* 36 (1993) 2175–2182.

[12] N. H. Saeid, I. Pop, "Transient free convection in a square cavity filled with a porous medium", *Int. J. Heat Mass Transfer* 47 (2004) 1917–1924.

[13] N. H. Saeid, I. Pop, "Natural convection from discrete heater in a square cavity filled with porous medium", *J. Porous Media* 8(1) (2005) 55–63.

[14] P. Rajeev, J. Kodikara, W. K. Chiu, T. Kuen, "Distributed Optical Fibre Sensors and Their Applications in Pipeline Monitoring", *Key Engineering Materials*. 558 (2013) 424–434.

[15] S. Sivasankaran, Younghae. Do, M. Sankar, "Effect of Discrete Heating on Natural Convection in a Rectangular Porous Enclosure", *Transp. Porous Med* 86 (2011) 261–281.

[16] Thierry Dubois, Rachid Touzani, "A numerical study of heat island flows: Stationary solutions", *Int. J. Numer. Meth. fluid* 59 (2009) 631–655.

[17] V. Prasad, F.A Kulacki, "Convective heat transfer in a rectangular porous cavity", *J. Heat Transfer* 106 (1984) 158–165.

[18] V. Prasad, F. A. kulacki, "Natural convection in horizontal porous layers with localized heating from below", *J. Heat Transfer* 109 (1987) 795–798.

[19] X. Zhao, W. Li, G. Song, Z. Zhu, J. Du, "Scour Monitoring System for Subsea Pipeline Based on Active Thermometry: Numerical and Experimental Studies", *Sensors, doi:10.3390/s130201490*. 13 (2013) 1490–1509.

[20] 49 CFR Part 195: "Transportation of Hazardous Liquids by Pipeline", *Code for Federal Regulations Title 49 (January, 2002)*.

Proceedings of the 25th CANCAM
London, Ontario, Canada, May 31 - June 4, 2015

BUOYANCY-DRIVEN FLOWS IN A CONFINED POROUS MEDIUM

Chunendra K. Sahu
Dept. of Mech. Engg.
University of Alberta
Edmonton, AB, Canada
chunendra@ualberta.ca

M. R. Flynn
Dept. of Mech. Engg.
University of Alberta
Edmonton, AB, Canada
mrflynn@ualberta.ca

ABSTRACT

Motivated by the voluminous literature on filling box flows with turbulent free plumes, we examine an analogue problem consisting of a laminar plume in a closed porous medium. A similarity solution is derived for Darcy plumes with Pe $>$ $\mathcal{O}(1)$ where Pe is the Péclet number. This solution is used to predict the plume volume flux and the reduced gravity as functions of the vertical distance from the point of injection, the source buoyancy flux, the permeability etc. Once the plume reaches the bottom of the closed control volume, it is discharged into an ever-deepening layer of contaminated fluid. We provide an estimate for the time-scale associated with the vertical motion of the first front, the interface separating the contaminated and ambient layers. Similitude laboratory experiments are performed with fresh-water and salt-water as the working fluids to verify the theoretical predictions. Generally good agreement is observed. Results from the study of porous media filling box flows may be used to understand the behavior of geological plumes, e.g. the plumes generated by injecting supercritical CO_2 into deep saline aquifers for the purpose of carbon sequestration.

Keywords: Buoyancy-driven flow, filling box model, laminar plumes, first front, porous medium

INTRODUCTION

The development of the MTT plume equations (Morton, Taylor & Turner [12]), which describe the dynamics of free buoyant plumes for Boussinesq fluids in an infinite unstratified ambient, motivated the subsequent development of the "filling box" equations, which describe the flow behavior when the plume is located inside a closed control volume or box. As explained by Baines & Turner [1], upon reaching the bottom of the control volume the (downwards-facing) plume first spreads laterally then forms an ever deepening layer whose upper surface (the "first front") approaches the plume source for large times. The inner solution of the filling box flow, adapted directly from the solution of the MTT equations, describes the vertical variation of the plume volume, momentum and buoyancy fluxes. The outer solution predicts the vertical advection speed for fluid parcels at the first front by equating the volume fluxes inside and outside of the plume. Knowing this advection speed, it is straightforward to compute the position of the first front as a function of time.

Since Baines & Turner's analytical and experimental investigation, there has been an extensive study of buoyancy-driven filling box flows and their application to mixing in chemical storage tanks, building ventilation systems, etc. (Manins [11], Kaye & Hunt [9], Germeles [5], Nabi & Flynn [13]). However, significantly less work has been conducted to describe filling box dynamics in a porous medium. Such considerations are relevant to numerous geological problems e.g. the buoyant plumes generated by the injection of supercritical CO_2 (sc-CO_2) into deep saline aquifers. A study of plumes and filling box models in porous media will help in understanding the behavior of sc-CO_2 plumes in such aquifers in connection with geological sequestration.

Flows in porous media are generally categorized based on the Reynolds number, $\text{Re} = d_0\bar{U}/\nu$, where d_0 is the mean grain diameter, $\bar{U}$ is the specific discharge i.e. the volume flux per unit flow area, and ν is the kinematic viscosity, i.e. into Darcy ($\text{Re} \lesssim \mathcal{O}(10)$) and non-Darcy regimes ($\text{Re} > \mathcal{O}(10)$) [2]. Moreover, the Péclet number, $\text{Pe} = d_0\bar{U}/D_d$ [7], in which D_d is the molecular diffusion coefficient, plays an equally important role in classifying such flows based on two different mass transport mechanisms, i.e. molecular diffusion (Pe $\lesssim \mathcal{O}(1)$) vs. mechanical dispersion (Pe $> \mathcal{O}(1)$). When Pe $> \mathcal{O}(1)$, it is, in fact, helpful to consider a dispersion coefficient, D, rather than D_d where

$$D = d_0 \bar{U} \left(1 + \frac{\omega}{\text{Pe}}\right), \tag{1}$$

and ω (< 1) is a parameter that depends on the tortuosity, i.e. the shape of the flow-path [7].

The dynamics of rectilinear line plumes in porous media were first studied by Wooding [16] for Darcy flow with Pe $\lesssim \mathcal{O}(1)$ i.e. when $D \simeq D_d$. Wooding derived the following formula for the plume volume flux, Q, as a function of the vertical coordinate, x:

$$Q = \left(\frac{36 D \phi F_0 k \Lambda^2 x}{\nu}\right)^{1/3}. \tag{2}$$

Here ϕ is the porosity, F_0 is the source buoyancy flux, Λ is the source width and k is the permeability.

Lai [10] and Chen & Ho [4] extended Wooding's work by studying plumes in the non-Darcy regime with Pe $> \mathcal{O}(1)$ and Pe $< \mathcal{O}(1)$, respectively. However, in both cases, similarity solutions could not be obtained. The authors therefore presented their solutions in terms of a non-similar variable, whose value depends on various physical parameters, e.g. k, ν etc.

Whereas all of the above analysis considered plumes rising or falling through an infinite ambient, Roes et al. [14] specifically investigated flow dynamics inside a confined geometry. More specifically, they incorporated Wooding's solution and developed a model for buoyant convection from a discrete Boussinesq source in a leaky porous medium, i.e. one that is closed except for the presence of one or more discrete fissures. Although the present work is an extension of the filling box model developed previously, it is different in two fundamental ways: (i) here we consider Darcy plumes with Pe $> \mathcal{O}(1)$ and (ii) our box is devoid of discrete fissures. Our problem therefore includes a continuous inflow of source (contaminated) fluid, but no outflow, and is consequently time-dependent by definition – see figure 1.

The rest of the article is organized as follows. Our plume solution is derived in the Theory section, which also outlines the details of the transient filling-box flow. Thereafter in the Experiments section, the complementary laboratory experiments are described; these have the purpose of corroborating the analytical predictions from Theory section. This discussion is followed by a comparison between theory and experiments in the Discussion and Conclusions section.

THEORY

We use the governing equations, previously presented by Wooding [16] and Roes et al. [14], based on mass continuity

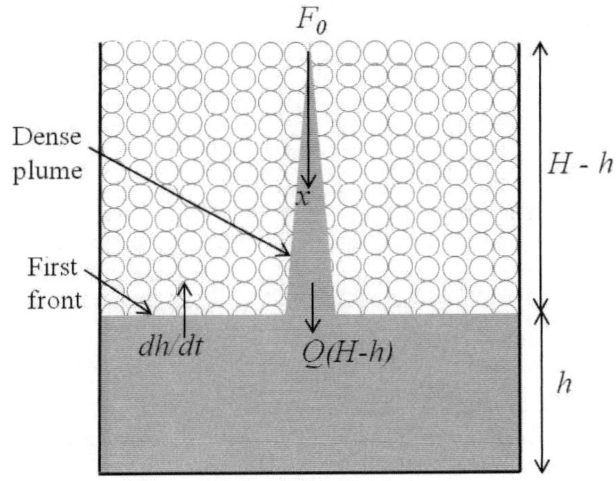

Figure 1: A filling box flow in porous media.

and Darcy's law for Boussinesq, two dimensional and rectilinear flows. However, in case of the solute transport equation, we assume a spatially variable dispersion coefficient, unlike the constant value presumed in these two previous investigations. Thus, on using (1), provided Pe $> \mathcal{O}(1)$ and $\omega < 1$, and introducing a dimensionless constant, α, such that $\bar{U} = \alpha u$, the dispersion coefficient becomes

$$D \simeq \alpha d_0 u \tag{3}$$

where u is the vertical transport velocity and the value of the fitting parameter, α, shall be determined based on experimental measurement. Hence the governing equations read

$$\frac{\partial u}{\partial x} + \frac{\partial v}{\partial y} = 0 \tag{4}$$

$$\frac{1}{\rho_0}\frac{\partial P}{\partial x} + \frac{\nu}{k}u = \frac{g\rho}{\rho_0} \tag{5}$$

$$\frac{1}{\rho_0}\frac{\partial P}{\partial y} + \frac{\nu}{k}v = 0 \tag{6}$$

$$\frac{1}{\phi}\left(u\frac{\partial S}{\partial x} + v\frac{\partial S}{\partial y}\right) = \frac{\partial}{\partial x}\left(D\frac{\partial S}{\partial x}\right) + \frac{\partial}{\partial y}\left(D\frac{\partial S}{\partial y}\right) \tag{7}$$

$$\rho = \rho_0(1 + \beta S) \tag{8}$$

Here P is the fluid pressure, v is the horizontal transport velocity, S is the solute concentration, β is the solute contraction coefficient, ρ is the plume fluid density and ρ_0 is the ambient fluid density which corresponds to a solute concentration of zero.

By adapting the above governing equations, and after much algebra (details presented in Sahu & Flynn [15]), we derive the following formula for the plume volume flux:

$$Q = \left[\left(\frac{16 F_0 k \Lambda}{\pi \nu}\right)^2 d_0 \phi \alpha (x + x_0)\right]^{1/4} \tag{9}$$

where x_0 is the virtual origin correction [16] and is given by:

$$x_0 = \frac{1}{d_0 \phi \alpha}\left(\frac{\pi \nu}{16 F_0 k \Lambda}\right)^2 Q_0{}^4.$$

Similarly, the plume momentum flux and the mean reduced gravity are determined to be, respectively,

$$M = \frac{F_0 k}{\nu} \text{ and } \bar{g}' = \left[\left(\frac{\pi F_0 \nu}{16 k \Lambda} \right)^2 \frac{1}{d_0 \phi \alpha (x + x_0)} \right]^{1/4}. \quad (10)$$

In the current study, we find that Q and $\bar{g}'$ are proportional to $x^{1/4}$ and $x^{-1/4}$, respectively, whereas in the previous investigations by Wooding [16] and Roes et. al [14], where $\text{Pe} \lesssim \mathcal{O}(1)$, these quantities are proportional to $x^{1/3}$ and $x^{-1/3}$, respectively. However, M is still given by the same equation, because our plumes plus those of Wooding satisfy the conditions for Darcy flow.

Once negatively-buoyant plume fluid reaches the impermeable bottom surface, it moves horizontally sidewards, and thus creates a pair of gravity currents. Further, after reaching the lateral sidewalls, the discharged plume fluid starts advecting upward. The interface separating this contaminated layer from the ambient layer above is initially curved, but becomes horizontal in the limit as $t \to \infty$ (a description of the gravity current and of the curved interface dynamics is presented in Sahu & Flynn [15]). The vertical advection speed of this horizontal interface can then be determined by simple volume flux balance, i.e.

$$\phi A U(h) = Q(H - h), \quad (11)$$

where Q is defined by (9), A is the cross-section of the box, which is assumed to be much larger than that of the plume and U is the velocity of the first front, whose elevation relative to the bottom of the box we denote by h. After applying (9), and with further integration and simplification, the change of first front elevation, $h_2 - h_1$ in a given interval of time, $t_2 - t_1$, can be determined from

$$h_2 = H + x_0 - \left\{ (H + x_0 - h_1)^{3/4} - \frac{3}{4} \frac{(t_2 - t_1)}{A} \left[\left(\frac{16 F_0 k \Lambda}{\pi \nu} \right)^2 \frac{d_0 \alpha}{\phi^3} \right]^{1/4} \right\}^{4/3}. \quad (12)$$

Qualitatively, (12) is similar to (B3) of Caulfield & Woods [3], who studied filling box flows for a box devoid of porous material.

EXPERIMENTS

Laboratory experiments were performed in an acrylic transparent rectangular box, 88.9 cm long × 7.6 cm wide × 50.8 cm tall, which was filled with either 5 mm or 3 mm Potters Industries A Series Premium glass beads and fresh water to a height of 40 cm. The porosity of the uniformly distributed spherical beads was considered to be $\phi = 0.38$ [6]. Saline water mixed with Procion MX dye was used as the negatively-buoyant source fluid. The density of the salt water was kept below 1.10 g/cm^3, so that the Boussinesq approximation was satisfied in all experiments. This source fluid was supplied from an overhead reservoir fitted with an internal weir to maintain a constant level. A Gilmont GV-2119-S-P flow meter along with a ball and a needle valves were used to measure and control the flow rate of dense fluid, which was subsequently discharged into the experimental box using a line nozzle. The design of the nozzle assures a uniform distribution of source fluid in the transverse direction. The nozzle was located at a height of 36 cm from the bottom of the box.

As summarized in table 1 of Sahu & Flynn [15], a significant number of experiments were performed by changing the bead diameter, reduced gravity, source volume flux and the fluid viscosities; these details are not reiterated here. Rather, we focus on four select experiments, chosen based on different bead diameter, reduced gravity and the source volume flux. A comparison between the experimental measurements for these four cases and the corresponding theoretical solutions is provided in the following section.

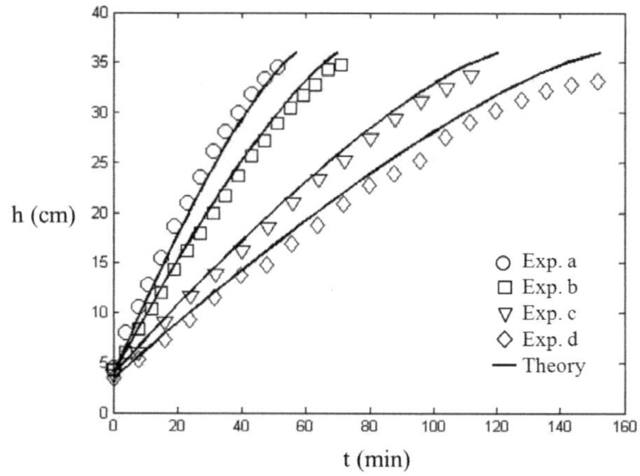

Figure 2: First front elevation vs. time. Experimental parameters are as follows: (a) source volume flux, $Q_0 = 0.9$ cm^3/s, bead diameter, $d_0 = 0.3$ cm, source reduced gravity, $g'_0 = 85.3$ cm/s^2, viscosity, $\nu = 0.01$ cm^2/s, (b) $Q_0 = 0.9$ cm^3/s, $d_0 = 0.3$ cm, $g'_0 = 54.9$ cm/s^2, $\nu = 0.01$ cm^2/s, (c) $Q_0 = 0.3$ cm^3/s, $d_0 = 0.5$ cm, $g'_0 = 21.0$ cm/s^2, $\nu = 0.01$ cm^2/s, and (d) $Q_0 = 0.17$ cm^3/s, $d_0 = 0.3$ cm, $g'_0 = 85.3$ cm/s^2, $\nu = 0.01$ cm^2/s.

DISCUSSION AND CONCLUSIONS

Based on the sum total of our laboratory data, the mean value of the fitting parameter, α, is found to be 0.05 with a standard and maximum deviations ± 3% and ± 5%, respectively [15]. Figure 2 shows the comparison of the theoretically predicted first front elevation vs. the corresponding experimental measurements. The curves and data sets do not start from the origin because the first front evolves from the gravity current whose mean height at the sidewall is prescribed by the y-axis intercept. In all cases, $\mathrm{d}^2h/\mathrm{d}t^2 < 0$, i.e. the contaminated layer depth increases at a decreasing rate. This behavior can be explained as follows: $Q \propto x^{1/4}$ (see equation 9), therefore the first front advects at a faster rate when it is close to the source.

The overall agreement between theory and observation in figure 2 is favorable, particularly in the case of experiments a, b and c. However, the discrepancies observed in experiment d is because of the deviations obtained in the error-minimizing value of α from its mean value [15], and as reported above this deviation is within $\pm$ 5%.

On a concluding note, it should be reiterated that the current study is performed under a set of limiting assumptions. Most notably, we consider Boussinesq and miscible fluids and Darcy flow in an isotropic porous medium, whereas in actual geological settings, e.g. in CO_2 sequestration, these assumptions may not apply. The goal of our next phase of this research will therefore be to examine flow dynamics in a medium having spatially-variable ϕ and k.

ACKNOWLEDGMENTS

The current research is supported by NSERC (Discovery Grant and RTI programmes) and Carbon Management Canada. The authors acknowledge the help provided by Mark Roes in setting up the experiment and in writing the image post-processing routines used in extracting the data shown in figure 2. The authors also wish to acknowledge the technical support kindly offered by the Department of Mechanical Engineering Machine Shop at the University of Alberta.

REFERENCES

[1] Baines, W. D. and Turner, J. S. 1969, Turbulent buoyant convection from a source in a confined region, *J. Fluid Mech.* Vol(37), pp. 51-80.

[2] Bear, J. 1972, *Dynamics of fluids in porous medium*, American Elsevier.

[3] Caulfield, C. P. & Woods, A. W. 2002, The mixing in a room by a localized finite-mass-flux source of buoyancy, *J. Fluid Mech.* Vol(471), pp. 33-50.

[4] Chen, K. S. and Ho, J. R. 1986, Effect of flow inertia on vertical natural convection in saturated porous media, *Int. J. Heat Mass Transfer* Vol(29), pp. 753-759.

[5] Germeles, A. E. 1975, Forced plumes and mixing of liquids in tanks, *J. Fluid Mech.* Vol(71), pp. 601-623.

[6] Happel, J. and Brenner, H. 1991, Low Reynolds Number Hydrodynamics: With Special Applications to Particulate Media, *Kluwer Academic* Vol(2).

[7] Houseworth, J. E. 1984, Longitudinal dispersion in non-uniform, isotropic porous media, PhD Thesis, California Institue Of Technology, Pasadena, California, Report No. KH-R-45.

[8] Huppert, H. E. and Woods, A. W. 1995, Gravity-driven flows in porous layers, *J. Fluid Mech.* Vol(92), pp. 55-69.

[9] Kaye, N. B. and Hunt, G. R. 2005, Overturning in a filling box, *J. Fluid Mech.* Vol(576), pp. 297323.

[10] Lai, F. C. 1991, Non-Darcy convection from a line source of heat in a saturated porous medium, *Int. Comm. Heat Mass Transfer* Vol(18), pp. 445-457.

[11] Manins, P. C. 1978, Turbulent buoyant convection from a source in a confined region, *J. Fluid Mech.* Vol(91), pp. 765-781.

[12] Morton, B. R., Taylor, G. I. and Turner, J. S. 1956, Turbulent gravitational convection from maintained and instantaneous sources, *Proc. R. Soc. Lond.* A 234, 123.

[13] Nabi, S. and Flynn, M. R. 2014, Influence of geometric parameters on the eventual buoyancy stratification that develops due to architectural exchange flow, *Building and Environment* Vol(71), pp. 33-46.

[14] Roes, M. A., Bolster D. T. and Flynn M. R. 2014, Buoyant convection from a discrete source in a leaky porous medium, *J. Fluid Mech.* Vol(755), pp. 204-229.

[15] Sahu, C. K. and Flynn, M. R. 2015, Filling box flows in porous media (to be submitted).

[16] Wooding, R. A. 1963, Convection in a saturated porous medium at large Rayleigh number or Péclet number, *J. Fluid Mech.* Vol(15), pp. 527544.

Proceedings of the 25th CANCAM
London, Ontario, Canada, May 31 - June 4, 2015

CARBON SEQUESTRATION MODELLING USING STABILIZED FINITE ELEMENT METHODS

Chris Ladubec
Dept. of Civil and Environmental Engineering
University of Waterloo
Waterloo, Ontario, Canada
cladubec@uwaterloo.ca

Robert Gracie
Dept. of Civil and Environmental Engineering
University of Waterloo
Waterloo, Ontario, Canada
rgracie@uwaterloo.ca

ABSTRACT

Carbon sequestration aims to reduce CO_2 emissions to the atmosphere by permanently storing CO_2 in deep saline (brine-filled) aquifers. Models of carbon sequestration systems involve a great deal of uncertainty related to system properties. Computationally efficient models are required to perform stochastic analyses. One such computationally efficient method is described in this paper. The model is based on the vertically averaged multiphase flow in a porous medium equations that describe the mass balance of the brine and CO_2.

In the solution methodology adopted in this paper the standard Galerkin Finite Element Method (FEM) can produce nonphysical oscillatory results in the saturation distribution. We compare the Galerkin FEM with three stabilized Finite Element Methods: Streamline Upwind (SU), Streamline Upwind Petrov Galerkin (SUPG) and the Least Squares Finite Element Method (LSFEM).

Keywords: Multiphase Flow, Carbon Sequestration, Galerkin, Finite Element Method, SUPG, SU, LSFEM

INTRODUCTION

It is believed that the release of CO_2 into the atmosphere contributes to global warming. As an alternative to releasing CO_2 from point sources such as coal-fired power plants, CO_2 may instead be captured and stored in deep geological structures, through a process known as carbon sequestration. An important class of potential geological structures are deep saline aquifers, which initially contain brine. The CO_2 is prevented from escaping the aquifer by a low permeability caprock overlying the aquifers.

The storage of CO_2 in deep saline aquifers involves a great deal of uncertainty stemming from the sparse availability of data and heterogeneity. Stochastic methods are necessary for site assessment and require numerical models that are computationally inexpensive [1]. The use of vertically averaged models reduce the dimensionality of the problem (from 3D to 2D or 2D to 1D) and thus reduce the computational cost [2, 3, 4]. In the solution approach adopted in this paper, one of the coupled governing equations contains an advection term. The standard Galerkin Finite Element Method (FEM) can produce nonphysical, oscillatory solutions when an advection term is present in the governing equations. To overcome this difficulty we examine three alternative FEMs and compare these solutions with the solution obtained from the Galerkin FEM.

In this paper we present a carbon sequestration model that is composed of vertically averaged multiphase flow equations. The coupled equations are discretized using the Galerkin FEM, the Streamline Upwind Finite Element Method (SU) [5], the Streamline Upwind Petrov-Galerkin Method (SUPG) [5] and the Least Squares Finite Element Method (LSFEM) [6]. The two coupled equations (pressure equation and saturation equation) are solved using a sequential solution strategy. An example demonstrates the capabilities of these stabilization methods.

GOVERNING EQUATIONS

The governing equations for this problem come from the vertically averaged mass balance of the CO_2 and brine phases [2].

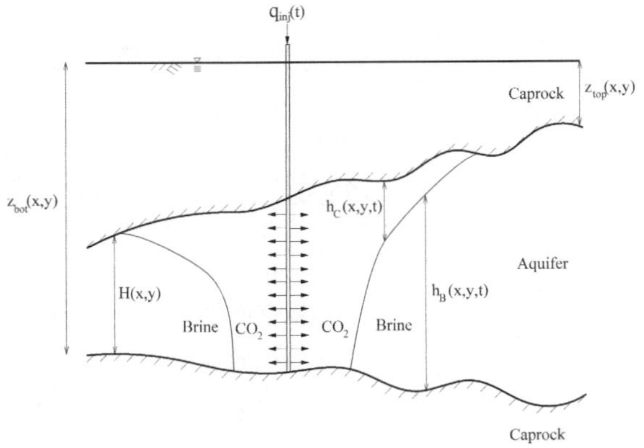

Figure 1: Carbon sequestration model parameters.

$$\phi(1-S_B)\frac{\partial h_C}{\partial t}+\frac{\partial \hat{q}_C}{\partial x}=Q_{inj}(t)\delta(x-x_{inj}) \quad (1)$$

$$\phi(1-S_B)\frac{\partial h_B}{\partial t}+\frac{\partial \hat{q}_B}{\partial x}=0 \quad (2)$$

where ϕ is the porosity of the aquifer, S_B is the residual brine saturation, h_C is the depth of CO_2, h_B is the depth of CO_2, $\hat{q}_C$ is the vertically averaged CO_2 flux, $\hat{q}_B$ is the vertically averaged brine flux, Q_{inj} is the injection rate of CO_2, and $\delta()$ is the Dirac delta function. The fluxes are determined by the multiphase extension of Darcy's law. Figure 1 illustrates the problem.

$$\hat{q}_C=-h_C\frac{kk_C^{rel}}{\mu_C}(\frac{\partial p_{bot}}{\partial x}-\rho_B g\frac{\partial H}{\partial x}+\Delta\rho g\frac{\partial h_C}{\partial x}+\rho_C g\frac{\partial z_{top}}{\partial x}) \quad (3)$$

$$\hat{q}_B=-h_B\frac{k}{\mu_B}(\frac{\partial p_{bot}}{\partial x}+\rho_B g\frac{\partial z_{bot}}{\partial x}) \quad (4)$$

where k is the permeability of the aquifer, k_C^{rel} is the relative permeability of CO_2, μ_C is the viscosity of CO_2, p_{bot} is the pressure at the bottom of the aquifer, ρ_B is the density of brine, ρ_C is the density of CO_2, $\Delta\rho=\rho_B-\rho_C$, g is the gravitational constant, H is the depth of the aquifer, z_{top} is the depth to the top of the aquifer, and z_{bot} is the depth to the bottom of the aquifer.

In order to arrive at the strong form of the problem, Equations (1) and (2) are added together to result in a pressure equation, where pressure is treated as the unknown. Equation (2) is termed the saturation equation, as the saturation (or average depth) of the brine is treated as the unknown. The resulting strong form of the pressure equation can be stated as:

$$\frac{\partial}{\partial x}\left(-h_C\frac{kk_{REL}^C}{\mu_C}(\frac{\partial p_{bot}}{\partial x}+\Delta\rho g\frac{\partial h_C}{\partial x}-\rho_B g\frac{\partial H}{\partial x}+\rho_C g\frac{\partial z_{top}}{\partial x})\right)$$
$$+\frac{\partial}{\partial x}\left(-h_B\frac{k}{\mu_B}(\frac{\partial p_{bot}}{\partial x}+\rho_B g\frac{\partial z_{bot}}{\partial x})\right)$$
$$=Q_{inj}(t)\delta(x-x_{inj}),\quad x\in\Omega \quad (5)$$

and

$$p_{bot}(x,t)=\overline{p}_{bot}(x,t) \text{ on } \Gamma_p \quad (6)$$

where $\overline{p}_{bot}(x,t)$ is the prescribed Dirichlet boundary condition applied to the domain boundary Γ_p. The pressure is approximated using linear two-node shape functions, $N_I(x)$, and the pressure at node I, p_I.

$$p_{bot}(x)=\sum_{I\in\mathcal{N}}N_I(x)p_I \quad (7)$$

The strong form of the saturation equation can be obtained by substituting (4) into (2), giving:

$$\frac{\partial h_B}{\partial t}=\frac{1}{\phi(1-S_B)}\frac{\partial}{\partial x}\left(h_B\frac{k}{\mu_B}(\frac{\partial p_{bot}}{\partial x}+\rho_B g\frac{\partial z_{bot}}{\partial x})\right),$$
$$x\in\Omega,\ \mathrm{t}\in[0,t_{end}] \quad (8)$$

$$h_B(x,t)=\overline{h}_B(x,t) \text{ on } \Gamma_h \quad (9)$$

$$h_B(x,0)=h_{Bo}(x) \text{ on } \Omega \quad (10)$$

where $\overline{h}_B(x,t)$ is the prescribed depth of CO_2 at the domain boundary Γ_h, and $h_{Bo}(x)$ is the initial brine distribution. The depth of brine is approximated using two node linear finite elements.

$$h_B(x,t)=\sum_{I\in\mathcal{N}}N_I(x)h_I(t) \quad (11)$$

where the depth of brine at node I, $h_I(t)$ is a function of time. The strong forms of the pressure and saturation equations presented in this section are converted to the weak forms, which are then discretized. The discretizations are presented in the next two sections. The coupled equations (5) and (8) are solved using a sequential solution technique described in [3, 4].

FINITE ELEMENT FORMULATION OF THE PRESSURE EQUATION

The pressure equation (5) is discretized using the Galerkin Finite Element Method. The system of equations are described by the matrices and vectors below. The system of equations shown in (12) is solved for p_{bot}, the pressure at the bottom of the aquifer.

$$\mathbf{K}^p\mathbf{p}_{bot}=\mathbf{F}^{inj}+\mathbf{F}^{buoy}+\mathbf{F}^{slope} \quad (12)$$

where $\mathbf{K}^p$ is the stiffness matrix, $\mathbf{F}^{inj}$ is the injection flux vector, $\mathbf{F}^{buoy}$ is the buoyancy flux vector, and $\mathbf{F}^{slope}$ is the slope flux vector. The matrices and vectors are:

$$\mathbf{K}^p=\int_{\Omega^e}\mathbf{B}^{e\mathsf{T}}\left(\frac{h_C k_C^{rel}}{\mu_C}+\frac{h_B}{\mu_B}\right)k\mathbf{B}^e d\Omega^e \quad (13)$$

$$\mathbf{F}^{inj}=\int_{\Omega^e}\mathbf{N}^{e\mathsf{T}}Q_{inj}(t)\delta(x-x_{inj})d\Omega^e \quad (14)$$

$$\mathbf{F}^{buoy} = -\int_{\Omega^e} \mathbf{B}^{\mathbf{e}\mathsf{T}} \frac{k k_C^{rel}}{\mu_C} \Delta\rho g h_C \frac{\partial h_C}{\partial x} d\Omega^e \tag{15}$$

$$\begin{aligned}\mathbf{F}^{slope} = \int_{\Omega^e} \mathbf{B}^{\mathbf{e}\mathsf{T}} (h_C \frac{k k_C^{rel}}{\mu_C} (\rho_B g \frac{\partial H}{\partial x} - \rho_C g \frac{\partial z_{top}}{\partial x}) \\ - h_B \frac{k}{\mu_B} \rho_B g \frac{\partial z_{bot}}{\partial x}) d\Omega^e\end{aligned} \tag{16}$$

where $\mathbf{N^e}$ is a matrix containing the shape functions for element e, and $\mathbf{B^e}$ is a matrix containing the derivatives of the shape functions.

FINITE ELEMENT FORMULATION OF THE SATURATION EQUATION

The saturation equation (8) is discretized using the Galerkin, SU [5], SUPG [5] and LSFEM [6] Finite Element Methods. A good presentation of stabilization methods can be found in [7]. The system of equations that needs to be solved has the form

$$[\mathbf{C}]\{\dot{\mathbf{h}}_\mathbf{B}\} + [\mathbf{K}]\{\mathbf{h_B}\} = \{0\} \tag{17}$$

where $\mathbf{C}$ is the storage matrix and $\mathbf{K}$ is the stiffness matrix. The Finite Difference Method (FDM) is used to solve the saturation equation in time, which provides us with the following system of equations which is solved for the brine depth at time step $n+1$.

$$\begin{aligned}\left([\mathbf{C}_\alpha]^{n+1} + \Delta t\theta[\mathbf{K}_\alpha]^{n+1}\right)\{\mathbf{h_B}\}^{n+1} = \\ (\left([\mathbf{C}_\alpha]^{n+1} - \Delta t(1-\theta)[\mathbf{K}_\alpha]^{n+1}\right)\{\mathbf{h_B}\}^{n}\end{aligned} \tag{18}$$

where α = G (Galerkin), SU, SUPG, or LSFEM. If $\theta = 1$ this is called the Backward Euler method.

The storage and stiffness matrices for each of the spatial discretization methods are given below.

$$\mathbf{K}_G = \int_{\Omega^e} \mathbf{B}^{\mathbf{e}\mathsf{T}} a \mathbf{N^e} d\Omega^e \tag{19}$$

$$\mathbf{C}_G = \int_{\Omega^e} \mathbf{N}^{\mathbf{e}\mathsf{T}} \mathbf{N^e} d\Omega^e \tag{20}$$

$$\mathbf{K}_{SUPG}^{stab} = \int_{\Omega^e} \mathbf{B}^{\mathbf{e}\mathsf{T}} \tau a^2 \mathbf{B^e} d\Omega^e \tag{21}$$

$$\mathbf{K}_{SUPG} = \mathbf{K}_{SU} = \mathbf{K}_G + \mathbf{K}_{SUPG}^{stab} \tag{22}$$

$$\mathbf{C}_{SUPG}^{stab} = -\int_{\Omega^e} \mathbf{B}^{\mathbf{e}\mathsf{T}} \tau a \mathbf{N^e} d\Omega^e \tag{23}$$

$$\mathbf{C}_{SUPG} = \mathbf{C}_G + \mathbf{C}_{SUPG}^{stab} \tag{24}$$

$$\mathbf{C}_{SU} = \mathbf{C}_G \tag{25}$$

$$\mathbf{K}_{LSFEM}^{stab} = \int_{\Omega^e} \mathbf{B}^{\mathbf{e}\mathsf{T}} \theta\Delta t a^2 \mathbf{B^e} \mathbf{d\Omega^e} \tag{26}$$

$$\mathbf{K}_{LSFEM} = \mathbf{K}_G + \mathbf{K}_{LSFEM}^{stab} \tag{27}$$

$$\mathbf{C}_{LSFEM}^{stab} = -\int_{\Omega^e} \mathbf{B}^{\mathbf{e}\mathsf{T}} \theta\Delta t a \mathbf{N^e} d\Omega^e \tag{28}$$

$$\mathbf{C}_{LSFEM} = \mathbf{C}_G + \mathbf{C}_{LSFEM}^{stab} \tag{29}$$

where τ is a stabilization parameter, a is the advective velocity and can be computed by the following equation.

$$a = -\frac{k}{\phi(1-S_B)\mu_B}(\frac{\partial p_{bot}}{\partial x} + \rho_B g \frac{\partial z_{bot}}{\partial x}) \tag{30}$$

The relation between the timestep size, Δt and the element size, Δx is given by the Courant number.

$$Cr = \frac{|a|\Delta t}{\Delta x} \tag{31}$$

EXAMPLE

To demonstrate the performance of the four formulations (Galerkin, SU, SUPG, and LSFEM) CO_2 is injected into an aquifer as shown in Figure 2. The system properties are given in Table 1. As the CO_2 is injected the brine is displaced. The system is symmetric, thus only half the model along with symmetric boundary conditions (no flux) are applied. The pressure equation and the saturation equation have Dirichlet boundaries at the end of the domain (x=1000). For the pressure equation, a hydrostatic pressure is applied at the domain boundary ($p_{bot}(1000) = p_{hyd}$). The saturation equation is solved with a boundary condition of $h_B(1000) = H$.

Backward Euler Finite Difference time integration results are illustrated in Figure 3. The Backward Euler method, which is implicit, gives realisitic, oscillation-free brine distributions for all four formulations when $Cr = 2$.

Figure 4 shows the brine distribution after 350 days of injection. The CO_2 plume has made contact with the domain boundary. Even though the Galerkin method gives oscillation-free results in early time before the CO_2 has made contact with the boundary (Figure 3), once contact is made with the Dirichlet boundary, oscillations contaminate the Galerkin solution, while the other formulations are unaffected.

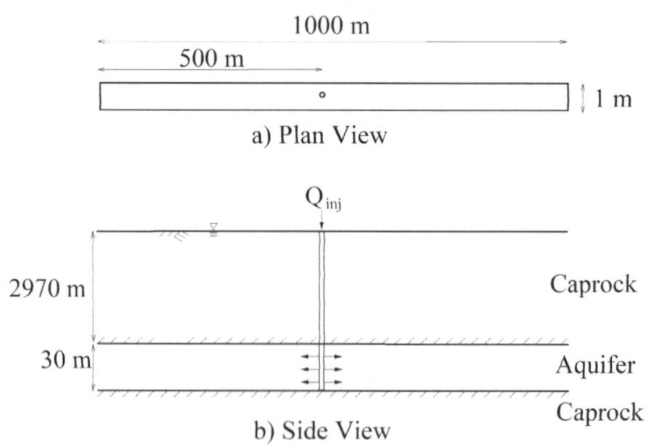

Figure 2: Injection into single aquifer system.

Table 1: System properties for example 1.

Property	Value	Units
μ_B	2.535e-4	Ns/m^2
μ_C	3.950e-5	Ns/m^2
ρ_B	1045	kg/m^3
ρ_C	479	kg/m^3
S_B	0	-
ϕ	0.15	-
k	2e-14	m^2
k_C^{rel}	1	-
Q_{inj}	13.34	m^3/day

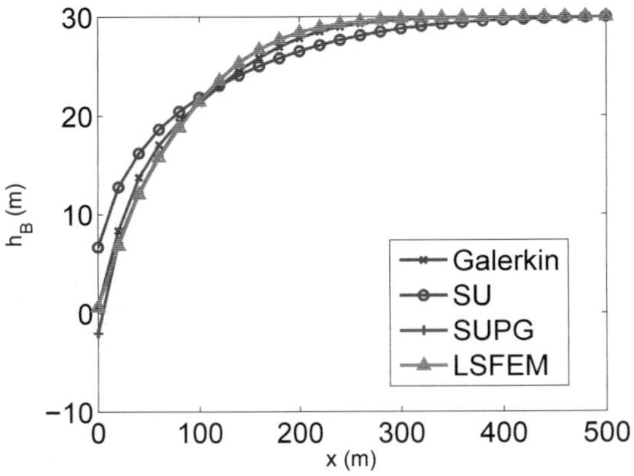

Figure 3: Saturation Distribution after 50 days of injection, $\theta = 1$, $Cr = 2$ using 25 linear finite elements.

CONCLUSIONS

Carbon sequestration can reduce emissions of CO_2 by permanently storing CO_2 in deep saline aquifers. A computationally efficient model is presented that is based on vertically averaged mass balance equations for CO_2 and brine.

In this paper we compared different Finite Element approximations (Galerkin, SU, SUPG, and LSFEM) of the saturation profile in the system. When Backward Euler time integration is used all four methods of approximating the saturation distribution avoid oscillations after 50 days of injection. However, when the CO_2 reaches the Dirichlet boundary (350 days), the Galerkin formulation is contaminated by significant nonphysical oscillations.

ACKNOWLEDGMENTS

The authors would like to acknowledge support from the Natural Sciences and Engineering Research Council (NSERC), NSERC-Network of Centres of Excellence- Carbon Management Canada and from NSERC Discovery Grant.

REFERENCES

[1] M. Sarkarfarshi, F. A. Malekzadeh, R. Gracie, and M. B. Dusseault, "Parametric sensitivity analysis for CO_2 geosequestration," *International Journal of Greenhouse Gas Control*, vol. 23, pp. 61–71, 2014.

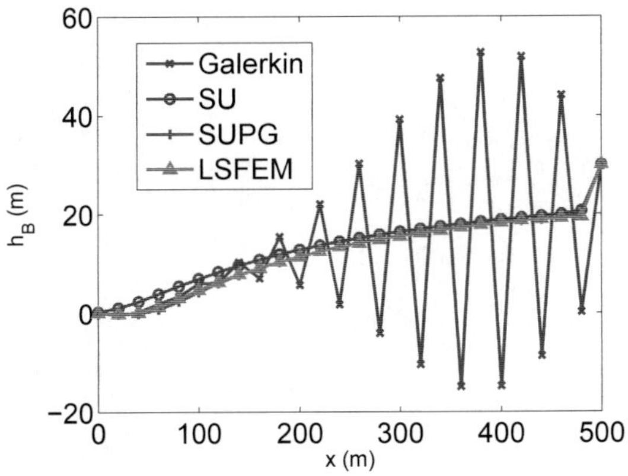

Figure 4: Saturation Distribution after 350 days of injection, $\theta = 1$, $Cr = 2$ using 25 linear finite elements.

[2] M. A. Celia and J. M. Nordbotten, "Practical modeling approaches for geological storage of carbon dioxide," *Groundwater*, vol. 47, no. 5, pp. 627–638, 2009.

[3] C. Ladubec, R. Gracie, and J. Craig, "An extended finite element method model for carbon sequestration," *International Journal for Numerical Methods in Engineering*, DOI: 10.1002/nme.4737, 2014.

[4] C. Ladubec and R. Gracie, "Stabilized finite element methods for vertically averaged multiphase flow in carbon sequestration," *Submitted to Computational Mechanics*, 2015.

[5] A. N. Brooks and T. J. Hughes, "Streamline upwind/petrov-galerkin formulations for convection dominated flows with particular emphasis on the incompressible navier-stokes equations," *Computer methods in applied mechanics and engineering*, vol. 32, no. 1, pp. 199–259, 1982.

[6] G. F. Carey and B. Jiang, "Least-squares finite elements for first-order hyperbolic systems," *International journal for numerical methods in engineering*, vol. 26, no. 1, pp. 81–93, 1988.

[7] J. Donea and A. Huerta, *Finite element methods for flow problems*. John Wiley & Sons, 2003.

Proceedings of the 25th CANCAM
London, Ontario, Canada, May 31 - June 4, 2015

REACTION-DRIVEN FLOW IN A POROUS MEDIUM

Jared Penney
Department of Applied Mathematics
University of Waterloo
Waterloo, ON, Canada
j2penney@uwaterloo.ca

Marek Stastna
Department of Applied Mathematics
University of Waterloo
Waterloo, ON, Canada
mmstastn@uwaterloo.ca

ABSTRACT

Chemically reactive flows in porous media occur in a number of geophysical and engineering processes. This paper presents a 2D numerical simulation of the flow of an incompressible, buoyant fluid through a square porous pile exposed to the atmosphere. The fluid contains a tracer that reacts exothermically with material contained in the porous medium. Spatial derivatives are calculated using Chebyshev spectral differentiation, and time-stepping is first order. The exothermicity of the reaction causes the flow to be dominated by buoyancy, which results in a region of strong tracer flux through the top of the pile.

Keywords: porous media flow, exothermic reaction effects, Darcy-Boussinesq equation, Chebyshev spectral differentiation, computational fluid dynamics

INTRODUCTION

In certain geophysical and engineering porous media flows, such as acid mine drainage in mine waste rock piles [1] and carbon sequestration [2, 3], there is often a concern with the transport of dissolved solutes. These processes often involve reactions that may occur between dissolved materials in the saturating fluid and the porous matrix. In addition to consuming the reactants, the reaction may also absorb or release heat, altering the buoyancy of the fluid and thereby affecting the overall flow.

This research examines the flow of an incompressible buoyant fluid containing an exothermically reactive dissolved solute, or tracer, through a pile of stationary, homogeneous, porous medium exposed to a steady, unidirectional wind. Equations and boundary conditions governing the flow are presented, and an appropriate nondimensionalization is introduced. General geometric properties of the flow fields are discussed, and a comparison of the advective and diffusive vertical tracer fluxes through the top of the pile is examined.

MODEL AND NUMERICAL METHODS

This model describes a two-dimensional rectangular pile of porous material exposed to the atmosphere with horizontal extent $x^* \in [-L, L]$ and vertical extent $z^* \in [0, H]$ resting on an impermeable surface (e.g., the ground) at $z^* = 0$. Note that asterisks are used to denote dimensional variables. The fluid is assumed to be incompressible,

$$\vec{\nabla}^* \cdot \vec{u}^* = 0, \tag{1}$$

with Darcy's law under the Boussinesq approximation governing the flow inside the pile,

$$\vec{u}^* = -\frac{\kappa}{\mu}\left(\vec{\nabla}^* p^* + \rho' g\hat{k}\right), \tag{2}$$

where $\vec{u}^*$ is the flow rate, κ is the permeability, μ is the dynamic viscosity, g is the acceleration due to gravity, $\hat{k}$ is the unit normal in the positive vertical direction, and p and ρ' are fluctuations from the base state of pressure and density, respectively. Applying incompressibility to Darcy's law with constant permeability and viscosity gives an equation for pressure,

$$\nabla^{*2} p^* = g\frac{\partial \rho'}{\partial z^*}. \tag{3}$$

The density fluctuation is assumed to be a linear function of temperature T only,

$$\rho' = -\rho_f \alpha (T^* - T_0), \tag{4}$$

with thermal expansion coefficient α, reference density ρ_f, and reference temperature T_0.

The tracer concentration C and temperature fields are governed by similar advection-diffusion equations,

$$\varphi \frac{\partial C^*}{\partial t^*} + \vec{u}^* \cdot \vec{\nabla}^* C^* = \varphi D \nabla^{*2} C^* - \varphi k^* (\vec{x}^*) C^*, \quad (5)$$

$$(\rho c)_m \frac{\partial T^*}{\partial t^*} + (\rho c)_f \vec{u}^* \cdot \vec{\nabla}^* T^* = \lambda_m \nabla^{*2} T^* - \varphi k^* (\vec{x}^*) \Delta H C^*. \quad (6)$$

The subscript m denotes an overall property of the medium, defined as $A_m = (1-\varphi) A_s + \varphi A_f$, where A is the given property, φ is the porosity of the medium, and the subscripts s and f denote properties of the solid matrix and the fluid, respectively. Volumetric heat capacity and thermal conductivity are given by (ρc) and λ, while D gives the molecular diffusion coefficient of the tracer. The spatially dependent first-order reaction coefficient is defined as

$$k^* (\vec{x}^*) = k_0 e^{-25\left(\frac{x^{*2} + (z^* - H/2)^2}{H^2}\right)^2}, \quad (7)$$

where k_0 has units of s^{-1}. This function represents a circular core of reactive porous material that increases in reactivity towards the centre of the domain. The reaction is assumed to occur in a single step between the tracer and the porous medium without affecting the permeability. ΔH is the enthalpy of the reaction of the tracer with the medium, which is negative for exothermic reactions.

The governing equations are nondimensionalized by introducing the typical length and velocity scales L and U, and choosing an advective time scale $\tau = L/U$. U is chosen by assuming that the typical fluid velocity inside the pile is set by the external deviation from atmospheric pressure Δp. By scaling with Darcy's law, this gives $U = \kappa \Delta p / \mu L$. In addition to the dimensionless variables $\vec{x} = (x, z) = \vec{x}^*/L$, $\vec{u} = (u, w) = \vec{u}^*/U$, $\vec{\nabla} = L\vec{\nabla}^*$, $\nabla^2 = L^2 \vec{\nabla}^{*2}$, $p = p^*/\Delta p$, and $t = t^*/\tau$, choosing appropriate reference values for temperature, concentration and the reaction coefficient gives dimensionless variables $T = T^*/T_0$, $C = C^*/C_0$, and $k = k^*/k_0$. Equations (1)-(6) can therefore be nondimensionalized as

$$\vec{\nabla} \cdot \vec{u} = 0, \quad (8)$$

$$\vec{u} = -\vec{\nabla} p + \beta (T - 1) \hat{k}, \quad (9)$$

$$\nabla^2 p = \beta \frac{\partial T}{\partial z}, \quad (10)$$

$$\varphi \frac{\partial C}{\partial t} + \vec{u} \cdot \vec{\nabla} C = \frac{1}{\mathrm{Pe}_C} \nabla^2 C - \mathrm{Da}\, k(\vec{x}) C, \quad (11)$$

$$\frac{\partial T}{\partial t} + \frac{1}{\sigma} \vec{u} \cdot \vec{\nabla} T = \frac{1}{\mathrm{Pe}_T} \nabla^2 T - \mathrm{Da}_T\, k(\vec{x}) C, \quad (12)$$

where $\beta = \frac{\rho_f g L \alpha T_0}{\Delta p}$ describes the effects of buoyancy relative to the applied pressure gradient, $\mathrm{Pe}_C = LU/\varphi D$ is the tracer Péclet number, $\mathrm{Da} = \varphi k_0 L/U$ is the Damköhler number, $\mathrm{Pe}_T = LU/(\lambda_m/(\rho c)_m)$ is the thermal Péclet number, and $\mathrm{Da}_T = \frac{\Delta H C_0}{T_0 (\rho c)_m} \mathrm{Da}$ is the thermal Damköhler number, the sign of which depends on the sign of ΔH. $\sigma = (\rho c)_m / (\rho c)_f$ is the ratio of overall heat capacity of the medium to the heat capacity of the fluid.

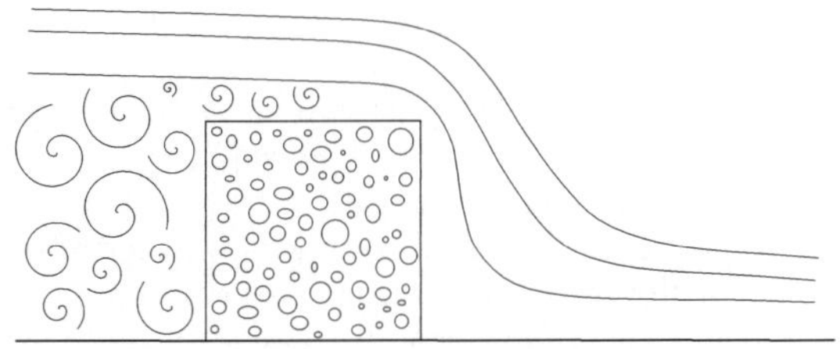

Figure 1: Hypothetical flow past a square pile.

The dimensionless pile is set to be a square with domain $x \in [-1, 1]$, $z \in [0, 2]$. The boundary conditions are defined by the exterior atmospheric conditions, while no flow occurs between the pile bottom and the ground. Therefore, Dirichlet boundary conditions are specified for the sides and top of the pile, and no flux conditions are specified along the bottom. It is assumed that the wind approaches the pile from the right with a high, constant far-field velocity, and that separation occurs near the top front corner of the pile, resulting in a turbulent wake along the top and back. This suggests a higher surface pressure at the bottom right corner of the pile that decreases linearly along its upwind face to a lower constant pressure wake along the top and downwind face,

$$p(-1, z, t) = 0.9, \quad p(1, z, t) = 1 - z/20, \quad (13)$$

$$p(x, 2, t) = 0.9, \quad p_z(x, 0, t) = 0. \quad (14)$$

Enough tracer is assumed to be present in the atmosphere that its concentration around the pile is constant,

$$C(-1, z, t) = 1, \quad C(1, z, t) = 1, \quad (15)$$

$$C(x, 2, t) = 1, \quad C_z(x, 0, t) = 0, \quad (16)$$

while it is assumed that there are no fluctuations in the atmospheric temperature,

$$T(-1, z, t) = 1, \quad T(1, z, t) = 1, \quad (17)$$

$$T(x, 2, t) = 1, \quad T_z(x, 0, t) = 0. \quad (18)$$

Inside the pile, the temperature and concentration initially match the atmospheric conditions ($T(x, z, 0) = 1$ and $C(x, z, 0) = 1$), while pressure and velocity satisfy Darcy's law.

The code designed to model this flow uses Chebyshev spectral differentiation for the spatial derivatives, with 61 grid points in both directions. First-order time-stepping is used in the conservation equations, with backwards Euler for the diffusion terms, forward Euler for the advection and reaction terms, and a dimensionless time step of 0.001. Once the concentration and temperature fields have been computed, the pressure and velocity fields are updated.

NUMERICAL SIMULATION

Devolatilization and drying are two short duration pre-steps in gasification. A part of those already took place during torrefaction. For comparing gasification of two feedstock (raw biomass and torrefied biomass) on a uniform base, both were devolatilized at a fixed temperature of 800°C, and then the product char is subjected to steam gasification. This part of the process is slowest and main part of the gasification process.

The kinetics of gasification of raw and torrefied biomass char in steam is studied in a Quartz Wool Matrix (QWM) reactor (Fig. 1) that consists of a stainless steel reactor 50 mm in diameter encircled by an electric heater. A temperature controller controls temperature inside the reactor. A precision balance sits on the top of the reactor to facilitate continuous measurement of mass of the substance being examined. A wire basket, whose one end is connected to the balance, holds the sample inside the reactor. The flow rate of gases is continuously measured with an electronic flow meter. But the steam flow rate is directly measured with its valve opening, which is first calibrated.

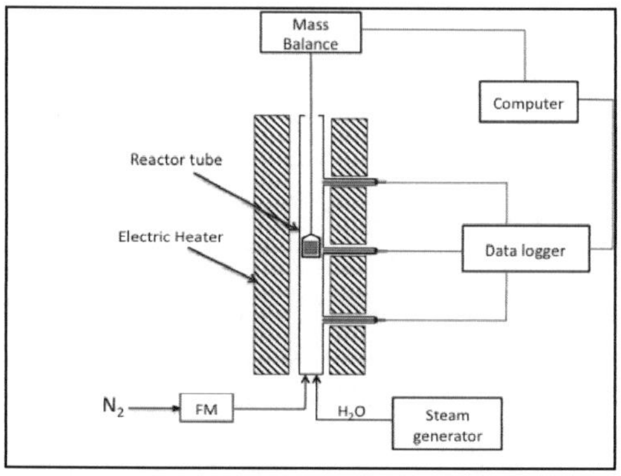

Figure 1: Schematic of QWM reactor

To produce a sample of torrefied biomass a 25 mm diameter and 75 mm long cylindrical piece of raw poplar is taken. For this, first the reactor is heated to the desired torrefaction temperature and then the above piece of raw biomass is placed in the bucket and dropped inside the reactor flushed with nitrogen. While the biomass sample gets heated and torrefied both temperature and the mass loss of the biomass is continuously measured. The biomass is placed for 1.0 hour and is taken out. For kinetics studies one sample of raw biomass and 2 samples of torrefied biomass (torrefied at 250°C and 300°C for 1 hour) are gasified at five different temperatures: 750°C, 800°C, 850°C, 900°C, and 950°C. Before gasification samples of both raw and torrefied biomass is devolatilized to produce char under identical conditions. For devolatilization, the reactor is heated to 800°C and then the biomass sample i.e. raw and torrefied biomass is dropped into the reactor. N_2 is supplied as a gaseous medium to drive off the volatiles, while providing an inert environment to supress occurrence of any other reactions. The mass loss of the sample is continuously measured, and as the mass becomes constant, the sample is taken out and cooled in a desiccator. The cooled sample is then ground to 300-450 microns size for gasification tests.

For gasification, the reactor is first preheated to the desired gasification temperature, and then the char samples obtained from the above devolatilization tests are placed in the bucket and lowered inside the reactor. The sample is gasified for 1.0 hour while steam is supplied from the bottom of the reactor. The mass of the char is continuously measured and its loss is used to calculate the char conversion rate for determining the kinetic rate. The Arrhenius form of equation (Eq. 1) is used for comparing the kinetics of gasification between raw and torrefied biomass char (Slopiecka et al., 2012).

$$k(T) = Ae^{\left(-\frac{E_A}{RT}\right)} \tag{1}$$

Where, E_A (kJ/mol) is activation energy, A (1/min) is frequency factor, T (K) is the absolute temperature and R (J/mol/K) is universal gas constant. To facilitate accurate collection of products of gasification (char, gas and tar), the reactor is closed from the top. The char sample is kept in the basket but the product gas is taken to the condenser and then collected in a gasbag, which is later analysed in gas chromatograph (GC). The condensed liquid is then evaporated to remove the moisture and the mass is measured for gravimetric tar.

RESULT AND DISCUSSION

Product Yield

Table 2 shows the mass yield of char, gas and tar obtained during torrefaction, devolatilization and gasification in terms of per unit of raw biomass weight. It is apparent that char yield during devolatilization is in on average 43% higher for torrefied biomass than that for raw biomass whereas tar yield is lower for torrefied biomass. The product gas yield of gasification is higher for raw biomass char than that for torrefied biomass char. Average value of gasification yield for char conversion is 82.07% for raw biomass compared to 47.21% for torrefied biomass. The average value of tar yield during gasification of raw biomass char is 79% higher than that for torrefied biomass char.

Kinetics Study

The Arrhenius plot for the gasification reaction of char produced from raw and torrefied biomass in presence of steam is shown in Figure 2. It shows that the kinetic rate of gasification of torrefied biomass is lower than that of raw biomass. This was surprising because both devolatilization and gasification processes were identical for both raw and torrefied biomass, and torrefied char showed higher internal surface area.

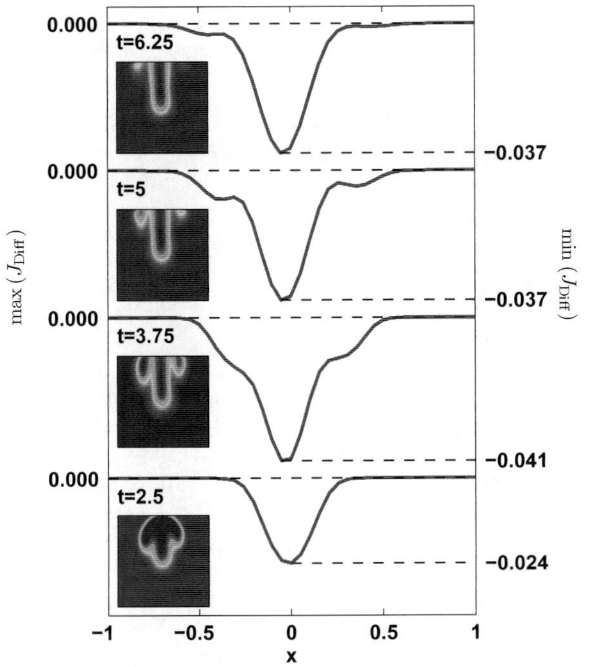

Figure 4: The vertical diffusive concentration flux along the top of the pile at $t = 2.5$, 3.75, 5, and 6.25, with corresponding concentration fields. The values on the left and right axes indicate the maximum and minimum flux at a given time, respectively.

greater advective flux along the top due to the applied pressure gradient along the windward face. The diffusive flux is essentially zero away from the tracer-poor region due to the constant C field throughout most of the domain. As the low C region rises and spreads, the region of diffusive flux expands, before the low concentration bulbs on the left and right are replenished with tracer from the rest of the pile, narrowing the region of inward flux. The gradient between the regions of high and low concentration is also much steeper than the gradient between the regions of high and low temperature, since the tracer Péclet number is ten times greater than the thermal Péclet number, indicating faster thermal diffusion than molecular diffusion.

Integrating the vertical boundary flux along the top of the pile over the entire horizontal domain gives the instantaneous rate of total tracer transfer through the boundary, which is presented as a function of time in Figure 5. The advective rate of tracer transfer out of the pile steadily increases to a local maximum at $t = 3.9$ before a small decrease. It then increases to a slightly faster steady state. Initially, no tracer passes through the top of the pile due to diffusion. By $t = 2.15$, the low concentration region has expanded close enough to the boundary to create a gradient and allow some inward diffusion. The rate of diffusive tracer transfer into the pile increases until $t = 3.15$ when the bulbs on the side of the tracer deficient region begin to separate and are replenished by tracer from elsewhere in the pile. Eventually, diffusive transfer reaches a slower steady state. During most of the flow's evolution, the advective transfer rate out through the top of the pile is at least an order of magnitude greater than the diffusive transfer rate into the pile, indicating that tracer is mainly released from the pile through the top.

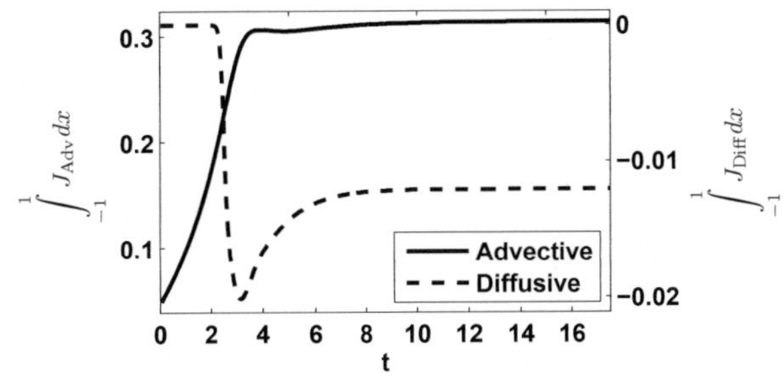

Figure 5: Instantaneous rate of total tracer transfer through the top of the pile. The solid line gives the rate of transfer due to advection, while the dashed line gives the rate of transfer due to diffusion, with corresponding vertical axes on the left and right.

CONCLUSIONS

A numerical simulation of fluid flow through porous pile containing an exothermically reactive tracer has been presented. For the dimensionless parameters selected, flow is initially driven by an external pressure gradient before being dominated by buoyancy effects resulting from the exothermic reaction. Advective tracer flux was observed to be significantly stronger than diffusive tracer flux.

ACKNOWLEDGMENTS

This research was performed with funding from the Natural Sciences and Engineering Research Council.

REFERENCES

[1] R. T. Amos, D. W. Blowes, L. Smith, and D. C. Sego. Measurement of wind-induced pressure gradients in a waste rock pile. *Vadose Zone J.*, 8(4):953 – 962, 2009.

[2] J. T. H. Andres and S. S. S. Cardoso. Onset of convection in a porous medium in the presence of chemical reaction. *Phys. Rev. E*, 83:046312, Apr 2011.

[3] T. J. Ward, K. A. Cliffe, O. E. Jensen, and H. Power. Dissolution-driven porous-medium convection in the presence of chemical reaction. *Journal of Fluid Mechanics*, 747:316–349, 5 2014.

MODELLING SOLIDS SEPARATION IN ROTATING BELT FILTERS USING A SEMI-EMPRICAL NUMERICAL APPROACH

Christopher T. DeGroot
Trojan Technologies
3020 Gore Road
London, ON, Canada
cdegroot@trojanuv.com

ABSTRACT

Rotating belt filter (RBF) technology has many benefits in comparison to conventional primary clarifiers for treating wastewater, including a smaller footprint, reduced construction and operational costs, and potential for energy savings and reuse. While computational fluid dynamics (CFD) methods are well-suited to modelling filtration processes, the combined sieve and cake filtration process encountered in RBF systems is particularly challenging to model since the properties of the cake layer dramatically impact the hydraulics and solids separation processes. This paper outlines a semi-empirical numerical model for analysis of RBF systems that uses experimental results for gravity filtration to supplement the CFD model of filtration on the moving filter mesh. Results are presented for an RBF in a channel, based upon wastewater properties derived from gravity filtration experiments on primary influent at a wastewater treatment plant (WWTP) in London, Canada.

KEYWORDS: Computational fluid dynamics, rotating belt filters, cake filtration, wastewater, multiphase flow

INTRODUCTION

RBF technologies operate by physically separating suspended particulate matter from wastewater using a mesh filter. They offer a smaller-footprint alternative with lower capital and operational costs when compared to primary clarifiers which operate on the principle of sedimentation (Franchi and Santoro 2015). Diversion of solids from the main treatment stream to side streams has a number of benefits related to a reduction of energy usage in aeration tanks due to lower oxygen demand and the potential for recovery of energy through anaerobic digestion (Batstone 2014).

RBF filtration processes operate on the principle of cake filtration which enables the removal of particles up to three times larger than the nominal pore size of the filter mesh (Tien 2012). The inclined mesh is continuously moving as it separates and retains solids from the influent wastewater that flows through the filter under the force of gravity. The solids that are retained on the filter mesh form a cake layer which transitions the filtration process from sieving to cake filtration. The filter mesh is cleaned with each revolution using compressed air, water spray, or mechanical scraping and the solids are collected. The filter speed is controlled dynamically in order prevent overflow by modulating the cake thickness to control the hydraulic loading.

Computational fluid dynamics (CFD) methods are well-suited to efficiently modelling the hydraulics of wastewater treatment unit processes (Laurent et al. 2014), including RBF systems, provided an appropriate semi-empirical model for the filtration and solids separation processes can be developed to avoid the need to simulate phenomena occurring at the scale of particle interactions with the filter mesh, which are computationally expensive. Recently, a CFD method for modelling RBF systems has been presented for the assessment of nutrient diversion potential, where the combined filter mesh and cake layer was considered a variable-permeability porous medium governed by Darcy's law (DeGroot et al. 2015).

In this paper, further details are provided regarding the experimental procedure that is used to generate empirical models for filter and cake permeability as well as the solids separation processes. It has been hypothesized that both the instantaneous, local values of permeability and removal efficiency can be directly correlated with the specific mass of the accumulated cake layer. These relationships are determined by gravity filtration experiments which will be described in detail in this paper.

The semi-empirical filtration model and the experimental procedure that is used to relate filter cake permeability and solids removal efficiency to the local specific cake mass is outlined first, followed by a description of the CFD numerical model. Finally, results will be presented for a basic configuration of a rotating belt filter placed in a channel using filtration properties obtained from gravity filtration experiments conducted on primary influent wastewater at a WWTP in London, Ontario.

METHODOLOGY

Theory

The flow of wastewater through the combined filter and cake layer is governed by Darcy's law in one dimension, given as (Ward 1964)

$$\frac{q}{A} = -\frac{k}{\mu}\frac{dp}{dx} \tag{1}$$

where q is the volume flow rate of fluid through the filter with area A, k is the filter permeability, μ is the dynamic viscosity of the fluid and dp/dx is the pressure gradient in the direction normal to the filter's surface.

The accumulation of solids on the filter mesh is governed by conservation of the solid-phase mass. For an influent total suspended solids (TSS) concentration of TSS_{in} and a removal efficiency η, the solid-phase mass balance may be expressed as

$$\Delta(TSS_A) = TSS_{in}\eta\frac{q}{A}\Delta t \quad (2)$$

where $\Delta(TSS_A)$ is the mass of solids accumulated per unit area of filter over the time interval Δt.

Both k and η are known to depend on TSS_A, which is the total accumulated mass of solids per unit area at a given location on the filter at a given time.

Gravity Filtration Experiments

To determine the relationships between the filter and cake layer permeability and instantaneous removal efficiency with the specific mass of solids retained on the filter mesh, experiments were conducted for filtration of wastewater in a vertical column under the driving force of gravity. The height of the water column was continuously monitored throughout the experiments using a custom-designed floatation device connected to a string potentiometer, while samples of the filtrate were taken at various intervals in order to compare the TSS concentrations with that of the influent to derive the removal efficiency. The TSS concentrations of samples were measured in a lab using gravimetric analysis.

Height measurements were used to calculate the instantaneous permeability of the filter and cake layer by recognizing the flow rate to be (Wells and Savage 1998)

$$q = -A\frac{dh}{dt} \quad (3)$$

where dh/dt is the derivative of height with respect to time.

The pressure difference across the filter is equal to the hydrostatic head of the water column, ie. $\Delta p = \rho gh$, where ρ is the fluid density and g is the acceleration due to gravity. This enables the definition of an effective form of Darcy's law, given as

$$\frac{q}{A} = k_{eff}h \quad (4)$$

where k_{eff} is the effective permeability, defined as

$$k_{eff} = \frac{\rho g}{\mu\Delta x}k \quad (5)$$

where Δx is the instantaneous filter and cake layer thickness. Through the definition of the effective permeability the effects of cake structure and thickness are combined into one effective parameter, eliminating the need to independently monitor or estimate cake thickness.

Combining Eqs. (3) and (4) results in the separable first order ordinary differential equation (ODE)

$$\frac{dh}{dt} = -k_{eff}h \quad (6)$$

Since the ODE is separable it may be rearranged integrated between the times t_i and t_{i+1}, corresponding to the heights h_i and h_{i+1}, resulting in

$$\ln\left(\frac{h_{i+1}}{h_i}\right) = -\int_{t_i}^{t_{i+1}} k_{eff}dt \quad (7)$$

Due to the non-constant coefficient, k_{eff}, the right side of Eq. (7) does not have a general solution. It can be considered as a reasonable approximation, however, to consider k_{eff} as being constant over a small time interval, Δt, in order to bring it outside of the integral. Following such an approach, the effective permeability for an interval can be calculated as

$$k_{eff} = -\frac{\ln\left(\frac{h_{i+1}}{h_i}\right)}{t_{i+1} - t_i} \quad (8)$$

The removal efficiency, η, over the same time interval was calculated as

$$\eta = 1 - \frac{TSS_{out}}{TSS_{in}} \quad (7)$$

where TSS_{out} and TSS_{in} are the TSS concentrations in the effluent and influent waters, respectively.

In order to relate the permeability and removal efficiency to the filter conditions, the mass of accumulated solids per unit area, TSS_A, was calculated as a summation of the $\Delta(TSS_A)$ values calculated using Eq. (2) over the appropriate sampling intervals.

Once the data has been calculated, a potential method of transferring the information into the CFD model would be via a curve-fitting procedure. This approach, however, has been found by the authors to be unsuitable because of the need to extrapolate beyond the range of available experimental data. For low values of TSS_A, this is due to the rapid initial flush of water that occurs before cake formation, which is challenging to capture. For high values of TSS_A, filter clogging occurs and it is difficult to measure the small differences in height. As an alternative to curve fitting, an approach to sort the data and generate step functions to capture the relationships of k and η with TSS_A is proposed, which will be further discussed later in the paper.

CFD Simulations

A schematic diagram of the two-dimensional CFD calculation domain is shown in Fig. (1), where primary wastewater enters from the left of the domain with depth $h_{upstream}$ and flows through the filter mesh which is at an angle θ to horizontal. Downstream of the filter, the effluent has a depth $h_{downstream}$ as it flows over a weir of height h_{weir} before exiting the domain to the right. The filter is moving in the direction of the arrowheads and is considered clean (TSS_A=0) at its lower left edge. As the filter moves, solids are retained and the cake layer increases in thickness.

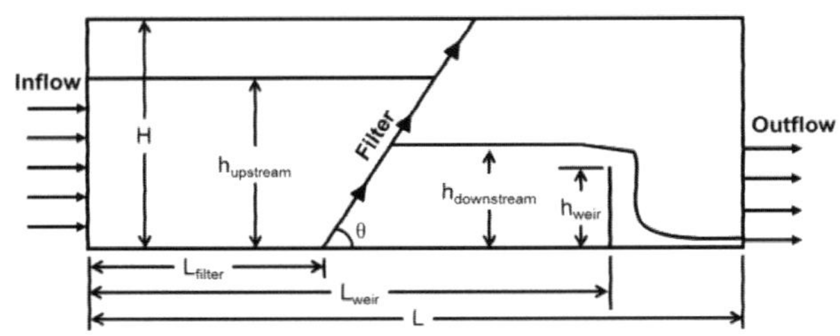

Figure 1: Schematic diagram of the CFD calculation domain.

Simulations were conducted using ANSYS Fluent (ANSYS, Canonsburg, PA) using a volume-of-fluid (VOF) model to account for the air and water phases. Turbulence was modelled using the k-ω SST model. The TSS concentration was modelled as a user defined scalar (UDS). To customize Fluent in order to solve the filtration problem, a custom User Defined Function (UDF) library was developed.

At the inlet, wastewater was introduced with constant velocity and constant TSS. The inlet velocity was dynamically adjusted using a PID control algorithm implemented in the UDF library in order to achieve a specified target upstream height, $h_{upstream}$.

The filter region was defined as a thin porous region with anisotropic viscous resistance. In the direction normal to the filter, the resistance was calculated from the experimental permeability values for a given accumulated mass of solids; in the direction parallel to the filter it was taken three orders of magnitude larger to restrict the flow as much as possible in this direction. The mass of solids accumulated on the filter was first calculated based on the local flow rate and removal efficiency and was then translated along the filter axis using a one-dimensional advection equation that was solved using an explicit Euler method for time integration and a second-order flux-limited interpolation for advection.

RESULTS AND DISCUSSION

Experimental Results

The filtration relationships were obtained by running five replicates of the gravity drainage experiment, taking five effluent samples and one influent sample for each replicate. Tests were conducted on primary influent wastewater at a WWTP in London, Ontario where the average TSS concentration was found to be approximately 290 [mg/L] on the day of the experiment.

The experimental results were processed using the procedure outlined previously to calculate the effective permeability, removal efficiency, and specific cake mass over the time intervals corresponding to the five effluent samples. In order to generate step functions of k_{eff} vs. TSS_A and η vs. TSS_A, each variable is sorted independently to ensure k_{eff} is a decreasing function of TSS_A and η is an increasing function of TSS_A. This is justified by the physical conditions that permeability must drop and removal must increase for cake layers of higher specific mass.

Figure 2 shows both the raw and sorted experimental data, along with fitted exponential curves for each. It is clearly observed that nearly identical fitting curves are generated whether or not the data is sorted, indicating that the sorting procedure does not induce any bias to the data. This is likely an indication that the measurement error is random, so that sorting has no impact on the overall trend. The advantage to sorting the data is that a step function can be generated which has constant values above and below the range of actual measurements. Physically, this can be justified based on the fact that for low TSS_A, a significant cake layer has not yet been formed and the dominant filtration mechanism is simply sieving by the filter mesh. For high TSS_A values, the obvious deviation of removal efficiency from the fitted exponential curve discourages the use of extrapolation and favors the step function approach.

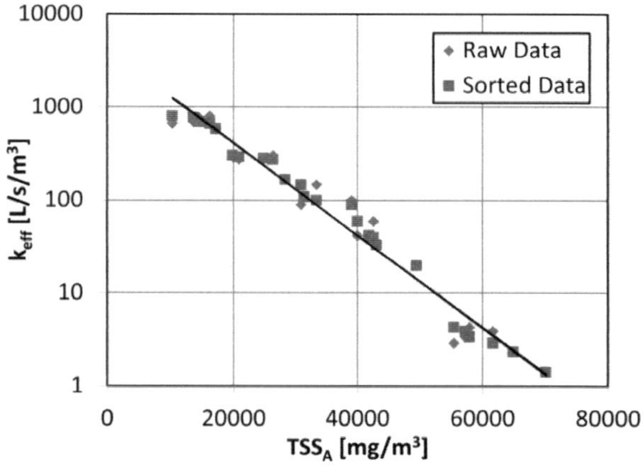

(a)

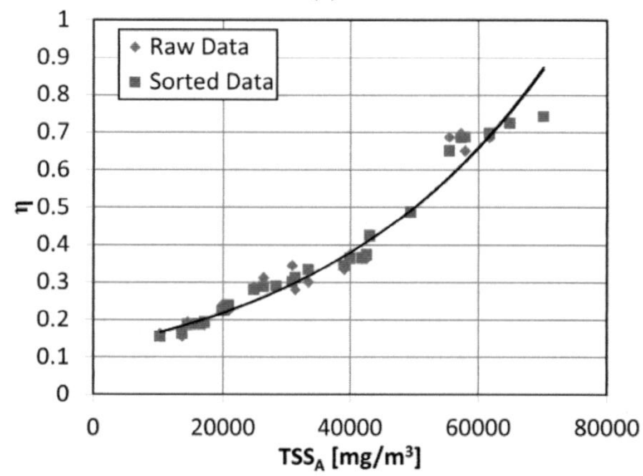

(b)

Figure 2: Results for effective permeability and removal efficiency as functions of specific cake mass, comparing the raw data with the sorted data used to generate step functions for the CFD model.

CFD Results

CFD results were obtained for the configuration shown in Fig. 1, with the parameters listed in Table 1. The mesh speed was varied to characterize the system's response for an influent TSS value of 300 mg/L. The key output variables are the total rate of wastewater flow through the system (hydraulic capacity) and the overall rate of solids removal compared to the rate of solids entering through the influent (removal efficiency).

Results for the hydraulic capacity per unit filter depth and overall removal efficiency as functions of the relative filter belt speed are plotted in Fig. (3). It may be observed that the hydraulic capacity increases with increasing belt speed; conversely, the removal efficiency decreases with increasing belt speed. In general, maximizing both hydraulic capacity and removal efficiency are competing objectives.

Increasing belt speed leads to a thinner cake layer which increases hydraulic capacity, but lowers removal efficiency. For the case considered, both hydraulic capacity and removal efficiency reach a particular limit after which the impact of increasing belt speed is diminished.

Table 1: List of geometric parameters used in the CFD calculations.

Parameter	Value
L	2 [m]
L_{filter}	0.5 [m]
L_{weir}	1.5 [m]
H	0.5 [m]
$h_{upstream}$	0.4 [m]
h_{weir}	0.25 [m]
θ	40°

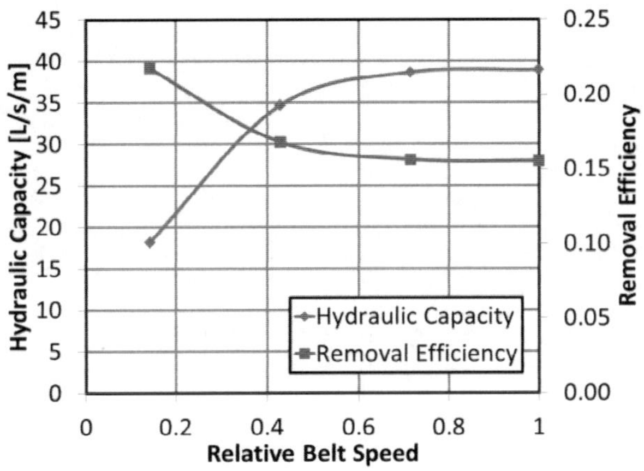

Figure 3: Results for hydraulic capacity and removal efficiency as functions of relative belt speed for different.

With the combined experimental and numerical procedure outlined in this paper, detailed calculations of RBF filter performance may be conducted in order to assess and optimize design for best performance. The experimental procedure allows for the characterization of the wastewater for a particular site/time which can vary quite significantly.

CONCLUSIONS

In summary, a semi-empirical CFD model has been presented for modelling the removal of suspended solids from wastewater using rotating belt filters. An experimental procedure was outlined in order to determine the filtration characteristics under gravity drainage and supply the numerical model with data to model the filtration process on a rotating belt. The numerical procedure was then given and results were presented for the key output variables of hydraulic capacity and overall removal efficiency as functions of filter belt speed, based on experimental filtration kinetics results obtained from primary influent wastewater.

ACKNOWLEDGMENTS

The author wishes to acknowledge the financial support received from the Natural Sciences and Engineering Research Council (NSERC) of Canada, as well as the support of Ehsan Sheikholeslamzadeh and Naghmeh Sharifi in conducting the gravity filtration experiments and laboratory measurements and Domenico Santoro for helpful discussions.

REFERENCES

1. Franchi, A., Santoro, D. 2015, "Current status of the rotating belt filtration (RBF) technology for municipal wastewater treatment", *Water Practice & Technology* (In press).
2. Batstone, D.J., Virdis, B. 2014, "The role of anaerobic digestion in the emerging energy economy", *Current Opinions in Biotechnology*, 27, pp. 142-149.
3. Tien, C. 2012, *Principles of Filtration*, Elsevier, Oxford, U.K.
4. Laurent, J., Samstag, R.W., Ducoste, J.M., Griborio, A., Nopens, I., Batstone, D.J., Wicks, J.D., Saunders, S., Potier, O. 2014, "A protocol for the use of computational fluid dynamics as a supportive tool for wastewater treatment plant modelling", *Water Science & Technology*, 70(10), pp. 1575-1584.
5. DeGroot, C.T., Sheikholeslamzadeh, E., Soleymani, A., Santoro, D., Batstone, D.J., Rosso, D. 2015, "Modeling rotating belt filter performance and nutrient diversion potential using computational fluid dynamics", *IWA Nutrient Removal and Recovery*, Gdansk, Poland.
6. Ward, J.C. 1964, "Turbulent flow in porous media", *J. Hydraulics Division, Proc. ASCE*, 90(HY5), pp. 1-12.
7. Wells, S.A., Savage, G.K., 1998, "Gravity drainage prior to cake filtration", *Advances in Filtration and Separation Technology*, 12, pp. 646-653.

Renewable Energy Systems

Proceedings of the 25th CANCAM
London, Ontario, Canada, May 31 – June 4, 2015

A NUMERICAL STUDY OF BASEMENT HEAT LOSSES

Lam Dang
Department of Mechanical and Industrial Engineering
Ryerson University
Toronto, Ontario, Canada
lam.dang@ryerson.ca

Wey H. Leong
Department of Mechanical and Industrial Engineering
Ryerson University
Toronto, Ontario, Canada
weyleong@ryerson.ca

1. ABSTRACT

This paper presents a study on heat losses from a basement of a house in a winter day. The study is based on a steady-state, two-dimensional numerical model. It was found that the heat losses depend much more on ground water content than the soil type and ground temperature. Also, with sufficient grid elements, the simulation time can be 2.45 times faster with constant thermal conductivity of the ground while the accuracy can still be maintained. In addition, adding insulation to the basement wall can save energy but not proportionally, so one should consider both the long-term energy saving with the additional cost of the insulation.

KEYWORDS: Numerical simulation, basement heat loss, 2D, COMSOL

2. INTRODUCTION

Many air-conditioned basements in houses and buildings have been providing comfortable and quiet living conditions for occupants. Basements also conserve land and make the house 10% more energy efficient [1]. However, if it is not properly designed, the basement can unnecessarily lose a great amount of heat to the surrounding cold ground. As a result, heat losses of houses and buildings from the below grade section can account to 10-40% of the heating load of a building [2].

An early study on basement heat loss was done by Latta and Boileau [3] who used a constant value of soil thermal conductivity to determine the average heat loss coefficients for insulated and uninsulated basement walls and floors. Also, some major textbooks (*e.g.*, McQuinston *et al.*[4]) still apply the heat loss coefficients to evaluate heat losses from basements.

However, investigations by many researchers have shown that the thermal conductivity of the ground, especially near ground surface, is generally not constant throughout a year. Rosen [5]evaluated the heat losses from partially buried and bermed heat storage tanks and realized that soil properties generally vary spatially. Consequently, Rosen [6] later developed a semi-empirical model to study how the properties of the soil berms could affect the heat loss of the partially buried heat storage tanks. The economics of using the berms in thermal energy storage systems was also examined [7]. In addition, Rosen *et al.* [8] evaluated how soils with spatially varying thermal properties affected underground thermal storages. Furthermore, Leong *et al.* [9] investigated the impact of spatially varying soil thermal properties on a ground thermal energy system for heating and cooling of dwellings.

Because computing powers have been increased vigorously, computer modeling techniques have become more advanced. As a result, it may soon be unacceptable to evaluate underground heat transfer without considering spatial variations of the ground thermal properties.

Therefore, the objectives of this paper are:

A. To study the heat losses from an uninsulated basement to the ground (as shown without rectangle EKLF in Fig. 1) with the ground thermal conductivity being constant and being a function of temperature and water content.
B. To study the heat losses from an insulated basement to the ground (as shown with rectangle EKLF in Fig. 1) with the ground thermal conductivity being constant and being a function of temperature and watercontent.
C. To compare the results in part B with the ones in part A.

3. NOMANCLATURE

c	Specific heat ($J/kg{\cdot}K$)
h	Heat transfer coefficient ($W/m^2{\cdot}K$)
k	Thermal conductivity ($W/m{\cdot}K$)
Q	Heat loss per unit depth into the page (W/m)
S	Degree of saturation (-) = $\frac{\theta}{\phi}$
t	Thickness (m)
T	Temperature (°C)
ϕ	Porosity (m^3/m^3) = $\frac{\forall_{pore}}{\forall_{total}}$
θ	Water content (m^3/m^3) = $\frac{\forall_w}{\forall_s}$
ρ	Density (kg/m^3)
$\forall$	Volume (m^3)

Subscripts

a	pertaining to air
avg	average
b	bulk
con	pertaining to concrete
dry	dry

FC	field capacity
FS	full saturation
in	pertaining to inside space of the basement
ins	pertaining to glass fiber insulation
out	pertaining to outside space of the house
s	pertaining to soil
w	pertaining to water

4. NUMERICAL MODEL AND GRID SENSITIVITY

The basement heat-loss problem for the present study is shown in Fig. 1. The added insulation is formed by EKLF. The segments AB, DE, EK, GH, and HI are adiabatic. The interior surfaces (EF or KL and FG or LG), the exterior surface (CD) of the basement and the ground surface (BC) have convective boundary conditions. The lower boundary (AI) is considered to have a deep ground temperature. The steady-state, two-dimensional numerical model is created in COMSOL [10] with dimensions of each segment and input parameters tabulated in Tables 1 and 2, respectively.

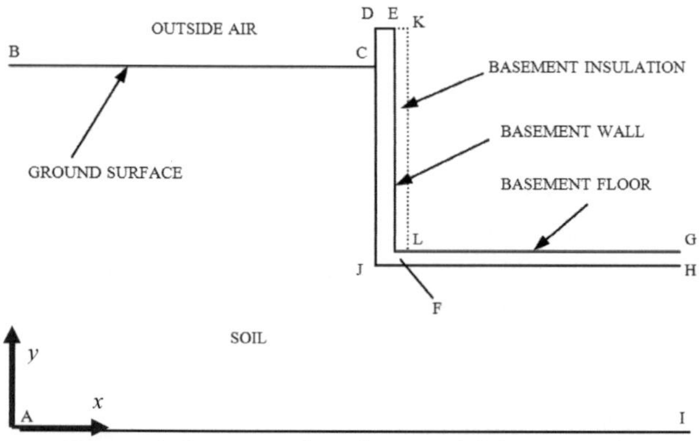

Figure 1. Basement heat-loss model (not to scale)

Table 1. Dimensions of segments in Fig. 1

Segment	AB	BC	CJ	DE	EF	FG	GH
Length(m)	3.0	5.0	2.0	0.2032	3.0	2.5	0.1016

Table 2. Input parameters for numerical simulations [10,11]

Parameter	ρ_{con} (kg/m^3)	c_{con} ($J/kg{\cdot}K$)	k_{con} ($W/m{\cdot}K$)	ρ_{ins} (kg/m^3)	c_{ins} ($J/kg{\cdot}K$)	k_{ins} ($W/m{\cdot}K$)
Value	2300	880	1.80	64.0	1400	0.0345
Parameter	T_{in} (°C)	T_{out} (°C)	T_{AI} (°C)	h_{in} ($W/m^2{\cdot}K$)	h_{out} ($W/m^2{\cdot}K$)	-
Value	20.0	0.00	10.0	4.00	4.00	-

The thermal properties of the ground as functions of T and/or θ are given as:

- Thermal conductivity: $$k_b = \frac{a_1 + a_2 T + a_3\theta + a_4\theta^2}{1 + a_5 T + a_6\theta + a_7\theta^2} \quad (1)$$
- Density: $$\rho_b = \rho_s(1-\phi) + \rho_w\theta + \rho_a(\phi-\theta) \quad (2)$$
- Specific heat capacity:

$$c_b = \frac{\rho_s c_s (1-\phi) + \rho_w c_w \theta + \rho_a c_a (\phi - \theta)}{\rho_b} \quad (3)$$

where a_1, ..., a_7 are listed in Table 3 for Ottawa sand (OT) and Richmond Hill (RH) fine sandy loam [12] for the temperature range of $2 \le T \le 92$°C and moisture content range of $0 \le \theta \le \phi$, and θ is assumed to be uniform for the entire ground. Three soil water contents are considered in this study, namely: dry, field capacity (FC) and full saturation (FS), as shown in Table 4.

Table 3. Correlation coefficients for k_b in Eq. (1) [12]

Coefficient	***Ottawa sand***	***Richmond Hill Fine Sandy Loam***[1]
a_1	0.3039	0.2079
a_2	-1.3300×10^{-3}	-9.4004×10^{-4}
a_3	-0.6934	-0.7811
a_4	42.9631	6.9407
a_5	-6.1050×10^{-3}	-5.8852×10^{-3}
a_6	-3.8360	-2.7028
a_7	17.6166	7.0293

Table 4. Water contents used in the study

Parameter	θ_{dry} (m^3/m^3)	$\theta_{FC} = \phi/2$ (m^3/m^3)	$\theta_{FS} = \phi$ (m^3/m^3)
Ottawa sand	0.0	0.183	0.366
Richmond Hill	0.0	0.290	0.571

The heat loss to the ground per unit depth Q (W/m) into the page is defined as:

$$Q = Q_{bw} + Q_{bf} = -\int_{y_J}^{y_C} k_s \left.\frac{\partial T}{\partial x}\right|_{x_J} dy - \int_{x_J}^{x_H} k_s \left.\frac{\partial T}{\partial y}\right|_{y_J} dx \quad (4)$$

Figure 2 shows the results of a grid sensitivity study of an insulated basement with 2.54-*cm* (1-*in.*) insulation at the interior side of the basement wall and dry Ottawa sand in the ground.

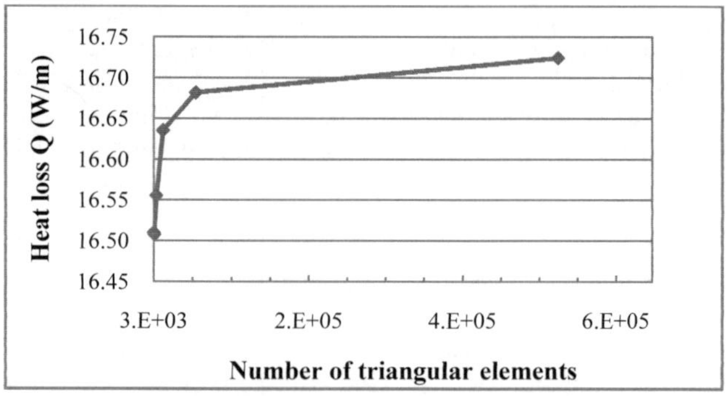

Figure 2. Results of grid sensitivity study

As shown in Fig. 2, using 5.27×10^5 triangular elements produces a slight difference of 0.25% in heat loss from using 5.73×10^4 elements. In terms of computational speed and accuracy, the mesh size of 5.73×10^4 elements is deemed to be warranted for the rest of the study.

[1]Nikolaev [12] originally mis-analyzed it to be a clay loam soil. But later it was correctly identified as a fine sandy loam [13].

5. RESULTS AND DISCUSSION

Figure 3 illustrates the heat loss through uninsulated basement with the ground thermal conductivities of $k_{b,T} = k_b(T,\theta)$ and $k_{b,T_{avg}} = k_b(T_{avg},\theta)$ from Eq. (1), where $T_{avg}= (T_{in} + T_{out})/2$ = 10°C. The percent difference between $Q(k_{b,T})$ and $Q(k_{b,T_{avg}})$ in Fig. 3 is defined as:

$$\%Difference=\frac{Q(k_{b,T})-Q(k_{b,T_{avg}})}{Q(k_{b,T})}\times 100 \quad (5)$$

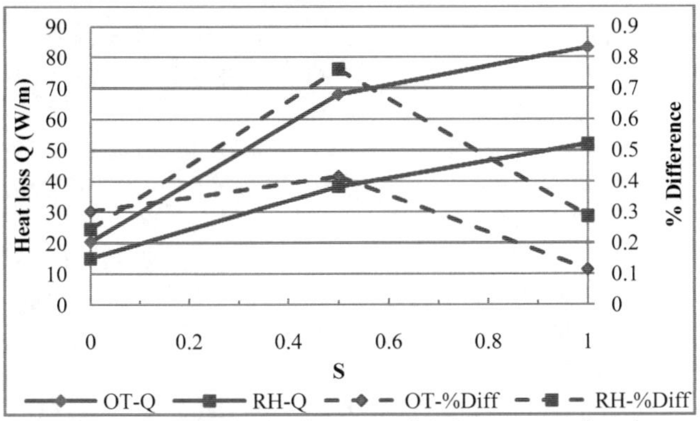

Figure 3. Q and %Difference of Q from uninsulated basement with $k_{b,T}$ and $k_{b,T_{avg}}$

As shown in Fig. 3, the more moisture the ground has, the more the heat from the basement is lost. It is due to increasing ground thermal conductivity as moisture content increases. In addition, using $k_{b,T}$ in the model results in insignificant differences (less than 1%) from using $k_{b,T_{avg}}$. Furthermore, the heat loss to the ground is the most significantly affected by the ground θ (up to 309% for OT between dry and FS), followed by the soil type (up to 78% between OT and RH at S = 0.5), and the least by the ground temperature (up to 0.76% for RH at S = 0.5) under unfrozen conditions (0 < T < 20°C). In other words, higher water content in the ground will cause more heat loss from the basement, while the ground temperature causes little effect in the heat loss and a constant $k_{b,T_{avg}}$ can be used in the model.

Also, by looking at Table 5, the time to run the simulations using $k_{b,T}$ is about 2.45 times longer than that using $k_{b,T_{avg}}$. Obviously, updating the thermal conductivities of finite elements in the ground during a simulation for the case of $k_{b,T}$ is computationally expensive. Although different computer configurations yield different simulation times, the time ratio in Table 5 should be very similar if the RAM memory of the computer is not exceeded. From Table 5, it is also interesting to point out that although there are about 11,500 more elements located in the insulation layer (EKLF in Fig. 1) of the insulated basement case (with constant thermal properties for the insulation), the simulation time is virtually the same as the uninsulated basement case. This seems to show that it is not computationally intensive to solve for 11,500 more elements out of 5.330×10^5 elements (*i.e.*, about 2.2%).

Table 5. Average simulation timeper case

Uninsulated Basement		
Ground thermal conductivity	**Average number of elements**	**Average simulation time per case**
Variable $k_{b,T}$	5.215×10^5	27.05 *s*
Constant $k_{b,T_{avg}}$	5.215×10^5	11.05 *s*
Insulated Basement		
Variable $k_{b,T}$	5.330×10^5	27.08 *s*
Constant $k_{b,T_{avg}}$	5.330×10^5	11.10 *s*

Figure 4 shows the percent energy saving by comparing the heat loss from insulated basements with uninsulated basements. The legend displays the soil type followed by the degree of saturation. The thickness of the insulation ranges from 2.54 *cm* (1 *in.*) to 15.24 *cm* (6 in.) with 2.54-*cm* (1-*in.*) increment. The %Energy Saving in Fig. 4 is defined as:

$$\%Energy\ Saving=\frac{Q(k_{b,T})_{NI}-Q(k_{b,T})_{WI}}{Q(k_{b,T})_{NI}}\times 100 \quad (6)$$

where NI means no insulation, WI means with insulation.

As illustrated in Fig. 4, thicker insulation reduces the heat loss from the basement to the surrounding ground, and hence resulting higher %Energy Saving with respect to the uninsulated basement. Although the energy saving ranges from 17-25% for 2.54-*cm* (1-*in.*) insulation to as much as 40-45% for 15.24-*cm* (6-*in.*) insulation, there is a diminishing return in terms of economics, as the rate of saving falls quickly for thicker insulation. For dry soils having low thermal conductivity, the added insulation to the basement wall provides lesser effect on the energy saving. On the other hand, when the soils are wet (FC and FS), the %Energy Saving is higher, because the insulation is more effective to prevent high heat loss due to higher ground thermal conductivity.

To further explore how much more energy saving due to addition of insulation, the extra percent energy saving per 2.54-*cm* (1-*in.*) addition of insulation is calculated as follows:

$$Extra\ \%\ Energy\ Saving = (Eq.6)_{t_{ins}+2.54\ cm} - (Eq.6)_{t_{ins}} \quad (7)$$

Average values of Extra %Energy Saving based on the two soils at the three levels of water content are plotted in Fig. 4 as a black curve. It can be seen that increasing the insulation thickness does not proportionally increase the energy saving.

For examples, the first 2.54-*cm* (1-*in.*) insulation has the best value, as it immediately provides 21% of %Energy Saving; then, using 5.08-*cm* (2-*in.*) insulation saves about 9% more energy than using 2.54-*cm* (1-*in.*) insulation while using 15.24-*cm* (6-*in.*) insulation saves only 2.1% more energy than using 12.7-*cm* (5-*in.*) insulation. As a result, it is worthwhile to consider both the energy saving and the additional cost of having a thicker insulation.

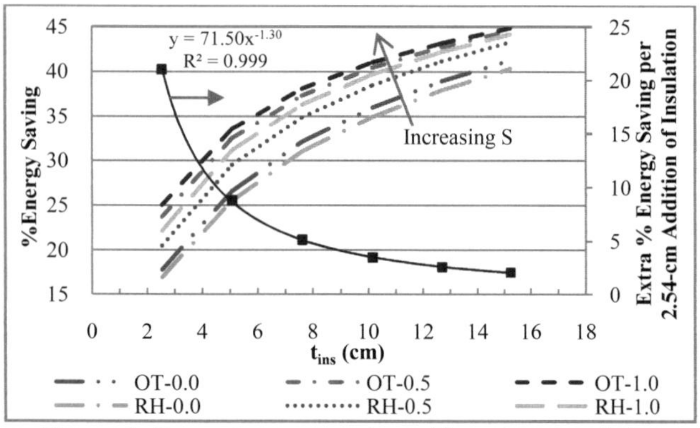

Figure 4. Percent energy saving with various insulation thicknesses

6. CONCLUSIONS

Heat losses from the basement depend on many factors, such as the ground thermal conductivity (which in turn depends on soil type, temperature and water content) and the thickness of insulation added to the basement wall. It has been shown that the basement would lose more heat to a wetter ground; therefore, adding insulation to the basement wall in this case can save more energy. Due to higher thermal conductivity, a coarse soil (OT) can result in higher heat loss from the basement than a fine soil (RH), especially when it is wet. Variation of ground temperature has negligible effect on heat loss under unfrozen conditions ($0 < T < 20$°C).Because of that, in terms of computational expense, it is acceptable to assume constant ground thermal conductivity in order to save the simulation time (2.45 times faster than the case with variable ground thermal conductivity) without sacrificing the accuracy of results. It is evident that adding insulation will save energy, but not proportionally; so, it is recommended to consider both the additional cost of adding more insulation to the basement and the long-term saving of energy with the additional insulation.

ACKNOWLEDGMENTS

Financial support provided by the Natural Sciences and Engineering Research Council of Canada for the present study is gratefully acknowledged.

REFERENCES

1. Hannah, J. 2004, "Eco-friendly Basements Provide Extra Living Space", *Concrete (London)*, Department of Residential Buildings, Concrete Center, Vol (38), No. 11, pp. 37-38.
2. Fournier, M. and Laurencelle, F. 2011, "Impact of Insulation Configuration on the Ability of a Basement Thermal Mass to Shape Heat Losses", *Building Simulation 2011, 12th Conf. of Int. Building Performance Simulation Assoc.*, Sydney, Australia, pp. 1104 - 1110.
3. Latta, J.K. and Boileau, G.G.1969, "Heat Losses from House Basements", *Canadian Building*, Vol (XIX), No. 10, pp. 39-42.
4. McQuiston, F.C., Parker, J.D., and Spitler, J.D. 2005, *Heating, Ventilating, and Air Conditioning — Analysis and Design, 6th Ed.*, Wiley, New Jersey, USA.
5. Rosen, M.A. 1990, "Evaluation of the Heat Loss from Partially Buried, Bermed Heat Storage Tanks", *Int. J. Solar Energy*, Vol (9), pp. 147-162.
6. Rosen, M.A. 1998, "A Semi-Empirical Model for Assessing the Effects of Berms on the Heat Loss from Partially Buried Heat Storage Tanks", *Int. J. Solar Energy*, Vol (20), pp. 57-77.
7. Rosen, M.A. 1998, "The Use of Berms in Thermal Energy Storage Systems: Energy-Economic Analysis", *Solar Energy*, Vol (63), pp. 69-78.
8. Rosen, M.A., Hanford, G.C., and Gunnewick, L.H. 1994, "Heat Loss Analysis of Bermed Thermal Storages with Spatially Varying Thermal Properties", *6th Int. Conf. on Thermal Energy Storage (Calorstock 94)*, Espoo, Finland, pp. 255-262.
9. Leong, W.H., Lawrence, C.J., Tarnawski, V.R. and Rosen, M.A., 2006, "Evaluation of a Ground Thermal Energy Storage System for Heating and Cooling of an Existing Dwelling", *Can. Conf. on Building Performance Simulation (eSim 2006)*, Toronto, pp. 141-148.
10. COMSOL Multiphysicsv4.3b, 2013, COMSOL Inc.
11. Incropera, F.P., DeWitt, D.P., Bergman, T., and Lavine, A. 2007, *Fundamentals of Heat and Mass Transfer, 6th Ed.*, John Wiley & Sons, New Jersey, USA.
12. Nikolaev, I. 2007*An Experimental Study of Soil Thermal Conductivity Using a Guarded Hot Plate Apparatus*, M.A.Sc. Thesis, Department of Mechanical and Industrial Engineering.
13. Nikolaev, I.V., Leong, W.H. and Rosen, M.A., 2013, "Experimental Investigation of Soil Thermal Conductivity Over a Wide Temperature Range", *Int. J. Thermophys.*, Vol (34), pp. 1110-1129.

Proceedings of the 25th CANCAM
London, Ontario, Canada, May 31 – June 4, 2015

SEASONAL PERFORMANCE INVESTIGATION OF A DUAL TANK SOLAR-ASSISTED HEAT PUMP SYSTEM

Carsen J. Banister
Mechanical and Mechatronics Engineering
University of Waterloo
Waterloo, Ontario, Canada
carsen.banister@uwaterloo.ca

Prof. Michael R. Collins
Mechanical and Mechatronics Engineering
University of Waterloo
Waterloo, Ontario, Canada
mike.collins@uwaterloo.ca

ABSTRACT

This work investigates the simulated performance of a dual tank solar-assisted heat pump (SAHP) system on the basis of domestic hot water (DHW) tank temperature and electricity consumption over a typical meteorological year for Ottawa, Canada. The modes of operation and model components were validated experimentally in a previous study, building confidence in the output results.

The DHW tank temperature results indicate that energy savings result from increasing the stored water temperature and enabling the SAHP system to operate more often. Despite the advantageous performance of this dual tank SAHP system, economic justification for DHW heating is difficult due to high capital costs relative to the scale of the load investigated. To improve justifiability, the feasibility of applying the system to space cooling during the summer months is explored.

Space cooling savings can be accomplished by storing chilled water in the second tank, enabling off-peak cooling. It was found that electricity consumption only increased marginally when disabling heating of the float tank during a period of the cooling season, suggesting that applying the load to space cooling is a feasible strategy to increase savings.

KEYWORDS: renewable energy, solar thermal, heat pump, TRNSYS, modelling, experimental

NOMENCLATURE

DHW – Domestic Hot Water
EC – Electricity Consumption
HP – Heat Pump
HX – Heat Exchanger
IEC – Increased Electricity Consumption
SAHP – Solar-Assisted Heat Pump
SDHW – Solar Domestic Hot Water
STC – Solar Thermal Collector
TMY – Typical Meteorological Year
YRF – Year-Round Float

INTRODUCTION

The negative impacts of fossil fuel consumption have motivated the development of energy systems that utilize renewable sources. Individually, solar thermal systems and heat pump systems are both attractive means for meeting thermal energy needs sustainably. The synergistic effects of these two systems prompted exploration of utilizing the technologies in tandem. A review of prior work was addressed in previous publications [1-3].

Performance improvements are generally found with solar-assisted heat pump (SAHP) systems, but the difficulty in justifying systems of this type lies in the added equipment costs. Many studies have either conducted performance analysis using numerical models or experimental results. Few studies have used validated models to analyze system performance under varying conditions and parameters.

This study investigates the annual performance of a dual tank SAHP system developed in previous work. A custom model was developed in TRNSYS and validated experimentally with a purpose-built test apparatus [1,2]. In previous work it was found that the system is difficult to justify for domestic hot water (DHW) production in single family homes [3]. The feasibility of modifying system operation during the summer months to enable space cooling capabilities is explored in this work.

METHOD

The dual tank SAHP system includes a heat exchanger (HX), hydronic pumps, and diverter valves in addition to solar thermal collectors (STCs), a custom heat pump (HP), and thermal storage tanks. A simplified schematic of the system is provided in Figure 1. The configuration allows for a variety of operating modes, such as Solar-HX-DHW, Solar-HP-DHW, Solar-Float, Float-HX-DHW, and Float-HP-DHW.

The dual tank SAHP system is investigated using TRNSYS by analyzing DHW tank temperature in comparison to a traditional solar domestic hot water (SDHW) system. On an annual energy basis, the solar thermal systems are also compared to a standard electric DHW system. The impact of disabling float tank heating for summer months is analyzed.

The water draw profile used for the investigations is CSA-C from Canadian Standard Association's "Packaged solar domestic hot water systems (liquid-to-liquid heat transfer)", which totals 300 L per day [4]. This draw profile is representative for a family of 5 individuals. The weather file used is typical meteorological year (TMY) data for Ottawa, Ontario, Canada [5]. The solar thermal collector

(STC) array modelled for all simulations is 7.5 m^2 of Viessmann Vitosol 200-F, Type SV2. The HP model is a custom type developed in previous work and has a nominal output of 2.3 kW [1].

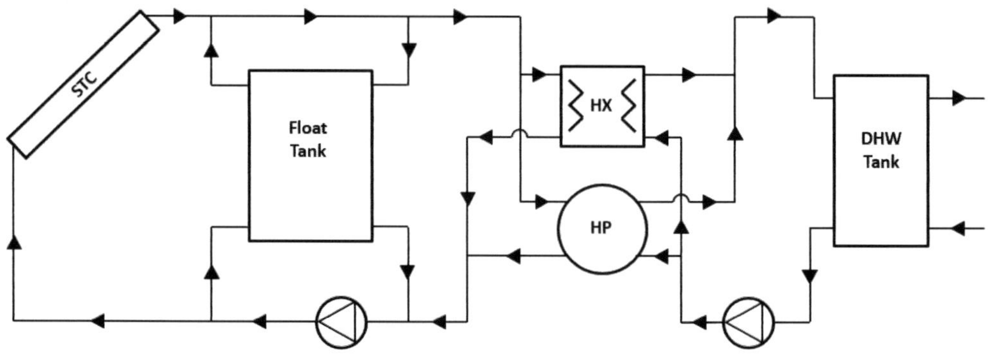

Figure 1: Dual tank SAHP system simplified schematic

RESULTS

The main goal of an SAHP system is to increase the amount of solar energy collection by reducing average collector temperature and boosting the temperature delivered to the load. Lower collector temperatures reduce losses to the environment, improving solar energy collection. Boosting the temperature of the fluid exiting the collectors allows solar energy to be obtained during periods of moderate insolation or low temperatures, when a SDHW system might be unable to operate.

The second storage tank provides additional thermal storage to the system for excess solar energy. The energy stored in the float tank can later be used to heat the DHW tank with either the HX or HP.

The dual tank SAHP, SDHW, and DHW systems were modelled for a calendar year to compare performance and assess the strengths of the proposed system. Weekly moving averages of DHW tank temperature and daily insolation for the SDHW and dual tank SAHP system are presented in Figure 2. Each period of 730 hours represents 1/12 of a calendar year; i.e. the length of an average month.

The variations in DHW tank temperature between these two systems highlights the performance improvements made by the dual tank SAHP system. From May thru August, the SAHP system significantly reduces the reliance on auxiliary heating. This is evident because the SAHP prevents the DHW temperature from falling below 60°C for the majority of this period. For the SDHW system during this time period, the DHW tank mainly oscillates between 57°C and 58°C. The performance improvement during April thru September can be considered 'summer energy increase'.

SDHW performance is at its worst outside May thru August, as overcast skies and cold temperatures in Ottawa prohibit effective solar energy collection [6]. However, the SAHP system provides noticeable performance improvements during September, October, February, March, and April. Performance improvements realized during September thru April can be considered 'winter operation increase'.

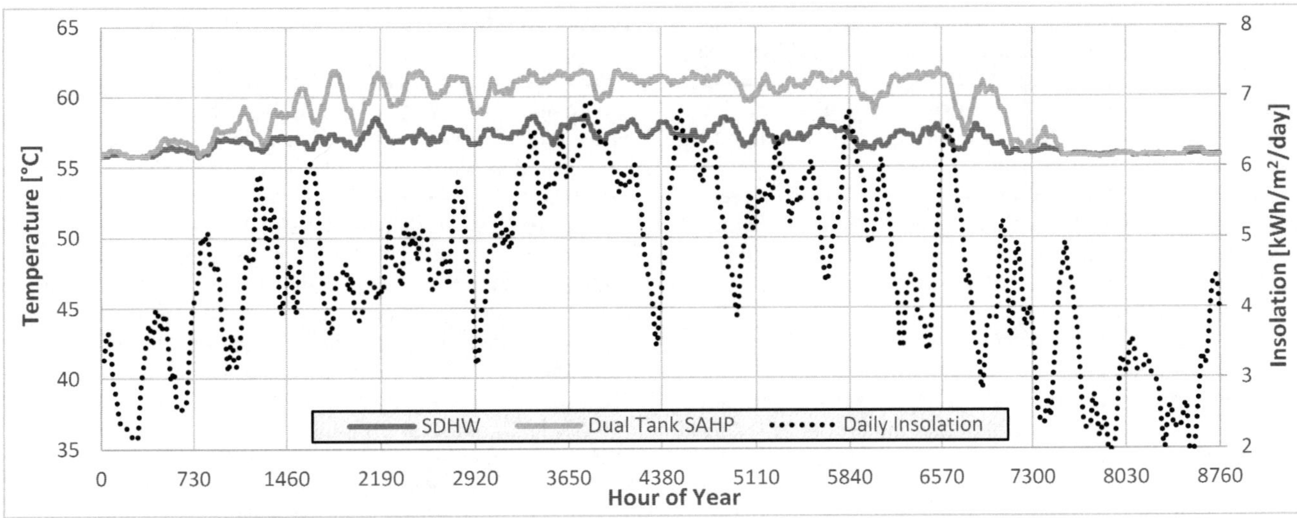

Figure 2: Weekly moving averages for DHW tank temperature and daily insolation

The dual tank SAHP system performance improvement is well depicted in a sorted annual DHW tank temperature plot, as shown in Figure 3. The increased area of the temperature-time plot for the dual tank SAHP illustrates the greater amount of energy storage and availability provided to the DHW tank, which reduces reliance on auxiliary heating.

The DHW tank temperature results, in combination with the weather data for Ottawa, show that a SAHP system of this type thrives on moderate insolation levels and is not severely impacted by very cold ambient temperatures. The coldest months of the year are January and February [6]; however, due to moderate insolation levels, the system performs better during February in comparison to December. December is much milder, but the very low insolation levels inhibit the operation of any solar energy system.

System performance during the summer months is improved by the heat pump and extra storage tank, but this period of the year is already the least vulnerable due to relatively abundant solar resources. For the investigated STC array and float tank size, the dual tank SAHP still occasionally relies on auxiliary heating; however, a control strategy modification may improve overall justification.

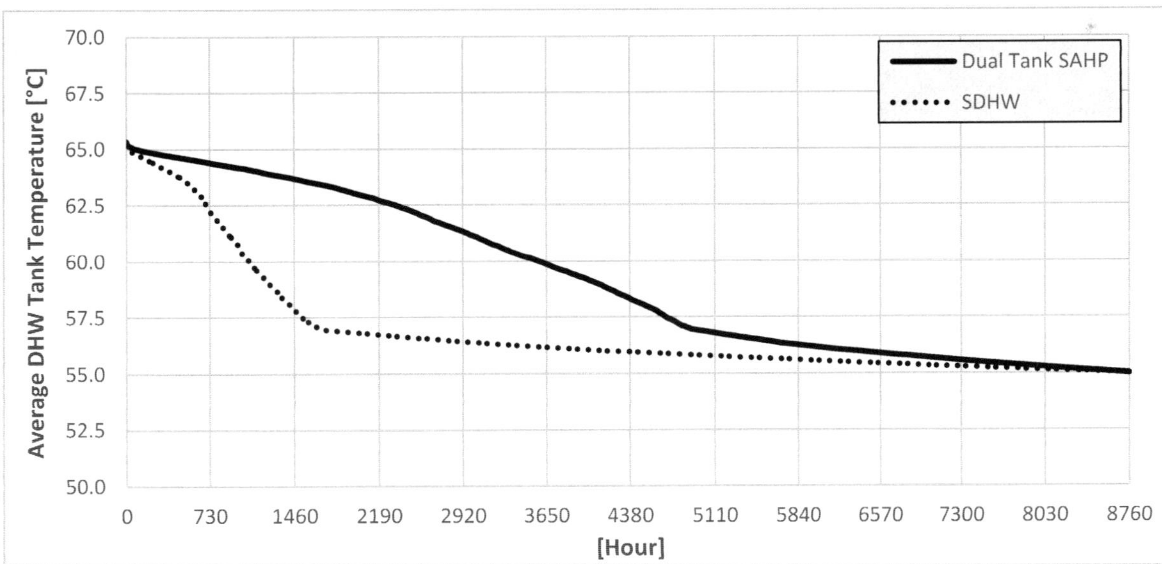

Figure 3: Average DHW tank temperature by hours of year, sorted descending

Since the dual tank SAHP system's only heat sources are solar energy and the auxiliary electrical element, the system relies heavily on resistance heating when solar insolation is low. This limitation suggests that the system could further benefit from an additional energy source.

To improve overall system performance and justifiability, it is worthwhile to consider adding an air-source heat exchanger. This additional energy source could reduce auxiliary energy usage throughout the year for DHW heating and also improve economic justification by allowing the system to meet the cooling load during hot and humid summer months.

Space cooling is an excellent additional load to meet with a dual tank SAHP. In Canada, about 52% of homes utilize air conditioning systems. For Ontario, this figure jumps to 80% due to a warm, humid summer [7]. The rates are even higher for homes built after 2000, meaning that most new residential construction in Ontario includes air conditioning. The installed residential air conditioning equipment in the province is underutilized, as it sits dormant for a large portion of the year. Air conditioning is generally only required in Ontario for the months of June, July, and August [6], meaning the compressor could be used as part of a SAHP system for the remainder of the year.

During the cooling season, the float tank could store chilled water rather than warm water, since solar resources are the greatest at this time. Less hot thermal storage is required to obtain a particular solar fraction for the system during this period. With a thermal storage medium for chilled water, the air conditioning portion of the system could operate during off-peak hours, when electricity rates are reduced and higher COPs can be achieved due to lower ambient temperature. Transitioning a substantial quantity of homes in Ontario to such a system could allow the summer midday electricity consumption peak to be noticeably decreased.

A further option using a dual tank SAHP system to meet the cooling load of a building is rejecting heat from chilled water to the DHW tank rather than air. Although at times the COP could be lower than exchanging to air due to higher temperature of DHW water, it is likely to use less energy than independently meeting these two loads.

To analyze the feasibility of producing chilled water with a dual tank SAHP during the cooling season, a modified operation strategy is considered. Rather than storing and extracting heat from the float tank for the whole year, disabling the capability during a portion of the summer is considered.

Table 1 compares the simulated performance of the new operating strategy to alternatives. The metric used for analysis is the increased electricity consumption (IEC) resulting from not using the float tank for a prescribed period in comparison to year-round utilization. IEC is calculated according to Eq. (1):

$$IEC\% = \frac{EC_{system} - EC_{SAHP\,(YRF)}}{EC_{SAHP\,(YRF)}} \times 100\% \qquad (1)$$

where EC_{system} is the annual energy consumption (EC) of the system being compared and *SAHP (YRF)* is the base case of the dual tank SAHP operating with year-round float (YRF) capability. (Note: 1 GJ = 278 kWh)

Table 1: Annual energy consumption of dual tank SAHP system with periods of unheated float tank in comparison to alternatives

System	**Electricity Consumption [GJ]**			**Increased Electricity Consumption**		
	HP	**Auxiliary**	**Total**	**Overall [GJ]**	**Days**	**Per Day [MJ]**
DHW	-	20.6	20.6	14.19 (221%)	-	-
SDHW	-	8.31	8.31	1.90 (29.6%)	-	-
SAHP, no float June 1 to Sept 30	1.32	5.79	7.11	0.70 (10.9%)	122	5.7
SAHP, no float June 1 to Aug 31	1.62	5.31	6.93	0.52 (8.1%)	92	5.6
SAHP, no float July 1 to July 31	2.09	4.56	6.65	0.24 (3.7%)	31	7.7
SAHP, year-round float (YRF)	2.28	4.13	6.41	-	-	-

The dual tank SAHP system with YRF decreased annual electricity consumption by 1.9 GJ or 528 kWh in comparison to the SDHW. Although the percent electricity savings are significant, the cost savings are not very significant due to the small scale load of single-family residential DHW.

The increased electricity consumption from disabling heating of the float tank during a period of the summer is noticeable, but not very large in comparison to the electricity savings resulting from utilizing the float tank for DHW heating during the remainder of the year.

For the main cooling period of June 1 to Aug 31 in Ottawa, the electricity savings dropped by 8.1%. This is in comparison to an overall drop of 29.6% for the SDHW system that has no float tank. As expected, the electricity consumption of the dual tank SAHP system increases as the duration the float tank is disabled increases.

CONCLUSIONS

Overall, it was found that a dual tank SAHP system provides significant annual energy savings in comparison to a SDHW system. This is accomplished through two main performance improvements: a summer energy increase and a winter operational increase. Despite the performance increase, economic justification for single family residential applications remains difficult due to relatively high capital costs. Therefore, a modified control strategy was evaluated which would enable the float tank to be used for space cooling during the hottest months of the year.

The results indicate that although the float tank improves energy savings during the summer months, the larger portion of savings is experienced during fall, winter, and spring. The minor increase in electrical energy usage by not heating the float tank from June 1 to August 31 could be offset by the savings from producing and storing chilled water to meet the building cooling load during this period. In an electricity market where peak summer load is dominated by space cooling, this alternative system provides an opportunity to reduce cooling costs and address regional concerns associated with peak electricity demand.

ACKNOWLEDGMENTS

This work was supported by: SNEBRN – Smart Net-Zero Energy Buildings Strategic Research Network, Canada; NSERC – Natural Sciences and Engineering Research Council, Canada; OGS – Ontario Graduate Scholarship, Ontario, Canada.

REFERENCES

1. Banister, CJ., Wagar, WR., and Collins, MR. 2013, "Solar-assisted heat pump test apparatus", *Solar Heating and Cooling Conference*, Freiburg, Germany.

2. Banister, CJ., Wagar, WR., and Collins, MR. 2013, "Validation of a single tank, multi-mode solar-assisted heat pump TRNSYS model", *Solar Heating and Cooling Conference*, Freiburg, Germany.

3. Banister, CJ., Collins, MR. 2014, "Dual tank SAHP configuration optimization", *Solar 2014 Conference*, San Francisco, USA.

4. Canadian Standards Association. 2009, "Packaged solar domestic hot water systems (liquid-to-liquid heat transfer)," Mississauga, Ontario, Canada.

5. TRNSYS: Transient System Simulation Tool. [Online]. http://www.trnsys.com.

6. Environment Canada. 2014, "Climate ID: 6158350", *1981 to 2010 Canadian Climate Normals*.

7. Office of Energy Efficiency, Natural Resources Canada. 2007, "Survey of Household Energy Use", Ottawa, Canada.

OPTIMIZED LOAD ALLOCATION OF BOILERS USING MIXED INTEGER NON–LINEAR PROGRAMMING FOR REDUCED NATURAL GAS CONSUMPTION

Altamash Ahmad Baig, PhD Student
Mechanical & Industrial Engineering
Ryerson University
Toronto, Ontario, Canada
altamashahmad.baig@ryerson.ca

Dr. Alan S. Fung, PhD, PEng, FCSME
Associate Professor
Mechanical & Industrial Engineering
Ryerson University
Toronto, Ontario, Canada
alanfung@ryerson.ca

ABSTRACT

Load management is a common practice in industrial plants that have similar equipment, having different operational efficiencies. Examples of such cases are boiler plants where boilers are arranged in parallel to meet the steam demand of the industrial plant. Greater load is assigned to the boiler which is considered to be more efficient, improving efficiency of the boiler plant and conserving energy. However, there is potential for improvement by optimizing load allocation. In this study an optimization technique is presented based on mixed integer non-linear programming (MINLP). Natural gas consumption versus steam demand curves were produced by conducting tests at various loading conditions. Gas consumption was expressed as a function of steam demand. Branch and Bound technique was used to reduce the mixed integer non-linear programming (MINLP) problem to a non-linear programming problem (NLP) at each node. The optimization technique was applied at two industrial plants and resulted in reduction of 2.6 % and 0.4 % in the natural gas consumption of the boiler room at plant A and plant B respectively. The savings in fuel cost associated with natural gas savings were $ 7,027 and $ 744 per year respectively while the reduction in greenhouse gas emissions were 63.8 and 6.8 metric ton equivalent of carbon dioxide per year respectively at plant A and plant B.

KEYWORDS: Energy Efficiency, Load Management, Optimization, MINLP.

INTRODUCTION

Optimized load allocation is a very useful and effective method of conserving energy in industrial plants that have similar equipment running at different operational efficiencies. Examples of such cases are boiler rooms that have more than one boiler sharing the steam demand load of the plant. Most of the studies undertaken on optimization of boiler load management have considered cost of running the boiler plant as the objective function [1-4]. This approach has generally been successful. However, the problem with using cost as the objective function is that the mathematical solution may result in optimal load for a boiler that corresponds to a firing rate below the minimum firing rate of a boiler. This may not be possible in reality as each boiler has a minimum firing rate at which it will run when it is 'fired' or turned 'on'. Hence, there is a need to set the optimization problem such that it eliminates the possibility of getting any solution of optimal loads corresponding to firing rates below the minimum firing rate of the boiler. In this study a novel optimization method based on mixed integer non-linear programming (MINLP) was developed with the objective of minimizing overall boiler plant energy (natural gas) consumption.

PROPOSED METHOD

The method is applicable to any boiler room with "n" number of boilers. Since, natural gas consumption varies with changing steam demand therefore, the steam load allocated to each of the boilers (represented by 'X') was taken as the independent variable. Also the decision to run a particular boiler was represented by the binary variable 'Y' which could be assigned values of "1" or "0" for "on" or "off" respectively. The natural gas consumption of each of the boilers along with the steam produced were noted at different loading conditions. Natural gas consumption per hour was expressed as a function, f(X, Y), of steam demand. It has been shown in [1] that observed boiler data fitted on a second order polynomials could be used to represent boiler efficiencies at different loads. In this study quadratic expressions were used for natural gas consumption of the boilers as shown by Equation (1).

$$f(X_k, Y_k) = Y_k * [a_k * (X_k)^2 + b_k * X_k] \quad (1)$$

$f(X_k, Y_k)$	is the natural gas consumption of the k-th boiler, as a function of the steam load
X_k	is the steam demand load allocated to the k-th boiler
Y_k	represents the decision to switch on k-th boiler ("1" for "on" and "0" for "off")
n	is the number of boilers

The steam demand of the boiler plant was taken as the constraint for the optimization problem. Furthermore, it was necessary to ensure that the solution of the optimization problem did not return a load allocation for boiler which corresponds to a firing rate less than the minimum firing rate of the boiler. This was achieved by adding an extra constraint that the steam load for a boiler could only be between the minimum and maximum load capacity of the boiler.

The optimization problem for 'n' number of boilers was set to minimize hourly natural gas consumption (Z) of the boiler room.

Minimize

$$Z = \sum f(X_k, Y_k)$$

Subject to

$$\sum X_k = \text{Steam load demand of the boiler room} \quad \text{(constraint 1)}$$

$$L_{k\,min} \leq X_k \leq L_{k\,max} \quad \text{(constraint 2)}$$

and

$$Y_k = \{1, 0\} \quad \text{(constraint 3)}$$

where

$L_{k\,max}$ is the maximum load capacity of k-th boiler when the boiler is running

$L_{k\,min}$ is the minimum load capacity of k-th boiler when the boiler is running

For the solution of the optimization problem Branch and Bound technique [5] combined with nonlinear programming (NLP) was used. At each node of the variable (Y) was assigned a value of either "1" or "0". This reduced the mixed integer non-linear programming (MINLP) problem to a quadratic programming (QP) problem of the form shown in Equation (2)

$$Q = \frac{1}{2}X^THX + F^TX \quad (2)$$

where

X is the vector of variables

H is the Hessian matrix for the objective function

F is the column vector comprised of coefficients of linear terms in the objective function

The problem was solved by using algorithm based on the solution of Karush-Khun-Tucker (KKT) conditions [6]. At each node the corresponding quadratic programming problem was solved to obtain the lower bound for the objective function. For generating branches, depth first technique was used (i.e., nodes on a branch were solved all the way to the last node). For finding candidate solution or fathoming nodes (i.e., "determining that it is not necessary to explore the descendants of a particular node in the search tree") [5], best lower bound method was used. If the optimal solution did not return a result better than the current lower bound, no further branches from that node were generated).

The boilers which were to be 'on' and the load allocation for those boilers were determined by solving the MINLP optimization problem. The baseline natural gas consumption was determined by using the load allocation in the boiler room at the time of the energy audit. The natural gas saving was determined as the difference between the baseline and the optimal.

$$\text{Energy Savings} = (\text{hr}) * (Z_{baseline} - Z_{opimal}) \quad (3)$$

hr hours of operation

$Z_{baseline}$ hourly natural gas consumption at the time of energy audit

$Z_{optimal}$ optimal hourly natural gas consumption

RESULTS

The method was applied for two industrial plants 'A' and 'B'. Natural gas consumption was plotted against steam production load. The coefficients 'a' and 'b' for demand vs. natural gas consumption curves of the boilers along with the coefficient of correlation (R^2) are shown in Table 1. The high R^2 value emphasize the strong correlation between steam load and natural gas consumption.

Table 1: Coefficients of equation representing the relationship between steam demand and gas consumption

Site	Boiler Number	Coefficients		Coefficient of Correlation
		a	b	R^2
A	1	-	0.0399	0.9997
	2	6.00E-07	0.0399	0.9999
	3	3.00E-08	0.0385	0.9999
B	1	6.00E-07	0.0395	0.9984
	2	1.00E-07	0.0379	0.9999
	3	5.00E-07	0.0375	0.9998

Natural gas consumption for the base case and optimal case were calculated. The savings for optimal case as a percentage of base case consumption are shown in Table 2.

Table 2: Natural gas savings by optimized load management

Site	Base Case Natural Gas Consumption of Boiler Room	Natural Gas Savings	Savings as Percentage of Base Case
	(m^3/yr)	(m^3/yr)	(%)
A	1,294,176	33,945	2.6
B	837,009	3,594	0.4

Fuel cost savings and greenhouse gas emission reduction were determined using the marginal cost of natural gas along with the greenhouse gas emission factor.

The change in cost per unit change gas consumption is called marginal cost of natural gas. Marginal cost was calculated through summation of all consumption dependent charges on the gas utility bill such as gas supply charge, cost adjustment charge, transportation charge, storage and delivery charges. Enbridge Gas' Rate-6 for commercial and industrial customers was used for this study. The applicable monthly charges under rate-6 are shown in Table 3.

Table 3: Marginal cost to commercial and industrial customers of Enbridge [7]

Charge	Rate
Gas supply charge	12.7 ¢/m^3
Transportation to Enbridge	3.15 ¢/ m^3
Cost adjustment	0.862 ¢/m^3
Delivery to Customer	3.98 ¢/m^3
Total Marginal Cost	20.7 ¢/m^3

The greenhouse gas emission factors in terms of grams equivalent of carbon dioxide per m^3 of natural gas in Canada based on the Canada's National Inventory Report (1990–2009) [8] are shown in Table 4. The greenhouse gas emission factor for Ontario was taken as 1879 grams equivalent of carbon dioxide per cubic meter of natural gas consumption.

Table 4: Carbon dioxide emission factors for natural gas [8]

Province	GHG Emission Factor (g eqCO_2/m^3)
Newfoundland and Labrador	1891
Nova Scotia	1891
New Brunswick	1891
Quebec	1878
Ontario	1879
Manitoba	1877
Saskatchewan	1820
Alberta	1918
British Columbia	1916
Yukon	2389
Northwest Territories	2454

Fuel cost savings and reduction in greenhouse gas emissions achieved by employing the proposed technique are summarized in Table 5.

Table 5: Fuel cost savings and reduction in greenhouse gas emissions by optimized load management

Site	Fuel Cost Savings	Reduction in Greenhouse Gas Emissions
	(\$/yr)	(tonnes CO_2/yr)
A	7,027	63.8
B	744	6.8

Boilers at site 'A', which had considerable differences in ages and efficiencies showed greater potential as compared to site 'B' which had boilers of the same age, which were installed at the same time and had very similar efficiencies. Site 'A' had three boilers having ages of 20 years, 40 years and 40 years while the boilers at site 'B' were relatively new and less than 10 years old. Therefore, the energy saving potential at site 'B' was considerably lower than the savings at site 'A'. Therefore, it can be concluded that this energy saving measure would be best suited to sites where there are considerable differences in the efficiencies and ages of the boilers in the boiler room

CONCLUSIONS

The optimization technique was successfully applied to optimize boiler loads at two industrial sites. The method ensured that the solution for optimized boiler loads did not correspond to firing rates below the minimum firing rate of the boilers. The optimization technique resulted in reduction of 2.6 % and 0.4 % respectively in the natural gas consumption of the boiler room at plant A and plant B respectiely. Optimization of boiler can potentially aid in conserving energy. However the potential of savings achieved would depend on the difference of efficiencies between the boilers.

ACKNOWLEDGMENTS

The authors would like to acknowledge the financial support provided by Enbridge Gas Distribution Inc, Connect Canada and MITACS Accelerate Ontario. In addition the authors also acknowledge Paul Morrison and Peter Goldman from Enbridge for as well as acting as liaison between their industrial customers and the authors.

REFERENCES

1. Cho, C.H., 1978, "Optimum boiler load allocation", *Instrumentation Technology,* Vol (25), pp. 55-58
2. Andreassen, O. Olsen, T.O., 1983, Optimal load allocation, *Modelling, Identification and Control,* Vol. (4), pp. 107-116

3. Dragićević, S., Bojić, M., 2009, "Application of Linear Programming in Energy Management", *Serbian Journal of Management,* Vol (4), pp. 227- 238
4. Dunn, A.C., Du, Y.Y., 2009, "Optimal Load Allocation for Multiple Fuel Boilers", *ISA Transactions,* Vol (48), pp 190 - 195
5. Lawler, E. L., Wood, D. E., 1966, "Branch-and-Bound Methods: A Survey", *Operations Research*, Vol (14), pp. 699 - 719
6. Kuhn, H.W., Tucker, A.W., 1951, Nonlinear Programming, *Berkeley Symposium on Mathematical Statistics and Probability*, University of California Press, pp 481 – 492
7. Enbridge Gas, *Rate 6 - Commercial and Industrial*, Retrieved January 1, 2014 from Enbridge Gas: https://www.enbridgegas.com/businesses/accounts-billing/gas-rates/large-volume-rates/rate-6.aspx
8. Environment Canada, (2014), Greenhouse Gas Emission Quantification Guidance - Fuel Combustion, Retrieved on January 1, 2014 from Environment Canada: http://www.ec.gc.ca/ges-ghg/default.asp?lang=En&n=AC2B7641-1#section1

IMPROVED GASIFICATION THROUGH TORREFACTION OF BIOMASS

Prabir Basu
Mechanical Engineering
Dalhousie University
Halifax, NS, Canada
prabir.basu@dal.ca

Manoj Kumar Raut
Mechanical Engineering
Dalhousie University
Halifax, NS, Canada
manojkumarraut@gmail.com

Bishnu Acharya
Greenfield Research Incorporated
Halifax, NS, Canada
bishnu.acharya@greenfieldresearch.ca

ABSTRACT

Biomass is a major renewable energy source. It has enormous potential as a feedstock for gasification, but its inherent characteristics limit the wide scale practical use of raw biomass. Properties such as high moisture content, high O/C ratio, low bulk density, heterogeneous nature, and lower grindability make its handling and gasification difficult. Furthermore, higher tar formation during gasification of raw biomass increases the downstream cost of gas separation. Pre-treatment of biomass through torrefaction could remove some of these limitations. This paper examines how and to what extent torrefaction removes these limitation of biomass gasification. Experiments in batch mode of steam-gasification show that torrefaction pre-treatment decreases the tar production and slightly increases the quality of product gas, but it reduces the gas yield when compared to raw biomass. Also, the reactivity for gasification of char from torrefied biomass is found to be lower than that from raw biomass. The detail experimental findings are discussed in the paper.

KEYWORDS: Torrefaction, Gasification, Kinetics of Gasification, Tar Yield

INTRODUCTION

Biomass is favoured as an alternative renewable fuel for heat, power and liquid fuel production (Svoboda et al., 2009). Gasification is one of the thermochemical routes for production of clean and quality heat, power and liquid fuels from biomass. However, some inherent properties of biomass like high moisture content, low-bulk density, fibrous nature makes its use difficult and additionally tar produced during gasification process hinders its downstream use (Devi et al., 2003). These limitations can be eliminated by pre-treating the with torrefaction, which is a mild pyrolysis process within the temperature range of 200-300°C in an inert environment (Basu, 2013). This pre-treatment makes the biomass more brittle, increases the bulk density and heating value, makes biomass hydrophobic, more stable and homogenous (Bergman et al., 2004) giving major boost to its practical use for energy. Furthermore, it helps to reduce the hydrogen to carbon (H/C) ratio and oxygen to carbon (O/C) ratio of biomass to values similar to that of coal. This reduction in H/C and O/C ratio further helps to improve the quality of the gas produced during gasification (Prins et al., 2006). Another major advantage of torrefaction pre-treatment is reduction in tar production during gasification. Because of these benefits there has been much interest in torrefaction and several studies has been done to understand the process. However, the application of torrefied biomass in gasification remains largely unexplored. This work examines the kinetics of steam gasification of raw and torrefied biomass using a batch reactor. The effect of torrefaction on gas and tar yield is also studied.

METHODOLOGY

The present work is based on woody biomass, Poplar wood. The proximate and ultimate analysis of the wood is shown in Table 1.

Table 1: Proximate and ultimate analysis of raw and torrefied poplar wood

	Proximate Analysis (%)				Ultimate Analysis (%)					Heating Value
	Moisture Content	Dry Basis			Carbon	Hydrogen	Nitrogen	Oxygen	Sulfur	HHV (MJ/kg)
		Volatile Matter	Ash	Fixed Carbon						
Raw Biomass	4.64	85.25	5.52	9.23	47.47	7.18	0.55	44.79	0.0	18.34
Torrefied Biomass @250 C, 1 hours	2.49	81.35	6.52	12.13	51.73	6.33	0.55	41.38	0.0	20.33

Devolatilization and drying are two short duration pre-steps in gasification. A part of those already took place during torrefaction. For comparing gasification of two feedstock (raw biomass and torrefied biomass) on a uniform base, both were devolatilized at a fixed temperature of 800°C, and then the product char is subjected to steam gasification. This part of the process is slowest and main part of the gasification process.

The kinetics of gasification of raw and torrefied biomass char in steam is studied in a Quartz Wool Matrix (QWM) reactor (Fig. 1) that consists of a stainless steel reactor 50 mm in diameter encircled by an electric heater. A temperature controller controls temperature inside the reactor. A precision balance sits on the top of the reactor to facilitate continuous measurement of mass of the substance being examined. A wire basket, whose one end is connected to the balance, holds the sample inside the reactor. The flow rate of gases is continuously measured with an electronic flow meter. But the steam flow rate is directly measured with its valve opening, which is first calibrated.

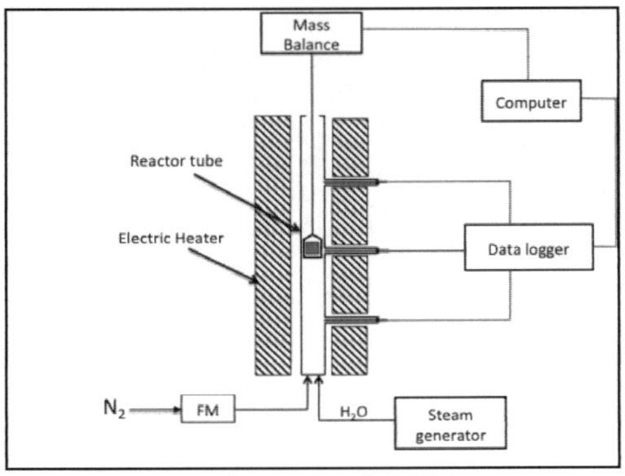

Figure 1: Schematic of QWM reactor

To produce a sample of torrefied biomass a 25 mm diameter and 75 mm long cylindrical piece of raw poplar is taken. For this, first the reactor is heated to the desired torrefaction temperature and then the above piece of raw biomass is placed in the bucket and dropped inside the reactor flushed with nitrogen. While the biomass sample gets heated and torrefied both temperature and the mass loss of the biomass is continuously measured. The biomass is placed for 1.0 hour and is taken out. For kinetics studies one sample of raw biomass and 2 samples of torrefied biomass (torrefied at 250°C and 300°C for 1 hour) are gasified at five different temperatures: 750°C, 800°C, 850°C, 900°C, and 950°C. Before gasification samples of both raw and torrefied biomass is devolatilized to produce char under identical conditions. For devolatilization, the reactor is heated to 800°C and then the biomass sample i.e. raw and torrefied biomass is dropped into the reactor. N_2 is supplied as a gaseous medium to drive off the volatiles, while providing an inert environment to supress occurrence of any other reactions. The mass loss of the sample is continuously measured, and as the mass becomes constant, the sample is taken out and cooled in a desiccator. The cooled sample is then ground to 300-450 microns size for gasification tests.

For gasification, the reactor is first preheated to the desired gasification temperature, and then the char samples obtained from the above devolatilization tests are placed in the bucket and lowered inside the reactor. The sample is gasified for 1.0 hour while steam is supplied from the bottom of the reactor. The mass of the char is continuously measured and its loss is used to calculate the char conversion rate for determining the kinetic rate. The Arrhenius form of equation (Eq. 1) is used for comparing the kinetics of gasification between raw and torrefied biomass char (Slopiecka et al., 2012).

$$k(T) = Ae^{\left(-\frac{E_A}{RT}\right)} \tag{1}$$

Where, E_A (kJ/mol) is activation energy, A (1/min) is frequency factor, T (K) is the absolute temperature and R (J/mol/K) is universal gas constant. To facilitate accurate collection of products of gasification (char, gas and tar), the reactor is closed from the top. The char sample is kept in the basket but the product gas is taken to the condenser and then collected in a gasbag, which is later analysed in gas chromatograph (GC). The condensed liquid is then evaporated to remove the moisture and the mass is measured for gravimetric tar.

RESULT AND DISCUSSION

Product Yield

Table 2 shows the mass yield of char, gas and tar obtained during torrefaction, devolatilization and gasification in terms of per unit of raw biomass weight. It is apparent that char yield during devolatilization is in on average 43% higher for torrefied biomass than that for raw biomass whereas tar yield is lower for torrefied biomass. The product gas yield of gasification is higher for raw biomass char than that for torrefied biomass char. Average value of gasification yield for char conversion is 82.07% for raw biomass compared to 47.21% for torrefied biomass. The average value of tar yield during gasification of raw biomass char is 79% higher than that for torrefied biomass char.

Kinetics Study

The Arrhenius plot for the gasification reaction of char produced from raw and torrefied biomass in presence of steam is shown in Figure 2. It shows that the kinetic rate of gasification of torrefied biomass is lower than that of raw biomass. This was surprising because both devolatilization and gasification processes were identical for both raw and torrefied biomass, and torrefied char showed higher internal surface area.

The difference in reactivity could be associated with some special changes to the biomass structure that takes place during torrefaction. Torrefaction is characterized by a slow heating rate. So, the volatile released from sample interior finds sufficient time to undergo condensation on pore walls of char around it. This condensation forms additional char known as secondary char. For this reason the char yield from torrefied biomass after devolatilization is around higher (43%) than that from the raw biomass, which confirms the formation of secondary char. This is more prominent in larger size biomass similar to that considered in this study (Dhungana, 2011). The characteristic of the secondary char is different from that of the primary char. Fisher et al. (2012) found that secondary char formed during torrefaction has a lower reactivity. So, gasification of torrefied mass involves gasification of less reactive secondary char and while that of raw biomass generally involves reactive primary char. Therefore, the overall reactivity of torrefied biomass during gasification is lower than the raw biomass.

Table 2: Char, volatile and tar yield during torrefaction, devolatilization and gasification

	Initial Mass	Torrefaction (250°C/1hr, N_2)			Devolatilization (800°C/1hr, N_2)			Gasification (800°C/15min, Steam)		
Sample		Biomass (g/g)	Volatiles (g/g)	Tar (g/g)	Char (g/g)	Volatiles (g/g)	Tar (g/g)	Char (g/g)	Gas (g/g)	Tar (g/g)
Raw1	5.7203				0.0813	0.9086	0.0101	0.0103	0.0695	0.0015
Raw 2	5.6949				0.1024	0.8885	0.0091	0.0203	0.0805	0.0016
Raw 3	5.7218				0.1073	0.8815	0.0112	0.0229	0.0831	0.0013
Torrefied 1	5.72	0.7596	0.2391	0.0013	0.1301	0.6248	0.0047	0.0673	0.0626	0.0002
Torrefied 2	5.7046	0.7321	0.2667	0.0012	0.1553	0.5731	0.0038	0.0784	0.0763	0.0005
Torrefied 3	5.7152	0.7499	0.2488	0.0013	0.1314	0.6138	0.0047	0.0738	0.0575	0.0002

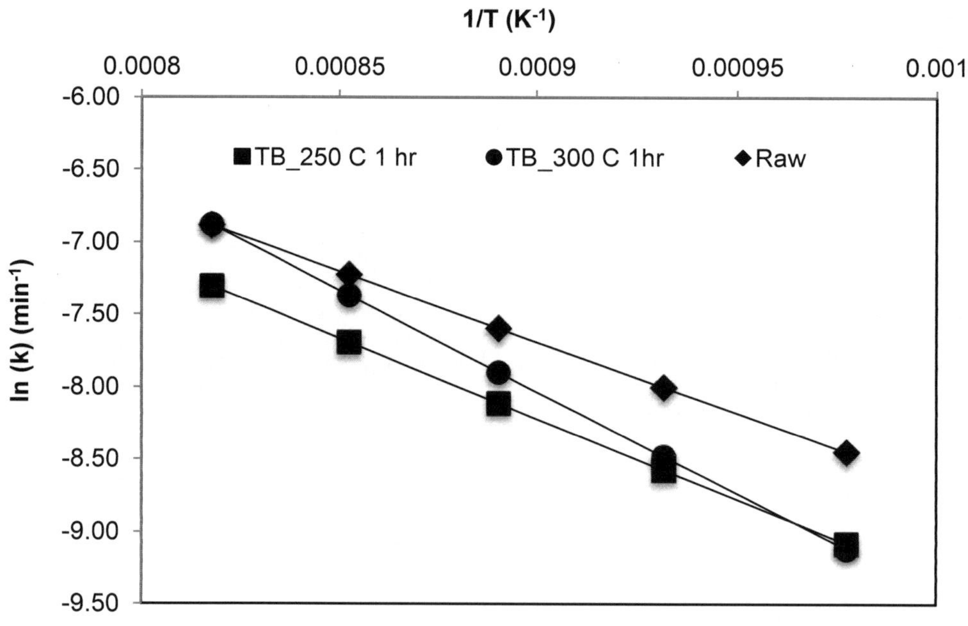

Figure 2: Arrhenius plot for gasification reaction of char obtained from raw and torrefied biomass

Tar Yield

Table 3 shows the total tar yields during gasification of raw and torrefied biomass (produced at 250°C & 300°C and 1 hour residence time) at 800°C in presence of steam. From Table 3, it is evident that the tar yield for torrefied biomass is lower than that of raw biomass.

Sweeney (2012) also found 66% reduction in tar yield of gasification for dark roasted torrefied biomass as compared to raw biomass when gasified in a bubbling fluidized bed gasifier.

Table 3: Total tar yield from gasification of raw and torrefied biomass chars at 800°C

	Raw biomass Char	Torrefied biomass char	
		250°C/1hr	300°C/1hr
Tar (mg/g)	10.12	6.39	3.08

The lignin and the hemicellulose components in a lignocellulose biomass make major contribution to the tar formation (Kiel et al., 2004). During torrefaction, most of the hemicellulose is devolatilized and lignin is partially devolatilized. Thus when the torrefied product is gasified less tar remains to be produced. Studies done on the pyrolysis of lignin showed that major decomposition of lignin occurs within the temperature range of 200-400°C (Beis et al., 2010). During this decomposition, aromatic hydrocarbons, phenolic, hydroxyphenolics and guaiacyl/syringyl-type compounds, which mostly have phenolic-OH groups, are formed. These constitute the bulk of the tar (Brebu and Vasile, 2009). During torrefaction, which occurs between 200-300°C, some of these tar components are released thereby reducing the tar produced during subsequent devoltilization of the torrefied biomass. Also as the heating rate is relatively low during torrefaction, the rate of volatilization of lignin fragments is not sufficiently high to avoid formation of secondary char (Brebu and Vasile, 2009). The secondary char is therefore formed, and since it is more stable during devolatilization and gasification, the tar yield is also less for torrefied biomass compared to that for raw biomass.

CONCLUSION

The present study of gasification of torrefied biomass highlights both positive and negative aspects of this pre-treatment process. The positive side is that torrefaction results in lower tar production during gasification. This can reduce the cost and effort needed for downstream gas cleaning. On the other side, the lower char conversion, which leads to lower gasification efficiency, could be a limiting factor for torrefied biomass. Kinetic studies showed that torrefied biomass has reduced gasification reactivity due to the formation of secondary char during the process. However, superior handling properties, hygroscopic and non-biodegradable nature, and an enhanced energy density could well compensate this limitation of torrefied biomass, and make it a better fuel for gasification.

REFERENCES

Basu, P., 2013. Biomass gasification, pyrolysis and torrefaction: Practical design and theory, Second edition. Academic press, an imprint of Elsevier, London NW1 7BY, UK.

Bergman, P.C.A., Boersma, A. R., Kiel, J. H. A., Prins, M. J., Ptasinski, K. J., and Janssen, F. J. J. G. 2004. Torrefaction for entrained flow gasification of biomass. The 2nd World Conference and Technology Exhibition on Biomass for Energy, Industry and Climate Protection, Rome, Italy.

Beis, H. S., Mukkamala, S., Hill, N., Joseph, J., Baker, C., Jensen, B. and Stemmler, E. A., 2010. Fast pyrolysis of lignins. Bio Resources 5, 1408-1424.

Brebu, M. and Vasile, C., 2010. Thermal degradation of lignin – a review. Cellulose chem. Technol., 44 (9), 353-363.

Devi, L., Ptasinski, K. J., and Janssen, F. J. J. G., 2003. A review of the primary measures for tar elimination in biomass gasification processes. Biomass and Bioenergy, 24, 125 – 140.

Dhungana, A. 2011. Torrefaction of biomass. M. A. Sc. Thesis, Dalhousie University, Halifax, Nova Scotia.

Fisher, E. M., Dupont, C., Darvell, L. I., Commandré, J.-M., Saddawi, A., Jones, J. M., Grateau, M., Nocquet, T. and Salvador, S. 2012. Combustion and gasification characteristics of chars from raw and torrefied biomass. Bioresource Technology, 119, 157–165.

Kiel, J.H.A., van Paasen, S.V.B., Neeft, J.P.A., Devi, L., Ptasinski, K.J., Janssen, F.J.J.G., Meijer, R., Berends, R.H., Temmink, H.M.G., Brem, G., Padban, N. and Bramer, E.A. 2004. Primary Measures to Reduce Tar Formation in Fluidized-Bed Biomass Gasifiers. ECN-C-04–014. Energy Research Center of the Netherlands.

Prins, M. J., Ptasinski, K. J., and Janssen, F. J. J. G. 2006. More efficient biomass gasification via torrefaction. Energy, 31, 3458-3470.

Svoboda, K., Pohorel, M., Hartman, M., and Martinec, J. 2009. Pretreatment and feeding of biomass for pressurized entrained flow gasification. Fuel Processing Technology, 90, 629–635.

Slopiecka, K., Bartocci, P., and Fantozzi, F. 2012. Thermogravimetric analysis and kinetic study of poplar wood pyrolysis. Applied Energy, 97, 491-497.

Sweeney, D. J. 2012. Performance of a pilot scale, steam blown, pressurized fluidized bed biomass gasifier. Ph.D. Thesis, Department Of Mechanical Engineering, The University Of Utah, USA.

THE ROLE OF HRW SYSTEMS IN REDUCING GREENHOUSE GAS EMISSION AND ENERGY SAVINGS OF ONTARIO-BASED PLANTS

Afarin Amirirad
Mechanical and Industrial Engineering
Ryerson University
Toronto, Ontario, Canada
afarin.amirirad@ryerson.ca

Alan S. Fung, PhD, PEng, FCSME
Mechanical and Industrial Engineering
Ryerson University
Toronto, Ontario, Canada
alanfung@ryerson.ca

ABSTRACT

In all industrial plants, the ventilation process is an essential component of Heating, Ventilation and Air-Conditioning (HVAC) procedure. Manufacturers need this process to reduce their exposure to extra heat and pollution which are produced in the environment. Our main focus in this research was on the ventilation sub-category of the seasonal energy consumptions in which we aimed to analyze a series of measured datasets from 15 Ontario-based industrial plants. In particular, we investigated whether the energy consumption and greenhouse gas (GHG) emission could be reduced by employing a Heat Recovery Wheel (HRW) technology and if they could, from financial side whether it was acceptable to use HRW in different plants with different characteristics.

KEYWORDS: Energy Saving, Greenhouse Gas Emission, Heat Recovery Wheel, Industrial Ventilation, Payback Period

INTRODUCTION

It is clear that decreasing the energy consumption is not only beneficial from a financial standpoint but is also valuable from environmental and global perspectives. The financial savings are persuasive for all sectors, in particular for the industrial sector which consumes approximately 37% of the world's total energy [1]. Furthermore, reduced energy consumption preserves the global natural resources while decreasing the amount of greenhouse gas emissions which is fundamental in decelerating the global warming process. Although at present the industrial sector is one of the world's largest energy consumers, it has been estimated that in the most "technology optimistic" perspective, the industrial energy consumption would in fact increase within the next 50 years [2, 3]. Consequently, there will be higher demand for industrial energy auditing in the near future which will require an investigation for alternative approaches in reducing the energy consumption.

Focusing on the industrial sectors, their energy consumption can be generally divided into two categories:

1. Process Consumption: The natural gas that is consumed in manufacturing product. This amount is assumed to be constant during one year.

2. Seasonal Consumption: The amount of natural gas which is used in ventilating and heating the environment during cold season and it is clear this parameter can change during one year. The seasonal consumption itself may also be subdivided into two different sub-categories, namely: 1-Space Heating, 2-Ventilation. Since a very large volume of air is required in industrial plants for ventilation purposes, the energy conservation and recovery should be considered to provide significant savings [4].

Previous research efforts on improving the industrial ventilation include various alternatives. Saman et al. [5] discussed the energy conservation opportunities in the area of industrial ventilation. They proposed several recommendations such as reducing the blower operating time/speed and static pressure for dust collectors, installing radiation shield on ovens and using outside air for cooling. Bhattacharjee [6] also discussed some opportunities in reducing the energy consumption of make-up air units. Some of these opportunities include: appealing proper insulation, the appropriate space temperature control, air balancing between the make-up air and the exhaust air systems in order to optimize the energy. Asiedu et al. [7] has formulated the payback period for the air to air heat and energy recovery systems based on a small HVAC configuration.

In this research, we have focused entirely on the ventilation process. As an alternative means for reducing energy consumption, we have employed the idea of recovering heat from the exhaust ventilation air. In particular, we will investigate the economical and environmental benefits of using a "Heat Recovery Wheel" unit.

METHODOLOGY

Based on ASHRAE Standard 62.1 [8], the required air flow for ventilation in industrial plants is calculated using plant area as shown in Eq. (1):

$$Ventilation\ Air\ Flow\ (CFM_{vent}) = 0.18 \times Plant\ Area\ in\ ft^2 \quad (1)$$

The employed plants descriptions for different parts of calculation have been shown in Table 1.

Table 1: Plants Description [9]

	Average Annual Natural Gas Consumption (m^3)	Total GHG Emission (tonnes/yr.)	Natural Gas Cost ($)	Area (ft^2)	Hours of Operation
Plant 1	3,263,751	6,133	734,344	190,008	6,240
Plant 2	676,090	1,270	152,120	60,000	5,616
Plant 3	987,794	1,856	222,254	186,026	6,240
Plant 4	544,200	1,023	122,445	40,000	7,488
Plant 5	964,854	1,813	217,092	186,500	6,240
Plant 6	628,339	1,181	141,376	270,000	8,400
Plant 7	545,359	1,025	122,706	70,000	2,500
Plant 8	340,017	639	76,504	100,000	3,640
Plant 9	458,033	861	103,057	46,609	2,210
Plant 10	492,795	926	110,879	65,000	2,080
Plant 11	302,332	568	68,025	110,270	2,000
Plant 12	1,283,047	2,411	288,686	213,668	2,000
Plant 13	397,663	747	89,474	71,722	3,000
Plant 14	158,235	297	35,603	10,573	3,120
Plant 15	779,754	1,465	175,445	230,453	8,000

The efficiency (η), power requirement and air flow of HRWs are given in their catalogues. Thus for the selected HRW (based on air flow rate CFM), the supply temperature of the ventilation system or the exhaust temperature of HRW to the indoor environment (i.e., T_{supp}) has been calculated using the HRW efficiency η which is defined as:

$$\eta = \frac{T_{supp} - T_{out}}{T_{in} - T_{out}} \tag{2}$$

In order to investigate and compare the feasibility of using HRW in severe weather conditions of Toronto this formula has been used for two different scenarios:

Scenario 1: T_{out} is equal to average temperature in winter time in Toronto

Scenario 2: T_{out} is equal to average temperature in January in Toronto

For the first scenario, T_{in} = 72°F and T_{out1} = 35.5°F. Note that T_{out1} is the average temperature in winter time (Oct. to May) during 1971 to 2000 [10]. The second scenario assumes T_{in} = 72°F and T_{out2} = 13°F. This time T_{out2} is the average temperature in January which was the coldest month during 1971 to 2000 [10]. T_{sup1} and T_{sup2} have been estimated using Eq. (2).

An important step of this project involves calculating the required amount of heat load, Q, in two different conditions: with and without the HRW system. Regarding calculation of heat loads, Eq. (3) has been employed:

$$Q = mC_p\Delta T \tag{3}$$

In order to convert mass flow unit to cubic feet per minute, we need to multiply it by air density $(0.075\frac{lb}{CFM})$, air specific heat $0.24(\frac{Btu}{lb°F})$ and $60(\frac{min}{hr})$. Finally, the Eq. (3) will transform to the following equation:

$$Q = 1.08 \times CFM \times \Delta T \tag{4}$$

Moreover, for each case we will do the calculation based on the aforementioned scenarios using Equations (5) and (6) and (7):

$$Q_{without\ HRV} = 1.08 \times CFM_{vent} \times (T_{in} - T_{out}) \tag{5}$$
$$Q_{with\ HRV-1} = 1.08 \times CFM_{vent} \times (T_{in} - T_{supp1}) \tag{6}$$
$$Q_{with\ HRV-2} = 1.08 \times CFM_{vent} \times (T_{in} - T_{supp2}) \tag{7}$$

For the next step, these two Q values were used in Equation (8) to estimate the total annual amount of energy that could be saved after adding an HRW system:

$$Annual\ Energy\ Saving(Btu) = (Q_{without\ HRV} - Q_{with\ HRV}) \times Annual\ Hours\ of\ Plant\ Operation\ in\ Winter \tag{8}$$

This energy saving has guided us to determine the amount of reduction in natural gas consumption which in turn has been used to infer the total savings in energy costs as well as the amount of reduction in the greenhouse gas emissions for different plants.

Every 1 m^3 of natural gas has Higher Heating Value (HHV) of 38 MJ [11], therefore (recall that 1 GJ = 947817 Btu):

$$HHV_{1\ m^3 Natural\ Gas} = 36017\ Btu \tag{9}$$

The above value of HHV together with the calculated annual energy saving due to HRWs would enable the estimation of the annual reduction in natural gas consumption, based on Eq. (10):

$$Annual\ Reduced\ Natural\ Gas\ Consumtion(m^3) = Annual\ Energy\ Saving(Btu)/36017 \tag{10}$$

According to the Canada's National Inventory Report (1990–2009) [12], the greenhouse gas emission factor in terms of grams of carbon dioxide per m^3 of natural gas in Ontario is 1879 g CO_2/m^3. Hence the total reduced GHG emission may be obtained as:

$$Annual\ Reduced\ GHG\ Emission(gr) = Annual\ Reduced\ Natural\ Gas(m^3) \times 1879 \tag{11}$$

RESULTS AND COMPARISON

The focal point of the study has been the reduction in natural gas consumption due to the application of HRW units. Figure 1 illustrates the reduced natural gas consumption (estimated by using Eq. (10) and the annual energy savings) in percentage (y-axis) along with the surface area of all the plants (x-axis). As shown in Figure 1, the natural gas consumption for ventilation relative to the total gas consumption can be reduced by 1.91% to 51.09% in all of the plants for both scenarios. Based on Figure 1 it may be concluded that, a colder outside temperature (e.g. in Scenario 2) increase the amount of natural gas reduction due to HRW. Therefore for industrial plants located in freezing weather conditions, using an HRW unit would be more efficient. The graph also shows that, in most cases the larger industrial plants have a higher percentage of natural gas reduction. However, for plants 1 and 8, this inference is not entirely accurate which can be a result of different ratios of winter hours of operation to the plant area. Another possible cause might be the variable ratio of the process to seasonal load which leads to plants 1 and 8 being outliers.

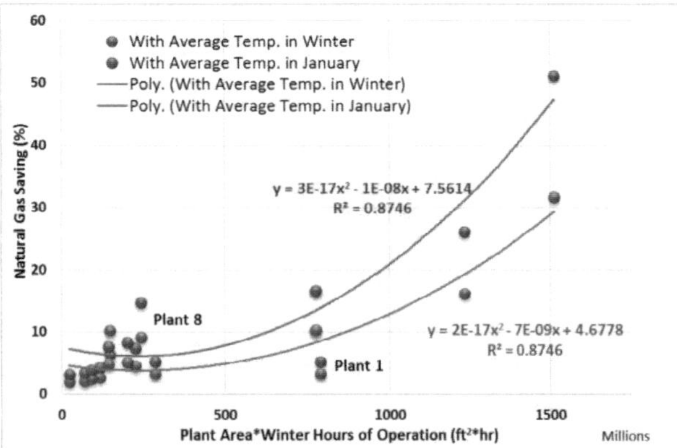

Figure 1: Natural Gas Saving and Plant Area Correlation for Scenarios 1 and 2

Furthermore the total reduced natural gas consumption in one year for all the 15 plants is equal to 902,156m^3 (representing 7.3% of total consumption) as it is shown in Table 2. This significant amount is about 3 days of marketable natural gas production in Ontario [13].

Table 2: Comparison of Energy Saving, Reduced Gas and Reduced GHG Emission for Scenario 1 and 2

	Annual Energy Saving Scenario1 (MMBtu)	Annual Energy Saving Scenario2 (MMBtu)	Annual Reduced Gas Scenario1 (m^3)	Annual Reduced Gas Scenario2 (m^3)	Reduced GHG Scenario1 (tonnes)
Plant 1	3,968	6,414	110,168	178,079	207
Plant 2	1,115	1,802	30,946	50,023	58
Plant 3	3,850	6,224	106,904	172,804	201
Plant 4	995	1,608	27,613	44,635	52
Plant 5	3,864	6,246	107,293	173,432	202
Plant 6	7,476	12,085	207,581	335,541	390
Plant 7	571	922	15,843	25,609	30
Plant 8	1,209	1,954	33,558	54,244	63
Plant 9	343	555	9,527	15,400	18
Plant 10	418	675	11,599	18,749	22
Plant 11	692	1,119	19,227	31,079	36
Plant 12	1,367	2,209	37,943	61,333	71
Plant 13	709	1,146	19,690	31,827	37
Plant 14	103	167	2,863	4,627	5
Plant 15	5,813	9,397	161,403	260,898	303
Total	32,493	52,523	902,156	1,458,280	1695

Another goal of this research has been to investigate the reduction in greenhouse gas emissions by applying HRW units. In this regard, Figure 2 shows that by using HRW in the ventilation system of a plant, the amount of GHG emission would be noticeably reduced. Although the amount of reduction in the GHG emission ranges from 1.9% to 31.6% for different plants, overall, the percentage of total GHG emission reduction for all 15 plants reaches approximately 7.3% of the total GHG emission without HRWs. This aggregate amount of GHG reduction cannot simply be neglected because this amount is equal to 1,695 tonnes of CO_2 as it is shown in Table 1 and is equivalent to the CO_2 emission from 155 residential buildings during one year, or equivalent to the total sequestered carbon from 43,462 tree seedlings grown for 10 years [14].

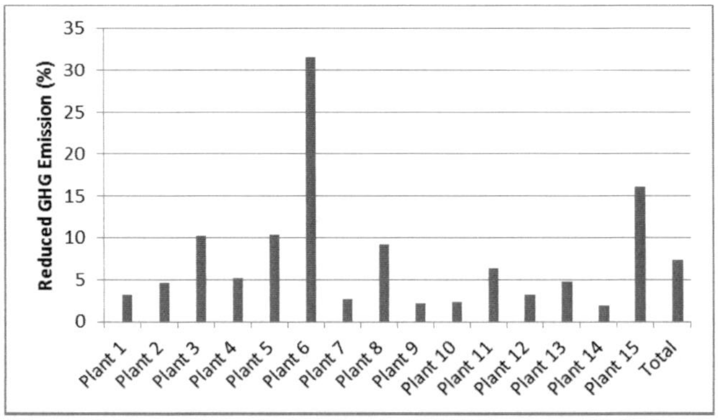

Figure 2: Percentage Reduction in GHG Emission Relative to the Total Emissions before Using HRW

It would be interesting to seek the primary parameter that determines the payback period. For this purpose, Figure 3 plots the payback period versus the plant area which illustrates that there exists a strong correlation between these quantities. As shown in Figure 3, larger plant size would result in shorter payback period. Consequently it can be concluded that for industrial plants with larger area it is more convincing to employ HRW and the business owners can receive back the money that they have invested in installing HRW much quicker. There are some exception for plant 2, 4 and 12 as it is shown in Figure 3. Although plants 2 and 4 have small area which must lead to higher payback period, but they have lower payback period compared to other plants with similar areas. The reason is because they have longer hours of operation which results in higher energy consumption in plants. Therefore, by using HRW more energy could be saved and payback period would be lower. For the plant 12 the situation is opposite and this plant has shorter hours of operation compared to plants with similar facility area, thus, lowering the potential for natural gas saving using HRW.

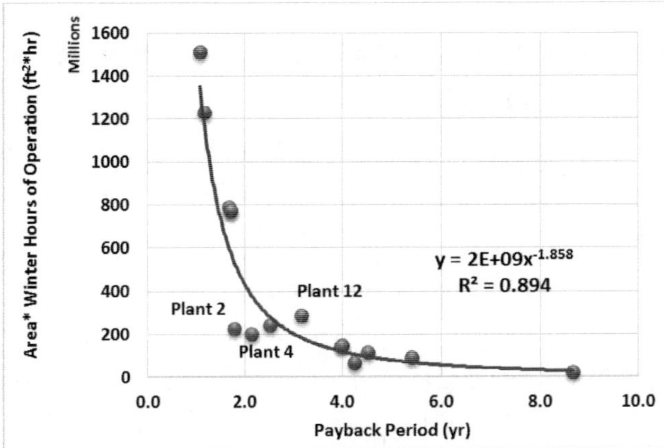

Figure 3: Payback Period vs. Plant Area and Hours of Operation Correlation

CONCLUSIONS

Based on the results presented in this report, it is concluded that using HRW in ventilation system can save energy, reduce natural gas consumption and reduce GHG emission for all the plants even in a severe January weather condition. Moreover the amount of energy that can be saved depends on the ventilation portion of the total natural gas consumption rather than process load or space heating load. Eventually, the larger the area is, the lower payback period can be attained by using HRW.

Although using HRW seems to be beneficial both from a financial and environmental perspectives, there are some issues that should be considered when this equipment is used. For instance, in cold weather conditions there might be higher demand for maintenance service which in turn would increase the overall HRW cost and subsequently leads to a longer payback period. One of the main drawbacks of HRW is a small amount of mixing between exhaust and supply air inside the wheel. Hence this equipment might not be an acceptable device for some plants with contamination in exhaust air.

ACKNOWLEDGMENTS

The authors would like to acknowledge the financial support provided by Enbridge Gas Distribution Inc. and MITACS Accelerate Ontario. The authors are grateful to Paul Morrison and Peter Goldman from Enbridge for as well as acting as liaison between their industrial customers and the Ryerson University team. In addition, the authors would like to thank their colleagues, Farzin M. Rad and Altamash Ahmed Baig, for conducting the energy audits and collecting the data.

REFERENCES

1. U.S Energy Information Administration, 2009, World Energy and economic Outlook; 2009, Retrieved on November 1, 2014, from http://www.eia.doe.gov/oiaf/ieo/world.html
2. Gielen, D., Taylor, M. 2007, "Modelling industrial energy use. The IEA's energy technology perspective", *Energy Econ*, Vol. (29), pp. 889–912.
3. Palm, J., Thollander, P. 2010, "An interdisciplinary perspective on industrial energy efficiency", *Applied Energy*, Vol. (87), pp. 3255–3261.
4. ASHRAE., 2007, "ASHRAE Handbook - Heating, Ventilating and Air-Conditioning Applications", Atlanta: American Society of Heating, Refrigerating and Air-Conditioning Engineers, Inc.
5. Saman, N., Nutter, D. 1994, "Energy Conservation Through Improved Industrial Ventilation in Small and Medium-Sized Industrial Plants", 16th Annual Industrial Energy Technology Conference.
6. Bhattacharjee, K. 2009, "Energy Conservation Opportunities in Industrial Ventilation Systems", *Energy Engineering*; 106, 3; ProQuest, pg. 73.
7. Asiedu, Y., Besant, R.W., Simonson, C. J. 2004, "Wheel Selection for Heat and Energy Recovery in Simple HVAC Ventilation Design Problems", *ASHRAE Transactions; ProQuest Science Journals*, Vol.(110), pg. 381.
8. ASHRAE., 2013, "ASHRAE Standard 62.1 - Ventilation for Acceptable Indoor Air Quality", Atlanta: American Society of Heating, Refrigerating and Air-Conditioning Engineers, Inc.
9. Ahmad Baig, A., 2014, " Analysis of Natural Gas Consumption and Energy Saving Measures for Small and Medium-Sized Industries in the Greater Toronto Area", Master's Thesis Report, Department of Mechanical & Industrial Engineering, Ryerson University, Toronto, ON
10. Climate Data and Information Archive. Daily Data Report. Retrieved from on November 1, 2014, EnvironmentCanada:http://climate.weather.gc.ca/ climate_normals/results_e.html?stnID=5097&lang =e&dCode=1&StationName=TORONTO&SearchType =Contains&province=ALL&month1=0&month2=12
11. Enbridge Gas Distribution, Retrieved from on November 1,2014, https://www.enbridgegas.com/corporate /environmental-commitment/ghg-reporting.aspx
12. Environment Canada, 2014, Greenhouse Gas Emission Quantification Guidance - Fuel Combustion, Retrieved on November 1, 2014 from Environment Canada: http://www.ec.gc.ca/ges-ghg/default.asp?lang=En&n=AC2B7641-1#section1
13. Marketable Natural Gas Production Archive, Retrieved from on November 12, 2014, National Energy Board Canada, https://www.neb-one.gc.ca/nrg/sttstc/ntrlgs/stt/ mrktblntrlgsprdctn-eng.html
14. US. Environmental Protection Agency, Retrieved from on November12,2014, http://www.epa.gov/cleanenergy/ energy-resources/calculator.html#results

Proceedings of the 25th CANCAM
London, Ontario, Canada, May 31 – June 4, 2015

A NEW DESIGN FOR THE IMPROVEMENT OF UNIFORM COOLING IN AIR-COOLED LI-ION BATTERY STACKS

Soroush Arghavan
Department of Mechanical and Materials Engineering
Western University
London, Ontario, Canada
sarghav2@uwo.ca

Mohammadhadi Akbari
School of Mechanical Engineering
Shiraz University
Shiraz, Fars, Iran
h-akbari@shirazu.ac.ir

ABSTRACT

This article presents a new design for thermal management of Li-ion batteries in electric vehicles through forced convection. This design carries the advantage of relatively uniform cooling of cells, while employing a simple geometry. In addition to a baseline case, a parametric study has been conducted on the contributing parameters such as the rate of heat generation in the battery, as well as cooling air velocity and temperature. It is shown that temperature distribution in the battery remains relatively unaffected although the value of temperature is changed with the operating conditions. While higher heat generation rate reduces temperature uniformity, increasing the inlet velocity is shown to lower cell temperature and increase temperature uniformity. Location of the maximum temperature will also be pushed downstream with higher velocities. Temperature of the battery can also be slightly reduced by an increase in the inlet turbulence intensity, however, temperature uniformity is compromised.

KEYWORDS: Li-ion batteries, thermal management, CFD modeling, air-cooled system.

INTRODUCTION

With the increasing popularity of electric vehicles, it is becoming progressively important to evaluate thermal management for mobile batteries. Power requirements of electric vehicles involve rapid charge and discharge rate cycles with high current densities during conditions such as quick acceleration, which cause considerable heat generation in return [1]. Heat generation, which is due to various internal chemical and electrochemical reactions, may cause serious damage to the battery. Common risks include overheating, loss of power and energy capability, reducing round-trip efficiency, explosion and high maintenance costs. For instance, every degree of temperature rise in an operating temperature of 30-40 °C causes the lifespan of a Li-ion battery to drop by two months [2]. Therefore, it is as much necessary to design an effective heat dissipation method as it is to increase the current density of the battery.

There are two main concerns regarding heat generation in a battery cell. The first is that the battery performance is highly affected in the case that temperature levels exceed permissible limits. Secondly, uneven temperature distribution in the battery pack results in localized deterioration. Uniformity of temperature within a cell and from cell to cell is important in maximizing performance of each cell and the whole stack.

Popular thermal management methods vary from tradition methods such as air thermal management systems using an electric fan and liquid thermal management with water or oil to more advanced methods such as thermal management using phase-change material.

When discussing thermal management solutions for electric vehicles, cost and simplicity of the system is crucial. The simplest approach that can be considered for electric vehicles is employing air cooling systems [3]. In order to maximize space efficiency, the required air can be supplied by the air conditioning unit of the vehicle. Among modern vehicles, Toyota Prius is one of the cars that uses conditioned air from the cabin for battery cooling. Zolot et. al. [4] used a system similar to Prius to cool (and heat) the battery. Lou [5] designed a cinquefoil battery pack with total voltage of 36 V to improve heat dissipation for Ni-MH batteries. Results revealed that while temperatures were well within the allowable range, cells near the fan would exhibit a larger temperature drop compared to those further from the fan. Moreover, Lou's structure was rather complex. For the purposes of reducing temperature drop variation among cells, Mahamud et al. [6] proposed a model consisting of two flip door valves that guide airflow through different channels and over the battery stack. The valves cause the flow direction over the stack to change with a specified period and therefore improve temperature uniformity.

While previous research shows that air-cooled systems are among the simplest methods that can be used in electric or hybrid vehicles, non-uniform distribution of temperature from cell to cell and hence, non-uniform wear and tear has always been an issue. In the present article, a design has been proposed in order to maintain uniformity among battery cells.

MODEL DESCRIPTION AND GOVERNING EQUATIONS

The proposed design consists of several parallel channels, each housing a battery cell. The channels are all placed perpendicular to an airflow channel which guides atmospheric air into a manifold and feeds each single channel through its 45-degree inlet. Air then cools the battery and is withdrawn from the structure through a common exit manifold. An advantage of this design is that the cold air entering each channel has the same temperature and therefore adjacent channels are cooled in the same manner and hence, uniform cooling of cells can be achieved. By placing a total of 40 Li-ion battery cells in the structure, enough power for an electric vehicle can be achieved which is about 50 kWh for a sports car. Fig. 1 shows a sketch of the proposed layout

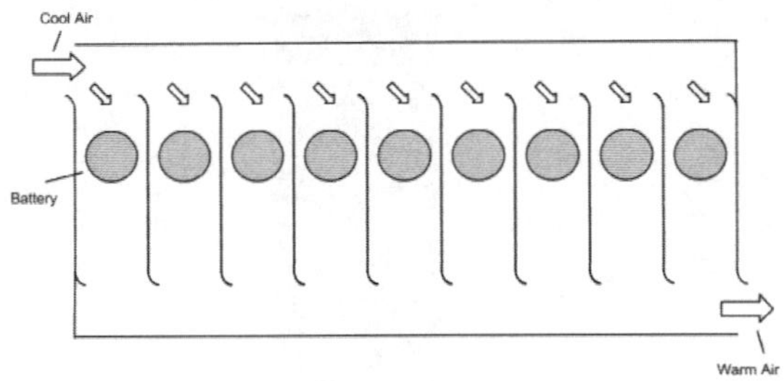

Figure 1: Proposed layout

The cylindrical Li-ion ($LiMn_2O_4/C$, capacity: 3.6 Ah, D × L = 42.4 mm × 97.7 mm) cell was considered for this study with an average density of 1400 $kg.m^{-3}$, specific heat capacity of 935 $J.kg^{-1}K^{-1}$ and thermal conductivity of 0.87 $W.m^{-1}K^{-1}$.
The geometry of the channel is assumed to be approximately 6.82 cm in width and 14.53 cm in length, long enough for the trailing vortices to damp. The battery is placed 3.83 cm below the inlet and both ends of the channel are 1-cm-radius arcs that form 45-degree angles with the airflow manifolds to improve guidance of airflow.
This case involves heat transfer by means of forced convection with air as the cooling fluid. The equations for continuity and conservations of momentum and energy, accompanied by transport equations of turbulence need to be solved throughout the domain in order to obtain the steady-state temperature distribution.
Keeping in mind that changes in air velocity and temperature are small, it can be assumed that the fluid is incompressible and the equation for conservation of mass can be written as follows [7]:

$$\nabla.(\vec{v}) = 0. \tag{1}$$

Conservation of momentum is given by the Navier-Stokes equation. In the present study, with air as the fluid, body forces can be omitted and the equation can be written as:

$$\frac{\partial}{\partial t}(\rho\vec{v}) + \nabla.(\rho\vec{v}\vec{v}) = -\nabla p + \nabla.(\overline{\overline{\tau}}), \tag{2}$$

where the stress tensor is described as Eq. 3:

$$\overline{\overline{\tau}} = \mu[(\nabla\vec{v} + \nabla\vec{v}^T) - \frac{2}{3}\nabla.\vec{v}I. \tag{3}$$

As the Reynolds number dictates at high inlet air velocities, the flow will become turbulent and transport equations for turbulence also need to be considered. The $k-\varepsilon$ model is chosen here which defines the transport equations as:

$$\begin{aligned}&\frac{\partial}{\partial t}(\rho k) + \frac{\partial}{\partial x_j}(\rho k u_j) = \\ &\frac{\partial}{\partial x_j}[(\mu + \frac{\mu_t}{\sigma_k})\frac{\partial k}{\partial x_j}] \\ &+G_k + G_b - \rho\varepsilon - Y_M + S_k,\end{aligned} \tag{4}$$

and

$$\begin{aligned}&\frac{\partial}{\partial t}(\rho\varepsilon) + \frac{\partial}{\partial x_j}(\rho\varepsilon u_j) = \\ &\frac{\partial}{\partial x_j}[(\mu + \frac{\mu_t}{\sigma_\varepsilon})\frac{\partial\varepsilon}{\partial x_j}] \\ &+\rho C_1 S\varepsilon - \rho C_2\frac{\varepsilon^2}{k + \sqrt{\nu\varepsilon}} \\ &+C_{1\varepsilon}\frac{\varepsilon}{k}C_{3\varepsilon}G_b + S_\varepsilon.\end{aligned} \tag{5}$$

Finally, the energy equation must be solved in the entire domain including the battery and the flow field in order to obtain the temperature distribution. The energy equation is given as:

$$\begin{aligned}&\frac{\partial}{\partial t}(\rho E) + \nabla.(\vec{v}(\rho E + p)) = \\ &\nabla.(k_{eff}\nabla T - \sum_j h_j\vec{J}_j + (\overline{\overline{\tau}}_{eff}.\vec{v})) + S_h,\end{aligned} \tag{6}$$

where $k_{eff} = k + k_t$ is the effective conductivity and k_t is the turbulent thermal conductivity calculated according to the $k-\varepsilon$ model. Also E is defined as function of sensible enthalpy such that:

$$E = h - \frac{p}{\rho} + \frac{v^2}{2}. \tag{7}$$

The two-dimensional model was designed using commercial software GAMBIT and analyzed by computation fluid dynamics commercial software FLUENT 6.3. The SIMPLE algorithm was used for coupling pressure and velocity fields. The Power Law discretization scheme was used for momentum and energy equations and second-order upwind scheme was implemented for the $k-\varepsilon$ model. The governing equations and employed schemes have been discussed in detail by Patankar [8].

BOUNDARY CONDITIONS

The proposed model is examined through a baseline case study and a parametric study. For the baseline case, the air inlet velocity at the top of the channel is assumed to be 1 $m.s^{-1}$ with a turbulent intensity of 10% and a hydraulic diameter of 3 cm which is based on the cell diameter. Battery properties such as discharge capacity and efficiency are based on the available literature (e.g. [9]). Assuming 3.6 Ah capacity and 90 percent cycle efficiency, a uniform heat generation rate of 12 $kW.m^{-3}$ throughout the cell would be a safe assumption.

MODEL VALIDATION

In order to make sure that the designed mesh used for the simulations provides accurate enough results, we shall compare temperature distributions predicted using three different grids varying from coarse to fine. Fig. 2 displays temperature distribution along the axial symmetry line of the cell for the three grids.

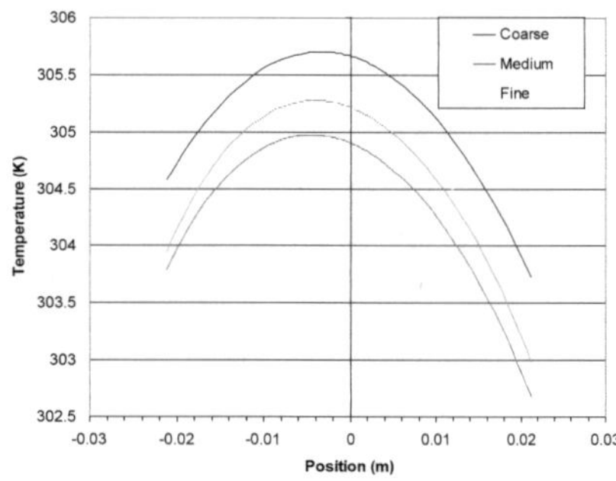

Figure 2: Grid study

It is observed that the difference in the peak temperature between the fine and medium grids is less than 0.5 K which itself is partly due to regression that is performed on the data points. Furthermore, it can be seen that the predicted cell temperature levels are consistent with previous research (e.g. [3], [4], [6] and [9]). In order to optimize computational load and achieve reliable results, the medium-sized grid is chosen for the analysis.

RESULTS AND DISCUSSIONS

Using the above mentioned boundary conditions, a baseline case is solved and temperature distribution and
flow field in the entire domain are obtained as shown in Figs. 3 and 4. It is observed in Fig. 3 that the maximum temperature point occurs near the center of the battery. Due to higher air velocities on the top compared to the cell wake where trailing vortices exist (See Fig. 4), the point of maximum temperature is closer to the bottom of the cell. Maximum temperature of the cell itself is 5 K higher than the upstream air temperature and maximum variation in temperature inside the cell is about 2 K which is relatively low.

It is also seen that air is heated by about 1 K which means that the withdrawn air from the batteries can be used as input for other components of the vehicle. Moreover, Fig. 4 shows that air velocity is higher on one side of the cell due to the 45-degree angle at the inlet. This results in uneven pressure distribution around the cell.

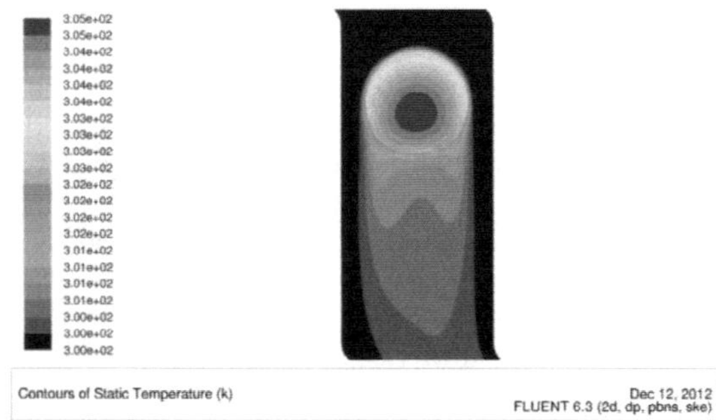

Figure 3: Temperature distribution in the cooling channel

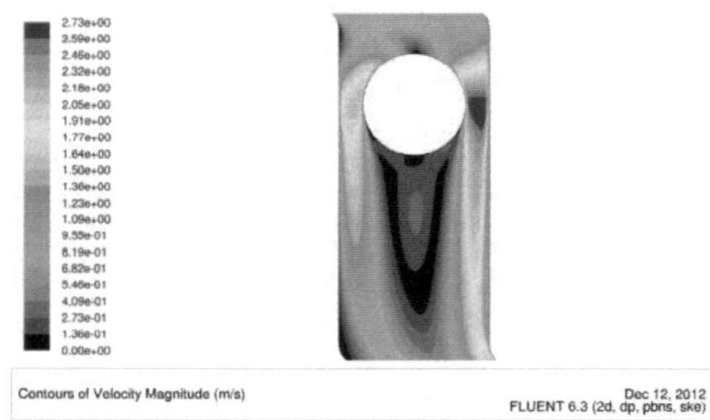

Figure 4: Contours of velocity magnitude

Having the general characteristics of the channel, a parametric study can be conducted to disclose the effects of influential parameters on temperature distribution throughout the cell.

One of the most important parameters is the rate of generated heat within the cell. The rate of generated heat depends on the cell load. Higher loads are exerted in quick vehicle acceleration and lighter loads are applied in cruise conditions. Fig. 5 exhibits temperature distribution along the axial symmetry line of the cell with various generated heat rates.

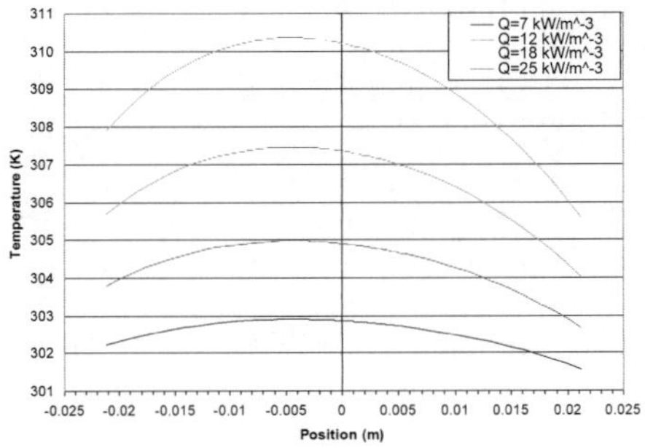

Figure 5: Temperature distribution along the axial symmetry line of the cell with various heat generation in the cell

As shown in Fig. 5, an increase in heat generation rates in the cell will result in higher cell temperatures as well as a more non-uniform temperature distribution. However, the position of temperature peak remains relatively unchanged. Doubling the generated heat rate will result in more than double the temperature rise as the amount of extracted heat through convection is limited. Also, maximum variation in temperature is nearly doubled. It should be noted that high heat generation rates occur for only short periods of time during high discharge loads and such non-uniform temperature distributions are not likely to happen in practical cases. The amount of withdrawn heat from the cell can be managed by modifying the inlet condition for the air flow. Temperature distribution for various magnitudes of inlet air can be seen in Fig. 6.

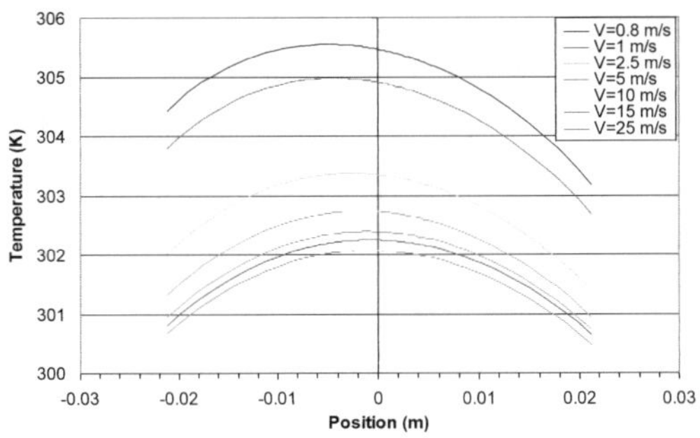

Figure 6: Temperature distribution along the axial symmetry line of the cell for various inlet air velocities

It is seen that increasing the airflow velocity can significantly enhance both temperature peak and temperature distribution in the cell. Furthermore, influence of air velocity becomes less significant with higher velocities which is due to the limited heat transfer capacity of forced convection. Pressure drop in the cooling channel is another important parameter. In the present model, average pressure drop can be anywhere between 2 Pa for inlet velocity of 1 $m.s^{-1}$ to about 700 Pa for 25 $m.s^{-1}$ which is relatively low for air-cooled systems.

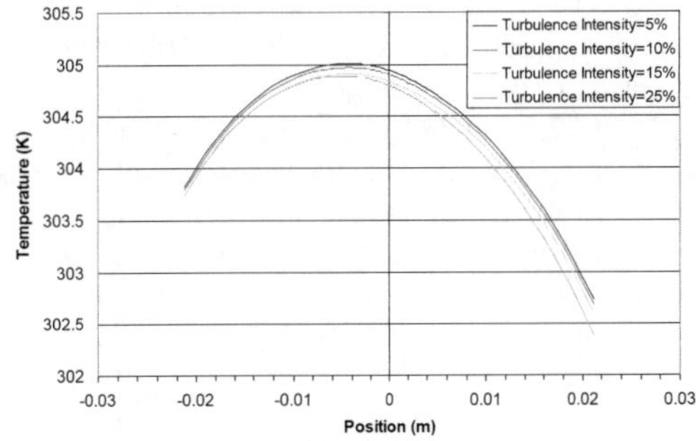

Figure 7: Temperature distribution along the axial symmetry line of the cell for various turbulent intensities

In addition to velocity, turbulence intensity at the inlet can be modified. Fig. 7 suggests that increasing turbulence intensity at the inlet does not result in considerable drop in cell temperature. However, maximum temperature variations along the cell is slightly increased.

CONCLUSIONS

A new design is proposed in order to enhance temperature uniformity between adjacent cells in air-cooled Lithium-ion battery stacks commonly used in electric vehicles. A structure consisting of two parallel airflow manifolds accompanied by several adjacent cooling channels which contain the batteries and guide air through the structure is suggested. It is seen that while this design results in uniform cooling between adjacent cells, it can be as effective as previous models in thermal management of a single cell with the advantage of considerable simplicity and minimum pressure drop. In addition, a parametric study is conducted which showed that air velocity can be increased at the inlet in order to improve thermal management. However, increasing air velocity beyond certain values will increase the costs due to higher pressure drops, while the change in temperature distribution would not be advantageous. Moreover, increasing turbulence intensity results in a small drop in cell temperature while uniformity of temperature distribution is slightly decreased. More importantly, high heat generation rates increase the overall cell temperature and reduce temperature uniformity.

REFERENCES

1. Rao, Z., and Wang, S., 2011, "A review of power battery thermal energy management". *Renewable and Sustainable Energy Reviews,* 15, pp. 4554-71.
2. Motloch, C.G., Christopheresen, J.P., Belt, J.R., Wright, R.B., Hunt, G.L., Sutula, R.A., Duong, T., Tartamella, T.J., Haskins. H.J., and Miller, T.J., 2002, *High-Power Battery Testing Procedures and Analytical Methodologies for HEV's*, SAE 2002-01-1950.
3. Jung, D.Y., Lee, B.H., and Kim, S.W., 2002, "Development of battery management system for nickel-metal hybrid batteries in electric vehicle applications". *Journal of Power Sources,* 109, pp. 1-10.

4. Zolot, M., Pesaran, A.A., and Mihalic, M., 2002, "Thermal evaluation of Toyota Prius battery pack". Hyatt Crystal City, Future Car Congress.
5. Lou, Y., 2007, "Nickel-metal hybrid battery cooling system research for hybrid electric vehicle". Shanghai, Shanghai Jiao Tong University.
6. Mahamud, R., and Park, C., 2011, "Reciprocating air flow for Li-ion battery thermal management to improve temperature uniformity". *Journal of Power Sources,* 196, pp. 5685-96.
7. Fluent, Inc., 2006, *Fluent 6.3 Users Guide.*
8. Patankar, S.V., 1980, *Numerical Heat Transfer and Fluid Flow*, McGraw-Hill Book Company.
9. Zhang, X., 2011, "Thermal analysis of a cylindrical lithium-ion battery". *Electrochimica Acta*, 56.3, pp. 1246-55.

AN AIR SOURCE HEAT PUMP FOR HYDRONIC HOME HEATING, APPLICATION OF AN AIR BASED PHOTOVOLTAIC/THERMAL COLLECTOR

Omid Nemati, PhD Student
Department of Mechanical Engineering
Ryerson University
Toronto, Ontario, Canada
omid.nemati@ryerson.ca

Dr. Alan S. Fung, PEng, FCSME
Associate Professor
Department of Mechanical Engineering
Ryerson University
Toronto, Ontario, Canada
alanfung@ryerson.ca

ABSTRACT

An energy model of a semi attached house was developed. The house is located in Toronto, and it is equipped with a cold climate, variable capacity, air to water heat pump (HP). A photovoltaic/thermal (PV/T) collector provides heated air, in place of outdoor air, to the heat pump in order to offset the compressor energy input during heating season. Previous hourly, coupled PV/T and HP models mostly took an optimization approach to modelling. They concentrated on a fixed air flow rate and instead engaged in detailed modelling of the thermal collection efficiency. However, this efficiency is known to be more or less constant for a given setup. This study was motivated by the fact that the heat pump performance during heating mainly depends on the flow rate requirement of its source coil. For this reason the thermal collection efficiency was assumed to be fixed, and an iterative scheme was devised to determine the heating load, flow rate requirement and compressor energy input concurrently. The addition of the PV/T collector was estimated to improve the heating COP by 10% or less. Including the electrical energy generation, 75% or more of the compressor energy input was estimated to be offset by the electrical gain. Other than Toronto, three cities of Montreal, Calgary and Vancouver were considered. The heat pump defrost cycle was neglected.

KEYWORDS: air source heat pump, photovoltaic/thermal, net zero energy, COP, thermal collection, eQuest.

INTRODUCTION

With increasing environmental concerns and tightening of building energy codes, the concept of net zero energy (NZE) housing is an ambitious and promising concept for residential new construction. This concept is based on aggressive envelope thermal performance, use of on-site renewable energy systems and recovering/storing as much residual heat as possible. The term, NZE, refers to a situation where all energy consumption is made up by local production. In Canada 12 NZE houses have been built through the eQuilibrium Housing Initiative, sponsored by the CMHC and NRCan. One other NZE construction is a set of twin houses located at the Toronto's Kortright Centre, referred to as the Archetype Sustainable House. This house was built in 1998 by the Toronto Region Conservation Authority (TRCA) and Building Industry and Land Development (BILD) Association. Renewable energy systems and unconventional thermal storage equipment are commonly used in these houses in order to achieve the NZE status. This paper presents a PV/T system in combination with air source heat pump for residential hydronic space heating application.

Suffering from high installation costs and inexpensive cost of competing fuel (natural gas), air source heat pumps have not had a significant market share in residential heating in North America (Minea, 2003). Canada also suffers from relatively colder winters than our neighboring country, which dramatically reduces the heat pump capacity, not to mention the energy performance. "Cold climate" variable capacity air source heat pump is a relatively recent development that tends to mitigate some of the drawbacks. It uses a larger volume of refrigerant than traditional heat pumps to its advantage, and it is also offered in two, three and variable speed compressor options. This system has been used effectively at TRCA House A (Safa et al., 2014). Hydronic systems (air to water heat pump) have the advantage over all air systems (air to air heat pump) that they likely eliminate a traditional occupant complaint about cool circulating air associated with all air systems (Minea, 2003).

Solar energy is a renewable source, which makes it attractive for NZE housing. Adding thermal collection to a solar PV setup is known to enhance the overall collection efficiency of the setup three- or four-fold. This is still "free"–it is only at an expense of slight (Candanedo et al., 2010) static pressure increase through the duct. All eQuilibrium houses, as well as the TRCA twin houses, have some form of solar thermal system. Air based systems have two main types: one type uses this preheated air as ventilation air, effectively bearing a task of dedicated outdoor air heat recovery, and the other type uses this air to achieve a better source temperature than ambient for heat pump operation. The first type has relatively small airflow rates, thus it is suitable for spandrel panels, which cannot allow high airflow to pass by a between-the-panes shading device. The second type has relatively high airflow rates, thus it is suitable for roof integration

(Candanedo et al., 2010; Liao et al., 2007). The House A considered in this project has a roof integrated PV/T. This type has the advantage over the other type that it does not have a dust collection issue. It is stipulated that a form of PV/Ts and air source heat pumps will continue to attract to-be-built NZE houses in Canada.

Comparison of air based building-integrated PV/T collectors reveals that the thermal collection efficiency has a small variation with operating conditions, including air flow rate. One model of a roof-integrated collector (Kamel, et al., 2014) only reported a 10% variation in this efficiency as the channel depth was increased from 1.5" (45% efficiency) to 3" (55% efficiency), while the flow rate (460 cfm) was fixed. One set of experimental data (Candandedo, 2010) showed that, with varying flow rate, the thermal collection efficiency also varied from 10% to 20%, however, the lower end of this range was attributed to free cooling mainly. Since only forced convection, likely at the highest fan speed, is expected to be encountered during the heating season, the thermal collection efficiency of that setup too was more or less constant for practical applications. Among other studies, Guiavarch et al. (2006) reported 7% efficiency for free cooling of curtainwall integrated semi-transparent PV/T. Gan et al. (2004) used 31% efficiency, at 111 cfm, to model a roof-integrated PV/T with fresh air introduced via an opening through the wall near the corner of an atrium. Nemati et al. (2010) reported 37% at 108 cfm for a curtainwall application. A summary of thermal collection efficiencies is presented in Table 1.

Author(s)	η_T (%)	Q (cfm)	Install. place	PV type
Guiavarch et al. (2006)	7	FC	CW	ST
Candanedo et al. (2010)	20	50	R	OP
Gan et al. (2004)	31	111	CW	OP
Nemati et al. (2010)	37	108	R	OP
Liao et al. (2007)	50	147	CW	OP

Table 1: Summary of thermal collection efficiency of building-integrated PV/T collectors (FC: free cooling, CW: curtainwall, R: roof: ST: semi-transparent, OP: opaque)

TRCA HOUSE A

The TRCA House A is a three storey (including attic), semi attached, single family house located in Toronto, Ontario. It was built in 1998. It is R2000 and LEED Platinum certified. Excluding the basement and attic, it has a floor area of approximately 1,900 ft^2. All windows (405 ft^2) are double glazed with a centre of glass U-value of 0.39 Btu/ft^2-°F (2.2 W/m^2-°C). Exterior wall insulation is provided by Roxul mineral fibre and 3" extruded polystyrene. The overall thermal resistance of the exterior walls is R-30. The roof is a Structurally Insulated Panel (SIP) with an overall thermal resistance of R40 (Barua, 2010). Space heating is provided by an air to water heat pump (36,000 Btu/h). This heat pump supplies hot water to the air handling unit (AHU), as well as infloor heating loop. Total supply air flow rate for the AHU is 1,650 cfm. Fresh air (100 cfm) is provided by a heat recovery ventilator, however, heat recovery was not modelled. Fig. 1 presents a schematic of the twin houses.

Figure 1: Schematic of the twin houses

PV SETUP AND PERFORMANCE

The rooftop PV array is a 5 × 5 arrangement. This arrangement is based on one possible arrangement analyzed by (Kamel et al., 2014). It is large enough to provide the required air flow (3,250 cfm) for the "outdoor" coil of the heat pump. Each array is 3 ft × 5 ft with a 1.5 in. air channel below it. The pressure drop was taken to be 0.5" water. The electric conversion efficiency (crystalline cells) was taken to be 15%. The thermal collection efficiency was taken to be 50%. The air channel from each PV module joins a common duct, which extends to the basement where the heat pump's "outdoor" unit is located. During the cooling season, the heated air is rejected to the exterior and does not reach the basement. Fig. 2 presents a schematic of the integration of the PV/T collector and heat pump.

HEAT PUMP PERFORMANCE

Performance results of two cold climate, air source heat pump manufacturers, Mitsubishi and Ecologix, were examined. The ARI 210/240 standard covers these heat pumps, similar to air to air heat pumps. It requires that the heat pump operation be tested at three ambient conditions: cold (-8.3°C), frost point (4.4°C) and warm (8.3°F). However, only the cold and warm setting results are presented by these manufacturers. The defrost temperature setting is missing entirely. The cold and warm settings were used to populate a graph of COP vs. outdoor air temperature (at 35°C hot water), along with a very cold setting (-25°C) and a few other settings in between (-3°C, 2°C and 7°C), all of which are outside the specified test settings.

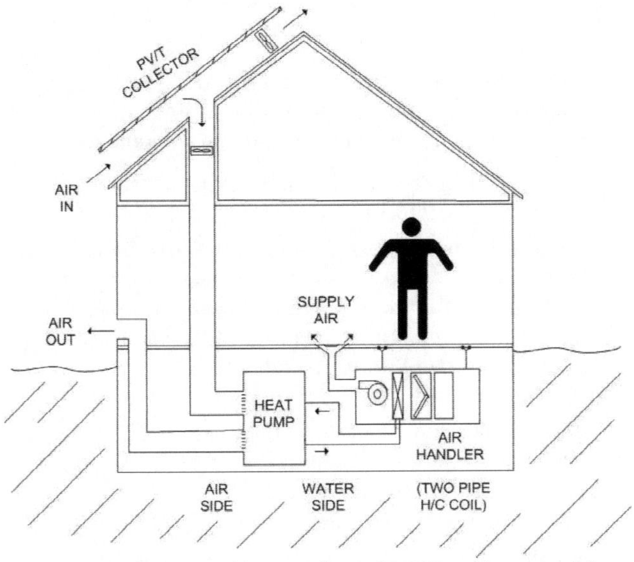

Figure 2: Schematic of integration of PV/T collector and heat pump

The results of both manufactures were amalgamated into one graph, presented in Fig. 3. A linear regression was made. Variation of COP with hot water temperature other than 35°C was not available, therefore, it was taken from a residential sized ground source heat pump (Safa, 2010) as a rough estimate. This variation was 0.2 for 5°C increase or decrease. To make this study more accurate in future, the COP at the defrost point should be added, and also the COP of any point below this point should be revised to include the effect of periodic defrost. In the product catalogues, the nominal heating capacity and operating, worst-case COP are provided at -25°C, as opposed to the -8.3°C or any other temperature given in the standard. Fig. 4 presents capacity data.

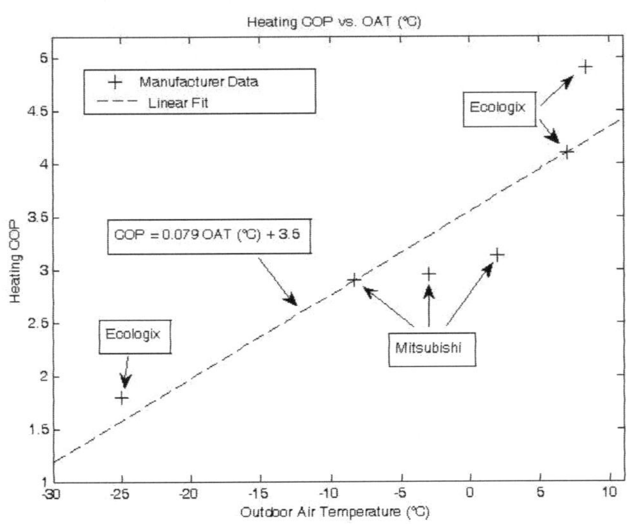

Figure 3: Heating COP vs. source temperature (Mitsubishi's PUZ–HA36NHA and Ecologix's A2W36-V, both 36,000 Btu/h nominal)

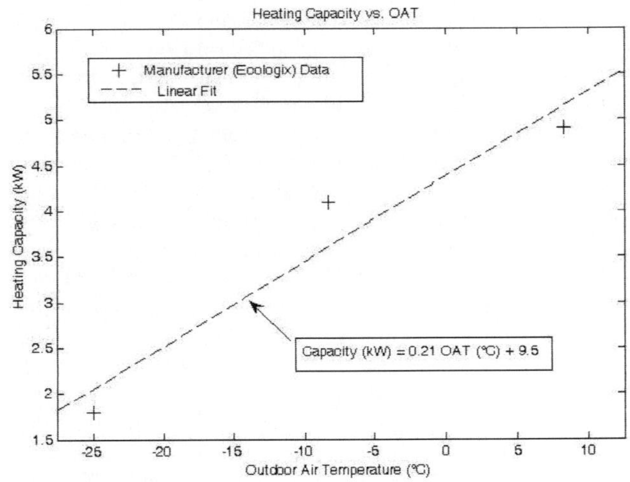

Figure 4: Heating capacity vs. source temperature

ENERGY SIMULATION

An hourly model was created in the software package eQuest 3.65. The lighting, hot water and appliance loads were based on the default values of eQuest. The occupancy schedule was set to high during early morning and also late at night, and it was set to low during night and mid-day. The space heating setpoint was 70°F. The plant was modelled as a chiller with this chiller defined as a "heat pump". The final peak heating load was 27,000 Btu/h (Toronto), consistent with the TRNSYS models of (Kamel et al., 2014) and (Barua et al., 2010).

The inter-relation between the flow rate of the air inside the channel and its temperature rise was accounted for via Eq. 1 (ASHRAE, 1997). This was an improvement over two previous simulations (Kamel et al., 2014, and Hailu et al., 2015) in TRNSYS of integrated air source heat pump and ducted, all-opaque PV/T system that used a fixed airflow rate year round. There is one previous study (Candanedo et al., 2010), through which a calibrated in-house model of a PV/T system was developed, but that study did not include heat pump operation. In this project at each hour, the heating load was estimated first, then a flow rate was determined accordingly, and finally the heating load was re-visited and simulation carried forward.

$$\Delta T = \frac{Btu/h}{1.08 \times cfm} \quad (1)$$

In this equation *Btu/h* is the rate of thermal collection efficiency. At 3,250 cfm, the maximum ΔT (at highest irradiation level) was 8°F (Toronto).

RESULTS

The energy performance (refer to Table 1) of the heat pump was estimated for the two cases of standalone operation and combined operation with PV/T collector. To present the results, the Heating Seasonal Performance Factor (ARI, 2008) was considered to be a suitable efficiency identifier. It is the ratio of the total heating delivered (in Btu/h) during an entire heating season to the total electrical energy input (in kWh) during this time. As a result, it provides a comprehensive assessment of the efficiency, as opposed to, for example, hourly and daily averages. Throughout this report, the HSPF is expressed in the units of COP (kWh over kWh) consistently. Electrical energy generation is summarized in Table 2. Two observations were made about Tables 1 and 2. First, the COP closely follows the ambient temperature and does not have a strong correlation with the amount of solar irradiation, as expected. Secondly, the incremental improvement in COP attributed to the PV/T collector is smaller than the electrical gain of this collector. For the cities considered, a maximum of 10% of the compressor energy input was shown to be offset by the PV/T collector, whereas a minimum of 75% of this input was shown to be offset by this collector. For Vancouver, all (100% of) compressor energy was offset.

City	COP_{HP}	$COP_{HP\&PV/T}$
Calgary	2.8	2.9
Montreal	3.0	3.1
Toronto	3.2	3.3
Vancouver	3.3	3.3

Table 1: Summary of seasonal COP, with and without PV/T

City	$C_{HP\&PV/T}$ (kWh)	$C_{HP\&PV/T}$ (kWh)	E_{PV} (kWh)
Calgary	878	851	(722)
Montreal	548	534	(402)
Toronto	878	851	(720)
Vancouver	493	492	(498)

Table 2: Summary of compressor energy input, with and without PV/T, and PV electrical energy generation

CONCLUSIONS

1) Neglecting defrost cycle, a cold climate, variable capacity heat pump has a seasonal heating COP of between 2.8 (Calgary) and 3.3 (Vancouver) for hydronic home heating.

2) Implementation of a roof integrated PV/T collector with fresh air introduced inside a channel and under the collector improves this COP by a maximum of 10%.

3) A minimum of 75% of the annual compressor energy input is readily offset by this collector.

ACKNOWLEDGMENTS

The authors would like to acknowledge the financial support of the Natural Sciences and Engineering Research Council (NSERC) of Canada Discovery Grant.

REFERENCES

1. ASHRAE/ANSI 210/240 (formerly ARI 210/240). 2008, "Standard for performance rating of unitary air conditioning and air source heat pump equipment".
2. ASHRAE Handbook of Fundamentals, "standard air calculations", 1997. pp. 28.15
3. Barua, R., Zhang, D., and Fung, A.S. 2010, "Analysis of energy performance of the sustainable archetype house at Kortright Centre", *International high performance buildings conference*, Purdue University, West Lafayetta, United States, pp. 265-278.
4. Candanedo, L.M., Athienitis, A.K., Candanedo, J.A., O'Brien, W., and Chen, Y. 2010, "Transient and steady state models for open loop air based BIPV/T systems", *ASHRAE Transactions*, 116, pp. 600-615.
5. Gan, G. and Riffat, S. 2004, "CFD modelling of air flow and thermal performance of an atrium integrated with photovoltaics", *Building and Environment*, 39, pp. 735-748.
6. Guiavarch, A. and Peuportier, B. 2006, "PV collectors efficiency according to their integration in buildings", *Solar Energy*, 80, pp. 65-77.
7. Hailu, G., Dash, P. and Fung, A.S. 2015, "Performance evaluation of an air source heat pump coupled with a building integrated PV/T system under cold climate conditions", *International Building Physics Conference*, 6, pending publication.
8. Kamel, R.S., and Fung. A.S. 2014, "Theoretical estimation of the performance of a PV/T system coupled with a heat pump in a sustainable house in Toronto", *ASHRAE Transactions*, 120 (1), pp. 179-193.
9. Liao, L, Athienitis, A.K., Candanedo, M.L., Park, K.W., Poissant, Y., and Collins, M. 2007, "Numerical and experimental study of heat transfer in a BIPV/T system", *ASME Journal of Solar Energy Engineering*, 116, pp. 423-430.
10. Minea, V. 2003, "North American test procedures and calculations for residential heat pumps with combined space and domestic hot water heating", *Hydro Quebec interim report*.
11. Nemati, O. Candanedo, L, Collins, M.R. and Athienitis, A. 2010, "Experimental and numerical investigations of a mechanically ventilated multiple skin façade with between the panes venetian blinds", *ASHRAE Transactions*, 116, Part 1.
12. Safa, A.A. 2012, "Performance analysis of a two stage variable capacity heat pump and a horizontal loop coupled ground source heat pump", *Master's thesis*, mechanical engineering, Ryerson University.

Safa, A.A., Fung, A.S., and Kumar, R. 2014, "Performance of two stage variable capacity air source heat pump: field performance results and TRNSYS simulation", Energy and Buildings.

COMPARISON ANALYSIS OF WEATHER NORMALIZATION FOR INDUSTRIAL PLANTS USING PRISM AND EXCEL

Md Maniruzzaman Akan
Mechanical and Industrial Engineering Department
Ryerson University
Toronto, Ontario, Canada
Email: makan@ryerson.ca

Dr Alan S Fung, P.Eng, FCSME
Mechanical and Industrial Engineering Department
Ryerson University
Toronto, Ontario, Canada
Email: alanfung@ryerson.ca

Tamima Ahmed
Mechanical and Industrial Engineering Department
Ryerson University
Toronto, Ontario, Canada
Email: tamima.ahmed@ryerson.ca

ABSTRACT

This research paper investigated result accuracy and user friendliness of calculating weather normalization by using PRInceton Scorekeeping Method (PRISM) and Excel. This paper analyzed the consumption of natural gas by sixteen small-and medium-enterprises (SMEs) within the Greater Toronto Area (GTA) to check comparison between PRISM and Excel analysis on weather normalization calculation. This research looked at accuracy and user friendliness of methods to calculate weather normalization because it is important for future energy benchmarking calculation and energy conservation. Plant can perform benchmarking analysis in order to look for potential energy savings by reducing calculating time and complexity. Normalized annual consumption (NAC), reference temperature, and coefficient of correlation are three major parameters can be determined by the custom-made Excel tools and commercially available PRISM software which is required to estimate weather normalization. Normalized weather consumption is required to perform benchmarking analysis. The custom-made Excel tool will be part of the integrated industrial energy benchmarking and conservation analysis tools, currently being developed as part of the overall Enbridge Gas Distribution Inc. University Partnership Program (UPP) for small and medium enterprises. It was found that the Excel based tool has comparable accuracy and was more compatible and flexible with any computer operating systems. PRISM took less time to process large amounts of data and to run analyses; though it required a 32-bit Windows XP operating system. Both methods estimated weather normalized annual consumption with a variation between 0.5% and 4.7%.

KEYWORDS: Small and medium enterprise, Normalized annual consumption, Heating degree days. Heating-and-Cooling (HC) PRISM, Heating-Only (HO), Cooling-Only (CO).

INTRODUCTION

Weather normalization is the process of removing abnormal or extreme weather impact from historical energy consumption data [1]. Normal weather conditions are the average weather experienced over the last thirty-one years. Forecasted normalized energy consumption is based on the weather-corrected base year.

Energy savings calculation in engineering does not always correlate with real world performance. Weather is one of the important differences between engineering energy savings calculation and real world performance [2]. Weather varies from year to year and continues to change. As a result, it is become difficult to forecast weather effect. Therefore, in energy savings calculation, it is vital to remove the energy consumption due to weather effect from total energy consumption. In order to calculate energy consumption due to weather relies on a reference temperature of individual buildings and can be normalized with historical weather data for realistic estimation. PRISM and Excel are two important methods to calculate those. PRISM is commercially available software based on statistical procedure that converts common utility billing data into useful weather adjusted estimates of annual energy use [3].

Sixteen SME industrial companies have been analyzed using the past three years of energy data collected from plant managers. These billing periods' weather data (temperature) has been taken from Environment Canada (1984-2013)[4].

Research analysis was performed in both Excel and PRISM software to verify results. PRISM calculates a reference temperature in order to achieve optimized linear regression [5]. This reference temperature obtained from the statistical approach is influenced by a few factors: facility envelopes, HVAC, production machinery, and appliances. Therefore, optimum reference temperature is a characteristic of a facility [6]. Higher reference temperatures in heat-dominated facilities indicate the higher space heating required at higher outdoor temperatures. PRISM runs on 32-bit Windows XP operating system, which became outdated in computer operating system.

METHOD AND METHODOLOGY

PRISM ANALYSIS

Obtaining the reference temperature in PRISM requires two important data files for each company: daily average outside temperature in Fahrenheit, and monthly energy consumption in any unit (in this analysis, m^3 of natural gas is used), showing respective billing dates. These files must be converted to windows Notepad format as described in the PRISM user manual, or prepared in a Excel or Notepad.txt and converted to Notepad file by PRISM through readable temperature (file ends with .TPS in PRISM) and meter file (file ends with .MTR), which are described in the PRISM Users' Guide: Reference Manual Chapter-1[3]. After the data file has been successfully prepared, PRISM runs in order to get heating and cooling reference temperature (TAU, τ), NAC and (coefficient of correlation) R^2.

PRISM calculated NAC value through an estimates consumption under average weather conditions [3]:

$$NAC = 365\,\alpha + \delta_h \beta_h H_0(\tau_h) + \delta_c \beta_c C_0(\tau_c) \quad (1)$$

(base level) + (heating part) + (cooling part)

Where

α = base-level consumption (m^3/day)[HO, CO, HC]
β_h = heating slop (m^3/°F-day) [HO,HC]
β_c = cooling slop (m^3/°F-day) [CO,HC]
τ_h = heating reference temperature (°F) [HO,HC]
τ_c = cooling reference temperature (°F) [CO, HC]

In Equation, δ_h= 1 for HO and HC models and δ_c= 0 while δ_c= 1 for the CO and HC models and δ_h= 0. Long-term heating and cooling degree days per year can be determined by H_0 (τ_h) and C_0 (τ_c), respectively. Therefore, base temperature τ_h and τ_c are determined by PRISM. M. Fels and C. Reynolds (1993) defined heating-only (HO), cooling-only (CO), and heating-and-cooling (HC) automated models in PRISM. A coefficient of correlation (R^2) of linear regression between energy consumption and heating degree days demonstrates the reliability of the model. A good model has an R^2 value close to 1.

EXCEL ANALYSIS

Analysis has been performed in Excel using simple mathematical and statistical methods. Heating degree days (HDD) is calculated by the Equation 2. ASHARE defines HDD are measures of how much in degrees, and for how long in days, the outside air temperature was below a certain level [11]

$$HDD = (\text{Reference temperature} - \text{Outdoor temperature}) * \text{Number of days in a month} \quad (2)$$

This analysis was performed several times using variable reference temperature data to determine the best R^2 value. Outdoor temperatures are monthly weather data from Environment Canada. HDD is determined for each month using the same Equation 2. A regression line was obtained by using the scatter plots in Excel where HDD is presented in the X-axis while monthly consumption is presented in the Y-axis. From this scatter plot, a R^2 value is found. Maximum R^2 value can be determined by varying the reference temperature in HDD Equation 2 and in the scatter plot. This maximum R^2 value and corresponding reference temperature is the base temperature of this analysis through Equation 2[7]. To calculate NAC in Excel, Equation 3 is used, where this reference temperature used as base temperature. The thirty one years' monthly average temperature was used instead of billing periods' monthly temperature [8].

$$NAC = MX + C \quad (3)$$

Where

M = heating or cooling slope obtained in the regression line in the scattered plot
X = HDD obtained from Equation 2 corresponds to maximum reference temperature
C = is the fix value in the regression line in the scattered plot

RESULT AND SUMMARY

Normalized weather consumption analysis performed for 16 industrial plants by using both PRISM and Excel, and estimated values are presented in the subsequent table and plot. The calculated reference temperatures found through Excel and PRISM analysis are in the Table 1. Company AAAM shows the highest reference temperature, although this company has a shortage of monthly data (Feb 2012 to Sep 2012) for eight consecutive months which, affects the regression analysis. It is assumed that this company might be shut down or has low production during those months. A similar situation was observed in AACF. There were only eighteen months of data in Excel and PRISM based on statistical analysis and historical weather data, which shows significant differences between companies. The overall percentage of error in PRISM analysis was 2.67 % higher than Excel analysis.

Name of the Company	Reference Temperature (°F)	
	EXCEL	PRISM
AAAM	90	90
D78	60	58
D22	53	53
AAAL	68	68
AABN	52	48
AACF	84	86
AAGF	60	64
AAKK	60	59
AAKK2	56	57
AAMP	64	63
AASN	57	46
AASU	54	47
AASP	69	65
AAWI	55	57
AAWR	65	64
AAKI	52	48

The value of coefficient of correlation observed in Excel analysis was 0.13 for AACF at the lowest and 0.9 for AAMP at the highest, which shown in Table 2. On the other hand, in PRISM analysis, AACF was 0.16, at the lowest and 0.9 for AAKK at the highest. Individual months' HDD and consumption analysis were checked in the PRISM and Excel analysis and some months in PRISM showed small HDD values, but Excel did not show this because of negative value, whereas consumption seems close.

After analysis, it was observed that PRISM runs using full one year's data while, Excel can perform analysis by fraction data of a year. Some companies showed a negative value in Excel, thus obtaining zero HDD value, but PRISM shows a positive value. The overall percentage of error in Excel analysis was 4.7% higher than PRISM analysis.

Company Name	R^2 Value	
	EXCEL	PRISM
AAAM	0.27	0.37
D78	0.78	0.81
D22	0.82	0.88
AAAL	0.70	0.71
AABN	0.68	0.69
AACF	0.13	0.16
AAGF	0.61	0.62
AAKK	0.89	0.90
AAKK2	0.64	0.67
AAMP	0.90	0.89
AASN	0.42	0.52
AASU	0.45	0.47
AASP	0.88	0.92
AAWI	0.65	0.63
AAWR	0.50	0.50
AAKI	0.45	0.51

Similar results were observed after NAC analysis using the same method shown in Table 3. This is because normalized consumption is based on statistical analysis. The overall percentage error in Excel analysis was 0.5% higher than PRISM analysis. Both the analysis showed feasible to estimate NAC value.

Company Name	NAC Value (m^3)	
	EXCEL	PRISM
AAAM	154339	155910
D78	969032	970041
D22	1320182	1329935
AAAL	586763	589732
AABN	357552	357134
AACF	437959	438817
AAGF	3373410	3381150
AAKK	1032623	1037489
AAKK2	649458	653247
AAMP	517909	522612
AASN	550606	547937
AASU	674445	705108
AASP	373975	374740
AAWL	301399	302647
AAWR	1061480	1057262
AAKI	1230668	1252789

CONCLUSION

Weather normalized energy consumption analysis is important to assess and estimate energy saving potential. Results of normalized energy consumption rely on accuracy of technique, quick computation of data, and user-friendliness of tools. This research paper analyzed the feasibility of computing technique in PRISM and Excel. The result found close agreement between PRISM and Excel. However, PRISM has limitation of computer operating system and run based on full one year slot base, which results the variation between findings. It was found that an Excel base computation tools can solve PRISM's limitation.

It is anticipated, future study to examine these findings in this research will help to develop more accurate and user-friendly weather normalization analysis tools.

ACKNOWLEDEMENTS

This work has been supported financially by the Enbridge Gas Distribution Inc. University Partnership Program (UPP), Connect Canada. The authors would like to thanks the participating industries in GTA for supplying their plant data. Special thanks to Farzin M. Rad and Altamash Baig for their help and support-as well as for conducting energy audits and data collection. The authors would also like to express their gratitude to Paul Morrison and Peter Goldman from Enbridge Gas Distribution Inc. for their help and assistance.

REFERENCES

1. Ontario Energy Board. (n.d.). Weather normalization for total utility load. Retrieved from http://www.ontarioenergyboard.ca/documents/cases/EB-2005-0317.
2. Avina, J. (2012). An introduction to utility bill weather normalization for energy contractors. Retrieved from htto://www.abraxasenergy.com/article/intro-weather-normalization-contructors.
3. Fels, M. F., Kissock, K., A, Michelle., Marean., & R, Cathey. (1995). PRISM (Advance Version 1.0) user guide. Princeton University: Centre for energy and environmental studies. New Jersey.
4. Environment Canada. (n.d.). Ontario-weather condition and forecast by location. Retrieved from http://www.weather.gc.ca/city/pages/on-143_metric_e.
5. Fels, M. F., Rachlin, J., & Socolow, R. H. (1986). Seasonality of non-heating consumption and its effect on PRISM results. Energy and Buildings. Pages 139-148.
6. Finch, G., Burnett, E., & Knowles, W. (2010). Energy Consumption in mid and high rise residential building in British Columbia. Portland.
7. Paulus, T. Mitchell., Claridge, E. David., & Culp, Charles. (2014). Algorithms for automating the selection of a temperature dependent change point model. Energy and Buildings. 87(2015)95-104.
8. Eto, H. J. (n.d.). A comparison of weather normalization technique for commercial buildings energy use. Retrieved from http://www.ornl.gov/sci/buildings/2012.
9. ASHRAE, Guideline 14-2002, Measurement of energy and demand savings, 2002.
10. Fels, M. F. (1086). PRISM: an introduction. Energy Build. 5-18.
11. 2009 ASHRAE Handbook- Fundamental. (n.d.). Climatic design information. Retrieved from http://www.ecop.rpi.edu/hartford/.

Proceedings of the 25th CANCAM
London, Ontario, Canada, May 31 – June 4, 2015

Energy Benchmarking Analysis Based on Ventilation Consumption of Small and Medium Powder Coating Industries in the Greater Toronto Area (GTA)

Tamima Ahmed
Department of Mechanical and Industrial Engineering
Ryerson University
Toronto, Ontario, Canada
Tamima.ahmed@ryerson.ca

Dr. Alan S. Fung, PEng, FCSME
Department of Mechanical and Industrial Engineering
Ryerson University
Toronto, Ontario, Canada
alanfung@ryerson.ca

ABSTRACT

For the past several years, energy benchmarking has become a very popular tool for the estimation of energy consumption and overall performance of buildings. Recently, industrial energy benchmarking has attracted attention all over the world due to its ever-increasing energy demand. Industrial facility ventilation is one of the most overlooked portion in terms of overall industrial sector energy consumption, and a large amount of energy could be reduced from the studied powder coating industrial plants if energy consumption is properly assessed and managed. Although several works have been previously reported on ventilation analysis of industrial facilities in Ontario, energy benchmarking has not been conducted so far on ventilation. Therefore, this paper aims to report a detailed energy benchmarking exercise based on ventilation energy consumption analysis. An energy benchmarking analysis was conducted in five different powder coating facilities in Greater Toronto Area (GTA) based on ventilation analysis. It was found from the analysis that, the worst performing facility could reduce up to 1-10% of its ventilation energy consumption compared to the top performing facility among the five facilities in this case study.

KEYWORDS: Benchmarking, process load, seasonal load, ventilation consumption, space heating and potential savings.

INTRODUCTION

"Energy benchmarking of a building provides the relative comparison of energy use of whole building with respect to a set of building alike" [1]. It contributes to the idea of target buildings for energy auditing and measuring the energy saving potentials. In recent years, industrial energy benchmarking has attracted attention all over the world because of its increasing consumption record. Energy benchmarking of an industrial facility refers to measure the energy consumption of an individual plant against the similar standard one [3]. Energy supplier companies and their dealers compare energy saving potentials with "typical" and "best practice" benchmarks. Both building managers and energy suppliers are always interested in comparing their energy performance with each other [1]. Benchmarking is also an excellent analytical tool to provide information about the energy consumption alone with the production capacity of the sector. In addition, benchmarking method can be widely used to determine the energy efficiency of an industrial facility and to measure the sectorial energy efficiency all over the country as well as the world [2]. Energy benchmarking helps energy auditors to understand the energy consumption pattern of buildings with energy saving potentials. It also contributes to improving energy economy [3].

In energy provider companies, energy saving potential is known as "best-practice" benchmark. The owners of buildings are always interested in achieving the best practice benchmark for their industries or buildings. Energy benchmarking is also used in residential premises, industries, schools, and hospital buildings to compare their energy pattern with one another [2].

LITERATURE REVIEW

A case study was conducted in a Toronto Coffee Industry by Bhattacharjee [4] in order to find energy saving opportunities in the ventilation system. The industry had three make-up air (MUA) units that were equipped with natural gas-fired heaters and refrigerator compressors for cooling in summer. The MUA unit was used to operate 24h/7 days a week. A decision was made to shut down the MUA unit on Sundays and to restart it on Mondays. Natural gas savings was calculated to be 52,363 m^3/year by executing this approach [4].

Another case study was conducted in 23 Dutch paper mills to benchmark specific energy consumption of similar process in different paper mills [5]. After introducing the energy benchmarking method, 5.4 PJ/year of potential energy savings were determined.

In a case study by Kabir [6], energy auditing in Portland cement production plants north-east Nigeria was conducted. Energy audit was conducted in several cement plants and major heat loss areas and equipment were identified. After conducting a detail thermal energy audit the study was concluded with 5.30MW of energy savings, which was equal to 10.4% of total input energy, and US$2318.18 per year of cost savings.

SIGNIFICANCE OF THE PAPER

Many research studies have been conducted in Ontario industrial energy management sector. However, there are still opportunities to improve the industrial energy sector to make energy usage more convenient and efficient. From the literature review, it was found that very few works have been done on energy benchmarking of small and medium-sized enterprises (SME) based industries in Ontario. Hence, a focused study is required in this sector in order to improve natural gas conservation. In addition, although several studies have been previously conducted on ventilation analysis of industrial facilities in Ontario, energy benchmarking based on ventilation energy consumption has not been reported so far. Therefore, to fill this gap, this paper reports a detailed energy benchmarking that is based on ventilation energy consumption analysis of five different powder coating industries in GTA.

METHODOLOGY

To conduct the analysis, walk through energy audit was performed on the selected five powder coating industries in GTA and utility bills, and other related data (for example, Cure Oven and Dry-Off Oven specifications, operational hours, Dry-Off Oven and Cure Oven inside temperature) were collected. Then, process load and seasonal load were separated from the utility bills. Process load was calculated from the average summer month consumption (according to Natural Resources Canada, June, July, and August are considered as summer months in Canada). It was found that during this period no space heating was required. Thus, the natural gas consumption was only used for the production purposes. The seasonal load was calculated by subtracting the processing load from the monthly energy consumption: To calculate the ventilation energy consumption, it is necessary to separate the space heating consumption from the seasonal load consumption. Space heating of a building is required for transmission loss, which includes the heat loss through walls, roof, window, floors, and doors. The ventilation heat loss includes infiltration loss, mechanical ventilation and indoor equipment air intake.

The transmission loss of the building was calculated using the Equation 1 [7]:

$$Q = U x A x \Delta T \tag{1}$$

Where U is an overall heat transfer coefficient [Btu/(hr-ft^2-°F) or W/(m^2-°C)], A is the net area of the (wall/roof) for a given component (ft^2), and ΔT is the difference between indoor and outdoor temperatures (°F).
Also, the overall heat transfer coefficient can be written in terms of thermal resistance of the components [7]:

$$U = \frac{1}{R'A} = \frac{1}{R} \tag{2}$$

Heat loss due to infiltration was calculated using air change method. Equation 3 [7] presents the infiltration loss:

$$Q = \frac{ACH\ x\ V}{C_T} \tag{3}$$

Where Q is infiltration rate (cfm or m^3/hr), ACH is the number of air change per hour, V is Gross space volume (ft^3 or m^3), and C_T is a constant, which is 60 for English units and 3600 for SI unit Also, infiltration loss is calculated using Equation 4 [7]:

$$Infiltration\ loss = CFM\ x\ 1.08\ x\ \Delta T \tag{4}$$

Here, CFM is infiltration rate and ΔT is the temperature difference between inside the building and outside weather. Now in the powder coating industries there are dry off oven and cure oven. So total oven air intake,

$$Oven\ air\ intake = Dry\text{-}Off\ oven\ air\ intake + Cure\ Oven\ air\ intake \tag{5}$$

Equation 6 [7] is used to calculate Dry-Off oven and Cure Oven air intake

$$Dry\ off/cure\ oven\ oven\ air\ intake = \frac{QC_p}{v}\ x\ (t_e - t_o) \tag{6}$$

Where, Q = volume flow rate of intake air [ft^3/hr], C_p = Constant pressure specific heat of air [Btu/(lbm-°F)], v = Specific volume of air, [ft^3/lbm], t_e = Outdoor temperature [°F], t_o = Indoor Temperature [°F]

Now, Equation 7 was used to calculate the mechanical ventilation consumption [4]

$$Mechanical\ Ventilation\ Consumption = \frac{CFM\ x\ 1.08\ x\ Operational\ Hours\ x\ \Delta T}{\eta_{make\ up\ air\ unit}\ x\ HHV_v} \tag{7}$$

Where CFM is ventilation rate, ΔT is temperature difference outdoor and indoor (°F), $\eta_{make\ up\ air\ unit}$ is thermal efficiency of make-up air unit, and HHV_v is higher heating value of natural gas in the volume basis (0.038 GJ from Enbridge website).

Ventilation rate (CFM) was calculated using Equation 7. After calculating ventilation rate of the building, benchmarking was performed on five different powder coating industries in Greater Toronto Area. Energy performance indicator is useful to understand the facility energy performance. Energy performance indicator is defined as the measurement of energy intensity to identify the energy management effectiveness. To develop an energy performance indicator, statistical correlation (R^2) was derived from ventilation consumption and area of the facility.

Ranking distribution is helpful to benchmark the industries. The ranking distribution of an industrial facility shows the standard energy performance of the plant within the database. The complete data set was divided into top performer, 75th percentile, 50th percentile and poor performer.

The potential savings of the industries were calculated after conducting the ranking distribution. The saving potentials were computed from the unit ventilation energy consumption of the top performer plant, meaning that per unit ventilation consumption of the facility was compared with respect to the plant area. Equation 8 and 9 and 10 [8] presents the potential saving analysis.

Potential Savings (%) = (Considered plant ventilation consumption (m^3/ft^2) – Top performer plant ventilation consumption (m^3/ft^2))/ Considered plant ventilation consumption (m^3/ft^2) x 100 (8)

Potential Savings (m^3) = Annual ventilation consumption of considered building (m^3/ft^2) x Annual saving potential (%) x Plant gross area (ft^2) (9)

Potential savings ($) = Annual ventilation consumption of considered building (m^3/ft^2) x Annual saving potential (%) x Plant gross area (ft^2) x per unit charge of natural gas ($/$m^3$) (10)

RESULT

Initially, the energy data were collected from the selected industrial plants. According to the methodology, process and seasonal energy consumptions were separated from the monthly energy consumption. It was found that 71% of total energy consumption was used as seasonal energy consumption. Figure 1 presents the breakdown of total energy consumption.

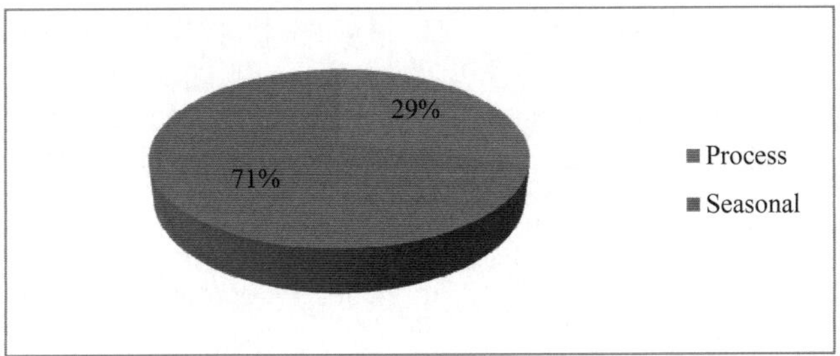

Figure 1: Percentage of seasonal and process energy consumption per year.

Seasonal load is responsible for ventilation consumption, space heating and mechanical ventilation consumption. Seasonal load was again divided into two parts: space heating and ventilation. Figure 2 shows the breakdown of the seasonal energy consumption.

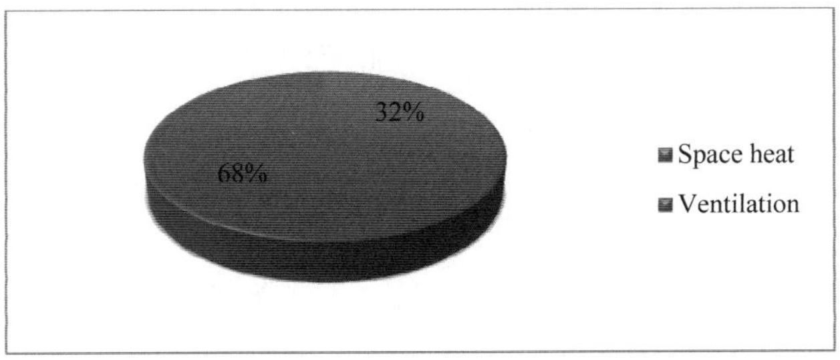

Figure 2: Percentage of space heating and ventilation consumption of the building

Ventilation consumption of an industrial plant is quite different from a residential building. In industrial buildings, some equipment, for example, Dry-Off Oven and Cure Oven intake air from the plant to heat them up. Thus, this intake air also increases the ventilation consumption of the plant. In this case, a calculation was made to find the intake air percentage of Dry-Off Oven and Cure Oven. Figure 3 presents the percentage of mechanical ventilation and oven air intake.

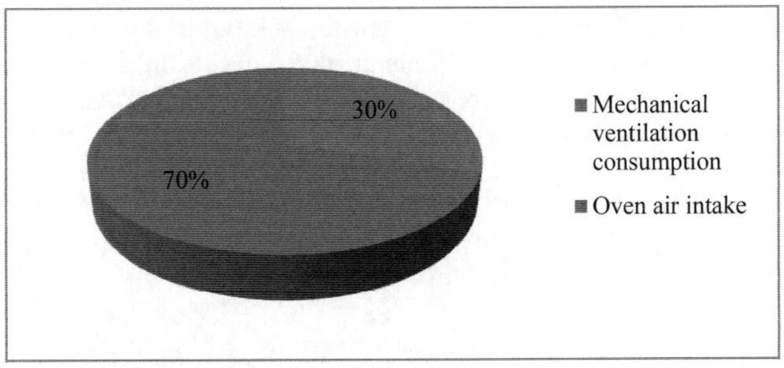

Figure 3: Percentage of mechanical ventilation consumption and oven air intake

In this analysis, it was found that only 32% of the seasonal consumption was used for space heating purpose and the rest was used for ventilation. That means that a tremendous amount of energy was used for ventilation purpose. Hence, there is a possibility to save energy from ventilation system. Figure 4 presents the ventilation energy consumption per unit plant area of the five powder coating industries.

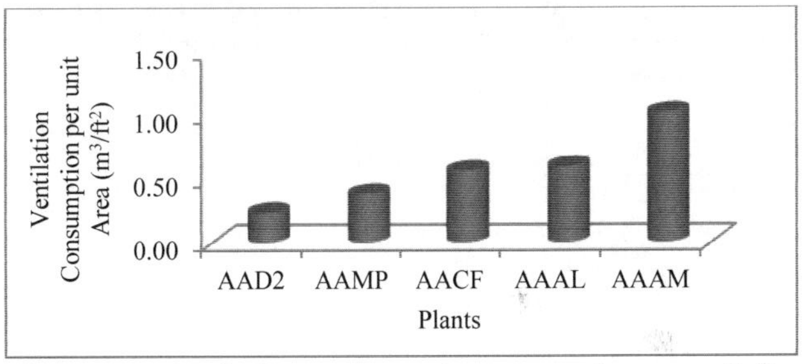

Figure 4: Ventilation consumption per unit area of five powder coating industries.

There is a strong statistical relationship between the annual ventilation energy consumption and the area as shown in Figure 5.

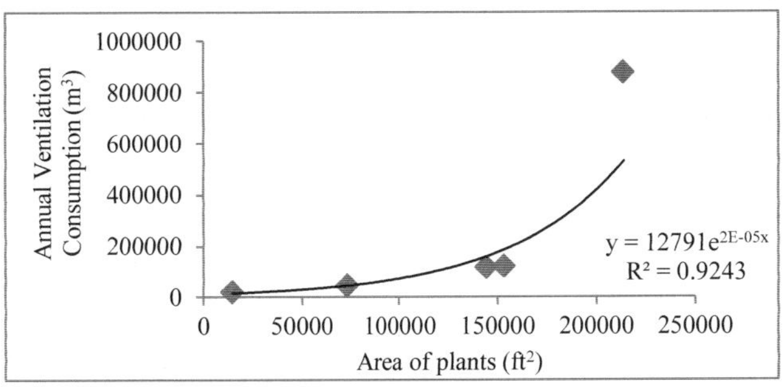

Figure 5: Scatter-plot of ventilation consumption vs. plant area

In addition, a performance-based ranking distribution of the industrial plants was conducted. In this analysis, the industry that consumed the least amount of energy per unit area was considered the top performing plant. Based on the energy performance per unit area, the industrial plants were classified into four percentiles top performer, 75th, and 50th and worst performer. Microsoft EXCEL ranking and percentile analysis were conducted to rank the plants. In this case, the plant that uses the energy that falls in the category above 75th percentile were considered the top performer and the energy consumption of 75th percentile plant was considered as the highest limit for the plants in the top performing category. In the same way all percentiles were categorized. Table 1 shows the ranking distribution of five audited industries.

Table 1: Energy Benchmarking of ventilation consumption Range for SMEs

Consumption Benchmark	Normalized Annual Consumption (m^3/ft^2)
Top Performer	0.22 or below
75th Percentile	0.23- 0.57
50th Percentile	0.58-0.81
Worst Performer	0.82 and above

Finally, probable energy consumption savings of the plants in 75th, 50th and poor percentile were calculated using Equation 8, 9 and 10 [8]. Table 2 presents the savings of natural gas and associated cost. The unit cost of natural gas was considered $0.25.

Table 2: Natural gas and cost savings

Benchmark	Saving Potential (%)	Natural Gas Savings (m^3)	Cost Savings ($)
Top Performer	0	0	0
75th Percentile	1 – 7%	1 – 5,294	1 – 1,323
50th Percentile	8 - 10%	5,295 - 15,735	1,324 - 3,933
Worst Performer	11 - 16%	15,736 - 25,917	3,934 – 6,479

CONCLUSION

In this paper industrial ventilation related natural gas consumption of five powder coating industries in GTA and the related potential energy and cost savings were analyzed. Detailed energy audit was conducted and in-depth natural gas consumption analysis was performed. From the ventilation analysis, it was found that 50 - 75% of the seasonal consumption was related to ventilation of the plants. The poor performing plants could save up to 1 - 16 % of their ventilation consumption which was equal to 5,294 - 25,917 m^3 of total natural gas consumption. The results from this study suggest that energy benchmarking is an excellent tool to identify energy saving opportunities in industrial plants especially in the ventilation system. This will allow better energy management practice in industrial sector for the potential reduction of considerable amount of energy and cost.

ACKNOWLEDGMENTS

The authors would like to acknowledge the financial support provided by Enbridge Gas Distribution Inc. and Connect Canada. The authors are grateful to Paul Morrison and Peter Goldman from Enbridge for liaising between their industrial customers and the Ryerson University team. In addition, the authors would like to thank their colleagues, Farzin M. Rad and Altamash Ahmed Baig, for conducting the energy audits and collecting the data.

REFERENCES

1. Wilde, P.D., Martinez-Ortiz, C., Pearson, D., Beynon, I., Beck, m., Barlow, N., (2013), "Building simulation approaches for the training of automated data analysis tools in building energy management", *Advanced Engineering Informatics*, Vol- 27, pp. 457–465.
2. Saygin, D., Worrell, E., Patel, M.k., Gielen, D.J., (2011), “Benchmarking the energy use of energy intensive industries in industrialized and in developing countries”, *Energy*, Vol-36, PP-6661-6673
3. Ke, J., Price, L., McNeil, M., Khanna, N.J., Zhou, N., (2013), "Analysis and practices of energy benchmarking for industry from the perspective of systems engineering." *Energy*, Vol-54, PP- 32-44.
4. Kaushik, B., (2009), "Energy conservation opportunities in industrial ventilation systems", *Energy Engineering*, Vol-106, pp- 73-79.
5. Ernst, W., Laurijssen, J., & Faaij, A., (2013), "Benchmarking energy use in the paper industry: a benchmarking study on process unit level", *Energy Efficiency*, pp- 49–63. DOI 10.1007/s12053-012-9163-9.
6. Kabir G., Abubakar, A.I., UA-EL-Nafaty, (2009), "Energy audit and conservation opportunities for pyroprocessing unit of a typical dry process cement plant", *Energy*, Vol- 35, pp-1237–1243.
7. Mcquiston, P., Spitler, (2005), Heating ventilating and air conditioning, ISBN-0-471-47015-5, USA.
8. Ghajarkhosravi, M., (2013), Utility benchmarking and potential savings of multi-unit residential building (MURBs) in Toronto, *Ryerson University*.

Proceedings of the 25th CANCAM
London, Ontario, Canada, May 31 – June 4, 2015

EXPERIMENTAL PERFORMANCE STUDY OF A VERTICAL AND A HORIZONTAL GROUND LOOPS COUPLED TO A GROUND SOURCE HEAT PUMP SYSTEM

Waleed S. Alzahrani, Alan S. Fung, and Wey H. Leong
Department of Mechanical and Industrial Engineering
Ryerson University
Toronto, Ontario, Canada
waleed.alzahrani@ryerson.ca, alanfung@ryerson.ca, weyleong@ryerson.ca

ABSTRACT

The performance of vertical and horizontal ground loops coupled to a ground-source heat pump (GSHP) was investigated under four different scenarios. For this purpose, an experimental set-up was designed and constructed at the Toronto and Region Conservation Authority (TRCA) Archetype Sustainable Houses (ASH) in Vaughan, Ontario, Canada. In the first two tests, the two vertical ground loops coupled to the GSHP were tested in heating, and cooling modes. In heating mode, the GSHP's COP ranged between 2.7 and 3.15. In cooling mode, the GSHP performed better than the heating mode with COP range of 3.75 to 5.4. In the last two tests, two scenarios were tested to compare the horizontal and the vertical ground loops in cooling mode. In the first scenario, the ground loop flow was divided equally between the loops and the GSHP's overall COP was 5.42. The last test used equal Reynolds number in both loops and the GSHP's overall COP was 5.36.

KEYWORDS: Ground-source heat pump, GSHP, horizontal loop, vertical loop, geothermal, COP

LIST OF ACRONYMS:

AHU	Air Handling Unit
ASH	Archetype Sustainable Houses
COP	Coefficient of Performance
DAQ	Data Acquisition System
DHW	Domestic Hot Water
ELT	Entering Load Temperature
EST	Entering Source/Sink Temperature
GHG	Greenhouse Gas
GSHP	Ground-Source Heat Pump
TRCA	Toronto and Region Conservation Authority

INTRODUCTION

Buildings account for the majority of the total energy consumption worldwide. In Canada, the increase of commercial and institutional facilities' energy demands grew by over 35% between 1990 and 2009. Also, due to the increase of the number of households and house sizes in the same period, energy consumption of the housing sector rose 37% [1]. Most of the energy in buildings is consumed to provide space heating, space cooling, and water heating. When these three major end-uses are combined, they account for 82% of the total residential building energy use and 60% of the commercial building energy use. In 2008, these sectors were responsible for 28% of the total greenhouse gas (GHG) emissions in Canada [2].

To lower energy consumption and its associated GHG emissions in buildings, a number of strategies can be employed not only to reduce the energy demands but also to make the most of the energy used by increasing the efficiency. One of the most energy efficient mechanical systems that can provide space heating, space cooling and domestic hot water (DHW) is the ground source heat pump (GSHP). However, GSHPs have not been commonly used due to their high cost of installation. Nevertheless, as the energy prices have increased and further industrial enhancements have been achieved, GSHPs are reconsidered as an alternative heating and cooling system.

A GSHP system has three major components: a heat pump, a ground loop, and a heating/cooling distribution system. The ground loop is where the heat transfer between the GSHP system and the soil occurs. The two most common types of ground loops are the horizontal and the vertical configurations. In the horizontal loop configuration, pipes are buried in horizontal trenches, whereas in the vertical loop configuration, the pipes go into vertical boreholes. Looking at the industry average in Canada over 2008-2010, 56.4% of residential systems had a horizontal ground heat exchanger (GHE) and 24.2% had a vertical GHE. Open loop systems that use wells follow with 13% of residential applications while pond and lake loops account for 6.4% [3]. The higher popularity of the horizontal loop configuration is due to the lower price of installing a horizontal ground loop. In 2010 for example, the average vertical GSHP system sold for $7886 per ton in Canada compared to $6116 per ton for a GSHP system with a horizontal GHE. The total average price for a 4-ton system is $31,544 for a GSHP system with a vertical GHE and $24,464 for a system with a horizontal GHE. Horizontal systems are roughly 25% less expensive per ton than the vertical option [3]. Although the horizontal GHE may require more piping in some cases, there are significant cost savings made on trenching versus drilling.

In this paper, the performance of vertical and horizontal ground loops coupled to a ground-source heat pump (GSHP) was investigated. For this purpose, an experimental set-up was designed and constructed at Toronto and Region Conservation Authority (TRCA) Archetype Sustainable Houses (ASH) located in Kortright Centre for Conservation, Vaughan, Ontario, Canada.

EXPERMANTAL SETUP AND METHODOLOGY

Constructed in the summer of 2008, the Archetype Sustainable House (ASH) is two semidetached houses; House A and House B. The technologies implemented into House A showcase current sustainable practices, while the technologies in House B present energy conservation practices and technologies expected to become prominent in the near future [4, 5]. The semi-detached House B investigated in this research is described briefly in Table 1. The GSHP system in House B was uniquely equipped with a fully instrumented horizontal and vertical ground loop (Table 2). The system has been configured to allow for both parallel and individual operation of the two loops and is powered by a WaterFurnace heat pump with 3.5 tons nominal capacity [4]. The heat pump charges a buffer tank used for radiant floor heating and forced air cooling. The buffer tank is used in between the GSHP and the in-floor system/AHUs to minimize short cycling. The house is equipped with over a 300 calibrated sensors and a data acquisition system (DAQ) which collected data every five seconds to monitor the performance of the electrical/mechanical systems and energy fluxes into and out of the house [5].

Table 1. Archetype House B summary [7]

Archetype House B	
Orientation	South facing
Stories	3
Floor Area	325 m^2/ 3500 ft^2
Peak Heating Load	7.94 kW/27.1 MBH
Peak Cooling Load	6.18 kW/21.1 MBH

Table 2. Horizontal and vertical loops specifications

Parameter	Vertical loops	Horizontal loops
Number of loops	2	2
Depth of loop	76.2 m (250 ft)	1.83 m (6 ft)
Length of each loop	152 m (500 ft)	366 m (1200 ft)
Nominal Diameter	1″	1.25″
Material	HDPE 4710	HDPE 3408
Loops total volume	193 L (51 US gal)	732 L (193 US gal)
Shape	U-loop	J-loop
Fluid	20% propylene glycol (PG) and 80% water with freezing point of -8°C	

A schematic of the GSHP in House B is shown in Fig. 1, depicting the ground loops lines and pump, the buffer tank line and pump, and the GSHP with the corresponding sensors and their locations [4,6]. To the left of the heat pump is the entering (T17) and leaving (T16) load temperatures (ELT and LLT), load circulator pump power consumption and load flow rate (FL6). To the bottom right of the heat pump in the schematic is the entering (T64) and leaving (T65) source temperature (EST and LST) and flow rate for both the vertical (V-loop) (FL 42 & 43) and horizontal (H-loop) (FL 44) loops as well as the circulator pump power consumption and the temperture sensors.

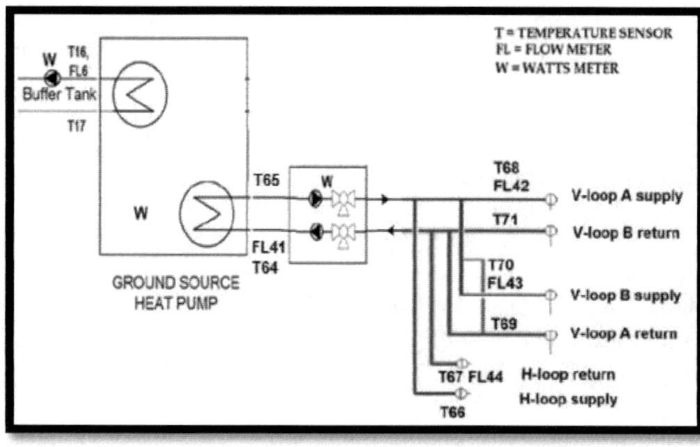

Figure 1: Schematic of performance test monitoring points [6]

In this study, the performance of vertical and horizontal ground loops coupled to GSHP was investigated under different scenarios. Both loops were tested in parallel and individually. In the first two tests, the two vertical ground loops coupled to the GSHP were tested in heating and cooling modes. Also, the GSHP with the vertical loop COP results were compared with a previous study when the system was coupled to the horizontal loop [7]. In the last two tests, two scenarios were tested to compare the horizontal and the vertical ground loops in cooling mode when connected in parallel. Table 3 summarized the tests conducted in this paper.

Table 3. Summary of the tests

The vertical loop tests (The two U-loops in parallel)	
Heating (May 14 – June 4, 2013)	**Cooling** (June 21 – July 8, 2013)
Vertical loops: Reynolds #: **8300** Flow: **48** LPM	Vertical loops: Reynolds #: **8300** Flow: **48** LPM
GSHP/buffer-tank loop flow rate: **50.4** LPM	
The vertical and horizontal loops tests (The two U-loops and the two J-loops in parallel)	
Cooling (July 19 – Aug. 7, 2013)	**Cooling** (Aug. 9 – Sep. 4, 2013)
Vertical loops: Reynolds #: **5540** Flow: **32** LPM	Vertical loops: Reynolds #: **4840** Flow: **28** LPM
Horizontal loops: Reynolds #:**4400** Flow rate: **32** LPM	Horizontal loops: Reynolds #: **4840** Flow rate: **35** LPM
GSHP/buffer-tank loop flow rate: **47** LPM	

RESULTS

The detailed analyses and findings in this study had been reported in the study by Alzahrani [6]. Therefore, this paper mainly discusses the operation characteristics and findings from the experimental work.

(1) In the first two tests, the GSHP operated using the vertical ground loops. The GSHP performed very well in the heating test period with a COP range of 2.7 (at an ELT of 45°C and EST of 0.5°C) to 3.15 (at an ELT of 40.5°C and EST of 3.5°C) as can be seen from Fig. 2. The heating test period COP was obtained to be 3.0. On the other hand, in the cooling mode, the GSHP performed much better than in the heating mode with test period COP range of 3.75 (at an ELT of 3°C and an EST of 19.5°C) to 5.4 (at an ELT of 10.5°C and an EST of 16.5°C) and a test period COP of 4.5. From the detailed analysis of both tests data, the GSHP showed a constant performance regardless of the weather conditions which can be explained by the relatively constant ground temperature.

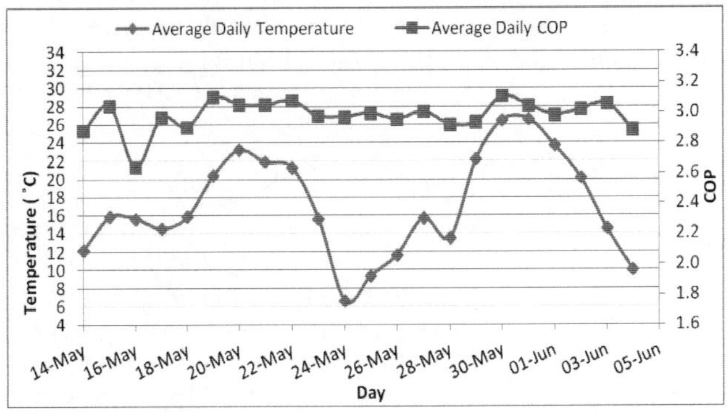

Figure 2: GSHP heating COP variation May 14 to June 4, 2013(2) The performance of the GSHP was not affected by the weather conditions during the tests and was shown to be dependent on the entering source temperatures (EST), entering load temperature (ELT) and the flow rates. The effects of the EST and the ELT on the COP during the GSHP heating mode test period using the vertical loops are shown in Fig. 3. As expected, heating season COPs increased as ELTs decreased and ESTs increased. The reverse occurred during the cooling season. The higher difference across the GSHP resulted in a lower COP. In the heating mode, the average difference between the EST and the ELT was around 41°C whereas in the cooling mode, the average difference was around 15°C. The higher difference between the EST and the ELT resulted in a higher power draw and lower COP. This change in the power draw is mainly due to the GSHP compressor which accounted for the majority of the GSHP system total power consumption since the power draw from the pumps is almost constant in all cases.

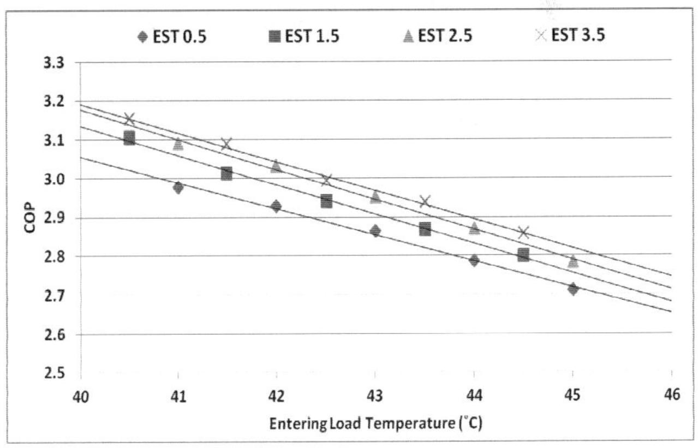

Figure 3: GSHP heating COP variation with EST and ELT

(3) Some characteristics were noticed when looking at the general operation cycles from the time the GSHP starts providing heating or cooling to subsequently turning off. The general trend in both the heating and the cooling cases is that the GSHP showed high COP at the beginning of operation

cycles. This is due to the fact that the temperature difference between the EST and the ELT always increased with time. This increase in the temperature difference resulted in higher compressor consumption which has affected the cycle COP. Within each of the tests, the shorter cycle time resulted in a higher COP. It is worth noting also that the GSHP averaged 32 minutes per cycle in the heating mode, whereas the cooling mode average was 20 minutes.

(4) The heating and cooling mode COPs associated with the vertical loop were 3.0 and 4.5, respectively, and those for the horizontal loop, on the same system, were 2.8 and 4.0, respectively [8]. Experimental results showed that the GSHP have a higher COP when using the vertical ground loop compared to the horizontal loop. Also, a vertical U-tube GHE configuration is preferred because it requires less ground area and offers better performance due to smaller seasonal swing in the ground mean temperature compared horizontal loop configuration. The fluid supply temperature from the vertical ground loop was not affected by the weather conditions during the tests. However, solar radiation near the ground surface, and warm weather conditions in the summer affected the horizontal ground loop. The fluid supply temperature from the horizontal loop increased at the end of the cooling season tests by almost 2°C.

(5) In the last two tests of this experimental study, the horizontal and the vertical loops have been used in parallel as a heat sink for the GSHP in cooling mode tests. When using both ground loops, the cooling COP of the GSHP system was 5.42 when both loops had equal flow and 5.36 when both loops had equal Reynolds number (equal flow characteristics). In the first test, the combined daily average heat rejected to the ground by the vertical and the horizontal loops was 15.5 kWh. Individually, the average heat rejected by the vertical and horizontal loops was 7.89 and 7.58 kWh, respectively. When taking into account the loops dimensions in Table 2, the vertical loop had better heat transfer per unit area of the pipe surface. This was mainly a result of the lower ground temperatures at greater depths and the higher Reynolds number due to the smaller diameter. When testing the same configuration with an equal Reynolds numbers by using a higher flow rate in the horizontal loop, the same conclusions were reached.

CONCLUSIONS

The performance of horizontal and vertical loops coupled to a ground-source heat pump was investigated experimentally. The application of the GSHP in buildings ensures a low operation cost. The results demonstrate that GSHP is a viable and energy efficient alternative to conventional systems for heating and cooling applications. In the study also, the GSHP system operated in reasonable agreement with the manufacturer performance specifications which helps to build confidence in the technology and the industry. During the heating season, the system delivered more than three units of heat energy for every unit of electrical energy consumed. This is further proof that GSHP systems require less energy than conventional systems to deliver an equivalent amount of heating. When comparing the vertical and the horizontal loop performances, the vertical loop configuration had a better heating and cooling COP and was not affected by the weather conditions.

ACKNOWLEDGMENTS

This experimental work has been supported by the Natural Sciences and Engineering Research Council of Canada (NSERC), GeoEnergy Solutions Inc. and the Toronto and Region Conservation Authority (TRCA). The authors are also thankful for the King Abdullah Foreign Scholarship Program and the Saudi Arabian Cultural Bureau in Canada for their support and guidance.

REFERENCES

1. NRCan. 2012. "Energy Use Data Handbook: 1990 to 2009", *Natural Resources Canada.*
2. NRCan. 2012. "Report to Parliament Under the Energy Efficiency Act For the Fiscal Year 2010-2011", *Natural Resources Canada.*
3. Canadian GeoExchange Coalition. 2011. "The State of the Canadian Geothermal Heat Pump Industry 2011: Industry Survey and Market Analysis", *Canadian GeoExchange Coalition.*
4. A. Dembo, A.S. Fung, K.L.R. Ng, A. Pyrka, The archetype sustainable house: investigating its potentials to achieving the net-zero energy status based on the results of a detailed energy audit, in: Proceedings of the 1st International High Performance Buildings Conference, 2010, pp. 1–8, 3247.
5. D.H. Zhang, R. Barua, A.S. Fung, TRCA-BILD archetype sustainable house—overview of monitoring system and preliminary results for mechanical systems, ASHRAE Trans. 117 (2) (2011) 597–612.
6. Alzahrani, Waleed Saeed, 2014. "*Experimental study of the performance of a vertical and a horizontal ground loops couples to a ground source heat pump system*". Master of Applied Science Thesis, Ryerson University, Toronto.
7. Safa, Amir Alizadeh, 2012. "*Performance analysis of a two-stage variable capacity air source heat pump and a horizontal loop coupled ground source heat pump*". Master of Applied Science Thesis, Ryerson University, Toronto.
8. A.A. Safa, A.S. Fung, R. Kumar, Comparative thermal performances of a ground source heat pump and a variable capacity air source heat pump systems for sustainable houses, Appl. Therm. Eng. 81 (2015) 279-287.

Turbulence

BOUNDARY LAYERS PROPERTIES FOR TURBULENT TAYLOR-COUETTE-POISEUILLE FLOWS IN A NARROW-GAP CAVITY

Poncet Sébastien
Department of Mechanical Engineering
Université de Sherbrooke
Sherbrooke, Québec, Canada
Sebastien.Poncet@USherbrooke.ca

Viazzo Stéphane
Laboratoire M2P2 UMR 7340
Aix-Marseille Université
Marseille, France
stephane.viazzo@univ-amu.fr

ABSTRACT

The present paper concerns Large-Eddy Simulations (LES) of Taylor-Couette-Poiseuille flows in a narrow-gap cavity using the Wall-Adapting Local Eddy-Viscosity model. The influence of the swirl parameter N is investigated in details. This coefficient, which compares the effects of the inner cylinder rotation and the axial flowrate, does not influence significantly the mean velocity field. Turbulence intensities are enhanced with increasing values of N with remarkably high peak values within the boundary layers. The two boundary layers exhibit the main characteristics of two-dimensional boundary layers. Thin coherent structures appearing as negative (resp. positive) spiral rolls are observed along the rotor (resp. stator) side. Their inclination depends on N, which fixes the intensity of the crossflow.

KEYWORDS: Large-Eddy Simulation, Taylor-Couette-Poiseuille, narrow-gap system, turbulent boundary layers

INTRODUCTION

Since Taylor's theoretical results in 1923, huge research efforts have been done to better understand the rich dynamical behaviour exhibited by the flow induced by the differential rotation of concentric cylinders. The present work has been motivated by the need of an aerospace company to better predict the heat transfer in the rotor-stator gap of an electrical motor, which can be modeled faithfully by considering a narrow-gap Taylor-Couette-Poiseuille system. A better knowledge of the hydrodynamic field is absolutely necessary before considering the heat transfer aspect as pointed out recently by Fénot *et al.* [3]. These last authors provided an exhaustive literature survey on that topic. They explained the difficulty to establish universal correlations for the heat transfer coefficients by the large numbers of parameters involved in the problem. Moreover, most of the previous studies were mainly concerned with temperature measurements without any idea of what the flow was.

Nouri and Whitelaw [8] and Escudier and Gouldson [2] provided experimental databases for a middle-gap cavity of large aspect ratio, later used to validate numerical approaches, which all assume the flow in the axial direction as being fully established. For example, Sung and co-authors have performed direct numerical simulations [1], then LES [6] for an infinite cavity of small to intermediate radius ratios. Jung and Sung [6] investigated the curvature effects on the near-wall coherent structures in the rotating case. Rotation effects are dominated by the inward axial flow and the boundary layers developed on each wall revealed some features of three-dimensional turbulent boundary layers (3DTBL) but not all of them because of transverse curvature effects. Chung and Sung [1] performed LES for three rotation rates corresponding to the experimental configuration of Nouri and Whitelaw [8]. They established the destabilization of the near-wall turbulent structures due to rotation of the inner wall giving rise to an increase of sweep and ejection events. Hadziabdic *et al.* [4] studied by LES a fully developed turbulent flow in a concentric annulus with an outer rotating wall while the inner one remains stationary. They focused their attention on the effect of the swirl parameter N on the coherent structures in the near-wall regions. A significant change in the turbulence structures is observed between N=2 and 2.8, which is attributed to the centrifuged turbulence prevailing over the axial inertial forces. For N= 4, the flow gets relaminarized with the appearance of spiral rolls close to the Taylor vortices encountered in enclosed Taylor-Couette flows.

As far as the authors know, there is no reference data for Taylor-Couette-Poiseuille flows with rotation of the inner cylinder in narrow-gap cavities for high values of the swirl parameter. The goal of the present work is then to investigate the influence of N with a particular emphasis on what happens in the boundary layers in real operating conditions.

NUMERICAL METHOD

The fluid is confined between two cylinders of radii R_1 and R_2 and height h (Fig.1). The inner cylinder rotates at a constant rate Ω, while the outer cylinder is stationary. The cavity is characterized by two geometrical parameters: its aspect ratio $\Gamma=h/e$ and its radius ratio $\eta=R_1/R_2$, with $e=R_2-R_1$. An axial throughflow is imposed within the gap at a constant bulk velocity W_m. The main flow parameters are the rotational $Re_\Omega=\Omega R_1 e/\nu$ and axial $Re_Q=W_m e/\nu$ Reynolds numbers (ν the fluid kinematic viscosity) gathered to form the swirl parameter $N=Re_\Omega/Re_Q$. The values of these parameters are in similarity with the real rotating machinery (Table 1).

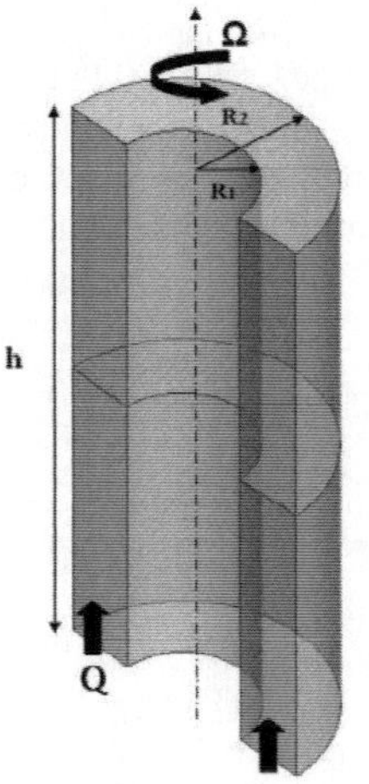

Figure 1: Sketch of the cavity

Case	A	B	C	D	E	F
Γ	10					
η	8/9					
Re_Q	3745			5617		
Re_Ω	8378	16755	25133	8378	16755	25133
N	2.24	4.47	6.71	1.49	2.98	4.47

Table 1: Geometrical and flow parameters

Concerning the numerical method, the time advancement is second-order accurate and is based on the explicit Adams-Bashforth scheme for the convective terms and an implicit backward Euler scheme for the viscous terms. The derivatives are approximated using fourth-order compact formula in the radial and axial directions. The time splitting scheme is an improved projection method, which ensures the incompressibility at each time step. The projection decomposition method is based on a direct non-overlapping multidomain Helmholtz/Poisson solver, which provides the solution of each Helmholtz/Poisson problem resulting in the velocity-pressure coupling. The multidomain solver based on the influence matrix technique ensures the continuity of the solution and its first normal derivative across the conforming interface. By assuming the periodicity of the solution in the azimuthal direction, the three-dimensional extension uses Fourier series. Periodic boundary conditions are applied in the axial and circumferential (angle of $3\pi/4$) directions and no-slip boundary conditions are imposed on the walls. The cavity is divided into 2 subdomains in the axial direction. For more details, the reader can refer to [9].

For cases A to C (resp. D to F), the mesh grid is composed of $65\times144\times130$ (resp. $91\times128\times182$) mesh points in the radial, tangential and axial directions respectively with a time step equal to 10^{-4}s (resp. 5×10^{-5}). It ensures a maximum radial wall coordinate r^+ equal to 0.86 (resp. 0.65).

The Wall-Adapting Local Eddy Viscosity (WALE) model developed by Nicoud and Ducros [7] is used as subgrid-scale model. No noticeable differences have been found in the present configuration compared to the Dynamic Smagorinsky model, while saving about 12% of calculation time as shown by Poncet *et al.* [9].

MEAN AND TURBULENT FLOW FIELDS

The mean flow field is mainly at two-component (the mean radial velocity being zero everywhere). It is divided into three flow regions: two thin boundary layers on each wall separated by a central core at mid-gap in almost solid body rotation at about $0.5\Omega R_1$. The radial distribution of the mean axial velocity component is close to the turbulent Poiseuille profile encountered in pipe flows. A slight asymmetry is observed with higher axial velocities along the stator [9].

A more interesting quantity is the intensity of the crossflow defined as the ratio between the mean tangential and axial velocity components and shown in Figure 2. This ratio strongly increases close to the inner cylinder because of the wall rotation. Contrary to the LES of Hadziabdic *et al.* [4] in the case of a stationary inner wall and an outer rotating one, it is nearly independent of N. When moving towards the outer cylinder, the ratio decreases from 0.6 (resp. 0.54) to 0.42 (resp. 0.47) for N=1.49 (resp. N=6.71). The crossflow is rather constant ($\approx$0.45) with the radial position for $r^*\geq0.4$ and $N\geq2.98$, while it decreases more significantly with r^* for $N\leq2.24$. One can thus expect to observe a better transport of turbulence from the inner wall region towards the gap and the outer wall for low values of the swirl parameter.

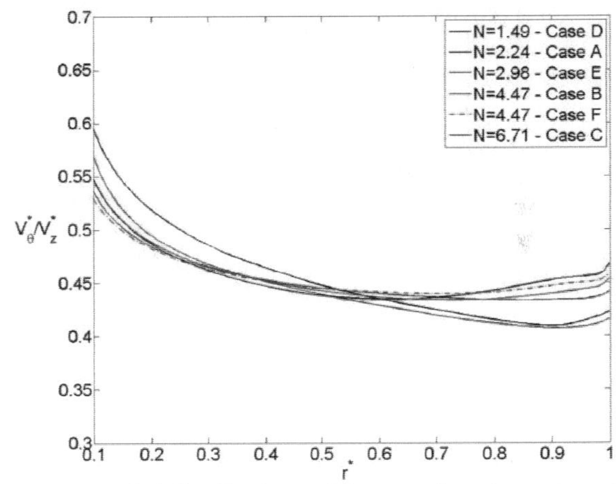

Figure 2: Radial distributions of the crossflow intensity with $r^*=(r-R_1/e)$, $V_\theta^*=V_\theta/(\Omega R_1)$ and $V_z^*=V_z/W_m$

The influence of the swirl parameter N on the tangential normal component of the Reynolds stress tensor is shown in Figure 3. All other components are displayed in [9] and the influence of N on them is quite similar. Turbulence is mainly concentrated in the boundary layers with remarkably high peak values. All profiles are slightly asymmetric. The rotor boundary layer appears to be more unstable than those of the stator highlighting the destabilizing effect of rotation for this range of N values. It may also be attributed to low curvature effects. Increasing the value of N leads to an increase of all turbulence intensities within the gap.

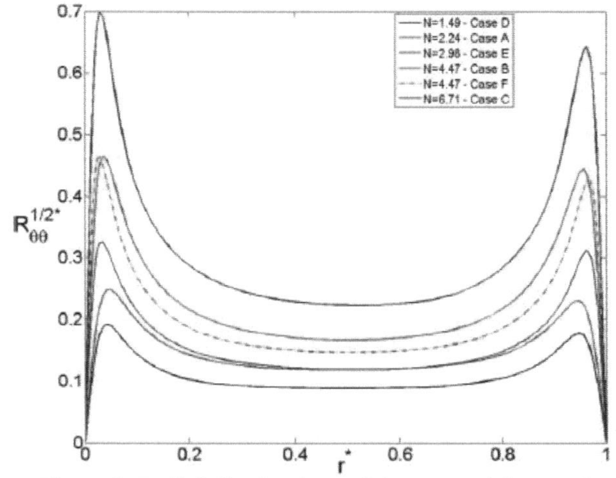

Figure 3: Radial distributions of the tangential normal component of the Reynolds stress tensor normalized by ΩR_1

NATURE OF THE BOUNDARY LAYERS

A 3DTBL is a boundary layer in which the mean flow direction changes with the distance from the wall. To fix the nature of the boundary layers, one can first evaluate the radial distributions of the mean velocity angle γ_s and the lag between the mean gradient velocity angle γ_g and, the turbulent shear stress angle γ_τ defined as: $\gamma_s=\arctan((V_\theta-V_w)/V_z)$, $\gamma_g=\arctan(\partial V_\theta/\partial V_z)$ and $\gamma_\tau=\arctan(\overline{v_r'v_\theta'}/\overline{v_r'v_z'})$, with V_w the tangential velocity of the wall.

One major feature of 3DTBL is the continuous change of direction of the mean velocity angle γ_s with the distance to the wall, which is not observed in Figure 4. Only the rotor side is shown but similar behaviours have been obtained along the stator. γ_s is nearly constant with the wall distance for all values of N and along both walls, highlighting the 2D nature of the rotor and stator boundary layers. Increasing the swirl parameter N leads to an increase, in magnitude, of γ_s.

Another feature of 3DTBL is that the direction of the Reynolds shear stress angle γ_τ in planes parallel with the wall is not aligned with the mean velocity gradient angle γ_g. Such a misalignment is observed $N\geq 2.24$ along both walls, whereas, for $N=1.49$, the lag is rather constant and close to 0. The misalignment increases with increasing values of N and is much larger at the edge of the boundary layers. Such large values of this lag make the assumption of eddy-viscosity isotropy to fail the prediction of such flows. Johnston and Flack [5] recalled that if this lag remains lower than 10°, the errors associated with the isotropic eddy-viscosity concept should not exceed 3%, which is the case here only for the lowest value of the swirl parameter $N=1.49$.

Another characteristic of 3DTBLs is the reduction of the Townsend structural parameter a_1 defined as the ratio between the shear stress vector magnitude τ to twice the turbulence kinetic energy k. As shown in Figure 5, one can clearly see a large reduction of a_1 in both boundary layers as the swirl parameter N is increased. The structural parameter remains above the limiting value a_1=0.15 (see in [5]) within the boundary layers except from case D. A small region at the edge of each boundary layer ($0.055\leq r^*\leq 0.11$ on the rotor and $0.9\leq r^*\leq 0.94$ on the stator) falls below 0.15. The boundary layers exhibit then another characteristic of 2DTBLs in the present case for the whole range of N values.

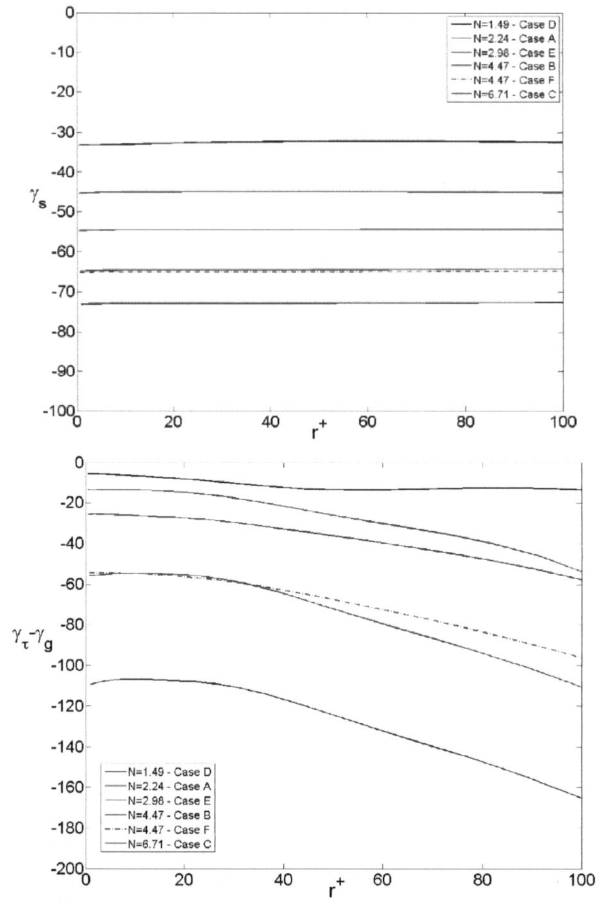

Figure 4: Mean velocity angle γ_s and the lag between the mean gradient velocity γ_g and the turbulent shear stress γ_τ angles against the wall coordinate r^+ along the rotor side

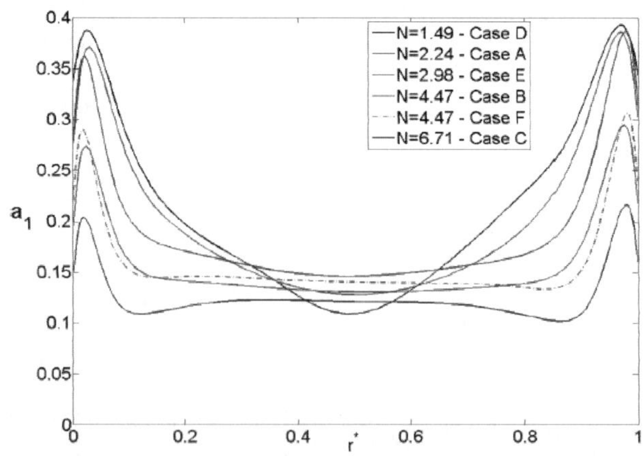

Figure 5: Radial distributions of the Townsend structural parameter a_1

For $1.49 \leq N \leq 6.71$, the boundary layers exhibit the main characteristics of 2DTBL except for the misalignment of the Reynolds shear stress and mean velocity gradient angles. They have then a more complex behaviour in this range of N values in the narrow-gap case. One may except that the two boundary layers may get 3D for decreasing N below 1.49.

COHERENT STRUCTURES

The isovalues of the Q-criterion are shown on the rotor and stator sides for case B in Figure 6. The LES highlights the presence of 3D unsteady coherent structures within the two boundary layers. They appear as thin negative spiral patterns along the rotor as they roll up in the opposite sense of the inner cylinder rotation. The same spiral network is obtained along the stator with a positive angle. When the axial throughflow is more important, these structures are more aligned with the axial direction. They get progressively inclined with the tangential direction when the rotation rate increases. As examples, along the rotor, the angle formed by the spirals with the tangential direction is equal to -45° for N=1.49, to -22° for N=4.47 and to -16° for N=6.71. These 3D unsteady structures appearing at a relatively small scale in the LES are responsible for the large peak values observed in the Reynolds stress tensor profiles. They may play a key role in the wall heat transfer process. They could explain why most of the RANS models, which assume the base flow as being steady and axisymmetric, fail to predict the right distributions of the heat transfer coefficient.

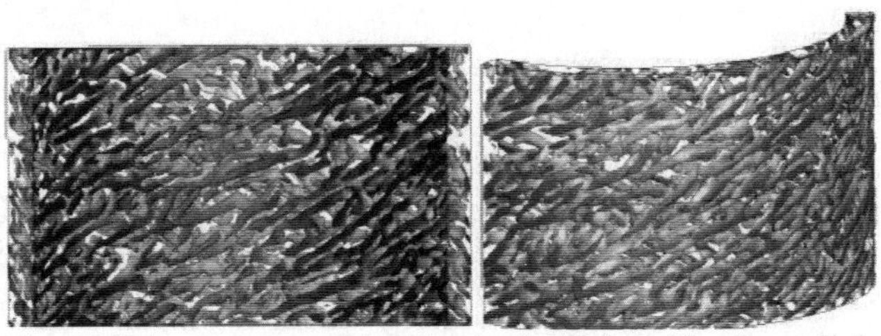

Figure 6: Isocontours of the Q-criterion colored by the radial coordinate (blue on the rotor and red on the stator) for case B (N=4.47) along the rotor (left) and the stator (right) sides. The inner cylinder is rotating counter-clockwise and the axial throughflow is from the bottom to the top

CONCLUSIONS

Accurate large-eddy simulations using the WALE model have been here performed in the case of turbulent Taylor-Couette-Poiseuille flows in a narrow-gap rotor-stator cavity (η = 0.89). Six different values of the swirl parameter N in the range [1.49–6.71] corresponding to real operating conditions of an electrical motor have been considered.

Turbulence is mainly concentrated within the boundary layers with remarkably high peak values compared to former studies in similar systems. Both boundary layers exhibit the main characteristics of two-dimensional turbulent boundary layers except for the misalignment of the Reynolds shear stress and the mean velocity gradient angles. Thin coherent structures appear within the boundary layers under the form of well-organized spiral networks, having a negative angle along the rotor and a positive one along the stator.

ACKNOWLEDGMENTS

This work was financially supported by Liebherr Aerospace Toulouse (project "Entrefer moteur") and by the LABEX MEC (ANR-11-LABX-0092) through the HYDREX program, which are gratefully acknowledged.

REFERENCES

1. Chung, S.Y., Sung, H.J. 2005, "Large-eddy simulation of turbulent flow in a concentric annulus with rotation of an inner cylinder", *Int. J. Heat Fluid Flow*, 26, pp.191-203.
2. Escudier, M.P. and Gouldson, I.W. 1995, "Concentric annular flow with centerbody rotation of a Newtonian and a shear-thinning liquid," *Int. J. Heat Fluid Flow*, 16, pp.156–162.
3. Fénot, M., Bertin, Y., Dorignac, E. and Lalizel, G. 2011, "A review of heat transfer between concentric rotating cylinders with or without axial flow," *Int. J. Therm. Sci.*, 50(7), pp.1138–1155.
4. Hadziabdic, M., Hanjalic, K. and Mullyadzhanov, R. 2013, "LES of turbulent flow in a concentric annulus with rotating outer wall," *Int. J. Heat Fluid Flow*, 43, pp.74–84.
5. Johnston, J.P. and Flack, K.A. 1996, "Review-advances in three dimensional turbulent boundary layers with emphasis on the wall-layer regions," *J. Fluid Eng.*, 118, pp.219–232.
6. Jung, S.Y. and Sung, H.J. 2006, "Characterization of the three-dimensional turbulent boundary layer in a concentric annulus with a rotating inner cylinder," *Phys. Fluids*, 18, 115102.
7. Nicoud, F. and Ducros, F. 1999, "Subgrid-scale stress modelling based on the square of the velocity gradient tensor," *Flow, Turbul. Combust.*, 62, pp.183–200.
8. Nouri, J.M. and Whitelaw, J.H. 1994, "Flow of Newtonian and non-Newtonian fluids in a concentric annulus with rotation of the inner cylinder," *J. Fluid Eng.*, 116, pp.821–827.
9. Poncet, S., Viazzo, S., and Oguic, R. 2014, "Large eddy simulations of Taylor-Couette-Poiseuille flows in a narrow-gap system", *Phys. Fluids*, 26(10), 105108.

VERY-LAGE-SCALE DIRECT NUMERICAL SIMULATION OF THE CANONICAL FLAT-PLATE BOUNDARY LAYER

Jean-Pierre Hickey
Mechanical and Aerospace Engineering
Royal Military College of Canada
Kingston, Ontario, Canada
Jean-Pierre.Hickey@rmc.ca

Xiaohua Wu
Mechanical and Aerospace Engineering
Royal Military College of Canada
Kingston, Ontario, Canada
Xiaohua.Wu@rmc.ca

ABSTRACT

Flat-plate boundary layer, an idealization of the flow over an aircraft wing, may transition from laminar to turbulent under infinitesimal disturbance (natural transition), or under weak yet finite amplitude disturbance (bypass transition). We define bypass transition in the narrow sense as the zero-pressure-gradient boundary layer beneath a continuous free-stream flow of grid turbulence decaying from an initial intensity between 1 to 4 percent. It has been widely believed in the literature that this type of transition always develops its own distinct primary instability in the form of streaks, as well as its own secondary instability. We show that bypass pass transition in the narrow sense may also proceed through the natural and sequential formation of quasi-spanwise structure, Lambda vortex, hairpin packet, infant turbulent spot, and hairpin forest. This is similar to the secondary instability and breakdown process of boundary layer natural transition. Streaks are present, but they are not dynamically important in our flow because they occur downstream of the Lambda vortices. Accuracy of the results is established by verifying the statistics agree with analytic laminar solution over an extended range prior to breakdown, and agree with accepted data in the fully-turbulent region after the completion of transition.

KEYWORDS: DNS, Turbulence, Transition, Boundary Layer

INTRODUCTION

One of the corner-stones of modern fluid mechanical science is the incompressible flow over a smooth flat-plate at zero pressure-gradient (canonical boundary layer), which forms the limiting case and calibration benchmark of many practical aeronautical engineering flows. Starting from a Blasius boundary layer, there are numerous ways of generating the downstream transitional and turbulent boundary layers. The most fundamental route, natural transition, is that the layer being subjected to infinitesimal disturbances in the sense of Schubauer & Skramstad (1947), who verified through experiments the existence of Tollmien-Schlichting wave predicted by linear stability theory. The second most fundamental route, bypass transition in the narrow sense, is a superposition of the canonical boundary layer, in its freestream, with the simplest possible turbulent flow – initially homogeneous isotropic turbulence of (mildly) finite amplitude. There is a pronounced plurality of this case compared to the case of natural transition since the length-scale and intensity of the isotropic grid turbulence participating in the superposition are the extra independent variables. Westin et al. (1994) pointed out the need for creating "standard" conditions in transition experiments, and mentioned that if the freestream turbulence intensity (FST) level is larger than 5% transition will be rapid and occurs at the minimum Reynolds number for self-sustained boundary layer turbulence to exist. It is also obvious that for boundary layers with very high FST level, e.g., 5%, the superposition will severely distort the basic boundary layer profile right at the inlet. For purposes of investigating mechanisms of bypass transition in the narrow sense, it has been accepted to only consider a narrow FST window between approximately 1% and 4% at the leading edge. The lower threshold 1% is chosen so that the Tollmien-Schlichting wave will not be the dominant mechanism. One important piece of experimental result from Westin et al. is confirming, for bypass transitional flows in the narrow sense, a quasi-linear dependence of squared peak streamwise turbulence intensity with streamwise distance. This linearity is an important result of the "algebraic/transient growth" theory, see Anderson et al (1999). Matsubara and Alfredsson (2001) extended the experiments of Westin et al. by varying leading edge freestream turbulent intensity from 1.5%, 2.2% to 6% and by performing flow visualization. They emphasized a plausible connection between their experiments and transient growth theory from three different angles: the aforementioned linearity, a collapse of y/δ^* versus $u_{rms}/u_{rms,peak}$ profiles at different streamwise locations onto the prediction from theory, and visualization images of high and low streamwise velocity streaks. Here, δ^* is boundary layer displacement thickness, y is wall-normal coordinate. In particular, they noted that interpretations of their results are guided by the algebraic/transient growth theory: receptivity giving rise to streaks, transient growth of streaks, breakdown of the flow owing to a secondary instability of the streaks. Though interesting, it seems that the experiments of Matsubara & Alfredsson (2001) should not be viewed as a full validation of the theory; especially considering that no causality was established between streak and breakdown from their visualization, and also considering that no causality was established between streak and the statistical properties of $u'^2_{rms,peak}/U^2$. Following the same line of thought, many authors such as Jacobs & Durbin (2001) also reported DNS studies on the problem of bypass transition in the narrow sense with a synthetic grid turbulence manufactured by superposing modes of the continuous spectrum of the Orr-Sommerfeld and Squire equations. They suggested that the breakdown into turbulent spots is related to local secondary instabilities of these streaks. Jacobs & Durbin (2001) observed that when upwelling lifts the streaks away from the wall they become a receptivity site for smaller scale freestream turbulence which initiates an instability near the top of the boundary layer and results in breakdown. Overall, it seems that most of these studies suggest that bypass transition in the narrow sense develops its own primary instability in the form of streaks whose algebraic/transient growth leads to secondary instability (streaks meandering or streak interaction with free-stream eddies), and eventually breakdown of the laminar layer.

Our idea is that there may exist the possibility that at least a sub-category of bypass transition in the narrow sense might be approximated with the secondary instabilities of boundary layer natural transition, implying a partial unification of these two branches of study. There is actually some literature support for this partial unification conjecture. Klebanoff et al. (1962) wrote "Increasing the ribbon amplitude had no significant effects on the character of the wave generated except to move the point of departure and breakdown further upstream. With sufficiently large amplitude, it was possible to by-pass the linear (TS-wave) range completely. This was graphically demonstrated by observing the behavior of a wave of such frequency that it should damp according to the linear theory. At low amplitudes the wave damped as expected, but at a sufficiently high amplitude it did not damp and very rapidly led to breakdown of laminar flow in much the same manner as a wave within the amplified zone." Their statement is remarkable and particularly relevant here because they semi-explicitly equated bypass transition with secondary instabilities of natural transition. The statement is further interesting because it seems to us that this was the first time the word "bypass" appeared in an archival journal within the context of boundary layer transition. In comparison, the often quoted first naming of bypass transition by Morkovin actually occurred seven years later in 1969. From the perspective of accurate boundary layer simulation, most of the previous DNS studies on bypass transition in the narrow sense were performed prior to the recent resurgence in boundary layer DNS. As a result, they made compromises due to hardware limitations, on streamwise domain size, boundary conditions, grid resolution and used synthetic inflow grid turbulence manufactured by superposing modes of the continuous spectrum of the Orr-Sommerfeld and Squire equations. The past few years have seen remarkable advances on very-large-scale (VLS) boundary layer DNS, in terms of both statistical accuracy and resolution of coherent structures, starting from the paper of Wu & Moin (2009) which provided unprecedented easy access to the details of boundary layer coherent structures. Prior to 2009, it had been quite difficult to detect these coherent structures from the then existing DNS databases (Adrian 2007). In light of this quasi step-change in boundary layer DNS capability since 2009, we are of the opinion that it is logical and imperative to revisit the simulation on bypass transition in the narrow sense so that this problem is re-examined using the new and improved more realistic approach. We are particularly interested in a resolution on our conjecture that a sub-class of bypass transition in the narrow sense might be viewed as the secondary instabilities of natural transition.

RESULTS

Two primary simulation cases were performed first. In case B2K, the domain size is $12750\theta_0$, $2250\theta_0$ and $562.5\theta_0$ in the streamwise, wall-normal, and spanwise directions, respectively. θ_0 is the inlet momentum thickness. The boundary layer grows from $Re_\theta = 80$ to 1974 by the exit. The grid size is 8192 × 500 × 256. In case B3K, the domain size is $21562.5\theta_0$, $2250\theta_0$ and $843.75\theta_0$. The boundary layer grows to $Re_\theta = 3018$ at $x = 21562.5\theta_0$. The grid size is 16384 × 500 × 512, yielding a total mesh points of 4.2 billion. Note at $Re_\theta = 3000$, the boundary layer thickness $\delta = 309\theta_0$, which implies that the computational domain is sufficiently high and wide. In both cases, the continuous free-stream flow above the boundary layer has an inlet FST level of 3%U, which was generated separately using a set of independent computations on the incompressible isotropic grid turbulence. Two auxiliary simulations were also performed by varying the inlet FST in case B2K from 3% to 1.3% (case B2Ka1), and to 0.01% (case B2Ka2), respectively. All other aspects of the case B2K were retained. Rex in these two auxiliary simulations is the same as that in case B2K, but Re_θ is different because the change of inflow FST modified the path of boundary layer development. At the exit of the computational domain ($x = 12750\theta_0$), Re_θ is 723 for case B2Ka1, and 682 for case B2Ka2. The reliability of the boundary layer simulation results is demonstrated by evaluating the skin-friction coefficient Cf in Fig.1. In the figure, the case B3K result falls almost exactly on to the Blasius solution over an extended region ($0 < x < 700\theta_0$, $80 < Re_\theta < 180$) prior to laminar layer breakdown. This suggests that the present free-stream perturbation near the inlet is mild and well-controlled; because otherwise Cf would show noticeable departure from the Blasius solution right away. This also suggests that the present simulation accuracy at the laminar end is good. In many previous bypass transition simulations, the inlet perturbations were so large that the near-wall flow was affected (as measured by ax sizable departure of Cf away from the Blasius) starting from the inlet, which would put a question mark on the applicability of linearized perturbation equations in those flows. During transition, the present Cf follows the trend of the T3A bypass transition experimental data of Roach & Brierley (1990). Even though in both cases the inlet free-stream intensity is 3%, an exact match between the present Cf and that of the T3A case is not expected since no efforts were made to match the length scales between the two cases. We also found that after the completion of transition, at approximately $Re_\theta = 800$, the case B3K and case B2K results start to converge to our previous results obtained using other types of inflow boundary conditions (wakes, and blocks of turbulence, not shown here to avoid making the figure too busy). Further downstream, our DNS skin-friction coefficient result in the fully-turbulent region agrees with classical experimental data. Our results are also in good agreement with the recent DNS results of Schlatter & Orlu (2010), Sayadi et al. (2013) with different inlet/free-stream boundary conditions. This suggests that the present flow eventually arrives at a typical flat-plate turbulent boundary layer state after first going through a representative bypass transition in the narrow sense. The accuracy of the present boundary layer DNS has therefore been verified in the lower laminar end before breakdown through the existence of an extended region over which the departure of skin-friction from the Blasius solution is negligibly small. At the higher turbulent end, the accuracy is demonstrated through comparisons of mean and second-order turbulent statistics with well-accepted datasets. It is thus reasonable and logical to accept the notion that the accuracy in the transitional region that is bracketed by these two verified ends is also good. Hence, after and only after these two verification, we are now ready to address the bypass transition mechanism issue. As a matter of fact, one important distinction of the present bypass transition study is that it is done after a thorough verification of accuracy at the two bounding ends. Overview for the case B3K boundary layer in the transitional region is presented using iso-surfaces of swirling strength λ_{ci} (a vortex identifier proposed by Adrian 2007) in Fig.2. Here $\lambda_{ci} = 4$ is used in Fig.2(a) to show the FST, and $\lambda_{ci} = 16$ is used in Fig.2(b) to remove the weak freestream vortex filaments blocking view into the boundary layer. Spanwise vortex filaments, obliquely oriented legs from detached lambda vortex, hairpin packet in front of a detached lambda vortex, and matured turbulent spot made of localized hairpin forest, are prominent vortex features in the figure. There is almost no evidence for the dominance of elongated streamwise vortices in the present bypass transition region. The figure also shows that in the early turbulent region, the landscape is blanked by hairpin forest. Streaks are not important in the present bypass transition.

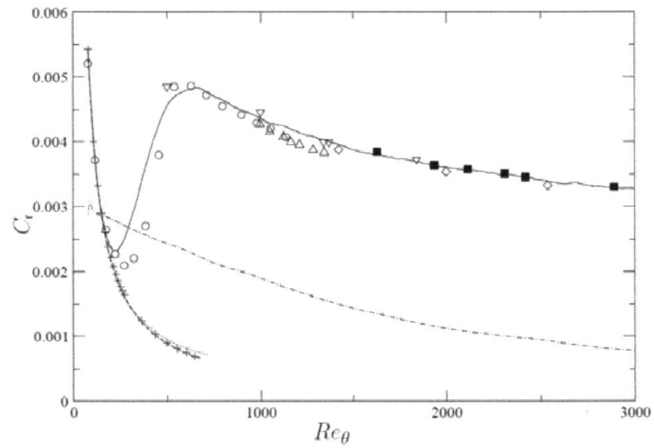

Figure 1: Variation of skin friction coefficient with Re_θ Present DNS: solid line: case B3K; dotted line: case B2Ka1 with 1.3% inlet FST; dashed line: case B2Ka2 with 0.01% inlet FST. Dash-dot-dashed line: one-tenth of the FST in case B3K. Plus: Blasius solution. Other symbols: previous experimental and DNS data.

REFERENCES

1. Schubauer, G. and Skramstad, H. 1947, "Laminar boundary layer oscillations and stability of laminar flow", J. Aero. Sci., Vol 14, pp. 69-78.

2. Westion, K. et al, 1994, "Experiments in a boundary layer subject to free-stream turbulence. I. Boundary layer structure and receptivity", *J. Fluid Mech.*, Vol. 281, pp.193-218.

3. Andersson, P., Berggren, M. and Henningson, D.S., 1999, "Optimal disturbances and bypass transition in boundary layers", *Phys. Fluids.*, Vol. 11, pp.134-150.

4. Matsubara, M. and Alfredsson, P.H. 2001, "Disturbance growth in boundary layers subjected to freestream turbulence", *J. Fluid Mech.*, Vol. 430, pp.149-168.

5. Jacobs, R.G. and Durbin P.A., 2001, "Simulation of bypass transition", *J. Fluid Mech.*, Vol. 428, pp.185-212.

6. Klebanoff, P.S., Tidstrom, K.D. and Sargent, L.M., 1962, "The three-dimensional nature of boundary layer instability", *J. Fluid Mech.*, Vol. 12, pp.1-34.

7. Wu, X. and Moin, P., 2009, "Direct numerical simulation of turbulence in a nominally zero-pressure-gradient flat-plate boundary layer" *J. Fluid Mech.*, Vol. 630, pp.5-41.

8. Adrian, R.J., 2007, "Hairpin vortex organization in wall turbulence", *Phys. Fluids.*, Vol. 19, 041301.

9. Sayadi, T., Hamman, C.W. and Moin, P., "Direct numerical simulation of complete H-type and K-type transitions", *J. Fluid Mech.*, Vol. 724, pp.480-509.

(a)

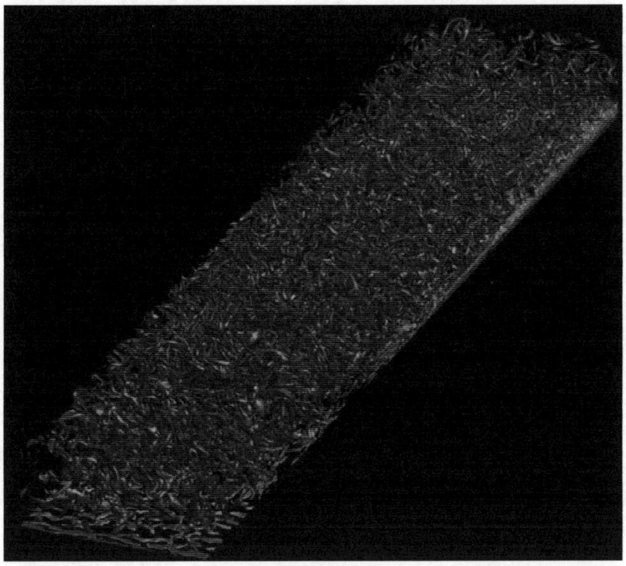

(b)

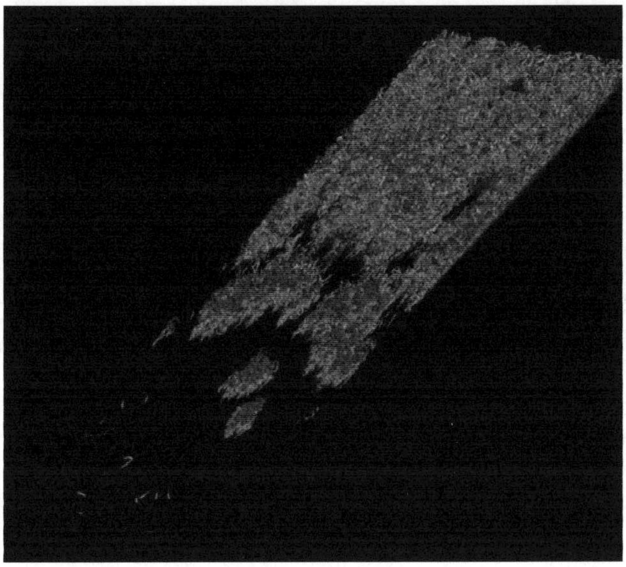

Figure 2: Visualization of boundary layer bypass transition in the first 25% of the streamwise domain of Case B3K. (a) λ_{ci} = 4; (b) λ_{ci} = 16. Note the absence of long streamwise vortices in the transitional region. Breakdown is due to lambda vortices and hairpin packet. Streaks are not dynamically important in the present bypass transition.

TURBULET FLOW BEHIND A PARTIALLY BLOCKING GRID

Fama Fouladi
Turbulence and Energy Lab
University of Windsor
Windsor, Ontario, Canada
fouladi@uwindsor.ca

Paul Henshaw
Turbulence and Energy Lab
University of Windsor
Windsor, Ontario, Canada
henshaw@uwindsor.ca

David S-K. Ting
Turbulence and Energy Lab
University of Windsor
Windsor, Ontario, Canada
dting@uwindsor.ca

ABSTRACT

Flow downstream of a 10 cm high and 65 cm wide diamond-shaped perforated plate suspended in the middle of a 76 cm high and 77 cm wide wind tunnel was examined. 1-D hotwire measurements were taken from Y/D = -13.2 to 13.2 over X/D = 10 to 30 at Z/D = 0; where X = streamwise distance, Y = vertical distance, Z = cross-stream distance from the center of the grid. The flow behind the 54% solidity ratio grid was reduced by approximately 50%. The shear layers generated by the upper and lower edges of the grid were clearly detectable and were characterized by turbulence intensity maxima and integral length and Taylor micro-scale minima.

KEYWORDS: Turbulence intensity, Integral length scale, Taylor micro-scale, Perforated plate, Hot-wire anemometer

INTRODUCTION

Grid-generated turbulence has been the subject of numerous investigations for over seventy years. Generating a nearly-isotropic turbulence using a perforated plate in a wind tunnel was first introduced by Taylor in 1935 [1]. At a short distance from the grid, the turbulence is neither homogeneous, nor isotropic. It becomes homogeneous in the lateral direction and quasi isotropic after some distance downstream of the grid [2]. This region is the initial developing region which has been determined to range from 10 ~ 15 D [2] to 20 D [3]. However, it has been pointed out by many researchers [4-9] that, even at larger distances, grid-generated turbulence can deviate from an isotropic and homogenous state and the amount of this deviation varies from study to study. After this stage there are the initial and final decay regions where the turbulent kinetic energy decays downstream [3, 10, 11].

Conventional grid-generated turbulence has been discussed extensively in recent years. Murzyn and Be´lorgey [12] experimentally investigated the time and length scales of grid-generated turbulence occurring in a flume with a free surface flow. In Liu *et al.*'s study [13], homogeneous, nearly isotropic turbulence was experimentally produced using an orificed, perforated plate in a wind tunnel. Llor [14] studied the decay of homogeneous isotropic turbulence theoretically. Liberzon *et al.* [15] investigated the turbulent dissipation rate in a decaying grid turbulent flow using particle image velocimetry and laser Doppler velocimetry. George [16] reviewed the experimental, theoretical and numerical studies for the asymptotic effect of initial and upstream conditions on turbulence. Horender [17] discussed the turbulent flow downstream of a large-solidity (96%) perforated plate at 1m/s bulk velocity.

Except for one study [18], in all of these previous grid turbulence studies, the perforated plate covered the whole cross section of the wind tunnel. In Judd *et al.*'s study [18] a wind-tunnel experiment was conducted on the flow around a single or multiple porous windbreak of finite height that was sheltering a model plant canopy with a lower height. In their work, the experimental emphases were on the interactions of flow with the underlying plant canopy [18]. A grid partially blocking the middle of a larger wind tunnel, however, has not been investigated. The somewhat symmetrical flow over and under (and around the two sides of) a partial grid is of fundamental interest. Furthermore, it provides a baseline assessment of the turbulent flow over a flat plate downstream of a partial grid. The intention is that the grid will enhance the heat transfer from the flat plate.

APPARATUS AND INSTRUMENTATION

The experiment was conducted in a closed loop wind tunnel with a 180 cm long test section (76 cm high and 77 cm wide). The maximum free stream velocity in the wind tunnel may be as high as 30 m/s when the test section is empty. The background turbulence level of the wind tunnel was measured to be less than 0.4%.

Fig. 1 presents an isometric view of the grid which was hung in the middle of the wind tunnel test section. The X, Y and Z directions are stream-wise, vertical and cross-flow, respectively, with the base point located at the center of the perforated plate. The perforated plate was cut from an aluminum sheet of 0.3 cm thickness by a computer numerical controlled machine. The outside dimensions of the perforated plate were 10 cm x 62 cm. After cutting diamond-shaped holes with a diagonal distance (D) of 1.4 cm, the resulting solidity was 0.545.

The velocity at different points downstream of the perforated plate was measured using a 1-D boundary layer hot-wire probe (type 55P15) as part of a Dantec streamline 55C90 constant-temperature anemometer (CTA). The probe has a platinum-plated tungsten wire with a diameter of 5 μm and a length of 1.25 mm. Signals were low-passed at 30 kHz and then sampled at 80 kHz over a period of approximately 25 s, resulting in 2 x 10^6 data points per measurement. A 12-bit PCI-6071E National Instrument data acquisition card was used to convert the analog signals into digital data. A computer-controlled 2-D traverse system was used to position the probe at the specified location (Fig. 1). The traverse system can move crosswise (Z) in the test section and vertically (Y) with a 3.2 mm (0.125 inch) resolution.

Measurements were performed at 75 points in the Y direction from Y = -18.5 to 18.5 cm (Y/D = -13.2 to 13.2; every 0.5 cm) at Z = 0 (Fig. 1) and repeated at three streamwise locations (X/D = 10, 20 and 30). Tests were performed at 6 m/s free stream velocity.

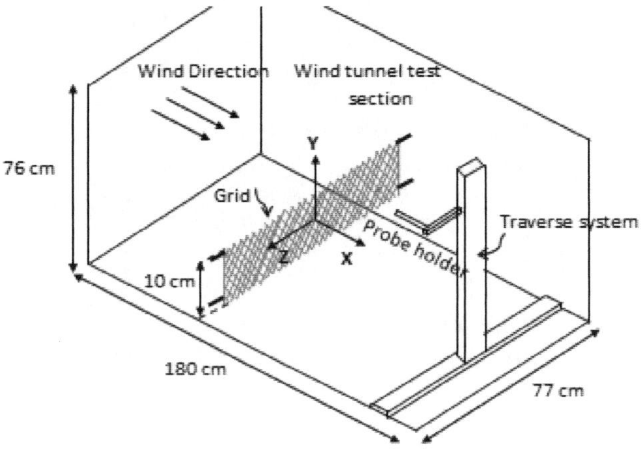

Figure 1: Test section isometric

RESULTS AND DISCUSSION

Velocity profiles

The overall flow behavior behind the grid can be detected from the streamwise velocity profiles. Fig. 2 shows the velocity profiles after a perforated plate at three different downstream distances. As can be seen, the perforated plate caused the flow to be retarded such that the velocity downstream of the grid was approximately one half the free stream value as expected due to the blockage effect of a 54% solidity ratio grid. This significant reduction in the flow velocity magnitude may not be a desirable outcome if our purpose is to enhance convection heat transfer of a flat plate perpendicular to the grid. Nonetheless, the mean velocity reduction should be considered along with other heat transfer alteration parameters which were affected, such as the turbulence intensity and length scale. Two shear layers at the edges of the perforated plate (Y ≈ ±5 cm) can be distinguished in these profiles. At X/D = 10, it can be seen that the velocity had its minimum value (~ 3.3 m/s) at Y = 0, increased slowly to almost constant values from Y = ±2.5 to Y = ±5 cm, and then the shear layers started at Y ≈ ±5 cm. At X/D = 20 and 30 for the area behind the grid (-5 cm ≤ Y ≤ 5 cm), the profile has smoother curvature but still has its minimum at Y = 0. The vertical growth of the shear layers with downstream distance can also be observed in these graphs. As can be seen, the shear layer started from the edges of the grid (Y ≈ ±5 cm) and extended to Y ≈ ±7.5, 9.5 and 10.5 cm in X/D =10, 20 and 30, respectively.

The total uncertainty associated with each parameter consists of bias and precision uncertainties [19, 20]. Bias or systematic uncertainty of the hotwire-measured instantaneous velocities has been estimated according to Ref. [19]. Precision uncertainty, which concerns the repeatability of the test, has been derived from Student's *t* distribution method with a 95% confidence interval [20]. As a result, the uncertainty in velocity (U) at point X/D = 10, Y = 0 in Fig. 2 is estimated to be around ±0.3 m/s, mostly due to the precision uncertainty.

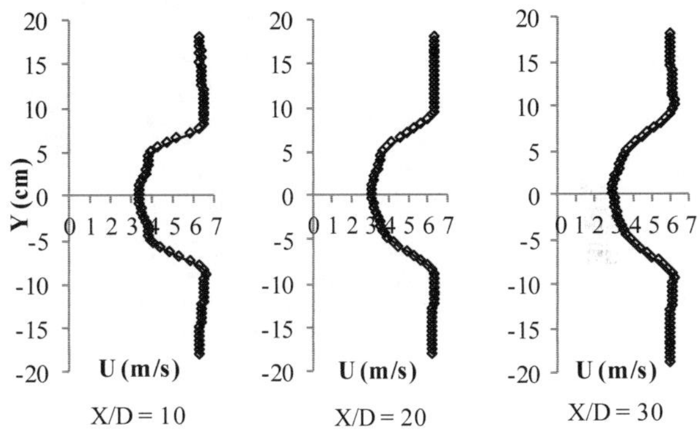

Figure 2: Velocity profiles after grid

Turbulence intensity

Other than flow velocity, turbulence intensity is known to play an important role in convection heat transfer. It is defined as the normalized standard deviation, or root-mean-square (rms), of the velocity fluctuations [19]:

$$Tu = \frac{u'}{U_\infty} \tag{1}$$

where U_∞ is the free stream velocity; and u' is calculated from [19]:

$$u' = \sqrt{\sum_{i=1}^{N} \frac{(u-U)^2}{N-1}} \tag{2}$$

where u is the instantaneous velocity, U is the time-averaged velocity and N is the sampling size (2 x 10^6).

Fig. 3 illustrates the turbulence intensity profile after the grid at the same three downstream distances. As is seen, Tu had almost a constant value in the area behind the grid, -5 cm ≤ Y ≤ 5 cm. At the level of the grid edges (Y = ±5 cm), Tu increased drastically to a peak value and then decreased with almost the same slope to the free stream turbulence intensity (~ 0.4%). By comparing these plots to the velocity profiles

(Fig. 2), it can be seen that the turbulence intensity increase happened at the beginning of the perforated plate edge's shear layer (Y = ±5 cm), as it extended outward from Y = 5, reached a peak at Y around 6.5 ~7.5 cm, and approached the free stream's Tu at the farthest edge of the boundary layer. This illustrates significant turbulence generation around the edge of the perforated plate. Tu for the mid-grid area decayed from 4.5% to 2.5% to 1.6% at X/D = 10, 20 and 30, respectively. This appears similar to the normal decay of grid turbulence [21]. The corresponding peak Tu in the shear layers decreased from roughly 9% to 7.5% to 6.3%. The uncertainty in Tu for the point X/D = 10, Y = 0 is around ±0.7% Tu.

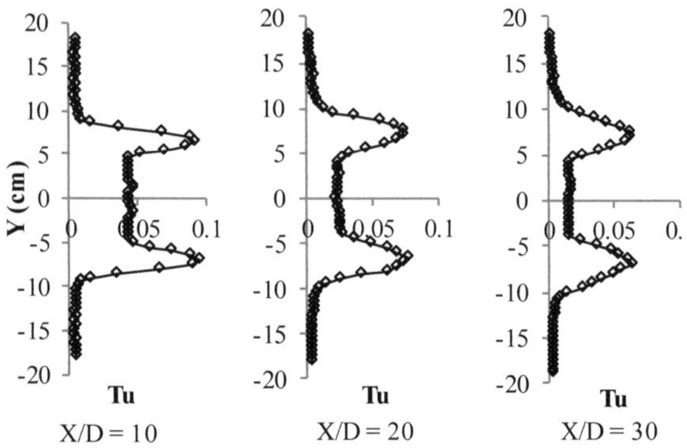

Figure 3: Turbulence intensity profiles after grid

As was observed, the level of Tu resulting from the shear flow around the edge of the grid was much higher than that behind the grid itself. Also, it seems that this higher level of disturbances lasted quite far downstream. Thus, a low height perforated plate at the leading edge of a surface can generate a high degree of disturbance near the surface and may be very effective for enhancing the heat transfer.

Turbulence length scales

Other than the fluctuation intensity, it is known that the turbulent length scales could also be effective parameters influencing momentum and heat transport [22, 23]. The most relevant turbulence scales for any application are not obvious. As a first step, one large and one small turbulent length scales were estimated. It is a common practice to pick the integral length scale to represent the large scale and the Taylor micro-scale to represent the smaller, dissipative eddies. For the detailed calculation procedure of the integral length scale and the Taylor micro-scale, see Ref. [24].

Fig. 4 presents the integral length scale profile at different downstream distances. Since the values of length scales, Λ, above the shear layer were ill-defined due to very low turbulence in the free stream, graphs are truncated beyond Y = ±8, 10.5 and 11.5 cm for X/D = 10, 20 and 30, respectively. As can be seen, the integral length scale has a peak at Y = 0 and decreased vertically upward and downward to a constant value at Y ≅ ±1, 2 and 3 cm for X/D = 10, 20 and 30, respectively, and remained at this value until reaching the outer edges of the shear layers. Within the shear layer where the velocity gradient is highest (see Fig. 2), the turbulence generation is at its maximum. Consequently Λ is smallest and it increases as we move away (upward/downward) from this maximum shear regime. The general trend of increasing Λ with downsteam distance confirms the decay of grid turbulence [4]. The unusually large increase in Λ around Y = 0 seems to be resulted in part by the expanding wake behind the grid, allowing the "domain confined" larger length to expand, in addition to normal decay and negligible turbulence production because the local velocity gradient is nil. The uncertainty in integral length scale value is estimated to be around 0.3 cm.

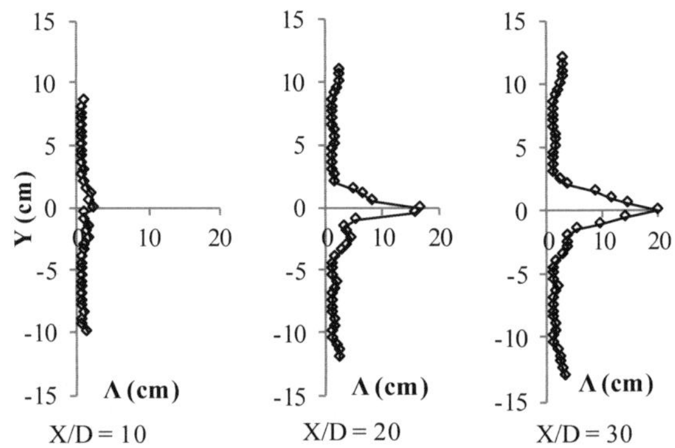

Figure 4: Integral length scale profiles after grid

Taylor micro-scales at some specific points for different free stream distances were calculated and are shown in Fig. 5. Again, it can be seen that the graphs have a peak value at Y = 0 cm but the difference between this value and the other points is not nearly as significant as that associated with the integral length. The peak values at Y = 0 cm, were 0.55 cm, 0.74 cm and 0.85 cm at X/D = 10, 20 and 30, respectively, showing a general trend of increasing Taylor micro-scales with streamwise distance, which demonstrates the decaying trend of turbulence moving downstream.

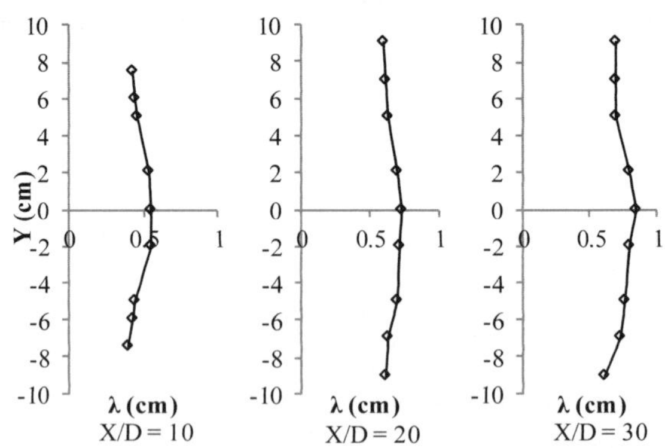

Figure 5: Taylor micro-scale profiles after grid

CONCLUSIONS

The flow downstream of a perforated plate (grid) partially blocking the middle of a wind tunnel test section has been studied to provide baseline data for enhancing the convection heat transfer via partial grid turbulence. The 54% solidity ratio grid retarded the flow to approximately 50% of its free stream value. The velocity profiles showed two shear layers at the grid edges. Significantly higher turbulence intensity and lower integral length and Taylor micro-scales distinguished the grid edges compared to the area behind the grid. These observations at the grid edge shear layers indicate great potential for furthering convection heat transfer enhancement well beyond grid turbulence

ACKNOWLEDGMENTS

This work was made possible by the Ontario Trillium Foundation and the Natural Sciences and Engineering Research Council of Canada.

REFERENCES

1. Taylor, G.I. 1935, "Statistical theory of turbulence," Proc. R. Soc. London, *Series A, Math. and Phys. Sci.*, 151, pp. 421–478
2. Hinze, J.O. 1975, *Turbulence*, 2nd edition, McGraw-Hill, New York, USA, ISBN 0-07-029037-7
3. Batchelor, G.K., and Townsend, A.A. 1948, "Decay of isotropic turbulence in the initial period," *Proc. R. Soc. London A.*, 193, pp. 539-558
4. Grant, H. L., and Nisbet, I.C.T. 1957, "The inhomogeneity of grid turbulence," *J. Fluid Mech.*, 2(3), pp. 263-272
5. Grant, H.L. 1958, "The large eddies of turbulent motion," *J. Fluid Mech.*, 4(2), pp. 149-190
6. Uberoi, M.S. 1963, "Energy transfer in isotropic turbulence," *Phys. Fluids,* 6(8), pp. 1048-1056
7. Uberoi, M.S., and Wallis, S. 1966, "Small axisymmetric contraction of grid turbulence," *J. Fluid Mech.*, 24(3), pp. 539-543
8. Comte-Bellot, G., and Corrsin, S. 1966, "The use of a contraction to improve the isotropy of grid-generated turbulence," *J. Fluid Mech.*, 25(4), pp. 657-682
9. Van Atta, C.W., and Chen, R. 1968, "Correlation measurements in grid turbulence using digital harmonic analysis," *J. Fluid Mech.*, 34(3), pp. 497-515
10. Batchelor, G.K. 1948, "Energy decay and self-preserving correlation functions in isotropic turbulence," *Quart. Appl. Math.*, 6, pp. 97-116
11. Batchelor, G.K., and Townsend, A.A. 1948, "Decay of isotropic turbulence in the final period," *Proc. R. Soc. London A.*, 194, pp. 527-543
12. Murzyn, F., and Be´lorgey, M. 2005, "Experimental investigation of the grid-generated turbulence features in a free surface flow," *Exper. Thermal Fluid Sci.*, 29(8), pp. 925–935
13. Liu, R., Ting, D.S-K., and Checkel, M.D. 2007, "Constant Reynolds number turbulence downstream of an orificed perforated plate," *Exper. Thermal Fluid Sci.*, 31(8), pp. 897–908
14. Llor, A. 2011, "Langevin equation of big structure dynamics in turbulence: Landau's invariant in the decay of homogeneous isotropic turbulence," *Euro. J. Mechanics B/Fluids*, 30(5) pp. 480–504
15. Liberzon, A., Gurka, R., Sarathi, P., and Kopp, G.A. 2012, "Estimate of turbulent dissipation in a decaying grid turbulent flow," *Exper. Thermal Fluid Sci.*, 39, pp. 71–78
16. George, W.K. 2012, "Asymptotic effect of initial and upstream conditions on turbulence," *ASME J. Fluid Eng.*, 134(6), pp. 061203-1-27
17. Horender, S. 2013, "Turbulent flow downstream of a large solidity perforated plate: near-field characteristics of interacting jets," *Fluid Dyn. Res.*, 45(2), pp. 025501-1-22
18. Judd, M., Raupach, J.M.R., and Finnigan, J.J. 1996, "A wind tunnel study of turbulent flow around single and multiple windbreaks, Part I: Velocity Fields," *Boundary-Layer Meteorology*, 80, pp. 127-165
19. Jørgensen, F. E. 2002, "How to measure turbulence with hot-wire anemometers - a practical guide," Dantec Dynamics
20. Figliola, R.S., and Beasley, D.E. 2011, *Theory and Design for Mechanical Measurements*, 5th edition, John Wiley and Sons, Inc., New York, USA
21. Patten, N., Griffin, P., and Young, T.M. 2013, "Effects of freestream turbulence on the characteristics in the boundary layer near the transition onset location," *ASME J. Fluid Eng.*, 135(7), pp. 071203-1-9
22. Carullo, J. S., Nasir, S., Cress, R. D., Ng, W. F., Thole, K. A., Zhang, L. J., and Moon, H. K. 2011, "The effects of freestream turbulence, turbulence length scale, and exit reynolds number on turbine blade heat transfer in a transonic cascade," *J. Turbomach.*, 133(1), pp. 011030-1-11
23. Sak, C., Liu, R., Ting, D.S-K., and Rankin, G.W. 2007 "The role of turbulence length scale and turbulence intensity on forced convection from a heated horizontal circular cylinder," *Exper. Thermal Fluid Sci.*, 31, pp. 279-289
24. Fouladi, F., Henshaw, P., Ting, D. S.-K., 2015, "Turbulent flow over a flat plate downstream of a finite height perforated plate," *ASME J. Fluid Eng.*, 137, 021203-1-12

THE TURBULENT STRUCTURE OF INNER LAYER OVER SUPERHYDROPHOBIC SURFACE

Babak Vajdi Hokmabad
Mechanical Engineering
University of Alberta
Edmonton, Alberta, Canada
vajdihok@ualberta.ca

Sina Ghaemi
Mechanical Engineering
University of Alberta
Edmonton, Alberta, Canada
ghaemi@ualberta.ca

ABSTRACT

Planar particle image velocimetry is applied to scrutinize the inner layer of the turbulent channel flow over a superhydrophobic (SHO) surface with random pattern in comparison to a smooth reference surface at Re = 9,600. To further investigate the wetting effects, a fully-wetted counterpart of the SHO surface with the same surface texture is also considered. No discernable difference in mean velocity is reported over the three surfaces whereas a reduction of normal and shear Reynolds stresses is observed in the buffer and logarithmic layers over the SHO surface. The wetted SHO surface demonstrates no effect on the mean velocity and normal Reynold stresses which suggests that the roughness size is in the range of hydrodynamically smooth wall. The reduction of turbulent Reynolds stresses over the SHO surface raises the potential application of randomly-patterned SHO surfaces for skin-friction reduction.

KEYWORDS: Turbulent channel flow, Superhydrophobic surface, Skin-friction, Particle image velocimetry

INTRODUCTION

The demand for high-efficiency transportation pipelines and marine vehicles has prompted interest in the reduction of skin-friction using various active and passive techniques. Modification of solid surface properties has been investigated as a viable passive method for the reduction of skin-friction. A recent technology, known as superhydrophobic (SHO) surface, utilizes fabrication of micro/nano size patterns on an intrinsically hydrophobic material. Due to the low surface energy of the SHO surface, the adhesion force cannot overcome the surface tension when in contact with a liquid phase. This phenomenon results in a partially or fully non-wetted surface in which microscale air pockets will be trapped in the pores of the surface forming a plastron [1]. The slip of the fluid flow over this plastron has been associated with the reduction of skin-friction in both laminar and turbulent wall-bounded flows [2]. In terms of surface topology, SHO surfaces are categorized into ordered and random structures.

The SHO surfaces with ordered pattern are mostly fabricated through photolithography and soft lithography processes which are typically limited to small surface areas (less than 100 cm^2). The direct numerical simulations (DNS) of Martell et al. [3] and Min and Kim [4] of SHO surfaces with ordered pattern of streamwise micro grooves in the turbulent regime showed reduction of skin-friction due to the attenuation of near-wall turbulent structures. The DNS of Min & Kim [4] and Martell et al. [5] showed that mean velocity profile near the SHO wall scales with the wall shear stress forming a log layer which is offset by a slip velocity. Despite the reduction of skin-friction, the application of SHO surfaces with ordered pattern is limited because of the complicated fabrication process, production in small-scale substrates, and physical durability [6].

SHO surfaces with random surface structures are fabricated by large-scales methods such as spray coating, spin coating, and dip coating methods [7] viable for industrial applications. The experimental study of Zhao et al. [8] in 2007 reported no appreciable decrease of skin-friction in turbulent flow over a flat plate with random SHO roughness. Later, Peguero and Breuer [9] conducted PIV measurements over two SHO surfaces with random patterns (sand-blast surface and random nanograss) and also did not observe any drag reduction. Most recently, Bidkar et al. [10] presented promising results of up to 30% drag reduction in a turbulent flow by means of randomly textured SHO surfaces. The apparent contradiction of the experimental results might be due to disruption of the non-wetted area (SHO longevity) in some cases and experimental uncertainties.

Aljallis, *et al.* in 2013 carried out a drag measurement on spray-coated SHO surface with micro/nano-scale pattern in a wide range of Re number [11]. The results showed a 30% skin-friction reduction in transition flow (10^5<Re<10^6) but at higher Reynolds (10^6<Re<10^7) skin-friction increased. They speculated that the reason is plastron depletion due to high shear flow resulting in direct contact of wall surface roughness with flow. However, Gad-el-Hak [12] in a reply argues that the characteristic length scale of the surface protrusions of this SHS (estimated about 1 μm from SEM images) is considered hydrodynamically smooth and should not cause increase of skin-friction. Haibao *et al.* [13] reported that the intensity of turbulent bursts decreases through hot-film anemometery measurement over a hydrophobic surface with a contact angle of 105.8°.

The present investigation aims at experimental evaluation of turbulent flow over a random patterned SHO.

The experiments are carried out using PIV in a channel flow over a non-wetted SHO surface, a wetted SHO surface and a reference smooth surface to shed light on the turbulent structures.

EXPERIMENTAL SETUP

The experiments are conducted in a closed-loop water channel facility at the Department of Mechanical Engineering of the University of Alberta with a rectangular test section of 680 mm wide, 470 mm deep and 5240 mm long. A 3 m flat-plate model made of cast acrylic with an elliptical leading edge and a sharp trailing-edge is placed parallel to the channel wall to generate a high-aspect ratio channel flow with $W = 50$ mm width as shown in Fig. 1. The measurement region is located 2 m downstream of the leading edge over a replaceable module incorporating the test surfaces.

To investigate the consequences of SHO surface failure and to study the flow over a similar roughness, a SHO surface with a similar texture was wetted using an aqueous solution of a surfactant (wetted SHO surface). In this case, the adhesion force overcomes the solution surface tension yielding in surface wetting. In total three surfaces of smooth, SHO and wetted SHO are applied.

Measurements are at average velocity of $U = 0.154$ m/s (Re = 9,600 based on W). Tripping wires are also installed 20 cm downstream of the leading edge on both channel walls to ensure a spanwise uniform flow. The mean velocity profile is symmetric at the measurement region with maximum velocity of $U_{max} = 0.177$ m/s at the centerline (smooth walls on both sides).

The applied SHO surface comprised of a layer of base micro-particles covered with nanoparticles which are both inherently hydrophobic (Ross Nanotechnology, LLC) [11]. To measure wetting characteristics of none-wetted SHO surface i.e. contact angle (CA) a goniometer system (CAM 100, KSV) was used, however, because of the very low surface energy of the SHO coating the droplet stuck to the dispenser needle (i.e. CA>165°) and made it impossible to measure exact CA. The surface roughness profile of the test plate was measured using Ambios XP-300 high resolution surface profilometer. The resolution of the measurement was set to 0.1 μm. The arithmetic roughness of the surface was found to be 7.9 μm.

A two-component PIV (2C-PIV) consisting of a dual cavity Nd-YAG laser (PIV-400-10, Spectra Phsyics) and an Imager ProX camera with a 14-bit CCD sensor of 2048×2048 px^2 (pixel size 7.4 μm × 7.4 μm) was applied to measure the flow field over a wall-normal streamwise field of view. The water channel has been seeded with 12μm hollow glass spheres (Potters Industries Sphericel® 110P8 Hollow Glass Spheres). The camera is set at a magnification of 0.29 (25.9 μm/pix) using a 105 mm Nikkor objective at f / 5.6 covering the channel width. The mean profiles are obtained by averaging the cross-correlation maps over an ensemble of 10,000 recordings with 12×12 pixels Gaussian windows elongated with 4:1 aspect ratio in the streamwise direction. For measurement of the turbulent intensities from instantaneous velocity vector fields, the images are processed using cross-correlation over square windows of 48×48 pixels and 75% overlap conducted by Davis 8.2.

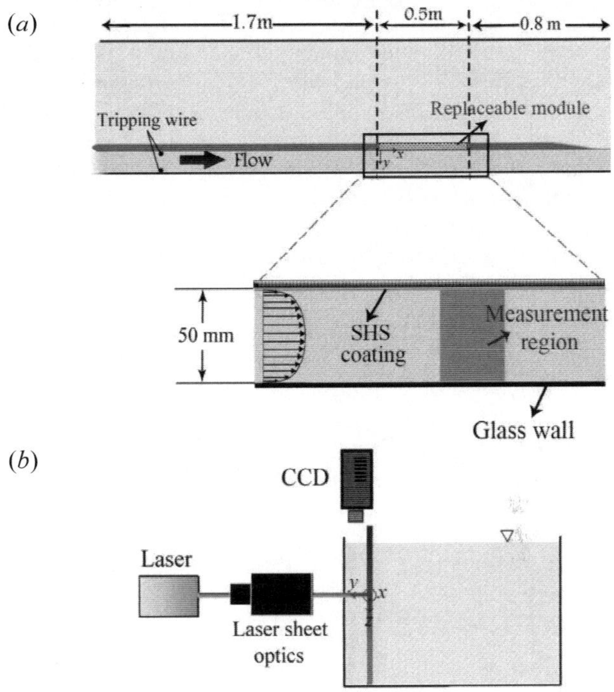

Figure 1. A schematic of the experimental setup showing (a) the top view of the channel formed between the flat plate and the wall and (b) frontal view of the channel and the planar PIV setup.

RESULTS AND DISCUSSION

The normalized mean flow velocity profiles of the turbulent boundary layers over the smooth surface, SHO surface and the wetted SHO surface are presented in in Fig. 2. No considerable velocity difference is observed between the velocity profiles of the three surfaces. The mean velocity profile of the wetted SHO is very close to the smooth wall indicating that the surface might be hydrodynamically smooth. The results are in contrast to the DNS of turbulent flow over SHO surfaces with patterned structures [3, 5] which demonstrate about 30% increase in maximum velocity and an slip velocity in the order of 80% of bulk velocity. Nonetheless, the experiments of Woolford et al. [14] over patterned SHO surface also exhibited no appreciable near-wall difference in velocity profiles which is in agreement with present results.

The magnified view of the normalized mean velocity in the vicinity of the wall is shown in Fig 3. The velocity profiles approximately collapse at $y^+ > 4$. There is no discernable difference in this region as the small variations are within the measurement uncertainty. The slope of the line passing through the center and the data points between $4 < y^+ < 8$ is used to estimate a reference wall shear stress (τ_0). The profiles exhibit approximately identical wall shear stress which might be due to insufficient measurement resolution and accuracy in sublayer.

The smaller velocity profiles of the SHO surface at $y^+ < 4$ relative to the other two surfaces is due to larger biased error of PIV caused by the strong reflection of the laser light. The SHO surface is covered with a thin air layer which acts as a mirror with an irregular surface causing high intensity glare points. The reflections are much weaker in the case of the smooth and wetted SHO as the laser light passes through the surface due to transparency.

The measurement of slip velocity and the possible difference in wall shear stress of the SHO surface would require a PIV experiment with higher spatial resolution (higher magnification and tracer density) along with fluorescent particles to avoid the intense and irregular reflection of the SHO surface.

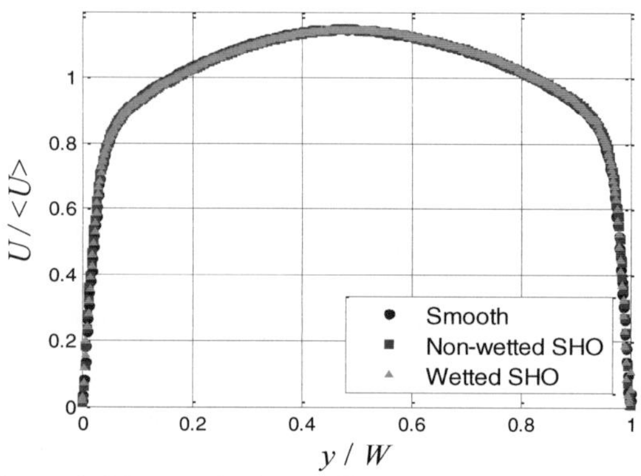

Figure 2. Mean velocity profiles vs wall-normal distance.

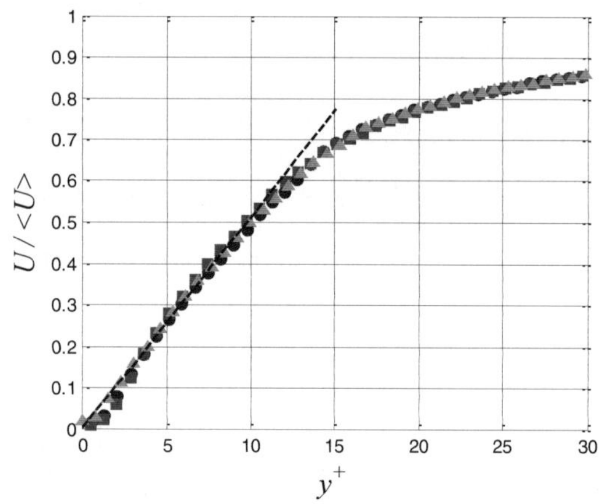

Figure 3. High magnification results of U vs y. Labels are the same as Fig 2.

The normalized profiles of streamwise Reynolds stress $\langle u^2\rangle/u_{\tau 0}^2$ are shown in Figure 4(a). The intensities of the smooth and the wetted SHO surface overlap which indicate that the micro-scale roughness (smaller than the viscous sublayer thickness) acts as hydrodynamically smooth wall with no effect on the flow [15]. It is also evident that the flow over the SHO wall experiences lower streamwise turbulent fluctuations compared to the smooth wall. The wall-normal Reynolds normal stress $\langle v^2\rangle/u_{\tau 0}^2$ profiles are presented in Fig. 4(b) showing a similar behavior to the $\langle u^2\rangle/u_{\tau 0}^2$ profiles. Reynolds shear stress over the smooth and wetted SHO walls show negligible difference whilst it is smaller over the SHO surface as observed in Fig 4 (c). The peak of turbulent intensities is also further away from the wall for the SHO surface implying a thicker viscous sublayer. The reduction of Reynolds stresses for flow over the current SHO surface with random pattern is consistent with the previous experiments [14] and DNS [3, 5] of flow over patterned SHO surfaces.

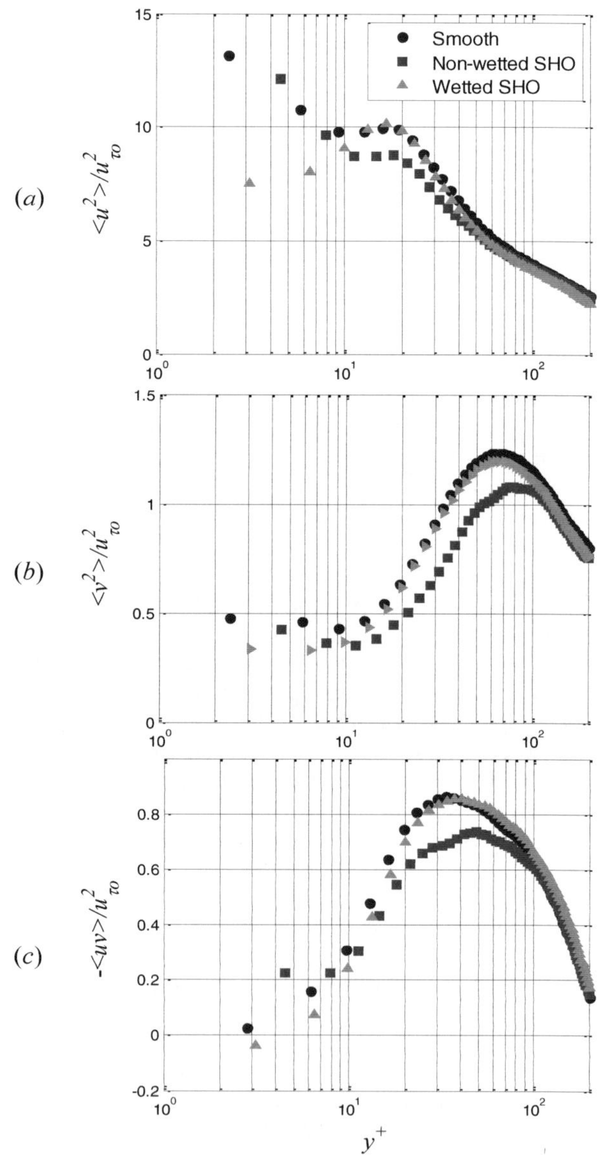

Figure 4. Normalized profiles of normal Reynolds stresses (a) $\langle u^2\rangle/u_{\tau 0}^2$, (b) $\langle v^2\rangle/u_{\tau 0}^2$ and shear Reynolds stress (c) $-\langle uv\rangle/u_{\tau 0}^2$ over the three surfaces.

The highest effect of surface slip is observed on the $<v^2>/u_{\tau0}^2$ which is not evident in the PIV measurement of Woolford et al. [14] over a pattered SHO. The turbulence attenuations over the SHO surface are not limited to the sublayer and buffer layer and are well extends into the log layer with the highest variation at about $20 < y^+ < 50$. In this region, turbulent fluctuations are the strongest and have a growing dominance over the viscous effects.

A significant reduction of the Reynolds stresses and in particular shear stress is observed over the SHO surface which indicates suppression of turbulence generation. This is in contrary to the mean velocity profile, which is almost the same over the three surfaces, except at $y^+ < 4$ which is not accessible by the current measurement. The lack of variation in the mean velocity profile over the SHO surface is associated with the smaller turbulent force ($-\rho\ d<uv>/dy$) which results in a smaller mean flow velocity in the region of $y < y_{<uv>\max}$ where $<uv>_{\max}$ is at about $y^+ = 50$ in Fig 4(c). This effect counteracts the effect of any possible slip velocity which would increase the near-wall velocity.

The reduction of Reynolds stresses suggests that SHO surfaces with random patterns can lead to skin friction drag reduction in a channel flow. This conclusion is in contrast to Aljallis et al. [11] who showed increase of skin-friction in a fully turbulent boundary layer. The results support Gad-el-Hak's [12] dubiety regarding measurement uncertainties and hydrodynamic smoothness of the possible wetted SHO surface in the work of Aljallis et al. [11].

CONCLUSIONS

2C-PIV was employed to measure the turbulent channel flow over a SHO surface with random pattern and a reference smooth surface. To investigate the wetting effects, the fully-wetted SHO surface with the same roughness was also considered. It was observed that the velocity profiles over the three surfaces are identical within the range of measurement uncertainty while Reynolds stresses reduce over the SHO surface. There is no significant change of Reynolds stresses over the wetted SHO in comparison with smooth surface. The effects on turbulent fluctuations are found to extend to logarithmic region and the turbulent intensity peaks happen at a larger wall-normal distance. The largest discrepancy in the intensities pertains to the wall-normal Reynolds stress. The results introduce randomly-patterned SHO surfaces as a promising method for reduction of wall turbulence.

ACKNOWLEDGMENTS

This work has been partially supported by the Natural Sciences and Engineering Research Council of Canada (NSERC RGPIN 1512 GHAEMI).

REFERENCES

1. A. Lafuma and D. Quéré 2003, "Superhydrophobic states.," *Nat. Mater.*, vol. 2, no. 7, pp. 457–60, Jul.
2. J. P. Rothstein, "Slip on Superhydrophobic Surfaces," *Annu. Rev. Fluid Mech.*, vol. 42, no. 1, pp. 89–109, Jan. 2010.
3. M. B. Martell, J. B. Perot, and J. P. Rothstein 2009, "Direct numerical simulations of turbulent flows over superhydrophobic surfaces," *J. Fluid Mech.*, vol. 620, p. 31.
4. T. Min and J. Kim 2004, "Effects of hydrophobic surface on skin-friction drag," Phys. Fluids, vol. 16, no. 7, p. L55.
5. M. B. Martell, J. P. Rothstein, and J. B. Perot 2010, "An analysis of superhydrophobic turbulent drag reduction mechanisms using direct numerical simulation," *Phys. Fluids*, vol. 22, no. 6, p. 065102.
6. T. Verho, C. Bower, P. Andrew, S. Franssila, O. Ikkala, and R. H. a Ras 2011, "Mechanically durable superhydrophobic surfaces.," *Adv. Mater.*, vol. 23, no. 5, pp. 673–8.
7. C. J. Brinker, 2008"Superhydrophobic Coating." Pending Patent–Application # 11/104,917.
8. J. Zhao, X. Du, and X. Shi 2007, "Experimental research on friction-reduction with super-hydrophobic surfaces," *J. Mar. Sci. Appl.*, vol. 6, no. 3, pp. 58–61.
9. Peguero, C., K. Breuer. 2009 "On drag reduction in turbulent channel flow over superhydrophobic surfaces." Advances in Turbulence XII. Springer Berlin Heidelberg, 233-236.
10. R. a. Bidkar, L. Leblanc, A. J. Kulkarni, V. Bahadur, S. L. Ceccio, and M. Perlin 2014, "Skin-friction drag reduction in the turbulent regime using random-textured hydrophobic surfaces," *Phys. Fluids*, vol. 26, no. 8, p. 085108.
11. E. Aljallis, M. A. Sarshar, R. Datla, V. Sikka, A. Jones, and C.-H. Choi 2013, "Experimental study of skin friction drag reduction on superhydrophobic flat plates in high Reynolds number boundary layer flow," *Phys. Fluids*, vol. 25, no. 2, p. 025103.
12. M. Gad-el-Hak 2013, "Comment on 'Experimental study of skin friction drag reduction on superhydrophobic flat plates in high Reynolds number boundary layer flow' [Phys. Fluids 25, 025103 (2013)]," *Phys. Fluids*, vol. 25, no. 7, p. 079101.
13. H. Haibao, D. Peng, Z. Feng, S. Dong, and W. Yang 2015, "Effect of hydrophobicity on turbulent boundary layer under water," *Exp. Therm. Fluid Sci.*, vol. 60, pp. 148–156.
14. B. Woolford, J. Prince, D. Maynes, and B. W. Webb 2009, "Particle image velocimetry characterization of turbulent channel flow with rib patterned superhydrophobic walls," *Phys. Fluids*, 21, p. 085106.
15. M. P. Schultz and K. a. Flack 2007, "The rough-wall turbulent boundary layer from the hydraulically smooth to the fully rough regime," *J. Fluid Mech.*, 580, p. 381.

EFFECT OF ASPECT RATIO ON THE NEAR FIELD DYNAMICS OF FREE SURFACE JET

Anuvrat Mishra
Department of Mechanical Engineering
University of Manitoba
Winnipeg, MB, Canada

Mark F. Tachie
Department of Mechanical Engineering
University of Manitoba
Winnipeg, MB, Canada
mark.tachie@umanitoba.ca

David C.S. Kuhn
Department of Mechanical Engineering
University of Manitoba
Winnipeg, MB, Canada
david.kuhn@umanitoba.ca

ABSTRACT

The interaction of a rectangular turbulent jet with a free surface for three jet-exit aspect ratios is experimentally investigated using Particle Image Velocimetry (PIV). The jet exits from a sharp edged rectangular orifice plate parallel to the free surface and has an equivalent diameter d_e=10 mm. Aspect ratios of 1, 2 & 4 are studied with a fixed offset of 3 d_e from the free surface. Reynolds and Froude numbers based on the bulk velocity are 6900 and 1.27 respectively. Detailed 2-D velocity fields are captured using the PIV in the central x-y plane for $0<x/d_e<23.5$. The center-line velocity decay, velocity profiles and turbulence intensities at several axial location in the central x-y plane are considered to investigate the effect of jet exit aspect ratio on jet – free surface interaction.

KEYWORDS: PIV, Free surface jets, aspect ratio

INTRODUCTION

The presence of a lateral free surface near a jet causes an unequal entrainment on the two sides of the jet. As shown in figure 1 a recirculation region is created in the fluid confined between the jet and the free surface which creates a local low pressure zone and deflects the jet towards the boundary. The jet attaches to the free surface and emerges as a free surface jet. This phenomenon is of particular interest since it has potential engineering applications ranging from remote sensing to the design of water dam spillways so as to mitigate the effects of hydraulic shock.

Walker et al [1] reported the effect of Reynolds and Froude numbers on circular turbulent jets discharged at a fixed offset from the free surface and found the high Reynolds number jets evolve more slowly in the presence of a free surface and at high Froude number violent surface deformation occurs and thus reduces the turbulence kinetic energy. The selective tendency of an offset jet to deflect towards the solid wall or the free surface depending upon the offset from the either boundary was investigated by Tsunoda et al [2] for a plane jet. A circular turbulent jet with a fixed offset of 5 from the free surface was investigated and compared against a free jet by Tian et al [3]. Wen et al [4] reported the dynamic structure of a submerged jet using Proper Orthogonal Decomposition (POD) analysis, they compared flows at 2 different Reynolds number and found the scale of vortical structures is much smaller in high Reynolds number jets. Most of the previous literature have focused on relatively simple configurations of plane or axisymmetric nozzles. There is recent interest in orifice type nozzles owing to its superior mixing properties. The present study focuses on the effect of rectangular orifice plate aspect ratio on jet – free surface interaction.

In the present study submerged turbulent water jets of different aspect ratios are discharged at a finite and fixed depth from the free surface and are investigated experimentally using a PIV system. The jets have a tendency to deflect towards the free surface. The effect of variation of aspect ratio is studied and mean velocity profiles, turbulence intensities and Reynolds shear stresses are reported at various axial locations, covering the deflecting jet/recirculation region, impingement point and the free surface jet region.

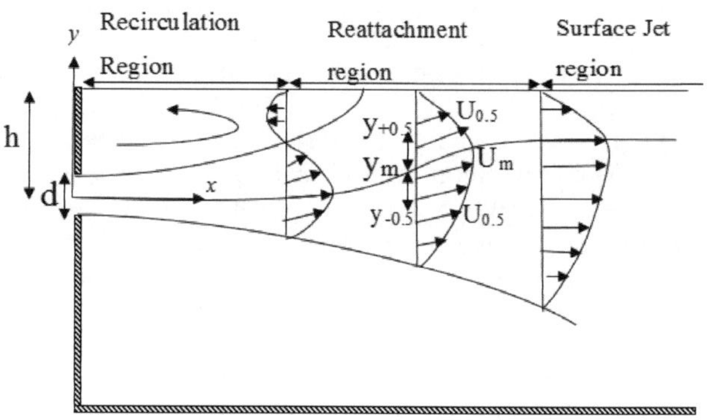

Figure 1: A typical free surface jet

EXPERIMENTAL PROCEDURE

Submerged turbulent water jets of different aspect ratios are discharged at a fixed depth and are investigated experimentally using a PIV system. In the present study, 'h' is the depth of centerline from the free surface, 'd' is the jet width and the aspect ratio is defined as the width to height ratio of the orifice plate opening. The water channel has a width and height of 200 mm and is made of acrylic glass for optical access. Orifice plates are constructed from 3 mm thick acrylic glass with sharp edge rectangular openings and are fixed flush to the vertical entrance wall. A single equivalent orifice diameter of d_e= 10 mm is studied and centered at height from the free surface 'h' = 3 d_e. The orifice aspect ratios studied are 1, 2 and 4. A constant flowrate is set to maintain a Reynolds number, based on the bulk jet velocity, of 6900 and a Froude number, based on the depth of the jet centerline, of 1.27. An equivalent area of all orifices is maintained to ensure a constant momentum flux through all the jets.

Three 2-D velocity fields are captured between the jet exit and x/d_e<23.5 using a high resolution Planar PIV system. Each plane has a field of view of 84.1×84.1 mm, with a 5 mm overlap between the first and second planes. The flow was seeded with 10µm silver coated hollow glass spheres having a specific gravity of 1.4. The flow field was illuminated with a New Wave Solo Nd: YAG double-pulsed laser that emits green light up to a maximum of 120 mJ/pulse at 532 nm wavelength and at 4 Hz repetition rate. The laser sheet was aligned with the mid-span of the channel, and a 12-bit high-resolution charge-couple device (CCD) camera with 2048 pixel×2048 pixel array and 7µm pixel pitch was used to image the flow field. A sample size of 5000 instantaneous image pairs was acquired in each plane and post-processed using the adaptive correlation option of Dynamic Studio. The flow images were post-processed using interrogation window size 32 pixels × 32 pixels with 50% overlap in both the *x* and *y* directions.

RESULTS

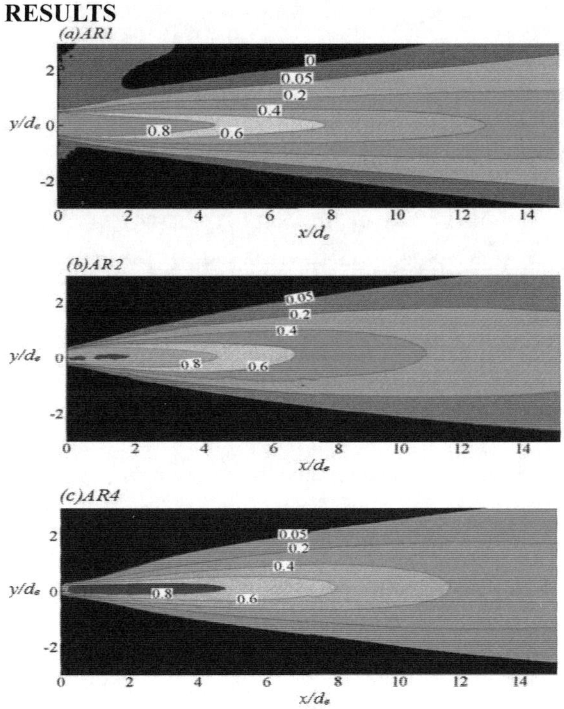

Figure 2: Normalized Mean streamwise velocity contours in *x*-*y* plane

The above figure shows the normalized mean streamwise velocity contours (0<x/d_e<15) in the *x*-*y* plane. The velocity is normalized by the maximum of centerline velocity U_{max} and the axial distance by the equivalent diameter d_e. The effect of the free surface is not pronounced in the near field and the profiles retain a typical Gaussian shape as that of a free jet. As the aspect ratio increases the jet expands less rapidly. The maximum jet exit velocities on the centerline for Aspect Ratio AR2 and AR4 are similar and greater than AR1. In this study the PIV yields dense enough instantaneous velocity fields to characterize higher order turbulence statistics such as Reynolds stresses or even their derivatives vorticity, swirling strength. These multi point statistics give a better understanding of the flow as they give an estimate of spatial and temporal extent to which the turbulence field is correlated.

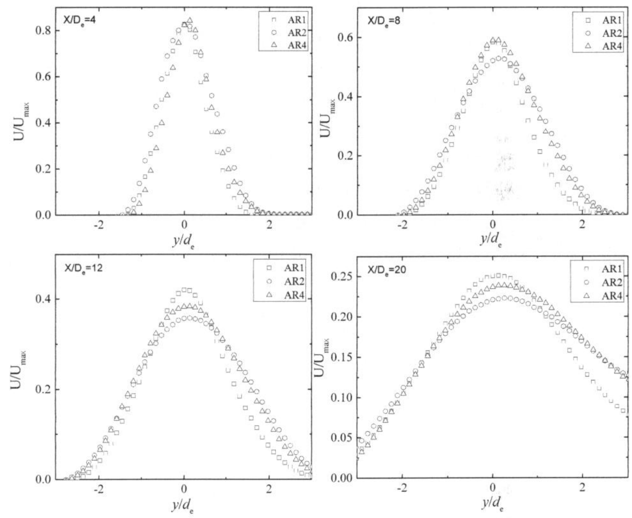

Figure 3: Mean streamwise velocity profiles in *x*-y plane

The mean streamwise velocity profiles at 4 different axial locations (x/d_e=4, 8, 12, 20) are plotted in Figure 3. At x/d_e=4 the profiles are symmetrical about the jet centerline similar to that of a free jet. As the axial distance increases the shear layer is increasing skewed towards the free surface and the jet finally attaches itself to the free surface and evolves as a free surface jet. For aspect ratios of 2 and 4 the jet maximum shifts towards the free surface but not for the case of aspect ratio 1. All the jets impinge the free surface between x/d_e=8 to 12. It can be observed from the plots the higher aspect ratio jets evolve more quickly as a free surface jet.

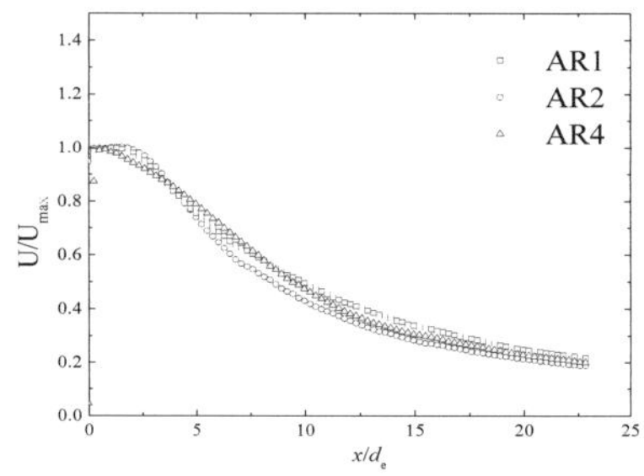

Figure 4: Centerline velocity decay in x-y plane.

The center-line jet velocity decay is shown in the figure 4. The mean streamwise velocity is normalized by U_{max} which is the maximum jet exit velocity on the center-line. The axial distance is non-dimensionalized by the equivalent diameter d_e. It is observed that the velocity decay rate of aspect ratio 2 is higher than for aspect ratio 1 in the region of $x/d_e = 3$ to 17. At the initial stage of jet development ($x/d_e<2.25$) the center-line velocity decay for the aspect ratio 4 is the highest. This can be attributed to the slender vortical structures near the jet exit plane in the aspect ratio 4 case, which interact strongly with the potential core and dissipate its momentum in the near field by entrainment of the ambient fluid thus reducing its length. Similar observations are reported by Quinn et al [5] for sharp edged rectangular free jets. In the far field ($x/d_e>12$) the mean stream wise velocity decays at a still higher rate for aspect ratio 2 while the decay rate for aspect ratio 4 approaches to that of aspect ratio 1. Similar results are reported for an elliptical free jet aspect ratio study by Lee et al [6]. In the 'recirculation region' (figure 1) the interaction with free surface is minimal and thus various analogies are seen with aspect ratio studies of free jets, Tian et al [3] showed the jet retains a profile closely similar to a free jet in this region.

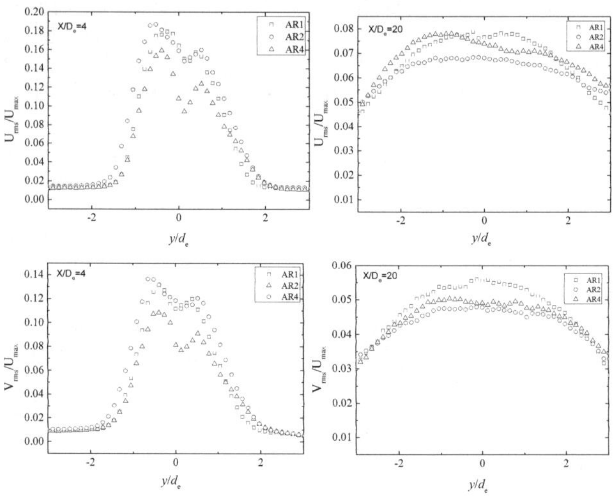

Figure 5: Streamwise and transverse turbulence intensities in the near and far field of the jet.

The streamwise and transverse turbulence profiles, figure 5, in the near field ($x/d_e<5$) have a typical double peak similar to observed in a free jet. At further downstream locations the peaks decay to form a broader peak indicative of increased turbulence penetration of the jet by the entrained flow. It is interesting to note that the magnitude of turbulence intensities for aspect ratio 2 is highest in the near field region but falls below that of aspect ratio 4 in the far field. This is similar to observations reported by Lee et al [6] for an elliptic jet study of aspect ratios of 1, 2, 4 & 8. The elliptic jet of aspect ratio 2 was reported to have the most intense turbulence characteristics in the near field region.

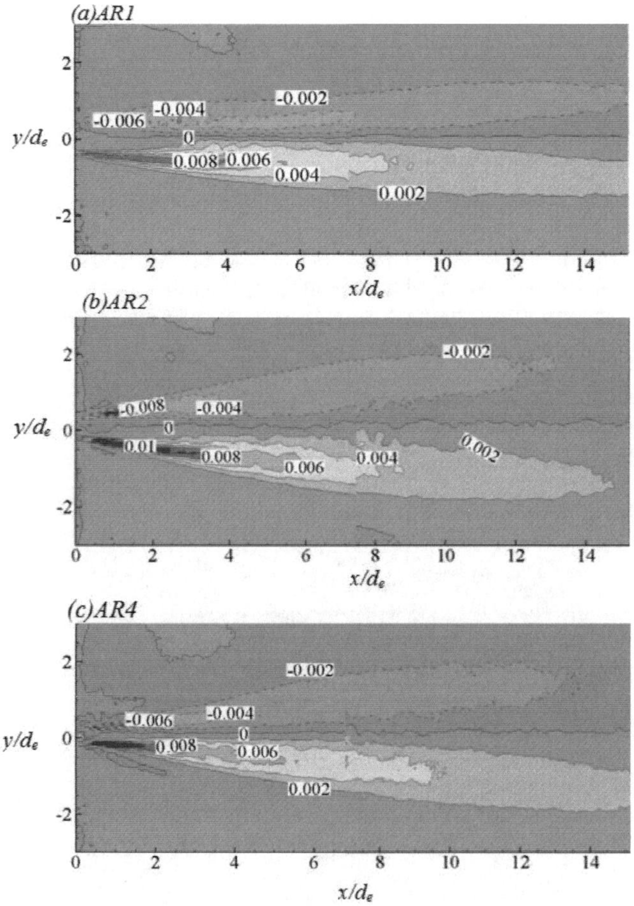

Figure 6: Streamwise development of normalized Reynolds shear stress.

Figure 6 shows the contour plots of normalized streamwise Reynolds shear stress normalized by the square of maximum centerline velocity. The peaks of the shear stresses are located in the shear layer regions but are not symmetric about the jet centerline in contrast to a free jet [3]. The smaller magnitude of shear stress near the free surface is attributed to the proximity of the free surface which diminishes the surface normal turbulent intensities [1] [3]. The shear stress decreases in the downstream direction as the jet expands.

The impingement point is the point where the offset jet finally impinges and attaches to the free boundary (air-water interface) and emerges as a free surface jet. The impingement point is calculated as the axial location where the mean streamwise velocity changes sign along at the free surface. It was found that aspect ratio 1 had the longest reattachment length of $10.87d_e$, while it is $10.48d_e$ and $9.69d_e$ for aspect ratios 2 & 4. The jet of aspect ratio 1 is expected to have the longest reattachment length since, despite having sharp corners, the orifice is most comparable to an axisymmetric nozzle. As the aspect ratio increases the rate of mixing increases in the near field [5] and thus the jet experiences a higher entrainment from the surrounding fluid and reattaches to the free surface earlier.

CONCLUSIONS

Mean flow in the near field is modestly modified by the presence of the free surface and the normalized mean streamwise velocity profiles retain shapes similar to that of a free jet. The centerline velocity decay and turbulence intensity results indicate the jet of aspect ratio 2 to have the greatest mixing in the near field region. The length of potential core for the jet of aspect ratio 4 is the smallest which can be attributed to needle like vortices which strongly interact with the potential core and dissipate its momentum to the adjacent shear layers. As the axial distance progresses the outer and inner shear layers skew progressively with a continuous shift of the maximum velocity to the free surface. The Reynolds shear stresses are suppressed in the outer shear layer due to the vertical confinement.

ACKNOWLEDGMENTS

This work was supported by research grants from National Science and Engineering Council of Canada.

REFERENCES

1. Turbulent structure in free-surface jet flows: D.T.Walker, C.-Y.Chen, W.W.Willmarth *Journal of Fluid Mechanics 1995, vol. 291, pp. 223-261*

2. Plane offset jet discharged into water of finite depth: Hiroyuki Tsunoda, Yoshihito Shimizu, Takeshi Kashiwagi *JSME International Journal 2006 Series B, vol. 49, No.4, 2006*

3. Characteristics of a jet in the vicinity of a free surface: Jiahao Tian, Vesselina Roussinova, Ram Balachandar. *Journal of Fluids Engineering 2012, vol. 134*

4. Dynamic structures of a submerged jet interacting with the free surface: Qian Wen, Hyun Dong Kim, Ying Zheng Liu, Kyun Chun Kim. *Experimental thermal and fluid science 2014 , 57(2014)396-406*

5. Turbulent Free Jet Flows Issuing from Sharp-Edged Rectangular Slots: The Influence of Slot Aspect Ratio W.R. Quinn. *Experimental thermal and fluid science 1992; 5:203-215*

6. The effect of aspect ratio on the near field turbulent structure of elliptic jets: Sang-Joon Lee, Seung-Jo Baek. *Flow Meas. Intrum. 1994, volume 5 Number 3*

Turbulence & Vortex Dynamics

Proceedings of the 25th CANCAM
London, Ontario, Canada, May 31 – June 4, 2015

Suppression of Vortex Shedding for Flow Past around a Circular Cylinder using Splitter Plate

Abdelkrem Eltaggaz
School of Engineering
University of Guelph
Guelph, Ontario, Canada
Email: aeltagga@uoguelph.ca

ABSTRACT

The vortex shedding phenomena usually occurs in flow induced vibration of bluff bodies. It consists typically of a pattern of swirling vortices in alternating directions. This paper presents an empirical study for the suppression of vortices shedding behind a circular cylinder. Experiments were conducted in a small duct using the passive control approach of vortex shedding by fixing a splitter plate behind a circular cylinder. This study investigated two cases: one with and one without splitter plate. The flow velocity range is 0.2 to 3.2 *m/sec* ($2200< R_e <12600$). The equation of motion is formulated using a mass spring system. The experimental and mathematical results gave the same critical speed (U_{VIV}= 1.1 *m/sec* at S_t= 1.8) of vortex-induced vibration (VIV). In addition, the effect of an attached the splitter plate on disappearance of vortex shedding and variation of Strouhal number are analyzed. Furthermore, the flows in both cases are simulated numerically using FLUENT software package.

KEYWORDS *flow induced vibration, vortex shedding, circular cylinder, splitter plate, FLUENT.*

INTRODUCTION

In fluid mechanics the flow around a bluff body has been an interesting topic of research with big challenges for engineers [3]. These bodies have been used widely in numerous engineering applications such as heat exchangers, marine technology, and buildings. In this technology there are some undesired phenomena which appear during a cross flow around bodies. The most common and significant phenomena is called vortex shedding. This phenomena appears behind bluff bodies in the wake region and it has serious effects on the body's structure. This phenomena has been studied systematically since Leonardo da Vinci's time (15 century) to Strouhal's time (1878) [7]. Recently, the vortex shedding suppression has been considered a significant problem because it can promote harmful effects such as excessive vibration, noise, and fatigue damage to the structure [9]. Therefore, many scientists have been investigating methods to suppress this phenomena. Examples of experimental testing to control vortex shedding are presented by Gad-el-Hak [11, 12] and Modi [13]. One of the passive methods of suppression vertex shedding that has been considered the most successful, is attaching a splitter plate behind a cylinder. This method was used in many experimental and numerical studies that mostly focused on the influence of the splitter plate on the vortex shedding behind a circular cylinder [9]. Roshko (1954), was the first to use a splitter plate in different positions behind a circular cylinder to suppress vortex shedding [9]. Aplet, West and Szewezyk (1973) performed experiments on a circular cylinder with a splitter plate behind at a high Reynolds number 104< Re < 5. 105. They investigated the effect of splitter plate length on vortex shedding, Strouhal number and form drag. In addition Kwon and Choi (1995) reported on the impact of splitter plates on vortex shedding of cylinders. Despite this have been many attempts to fix the splitter plate behind a circle cylinder. Few studies have used this technology to suppress vortex shedding behind different cross sections such as square and rectangular. Ozono (1999) investigated numerical flow control of vortex shedding around a rectangular cylinder using a short splitter plate. Although many researchers have investigated experimentally the effect of the splitter plate on the vortex shedding behind bluff bodies, the understanding of the flow mechanism of vortex shedding is still complex [2].

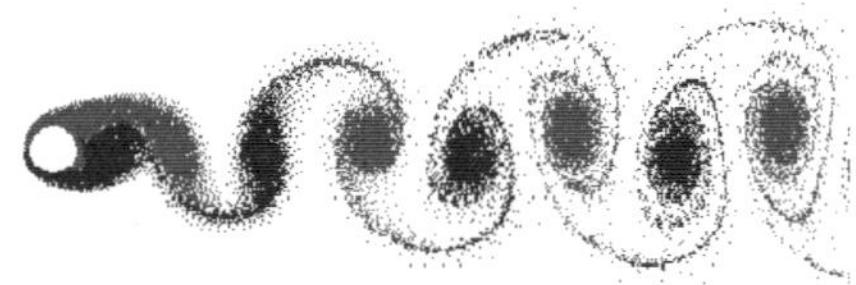

Figure 1: Vortices shedding behind a circular cylinder.

This paper presents a passive control method for suppress vortex shedding behind cylinders experimentally. The wake splitter plate was fitted behind a cylinder to stop the interaction between two separate shear layers. This means reducing drag amplitude and the negative effect of vortex shedding as well. The vibratory system was represented by a single mass connected to a spring. Therefore, the equation of motion that was used is for single degree of freedom (SDOF). The experiment started by testing the flow behind a cylinder without a splitter plate at various velocities. Next, splitter plate was connected behind the model and the experiment was conducted at the same velocities. From the comparison of the two flow fields we utilized the results to investigate the effect of a splitter plate on creating vortex shedding, Strouhal number, and drag coefficient. For both cases the numerical analysis was carried out for turbulent viscosity, pressure distribution, and velocity magnitude by using FLUENT software

I. THEORY MATHEMATICAL MODEL:

This part shows the theory of vortex induced vibration (VIV) for elastic cylindrical model with one degree of freedom in a mass-spring system.

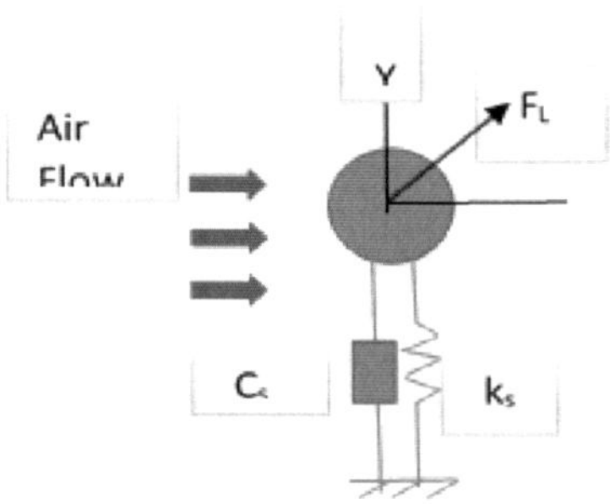

Figure 2: Mass spring model.

The test model was considered as a simple mass spring and dashpot system. The equation of motion of the flexibly mounted cylinder is [1]:

$$m_s\ddot{y} + c_s\dot{y} + k_s y = F_{vs}(t) \quad (1)$$

We can rewrite the equation,

$$\ddot{y} + 2\zeta_s\omega_n\dot{y} + k_s y = \frac{F_{vs}(t)}{m_s} \quad (2)$$

Where:

$$F_{vs} = \frac{1}{2}\rho U^2 D C_L \sin(2\pi\ f_v\ t) \quad (3)$$

And, $f_v = \frac{U S_t}{D}$,and $\omega_n = \sqrt{\frac{k}{m+m_a}}$, $f_n = 3.2609$ Hz experimentally.

The Strouhal number is assumed [8]: $St = 0.18$ (see appendix: Chart Strouhal No. with Reynolds No.)

Hence, the equation of motion becomes:

$$\ddot{y} + 2\zeta_s\omega_n\dot{y} + k_s y = \frac{1}{2\text{m}}\rho U^2 D C_L \sin(2\pi\ f_v\ t) \quad (4)$$

The standard solution of this equation is:

$$y(t) = A\sin(2\pi f_v t + \emptyset) \quad (5)$$

Hence, A is the response amplitude [6],

$$A = \frac{\rho D c_L U^2}{2m\omega_n}\frac{1}{[(1+r^2)^2 + (2\xi r^2)^2]^{0.5}} \quad (6)$$

Where, $= \frac{f_v}{f_n}$, and $\phi = tan^{-1}\left(\frac{2\xi r}{1-r^2}\right) \quad (7)$

II. EXPERIMENTAL SETUP

The model test consisted of a circle cylinder mounted on a thin AISI 1045 steel. The dimensions of the cylinder are: diameter = 0.0618 *m*, length = 0.32 *m*, and total mass=0.01945 kg. The model was fixed inside a square section wind tunnel (0.82 *m* x 0.82 *m* x 1.2 *m*) and one meter in front of a fan. A flow meter was used to determine airflow velocity. A laser sensor was installed close to steel plate base. An electric fan faced the model in order to get various flow velocities, as shown in the figure below.

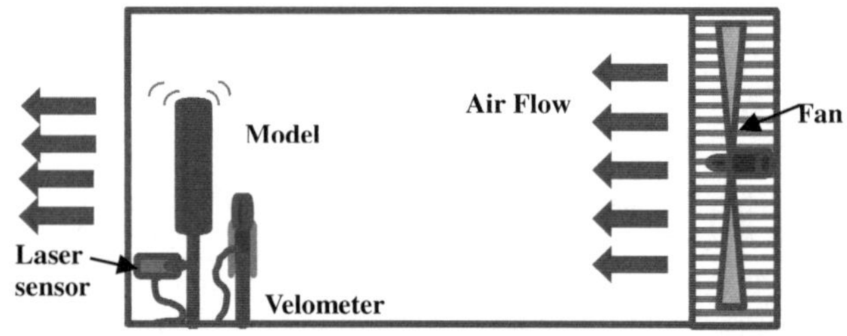

(a)

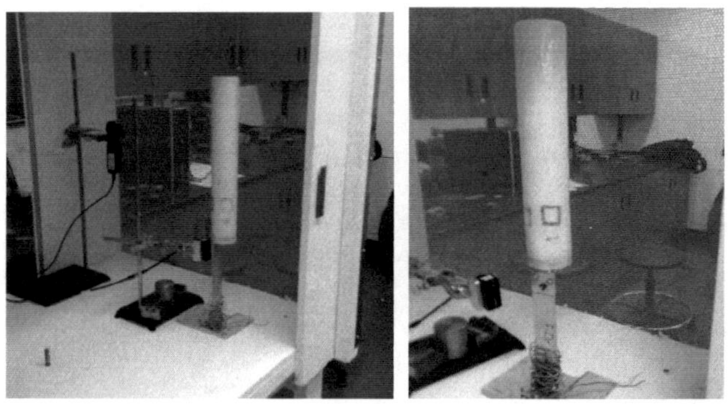

(b)

Figure 3: (a) - Experimental setup. (b)- The model stands inside the duct.

III. EXPERIMENTAL WORK

The natural frequency of the model without a splitter was estimated from the average of four pluck test readings (f_n = 3.2609 Hz). An experiment was then performed on a cylinder model without a splitter at first. The upstream face cylinder was located a distance of 1*m* from the fan. The next step, was to start the fan until flow meter reading reached 0.5 *m/sec*. Then we waited between 20 to 30 seconds in order to get steady flow rate. After that the reading was recorded and saved. The procedure was repeated at various flow velocities from 0.5 to 3.2 *m/sec*. The natural frequency was calculated for the model with splitter plate attached f_n = 2.91 Hz. While an experiment was conducted under the same conditions, the cylindrical model was fixed in the same position inside the duct.

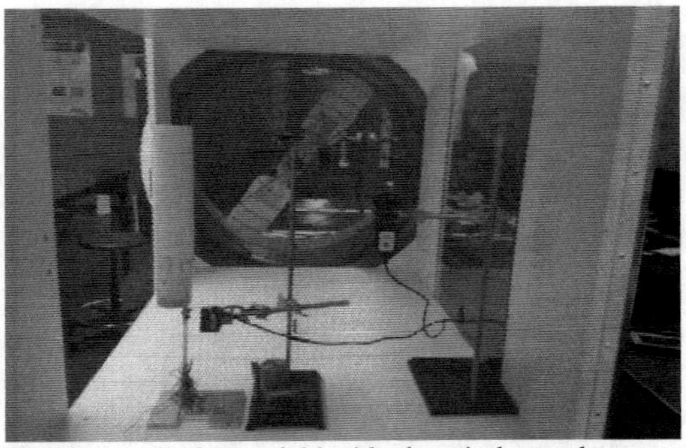

Figure 4: The circle cylinder model inside the wind tunnel.

IV. NUMERICAL SIMULATION

Fluent was used for the numerical analysis. Analysis of two cases was carried out to compare to experimental results. A cylinder diameter of 0.0618*m* and length 0.32 *m* with and without a splitter plate (0.02 *m*) passed through laminar airflow ranging from 0.2 to 3.5 *m/sec*. Simulation plots of velocity magnitude, pressure distribution, and turbulent viscosity around the cylinder for both cases were plotted and compared to experimental results.

V. RESULTS AND DISCUSSION

The results showed the critical speed for vortex induced vibration (VIV), for the model without splitter plate, is $U_{VIV} = \frac{f_n D}{S_t} = 1.1\ \mathrm{m/sec}$, at St=0.18 and Re= 4.729x103. That means the expected lock-in zone phenomenon has to be between 0.88*m/sec* to 1.32*m/sec.* Figure (4) illustrates clearly that the lock-in zone occurred from 0.8 *m/sec* to 1.2 *m/sec* which proves the mathematical calculation of U_{VIV}. Moreover, figures 5-8 clearly show the significant attenuation in amplitude (approximately 50%) due to the connected splitter plate in the wake region. Furthermore, it eliminates the lock-in zone phenomenon. The result also supports the importance of the flow control behind the cylinder particularly in the wake region and makes the bluff body more stable.

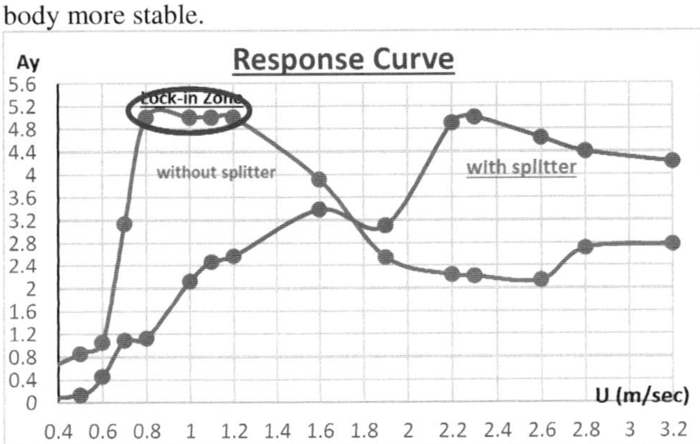

Figure 5: Response curve between amplitude and flow velocity.

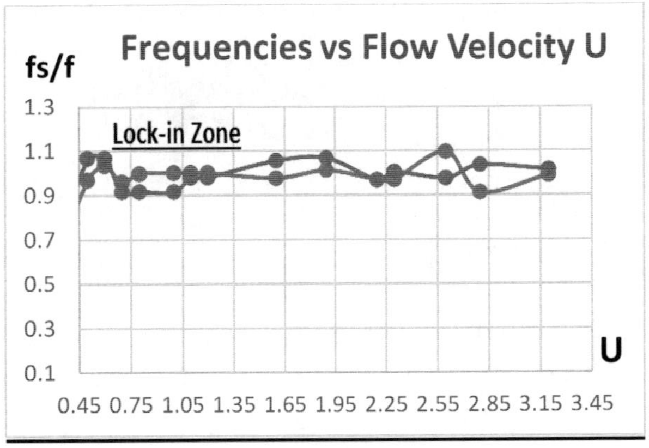

Figure 6: Variation of frequency ratio with flow velocity.

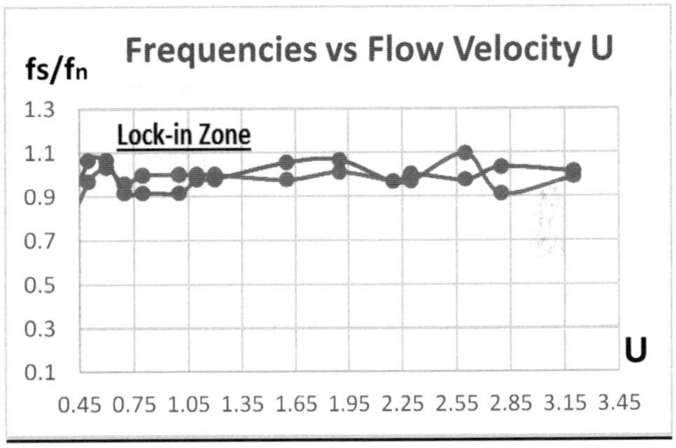

Figure 6: Variation of frequency ratio with flow velocity.

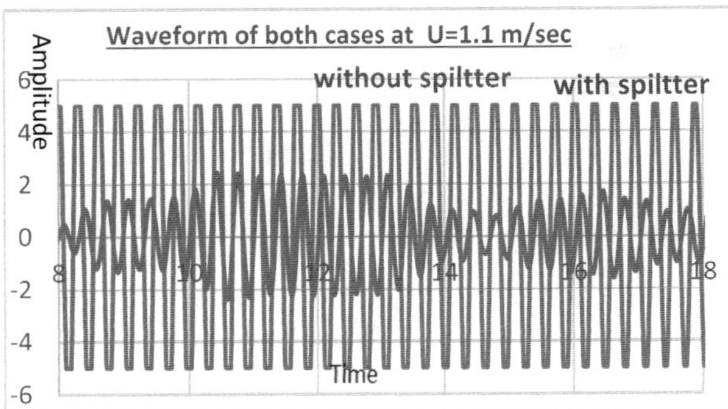

Figure 7: Response amplitude comparison (attention 50%) between two cases at critical speed of vortex-induced vibration of case one.

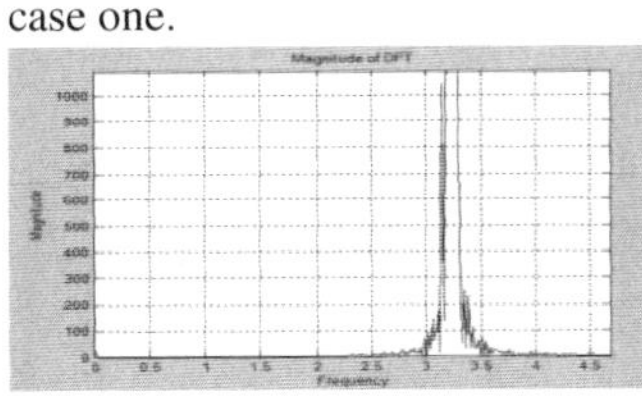

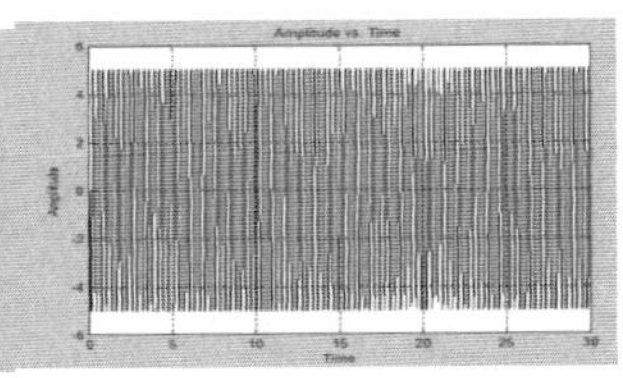

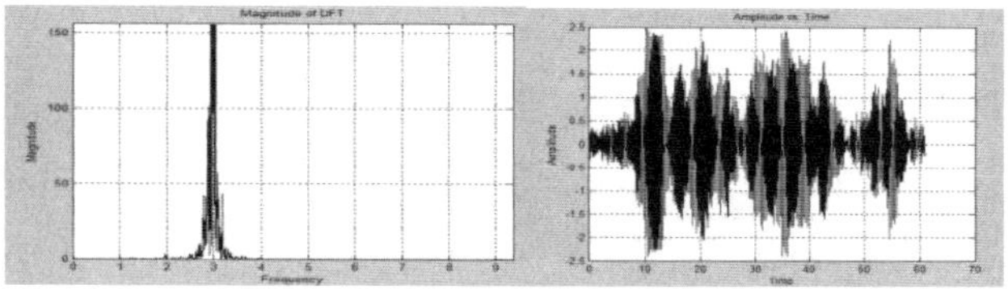

Figure 8: Power spectrum and waveform of model response with and without splitter plate at flow velocity =1.1 m/sec.

Adding the splitter plate to the cylinder model led to decreasing the RMS values by 60% as shown in figure (9). That means the splitter plate has a great effect on supersession of vortex shedding. When flow velocity reaches to 1.8 m/sec, the RMS curve of cylinder with the plate overtakes the other curve due to the smallness of the model diameter. In addition, insertion of a splitter behind the cylinder model shows a slight effect in variation of the Strouhal number with Reynolds number (Fig.10). This is because vortex shedding frequencies values are close to each other.

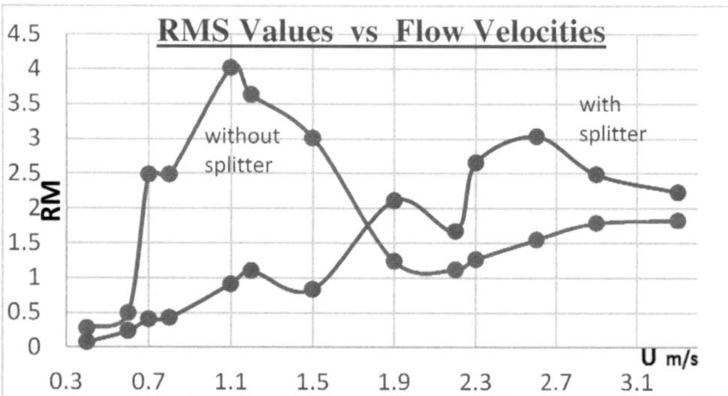

Figure 10: Variation of RMS value with flow velocity.

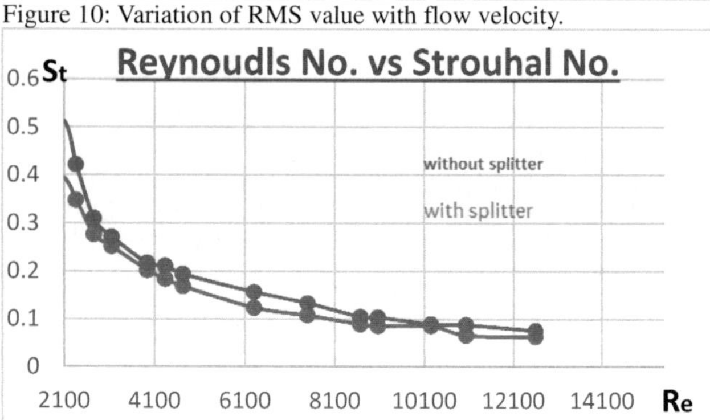

Figure 11: Variation of the Strouhal number with Reynolds number.

On the other hand, the Figure (11) shows the simulation result obtained from fluent software. Figure (11-a) describes pressure contours of air flow past a cylinder with and without a splitter plate. The pressure contours appeared around the model with a splitter more density than the other model. This result shows that putting a splitter in the wake region is helpful to reduce a negative pressure gradient effect.

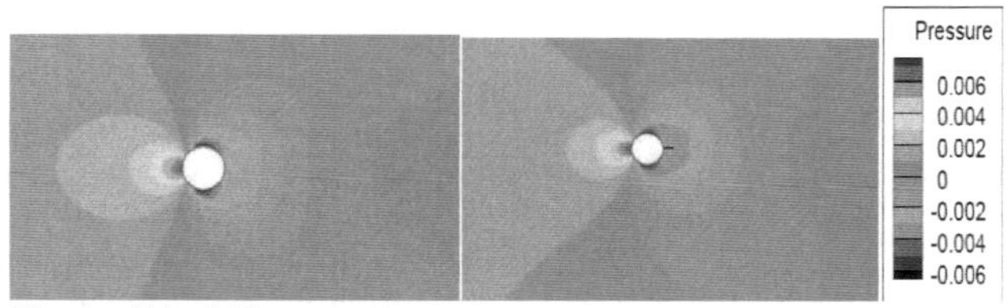

Figure (11-a) Pressure contours

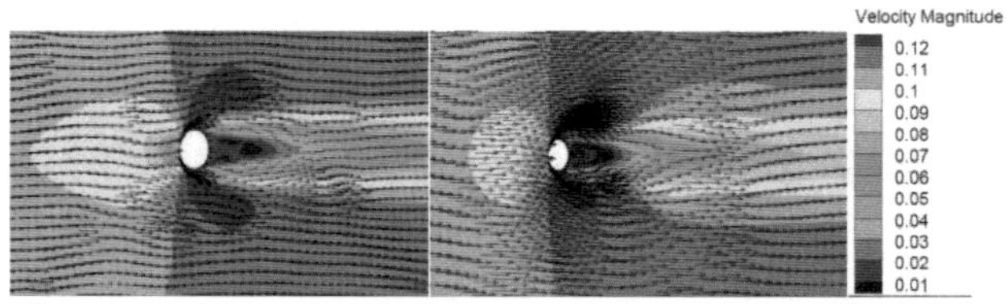

Figure (11-b) Stream traces of model with and without splitter.

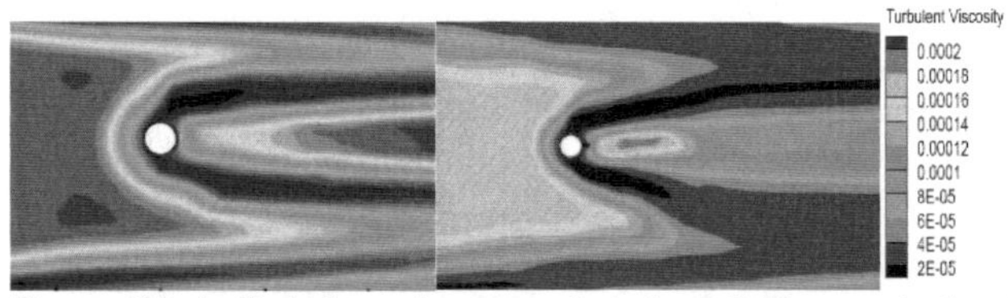

Figure (11-c). Turbulent viscosity of circle cylinder attached and detached splitter plate.

Theses figures show that flow behaviour is different in the wake region behind the cylinder for two cases. Its clearly seen that when the splitter plate is placed in the wake region, the trubelunate flow area becomes small. It was also observed that flow is more attached in upstream and downstream for a splitter plate case.

VI. CONCLUSION

This work introduced experimental and numerical investigation to test passive control of air flow using a splitter plate behind a cylinder, inside a small wind tunnel duct. The results gained in this research can be summarized as follows:

- When the critical speed for vortex induced vibration (VIV) is around 1.1 m/sec, the bluff body oscillates close to its natural frequency.
- Passive control of flow by a fixed splitter plate in the wake region reduces the effect of vortex shedding.
- The simulation result illustrated in figure 11-c shows different flow behaviors, when splitter plate is placed behind the cylinder. Thus, the turbulent flow area shrank, as well as the negative pressure gradient effect was decreased.

REFERENCES

1. Dr. Marwan Hassan, 2014. "Flow Induced Vibration". Lecture No.5, School of Engineering, University of Guelph. Guelph, Ontario, Canada.
2. K. Kwon and H. Choi, 1995. "Control of Laminar Vortex Shedding behind a Circular Cylinder using Splitter Plates". Department of Mechanical Engineering, Seoul National University, Seoul 151-742, Korea.
3. K. W. Chang & J.H. Chen, 2014. "Vortex Shedding and its Suppression for a Shear Flow Past a Circular Cylinder ", 11th. World Congress on Computational Mechanics (WCCM XI, Barcelona, Spain.
4. M.T. Asyikin, 2012. "CFD Simulation of Vortex Induced Vibration of a Cylindrical Structure", Department of Civil and Transport Engineering, Norwegian University of Science and Technology.
5. Mohammed Hosny Eid, 2004."Vortex Shedding from Single and Tandem Finned Cylinders". Master Degree Thesis, McMaster University, Ontario, Canada.
6. P. Badami, V. Shrivastava, Saravanan V., N. Hiremath, K.N. Seetharamu, 2012. "Numerical Analysis of Flow past Circular Cylinder with Triangular and Rectangular Wake Splitteer". World Academy os Science, Engineering and Technology. Vol: 62012-09-20.
7. P. J. Strykowski & K. R. Sreenivasan, 1990. "On the Formation and Suppression of Vortex Shedding at Low Reynolds Number", J. Fluid Mechanical, volume, 218, pp. 71-107. Great Britain.
8. Prof. A. H. Techet, 2005. "Vortex Induced Vibration". 13.42 lecture.
9. Said Turki, 2008. "Numerical Simulation of Passive Control on Vortex Shedding behind Square Cylinder using Splitter Plate". Journal of Engineering Applications of Computational Fluid Mechanics. Vol. 2, No.4, pp. 514-524.
10. Gad-el-Hak M. "Modern developments in Flow Control". Applied Mechanics Review, 1996; 49:365-379
11. Gad-el-Hak M. "Passive, Active and Reactive Flow Management". Cambridge University. Press: Cambridge, U.K., 2000; 421.

Appendix:

Strouhal Number vs. Reynolds Number

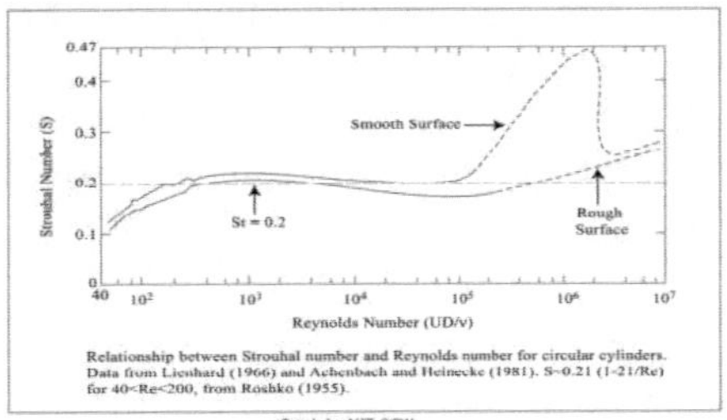

Graph by MIT OCW.

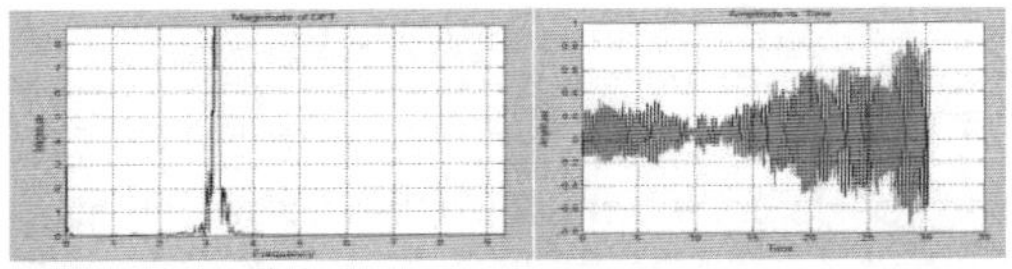

without (U=0.5 m/sec)

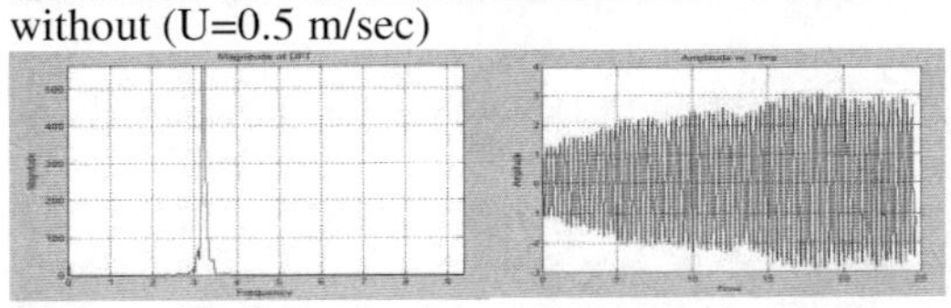

Without (U=0.7 m/sec)

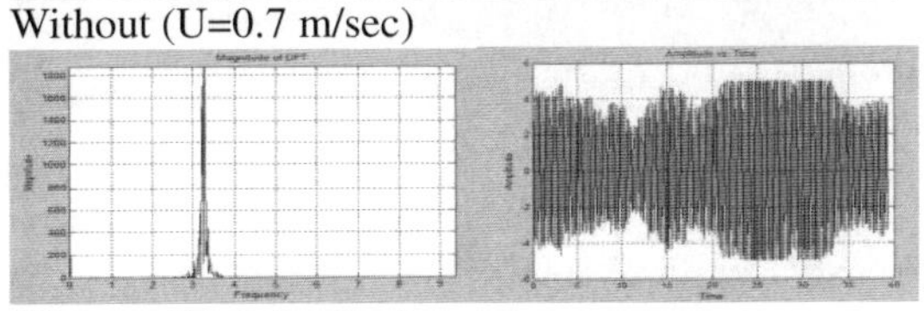

Without (U=1.6 m/sec)

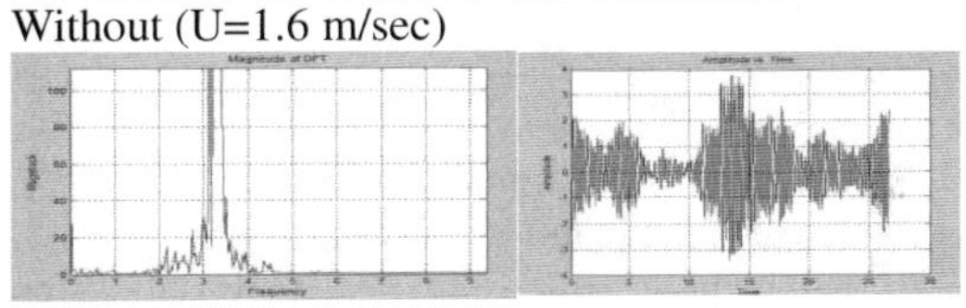

Without (U=2.2 m/sec)

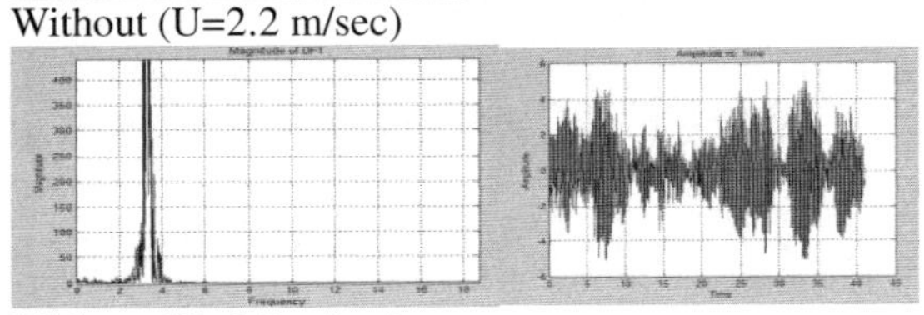

Without (U=3.2 m/sec)

QUASI-STEADY HYPOTHESIS AND CONDUCTOR GALLOPING

Karine Lavertu
Mechanical Engineering Department
Université de Sherbrooke
Sherbrooke, Québec, Canada
Karine.Lavertu@usherbrooke.ca

André Laneville
Mechanical Engineering Department
Université de Sherbrooke
Sherbrooke, Québec, Canada
Andre.Laneville@usherbrooke.ca

Hachimi Fellouah
Mechanical Engineering Department
Université de Sherbrooke
Sherbrooke, Québec, Canada
Hachimi.Fellouah@usherbrooke.ca

ABSTRACT

The goal of this research project is to find the application limits of the quasi-steady hypothesis used in the analysis of the galloping phenomenon. This paper presents the results from static tests, which consist of measuring, in a wind tunnel, the steady force coefficients on a D-section cylinder, representing an iced conductor. Strouhal numbers have also been calculated at different angles of attack. The effect of the azimuth angle was studied by conducting static tests where the conductor was inclined within the plane parallel to the flow. Finally, dynamic tests were conducted where the conductor was forced to oscillate in rotation. Rotational speed and amplitude of forced oscillations were varied to observe the effect of the reduced frequency on unsteady lift, drag and moment coefficients.

KEYWORDS: galloping, D-section, quasi-steady hypothesis, conductor, force coefficients, wind tunnel.

INTRODUCTION

The wind, by its intensity, direction and unsteady content of turbulence, may cause structural damage on overhead transmission lines that are subject to wind-induced motion such as conductor galloping.

Galloping arises from aerodynamic and/or aeroelastic instability. This phenomenon occurs when the lift, resulting from the asymmetry of the cross-section, is modulated by the periodic motion of the conductor. Small-amplitude oscillations generate forces which increase the amplitudes to large values [Parkinson, 1971]. In the case of transmission lines, galloping usually occurs when ice adheres to the conductors.

The analysis of the galloping phenomenon and, as a result, the design of countermeasures suitable for transmission lines, are usually based on the quasi-steady hypothesis. For a conductor moving downward in its oscillation cycle at a transverse velocity ($\dot{y}$), in a flow of speed (V), this hypothesis assumes that its cross-section experiences the same instantaneous aerodynamic force that it would if held stationary in a flow of speed $V_{rel} = \sqrt{\dot{y}^2 + V^2}$ coming upward at an angle of attack $\alpha = \tan^{-1}(\dot{y}/V)$[Laneville and Parkinson, 1971].

Applying aerodynamic coefficients measured on a stationary model to evaluate the loads on an oscillating conductor is an approach for which the limits of its validity have not been well established. Consequently, the goal of this research project is to determine experimentally these limits in the case of a D-section cylinder, a shape frequently adopted to simulate the galloping of conductors. The project includes several phases and this paper deals with both the static and dynamic tests.

EXPERIMENTAL SET-UP AND METHODOLOGY

The experimental tests were carried out in the wind tunnel of the Université de Sherbrooke (Sherbrooke, Quebec). The conductor was modeled using a circular steel tube on which a D-section was fixed to simulate the ice effects. The aerodynamic forces and moment were captured through a "floating" portion of the D-section that was only attached to an internal balance (Figure 1). This "floating" section was built from foam and balsa (166.7 g) to minimize the effects of inertia during the dynamic tests. Static tests were conducted before proceeding to dynamic tests where the conductor was forced to oscillate in rotation. The effect of the azimuth angle was also studied by inclining the conductor. More details on the experimental methodology were presented in [Lavertu et al., 2013].

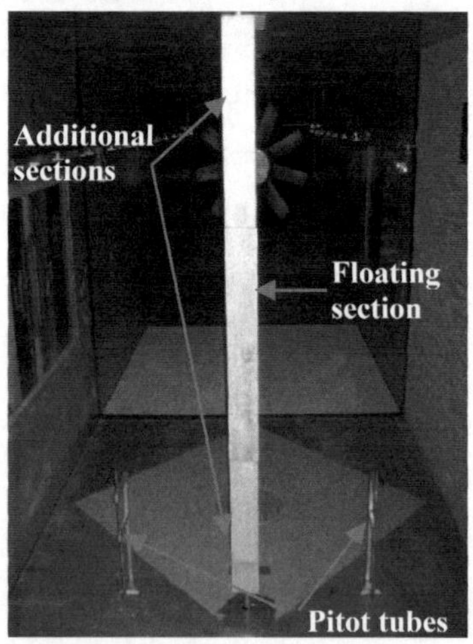

Figure 1: Floating section in foam and balsa

STEADY FORCE AND MOMENT COEFFICIENTS

Figure 2 shows the measured steady forces and moment coefficients. The axis of the model was in a vertical plane parallel to the flow. The reference lengths and velocity used to calculate these coefficients are respectively the D-section diameter, the length of the floating section and the velocity measured halfway between the model and the tunnel wall. The results are consistent with those of previous studies [Cheers, 1950] [Chadha and Jaster, 1975]. At an angle of attack of 0 degrees, the flow separates symmetrically from the two corners of the D-section. At this angle, the drag is maximal and the lift is zero because of the symmetrical pressure distribution. In the range 40° < α < 95°, these coefficients vary substantially, especially the moment coefficient. The lift coefficient reaches its maximum just before the reattachment of the shear layer and starts to decrease as the reattachment point moves forward.

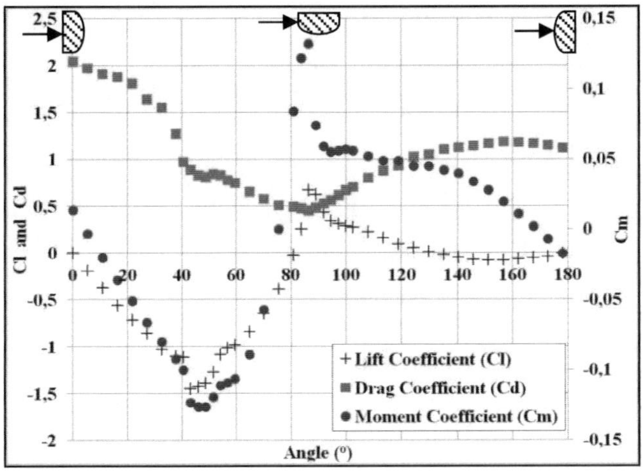

Figure 2: Lift, drag and moment coefficients for a vertical D-section as a function of the angle of attack (Re = 39 387)

EFFECT OF THE AZIMUTH ANGLE

The azimuth angle has the effect of changing the shape of the section seen by the flow. Few papers have presented results on an inclined D-section. Nakamura and Tomonari (1981) studied the effect of the aspect ratio of the D-section on the pressure distribution. Pierre Van Dyke conducted galloping measurements on a high-voltage overhead test line equipped with a configuration of three conductors with a D-section and with interphase spacers. One of the results of his study is that wind azimuth may strongly influence the galloping amplitude. Maximum amplitudes were attained at angles close to 50 degrees from perpendicular to the conductor and decreased to a minimum when the wind was perpendicular or parallel to the conductor [Van Dyke and Laneville, 2008].

To study this effect, the D-section was inclined at an angle of 40 degrees with respect to the horizontal (50 degrees from perpendicular to the conductor).

Figure 3: Inclined D-section

Forces of gravity that the "floating" portion of the D-section applied to the internal balance were first measured at all angles with no wind. They were subtracted from the total forces measured with wind to obtain the aerodynamic forces. Figure 4 shows the measured steady forces and moment coefficients. The reference length and velocity used to calculate these coefficients are respectively the D-section diameter and the velocity, parallel to the wind tunnel, measured halfway between the model and the tunnel wall. The other reference length was the length of the floating section for the calculation of the lift and moment coefficients, and the length of the floating section divided by sin 40 (projected length) for the drag coefficient. In the range 40° < α < 95°, steady forces and moment coefficients vary more smoothly than in the vertical case. Also, the maximum drag, lift and moment coefficients are smaller.

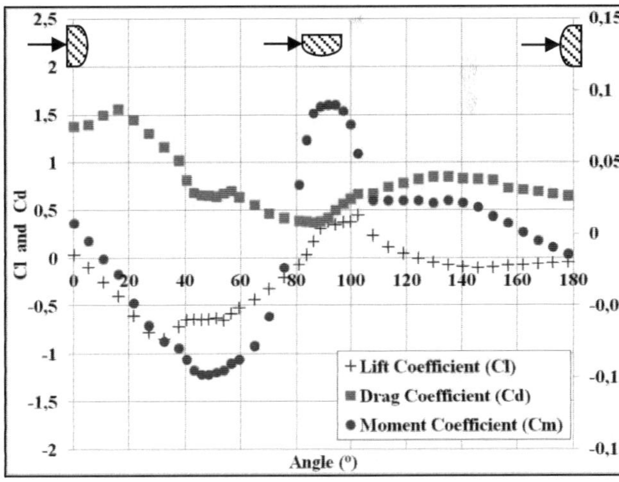

Figure 4: Lift, drag and moment coefficients for an inclined D-section at 40° with respect to the horizontal (Re = 39 190)

STROUHAL NUMBER FOR A VERTICAL D-SECTION

Given the very sensitive internal balance, it was possible to obtain the frequencies associated with vortex shedding by conducting a Fourier transformation on the signals measured at each angle during static tests. Figure 5 shows the Strouhal number as a function of the angle of attack for a vertical D-section and a flow speed of 5.8 m/s (Re = 39 387). Few papers have presented Strouhal numbers of a D-section at different angles of attack [Weaver and Veljkovic, 2005]. For angles smaller than 90 degrees, the flat face of the D-section faces the wind, and the flow separation points are the corners of the section. In the case of a rectangular section at an angle of 0 degrees, the Strouhal number ranges between 0.13 and 0.15 depending on the length of the afterbody [Knisely, 1990]. For angles greater than 90 degrees, the curved surface faces the wind, and it can be observed that the Strouhal number is similar to that of a circular cylinder (St = 0.21).

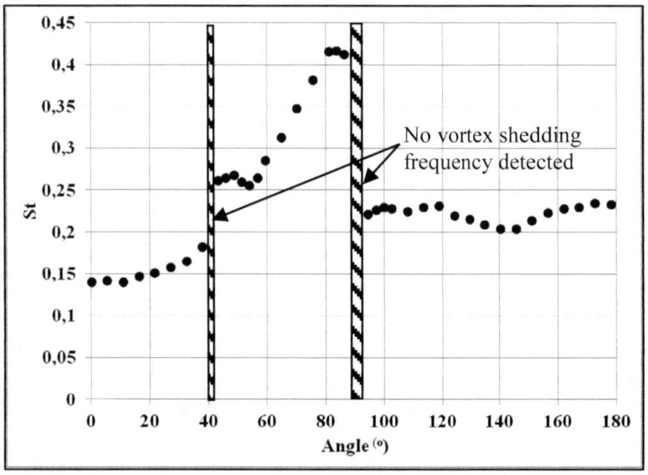

Figure 5: Strouhal number for a vertical D-section as a function of the angle of attack (Re = 39 387)

The amplitudes of the frequencies associated with vortex shedding (Figure 6) decrease at specific angles, then reappear. Furthermore, the vortex shedding ceases to be organized, as no vortex shedding frequency has been detected at 40.5 degrees, 89.1 degrees and 92.8 degrees.

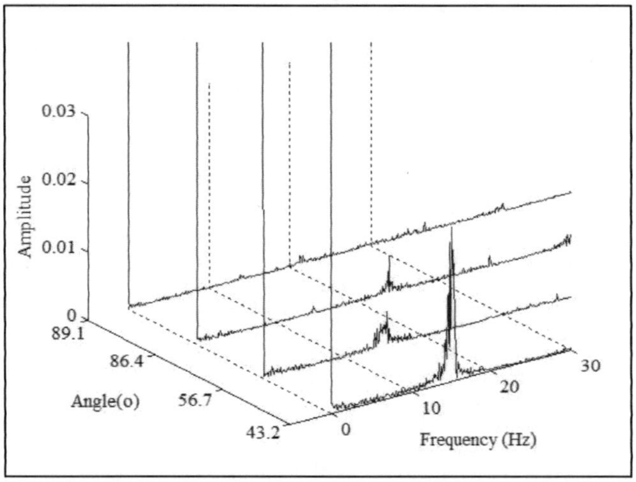

Figure 6: Amplitudes of the frequencies associated to vortex shedding (vertical D-section)

The authors' hypothesis is that it may have a connection between the shear layer fluctuations on the upper and lower surfaces of the D-section: the reattachment may be intermittent and the position of reattachment may vary in a random manner so no specific frequency can be detected. Figure 5 shows clearly that the vortex shedding frequency more than doubles as the angle is varied. This indicates that the vortex-induced vibration range is much larger than presumed previously in the case of the D-section.

STROUHAL NUMBER FOR AN INCLINED D-SECTION

Fourier transformations on the signals measured by the internal balance for the inclined D-section were conducted at each angle. Results are presented in Figure 7.

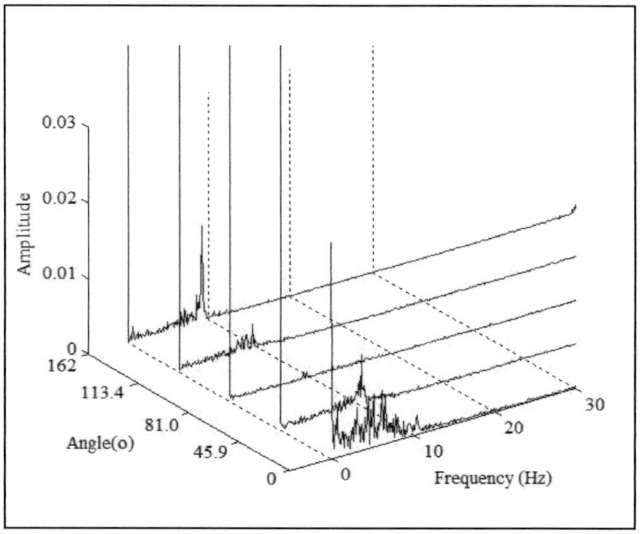

Figure 7: Amplitudes of the frequencies associated to vortex shedding (inclined D-section)

From 0 to 40.5 degrees, many low frequencies were detected as shown in Figure 6 for the angle of 0 degree. From 43.2 to 75.6 degrees, the vortex shedding frequency ranges from 9.5 to 10 Hz, however the amplitude is generally small. These frequencies correspond to Strouhal numbers ranging from 0.16 to 0.17, using the D-section diameter and a flow of 5.8 m/s. The range of angles where the vortex shedding ceases to be organized is larger for the inclined tube than it is for the vertical tube. Between 81 and 102.6 degrees no vortex shedding frequency has been detected as shown in Figure 5 for the angle of 81 degrees. At 113.4 degrees, frequencies are detected again; they become clearer and their amplitude increases as the angle increases as shown in Figure 5 for the angles 113.4 and 162. From 113.4 to 178.2, the vortex shedding frequency ranges from 9.0 to 9.4, corresponding to a Strouhal number of 0.16. For the inclined D-section, the vortex shedding frequency doesn't vary as it was the case for the vertical D-section.

DYNAMIC TESTS

Dynamic tests have been conducted in the wind tunnel by forcing the conductor to oscillate in rotation in order to vary the angle of attack of the D-section. The tube was mounted vertically and both extremities were supported by ball bearings which were attached to a frame independent from the wind tunnel to avoid transferred vibrations. A stepper motor and a Labview program were used for oscillating the tube. The motor was fixed to a structure independent from the D-section and was coupled to the D-section with a timing belt. The number of pulses sent to the drive allows the calculation of the angle. To make sure that the correct command was sent to the drive, a bi-directional incremental encoder was also used to measure the angle.

Rotational speed and amplitude of forced oscillations were varied to observe the effect of the reduced frequency on unsteady lift, drag and moment coefficients. Forces of inertia that the "floating" portion of the D-section applied to the internal balance were first measured with no wind while the tube was oscillating. For this measurement, a sheath in aluminum covered the floating section, without touching it, in order to not measure any aerodynamic forces. Forces of inertia, measured with no wing, were subtracted from the total forces measured with wind to obtain the aerodynamic forces. Results are being analyzed and will be presented at the conference.

CONCLUSIONS

Static force coefficients obtained from wind tunnel tests are consistent with previous studies conducted on D-sections. The D-section has been observed to have a peculiar type of vortex shedding, its frequency varying substantially from more than 200% at some angles of attack to collapsing at others. The effect of incidence was observed to reduce the aerodynamic coefficients defined in the flow axis. The vortex shedding ceases to be organized on a larger range of angles when the tube is inclined. Also, the vortex shedding frequency doesn't vary as was the case for the vertical D-section.

ACKNOWLEDGMENTS

This research project has received financial support from NSERC through a grant to the Research Director. Authors would like to thank the COOP-students that have participated to this research project: Dominique Magnier, Ugo Masarotti, François Boisguerin, Éric Blais-Desgagnés, Emmanuel Côté and Dimitri Cuesta Lavoie.

REFERENCES

1. Chadha, J., Jaster, W., 1975, "Influence of turbulence on the galloping instability of iced conductors", *IEEE Trans. Power Apparatus and Systems*, PAS-94 (5), 1489- 1499.

2. Cheers, F., 1950, "A Note on Galloping Conductors," *Rapport no.MT-14*, National Research Council of Canada.

3. Knisely, C.W., July 1990, "Strouhal numbers of rectangular cylinders at incidence: A review and new data," *Journal of Fluids and Structures*, Volume 4, Issue 4, p. 371-393.

4. Lavertu K., Laneville A., Fellouah H., 2013 "Strouhal number and aerodynamic force coefficients of a D-section", *24th Canadian Congs App Mech*, Saskatoon, Canada.

5. Nakamura, Y., Tomonari, Y., 1981. "The aerodynamic characteristics of D-section prisms in a smooth and in a turbulent flow", *Aeronaut.* Q. 32 (2), 153–168.

6. Parkinson, G.V. 1971, "Wind-induced instability of structures," Phil. Trans. Roy. Soc. Lond., A.269, p.395-409.

7. Van Dyke, P., Laneville, A., 2008, "Galloping of a single conductor covered with a D-section on a high voltage overhead test line", *Journal of Wind Engineering and Industrial Aerodynamics*, 96, 1141-1151.

8. Weaver, D.S., Veljkovic, I., 2005, "Vortex shedding and galloping of open semi-circular and parabolic cylinders in cross-flow", *Journal of Fluids and Structures*, 21, 65-74.

Proceedings of the 25th CANCAM
London, Ontario, Canada, May 31 - June 4, 2015

LARGE EDDY SIMULATION OF SUSPENDED SEDIMENTS IN TURBULENT FLOW

M. Dallali
Dipartimento di Ingegneria e Architettura
Universita degli Studi di Trieste
Trieste, Italy
maryam.dallali@phd.units.it

V.Armenio
Dipartimento di Ingegneria e Architettura
Universita degli Studi di Trieste
Trieste, Italy
Armenio@dica.units.it

ABSTRACT

A resolved large eddy simulation is carried out to study the sediment suspension that occurs in turbulent open- channel flow. Transport of suspended sediments in turbulent channel flow is usually assumed in passive form in which the feedback effect of sediment suspension on near-wall momentum transport is negligible. The main purpose of present study is to clarify the effect of suspended sediment concentration on dynamics of turbulent flow by using the single-phase approach. The sediment-induced buoyancy term is considered in the Boussinesq form of the 3D Navier-Stokes equations. The contribution of the small scales is parametrized by a dynamic Smagorinsky model. Results reveal that in presence of large suspension, the buoyancy effect modifies the dynamics of flow field.

Keywords: Large Eddy Simulation, Suspended Sediment transport, Stratification effects

INTRODUCTION

In alluvial channels, the sediment load transported by the flow may move either as bedload or as suspended load. The suspended load is fine material entrained from erodible bed and is carried downstream. In most of literatures, the problem of suspended sediment transport has been studied as a single mixed phase approach or two-phase approach in Eulerian framework. In the former approach a single set of momentum equations for the mixture of flow and sediments is employed. The feedback effect of sediment transport on flow properties is modeled through a correction term in equations of motion (Guo and Julien (2001)). In the latter model the governing equations are derived for both phases and they are coupled through their interaction forces (Teisson et al. (1992)).
Several experimental studies have been reported on the effect of suspended non-cohesive sediments, i.e. sand, on the dynamics of turbulent open-channel flow from that of Vanoni (1946), Enstein and Chien (1955) and Elata and Ippen (1961) to experiments by Muste and Patel (1997), Gelfenbaum (1986) and Muste et al. (2005). According to their studies the von Kàrmàn constant (κ) decreases with increasing suspended sediment concentration (C) inside the flow domain. Sediment suspension seems also correlated with large turbulent eddy motion. It has found that particles extract turbulent kinetic energy from the flow and suppress the turbulence (Muste et al. (2005); Guo and Julien (2001)).
Nevertheless, there certainly have been attempts to understand the turbulence-sediment interaction, few numerical studies addressed the effect of suspension on the both mean flow and concentration profile. In the present study, the writers focus on the effect of sediment concentration and consequently the density variations on the dynamics of the flow field by Large eddy simulation. The comparison of numerical results for two Eulerian single-phase models with and without considering the sediment-induced buoyancy term is presented.
This paper is organized as follows. First, the governing equations and boundary conditions for flow and dispersed phase simulations are presented in details. Then the numerical results are analyzed. Finally, conclusion remarks are given at the end of the paper.

GOVERNING EQUATIONS and BOUNDARY CONDITIONS

In present simulations, a LES code has been used to solve the 3-dimensional Navier-Stokes equations for incompressible flows and the advection-diffusion equation for suspended sediment concentration. The Boussinesq approximation for buoyancy effects is applied. The small dissipative scales are modelled by means of a subgrid-scale (SGS) closure. Scale separation is mathematically carried out applying a low-pass filter to the governing equations. Finite−difference space−discretization of the governing equations implies the use of a top-hat filter whose width is twice the cell size. Application of a filter (denoted by the over-bar) to conservation equations gives:

$$\frac{\partial \overline{u}_i}{\partial x_i} = 0, \tag{1}$$

$$\frac{\partial \overline{u}_i}{\partial t} + \frac{\partial \overline{u}_i \overline{u}_j}{\partial x_j} = \frac{1}{\rho_0}\Pi - \frac{1}{\rho_0}\frac{\partial \overline{p}}{\partial x_i} - g_m \delta_{i2} + \nu \frac{\partial^2 \overline{u}_i}{\partial x_j \partial x_j} - \frac{\partial \tau_{ij}}{\partial x_j}, \tag{2}$$

where u_1, u_2, u_3 are the velocity components in the streamwise, vertical upward, spanwise directions, Π is an imposed pressure gradient driving the flow, p is the hydrodynamic pressure, ρ_0 is the density of the fluid, g_m is the modified gravity (defined later on) and ν is the kinematic viscosity of the fluid. The effect of unresolved velocity field to the resolved large scales of motion is accounted for in term τ_{ij}.
Similarly, filtering the scalar transport equation for suspended sediments yields:

$$\frac{\partial \overline{C}}{\partial t} + \frac{\partial}{\partial x_j}[(\overline{u_j} - v_s \delta_{j2})\overline{C}] = \frac{\nu}{\sigma}\frac{\partial^2 \overline{C}}{\partial x_j \partial x_j} - \frac{\partial \chi_j}{\partial x_j}, \tag{3}$$

where C is the sediment concentration, σ is the Schmidt number assumed to be 1 as in Kraft et al. (2011) and v_s is the sediment settling velocity (Zanke (1977)); The term χ_j represents the contribution of the unresolved scales to the resolved concentration fluxes.
To account for the effects of sediment concentration on the flow field, the modified gravity term is considered in Navier-Stokes equations. We account for weak density variations in buoyancy effects through the Boussinesq approximation. Therefore, we retain density variations in the buoyancy term in the vertical direction of the momentum equation. We define the buoyancy coefficient as:

$$c_b = \frac{\rho_s - \rho_0}{\rho_0} C. \tag{4}$$

where ρ_s and ρ_0 are sediment and clear-water density, respectively. Thus, the modified gravity is obtained as (Teisson et al., 1992; Winterwerp, 2001) :

$$g_m = g c_b \tag{5}$$

in which g is gravitational acceleration.

The SGS stresses $\tau_{ij} = \overline{u_i u_j} - \overline{u}_i \overline{u}_j$ in Equation (2) are modelled using a dynamic eddy-viscosity model where $\tau_{ij} = -2\nu_{sgs}\overline{S}_{ij}$ with ν_{sgs} the SGS momentum diffusivity and $\overline{S}_{ij}$ the resolved strain rate tensor. The model is formulated according to Armenio and Piomelli (2000). The SGS turbulent fluxes for sediment concentration ($\chi_j = \overline{u_j C} - \overline{u}_j \overline{C}$) are expressed as:

$$\chi_j = -k_{sgs}\frac{\partial \overline{C}}{\partial x_j}. \tag{6}$$

The boundary conditions for the fluid are no-slip on the bottom bed, free slip on the top and periodic on all lateral boundaries. The flow is driven by a constant pressure gradient that yields a friction velocity $u_\tau = \sqrt{\delta\Pi/\rho_0} = 0.01 m/s$ and friction Reynolds number $Re_\tau = u_\tau \delta/\nu = 600$ based on the channel height. The boundary conditions implemented for the dispersed phase is no net flux for the top surface and the reference concentration for the bottom boundary given by Smith and Mclean (1977) as:

$$\overline{C}_{ref} = C_0 \frac{\gamma_0 \overline{S}_0}{1 + \gamma_0 \overline{S}_0}, \quad \overline{S}_0 = \frac{\overline{\tau}_w - \tau_{cr}}{\tau_{cr}} \tag{7}$$

where $\gamma_0 = 2.4 * 10^{-3}$ is a constant and $C_0 = 0.65$ is the maximum permissible concentration for single-grain spherical sediments. The critical bed shear stress (τ_{cr}) is calculated from the Shields curve (Vanoni (1975)) for corresponding flow conditions. The channel has dimensions 5δ and 2.5δ in streamwise and spanwise directions, respectively, where $\delta = 0.06\ m$ is the channel height. The spatial resolution of grid cells is $N_x * N_y * N_z = 64 * 32 * 64$. The grid is stretched in vertical direction and $\Delta y^+ = 1$ at the first grid point. In order to assess the effects of buoyancy, the non-dimensional grain sizes are chosen as $d^+ = u_\tau d/\nu = 1, 3, 5, 6.5$ and the sediment density is $\rho_s = 2650 kg/m^3$.

NUMERICAL RESULTS

The simulation analysis accordingly entails two main features: quantify the effects of particle size on sediment concentration distribution; and characterize the turbulence-particles interaction through the buoyancy term. We first discuss the sediment concentration profiles for each particle size with and without buoyancy term. The Reynolds averaged form of the sediment concentration equation reads:

$$k\frac{\partial < C >}{\partial y} - < C'v' > + < C > v_s = F_c, \tag{8}$$

where F_c is the concentration flux at the wall. On the left hand side (LHS) of Equation (8) the concentration flux is composed of the sediment diffusive flux ($k\frac{\partial <C>}{\partial y}$) and the turbulent one $< C'v' >$. Basically, the sum of the absolute values of the these two terms with the deposition contribution $< C > v_s$ on the LHS balances the absolute value of the term on the right hand side (RHS). Thereby the larger the deposition flux leads to the smaller the sum of the diffusive and turbulent fluxes. Close to the wall, the turbulent flux is negligible compared to the diffusive one and the steep gradient of concentration profiles is associated with the large value of deposition flux and diffusive term. Moving toward top boundary, the sediment concentration is affected more by turbulent flux. The no-flux boundary condition prevents the upward flux at free surface. Same behavior is observed for all concentration profiles depicted in fig. 1.
The deposition flux is a function of settling velocity which varies by particle size. Comparing the sediment concentration profiles, from the biggest particle ($d^+ = 6.5$) to the smallest

one ($d^+ = 1$), the available sediment concentration in the water column increases which is a result of smaller deposition term and larger available reference concentration at bottom boundary. In summary, the concentration profile is affected by the particle size variation and a smaller value of the particle diameter results in higher concentration level and larger suspended load.

By considering the turbulence-particles interaction through the buoyancy term in eq. (2) (two-way coupling), the concentration profiles change compared to the previous (one-way coupling) cases. For each particle size, the two-way coupling profiles are shown in fig. 1 by delta symbols. Equation (8) remains unchanged although the redistribution of the three terms ruling the vertical profile of concentration and the concentration wall fluxes are affected by stratification. Overall the suspension amount appears reduced compared to the non-stratified case. This effect is well observable when the non-dimensional diameter of sediments is smaller than 3.

In classical studies of sediment transport, the vertical profile of mean concentration is expressed though the theoretical Rouse profile:

$$\frac{C}{C_{ref}} = \left(\frac{\delta - y}{y} \cdot \frac{a}{\delta - a}\right)^{\alpha}; \ \alpha = \frac{v_s}{\beta \kappa u_\tau} \tag{9}$$

where the reference concentration C_{ref} is calculated at elevation a and α is the Rouse number. The parameter β, which relates sediment and momentum diffusion coefficients is generally assumed to be 1. The comparison of two-way coupling approach with Rouse fitted profile (see Muste et al. (2005) for more details) is presented in fig. 2. The reference height a is taken at the first centroid off the wall and the parameter β varies between 0.1 and 0.6. In presence of large amount of suspension in water column (smaller sediments), the Rouse profile is in close agreement with numerical result. As sediment concentration decreases in the water depth ($d^+ = 5$) the Rouse fitted profile substantially departs from simulated data. Note that since the larger particles are more likely transported in the bed load mode, the Rouse formula, which describes suspended mode transport may not be appropriate analytical formula to qualify the numerical result. The need to use values of β smaller than the canonical one may be related to Reynolds number effects and to the assumption of constant turbulent diffusivity of sediment concentration along the depth (Muste et al., 2005).

To quantify the level of stratification in presence of sediment suspension, we also discuss the buoyancy effects on the velocity field. The comparison of streamwise mean velocity profiles is shown in fig. 3. In the two-way coupling case, it is observed that in presence of sediments, the vertical profiles of streamwise velocity are modified by buoyancy effects. From studies of stratified wall-bounded turbulence, it is known that turbulence is damped by density gradients (Vanoni, 1946; Muste and Patel, 1997; Cellino and Graf, 1999; Armenio and Sarkar, 2002) and vertical momentum transport is suppressed with respect to the passive scalar case. To keep the total shear stress in balance, the mean velocity gradient

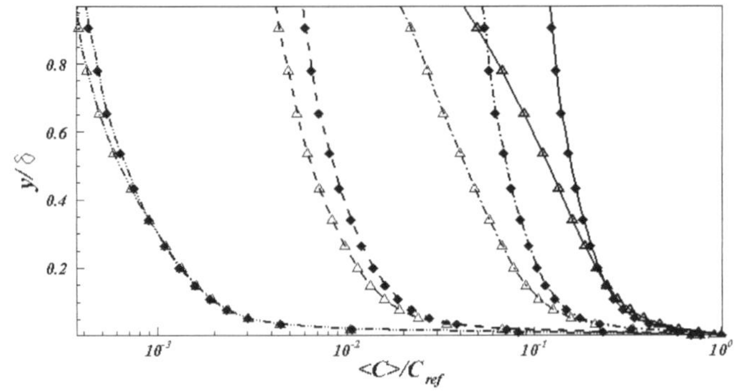

Figure 1: The normalized vertical profiles of averaged sediments concentration for 4 different particle sizes. $d^+ = 1$: solid line, $d^+ = 3$: dashed dot line, $d^+ = 5$: dashed line, $d^+ = 6.5$: dashed dot dot line; One-way coupling simulation (diamond); two-way coupling simulation (delta).

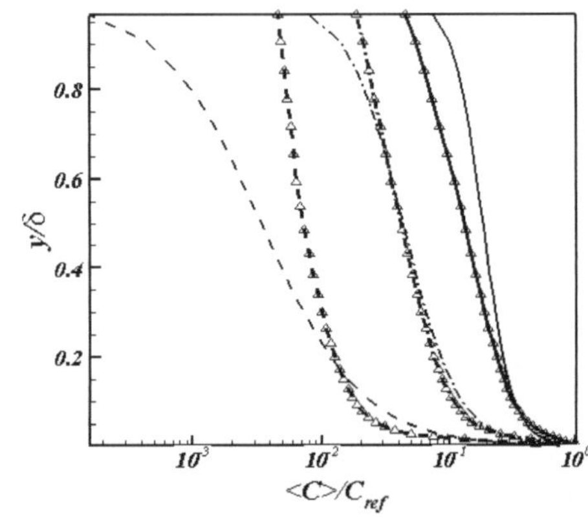

Figure 2: Two-way coupling sediments concentration profiles (delta) and Rouse equation (no symbol) for different particle sizes. $d^+ = 1$: solid line, $d^+ = 3$: dashed dot line, $d^+ = 5$: dashed line ; Rouse fitted profile, $\beta = 0.1$:solid line, $\beta = 0.38$:dashed dot line, $\beta = 0.6$:dashed line.

must increase when the Reynolds shear stress decreases for stratification effects. This increase is more significant for the smallest particle, namely for large buoyancy effects. For larger particles, the suspension amount is small along the fluid column and hence the mean velocity profile does not show any observable change comparing to one-way coupling model. To sum up, the interaction of suspended particles with the underlying flow suppresses the vertical turbulent fluctuations and consequently reduces the von Kàrmàn constant (κ).

CONCLUSIONS

This study investigates the buoyancy effect of suspended sediment concentration on flow field properties by the large-eddy simulation methodology. The results are analyzed for one-way (passive) and two-way coupling models where the sediment concentration adds on a buoyancy contribution to momentum in two-way coupling model. It shows that the

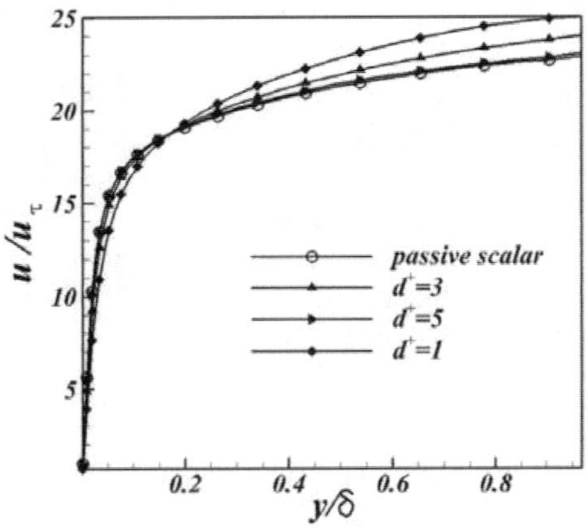

Figure 3: Vertical profiles of averaged streamwise velocity for three stratification levels

combination of large near-wall concentration and small deposition term, enhances the suspension concentration in case of small particles. Considering the buoyancy effect (two-way coupling model), the suspension concentration is reduced compared to passive case. The difference is more relevant for small particles where the sediment-induced buoyancy can not be neglected.

It is also shown that the averaged streamwsie velocity profile modifies by employing the buoyancy term in numerical simulations. To summarize, for small particles, large suspension amount is observed in the water column which causes stratification effects in the flow field. The results clearly show that in presence of large suspension, the two-way coupling approach should be applied to represent the sediment-turbulence interaction.

ACKNOWLEDGMENTS

This research has been carried out using computational facilities procured through the European Regional Development Fund, Project ERDF-080 'A Supercomputing Laboratory for the University of Malta' (`http://www.um.edu.mt/research/scienceeng/erdf_080`). The present study has been financially supported by the University of Trieste, through the project FRA 2013: Modellistica numerica e sperimetale di fenomeni di erosione localizzata in alvei fluviali.

REFERENCES

Armenio, V. and U. Piomelli (2000). Lagrangian mixed subgrid-scale model in generalized coordinates. *Flow, Turbulence & Combust 65*, 51–81.

Armenio, V. and S. Sarkar (2002). An investigation of stably stratified turbulent channel flow using large-eddy simulation. *J.Fluid Mech 459*, 1–42.

Cellino, M. and W. H. Graf (1999). Sediment-laden flow in open-channels under noncapacity and capacity condition. *J. Hydraul. Eng 125*(5), 455–462.

Elata, E. and A. T. Ippen (1961). The dynamics of open channel flow with suspensions of neutrally buoyant particles. Technical report, Hydrodyn Lab. Mass.Inst of Technol., Cambridge.

Enstein, H. A. and N. K. Chien (1955). Effect of heavy sediment concentration near the bed on the velocity and sediment distribution. *M. R. D. Sediment Series* (8). Univ. of California at Berkeley , USA.

Gelfenbaum, G.and Smith, J. D. (1986). Experimental evaluation of a generalized suspended-sediment transport theory. *Shelf sand and Sandstones 11*, 133–144.

Guo, J. and P. Y. Julien (2001). Turbulent velocity profile in sediment-laden flows. *J. Hydraulic Research 39*(1), 11–23.

Kraft, S., Y. Wang, and M. Oberlack (2011). Large eddy simulation of sediment deformation in a turbulent flow by means of level-set method. *J. Hydraul. Eng 137*(11), 1394–1405.

Muste, M. and V. C. Patel (1997). Velocity profiles for particles and liquid in open-channel flow with suspended sediment. *J. Hydraul. Eng 123*(9), 742–751.

Muste, M., K. Yu, I. Fujita, and R. Ettema (2005). Two-phase versus mixed-flow perspective on suspended sediment transport in turbulent channel flows. *Water Resour. Res. 41*(10). doi:10.1029/2004WR003595.

Smith, J. and S. Mclean (1977). Spatially averaged flow over a wavy surface. *J. Geophys. Res 82*(12), 1735–1746.

Teisson, C., O. Simonin, J. Galland, and D. Laurence (1992). Turbulence and mud sedimentation: A reynolds stress model and a two-phase flow model. In *23rd International Conference on Coastal Engineering*, pp. 2853–2866.

Vanoni, V. A. (1946). Transportation of suspended sediment by water. *Trans Amer. Soc. Civil Eng. 111*, 67–133.

Vanoni, V. A. (1975). *Sedimentation Engineering*. ASCE Task Committee for the Preparation of the Manual on Sedimentation of the Sedimentation Committee of the Hydraulics Division.

Winterwerp, J. C. (2001). Stratification effects by cohesive and noncohesive sediment. *J.Geophysical Research 106*(10), 22559–22574.

Zanke, U. (1977). Berechnung der sinkgeschwindigkeiten von sedimenten. Technical Report 42, Mitteilungen des Franzius-instituts fur wasserbau, Technische Universitat Hannover.

Proceedings of the 25th CANCAM
London, Ontario, Canada, May 31 – June 4, 2015

SPATIAL DEVELOPMENT OF THE ZONE OF NEGATIVE TURBULENCE PRODUCTION IN AN ASYMETRY CHANNEL FLOW

Ali Esmaeilnejad
Department of Mechanical Engineering
University of Alberta
Edmonton, Alberta, Canada
aliesmaeilnejad@ualberta.ca

Sina Ghaemi
Department of Mechanical Engineering
University of Alberta
Edmonton, Alberta, Canada
ghaemi@ualberta.ca

ABSTRACT

The spatial development of the turbulent structure of an asymmetry turbulent channel flow at Reynolds number of *Re* = 9200 is investigated using particle image velocimetry (PIV). The asymmetry in the channel flow is introduced using a *k*-type rough wall on one side while the other wall is kept smooth. The ratio of shear stress at the rough wall to the smooth wall (τ_r/τ_s) is about 1.3. Measurements are carried out using two simultaneous PIV planes located at x/H = 59 and 65 where x is the streamwise distance from the channel inlet and H is the height of channel from tip of the roughness elements to the smooth wall. The results show a small region of negative energy production at x/H = 59 (the upstream measurement location) in between the location of maximum streamwise velocity (y_U) and zero Reynolds shear stress (y_{uv}). At the downstream measurement location of $x/H = 65$, y_U and y_{uv} overlap and the region of negative turbulent production disappears. The results suggest that the non-coincidence of y_U and y_{uv} and the region of negative turbulent production in asymmetric channel flows are associated with slight deviation from the fully developed state.

KEYWORDS: Asymmetry turbulent channel flow, Negative production of turbulence, Particle image velocimetry (PIV).

INTRODUCTION

Asymmetric turbulent channel flows are of fundamental importance due to the interaction between two wall flows with turbulent structures of different spatio-temporal scales. Most of the interactions occur over a relatively wide region in the midsection where a tangled structure of ejection and sweeps of the two opposing walls evolves (Ghaemi *et al.* 2015). In this midsection of an asymmetric channel flow, of particular interest is the non-coincidence of the location of the maximum mean streamwise velocity (y_U) and the location of zero Reynolds shear stress (y_{uv}). The narrow region between these two points is associated with negative production of turbulence (negative $-\langle uv\rangle\ d\langle U\rangle/dy$) where the mean flow extracts energy from the turbulent field (Maubach & Rehme 1972). This is a consequence of $\langle uv\rangle$ and $d\langle U\rangle/dy$ having the same sign in this narrow region. The formation and the role of this narrow region in the interaction of the two wall flows and the development of the asymmetric profile require further scrutiny.

The non-coincidence of y_U and y_{uv} has motivated several investigations on the turbulent structure of the central region of asymmetric channel flows produced by introducing roughness on one of the walls. Hanjalić & Launder (1972) investigated the turbulence structure of an asymmetric channel flow for a range of *Re* = 18,700 to 56,000 (based on full channel width) using hot-wire technique. The *k*-type roughness element consisted of square-sectioned transverse bars resulted in τ_r/τ_s of about 4. They observed a significant non-coincidence (y_U - y_{uv}) of about 10% of the channel width. Nagano *et al.* (2004) investigated the effects of a number of roughness patterns including the pattern investigated by Hanjalić & Launder (1972) on the statistical quantities of the turbulent and thermal fields using direct numerical simulations (DNS). The simulation was at Reynolds number of $Re_{\tau 0}$ = 150 (based on friction velocity and channel half width) with τ_r/τ_s of about 2. The results demonstrated a considerably lower (y_U - y_{uv}) of about 2 % which was associated with the smaller *Re* relative to Hanjalić & Launder (1972) experiment. Burattini *et al.* (2008) carried out both measurement and DNS of an asymmetric channel flow with *k*-type roughness over *Re* of 7,200-26,000 (based on full channel width) with τ_r/τ_s of about 1.17 to 2.96 for *Re* = 7,200 to 26,000, respectively. They concluded almost indistinguishable (y_U - y_{uv}) and associated the approximate overlap of y_U and y_{uv} with the difference in $Re_{\tau r}$ (based on friction velocity at the rough wall). Burattini *et al.* (2008) proposed that there may be a range of $Re_{\tau r}$ in which the two locations would not overlap, yielding a region of negative energy production.

The non-coincidence of y_U - y_{uv} is not only limited to one-sided rough channel flows and has also been observed in asymmetric profiles of turbulent annuli flow. Rehme (1975) proposed that turbulence diffusion is higher in asymmetric profiles causing the non-coincidence, although no experimental evidence was provided. Ghaemi *et al.* (2015) speculated that the non-coincidence is due to slight deviation from fully developed one-dimensional flow field. They associated the region of negative turbulence production with smoothing the mean velocity profile at the joint of the two wall flows by equalizing its curvature on the two sides of y_U.

The objective of this study is to experimentally investigate the effect of spatial development of the flow field on the location of y_U and y_{uv} in an asymmetric turbulent channel flow using particle image velocimetry (PIV). The asymmetry was introduced by increasing the shear stress at one of the walls using a *k*-like type roughness while the other was left smooth. The locations of y_U and y_{uv} are evaluated at two streamwise locations of x/H = 59 and 65 to investigate the effect of streamwise development on the possible non-coincidence of y_U and y_{uv}.

EXPERIMENTAL SETUP

The experiments are performed in the open surface water channel facility located at the Laboratory of Turbulence of the University of Alberta. As it is seen from Figure 1, the closed channel flow was formed between the bottom wall of the channel and a top flat plate of 3 meters length. This flat plate has an elliptical leading edge and a sharp trailing-edge and is located 54 mm above bottom of channel.

The asymmetric profile is developed by placing a rough surface of 3.05 m long and 3 mm thick (h_2) on the lower wall of the water channel as detailed in Figure 2. The height of roughness (h_1) is 1 mm and their spacing is w = 3 mm. The roughness elements are *k*-like type two-dimensional spanwise elements with triangular cross-section. The distance between the smooth wall and the tip of the roughness elements is 51 mm (H = 51 mm). The Reynolds number based on the mean velocity across the channel $\langle U_b \rangle$ and the channel width H is Re = 9200.

Figure 1: The PIV setup consists of two CCD cameras imaging perpendicular to the laser sheet. The flat plate is located parallel to the bottom wall of the water channel with a distance of 54mm.

The PIV measurements are carried out at two streamwise positions along the channel at x/H = 59 and 65 (x is streamwise distance from the inlet of channel). The centers of the two FOVs are 320 mm apart (6H). The y axis is in the wall normal direction starting from smooth wall (y = 0). The streamwise and wall normal velocities are specified by (U_1, U_2) = (U, V) while their fluctuations are denoted by (u, v), respectively. The PIV setup consists of a double-pulse laser (Spectra physics, PIV-400-10) and two CCD cameras (Imager Intense, LaVision) with 2048×2048 pixel. Two sets of 8,000 double-frame recordings with 1000 μs time interval is recorded by the two cameras. The two upstream and downstream fields of views are referred to as FOV1 and FOV2, respectively. The laser beam is expanded to a laser sheet of about 0.6 m wide and 1-2 mm thick to cover both FOV1 and FOV2. The spatial resolution of the upstream camera is 26.08 μs/pix while the spatial resolution of the downstream camera is 26.48 μs/pix. The recorded images are processed in Davis 8.2 (LaVision) with a final interrogation area of 32×32 pixels (0.8367×0.8367 mm) with 75% overlap.

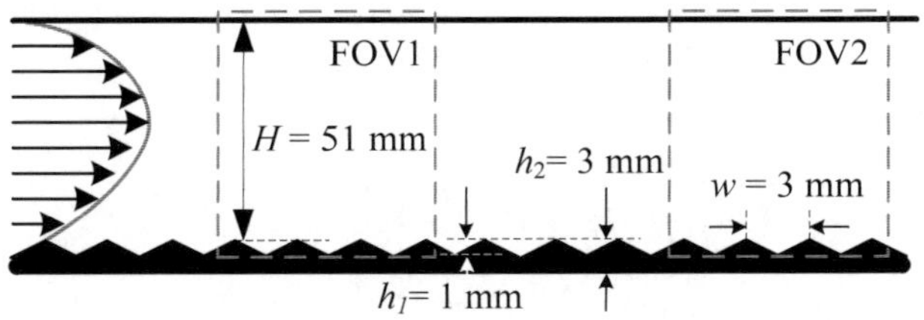

Figure 2: A schematic of the channel with the smooth and rough walls. The rough wall cross section consists of spanwise oriented elements with triangular cross-section.

RESULTS AND DISCUSSION

The profiles of mean streamwise $\langle U \rangle$ and wall normal velocity $\langle V \rangle$ for FOV1 and FOV2 are shown in Figure 3. The streamwise velocity is multiplied by $1/\langle U \rangle_{max}$ while the wall normal velocity is multiplied by $-100/\langle U \rangle_{max}$. The $\langle U \rangle$ profile is asymmetric with lower velocity on the rough-wall side while $\langle U \rangle_{max}$ is closer to the smooth wall. There is a slight difference between $\langle U \rangle$ profiles of FOV1 and FOV2 in the near wall regions, which suggests development of the profile with streamwise distance. In addition, it is observed from Figure 3 that these two profiles have a smaller discrepancy at the center of channel.

In addition, Figure 3 shows the profiles of average wall-normal velocity. Considering that the data is multiplied by a factor of -100, it is clearly negligible in comparison with $\langle U \rangle$. However, $\langle V \rangle$ for FOV1 is about 4-5 times larger than the values at FOV2. This shows slight deviation from the fully developed state at FOV1 relative to FOV2. It may come from this fact that FOV1 is still located in a developing region while FOV2 is a more developed towards a one-dimensional fully-developed profile with $\langle V \rangle = 0$.

Figure 4 shows a magnified view of $\langle U \rangle$ profiles of FOV1 and FOV2 in the vicinity of $\langle U \rangle_{max}$ at the mid-section of the channel. The maximum streamwise velocity $\langle U \rangle_{max}$ is located at y/H = 0.3895 and y/H= 0.4025 for cameras #1 (FOV1) and #2 (FOV2), respectively. The displacement of $\langle U \rangle_{max}$ is about 0.013H over a streamwise distance of 6H. Therefore, the rough wall boundary layer is becoming slightly thinner with streamwise development.

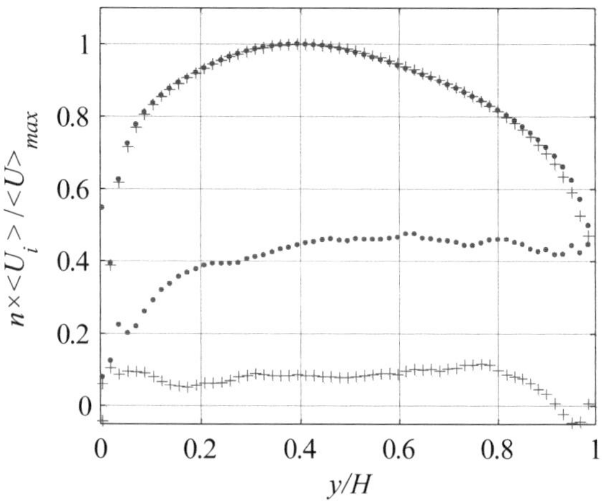

Figure 3: Distribution of $<U>/<U>_{max}$ (upper plot in blue) and $-100\times<V>/<U>_{max}$, (two lower plots in red). Symbols: ., FOV1; +, FOV2. One out of every four data points is shown for clarity.

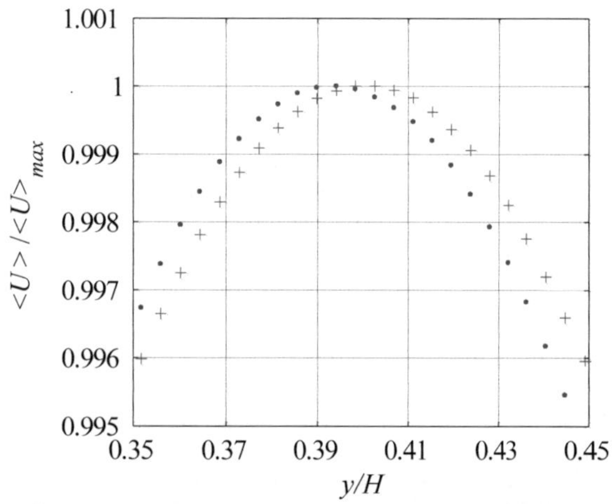

Figure 4: Profiles of $<U>/<U>_{max}$ at the mid-section across the channel. Symbols: ., FOV1; +, FOV2.

The corresponding distributions of normal and Reynolds shear stress normalized by using $<U>_{max}$ are shown in Figure 5. It is observed that the profiles are asymmetric while they are skewed towards the smooth wall. The Reynolds shear stress $<uv>$ varies linearly in the central region while the highest intensities occur near the walls. It is observed that the minimum of $<u^2>$ for FOV1 is at y/H=0.3559 while it is located at the y/H=0.3602 for FOV2. The location of $<uv>$ = 0 is closer to the smooth wall and at about y/H=0.3856 and y/H=0.4025 for FOV1 and FOV2, respectively. In the vicinity of the walls, the local maximums of the $<v^2>$ are at y/H = 0.1356 and y/H = 0.911 for FOV1 and y/H = 0.1271 and y/H = 0.9025 for FOV2. The exact location of maximum $<u^2>$ in the near wall region is not visible due to lack of spatial resolution in the vicinity of the rough wall. The results show that the profiles of turbulent intensities are slightly developing with streamwise distance. The displacement of the local minima of $<u^2>$ and $<v^2>$ and also $<uv>$=0 from FOV1 to FOV2 shows that the rough wall boundary layer becomes thinner with increase of streamwise distance.

The distribution of the Reynolds shear stress ($<uv>$) profiles of the two cameras in the midsection of the channel also confirms the fully developed state of FOV2 relative to FOV1. This is due to perfectly linear profile of $<uv>$ for FOV2 while the $<uv>$ profile of FOV1 is slightly curved.

The magnified views of $<uv>$ profiles in the midsection are shown in Figure 6. A shift towards the rough wall is observed in FOV2 in comparison with the FOV1 for $<uv>$=0 across the channel. The y_{uv} for FOV1 and FOV2 is at y/H = 0.3856 and 0.4025, respectively. The comparison of Figures 4 and 6 shows that (y_U - y_{uv})=0.004H is positive at FOV1 confirming their non-coincidence while y_U and y_{uv} for FOV2 coincide.

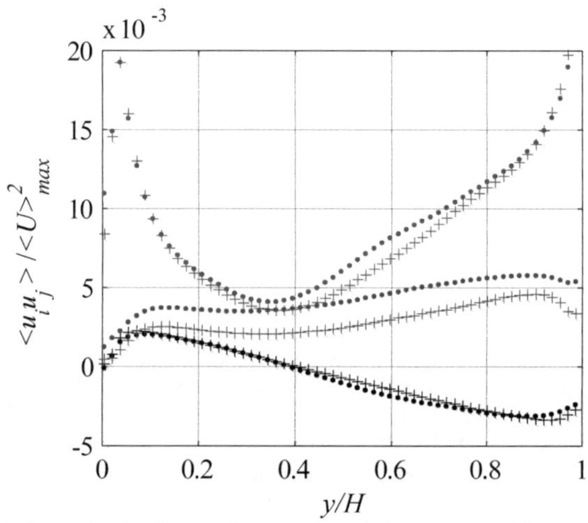

Figure 5: Profiles of normal and shear Reynolds stresses across the channel. $<u^2>$ is the upper plot in red, $<v^2>$ is the middle plot in blue and $<uv>$ is the lower plot in black. Symbols: ., FOV1; +, FOV2. One out of every four data points is shown for clarity.

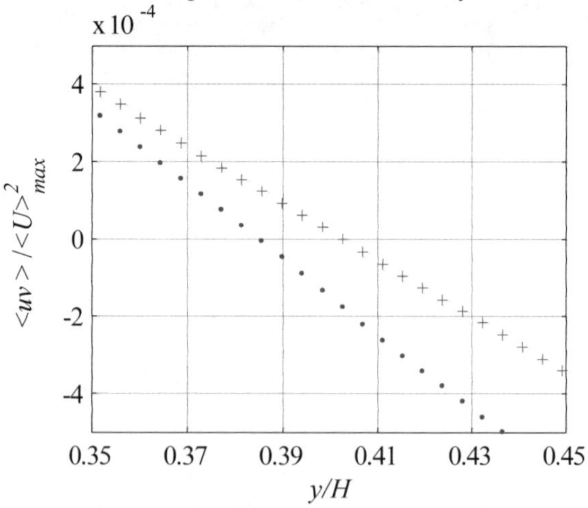

Figure 6: Profiles of $<uv>$=0 at the mid-section across the channel. Symbols: ., FOV1; +, FOV2.

The results demonstrate that in FOV1, where the flow is still developing, there is a slight non-coincidence of y_U and y_{uv} ($0.004H$). With increase of streamwise distance from FOV1 to FOV2, y_U shifts $0.013H$ towards the rough wall while y_{uv} shifts $0.0169H$ causing the coincidence at this streamwise location.

The measured distance between y_U and y_{uv} at FOV1 is about 0.4% of H which is comparable with the 2% observed by the DNS simulation of Nagano *et al.* (2004) and undistinguishable distance reported by Burattini *et al.* (2008). However, the results do not agree with 10% discrepancy of y_U and y_{uv} predicted by Hanjalić & Launder (1972). This might be an over-prediction due to experimental uncertainties as scrutiny of Figure 3 of Hanjalić & Launder (1972) shows lack of statistical convergence. In particular close the $<U>_{max}$, $d<U>/dy$ is small and any noise in the data would significantly shift the location of the measured $<U>_{max}$.

In the thin layer between y_U and y_{uv} in developing region of channel (FOV1), the production of turbulent kinetic energy is negative as both $<uv>$ and $d<U>/dy$ are negative. This negative production of turbulence is speculated to equalize the curvature of the mean velocity profile on the two side of y_U. It is clear that this thin region of negative production of turbulence is not observed in the FOV2 for fully developed region as the mean velocity profile has reached the asymptotic state.

CONCLUSIONS

An asymmetric turbulent channel flow produced by a rough wall and a smooth wall is experimentally investigated at Reynolds number of 9200 by PIV measurement. The main goal of this experimental investigation was to scrutinize the effect of streamwise flow development on the coincidence of the location of maximum velocity (y_U) and zero Reynolds shear stress (y_{uv}). The measurements were carried out at two streamwise regions of FOV1 and FOV2 along channel located at x/H=59 and 65. Measurement at FOV1 showed small $<V>$ residue along with small curvature in $<uv>$ demonstrating developing condition of the channel flow at x/H=59. The comparison of $<U>$ and $<uv>$ profiles also showed non-coincidence of y_U and y_{uv} at this upstream location. At FOV2, $<V>$ residue was 5 times smaller than FOV1 while $<uv>$ had a linear distribution. Therefore, FOV2 shows fully developed characteristics while the location of y_U and y_{uv} also overlap.

The local minimum values of $<u^2>$ and $<v^2>$ are observed to coincide with $<uv>=0$ while they are apart from $<U>_{max}$ location at FOV1. However, the local minimum values of $<u^2>$ and $<v^2>$, $<uv>=0$ and $<U>_{max}$ overlap at FOV2 where the asymmetric channel flow is fully developed.

REFERENCES

1. Burattini, P., Leonardi, S., Orlandi, P., and Antonia, R.A. (2008), "Comparison between experiments and direct numerical simulations in a channel flow with roughness on one wall", *Journal of Fluid Mechanics*, 600, pp. 403–426.
2. Ghaemi, S., Bizhani, M., and Kuru, E. (2015), "Turbulent structure at the midsection of an annular flow", *Physics of Fluids*, under review.
3. Hanjalić, K., and Launder, B.E. (1972),"Fully developed asymmetric flow in a plane channel", *Journal of Fluid Mechanics*, 51, pp. 301-335.
4. Maubach, K., & Rehme, K. (1972), "Negative eddy diffusivities for asymmetric turbulent velocity profiles", *International Journal of Heat and Mass Transfer*, 15(3), pp. 425-432.
5. Rehme, K. (1975), "Turbulence measurements in smooth concentric annuli with small radius ratios", *Journal of Fluid Mechanics*, 72(01), pp. 189-206.

Vibration of Structures

Proceedings of the 25th CANCAM
London, Ontario, Canada, May 31 – June 4, 2015

RECREATION OF RANDOM VIBRATION DURING TRUCK TRANSPORTATION

Enayat Mahajerin
Mechanical Engineering Department
Saginaw Valley State University
University Ctr., Michigan 48710, USA
Email: *mahajeri@svsu.edu*

Gary Burgess
School of Packaging
Michigan State University
East Lansing, MI 48824
Email: *burgess@msu.edu*

ABSTRACT

The vibration that occurs during truck transportation has been studied using accelerometers mounted on the floor of the trailer. The recorded accelerations are used to produce a power spectral density plot, which shows the frequencies present and their contributions to the measured accelerations. This plot is used to drive a vertical vibration table to recreate the ride. The table does not reproduce the motion of the trailer floor for various reasons. In this paper, an analytical solution for a truck traveling over an arbitrary road consisting of straight line segments is developed for the purpose of driving a vibration table in real time. This eliminates most of the problems with PSD driven tables. The technique is validated by comparing the rms acceleration to known values. Various numerical simulations are presented. MATLAB® is used to expedite the computations.

KEYWORDS: ASTM D4728, D999, random vibration, rnd, PSD, RMS, transportation simulation.

INTRODUCTION

Products in packages are subjected to complex dynamic forces during transportation. Resonant responses caused by vibration can cause damage. Simulating this involves testing the product on a vibration table. Figure 1 shows a TV being tested this way. The standard test procedures for this are ASTM D999 [1] and ASTM D4728 [2]. These tests advocate the use of different methods depending on the equipment available to do the test. ASTM D999 uses sinusoidal vibration. The frequency of vibration is adjusted until the package begins to bounce on the table. The package is then vibrated at this frequency for 1 hour.

The newer test, ASTM D4728, uses random vibration. The truck transportation environment is first recorded by mounting vibration recorders on the trailer floor. These recorders house tri-axial accelerometers, a power supply, and electronic hardware (memory, filters, processors, etc.). They sample the acceleration in g's at a user specified sampling rate and can collect this data for several weeks. The recorded accelerations are then used to produce a power spectral density (PSD) plot. The PSD plot shows the frequencies present and

their contributions to the measured accelerations. This information is used to drive a vertical vibration table.

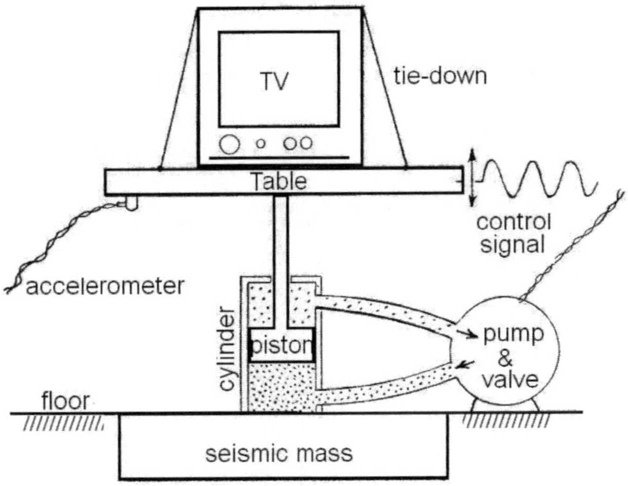

Figure 1. Vibration Table.

The resulting motion of the table has the same frequencies and average amplitudes as the recorded motion, but is not the same as that of the trailer floor for the following reasons:

1) The table is never quiet. When the trailer wheels hit a bump, there is an initial shock followed by a quiet down period. This doesn't happen on the vibration table.
2) Large shocks from the trailer going into potholes and over railroads tracks are averaged out in the creation of the PSD plot. These are infrequent but often the cause of damage.
3) It is not a simple matter to adjust the PSD plot for different roads, different trailer weights, and different truck speeds.

The objective of this paper is to remove these drawbacks by driving the vibration table with a real time acceleration vs. time signal representing the road. This has the following advantages:
1) The table responds to each bump as it occurs.

2) Railroad tracks and potholes can easily be accounted for.
3) The road profile can be easily adjusted.

This paper suggests an algorithm for generating the displacement vs. time drive signal. The technique is validated by comparing the response of a mathematical model of the truck trailer traveling over a road with published rms g values. Various numerical simulations are presented and MATLAB is used to expedite the computations.

DYNAMIC MODELS

The truck trailer is modeled as the single degree of freedom spring-mass-dashpot system shown in Figure 2.

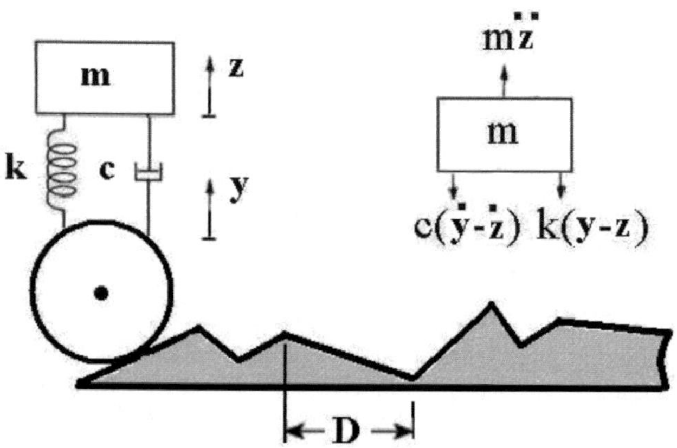

Figure 2. Truck Trailer Model.

The mass of the trailer is m and the stiffness and damping constant of the suspension system are k and c. As the trailer travels over the road, its motion is governed by

$$m\ddot{z} + c\dot{z} + kz = c\dot{y} + ky \quad (1)$$

Subjected to the initial conditions,

$$z(0) = z_0, \dot{z}(0) = v_0 \quad (2)$$

The road is modeled by straight line segments as shown in Figure 4.

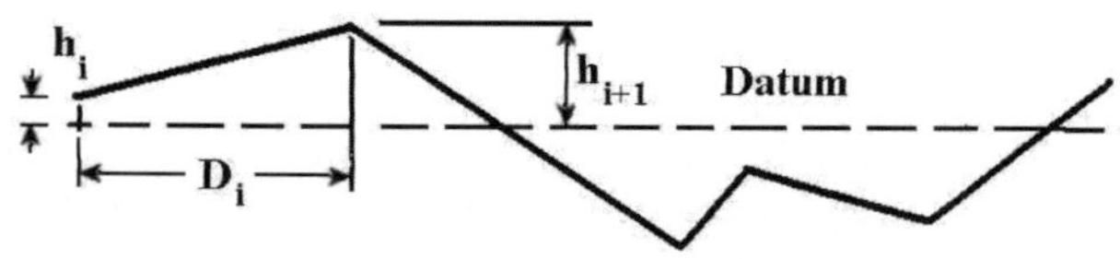

Figure 4. Road Segment.

The length of the ith segment is D_i and the starting and ending heights are h_i and h_{i+1}. The lengths and heights are random but are bounded by

$$D_{min} < D < D_{min} \quad (3)$$

$$-h_{max} < h < h_{max} \quad (4)$$

Where D_{max} and h_{max} are known road parameters and

$$D_{min} = \sqrt{2Rh_{max}} \quad (5)$$

A lower bound D_{min} on segment length exists because the position coordinate y in Figure 2 refers to the motion of the axle, not the road. Even for the step bump in Figure 3, the axle must travel the length D_{min} to rise to the height of the bump. Solving for D_{min} using the triangle in Figure 3 and using h<<R gives the result in Equation (5).

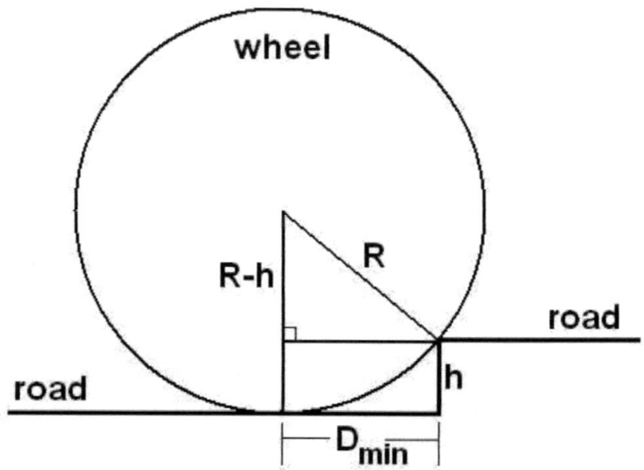

Figure 3. Step Bump.

The goal is to generate D's and h's consistent with known road profiles and use them to compute the floor accelerations as the truck travels over the road at speed V. For any segment i, we can write

$$D_i = D_{min} + (D_{max} - D_{min})rnd \quad (6)$$

$$h_i = h_{max}(2rnd - 1) \quad (7)$$

Where *rnd* is a random number between 0 and 1 from a uniform distribution. For the road segment i,

$$\dot{y} = \frac{h_{i+1} - h_i}{D_i/V} \quad (8)$$

$$y = h_i + \dot{y}t \quad (9)$$

Where V is the truck speed and t=0 corresponds to the beginning of segment i. The exact solution to Equations (1) and (2) can be used to find the acceleration versus time for segment i. Using standard solution techniques, the acceleration will be:

$$a_i(t) = e^{-ct/2m}[P_i \sin(\omega t) + Q_i \cos(\omega t)] \quad (10)$$

With

$$P_i = [(\tfrac{c}{2m})^2 - \omega^2]\,C_1 + \tfrac{c\omega C_2}{m} \quad (11)$$

$$Q_i = [(\tfrac{c}{2m})^2 - \omega^2]\,C_1 - \tfrac{c\omega C_1}{m} \quad (12)$$

The starting position z_0 and velocity v_0 for segment i are the ending position and velocity for segment i-1. The *root mean square* (rms) acceleration over any time interval T is [4, 5].

$$C_2 = z_0 - h_i \quad (13)$$

$$C_1 = \left[\, v_0 - b + \tfrac{cC_2}{2m} \right] / \omega \quad (14)$$

$$b = V(h_{i+1} - h_i)/D_i \quad (15)$$

The rms g is considered to be a measure of the overall ride severity. The rms g can be determined exactly using Equation (10) for each segment. To expedite the Computational process, MATLAB [6] programing was used.

$$g_{rms} = \frac{1}{g}\sqrt{\frac{1}{T}\int_0^T a(t)^2 dt} \quad (16)$$

EXAMPLES

The following examples demonstrate how this approach can be used to determine the effect of truck speed and trailer weight on roughness of the ride. This cannot be done using a PSD approach.

Consider a loaded truck trailer weighing $2\text{x}10^4$ lbf ($8.896\text{x}10^4$ N) traveling at speeds V=25-90 mph (11.2-40.2 m/s) on a road having D_{max}=20 ft (6.1 m) and h_{max}=0.5 in (12.7 mm). The radius of the trailer wheel is taken to be R=1.5 ft (0.457 m). For a trailer of this weight, the up/down natural frequency is about 5 Hz and the damping ratio is 0.2 0.2 [1, 2]. Based on this, the following truck suspension properties were used:

k =$6.4\text{x}10^4$ lbf/ft ($9.3396\text{x}10^5$ N/m)
c =$8\text{x}10^3$ lbf-s/ft ($1.167454\text{x}10^5$ N-s/m)

Using zero initial values (z_0=0, v_0=0) for the 1st segment, we computed accelerations analytically from Equation (10) and the corresponding RMS acceleration from Equation (16). The maximum acceleration in g's was also computed. A typical rms g for a trailer traveling over expressways is ½ g [3].

Figures 5-6 show the effect of truck speed on rms g's and maximum g's for a section of road consisting of 1000 segments. As truck speed increases, both the rms and maximum g's increase. This was expected since the wheels have a shorter time to rise the same height when going over a bump at a greater speed.

Figures 7-8 show the effect of trailer weight. As weight increases, the rms and maximum g's decrease. This was also expected. The greater the weight, the greater the inertia. The trailer tends to move less in response to the same bump.

The relationship between acceleration and speed appears to be linear, $a=c_1+c_2V$, and the relationship between acceleration and weight appears to be of the form $a=c_3/W$, where c_1, c_2, and c_3 are constants. This fits with the approximate solution to the much simpler case of a truck trailer traveling over a flat road approaching a step bump of height h. When the wheels hit the bump, the spring and dashpot are compressed an amount h, at speed h/t_h, where t_h is the time required for the wheels to rise to height h. From Figure 3, $t_h=D_{min}/V=(2Rh)^{½}/V$. The unbalanced force up on the trailer is the sum of the spring and dashpot forces, $kh+ch/[(2Rh)^{½}/V]$. Equating this to the product of mass and acceleration gives

$$a(t) = [kh \; + cV(\tfrac{h}{2R})^{1/2}]/\text{W} \quad (17)$$

Using $h=h_{max}$=0.042 ft (0.0128 m), W=20,000 lbf ($8.896\text{x}10^4$ N), and the trailer data: V=55 mph=80 ft/s (24.384 m/s), k=640,000 lbf/ft ($9.3396\text{x}10^5$ N/m), c=8,000 lbf-s/ft ($1.167454\text{x}10^5$ N-s/m), R=1.5 ft (4.572 m) obtain a=5.1 g's. This is much lower than the maximum acceleration reported for these conditions in Figures 6 and 8 because the trailer does not enter every bump from a flat road. The trailer could be moving down as it hits the bump, which would increase the g's. This demonstrates why it is necessary to consider the motion in real time.

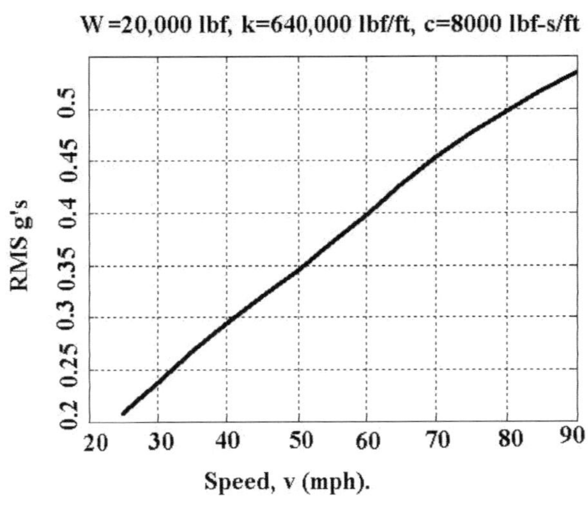

Figure 5. RMS g's vs. Truck Speed.

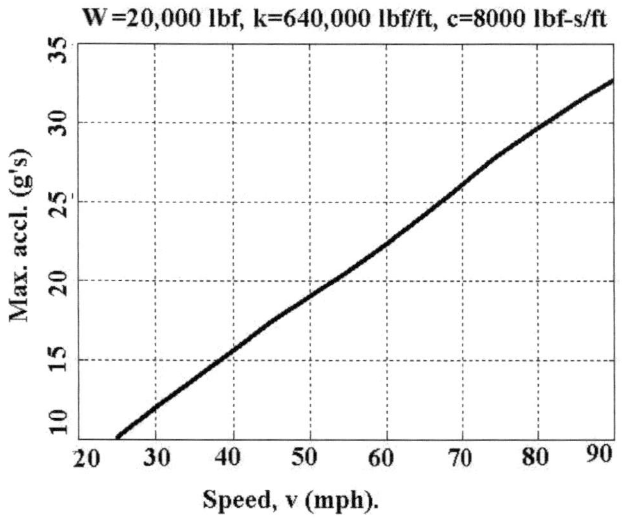

Figure 6. Max. Accel. Vs. Truck Speed.

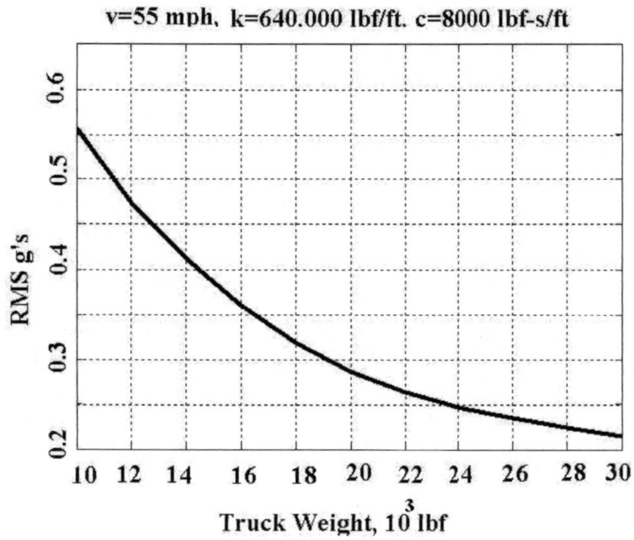

Figure 7. RMS g's vs. truck weight.

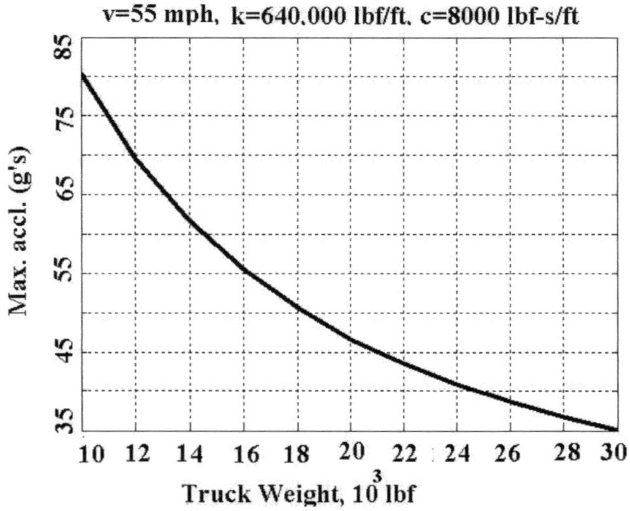

Figure 8. RMS g's vs. truck weight.

CONCLUSIONS

A method for generating trailer floor accelerations has been developed. With most commercial vibration tables, it is possible to input this acceleration directly into the controller. The table should replicate this motion if the transfer function is unity. Therefore, this method eliminates the need to record accelerations and drive the vibration table with PSD plots derived from these recordings. The results indicate that the method can be used with trucks of arbitrary weights traveling at variable speeds on roads having any profile.

REFERENCES

1. ASTM D999-08, "Standard Test methods for Vibration Testing of Shipment Containers."
2. ASTM D4728 - 06(2012), "Standard Test Method for Random Vibration Testing of Shipping Containers. "
3. *Manuel-Alfredo, et al,* "Measurement and Analysis of Vibration Levels," Packaging Technology and Science, *Package. Technol. Sci.* 2008; 21:439–451, Published online, 19 November 2007 in Wiley Inter-Science. Measurement and Analysis of Vibration Levels for Truck Transport in Spain as a Function of Payload, Suspension and Speed.
4. Palm III, W. ,J., "Mechanical Vibration," J. Wiley 2007.
5. Meirovich, J., "Elements of Vibration Analysis," McGraw-Hill, (1986).
6. MATLAB®, The Mathwork Inc., (2010),3 Apple Hill Drive, Natick, MA 01760-2098 USA.

Proceedings of the 25th CANCAM
London, Ontario, Canada, May 31 – June 4, 2015

FOURTH ORDER APPROXIMATE SOLUTIONS FOR THE UNDAMPED FREE VIBRATION OF STRONGLY NONLINEAR OSCILLATORS SUBLECT TO INITIAL DISPLACEMENTS

Vinicius B. F. Rabelo
Department of Mechanical Engineering
Lakehead University
Thunder Bay, ON, Canada
vbelemf@lakeheadu.ca

M. L. Liu
Department of Mechanical Engineering
Lakehead University
Thunder Bay, ON, Canada
mliu@lakeheadu.ca

ABSTRACT

In this paper, the approach for obtaining fourth order approximate solutions of a class of strongly nonlinear oscillators is presented. The procedure combines the Newton's method for solving a nonlinear algebraic equation and the harmonic balance method, and is applied successively to achieve approximations of higher and higher orders. It is an extension of the analytical approximate technique developed by Wu, Sun and Lim. An example is included to demonstrate the effectiveness and accuracy of the proposed technique.

KEYWORDS: Large amplitude oscillation, fourth order approximate solutions, free vibrations due to initial displacement

INTRODUCTION

Nonlinear oscillation has long been an important area of investigations. This is especially the case in mechanical engineering since many mechanical systems are inherently nonlinear. As demonstrated in Refs. 1 and 2, many methodologies have been developed to solve problems involving nonlinear oscillation. They range from methods of straightforward expansion and Lindstedt-Poincarè perturbation, to methods of multiple scales, harmonic balance and averaging.

With particular reference to undamped free vibration due to initial displacement only, for example, Lin (Ref. 3) investigated the Duffing oscillator with fifth order nonlinearity by the parameter iteration method. In 2006, Lai and Lim developed analytical approximate solutions to strongly nonlinear third order Duffing oscillator (see Ref. 4). These analytical approximate solutions were obtained by successively applying Newton's method (for solving nonlinear algebraic equations) in combination with the harmonic balance method. The methodology was then generalized and applied to a class of odd-functioned nonlinear oscillators in Ref. 5. It is noted that, in both Refs. 4 and 5, the methodology and resulting approximate solutions were up to the third order only.

In terms of undamped free vibrations due to arbitrary initial conditions (that is, initial displacement and velocity are both non-zero), Ref. 6 reported approximate solutions for the third order Duffing oscillator and Duffing-harmonic oscillator. Finally, for undamped vibrations in general, Ref. 7 presented jump frequencies and amplitude-frequency diagrams of a restrained cargo system which was modeled as a Duffing oscillator with various orders, in addition to natural frequencies (which are amplitude dependent) of the oscillator when in free vibration.

The investigation reported here is focused on obtaining the fourth order approximate solutions for the undamped free vibrations of strongly nonlinear oscillators due to initial displacements. The proposed methodology is an extension of that presented in Ref. 5. The motivation behind such an extension is twofold. First of all, when applying the approach presented in Ref. 5, that is, the third order approximation, to the restrained cargo system of Ref. 7, the authors found that the amplitude-dependent natural frequencies did not converge in a monotonic fashion; therefore fourth order (or even higher order) approximation would be deemed necessary. Secondly, the fourth order extension lays the foundation for higher order approximation and for investigating undamped free vibrations due to arbitrary initial conditions.

The organization of the paper is as follows. The current section is Section 1, introduction. The proposed methodology will be presented in Section 2, which is followed by an illustrative example given in Section 3. Finally, conclusions will be given.

METHODOLOGY

In this section, a fourth order expansion to the analytical approximate technique of Ref. 5 will be presented. For completeness in presentation, some details of first, second and third order approximation will be given. Readers can refer to Ref. 5 for more details.

As mentioned in the introduction, the approach to the analytical approximate solutions in Ref. 5 consisted of successive applications of Newton's method in conjunction with the harmonic balance method. Due to linearization by using Newton's method prior to harmonic balancing, linear algebraic equations were obtained instead of a system of complex nonlinear algebraic equations that was typically the outcome of harmonic balance method. The approach was applicable over a large range of parameters, and was valid for small as well as large amplitude of oscillation, as demonstrated by Refs. 4 and 5.

The contribution of the present work is in the extension to the fourth order. Furthermore, this extension lays the foundation for approximate solutions of even higher orders, and for incorporating non-zero initial velocities.

A class of strongly nonlinear oscillators

Consider a conservative nonlinear oscillator described by the following second order differential equation:

$$\frac{d^2u}{dt^2} + f(u) = 0 \tag{1}$$

with initial conditions of

$$u(0) = A, \frac{du}{dt}(0) = 0 \tag{2}$$

where the nonlinear function $f(u)$ is odd and satisfies $u{\cdot}f(u) > 0$ for $u \in [-A, A]$, $u \neq 0$. The period and corresponding periodic solution are dependent on the oscillation amplitude A.

Introducing a new independent variable $\tau = \omega t$ where ω is the (unknown) circular natural frequency of the oscillator. Eq. (1) can be rewritten as

$$\Omega u'' + f(u) = 0 \tag{3}$$

where $\Omega = \omega^2$ and the prime denotes differentiation with respect to τ. The initial conditions associated with Eq. (3) are

$$u(0) = A, \ u'(0) = 0 \tag{4}$$

First order approximation

The approximated periodic solution is

$$u_1(\tau) = A\cos\tau, \ \Omega_1(A) = \frac{a_1}{A} \tag{5}$$

where a_1 is a Fourier coefficient determined by,

$$a_{2i-1} = \frac{4}{\pi}\int_0^{\frac{\pi}{2}} f\big(u_1(\tau)\big)\cos[(2i-1)\tau]\,d\tau \tag{6}$$

where $i = 1, 2, 3, \ldots$

Second order approximation

The approximated periodic solution now takes the form:

$$\begin{aligned} u_2(\tau) &= u_1(\tau) + \Delta u_1(\tau), \\ \Omega_2(A) &= \Omega_1(A) + \Delta\Omega_1(A) \end{aligned} \tag{7}$$

with

$$\Delta u_1(\tau) = x_1(\cos\tau - \cos 3\tau) \tag{8}$$

where x_1 and $\Delta\Omega_1$ are determined by

$$\begin{bmatrix} b_0 - b_4 - 2\Omega_1 & -2A \\ b_2 + b_4 - b_0 - b_6 + 18\Omega_1 & 0 \end{bmatrix} \begin{Bmatrix} x_1 \\ \Delta\Omega_1 \end{Bmatrix} = \begin{Bmatrix} 0 \\ -2a_3 \end{Bmatrix} \tag{9}$$

In Eq. (9), the Fourier coefficients a_{2i-1} are as given by Eq. (6). On the other hand, the coefficients $b_{2(j-1)}$ are,

$$b_{2(j-1)} = \frac{4}{\pi}\int_0^{\frac{\pi}{2}} f_u\big(u_1(\tau)\big)\cos[(2(j-1)\tau]\,d\tau \tag{10}$$

where f_u denotes df/du and $j = 1, 2, 3, \ldots$. Once x_1 is determined, $u_2(\tau)$ becomes:

$$u_2(\tau) = X\cos\tau + Y\cos 3\tau \tag{11}$$

where

$$X = A + x_1, Y = -x_1. \tag{12}$$

Third order approximation

In a similar fashion, the approximated periodic solution is

$$\begin{aligned} u_3(\tau) &= u_2(\tau) + \Delta u_2(\tau), \\ \Omega_3(A) &= \Omega_2(A) + \Delta\Omega_2(A) \end{aligned} \tag{13}$$

with

$$\Delta u_2(\tau) = y_1(\cos\tau - \cos 3\tau) + y_2(\cos 3\tau - \cos 5\tau) \tag{14}$$

The unknowns y_1, y_2 and $\Delta\Omega_2$ are determined by

$$\begin{bmatrix} \alpha_1 & \alpha_2 & -2X \\ \alpha_3 & \alpha_4 & -18Y \\ \alpha_5 & \alpha_6 & 0 \end{bmatrix} \begin{Bmatrix} y_1 \\ y_2 \\ \Delta\Omega_2 \end{Bmatrix} = -\begin{Bmatrix} 2c_1 - 2X\Omega_2 \\ 2c_3 - 18Y\Omega_2 \\ 2c_5 \end{Bmatrix} \tag{15}$$

The constants α_1 through α_6 are

$$\begin{aligned} \alpha_1 &= d_0 - d_4 - 2\Omega_2 \\ \alpha_2 &= d_2 - d_6 \\ \alpha_3 &= d_2 + d_4 - d_0 - d_6 + 18\Omega_2 \\ \alpha_4 &= d_0 + d_6 - d_2 - d_8 - 18\Omega_2 \\ \alpha_5 &= d_4 + d_6 - d_2 - d_8 \\ \alpha_6 &= d_2 + d_8 - d_0 - d_{10} + 50\Omega_2 \end{aligned} \tag{16}$$

and

$$\begin{aligned} c_{2i-1} &= \frac{4}{\pi}\int_0^{\frac{\pi}{2}} f\big(u_2(\tau)\big)\cos[(2i-1)\tau]\,d\tau \\ d_{2(j-1)} &= \frac{4}{\pi}\int_0^{\frac{\pi}{2}} f_u\big(u_2(\tau)\big)\cos[(2(j-1)\tau]\,d\tau \end{aligned} \tag{17}$$

Again, once y_1 and y_2 are determined, $u_3(\tau)$ becomes

$$u_3(\tau) = X_1\cos\tau + Y_1\cos 3\tau + Z_1\cos 5\tau \tag{18}$$

with

$$X_1 = X + y_1, \ Y_1 = Y - y_1 + y_2, Z_1 = -y_2. \tag{19}$$

Fourth order approximation

Starting with the following approximated periodic solution

$$\begin{aligned} u_4(\tau) &= u_3(\tau) + \Delta u_3(\tau), \\ \Omega_4(A) &= \Omega_3(A) + \Delta\Omega_3(A) \end{aligned} \tag{20}$$

where

$$\begin{aligned} \Delta u_3(\tau) = z_1(\cos\tau - \cos 3\tau) + z_2(\cos 3\tau - \cos 5\tau) \\ + z_3(\cos 5\tau - \cos 7\tau) \end{aligned} \tag{21}$$

The four unknowns z_1, z_2, z_3 and $\Delta\Omega_3$ are determined by

$$\begin{bmatrix} \beta_1 & \beta_2 & \beta_3 & -2X_1 \\ \beta_4 & \beta_5 & \beta_6 & -18Y_1 \\ \beta_7 & \beta_8 & \beta_9 & -50Z_1 \\ \beta_{10} & \beta_{11} & \beta_{12} & 0 \end{bmatrix} \begin{Bmatrix} z_1 \\ z_2 \\ z_3 \\ \Delta\Omega_3 \end{Bmatrix} = R \quad (22)$$

where

$$R = -\begin{Bmatrix} 2e_1 - 2X_1\Omega_3 \\ 2e_3 - 18Y_1\Omega_3 \\ 2e_5 - 50Z_1\Omega_3 \\ 2e_7 \end{Bmatrix} \quad (23)$$

and

$$\begin{gathered} \beta_1 = f_0 - f_4 - 2\Omega_3 \\ \beta_2 = f_2 - f_6 \\ \beta_3 = f_4 - f_8 + 98\Omega_3 \\ \beta_4 = f_2 + f_4 - f_0 - f_6 + 18\Omega_3 \\ \beta_5 = f_0 + f_6 - f_2 - f_8 - 18\Omega_3 \\ \beta_6 = f_2 + f_8 - f_4 - f_{10} \\ \beta_7 = f_4 + f_6 - f_2 - f_8 \\ \beta_8 = f_2 + f_8 - f_0 - f_{10} + 50\Omega_3 \\ \beta_9 = f_0 + f_{10} - f_2 - f_{12} - 50\Omega_3 \\ \beta_{10} = f_6 + f_8 - f_4 - f_{10} \\ \beta_{11} = f_4 + f_{10} - f_2 - f_{12} \\ \beta_{12} = f_2 + f_{12} - f_0 - f_{14} \end{gathered} \quad (24)$$

The coefficients e_i and f_i are,

$$\begin{aligned} e_{2i-1} &= \frac{4}{\pi}\int_0^{\frac{\pi}{2}} f(u_3(\tau)) \cos[(2i-1)\tau]\, d\tau \\ f_{2(j-1)} &= \frac{4}{\pi}\int_0^{\frac{\pi}{2}} f_u(u_3(\tau)) \cos[(2(j-1)\tau]\, d\tau \end{aligned} \quad (25)$$

Finally,

$$u_4(\tau) = X_2\cos\tau + Y_2\cos3\tau + Z_2\cos5\tau - z_3\cos7\tau \quad (26)$$

with

$$X_2 = X_1 + z_1, Y_2 = Y_1 - z_1 + z_2, Z_2 = Z_1 - z_2 + z_3 \quad (27)$$

In passing, it should be mentioned that Eqs. (22) through (24) have been verified both manually and with the aid of symbolic computation package MAPLE.

ILLUSTRATIVE EXAMPLE

In this section, an example will be presented to demonstrate the effectiveness and accuracy of the proposed methodology. The nonlinear oscillator considered here is an oscillator with fractional-power restoring force:

$$\frac{d^2u}{dt^2} + \text{sign}(u)|u|^{\frac{1}{3}} = 0 \quad (28)$$

where sign(·) being the signum function. This oscillator is subject to the same initial conditions as given in Eq. (2). The analytical (exact) period of this oscillator is

$$T_e = \sqrt{6\pi}\frac{\Gamma(\frac{3}{4})}{\Gamma(\frac{5}{4})}A^{\frac{1}{3}} \quad (29)$$

with Γ(·) denoting the gamma function.

For the first order approximation, setting $f(u) = u^{\frac{1}{3}}$, $\frac{df}{du} = f_u(u) = \frac{1}{3}u^{-\frac{2}{3}}, u_1(\tau) = A\cos\tau$ and applying Eq. (6) gets

$$a_1 = \sigma A^{\frac{1}{3}} \quad (30)$$

where

$$\sigma = \frac{\sqrt{\pi}}{\Gamma(\frac{2}{3})\Gamma(\frac{5}{6})} \quad (31)$$

Subsequently the period by first order approximation is

$$T_1 = \frac{2\pi}{\sqrt{\sigma}}A^{\frac{1}{3}} \quad (32)$$

For the second order approximation, the required Fourier coefficients are, by applying Eq. (6) and Eq. (10)

$$\begin{aligned} a_3 &= -\tfrac{1}{5}\sigma A^{\frac{1}{3}} \\ b_0 &= \tfrac{4}{3}\sigma A^{-\frac{2}{3}} \\ b_2 &= -\tfrac{2}{3}\sigma A^{-\frac{2}{3}} \\ b_4 &= \tfrac{8}{15}\sigma A^{-\frac{2}{3}} \\ b_6 &= -\tfrac{7}{15}\sigma A^{-\frac{2}{3}} \end{aligned} \quad (33)$$

Solving Eq. (9) yields,

$$x_1 = \frac{2}{85}A,\ \Delta\Omega_1 = -\frac{6}{425}\sigma A^{-\frac{2}{3}} \quad (34)$$

such that

$$\begin{aligned} u_2(\tau) &= \frac{87}{85}A\cos\tau - \frac{2}{85}A\cos3\tau \\ \Omega_2 &= \frac{419}{425}\sigma A^{-\frac{2}{3}} \\ T_2 &= 2\pi\sqrt{\frac{425}{419\sigma}}A^{\frac{1}{3}} \end{aligned} \quad (35)$$

For the third and fourth order approximations, analytical solutions could not be obtained due to the integrals defined by Eq. (17) and Eq. (25). Numerical integration is adopted instead. Specifically, the adaptive Simpson quadrature available in Matlab® is employed in the present study.

Table 1 below shows the ratios of approximated periods to the exact period, or T_e of Eq. (29). Amplitude A has been set to unity for easy comparison with results in Ref. 5. It should be noted that, for the oscillator defined by Eq. (28), Ref. 5 did not follow strictly the approach it had outlined for the third order approximation. Specifically, Z_1 was given an assumed form (see Eq. (67) in Ref. 5) instead of setting $Z_1 = -y_2$ as indicated by Eq. (19). The present approach is systematic, and is able to handle any complex odd function $f(u)$, albeit employment of numerical integration when it is called for.

Table 2, on the other hand, lists the exact periods, and the ratios of the computed approximate periods to their respective exact ones, over a wide range of A. It is seen that, (1) the ratio of approximated period to exact period remains unchanged, over the entire range of A that was investigated, for the first and second approximations; (2) the computed approximate periods converge to the true period in an oscillatory pattern similar to that of a damped free vibration; that is, the first order approximation yields a lower bound estimation on period, followed by an upper bound value as given by the second order approximation, and so on and so forth.

The time history plots of u_1, u_2, u_3 and u_4 versus real time t are given in Fig.1 for $A = 1$ and in Fig. 2 for $A = 250$.

Table 1: Ratios of Approximated Periods to T_e ($A = 1$)

Method	T_1/T_e	T_2/T_e	T_3/T_e	T_4/T_e
Present study	0.994062	1.00115	0.999691	1.00014
Ref. 5 (Example 3)	0.994062	1.00115	0.999692	N/A

Table 2: Ratios of Approximated Periods to T_e

A	T_e	T_1/T_e	T_2/T_e	T_3/T_e	T_4/T_e
0.005	1.00370	0.994062	1.00115	0.999686	1.00013
0.01	1.26458	0.994062	1.00115	0.999686	1.00013
0.1	2.72446	0.994062	1.00115	0.999691	1.00014
1	5.86966	0.994062	1.00115	0.999691	1.00014
10	12.6458	0.994062	1.00115	0.999692	1.00014
100	27.2446	0.994062	1.00115	0.999692	1.00014
500	46.5876	0.994062	1.00115	0.999692	1.00014

CONCLUSIONS

In this paper, a fourth order approximate solution for a class of strongly nonlinear oscillators was presented. The methodology can be seen as the extension of the analytical approximate technique developed by Wu, Sun and Lim in Ref. 5. The methodology consisted of successive applications of Newton's method in conjunction with the harmonic balance method. Due to linearization prior to harmonic balancing, linear algebraic equations were obtained instead of a system of complex nonlinear algebraic equations. The approach was applicable over a large range of parameters, and was valid for small as well as large amplitude of oscillation. The illustrative example demonstrated the effectiveness and accuracy of the proposed technique. Furthermore, this methodology lays the foundation for approximate solutions of even higher orders, and for incorporating non-zero initial velocities.

ACKNOWLEDGMENTS

The first author gratefully acknowledges the financial support by the Brazilian Government through CAPES, Proc. 88888.020471/2013-00, a Science Without Borders Scholarship which enabled him to spend one year in Canada. He is currently at Universidade Federal do Ceará, Brazil.

REFERENCES

1. Nayfeh, A.H. 1973, *Perturbation Methods*, Wiley, New York.
2. Nayfeh, A.H., and Monk, D.T. 1979, *Nonlinear Oscillations*, Wiley, New York.
3. Lin, J. 1999, "A new approach to Duffing equation with strong and high order nonlinearity", *Communications in Nonlinear Science and Numerical Simulation*, Vol(4), pp. 132-135.
4. Lai, S.K., and Lim, C.W. 2006, "Higher-order approximate solutions to a strongly nonlinear Duffing Oscillator", *International Journal for Computational Methods in Engineering Science and Mechanics*, Vol(7), pp. 201-208.
5. Wu, B.S., Sun, W.P., and Lim, C.W. 2006, "An analytical approximate technique for a class of strongly non-linear oscillators", *International Journal of Non-Linear Mechanics*, Vol(41), pp. 766-774.
6. Chen, Y.M., and Liu, J.K. 2007, "A new method based on the harmonic balance method for nonlinear oscillations", *Physics Letters A*, Vol(368), pp. 371-378.
7. Lesage, J. and Liu, M.L. 2008, "On the investigation of a restrained cargo system modeled as a Duffing oscillator of various orders", *2008 ASME Early Career Technical Conference*, Miami, U.S.A., paper number 108.

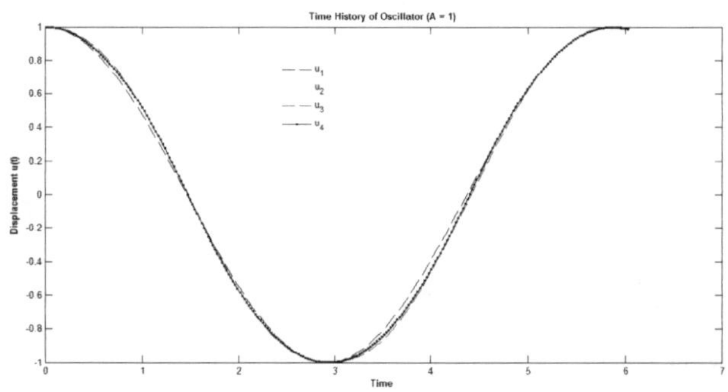

Figure 1: Time History Plot of Oscillator when $A = 1$

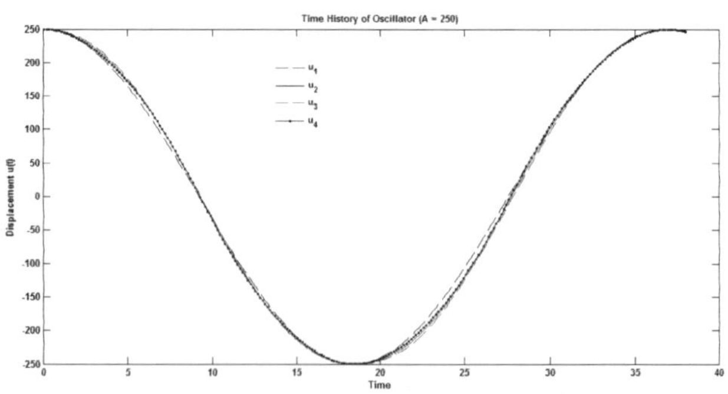

Figure 2: Time History Plot of Oscillator when $A = 250$

Proceedings of the 25th CANCAM
London, Ontario, Canada, May 31 – June 4, 2015

SEISMIC RESPONSE INVESTIGATIONS OF NONLINEAR SINGLE-DEGREE-OF-FREEDOM SYSTEMS EQUIPPED WITH GYROMASS DAMPERS

Reza Mirza Hessabi
Department of Civil Engineering
University of Toronto
Toronto, Ontario, Canada
reza.hessabi@mail.utoronto.ca

Jiejing Zhou
Division of Engineering Science
University of Toronto
Toronto, Ontario, Canada
jiejing.zhou@mail.utoronto.ca

Oya Mercan
Department of Civil Engineering
University of Toronto
Toronto, Ontario, Canada
oya.mercan@utoronto.ca

ABSTRACT

In this paper, the effectiveness of gyromass dampers (GMDs) in vibration control of nonlinear single-degree-of-freedom (SDOF) systems under two historical ground motions is evaluated. A single lumped mass model is employed to represent the main structure and a bilinear material behaviour is assumed for this SDOF system. The damping and accumulated input energies modified by the introduction of the GMD are examined and the peak displacement and acceleration ratios in the systems with and without GMDs are compared as the performance measures. Performance of a GMD-nonlinear SDOF system could be influenced by several parameters such as the inertance ratio of the GMD, dynamic properties of the main structure such as the natural period and post-yielding stiffness ratio, and properties of the input ground motion. To assess the effects of each of these parameters, numerical models are used and time history analyses are carried out. It is shown that for the ground motions considered, GMDs have the potential to be effective in improving the seismic performance of the nonlinear SDOF systems.

KEYWORDS: Gyromass dampers, seismic protection, passive control, bilinear behaviour, single-degree-of-freedom systems

INTRODUCTION

Some of the conventional seismic design methods assume that the undesirable earthquake energy should be dissipated by inelastic behaviour of structural elements. However, the damage occurred in the structure during an earthquake may render it nonfunctional. As the resilience of civil structures, such as buildings and bridges, becomes more important, the focus has shifted towards the development of structural control systems. In recent years, considerable attention has been given to research and development of new devices that can modify the dynamic characteristics of structures and consequently improve the seismic performance of these systems. One novel passive control strategy is the application of gyromass dampers (GMDs) which will be the subject of this paper. GMDs are newly developed mechanical devices that generate a resisting force, which is proportional to the relative acceleration applied at their end nodes. This is possible through the transformation of the rotational inertial forces of the rotary components of a GMD to a resultant translational force.

Early applications of GMDs have been adopted to mitigate the vibrational response of car, motorcycle and train suspension systems [1-3]. Despite the considerable amount of research focused on the use of GMDs in these fields, their application in civil engineering is relatively new. In 2004, Saito et al. numerically investigated the effects of a new type of viscous damper where an intrinsic rotary mass subsidiary behaved as a gyromass [4]. Ikago et al. [5] investigated the effectiveness of these modified viscous dampers in the seismic control of single-degree-of-freedom (SDOF) structures and derived closed-form optimum design equations. They consequently verified their numerical simulation by conducting shake table tests. Based on the results of their studies, these devices have been installed in a steel structure built in Japan [6]. Other researchers have also considered SDOF systems to evaluate the effects of GMDs on various types of structures. For instance, Saitoh [7] used a GMD to mitigate the lateral displacements of base isolators.

Yet despite all these efforts, it can be seen that most of the studies listed above have investigated the interaction between a GMD and a linear primary structure. Thus the understanding of the dynamic effects of GMDs on nonlinear structural systems has still not been well developed. With the objective of explaining the performance benefits of GMDs in real-life structures, the present article investigates the dynamic behaviour of GMD-nonlinear SDOF systems.

GYROMASS DAMPERS

A GMD is a control system that generates a reaction force proportional to the relative translational acceleration between its end nodes (Fig. 1). This force (F) can be expressed as:

$$F = b\ddot{x} \tag{1}$$

where, b in Eq. 1 is called the inertance and $\ddot{x}$ is the transitional acceleration of the damper.

When a structure is equipped with a GMD, its dynamic response is greatly modified. Due to the complexity of the dynamic response of structures to earthquake ground motions and the uncertainty associated with ground motions, a thorough understanding of the impact of adding a new device to a structure is necessary to ensure the effectiveness of the system. Considering a general SDOF system that is excited by an earthquake ground motion, the governing differential equation of motion can be written as shown in Eq. 2:

$$m\ddot{x}(t) + c\dot{x}(t) + F_s(t) + b\ddot{x}(t) = -m\ddot{x}_g(t) \quad (2)$$

where, m and c are the mass and damping of the main primary SDOF system; $x(t)$, $\dot{x}(t)$ and $\ddot{x}(t)$ are the displacement, velocity and acceleration of the mass, respectively and F_s is the force developed in the primary structure which can exhibit nonlinear behavior. Finally, $\ddot{x}_g(t)$ is the ground motion accelerations at time t.

Figure 1: A small-scale prototype of gyromass damper

According to [8], where the effects of the GMD on the response of linear SDOF systems have been discussed, with the introduction of these dampers, the fundamental vibrational period of the structures is elongated. This can improve the seismic performance of the system while decreasing the input seismic energy. On the other hand, GMDs reduce the damping energy of the linear SDOF systems. In the following sections, the efficiency of the GMDs in improving the seismic nonlinear response of structures is investigated and compared to the results from the uncontrolled systems.

NUMERICAL SIMULATIONS AND RESULTS

In order to examine the proposed control system, a bilinear material behaviour is considered for the SDOF systems and numerical simulations are carried out. The structure is represented by a single lumped mass (equal to 60 tons) and the initial natural period of vibration of the primary structure is 0.8 s. Also, the damping ratio of the main system is taken as 5%. In order to idealize the behaviour of the nonlinear system by the bilinear piecewise linear model, the characteristic parameters (i.e., yield strength and post-yield stiffnesses ratio) have to be defined. The value of 0.10 is chosen for the post- to pre-yield stiffness ratio (α), which can be considered as an average of values suggested in the literature. The yield strength may be selected within a suitable range depending on the value of the vertical force due to the weight of the structure. Based on values proposed in the literature, 0.05 ~ 0.20 of the weight can be used. In this study, the value of 0.05 is chosen. Due to the similarities between GMDs and tuned mass dampers, it is common in the literature to determine the inertance of the GMD in terms of the mass of the SDOF system. Here, the inertance ratio (i.e., inertance over mass ratio) of 0.50 is considered. For input earthquake ground motions, two different historical ground motions were selected from the PEER-NGA database [9], and time history analyses were performed. The details of each of these ground motions are reported in Table 1.

Table 1: List of the earthquake ground motions considered in this study [9]

	Event	NGA #	Magnitude	PGA (g)	PGD (cm)
1	Northridge	1044	6.69	0.58	17.98
2	Imperial Valley	0183	6.53	0.60	32.21

For the above-described parameters of the GMD-nonlinear SDOF system, the response time history under the Northridge earthquake is shown in Fig. 2. As it can be seen in this figure, the introduction of the GMD affects the response of the system in time domain where a reduction of the displacements of the nonlinear SDOF system with the GMD can be observed.

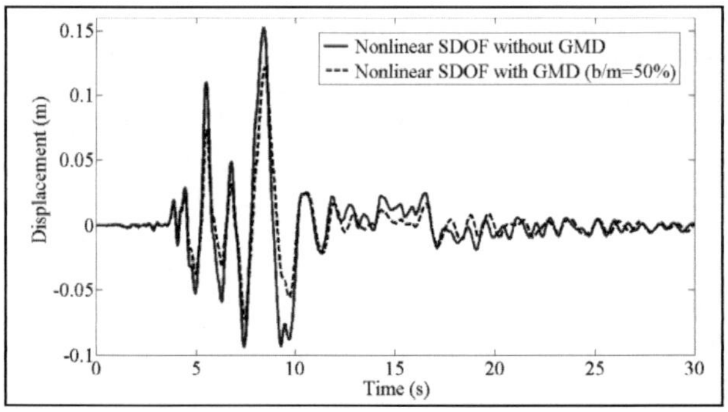

Figure 2: Comparison between the responses of the nonlinear models with/without GMDs (Northridge record)

In order to compare the effects of GMDs on the energy dissipation of the nonlinear systems, hystereses are plotted and compared in Fig. 3.

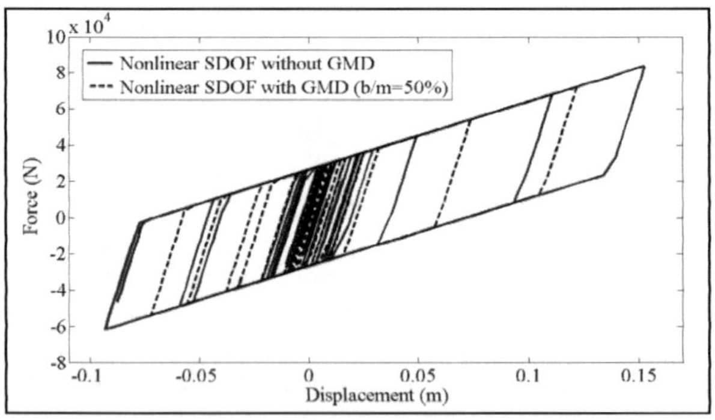

Figure 3: Hystereses of models with/without GMDs

A closer look at Fig. 2 and 3 shows the benefits of the GMDs as the peak and root mean square of displacement are decreased by 20 % and 28 %, respectively. However, similar to linear systems, this decrease is achieved at the expense of the reduction of the system energy dissipation capacity.

The reason for the decrease in the GMD-nonlinear system displacements can be attributed to the reduction of the input energy of the system. Seismic input energies of the models with/without GMDs are compared in Fig. 4. The maximum input energy is reduced by 37 % in Fig. 4.

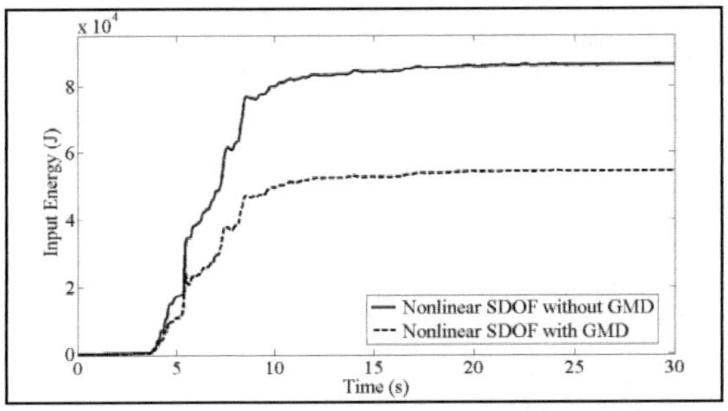

Figure 4: Effects of the GMDs on the cumulative input energies of the structures (Northridge earthquake)

To investigate the effects of the vibrational period of the original system on the response of the controlled system, energy spectra are used here. The energy spectra are generated by changing the natural period of the primary system and plotting the corresponding maximum value of the input energy terms. As it can be seen in Fig. 5, GMDs are effective throughout the applicable civil engineering period range. Moreover, there is a higher reduction in the seismic input energy for lower vibrational periods.

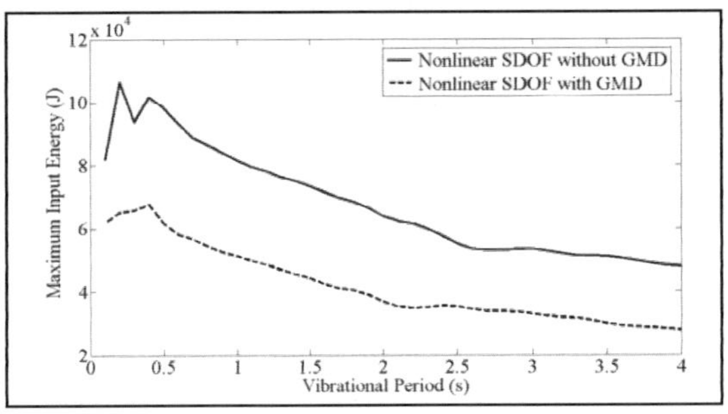

Figure 5: Effects of the vibrational period of the original structural on the damping energy spectra

The sensitivity of the response to the nonlinear parameters is examined by changing the yield strength over weight and post-to pre-yield stiffnesses ratios. Peak displacements of the nonlinear models with GMDs are normalized by the corresponding values from the models without the GMD. This process is repeated for the earthquake records shown in Table 1. It should be noted that the GMDs are effective when the vertical axis values are less than unity in Fig. 6 through 9. Based on the results in Fig. 6 similar effects were observed under both earthquakes that were used in this study and but it is clear that the control strategy is more sensitive to the nonlinear parameters when the system is subjected to the Northridge record.

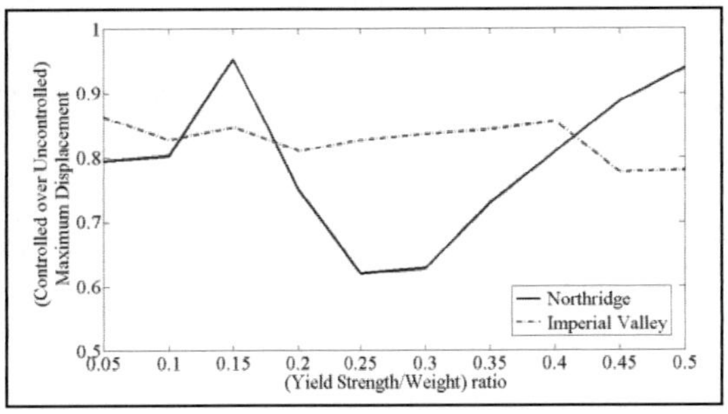

Figure 6: Effects of the yield strength over SDOF weight ratio on maximum displacements

For results shown in Fig. 7, for both of the earthquakes, GMDs are effective when the pre- to post-yield stiffness ratios are lower. More details about the performance of GMD-linear SDOF systems (i.e., when the ratio is close to one) can be found in [8].

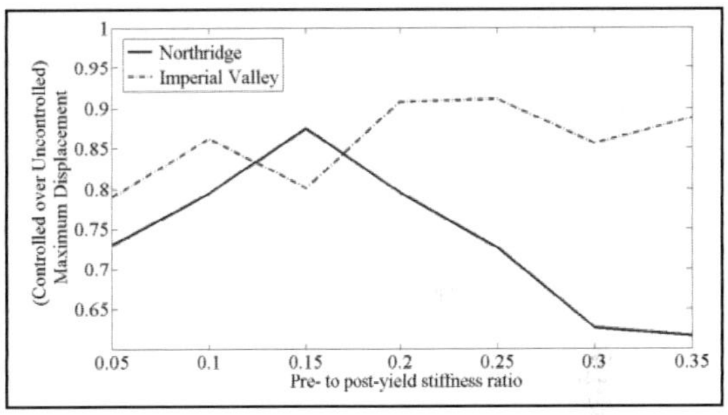

Figure 7: Effects of the post-to pre-yield stiffnesses ratio on maximum displacements

Finally, all of the results shown in the previous sections are obtained for an inertance of 6 tons. This value corresponds to a total inertia over total mass ratio of 10%. This ratio is plotted against the normalized maximum displacements and accelerations in Fig. 8 and 9, respectively.

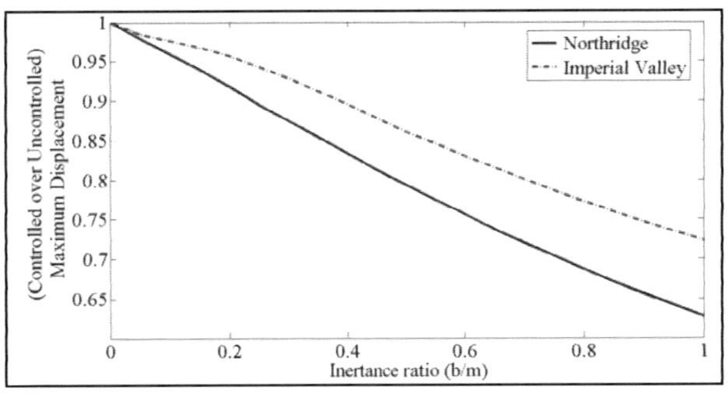

Figure 8: Effects of the inertance ratio on displacements

Fig. 8 shows that the effectiveness of the GMDs depends on the inertance ratio. Both displacements and accelerations decrease monotonously with the increase of the inertance. However, since physical limitations should also be considered in selecting the inertance, future studies are necessary to determine the values of the inertance over mass ratio to attain optimum performances.

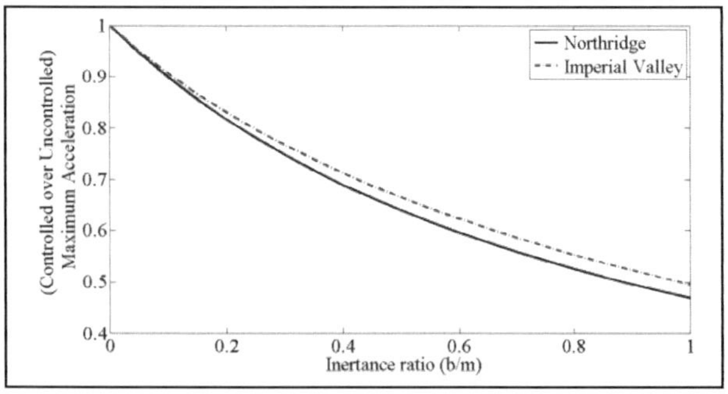

Figure 9: Effects of the inertance ratio on maximum accelerations

Fig. 9 also reveals that, as expected, GMDs are more effective in reducing the peak accelerations. This is especially useful for structures that contain acceleration sensitive equipments and components.

CONCLUSIONS

In order to consider the effects of GMDs on SDOF systems with nonlinearities, a bilinear model is used and the results are compared to the corresponding uncontrolled GMD-SDOF system. By using a GMD device, peak response can be reduced significantly. This can be attributed to the reduction of the input energy. The efficiency of the entire structural control system may vary for different input earthquake ground motions as the frequency content and intensity of the earthquake can force the system to resonate. Moreover, through several numerical simulations, it has been shown that the combination of yield strength and post- to pre-yield stiffnesses ratio can also result in different responses. Appropriate selection of a combination of these variables should be made during the design procedure of the GMD-nonlinear SDOF systems. It has been found that consistent with the observations made in the cases where the SDOF system remained linear, the GMD-nonlinear SDOF systems with higher inertance ratios exhibits lower peak displacements and accelerations. In addition, the acceleration is more sensitive to the choice of inertance ratio. For the ground motions considered, GMDs are found to be effective in improving the seismic performance of the SDOF systems with bilinear behaviour. This is of practical significance in the view of the current trend toward performance-based design. Future studies should be carried out to investigate the performance of nonlinear multi-degree-of-freedom structures with GMDs.

REFERENCES

1. Smith, M.C., and Wang, F.C. 2004, "Performance benefits in passive vehicle suspensions employing inerter", *Vehicle System Dynamics*, 42(4), pp. 235-257.
2. Evangelou, S., Limebeer, D.J.N., Sharp, R.S., and Smith, M.C. 2007, "Mechanical steering compensators for high-performance motorcycles", *Journal of Applied Mechanics-Transactions of the ASME*, 74(2), pp. 332-346.
3. Jiang, J.Z., Matamoros-Sanchez, A.Z., Goodall, R.M. and Smith, M.C. 2012, "Passive suspensions incorporating inerters for railway vehicles", *Vehicle System Dynamics: International Journal of Vehicle Mechanics and Mobility*, 50 (Supplement 1), pp. 263-276.
4. Saito, K., Yogo, K., Sugimura, Y., Nakaminami, S., and Park, K. 2004, "Application of rotary inertia to displacement reduction for vibration control system", *13th World Conference on Earthquake Engineering*, Vancouver, BC, Canada, Paper No. 1764.
5. Ikago, K., Saito, K., and Inoue, N. 2012, "Seismic control of SDOF structure using tuned viscous mass damper", *Earthquake Engineering and Structural Dynamics*, 41 (3), pp. 453-474.
6. Sugimura, Y., Goto, W., Tanizawa, H., Saito, K., and Nimomiya, T. 2012, "Response control effect of steel building structure using tuned viscous mass damper", *Proceedings of the 15th World Conference on Earthquake Engineering*, Lisbon, Portugal.
7. Saitoh, M. 2012, "On the performance of gyro-mass devices for displacement mitigation in base isolation systems", *Structural Control and Health Monitoring*, 19(2), pp. 246-259.
8. Mirza Hessabi, R., and Mercan, O. 2015, "Application of Gyro-Mass Dampers to Mitigate the Seismic Failure in Soft First Story Buildings", *Structures Congress 2015 - ASCE*, Portland, Oregon, USA.
9. Pacific Earthquake Engineering Research Center. 2014, *Next Generation Attenuation Relationships – Strong motion Database*, Regents of the University of California, Berkeley, California, USA. Retrieved from http://peer.berkeley.edu/nga.

Proceedings of the 25th CANCAM
London, Ontario, Canada, May 31 – June 4, 2015

HIGH-CYCLE FATIGUE TESTING ON AERO-ENGINE BLADE

Nima Bahrani
Mechanical and Industrial Engineering
University of Toronto
Toronto, ON, Canada
nima.bahrani@mail.utoronto.ca

Kamran Behdinan
Mechanical and Industrial Engineering
University of Toronto
Toronto, ON, Canada
behdinan@mie.utoronto.ca

ABSTRACT

An appropriate high-cycle fatigue testing methodology is described to investigate the fatigue behavior of the material and generate the S-N curve for aero-engines blades. The sine-dwell excitation approach is employed to accelerate fatigue failure of the blade by applying a base excitation at the resonance frequency of the blade using an electrodynamic shaker. Two criteria are discussed to track the resonance frequency and update the frequency of the excitation in order to obtain maximum cumulative fatigue in the structure. A computerized feedback control system is connected to the shaker to implement the dwell criterion and update the excitation frequency. The results of proposed high-cycle fatigue testing on three axial-fan blades of an aircraft engine are presented. The vibration parameters of the blade at the resonance before fatigue failure are compared with corresponding parameters after crack initiation to conclude the effects of the crack initiation on the structure.

KEYWORDS: High-cycle fatigue, Vibration, Sine-dwell excitation, Crack initiation

INTRODUCTION

High-cycle fatigue (HCF) had been identified as the most significant cause of blade failure in gas turbine engines [1]. High-cycle fatigue tests are employed to investigate the fatigue behavior, generate the S-N curve, and determine the stress or strain limit for design. High-cycle fatigue testing is a type of fatigue test which evaluates the fatigue behavior in high-cycle regime approximately below 10^7-10^8 cycles [2]. Fatigue behavior of a design essentially depends on material properties. However, geometrical effects can alter fatigue behavior of the design. Therefore, the most reliable method to determine fatigue behavior of the blade design is to perform a high-cycle fatigue test on several specimens which have exact same blade design.

Traditionally, low-frequency servo-hydraulic testing machines are used to evaluate the fatigue behavior of materials. These machines are not suitable to handle HCF testing since a high-cycle test takes several days in a low-frequency servo-hydraulic machine. Several fatigue testing systems have been developed to handle high cycles in recent two decades including ultrasonic fatigue testing systems [3], high-frequency servo-hydraulic machines [4], piezoelectric actuators [5], electro-magnetic resonance machines [6], and electrodynamic shakers [7]. Depending on test case and study purposes, each of these testing systems has advantages and disadvantages in comparison with alternatives. For instance, performing ultrasonic tests at high frequencies (typically 20 KHz) result in unusual physical effects such as internal friction and self-heating [8], while aero-engines blades experience vibration rates in range of 100 to 4000 Hz.

Electrodynamic shaker is a well-suited machine for studying fatigue behavior of aero-engines blades since fatigue testing can be performed on original blade specimens using shakers, unlike other systems which are suitable for standardized material test specimens. Furthermore, electrodynamic shakers provide vibrating force in wide frequency range. This paper describes a methodology for HCF test using electrodynamic shakers and the sine-dwell excitation approach. The experimented results of the HCF testing on an aero-engine blade are also presented.

SINE-DWELL EXCITATION

Sine-dwell excitation accelerates fatigue failure by applying a vibrating force on a part at its resonance frequency due to the vibration amplitude of the part being at maximum at the resonance frequency which results in higher stress or strain levels. In addition, the testing time will be reduced if the resonance frequency is selected at higher frequencies especially after 100 Hz. In other words, choosing higher resonance frequency makes the high-cycle fatigue testing practical where 10^7 cycles of vibration is achieved in lower testing time at higher frequencies.

The excitation force is applied to the support points known as base. A generalized base excitation problem has been shown in Fig. 1 for a single degree-of-freedom (SDOF) system.

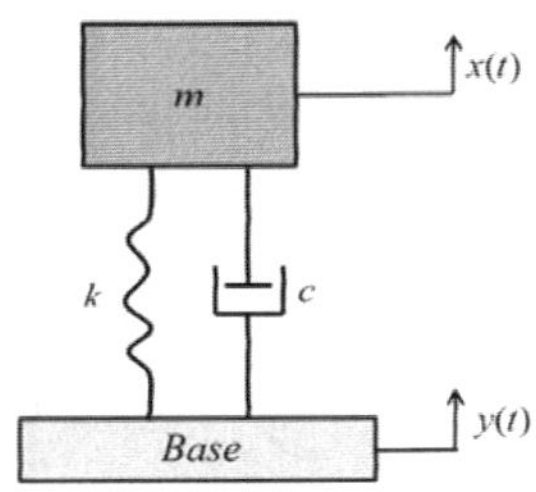

Figure 1: Base excitation of a SDOF system

The base excitation $y(t) = Y \sin \omega t$ causes the mass m to vibrate as a function of excitation frequency and amplitude $x(t) = X\sin(\omega t + \phi)$. By solving the differential equation of the system the ratio of the response amplitude X to base excitation amplitude Y known as the displacement transmissibility can be expressed as [9]

$$\frac{X}{Y} = \sqrt{\frac{1+(2\zeta r)^2}{(1-r^2)^2+(2\zeta r)^2}}, \qquad (1)$$

and the phase difference ϕ is calculated according to Eq. (2)

$$\phi = \tan^{-1}\left[\frac{2\zeta r^3}{1+(4\zeta^2-1)r^2}\right], \qquad (2)$$

where ζ is given by $\zeta = c/2m\omega_n$, r is given by $r = \omega/\omega_n$, ω and ω_n represent the excitation and the natural frequency, respectively.

Sine-dwell tests excite structural resonances with a sinusoidal force which is applied at a fixed frequency or at a tracking resonance frequency with respect to vibration amplitude or phase. In the fixed frequency method, the vibrating force is applied at the predetermined resonance frequency which is fixed throughout the test while in the tracking mode the resonance frequency is updated. The sine-dwell test is utilized to promote fatigue failure. Damage progression and thermal effects can cause variations in mechanical properties such as stiffness, which results in changes in resonance frequency [10]. Consequently, the resonance frequency of the structure shifts progressively during the test. Therefore, the resonance frequency should be tracked and modified during the test, in order to obtain maximum cumulative fatigue in the structure. In the tracking method, the frequency of the vibrating force is updated to instantaneous resonance frequency using a computerized feedback control system which updates these frequencies according to amplitude criterion or phase criterion.

The amplitude criterion updates the frequency of the base excitation based on the response amplitude as shown in Eq. (3) [11]

$$\begin{cases} \omega_{n+1} = \omega_n + \partial\omega_s & \text{if } x_n > x_{n-1} \\ \omega_{n+1} = \omega_n - \partial\omega_s & \text{if } x_n < x_{n-1} \end{cases}, \qquad (3)$$

where $\partial\omega_s$ is calculated by $\partial\omega_S = \Delta\omega/S$, $\Delta\omega$ is the frequency bandwidth which is predetermined manually, S is the sine-dwell tracking step factor, ω_n demonstrates the excitation frequency at n cycles, and x represents the response amplitude. The frequency is updated depending on the vibration amplitudes of previous steps. The purpose of this criterion is to maintain the amplitude at maximum as it is imposed in the Eq. (3).

The phase dwell criterion updates the excitation frequency to maintain the input-output phase difference at a targeted value. The control system compares the phase difference with the target phase and updates the frequency with respect to Eq. (4) and Eq. (5) [11]

$$\begin{cases} \omega_{n+1} = \omega_n + \partial\omega_s & \text{if } \omega_T - \omega_n > \partial\omega_s \\ \omega_{n+1} = \omega_T & \text{if } \omega_T - \omega_n \le \partial\omega_s \end{cases}, \qquad (4)$$

$$S_p = \frac{\phi_n - \phi_{n-1}}{\omega_n - \omega_{n-1}}, \quad \omega_T = \frac{1}{S_p}(\phi_T - \phi_{n-1}) + \omega_{n-1} \qquad (5)$$

where $\partial\omega_s$ is calculated by $\partial\omega_s = \Delta\omega/S$, $\Delta\omega$ is the frequency bandwidth which is predetermined manually, S is the sine-dwell tracking step factor, S_p is the slop of the plot of ϕ versus ω, ϕ_T is the target phase, ω_n demonstrates excitation frequency at n cycles, and ϕ represents input-response phase difference.

TEST PROCEDURE

The first step to perform a high-cycle fatigue test is to set up the required equipment. A strain gauge is placed in the maximum stress spot of the blade which is determined before the testing by finite element analysis as the maximum strain level should be monitored for plotting the S-N curve. The specimen is attached to a fixture which is mounted on the shaker that simulates the fixed-free (cantilever) boundary condition. An accelerometer is attached to the fixture in order to evaluate and control the excitation force. The laser beam of the vibrometer, which is mounted on a tripod and placed at an optimum stand-off distance from the specimen, is pointed on the specimen as close as possible to its tip. The vibrometer measures the velocity or displacement response of the forced excitation for the specimen. A schematic for the testing setup has been shown in Fig. 2. All sensors including the accelerometer, the strain gauge, and the vibrometer are connected to software through a data acquisition system to process the measured signals. The software is also connected to an amplifier, which provides power for the shaker, through a controller. The software controls the frequency of excitation by changing the input signal of the amplifier using frequency tracking criterion. The shaker is bolted down to an optical table and supported with a cooling unit.

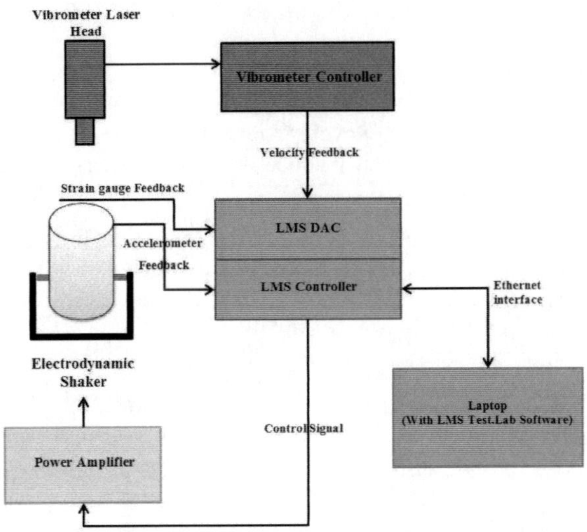

Figure 2: HCF testing configuration

- **Calibration Testing:** Occasionally, at high g-level, strain gauges can break during the HCF fatigue test. As a result, a calibration test is required to establish the relationship between strain level, displacement, and excitation g-level by performing several sine sweep tests at different g-levels. FRF, strain level and corresponding g-levels are recorded in order to plot g-level versus strain graph.
- **Sine Sweep Testing:** To search for resonance frequency, a sine sweep test is required to be performed through a range of frequencies which includes the resonance mode of interest. The g-level is determined with respect to the required g-level for the sine-dwell test.
- **Sine Dwell Testing:** The specimen is excited by the shaker at resonance frequency which is determined by the sine sweep test in previous steps. The resonance frequency is tracked by phase dwell criterion until the blade cracks or 10 million cycles is reached. Blade cracking will result in a substantial shift in the resonance frequency because of the change in the structure stiffness after crack initiation. The test is stopped either by detecting crack on the surface or by observing a considerable shift in the resonance frequency. Consequently, the software is set to stop exaction when the frequency shift exceeds 1.5%. The accelerometer, strain gauge and vibrometer are used throughout the test to enable the controller to track the resonance frequency and evaluate the strain level. As soon as the specimen cracks, number of cycles is recorded in order to use it for plotting the S-N curve.

RESULTS

Three specimens have been tested to evaluate the fatigue behavior of the material. These specimens are axial-fan blades of an aircraft engine which were fabricated from aluminum.

- **Calibration Results:** The calibration test was performed before the main high-cycle fatigue test to correlate the results obtained from different g-levels. The velocity and displacement of a point near to the tip was also measured to evaluate the magnitude of the vibration by the vibrometer. The g-level was increased from 5g to 35g. The values for measured velocity, displacement, and strain levels are related to the amplitude of these parameters at the fundamental frequency. As the amplitude of the base excitation (g-level) was increased, displacement, velocity, and strain value were increased with respect to the g-level. The correlation between the acceleration of the applied excitation in terms of g-level as the input and maximum strain level of the blade as the response has been presented in Fig. 3. The calculated correlation coefficient is relatively close to 1 which implies that the g-level and strain levels have a linear correlation.

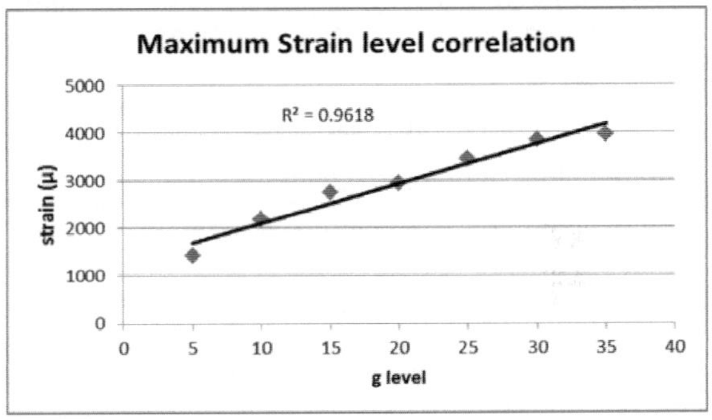

Figure 3: Correlation between g-level and maximum strain level of the blade

- **Dwell Test Results:** The dwell tests were conducted at different g-levels. Specimens were excited at their first resonance frequency (bending mode) at 10g, 13g, and 15g until the blade cracked. The resonance frequency of the blade was observed to a shift away from its initial position when the blade cracked. The results summary which is required for plotting the S-N curve has been presented in Table 1.

g level	Strain level (µ)	Number of Cycles (cracking)	Resonance Freq. difference	Displacement Amplitude difference
10	2157.1	9.148620×10^6	-1.79 %	+62 %
13	2395.6	1.299052×10^6	-2.84 %	+72 %
15	2714.6	0.345439×10^6	-2.30 %	+49 %

Table 1: Dwell test Results

Fig. 4 represents the comparison of displacement spectrum before and after crack initiation when the excitation was applied at 10g. These spectrums were obtained by performing the sine sweep test before and after the crack initiation. The resonance frequency of the specimen has been shifted due to the crack initiation in the blade. The crack initiation causes reduction in the stiffness of the blade which essentially results in a frequency shift and a higher level of the vibration as shown in Fig. 4.

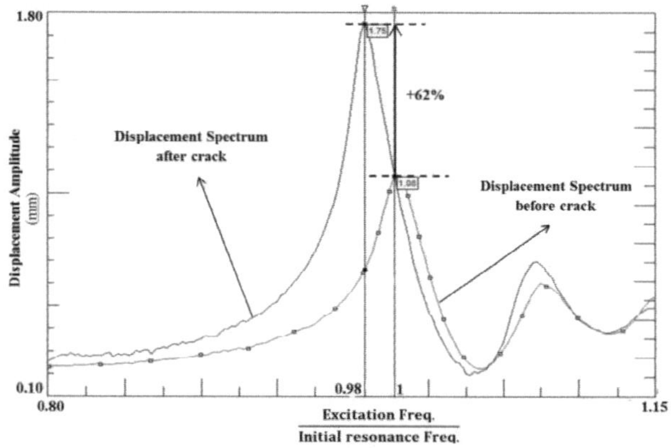

Figure 4: Displacement spectrums at 10g

- **S-N Curve:** The S-N curve was obtained based on the results of three sine-dwell tests at different g-levels as presented in Fig. 5.

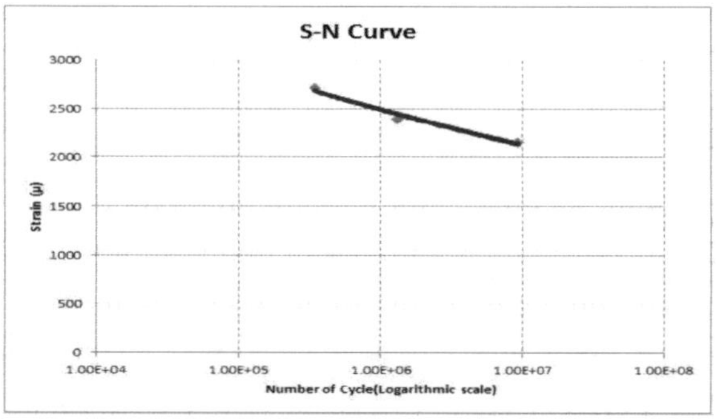

Figure 5: S-N curve

CONCLUSIONS

An appropriate methodology has been proposed to perform HCF testing on aero-engines blades. Based on observations of the high-cycle fatigue of three aluminum blades, the following conclusion can be made:

1. In order to obtain the maximum cumulative fatigue in sine-dwell excitation, the resonance frequency of the part should be tracked and the excitation frequency should be updated to the instantaneous resonance frequency using a computerized feedback control system.
2. The calibration results represent a linear correlation between the level of the base excitation and the response of the part as the theoretical foundations indicate (Eq. (1)).
3. Based on comparison of displacement spectrums before and after crack, the crack initiation results in stiffness reduction which causes higher vibration levels as well as a decrease in the resonance frequency as the sine-dwell experimental results represent (Fig. 4).

ACKNOWLEDGEMENT

The authors would like to acknowledge the funding and support for this project received from the University of Toronto Advanced Research Laboratory for Multifunctional light Weight Structures (ARL-MLS), CFI, and NSERC.

REFERENCES

1. B. A. Cowles, "High cycle fatigue in aircraft gas turbines—an industry perspective," *Int. J. Fract.*, vol. 80, no. 2–3, pp. 147–163, 1989.
2. H. Mayer, "Influence of cyclic loads below endurance limit or threshold stress intensity on fatigue damage in cast aluminium alloy 319-T7," *Int. J. Fatigue*, vol. 27, no. 2, pp. 129–141, Feb. 2005.
3. A. Shyam, C. J. Torbet, S. K. Jha, J. M. Larsen, M. J. Caton, C. J. Szczepanski, T. M. Pollock, and J. W. Jones, "Development of ultrasonic fatigue for rapid, high temperature fatigue studies in turbine engine materials," in *10th International Symposium on Superalloys, Champion, PA, September*, 2004, pp. 19–23.
4. J. M. Morgan and W. W. Milligan, "A 1 kHz servohydraulic fatigue testing system," *High cycle fatigue Struct. Mater. TMS, Warrendale, PA*, pp. 305–312, 1997.
5. C. Fischer, R. Wagener, A. Friedmann, C. Axt, M. Matthias, T. Melz, and H. Kaufmann, "Piezoelectric Driven Testing Facilities to Research the Very High Cycle Fatigue Regime*," *Mater. Test.*, vol. 54, no. 11–12, pp. 746–749, 2012.
6. M. Hirao and H. Ogi, "Electromagnetic acoustic resonance and materials characterization," *Ultrasonics*, vol. 35, no. 6, pp. 413–421, Sep. 1997.
7. T. J. George, J. Seidt, M.-H. Herman Shen, T. Nicholas, and C. J. Cross, "Development of a novel vibration-based fatigue testing methodology," *Int. J. Fatigue*, vol. 26, no. 5, pp. 477–486, May 2004.
8. T. E. Owen, D. L. Davidson, and A. Nagy, "High-cycle fatigue test machine." U.S Patent No. 6023980, 15-Feb-2000.
9. S. S. Rao and F. F. Yap, *Mechanical vibrations*, vol. 4. Addison-Wesley New York, 1995.
10. V. Rouillard and M. A. Sek, "A frequency domain technique for maintaining resonance condition during sine dwell vibration testing of packages," *Packag. Technol. Sci.*, vol. 13, no. 6, pp. 227–232, Nov. 2000.
11. LMS International Co., "Enviromental Testing Theory," LMS Instruments, Belgium, 2009.

A STRUCTURED APPROACH FOR SOLVING THE INVERSE EIGENVALUE PROBLEM OF A BEAM WITH A GROUNDED MASS-SPRING ATTACHEMENT

Farhad Mir Hosseini
Mechanical Engineering
University of Ottawa
Ottawa, ON, Canada
fmirh084@uottawa.ca

Patrick Dumond
Mechanical Engineering
University of Ottawa
Ottawa, ON, Canada
pdumond@uottawa.ca

Natalie Baddour
Mechanical Engineering
University of Ottawa
Ottawa, ON, Canada
nbaddour@uottawa.ca

ABSTRACT

The effect of adding lumped elements to a beam on its natural frequencies and corresponding modeshapes is a problem which has been extensively studied. A good knowledge of these effects is required when considering the inverse problem of imposing the natural frequencies of a beam and subsequently determining the characteristics of the added lumped attachments such as the stiffness of a spring or the magnitude of a mass.

A previous investigation focused on the effects of adding lumped masses to a Euler-Bernoulli beam on its natural frequencies and corresponding modeshapes and proposed the determinant method in order to solve the inverse problem. The inverse problem imposed up to four natural frequencies on the Euler-Bernoulli beam with either simply-supported or clamped-free boundary conditions by modifying the position and value of the masses. The current investigation widens the scope of the previous work by analyzing the effect of adding a grounded mass and spring system on the beam's natural frequencies and corresponding modeshapes. The inverse problem, using the determinant method, is then considered for this system by imposing two desired natural frequencies on the Euler-Bernoulli beam with simply-supported boundary conditions to which the grounded mass-spring system is attached at a specified location along the beam. It is then solved for the magnitude of the mass and the stiffness of the spring.

KEYWORDS: Inverse eigenvalue problem, Euler-Bernoulli beam, vibration analysis, mass-spring system

INTRODUCTION

Determining the eigenvalues (natural frequencies) and eigenvectors (modeshapes) of multi-component dynamic systems has been the subject of extensive study. In most cases, no closed-form solutions exist. One such system, whose vibrational analysis is of great interest due to its wide application, is a beam to which several lumped elements are attached. These lumped elements can take different forms. These include point or rotary masses, translational as well as torsional springs and translational as well as torsional dampers.

Most of the research done in this area involves the determination of the system's natural frequencies and corresponding modeshapes, assuming that the characteristics of the systems are known (forward eigenvalue problem). Kukla and Posiadala in [1] and Nicholson and Bergman in [2] used Green's function to determine the exact eigenvalues for the combined dynamic system. Low in [3], [4] used Lagrange multipliers to calculate the eigenvalues and then compared it to Rayleigh's method. Cha in [5] proposed a method for calculating the natural frequencies (eigenvalues) of a beam with multiple miscellaneous lumped attachments by reducing the order of the matrices whose determinants are to be solved from N (number of assumed modes) to S (number of attachments).

Conversely, few researchers have studied the inverse eigenvalue problem for combined dynamic systems. Cha and Pierre in [6] used the assumed modes method in order to develop a forward solution for determining the frequencies of a linear elastic structure (beam) to which a system of masses and springs were attached at particular points. This approach then provided a method for solving the inverse problem of imposing nodes at a specific location. For this to happen, the frequency of the combined system must be equal to the frequency of the grounded mass-spring system that is, the natural frequency of the isolated mass-spring attachment ($\omega = \sqrt{k/m}$) must exactly match that of the combined beam-attachment system. Mir Hosseini and Baddour in [7] considered the effect of adding multiple lumped masses to a Euler-Bernoulli beam on the beam's natural frequencies and corresponding modeshapes. Based on this forward modeling investigation, a new method was proposed in order to solve the inverse eigenvalue problem of imposing up to four desired natural frequencies on the beam with either simply-supported or cantilever boundary conditions by adding and modifying the position and value of two lumped masses. Dumond and Baddour proposed two separate methods for solving the inverse eigenvalue problem of discrete vibrating systems. The first method makes special use of the Cayley-Hamilton theorem [8], while the second method solves the determinant of the generalized eigenvalue problem for specified variables [9]. The determinant method is computationally more efficient for the partially described problem described above. In either case, inverse eigenvalue problems are ill-posed, meaning they have multiple solutions, which may or may not be physically realistic. Thus, a comprehensive study of the forward problem is an indispensable prerequisite to considering the inverse problem. It paves the way for introducing and applying educated and quasi accurate inputs to the inverse problem, enhancing the chance of obtaining accurate solutions. Hence, the importance of understanding the effects of changes made to the forward model on the natural frequencies and modeshapes of the system is emphasized.

In this paper, the authors extend the scope of the previous work by considering the effects of adding a grounded mass-spring system, without adding degrees of freedom, on the simply supported Euler-Bernoulli beam's natural frequencies as well as the corresponding modeshapes. Furthermore, the inverse problem of imposing two desired natural frequencies is considered.

THEORY

Forward Problem

In this paper, the assumed modes method is used as a forward model in order to discretize the problem, using the principle of modeshape superposition. Thus, the transverse displacements of the beam can be modeled as:

$$w(x,t)=\sum_{j=1}^{N}\phi_j(x)\eta_j(t) \tag{1}$$

where, $w(x,t)$ is the transverse displacement of the beam, $\phi_j(x)$ are the space-dependent eigenfunctions, $\eta_j(t)$ are the generalized coordinates and N is the number of modes used in the problem.

Expressions for the kinetic and potential energy can be written as:

$$T=\frac{1}{2}\sum_{j=1}^{N}M_j\dot{\eta}_j^2+\frac{1}{2}m_1\dot{w}^2(x_1,t) \tag{2}$$

and

$$V=\frac{1}{2}\sum_{j=1}^{N}K_j\eta_j^2(t)+\frac{1}{2}k_1w^2(x_1,t) \tag{3}$$

where M_j and K_j are the generalized masses and stiffnesses of the beam respectively, and m_1 and k_1 are the mass and stiffness of the attachments.

Once Eq. (1) is substituted into Eq. (2) and (3), Lagrange's equations can be used to determine the dynamic system's equations of motion:

$$\frac{d}{dt}\left(\frac{\partial T}{\partial\dot{\eta}_i}\right)-\frac{\partial T}{\partial\eta_i}+\frac{\partial V}{\partial\eta_i}=0 \qquad i=1,2,..,N \tag{4}$$

The equations of motion can be written in matrix form such that

$$\mathbf{M}\ddot{\underline{\eta}}+\mathbf{K}\underline{\eta}=0 \tag{5}$$

where $\mathbf{M}$ and $\mathbf{K}$ are the system's mass and stiffness matrices respectively and are given by

$$\mathbf{M}=\mathbf{M}^d+m_1\cdot\underline{\phi_1}\cdot\underline{\phi_1}^T \tag{6}$$

$$\mathbf{K}=\mathbf{K}^d+k_1\cdot\underline{\phi_1}\cdot\underline{\phi_1}^T \tag{7}$$

where $\mathbf{M}^d$ and $\mathbf{K}^d$ are diagonal matrices constructed using the generalized masses and stiffnesses of the beam, whose diagonal elements are defined as

$$M_i=1 \tag{8}$$

and

$$K_i=\frac{(i\pi)^4 EI}{\rho L^4}, \tag{9}$$

respectively and where

$$\underline{\phi_1}=\left[\phi_1(x_1)\ \ .\ \ .\ \ \phi_N(x_1)\right]^T \tag{10}$$

is an N-dimensional column vector consisting of the N eigenfunctions evaluated at the attachment point x_1.

Inverse Problem

The inverse problem, considers the Euler-Bernoulli beam with simply-supported boundary conditions as described above. However, in this case, two desired natural frequencies, as well as the mounting position of the grounded mass-spring system, are imposed on the system. The variables considered as unknowns are the magnitude of the mass m_1, as well as the stiffness of the lumped spring k_1. This can be described as a system of two equations and two unknowns

$$\begin{cases}\det(\mathbf{K}-\lambda_a\mathbf{M})=0\\ \det(\mathbf{K}-\lambda_b\mathbf{M})=0\end{cases} \tag{11}$$

where $\mathbf{M}$ and $\mathbf{K}$ are the overall mass and stiffness matrices as defined by Eq. (6) and (7), respectively and λ_a and λ_b are the square of the imposed (desired) natural frequencies.

RESULTS

In this paper, the effects of adding a lumped mass-spring system with no extra degrees of freedom (grounded) to a Euler-Bernoulli beam on the natural frequencies and corresponding modeshapes of the beam are considered and the results are tabulated.

Table 1 depicts the first five modeshapes of the Euler-Bernoulli beam with simply-supported boundary conditions, to which the grounded mass-spring system is attached at a position of $x=0.3L$ from the end of the beam. This position is chosen because it allows for the verification of the Maple code by comparing results with the frequency results obtained by Cha in [5]. In this case excellent agreement was observed. Each plot shows the results for five different masses mounted on a specified spring. The mass is defined as $m_1=c\rho L$ and the stiffness is defined as $k_1=kEI/L^3$, where c and k represent variable constants.

The inverse problem considers the forward model described above. In this case, the two imposed natural frequencies (2nd and 5th) squared are

$$\begin{cases} \lambda_2 = 886.0514215 \\ \lambda_5 = 44133.26578 \end{cases} \quad (12)$$

Table 1: Modeshapes of the simply-supported Euler-Bernoulli beam with a grounded mass-spring at $x = 0.3L$

Stiffness	Modeshape1	Modeshape2	Modeshape3	Modeshape4	Modeshape5
$k_1 = 0$	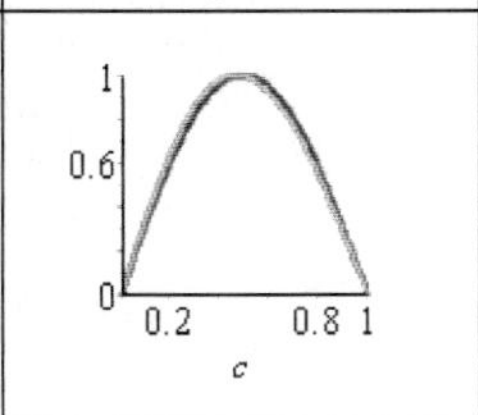	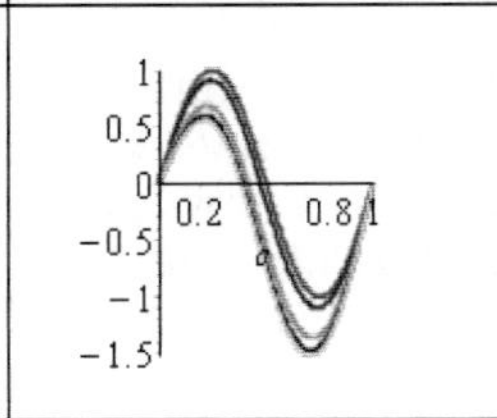	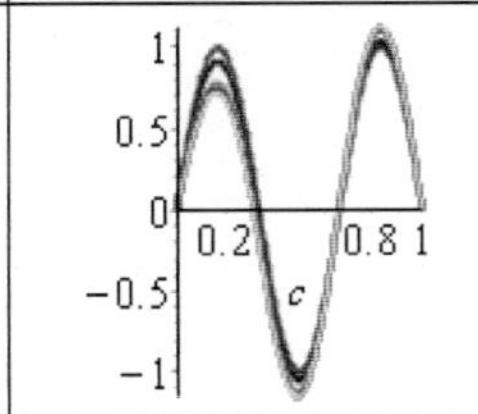	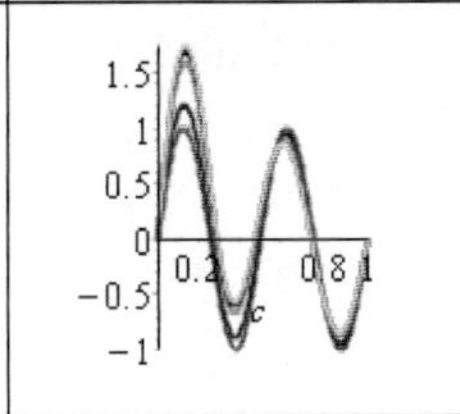	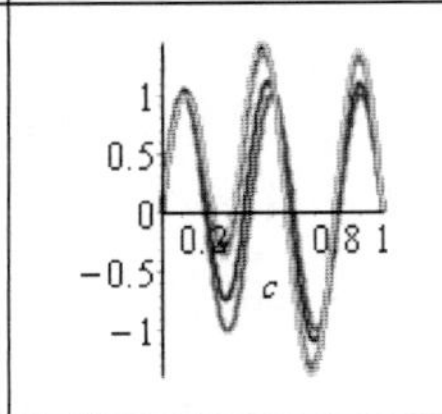
$k_1 = \frac{3EI}{L^3}$	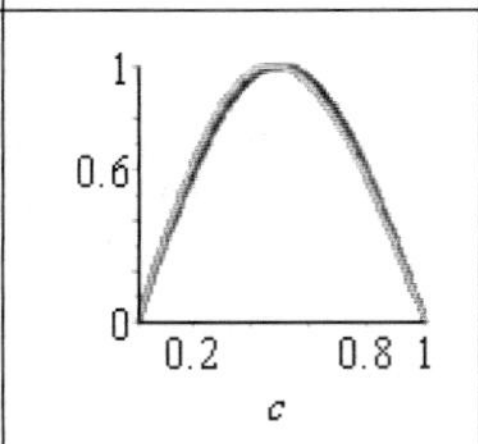	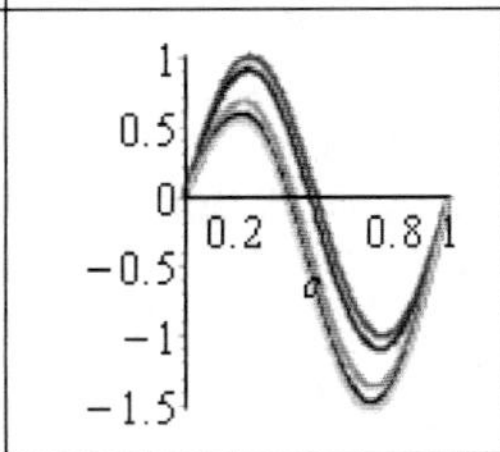	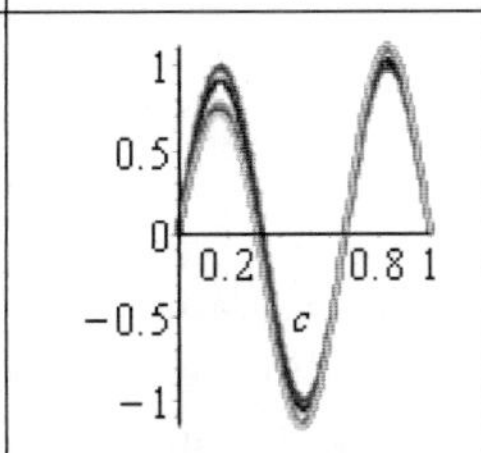	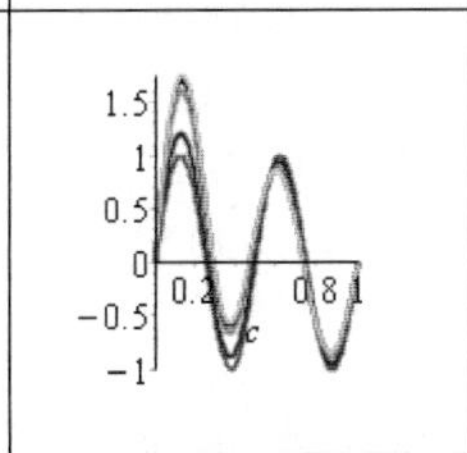	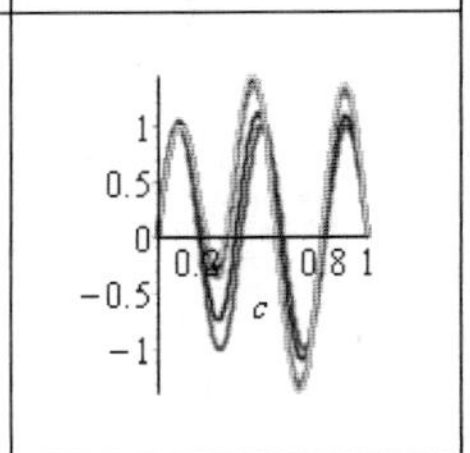
$k_1 = \frac{200EI}{L^3}$	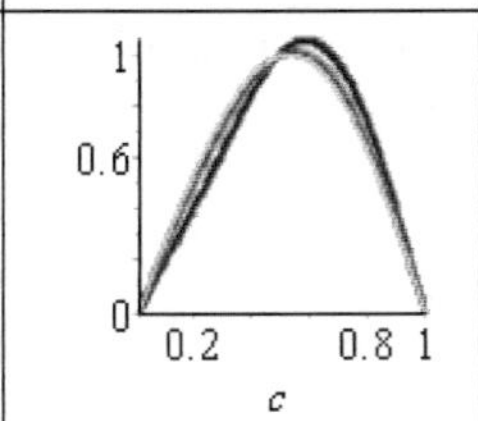	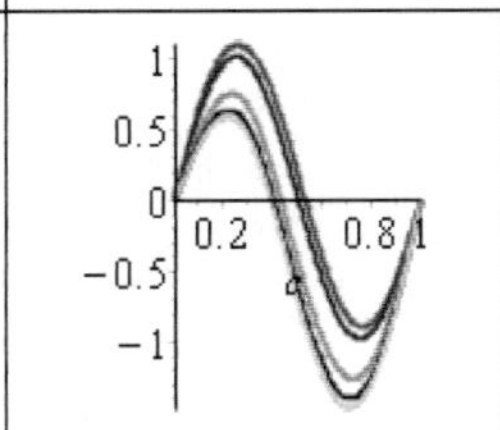	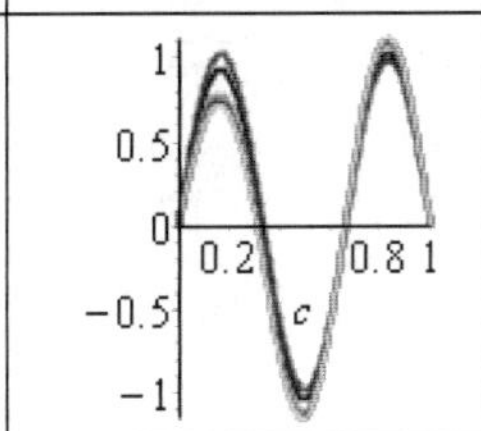	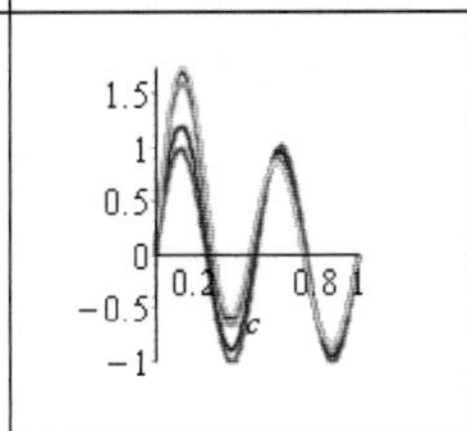	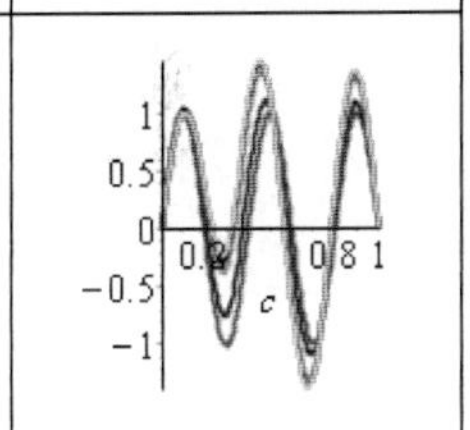
$k_1 = \frac{800EI}{L^3}$	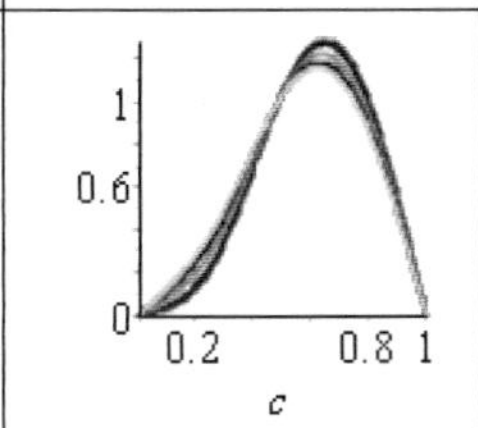	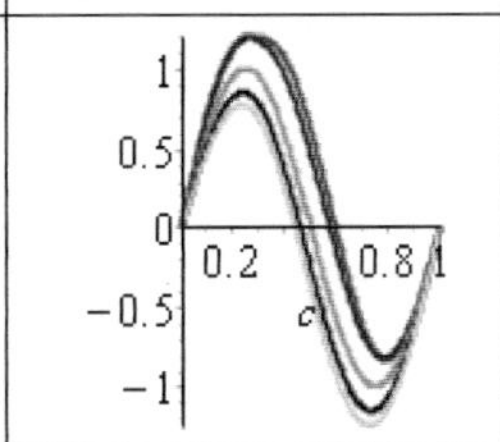	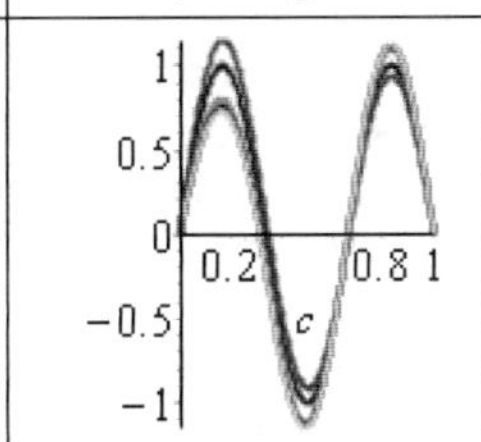	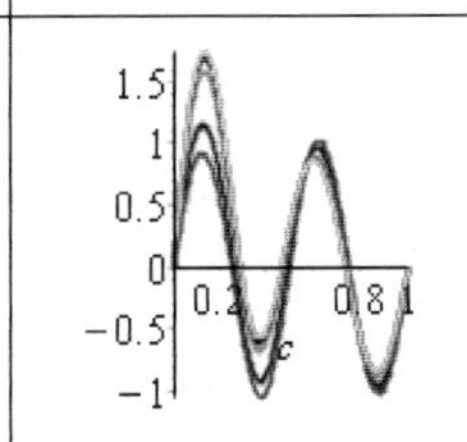	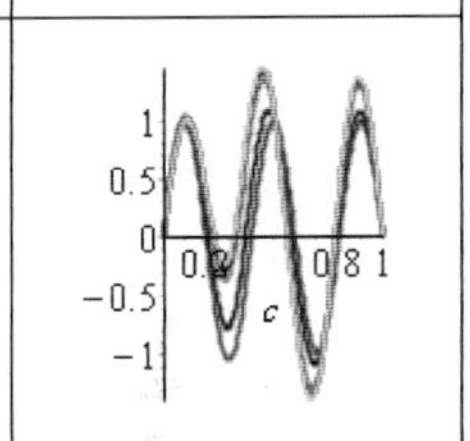
$k_1 = \frac{1000EI}{L^3}$	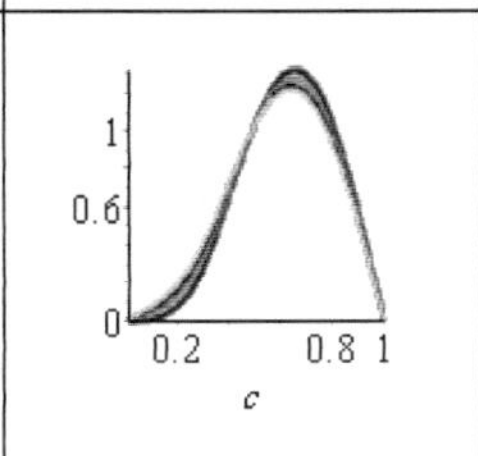	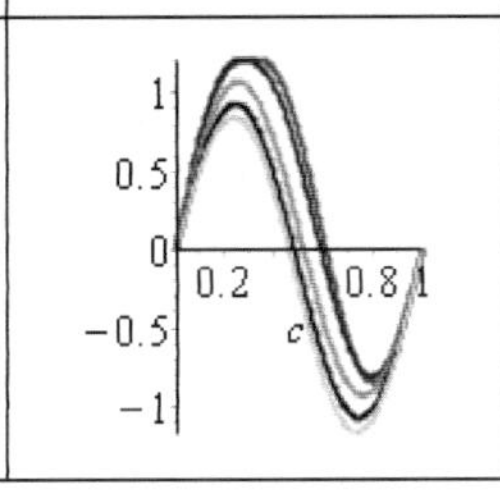	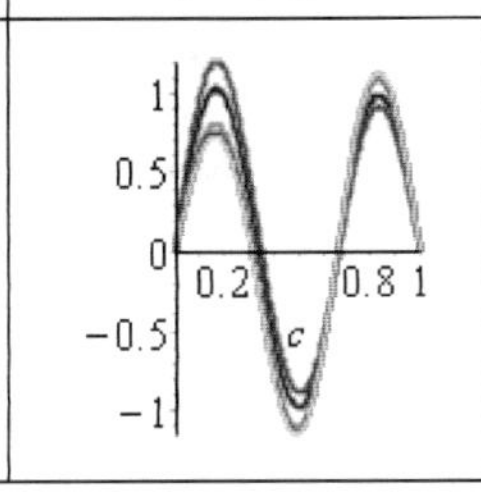	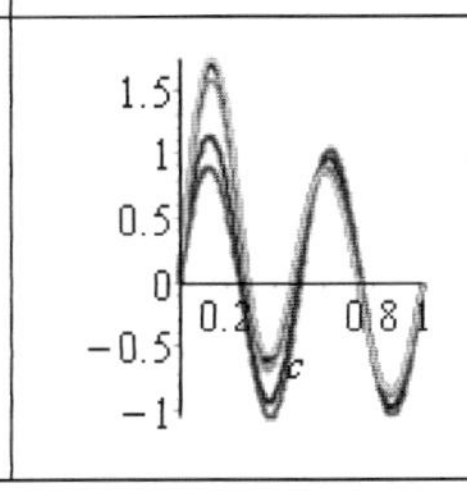	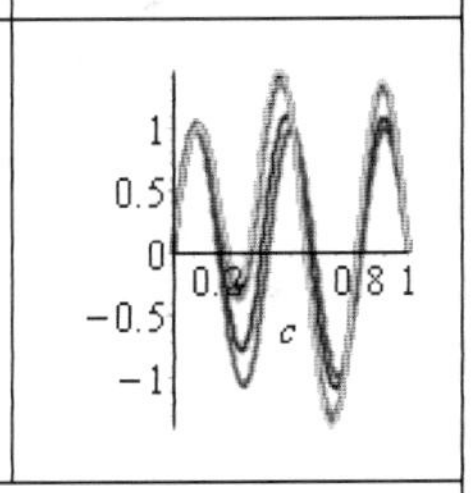
—— Beam only —— $0.1\rho L$ —— $0.5\rho L$ —— $0.8\rho L$ —— ρL					

which were obtained using the forward problem when $m_1 = \rho L$, $k_1 = 3EI/L^3$ and using a degree of discretization of $N=10$.

Using the determinant method and the same number of assumed modes ($N=10$), the exact solution of ($c=1$ and $k=3$) is obtained using both *fsolve* and *DirectSearch* commands in Maple. This is the sole plausible solution obtained by both solvers. Due to the round off errors caused by the very large coefficients in the resulting determinant polynomials, it is important to use the same number of assumed modes for both the forward and inverse problems. Otherwise, it proved impossible to retrieve the original solution.

This issue is illustrated by employing a greater number of assumed modes for the forward problem ($N=20$) while retaining $N=10$ for the inverse problem. This approach has previously been shown to yield good results [7]. However, in this case the only plausible result obtained is

$$m_1 = 1.1512\rho L, k_1 = \frac{134.65EI}{L^3} \quad (13)$$

As can be seen, the results in (13) are not the same as the original results used in the forward problem. To check whether the results in (13) produce the same order of frequencies as those of the original problem, these results are substituted into the forward problem. The first eight frequencies were then obtained such that

$$\Lambda = \begin{bmatrix} 109.1083 \\ \mathbf{885.2803} \\ 7511.598 \\ 20392.22 \\ \mathbf{43612.37} \\ 115896.8 \\ 225990.8 \\ 225990.8 \\ 316533.7 \\ 225990.8 \end{bmatrix} \qquad (14)$$

As can be seen in (14), the order of frequencies has been preserved and the obtained frequencies (in bold) are in good agreement with the desired frequencies $\lambda_2 = 886$ and $\lambda_5 = 44133$. The higher frequency shows greater discrepancy with the desired frequencies when compared to the lower one, which is to be expected when the number of assumed modes is reduced.

CONCLUSIONS

In this paper, the inverse problem of imposing two desired natural frequencies on a dynamic system, consisting of a Euler-Bernoulli beam with simply-supported boundary conditions to which is attached a grounded mass-spring system, is considered.

In order to properly describe the inverse problem, it is crucial that the effects and limitations of adding a grounded mass-spring system on the natural frequencies and corresponding modeshapes of the Euler-Bernoulli beam, with simply-supported boundary conditions, be thoroughly understood. In this study, the grounded mass-spring system is positioned at $x = 0.3L$ from the end of the beam, where L is the length of the beam. This position is chosen because it enables the verification of the code by comparing the forward modeling results with those of Cha [5]. Having done this verification, complete agreement was observed. The first five modeshape plots for the dynamic system were generated and tabulated. It is shown that an increase in stiffness at the location of the attachment significantly reduces the amplitude of the modeshape in the region of the attachment. However, this effect is intuitively reduced when the mass is increased accordingly.

This investigation was an indispensable preamble for considering the inverse problem of imposing desired frequencies by the addition of the grounded mass-spring attachment, as it provides the necessary insight for making educated assumptions about the inverse problem.

In this study, the specific inverse problem considered two desired natural frequencies to be imposed on the system, as well as the given mounting position of the grounded spring-mass attachment as inputs. Then, using the determinant method, a system of two non-linear equations was obtained from which possible values of the lumped mass and spring elements were found. Due to the ill-posed nature of the inverse problem and the round off errors resulting from the great discrepancies in the determinant polynomial coefficients, using the same degree of discretization for the inverse and forward problems is recommended.

ACKNOWLEDGMENTS

The authors would like to acknowledge the generous support provided by the Natural Sciences and Engineering Research Council of Canada.

REFERENCES

[1] S. Kukla and B. Posiadala, "Free Vibrations Of Beams With Elastically Mounted Masses," *J. Sound Vib.*, vol. 175, no. 4, pp. 557–564, 1994.

[2] J. Nicholson and L. Bergman, "Free Vibration of Combined Dynamical Systems," *J. Eng. Mech.*, vol. 112, no. 1, pp. 1–13, 1986.

[3] K. H. Low, "On the Eigenfrequencies for Mass Loaded Beams Under Classical Boundary Conditions," *J. Sound Vib.*, vol. 215, no. 2, pp. 381–389, Aug. 1998.

[4] K. H. Low, "Comparisons of experimental and numerical frequencies for classical beams carrying a mass in-span," *Int. J. Mech. Sci.*, vol. 41, no. 12, pp. 1515–1531, Dec. 1999.

[5] P. D. Cha, "A general approach to formulating the frequency equation for a beam carrying miscellaneous attachments," *J. Sound Vib.*, vol. 286, no. 4–5, pp. 921–939, 2005.

[6] P. D. Cha and C. Pierre, "Frequency analysis of a linear elastic structure carrying a chain of oscillators," *J. Eng. Mech.*, vol. 125, no. 5, pp. 587–591, 1999.

[7] F. Mir Hosseini and N. Baddour, "A Structured Approach to Solve the Inverse Eigenvalue Problem for a Beam with Added Mass," *Math. Probl. Eng.*, vol. 2014, p. e292609, Mar. 2014.

[8] P. Dumond and N. Baddour, "A structured approach to design-for-frequency problems using the Cayley-Hamilton theorem," *SpringerPlus*, vol. 3, no. 1, p. 272, May 2014.

[9] P. Dumond and N. Baddour, "A structured method for the design-for-frequency of a brace-soundboard system using a scalloped brace," *Appl. Acoust.*, vol. 88, pp. 96–103, Feb. 2015.

Proceedings of the 25th CANCAM
London, Ontario, Canada, May 31 – June 4, 2015

STOCHASTIC DYNAMICS AND STABILITY OF A CLASS OF MEMS GYROSCOPES

Soroush Arghavan
Department of Mechanical and Materials Engineering
Western University
London, Ontario, Canada
sarghav2@uwo.ca

Mohamed Bognash
Department of Mechanical and Materials Engineering
Western University
London, Ontario, Canada
mbognash@uwo.ca

Samuel Asokanthan
Department of Mechanical and Materials Engineering
Western University
London, Ontario, Canada
sasokant@uwo.ca

ABSTRACT

Effect of stochastic fluctuations in angular velocity on the stability of two DOF ring-type MEMS gyroscopes is investigated. The governing Stochastic Differential Equations are discretized using the higher-order Milstein scheme in order to numerically predict the system response assuming the fluctuations to be white noise. Simulations via Euler scheme as well as a measure of Largest Lyapunov Exponents are employed for validation purposes due to lack of similar analytical or experimental data. The stability investigation predicts that the threshold fluctuation intensity increases nonlinearly with damping ratio. Under typical gyroscope operating conditions, nominal input angular velocity magnitude and mass mismatch appear to have minimal influence on system stability.

KEYWORDS: MEMS gyroscope, stochastic differential equation, dynamic stability, white noise.

INTRODUCTION

Gyroscopes can generally be categorized into three grades based on their accuracy and performance, namely rate, tactical and inertial grade, where the inertial grade provides the highest accuracy possible while the rate grade is used for low-accuracy demanding applications. Micro Electro Mechanical Systems (MEMS) based sensors primarily fall in the rate grade. Improvements in accuracy as well as their drift performance are warranted before they can be accepted as a true tactical or inertial grade device. To this end, several recent research as well as development efforts are underway.
One type of currently rate grade vibratory sensors, the ring-structure gyroscopes, gained popularity for their minimal drift and low sensitivity to environment vibrations [1]. Dynamics and stability of ring-structure gyroscopes have been of interest in the recent past. A method for investigating the stability of a linear gyroscopic system via a model of a rotating beam has been developed in [2]. Recently, stability of a ring-based gyroscope subjected to harmonic perturbation in angular rate has been studied via the method of averaging in [3]. Although harmonic vibration may be a simple approximation useful for modeling angular speed fluctuation emanating from the environment, white noise, which covers a wide range of frequencies, may be a more realistic representation. Hence, in the present study, extending the work performed in [3], introduction of the white noise is envisaged to aid in a more accurate prediction of the dynamic behavior of these devices as well as the physical systems they are mounted on. The present investigation focuses on the dynamic stability of ring-structure gyroscopes based on their dynamic response, considering random perturbation in input angular velocity.
For this purpose, considering the random fluctuation, the governing equations of motion of the gyroscopic system under investigation are written in the form of a system of standard Stochastic Differential Equations (SDE). It is known that closed-form analytical solutions cannot be obtained for this class of systems owing to their highly non-differentiable character of the realization of the Wiener process [4]. Therefore, in the present study, the higher-order Milstein numerical scheme is employed to simulate the time response so that the stochastic response of ring-based gyroscopes can be quantified for certain parameters of interest. The stability analysis is then performed based on the simulated responses so that the stability behavior of this class of gyroscopes can be predicted. To this end, an algorithm for computing the characteristic Lyapunov exponents of the response have been employed for validating the stability predictions via the stochastic response. Effects of damping and angular speed fluctuation magnitude on system stability have been quantified for different input angular velocities in addition to a parameter representing the non-uniformity in ring mass distribution.

GOVERNING EQUATIONS

The governing equations for the system developed in [3] based on previous work in [1] have been employed to study the effect of stochastic fluctuation of input angular rate. The material chosen for the ring is assumed to be isotropic and homogeneous while the transverse shear deformation and the rotary inertia of the ring is considered negligible since the mathematical model has been based on the Euler-Bernoulli theory.
It is known that for a typical ring-based gyroscope operation, one of the second flexural modes is harmonically excited while the measurement of angular shift of this mode towards the corresponding degenerate mode is employed in

quantifying the angular rate (see, e.g., [5]). Generalized coordinate vector $\mathbf{q} = [q_1 \quad q_2]^T$ that represents radial displacements associated with the pair of modes is introduced to derive the governing equations of motion for the gyroscope [3]:

$$\begin{bmatrix} 1 & 0 \\ 0 & 1+\delta m \end{bmatrix} \ddot{\mathbf{q}} + \begin{bmatrix} 2\zeta\omega_{01} & -2\Omega\gamma \\ 2\Omega\gamma & 2\zeta\omega_{02} \end{bmatrix} \dot{\mathbf{q}} + \begin{bmatrix} \kappa_1 + \kappa_2\Omega^2 & -\dot{\Omega}\gamma \\ \dot{\Omega}\gamma & \kappa_1 + \kappa_2\Omega^2 \end{bmatrix} \mathbf{q} = \begin{bmatrix} f\cos\omega t \\ 0 \end{bmatrix}, \quad (1)$$

where γ, κ_1 and κ_2 are parameters that contain the geometrical information of the ring. The coordinates q_1 and q_2 represent the driving and the sensing directions respectively. Also, the excitation force f is applied with a harmonic frequency ω close to the undamped natural frequencies of the system ω_{01} and ω_{02} in order to resonate the driving motion of the ring. Focusing on the system stability where the steady state response of the system is not of interest, the homogenous form associated with Eq. (1) in the absence of the excitation force is considered for the present analysis.

The mass matrix in Eq. (1) incorporates the assumption that mass is not uniformly distributed along the circumference of the ring and the term δm represents the mass mismatch of the ring. It may me noted that the stiffness matrix includes the centrifugal force term that depends on a factor κ_2 which takes a negative value for the present system. Hence, overall system stiffness decreases with higher angular velocity which may lead to lower system stability. The damping matrix, apart from representing viscous dissipation via damping ratio ζ , includes the gyroscopic coupling term $2\Omega\gamma$ which is dependent on the input angular velocity Ω . It is worth noting that owing to the gyroscopic as well as the centrifugal effects the two undamped system natural frequencies ω_{01} and ω_{02} vary with the input angular velocity. However, at typical low input angular velocities of about 2π rad/s, these two frequencies take nearly identical values [3].

Moreover, for constant angular rate, the term $\dot{\Omega} = 0$. This case is not practical in the presence of fluctuations in the angular rate. However, for the system under investigation, the contributions of the associated terms, $\dot{\Omega}\gamma q_1$ and $-\dot{\Omega}\gamma q_2$ are negligible when compared to $2\Omega\gamma\dot{q}_1$ and $-2\Omega\gamma\dot{q}_2$ at high angular rates where instability becomes an issue, it is sufficient to approximate $\dot{\Omega}\gamma q_1 = -\dot{\Omega}\gamma q_2 = 0$ for the purpose of stability analysis. Also, for the case of uniform mass distribution of the ring, mass mismatch $\delta m = 0$. However, for natural frequency computations, the mass mismatch is taken into account.

It is known that numerical schemes developed for stochastic differential equations are limited to first order differential equations. Therefore, prior to applying the numerical methods, Eq. (1) is converted to four first order differential equations by defining the state variables $x_1 = q_1$, $x_2 = \dot{q}_1$, $x_3 = q_2$ and $x_4 = \dot{q}_2$. In order to represent the random fluctuation in the input angular velocity, these fluctuations are assumed to take the form of a white noise process. Understanding that the first time derivative of a Brownian motion process is Gaussian white noise, a Brownian motion function $W(t)$ is employed to simulate the random fluctuations (see, e.g., [6]).

Introducing the random fluctuation to a nominal input angular velocity Ω_0 , the input angular velocity is written as $\Omega = \Omega_0 + \mu_0\xi(t)$, where $\xi(t) = dW/dt$ is white noise and μ_0 is the noise intensity magnitude. Hence, using Eq. (1), a system of SDEs that represents the motion is obtained as:

$$\begin{bmatrix} dx_1 \\ dx_2 \\ dx_3 \\ dx_4 \end{bmatrix} = \begin{bmatrix} 0 & 1 & 0 & 0 \\ -\kappa_1 - \kappa_2\Omega_0^2 & -2\zeta\omega_{01} & 0 & 2\Omega_0\gamma \\ 0 & 0 & 0 & 1 \\ 0 & -2\Omega_0\gamma & -\kappa_1 - \kappa_2\Omega_0^2 & -2\zeta\omega_{02} \end{bmatrix} \begin{bmatrix} x_1 \\ x_2 \\ x_3 \\ x_4 \end{bmatrix} dt + \begin{bmatrix} 0 & 0 & 0 & 0 \\ -2\kappa_2\mu_0\Omega_0 & 0 & 0 & 2\mu_0\gamma \\ 0 & 0 & 0 & 0 \\ 0 & -2\mu_0\gamma & -2\kappa_2\mu_0\Omega_0 & 0 \end{bmatrix} \begin{bmatrix} x_1 \\ x_2 \\ x_3 \\ x_4 \end{bmatrix} dW, \quad (2)$$

where the resultant 4×1 coefficient matrix associated with the term dt is called the drift matrix, $\mathbf{a}[\mathbf{X}(t)]$ and the resultant 4×1 coefficient matrix associated with the term dW is called the diffusion matrix, $\mathbf{b}[\mathbf{X}(t)]$, where the state vector $\mathbf{X}(t) = [x_1 \quad x_2 \quad x_3 \quad x_4]^T$ (see e.g. [4, 6]).

In order to better characterize the fluctuation magnitude μ_0 , a parameter μ that represents the noise intensity ratio is introduced as:

$$\mu = \frac{\mu_0 \max(\xi(t))}{\max(\Omega_0)}. \quad (3)$$

In the present numerical study, a smooth increase of the nominal input angular rate Ω_0 from zero to 2π rad/s is employed for the purposes of response predictions. Conforming to the goal of the present study, namely the stability investigation, free vibration of the gyroscope when subjected to an initial disturbance is examined.

Density	$\rho = 8800\, kg/m^3$
Young's Modulus	$E = 210\times10^9\, N/m^2$
Mean Radius	$r = 500\,\mu m$
Radial Thickness	$\bar{h} = 12.5\,\mu m$
Axial Thickness	$\bar{b} = 30\,\mu m$

Table 1: Geometric and material properties.

For the purposes of numerical simulations, the geometric and material properties of the ring as shown in Table (1) have been used. In particular, these properties are employed to calculate the ring parameters γ, κ_1 and κ_2 as defined in [3].

NUMERICAL SIMULATIONS

It is known that an exact analytical solution does not exist for this class of equations that represents the motion of a gyroscope. A numerical simulation of the system seems to be the most efficient method. For this purpose, the vector form of the higher-order Milstein scheme which takes advantage of the Itô-Taylor expansion for discretization of the SDE [6] has been chosen for this study instead of the simpler Euler approach in order to achieve higher accuracy for the numerical simulations. Using Milstein scheme, the uth component of the state vector for a four-dimensional system takes the form:

$$\mathbf{X}^u(t_{i+1}) = \mathbf{X}^u(t_i) + \mathbf{a}^u[\mathbf{X}(t_i)]\Delta t + \mathbf{b}^{u,1}[\mathbf{X}(t_i)]\Delta W_i + \sum_{k=1}^{4} \frac{1}{2}\mathbf{b}^{k,1}[\mathbf{X}(t_i)]\frac{\partial \mathbf{b}^{u,1}[\mathbf{X}(t_i)]}{\partial \mathbf{X}^k}\left\{(\Delta W_i)^2 - \Delta t\right\}, \quad (4)$$

$u = 1, 2, \cdots, 4.$

Considering the four state variables of Eq. (2), the resulting equations are employed in the prediction of gyroscope response.

RESULTS AND DISCUSSIONS

Simulations performed are based on an assumption that an initial displacement of $10^{-5}\,m$ being exerted on the driving coordinate q_1. The free vibration of the sensing coordinate q_2 is then quantified for the purposes of characterizing the gyroscope response. In order to ascertain the accuracy of the developed model, the convergence of the simulation algorithm is tested with four different damping ratio values, namely 0.03, 0.05, 0.09 and 0.13. The numerical convergence is reached in 150000 or more time steps for a duration of 0.009 seconds of simulation time and the number of time steps is considered adequate for the remainder of the study.

System time responses generated by the two methods were compared for the purposes of verifying the response predictions. Maximum relative noise intensity, as defined in Eq. (3) has been used as a magnitude measure for representing environment fluctuation.

Analysis of the simulated time responses reveals that with sufficient damping, the system is observed to remain stable. However, as noise intensity is increased to a sufficiently high value to cause noticeable disturbance in the system, yet low enough to be damped, the system shows oscillatory motion until complete decay. It may be noted that a certain threshold intensity measure for each damping ratio is associated with transition to instability and this measure can be computed using the time responses. For damping ratios less than the threshold value or, alternatively with sufficiently high noise measure, the system response becomes unstable. Furthermore, it may be noted that Euler and Milstein schemes both predict similar time responses, with Milstein scheme which accounts for additional higher-order terms showing slightly larger values for the response compared with those predicted via the Euler scheme. Further, the Euler scheme which does not take the higher order terms into account results in an under-predicted response, hence exhibits increased system stability. As a result, under certain conditions where noise intensity is close to the threshold values, it appears that Euler scheme can predict a stable system behavior while Milstein scheme predicts an unstable behavior. An example of such conditions is shown in Fig. 1.

Therefore, due to lack of any exact analytical or experimental results to verify the acquired data against, it is necessary that the accuracy of the employed model is assessed before proceeding to the parameter sweep for characterizing the system behavior under changes in pertinent system parameters. For this purpose, dynamic stability of the system is studied employing the two schemes using the Largest Lyapunov Exponent (LLE) method as a stability measure (see, e.g., [7]). Any system with at least one positive Lyapunov exponent will inevitably become unstable, with the magnitude of the exponent reflecting on the time scale that system dynamics will diverge. Therefore, it is sufficient to calculate the largest Lyapunov exponent for characterizing system stability. In the present study, a practical implementation of the LLE search algorithm, based on the method of time delay as presented in [7], is employed.

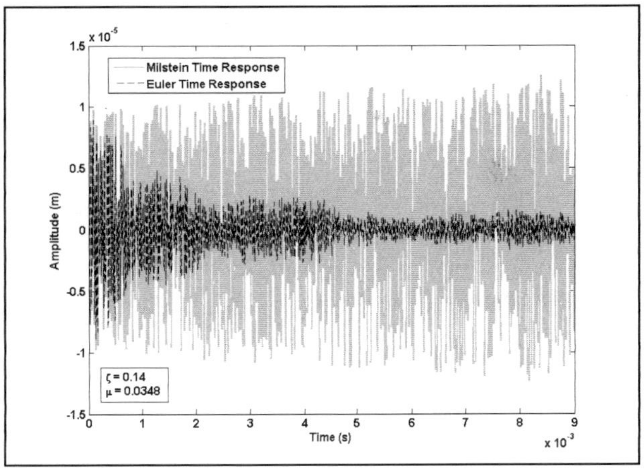

Figure 1: Under-prediction of results by Euler scheme

Fig. 2 shows the values of threshold noise intensity factors obtained at which the system starts to exhibit instability. For the purposes of confirming that the results from Milstein and Euler schemes are compatible, and also for studying the error introduced by neglecting higher-order terms in Euler scheme, a number of points have been chosen for the LLE study; three highly stable points and six marginal points. LLE values evaluated at the highly stable points resulted in the same sign, confirming that the Euler and Milstein schemes predict the stable and unstable conditions similarly. However, LLE values predicted by Euler's scheme have been found to be lower than those predicted by Milstein Scheme by 3 to 8.5 percent in highly stable cases. Further, comparison of the time responses predicted by these two schemes reveals that the Euler scheme tend to under-predict the response for the present gyroscopic system. Hence, it may be concluded that contribution from the higher-order terms is significant and the use of Milstein scheme is beneficial in predicting response of this class of gyroscopes.

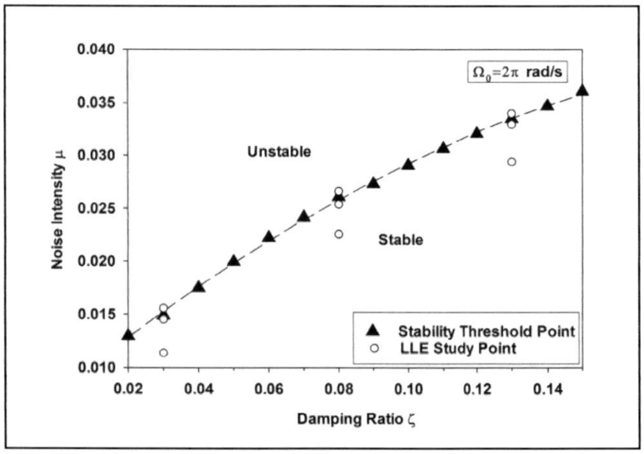

Figure 2: Stability Boundary in the $\mu - \zeta$ space

($\Omega_0 = 2\pi$ rad/s)

It is known that external noise is one of the most important parameters that affect the gyroscope dynamic behavior. Such noise usually results from environment factors and the nature of the system operation and can be exerted on the system at any frequency range depending on the source. Therefore, a parametric study is performed to assess the noise intensity stability threshold for a number of damping ratio values.

Fig. 2 shows that noise intensity threshold is increased at higher damping ratios. In addition, studies performed on varying input angular rates of practical significance for typical gyroscope applications revealed that changes in angular velocity does not affect noise threshold significantly. However, under unrealistic high angular velocities, namely values close to system natural frequencies, where the first natural frequency tends to increase while the second natural frequency decreases, noise thresholds seem to increase. This behavior may be attributed to the centrifugal force component present in the stiffness matrix which increases with angular velocity, affecting overall system stiffness. Further, it has been found that moderate changes in mass mismatch do not affect the noise threshold. However, unrealistic mass mismatch values close to 20 percent cause a reduction in the second natural frequency and as a consequence appear to contribute to a reduction in the noise threshold. It is worth noting that such values for mass mismatch and angular velocity are not of practical significance under typical gyroscope operating conditions.

CONCLUSIONS

Stability of ring-based gyroscopes subjected to random fluctuation in the angular velocity is studied using a two DOF dynamic model developed for flexible circular rings. Resulting equations of motion have been shown to predict the response in the driving as well as the sensing directions of the gyroscope. The fluctuation has been assumed to represent a standard white noise and the associated Stochastic Differential Equations have been formulated employing the Milstein scheme to predict the system response. Due to lack of analytical or experimental data, response predictions via Euler scheme as well as the Largest Lyapunov Exponents have been used for validation. The predicted threshold values for noise intensity at different values of damping exhibit a nonlinear increasing trend. Variation in input angular velocity and mass mismatch have been found not to impose significant influence on system stability under typical gyroscope operating conditions.

ACKNOWLEDGEMENTS

This research was supported in part by the Natural Science and Engineering Research Council (NSERC) of Canada discovery grant, Western Engineering Scholarship (WES) scheme from Western University, London, Ontario, Canada and Libyan-North American Scholarship program.

REFERENCES

1. Asokanthan, S.F., Ariaratnam, S.T., Cho, J., and Wang, T. 2006, "MEMS vibratory angular rate sensors: Stability considerations for design", *Structural Control and Health Monitoring, 13*(1), pp. 76-90.
2. Kammer, D.C., and Schlack Jr, A.L. 1987, "Effects of nonconstant spin rate on the vibration of a rotating beam", *Journal of Applied Mechanics,* 54, pp. 305–310.
3. Asokanthan, S.F., and Cho, J. 2006, "Dynamic stability of ring-based angular rate sensors", *Journal of Sound and Vibration,* 295, pp. 571-583.
4. Higham, D.J. 2001, "An algorithmic introduction to numerical simulation of stochastic differential equations", *SIAM Rev. 2001,* 43, pp. 525–546.
5. Cho, J. 2009, *Nonlinear Instabilities in Ring-based Vibratory Angular Rate Sensors* [Order No. NR50210], The University of Western Ontario, London, Canada.
6. Kloeden, P.E., and Platen, E. 1999, *Numerical Solution of Stochastic Differential Equations,* Springer.
7. Kliková, B., Raidl, A. 2011, "Reconstruction of Phase Space of Dynamical Systems Using Method of Time Delay", *WDS'11 Proceedings of Contributed Papers: Part III – Physics (eds Safrankova J and Pavlu J)*, Prague, Czech Republic, pp. 134–137.

Proceedings of the 25th CANCAM
London, Ontario, Canada, May 31 - June 4, 2015

TRANSIENT RESPONSE OF A SANDWICH STRUCTURE DAMPED WITH A FIBROUS CORE MATERIAL

Elsa Piollet
Mechanical Engineering Dept.
Polytechnique Montréal
Montréal, Québec, Canada
elsa.piollet@polymtl.ca

Guilhem Michon
ICA, ISAE
Université de Toulouse
Toulouse, France
guilhem.michon@isae.fr

Annie Ross
Mechanical Engineering Dept.
Polytechnique Montréal
Montréal, Québec, Canada
annie.ross@polymtl.ca

ABSTRACT

This paper investigates the effect of a fibrous core material on the transient response of a sandwich structure. The core material studied is made of entangled carbon fibers cross-linked with epoxy resin. In order to understand its complex behavior, the material is characterized in shear oscillation, and the obtained shear-stress loops are described using Dahl's dynamic hysteresis model. Then, the transient response of a single-degree-of-freedom system including this material is simulated and the amplitude dependency of the response is analyzed. Sandwich beams with the fibrous core material are tested experimentally with a impact hammer and compared with classical honeycomb and foam cored sandwich beams. The entangled cross-linked fibers are shown to provide high damping resulting in a fast return to the equilibrium position. Both theoretical and experimental studies showed nonlinear damping. In the amplitude range studied, the material is more interesting at high impact amplitude.

Keywords: Vibration damping, sandwich structures, nonlinear behavior, transient response

INTRODUCTION

Aerospace and automotive structures need to be as light as possible in order to limit fuel energy consumption, while maintaining high stiffness and strength properties. Sandwich structures, composed of two facesheets separated by a light core, are widely used in such applications as they provide high stiffness-to-weight ratios. However, the use of stiff lightweight structures can lead to noise and vibration amplification. This can be detrimental to user comfort and structural durability. One way to reduce noise and vibration is to investigate new core materials with high energy dissipation properties.

Studies on the static behavior of entangled fibers indicate that energy is dissipated by friction between the fibers when they are deformed [1]. This has led to the development of a new sandwich core material in which entangled fibers are cross-linked with epoxy resin [2]. Resin cross-links ensure cohesion and stiffness of the material while fibers can provide energy dissipation. The present study investigates the effect of this material on the damping properties of sandwich structures through a transient analysis.

In this paper, the material behavior is first measured and identified with a hysteresis model. Then, the transient response of a single-degree-of-freedom system is studied numerically. The consequences of the material behavior on this simple structural response is analyzed. Finally, the transient response of a sandwich beam with entangled core material is compared experimentally with the response of classical honeycomb and foam sandwich beams.

MATERIAL CHARACTERIZATION

The material is made following the fabrication process described by Mezeix [2]. Different types of fibers can be used. Here, carbon fibers are chosen, since carbon-epoxy materials are commonly used in aerospace applications. The fibers are 7 μm in diameter, with a Young's modulus of 240 GPa and a density of 1770 kg/m^3. They are first cut to a length of 31 mm, and are then entangled in a blower room under an air flux at a pressure of 5 bar.

Epoxy resin is then sprayed on the entangled fibers with

Figure 1: Scanning electron microscopy of the entangled cross-linked fibers after polymerization.

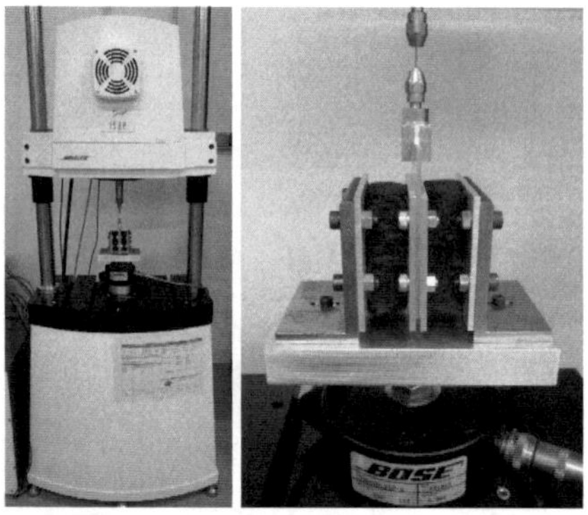

Figure 2: Experimental set-up for material characterization in shear.

a commercial paint gun at a pressure of 2 bar. The small droplets create links between the fibers at several contact points. Finally, the entangled cross-linked fibers are polymerized in a mold at 70° C during 8 hours.

The final density of the material is 180 kg/m^3 (150 kg/m^3 of carbon fibers and 30 kg/m^3 of epoxy resin). Figure 1 shows a scanning electron microscopy of the polymerized material. It can be seen that some contacts are bonded by resin while others are free to slip.

As the material is intended to be used as a core material in sandwich structures, shear is the most important mechanism to be analyzed [3]. Figure 2 shows the set-up for the shear test on a BOSE ElectroForce® 3330. Two samples of dimensions 20 mm × 40 mm × 60 mm are tested together to ensure shearing only. The samples are held by two steel brackets clamped on the bottom plate. A vertical displacement is imposed between the samples by an aluminum plate attached to the vertical motor. The resulting force is measured. From the displacement and force, the shear strain γ and the shear stress τ in the samples are deduced by assuming a constant shear through the thickness.

Measurements are made at 1 Hz, 20 Hz, 40 Hz, 60 Hz and 80 Hz, and the material behavior is found to be frequency independent in the tested frequency range. The following study is based on measurements made at 20 Hz, as they were found

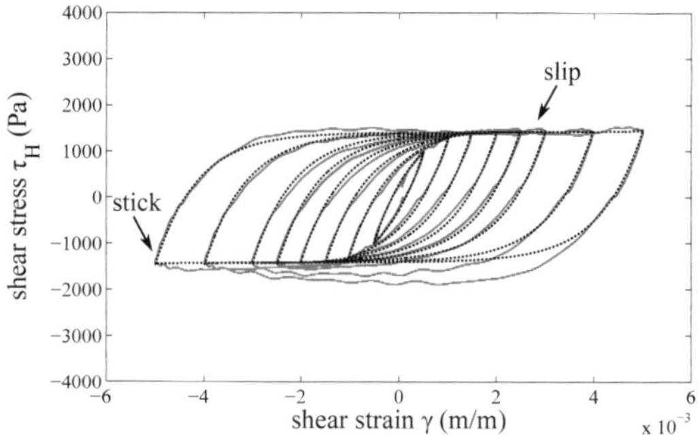

Figure 3: Shear stress-strain hysteresis loops obtained by removing the linear part of the shear stress: $\tau_\mathrm{H} = \tau - G_1\gamma$ with G_1 identified in Eq. (3). Loops measured at 20 Hz (grey line) and identified (black dotted line).

to be less sensitive to noise coming from the set-up.

The measured shear stress can be decomposed as a linear part of modulus G_1 and a hysteresis part defined as:

$$\tau_\mathrm{H} = \tau - G_1\gamma \tag{1}$$

The measured shear stress without the linear part, τ_H, is represented against the measured shear strain γ on Fig. 3 (grey line). The physical interpretation of the observed hysteresis is the following: after each change in the strain direction, the strain rate is null, and all unbonded contacts between fibers are blocked in a "stick" configuration, which creates an initial stiffness; then, stiffness decreases as contacts begin to slip individually, until they are all in a "slip" configuration. As for contacts cross-linked with epoxy resin, they never enter the "slip" configuration, and they are responsible for the linear behavior $G_1\gamma$ which does not appear on the figure.

The shape of the hysteresis loop, combined with the frequency independent behavior, indicates that dissipation comes from the friction between the fibers.

The hysteresis loops are described using Dahl's dynamic hysteresis model [4]:

$$\frac{\mathrm{d}\tau_\mathrm{H}}{\mathrm{d}\gamma} = \frac{\sigma}{\tau_\mathrm{C}}\left(2\tau_\mathrm{C} - (\tau_\mathrm{H} - \tau_\mathrm{m})\,\mathrm{sgn}(\dot{\gamma})\right) \tag{2}$$

where $\dot{\gamma}$ is the strain rate, σ and τ_C are constants of the model and τ_m is the value of the stress at the last change in the strain direction.

Figure 3 shows the identified hysteresis loops for one set of samples (black dotted line). Two sets of two samples each are tested, leading to the following parameters for the model:

$$G_1 = 6.02 \cdot 10^6 \text{ Pa} \quad \sigma = 1.49 \cdot 10^6 \text{ Pa} \quad \tau_\mathrm{C} = 1.39 \cdot 10^3 \text{ Pa} \tag{3}$$

SINGLE-DEGREE-OF-FREEDOM SYSTEM

A theoretical single-degree-of-freedom system is now simulated in order to understand the consequences of the measured material properties on the transient response of a structure.

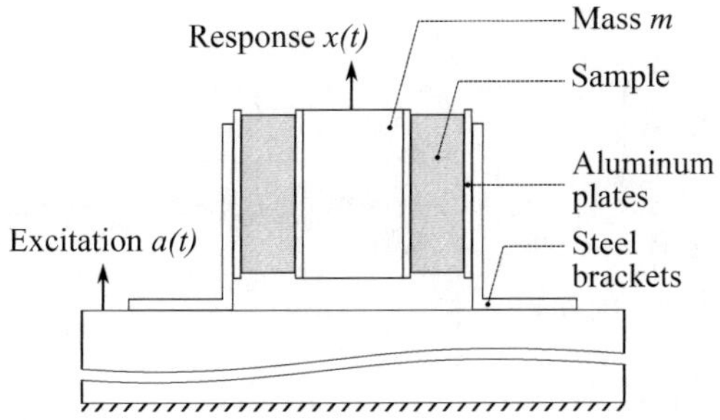

Figure 4: Principle of the studied single-degree-of-freedom system, including the entangled fiber samples in grey.

The structure studied is composed of two samples in shear configuration, as tested previously, but separated by a rigid mass. Figure 4 shows the principle of the system. The mass is taken to be $m = 1$ kg, and is assumed to move only in the vertical direction. The parameters of the material are as given in Eq. (3). A base acceleration is applied at the bottom of the structure. The shock acceleration is a half sine of duration 1.5 ms, as shown on Fig. 5(a). Transient response is computed for seven amplitudes ranging from 1 g to 25 g.

Figure 5(b) shows the simulated transient response from 0 s to 0.2 s. The response to the 25 g shock shows three decay regimes. First, at high oscillation amplitudes, fast linear decay typical of friction is observed (regime 1). Then, the decay seems to become exponential, which is more representative of viscous damping. However, the slope decreases with the oscillation amplitude, from relatively fast (regime 2) to very slow decay (regime 3). Thus, while most of the oscillation is damped before 0.1 s, the remaining oscillation is dissipated slowly.

For lower shock amplitudes, the response enters directly the "exponential" decay regimes 2 and 3, as shown on Fig. 5(c) which spans 0.5 s. The resulting decay rate is slower for those low amplitude oscillations, and the remaining oscillations when entering regime 3 represent a large part of the initial oscillation amplitude.

This simulation shows that the material behavior leads to a transient response that depends on the shock amplitude. For this simple structure and in the amplitude range studied, the entangled material leads to higher damping for higher excitation levels. The following part presents the behavior of a real structure studied experimentally in order to see if the same phenomena are encountered.

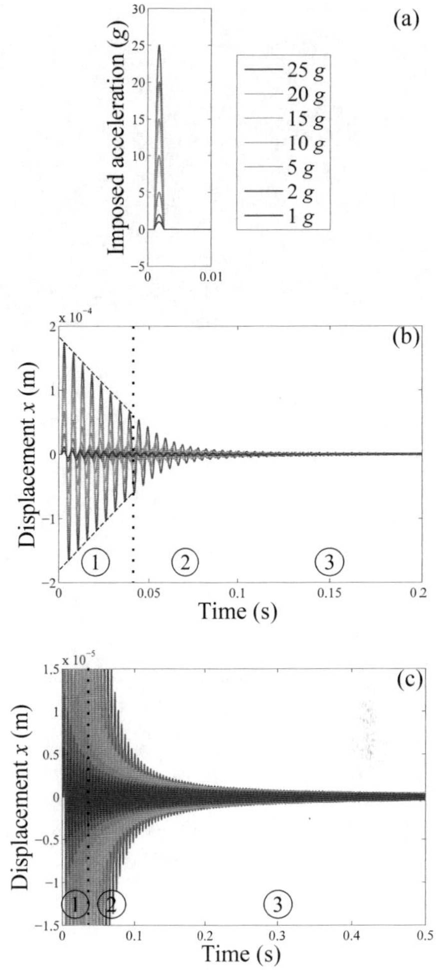

Figure 5: Simulated transient response of the single-degree-of-freedom system of Fig. 4: (a) imposed acceleration, (b) response for all amplitudes during 0.2 s and (c) zoom on the low response amplitudes during 0.5 s. Numbers indicate decay regimes: linear decay (1), and fast (2) and slow (3) "exponential" decay.

TRANSIENT RESPONSE OF SANDWICH BEAMS

The transient responses of three sandwich beams are studied experimentally. For the purpose of comparing with standard materials, the core materials of the beams are the following, along with their density ρ and linear shear modulus G:

- entangled cross-linked carbon fibers ($\rho = 180$ kg/m^3, $G_1 = 6.02$ MPa)
- Nomex honeycomb ($\rho = 48$ kg/m^3, $G = 44.8$ MPa)
- PMI foam ($\rho = 52$ kg/m^3, $G = 19$ MPa).

The beam length is 253 mm and the width is 40 mm. The skins are made of seven plies of prepreg carbon-epoxy fabric used in the aeronautic industry, for a total thickness of 2 mm. The core thickness is 20 mm.

The set-up is shown on Fig. 6. The sandwich beams are tested in cantilever configuration. Excitation is applied with an impact hammer close to the clamped end, and the response is measured at the free end with a laser vibrometer.

Figure 7 shows the measured impacts and transient response for two impact amplitudes of 10 N and 100 N. The durations of the impacts were 1.7 ms and 1.4 ms respectively. The response shown was obtained as the average of five successive tests.

Figure 6: Experimental set-up for transient testing.

These measurements show that the entangled cross-linked fibers lead to a large improvement of damping as compared to classical core materials. While the PMI foam dissipates more than the honeycomb, the entangled cross-linked fibers are far more dissipative. The difference increases with impact amplitude, and the material is thus more interesting for high impact amplitudes.

As expected from the previous simulation, the initial slope of the response is steep, particularly at 100 N. Then, the decay rate decreases, and at low oscillation amplitude the material provides little dissipation.

While this study demonstrates very good damping properties, it should be noted that the entangled cross-linked fibers tested have a higher density and a lower stiffness than the two other core materials.

CONCLUSIONS

The effect of a fibrous material on the transient response of a structure has been studied. Material study showed a hysteretic behavior indicating dissipation coming from friction between the fibers. Dahl's dynamic hysteresis model was used to model the measured hysteresis. The simulated transient response of a single-degree-of-freedom system showed three decay regimes depending on the amplitude. The first regime, typical of friction, corresponds to the higher decay rate, and is thus the most interesting. For the amplitude range studied, the entangled material is more dissipative for higher excitation amplitudes.

The transient response of a sandwich beam with entangled core material showed high added damping as compared to two classical core materials (Nomex honeycomb and PMI foam). Moreover, as predicted by the single-degree-of-freedom simulation, the added damping was increased at higher impact amplitude.

This study shows that the entangled-cross-linked material is a good candidate for vibration damping of sandwich structures. It should be noted however that its lower stiffness and higher density than classical core materials prevents a total replacement of classical materials. Future studies should be conducted to investigate possible applications and analyze the effect of the fiber density and fiber type on damping.

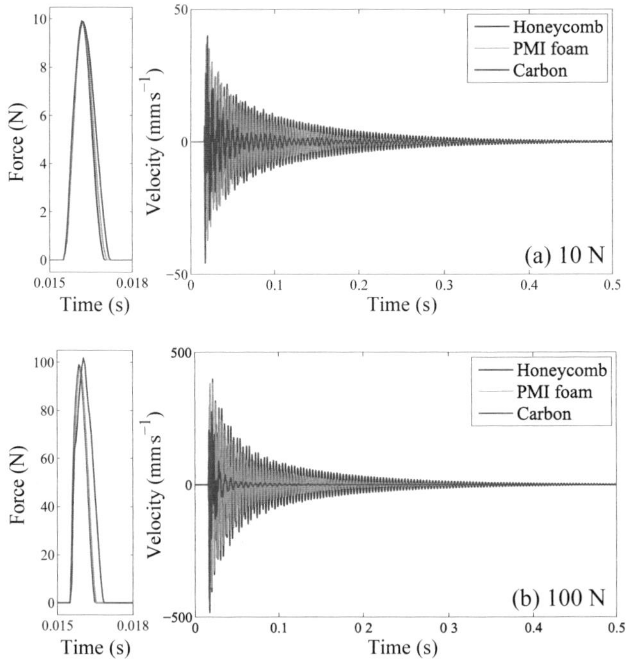

Figure 7: Transient response of sandwich beams with three different core materials to two impacts: (a) 10 N, 1.7 ms and (b) 100 N, 1.4 ms.

ACKNOWLEDGMENTS

This study was carried out during the first author's PhD work at Institut Supérieur de l'Aéronautique et de l'Espace (ISAE) in Toulouse, France. Midi Pyrénée region, Université de Toulouse and Ecole Doctorale Aéronautique et Astronautique are gratefully acknowledged for their financial support.

REFERENCES

[1] Poquillon, D., Viguier, B. and Andrieu, E. 2005, "Experimental data about mechanical behaviour during compression tests for various matted fibres," *Journal of Materials Science*, Vol(40), pp. 59635970.

[2] Mezeix, L., Bouvet , C., Huez, J., and Poquillon, D. 2009, "Mechanical behavior of entangled fibers and entangled crosslinked fibers during compression," *Journal of Materials Science*, Vol(44), pp. 3652-3661.

[3] Mead, D. J., and Markus, S. 1969, "The forced vibration of a three-layer, damped sandwich beam with arbitrary boundary conditions," *Journal of Sound and Vibration*, Vol(10), pp. 163-175

[4] Dahl, P. R. and Wilder, R. 1985, "Math model of hysteresis in piezo-electric actuators for precision pointing systems," *Guidance and control*, pp. 61-88

Proceedings of the 25th CANCAM
London, Ontario, Canada, May 31 – June 4, 2015

VIBRATION CONTROL OF TALL BUILDINGS USING AERODYNAMIC OPTIMIZATION

Ahmed Elshaer
Civil and Environmental Department
Western University
London, Ontario, Canada
aelshae@uwo.ca

Girma Bitsuamlak
Civil and Environmental Department
Western University
London, Ontario, Canada
gbitsuam@uwo.ca

Ashraf El-Damatty
Civil and Environmental Department
Western University
London, Ontario, Canada
damatty@uwo.ca

ABSTRACT

Traditionally the shape and orientation of tall buildings are driven by architectural considerations, functional requirements, and site limitations rather than aerodynamic considerations. As a result, they are bluff bodies characterized by high wind-structure interaction induced motion. Vibration control is becoming an important aspect of tall building design in order to keep the building's motions within comfortable limits. In this study, an optimization procedure is presented for minimizing the wind-induced vibrations of tall buildings by aerodynamic mitigation of corners. Computational Fluid dynamics (CFD) models are used to simulate the aerodynamics of wind around a building. An optimization problem is presented to demonstrate the developed procedure.

KEYWORDS: Tall Buildings, vibration, Optimization, CFD, Wind-induced responses

INTRODUCTION

Tall buildings attract people and businesses because of their elegant shape and being considered landmarks. However, this appeal could be affected due to vibrations resulting from lateral loads such as wind and earthquake. For super-tall buildings, wind-induced motion will be even more significant. This will be even more significant in the recent generation of tall buildings owing to being elaborate and slender. Tall buildings can be susceptible to excessive motion during wind events that can cause occupant discomfort and reduce the overall appeal of the structure [1]. Furthermore, keeping the motions of a tall building within comfortable limits could sometimes govern the design rather than the strength requirements. Improving the aerodynamics of a building through corner modification can result in a significant reduction in the lateral load and limit vibrations caused by the wind leading to a more economic and comfortable tall building.

One of the most significant vibrations that buildings experience is the across-wind motion resulting from wind vortex shedding. Some cross-sections such as lens shape are more prone to across-wind buffeting because their streamlined shape causes them to act somewhat like a vertical airfoil. This results in generating high across-wind force variations for relatively small changes in wind angle of attack caused by turbulence. Shape changes that make them less like an airfoil can be effective in this situation [2].

LITERATURE REVIEW

The aerodynamic performance improvement for tall buildings using shape modifications has prompted many scholars to study the relationship between the aerodynamic characteristics of a structure and the resulting wind–induced responses. These include Dagnew [3], Kawai [4], Irwin and Breukelman [5], Tamura and Miyagi [6], Bitsuamlak et al. [7], Elshaer et al. [8], and Kareem et al. [9]. Rather than the classical sharp edged corner of most buildings, building corners can be chamfered, rounded, recessioned, fined, or any other non-uniform shapes. Tse et al. [10] investigated the effect of chamfered and recessed corners on tall building responses. Kwok et al. [11][12] studied the effects of chamfered corners and slotted corners of prismatic tall buildings on the wind-induced responses. The latter reported that chamfering 10% of the building's width may result in a 30% to 40% reduction in both alongwind and crosswind induced motions. Also, the corners on "Taipei 101" were stepped in order to reduce across-wind responses and drag, resulting in 25% reduction in base moment [2]. Most of these studies primarily discuss the advantage of group of modifications compared to others. However, they fall short in advising what type of mitigations provide a "near optimal" solution rather than an "ad-hoc" solution that may or may not be "optimal".

OPTIMIZATION PROCEDURE

The Aerodynamic Shape Optimization (ASO) for tall buildings' corners procedure starts by choosing the objective function and the design variables. The objective function is an aerodynamic characteristic that is intended to be minimized or maximized. The design variables are geometric parameters controlling the shape of the corner. Different combinations of design variables will form different candidates in the optimization problem. Those candidates were ranked according to their fitness (achieving of the objective function). Values of the objective function for different candidates depend on the values of the design variables for each candidate. In the current ASO, the values of the objective function were evaluated either using Computational Fluid dynamics (CFD) models or surrogate models such as neural networks (NN) trained by the data coming from CFD simulations. The values of the objective function were evaluated for different wind angles of attack A(i) and return to the optimization algorithm with the highest value. The procedure of evaluating the objective function values is summarized in Figure 1. The role of the optimization algorithm is to search for the optimal solution through an iterative procedure.

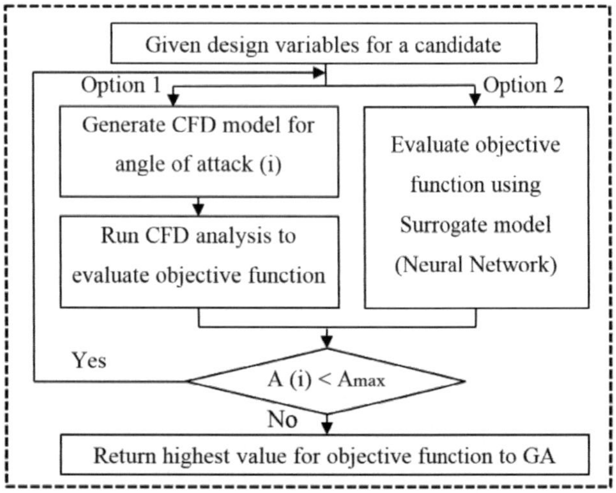

Figure 1 Objective function evaluation

In the current optimization problem, v_1 and v_2 were selected to be the design variables. c and a are calculated according to Eq. (1). The objective function is the standard deviation of the lift coefficient, where the lift coefficient is calculated according to Eq. (2).

$$c = v_1 L \quad , \quad a = v_2 c \tag{1}$$

where c is the modification length of the corner, L is the side length of the square section of the building, and a is the distance measured from the edge of the building to the corner tip (Figure 2).

$$C_L = \frac{F_L}{\frac{1}{2}\rho v_R^2 A_p} \tag{3}$$

where F_L is the lift force acting laterally to the wind direction, ρ is the air density, v_R is the reference velocity, and A_p is the projected area of the building section.

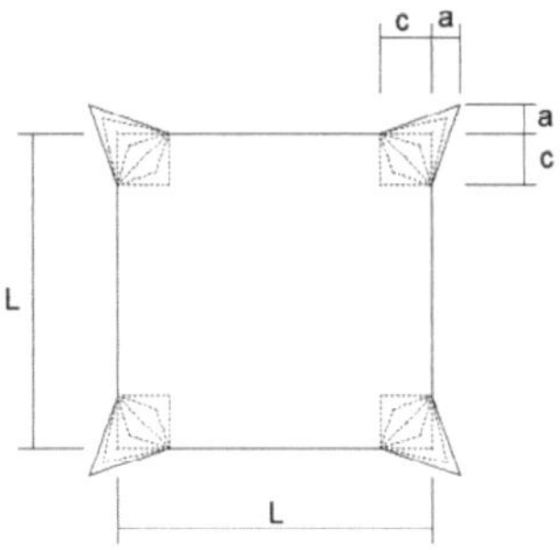

Figure 2 Geometric design variables

OPTIMIZATION ALGORITHM

The Genetic Algorithm (GA) was adopted in the current research because of its reliable ability to find the global extreme value of the objective function without being trapped in a local extreme value. In addition, GA was found to be efficient in solving similar optimization problems [13][14]. In GA, the optimization process starts by evaluating the objective function value for the initial population candidates. Crossover and mutation operators were applied on the current population to produce new candidates to be tested. The role of those operators is to produce new candidates of higher fitness (minimizing the objective function) and investigating areas of the search space that had not been explored. After that, the value of the objective function will be evaluated for the new generated candidates. The step of applying the operators and evaluating the objective function value is iteratively repeated until there is no significant improvement in the optimal solution. Each repetition of that step is considered a new generation in the optimization procedure.

In the current problem, the initial population was formed of *70* candidates with different combinations of v_1 and v_2. The optimization analysis was continued for *50* generations. Three types of crossover operators and four types of mutation operators were applied in each generation on each of *23* candidates of the existing population. The objective function values were evaluated using the NN surrogate model, which was trained by the data resulting from the CFD numerical simulations. NN and CFD will be explained in the following sections.

CFD MODEL PROPERTIES

Large eddy simulations (LES) for the flow around a tall building were conducted to obtain the time history of the lift coefficient for different corner shapes, as shown in Figure 3. The simulations were conducted using a length scale of *1:500,* time scale of *1:100,* and a uniform inlet velocity *10 m/s*. Commercial CFD package (STAR CCM+) [15] was utilized to solve the LES. The dynamic Sub-Grid Scale model by Smagornisky [16] and Germano et al. [17] is used to account for the turbulence. The total number of mesh cells in each model was more than *200,000* cells. The polyhedral mesh size was less than (*D/20*), where *D* is the width of the building. The time step was set to be equal to *5E-4* sec to maintain Courant Friedrichs-Lewy (CFL) lower than (*1.0*). The total physical time was *0.5* sec, which produces from *10* to *18* vortex-shedding cycles. The dimensions of the employed computational domain followed the recommendation of COST [18] guidelines. Figure 4 shows the generated mesh and the velocity contour for different corner shapes and angles of attack.

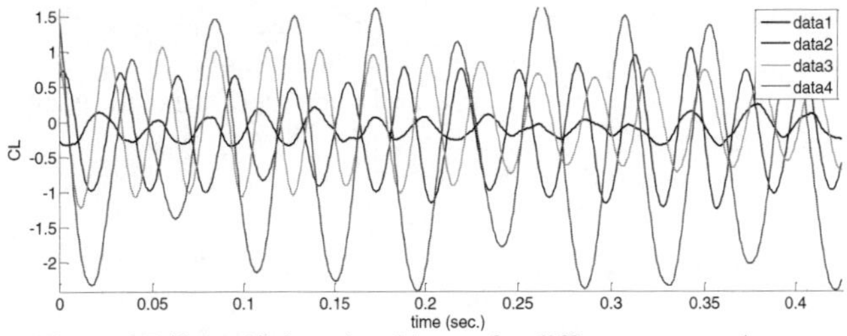

Figure 3 Lift coefficient time history for different corner shapes

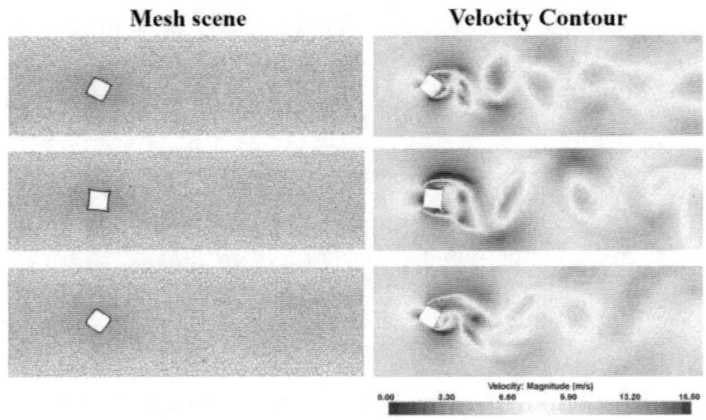

Figure 4 Mesh scene and Velocity contour plan view

NEURAL NETWORK

NN was used as an alternative model to estimate the objective function value as it is considered an efficient data-fitting tool that is capable of estimating outputs values based on input values after being trained using a database of different inputs and outputs. A database of *200* samples was selected randomly for different combinations of v_1, v_2, and angle of attack, as shown in Figure 5.

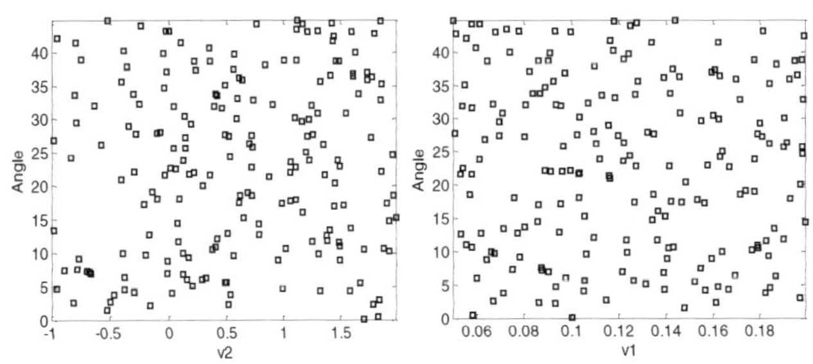

Figure 5 Selected candidates to train NN

The objective function values for all samples were evaluated using CFD models. NN was trained using v_1, v_2, and angle of attack as input data, and corresponding C_L' values as output data. As shown in Figure 6, the regression coefficient of the generated NN model was *0.9964*, which reflects the reliability of the NN as a surrogate model in this application.

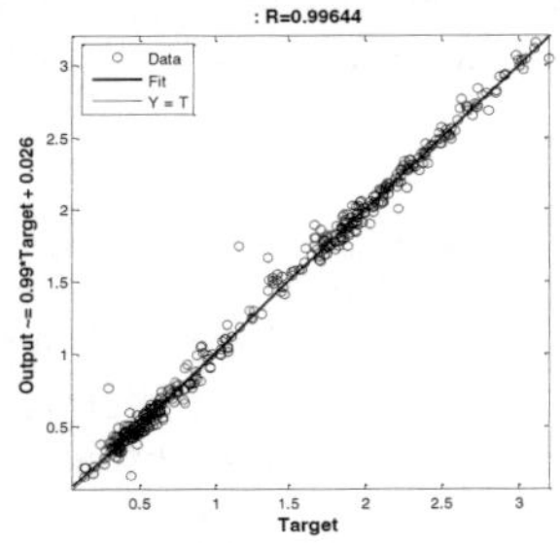

Figure 6 Fitting model (NN) vs target data (CFD)

OPTIMIZATION PROBLEM RESULTS

After running the analysis, the best fitness-generations curve and the optimal corner shape was acquired, as shown in Figure 7. The latter curve shows the aerodynamic improvement over generations resulted from the optimization process. The optimization analysis was repeated four times to ensure reaching the optimal solution without being trapped in local minima. Figure 7 shows that after running the analysis more than once all optimal solutions converges to the same shape, which indicates reaching the global minima of the search space. Optimal solution acquired from the analyses was of $v_1 = 0.127$, $v_2 = -0.94$, and $C_L' = 0.503$. The resulted optimal solution was found to be of less fluctuating lift coefficient by *24%* than that of square cross-section.

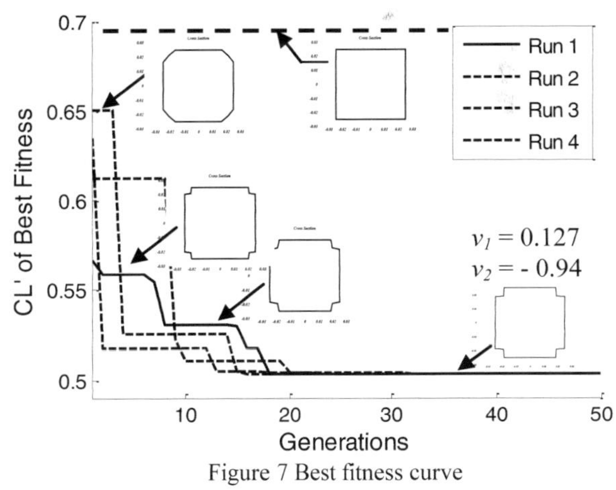

Figure 7 Best fitness curve

CONCLUSIONS

In the current study, the idea of controlling the vibration of tall buildings by corner shape modifications has been discussed. An optimization procedure was introduced by adopting the Genetic Algorithm (GA) to minimize the standard deviation of the lift coefficient of tall buildings by mitigating their corner shape. A demonstration example was presented to show the capability of the developed procedure. NN was trained using the data resulted from CFD models, which simulate the aerodynamics of the tall buildings. NN was found to be a reliable surrogate model for objective function estimation. The fluctuating lift coefficient of the resulted optimal solution is 24% less than that of square cross-section.

ACKNOWLEDGMENTS

The authors would like to acknowledge SharcNet [19] for providing access to their high performance computation facility and CD-adapco for their powerful numerical tool [15].

REFERENCES

[1] A. Kareem, "Dynamic response of high-rise buildings to stochastic wind loads," *J. Wind Eng. Ind. Aerodyn.*, vol. 42(1–3), pp. 1101–1112, 1992.

[2] P. A. Irwin, "Wind engineering challenges of the new generation of super-tall buildings," *J. Wind Eng. Ind. Aerodyn.*, vol. 97, no. 7–8, pp. 328–334, Sep. 2009.

[3] A. K. Dagnew and G. T. Bitsuamlak, "Computational evaluation of wind loads on a standard tall building using LES," *Wind Struct.*, vol. 18, no. 5, pp. 567–598, 2014.

[4] H. Kawai, "Effect of corner modifications on aeroelastic instabilities of tall buildings," *J. Wind Eng. Ind. Aerodyn.*, vol. 74–76, pp. 719–729, Apr. 1998.

[5] P. A. Irwin, B. Breukelman, C. Williams, and M. Hunter, "Shaping and orienting tall buildings for wind," in *Proceedings of the Structural Engineers World Congress, ASCE, San Francisco*, 1998, pp. 19–23.

[6] T. Tamura and T. Miyagi, "The effect of turbulence on aerodynamic forces on a square cylinder with various corner shapes," *J. Wind Eng. Ind. Aerodyn.*, vol. 83, pp. 135–145, 1999.

[7] G. Bitsuamlak, W. Warsido, E. Ledesma, and A. Gan Chowdhury, "Aerodynamic mitigation of roof and wall corner suctions by using simple architectural elements," *J. Eng. Mech., Spec. ISSUE Exp. Methods Damage Detect. Wind Eng.*, pp. 396–408, 2013.

[8] A. Elshaer, G. Bitsuamlak, and A. El Damatty, "Wind Load Reductions due to Building Corner Modifications," in *22nd Annual Conference of the CFD Society of Canada*, 2014.

[9] T. Kijewski and Ahsan Kareem, "Mitigation of motions of tall buildings with specific examples of recent applications," *Wind Struct.*, vol. 2, no. 3, pp. 201–251, 1999.

[10] K. T. Tse, P. A. Hitchcock, K. C. S. Kwok, S. Thepmongkorn, and C. M. Chan, "Economic perspectives of aerodynamic treatments of square tall buildings," *J. Wind Eng. Ind. Aerodyn.*, vol. 97, no. 9, pp. 455–467, 2009.

[11] K. C. S. Kwok and P. A. Bailey, "Aerodynamic devices for tall buildings and structures," *J. Eng. Mech.*, vol. 113, no. 3, pp. 349–365, 1987.

[12] K. C. S. Kwok, P. A. Wilhelm, and B. G. Wilkie, "Effect of edge configuration on wind-induced response of tall buildings," *Eng. Struct.*, vol. 10, no. 2, pp. 135–140, 1988.

[13] A. M. El Ansary, A. A. El Damatty, and A. O. Nassef, "Optimum Shape and Design of Cooling Towers," *World Acad. Sci. Eng. anf Technol.*, vol. 9, pp. 4–13, 2011.

[14] A. Oyama, "Real-Coded Adaptive Range Genetic Algorithm And Its Application to Aerodynamic Design," *JSME Int. J. Ser. A*, vol. 4, 2002.

[15] "STAR CCM+." CD-ADAPCO Product, www.cd-adapco.com/products/star-ccm, 2015.

[16] J. Smagorinsky, "General circulation experiments with the primitive equations," 1963.

[17] M. Germano, U. Piomelli, P. Moin, and W. H. Cabot, "A dynamic subgrid-scale eddy viscosity model," *Phys. Fluids*, vol. A 3(7), pp. 1760–1765, 1991.

[18] J. Franke, A. Hellsten, H. Schlünzen, and B. Carissimo, Eds., *COST . Best practice guideline for the CFD simulation of flows in the urban environment*, vol. Action 732. Hamburg: University of Hamburg, 2007.

[19] "SHARCNET is a consortium of colleges, universities and research institutes operating a network of high-performance computer clusters across south western, central and northern Ontario," 2015. [Online]. Available: www.sharcnet.ca.

BOUNCING DYNAMICS CONSIDERATIONS FOR MICRO SWITCH DESIGN

Mohamed Bognash
Department of Mechanical and Materials Engineering
Western University
London, Ontario, Canada
mbognash@uwo.ca

Tianfu Wang
Global R&D
Arcelor Mittal
1330 Burlington Street E
Hamilton, Ontario, Canada
tianfuwang@hotmail.com

Samuel Asokanthan
Department of Mechanical and Materials Engineering
Western University
London, Ontario, Canada
sasokant@uwo.ca

ABSTRACT

The transient dynamic behavior of cantilever type electro-statically actuated MEMS switch has been studied. Particular emphasis has been placed on gaining an understanding of bouncing dynamics via numerical simulations. An asperity-based contact model which represents the nonlinear contact mechanics is incorporated for capturing the bouncing dynamics together with models of electrostatic actuation forces and squeeze film damping forces. Euler-Bernoulli beam theory and an approximate analytical approach based on Galerkin's method have been employed for predicting transient dynamic responses. The obtained numerical results performed on a micro-switch have captured the main switching features of transient response. In particular, these results indicate good match in the case of certain physical phenomenon during switch operation and the variation of some key switch design parameters have been studied.

KEYWORDS: micro-switch, MEMS, transient response, bouncing dynamics, contact asperity

INTRODUCTION

Micro-mechanical switches have been recognized as primary building blocks for the design of MEMS applications [1]. Cantilever-based switch is one of the most often used configurations in this class of applications. Although improvements have been achieved in their design, analysis and fabrication, the dynamic behavior has not been adequately understood, in particular the transient response. More specifically, during the switch closure stage, the beam bounces several times before making a permanent contact with the drain. Significant performance degradation in the switch is likely to occur due to excessive bouncing. For instance, an increase in the transition time or a complete breakdown of the switch can be a result of bouncing. In order to understand the dynamic characteristics of such switches, a model which includes the interacting forces has been considered.

A detailed study of the transient dynamics of an electrically actuated micro-cantilever has been carried out by Krylov, et al. [2] employing Galerkin's method. Dynamic pull-in stability has been investigated and nonlinear damping has been taken into account. The simulation results have been confirmed by experiments and this study suggested that a significant increase in damping when the structure approaches the substrate can be beneficial for reducing the bouncing in micro switches. However, the bouncing effect was not considered in Krylov's study.

A finite element model using ANSYS software was proposed by Ostasevicius, et al. [3]. In this model, an electro-statically actuated cantilever-type switch was considered. The effects of interaction between electrostatic actuation, squeeze-film damping effects as well as the impact interaction have been taken into account in this model. The study has only emphasized the pull-in voltage and has not emphasized the bouncing in the switch.

Contact bounce phenomenon has been considered in several previous studies. McCarthy et al. [4] included contact bounce effects in the dynamic model of an electronically actuated micro-switch. A simple linear spring model for the contact tips, the electrostatic actuation force, and squeeze-film damping were included. The finite difference method was employed to predict the dynamic behavior of the cantilever micro-switch. Later, a simplified model was also proposed by Granaldi and Decuzzi [5] in order to predict the switching time and bouncing. In another study, Decuzzi et al [6] examined the bouncing dynamics of a cantilever micro-switch with the consideration of adhesive force at the tip. It may be noted that, in the studies [4] – [6], only a simple linear model was used in representing the contact bouncing.

Greenwood and Williamson [7] have proposed an asperity-based model for surface contact. The model has been used widely, and in this approach the rough surface is represented by a collection of spherical asperities with identical end radii. Several researchers have improved the above model to account for elastic, elastic-plastic contact of asperities. Recently, Wang [8] adopted an asperity contact model in MEMS switch to simulate the contact between the beam tip and the substrate.

Based on the previous research carried out, it appears that the micro-switch dynamics that takes into account the bouncing phenomena consider more realistic contact model has gained a little attention; therefore, the present study aims to investigate the transient dynamic behavior of cantilever-type micro-switch when a switch undergoes contact bouncing using uniformly distributed asperity based approach.

To this end, asperity-based contact mechanics, the dynamic coupling effects of electrostatic forces as well as squeeze-film damping forces are to be taken into account. For response predictions, an approximate analytical approach based on Galerkin's method is employed, since closed-form solutions are not achievable in such cases. The predicted transient response employing a micro-beam model has been compared to a pervious experimental work performed [8].

MODELING

A micro switch model developed by Wang [8] is employed to study the transient dynamics. Figure 1 shows a typical micro switch device which has a cantilever configuration of length L, tip height d_T and initial gap d. Electrostatic force is used to actuate this switch. If a voltage is applied via electrode of length $L_2 - L_1$, an electrostatic force will be generated and beam can be pulled down to make contact between the free end tip and the drain. Due to scaling effect, the air damping force cannot be ignored, and hence, a modified Reynold's equation is employed to formulate the damping force. Further, the contact boundary condition is also established via an asperity-based contact model.

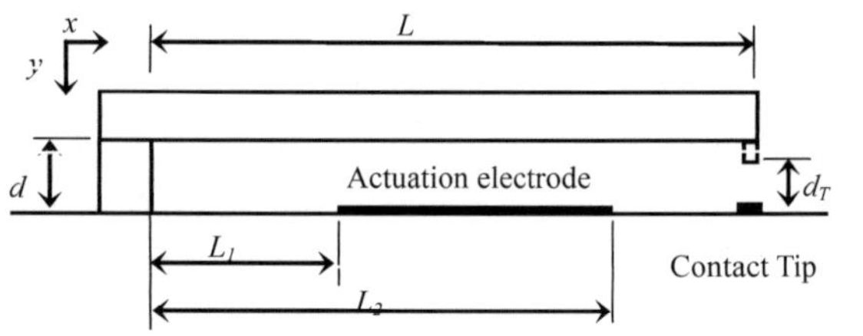

Figure 1: Typical cantilever beam type micro-switch.

Governing Equation

Euler-Bernoulli beam theory is used to develop the equation of motion that governs the flexural dynamics of the switch. When the electrostatic force, air damping force, and the contact force are incorporated in the modeling, the equation takes the form:

$$EI(x)\frac{\partial^4 y(x,t)}{\partial x^4} + \rho A(x)\frac{\partial^2 y(x,t)}{\partial t^2} = f_E + f_D - f_C \ , \quad (1)$$

where x, represents the coordinate along the beam, t represents time, y represents the downward deflection of the beam, E represents the Young's modulus, and ρ represents the material density, while A , I, respectively, denote the area and moment of inertia associated with the beam's cross-section. The force f_E represents the actuation electrostatic force per unit length of the beam, while f_D denotes the dissipating damping force due to the squeezed air film damping effect between the beam and the substrate. The force f_C represents the contact force. The detailed expressions for these forces are discussed in the following sections.

Electrostatic Force

The electrostatic force per unit length of the beam is denoted by f_E and it takes the following form [9]:

$$f_E = \frac{\varepsilon_0 V^2 a}{2h^2}(1+\beta)H(x-L_1)H(x-L_2). \quad (2)$$

The parameter ε_0 represents the permittivity of air, V represents the applied voltage between the electrode and the beam, a represents the beam width, $h = d - y$, and β represents the fringe effect correction coefficient [9]. The function, $H(x)$, denotes the Heaviside step function, and the electrostatic force is considered to act on the beam segment between lengths L_1 and L_2.

Squeeze-Film Damping Force

The damping effects of air in micro-scale have been included in the form of the squeeze-film air damping. Li, et al. [10] have utilized the modified Reynold's equation to represent the squeeze-film air damping force. Based on the above, the squeezed air film damping force per unit length of the beam, f_D, can be expressed as:

$$f_D = -\frac{\mu a^3}{h^3}\frac{\partial h}{\partial t} \ , \quad (3)$$

where μ denotes the air viscosity.

Contact Force

Wang [8] has proposed a simplified elastic modified model based on the model proposed by Greenwood-Williams. This model utilizes a uniformly distributed height for the asperities and the total contact force is given by

$$f_C = \eta A_n P_0(w) \ , \quad (4)$$

where η is the area density of asperities, A_n represents the nominal contact area and $w = z - (y - d_T)$, where z denotes the asperity height. The applied load for a single asperity based on Hertzian contact theory is represented by $P_0(w)$:

$$P_0(w) = \frac{4}{3}E^* R^{\frac{1}{2}} w^{\frac{3}{2}}, \quad \frac{1}{E^*} = \frac{1-v_1^2}{E_1} + \frac{1-v_2^2}{E_2}, \quad (5)$$

where P_0 represents the applied load while R and w, respectively, denote the radius of the sphere asperity and the contact interference. The equivalent Young's modulus E^* is based on the Young's moduli E_1, E_2 and the Poisson's ratios v_1, v_2 of the two contact surfaces.

NUMERICAL SIMULATIONS

Galerkin's approach is used to develop approximate analytical solution since closed-form solutions of the governing equations are not achievable. For this purpose, the natural frequencies and the mode shapes of the cantilever beam are evaluated first. By employing the Galerkin's method, the continuous system can be discretized to form a system of ordinary differential equations which is solved analytically. The solutions of these ODE's are then used to predict the response. To this end, the transverse displacement of the beam is written in the following form:

$$y(x,t) = \sum_{r=1}^{\infty} Y_r(x)\eta_r(t). \tag{6}$$

For discretizing the system equation, normal modes are employed to obtain the response. Substituting Eq. (6) into the beam equations of motion presented in Eq. (1) and considering the orthonormality conditions, the following ODEs representing the motion are obtained:

$$\ddot{\eta}_r(t) + \omega_r^2\eta_r(t) = N_r, \qquad r = 1,2,\dots, \tag{7}$$

where N_r represents the modal forces given by the expression

$$N_r = \int_0^L Y_r(x) f dx\ , \quad r = 1,2,\dots \tag{8}$$

It may be noted that the modal forces capture the effects of all forces that influence the system dynamics. The equations of motion in the normal co-ordinates as given by Equations (7) are decoupled and hence a single degree of freedom system is used to represent the individual equation. If system is subject to the initial displacement, x_0, and initial velocity, v_0, and if the modal force is assumed to be constant, the solution can be expressed as follows:

$$x(t) = \frac{N_r}{\omega_n}[1 - \cos(\omega_n t)] + x_0\cos(\omega_n t) + \frac{v_0}{\omega_n}\sin(\omega_n t)\,. \tag{9}$$

If an assumption is made that during a finite small time step, only a finite small displacement is obtained, the modal force N_r can be considered to take a constant value during this finite small time step. Hence, Equation (9) has been used in the computation of the system response during each time step. Figure 2 summarizes the overall computation procedure used for predicting the system response.

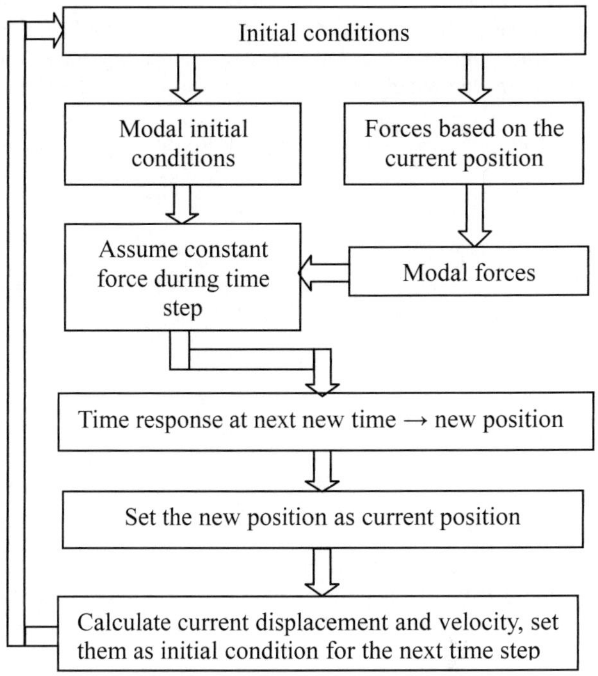

Figure 2: Response computation process.

RESULTS AND DISCUSSION

Transient response simulation of micro switch has been performed to quantify important switch design parameters based on the predicted response. For the purposes of comparison, the material and geometric properties of micro-switch used in [8] are employed in this work. Figure 3 shows some of the important parameters that characterize the switch performance which can be defined as follows:

- Initial contact time, t_i, is the time taken for the first contact.
- Permanent contact time, t_p, is the time taken for the switch to maintain permanent contact.
- Bouncing time, t_b, is the time from the instant of making first contact to the instant when the single significant bounce ends.
- Major bounce height, H.

This model has been used to evaluate the dynamic pull-in voltage, V_{th}, at which the switch collapses when the electrostatic force overcomes the elastic restoring force with the consideration of inertia effect and it takes a value of 64.5 volts for the present switch configuration.

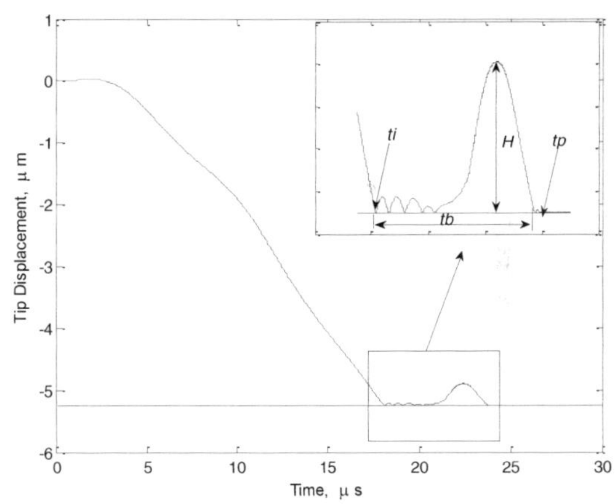

Figure 3: Typical microswitch tip-end response for $V = 1.25\ V_{th}$

The simulation results have been compared with previous experimental work [8] in which the beam motion is measured using a micro scanning Laser Doppler Vibrometer. The numerical predictions have been found to match the nature of the experimental results, hence providing adequate confidence in using the theoretical model as a basis for understanding switch dynamics. The variation of the switch design parameters at different activation voltages are shown in Figure 4 and 5. An overall reduction in the switching time with increasing actuation voltages can be observed from Figure 4 which indicates performance improvement can be achieved with higher actuation voltages. However, this improvement is seen to be obtained at the expense of greater major bounce heights as seen in Figure 5 which predicts that the major bounce heights increase with increasing activation voltages, hence limiting the achievable activation voltages.

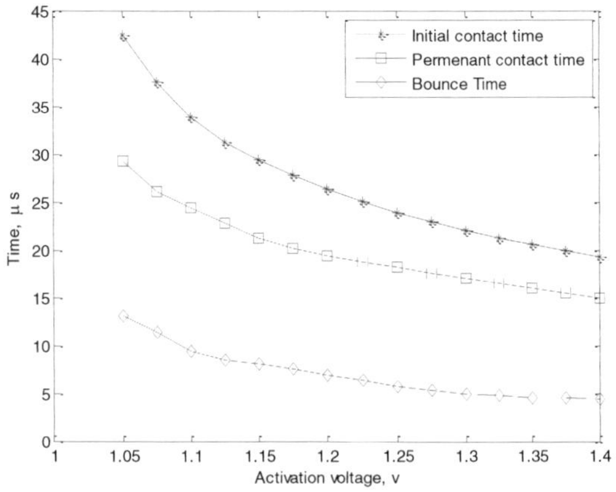

Figure 4: Variation of initial contact time, permanent contact time, and the bounce time versus activation voltages.

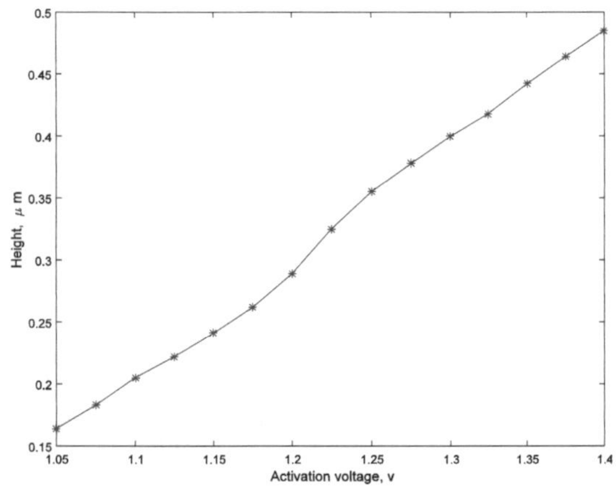

Figure 5: Major bounce height *vs* with activation voltage.

CONCLUSIONS

A micro switch model that includes the relevant forces and a suitable numerical scheme for providing an insight into the switch nonlinear transient dynamics is employed. An asperity-based contact model which represents the nonlinear contact mechanics is included in the model. Euler-Bernoulli beam theory and an approximate analytical approach based on Galerkin's method have been employed for predicting transient dynamic responses. This model and the scheme are capable of predicting responses, and they are envisaged to aid the design of micro switches. The predictions of the switch closure parameters are performed. These results provide insight on improve the switch dynamics. The physical phenomenon during switch operation and the variation of some key parameters are found to have good match with the previous experimental work.

ACKNOWLEDGMENTS

This research was supported in part by the Natural Science and Engineering Research Council (NSERC) of Canada discovery grant, Western Engineering Scholarship (WES) scheme from Western University, London, Ontario, Canada and Libyan-North American Scholarship program.

REFERENCES

1. Ramezni, A., Alasty, J., and Akbari, J. 2006, "Closed-form solutions of the pull-in instability in nano-cantilevers under electrostatic and intermolecular surface forces", *International Journal of Solids and Structures,* Vol(44), pp. 4925–4941.
2. Krylov, S., and Maimon, R. 2004, "Pull-in dynamics of an elastic beam actuated by continuosly distributed electrostatic force", *Journal of Vibration and Acoustic, Transaction of the SAME,* Vol(126), pp. 332-342.
3. Ostasevicius, V., Dauksevicius, R., Tamulevicius, S., Grigaliunas V., and Pallecvicius, A. 2005, "Design, fabriction and simultion of cantilever-type electrostatic micromechanical switch", *International Society for Optical Engineering, Smart Structures and Materials,* Vol(5763), pp. 436-445.
4. McCarthy, B., Adams, G., McGruer, N., and Potter, D. 2002, "A dynamic model, including contact bounce, of an electrostatically actuated microswitch", *Journal of Microelectromechanical systems,* Vol(11), pp. 276-283.
5. Granaldi, A., and Decuzzi, P. 2006, "The dynamic response of resistive microswitches: Switching time and bouncing," *Journal of Micromechanics and Microengineering,* Vol(16), pp. 1108-1115.
6. Decuzzi, P., Demelio, G., Pascazio, G., and Zaza, V. 2006, "Bouncing dynamic of resistive microswitches with an adhesive tip", *Journal of Applied Physics,*Vol(100), pp. 024313.
7. Greenwood J. A., and Williamson, J. P. 1966 "Contact of Nominally Flat Surfaces," *Proceeding of the Royal Society of London, Series A, Mathematical and Physical Sciences,* Vol(95), pp. 300-319.
8. Wang, T., "Nonlinear and Stochastic Dynamics of MEMS Based Angular Rate Sensing and Switching Systems," University of Western Ontario, London, Ontario, Canada, 2009.
9. Bao, M. 2001, *Analysis and Design Principles of MEMS Devices*, Elsevier, 2001, p. 179.
10. Li, Y., Meguid, S., Fu, Y., and Xu, D. 2013, "Unified nonlinear quasistatic and dynamic analysis of RF-MEMS switches", *Acta Mechanica,* Vol(224), pp. 1741–1755.

Proceedings of the 25th CANCAM
London, Ontario, Canada, May 31 – June 4, 2015

HORIZONTAL MODELS OF UNBONDED FIBER REINFORCED ELASTOMERIC ISOLATORS

Niel Van Engelen
Department of Civil Engineering
McMaster University
Hamilton, Ontario, Canada
vanengn@mcmaster.ca

Michael Tait
Department of Civil Engineering
McMaster University
Hamilton, Ontario, Canada
taitm@mcmaster.ca

Dimitrios Konstantinidis
Department of Civil Engineering
McMaster University
Hamilton, Ontario, Canada
konstant@mcmaster.ca

ABSTRACT

Base isolation is an effective method of decreasing the overall seismic demand of a structure during an earthquake event. The conventional and commonly used steel-reinforced elastomeric isolator is heavy and expensive. The weight and cost of this type of bearing have been perceived as barriers to widespread base isolation application. The use of fiber reinforcement in lieu of steel can significantly reduce the weight and cost while still achieving similar performance. The combination of fiber reinforcement and an unbonded application to the isolator supports results in a desirable non-linear horizontal hysteretic response. The horizontal force-displacement relationship is characterized by softening and subsequent stiffening at large displacements. In this paper, four proposed models on the horizontal force-displacement relationship of unbonded fiber-reinforced elastomeric isolators are reviewed and compared. Of these models, only one predicts and quantifies the stiffening that occurs at large displacements.

KEYWORDS: base isolation, fiber reinforced, elastomeric isolator, earthquakes, horizontal stiffness

INTRODUCTION

Earthquakes continue to be the source of large economic loses and the tragic loss of thousands of lives. A recent study of potential earthquakes in Canada indicated that Canadian cities are not immune from the substantial risks that earthquake hazards present [1]. The risk presented by earthquakes is in part due to the unfortunate harmony that exists between the fundamental frequency of structures and the critical high energy frequency range of a typical earthquake event. Life safety is the primary concern of most modern building codes. Thus, the conventional approach to designing structures to withstand earthquakes does not prevent damage to the structure or the nonstructural components and systems (NCS), focusing primarily on preventing collapse. From an economics perspective, this is especially concerning since the value of NCS often dominates the total value of the structure.

The concept of base isolation requires the introduction of a flexible horizontal layer at the foundation level of the structure. The flexible layer decouples the structure from ground motions. The resulting large displacements that occur at the flexible isolation layer dominate the response of the structure, effectively shifting the frequency response out of the critical high energy range. The structure ideally translates as a near rigid body on-top of the isolation layer. This process minimizes the damage to the structure and the NCS, even after large earthquake events. The idealized response of a fixed base and base isolated structure to ground motions is illustrated in Fig. 1.

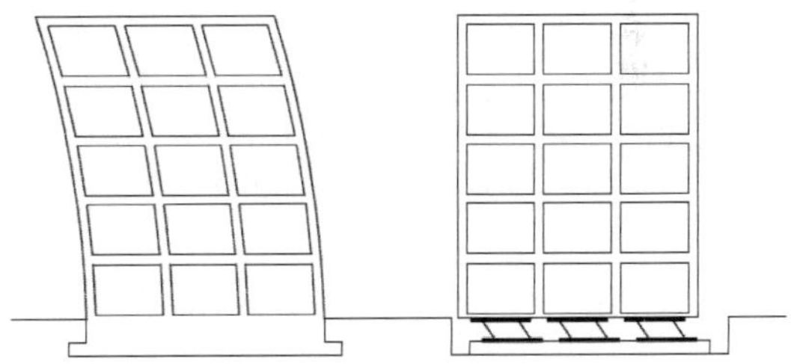

Figure 1: Idealized comparison between a fixed base (left) and base isolated (right) structure

It is necessary that the isolators that make up the flexible layer retain a large vertical stiffness to suppress potential rocking modes. In this respect, the low shear modulus and near-incompressible nature of elastomers render them ideal for application as base isolators. Conventional elastomeric isolators are composed of alternating horizontal layers of steel reinforcement and elastomer. The steel reinforcement restrains the lateral bulging of the elastomer under an applied vertical compressive stress. Combined with the near incompressibility of the elastomer, the restraint of the lateral bulging develops a large vertical stiffness while leaving the low horizontal stiffness relatively unaltered.

The large weight and substantial cost of steel-reinforced elastomeric isolators have been perceived as barriers to widespread base isolation application. To overcome these barriers, it has been proposed that the steel reinforcement can be replaced with fiber reinforcement of similar properties in tension [2]. Furthermore, the large steel end plates required to mechanically fasten the isolator to the upper and lower supports can be removed, relying on friction between the elastomer and the supports to transfer lateral loads. The resulting bearing, denoted as an unbonded fiber-reinforced elastomeric isolator (FREI), is significantly lighter, and has considerable advantages in the manufacturing process. Experimental and numerical studies of unbonded FREIs have indicated that comparable performance can be achieved when compared to similar bonded FREI or SREI designs [3, 4].

Unbonded FREIs of certain designs exhibit a characteristic softening and stiffening response, as shown in Fig. 2. This response, similar to other adaptive isolation devises, is considered advantageous to achieve multiple performance objectives at different earthquake hazard levels. In this paper, methods of modeling the horizontal force-displacement relationship of unbonded FREIs are reviewed and discussed, identifying the unique challenges that originate as a consequence of the fiber reinforcement and unbonded application.

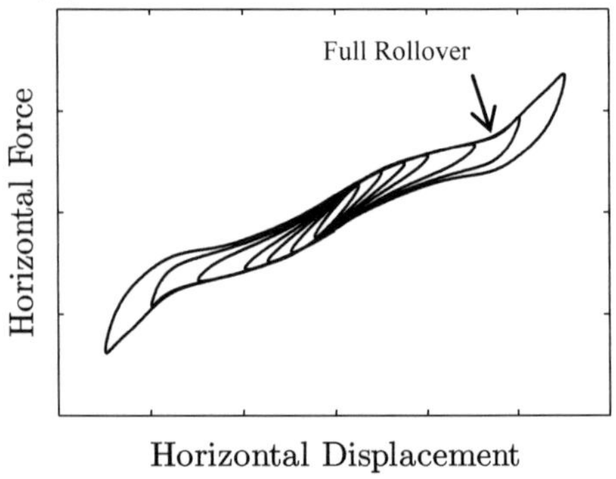

Figure 2: Generalized hysteresis loops of a stable unbonded FREI

ROLLOVER AND FULL ROLLOVER

The combination of the unbonded application and lack of flexural rigidity of the fiber reinforcement results in a unique *rollover* deformation. Rollover occurs as the initially horizontal surfaces of the isolator rotate and lose contact with the supports with horizontal displacement. This process continues until *full rollover*, when the initially vertical faces contact the upper and lower supports. Rollover and full rollover are illustrated in Fig. 3.

Full rollover prevents further rollover from occurring. The horizontal resistance to displacement is greater in simple shear than an equivalent rollover section, the size of which is proportional to the horizontal displacement. Thus, as rollover occurs, the effective horizontal stiffness softens. The effective horizontal stiffness continues to decrease until full rollover, when the response begins to stiffen (see Fig. 2). These adaptive characteristics are considered advantageous for seismic isolation. The softening increases the shift of the fundamental period of the isolation system, further out of the critical high energy range of a typical earthquake event, whereas the stiffening is believed to act as a self-restraint against excessive displacement during large earthquake events. If the tangential stiffness remains positive over all levels of imposed horizontal displacement the isolator is considered stable. Retaining horizontal stability is a requirement for isolation devices over all levels of imposed displacement.

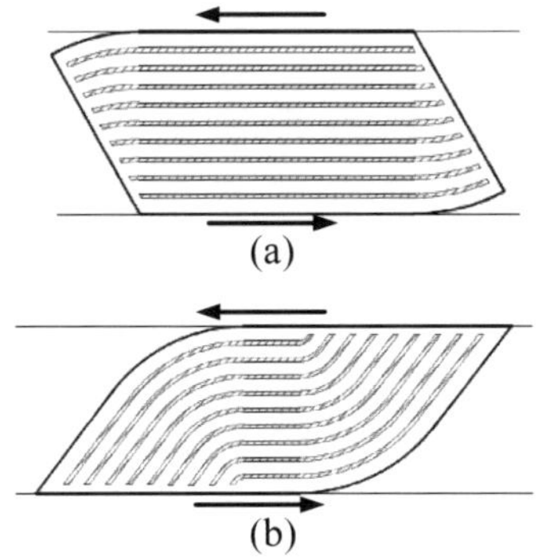

Figure 3: An unbonded FREI showing (a) rollover and (b) full rollover.

HORIZONTAL MODELS

It is convenient to analyze the isolator by dividing it into three sections, shown in Fig. 4. The central section refers to any part of the bearing that remains in contact with the upper and lower supports. This is generally determined by a vertical line drawn from the opposing edge of the isolator (see Fig. 4). The remaining two sections are identical and represent the rollover of the bearing. Note that the size of these sections will change depending on the horizontal displacement.

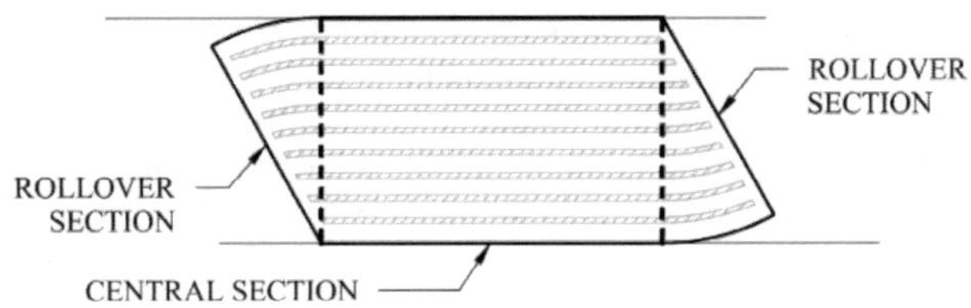

Figure 4: Idealized division of an unbonded FREI into sections

Lower Bound (Model M1)

A lower bound force-displacement relationship can be determined by neglecting the rollover sections, assuming that they are stress free and contribute no resistance to horizontal displacement [5]. In the model, only the central section of the isolator is considered. If the length of the rollover surface is assumed equal to the horizontal displacement, x, than the width of the central section also decreases by x as rollover occurs. Therefore, the horizontal force, F, is

$$F = \frac{2bGx}{t_r}\left(2a - x\right) \quad (1)$$

where t_r is the total thickness of the elastomeric layers, $2b$ is the length (perpendicular to the direction of loading), $2a$ is the width, and G is the shear modulus of the elastomer.

Equation (1) gives a parabolic force-displacement relationship. As the horizontal displacement increases, F reaches a maximum at a displacement equal to half the width of the isolator. Further horizontal displacement results in a negative tangential stiffness, becoming horizontally unstable. Horizontal instability as a result of full rollover is common to isolators with low width-to-total height aspect ratios, where the area of the central section decreases rapidly in comparison to the initial loaded area. It should be noted that this model is not intended to predict the relationship past the point of instability.

Including Vertical Deformation (Model M2)

Russo et al. [6] proposed an equivalent linear force-displacement relationship determined from an average area in simple shear. It was proposed that the vertical deformation of the bearing, which reduces the total height of the bearing, delays the loss of contact between the isolator and the upper and lower supports. The region of the bearing in contact with the supports was assumed to behave in simple shear, and, as with the Konstantinidis et al. [5] model, the rollover sections were assumed to be stress free.

The average area of the central section was determined from two points: the area at zero horizontal displacement and at full rollover. It should be noted that the model requires a vertical deflection and a full rollover displacement to determine the average area of the central section, which in [6] were determined from experimental results. The horizontal force is

$$F = \frac{GA_{ave}}{t_r} x \tag{2}$$

where

$$A_{ave} = \frac{A_c(0) + A_c(x_r)}{2} \tag{3}$$

is the average area of the central section, A_c, determined at zero horizontal displacement and full rollover, x_r.

Equivalent Area Method (Model M3)

Based on observations made through finite element analysis, Toopchi-Nezhad [7] acknowledged that the shear strain in the rollover sections was considerably less than the central section. The horizontal resistance of the rollover section was approximated by introducing an effective area, which accounted for the area of the central section, in addition to a reduced area representing the contribution of the rollover sections. In effect, this model is similar to Russo et al. [6] in that it increases the size of the central section, but otherwise is identical to Konstantinidis et al. [5]. In this representation, the increase in the area of the central section is intended to account for the rollover section, not the vertical deflection.

Two models were proposed depending on the assumed contribution of the rollover sections. The first, presented herein, assumed that the rollover sections could be represented by simple shear assuming 25 % of the total area of the rollover section contributed. The second model required the effective area of the rollover section to be adjusted based on the full rollover displacement. According to the former, the horizontal force is

$$F = \frac{2bGx}{t_r}\left(2a - \frac{3}{4}x\right) \tag{4}$$

Note that Eq. (4) is identical to Eq. (1) aside from the coefficient of the second term in the parenthesis.

Including Full Rollover (Model M4)

Van Engelen et al. [8] developed an analytical model based on a cantilever representation of the full rollover sections using large deflection theory. This model is the most comprehensive available for unbonded FREIs as it predicts the occurrence of full rollover and the horizontal force-displacement relationship up to and exceeding full rollover.

The model divided the isolator into three sections, as shown in Fig. 4. The central section was assumed to follow Eq. (1) and the total horizontal force was considered to be the sum of the three sections.

The rollover sections were assumed to be the combination of shear and bending displacements, which are both the result of the horizontal resistance of the rollover sections. A relationship between the horizontal force and these displacements was determined based on a cantilever representation of the rollover sections where the length was assumed to be equal to the total horizontal displacement. Based on large deflection theory, an iterative process was applied to determine the deflection of the cantilever and the horizontal force contribution of the rollover sections. The deflected shape of the rollover section was determined from the cantilever deflection and used to predict full rollover.

Full rollover restrains further rollover from occurring, thus, the assumptions made for horizontal displacements up to full rollover are invalid due to the changing boundary conditions. It was proposed by Van Engelen et al. [8] that the restraint of further rollover could be idealized as an isolator in pure shear deformation. Therefore, a linear force-displacement relationship was selected for displacements exceeding full rollover.

MODEL COMPARISON

The horizontal force-displacement relationship generated by the four models are compared with experimental data in Fig. 5, where F has been normalized by G and the total plan area, A. The models are represented in the figure in the order presented herein. For the quarter scale isolator considered, t_r = 19 mm, $2a$ = 76 mm, $2b$ = 52 mm, and G = 0.35 MPa [9]. Full rollover was assumed to occur at a horizontal displacement of 1.96 t_r and the vertical compressive strain was assumed to be 0.05.

All four models make similar predictions of the horizontal force up to 0.75 t_r. Since the softening increases with increasing horizontal displacement, initially the size of the rollover section is small. Consequently the influence of the rollover section on each model is also small. As the displacement increase, the models begin to diverge. M1[5] and M3 [7] both share a similar form. M1 [5] closely follows the experimental data but incorrectly predicts an instability. M2 [6] initially under predicts the horizontal force. As substantial softening occurs, the model begins to substantially over predict the horizontal force. For the isolator considered, the magnitude of the softening is too substantial to be accurately predicted by a linear model. M4 [8] over predicts the force in the intermediate range. Full rollover can be observed at approximately 2.0 t_r. At displacements exceeding full rollover, M4 [8] accurately captures the force-displacement relationship.

It should be noted that none of the proposed models take into account the interaction between the vertical load and the horizontal stiffness. Depending on the design of the bearing, this interaction can substantially reduce the effective horizontal stiffness. Thus, although M1 [5] appears to accurately follow the data up to full rollover, if the effect of the vertical force on the horizontal behaviour were included, the accuracy of the model prediction would decrease [10].

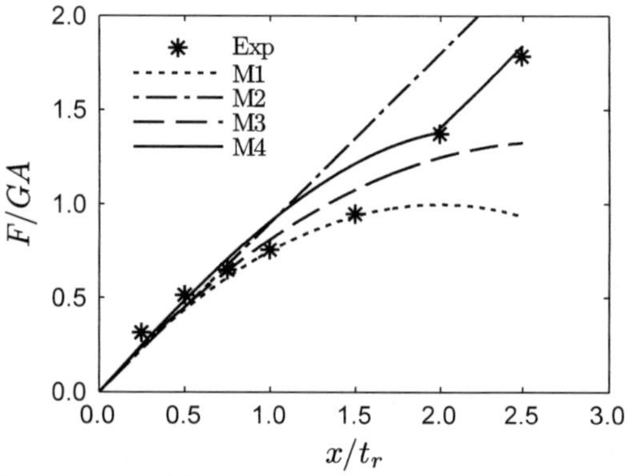

Figure 5: Model comparison with experimental data

CONCLUSIONS

The occurrence of rollover and full rollover are responsible for the adaptive softening and stiffening characteristics in unbonded FREIs. Although considered beneficial, the non-linear response presents a challenge for designers to accurately predict and model. In this paper, four models from the literature were presented. Each model accounts for rollover in a different method and only one considered the stiffening due to full rollover. When compared to experimental data, all of the models accurately predict the force-displacement relationship at small displacements. As full rollover occurs and softening increases, the accuracy of the models begin to diverge. Without correcting for full rollover, the models are unable to accurately capture or predict this beneficial feature.

ACKNOWLEDGMENTS

Financial support for this study was provided by the McMaster University Centre for Effective Design of Structures (CEDS) funded through the Ontario Research and Development Challenge Fund (ORDCF) as well as an Early Researcher Award (ERA) grant, both are programs of the Ministry of Research and Innovation (MRI). Support was also provided through the Natural Sciences and Engineering Research Council (NSERC) of Canada. The support of a Vanier Canada Graduate Scholarship is also gratefully acknowledged.

REFERENCES

1. AIR Worldwide. 2013, *Study of Impact the Insurance and Economic Cost of a Major Earthquake in British Columbia and Ontario/Quebec*, AIR Worldwide, Boston.
2. Kelly, J.M. 1999, "Analysis of fiber-reinforced elastomeric isolators", *Journal of Seismology and Earthquake Engineering,* 2(1), pp. 19-34.
3. Toopchi-Nezhad, H. Tait, M.J. and Drysdale, R.G. 2011, "Bonded versus unbonded strip fiber reinforced elastomeric isolators: finite element analysis", *Composite Structures,* 93(2), pp. 850-859.
4. Naghshineh, A.K., Akyuz, U., and Caner, A. 2014, "Comparison of fundamental properties of new types of fiber-mesh-reinforced seismic isolators with conventional isolators", *Earthquake Engineering and Structural Dynamics,* 43, pp. 301-316.
5. Konstantinidis, D., Kelly, J.M., and Makris, N. 2008, *Experimental Investigations on the Seismic Response of Bridge Bearings,* EERC Report No. 2008/02. Earthquake Engineering Research Center, Berkeley.
6. Russo, G., Pauletta, M., and Cortesia, A. 2013, "A study on the experimental shear behavior of fiber-reinforced elastomeric isolators with various fiber layouts, elastomers and aging conditions", *Engineering Structures,* 52, pp. 422-433.
7. Toopchi-Nezhad, H. 2014, "Horizontal stiffness solutions for unbonded fiber reinforced bearings", *Structural Engineering and Mechanics,* 49(3), pp. 395-410.
8. Van Engelen, N.C., Tait, M.J., and Konstantinidis, D. 2014, "Model of the shear behavior of unbonded fiber-reinforced elastomeric isolators", *Journal of Structural Engineering.*
9. Van Engelen, N.C., Tait, M.J., and Konstantinidis, D. 2012, "Horizontal behaviour of stable unbonded fiber reinforced elastomeric isolators (SU-FREIs) with holes", *15th World Conference on Earthquake Engineering,* Lisbon, Portugal.
10. de Raaf, M.G.P., Tait, M.J., Toopchi-Nezhad, H. 2011, "Stability of fiber reinforced elastomeric bearings in an unbonded application", *Journal of Composite Materials,* 45(8), pp. 1873-1884.

Proceedings of the th CANCAM
London, Ontario, Canada, May 31 - June 4, 2015

NONLINEAR THERMO-ELECTRO-MECHANICALLY INDUCED VIBRATION OF FGPM BEAMS

M. Komijani
Civil and Environmental Engineering Department
University of Waterloo
Waterloo, Ontario, Canada
mohammad.komijani@uwaterloo.ca

Dr. R. Gracie
Assistant Professor
Civil and Environmental Engineering Department
University of Waterloo
Waterloo, Ontario, Canada
rgracie@uwaterloo.ca

ABSTRACT

The presented article investigates nonlinear thermo-electro-mechanically induced vibration of Functionally Graded Piezo-electric Material (FGPM) actuators. Timoshenko beam theory with various inplane and out-of-plane boundary conditions are considered. The beam is assumed to be under mechanical, thermal, and electrical dynamic excitation. All thermo-electro-mechanical properties of the beam are graded through the thickness (i.e., height) of the beam, according to a power law distribution. The von Kármán geometric nonlinearity is used to account for the large deflection response of the beam. A displacement-based finite element formulation is developed. The resulting system of nonlinear algebraic equations is solved using a Newton-Raphson iterative technique, based on the Newmark's scheme for time integration. The numerical results of the lateral deflection as a function of time under dynamic loading reveals induced vibrational response of the structure. The effects of load type, boundary conditions, beam geometry, composition rule of the constituents, and the applied actuator voltage are discussed through parametric studies. It is demonstrated that the applied voltage has a significant effect on the themo-electrio-mechanically induced vibration of the structure.

Keywords: Nonlinear FEM, Functionally graded piezoelectric, Timoshenko beam theory, von Kármán nonlinearity.

INTRODUCTION

Piezoelectric structures possess excellent thermo-electro-mechanical features and characteristics to convert electrical energy to mechanical one and vice versa. Hence, this kind of structures spans a wide range of applications that embrace surpassing vibrational motion, controlling systems, suppression of structural instabilities, noise attenuation, damage detection, etc. [1, 2]. Traditional piezoelectric sensors and actuators have usually been made of several mono-layers of various piezoelectric substances. Several theoretical and mathematical models were proposed for laminated composite structures with sensors and actuators made of piezoelectric materials [3]. Due to abrupt changes in both their thermo-electro-mechanical properties from one layer to another, these sandwich devices is that high stress concentrations at the interfaces between bonded piezoelectric layers. These stress concentrations may pose problems such as initiation and propagation of micro-cracks [4]. Owing to these limitations, the use of piezoelectric layered structures in application where high reliability is needed is limited. In order to tackle these disadvantages of conventional layered piezoelectric elements, Wu et al. [5] extended the concept of functionally graded materials (FGMs) to piezoelectric structures. FGPMs are a kind of piezoelectric material in which thermo-electro-mechanical properties change smoothly and continuously in one or more directions. FGPM actuators not only can supply large displacements but also decrease the internal stress concentrations, which enhances significantly the lifetime of the actuators.

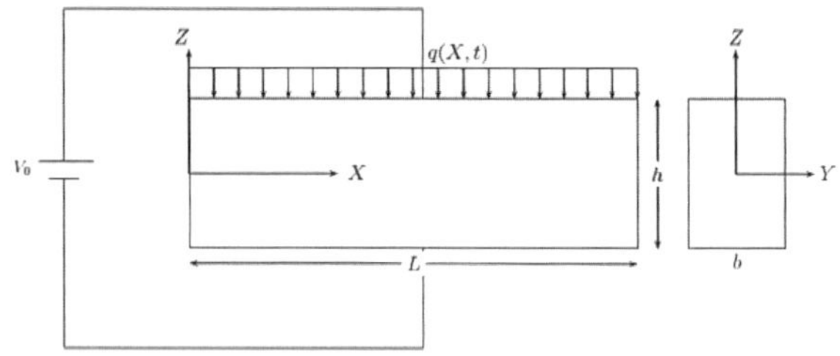

Figure 1: Geometry and coordinate system of an FGPM beam.

Governing equations

A rectangular cross section beam made of a functionally graded piezoelectric material (FGPM) of length$L$, width$b$, and thickness$h$is considered. The beam is subjected to lateral distributed force$q$, uniform temperature riseΔT, and external applied electric voltageV_0, as shown in Fig. 1.

All material properties are considered to vary continuously across the thickness according to the power law distribution function (see [6]).

In the present work, Timoshenko beam theory is used with the following displacement field

$$U(X,Z,t) = U_0(X,t \quad Z \quad X,t), \quad W(X,Z,t) = W_0(X,t) \tag{1}$$

where(U_0, W_0)represent the displacement components of a point on the mid-plane of the beam along axial and transverse coordinates, respectively, and the corresponding rotation of the cross section is denoted byΨ.

In this study, the electric potential$V$ is assumed to have the following distribution function that includes both reverse and direct piezoelectric effects.

$$V(X,Z,t) = \quad \beta Z(\quad X,t) + \frac{V_0}{h} Z \tag{2}$$

where$\beta = \frac{\pi}{h}$, and $X,t)$is the spatial function of the electric potential and the second term represents the external applied electric voltage to the upper and lower surfaces.

The constitutive model for an FGPM beam under thermo-electro-mechanical loads may be expressed as follows

$$\sigma_X = Q_{11}\varepsilon_X - Q_{11}\alpha_1 \Delta T - e_{31} E_Z, \tag{3a}$$

$$\tau_{XZ} = Q_{55}\gamma_{XZ} - e_{15} E_X \tag{3b}$$

$$D_X = e_{15}\gamma_{XZ} + k_{11} E_X, \tag{3c}$$

$$D_Z = e_{31}\varepsilon_X + k_{33} E_Z + p_3 \Delta T \tag{3d}$$

whereε_X,γ_{XZ},σ_X,τ_{XZ},D_i, andE_i state respectively the axial strain, shear strain, axial stress, shear stress, dielectric displacements, and electric field components;e_{ij},k_{ij},α_1,p_3 are the piezoelectric, dielectric, thermal expansion, and pyroelectric coefficients, respectively;$\nu$is Poisson's ratio, and Q_{11} andQ_{55} represent the elastic stiffness coefficients.

The von Kármán type nonlinear strain–displacement relations are obtained using Eq. (1) as

$$\varepsilon_X = U_{0,X} + \tfrac{1}{2}W_{0,X}^2 + Z\Psi_{,X}, \quad \gamma_{XZ} = W_{0,X} + \Psi \tag{4}$$

It is noted that the electric field is the negative gradient of the total electric potential function.

The motion equations are obtained on the basis of the principal of virtual work :

$$\delta \int_t (K - H + R)dt = 0 \tag{5}$$

where the variation of the electric enthalpyδH, and the variation of the kinetic energyδKcan be represented respectively as:

$$\delta H = b \int_0^L \int_{-\frac{h}{2}}^{\frac{h}{2}} \Big(\sigma_X \delta\varepsilon_X + K_s \tau_{XZ} \delta\gamma_{XZ} - D_X \delta E_X - D_Z \delta E_Z\Big) dZdX,$$

$$\delta K = b \int_0^L \int_{-\frac{h}{2}}^{\frac{h}{2}} \rho(U_{,t}\delta U_{,t} + W_{,t}\delta W_{,t}) dZdX \tag{6}$$

whereK_s is a shear correction factor and is taken as$K_s = \pi^2/\ $.

The virtual workδRdue to the external force is:

$$\delta R = b \int_0^L \Big[q\,\delta W_0\Big] dX \tag{7}$$

where$q$is transverse distributed force that is measured as per unit area of the beam.

Making use of the associated constitutive relations in the variational form of the energy (Eq. (5)), and applying the finite element interpolation function for each variable, the final FE model can be expressed in a compact form as:

$$[M]\left\{\ddot{\Delta}\right\} + \quad K_L] + [K_{NL1}] + [K_{NL2} \quad \{\Delta\} = \{F\} \tag{8}$$

Results and Discussion

As an example, figure 2 depicts the midpoint deflection versus time curves of a$C-C$FGPM beam ($L/h \quad = \quad , V_0 = 0V$) subjected to a suddenly applied lateral distributed load of$q = \quad KNm^{-2}$. The beam is made from PZT-4/PZT-5H with various values of power law index. The results obtained using the nonlinear numerical solution are compared with the linear dynamic analysis results of [7]. As seen, the nonlinear results of this work are in good agreement with those reported for linear case in [7], when deformations are small, but deviate significantly as deformantions become finite.

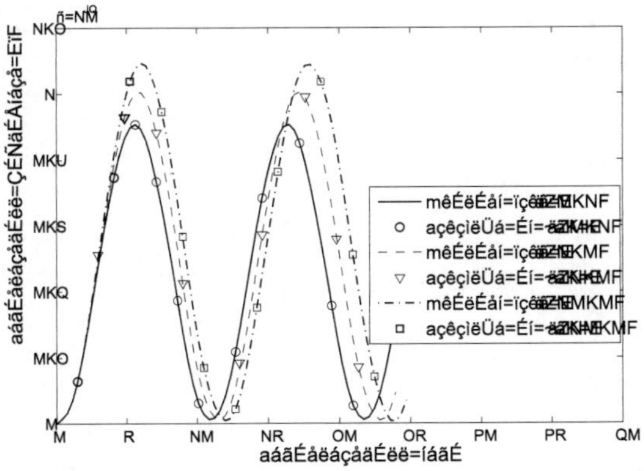

Figure 2: A comparison on transient response of a C-C FGPM beam(L/h =) with various volume law indices under a suddenly applied lateral distributed load(q= KNm^{-2}).

CONCLUSIONS

To analyze the effects of structural parameters on the induced vibration of an Functionally Graded Piezoelectric Material (FGPM) beam under thermo-electro-mechanical dynamic loads, a nonlinear displacement finite element model is developed. First-order shear deformation theory (i.e., the Timoshenko beam theory) combined with the von Kármán geometrical nonlinearity are considered to develop the model. The Newton-Raphson iterative scheme together with the Galerkin form of Newmark's approach is used to obtain the solution of the system of nonlinear equations for each time step. It is concluded that the applied external electric voltage has a great effect on the induced vibration of actuators, and hence, the vibration behavior of the structure may be noticeably controlled by applying a suitable external voltage.

Acknowledgements

Support form an NSERC Discovery Grant is gratefully acknowledged.

REFERENCES

[1] H.S. Tzou, Piezoelectric shells: distributed sensing and control of continua.*Dordrecht: Kluwer Academic Publisher*, 1993.

[2] J.S. Yang, Introduction to the theory of piezoelectricity. *New York: Springer*, 2005.

[3] N.J. Pagano, Exact solutions for rectangular bidirectional composites and sandwich plates, *Journal of Composite Materials*, 1970, 4, pp. 20-34.

[4] X.H. Zhu, Z. Meng, Operational Principle, Fabrication and Displacement Characteristics of a Functionally Gradient Piezoelectric Ceramic Actuator,*Sensors and Actuators A: Physical*, 1995, 48, pp. 169-176.

[5] C.C.M. Wu, M. Kahn, W. Moy, Piezoelectric Ceramics With Functional Gradients: A New Application in Material Design,*Journal of American Ceramic Society*, 1996, 79, 809-812.

[6] Komijani, M., Reddy, J. N., and Eslami, M. R., Nonlinear analysis of microstructure-dependent functionally graded piezoelectric material actuators,*Journal of the Mechanics and Physics of Solids,*DOI: 10.1016/j.jmps.2013.09.008, 2013.

[7] A. Doroushi, M.R. Eslami, A. Komeili, Vibration analysis and transient response of an FGPM beam under thermo-electro-mechanical loads using higher-order shear deformation theory, *Journal of Intelligent Material Systems and Structures*, 2011, 22, pp. 231-43.

Proceedings of the 25th CANCAM
London, Ontario, Canada, May 31 – June 4, 2015

THE INVERSE EIGENVALUE DETERMINANT METHOD FOR JACOBI AND TRIDIAGONAL MATRICES

Patrick Dumond
Mechanical Engineering
University of Ottawa
Ottawa, ON, Canada
pdumond@uottawa.ca

Natalie Baddour
Mechanical Engineering
University of Ottawa
Ottawa, ON, Canada
nbaddour@uottawa.ca

ABSTRACT

This paper investigates the use of the inverse eigenvalue determinant method for solving problems involving Jacobi or tridiagonal matrices. Specifically, this paper demonstrates this method's application to oscillating mass-spring systems having $2n$ degrees of freedom and fixed-free boundary conditions. The method demonstrates how to reconstruct the mass-spring system from knowledge of its eigenvalues and various physical parameters. Three algorithms are proposed and examples are given. The solvability of the method is discussed and its advantages over other numerical methods is considered.

KEYWORDS: Inverse eigenvalue determinant method, Jacobi matrices, tridiagonal matrices, mass-spring system

INTRODUCTION

When faced with the problem of specifying a dynamic system's natural frequencies (whether structural, mechanical or acoustic), engineers will typically design the system based on past experience and then verify how close they are to their specified target frequencies (eigenvalues). Inverse eigenvalue methods present an interesting alternative to the typical iterative procedure used by most engineers when faced with such a vibration problem. Inverse eigenvalue theory stipulates that a system can be reconstructed directly from knowledge of its spectral data (eigenvalues). Chu and Golub give a good summary of inverse eigenvalue theory for many types of systems [1], [2]. In most cases, complex systems must be modeled and analyzed discretely. This generally leads to a matrix representation of the system.

A $2n \times 2n$ Jacobi matrix is a specific type of structured matrix which is typically written as

$$J_{2n} = \begin{bmatrix} a_1 & b_1 & & & \\ b_1 & a_2 & \ddots & & \\ & \ddots & \ddots & \ddots & \\ & & \ddots & a_{2n-1} & b_{2n-1} \\ & & & b_{2n-1} & a_{2n} \end{bmatrix} \tag{1}$$

where a_i and b_i are all real positive values. This matrix can also be described as a tridiagonal matrix, which only has non-zero elements on the diagonal and the two first off-diagonals. A good survey of Jacobi inverse eigenvalue problems is given by Boley and Golub [3]. The solution of inverse eigenvalue problems involving Jacobi matrices is of great interest in vibration analysis, because they represent a large range of typical dynamic systems. One such system is the oscillatory mass-spring system shown in Figure 1.

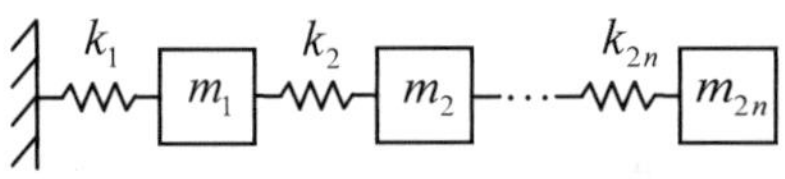

Figure 1: 2n fixed-free mass-spring system

In this particular case, the Jacobi matrix can be constructed from the mass and stiffness elements as

$$a_i = \frac{k_i + k_{i+1}}{m_i},\; b_i = \frac{k_{i+1}}{\sqrt{m_i m_{i+1}}},\; i = 1,2,...,2n-1 \tag{2}$$

and

$$a_{2n} = \frac{k_{2n}}{m_{2n}} \tag{3}$$

For the inverse problem, the Jacobi matrix is constructed from spectral data, then the mass and stiffness elements are recovered using an algorithm proposed by Gladwell [4]. With knowledge of the total mass, m_{tot}, and vector $\vec{e}_1 = \begin{bmatrix}1 & 0 & \cdots & 0\end{bmatrix}^T$, find $\vec{x}$ such that

$$\vec{x} = J_{2n}^{-1}\vec{e}_1 \tag{4}$$

Then calculate c from

$$c = \frac{\sqrt{m_{tot}}}{\|\vec{x}\|} \tag{5}$$

Vector $\vec{d}$ can then be calculated from

$$\vec{d} = c\vec{x} \tag{6}$$

Making the entries of vector $\vec{d}$ the diagonal entries of matrix D such that $D = \operatorname{diag}(d_1, d_2, \ldots, d_{2n})$, then the system mass matrix can be recovered from

$$M = D^2 \tag{7}$$

and the system stiffness matrix from

$$K = DJ_{2n}D \tag{8}$$

Two inverse eigenvalue problems are typically discussed when considering Jacobi type systems. The first was posed by Hochstadt [5] and was described as follows.

Problem 1. Given an $n \times n$ Jacobi matrix J_n, and a set of distinct real values $\{\lambda_i\}_{i=1}^{2n}$, construct a $2n \times 2n$ Jacobi matrix J_{2n} such that $\{\lambda_i\}_{i=1}^{2n}$ are its eigenvalues and J_n is its leading principal submatrix.

The second inverse eigenvalue problem can also be considered directly from the mass-spring system shown in Figure 1. In this case the problem can be described as follows.

Problem2. Given $\{m_i\}_{i=1}^{n}$, $\{k_i\}_{i=1}^{n}$ and a set of distinct real values $\{\lambda_i\}_{i=1}^{2n}$, determine $\{m_i\}_{i=n+1}^{2n}$, $\{k_i\}_{i=n+1}^{2n}$ such that $\{\lambda_i\}_{i=1}^{2n}$ are the eigenvalues of the mass-spring system of Figure 1.

In this case, Eq. (2) and (3) can be used to convert $\{m_i\}_{i=1}^{n}$ and $\{k_i\}_{i=1}^{n}$ into J_n and then problem 2 can be turned into problem 1. From there, Gladwell's algorithm can be used to recover the remaining mass and stiffness values. It is important to note however, that in this case, a_n cannot be obtained, since $a_n = (k_n + k_{n+1})/m_n$ and k_{n+1} is unknown. Thus, problem 2 contains less information than problem 1.

Finally, following a similar approach, a third problem can also be described as follows.

Problem 3. Given $\{m_i\}_{i=1}^{n}$, $\{k_i\}_{i=1}^{n}$ and a set of distinct real values $\{\lambda_i\}_{i=1}^{2n}$, construct a mass matrix M and a stiffness matrix K such that $\{\lambda_i\}_{i=1}^{2n}$ are the eigenvalues of the mass-spring system of Figure 1.

Wei and Dai give a good account of the methods typically used for solving problems 1 and 2 in [6]. Problem 3 is more complex and can be solved using methods developed by Dumond and Baddour. The first method uses a result of the Cayley-Hamilton theorem [7], while the second method solves the determinant of the generalized eigenvalue problem [8].

In this study, the determinant method is used to investigate the solvability of all three problems.

THEORY

In order to solve the determinant of the Jacobi eigenvalue problem of the first two problems, a set of equations must be written such that

$$\det(J_{2n} - \lambda_i I) = 0, \; i = 1,2,...,2n \tag{9}$$

Depending on the problem being solved, given information is included in the equation and unknown information is left as variables in the Jacobi matrix.

For the third problem, the equation of motion of an oscillating mass-spring system as shown in Figure **1** can be written as

$$M\ddot{\vec{x}} + K\vec{x} = 0 \tag{10}$$

where $\vec{x}$ corresponds to the coordinate vector, M the mass matrix defined as

$$M = \begin{bmatrix} m_1 & & & & \\ & m_2 & & & \\ & & \ddots & & \\ & & & m_{2n-1} & \\ & & & & m_{2n} \end{bmatrix} \tag{11}$$

and K the stiffness matrix defined as

$$K = \begin{bmatrix} k_1 + k_2 & -k_2 & & & \\ -k_2 & k_2 + k_3 & -k_3 & & \\ & -k_3 & \ddots & \ddots & \\ & & \ddots & k_{2n-1} + k_{2n} & -k_{2n} \\ & & & -k_{2n} & k_{2n} \end{bmatrix} \tag{12}$$

The determinant of the generalized eigenvalue problem can then be written as

$$\det(K - \lambda_i M) = 0, \; i = 1,2,...,2n \tag{13}$$

In this case, given values of m and k are included in the matrices and the rest are found numerically.

In all three problems, the determinant of a $2n$ tridiagonal matrix must be solved. The easiest numerical method for doing this involves solving the $2n^{\text{th}}$ order extended continuant such that

$$\begin{aligned} C(0) &= 1 \\ C(1) &= a_1 \\ C(2n) &= a_n C(n-1) - b_{n-1}^2 C(n-2) \end{aligned} \tag{14}$$

where $C(2n)$ represents the solution of the matrix determinant.

NUMERICAL METHODS

Algorithms for solving all three problems are presented. In all cases, it will be assumed that the eigenvalues $\{\lambda_i\}_{i=1}^{2n}$ used as input information for the examples below, are computed for the case where $m_i = 1$ and $k_i = 1$ for $i = 1,2,...,2n$.

The first problem can be solved using the following algorithm.

Algorithm 1

Given J_n and $\{\lambda_i\}_{i=1}^{2n}$

1. Populate J_{2n} of Eq. (1) from the values of J_n.

2. Compute $p_i = \det(J_{2n} - \lambda_i I)$ for $i = 1, 2, ..., 2n$ using the extended continuant.
3. Solve for the remaining a_i and b_i by solving the set of $p_i = 0$ non-linear equations.

Example 1. Let $n = 2$, $\{\lambda_i\}_{i=1}^4 = \{0.12, 1, 2.35, 3.53\}$ and

$$J_n = \begin{bmatrix} 2 & 1 \\ 1 & 2 \end{bmatrix} \quad (15)$$

Since J_n is the leading principal submatrix of J_{2n}, then

$$J_{2n} = \begin{bmatrix} 2 & 1 & 0 & 0 \\ 1 & 2 & b_2 & 0 \\ 0 & b_2 & a_3 & b_3 \\ 0 & 0 & b_3 & a_4 \end{bmatrix} \quad (16)$$

Computing the $p_i = 0$ for $i = 1, ..., 4$ requires solving

$$\text{Det}\begin{bmatrix} 2-\lambda_i & 1 & 0 & 0 \\ 1 & 2-\lambda_i & b_2 & 0 \\ 0 & b_2 & a_3-\lambda_i & b_3 \\ 0 & 0 & b_3 & a_4-\lambda_i \end{bmatrix} = 0 \quad (17)$$

from which 4 nonlinear equations are obtained. Solving these equations gives $b_2 = 1, b_3 = 1, a_3 = 2, a_4 = 1$, the exact solution to the problem.

The algorithm for the problem 2 is similar to that of problem 1.

Algorithm 2

Given $\{m_i\}_{i=1}^n$, $\{k_i\}_{i=1}^n$ and $\{\lambda_i\}_{i=1}^{2n}$

1. Calculate known $a_i = \dfrac{k_i + k_{i+1}}{m_i}$, $b_i = \dfrac{k_{i+1}}{\sqrt{m_i m_{i+1}}}$.
2. Populate J_{2n} of Eq. (1) from known a_i and b_i.
3. Compute $p_i = \det(J_{2n} - \lambda_i I)$ for $i = 1, 2, ..., 2n$ using the extended continuant.
4. Solve for the remaining a_i and b_i by solving the set of $p_i = 0$ non-linear equations.
5. Calculate $\{m_i\}_{i=n+1}^{2n}$, $\{k_i\}_{i=n+1}^{2n}$ using Gladwell's method

Example 2. Let $n = 2$ and $\{\lambda_i\}_{i=1}^4 = \{0.12, 1, 2.35, 3.53\}$. The $2n$ Jacobi matrix can be written as

$$J_{2n} = \begin{bmatrix} 2 & 1 & 0 & 0 \\ 1 & a_2 & b_2 & 0 \\ 0 & b_2 & a_3 & b_3 \\ 0 & 0 & b_3 & a_4 \end{bmatrix} \quad (18)$$

Computing the $p_i = 0$ for $i = 1, ..., 4$ requires solving

$$\text{Det}\begin{bmatrix} 2-\lambda_i & 1 & 0 & 0 \\ 1 & a_2-\lambda_i & b_2 & 0 \\ 0 & b_2 & a_3-\lambda_i & b_3 \\ 0 & 0 & b_3 & a_4-\lambda_i \end{bmatrix} = 0 \quad (19)$$

from which 4 nonlinear equations are obtained. Solving these equations would require solving 4 equations for 5 unknown values. Thus, a further method would need to be described for defining the extra parameter. Methods for doing this have been discussed in [2], [6].

Problem 3 can be solved using a modified version of the previous algorithms.

Algorithm 3

Given $\{m_i\}_{i=1}^n$, $\{k_i\}_{i=1}^n$ and $\{\lambda_i\}_{i=1}^{2n}$

1. Populate M of Eq. (11) and K of Eq. (12)from known m_i and k_i.
2. Compute $p_i = \det(K - \lambda_i M)$ for $i = 1, 2, ..., 2n$ using the extended continuant.
3. Solve for the remaining m_i and k_i by solving the set of $p_i = 0$ non-linear equations.

Example 3. Let $n = 2$ and $\{\lambda_i\}_{i=1}^4 = \{0.12, 1, 2.35, 3.53\}$. The mass and stiffness matrices can be written as

$$M = \begin{bmatrix} 1 & 0 & 0 & 0 \\ 0 & 1 & 0 & 0 \\ 0 & 0 & m_3 & 0 \\ 0 & 0 & 0 & m_4 \end{bmatrix} \quad (20)$$

and

$$K = \begin{bmatrix} 2 & -1 & 0 & 0 \\ -1 & 1+k_3 & -k_3 & 0 \\ 0 & -k_3 & k_3+k_4 & -k_4 \\ 0 & 0 & -k_4 & k_4 \end{bmatrix} \quad (21)$$

Computing the $p_i = 0$ for $i = 1, ..., 4$ requires solving the determinant of

$$\text{Det}\begin{bmatrix} 2-\lambda_i & -1 & 0 & 0 \\ -1 & 1+k_3-\lambda_i & -k_3 & 0 \\ 0 & -k_3 & k_3+k_4-m_3\lambda_i & -k_4 \\ 0 & 0 & -k_4 & k_4-m_4\lambda_i \end{bmatrix} = 0 \quad (22)$$

from which 4 nonlinear equations are obtained. Solving these equations gives $k_2 = 1, k_3 = 1, m_3 = 1, m_4 = 1$, the exact solution to the problem.

DISCUSSION

Three inverse eigenvalue problems have been presented for describing the reconstruction of the mass-spring system of Figure 1. The determinant method was applied to all three problems. Problems 1 and 3 were solved efficiently for low order systems. Since the use of the determinant method requires the solution of nonlinear equations, it depends heavily on the availability of nonlinear equation solvers. In most cases, these solvers cannot handle higher order equation sets. Problem 2 cannot be solved without an additional input source.

Although problems 1-3 specify the input information in each case based on previous studies, the determinant method is able to generate the $2n$ nonlinear equations regardless of the input information. Necessary input information for the determinant method consists of any information which leaves exactly $2n$ variables to be solved in the Jacobi matrix or mass and stiffness matrices being considered. This can be demonstrated in the following example.

Example 4. Let $n=2$, $\{\lambda_i\}_{i=1}^{4}=\{0.12, 1, 2.35, 3.53\}$ and $\{m_i\}_{i=1}^{n}=1$. The mass and stiffness matrices can be written as

$$M=\begin{bmatrix}1&0&0&0\\0&1&0&0\\0&0&1&0\\0&0&0&1\end{bmatrix} \quad (23)$$

and

$$K=\begin{bmatrix}k_1+k_2&-k_2&0&0\\-k_2&k_2+k_3&-k_3&0\\0&-k_3&k_3+k_4&-k_4\\0&0&-k_4&k_4\end{bmatrix} \quad (24)$$

Computing the $p_i=0$ for $i=1,\ldots,4$ requires solving the determinant of

$$\text{Det}\begin{bmatrix}k_1+k_2-\lambda_i&-k_2&0&0\\-k_2&k_2+k_3-\lambda_i&-k_3&0\\0&-k_3&k_3+k_4-\lambda_i&-k_4\\0&0&-k_4&k_4-\lambda_i\end{bmatrix}=0 \quad (25)$$

from which 4 nonlinear equations are obtained. Solving these equations gives $k_2=1, k_3=1, m_3=1, m_4=1$, the exact solution to the problem.

Uniqueness is not always guaranteed. Although the expected solution to example 4 was obtained, one of many alternate valid solutions includes $k_1=1.72$, $k_2=0.64$, $k_3=1.30$, $k_4=0.70$. Previous studies have developed algorithms for solving problems 1-3 requiring strict constraints on acceptable input information. Furthermore, although uniqueness is not guaranteed, solutions using other algorithms typically provide singular solutions. Thus, the determinant method represents a much more flexible alternative to solving these problems for low order systems.

CONCLUSIONS

This study considered the use of the inverse eigenvalue determinant method for solving Jacobi and tridiagonal matrices. These problems arise from the inverse vibration problem of an oscillating mass-spring system with $2n$ degrees of freedom and fixed-free boundary conditions. Three numerical algorithms were given and their solvability discussed. The inverse determinant method is found to be suitable for low order systems and its main advantage is the flexibility of suitable input information compared to the strict information required by other methods.

ACKNOWLEDGMENTS

The authors would like to acknowledge the generous support provided by the Natural Sciences and Engineering Research Council of Canada.

REFERENCES

[1] M. T. Chu, "Inverse eigenvalue problems," *SIAM Rev*, vol. 40, pp. 1–39, 1998.

[2] M. T. Chu and G. H. Golub, *Inverse Eigenvalue Problems: Theory, Algorithms, and Applications*. Oxford University Press, USA, 2005.

[3] D. Boley and G. H. Golub, "A survey of matrix inverse eigenvalue problems," *Inverse Probl.*, vol. 3, no. 4, p. 595, 1987.

[4] G. M. L. Gladwell, *Inverse problems in vibration*. Kluwer Academic Publishers, 2004.

[5] H. Hochstadt, "On some inverse problems in matrix theory," *Arch. Math.*, vol. 18, no. 2, pp. 201–207, 1967.

[6] Y. Wei and H. Dai, "An inverse eigenvalue problem for Jacobi matrix," *Appl. Math. Comput.*, vol. 251, pp. 633–642, Jan. 2015.

[7] P. Dumond and N. Baddour, "A structured approach to design-for-frequency problems using the Cayley-Hamilton theorem," *SpringerPlus*, vol. 3, no. 1, p. 272, May 2014.

[8] P. Dumond and N. Baddour, "A structured method for the design-for-frequency of a brace-soundboard system using a scalloped brace," *Appl. Acoust.*, vol. 88, pp. 96–103, Feb. 2015.

Proceedings of the 25th CANCAM
London, Ontario, Canada, May 31 – June 4, 2015

Vibration Analysis of Coupled Axial and Transverse Vibrations of Oil Well Drill Strings without Effect of Drilling Mud

Mohammad Damshenas
Department of Mechanical Engineering
Ms.c in Mechanical Engineering; Applied Mechanics & Design
Damshenas1987@yahoo.com

Mohammad Taghi Ahmadian
Department of Mechanical Engineering
Sharif University of technology
Tehran, Iran
Ahmadian@sharif.edu

ABSTRACT

Vibration is one of the most influential factors in the design, optimum performance and lifetime of all machinery. Analysis of the vibrations created in drill string like other fields such as rotary drilling is important and due to the length of the drill string whirling is probably one of the most important issues that leads to reduce Penetration rate therefore increases the time and cost of drilling operation. The purpose of this article is the analysis of the linear or nonlinear of inclined drill string and the effect of stick-slip in starting whirling. For this purpose, a series of kinetic and potential energy equations for rotary drill string, considering the longitudinal and transverse vibrations plain and Lagrange method to achieve equations of motion are used.. The modeling is done by using the finite element software ANSYS™ and drill string weight and the effect of stick-slip phenomenon is well taken into boundary conditions. Finally, the method of nonlinear analysis because of the nonlinear terms in the equations and also due to the stick-slip phenomenon and momentum changes indicates importance of sentences nonlinear natural frequency and time response shows. The results without injection of the drilling mud is considering the effect stick-slip show that this phenomenon is beginning to move in the longitudinal vibrations and then will effect in the transverse vibration state.

KEYWORDS: drill string, non-linear vibrations, weight on bit, angular velocity, finite elements, stick-slip phenomenon

INTRODUCTION

Vibration is one of the factors affecting design, performance and the life of the drill string. Investigation of the vibrations is generated in the drill string, as other rotating machines have special importance and due to the high length of drill string, the necessity of investigation of the vibration increases [1 - 3]. Drill string vibration can reduce *Rate of Penetration* (ROP) and thus increases the time and cost of drilling. It can also lead to permanent damage to the wall well and cause premature failure of drilling and other drilling equipment [4, 5]. High cost and difficulties to extract broken drill string from the well, due to drilling companies, are paid much attention to investigate the string vibration.

In drill strings, transverse, longitudinal and torsional vibrations occur simultaneously [1]. The integration between these three types of vibrations caused the complexity of drill string vibration investigation and caused it not to be completely modeled [6]. In addition to contact drill string with the walls of the well, the high length of drill string and possibility of buckling in the end, the existence of mud in and around the drill string, stick-slip of drill bit in the stone, and the torsion impact on the drill string, drill bit off released contact from the stone and imposed hitting to the drill string and etc. It caused doubling the complexity of the drill string vibration investigation [5]. Therefore, for drill string vibration modeling, simple drop assumptions are used that are discussed in the next section. Figure 1 is a representation of the types of vibrations applied to the drill string.

According to the description provided, various factors can cause modes of vibration. Table 1 classifies the factors that cause vibrations in the drill string. The various modes of vibrations are also presented [6 , 7]. As can be seen in this table, each of these factors excited various modes of vibration. The primary excitation modes are the modes that are due to that factor which is excited and it begins to vibrate. However, the secondary excitation modes are that may excite it or may not excite it.

The diameter of the drill string is less than the diameter of the well; the drill string can be made by lateral vibration. According to table 1, this vibration mode is considered as the most important failure factor of drill string (especially in bottom hole assembly) [7]. The system damping caused this vibration is not detectable on the surface and this makes the analysis more complicated. All these have led to the lateral vibration of drill string which is the focus of research in recent years. Due to the coupling transverse-axial, these effects caused lateral instability to increase [8]. Drill bit plays a significant role in coupling and the axial vibrations can become transverse vibrations.

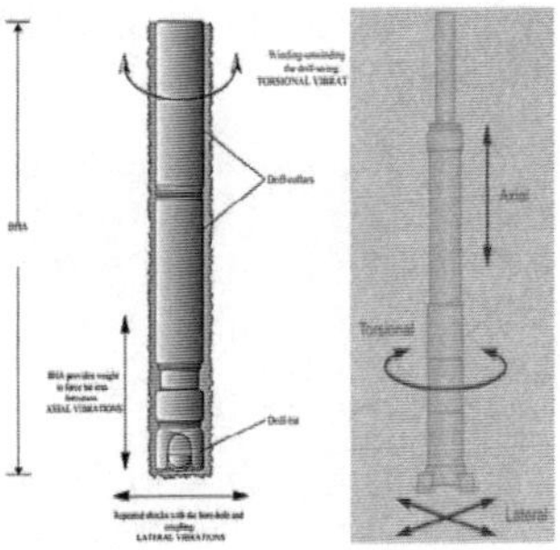

FIGURE 1: TYPES OF VIBRATIONS APPLIED ON THE DRILL STRING

Table **Error! No text of specified style in document.** Factors of vibration in drill string and excitation modes

Physical mechanism	Primary excitation	Secondary excitation
Mass imbalance	lateral	Axial-Torsional-Lateral
Misalignment	Lateral	Axial
Three cone Bit	Axial	Torsional-Lateral
Loose drill string	Axial-Torsional-Lateral	----------------------
Rotational walk	Lateral	Axial-Torsional
Asynchronous walk or whirl	Lateral	Axial-Torsional
Drill string whip	Lateral	Axial-Torsional

The Equations of Vibrations and Their Simplification

In this section the governing assumptions of problem are examined and achieved vibrational governing equations of drill string will be brought. To calculate the equations energy method is used and finally it has come to the conclusion that the analysis of drilling complex in a finite element program analysis such as ABAQUS ™ or ANSYS ™ linear assumption is better or nonlinear analysis is more close to the fact of the problem. To simplify the modeling of drill string vibrations and obtain the equations governing on it, reasonable assumptions have been considered as discussed below.

- Drill string vibration is studied in all three spatial directions (a longitudinal direction (u) and two transverse directions (v, w)). Therefore, evaluation of vibrations in this direction would be continuous and nonlinear.
- The governing equations of transverse vibrations of the drill string, and using of the Euler Bernoulli beam equations are obtained show that, according to large length of the drill string are appropriate assumptions.
- Review of torsional vibrations of drill string has been refused and drill has been considered as a beam.
- Drill rotation is modeled with constant angular velocity.
- The distance (Clearance) between the stabilizer and the well's wall is ignored.
- For the initial conditions of inclined drilling, transverse effect of weight will be neglected and drill string without whirling is assumed. However, the increasing strain in the axial direction (due to the axial component of the weight force) is considered.

Derive Equations

Equation for kinetic energy and potential energy of the drill string in rotating frame (with constant angular velocity Ω around x axis) for an element from drill string are given to its free diagram that obtains [8-12]:

$$U_e = \frac{1}{2}\int_{x_1}^{x_1+L_e}\Big(EI_{ez}(v'')^2 + EI_{ey}(w'')^2 + EA_e(u' + \frac{1}{2}(v')^2 + \frac{1}{2}(w')^2)^2\Big)dx \qquad (1)$$

$$T_e = \frac{1}{2}\rho_s A_e \int_{x_1}^{x_1+L_e} (\dot{u}^2 + \dot{v}^2 + \dot{w}^2)dx + \frac{1}{2}\int_{x_1}^{x_1+L_e} (J_{ey}(\dot{w}' + \Omega v')^2 + J_{ez}(\dot{v}' - \Omega w')^2)dx \qquad (2)$$

In Equation 1, v and w are the displacement of the drill string to the axes y and z, u longitudinal is displacement along the x-axis, E is Young's modulus of elasticity, I_e is moment of inertia of the cross section of drill string, A_e is Cross-sectional area, x_e is global coordinate of the first node, L_e is length of element and $()'$ represents the partial derivative with respect to x. Also, index of e represents corresponding element.

The second part of Equation 1, according to Figure 2, is calculated as follows [8-12]:

$$dL = ds - dx = \sqrt{(dx + du)^2 + (dv)^2 + (dw)^2} - dx = \left(\sqrt{\left(1+\frac{du}{dx}\right)^2 + \left(\frac{dv}{dx}\right)^2 + \left(\frac{dw}{dx}\right)^2} - 1\right)dx \cong \left(1 + \frac{1}{2}\left[2\frac{du}{dx} + \left(\frac{du}{dx}\right)^2 + \left(\frac{dv}{dx}\right)^2 + \left(\frac{dw}{dx}\right)^2\right] - 1\right)dx = \frac{du}{dx} + \frac{1}{2}\left[\left(\frac{du}{dx}\right)^2 + \left(\frac{dv}{dx}\right)^2 + \left(\frac{dw}{dx}\right)^2\right] \cong \frac{du}{dx} + \frac{1}{2}\left(\left(\frac{dv}{dx}\right)^2 + \left(\frac{dw}{dx}\right)^2\right) \qquad (3)$$

According to the nonlinear potential energy presented in Equation 1, the dependence between longitudinal and transverse vibrations can be realized. It should be noted in this Equation, the potential energy of the drill string resulting from weight force and force from bottom of well has not been considered and how to apply the energy from these forces in the equations will be explained in Section 3.2. The kinetic energy of an element of the drill string by considering longitudinal and transverse vibrations is expressed by Equation 2. In this Equation, the density of the drilling string isρ_s, drill string rotational speed is Ω, Mass moment of inertia is J_e and (^{o}) means the partial derivative with respect to time t.

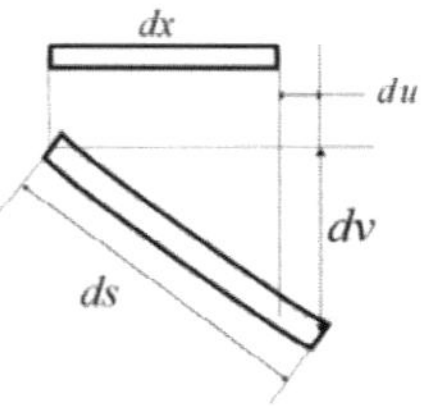

Figure 2 Diagram of a beam element

Simplification of Equations

Due to the non-linearity of the equations for potential and kinetic energy, it cannot be solved analytically, and numerical methods must be used. Therefore the finite element method to convert a non-linear integral energy to nonlinear scalar is used

Weight Force

The coordinate system attached to the drill string and the commitment that the x-axis along the axis of the drill string is placed, the weight force is decomposed into the two components in the longitudinal and transverse direction.

$$(W_x)_e = (\rho_s - \rho_f)A_e g \cos(\alpha) \tag{4}$$

$$(W_y)_e = (\rho_s - \rho_f)A_e g \sin(\alpha)\sin(\Omega t) \tag{5}$$

$$(W_z)_e = (\rho_s - \rho_f)A_e g \sin(\alpha)\cos(\Omega t) \tag{6}$$

The above equation represents the weight per unit length along the longitudinal and transverse components of the forces there.

α is an angle of inclined drill string, Ωt is the phase difference between the drill strings with its initial state and ρ_f is mud density. Axial component of weight force will be considered as a static deflection and transverse components of weight force which are calculated from the external forces and added to the equation. Given to static force in the longitudinal direction is the force obtained from drill string weight and the force on drill bit (*P*), we can calculate function $u_s(x,t)$ in equation 3.8 [13]:

(7)

That the first term represents the deflection of the drill string in response to $P(t)$ force and the second term represents displacement obtained from drill string weight force. Finally, axial and static displacement of drill string for both drilling pipe and drill collar is explained.

$$u_s(x,t) = \begin{cases} -\dfrac{P(t)x}{EA} + \dfrac{\rho g}{E}x\left(L - \dfrac{x}{2}\right) & 0 \le x \le L_1 \\ -\dfrac{P(t)}{E}\left(\dfrac{L_1}{A_1} + \dfrac{x - L_1}{A_2}\right) + \dfrac{\rho g}{E}x(L - \dfrac{x}{2}) & L_1 < x \le L \end{cases} \tag{8}$$

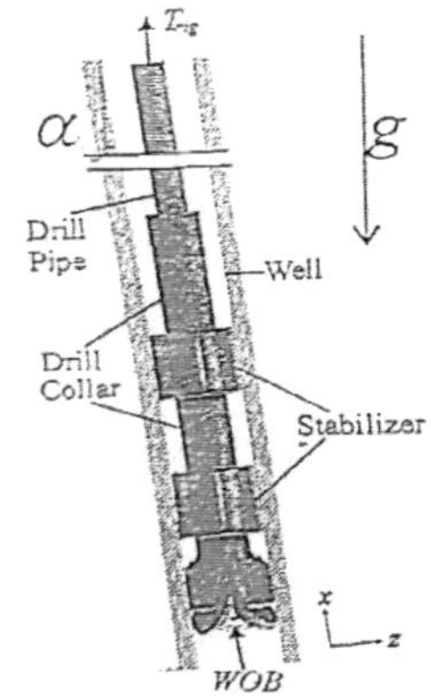

Figure **Error! No text of specified style in document.** a shematic of the drill string in inclined well

In this equation, $P(t)$ is force on the drill bit, g is acceleration of gravity and $\rho = \rho_s - \rho_f$ where ρ_s and ρ_f are drill string density and drill mud density, respectively. According to the Figure 3.3 force on bit drill is explained as follows where $H(t)$ is the height of the neutral point.

Simplification of the Kinetic Energy

For simplification by placing equations in kinetic energy equation (equation 2) and integration, we can write:

$$\begin{aligned} T_e = {} & \frac{1}{2}M_e^{PP}\dot{P}^2 + \dot{P}\{M^{Pu}\}_e^T\{\dot{u}\}_e + \frac{1}{2}\{\dot{u}\}_e^T[M^{uu}]_e\{\dot{u}\}_e \\ & + \frac{1}{2}\Omega^2\{v\}_e^T[M^{vvJ1\Omega}]_e\{v\}_e + \frac{1}{2}\Omega^2\{w\}_e^T[M^{wwJ2\Omega}]_e\{w\}_e \\ & + \Omega\{\dot{w}\}_e^T[M^{wvJ1\Omega}]_e\{v\}_e - \Omega\{\dot{v}\}_e^T[M^{vwJ2\Omega}]_e\{w\}_e \\ & + \frac{1}{2}\{\dot{w}\}_e^T([M^{ww}]_e + [M^{wwJ1}]_e)\{\dot{w}\}_e \\ & + \frac{1}{2}\{\dot{v}\}_e^T([M^{vv}]_e + [M^{vvJ2}]_e)\{\dot{v}\}_e \end{aligned} \quad)9)$$

As equation associated with the kinetic energy, replacement relations in the potential energy equation can be written:

$$\begin{aligned} U_e = {} & \frac{1}{2}K_e^{PP}P^2 + K_e^{Pg}Pg\cos(\alpha) + \frac{1}{2}K_e^{gg}g^2\cos^2(\alpha) \\ & + P\{u\}_e^T\{K^{Pu}\}_e + g\{u\}_e^T\{K^{gu}\}_e\cos(\alpha) + \frac{1}{2}\{u\}_e^T\{K^{uu}\}_e\{u\}_e \\ & + \frac{1}{2}\{v\}_e^T\{K^{uu}\}_e\{v\}_e + \frac{1}{2}P\{v\}_e^T\{K^{Pvv}\}_e\{v\}_e \\ & + \frac{1}{2}g\{v\}_e^T\{K^{gvv}\}_e\{v\}_e\cos(\alpha) + \frac{1}{2}\{v\}_e^T\{u\}_e^T\{K^{uvv}\}_e\{v\}_e \\ & + \frac{1}{8}\{v\}_e^T\{v\}_e^T\left[[[K^{vvvv}]]\right]_e\{v\}_e\{v\}_e + \frac{1}{2}\{w\}_e^T\{K^{ww}\}_e\{w\}_e \end{aligned} \tag{11}$$

$$+\frac{1}{2}g\{w\}_e^T\{K^{gww}\}_e\{w\}_e\cos(\alpha)+\frac{1}{2}P\{w\}_e^T\{K^{Pww}\}_e\{w\}_e$$
$$+\frac{1}{2}\{w\}_e^T\{u\}_e^T[[K^{uww}]]_e\{w\}_e$$
$$+\frac{1}{8}\{w\}_e^T\{w\}_e^T\left[[[K^{wwww}]]\right]_e\{w\}_e\{w\}_e$$
$$+\frac{1}{4}\{w\}_e^T\{v\}_e^T\left[[[K^{vvww}]]\right]_e\{v\}_e\{w\}_e$$

The General Equation of Energy

By considering the boundary conditions with respect to equations 4 and 10 total kinetic energy and potential for drill string are derived as follows:

$$T=\frac{1}{2}M^{PP}\dot{P}^2+\dot{P}\{M^{Pu}\}^T\{\dot{u}\}+\frac{1}{2}\{\dot{u}\}^T[M^{uu}]\{\dot{u}\} \tag{11}$$
$$+\frac{1}{2}\Omega^2\{v\}^T[M^{vvJ1\Omega}]\{v\}+\frac{1}{2}\Omega^2\{w\}^T[M^{wwJ2\Omega}]\{w\}$$
$$+\Omega\{\dot{w}\}^T[M^{wvJ1\Omega}]\{v\}-\Omega\{\dot{v}\}^T[M^{vwJ2\Omega}]\{w\}$$
$$+\frac{1}{2}\{\dot{w}\}^T([M^{ww}]+[M^{wwJ1}])\{\dot{w}\}$$
$$+\frac{1}{2}\{\dot{v}\}^T([M^{vv}]+[M^{vvJ2}])\{\dot{v}\}$$

$$U=\frac{1}{2}K^{PP}P^2+K^{Pg}Pg\cos(\alpha)+\frac{1}{2}K^{gg}g^2cos^2(\alpha) \tag{12}$$
$$+P\{u\}^T\{K^{Pu}\}_e+g\{u\}^T\{K^{gu}\}\cos(\alpha)+\frac{1}{2}\{u\}^T\{K^{uu}\}\{u\}$$
$$+\frac{1}{2}\{v\}^T\{K^{vv}\}\{v\}+\frac{1}{2}P\{v\}^T\{K^{Pvv}\}\{v\}$$
$$+\frac{1}{2}g\{v\}^T\{K^{gvv}\}\{v\}\cos(\alpha)+\frac{1}{2}\{v\}^T\{u\}^T\{K^{uvv}\}\{v\}$$
$$+\frac{1}{8}\{v\}^T\{v\}^T\left[[[K^{vvvv}]]\right]\{v\}\{v\}+\frac{1}{2}\{w\}^T\{K^{ww}\}\{w\}$$
$$+\frac{1}{2}g\{w\}^T\{K^{gww}\}\{w\}\cos(\alpha)$$
$$+\frac{1}{2}P\{w\}^T\{K^{Pww}\}\{w\}+\frac{1}{2}\{w\}^T\{u\}^T[[K^{uww}]]\{w\}$$
$$+\frac{1}{8}\{w\}^T\{w\}^T\left[[[K^{wwww}]]\right]\{w\}\{w\}$$
$$+\frac{1}{4}\{w\}^T\{v\}^T\left[[[K^{vvww}]]\right]\{v\}\{w\}$$

According to the Lagrange principle equations of motion will have to:

$$\frac{d}{dt}\left(\frac{\partial T}{\partial\dot{\bar{u}}_j}\right)-\frac{\partial T}{\partial\bar{u}_j}+\frac{\partial U}{\partial\bar{u}_j}-Q_{\bar{u}_j}[M^{uu}]\{\ddot{u}\}+[K^{uu}]\{v\} \tag{11}$$
$$+\ddot{P}\{M^{Pu}\}+P\{K^{Pu}\}+g\cos(\alpha)\{K^{gu}\}$$
$$-Q^2(\{K^{uFL}\}+\{K^{uFP}\})$$
$$+\frac{1}{2}\left[[[K^{uvv}]]\{v\}\right]\{v\}+\frac{1}{2}\left[[[K^{uww}]]\{w\}\right]\{w\}=\{0\}$$

$$\frac{d}{dt}\left(\frac{\partial T}{\partial\dot{\bar{v}}_j}\right)-\frac{\partial T}{\partial\bar{v}_j}+\frac{\partial U}{\partial\bar{v}_j} \tag{14}$$
$$-Q_{\bar{v}_j}([M^{vv}]+[M^{vvJ2}]-[M^{vvin}]-[M^{vvout}])\{\ddot{v}\}$$
$$-Q\begin{pmatrix}[C^{vvin}]+[C^{vvout}]+[C^{vvFN}]+\frac{1}{Q}K[C^{vvKFN}])\{\dot{v}\}\\-\Omega^2[M^{vvJ1\Omega}]\{v\}-\Omega\{\dot{w}\}^T[M^{wvJ1\Omega}]+\Omega[M^{vwJ2\Omega}]\{\dot{w}\}\end{pmatrix}$$
$$+\begin{pmatrix}[K^{vv}]-Q^2[K^{vvin}]\\+[K^{vvout}]+[K^{vvFN}])\end{pmatrix}\{v\}$$
$$+(P\{K^{Pvv}\}+gcos(\alpha)\{K^{gvv}\})\{v\}$$
$$+\{u\}^T[[K^{uvv}]]\{v\}+\frac{1}{2}\{v\}^T\left[[[K^{vvvv}]]\right]\{v\}\{v\}$$
$$+\frac{1}{2}\{w\}^T\left[[[K^{vvww}]]\right]\{v\}\{w\}-g\sin(\alpha)\sin(\Omega t)\{G^{gv}\}$$
$$=\{0\}\begin{pmatrix}[C^{vvin}]+[C^{vvout}]+[C^{vvFN}]\\+\frac{1}{Q}K[C^{vvKFN}]\{\dot{v}\}+Q^2[K^{vvin}]\\+[K^{vvout}]+[K^{vvFN}]\end{pmatrix}\{v\}$$

$$\frac{d}{dt}\left(\frac{\partial T}{\partial\dot{\bar{w}}_j}\right)-\frac{\partial T}{\partial\bar{w}_j}+\frac{\partial U}{\partial\bar{w}_j}-Q_{\bar{w}_j}[M^{ww}] \tag{15}$$
$$+[M^{wwJ1}]-[M^{wwin}]-[M^{wwout}]\{\ddot{w}\}$$
$$-Q\left([C^{wwin}]+[C^{wwout}]+[C^{wwFN}]+\frac{1}{Q}K[C^{wwKFN}]\right)\{\dot{w}\}$$
$$-\Omega^2[M^{wwJ2\Omega}]\{w\}+\Omega[M^{wwJ2\Omega}]\{\dot{v}\}-\Omega\{\dot{v}\}^T[M^{vwJ2\Omega}]$$
$$+([K^{ww}]-Q^2[K^{wwin}]-[K^{wwout}]-[K^{wwFN}])\{w\}$$
$$+(P\{K^{Pww}\}+gcos(\alpha)\{K^{gww}\})\{w\}$$
$$+\{u\}^T[[K^{uww}]]\{w\}+\frac{1}{2}\{w\}^T\left[[[K^{wwww}]]\right]\{w\}\{w\}$$
$$+\frac{1}{2}\{v\}^T\left[[[K^{vvww}]]\right]\{v\}\{w\}-g\sin(\alpha)cos(\Omega t)\{G^{gw}\}$$
$$=\{0\}$$

Linear and non-linear equations are from omitting and not omitting of K^{uvv} and K^{vvvv} and also from K^{uww} and K^{wwww} of governing equations on drill string vibration. So, because of magnitude of these matrixes nonlinear, analysis is nearer to fact of the problem.

Results

Analyzing the results of the vibration governing equations of the drilling string has occurred in several parts. In the first section, we have talked about the boundary conditions that include how to obtain axial forces from the up, applied harmonics rotational forces and how to consider stick-slip phenomena.

In the second section we discussed the the stability criteria, factors affecting the instability and the neutral point.

In the third section, the longitudinal and transverse frequencies were obtained and finally in the fourth section, the time response curves were obtained and a general conclusion can be deduced. In table 2, Values used for obtaining the results are presented. These data are related to the model presented by Khulief (except length of drill collar) [2].

Drilling pipe Specifications		Material specifications	
length	*1000 m*	**drilling string density**	*7850 kg/m³*
Internal diameter	*0.095 m*	**Drilling mud density**	*1500 kg/m³*
External diameter	*0.127 m*	**Modulus of elasticity**	*210 GPa*
Collar drilling specifications		**General Specifications**	
length	*50 m*	Wall wells diameter	*0.3 m*
Internal diameter	*0.0762 m*	Acceleration of gravity	*9.81 m/s²*
External diameter	*0.2286 m*	Slope angle of wells	*0.0524 rad*

Part one: The amount of rotational force applied to the developed model is a harmonic force equal to *H=10+10sin*

(10πt/3) that was used in the previous work. Angular velocity is equal to *120 rpm*.

A tensile axial force from top to the drill string is equal to the drill string that its total weight is obtained according to table 2.

Finally, for the boundary conditions of the bottom of the well, the stick-slip equation is used and programming by micro writing in ANSYS™ is applied. The accuracy of model is confirmed based on mesh analysis and the elements are some types of pipe.

Usually for diagonal drilling, deviation from 3° to 5° is considered. This work only has 3° deviation from the vertical line and the rate of penetration of drill bit is 20 cm per minute. In the next section, stick-slip modeling is talked completely.

1	2	3	4	5	Flow rate
1.197727	1.17127	0.117155	1.411661	1.498512	0

Part two: The main reasons for the instability to vertical direct drilling are three principles:

- High velocity (or in other words, flow rate) of drilling mud
- Drill string buckling under the weight pressure force of **P**
- Drill string angular velocity

For the first case in this study that are not injected drilling mud and the total weight of the drill string has been inhibited by tensile force from top of complex. According to previous research, the drill string is unsustainable because of high angular velocity more than 500rpm and while working angular velocity is between 100 to 300 rpm and in this work the angular velocity is constant and equal to the 120 rpm [14]. As with the definition of neutral point stated weight is collar part that is not drawn from top and this point is not in tension and pressure. Here the total weight censored from up because of two reasons: First, the research on the field before drilling locations on the neutral point 13 and 26 meters is done and second, in fact as a drilling mud weight is always on the drill bit.

Figure **Error! No text of specified style in document.**

Meshed part model from Ansys™

Part three: It is clear there is no association among frequencies of longitudinal vibrations of the drill string, the weight force, and drilling mud. This frequency depends on the physical properties of the drill string. According to this explanation, the natural frequency of longitudinal vibrations of drill string for the data presented in Table 2 is obtained from the following table 3. Also for the first 5 transverse vibrations mode dependent on loading and boundary values is given in table (4).

Table 3 Natural frequency of longitudinal vibrations of drill string with the specifications provided in Table 2.

	1	2	3	4	5
longitudinal natural frequency*(rad/s)*	15.98	31.08	40.63	51.33	66.81

Table **Error! No text of specified style in document.** natural frequency of drill string by assumption of H=0

Part four: The reason was explained in the previous section using nonlinear approach and all results presented in this section are related to the displacement of a point of drilling pipe (*x=1000m*). Physical properties provided in Table 2 are based on khulief.

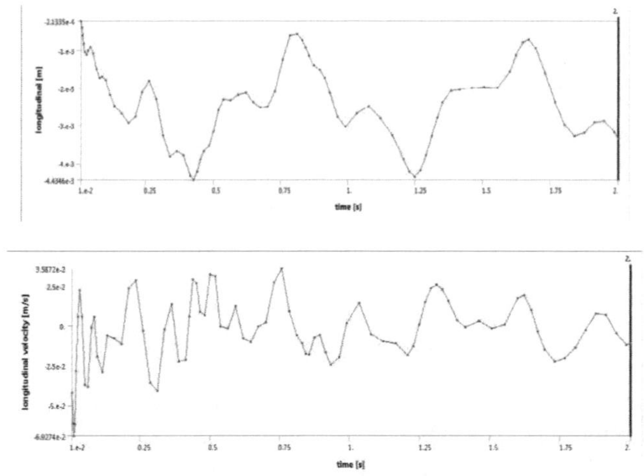

Figure 3 longitudinal vibration for X=1000m, Ω=120rpm

As seen in longitudinal vibration effect on the movement pipe and collar point is softer as well as speed reduction, but not zero, while achieve to the amount will be determined that penetration rate is equal to 20 centimeter per minute.

$$20\frac{Cm}{min}\times\frac{1min}{60sec}\times\frac{1m}{100cm}=0.33\times10^{-2}$$

This seems quite reasonable.

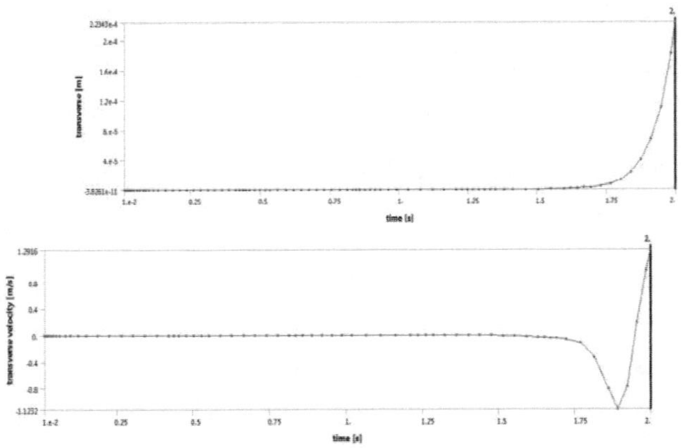

Figure 4 Transvers vibration for X=1000m, Ω=120rpm

As can be seen, the vibration damping system is low and these lead to increase the amplitude because weight divided into two components with respect to the vertical drilling is causing decrease in frequency and increase inbthe amplitude.

CONCLUSIONS

The damping effect of drilling mud is extremely important to the system over time to converge on the equilibrium point that here in diagrams of the point X = 1000m there is no convergence. However, such analysis is transient and only 2 seconds to cover this is clearly visible. Also drilling mud has so important tasks that it is impossible to remove them from the drilling cycle some important points are listed below:

- Cooling and lubrication the drill bit
- Lubrication and cooling the drill string
- Control of ground formation pressure
- Transport finely ground particles out of the well
- Well stability and prevent its falling
- Helping to well logging

REFERENCES

1. Dykstra M. *Nonlinear drill string dynamics*, 1996, University of Tulsa, The Graduate School.
2. Khulief, Y.A. and H. Al-Naser, *Finite element dynamic analysis of drill string*. Finite Elements in Analysis and Design, 2005. **Volume 41**(13): p. 1270-1288.
3. Jansen, J.D. and L. van den Steen, *Active damping of self-excited torsional vibrations in oil well drill strings*. Journal of Sound and Vibration, 1995. **Volume 179**(4): p. 647-668.
4. Mason, J.S. and M. Sprawls, *Addressing BHA Whirl – The Culprit in Mobile Bay*, in *SPE/IADC Drilling Conference* 1996: New Orleans, Louisiana.
5. Jansen, J.D. *Non-linear rotor dynamics as applied to oilwell drill string vibrations*. Journal of Sound and Vibration, 1991. **Volume 147**(1): p. 115-135.
6. Tucker, W.R. and C. Wang, *An Integrated Model for Drill-string Dynamics*. Journal of Sound and Vibration, 1999. **224**(1): p.123-165.
7. Mitchell, R.F. and M.B. Allen, *Lateral vibration: the key to BHA failure analysis*. Journal Name: World Oil; (United States); Journal Volume: 200:4, 1985: p.Medium: X; Size: Pages: 101-104, 106.
8. Yigit, A.S. and A.P. Christoforou, *Coupled Axial and Transverse Vibrations of Oil Well Drill Strings*. Journal of Sound and Vibration, 1996. **195**(4): p. 617-627.
9. Christoforou, A.P. and A.S. Yigit, *Dynamic Modeling of Rotating Drillstring with borehole interactions*. Journal of Sound and Vibration, 1997, **Volume 206**(2): p.243-260.
10. Fung, R.F. And J.P. Lee, *Parametric Variable Structure Control of a Spinning Beam system via Axial Force*. Journal of sound and vibration, 1999. **F. B. 227**(3): p. 545-554.
11. Gmur, T.C. and J.D. Rodrigues, *Shaft Finite Elements for Rotor Dynamics Analysis*. Journal of Vibration and Acoustics, 1991. **F. B. 113**(4): p. 482-493.
12. S. H. Choi, C. Pierre, and A. G. Ulsoy, *Consistent modeling of rotating Timoshenko shafts subject to axial loads*. Journal of Vibration and Acoustics, 1992. **114**: p. 249-259.
13. R. P. beer and E.R. Johnston, *Mechanics of Material* 1992, McGraw Hill: United States.
14. Taheran, F. *Coupled vibration analysis of directional drilling string considering effect of drill mud,* 2012, *M.* Sc. Thesis; Department of Mechanical Engineering, Sharif University of Technology.
15. Faraji Mahyari, M. *lateral vibration analysis of drill string, considering drill mud around it,* 2007, M. Sc. Thesis; Department of Mechanical Engineering, Sharif University of Technology.
16. Damshenas, M. *Vibration Analysis of Drill String under Inclined Drilling Including Stick-Slip Phenomenon,* 2014, M. Sc. Thesis; Department of Mechanical Engineering, Sharif University of Technology, International campus.

SEISMIC ANALYSIS FOR COMPOSITE CONICAL TANKS

Ahmed Elansary
Department of Civil and Environmental Engineering
Western University
London, Ontario, Canada
aelansa2@uwo.ca

Ashraf El Damatty
Department of Civil and Environmental Engineering
Western University
London, Ontario, Canada
damatty@uwo.ca

ABSTRACT

Composite steel/concrete conical vessels can be used for liquid storage in elevated tanks. These vessels consist of a steel wall and a reinforced concrete wall connected through studs. Analysis of composite tanks under seismic loading is not well studied in the literature. A coupled finite/boundary element model is developed in this study to perform time history analysis for composite tanks. The study focuses on a composite conical elevated tank recently constructed at a high seismic area in Mexico. The analysis considers the effect of both the horizontal and vertical components earthquakes. Time histories for the displacements along the vessel's height and the forces at the vessel's base for the studied tank are presented. The resulted straining actions at the base are compared with their counterparts from the response spectrum analysis. The comparison revealed that the response spectrum analysis provides a conservative solution with safety factors of 1.1, 1.4 and 1.2 for the base shear force, moment and normal force respectively.

KEYWORDS: Composite, Conical tank, Time history, Finite element, Reinforced concrete, Steel, Studs.

INTRODUCTION

Strong earthquakes are often followed by fire events that can cause catastrophic destruction. As such, water storage structures must be well designed to withstand severe earthquake events in order to remain functional post-earthquakes to supply water needed to extinguish fire. Elevated conical tanks are subjected to both lateral and vertical forces during earthquake events. These tanks typically consist of a steel, concrete or composite vessel supported on a tower. Composite conical elevated tanks are considered an efficient system to store large amount of water. There are no guidelines in the current codes of practice for the analysis or design of composite tanks.

Haroun and Housner (1981) studied the behaviour of cylindrical steel tanks under seismic excitation. They showed that the flexibility of the tank's wall has a significant effect on the magnitude of forces resulting from earthquakes. Haroun and Ellaithy (1985) studied seismically induced fluid forces on the supporting towers of elevated tanks. They studied the seismic forces on two different supporting systems: cross braced frames and pedestal tower. They developed a simplified design procedure to calculate the forces on the supporting system. They concluded that wall's flexibility has a significant effect on the dynamic response of large capacity tanks more than its effect on small capacity tanks. El Damatty et al. (1997) studied the stability of conical tanks under seismic loading. They considered the fluid – structure interaction to estimate the hydrodynamic pressure forces on the walls of the tanks. El Damatty and Sweedan (2006) developed an equivalent mechanical model for the dynamic analysis of pure conical tanks under horizontal ground excitations. Their model accounts for both the impulsive and sloshing earthquake components.

Regarding the effect of vertical acceleration, Haroun and Tayel (1985) studied the response of cylindrical steel tanks under vertical earthquake excitations. They applied two types of ground earthquake excitations on cylindrical ground-supported tanks: step forcing loads and different earthquake histories. In their study, the tanks' walls were modelled using the finite element, while the liquid inside the tank was modelled analytically. Sweedan and El Damatty (2005) proposed an equivalent model for pure conical tanks under vertical ground excitation. In their study, they developed a simple procedure for estimating the fundamental natural frequency and the seismic forces on steel conical tanks subjected to vertical seismic excitations.

The objective of this paper is to evaluate the straining actions at the vessel's base of composite conical tanks subjected to an earthquake and compare them with their counterparts from the response spectrum analysis.

FINITE ELEMENT MODEL

The concrete and steel walls are modelled using a 13-node triangular element, while the studs between the two walls are modelled using a 26-node contact element. The 13-node element was first introduced by Koziey and Mirza (1997). In the formulation of this element, cubic and quadratic polynomials are assumed for the approximation for the displacements and rotations respectively. This formulation ensures avoiding the spurious transverse shear strain modes that were found in the isoparametric shell elements. The 26-node contact element was first introduced by Siddique and El Damatty (2012) to model the adhesive between steel and GFRP plates. This element consists of two triangular 13-node elements. Each node on each triangular element is connected to a node on the adjacent element using three-dimensional springs. In the current study, the 26-node element is extended to model the studs between the steel and concrete walls for composite conical tanks.

The reinforced concrete and steel walls are modeled using 32 shell elements per wall where eight elements are used at each row and eight elements at each column as shown in Fig. 1. Only one half of the vessel is modelled due to the symmetry about the axis in the excitation direction.

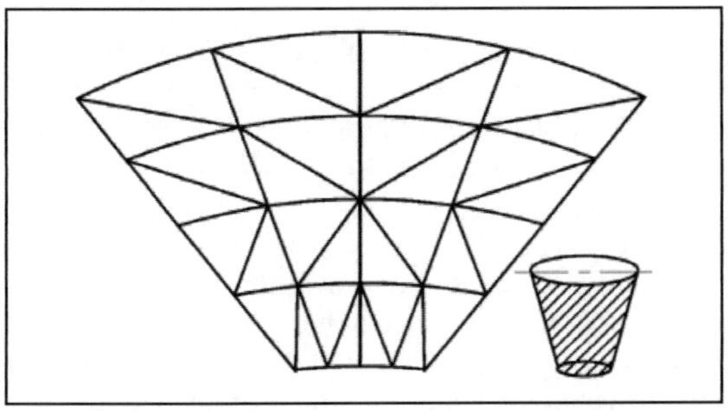

Figure 1: Finite element mesh for each wall

One mesh of 32 contact elements is used to connect the reinforced concrete and steel walls. This mesh simulates the studs that are connecting the two walls. The reinforced concrete shaft is modelled using two springs representing its vertical and horizontal stiffnesses with values of 11×10^9 N/m and 1.4×10^9 N/m respectively. The vessel is assumed to be free at the top edge as there is an aluminum roof which has a relatively smaller stiffness than the vessel's wall. At the lines of symmetry, the horizontal and vertical displacements are allowed while the displacements perpendicular to the line of symmetry are prevented.

GEOMETRY AND MATERIAL PROPERTIES

The vessel's height, radius at bottom and inclination angle are 9 m, 4 m and 51.6° respectively. This vessel is composed of a 113 mm thick reinforced concrete wall and an 8 mm steel wall. The two walls are connected together with studs with a diameter and spacing of 13 mm and 400 mm respectively. The vessel is supported on a 15 m high reinforced concrete shaft.

The concrete properties are as follows: characteristic strength $f_c^{'} = 24.5$ MPa ; Poisson's ratio $\upsilon_c = 0.2$; and modulus of elasticity c , 5 P a. Material properties of steel are as follows: Yield stress $\sigma_y = 248\ MPa$; ultimate stress u 4 Pa ; Poisson's ratio s . ; and modulus of elasticity s , Pa .

TIME HISTORY ANALYSIS

The water-filled composite tank in Mexico is analyzed under both free horizontal and vertical vibrations using a coupled finite/boundary element program. The frequencies from the horizontal free vibration analysis for the first four modes are found as 1.94, 2.23, 3.11 and 6.70 Hz, while the frequencies from the vertical free vibration analysis for the first four modes are found as 6.74, 12.58, 16.67 and 18.38 Hz. Time history analysis is performed for this tank under the horizontal and vertical components of the Victoria earthquake. This earthquake occurred in Mexico in 1980 with a magnitude of 6.1.

One record for Victoria earthquake is used in the time history analysis for the studied composite tank. This record is chosen because the frequency for the first four modes for the studied tank are within the frequency content of this record. The history for the strongest ten seconds in this record is chosen with a time step of 0.02 second. This time step is much smaller than that of the fundamental horizontal and vertical periods for the studied tank. The time histories for the vertical and horizontal components for the chosen earthquake are shown in Figures 2 and 3. The response spectrum for the chosen earthquake at a damping ratio of 5% is plotted in Fig. 4.

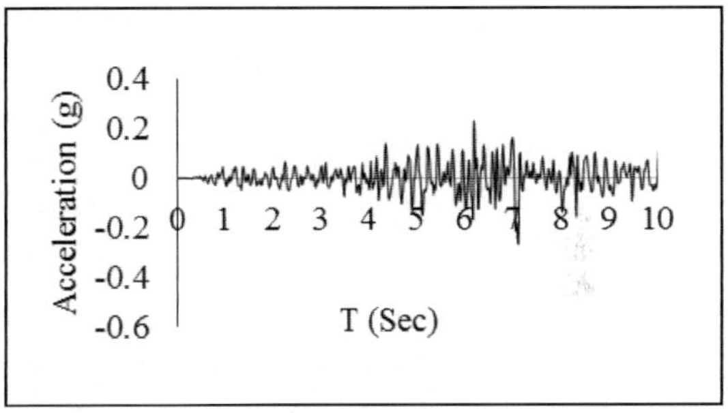

Figure 2: Time history for vertical component of Victoria earthquake

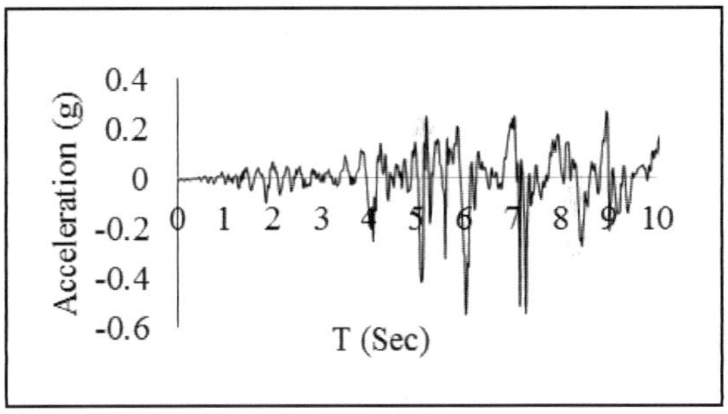

Figure 3: Time history for horizontal component of Victoria earthquake

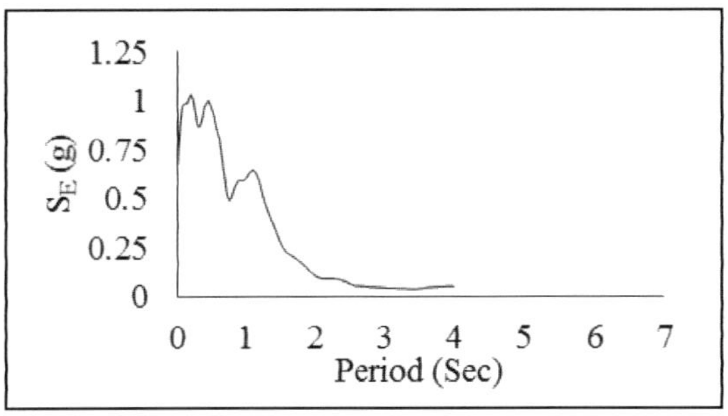

Figure 4: Response spectrum for horizontal component of Victoria earthquake

The elastic response spectrum for the horizontal excitation is plotted from the Mexican Manual of Civil

Structures, MOC (2008) at a damping ratio of 5% as shown in Fig. 5. For each vibration mode, a scale factor is calculated by dividing the acceleration from the Mexican manual spectrum by the acceleration for Victoria earthquake spectrum. The scale factors for the first four vibration modes, spectral accelerations for Victoria earthquake, S_E and spectral accelerations from the Mexican manual, S_M are listed in Table 1. An overall scale factor of 0.24 is chosen such that the horizontal acceleration of the earthquake record is equal to or larger than the zonal horizontal acceleration. Both the horizontal and vertical components for the chosen earthquake record are scaled using the calculated overall scale factor.

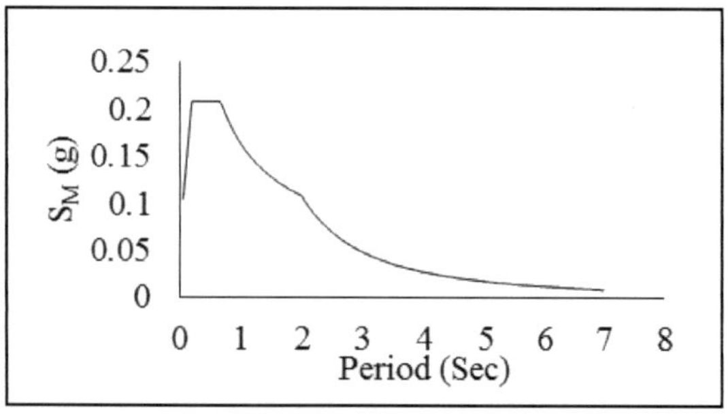

Figure 5: Elastic response spectrum for the horizontal component from MOC (2008)

Table 1: Scale factors for earthquake history scaling

Mode	1st	2nd	3rd	4th
T_F (Sec)	0.51	0.45	0.32	0.15
$S_M(T)$ (g)	0.21	0.21	0.21	0.14
$S_E(T)$ (g)	0.98	0.99	0.87	0.99
Scale factor	0.21	0.21	0.24	0.14

RESULTS

Time histories for the displacements at the middle height of the vessel and straining actions at the base are presented. The values at T = 0 correspond to the response under hydrostatic water pressure that is applied before the seismic loading. Time histories for the vertical and horizontal displacements at the middle height of the vessel at angles Θ ° and 9 ° are plotted in Figures 6 and 7.

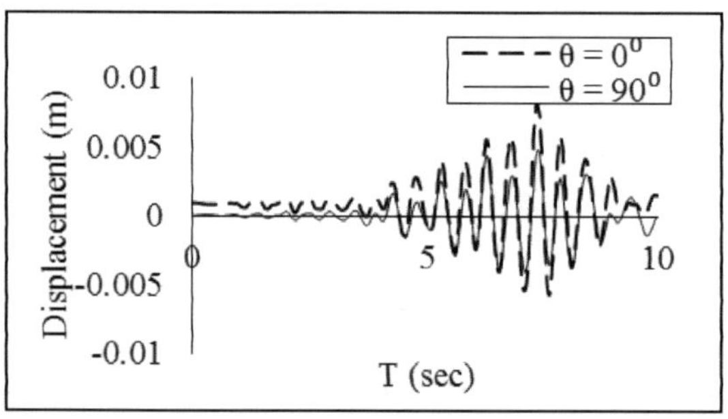

Figure 6: Time history for horizontal displacement at middle height of the vessel

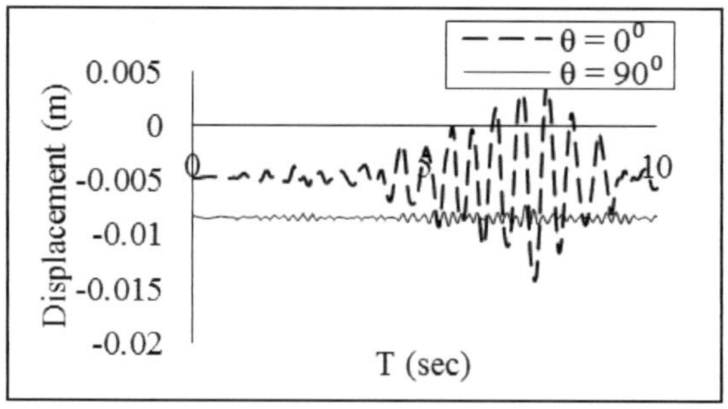

Figure 7: Time history for vertical displacement at middle height of the vessel

From these figures, it can be noted that the ratio between horizontal displacement due to the seismic load and horizontal displacement due to the hydrostatic load at the vessel's middle height is 7.5. It can also be noted that the response at Θ 9 ° results from the vertical excitation only, while the response at Θ ° results from both the horizontal and vertical excitations. Fig. 7 shows that the upward displacement due to the vertical excitation exceeded the downward displacement due to the hydrostatic load.

Time histories for the normal force, shear force and bending moment at the base of the vessel are shown in Figures 8, 9 and 10 respectively. The normal force results from the vertical component of the earthquake excitation, while both the base shear and moment result from the horizontal component of the earthquake excitation.

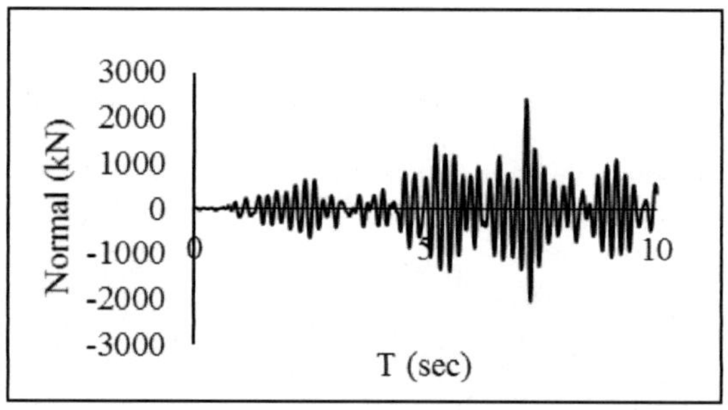

Figure 8: Time history for base normal force

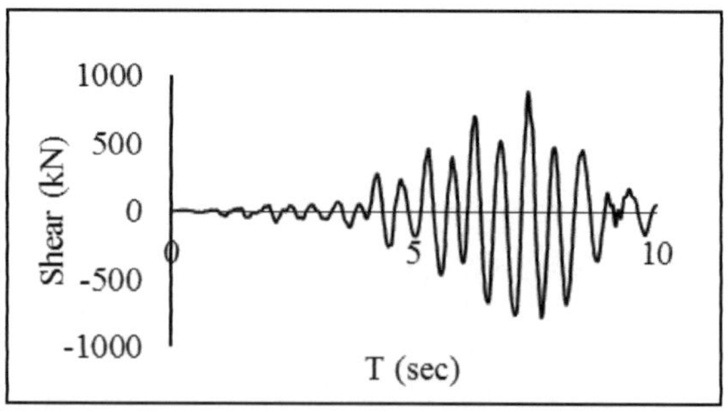

Figure 9: Time history for base shear force

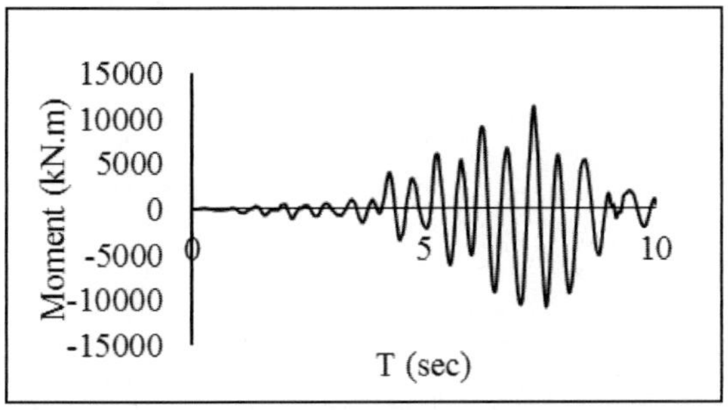

Figure 10: Time history for base moment

From these figures, it can be observed that the maximum normal force, base shear force and base moment for half of the tank are 2430 kN, 885 kN and 11300 kN.m. These values occur at times 7.2, 7.4 and 7.4 respectively. It can be noted that the maximum base shear and moment occur exactly at the same time, while the maximum normal force occurs at a slightly different time. Knowing that the mass of half the fluid inside the tank and the PGA for the vertical and horizontal components are equal to M_F = 1.5 1 ^{6} kg, a_V = . 65 g and a_H = 0.13g respectively, the following two relations can be deduced:

$$N_b = 2.6\ M_F a_V \quad (1)$$

$$Q_b = 0.46\ M_F a_H \quad (2)$$

Where N_b and Q_b are the normal force and shear force at the vessel's base. Therefore, a factor of 2.6 of the fluid mass inside the tank acts as rigid mass connected to the tank's wall and affects the vertical normal force. Also, approximately half of the fluid mass inside the tank acts as rigid mass connected to the tank's walls that affects both the base shear force and the base moment.

The staining actions from the time history analysis are compared with their counterparts from the response spectrum analysis that is performed using the mechanical model by El Damatty and Sweedan (2005) and (2006). The base shear force, moment and normal force from the response spectrum analysis are 951 kN, 15670 kN.m and 2900 kN respectively. It can be observed that the response spectrum analysis provides conservative values for both the base shear force, moment and shear force with a safety factors of 1.1, 1.4 and 1.2 respectively.

CONCLUSIONS

Based on the time history analysis of the composite conical tank in Mexico under Victoria earthquake, it can be concluded that the displacements along the vessel's height increase significantly due to the seismic loading such that it reaches 7.5 at the middle height of the vessel.

It is found that a factor of 2.6 of the fluid mass affects the base normal force, while a factor of 0.46 of the fluid mass affects the base shear and base moment. It also found that reasonable values for staining actions can be obtained using the response spectrum analysis. This analysis provides conservative values for the base shear force, moment and normal force with safety factors of 1.1, 1.4 and 1.2 respectively.

ACKNOWLEDGEMENTS

The authors would like to express their gratitude to the government of Ontario for their generous support through the Ontario Trillium Scholarship.

REFERENCES

1. Haroun, A., and Housner G. 19 1, "Seismic design of liquid storage tanks", *Journal of the Technical Councils of ASCE*, 107, pp. 191-207.
2. Haroun, A., and Ellaithy, H. 19 5, "Seismically induced fluid forces on elevated tanks", *Journal of Technical Topics in Civil Engineering,* 111, pp. 1-15.
3. Haroun, A., and Tayel, A. 19 5, "Response of tanks to vertical seismic excitations", *Earthquake Engineering and Structural Dynamics,* 13, pp. 583-595.
4. El Damatty, A., Korol, R., and Mirza, F. 1997, "Large displacement extension of consistent shell element for static and dynamic analysis", *Computers and Structures,* 62(6), pp. 943–960.
5. Koziey, L., and Mirza, F. 1997, "Consistent thick shell element", *Computers and Structures*, 65(4), pp. 531-549.
6. El Damatty, A., Korol, R., and Mirza, F. 1997, "Stability of elevated liquid filled conical tanks under seismic loading, Part I-Theory", *Earthquake Engineering Structure Dynamics*, 26, pp. 1191–208.
7. Sweedan, A., and El Damatty, A. 5, "Equivalent Models of Pure Conical Tanks under Vertical Ground Excitation" *Journal of Structural Engineering* ASCE pp. 725-733.
8. El Damatty, A., and Sweedan, A. 6, "Equivalent mechanical analog for dynamic analysis of pure conical tanks", *Thin-Walled Structures*, 44(4), pp. 429-440.
9. MOC- , , "Manual of Civil Structures", Section 3.1: Spectra of seismic design for Mexican territory at Mexico, Electrical research institute, Federal electricity commission.
10. Sweedan, A., and El Damatty, A. 2009, "Simplified procedure for the design of storage combined conical tanks", *Thin-walled structures,* 47, pp. 750-759.
11. Siddique, M., and El Damatty, A. 2012, "Enhancement of buckling capacity of steel plates strengthened with GFRP plates", *Thin-Walled Structures,* 60, pp. 154–162.

LOW-FREQUENCY VIBRATIONS OF INHOMOGENEOUS ELASTIC STRUCTURES WITH CONTRAST PROPERTIES

Dr Danila Prikazchikov
School of Computing and Mathematics
Keele University
Keele, Staffordshire, UK
d.prikazchikov@keele.ac.uk

Olga Sergushova
School of Computing and Mathematics
Keele University
Keele, Staffordshire, UK
o.sergushova@keele.ac.uk

ABSTRACT

Time-harmonic free vibrations of an elastic composite rod containing three inhomogeneous parts are considered. The attention is drawn to the case when the inner and outer parts have contrast parameters. The global low-frequency regime is first investigated for a symmetric composite rod with constant properties of the inner and outer parts. Relations between the material properties allowing existence of low-frequency modes are derived, revealing the contrast between the parameters as essential condition. Finally, the consideration is extended to the case of variable elastic parameters depending on the longitudinal coordinate.

Keywords: rod, bending, low-frequency, composite, contrast parameters.

INTRODUCTION

The problem of vibrations of inhomogeneous structures has numerous applications in modern engineering, see e.g. [1, 2] among other contributions. The component parts may often have contrast material properties, which occurs for example in sandwich plates [3, 4]. A related area of homogenizing these structures should also be noted [5].

This paper is concerned with analysis of low-frequency vibrations in such structures within 1D setup. A non-uniform elastic rod containing three components is considered subject to free ends boundary conditions. It is demonstrated that the contrast material properties of the components may cause an unexpectedly low natural frequency, which may get rather close to zero, corresponding to rigid body motion.

First, we derive the exact dispersion relation for the case of a symmetric rod, with the material parameters of the outer and inner parts been independent of the longitudinal variable. The global low-frequency regime is then investigated, and the conditions on the contrast material parameters allowing the first natural frequency close to zero are derived.

Then the analysis is extended to a more general case of a composite rod with non-homogeneous components, with the material parameters dependent on the longitudinal variable. A perturbation procedure is implemented, leading to approximations for the first natural frequency along with the corresponding displacement profile.

PROBLEM STATEMENT

Consider time-harmonic vibrations of an elastic composite rod consisting of three inhomogeneous parts. The axis Ox is chosen such that the origin O is located in the middle of the inner part. The outer parts of the rod have free ends, and continuity is assumed between the components. The inner part of the rod occupies the region $|x| \leq h_1$, with the outer parts specified by $-h_1 - h_3 \leq x \leq h_1$ and $h_1 \leq x \leq h_1 + h_2$.

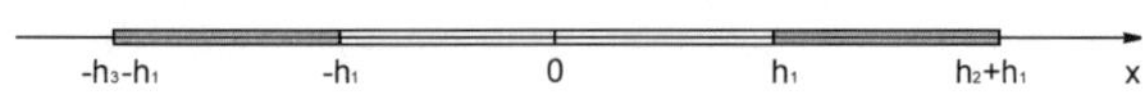

Figure 1: Composite rod

The equations of motion are written in the form

$$\frac{d}{dx}\left(E_i \frac{du}{dx}\right) + \rho_i \omega^2 u = 0, \qquad i = 1, 2, 3, \tag{1}$$

where $u(x)$ is the displacement, $E_i(x) = E_i^* \tilde{E}_i(x)$ are the Young's moduli, $\rho_i(x) = \rho_i^* \tilde{\rho}_i(x)$ are the material densities and ω is the vibration frequency.

Note that $c_i(x) = \sqrt{E_i^* \tilde{E}_i(x) / \rho_i^* \tilde{\rho}_i(x)}$ are the longitudinal wave speeds with $c_i^* = \sqrt{E_i^* / \rho_i^*}$. Here the indices 1, 2 and 3 correspond to the inner, right outer and left outer parts, respectively.

The continuity conditions at the interfaces are written as

$$\begin{aligned} \lim_{x \to \pm(h_1 - 0)} u &= \lim_{x \to \pm(h_1 + 0)} u, \\ \lim_{x \to h_1 - 0} E_1 u' &= \lim_{x \to h_1 + 0} E_2 u', \\ \lim_{x \to -h_1 + 0} E_1 u' &= \lim_{x \to -h_1 - 0} E_3 u', \end{aligned} \tag{2}$$

with free ends boundary conditions given by

$$\begin{aligned} u' &= 0 \quad \text{at} \quad x = h_1 + h_2, \\ u' &= 0 \quad \text{at} \quad x = -h_1 - h_3. \end{aligned} \tag{3}$$

Let us introduce the following dimensionless quantities

$$E_{ij} = \frac{E_i^*}{E_j^*}, \quad \rho_{ij} = \frac{\rho_i^*}{\rho_j^*}, \quad c_{ij}^* = \frac{c_i^*}{c_j^*}, \quad h_{ij} = \frac{h_i}{h_j},$$

where $i, j = 1, 2, 3$, along with the dimensionless variable

$$\chi = \frac{x}{h_1},$$

and scaled frequencies

$$\lambda_i = \frac{\omega h_i}{c_i}, \qquad i = 1, 2, 3.$$

VIBRATIONS OF A SYMMETRIC ROD

Let us begin with a simple problem for a symmetric composite rod with constant material parameters. In this case due to symmetry of the rod

$$E_2 = E_3, \quad \rho_2 = \rho_3, \quad c_2 = c_3, \quad h_2 = h_3, \quad \lambda_2 = \lambda_3,$$

and

$$E = \frac{E_1}{E_2}, \qquad \rho = \frac{\rho_1}{\rho_2}, \qquad c = \frac{c_1}{c_2}, \qquad h = \frac{h_1}{h_2}.$$

Below we consider antisymmetric vibrations allowing low-frequency regime. The displacement of the inner part is given by

$$u = A \sin \lambda_1 \chi \quad \text{at} \quad 0 \le |\chi| \le 1, \tag{4}$$

following from (1). The displacement of the outer part may be written in general form

$$u = C \cos h\lambda_2 \chi + D \sin h\lambda_2 \chi \quad \text{at} \quad 1 \le |\chi| \le 1 + h^{-1}. \tag{5}$$

Substitution of (4) and (5) into equations (2) and (3) gives a linear algebraic system, which possesses non-trivial solutions provided that the associate determinant vanishes, leading to the frequency equation

$$\tan \lambda_1 \tan \lambda_2 = \frac{E}{c}. \tag{6}$$

It is then possible to represent the displacement profile as

$$u = \begin{cases} \sin \lambda_1 \chi, & 0 \le |\chi| \le 1, \\ \sin \lambda_1 \cos h\lambda_2(\chi - 1) + \\ \dfrac{E}{c} \cos \lambda_1 \sin h\lambda_2(\chi - 1), & 1 \le |\chi| \le 1 + h^{-1}. \end{cases} \tag{7}$$

The studied low-frequency motion may be classified into cases of global low-frequency regime ($\lambda_1 \ll 1$, $\lambda_2 \ll 1$), and local low-frequency behaviour ($\lambda_1 \ll 1$, $\lambda_2 \gtrsim 1$ or $\lambda_1 \gtrsim 1$, $\lambda_2 \ll 1$). In this paper the consideration is restricted to global low-frequency regime only.

In case of $\lambda_1 \ll 1$ and $\lambda_2 \ll 1$ the frequency equation (6) reduces to the following approximate form

$$\lambda_1 \lambda_2 = \frac{E}{c}, \tag{8}$$

from which it may be deduced that

$$\lambda_1 = \sqrt{\rho h}, \qquad \lambda_2 = \sqrt{\frac{E}{h}}. \tag{9}$$

The latter, combined with strong inequalities $\lambda_1 \ll 1$, $\lambda_2 \ll 1$, provide the conditions for material parameters allowing global low-frequency regime

$$E \ll h \ll \rho^{-1}. \tag{10}$$

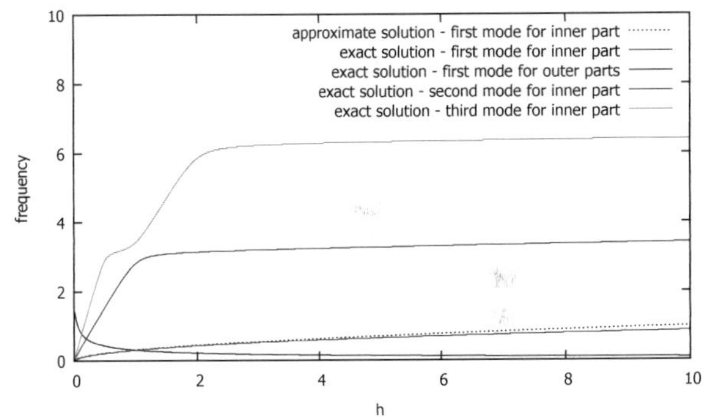

Figure 2: Frequency vs h ($E = 0.1$, $\rho = 0.1$)

Figure 2 contains frequency curves depending on the relative thickness h, including the decreasing curve of the first mode for λ_2 defined by (6), and the increasing curves for the first three modes for λ_1 along with the asymptotic approximation for the first mode (9). It can be seen from Figure 2 that varying h between E and ρ^{-1}, it is possible to find a range of values of h, such that λ_1 and λ_2 are both small. Good agreement between the first harmonic for λ_1 defined from (6) and its approximation (9) is also observed.

It may be noticed that the distance between the first and second modes is substantial. This may have an implication for the dispersion analysis of bending of sandwich plates composed of layers with contrast properties. Indeed, the frequencies studied within 1D framework correspond to the cut-off-frequencies of the plate [6]. Therefore, one may expect that the first cut-off-frequency being small leads to the phenomena of weak coupling between the fundamental mode and the first harmonic, requiring special plate theories of Timoshenko-type, see e.g. [7, 8].

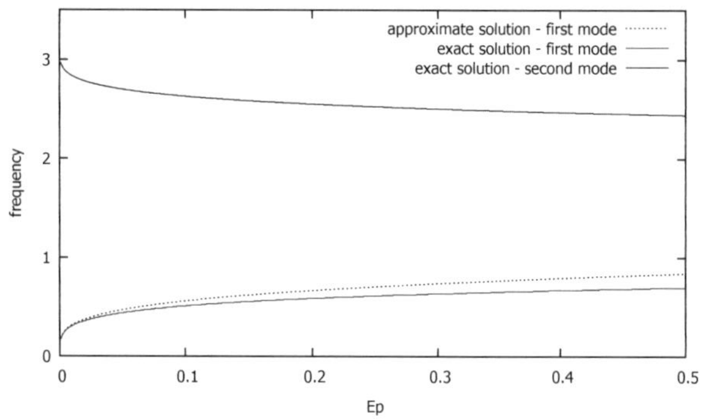

Figure 3: Frequency λ (11)-(12) vs the parameter $E\rho$

It is also worth noting that in the vicinity of intersection of curves for λ_1 and λ_2 in Figure 2, when $\lambda_1 = \lambda_2 = \lambda$, the frequency equation (6) reduces to

$$\tan^2 \lambda = \sqrt{E\rho}, \tag{11}$$

hence, in the low frequency region

$$\lambda = (E\rho)^{1/4}. \tag{12}$$

The low-frequency regime corresponds to $E\rho \ll 1$, which may be interpreted physically as having a soft inner part and firm outer parts.

It may be observed from Figure 3 that the first harmonic for λ defined by (11) decreases as $E\rho \to 0$, with the distance between the first and second modes increasing. Predictably, the accuracy of the approximation (12) increases as $E\rho \to 0$.

Noting from (8) that $\dfrac{E}{c} \ll 1$ and introducing the scaled displacement $U = \dfrac{u}{\lambda_1}$, we arrive at the following approximate profile of the eigenform (see Figure 4)

$$U = \begin{cases} \chi, & 0 \le |\chi| \le 1, \\ 1, & 1 \le |\chi| \le 1 + h^{-1}, \end{cases} \tag{13}$$

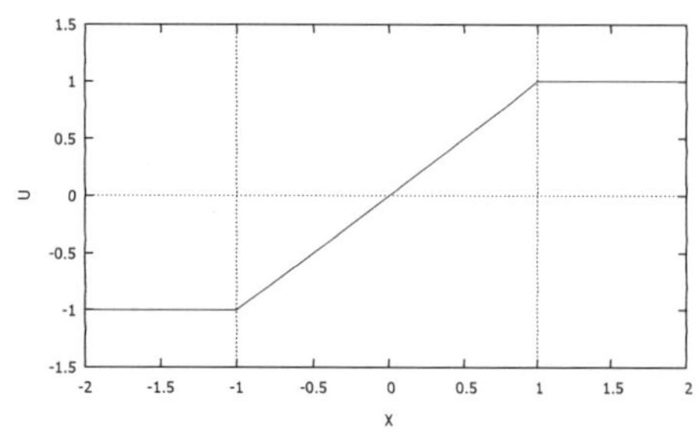

Figure 4: Displacement profile (13) for $h = 1$

VIBRATIONS OF AN ASYMMETRIC ROD

Now the analysis is extended to an asymmetric rod with free ends where the densities and the Young's moduli of the components are dependent on χ. Since it is not possible to find the exact solution in this case, we implement the perturbation technique.

Let us introduce a small parameter $\varepsilon = \dfrac{E_{12}}{h_{12}} \ll 1$ associated with contrast material properties of the components and consider the global low-frequency regime of the form $\lambda_1 \sim \lambda_2 \sim \lambda_3 \sim \varepsilon$. The frequencies and the displacement are expanded as asymptotic series

$$\begin{aligned} \lambda_i^2 &= \varepsilon(\lambda_{i0}^2 + \varepsilon\lambda_{i1}^2 + \varepsilon^2\lambda_{i2}^2 + \ldots), \\ u &= u_0 + \varepsilon u_1 + \varepsilon^2 u_2 + \ldots\ . \end{aligned} \tag{14}$$

The equation of motion (1) can be rewritten as

$$\frac{d}{d\chi}\left(\tilde{E}_i \frac{du}{d\chi}\right) + \tilde{\rho}_i h_{1i} \lambda_i^2 u = 0, \qquad i = 1, 2, 3. \tag{15}$$

On substituting the expansion (14) into the equation of motion (15), at leading order ε^0 for the displacement of the right outer part we get

$$\frac{d}{d\chi}\left(\tilde{E}_2 \frac{du_0}{d\chi}\right) = 0, \qquad 1 \ll \chi \ll 1 + h_{21}. \tag{16}$$

From the continuity condition at $\chi = 1$

$$\lim_{\chi \to 1-0} \varepsilon \frac{du}{d\chi} = \lim_{\chi \to 1+0} \frac{du}{d\chi}, \tag{17}$$

it may be deduced $\dfrac{du_0}{d\chi} = 0$ at $\chi = 1$, hence $\dfrac{du_0}{d\chi} = 0$ at $1 \ll \chi \ll 1 + h_{21}$. It should be noted that the boundary condition (3) at $\chi = 1 + h_{21}$ is satisfied automatically. Thus,

$$u_0 = U_R, \qquad 1 \ll \chi \ll 1 + h_{21}. \tag{18}$$

Similarly, the displacement of the left outer part is

$$u_0 = U_L, \qquad -1 - h_{31} \ll \chi \ll -1. \tag{19}$$

Here U_R and U_L are constants.

Now we proceed with analysis of the displacement u_0 in the inner part. From (15), using (14) we get

$$\frac{du_0}{d\chi} = \frac{C}{\tilde{E}_1}, \qquad -1 \ll \chi \ll 1. \tag{20}$$

Integrating the latter and employing continuity conditions (2) along with (19)-(20) we result in

$$u_0 = \int_{-1}^{1} \frac{U_R - U_L}{\tilde{E}_1 \theta_{E_1}}\, d\chi, \tag{21}$$

where

$$\theta_{E_1} = \int_{-1}^{1} \tilde{E}_1^{-1} d\chi.$$

It is worth noting that the leading order displacement (18, 19, 21) correlates with (13) obtained for a symmetric rod in the previous section, see Figure 4.

At next order analysis of (15) gives for the displacement of the right outer part

$$\frac{d}{d\chi}\left(\tilde{E}_2\frac{du_1}{d\chi}\right)+\tilde{\rho}_2 h_{12}\lambda_{20}^2 u_0=0, \quad 1\ll\chi\ll 1+h_{21}. \tag{22}$$

Integrating this equation one arrives at

$$\tilde{E}_2\frac{du_1}{d\chi}|_{\chi=1+h_{21}}-\tilde{E}_2\frac{du_1}{d\chi}|_{\chi=1}=-\lambda_{20}^2 U_R h_{12}\theta_{\rho_2}, \tag{23}$$

where

$$\theta_{\rho_2}=\int\limits_{1}^{1+h_{21}}\tilde{\rho}_2 d\chi, \qquad \theta_{\rho_3}=\int\limits_{-1-h_{31}}^{-1}\tilde{\rho}_3 d\chi.$$

Employing the continuity condition (17) and the boundary condition for free faces (3) at $\chi=1+h_{21}$, we deduce

$$\lambda_{20}^2=\frac{\tilde{E}_2(1)(U_R-U_L)h_{21}}{\tilde{E}_1(1)U_R\theta_{E_1}\theta_{\rho_2}}. \tag{24}$$

Similarly the expression for λ_{30}^2 is found as

$$\lambda_{30}^2=\frac{\tilde{E}_3(-1)(U_L-U_R)h_{31}h_{32}E_{23}}{\tilde{E}_1(-1)U_L\theta_{E_1}\theta_{\rho_3}}. \tag{25}$$

Using the relation $\lambda_2=h_{23}c_{32}\lambda_3$ and (25), it is possible to express λ_{20}^2 as

$$\lambda_{20}^2=\frac{\tilde{E}_3(-1)(U_L-U_R)h_{21}\rho_{23}}{\tilde{E}_1(-1)U_L\theta_{E_1}\theta_{\rho_3}}. \tag{26}$$

Therefore from (24), (26) we obtain

$$\lambda_{20}^2=\frac{h_{21}}{\theta_{E_1}}\left(\frac{\tilde{E}_2(1)}{\tilde{E}_1(1)\theta_{\rho_2}}+\frac{\tilde{E}_3(-1)\rho_{23}}{\tilde{E}_1(-1)\theta_{\rho_3}}\right) \tag{27}$$

and

$$U_L=-U_R\rho_{23}\frac{\tilde{E}_1(1)}{\tilde{E}_2(1)}\frac{\tilde{E}_3(-1)}{\tilde{E}_1(-1)}\frac{\theta_{\rho_2}}{\theta_{\rho_3}}. \tag{28}$$

In case when E_i and ρ_i are independent of χ the formulae (27)-(28) reduce to

$$\lambda_{20}^2=\frac{1}{2}(1+\rho_{23}h_{23}), \qquad U_L=-U_R\rho_{23}h_{23}. \tag{29}$$

CONCLUSIONS

In this paper the low-frequency vibrations has been analyzed in respect of a composite non-homogeneous rod with contrast material properties of the components. It is shown that the existence of contrast may lead to the first natural frequency being close to zero. The underlying physics may therefore be described as perturbation of the rigid body motion, similarly to ideas developed in [9].

It should be emphasized that the developed conditions (10) provide the range of parameters, for which the first natural frequency is small, thus providing a basis to interaction between the two close eigenvalues and justifying development of Timoshenko-type theories for plates and shells.

Finally, we note that the presented approach may be extended further to infinite periodic rods, and also to composite plates and shells.

ACKNOWLEDGMENTS

Fruitful discussions with Prof J. Kaplunov are very gratefully acknowledged.

REFERENCES

[1] Elishakoff, I. 2004, *Eigenvalues of inhomogeneous structures: unusual closed-form solutions*, CRC Press, Boca Raton.

[2] Horgan, C. O., Chan, A. M. 1999, "Vibration of inhomogeneous strings, rods and membranes," *Journal of Sound and Vibration*, Vol(225.3), pp. 503-513.

[3] Nayak, A. K., Moy, S. S. J., Shenoi, R. A. 2002, "Free vibration analysis of composite sandwich plates based on Reddy's higher-order theory," *Composites part B: engineering*, Vol(33.7), pp. 505-519.

[4] Chapman, C. J. 2013, "An asymptotic decoupling method for waves in layered media," *Proc. R. Soc. A*, Vol(469), 20120659.

[5] Cherednichenko, K. D., Smyshlyaev, V. P., Zhikov, V. V. 2006, "Non-local homogenized limits for composite media with highly anisotropic periodic fibres," *Proc. R. Soc. Edin.*, Vol(136), pp. 87-114.

[6] Kaplunov, J. D., Kossovitch, L. Y., and Nolde, E. V. 1998, *Dynamics of thin walled elastic bodies*, Academic Press, San-Diego.

[7] Goldenveizer, A. L., Kaplunov, J. D., Nolde, E. V. 1993, "On Timoshenko-Reissner type theories of plates and shells," *International Journal of Solids and Structures*, Vol(30.5), pp. 675-694.

[8] Le, K. C. 1999, *Vibrations of shells and rods*, Springer, Berlin.

[9] Kaplunov, J., Shestakova, A., Aleynikov, I., Hopkins, B., Talonov, A. 2015, "Low-frequency perturbations of rigid body motions of a viscoelastic inhomogeneous bar," *Mechanics of Time-Dependent Materials*, doi:10.1007/s11043-015-9256-x, pp. 1-17.

SEISMIC BEHAVIOUR OF STEEL LIQUID-FILLED CONICAL TANKS DESIGNED UNDER HYDROSTATIC PRESSURE

A. Musa
Civil and Environmental Engineering
Western University
London, ON, Canada
amusa5@uwo.ca

A.A. El Damatty
Civil and Environmental Engineering
Western University
London, ON, Canada
damatty@uwo.ca

ABSTRACT

Steel conical tanks are widely used for liquid storage in North America and elsewhere. A number of those tanks collapsed in the last decades as a result of instability of the steel shells. Despite being widely used, no specific design procedure is available for conical tanks either under static or dynamic conditions. In this study, an assessment is made for the capacity of tanks designed under hydrostatic pressure only to resist the hydrodynamic pressure resulting from horizontal and vertical ground excitations. The assessment is done using non-linear time history analysis based on a set of natural ground motions scaled to different seismic zones in Canada. The study is conducted numerically using an in-house non-linear finite element model. Finally, the adequate design of the steel conical tanks required to sustain the ground motions is determined.

KEYWORDS: Steel conical tanks, hydrodynamic pressure, time-history analysis, buckling, finite-element

INTRODUCTION

Steel conical-shaped liquid storage tanks are commonly used for the purpose of storing different kinds of liquids. A conical tank might consist of a pure truncated cone only as shown in Fig. 1a or might be capped with a cylindrical part and is referred as combined conical tank as shown in Fig. 1b. Although such tanks are widely constructed, no specific design procedure is found for conical tanks in most liquid-tanks specifications. The only guidelines, found in some design provisions are based on treating a conical-shaped tank as an equivalent cylinder.

Figure 1: a) Pure conical tank, b) Combined conical tank

The main difference in the structural behaviour between conical and cylindrical tanks is due to the inclination of the conical tank wall. The volume of the contained liquid can be divided into two portions: vol. 1 and vol.2 as shown in Fig. 2. The first one is transferred directly to the supporting concrete shaft through the tank base, while vol.2 is resting on the inclined wall of the tank. The latter volume increases as the base of the tank is approached and is associated with a decrease in the vessel radius. This leads to the development of high compressive meridional stresses σ_m in this region making those steel tanks vulnerable to collapse due to buckling if not properly designed.

For a liquid-filled conical tank subjected to an earthquake excitation, the walls are affected by the inertial forces of the wall in addition to the hydrodynamic pressure of the contained liquid. The contained liquid can be divided into two parts: the first part is mainly the lower amount of liquid which moves in synchronism with the walls of the tank and is called the impulsive liquid mass, while the second part is mainly the upper amount of liquid and is called the convective liquid mass which undergoes sloshing due to vibration. The ratio of the convective mass to the impulsive one is small in general for practical dimensions for steel conical tanks but it increases as the tank becomes shallower. Also, the frequency of the impulsive vibration mode is much higher than this for the convective vibration mode.

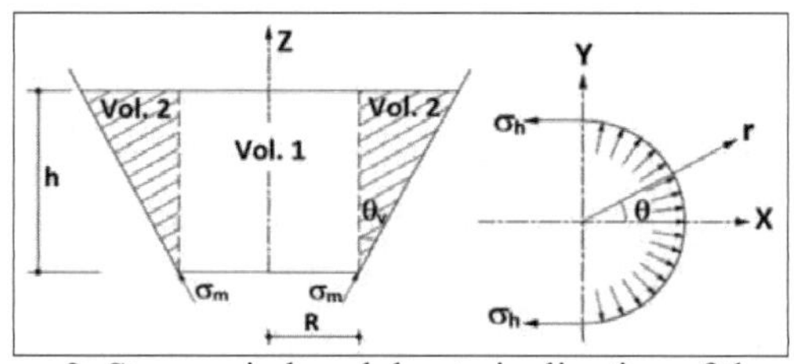

Figure 2: Stresses induced due to inclination of the wall

The first study related to steel conical tanks was conducted by Vandepitte et al. (1982) after the collapse of a conical steel water tower in Belgium. In this study, a large number of small-scale conical tank models were gradually filled with water till buckling occurred. A set of design curves for different base restraining conditions were developed. In 1990, a steel conical water tower collapsed in Fredericton, Canada when it was filled with water for the first time. Vandepitte (1999) concluded that the main cause of failure was related to the inadequate thickness of the tank walls at the base as the designers underestimated the amplitude of the initial geometric imperfections.

The first study conducted to assess the behavior of conical tanks under seismic loading was done by El-Damatty et al. (1997b). They developed a coupled shell element-boundary element formulation to simulate the fluid-structure interaction. A fluid-added mass matrix was obtained from the above formulation and is added to the mass matrix of the structure and is incorporated into a nonlinear dynamic analysis routine. The formulation was conducted for both horizontal and vertical excitations.

HYDRODYNAMIC PRESSURE

Hydrodynamic pressure is induced on the tank wall during seismic excitation acting on a conical tank. The hydrodynamic pressure can be divided into impulsive pressure P_I and convective pressure due to sloshing of the water surface P_S. The sloshing component is a long period component relative to the impulsive one and hence the two components can be decoupled in their derivations.

The impulsive component associated with the hydrodynamic pressure for a conical tank containing an ideal fluid can be obtained by solving the following differential equation inside the fluid volume:

$$\nabla^2 P_I(r,\theta,z,t) = 0 \qquad (1)$$

where the coordinates are described in Fig. 2 and t is the time. The solution of Eq. (1) was done using the boundary element method by Haroun (1980) for cylindrical tanks and El Damatty et al. (1997b) for conical tanks. Based on the latter study, the impulsive component of the hydrodynamic pressure resulting from horizontal and vertical excitations are expressed as shown in Eqs. (2,3), respectively.

$$P_{IH}(r,\theta,z,t) = \sum_{n=1}^{N_2}\sum_{i=1}^{N_1} A_{in}(t) I_n(\alpha_i r)\cos(\alpha_i z)\cos(n\theta) \qquad (2)$$

$$P_{IV}(r,\theta,z,t) = \sum_{i=1}^{N_1} A_{io}(t) I_o(\alpha_i r)\cos(\alpha_i z) + \ddot{G}_v(t)\left(1 - \frac{z}{h}\right) \qquad (3)$$

$$\alpha_i = (2i-1)\,\pi/2h \qquad (4)$$

where $A_{in}(t)$ is the amplitude function of time, I_n are the modified Bessel's functions of the first kind, $\ddot{G}_v(t)$ is the ground vertical excitation, and n is the wave number.

The sloshing component P_S can be evaluated considering the tank walls are rigid with a reasonable accuracy as noted by Haroun (1980). Based on this assumption, El Damatty et al. (2000) derived an expression for the fundamental sloshing component of the hydrodynamic pressure as follows

$$P_S(r,\theta,z,t) = B(t)\,\rho_F J_1(k_1 r)\cosh(k_1 z)\cos(\theta) + \rho_F \ddot{x}(t) r\cos(\theta) \qquad (5)$$

where $J_1(k_1 r)$ is the Bessel's function of the first kind of order one, B(t) is arbitrary function of time, ρ_F is the fluid density and $\ddot{x}(t)$ is the base acceleration acting on the tank. A procedure to evaluate the constant k_1 is discussed in details by El Damatty et al. (2000).

GEOMETRIC IMPERFECTIONS

The steel vessel of a conical tank is constructed from curved panels welded together. In fact, it requires a superior quality control to get perfect straight thin steel plates which is almost impossible. As a result, geometric imperfections of the conical tank walls will exist and define the buckling capacity of the tanks. A commonly used model for simulating the geometric imperfections W(s) is

$$W(s) = w_0 \sin(2\pi s/L_I)\cos(n\theta) \qquad (6)$$

where w_0 is the imperfection amplitude, L_I is the imperfection wavelength, s is a coordinate measured along the generator of the vessel, and n is an integer defining the circumferential wave length of the imperfection shape.

According to Vandepitte (1982), a conical tank with the ratio w_0/L_I less than 0.004 is considered a good cone while a conical tank with the ratio w_0/L_I ranging from 0.004 to 0.01 is considered a poor cone. For the case of steel conical tanks under the effect of hydrostatic pressure only, Vandepitte (1982) used experimental results to get an expression for the critical buckling wave length and El Damatty et al. (1997a) have shown that an imperfection shape with axisymmetric distribution along the surface of the vessel leads to minimum buckling capacity of pure conical tanks.

For the case of combined hydrostatic and hydrodynamic pressure, non-linear static analysis is used to determine the critical buckling wave length. A pressure distribution following first mode of Eq. (2), i.e., n=i=1 is applied on the tank walls incrementally till buckling takes place. The procedure is repeated for 60 tanks with several analyses performed for each tank with different imperfection wave length and the one leading to minimum buckling capacity is considered the critical imperfection wave length L_{cr}. It is found that an axisymmetric distribution is still the critical one and based on regression analysis L_{cr} is given by

$$L_{cr} = 4.03\sqrt{R\,t/\cos\theta_v} \qquad (6)$$

where R is the tank bottom radius, t is the wall thickness, and θ_v is the inclination angle with the vertical as in Fig. 2.

TIME-HISTORY ANALYSIS

In this study, non-linear time history analysis is used in order to assess the behaviour of pure steel conical tanks under the effect of both horizontal and vertical excitations. Using a coupled finite-boundary element approach, El Damatty et al. (1997b) derived a fluid added-mass matrix that simulates the effect of the impulsive component of the hydrodynamic pressure for both horizontal and vertical excitations which is then added to the structure mass matrix and incorporated in the dynamic analysis.

The selection of the ground motions to be used in this study is based on the deaggregation of seismic hazard procedure proposed by Halchuk and Adams (2004) for different locations in Canada. The seismic hazard is deaggregated in terms of distance and magnitude of the ground motion. Three Canadian seismic zones are considered: Toronto, Montreal, and Vancouver. The chosen natural ground motion records are summarized in table 1.

Table 1. Selected ground excitation records corresponding to different seismic zones

Seismic Zone	Record Name	Date	Duration (sec)
Toronto	Chalfant Valley-02	1986	20
	Whittier Narrows-01	1987	20
	Northridge-04	1994	20
	Chi-Chi, Taiwan-02	1999	30
Montreal	Big Bear-01	1992	25
	Chi-Chi, Taiwan-06	1999	35
	Northridge-01	1994	20
	San Fernando	1971	20
Vancouver	Big Bear-01	1983	25
	Chi-Chi, Taiwan-05	1992	20
	Coalinga-01	1999	30

The scaling procedure is based on matching the spectral acceleration of the record under consideration with the spectral acceleration of the response spectrum obtained from NBCC 2010 at the natural frequency of the first impulsive mode. This scaling is applied separately for each of the horizontal and vertical components of the ground motion record and differs from tank to another. The natural frequencies for the impulsive vibration modes for both horizontal and vertical vibrations are obtained through free vibration analysis using the previously discussed fluid added matrix. A value of 2% and 0.5% damping are used for impulsive and sloshing vibration modes, respectively. During the time-history analysis, hydrostatic pressure is applied prior to the application of the ground accelerations at each time increment with a load factor equals to unity.

A group of 25 tanks of practical dimensions are chosen for this study with the tank height h ranging from 5.0 to 9.0m with a step of 1.0m and the tank bottom radius R ranging from 4.0 to 6.0m with a step of 0.5m. The tanks are analyzed with three values of the angle θ_v: 30°, 45°, and 60° leading to a total number of 75 tanks. The tanks were preliminary designed under hydrostatic pressure only based on the simplified method proposed by Sweedan and El Damatty (2009).

FINITE ELEMENT MODEL

Three-dimensional numerical models are developed for the steel conical tanks using the finite element method. The numerical model is based on a consistent subparametric triangular shell element shown in Fig. 3 that was developed by Koziey and Mirza (1997) and extended later on by El Damatty et al. (1997c) to include the large deformations and geometric nonlinear effect. Such a consistent subparametric formulation eliminated the spurious shear modes and locking phenomenon observed in isoparametric shell elements when used in modeling thin shell structures.

Due to symmetry about the horizontal axis in both loading and geometry, only half of the cone is modelled and used in the analysis. A mesh of 52 triangular elements is used for the model as shown in Fig. 3. The tanks are assumed to be hinged at the base along the circumference. Finally, the horizontal and vertical degrees of freedom at the tank base level are constrained in order to have the same deformations due to the existence of a concrete slab underneath the tank.

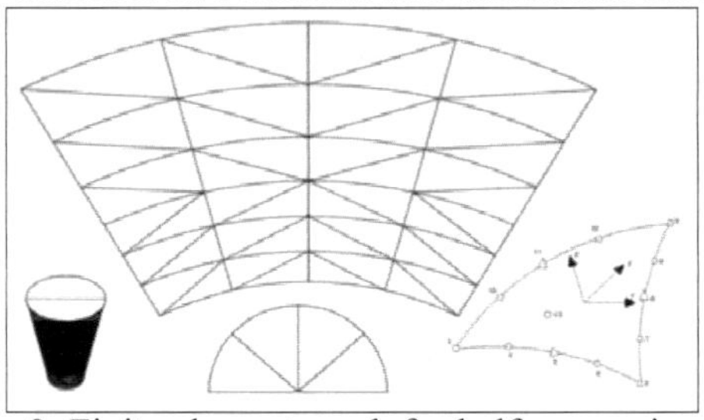

Figure 3: Finite element mesh for half cone using the consistent shell element

RESULTS OF ANALYSIS

The set of the steel conical tanks chosen for this study are subjected to the eleven ground motion records in the form of horizontal and vertical ground accelerations. The deformed shape of one of the conical tanks walls is shown in Fig. 4a at two stages; (1) Just after the application of the hydrostatic pressure, (2) At maximum displacement throughout the time history of the ground motion. Due to the stress concentration near the base of the steel conical tank, buckling waves starts to initiate during the hydrostatic pressure stage and increase with the application of the hydrodynamic pressure. Such deformation pattern is commonly known as elephant-foot buckling and was noted in several liquid storage tanks after some earthquake events as shown in Fig. 4b.

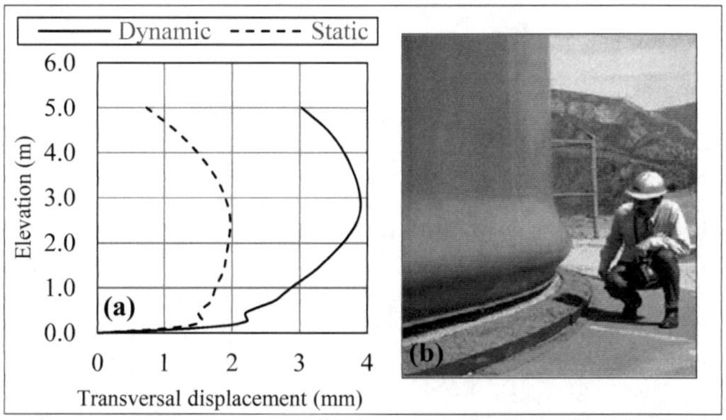

Figure 4: (a) Transversal displacement for a steel conical tank (R=4m, H=5m, θ_v=45°) subjected to Northridge-01, (b) Elephant-foot buckling phenomena

As discussed earlier, the finite element model used in this study takes into consideration both geometric and material nonlinearities. As a result it is capable of capturing any localized buckling taking place before yielding of the tank material i.e., elastic buckling or after yielding i.e., inelastic buckling. The aim of this study is to assess the ability of the chosen steel conical tanks designed under hydrostatic pressure with a factor of safety of 1.4 to resist the set of ground motions discussed earlier. The effect of degree of the geometric imperfection is studied with two levels: the first one with $w_0 = 0.004L_{cr}$ to represent good tanks while the second one with $w_0 = 0.01L_{cr}$ representing poor tanks.

For the set of steel conical tanks considered in this study, one of three final states can take place; (a) Safe Elastic, if the tank survived the scaled components of the ground excitation without yielding at any location, (b) Safe Plastic, if the tank survived the scaled components of the ground excitation with yielding taking place, (c) Unsafe Plastic, if the tank failed under the scaled components of the ground excitation due to local buckling.

Time history analyses show that all the perfect steel tanks behaved elastically and survived ground motions corresponding to the three seismic zones. The same observation is found for the case of the good conical tanks corresponding to the seismic zones of Toronto and Montreal. For Vancouver seismic zone, majority of the tanks survived the ground excitations either elastically or inelastically.

Regarding poor tanks, over 90% of the steel conical tanks with θ_v=30°, 45° survived the ground excitations corresponding to Toronto seismic zone elastically while the remaining tanks survived inelastically. For tanks with θ_v=60° corresponding to Toronto seismic zone, the majority of the tanks survived the ground excitations inelastically with some tanks failing due to buckling.

For ground excitations corresponding to Montreal seismic zone, majority of poor conical tanks with θ_v =30° did not survive the ground excitations elastically while none of the tanks with θv=45°, 60° survived the ground excitations elastically. Finally for Vancouver seismic zone, all poor conical tanks did not survive the ground excitations elastically with more than 90% of the tanks failing due to buckling. The ranges for the thicknesses adequate to sustain the ground motions elastically for different seismic zones are summarized in table 2, where t_s is the initial hydrostatic thickness.

Table 2. Adequate tank thicknesses range for different seismic zones

Zone	θ_v	Perfect	Good	Poor
Toronto	30	$1.0t_s \sim 1.0t_s$	$1.0t_s \sim 1.0t_s$	$1.0t_s \sim 1.0t_s$
	45	$1.0t_s \sim 1.0t_s$	$1.0t_s \sim 1.0t_s$	$1.0t_s \sim 1.2t_s$
	60	$1.0t_s \sim 1.0t_s$	$1.0t_s \sim 1.0t_s$	$1.0t_s \sim 1.2t_s$
Montreal	30	$1.0t_s \sim 1.0t_s$	$1.0t_s \sim 1.0t_s$	$1.0t_s \sim 1.6t_s$
	45	$1.0t_s \sim 1.0t_s$	$1.0t_s \sim 1.0t_s$	$1.2t_s \sim 1.6t_s$
	60	$1.0t_s \sim 1.0t_s$	$1.0t_s \sim 1.0t_s$	$1.4t_s \sim 1.6t_s$
Vancouver	30	$1.0t_s \sim 1.0t_s$	$1.0t_s \sim 1.4t_s$	$1.2t_s \sim 2.0t_s$
	45	$1.0t_s \sim 1.0t_s$	$1.0t_s \sim 1.4t_s$	$1.4t_s \sim 2.0t_s$
	60	$1.0t_s \sim 1.0t_s$	$1.0t_s \sim 1.4t_s$	$1.4t_s \sim 2.2t_s$

CONCLUSIONS

The aim of this study is to assess the ability of pure steel conical tanks designed to resist hydrostatic pressure with a certain factor of safety to resist ground excitations scaled to match three seismic zones in Canada. The study is done using non-linear time-history finite element code that is able to capture the induced fluid-structure interaction. The conclusion of the study is that design under hydrostatic pressure only with a factor of safety of 1.4 is not adequate in all cases as some tanks suffered buckling when subjected to the ground excitations. The thickness required to resist the ground motions reached 120%, 160%, and 220% of the hydrostatic thickness for Toronto, Montreal, and Vancouver, respectively, based on the level of imperfections

ACKNOWLEDGMENTS

The authors would like to akcnowledge the Boundary layer wind tunnel at Western University for their funding through the NSERC IPS program. Also, this work was made possible by the facilities of the Shared Hierarchical Academic Research Computing Network (SHARCNET: www.sharcnet.ca).

REFERENCES

[1] Vandepitte, D., Rathe, J., Verhegghe, B., Paridaens, R., and Verschaeve, C. 1982, "Experimental investigation of hydrostatically loaded conical shells and practical evaluation of the buckling load", Proc. State of the Art Colloquium, Universitat Stuttgart, Germany, pp.375-399.

[2] Vandepitte, D. 1999, "Confrontation of shell buckling research results with the collapse of a steel water tower", Journal of Constructional Steel Research, Vol(49),pp. 303–314.

[3] El Damatty, A. A., Mirza, F.A., and Korol, R.M. 1997b, "Stability of elevated liquid-filled conical tanks under seismic loading, Part I-Theory", Earthquake Engng. Struct. Dyn., Vol(26), pp. 1191-1208 .

[4] Haroun, M. A. 1980, "Dynamic analyses of liquid storage tanks", EER Report No. 80-04, Earthquake Engineering Research Laboratory Report, California Institute of Technology, Pasadenas, CA.

[5] El Damatty, A. A., Korol, R. M., and Tang, L. M., 2000, "The sloshing response of conical tanks", Proceedings of World Conference of Earthquake Engineering, Paper No. 966, Topic No. 7, New Zealand.

[6] El Damatty, A. A., Korol, R. M., and Mirza, F. A., 1997a, "Stability of Imperfect Conical Tanks under Hydrostatic Loading, Journal of Structural Engineering", ASCE, Vol(123-6), pp. 703-712.

[7] Halchuk, K.S., and Adams, J., 2004, "Deaggregation of Seismic Hazard for Selected Canadian Cities", 13th World Conference on Earthquake Engineering, Vancouver, B.C., Canada, paper No. 2470

[8] National Building Code of Canada 2010, National Research Council of Canada, Ottawa.

[9] Sweedan, A.M., El Damatty, A. A. 2009, "Simplified procedure for design of liquid-storage combined conical tanks", Thin-Walled structures, Vol(47),pp. 750-759.

[10] Koizey, B., Mirza, F.A. 1997, "Consistent thick shell element", Computer & Structures, Vol(65-12), pp.531-541.

[11] El Damatty, A. A., Korol, R. M., and Mirza, F. A., 1997c, "Large displacement extension of consistent shell element for static and dynamic analysis", Computers and Structures, Vol(62-6), pp. 943-960.

An algorithm for optimal seat selection for heavy mobile machinery in different industrial vibration environments

Xiaoxu Ji, James P. Dickey
Kinesiology/Faculty of Health Science Western University
London, Ontario, Canada
xji23@uwo.ca, jdickey@uwo.ca

Tammy R. Eger
School of Human Kinetics
Laurentian University
Sudbury, Ontario, Canada
teger@laurentian.ca

ABSTRACT

Low back pain is a major health issue among heavy machine operators. Seat selection is important for reducing vibration exposure. An efficient neural network approach modeled the vibration attenuation properties for five suspension seats. We implemented the models to predict seatpan vertical accelerations for a variety of skidders from the forestry sector and load-haul-dump (LHD) vehicles from the mining industry. Our results demonstrated that seat selection is not universal. The performance of industrial seats varies between vibration environments.

KEYWORDS: vibration attenuation, whole-body vibration, seat selection, neural network, *A*(8)

INTRODUCTION

Low back pain (LBP) is a major health issue among heavy machine operators over the course of their careers (Bovenzi, 1996; Jack et al., 2008). Whole-body vibration (WBV) is a main factor to cause low back pain (LBP) while operators are exposed to vibrations with a sitting posture (Johanning, 2011). Forestry and underground mining are two industrial sectors that have a high rate of low back injuries (Hagen et al., 1998; Burgess-Limerick, 2005).

Seats are one of the easiest engineering controls to influence the vibration exposures transmitted to the heavy machine operators (Griffin et al., 2006; Gunaselvam and Van Niekerk, 2005). Due to the frequency-dependent properties of the suspension system, industrial seats can amplify the vibration exposures if they are adjusted incorrectly or selected inappropriately (Paddan and Griffin, 2002; Gunaselvam and Van Niekerk, 2005). A seat selection algorithm would be helpful for optimizing the choice of industrial seats for the specific vehicles and workplaces.

Neural networks (NN) can efficiently model complex nonlinear systems by learning from the system's input (e.g. chassis acceleration data) and output signals (e.g. seatpan acceleration data) to optimize the predictions. They do not require the parameters (e.g. stiffness, damping coefficient, etc) like some other biomechanical models (Cho and Yoon, 2001). Therefore the purpose of this paper was to develop and validate a NN model that can be used to select seats for forestry and underground mining vibration environments.

METHODOLOGY

Five industrial seats (Access; Amobi; CAT; KAB 301; KAB 525) were tested in this project. Each seat was mounted to the top surface of a 6-degree of freedom (df) robotic platform (R3000, Mikrolar Inc. Hampton, NH, USA) and the suspension was appropriately adjusted to its mid-travel position. Two 6-df inertial measurement units (IMUs; MechTrack – Analog version, Mechworks Systems Inc., Vancouver, BC, Canada) were used to record translational accelerations and rotation velocities at the seatpan/operator interface and the center surface of the robotic platform (chassis). Ten subjects with a range of anthropometrics were tested to develop the NN model for the industrial seats.

The 5-layer neural identification network model was developed for this application and is shown in Figure 1. The inputs (x_{xi}, x_{yi}, x_{zi}) were the recorded time series translational chassis acceleration data and the *BMI prime* for each subject. The *BMI prime* is a normalized parameter reflecting the subject's anthropometrics (1/25 * body mass/height^2; Gadzik, 2006). In the 2nd layer, each neuron acted as a band pass filter. The root mean square (r.m.s.) values of the time series chassis acceleration data were calculated in the 3rd layer. The vertical r.m.s. accelerations of seatpan were predicted in the 4th layer. All the optimal weighting parameters (w_j, w_c, w_{bp}, b_1, b_2, b_{bp}) were determined through system identification process by matching the predicted outputs to the corresponding measured values in the 5th layer.

Random broad-band vibration signals were generated with r.m.s. amplitudes between 0.2 and 2.0 m s^{-2} and with uniform flat spectra between 0.5-20 Hz on all three translational axes, and were used to develop the NN model for each seat. Thirty selected field vibration profiles, representing the most common vibration characteristics of the field exposures (Dickey et al., 2010), were used to validate each NN seat model. Each participant was tested using seven random vibration exposures, and ten randomly selected field exposures from the set of 30. Each seat model was evaluated by correlating the observed and predicted seatpan accelerations.

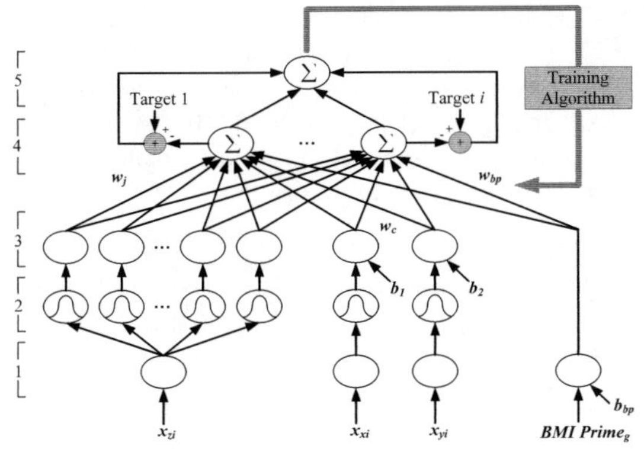

Figure 1: 5-layer neural network architecture with *BMI Prime* values and chassis r.m.s. accelerations in X,Y and Z directions as inputs, and the predicted seatpan r.m.s. accelerations as outputs.

We implemented each of our seat models to predict the daily 8-hour equivalent frequency weighted accelerations (ISO 2631-1, 1997), *A*(8), for eight skidders and ten LHD vehicles (Jack et al, 2010; Eger et al, 2011). Three specific *BMI Prime* values (0.74, 1.00 and 1.20) were selected to reflect the influence of machine operator anthropometrics. The weighted r.m.s. accelerations of seatpan in the vertical direction were used to evaluate the health risk for operators and to identify the most suitable seat for different working environments.

RESULTS

The attenuation characteristics of five industrial seats were successfully modeled using a NN algorithm applied to broadband random vibration exposures, and validated the models using field profiles (Figure 2). The coefficients of determination (r^2) between the measured and predicted seatpan r.m.s. accelerations for both the training and validation profiles ranged between 0.93-0.99. The slopes of linear regression equations were also close to 1.0 and the intercepts were close to zero (Figure 2).

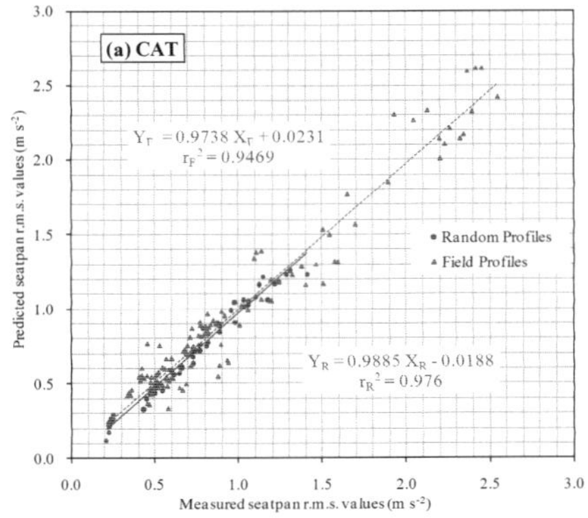

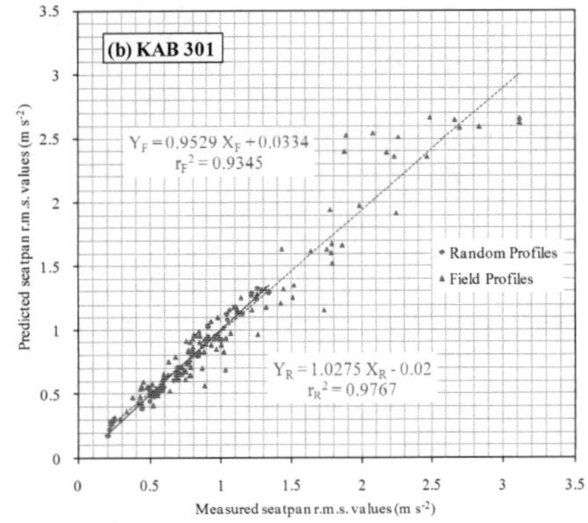

Figure 2: Relationship between the measured and predicted seatpan weighted r.m.s. acceleration from random profiles that were used to generate the NN models (blue circle symbols), and from field profiles that were used to validate the NN models (red triangle symbols). The regression equations and coefficients of determination are showed for the random and field trials for the CAT (a) and KAB 301 (b) seats. The other seats showed similar goodness of fits.

The predicted *A*(8) accelerations varied substantially between industrial seats. In forestry environment, only one of the industrial seats (CAT) attenuated 96% (23/24) of the vibration exposures below the upper limit of the ISO 2631-1 HGCZ (0.9 m s^{-2}; Table 1). In mining environment, the KAB301 seat reduced the vibration magnitude (14 of the 30 vehicle and driver combinations, Table 2) to levels that could be tolerated for more than eight hours (ISO 2631-1, 1997). This proportion of exposures below the eight hour threshold was greater than the other four seats.

Table 1. $A(8)$ results for five industrial seats with different driver anthropometrics (*BMI primes*) in the forestry environments. Values of A(8) that are below the upper limit of the ISO 2631-1 HGCZ (0.9 m s^{-2}) are identified in bold, and represent exposures that can be tolerated for at least eight hours.

Skidders	*BMI Prime*	$A(8)$ (m/s^2)				
		Access	Amobi	CAT	KAB301	KAB525
S1	0.74	**0.8**	0.99	**0.68**	**0.81**	**0.87**
	1	**0.76**	0.9	**0.61**	**0.75**	**0.83**
	1.2	**0.73**	**0.83**	**0.55**	**0.7**	**0.8**
S2	0.74	**0.8**	0.95	**0.7**	**0.79**	**0.88**
	1	**0.76**	**0.86**	**0.63**	**0.73**	**0.84**
	1.2	**0.73**	**0.79**	**0.57**	**0.69**	**0.81**
S3	0.74	**0.79**	0.98	**0.81**	0.9	**0.86**
	1	**0.75**	**0.89**	**0.74**	**0.84**	**0.82**
	1.2	**0.72**	**0.82**	**0.69**	**0.8**	**0.79**
S4	0.74	1.34	1.63	0.97	1.37	1.58
	1	1.3	1.54	0.9	1.32	1.54
	1.2	1.27	1.48	**0.84**	1.27	1.51
S5	0.74	1.11	1.33	**0.86**	1.13	1.3
	1	1.07	1.24	**0.79**	1.07	1.26
	1.2	1.04	1.17	**0.73**	1.02	1.23
S6	0.74	1.03	1.3	**0.69**	1.08	1.21
	1	0.99	1.21	**0.62**	1.02	1.17
	1.2	0.95	1.14	**0.57**	0.98	1.14
S7	0.74	**0.87**	0.99	**0.77**	**0.84**	0.94
	1	**0.83**	0.9	**0.69**	**0.78**	0.9
	1.2	**0.8**	**0.83**	**0.64**	**0.74**	**0.87**
S8	0.74	0.94	1.17	**0.66**	0.99	1.08
	1	0.9	1.07	**0.59**	0.93	1.04
	1.2	**0.87**	1	**0.53**	**0.88**	1.01

Table 2. $A(8)$ results for five industrial seats with different driver anthropometrics (*BMI primes*) in the underground mining environments. Values of A(8) that are below the upper limit of the ISO 2631-1 HGCZ (0.9 m s^{-2}) are identified in bold, and represent exposures that can be tolerated for at least eight hours.

LHDs	*BMI Prime*	$A(8)$ (m/s^2)				
		Access	Amobi	CAT	KAB301	KAB525
M1	0.74	1.10	1.20	1.18	**0.80**	0.96
	1.00	1.06	1.11	1.10	**0.74**	0.92
	1.20	1.03	1.04	1.05	**0.70**	**0.89**
M2	0.74	1.04	1.15	1.21	**0.80**	0.92
	1.00	1.00	1.06	1.14	**0.74**	**0.88**
	1.20	0.97	0.99	1.09	**0.69**	**0.85**
M3	0.74	1.11	1.22	**0.89**	0.92	1.16
	1.00	1.08	1.14	**0.82**	**0.86**	1.13
	1.20	1.05	1.07	**0.77**	**0.82**	1.10
M4	0.74	**0.60**	**0.69**	**0.50**	**0.66**	**0.69**
	1.00	**0.56**	**0.60**	**0.44**	**0.60**	**0.65**
	1.20	**0.53**	**0.53**	**0.39**	**0.56**	**0.62**
M5	0.74	1.89	1.88	1.46	1.39	1.84
	1.00	1.85	1.79	1.39	1.33	1.81
	1.20	1.82	1.73	1.34	1.29	1.78
M6	0.74	1.54	1.58	1.39	1.15	1.43
	1.00	1.50	1.49	1.32	1.09	1.40
	1.20	1.47	1.42	1.27	1.05	1.37
M7	0.74	1.69	1.73	1.60	1.28	1.52
	1.00	1.65	1.64	1.53	1.22	1.49
	1.20	1.63	1.58	1.48	1.18	1.46
M8	0.74	1.24	1.30	1.15	**0.87**	1.12
	1.00	1.20	1.21	1.08	**0.81**	1.08
	1.20	1.16	1.14	1.03	**0.77**	1.05
M9	0.74	1.80	1.81	1.46	1.49	1.75
	1.00	1.76	1.73	1.39	1.43	1.71
	1.20	1.73	1.66	1.34	1.39	1.68
M10	0.74	1.56	1.63	1.58	1.45	1.47
	1.00	1.52	1.54	1.51	1.39	1.44
	1.20	1.49	1.47	1.45	1.35	1.41

DISCUSSION

The attenuation characteristics of each industrial seat were successfully identified based on a NN concept model. This enabled us to predict the r.m.s. magnitude of the seatpan vertical acceleration for different chassis accelerations and different subject anthropometrics. The strong fits in the both random and field data, with slopes of linear regression equations that are close to 1.0 and intercepts close to 0.0, enhance our confidence that the NN models are robust and can be applied to a broad variety of field vibration exposures. We evaluated the performance of five industrial seats based on the predicted seatpan r.m.s. accelerations for each of eight specific skidders and ten LHD vehicles, and three variations of driver anthropometrics. Among the seats, our predictions indicate that the CAT seat was the best choice for the forestry skidders, and the KAB301 seat was the best choice in the LHDs in the mining environment because they usually had lower *A*(8) values than the other seats. Also, they will allow heavy machine operators to work more than 8 hours to reach the upper border to the HGCZ (ISO 2631-1, 1997). These findings affirm that seat selection is not universal - the performance of industrial seats varies between vibration environments. It is helpful to evaluate seat performance in specific environments to provide design information for the seat manufacturers (such as selecting the stiffness, the damping coefficients etc). In terms of limitations, we acknowledge that these findings are based on a relatively small number of subjects for the human testing, specific vibration profiles from two industrial sectors, and a relatively small number of industrial seats. We recommend continued efforts to develop appropriate tools for the important goals of evaluating the responses and health risks of heavy machine operators with multi-axis vibration, and the development of seat selection algorithms.

ACKNOWLEDGMENTS

We acknowledge funding from the Ontario Ministry of Labour, the Workplace Safety and Insurance Board of Ontario and the Natural Science and Engineering Research Council of Canada. We acknowledge infrastructure support from the Canadian Foundation for Innovation.

REFERENCES

1. Bovenzi, M. 1996. Low back pain disorders and exposure to whole-body vibration in the workplace. Elsevier

2. Burgess-Limerick, R. 2005. Reducing injury risks associated with underground coal mining equipment. *Ergonomics Australia*. 19, 14-20.

3. Cho, Y., Yoon, Y. S. 2001. Biomechanical model of human on seat with backrest for evaluating ride quality. *Int.J.Ind.Ergon.* 27, 331-345.

4. Dickey, J. P., Eger, T. R., Oliver, M. L. 2010. A systematic approach for studying occupational whole-body vibration: A combined field and laboratory based approach. *Work* 35, 15-26.

5. Eger, T., Stevenson, J. M., Grenier, S., Boileau, P. M. and Smets, M. P. 2011a. Influence of vehicle size, haulage capacity and ride control on vibration exposure and predicted health risks for LHD vehicle operators. *Journal of Low Frequency Noise, Vibration and Active Control*. vol. 30, pp. 45-62.

6. Gadzik, J. 2006. " How much should I weigh?"--Quetelet's equation, upper weight limits, and BMI prime. *Connecticut medicine* 70, 81-88.

7. Griffin, M. J., Howarth, H. V. C., Pitts, P. M., Fischer, S., Kaulbars, U., Donati, P. M., Bereton, P. F., 2006. Guide to good practice on whole-body vibration. European Commission Directorate General Employment, Social Affairs and Equal Opportunities (contract VC/2004/0341), Brussels.

8. Gunaselvam, J. and Van Niekerk, J. L. 2005. Seat selection guidelines to reduce whole-body vibration exposure levels in the SA mining industry. *Journal-South African Institute of Mining and Metallurgy*. 105, 675-686..

9. Hagen, K. B., Magnus, P., Vetlesen, K. 1998. Neck/shoulder and low-back disorders in the forestry industry: relationship to work tasks and perceived psychosocial job stress. *Ergonomics* 41, 1510-1518.

10. ISO 2631-1. 1997. Mechanical Vibration and Shock: Evaluation of Human Exposure to Whole-body Vibration. Part 1, General Requirements: International Standard ISO 2631-1: 1997 (E). ISO

11. Jack, R. J., Oliver, M., Cation, S., Dony, R. 2008. Tri-Planar Trunk Motion in Northern Ontario Skidder Operators. *International Journal on Industrial Risk Engineering* 1, 1-20.

12. Jack, R. J., Oliver, M., Dickey, J. P., Cation, S., Hayward, G., Lee-Shee, N. 2010. Six-degree-of-freedom whole-body vibration exposure levels during routine skidder operations. *Ergonomics* 53, 696-715.

13. Johanning, E. 2011. Diagnosis of whole-body vibration related health problems in occupational medicine. *Journal of Low Frequency Noise, Vibration and Active Control*. 30, 207-220.

14. Paddan, G. S. and Griffin, M. J. 2002. Effect of seating on exposures to whole-body vibration in vehicles. *Journal of Sound and Vibration*. 253, 215-241.

Proceedings of the 25th CANCAM
London, Ontario, Canada, May 31 – June 4, 2015

ASYMMETRIC INFLUENCES ON THE ENERGY DISSIPATION PROPERTIES OF A METASTABLE MODULE EXCITED NEAR RESONANCE

Narayanan Kidambi*, R.L. Harne, and K.W. Wang
Department of Mechanical Engineering
University of Michigan
Ann Arbor, MI, USA
***email: kidambi@umich.edu**

ABSTRACT

Recently, researchers have explored the development of structural/material systems with large and adaptable damping realized by strategically designed negative stiffness inclusions or bistable elements. From the viewpoint of system integration, bistable, negative stiffness elements often interface with positive stiffness elastic members. In such cases, the system may exhibit coexisting metastable states, where the macroscopic displacement remains fixed while the reaction force varies due to internal change. Much remains to be understood regarding the dynamics and energy dissipation characteristics of such systems when excited near resonance, where nonlinear dynamics are more easily activated and damping design is particularly important. Thus, to explore the versatility of damping properties afforded by exploiting negative stiffness inclusions in structural/material systems, this research investigates a mechanical module which leverages a coexistence of metastable states: an archetypal building block for system assembly. The studies undertake numerical and experimental efforts to examine how the static load on the module, an asymmetry-inducing feature, influences the activation of distinct dynamic response regimes each resulting in distinct energy dissipation characteristics. Orders of magnitude in energy dissipation adaptability are observed by transitioning amongst the dynamic regimes induced in consequence to the near resonant excitations.

INTRODUCTION

Many modern engineering applications call for structures and materials exhibiting enhanced performance, including high and adaptable levels of damping and energy dissipation. For example, large energy dissipation may be required when a structure undergoes sudden, extreme loading, while low damping during regular operating conditions is beneficial to effectively transmit vibrational signals for system identification and damage detection. Recently, researchers have investigated the potential for negative stiffness structural/material inclusions to achieve large damping and energy dissipation, by exploiting energetic transitions between stable equilibria; a phenomenon known as *snap-through*. It has been shown that the integration of positive and negative stiffness elements may be tailored to effect variation in macroscopic structural/material properties [1]. Stability criteria for viscoelastic composite media having a negative stiffness phase have been derived, providing guidelines for material selection to achieve extreme damping [2]. A system of parallel columns under low frequency harmonic loads has been shown to result in large damping when the weaker columns buckle [3]. Researchers have also demonstrated that a bistable von-Mises truss exhibits large change in damping as harmonic excitation frequency varies [4] and energy dissipation adaptation in a compressed mass-spring bistable device was shown to be influenced by initial conditions of displacement and velocity [5].

The previous works have highlighted the versatility in energy dissipation afforded by bistable, negative stiffness elements. However, from a system integration standpoint, these bistable elements are often connected to elastic, positive stiffness members [2]. Hence a fundamental, one-dimensional, model of the integration may be of a bistable spring connected to a linear positive stiffness spring in series. This module exhibits distinct mechanical properties as compared to an individual bistable device. An individual bistable element may exhibit one of two unique displacements for a given reaction force, but only one possible reaction force for a given displacement. In contrast, a bistable element in series with a linear spring may exhibit two unique global displacements for a given reaction force, as well as two possible reaction forces for a given global displacement when the linear spring stiffness is below a critical value [6]. The range of global displacements for which there exists two distinct reaction forces may be referred to as the metastability range, and a module exhibiting this behavior is known as a metastable module in this research. While the static and low-frequency dynamic responses of such modules have been explored [1], much remains to be understood when the modules are excited near resonance. These are excitations at frequencies near the linearized resonances of the locally stable states where snap-through oscillations may be more easily influenced and damping change has great importance. Additionally, the role of asymmetries in the loading remains unclear and is in need of focused investigation. Therefore, this research explores an experimental metastable module and develops a one degree-of-freedom (DOF) model to examine its dynamic characteristics under near resonant excitations with and without asymmetric effects.

EXPERIMENTAL SETUP, MODEL FORMULATION AND GOVERNING EQUATION

The experimental metastable module is shown in Fig. 1. A thick polycarbonate base supports arms connected to a pair of thin spring steel suspension beams. At the middle of each of these suspension beams is a bearing that connects to a rotating arm. The other ends of the arms are pinned together by an additional bearing. The distance between suspension beams is less than the overall length of the two rotating arms, resulting in a bistable element with two statically stable configurations. The device is able to snap between its two stable states due to the compliance provided by the suspension beams. A pair of antagonistic linear springs joins the pin connecting the rotating arms to two bolts protruding from the bottom of a U-channel. The U-channel is rigidly attached to a controlled APS-400 electrodynamic shaker whose motion is measured by a PCB 352 C33 accelerometer. The angular displacement of the rotating arms is measured by a Novotechnik RFC 4801 Hall effect sensor.

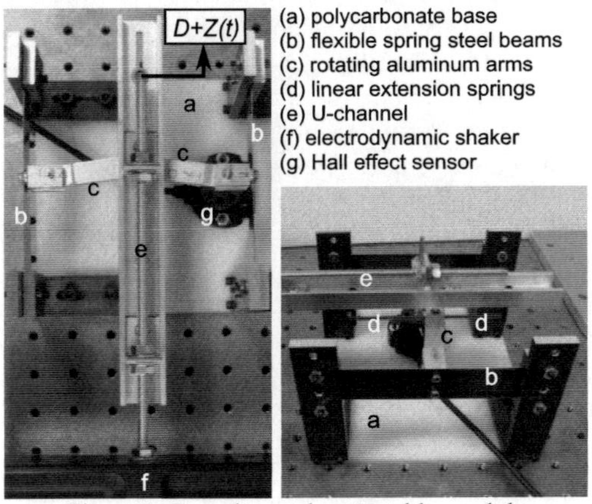

Figure 1. Experimental metastable module

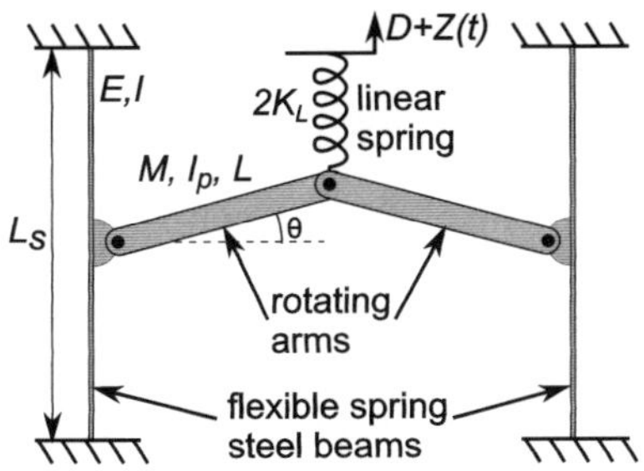

Figure 2. Schematic of experimental metastable module.

The governing equation of motion for the moving internal mass of the metastable module is derived according to the schematic representation of the experimental system shown in Fig 2. Assuming all components except for the rigid rotating arms are massless, the total kinetic energy is written to be

$$T = 2\left(\frac{ML^2}{8} + \frac{I_p}{2}\right)\dot{\theta}^2 \qquad (1)$$

Due to geometric constraints, both arms share the same angular displacement $\theta(t)$. M is the mass of each rotating arm with length L and moment of inertia I_p. The potential energy stored in the system may be expressed as

$$U = 2\frac{192EI}{L_s^2}\left[1 - \theta\sin(\theta) - \cos(\theta) + \frac{1}{2}\theta^2\cos(\theta_0)\right] + K_L[Z + D - L\sin(\theta)]^2 \qquad (2)$$

where E is the Young's modulus of the suspension beam, I is the area moment of inertia about the neutral bending axis, and L_s is the length of each beam between clamps. θ_0 is the stable equilibrium angular position of the rotating arms in the absence of a connection to the linear springs. K_L is the stiffness of each linear tension spring, while D refers to the static offset of the periodic excitation $Z(t) = Z_0\cos(\Omega t)$ from the neutral position of the bistable device. The offset D is analogous and mathematically equivalent to a static load added upon the metastable module, for example if serving as a vibration isolator. Consequently, non-zero offsets D induce asymmetry in the loading/excitation conditions on the module. Dissipation is approximated by viscous damping on $\dot{\theta}$ with coefficient B, which is determined by fitting numerical results to experimentally obtained data of a free decay response. Lagrange's equations yield the governing equation for $\theta(t)$ due to end displacement motions $Z(t)$:

$$\left(\frac{ML^2}{2} + 2I_p\right)\ddot{\theta} + B\dot{\theta} + 2K_L L^2\sin(\theta)\cos(\theta) - \frac{384EI}{L_s^2}\theta[\cos(\theta) - \cos(\theta_0)] - 2K_L L\,(Z + D)\cos(\theta) = 0 \qquad (3)$$

In this study, the energy dissipated per excitation cycle is the measure of interest, and is computed by calculating the area enclosed by hysteresis loops of end displacement and reaction force. The reaction force due to elastic and dissipative effects is captured by the force due to the deflection of the linear springs.

$$F(t) = 2K_L[Z + D - L\sin(\theta)] \qquad (4)$$

EXPERIMENTAL RESULTS

To investigate the influence of excitation offset D (i.e. different static loads) and the near resonant excitation frequencies, experiments are conducted using very slow excitation frequency sweep rates of ± 0.05 Hz/s. The system parameters in Table 1 are also used to numerically integrate the governing equation (3) and reconstruct the dynamic reaction force, eq. (4), for comparison to measured trends.

Table 1. Experimental system parameters used in numerical simulations.

M (g)	I_P(kg.m^2)	L (cm)	E (GPa)	I (m^4)
12	1.05	5.08	180	5.46×10^{-15}
L_s (cm)	**θ_0 (°)**	**K_L(N.m^{-1})**	**B(N.s.m^{-1})**	**Z_0**
10.02	±10	84	1.5×10^{-4}	300 μm

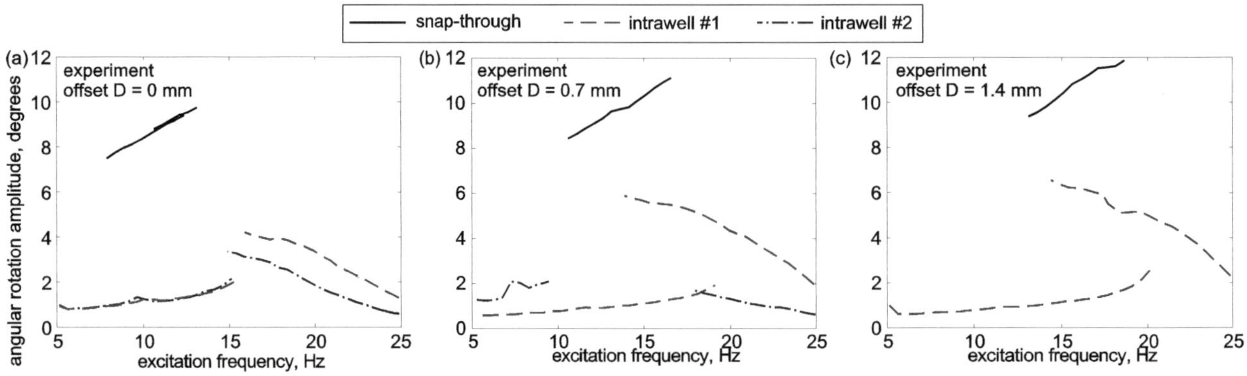

Figure 3. Experimentally measured angular rotation amplitude of the rigid arms as the excitation frequency is varied, for excitation offsets (a) $\boldsymbol{D} = 0$ mm, (b) $\boldsymbol{D} = 0.7$ mm, and (c) $\boldsymbol{D} = 1.4$ mm.

Fig. 3 presents measurements carried out for fixed excitation amplitude of $Z_0 = 300$ μm across the range of frequencies considered. Three excitation offset configurations are examined as shown in Fig. 3(a): $D = 0$ mm, (b) $D = 0.7$ mm, and (c) $D = 1.4$ mm. For the offsets $D = 0$ mm and 0.7 mm and with no harmonic excitation, the bistable constituent of the metastable module yields a double-well potential energy profile and thus induces two statically stable internal configurations of the module (metastable states). For clarity, the two local wells of potential energy are referred to as *well #1* and *well #2.* Two distinct types of oscillations are observed when the system is subjected to harmonic excitation: *intrawell* and *snap-through. Intrawell* responses are those for which the internal mass oscillates around a statically stable metastable state, and are referred to as *intrawell #1* or *intrawell #2* according to the local potential well in which the responses are observed. Due to the near-resonant excitation frequencies, two different intrawell-type responses are uncovered for each potential well in Fig. 3(a) when the offset is $D = 0$ mm: *low* and *high amplitude* intrawell. For some excitation frequencies, *snap-through* dynamics are revealed, wherein the internal mass of the metastable module vibrates across the unstable equilibrium of the internal bistable spring, thus leading to greater amplitude internal motions than the intrawell-type responses. For the excitation offset $D = 0$ mm shown in Fig. 3(a), both local potential wells are statically stable and lead to intrawell dynamics within one or the other well across the full frequency spectrum considered. On the other hand, snap-through vibrations are achievable in the range of frequencies of approximately 7 to 13 Hz as shown in Fig. 3(a).

With a nonzero offset $D = 0.7$ mm, the metastable module is biased towards well #1. This behavior is seen in Fig. 3(b) where the asymmetry results in a greater bandwidth of frequencies across which the intrawell #1 responses occur than those oscillations remaining in well #2. When the offset is further increased to $D = 1.4$ mm, well #2 is no longer statically stable, thus no intrawell type oscillations are observed in well #2. In fact, under this condition the module is comparable to a strongly asymmetric Duffing oscillator which also exhibits at most three dynamic regimes including low- and high-amplitude intrawell behaviors as well as a still higher amplitude dynamic similar to a snap-through feature [7]. Snap-through type oscillations shown in Fig. 3(c) are those for which the bistable element crosses its neutral position. The trends exemplified in Fig. 3 clearly demonstrate that the existence of dynamic regimes depends strongly on excitation frequency and offset conditions. The activation of numerous distinct dynamic responses from the near-resonant excitations suggests that the energy dissipation characteristics are also distinct and thus a transition from one regime to another -- all other factors the same -- would serve as a useful strategy for adaptable energy dissipation.

Fig. 4 plots the dynamic reaction force $F(t)$ as defined in eq. (4) against global end displacement $Z(t)$ for measured and simulated results taken at an excitation frequency of 16 Hz and range of excitation offsets considered above. Since the excitation and responses are periodic, the force-displacement trajectories form loops. The area enclosed by each of the loops is the work done by the reaction force and is equal to the energy dissipated by the module over a single excitation period. The hysteresis loops in Fig. 4 shown by solid lines are plotted over the static force displacement curves shown by dashed lines. Figs. 4(a) and (d) show experimental and simulated results, respectively, obtained at offset $D = 0$ mm. From Fig 3(a), two high amplitude intrawell responses are observed at the 16 Hz excitation frequency, one in each potential well. This is reflected by the hysteresis loops in Figs. 4(a) and (d), which exist near the two static equilibrium positions. Simulations predict smaller degrees of energy dissipation than those experimentally measured, a discrepancy attributed to unmodeled stick-slip friction in the experimental module, which has a more pronounced effect on lower amplitude intrawell oscillations than on the higher amplitude snap-through responses. Figs. 4(b) and (d) present experimental and simulated hysteresis loops, respectively, using the offset $D = 0.7$ mm. Stable low and high amplitude oscillations in well #1 are uncovered by the hysteresis loops, as well as a snap-through response which is the loop enclosing a substantially greater area. In fact, there is a 44 times difference in energy dissipation from the low amplitude intrawell to the snap-through responses. At the excitation offset $D = 1.4$ mm, the frequency domain profiled in Fig. 3(c) shows the same qualitative range of

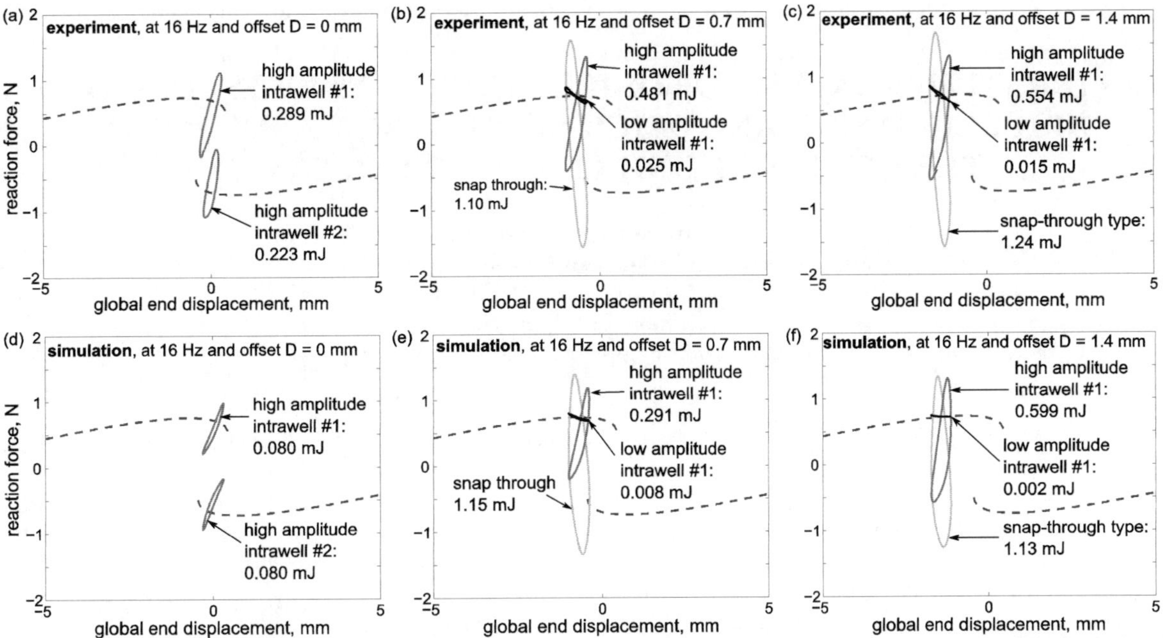

Figure 4. Hysteresis loops showing energy dissipated per cycle computed as the area enclosed by the loops, at 16 Hz frequency. (a-c) Experimental and (d-f) simulated results are shown. Loops plotted over static force-displacement profile (dashed curves)

dynamic behaviors as those observed when D = 0.7 mm, which is reflected in the loops characterized in Figs. 4(c) and (e). Under these excitation conditions and offset-induced asymmetries, a similarly large adaptation in energy dissipation performance is available, based on the substantial range in areas enclosed by the hysteresis loops.

The results in Fig. 4 show that the static excitation offset has significant influence on the energy dissipation adaptability. For a given excitation amplitude, different asymmetric loading conditions induced by the offset affect the bandwidth of frequencies for which specific response regimes are observed as well as the magnitude of the energy dissipation each regime produces. Modulation of the excitation offset also enables smooth variation in the energy dissipation as shown amongst the various high amplitude intrawell loops in Figs. 4(a-c). Consequently, the change in excitation offset is found to critically govern the energy dissipation adaptability, which helps guide effective deployment of such metastable modules whether employed individually or assembled into adaptive structures/materials.

CONCLUSION

This paper explores the unique dynamic and energy dissipation characteristics of a metastable module excited at frequencies near resonance, with specific focus on the influence of a static offset on energy dissipation performance. The archetypal module considered here represents a fundamental unit for the integration of negative stiffness bistable elements within an overall structural/material system for enhanced damping characteristics. It is found that tailoring the excitation offset has significant effect on energy dissipation characteristics, due to inducing asymmetric static mechanical properties that govern the realization of particular dynamic energy dissipation regimes. Modulation of static offset may also allow for smooth variation in energy dissipation if a given response regime is maintained. By understanding these sensitivities, large adaptation of damping features may be leveraged in individual metastable modules or structures/materials assembled from such building blocks.

REFERENCES

1. Klatt, T. and Haberman, M.R., 2013, "A nonlinear negative stiffness metamaterial unit cell and small-on-large multiscale material model," Journal of Applied Physics, **114**, p. 033503.
2. Kochmann, D.M., 2014, "Stable extreme damping in viscoelastic two-phase composites with non-positive-definite phases close to the loss of stability," Mechanics Research Communications, **58**, pp. 36-45.
3. Dong, L. and Lakes, R.S., 2013, "Advanced damper with high stiffness and high hysteresis damping based on negative structural stiffness," International Journal of Solids and Structures, **50**, pp. 2416-2423.
4. Barbarino, S., Pontecorvo, M.E., and Gandhi, F.S., 2012, "Energy dissipation of a bi-stable von-Mises truss under harmonic excitation," *54th AIAA/ASME/ASCE/AHS/ASC Structures, Structural Dynamics, and Materials Conference*, Honolulu, HI, USA, pp. AIAA 2012-1712.
5. Johnson, D.R., Thota, M., Semperlotti, F., and Wang, K.W., 2013, "On achieving high and adaptable damping via a bistable oscillator," Smart Materials and Structures, **22**, p. 115027.
6. Wu, Z., Harne, R.L., and Wang, K.W., 2014, "Muscle-like characteristics with an engineered metastructure," *ASME 2014 Conference on Smart Materials, Adaptive Structures, and Intelligent Systems*, Newport, RI, USA.
7. Kovacic, I., Brennan, M.J., and Lineton, B., 2009, "Effect of a static force on the dynamic behaviour of a harmonically excited quasi-zero stiffness system," Journal of Sound and Vibration, **325**, pp. 870-883.

VIBRATION ANALYSIS OF DRILL STRING UNDER INCLINED DRILLING INCLUDING STICK-SLIP PHENOMENON WITH THE EFFECT OF DRILLING MUD INJECTION

Mohammad Damshenas
Department of Mechanical Engineering
Sharif University of technology
International campus, Iran
Damshenas1987@yahoo.com

ABSTRACT

Vibration is one of the most influential factors in the design, optimum performance and lifetime of all machinery. Analysis of the vibrations created in drill string like other fields such as rotary drilling is important and due to the length of the drill string whirling is probably one of the most important issues that leads to reduce Penetration rate therefore increases the time and cost of drilling operation…The purpose of this article is to find the effect of variable injection of drilling mud around inclined drill string with Stick – Slip phenomenon and deformation because of whirling. For this purpose, a series of kinetic and potential energy equations for rotary drill string, considering the longitudinal and transverse vibrations plain and Lagrange method to achieve equations of motion are used.. The modeling is done by using the finite element software ANSYS™ and drill string weight and the effect of stick-slip phenomenon is well taken into boundary conditions. Finally, the results with considering drilling mud injection and the effect of Stick – Slip show that this phenomenon is beginning to move in the longitudinal vibrations and then effected in the transverse vibration state.

KEYWORDS: drill string, non-linear vibrations, weight on bit, finite elements, stick-slip phenomenon

INTRODUCTION

The set of equipment for drilling is called drill string. Drill string consists of the hollow cylindrical pipes and various parts such as drill pipe, drill collar, stabilizers, drill bits and hoisting system on the top of the well that its tasks are transmitted torque to the bottom of it (bit blades).

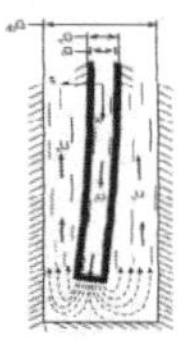

Figure 1: a view from inside of the drill string

DERIVE THE GOVERNING EQUATION OF FLUIDES

Forces exerted from the mud on the drill string can be divided into three categories: viscous, inviscid and friction (between bit and ground) forces. Continuation of investigation into the relationship of each of these forces will be discussed.

Internal Flow Drilling Mud Force: To obtain the relations, a small element of the drill string and internal mud elements which have length of δx, are considered (figure 2).

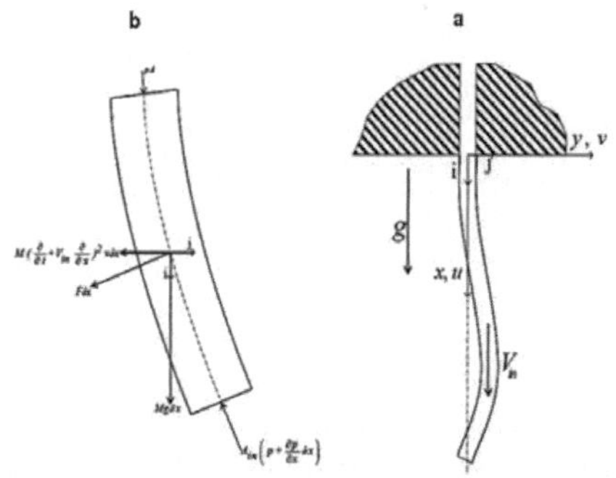

Figure 2 a: Overview of internal mud flow, b: Elements of mud into the drill string

By assuming that the volume of this element is considered to beδD, change of linear momentum within this element is [2]:

$$\frac{dM}{dt} = \iiint_{\delta D} \left[\frac{\partial V}{\partial t} + (V_{in}.\nabla).V_{in}\right]\rho_f dD \qquad)1($$

$$V = V_{in}\hat{\mathrm{I}} + \frac{\partial v}{\partial t}\hat{\mathrm{J}} + \frac{\partial w}{\partial t}\hat{\mathrm{K}} \qquad)2($$

By assuming small mud velocity changes in the radial direction, velocity of mud can be considered as equation 2 [2] that V_{in} mud velocity in element at each moment of the, ρ_f is the density of drilling mud and $\boldsymbol{M}$ is Linear Momentum of mud and in equation 2, V_{in} is mud velocity in drill string and v and w are the movement of drill string in both directions. It can be assumed v, $\frac{\partial v}{\partial t}$ and $\frac{\partial w}{\partial t}$ to be small (also for W) by combining equation 1 and 2 that obtained 3 [2].

$$\frac{d\bar{M}}{dt} = M_{in}\frac{\partial V_{in}}{\partial t}\delta x\hat{\imath} + M_{in}\left(\frac{\partial}{\partial t} + V_{in}\frac{\partial}{\partial x}\right)^2 (v\hat{\jmath} + w\hat{k})\delta x\hat{\jmath}$$)3(

In this equation, M_{in} is the mass per unit length inside the drill string and is equal to:

$$(M_{in})_e = \frac{\pi}{4}\rho_f (d_{in})_e^2$$ 4

About drill string by assuming no change in mud flow rate with time, the first term in right side of equation 2.3 is neglected to obtain the following simplified equation:

$$(F_{in})_e = -(M_{in})_e\left(\frac{\partial}{\partial t} + (V_{in})_e\frac{\partial}{\partial x}\right)^2 (v\hat{\jmath} + w\hat{k})$$)5(

F_{in} is a restoring force that is on the unit of length and by the mud forces to drill string.

External Flow Force of Drilling Mud: Forces acting on the drill string from the external flow mud are divided into 3 categories that two categories in the transverse direction and the third category in the longitudinal direction affect on the drill string [2-4].

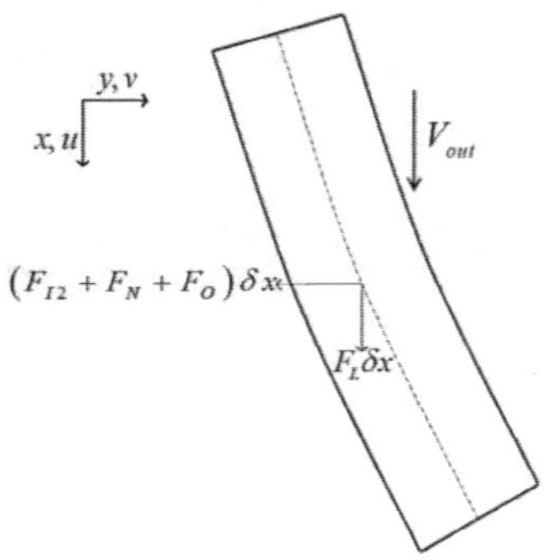

Figure 1: Free body diagram of a drilling string elements and external mud flow and acting forces

In this figure, F_{out} is restoring force in the transverse direction (v) and is on the unit of length that is as acted by mud flow to the drill string and its value is calculated by the equation 6.

$$(F_{out})_e = -(M_{out})_e\left(\frac{\partial}{\partial t} + (V_{out})_e\frac{\partial}{\partial x}\right)^2 (v\hat{\jmath} + w\hat{k})$$)6(

$$(M_{out})_e = \frac{\pi}{4}\rho_f (C_M)_e (d_{out})_e^2$$)7(

In this equation, V_{out} is the external mud flow around the drill string and M_{out} is the extra equivalent mass to mud that is calculated by equation 7 and C_M is the correction coefficient equivalent to mass of external mud.

$$(C_M)_e = \frac{\left(\left(\frac{d_B}{(d_{out})_e}\right)^2 + 1\right)}{\left(\left(\frac{d_B}{(d_{out})_e}\right)^2 - 1\right)}$$)8(

The second force on the drill string by the external mud flow is a drag force in the transverse direction and is on a unit length that is created due to viscosity [2-4]. This force is indicated by F_N and its value is determined by the equation 9[4].

$$(F_N)_e = \frac{1}{2}C_f\rho_f (d_{out})_e (V_{out})_e{}^2 (C_f \sin\bar{\alpha} + C_D(\sin\bar{\alpha})^2)$$)9(

That $\bar{\alpha} = \sin^{-1}\left[\frac{1}{V_{out}}\left(\frac{\partial}{\partial t} + (V_{out})_e\frac{\partial}{\partial x}\right)\right]$ and

$C_f = 0.0125$ 11

$$k = \frac{2\sqrt{2}}{\sqrt{S}}\frac{(1+\bar{\gamma}^3)}{(1-\bar{\gamma}^2)^2}\rho_f A_{out}$$ 11

$$\bar{\gamma} = \frac{d_{out}}{d_B}$$ 12

$$A_{out} = \pi (d_{out})_e L_e$$ 13

The third force from flow mud to drill string is the drag force in longitude direction that is

$$F_{L_i} = \frac{1}{2}C_f\rho_f (d_{out})_e (V_{out}^2)_e \cos\bar{\alpha}$$ 14

The force of mud pressure can be separated into two parts. First, the force is the hydrostatic pressure caused by the weight, so the force obtained from hydrostatic pressure can be modeled by subtracting the buoyancy force from the weight force of the drill string. Ultimately, axial force acting on the drill string in the longitudinal (axial) direction can be expressed as:

$$F_{ax} = (\rho_S - \rho_f)gA$$)16(

However, for cases where flow mud is to have a pressure drop, the resultant force of the pressure drop equation is modeling and has expressed in the equations. For this purpose, used pump power that is injecting drilling mud, since required force to move flow mud and overcome the pressure drop, is supplied by the injector pump.

DERIVE THE GOVERNING VIBRATIONAL EQUATION

In my previous article, the general equations of motion based on energy and Lagrange method was examined that Due to the lack of drilling mud, working force of mud was not calculated. In the present work, according to the drilling mud in the variable flow rate of these forces are of extraordinary importance, therefore, it is necessary to examine the forces that are to be addressed in detail below.

Internal Flow Force of Drilling Mud: To calculate the work done by the transverse force exerted by the mud flow inside the drill string, can be written as [5-7]:

)17(

$$\delta W_{F_{in_y}} = \int_{x_1}^{x_1+L_e} F_{in_y}(\delta v)dx$$

)18(

$$\delta W_{F_{in_z}} = \int_{x_1}^{x_1+L_e} F_{in_z}(\delta w)dx$$

to convert the fluid velocity in the drill string through flow

$$W_{F_{in_y}} = \delta\{v\}_e^T\begin{pmatrix}[M^{vvin}]_e\{\ddot{v}\}_e + Q[C^{vvin}]_e\{\dot{v}\}_e \\ +Q^2[K^{vvin}]_e\{v\}_e\end{pmatrix}$$)19(

)21(

$$\delta W_{F_{in_z}} = \delta\{w\}_e^T\begin{pmatrix}[M^{wwin}]_e\{\ddot{w}\}_e + Q[C^{wwin}]_e\{\dot{w}\}_e \\ +Q^2[K^{wwin}]_e\{w\}_e\end{pmatrix}$$

Restoring Force Work of External Flow: Calculating the work done by the transverse restoring force of flow mud out of drill string is quite similar to the force work of internal flow. So we can write [5-7]:

$$\delta W_{F_{outy}} = \int_{x_1}^{x_1+L_e} F_{out_y}(\delta v)dx \qquad)21($$

$$\delta W_{F_{outz}} = \int_{x_1}^{x_1+L_e} F_{out_z}(\delta w)dx \qquad)22($$

to convert the fluid velocity in the drill string to the flow transmission, it can be concluded:

$$\delta W_{F_{outy}} = \delta\{v\}_e^T \begin{pmatrix} [M^{vvout}]_e\{\ddot{v}\}_e + Q[C^{vvout}]_e\{\dot{v}\}_e \\ +Q^2[K^{vvout}]_e\{v\}_e \end{pmatrix} \qquad)23($$

$$\delta W_{F_{iouty}} = \qquad)24($$

$$\delta\{w\}_e^T \begin{pmatrix} [M^{wwout}]_e\{\ddot{w}\}_e + Q[C^{wwout}]_e\{\dot{w}\}_e \\ +Q^2[K^{wwout}]_e\{w\}_e \end{pmatrix}$$

Longitudinal Forces: Work done by the forces in the longitudinal direction of F_L and F_P longitudinal displacement of the various points of drill string (δu), is calculated as follows [5-7] :

$$\delta W_{F_x} = \int_{x_1}^{x_1+L_e} F_x(\delta u)dx = \delta\{u\}_e^T(Q^2\{K^{uFL}\}_e + Q^2\{K^{uFP}\}_e) \qquad)25($$

Drag Force Work of External Flow: $F_N$is the force from drag due to mud viscosity and is in the transverse direction of the drill string which is given

$$\delta W_{F_{Ny}} = \int_{x_1}^{x_1+L_e} F_{N_y}(\delta v)dx \qquad)26($$

$$\delta W_{F_{Nz}} = \int_{x_1}^{x_1+L_e} F_{N_z}(\delta w)dx \qquad)27($$

The relationship according to equation 9 is obtained as follows:

$$\delta W_{F_{Ny}} = \delta\{v\}_e^T \begin{pmatrix} [M^{vvN}]_e\{\ddot{v}\}_e + Q[C^{vvN}]_e\{\dot{v}\}_e \\ +Q^2[K^{vvN}]_e\{v\}_e \end{pmatrix} \qquad)28($$

$$\delta W_{F_{Nz}} = \delta\{w\}_e^T \begin{pmatrix} [M^{wwN}]_e\{\ddot{w}\}_e + Q[C^{wwN}]_e\{\dot{w}\}_e \\ +Q^2[K^{wwN}]_e\{w\}_e \end{pmatrix} \qquad)29($$

The present work is done by the ANSYS ™ software that due to the presence workbench such as CFX and FLUENT does not have to use equal pressure of such ABAQUS ™ software. Using equal pressure interchangeably mass flow rate in ABAQUS ™ software is because of ALE and it is because solids have Lagrangian treat but fluids have Eulerian treat, wherever we have the interaction between those an advanced topic or the name of the Arbitrary Lagrangian Eulerian Finite Element Method (ALE) comes.

RESULTS

Analyzing the results of the vibration governing equations of the drilling string has occurred in several parts. In the first section, we talked about the boundary conditions that include how to obtain axial forces from the up, applied harmonics rotational forces and how to consider stick-slip phenomena.

In the second section we discussed the stability criteria, factors affecting the instability and the neutral point.

In the third section, the longitudinal and transverse frequencies were obtained and finally in the fourth section, the time response curves were obtained and a general conclusion can be deduced. In table 1, Values used for obtaining the results are presented. These data are related to the model presented by Khulief (except length of drill collar) [2].

Table 1 data of drill string, drilling mud and wells [1]

Drilling pipe Specifications		**Material specifications**	
length	*1000 m*	**drilling string density**	*7850 kg/m³*
Internal diameter	*0.095 m*	**Drilling mud density**	*1500 kg/m³*
External diameter	*0.127 m*	**Modulus of elasticity**	*210 GPa*
Collar drilling specifications		**General Specifications**	
length	*50 m*	Wall wells diameter	*0.3 m*
Internal diameter	*0.0762 m*	Acceleration of gravity	*9.81 m/s²*
External diameter	*0.2286 m*	Slope angle of wells	*0.0524 rad*

Part one: The amount of rotational force applied to the developed model is a harmonic force equal to *H=10+10sin (10πt/3)* that has been used in the previous work. Angular velocity is equal to *120 rpm*.

A tensile axial force from top to the drill string is equal to the drill string that its total weight is obtained according to table 1.

Finally, for the boundary conditions of the bottom of the well, the stick-slip equation is used and programming by micro writing in ANSYS™ is applied. The accuracy of model is confirmed based on mesh analysis and the elements are some types of pipe.

For the boundary conditions of the bottom of the well, the stick-slip equation is used and programming by micro writing in ANSYS™ is applied.

$$\mathrm{WOB} = W_0 + k_f x_0\,(1 - \sin 2\ f\ t) \qquad)31($$

Where f is the frequency of the fluctuation and x_0 is the depth of cut in one bit revolution. The frequency f is related to the depth of cut x_0 and the rate of penetration (ROP) $\dot{x}$, as $f = \dot{x}/x_0$ and WOB is weight on bit. Usually for diagonal drilling, deviation from 3° to 5° is considered. This work only has 3° deviation from the vertical line and the rate of penetration of drill bit is 20 cm per minute and no stabilizer is on collar.

Part two: The main reasons for the instability to vertical direct drilling are three principles:

- High velocity (or in other words, flow rate) of drilling mud
- Drill string buckling under the weight pressure force of **P**
- Drill string angular velocity

For the first case that according to research before drilling mud injected at a rate of several hundred liters per second will cause instability but in actual conditions this amount is less than one hundred liters per second and has an optimal value. The total weight of the drill string has been inhibited by tensile force from top of complex. According to previous research, the drill string is unsustainable because of high angular velocity more than 500rpm and while working angular velocity is between 100 to 300 rpm and in this work the angular velocity is constant and equal to the 120 rpm [6]. As with the definition of neutral point stated weight is collar part that is not drawn from top and this point is not in tension and pressure. Here the total weight censored from up because of two reasons, First, the research on the field before drilling locations on the neutral point 13 and 26 meters are done and second, in fact as a drilling mud weight is always on the drill bit.

Part three: As equation 11, it is clear there is no association among frequencies of longitudinal vibrations of the drill string, the weight force, and drilling mud. This frequency depends on the physical properties of the drill string. According to this explanation, the natural frequency of longitudinal vibrations of drill string for the data presented in Table 2 is obtained from the following table 1. Also for the first 5 transverse vibrations mode dependent on loading and boundary values is given in table (2).

Table 2 natural frequency of drill string by assumption of H=0

		Number of lateral vibration modes(Hz)				
		1	2	3	4	5
Flow rate of drilling mud(m^3/sec)	0	0.097727	0.17127	0.317355	0.433661	0.498512
	0.01	0.097716	0.17125	0.317328	0.433622	0.498468
	0.02	0.0976682	0.171120	0.317249	0.433506	0.498335
	0.03	0.0976627	0.17111	0.317116	0.433311	0.498114
	0.04	0.097549	0.17099	0.31693	0.433039	0.497803
	0.05	0.097449	0.17083	0.316691	0.432689	0.497404

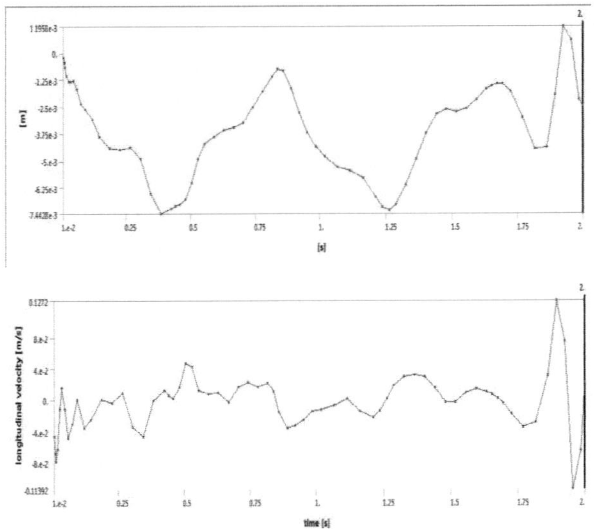

Figure4: longitudinal vibration for X=1000m, =120rpm, Q=10lit/sec

As seen in longitudinal vibration effect on the movement pipe and collar point is softer as well as speed reduction, but not zero, while achieve to the amount will be determined that penetration rate is equal to 20 centimeter per minute.

$$20\frac{Cm}{min}\times\frac{1min}{60sec}\times\frac{1m}{100cm}=0.33\times10^{-2}$$

This seems quite reasonable.

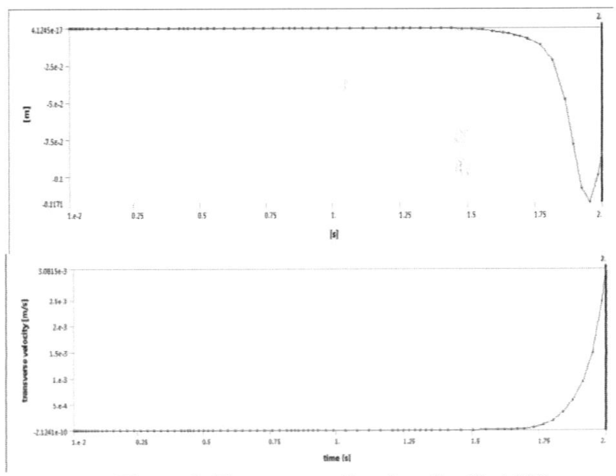

Figure 5: Transverse vibration for X=1000m, =120rpm, Q=10lit/sec

As seen in the presence of the drilling mud transverse vibrations will converge but without injecting drilling mud seemed to diverge.

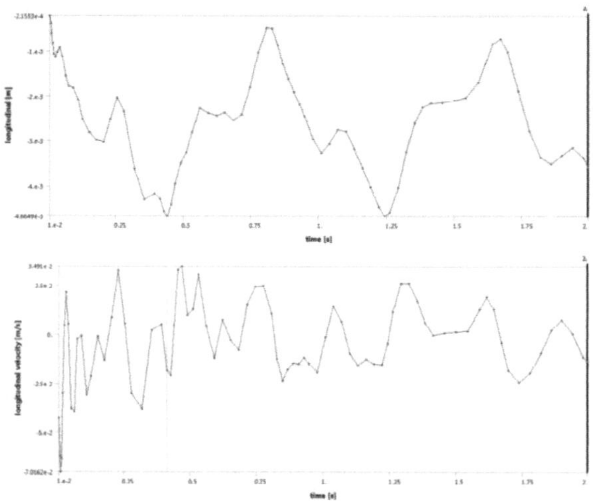

Figure 6: longitudinal vibration for X=1000m, =120rpm, Q=20lit/sec

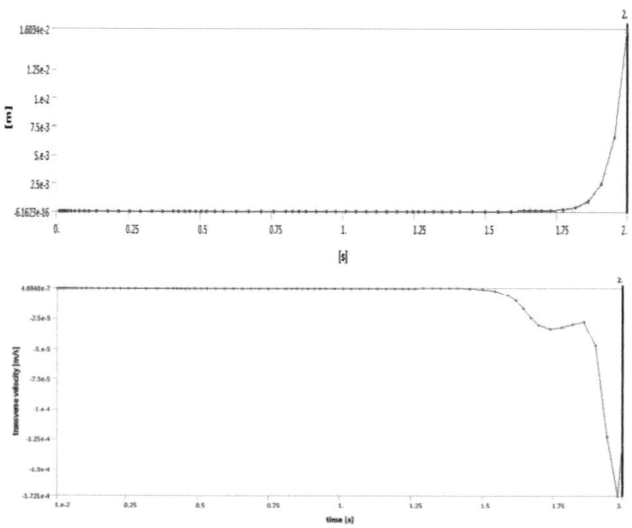

Figure 7: Transverse vibration for X=1000m, =120rpm, Q=10lit/sec

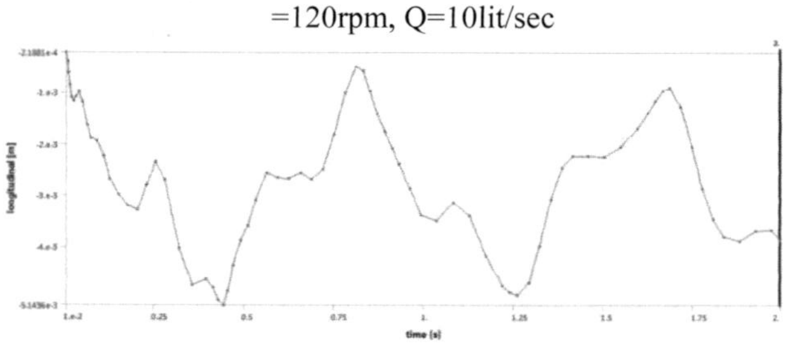

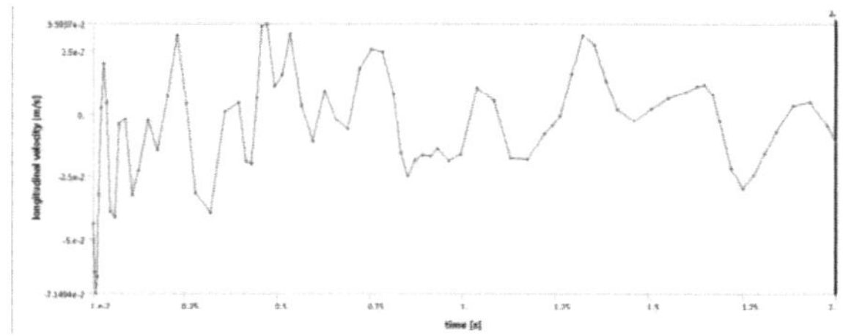

Figure 8: longitudinal vibration for X=1000m, =120rpm, Q=50lit/sec

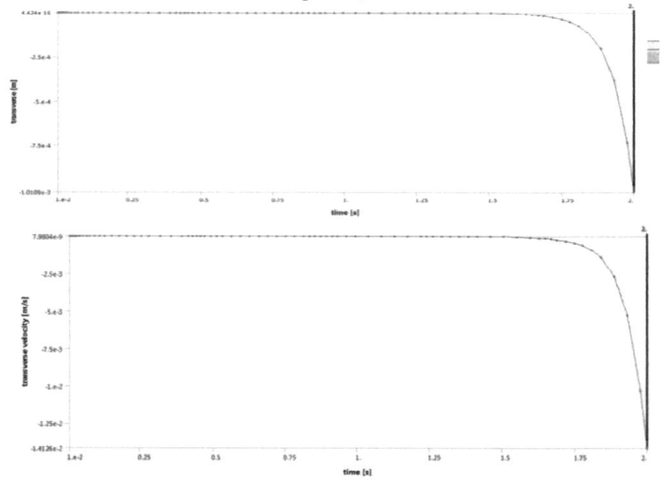

Figure 9: Transverse vibration for X=1000m, =120rpm, Q=50lit/sec

CONCLUSIONS

Drilling mud is extremely important to the damping effect of the system over time to converge on the equilibrium point is that without the injection of drilling mud near the divergence. With 10lit / sec it was seen that the transverse displacement of the point X = 1000m of drill string is returned, and this shows a lack of divergence. The 20 lit / sec of flow rate decreases transverse displacement that the most destructive factor in vibrations and optimal flow of current is optimal, but it is not absolute state and the most important factor for adjusting the flow rate of the drilling is formation pressure. The drill mud flow rates Q=50lit/s or more, we show similar behavior as same as zero mass flow rate (Q=0) that indicates we have optimum mass flow rate and after it increases mass flow rate, it leads to instability. According to table 4 the first transverse mode without injection of drilling mud starts from 0.097727 that the same value for 50 lit/sec decreases to 0.097479. This value is important when we know that the working frequency is less than half (sometimes ⅓ of first mode) of the first mode and this reason causes limitation in working range.

REFERENCES

1. Khulief, Y.A. and H. Al-Naser, *Finite element dynamic analysis of drill string*. Finite Elements in Analysis and Design, 2005. **Volume 41**(13): p. 1270-1288.
2. Paidoussis, M.P. and N.T. Issid, *Dynamic Stability of pipes conveying fluid*. Journal of sound and Vibration and, 1974. **33**(3): p. 267-249.
3. Paidoussis, M.P., *Dynamics of flexible slender cylinders in axial flow Part 1. Theory*. Journal of Fluid Mechanics, 1966. **26**(04): p. 717-736.
4. Paidoussis, M.P., *Dynamics of flexible slender cyliders in axial flow Part 2. Experiments*. Journal of Fluid Mechanics, 1966. **26**(04): p. 737-751.
5. Taheran, F. *Coupled vibration analysis of directional drilling string considering effect of drill mud,* 2012, *M.* Sc. Thesis; Department of Mechanical Engineering, Sharif University of Technology.
6. Faraji Mahyari, M. *lateral vibration analysis of drill string, considering drill mud around it,* 2007, M. Sc. Thesis; Department of Mechanical Engineering, Sharif University of Technology.
7. Damshenas, M. *Vibration Analysis of Drill String under Inclined Drilling Including Stick-Slip Phenomenon,* 2014, M. Sc. Thesis; Department of Mechanical Engineering, Sharif University of Technology, International campus.

Wind Engineering

Proceedings of the 25th CANCAM
London, Ontario, Canada, May 31 – June 4, 2015

APPLICABILITY OF QUASI-STEADY THEORY ON BUILDING POINT PRESSURES

Chieh-Hsun Wu
Boundary Layer Wind Tunnel Laboratory
University of Western Ontario
London, ON, Canada
cwu283@uwo.ca

Gregory A. Kopp
Boundary Layer Wind Tunnel Laboratory
University of Western Ontario
London, ON, Canada
gakopp@uwo.ca

ABSTRACT

This paper presents an assessment of the applicability of quasi-steady theory in relating the instantaneous wind speed and point pressures near the corner of a low-rise building. Synchronized measurements of pressure and velocity were conducted. The results showed that the pressures respond more significantly to the instantaneous wind velocity and azimuth, and less to the wind elevation angle.

KEYWORDS: Building aerodynamics, wind loads, quasi-steady theory.

INTRODUCTION

Extreme wind speed is the most important parameter affecting the risk level of houses during severe windstorms. Thus, it is useful to understand the relationships between the magnitude of wind speed and corresponding levels of potential damage. Because the full relationship between wind speed and damage relates to many aspects of knowledge (including meteorology, building aerodynamics and construction integrity), the current research mainly focused on effect of local wind environment on the wind loading.

In storm events such as tornadoes, buildings undergo intense wind that changes rapidly in magnitude and direction within a short period of time. The best approach is to relate the instantaneous wind field to the instantaneous surface pressures. The quasi-steady theory is proposed to be used to relate the instantaneous wind speed and building surface pressure. This theory basically indicates that the instantaneous variation of building surface pressure is caused by instantaneous variation of wind speed magnitude and direction. Before applying the quasi-steady theory to predict building surface pressures for assumed wind speed variations, the first stage of analysis to be carried out relates to the applicability of this theory. This paper presents the work done in this regard.

QUASI-STEADY THEORY

In quasi-steady vector model, the surface pressure, p, is related to the instantaneous local velocity, V, through the instantaneous pressure coefficient, $C_{p,inst}$, as

$$C_{p,inst} = (p - p_0)/(0.5\rho V^2) \tag{1}$$

where p_0 and ρ are the static pressure and the air density, respectively. The total wind velocity is defined as $V^2 = u^2 + v^2 + w^2$ where u, v, and w are wind speeds in longitudinal, transverse and vertical components. The direction of the instantaneous winds is represented in terms of azimuth, $\Theta = \tan^{-1}(v/u)$, and elevation, $\beta = \tan^{-1}[\beta/(u^2+v^2)^{0.5}]$. If the wind azimuth is considered only, the $C_p(\Theta)$ is defined as the average of all instantaneous pressure coefficients measured when the corresponding wind azimuths equal to Θ, such that,

$$C_p(\theta) = E[(p - p_0)/(0.5\rho V^2)|\theta] \tag{2}.$$

Instead of the $C_{p,inst}$ expressed in Eq. (1), the conditionally-averaged $C_p(\Theta)$ defined in Eq. (2) is referred to as the instantaneous pressure coefficient throughout this paper.

Under the assumption of quasi-steady flow, the building surface pressure pressures are assumed to respond to the instantaneous local wind in a manner that is the same as the way that the mean pressure responds to the mean flow. If the relation between wind flow and pressure can be described by the quasi-steady theory, then the instantaneous pressure coefficient in Eq. (2) can be found by [1]

$$E[C_p(\theta)] = \int C_p(\theta) f(\theta - \bar{\theta}) d\theta \tag{3}$$

where f is the probability density function of the wind direction. Note that the exact azimuth Θ is equal to the sum of the mean, $\bar{\theta}$, and fluctuating azimuth, $\theta - \bar{\theta}$. The left hand side of Eq. (3) is the mean pressure coefficient,

$$E[C_p(\theta)] = E[p - p_0]/E[0.5\rho V^2] \tag{4}$$

which can be obtained from wind tunnel testing. Once the mean pressure coefficient, $E[C_p(\Theta)]$, has been obtained for a range of wind azimuths, the instantaneous pressure coefficient can be obtained by using either an analytical method [1] or iterative numerical method [2] to solve Eq. (3).

EXPERIMENTAL SETUP

Simultaneous measurements of surface pressures and flow velocities were conducted in Boundary Layer Wind

Tunnel II at UWO. Figure 1 shows the 1/50 scale-model of the TTU building used for the pressure measurements, which is accompanied by the cobra probes placed near the building for velocity measurements. A total of 204 pressure taps were uniformly distributed on the building model. However only data measured at tap #53 are discussed in this paper. The tubing system and frequency response are described in [3]. Measurements were conducted from mean azimuth wind angles varied from 0° to 90° with an increment of 5°.

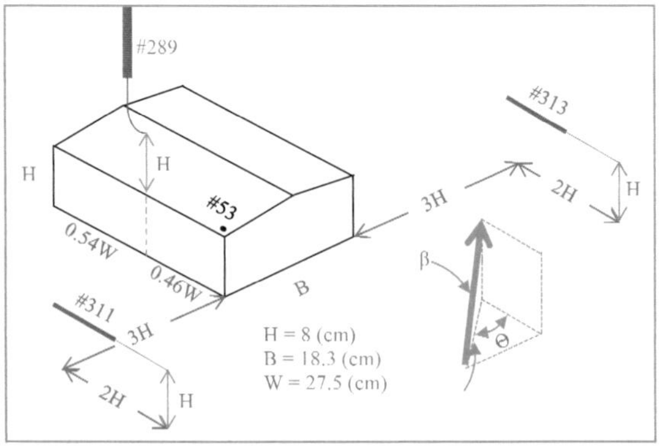

Figure 1: TTU building model and cobra probes

The locations of the cobra probes were selected to obtain the velocities representative of the flow at the building location, while minimizing the effects on the building pressures. However, the correlation of signals measured between probes decreases significantly as the transverse separation distance increases. So the velocity signal measured from cobra #289 is used for analyses throughout this paper. Furthermore, the cobra #289 was fixed to the position with respect to the building for each mean wind azimuth angle in order to analyze the velocity-pressure data using quasi-steady theory. The synchronized pressure and velocity time series were sampled at 625 Hz for 200 seconds. The definition of velocity vector is also shown in Figure 1.

A "suburban" turbulent boundary layer wind environment was created by distributing the roughness elements on the tunnel floor between the inlet and the building site. Three 122-cm tall spires and a 38-cm tall barrier were also placed at the inlet of the tunnel. The mean values of the velocity vector measured in this terrain by cobra #289 without the building in place are: $\overline{V} = 10.5\,\text{m/s}$, $\overline{\theta} \approx \overline{\beta} \approx 0^\circ$. The probabilities of the velocity vector are nearly Gaussian distributed with the standard deviations: σ_V= 2.2 m/s, σ_Θ= 9.2° and σ_β= 7°.

THE MEAN AND THEORETICAL INSTANTANEAOUS C_P

The first step in solving the theoretical instantaneous C_p is to establish the mean C_p values as a function of the mean wind azimuth. Because of the symmetric shape of building and pressure tap layout, the measurements between $\overline{\theta} = 0^\circ$ and 90° can be extended to entire 360° mean wind azimuth. The mean C_p's are then fitted by Fourier series with 9 orders and shown in Figure 2. The choice of Fourier order should be based on minimizing the error in fitting ($R^2 \geq 99.7\%$), while minimizing the number of order. The theoretical instantaneous C_p can be obtained by using Eq. (3) with the probability density function of the fluctuating wind azimuth and mean C_p curve. An iterative method was used to solve Eq. (3) for each angle within $\overline{\theta} = 0^\circ$ and 360°. The solved theoretical instantaneous C_p values for tap #53 are also shown in Figure 2.

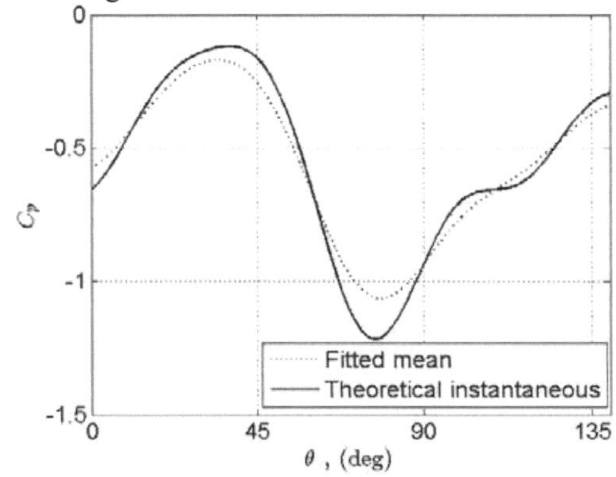

Figure 2: The Fourier-fitted mean and theoretical instantaneous C_p for tap #53.

The instantaneous C_p curves were found to have small curvature at peak regions, which lead to lower or higher peaks when they are compared to the mean C_p. This is because when the wind sways around a band of azimuth angle, the peak instantaneous C_p is averaged by the adjacent C_p's, leading to a smoothed mean C_p curve. This averaging process is also described and presented in [1] and [2].

EFFECT OF WIND DIRECTION ON C_P

In order to see the effect of wind direction on the building pressures, the instantaneous C_p values were investigated for various states of the wind elevation angle. This was done by first separating the possible wind elevation angles into three categories: (i) downward-acting winds ($-50^\circ \leq \beta < -2.5^\circ$), (ii) horizontal winds ($-2.5^\circ \leq \beta < 2.5^\circ$), and (iii) upward-acting winds ($2.5^\circ \leq \beta < 50^\circ$), for each differential range of azimuths, $\Theta - d\Theta/2 \leq \Theta < \Theta + d\Theta/2$. For wind directions falling into each of the differential ranges, the corresponding $C_{p,inst}$ values obtained using Eq. (1) were averaged to form an instantaneous C_p measurement. These results measured at taps #53 are presented in Figure 3. The symbols (dots) in different colors are associated with the three different elevation angle ranges. These colors are further categorized by gradient to denote number of samples used for averaging, which is further shown in terms of equivalent time (the right color bars) in model scale. Note that each data point was obtained by taking measurements less than 3 seconds. The solid lines used for comparison in Figure 3 is the theoretical C_p function obtained in Figure 2.

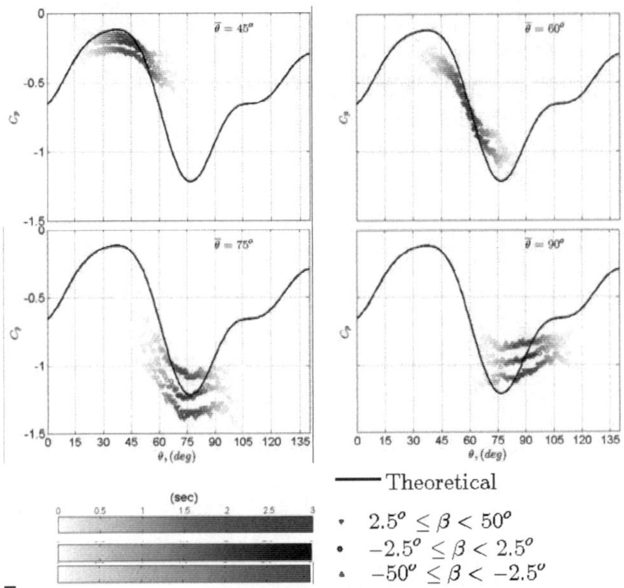

Figure 3: Theoretical and measured instantaneous C_p obtained at tap #53.

The peak near Θ = 75° indicated by the theoretical line is due to the conical vortices forming on the roof near this tap. For most mean wind azimuths, the variation of the measured C_p's followed the trend suggested by the theoretical line but with milder slopes and curvatures. Banks and Meroney [2] attributed the differences between measured and theoretical C_p's to a 'hysteresis effect'. Generally this means that, when the wind sways from $\bar{\theta}$ to $\bar{\theta}+\Delta\theta$, the instantaneous flow near the tap is not developed fully compared to the vortical structure established by the mean flow in for the particular direction of mean azimuth, $\overline{\theta+\Delta\theta}$. For example, consider the flow separation formed above the tap which governs the pressures for a mean wind azimuth, $\bar{\theta}=90^\circ$. The instantaneous change of wind to Θ = 75° did not establish the conical vortices above the tap as the $\bar{\theta}=75^\circ$ wind did, and therefore the tap pressure generated by this momentary shifted wind direction was not as low as the pressure generated by the mean wind. The reverse situation is true for wind flowing in mean azimuth, $\bar{\theta}=75^\circ$, which means the conical vortices did not disappear and become a separation bubble when the instantaneous azimuth changed to Θ = 90°. The mechanism described above is thought to be related to the effect of turbulence length scale which controls the duration of the gusts. As the turbulence becomes larger in length scale, the wind gusts become more similar to mean wind flow with respect to the building

The variation of elevation angles shows significant effect on measured C_p values. Generally the magnitude of the coefficients go up as the wind elevation becomes more positive. Also, the coefficients for the horizontal wind matches more to the theoretical line. Thus, upward-acting winds lead to more severe wind loading on the roofs of buildings. However, these events are relatively rare because of the negative correlation in the Reynolds shear stress.

Since the variation of C_p due to the various ranges of elevation angles were evaluated, there is a need to define the exact value of elevation angle within each range in order to quantify the gradient, $C_p/d\beta$. Given a differential azimuth, $\Theta \equiv \{\Theta - d\Theta/2 \leq \Theta < \Theta + d\Theta/2\}$, the exact elevation angle is the conditional average of wind elevation within the range, $\mathrm{B} \equiv \{\beta_1 \leq \beta < \beta_2\}$,

$$E[\beta|\Theta,\mathrm{B}] = \int_{\beta_1}^{\beta_2} \beta \cdot f(\beta|\Theta,\mathrm{B})d\beta$$

where β_1 and β_2 are the bounds for an elevation band and $f(\beta \mid \Theta, \mathrm{B})$ is the probability of β given the condition of Θ, B. By plotting and connecting the exact elevation angles for each elemental azimuth, the 'locus' of the exact elevation along the azimuth can be obtained. Figure 4(a) first shows the joint probability between wind azimuth and elevation fluctuation in the shaded background, with the color bar on the right denoting the density. Note that the nearly concentric shaded contour denotes the uncorrelated wind azimuth and elevation. The joint probability was then used to calculate the conditional probability in Eq. (5). Eventually the 'locus' of wind elevation for each of the three bands are plotted on the same figure. It is clear that the loci are nearly uniform with respect to azimuth axis, $\beta \approx$ -7°, 0° and 7° for downward-acting, horizontal and upward-acting winds, respectively. These elevation loci are associated with the measured instantaneous C_p loci in Figure 3 by using the same color.

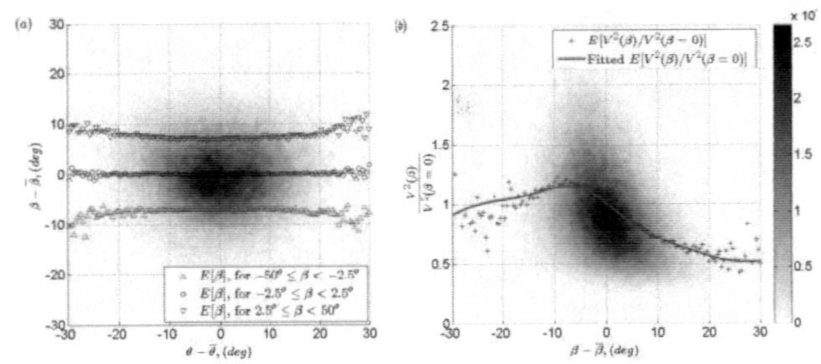

Figure 4: (a) The joint probability distribution between fluctuating azimuth and elevation (the shaded background), and $E[\beta]$ loci for given ranges of β. (b) Joint probability between $V^2(\beta)/V^2(\beta = 0^o)$ and fluctuating β, and $E[V^2(\beta)]/E[V^2(\beta = 0^o)]$ locus, obtained on the condition, $-2.5^\circ \leq \theta - \bar{\theta} < 2.5^\circ$.

The discussions on the wind elevation angle effect on building pressures are usually related to Reynolds stress effects [3]. The Reynolds stress means that the wind elevation angle is prone to be more negative as the velocities fluctuate to the higher magnitudes. This adds complexity when the results shown in Figure 3 are being interpreted. Because the instantaneous C_p is defined as the ratio of surface pressure to instantaneous dynamic pressure, the variation of instantaneous C_p can be attributed one of the following assumptions: (i) aerodynamic effects, (ii) the reduction in dynamic pressure while the surface pressure remains unchanged, or (iii) some combination of assumptions (i) and (ii). The aerodynamic effects of elevation wind angle were found by tilting the building model forward or backward with respect to incident wind in [3]. Since there are no experimental data available for non-zero mean elevation angles, our approach is to make the assumption of second argument and check the applicability. The second argument can be sensed based on the conclusion that have been made from Figures 3 and 4(a), and the concept of Reynolds stress: Higher magnitude instantaneous C_p are usually associated with more positive wind elevation angles, and the instantaneous dynamic pressured is negatively correlated to the wind elevation angle. Furthermore, Banks and Meroney [2] addressed that the surface pressures respond more to wind fluctuations of larger length scale. The preliminary comparison showed the fluctuating wind elevation has the smallest length scale as compared to velocity and azimuth fluctuations. So, the assumption that the surface pressures are dominated by magnitude and azimuth fluctuation of the wind, while the fast- and small-changing wind elevations do not affect the surface pressure so much, may hold.

By following the assumptions made earlier, the instantaneous $C_p(\Theta,\beta)$ along a range of elevation angle can be represented by $C_p(\Theta,\beta = 0)$. The reason $C_p(\Theta,\beta = 0)$ being selected is that its theoretical function is already obtained (see solid lines in Figure 3). If a small band of azimuth is selected, $\Theta - d\Theta/2 \leq \Theta < \Theta + d\Theta/2$, the instantaneous C_p can be obtained by assuming that the pressure is not responding to β, $C_p(\beta) = E[\Delta p(\beta = 0^\circ)/(0.5\rho V^2(\beta))]$. Expanding the right hand side by substituting $\Delta p(\beta = 0^\circ)$ by $C_p(\beta = 0^\circ)\bullet\rho V^2(\beta = 0^\circ)/2$, the instantaneous C_p becomes,

$$C_p(\beta)=C_p(\beta=0^\circ)\cdot E\left[V^2(\beta=0^\circ)\right]/E\left[V^2(\beta)\right] \quad (6)$$

The joint probability between the ratio, $E[V^2(\beta)]/E[V^2(\beta = 0^\circ)]$, and fluctuating wind elevation, $\beta-\overline{\beta}$, is first obtained and plotted in the shaded background in Figure 4(b), for a 5 degree band of fluctuating azimuth, $-2.5^\circ \leq \theta-\overline{\theta} < 2.5^\circ$. The conditional average of this ratio was then obtained in the same way as for obtaining elevation loci in Figure 4(a), which displayed an increasing trend starting from 0.5 at $\beta-\overline{\beta} \approx 30^\circ$ to 1.2 at $\beta-\overline{\beta} \approx -8^\circ$, and slightly decreasing trend for further negative fluctuating β values. A fitted curve of these conditionally-averaged ratios was also obtained and used in calculating $C_p(\beta)$ in Eq. (6).

The $C_p(\beta)$ measured within 5 degree azimuth band around various mean wind directions for the tap are presented in Figure 5. Again the symbols (dots) were further categorized by the gradient color bar corresponding to the equivalent sampling time in model scale. The assumption of $C_p(\beta)$ represented in Eq. (6) is also shown in these figures by taking the measured $C_p(\beta = 0^\circ)$ and dividing it by the fitted velocity square ratio. The assumption seems to hold only for $\overline{\theta} = 45^\circ$ case, while partially matches the trend for the rest of cases. Therefore, the surface pressures were found to be able to respond to the wind elevation angles to some extent, with the degree of sensitivity depending of the wind azimuth.

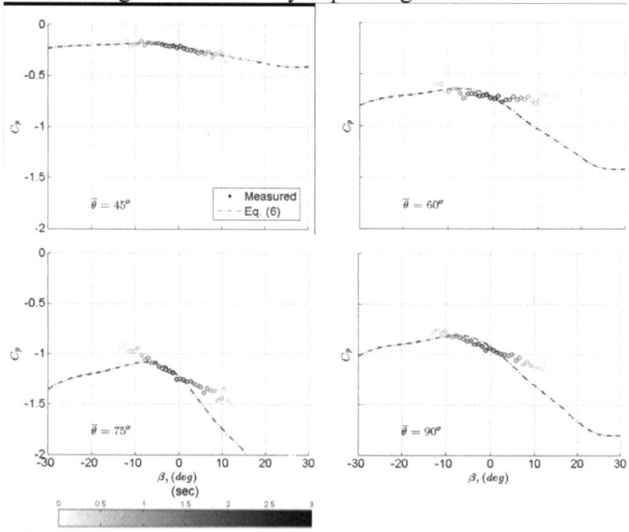

Figure 5: Instantaneous C_p's versus elevation angle obtained on the condition, $-2.5^\circ \leq \theta-\overline{\theta} < 2.5^\circ$.

CONCLUSIONS

The applicability of quasi-steady theory in relating the wind-speed and surface pressures at a point near the roof corner was assessed. The theoretical instantaneous C_p function was obtained by using methods developed in [1] and [2], and compared with measured results. Good agreement was found for horizontal winds swaying between 15° and 75° in azimuth. Poorer agreement was observed between 75° and 90°, which was attributed to the effect of building generated turbulence and the length scale of the incident wind turbulence. Although the change in wind elevation angles was found to affect the instantaneous C_p magnitude significantly, the excitation of pressure due to the fluctuating wind elevation angles was dampened through aerodynamic effects.

REFERENCES

1. Richards, P.J., Hoxey R.P., Wanigaratne, B.S., 1995. "The effect of directional variation on the observed mean and rms pressure coefficients", *J. Wind Eng. Ind. Aerodyn.*, 54/55, 359-367.
2. Banks, D., Meroney, R.N., 2001. "The applicability of quasi-steady theory to pressure statistics beneath roof-top vortices", *J. Wind Eng. Ind. Aerodyn.*, 89, 569-598.
3. Ho, T.E.C., Surry, D., Morrish, D., Kopp, G.A., 2005. "The UWO contribution to the NIST aerodynamic data base for wind loads on low buildings: Part 1. *J. Wind Eng. Ind. Aerodyn.*, 93, 1-30.

4. Sharma, R.N., Richards, P.J., 1999. "The influence of Reynold stresses on roof pressure fluctuations. *J. Wind Eng. Ind. Aerodyn.*, 83, 147-157.

LARGE SCALE PARTICLE STREAK VELOCIMETRY IN THE WINDEEE DOME

Dan-Stefan Parvu
WindEEE Research Institute
Western University
London, Ontario, Canada
dparvu@uwo.ca

Adrian Costache
WindEEE Research Institute
London, Ontario, Canada

Horia Hangan
Professor and Director
WindEEE Research Institute
Western University
London, Ontario, Canada

ABSTRACT

Flow visualization is a technique that enables researchers to observe the physical properties of fluid flows. Fluid flows have been characterized, quantitatively, using a variation of different methods, which can be either intrusive or non-intrusive. Non-intrusive techniques include particle-tracking velocimetry, molecular-tracking velocimetry and particle-image velocimetry.

An experiment has been set up in the recently constructed WindEEE Dome, at Western University, in London, ON, Canada. A downburst flow with a jet diameter of 5 m was simulated and then investigated using the proposed measurement technique. Downbursts are strong ground-level wind systems, originating from a thunderstorm cloud base, directed downwards, which generate short duration high intensity winds close the ground. Velocity proved to be difficult to obtain for such a phenomenon using traditional means. For this purpose a Large Scale Particle Streak Velocimetry (LSPSV) setup has been designed, built and tested.

KEYWORDS: WindEEE; Downburst; Particle Streak Velocimetry, physical simulations

INTRODUCTION

Flow visualization is a technique that enables researchers to observe the physical properties of fluid flows. Fluid flows have been characterized, quantitatively, using a variation of different methods, which can be either intrusive or non-intrusive. Non-intrusive techniques include particle-tracking velocimetry, molecular-tracking velocimetry and particle-image velocimetry [1] [2].

The non-intrusive method of Particle Image Velocimetry (PIV) is typically used to analyze the velocity field of gasses or liquids. The standard PIV method can usually resolve two-dimensional flow fields. In order for the method to function, the studied medium needs to be seeded with tracer particles and illuminated periodically with a high power light source. PIV is most suitable for a relatively small region of interest. There are experiments in which the region of interest exceeds the capabilities of the standard PIV method – for cases such as the one presented herein, other flow measurement techniques have been developed. These methods include LSPIV (Large Scale PIV) [3], LFPIV (Large Field of View PIV) [4] and PTV (Particle Tracking Velocimetry). While the standard PIV method has been refined over the years, LS- PIV further complicates experimental setups. Usually, the number of cameras used to capture the required region of interest is increased and a much more powerful light source is needed in order to illuminate the larger area.

Particle Tracking Velocimetry is also known as Low Image Density PIV [5] [6], in which the seed particles are not as abundant as in the case of standard PIV. The PTV method measures the movement of a particle in a known time period and evaluates its velocity. PTV methods can be applied to both 2D and 3D flow cases. A benefit of using PTV is the reduced number of images that need to be processed. The main difference between PTV and PIV is that the former cannot provide a high density velocity field because of the different seeding procedure – the particles are randomly located.

Particle Streak Velocimetry (PSV) has been used traditionally for small scale experiments with only a few studies of medium to large scale applications. The main advantage of this technique is that it can be used to analyze a large area and it is easier deploy and use than PIV or PTV methods. The earliest usage of this technique seems to have been in 1932, by Fage and Townend [7] in an attempt to characterize turbulent flow in both a square and circular brass pipe. Their source of light was a small 5 ampere arc lamp focused using a single condensing lens. Measurements were done using a microscope over an area of 1.78 mm in diameter and printed on celluloid. At one end of the range, Dixon et al. [8] performed microscopic measurements of colloidal spheres approximately 1.5 μm in diameter in a 100 μm microfluidic channel and Sinha and Kuhlman [9] captured and analyzed the streaks produced by 60 μm particles in a 2D Couette flow. While at the opposite end of the range, Bosbach et al. [10] used a combination of PIV and PSV to evaluate the convective air flow pattern in an Airbus A380 cabin. The tracer particles used were neutrally buoyant helium filled soap bubbles (HFSB) and the light source was a high powered laser needed to illuminate a field of view of 1.5 x 2 m^2. Voss et al. [11] benchmarked a novel technique for 3D flow measurement of interfacial transport processes. They designed a complicated system of bichromatic illumination with intensity modulation of two LED arrays. Their system allowed for the extraction of tracer particle trajectories in a volume of 415 x 405 x 50 mm^3. There are several occasions when a quick estimation of the flow characteristics, such as velocity and direction is needed. Having this in mind, this paper presents a method of obtaining flow characteristics for very large areas, in the order of 10 m^2, by means of Large Scale Particle Streak Velocimetry (LSPSV), a combination between traditional PTV and PSV.

An experiment has been set up in the recently constructed WindEEE Dome, at Western University, in London, ON, Canada. A downburst flow with a jet diameter of 5 m was simulated and then investigated using the proposed measurement technique. Downbursts are strong ground-level wind systems, originating from a thunderstorm cloud base, directed downwards, which generate short duration high intensity winds close the ground. Velocity proved to be difficult to obtain for such a phenomenon using traditional means. For this purpose an LSPSV setup has been designed, built and tested.

TEST FACILITY

The WindEEE (Wind Engineering, Energy and Environment) Dome is a new three dimensional wind tunnel. In the form of a hexagon, the wind testing chamber has a diameter of 25 m, shown in Figure 1, and a return circuit of the same shape with a larger diameter of 45 m. High intensity wind systems can be physically simulated inside the WindEEE test chamber, such as tornados, downbursts, gust fronts [12]. In the case of downbursts the upper fans are pushing air down into the test chamber through the bell-mouth. The column of air created impinges on the test chamber floor and afterwards the air is evacuated through the peripheral walls.

Figure 1: Test chamber in the WindEEE Dome.

EXPERIMENT SETUP AND PLAN

Experimental setup

The experiment was set up in the WindEEE dome. A downburst flow was investigated in order to assess the applicability of the proposed velocimetry method. Downbursts are axisymmetrical phenomena and based on this assumption, the experiment was set up in the test chamber, with the light source pointing towards one of the walls. Using this setup, it was possible to analyze a two-dimensional vertical slice of the flow which covered an area of 5 x 3.8 m^2.

Seeding particles

Particle tracking techniques rely on discrete particles that follow the flow. These particles also need to have good reflectivity so that their motion can be captured by the camera. In the case of the PTV and PSV methods, particle seeding density is not a major issue [10]. Since in these methods velocity vectors are determined based on single particle recognition and tracing, neutrally buoyant HFSB are ideal candidates for seeding particles. A commercially available HFSB generator was used to seed the test chamber prior to running the fans. The seeding density was chosen so that the camera could easily acquire images and capture the particles even in low light conditions, without having to increase the exposure time to a large value. Another limiting factor was the amount of particles that populated the light sheet. The challenge here was to have a sufficient number of particles to analyze while at the same time not overcrowding the light sheet volume. This was accomplished by vertically separating the seeding outlets using a stand on which the two outlet tubes were placed.

Illumination

The majority of particle tracking techniques require a high powered light source, usually in the form of a laser. Lasers are expensive and difficult to perform maintenance on and they require trained staff to set up and operate as well. In the case of this experiment a stroboscopic lamp was chosen as the main (and only) light source due to its simplicity to operate and low costs compared to a laser system. A pulsating light sheet was needed to obtain different illumination effects on a particle streak that could be used afterwards to obtain velocity data. In other words, the particle streak has segments that are lighter or darker, dark breaks in the complete segment. The strobe itself was placed close to the center of the turntable, as shown in Figure 2. A metallic slit was placed in front of the strobe optics to create a thinner light sheet.

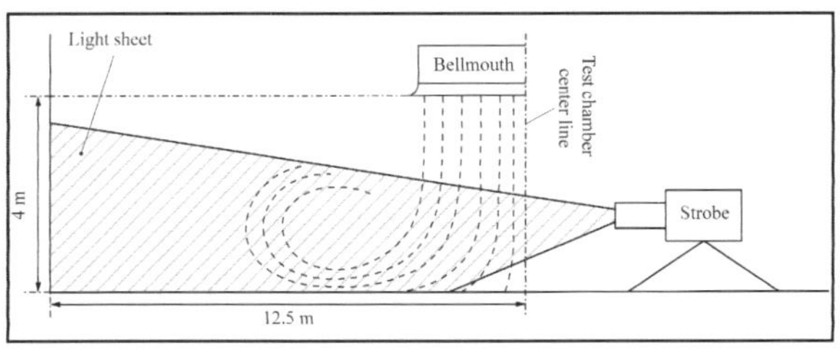

Figure 2: Schematic of the test setup showing strobe lamp position and light sheet position relative to the flow.

Camera

A commercially available camera with a 7360 x 4912 pixel CMOS sensor (effective resolution: 12 megapixels) was used to capture images. The camera was placed at two different positions, as shown in Figure 3. These two positions were required to generate a wider field of view that would cover the entire distance from the center of the turntable to the wall of the test chamber.

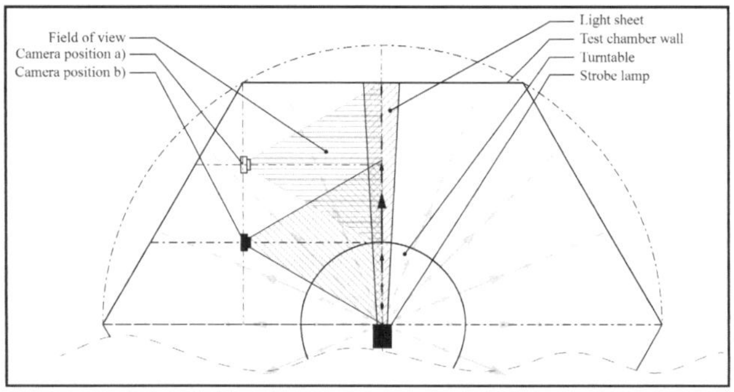

Figure 3: Top down view of the experimental setup. The vectors represent the theoretical wind velocity of a downburst.

Alignment and calibration

The strobe lamp was placed on one of the center lines of the turntable as shown in Figure 3, and the camera was placed so that the field of view would be perpendicular to the light sheet. Two adjustable height stands were placed on the center line at a known distance from each other (5 m) in order to calibrate the pixel to physical length ratio, as shown in Figure 4. Using the camera, a few calibration images were taken which cover the area highlighted in Figure 4 as "Camera frame".

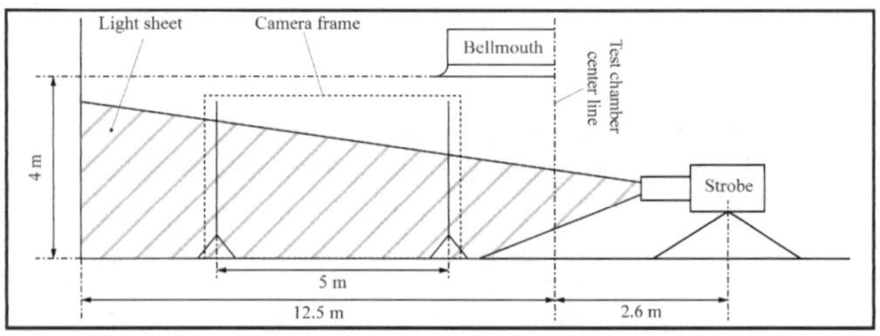

Figure 4: Camera calibration relative to light sheet for camera position a).

Experimental plan

After the test chamber was sufficiently seeded with HFSB, a downburst flow is created, that is impinging on the test chamber floor, using the top fans. Images were captured by the camera in burst mode until no more HFSB were in the light sheet. Images were acquired for seven different speeds of the top fans in the WindEEE dome, in 5% increments from 10% to 40% of fan capacity. Since two camera positions were used, the image sets are split in two categories: a set of images for the center of the downburst (camera position b) in Figure 3) and another set of images for the extremity (camera position a) in Figure 3).

RESULTS AND DISCUSSION

Figure 5 and Figure 6 present the images obtained for the downburst test at the extremity and the center, respectively, using the top fans at 15% capacity.

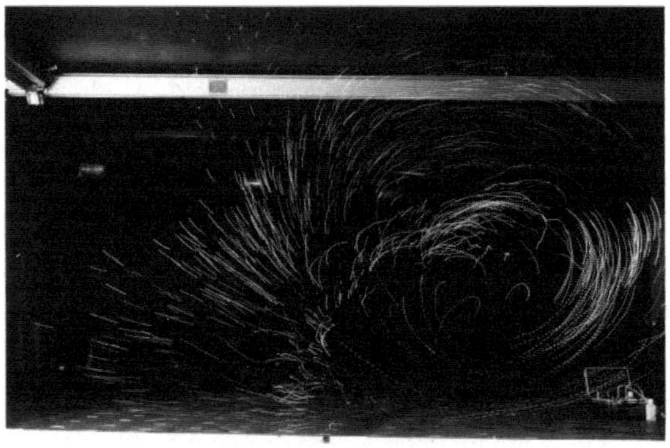

Figure 5: 15% fan speed, extremity.

Figure 6: 15% fan speed, center.

The acquired images were processed using an algorithm developed in MatLab. The background image is subtracted from each of the leaving just the particle streaks visible. The resulting image is then thresholded to grayscale and then binarized, which essentially converts the image into a logical 2D array.

A single sample streak is processed and shown in Figure 7. The position and velocity of a single particle is evaluated using one streak. Based on the camera exposure, streak length as well as the number of breaks, the velocity and position of the tracked particle are calculated. Vectors are obtained by essentially counting the number of breaks in a streak and measuring how long they are, in pixels. Pixel values are then converted into length units (millimeters) using camera calibration values. Velocity data is generated based on the ratio of the individual break length and camera exposure time. The complete streak is used as orientation information – each vector starts on, and is tangent to, this curve. Vectors are then plotted, as shown in Figure 8. The procedure should then be repeated for each streak identified in the images in order to obtain a vector field.

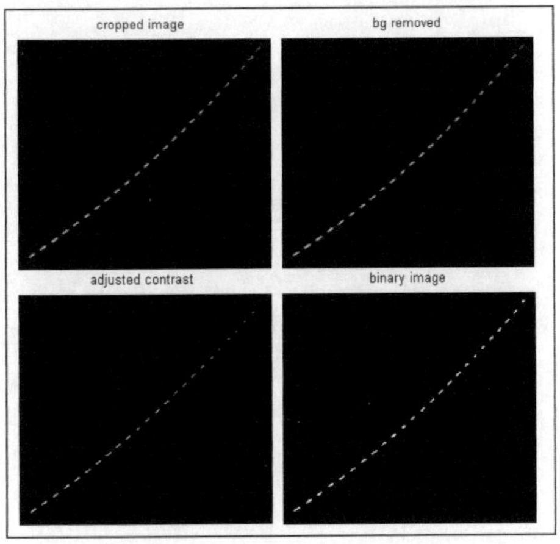

Figure 7: Image processing.

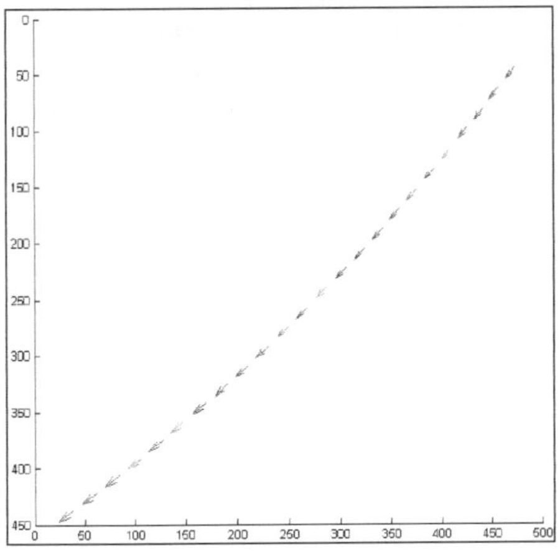

Figure 8: Computed velocity vectors. Axes show pixel number.

CONCLUSION

A large scale downburst experiment has been set up in the WindEEE dome at Western University, in London, ON, Canada. The LSPSV setup proposed and tested herein shows the potential to be a less costly and easier to use alternative to large scale PIV.

Using this method a very large area of 5 x 3.8 m^2 was analyzed with minimal preparation and equipment. The setup time is considerably less than the time needed to assemble a traditional PTV or PIV configuration. A pulsating light sheet covering the region of interest, in which HFSB particles were placed, allowed the extraction of velocity vectors as well as position information from the images acquired. The difficulty of this method consists in having good HFSB seeding and the post processing of the obtained images. The obtained data will be validated against measurements taken using equipment such as pressure probes and PIV.

REFERENCES

1. R.B. Miles, J.J. Connors, E.C. Markovitz, P.J. Howard and G.J. Roth, Instantaneous Profiles and Turbulence Statistics of Supersonic Free Shear Layers by Raman Excitation plus Laser-Induced Electronic Fluorescence (Relief) Velocity Tagging of Oxygen, Experiments in Fluids 8, 17-24, 1989.
2. W. Lauterborn and A. Vogel, Modern Optical Techniques in Fluid Mechanics, Ann. Rev. Fluid Mech. 16: 223-44, 1984.
3. I. Fujita, M. Muste and A. Kruger, Large-scale particle image velocimetry for flow analysis in hydraulic engineering applications, Journal of Hydraulic Research, 36:3, 397-414, 1998.
4. S. Pol, B. Balasubramaniam, Large Field of View Particle-Image Velocimetry (LF-PIV): Design and Performance, 50th AIAA Aerospace Sciences Meeting including the New Horizons Forum and Aerospace Exposition, Aerospace Sciences Meetings, American Institute of Aeronautics and Astronautics, 2012.
5. Y. Stegeman, Particle Tracking Velocimetry, 1995.
6. R. J. Adrian, Particle-imaging techniques for experimental fluid mechanics, Annual review of fluid mechanics, 23(1), 261-304, 1991.
7. A. Fage and H. C. H. Townend, An Examination of Turbulent Flow with an Ultramicroscope, Proceedings of the Royal Society of London. Series A, Containing Papers of a Mathematical and Physical Character, Vol. 135, No. 828, pp. 656-677, 1932.
8. L. Dixon, F.C. Cheong, and D.G. Grier, Holographic particle-streak velocimetry, Optics express, 19(5), 4393-4398, 2011.
9. S.K. Sinha and P.S. Kuhlman, Investigating the use of stereoscopic particle streak velocimetry for estimating the three-dimensional vorticity field, Experiments in fluids, 12(6), 377-384, 1992.
10. J. Bosbach, M. Kühn and C. Wagner, Large scale particle image velocimetry with helium filled soap bubbles, Experiments in fluids, 46(3), 539-547, 2009.
11. B. Voss, J. Stapf, A. Berthe, and C.S. Garbe, Bichromatic particle streak velocimetry bPSV, Experiments in Fluids, 53(5), 1405-1420, 2012.
12. H. Hangan, The Wind Engineering Energy and Environment (WindEEE) Dome at Western University, Canada, Wind Engineers, JAWE, Vol. 39, No.4, October 2014

EFFECT OF ATMOSPHERIC WIND ON A GROUND MOUNTED SOLAR PANEL ARRAY

Chowdhury Jubayer
WindEEE Research Institute
Western University
London, Ontario, Canada
cjubayer@uwo.ca

Horia Hangan
WindEEE Research Institute
Western University
London, Ontario, Canada
hmhangan@uwo.ca

ABSTRACT

Aerodynamic loads and wind flow field for a ground mounted solar panel array immersed in the atmospheric boundary layer (ABL) are investigated using Reynolds-Averaged Navier-Stokes (RANS) approach. A full scale three-dimensional (3D) unsteady solver with Shear Stress Transport (SST) k-ω turbulence closure is employed. Four different approach wind directions (0°, 45°, 135° and 180°) for a Reynolds number of $3x10^6$ are considered. Detail analysis of wind loads on the array in relation to the flow field around the array is provided. For the array configuration the results show that all trailing rows are in the complete wake of the leading row for 0° and 180°, but not for 45° and 135° wind directions. In terms of maximum drag and lift, the leading row experiences the critical loading for all four wind directions.

KEYWORDS: Solar panel array, wind load, CFD, ground mounted

INTRODUCTION

Utility scale photovoltaic (PV) power stations or solar farms are emerging as a significant contributor to the electricity generation. As of 2014, at least 53 solar farms with capacity more than 50MW are operating in 13 countries (REN 21, 2014). Typically, solar farms are sited in a large open field and employ arrays of ground mounted solar PV panels designed to supply electricity to the commercial power grid. Optimizing PV panel support structures to withstand aerodynamic forces is one of the challenges experienced by the solar panel manufacturer and installer. Both experimental (Bechtel National Inc., 1980; Miller and Zimmerman, 1981; Kopp et al., 2012; Warsido et al., 2014) and numerical studies (Bitsuamlak et al., 2010; Shademan et al., 2014) have been performed to investigate wind loads on ground mounted solar panel array.

Both studies by Bechtel National Inc. (1980) and Miller and Zimmerman (1981) were conducted as commercial industrial projects to initiate an effort towards reducing the cost of solar panel arrays. Bechtel National Inc. (1980) reported the wind loads on ground mounted solar array using strain gages in a boundary layer wind tunnel. Aerodynamic coefficients were independent of the scales used (1:24 and 1:12). In addition to the scales, solar panel height above the ground (lower edge height of 0.61 and 1.22 m), porosity (2.5%, 5% and 10%) and aspect ratio (Width/Breadth, W/B of 2, 3 and 4) did not have a significant effect on the aerodynamic coefficients of the panel. Unlike Bechtel National Inc. (1980), Miller and Zimmerman (1981) utilized pressure taps to measure wind loads on the panel and was able to provide pressure distribution on the solar panel surface. For the wind tunnel speed of 15.24 m/s, between 1:12 and 1:24 scales, the dimensionless pressure and force coefficient were independent of Reynolds number.

A wind tunnel study on a ground mounted solar panel array was performed by Kopp et al. (2012) to investigate the effect of buildings on roof mounted arrays. For the ground mounted array, wind load was found maximum for the first windward row with minimum for the third or fourth row. Wind tunnel tests were also conducted by Warsido et al. (2014) to investigate the effect of row spacing on wind loads for a solar panel array mounted on the flat roof of a building and also on the ground. Minimal influence of lateral spacing between panels was found on the inner panel columns of the ground mounted system. Also, increasing longitudinal spacing between panels increased wind loads on the panels.

Using numerical simulations, one case of arrayed ground mounted solar panel was studied by Bitsuamlak et al. (2010). It was found that the sheltering effect from the upwind panel reduced wind loading on the downstream panels significantly. Shademan et al. (2014) also employed numerical simulations and investigated the effect of row spacing for a ground mounted solar panel array for 45° inclination angle and for only 0° wind direction. It was observed that the row spacing did not have an effect on wind loading of the first windward row of panels. However, for the rest of the rows, drag increased and lift decreased with the increase of row spacing.

North American building standards (ASCE 7-10, 2010 and National Building Code of Canada, 2010) do not cover any information regarding minimum design wind load for solar panel arrays, either ground mounted or roof mounted. However, Australia (AS/NZS 1170.2, 2011) and California (SEOAC, 2012) have some standards for roof mounted solar panels but not for ground mounted installation.

The present study is performed to investigate not only the wind load but also the underlying aerodynamic mechanism responsible for wind loads on a ground mounted solar panel array. Four wind directions (0°, 45°, 135° and 180°) are considered. Computational Fluid Dynamics (CFD) approach is undertaken to satisfy the objective of this study.

METHODOLOGY

Details of the numerical modelling setup are discussed here. The dimension of the individual solar panel module used for the solar panel array is 1.2 m (length) × 0.6 m (width) × 0.007 m (thickness). One row of the array field has 72 aforementioned individual solar panel modules arranged in 4 (row) × 18 (column) with 25° inclination. The whole array field has 5 rows. The overall dimension of one array is 2.47 m (B) × 21.92 m (W) × 1.65 m (H) (Fig. 1).

The COST guideline for CFD simulation in the urban environment developed by Franke et al. (2007) is followed to create the computational domain and mesh. The dimension of the domain is 30.2H (X) × 6.0H (Y) × 48.4H (Z). Three different grids with 1,492,982 (G1), 2,209,925 (G2) and 2,780,048 (G3) cells are created for 0° wind direction to check grid sensitivity. Based on the grid sensitivity analysis on the surface average pressure coefficients, G2 (Fig. 2) is chosen as the final grid for simulation.

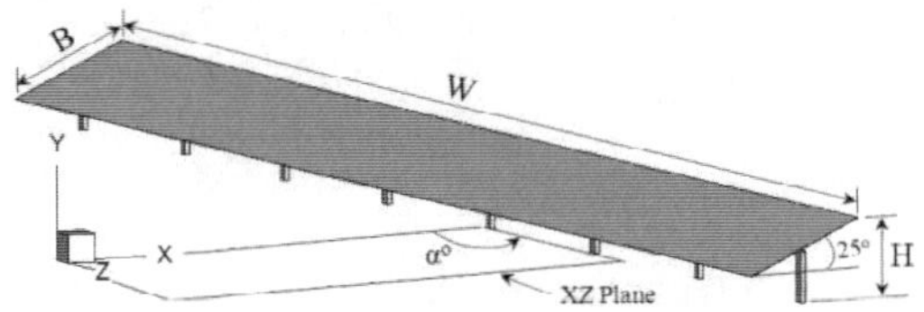

Figure 1: Computational model of a single array with wind direction (α°) shown in the XZ plane

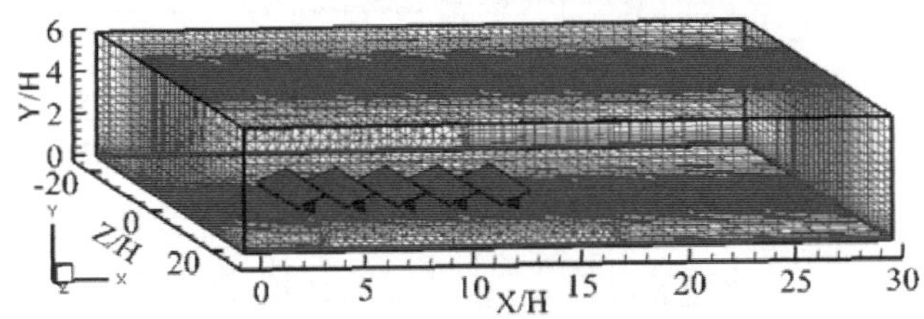

Figure 2: Isometric view of the computational domain and mesh for 0° wind direction

Three dimensional RANS simulations are carried out using a transient solver in the open source CFD software package OpenFOAM® 2.1.0 (ESI Group). The transient solver pisoFoam, which applies PISO (Pressure Implicit with Splitting of Operators) algorithm to solve the Navier-Stokes equations, is used. The SST k-ω turbulence model is employed in this study. Boundary conditions and wall-functions are similar to the study by Jubayer and Hangan (2014) with ESDU (1982, 1983) velocity and turbulence intensity (TI) profiles for open terrain roughness (y_0 = 0.03 m) at the inlet, fixed uniform zero gauge pressure at the outlet, slip walls at sides of the domain, fixed values of velocity and turbulence properties from the top of the inlet boundary throughout the top of the domain, no slip rough wall at the bottom of the domain and no slip smooth walls at the surface of the solar panel. The Reynolds number based on the incident wind speed at H and the length scale B is 3×10^6. Second order discretization schemes and linear interpolation scheme are chosen for this study. For each wind direction, simulation is run for a non-dimensional time, t^* (= $t\overline{U_H}/B$, here t is the flow time (s)) of 144. Before running simulations with the solar panel array inside the domain, simulations are performed inside an empty domain to ensure that the equilibrium ABL flow (Blocken et al., 2007) is achieved. Between the lower edge (0.36H) and the upper edge height (H) of the panel, inlet and incident profiles match with each other within 3% difference for both mean wind speed and turbulence intensity.

RESULTS AND DISCUSSION

Normalized mean velocity contours with streamlines at a vertical plane as well as at a horizontal plane through the center of the solar panel array are shown in Fig. 3 and 4 respectively. For all results presented here, Row 1 is always the leading row irrespective of the wind direction. For 0° wind direction (Fig. 3 a) at Row 1, flow is attached on the upper surface of the row of panels while on the lower surface, it separates at the leading edge forming a separation-reattachment bubble. Rows 2 to 5 are completely in the wake of the first row of panels. For 45° wind direction (Fig. 3 b), a very small vortex is observed close to the leading edge on the lower surface for all rows. This small vortex may be associated to the formation the corner vortices at this oblique wind direction. For 135° wind direction (Fig. 3 c), flow impingement can be seen on lower surfaces and close to the leading edges for all rows. Fig. 4 (a, d) clearly shows that Rows 2 to 5 are completely in the wake of Row 1 for the straight wind directions (0° and 180°). Also, the flow pattern and velocity magnitudes are very symmetric about an axis through the center of a row and parallel to the X axis for the two straight wind directions. For 0° wind direction, the flow on both sides (along the Z axis) of the array has a tendency to move closer to the array whereas for 180°, it is the opposite. This behaviour of the flow may have resulted in the vortices on the leeward sides of Rows 3 to 5 in 0° wind direction but not for 180° wind direction. On the other hand, for the oblique wind directions (45° and 135°) (Fig. 4 b, c), the wind flows in between the rows resulting in higher wind speed around Rows 2 to 5 compared to 0° and 180° wind directions. Out of all these four wind directions, the maximum wind velocity magnitude is observed around Rows 2 to 5 for 135° wind direction.

Surface pressure distributions in terms of mean pressure coefficients (C_p) on the upper and lower surfaces for every row of panels are provided in Fig. 5 for all four wind directions. For 0° wind direction, Row 1 experience higher wind loads compared to the other four rows (Fig. 5 a). Especially, surface area close to the leading edge for Row 1 sees the maximum wind load since the incoming flow impinges in the region close to the leading edge on the upper surface (Fig. 3 a). For Rows 2 to 5, major portions of the

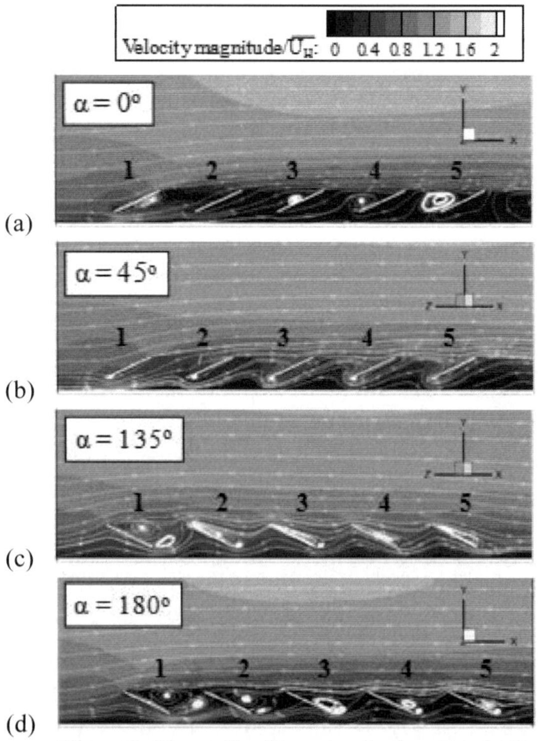

Figure 3: Normalized mean velocity magnitude contours with streamlines at the plane through the center of the panel and parallel to the side edge of the panel for (a) 0°, (b) 45°, (c) 135° and (d) 180° wind directions

surface area on upper and lower surfaces have negative C_ps and since these rows (Rows 2 to 5) are in the wake of Row 1 (Fig. 3 a, 4 a), these rows show much lower net mean C_p than Row 1. For 45° wind direction, the critical region, in terms of maximum net mean C_p, is the region close to the leading shorter edge for all rows (Fig. 5 b). Conical shape C_p distributions on the leeward surface at the leading edge corner of each row suggest the presence of corner vortices (Fig. 5 b). Comparable to the 45° wind direction case, lower C_p values in a conical shape distribution are observed on the leeward side of the panel for 135° wind direction for all five rows (Fig. 5 c). Also, the gradual decrease in C_p values from the leading shorted edge towards the trailing shorter edge on the windward surfaces for 45° and 135° wind directions can be related to the flow acceleration in between two consecutive rows of panels. Similar to 0° wind direction, the maximum net mean C_p is found close to the leading edge of Row 1 (Fig. 5 d) which is the location of flow stagnation for this row (Fig. 4 d). Again, since Rows 2 to 5 are completely in the wake of Row 1 (Fig. 4 d), very low C_p values are found on both upper and lower surfaces of these rows for 180° wind direction (Fig. 5 d).

Mean drag (C_D) and lift coefficients (C_L) for all four wind directions are shown in Fig. 6. Fig. 6 shows that for all four wind directions, the first windward row (Row 1) is the critical row compared to the rest of the rows (Rows 2-5) based on higher drag and lift. In terms of individual loading

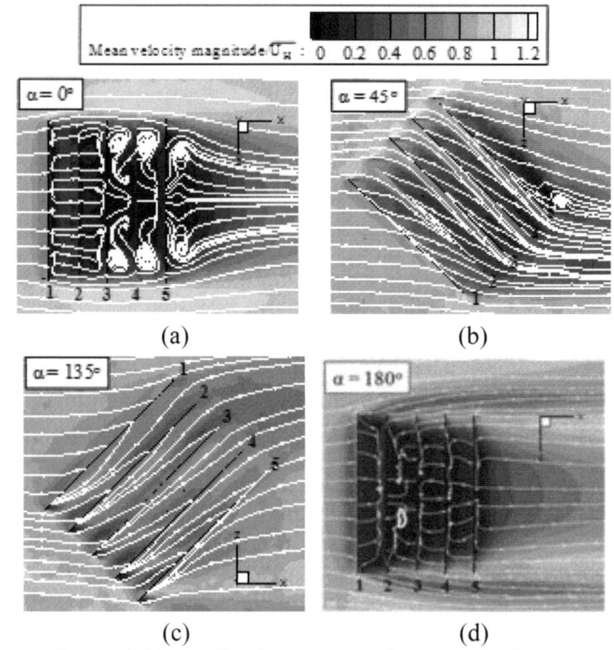

Figure 4: Normalized mean velocity magnitude contours with streamlines at the horizontal plane through the center of the panel for (a) 0°, (b) 45°, (c) 135° and (d) 180° wind directions

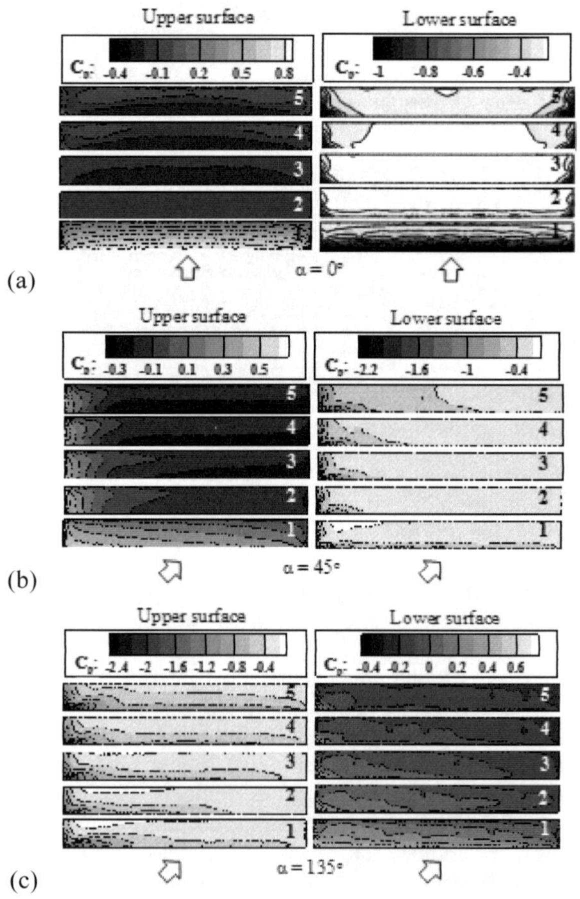

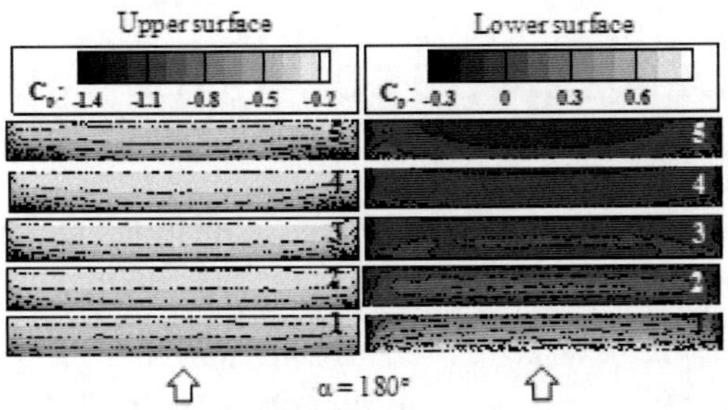

(d)

Figure 5: Mean C_p distributions on the upper and lower surfaces of the solar panels for (a) 0°, (b) 45°, (c) 135° and (d) 180° wind directions

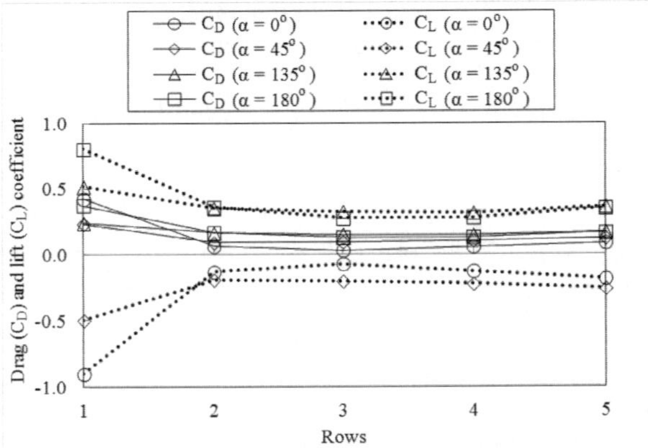

Figure 6: Mean drag (C_D) and lift coefficients (C_L)

component, 180° wind direction is critical for maximum uplift and 0° wind direction is for maximum drag. Also, minimum wind loads (both drag and lift) are found at Row 3 for 0° and 180° wind directions, at Row 2 for 45° wind direction and at Row 4 for 135° wind direction.

CONCLUSIONS

The current study investigated the wind loads and the wind flow fields around an array of ground mounted solar panels using computational fluid dynamics (CFD) approach. For 0° and 180° wind directions, Rows 2 to 5 are completely in the wake of the first windward row (Row 1). Corner vortices are observed on the leeward sides of all five rows of panels for 45° and 135° wind directions. Row 1 experiences the maximum wind loads (drag and lift) for all four wind directions (0°, 45°, 135° and 180°). For Rows 2 to 5, the wind loads (lift and drag) are higher for oblique winds (45° and 135°) than for straight winds (0° and 180°). Overall, this study provides a very detail analysis of wind effects on ground mounted solar panel array by considering not just straight wind directions (0° and 180°) but also oblique wind directions (45° and 135°).

REFERENCES

1. ASCE 7-10, 2010, "Minimum Design Loads for Buildings and Other Structures", *ASCE 7-10*, Virginia, USA: American Society of Civil Engineers.
2. AS/NZS 1170.2, 2011, "Australian/New Zealand Standard Structural Design Actions, Part 2: 2011", Standards Australia International Ltd., Sydney, Australia.
3. Bechtel National Inc. 1980, "Wind design of flat panel photovoltaic array structures", *Final Report, Report SAND79-7057*, Bechtel National Inc., California.
4. Bitsuamlak, G. T., Dagnew, A. K., and Erwin, J. 2010, "Evaluation of wind loads on solar panel modules using CFD", *The Fifth International Symposium on Computational Wind Engineering,* Chapel Hill, USA.
5. Blocken, B., Stathopoulos, T., and Carmeliet, J. 2007, "CFD simulation of the atmospheric boundary layer: wall function problems", *Atmospheric Environment,* Vol (41), pp. 238-252.
6. ESDU, 1982, "Strong winds in the atmospheric boundary layer. Part 1: Mean hourly wind speeds", *Engineering Sciences Data Unit Number 82026.*
7. ESDU, 1983, "Strong winds in the atmospheric boundary layer. Part 2: Discreet gust speeds", *Engineering Sciences Data Unit Number 83045.*
8. Franke, J., Hellsten, A., Schlunzen, H., and Carissimo, B. 2007, "Best practice guideline for the CFD simulation of flows in the urban environment", *COST Action 732; Quality assurance and improvement of microscale meteorological models.*
9. Jubayer, C. M., and Hangan, H. 2014, "Numerical simulation of wind effects on a stand-alone ground mounted photovoltaic (PV) system", *Journal of Wind Engineering and Industrial Aerodynamics,* Vol (134), pp. 56-64.
10. Kopp, G. A., Farquhar, S., and Morrison, M. J. 2012, "Aerodynamics mechanisms for wind loads on tilted, roof-mounted, solar arrays", *Journal of Wind Engineering and Industrial Aerodynamics*, Vol (111), pp. 40-52.
11. Miller, R. D., and Zimmerman, D. K. 1981, "Wind loads on flat plate photovoltaic array fields", *Phase III, Report DOE/JPL/954833-81/3*, Boeing Engineering and Construction Company, Seattle, Washington.
12. National Building Code of Canada, 2010. National Research Council of Canada, Ottawa, Canada.
13. REN 21 2014, "Renewables 2014 Global Status Report", *REN 21 Secretariat*, Paris, France.
14. SEAOC 2012, "Wind loads on low profile solar photovoltaic systems on flat roofs", *Report SEAOC-PV2-2012.* Structural Engineers Association of California.
15. Shademan, M., Barron, R. M., Balachandar, R., and Hangan, H. 2014, "Numerical simulation of wind loading on ground-mounted solar panels at different flow configurations", *Canadian Journal of Civil Engineering*, Vol (41), pp. 728-738.
16. Warsido, W. P., Bitsuamlak, G. T., and Barata, J. 2014, "Influence of spacing parameters on the wind loading of solar array", *Journal of Fluids and Structures*, Vol (48), pp. 295-315

Proceedings of the 25th CANCAM
London, Ontario, Canada, May 31 – June 4, 2015

AERODYNAMIC AND ACOUSTIC EFFECTS OF A SERRATED TRAILING EDGE ON WIND TURBINE BLADES

Nicholas Tam
Mechanical Engineering
University of Waterloo
Waterloo, ON, CA
ntam@uwaterloo.ca

Kobra Gharali
Mechanical Engineering
University of Waterloo
Waterloo, ON, CA
kgharali@uwaterloo.ca.

Professor David A. Johnson
Mechanical Engineering
University of Waterloo
Waterloo, ON, CA
david.johnson@uwaterloo.ca

ABSTRACT

The problem of noise with wind turbines is not exclusive to large machines alone. Small scale wind turbines, operating in laminar or transitional regimes, can emit tonal noise which can be a nuisance. Serrations, which have been used on large wind turbines, are examined under controlled and dynamic conditions at a chord based Reynolds number of $4x10^4$ to evaluate their effectiveness for small scale wind turbines.

Phased averaged PIV measurements showed an increase in lift coefficients near dynamic stall points when an SD-7037 airfoil was pitched sinusoidally. Little effect was observed at lower angles of attack. The stall point was also delayed with the extension of the trailing edge through attachment of serrations.

Acoustic measurements taken under dynamic conditions exhibited tones that can also be found under static conditions at low angles of attack. Reduction was found for one of the tones emitted by the airfoil at the frequency of 4 kHz.

KEYWORDS: Wind turbine, acoustics, serrations, aerodynamics, fluid mechanics

INTRODUCTION

While wind turbine technology is well-merited in being a clean and viable manner to produce electricity, its adoption is often hindered by concerns of the nuisance from the noise that it generates. This is a problem well associated with large wind turbines in commercial wind farm projects. However, small scale wind turbines are also afflicted with this issue and should not be overlooked.

The noise source of wind turbines of any scale is predominantly aerodynamic in nature. However, with turbine blades operating under chord Reynolds numbers from 10^4 to 10^6, mechanisms for noise generation with small and large wind turbines differ. For large scale turbines, there is interest in reducing a broadband noise characteristic from the trailing edge caused by turbulent boundary layer flow. The addition of triangular "saw tooth" serrations have been found effective in this case.

Small scale turbines experience laminar and transitional boundary layer flow; with its trailing edge noise being tonal in character. Saw tooth serrations have also been investigated in this regime and are found effective in reducing noise as well; with notable studies by Moreau et al [1], and Chong et al [2].

The following study examines the effects of serrations in the low Reynolds number regime for its effectiveness with small scale wind turbine applications. Because wind turbines operate in conditions where wind shear, turbulence and yaw misalignment causes fluctuating inflow conditions, it is beneficial to examine the serrations under dynamic stall conditions. This can provide further insights into how serrations perform in field conditions; allowing a better prediction of their performance.

EXPERIMENTAL SET-UP

Experiments were conducted on a 25mm chord SD-7037 airfoil specimen with a closed circuit wind tunnel of the wind energy group. Trailing edge attachments made from 5 mil polyester shim stock were glued to the pressure side of the airfoil using cyanoacrylate adhesives. Cases examined are summarized in Table 1.

Table 1- Trailing Edge Cases

Trailing Edge Cases	Description
Original Trailing Edge Case (OTE)	Unmodified airfoil with SD-7037 profile
Flat Plate Case (FPL)	Extends the trailing edge of the SD-7037 airfoil with a 2mm flat plate (see figure 1a). Shim has a 2mm overlap for gluing to the pressure side of the airfoil specimen.
Narrow Serrations Case (NSR)	Extends the trailing edge of the SD-7037 with a 4mm serrated plate of aspect ratio $\lambda/h = 0.67$ (see figure 1a and 1b). Shim has a 2mm overlap for gluing to the pressure side of the airfoil specimen. A wider case ($\lambda/h = 1.33$) is planned to be studied but is not discussed here.
Zero Length Attachment Case (ZLA)	Does not extend the length of the airfoil. This case is to examine if the attachment method has any effect to the acoustics of the airfoil.

The serration aspect ratio was chosen to be similar as those found effective by Moreau et al [1] at low to moderate Reynolds number flow. FPL case, which only extends half the length compared with serrated case, was included in this study for comparison with the NSR case as an airfoil with regular trailing edge of the same wetted area.

Experiments were conducted for a chord Re of 4×10^4, based on the airfoil chord according to Eq. (1)

$$Re = \frac{\rho u c}{\mu} \qquad (1)$$

where ρ and μ are air density and viscosity, u is free stream velocity, and c is the airfoil chord.

The airfoil operated under a sinusoidal pitching oscillation as defined in Eq. (2)

$$\alpha = \alpha_{mean} + \alpha_{amp} \sin(2\pi f t) \qquad (2)$$

with mean angle of attack, α_{mean}, of 11°, and amplitude of oscillation, α_{amp}, of 11°.

The airfoil was oscillating at a frequency, f, of 25Hz, with the reduced frequency, k, of 0.08. (see Eq.(3)).

$$k = \frac{\pi f c}{u} \qquad (3)$$

Lift measurements were measured by phased average PIV measurements. Further details of the setup and post-processing technique can be found in Gharali et al. [3].

Acoustic experiments were conducted using a B&K 4192 microphone that is recess-mounted in the wind tunnel on the pressure side of the airfoil. Setup and processing technique can be found in McPhee [4]. Additional experiments for noise source verification were conducted with University of Waterloo's blow down open jet wind tunnel. Single microphone measurements were taken with a B&K 4189 microphone using a B&K 2250 handheld analyzer.

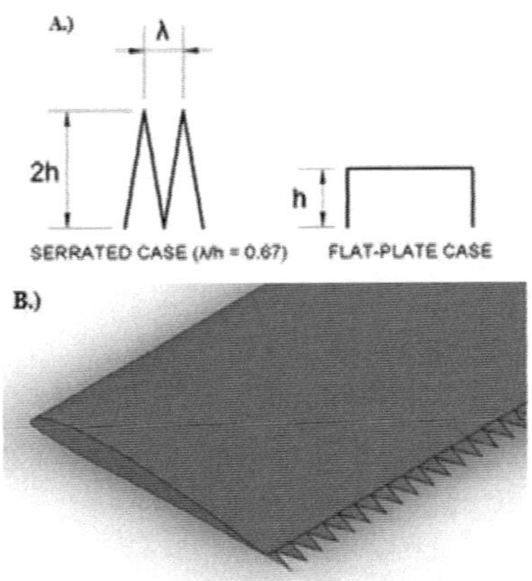

Figure 1A.) Diagram depicting NSR case (left) and FPL Case (right). Variables defined by Howe [5]. 1B.) Depiction of a serrated airfoil (NSR case) Source: Gharali et al. [3]

RESULTS AND DISCUSSION: AERODYNAMIC

Lift coefficients of the airfoil were calculated along various angles during the upstroke and downstroke movements in the sinusoidal pitching. The lift curve for different trailing edge configurations can be seen in figure 2.

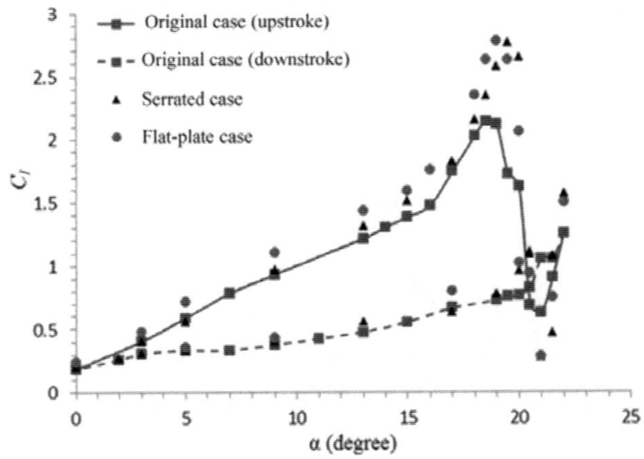

Figure 2 Lift coefficient vs. angle of attack for SD-7037 with various TE configurations. Serrated case represents NSR configuration. Flat-plate case represents FPL configuration. Source: Gharali et al [3]

Both FPL and NSR case did not show altered lift behaviour at low angles of attack for upstroke and downstroke movement. However, for upstroke angles close to the dynamic stall point, the lift coefficient was found to increase for the modified trailing edge cases. In examining the leading edge vortex (LEV) for the different cases, it was found that the FPL showed increased circulation at these higher angles of attack while NSR values were similar to the OTE values (except near the stall point). This suggests another mechanism contributes to the increased lift observed in the range between the static stall point and dynamic stall point.

Another difference was found with the stall point itself. The OTE case showed the airfoil to dynamically stall at 18.5 degrees. This stall was delayed to 19 degrees with the FPL case and to 19.5 degrees for NSR case. It is speculated that the added length of the airfoil allows the LEV to continue developing to a higher angle. Continuing growth of the LEV delayed stall as a result.

Further discussion on the aerodynamic effects of the serrations under dynamic conditions can be found in Gharali et al [3].

RESULTS AND DISCUSSION: ACOUSTIC

Un-weighted sound pressure level with respect to the sound frequency for the different trailing edge cases for a sinusoidally pitched airfoil can be seen in figure 3. Plots were generated from FFT analysis with a frequency resolution of 0.75Hz. As the airfoil was found to produce noise above background levels between 2.5 kHz to 5 kHz, only this range is shown.

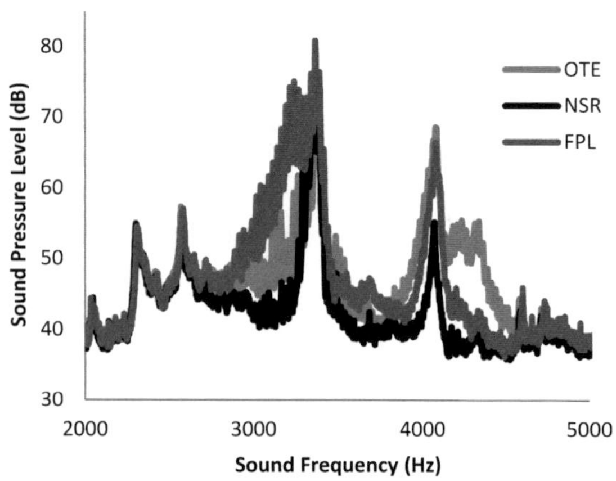

Figure 3- FFT results of un-weighted sound pressure levels with SD-7037 airfoil under sinusoidal pitching with different trailing edge configurations

Tones were found centred at 3.4 kHz and 4.1-4.3 kHz. They were found in all the different trailing edge cases, albeit with different sound pressure levels. The same tones were observed with the airfoil under a static condition, independent of trailing edge case, at low angles of attack.

Before discussion of the effects of the different trailing edge configurations, it is important to determine if the tones observed were caused by laminar boundary layer conditions or some other factors. To aid with this effort, the program NAFnoise [6], developed by NREL, was used. Testing conditions were entered with the program set to calculate noise levels using models developed by Brooks et al [7] (BPM model).

The dominant noise source found by the program at the static angle of attack of 0 degrees was laminar boundary layer noise. Blunt trailing edge noise and inflow turbulence noise were not expected to be an issue with the current setup and therefore excluded from analysis. The results were given as un-weighted sound pressure levels in third octave bands and were compared with experimental results of the OTE case at 0 degrees angle of attack.

As seen in figure 4, there were some differences between the results. This was expected, however, since the BPM model uses empirical data of NACA 0012 airfoil in its calculations. Nevertheless, the NAFnoise results suggest the observed tones are likely caused by a laminar boundary layer source. The estimated peak in noise is within a third octave band of the experiment results. The difference in peak value between NAFnoise and experimental results is 3.3dB.

In order to ensure the tones observed are not caused by resonance in the wind tunnel, the airfoil specimen under the OTE case was examined with the open jet blow down wind tunnel. Under same wind speed conditions, tonal noises were found at the same frequencies. Since there were no walls under this test setup, the tones found can therefore be attributed from the airfoil and independent of the tunnel.

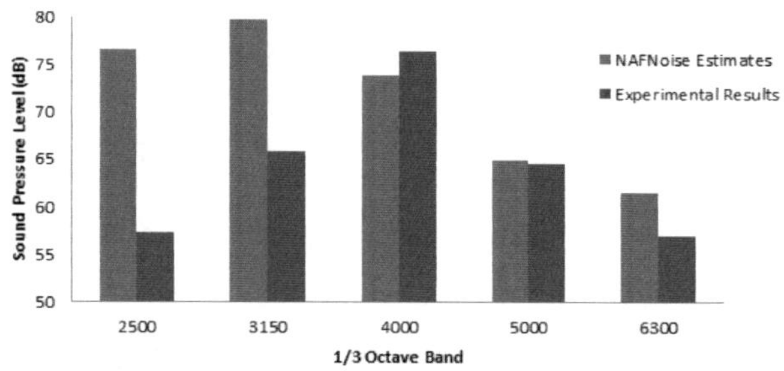

Figure 4- Comparison of un-weighted sound pressure levels computed from NAFnoise [6] with experimental results at 0 degrees angle of attack under static conditions

The effect of the attachment method should also be taken into consideration. Noise level for the ZLA case was compared with the OTE case in figure 5. No additional tones were found to be generated from this method. Noise levels did not appear to be affected significantly at 3.4 kHz but there was an increase in noise levels for the tone at 4.1-4.3 kHz. Third octave band levels centred at 3.15 kHz showed a difference between the two cases to be 1.16dB while there was a 3.67dB difference at the 4 kHz band.

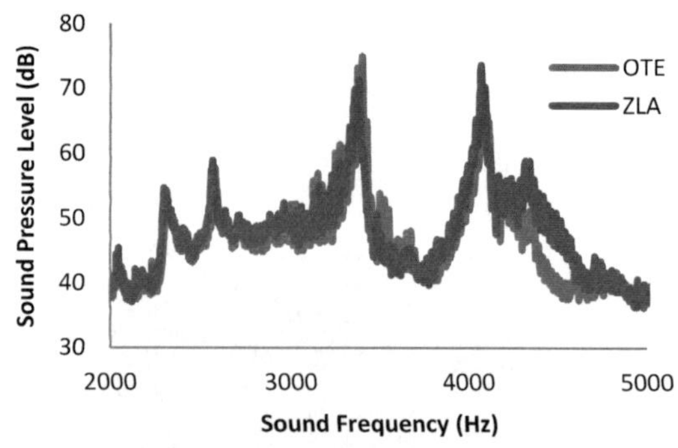

Figure 5- Comparison of un-weighted sound pressure level with ZLA case and OTE case under sinusoidal pitching conditions

Given a more thorough understanding of the nature of the noise source observed, results shown figure 3 can be compared. Notable reduction was seen in the NSR case at the 4.1-4.3 kHz tone. For the third octave band centred at 4 kHz, there was a 10.9dB reduction between the NSR case and OTE case.

For the FPL case, there was a notable increase in sound pressure level at the 3.4 kHz tone. A 7.16 dB increase was found at the third octave band centred at 3.15 kHz between the FPL case and the OTE case.

CONCLUSIONS AND FUTURE WORK

Results thus far have shown serrations under dynamic cases to reduce some tonal components produced by the airfoil. Aerodynamic performance has also been altered, with increased in lift coefficient near stall angles and a delay in stall.

To further understand the effects of dynamic flow conditions on the performance of serrations, a more detailed comparison of the acoustic results between the dynamic cases with static cases is needed. Further examination into the flow conditions around the airfoil may also provide insights into the noise generation mechanism.

In terms of its aerodynamic effects, further investigation is needed to explain the increased lift phenomenon between static stall angle and dynamic stall angle for the serrated trailing edge.

ACKNOWLEDGMENTS

The authors would like to thank NSERC for their support with this project.

REFERENCES

1. Chong T.P and Joseph P.F. "An Experimental Study of Airfoil Instability Tonal Noise with Trailing Edge Serrations", J. of Sound and Vibration. **332**, 6335-6358, 2013.

2. Moreau B.J., Brooks L.A. and Doolan C. "On the Noise reduction Mechanism of a Flat Plate Serrated Trailing Edge at Low-to-Moderate Reynolds Number", AIAA. **51(10),** 2513-2522, 2013.

3. Gharali K, Tam N and Johnson D.A "A PIV Load and Flow Structures Study of A Serrated Dynamic Airfoil", 17th International Symposium on Applications of Laser Techniques to Fluid Mechanics, 2014.

4. McPhee A. "The Development of A Research Technique for Low Speed Aeroacoustics", MASc thesis, University of Waterloo, 2008

5. Howe M.S. "Aerodynamic Noise of a Serrated Trailing Edge" J of Fluids and Structures. **5**, 33-45, 1991.

6. Moriarty P. "NAFNoise User's Guide" National Renewable Energy Laboratory. 2005

7. Brooks T.A, Pope D.S and Marcolini MA. "Airfoil Self-Noise and Prediction" NASA Reference Publication. **1218**, 1989

LARGE EDDY SIMULATION OF WIND-INDUCED PRESSURE ON A LOW RISE BUILDING

B. Hajra,
Postdoctoral fellow,
Western University, Canada
email: bhajra@uwo.ca

H. Aboshosha,
Postdoctoral fellow,
Western University, Canada
email: haboshos@uwo.ca

G. T. Bitsuamlak,
Associate Professor,
Western University, Canada
email: gbitsuam@uwo.ca

ABSTRACT

The wind-induced pressure on the roof of a low-rise building was assessed computationally using Large Eddy Simulation (LES) for wind angle of zero degrees. The low-rise building consists of a complex roof system with varying slopes. Turbulent inflow data used for the Computational Fluid Dynamics (CFD) simulations was generated based on the wind tunnel measured data. Furthermore, the grid size was changed from a coarser mesh at the computational domain to a finer mesh at the building of interest for better accuracy of results. The mean pressure coefficients obtained from LES were compared to Wind tunnel measurements that were carried out at the Boundary Layer Wind Tunnel (BLWT) II at Western University, Canada. Results show good comparison (within a factor of 1.2) of mean pressure coefficients between LES and wind tunnel data at points away from the roof-ridge of the building. However, there is a departure between LES and wind tunnel data at locations closer to the ridge of the roof, although the overall trends of data are similar. Results from LES simulation are encouraging and can be useful for wind load estimation and subsequent design of low-rise buildings.

KEYWORDS: Low-rise building, Complex roof, LES, Turbulent inflow data, Pressure coefficient.

INTRODUCTION

Wind induced structural damage to low-rise buildings is not uncommon especially in coastal areas like Florida which is often affected by high wind speeds and hurricanes (Pinelli et al., 2004). Interaction between wind and buildings produces a high positive pressure on the walls and suctions on the roof. The roof of a low-rise building is more likely to get damaged due to these high suctions, leading to partial or complete failure of the structure (Dagnew and Bitsuamlak, 2014). Most available building codes such as the NBCC 2005 and ASCE 7, 2010 provide formulations to assess wind-induced pressures on low rise buildings of regular sloped roofs and recommend wind tunnel measurements for complex roof systems. In most urban areas, buildings do consist of complex roof systems with varying slopes. Also, due to time and financial constraints it is not always possible to carry out wind tunnel measurements for such cases (Dagnew et al., 2009; Dagnew and Bitsuamlak, 2013). According to Paneerselvam, 1997, "*In LES, the fluctuating motions of turbulence can be computed exactly except for eddies that are smaller than the grid size. The smaller eddies are modelled using eddy viscosity models. Because of this, LES can compute fluctuating pressures on buildings due to turbulence*". Therefore, the present study focuses on an actual low rise building located in suburban terrain conditions in Florida, USA for which wind tunnel measurements were carried out at the BLWT II at Western university, Canada. The results from the wind tunnel data were simulated using LES for which the turbulence inflow characteristics were developed based on the approach of Aboshosha et al., 2015. The wind induced mean pressure coefficients on the building roof obtained from wind tunnel and LES compared reasonably well, particularly at locations away from the roof ridge. The results from this study can be very useful for designers, since current codes do not provide formulations for such cases.

WIND TUNNEL MEASUREMENTS AT WESTERN UNIVERSITY, CANADA

Wind tunnel measurements were carried out at the BLWT II at Western University, Canada. Spires and roughness elements (see Figure 1) to simulate a suburban terrain (exposure B, as per ASCE-7, 2010) were used for the experiments because the building (hereafter referred as FL27) is located close to the wooded areas, not far from the Atlantic coast.

Figure 1: Upstream terrain simulation (Kopp and Gavanski, 2011)

Figure 2 shows the building model used for wind tunnel experiments. The mean and peak pressure coefficients were measured using electronic ESP-16TL pressure scanners. The sampling rate of the pressure measurements was 400Hz, while the sampling time for each measurement was 360 seconds. The roughness length (Z_o) was 0.23 m (full scale), power law exponent (α) was 0.165 and length scale was 1:50. The wind velocity measured in the wind tunnel at the building height of 4.26 m (full scale) was 8 m/s.

Figure 2: Wind tunnel model of FL27 (Kopp and Gavanski, 2011)

496 pressure taps were located on the building roof and wind directions for 0° to 350° were tested for every 10° increment, although the present study focuses only on the 0° wind direction case. For additional experimental details, see Kopp and Gavansky, 2011.

CFD MODELLING

The Computer aided design (CAD) model was prepared using STAR CCM (CD-Adapco, 2012). While meshing the model, a nested meshing approach (see Nozu et al., 2008) was used. The computational domain (CD) was of 227m X 103m X 37m (full scale) and a 'base size' of 1.5m was used. Totally 3 inner blocks were constructed that engulfed the building model within the CD. From outer to the inner block, mesh sizes (polyhedral and surface mesh) were reduced gradually from coarser to finer as: 65%, 40% and 15% of the base size respectively (see Figure 3).

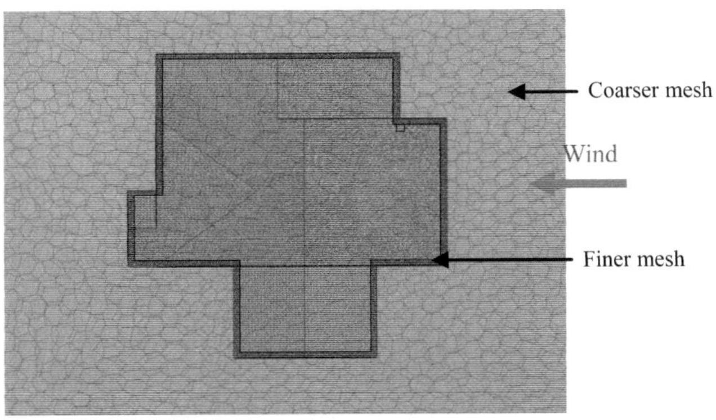

Figure 3. Top view of low-rise building (FL27) with meshing

The building roof surface was provided a finer surface mesh of 10% of base size. This nested meshing approach has been utilized earlier by Nozu et al., 2008 in successfully simulating the wind flow conditions in Tokyo city using LES. The meshed model contains 1145574 cells, 8205065 faces and 7146163 vertices. The use of LES has primarily improved due to better turbulence inflow generation (Lund et al., 1998), efficient sub grid scale modelling technique and better discretization techniques for complex geometry (Tamura et al., 2008). Figure 4 shows a side view of the meshing, and it can be clearly seen that the size of the mesh is finer closer to the region of interest (building roof) than other regions of the CD.

Figure 4. Side view showing the meshing of the building

A symmetry boundary condition was used for the top and lateral surfaces while the solid walls had a 'no-slip' condition (COST, 2007). The blockage ratio, which is defined as the ratio of the projected area of the building surfaces in the flow direction to the area of inlet boundary was less than 5 % in the present case. In order to generate an inflow condition equivalent to that produced in the wind tunnel, the technique developed by Aboshosha et al., 2015 was used. According to Aboshosha et al., 2015, "*Ground surface roughness is simulated using fractal surfaces generated by random Fourier modes (RFM) and scaled to match a targeted aerodynamic roughness z_o. Simulated wind velocities are averaged spatially and temporally to extract the mean and turbulent components*". The Dynamic Smagornisky-Lilly subgrid-scale (SGS) model based on Germano et al. (1996) and Lily (1992) have been employed. In order to have Courant number less than 1, a time step of 0.01 s was used to estimate the pressure coefficients at various locations on the roof. MATLAB was used to process the data (instantaneous pressure coefficients) obtained from LES simulations for different time steps in order to obtain the average pressure coefficients. The computations were carried out using the facilities provided by SHARCNET, Canada using more than 20 processors.

RESULTS AND DISCUSSION

Various surfaces on the roof were selected to estimate the mean pressure coefficient. The mean pressure coefficient (Cp) is defined as:

$$Cp = 2(P-Po)/\rho V^2 \quad (1)$$

P is the pressure on a particular point on the roof (Pa);

Po is the free stream static pressure (Pa);

ρ is the density of air (1.22 kg/m^3);

V is the wind velocity at the building height (V = 8 m/s)

Figure 5 shows mean pressure coefficients estimated from LES on the roof of FL 27.

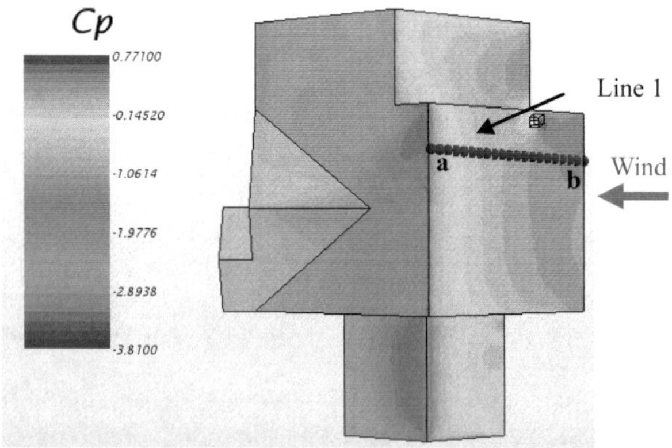

Figure 5. 'Mean Cp' values on the building roof from LES

The range of mean pressure coefficient (Cp) generally lies between 0.771 to -3.810, with predominantly negative pressures (suction) on the roof as expected. In fact, the values of Cp are generally higher (between -2 to -3.81) closer to the ridge (centre of the roof), indicating high suction values, while gradually away from the ridge there is less suction. A close inspection of the lower roof on the left of the wind flows indicates higher suctions (close to -3) immediately downwind of the ridge because of the flow separation. Similar trends are observed from wind tunnel data as shown in Figure 6, although the values ranging from 0.45 to -1.12, are somewhat lower than those produced by LES.

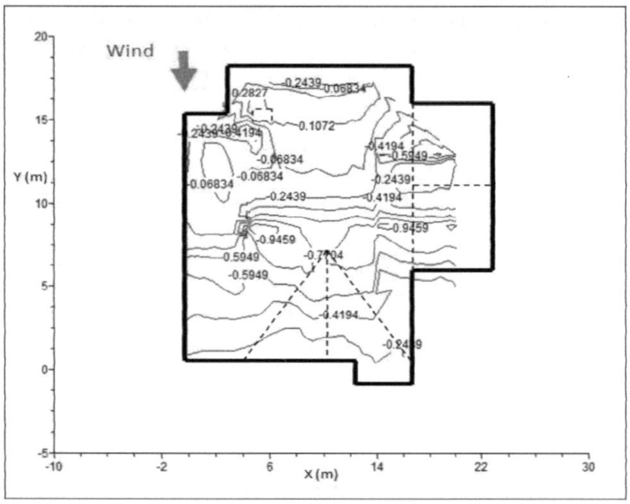

Figure 6. 'Mean Cp' contour plot from wind tunnel measurements (Kopp and Gavanski, 2011)

To get a clearer indication of the values, the mean pressure coefficients obtained from wind tunnel and LES are plotted for a straight line along the main roof (see Line 1 on Figure 5) of the building, as shown in Figure 7.

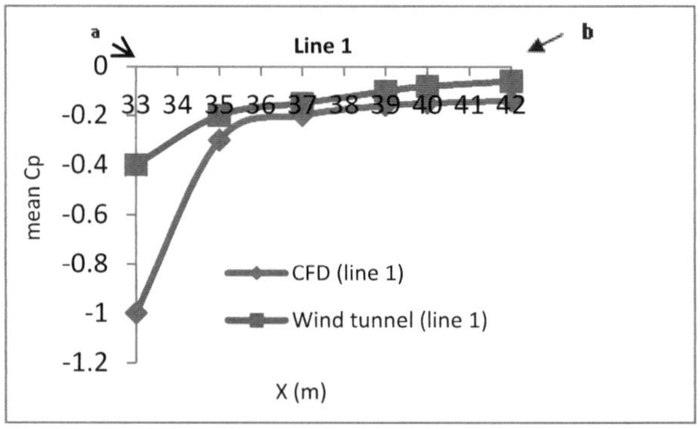

Figure 7. Comparisons of mean pressure coefficient (Cp) between LES and wind tunnel

Results show that Cp obtained from wind tunnel measurements are generally higher (about a factor of 1.2) than LES simulations at locations away from the ridge line. However, at locations on the roof-ridge the values of Cp predicted by CFD are more negative (higher suctions) than wind tunnel by a factor of 2.5. Although the overall trends predicted by CFD are similar to wind tunnel data, the former generally predicts higher suctions than the latter at all locations. Also, close to the ridge where flow separation occurs, resulting in higher suctions, LES modelling may require some more refinement (e.g. by using finer mesh at the ridge compared to other roof locations). Despite the discrepancies, the results are quite encouraging and this study shows that in the absence of experimental measurements, LES modelling can certainly assess wind induced mean pressures on buildings reasonably well. Estimation of peak pressure coefficients on the roof of FL 27 is currently underway. Future research must be focussed on improving CFD simulations by comparing with field measurements for more realistic 'multiple-building' urban scenarios.

CONCLUSIONS

This paper presents comparisons of mean Cp between LES and wind tunnel data for wind azimuth of 0 degrees, on the roof of a low rise building (FL27) located in Florida, USA. Wind tunnel measurements were carried out at the BLWT II at Western University (Kopp and Gavanski, 2011), while CFD simulations using LES were carried out using turbulence inflow generating technique of Aboshosha et al., 2015. Results indicate good comparisons between LES and wind tunnel at locations away from the ridge of the roof (factor of 1.2), while at the ridge, LES predictions are more negative (higher suctions) than wind tunnel data (factor of about 2.5). The overall trends between LES and wind tunnel are similar, despite the discrepancy at certain locations. It may be possible to obtain better correlation between wind tunnel and LES, by having finer mesh closer to the locations of separation (e.g. ridge) compared to other roof locations. In general, the results are quite encouraging, and with suitable adjustments, LES modelling may be a useful tool for the future for assessing wind induced pressures on buildings in the absence of experimental facilities. Greater collaborations between experimentalists and numerical modellers can result in the improvement of the latter, especially for studying realistic (multiple-building) urban scenarios.

ACKNOWLEDGMENTS

The authors are grateful to Dr. K. Gurley, Dr. D. Prevatt, Dr. F. Masters (University of Florida, USA) and Dr. G. Kopp (Western University, Canada) for providing the wind tunnel data used for the complex roof comparison (NSF Grant CMMI-0928563). The computing facility provided by SHARCNET, Canada is gratefully acknowledged.

REFERENCES

1. ASCE-7. 2010. Minimum design loads for buildings and other structures, American Society of Civil Engineers, Reston, VA, USA.
2. Aboshosha, H, Bitsuamlak, G, El Damatty, A. 2015, "Turbulence characterization of downbursts using LES", *Journal of Wind Engineering and Industrial Aerodynamics*, 136, 44-61.
3. CD-Adapco, 2012. "User Guide: Star CCM+". version 7.06, USA
4. COST. 2007. "Best practice guideline for the CFD simulation of flows in the urban environment", COST Action 732.
5. Dagnew, A.K, Bitsuamlak, G.T, Ryan, M. 2009, "Computational evaluation of wind pressures on tall buildings", Proceedings of the 11th American conference on wind engineering, San Juan, Puerto Rico, USA.
6. Dagnew, A.K. Bitsuamlak, G.T. 2013, "Computational evaluation of wind loads on buildings: a review", W*ind and Structures*, 16(6), 629-660.
7. Dagnew, A.K. Bitsuamlak, G.T. 2014, "Computational evaluation of wind loads on a standard tall building using LES", *Wind and Structures*, 18(5), 567-598.
8. Germano, M.U, Piomelli, P, Moin, P, Cabot, W.H. 1996, "Dynamic Sub-grid scale eddy viscosity model", In Summer workshop, Centre for turbulence research, Stanford, CA, USA.
9. Kopp, G.A, Gavanski, E. 2011, "Wind tunnel pressure measurements on two FCMP houses in Florida", Report-BLWT-5-2010, Boundary Layer Wind tunnel Laboratory, Faculty of Engineering, Western University, London, Canada.
10. Lily, D.K. 1992. "A proposed modification of the Germano subgrid scale closure model", *Physics of Fluids*, 4, 633-635.
11. Lund, T.S, Wu, X, Squires, K.D. 1998. "Generation of turbulent inflow data for spatially developing boundary layer simulations", *Journal of Computational Physics*, 140, 233-258.
12. NBCC. 2005. "National Building Code of Canada", National Research Council, Ottawa, Canada
13. Nozu, T, Tamura, T, Okuda, Y, Sanada, S. 2008, "LES of the flow and building wall pressures in the centre of Tokyo", *Journal of Wind Engineering and Industrial Aerodynamics,* 96, 1762-1773.
14. Paneer Selvam, R. 1997. "Computation of pressures on Texas tech university building using large eddy simulation", *Journal of Wind Engineering and Industrial Aerodynamics*, 67, 647-657.
15. Pinelli, J. P., Simiu, E., Gurley, K., Subramanian Zhang, L., Cope, A, Hamid, S. 2004. "Hurricane damage prediction model for residential structures", *Journal of Structural Engineering*, 130(11), 1685-1691.
16. Tamura, T, Nozawa, K, Kondo, K. 2008. "AIJ guide for numerical prediction of wind loads on buildings", *Journal of Wind Engineering and Industrial Aerodynamics*, 96, 1974-1984.

Author Index

A

Abdollahi, Mehrzad 558
Aboshosha, H. 844
Abulkhair, Hani 125
Acharya, Bishnu 678
Adamiak, Kazimierz 202
Afriyie, Yaw Y. 397
Afshari, Hamed Hossein 509
Agelin-Chaab, Martin 392,409
Aguirre, Rogelio Hernández 611
Ahmadian, Mohammad Taghi 794
Ahmed, Ryan 509
Ahmed, Tamima 695, 699
Akafuah, N.K. 219
Akan, Md Maniruzzaman 695
Akbari, Mohammadhadi 686
Akhtar, Ali 69
Akon, Abul Fahad 190
Alam, J. M. 643
Alam, Md. Mahbub 104
Al-Athel, Khaled S. 321
Alba, K. 245, 621, 630
Alemohammad, Hamid 496
Ali, Shady 206
Alibabaei, Nima 513
Al-Qaysi, Zaid 388
Al-Rawahi, Nabeel 522
Alvarado, Fabian 121
Aly, Shahzada P. 321
Alzahrani, Waleed S. 703
Amon, Cristina H. 121
Anderson, Shammawi A. 211
Annamalai, Mani 265
Araghi, Orang Vahid 541
Arghavan, Soroush 767
Arif, A. F. M. 321, 583
Armenio, V. 738
Asgari, Meisam 152, 162
Asokanthan, Samuel 767, 779
Attarzadeh, S.M.Reza 273
Avari, Hamed 147

B

Baddour, Natalie 763, 790
Bahrani, Nima 759
Bahreyni, Behraad 500
Baig, Altamash Ahmad 674
Balachandar, R. 269, 423
Banerjee, Joydeep 541, 546
Banister, Carsen J. 669
Barron, R.M. 269, 423
Barthelat, Francois 160, 161, 459
Bashar, Mohammad 334
Bassom, A. 223
Basu, Prabir 678
Behdinan, Kamran 309, 313, 759
Benmokrane, Brahim 19
Bergstrom, Donald J. 277
Bhakhri, V. 376, 380
Bhuiyan, A. S. 643
Birhane, Tibebu H. 482
Biria, Aisa 152
Bitsuamlak, G. T. 440,482,775,844
Blackman, Karin 474
Blaine, Sitarz 575
Bognash, Mohamed 767,779
Bordatchev, Evgueni 36, 48, 60
Bourgeois, N.R. 185
Brinkerhoff, Joshua 194
Brooks, Kevin 286
Bruneau, Steve 56
Bruyn, John R de 129
Bubshait, H.M.H. 32
Burgess, Gary 293, 747
Bustamante, Carlos A. 241, 253

C

Cai, Xiaobing 593, 608
Campbell, Gord 166
Campbell, Robert J. 156
Carrick, Roger 417
Castle, G. S. Peter 202
Chanda, Sourayon 455
Chaudhry, Muhammad Salman 417
Chekurthi, Shailesh 575, 579
Chin, Melanie 166
Chintapalli, Ravi 160
Chishty, Wajid A. 363
Choi, Steve 86
Chow, Jacky 597
Chtchetinina, Anna 113
Cohen, J.D. 219
Colborne, Bruce 56
Collins, Michael R. 669
Cooper, Scott 144
Costache, Adrian 832
Csernyei, Chris 342
Cuillière, Jean-Christophe 24, 297
Czekanski, Alex 417

D

Dallali, M. 738
Damatty, A.A. El 800,808
Damshenas, Mohammad 794, 820
Dang, Lam 665
Das, Sreekanta 460, 464
Dastjerdi, Ahmad K. 459
DeGroot, Christopher T. 660
DiCarlo, Amanda L. 140
Dickey, James P. 812
Dolatabadi, Ali 273
Dong, Chenying 413
Doxtator, Andrew 69
Doyle, Matthew G. 121
Dumond, Patrick 763, 790
Dutta, Joydeep 522

E

Egberts, Philip 619
Eger, Tammy R. 812
Ekmekci, Alis 90
Elansary, Ahmed 800
Elhalwagy, Mahmoud 215
Elkaseh, Husam 156
Ellis, Christopher G. 137
Elshaer, Ahmed 775
Eltaggaz, Abdelkrem 729
Engelen, Niel Van 783
Epp, Tom 397
Eslami, A. 245
Esmaeilnejad, Ali 742
Essel, Ebenezer E. 401

F

Fartaj, Amir 330
Feero, Mark 74
Fellouah, Hachimi 734
Feltham, Graham 90
Ferreira, Louis 181
Finney, M.A. 219
Floryan, J.M. 223, 228, 232, 237, 281, 338
Flynn, M. R. 648
Fortais, Adam J 129
Forthofer, J.A. 219
Fotowat, Shahram 330
Fouladi, Fama 716
François, Vincent 24, 297
Fraser, Graham M. 137
Frigaard, I.A. 245
Fung, Alan S. 513, 674, 682, 691, 695, 699, 703

G

Gadallah, Aly H. 363
Gadsden, Andrew 509
Gallant, Tyler E. 65, 82
Gashti, M. Parvinzadeh 625
Ghaednia, Hossein 460, 464
Ghaemi, Sina 355, 720, 742
Ghannoum, AbdulRahman 496
Gharali, Kobra 840
Ghasemi, Abbas 347
Ghazanchaei, Mohammadreza 202
Ghorbanishohrat, Faegheh 257
Ghosh, Suvojit 567
Godaba, Hareesh 15
Goldman, Daniel 137
Gollner, M.J. 219
Golnaraghi, Farid 500
Gómez, Natalia 253
Gorham, D.J. 219
Gracie, Robert 301, 317, 652,787
Greener, J. 625
Grosse-Wortmann, Lars 121
Guo, Qiuquan 608
Gupta, Ujjaval 15

H

Habibi, Saeid 509
Haghshenas, M. 380
Hajiani, P. 638
Hajra, B. 844
Hamilton, Benjamin 48,60
Hamouda, Ouajih 436
Hangan, Horia 491, 832, 836
Harne, R.L. 816
Hashemi, Seyed M. 615
Hassan, Mohamed 19
Hassan, Mohamed T.Z. 554
Haya, Laura 109
He, Yuping 537
Henshaw, Paul 351, 716
Heris, Hossein K. 162
Hessabi, Reza Mirza 755
Hickey, Jean-Pierre 712
Hokmabad, Babak Vajdi 720
Honjo, Osami 121
Hossain, M. A. 643
Hossain, M. Z. 223, 338
Hosseini, Farhad Mir 763
Hu, Fei 558
Hu, Mingjun 608
Huang, Yifan 451
Hussein, Sama 48, 60

I

Ide, Haruki 121

Iftekhar, Hassan 392, 409
Islam, AZM Ariful 376
Ismail, Mohammed 330
Iyer, Krishna 496

J

Jafari, M. 269, 423
Jahed, Hamid 372, 405
Ji, Xiaoxu 812
Jiang, Liying 11,413, 532, 563
Jobehdar, Mona Hassanzadeh 363
Johnson, David A. 65, 82, 257, 840
Jubayer, Chowdhury 836

K

Kadem, Lyes 113, 125
Kalbfleisch, Alan 359
Karazis, Kostas 432
Karganroudi, Sasan Sattarpanah 24
Kenno, Sara 464
Kettlewell, Brian Q. 117
Khajepour, Amir 611
Khalilollahi, Amir 293
Khan, Ishan Ali 615
Khayat, Roger E. 11
Kheiri, Mojtaba 444
Khosravipour, Ida 177
Kiani, AmirKianoosh 28, 174
Kidambi, Narayanan 816
Kilpatrick, Ryan 491
Klassen, R.J. 368, 376,380
Knopf, George K. 601
Komijani, M. 787
Konstantinidis, Dimitrios 783
Kopp, Gregory Alan 6,190,827
Koupriyanov, Mikhail 397
Kuhn, David C.S. 724
Kunwar, Sumesh 44
Kusins, Jonathan 181
Kwon, H. J. 40

L

Ladubec, Chris 652
Lai, Yong Jun 156, 597
Lajimi, S. Amir Mousavi 500
Lambert, Andrew 198
Laneville, André 734
Langlois, R.G. 185
Larachi, F. 638
Larson, R.G. 249
Lavertu, Karine 734
Lavoie, Philippe 74
Leask, Richard L. 144
Lee, Eungkil 537
Lee, Tim 86
Leong, Wey H. 665, 703
Lewalle, Jacques 100
Li, LePing 171
Li, Xianguo 347
Li, Yang 563
Liang, Yiteng 541
Libii, Josué Njock 469
Lightstone, M. F. 261
Lin, W.E. 478
Liu, Changli 517
Liu, Huang 532
Liu, Hui 563
Liu, M. L. 751
López, Tatiana 241
Luo, Yunhua 177

M

Mahajerin, Enayat 293, 747
Mali, Sarwesh 401
Martini, Roberto 160
Maryniuk, Rhyse 504
Marzbanrad, Bahareh 405
Marzouk, Amr 500
Massri, Maha Al 427
Mayemba, Bob Munlemvo 384
Mayer, J.R.R. 52
McAllister, S.S. 219
McClure, Jeff 96
McDonald, Andre G. 211
McMeeking, Robert M. 5
McMillan, Brian D. 575
McPhee, John 541, 546
Meguid, S.A. 554, 589
Mercan, Oya 755
Michon, Guilhem 771
Minaker, Bruce P 550, 558
Mirkhalaf, Mohammad 161, 459
Mishra, Anuvrat 724
Moallemi, Nima 194
Mohajerani, Shirzad 36
Mohammadi, A. 228, 232
Mollaabbasi, R. 621,634
Mongeau, Luc 162
Moradi, H. V. 281
Moss, Eddie B. 156
Mostaghimi, Javad 321
Murray, S. A. 261
Musa, A. 808
Myint, Myo Tay Zar 522

N

Nana, Alexandre 297
Nasir, Zoheb 440
Nemati, Omid 691
Nematollahi, Ali 401
Newson, Timothy A. 166,427
Ng, Hoi Dick 113
Nie, Weirong 532
Nieto, César 241, 253
Nieva, Patricia 496, 611
Nikumb, Suwas 601
Nobes, David S. 355
Noie, S.H. 634
Noori, Navid 500

O

Onaizah, Onaizah 133, 140
Oreskovic, Chris 487
Orf, L.G. 478
Oviasuyi, R. 380
Ozyoruk, A. 478

P

Païdoussis, Michael P. 432
Pang, Kun 413
Parchei-Esfahani, Matin 317
Parfilko, Yevgeniy 290
Parvu, Dan-Stefan 832
Pejhan, Shabnam 177
Penney, Jared 656
Perera, Sharman 69
Perret, Laurent 474
Peterson, Sean D. 117
Piollet, Elsa 771
Poepping, Tamie L 129, 133,140
Porto, Juliette 487
Prikazchikov, Danila 804
Puri, Ishwar K. 567

Q

Qian, Yanjun 40
Qin, Yanzhou 347

R

Raahemifar, Kaamran 513
Rabelo, Vinicius B. F. 751
Radmanesh, Mitra 28, 174
Raghavan, J. 575, 579
Rahman, Md Mizanur 52
Rahman, Mohammad Anisur 309, 313
Rajakumar,H. 368
Raut, Manoj Kumar 678
Ray, Rajan 351
Riznic, Jovica 436
Roberto, Matthew 487
Rodriguez, Marcel 171
Roostaei, Ali 372
Ross, Annie 771
Ru, C.Q. 449, 455

S

Sadrekarimi, Abouzar 388
Sadri, Sobhan 504
Sahu, Chunendra K. 648
Saito, K. 219
Sakib, Md. 237
Sanders, R. Sean 355
Savory, Eric 147, 206, 474, 478,487
Sébastien, Poncet 708
Sellmann, Asaad 541
Sergushova, Olga 804
Shahbazian, Negin 309, 313
Shao, Weijie 392
Shokri, Rouholluh 355
Shrivastava, Suresh 384
Siddiqui, Kamran 334, 359, 363, 491
Siddiqui, M. U. 583
Simmons, Craig A. 528
Sinar, Dogan 601
Skiba, Oxana 301
Spence, Allan D. 32
Srivastava, Arpita 265
Stastna, Marek 656
Stéphane, Viazzo 708
Straatman, Anthony G. 342
Su, Ying Yu 86
Su, Yu 305
Sullivan, Ashley 56
Sullivan, Pierre 74, 78
Sumaiya, Syeda Noor E 575

T

Tachie, Mark F. 397, 401, 724
Taghavi, S. M. 245, 249, 621, 630
Tahan, Souheil-Antoine 24
Tait, Michael 783
Tallman, T. N. 571
Tam, Nicholas 840
Tavoularis, Stavros 8, 109
Tembely, Moussa 273
Teng,Ming 277
Tenney, Andrew S. 100
Ting, David S-K. 716
Tiwari, Shaligram 265
Toyserkani, Ehsan 405

Tsai, Peiying Jennifer 567
Tse, Leonard W. 121
Tullis, S. 261
Tutunea-Fatan, O. Remus 36, 48, 60, 181

V

Varela, Benjamin 290
Varty, Robert L. 326

W

Waller, Edward 69
Walsh, Paul 69
Wan, Alwin M. D. 526
Wang, Kon-Well 7, 571, 816
Wang, Longjun 104
Wang, Tianfu 779
Wang, Xuejun 517
Wang, Y.J. 449
Weaver, David S. 2, 436
Weisel, Richard D. 121
White, Paul 504
Wong, Jeremy 528
Wu, Chieh-Hsun 827
Wu, Christine 504
Wu, Peidong 567
Wu, Xiaohua 712
Wyss, Urs P. 177

X

Xie, Kongying 156

Y

Yang, Jun 9, 593, 605, 608
Yao, Zheng 550
Yarusevych, Serhiy 96, 198
Young, Edmond W. K. 526
Yuan, Fang 413

Z

Zamani, Nader 286
Zamir, Mair 133, 140
Zandi, S. 232
Zdravkova, Liliana 496
Zeinali, Meysar 44
Zeng, Danielle 563
Zhai, Jianhui 305
Zhang, Chao 206, 215
Zhang, Dongxing 605
Zhang, Jin 589
Zhang, Wanyu 121
Zhang, Wenjun 517
Zhang, Zetian 56
Zhou, Jianyou 11
Zhou, Jiejing 755
Zhou, Wenxing 451
Zhou, Yu 104
Zhu, Jian 15
Ziadé, Paul 78
Zohrehheydariha, Jamshid 460, 464